1 MONTH OF
FREE
READING

at
www.ForgottenBooks.com

ISBN 978-0-332-70922-2
PIBN 11227388

This book is a reproduction of an important historical work. Forgotten Books uses
state-of-the-art technology to digitally reconstruct the work, preserving the original format
whilst repairing imperfections present in the aged copy. In rare cases, an imperfection in
the original, such as a blemish or missing page, may be replicated in our edition. We do,
however, repair the vast majority of imperfections successfully; any imperfections that
remain are intentionally left to preserve the state of such historical works.

CANADIAN MACHINERY
AND
MANUFACTURING NEWS

Vol. XXIV., No. 1

July 1, 1920

Common Sense and Sweat Can Do It

CANADA has seen twenty months of business and trade since the signing of the armistice, and in those twenty months Canada has demonstrated a resourcefulness that is remarkable, and a business stability that carries with it a tremendous degree of assurance for the future.

Business prophets, prior to the end of the war, when it appeared likely that munition contracts would soon cease, predicted that the final chapter of Canada's industrial war effort would end in a blur of disaster.

But they reckoned without taking into account the same initiative and determination that made Canada one of the greatest shell and fuse shops in the world.

And to-day almost every industrial centre of the Dominion has a thriving industry of some sort—a monument to the desire of our industrial leaders to hold their war-work organizations intact and make them count in peace-time development.

It is nonsense to claim that everything is rosy, and it is unfair to indulge in booster talk to the exclusion of recognizing adverse facts that are very squarely facing us.

The much-heralded adjustment of values is upon us. So far machinery, iron, steel and metals in general, from their essential character, have not figured in any spectacular deflation. There have been forces and influences at work all over the world that, so far at least, have prevented supply and demand becoming equalized, and it is due to these forces that deflation has not touched this field.

The process, sooner or later, will become operative. The first step may be the elimination of the premium market, and there will be no sorrow when this happens. It is highly desirable that we arrive, as soon as possible, consistent with safety, at a point where a definite market price can be quoted for staple lines of raw, semi-finished or finished material.

The organization that is giving a real service, that is performing some useful function, that is not taking out of our industrial pocket something for which it is not giving value, has nothing to fear.

The hanger-on, the gouger and the schemer —whether in the form of an individual or a company—is apt to be in for a rough passage, and a few funerals in this category will be as tonics and restoratives to legitimate effort.

To men in the control of industry, and to men who wield influence in the workshops, there is a plain message—Don't rock the boat.

Now that's an old phrase and a plain one and its meaning is clear.

Many countries, not yet recovered from the ravages of war, not yet fitted to meet new conditions, are looking to Canada for supplies.

Canada has her chance NOW—not some other time—but NOW—to make advances in a few years that under ordinary circumstances she could not make in two decades.

The man who can do damage now is the labor leader who teaches the suicidal doctrine that by keeping down production it is possible to create more employment. It is a vicious belief, and prevents any man from giving wholehearted and loyal service to his employer.

Not far removed from him is the employer— fortunately his kind is disappearing—who runs his business for the sole object of paying dividends. His responsibility to his community, to his employees and to the furtherance of decent business practices, is conveniently ignored.

In our post-war experience we have had an avalanche of strikes, and we should by now be at the stage where we have learned that the strike hurts the strikers, especially where it is used in defiance of signed and accepted contracts.

Common sense and a little honest sweat can work wonders for this country in the coming months.

Canadian *Exporter* Has His Own Problems

STRIKES, NEW TARIFFS, FREIGHT INCREASES and EXCHANGE

CANADIAN exporters of agricultural implements have had their good spots and their bad ones. Business has been carried on and developed in many cases in spite of conditions rather than with their assistance. Strikes, high freight and ocean rates, lack of certainty in shipping, exchange rates, changing tariffs, and the uncertainty of the war-torn countries of Europe have all been on hand to see to it that Canadian firms had plenty of things to attend to beside making implements and putting them on the cars at the factory.

Massey-Harris Co. have a world-wide organization, and export business is bread and butter to them. "Things are different now to what they were in days before the war," stated Mr. J. N. Shenstone, first vice-president of the company, to this paper. "Then it was simply a case of turning the material out and putting shipping tags on it, and away it went, and the chances always were that it would reach its destination at the earliest possible time. Right now, and starting closest to home, the dock strike in New York is one of the worst things with which we have to contend. We have to go at it about the same as we did when the war was on, viz., advise the people down there that we have certain shipments, and wait until they tell us that they can take them on. In many cases they have to be sent by rail to Portland, Boston, or some other port on that coast, or even as far as Montreal."

Troubles With Material

While the making of agricultural machinery is not exactly a part of the export, it is interesting to note that there is trouble enough in getting supplies of raw material from United States points. Much of the steel that it brought in now is not, as usual, in multiples. It usually comes, say in even lengths of ten feet, so that a quantity of it can be cut up into five two-foot lengths. The material coming now is odd lengths, and often bent, so that operations are interfered with and there is more than usual waste. The manner in which the supplies are coming in makes it almost impossible to get the shops up to speed on any one line, as they have to keep switching from one to another as there is not sufficient raw material to operate on them all at once. Like many another Canadian company, the Massey-Harris people have their own men tracing cars in many United States districts, trying to get in supplies of coal and steel. According to reports from the company it makes a lot of difference who goes after the cars. One of the Massey-Harris men is an ex-railroader, and he knows the business thoroughly, and understands the handling of men. He has that thing most necessary, a personality, and he can get results where others fail

down. He has done a big work in keeping cars moving toward the Massey-Harris works.

The Exchange Problem

"Undoubtedly exchange has been the biggest problem that we have had to deal with this year," stated Mr. Shenstone. "It is the cause of an unending stream of correspondence. "In spite of it, we are doing a big French business. France is coming back in a remarkable way. Our French office will buy $100,000 worth of francs and send them here. We deal direct with the dealers in France, and sell the machines at the difference between the value of the Canadian money and the French currency. There is no other way in which we could carry on there. It really has meant that our prices have fluctuated with the changes of exchange, which have been frequent as you know and it has kept us busy following the changes in money values as between Canada and France.

Nothing In Germany

"There is no business to be done with Germany yet in agricultural implements as the value of the German mark is still too low. Foreign offices believe that there will be a recovery in Germany if the country holds together."

One of the correspondents of the Massey-Harris Company who has recently finished a survey of the German situation for the firm, has submitted a long report. Among other things he draws attention to the human element in the situation:

Consideration as to what should be done was, of course, wrapped up with judgment as to the future. There was great need, he pointed out, in view of the heavy discount against German money, of the German people trading within their own limits and employing their own labor and materials as much as possible. They could not afford to do business with the outside world at the existing exchange, and if the political misfortunes of other countries should overtake Germany, it would be difficult to say what might happen. There was always the possibility that the country might break up into various divisions, or that revolt might occur. "Dull despair has settled over every class. They see their lives now chained to a servitude of taxation. . . . Whether this despair will give way to untoward action in an endeavor to avoid the heaviness of the burden, is a matter for speculation."

According to information reaching the Massey-Harris company, there would not be much business done with Germany in any line until such time as the mark had reached five or six cents. Some of those who had studied

the field were rather optimistic in regard to this, and were certain that it would not be long before the mark would have a standing of at least ten cents. In this same connection, it might also be remarked that there are students of the situation abroad who predict that the British pound will be at a premium inside of six months.

An Ingenious Representative

The Massey-Harris Company have a representative in charge of their German business who has displayed no small amount of ability in holding the business of the company during the war. Of course all communication with Germany was cut off when war started, and there were no parts shipped for repairs. When this service was stopped, the German agent took parts of machines he had in stock, which were called for most frequently, and had patterns made, from which castings were turned out. In this way he was able to go ahead and give service to the users of Massey-Harris machines, and in the end had built up a business on which he was actually making money.

The Massey-Harris company now own their German head office, the building being worth about $275,000, but the purchase price at the exchange working out something like $47,000. Other reports on the situation make it plain that it will be necessary for Germany to import agricultural machinery for some time, as the country has neither the raw material nor the facilities for turning out sufficient to keep up the production of the soil.

Russia Still a Closed Market

Nothing has been done to open up trade with Russia, nor does it seem that this will be done in the very near future. The Massey-Harris company had large premises in Moscow, but of course their representatives left there some time ago, and it is not known what has become of the property. Some propositions have been put forward to the London office to send shipments of implements to Russia, but they are usually somewhat involved, and after some investigation, are dropped as too dangerous. Many of the firms that are trying to secure implements to sell to the Russian people want to do so on the barter system —taking raw materials and Russian goods in exchange. "When the banks come to the point where they will loan money on that," stated Mr. Shenstone, "we will probably resume the business, but not before."

The Australian Tariff Changes

The tariff wall which the people of Australia intend to build at the first of the year will be a serious problem for the Canadian firms at present shipping into that territory. Of course many of them are sending large stocks in there ahead of the coming into effect of the tariff, so that it may be some time before it actually begins to influence their shipments.

Massey-Harris have not decided just what they will do to deal with the new Australian situation. They hardly see enough business there to warrant starting a branch factory, which is evidently the object of the Australian tariff. Australian trade has been very satisfactory for some time now, and the tariff that is coming into effect is of the real Haman's gallows type. Canada will apparently come under the intermediate or general tariff, but not under the British preference, at least that is the understanding of the Massey-Harris officials.

Reapers and binders, under the new ruling, pay under the British preference £6 10s; intermediate tariff, £9 10s and general tariff £10, or an ad valorem duty of 30%, 40%

or 45%, depending on which will make the highest revenue.

Mowers will pay under the British preference £2 8s; under the intermediate tariff £3 15s, and under the general tariff £4, or an ad valorem of 30%, 40% or 45 %, depending on which ruling will return the greater revenue to the government.

Cream separators, free under the British preference; 5% under the intermediate and 10% under the general.

The Australian government go the whole distance, however. They are giving this protection, and they are also setting the price at which machinery under it shall be sold to the purchaser, and the wages which shall be paid to the mechanics making it.

Other items in the Australian tariff indicate that it is the desire to build up a steel and iron industry. Pig iron is to be taxed under the British preference, 20s per ton, under the intermediate 30s per ton and under the general 40s per ton. Ingots, blooms, slabs, billets, puddled bars less finished than iron or steel bars, but more advanced than pig iron, except castings per ton, British preference 32s, intermediate 52s and general 65s.

Bar, rod, angle, tee, per ton 44s, 65s and 80s.

Plate and sheet up to and including 1-16, under British preference, free; intermediate, 5%; general, 10%; on and after January 1, 1922, British preference, 65s; intermediate, 82/6, and general 100s.

Wire, ton, British preference 52s, intermediate 72/6, and general 90s.

As a general thing machinery will be taxed, British preference 27½%, intermediate 35%, and general 40%, while automobiles come under the following, 35%, 45% and 55%.

The Recovery of Spain

In the opinion of Mr. Shenstone, Spain is developing into a good market for the Canadian firms with anything to sell to the trade there. "In the last five years there has been a great change in that country. Formerly we did not regard it as a good customer. We are shipping implements there now, and the trade is developing in a way that has been a pleasant surprise."

Exchange has been against Italy rather hard, and the uncertain conditions there have not been conducive to attracting trade. Denmark and Sweden are also in the class where exchange and high freight rates are retarding factors.

Speaking of freight rates Mr. Shenstone recalled that there had been rates all the way from 10 to 50 shillings per ton to Australia. Things work in a circle. The ship owners come together and fix rates at a high figure, so high that it induces some other lines to come in at a competing price, until they are taken in, and the operation begins all over again. The rate to Liverpool used to be about 8s to 10s, and is now around 25s. Shippers, though, very often get some measure of relief from the tramp steamers in getting their goods away at a fair price. They may have some space left, and a shipper may have the goods to fill it. It costs these tramps money every day they tie up, so there is often a bargain made at quite a few shillings per ton under the prevailing rate. The tramp fills up her space and gets away, and the shipment gets that much start in the matter of delivery, and there is money saved as well.

With the clearing up of conditions in the war-ridden countries of Europe, Massey-Harris look for a continued development of their export trade.

Handling Export Business

In preparing goods for export there are
certain points one must watch carefully.
They are stated herein.

By J. H. Moore

HOW do you pack your product?
Do you slap some boards around
it, stick a brace here and there,
stencil the proper name and destination, then trust
to luck for the rest of the journey. If
you do, you've got the wrong idea. No
matter what the distance, your product
should be carefully packed. Should the
distance be short, such as local shipment by team or truck, the work entailed is not so great. This also applies
to domestic shipment, as generally a
good, stout openwork crate will suffice
for such purposes, but when you ship
to the export trade, you are up against
an entirely different proposition.

The damages caused by inadequate
packing are already too well known to
require comment. Breakage, rust and
other attendant evils, even to thieving,
generally occur. Many methods have
been adopted to overcome these various
difficulties. and each firm, adopting what
in their opinion is the best practice.
Believing that an exchange of these
methods would be a good thing for all
concerned, we have prepared in this issue various systems adopted. Let us
first consider how the Brown & Sharpe
Mfg. Co., Providence, R. I., prepare their
shipping cases.

Four Important Rules

This concern has four rules to which
they strictly adhere. First, they prevent
breakage by so boxing and bracing as to
protect against, not only ordinary handling, but also against rough usage. Secondly, they see that their product is
thoroughly protected against rust, and
thirdly, they so arrange their packing
as to ensure protection against thieving.
Last, but not least, they are very particular regarding invoicing and marking,
and at all times aim to meet the many
varying requirements of countries to
which their goods are shipped.

As we are particularly interested in
export shipping, we will pass over their
methods of packing for local and domestic shipment and consider only the export trade.

In boxing machinery, there are certain
principles which are important to observe. These could be placed as follows: 1. The platform and skids must be
strong enough to support the entire machine, and the platform should be attached to the skids by using heavy
spikes toed in on a slant both ways
to keep the boards from lifting up when
barring.

2. The size of the box should be such
that when squared up for length, width,
and height, there will be a clearance of
at least half an inch.

3. The machine should be so blocked
in position that if tipped over the weight
will be supported, and all delicate parts
protected. All blocking should be placed
lengthwise of the grain so as to be less
affected by swelling and shrinkage.

4.—The top should be sufficiently
strong, so as to support any crushing
strain which comes on account of the
use of a sling in handling when loading
on vessels, etc.

5: All small parts, such as wrenches,
etc., should be packed in small boxes,
and attached to the main box in a substantial manner.

Some years ago, a Mr. J. F. Monaghan, who was at that time counsel at
Chemnitz, Germany, sent in a report on
forty cases of machine parts shipped
from America. These were badly damaged in transit, and in his report he called
attention to the fact that this serious
breakage had occurred in spite of the
case being of rugged construction. He
commented as folows:,

*"The main fault was that the cases
were much too spacious for their contents. The heavy material was given
the opportunity to flop from side to
side, and hence was subjected to tremendous strain when being handled on
a rough, rolling sea. The more delicate
parts, such as slides, glass plates, and
reflectors, should have been packed in
separate boxes, with a liberal use of
excelsior, or other soft material.
Wrenches, latches, bolts, etc., should
also have been placed in small boxes.*

When assembling a case, make it of
sufficiently heavy material. Remember
that a box containing a valuable machine of great precision is of no more
interest to the laborer or longshoreman
handling it, than would be a box of rough
building material. Some claim if you
mark the case "Handle with Care" that
it is assured of careful treatment. As a
general rule, little if any attention is
paid to such a marking, no matter how

FIG. 5—ONE COPY OF THE PACKING LIST SHOWN GOES WITH
INVOICE AND ANOTHER SIMILAR IS PACKED WITH THE MACHINE.
FIG. 6—ILLUSTRATING HOW THE BOX IS MARKED.

prominently it is displayed, or in what
language.

Use paper between the blocking and
painted surfaces in order to avoid, as
far as possible, bruising of the paint.
Cover all ways or sliding surfaces with
waterproof burlap paper, and wrap all
shafts in paper of similar nature. Protect all holes, if practical, with wooden
plugs, and always place under the top
cover a layer of burlap waterproofed
paper. Of course, no matter how careful you may be, accidents can happen.

but if preventative measures are adopted the damage toll will be reduced to almost nil.

In order to present by photograph an actual example of boxing for foreign shipment, we refer readers to Figs. 1 to 4. These pictures illustrate various steps in the packing of a Brown & Sharpe Universal grinding machine. In this case, the platform and skids are arranged to support the base of machine and other small parts. It will be noted that paper is placed between the base and platform for the purpose already spoken of. The pulleys rest on blocking and are firmly held in position. A wooden plug is passed through the holes in these pulleys, then held by the blocking shown. The countershaft hangers are held down with not only blocking, but lag screws as well.

All loose parts, or parts which might become loose are wired in position, or packed separately. It might be well to add that the cupboard of a machine is a mighty handy place in which to pack small parts. The table is lifted from the ways, which are slushed with thick lubricating grease, as is also all interior finished mechanism. The outside is slushed with a good heavy compound. Pine strips are inserted between the table and its ways, and on these the table rests. This arrangement separates the teeth of the table rack and pinion

and ensures that shock in transit will not break the teeth.

At Fig. 2 in the foreground can be seen the box in which the grinding wheels are packed in sawdust. The cover of this box is not nailed, but screwed on, to reduce the danger of breakage. This box is so located that it can be readily removed without disturbing other parts of the packing in case the machine is ordered without grinding wheels. This is a provision well worth observing in any type of packing. The cone pulley shown at the right has projecting hubs, and can thus be held in position by the hubs fitting into holes in the blocking.

Fig. 3 illustrates four men engaged at further stages of the case. One of these men is always a gang boss, the others working with him. In the B. & S. plant, the packing of all regular machines is standardized, and after once having proved satisfactory, a card record is made of all the details of the boxing, and a stick marked with important dimensions is also kept on the job. This guarantees that should a new gang of men, who have not boxed the particular machine before, but are familiar with the B. & S. method of boxing, can, from instructions at hand, box the machine without difficulty.

The remaining photo, Fig. 4. illustrates the use of metal box strapping bands

to further strengthen after boxing, also the method of using a sling in handling the boxed machine for storage or shipment.

A very important point to watch before boxing is that the machine be handled in a way to avoid danger, both to the workman and to the machine. Always look out for the possibility of chafing or cutting of the rope, and where necessary insert a block of wood to overcome any such tendency. Should you sling a rope round a shaft or some finished part, protect that portion with a cloth or some such article. Watch also the slipping tendency, for the slipping of a rope has caused many an accident.

We have already stated the necessity of slushing the inside finished parts will thick lubricating grease, and the outside parts with a heavy compound. In the plant we are speaking of, this compound was selected by exposing to the weather during long periods of time, various samples coated with different preparations, and, of course, the preparation showing the best results was adopted.

Several kinds of specially prepared papers are used in packing, as it has been found that newspaper, or in fact any ordinary white paper, even when used over the slushed surfaces, encourages rust. A special paraffin paper,

FIGS. 1 TO 4—ILLUSTRATING VARIOUS STEPS IN THE PACKING OF A UNIVERSAL GRINDING MACHINE.

coated on both sides, has been found very satisfactory for this purpose. The **waterproofing** paper used inside the cover of the boxes is made of prepared tar with burlap rolled in. This style paper is also put on as an additional covering on the entire box should there be any possibility of its being exposed for a long period to the weather. This condition especially applies to large cases which cannot be placed in box cars, and where, due to congestion of shipping, boxes may stand in exposed places for long periods. For small tools, they use what is known as the "Puritan" brand of oil to protect all finished surfaces. This serves both the purpose of a lubricant, and a rust preventative. For general packing of small tools, a heavy waterproofed paper (15 lb. weight) is used. It is not so heavy or stiff as the burlap paper.

A very interesting example of unexpected exposure with no serious results is given by this concern. This was a case of two vertical millers that had been shipped to go to Russia, and were returned over a year afterwards having been in continuous transit. The cases showed the effects of the weathering and rough handling, but on unpacking, the machines were found to be in perfect shape for reshipment.

Other incidents have occurred where cases have fallen into the salt water, and been under water for several days with no ill effects. For some foreign shipments, especially for small tools, a complete zinc lining is placed in the box. The box, when enclosed and soldered, make a complete protection against rust from outside causes.

A difference of opinion exists as to whether in some cases waterproofing covering may be a detriment rather than a help in preventing rust, because some claim that when dampness once penetrates a case, if it is covered, it holds the dampness longer, and therefore does more damage. This, of course, all depends on how the case is protected, and

the B. & S. people are of the strong belief that the more thorough the provision for keeping out the dampness, the better the condition in which the machine will arrive.

Unpacking Should be Watched

Is is a peculiar fact that the more thoroughly a machine is packed, the more care is needed in unpacking, to see that none of the parts are lost. One illustration was cited where parties receiving a machine wrote back and stated certain parts mentioned on the packing list were missing. After considerable correspondence the parties receiving the machine decided to look at the portions of the case, to make dead sure the missing parts were not there. A search in the boiler room, where the boxing had been burned, disclosed the remnants of the missing parts which had been through the fire. The portions under discussion were long shafts packed in a long box, and which were mistaken for blocking timber.

In another case the foreman of the packing department travelled quite a distance to prove that the parts which were claimed to be missing had actually been sent. It was treated rather as a joke that he should imagine he could discover the parts after all the search that had been made, but he did just the same, among some packing from another machine.

It is in such cases that the value of a packing list becomes apparent. By having a list such as shown at Fig. 5, signed by the inspector, and verified by the man who boxed the machine, there should be little danger of parts being overlooked.

In the case of very small parts, it is sometimes the practice to make the box larger than the size of the parts in order to draw attention to them. The box has even been painted red in some cases in order to attract attention. The more a machine is dismantled in shipping, the greater the danger of parts being lost, or of difficulty in assembling.

Or in such cases special instructions, in addition to the regular printed matter, should be supplied.

The concern spoken of follow out certain rules for marking their boxes or crates. Fig. 4 and Fig. 6 are good examples of this. The object of marking on both ends and sides is so that whenever boxes are stored, either at the plant, or in transit, whichever side is exposed, always designates its contents. Special markings in foreign languages are very seldom called for, the English weights, etc., being easily understood. Much trouble in shipping is caused by putting the marking on tags, or cards, instead of painting directly on the crate. Never mark a tag for quickness' sake, for the minute or so saved, may mean weeks' delay. Another careless habit is the leaving of some old marking, if using second-hand lumber. The motto of careful shippers should be: "Make sure you are right, then make sure again."

Always get proper information as to the conditions and requirements of the country to which you are shipping. By doing so you will avoid both expense and delay. If at all possible keep the dimension of your case so that it will enter the door of a box car. In shipping small tools some style of manifold form will be found convenient. A good plan is to have six copies, one for the office, one for the stock room, one to be packed and shipped with the goods, one for a numerical record of the shipment, one for an alphabetical record, and one for the correspondence department. Different colored inks can be used for these different sheets.

The main point to be remembered in packing of goods for export is the long and rough journey ahead of them. If you are going a long journey you would make all due preparations to arrive at your destination in as good condition as your left. How much more so then should you see to it that your product, which is after all a part of you, should be so packed as to arrive in perfect condition.

Using All the Shipping Space You Pay For

D URING the past few years much has been written about exporting. Especially has the manner in which Canadian and United States manufacturers box their merchandise for export shipment come in for reams upon reams of adverse criticism. No doubt some of this criticism is quite justifiable, but we believe poor packing is the exception rather than the rule among the North American manufacturers, who are making a real effort to gain a foothold in foreign markets. So comments Mr. W. W. Clark, export manager of the Hart-Parr Company of Charles City, Iowa, in answer to our request for information regarding their system of packing their well-known line of tractors. He proceeds as follows:

THE CRATE AS IT APPEARS BEFORE THE COVERING GOES ON.

A few months ago the writer enjoyed the privilege of standing on a foreign

wharf and of watching a cargo of American merchandise unloaded. Case after

care of merchandise was swung from the hold by the mammoth winches, to be deposited on the pier. Hundreds of manufacturers in Canada and the United States were represented, and there were

NOTE THE COMPACT AND SOLID LOOK OF THE COMPLETE CASE.

all kinds of articles ranging from condensed milk to heavy plantation machinery.

During the several hours that this work was watched, just two breakages occurred. One flimsy case of hardware buckled in the middle and pots and pans rolled into the ocean with a great splash. The second breakage was of a number of boxes of a North American medicinal water, which was put up in boxes jointed at the end but not nailed. They were not sturdy enough to stand the strain put on them in the swing and as a result the fish in the bay received a dose large enough to cure all their ailments for the next fish generation.

We might hold these two examples of careless packing up for our readers as a testimonial to the glaring lack of understanding on the part of our manufacturers of what is necessary in the way of packing service to overseas customers. Most of our contemporaries are doing that. But we don't think that such a policy is right or just. We were very much more impressed by the evidence that ninety-nine per cent. of our manufacturers do know how to pack and that their goods to arrive at foreign ports in just as good condition as the day they left the factory.

The problem of the hardware or condensed milk manufacturer is tame compared with that of the manufacturers who are introducing into overseas markets the agricultural machinery which has made our country famous. The tractor manufacturers especially have had to solve a big problem in the matter of crating their machines in such a way that the minimum of space will be used and that at the same time tractors may arrive at the foreign destination ready to go into the field with no parts broken or missing.

For the benefit of readers who may not be familiar with ocean freight tariffs, we will explain that all merchandise pays either by weight or measurement at the option of the transportation company. The tariff on tractors to a foreign port is so much per ton or so much per cubic foot and in practically all cases the cubic foot rate is the one which is applied. Under these circumstances it behooves the manufacturer to protect his foreign customers in the matter of freights by reducing the number of cubic feet in the case to the smallest possible figure.

As an example of the systematic carefulness tractor manufacturers in general are showing in connection with their export business we will cite our own case, as naturally we are more familiar with it, than others. During the past year we have shipped several hundred of our model "30" tractors abroad and among all those tractors which have been subjected to all the hazards of five oceans, only two small castings have been reported broken.

This excellent record is not the result of a run of good luck, but a result of the same specialized effort and rigid inspection that has made our tractors one of the leaders in the domestic markets. Before a tractor was ever exported, we tried out numerous styles of crates. To gain compactness and therefore economize on freight for our customers we found it desirable to partially dismantle the machine. That we have succeeded in utilizing every available inch of space in our case is shown by the photographs reproduced. The customer is not paying for any waste space on a case like this.

Space reduced to a minimum, the problem of absolute security had next to be solved. A study of the photographs showing the design of the case will be well worth while. Note the double sill of two by fours on which the tractor rests, the heavy reinforcements in the construction of the frame, the corner straps of steel. In the case, every part is securely bolted, wired, or pinned to the rugged case floor. Nothing can roll around. No one part may chafe against another and before the case is sealed, a waterproof paper is put inside the tongue and grooved lumber siding so that the tractor reaches its buyer just as bright and shiny as the day it left the factory.

Another of the photographs accompanying the article illustrates our busi-

NOTE HOW EVERY CUBIC INCH OF SPACE IS UTILIZED IN THIS ARRANGEMENT.

EACH TRACTOR IS DISASSEMBLED AND PACKED IN A SEPARATE STALL.

ness-like methods followed out in boxing our tractors. A portion of our plant is given over entirely to the export boxing department. As the tractors destined to be shipped to overseas territory leave the paint shop a giant crane swings them into place in the boxing department. Each tractor to be boxed is set in a separate stall. There are eight of these stalls, each partitioned off from the other so that when the parts are taken off the tractor preparatory to being put into the case, there is no chance for them to become lost or mixed with parts from another machine.

In this department, as in all others, we follow our theory of massed · unit production. One group of men go from stall to stall dismantling the tractors. Close on their heels follow the packers who place the dismantled machine on the floor of the case, leaving it in the condition shown in photograph. The result of the same packing specialists doing all of this work, is that every case is packed in the same manner, and because these men do nothing but pack export cases, not only can they accomplish a great deal more work in a given time, but, what perhaps is more important, their long experience in this class of work has taught them how to get every piece in the place which the experiments of the company have proven

to be most secure. The carpenters next set up the sturdy framework, envelop the machine in waterproof paper and nail on the sides. Finally the marker, with instructions in hand, stencils each case with the proper shipping marks, thus forming the last link in the long chain of workers, each of whom does his little bit toward preparing the case in the proper manner.

Do not think that this system of packing is the result of guesswork on our part or that we were lucky in hitting on a good design. We have spent a good many dollars in experimenting and knowingly breaking up valuable machinery before we were satisfied that our export packing was as close to perfection as it could be made. Case after case was packed with a tractor. Not until a method of packing was found which would permit the tractor being picked up by a crane and dropped several feet, rolled over several times side-wise and end for end, in short, submitted to worse strains than any tractor could normally be expected to undergo in ocean shipment—not until the crated tractor could pass through such a test and come out unscathed were we satisfied as to overseas packing.

We believe our case as we have cited it is far more typical of the service foreign buyers are receiving from North American manufacturers than are the isolated cases of poor packing concerning which a woeful howl has echoed and re-echoed undiminished through journalistic pages for the past decade. Let's quit giving our foreign competitors sales arguments to use against us. If there are still manufacturers who do not know how to pack, let's try to teach them by example rather than by scathing censure which casts a reflection on the other ninety-nine per cent. who are blameless. In short, when it comes to North American export packing, let's throw away our hammer and buy a horn.

Applying System to the Packing of Millers

"IN preparing for export shipment, just what procedure do you adopt?" This was the question we put to the Ford-Smith Machine Co., Hamilton, Canada, and following is the gist of the information we received. It might be well to add that this concern ship their millers and other products to India, Australia, South Africa, Belgium, England, Russia, Japan, and other countries, so that their method of packing and shipping is worth while studying.

The big secret in making export business a success, said this concern, is to thoroughly understand what you are undertaking. Every country has its own peculiar requirements as to methods of shipping, and these should be thoroughly studied. Some people run away with the idea that the export business is chuck full of red tape and technicalities. This is not the case, for if you proceed in the correct manner, the handling of export trade is just as easy as domestic business.

In our particular case, we make it a point to handle export business in one way only, in as far as payment is concerned. We have an arrangement whereby we receive payment as soon as our goods leave the factory. Of course, readers will understand that we simply present the draft to the bank and they pay on sight.

Suppose we take an imaginary case. We will suppose that someone in South Africa has placed an order for 24 millers and 36 grinders, but has placed this order through our New York agent. What happens? He notifies us as to his receiving an order for such and such a quantity of machines, and requests that we wire him when we can ship these goods. We do so, and also mail him a pro-forma invoice confirming our wire. This invoice is simply

a verification of our wire, and shows when we expect and hope to ship the machines.

As soon as he receives our wire, or pro-forma invoice, as the case may be, and taking it for granted the same is satisfactory, he arranges for what is known as a G.O.C. permit from the United States Traffic Control Commission at 59 Broadway, New York. This commission handles and controls all traffic through the States at the present time, and anyone shipping goods to other countries from United States ports must have such a permit.

The agent having secured the permit mails the same to us with all necessary shipping instructions. It might be well to add that this permit generally allows a latitude of some two weeks from approximated shipping date, so that you are by no means tied down to a specified date.

On the back of the official order, which is sent at the same time as the G.O.C. permit, there is usually printed all necessary shipping instructions. To give a concrete example we refer the reader to the following instructions, which were printed on the back of such an order:

Instructions

No. 1 Orders. Acknowledge receipt of order by filling out and returning the acknowledgment card we send attached.

Invoicing

No. 2. Invoices should show order number, mark, gross and net weights, in lbs. and kilos (kilo—2.20 lbs.), contents and measurements of each package. Also kind of package used, whether crate, case or bundle. Send invoices to us at latest on the day following shipment of goods from factory.

No. 3. State net weight (and size of electros), of printed matter separately and case in which packed.

Explanation of Weights

No. 4. Net weight is the weight of the goods alone without any packing whatsoever.

Legal weight is the weight of the goods, plus the weight of interior packages, such as glass, tin or cardboard receptacles, in which the goods are contained; but not including the outside protecting case.

Example: The legal weight in any article in glass, packed in boxes, would be the weight of the article plus the weight of the bottle, but excluding the outside case. When there is no interior packing, the legal weight is the same as the net weight.

Gross weight is the combined weight of the goods and all interior and exterior packing.

Packing

No. 5. Each order must be packed separately. Use strong substantial cases and do not make any package excessively heavy. Iron strap all cases. Pack in as small a space as possible. The condition in which the goods arrive governs the customer's judgment of the goods, and it depends entirely on the packing.

No. 6.—Goods, cartons and packing cases must be marked "Made in Canada."

Shipping

No. 7. · Ship promptly. A day's delay at the factory may mean that we miss a steamer, and thus the delay here may be a month or more. Slow ship-ment means limited consumption of the goods; besides creating a bad impression on the buyer as to your facilities for executing orders.

No. 8.—Ship via fastest freight line.

No. 9.—Prepay freight. Show export marks and the words, "For Export" on B-L. Send R.R.B.L. to us the day after the goods are shipped, at latest. Do not fail to show on B-L. our address, 44 Whitehall St., New York City.

Marking

No. 10. Mark each package with stencil—on four adjacent sides. Marks must be clear. Our name and address should appear on a tag tacked on the case. No other marks should appear on cases except the export mark we give you and the gross and net weights in lbs. and kilos.

Failure to carry out instructions will oblige us to hold you responsible for consequences.

PASS THESE INSTRUCTIONS TO THE PROPER DEPARTMENTS

In this case quoted the instructions are very simple, but, of course, in other cases the instructions are more varied. It depends entirely on the nature of the goods, and to what country they are being shipped. For example, any goods shipped to South Africa must have the

THIS ILLUSTRATES HOW THE MILLER LOOKS FROM THE SIDE.

AS IT LOOKS FROM THE END.

price of the machine marked on the invoice, also the cost of the case, and as noted in the previous instructions, catalogues, electros, etc., must be stated clearly and separately from the rest of the contents.

We make an unfailing practice of stencilling the cases on all four sides, that is, the two sides, and the ends. On every order we get some sort of a simple shipping mark to go by, and in one particular case this took the form of the letters A D S inside of a diamond shaped frame. This shipping mark is easily remembered, and the loaders at the boat can quickly pick it out. Another simple mark we have had was simply the name McLeod, 56, Calcutta. From these two cases readers can no doubt grasp the principle.

On the upper left-hand side of the case we stencil the gross weight, the net weight, and the case number. At the lower left-hand side we stencil the size of crate in cub. ft., and at the lower right-hand side we state the destination and the fact that the goods are Canadian products.

Leaving the actual packing of the product to the last, let us suppose we have completed our order, packed the goods, and notified the railroad to send their motor truck for the same. We now make out certified invoices . and send them as follows: We forward six certified invoices and packing lists to customers New York agent. We send two copies to the transportation company, also two copies of the custom entry papers. The bills of lading are, of course, filled out in the usual manner, with the exception that more are made

out than in the case of domestic shipment.

Away goes the goods to the New York port, where the customers shipping agent attends to them and sees they are placed on the boat at the proper time. We know everything is all right as soon as the goods leave our plant. The main point is to state on your proforma invoice a shipping date that you can actually meet. Ship as near that date as possible and avoid confusion and congestion. We believe a great deal of success on our part is due to watching these points.

Regarding Packing

Of course, we know that in all respect to watching the shipping instructions, one could watch them for all they were worth, but their goods would never reach their destination if not properly packed. Packing is the stumbling block in export shipping. Some believe any old thing is good enough, but this is an erroneous impression. An export case must be perfect, for it stands, or should stand unexpected strains. A workman at the docks does not care what is in your case. He knows he has to get it into the hold quickly, and quickly it goes. If you have not built your case sufficiently strong, then it's up to you.

It pays to spend a bit of thought on how to pack your product. Personally, we devoted considerable study to the matter, with the result that we have our cases standardized. Fig. 1 illustrates how our No. 2 miller looks in its case before the outside covering goes on. Note how well every part is braced. Nothing can rock or chafe on another

part with the result that the machine arrives in perfect condition. We paint all polished and machined surfaces with a 'special rust preventative, and, of course, use waterproof paper inside the case before the outside boards go on.

We would suggest that readers do not merely casually glance at this photo, but study all the details, for it is really the small details that count. Fig. 2 illustrates how the machine appears from the end. It will be noted that provision has been made for hoisting hook rings to go through the case and we do this on all our miller cases. To illustrate to what an extent we have gone in standardizing our cases, we refer you to Fig. 3, which gives the details of our No. 2 miller case. Our material list states all particulars as to size, etc., and the detailed view illustrates clearly the construction, so that even a new worker can be sure of preparing a perfect case. We always place band irons around the outside of our export cases, and we cannot too strongly emphasize the importance of proper packing.

Other Features to Watch

There are numerous other points to be watched, but it is, after all, a case of practice makes perfect. The more experience one has in export business, the more simple it becomes. For example, the usual form of f.o.b., New York, is not good practice in stating price or export business. F.O.B. New York can mean both f.o.b. the railroad cars to New York, or transportation to the hold of the ship. Always be specific, otherwise you will lose some

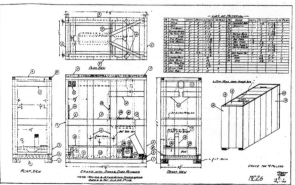

A DETAILED DRAWING OF THE CASE SHOWING ALL PARTICULARS.

money. We usually ship our product f.o.b., railroad cars to point of shipment. Another case is the C.I.F. mark, bourne.

We have shipped goods C.I.F. Melbourne. This, of course, means that we covered the cost of both insurance and freight to that point.

In making up a shipping package remember that the steamship companies have the privilege of charging you by weight or by cubic measure. 40 cubic feet equals 2,240 lbs., as far as the shipping companies are concerned, so that it behooves you to pay attention to this

point, for by careful packing you can save yourself some more money.

Never guess that you are right. Be sure. The Canadian Trade Commission at Ottawa get out a handbook for exporters, and valuable information is contained in these books, which are issued regularly. As they state themselves, "Exporting firms are advised to communicate with the Canadian Trade Commission regarding any matters on which they require specific information."

The different express companies also issue various books of shipping suggestions to the different countries, and they

are well worth securing. The Canadian Manufacturing Association issue bulletins of interest, and there is no reason why those desiring to cater to export business cannot do so without any difficulty. After all is said and done, your forwarding agent is really the one who arranges the shipping route. Your best plan is to place your forwarding business in the hands of a reliable firm, forget about that end of the business, and manufacture your goods as speedily and as perfectly as possible. Remember that it depends on how you treat your export business, how it will, in turn, treat you.

Export Shipping is a Standardized Science

WHEN you start to consider an export system capable of being applied to all the plants under the direction of the International Harvester Co., Ltd., then you enter upon a method which has been standardized only after long years of experience. The system as adopted by the Hamilton plant, and which we will describe herein is of course the same as adopted in all other branch plants. The Chicago office, which is the headquarters of the company mentioned, is the pivot point of the complete system.

As they ship to all parts of the world, they have at each plant various traffic experts, men who thoroughly understand their business, and at the Hamilton plant they have a self-contained traffic dept., which takes care of all inland lake shipping. As we are however chiefly interested in the export system, we will ignore the domestic method altogether.

The Question of Packing

In practically every case the weight of their products is so small, that the shipping companies charge them by cubic measurement, and not by weight. This necessitates great care in the arrangement of packing, as every cubic inch wasted means added expense. They ex-

port a great amount of material every year, and the total loss by careless packing would be very great indeed were not this point so carefully watched.

FIG. 2—THE STYLE OF RECORD AND SHIPPING INSTRUCTION FORM USED.

Standard styles of cases, crates, etc are made, depending upon the article or machine being packed. This packing, of course, is handled by the packing dept.

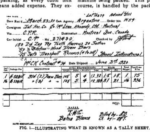

FIG. 1—ILLUSTRATING WHAT IS KNOWN AS A TALLY SHEET.

only. Band iron is used to protect export cases from damage, and inch lumber is as a general rule used for the outside casing, although on plow and some such

heavy machines, stronger lumber is used. In all cases, however, the lumber is made a little over the required strength.

To thoroughly understand the procedure adopted by this concern, we will take an imaginary case to illustrate the system. As stated previously, Chicago is the head office, and all orders of course come from there.

We will suppose we are following an order which has just been issued from the Chicago office to the Hamilton office. This order on reaching Hamilton arrives at the proper clerk on this work who enters the order. This might read as follows: "Make 500 eastern drills, type so and so." No destination is stated, nor is any customers name mentioned. The number of machines to be manufactured, with whatever special attachments may be ordered, is all that is given. This order is what is termed a manufacturing order.

Specification lists, which are already prepared for all machines made by this concern, are issued by the experimental department, and these are consulted.

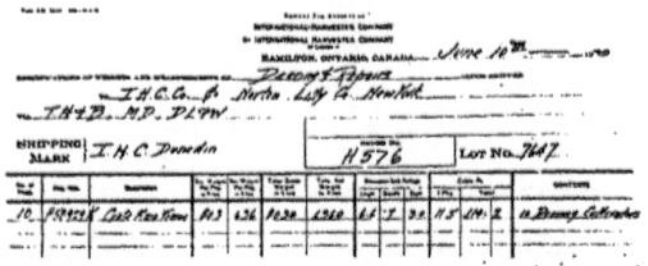

FIG. 3—SPECIFICATION AND MEASUREMENT SHEET AS ADOPTED BY THIS FIRM

These lists not only show what the machine consists of, but what parts are peculiar to the different countries. This point can be readily understood as agricultural machinery can differ considerably to suit the various soils.

An order for the parts as stated on these lists is now issued on the factory. The various departments get busy and eventually the machines are completed. They now go to the packing department, but before this dept. can pack the goods they must receive from the planning dept. as to how the goods must be packed. Special packing lists are made out for this purpose which the packers must follow. In previous issues we illustrated some of the packing systems as adopted by this company.

Now comes a shipping order from Chicago. On this order is placed the country to which the goods are to be shipped. The factory order, and the foreign order numbers are also placed on this shipping instruction order. The receiving of this order is entered in a record book for the purpose, and the order is passed over to a ledger keeper who is acquainted with progress of the work.

The tally card as shown at Fig. 1, is now partially filled out. All the upper portions of this card possible are filled out and the card is sent to the inspection department to be checked.

Advice notes are next issued, and copies of these are sent to the different parties interested. In this case, six are sent out, one to their own traffic manager, one to their agent in New York, one to the foreign sales dept., and three to the traffic dept. in Chicago. This procedure would differ of course in other plants, so that six is by no means a standard number to use.

The ledger keeper now looks up to see if the goods are packed, and if so the tally cards as shown at Fig. 1 are sent up to the shipping dept. where they are filled out as the cars are loaded.

The bills of lading are next in order. Five copies are made out, the first, the original, the second, third and fourth for company record purposes, and the fifth for the railroad for custom purposes only. As the standard bill of lading is used there is no need to reproduce this already well-known form of bill.

The record of shipping instructions form as shown at Fig. 2, is the style used by this concern. For illustrative purposes we have filled in certain sections of this form., Four copies are issued, one for their files, one to Chicago office, one to foreign sales dept. of the same office, and the remaining one to the agent at New York, or whatever port it may be.

The invoice form as used by this concern is of the usual design, so no mention need be made.

The customs forms are also of standard type, so we will take it for granted that readers are familiar with this procedure. There are other numerous forms which are used but are only peculiar to this concern, so in order to avoid confusion we will ignore them altogether.

A very interesting sheet is that known as the specification and measurement sheet. This is illustrated at Fig. 3. For the purpose of example we have filled out a portion of the sheet. A very handy method is employed to find the total cubic ft. of each case, a set of tables known as Nash's Merchant and Seaman's expeditious measure being used. This set of tables gives at one reading the solid cubic contents from the length, breadth, and depth dimensions Baker and Taylor Co., 354 Fourth Avenue, New York, are publishers of this book, and a copy is well worth having.

In a special file in the traffic dept. is kept a list of shipping instructions as standard for the standard for the different countries. For example, refer to Fig. 4. Here we see all particulars as to procedure adopted for shipping goods to Dunedin, New Zealand. Other forms for the various countries are made out in similar style.

To go into the multitudinous methods as practised by this concern would not accomplish anything definite, as ma steps taken by them would not apply to the average case, so that we will leave the matter o the various forms used at this point and conclude by generally summarising up what, in their opinion, constitutes sensible export shipping.

First, leave the shipping arrangements to those who understand such a business. Every firm cannot afford to have their
Continued on page 15

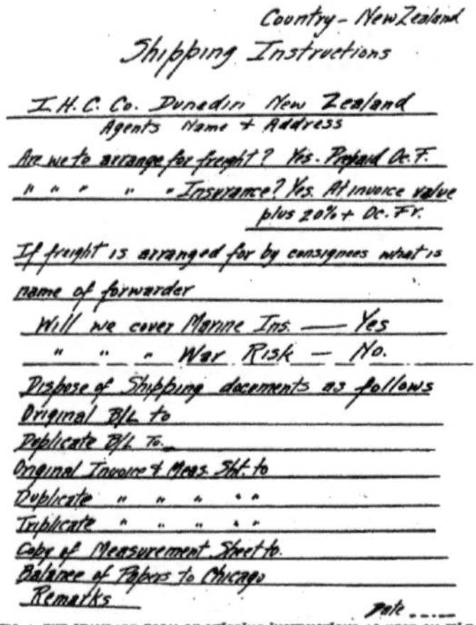

FIG. 4—THE STANDARD FORM OF SHIPPING INSTRUCTIONS AS KEPT ON FILE.

Canada is by no Means a Midget of a Country

We Can Lay Claim to Some of the Largest Engineering Achievements.—We possess the greatest pulp resources, and our industrial Development Has Been Very Rapid Indeed.   Read These Facts.

DID you ever pause to consider what a wonderful country this Canada of ours really is? Do you know that we can lay claim to some of the largest agricultural and engineering achievements in the world These facts should be on everybody's tongue, for every true Canadian should be a booster, especially as we have good reason to be proud of the country in which we live.

For instance, we have the largest grain mills, the largest lift-lock, and the largest and richest nickel mines in the world. Our Quebec bridge has the largest span of any bridge in the world, and at our harbor in Montreal is the largest grain conveying system known.

We also possess the greatest pulp resources of any country, and have one of the largest single canal locks at Sault Ste. Marie. At Port Arthur we have the largest elevator in the world, having a capacity of nearly ten million bushels. There is a huge dam at Bassano, Alberta, that is the largest project of its kind on the American continent. Our mines are also noted for their production, and one of our coal mines at Stallarton, Nova Scotia, has the honor of having shown a coal seam measuring 47 feet. This is said to be the largest ever discovered.

We are no midget of a country as the following figures will show. Our area is actually one-third the area of the British Empire, and we have approximately 13,000 miles of coast line. This distance is almost half the circumference of the earth. We can also lay claim to the largest body of fresh water in the world, that being Lake Superior, which has an area of 31,800 square miles.

Our fishing industry is also well known, for we have some 12,000 miles of coast line fishing, and some 220,000 square miles of fresh water · fishing. Would you believe that we produce over 80·per cent. of the world's output of nickel, and that on a basis of populati! we have the highest ratio of railroad mileage of any country. Take the St. Clair tunnel as an example of efficient engineering. Here is one of the largest tunnels of its kind in the world, the trains being carried through by electric locomotives. One point of particular interest is that this tunnel is actually a passageway between Canada and the United States.

In scenery we take no back seat, for every province is noted for its special style of beauty. We have fishing for those who like to ply the rod, we have mountains for those who like·to climb, and last, but not least, big game for those who care to use the gun.

Our water power facilities are ·simply enormous, and more use of this source of energy is being adapted every day. Our potential water power is 'approximately twice that of the United States, and equal, if not greater, to any other in the world, over 25,000,000 H.P. are available from this source.

Our agricultural record is something of which we can be justly proud We exported alone during 1919, 35,223,983 dollars worth of dairy produce. We have colleges, agricultural and otherwise, that ensure the correct type of Canadian citizens, and our educational system is general needs no championing. In· Ontario alone, we have an approximate forest area of 260,000 square miles, so one can judge our enormous timber limits.

We can also point with pride to our waterway system of transportation. This

DOMINION BUREAU OF STATISTICS

Ottawa, May 6, 1920

1. Summary of the Trade of Canada.

	Twelve months ending March,		
	1918.	1919.	1920.
IMPORTS FOR CONSUMPTION—			
Dutiable goods	$ 542,319,623	$ 526,475,717	$ 693,543,211
Free goods	421,191,056	393,230,085	370,872,958
Total imports, mdse...	$ 963,510,679	$ 919,705,802	$1,064,516,169
Duty collected	161,588,465	158,944,436	187,520,613
EXPORTS—			
Canadian	$1,540,027,788	$1,216,443,806	$1,239,492,006
Foreign , ...	46,142,004	52,321,479	47,166,611
Total exports, mdse.	$1,586,169,792	$1,268,765,285	$1,286,658,709
IMPORTS BY COUNTRIES—			
United Kingdom	$ 61,302,403	$ 73,024,916	$ 126,269,274
Australia	2,356,688	4,963,446	1,569,218
British East Indies	16,434,256	13,223,434	16,293,738
British Guiana	5,716,647	6,747,072	7,412,931
British South Africa	553,392	1,300,259	676,070
British West Indies	10,550,550	9,437,925	12,093,144
Hong Kong	1,505,015	2,321,909	3,194,969
Newfoundland	2,947,527	3,096,834	2,139,614
New Zealand	3,735,559	7,855,436	3,415,096
Other British Empire	1,611,037	688,307	1,267,322
Argentine Republic	984,955	1,139,267	3,057,369
Belgium	12,973	6,270	911,407
Brazil	990,777	1,156,232	1,273,768
China	1,236,690	1,954,469	1,291,579
Cuba	1,065,547	3,040,953	17,585,526
France	5,274,053	3,641,244	10,604,357
Greece	20,296	33	700,896
Italy	771,187	505,112	999,040
Japan	12,235,519	13,618,122	13,635,680
Netherlands	1,054,176	495,469	2,222,434
United States	792,804,959	750,199,879	802,101,187
Other foreign countries	13,796,066	30,238,277	35,448,655
EXPORTS BY COUNTRIES—			
United Kingdom	$ 845,480,069	$ 840,750,977	$ 489,151,806
Australia	8,655,635	14,019,629	11,415,623
British East Indies	3,774,475	3,831,741	6,762,259
British Guiana	1,975,223	2,646,160	3,109,381
British South Africa	6,065,658	11,992,299	8,649,726
British West Indies	6,839,563	10,200,582	10,868,693
Hong Kong	1,003,900	995,116	1,343,867
Newfoundland	10,191,564	11,325,618	16,176,443
New Zealand	4,069,623	5,227,509	6,587,908
Other British Empire	1,712,366	3,170,149	7,923,323
Argentine Republic	1,205,142	4,603,130	6,126,417
Belgium	4,909,453	950,318	29,463,855
Brazil	974,385	4,088,584	2,703,488
China	1,954,095	2,856,933	8,659,805
Cuba	4,015,940	5,035,975	6,329,783
France	291,033,576	96,163,142	61,106,938
Greece	4,262	16,902	29,588,984
Italy	3,336,059	13,181,514	16,961,312
Japan	4,861,244	12,245,439	7,732,514
Netherlands	2,462,574	198,365	5,653,218
United States	417,812,807	454,873,170	464,029,914
Other foreign countries	8,651,832	17,129,975	42,349,571

method has been a great help in reducing the cost of transportation.

And what of the industrial developments of this Canada of ours We do not need to point out any specific case, for our plants and their products are known all over the world. In that dangerous period of uncertainty, when blood was flowing freely in Europe, Canada not only sent her boys to help, but her industrial efforts as well. That record stands to-day as something unique in the annals of history. The munition business was something entirely new to

us, yet we turned out the different styles shells, etc., in record time.

We have over 15,000 factories in Ontario alone, and the total imports for this province up to March, 1919 were $470,650,679, while the total exports were $298,270,478. To state the nature of factories throughout Canada would be to name every conceivable product imaginable. Iron and steel, foundry products, heating appartus, agricultural implements, motor vehicles, and in fact every line of industry is represented.

There is a saying that facts and fig-

ures carry more weight than mere words, so lest we should be accused of idle baosting, look at the folliwing items:

In 1879 we done a total trade of $149,489,188; in 1889, $196,300,107; in 1899, $304,227,339; in 1909, $548,139,881; in 1919, $2,185,194,620; and in 1920, year ending March 31st, $2,351,174,886. Note the steady and rapid increase. These figures show that we grew from $35.60 to $261 per capita during the last 41 years. In no department of Canadian industry has development been more rapid than in that of manufacturing, the value of annual production having increased from $71 to $360 per capita from 1881 to 1917.

Export figures show us that in 1920 we exported $1,228,359,325 worth of goods; and in May, 1920, we were $12,-000,000 over the mark of the similar month in 1919.

Everthing in Canada is on the upward trend. Our banks paid up capital and reserve is greater than ever beford. Our crop situation is better than ever, our insurance companies report more insurance than ever before, and all branches of industry are digging their toes in and intend making 1920 a banner year in every way. Water power development in Canada is proceeding very rapidly. According to a recent computation, the water power resources of the British Empire have been placed at 50 to 70 millions horse power. To this total Canada contributes in the neighborhood of 20 million horse power, and continued investigation will no doubt add to the figure. It is supposed that after recent installations are complete the total horse power will reach 3,385,000. We could go on quoting figures indefinitely but, why go further. These figures need no comment, they speak for themselves as to what Canada has been accomplishing, and there is no reason why 1920 should not be a still greater year, a period memorable in the trade annals of our country. If you meet a pessimist, knock him cold with come of these figures. Boost your country for all you are worth and let your sloggan be "All together for a bigger and better Canada."

DOMINION BUREAU OF STATISTICS

Ottawa, May 6, 1920

3. Principal Articles of Canadian Produce Exported from Canada

		Twelve months ending March.		
		1918.	1919.	1920.
Animals, living	$	19,707,242	$ 35,278,269	$ 50,226,158
Butter	Lb.	4,926,154	13,659,157	17,612,505
	$	2,000,467	6,140,864	9,844,259
Cheese	Lb.	169,330,753	152,207,057	126,295,777
	$	36,602,504	35,222,983	36,336,883
Clothing	$	9,702,207	13,426,235	8,928,906
Coal	Ton	1,902,019	1,826,639	2,130,138
	$	8,684,038	10,149,722	13,183,666
Cartridges	$	351,342,138	212,432,531	7,366,733
Other explosives	$	34,997,136	37,506,294	4,875,047
Fish	$	31,782,258	36,392,626	40,597,172
Furs	$	8,199,312	13,737,621	20,921,971
Grain—Oats	Bush.	54,877,882	17,879,783	10,768,872
	$	37,644,293	13,193,527	9,349,455
Wheat	Bush.	150,392,037	41,808,997	77,978,037
	$	366,341,565	96,985,050	185,944,806
Other grain	Bush.	8,139,867	4,706,232	17,139,122
	$	11,027,411	8,457,082	25,511,421
Hides and skins	$	8,928,063	7,790,048	19,765,646
Leather	$	10,986,521	12,437,712	18,057,152
Meats—Bacon and ham	Lb.	207,833,118	124,688,741	223,642,600
	$	60,082,494	40,242,195	70,123,580
Beef	Lb.	86,565,104	127,810,294	110,047,800
	$	13,016,278	26,594,814	19,637,656
Canned meats	Lb.	13,422,624	14,140,717	2,812,706
	$	3,695,384	5,701,310	1,102,842
Pork	Lb.	7,909,803	37,318,106	6,682,300
	$	2,052,192	11,711,024	1,541,570
Other meats	$	1,053,257	1,340,738	3,655,586
Metals—				
Aluminum, ingots, etc.	Cwt.	215,740	202,839	192,069
	$	7,581,838	6,712,053	5,680,871
Asbestos	Ton	141,099	149,244	129,202
	$	5,693,153	9,159,022	8,797,856
Brass, old and scrap	Cwt.	521,108	72,685	91,512
	$	8,083,864	1,148,819	1,217,940
Copper	$	21,876,611	20,991,179	13,879,332
Gold	$	13,688,700	9,302,033	5,974,334
Iron and steel	$	45,810,267	58,854,318	61,912,659
Nickel	Cwt.	830,499	791,644	441,407
	$	9,029,535	11,170,389	9,039,221
Silver	Oz.	21,960,827	19,759,478	12,379,642
	$	18,428,571	19,519,642	14,255,601
Milk and cream	$	5,862,976	7,882,799	10,216,861
Paper—Printing	Cwt.	12,101,865	13,248,542	14,272,513
	$	33,978,347	40,718,021	53,203,792
Other paper	$	3,886,983	8,447,774	10,049,627
Rubber	$	2,911,505	5,629,590	10,069,963
Seeds—Flax	Bush.	6,424,550	1,890,978	1,127,986
	$	19,764,255	7,759,852	5,396,675
Textiles	$	21,103,575	14,819,058	22,063,620
Vegetables	$	19,034,528	12,841,422	11,656,483
Vehicles—Autos	No.	8,447	14,180	24,506
	$	3,807,278	7,303,678	14,883,607
Auto parts	$	1,557,712	1,552,296	3,097,466
Wheat flour	Brl.	9,931,148	9,266,439	8,863,068
	$	95,896,492	99,931,650	94,262,928
Wood—				
Unmanufactured	$	51,829,121	70,487,288	105,236,768
Manufactured—Wood pulp	Cwt.	9,696,704	11,841,656	15,359,582
	$	25,620,892	34,706,771	41,383,482
Other manufactured	$	777,053	1,055,029	3,449,576
Total principal and other articles exported		$1,540,027,788	$1,216,443,806	$1,239,492,098

CHOOSING THE RIGHT BELT GUARD.

There is more to guarding a moving belt than the mere act of placing a guard about it. In fact, many instances are recorded where guards have turned the popular slogan to "Safety Last." Assume, for example, that a pipe rail guard is thrown about a vertical or slanting driving belt from the main shaft on the floor below to a machine on the floor above. If close to the belt the guard may be the means of a workman's hurting himself.

This is easily imaginable should the employee lean against the guard on the floor through which the moving belt passes. He is quite likely to have some part of his clothing drawn into the narrow slit in the floor by air suction, or by the belt lacing, and perhaps even by the belt clips as they go by. If the

clothing does not tear, a serious accident is probable.

For that reason, in safeguarding belting it is advisable to consider each particular case, and not hope to settle upon one style of guard to meet every condition. One rule which fits a great many cases provides that all vertical and inclined belts, and also rope drives, be completely enclosed to a height of 6 feet above the floor, if the guards have to be less than 15 inches from the belts. It is better to omit a would-be guard entirely than to make that guard itself a hazard.—Belts.

ACETYLENE WELDING OF ALUMINIUM.

The chemical factory Griesheim-Elektron, Frankfort am Maine has introduced a new process which entirely prevents the formation of oxide of aluminium, and produces a weld which is of the same strength as the metal itself, and can be rolled and hammered.

The method is applicable to castings, sheet metal, etc., and depends chiefly on the admixture of a small quantity of fluorine to the ordinary flux made with 60 parts of chloride of potash, 12 parts of chloride of soda and 4 parts of sulphate of potash. This flux, which may be used as powder mixed with water to a paste, melts under the acetylene flame sooner than the aluminium and fully protects the heated surfaces from the air. A perfectly homogeneous joint is obtained with the greatest ease and without risk of impurities entering. The flux can also be used in crucibles for melting scrap aluminium, as it effectually prevents any loss of metal by oxidation.

The latest pattern of night landing light for aerodromes, says Acetylene & Welding Journal, is a complete oxyacetylene unit in itself. The light and all the necessary apparatus for its maintenance are mounted on a trolley fitted with aeroplane type wheels. All conditions of the landing grounds have been taken into consideration, and the high mounting of the axle, etc., enables the light to be taken to any part of an aerodrome. As the units are each complete in themselves, and the light can be got going in two or three minutes; emergency landings can always be assisted.

EXPORT SHIPPING IS A STANDARDIZING SCIENCE

Continued from page 12

own traffic staff, they can at least employ expert advice at a reasonable figure. Money spent on such advice is not money lost, but money gained, for by the advice given you will more than likely save both money and delay.

Always be sure of your shipping package. Use the best of material and pack intelligently, not haphazardly. We have brought all our cases and crates to a standard which we know gives ideal results. Having arrived at this point we can safely advise. Make your case heavier than just heavy enough, if you get the idea. Rough usage is one of the things an export case has to contend with, so prepare it for such treatment.

Having manufactured and packed your product, follow out the shipping instructions implicitly. Do not deviate from them one iota. You may not see the sense in them, but your forwarding agent does.

Last, but not least, treat your export business as you would have them attend to your wants were you in the reverse position, for if this is done the relations will not only be friendly but profitable.

One of the reasons why wireless telegraphy has not been made compulsory on smaller vessels is that its value has depended upon an operator being continuously on duty. Important calls might come at any moment, night or day, and unless there was an operator always with his ear to the telephone, the value of the installation would be enormously reduced. This drawback has been removed by a recent invention by a British wireless telegraph expert. It is described as an "automatic call device" which rings a bell when messages of a certain kind, such as the ship's special call signals or the S.O.S. signal for help, are being sent out. When this device is installed, a ship does not need relays of operators continually in the wireless cabin. It is enough to train two or three officers in the use of the wireless telegraph instrument. The bell calls them when they are likely to be required—as in the case of the ordinary telephone.

DOMINION BUREAU OF STATISTICS

Ottawa, May 6, 1920

2. Principal Articles Imported into Canada for Consumption

	Twelve months ending March.		
	1918.	1919.	1920.
Animals, living	$ 2,764,371	$ 1,827,849	$ 2,568,307
Articles for army and navy	130,773,478	50,704,709	1,679,079
Asphaltum and asphalt	374,997	440,722	446,587
Books and printed matter	6,358,839	7,824,499	11,240,814
Breadstuffs	16,941,610	26,717,140	26,519,958
Bricks, clays and tiles	4,302,884	4,298,745	2,470,812
Butter	136,269	718,671	176,994
Cheese	114,635	64,867	206,960
Chemicals	26,622,172	32,786,794	19,785,974
Clocks and watches	2,248,934	2,448,449	3,125,267
Coal—Anthracite	28,047,226	26,191,798	32,647,780
Bituminous	46,277,715	44,411,207	27,424,870
Cocoa and chocolate	2,008,427	3,783,426	7,626,745
Coffee	2,122,058	1,865,612	5,077,103
Cotton	36,952,806	73,377,554	89,367,984
Curtains and shams	357,328	367,320	474,779
Earthenware and chinaware	3,562,776	2,256,600	3,511,447
Eggs	1,504,234	681,849	2,837,442
Fish	2,316,629	2,497,054	3,491,579
Flax, hemp and jute	11,511,778	13,513,912	15,543,245
Furs	3,967,470	4,623,037	13,511,205
Hides and skins	8,794,289	5,426,908	23,020,976
Jewelry	871,816	756,771	1,242,010
Lard	758,142	554,867	2,220,413
Leather	8,916,611	11,468,787	17,102,702
Meats	24,418,720	5,905,271	22,100,333
Metals—Brass	5,410,881	5,231,230	4,565,756
Copper	6,545,300	5,997,626	8,568,035
Gold and silver	292,810	247,870	704,938
Iron and steel	159,309,323	181,619,059	149,846,502
Lead	1,384,082	966,982	937,312
Tin	15,241,179	15,131,175	11,419,016
Zinc	1,993,356	1,180,412	835,596
Musical instruments	3,707,407	3,164,227	4,329,063
Paints, colors and varnish	3,196,502	3,603,231	4,121,681
Paper	7,516,389	9,044,390	9,970,656
Pickles and sauces	449,532	418,503	819,630
Ribbons	1,560,695	1,885,632	2,899,429
Rubber	12,864,355	12,065,693	17,655,992
Seeds	1,887,697	2,038,006	4,208,845
Settlers' effects	6,367,291	5,691,039	10,181,034
Silk	14,942,206	21,182,992	34,432,789
Soap	1,156,953	1,267,868	1,534,082
Stone, marble and slate	1,857,421	2,194,247	3,277,420
Sugar and molasses	39,484,978	39,493,078	73,618,354
Tea	13,713,427	3,793,724	9,336,165
Tobacco	7,975,796	11,813,350	4,673,550
Tobacco, pipes, etc.	831,804	754,757	1,000,023
Vegetables	4,621,555	3,896,463	5,602,017
Vehicles	22,308,227	18,423,384	32,635,350
Vessels	2,202,740	3,901,424	6,186,391
Wood	14,615,607	18,925,455	22,431,670
Wool	25,086,969	40,167,925	63,493,535
Total principal and other articles imported	**$ 963,510,679**	**$ 919,705,802**	**$1,064,316,169**

The Making
OF
Morrow Drills

BY
J. H. Moore

I N preceding article dealing with the screw producing portion of the John Morrow Screw & Nut Co., we commented on the splendid lighting provided, proper aisle space and other features that go to make up ideal working conditions. The same holds good in this the second portion of their plant, which is devoted to the manufacture of drills of all kinds.

The number of operations necessary to complete a drill would surprise the uninitiated, there being more work on the same than one would imagine. For instance, before even the first operation is performed, the question of proper proper steel must be gone into. Without the correct quality of steel no amount of careful workmanship will make an efficient drill. This portion of their business is perhaps

the most closely watched of all, for they pride themselves on the fact that only steel of proven quality goes into their drills.

After a proper test has been made to ensure that the steel is of correct grade, the drill goes through the following operations:

The stock is first cut off to length, the cut-off bits are snagged on an emery wheel, then the lengths are pointed to shape. Next comes the centering operation, followed by turning to size, that is the turned, and not the ground size of drill.

The tangs are next slab milled, after

which the grooves or lips are milled. Next in order comes the filing of burrs, etc., folowed by the soft grinding of the grooves. They are now soft cleared and the points soft ground, after which they are stamped as to size, style, etc.

They are next hardened and tempered, hard straightened and the grooves polished, after which they go to the sand blast. Now comes the grinding to size, hard clearing, and drawing up operations.

The shanks are next polished, also the margins and the points. This completes the drill with the exception of inspection, which makes a total of 23 different operations.

Of course these operations vary somewhat, depending upon the size and nature of the drill, but our idea was mainly to

This photograph of course shows only a small portion of the department.

FIG. 1—A GROUP OF BRADLEY HAMMERS FORGING THE TOUGHNESS INTO THE DRILLS.

illustrate to readers the many operations through which the average drill passes. All high speed drills over ¾ in. diameter do not follow the routine already given, but are forged in place of milled. The sequence of operations are also varied for different style drills, but on the whole the list as given will allow readers to form a fair idea of the routine adopted.

Method of Storing Steel

The method of storing steel in this plant is well worthy of comment. Special racks with sections of about 6 in. to 12 in. square are installed. Each section is labelled plainly as to the quality of steel contained in that section, what it is to be used for, and so on. A card system is in operation which ensures an accurate knowledge of stock on hand, amount used, etc., so that no confusion arises when material is required.

The first few machine operations are so widely known that we make no comment on the same, but for illustrative purposes we will suppose it is a forged drill that is about to be manufactured.

After cutting off, and welding stock to the shank if necessary, the next operation is that of flattening to proper shape for twisting. This work is done on Bradley cushion hammers, and Fig. 1 will give a good idea of a small portion of this hammer and forging department. To see the operators at work is a pleasure, for they are so accustomed to their duties that not a single blow of the

hammer is wasted. When flattened out to shape, the stock is placed in a very simple but ingenious twisting fixture, and the drill now assumes more of the finished product appearance.

After twisting, the drill is annealed, then pointed and centered, after which it is soft straightened. These few remarks will show readers what we have already mentioned, namely, that although the operations do not occur in the same sequence, they are almost identical in style.

The Milling Operation

Omitting the next few operations, let us consider the milling of the grooves. At Fig. 2 we illustrate the machines used for drills from No. 80 up to ¼ in. diameter.

This type machine is known as a horizontal drill miller, and one girl attends to as many as six to a dozen machines. When we state that the No. 80 drill is only 13 thousandths in diameter, and that it takes 30,000 to make 100 lbs. weight, readers can no doubt realize the delicacy of the milling operation on a drill of this size. It is the intention of the firm to build a larger size machine of the horizontal type to accommodate up to ⅛ in. diameter, but at present anything over ¼ in. is milled in a vertical style machine. On the horizontal type only one groove is milled at a time, while on the vertical machine the two grooves are milled at once. Of course

the advantage of the smaller type machine is that they are automatic in action, and one operator can attend to a battery of machines.

At Fig. 3 we show a general view of one of the vertical style machines, while at Fig. 4 we illustrate a close-up view of the same type machine busily engaged at work. Note the angle to which the cutter is set, and of course there is another similar cutter on the other side milling the second groove.

The complete machine is operated through gearing, the action being as follows: The central spindle, which holds the drill in position, turns slowly at a predetermined rate while the cutters mill the grooves. As the spindle turns, it rises, until the drill grooves are completed. It is no doubt a well-known fact that a properly milled drill is thicker in section between the grooves at the back, than at the point. This is also allowed for in the setting up of the machine.

Miscellaneous Operations

The grinding operations are very important and these are performed on high-grade grinders, both the Landis and Norton style being used. Fig. 5 illustrates a small group of these machines, the general style of grinding practice being adopted. We need not go into any special comment in this regard.

The grinding of the clearance on the drills is accomplished by means of spe-

FIG. 1.—GIRLS OPERATE THESE SMALLER TYPE HORIZONTAL   FIG. 3.—A GENERAL VIEW OF THE LARGER STYLE DRILL DRILL MILLERS.   MILLING MACHINE.

cial shaped wheels, mounted on small grinder frames as shown at Fig. 6. The operators get very expert on this work and can produce the drills at a surprising rate of speed. Starting the drill at the point, they can, with a twisting movement of the hand, follow the groove to a nicety, even to the difference in thickness between the grooves already spoken of.

The sand blasting operation is also a very interesting one to watch. The best of silica sand is used for sand blasting purposes, the result being a perfectly clean and smooth finish drill.

The Heat Treating Department

This department is the place par excellence as far as the plant is concerned. It is here that the product can be made or marred, for without proper heat treatment all the previous operations is time wasted. There is no danger, however, of the product being marred, for every precaution has been taken to ensure against such a possibility. Even the pyrometers are placed to a considerable depth in the ground, to insure accuracy and a double check is kept on all furnaces and baths. Special handling tongs are adopted and nickel chrome tubes are used in many instances, especially in the hardening of the smaller size drills.

After the drills are tempered, they go to what is known as the hard straightening department. This is a room specially constructed for the purpose and everything is painted green, even to the windows, which might cast a confusing glare on the workman's

FIG. 4.—GRINDING THE CLEARANCE ON THE DRILLS.

eyes. In separate compartments, these men painstakingly see that every drill is straight before passing on to the next

operation. If not, they proceed at this point to work with it until perfect, This is a very important operation and con-

FIG. 4.—A CLOSE-UP VIEW OF THE DRILL MILLING MACHINE.   FIG. 5.—A GROUP OF GRINDERS FINISHING THE DRILLS.

siderable trouble is taken to be sure that only first-class work passes through their hands.

The Inspection Department

Having taken readers through the general procedure of the making of a drill, it would not be fair to leave the subject without discussing the most important operation of all, namely, inspection. This department is pointed to by the Morrow Co. with pride, and justly so, because rather than pass a drill with the slightest flaw, they scrap it immediately. There are no such things as seconds. They must be first quality or nothing. The men in charge of this department are old-timers, who believe in taking a personal pride in their work,

and are, I believe, actually delighted if they discover a flaw. This, however, is the exception rather than the rule, but they nevertheless look for flaws with painstaking zeal.

When the drills have been inspected they are dipped in a non-rusting compound before being packed and shipped.

Card systems are in vogue in this section of the plant, that keep track of the customers' requirements, the progress of work, material used, time spent on the work, etc., etc. In this way all orders can be carefully checked as to costs and so on. In other words, if a customer is promised an order on a certain date, it is so marked on the card, and in as far as is humanly possible, the order is shipped in plenty of time to

reach its destination on the promised date.

In conclusion, we might add that anyone imagining the making of a drill an easy proposition had better take a walk through a drill-making plant such as described, then change their mind, for the main object throughout is absolute accuracy. If a drill is not accurate, it is worthless. If a drill has the slightest flaw, it is likewise of no use, and if a drill is not properly heat treated, you might as well use a chunk of iron in place of the more expensive steel. The art of drill-making is a study in itself and it is hardly necessary to add that the company whose plant we have described are well equipped and experienced in that art.

Proposed Network of Wireless Communication

This Plan, as Proposed by the Marconi Wireless Telegraph Company to the British Government, is Hoped to Relieve the Abnormal Congested Condition of Submarine Cable Service

By J. H. RODGERS

THE proposed network of wireless communication covered by the plan submitted by Marconi's Wireless Telegraph Company to the British Government, would appear to provide the much desired Imperial chain, and as such it is hoped the scheme will be approved and adopted, either in its entirety or with some slight modifications, where necessary changes may seem advisable for the best interests of the system and the various countries that it will serve.

Realizing the abnormal congested con-

ditions of submarine cable service, the Marconi Company has developed and perfected a plan that will not only give relief to existing systems, but will eventually prove of inestimable value to Great Britain, the overseas Dominions, and outlying British colonies. The late war has shown the imperative need of direct and uninterrupted communication between governments and operating field forces, and in no instance has wireless been equalled for reliable and effective service. The wireless method does not give the

enemy the opportunity of severing communication, and therefore establishes an independent and effective means of creating a network of rapid communication, capable of embracing, not only all British territory, but also those equally important units of the Empire: ships at sea.

The terms upon which the proposal is submitted may be stated, briefly, as follows: The Company agrees to bear the entire cost of construction, maintenance, and operation of the complete network of wireless system, and pay a sum equal to

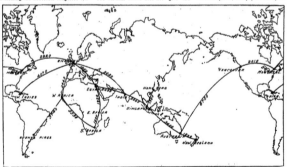

MAP SHOWING PROPOSED TRUNK ROUTES.

25 per cent. of the net profits . of each station or stations to the Government in whose territory such stations are located.

On the expiration of thirty years of service of any portion of the network, the system affected may become, if so desired, the property of the Government concerned, free of any payment. Special terms are provided should the Government, before the expiration of the thirty years, desire to acquire the system. The Government will have the right to take over the control of the stations during any period of war or national emergency, under arrangements agreed upon between the Company and the military authorities of the Government.

The proposal requires that 'the Government concerned issue all requisite licenses during the term of agreement, and grant every facility for the acquisition of sites, and that the Government construct and maintain the necessary underground and overhead cable systems that would be required for the telephonic and telegraphic feeder services. Special provision is made in the proposal to take care of the world trade in general, but the fundamental principle of Imperial preference is the keynote of the system.

The construction of the network would call for the erection of one or more main trunk stations in each country, for the carrying on of communications over long distances. In a country which forms a terminal of more than one route, and would necessitate a trunk station for each line of communication, the trunk stations would be grouped as close together as avoidance of mutual induction would allow, so as to maintain only one trunk transmitting area in any country. These areas would also include smaller transmitting stations, which, in turn, would be grouped to form a main feeder transmitting area. The receiving areas for the various services would be similarly arranged for economical and efficient service. The entire network would be under the control of a central office situated in a convenient telegraph centre of the country, with the transmitting and receiving areas located as nearby as economic condition would permit.

In proposing a network of this character, the Company has in mind, not only the commercial value of the enterprise, but the stategic value as well. As range is not restricted to definite distances the choice of sites would be a very important factor, and selection of location would be decided by joint conference between the Company's engineers and officers of the various fighting forces of each country.

The proposed main trunk routes are shown on the map, the necessary number of main trunk stations for this tentative network being twenty-six, five of which would be located in England. Two of the twenty-six would be located at Montreal, and two others at Vancouver,

with feeder stations for Western and Eastern communication.

While the system of network has been developed on the best experience of wireless engineers, the final details and

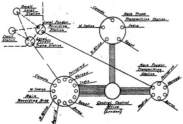

LAYOUT SHOWING PROPOSED ARRANGEMENT OF STATIONS AND THEIR CONTROL.

modifications, if any, would be agreed upon between engineers of the Marconi Company and representatives of the different countries. While the primary object of the proposal is to advance the commercial interests of the Empire, the future possibilities of war have been given due consideration, and with this in view, the personnel of such a network would become members of naval or military reserve forces, and capable of immediate mobilization in ranks comparable with the position they hold at their different stations.

The advantages of such an Imperial scheme of communication cannot be over-estimated and it is hoped that the co-operation of the various Governments will enable the proposed network to become an integral part of British enterprise.

PROMET BEARING METAL

HEAT treatment as a means of increasing the efficiency of steel has become a universal factor in engineering problems, but little knowledge has been acquired as to the heat treatment of non-ferrous metals for adding to their strength and durability. In Promet metal the manufacturers, the American Crucible Products Company of Elria, Ohio, have obtained a metal that possesses all the advantages that has been accorded to the bronze and other non-ferrous metals of olden days, when such metals contained all the tempering characteristics now possible only in steel tools. While the actual process used by the Egyptians of olden times will probably always remain a mystery, the results obtained in the making of Promet are in many respects identical with ex-

isting specimens of those pre-historic bronzes and other metals.

While the analysis of Promet is the same before and after heat treatment, the process undoubtedly changes a re-

latively good bearing metal into one of incomparable superiority in both bearing and lubricating qualities. The accompanying cuts will show the remarkable change in structure and the marked similarity to steel before and after treatment. It is claimed that the properties of this metal will remain permanent under any and all conditions; heating up to 2,000 degrees F., time or exposure, conditions of storage or long and exacting service, have no deteriorating effect upon the metal, and the metal can be remelted and re-used over and over again, and still retain all the advantageous qualities of the virgin metal.

It is also claimed that it has many times the bearing properties of any tin base babbit, and many times the lubricating properties of any lead base babbit. Many working tests have been made with material, and loads of 9,700 pounds per sq. in. have been carried without distorting the metal, and the co-efficient of friction is remarkably low, particularly where lubrication has been seriously neglected. It will melt at 600 degrees F., and can be heated to 2,000 degrees without the least injury. The recommended temperature for pouring is between 700 and 850 degrees F., and it is claimed that the metal is non-shrinkable when worked at this heat. An important feature of the metal is that it will not stick to the shaft when the bearing becomes overheated. Hot slag and molten metal cars, fitted with Promet, have shown no signs of wear after four or five months' service. Promet is equally adapted to high speeds and heavy loads. The Engineering Sales and Service Company, 55 St. Francis Xavier Street, Montreal, are the Canadian distributors.

Strike Did Not Stop the Bonus System

McKinnon Industries at St. Catharines Are Working Out Several Plans in Their Plant

Cafeteria is a Good Centre for Social Side of Shop Life to Work From—Their Own Orchestra Too

ONE might imagine that a firm, after experiencing the inconvenience and loss of production attendant upon a strike, might not be very keen about putting into effect such things as bonuses for efficiency or long service, cafeteria service, etc. But that is just what is taking place at McKinnon Industries, Ltd., at St. Catharines. No resentment is held on the part of the company that the men should have gone on strike, as they did a short time ago, the belief being that the men quit work in order to retain membership in their international bodies, and that many of them were reluctant to walk out. Of course, a man who went out forfeited his length of service bonus, and has to start from the bottom again to regain his standing.

Officials of the McKinnon organization are not anxious to talk of what has been done, nor of what they intend to do. Above all else they seem to fear that the impression may get out that they are "doing something" for the men in the shops, whereas, as far as can be gathered, they are simply aiming to make working conditions favorable to the man who really wants to turn in a good day's production, and to reward service in a tangible way. The social side of the business is being developed and encouraged, but what leadership there may be in this is coming from the men themselves, and in this the company is ready to give every encouragement.

McKinnon Industries, Ltd., is situated on the outskirts of St. Catharines, in splendid surroundings, and the construction of the buildings is such as to give plenty of light and air to the men in the plant. The firm consists of a group of industries that have grown, one after the other, until now they have a plant where some 1,000 hands find employment. Roughly there is the malleable iron and hardware shops, then the plant turning out automobile differentials and transmission, and also automobile radiators. The McKinnon Columbus Chain Co., where electric-weld and fire-weld chain is manufactured, is also situated in the same group. The present company is really the outcome of the business established some forty years ago when the making of dashboards and fenders was the extent of the lines developed. Now the company makes mal-

leable castings, drop forgings, carriage, wagon and saddlery hardware, automobile accessories, etc. L. E. McKinnon is president and general manager; J. C. Notman, vice-president and sales manager, and W. A. McKinnon, assistant treasurer and in charge of export sales.

Like many other Canadian plants, McKinnon Industries were engaged on war contracts, turning out large shipments of shells and fuse. On the completion of these contracts part of this war plant was utilized in developing the making of automobile parts, and a substantial peace-time line has grown out of the shut-down munition plant.

Making a Social Centre

The need for a cafeteria had been felt for some time, especially in the winter months. Many of the employees are in that enviable position where they can go home for dinner at noon, but bad weather often interferes with this. The company built the cafeteria, and then turned it over to one of the employees of the firm, who had had previous experience in this line of work. A very substantial meal can be secured for 30 to 35 cents, and those who bring their lunch are invited to use the place and purchase what they care to. In the winter season the company seeks to combine entertainment with instruction in the noon hours, so they bring in moving pictures or lantern lectures. Of course, they keep these events from being dull or academic so they never lose their interest, but at the same time always strive to keep them educational as well as entertaining. Song sheets are also used, containing a lot of popular and patriotic airs. But where do they get the music for all this? A lot of it from the shop orchestra. Yes, that's just what we said—shop orchestra. Never had one in your plant? Well, neither did McKinnon Industries until they got busy and organized it, and it's a success. A man in the office had a good knowledge of music and was interested in orchestral work in the city. Notices were posted through the works, calling a meeting of all who would be able to assist by playing any instrument. The result is an organization of piano, drums, three violins, saxaphone, flute and cornet.

One of the latest ventures is the publication of the McKinnon Bulletin, a

shop paper. St. Catharines recently lost one paper, so this shop is seeking at an early date to keep the number of publications up to normal. Suggestion boxes are placed in the works, and the men in the various departments furnish most of the material for the paper. The employment manager is at present the editor of the Bulletin, but as the plans work out it is the intention to have a committee from the shop elect their own editor. Bowling and baseball leagues in season are encouraged.

The Office Viewpoint

Through the paper the office also takes advantage of putting certain phases of business before the men. The pay envelope is also used to bring out any points that it is considered advisable to emphasize in order to keep the shop organization up to a good standard. For instance, here is a little "stuffer" that was put in the envelopes recently:

"How long has that man worked for you?" asked the Caller.

"About four hours," replied the Boss.

"I thought he had been here longer than that," said the Caller.

"He has," said the Boss. "He has been here for four months."

The above little story is just to attract your attention, and, by the way, that reminds us of the man who was taking a course in short story writing. He was advised to use an unusual beginning to his story to startle the reader into immediate attention, and on being instructed to write a story illustrating this principle, the introduction to his story read like this "Oh Hell, said the Duchess, as she spat on the floor."

What we really want to talk to you about is "Loafing on the Job." Bill passing Fred in the Plant stops for a five-minute chat—ten minutes lost. Along comes Sam, and stops to hear the latest—fifteen minutes lost—and so it goes on. Now try to realize this. Under an eight-hour working schedule, "loafing on the job" one hour a day would mean the defrauding of more than three full days out of every month, about six weeks out of every year. If a man draws $100 a month, it would mean that he accepts $150.00 a year without giving anything in return. If other employees be led by his example to do the same, he touches his employer for a great deal more. "Loafing on the job" one hour a day means that eight men are required to do seven men's work. If one thousand people be employed in a given industry, and all waste one hour a day, the employer is put to the expense of keeping one hundred and twenty-five more people at work than would be necessary. If these one hundred and twenty-five men, who would otherwise not be needed in that

Plant, draw an average of $1,200 a year, it would mean that the operating expenses are $170,000 a year more than they should be.

The next time you find yourself tempted to stop for a chat, or see others doing it, remember the figures given above, and be fair to the employer, giving a "fair day's work for a fair wage."

The Shop Bonus Idea

McKinnon Industries have two bonus plans, one of them based on the production and efficiency of the employee, and the other on the length of service and his earning power combined. The bonus rate table, part of which is shown in this article, is posted in various parts of the works in order that the employees may see just what can be done when they attain certain degrees of efficiency. The average rate in the McKinnon Industries is 50 cents per hour. According to the table the employee at that rate is supposed to be 65 per cent. efficient. As he leaves that point and moves toward 100 per cent. his wages automatically increase as his production increases, until at 100 per cent. efficient he has come up to 75 cents per hour.

Do most of the men get in on the bonus? They do. In order to have this plan working satisfactorily it is necessary to have a standards department, where special attention is given to time studies. In this department a group of men, independent of the foremen or superintendent, analyze each operation that is to be put on a bonus plan. They find out what should be reasonable time under average conditions to complete the work, and on this the price is adjusted. It is not meant to suggest that the work is done over the superintendent's head, as the findings are submitted to him.

Canadian Machinery put the question that at once suggests itself: Will these rates stand if an operator gets to the point where he can make well above the average rate? The answer was that as a general thing the rates stood for a year, and the men were encouraged to go ahead and get as far along the bonus line as they could. There have been few complaints about the time set as standard for the operation. Of course, there will always be some who will object, but the average find them fair and quite within reach.

Every effort is made on the part of the company to allow men to get to the operation where their special talent or bent will allow them to make the bonus. They prefer bonus employees. If a man cannot make good in one department there may be another where he will get along much better. Before an employee is dismissed as not being able to fit into the organization, he is given all possible chance to find the position where he can make good and be of service in the plans of the organization as a whole.

The Inspection Department

Any tendency on the part of a worker to crowd material through in order to run up a day's work is checked by the inspection department, which is well organized and maintained in this shop. That the men are not trying to take advantage is shown by the small percent-

age of rejects, which, according to the records of the shop, show around 2 per cent.

There is no trouble now in getting all the employees needed. Replacement figures in recent months show the following:

February, 8 per cent.; March, 15 per cent.; April, 14 per cent.; May, 12 per cent.

The Service Bonus

Three years ago McKinnon Industries commenced the payment of a yearly bonus, calculated on the earnings and the length of service. It starts at the end of the first year's service at two per cent., and increases until at six years the employee gets ten per cent. This is paid in a lump sum, and is quite separate from any other wage bonus paid in the shop. If a man leaves or goes on strike he automatically loses his standing as an employee of so many years' standing. This is what happened when the strike of moulders and machinists took place at the plant, and on their return after being out some ten or twelve weeks they had to start in again to qualify for the service bonus.

"Did the strike change the attitude of the company toward this form of work?"

"It did not," was the answer of Mr. R. A. McKinnon, who has had considerable to do with the operation of the various plans outlined above. "We considered it good business before the strike and we consider it good business now. Many of the men did not want to go out, and went only through fear of losing their membership in their union. We try to avoid anything that looks like giving the men favors, but rather aim to put something worth while before

them and make conditions so that they can go ahead and make good themselves."

A recent paper presented to the Institution of Automobile Engineers deals with "The Electro-deposition of Iron as Applied to Motor Vehicle Repair Work," Mr. B. H. Thomas describes work of this class which was carried out during the war in a heavy repair shop. The parts reclaimed included stub axle arms, steering swivel pins, brake and clutch shaft ends, change-speed lever shafts, insides of wheel hubs, outsides of axle tubes, and universal joint pins. Putting the case generally, the author states it was found possible to deposit a layer of iron up to about 2 mm. in actual thickness on any simple cylindrical surface of wrought iron or steel, mild or cast. If properly done, it is practically impossible to chip this layer away from the basis metal with hammer and chisel. It is deposited direct on the surface without "coppering" first, and can be subjected to red heat without apparent deterioration. It can be carbonized and hardened in the ordinary way and can be filed or ground, and takes a high finish. Its wearing qualities on a fast-running journal are, as far as the author knows, untried. The chief limitation of the process is, in the opinion of the author, the inability to deposit satisfactorily on such materials as cast iron or aluminium. If it were possible to deposit an adherent coating of iron on such parts as worn gear-box ball race hosings, etc., the scope and usefulness of the process would be greatly increased.

Bonus Rate Table

Rate Per Hour	65%	70%	75%	80%	85%	90%	95%	100%	105%	110%	115%	120%	125%
20	.20	.205	.21	.22	.23	.25	.27	.30	.315	.33	.345	.36	.375
22	.22	.225	.23	.24	.26	.28	.30	.33	.345	.36	.38	.395	.41
24	.24	.245	.25	.26	.28	.30	.32	.36	.38	.395	.415	.43	.45
26	.26	.265	.27	.28	.30	.32	.34	.39	.41	.43	.45	.47	.49
28	.28	.285	.29	.30	.32	.35	.38	.42	.44	.46	.48	.50	.525
30	.30	.31	.32	.33	.35	.38	.41	.45	.47	.495	.52	.54	.56
32	.32	.33	.34	.36	.38	.41	.42	.48	.50	.53	.55	.58	.60
34	.34	.35	.36	.38	.41	.44	.47	.51	.535	.56	.59	.61	.64
36	.36	.37	.38	.40	.44	.46	.50	.54	.57	.595	.62	.65	.675
38	.38	.39	.40	.42	.45	.48	.53	.57	.60	.63	.65	.685	.71
40	.40	.41	.42	.44	.47	.50	.55	.60	.63	.66	.69	.72	.75
42	.42	.43	.44	.47	.49	.53	.58	.63	.66	.69	.725	.755	.79
44	.44	.45	.46	.51	.52	.56	.60	.66	.69	.725	.76	.79	.825
46	.46	.47	.48	.53	.54	.58	.62	.69	.725	.76	.79	.83	.86
48	.48	.49	.50	.53	.56	.60	.65	.72	.755	.79	.83	.86	.90
50	.50	.51	.52	.55	.58	.63	.68	.75	.79	.825	.86	.90	.94
52	.52	.53	.55	.58	.61	.66	.71	.78	.82	.86	.90	.93	.97
54	.54	.55	.57	.60	.64	.69	.74	.81	.85	.89	.93	.97	1.01
56	.56	.57	.59	.62	.66	.71	.77	.84	.88	.92	.97	1.01	1.05
58	.58	.59	.61	.64	.68	.73	.80	.87	.91	.96	1.00	1.04	1.09
60	.60	.61	.63	.66	.70	.76	.83	.90	.95	.99	1.03	1.08	1.13

Modern Methods
of
Production

The Automatic is herein
discussed as one method
of increasing production

By
J. M. Harrison

"WE want more production." In these four words is echoed what is on the lips of practically every manufacturer at the present time. Ignoring the labor and material phase of the situation, let the writer suggest that the tool equipment be looked over thoroughly. Is it up-to-date? More. important still, is it the proper kind of equipment to suit your product? All these factors enter into the solution of added production. Suppose we had 50,000 screws per week to make. Would we think of producing these on ordinary lathes, modern though they might be. Of course not, for that is not the proper field for that style of machine tool.

True, the average manufacturer would not think of using an ordinary lathe for rapid production of standard screws, still there are many cases where the wrong type of equipment is used for other classes of work. A more common danger existing than the one already spoken of, is the hanging on to, and continuing using of, out-of-date machinery.

If we had a long journey to cover in a hurry we would hardly take a mule as a means of transportation, but would pick on an express train, or better still, an aeroplane if possible. In other words we would use the quickest method to get there. This same spirit of "get there" can be used to advantage in the manufacturing plant of to-day. Added production necessitates some means of speeding up, and a study of your tool equipment will often produce surprising results.

It is not our intention, however, to offer any definite solution to any one class of manufacturer, but rather to present in article form different modern methods of production. Various classes of machine tools will be shown, together with interesting set-ups, and the readers will be left to decide for themselves if such style of equipment meets with their own particular requirements.

The Automatic

When we speak of automatic machinery, we realise that the term covers a very wide field. For our purpose, we will confine ourselves in this article to one class of automatic machine, namely, the style of equipment known as the Acme Multiple-Spindle Automatic Screw Machine.

In order that readers can become familiar with the general appearance of this class of machine, we illustrate at Figs. 1 and 2, two different styles. At Fig. 1 is shown what is known as the smaller standard plain type. This type of machine is built in capacities from 9-16 in. to 2 1-4. At Fig. 2 is illustrated the 3 in. to 4 in. machine, and as can be noted this is a much heavier and more rigid machine in every way, being designed for speed, heavy work and accuracy.

Digressing from the subject proper for a moment, and delving into the history of the manufacture of screws, we find that they started to make screws about the time Rome was built. For eighteen hundred years the progress in screw-making was very slow, but crudely made screws of bronze and iron came into use about the fifteenth century.

FIG. 1.—A GENERAL VIEW OF AN ACME MULTIPLE SPINDLE
AUTOMATIC.

FIG. 2.—THIS IS THE NEW TYPE OF ACME MULTIPLE SPINDLE
AUTOMATIC HAVING A CAPACITY FROM 3" TO 4".

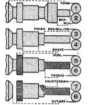

FIG. 3.—AN EXAMPLE OF WORK, AND THE TOOLS USED ON THE SAME

FIG. 5.—EACH TOOL STEP CAN BE CLEARLY FOLLOWED.

In 1775 came the invention of the ordinary lathe, although a poor example compared with the lathe of to-day. Of course this innovation made screw-making somewhat simpler, but in 1835 another type of machine was invented that still reduced the labor in making screws. In 1873 another machine was built that did practically all the work automatically.

Progress of real worth occurred from this date. The multiple spindle idea still further increased production, as it allowed a number of bars to be used in place of one. Leaving the history at this point, let us consider the principle of design in the modern machine we now have under discussion.

Principle of Design

In principle, these machines comprise four work spindles, which are held in a

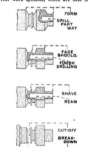

FIG. 4.—A GOOD EXAMPLE OF TOOL LAYOUT.

cylinder and rotated by gears, the cylinder being indexed to bring these work spindles consecutively into position in front of the side-working, and end-working tools. It requires one-quarter cycle of the cylinder to complete one piece, this being accomplished in the time required to perform the longest cut, which is usually a box turning, forming, or box tool turning operation.

In many cases it is possible to complete a piece in less time than the time required for the longest cut. Take for example, a long milling cut, which can be divided between the box tools in the first and second positions. In that case the time for the completion of the piece is considerably less than the usual time for box turning, being slightly more than half.

In drilling, three tools are frequently used in this manner, so that only a trifle over one-third of the time is required to complete the piece. The feeding of the bar takes place between the first and fourth positions, when the cylinder is indexing (on the smaller machines) and in the first position on the larger machines, the stop swinging out of the way before the tools approach the work.

By referring to Fig. 1, it will be seen that there are eight standard tool positions, namely, four end working tools, two horizontal forming and cut-off slides, and two stop slides working in the second and third positions. With this arrangement it is possible to have two tools working on each of the four bars at one time. This is, of course, a great advantage from production standpoint. The only idle, or non-productive moments in the style machine under discussion, are the withdrawing of tools, and indexing the cylinder carrying the work spindles. These moments occur at the completion of each piece.

No general rule can be stated as to tooling, or the use of same, as that depends entirely on the nature of the piece. As a general rule the operations performed at first position are forming, (with a tool held on forming side), turning, drilling, facing, counterboring, etc., with tools working from the end and held in the main tool slide.

The second position is sometimes as follows: From the end we might have turning with box tools, drilling, reaming, countersinking, counterboring, facing, etc. We might also have such operations as shaving, light forming, burling, and frequently thread rolling can be performed at this position.

In the third station, such operations as threading (with dies, taps, or chasers), drilling, reaming, counterboring, etc., can be performed, while turning with a box tool can be accomplished from the end. This position also lends itself admirably to the use of special attachments for milling, cross drilling, etc., their use being made possible because of the fact that the rotation of work spindle can be stopped in this position for the threading operation. Shaving, thread rolling, and knurling operations are often performed in this position, the tools being held on the top slide, and operating in this position.

The fourth and last position is used in many ways. The slide working tool cuts the piece off, in some cases after a forming operation has been performed. Tools from the end can counterbore, countersink, ream, recess, drill or turn. In many cases a variety of these stated operations are combined, in order to still further speed up production. The study of proper tooling on any style machine is a very important one, and the same condition holds good on these machines. By a careful study of the product, maximum production tooling can be installed. In this regard, the company whose machine we are describing lends valuable aid and suggestions.

Take for example, the product and its tools shown at Fig. 3. This set of eight tools was used for the production of the piece shown at lower right hand of picture. In the first position, the piece was formed, and a large hole drilled. In the second position the already formed piece was shaved on the shaving tool shown, while at the same position a smaller hole was drilled. The third position devoted to the rolling of the two threads, while the fourth position was used to drill another smaller hole completely through the piece, at the same time cutting it off, thus completing.

Other interesting examples illustrating

FIG. 6—ANOTHER EXAMPLE OF TOOL LAYOUT.

the different positions and tooling, are shown at Figs. 4, 5 and 6. As these diagrams are self-explanatory no mention need be made.

Various Attachments

On certain classes of work it is advisable to use other attachments in addition to the regular tool equipment. For example, suppose you desire to mill some piece from the side without removing from the machine. This can be accomplished by the use of the attachment shown at A, Fig 7. It is used while the other tools are working, therefore the piece is finished in the same length of time as without the milling. In other words you can perform an extra operation any additional time.

Should special reaming be necessary, the attachment at B, Fig. 7, will accomplish the work, as it is so constructed to feed farther and faster than the tool slide or drills. At C is shown another attachment for slabbing and slotting. One, two or three slitting saws or milling cutters can be used, and may be set horizontally or vertically, or may be swung at any desired angle in between.

At D, we see a combination arrangement for drilling. In this case the attachment is fitted with a gear drive, and telescopic ball-jointed shaft. At E is illustrated an end milling fixture, with

belt drive attachment. Such a fixture is very easily taken off or placed on the machine. Lastly, at F, is shown a cross drilling attachment which is very often used.

Every attachment illustrated is really a standard attachment, but in addition to these shown, other special fixtures can be used, such as auxiliary forming, special threading, self-opening die, left-hand threading, and so on. The beauty of a machine such as we are describing is the fact that you can suit your tooling to the nature of your product.

To illustrate all the attachments described would entail more space than we can devote to it, but at least we have spoken of the possibilities of such attachments.

It has been said that one thing at a time (as far as the making of screws

FIG. 12—VARIOUS STEPS IN A TYPICAL SCREW JOB.

been possible not only because of the four spindles, but because of the other attachments as well. Complicated work

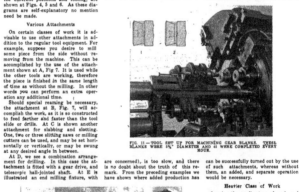

FIG. 11—TOOL SET UP FOR MACHINING GEAR BLANKS. THESE BLANKS WERE 3⅞" DIAMETER AND 48 WERE COMPLETED EVERY HOUR.

are concerned), is too slow, and there is no doubt about the truth of this remark. From the preceding examples we have shown where added production has

FIG. 7—VIEWS OF VARIOUS ATTACHMENTS USED IN CONJUNCTION WITH SUCH MACHINES.

can be successfully turned out by the use of such attachments, whereas without them, an added, and separate operation would be necessary.

Heavier Class of Work

Leaving the smaller type of work, and entering into the use of the larger machine, we feel the best method of explaining the use of such machine is to show actual examples of work being performed. At Fig. 8 is illustrated a typical tool set up on this machine. The completed product can be seen to the lower left-hand side of photograph. while the other work spindles and tooling are clearly shown. Note the compact arrangement of these tools, and how they have been so placed so as to attain maximum production.

A threading attachment is illustrated at Fig. 9, and again the nature of work performed can be seen on the photo. It will be remembered that we stated that the work spindle, in the third position, stopped rotating if necessary, and it is,

of course, at this position that the operation shown is performed. The die being used is one made by the concern itself and is known as the Namco self-opening die head. This same head can be used as a knurl holder, if so desired.

At Fig. 10, shown in heading of article, is illustrated an example of accelerated reaming. The special reaming attachment is clearly shown, and the method of taking the work from the machine, also the nature of the product can be clearly noted.

Fig. 11 illustrates an interesting tool set up for machining gear blanks. These blanks are 3 3-4 ins. diameter, made from cold rolled steel, and with tooling shown 45 pieces are completed an hour. Every detail being so clear, no further mention is necessary.

To show in actual photographic form the steps in the making of a typical screw job, we have prepared Fig. 12. We have not illustrated the completed portion, simply because the piece is completed with exception of cut-off in the lower position on photo. A picture such as this speaks louder than words, and while only a common example, it serves the purpose of bringing to the eye the procedure adopted. The piece, as shown, was slightly over 4 ins. in length, the threaded portion being about 2 1-4 in.

In order to ensure that readers are not running away with the erroneous impression that a machine of this nature is useful for screw work only, we might say, that not only is such machine used for all kinds of screws and nuts, but also for all classes of complicated bar work.

Study Your Product

Once more we emphasize the importance of studying your product. Where duplication work is concerned the first thing to do is to turn over in your mind what type of machine is best suited to produce the piece. Supposing you have

decided on an automatic, such as we are describing, the first thing to consider is the best method of applying the various tools, dividing the cuts, etc. Make sure you get the greatest possible production without crowding any of the tools with a heavier feed than they will stand under continued service. Another point to bear in mind is to cover the work as much as possible with top and side working tools, to avoid the necessity of making elaborate or expensive gauges for inspecting, where, if forming tools were properly designed, a simpler gauge would serve the purpose as well.

Parts which should be accurate as regards diameter should be covered with a shaving tool and every possible advantage should be taken to produce the work accurately. Experience is of course the best teacher, and to secure the advantages from such a machine, a clear understanding of its possibilities is absolutely necessary. Actual use brings that experience and operators are naturally always keen to devise, or find out quicker means of production.

As it is not our intention to go into the actual operation of the machine, but rather to discuss its possibilities, we leave the matter at this point, allowing readers to form their own judgment as to whether automatic machinery of such a nature as describing will solve their production difficulty. In our next article of this series we will discuss another machine tool of interest.

The Hugh Park Foundry Co., Ltd., Oshawa, Ont., has been incorporated with a capital stock of $200,000 by Frank A. Park, Oshawa; James Parker, 157 Bay Street; Maurice Crabtree and others of Toronto, to manufacture castings, forgings, tools, etc.

CHANGE NAME OF LATHES

An effort is being made in the machinery manufacturing trade to change the name of the lathes now commonly called screw machines, says a Cleveland despatch. The development in turret lathes and broadening of their uses has made earlier designation inappropriate. General support of the change is indicated from manufacturers in all parts of the country.

INSURANCE POLICY FOR EMPLOYEES

An insurance policy for the sum of five hundred dollars is now given entirely without cost to every employee of the American Westinghouse Electric and Manufacturing Company who has been in the service of the company for a period of six months or more. Furthermore all employees who have been in the company's service for six months or longer and who deposit a sum each pay-day in the Employees' Saving Fund equal to 2 per cent. or more of their earnings, will not only receive 4½ per cent. interest compounded semi-annually on such deposits, but, in addition, will automatically have their insurance increased to amounts up to 2,000 dollars, depending on the length of time they have been with the company. For example, a man who has been in the service of the company for at least fifteen years, and has regularly deposited in the Employees' Saving Fund 2 per cent. or more of his salary, on which he regularly receives 4½ per cent. compound interest, is presented with an insurance policy in one of the established companies for the sum of 2,000 dollars at no cost whatever to him.

FIG. 8—NOTE THE COMPACT ARRANGEMENT OF THESE TOOLS.   FIG. 9—SHOWING THE MACHINE IN ITS THREADING POSITION.

What Goods *Must Stand* on the High Seas

By T. H. FENNER, Editor Marine Engineering

SINCE the galleys of the Phoenicians cume over the seas to take tin from the mines of Ancient Britain, vessels and methods have improved, but the perils of the old ocean remain much the same. Science has provided us with stronger weapons to overcome the treachery of wind, sea and fog, but ships laden with valuable cargoes still leave port to be heard of no more, or to arrive battered wrecks, with cargoes water sodden, fit only for sale by auction to the hangers-on of commerce.

The day of the supercargo has passed, but it might be worth while in many cases for a manufacturer who is selling goods for export to far countries to send a representative for a voyage to see for himself what his goods have to go through before they arrive at their port of destination. He would come back with stronger impressions of the necessity for expert packing than the reading of a thousand articles would convey to him. Even when goods are crated and packed with the most meticulous care there are many ways in which they can be damaged during their voyage over which the shipper has very little chance to prevent or control.

One of the most annoying sources of loss, and one which seems almost impossible to eradicate, is that of pilfering by longshoremen and others who contrive to find their way to the docks where ships are unloading. Sometimes this is aided and abetted by the crews of the ships themselves, and so serious has it become that shipowner and forwarding agents, together with underwriters, are getting together to find some effective means of combatting it. Of course the exporter of heavy machinery is not much bothered with this class of loss, but manufacturers of goods that are exported in packages are the victims. Such articles as books, shoes, textile products, cigarettes and tobacco, which are readily hidden away, are favorites with the dockside robber. The packages may be rifled at the port of shipment while they are being loaded, as well as at the port of destination. Some ports are notorious for this class of thieving, while others having more efficient supervision by the dock police keep it within bounds. When the stealing is done on the dock side it is out of control of the ship's officers, but when it is done in the hold of the vessel it will depend upon the calibre of the officer in charge of that hold how

much can be got away with. If he is a man of long experience and some personal courage he can prevent a great deal, but if he is a young chap and susceptible to intimidation, it will be carried on almost openly.

The favorite method of ascertaining what class of goods is in a case, and of making it accessible at the same time, is to have it slip out of the slings when lowering. This will be done at a height just sufficient to burst the case open. If the packages are being lowered in a cargo net where it is not very convenient to slip them out, the signal to the winchman to stop lowering is delayed long enough to let the net reach the bottom of the hold with a good sharp bang, which does the trick very nicely. The broken case is then put over in a wing of the hold and left strictly alone for a while. Then an opportunity comes when the officer is in some other part of the hold to abstract a few pairs of boots or boxes of tobacco, suits of underclothes, or whatever it may contain, and these are rapidly distributed and cached till the time comes to get them ashore. If the officer who is looking after the loading happens to catch them, it will depend on his ability to handle men whether they are put back or not. While as a rule these men will not resort to violence, they can act and look ugly enough to get the average man to look the other way. The officer usually consoles himself with the thought that his business is only to see that the cargo is stowed properly, and it is up to the dock police to see that nothing gets out of the gates. There is considerable stealing done also at night time by men who come alongside in boats, clamber up the side by means of a handy rope, and abstract goods wholesale. They are usually aided by the fact that the nightwatchman is stationed at the side of the ship next the wharf to watch the gangway. The thieves are not only bold, but possess a certain sense of humor. The writer was in a ship lying in Genoa, a notorious port for thieving, and the skipper had made a vow he would get some of them as they had not been content with cargo but had been stealing the bo's'n's stores. The old man had spoken pretty freely of his intentions, and word had evidently reached some of the gentry ashore. The ship lay moored to a buoy at the bow and the stern on to the wharf, and a ladder was over the stern to give access

to and from the ship. Just at noon one day, when practically all hands were on deck, the red ensign (affectionately known among sailors as the "blood and guts of Old England"), which floated lazily from the jack staff at the stern, slowly disappeared, and just as the rush of angry men reached the poop, the communicating ladder was wafted away. Three brown-skinned rascals laughingly put our ladder up against the freight shed, mounted it, pulled it up after them, laid flag and ladder on the roof of the shed, and disappeared over the other side with a wave of the hand and a parting smile. It was many days before our skipper heard the last of that.

Just to show that ingenuity in thieving is not confined to any particular place, a case occurs to my mind that happened in Liverpool. This was not connected with cargo, but doubtless the perpetrators had gained experience in that field of endeavor. A large valve, made of bronze and weighing about 600 pounds was to be sent ashore for repairs. Three men arrived at the dock with a hand cart, one from the shipowner's own repair shop, and stated they have come for the valve, which was promptly given to them, and pushed their cart with the valve in it out of the dock with the accustomed stately deliberateness of the British working-man. It was never seen again, for by the time the hoax was discovered all traces of it had disappeared. It was so common a sight that no one took a second glance at the men, and they no doubt had a convenient foundry to dispose of it without loss of time.

Damage by Salt Water

The foregoing, while showing a very prolific source of loss, does not really come under the definition of perils of the sea. When everything is safely stowed, hatches battened down, and the vessel squared away for her run, there are many opportunities for damage to cargo.

An instance occurs to me where a cargo was carried safely for 12,000 miles through fair weather and foul, only to have a very valuable portion of it almost irretrievably ruined while lying calm and peaceful in harbor. Modern cargo vessels are equipped with what are called deep tanks. There are situated in the main hold, just abaft the engine room bulkhead, and are carried

up to the twin decks and provided with watertight covers. When the ship is loaded they are used for cargo, but when the ship is in ballast they are filled with water, and serve the purpose of making the ship trim by the stern. This gives the propeller a better hold on the water and reduces the metacentric height, both desirable purposes. The deep tanks in this case had been utilized for a shipment of pianos and organs, and the vessel was at her final port of discharge in New Zealand, and was to proceed to Newcastle, N.S.W., in ballast after discharging. Everything was out of the holds except the lower tier of pianos in the deep tank, and these were expected to be finished late that night. The ship's carpenter had instructions to close up the tank when the cargo was finished and pass the word to the engine room to fill the tanks. Now there is always supposed to be one officer and one engineer aboard the ship, but this rule, like many others, is often broken. In this case the fourth engineer and third officer, who were the respective deck and engine-room officers on duty, decided that everything would be alright for a while aboard the ship, while ashore the girls were pretty and the wine was good. "Chips," which is the carpenter's official name, promised the third officer that all would be well, and the fourth engineer relegated his duties to the donkeyman. For some unexplained reason the longshoremen decided they would not finish the job that night, and left about 9 o'clock. "Chips," who had spent the intervening time communing with the spirits in his room, sensed the silence which ensued, and the spirit of duty, among others stirring within him, roused out his mind, and proceeded to close up the tank. This being done, the donkeyman was notified, and opened up the valves, letting the sea water rush merrily in. The third officer, coming aboard shortly after, was notified by the attendant "Chips" that all was shipshape and Bristol fashion; The fourth engineer was duly informed by the donkeyman that the tank was filling up. Everybody was happy with that contented feeling that comes with a duty well done. It was only in the morning when the crew were putting the finishing touches to battening down and making all secure, ready for sailing, that a cynical and sarcastic boss longshoreman arrived and delicately conveyed to the skipper the news that some thirty pianos were enjoying the benefits of a salt water treatment. As it is impossible to convey by cold type the impassioned utterances of the old man to the third mate, we will draw a veil over the harrowing scene. Suffice it to say that the third mate, like a good shipmate, kept quiet about the fourth engineer's absence. The engine room staff were clear, anyway, as they only carried out orders, but it would have been decidedly unpleasant for the fourth had it been known he was ashore also.

Another case was in carrying a cargo of soap, lye, and stuff of that nature down the Mediterranean. This cargo was in the after hold, and crossing the Bay of Biscay the vessel got a severe trouncing. The ship was pretty old and the straining caused the shaft tunnel to draw away from the tank top in places, besides which the after peak tank commenced to leak. The stern gland, which prevents, or should prevent the water leaking through the stern tube into the tunnel, was leaking badly, and the water found its way through the tunnel to the after hold. Some of the cases on the bottom of the hold got water soaked, and the contents disintegrated, so that the cases got smashed up, and the weight above them settled down, loosening up the whole stowage. The hold bilges naturally got choked up, and as these bilges act as a receptacle for all drainage being pumped out from the engine room, the water rose in the hold.

The result was a greasy, slimy mess sweeping through into the tunnel, which made the tour of inspection necessary while on watch, a thing of peril to the unfortunate engineers. The leakage made it necessary to keep one of the bilge pumps continually pumping in the tunnel, and when the hold was finally opened up the whole of the cargo in the lower portion was a beastly mess, fit for nothing but pumping overboard.

Even such sturdy things as barbed wire and steel rails are not immune. It is usual to provide ventilators from the holds to the deck to use when the cargo makes it necessary. When not in use these ventilators are made so that the top part can be unshipped and the opening of the pipe, about two feet above the deck, is covered with tarpaulin to make them tight.

On the occasion I have in mind we were bound from Cardiff to Buenos Ayres with corrugated iron, barbed wire, and steel rails. The ship ran into very heavy weather while in the Bay of Biscay, that famous old bay. She had been hove to for about twenty-four hours, when, the weather moderating, it was decided to go ahead. The sea was still very heavy, and she made bad weather of it. finally shipping a sea that stove in the starboard bulwarks and swept the ventilators clean with the decks. The result was that tons of water found their way into the holds, and a considerable amount of barbed wire was so badly rusted on arrival that the consignees refused to accept it.

There is of course always the danger of fire and collision, resulting in total loss, and damage through hatches being stove in as a result of bad weather. Cargo is still damaged through inefficient stowage, though this is rare. These few instances will serve to show that the merchant with goods on the high seas is still liable to the anxieties which best Bassanio and gave Portia the opportunity to show that a lawyer may still be a lover.

Extensive deposits of graphite exist in North-western Siberia on the left bank of the river Kureika, near its junction with the river Yenisei, ninety miles from the mouth of the latter river, says Engineering. The graphite area, forming a horizontal plateau, contains two layers of graphite, which is of a solid, steel grey colour, soft, and of an excellent quality. The carbon constituent found in graphites in other parts of the world is said to be superior in quality to that world. The graphite is not inflammable, and is very plastic. It is believed that in the future the graphite from these mines will supply Russian demands, and that large quantities will be available for export. The chief sources of graphite have hitherto been Ceylon, Bohemia, Germany, France and the United States. The annual world production has been, approximately, 120,000 short tons.

Mechanical Features of the Modern Tannery

Various Details of Belt Splitting Machines Are Shown—Other Mechanical Problems Encountered in a Modern Tannery Are Discussed and Construction of the Building in Itself is of Interest

By DONALD H. HAMPSON

IF we speak of leather manufacture, nine out of ten hearers will unconsciously tilt their nose to escape the odor of tanning and hides they fear will accompany the words. But the manufacture of hides through all the processes up to the shipping room is full of interest to every man, while the odor—always a healthy one—tapers off toward the end into a fascinating aroma.

It is not the purpose of this article to describe the manufacture of leather in detail, but rather to set forth some of the mechanical features of the special machinery, its construction and repair, and the shop practice in the plants. Unwarned, the machinist would be apt to remark: "A tannery, huh, there's nothing there but a lot of tubs and sliminess." But ask that machinist how they cut a sheet one-sixteenth thick off a hide roughly eight feet square, how they get a micrometer measurement at any part of that sheet when cut, and he will straightway "register interest," as they say in the movies.

The "splitting" machines are the most important ones in the plant—and the most complicated. A hide as tanned is approximately ¼ inch thick; except for sole leather and belting and a few similar purposes, all hides are split, dividing this ¼ inch thickness into several sheets called "splits," which are then treated to various finishing processes according to uses, whether automobile seat covers or shoes or bags or the thousand and one other demands for leather goods. That part of the hide next the outside of the animal is best in every respect, and a section about three by four feet at the middle of the back possesses to the greatest degree those desirable qualities of toughness, pliability, and close texture.

With the upholstering of the automobile there came a demand for vast numbers of the largest and best hides, and this work furnished new problems for those who man the splitting machines. The cattle from South America

furnished hides which met the requirements very nicely—hides of the largest sizes—and incidentally their weight added much to the physical exertions of the men in the leather plants.

There are in use two styles of splitting machines; one a hand-operated machine for the finest class of work, where two operators pull and roll the hide against a blade of razor sharpness and the other, the production machine, outlined in Fig. 1, being a power-driven machine, massive and complex, and manned by four operators, who are able to split a hundred hides a day into the requisite thicknesses. Those who are familiar with special handling machinery are accustomed to parts which are regular in shape—they will appreciate more than others the difficulties of handling hides which are irregular in outline and bear the "projections" left by including short sections of the legs and neck in the hide, besides being a material which is heavy, damp, and limp at the time of working.

This belt knife splitting machine bears some resemblance to a band saw, though it uses a "saw" which has no teeth and its wheels are set horizontally instead of

one over the other. Referring to Fig. 1, the blade passes over two wheels 30 ins. in diameter, one of which is adjustable to take up any wear or stretch in the parts. The blade is of spring steel, without joints, tempered so it marks a file and having a section when new of 1/16 in. x 2½ ins. with a double bevelled

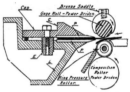

FIG. 2—A SECTION THROUGH THE MACHINE KNIVES.

cutting edge whose included angle is 10 degrees to 15 degrees.

The machines are built with a working section from 100 ins. to 120 ins. long. On the upper of these tangents, the knife does the cutting; on the lower, it is kept constantly sharp by the two emery wheels as shown. Naturally, to do accurate work the knife must be well supported. It runs between hardened steel surface of undoubted straightness, which guide the knife close to the cutting edge. These surfaces are blades similar to machine knives; they are fastened to massive lower and upper jaws, marked L and C respectively in Fig. 1. A section is shown by Fig. 2.

It will be noticed in Fig. 1 that the lower jaw is suspended on end keys between the frame of the machine; this always locates the jaw in the same relation for bolting fast. Portions of the trussed top, A and B, which are adjustable, are set by tissue paper feelers before bolting fast as additional supports for the jaws. The frame members also carry the bearings and supports for the various parts shown in the sectional view.

Referring to this section, the lower jaw L is a ribbed casting of a thousand pounds weight. The upper jaw C is known as a cap. Between the hardened surfaces P and P' there runs the knife shown in heavy black. The jaws are shown separated for the sake of clearness; there is a means of adjustment when drawing down the cap bolts by setting up the special tapped screws E from

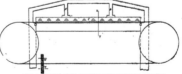

FIG. 1—GENERAL VIEW OF BELT KNIFE SPLITTING MACHINE.

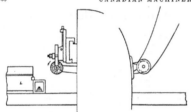

FIG. 2—THE SET UP ON THE PLANER TO GRIND JAW.

below—the cap is tipped very slightly to make a close bearing near the front of the knife. The hardened faces of the jaws have to be ground very accurately; when this is done and the machine properly set up, there is provided a keen-edged knife, true within a thousandth of an inch and rigidly held so. A series of steel plates (not shown) back up the knife, taking the thrust and keeping the cutting edge in a known relation to the hard blades.

Method of Feeding

The means for feeding and gaging the stock are shown in Fig. 2. A driven roller, called the "gage roll," is fixed directly over the knife; it is adjustable in a vertical position to allow for wear and the taking of splits of different thicknesses. It is adjusted by tissue paper feelers from the knife, and owing to its length and small diameter two or more bearings called saddles must be provided about forty inches apart. With these saddles the upward pressure is supported and the sag is kept out while setting.

Beneath is a heavy rubber composition roller driven by power and bearing upward against a roller made up of a series of brass rings shaped like ferrules which are loosely strung on a rod. This soft roller and the sectional one above make it possible to counteract any difference in thickness to any point in the hide—an important consideration in

FIG. 4—HOW JAWS WERE HANDLED.

maintaining the split to an even thickness.

This description will enable the reader to see from Fig. 2 just how a hide·· is split. The finished piece is taken off over the top of the knife, while the surplus passes down below to be again split up when ready. Setting the parts of

the machine is as close a job as anything in machine shop work. After that, operation is a matter of speed in handling the hides. The rolls above and below the knife act as feed rolls—they travel at about forty feet a minute surface speed while the knife travels around a thousand feet per minute.

Besides the general interest, much of this description would be necessary

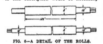

FIG. 6—A DETAIL OF THE ROLLS.

to enable one to grasp the machine shop maintenance problem. Accurate jaws and gage rolls are essential. Most tanners send their equipment to shops who specialize on this work, usually located in leather centres; for instance, in Eastern United States, Newark is the leather centre and there exists the feeling

FIG. 5—SPECIAL ANGLE PLATE TO HOLD CASTINGS FOR REBORING.

among manufacturers within a thousand miles, that they must send their repair work to Newark to get it properly done.

Some years ago the shop with which the writer was connected decided to make a try for this business as found in a large local manufactory and several smaller ones within a fifty mile radius. At

first, the manufacturers were loath to believe that our shop could handle the work, but after much harping on the point that every metal working job naturally falls in a certain class regardless to the use of the piece outside, we were permitted to go ahead on one set. Once the requirements were thoroughly understood and the best method of doing the work settled, there was no trouble at all in turning out as good or better work than the manufacturer had been in the habit of getting. And that at a somewhat lower figure—coupled with no freight charges, and no long weights for work in transit and awaiting its turn in a distant shop. These latter points have proven especially valuable during the unsettled conditions of the past six years.

Our policy was to attack the hardest job first—this being the re-facing of the hard plates P and P. Following a year's running, the plates get worn bell mouthed, when they have to be loosened up, papered under them, and ground off on top. Enough paper has to be put under to keep the newly ground surface 1/64 ins. above the level of the iron just back of them.

The weight and the near-sharpness of the blades make some handling arrangement necessary. The cap may be handled by two men, but not so with the lower jaw, which is also "tippy" when set down right side up. We made the fork shown in Fig. 4 of 1 inch steel to span three holes, and by removing two of the adjusting screws, E E, we could put the fork in their place and lift the lower jaw in perfect balance, handling it this way from truck to the planer and avoiding all risk.

Fig. 3 shows the set-up on the planer. Five-inch parallels are clamped at each end and on these rest the keys in the end of the jaw—this holds the jaw just as it is held in its own frame. A grinding arrangement was fixed up as may be seen, taking power from the countershaft down to a jack shaft bolted to the planer housing and driving from here to the shaft held in a fork clamped in the tool block. On the jack shaft the driving pulley has eight inches of face, which permits of side feeding without the belt running off.

This makeshift has ground scores of jaws to a most satisfactory degree of accuracy. Though not shown, the emery wheel is fitted with a hood and the clapper is secured against lifting. The direction of rotation is such that during the return stroke of the planer, the wheel cuts against the work, an arrangement found by trial to give the best all-round results. A Norton 80 M wheel is used, the size being 8 inches by 1-2 inch by 3-4 inch. Sometimes when new blades are put on, there is as much as a thirty-second of metal to remove. This is ground entirely, both the hard front and the soft metal toward the back of the blade.

In re-grinding a set of jaws, they are set up on the planer and clamped. Then

the blades are loosened up to permit the insertion of strips of paper along the edge of the jaw, which raise the blade so the worn top is up about the wearing level for properly ground blades. To get the right height, a surface gage carrying an indicator instead of a pointer is used—this gives a direct reading at

FIG. 7—HOW THE HIDES ARE MEASURED.

any spot and makes it easy to re-set for a minimum of grinding and to get the maximum wear from a set of blades. It takes about two nine-hour days to thus set up and finish the old blades on one jaw.

Our practice was to finish the cap first. Then do the lower jaw. Next to place the cap on the lower jaw and bolt it down in working position. The emery wheel would then be lowered and run along the edge of the jaw, making one directly above the other in a continuous straight line from end to end, thus affording equal support both sides of the knife as well as a true surface for measuring purposes. If this edge grinding showed up thick or stubby portions, the wheel was tipped and a bevel cut run over the surface, for it has been found that a thickness of a sixty-fourth is all that may be left without catching the leather as it passes through.

Turning of Rolls

Gage roll turning is quite similar to shafting turning, a job that modern machinists know but vaguely, because of shaft-turning lathes and cold-drawing processes. These rolls when new are 2 inches in diameter and there are adjustments on the machine allowing a decrease to 1 7-8 ins. before scrapping. As made, these rollers are turned on centres, but we didn't have any lathe long enough to take them that way. The rollers are made of maganese steel and must run very true.

FIG. 8—PUTTING THE POSTS IN CAST IRON SHOES.

Fig. 5 gives the details of the rolls. In turning, we put one end in an independent chuck, grasping it by the journal and truing by that portion also. With the tailstock removed there is an overhang of three feet beyond the steady rest set at the end of the lathe. Between this rest and the chuck, we put another rest to absorb the vibration and avoid chatter. It is a well-known fact that a heavy cut is less apt to chatter than a light one; on this work, all that was permitted was a light cut and to avoid chatter on this relatively slender piece, a narrow round-pointed tool was used and as coarse a feed as could be worked with a fairly smooth cut. With but one steady rest, it was impossible to avoid chattering.

The belt speed of the lathe was resorted to instead of the gear speed, not so much to save time as to produce a smoother cut—the lathe we had was not of the heaviest type and it was found that there were a series of longitudinal marks, one for each tooth in the back gears, and the marks, though not measurable, were sufficient to impair the finish of the roll.

Our plan of turning was to first micrometer the worm spot of smallest dia-

FIG. 9—FITTING AND CALKING THE BLADES

meter; then start a cut about five thousandths smaller than this. We would turn as far as a steady rest would allow, then shift and carefully re-set the rest, and turn some more. We would keep our cut between .001 in. over and .001 in. under the set diameter. The difference was wiped out in a careful file finish. After turning the seven feet within the travel of the carriage, the roll was reversed and chucked by the opposite journal and trued up by it and the nearest point of the newly finished cut. The remainder was turned as before. The points of greatest wear are always where the saddles bear, for these resist considerable thrust during the leather splitting.

Fig. 5 illustrates well-known points in machine design. Practically all machines come through with rollers like the lower drawing, with sharp corners at the inner end of the journal and with a driving gear at both ends. During the operation of changing rolls, a man might drop one a foot or so—which invariably meant a broken end; sometimes one of these journals would twist off in the machine while running. There is the same drive to each end of the rolls and the only way we could account for the twisting off when working was that the leather caused more drag one place than another, setting up just so much torsion in the roll that could not relieve itself because tied to gears at the end.

FIG. 10—PECULIAR CONSTRUCTION OF BUILDING.

By doing away with the double drive, as shown by the upper drawing, the breakage in the machines ceased entirely.

Rolls with broken journals were repaired by drilling out the end and setting in a new piece slightly larger and two inches longer than needed. This piece was chucked, trued up by the body of the roll, and turned a new journal with the gear fit beyond—the extra two inches was sawed off after the turning was over. In drilling, the roll was caught in the two steady rests and the drill in a universal chuck, then with a dog on the roll the carriage feed was used to feed the work on the drill. All this details one way of doing long work on a short lathe and doing it accurately too.

The upper drawing Fig. 5, shows filleted corners, which were adopted after it was shown that the rolls themselves were amply strong for the work. Likewise the diameter was increased one-eighth—then no chance was could snap off a journal by dropping it.

The roll saddles that have been mentioned are shown both in Fig. 2 and Fig. 6. They are bronze castings and would be made to encircle the roll entirely but for the clearance needed toward the bottom for the passage of the leather. Along with the wear on the roll, the saddles themselves wear, making a loose fit. When the rolls are turned up, the decrease in diameter makes a still looser fit. Then the metal is bent or hammered into a smaller size, which, of course, is only approximately round.

We made the special angle plate shown in Fig. 6 to hold these castings for reboring. There will be noticed a counterbore in which the hub sits, while a bolt from the back holds it firmly in position. Then the four set-screws set on the angle are brought up to support the open sides of the piece—and this they do very nicely. The rig provides a quick and most satisfactory means of holding while the bronze is bored out. Many places scrap the saddles when worn because they have no cheap way of boring out to a close fit—the fixture shown is

a good example of one that will save its cost on the first job.

There are in use saddles made with two swinging sides, pivoted at the top and drawn together by a screw. These find favor because they may be taken up from time to time—there is an objection to them in that there is a tendency to

FORM OF EMERY WHEEL AS SOMETIMES USED.

keep a worn roll in longer than is right, because the saddles may be kept up close and in doing this, the greater evil is encountered of running a worn roll too long and this is worse than a loose saddle. Rollers have been fitted to the wearing ends of these adjustable saddles, but they are generally discarded because of their short life due to a small size and the fact that they seem to wear the roll as much as a plain bearing.

Accurate splitting has already been mentioned. The task of measuring thickness in the centre of a hide brings visions of loner-jawed calipers and gages with a pointer running over a dial something like the direct reading calipers that appear on the market every so often. None of these. Just a micrometer of the type with broad measuring faces for thin materials and soft stock. Referring to Fig. 7, we see how this well-known tool is used—the hide is doubled at any desired point and a measurement taken over two thicknesses. If there is any doubt as to the truth of the reading, two or three measurements in the vicinity will dispel it.

Varied Uses For Leather

There are many other jobs and machines about a leather plant of interest. Workmen there learn to "strike from where your hand is," which interpreted means to use leather for the purposes that the average machinist would think metal or pasteboard or wood should be used. Thus, in babbitting ordinary bearings, a hole is cut in a piece of scrap leather and it is used over the shaft as a cap instead of pasteboard or putty.

Leather scraps are burned under steam boilers. It requires more air, but when mixed with a small amount of coal it does very well as fuel. One plant saves one-fifth of its fuel bill in this way.

Shaft couplings are constructed with leather—making couplings of the flexible type. This has been the means of some power saving, too, for in many buildings the floors and overheads are none too

rigid so that a regular shaft and hanger equipment does not stay in shape very long, with the consequence that it runs hard in a short time. It has been customary to move the regular couplings back an inch or less from the end of the shaft length, slip in a circle of sole leather, and bolt to each flange.

The vats in which the hides are tanned are located on the ground floor. Sometimes in a one-story building and sometimes in a building of more height. The construction of the vats precludes the use of piers or other means of support for floors above, this requiring a different type of construction where land is at a premium. The building that is erected is virtually a shell, having a heavy foundation and substantial side walls surmounted by a roof structure having a high truss. From this truss, rods are dropped to support the floors which are put in below, these floors having no connection with the side walls except as the beams are set in at the ends and for the horizontal tie rods that are applied. (See Fig. 10).

Another structure is the bark grinding mill, usually a building resembling a frontier block house. Supported on four 12 in. x 12 in. posts, with a stone or concrete floor, wagons may drive right under to receive and discharge their loads. But this level-with-the-road floor works havoc with the posts, which are exposed to the elements and rot out at the bottom in a comparatively short time. The management of one plant adopted the scheme shown by Fig. 8 of cutting off the posts a foot and putting them in cast iron shoes. Now the water does not remain about the foot and the posts look good for twenty years

—the shape of the shoes (tapered in, with a small bearing on the concrete) also makes a neat job—one that lends itself to washing down the floor with a hose.

A split bearing in a tannery is never shimmed with pasteboard or metal liners. Leather is used. The variety of thicknesses found in the scrap invites this.

The hides in the vats are mechanically agitated and there is more or less slowmoving shafting and gearing about the vat building. Sometimes the lines are two or three hundred feet long. Spur gear reductions are used between these lines and the motor or line shaft drive. By all precedent, cast (tooth) gears are correct for such drives and they are extensively used. The fact that some of them had to be renewed every four months was accepted as a matter of course. In one case, however, the writer got permission to put in a cut gear and pinion; this ran twenty-nine months before discarding, as against four months, and now they are saving money with cut gears all around.

There is an exception to the wide use of leather in leather plants. Woven belts are quite extensively used. Not only do they wear as well and have the pulling power from the day of installation, but they cost less than leather belts. Belting is a branch by itself and only a small number of the plants make leather belts. It would not pay to make up a few any more than similar maintenance or tool propositions in a manufacturing machine shop.

One machine that saves much hand labor is employed to remove hair and fleshings from the hides. A drum about six inches in diameter and two feet long has a series of helical blades projecting an inch from the surface; the appearance is similar to a herringbone pinion in a turbine reduction set. As this drum revolves at a high rate, the hide is passed over the top somewhat as a board over a jointer head. In time the blades wear and are ground down, calling for renewal. They are calked into place in slots

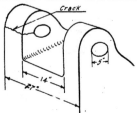

FIG. 12—A PORTION OF THE BASE OF A HYDRAULIC PRESSING MACHINE.

in the drum, using soft iron strips for calking. Fig. 9 gives an idea of this. It is no easy task to fit the ends and calk a set of these spring steel blades.

The knives which are used on the hand-splitting machines have been mentioned. These are about eight feet long and have a thin cutting edge that must be ground with the nicety of a

precision tool; it is hollow ground on both sides. Knife grinders of the familiar reciprocating-table kind are used for this work except in the smaller plants, where some home-made rig is employed or the blades sent out to be re-ground.

Fig. 11 shows a form of wheel used in some plants. A 20-inch pulley of heavy type is selected and has a number of 3,4 inch holes drilled through the rim and countersunk on the inside. A rough form is made and a ring of lead about two inches thick poured around the outside. This is turned up true in the lathe on its own spindle and it becomes the grinding wheel. It is charged with emery when it is ready to use. The emery gets bedded in the lead and makes a quite satisfactory wheel, though not to be compared with a manufactured abrasive wheel. Guides are provided to rest the knives in and bring them against the wheel. This improvision does excellent work.

It is obvious that one of the layers into which the hide is split can have the natural grain on its surface. This one is the outside split. But for many purposes, the outside of cattle hides does not have grain enough to meet the demands required of it. Belting and shoes and travelling bags may have a smooth surface, but purses and auto covers and belts must have a more distinctive finish, which may run all the way up to that imitating the skin of the alligator.

According to the finish, this grain is rolled or pressed in—the processes being known as pebble rolling or embossing. The embossing is done largely on hydraulic operated machines, which bring two flat surfaces together with the leather between them, one of the surfaces being the die for the particular grain being run. Fig. 12 gives an idea of the base of one of these machines. It is a massive steel casting. In 1916 one one of these cracked as shown. It was Thermit welded and put in service, breaking again in a week. This was repeated and it again broke. Then the welders filled the entire hole in making the weld, on the theory of equal heating and contraction of the mass. The hole was bored and the machine is running to-day as sound as ever.

HANDY RULES FOR BELTING

Rules for finding length of crossed belts:—First, find the length of straight belt. Square each the diameter of the large pulley and distance between centres. Add together and extract square root of sum. Subtract from this the distance between centres. Multiply the remainder by two and add to length of straight belt as previously found. The result will be the length of crossed belt.

As a belt depends entirely on its power of adhesion to the pulleys to perform its function, see to it that its power of adhesion is maintained at the maximum by not overloading, and by the consistent use of suitable belt dressing, sparingly but regularly applied. Do not allow the dressing to clog on the pulleys. As belt slip is such an insidious danger—it makes no noise—and may be going on unsuspected until irretrievable damage has been done to the belt. Begin the application of belt dressing immediately the belt is put into use.

Rules for Finding Length.

When it is not convenient to measure with the tape line the length required, the following rule will be found of service: Add the diameter of two pulleys together, divide the result by two, and multiply the quotient by 3 1-7; add the product to twice the distance between centres of the shaft and you have the length required, substantially. If one pulley is considerably larger than the other, a little extra allowance should be made, because the distance from the centre of the top of one to the centre of the top of the other is a little greater than the exact distance between the centres of the shafts.

To ascertain the number of feet in a roll of belting: Add diameter of roll in inches to diameter of hole in centre. coils in roll, and then by 132; the three left-hand figures being the number of feet in roll.—Belts.

INTERNATIONAL STANDARDIZATION OF WIDTHS ACROSS FLATS ON NUTS AND BOLT HEADS

The Swiss Standards Association has addressed a communication to the national engineering standardizing bodies of the various countries, proposing the international standardization

---

MAKE USE OF THE STAFF ROUTING LIST

A decided innovation has been started in this issue in the form of a convenient staff routing list. This is placed directly on the front cover, and we strongly advise our readers to make good use of the same.

It is hardly necessary to explain its use, but what we can emphasize is the advantage of using such a list. Your magazines multiplies its usefulness, and does not merely rest on your desk after you have perused its contents. Simply mark the list saying where you want it passed on to, and if in your opinion there is an article of special interest, check it off and refer to the page number. All this takes but a moment, but the benefits derived are surprising.

Note that a place is left on the list to ensure the magazine being returned to the proper party for filing purposes. Use this list and emphasize to the other parties concerned the importance of studying and marking articles of real merit. The success of this venture lies with the readers themselves.

---

of the widths across flats on nuts and bolt heads.

The proposal covers the range of ¼ in. (6 mm), 3 in. (80 mm), diameter of bolts. The numerical values proposed are a compromise between the United States Standard, the British, or Whitworth, and the metric "Système Internationale." The communication states as follows:

"If at this moment the parties concerned were willing to do their best to bring about an agreement, a satisfactory international standard could be attained, while, when nationally individual standards would be introduced into practice, an attempt to reach an agreement could hardly be expected to be successful. We hold it advisable to subject the matter to a referendum of widest possible scope. It seems to us that by generous co-operation we ought to be able to find some system, which could induce France to agree upon, and at the same time form a foundation which would render possible the adoption by America and England.

The V. S. M. Committee of Standards being ardently urged by the Swiss industry, is compelled to decide upon a system of widths across flats in the near future, but doesn't care to do this unless one more attempt to bring about an international agreement in this matter be made by the V. S. M. Bureau of Standards." The communication was addressed to the standardizing bodies of America, Belgium, England, France, Germany, Holland and Sweden.

DUPLICATE BELTS

In a color and paint factory there were in service a number of grinding and pulverizing machines. These machines were belt-driven, and because of the excessively hard service to which they were subjected, as well as the unfavourable working conditions, difficulty belts. The belts often broke open at the joints, when an abnormal stress was produced. As a rule, this resulted in the shutting down of the machine for a half-hour or so while the millwright was sought out and the belt fixed up and put back in service. These breakdowns usually tied up other operations, too.

The millwright finally hit upon the following scheme: He made up duplicate belts for all of the machines that gave trouble, fitted these belts upon the pulleys, and then took them off and hung them in closets adjacent to each machine. Then when the belt in service broke the machine operator had only to go to the closest, get out the duplicate belt and slip it onto the pulleys, involving a tie-up of but a few minutes. The broken belt was promptly fixed up by the millwright, and put back into the spare belt closet for future use.

While this plan ties up a little money, it eliminates delays in connection with the grinding operations, and can well be duplicated in other establishments where troublesome belts are found.—Factory.

WELDING
AND CUTTING

Lessons on the A.B.C. of Good Welding

This is the Seventh Lesson by an Authority on the Subject — The Remainder of the Series Will Follow

By W. B. Perdue*

ALL the welds described in previous lectures have been made by forward movement of the torch. By this is meant that the progressive motion of the torch is in the same direction as the progressive work on the weld. When the weld is begun at the left and worked to the right it will be seen by reference to Figures 25 and 26 that the torch must necessarily back away from the finished portion of the weld, hence the name "Welding Backward."

This method of welding dates from the birth of the oxy-acetylene industry, and was the first to be used in France with the pioneer blow pipes for high pressure acetylene. That is with acetylene under pressures from 5 to 15 pounds. By using the acetylene under pressures a trifle higher than that of the oxygen, these pioneers eliminated the tendency to flash that is ever present when the heat of the flame is thrown directly backward upon the welding tip. They were not able to secure proper mixtures of the gases or a stabilized flow of same, consequently the process was superseded by the method of "Forward Welding," which has been outlined in previous lectures.

Many of the pioneer French generators were of the low pressure type which necessitated the use of injector type torches. The tendency of these torches to flash upon being brought too

*Welding Instructor, Heald's Engineering School, San Francisco, Cal.

close to the metal mitigated against the use of this form of welding, with the result that it was not brought seriously to the attention of the welding public

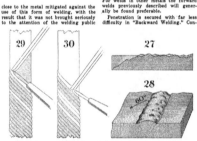

Fig. 27.—Section through centre line of bevel showing appearance of "backward" or "Vertical" weld improperly executed.
Fig. 28.—Improperly executed "backward weld." There is very likely to be a thinning of the edges at the point of contact with the bevel.
Fig. 29.—Showing start of fill, working Vertical weld from the bottom up.
Fig. 30.—Vertical weld. The metal is held in place partly by the action of the flame, partly by the manipulation of the rod.

until recent developments in methods for stirring and stabilizing the flow of gases from the torch eliminated the possibilities of the flash and at the same time

made it possible to execute a weld of this type without danger of burning out any of the essential elements of the metals welded, or impregnating them with impurities from the welding flame.

For welding heavy sections of steel or steel casting this method of welding will be found to possess many advantages. For welds in other metals the forward welds previously described will generally be found preferable.

Penetration is secured with far less difficulty in "Backward Welding." Con-

sequently the welder who first learns backward welding seldom if ever secures penetration when he finds it necessary to do "Forward Welding." For this reason the student is first required to master "Forward Welding." With very little practice he can then become proficient in "Backward Welding," after which time he is thoroughly competent to use his own discretion as to the means to be employed in handling any complicated piece of work.

Position of Rod and Torch

Reference to Figure 3, with the first lecture of this series (our issue of April 22), shows the proper position to be assumed for either forward or backward welding. Figures 25 and 26 are intended to show a cross section of the weld through the centre line of the

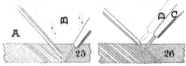

Fig. 25.—Welding backward. Showing position of rod and, torch at the beginning of a fill and in normal work.
Fig. 26.—Welding backward, showing position of rod at finish of ripple and positions of torch to secure finish of same.
These figures also show positions used in working vertical weld from the to pdown (see text).

bevel. The position of the torch in Figure 25 is that ordinarily assumed for thin sections; the position "C" in Figure 26 shows the highest point on the fill reached by the torch in welding heavy sections except the "finishing swing" as shown at "D," which serves to give finish to and melt the upper edge of the ripple into one homogeneous

of 30 degrees on the edge of each of the pieces to be joined.

No channels or depressions which represent a thinning of the edges will be in evidence at the edges of a good weld, neither will there be any lumps, hills or valleys present, or any side lumps formed by an adhesion of the melted metal from the rod on the un-

operator who makes frequent test welds and subjects them to bending, shock and etching tests to determine his skill in handling the torch. Such tests should be made even by the experienced operator before undertaking any weld of extreme importance, the failure of which might endanger life or property.

Welding Rod

The symmetry of the weld depends largely upon the motion imparted to the welding rod, as does the elimination of impurities and the grain of the metal within the weld. The motion imparted to the torch should not be rapid, but that of the rod must be as vigorous as can be continuously maintained by the operator. This motion for materials in excess of 1-8 inch should be the same as that indicated for the torch in the small diagram below Figure 1 in the first lecture of this series. For thinner sections the rod is alternately driven to the bottom of the melted pool and lifted to the surface.

FIG. 62.—CLOSE UP OF VERTICAL WELD. IT IS QUITE IMPORTANT THAT THE RIPPLES EXTEND ALMOST STRAIGHT ACROSS.

mass. Figures 27 and 28 show the rough and irregular appearance that is often obtained by inattention to the proper formation of the ripple or from an attempt to add too much metal at any one time. The side motion imparted to the rod and torch is the same as this illustrated by the small outline below Figure 1 (April 22 issue).

The angle of inclination of the torch can be checked up at any time by the operator, but it is quite essential to the smooth appearance of the weld that this angle of inclination remain constant. Holding the torch too near the horizontal will disturb the even regular appearance of the ripple and further tends to melt the rod too rapidly. This, too, may cause oxidation of the added metal by sucking in too great an amount of oxygen from the nearby atmosphere.

In backward welding particular attention must be given the selection of tips of the proper size. The end of the white cone must at all times just lick the metal.

Unless the cone is forced all the way to the bottom of the bevel the flowing together of the edges of the weld may form adhesions at or near the bottom. The size of the tips selected will depend to some extent upon the experience of the operator. Beginners will do well to practice with smaller tips than those used by the more experienced operator, while experts can use much larger sizes to advantage.

Execution of Sound Welds

Sound welds will be quite similar to those made by forward welding, the principal difference being a slight variation in the form of the ripple and the width of the weld. Forward welding requires a bevel of a full 90 degrees, backward welds may be executed with a bevel of 60 degrees; that is with a bevel

melted edges. In the work of beginners these faults will be found quite common. The reverse side must be similar in appearance to the forward weld with the exception that beads are generally in evidence instead of the better defined ripple. In any case penetration must be thorough and complete in order that there may be no starting points for cracks.

The tendency to hold the torch dead and burn or slick away the metal from the sides of the bevel is greater than with forward welding, but the tendency to form adhesions is lessened. In either case many regrets will be saved the

It is not necessary at any time to lift the rod from the melted bath or pool of metal in which it is being manipulated.

The choice of a steel rod of the proper grade and size merits some attention. A rod that is too small may cause oxidation of the weld, while one that is too large absorbs too much heat from the melted bath and retards the speed of welding. Either extreme may seriously influence the quality of the work.

A very simple rule to determine the approximate size of rod for any steel weld is to take ⅜ the thickness of the metal and to the result add 1-16 inch. This will work out in all cases except thickness under ¼ inch for which due allowance must be made. In using this rule whenever fractions of an inch

Continued on page 160

FIG. 31.—SHOWING PROPER METHOD AND POSITION FOR EXECUTION OF VERTICAL WELD.

DEVELOPMENTS IN SHOP EQUIPMENT

ELECTRICALLY CONTROLLED

THE Albert Herbert Co., 31 Yonge Street, Toronto, Canada, who are Canadian agents for the Stirk line of planers in Canada, have supplied us with some very interesting information relative to this novel planing machine.

The machine, which is called a "Hilo-plane," and is illustrated in Fig. 1, contains a number of interesting features including a reversing motor drive (current to which is supplied through a special generator set which is part of the equipment), magnetic feeds, the introduction of cross planing, the simplicity of setting table stroke on a graduated dial, no table dogs being employed for this purpose, but only in cases where one desires to skip gaps, as is often necessary.

In its mechanical details the box section bed is of unusually substantial proportions. The driving pinions are forged solid with their shafts, as the makers contend that at the speeds at which they run this machine solid pinions are safer in every way. One change of speed for slow speed work is obtainable in the table gearing. The entire absence of belts, and the convenience of quick power traverses in all directions may be noted.

The most striking features, however, are the flexibility and ease of control obtained by the special system of driving and the magnetic feeds. The essential

FIG. 2.—DETAIL OF MAGNETIC FEED CONTROL.

principle of the drive is the Ward-Lennard System of using a special generator to supply current to an individual motor, the control of the latter being principally accomplished by manipulation of the fields of the former. The driving motor (which may be an alternating or direct current motor) is directly coupled to the special generator which in common with the final reversing motor must be a direct current machine, with independently excited fields. Variation in strength of the generator field provides a variable voltage with the corresponding

variation of speed in the final motor. This gives an infinite range of speed downwards from a given normal by reducing the generator voltage to zero, although in practice the speeds are never reduced below 25 per cent. of the normal on account of the reduced torque. The overall range of the final reversing motor is usually increased from 8 to 1 by the variation upwards from the normal, by weakening the motor shunt field, the speed variation being accomplished by adjustment of resistance values in the generator and motor field. It is a simple matter to secure sudden alterations of speed at any part of the stroke and an accelerating switch is fitted on all hilo-planes, which enables cutting speed to be increased after the tool has entered the metal, or table speed increased to return speed, in order that gaps between surfaces may be quickly bridged.

An exceptionally wide range of feeds is obtained by the special magnetic feed control which is operated by solenoid and small separate motor, illustrated in Fig 2. Light feeds are obtained by solenoid control only, broad feeds are actuated by the motor controlled by the solenoid working in synchronism with it, the feed obtained being according to the position of the handle in the horizontal slot "a" shown in Fig 3. The motor is only energized for a few seconds at the beginning of the cut and is fitted with an adjustable slipping clutch. The feed motor mentioned is also used for quick power traverses to the heads, which are obtainable in all directions, also for

FIG. 1.—GENERAL VIEW OF THE HILO-PLANE.

FIG. 3—ILLUSTRATING THE FEED BOX

cross planing, a unique feature very valuable for short bosses on large pieces, and for jig work and vertical planing or slotting.

The control of table stroke, or cross stroke to head for, cross planing is exceedingly simple—this is clearly illustrated in Fig 3. The length of stroke is obtained by setting the dogs to the necessary graduations, which are direct reading in feet and inches. The reversal of stroke is obtained by the action of the dogs on the master reversing switch shown in the same illustration. The combination enables a stroke of only 4 ins. to be obtained on a twelve-ton table. This stroke control entirely does away with table dogs and enables the operator to set both ends of the travel accurately from a given point. The table may be started or stopped at the master switch or from stations on either side of the bed; a hanging pear switch is also provided for convenience in setting up—it acts as an inching control; a safety switch is also provided to prevent the table running off the rack.

The simple type of switch gear employed is to be noted. The disk and master switch mentioned in the previous paragraph, and illustrated in Fig. 3, actuate two contractors, which, without any other switch gear whatever, provide the quick return reverse to the table motor. This apparent impossibility is accomplished by the division of the shunt field winding of both generator and final motor into two sections.

In the case of the generator these sections are connected to oppose each other, so that alternative polarity results from the alternative excitation of the two sections.

The two sections of the motor field are connected to agree, and one is permanently excited. The second section is excited during the cut stroke only, and a sole cutting speed and a quick return are thereby provided.

The Hiloplane is not made in smaller sizes than five feet by five feet in the housings, but Messrs. J. Stirk & Sons,

Limited, Halifax, England, manufacture a very complete line of planer machines, from the smallest size upwards, in standard types with various drives.

SMALL TOOLS

The Jones & Shipman, Ltd., Leicester, England, who now have a Canadian agency in the Page Bldg., King and Jarvis Sts., Toronto, are introducing to the Canadian market their varied line of small tools. It is perhaps not correct to say introducing, as this line is already known to some manufacturers in Canada, but they are now out to make every plant manager know of their line. For the benefit of those not familiar with this line we might state that it includes tool holders of all styles, cut-off tool holders, turning tool holders, etc.; knurling holders, threading tools, boring bars and holders, lathe dogs, heavy and light duty; drill sleeves, taper sockets, taper gauges, drill and lathe chucks, and in fact all small tool lines, including, of course, their well-known centre drill.

This style centre drill was placed under a very severe test, being placed by comparison with other makes. The result was that their drill completed 359

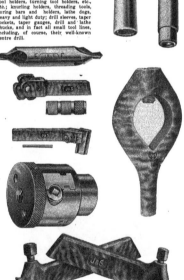

A GROUP OF VARIOUS J. & S. SMALL TOOLS.

GENERAL VIEW OF SLOTTING ATTACHMENT

holes before breaking against 44 and 153
of other makes. The claim that this concern makes for the long lasting qualities
of these drills are as follows: The drill
is completely ground all over, and is
ground after hardening, which, of
course, eliminates all distortion. The
tip being ground, also the lip, means
that every hole drilled is true to size,
smooth and free from ridges. The
diameter of the tip is also slightly
greater at the point than at the root.
The cutting edge of the countersink is
at an angle, which eliminates any tendency to chatter, and after hardening
every drill is backed off on machines of
their own design to absolute exactitude.

Their various tool holders are made
from steel drop forgings, and the cutters or tools supplied are made from the
finest grade high speed steel. On their
knurls, they make a point of generating
the teeth on the knurl discs, which ensure not only accuracy, but a true knurl.

Safety setscrews are used on all lathe
dogs, and both straight and bent tails
are handled. The sleeves and taper
sockets are made to absolute standards
and are of splendid workmanship
throughout.

Every tool passes rigid inspection before being sent out to the trade, and, as
stated to our representative, all tools
are guaranteed to be correct.

They also make a new two-jawed
chuck, which is deserving of close attention, and in a later issue we will go
more deeply into the individual merits
of their respective lines.

MILLER ATTACHMENTS

The Ford-Smith Machine Co., Hamilton, Canada, are now manufacturing a
full line of miller attachments in addition to their regular line of grinders
and millers. Any of the usual style attachments can be supplied, and two of
these lines are illustrated herewith.

Fig. 1 shows a clothing attachment
which has a 3¼ in. maximum stroke.
All parts are of sufficient strength to
withstand heavy duty, and the machine
work put into this attachment is up to
the usual Ford-Smith standard.

The Universal milling attachment illustrated at Fig. 2 has a wide range and
covers all general requirements of such
an attachment.

The slotting attachment receives its
motion by means of a special nose sleeve
which enters the main spindle. This
drives a disc, on which is a toe slot. The
connecting rod of the slotter ram is fastened to this disc, and, of course, the
length of stroke can be changed in an
instant.

The universal attachment is driven by
gearing, which, of course, first receives
its motion from the miller spindle. Anyone interested in either of these fixtures
or other miller attachments can get additional information from the makers.

HYDRAULIC PRODUCTION PRESSES

The Eastern Production Co., Detroit,
Mich., manufacturers of hydraulic machinery, have developed a full hydraulic

horizontal press for sizing the ends of
axle and pinion tubes used in motor car
construction. The sizing of these tubes
is necessary for pressing into housings
and flanges on either side. The time required for the sizing of the two ends
of an axle tube and within a limit of
.001 of an inch in diameter and removing from the machine, is thirty seconds.

Up to the present time the method of
sizing the two ends of an axle tube is
by grinding or turning. By this method
the sizing is left entirely to human element, resulting in a loss by reason of
turning or grinding under size from 5
to 20 per cent. The machine which this
company has developed receives the tube
in its receiver, presses it into a die, both
ends at the same time, and removing it,
therefore the diameter of the tube is
fixed for the size of the die and is
easily held within fixed limits with no
scrap.

In addition to this the machine is used
for reclaiming of tubes which have been
turned or ground under size. The operation of this machine is by hydraulic
pressure developed by a 3-cylinder hydraulic pump. The ram is driven with a
speed of 1¾ ins. per second and is returned by hydraulic pressure with the speed
of 3 ins. per second. Sizing dies can be
replaced readily and the machine carries
a built-in motor. It is made in 35, 50,
75 and 100-ton capacity.

Among other developments of the
Eastern Production Company in production equipment are vertical hydraulic
presses from 10 to 2,000 tons standardized on 20, 35 and 50 tons production
basis; horizontal presses standardized on 20, 35, 50, 75, 100 and 200-ton
basis; hydro-pneumatic presses of 10, 15,
20 and 30 tons; extra heavy presses for
assembling 5-ton truck axles; special
broaching machine, 20-ton capacity, 30-
inch stroke for 20-inch broaches, used in
broaching universal joint yokes and
spline gears.

DIFFERENT STYLE TOOL HOLDERS AND CUTTERS.

THIS ILLUSTRATES THE UNIVERSAL MILLING
ATTACHMENT.

NEW THINGS IN MACHINE TOOLS

WHIPP CRANK SHAPER

The Whipp Machine Tool Company, of Sidney, Ohio, has brought out an improved type of shaper. Quick and accurate adjustment is provided for the swivel head, by means of an eccentric lock. Micrometer graduation is fitted to the down feed screw. The automatic cross feed may be adjusted while the machine is in motion, likewise the stroke of the ram. The ram has a stroke of 16½ inches. Table has a vertical travel of 13 inches and a horizontal movement of 20 inches. Vertical feed of tool head, 6 inches. Table dimensions, 10 x 16 inches top surface, by 12 inches deep.

INSERTED LATHE CENTRE

The Federal Steel Company of Detroit, Mich., is placing on the market a special inserted centre for use on lathes, milling machines and grinders. The small inserted portion is made of high-speed steel, the shank is made of carbon steel, and the fit is made with a Morse standard taper.

DOUBLE-SPINDLE GRINDING MACHINE

A grinding machine that is intended for the finishing of both sides of parallel pieces has been developed by the Badger Tool Company, of Beloit, Wis. The disc wheels are 20 inches in diameter, and are carried on spindles 2 3-16 inches in diameter. Special facilities are provided for preventing the grit from getting under the sliding head. The sliding head is equipped with positive micrometer stops. The machine can be supplied for either belt or motor drive.

SCREW PRESS

The Manhattan Machine and Tool Works of Grand Rapids, Mich., are now manufacturing a special screw press intended for the testing of dies and other experimental work. The punch head is operated on four posts and is operated by means of a central screw that carries a large ratchet wheel, operated by a pawl that is fitted to a lever fulcrumed to one of the corner posts. This gives a much more powerful drive than where the screw is operated direct.

SPECIAL SCREWDRIVER

The Ackland Specialty Company of Springfield, Mass., have placed on the market a screw-driver and attachment for the specific purpose of removing stubborn screws that have become rusted in position. The screw-driver is operated by means of a ratchet wrench and pressure is applied by means of a lever which is fulcrumed on a shaft that is placed in a convenient position. The ratchet wrench is double ended, one end fitting the Ackland socket and the other adapted for the standard sockets.

PRODUCTION DRILLING MACHINE

A drilling machine that has been designed for quantity production has just been placed on the market by the Minster Machine Company of Minster, Ohio. The power is transmitted through hardened stub-tooth gears. Two sets of speed change, and one set of feed transposing gears are fitted to the machine. Provision is made for the disengagement of the feed when the desired depth is attained. Spindle is fitted with a ball-thrust bearing. The machine is primarily intended to be belt driven, but may be fitted with a motor if desired.

SAFETY PRESS GUARD

The D. & M. Guard Company of Rochester, N.Y., has recently designed a guard for power presses that provide three distinct guards that are arranged to swing independently and in proper time with each operation of the press. One of these guards is operated by means of the foot treadle, swinging across the front of the die; the second one has the same movement but is operated by the downward motion of the ram. The third "safety" prevents the feeding of the press from the side. The bridge carrying the three guards is hinged and fastened by means of a pin and padlock.

BAR STRAIGHTENING MACHINE

The Whitney Metal Tool Company of Rockford, Ill., are now manufacturing a straightening machine that can be placed on the bench or a machine for handling shafts that require to be straightened before working. The pressure head that carries the screw is fitted to a slide and may be located in any position along its length. A pair of V blocks are provided, and to the side of these is fitted a set of disc rollers on which the shaft is revolved when determining its accuracy. When pressure is applied the wheels depress so that the shaft rests upon the vee's in the block.

MULTIPLE-SPINDLE PROFILING MACHINE

The Automatic Machine Company of Bridgeport, Conn., has recently placed on the market a special machine of the multiple type, for the rapid profiling of irregular contours. The machine can be designed with any desired number of spindles and the construction modified to suit almost any requirements. The spindles are driven by means of helical gearing, the gear on the main driving shaft being long enough to engage with all the gears on the different spindles. The lateral movement of the spindle heads is obtained by means of a cam fitted to one end of the machine; the table movement is derived from another cam located at the rear of the machine. The combined motion of these two cams provides the path of the cutter, and these cams may be made to suit any desired contour within the range of the machine. The two cam shafts are driven by worm gearing and timed to act in unison, the drive being obtained from the spindle drive shafts, so that overload on the cutters will not affect the movement in a horizontal direction. A timed clutch is provided to control the feed, so that the feeding of the cutters will always stop at the point from which they were started. The head carrying the spindles may be raised or lowered, and a micrometer adjustment is fitted for accurate setting. The spindle design is of special construction for taking up the wear in the bearings.

ADJUSTABLE LIMIT GAGE

The John M. Rogers Works of Gloucester City, N.J., are now making a special line of snap gages. The malleable iron frames are designed for strength and rigidity and are well seasoned before machining to guard against the danger of warping or twisting. The anvils are made of alloy steel, hardened, ground and lapped, and designed so that they will not turn in the frame. The anvils may be set to any required dimension within the range of the gage, and sealed after locking, to avoid the possibility of tampering. They are made in sizes up to 8½ inches.

CENTRELESS GRINDING MACHINE

An improved type of centreless grinding machine has just been brought out by the F. C. Sanford Manufacturing Company of Bridgeport, Conn. The grinding wheel, which is 20 inches in diameter with 4 inch face, is carried on a high-carbon heat-treated 3¼ in. steel shaft. The machine is intended for grinding cylindrical work from 3-16 inch to 6 inches in diameter, with lengths up to 20 inches. When large quantities of work are to be produced it is intended that several of these machines will be installed so that the work can be passed from one machine to another, each taking off a certain amount of the stock, to avoid the necessity of sizing adjustment, which would be required if only one machine were used.

VERTICAL DRILLING MACHINE

The Hoosier Drilling Machine Company of Goshen, Ind., has added a new type of 20-in. vertical drilling machine to their standard lines. Several sizes are made to meet the varying requirements of the trade. The larger sizes are fitted with back gear drive and power feed, and are especially designed for heavy service. Ball thrust bearings are fitted to the spindle, which is counterbalanced by a weight in the column. The machine has eight changes of speed, four with and four without the back gear. The graduated spindle sleeve is provided with an automatic stop. The centre hole of the table is bored in alignment with the drill spindle.

The MacLean Publishing Company

LIMITED
(ESTABLISHED 1887)

JOHN BAYNE MacLEAN, President.   H. T. HUNTER, Vice-President
H. V. TYRRELL, General Manager.

PUBLISHERS OF

CANADIAN MACHINERY
AND MANUFACTURING NEWS

A weekly journal devoted to the machinery and manufacturing interests.

B. G. NEWTON, Manager.   A. R. KENNEDY, Managing Editor.

Associate Editors:

J. H. MOORE   T. H. FENNER   J. H. RODGERS (Montreal)
Office of Publication: 143-153 University Avenue, Toronto, Ontario.

VOL. XXIV.   TORONTO, JULY 1, 1920   No. 1

The Fuel Problem

AN item in the news from Montreal discloses the fact that a shortage of fuel is causing great inconvenience there. Montreal uses large quantities of soft coal, which in the years before the war was shipped up the St. Lawrence in ample quantities to supply the needs and furnish a substantial reserve. In those days the methods of using coal which have largely contributed to the present shortage were very much in evidence, not only in that particular district, but all over the country where coal was used. During the strenuous days of the war a great cry for conservation was raised, and our Government came very nearly achieving something in the direction of teaching fuel users better methods.

With the passing of the high productive period of the war, the Government enthusiasm waned, and little more was heard of conservation from an official source. The mounting cost of coal and the difficulty of getting it is causing interest in this question again. The necessity for education of the men who are responsible for the proper burning of coal is as great as ever.

It is as true now as it was five years ago that the average efficiency of the steam plants in Canada would not reach the 50% mark. In some places it is feasible to use oil fuel, and the owner then comforts himself with the thought that he is getting good combustion efficiency. Perhaps he is, but it is just as easy to burn oil inefficiently as it is coal.

From the point of view of economy, and also the conservation of natural resources, it is time that a really comprehensive effort was made to make the problems of combustion better understood among the rank and file of power plant operators. It would well repay a group of manufacturers in a district to pool together and get a combustion expert into the district to personally visit each plant and set them in the right way. The saving that can be made in a steam plant by the exercise of a little care with no expense at all for instruments is enormous. We speak of the average plant, not of course of large installations where the best of scientific methods are in force. The average power house operator is anxious to do the best by the plant, and it is up to the owner of the plant to give him every facility.

If half the energy that is devoted to other forms of propaganda were to be put into this question of fuel efficiency, it would not be long before the country would be getting along with half its present consumption of coal. What is needed is a concerted effort, and representative bodies of manufacturers and engineers should get together and devise a programme of education that would cover the country. The question of expense need not deter them, bcause they will get it back a hundred fold. It will be a long long time before we can do without coal or oil fuel in this climate, and it will be a considerable time before the hydro-electric power entirely supplants steam, if it ever does. The more general use of the internal combustion engine will no doubt relieve the situation to some extent, and the installation of more economical steam engines is a factor in the right direction. The greatest source of waste is still the boiler furnace, and that is the place where the remedial measures should be applied in the first instance.

In the British Market

FOR the most part Canadian goods are well thought of in the Old Country markets. Britain has been for some time, and continues to be, our best customer when it comes to selling. In 1919 Canada sold to Old Country buyers $528,000,000 worth of goods, or 43 per cent. of the total amount of our sales.

Formerly the bulk of our trade to Britain was in raw materials or farm produce, but more and more we have been sending in the finished material, and it is getting a good reception in the British market. One firm recently went to a number of the larger stores in Britain to find out how the Canadian goods were received. In groceries, provisions, flour, rolled oats, bacon, cheese and salmon it was found that the Canadian article was considered as the standard of the trade. Strange as it may seem, our spaghetti and macaroni are competing with the article made in Italy for the London trade. It was hard to get a British dealer to admit that Canada, or any other country for that matter, could make jams to equal the Old Country product. Canned fruits were severely criticized in many particulars, although Canadian apples are acknowledged as being the best reaching the market.

Canadian made garden utensils, tools and garden furniture sell well and are popular. The Britisher is wedded to white enamelware, and in that essential the Canadian product, which is often finished with a grey or brown speckle, is lacking. They consider our tinware rather clumsy, but oil stoves compete with anything sent in from American makers. The Canadian kitchen range is perhaps a little too heavy and large for the British requirement.

For electrical equipment there seemed to be a real opening for Canadian-made goods. Only a few lines appeared in the stores visited, but these had sold well against all competition. It must be remembered that there are many voltages met with in Britain, and electrical goods going in there should be of the universal type. Woodenware, kitchen cabinets, fibreware, etc., from the Dominion are acknowledged to be the best reaching the British buyer.

In spite of all that was said during the early stages of the war about Canadian-made shoes, there is a good sale for the Canadian article.

One thing that the Canadian exporter must bear in mind—no matter whether his product is great or small, metal, leather or wood—or anything else—he can offend the British market by a poor shipment, and his poor stuff can prejudice the minds of British buyers against anything and everything that comes from Canada. There is a great big responsibility attached to each Canadian seeking to enter the British or any other market abroad. The customer may form his idea of all Canadian goods from the quality, or lack of it, that is put into any one shipment.

The Week in Review

F. M. RIGGS, vice-president of the Desmond-Stephan Co., Urbana, Ohio, was in Toronto during the week. He had with him an interesting collection of diamonds for the trueing of abrasive wheels. Stones, even for this purpose, are expensive at present: They are uncuttable diamonds, and nothing gets past now that can be utilized for a cut stone. The increased earnings of a great many people during and since the war has caused them to become buyers of diamonds, which has driven the prices to new levels. It used to be quite common to find a cuttable stone in a shipment of supposedly uncuttable ones, but such an event now is a rarity.

*　*　*

The new Australian tariff, which is to become operative at the start of 1921, gives Canadian exporters who have been selling in that market something to think about. The Australian Government is putting on a real tariff, fixing the price at which articles shall be sold to the home trade and fixing the rate that shall be paid to the men engaged in the industry. Canadian implement makers have built up a good trade with Australia, but they will find it a hard matter to get through the tariff wall, which is about a forty per cent. structure, and do a profitable business. Castings are allowed in for 10 per cent, apparently with the idea of getting some outside firms to come over and start business by doing their assembling there. Canadian firms have not yet announced how they will meet the new conditions surrounding Australian trade.

*　*　*

The closing of the finishing departments of the Steel Company of Canada is an unfortunate occurrence, for Canadian industry, as well as for the company. A steel mill, all through, is of necessity a continuous operation, and a break is a hard thing to overcome. Going back to first causes it is quite true to say that the Steel Co. of Canada is closed to-day because a gang of outlaw switchmen decided to paralyze traffic on United States lines, and succeeded to a degree that the authorities do not care to recognize or admit. The coal supply from U. S. mines was cut off from the Hamilton plant, and a steel mill can no more run without coal than a blast furnace can operate without ore. The strike of United States, outlaw or recognized; are to all intents and purposes strikes against Canadian industry if they touch lines for which we depend on United States for our supply.

*　*　*

The fact that their men went on strike a few months ago has not made any difference in the shop bonus and service bonus plans of the McKinnon Industries at St. Catharines. In the natural course of events it might have been thought there would be a tendency on the part of the company to retrench in this line of work following the strike. "In spite of the strike, we consider it good business," was the way one of the company officials stated the case to this paper. The company claim that many of their men went out through pressure from their unions, fearing that if they stayed they might lose their membership.

*　*　*

After all, a personality counts for a lot. One large Canadian firm of implement makers have several men on the roads of United States keeping cars moving in the direction of the plant in Toronto. "One of our men in an old railroader, a big chap who knows how to meet men and talk to them. He can get more cars moving in a shorter time than all our representatives put together." That is the company's estimate of this man. And it's the wise company that watches out for these men and develops them. The man who has magnetism and personality, if properly treated, can be made a wonderful asset in the organization. On the other hand, if he is ignored or

snubbed he can become a dangerous man. He can attract men and influence them, and the company that is not on the alert for this ability has not learned the first lesson in real organization work and team play.

*　*　*

The fact that the steel companies and some of the other large melters are not buying scrap now may cause the prices to drop on such lines as heavy melting steel, machinery scrap, etc. If you can afford to do so, it might be better to hold material, for a recovery is almost certain to come in one way or another. An improvement in the coal and coke situation will carry with it a call for several lines of scrap material, especially of the iron and steel kinds, and prices will be stronger.

*　*　*

Writing to this paper from Glasgow, Scotland, James Johnstone & Son, iron and steel merchants, 212A St. Vincent St., remark on the low prices quoted in Canadian scrap metal trade. A comparison shows the following:

	Canada	Scotland
Heavy melting	$18.00	£11/10
Rails	18.00	£11
Malleable scrap	25.00	£12
No. 1 machine cast iron	33.00	£13
Steel axles	20.00	£15

Export duties on iron and steel have been fixed by a Luxembourg Grand Ducal Decree as follows, all duties are per metric ton:—Pig iron, 40f.; ingots, bars, billets, shapes, wires, plates, sheets, &c., 60f.; scrap, drillings, turnings, &c., of iron and steel, 25f.; iron ore, mill and other scale, 1f. The duties were to be levied as from May 1st, and may be revised every month.

The electrification which has been adopted for the Paulista Railway, a line of about 28 route and 76 track miles, in the State of Sao Paulo, is to be worked on the 3,000-volt continuous current system, with overhead contact lines and geared locomotives. The arrangements will be very similar to those of the Chicago. Milwaukee and St. Paul line in America. The electrical supply for the railway is to be obtained from the Sao Paulo Light and Power Company. This company obtains most of its energy from hydraulic stations, and will provide the railway with a three-phase, 60-cycle, 88,000 volt supply.

Bernhill in *Punch*

HIS OWN BUSINESS.

Uncle Sam—"If I wasn't so preoccupied with Ireland I might be tempted to give myself a mandate for this."

MARKET DEVELOPMENTS

Just 12 Weeks Coming from Pittsburgh

Some Shipments Are Arriving Now that Have Had a Long Wait to Get Through—Scrap Market Sags When the Large Buyers Drop Out—Machine Tools Are Quiet.

THERE has been a quiet period during the week in machine tool circles, and the same is true of several kindred lines. There is nothing unusual about a dull period at this season, and a revival of machine tool buying is looked for as soon as there is something more definite in regard to the securing of material.

There is little, if any, improvement in the manner in which shipments are coming across the line. Some warehouses have been fortunate in getting shipments through, but in one case a car had been on the road for twelve weeks from the Pittsburgh district. Canadian firms in increasing numbers are sending their own men across the line in an attempt to get cars moving in their direction, and some of them are meeting with pretty fair success in this work.

It was thought that there might be some revision in the prices of small tools, particularly drills, but Canadian makers find that they are paying a price for their high speed steel that keeps them from coming to lower levels. The idea of a cut would be to keep on a level with prices that are being quoted on Old Country material, although some firms claim the latter will have to come to higher levels, which will practically equal the Canadian price.

The scrap metal market, with the steel companies and other large melters out of the market, is in for a dull season, and dealers say that prices will likely be marked down on such lines as heavy melting steel, cast scrap, etc. The prices paid for cast scrap in Scotland yards now is away above the prices here, and reports have it that United States interests have been shipping there to take advantage of the high prices that are offered.

Dealers selling motors report that, with the shortage of power in the Niagara district, there has come quite a falling off in the sale of electrical equipment, while in other sections, where power is available in larger quantities, the sale of motors is very brisk.

UNSETTLED CONDITIONS PREVENT A LARGER TRADE IN MONTREAL FIELD

Special to CANADIAN MACHINERY.

MONTREAL, July 1.—The market is bearing up under the continued difficulty that features the transportation of materials at the present time. It cannot be said that that any pronounced improvement has developed in connection with delivery, although some jobbers claim that there has been a slight betterment in getting commodities through from the States. The threatening conditions that hover round the steel and railroad activities do not favor a speedy return to normal, and many dealers here fear that further disturbances in the States will be reflected in this district, and perhaps have a serious effect upon many lines of Canadian activity. Already the supply of soft coal has fallen off considerably, and many manufacturing plants have suspended operations, while others are contemplating shutting down or curtailing operations as a means of conserving fuel now on hand. Machine tool movement has shown a slight easing off, the demand being lighter and the delivery uncertain. Good business is still reported in small tools and accessories, with prices well maintained. The metal situation, while not active, shows improved signs of settlement. Demand is comparatively light but fair business is reported. Old materials are moving slowly, with accumulation of yard supplies increasing.

Local Supplies May be Affected

"The gathering strike clouds in the States may again set us back to where we were six months ago. Should the trouble assume any magnitude the present situation, which is far from being satisfactory, would become increasingly acute, and we here in Canada would be confronted with difficulties that could only be relieved by early adjustment of American labor troubles." This statement by a local dealer is some indication of the situation that may be created by drastic action of steel and railroad employees in the States. The coal conditions here in Canada are none too promising, and reduced American supplies would seriously affect Canadian industrial operations. Increasing quantities of Nova Scotia coal have been diverted for bunker purposes, so that industrial supplies from these fields have shown a marked falling off.

"Our mills in the States," continued the dealer, "have been showing slight increases in tonnage output, and receipts at this end have been better, but while endeavoring to meet the requirements of our customers we are not over anxious to carry heavy warehouse supplies, as much of the stuff coming through just now carries a premium, and this would not be a profitable factor on large stock accumulation. At present we are in a position to place the material in the hands of waiting consumers, but this condition may not always prevail. Further disorganization of the steel industry, and railroad inactivity, would tend to reopen a wound that was beginning to show signs of healthy convalescence." The sheet and light plate demand exceeds the supply coming in and dealers frequently find it a difficult matter to satisfy the needs of customers, as shipments coming in are, invariably, allotted so as to accommodate several of those constantly crying for material. The sizes in general demand are the hardest to get. Tubes are coming in slowly, but the demand is lighter.

Slight Improvement in Metals

The quiet condition of the metal market, which has been practically feature-

less for many weeks, has shown slight improvement, and dealers believe that the situation will take on a firmer tone shortly. Prices have been lower than those quoted but the firming tendency gives strength to the possibility of maintaining quotations. Tin is the only metal to show weakness and local dealers are asking the average price of 62 cents per pound.

Activity in scrap continues to be far from satisfactory and trade expectations do not favor any marked change unless it be for additional dullness. "Extremely quiet, with little hope of early change," said one dealer. "We were in the hopes that the worst was over, but the possibilities of trouble in American steel districts have not given a hopeful outlook for the disposal of scrap." All lines are listless and what little movement is going favors machinery cast.

JUNE HAS BEEN VERY GOOD MONTH

Warehouses Have Had a Good Trade— No Improvement in the Railroad Situation

TORONTO.—Occasionally an importer gets a deluge of cars from points across the line, but the occasions are not frequent. Improvement in the traffic situation may have taken place in theory, but in actual practice, as far as this district is concerned, it is not felt.

Dealers in machine tools agree that it is not unusual to have a quiet period at this time of year, and they are experiencing that now. The railroads are taking on some material now, but that is pretty well split between Canadian makers, and the dealers do not draw much from it.

Some of the dealers look to see a revival of machine tool buying in the early fall. They point out that by that time it ought to be possible to secure material for the shops to work with, and if the prospects of a good crop materialize, there should be enough backing in the country to warrant going on with several undertakings that are for the present tied up.

The power situation—or rather the power shortage—is having its effect on industrial life. In the Niagara belt dealers report that the sale of motors has dropped off very considerably, although in other parts, where power is to be had, the sale is very satisfactory.

The shortage of coal and the need of making the most out of a power or heating plant is making the call for tubes more insistent, and there it ends, for it is almost impossible to get through a shipment of boiler tubes.

The business being done in small tools is smaller in volume, according to a summing up of reports from several sources. There was some talk of a change being made in the relative prices of high speed and carbon drills, but on going over the figures the makers found that they were paying such a price for their high speed steel to make the drills

POINTS IN WEEK'S MARKETING NOTES

Coal companies have a preference now in getting cars up to a full 100 per cent of the mines rating.

—

Pittsburgh bankers do not want to see call prices go any higher as many of their clients are in business where they cannot pay excessive material charges.

—

Both Bessemer and basic pig iron have advanced during the past week, although it was expected that there would be a recession in these prices.

—

The steel mills in Ontario are largely out of the market for scrap material, and a slump in prices would not cause any surprise.

—

Some shipments of sheet came to Toronto this week that were placed on the cars at Pittsburgh twelve weeks ago.

—

The shortage of power in this district is having the effect of slowing down the sale of motors.

—

Importers fail to see any improvement in the matter of getting machine tools from American points.

that it would be impossible to make any drop in the selling prices.

Shipments Come in Spots

One Toronto jobber got through a dozen cars in about two days during the week, so many in fact that they could not be handled. But that is the exception and not the rule. Delayed arrivals are still the order of the market. Some of the material—mostly sheets—that arrived in the above lot had been on the road for 12 weeks. Fortunately there was no loss either way, to the jobber or consumer, as the price arranged was to be the price at the time of delivery. There is still necessity to appeal direct to Ottawa for rulings regarding the operation of the new tariff, as there are a number of cases where a special finding has to be made before there is anything like a precedent to go by. One line that causes trouble is where the boiler tubes (bent) have to be imported for a certain type of boiler. Local collectors hold that these are manufactured and therefore subject to duty, while Ottawa thinks that they are, as far as the use is concerned, simply boiler tubes and therefore entitled to come in free.

Warehouses are winding up their month, and contrary to expectations it has been, in most cases, a good one. It had been thought that with the trouble in getting shipments, a falling-off would be noticed in a good many lines. It may be that the tonnage is less

in many places, for the high price keeps the value of the business from going very far down.

Very little relief is to be had for the Canadian manufacturer, even from the Canadian mills. The shutting down of the finishing department of one mill, even for a short time, takes away material that is badly needed. Complaint is also made by some agricultural firms in this country that the material they are receiving has not the uniformity that characterized the shipments some months ago. Bars, for instance, do not always cut in units of two-foot lengths, and as a result there are more than usual short ends left over for the scrap pile. Firms needing special steels and alloys are finding it hard to get these, more so than the usual run of mill stuff.

Scrap Metals Are Quiet

The recovery of the scrap metal market in United States, which is spoken of as a possibility, may help conditions here, but there is nothing much else in sight that will do this. Dealers frankly say that they look for a drop in the prices now quoted.

The big melters that are usually taking on material are out of it for the present and another is well supplied for some time to come. So it may be that heavy melting, agricultural scrap, and perhaps wrought iron will come down from their rather lofty positions. The dealer from whom Canadian Machinery gets prices each week frankly states that the schedule quoted is nominal, and that he will not guarantee to take on unlimited amounts at these figures. Buying in many cases, would be necessary to keep up shipments against contracts. The reds and yellows are also dull. The consumers in Canada cannot take on half of the material that is stacked up in the yards now. The change that is taking place in British sterling may make a change in the scrap export business soon.

HIGH PRICES FOR SCRAP IN GLASGOW

Rumored That U.S. is Seeking to Sell in That Market at Attractive Figures

James Johnstone & Son, 212A St. Vincent Street, Glasgow, writing to Canadian Machinery under date of June 16, draws attention to the high prices being paid in Scotland:

"We are obliged for the June number of your paper just received. One thing that surprises us is the low prices of scrap material ruling with you. It may interest you to know the following prices are paying delivered into their works:

Heavy steel melting scrap, £11 10s. per ton; scrap iron rails, £12; scrap steel rails, £11; heavy malleable iron scrap, £12; No. 1 machinery cast iron scrap, £13; heavy steel turnings, £9 12s. 6d.; clean cast iron borings, £9 7s. 6d.; wrought iron railway axles, £18; steel railway axles, £15.

PITTSBURGH REPORTS CONDITIONS AS UNSETTLED AND PRICES UNCERTAIN

Special to CANADIAN MACHINERY.

PITTSBURGH, July 1.—Transportation conditions as affecting the shipment of pig iron and finished steel products have not improved during the past week or two, and on a precise showing would probably be seen to have grown a shade worse. Supplies of raw materials, however, appear to have been kept up, whereby production is not decreased. The rate of output is indeed quite high considering the general financial and industrial situation, which is one of doubt and uncertainty rather than of full activity and confidence.

Coal Movement Improving

Effective Monday of last week the Interstate Commerce Commission issued an order requiring that full preference be given the coal mines in the matter of car supplies, up to 100 per cent. of mine ratings. The previous rule had been for preference up to 50 per cent. Already there is a materially heavier movement of coal as a result of the order. Quite typical of the fault-finding spirit so generally prevalent, which is perhaps the main retarding influence now to real progress and prosperity, complaints regarding the coal car order are made in various quarters, and newspaper space is generously given to the complaints. For instance, the Connellsville coke operators complain that the order is going to take cars away from them, with the intimation that nothing could be worse than to restrict the movement of Connellsville coke. Now coke is used mainly for making and melting pig iron and the by-product ovens have been very short of coal, so that a better coal movement will help them. As to the relative importance, last year the Connellsville region produced 10,000,000 tons of coke and the by-product ovens 25,000,000 tons. After pig iron has been made, the steel mills need coal, and the better coal movement will help the steel mills also. Nevertheless, what one sees in the average daily or trade paper is not that the steel mills are going to get more coal, which they need badly, but that their car supplies for shipping steel may be decreased a trifle. Even then, however, it is improbable that any such thing will occur. Pittsburgh bankers, who are interested in seeing all the wheels of industry move freely, have been expressing grave apprehensions over the coal market situation, Pittsburgh coal having brought $10 per net ton at mine, a price some industries cannot afford to pay so that they are forced to close. If the industries cannot run there is no use making steel or shipping it.

American Federation of Labor

By reason of the action taken at the Montreal convention of the Federation of Labor it is thought in the iron and steel industry that no attempt will be made by the Federation to call a strike in the industry in the near future. The Federation directed that the committee which handled the strike of last September be disbanded, its field workers be called in by July 1, and its funds turned over to the Federation's executive committee. A meeting of the heads of the international unions is to be held in Washington, to appoint a new committee and lay plans for a fresh general campaign to organize the industry. No doubt the Federation will attempt to profit by the experience of the last strike, which had too much of the "red" element in it. Doubtless the Federation management did not like this, but it was not strong enough to curb the Fitzpatrick-Foster element that dominated the committee now disbanded, Fitzpatrick and Foster themselves having resigned several months ago.

Any activity of the Federation will be entirely apart from the Amalgamated Association, which broke with the union sheet and tin plate mills as detailed in last issue, and which will have a strike this summer unless there is a last-hour recession from the impossible demands it made upon the mills.

Coke and Pig Iron

Connellsville coke for spot shipment has advanced further, being now fully $16 for furnace and $16 to $17 for foundry. Many consumers will not pay anything like such prices, preferring to curtail operations, but the offerings are so limited that everything is absorbed by those who are willing to pay practically any price asked. Production in the Connellsville and Lower Connellsville region is about 70 per cent. of the rate in May before the rail strike and the consequent traffic congestion.

Before the traffic difficulties became pressing a moderate amount of furnace coke had been put under contract for the second half of the year, at what then seemed to be very high prices, but prices lately asked are still higher, and the furnaces are placed in a particularly awkward position because they are called upon to buy their coke for the half year and they have sold only a part of their output, perhaps less than half, for the six-month period, while the market is too dull now to sell the remainder, and no one knows whether pig iron is going to advance or decline on the next buying movement. The coke operators say it is going to advance. Some business was done, as noted in previous reports, on a 4-to-1 ratio basis, which would make the coke cost $11, if basic pig iron stays at its present level of $44, valley, but would help the furnace in case pig iron declined. The lowest flat price quoted seems to be $12, and furnaces do not care to risk paying that price.

While the pig iron situation generally is quiet and the balance of probability seems to be that before the end of the year iron will decline rather than advance, the bare fact is that Bessemer and basic have advanced in the past week. The only market to quote is the prompt market and deliveries are so scant that higher prices are obtainable. Two lots of 2,000 tons each of Bessemer and basic have been sold, for early delivery, establishing the market at advanced prices, $44 for Bessemer, or a $1 advance, and $44 for basic, a 50c advance, while foundry iron remains at its old price of $45, and strong at the figure, all prices here mentioned being f. o. b. valley furnaces.

Unfinished Steel

The billet and sheet bar market presents an easier tone, though prices can hardly be quotable definitely at lower levels than formerly There was a sale a few days ago of over 5,000 tons of sheet bars at $75, the old price, but it seems the buyer was influenced by ulterior motives. Reports have it that prospects of idleness at some of the sheet and tin mills, hitherto union, led to the offering of some sheet bars by these mills for resale, at between $65 and $70. Then there was a sale of a few thousand tons of small billets at $67.50, lower than the price lately regarded as the market in the absence of definite sales.

Prices on finished steel products for early shipment continue to soften, but the offerings in most cases are very light. Sheets can hardly be obtained at all. Plates, however, are more readily obtainable for reasonably prompt shipment, and 3.50c can be done, if not 3.25c, while a couple of months ago prompt plates were bringing 4c readily, and up to 4.50c in a few cases.

MAY SELL THE ROAD FOR SCRAP

Railway in British Columbia May Come to Its Last End in Scrap Yard

Toronto.—The Spokane and British Columbia Railroad, in the State of Washington, constituting the principal asset of the bankrupt Dominion Permanent Loan Company, may be sold for scrap, it was announced. The loan company recently failed with heavy liabilities to depositors and shareholders after payments of more than $2,000,000 in dividends from the capital, according to the liquidator.

Chas. A. Strelinger Co., who recently established their Canadian headquarters at Windsor, Ont., have made some considerable addition to the floor space of their warerooms and are now enabled to make effective display of their machinery, tools, and shop supplies. "Our parent house in Detroit has a 36-year start on us," said Mr. Doherty, the Canadian manager, "but watch us grow."

NEW MACHINE HAS BEEN SHOWN
AT THE BUILDING TRADES EXHIBITION

THE following information has been furnished by the British Trade Commissioner at Toronto, based on notes forwarded by the British Government's Department of Overseas Trade, London:—

Builders & Contractors' Plant, Ltd., 17 Victoria Street, London, S.W.1. exhibited a new portable hoist at the Building Trade Exhibition which would appear to have created a great deal of interest. The frame consists of angle-iron guides, and these are attached to the scaffold or brick-work. The carriage is raised to the top by means of a cable and will take either a barrow or other conveyance. At the head is the revolving apparatus by means of which the carriage is turned inwards and the barrow with its contents wheeled away.

This is operated by means of a geared friction winch which weighs in itself 330 lb. The whole apparatus weighs only 650 lb. The overall space required for the winch being 3 ft. by 3 ft. 6 in., it can be erected between two houses or in the middle of a block, and the material on arrival at top, fed in any direction. The rails are adjustable to any height. The winch can be driven by either a belt drive or a combined electric motor or petrol engine, and the power required to raise the load at a speed of 200 ft. per minute is only 3 h.p.

A larger size on similar lines for use on large buildings is also shown.

Clayton and Shuttleworth, Ltd.—At the nineteenth annual meeting, the chairman, in referring to the new programme of extension, said that favorable arrangements had been made with the Government for the erection of the Abbey Works and the Clayton Forge. Both were constructed primarily for war interests, but they had proved to be admirably suitable for normal purposes, and were amongst the finest works of their kind for the manufacture of rolling stock and dropped forgings, for both of which there was a great demand. As these works were brought into full production their output would be very large. At the Abbey Works they were now constructing Pullman cars, the very highest class of railway work. The demands for the Clayton steam wagon, introduced in 1912, had been so great that it required the provision of special manufacturing facilities. The Titanic Works were being prepared for that purpose, and they hope soon to have the larger, and best equipped works in the country for the exclusive manufacture of steam and electric commercial motors.

Two large foreign orders have been placed over the last half and general indications of the market tend to show that the last quarter should be a good buying period. Basic users not covered for the last half and now desiring to place orders are unable to do so in the Chicago district, and have been forced to order from Detroit firms at slightly higher rates.

On the whole conditions are not normal generally at the steel plants in the Chicago district, due to lack of coke and difficulty in procuring coal owing to transportation conditions. Dealers and selling agencies hold that the last two weeks have been exceptionally quiet. There is a certain timidity about buying, due to the scarcity of coke, and the general attitude adopted by the banks discouraging outlay on anything but absolute essentials. Transportation conditions are still the most prominent factor in determining the policies of both producers and buyers.

PRIZES FOR AEROPLANES

The progress of aircraft design and construction in Great Britain is being stimulated by the offer of valuable prizes by the Air Ministry. Three prizes will be given in each of the following classes; Small-type aeroplanes for six passengers; large-type aeroplanes for seven or more; and amphibious seaplanes for two passengers. In all cases the accommodation is to be exclusive of the crew. The prizes for the large-type aeroplanes are approximately $100,000, $40,000, and $20,000; and those for the small-type seaplanes and aeroplanes are $50,000, $20,000, and $10,000. The competition is open to British subjects for machines designed and manufactured anywhere within the British Empire. Each competing machine will be subjected to exhaustive tests drawn up in consultation with the Society of British Aircraft Constructors and designed to ensure all-round excellence in the successful machines. This competition is expected to have a very marked effect on the development of civil aviation.

A Gloomy Picture.—City Clerk Kent, of Hamilton, who returned from a trip to the mines, where he made an effort to secure a civic coal supply, gave a gloomy picture of the coal situation. He urges citizens to get what coal they can as soon as possible as he fears a great shortage this winter, and thinks it not improbable that many may go cold. Labor troubles and car shortage are the causes for the sub-normal supply. He knew of an instance where a manufacturer paid $11 a ton for soft coal at the mine, and after he had closed the deal the coal was sold to another for $11.50 per ton.

"It is reported that the United States firms have sold a very large quantity of melting steel scrap for delivery c.i.f. Glasgow."

FAVOR PHYSICIAN
ON THE BOARD NOW

Hinted That Some Doctors Have Been in the Habit of "Working" the Board

The Toronto "Globe" says: "Considerable controversy is going on in political circles over the vacancy on the Workmen's Compensation Board. A claim is advanced on behalf of labor that the post should be filled by one who is identified with the Labor Government. The employers would like to have a direct representative on the board. The 'Globe' understands that one member of the board at least is in favor of appointing a physician. The reason given for this is that the board is often the victim, if not of fraud, at rate of connivance on the part of physicians in misrepresentation. It sometimes happens that a doctor is attending a poor patient and finds it difficult to collect his bills. Through an accident the patient comes under the jurisdiction of the Workmen's Compensation Act, and the case is exploited by the attending physician to include a part, if not all, of his uncollected account. It is claimed that a physician sitting directly on the board could offset this to a greater extent than is possible by the medical referee. The 'Globe's' informant stated that the board's experience was that these entitled to compensation invariably were honest, and it was not from the laboring class that this difficulty arose, but there were cases where it was apparent that some physicians had attempted to 'work' the board, and it is to overcome this that a physician has been suggested to fill the vacancy."

U.S. SCRAP METAL

The scrap market at U. S. points is showing a better trend, though buying has been light. Conditions are generally weak. Transportation facilities for both fuel and material are greatly determining the market conditions. Further complications were caused by the embargo placed by the Government on the use of open cars for conveying scrap which became effective June 16. This measure was adopted in order to make efforts to restore the fuel supply to as near normal as possible. A consumer in the Buffalo district is understood to have secured 10,000 tons of heavy melting steel recently at a reported price of $26. For the most part the tendency of prices is to remain set, although there have been indications of a revision upwards in the dealers' price lists.

PIG IRON TRADE

There is a tendency for basic pig iron to be scarce, which causes certain fluctuation in market prices. One sale is reported as high as $44.70 valley, but the market can be actually quoted at $43.50 to $44.

SELECTED MARKET QUOTATIONS

Being a record of prices current on raw and finished material entering into the manufacture of mechanical and general engineering products.

PIG IRON

Grey forge, Pittsburgh	$42 40
Lake Superior, charcoal, Chicago	57 00
Standard low phos., Philadelphia	50 00
Bessemer, Pittsburgh	43 00
Basic, Valley furnace	42 90
Toronto price:—	
Silicon, 2.25% to 2.75%	52 00
No. 2 Foundry, 1.75 to 2.25%	50 00

IRON AND STEEL

Per lb. to Large Buyers	Cents
Iron bars, base, Toronto	$ 5 50
Steel bars, base, Toronto	5 50
Iron bars, base, Montreal	5 50
Steel bars, base, Montreal	5 50
Reinforcing bars, base	5 00
Steel hoops	7 00
Norway iron	11 00
Tire steel	5 75
Spring steel	10 00
Band steel, No. 10 gauge and 3-16 in. base	6 00
Chequered floor plate, 3-16 in.	8 40
Chequered floor plate, ¼ in.	8 00
Bessemer rails, heavy, at mill	....
Steel bars, Pittsburgh	3 00-4 00
Tank plates, Pittsburgh	4 00
Structural shapes, Pittsburgh	3 00
Steel hoops, Pittsburgh	3 50-3 75
F.O.B., Toronto Warehouse	
Small shapes	4 25
F.O.B. Chicago Warehouse	
Steel bars	3 62
Structural shapes	3 72
Plates	3 67 to 5 50
Small shapes under 3"	3 62

FREIGHT RATES

Per 100 Pounds.

Pittsburgh to Following Points	C.L.	L.C.L.
Montreal	33	45
St. John, N.B.	41½	55
Halifax	49	64½
Toronto	27	39
Guelph	27	39
London	27	39
Windsor	27	39
Winnipeg	89½	135

METALS

	Gross. Montreal	Toronto
Lake copper	$25 00	$24 00
Electro copper	24 50	24 00
Castings, copper	24 00	24 00
Tin	66 00	65 00
Spelter	12 00	12 00
Lead	11 50	11 00
Antimony	14 50	14 00
Aluminum	34 00	36 00

Prices per 100 lbs.

PLATES

Plates, 3-16 in.	$ 7 25	$ 7 25
Plates, ¼ up	6 50	6 50

PIPE—WROUGHT

Price List No. 44—April, 1920.

STANDARD BUTTWELD S/C

	Steel Black	Steel Galv.	Gen. Wrot. Iron Black	Gen. Wrot. Iron Galv.
⅛ in.	$6 50	$ 8 50	....	....
¼ in.	5 13	7 26	$ 5 43	$ 7 56
⅜ in.	5 13	7 26	5 43	7 56
½ in.	6 84	8 42	7 27	8 84
¾ in.	8 45	10 58	9 03	11 16
1 in.	12 50	15 64	13 35	16 49
1¼ in.	16 91	21 16	18 06	22 31
1½ in.	20 21	25 30	21 69	26 68
2 in.	27 20	34 04	29 05	35 89
2½ in.	43 00	53 82	....	....
3 in.	56 23	70 38	....	....
3½ in.	71 30	88 32	....	....
4 in.	84 48	104 64	....	....

STANDARD LAPWELD S/C

	Steel Black	Steel Galv.	Gen. Wrot. Iron Black	Gen. Wrot. Iron Galv.
2 in.	$30 90	$37 74	$34 60	$41 44
2½ in.	45 34	56 16	51 19	62 01
3 in.	59 29	78 44	66 94	81 09
3½ in.	73 14	90 16	82 84	99 36
4 in.	86 66	106 82	97 56	117 72
4½ in.	0 98	1 23	1 24	1 49
5 in.	1 15	1 44	1 44	1 73
6½ in.	1 49	1 86	1 87	2 25
7 in.	1 94	2 43	2 42	2 90
8-L in.	2 04	2 55	2 54	3 05
8 in.	2 35	2 94	2 92	3 51
9 in.	2 81	3 52	3 50	4 21
10-L in.	2 61	3 26	3 25	3 90
10 in.	3 36	4 20	4 18	5 03

Prices—Ontario, Quebec and Maritime Provinces

WROUGHT NIPPLES

4" and under, 60%.

4½" and larger, 50%.

4" and under, running thread, 30%.

Standard couplings, 4-in. and under, 30%.

Do., 4½-in. and larger, 10%.

OLD MATERIAL

Dealers' Average Buying Prices.

	Per 100 Pounds. Montreal	Toronto
Copper, light	$15 00	$14 00
Copper, crucible	18 00	18 00
Copper, heavy	18 00	18 00
Copper wire	18 00	18 00
No. 1 machine composition	16 00	17 00
New brass cuttings	11 00	11 75
Red brass cuttings	14 00	15 75
Yellow brass turnings	8 50	9 50
Light brass	7 00	7 00
Medium brass	8 00	7 75
Scrap zinc	6 50	6 00
Heavy lead	7 50	7 75
Tea lead	4 50	5 00
Aluminum	19 00	20 00
	Per Ton	Gross
Heavy melting steel	18 00	18 00
Boiler plate	15 50	15 00
Axles (wrought iron)	22 00	20 00
Rails (scrap)	18 00	18 00
Malleable scrap	25 00	25 00
No. 1 machine east iron	32 00	33 00
Pipe, wrought	12 00	12 00
Car wheel	26 00	26 00
Steel axles	22 00	20 00
Mach. shop turnings	11 00	11 00
Stove plate	26 50	25 00
Cast boring	12 00	12 00

BOLTS, NUTS AND SCREWS

	Per Cent.
Carriage bolts, ⅝-in. and less	10
Carriage bolts, 7-16 and up	Net
Coach and lag screws	25
Stove bolts	55
Wrought washers	45
Elevator bolts	Net
Machine bolts, 7/16 and over	Net
Machine bolts, ⅜-in. and less	15
Blank bolts	Net
Bolt ends	Net
Machine screws, fl. and rd. hd., steel	27½
Machine screws, o. and fil. hd., steel	10
Machine screws, fl. and rd. hd., brass	net
Machine screws, o. and fil. hd., brass	net
Nuts, square, blank ... add	$2 00
Nuts, square, tapped ... add	2 25
Nuts, hex., blank ... add	2 25
Nuts, hex., tapped ... add	2 50
Copper rivets and burrs, list less	15
Burrs only, list plus	25
Iron rivets and burrs	40 and 5
Boiler rivets, base ¾" and larger	$8 50
Structural rivets, as above	8 40
Wood screws, O. & R., bright	75
Wood screws, flat, bright	77½
Wood screws, flat, brass	55
Wood screws, O. & R., brass	55½
Wood screws, flat, bronze	50
Wood screws, O. & R., bronze	47½

MILLED PRODUCTS

(Prices on unbroken packages)

	Per Cent
Set screws ... 25 and	5
Sq. and hex. hd. cap screws	22½
Rd. and fil. hd. cap screws ... plus	17½
Flat but. hd. cap screws ... plus	30
Fin. and semi-fin. nuts up to 1-in.	20
Fin. and Semi-fin. nuts, over 1 in., up to 1½-in.	10
Fin. and Semi-fin. nuts over 1½ in., up to 2-in.	Net
Studs	15
Taper pins	40
Coupling bolts	Net
Planer head bolts, without fillet, list	10
Planer head bolts, with fillet, list plus 10 and	net
Planer head bolt nuts, same as finished nuts.	
Planer bolt washers	net
Hollow set screws	net
Collar screws ... list plus 20,	30
Thumb screws	40
Thumb nuts	75
Patch bolts ... add	20
Cold pressed nuts to 1½ in. ... add	$1 00
Cold pressed nuts over 1½ in. ... add	2 00

BILLETS

	Per gross ton
Bessemer billets	$60 00
Open-hearth billets	60 00
O.H. sheet bars	76 00
Forging billets	56 00-75 00
Wire rods	52 00-70 00

Government prices.

F.O.B. Pittsburgh.

NAILS AND SPIKES

Wire nails	$5 70
Cut nails	5 85
Miscellaneous wire nails	60%
Spikes, ⅜ in. and larger	$7 50
Spikes, ¼ and 5-16 in.	8 00

ROPE AND PACKINGS

Drilling cables, Manila	0 39
Plumbers' oakum, per lb.	0 10¼
Packing, square braided	0 38
Packing, No. 1 Italian	0 44
Packing, No. 2 Italian	0 36
Pure Manila rope	0 35½
British Manila rope	0 28
New Zealand hemp	0 28
Transmission rope, Manila	0 47
Cotton rope, ¼-in. and up	0 88

POLISHED DRILL ROD

Discount off list, Montreal and Toronto ... net

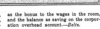
SAVING IN BELT TRANSMISSION

At a foremen's meeting the other
evening, one of the questions discussed
was, "Why do we lose so much time in
getting the full product from a machine
with a new belt?"

One man asked: "Has it anything to
do with the way the belt hugs the pulley?" The reply came quickly from two
or three: "The trouble is the belt was
not properly stretched before it was put
in place."

The next man who rose, said: "All
my belts jump when the belt hooks go
over either the receiving or driving pulley."

Finally the manager said: "Well, let
us take this up as we did the other power
problems. We will have the engineering
department take full charge of all the
belts that are put on new in two or three
rooms, for the next two weeks, and then
hear their report."

When the tests were made, and the
report read, the following facts were
brought out:

First: All the belts, even those as narrow as 1¼ inch, showed the least
trouble when stretched on the floor in
the roll before being measured for the
length required. This was done by securing the ends of the belt by nails at
one end, the other being put in the regular clamps and drawn up a little every
day before cutting to length.

Second: All the belts of whatever
width that were used showed a loss in
power when they were put together,
either with lacing, or any other style of

belt hook. Those that were cemented,
with no extra thickness at the joint,
showed regular transmission, no lost motion on machine or shaft, and the production very soon, after the new belt
was put on the machine, came up to
full product.

The tests showed such favorable results that the following rules were adopted after the points had been thoroughly
discussed:

Rule 1.—All belts shall be given out
only in lengths required for applying
to the drive, and only after same have
been subjected to a regular process of
stretching, the roll being secured to the
floor at one end, and the clamps applied
to the other, the amount taken up being
determined by tests on the exact width
of the belt being kept, one inch not being reduced by the pull to three-quarters,
and so forth.

Rule 2.—Any roll of belting, so
stretched, that shows uneven width shall
be laid to one side, and shown to the
purchasing agent, to be taken up with
the firm that supplied the belting.

Rule 3.—The putting on of the belts,
after the order is received from the foreman of the department, shall be turned
over to the belt man, who will apply the
same.

Rule 4.—All the rooms will have, as
standard the amount of belting used in
the last twelve months, and will be credited with all the saving made by the
above rules for the next two months.
This to be divided at the end of each
two months' period, giving 75 per cent.

as the bonus to the wages in the room,
and the balance as saving on the corporation overhead account.—*Belts*.

———

Many industrial research associations
have been formed in Great Britain and
are working in conjunction with the Department of Scientific and Industrial Research. The expenses of the research
are met partly by the industries concerned and partly from State funds. Co-
operative work of this kind has not, however, discouraged private enterprise in
research. Many firms are extending their
research laboratories and prizes are being offered for success in special directions. Sir Robert Hadfield, for example,
has deposited £200 with the Institution
of Mechanical Engineers to be awarded
as a prize for a new and accurate
method of measuring the hardness of
metals. Present methods are not very
satisfactory, especially with very hard
metals.

———

A paper entitled "Blowholes, Porosity
and Unsoundness in Aluminium Alloy
Castings," and issued by the United
States Bureau of Mines, discusses the
magnitude of the light alloy industry,
gases in aluminium, solidification of
metals, analogy with steel, effects of various processes in casting, melting, moulding, etc., as well as the metallography
and radiography of the subject. The
author is Mr. R. J. Anderson of the
Bureau's branch at Pittsburgh.

MISCELLANEOUS

Solder, strictly	$ 0 40
Solder, guaranteed	0 43
Babbitt metals	18 to 70
Soldering coppers, lb.	0 62
Lead wool, per lb.	0 16
Putty, 100-lb. drums	8 30
White lead, pure, cwt.	20 00
Red dry lead, 100-lb. kegs, per cwt.	16 50
Glue, English	0 40
Tarred slater's paper, roll	1 30
Gasoline, per gal., bulk	0 35
Benzine, per gal., bulk	0 35
Pure turp., single bbls., gal.	3 60
Linseed oil, raw, single bbls.	3 00
Linseed oil, boiled, single bbls.	3 03
Sandpaper, B. & A.	List plus 43
Emery cloth	List plus 37½
Sal Soda	0 03½
Sulphur, rolls	0 05
Sulphur, commercial	0 04½
Rosin, "D," per lb.	0 14
Boray crystal and granular	0 14
Wood alcohol, per gal.	2 70
Whiting, plain, per 100 lbs.	2 75

CARBON DRILLS AND REAMERS

S.S. drills, wire sizes	32½
Can. carbon cutters, plus	20
Standard drills, all sizes	32½
3-fluted drills, plus	10
Jobbers' and letter sizes	32½
Bit stock	40
Ratchet drills	15
S.S. drills for wood	40
Wood boring brace drills	25
Electricians' bits	30
Sockets	50
Sleeves	50
Taper pin reamers	25 off
Drills and countersinks	net
Bridge reamers	50
Centre reamers	10
Chucking reamers	net
Hand reamers	10
High speed drills, list plus 20 to	40
Can. high speed cutters, net to plus	10
American	plus 40

COLD ROLLED STEEL

[At warehouse]

Rounds and squares	$7 base
Hexagons and flats	$7.75 base

IRON PIPE FITTINGS

	Black	Galv.
Class A	60	75
Class B	27	37
Class C	18	27

Cast iron fittings, 5%; malleable bushings, 22½%; cast bushings, 22½%; unions, 37½%; plugs, 20% off list.

SHEETS

	Montreal	Toronto
Sheets, black, No. 28	$ 8 50	$ 9 50
Sheets, Blue ann., No. 10	8 50	9 00
Canada plates, dull, 52 sheets	8 50	10 00
Can. plates, all bright	8 60	9 00
Apollo brand, 10¾ oz. galvanized	....	....
Queen's Head, 28 B.W.G.	11 00	....
Fleur-de-Lis, 28 B.W.G.	10 50	....
Gorbal's Best, No. 28	....	....
Colborne Crown, No. 28	....	....
Premier, No. 28, U.S.	11 50	10 50
Premier, 10¾-oz.	11 50	10 90
Zinc sheets	16 50	20 00

PROOF COIL CHAIN

(Warehouse Price)

B

¼ in., $13.00; 5-16, $11.00; ⅜ in., $10.00; 7-16 in., $9.80; ½ in., $9.75; ⅝ in., $9.20; ¾ in., $9.30; ⅞ in., $9.50; 1 in., $9.10; Extra for B.B. Chain, $1.20; Extra for B.B.B. Chain, $1.80.

ELECTRIC WELD COIL CHAIN B.B.

⅛ in., $16.75; 3-16 in., $15.40; ¼ in., $13.00; 5-16 in., $11.00; ⅜ in., $10.00; 7-16 in., $9.80; ½ in., $9.75; ⅝ in., $9.50; ¾ in., $9.30.

Prices per 100 lbs.

FILES AND RASPS

	Per Cent.
Globe	50
Vulcan	50
P.H. and Imperial	50
Nicholson	32½
Black Diamond	27½
J. Barton Smith, Eagle	50
McClelland, Globe	50
Delta Files	20
Disston	40
Whitman & Barnes	50
Great Western-American	50
Kearney & Foot, Arcade	50

BOILER TUBES.

Size.	Seamless	Lapwelded
1 in.	$27 00	$.....
1¼ in.	29 50	.....
1½ in.	31 50	29 50
1¾ in.	31 50	30 00
2 in.	35 00	30 00
2¼ in.	35 00	29 00
2½ in.	42 00	37 00
3 in.	50 00	48 00
3¼ in.	.. ..	48 50
3½ in.	63 00	51 50
4 in.	85 00	65 50

Prices per 100 ft., Montreal and Toronto

OILS AND COMPOUNDS.

Castor oil, per lb.	
Royalite, per gal., bulk	24½
Palacine	27½
Machine oil, per gal.	43½
Black oil, per gal.	18½
Cylinder oil, Capital	82
Cylinder oil, Acme	70
Standard cutting compound, per lb.	0 06
Lard oil, per gal.	$2 60
Union thread cutting oil, antiseptic	88
Acme cutting oil, antiseptic	37½
Imperial quenching oil	39½
Petroleum fuel oil, bbls., net	13½

BELTING—No 1 OAK TANNED

Extra heavy, single and double	10%
Standard	10%
Cut leather lacing, No. 1	2 75
Leather in side	2 40

TAPES

Chesterman Metallic, 50 ft.	$2 00
Lufkin Metallic, 603, 50 ft.	2 00
Admiral Steel Tape, 50 ft.	2 75
Admiral Steel Tape, 100 ft.	4 45
Major Jun. Steel Tape, 50 ft.	3 50
Rival Steel Tape, 50 ft.	2 75
Rival Steel Tape, 100 ft.	4 45
Reliable Jun. Steel Tape, 50 ft.	3 50

PLATING SUPPLIES

Polishing wheels, felt	$4 60
Polishing wheels, bull-neck	2 00
Emery in kegs, Turkish	09
Pumice, ground	06
Emery glue	30
Tripoli composition	09
Crocus composition	12
Emery composition	10
Rouge, silver	60
Rouge, powder, nickel	45

Prices per lb.

ARTIFICIAL CORUNDUM

Grits, 6 to 70 inclusive	.08½
Grits, 80 and finer	.6

BRASS—Warehouse Price

Brass rods, base ½ in. to 1 in. rod	0 34
Brass sheets, 24 gauge and heavier, base	$0 42
Brass tubing, seamless	0 46
Copper tubing, seamless	0 48

WASTE

XXX Extra	.24	Atlas	.20
Peerless	.22½	X Empire	.19½
Grand	.22½	Ideal	.19
Superior	.22½	X Press	.17½
X L C R.	.21		

Colored

Lion	.17	Popular	.13
Standard	.15	Keen	.11
No. 1	.15		

Wool Packing

Arrow	.35	Anvil	.22
Axle	.28	Anchor	.17

Washed Wipers

Select White	.20	Dark colored	.09
Mixed colored	.10		

This list subject to trade discount for quantity.

RUBBER BELTING

Standard ... 10% Best grades... 15%

ANODES

Nickel	.58 to	.65
Copper	.38 to	.45
Tin	.70 to	.70
Zinc	.18 to	.18

Prices per lb.

COPPER PRODUCTS

	Montreal	Toronto
Bars, ½ to 2 in.	$42 50	$43 00
Copper wire, list plus 10		
Plain sheets, 14 oz., 14x60 in.	46 00	44 00
Copper sheet, tinned, 14x60, 14 oz.	48 00	48 00
Copper sheet, planished, 16 oz. base	46 00	45 00
Braziers', in sheets, 6x4 base	45 00	44 00

LEAD SHEETS

	Montreal	Toronto
Sheets, 3 lbs. sq. ft.	$10 75	$14 50
Sheets, 3½ lbs. sq. ft.	10 50	14 00
Sheets, 4 to 6 lbs. sq. ft.	10 25	13 50

Cut sheets, ½c per lb. extra.
Cut sheets to size, 1c per lb. extra.

PLATING CHEMICALS

Acid, boracic	$ .23
Acid, hydrochloric	.03½
Acid, nitric	.10
Acid, sulphuric	.03½
Ammonia, aqua	.15
Ammonium, carbonate	.20
Ammonium, chloride	.22
Ammonium hydrosulphuret	.75
Ammonium sulphate	.30
Arsenic, white	.14
Copper, carbonate, annhy	.41
Copper, sulphate	.16
Cobalt, sulphate	.20
Iron perchloride	.62
Lead acetate	.30
Nickel ammonium sulphate	.08
Nickel carbonate	.32
Nickel sulphate	.19
Potassium sulphide (substitute)	.42
Silver chloride (per oz.)	1.25
Silver nitrate (per oz.)	1.20
Sodium bisulphate	.11
Sodium carbonate crystals	.06
Sodium cyanide, 127-130%	.38
Sodium hyposulphite per 100 lbs	8.00
Sodium phosphate	.18
Tin chloride	1.00
Zinc chloride, C.P.	.30
Xinc sulphate	.08

Prices per lb. unless otherwise stated

Sharp Chasers
Cut Clean Threads

Accurate, uniform threads result only from dies which are maintained in the highest state of cutting efficiency. This means that chasers must be kept sharp, and ground uniformly.

Even if just touched up from time to time, the chasers respond splendidly, with clean threads. And with this machine — the Geometric Chaser Grinder—the matter of keeping threading tools up to scratch is a comparatively simple matter.

Various makes of chasers can be ground on this adaptable machine. The two wheels permit the easy grinding of both milled and tapped chasers. In addition, the plain wheel lends itself readily to various kinds of tool grinding.

Uniform grinding of a set of chasers is purely a mechanical matter through the use of adjustments which can be accurately set to govern the grinding of an entire set of chasers.

The Catalogue describing this machine is a mine of information on chaser grinding. Write for it.

THE GEOMETRIC TOOL COMPANY
NEW HAVEN   CONNECTICUT

Canadian Agents: Williams & Wilson, Ltd., Montreal; The A. R. Williams Machinery Co., Ltd., Toronto, Winnipeg, and St. John, N.B.; The Canadian Fairbanks-Morse Co., Ltd., Manitoba, Saskatchewan, Alberta.

If interested tear out this page and place with letters to be answered

Production of Pig Iron and Steel in 1919

THE total production of pig iron in Canada in 1919 excluding the production of ferro-alloys was 917,346 short tons (819,059 gross tons), having a value of $24,536,432 as compared with a total production in 1918 of 1,195,551 short tons (1,06,456 gross tons) valued at $33,495,171 showing a falling off of 278,205 tons, or 23 per cent. Of the 1919 total 910,080 tons were made in blast furnaces and 7,266 tons (subject to revision) were made in electric furnaces from scrap metal, chiefly shell turnings. In 1918 the blast furnace production was 1,163,510 tons and the electric furnace production from scrap steel was 32,031 tons.

The production of blast furnace pig iron in Nova Scotia in 1919 was 285,087 tons as against 415,870 tons in 1919 and with the exception of 1914 was the smallest production in that province since 1905. In Ontario the production of blast furnace pig iron was 624,993 tons, as against 747,650 tons in 1918. Although less by 16 per cent. than in the previous year, the 1919 production in Ontario was exceeded in only four previous years.

Less than one quarter as much pig iron was made from electric furnaces from scrap steel as in the previous year the output being derived from six furnace plants in 1919 as compared with ten plants operated in 1918.

By grades the 1919 production included: Basic, 580,426 tons; Bessemer, 7,637 tons; foundry and malleable, etc., 322,017 tons; low phosphorous iron (electric furnace), 7,266 tons. The 1918 production included: Basic, 966,409 tons; Bessemer, 15,415 tons; foundry and malleable, etc., 181,696 tons; low phosphorous iron (electric furnace), 32,031 tons.

The blast furnace plants operated included those of the Dominion Iron and Steel Company at Sydney, N.S.; The Nova Scotia Steel and Coal Company at North Sydney; The Standard Iron Company at Deseronto, Ont.; The Steel Company of Canada at Hamilton, Ont.; The Canadian Furnace Company at Port Colborne, Ont.; The Algoma Steel Corporation, Ltd., at Sault Ste. Marie, Ont.; The Midland Iron and Steel Company at Midland, Ont., and the Parry Sound Iron Company, Ltd., at Parry Sound, Ont.

Electric furnaces were operated for the production of pig iron from scrap at Hull and Shawinigan Falls in Quebec, at Belleville and Welland in Ontario, and at Vancouver, British Columbia.

The production of ferro-alloys in Canada in 1919 including ferro-silicon, silico spiegel, spiegeleisen and ferro-phosphorus, all with the exception of the speigeleisen being made in electric furnaces, was about 48,579 tons valued at $1,998,779. In 1918 the production was 44,704 tons valued at $4,731,521. Over

one-half the tonnage made in 1919 was speigeleisen made by Algoma Steel Corporation for the company's own use.

The exports of pig iron during 1919 were 63,605 tons valued at $1,820,260, or an average of $28.62 per ton and of ferro-alloys 22,449 tons valued at $1,229,341, or an average of $54.76 per ton. The exports of pig iron included 57,845 tons to the United States; 783 tons to Chili; 7 tons to Japan, and 4,970 tons to other countries. The ferro-alloy exports included 2,564 tons to United Kingdom; 15,371 tons to United States; 4,514 tons to other countries.

The imports during 1919 included 35,-800 tons of pig iron valued at $1,022,871, or an average of $28.80 per ton, and 16,-221 tons of ferro-alloys, valued at $901,-678, or an average of $55.58 per ton, making a total import of pig iron and ferro-alloys of 52,021 tons valued at $1,924,549. The United States trade records show exports to Canada during 1919 of pig iron and ferro-alloys amounting to 33,751 gross tons (37,801 short tons), valued at $1,052,103.

Steel

The total production of steel ingots and direct steel castings in 1919 subject to possible slight revision, was 1,031,329 short tons (920,844 long tons), of which 904,349 tons were ingots and 39,980 tons direct steel castings.

The total production in 1918 was 1,-873,708 short tons (1,672,946 long tons), of which 1,800,171 tons were ingots and 73,537 tons were castings.

The 1919 production included: Open-hearth steel, 1,008,540 tons; electric steel, 15,467 tons; crucible and converter steels, 7,322 tons. The 1918 production included: Open-hearth steel, 1,746,334

tons; electric steel, 119,130 tons; crucible and converter steels, 8,244 tons.

The total production of electric furnace of which 1,800,171 tons were ingots and in 1916, 19,639 tons.

The total production of pig iron, ferro-alloy and steel in electric furnaces was about 43,540 tons in 1919 as compared with 191,869 tons in 1918 and 101,031 tons in 1917.

The exports of steel during 1919, as per Customs Department records included billets, blooms and ingots 29,087 tons valued at $1,731,629, or an average of $61.65 per ton; bars and rods, 52,191 tons valued at $3,304,894, or an average of $65.05 per ton; steel rails, 30,737 tons valued at $1,297,836, or an average of $42.22 per ton; wire and wire nails valued at $5,745,773; structural steel, 5,515 tons valued at $465,989, or an average of $84.49 per ton; scrap iron and steel, 245,214 tons valued at $3,779,179, or an average of $15.41 per ton, together with a large quantity of manufactured iron and steel goods.

The production of rolled iron and steel products in 1919 included: Steel rails, 316,304 short tons; plates and sheets, 25,406 short tons; wire rods, 153,723 short tons and structural shapes, 29,295 short tons, and a large tonnage of iron and steel bars, rods, etc., for which returns are not yet complete. The total production in 1919 of finished rolled products was 1,146,610 short tons, which included steel rails, 162,747 tons; wire rods, 154,789 tons; merchant bars and rods and structural shapes, 415,017 tons; plates and sheets, 26,413 tons; rolled blooms and billets for forging purposes and rolled blooms, billets, or slabs sold for export, 395,644 tons.

CONSOLIDATION OF DROP-FORGING PLANTS HAS BEEN ANNOUNCED

ON April 2nd the stockholders of J. H. Williams & Co., manufacturers of drop-forgings and drop-forged tools at Brooklyn and Buffalo, N.Y., and of the Whitman & Barnes Manufacturing Co., makers of twist drills, reamers, wrenches and drop-forgings at Akron, Chicago and St. Catharines, Ontario, ratified the agreement providing for the consolidation of the wrench and drop-forging plants and business of Whitman & Barnes at Chicago and St. Catharines with J. H. Williams and Co. The Whitman & Barnes Manufacturing Co. retain their twist drill and reamer business and will continue, as an entirely separate organization and on an extended scale, the manufacture of these tools at Akron, Ohio.

History of the Company

J. H. Williams & Co., makers of drop-forgings and drop-forged tools, was one

of the first to enter this field commercially. The business was founded in Flushing, L.I., as a partnership in 1882 by James H. Williams and Matthew Diamond under the style of Williams & Diamond, later Williams & Brock. It located in Brooklyn in 1884, assumed its present name in 1887 and was incorporated in New York State in 1895.

The Brooklyn Works employ about 750 persons, occupy two adjoining square blocks of the city property, including the street between, which is closed, and contain about 197,000 sq. ft. of floor space. In 1914 the Buffalo Works, an entirely new plant, began operation; they now employ over 1,000 persons and comprise 47 acres, with about 233,000 sq. ft. of floor space.

The Whitman & Barnes Manufacturing Co. was founded as a partnership in 1846 by Augustus Whitman and Al-

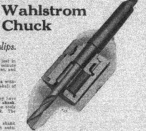

fred G. Page under the title, Page, Whitman & Co. In 1868 the Akron factory was built, and, in 1877, the business was combined with that of George Barnes & Co. of Syracuse, N.Y., the enlarged concern assuming its present name—The Whitman & Barnes Mfg. Co. The St. Catharines plant was acquired in 1882 and eleven years later the Chicago works were built. They now consist of 45 acres of land, with 320,000 sq. ft. of floor space, and are connected by private trackage with four railroads. They employ about 800 people.

The St. Catharines works, manufacturing special drop-forgings to order and standard lines of drop-forged wrenches, pipe wrenches, adjustable wrenches of various styles, pliers, hammers, hatchets, swedges, etc., are located on the Welland Canal and obtain power therefrom. They occupy about 2½ acres, with floor space of 35,000 sq. ft. The company owns some 6 acres of additional property, which may be utilised in the near future for the erection of larger and improved facilities to care for the Canadian trade.

Through this consolidation of the four plants—Brooklyn, Buffalo, Chicago and St. Catharines—the ultimate geographical ideal for the service of customers has been attained.

Relations With Employees

The company has been particularly fortunate in its relations with its men. Operating throughout its history on the principle of the highest wages in the industry for the most efficient work, supplemented (but not substituted) by expenses liberally incurred to provide the best possible working conditions, it has succeeded in establishing a relation of mutual confidence that has frequently been the subject of comment in the industrial and social press—in 1912 this was the cause of an unsolicited visit from Theodore Roosevelt, followed by a special article by him in "The Outlook" on the subject.

The Brooklyn factory, in 1893, had the distinction of being the first factory in this country to provide shower baths for its workmen, and again, in 1914, of leading its industry in the establishment of separate pension and life insurance systems for the direct benefit of its employees, and, of course, for the indirect benefit and satisfaction of its stockholders.

The sales for the first five months of 1920 indicate that this will be the largest year in the company's history. All of the factories are far behind their orders in spite of increased production, and permanent night shifts have been organized to help meet the demands of the trade.

Management

The business will continue to be operated by those who have been continuously identified heretofore with the business of J. H. Williams & Co. and with the Chicago and St. Catharines plants of the Whitman & Barnes Manufacturing Co., the organization being as follows:

President and managing director, J. Harvey Williams.
Vice-president, A. D. Armitage.
Secretary and treasurer, W. A. Watson.

Controller, R. S. Baldwin.
General sales manager, F. W. Trabold.
Eastern Dist. sales mgr., A. S. Maxwell.
Western Dist. sales mgr., W. E. Rowell.
Buffalo Dist. sales mgr, J. C. Cotter.
Detroit Dist. sales agent, A. M. Thompson.
Publicity manager, Hugh Aikman.
General works manager, Capt. W. N. McMunn.
Chief engineer, Willard Doud.
Supt. Brooklyn division, R. J. Smith.
Supt. Buffalo division, T. F. Du Puy.
Supt. Chicago division, Barney Nelson.
Gen'l purchasing agent, J. C. Scanlon.
Gen'l traffic manager, J. B. Payne.
Canadian manager, W. J. Elliott.

BIG CONTRACT WITH THE SOVIET

Montreal Firm Admits That More Than $5,000,000 is Involved

New York.—F. W. Boyer and W. W. Sloan, of the firm of Boyer, Sloan & Co., Montreal, are at present in New York. In conversation with the Canadian Press they confirmed the report that they had signed a contract with Ludwig C. A. K. Martens, chief of the Russian Soviet Bureau here, to trade with Soviet Russia.

Messrs. Boyer and Sloan refused to give the amount involved in the contract but intimated that it was in excess of $5,000,000. They said that until the details of the deal had been fully worked out they would not be in a position to go into particulars.

MUST DEVELOP OUR RESOURCES

Canada Cannot Continue to Look for Coal From United States Mines

Ottawa.—That Canada need not look to the United States for its future supply of coal, and that the present shortage serves as a good lesson for the Dominion to develop its own resources was the gist of an interview given by H. A. Harrington, fuel controller for the Province of Ontario.

Mr. Harrington was very reticent when questioned as to the probabilities of the United States supplying Canada's need during the coming winter. He stated the strong probabilities were that the United States would need all of her own output, especially the anthracite coal. There was all kinds of coal in the Dominion, but the people of the Dominion were not "digging" for it. Rather they were depending on the people to the South. They had to learn to be independent and "dig" for themselves.

—Marcus in New York "Times."
WHEN YOU WAVE A RED FLAG, LOOK OUT FOR THE BULL.

Quality

THE
STEEL COMPANY
OF
CANADA
LIMITED
HAMILTON MONTREAL

Service

Some General Principles of Tool Hardening

There Are Over Two Hundred Metallurgical Factors That Influence the Efficiency of a High Speed Cutting Tool—The Main Principles to Observe Are Stated Herein.

By MAJOR A. E. BELLIS

IT has recently been determined that there are over two hundred metallurgical factors (or variables) that influence the efficiency of a high speed cutting tool. This means that there are two hundred reasons why a tool may not give performance up to standard, and two hundred possible excuses when a tool fails. Let us, however, regard this fact as a challenge to systematically attack the problem of tool failures in order to find the best means of controlling these two hundred variables.

The fact that there are so many variables that affect the successful hardening of a tool accounts for the many traditions and secret processes of the hardening room. One of the early processes in the manufacture of the steel may be the real cause of failure of a tool, but if the hardener used a slightly different means of heating or quenching than usual he cannot help feeling that this is connected with his result. He therefore superstitiously avoids this means ever after, if the result is unsatisfactory, or clings to it if result is good. This fact also explains the reasons why the shop man can generally locate and correct troubles more quickly than the laboratory theorist. There are so many little details that the hardener is familiar with that influence results, that he seems to "sense" a cause of variation in results that any one not so familiar with his process would take a long time to discover. The fact that there are so many factors involved in successful hardening also explains why there is always a "come back" on a rejection. The "defendant," by emphasizing some other variables, can generally prove an "alibi" or show a way of making use of material not up to the letter of the specification. From a purely theoretical standpoint, in order to study the effect of any one of these two hundred variables on the hardened tool, we would have to make all the others constant. An algebraic equation expressing this would contain two hundred factors (the letters of eight alphabets would be required for symbols) and in order to study the effect of varying one, all of the others would have to be kept constant.

The Vital Principal

This theoretical conception is very useful, for it leads at once to the vital principle of success in all tool steel work. This can be expressed in one word, uniformity. The practical means of controlling results from the foundry, through rolling, annealing, forging, and hardening, is to maintain uniformity.

*Presented before the American Steel Treaters' Society.

Uniformity of composition and uniformity of heat treatment will solve most tool steel problems.

Consideration of the practical problem of controlling these two hundred variables, leads to an appreciation of the secret of the tool hardener's skill. While he directly controls some of the factors affecting results, he has it in his power to compensate to some degree for the shortcomings of all the others who have had a hand in the manufacture of the tool. He directly controls some variables and indirectly a great many others. In order to advance in his profession he has to consider and weigh all of them. Stated more concretely, the hardening room can often make successful tools from improperly manufactured material. During the rush of war production about ten per cent. of material tested by the writer could have been theoretically rejected. Actual rejections were less than one per cent. After the resulting gain was much greater than the value of the material saved because a loss of actual rifles produced could never be made up. Without, in any way, sacrificing quality, a great saving and increase in production was made by getting everything possible out of the steel in the heat treating department. The saving in quality and quantity was incomparably greater than the cost of good heat treatment work.

The Study of Failures

The systematic study of failures should begin with the melting of the steel, but as we are more interested in the failures that show up in, or after, hardening, we will start at the wrong end first. When the shop complains of bad stock or poor tools the hardening room has a chance to show its efficiency, to show its spirit of co-operation with other units of the organization, and to show its ability in conducting a quick, systematic and thorough investigation. An inefficient organization frantically starts making changes; more treatments, methods and conditions are changed in half an hour than can be corrected in weeks. Probably only one little variable needs correction. The systematic location of this is the real test of the efficiency of the department. The trouble may not be in the hardening room at all, and, here is where the personality of the head of the department is important. He should be given enough authority to be able to absolutely stop production if serious trouble occurs until the cause is found, because this is always cheaper than random experimenting and continued failures. Too much emphasis cannot be put on finding what factor has chang-

ed, what has gone wrong, before making any changes of treatment or methods.

There are a few quick tests that are invaluable and should be made at once when trouble occurs. Assuming that the method and treatment are the best for the work and have been properly carried out, quick information can be obtained from tests of composition, fracture and hardness. With a little experience and under uniform conditions the spark test is a quick guide for composition. If a carbon steel or a "semi" high speed has been mixed with the high speed the spark test will show it at once. An estimate of the carbon content can be made from the spark by comparison with known standards. A metallographic test can give a quick estimate of carbon content, or more exact information can be obtained in fifteen minutes or half an hour from the combustion method of carbon analysis. These three methods can be used to check each other.

The most valuable quick test is that of fracture. It is surprising how little advantage is taken of this most rapid and direct source of information. If the habit is formed of observing fractures, it is surprising how much information can be gleaned regarding the properties of steels and their variations with treatment and how thorough a check can be quickly made by simply breaking one piece from a lot. More information can be obtained by fracturing after hardening than after tempering. This test can be made quantitative though less speedy by using an impact machine of standard make such as the Charpy or Izod type. After a little experience an excellent check can be made of the two hundred variables that have influenced the structure of a piece of steel by comparing its fracture with standard pieces.

These quick tests should always be supplemented by more complete and thorough investigations, for, though the worst cause of trouble may have been found in these quick tests, there are no doubt contributory causes of failure that can be more easily found and more effectively corrected while interest is aroused over the serious failures.

Establish Performance Standards

In order to establish standards of performance it is necessary to periodically make work tests to destruction. Unless a test is carried to the limit to show how a piece fails, it is of very little real value. In connection with such tests, variation should be tried out to establish the heat treating ranges that give satisfactory results with a particular steel. It is often surprising to

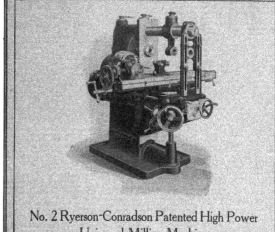

find the great increase in performance resulting from a slightly tougher piece produced by a higher drawing heat, or the better wearing qualities from a lower draw. Instead of depending upon the opinion of a shop foreman, the heat treating can be scientifically placed on a basis of established facts. In connection with routine performance tests, it is well to carry out trials of recommended new steels under these same conditions. Unless a new steel shows an exceptional and consistent improvement it should not be adopted, because its adoption will immediately introduce a number of variables into the heat treatment of the part. Thus, the decision for adoption of new steels should rest with the man responsible for heat treating, and not with a purchasing agent who has no means of understanding the effect of such a change.

An example of the improvement in a tool from carefully studied performance tests is that of the drills used on rifle barrels. It used to be customary to grind a drill after each barrel. The barrel is 24 inches long, of heat treated .55 carbon steel, and is drilled at a feed of 1 inch per minute. By standardizing the best steel (not high speed) and best treatment, it was possible to produce a drill that without grinding would go through 34 barrels, 78 feet of heat treated barrel steel.

Careful work tests always pay richly in increased tool life and increased production. Often a lower grade steel properly treated, will give as good or better performance than a higher grade or higher priced one. The steel salesman "cure all" is a higher priced alloy, but this is frequently carried too far. Thousands of fine finished cuts, as reaming, splinting and drilling, that could be performed much more efficiently and economically with carbon or a "semi high-speed" steel, are being performed with high speed steel simply because of a steel salesman's recommendations unverified by proper work test.

A Laboratory Analysis Pays

It pays to have a laboratory analysis made to check the composition of each lot of steel, not only for the direct information to guide the heat treatment, but for the insurance value that results from the psychological effect on the steel maker. Errors of composition directly affect the physical properties, critical range, fatigue, and corrosion resisting properties and are easily avoided by proper chemical test.

Failures occurring from errors of casting or working are not so easily avoided. Too high casting temperature leads to ingotism, banding and segregation, while too low a casting temperature results in "cold shuts," "pipes" and "blow holes." Foreign matter can be included by the casting operation with resulting slag inclusions or seams. Metallographic tests will usually show these defects.

Working or rolling at improper temperature can result in laps, decarburization, and strains leading to distortion

troubles. All of these points should be borne in mind before concluding that the "wrong hardening temperature" was the only cause of failure. The signs of overheating are so obvious and familiar that they are often considered as entirely accounting for a failure that was affected primarily by other causes.

The means of carrying out the heat treating operations determine largely the degree to which the desired uniformity can be maintained in hardening, tempering, quenching or annealing. This is why the subject of equipment is such a vital and interesting one.

There has been so much good development work in recent years in this line, and such excellent papers have already been presented to you on the subject, that the writer hesitates to touch upon it, but one or two guiding principles can be stated, though particular problems are always making exceptions to these. For heating for hardening the open fire is the poorest method possible. This outgrowth of the old blacksmith's forge is incapable of control with any tolerable degree of uniformity. Much better results can be obtained with the lead bath. The writer has successfully used salt baths of special composition and low melting point more satisfactory than lead. Where uniformity of heating, temperature control, and furnace atmosphere regulation are desired, a well designed gas system can give better results than any other fuel. Where a large volume of packed work is being treated, oil is no doubt the cheaper fuel. A well-designed annealing or pack carbonizing furnace should have a continuously heated chamber with directly connected cooling chamber for the sake of fuel economy. The dimensions of the pots for pack work are important, the least dimension of the pot determining the heating time, so that a narrow pot of the same capacity heats very much more quickly than a wider one.

The writer has found illuminating gas a satisfactory carbonizing material for case hardening to two or three-tenths of an inch depth. The economy in its use results from the saving in not having to handle or heat pots or packing material. Much more heat is required for these materials than for the actual pieces being treated in the ordinary pack hardening method.

In order to maintain the proper uniform conditions for quenching, it is necessary to have large enough quenching tanks and circulating system, a large, rapidly circulating and well-cooled supply of the very cheapest mineral oil is more satisfactory than any specially prepared oil used in too small amount. When a water quench is used it is economical to run water into quenching tank in great enough volume to maintain a uniform temperature. Applying the principle of uniformity to the use of brine, it should be remembered that for a given temperature a saturated solution always has the same strength, and uniformity can be attained by keeping the temperature constant and an excess of salt in the tank.

Enough has been said to show that a thoughtful application of the principle of "uniformity" will result in a high standard of production in all tool hardening work.

THE SMALLEST WIRELESS RECEIVER

A late British wireless invention is a portable wireless telegraph receiver. The whole apparatus—receiving coil, tuning condenser, two valve detectors, accumulator dry battery, transformers and reaction coil—goes into a box which is only fourteen inches long, thirteen inches broad, and five inches deep. The cover of the box lifts off its hinges and forms a base with a pivot on which the receiver can be rotated so as to get the strongest signals according to the direction in which the waves are coming. A compass card on the cover enables the receiver to be set in the correct position for receiving from certain stations. With this little appliance, which weighs only 20 lbs., signals can be read in Great Britain from the largest stations in France and Germany as well as in England, without any additional wires. The signals, which are read in a telephone, are clearly audible in a reasonably quiet room.

STANDARDIZATION OF SHAFTING

The American Engineering Standards Committee has invited the American Society of Mechanical Engineers to act as sponsor for the standardization of shafting. The Society has already done a considerable amount of work on a set of standard diameters for transmission and machinery shafting. It is proposed that the work, which will be carried out by a sectional committee working under the rules of procedure of the A. E. S. C., shall be broadened to include the standardization of the method of determining what diameters of transmission shafting should be used for given loads, the dimensions of shafting keys and keyways; and the setting of dimensional tolerances.

STANDARDIZATION OF PIPE FLANGES AND FITTINGS

The American Society of Mechanical Engineers has been requested by the American Engineering Standards Committee to assume sponsorship of the standardization of pipe flanges and fittings. In 1914, the Society issued a report covering a schedule of pipe flanges and fittings, for diameters from 1 in. to 100 ins. for 125 lbs. pressure, and also a schedule for extra heavy pipe, covering a range from 1 in. to 48 in. diameter, and for 250 lbs. pressure. In 1918, a supplementary report was published for working pressures of 50, 800, 1,200, and 3,000 lbs.

While the work has not been active during the last year, it is now proposed that it be continued, the Society being formally recognized as sponsor, under the rules of procedure of the A. E. S. C.

GRINDING HARDENED
BEARINGS IN PLACE

STARTING and finishing this job with one set-up—from machining the casting to grinding the hardened bearing in place—is a good illustration of the **DUMORE** grinder on production. This method positively aligns each bearing with the rest of the machining, by tempering. And the danger of chatter-marks, taper or bell-mouth is eliminated because the **DUMORE** grinder is perfectly balanced throughout and gives correct cutting speeds to even small emery wheels. Limits of .0001" are easily maintained.

There are any number of jobs in your shop for the **DUMORE** grinder which this illustration will suggest. Give your workmen the use of a tool that will speed their work, turn out better tools and improve the quality of your product. The **DUMORE** grinder is light and portable—may be carried anywhere in the shop and quickly set up.

Ask your dealer for a demonstration. If he is not able to supply your needs, write us for further information and prices.

WISCONSIN ELECTRIC COMPANY
2925 Sixteenth Street   Racine, Wis.

DUMORE HIGH SPEED GRINDERS

If interested tear out this page and place with letters to be answered

A Drinking Water Service that will last a LIFETIME

What care and a few ordinary drinking cups still save the lives of the

PURO
SANITARY
DRINKING
FOUNTAIN
(MADE IN CANADA)

MACHINE AND BENCH VISES
Write for Prices
Superior Machinery Co., London, Canada

A. B. C. OF WELDING
Continued from page 34

less than 1-16 inch ocour they are dropped as their addition might cause the selection of too large a filler.

Vertical Welds—How Executed

Vertical welds in steel may be worked either from the bottom up or from the top down. Figure 31 shows the proper position of operator, rod and torch in either case. Vertical welds in cast iron are possible only by working from the bottom up. Even then they are not recommended except in cases of extreme emergency. Similarly malleable iron or other metals may be welded with bronze if the necessary caution is employed.

Figures 29 and 30 are intended to show the position of fills through a cross section from the centre of the bevel. The position in Figure 29 is that assumed at the beginning of the fill. As the edges and sides are brought to fusion the metal is held in place partly by the action of the flame, partly by the manipulation of the rod, and in even greater measure by its own tenacity and the peculiar tendency of the heated metal to cling to any supporting surface.

Choosing a tip of the correct size enables the operator to keep the metal under proper control, so that with very little practice he should be able to support the fill in the manner shown in Figure 31, and to execute a ripple as uniform and regular as in horizontal work.

In working a vertical weld forward the individual fills are worked backward. The metal is brought to fusion at the bottom of the bevel, the rod plunged in and added continuously by vigorously working it across the weld in the manner shown in Figure 1 with the first lecture of this series. The torch is moved slowly across and backs up slowly as fusion is obtained with the preceding fill, the sides and the welding rod.

It is quite important that the ripples extend almost straight across the weld, as a pronounced curve tends to spill the melted metal out from the centre of the curve at the bottom. When the bottom of the ripple is reached the weld can be reinforced as much as is desired, bearing in mind that too great a reinforcement is apt to produce too much stiffness of the parts and thus invite fracture. Other ripples are formed by repetition of this process. The properly executed weld will be similar in appearance to the close-up photographic illustration in Figure 32. If improperly executed it is quite likely to assume the appearance shown in Figures 27 and 28.

By turning Figures 25 and 26 on the side so that the letters B and D are right way up we get an illustration of the centre line of the bevel in working a vertical weld from the top down. The beginning is made with the torch and rod in the position D, Figure 26, and the positions shown at C and at B (Figure 25), taken as the fill progresses. The motion of the rod and torch across the weld is the same as in working from the bottom up.

The operator who is sufficiently care-

ful to secure proper fusion, and observes that none of the melted metal precedes the welding flame may at times find this method of welding a great advantage.

TRADE GOSSIP

Making New Line.—"The Wilt Twist Drill Company of Canada, Limited, Walkerville, Ontario, are now manufacturing a complete line of specially designed Ford reamers, catalogues of which are now ready for distribution and will be furnished on request."

Hamilton Has New Industry.—The Peterson Core Oil and Manufacturing Co. have occupied their new factory at the foot of Harmony Avenue. The new factory will be able to supply the local foundries with supplies that they formerly had to go out of town to procure. The products of the new company will include core oil, high-grade parting and sea coal.

Industry Extends at Cobourg.—The ammunition plant formerly occupied by the Cohoes Steel Co., Ltd., is to be taken over by the Bird, Archer Co., of New York, who have been operating a plant for the last three years manufacturing boiler compound. The plant will be in operation about July 1 and will be devoted to the manufacture of high-speed cast steel tools and foundry chemicals.

Coal Shortage in Montreal.—There is a dire lack of soft coal in Montreal and district, and production in some cases has been curtailed 50 per cent. The Canada Cement Co. is a case in point, they having been compelled to close down two of their mills. Coal merchants have notified customers that only 50 per cent. of their orders can be filled, and it is expected that the Government will be approached for aid if the situation does not improve soon.

Sells Invention to Steel Industry.—Frank B. McKune, of 179 Herkimer St., Hamilton, superintendent of the open-hearth furnaces of the Steel Co. of Canada has just returned from the United States where he has been disposing of the U. S. rights of his invention for feeding open hearth furnaces with by-product gas. The invention is in use at the Steel Co.'s plant in Hamilton, and it is understood that Mr. McKune received a large sum for the United States rights to it.

Steel Plant Closed Down.—The Steel Company of Canada, at Hamilton, has announced that its plant will have to stand idle for a short time owing to coal and fuel oil shortage. The close-down at this time means considerable to the company as many orders are on hand to he filled. Every effort will be made to secure a supply of fuel, but it is feared the plant may have to lie idle for a couple of weeks. In the neighborhood of 2,500 employees will be affected. This is the first occasion in many years upon which the company has found it necessary to close down its plant.

Must Pay Two Per Cent.—A definite and final ruling has been received by the

The Publisher's Page

Improving Our Service

THE Directors of the Canadian Press Association decided in 1919 that the MacLean Publishing Company was the largest newspaper organization in Canada and must therefore pay the highest annual fee to the Association. The Montreal "Star," with its two big weeklies, comes second, but the "Star" group had been doing business for 18 years before the MacLean Company began. While we have not the details upon which the Press Association based their decision, we understand they figured that the MacLean Company had a larger advertising revenue. This is not the case. Lord Atholstan's properties probably carry a third more business, use more paper and have a greater total circulation. The MacLean papers on the other hand get more revenue from circulation; have to pay perhaps three times as much in salaries and wages and show considerably less profit.

Good service to readers first and advertisers next has been the cardinal principle upon which the MacLean Company has been built. In doing this, every man, woman and junior apprentice on our staff has had a share. We have tried to gather about us and train the best experts in the country. Here is an example of what we are doing:

In the recent annual examinations of the Toronto Technical School, session 1919-20, in the Typography branch—that is the department where they learn how to set type—only 27 boys out of the whole city passed, of whom 10, or nearly 38 per cent., were boys of our own composing room, while in two out of the three classes, MacLean boys stood at the head.

Several of our papers are the best of their class in the world, but we are still far from satisfied with the service we are giving. As we can afford it, as the country grows, as our business grows, the service will be improved. We ask our readers to take these Technical School results as an evidence of our efforts to serve them better and better, for by giving a superior training to these boys we are laying the foundation for the still better work we hope to do in the future.

Readers can help us to give them a better service by sending in—direct to the President, Vice-President, or General Manager—criticisms and suggestions.

secretary of the Canadian Manufacturers' Association at Montreal regarding the 2 per cent. sales tax on material sold by the manufacturer to the consumer or retailer to the effect that all railway companies, power companies, steamship lines and public utility corporations are ruled by the Department of Inland Revenue to be consumers, and must, therefore, be charged by the manufacturer the two per cent. sales tax on each invoice. This ruling has been given officially by George B. Taylor, Assistant Deputy Minister of Inland Revenue. The impression had gone abroad that a one per cent. tax only applied, but this is in error, and the department desires manufacturers to be governed thereby.

MARINE

Toronto.—The new racing hydroplane Miss Toronto II, built by F. C. Ericson & Co. for a local syndicate, was successfully launched recently. The little vessel is 22 feet long, and has a motor of 450 h.p. installed, which the builders expect to drive her at 72 miles per hour. The Miss Toronto II was sent to Burlington, Mass., to take part in her first race.

Toronto. — Some interesting figures anent Toronto Harbor were given in the Federal House recently. There are four and a half miles of docks in the present scheme. The depth of water will be 24 feet; 2,033 vessels, aggregating 1,408,-465 tons, arrived during the fiscal year ending March 31st. The vessel, leaving were 1,940, their tonnage being 1,193,-318. The Dominion Government has spent $3,303,384 on the harbor since the inception of the plan.

Genoa.—The absence of any delegation from the United States called forth strong protests at the meeting of the International Seamen's Conference when the question of regulating conditions of inland navigation came up. Mr. Thomas Robb, secretary of the Shipping Federation of Montreal, who represents the employers, said: "In behalf of Canadian shipping interests I protest strongly against any recommendation being made by this conference without hearing from our friends from the United States. It is a very serious thing to make recommendations concerning such a large inland sea without both parties being present. Regulations on these lakes are not international, each country adopting its own regulations."

Tenders will be received until 12 noon on July 13, 1920, for the construction of alterations and additions over the boiler room of the Printing Bureau at Ottawa. R. C. Desrochers, Secretary of the Dept. of Public Works.

Classified Opportunities

Next Export Number
August 5th.
Reserve Space At Once!

If what you need is not advertised, consult our Buyers' Directory and write advertisers listed under proper heading.

CANADIAN MACHINERY

AND

MANUFACTURING NEWS

Vol. XXIV., No. 2.   July 8, 1920

Making Up Part Lists for Machines in Sections

The Preparing of Part Tests for Standard Machines is Simple, but When Machines Are Built in Sections the Proposition Becomes More Complex — One Plant Prepared Their Tests as Shown

By F. SCRIBER

ONE of the quite prevalent customs in the building of machinery is to furnish for each machine or group of machines to be built a list of parts which go to make up the machine or group. This usually originates in the engineering department and is known by such names as "Lists of Parts," "Sched. ule," "Bill of Material," etc.

When the machines built are all simi. lar the making of this list by one familiar with the particular machine is a fairly simple matter, but in some types of ma. chines, most notably those made in sec. tions, having variant numbers of units in a section or variant numbers of sec. tions in a machine, the making of a suit. able list is a much more complicated pro. position. Types of machines in this class are multiple spindle drills and some forms of textile machinery.

The purpose of this article is to illus. trate some suitable forms which have been found convenient for listing parts which must be included to make up ma. chines of the latter class, namely, "ma. chines requiring varying quantities to make one complete machine to order as regards size or number of units wanted.

The first list shown, Fig. 1, is a list complete with the exception of sundries such as cotter pins, screws, etc., for get. ting out parts to make a complete ma. chine, and the first column at the left gives the piece number or drawing num-

ber of the various parts, the second column is for the number of parts want. ed per machine, and in this connection it will be noted that where the quantity is common to all sizes of machines this is filled in, while a blank space is left otherwise, this to be filled in by the person who issues each individual list; the third column gives the name of each part, while the remaining three columns are a repetition of the others. This list is issued primarily for the finished stock department to issue stock to the as. semblers, consequently no columns giv. ing kind of material or size of stock are needed, although such is usually given in lists for the order and rough stock de. partments.

FIG. 2.—A PORTION OF SUNDRY LIST.

FIG. 1.—PORTION OF COMPLETE LIST WITH EXCEPTION OF SUNDRIES.

Referring to Fig. 2 this shows a list of sundries, the column at the left being the symbol number, or in the case of screws or other purchased parts the size of same, the second column gives the material used for making the various parts where this is desirable, and the third column gives the names of the var. ious parts; at the head of the sheet the part list number is given and also the name of the machine and other general data of a quite obvious nature. At the right of this illustration are six columns, each of which are headed by two letters, these letters being symbol letters designating different types of machines, and in these columns are noted the neces. sary quantities for making the various machines, this type of list being one which can be made to cover quite a num. ber of different machines having the same general characteristics.

A somewhat different type of sundries list, and one having some very com. mendable features is shown by Fig. 3. The first line at the top of this list gives the name of the machine, the second line is a general description giving the sub. ject matter of the list, while the columns are divided into three groups each giving the following information, size of screw or part, "Number Wanted" column and column giving part name and purpose for which used.

Taking for instance the column headed A, at intervals along this column a sketch

of the type of screw wanted is made, and below these sketches are listed the size and length of screws wanted. In column C opposite each sketch is given the name of each type of screw, below the names in this column, are noted the screw each screw is used. Column B, which is left blank, is filled in with the quantity number required to make the particular group of machines wanted, while the blank space at the head of the sheet is for indicating the number of machines being made.

This last list makes a very vivid presentation of what is wanted in the nature of screws, washers, etc., while all forms shown may be suggestive of an adaptation suitable for some of the readers' needs.

FIG. 3—A VERY HANDY FORM OF SUNDRY LIST.

Stellite, Festel Metal and Stainless Steel

The Composition of Stellite, Together with the History and Properties of Stainless Steel, Are Presented in a Very Able Manner in This Article—Discussion of the Paper is Also Given.

STELLITE and stainless steel were the subjects of an interesting paper recently presented before the Engineers' Society of Western Pennsylvania, Union Arcade Building, Pittsburgh, by Elwood Haynes, president Haynes Stellite Co., Kokomo, Ind. The paper is given largely in full below:

The three metals, iron, nickel, and cobalt, are termed by chemists the "metals of the iron group." The reason for classifying them thus is the fact that their respective properties are all quite similar.

1. They are all distinctly malleable.
2. They are all distinctly magnetic.
3. They possess high tensile strength and high modulus of elasticity.
4. When pure they take a high polish and show a distinct metallic lustre.

They also resemble one another in their chemical properties.

1. Each is readily soluble in nitric acid.

2. Each form a monoxide with oxygen, as FeO, NiO, and CoO. Each also forms a sesquioxide, Fe_2O_3, Ni_2O_3, and Co_2O_3.

3. Aqueous solutions of their chlorides, when evaporated to dryness, are transformed into oxides.

4. Their oxides are all readily reduced by either carbon monoxide or hydrogen.

5. Their melting points coincide quite closely.

6. Their atomic weights are quite close together, that of iron being 56, and those of cobalt and nickel approximating 59.

When solutions of cobalt and nickel are mixed, it is difficult to separate the metals one from the other, owing to the fact that their behavior under most precipitants is practically the same.

Composition of Stellite

In 1899 the writer produced an alloy consisting of practically pure nickel and pure chromium by heating their mixed oxides with aluminum. This alloy, when polished, retained its lustre, even in the atmosphere of a chemical laboratory, and proved to be practically insoluble in nitric acid, even when boiling. It is also malleable when cold, and under proper annealing can be worked into sheets and wire.

Shortly afterward an alloy of cobalt and chromium was produced, which not only showed the same untarnishable properties as the nickel-chrome alloy, but possessed much greater hardness. The alloy could not be worked to any extent cold, but was found to be malleable at a bright orange heat.

It was not until 1906 that the alloy was produced in sufficient quantity to determine its properties fully. In 1909 a cutting blade was made of the alloy, which took an edge comparable to that of tempered steel. Later, tungsten or molybdenum was added, and the alloy thus produced was sufficiently hard to turn iron and steel on the lathe. Later experiments demonstrated that such alloys, when properly formed, would scratch any steel, and would stand up under much higher speeds on the lathe than the best high-speed steel tools. This fact gave the cobalt-chromium-tungsten alloy termed stellite (from the Latin word, stella—a star) a field of its own, and placed it in a class by itself as a material for high-speed tools.

Generally speaking, the cobalt-chromium alloys possess three distinctive properties, namely:

1.—They are untarnishable under all atmospheric conditions, and immune to nearly all chemical reagents.

2.—They possess great hardness.

3.—They retain their hardness up to visible redness.

Festel Metal

Some of the stellite articles for ordinary use are formed from alloys of cobalt and chromium only. This alloy answers well for table knives, spoons, etc. The harder edged tools, such as pocket knives, surgical instruments, etc., contain in addition to cobalt and chromium a certain amount of tungsten to give them greater hardness, while in other instances a certain amount of iron is introduced into the alloy to soften it so that it may be more readily worked. Such articles include table-knife blades, pocket-knife handles, certain dental instruments, etc. When iron is added to the alloy, the resulting mixture is termed "Festel metal," being made up from the chemical symbol for iron (Fe) and the first syllable of stellite.

This beautiful and easily workable alloy is well adapted to the manufacture of fine door latches, door-knobs, and high class sanitary fittings for bathrooms, lavatories, etc. It is not malleable except at a bright red heat, but when a certain portion of nickel is added it may be worked cold on the lathe or under the file. By suitable means it can be given a beautiful stippled surface, resembling that of frosted silver, and it is needless to add that under all conditions it retains its lustre in the most satisfactory manner.

Some of the later stellite alloys have shown most remarkable resistance to chemical reagents. One of these, possessing quite high chromium, takes a burnished silver. This alloy retains its lustre perfectly in boiling aqua regia, and is not affected in the slightest degree after immersion in that liquid for a period of 14 days. It is slowly attacked by cold hydrochloric acid, but is practically immune to cold, strong sulphuric acid, and nearly immune to the same acid in the diluted form. It is strictly immune to nitric acid of all strengths.

Balance weights made of this material retain their lustre under the most trying conditions. They present a beautiful appearance, owing to their superb lustre, and are so hard that their loss from ordinary wear will be perhaps unweighable for several years.

History of Stainless Steel

In the year 1911, I made some experiments on alloys of iron and chromium with a view to ascertaining definitely their properties. I quote from my notes as follows. [Mr. Haynes gives memoranda from his notebook, made in 1911 and 1912, of tests of alloys of varying quantities of iron and chromium.]

The experiments recorded above distinctly show that the non-corrosive qualities of chrome-iron and chrome-steel alloys were not only discovered by the writer at the time specified, but that their physical properties were also fully demonstrated. The discovery rests not on the possibility of adding to the steel other elements which may render it more or less immune to corrosion, more or less easily workable, but upon the fact that immune chrome-steels must contain more than 8 per cent. chromium, though for certain purposes they may contain much more than that amount, even up to 60 per cent.; that such steels are distinctly workable and useful, whether subjected to heat treatment or not; furthermore, that the proportion of carbon may be raised as high as 2 per cent. without materially interfering with the untarnishable qualities of the alloy, though such alloys are, generally speaking, more easily worked if the carbon is below 1 per cent.

Numerous metals may be added to stainless or rustless steel, and some of these may contribute slight benefit, while others may be slightly detrimental. Among these are nickel, cobalt, vanadium, silicon, boron, tungsten, molybdenum, titanium and tantalum. It is evident that an indefinite number of alloys could be thus formed, some with and some without the above elements, but none would be stainless unless it contained the proper amount of chromium, which is the essential element to be added to nickel, cobalt, or iron to produce a stainless alloy.

About two or three years after the discoveries recorded above Harry Brearley, of Sheffield, England, discovered practically the same properties in chrome-steel, and I am practically certain that the melting point, even in the presence of any discoveries made by me.

Immediately after making the discoveries recorded above, I applied for a patent, but my application was not at first granted, on the ground that chrome-steels were not new. Without going into details, I will say that I later made a second application, and that about fifteen days later. Mr. Brearley filed an application for practically the same thing. The United States Patent Office granted a patent to Mr. Brearley on the ground that his application contained a provision that it was necessary to polish and harden the steel in order to render it immune. This, however, was later found not to be correct, and in May of this year practically all of the claims in the application of the writer were granted by the Patent Office.

A personal service corporation was formed in this city, to which both the Haynes and Brearley patents were assigned, and licenses have now been granted to the principal steel makers for the manufacture of stainless steel under these patents. This corporation is the American Stainless Steel Co., with offices at present in the Oliver Building, Pittsburgh.

Properties of Stainless Steel

Stainless or rustless steel consists essentially of an alloy of iron and chromium, containing usually from 0.1 to 1 per cent. of carbon, though the latter element may be present up to nearly 2 per cent. without interfering seriously with the working qualities of the steel.

Owing to the high percentage of chromium and its tendency to oxidize at the melting point, even in the presence of carbon, it has been found advisable to melt the steel either in crucibles or in the electric furnace. After melting, the metal may be poured into ingot molds in the usual manner, and the ingots thus obtained may be forged or rolled into bars or sheets. If the ingots are of comparatively small size, they will be found to be very hard after casting, especially if they have been stripped hot and allowed to cool rather rapidly in the air. Indeed, small bars thus produced are likely to be almost file hard.

If a small piece of the steel thus produced be placed in a beaker with a piece of ordinary steel and covered with nitric acid, the ordinary carbon steel will be dissolved with great violence, while the chrome steel will remain utterly unchanged, thus proving that its immunity is primarily due to its composition. This is true whether the steel contains carbon in large or only minute quantities.

Cold chisels cast in iron or graphite molds are sufficiently hard, without tempering to cut ordinary iron or steel.

By heating cast bars to a bright orange temperature, they can be forged pretty readily into various forms. After the forging is completed, the metal may be allowed to cool in the air, and will be found to possess remarkably fine grain and good cutting qualities.

Quenching in water enhances the hardness to a considerable degree, particularly if the steel contains more than 0.4 per cent. carbon. It is best, however, to use oil for quenching, in order to avoid local contraction stress in the finished article, which might cause it to break under a slight shock or jar.

Notwithstanding the comparatively high temperature of working this steel, the bars show almost no scale during the forging operation, and when finished are covered with a blue-black "skin" consisting of a thin film of oxide.

Owing to the absence of deep oxidation and resistance to deformation at comparatively high temperatures the alloy is admirably suited for casting engine valves and distilling apparatus, and for many other purposes of like nature. When ground and polished, the alloy resists tarnish to a remarkable degree. It

is superior in this respect to brass, copper, and nickel plate, and far superior to any other steel yet produced. Axes, hatchets, saws or chisels made from it, not only will not rust in the atmosphere, but are unchanged when exposed to salt water or salt air. It will likewise doubtless find a large use in the manufacture of propeller blades for steamers, since its modulus of elasticity is much higher than that of bronze; and it resists the action of both fresh and salt water perfectly. Its great strength and comparatively high elastic limit are likewise in its favor. It will doubtless also have a large application in the manufacture of pump-rods, cylinder linings, pump valves, etc.

It is slowly attacked by dilute or strong sulphuric acid, and also by hydrochloric acid; but nitric acid has little or no effect upon the polished surface of the metal. When properly made it is impervious to practically all the fruit acids, including strong vinegar.

The alloy will fill a long-felt want among carpenters and others using wood cutting tools, since its freedom from rust, together with its capability of taking a keen cutting edge, renders it admirably suited for wood working tools. As noted above, it has been made into auger bits, and these have remained bright for years under all sorts of atmospheric influences.

Discussion Then Followed

In answer to a question as to the tensile properties of stellite, Mr. Haynes stated that "a good representation of malleable stellite would be 80,000 lb. elastic limit, 110,000 tensile strength, and with an elongation of 9 per cent."

Relative to the use of stellite during the war, he said: "We have no definite record, but our men claim that stellite was used in the rough turning of three-fourths of all the shrapnel made in this country during the late war. The last year of the war we delivered about three million dollars' worth of stellite for war purposes principally."

In answer to a question as to erosion on stellite from the effects of abrasion, such as the wear from water rapidly passing over the surface of the parts of a pump runner, Mr. Haynes explained that "it is very satisfactory. In the first place it is harder to grind than any steel, because steels do not resist erosion or abrasion to any remarkable extent. The hard alloys do not show as high scleroscopic or Brinell tests as the high-speed steels, but they work very much better because their abrasive hardness is very great. These two properties are not always fully understood. Some of us have thought that if we put a bar of steel under the scleroscope and it shows a high test, it is an extremely hard steel which would resist abrasion. A steel of that kind has a high elastic hardness, but not necessarily a high abrasive hardness. A vanadium steel of moderately high carbon and hardness will throw the hammer very high and the scleroscope test may show a hardness of 90, but at the same time you can take a file and file it with ease. Stellite alloys are just the reverse. They show a low elastic hardness and a very high abrasive hardness."

Power Transmission

The start of a series covering various styles of transmission. The chain drive and examples of its installation are shown.

By
J. H. Moore

THE question of proper power transmission is perhaps one of the most important met with in the modern industrial plant of to-day. In order that readers can have an opportunity of studying this matter for themselves, we are preparing a series, of which this is the first, on the different styles, and phases of transmission.

Through the courtesy of the Link-Belt Co., of Chicago, we are able to present the following illustration on chain drives as applied to transmissions. We feel some readers will appreciate the opportunity of noting where this style drive can be used to advantage, so suppose we ask ourselves the question; "What style drive shall we use, and what requirements shall we look for in the drive we decide upon?"

Points to Be Remembered

There are numerous points to be considered before deciding on the style of drive to be used, but perhaps most important of all is the question of loss of power due to slippage. One can readily understand that the moment any slip occurs, a certain percentage of power is going to waste, so that whatever system is adopted, ascertain if that system will ensure against such loss.

We are all aware that gearing is positive in its action, for, as the teeth mesh with each other, there is a positive drive which cannot slip. It is this fact which makes gearing the ideal driving power for automobiles, and so on.

We can readily realize, however, that it is quite impossible to use gearing for all transmission purposes. Apart from the high cost of cutting the teeth, the space the gears would occupy, the difficulty of securing proper ratio, and the noise accompanying the running of a

huge train of gears, especially at high speed, would soon prove its not being the proper methd. Having been convinced upon these basic points, suppose we ascertain if the chain belt system will meet in with the non-slip requirement.

In referring to its construction, we note that the chain consists of a series of leaves or plates, the joints of which consist of segmental case-hardened liners, or bushings, and case-hardened steel pins. These liners, which are removable, extend across the entire width, which makes for a double bearing surface, and halves the bearing pressure on the joints.

This, of course, means that the chain wears uniformly and keeps round. It runs upon a sprocket with teeth somewhat after the style of a gear, although they are shaped different, owing to the construction of the links. From these few points, we find that by transmitting power by means of a chain we are practically accomplishing the same result as if we were driving direct by gears, with the added advantage that we get away from the difficulty of ratios.

If we look at a chain drive closely, we will discover that there is actually more working contact in a chain drive than

FIG. 2—5 H.P. CHAIN DRIVE OPERATING 24" BACK GEARED SHAPER.

even in a gear drive, so that we can, with safe assurance, look forward to a longer life drive, together with smoother action. This smooth action would tend to answer the unspoken question regarding what noise can we expect from a drive of this nature, for if we have smooth action we naturally overcome throbbing, uneven motion, vibration, and other disadvantages which go to make up that one word, noise.

Another big point to be looked into is the matter of space, or in other words, what space is necessary to install a chain drive? Once more the answer is satisfactory, for the distance between centres can be made to suit your individual conditions. Gears require fixed centres, even to the thousandth of an inch, but with a chain installation you can make your centre distance meet in with your requirements.

There is also the question of ratios, or to put it more plainly, speed reduction. In gearing it is not always practical to step down at once to the desired speed, which means that intermediate reductions are necessary, but with the chain drive this becomes unnecessary, which naturally is a big point in its favor.

So far it would appear as if the chain drive was coming up to our requirements, so next we ask ourselves the question, "How about the power required to operate any machine? Can we, by using a chain drive, save any of that power?"

The practical mechanic knows that it is impossible to avoid some variations in the cutting and setting of gears, also that in proper meshing and wear, will increase the power required to operate a machine. These drawbacks are overcome in using a chain drive, as for example:

A certain machine required 7½ horsepower to drive it by the regular method,

FIG. 3—16 H.P. CHAIN DRIVE OPERATING PATTERN MAKERS' DISC GRINDER.

but on it being equipped with individual motor, and chain drive, 5 horsepower proved efficient. This shows a clear saving of 2½ horsepower in this particular case.

Other Features of Interest

There is also the question of tool upkeep to be considered. From tests made, it has been proved that where conditions were exactly alike, with the exception of the drive, a greater and a better output was possible by the use of the chain drive. The effect on the tools was very noticeable, one case showing that only two sharpenings of a drill was neces-

FIG. 4—15 H.P. DRIVE WITH SPECIAL SLIP PINION TO TAKE CARE OF OVERLOAD.

sary in place of ten, after the chain drive had been installed.

This feature of longer life of tools means less expense, not to speak of the time saved, without question. It is a serious matter, the removing, sharpening, and replacing of the various tools. The driving force of a chain is positive, yet elastic, and is equal at each instant to the force required to do the work at the rate of feed and speed as set. The result is a uniform character of work. There is no tearing or chattering, and a better and more uniform product is produced.

It is claimed by manufacturers that in the average case a chain drive can help to obtain 20 per cent. greater output, save 20 per cent. in power, give from four to ten times the service on cutting tools, with the consequent saving of time of operation for grinding, re-setting, etc., give a better finish to work, make less wear on the bearings and rotating parts, and larger and heavier work can be accomplished on the same machines.

To suggest where chain drives can be used to good advantage, is almost impossible, for their field of usefulness is practically innumerable. Cement plants, paper making plants, textile mills, rubber mills, machine tool plants, in fact every industry can use them to good advantage.

Another point which we have so far omitted to mention is that of the safety feature of a drive of this nature. A chain installation is always easily protected, and if enclosed in a dust-proof oil-tight casing, the result is, of course, perfect, for such a procedure guarantees ample lubrication and protection from dirt.

In looking over the various illustrations in this article we suggest you not only look, but study them. Run over, in

FIG. 3—A 7½ H.P. CHAIN BELT OPERATING MILLER PREVIOUSLY OPERATED BY MECHANICAL DRIVE.

FIG. 4—A 5 H.P. DRIVE ON A STERILE TURRET LATHE.

FIG. 1—3 H.P. DRIVE OPERATING MULTIPLE AUTOMATIC.

FIG. 2—7½ H.P. DRIVE OPERATING LATHE. CHAIN SHOWN IS ¾" PITCH AND 4" WIDE.

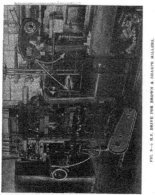

FIG. 9.—A H.P. DRIVE FOR BROWN & SHARPE MILLER.

FIG. 10.—CLOSE UP VIEW OF DRIVES SHOWN AT FIG. 11.

FIG. 11.—A BATTERY OF GEAR CUTTERS DRIVEN BY CHAIN FROM LINE SHAFT UNDER FLOOR.

FIG. 12.—A 2 H.P. DRIVE OPERATING A JONES & LAMSON TWO-SPINDLE LATHE.

your mind, if there are speeds, feeds, or drives in your plant, which could be improved by such a change. If you follow out this suggestion, then the article will really have accomplished its purpose, namely, proved itself of practical value.

As this is strictly a machinery journal we have confined ourselves to that field, believing that readers will be more interested in machine tool installations than in any others.

Some Interesting Installations

Take for example, Fig. 1. Here we have a main drive of 400 horsepower. The installation shown is that of one of the main drives at the Latrobe Electric Steel Co., Latrobe, Pa. It will be noted that the casing has been raised in order to show the chain. One cannot help but remark on the compactness and business-like look of this drive. It has been said that a chain drive of the nature shown is flexible as a belt, and positive as a gear. These claims are readily understood after a glance at the photograph.

Our next illustration, Fig. 2, shows a Kelly shaper equipped with individual motor and chain drive. Again we note the compactness of the arrangement. The drive is 5 horsepower and the chain 5-8 of an inch pitch, by 1½ inches wide.

Fig. 3 demonstrates once more the ease with which one can connect a machine by chain drive. The machine is of course recognised as a Besly Pattern Makers disc grinder, and the drive is 15 horsepower.

The installation shown at Fig. 4 is somewhat out of the ordinary. The machine is an Espen-Lucas saw equipped with a 15 horsepower chain drive, with a special slip pinion to take care of overload. The need of the slip pinion is of course obvious, so we need not com-

ment on this point, except to say that the possibilities of chain drives are wider than often realized.

The next installation, Fig. 5, shows an adapted drive on the spindle of a Lincoln type miller which was previously driven by mechanical drive.

The drive shown is 7½ horsepower. The motor is run at 1,150 r.p.m., and the shaft centre is 31¾ inches. Here is a case, not only compact, but ingenious in its adaptation.

Next in order comes Fig. 6. Here we have a compact drive installed on a Steinle turret lathe. The neatness of

FIG. 13—5 H.P. DRIVES OPERATING PUNCH PRESSES. NOTE THAT TEETH ARE CUT DIRECTLY IN FLY WHEEL.

the arrangement is at once apparent, and following are the data of installation: The motor runs at 1,150 r. p. m., the driven shaft at 285 r. p. m., and the centre distance is 18 inches. This latter figure emphasises the advantages of such a drive.

The multiple automatic opens up a good field for chain drives. The machine shown at Fig. 7 is one made by the Davenport Machine Tool Co., New Bedford. Mass., and the motor is 3 horsepower. Note how the motor is placed underneath the bed, and how compact and dependable the drive looks. Installations such as shown speak best for themselves.

Fig. 8 illustrates a 7½ horsepower chain drive operating a large sized lathe. The chain used is 5-8 inch pitch, and 4½ inches wide. At Fig. 9 is shown a No. 2B Brown & Sharpe miller equipped with chain drive. A 5 horsepower motor provides the power and the neatness of the installation is obvious. It will be noted that various machines, part of which are shown in the photograph, are likewise equipped.

A Barber-Coleman gear hobber next demands our attention. This is illustrated at Fig. 10, and the casing has been removed in order to show the drive to the best advantage.

Fig. 11 illustrates a battery of Brown & Sharpe gear cutters driven by chain from a line shaft under the floor. It will be noticed that the operating levers come up through the floor so as to be handy to the operator.

At Fig. 12 is shown a 3-horsepower drive on a two spindle Jones & Lamson lathe. Once more one cannot help but note the compactness possible with a drive of this nature. The motor runs at 1,150 r.p.m., the driven shaft at 800 r. p. m., and the centres are 2½ inches.

FIG. 14—LEFT VIEW SHOWS 5 H.P. DRIVE OPERATING A BLISS PRESS, WHILE RIGHT HAND VIEW ILLUSTRATES A 5 H.P. DRIVE TO A BORING MILL.

It will be noticed that a neat motor bracket is arranged on top to suit the motor.

Fig. 13 illustrates some very compact and serviceable installations of chain drives operating punch presses. The motors are 5 horsepower in each case. Note the fact that the teeth for the chain to engage with, are cut directly in the flywheel.

At Fig. 14 is shown two different installations. The one to the left is a 5 horsepower drive on Bliss presses. The motors run at 1,150 r.p.m., the driven shaft at 230 r.p.m., and the shaft centres are 42 inches. The view to the right is a 3 horsepower drive operating a King boring mill. The motor also runs at 1,150 r.p.m., the driven shaft at .185 r.p.m., and the centre distance of shafts is 18 inches.

Our last illustration at Fig. 15 shows an extremely neat and efficient drive system on a group of twelve Barber-Coleman gear hobbers. Note the compact drive, and the safe chain guards, and the general neat appearance of the complete installation. Everything of a moving nature is enclosed. No danger exists from getting caught in moving parts, and in fact from all points of view the advisability of an arrangement as shown is self-apparent.

The various illustrations which have been shown and explained do not by any means cover the field of the chain drive. Volumes could be written on the subject, but we have at least covered a certain amount of ground in the machine shop

FIG. 15—A GOOD ILLUSTRATION OF COMPACT CHAIN DRIVES. THESE DRIVES OPERATE A GROUP OF GEAR HOBBERS.

field. When considering a transmission drive remember that you should look for the following requirements in the drive:

First, it must be efficient with the best economy of power. Your speed, reduction, etc., should be carefully considered, also the fact of protection against overloads. Do not forget that space saved, means more room for additional equip-

ment, which in turn means added production. The safety feature should also be considered carefully, for this is a very important point in these days of speeding up production.

In the next article we will present the belt system of transmission, and discuss drives of interest in this style of transmitting motion.

Some General Observations on Valve Motion

This Paper, Read at the Canadian Railway Club Meeting, Should Be of Especial Interest to Our Railroad Readers—It Gives Some Valuable Hints on the Economical Distribution of Steam.

By F. WILLIAMS

THIS subject was ably presented by the above author, who is mechanical designer of the Canadian National Railways, Moncton, N.B., at a recent meeting of the Canadian Railways Club, Montreal. To anyone interested in the construction and adjustment of locomotive valve gears, we can readily suggest a careful perusal of this material.

Mr. Williams commences as follows:

"The duties which a locomotive valve gear has to perform are exacting in the extreme, as it has to control the distribution of steam to the cylinders with almost perfect precision through a wide range of cut-offs in forward and reverse direction. There is no apparatus on a locomotive upon which the economical working depends so largely, and when we consider that at diameter-speed the movement of the distribution valve is

reversed 672 times per minute, we can appreciate with what care the design must be undertaken.

"From the point of view of economical steam distribution, valve motion design has to-day reached a point where it cannot be greatly improved upon, and the chief attention of the designer has for the last few years been taken up with questions of accessibility and low maintenance cost, his aim being to apply

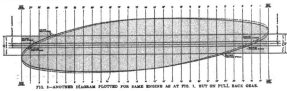

FIG. 2—ANOTHER DIAGRAM PLOTTED FOR SAME ENGINE AS AT FIG. 1. BUT ON FULL BACK GEAR.

a gear which would run and keep square from shopping to shopping' with the minimum of attention.

"Considered from the standpoint of steam distribution alone, I doubt if a well-designed and properly set Stephenson gear has ever been excelled, but owing to inaccessibility, high maintenance cost and its great liability to get out of square due to the springing of parts and development of lost motion, the Stephenson gear has become a back number, and I shall only refer to it for purposes of comparison.

"For several years past practically every locomotive built in this country has been equipped with an outside gear, the vast majority with Walschaerts', and to this gear I shall devote most attention.

"The simplicity of the gear lies in the fact that the valve receives its motion from two sources, first from the crosshead through the combination lever, and second from the eccentric through the link, and each of these sources of motion can be dealt with separately without considering the influence of the other in designing and setting. The motion derived from the combination lever is equal to the steam lap plus the lead and it attains its maximum travel when the engine is on the dead centre. It is not affected in any way by the reverse gear, but remains the same in all positions of the lever.

"The motion derived from the link is simply a symmetrical motion front and back of the centre line, and is increased or decreased according to the distance of the link block from the centre of the link.

"When the link block is exactly in the centre of the link there is, of course, no motion from this source and as the block gets by the centre the motion is reversed. When the engine is on the front or back dead centre the link assumes such a position that the reverse lever

can be moved backward and forward through the entire travel without imparting any motion to the valve, and the distance that the valve is off the centre is entirely due to the position of the combination lever which is at its maximum travel at these points.

"The length of the combination lever from the radius bar connection to the union link connection must bear the same proportion of its length from the radius bar connection to the valve steam crosshead connection as does half the stroke of the piston to the lap plus the lead plus 1/64 in. The 1/64 in. is added to the lap plus the lead to take care of the lost motion. Care must be taken that the length of the combination lever adopted will bring the lower end of the lever to the correct level to connect up with the union link, especially if the union link is connected directly to the wrist pin, which is the practice generally adopted.

"The radius of the link slot centre line is, of course, determined by the length of the radius bar, and the preferred location of the link support bearings is such that the horizontal centre line is on a level with the radius bar connection to the combination lever. This location may be varied within reasonable limits without affecting the valve events to any appreciable extent; for instance, on a locomotive with a very large cylinder the steam chest centre line and the cylinder centre line are of necessity quite a distance apart and in this case the link support is sometimes lowered an inch or two to bring the link tail nearer to the horizontal centre line of the axle.

"The angle through which the link rocks should not exceed 45 deg., and if it can be kept lower, so much the better. The eccentric rod connection to the link tail should be kept within 3 or 4 ins. of the horizontal centre line of the axle in order to keep the angularity of the eccentric rod within limits and owing to

this angularity of the rod, it will be found necessary to offset the tail connection of the link in order to give it the same angular travel on either side of the central position. I have heard men with a good deal of experience state that an approximately correct offset is all that is required, but as it is just as easy to make this offset correct as otherwise, I always prefer to make it dead right.

Points to Watch

"The eccentric crank must be set so that it brings the link dead on its central position when the engine is on either front or back dead centre and the throw of the eccentric pin must be such that, acting in combination with the radius of the link tail, it will give the required angular travel to the link.

"The reverse shaft location, length of arm, and swing link are very important considerations, and unless great care is exercised in the arrangement of these details the efficiency of the motion may be considerably reduced. The arc which the reverse shaft arm describes should be so arranged as to reduce the link block slip to a minimum in all positions of the reverse lever, special attention being paid to the running position in fore gear.

"It is impossible to avoid link block slip altogether, but it can be kept pretty low, and if this is not carefully looked after the effect will be seen in the valve events and also in the wear on the link and link blocks. The steam chest centre line should be outside the cylinder centre line far enough to permit of bringing the whole motion into practically a straight line, thus eliminating the necessity for rockers and doing away with the twisting effect and lost motion which the use of rockers involves.

"Engines equipped with the Walschaerts valve gear should be so arranged that the link block is in the bottom half of the link for forward gear, the

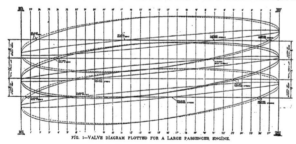

FIG. 1—VALVE DIAGRAM PLOTTED FOR A LARGE PASSENGER ENGINE.

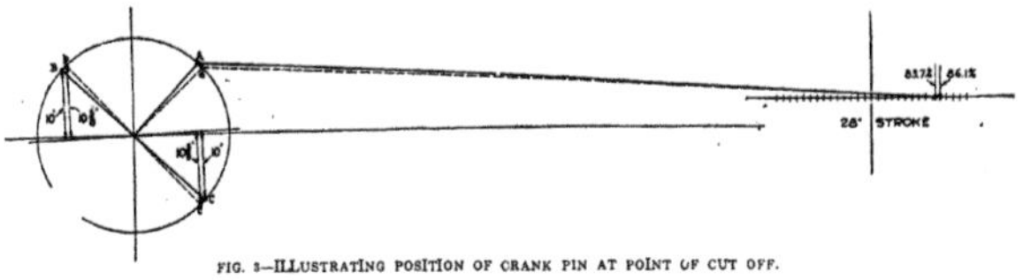

FIG. 3—ILLUSTRATING POSITION OF CRANK PIN AT POINT OF CUT OFF.

eccentric, of course, following the crank pin. The advantages of this arrangement are that the wear in the link support bearings is diminished and the link block slip in running position may be kept very small, as the swing link describes an arc which is very similar to the arc struck by a point in the bottom of the link, the concave side of both these arcs being uppermost.

"It is very important that the design of this gear should be as good as it is posible to make it, for if it is faulty, it is impossible for the valve-setter to correct its faults. If the design is good all the valve events naturally come within very close limits of being square, but if they do not there is practically nothing that the valve-setter can do to correct them. A perfect steam distribution will, of course, give a perfect exhaust, but a perfect sounding exhaust does not necessarily mean a perfect steam distribution. The steaming properties of the boiler and the fuel economy depend very greatly on the evenness of the exhaust, and if the exhaust is ragged the vacuum in the smoke-box is unsteady and the fire is soon pulled in holes, resulting in a great waste of fuel. The designer therefore endeavors to obtain a perfect exhaust by giving as nearly as possible a perfect steam distribution, the valve-setter has to get an even exhaust at all costs.

"The usual setting of a Walschaerte gear is square on dead centres, with a constant lead in all positions of the reverse lever, but the main object of lead is to give an unrestricted supply of steam to the cylinder when the piston begins its stroke, and with the preadmission down to about 1/64 in. it is impossible that the steam admitted to the cylinder can exert any appreciable turning movement on the axle until the crank pin has gone over the centre.

Various Diagrams

"Fig. 1 shows a valve diagram plotted for one of our large passenger engines. This engine has 24 in. by 28 in. cylinders, a 14 in. valve, 6 in. steam travel, ¼ in. constant lead, 1/16 in. steam lap and ¼ in. clearance. The broader ellipse in the centre shows the valve travel in relation to the piston travel on full forward gear, and the narrow ellipse inside it shows the same thing with the lever notched up to 25 per cent. cut-off. The distance from the steam edge to the exhaust edge on the valve over the packing rings is 2¾ in., therefore the similar ellipses which are plotted 2¾ in. above and below the centre ellipse with lighter lines must represent the movement of the exhaust edges of the valve. The three ellipses shown in dotted lines represent the movement of the valve set with no lead in full fore gear. Picking out the valve events we find that with the ¼ in. lead setting we have the cut-off at 23 in. and 23⅞ in., the release at 25 9/16 in. and 26¼ in., and the closure at 26 13/16 in. and 27¼ in. Set with no lead, however, we have the cut-off at 23 9/16 in. and 24⅝ in., the release at 26⅛ in. and 26¾ in., and the closure at 27¼ in. and 27⅝ in., so that the net result of adopting this latter setting is to delay the cut-off from 83.7 per cent. to 86.1 per cent., the release from 92.6 per cent. to 94.5 per cent., and the closure from 96.6 per cent. to 98 per cent., an improvement in the starting position of 2.4, 1.9 and 1.4 per cent. of the stroke respectively.

"The valve diagram shown in Fig. 2 is plotted for the same engine on full back gear, the ellipse shown in dotted lines representing the valve movement with a variable lead setting, which setting is made possible by the lead slightly increasing in the forward motion, with a corresponding decrease on the backward motion. In this case there are two lines ¼ in. above and below the centre line representing the amount of the exhaust clearance. These lines will determine the release and closure points in the same way that the outside edges of the steam parts did in Fig. 1.

"Fig. 3 shows the position of the crank pin of this engine at the point of cut-off, the full line indicating position with normal setting and the dotted line with variable lead setting. It is evident that when the engine is standing in this position we shall have the minimum starting effort, as all the turning movement has to come from the other crank, which will be either at Bb or Cc, according to whether Aa represents the right hand crank or the left. The effective length of the crank, which is doing the work, is 10 in. for normal setting, and 10⅝ in. for the variable lead setting, or a difference in favor of the variable lead of 6.25 per cent., so that the minimum starting effort of this engine is increased by 6.25 per cent. by this setting. The maximum traction effort is not affected in any way. As soon as the engine has turned a wheel the advantage almost entirely disappears.

"Fig. 4 shows a diagram plotted for one of our Mikado engines. This engine has cylinders 27 by 30 in., 14 in. valve, 6½ in. valve travel, ⅛ in. constant lead, 1 in. steam lap, and no exhaust clearance. The reduction or total

Contiued on page 59

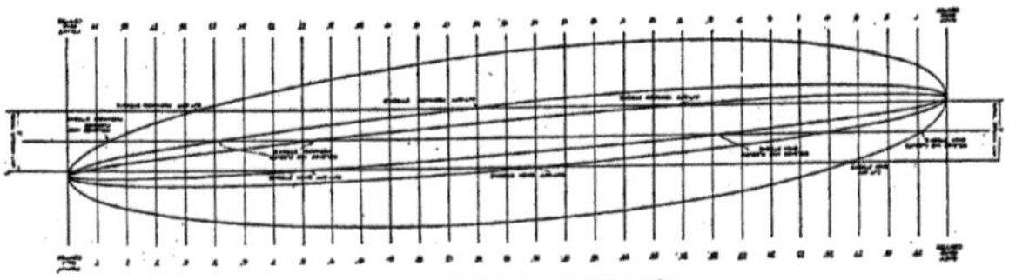

FIG. 4—DIAGRAM PLOTTED FOR A MIKADO ENGINE.

WHAT OUR READERS THINK AND DO

HANDY MILLING FIXTURE
By G. Blake

The sketch illustrates a milling fixture designed by the writer for slotting the piece of work (brass) shown in the sketch. A is the body with a tongue let in for locating in the slot of the milling machine table. The clamping bar, B, is pivoted with a pin G. The sectional view shows the method of locking éccentric shaft E, with a lever D pinned on, accomplishing the work The fixture is open when the lever D is in a vertical position .

The spring F $_{is}$ for raising the clamping bar B to allow the work to be placed in position, a stop piece (not shown in the sketch) locates the work so that the slot is in exactly the same position on every piece milled.

By pressing the lever D down in a horizontal position, it locks the work securely by means of the cam on the shaft E. The setting plate C, (cast steel hardened), is for setting the milling cutter to the required depth of the slot. The shaft E, and bar B, are of mild steel casehardened at the working parts. The body A, is of cast iron.

AN EXCEPTION AND ITS ANSWER

In our volume XXIII., No. 22, we published an article on Repairing a Double Throw Crankshaft, by W. R. Green. One of the readers, G. Durham, on perusing this article, felt he had a few things to say on the subject himself and wrote us as follows:

"I wish to take exception to some of the figures in Mr. Green's article. His workmanship in turning shaft and sleeve

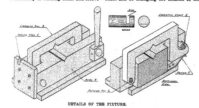

DETAILS OF THE FIXTURE.

must have been excellent and I believe the best method possible, outside of electric and oxy-acetylene welding, but why the long 1½-in. hole, or how long was the hole? Then he says for shrinkage was given 5-32-ins. on a 7-in. diameter shaft. This figure is, I believe, impossible. It is about all the steel would expand at a red heat, and would burst when cooling. 13-64 for the 10-in. is also too much, to my mind. I have found from practical experience that 1-80 part of 1 inch of 1 ft. is far better than guessing the limit of expansion that the metal will allow. I will admit that machine steel is quite elastic and will stretch quite a large amount, but you are not getting any benefit by so doing as you are only straining the metal.

"For comparison I give the following figures, based on 1-80 part of 1 inch of 1 ft:

Mr. Green's figures.   My conception. .
7″ diam. 5/32″=.156″........ =.007″
10″ diam. 13/64″=.003″....... =.01″

"By using the 1-80 basis it is not necessary to get the shrink so hot, consequently it is easier to handle. I might also add that any good handbook will give about half of my figures for shrinking as standard."

On receiving this data we forwarded it to our correspondent, Mr. Green, and asked his viewpoint, after reading the letter. Here is what we received:

June 8th, 1920.

"Your letter with enclosure received to-day. I will answer same to the best of my ability. First: the long hole was drilled to stop or head-off a further possible crack, which might have been caused by turning off the outer part of shaft and so changing the tension of the

steel. Whether it did any good or not, I cannot say. It was the order of the foreman. The hole extended beyond the outside crack limits.

"Second: As far as I can remember about the shrinkage problem, the smith and myself consulted the head moulder with regard to the necessary amount to allow for shrinkage. At that time, what books on engineering I possessed only gave tables for castings shrinkage and for lineal expansion for steel and iron (rolled). These were no good for what I wanted to know, but I may possibly have made a mistake in my diary when putting down the figures, but I do not think so. I do know that the flange buckled a little when cooling, but there was no sign of a burst sleeve, which would surely have been the case when being finish turned if the expansion had been excessive.

"I wonder if Mr. Durham knows that a shaft with a bored hole in it will very seldom warp or twist when working. It is a fact, and if a large shaft shows spline cracks in angle of spline or keyway, a small hole bored in centre of shaft will frequently save changing the shaft. I have learned these tricks of the trade in out-of-the-way places, where new material was generally hard to get, and the average machine tender was NOT a tradesman and therefore would frequently tighten down bearings till they squealed, thereby putting great torsion on the shafting and causing cracks to start.

. "I will try and find out from a friend of mine who is working near the place where this engine was, whether the same shaft is still at work or not. If he writes to me in answer I will pass his letter on to you."

STANDARDIZATION OF PLAIN LIMIT GAGES

A sectional committee of the American Engineering Standards Committee has just been organized to undertake the standardization of plain cylindrical gages for general engineering work, under the sponsorship of the American Society of Mechanical Engineers. The immediate occasion for undertaking the work was a request of the British Engineering Standards Association for co-operation on the subject. The committee held its organization meeting on June 11th. It is understood that this committee will recommend to the American Engineering Standards Committee that the scope of the work should be broadened so as to

cover all plain limit gages for general engineering work.

The present personnel of the committee is as follows:

E. C. Peck, chairman, gen. supt. Cleveland Twist Drill Co.; L. D. Burlingame, chairman, industrial supt. Brown & Sharpe Manufacturing Co.; H. W. Bearce, secretary, Gage dept. Bureau of Standards, sec'y National Screw Thread Commission; P. W. Abbott, Lincoln Motor Company; John Bath, president John Bath & Co., Inc., engineer of standards, Pratt & Whitney Company; Fred H. Colvin, assoc. editor, "American Mechinist"; W. A. Gabriel, chief draftsman and designer Elgin National Watch Co.; F. O. Hoagland, vice-pres. and works manager, The Bilton Machine Tool Co.; Edward H. Ingram, works manager, The Cleveland Drilling Machine Co.; J. O. Johnson, office of Chief of Ordnance War Department; A. W. Schoof, gage enginer, Greenfield Tap & Die Corp.; G. T. Trundle, consulting engineer, Engineers Building, Cleveland, Ohio; H. L. VanKeuren, The VanKeuren Company.

VALVE MOTION

Continued from page 57

elimination of the exhaust clearance lengthens the period of expansion by delaying the release, and this in itself is a good feature, but it has also the effect of advancing the closure point and the question naturally arises as to why it should be permissible to eliminate exhaust clearance and thus advance the closure point on freight engines and not on passenger engines. The first reason is that in running position the passenger engine is generally notched up to a much earlier cut-off than the freight engine,—about 25 per cent. of the stroke instead of 50 per cent.—and the second that the piston speed of the passenger engine averages much higher than that of the freight engine.

"The passenger engine under consideration has a piston speed of 1,136 per hour, or over 40 per cent. higher than that of the freight engine at 30 miles per hour, and the higher the speed the higher the compression will be, provided that all other conditions are equal. By giving this engine exhaust clearance we not only delay the closure point, but also give a greater exhaust part opening, thus allowing the exhaust freer access to the atmosphere, and the result is a freer running engine. It may be imagined that when the engine is working at a short cut-off less steam is admitted to the cylinder, and, therefore, the piston has less to sweep out on the return stroke, but the exhaust begins with the release, and by the time the return stroke has begun there is very little difference in the amount of steam left in the cylinder, whether running on long or short cut-off.

"From the point of view of economy the question of steam consumption per unit of power developed, the higher the compression the greater the economy, provided we do not run the compression higher than boiler pressure. This is on account of the clearance volume, and we can readily understand that the higher the compression the less steam has to be supplied from the boiler to build up the initial pressure, and if the compression reaches boiler pressure there is no steam drawn from the boiler until the piston actually starts its working stroke. The next thing to consider is the power required, as it is no use trying to run on a very fine thread of steam if we cannot get the tractive effort necessary to keep the load moving.

"If we compare Figs. 1 and 4 we find that the closure takes place at 76 per cent. of the stroke for the passenger engine in running position; with the freight engine running at the same cut-off the closure takes place at 73 per cent., but if we eliminate the exhaust clearance on the passenger engine we shall advance the closure point from 76 per cent. to 68 per cent. of the stroke, which is a marked advantage in point of economy.

"This goes to show how much the closure point is affected by the amount of exhaust clearance. For my own part I do not think the actual closure point is of very much importance, but that the exhaust port opening has a great deal more influence on the compression than the actual position of the closure point. I contend that if we pay proper attention to the exhaust port opening the closure point will take care of itself. There is no purely mathematical means of determining the most desirable exhaust port opening, and this, like to many other problems in locomotive work, has had to be determined by practical experiments.

Passenger Versus Freight

"It is here that we find the chief difference between passenger and freight engine setting, and, still referring to Figs. 1 and 4, we see that the maximum exhaust port opening in running position for the passenger engine is just over $1\frac{5}{8}$ in., whereas that of the freight engine is just over 1 9/16 in. when running at a 50 per cent. cut-off, while if we notch up the freight engine to the same cut-off as the passenger engine we have a maximum exhaust port opening of only $1\frac{1}{4}$ in.

"This maximum port opening is only maintained for a few inches of the stroke, and it is easy to understand that when this port opening begins to narrow down it will form quite a choke for the exhaust at a high piston speed, and will build up quite a little compression before the closure point is reached. We all realize that a locomotive exhaust has to be choked to a certain extent to obtain a high velocity jet up the stack, which will induce a proper draft through the grates, but this choking should be done by the exhaust pipe tip and not by the valve. The area of the exhaust pipe on the passenger engine under consideration is about 23 sq. in., and on the freight engine 29 sq. in., and the valve displacement necessary to give a part opening equal to the area of the tip will be approximately 11/16 in. for the Pacific engine and $\frac{7}{8}$ in. for the Mikado. This 11/16 in. port opening is maintained for 53 per cent. of the stroke on the Pacific engine, but on the Mikado the $\frac{7}{8}$ in. port opening is only maintained for 36 per cent. of the stroke when notched up to the same cut-off as the Pacific. When the Mikado is running at 50 per cent. cut-off, which is approximately the running position, the $\frac{7}{8}$ in. exhaust port opening is maintained for 58 per cent. of the stroke, which compares favorably with the Pacific. This gives us the chief reason why the Pacific setting is found to be more suitable for high speeds and short cut-offs, while the Mikado setting is better on the slower speeds and long cut-offs. From these comparisons we can see that the passenger engine setting has been developed to give a smart and free running engine at high speeds and short cut-offs, while the chief consideration with the freight engine setting is to obtain the greatest possible traction effort at moderate speeds and to run on comparatively long cut-offs with the greatest economy by delaying the release point as far as possible."

British engineering firms are showing remarkable skill and enterprise in developing new types of internal combustion engine. One of the latest and most interesting of these types has been designed to meet the demand in steel works for large slow-speed gas engines. Continental practice has favored the horizontal engine, which is huge enough to fill a church, but the new British engine referred to is of the vertical design, which takes up only a fraction of the space filled by the horizontal engine. Most vertical engihes have previously been high-speed machines, but this one runs at only 125 revolutions per minute. This result is achieved by arranging the cylinders in pairs, one above the other, and placing a pair on each side of the crank. Thus four cylinders combine in driving a single crank. The mechanical arrangements are so admirable that although an engine of 1,000 horse power is about 20 feet high, hardly a tremor can be noticed at the extreme top even when the engine is running on an overload. An engine of this power is now working successfully in a leading British ironworks, being used for generating electricity.

According to Wireless World there are three wireless direction-finding stations being erected at the mouth of the Mississippi River for the purpose of directing ships to New Orleans. Eight stations of this nature are also in course of construction along the Pacific Coast from San Diego to Alaska. One station, that on the Farralone Islands, is nearly completed. The sites of the other seven are at Pt. Reyes, Bird Island, Pt. Montara, Pt. Hueneme, Avalon, Pt. Firmin, and Imperial Beach.

DEVELOPMENTS IN
SHOP EQUIPMENT

HEAVY DUTY SHAPERS

The Columbia Machine Tool Co., Hamilton, Ohio, have placed on the market a varied line of heavy duty shapers. These machines are made in various sizes ranging from 16 ins. to 28 ins., and are of complete new design, having certain features which ensure exceptional efficiency in operation.

The outstanding feature is their strong, rugged construction. All parts are carefully fitted, practically all parts being made to jigs and templates, on the interchangeable basis. The alignment of vise and table with ram is held within a limit of .001, within their full length. This is only one example of how they are kept to close limits.

The various movements of these machines are controlled from the front or working side. There is a quick change feed, and a single lever for locking cross rail to the column. The table support is placed at front of table, where it is most accessible, and ring-oiling bearings are provided.

The quick change feed, as already mentioned, is particularly valuable as it adjusts itself automatically to position of cross rail, and is graduated to show the amount of feed. It also has bevel gear reverse which automatically insures feed taking place on return stroke. Two levers control all feed changes, one for amount of feed, the other for direction of feed. Changes can be made instantly while machine is running.

The illustrations depict the 20 in. extra heavy type with speed box, friction clutch and brake.

The speed box is of the selective gear type, strong, substantial and convenient to operate. The gears are semi-steel, affording strength for the heaviest service. Shafts are ground and run in ring-oiling and bushed bearings. The cover is marked to show the proper position of the two operating levers for each speed. The hand wheel on side of speed box is for making hand adjustments when required.

A powerful friction clutch and brake is placed between the driving pulley (or motor) and the speed box, operating by the curved lever at side of column, within convenient reach of the operator, permitting quick starting and stopping at any point of the stroke.

The crank gear is very liberally proportioned and is carried close to the rocker arm, in bearing of extra large diameter, preventing torsional strain. The machine is powerfully back geared, which, in connection with the four-change speed box affords eight changes of speed; the speeds have been carefully calculated to afford the proper range and increments, being in geometrical progression as required by the best practice. The rocker arm is exceptionally heavy, the sides being joined by a strong central tie. It is connected to the ram through a heavy link, which affords a direct pull on the ram.

The rocker block is a steel casting, with large bearings, provided with taper gib for adjustment. The driving shaft bearings are ring-oiling, with renewable cast-iron bushings. Other cylindrical bearings are bronze bushed, all shafts are ground, and every provision is made for strength and durability. Liberal oiling facilities are provided.

The feeds are produced by a substantial feed box. It will be noted that this is designed as part of the machine, not as an attachment. A range of feeds is provided sufficient for all classes of work, eight on the 16- and 20-inch shapers, twelve on the 24- and 28-inch. They are operated by merely shifting a small handle, conveniently located, and can be made while the machine is running. The arrangement is such that the feed takes place only on the return stroke and the mechanism automatically adjusts itself to the position of the cross rail. A small lever is provided—located just back of the cross rail—which controls and in-

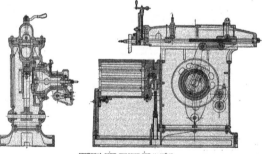

SECTIONAL VIEW THROUGH THE MACHINE.

dicates the direction of the feed; this lever can also be operated without stopping the machine. A safety friction is provided to guard against breakage. The feed gears are of steel, hardened and running in oil, and the arrangement throughout is compact, durable and extremely convenient.

The other parts of these machines are in like proportion of strength and convenience, and further details can be procured by those interested.

GENERAL VIEW OF THE SHAPER.

NEW THINGS IN MACHINE TOOLS

PNEUMATIC TOOLS

The Ingersoll-Rand Company have added several new tools to their pneumatic lines, and will be known as the "Little David." Included in these tools will be a small size of close-quarter drill, two types of small high-speed grinders, and a lightweight drill in two sizes.

DRILL CHUCK

The Scully-Jones Company of Chicago, Ill., are now manufacturing a new drill chuck or collet that is designed to hold a straight shank drill in a taper hole of the drilling machine. The only operation necessary to use an ordinary straight shank drill in this chuck is to grind a flat on one side of the shank. One advantage claimed for this collet is that closer spacing can be made when used in connection with multiple drilling. These collets are made in a wide range of sizes

MULTIWHIRL OIL COOLER

For the cooling of oil used in lubrication of turbine bearings and reduction gears, or the quenching oil used in heat treatment, the Griscom-Russell Company of New York has manufactured a special apparatus, that provides for constant circulation of the oil through the cooler, and maintains the oil at a fixed temperature, and permits the continual use of the original quantity of the oil and its maintenance at the proper viscosity. Helical baffle plats cause the oil to take a whirling path as it passes through the machine. The shell is of cast iron and the tubes of seamless brass or copper tubing.

CYLINDER GRINDING MACHINE

The Sutherland Machine Shop, of Omaha, Neb., are now making a machine of special design for the boring and grinding of automobile cylinders. The device is made in the shape of an attachment that may be secured to the ordinary lathe, the head being independent of the head of the lathe. The fixture that carries the cylinders is fastened to the carriage of the lathe. The holding brackets may be adjusted to support cylinder blocks of various sizes. Micrometer adjustment is provided for bringing the different cylinders into alignment. The wheel spindle is carried within the outer slow-running spindle which is used for operating the boring tools. Micrometer adjustment is provided for regulating the eccentricity of the tool or wheel movement.

GAGING DRILL RACK

The Feerless Machine Company, of Racine, Wis., are placing on the market a drill rack of interesting construction. It provides for the placing of the drill with their points downwards, the holes in the stand being drilled clear through, each hole having two diameters. The holder acts as a limit gauge, inasmuch as a drill of the size intended for the hole will enter the top portion but will not pass through the second, and a drill that goes into the lower diameter is obviously in the wrong location. These racks are made in seven sizes, the largest size being used for drills from 41-64 to 1¼ inches in diameter.

RIGHT-LINE RADIAL DRILL

The Niles-Bement-Pond Company, of New York, are placing a new type of drilling machine on the market. This machine is of the radial type and is built in 5 ft. and 6 ft. sizes. The special feature is the design of the column, which permits of a simplified drive. The entire mechanism is controlled from the head, and the clamping of the column is performed electrically. The column is of special box construction and swings with the arm so that all bending stresses due to the drilling operations are always taken in the same plane. The base of the column is supported on ball thrust bearings and is likewise provided with roller bearings for the rotating motion. When in a position for rotating, the column is raised slightly from the pedestal by means of flat springs, but when in a working position the action of the clamp pulls the column down on to the flange of the base. This clamping of the column is accomplished by a ring that is operated through a worm-wheel driven by a small electric motor. An automatic stop is provided that prevents accident should the arm be run beyond the limit of its travel in either direction. The motor for driving is located on the back of the arm saddle. The horizontal shaft passing through the column and giving a direct drive. A maximum distance of 68 inches is provided between the spindle and the base. The spindle has a traverse of 18 inches, with 28 speeds ranging from 20 to 400 revs. per min.

PISTON BLASTING MACHINE

A blasting machine that has been adapted for the cleaning of the insides of pistons is now being manufactured by the Gray Machine Tool Company of Buffalo. The general principle is the same as that adopted for the cleaning of shells. The piston is located in the machine with the open end downwards, and when lowered to a certain position starts revolving. The sand blast is operated by a quick-acting valve and the nozzles can be placed in any position desired. The machine is of the duplex type, so that one piston may be placed in and removed while the blast is operating on the other. The air consumption will approximate 40 cu. ft. per min. at a pressure from 70 to 90 lbs. per sq. in.

The MacLean Publishing Company
LIMITED
' (ESTABLISHED 1887)
JOHN BAYNE MACLEAN, President. H. T. HUNTER, Vice-President
H. V. TYRRELL, General Manager.
PUBLISHERS OF

CANADIAN MACHINERY
~ MANUFACTURING NEWS ~

A weekly Journal devoted to the machinery and manufacturing interests.
B. G. NEWTON, Manager.   A. R. KENNEDY, Managing Editor.
Associate Editors:
J. H. MOORE   T. H. FENNER   J. H. RODGERS (Montreal)
Office of Publication: 143-163 University AVenue. Toronto, Ontario.

VOL. XXIV.   TORONTO, JULY 8, 1920   No. 2

Leaning on United States

FOR several months now Canada has been witnessing the spectacle of fuel controllers and various civic officials from this country making sporadic excursions to the coal-bearing areas of United States. And following these excursions there invariably comes the statement that the "situation is very serious," "Canadians should make every effort to get as much coal as possible from whatever source is available to them," etc.

Now this has been going on for some years. There is nothing new in the situation, and it varies only with the varying ability of the different official tourists from this country to tell a hard-luck and cold winter story on their return.

And the very next day we can probably turn to some proxy Government report, or some peppery booster material from Western municipalities, and read that Canada has some sixteen or seventeen per cent. of the known coal reserves of the world.

Where is the meeting point between these two situations? There must be some plane where these conditions find common ground. If there is not, and if there is no prospect of such in the near future, the sooner we put the everlasting quietus on our talk about owning coal reserves, the sooner will we cease to be laughing stock.

In round figures we produce in Canada thirteen million tons of coal per annum, and in that same time we want about thirty million tons, and so we send our tourists scurrying all over United States to see if they can scratch around and find us that other seventeen million tons.

Right now United States wants every ton of coal it can secure. Industries and householders there are short away short.

Were it not for the attendant loss, inconvenience, and suffering it might be a good thing for Canada to spend a winter of cold feet and frozen ears. We have been told for years that the haul is too great to bring coal from either West or East to the central portion of the Dominion. In such a policy the railroads are short-sighted. It is time we began to see that if coal is not forthcoming in plentiful quantities there will be nothing to take away from the factories and the workshops of the industrial centres. It is a matter of self-preservation for the railroads of the country that there should be a good supply

of coal in every workshop of the Dominion. Railroad figures of the distance East or of the distance West must not scare our people into throwing up their hands in that miserable attitude of "It can't be done."

The people of this country should also get it firmly established in their minds that United States is using coal at the rate of 700,000,000 tons annually, and her supplies are not inexhaustible. Then they must recognize that every industrial squabble that stirs in United States, is felt in Canada in a very short time. A coal strike in United States is in effect, if not in reality, a coal strike in Canada, and more and more we are becoming the victims of all the strikes that outlaw leaders and loud-mouthed agitators are pitchforking onto the industrial life of the Republic.

A self-contained Canada is the solution and the only way out. It is time that we advanced from the academic discussion of this question to putting some real national sweat into its solution.

The Week in Passing

REPORTS from London, Eng., indicate that the Canadian Exhibition there was not a success, and that it did not attract very large crowds. Members of the Krassin party from the Soviet Government in Russia were visitors. One London manager of a Canadian firm offers the following as an explanation of the lack of success: "There is not enough of any one class of merchandise in the exhibition to attract anyone to it. Those Canadian firms wanting to exhibit should send their hardware to our hardware exhibitions, their engineering plant to our engineering exhibit, their grocery goods to the foodstuffs exhibition, and their agricultural machinery to our agricultural exhibition. A mixed exhibition such as this was bound to be a failure no matter how well it was worked." The whole thing was a bid on Canada's part for export trade. Experimenting is one sure way—perhaps a bit costly—to find out the best way to get our goods before the British or Continental buyer.

LAST week reference was made in Canadian Machinery to the high prices that were being paid in Scotland by the ironmasters there for several grades of steel and iron scrap. The prices were very much higher than those quoted locally or in United States market. The suggestion that Canada should sell into that market came readily. Frankel Bros., of Toronto, have made several attempts to ship in to this market and have offered as high as $20 per ton for shipping space, but have been unable to secure accommodation. So far as any advantage to Canadian trade is concerned, it will have to be extracted from the rather unsatisfactory reflection that the thrifty Scot can turn in his old steel axles at £15 per ton, against $20 that is offered here by dealers, who are none too anxious to buy.

COMPLAINT is made, and apparently with good reason, of the slowness attending the return of our freight cars sent with shipments to United States. At the first of February this shortage was up to 16,000 cars, and on April 15 the figures stood at 21,000. While these cars are across the line they are of no use to the shippers here, and they are doing poor service there as far as Canadian trade is concerned unless they are being used to bring back to this country some shipments of much-wanted raw or semi-finished materials.

THE Niagara district is taking a new way of getting its advantages before manufacturers looking for sites. The association is holding a three-day congress, when a tour of the peninsula will be made and its advantages pointed out. There is something to commend this idea of publicly going about this work in an open way. In too many cases municipal promotion work has consisted of a poorly-concealed auctioning of a much pedaled proposition, the municipalities being the willing and apparently open-eyed victims, with open-mouthed simplicity swallowing all sorts of bonused and doctored cripples in the hope of some day grabbing a real one out of the deluge.

MARKET DEVELOPMENTS

Believe Premium Market is Near an End Here

Some Sales Being Made at Better Prices From the Buyer's Standpoint—Can't Ship Canadian Scrap Metal to the Old Country Market—Railroads Looked to as Best New York Buyers.

CERTAIN of the Canadian steel merchants are firmly of the opinion that 'we are facing a better steel market from the standpoint of the consumer, and they look for the disappearance of the premium dealer as the first move in this direction. They point to actual cases where they are buying from the same firms at better prices than they could hope for a few weeks ago. Sheets, for instance, were held up for $8 per hundred, while now $6.50 is gladly taken. The protracted helplessness of the American railroads to tackle the traffic situation across the line and straighten it out helps to keep the market at a higher level, and there are good reasons to believe that with good transportation facilities we would be now dealing with a normal steel market.

Power and fuel problems are troubling Canadian industry in no small degree. In a plant where there are hundreds of employees, the great majority of them depending on power to get reduction, it is a serious matter Shops in many cases are finding that they have to pay for a good many hours in the course of a month for which they receive no return.

Old Country ironmasters are paying high prices for scrap material, much higher than are quoted in Canada or United States. Canadian dealers would like very much to get a chance to ship to that market, but they cannot get shipping space. One Toronto firm has offered as high as $20 per ton—as much as some of the metal can be sold for here—for shipping space to reach the Old Country market, but there has been no response, and there the matter rests.

Bar mill material is much needed, especially in the industrial centres of Ontario at present. The agricultural trade needs flats, while the building and contracting interests are after rounds for reinforcing purposes.

MONTREAL PLANTS NEED COAL AND LACK OF IT KEEPS BACK BUYING

Special to CANADIAN MACHINERY.

MONTREAL, Que., July 8.—A condition that may have an early and serious effect upon industry in this district, is the increasing difficulty experienced by many manufacturing concerns in getting delivery of coal in sufficent quantities to maintain normal operations. Not only do these adverse circumstances have a direct bearing on plants affected, but the curtailment is naturally reflected in other directions. The general relaxation that becomes effective during the midsummer season makes the shortage of coal less noticeable, but unless adequate relief is given to the coal situation during the next two months, the outlook for the fall and winter is none too promising. One would suppose that with less cars being used for the transportation of fuel that better delivery of other commodities would be in order, but the reports do not indicate that any material improvement has been made in this regard. Coal requires cars of special construction and it is just possible that few of them could be utilized for any other purpose.

Increased activity is noted in some of the local metal working plants, and this is largely due to many of the district foundries resuming operations after a period of enforced idleness. It was stated by one manufacturer here, that a few weeks back he had to refuse orders owing to the inability to obtain the necessary castings. Metal-working shops are quite busy, but seldom in a position to give definite promise of completion of orders, owing to the prevailing uncertainty of material shipments.

Little Change in Steel

"No change to speak of. We were hoping for some relief but this has not developed. Production has not been aided by the lack of coal, so that we are still in a position of watchful waiting." This statement by a dealer seems to reflect the general situation, as the quiet tone of the market would appear to indicate a decline of industrial activity that is more than seasonable. The midsummer period is frequently marked by business depression, but in comparison with previous activity, the present is a time that carries with it more than the usual uncer-

tainty. The volume of business is not so pronounced but dealers seem quite satisfied with the average turnover, as this, invariably, is well balanced by the prices that prevail. Apart from the possible effect of fuel scarcity, the impression is growing that the return to normal conditions will only be a matter of a few months. The fact that some domestic industries are becoming affected by the falling off in demand, has acted as a sort of brake to other activities, and steel dealers reluctantly assume that these conditions will eventually react in a general way.

Machine Trade Quieter

The customary seasonable falling off in the machine tool movement is becoming a feature, as reports would indicate that the demand is lighter and requirements less pronounced, or buying deferred until after the holiday period. There continues to be a fair volume of supplies passing through the hands of the dealers, and the same applies to used equipment, although sales of the latter are less active than formerly.

Downward Tendency in Steel Scraps

The listless condition of the old material market continues and dealers report a quiet week. There has been a slight betterment in the trading of

machinery scrap, owing to the renewed activity of local foundries. This, however, has not been marked, but small as it is, it retains the feature position. "There is little demand at present," remarked a dealer, "and the trend is apparently moving in an easier direction. Non-ferrous scraps are moving quietly and prices are comparatively firm, but dullness in irons and steels may eventually develop a weaker condition and a decline in quotations," Little scrap is coming in, and this is partly due to the attitude of dealers, as they are evidently reluctant to carry any more yard and warehouse stock than conditions warrant.

OFFERED $20 TON FOR OCEAN RATES

Local Dealers Would like to Sell Into the Old Country Scrap Market at High Prices

TORONTO—A great deal depends on power, and supplies, as far as the industrial life of this district is concerned. Many of the shops are constantly in trouble with the shutting off of their power. When this takes place the shop generally has to pay the men, as it is no fault of theirs that there is no juice on hand. Expenses go on just the same but there is no output. When this takes place many times in a plant where there are some hundreds of employees it becomes a serious matter. Add to this the shortage of material, and the shortage of coal, and it makes a condition that is none too favorable to getting ahead with the work.

But in spite of all these things there are buyers in the market, and there are inquiries and prospects, which, if only a fair percentage of them turn out to be good, will mean a nice booking for the early fall.

One of the leading makers of machine tools in the United States has announced a further increase this week. It has been some days since any advances have been made, and dealers have come to regard the machine tool market as more stable, and if anything have been anticipating some marking down.

Deliveries are reported by several firms as having shown quite a decided improvement during the week, and shipments are coming in from points now that have been particularly affected by the strike on United States roads.

Old Country firms are also getting better treatment in this matter of getting their goods over here, and one firm has opened a warehouse with the assurance that it will have a good line of material to show under power.

Several sales have been made recently owing to the placing of better machinery in certain plants, and it is likely that this will account for several more sales. Where the salesman can show how a shop operation can be turned out more efficiently, and where time and expense can be saved by the placing of new machinery, he can always get an audience, and there is undoubtedly a disposition on the part of plant executives

POINTS IN WEEK'S MARKETING NOTES

Some steel merchants believe there is an easier tone to the steel market, and that if railway traffic were able to handle the material piled at the mills, we would be facing a normal steel market.

Bar mill material is in demand, but the supplies forthcoming are not showing any signs of improving. Canadian mills are being pressed by their customers for these shipments.

Although prices for scrap metal are very high in the Old Country now Canadian exporters cannot get into the market, claiming that they cannot get shipping space, although offering $20 per ton.

The New York machine tool market is looking to the railroads to come into the situation there as large buyers of equipment.

Old Country firms report that they are getting better shipments of their orders for the Canadian market. Heavy duty machinery, such as boring mills and heavy planers are on a more extended schedule.

One maker of machine tools advised the Canadian trade of an increase in price this week.

It is estimated that United States steel mills now have about three million tons of material, finished and semi-finished, piled at their plants, waiting shipment.

Prices More Reasonable

There is every reason to believe that any price movement that takes place in the steel market will be toward a downward revision.

That is not a guess, but the statement of one of the largest warehouse firms in this district. One bit of evidence is given in the experience of this firm in the matter of sheets from the premium mills. "A short time ago the only basis on which they would talk business to us," stated this firm, "was $8.50 base, while to-day the same people will gladly take on business at $6.50 base, and give us just as good service in the matter of delivery at the lower price as they were doing at the former price of $8. There are other lines too where we find that there is quite a marked tendency to give us better price treatment, and a greater desire to do business than there has been for some time past." A good many sheets are selling this week around the basis of $10.50 per hundred for 28 gauge.

The matter of securing a good supply of bar mill material is becoming a question that is not easy to handle. The Canadian mills have a large tonnage on their books, much of it very urgent owing to the fact that it must be secured for certain seasons, and it is almost impossible for the mills here, owing to shortage of materials, to get the rollings through to satisfy the trade. Bars for some time were one of the best classes in the trade, but recently they have become one of the worst.

Old Country dealers are getting in shipments now that were ordered last November, and are getting fairly good satisfaction in the matter of future deliveries, unless they go into the line of heavy duty machinery, when the time gets extended by some months. The supposition is that many of the large steel and iron plants on the continent are getting equipment ready to start operating again, and in this way there is an unusual demand for this class of equipment. Boring mills and heavy planers, for instance, are hard to secure unless the buyer is prepared to wait his turn.

The Scrap Market

Canadian dealers are experiencing a dull season in scrap metal in almost all their lines. Last week this paper published a letter from a Scottish iron and steel merchant, calling attention to the prices that were being paid by the ironmasters of that country for scrap lines, and noting the difference in price between their figures and what is offered here. Canadian Machinery took this matter up with one of the large yards here to-day, and inquired why it was that Canadian scrap could not be sold profitably in Scotland under these circumstances.

"The explanation is simply this," was the reply. "We cannot ship the bottoms to take our stuff. We have offered as high as $20 per ton to carry shipments of steel and iron scrap, but we can get no takers. So you see the price may as well be anything as far as we are concerned. It is a nice market, but we have no way to reach it now."

Coppers for a day or so seemed to show some strength, but the demand locally is weak. Any of these metals may seem to be worth more than the figures quoted at times, but the real touchstone as far as this market is concerned seems to be, "Do you want it, or don't you?" If there are no buyers wanting metals then the chances are that prices will stay down.

Can Get Started. — The Hamilton Chamber of Commerce was advised that a Western Canada coal company can send 1,000 tons of coal per day East for the next eight weeks. Local manufacturers have been advised of this.

Steel Co. Starts.—The Steel Company of Canada has secured a supply of coal and oil again, and will start part of its plant next week. It has not been able to get full supplies yet.

STEEL MILLS CONTINUE TO PILE
AS CAR SUPPLY IS TAKEN FROM THEM

Special to CANADIAN MACHINERY.

PITTSBURGH, July 8.—The sheet and tin mill wage matter has been settled, and there is no strike. As noted in our report of a fortnight ago, the Amalgamated Association broke with the sheet and tin plate manufacturers who had formerly recognized the union, the issue between an amendment proposed by the union which would enable it to use its hold on the hot mill labor to induce workmen in other departments to join by being able to promise such men, only ten being required to form a new lodge, that they would use their previous strength to secure signatures for new and additional scales. When the Atlantic City conference broke up, June 14, President Tighe of the Amalgamated Association stated that if an adjournment were had it would be to a date not later than the first Tuesday in September, which of course meant that the men would walk out July 1st, as the existing scale agreement expired June 30. However, probably on account of pressure of the members, the Amalgamated Association officials reconsidered the matter and called for another conference, to which the mills acceded. The conference was in session at Columbus, June 30, when the Amalgamated Association officials telegraphed the lodge that the men should continue at work, and an agreement was reached late last Thursday. At the Standard Tin Plate Company plant, Canonsburg, Pa., the men walked out as a local president refrained from acting on his telegram. A few mills closed because the management had made arrangements in accordance with the common expectation that the walkout would be attempted. Otherwise the mills are working the same as formerly.

Besides the amendment to the preamble, the men had demanded a 20 per cent. increase in basis rates, a six-hour shift instead of the eight-hour shift that has been in force for many years, and a great many other things, but the mills made only small concessions. Outside of some trivial matters the changes were as follows: In the tin mill scale a spell hand is introduced, not a regular member of the crew, who will help the catcher part of the time and help in other ways, the man to be paid about 50 per cent. as much as the catcher, while the catcher's rate is reduced 10 per cent. The increments in the sliding scale rates, when tin plate is selling above the $3.50 base, are to be 1 1-3 per cent. instead of one per cent. Thus when the realized average on tin plate shipments is $7.40, as it was at the last bi-monthly settlement, the men would receive 87% per cent. above the base rates instead of 78 per cent. above. In the spell scale, an extra spell 'hand' is also introduced, to be paid also by the company, but not to be considered a regular member of the crew. In each case, the extra man will conduce to greater output, this benefiting the men. The wage cost per ton will be increased between $1 and $2 per ton, but this may be made up to the management by increased output.

Not Much Organizing Expected

The difference between the American Federation of Labor, which contemplated by its action at the Montreal convention a fresh effort to organize the iron and steel industry, and the Amalgamated Association, which held that it should have sole charge of any organizing of the iron and steel industry, was recently settled between the two bodies on the compromise basis that if the Federation committee were formed by the various unions, the Amalgamated Association would be accorded a 51 per cent. representation on it. The manufacturers believe, however, that this arrangement is made simply to save the faces of the Federation management, and that the committee is not likely even to be formed. The present outlook, therefore, is that little, if any, attempt at organizing will be done and that the iron and steel industry will remain non-union.

Even the hold the Amalgamated Association has had, by controlling some 30 to 40 per cent. of the hot mills in the sheet and tin plate industries, and the tonnage men at most of the iron mills, may possibly disappear in time, for many of the mills that have hitherto been quite willing to sign Amalgamated Association scales year after year are now disposed to embrace the next favorable opportunity to eliminate the Amalgamated Association entirely. It is observed the non-union sheet and tin mills, which include all those of the Steel Corporation, get along very nicely, and thus the Amalgamated Association is not useful as it once was in preventing minor disturbances from day to day or week to week.

Car Shortage

In connection with the Interstate Commerce Commission order that the railroads give full preference to coal mines in furnishing coal cars, the curious situation develops that the coal operators claim they are not getting many more coal cars than formerly, while the iron and steel interests claim their car supplies have been greatly diminished, through gondola cars, particularly high side gondolas, being transferred from their service to the coal service. Any increase in coal movement would benefit the iron and steel industry, by supplying more coal to the by-product coke ovens and more coal to the steel mills, but the industry wants to ship more material rather than to make more, since right along it has been accumulating stocks.

Some estimates of the stock of steel accumulated at mills on account of car shortage now run up to 3,000,000 gross tons, while the lowest guess is about 2,000,000 tons. Thus the accumulation represents something like from three to five weeks of production at the rate recently obtaining.

Thus the accumulations of steel have become a much more serious matter, a great difficulty being that some of the steel will be shipped so much later than expected that the customers may not require it at that time. It would at best require a long time to work off the extra steel, given the best transportation conditions that can be expected, and thus far even the mills are still shipping less steel than they produce. At a number of plants rolling schedules have been changed so that finishing operations are decreased, without the production of steel itself being decreased, and thus more steel is being accumulated in the intermediate forms of blooms, billets, slabs and sheet bars. Steel in this form can be piled conveniently and can later be finished into the forms then desired by customers.

Markets Quiet

The markets continue quiet all along the line, but the situation attracts less attention than formerly, as the traditionally dull midsummer period has been entered upon. Premiums for prompt shipment of steel products continue to sag, but the large independents have not changed their prices, which cover moderately early delivery, and the Steel Corporation of course continues its adherence to the Industrial Board schedule. June is expected to show a further increase by bookings during the month exceeding the shipments.

LOOK TO RAILS TO
KEEP TRADE UP NOW

Otherwise There Are Not Many Large Buyers in the New York Tool Market

Special to CANADIAN MACHINERY.

NEW YORK, July 3.—The machine-tool trade is centering its attention on railroad business, most of which is still in a prospective stage. It is apparent, however, that the railroads will buy a great deal of shop equipment when funds become available. The Norfolk & Western, which issued a list of about 70 machines several weeks ago, is now about to place orders. These orders will probably come through this week. The Delaware, Lackawanna & Western has bought a driving wheel lathe and a wheel press. The Seaboard Air Line is expected to be a fairly large purchaser. Some small inquiries have been issued by the New York Central Railroad, but a large list is being held up pending the approval of the necessary appropriation.

Conditions as a whole in the machine-tool trade continue pretty much as they have been during the past six or eight weeks. The credit situation and the freight congestion have pretty effectually stopped all buying except for immediate requirements. A great many manufacturers will not purchase tools

unless they can obtain immediate delivery, and as there are few lines of new tools on which prompt delivery can be promised the sale of used tools is stimulated. In the past week inquiry has been slightly better than in the two weeks preceding. The General Electric Co., Schenectady, N. Y., has issued an inquiry for about 20 machines, half metal work-

ing and half woodworking, for its Baltimore, Md., plant. The E. W. Bliss Co., Brooklyn, N. Y., has issued a list of about 15 tools, most planers, slotters and lathes, for a prospective plant at Birmingham, England. The Wasson Piston Ring Co., Plainfield, N. J., and the Edison Lamp Works, Harrison, N. J., have bought a few machines each.

COAL STRIKE IN THE EAST WAS
AVERTED BY NARROW MARGIN

Sydney, N. S.—That a general strike in the Nova Scotia coal fields was only just averted by the arrival of Hon. Gideon Robertson's telegram announcing the appointment of a Royal Commission on wages, was the statement made by J. B. McLachlin, District Secretary, at the United Mine Workers' headquarters.

The strike order had been voted upon and was on the table ready for signature when the all important message came in. Had it been delayed its arrival would have found twelve thousand miners idle, and not a colliery working in the Provinces of Nova Scotia and New Brunswick.

In view of the coal shortage in Upper Canada and the recent Government proposal for an embargo on foreign shipment of Cape Breton coal, the miners say

they consider the Robertson message the most important that has reached this part of the country for many years.

Senator Robertson notified the District Board that Royal Commission demanded by them has been authorized by Order-in Council, that the names of the members will be announced in a day or two, and that the first session of the Commission will take place at Sydney or Halifax on July 12. The U. M. W. Board immediately wired acceptance of the Government's offer.

The miners have been clamoring for a commission since the first of May. Continued delay left them in an exceedingly ugly mood, practically forcing the Executive to the preparation of a general strike order, which happily will not now be needed.

FARM MACHINERY WANTED IN
RUSSIA

Great Shortage of Agricultural Equipment to be Supplied in Near Future

'Agricultural machinery in Russia is almost non-existent at the present time,' said the London manager of "Selskosoyus," the All-Russian Purchasing Union of Agricultural Co-operation, to a London, England, reporter.

"Until the former stock, destroyed or worn out during the war, can be replaced, there is little hope of Russia resuming its place as a great grain-producing country.

"We are here as representatives of the Russian co-operative societies, to export agricultural machinery, chemicals, and seeds, for which we shall pay in goods imported from Russia or by long credit. So far we have been carrying on a very uphill undertaking. and our exports have only amounted to £300,000. The difficulties of communications have been almost insuperable, and are now beginning to be surmounted. Even to-day we are afraid that a great quantity of our goods may have been endangered by the British evacuation of Batoum.

"The 'Selskosoyus,' or 'Village Union,' is the great association of Russian co-operative societies. It is divided into three grades—the village society, the association of villagers, and the union of districts. The whole movement is under the aegis of the Moscow Narodny Bank, and is entirely non-political.'

POWER PROBLEM
GROWS SERIOUS

Plants Have to Pay for Time and They Get No Production to Balance It

At a recent meeting of Canadian manufacturers the question of power was discussed, and it was found that real difficulty was being experienced by firms which were being cut off.

Mr. F. G. Perrin of the Willys-Overland Co., in stating the case for his plant, gave the experience of a number of other firms in the district. Mr. Perrin said:

"When our plant shuts down we have the choice of paying the men off, or keeping them on and paying them. When it is shut down, it is always difficult to know when it will be on. Consequently we lay men off for a certain length of time and pay them, so we are paying them for producing nothing, thus doubling our loss. It is a condition that we are all up against. It is not only our own plant that is shut down, but all the plants that supply us with materials. Along the Windsor border our representatives telephoned us that all three plants upon which we were depending for our raw material for our plant, were closed down through lack of power. In our own case we have kept a record of all which registers all current coming into the factory, shows stoppages aggregating 38¼ hours' duration from December 2 to April 27, which is a very serious matter."

Another embarrassing factor in the pig iron market is the embargo placed by the Inter-State Commerce Commission on the use of open cars for the conveyance of anything but fuel. This is now hitting the industry through the limestone supply as well as regular transportation facilities for iron, and in consequence quite a large number of stacks have banked down.

The current market ranges from $44 to $45.50 delivered in the East.

The lower price of bessemer has tended towards an equalization in the price of the two grades. In fact nearly all grades of iron are now quotable at a uniform price of $45 valley furnace.

Some iron and steel works are so pressed for pig iron that a large percentage of their melt is scrap iron.

New England foundries are nearly without exception reported with orders booked up for eight months ahead.

The bulk of present activity is for prompt iron.

There have been some large requests for export recently, one export house inquiring for 80,000 tons basic and 20,000 tons No. 1 foundry for Italy was only able to get quotations for 10,000 tons basic and 20,000 tons of foundry. The exporter stated that the prices quoted were too high to interest the prospective buyer. There were also numerous other inquiries, some of which call for 20,000 to 30,000 each of various grades of iron.

The trade is mostly concerned with transportation facilities and fear that unless some modification in the embargo on the use of open cars is made that a high percentage of furnaces will be forced to bank down.

U.S. SCRAP METAL

In spite of the dullness of the scrap metal market generally the demand for melting steel has been strongly held by the large foreign sales during the last month, England and Scotland having purchased 150,000 to 200,000 tons, delivered seaboard at between $24 to $30.

The market is generally dull everywhere, owing to the difficulties of transportation and scrap metal. Consumers who have not covered for future requirements are beginning to fear that their indifference in this matter may cost them dearly.

There is a general feeling amongst dealers that there will be a rise in the scrap market, as steel grades have advanced about 50 cents per ton above last week's level, and they are covering to protect sales already effected.

On account of the large demand for the best quality grades, for export, is the general impression that domestic buyers will have difficulty when they enter the market.

CANADIAN MACHINERY
AND
MANUFACTURING NEWS

Vol. XXIV. No. 3.   July 15, 1920

The Manufacture of Iron by the Use of Rolls

This Interesting Branch of Industry is Not as Well Known as It Should Be—Methods of Grading, Piling, Refining and Other Processes to Completion Are Discussed Herein

By W. S. STANDIFORD

THE manufacture of iron by the use of rolls occupies a position in the industrial world that is of the utmost importance to the human race. The comforts that generations of people have enjoyed, have been largely due to the workers in steel and iron. As this article deals with the making of iron bars, no attention will be given to steel rolling—except to state that the drafts and designs of rolls should be different if efficient results are desired. The best iron is from Norway, it being of unusual purity, which is due to lack of sulphur and phosphorus in the ores, these chemicals not being desired on account of their bad effects on the finished metal; sulphur making iron brittle when hot (called red-short), phosphorus causing it to break easily when cold (termed cold-short). The grades of iron as usually marketed are classified as common, refined, and double-refined.

Common iron is made by using scrap consisting generally of old horse shoes, bolts, pieces from machines of various kinds and junk of every description. This mass is piled on boards and put into a heating furnace; it is then heated to a white-heat, run through the rolls, and made into flat bars of varying widths, thicknesses, also lengths, all depending on what size of mill they are to be used upon, also the finished length and diameters of the iron bars that are required.

They are then cut up into desired lengths and one bar is laid on the bottom of a pair of U-shaped irons (these are used to hold the box together until it is finished); bars are also placed on each side, thus making a box pile. The latter is then filled with scrap, getting it as full as possible, then another bar is laid on top and the side of box bound tightly with band iron to hold it together. It it then technically termed a "box pile"

in some mills and "faggots" in others. They are then put into the furnace, heated to a white-heat, and run through the rolls. Fig. 1 shows what these box piles look like. The flat bars forming the sides having been rolled once, before being made into box piles, have a more denser grain than the centre composed of rusty and dirty scrap iron; the resulting finished material, while having a nice smooth appearance—if the rolls are in good shape—will not stand much bending, as it splits through the centre.

As scrap metal contains anything from Norway iron to tool steel, the bars made from it cannot be expected to be very good. However, in places where common iron can be used, such as for making fences, it comes in handy, as it would be folly to use a high-priced metal of great tensile strength for the straight bars of fences where there is no strain. Refined iron is made by melting pig iron

FIG. 1—BOX PILES.   FIG. 2—BAR PILES.

FIG. 3—EXTERIOR OF A PUDDLING FURNACE.

and puddling it in a puddling furnace, which process removes some of the carbon and converts it into wrought iron in an impure condition. It is then run through rolls and made into flat bars. If the latter are examined at this stage they will be found to be porous on the surface like a sponge. This is caused by the impurities such as slag, etc., which they contain. The bars are then cut up into the required lengths, piled upon each other until a pile of the desired height and weight is obtained. It is then put into the heating furnace until at the proper heat, next being run through the rolls, making finished bars ready for sale.

Another method of manufacturing refined iron that is in use by some mills is to buy 12 or 16 pound to the yard old mine rails of wrought iron and then cutting them up to desired lengths and tying two or more sections together with iron wire, heating them in the furnace and rolling into the desired sizes and shapes. This method of working makes a very strong iron that can be used for almost any purpose, except for bars that are to be used for bracing the bottoms of freight and passenger cars — when an iron of the utmost strength and ductility is needed. The rust and dirt on the surfaces of the old rails is pretty well worked out by the pressure exerted by the rolls, the result being an iron that has a fair average strength and one that makes a good material for general use in machine construction. Double-refined iron is also made from pig iron, each mill having a special formula of its own. It goes through the same process in the puddling furnace and rolling through the puddle rolls as the refined iron does, but with this exception.

The iron, after going through the puddle rolls is cut up into desired lengths, then taken over to the finishing mill furnace, the cut lengths being piled upon each other to the required height, heated to a white heat in furnace and run through the finishing rolls. The bars are then cut up into suitable lengths, piled one upon another, reheated and given the final rolling to the desired section of finished iron. Thus it will

be readily seen that the double-refined metal has received three rollings, one in the puddle mill and two in the finishing mills. The triple rolling that the iron has gotten renders it very dense and fibrous, the dirt and impurities being well worked out. It is an axiom that the more iron is worked the more fibrous it gets. Thus, contrasted with the refined article, it is far stronger, which, other things being equal as regards quality of metal, the extra rolling imparts greater strength. It may be thought that if three rollings improved the metal as regards tensile strength and ductility, that eight or ten rollings would make the iron have tremendous strength. Theoretically this would be the case, as the impurities would be practically driven out, but laying aside the question of expense involved in such an undertaking, experience shows that with the present methods of heating iron by means of coal or gas that the iron absorbs too much sulphur from the fire; it also suffers in its physical characteristics by becoming brittle, which is, no doubt, due to its getting burnt by the frequent re-

heatings. Fig. 2 depicts a bar pile ready for the final rolling into the finished product.

Heat Piles Evenly

This double-refined iron is a splendid material if the puddling has been carefully done and care taken in regard to heating the piles evenly in the furnace. The temperature at which the metal is rolled also has a great influence on the strength of the finished bars. Double-refined iron is a fine metal, it having a breaking strength of over 50,000 pounds per square inch with an elongation of 27 per cent. The metal is well liked by blacksmiths on account of its fine working qualities, such as being easy to weld, etc. The strength of this iron may seem small when compared with the strength of steel bars of equal diameter, but when subjected to constant vibration steel has been found to lose its tensile strength and ductility, becoming very brittle; it is said to "crystallize," and shows a granular grain on fracture. The process of rolling makes both metals fibrous in nature, but iron seems, for some reason or other, to keep its fibrous character, and stands vibration better. Double-refined iron is used extensively under freight and passenger cars. When it is realized that a car weighing 45,000 pounds and carrying a load of 140,000 pounds of coal is subjected to continuous shock as it passes over rail-joints of uneven roadbeds, besides receiving a severe shaking during transit from one locality to another, it will readily be seen that a fine grade of reliable metal is needed for such service. Iron also withstands the action of the weather better than steel.

A rough test of the quality of any iron can be made as follows: If fracture gives long, silky fibres of leaden-gray hue, the fibres cohering and twisting together before breaking, it may be considered a soft, tough iron. An iron of this character will stand being twisted and knotted like a pretzel. The specimen that the writer saw showed no cracks or breaks whatever. An iron having a medium, even grain mixed with fibres,

FIG. 5—PUDDLE ROUGHING ROLLS IN HOUSINGS.

denotes a good iron. A short, blackish fibre indicates badly refined iron. A very fine grain shows a hard, steely iron, apt to be cold-short and hard to work with a file.

Coarse grain with brilliant crystallized fracture, containing yellow or brown spots, indicates a brittle iron, which is cold-short, working easily when heated and also welding easily. Cracks on the edges of bars is a sign of hot-short iron. Good iron is readily heated, soft under the hammer and it throws out few sparks when being hammered hot. It should also be low in phosphorus as possible, the lower the better, as only 0.5 to 0.8 per cent. is sufficient to produce cold-short-ness in iron. Carbon also has the effect of making iron hard, and should therefore be kept as low as practicable, 0.20 per cent. being a fair percentage, as high carbon prevents good welding. The qualities of various grades of merchant iron having been described, the manu-facture of a 1¼-inch round iron bar will be described along with the drawings of rolls illustrating the passes used by the workers to make double-refined metal. This grade is made from pig iron, and as the latter contains anywhere from 3 to 1-3 per cent. of carbon, the excess will have to be gotten rid of by the puddling process before it can be con-verted into wrought iron. The decar-bonization of pig iron in a puddling fur-nace is effected by the joint action of oxidizing fluxes and a current of air, the impurities passing out of the fur-nace in the form of cinder or slag.

Fig. 3 illustrates what the outside of a puddling furnace looks like. Over it can be seen the brickwork enclosing the boilers which supply steam (generated by waste heat) to the engines operating the rolls, etc. Fig. 4 shows a sectional and plan view of a puddle furnace. The fireplace is rectangular and separated from the hearth by a low bridge wall; the roof is arched and slopes down toward the flue, which causes the flames to beat down upon the metal. The grate bars are removable to allow the clinkers and ashes to be easily removed. The amount of air passing through the furnace, it being operated by forced draft, is regu-lated by the damper located on top of the stack. The furnace's outside walls are enclosed in cast iron plates and the bottom of the bed is also made of cast iron plates supported on pillars of the same material. Other mills have the bottom of furnace plates laid on a fire-brick bed overlaid with a layer of sand, the sides of the bed being constructed to suit the method of cooling adopted in that mill. Frequently there are made of hol-low cast iron blocks through which a current of air passes.

The place where the metal is melted is called the "hearth." and at both ends of the latter is a straight wall of brick called the "fire-bridge" and the "flue-bridge." The brickwork (fire-brick be-ing used) overlaps the tops of the side frames so as to form a recess for the fettling or fix with which it is lined. This fettling is a mixture of oxide of iron and sand from the bottom of the hearth. Under the great heat generated in the furnace some of this sand melts with the pig iron and forms with the cinder what is called a "bath," in which the puddling process is done. The silica in the sand unites with the iron and makes a slag. The latter protects the iron from oxidizing, so that large puddle balls can be made from small ones. This slag also makes the iron weld easily, it being a necessary impurity in the first stage of manufacturing puddle iron, a large percentage of the slag being worked out in the further refining which the metal undergoes.

The method of working is as follows: The furnace having been thoroughly heated is charged with pig iron, and all bad joints are closed with fire-clay. The grate is stirred and fresh fuel added, the fire-door being stopped to prevent air from entering except through the ashpit. As the temperature increases the metal soon melts, and when in a liquid condition it is stirred briskly with a long bar of iron called a "rabble," which the puddler inserts through a notch in the working door. The tem-perature is next lowered gradually by closing the damper at the top of stack. The metal finally becomes covered with melted slag which protects it from fur-ther oxidation. As soon as the metal becomes pasty the reactions of the car-bon with the sand and oxides become apparent, blue colored flames in copious amounts making their appearance. Then the damper is opened and the tempera-ture becoming higher the surface begins to boil. The puddler then pushes some of the slag off with his rabble, some being pushed toward the neck or flue of fur-nace, and also through the working door into an iron two-wheeled buggy, the boiling being accelerated by stirring.

As the carbon is gradually eliminated the mass of iron becomes stiff and comes to "nature," as it is technically termed. As soon as the iron has been thoroughly worked, it is made into balls of from 60 to 80 pounds each. The working door and stopper hole are now closed and the last heat given. The balls are then withdrawn through the working door with a large tongs and conveyed on a small two-wheeled truck to the squeezer. There are several forms of this device, the rotary one being the best and most generally used. It consists of a cylindrical casting with corrugated surfaces on the inner side; within this is a corrugated cast iron cylinder which revolves. The cylinder has projecting studs inserted at intervals on its sur-face, the purpose of these being to help drag the puddle ball into the squeezer

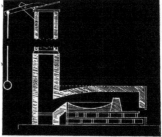

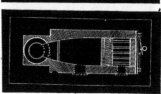

FIG. 4—SECTIONAL AND PLAN VIEW OF A PUDDLE FURNACE.

or "roundabout" as it is called by the workmen, The fixed circular casting forms three-fourths of a circle and the revolving drum is placed eccentrically on its axis. As the surfaces are parallel the distance between them gradually decreases in the direction of rotation.

The puddle balls entering at the widest part are pulled into the machine by the studs on the revolving cylinder, being gradually reduced in size and shape until they pass out at the narrower end in the form of a cylindrical bloom, ready to be put through the first pass in the roughing rolls. The making of wrought iron by the puddling process is so exceedingly severe on the workman or "puddler" as he is called, who is subjected not only to an intense heat radiated from the hot metal and furnace, but also has to do hard physical work in moving the heavy lumps of metal; and in the necessary stirring required to work the carbon out, that many expen-

the bloom undergoes on leaving squeezers is rolling through the puddle roughing rolls. Fig. 5 depicts the puddle roughers in their housings at working order, the pictures being taken just after the rolling was over and the mill shut down for invoicing and repairs.

This is known as a "two-high" train of rolls, which is the type most generally used throughout the United States and Canada. The rolls consist of long cylinders of iron or steel having journals on each end, the end of each journal having a driving device called a "wobbler" to rotate the rolls. The wobbler in reality is a continuation of the journal, only it is made one-half an inch smaller in diameter than the journal or "neck" to use rolling mill workers vernacular. Wobblers contain four grooves called "ways" spaced at equal distances apart and parallel to the axis of the roll. These ways are about 3 inches wide at their tops and 1½ inches deep.

FIG. 6—PUDDLE FINISHERS.

sive machines have been devised to do the work for him—but all have been failures from a commercial standpoint, as well as from a mechanical one. The puddling of iron is not entirely a mechanical process; but it requires a certain amount of thinking and judgment, as the conditions in which the puddler works are variable; which is due to the varying qualities of the materials that he works with.

Puddle Furnace Points

Puddle furnaces are generally constructed to hold from 1,000 to 6,000 lbs. of iron at one time; the latter being made by large modern furnaces, the iron in this case having been melted outside of the furnace and then conveyed to it for puddling. Years ago, it was thought that steel was going to drive iron from the market in this country; many rolling mills manufacturing steel instead of iron. The result was, that large numbers of puddlers drifted into other occupations. It was found out by experience, that there was a place in the industrial world for iron, which steel could not displace, the making of iron was then resumed, but there was a shortage of puddlers for a time; however, there is a fair supply of the workers now. The next process that

To drive the rolls, a cylindrical cast iron box is used, which is about one-eighth of an inch larger in inside-diameter than the wobbler of the rolls. It also has ways lengthwise of its axis, these ways being slightly smaller than those on the roll. In working, the box is pushed half-way onto the wobbler, the other half of the box being resting upon a device called a "spindle." This is a long cast iron piece having ways lengthwise on its surfaces to match those on the box. Fig. 6 of the puddle finishers shows the construction of the journals and wobblers on the ends of the rolls. One end of the spindle being connected to the engine shaft, and the box partly on roll wobbler and also spindle.

When the engine shaft revolves, the spindle and box turns around until the projections on the inside of the box engage the sides of the slots in roll wobbler, the roll then turning around. There are various shapes of roll wobblers made for roll driving, but this style is about as good as any and it is used all over the world in quite a number of mills. Two-high rolls are driven by two cog wheels placed in a separate set of bousing; this causes the rolls to revolve in opposite directions; the top roll in a two-high mill turns from left to right

and the bottom one in a reverse direction. The rolls are placed one on top of the other, between strong iron frames called "housings," which are securely bolted to the foundation of the mill. Each housing is fitted with a screw placed in the centre of the top. This allows the top roll to be raised or lowered, as shown in photograph Fig. 5; the rolls have grooves cut into their peripheries; These grooves, or passes as they are termed, being opposite each other when they are in the housings. It will be readily seen that if a white-hot box pile, bar pile or a puddle bloom of a size somewhat larger than the groove in the rolls, is pushed against the pass, it will be pulled through the space and have a section given to it precisely of the form and size of the pass.

Rolls are divided into two classes: Roughing and finishing. They are in construction and principles of working, precisely the same; the only practical difference being that the roughing rolls give larger sections than those produced by the finishers. This is necessary in order to allow the metal in finishing rolls passes—which are smaller in sectional area—to help work out the impurities in the iron and at the same time gradually to bring it to the desired size. The puddle mill roughing rolls illustrated in photograph No. 5 are designed to make flat puddled bars of 2-inch, 3-inch, 4-inch and also 2¾-inch square billets. They will also make wider iron, such as 6 inches. However, as this article deals with the manufacture of finished bars of 1¼-inch round sections of double refined iron, the rolling of puddle bars of 4 inches. wide and ⅝ of an inch thick will be described. This is the size of iron to be used on the "guide mill" or finishing mill to make the 1¼-inch round bar of iron.

The bloom from the squeezers is shoved through the largest groove in the puddle roughers, then it is lifted up with a tongs from the other side, being put on top of the top roll and opposite the next pass that it is to go through. The rotation of the roll carries it over to the other side, then the worker—called rougher—pushes it into the pass next to the one first used. Each groove is made smaller in sectional area than the one previously used. The bar is gradually worked into a rectangular shape, from a short thick bloom; it emerges into a bar about ten feet long. and 3¾ inches wide, by 2¾ inches thick. It is then ready for the roughing pass in the puddle finishers. The top roll of the puddle roughers has the first four grooves "ragged" or made rough on its periphery by having pieces cut out of the surface with a cold chisel. These cuts are about ¼ inch deep and are spaced fairly close together. This is done to make the iron enter the rolls easier, as the passes get smooth and slippery when the rolls are heated by repeated contact of hot bars. Steel

rolls being worse than iron ones in this respect.

When the metal first enters the passes, it is necessary that its particles be welded together while it is at a high temperature, the idea being to convert the metal into a solid, homogeneous bar. This can only be done while the metal is at a white heat or near to it. To weld the iron thoroughly together, the rolls require to be drafted very heavy; in other words, the metal should be squeezed to the utmost limits that the rolls and engine power will stand; hence the necessity of ragging the first few passes enables them to take hold of the iron quickly while it is hot. After it is welded together and has gone through four passes, it then becomes sufficiently shaped and compressed so that it will go through the remaining grooves without ragging them. In these days of efficiency methods, it is of importance that the drafts should be as heavy as practicable, not only from a welding standpoint, but also from a production one, the idea being to make the material into finished bars as quickly as possible; thus a large output is obtained. This requires very powerful engines or electric motors, as the case may be, in order to drive the train of rolls at a decent speed, so that both the roughing and finishing rolls can be in action at the same time; it being advisable to have both sets working continuously, instead of the roller who works on the puddle roughers having to wait until the finished bar was made and then putting his iron into the roughers while the other fellow waited for him.

This was the case of a mill making a very high grade of iron. In rolling puddle iron, two bars could be rolled on the mill at a slow speed, but on making re-heated metal with one bar in the finishers and a box pile in the roughers, the engine would stop when the bar in the finishing pass was half-way through. This was done to a lack of steam, which is remedied now. Having described the reasons for using heavy drafts in the roughing rolls, the design of the finishers will next be considered. These are made of the same diameters as the puddle roughers, the diameters of the rolls being 18 inches—the mill being termed an 18-inch mill—the shape of the grooves in the finishers are different as contrasted with those of the roughers as will be seen by reference to puddle finisher Fig. 6, which shows the bottom roll as having deep passes, while the top roll has its iron cut out in such a manner so as to form tongues that project into the passes, the sides of the latter called "collars" preventing the iron from spreading sidewise; thus forcing the metal while it is being rolled, to lengthen at a right-angle to the axis of the rolls.

These finishers make iron 2 inches, 3 inches and 4 inches wide, ⅝ of an inch thick, although, some rolling mill superintendents prefer ½ or ¾ inch thickness for their puddle bars. The different sizes are gotten by raising or lowering the top roll by means of the housing screws. It will be noticed that the rolls contain an extra 4-inch pass, as there is more demand for the four-inch size; the finishing groove wears out more quickly than does the finishing 2- and 3-inch passes. By having the extra finishing pass, the wear is divided more equally so as to make the three grooves wear uneven and play out at the same time, before rolls require dressing. In order to prevent the hot metal from sticking to the top roll, it is customary to make the top roll passes one-fourth of an inch larger in diameter than those in the bottom roll. The peripheral speed being greater, it throws the iron downwards—assisted by gravity—and away from the surface of the top roll. In order to keep the iron from sticking to the grooves in the bottom roll, a bar called a "guide" is used. This is a close-fitting piece of iron of the proper width and length, etc.; one end being made with a knife-like edge, and rests upon the roll; the other end is fastened to a device called a "clamp-bar," which is held in place by wedges to a slot in the housings. As the bar comes through the rolls, the end turns down, but striking the guide, is deflected from its course and delivers in a straight line.

Sometimes a sliver on a bar knocks a guide out of place. The iron then sticks to the pass, going around with the roll until it meets the rest of the oncoming bar; then it welds together in an endless ring. The pass is completely filled with metal and it has to be cut through with cold chisels and pried off the roll. The bar from the puddle roughers is put three passes in the finishers to get it to the desired size, viz., 4 inches wide by ⅝ of an inch thick. On leaving the finishing rolls, the bars are straightened out on a place called a "hot-bed" and then cooled by a stream of water, so that they can be handled. Then they are cut into short lengths of the required size and made into piles of suitable height and width by laying them on each other. After that, they are put into a heating furnace and run through the rolls again—the size and thickness of the bar being the same as previously described.

Double Rolling

The double rolling makes the metal very dense and tough. Some mills have a heating furnace installed in their puddling departemnt, and the work is so arranged that, while the puddlers are charging their furnaces, the iron from the heating furnace is being rolled. This allows the same rolls to be used for both puddle and re-heated iron. This method of operating is all right for a small mill having a few puddle furnaces; and a limited demand for a first-class iron, but when it is used in a mill having a dozen or more puddle furnaces, it is a nuisance, as it interferes with the quick working of both heating and puddle furnaces. The conditions of puddling requiring such careful work and the charge varying in composition, it is a difficult matter to keep a number of puddlers working their furnaces in rotation, so that one will be ready to send iron to the rolls, after another has discharged the contents of his furnace. After the iron has been re-rolled, it is next taken to the guide mill side and cut up into bars by means of shears which latter is run by an electric motor of 7½ h.p. The bars are piled upon each other so as to make a pile 4⅝ inch high by 4 inches wide, these are then ready to be heated in the heating furnace. As the heating furnace is very similar in outside appearance to a puddling furnace, no picture is shown of it. The hearth of the heating furnace differs only from the puddling one in being nearly level, its hearth is built solid with firebrick and overlaid with a bed of sand for the iron to rest upon while the heating is being done. Toward the rear is the fuel door of the furnace, with the exception noted, it being substantially the same in construction as a puddle furnace.

The craftsman who operates a heating furnace is called a "heater.". The furnace is charged with the iron piles by the heater on a long tool having a flat end, care being taken so as not to disturb the arrangement of the bars; if this should be done and one projected out from a pile, that part would be burnt, thus rendering the iron weak in structure. The heating of the piles calls for great care and skill on the part of a heater; the main object being to have the iron evenly heated all through. It may be thought that as they are all the same size and weight, that this would be an easy thing to do. But although the furnace is supplied with piles of a uniform size and weight, so as to roll a lot of bars having practically the same length—it by no means follows that the heating of any one pile will take the same time as another. The heating of a furnace with coal precludes the possibility of securing an even heating of all of the piles at the same time. As a general rule, there will be a sufficient number of them heated to start the rolling in the guide mill roughing rolls, and by the time these are all sent to the rolls, the others are ready. This is the system adopted by most rolling mills in the United States, Canada and England. In the days of natural gas, when it was plentiful in supply, the mills in the Youngstown district used that fuel for heating iron and steel, the result being that a more uniform heat could be secured and more piles be ready for the rolls at one time, thus allowing the rolling process to be started and finished quickly. The gas supply has long been exhausted and taken away from the mills by the gas companies, it has forced the former to fall back on coal, producer gas or powdered coal.

Continued in next issue

What is a Hack Saw and How Should It Be Used?

One Chap Said It Was a Thing With Teeth In—This Article Not Only Tells What a Hack Saw is, but How It Should be Used

What do we actually know regarding the hack saw? Ask the average person what is a hack saw and, while he will tell you what it is, he will not, and cannot, tell you why it is. By why it is, we mean why one style of blade is better for a certain class of work than another. To him a hack saw is a hack saw irrespective of its make of teeth.

The L. S. Starrett Company, Athol. Mass., evidently realized this fact for they went to the trouble of preparing a book on the subject, and which is of such real merit that for our readers' benefit we reproduce certain portions of it. Some of the data given in this book is based, of course, on their own experience, but wherever possible these statements and deductions have been carefully checked, and in all cases the statements are made to cover hack saws in general, not so much as to their make alone. Following is what they say:

Hack Saw Economy

The beginning of all wisdom in the practice of hack saw thrift is to comprehend fully the many factors that enter into the production, purchase and use of hack saws. Whether or not we appreciate them, they all have their influence as regards the result. It is, therefore, well that we first consider what these elements are and how they are related, an undertaking in which we will be aided by a study of the following table, through which we may recognize the individual degrees of control exercised by the manufacturer and the purchaser, and the influences exerted by the material to be cut.

A.—The Saw

1. Composition. 2. Heat treatment and tempering. 3. Dimensions; length, width, thickness. 4. Teeth; number, accuracy, set. A different combination of 1, 2, 3, and 4 is essential for each important difference in material or conditions under which the blade is used. These are, of course, determined by the saw manufacturer.

B.—The Material Cut

1. Composition—various grades of steel, iron, brass, lead, etc., conduit, slate and like materials. 2. Shape—round, square, structural, plate, tube, etc. 3. size. These are, of course, fixed by the requirements of the work.

C.—The Operation

1. Hand or machine. 2. Tension. 3. Pressure. 4. Speed. 5. Lubrication. 6. Length of time saw is used. 7. Labor cost and overhead. From all of which may be summarized:

1. Life of saw, or 2, total cuts during life of saw, or 3, average cuts per hour during life of saw, or 4, square inches cut during life of saw, or 5, number of strokes for one cut, or 6, square inches

cut for one cut, or 7, average cost per cut for saw, labor and overhead.

These are determined by proper combination of A and C in connection with B.

The Measure of Hack Saw Efficiency

Efficiency in the case of hack saws is measured both by the length of service and by the speed with which they cut. No purchaser, except one using an enormous number of blades, would ever dream of having his saws made to order, and so for practically all cases, the manufacturer determines what the saw shall be. He has the experience, the knowledge and the equipment which make possible the manufacture of saws in the quality and quantity that are necessary to meet the many and diverse needs of industry. But his service to the purchaser is not complete unless he makes it possible for the latter to select instantly the particular saw best suited to the work in hand. Experience has proved that this may be most satisfactorily accomplished through the medium of a hack saw chart based, not upon the characteristics of the saws, but upon the different materials which have to be cut in ordinary work, thus enabling the machinist to select always the proper blade by looking, not for the number of a blade, but for the

name of a material he has to cut and employing the blade indicated under that heading on the chart.

As materials differ in texture or fibre, hardness and shape, so must the saws vary, if we are to attain economy in the cutting-off department. When the right combination of saw and material has been made, the burden of responsibility rests upon the operator. It is his task to see that the greatest output is secured for the least cost for saws, labor, and overhead.

Economy is more than a mere reduction of first costs. As a matter of fact, consideration of first cost only usually results in inefficiency and an increased ultimate cost.

Knowledge, training, and experience are as necessary to the efficient use of hack saws as of other cutting tools. Failure to appreciate and profit by this fact leads to unnecessary waste of saws, material, time and labor—to an amount far larger than is generally realized. Because of the universal use of the hack saw, both as a hand and a power driven tool, and because of the great variety and the lack of uniformity in the work it is called upon to do, regulation and control in its use have been somewhat difficult.

Taylor, and other exponents of scien-

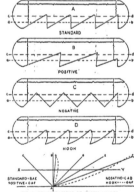

FIG. 1.—EXAGGERATIONS OF TOOTH SHAPES.

tific management, have taught us much regarding the efficiency of cutting tools and we have profited by it; but until hack saw practice was standardized by the establishment of a practical hack saw chart, as given with this article, there was no definite basis for economic progress in this field.

Selecting the Right Make of Saw

As the saw, so is the reputation of its maker. Nothing but the best in composition of material, heat treatment and the greatest accuracy in cutting, spacing and setting the teeth will give the best results, no matter how skillfully the blade may be used. And, vice versa, no matter how high the quality of the saw, economy of or in operation cannot be secured if the blade is put to uses other than those for which it was intended by the maker, or if certain fundamental facts are overlooked.

In selecting a hack saw, three objects should be kept in mind: The blade must cut fast; the blade must do the maximum amount of work at the minimum combined cost of the saw itself and the labor and overhead chargeable to the operation; the blade must be suited for use on the largest variety of materials possible without interfering with above.

Any system of blade selection to be really practical must be based wholly on the material to be cut, rather than on the classification of blades according to a list of trades or an arbitrary classification by number. Were it possible to say that in any particular trade only certain classes of material were cut, such a classification would be ideal; but, dividing the metals according to their influences on cutting speed, saw life, etc., which is the only true basis to use, it is practically impossible to say that any given trade will have occasion to cut only a definite portion of the list. Efficiency is best served by classifying the various materials, the one group of factors with which every operator, skilled or unskilled, is familiar, and putting opposite the name of each metal the catalog number or designating letter of the saw which should be used.

The Modern Hack Saw

The hack saw, as we know it to-day, is an expensive tool to manufacture, requiring very costly and most ingenious machinery, much of which is of special design, necessitating expert workmen and the most careful and rigid system of inspection.

To cut one metal by another is obviously not only a comparatively slow process, but essentially a difficult one. The saw blade must not only be of sufficient hardness to enable the teeth to cut the opposing metal and to withstand the constant wear, but the metal of which the saw is made must have sufficient toughness and ductile strength to reduce to a minimum the breakage of saws and blades.

The carbon content of the steel from which the blade is made determines its hardness and, to a great degree, the speed at which it will cut. An excess of

carbon, however, as every machinist knows, makes steel brittle. Careful tests have proved that a carbon content of over 1 per cent. not only adds little or nothing to the wearing quality of the tooth or blade, but, on the other hand, adds enormously to the possibility of the teeth cracking during the hardening process, or "spalling" when in use. Speed in cutting, however, requires the use of high carbon steel and to overcome its brittleness other elements are added. These include tungsten, chromium and vanadium. Manganese is often added because of its effect during the hardening process.

Differences in Grades of Steel

Hack saw manufacturers recognize three grades of steel. No. 1, the highest, contains from 0.90 per cent. to 1.25 per cent. carbon; tungsten 0.55 per cent. to 2 per cent.; manganese 0.20 per cent. to 0.50 per cent.; chromium 0.20 per cent. to 0.80 per cent.; and vanadium about 0.25 per cent.

The best hack saw blades are made from No. 1 steel containing 0.90 per cent. carbon and 1.25 per cent. tungsten, with manganese, chromium and vanadium present to the extent of approximately 0.25 per cent. each. Repeated tests, under working conditions and in competitions between various makes of blades, have shown that 0.90 per cent. carbon steel is sufficiently hard to give maximum cutting speed without excessive brittleness, while the tungsten content, given above, insures the requisite ductile strength. The percentage of manganese has a direct effect upon the depth to which the "hardening" penetrates through and beyond the teeth of the blade, and for that reason is an especially important factor in the manufacture of "flexible" or "soft-back" hack saws.

The unseen element in a saw is the quality given to it by heat treatment and quenching. Heat treatment may ruin the best of steel, or it may, when properly done, give to each blade the exact quality that best suits it for the conditions under which it is to be used. A hack saw must have not only hardness, but also toughness and flexibility, even in blades tempered throughout. Naturally the degree of "temper" necessarily varies with the material to be cut, but a uniformly tempered blade is always hard and tough, and has a surface which will not shell off. A soft blade is tougher and the teeth will stand greater strain without breaking; and when used in the softer metal, for which it is designed, it cuts more freely but requires more clearance and a coarser pitch to clear the "kerf" of chips. Naturally, however, a hard blade will outwear a soft one.

Effect of Abuse

More saws break than wear out. As a rule this is not so much because the saws are not properly "tempered" as because they are abused or misused. In fact, most of the success of a hard hack saw,

depends upon whether it is used by skilled or unskilled labor. Through ignorance, a hard tempered saw may be easily twisted, bent and broken. For this reason saws should not be too readily condemned on account of breakage. In other hands it might give long and satisfactory service. In case of an unusual amount of breakage in a lot of saws, as compared with their previous performance, there is quite as much reason to suspect a lack of uniformity, or change in the physical condition of the material being cut, as any deterioration in the quality of the saw. In annealed steel it not infrequently happens that opposite ends of the same bar will have had very different degrees of annealing and can consequently be cut at markedly different rates.

The most important factor in the efficiency of a hack saw, next to the composition of the steel from which it is made and the hardening or "tempering" of the blade, is the shape of the teeth. The great majority of hack saws to-day have their teeth milled perfectly straight on the face, and are cut to a depth of little less than the distance between two teeth. Experiments and experience show this proportion gives not only the most serviceable, but, taking the life of a blade as a whole, the fastest cutting saw.

The slope of the back face of the tooth has a direct effect upon the durability of the blade and its total output of work. When a saw becomes worn, the ends of its teeth are flattened or rounded. As the area of contact of the teeth with the work increases, the pressure necessary to make the saw "bite" increases, and consequently there is a tendency either to strip the teeth from the blade or to crush them, and it is this strain which must be borne largely by the back face of the tooth. Eventually, this flattened area will become so great that it is impossible to make the saw cut without breaking the blade. Consequently, the amount of work that can be got from a saw depends largely on the volume of tooth available for wear before this limiting area of contact is attained. The angle of the forward or cutting face of the tooth is even more important. The illustration, Fig. J, shows angles at which this face may be cut.

How Different Shaped Teeth Cut

To understand what this means let us take the different styles of teeth and exaggerate them as shown at Fig. 1. This shows exactly what occurs as each becomes worn through use.

The standard type of tooth, A, with its forward face vertical, strikes the metal to be cut at the same angle, regardless of the amount of wear. C, which has a negative rake, scrapes the metal from the very start, rather than cuts it. As the wear increases, as shown by lines a b and c d, the area of the tooth in contact with the work increases much more rapidly than is the case where the standard type of tooth is used. On B, the positive rake necessitates an abnormally

obtuse angle on the back face of the tooth, and consequently a disproportionate increase in the contact area as the saw becomes worn. In addition to this undesirable feature, the forward slope of the teeth tends to make them dig into the metal, increases the strain on the sharp cutting edge of the tooth and causes the saw to stick and break. D, which is distinctly hook-shaped, combines most of the bad features of B and C and has also a tendency to crush.

It has been found by test that while the hook-shaped tooth may, in some cases, cut slightly faster on the first two or three cuts, yet once the first keenness is gone, there is an unusual tendency to slip or slide over the work, due to the extremely rapid increase in contact area and the angle at which what is left of the "hook-tooth" meets the metal that is being cut. One contributory cause to the rapid loss of efficiency in the "hook-toothed" blade, when used in power machines, is the fact that it always is extremely difficult to avoid a slight dragging of the blade on the return stroke. This dragging not only takes off the first keenness of any blade, but on saws of this type seems to tend to turn the forward edge of the tooth up, an effect which, if exaggerated, would produce a ball-pointed "tooth." Referring, however, only to the normal wear of the teeth by cutting, it is evident that the "Contact Area" increases more rapidly as the "hook-tooth" is worn down, than in the "standard" type of tooth.

The pitch of the blade, or number of teeth to the inch, is of almost supreme importance in determining the efficiency of a saw, and it is a subject so directly related to and scarcely separable from the hack saw chart, or classification of blades according to the work they are to do, that its discussion will be reserved for a later page.

The "Set" of a Saw

The milling of the teeth by multiple milling cutters alone insure absolute uniformity in their spacing, shape and sharpness—conditions most essential to quick cutting and endurance. The cutting points must be square and should be set in such manner that proper clearance is provided and that each tooth is given an opportunity to do its full share of the work. The standard "set," as indicated by general practice, is "Regular Alternate"—that is, one tooth is slightly turned to the right and the next to the left—just enough to insure a free, smooth, rapid cut in a slot a little wider than the blade itself, removing no more stock than is necessary. In certain fine-toothed saws, every pair of teeth is "set" alternately right and left, a style of setting known as "Double Alternate."

While the majority of manufacturers have, by tests and experience, proved to their own satisfaction the advantages of the "Regular Alternate Setting," some makers "set" one tooth to the left, the next to the right, while the third is left straight to serve as a sort of "Raker," its purpose being to clear the cut of chips as well as to deepen it. This style of "setting" is known as "Alternate and Centre."

The third style of "setting" is known as "Alternate Full and Half," in which the teeth are "set" alternately left and right, but in different degrees. That is, the first two teeth will have more "set" than the next two, and so on. Another style of "setting" is the "Double Alternate" in which the teeth are "set" right and left in pairs.

The Effect of Pitch

The proper pitch of the teeth is determined by the power which may be applied, as well as by the hardness, texture and shape of the material to be cut. For all-round work with a general purpose blade used in a hand frame, 18 teeth per inch best meets conditions in the case of all solids, tool steel, wrought iron, cast iron and slate. But for cutting cold rolled stock and machine steel in a light machine a coarser pitch (of 14 teeth to the inch) is better; while with an extra heavy machine, as low as 10 teeth per inch is better still. Such a saw, with its big teeth and coarse pitch, will naturally "bite" into the metal and cut much faster than one with fine teeth operated by less power. But a 14-tooth saw, for instance, would not be serviceable for cutting piping, soft metal tubing, or sheet metal, because it will "chew" in too much on each cut and will have a tendency to bind or choke, because there are not enough teeth in contact at any period in the cut.

For that reason, a blade with 24 or 32 teeth to the inch is best suited to sawing tubing or thin copper and sheet metal, light structural iron, wrought iron pipe, electrical casing and brass. The teeth are not cut so deeply and naturally take the proper depth at each stroke. A coarser pitch blade used upon sheet metal gives the teeth an opportunity to "straddle" the work so that only one tooth can cut at a time, and renders the saw liable to stripping. On thin metals, brass tubing and sheet steel, a pitch as fine as 32 to the inch, giving two or three or more teeth the chance to cut at the same time, is necessary to avoid this danger.

Broadly speaking, the length of a blade should be in proportion to the size of the stock to be cut, though the shorter the blade, the greater the pressure it will withstand. Where the blades are selected according to diameter of the work to be done, it is possible to maintain a constant cutting speed, whereas if the strokes per minute be adjusted according to the size of the blade, there is a different speed with each length of blade used.

How Material to be Cut Determines Type of Saw

As the cutting of material is the obvious purpose for which the saw is made, it is manifest that the character

MATERIAL TO BE CUT	BLADES FOR HAND FRAMES										BLADES					
	All Hand					"Flexible" or Soft Back					Light Machine					
	Cat. No.	Pitch	Gage	Width"	Lengths"	Cat. No.	Pitch	Gage	Width"	Lengths"	Cat. No.	Pitch	Gage	Width"	Lengths"	St. Wt. lbs.
Light Angles Channels Tee Iron Ornamental Iron	102	24	23	1/16 1/2	6–9 10–12	252	24	23	1/16 1/2	6–9 10–12	115	24	21	5/8	10–14	24
Heavy Angles Channels Tee Iron	103 112	18 18	23 22	1/16 1/2 5/8	6–9 10–12 8–14	250	18	23	1/16 1/2	6–9 10–12	115-B	18	21	5/8	10–14	24
Light Structural	112-B	14	22	5/8	8–14	250-B	14	22	5/8	12–14	115-B	18	21	5/8	10–14	24
Heavy Structural	112-B	14	22	5/8	8–14	250-B	14	22	5/8	12–14	114	14	21	5/8	10–17	24
Rails	112-B	14	22	5/8	8–14					...	114	14	21	5/8	10–17	24
Steel and Iron Pipe Conduit Brass Pipe	102	24	23	1/16 1/2	6–9 10–12	252	24	23	1/16 1/2	6–9 10–12	115	24	21	5/8	10–14	24
Solid Stock Cold Rolled Machine Steel	103-B 112-B	14 14	23 22	1/16 1/2 5/8	6–9 10–12 8–14	250 250-B	18 14	23 22	1/16 1/2 5/8	6–9 10–12 12–14	114	14	21	5/8	10–17	24
Tool Steel Cast Iron	103 112	18 18	23 22	1/16 1/2 5/8	6–9 10–12 8–14	Not recommended					114	14	21	5/8	10–17	24
Brass	103 112	18 18	23 22	1/16 1/2 5/8	6–9 10–12 8–14	Not recommended					115	24	21	5/8	10–14	24
Sheet Metal And Tubing Less than 18 gage	253	32	23	1/16 1/2	6–9 10–12	258	32	23	1/16 1/2	6–9 10–12	Not ordinarily cut on power machine					
Sheet Metal And Tubing Over 18 gage	102	24	23	1/16 1/2	6–9 10–12	252	24	23	1/16 1/2	6–9 10–12	Not ordinarily cut on power machine					
Aluminum Copper Tin Babbitt						250 250-B 252 258	See above									

FIG. 2—THE CHART AS SPOKEN OF IN TEXT.

of the material to be cut, and its shape, must be most carefully considered in the design, manufacture and selection of the blade, if the highest efficiency is to be obtained. This is becoming increasingly important with the rapid growth of the practice of power saving.

One manufacturer, in the same spirit of efficiency that pervades the business upon which a world-wide reputation has been built—the manufacture of Precision Tools—has, through painstaking investigation and experiment, determined, step by step, just what features in the saw give greatest efficiency in the cutting of each particular material under various working conditions of hand and machine sawing.

The Hack Saw Chart

Which saw will insure the most economical cutting of any particular material? One has only to consider the multiplicity of factors as represented by the material, the saw and its operation, to realize the difficulties encountered in arriving at the correct solution of this problem.

Practically every saw manufacturer has made some attempt to answer the question. A number have gone no further than to suggest the obvious—that is, that the saw used for cutting brass or sheet metal should have more teeth per inch than that used when cutting steel or round stock. Some, however, go further than this, though not necessarily to the extent of conducting careful tests and the compilation of records over a series of years as one or two have done, but yet far enough to recommend with some degree of definiteness, the employment of blades of certain physical characteristics, to cut certain classes and shapes of metal.

One manufacturer, as the result of experiments and tests, has divided materials and shapes into nine such classes, the distinction being based on observation of the influences exerted by the hardness, ductility, crystalline structure, and chemical composition of the various metals, and the diameter and shape of the material to be cut. For these different classes of materials, different blades have been designed and manufactured, and the classifications, with the blade recommended for each have been embodied in the hack saw chart, which is reproduced at Figs. 2 and 3.

In a more elaborate and technical way this chart might have been made up so as to show the composition and temper of each blade, but for the fact that saw numbers on the chart supply all necessary information where saws of this make are employed.

Getting the Most Out of the Right Saw

In the small shop, where hand sawing is done and each man owns his own frame, and where but few pieces of different materials are cut from time to time, a change in blades is naturally out of the question for each change in the material. Hence the convenience of a blade like the No. 103 (see chart), that

will in a degree meet all ordinary requirements. But there are few mechanics, who, under even these conditions, cannot at times work to better advantage with a flexible back blade, like the No. 250, which has 18 teeth, or who do not sometimes require a finer tooth blade, like the No. 258, with 32 teeth to the inch, for cutting thin sheet metal. In the end, time will be saved and inconvenience avoided if these saws are kept strained up in separate frames, so that one can be picked up and used instantly. The man who will thus equip himself may generally be relied upon to use with discretion the various blades furnished to him, but wisdom should be exercised in their selection and purchase.

Whatever saws are adopted as the standard, a hack saw chart, based on that make of saw, should be established as the basis of selection for the work in hand. Not only should the tool or stockroom from which the saws are issued, be provided with these charts, but definite control should be maintained over the delivery of saws of a given number, according to the requirements of the work and the machine, where power saws are used. Each workman, in turn, should have a copy of the chart, perhaps somewhat modified, so that he too may know the blade which is best suited for each class of material.

To meet the requirements of different classes of work by the selection of the blade designed for that work, to increase thereby the individual efficiency of each saw and reduce the total number

of blades required, it is obvious that a complete line should be carried. It is better to carry and issue exactly the right saw for unusual and intermittent requirements, than to do the work inefficiently for lack of the proper blade.

The output with all power blades should be carefully watched and the proper steps, taken to insure that a blade is discarded as soon as it begins to lose efficiency to an extent that it cannot be overcome by an increase of pressure.

Testing Hack Saws

Every hack saw test has two distinct objectives. The first point to be determined is, what make of blade shall be selected as most efficient for the work of the shop as a whole? The second point, which is no less important, is—what particular blade made by the manufacturer whose brand has been selected, is best adapted to the conditions governing any particular piece of work?

The difficulties in the way of making sufficiently exhaustive tests to determine the all-round efficiency of various makes of saws has already been mentioned. To reach conclusions that are beyond doubt, every factor must be taken into consideration—the composition, temper, pitch of teeth, "set," gage, depth, and cost of the blade, the tension, pressure, speed and lubrication, and the characteristics of the material cut. As it is a fundamental in comparative tests that only one variable should be changed at a time, the magnitude of the task of making complete comparisons is mani-

FOR						POWER						MACHINES					
Medium Machine						**Heavy Machine**						**Extra Heavy Machine**					
Cut No.	Pitch	Gage	Width"	Length"	St. Wt. Lbs.	Cut No.	Pitch	Gage	Width"	Length"	St. Wt. Lbs.	Cut No.	Pitch	Gage	Width"	Length"	St. Wt. Lbs.
262	18	18	¾	12-14	24	Not cut economically on this class of machine						Not cut economically on this class of machine					
255	14	18	¾	10-20	44	Not cut economically on this class of machine						Not cut economically on this class of machine					
255	14	18	¾	10-20	44	254	12	18	1	12-24	44	256	12	16	1	14-24	64
255-B	12	18	¾	10-20	40	254-B	10	18	1	Same as above	34	256-B	10	16	1	14-24	53
255-C	10	18	¾	17-20	35	254-B	10	18	1	12-24	34	256-C	8	16	1	14-24	43
262	18	18	¾	12-14	24	259	18	18	1	14-17 only	44	Not cut economically on this class of machine					
255-B	12	18	¾	Bth. Blds.		254-B	10	18	1	12-18	34	256-B	10	16	1	14-24	53
255-C	10	18	¾	10-20	35					12-24		256-C	8	16	1	14-24	43
255	14	18	¾	10-20	44	254	12	18	1	12-24	44	256	12	16	1	14-24	64
262	18	18	¾	12-14	24	259	18	18	1	14-17 only	44	Not cut economically on this class of machine					

All recommendations on this chart and throughout the book must be taken in a general rather than a specific sense and should be applied as the common sense of the user indicates.

NOTE: When testing hack saws for which the manufacturers have made no specific recommendations, it may be convenient to identify them with the number of the Starrett Blade most nearly similar and use according to the recommendations on this chart.

FIG. 3—STILL ANOTHER CHART OF DECIDED INTEREST.

fest. The influence exerted by a variance in the composition and temper of the blade and the pitch and "set" of the teeth has already been indicated; these are factors beyond the control of the user of the saw.

The gage of the blade, another of the elements determined by the maker, directly affects the cutting speed. Tests have shown that if two saws differing only in gage are used to cut the same metal, the first few cuts are usually made by both blades in the same time, but about the fifth or seventh cut, the thinner saw will be found to be cutting faster. This is, of course, assuming that both saws are, except for the matter of gage, equally well adapted to use on that particular kind or shape of metal. The first half dozen or so cuts remove the extreme sharpness of the teeth and permit the element of "contact area" to come into play. Obviously, the thicker gage of one saw simply means more friction to be overcome and necessitates greater pressure to make the teeth cut. To make a 14 gage saw cut as fast as one of 20 gage, throughout a long series of cuts on the same material, probably double the amount of weight would have to be put on the thicker blade toward the last of the test.

In selecting a saw to use on any particular job it is worth while remembering that a short blade, though of lighter gage, is stiffer in proportion, and consequently can carry a coarser tooth.

Factors Under Control of Operator

Tension, pressure, speed of strokes per minute, and lubrication, or the use of compound, are factors directly under the control of the operator, and have a direct bearing on the worth of any test. Their influence will be discussed elsewhere later. Time is another factor under the control of the operator, and is a frequent source of abuse through crowding saws far beyond their normal speed, so that they are destroyed in perhaps half a dozen cuts. The unreasonableness of such treatment will be clearly indicated later.

The general consensus of opinion is that the most practical method of testing blades, in order to determine what make or kind of saw, as regards "pitch," gage and "set," is best adapted to the requirements of the purchaser, especially where there is a fairly uniform class of work, is, after all, to compare the run-of work output of different kinds or makes of saws of exactly the same characteristics, operated under identical conditions. To make such comparisons reliable, each must represent the average of results obtained from a number of saws of the same make and kind.

A great deal has been said and written regarding the testing of hand hack saw blades, and various means have been devised from time to time, the majority of which necessitate the use of a power machine. It is extremely doubtful whether or not any useful purpose is served by testing a blade under conditions which will never be met in actual

work, and for which it is not designed. Nevertheless, the practice is fairly common. Of course, the object of using the power machine for testing hand blades is to remove the human element so far as is possible, although this element is one of the factors directly allowed for in the design of this class of blades. Hand hack sawing, like filing, is one of the most difficult operations. The hand blade, unlike the saw fixed in a power machine, must be guided by hand alone, and the accuracy with which this is done, both with regard to direction and evenness of pressure, will directly affect the efficiency of the blade.

Power sawing differs from hand work in that the regulation of the weight or pressure is purely mechanical, rather than instinctive, as when the hand frame is used. The natural tendency of the workman with a hand frame is to use less pressure on the first few strokes, thereby preventing the breaking or dulling of the teeth, and to increase the pressure gradually as the work progresses and the teeth lose their first keenness. The workman's "feel," when cutting with a hand frame, gives him a decided advantage over the machine in this respect. It is, therefore, necessary, if efficient work is to be done with the power saw, that the principle of gradually increasing the weight, as the work progresses, be understood and correctly applied.

Testing Power Blades

One method of testing power blades, which is at least fair to all, is to purchase several blades of each of such makes or brands as it is desired to test, buying one blade of each kind tested from at least three different merchants if possible, in order to get an average run of the product. The blades tested should be of the same length, gage, depth, and "pitch" for each of the different makes, unless one or more of the manufacturers makes some specific recommendations as to the type of blade to use on the material on which the test is to be conducted. Where such recommendations are made they should be

scrupulously observed. The blades should all be tested on the same grade of material and in the same power machine, preferably one with gravity feed. The reason for testing all of the saws on the same kind, size and shape of material is obvious. Should the test be made, as is usually the case, on unannealed tool steel, the blades should be alternated, so that should the steel not be uniform no saw will be placed under a handicap by a hard spot.

One point, however, must not be overlooked. Some manufacturers have given the question of hack saw efficiency sufficient thought to evolve certain specific recommendations as to the blades, strokes per minute and weight to be used on certain classes of materials and with certain machines. If any of the manufacturers of any of the saws to be tested makes any definite recommendations as to what blade to use under the conditions of the proposed test, they should be rigidly adhered to. Otherwise, ignorance of some of the factors that entered into the manufacture of the blade, and the consequent ignoring of the purposes for which it was intended, will make the test, if not directly misleading, at least of doubtful or little value so far as that particular blade is concerned and will result in conclusions being drawn which are unfair to the maker of the saw. Where definite recommendations as to the use of a saw are not made by its manufacturer, some confusion as to the conditions under which it should be tested may be avoided by identifying it by means of gage and pitch, with a blade whose proper use is known, and testing it according to the instructions for that blade, as given in the hack saw chart.

Careful records should be kept as regards the saw used, the material cut, the speed, pressure and work accomplished. For exactness, the time taken for each individual cut should be recorded in minutes and seconds. It is simplest to read the exact time of day at which each cut is completed, and from it subtract the time at which the previous cut was fin-

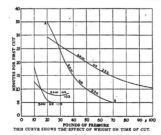

THIS CURVE SHOWS THE EFFECT OF WEIGHT ON TIME OF CUT.

ished. Of course the elapsed time then includes that required for feeding up and putting in new pieces.

A special form for recording the results of power tests has been prepared by The L. S. Starrett Company of Athol, Mass., for the convenience of hack saw users who wish to make run-of-work tests.

The Effect of Weight or Pressure

The demonstrated advantage that has resulted through the establishment of specific cutting speeds and feeds as most economical in the case of the lathe tool, points to the necessity of likewise determining the best working conditions for efficient hack saw work, as regards speed, pressure and lubrication for each type of saw and class of material.

The most vital variable under ready control is the pressure upon the saw,— which, other things being equal, determines the rate of cutting and the endurance of the blade. The proper weight to be applied on starting the cut, and when and how much it should be increased, varies with different makes of saws, and in the absence of recommendation by their makers can be estimated only by trial and by comparison of results.

The effect of pressure on the life of the saw and how it affects cutting costs is very clearly shown in the succeeding tables and curves. From all of these, it will be at once evident that too light a pressure will allow the saw to slide and wear without doing effective cutting. This may show long life in hours, but also requires long periods per cut, and results in an ultimate low efficiency. On the other hand, excessive pressure is unnecessarily destructive. The saw may cut much faster while it lasts, but it doesn't last long enough to justify the practice. Between these extremes lies the happy mean which represents the acme of hack saw economy.

The actual pressure, in pounds per square inch of "contact area" of the tooth, which has been found to give the most satisfactory results as regards both the time per cut and the number of cuts per saw, varies from 20 to 30 pounds. It has been determined by careful tests that pressures within these limits, while not overloading the saw, are sufficient to avoid any possibility of the blade slipping and sliding over the work, and thereby becoming dull without having done more than a portion of its rated capacity. It has also been found that the basic weight or pressure is directly proportional to the gage of the saw. Standard practice indicates using the 20-lb. unit of pressure for blades not over 0.040-in. thick, 25 lbs. for saws between 0.040-in. and 0.060-in., and 30 lbs. for those whose thickness is 0.060-in. and over. The so-called unit or basic pressure must not be confused with the weight actually resting on the blade when in use, but must be taken as a constant, by the use of which the actual pressure on the saw may be calculated.

This weight, or pressure, is measured by attaching a spring balance to the forward end of the saw-frame when the blade is in mid-stroke, and lifting. The amount of "pull" indicated by the needle of the scale is the weight on the blade.

In setting the weight for a new blade it is evident that a higher initial pressure will be used when cutting heavy stock than for light stock. In practice, however, the figures given in the hack saw chart have been worked out for an average size of bar to be cut, and it has become customary to use the weights thus determined for a given saw as the proper pressure at which to commence a series of cuts with that particular blade, regardless of the dimensions of the material to be cut.

The necessary adjustment of the weight, as affected by the diameter of the work, can easily be made after the first cut, or even after a few strokes have been made, while the weight indicated is sufficient to prevent the saw from slipping or sliding, yet not so excessive in any case as to incur the danger of stripping.

The weights given have been figured out for cutting the more common materials, such as machine steel, etc., and must be taken as approximate, rather than absolute figures. For other materials the proper weight to be used in actual practice may vary somewhat, according to the material and the speed at which it was to be cut. The figures are, however, sufficiently accurate for all materials to afford a working basis from which to start.

In using the weights recommended in the hack saw chart, one point which must

---

INTERESTING FACTS TO BE FOUND IN THIS WEEK'S ADVERTISING SECTION

How to obtain a steady volume of work.

A hammer noted for its blow.

A chemically pure product and where to get it.

What to use in repairing broken machinery.

A chuck that has reversible screws.

A new stirring and stabilizing process.

How to put efficiency into your shop.

A device that is said to increase production, prevent accidents, and reduce costs.

A battery of machines that are giving perfect satisfaction.

An announcement of interest.

What to think of when purchasing belting.

That it never pays to be prejudiced.

How to choose the proper grinding wheel.

A concern that has been a factor in helping Canada's railways.

---

not be lost sight of is that, to obtain efficiency in hack-sawing, is it just as important to keep the proper pressure on the blade throughout the entire series of cuts, as it is to secure it at the outset. Remember, that a saw wears as it is used, and that the area of the teeth in contact with the work becomes greater as the number of cuts increases, and therefore in order to maintain the same cutting speed, the pressure must be increased from time to time. It may be taken as a maximum of hack saw thrift that, for efficiency, the weight must be increased as the number of cuts progresses.

While speaking of the necessity for the proper adjustment and increase of weight, it may be worth while to note that equal diameters of work of the same material will be cut at about the same rate, without regard to their comparative shapes. That is, 1½-ins. square, round or elliptical bars of the same grade of steel would cut at the same rate. Angle irons, channels and tees of the same greatest dimensions, will cut alike, provided they are properly placed in the vice.

Cutting speed on rounds of any given material increases as the diameter of the work and varies in the same proportion as the saw goes through any particular cut. Theoretically, the weight during the progress of each cut should vary directly as the diameter of the work, though this is, of course, impractical. However, it is essentially practical that, when the time required for any single cut becomes greater than the average time required for cutting that particular class of material, with that particular blade, the weight or pressure on the blade should be increased a few pounds before beginning the next cut.

Continued in next issue

In a lecture recently delivered before the Institution of Civil Engineers in London, England, Sir Dugald Clerk revived an interesting proposal made by the late Lord Kelvin for the heating of rooms. This proposal is not easy to explain without diving into mathematics and the abstruse laws of heat, but it may be expressed as a process of using an engine to extract some of the heat from the cold air outside a room and adding it to the heat of the air inside the room. The curious and puzzling thing about this process is that the heat so added is, under certain conditions, much greater than the heat equivalent of the work done by the engine. In theory, therefore, an electric motor may be used very efficiently to warm a room. Whether the notion will work out satisfactorily, in practice remains to be seen, but in view of the ever-increasing cost of fuel this fascinating problem is likely to be soon attacked by British engineers.

Two Efficiency Forms of Decided Interest

It is Only by Exchange of Ideas That We Can Improve Our Present Methods—Perhaps the Suggestions Given Herein May be Suitable for Use in Your Plant

By J. H. MOORE

ANY form, which on being filled out day by day depicts progress throughout a plant, might well be termed an efficiency report. Proceeding on this assumption, we illustrate two very useful forms as adopted by the International Harvester' Co., Hamilton, Canada.

At Fig. 1, we show a portion of what is known as their daily report sheet.

This style form is adaptable to any department, but for illustrative purposes we depict the sheet filled out for the

First comes the filling out of the proper department, together with the date. Following comes a note of importance which reads that the report must be sent to superintendent's office at 7 a. m. daily, whether or not a department was delayed for any reason. The remainder of this notice should be read by every reader, as it covers quite a range of possible, and sometimes unavoidable delays.

The remaining headings are self-explanatory and are not filled out in this case, simply because the delay described

These reports on arriving at the superintendent's office act as an illustrative chart, showing every detail of progress throughout the entire plant. All delays are investigated, and repairs to crippled equipment made at once. It is this fact that allowed us to call these forms efficiency reports, for they make for systematic handling of any product.

The returns issued by Lloyd's Register for the quarter ended March 31st, 1920, show that Britain has regained its lead as the world's greatest shipbuilding nation. The tonnage under construction in Great Britain exceeds that of the United States by over 800,000 tons. New British vessels put in hand during the quarter numbered 229, and aggregated nearly 710,000 tons. There is a notable increase in the number of large ships building, there being 62 vessels of over 10,000 tons under construction, as compared with 55 at the end of 1919.

DAILY REPORT OF DELIVERIES			
FROM **SHEET METAL** DEPT.		TO **ERECT No 1** DEPT	May 14th 1920

CAT. No	DELIVERIES	PIECES	WEIGHT
H 5373	PAINT DEPT	16 & 30	26
F 7639	" "		1000
B 1101	" "		205
M1-123	ERECT No 2		2000
3/16 X 1 3/4	Washer B & R DEPT		3000
1439 N	PAINT DEPT		15

FIG. 1—SHOWING A PORTION OF WHAT IS TERMED THE DAILY REPORT SHEET.

sheet metal department. As will be noted, the first item to be filled in is that of the date. Next comes the name of department, also where the material has to be trucked to.

In other words, the chart forms a double purpose. First, it records the amount of material passing through the different departments each day, while it also informs those interested the destination of the parts. The symbol number is of course always given.

Each department has its own truckman, or truckmen (if the trucking demands more than one man), who looks after every load leaving his department. He fills out the form as he delivers each load, when at the end of the day the timekeeper enters these records into a permanent record book.

At the end of each month the timekeepers total up the production of their own particular departments, which shows immediately how the work in all departments is progressing.

The second of these forms (shown at Fig. 2) comes under the name of a manufacturing delay report, and as the name implies, the form is used as a means of recording delays, and the reasons for these occurring.

Again, we will take the sheet metal department as an illustrative example.

was of another nature to any of the headings shown.

As will be noticed, the piece 1030N did not arrive at the proper time in the sheet metal department. This delay was caused by the breaking down of a press in the forge shop. This may seem like a strange excuse, but when we explain that the routing of this particular part is from the forge, to sheet metal department, the cause of delay is understood. If a press accomplishing a previous operation on any part breaks down, it stands to reason the parts cannot be forwarded on time.

Important trials with pulverized coal and "colloidal fuels" have been carried out recently on the Great Central Railway, England. Colloidal fuel is powdered fuel suspended in thick oil, and pulverized coal is dry coal reduced to a fine dust. The trials, which were made in comparison with ordinary coal, showed that the special fuels could easily maintain a full head of steam with a high degree of superheat, even on heavy gradients and sharp curves. Locomotives adapted for burning pulverized coal or colloidal fuel show an economical freedom from ashpan cleaning, smoke-box cleaning and repairs, fire cleaning, and so on. Further, when the locomotive is delayed in a siding there is much less waste of fuel than in the case of ordinary engines.

MANUFACTURING DELAY REPORT				

Sheet Metal DEPT   May 14th 1920

This Report must be sent to Supts office at 7AM daily whether or not a department was delayed for any reason. Report Shortage of Material, Breakage of Machinery, Stoppage of Power, Mistake in Ordering, or anything that interferes with getting out a full day's product.

Number of Employees Lost Hours	No of Employees Transferred to other Dept	Number of day each hours kept idle	Number of day each hours allowed	CAUSE OF DELAY
				1030 N from Press in Forge
				Route Forge to Press
				to Sheet Metal
				Press in Forge Shop broke down

APPROVED   A. L. Blank FOREMAN.

FIG. 2—THIS DELAY REPORT HAS SAVED MANY AN EMBARRASSING QUESTION.

WELDING AND CUTTING

Welding is Steadily Gaining Favor in Boiler Industry

By ARTHUR MALET

THERE is no question but that welding in general and oxy-acetylene welding in particular is steadily gaining in favor in the boiler industry, as is evidenced by the number of shops in which its use is gradually being extended in regular production. This interesting statement is made by the above author in a recent issue of Acetylene Journal; and he goes in the following manner:

Although it has been condemned by some individuals, the causes of failures of oxy-acetylene welding are generally traceable to the inexperience or poor workmanship of the operator. To become a good welder or to be efficient in any line of work, it is necessary that a man must like that work and try his hardest at all times to produce the best results of which he is capable.

In the case of oxy-acetylene welding, the operator must study the action of the metal when subjected to the intense heat produced by the flame and understand why the welding progresses as it does. He must watch his flame and keep it neutral and be sure that he is actually welding and not simply sticking the metal from the filler rod on to the piece being welded. He must learn how to handle his torch and the filler to the best advantage. He must also be a cold-blooded sort of chap to be able to stand the intense heat in which he must sometimes work if he is welding a piece surrounded by a bed of red-hot coals, as happens when preheating is necessary.

In the shops where the writer is located the torch is put to every possible use in the work. In its cutting capacity it is used to burn out all staybolts and old sheets and to cut all rivets. If the operator is careful, he can bevel the sheets to be welded with the cutting torch and save time and labor that would be required to do the same work with a chisel and air hammer.

Ash pans are in a great many cases welded now and seem to give as good service as pans with angle irons and rivets. In building up ash pans the pieces of plate may be riveted to the casting before or after welding. The sides of the pan are held together at the top by a wire clamp. Before the welding is started, thin wedges are driven between the edges of the plates, forcing them apart about ¼ inch.

All tool boxes used by the engine crews are welded. These boxes are built up of No. 12 gauge sheet iron and are 18 inches long, 8 inches wide and 8 inches deep. Welding is done from the inside, the outside edges being smoothed after the operation is completed.

The great speed that can be accomplished in doing any of this shop work by means of welding is very much in favor of the process.

Recently it was necessary for the men in the shop to put a patch in the left back mudring corner of an engine. The patch was to cover an area which included two staybolts, eight mudring rivets on the door sheet, and three mudring where it lapped on to the side sheet, and three rivets in the running seam. It took just eight hours for a boiler maker, a helper and a welder to remove the old piece and weld the patch into place. All that remained after this was to drill the holes for the mudring rivets, drive the rivets, apply the two staybolts, and chip and caulk the patch.

Incidentally, the cutting torch was used to burn off the staybolts and the old sheet, as well as to preheat the patch when it was being laid up.

As yet we have not had to install any new side sheets or flue sheets, but probably will in the near future. When this becomes necessary we intend to try cutting out the top of the back flue sheet just below the braces and weld in a new top piece, as several of the flue sheets are badly worn in the tube area, but perfectly sound below.

Can You Weld Successfully with Ethylene?

THE rapid and successful growth of the oxy-acetylene industry has led, from time to time, to attempts to replace it by some other combustible, often introduced with advertising boom and supported by statements which will not withstand criticism. We have frequently dealt with processes which on paper have appeared to be serious rivals of acetylene welding, but which have failed to survive the formidable test of industrial application. A search in the advertisement columns of technical journals for these processes shows invariably that "they have had their day and ceased to be." Thus comments the Acetylene and Welding Journal in a recent issue.

At the present moment welding with ethylene is receiving a great amount of attention by the technical press, due to the glowing description of its possibilities given by Mr. R. B. Tunison before the American Chemical Society. We cannot do better than quote his statement regarding this process.

Mr. R. Tunison maintained that in many respects the ethylene flame is far superior to the acetylene flame for cutting and welding metals. As far as the heat of combustion is concerned, ethylene has a slightly higher value. It has a rapid rate of combustion, preventing loss of temperature from absorption and conduction of heat by the metal being cut, and it will produce a rapid rate of cutting with a low consumption of gas, making a high efficiency possible from the regular operators of the blowpipe. It is a non-poisonous and safe gas without objectionable odor, and has a clearly-defined cone in the flame, without which a workman would be badly handicapped. In the working of copper it has been impossible to make a satisfactory weld with acetylene because of the formation of carbon and the consequent blistering in the weld. Perfect copper welds have been obtained with ethylene. In aluminium welding and lead burning, ethylene is infinitely better than acetylene. There is no carbon formation, and no kick back in the burner, while a terrific heat may be maintained. Ethylene is much safer to handle than acetylene, since it does not explode spontaneously. It can be compressed and handled in the standard form of gas cylinder with-

out either packing or acetone. For this reason it is possible to make a saving of half the cost of cylinders and about half the cost of freight. Furthermore, ethylene can be compressed so that a cylinder will hold over 200 cubic feet of ethylene. The weight of ethylene per 100 cubic feet, including the container, is only 40 lbs., as compared with 90 lbs. for acetylene. Ethylene is sold at the same price as acetylene. It may be used in an ordinary welding blowpipe, but preferably one with a mixing chamber. A change in the size of tip is also desirable.

A first perusal of this statement might lead to the assumption that at last a process which must displace acetylene has arrived. But manufacturers and those familiar in detail with oxy-acetylene welding will have no apprehensions as to it arresting or displacing the acetylene process. Ethylene belongs to the class of compounds known as hydrocarbons, of which there are hundreds, all more or less thoroughly studied. Although ethylene has been known for a very long time it has never been considered as a combustible for welding purposes. It is most conveniently prepared by the action of sulphuric acid upon alcohol. The statement regarding the great superiority for cutting purposes on the grounds of combustion will be found to be mythical. The fact that one volume of the gas requires three volumes of oxygen for complete combustion shows that for comparative purposes of costs this property would become important. The question of safety, carbon deposit, and back-firing made by the writer, are open to criticism. It should be clearly understood that acetylene cannot be exploded, detonated, or otherwise set off, in any manner that could warrant it being called explosive. The mixture of ethylene with the quantity of oxygen required for complete combustion is a powerful explosive, and carbon formation and backfires are just as probable with ethylene as with acetylene. The writer's references to copper and aluminium welding, and lead burning are, as most of our readers will know, incorrect. Ethylene burns in air with a luminous smoky flame, and the statement that it should be used preferably in a blowpipe with a mixing chamber is distinctly good. Enough has been said to show the value of the comparisons made between ethylene and acetylene, and it remains for the supporters of the process to produce the necessary industrial judgment to place it in its proper position as regards being a rival of acetylene welding.

The Large Locomotive Boiler

By F. W. BREWER

THE boiler is essentially the most important part of a locomotive. It must be carefully designed, well made, and constructed of the best material, since, besides the stresses due to steam pressure, it has to withstand the strains caused by expansion and contraction, and also those arising from the severe conditions under which locomotives ordinarily work. Moreover, continues "Locomotive," as it is difficult to determine with absolute accuracy all the stresses to which it is subject, it must be of such strength as to provide an adequate margin or factor of safety. Whatever its defects may be there is no doubt that the ordinary multitubular locomotive boiler, on the whole, is admirably adapted to fulfill the duties peculiar to railway traction. It has one conspicuous advantage in that it enables a large amount of heating surface to be got within a comparatively small compass. This is of the utmost importance, because, other things being equal, the hauling efficiency of the engine is entirely dependent upon the capacity of the boiler, which should be large enough to supply more than sufficient steam for ordinary requirements in order that the engine may be able to respond to special demands upon it for greater power. It being essential, therefore, that steam should be generated as rapidly as may be necessary at all times while the engine is running, the whole heating surface must be arranged so as to effect this object, and this is best accomplished by allowing both steam and water to circulate freely throughout the boiler.

Heating Surface Versus Power

It must not be supposed that a large amount of heating surface necessarily means a high evaporative power, or that a nominally small heating surface invariably implies a low evaporative efficiency. The boiler barrel, for instance, may be so overcrowded with tubes as to render ineffectual the greater proportion of the heating surface presented by them, due solely to the absence of a proper circulation. The disintegration of the steam particles from the water is then consequently retarded, so that the water is apt to rise with the steam and produce what is technically known as priming. Similarly, if the water spaces surrounding the firebox are too narrow, and if the sides of the box are vertical instead of being inclined towards the top, the circulation of the water is again impeded, and the production of steam is consequently injuriously affected. From this it follows that for a boiler of a given diameter and length (including the firebox casing) it is better to employ a relatively moderate number of tubes and to sacrifice some of the width and length of the inner firebox, the alternative advantage being a much better circulation of the water by reason of which a greater quantity of steam is generated in a given time, while the formation of scale is lessened and the tendency to priming reduced. Again,

the proportion of firebox surface to tube area, the length and diameter and arrangement of the tubes, apart from their number, and the diameter of the boiler barrel itself are all matters which must be taken into consideration as affecting steam production. Thus the firebox, which is exposed to the incandescent fuel and heated gases, has a much higher evaporative efficiency per square foot than the tubes, so that, within limits, it is always preferable to use as large a firebox as possible, consistent with the work to be done and the quality of the fuel employed. The tubes, however, play an important part in the economy of the engine, since they serve to utilize the heat as it leaves the fire box, and by imparting the heat to the main body of water in the barrel, though necessarily less effectively than the directly heated plates of the firebox itself, they provide, nevertheless, an indispensable contribution to the total heating surface.

Length of Tubes

As regards the length of the tubes, practice teaches us that the minimum for modern express engines should be 12 ft. between tube plates, while this length can often be advantageously increased, because in such engines the velocity of the escaping gases is usually very high, and long tubes, therefore, abstract a greater proportion of heat than do short tubes. The tubes are arranged either in vertical or in diagonal rows. Vertical rows are generally considered preferable as it is thought that they promote circulation and assist the rising of the steam bubbles, at the same time allowing solids in the feed water to pass to the bottom of the boiler. To compensate the lower evaporative power of the tubes, their aggregate area in contact with the water is very large compared with that of the firebox, and may be anything from ten to fifteen times greater.

At one period the size of the boiler as a whole was mainly governed by the mean theoretical tractive force which an engine was calculated to develop with a given load at a given speed. That is to say, a boiler of minimum dimensions was generally provided, and this sufficed for moderate needs. Modern requirements, however, have by degrees reversed this order of things until to-day it may more correctly be said that the hauling power is only limited to the extent to which the size of the boiler can be increased. It is now fully recognised that the actual—as distinct from the nominal—tractive force is appreciably augmented by enlarging the steaming capacity for a given cylinder volume. The dimensions, therefore, can only be said to define the minimum size of boiler necessary for a given theoretical tractive power. They cannot of themselves be taken as a general indication of the absolute power, for that is inseparable from the boiler capacity, and will be greater proportionately to the increased facilities for steam production. Most British locomotives, it is true, have done exceptionally good work, hav-

(Continued on page 64)

DEVELOPMENTS IN SHOP EQUIPMENT

"LETTGO" MECHANICAL OVERLOAD RELEASE

An effective "safety-first" mechanical device that will instantly disengage a drive when the load exceeds a predetermined point, has been developed by the Link-Belt Company, 910 S. Michigan Ave., Chicago, Ill. It is known as the "Lettgo" Mechanical Overload Release, and is especially adaptable for elevating, conveying, and power transmission machinery.

This release will automatically disengage the driving from the driven machinery if the load exceeds the fixed amount, thus allowing the driving motor or other source of power to run free and prevent damage, due to the inertia of the motor armature or other high-speed moving parts.

The construction of this device is such that it will release whether the load is gradually or suddenly applied, but it can

FIG. 1.—"LETTGO" OVERLOAD RELEASE WITH PARTS IN POSITION FOR DRIVING.

be set so that it will not trip from jars or shocks.

It is symmetrical, and can be assembled to operate in either direction, and can be adjusted for tension, so that it will operate for any desired overload. The mechanism is entirely inclosed and can be packed with grease for lubrication purposes.

Following is the action of the release:

The spider A, keyed to the shaft B, has triggers C pivotally mounted on the links D, with the ends engaging inside notches in the rim of the drum F, and rollers K.

The springs E, regulated to any desired pressure by the adjusting nuts H,

hold the ends of the triggers on the rollers K, under normal conditions, but when the drive is over-stressed, the

FIG. 2.—THE OVERLOAD RELEASE WITH THE PARTS IN RELEASED POSITION.

compression of the springs will permit the ends of the triggers to drop into the position shown in the enlarged section, releasing connection with the rim F, and allowing the driven machine to stop immediately.

To set the triggers again in the driving position, the collar J is provided with fingers that engage the pins on the lower ends of triggers C. By turning this collar with a spanner wrench, the triggers are moved to the original position, and the outer ends at the same time enter notches in the drum F, thus re-establishing the transmission connection. The cover G, fitting the end of the drum F, incloses and protects the entire mechanism. The hub of the drum F

may be extended to receive a wheel or gear, and have a bushing for running loose on the shaft; or may be keyed directly to a separate shaft, thus forming a coupling device, in which either element may be the driver.

DIRECT-CURRENT SERIES-WOUND MOTORS

For severe intermittent varying speed service where heavy starting torque is required, such as for cranes and hoists, the Westinghouse Electric & Manufacturing Company has recently brought out the type "HK" direct-current, series-wound motors. These motors are especially designed for use where the load consists of a series of starts, stops and reversals, the motor being idle for only short periods of time.

These motors are of enclosed construction with small openings in the lower part for ventilation. Covered openings in the top half of the frame give access to the brushes and the commutator. The most prominent feature of this motor is its compact construction, giving small overall dimensions, light weight and great mechanical strength.

The motor has a forged open-hearth steel frame and solid forged-steel feet. The motors above three h.p. rating are equipped with commutating poles so that high momentary loads can be carried without series sparking, and long life of commutator and brushes thereby insured. All field coils are thoroughly insulated and impregnated. The brackets are strong and rigid, each complete with oil well and bearing housing. Oil ring

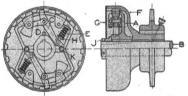

FIG. 3.—DIAGRAMMIC VIEW SHOWING THE CONSTRUCTION OF THE DEVICE.

lubrication is used. The brush holders are clamped to insulated rods mounted on the front bracket which are dowelled in place. The position of the brushes is thereby fixed for both directions of rotation.

Armature coils are form-wound and thoroughly insulated and impregnated

GENERAL VIEW OF MOTOR.

before being placed in the slots. A blower is placed in the rotor which effectively ventilates both armature and field windings. This allows a smaller diameter armature to be used, resulting in low flywheel effect, consequently little energy is required to start and stop the motor, thereby reducing to a minimum, wear and tear on bearings and brakes.

Electrically-operated shoe-type brakes are supplied when ordered. They are bolted to the machined lugs on the motor bracket, making a compact unit of motor and brake. The brake is simple and rugged, and provision is made for adjusting for wear of parts.

The company states that the ability of any motor to do its daily work, as

far as its temperature rise is concerned, is determined by the continuous rating it will carry at a safe temperature. A ventilated motor has a larger continuous rating than the non-ventilated motor of the same short time rating. A ventilated motor rated 7½ h.p., one-quarter hour, has about 30 per cent. more continuous capacity than the non-ventilated motor rated 7½ h.p. one-half hour.

DIRECT MOTOR-DRIVEN PLANER

The Oliver Machinery Co., Grand Rapids, Michigan, U.S.A., have forwarded us some interesting data regarding their new direct-coupled, motor-driven, ball-bearing hand planer and jointer. They comment as follows regarding this machine, which is illustrated herewith:

"For some time we have put out a machine with the motor built right into the machine, but being limited to three phase, 60 cycle, 220 volt. We decided to bring out this new type of jointer which would permit the use of any current motor providing the speed is approximately 3,600 r.p.m. That is, we can furnish this type for direct current or for alternating current or for one, two or three phase in the various voltages.

This machine has all the advantages of our No. 166 Ball-Bearing Jointer, at the same time eliminating bothersome belting and the danger that accompanies such belting. There is no necessity for belt guards; it saves room on the floor. It operates at a greater efficiency because of there being less bearings and no belting.

This machine is equipped regularly with three high speed knives in the head, which is of the Oliver round or safety type. The machine above described does especially smooth work and is highly recommended for furniture

VIEW SHOWING DIRECT MOTOR DRIVE.

plants, cabinet shops, pattern shops and other institutions requiring a smooth and accurate cutting jointer,

AUTOMATIC HOB-GRINDING MACHINE

The H. E. Harris Engineering Company of Bridgeport, Conn., have placed on the market a new type—No. 815-B—automatic hob-grinding machine that will grind hobs with any number of flutes up to 26, not exceeding 8 inches in diameter and 10 inches long. The adjustment that controls the rotation of the hob can be made within one-half minute of the arc, and can be changed while the machine is in motion. Indexing is automatic and is made while the machine is in one position only. The grinding head is so mounted that it may be swung to the required angle for grinding the flutes of helical hobs. An automatic feed is provided, and this can be disengaged when the desired amount of metal has been removed.

BENCH LATHE PLANING ATTACHMENT

An attachment for the cutting of keyways, small internal gears, slotting, etc., has been brought out by the Elgin Tool Works, of Elgin, Ill., for use on their small bench lathes. The stroke is adjustable from ⅛ inch to 1 inch, and the feed is regulated by a large knurled knob. A clapper is provided for tool relief on the return stroke. Either round or square tools may be used.

FAIRBANKS TRACK SCALE

Several new features have been embodied in the track scale recently developed by the E. & T. Fairbanks Company. In the new construction a special web section replaces the customary section levers. Unequal loads are more evenly distributed by the adoption of the web type of lever, as the initial load is suspended from the centre of the main lever bearing, and from there is transmitted through the true centre line of the lever to each succeeding bearing. The construction is such that the load on each through the true centre line of the lever are so designed, that there is neither torsional strain in the levers nor any tendency to displace the levers from their bearings, nor to produce unequal wear upon different parts of the same bearings. These scales are built in four sections. and in two capacities of 60 tons and 75 tons per section. The difference between the light and heavy duty scales is mainly in the loading per linear inch on the knife edge. this being 6,000 lbs., and 5,000 lbs. respectively.

The Brush Pilot Gauge Company, of Springfield, Mass., are now making plug gauges with a brush attachment on the nose, that cleans out the hole before the gauge enters, thereby minimizing the wear and assuring greater accuracy in gauging.

The MacLean Publishing Company

LIMITED

(ESTABLISHED 1887)

JOHN BAYNE MACLEAN, President. H. T. HUNTER, Vice-President
H. V. TYRRELL, General Manager.

PUBLISHERS OF

CANADIAN MACHINERY
~ MANUFACTURING NEWS ~

A weekly journal devoted to the machinery and manufacturing interests.

B. G. NEWTON, Manager.   A. R. KENNEDY, Managing Editor.

Associate Editors:

J. H. MOORE   T. H. FENNER   J. H. RODGERS (Montreal)

Office of Publication: 143-153 University Avenue, Toronto, Ontario.

VOL. XXIV.   TORONTO, JULY 15, 1920   No. 3

PRINCIPAL CONTENTS

The Week in Passing

NEW stories come in almost every week regarding the manner in which Canadian manufacturers are chasing material across the line. They quit months ago depending on letters, wires, telephone calls, or any such thing, and have substituted the more costly, but more certain, method of placing their own men at various points at the border and at the mills. In some cases the reliable truck is doing great service in getting material direct from the mills to the tracks where freight is moving in the direction of the Canadian border. It is an expensive way of doing business, and it is adding considerable to the cost in Canadian plants, but there is the poor option of doing this or trying to get along with dribbles coming in at the receiving end of the plant. The Canadian manufacturers are showing a spirit of determination that is commendable, and it speaks well for the future development of business in this country.

IF transportation facilities were normal, or anything approaching it, it is the conviction of many conservative business men, that the machine tool trade would be in the midst of one of the greatest business seasons it has ever experienced in Canada. Shortage of material, delay in shipping machines, and other well-known conditions have tended to keep the brakes on for some months past. There is a big volume of this trade that is going to stay good. It is not going to pass out simply because it could not be attended to at once. When conditions become such that it is possible to get supplies, a good deal of buying will begin. Building costs have also been responsible for spoiling a considerable amount of business for the machine tool trade. When equipment is purchased to any extent, the chances are that a new house is required to properly care for it in such a way as to secure maximum results. Several propositions that would have called for a nice line of tool equipment have been scared off the market, but not out of existence. Moderating costs will bring them back to life.

THE Department of Overseas Trade in Canada, in the yearly review of business conditions in the Dominion, draws attention to the fact that "there is room for much more advertising by the majority of United Kingdom firms doing business here. The local agents among others, should be consulted as to the best methods of expending the advertising appropriation." This matter has been brought up before, and firms in Canada trying to push British-made goods find that they are handicapped in that they do not get the aggressive and whole-hearted support that American firms put at the disposal of their representatives in this country. The British manufacturer ought to remember that when he comes into the Canadian market he has got to get out and scratch for the business. The printed page often gets a hearing while the salesman is warming his shins on the waiting bench outside the office door.

THOSE who are engaged in the controversy about building or taking over transportation lines in Ontario for the purpose of electrification, should think seriously and well about the changing aspect of transportation. It is folly to overlook the coming of the truck, as well as the passenger car. Banks are recognizing this in many places, and are financing the making of trucks and tractors, but not passenger cars. When well-built highways are put through the country, and the use of trucks is made possible and profitable, there is bound to be a falling-off in the call for electric or steam road service. The form of locomotion that will take the load from the barn direct to the city market is by all odds the most desirable, as the start, the route and the terminal are all capable of being adjusted to the convenience of the truck owner.

ONE writer draws attention to the fact that a number of Canadian firms, which have formerly confined their efforts to home business, do not study all the tariff regulations in order that they may take advantage of them. He states that some firms are not aware of the fact that they are entitled to a rebate on duty paid on material brought in from United States, made up in Canada, and exported. He goes on to cite a case where, when this was found out, the Canadian exporter was able to make a considerable reduction in his export price, which put him in a very favorable position to meet other competition. The Department of Trade and Commerce could well afford to see to it that every manufacturer was advised of these regulations. It is generally taken for granted that all this information is thoroughly known to the trade, but it is possible that firms entering the export field for the first time may not have gone fully into the regulations. A Canadian firm entering the foreign field should be armed with every weapon its government can hand it.

ARE we facing a better steel market from the standpoint of the consumer? There are some who state openly that we are, and give actual cases to back up their opinion. Just how this will come about cannot be stated precisely, but it would be reasonable to suppose that the disappearance of the premium market would lead the way, and there are indications of the shrinkage of this unpopular thing. If there were enough cars at the mills now to take away the material that is piled there, we might find that we would be very close to a normal steel market. Traffic troubles, strikes, and rumors of strikes have been as spring tonics and summer stimulants to the fagging premium market. A re-established freight service would be a real steam roller for the premium business.

MARKET DEVELOPMENTS

Hustling Now to Secure Their Material

Canadian Firms Are Keeping Their Own Men at U.S. Points of Supply in Order to Secure Shipment—New York Market Reported Very Quiet—Scrap Situation Dull and Uninteresting.

CANADIAN industries are displaying no small amount of enterprise and ingenuity in trying to keep a sufficient amount of material coming to their plants. The old order of things is quite reversed, and it is now a case of a firm not only ordering material, but spending money, and lots of it too, in making sure that it comes to their premises. Many of the establishments in this country have men at Buffalo, Toledo, Pittsburg, Youngstown, and other points, as well as between the other producing centres and the border. Their business is to get the material their plants must have to keep going. One firm has two crews with large trucks taking steel from several of the mills to cars on lines that are moving freight with fair rapidity, and making certain that they get started toward the Canadian border. Of course all this adds considerable to the cost of the finished product.

There is a fair volume of business being done in machine tools, although none of the dealers have anything in the line of large orders for equipment. Several firms hold to the belief that an easing off in the price of building will be a good thing for the sale of machinery, as an order for a large amount of equipment generally means new housing for it. Several fairly large expenditures in machine tools are hung up now on account of the too high price of the necessary building.

There is a dull market in the scrap metal trade. Some dealers have had to come out and buy this week, but they have done it at a fraction under the quoted prices. As a general thing those having material to sell are advised to hold, as there may be an improvement in the market.

MONTREAL IS GETTING MORE MATERIAL FROM STEEL MILLS

Special to CANADIAN MACHINERY

MONTREAL, July 15.—Combinations of circumstances at the present time tend to retain the industrial situation in a more or less disorganized condition. The irregular receipt of material, including fuel, and the general slowing up of activity at this period of the year, are a few of the factors that add to the difficulties of carrying on normal business. Many plants have experienced the troubles on ineffective operation as a result of the inadequate supply of coal and raw materials, some factories to the point of closing down. Shortage of cars has frequently been given as being the cause of delayed shipments, but this is very often counteracted by the appearance of empties standing idle at different points. At the present time about 300 empties are awaiting movement from a siding on the south shore of the river. It would seem that the supply of cars in the aggregate is equal to the situation, but district shortage is pronounced owing to the periodical congestion at terminals, particularly at ocean ports. The application of the railroads to the Canadian Railway Board, for a 30 per cent. increase in freight rates, has

created considerable interest in trade circles, and while some time will probably elapse before any definite action is taken, the possibility of higher charges, though no doubt justified, will nevertheless add to the difficulties of trading in general, and will probably mean further adjustments in other directions. Some of the engineering plants in this locality report more normal activity owing to the renewal of regular supplies of castings. The local foundries are quite busy trying to overcome the handicap of a six week's suspension.

The steel situation has not shown any marked improvement, although to some extent the mill output has been increased, but railroad facilities are inadequate to supply manufacturers with needed requirements. The metal situation, with the exception of copper, seems to foretell a buying movement. Machine sales are falling off with general supplies about normal.

Betterment in Steel

Steel dealers in this district report a fair volume of material coming in from the mills, but not more than can be placed in the hands of waiting consumers.

It would seem that the general demand for steel lines is not quite so brisk just now, and the impression is that production is gradually getting to the stage when mill and warehouse supplies will show steady accumulation, despite the fact that midsummer operations tend to lower the normal output. There is considerable dissatisfaction expressed by some dealers and manufacturers, regarding the non-delivery of material, both in connection with non-shipment and delay in transit. It is claimed that considerable rolling stock has been diverted for the transportation of coal and raw materials, the result being to assist the producing end of primary industries at the expense of other manufacturers. However, it is anticipated that the improved conditions now developing in the mill districts will be reflected in other directions, so that the early fall will experience a country-wide improvement in every branch of industrial activity. While no actual change is announced in warehouse conditions, the reports of dealers would imply that better inducement is being offered from the sheet mills, and it would seem that lower prices and more favorable shipments would probably rule in the early future. Dealers are still able to place all material coming through, but prospects are for larger surpluses for warehousing. Tubes are still in demand and hard to obtain.

Metals Show Weakness

Recent inquiries for copper for export have been quite encouraging and it is believed that the market will take on some strength, or at least retain the firm position it now holds. The local market is well maintained and sales are reported as seasonable. Other metals are inclined to be easier and the market is reflected in the lower quotations on tin, which is quoted at 62 cents; spelter at 11½, is one cent lower; lead has declined on weaker demand, and the price asked is 10¾ cents, a drop of ¾ cent on the previous quotation. Antimony is quoted at from 13 to 14 cents per lb.

Supply Demand Fair

Apart from the slight betterment reported in supply movement, the machine tools situation shows little change over last week. The summer months, invariably, are a period of lesser activity, and this, along with other conditions, such as curtailment of industrial enterprise, due to insufficient material and fuel supply, tend to put a damper on business, so that sales are usually reported on a lower basis. Orders on heavy tools are less in evidence and inquiries for used machinery have fallen off. Supplies, while quiet, are the feature of the present business, and dealers are apparently well satisfied with accessory requirements. Old country goods are coming to hand readily and delivery from this source is well within the time limits, with few exceptions.

Scrap Movement Restricted

The market is still devoid of any condition with feature characteristics. Some hope was entertained that considerable trade could be developed with Britain, following the reported need there, but the inability to get transportation for this class of freight, and the high rate prevailing, makes the prospect very doubtful, if not prohibitive. While the local movement of machinery scrap continues in fair volume, the business in old materials, generally, is below normal. Non-ferrous metals have their ups and downs, fluctuating both in demand and price, so much so, that quotations are based on sales rather than on actual prevailing condition. Prices quoted are a fair average of what dealers are paying, with the tendency lower.

CANADIANS TRUCK THEIR OWN STUFF

In An Endeavor to Get Material From Mills to the Roads That Are Moving

TORONTO.—"We have no complaint about the business we are getting. Of course at this season of the year we expect there will be a little dropping off, especially if the call for the equipment is not pressing. But we are looking for better business, and it is going to get better just as quickly as the railroad

POINTS IN WEEK'S MARKETING NOTES

Canadian firms are trying many unusual ways of getting material from U. S. points, and are keeping their plants going fairly well in spite of this trouble.

—

New York machinery market claims that in several cases large buyers are staying out, waiting for a reduction in prices.

—

Cancellations in some lines are not because firms do not want the material, but the belief grows that an attempt is being made in this way to batter down existing price schedules.

—

Warehouses are getting few deliveries this week. This refers to sheets, plates, tubes and bars

—

Large dealers in scrap are advising their customers to hold their material. Some buying was done during the week to cover contracts, but the material was taken in at a fraction under current prices.

—

U. S. Steel Corporation's unfilled obligations increased during the month of June by 34,351 tons. The total on their books now is 10.978,-817 tons, meaning capacity operation for at least eight months.

situation gets to the point where it can take care of it. It is our firm belief—and there are others who hold the same view—that had there been good railroad service the Canadian trade would never have seen a greater season for the selling of machine tools than right now. A great deal of this business is still good. One of the things that will help in bringing it into the market is a more reasonable and stable level in the figures for building. From what we can judge, conditions in Canada are fundamentally sound, and early fall will bring all the trade we can possibly get the material to accommodate."

The Scrap for Material

Canadian firms are getting to the point where they know what it means to get out and dig for their material. It used to amount to getting out to dig for the orders, but it is not yet a case of that so much as chasing material. For instance, one well-known Canadian firm has a number of men at one of the Pittsburg mills while another group works around Youngstown, in the steel district. They have a number of heavy trucks, and as soon as they locate any material belonging to their mill they grab it, and run it away to a line that is operating fairly well, fill and seal the cars and see that they get headed out pointing to the Canadian border. This performance is costing them quite a penny, but it means that they are able to get production instead of being at the point of closing down.

One Toronto dealer stated that he had instructions from several firms in this district to get material, no matter how it was sent. They are all willing to stand the added expense of having stakes put in as supports around the sides of flat cars, an unusual way of shipping many lines of steel. The mines are getting the preference on anything that looks like coal-carrying cars now. One Toronto dealer was in Chicago a few days ago at one of the mills there. This mill had a shipment of light rails for one of the mines that was shipping coal to them. They could not hold the cars long enough to get this shipment on board and send it on the return trip to the mines. The claim is made that were an exception made in this case it would create a precedent, and in the end defeat the drastic order that has been put into force to deal with the coal situation.

Little Arriving Here

The warehouses here are getting little assistance from any quarters this week. Shipments are dwindling away from the district. Reports of cars coming in after having been on the road six weeks and two months are quite common, but it is a healthy sign that practically all this business remains good, and cancellations on account of slow delivery are not often met with.

One view was expressed in the steel market this morning, viz., that many of the cancellations in other lines were not due to the fact that people did not want the goods, but rather that they were using the cancellation option as a club to batter down prices from their present high levels, and that once this were accomplished the orders would come in again. The fact that there is a demand for all sorts of material is one of the best indications of the fundamental strength of the market.

Hardly any class of material owns up to getting good treatment now. One can go over the list of material coming in, or rather not coming in, and be assured that sheets, bars, tubes, plate, etc. are all very slow in arriving, although the trade is getting a little more assistance now in the line of bars from the Canadian mills.

There is a scarcity of labor in some of the local steel warehouses. The work is fairly heavy, and apparently the available men can find easier jobs elsewhere.

No Recovery in Scrap

The scrap market is quiet. Some of the larger dealers had to come into the market a few days ago and take on quite a tonnage in order to cover themselves on promised shipments, and this they were able to do at half a cent under the market. Although it is possible to take on material at fractions off the recognized

price, the yards show no tendency to buy as they do not want to take on any more stock. In most cases they are advising their customers to hold their material. There are very few buyers in the market now, and the demand is weak.

INDEPENDENTS KEEP GOING ON ORDERS WHERE DELIVERY IS OF BIG CONCERN

Special to CANADIAN MACHINERY.

PITTSBURGH, July 15,—Production of steel ingots in June by the 30 companies making monthly reports to the American Iron and Steel Institute was 2,980,690 gross tons, against 2,885,164 tons produced in May by the same companies, which in 1918 made 84.03 per cent. of the total ingot output. Allowing for producers not making monthly reports, and taking account of the number of actual working days, the rates of steel ingot production for the whole industry appear to have been about 41,050,000 tons a year in May and 42,450,000 tons a year in June, showing an increase in the rate, from May to June, of about 3 per cent.

The production by the 30 companies reporting for the whole half year is 17,634,434 tons, equivalent to a rate for the whole industry of about 42,100,000 tons a year. This compares with estimated production in 1919 (official statistics not being available as yet) of 33,000,000 tons, and reported production of 43,081,022 tons in 1918 and 43,619,200 tons in 1917, the record year for production thus far. In each of the two best years before the war, 1912 and 1913, production was about 30,280,000 tons, so that production in the half year just ended may be said to have been almost 40 per cent greater than the best average rate before the war.

The half-year's production was, however, far short of capacity, for capacity is between 50,000,000 and 55,000,000 tons a year, given reasonably fair operating conditions. Throughout the half-year there were serious manufacturing difficulties of one sort or another, except that conditions were moderately fair in March. In January there was a serious shortage of coke and this condition was not corrected entirely in February or March. In April came the outlaw rail strikes, greatly decreasing the movement of coal and coke, and after the strikes, as such, became of small moment there was a traffic congestion that greatly interfered with operation. The recovery in production since April has really been greater than might be expected from the loud complaints heard everywhere of insufficient car supplies for moving raw materials. The month of April as a whole shewed a rate of ingot production of about 37,550,000 tons a year, but at the worst time in the month the rate was probably down to about 33,000,000 tons, while the rate at the close of June was probably about 43,000,000 tons, showing an improvement from the worst of fully 30 per cent.

Production of pig iron in June was at the rate of about 37,400,000 tons a year, showing an increase over the May rate of about 5 per cent. The production of pig iron increased more than the production of steel, indicating perhaps that the steel works piled pig iron to some extent during June, although it is possible that in May they were consuming some stocks previously accumulated.

In both pig iron and steel the June production was well in excess of the shipments. Pig iron was piled by a number of merchant furnaces, which had plenty of orders but were lacking in cars, while a number of consumers bought pig iron in the open market because they were not getting sufficient deliveries from their regular sources of supply. It has been the common remark of late that buyers of pig iron are really buying deliveries rather than pig iron. What they want is delivered pig iron. No one would at this time buy any pig iron that was to remain in the furnace yard for any length of time.

A rough guess would be that of the steel made in June between 75 and 80 per cent. was actually shipped, the remainder being added to the already large stocks in mill warehouses and yards. Assuming capacity in steel ingots to be 52,500,000 tons a year, the June production was at the rate of 81 per cent. capacity, while if shipments were 75 to 80 per cent. of production they were 60 to 65 per cent. of capacity. Thus whatever scarcity of steel there may be, and there is not so much scarcity apparent as a month or two ago, is to be attributed entirely to the difficulties in making shipments. If the mills could ship at their productive capacity the shipments would be 50 or 60 per cent. greater than they have been of late. A rough guess is that the accumulations of steel at mills now amount to the equivalent of a month's production at the rate lately obtaining.

Corporation Tonnage

The United States Steel Corporation's unfilled obligations increased during June by 38,351 tons, against increases averaging nearly 500,000 tons a month in the three preceding months. The increase in unfilled obligations is, of course, the month's bookings minus the month's shipments, not the month's production, as when the corporation rolls an order to customer's specifications and does not ship the steel it is not a filling of the obligation, the steel being invoiced only as actually shipped. The June increase in unfilled obligations was about 5 per cent. of the month's capacity, while the June shipments may be estimated at about 75 per cent. of capacity (actual production in June may be estimated at about 90 per cent.) so that the bookings in a month appear to have been about 80 per cent. of capacity, this comparing with 112 per cent. and 94 per cent. for April.

The June bookings, while greatly decreased, were still quite large considering that the steel market as a whole was very quiet. It was only because the corporation has held its prices down to the Industrial Board level of March 31, 1919, while all the independents have advanced their prices, that the corporation's customers are willing, in this quiet market, to place contracts for such indefinite delivery as is now involved in the Steel Corporation's business. The 10,978,817 tons of unfilled obligations on books July 1 is equal almost precisely to the tonnage that would be made in eight months, operating at full capacity,

Montreal Notes

Jas. Coleman, formerly manager of the Boss Lock Nut Factory, of Montreal, has become associated with the Consolidated Equipment Company, with offices in the Transportation Building.

The prospect of a large modern hotel in the uptown district of Montreal, has again been revived by the visit of F. A. Dudley and G. H. O'Neill, president and manager respectively, of the United Hotels Company, to this city. Several sites are under consideration, but no definite action has yet been taken. It is more than likely that further negotiations will take place in an effort to have the limited height restrictions removed.

As an indication of what has been done by the Canadian Pacific Railway in re-establishing ex-service men, it has been stated that of the eleven thousand that enlisted from the ranks of the company over seven thousand has been re-employed, many of them in the same capacity as formerly. In addition to the re-instatement of ex-C. P. R. men, the railway company has given employment to over eleven thousand men, making a grand total of returned soldiers now at work of over 18,000.

Last Saturday about twenty prominent American manufactnrurers various points in the United States were in Montreal and spent the day as guests of the Montreal branch of the Canadian Manufacturers' Association. A sight-seeing trip along the harbour front, on a special train provided by the Harbour Commissioners, was the feature of the morning. After luncheon the guests were taken for a trip around the city. Heading the party from the States was Stephen C. Mason of Pittsburg, president of the U. S. National Association of Manufacturers.

or to March 1, 1921, as an average date, while operating at 89 per cent. of capacity the unfilled tonnage would carry the corporation to April 1 as an average date. Some departments have more orders than orders, in proportion to capacity, but the general statement can be made that the average customer of the Steel Corporation now placing a contract

cannot expect delivery before the first or second quarter of next year. Customers of the independents, on the other hand, are buying for delivery any time from immediate to say three or four months hence. Beyond such a time they do not care to commit themselves, on account of the difference between Steel Corporation and independent prices.

lv understood, a special meeting of the council was called for Wednesday, when it was expected that Mr. J. Birnie, K.C., would present a legal opinion, this being done upon request of the council through the mayor.

NEW YORK SHOPS LAY OFF MEN
AND GET THE SAME PRODUCTION

Special to CANADIAN MACHINERY.

NEW YORK. July 15.—The Norfolk & Western Railroad within the past week has closed for about $150,000 to $200,000 worth of shop equipment, including planers, punches and shears, lathes, wheel presses and other special railroad tools. These orders were placed against an inquiry for. 72 machines, which was issued several weeks ago. Not all of the list was bought, owing to lack of sufficient appropriation. The Chesapeake & Ohio Railroad, which issued a list of its requirements some weeks ago, is expected to close soon. Other railroad business is in sight, but it is only prospective at this time, and there is nothing definite as to when inquiries will be issued. Some of the railroad purchasing departments are delaying action because of failure of the directorates to provide finances. Probably there will be greater railroad buying as soon as the new railroad freight rates,

on which the Interstate Commerce Commission is now working, are announced. These rates will take effect Sept. 1 next, and will approximate 25 to 30 per cent. advance.

Buying by industrial companies is at low ebb. In many instances manufacturers quite frankly say they are waiting for prices to come down. They are also taking advantage of the present lull to work out a readjustment of the labor situation. Labor has had the upper hand for a considerable time, and now there are evidences of a change. Some plants have deliberately laid off men for the psychological effect it might have upon the workmen who are retained. In certain cases the reduction of working forces has worked like magic. The forces remaining at work have speeded up production so that the loss of men has not had much effect, if any, on the gross output.

PIG IRON TRADE

There appears to have been a more active demand for prompt shipment in the pig iron industry in general recently.
Recent transactions in the East totalling 40,000 tons were negotiated at mostly $45.
There is an increasingly active demand for prompt iron, but while prices are holding steady this grade iron is increasingly difficult to get.
There has been some effort to secure delivery for first quarter, but most furnaces are unwilling to accept orders so far ahead.
The Chicago market is brighter but the new embargo on shipping without special permit within the Chicago switching district is having a detrimental effect.
The transportation question is not quite so acute, some shippers reporting a better supply of cars, which may be due to accumulation over the holidays.
There is great complaint of the time occupied in transit, it being reported that shipments which used to take ten to twelve days now occupy twice as long.
The average market may be held at $45, small lots having been sold at $45.75 furnace.
Transportation difficulties and the shortage of coke and limestone are still the predominating factors in determining the condition of the pig iron market.

COLLINGWOOD MAY ENFORCE THE
PROVISIONS OF OLD AGREEMENT

Collingwood Bulletin : — This week there are some discomforting rumors going around in respect to the steel plant which it was recently reported as sold to the Baldwins, of Toronto.
The latest report is that the company propose to scrap the plant and that next week the work of removing it will be commenced.
The presence in town of a couple of strangers whose time was spent around the plant in conjunction with reported hiring of men gave rise to the report. Letters to the firm in Toronto are said to remain unanswered, in fact not acknowledged.
Taking it for granted that something more real was behind the rumors of removal, the town council have been looking into the agreements respecting the plant, with the result that there is ground for believing that removal is guarded against and that under certain conditions the town may step in and enforce payment of certain annual amounts.
It apears that the original agreement was entered into on March 5th, 1900, by Cramp Co., with Charles D. Cramp as president and J. A. Currie, secretary. By this a bonus of $115,000 was voted. On November 10, 1903, a new agreement

was made between the Cramp Co. and the town of Collingwood, this being ratified by the act of the Ontario Legislature and assented to on April 26, 1904.
This agreement is signed by Dan Wilson, mayor; J. H. Duncan, clerk; J. Wesley Allison, president; J. A. Currie, secretary, with the seal of the corporation of the town of Collingwood and the Cramp Steel Co., Ltd.
By clause 7 of this agreement, the company covenants and agrees to properly maintain and operate the said works, and such works as shall hereafter be erected upon the said land at the town of Collingwood for a period of thirty years from and after July 1, 1904, and should the said works not be properly maintained and operated within the time intent and meaning of the agreement for a period of twelve months, the company shall, in addition to said taxes, repay to the corporation annually the sum of $2,000 for each year and every year remaining of the thirty years that the said company shall fail to operate the said plant as aforesaid.
A further clause provides that by this agreement the parties are free from the conditions and terms of the agreement of March 5, 1900.
In order to have the matter thorough-

U.S. SCRAP METAL

Transportation difficulties are still acting as a strong deterrent to trading.
Some fluctuations in prices are noticeable, but the general situation is steady.
There was a heavy purchase recently of 50,000 tons by a Buffalo mill, half heavy melting steel and half hydraulic compressed; at an average of $25.00.
It is understood that heavy melting steel can be generally procured from $25.00 to $25.50.
Consumers in the middle west are confining purchases to materials on which they can get certain and quick delivery.
The dealers of the St. Louis district forwarded a protest to the Interstate Commerce Commission which resulted in a modification of the embargo on the use of open cars. This did not have the beneficial effect hoped for due to strikes and traffic congestion.
On the whole the general condition of the scrap market is steadier than last week.

Geometric Solid Adjustable Die Heads

When one uses the word "ONLY," he needs to be sure of what he is talking about. With confidence we use the word in saying that a Geometric Solid Adjustable Die Head, fitted with a set of milled dies, is the only solid die head that will back off without stripping the thread.

Geometric Solid Adjustable Die Heads may be fitted with either a releasing shank, plain shank, or with special shank for the Gridley Automatics. The releasing shank permits the head to disengage from the shank upon completion of the thread. The plain shank fits the head for use on the turret of a lathe or on a live spindle, such as a drill press.

Apart from the fact that it is not equipped with the self-opening feature or the roughing and finishing attachment, the Geometric Solid Adjustable Die Head is the same in principle and construction and will do equally as accurate work as the other styles of Geometric Die Heads.

Whatever the requirements, there is a type of Geometric Die Head best adapted to the work.

THE GEOMETRIC TOOL COMPANY
NEW HAVEN   CONNECTICUT

Canadian Agents:

Canadian Agents: Williams & Wilson, Ltd., Montreal; The A. R. Williams Machinery Co., Ltd., Toronto.
Winnipeg, and St. John, N.B.; The Canadian Fairbanks-Morse Co., Ltd., Manitoba, Saskatchewan, Alberta.

If interested tear out this page and place with letters to be answered.

SELECTED MARKET QUOTATIONS

Being a record of prices current on raw and finished material entering into the manufacture of mechanical and general engineering products.

PIG IRON

Grey forge, Pittsburgh	$42 40
Lake Superior, charcoal, Chicago	57 00
Standard low phos., Philadelphia	50 00
Bessemer, Pittsburgh	43 00
Basic, Valley furnace	42 90
Toronto price:—	
Silicon, 2.25% to 2.75%	52 00
No. 2 Foundry, 1.75 to 2.25%	50 00

IRON AND STEEL

Per lb. to Large Buyers	Cents
Iron bars, base, Toronto	$ 5 50
Steel bars, base, Toronto	5 50
Iron bars, base, Montreal	5 50
Steel bars, base, Montreal	5 50
Reinforcing bars, base	5 00
Steel hoops	7 00
Norway iron	11 00
Tire steel	5 75
Spring steel	10 00
Band steel, No. 10 gauge and 3-16 in. base	6 00
Chequered floor plate, 3-16 in.	8 40
Chequered floor plate, ¼ in.	8 00
Bessemer rails, heavy, at mill	....
Steel bars, Pittsburgh	3 00-4 00
Tank plates, Pittsburgh	4 00
Structural shapes, Pittsburgh	3 00
Steel hoops, Pittsburgh	3 50-3 75
F.O.B., Toronto Warehouse	
Small shapes	4 25
F.O.B. Chicago Warehouse	
Steel bars	3 62
Structural shapes	3 72
Plates	3 67 to 5 50
Small shapes under 3"	3 62

FREIGHT RATES

Per 100 Pounds.

Pittsburgh to Following Points

	C.L.	L.C.L.
Montreal	33	45
St. John, N.B.	41½	55
Halifax	49	64½
Toronto	27	39
Guelph	27	39
London	27	39
Windsor	27	39
Winnipeg	89½	135

METALS

	Gross. Montreal	Toronto
Lake copper	$25 00	$24 00
Electro copper	24 50	24 00
Castings, copper	24 00	24 00
Tin	62 00	65 00
Spelter	11 50	12 00
Lead	10 75	11 00
Antimony	13 00	14 00
Aluminum	34 00	36 00

Prices per 100 lbs.

PLATES

Plates, 3-16 in.	$ 7 25	$ 7 25
Plates, ¼ up	6 50	6 50

PIPE—WROUGHT

Price List No. 44—April, 1920.

STANDARD BUTTWELD S/C

	Steel Black	Steel Galv.	Gen. Wrot. Iron Black	Gen. Wrot. Iron Galv.
⅛ in.	$6 50	$ 8 50	....	....
¼ in.	5 13	7 26	$ 5 43	$ 7 56
⅜ in.	5 13	7 26	5 43	7 66
½ in.	6 84	8 42	7 27	8 84
¾ in.	8 45	10 58	9 08	11 16
1 in.	12 50	15 44	13 55	16 49
1¼ in.	16 91	21 16	18 06	22 31
1½ in.	20 21	25 30	21 59	26 68
2 in.	27 20	34 04	29 05	35 89
2½ in.	43 00	53 82	....	....
3 in.	56 23	70 38	....	....
3½ in.	71 30	88 32	....	....
4 in.	84 48	104 64	....	....

STANDARD LAPWELD S/C

	Steel Black	Steel Galv.	Gen. Wrot. Iron Black	Gen. Wrot. Iron Galv.
2 in.	$30 90	$37 74	$34 60	$41 44
2½ in.	45 84	56 16	51 19	62 01
3 in.	59 29	73 44	66 94	81 09
3½ in.	73 14	90 16	82 34	99 36
4 in.	86 66	106 82	97 56	117 72
4½ in.	0 98	1 23	1 24	1 49
5 in.	1 15	1 44	1 44	1 73
6½ in.	1 49	1 86	1 87	2 25
7 in.	1 94	2 43	2 42	2 90
8-L in.	2 04	2 55	2 54	3 05
8 in.	2 35	2 94	2 92	3 51
9 in.	2 81	3 52	3 50	4 21
10-L in.	2 61	3 26	3 25	3 90
10 in.	3 36	4 20	4 18	5 03

Prices—Ontario, Quebec and Maritime Provinces

WROUGHT NIPPLES

4" and under, 60%.

4½" and larger, 50%.

4" and under, running thread, 30%.

Standard couplings, 4-in. and under, 30%.

Do., 4½-in. and larger, 10%.

OLD MATERIAL

Dealers' Average Buying Prices.

	Per 100 Pounds. Montreal	Toronto
Copper, light	$15 00	$14 00
Copper, crucible	18 00	18 00
Copper, heavy	18 00	18 00
Copper wire	18 00	18 00
No. 1 machine composition	16 00	17 00
New brass cuttings	11 00	11 75
Red brass cuttings	14 00	15 75
Yellow brass turnings	8 50	9 50
Light brass	7 00	7 00
Medium brass	8 00	7 75
Scrap zinc	6 50	6 00
Heavy lead	7 50	7 75
Tea lead	4 50	5 00
Aluminum	19 00	20 00
	Per Ton	Gross
Heavy melting steel	18 00	18 00
Boiler plate	15 50	15 00
Axles (wrought iron)	22 00	20 00
Rails. (scrap)	18 00	18 00
Malleable scrap	25 00	25 00
No. 1 machine cast iron	32 00	33 00
Pipe, wrought	12 00	12 00
Car wheel	26 00	26 00
Steel axles	22 00	20 00
Mach. shop turnings	11 00	11 00
Stove plate	26 50	25 00
Cast boring	12 00	12 00

BOLTS, NUTS AND SCREWS

	Per Cent.
Carriage bolts, 7-16 and up	+10
Carriage bolts, ⅜-in. and less	Net
Coach and lag screws	—15
Stove bolts	55
Wrought washers	—25
Elevator bolts	+10
Machine bolts, 7-16 and over	+10
Machine bolts, ⅜-in. and less	+10
Blank bolts	Net
Bolt ends	Net
Machine screws, fl. and rd. hd., steel	27½
Machine screws, o. and fil. hd., steel	+25
Machine screws, fl. and rd. hd., brass	net
Machine screws, o. and fil. hd., brass	net
Nuts, square, blank .... +25 add	$2 00
Nuts, square, tapped ... add	2 25
Nuts, hex., blank ... add	2 50
Nuts, hex., tapped ... add	3 00
Copper rivets and burrs, list less	15
Burrs only, list plus	25
Iron rivets and burrs	40 and 5
Boiler rivets, base ¾" and larger	$8 50
Structural rivets, as above	8 40
Wood screws, O. & R., bright	75
Wood screws, flat, bright	77½
Wood screws, flat, brass	55
Wood screws, O. & R., brass	55½
Wood screws, flat, bronze	50
Wood screws, O. & R., bronze	47½

MILLED PRODUCTS

(Prices on unbroken packages)

	Per Cent.
Set screws .... —20%, 25 and	5
Sq. and hex. hd. cap screws	12½
Rd. and fil. hd. cap screws ..plus	25
Flat but. hd. cap screws ....plus	50
Fin. and semi-fin. nuts up to 1-in.	12½
Fin. and Semi-fin. nuts, over 1 in., up to 1½-in.	—5
Fin. and Semi-fin. nuts over 1½ in., up to 2-in.	+12½
Studs	+5
Taper pins	—12½
Coupling bolts	+40
Planer head bolts, without fillet, list	+45
Planer head bolts, with fillet, list plus 10 and	+55
Planer head bolt nuts, same as finished nuts.	
Planer bolt washers	net
Hollow set screws	+60
Collar screws ......list plus 20,	30
Thumb screws	40
Thumb nuts	75
Patch bolts ......add	+85
Cold pressed nuts to 1½ in...add	$1 00
Cold pressed nuts over 1½ in..add	2 00

BILLETS

	Per gross ton
Bessemer billets	$60 00
Open-hearth billets	60 00
O.H. sheet bars	76 00
Forging billets	56 00-75 00
Wire rods	52 00-70 00

Government prices.

F.O.B. Pittsburgh.

NAILS AND SPIKES

Wire nails, base	$6 10
Cut nails, base	7 00
Miscellaneous wire nails	50%

ROPE AND PACKINGS

Plumbers' oakum, per lb.	0 10¼
Packing, square braided	0 38
Packing, No. 1 Italian	0 44
Packing, No. 2 Italian	0 36
Pure Manila rope	0 35½
British Manila rope	0 28
New Zealand hemp	0 28

POLISHED DRILL ROD

Discount off list, Montreal and Toronto	net

MISCELLANEOUS

Solder, strictly	$ 0 35
Solder, guaranteed	0 39
Soldering coppers, lb.	0 62½
White lead, pure, cwt.	20 35
Red dry lead, 100-lb. kegs, per cwt.	16 00
Gasoline, per gal., bulk	0 38
Pure turp., single bbls., gal.	3 15
Linseed oil, raw, single bbls.	2 37
Linseed oil, boiled, single bbls.	2 40
Wood alcohol, per gal.	4 00
Whiting, plain, per 100 lbs.	3 00

CARBON DRILLS AND REAMERS

S.S. drills, wire sizes	32½
Can. carbon cutters, plus	20
Standard drills, all sizes	32½
3-fluted drills, plus	10
Jobbers' and letter sizes	32½
Bit stock	40
Ratchet drills	15
S.S. drills for wood	40
Wood boring brace drills	25
Electricians' bits	30
Sockets	50
Sleeves	50
Taper pin reamers	25 off
Drills and countersinks	net
Bridge reamers	50
Centre reamers	10
Chucking reamers	net
Hand reamers	10
High speed drills, list plus 20 to	40
Can. high speed cutters, net to plus	10
American plus	40

COLD ROLLED STEEL

[At warehouse]

Rounds and squares	$7 base
Hexagons and flats	$7.75 base

IRON PIPE FITTINGS

	Black	Galv.
Class A	60	75
Class B	27	37
Class C	18	27

Cast iron fittings, 5%; malleable bushings, 22½%; cast bushings, 22½%; unions, 37½%; plugs, 20% off list.

SHEETS

	Montreal	Toronto
Sheets, black, No. 28	$ 8 50	$ 9 50
Sheets, Blue ann., No. 10	8 50	9 00
Canada plates, dull, 52 sheets	8 50	10 00
Can. plates, all bright	8 60	9 00
Apollo brand, 10¾ oz. galvanized	....	....
Queen's Head, 28 B.W.G.	11 00	....
Fleur-de-Lis, 28 B.W.G.	10 50	....
Gorbal's Best, No. 28	....	....
Colborne Crown, No. 28.	....	....
Premier, No. 28, U.S.	11 50	10 50
Premier, 10¾-oz.	11 50	10 90
Zinc sheets	16 50	20 00

PROOF COIL CHAIN

(Warehouse Price)

B

¼ in., $13.00; 5-16, $11.00; ⅜ in., $10.00; 7-16 in., $9.80; ½ in., $9.75; ⅝ in., $9.20; ¾ in., $9.30; ⅞ in., $9.50; 1 in., $9.10; Extra for B.B. Chain, $1.20; Extra for B.B.B. Chain, $1.80.

ELECTRIC WELD COIL CHAIN B.B.

⅛ in., $16.75; 3-16 in., $15.40; ¼ in., $13.00; 5-16 in., $11.00; ⅜ in., $10.00; 7-16 in., $9.80; ½ in., $9.75; ⅝ in., $9.50; ¾ in., $9.30.

Prices per 100 lbs.

FILES AND RASPS

	Per Cent.
Globe	50
Vulcan	50
P.H. and Imperial	50
Nicholson	32½
Black Diamond	27½
J. Barton Smith, Eagle	50
McClelland, Globe	50
Delta Files	20
Disston	40
Whitman & Barnes	50
Great Western-American	50
Kearney & Foot, Arcade	50

BOILER TUBES.

Size.	Seamless	Lapwelded
1 in.	$27 00	$.....
1¼ in.	29 50	.....
1½ in.	31 50	29 50
1¾ in.	31 50	30 00
2 in.	35 00	30 00
2¼ in.	35 00	29 00
2½ in.	42 00	37 00
3 in.	50 00	48 00
3¼ in.	... ..	48 50
3½ in.	63 00	51 50
4 in.	85 00	65 50

Prices per 100 ft., Montreal and Toronto

OILS AND COMPOUNDS.

Castor oil, per lb.	.....
Royalite, per gal., bulk	27½
Palacine	30½
Machine oil, per gal.	54½
Black oil, per gal.	28
Cylinder oil, Capital	82½
Petroleum fuel oil, bbls., net	18

BELTING—No 1 OAK TANNED

Extra heavy, single and double		6½
Standard		6½
Cut leather lacing, No. 1		2 00
Leather in side	2 40	3 00

TAPES

Chesterman Metallic, 50 ft.	$2 00
Lufkin Metallic, 603, 50 ft.	2 00
Admiral Steel Tape, 50 ft.	2 75
Admiral Steel Tape, 100 ft.	4 45
Major Jun. Steel Tape, 50 ft.	3 50
Rival Steel Tape, 50 ft.	2 75
Rival Steel Tape, 100 ft.	4 45
Reliable Jun. Steel Tape, 50 ft.	3 50

PLATING SUPPLIES

Polishing wheels, felt	$4 60
Polishing wheels, bull-neck	2 00
Emery in kegs, Turkish	09
Pumice, ground	06
Emery glue	30
Tripoli composition	09
Crocus composition	12
Emery composition	11
Rouge, silver	60
Rouge, powder, nickel	45

Prices per lb.

ARTIFICIAL CORUNDUM

Grits, 6 to 70 inclusive	.08½
Grits, 80 and finer	.6

BRASS—Warehouse Price

Brass rods, base ½ in. to 1 in. rod	0 34
Brass sheets, 24 gauge and heavier, base	$0 42
Brass tubing, seamless	0 46
Copper tubing, seamless	0 48

WASTE

XXX Extra	.24	Atlas	.20
Peerless	.22½	X Empire	.19½
Grand	.22½	Ideal	.19
Superior	.22½	X Press	.17½
X L C R	.21		

Colored

Lion	.17	Popular	.13
Standard	.15	Keen	.11
No. 1	.15		

Wool Packing

Arrow	.35	Anvil	.22
Axle	.28	Anchor	.17

Washed Wipers

Select White .20 Dark colored .09
Mixed colored .10

This list subject to trade discount for quantity.

RUBBER BELTING

Standard ... 10% Best grades... 15%

ANODES

Nickel	.55 to .60
Copper	.38 to .40
Tin	.70 to .70
Zinc	.16 to .17

Prices per lb.

COPPER PRODUCTS

	Montreal	Toronto
Bars, ½ to 2 in.	$42 50	$43.00
Copper wire, list plus 10..		
Plain sheets, 14 oz., 14x60 in.	46 00	44 00
Copper sheet, tinned, 14x60, 14 oz.	48 00	48 00
Copper sheet, planished, 16 oz. base	46 00	45 00
Braziers', in sheets, 6 x 4 base	45 00	44 00

LEAD SHEETS

	Montreal	Toronto
Sheets, 3 lbs. sq. ft.	$10 75	$14 50
Sheets, 3½ lbs. sq. ft.	10 50	14 00
Sheets, 4 to 6 lbs. sq. ft.	10 25	13 50

Cut sheets, ½c per lb. extra.
Cut sheets to size, 1c per lb. extra.

PLATING CHEMICALS

Acid, boracic	$ .23
Acid, hydrochloric	.04
Acid, nitric	.11
Acid, sulphuric	.04
Ammonia, aqua	.15
Ammonium, carbonate	.23
Ammonium, chloride	.22
Ammonium hydrosulphuret	.75
Ammonium sulphate	.30
Arsenic, white	.14
Copper, carbonate, annhy.	.41
Copper, sulphate	.15
Cobalt, sulphate	.20
Iron perchloride	.62
Lead acetate	.30
Nickel ammonium sulphate	.20
Nickel carbonate	.32
Nickel sulphate	.22
Potassium sulphide (substitute)	.42
Silver chloride (per oz.)	1.30
Silver nitrate (per oz.)	1.25
Sodium bisulphate	.14
Sodium carbonate crystals	.06
Sodium cyanide, 127-130%	.38
Sodium hyposulphite per 100 lbs	8.00
Sodium phosphate	.15
Tin chloride	.80
Zinc chloride, C.P.	.30
Zinc sulphate	.10

Prices per lb. unless otherwise stated

Quality

CONFIDENCE

THE GREATNESS OF BRITAIN IS FOUNDED ON HER INTEGRITY.

HER STRENGTH SHE DERIVES FROM WORLD COMMERCE—HER ENDURANCE FROM CONFIDENCE.

THE WAR TEMPORARILY RETARDED HER CONSTRUCTIVE EFFORTS, BUT THE POWER OF THE CONFIDENCE OF THE PEOPLES OF THE WORLD WILL RESTORE TO HER IN MORE BRILLIANT FORM THE LEADERSHIP SHE HAS JUSTLY EARNED AND PROUDLY HELD.

CONFIDENCE IS THE FOUNDATION OF OUR POSITION IN THE WORLD OF IRON AND STEEL— OUR PROGRESS DEPENDS ON IT.

WE HAVE BEEN SUCCESSFUL IN THE PAST— THE FUTURE IS BEFORE US. TO ADVANCE, WE MUST CONTINUE TO MERIT THE CONFIDENCE OF THE BUYERS OF STEEL AND IRON PRODUCTS OF EVERY DESCRIPTION IN CANADA, AND WE ARE DETERMINED TO ADVANCE.

THE
STEEL COMPANY
OF
CANADA
LIMITED
HAMILTON   MONTREAL

Service

If interested tear out this page and place with letters to be answered.

INDUSTRIAL NEWS

NEW SHOPS, TENDERS AND CONTRACTS
PERSONAL AND TRADE NOTES

TRADE GOSSIP

Want a Machine.—The Telfer Biscuit Co.. Limited, 139 Sterling Road, Toronto, are in the market for a wood working machine, similar to the Elliott woodworker, made in Belleville.

Canadian Plant.—A new factory is being erected by the Coleman Lamp Company of Wichita, Kansas, at the corner of Queen East and Don roadway, for the manufacture of all kinds of portable lamps.

Sold Outfit.—The Hanson Van Winkle Co. have just sold to the Kelsey Wheel Co., of Walkerville, a large generator plant, including a 2,000 amp. dynamo, which they are using for electro-plating automobile wheels and accessories.

Extending at Preston.—The McKenzie Machinery Co., Preston. have plans out for extending the main shop an additional 125 feet. It is expected that work will be started at once as the extra space is urgently required.

Amalgamation of Plants.—An important tool company merger has been announced in the affiliation of the Mann Axe & Tool Co., Ltd., of St. Stephen, N.B., with the James Smart Manufacturing Company of Brockville.

Sold Property.—The Gurney Foundry Company have sold No. 740 and 742 Yonge Street, Toronto, but as yet the name of the purchaser is not disclosed. The sale price is reported to be around $90,000 and the lot has a frontage of 33 feet by a depth of 325 feet to Balmuto Street.

Selling Stock.—The Dillon Crucible Alloys, Ltd., are advertising in the Welland paper, giving to citizens an opportunity to buy a small block of stock, which will pay dividends of not less than eight per cent., at the price of $100 per share, cumulative preferred, with bonus of 50 per cent. common at $5 per share; for instance, $1,025 would buy ten shares preferred and five shares common.

Need Crane Runway.—The Cincinnati Iron and Steel Co., are in the market for a self-supporting steel crane runway, carrying a bridge, with two five-ton lifts. The crane runway is to be arranged so that later on a roof and siding could be put on, making it a completely enclosed warehouse. Length of crane runway to be either 500, 600 or 700 feet, height approximately 25 feet, span 85 feet, quoting a price on each.

Shipments Up. — The shipments of feldspar from Frontenac are now double what they have been in past years. The centre of the feldspar mining industry is now at Tichborne. Feldspar is being found at other places in this district, but the Frontenac County rock is said to be much better in quality. An American company has been developing an iron pyrites mine at Flower, Ontario, in Frontenac. The material is mostly sent to Hamilton.

Chatham's Water.—At a meeting of the Chatham Water Commissioners it was decided to submit a referendum to the ratepayers of Chatham at the earliest possible moment asking them to decide between Lake Erie and the River Thames as the permanent source of a civic supply of water, with the understanding that the city council will apply to the Railway and Municipal Board for permission to issue debentures for raising money to carry into effect the scheme selected.

Paid for One Week.—The Renfrew Machinery Company on Saturday last closed down for a period of two weeks, partly in order to provide their em-

TREATING IRON ORE BY NEW METHOD

Big Future is Assured for the Low-Grade Ores of the Dominion

Ottawa.—Satisfactory results have been obtained from the investigations carried out by Prof. Alfred Stansfield of McGill University into the reduction of iron ores by gases at low temperatures and with the electric furnace.

This research was assisted by a grant of $1,200 from the Honorary Advisory Council for Scientific and Industrial Research. On some of the findings, the council states, patents have been applied for by permit of the commercial development in Canada of the methods which will make the low-grade ores of Canada utilizable and of the greatest value.

Iron ore, iron and steel and their products are annually imported into Canada to the value of more than $150,000,000. Such methods of utilising Canada's low-grade ores would render the greater part of this importation unnecessary, in the opinion of the council, which, it states, if it had the funds available to do so, would make provision for such investigations as the importance of the subject warrants being undertaken.

ployees with the opportunity of taking a summer vacation and partly to take stock. In making the announcement with reference to closing down, the firm also announced that any of their employees who have been with the firm for three years or over would receive one week's pay during the holiday.

Take Toronto Lease.—The Keasbey Mattison Company, one of the largest manufacturers of asbestos and magnesia products in the United States, have leased for a long term of years the entire building at No. 25 Front Street East, Toronto, formerly occupied by the Colville Cartage Company. The parent plant of the lessors is at Ambler, Pa., and they have branches in eleven large American cities and a foreign branch at London, Eng. No manufacturing will be carried on here, the building being taken over for a Canadian distributing depot, under the resident managership of B. A. Simon.

Destroyed by Fire.—New Glasgow, N.S., had one of the most serious fires of years, when in a few minutes flames converted the main workshop of the Maritime Bridge Company into a mass of ruins. The fire started in the northern end of the large building where the riveting shop is located in a wooden annex. A strong wind was blowing from the south, and the flames worked back against the wind through the long shop. The roof of the whole structure was of wood, but the walls for the greater part were concrete. So quickly did the flames spread that almost before the alarm had ceased sounding the main building was doomed.

Increasing Plant.—Butterfield & Co., Division of the Union Twist Drill Co., Rock Island, Quebec, are increasing their office and shipping room capacity by the addition of a 2-storey brick building, 160x40 feet. J. J. Powers, of Worcester, Mass., who has the contract for their new United States factory, just across the International Boundary line in Derby, Lim, Vt., has received the contract for this building and expects to have it roofed over before snow flies. The necessity for the increase in both plants of the Butterfield Co. is due not only to the demands of domestic trade, but fully as much to the influx of orders from foreign buyers.

London's New Scheme.—The Public Utilities Commission, of London, Ont., in an endeavor to maintain the city's spring water supply, has secured a drill-

ing outfit and has commenced prospecting for new artesian wells on the pipe line road near the second cove. It is reported that there are large quantities of pure water available there, and if the claim is substantiated it is the intention of the commission to forthwith install deep well turbine pumps on each well. No separate pumping station would be maintained at the field, the machines being equipped with remote control apparatus, the scheme having worked out in a satisfactory manner at the Foster wells.

Departments Combined. — Announcement is made of a combination of the sales, purchasing, accounting and executive departments of the Reed-Prentice Co., Worcester, Mass.; Becker Milling Machine Co., Hyde Park, Mass., and the Whitcomb-Blaisdell Machine Tool Co., Worcester. This arrangement took place a week or so ago, and the main offices are now permanently located at No. 53 Franklin Street, Boston, Mass. The Canadian agencies of these firms remain as formerly, viz., Reed-Prentice Co., represented by the Canadian-Fairbanks-Morse; Becker Milling Machine Co., represented by the Rudel-Belnap Machinery Co., of Montreal, and the A. R. Williams Machinery Co., of Toronto; Whitcomb-Blaisdell Machine Tool Co., represented by the A. R. Williams Machinery Co.

Put in New Hammers.—Two 6,000-pound hammers and their equipment arrived at Canada Forge Companies plant in Welland. All preparations had been made for their arrival, work on the beds having been in progress for some weeks, and it is expected that this new section of the plant will be in operation by August 1. Automobile forgings will be manufactured, and this output, combined with that of the Billings & Spencer plant will bring Welland well up to the top in this particular line of industry. With

CZECHOSLOVAKIA AGENTS VISIT CANADA TO SEE WHAT TRADE CAN BE DEVELOPED

MEMBERS of the Czechoslovakia party, now in Canada, are receiving a good impression of this country as a place where trade can be developed. Many of the Canadian centres have groups of these people in them now. Marcel Zoltan, M.E., who has been making his headquarters in Toronto for some days, believes that there should be a good business developed between this country and the new kingdom that has been brought into being since the close of the war. M. Zoltan, speaking to

GETTING A BIT RESTLESS

Canadian Machinery to-day, stated that he had been much impressed with what he had seen in Canada, and also with the kindness and courtesy that had been shown to him wherever he went.

"The new country, Czechoslovakia," he said, "is made up of two peoples. The Bohemians, while to a certain extent an agricultural people, can more properly be described as an industrial race, while the Slovaks, who were in Hungary as it existed prior to the war, are more of an agricultural force. Before the war we made a lot of machinery, especially for textile purposes, heavy special machinery, and the chemical industry was also in a high state of development. Then we have the largest deposits so far known of radium. The coal mines are very important, also the deposits of gold and silver. What we must do now is to take care of the trade that used to go to Germany, and that is why we have been looking all over Canada to see what there is here. One of our parties is at Vancouver, another at Calgary, and one at Winnipeg. We have been gathering information and will report back to the home government, and after that I expect that some of us will come back to this country to see what can be done in the way of making permanent trade connections."

"What can Canada send you?" inquired Canadian Machinery.

"Well, that is something that I cannot answer right offhand," answered M. Zoltan, "and it is for that reason that we are looking around and asking questions, but there are certain kinds of salt for paper making, your cobalt and nickel, many lines of farm implements, wires and cables, that we have found here. I have had the opportunity of calling on several of the plants and found them all very anxious to do everything they could for our nation. In a short time we should be in a position to have for export from our own country special machinery, especially for metal working industries, special hoists, chemicals and dyes, textiles, etc."

The Czechoslovakia country is much the same as Canada, as far as climatic conditions are concerned, although it is hardly likely that they get the same extremes in the winter.

"There is one thing," stated M. Zoltan in conclusion, "and that is that the people of Canada do not appreciate the safety and security in which they live—the security of their homes and of their business places. In Russia it is very uncertain, although we believe that conditions will work out there for the best, and that political recognition will be given to the government there. But it is so different in this country. It is quiet, safe and pleasant, and the people of Canada should appreciate it."

The average banker is over forty. The hustling business man who borrows is usually under forty. Nature gives the young man ambition, ability, and willingness. Nature gives the middle aged man judgment, experience and conservation.

I Want Every Large Shop to Use "MORROW" Forged High Speed Drills

"The Ad-Writer"

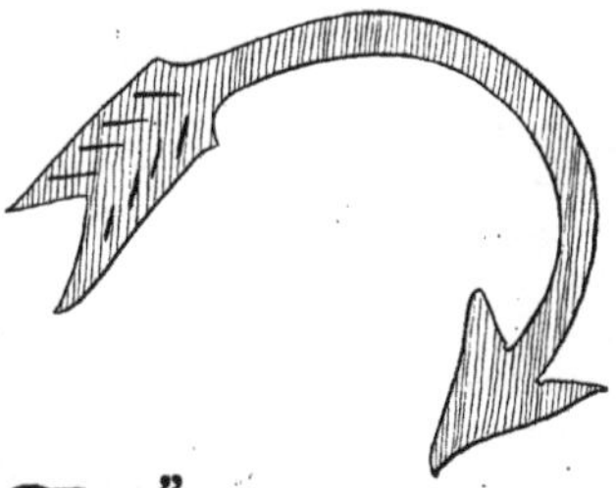

We claim "Morrow" Drills will give more holes with less regrinding—try out this statement at our risk. If *you* are not satisfied after you try them, you get your money back.

Buy from the Jobber

No Reliable Jobber Will Substitute

The John Morrow Screw & Nut Co., Limited
INGERSOLL, CANADA

Montreal
131 St. Paul St.

Winnipeg
Confederation
Life Building

Vancouver
1290 Homer St.

7 Hop Exchange, Southwark St., London, England

If interested tear out this page and place with letters to be answered.

the ever-increasing demand for cars, a ready market will be available. A good number of orders are already on hand and will be started on as soon as possible. Mostly skilled men will be employed.

MARINE

Kingston:—Owing to delay in dredging considerable trouble is being encountered by vessels in Kingston harbor. One result is that vessels loading at Oswego and compelled to go with only part cargo to avoid trouble.

Sarnia:—The "Harmonic" left for Detroit under her own steam to be dry docked for examination after being aground. The "Harmonic" was not seriously damaged, but she will be docked and the hull examined, also the propeller and tail-end shaft. Sahe will soon be in commission again.

Kingston.—The big dredge, "Kennqubair," which was bought from the Great Lakes Dredging Co. by the Dominion Government back in 1917, has been made ready for a voyage. Originally bought for work on the Hudson Bay Terminals, she has been lying in Cornwall dry dock for nearly three years. She will now be employed at Cape Tormentine, P.E.I., on the terminal there for the

Toronto:—The Dominion Shipbuilding Co., successfully launched the steamer "Floraba" from their yard at the foot of Spadina Ave. This is the second vessel for the Gulf Navigation Co., of New Orleans, the first having been launched about three weeks ago, and now ready for delivery. The vessel is of the usual dimensions for passing the Lower St. Lawrence Canals. A large and influential gathering witnessed the launching ceremony.

INCORPORATIONS

"The Jolliette Castings and Forgings, Ltd., is incorporated with effect from July 3rd, for the purpose of the manufacture in steel and iron products, and to act as importers and exporters for same. Authorised capital stock of $1,000,000, divided into 10,000 shares of $100 each. The chief place of business to be Montreal, P.Q.

Authority has been granted with effect from June 19th, for "The Anglo-Canadian Malleable Steel Manufacturing Co., Ltd.," to increase their capital from $50,000 to $200,000, such increase to consist of 15,000 shares of $100 each, and sub-dividing six unissued shares of the company now of the par value of

$1,000 each, into sixty shares of the par value of $100 each.

Notice is received of the incorporation of "The Belanger Foundry, Limited," with effect from June 11th, for the purpose of acquiring the firm of O. Belanger, Inc., and to engage in the manufacture of iron and steel in all its branches. The company to have a capital stock of $100,000, divided into 1,000 shares of $100 each. The chief place of business to be Montreal, P.Q.

PERSONAL

Mr. Harvey C. Rose, son of City Treasurer J. O. Rose, has been appointed as resident engineer, with headquarters in Guelph, in connection with the construction of Provincial Highways in Western Ontario. Mr. Rose will have as his section of the highway the road between Guelph and Hespeler, which has been designated as a provincial highway, and the road from Hespeler to Kitchener, by way of Kossuth; also the road from Kitchener to Galt, via Preston, and from Guelph to Puslinch Station, on the Brock road. The road from Guelph to Mount Forest and Owen Sound, has also been designated as a provincial highway and some construction work will be done on it this year.

One of the latest British patents applied for is a 10-ton hydraulic jack made with a 4.5 high explosive shell and an 18-pounder shell as the principal parts. The big shell forms the body and the other the ram. The pump plunger is out from a length of .303 rifle barrel. As a means of using up old war stock this invention is very ingenious and interesting.

THE LARGE LOCOMOTIVE BOILER
(Continued from page 80)

ing regard to their nominal tractive force and heating surface, and while there are many other conditions which have their influence on the hauling efficiency (such as the setting of the valves, the diameter of the blast-pipe, the weight on the driving wheels, and so on), the fact remains that it is now generally advisable to use a much larger boiler than that determined by the former methods of calculation. The most that can be said is that cylinder volume is an approximate guide, while the maximum size of boiler possible is determined by considerations which lie quite outside. These are the restrictions due to height, width, length, and weight. Generally speaking, the height of the chimney-top above rail-level must not exceed 13 ft. 6 in.; the width, or space available transversely between the tires of the driving wheels, so that it will be seen that these two limitations are the primary ones, and being interdependent are alone sufficient to fix the size of the boiler diametrically. Owing to difference in the vertical loading gauge, what

might be possible on one line might be out of-question on another, but it may be taken as a general rule that large-diameter boilers and large driving wheels cannot go together. The diameter of the one or the other must be kept down, and since the boiler capacity is of paramount importance, it may be preferable to use relatively small driving wheels, more especially as the tractive power for a given cylinder diameter and piston stroke will then be correspondingly greater.

As regards the length of the barrel and firebox combined, this must be such as to permit of the rigid wheelbase being short enough to enable the engine to traverse safely the shortest curve on the line over which it will be required to run. There is also an assumed limit to the length to which tubes can be usefully extended. The same may be said respecting the length of the firebox, as, although the objection to the use of long coupling rods is removed by placing all the driving wheels in front of the box and employing a supplementary carrying-axle underneath it, yet very long grates are difficult to fire efficiently. The condition as to weight depends upon the strength of the permanent way and bridges, but on all the more important main-line railways a reasonable allowance is now made for the probable development of the locomotive in this respect.

But it will be understood, no doubt, that there are certain other factors which have to be reckoned with in connection with the increase in boiler capacity. For example, in the case of express engines the limit to which a boiler can be advantageously enlarged for a given cylinder diameter is reached when the cylinders, owing to back pressure and the necessarily reduced port opening high speeds, can use only a comparatively small quantity of the steam generated. This limit will be determined by the particular type and arrangement of the valves and valve gear, by the respective areas of the steam and exhaust passages, and by the diameter of the blast-pipe, which details ought to be planned with a view to increasing the positive pressure on the pistons, without, however, reducing the compression below the amount necessary to absorb the momentum of the moving parts at the end of each stroke. It is not easy to adjust these matters to a nicety, but the big boiler, if provided with adequate grate area, is of itself a help in this connection, since it allows of a larger blast nozzle being used; and the latter, by diminishing back pressure, permits the cut-off to be later, and this, in turn, enables the engine to haul a heavier load at a given speed, or to take a given load at a higher speed, as may be required.

Nothing has been said about superheating, for the reason that it is not, as a rule, now deemed advisable to reduce the size of the boiler in consequence of the adoption of that excellent feature, the advantages obtained being by way of relative economy -in coal and water consumption.

Classified Opportunities

CANADIAN MACHINERY
AND
MANUFACTURING NEWS

Vol. XXIV. No. 4.   July 22, 1920

Electric Rocking Furnace of Novel Design

The Melting of Non-Ferrous Metals by Means of the Electric Furnace is a More Difficult Task that the Average Person Realizes —One Such Type is Herein Described.

By W. F. SUTHERLAND

WHILE the electric furnace has been used successfully for many years for the melting and refining of steel and iron, it is only of comparatively recent date that its use in the non-ferrous field has been attended with that success that one might reasonably expect. At first glance the problem of adapting electric heat to the brass foundry and other allied industries would appear simple and easily solved. This is not the case for the securing of a proper furnace design, one which will conserve the more volatile elements in the charge, and one which will melt economically, is not easily secured.

Although this is the case much research work has been done and the electric furnace, for brass and non-ferrous alloy melting, is rapidly replacing many other types of furnaces fired with some form of fuel. The reason for the present rapid growth of electric brass melting lies principally in the important saving of metals, especially zinc and lead, which the use makes possible. The electric furnace has certain other important advantages over older methods of melting.

H. M. St. John states in the Electric Journal for September, 1919, page 374, that: 'It has been found that a more uniform quality of metal can be produced in the electric furnace than in fuel fired furnaces operating under similar conditions, and that it is easier to produce an alloy of closely specified composition. The molten metal is also much cleaner and can be poured free from metallic drosses without the use of charcoal or flues of any kind. In general, these advantages are to be considered as inherent in properly conducted electric furnace operation, as a result of the greatly reduced loss of volatile metals and the elimination of contaminating combustion gases.

The saving of metal, otherwise unavoidably lost, is the principal economic advantage which the electric furnace is required to show in the melting of copper alloys, particularly yellow brass. It has been demonstrated, both in the laboratory and in the plant, that such a

saving is possible by virtue of the fact that the electric furnace can be tightly closed during the melting period and a neutral or reducing atmosphere maintained.

The production of perfect castings or billets, with the least possible number of defectives, depends in a large degree upon the use of metal at a temperature which conforms closely with that known to be most favorable for the work in hand. The electric furnace lends itself readily to exact temperature control and thereby enjoys an important advantage.

In general, the speed of melting is greater with electric furnaces than with fuel fired furnaces, because of their higher operating temperature and greater efficiency. The electric furnace can also be used in larger units than is commonly the case with fuel fired furnaces.

The cost of crucibles has always been a considerable item in the brass foundry. The electric furnace eliminates this item of expense, and while large fuel fired furnaces also effect the same saving

from a metallurgical point of view they are seldom as satisfactory as fuel fired crucible furnaces.

More favorable working conditions result from the use of the electric furnace, this tends to increase their efficiency as well as their comfort.

Much depends upon the wisdom with which metallurgical requirements have been met in the design of a furnace. The non-ferrous metal melting furnace is a different type of design altogether as compared with the steel furnace and has different obstacles to overcome. Steel melts at a high temperature and may be heated quite rapidly while even the apex in brass is somewhat volatile and oxidizes more readily than steel. Lead is more volatile than copper and oxidizes very easily. Tin is exceedingly volatile at molten brass temperatures. All copper alloys must be treated carefully during the melting process in order that loss of metal by oxidation and volatilization may be kept low.

Yellow brass for thin castings must be poured at a temperature not far be-

VIEW OF POURING SIDE OF FURNACE.

low its boiling point in order that the metal may be sufficiently fluid. At this temperature zinc which comprises 30 to 40 per cent. of the alloy, has a tendency to vaporise rapidly. So long as this metal is contained in a tightly closed furnace chamber, which can easily be done in the electric furnace, this tendency is counterbalanced by the vapor pressure of the metal which has already been vaporised and with which the furnace atmosphere is saturated. When the furnace is opened for pouring the metal, or for any other purpose, the vapor pressure is released and further quantities of zinc will escape from the alloy without restraint. If the heating has been perfectly uniform and all portions of the melt are at approximately the same temperature, the loss of zinc which ensues will constitute an unavoidable minimum. If the heating has not been uniform some portions of the melt will be at a temperature higher than the desired pouring temperature and such portions will lose zinc at a higher rate. If the lack of temperature uniformity is very great the loss which occurs after the furnace is opened and during pouring, will be decidedly excessive. In fact, some portions of the metal may be so seriously overheated during melting that the high vapor pressure generated within the furnace will force considerable quantities of zinc through crevices in the furnace structure. In some cases it may be practically impossible to keep the furnace chamber closed even to a reasonable degree.

Since electric heat is more costly than that derived directly from fuel, it is important that the thermal efficiency should be as high as can be obtained consistent with other requirements. A high thermal efficiency in electric melting, unless the heat is generated in the metal itself, requires a high temperature heat source, located as close as may be to the metal, under conditions which offer the least possible opposition to the flow of heat from the source to the metal. In addition, the walls of the furnace must be sufficiently thick and of high heat-insulating quality in order ‭ a heat may not be dissipated useless-‬ ‭ly. t‬

The electric furnace, to reap the full benefits of its economic possibilities, must operate in large units and must not use crucibles. The higher its speed of melting the better, so long as speed is not detrimental to metallurgical results.

The electrical characteristics of the furnace must be such as to make it a desirable load for the central · station company or the factory power plant. Its power factor must not be abnormally low and its power fluctuations must not be so violent as to endanger transformers and other electrical equipment, or to interfere with satisfactory service to other customers of the central station company who may be connected to the same power line.

In order to be thoroughly satisfactory the electric brass melting furnace should be highly flexible. That is to say it

should be able to operate under any desired foundry conditions, and to handle a charge of any ordinary nature and composition. The furnace should be readily applicable to any schedule; one-shift, two-shift, or three-shift daily operation. The melting of new metal, or composition ingot or course scrap or fine scrap or a charge with a high non-metallic content, should all be practicable in the same furnace, and it should be possible to change from one composition to an entirely different one in successive heats, without affecting adversly the operation of the furnace or the quality of the metal produced. With reference to these features the electric furnace must at least equal the performance of fuel fired furnaces. As a matter of fact, the electric furnace can be made more flexible than other furnaces and thereby gains a material advantage.

From an inspection of the above considerations as outlined by H. M. St. John, it will be seen that the proper design of an electric furnace for the melting of a non-ferrous metal is a somewhat difficult task and has occupied the minds of investigators for some little time. Several designs are now on the market, and it is the purpose of this article to describe the rocking furnace now made by the Detroit Electric Furnace Company, of Detroit, Michigan, who are licensed manufacturers under the U.S. Bureau of Mines Patents.

This furnace consists essentially of a cylindrical shell lined with refractory brick and provided, with a combined charging door and pouring lip. The ends are provided with electrode gear which serves to support the electrodes in position and to advance them into the interior of the furnace. Graphite electrodes are used and they pass through stuffing boxes located in the ends of the furnace so as to make a gas-tight joint. Water cooling is provided for the electrode clamps on the larger sizes of furnaces.

The shell is provided with bearing tracks for the supporting rollers and also with two gear rings which engage with the rocking mechanism. This is motor driven and the amplitude of the rocking motion is adjustable according to the requirements of the melt.

The efficiency of this type of furnace is high. This is due to the heat reclaimed from the walls of the furnace by a sort of regenerative action. In any stationary arc furnace all parts of the furnace chamber which are located above the metal line are superheated by the arc to a temperature considerably above that of the metal and the radiation losses from the roof and walls are high.

In the rocking furnace about four-fifths of the area of the furnace exposed to direct radiation from the arc is washed twice each minute by the cooler metal, which absorbs the excess heat in the brick work and thus increases the percentage of heat usefully applied.

The vigorous mixing which the metal in the furnace undergoes results not only

in uniformity of temperature but also in uniformity of composition. This is a particularly important feature in the melting of alloys high in lead, which are notoriously difficult to produce in homogeneous mixtures. In this type of furnace alloys containing as much as 25 per cent. of lead have been poured into homogeneous ingot and castings without any mechanical stirring other than that offered by the furnace itself.

Although rocking of the furnace is unnecessary during the first few minutes of the heat, while the metal is entirely solid, it is necessary to begin the mixing process before all the metal is melted since otherwise excessive volatilization of zinc will begin. The initial rock begun at this time is moderate in degree for otherwise pieces of solid metal will fall against the electrodes and break them. This initial rock is gradually increased until, when the charge is entirely molten, the full rock is employed. In this way the degree of agitation is always suited to the condition of the charge and metal losses are avoided without endangering the electrodes.   ‭ ‬

The makers state that as compared with combustion furnaces, the Detroit furnace reduces the melting cost by approximately 50 per cent., saves at least 75 per cent. of the metal wasted by oxidation and volatilization, and effects a very substantial improvement in foundry conditions and in the quality of the metal produced.

Production.

It also produces metal very rapidly because the use of the indirect arc makes possible a more rapid and efficient application of heat to the metal, without injury to the refractories, while the rocking motion of the furnace absolutely prevents injury to the metal by stirring it vigorously and thus maintaining it at a uniform temperature.

It is stated that a large rolling mill, melting 60-40 yellow brass in Detroit furnaces, has attained an average, over 500 consecutive heats of 50 minutes melting time per heat, accompanied by a net metallic loss of less than one per cent., while a smelting plant reports that it has melted 44,000 lb. of · red brass -borings per 24 hours, in one 2,000 lb. Detroit furnace, averaging 20 2,200 lb. heat.

In many cases yellow brass has been melted at the rate of 2,000 lb. in 40 minutes without the slightest injury to the metal.

Quality

The use of the indirect arc lends itself to exact temperature control, while the rocking motion of the furnace keeps the metal thoroughly mixed and results in remarkable uniformity of composition.

A foundry in a large automobile plant gives the following analysis as the average result from 13 consecutive heats of phosphor bronze :

	Copper	Tin	Lead	Zinc	Phos.
Av. composition	87.05	8.24	1.27	3.42	.017
Av. deviation	0.22	0.26	0.10	0.23	.004

Max. devia-
tion .... 0.75 0.60 0.36 0.54 .008

In addition to the uniformity of analysis, this plant found that the tensible
strength of the electrically melted metal
was 40,000 lbs. as compared with 35,-
000 lbs. for the same metal melted in
their combustion furnaces.

A smelting plant, melting a high-lead
bearing bronze, took analysis on four
consecutive 2,000 lb. heats from the
first 25 lb. ingot poured from the first
ladle of the heat in each case, as compared with the last ingot poured from
each heat, and reported the following
deviation in composition:

　　　　　　　Copper Tin Lead Zinc
Composition of al-
loy 66.00 4.50 26.0 3.0
Av. deviation. . 0.63 0.18 0.34 0.43

It has been found possible to use
cheap scrap, such as borings, chips and
grindings, in place of a large portion of
the new metal and composition ingot
used, without in any way impairing the
quality of the finished product.

Economy.

The rolling mill previously referred
to melted 500 consecutive 2,000 lb. heats
of 60-40 yellow brass at an average
electric energy consumption of 224 kw.
hrs. per ton: 50 consecutive heats showed an average of 215 kw. hrs. per ton.
These figures include preheats and re-
heats. The entire consumption of electric energy during the periods named.
On several occasions five or more consecutive heats averaged less than 200
kw. hrs. per ton. During this time the
net metallic loss from this high-zinc alloy varied from 0.7 per cent. to 1 per
cent.

A large smelting plant has melted red
brass (80 copper, 4 tin, 8 lead, 8 zinc)
in 2,500 lb. charges at 246 kw. hrs. per
ton.

Another smelter reports 256 kw. hrs.
per ton for red brass (83 copper, 3 tin,
4 lead, 10 zinc), made up from a charge
containing 2.0 per cent. oil and moisture:
the metal loss was 0.7 per cent. despite
the fact that oily yellow borings made
up 30 per cent. of the charge. The same
firm reports an energy consumption of

only 215 kw. hrs. per ton in melting borings of another red alloy (85 copper, 5
tin, 5 lead, 5 zinc).

It is stated that the refractory lining
of the furnace lasts for 600 heats or
more before extensive patching or renewals become necessary, a performance
which reduces this element of cost to
less than 25 cents per ton of metal melted.

Under favorable conditions the following cost figures have been attained with
the 2,000 lb. furnace:

　　　　　　　　　　　　Cost per ton
Electric energy (225 kw. hrs. @ 1c) $2.25
Refractories .... .25
Electrodes . ... .63
Labor 1.42
Incidentals . ... .50
Interest and depreciation .... .43
　　　　　　　　　　　　　　　　　————
Total melting cost $5.48

The two following tables giving data
obtained on actual runs may be of interest in the estimation of the furnace
performance:

Furnace No. 1.—General Aluminum
and Brass Co.:
Alloy melted—Copper, 85%; Tin, 5%;
Lead, 5%; Zinc, 5%.
Amount metal melted, 68,401 lbs.
No. of heats, 35.
Average heat, 1,954 lbs.
Total working time, 59.67 hrs.
Heats per day, six.
Av. kw. hrs. per ton, 280.
Av. electrode consumption per ton, 3.47
lbs.
Av. melting time per ton, 1 hr. 1min.
Gross metal loss, 0.44 per cent.
Av. pouring temperature, 2,084° F.
Character of charge—
Copper ingot .... 29.1 p.c.
Brass ingot .... 4.4 "
Foundry scrap .... 38.2 "
Borings .... 24.0 "
New lead .... 3.4 "
New zinc .... 0.1 "
New tin .... 0.9 "
　　　　　　　　　　　　　　————
　　　　　　　　　　　　　　100.1
Furnace No. 2.—Cleveland Brass and
Copper Rolling Mills:

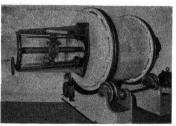

THE REAR END SHOWING ROCKING MECHANISM.

Alloy—Copper, 60%; Lead, 2.5%;
Zinc, 37.5%.
Amount of metal melted, 66,500 lbs.
No. of heats, 35.
Av. heat, 1,900 lbs.
Heats per 9-hr. day, 7.
Av. kw. hrs. per ton, 210.
Av. melting time per ton, 43 min.
Gross metal loss, 1.0 per cent.
Character of charge—
Light scrap .... 50 p.c.
Heavy scrap .... 50 p.c.

The manufacturers of concrete pipes
by the "Hume" centrifugal process—
Hume Pipe Company, South Africa, Limited—is described in a paper by Mr. W.
Wolstenholme recently read before the
South African Institution of Engineers.
The materials employed with the cement
are crushed stone from the mines in the
Witwatersrand area, and, for reinforcement, discarded winding ropes, of which
large quantities are available. The ropes
which are slowly rotated, are cut into
suitable lengths, annealed, and unstrand-
ed, and then woven into cages which are
placed in the pipe moulds first. The
concrete is thrown into the moulds until
the required amount is furnished, the
machine is gradually speeded up, and the
centrifugal force causes the material to
be evenly distributed throughout the
length of the moulds; water is also driven
out by this method. Six 4-in. or 6-in.
diameter pipes can be completed in eight
or nine minutes, and four 15-in. or 18-in.
pipes in fifteen minutes. Larger sizes
up to 60 in. diameter can be finished
in about twenty minutes. Sizes up to
9 in. diameter will withstand a working
pressure of from 300 to 350 lb. per square
inch, while larger sizes have been tested
up to 210 lb. per square inch. When subjected to excessive pressure these pipes
become a mass of pores, and have cracks
which allow the pressure to escape. The
pipes do not, therefore, fail in the manner of a cast iron pipe.

A recent communication to the
French Academy of Sciences, states that
coal has been discovered in the region of
Port Gueydon, French North Africa. This
coal appears to belong to very extensive
beds lying in the region mentioned.
Analysis taken have revealed the following composition:—Moisture, 1.05 to 1.15
per cent.; ash, 37.15 to 53.65 per cent.;
volatile matter, 8.05 to 12.27 per cent.;
fixed carbon and sulphur, 37.25 to 49.45
per cent; the calorific value varies from
3960 to 5390 calories.

According to preliminary figures published by the United States Geological
Survey, the total output of crude petroleum in the United States in 1919 was
about 377,719,000 barrels, as compared
with 355,927,716 barrels in 1918. The
home consumption last year was some
375,559,000 barrels, as compared with
380,242,153 barrels in 1918. Some 52,-
000,000 barrels of crude petroleum were
imported from Mexico out of a total output of 80,556,228 barrels in that country
last year, and handed to the refineries on
the coast.

Ideas of Interest to the Practical Mechanic

DIE FOR MAKING ROOF LOCKS

By John S. Watts

An interesting device for making the U-shaped Chicago roof locks, used in the construction of steel frame box cars, is shown in the accompanying sketch. The die here illustrated was designed to make the roof lock in one operation—cutting to length from the coil and bending to shape—at one stroke of the machine. The fixture is constructed as an attachment for a shaper, but may be modified and adapted to any other suitable machine.

The female holder A is made to bolt to the head of the shaper ram, after first removing the tool post and the clapper box. On the lower left side of the piece A, a circular shear block B is fitted, extending a little clear of the edge, and provided with six grooves, so that by reversing the block and rotating twelve cutting points are available. At the opposite or right-hand side an extension is provided to carry two special shaped pin blocks C, C, which have suitable grooves for guiding the wire when forming the roof locks.

The male die or stationary portion D is firmly secured to the shaper table and kept in alignment by the tongues shown, which fit into the slots in the table top. At the left of the piece D, and the forming block F is bolted in a slot that is located in a central position below the pin blocks C, C. On the entering end, the forming block is provided with a tapered slot that facilitates the entrance of the wire G. The gauge H is bolted to the side of the base, and is made from strap iron.

In operation, the wire is fed from the coil through the hole in the shear block E and up against the gauge H. The cutting die is set a little ahead of the forming block so that the wire is cut off slightly in advance of the bending operation. In this particular instance the shaper was operated at a speed of 20 strokes per minute, giving an output of about 1,200 pieces per hour. The spring of the wire, after forming, was sufficient for the piece to be carried back with the movable head, from which it was automatically ejected by an extending finger.

The report of the Lackawanna Steel Company for the year's second quarter, shows a profit of about $1,882,000, against a deficit of $233,000 in the same quarter of 1919.

MILLING A SLOT WITHOUT CLAMPING THE WORK

By F. M.

A rather interesting job is shown by the accompanying illustration, the pro-

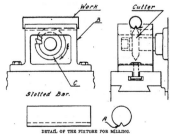

DETAIL OF THE FIXTURE FOR MILLING.

position being to mill the slot A along the rod without clamping the rod in any manner, it being understood that the groove is very shallow, about 1-13 inch wide by the same depth.

To accomplish this, a bracket B was made and the cutter C is caused to revolve beneath the upper face of this bracket in the manner indicated. In this bracket, a vee groove is cut, and the work is held in and pushed along this groove by hand in much the same manner as a board would be split with a circular saw.

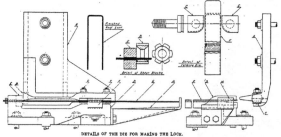

DETAILS OF THE DIE FOR MAKING THE LOCK.

What is a Hack Saw and How Should It Be Used?

Continuation of Article Dealing with Above Subject from Last Issue—One Should Be Able to Choose Wisely as Far as Hack Saws Are Concerned, After Perusal of this Matter.

The most important factor in determining the efficiency of a properly selected saw is the regulation of pressure so that the blade will give the greatest number of cuts in the shortest time and for the least money. Obviously, mere endurance without cutting effect, as exemplified in a saw that is worked too "gingerly," does not represent efficiency.

The cost of the saw is so small in comparison with that of the incidental labor, that even a slight increase in the cutting time represents a cost far in excess of what is saved by using the saw for a few cuts more.

The effect of the regulation of weight on the time per cut is clearly shown by the curves in Fig. 4.

On these curves each point marks the time required by a new saw to make the first cut. In each case the material cut was such as the blade was adapted for by design and the weight used is indicated by the figures at the bottom of the diagram.

To follow the performance of a single saw, let us take the curve for saw No. 254. In this experiment a new blade was properly placed in a Starrett Hack Saw Machine, a piece of three-inch machine steel put in the vise, the weight set at 20 pounds and a cut was made. The time required was 35 minutes, which gives us the first point on the graph. Another blade was then placed in the machine, the pressure increased to 25 pounds and another cut made. The time required was 30 minutes. With 30 pounds weight a new saw completed the first cut in about 24 minutes. Another saw was inserted, the weight increased to 40 pounds, and the time of the first cut was reduced to 18 minutes. The next increase was to 44 pounds, which is the weight recommended on the Hack Saw Chart for this particular saw, and a new blade made its first cut in 11 minutes and 30 seconds, which is slightly better than a good average time for a first cut on this size and class of material. The weight was then increased to 50 pounds and subsequently in increments of 10 pounds to 70 pounds, a new blade being used each time, and the time of the first cut noted. Connecting these points gives us the curve AB.

While the time per cut continued to decrease as the weight increased, the saw in each case was actually cutting at a destructive rate when the weight was increased beyond the amount recommended for that particular blade. For instance, the curve for saw No. 256 shows a decrease in time for all weights from 65 to 100 pounds, but, had any of the saws used been tested for the number of cuts per blade and general efficiency, as well as for the time per cut, it would have been seen that when the weight rose above 70 pounds on the first cut, the life of the saw was considerably shortened.

It should be borne in mind, that, as stated before, the ability to cut in the shortest time is but one of three objects that determine the value of a Hack Saw, and, therefore, when time only is considered as of value, the chances for a loss of efficiency are exactly two out of three. In other words, to test a Hack Saw by cutting with it at a rate that is known to be destructive, is to disregard deliberately two of the three factors that determine economy in this field of shop work. For this reason, any destructive cutting speed, is not only misleading in character, but dangerous in the selection of saws.

The maximum efficiency is to be found in the saw that "cuts quickest and lasts longest"; that combines cutting efficiency with endurance.

The gradual lowering of efficiency as a saw is used and the effect of successive increases in the pressure are very clearly displayed in Fig. 5. This is not presented to show just when increases should be made, but rather to emphasize the rate of increase in time required per cut, or corresponding decrease in efficiency, when the pressure is permitted to remain constant, and the immediate effect of increasing the pressure. The curve for 44 pounds shows how rapidly the saw was running beyond the limit of efficiency, while that for 64 pounds makes clear how the tendency was immediately restrained by the application of more weight. An example of efficient control of pressure is shown in the curves representing the entire life history of saws shown in Fig. 8.

Fig. 5 further illustrates the effect of weight or pressure on the rate of cutting and the efficient endurance of the saw, identical blades being used on the same material in this test and different

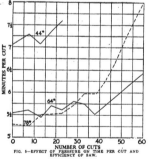

FIG. 5—EFFECT OF PRESSURE ON TIME PER CUT AND EFFICIENCY OF SAW.

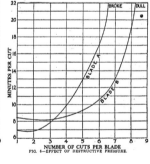

FIG. 6—EFFECT OF DESTRUCTIVE PRESSURE.

weights employed, the starting weight remaining constant throughout the test in each case.

In this experiment one of the weights used was below that recommended for the blade employed and the result is indicated by the upper curve in the chart, marked "44-lb. weight." The second solid line shows the performance of a similar saw using the proper weight of 64 pounds, and the dotted line, the result when a weight of 78 pounds, or a slightly excessive pressure, was employed.

Insufficient weight resulted in an excessive time per cut, 7 minutes and 15 seconds being required for the first cut, an amount of time in excess of what is considered a good average cutting speed on the class of material used in the test.

Better Too Much Weight Than Too Little

From this chart two conclusions may be drawn; First, it is better to exceed the proper weight a little at the outset than to use too little pressure: Second, no matter how nearly correct the weight is at the outset, after a certain number of cuts have been made, the pressure must be increased, not only for the sake of reducing the time per cut to a point within the limits of efficiency, but also to prolong the life of the saw. The effect of a slight excess in weight, not only on the life of the saw, but on the average time of all cuts made before the blade is dulled beyond all use, is clearly shown by the two lower curves in Fig. 5. For the first 15 cuts the advantage lies clearly with the heavier weight, but from that point on, it is apparent that the first saving in time was made at the expense of the general average for the entire series of cuts.

An example of what an excess of pressure will do is shown in Fig. 6. In this case, both saws were pushed to the limit from the very start by applying a pressure far in excess of that recommended, and one which it was known in advance would force the blades to cut at about twice their normal rate. Blade "A" was weighted 10 pounds more than "B" and was destroyed in one and a half less cuts. In each case there was little or no increase of time for the first three or four cuts. Beginning with the fifth cut, however, the time per cut mounted rapidly, with the result that one saw broke after six and one-half cuts, while the other was worn smooth in slightly more than eight cuts. While a hack saw must be made to withstand a great amount of abuse, there are, nevertheless, limits beyond which it will not go; and where a saw is forced to cut at a greatly excessive speed, the user must decide for himself whether or not the gain in time per cut offsets the loss in saws, spoiled stock, etc. So far as its value as a test is concerned, however, the practice of using excessive or destructive pressure not only has nothing to recommend it, but is actually dangerous and misleading.

Effect of Insufficient Pressure

The conditions in the tests charted in

Fig. 7, were exactly the reverse of those shown in Fig. 6, a weight being applied in each case which was considerably less than that recommended for the particular blade used in the test. Not only was the time per cut far in excess of what should have been required to cut the class of material on which the test was made, but the life of the saw was destroyed almost as rapidly as when too much weight was used. In each case the teeth of the blades were destroyed by slipping and sliding over the work, rather than by cutting. Blade "A," which had the greater weight (and shortest life) in the preceding test, had the least weight here and was worn dull in about five less cuts than were made by blade "B." These results make it evident that using too little pressure is almost as inefficient and costly as using too much, while the practice of using insufficient pressure has not even the doubtful advantage of saving time at the expense of the blade and stock, as is the case where too much pressure is employed.

Effect of Proper Regulation of Weight

In this test, Fig. 8, each of the blades was started with the weight recommended for that saw by the Hack Saw Chart. With the exception of Saw No. 256-C, the time required for the final cut, before the saw was completely ruined, was less than that for the first cut, and, in practically every case, an increase of weight was followed by a recession of the curve to a new low level. While the number of cuts per blade was unusually high for each test, the curve is typical of the performance of a Starrett saw under proper regulation of weight or pressure when cuting a material for which it was designed.

There is a slight tendency in some shops and by some manufacturers to recommend the use of a flexible blade in power saws, for the same reasons that are advanced in hand sawing. The two

bottom curves on the chart show the comparative efficiencies of the all-hard and the flexible, or "soft-back," blade on the same material.

Performance of Flexible and All-hard Blades Compared

Saw No. 250-B is a 14-pitch, 22-gage, flexible or "soft-back" blade. The starting pressure used was 24 pounds. Saw No. 255-B is a 12-tooth, 18-gage, all-hard saw, designed for use in a medium power machine under a starting pressure of 44 pounds. Both are designed to cut machine steel. The comparative efficiencies, under the conditions of the test, and the conclusions to be drawn regarding the use of the flexible blade in a power frame, are clearly indicated by the two curves. (See Fig. 8).

From the very first cut, the "all-hard" blade, designed for use in machines, required less time per cut than was needed by the flexible blade; while the total number of cuts made by each blade was practically identical and both saws ultimately failed in almost the same manner. The "hump" in the curve for the flexible saw might have been avoided by the use of greater pressure were it not for the fact that its very flexibility militated against such a step, for the reason that had extra pressure been put on while the teeth still retained something of their first keenness, the blade would have in all probability "buckled" to such an extent that it would have been impossible to make a smooth, straight cut. It, therefore, appears that the use of flexible saws in power machines is unavoidably attended by a loss of time in cutting speed without any corresponding gain in the life of the blade.

Improper vs. Proper Starting Weights

Fig. 9 is especially interesting because it compares the results of starting with too light a weight, combined with proper subsequent regulation, with those obtained by using a slightly ex-

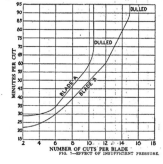

FIG. 7—EFFECT OF INSUFFICIENT PRESSURE.

cessive weight to start with and then regulating it properly as the number of cuts progresses.

The first blade tested was a No. 255-C, used in a medium power machine on 1¾ x 3 inch machine steel, the weight being taken as 24 pounds, which is less than that recommended on the chart. Almost from the first, the time per cut was excessive until about the 30th cut, when the weight was increased to 32 pounds, with a resultant drop in time per cut from 13 minutes and 35 seconds to 10 minutes and 35 seconds. However, the mischief had been done by cutting with too light a pressure, and the blade dulled so much by sliding over the work that it was not until the weight had been increased to 48 pounds, after nearly 120 cuts had been made, that the time per cut dropped to within the limits of efficiency. After 180 cuts, the time again rose to an excessive amount, and the blade failed after making the remarkable total of nearly 260 cuts.

In the second test, the weight recommended for the saw used, No. 256-C, is 43 pounds, whereas, that recommended for No. 256, or 64 pounds, was employed. The blade was used in a heavy machine to cut 2½ x 2½ inch machine steel, the square area cut being a little greater than that cut by No. 255-C.

In this case the blade, while over-weighted at the start, cut well within the limits of efficiency for nearly 40 cuts, rising during the next 20 cuts to a maximum time of 8 minutes and 45 seconds. At this point the pressure was increased to 78 pounds, and the time per cut reduced to about 5 minutes and 30 seconds. At no time during the next 240 cuts did the time rise above 8 minutes. When the blade failed it had cut a total area of 1742 square inches, or a trifle over twelve square feet as compared with the 1365 square inches, or nearly nine and a half square feet, which were cut by the No. 255-C—both of which are truly remarkable performances.

It may be worth while, at this point, to call attention to the fact that it is the number of square inches cut by two saws, rather than merely the number of cuts made by each, that determines their relative efficiencies. It often happens in actual practice that it is inconvenient to test all saws on exactly the same size of material and when this is the case, so long as the different stocks cut are near enough in size and shape to require the same type of blade, according to the recommendations in the Hack Saw Chart, the purpose of the test may be served by converting the number of cuts made by each blade into the total square inches cut, and making deductions or conclusions on that basis.

Comparing the average time per cut for each of the series, and noting especially the comparative "flatness" for the curve for Saw No. 256-C and the greater percentage of cuts it made at a less time than that required by No. 255-C, it is evident, as previously stated, that, of two evils, it is far preferable to use a slightly excessive weight at the outset than to use too little pressure.

The graphs in Fig. 10 above show characteristic curves for average performances, as regards the number of cuts, and also show in two cases the effect on the time per cut of using a saw under conditions for which it was not designed, and with an excessive starting pressure. Saw No. 112 is intended for use in a hand frame, and in any case should be started with an actual weight, or pressure on the blade, of 12 pounds. The result of starting with a 24-pound weight is indicated by the curve. Saw No. 256 was started with the proper weight, but the pressure was increased at an unnecessary rate and the life of the saw greatly shortened.

Effect of Lubricant

Fig. 11 shows vividly the comparative efficiency of a blade when used with compound and when cutting dry. The blade used in each case was a Starrett No. 250 and the entire series of cuts for all blades was made on the same bar of metal. A flexible back saw was chosen because, in the test, the abuse of the blade was intentional in the first two instances; and had an all-hard saw been selected, the probabilities are that it would have failed even sooner, especially as the strokes per minute, when cutting dry, were excessive. All these blades were started under the same pressure of 24 pounds. No. 1 saw, whose performance is shown by the line AB, was run without lubrication at 65 strokes per minute, an excess of 15 strokes over the recommended speed and failed when partly through the fifth cut. No. 2, curve CD, was run "dry" at 100 strokes per minute, or double the proper speed, and failed when about half way through the third cut. No. 3, used with compound at 100 strokes per minute, completed 50 cuts at a fair average speed, and was cutting at its apparent maximum efficiency under 56 pounds of pressure, when the test was discontinued.

The importance of lubrication, in its effect on the time per cut and the number of cuts per blade, is clearly shown here. Just to what extent the excessive speed employed with the blades that were run "dry" contributed to their early failure is none the less unmistakable. It may be said in this connection that, when cutting dry, an excess speed of as little as ten strokes per minute will ruin a saw by drawing the temper.

Cutting Costs

Naturally, all comparisons of saws or methods of use of saws, resolve themselves into a consideration of dollars and cents — the ultimate measure of Hack Saw economy. The cost of the saw, its cutting efficiency and endurance, the labor cost and overhead chargeable to its use, must all enter into the comparisons. Attention must be centered upon the reduction of labor cost through the use of machines and the maintenance of saw efficiency and endurance. All this has been carefully considered by manufacturers in the development of hack saw machines and in the design of various types of blades, but perhaps the most practical step was the standardization of hack saw practice through the medium of the hack saw chart.

The cost of the saw is always small —practically negligible—as compared with the value of the time of a man to use or tend it. Seldom, in fact, does the saw cost as much as an hour's ordinary wage. The total difference in cost between the best and the poorest saw is so slight as to be hardly discernible when reduced to the cost per cut, or even to the cost per hour of its use-

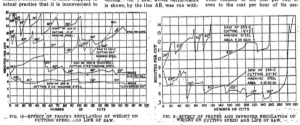

ful life. For comparison with hourly labor cost, it is convenient to divide the cost of the saw by the hours it endures and thus get an hourly rate.

From the accompanying table, page 48, which gives the "Cost per Hour for Saws of Different Costs and Endurance," it will be noted, for instance, that a blade costing 20 cents and cutting for 20 hours, would be chargeable at only 1 cent per hour. At a cutting rate of, say 10 cuts per hour, this would make the cost per cut for the blade but one-tenth of one cent. Even though a blade costing 20 per cent. less could be found that would show exactly the same efficiency—an obvious impossibility — the saving per cut would be only one-fifth of one-tenth of a cent, or one fiftieth of a cent. The chances are, however, that a 20 per cent. reduction in the cost of the blades would show as much or even greater reduction in efficiency, so that the cost per cut would actually be increased.

The most careless comparison of saw

of 5 hours, or a total of 35 cents, the labor and overhead charge being 30 cents as before, would cost 5.1 cents per cut. From these two examples it is again made manifest that the cost of the saw is a relatively small factor, and that in order to keep the labor and overhead down to the minimum, it is of the utmost importance to use only the best blades. With present labor costs it is better to discard a blade too soon than run it too long.

The sum and substance of all hack saw efficiency is—how long should it take to cut any particular piece of work, and how many cuts should a blade be expected to make before it is dulled?

As has been shown already, the characteristics of the metal which is being cut is the most important factor in determining how long it will take to cut it and, as every machinist knows, no two pieces of steel or iron are exactly alike as regards hardness, crystalline structure, etc. Therefore, no hard and fast limits regarding the cutting time

of any individual blade may be, the average time required to cut any given material, and the average number of cuts any blade will deliver, will in the course of the year's work be found to be very near the figures given. In power sawing, as in hand sawing, though in a lesser degree, the skill and attention of the operator will affect the efficiency of the blade. A careful man, who observes the recommendations of the manufacturers, will have little difficulty in obtaining better than the average time per cut and a greater number of cuts per saw, while a careless or indifferent operator may lag far behind the average.

Extended Tests Necessary

Comparisons, while odious, are nevertheless often profitable in the deductions which may be made from them. The following test, for instance, shows how necessary it is that any test be carried over an extended period and a large number of cuts before any final conclusion is drawn.

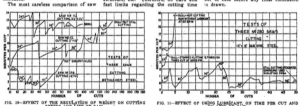
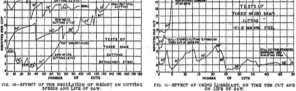

FIG. 10—EFFECT OF THE REGULATION OF WEIGHT ON CUTTING SPEEDS AND LIFE OF SAW.

FIG. 11—EFFECT OF USING LUBRICANT ON TIME PER CUT AND ON LIFE OF SAW.

costs with prevailing labor costs emphasizes most forcefully the fact that it is "penny wise and pound foolish" to buy anything but the best in the way of hack saws. It stands to reason that any standard product that is cheaper than the dearest is poorer than the best. Therefore, it is actually economy to get the best by paying the price.

Cost Per Cut Only True Basis

Ultimately, all expenses must be reduced to the cost per cut, as the true measure of efficiency and output. This may be readily determined from the accompanying tables for "Combined Cost per Hour for Saw, Labor and Overhead" on page 49. Thus, for instance, from Table B, 7 cuts per hour for a period of five hours at a labor and overhead charge of 30 cents per hour, the blade cost being taken at 15 cents, represented a cost for these items of 4.7 cents per cut. This figure is derived as follows: From Table B the cost per hour, for a five-hour period, of a saw costing 15 cents and a labor and overhead expense of 30 cents per hour, is 33 cents. Multiplying this by 5, the number of hours, we get $1.65; dividing by 35, the total number of cuts made during the period, we get a cost 4.7 cent per cut. A blade costing 30 cents, making an average of 7 cuts per hour for a period

of various saws nor their life can be given nor can it be said with certainty, just how long it will or should take to cut any given piece of material. We can, however, take the average of as many cuts of each class of materials as possible and pay that, if such of the factors as are under the control of the operator are made to conform to the recommendations given in this treatise, this average represents approximately the time in which a certain material should be cut and, in the same manner, we can say that it is reasonable to expect that a certain blade will deliver approximately so many cuts before becoming dull.

Some Average Performances

It must be remembered that the following figures are only averages, and because of differences in the material being cut. differences which only a chemical analysis could reveal, the very first saw that is compared with them may show a great many more cuts during its life, or may fail completely before it has made half the "average" number of cuts. On any given material, it may require half as long again per cut, or it may make a great many more cuts than the "average" blade and in half the time. But however far from the "Tables of Averages" the performance

Both makes of blades were run at 92 strokes per minute with a starting pressure of 44 pounds and a maximum pressure of 132 pounds. The greatest number of cuts was made with an average weight of 80 pounds.

On the face of it, this test clearly demonstrated the superiority of Blade "A," but following the test through, step by step, shows that Blade "B" put up a much more creditable performance than the final result would indicate, and that if the test had been stopped, after two blades of each manufacturer had been used, the result would have been reversed. When four saws of each brand had been tested, the results were in favor of Blade "A" to a much greater degree than after nine of that make had been used, but had the test stopper there, any conclusion as to the comparative merits of the blades would have been decidedly misleading. In other words, it was only by carrying out the test for a considerable number of cuts that a fair conclusion could be drawn.

Hack Saw Practice

Because a hack saw is so simple a tool, it is one of the most abused. Most of the waste in breakage, and most of the blame for inefficient cutting, is due to the inexperience of the user rather

Total Hours Used	Net Cost of Blade							
	$0.05	$0.10	$0.15	$0.20	$0.25	$0.30	$0.35	$0.40
5.	.01	.02	.03	.04	.05	.06	.07	.08
7.5	.0067	.0133	.02	.0267	.033	.04	.0467	.0533
10.	.005	.01	.015	.02	.025	.03	.035	.04
12.5	.004	.008	.012	.016	.02	.024	.028	.032
15.	.0033	.0067	.01	.0133	.0167	.02	.0233	.0267
17.5	.0028	.0057	.0086	.0114	.0143	.0171	.02	.0228
20.	.0025	.005	.0075	.01	.0125	.015	.0175	.02
22.5	.002	.004	.006	.008	.01	.012	.014	.016
25.	.0017	.0033	.005	.0067	.0083	.01	.0117	.0133
30.	.0014	.0029	.0043	.0057	.0071	.0086	.01	.0114
35.	.0013	.0025	.0038	.005	.0063	.0075	.0088	.01
40.	.0011	.0022	.0033	.0044	.0055	.0067	.0078	.0089
45.	.001	.002	.003	.004	.005	.006	.007	.008

TABLE A—COST PER HOUR FOR SAWS OF DIFFERENT COSTS AND ENDURANCE.

than to a lack of quality in the blade. Even a little instruction will be thoroughly worth while in the interests of hack saw economy.

In hand work, see first of all that the blade is well strained in the frame, with the rake of the teeth such that it will cut on the forward stroke. A flexible-back blade, because it lacks natural rigidity, should be strained tighter than an "all-hard" saw. A slack or a light blade has a much greater tendency to buckle and break than a taut or heavy saw, and will prevent the user from doing as much or as good work. The tendency of a saw to "drift," that is, to cut at an angle, depends on its gauge, on the skill of the mechanic and on the tension of the blade. The saw should be taut, but not over-strained. A properly strained blade, when "thumbed," gives a clear, humming note, which, once heard, is easily remembered. In using a flexible blade, the tension should be increased while cutting, as the blade stretches while in use.

Rigidity of Work Essential

Rigidity in the material to be cut is scarcely less important than proper tension on the blade. If possible, lock the work securely in a vise. In cutting sheet metal, if the gauge and size will permit, place the work in the vise so that a flat surface, rather than an edge, is in contact with the saw. If this is impractical it is well to put a light strip of wood on either side of the sheet of metal and cut through both the wood and metal at the same time. The angle at which any work is placed in the vise has a direct effect, not only on the time per cut but on the number of cuts the saw will make. This is a subject, the importance of which is often overlooked, and it may be worth while to state here that the proper "cornering" of the material to be cut is essential to economy in this class of work. Always place the work in the vise so as to provide as great a bearing for the saw as possible. That is, set the work, if a piece of structural steel, channel or similar material, so as to engage the maximum number of teeth throughout the cut. The diagrams as shown illustrate the proper manner in which to "corner" certain common shapes of material. When placing work in a vise, the tendency of coarse teeth to "straddle" must be borne in mind, and also that the fewer the teeth engaged at any particular moment during the cut, the greater the strain on each tooth and consequently the greater likelihood of stripping the blade.

Three Fundamentals

There are three fundamentals to be observed if efficiency is to be obtained in sawing with a hand blade. Be sure that the blade cuts, rather than slides or rubs. Always lift the saw on the return stroke. Don't crowd the work—40 to 50 strokes per minute should be the maximum.

Start the cut slowly, using the same motion as in filing. Put on enough pressure to make the blade cut and not slide or slip over the metal. This point is most important. Too little pressure in the first stages of a cut will take more out of a saw than fifty cuts made under proper conditions. By starting with a very light pressure, allowing the saw to rub rather than cut the work, and then increasing the weight at a destructive rate, once the first sharp edge is worn off the blade, it is possible to so completely ruin a saw that it will not make one full cut on a 2-inch round. Slipping or sliding the blade over the work simply glazes the cutting edges of the teeth and makes the saw blunt. If there is scale on the material to be cut, more pressure will be required than if the surface is smooth and clean. On the other hand, care must be taken not to use so much pressure that the teeth will engage the work too rapidly, or stripping of the blade may result. At the end of the forward stroke lift the blade slightly to avoid dragging or rubbing the teeth on the stock during the return stroke. Pressure during the return stroke is a frequent cause of premature loss of efficiency in hand hack saw blades.

Generally speaking, hand hack saw blades should not travel faster than 50 strokes per minute; thirty-five to forty strokes are even better; Blades for hand frames are comparatively thin and their teeth small; therefore, care must be exercised to avoid overheating the blade by too rapid cutting, and so drawing its temper. Too fast a stroke also almost inevitably results in dragging on the return stroke. In hand sawing it will generally be found worth while, after the first few strokes, to retighten the blade in the frame. This takes up any stretch that may have occurred, prevents drifting, and tends to produce better cutting. When using flexible or "soft-back" blades it is absolutely essential that the saw be re-tightened during use.

When starting on the corner of a square bar, it is worth while to notch it with a file, so that at least two teeth can engage at one time. Otherwise the teeth may straddle the metal, in which case there is a great tendency for them to strip. This danger also exists in sawing through tubing or conduit, hence a fine pitch blade should be used. When sheet metal is being cut, it is preferable, where possible, to saw along the flat surface rather than across the edge, thus preventing vibration as well as

Hours Used	Cost of Labor and Overhead per Hour Chargeable to Operation of Sawing						
	.10	.20	.30	.40	.50	.60	.70
5	.13	.23	.33	.43	.53	.63	.73
7.5	.12	.22	.32	.42	.52	.62	.72
10	.115	.215	.315	.415	.515	.615	.715
12.5	.112	.212	.312	.412	.512	.612	.712
15	.11	.21	.31	.41	.51	.61	.71
17.5	.1186	.2186	.3186	.4186	.5186	.6186	.7186
20	.1175	.2175	.3175	.4175	.5175	.6175	.7175
25	.116	.216	.316	.416	.516	.616	.716
30	.115	.215	.315	.415	.515	.615	.715
35	.1143	.2143	.3143	.4143	.5143	.6143	.7143
40	.1138	.2138	.3138	.4138	.5138	.6138	.7138
45	.1133	.2133	.3133	.4133	.5133	.6133	.7133
50	.113	.213	.313	.413	.513	.613	.713

TABLE B—COMBINED COST PER HOUR FOR SAW, LABOR AND OVERHEAD (Net Cost of Blade .15).

Hours Used	Cost of Labor and Overhead per Hour Chargeable to Operation of Sawing						
	.10	.20	.30	.40	.50	.60	.70
5	.14	.24	.34	.44	.54	.64	.74
7.5	.1266	.2266	.3266	.4266	.5266	.6266	.7266
10	.12	.22	.32	.42	.52	.62	.72
12.5	.116	.216	.316	.416	.516	.616	.716
15	.1133	.2133	.3133	.4133	.5133	.6133	.7133
17.5	.1114	.2114	.3114	.4114	.5114	.6114	.7114
20	.11	.21	.31	.41	.51	.61	.71
25	.108	.208	.308	.408	.508	.608	.708
30	.1067	.2067	.3067	.4067	.5067	.6067	.7067
35	.1057	.2057	.3057	.4057	.5057	.6057	.7057
40	.105	.205	.305	.405	.505	.605	.705
50	.1044	.2044	.3044	.4044	.5044	.6044	.7044

TABLE C—COMBINED COST PER HOUR FOR SAW, LABOR AND OVERHEAD (Net Cost of Blade .20).

avoiding stripping by bringing the greatest possible number of teeth in contact with the work at one time. For work which is not rigid or has a spring to it, the flexible back is the logical blade to use. It is particularly valuable on intricate work, or on work which is difficult o' access.

Cornering Work

In either hand or power work, the efficiency of the saw, when cutting angles, shapes and similar material, is directly affected by the way the material is, "cornered," or placed in the vise. A fundamental is to always start the work with the least angle facing the thrust of the saw teeth. Disregarding this point increases the tendency of the saw to strip. Too much stress cannot be laid upon the necessity of keeping the work rigid. In cutting small work with a hand saw, do not lay it across the knee, or merely hold one end on a bench. Put it in a vise. If there is no vise obtainable, drive a couple of nails into the bench and jam the piece to be cut between them, cramping it so that it will be held firmly.

To tire is human. It is, therefore, natural that as a man's efficiency is diminished by continued effort, neither speed nor quality of work is likely to be maintained. Hand work will not be so uniform nor will a hand blade last so long as one used in a power machine. More important, however, than the difference in quality or quantity of work produced by the two methods, especially while the present high wages prevail, is the fact that no man can be thriftily employed in hand sawing except on transient and odd jobs.

The machine never tires, assures uniformity in cutting, an output in excess of human effort; and, because it never requires the undivided attention of a man, costs less to operate where reasonably continuous work is to be done.

Importance of Proper Blade Selection

The hack saw chart naturally plays a very important part in the successful operation of saws in a machine. The careful selection of blades for different types of machines, and for different materials, makes it easy, where the recommendations of the chart are followed, to avoid picking too heavy a blade for a light machine or too light a blade for a heavy machine, and insures the proper saw being used for the material to be cut. If a blade is too light, it will tend to "drift," and will break if the pressure is not carefully regulated.

In power work, as with the hand frame, the same fundamentals must be observed if efficiency and economy are to be obtained. Most machines are made to cut upon the "draw stroke." This should be remembered when placing the saw in the machine; the rake of the teeth should always be in the direction the cut is to be made. In a "push stroke" machine, made to cut in the reverse direction, the blade should be placed with the rake of the teeth forward, while in machines cutting on the "draw stroke," they should point toward

the shaft. Retighten the blade after a few strokes have been made.

As in hand sawing, the cut should be started with comparatively light pressure, so that the teeth will not "bite" too rapidly and thereby be stripped. As has already been stated, too much weight put upon a new saw will destroy it quickly—while pressure insufficient to prevent the saw from slipping or sliding without cutting will dull the blade rapidly. It is good practice to relieve the blade of part of the weight of the frame during the first stroke or two, by holding it up with the hand, after which it may be released. Otherwise the teeth may "straddle" the metal and there is a great tendency to strip the saw. In sawing tubing or conduit, a fine-tooth blade should be used, not only because of the comparative softness of the metal, but because the small cross-sectional area will permit the teeth of a coarse saw to "straddle" the walls of the tube.

Increase of pressure to compensate for the wear of the blade is absolutely essential to economy in hack saw work. The common tendency among machinists is, not only to use too light an initial pressure or weight, but to practically ignore the necessity for increasing the weight as the work progresses and the saw is dulled by successive cuts.

The exact speed which will give the best results with each type of saw and kind of material can be determined for the individual machine only by comparison. In a general way, however, in power cutting, 50 to 60 strokes may be considered a maximum when water or compound are not used. With the use of lubricant, however, this speed can be increased to 100 strokes per minute on soft steel; 65 to 80 strokes on annealed tool steel; and 60 to 80 strokes on unannealed tool steel.

Compound should always be used except when cutting iron castings, as it greatly increases the output, doubling it

at least. Either water or "compound" may be used, but never oil.

Whatever the liquid employed, its purpose is not so much to lubricate the work as to cool the blade, since the high speed of cutting generates considerable heat which draws the temper of the blade. The work should be kept flooded, for a scanty flow of compound simply increases the likelihood of chips sticking in the cut and breaking the blade.

Formula for Compound

A formula for making compound is as follows: To a quart of sal soda thoroughly dissolved in ten gallons of cold water, add four quarts of equal parts of mineral and lard oil, and mix thoroughly. The compound will be ready to use ten or twelve hours after it is made.

The necessity of increasing the weight as the saw is dulled must not be overlooked, and is a factor for which no absolute rule can be given. The most that can be said is, knowing approximately the time in which a certain blade should cut an average specimen of a certain kind and size of material, the weight should be increased by increments of from 5 to 10 pounds as often as the time of three successive cuts is above the average. Remember that the weight on a heavy gauge saw must be increased faster, in proportion to the extra amount of dulled surface or contact area, than for a thinner gauge saw, regardless of the difference in weights on the first cut with each. In hack sawing, as in any other form of work, common sense, attention to detail and observation of manufacturers' instructions, are great assets.

To sum up: The attainment of real efficiency in cutting metals with either hand or power hack saws depends upon:

1 The selection of the proper make of saw.

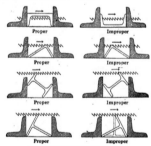

FIG. 25—SHOWING HOW TO CORNER WORK.

2.The selection of the proper saw of that make for the material that is to be cut.

3. The proper use of the saw selected. It is possible for any user of hack saws, at the expense of time and money, to evolve for himself a system for se-

lecting and using hack saws to the best advantage. It is likely there are many machinists whose experience is sufficiently wide to warrant their making their own selection of saws and establishing their own code of usage. To this class of men, this article will have been

but little more than a confirmation of facts they already know. To the man who has not had the advantage of such wide experience, it is hoped that this matter will prove of value in his search for the means of more efficient work and methods.

Making Cold Punched and Machine Made Nuts

Choice of Material is a Very Important Feature in the Making of Nuts of Either Variety, and the Press Operations in the Cold Press Type Will No Doubt be of Interest.

By J. H. MOORE

IN this the third, and concluding article dealing with the John Morrow Screw and Nut organization, we will discuss the method of manufacturing their varied line of nuts. A full line of cold-punched, and machine-made nuts is produced, also special nuts for any requirement; but first let us consider the routine of operations adopted in the making of a standard cold punched nut.

The first point considered is the choice of raw material. This is a very important matter and one that receives careful attention. Only first quality stock is used, and all material used undergoes a test of its ability to stand up under pressure, etc.

Having decided on the grade of stock to be used, the material is cut off into certain lengths, depending upon the style and size of nut to be made. Next comes the punching operation, which consists of punching the centre hole. This operation is depicted at Fig. 1. It will be noted that owing to the thickness of stock, together with the close proximity of the holes, the material takes a peculiar curl, really bending into the form of an arc. This, of course, does not injure the stock in any way, and

the operator gets so used to handling the material in this shape, that it is no disadvantage to him.

After coming from this press it goes through what is known as the cutting-out operation. This really consists of the punching out of blanks from stock shown at Fig. 1, at the same time rounding the corners. This makes the blanks look like A at Fig. 2. The nut now passes to another press where it is rough cornered, or made as shown at B, Fig. 2.

After this comes the chamfering operation as shown at Fig. 3. It will be noted that the operator feeds the blanks down a hopper at the side of press. The trip action shown at front of press actuates a slide that pushes a small ram out and in at the proper time. As each stroke ascends, this ram goes forward, pushes the completed nut out of the way, allowing another nut to slip down the hopper into place under the chamfering punch.

A trimming operation is next in order, and as this is a very simple one, we will pass on to the next, which is re-trimming, or final trim as it is termed. After the final trimming punch has per-

formed its duty the nut looks like C at Fig. 2.

When this is completed the nut is reamed on multiple spindle machines of both horizontal and vertical type, after which they are tapped on special machines for the purpose. Two of these machines are shown at Fig. 4, and they

FIG. 2—THE NUT IN VARIOUS STAGES OF COMPLETION.

are of the multiple spindle type. The spindles raise and lower at predetermined rates, taking their action from cams and rollers, so that one operator can tap an enormous quantity of nuts in one day.

After tapping, the nuts are faced on machines of both vertical and horizontal design. This practically completes the product with the exception of inspection.

The Bar Variety

Making nuts from the solid bar is an entirely different proposition, and briefly the procedure is as follows: The hexagon bar is placed in the turret lathe, and with proper tooling, the nuts are turned to proper thickness, chamfered, and cut off.

Next comes the operation of drilling, which is accomplished as shown at Fig. 5, on a battery of Colburn drills. Of course lighter nuts are drilled on lighter machines, but the operation shown is that of the heavier variety.

After drilling, comes the regular tapping operation, and as the shaping of the nut was accomplished in the turret lathe, this completes the work with the exception of removing of burrs, and final inspection.

In the inspection department every nut is gone over. Plug and screw gauges are used and unless a nut is perfect it is rejected. At various stages of man-

FIG. 1—PUNCHING THE CENTER HOLE. NOTE HOW STOCK CURLS.

ufacture, sub-inspection takes place, but the final inspection is most exacting. Between facing and inspection operations, the nuts, are thoroughly washed in order that no grit remains when entering inspection department.

Some Points of Interest

This concern has a rather good method of storing completed stock which is worth calling attention to. The stock bins are arranged in pillar form, being approximately 12 ft. in height. An elevator takes the completed product up to the top and there the nuts are dropped into the bins. When needed they are let out of a sliding door at the bottom. Should any overflow occur on the opening of this door, a hopper, about the height of a man's shoulder, is provided in which any such overflow is placed, these working their way gradually to the bottom. The arrangement, as described, is very neat and helps the attendant to keep track of his stock supply, which he does in the following manner: A card is attached to each bin in a sheet metal holder, and on the card is marked the number of nuts contained. As they know definitely how many nuts of a certain size there are to 100 pounds, this totalling is an easy matter.

It might be mentioned at this point that after every operation in the making of the nuts, they are rumbled in barrels to take off all rough edges.

A cost system is also in force that enables them to keep track of all stock going through their plant. They not only know the cost, but the actual output of each workmen for every day.

They have their own tool room in which are made their dies and other fixtures, in fact in as far as possible they accomplish everything themselves, even to the castellating of special nuts. This operation is accomplished on various styles of machines depending upon the size of the nut.

The plant is particularly lucky in having a splendid location and good light,

FIG. 5—DRILLING NUTS FROM THE SOLID COLD STOCK.

which naturally makes for ideal working conditions, and maximum production.

Over 752 tons of nuts are produced every year so that readers can see that efficiency and production has been developed to a high degree.

————

Much interest has been taken throughout England in a new plan for the building of a city about 20 miles from London on model lines. Under this Welwyn Garden City scheme, it is intended that a town of ultimately about 50,000 inhabitants shall be gradually constructed as a complete unit, with its own factories, warehouses, shops, and residences, in contrast to the many districts which are almost exclusively confined to dwelling houses for people who go into the metropolis daily in pursuit of their livelihood. Such places have come to be known as dormitory suburbs, and the new scheme is based on the theory that the only solution of the whole problem of comfortable housing and convenient transport is to set up instead, what are

described as satellite cities, in the sense that London itself is the centre of all things, but that within a radius of 20 to 25 miles there should be a ring of these almost self-contained communities. An interesting feature of the scheme is that the capital provided will only receive a maximum interest of 7 per cent., and as the town is gradually built, any increment of values arising from the settlement of the people will be conserved for their own social advantage.

————

To Consider Application.—The Board of Railway Commissioners has acted upon the application of the Railway Association of Canada filed on July 10, asking for a general increase in freight tariffs of 30 per cent. over those charged at the present time. The board has decided to consider the application at Ottawa, Aug. 10, at which time and place all interested parties, both for and against the application, will be expected to appear ready to proceed.

FIG. 3—CHAMFERING THE NUTS. NOTE THE   FIG. 4—TAPPING THE NUTS ON MULTIPLE SPINDLE AUTOMATICS.
HOPPER.

The Use and Abuse of Arc Welding Equipment

Welding Equipment of Any Kind Should be Used Intelligently—
A Thorough Understanding of the Equipment is Essential if Good
Work is to be Performed.

By H. L. UNLAND *

THERE are several simple precautions to be observed in the use of electric arc welding equipment whatever the nature of the apparatus may be.

Many of the accidents which occur are generally the result of a misconception of the nature of the equipment and its proper use. This applies more particularly to the auxiliary apparatus.

The eyes should be thoroughly protected from the light of the arc. Painful and more or less serious burns to the interior of the eye will certainly result from carelessness in this respect. No chinks or holes in the mask should be permitted, since only a brief exposure of the eyes is required to bring on painful results. The inside of the mask should be kept painted dull black to prevent reflection of the light from behind.

The mask is principally used where carbon electrode welding is being done. It consists of a thin sheet of aluminum formed to the proper shape and provided with an adjustable band for supporting it from the operator's head. An opening in the front of the mask is provided for a window of glass. This glass may be either a number of individual sheets of different colors or a single compound sheet of glass may be used.

The colored protective glass should be sufficiently dense to reduce the light intensity to a value not objectionable to the eye and at the same time the definition of the area immediately around the arc should be clear to enable the operator to properly follow the work. Different color combinations are used but the most general seems to be a combination of red and green glass.

The glass is held in a recess in the front of the mask by means of a clamping frame secured by four small bolts. By this means it is rendered impossible for light from the arc to pass through joints or cracks around the edge of the glass. The bolts should always be in place when the mask is used, as the small amount of light coming through one of these openings would in a short time affect the eyes of the operator.

To change the glass, the bolts and clamping frame are removed, the glass changed and the frame and bolts replaced. It is advisable to keep a piece of clear glass on the outside, since in welding, this outside surface will be struck by numberless particles of molten metal, and will become roughened to such an extent that it becomes useless and must be replaced.

The Hand Shield

The hand shield is principally used in doing metallic electrode welding. It consists of a light wooden frame with provision for a protective glass window. The protective glass is the same as used in the mask. The shield is also used by inspectors and others who require the protection only for short periods and at infrequent intervals. A light box frame surrounding the window is fitted to the operator's face, preventing light from the side or rear reaching the operator's eyes, thus eliminating any interference of a number of operators due to the light from the arcs. The protective glass of the hand shield is supported in guides on the front of the shield and is clamped in place by a wooden wedge driven through openings in the guides.

Electrode Holders

The function of the electrode holder is to electrically connect the electrode used to the cable connected to the welding equipment. The requirements of this service are:

First: It must securely grip the electrode so the welder can operate it without play in the mechanism or without the electrode becoming loose in the holder while being used.

Second: The clamping arrangement should be such as to facilitate changing electrodes.

Third: It should be so constructed that the minimum heat reaches the operator's hand.

Fourth: The weight should be as low as possible and the balance such as to facilitate manipulation by the operator.

Fifth: The construction should be such that the operating parts are protected from accidental contact to avoid injury by burning or by being struck.

Sixth: The general construction should be substantial to avoid light or flimsy parts subject to bending or jamming.

The carbon electrodes should be rods of hard, homogeneous uncored and uncoated carbon. The diameter used will vary with the current to be used and this information is given elsewhere. The length depends on the particular class of work to be done. Long carbons reduce the percentage of short ends thrown away, but are more liable to breakage. The average lengths range from 9 to 12 inches.

For welding iron and steel the metallic electrode should be a high grade of low carbon steel wire. A large number of tests were made by the Emergency Fleet Corporation to determine the best chemical analysis of wire for this purpose, and the wire now made by a number of manufacturers meets these requirements. This material can be purchased either direct from the makers or through jobbers and can be obtained either in rolls or in short lengths, cut and straightened. In ordering, "Electric Welding Wire" should be specified, since wire for acetylene welding is often treated in such a way as to render it unsuitable for electric welding.

The electrode wire should be cut into pieces convenient for the operation. A length of 18 inches is satisfactory, since it is about the greatest length an operator can handle; at the same time it reduces the number of times the electrode is changed, and consequently the wastage, to a minimum.

Cables

On account of the intermittent nature of the work, it is possible to use smaller cable for the welding circuits than is standard for the current capacities. In this way, there is also a gain in flexibility which permits better control of the welding arc, by facilitating the manipulation of the electrode holder.

In metallic electrode welding a length of at least fifteen (15) feet of extra flexible cable should be connected to the electrode holder to allow the operator to fully control the arc through manipulation of the holder. For the ground or return cable the standard extra flexible apparatus or dynamo cable insulated with varnished cambric for low voltage circuit and covered with double weatherproof braid has been found suitable.

The carbon electrode welding arc is not as unstable as the metallic arc and therefore the manipulation of the electrode is not so important. For this reason the standard extra flexible dynamo cable referred to above may be used for connection to the electrode holder, as well as for the return circuit.

It is difficult to give universally applicable figures covering amperes, speed, etc., for electric arc welding due to the effect of conditions under which the work is done, the character of the work, and to a very large extent the skill of the operator.

The following figures are based on favorable working conditions and a skilled operator. However, they are approximations only and are given merely as a general guide.

Metallic Electrode Welding

Light work ....25 to 125 amperes
Heavy work ....up to 225 amperes
                                        Correspondence

Electrode Diameter Inches	Amperes	Plate Thickness Inches
1/16	25-50	up to 3/16
3/32	50-90	up to 1/4
1/8	80-150	1/8 to 3/8
5/32	125-200	1/4 up
3/16	175-225	3/8 up

The same size electrode may be used

Continued on page 67

*Power and Mining Engineering Department, General Electric Co.

WHAT OUR READERS THINK AND DO

GAUGE FOR USE IN CUTTING-OFF LEAD
By Tyke

One of the most rapid operating devices that the writer has ever made is shown in the accompanying sketch. A large number of brass rods had to be cut off in the lathe and one end tapered. Practical accuracy had to be maintained but not to extremely fine limits, and the device shown speeded up things very materially and at the same time showed no variation from the desired dimensions in the finished work. A machine piece A was turned to fit the tailstock sleeve, slotted and drilled as shown, a pin of cold-rolled steel ¼ x ¼ x 7 inches long was inserted in the slot, drilled in position and pinned. The piece was a very close fit so that it would remain in the position desired. A parting tool was set to approximate position and carriage clamped. The stock to be cut was put through the lathe spindle and advanced just the right amount for cutting off. Rod B was brought in line and the tailstock spindle advanced until the rod B touched the end of the stock W, then the tailstock spindle was clamped and the rod B tapered down slightly as shown by dotted line. The piece W was cut off, rod B raised to centre alignment, and stock advanced till it touched the end of the rod and the chuck tightened. After cutting off the compound, rest was set for the proper taper, and one roll held in the chuck and rod brought into position. A trial cut was taken, adjustments made, then as before the tailstock was clamped and the tapers finished with a surprising degree of ac-

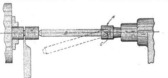

DETAILS OF THE CUT OFF GAUGE.

curacy. This device can be used to advantage on any cutting-off job where even but a few pieces are wanted, the big feature being that the bar is always within easy reach, and a light tap with the chuck wrench just clears it from

the work. No scale is required after setting the first one correctly, and greater accuracy can be obtained, to say nothing of the increased output, which is very marked.

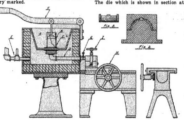

SECTIONAL AND GENERAL VIEW OF DIE-CASTING MACHINE.

DIE CASTING MACHINE
By G. Barrett

The illustration depicts a die casting machine as used to cast connecting rod bearings.

Fig. 1 shows a sectional view of the melting pot, and die set in place. The big feature of this pet is that the hottest metal is always nearest the discharge to the mould.

The main chamber is shown at A, Fig. 1. This carries the bulk of the metal; also a cylinder B which is connected by two flanges shown. The opening C acts as a valve. When the metal begins to melt, it falls through the opening C into the pipe D and is kept at correct

heat by the action and location of the burner E. The metal is then forced into the die by a plunger which is connected to lever F.

The die which is shown in section at-

Fig. 2 is bolted to a carriage, and is placed into position by the hand wheel H, Fig. 2. The operation of the machine is to turn hand wheel H until the die comes close to mouth of melting pot, then lift the gate valve G, and pull down lever F. Close the gate valve to stop the metal from running out, then turn back hand wheel and separate mould by releasing clamp screw I.

The small sketch, Fig. 3, shows the bearing which is made from hard babbitt, but may be made of different kinds of metal. Various shapes of dies may also be used on the same machine to advantage, and this method is well worth the expense where there is any quantity of casting to be made.

INTERESTING THREAD DATA

Some very useful thread data is published in the current issue of threads in the "He wants to know section," and we reproduce the same for readers' perusal.

The first question asked is this: "I have formulas for measuring internal threads by the three wire system, but, have not been able to check these correctly. Can you send me correct formulas for U.S.S. and V internal threads?" Here follows the answer:

As to the three wire system for measuring internal threads, we believe that

this system cannot be used satisfactorily on internal threads, for the reason that the wires will not set into the internal thread as they do on the external.

The wires on the external thread will properly rest on the point of tangency, but on internal threads this is not so, as the wire would have to be made to conform perfectly to the internal diameter and the point of bearing.

We know of no satisfactory method of measuring internal threads except with standard gauges.

The second question is equally interesting:

"Is the A. S. M. E. thread another name for the A. L. A. M. thread?" The answer follows:

The thread standards of the A. S. M. E. (American Society of Mechanical Engineers), and that of the A. L. A. M. (Association Licensed Automobile Manufacturers), are not the same.

The A. S. M. E. and the A. L. A. M. threads have the same form—the United States standard form—but have their own standard diameters and pitches.

The A.S.M.E. sizes are given by numbers, and are the same in diameter as the machine screw sizes.

In 1911 all A. L. A. M. standards were adopted by the S. A. E. (Society of Automotive Engineers).

The name of the association of Licensed Automobile Manufacturers has been changed to the National Automobile Chamber of Commerce (N. A. A. C.)

The present S. A. E. standard for screw threads gives the present thread standard of the automobile industry.

The third and last question should prove useful: "What is the reason that a thread is tapered on the back end of a stud six inches long?"

Should this be a matter of the thread tapering smaller at the end of the cut, it is then quite likely that the machine on which the die head is used is out of alignment wiht the work.

New machines have sometimes been proven out of alignment, and so it would be well to test the machine alignment. Prove alignment of the machine by fitting centres in head of machine and in the turret hole. Or, by turning a piece of stock held in the chuck or collar, to the diameter of turret hole, and then bring the turret up to the piece.

If the turret is out of alignment, this test will show which way the machine is out of line.

Should the thread taper larger at the end of the cut, the thread chasers are probably cutting either ahead or behind. All cutting should be done on the chamfer, and not back on the chasers.

If the lead is short, it can be corrected by stoning the back edge of the first tooth on each chaser.

If the lead is long, it can be corrected by stoning the front edge of the first tooth on each chaser. To anyone interested in thread work these questions and answers should prove well worth while.

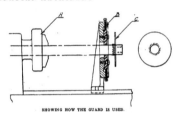

SHOWING HOW THE GUARD IS USED.

SHEET METAL CHIP GUARD
By F. M.

While boring a hole in the end of a bar considerable trouble was caused by small chips lodging between the steady rest jaws and the work, the work in this instance being gripped in a chuck A, Fig. 1, with a rest B, at the outer end. To overcome the trouble mentioned a circular disc, C, was employed.

This is flanged inward and the flange is prong shaped as shown, and made so it will spring on the bar and revolve with it, thereby forming a protecting hood against which the chips are thrown.

INTERESTING GRINDING FACTS
By H. W. DUNBAR

Many times one is asked why they use hardened steel shoes in some cases and wood shoes in other cases.

Normally, in the production of commercially ground parts in large numbers, the work shoe must be of such shape, size and proportions as to remain a constant factor in the grinding operation, because the operator is constantly turning out piece after piece at short intervals. If he had to take into consideration variation in size due to change in the work shoe, his production would be cut down materially. Therefore, the hardened steel work shoe of the size of the finished work for finishing, and of the size of the rough ground work for roughing, provides a means of supporting the work and removing one of the variables in the grinding operation, because the hardened steel work shoe retains its size longer and can be depended upon to produce parts of the same size, piece after piece, without consideration on the part of the operator.

When a machine is used for general all-round grinding, when there are seldom more than one half-dozen pieces to be ground at one given time, it would be manifestly unreasonable to hope to have hardened steel shoes of the proper size for all the various pieces being finished. Therefore, a steadyrest with adjustable wood shoes is provided which has a range within the capacity of the size of machine being used, and can be readily adjusted to the diameter of work being ground.

Much discussion has been met with in connection with the use of hardened steel shoes, because, it is argued, the work being ground is liable to be scratched. This has proved to be a false line of reasoning, and it is readily understood when one considers that steadyrest shoes of soft material will pick up particles of dirt, abrasive, chips, etc., and retain them, really converting the shoe into a lap, and consequently the hardened steel shoes will not become imbedded with these particles so easily. Therefore, if it fits the work properly and is in line with the work, hardened steel is less apt to scratch than the soft materials.

In the case of hardened steel shoes it is very important that they be properly constructed and properly mounted in the steadyrest. The shoe must present a considerable portion of its bearing surface to the work being ground. It must be in line with the axis of rotation of work, securely fastened in the steadyrest, and properly hardened and smooth.

The other forms of steadyrest do not need such careful attention as the hardened steel, for, since they are soft by nature, they quickly wear to a point where they meet all of these conditions. The fact that the hardened steel shoes do not wear so rapidly is the reason why they must have more attention in the making.—Grits and Grinds.

In the course of a paper on the "Airlift System of Raising Oil," read before the Institution of Petroleum Technologists, Mr. R. Stirling mentioned that the arguments against the use of the air lift are that the inflow of the water is induced thereby, and that with very heavy viscious oils, if over 5 per cent. and under 20 per cent. of water be present with the oil an emulsion may be formed from which it is difficult to separate the water. In a well with 10 to 15 per cent. of water, the difficulty was overcome by running into the well an additional 20 per cent. of water. No serious emulsion was then formed, and the discharge was much facilitated and increased.

DEVELOPMENTS IN SHOP EQUIPMENT

CARBURIZING COMPOUND MIXER

The Kent Machine Company, Kent, Ohio, is now manufacturing a machine for mixing bone meal and the various compounds used for pack hardening and carburizing steel. We believe it to be of great interest to all who do this class of work, and in particular, to the manufacturers of roller bearings, etc.

After each heat it is customary to replenish the used compound with a certain amount of new, usually about 25%, which, to insure a uniform depth of penetration, should be thoroughly mixed with the old. The machine as designed is adapted to this specific work.

The operation of the mixer is continuous and it automatically proportions, feeds, mixes and discharges the mixed material without requiring any man labor beyond the loading operation. This machine has been used for a number of years in the Canton, Ohio, plant of one of the largest manufacturers of roller bearings in the world. This same company recently ordered another mixer for their new plant at Columbus, Ohio.

Referring to the cut of this machine, the large hopper holds the used material, and the smaller the new bone meal or compound that is to be mixed with it. On the throat of the small hopper there is a gauge or gate which can be adjusted to restrict the opening through which the material is carried by the reciprocating action of the feed plate. This adjustment makes it possible to obtain any proportion of the new and used material desired.

The action of the feed plate is as follows: The material in the hoppers rests upon the feed plate and at each forward stroke a layer of the compound or bone meal is carried into the throat, the material in the hopper settling down as

ILLUSTRATING THE BELT TYPE MACHINE.

the plate moves forward. During the backward motion of the plate this layer, which cannot be carried back into the hopper on account of the material behind it, is dropped off the edge of the plate into the mixing trough.

Both new and old materials drop into the mixing trough at approximately the same place. The mixing is done by a long horizontal shaft on which there is a series of specially designed paddles or blades which thoroughly mix the ma-

terials, and at the same time convey them toward the discharge end of the trough. Careful experiments were made to determine the length of time required to obtain a thorough and uniform mix and the mix-

ing shaft is so designed that it is impossible for the material to be discharged from the trough before complete and thorough mixing has taken place. The mixing trough is provided with a cover which prevents the excessive amount of dust which is usually found where bone meal or compound is being mixed.

THE MOTOR DRIVEN MACHINE.

A particular feature of the machine which will appeal to those who are operating plants where large quantities of bone meal or compound are used in the possibility of building large bins over the present hoppers. These can be used as storage bins and it is not necessary to mix more material than can be used in any particular heat. In some plants these bins are carried up to the floor

GENERAL VIEW OF COMPOUND MIXER.

above and the material is dumped into them through a trap door, while in others a small belt conveyor or bucket elevator is used to elevate the material into the bins.

GRINDER FOR VALVE FACING

D. F. Dunham, 830 West 37th Street, Los Angeles, Cal., supplies us with the following information of his valve grinder which he has placed on the market:

The machine, as illustrated, is arranged to face valves of any size and angle, and if the valve centre has been mutilated it makes no difference, as an adjusting screw with a ball and socket joint overcomes this defect. The valves are ground true with the stem and not with the centre.

A spring stop can also be arranged for grinding valve seat reamers.

The valve is worked back and forth on the face of the wheel, the valve being revolved meanwhile. The arm on the machine is graduated from 30 to 60 degrees, and the strong, rigid construction can be clearly noted. Three models are made Nos. 1 to 3, the first a motor grinder, the second a belt-driven grinder, and the third a grinding attachment, as shown.

GUARD FOR PUNCH PRESS

The D. & M. Guard Co., Rochester, N. Y., have placed on the market what is known as their D. & M. guard for punch presses.

This guard has two vanes or guards, both swinging as pendulums from a stud on the bridge of the guard. The guard is really three-fold, and as the makers term it, Safety First, Second and Third. The first guard swings outside of the bridge; the second, inside the bridge, and the third is a blinker or side guard. The first guard is moved by every stroke of the treadle, the second by every stroke

of the ram, and the third swings from the bridge and prevents the operator feeding press from the side.

The feature of the inner or second guard is, of course, the fact that if the ram repeats unexpectedly after the treadle has been raised, the guard does its duty as soon as the ram starts to descend. This, of course, kicks the operator's hand out of danger, as the guard moves faster than the ram.

The swinging bridge which we have been speaking of carries all three guards. This bridge is hinged at the left and is

locked by a pin and padlock on the right. The pin can be unlocked and quickly withdrawn and the bridge can then be swung to the left, thus removing bridge and guards from front of press. The pitman, press head, etc., are then readily accessible. A few seconds completes the operation just described.

This style guard has been thoroughly tested before placing on the market, and the Eastman Kodak Co. have over 140 in operation with great success. The illustration will clearly illustrate the features spoken of.

NEW THINGS IN MACHINE TOOLS

DOUBLE END TAPPING MACHINE

A special machine for simultaneous tapping both ends of such articles as turn-buckles, etc., with right and left-hand threads, has been designed and built by the Cadillac Tool Company, of Detroit, Mich. The machine is built with beds of different lengths to accommodate work from 12 inches to 60 inches long. The maximum size tap used is ⅝ inch. The machine is fitted with adjustable steps that operate the switch for reversing the motor when the desired depth of thread is attained. A connecting rack, with pinions on either head, provided equal movement to tapping spindles when starting the thread by hand.

MULTI-SPEED PLANER

The Conradson Machine Tool Company have added a new type of planer to their machine lines. The design was primarily intended for motor drive, but equally efficient service may be had when arranged for constant speed drive with belt pulley. When a motor is used the

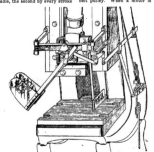

THIS VIEW SHOWS THE ACTION OF GUARD CLEARLY.

drive is made direct to the main shaft through a Clark flexible coupling. Reverse mechanism is operated by means of pneumatic clutches, which, it is claimed, provides self-compensation for wear, and requires little attention. Reverse clutch levers are directly connected, so that air is admitted to one while being discharged from the other. There are four operating speeds ranging from 25 to 45 ft. per min., with a return speed of about 120 ft. per min. These speeds may be varied to suit special conditions. All small gears and pinions are of chrome-nickel steel, heat-treated, large gears and rack being of cast steel with herringbone teeth. Splash system is used for lubrication. Rack feed is operated pneumatically, and this is controlled automatically from the table trips. Where shops are not provided with compressed air, a small compressor may be provided and driven from the main shaft of the machine. These planers are built in five sizes, ranging from 24 x 24 to 48 x 48 inches, the working length of table being 6, 8 and 10 feet. Net weight of the complete machine will approximate 8,000 lbs. for the small machines and about 32,000 lbs. for the heaviest.

RECIPROCATING TYPE MILLING MACHINE

The Ingersoll Milling Machine Company, of Rockford, Ill., has developed a special type of reciprocating milling machine that may be designed to meet specific requirements, and for this reason no attempt has been made toward this particular type of machine. These millers are intended for working on duplicate pieces, where setting up may be carried at one end while the cutter is operating on the other piece. When working cut is finished, the feed is automatically reversed and the table is rapidly traversed until it approaches the new piece, when the normal feeding speed is again resumed. The release, reversal of feed, rapid traverse, and reduction to normal cutting speed, continues indefinitely, while the operator is removing and replacing the work. These machines may be designed for any class of work. All cutter spindles are adjustable and all feeds controlled by the one lever located at the front of the machine.

The MacLean Publishing Company.
LIMITED
(ESTABLISHED 1887)
JOHN BAYNE MACLEAN, President. H. T. HUNTER, Vice-President
H. V. TYRRELL, General Manager.

PUBLISHERS OF

CANADIAN MACHINERY
and MANUFACTURING NEWS *and*

A weekly journal devoted to the machinery and manufacturing interests.

B. G. NEWTON, Manager.   A. R. KENNEDY, Managing Editor.

Associate Editors:
J. H. MOORE   T. H. FENNER   J. H. RODGERS (Montreal)
Office of Publication: 143-153 University Avenue, Toronto, Ontario.

VOL. XXIV.   TORONTO, JULY 22, 1920   No. 4

THE WEEK IN PASSING

IT is now costing the English mine owner as much to produce a ton of coal as it costs in United States, and in that some of the U. S. coal interests see a chance for world conquest for the coal trade. They count on the increasing costs in Britain, the coal-producing country of the world, to make it possible to crowd in on the market, providing they can secure ocean transport at fair prices. Now, this is all very well, until one stops to read that the Western States are apt to have their feet frozen this winter because they can't get coal enough to keep them warm, and the Canadian market continues to take all the coal U. S. cares to ship in this direction. Does it seem reasonable that United States is going to peddle coal all over the world, when her own people and her nearest market are sifting the cinders and wondering where the next shipment is coming from?

* * *

At the present time the shops of the Government roads, located at Moncton, are advertising for tenders for a long list of small tools, and the chances are that they will get a number of answers to their specification list. Dealers and jobbers like to deal with the Moncton office better than with any other branch of the Government-owned railroads, because they get their settlements more promptly. Some of the firms claim that the Moncton authorities are the old order, and have been running their business for a long time, while the branches coming under Government control recently have not acquired the habit of paying quickly. There is a lesson for them in the attitude of the dealer. There is a desire to get the Moncton business, and to give the very best service in the matter of shipment, whereas in the other branches of the Government service dealers are not nearly so keen to get the business, nor do they get as prompt treatment in the settlement of debts.

* * *

A freight car rolling into a Toronto yard with over sixty tons of steel shows the way United States is dealing with the car shortage. The trouble is that many of the cars on the track these days are not physically able to take care of such a tonnage, and were they put to the task would be returned as cripples. Thirty per cent. of the car miles are computed to be empties, as loading to capacity has not yet become a real art in the transportation business in this country: United States figures show that the best movement ever attained in that country was in 1916, when 26.9 car-miles per day was attained. Of course, it is much below that figure to-day. To reach anything like a working efficiency that will meet the present congestion it would be necessary to raise the standard to at least 32 miles per day. To do this will call for something approaching real honest-to-goodness sweat, and the trouble seems to be that fewer people every day are inclined to indulge in any form of labor calling for this accompaniment.

* * *

Mayor Church wants to spend millions to build radials to develop Toronto to a population of one million. The representatives of the surrounding towns and villages want them for the same reason. Neither explain from where the population will come. If the big cities are to be built up the population must come from the towns and villages. If the latter are to grow to the figures in Sir Adam Beck's fairy tales, the people must move from the farms. This is exactly what a Farmer-Labor Government does not want. The farmers want to keep the people on the farms and the city laborers do not want to increase local competition. The back-bone of the towns and villages is the local business people and they certainly do not want to be taxed to send villagers to the city to do their shopping, or to reduce their local field by moving their population to the big cities. E. M. Trowern, manager of the Retail Merchants' Association, makes this view very plain. He might have gone further and shown that having numerous happy and prosperous business centres near by makes rural life bearable. The absence of it is what makes the Western conditions so hard.

PROSPERITY!"
—Alley in the Memphis "Commercial Appeal."

MARKET DEVELOPMENTS

Holiday Season Dulls Machine Tool Market

Buying that is Done is Not on a Large Scale—New York Reports that Little is Being Done There—Montreal Needs Cars and Coal —Scrap Market Continues Dead.

MACHINE tool dealers report that trade has been quieter for the past week or so. Buying that is going on now is not in large volume. The New York market is dull now, and there have been some cancellations. But the makers are well booked with orders and any dullness or cancellations will not likely have the effect of forcing down prices, as there has been practically no letting off in the cost of building.

Importers in this district have been getting indifferent service in the matter of delivery. Some of the yards have received quite a tonnage, while in other cases there has been nothing at all. Some of the cars that do get across from American points are loaded to the hilt, much the same as they were doing in the strenuous days of the war. One car carrying steel brought over a load weighing better than sixty tons this week. This is not always possible owing to the poor physical condition of much of the roll-

ing stock that is in use. The arrival of plate has relieved some anxious times in the shipyards, which were getting close to the point of closing down for want of material. In many lines, such as sheets, blue or black, tubes, bars, etc., there is a scarcity that in many cases can be described as acute.

Scrap yards report very little business as being done. There are no large buyers in the market. The steel mills, considering the difficulty they have in securing sufficient supplies of coal in the first place, are not buying much scrap, and their absence from the list of buyers leaves matters rather flat.

Some of the dealers intimate that they consider that trade will be resumed at a fairly early date, but with this guess they insist on coupling up the provision that the resumption will be on a lower price basis.

SHORTAGE OF COAL AND CARS IS BEING KEENLY FELT IN MONTREAL

Special to CANADIAN MACHINERY.

MONTREAL, July 22.—Acute coal conditions and the strenuous efforts to relieve the tension in the meat industry, are the features of the present activities. The Dominion Board of Railway Commissioners are taking immediate steps to conserve the available supply in anticipation of a scarcity that is apparently developing for the coming winter requirements. One of the first moves will probably be the restrictions placed on coal going out of the country. It is likely that the Board will prohibit the exportation of Canadian coal, except to Newfoundland and the United States. Further bunkering of ocean-going vessels at Montreal with American coal will be curtailed, if not entirely prohibited. This practice of coaling Atlantic steamers requires large quantities of American coal that is intended for consumption in this country, so that Canada is deprived of considerable coal that is delivered here from the United States mines. In future, these vessels will be forced to take on coal produced from the Canadian mines. It is also given out that additional measures will be enforced if coal conditions warrant it. There is little doubt that the business outlook in Canada is

temporarily clouded by the evident scarcity of coal, as the situation has been so acute in certain quarters as to necessitate the closing down of some industrial plants, which in normal times, use, and at present are unable to get large quantities of coal. It is stated that the Montreal Light, Heat and Power Company, who use about 800 tons of coal per day in the operation of their gas plant, and at present time have only a few thousand tons ahead, are threatened with serious curtailment if coal supplies are not forthcoming in normal quantities.

Car Regulations Affect Shipments

The steel situation has experienced little relief during the past week, and dealers report themselves in very much the same position, where difficulty in obtaining material dominates the market. Dealers here state that shipments from the Canadian mills are relatively better than where the material is coming in from the States, and as the greater amount is from the latter source, the industry is not much benefitted by the local improvement. While other factors combine to affect business, that of car shortage is probably the chief reason for present

disruption. The diverting of cars for coal carrying purposes has created a partial famine in other directions, and frequently the reported improvement in production figures is counteracted by the fact that this additional output must be stored at the mills awaiting better transportation facilities to get it into the hands of dealers and consumers. Asked as to the possible result of the recent order of the Interstate Commerce Commission, to the effect that all open-top cars would be used for coal transport, one dealer stated that: "It would, perhaps, curtail the shipment of ore to the blast furnaces, and also other raw materials to the mills, but stocks now on hand would prevent any immediate shortage of finished material. The apparent difficulty at the present time is in getting this completed material from its stored position at the mill to its intended destination."

Seasonable Quietness in Tools

The machine tool situation offers little of interest just now, owing to seasonable quietness in the market. This is the general tendency, but isolated cases of good demand are now and then reported. Dealers who are awaiting shipment of equipment from the States have stated that delivery is still very uncertain owing to transportation difficulties. Several industrial firms here are temporarily held up awaiting the coming of some

heavy tools. Supplies are moving normally, with advances reported on many lines. This condition will be further emphasized should increased freight rates become effective.

Little Doing in Scrap

The market here is a repetition of continued dullness and reported sales are on a low standard. Local mills and foundries are regular buyers but the volume is not excessive. The activity of grey iron foundries, since the strike, has aided in the movement of machinery scrap, but apart from this the situation is devoid of even normal demand. Dealers are in a conservative mood and are not anxious to stock more than they require for immediate needs.

CAN GET GOODS IF THEY GO AFTER THEM

Harder to Get Material—Government Roads Have a List of Small Tools Out Now

TORONTO.—A tendency to load cars to the limit seems to be growing if a few of the cars that have crossed the line recently from U.S. points is anything to go by. One of the steel warehouses had a car placed a day or so ago and the papers showed that it had a few pounds better than sixty tons. Some of the dealers report that they are getting good service in the matter of said delivery, while in others practically nothing at all has come through. There seems to be no way of finding out when cars are likely to arrive, as some of the shipments that have slipped across are well ahead of the schedule of expectations.

There is a fair amount of machine tool business being done, but orders are not large, and the holiday season seems to have been quite an effective stop on a good deal of the trading. Several of the inquiries that are in the market now are for machines to turn out a special piece of work, preferably by automatic operation, if at all possible. Very often, in this way, if the firm is able to give good suggestions, business results in what would otherwise be a dull season. The firms that seem to be getting quite a number of inquiries are those that are able to take a man's blue prints and show him the best and the most effective way of turning out the work he requires.

Very Small Shipments

There is no general improvement in the steel market. Now and then some lucky dealer comes out with the information that he has a number of cars placed, but against this there are firms that have received nothing at all to place against the big list of orders in the office.

Bars took an increase on July 13 to $5 per 100 at some of the Canadian mills. Some of these mills, especially in the Montreal district, have been able

---

POINTS IN WEEK'S MARKETING NOTES

Pittsburgh says that the shipments of steel have fallen behind last month. One large firm says the decrease is from 20 to 25 per cent. in its case.

—

When mills cannot ship, they prefer to run and pile semi-finished material, fearing that some of the finished article may not be acceptable when it gets through the rollers.

—

The Government-railroads have a requisition out now for small tools for the Moncton shops.

—

New York reports state that the machine tool market there is very dull, but makers have so much business on their hands that there is not much chance of prices dropping.

—

The delivery of many lines of machine tools would be much improved were the shipping facilities better.

—

Montreal needs more cars and more coal. The lack of these hinders business in several lines.

—

The scrap metal market is dull. Dealers believe there may be a resumption of business later on but hold that it may be at a lower level.

—

One car of steel which came into a Toronto yard this week carried over 60 tons of steel.

—

Officials claim that cars should be loaded much nearer capacity than at present if traffic congestion is to be relieved.

---

to ship quickly, but it is unlikely that this advantage will be continued, as some of them are having labor troubles which threaten to interfere with their business. Of late the Canadian mills have not been able to do very much for the trade here in the way of bar mill products, and there are many lines that are feeling the effects.

The lack of good shipping keeps buyers guessing as to the best way to handle the material. One Hamilton firm, using trucks that will accommodate five tons, has moved as much as 150 tons now. The firm bought this material, and it was drawn some time ago when prices were less than at present, so it is really a paying proposition to use the trucks as prices have gone up considerably since the order was ready for shipment.

In one yard a car came in this week that weighed better than sixty tons. There are cars getting over now that

---

have never carried steel before. Some of them are in poor physical shape, and the right place for them would be in the repair shops of the roads handling them.

Looking for Material

A man from the head of the Great Lakes with a penalty job on his hands was in Toronto a few days ago looking for reinforcing material. He was quoted out of stock here on goods that had been taken in at a fairly high price. He went on to Montreal and left an open order there hoping to beat out the price he had received in Toronto. The chances are that he will be a long time getting his material, as many of the places turning out what he wants in the east are not supplying it now. Buyers are constantly on the road between mills and warehouses seeing what they can find. Very often a stiff warehouse price does not represent any more actual profit to the dealer than the old price. He has to pay high in the first place for his stock and it takes more money to keep a warehouse up to a well-assorted stock.

The attitude of some of the mills can best be shown by the following letter which was sent out this week by one of the large companies, showing the conditions under which they would accept business. Here are some of the conditions named:

New specifications will be accepted only on the basis of rolling and piling the material for shipment at a later date, except:

(a) Where a customer can haul material from our plant or supply barges for river shipment.

(b) Where a customer can furnish us with a privately owned car not controlled by the railroad company.

(c) Where a permit can be secured to load cars which it is not possible to load under the ruling of the Interstate Commerce Commission.

All tonnage except covered by the above will be taken with the understanding:

Shipments subject to our car supply at the time of rolling, railroad conditions in general and other conditions beyond our control. Will you kindly note on your orders whether if cars are not available at time of rolling, we shall roll, order and pile, or withdraw from schedule and roll at a later date. All orders accepted are understood as not being subject to cancellation.

The arrival this week of some cars of plates, bars and sheets, has meant relief for several industries, as some of the shipyards were more or less embarrassed in their operations owing to the scarcity of material.

Railroads Have List Out

The National Railroads have a long list out for small tools for the Moncton shop. Strange as it may seem, dealers are more eager to do business with this shop than with others in the government ownership ring. One dealer explained it this way to Machinery: "We like to do

business with the Moncton shops because they are very quick at settling up their bills. They seem to be under separate management from the government shops that have been taken on in this district, where they keep us waiting a long time to straighten up accounts. When Moncton sends out a requisition we want the business as it is quick pay."

The last few days has been better for small tools than for some before.

Scrap Market is Poor

There is little in the scrap market. Business is very slow, and buyers are few and far between. Some of the yards fear to buy as they don't want to stack stuff. Some of them intimate that they look for a resumption of business a little later on, going so far as to couple with this belief that the resumption will be on a lower level than the one they are trying to retain at the present time.

NEW YORK MARKET IS VERY DULL NOW

But Makers of Tools Will Have to Need Business Before Prices Come Down

NEW YORK, July 22.—Machine tool business in this market has dropped to about the lowest point in nearly a year. Most of the current orders are for single tools, and the aggregate of such orders is not very large. There is, of course, a mid-summer lull every year, but the situation is aggravated this year by the transportation conditions and the tightness of money. Moreover, many large buyers of machine tools seem to be convinced that there will be some price reductions, but it is not apparent that machine tool builders will make any move toward reducing prices until they are in greater need of orders than is the case at present. Nearly every plant is comfortably well off for two or three months, at least, and some have orders booked that will keep them busy from six months to nearly a year. Certain types of tools, however, are now to be had for earlier delivery than could have been obtained 60 days ago. There have been cancellations, and the machines cancelled have become available for sale to other customers.

The shipping situation shows improvement. It is still far from normal, but many plants are making better shipments than 30 or 60 days ago.

Most of the large industrial companies in the East have declared to sellers of tools that they are definitely out of the market, for the immediate future, at least. For example, the General and Electric Co. will make no further purchases for its old established plants, but will continue to buy such equipment as is urgently required for its new plants at Baltimore, Md., and Bridgeport, Conn. A new inquiry for one of these plants

calls for fuel oil furnaces, a blower, six moulding machines and one or two machine tools.

The Willys Corporation, Elizabeth, N. J., as bought a number of tools to balance up the equipment of the new automobile plant now under construction. This plant will be in production some time in the early fall, but it will be a year before full production is reached.

BUYERS NOT URGENT, KNOWING THAT THEY CANNOT SECURE DELIVERY

Special to CANADIAN MACHINERY.

PITTSBURGH, July 22.—The iron and steel markets are extremely quiet. Consumers are buying only what they absolutely must have, and that is not a great deal. As sellers are difficult to find when it is so hard to ship it does not require a great deal of demand to give the markets an appearance of strength.

Pig iron and steel producers continue to complain of railroad conditions and it really looks as though they felt the restriction in freight movement more than the consumers. A very awkward market situation is presented. Whether values will advance or decline is almost as uncertain a question as to which way the cat will jump. In pig iron it is almost impossible to buy prompt deliveries, while pig iron is piling up in furnace yards and buyers are consuming less, partly from not having as much business, partly from not getting the pig iron and partly from not getting coke or other supplies. In steel there is the situation of assured prompt deliveries still commanding premiums over deliveries in say two to four months, if the mills really succeed in making those deliveries, and such deliveries in turn being $10 or $20 a ton higher than the Steel Corporation prices, which are for delivery in rotation, the material to be shipped after present obligations to the customer are filled, whenever that may be. As the independents have been catching up with their order books for months past while the Steel Corporation has been falling farther behind, on account of its heavy bookings, there is easily a possibility of advancing prices by independents on account of the much earlier deliveries they will be able to make, while just along the same line there is a possibility of their prices declining through their orders running out.

Shipping Conditions

It is illustrative of the spirit of the times, which is one of criticism and complaint rather than one of enthusiasm and progress, that Service Order No. 7 of the Interstate Commerce Commission, requiring that priority be given to coal cars for coal mines, seems to provoke nothing but criticism and complaint, when obviously the order is necessary and is calculated to provide the greatest good for the greatest number. The coal operators insist that it does not give them any more cars, or at any rate not many more, while the blast furnace and steel interests complain that it has taken nearly all their cars away from

them. The newspaper reports put it in the form that closings of steel works are "threatened" on account of the car situation, the circumstances being that the mill warehouses and yards are filled with steel, about 2,500,000 tons, while the northwest is threatened with freezing next winter.

There is little doubt but that in general the shipments of steel thus far this month have fallen behind those in the same period of last month. One large interest reports tonnage figures that indicate a decrease of 20 to 25 per cent. So large a decrease can hardly be typical, but it is illustrative of the general situation.

Improvement in railroad conditions is all for the future, it being certain that no definite improvement has yet begun. Executive officials of the railroads seem now to be alive to the situation. One of the most prominent has admitted very publicly that the officials have not been co-operating as they should. One may infer that the lack of co-operation is due in part to the government guarantee of earnings, which was extended for six months after the return of the roads to their owners, or to September 1. It is common talk in steel circles now that the guarantee has not worked out well and hopes are entertained that when the railroads have to make their money themselves they will do better. Practically no one doubts that if the railroads fail in the test it will be impossible to oppose government ownership and operation of the roads. Shippers are, of course, utterly opposed to that, and they wish they could be sure that the railroad officials are equally opposed. The problem is to get more efficiency out of the physical equipment now existing. Buying cars would not help, the difficulty now being that there are more cars than are being moved, and as to enlargement of yards and terminals, which seems absolutely necessary, that will require years of time. Daniel Willard, president of the Baltimore & Ohio and chairman of the executive committee set up July 1 by the railroad officials, says the best average freight car movement was in 1916, 26.9 miles a day (while of late of course the movement has been vastly less) when the roads ought to and probably can move cars 30 miles a day, that 32 per cent of the movement is empty, when this proportion could be reduced, and that cars are loaded to only 70 per cent of capacity, a proportion that must be

increased. If the other railroad officials, and the shippers, will co-operate the traffic situation can be greatly improved.

Mill Closings

While much has been said of closing of mills there is no definite evidence that the production of steel has declined from the June rate, which was very good. The difference is in finished product, as finishing departments have been closed or have been running at reduced rates, semi-finished steel therefore accumulating. The idea is that the semi-finished steel can be piled more rapidly than the finished product, and that steel now finished to customers' specifications might prove quite unacceptable months hence when the material would be shipped. While this procedure has some objectionable features the influence upon labor is very satisfactory. The more steel that is piled the more the men value their jobs. When finishing departments are closed the mills always endeavor to arrange closings in rotation so that the men will not lose time for long periods. The labor situation is described as fairly satisfactory, but eminently satisfactory as compared with conditions in some other industries.

Pig Iron

Both Bessemer and basic pig iron have sold at $46, valley, or $1 advance, while foundry iron remains quotable at $45, valley. This, however, is almost a nominal quotation, as prompt deliveries only are inquired for and prompt deliveries are scarcely to be had. One large interest has made a few sales at $45 in the past few days, but states that the sales have been made to old customers only, inquiries from new sources being turned down flat. The question is which will come first, a buying movement or a selling movement. Some consumers of foundry iron cover-

ed only for the first half of this year but as deliveries were delayed, they still have iron due them enough to run into September, as an average period, and some of this iron is at $28.75, valley, so that it would be quite a jump for such consumers to have to pay $45 on their next purchases. Connellsville coke for spot shipment is about $18 to $18.50 for furnace and $19 to $19.50 for foundry, per net ton at ovens, when 14 months ago furnace coke sold at $3.50. Thus while pig iron prices are high a blast furnace may easily be in the position of losing money.

U.S. SCRAP METAL

The scrap market is decidedly slow with indications of softening; however prices remain the same as quoted a week ago with a few exceptions. The car shortage is still acutely felt and the special switching regulations relative to the Chicago district are a great deterrent to business there. There is a reason to believe that other mills will come into the market shortly for old material, and dealers stocks are being held to await this movement.

PIG IRON TRADE

The pig iron market continues to remain irregular on account of the fuel and transportation situation. Sales of fairly large quantities of standard basic iron have just been concluded in the Pittsburgh area for $46. Valley furnace basis and another advance of $1 per ton has been established. There is practically no change in the summer dullness.

and there is no buying in great bulk. Prices hold firm on account of lack of fuel and transportation difficulties. $46 furnace is being quoted for the remainder of this year and first half of next by the Virginia Iron, Coal and Coke Co. This is $1 lighter than the average price for the remainder of the year. There has been active enquiry for export but shipping conditions are too discouraging for anything to materialize.

INQUIRE ABOUT CANADIAN TRADE

Several Firms Want to Get Material and Manufactured Lines From This Country

Enumerated under are a list of foreign firms wishing Canadian trade, Canadian industries interested may obtain full particulars on quoting the identification numbers to the Intelligence Branch of the Department of Trade and Commerce at Ottawa.

No. 1289. A firm in Roumania desires to receive quotations and other particulars from Canadian firms in a position to export iron.

No. 1290. A young Brazilian firm of general import agents in Rio de Janerio, wish to hear from manufacturers and exporters of paints and varnishes, motor cars, lorries and accessories, pumps (centrifugal, etc.), equipment, machine tools, shaft hardware, brass goods, saw mill machinery, dairy equipment, shovels and tools, bicycles.

No. 1291. A Brazilian firm of general importers in Rio de Janerio wish to import agricultural machinery, metal working machinery, hardware of all kinds, copper and brass sheets, brass and pipes, lubricating oils, paints and varnishes and textiles from Canada.

The Week's Events in Montreal Industry

Jenkins Bros., of Montreal, have just completed a new addition to their plant, and this section will be used for the machining of iron valves which they will now manufacture. This will increase the factory floor space by about 16,000 sq. ft. The opening was inaugurated by a social evening to all the employees and the installation of the necessary equipment will be proceeded with immediately.

Following the developments that will eventually give Montreal additional electrical equipment for the pumping of water at the low level pumping station, the League of Proprietors, at a recent meeting, passed a resolution condemning the proposed contract that had been contemplated for the supplying of electrical energy. The objection is based on the belief that the term of twenty-five years is too long, as changing conditions

may make it possible for the city to obtain power at a cheaper rate. The commissioners have decided to call for tenders in connection with these power requirements.

H. T. Bates of London, Eng., is at present in Canada negotiating with different Canadian manufacturing firms with a view of getting some firm to take on the production of certain portions of electric welding equipment. Mr. Bates states that many British firms are reaching out to new fields, both in the way of direct trade and also in the establishment of branch factories to manufacture their products. Mr. Bates is representing the Alloy Welding Process Ltd.; Fuller's United Electric Works, S. Wolf and Company, and the Twin Engineering and Electric Transmission Company, all of London.

In view of the feeling that has existed for some time among business men in Canada it was expected that any strenuous objection will be made to the proposed advance of 30 per cent. in freight rates. Should this increase become effective the rate will be practically double what it was before the war. Speaking at the Montreal Board of Trade one of the members stated: "We do not look for any strenuous opposition to the present demand for an increase in freight rates. We have known, for months past that such an advance must come. On the last advance of 25 per cent., which was made in August, 1918, the Montreal Board of Trade took the ground that it would not oppose such an advance if the Board of Railway Commissioners found it warranted. That will be the probable attitude taken by the board now."

SELECTED MARKET QUOTATIONS

Being a record of prices current on raw and finished material entering into the manufacture of mechanical and general engineering products.

PIG IRON

Grey forge, Pittsburgh	$42 40
Lake Superior, charcoal, Chicago	57 00
Standard low phos., Philadelphia	50 00
Bessemer, Pittsburgh	43 00
Basic, Valley furnace	42 90
Toronto price:—	
Silicon, 2.25% to 2.75%	52 00
No. 2 Foundry, 1.75 to 2.25%	50 00

IRON AND STEEL

Per lb. to Large Buyers	Cents
Iron bars, base, Toronto	$ 5 50
Steel bars, base, Toronto	5 50
Iron bars, base, Montreal	5 50
Steel bars, base, Montreal	5 50
Reinforcing bars, base	5 00
Steel hoops	7 00
Norway iron	11 00
Tire steel	5 75
Spring steel	10 00
Band steel, No. 10 gauge and 3-16 in. base	6 00
Chequered floor plate, 3-16 in.	8 40
Chequered floor plate, ¼ in.	8 00
Bessemer rails, heavy, at mill	....
Steel bars, Pittsburgh	3 00-4 00
Tank plates, Pittsburgh	4 00
Structural shapes, Pittsburgh	3 00
Steel hoops, Pittsburgh	3 50-3 75
F.O.B., Toronto Warehouse	
Small shapes	4 25
F.O.B. Chicago Warehouse	
Steel bars	3 62
Structural shapes	3 72
Plates	3 67 to 5 50
Small shapes under 3"	3 62

FREIGHT RATES

Per 100 Pounds.

Pittsburgh to Following Points

	C.L.	L.C.L.
Montreal	33	45
St. John, N.B.	41½	55
Halifax	49	64½
Toronto	27	39
Guelph	27	39
London	27	39
Windsor	27	39
Winnipeg	89½	135

METALS

	Gross. Montreal	Toronto
Lake copper	$25 00	$24 00
Electro copper	24 50	24 00
Castings, copper	24 00	24 00
Tin	62 00	65 00
Spelter	11 50	12 00
Lead	10 75	11 00
Antimony	13 00	14 00
Aluminum	34 00	36 00

Prices per 100 lbs.

PLATES

Plates, 3-16 in.	$ 7 25	$ 7 25
Plates, ¼ up	6 50	6 50

PIPE—WROUGHT

Price List No. 44—April, 1920.

STANDARD BUTTWELD S/C

	Steel Black	Steel Galv.	Gen. Wrot. Iron Black	Gen. Wrot. Iron Galv.
⅛ in.	$6 60	$ 5 50	....	....
¼ in.	5 13	7 26	$ 5 43	$ 7 56
⅜ in.	5 13	7 26	5 43	7 56
½ in.	6 84	8 42	7 27	8 84
¾ in.	8 45	10 68	9 08	11 16
1 in.	12 50	15 64	13 35	16 49
1¼ in.	16 91	21 16	18 06	22 81
1½ in.	20 21	25 30	21 59	26 68
2 in.	27 20	34 04	29 05	35 89
2½ in.	43 00	53 82	....	....
3 in.	56 23	70 38	....	....
3½ in.	71 30	88 32	....	....
4 in.	84 48	104 64	....	....

STANDARD LAPWELD S/C

	Steel Black	Steel Galv.	Gen. Wrot. Iron Black	Gen. Wrot. Iron Galv.
2 in.	$30 90	$37 74	$34 60	$41 44
2½ in.	45 34	56 16	51 19	62 01
3 in.	59 29	73 44	66 94	81 09
3½ in.	73 14	90 16	82 34	99 86
4 in.	86 66	106 82	97 59	117 72
4½ in.	0 98	1 23	1 24	1 49
5 in.	1 15	1 44	1 44	1 73
6½ in.	1 49	1 86	1 87	2 25
7 in.	1 94	2 43	2 42	2 90
8-L in.	2 04	2 55	2 54	3 05
8 in.	2 35	2 94	2 92	3 51
9 in.	2 81	3 52	3 50	4 21
10-L in.	2 61	3 26	3 25	3 90
10 in.	3 36	4 20	4 18	5 03

Prices—Ontario, Quebec and Maritime Provinces

WROUGHT NIPPLES

4" and under, 60%.

4½" and larger, 50%.

4" and under, running thread, 30%.

Standard couplings, 4-in. and under, 30%.

Do., 4½-in. and larger, 10%.

OLD MATERIAL

Dealers' Average Buying Prices.

Per 100 Pounds.	Montreal	Toronto
Copper, light	$15 00	$14 00
Copper, crucible	18 00	18 00
Copper, heavy	18 00	18 00
Copper wire	18 00	18 00
No. 1 machine composition	16 00	17 00
New brass cuttings	11 00	11 75
Red brass cuttings	14 00	15 75
Yellow brass turnings	8 50	9 50
Light brass	7 00	7 00
Medium brass	8 00	7 75
Scrap zinc	6 50	6 00
Heavy lead	7 50	7 75
Tea lead	4 50	5 00
Aluminum	19 00	20 00
	Per Ton	Gross
Heavy melting steel	18 00	18 00
Boiler plate	15 50	15 00
Axles (wrought iron)	22 00	20 00
Rails (scrap)	18 00	18 00
Malleable scrap	25 00	25 00
No. 1 machine cast iron	32 00	33 00
Pipe, wrought	12 00	12 00
Car wheel	26 00	26 00
Steel axles	22 00	20 00
Mach. shop turnings	11 00	11 00
Stove plate	26 50	25 00
Cast boring	12 00	12 00

BOLTS, NUTS AND SCREWS

	Per Cent.
Carriage bolts, 7-16 and up	+10
Carriage bolts, ⅜-in. and less	Net
Coach and lag screws	—15
Stove bolts	55
Wrought washers	—25
Elevator bolts	+10
Machine bolts, 7-16 and over	+10
Machine bolts, ⅜-in. and less	+10
Blank bolts	Net
Bolt ends	Net
Machine screws, fl. and rd. hd., steel	27½
Machine screws, o. and fil. hd., steel	+25
Machine screws, fl. and rd. hd., brass	net
Machine screws, o. and fil. hd., brass	net
Nuts, square, blank .... +25 add	$2 00
Nuts, square, tapped ....add	2 25
Nuts, hex., blank ....add	2 50
Nuts, hex., tapped ....add	3 00
Copper rivets and burrs, list less	15
Burrs only, list plus	25
Iron rivets and burrs	40 and 5
Boiler rivets, base ¾" and larger	$8 50
Structural rivets, as above	8 40
Wood screws, O. & R., bright	75
Wood screws, flat, bright	77½
Wood screws, flat, brass	55
Wood screws, O. & R., brass	55½
Wood screws, flat, bronze	50
Wood screws, O. & R., bronze	47½

MILLED PRODUCTS

(Prices on unbroken packages)

	Per Cent.
Set screws ....—20%, 25 and	5
Sq. and hex. hd. cap screws	12½
Rd. and fil. hd. cap screws ..plus	25
Flat but. hd. cap screws ....plus	50
Fin. and semi-fin. nuts up to 1-in.	12½
Fin. and Semi-fin. nuts, over 1 in., up to 1½-in.	—5
Fin. and Semi-fin. nuts over 1½ in., up to 2-in.	+12½
Studs	+5
Taper pins	—12½
Coupling bolts	+40
Planer head bolts, without fillet, list	+45
Planer head bolts, with fillet, list plus 10 and	+55
Planer head bolt nuts, same as finished nuts.	
Planer bolt washers	net
Hollow set screws	+60
Collar screws ......list plus 20,	30
Thumb screws	40
Thumb nuts	75
Patch bolts ....add	+85
Cold pressed nuts to 1½ in...add	$1 00
Cold pressed nuts over 1½ in..add	2 00

BILLETS

	Per gross ton
Bessemer billets	$60 00
Open-hearth billets	60 00
O.H. sheet bars	76 00
Forging billets	56 00-75 00
Wire rods	52 00-70 00

Government prices.

F.O.B. Pittsburgh.

NAILS AND SPIKES

Wire nails, base	$6 10
Cut nails, base	7 00
Miscellaneous wire nails	50%

ROPE AND PACKINGS

Plumbers' oakum, per lb.	0 10¼
Packing, square braided	0 38
Packing, No. 1 Italian	0 44
Packing, No. 2 Italian	0 36
Pure Manila rope	0 35½
British Manila rope	0 28
New Zealand hemp	0 28

POLISHED DRILL ROD

Discount off list, Montreal and Toronto .... net

MISCELLANEOUS

Solder, strictly	$ 0 35
Solder, guaranteed	0 39
Soldering coppers, lb.	0 62½
White lead, pure, cwt.	20 35
Red dry lead, 100-lb. kegs, per cwt.	16 00
Gasoline, per gal., bulk	0 38
Pure turp., single bbls., gal	3 15
Linseed oil, raw, single bbls.	2 37
Linseed oil, boiled, single bbls.	2 40
Wood alcohol, per gal.	4 00
Whiting, plain, per 100 lbs.	3 00

CARBON DRILLS AND REAMERS

S.S. drills, wire sizes	32½
Can. carbon cutters, plus	20
Standard drills, all sizes	32½
3-fluted drills, plus	10
Jobbers' and letter sizes	32½
Bit stock	40
Ratchet drills	15
S.S. drills for wood	40
Wood boring brace drills	25
Electricians' bits	30
Sockets	50
Sleeves	50
Taper pin reamers	25 off
Drills and countersinks	net
Bridge reamers	50
Centre reamers	10
Chucking reamers	net
Hand reamers	10
High speed drills, list plus 20 to	40
Can. high speed cutters, net to plus	10
American plus	40

COLD ROLLED STEEL

[At warehouse]

Rounds and squares	$7 base
Hexagons and flats	$7.75 base

IRON PIPE FITTINGS

	Black	Galv.
Class A	60	75
Class B	27	37
Class C	18	27

Cast iron fittings, 5%; malleable bushings, 22½%; cast bushings, 22½%; unions, 37½%; plugs, 20% off list.

SHEETS

	Montreal	Toronto
Sheets, black, No. 28	$ 8 50	$ 9 50
Sheets, Blue ann., No. 10	8 50	9 00
Canada plates, dull, 52 sheets	8 50	10 00
Can. plates, all bright	8 60	9 00
Apollo brand, 10¾ oz. galvanized	....	....
Queen's Head, 28 B.W.G.	11 00	....
Fleur-de-Lis, 28 B.W.G.	10 50	....
Gorbal's Best, No. 28	....	....
Colborne Crown, No. 28.	....	....
Premier, No. 28, U.S.	11 50	10 50
Premier, 10¾-oz.	11 50	10 90
Zinc sheets	16 50	20 00

PROOF COIL CHAIN

(Warehouse Price)

B

¼ in., $13.00; 5-16, $11.00; ⅜ in., $10.00; 7-16 in., $9.80; ½ in., $9.75; ⅝ in., $9.20; ¾ in., $9.30; ⅞ in., $9.50; 1 in., $9.10; Extra for B.B. Chain, $1.20; Extra for B.B.B. Chain, $1.80.

ELECTRIC WELD COIL CHAIN B.B.

⅛ in., $16.75; 3-16 in., $15.40; ¼ in., $13.00; 5-16 in., $11.00; ⅜ in., $10.00; 7-16 in., $9.80; ½ in., $9.75; ⅝ in., $9.50; ¾ in., $9.30.

Prices per 100 lbs.

FILES AND RASPS

	Per Cent.
Globe	50
Vulcan	50
P.H. and Imperial	50
Nicholson	32½
Black Diamond	27½
J. Barton Smith, Eagle	50
McClelland, Globe	50
Delta Files	20
Disston	40
Whitman & Barnes	50
Great Western-American	50
Kearney & Foot, Arcade	50

BOILER TUBES.

Size.	Seamless	Lapwelded
1 in.	$27 00	$.....
1¼ in.	29 50	.....
1½ in.	31 50	29 50
1¾ in.	31 50	30 00
2 in.	35 00	30 00
2¼ in.	35 00	29 00
2½ in.	42 00	37 00
3 in.	50 00	48 00
3¼ in.	.. ..	48 50
3½ in.	63 00	51 50
4 in.	85 00	65 50

Prices per 100 ft., Montreal and Toronto

OILS AND COMPOUNDS.

Castor oil, per lb.	
Royalite, per gal., bulk	27½
Palacine	30½
Machine oil, per gal.	54½
Black oil, per gal.	28
Cylinder oil, Capital	82½
Petroleum fuel oil, bbls., net	18

BELTING—No 1 OAK TANNED

Extra heavy, single and double		6½
Standard		6½
Cut leather lacing, No. 1		2 00
Leather in side	2 40	3 00

TAPES

Chesterman Metallic, 50 ft.	$2 00
Lufkin Metallic, 603, 50 ft.	2 00
Admiral Steel Tape, 50 ft.	2 75
Admiral Steel Tape, 100 ft.	4 45
Major Jun. Steel Tape, 50 ft.	3 50
Rival Steel Tape, 50 ft.	2 75
Rival Steel Tape, 100 ft.	4 45
Reliable Jun. Steel Tape, 50 ft.	3 50

PLATING SUPPLIES

Polishing wheels, felt	$4 50
Polishing wheels, bull-neck	2 00
Emery in kegs, Turkish	09
Pumice, ground	06
Emery glue	30
Tripoli composition	09
Crocus composition	12
Emery composition	11
Rouge, silver	60
Rouge, powder, nickel	45

Prices per lb.

ARTIFICIAL CORUNDUM

Grits, 6 to 70 inclusive	.08½
Grits, 80 and finer	.6

BRASS—Warehouse Price

Brass rods, base ½ in. to 1 in. rod	0 34
Brass sheets, 24 gauge and heavier, base	$0 42
Brass tubing, seamless	0 46
Copper tubing, seamless	0 48

WASTE

XXX Extra	.24	Atlas	.20
Peerless	.22½	X Empire	.19½
Grand	.22½	Ideal	.19
Superior	.22½	X Press	.17½
X L C R.	.21		

Colored

Lion	.17	Popular	.13
Standard	.15	Keen	.11
No. 1	.15		

Wool Packing

Arrow	.35	Anvil	.22
Axle	.28	Anchor	.17

Washed Wipers

Select White	.20	Dark colored	.09
Mixed colored	.10		

This list subject to trade discount for quantity.

RUBBER BELTING

Standard ... 10% Best grades... 15%

ANODES

Nickel	.55 to .60
Copper	.38 to .40
Tin	.70 to .70
Zinc	.16 to .17

Prices per lb.

COPPER PRODUCTS

	Montreal	Toronto
Bars, ½ to 2 in.	$42 50	$43 00
Copper wire, list plus 10.		
Plain sheets, 14 oz., 14x60 in.	46 00	44 00
Copper sheet, tinned, 14x60, 14 oz.	48 00	48 00
Copper sheet, planished, 16 oz. base	46 00	45 00
Braziers', in sheets, 6 x 4 base	45 00	44 00

LEAD SHEETS

	Montreal	Toronto
Sheets, 3 lbs. sq. ft.	$10 75	$14 50
Sheets, 3½ lbs. sq. ft.	10 50	14 00
Sheets, 4 to 6 lbs. sq. ft.	10 25	13 50

Cut sheets, ½c per lb. extra.

Cut sheets to size, 1c per lb. extra.

PLATING CHEMICALS

Acid, boracic	$ .23
Acid, hydrochloric	.04
Acid, nitric	.11
Acid, sulphuric	.04
Ammonia, aqua	.15
Ammonium, carbonate	.23
Ammonium, chloride	.22
Ammonium hydrosulphuret	.75
Ammonium sulphate	.30
Arsenic, white	.14
Copper, carbonate, annhy	.41
Copper, sulphate	.15
Cobalt, sulphate	.20
Iron perchloride	.62
Lead acetate	.30
Nickel ammonium sulphate	.20
Nickel carbonate	.32
Nickel sulphate	.22
Potassium sulphide (substitute)	.42
Silver chloride (per oz.)	1.30
Silver nitrate (per oz.)	1.25
Sodium bisulphate	.14
Sodium carbonate crystals	.06
Sodium cyanide, 127-130%	.38
Sodium hyposulphite per 100 lbs	3.00
Sodium phosphate	.15
Tin chloride	.80
Zinc chloride, C.P.	.30
Zinc sulphate	.10

Prices per lb. unless otherwise stated

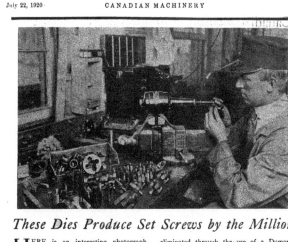

These Dies Produce Set Screws by the Million

Canadian Exhibition in London Not a Success

Causes Working to This End Set Forth in a Report from One of the Trade Commissioners—Many of the Exhibits Were Slow in Arriving—The Season Was Not a Suitable One.

By HARRISON WATSON, Trade Commissioner

LONDON.—It is to be feared that the Canadian Industries Exhibition, which was held at the Royal Agricultural hall, London, from June 7 to 19, failed to realize the hopes of many of the Canadian exhibitors, because the attendance both of the trade and the public was disappointing, although it improved somewhat towards the close of the exhibition. This was particularly regrettable because, while the display was not representative of all Canadian industries, several important Canadian manufacturers exhibited, and the general display was quite attractive.

The inception of the Canadian Industries Exhibition, which was actually the first of its kind held in this country—although Canadian industries had been represented to some extent in previous Canadian Government exhibitions—was due to an offer made last summer to Mr. Lloyd Harris, chairman of the Canadian Mission in London, by Mr. H. Greville Montgomery, an experienced manager of London trade exhibitions, who, having secured the Agricultural hall for a fortnight in June, suggested that the holding of an exhibition of the products of Canadian industries would advantageously supplement the propaganda work which the mission was carrying on.

Unfortunately, however, ill-fate has dogged the venture almost throughout. By the time the organiser arrived in Canada last October, many of the leading manufacturers had become extremely busy owing to the revival of an active home market and saw no prospect of being able to deal with export trade; and although sufficient exhibitors were secured to warrant the carrying out of the project, a number of manufacturers who originally promised to make displays subsequently found it impossible to do so.

A further misfortune, and indeed a severe blow to the prospects of the exhibition, was that resulting from delay in transportation and delivery; the bulk of the exhibits were lying at the London docks at the time originally advertised for the opening of the exhibition, which necessitated its postponement for several days and naturally caused general disorganization.

In any case, the wisdom of holding a trade exhibition—and a new venture at that—during the height of the summer and of the London season is extremely questionable.

The original intention was to hold an exclusively trade exhibition, admission to which would be limited to invitations, but it was subsequently decided that, owing to the participation of the Canadian railways and similar features, the public should be admitted by payment during the afternoons and evenings, and this certainly led some of the exhibitors to expect a much larger attendance of the public than resulted, and for this reason doubtlessly some of the exhibitors neglected to personally circularise firms in their trade, as is customary in trade exhibitions in this country. Upon the other hand it should be emphasized that several of the manufacturers who made comprehensive displays and personally invited prospective customers to call and inspect samples, reported excellent results, and many of them stated that they could have secured very large orders if they had been in a position to fill them.

While there was a certain amount of justification in the complaints made by the exhibitors generally that the exhibition had not been sufficiently advertised, the results of subsequent action taken in this direction was rather to attract persons wishing to view the free displays of Canadian films made by the Canadian Pacific Railway Company than business men wishing to inspect the manufacturers' stalls.

As a matter of fact the management stated that they had issued invitations to some 30,000 business firms, including several thousand names furnished by the Canadian Government Trade Commissioners in this country as known to be interested in Canadian products. The management also reported upon the conclusion of the exhibition that the number of persons who paid for admission at the turnstiles aggregated about 12,000; and while no exact check was kept of the number of visitors presenting complimentary tickets, it was estimated that they approximated between 25,000 and 30,000.

It must also be borne in mind that while the trade exhibitions which are annually held in London and elsewhere are established events, with perfected organizations, the Canadian Industries Exhibition was an entirely new departure, and as such bound to encounter initial difficulties.

A subject of criticism was that the scope of the exhibition was too general, except for propaganda purposes, because the regular series of trade exhibitions referred to above limit themselves to particular branches of trade which appeal to persons engaged in that trade or industry; and it was frequently stated that no particular trade was sufficiently represented at the Canadian Industries Exhibition to repay the ordinary busy buyer for the loss of time involved in a special visit to Islington.

It might be suggested, moreover, that should further exhibitions of this kind be held, the exhibitors should beforehand organize a committee of Canadian representatives to supplement the efforts of the management, which would tend to better co-operation and unification.

To return, however, to the exhibition itself. Although it was a private venture with which the Canadian Government was in no way connected, Sir George Perley, the High Commissioner for Canada, had accepted an invitation to perform the opening ceremony, but owing to other pressing engagements found it impossible to be present upon the postponed date, so the inauguration was kindly undertaken, at short notice, by Sir George McLaren Brown, European manager of the Canadian Pacific Railway Company.

The exhibits, although insufficient to necessitate the utilization of the galleries and outside halls, comfortably filled the whole of the ground floor, and included much that was interesting and attractive.

While, as will be observed from the list of exhibitors which is given later on, several industries, notably heavy iron and steel, and confectionery, were well represented, the display was far from representative of Canadian resources, a notable omission being the pulp, paper and associated industries, with the exception of wall-paper.

It will be noticed, however, that a number of displays were made by the resident agents and distributors in the United Kingdom, several of whom reported being disappointed at the last moment of exhibits which they had confidently expected.

The largest and most comprehensive exhibit of all was made by the Canadian Pacific Railway Company, which also provided the free displays of films already alluded to and rendered every possible support to the management by liberally advertising the exhibition. The Canadian National Railways also made a tasteful display.

It should also be mentioned that the newspapers The Canadian Gazette and Canada both had stalls at the exhibition and indeed did everything in their power to contribute to the success of the undertaking.

In view of the original close connection of the project with the Canadian Mission in London, its successors, the Overseas Branch of the Department of Trade and Commerce, agreed to maintain an office in the exhibition, where an official was in constant attendance. The

three Canadian Trade Commissioners who are at present in the United Kingdom—the Bristol and Liverpool Commissioners being at present in Canada—spent several days at the exhibition, interviewing the exhibitors in the endeavour to supply information about export openings, and to render all possible assistance.

List of Exhibitors

Iron, Steel, Hardware, etc.—Dominion Iron and Steel Co., Ltd., Sydney, N. S.; Steel Company of Canada, Hamilton; Nova Scotia Steel and Coal Co., Ltd., New Glasgow, N. S.; Sheet Metal Products Co., Ltd., Toronto; Dominion Steel Products Co., Ltd., Brantford, Ont.; British Smelting and Refining Co., Ltd., Montreal; Champion Spark Plug Co., of Canada, Ltd.; C. A. Dunham Co., Ltd., Toronto; Gillette Safety Razor Co., of Canada, Ltd.; Hoover Suction Sweeper Co., Ltd., Hamilton; Canada Cycle and Motor Co., Ltd., Weston, Ont.; Machine and Stamping Co., Ltd., Toronto; Maxwells, Ltd., St. Mary's, Ont.; J. H. Connor & Sons, Ltd., Ottawa; Whitman & Barnes Manufacturing Co., St. Catharines, Ont.

Machinery—Spramotor Co., London, Ont.; Massey-Harris, Co., Ltd., Toronto; London Gas Power Co., Ltd., London, Ont.; Chase Tractors Corporation, Ltd., Toronto; Acadia Gas Engines, Ltd., Bridgewater, N. S.

Chocolates and Confectionery—Willard's Chocolates, Ltd., Toronto; Canadian Biscuit and Confectionery Export Co., Ltd., Toronto, exhibiting for: Cowan & Co., Ltd., Toronto; McCormick Mfg. Co., London, Ont.; C. J. Bodley Co., Ltd., Toronto; Wm. Neilson, Ltd., Toronto; Patterson Candy Co., Ltd., Toronto; Maple Tree Producers Association, Ltd., Montreal.

Pianos, etc.—Bell Piano and Organ Co., Ltd., Guelph; Williams Piano Co., Ltd., Oshawa; Sherlock-Manning Piano and Organ Co., Ltd., London; Doherty Piano Co., Ltd., Clinton, Ont.

Furniture, etc.—McLagan Furniture Co., Ltd., Stratford, Ont.; Simmons, Ltd.; Arnprior Cabinet Co., Arnprior, Ont.; Kindel Bed Co., Ltd., Stratford, Ont.

Chemicals, etc.—Shawinigan, Ltd., representing:—Canadian Electric-Products Co., Ltd., and Canada Carbide Co., Ltd.; Canadian Bronze Powder Works, Ltd., Montreal; Palmolive Co., Ltd., Toronto; Channell Chemical Co., Ltd., Toronto; Wall-paper—Stauntons, Ltd., Toronto; Reg. N. Boxer Co., Ltd., Toronto.

Leather—Breithaupt Leather Co., Ltd., Kitchener, Ont.; C. H. Peters' Sons, Ltd., St. John, N. B.

Miscellaneous — Canadian Polishes, Ltd., Hamilton; Imperial Varnish and Colour Co., Ltd., Toronto; Canadian Bee Supply and Honey Co., Ltd., Toronto; E. D. Smith & Son, Ltd., Winona, Ont.

United Kingdom Merchants and Distributors handling various miscellaneous lines—John B. Keeble & Co., Ltd., acting for—Meakins & Sons, Ltd.; Stratford Mfg. Co., Ltd.; Schultz Bros. Co.; Canadian Veneering Co.; Springer Lock Mfg. Co.; Gavenite Products, Ltd.; Canadian Woodenware Co.; Megantic Broom Mfg. Co.

Export Associating of Canada, Ltd., representing—A. Ramsay & Son Co.; C. O. Clark & Bro.; Martin-Orme Piano Co.; McClary Mfg. Co.; Eaton & Sons, Ltd.; Windsor Phonograph and Record Co.

British Canadian Export Co., Ltd., representing — The Barnet-Canadian Kitchen Cabinet and other Renfrew industries.

Canada Overseas Trading Co., Ltd. showing British Columbia salmon, timber, also canned fruits, jams, etc.

A. H. Parker & Sons, Ltd., Bristol, acting for—Maritime Fish Corporation; Eastern Canneries, Ltd.; Saskatchewan Co-operative Creameries, Ltd.; Canadian Cereal and Flour Mills Co.

C. H. Dudemore & Co., Ltd., exhibiting models of British Columbia wooden houses; Dominion Machinery Co., Ltd., representing the "Ellot" woodworking machine and a variety of Canadian machinery; S. D. Simond & Co., Ltd., showing—Maple products, mincemeat, and various canned and preserved goods.

C. H. Baber: Boots and shoes of—Blachford Shoe Mfg. Co., Ltd.; Scott-Chamberlain, Ltd.; Getty & Scott; Nursery Shoe Co.

Dominion Industries, Ltd., exhibiting a wide range of Canadian manufactured products.

Sydney Smith: flour, grain and feed products.

Hill, Seddon & Co., representing—Sales, Ltd., H. Levy & Sons, Ltd., Basque Chemical Co., Non-Such Mfg. Co., Satinette Products Mfg., Co., Normandy Tire and Rubber Co., Smalls, Ltd.

T. M. Stevens & Co., showing different grades of British Columbia salmon and fruits.

NEW MOTOR FUEL NOW USED IN NATAL

Report to Admiralty Says it Has Been Found Possible to Produce It in Large Quantities

London, July 1.—After a series of trials in motor boats, the British Admiralty has just made a confidential report on the far-reaching possibility of using a new motor fuel—"Natalie"—in naval vessels.

This substitute for gasoline, which has been used successfully three years in Natal, South Africa, the fuel manager of an automobile association said to-day, is a mixture of alcohol and ether and other things, and is made cheaply in South Africa, where there are large tracts of waste vegetation, from which power alcohol can be extracted.

It is manufactured in some places as a by-product of the sugar industry.

When a Motorist Loves a Horse

The Ryerson-Conradson
Four Purpose Radial

U.S. AND FOREIGN PATENTS

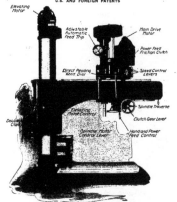

Drilling   Tapping   Boring   Reaming

can be performed on this machine with equal efficiency.

Power consumption practically 40% less than that of other radial drills.

All shafts concentrated in head.   Splash oiling system.

All bevel gears and friction clutches for tapping eliminated.

Exceptional feed range. Sixteen feeds from .005" to .370".

*The new Bulletin No. 4001 describes completely the special features
and advanced design of the Ryerson-Conradson Radial. Write for it.*

ESTABLISHED 1842   INCORPORATED 1898

JOSEPH T. RYERSON & SON
CHICAGO, ILL., U.S.A.
MACHINERY

Canadian Representatives:

GARLOCK-WALKER MACHINERY CO., 32-34 Front Street, TORONTO

MONTREAL   WINNIPEG

If interested tear out this page and place with letters to be answered.

MR. A. E. JUHLER HAS RESIGNED
AS RUDEL-BELNAP MANAGER

MR. A. E. JUHLER, who for some years has been Toronto manager of the Rudel-Belnap Co., has resigned from that position. Mr. Juhler has a wide connection with the Canadian machine tool field, especially during munition making in this country, coming in contact with a large number of manufacturers. His early training fitted him admirably for a machine tool salesman, and enabled him them to Europe, where he joined the firm of V. Lowener in Copenhagen, handling machine tools. About 1907 he returned to Canada, and joined the Imperial Oil Co. engineering forces, later taking charge of the Rudel-Belnap Toronto business. Mr. Juhler has not decided definitely on his future course, as he is considering several propositions at present.

Mr. J. H. Ryder, who has been con-

A. E. JUHLER

as well to give valuable advice to his customer. For some years after leaving college Mr. Juhler was with the Bethlehem Steel Corporation, going for nected with the Toronto office of Rudel-Belnap for some time, will, we understand, succeed Mr. Juhler in the position of Toronto manager.

CLYDE SHIPBUILDING FEELS THE
LACK OF MATERIAL AT PRESENT

THE Clyde shipping output for May is considered very satisfactory, being higher than any other May with two exceptions: May, 1911, and 1907.

For the first time in the history of the Clyde a shipload of 1,500 tons of plates, from Port Talbot, South Wales, had to be reported. If there was a surplus of ship plates available for export, it will then readily be seen that a good many thousand tons could be disposed of easily at a handsome figure.

Engineering departments are very short of material, particularly boiler and ship plates.

There is a feeling that very serious labor troubles may develop during the next two or three months, and result in seriously hampering the industry.

At an exhibition held in Glasgow, Scotland, recently, under the auspices of the Department of Overseas Trade of Great Britain, and the Glasgow Chamber of Commerce and Manufacturers, it was most noticeable that the goods were chiefly of German, United States and Japanese manufacture. This exhibition attracted considerable attention from local manufacturers and merchants, and the exhibits covered the following range of products: electrical smallware and fittings, cycle accessories, leather goods, brushes, toys and games, dyes, soaps, scents, and pomades, stationery requisites, glass bottles, surgical instruments, tools and hardware, woollen undercloth-ing and woollen piece-goods, hosiery, hats and caps, and cotton prints.

Included in the "foreign manufactured" goods there were a few Canadian commodities including two samples of woollen undergarments, two or thee toys, two hand-bags, etc.

One set of samples exhibited should be of particular interest to Canadian firms manufacturing agricultural implements. This set included half-a-dozen hay rakes of various sizes, different from any which are manufactured in Canada at the present time. This rake was made of galvanized iron, and the teeth were hollow. The lower part of the rake was from 13 inches to 20 inches in length, and the teeth were about 3 inches long on the average. One rake, which is 19½ inches wide, had 12 of these hollow iron teeth. This is a German product, and is sold very cheaply on the continent. It is claimed that its lasting qualities are much superior to the ordinary wooden rake, and that it is much lighter than the ordinary Canadian iron rake, and just as durable.

ECHO OF SHELL DAYS

Action brought by the United States of America against Motor Trucks, Limited, for the recovery of certain land and buildings at Brantford which the plaintiff alleges are the rightful property of the United States Government in consequence of a post-armistice agreement, is proceeding at Osgoode Hall before Mr. Justice Kelly. The properties in dispute were obtained by Motor Trucks, Limited, to enable the firm to carry through a contract with the American Government to manufacture 125,000 high-explosive shells, and for the purpose of financing the project the sum of $937,000 was borrowed by the defendant company. But before the manufacture of shells commenced the armistice was signed, the contract was withdrawn, and the parties submitted settlement to the Imperial Munitions Board. On its claim of $2,000,500 Motor Trucks, Limited, was awarded $1,648,000, which sum was paid over in consumation of an agreement by which the company transferred to the United States all properties mentioned in the schedule accompanying it. The action resulted because, the plaintiff claims, description of the company's land and buildings was omitted from the schedule by error. The United States seeks possession of the property on these grounds, while the defendant company claims right of ownership since the schedule did not specifically mention the properties.

TRADE GOSSIP

Sir Lomer Gouin and Sir Clifford Sifton have both consented to join the board of the British Empire Steel Corporation.

The shareholders of the Dominion Steel Corporation at a special general meeting ratified the proposals to merge their company in the $300,000,000 British Empire Steel Corporation.

The H. B. McCarthy Co., Ltd., of Port Hope, have rented the "Helm Mill" property for 3 years on option from the Board of Trade. This move was necessary on account of steadily expanding business.

To Make Briquettes.—The $400,000 appropriation jointly voted by the Federal and Saskatchewan and Manitoba Governments for the construction and operation of a lignite briquetting plant at Estavan has been increased to $600,000.

Windsor is Buying.—Acting on the advice of the Fuel Controller, the Windsor City Council have placed orders with mine operators in Alberta for large supplies of coal to be shipped immediately. Information gathered shows that this coal can now be transported here as cheaply as the coal from the United States.

Error in Advertisement.—It is regretted that in our issue of July 8th, an error occurred in the advertisement of the Dominion Foundries and Steel Ltd., of Hamilton, Ont., which read Drake Drums instead of Brake Drums in one line of the advertisement. This was detected after a few copies were off the press and was corrected for the rest of the run.

ARC WELDING EQUIPMENT

Continued from page 99.

with various thicknesses of plate, the heavier plate will require the use of the heavier currents.

Approximate speeds of welding sheet metal with the metallic electrode are given in the following table:

Thickness Plate	Speed Feet per hour	Cost per foot
1/16	20	2.12
1/8	16	3.12
1/4	10	7.13
3/8	6.5	12.3
1/2	4.3	19.8
3/4	2.6	41.7
1	1.4	61.3

The above figures are based on average figures for materials and labor.

The carbon electrode can be used for welding and for building up metal in a large number of cases where the metal is not subjected to high strains or where it is under compression only.

The average current ranges for different types of work are as follows:

Light welding ....150 to 250 amps.
Medium welding ....250 to 350 amps.
Heavy welding and medium cutting
...................400 to 600 amps.
Very heavy welding and heavy cutting ....600 to 1,000 amps.

The maximum values of current permissible for the carbon electrodes are as follows:

Diameter of Electrode	Maximum Amperes
1-4 inch	100
1-2 inch	300
3-4 inch	500
1 inch ..,.,................	1,000

Graphite electrodes permit the use of somewhat higher current densities but the higher cost of graphite electrodes is a serious handicap to their use. Lower currents than the above may be used, but higher values will result in undue burning of the electrode.

Classified Opportunities

If what you need is not advertised, consult our Buyers' Directory and write advertisers listed under proper heading.

CANADIAN MACHINERY

AND
MANUFACTURING NEWS

Vol. XXIV. No. 5   July 29, 1920

Making Connecting Rods for Overland Four

Have You a Manufacturing Problem? Ten Chances to One the Solution is This, "Make Sufficient Efficient Fixtures to Do the Work—Then Go to It."

By J. H. Moore

IN previous articles dealing with the plant of the Willys-Overland Co., Ltd., we have described and illustrated several general features in connection with the making of the Overland Four, but in this case we will touch on only one part of the car, namely, the making of the connecting rod.

In order that readers can be familiar with all details regarding the machinery of this rod we are about to describe, the sketch at Fig. 2 has been prepared. We would not only suggest a study of this drawing as to sizes, but also as to the careful routine of operations. It is the policy of this plant to issue a set of operation and tool record sheets with every piece that goes into the car, and these sheets are blue-printed and used in special operation books, where they can be looked up for reference at any time. At a later date we intend showing this sheet in detail.

Suffice for the present to state that on these sheets are mentioned all data regarding tools used and the sequence of operations. Referring to such sheets we find that the connecting rod is made from a steel forging and passes through the following 18 operations:

First the rod is straightened on a special straightening fixture on an arbor press, and a step gauge is used to ascertain if the correct height between the large and small bosses has been obtained. Next comes the operation shown at Fig. 1. This consists of the straddle milling of both bosses. The operation is accomplished on a special Becker miller and a rotary fixture is installed. The holding stations are really separate vises and are worked from worm screws. Two small vee blocks are made to grip the bosses securely, and the screw pushing these out tightens the work. Twenty rods are milled at one time, and 250 rods are completed every 8¾ hours. Two sets of cutters are used and the two large cutters mill the large end, while the two small cutters mill the other end. The method of setting these cutters is very simple. Special gauges are made the exact width which the cutters should be apart, and they are set from these gauges. After milling the rods are tested by both step and snap gauges.

The third operation is shown at Fig. 3. This consists of the drilling of a 1 19-32" hole in the large boss on rod. A four-spindle Foote-Burt drill is used for this purpose, and four duplicate jigs are placed on the table as shown in the photograph. Note the type of screw bushings used on these jigs. They are very quick in action and the bottom is bell shaped to centralize the large boss on the rod. The other side of this boss also rests on a stationary bell-shaped bush. The rod itself simply hits up against the jig as shown, this preventing it from turning. Plug gauges are used after the operation to ensure accuracy.

The fourth operation consists of chamfering the hole which has just been drilled, and the work is done on a 20" McDougall drill by means of a 90° countersinking tool. Next comes the fifth

FIG. 3—USING ALL FOUR SPINDLES TO SPEED UP PRODUCTION.

FIG. 4—A SIMPLE TYPE OF DRILLING JIG USED FOR THIS WORK.

operation shown at Fig. 4. The 23-32"
hole in the small boss is now being drilled, and the method of holding the same
is self-apparent. The jig is so arranged
that the large hole in the rod goes over,
and is centered by an oval-shaped pin,
whose major axis is the correct diameter
to suit the hole. The small boss of the
rod fits up against a vee block, and the
tightening screw, set at an angle,
squeezes the rod both down and into
the vee block. This work is accomplished
on two 22" Barnes drills, the one operator attending to both machines, and
sometimes also the chamfer operation
previously mentioned. Plug gauges are
used after this operation to make sure
the hole is keeping to correct size.

Broaching the Rods

Next comes the sixth and broaching
operation. The work is performed on a
No. 3 Lapointe broaching machine, with
two broaches, one the size of the large
hole, the other the size of the small
hole. Fig. 5 illustrates this machine at

FIG. 1.—MILLING BOTH ENDS OF CONNECTING RODS ON A
BECKER DOUBLE SPINDLE MILLER.

work, and the method of holding the
rods can be clearly noted. Oval-shaped

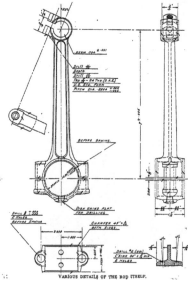

VARIOUS DETAILS OF THE ROD ITSELF.

pins locate the one end of the rod, and,
of course, the broaches locate the other.
The operator can keep both broaches
going very easily and the production
figure is all that could be desired. The
broaches are of sectional design, this allowing to regrind and push up the sections as required. The size of broaches
are 1⅜" and .734", and a combination
plug gauge is used after the operation
to test both holes.

The seventh operation is that of drilling the bolt and oil scoop holes. This
is accomplished on a seven-spindle Natco
drill, but only five spindles are used.
A special indexing jig is made as shown
at Fig. 6, and briefly, the drilling performed is as follows. The two bolt
clearance holes are drilled half way down
in the rod, then the two tapping size
holes are drilled the remainder of the
distance. Lastly a 11-32" hole is drilled
on top of the rod to suit the pipe tap
which is later tapped into this portion.
The jig as shown at Fig. 6 illustrates
the style of locating pins used. One of
these can be seen at the small end of the
loading station A. The jig locates every
90 degrees by means of a spring handle
as shown.

The next operation is termed operation 7A, and is merely the minor operation of reaming out the bolt holes. This
is done on a .25" Barnes drill, and a
special Hoefer drilling head is used on
the single spindle in this way, making
the drill a multiple spindle machine, and
both ⅝" reamed holes are finished at
one time.

Operation No. 8 consists of tapping
the ¼" pipe tap hole on top of the rod.
This is done on a 20" McDougall drill,
an Errington tap chuck being used. As
this operation is so simple we need make
no further comment. The ninth operation is that of spot facing the two bolt
holes. This is also a very simple operation, so we pass on to the tenth step,
namely the milling of bolt head surface.
This is done on a No. 3 Kempsmith, with
a fixture for the purpose, but as the
operation has no special feature we will
pass on to the eleventh operation, depicted at Fig. 8.

FIG. 5—THE BROACHING FIXTURE USED. NOTE THAT ONE BROACH IS SMALL, THE OTHER LARGE.

Using Four Spindles to Advantage.

This is really a combination of operations, Nos. 11, 12, 13 and 14, and these steps are accomplished on the same drill, this being a four-spindle Leyland-Gifford type. The first step is to drill an 11-92" hole as shown at A, the second step the drilling of a 17-64" hole as shown at B, while the third step is the tapping of a 5-16 x 24 SAE tap hole as shown at C. This is, of course, the natural sequence of operations, as first the hole at A was a bolt clearance hole, the second hole, at B, the tapping size hole, and the operation at C the actual tapping. The fourth step in this series of operations is the drilling and countersinking of the two oil holes in the lower boss of the connecting rod. A combination drill and countersink is used for this purpose. Note the style of sliding clamps used on these jigs, and in order that readers can see the spring used behind these clamps we have left the clamp off altogether at C. This style clamp can be adopted in many styles of jigs, and should be used as it is exceedingly quick and handy.

The fifteenth operation consists of the sawing of a ¼" slot and facing of bolt head surface on the small boss end of the rod. The work is accomplished on a Toledo hand miller, and needs no special comment.

The sixteenth operation, shown at Fig. 9, is that of sawing off the cap. This is performed on a No. 3 Kempsmith miller, and a special milling fixture is used. Referring to the photograph we note that the fixture is of the revolving type, the handle A locating the fixture by means of a pin entering a hole similar to that at B. In fact, when the jig is reversed from the position shown in photograph the pin is engaged with the hole B.

An explanation of the action and loading of this fixture will be time well spent. First note the shape of locating pin at C. This shape of pin is used universally throughout the plant, having proved itself to be much better than the regular round pin. The plug D is made of correct diameter to suit the hole in the rod, and this plug is removed after the cap has been sawn off. The plug is held in place by the screw E, and this screw is tightened by the wrench F. The construction of the tightening screw, as shown, is well worth duplicating in any jig, especially where a good pressure is needed in a speedy manner. A wrench such as shown at F is easily made and accomplishes its purpose admirably. The jig itself revolves 180 degrees, and while the cutter is slitting or sawing off the one cap, the operator is unloading the completed one, and loading on a fresh one.

The seventeenth operation is a bench operation and consists of the removing of all burrs or rough spots of any kind on the rod. The last, and eighteenth operation, is that of final inspection.

Having thus completed our journey of operations, and left the rod in a completed state in as far as machinery is concerned, we need not consider the babbitting or assembling of the rod to the engine, but had better summarize what we have discovered during our tour and discussion.

To the writer's mind we have discovered various things of interest, chief among them being the need and advantage of systematic planning of operations. It will be noted that every operation has been carefully thought out and planned step by step. There is no lost motion, and in various cases the one operator performs several operations unassisted. The machinery is so laid out that the rods pass directly down a line and do not have to be trucked here, there

FIG. 6—A CONTINUOUS DRILLING FIXTURE FOR TOP OF RODS.

FIG. 8—A GOOD EXAMPLE OF FOLLOW UP STEPS IN DUPLICATE MANUFACTURE

and everywhere. In every case the rods are handled on special trucks which have wooden pins in them on which the rods hang. Every operation has its gauge or gauges so that there is no guessing as to the product being correct. Each operator is a specialist in his own particular operation and this obtains maximum production. The benefits derived from special fixtures are fully realized and the most is made of this fact. By such a statement we do not mean that they spend money widely on intricate fixtures, but rather that they make the fixtures as simple and fool proof as possible, yet in every case efficient. This combination of simplicity, efficiency and fool proofness is an ideal one and should always be adhered to, for it is not the intricate or expensive jig that always turns out the best or the most work.

All those points are well worth remembering, and if such a concern as the one we have been describing finds it good practice to adopt such a procedure, it is surely a safe and sane policy for every reader to follow out.

CONE PULLEY BELT SHIFTERS

Speaking on the above subject before the summer conference of the National Safety Council (Engineering Section), Mr. H. E. Somes, electrical engineer of the Chevrolet Motor Co., made the following statements:

At your last meeting, I volunteered to obtain time studies on mechanical belt shifters for cone pulley belts and on the hand method. After going through a number of large machine shops, I was very much surprised to find that, with the exception of an isolated instance, belt shifter devices for cone pulley belts were not used, and due to that fact, and also the short period of time left to investigate this subject, I have been unable as yet to obtain the time studies. One of our factory managers who has a successful belt shifter installed expressed it as his opinion that a green workman can shift cone pulley belts by means of the belt shifter as rapidly as an experienced workmen can shift them by hand —with the additional advantage that in

doing so he is safeguarded from injury. However, in my investigations on this subject, I have been convinced of the fact that mechanical belt shifters can be developed that will shift belts on cone pulleys more rapidly than by means of the hand shifting method—much more safely and efficiently and without serious injury to the belt itself. Possibly such shifters have already been developed.

The problem, however, seems to be where should cone pulley belt shifters be required? Such a shifter has its maximum value when it is installed on a machine where belts are shifted many times per day. If a saving can be realized in the workman's time, due to an improved method of shifting belts and the machine itself has a greater capacity, these advantages will become more apparent as the number of belt changes increases. Usually the advantages become of lesser importance as the number of shiftings decreases.

I would submit, therefore, at this time that mechanical belt shifters for cone pulleys are particularly adapted to shop installations where frequent shiftings of the belts from one step cone to another is necessary. Among such installations we find the tool room, the maintenance or the repair shops, and many other types of manufacturing machine shops which can possibly be classified wherein belt shifting is a necessity.

Considering the modern manufacturing machine ship, such as the automobile manufacturing plant, we are struck at once by the fact that belt shifting is very uncommon, and in fact in many cases, if the manufacturer could obtain them, single purpose machines would be purchased.

In a shop of the type commonly called production machine shop, we find that an individual machine is tooled up for a single job, and that frequently this machine runs several years without having any change made in the method of manufacture or the speed of either feed or spindle.

Illustrating this point, I was in Flint, Mich., recently and went to our plant to see a screw machine which had been fitted with a mechanical belt shifter. I asked the operator what he thought of the shifter in question and how he found he could shift the belt by means of the shifter. He replied that he had been employed on that particular job for approximately eight months; he thought the belt shifter was very satisfactory, but that he never used it to shift belts with, excepting to satisfy himself that the shifter would work. This condition exists throughout that entire plant.

The field where the belt shifter would give greater service would undoubtedly be in a shop not engaged in production manufacturing, but rather where on account of the special nature of the differ-

FIG. 9—THE ROD SLITTING FIXTURE. NOTE ITS REVOLVING ARRANGEMENT.

ent jobs, wide variations in the spindle speeds are necessary. This would also apply to drill press operations where the step-cone pulleys are not on the machine itself and where one is exposed within twelve or eighteen inches of the floor. Drill presses, excepting where they are used over long periods of time on a single operation, require a large number of changes throughout the working day, owing to the different sizes of drills which may be used on an individual job.

While observing a belt shifter at Flint, several days ago, I noticed that it seemed to be very satisfactory in shifting from high to low; but in shifting from low to high speeds, I noted that the belt curled and that it was very difficult to make this shifting. The plant engineer advised me that the shifter had been thoroughly satisfactory until the belt became oil soaked from long use.

In order, however, that we may obtain accurate information as to the advant-

ages of using mechanical belt shifters over the hand shifting method, and also in order that we may know definitely whether a satisfactory shifter has been developed, our company has arranged to conduct a test on several types of shifters. This test will be strictly impartial, and as a result, we hope to learn whether by using a mechanical shifter the belt is injured, and whether the shifters are satisfactory after the belt has become thoroughly oil soaked and very pliable.

The Manufacture of Iron by the Use of Rolls

This Concludes the Article which Appeared in July 15th Issue— The Design of Rolls is Next Discussed, Also Various Types of Gases Used.

By W. S. Standiford

Producer gas, unfortunately, has only one-half of the heat calories contained in the natural article, it being therefore more expensive to use. As regards powdered coal, the writer must say that he is prejudiced against that material for use in heating furnaces, for the reason that the moisture or some other cause made the powdered coal clog the blast pipes where it entered the furnaces, the result being that—in one mill where the writer saw it used—the heating furnaces had to be shut down for a time, while the machinists cleared the pipes, it sometimes taking two to three hours. This would occur at different intervals during a month's time, it sometimes taking longer for the pipes to clog up.

New Type of Gas

There is a new gas that has been invented by a man named Elliott, that is far superior in cheapness and heating power to the natural gas, which it would seem has a bright future ahead of it for the melting and heating of iron or steel, etc. The Struthers Furnace Company, of Struthers, Ohio, are using the new gas to operate blast furnaces, also the city of East Palestine Ohio, has discarded natural gas, which latter has gotten to be unsatisfactory for use in cooking stoves owing to its weak pressure

and uncertain supply. East Palestine having bought the mains from the natural gas company, is installing the necessary apparatus to make the new gas. The problem of fuel in these modern times in our days of keen competition has become a most important item in the overhead costs of article production. The heater of iron or steel, by his handling of the furnace, can spoil the best metal ever produced—too quick heating burns the outside, while the interior of the pile is not hot enough to weld. If they are sent to the rolls in this condition, the resulting bars will be burnt on the outside and not welded together on the inside. When bent, such a bar will break easily and split lengthwise through its centre. If the iron is, on the other hand, left too long in the furnace, it loses its quality by the absorption of gases given out by the coal, the result being that the metal is brittle and it has little tensile strength. The guide or finishing mill comes next in description.

The rolls in this mill are smaller and lighter than those used on the puddle mill, the latter weighing 5,000 lbs. and over, while the heaviest ones operated in the guide mill weigh about 1,000 pounds apiece. There are various sizes of rolls used on guide mills, some being 5, 6, 7, 8, 9, 10 or 12 inches in diameter.

The smallest sized rolls, such as those of 5 or 6 inches diameter, are generally used to make the smaller sizes of rounds, such as 3-16 of an inch, etc.; the smaller rolls require to be driven at a very high speed, as the iron or steel gets cold very quickly before it reaches the finished size. Small rolls wear out more rapidly and require dressing oftener, which is due to the sharper peripheral surface and contact points rubbing against the metal rolled. Some persons in rolling mills seem to think that only 5 or 6-inch diameter rolls should be used for making 3-16 or ¼ inch metal. But it has been found out by experience better to make small sizes of iron and steel on larger rolls as they do not wear out so rapidly. This is especially true when hard manganese steel rods are to be rolled. Wire rod mills use rolls of 12½ inch diameter, the rod made, a number 5, being about 7-32 of an inch thick. The mills that we are to make these 1½-inch round on uses rolls ten inches in diameter and consists of the roughing, strand, ovals and finishing rolls. Each set has its own pair of housings securely fastened to the bed, and all rolls are connected together by means of boxes and spindles as described in the puddle mill. The roughers are three-high and are driven at the middle, the

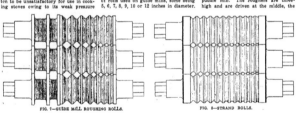

FIG. 7—GUIDE MILL ROUGHING ROLLS.   FIG. 8—STRAND ROLLS.

centre roll revolving in an opposite direction to that of the top and bottom rolls. This arrangement enables the iron to be rolled between the grooves of the bottom and middle rolls and then passed between those of the top and middle rolls. It will readily be seen that by using a three-high mill, much time is saved over that of the two-high design as well as a greater output of rolled metal being obtained.

The passing of the iron over the top roll in order to enable the worker to put it into the next groove smaller in area than the one that preceded it, is avoided by the three-high system. The worker also at the end of his day's work is less fatigued by using the three-high rolls, as a minimum of heavy lifting is done. The average quality of the iron rolled is also better—which is due to the fact that it has a better chance of its component parts being welded together. It is absolutely necessary that the pile should be welded into one solid mass while it is at a white-heat, and this is best done with three-high rolls, as no time is lost dragging the pile over the top roll as is done in the two-high mill.

Fig. 7 illustrates a good working set of guide roughers, that is made to take a box or bar pile 4 inches wide. This style is used extensively and they are called "box and edging" roughers. The use of square grooves in rolls of this description allows very heavy drafts to be used to weld the metal together. After the pile is heated in the furnace, it is pulled to the front by the heater, who uses a long rod for the purpose. It is then gripped between the jaws of a long pair of tongs, which latter has a ring through it. Overhead is a curved track containing a two-wheeled trolley traveller having a chain of suitable length

rolls is on the top and bottom of the pile, the iron getting welded together in that direction. In the edging grooves, the sides are well compacted together due to the pile being turned over and worked on its sides. After it goes through the top and bottom edging passes, it emerges in the shape of a long bar having sides of equal width.

By reference to Fig. 7, of the guide roughers, it will be seen that all remaining grooves are put in roll on an angle. This causes the metal to be rolled on all four sides and also on its top and bottom but not the corners opposite the joint between the rolls. These passes are called "square grooves," although in reality, their angles are five degrees more open than a square which has a 90 degrees angle. In general, roughing roll angles vary from 95 to 100 degrees, the latter being called a "diamond." The variation in angles in various mills is due to the personal preference of the roll designer, some believing that one angle works the impurities out of the iron better than another. Square and diamond grooves do not allow of such heavy drafts as can be gotten by using box and edging passes; the latter are also safer to work with in regard to the flashing of melted metal of which there is always a certain amount in the interior of a white-hot pile. Some roll designers use a diamond groove for the first pass instead of the box variety. While a diamond groove reduces the metal quickly, it is a most dangerous form to use on a pile that contains melted metal in its interior as the violent change in shape from a square to a diamond, causes the molten iron to spurt out viciously which the rougher tries to dodge, but often in vain. Any visitor to a rolling mill will

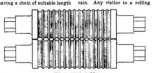

FIG. 9.—OVAL ROLLS.

fastened to one end; the other end of chain is attached to the tongs. This takes the weight of the pile off of the worker. The latter, seizing the white-hot pile with tongs, runs at full speed to the rolls, the object being to get it there as quickly as possible in order to avoid as much loss of heat as possible during transit. Then he pushes it into the largest pass in bottom roll and returns to his furnace. Stationed at the rolls are two roughers, one on each side. After the pile goes through rolls, the rougher on other side returns it through the top roll groove. The pile is then turned over on its side and sent through the next pass in the bottom roll called "edging" pass. In going through the first two passes the work done by the

notice that the roughers' clothes are pretty well shot to pieces by flashes from the rolls. The writer has seen a man who had one eye put out by a flash. The companies usually provide goggles for their roughers, but the men get careless and leave them in their cupboard. There would not be such violent flashing in using a diamond pass, if the pile was put in the pass on an angle, but this would not weld the material together as effectively, as when the pile is placed with its flat side facing upwards.

After the bar goes through the last edging pass, it is turned over on its corner and run through the first square groove in the bottom roll; the rougher on the other side then turns the bar over to a right angle to the direction in

which it came from the first pass and inserts it into the next smaller groove between the middle and top rolls, each rolling making it reduced in section and also longer. The roughers on both sides continue the process of passing the bar back and forth through the various sized passes until it is of the required size to make the 1½ inch round, which in this case is 1 1-2 inches square.

Now to the Strand Rolls

At this stage, the bar is then ready to go into the strand rolls. These are three-high and are made shorter in length than the roughing rolls. The grooves are also put in on an angle, the latter being 92 degrees. They are made out of chilled cast iron, which wears better than the soft variety. Chilled rolls are made with an intensely hard surface by casting in a metal jacketed mold of the shape desired. When the hot iron comes into contact with the cold sides, the latter cools suddenly, the result being a fine and close-grained material that will stand a lot of wear. The chilling of iron also makes it very brittle; so that the rolls have to be handled very carefully in putting them in the mill. Strand rolls usually contain a number of duplicate grooves. This is done to keep the number of rolls on a mill to be changed down to a minimum. By arranging them in this manner, the strands will last from 3 to 5 months in a mill before they require dressing. The bar from the roughing rolls goes into the 1 3-8-inch pass between the middle and bottom rolls; then it is put through the 1¼-inch grooves between the middle and top rolls. It is then turned over at a right angle to its previous position and put into the same pass between the middle and bottom rolls. This is done so, that all the corners will measure the same in diameter as the metal is getting near the finishing rolls, and it is necessary to have it as perfect as possible.

Fig. 8 shows what strand rolls look like. The next set to be described are the oval rolls illustrated in Fig. 9. These are made two-high as only one pass is needed to convert the square strand bar into an oval one. The length of body is 30 inches, not including necks and wobblers. They also have duplicate grooves. This set contains the following grooves: one¾, one 13-16, two 7-8, two 15-16, three 1 inch, two 1 1-16, two 1 1-8, one 1 3-16, and one 1¼ inch sizes. From the foregoing, it will be seen that there is a sufficient variety to keep the rolls in the housings for quite a while before they need dressing. They are also made out of chilled iron and wear well. The width of the oval to make a 1 1-8-inch round bar is 1 7-16 inches with a total thickness of 15-16 inch. This thickness is divided, half of the depth being put in each roll when they are turned; allowance is also made for wear and springing apart of the rolls which ensues when a bar enters. The reduction of the square strand bar to an oval-shaped one allows heavier drafts to be used, the metal thus being rolled quicker. By the time the strand bar is ready to go

into the oval pass, though the reduction is heavy, there is no flashing, as practically all of the impurities are worked out of the metal by this time. In the manufacture of small rounds such as those of ¼ or ⅜ of an inch sizes, the metal from the strand rolls is reduced by passing the strand bar into an oval groove—the oval bar then going into a square pass. The process is repeated alternately until the final sized oval is

inches, it will be seen that the strand bar is spread out to the oval width, the reduction on top and bottom sides doing the work, and filling the area of the oval pass completely. There are different styles of oval grooves put in rolls; some are very wide and not very thick; others approach the size of round to be made, the thickness being nearly the same diameter as the round and the width only 3-16 of an inch larger. Such

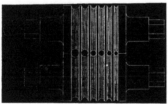

obtained, ready for the finishing rolls. If this were not done, the iron would get cold too quickly, long before the finishers were reached, as small iron cools very rapidly.

The manufacture of iron while it is hot is also easier on the machinery and does not put such heavy strains upon it; as well as on the rolls. In fact, nothing will wear out rolls more quickly than running cold iron into them. It must be understood by this, it is meant that the metal is black-hot and not cold as defined in the absolute sense of the term commonly used. The strand bar has its end cut off while it is hot, by a small shears placed in front of the rolls and operated by suitable means. The ends of any bar are colder than other parts, which is due to contact with the iron floor plates, so in order to make strand bar enter the oval groove easily about six inches is cut off of it. Only occasionally do the other bars from the roughers need one end cut off, as they are hotter than the bars which are nearly finished. After the strand bar's end is removed, it is shoved into the oval—flat side facing upward, and held in this position while passing through the groove by side guides.

The 1¼-inch strand bar is in actual practice made slightly larger than the sizes given, it generally being about 23/1000 of an inch oversize; this allows for the scraping action of the guides, which have to be made sufficiently tight to prevent the oval bar from turning over; which would cause a jam in the pass with a probable breaking of the roll. The roller in charge makes the bars slightly larger by means of the housing screws. As the width of the strand bar is 1¼ inches and oval 1 7-16

an oval will not cause the rolls to wear as heavy as the thinner and sharper ovals do; therefore they last longer in the mill before requiring dressing. There is one drawback to the round shape ovals, viz., they take more force and are harder to get into the finishing round groove, especially when the speed of the mill is high. On the other hand, the sharper edged ovals enter a finishing groove very quickly, whether the speed of the mill is high or low, hence they are extensively used by a great many mills throughout the country.

Design of Finishing Rolls

We now come to the resign of the finishing rolls depicted in Fig. 10. As this is a 10-inch train the rolls are ten inches in diameter. The top roll is made of 1-8 of an inch larger so as to throw the iron down. Their length is 12 inches, not including that of the necks and wobblers. They are two-high as only one pass is required to convert the oval bar into a round one. Rolls contain the following passes: One ⅝, one 15-16, two 1 inch, one 1 1-16 inch and one 1¼ inch groove. As a general rule, there is a heavier demand for one inch iron than for some of the other sizes, thus an extra pass is put in for the former. This enables these rolls to be kept in the housings much longer, the idea being to have all of the passes equally worn before they are sent to the lathe for dressing. Dressing consists of taking just enough metal off rolls so as to make them smooth. Turning rolls is the making of new ones from rough castings as they come from a foundry. When the bar comes from the oval rolls, it is taken by the workman (called "the finisher") and inserted in the guides with sharp

edge of oval facing upwards. The finisher then waits until the iron is at the right heat—which is between a cherry-red and a yellow color, and then pushes it all the way through the guides and into the grooves, the rolls then pulling it through by friction. In front of the pass used, on discharge side, a rail is laid on its side or piece of angle iron as preferred by the roller. This guides the hot bar straight on its way to the place to be cooled, called "the hot bed." The temperature at which iron is rolled in the guide roll is most important; too high a heat causes the oval to over-fill the finishing round groove, thus making it mark the iron bar at joints between top and bottom rolls.

With too low a heat, the metal will not fill out the entire area of pass in rolls, thus making a bar of uneven diameter. The temperature of finishing iron and steel is of the greatest importance to the appearance of the finished product. Too high a temperature will cause a thin layer of scale, which will fall off in spots, especially under strain of bending and spoil the looks of the bar. Too low a finishing temperature will yield a spotty-reddish appearance, which is prone to oxidize or rust. To prevent the latter in damp warehouses, some mills pass their iron as it comes hot from the rolls, through a trough containing cuttings of fibres or leather saturated with tar—the vapor of which will give the metal a thin, glossy coat. Others paint their bars when cold; while some mills leave them in their natural color as they are rolled.

The setting of guides in the finishing rolls is also a matter of great importance to the appearance of the finished bar. If one guide is set farther over on one side of the groove, the iron will twist from that guide and have a flat side, and also mark the iron on the opposite side of bar. Should one guide be higher than the other, there will be a twist from the higher guide. If the oval is not quite large enough to fill the area of the round groove, there will be flat sides to the round. Should the grooves in top and bottom rolls not be exactly opposite each other, the iron will be marked on its sides. It will be seen from the foregoing, that both rolls and guides have to be set carefully in order to make as perfectly shaped bars as possible. The straightening of the metal comes next. This is usually done by boys. After a bar leaves the rolls, the straighteners take hold at each end with tongs, pull on it and then throw bar down on hot-bed and roll it against the other bars. After the first one is straight, it is an easy matter to make the others the same. They are next cut to desired lengths and put on a small car and weighed and put in warehouse.

Where mills make a specialty of supplying railroads and car builders with iron (which is tested by both the makers and railroad inspectors), it is absolutely necessary to keep the chemical constituents and rolling factors controlling manufacture as uniform as possible, in

order to keep the tensile strength and elongation of metal testing highly, if the companies' reputation for manufacturing a high-grade product is to be maintained. So it will be seen that all persons engaged in making the product have to work together if high-grade iron is desired. To readers who are not familiar with rolling mill work, the methods of lubricating the hot necks of the rolls will be interesting. The contact of hot iron with the rolls heats them through so that they have a very high temperature. This is kept down to a certain extent by means of streams of water, used on both rolls and necks. The latter have pieces of beef suet placed against them; the heat causes the suet to melt

gradually, and the water carries the melted suet into the bearings. On some journals, a thick black grease is used. But the suet is superior on account of its great stiffness. The lubrication of the necks is very effective when either of the above-mentioned materials are used, it being renewed from time to time as needed. The roll turners' part of the work, in designing and turning the rolls, is most important and complex. Rolls that work well in some mills will not do so in others, but require to be altered. This is due to the material used and the engine power. Attempts have been made to find quicker methods of making iron bars than by rolling.

One of these consisted of a large cylinder, open at one end; the other had a removable die with a 1-inch round hole in its centre. The cylinder was charged with hot iron. The piston being worked by hydraulic power was inserted and operated, the metal emerging in the form of a long round bar. On testing, the quality was found to be very good —but the dies wore out rapidly and would not stand the heavy pressure combined with the heat imparted by the metal. It is most likely, that nothing better can be devised to take the place of making sections in iron and steel, etc., than the rolling process, which is used extensively in all civilized countries.

Grinding as Applied to Machine-Shop Work

Australia is Vitally Interested in the Art of Grinding as This Article, which Appeared in "Commonwealth Engineer," Clearly Shows—A Perusal Will Prove Worth While.

HISTORICALLY, the use of revolving wheels and also abrasive powders for shaping, forming and polishing materials, is one of the oldest mechanical practices. In modern engineering works, however, it is only comparatively recently that their use has become general as a machining process. The general principles of abrasive wheels and grinding processes are still new to many Australian shops.

There are two distinct machine operations in which an abrasive material is employed, viz., grinding and lapping or polishing. These found a very wide use in munition manufacture during the war; in after-war plants they are being very extensively employed. Grinding is applied to the forming or finishing of parts by the action of a rigid revolving wheel composed of a hard abrasive substance. Lapping or polishing is employed only when the highest accuracy or finish is desired, such as in finishing gun tubes, precision tools, gauges, etc.; the abrasive substance used for this purpose is in the form of a powder which is carried embedded in some soft medium such as lead, or gummed to a disc or belt. Lapping requires much hand work and is expensive. In Australian workshops this process is very little called for.

Action of Grinding

Grinding compared with other machining processes, such as turning, planing, etc., in which a single point tool is employed, enables more accurate work to be carried out, it produces a finer finish, and the hardest metals may be dealt with. On the other hand, it requires greater motive power and more skill in keeping the wheel and machine in proper adjustment for turning out good class work. The reason for these differences can be understood by comparing the action of a grinding wheel with that of a single point tool, if we regard the wheel as consisting of a large number

of single point tools, or as a multiple tool milling cutter.

Application of Grinding

It is obvious that each grain of abrasive material on the surface of a wheel offers very inefficient cutting angles. The top rake on the points being invariably in the wrong direction and also on a blunt grain, the bottom rake or clearance is practically non-existent. Truing up the wheel with a diamond improves the bottom rake, but has little effect on the cutting angle or top rake. Further, in grinding the length and depth of cut and the feed are all small compared with similar qualities where a single pointed tool is employed. The result is that the material is removed in very fine particles.

This method of removing metal obviously requires more power to be expended than when heavier turnings are cut, and, coupled with the inefficient cutting angles, it accounts for grinding requiring a great deal more power than if the same amount of material were removed by a single pointed tool. For this reason grinding is rarely employed as a roughing process.

Certain firms specializing in grinding now claim that they are producing wheels and machinery which make grinding efficient for removing mass material. However, a study of up-to-date shops shows that this claim is only justifiable for special cases. In repetition work it is sometimes possible to finish a part with sufficient accuracy from the rough by one grinding cut, and also to achieve this by removing a minimum of metal, possibly one-thirty-second of an inch only. Turning, on the other hand, would require to take, perhaps, a 1-8 inch cut to allow the tool to cut well under the hard skin of the casting or forging. This obviously means a considerable saving in material in favor of grinding.

In general, grinding is used to supplement the usual machining processes where accuracy is essential. The bulk of the metal is removed, leaving two or three thousandths of an inch for the grinding operation. Under these conditions it is not unusual to attain an accuracy on a finished product of two-ten thousandths of an inch. With a finishing cut amounting to only a few thousandths, the effect of spring in the grinding machine is practically eliminated. In precision grinding, by reversing the direction of the cut or feed, the effect of wear on the grinding wheel itself is inappreciable.

Grinding Wheels and Abrasive Materials

For machining materials of different toughness and hardness, wheels of varying coarseness, hardness, and toughness, are necessary. The coarseness or fineness of the material in the wheel is called the grit. The hardness which depends on the quality and quantity of the material cementing the grains in a wheel is termed the grade, while the hardness or toughness of the abrasive itself is varied by employing different kinds of abrasive material.

In manufacturing the abrasive material the artificial or natural mass is crushed and sieved into sizes designated by numbers ranging from 6 to 250. These mean the number of spaces or wires per inch in the sieve, e.g., No. 30, has passed through a sieve with 30 wires to the inch run. Sizes finer than 250 are separated by making use of the fact that a small particle takes longer to settle in water than a larger one. The powder from No. 250 sieve is mixed with a slowly moving stream of water which classifies it on the principle of riffles in tin or gold concentration into sizes known as F, FF, and FFF, etc. There are other methods of designating this, such as the time taken for the powder to settle

in water, but the one mentioned above is that in most general use.

The most common grits for wheels used in engineering shops are Nos. 24 to 80. A coarse grit cuts larger chips, and is therefore used where a considerable amount of material is to be removed; the coarseness of the grit is governed by the toughness of the material to be cut. A satisfactory finish for ordinary purposes can be obtained from a fairly coarse grit, provided the work-speed and feed are not excessive. The wheels composed of fine grit are stronger than those of coarser grain and may be run at higher speeds. At the same time they are more liable to glaze or load, especially in tough metals such as mild steel, than coarser wheels, which are of a more open texture.

Grade

This term is used to denote the hardness of the wheel, and depends on the amount and nature of the cement or bond holding the particles of the wheel together, also, to some extent, on the abrasive material of which the wheel is composed. Grade is not a definitely measurable quantity. It is estimated by the resistance offered to a steel pointed tool, pressed in by hand. Mechanical tests to measure as a skilled tester working in the above manner.

The grade of a wheel requires to be regulated to the work so that as a grain of abrasive becomes dull and bruises material away instead of cutting it, the frictional resistance should tear the blunt grain out of the wheel. Further, in turning, the chips easily fall clear, but in grinding this may not be so, and the bond should be porous enough to obviate chips being driven into the wheel and causing glazing. Another requirement of a good bond is that it should not be affected by water, oil, cutting compound, or deterioration with age.

The three classes of bond in most common use are known as vitrified, silicate, and elastic. In vitrified wheels the abrasive is thoroughly mixed with clay or kaolin, pressed into a mould of the desired shape, dried and burned in a kiln until the bond material frits or runs. It is then cooled slowly, and when cool the lead centre is run in and the wheel face trued up with a diamond. The wheel is then spun and properly balanced. Later it is tested for safety by being revolved at about 50 per cent. above its working speed. Vitrified bond is not as strong as elastic bond, but is stronger than silicate bond. The vitrified type of wheels is generally employed in engineering shops, it is porous, not liable to load or glaze, is very fast cutting, and is unaffected by liquids or age.

In elastic wheels, as the name signifies, the bond is rubber. They are much stronger than wheels of other bonds, and are consequently used for very thin wheels, or where a side thrust may come into play. They are affected by some cutting compounds and may deteriorate with age.

Silicate bond employs sodium silicate as the cementing material. Wheels of special shape and of large size can be made by this process, such as would be difficult to manufacture by the vitrified method. Silicate bond is weaker than either of the above, but can be made more uniform in quality. It is employed where a soft wheel is required, being especially suitable for surface grinding with cup wheels.

Choice of Grade

When grinding hard materials the bond should be soft enough to allow the particles being torn from the wheel as they are moderately dulled. The softer the material, on the other hand, the stronger should be the grade and the longer the grains of abrasive should be retained in the wheel before they are torn out. Dull grains in soft material produce a better finish than sharp grains, the latter being liable to give a scratchy finish.

Different makers have various methods of designating grade. That adopted by the British Abrasive Wheel Co., and the Norton Co., is probably the best known. Vitrified and silicate wheels are lettered from E to Q—the lower letters of the alphabet signifying soft, wheels, and the later letters hard wheels. In general, it may be said that E, F and G are very soft, and used on surface grinding with cup wheels; H to K are medium soft and suitable for work on hardened steel or tool steel; K to M are for general work on mild steel, or when made, with carborundum for cast iron. Internal grinding requires slightly softer wheels than external grinding, and for surface work even softer still.

Abrasive Materials

Passing over quartz or silicate, well-known on sand belts and in grindstones, the oldest and probably best known abrasive is emery. This was mined by the ancients in the Grecian Islands, and consists of a mixture of oxide of aluminium and iron. It is dark in color, moderately hard and tough, and breaks with an irregular fracture, giving a good grip for the bond on a wheel, or the glue in the case of emery cloth. Emery offers a cheap wheel suitable for steel and for the intermittent work in small general shops.

Corundum

Is nearly pure oxide of aluminium. It is yellow or almost colorless, and possesses the qualities enumerated for emery, but in a more marked degree. It is harder, faster cutting, and very suitable for general work, particularly steel, but is not adapted for cast-iron grinding.

Artificial Abrasives—Carborundum

Is a chemical compound of carbon and silicon, produced in the electric furnace by heating together coke, sand, sawdust, and salt to a temperature of 7,000 deg. Fah. for 30 to 40 hours. As mass of irridescent, usually black crystals is produced, which is broken up, ground, and classified into the various grits. Carborundum is extremely hard, being next

to diamond on Mohs scale, and breaks with a smooth, glassy fracture. It is most effective on hard and brittle materials, such as glass and cast iron, but is not efficient on steel. The reason for this is probably that the grains having a glassy surface, do not offer a sufficient grip for the bond, and are thus torn out by the resistance of tough metal.

Artificial Corundum

Is sold under various trade names—alundum, aloxite, electric, etc.—and is an electric furnace product made of bauxite, a hydrated oxide of aluminium. By regulating the temperature, rate of cooling, and working up the cooling mass, and the admixture of small quantities of iron and chromium, various degrees of hardness and toughness are obtained. This gives a range of abrasive which are eminently suited for grinding and finishing all classes of steel, bronze and general work.

The Canadian Engineering Standards Association have informed us that a stock of the most important publications of the British Engineering Standards Association are now available for distribution. Of course a nominal price must be charged for these publications. Below is a description of some of these books.

Requests for copies should be addressed: Secretary Canadian Standards Association, Room 112 West Block, Ottawa, Ont.

10-1904. British Standard Tables of Pipe Flanges.—This report gives the British standard dimensions for pipe flanges for steam and water piping for low pressures and high pressures, dimensions of welded-on flanges for pipe lines, for working steam pressures of 125, 225, and 325 lbs. per square inch, dimensions for short flanged flanges and tees of cast metal for pressures up to 325 lbs. per square inch, and dimensions for long bends of wrought iron and steel. 25 cents net.

46-1909. British Standard Specifications for Keys and Keyways.—The specification covers material, tests, definitions and tables of dimensions for three classes of key: (a) Parallel Sunk Key; (b) Taper Key; (c) Taper Sunk Key. 25 cents net.

21-1090. Report on British Standard Pipe Threads for Iron or Steel Pipes and Tubes. (Revised November, 1909). —This report gives definitions and tables of dimensions for British standard pipe threads. In this system the Whitworth form of thread is employed, but fine pitches are used, and both parallel and conical screw ends are provided for. 25 cents net.

95-1919. British Standard Tables for Use in Engineering Workshops. Giving Corrections to Effective Diameter Required to Compensate Pitch and Angle Errors in Screw Threads of Whitworth Form.—This important paper gives the necessary information in convenient form for correcting threads of Whitworth form from 28 per inch to 2¼ per inch. 25 cents net.

Investigations Pertaining to Heat Treating

This Paper, which Was Presented Before the American Steel Treaters' Society, States in an Able Manner the Results of Various Investigations in the Heat Treating Field.

By Frederick J. Griffiths[*]

IN the past a great deal has been written and said pertaining to the heat treatment of both carbon and alloy steels. This has contained a great deal of valuable data which, when properly applied, no doubt has been of material value in furthering the art of heat treatment and bringing it up to its present day standard. But in a great many plants, especially those which are not large enough to employ the expert service of metallurgist, there are still a great many antique and injurious methods used in heat treating and many erroneous ideas entertained in regard to the thermal treating of steel. Because of the necessarily brief character of this paper, we will confine our remarks mostly to alloy steels used for structural purposes disregarding the refinements of procedure required by the high-priced tool and kindred steels.

The time that is given over to the consideration of fuels before furnace equipment is installed is time well spent. Very satisfactory results have been obtained in using natural or artificial gas, oil, coal, coke and electricity in different localities. The controlling feature of the decision upon fuels is the cost of obtaining same, and the question of the constant source of supply as well as the availability of cheap labor to handle the same, the labor consideration being especially important in the case of the solid fuels. Coal, coke and oil also involves the consideration of storage space. In tool work where ease of regulation and accuracy of heat control are very important and only small quantities of fuel are used and where high-priced steels are always handled the higher priced fuel often shows the highest ultimate economy.

Low Furnaces

Too much attention cannot be paid to the proper furnace equipment. Almost every special class of work requires some modification of the general types of heating furnaces that are now on the market. A great many of the smaller parts when produced in quantities can be cheaply and properly handled in continuous furnaces. These usually consist of a chamber in which the heat is generated and applied, and a revolving drum, the drum being equipped with an automatic charging and discharging device. In some cases the heat may be directly generated in the revolving drum. With this type of furnace very uniform results may be obtained with very little attention and with low production costs.

Crankshafts, camshafts, leaf springs,

[*]Central Steel Co., Massillon, Ohio.

axle and propellor shafts as well as connecting rods can be advantageously handled in flat hearth continuous furnaces, when produced in quantity. These are usually constructed with some sort of guides and runners on the hearth in the shape of rails or flat strips. With this type of furnace one or several of the parts to be heat treated are introduced periodically into the furnace with some mechanical device pushing them along the hearth until they are dropped at the other end of the furnace into a quenching medium. These furnaces are usually so constructed as to permit a regulation of temperature as may be required by the different analyses of steels used. The charge end is usually of a low temperature, the furnace gradually becoming hotter to where the proper temperature of quenching is maintained.

In selecting a furnace for special work where quantity production is not possible, it is well to bear the following facts in mind: The furnace should be of the proper size to allow the uniform heating of the steel without direct contact with the flame. It is usually safer to use a large furnace rather than one just large enough to accommodate the work to be handled. For annealing operations of any extent and quality, the car type furnace probably gives the most satisfactory results considering the amount of time taken in loading and unloading the furnace.

If the furnace is so located as to permit of charging at one end and discharging at the other, two cars may be used permitting the loading and unloading of one while the other is in the furnace. This also permits the utilizing of the heat retained in the furnace walls when the car is drawn by a rapid introduction of the other car, thereby preventing an excessive cooling of the furnace.

Pyrometery

In the last few years with the introduction of alloy steels for most stressed parts in automotive engineering, and machine manufacture, high physical properties have been required by manufacturers' specifications. These properties were not only required in test pieces but were uniformly demanded in every part combined with a great volume of output. This accurate duplication of the best qualities that the steel contained can only be obtained by the most accurate control of the time and temperature of the heat application. In order to constantly obtain this, a system of reliable heat measuring instruments must be installed in such a way as to at all times show a representative record of the ther-

mal condition of the interior of the furnace as close to the steel as is practically possible.

It is not sufficient to merely put a thermo couple in one end of the furnace and take temperature regardless of how the furnace is loaded or being operated. A couple inserted in the top of an under-fired furnace will usually show from 15 to 40° higher temperature than an instrument placed near the work. There is also some difference in temperature across the width of the furnace. This is in well-constructed furnaces not more than 20° to 25° and does not seriously affect heat treating operations. Constant checking and replacing of defective and worn-out thermo couples is essential to reliable heat control.

It is very important to take into consideration the difference between the temperature of cold end junction of the thermo couple and the room temperature, as the measuring instrument only shows the difference in temperature between the hot and cold end junction. It has very often been found that the cold end junction of the thermo couples that was inserted through the top of the furnace would have a temperature of 175° while the registering instrument would be calibrated to take care of a cold end temperature of 75°. Such practice of course makes it impossible to even gain a close approximation to accurate heat treating. In order to eliminate any worry in this regard two very simple methods of automatically controlling cold end temperatures are in use at present. The first, which is probably the more advocated at present, is accomplished by running leads of the same composition as that of the thermo couple wires directly to the measuring instrument. The other is by sinking a hole down 10 to 20 feet into the ground and running leads of the same composition as that of the thermo couples into this hole and there connecting any convenient leads to the indicating instrument. Some other methods of controlling cold junction temperatures are in use such as taking temperatures by thermometer and calibrating the measuring instrument to this. Another is by encasing the cold end with a cold water jacket keeping a constant flow of cold water in this jacket and calibrating the instrument to the overflow water temperature.

A great amount of breakage in thermo couples can be avoided by using care in charging and discharging the furnace. When the couple has attained a high heat very little shock is sufficient to often times rupture the protecting casing, allowing the furnace gases to have free

access to the couple wires, rapidly oxidizing them.

Handling Apparatus

The proper heat treatment of many irregular shaped pieces depends largely upon the way they are handled after the proper temperatures have been applied for a sufficient length of time. It is well for a hardener to study the proper handling of every new shape to be heat treated and in this way much scrap due to warpage and breakage can be eliminated. Small gears can be usually successfully handled by quenching along the axis of the hole. Flat gears of small face and considerable diameter will usually warp less when quenched perpendicular to the axis of the hole. Machine quenching by having dies that firmly grasp the web of the gear and so constructed that proper oil circulation is obtained has been successfully used on many gears which by other methods of quenching were warped beyond possibility of straightening.

Straight shafts should be quenched parallel to the axis, and when of considerable length can be kept straight by rolling into and through the quenching medium by a series of screens of guides. Small saws such as screw slotting saws can easily be kept straight in hardening by clamping them between two internally water-cooled cast iron disks. Leaf springs are hardened by placing the straight hot leaf between curved dies which come together shaping the spring and then running in the quenching medium. Bolts, valves, universal joint bows and similar small parts can be hardened in quantity by making sheet metal retainers through which holes are drilled to receive the parts. The loaded retainer is inserted in the furnace and brought up to heat and quenched. Small parts are often laid in sheet metal pans and heated and quenched. Special shaped tongs are very useful in handling gears, the tongs being so constructed so that the jaws open out while the handles are pressed together, enabling the handling of gears by expanding in the hole, thus eliminating strains set up by unequal cooling at the contact point of the tongs.

When low furnaces are used, having the hearth of the furnace on the floor level or level with a small platform, a device consisting of a flat plate mounted on two wheels and equipped with a long handle can very advantageously be used for charging and discharging the furnace. This device will equally well handle carbonizing boxes, automobile front axles, heavy forgings and the like.

There is a considerable number of parts used in construction to-day which require that certain sections be left softer than others, sometimes due to the fact that a cotter pin or taper pin hole is drilled in the part after assembling or to avoid chipping and cracking of threads or to give the section a greater amount of ductility in order to resist shock at that one point. Threads on shafts can usually be protected by slipping on a loose nut before hardening. In rivet sets when it is desirable to have the cen-

tre section ductile to resist shock, and the shank and cup hard to resist wear a simple jig fixture can be made covering the part desired to be left soft and quenching the jig and sets clamped in place. By tapering off the edges of the jigs a sharp line of demarcation of hard and soft sections can be avoided. Designs of this kind easily suggest themselves to hardeners for the particular shape they may have to treat.

Forging Operations

A very considerable portion of the steel which goes through the heat treating room has had a previous forging operation which has a very direct and important bearing on all subsequent operations. The amount of care which is expended on the heating operations of forging is well repaid by the elimination of trouble and extra operations necessary to rectify careless forge shop heating in the hardening room. There is no doubt but that there is a considerable amount of steel ruined by being overheated and burned in the forge shop, where most of the furnaces used are constructed to allow direct impinging of the flame upon the work to get maximum speed of heating and where no temperatures control except the eye of the heater is maintained. Due to the fact that greater ease of working the steel is obtained as the temperature is increased many forgers work dangerously close to the burning point of steel to get the maximum flow of metal, especially in difficult forgings where great care must be taken in filling up the dies. After a piece of steel is once burned it is impossible to put it in condition where it will again even approximate the physical properties of a sound piece of steel and its only value is that of scrap. Even though a piece of steel is not actually burned it may have been overheated to such a degree that considerable grain growth has been induced. This makes a piece of steel brittle, low in shock resisting qualities, giving very little elongation and reduction of area. To bring a piece of overheated steel to its normal structure expensive heat treating operations must be followed in the form of normalizing and annealing which will be discussed later. It is advisable for every forgeman to study his product and to determine the lowest temperature at which a reasonable production from his dies can be obtained and not to exceed this temperature, taking into consideration that different classes of steel will require different forging temperatures.

Annealing and Normalizing

All steel which is to be used for any part subjected to stresses of any amount should be first put into a proper condition for complete response to heat treating operation, which is required to bring out the necessary physical properties. This is accomplished by an annealing or normalizing operation or both. Too little attention has in the past been given to the value of the normalizing treatment.

The high temperatures used in rolling and forging of steel causes a considerable grain growth, leaving the steel brittle and hard and in a condition where it is not susceptible to steel hardening treatments. It, therefore, becomes necessary to heat the steel to varying degrees of temperature dependent upon the grain growth, but always above the critical point in this coarse structure. Where a very coarse grain size is dealt with temperatures as high as 1,800° F. are required to break this up. It is then advisable to quench, as on slow cooling from this high temperature to the critical point grain growth again sets in and the same condition which was present in the original piece is set up, but to a much less extent.

After this normalizing treatment an annealing treatment for machinability is required. This can be either to a point closely above or below the critical point and should have as slow cooling as is consistent with the facilities of the plant and the urgency of the work, down to 800° and 900° F., after which rate of cooling is practically immaterial to the final condition of physical properties.

Annealing to remedy structure should always be carried out before machining operations are started. The possibility of warping and shrinking of steel is reduced to a minimum if a proper annealing or normalizing has been used prior to the quenching and drawing operation. When heavy machining operations are performed on steel a short anneal before hardening to avoid warping, due to the machine strains, is beneficial. The finished machining operation should be performed after this light anneal.

Hardening—Oil Quenching

Alloy steel should never, except in very exceptional cases, be used without heat treating. It is a useless expense to buy alloy steels and not develop their maximum possibilities. In order to intelligently handle any class of steel it is necessary in order to determine the proper heat treatment to know the chemical analysis of the steel. Reliable information of this nature can usually be obtained from the invoice of the manufacturer to the user. Since the proper heat treating data for all analysis of structural steels have been fairly well determined upon, this information can always be secured from the steel manufacturer, if he is told what application is to be made of the steel. Probably the safest way to handle alloy steel is to use oil as a quenching medium. This reduces the danger or warpage and crackage to a minimum, and leaves the steel in a tough, fibrous condition. After about 35 points of carbon is used in alloy steels oil quenching becomes almost a necessity except as mentioned later. For camshafts, piston pins, gears, and in the larger shafts, high carbon oil quenched steels are preferable, due to the increased strength and ease of manipulation, freedom from spauling or surface chipping, decreased warpage, cheapness of treating, greater uniformity of results and

increased production as compared with carbonizing steels. Oftentimes sufficient hardness is not obtained with oil quenching due to deficiencies of the oil cooling and circulating system and often causes good steel to be blamed.

Water hardening of alloy steels should not be attempted with steels above 35 carbon unless facilities are at hand to keep hot water at constant temperatures, except in cases when the alloy content in the steel is very low. Water quenching steel may well be used for certain classes of work such as small shafts, nuts and bolts, steering arms and knuckles, crankshafts, connecting rods and similar parts. Water quenching usually causes hardening to penetrate deeper than oil for the same section. Water hardening lends itself to many different methods of quenching. When very drastic cooling is required a spray of water under pressure gives very desirable results. Almost any physical properties from extreme hardness to those closely approximating oil quenching can be obtained by regulating the temperature of the water and its circulation and renewal, but this usually involves more or less mechanical equipment to regulate the supply of water at a constant temperature, therefore, it is usually customary that when the physical properties which are required can not be readily obtained by quenching in water as supplied from the main and using a subsequent draw, or in cases when the carbon is so high that danger of cracking or warping is imminent, to use an oil quench.

Brine and Solution

Brine and other solutions play a very small part in alloy steel hardening. The object of using brine or solution in quenching is that of increasing the speed of cooling and thereby gaining greater hardness. It seems plausible enough to use this method when an off analysis of steel or low carbon alloy content is being used to save the steel, but if steel is ordered for a particular purpose, a proper combination of carbon and alloys can in most cases be obtained so that a special solution is not necessary.

Double Treatment

In occasional instances such as in aeroplane and in some motor car construction when lightness and strength is of considerable moment and where cost is a secondary consideration, a double treatment of steel has been extensively advocated and used. Beyond a doubt a considerable increase in physical properties can be obtained, in exceptional cases having obtained 10 per cent. increase. The greatest value of this treatment seems to lie in the fact that a uniformity of results is secured by this method. Any refinement of grain structure which is not obtained by the first treatment is secured by the second. The first treatment is considerably above the critical point, usually 75-100°, while the second is ordinarily very little above the critical point.

Corbonizing

A great deal more importance has been attached to the carbonizing treatment in the past than is justified by more recent developments in the handling of high carbon alloy steels. Carbonising at best is always an expensive, uncertain and dirty process, necessarily tying up a great deal of furnace equipment and labor causing a considerable expenditure on carbonizing boxes and carbonizing materials. It also takes up a very large amount of floor space due to the packing and storage of the boxes and material and the necessarily slow cooling of the boxes after the carbonizing operation. The process itself is incongruous to the opinion on the handling of steel entertained by the best informed men in the country. No metallurgist to-day would consider it good practice to overheat a piece of steel at long intervals and then subject it to several different heat treatments in order to again regain its proper structure when he could easily attain the same or better results by a simple quench and draw. In the author's opinion this is being done daily in carbonizing.

There are still some isolated cases where it has not as yet been found practical to supplant carbonizing steel by high carbon steel. Successful results have been obtained in making piston pins, camshafts, practically all sizes and shapes of gears, valve tappets, shafts and numerous other similar parts from high carbon heat treated steels.

Lead and Cyanide

Lead and cyanide has been extensively used as a heating medium in hardening materials which would be spoiled by excessive scaling. With present day construction of muffle furnaces it is not altogether necessary to go to expense of a costly lead and cyanide pot installation.

Tempering

There is probably no more important phase of heat treating than the drawing operation. Oftentimes steel that has been rejected as absolutely worthless for certain parts has proven itself ideal after the proper drawing conditions have been established. Time plays as important a part in the drawing operation as the temperature itself, and it is safe to say that in practically half the cases where a drawing is used, were the time to be doubled at which the part is maintained at temperature, considerably better results would be obtained. The opinion has been entertained in the past that in sliding and wearing parts such as gears on spline shafts, gear teeth meshing with gear teeth, ball and roller bearings working on shafts or races and rollers working on camshafts, absolute or glass hardness must be obtained in order to make the life of the part consistent with its application. This has proven itself erroneous in the majority of cases. It has been found that high carbon heat treated alloy steel, and this in particularly true of properly refined carbon chrome

steel, when given a sufficient draw to bring the scleroscope hardness up to between 60 and 70, possesses a satisfactory wearing surface, although it may be readily touched with a file. After a short time in actual operation such steel takes a very high surface burnish with a negligible amount of wear.

The Canadian Engineering Standards Association have informed us that a stock of the most important publications of the British Engineering Standards Association are now available for distribution. Of course a nominal price must be charged for these publications. Below is a description of some of these books.

Requests for copies should be addressed: Secretary Canadian Standards Association, Room 112 West Block, Ottuwa, Ont.

C. L. 3750. Interim Memorandum on French Metric Screw Threads for Aircraft Purposes.—This memorandum describes the system of screw threads for aircraft purposes used by the French military authorities, and is accompanied by tables showing limits of size, tolerances, etc., for two grades of fit. The form of thread is that of the Systeme International, in which the crest is cylindrical, while the root of the thread is curved in section. The finer tolerances are provided for cases where great accuracy is required. The second grade tolerances are suitable for ordinary bolts and nuts. 15 cents net.

C. L. 7270. Interim Report on British Standard Whitworth (B.S.W.) Screw Threads and Their Tolerances (¼ inch to 6 inches diameter). (Superseding Reports Nos. 20 and 38.)—This important report gives the British standard nomenclature and definitions referring to screw thread work, also standard dimensions of B.S.W. threads, followed by tables of standard sizes and tolerances for both bolts and nuts, on pitch, angle, full diameter, effective diameter, and core diameter. 25 cents net.

C. L. (M) 7271. Interim Report on British Association (B.A.) Screw Threads With Tolerances for Nos. 0 to 15 B.A. Superseding Report No. 20.) — Gives similar information to that contained in C. L. (M) 7270, but applying to B.A. threads from .010 to .236 inch full diameter (0.25 to 6.0 mm.). 25 cents net.

84-1918. Report on British Standard Fine (B.S.F.) Screw Threads and Their Tolerances.—This report gives revised tables of dimensions for British standard fine screw threads and covers theoretical dimensions and standard sizes and tolerances of bolts and nuts for two grades of fit. The report also contains an appendix dealing with methods of determining and compensating for errors in pitch, form of thread, and diameter. Much information is given regarding methods of gauging screw threads. 25 cents net.

WELDING AND CUTTING

BLOWPIPE MANIPULATION

POSITION of blowpipe.—The flame of the blowpipe should be given a definite inclination. In certain cases and especially with expert welders, familiar with the new method of executing welds, this inclination of the flame to the plane of the weld is very large; in other words, the flame is almost perpendicular to the weld. We have, however, made a series of experiments to discover the best angle and have found that an angle of 20 degrees gives the best results, that is to say, the angle between the nozzle of the blowpipe and the perpendicular should be 20 degrees, the flame being turned backwards as shown in Fig. 6. Beginners should pay particular attention to obtaining and practically working at this angle of inclination.

We have previously stated that in welding backwards it is the welding rod that is given a movement and not the blowpipe. The blowpipe is therefore held in such a manner that the flame advances along the bevelled faces with as great a regularity as possible, the rate of movement being in proportion to the speed of welding. A very slight transversal movement may be given to the blowpipe to produce more rapid fusion of the two bevelled faces.

A series of welds have been made by this new method with a blowpipe fitted to the loose headstock of a lathe, the advancing movement being thus mechanical. In spite of the absence of the slight transversal motion we are convinced that this basis of construction can be adapted for welding machines constructed for welding backwards.

However, it is interesting to note that the hand of a welder holding a blowpipe, even without movement other than that of advancement, seems to be much preferable to a machine because, without giving precisely the lateral movements to the flame, the hand can constantly direct the welding part of the flame to the requirements and thus advance or retard the movements in such a way that the bevelled faces are executed in conjunction with the proper handling of the welding rod.

(Position of the Flame.)—The white cone of the flame should penetrate very deeply into the angle of the V as shown in Fig. 7. If held too high as shown in Fig. 8 the melting at the bottom of the V is not sufficient, the size of the weld is unnecessarily increased, the metal near the surface is overheated and the speed of welding is diminished.

The penetration of the white cone should be carefully observed if the advantages, economy and quality of welds which can be obtained by welding backwards are required.

Position of the Welding Wire.—The melting of the metal is produced, as previously explained, behind the blowpipe and not in front as is the common practice.

This melting is no longer obtained by the welding cone of the flame, but by the additional heat contributed by the envelope of the flame, the blowpipe being inclined towards the rear, in other words, towards the welding wire, as shown in Fig. 6.

The position of the welding wire and its movement should be closely followed. The wire is inclined to the line of welding, in the advancing direction, that is to say, in the opposite direction to the inclination of the flame. The best angle of inclination, between the weld and the wire, has been found to be 45 degrees for material about ¼-inch thick, and for thinner material, say ⅛-inch, an angle of about 30 degrees. This inclination is maintained whilst the welding wire is given its proper movement in the line of welding. This movement for the thicker material, say, about ¼-inch, consists in alternately moving the molten extremity of the wire from one side to the other of the line of welding, as shown in the small illustration in Fig 6. The movement for material less than this thickness becomes first of all elliptical or gyratory, and then for material about ⅛-inch, and especially when the material is about 1-16-inch the movement is translated into a reciprocating one without any transversal movements, as shown in Fig. 9. In both these cases the extremity of the wire remains continually in the molten bath.

Execution of welds.—In order that the line of welding should present a homogeneous appearance, it is advisable to operate in the manner already laid down and with the same speed at the commencement and completion of the execution. If, say, one of the extremities of the weld is attacked too soon with the blowpipe, free fusion and regular advancement are not obtained until after a certain time, with the result that irregularities are noticeable at the beginning of the weld. To avoid this the plates should be preheated for a length of a few inches with the blowpipe so as to obtain, at the beginning, regularity, and a normal rate of welding. The blowpipe and the welding wire being held in the manner indicated, the cone of the flame is directed so as to well penetrate into the angle of the bevel, and the first molten bath is obtained by giving the blowpipe a slight gyratory movement, immediately after which the extremity of the welding wire

FIG. 6.—POSITION OF BLOWPIPE AND WELDING WIRE.   FIG. 7.—POSITION OF THE FLAME.   FIG. 8.—WRONG POSITION OF THE FLAME.

is introduced into the molten bath and the blowpipe is then given its regular advancing movement.

The welding wire, on the other hand, follows immediately after the flame, and describes a reciprocating or a more or less elliptical and longitudinal motion, according to the thickness of the metal, as indicated and shown in Figs. 6 and 9, taking care to always maintain the given

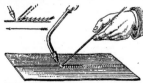

FIG. 9.—MOVEMENT OF WELDING WIRE FOR THIN MATERIAL.

angle of inclination The weld is thus obtained in a normal and very continuous manner. Care must be taken to use a welding wire which satisfactorily fills the lines of welding, without excess or insufficient addition of material. If necessary, the position of the blowpipe is changed when the extremity of the weld is reached in order to obtain a clean finish as is usual with the ordinary method of welding.

The melted metal being attacked in the rear, as a result of the inclination of the flame, the bevelled faces are always well melted; the weld is what is commonly said to be well penetrated and the defect of adhesion is almost impossible. However, it is advisable not to travel too fast so as to give the beveled surface sufficient time to melt freely, otherwise candles of molten metal will appear on the underside of the weld as a result of the addition of too much heat at the bottom of the V.

(Defects in Welds).—If the weld is badly executed, as is common with beginners, its general appearance is irregular. The surface does not consist of a series of overlapping ripples all practically in the same plane but consists, as a section along the line of welding will show in Fig. 10, of a series of hills and valleys.

The most serious defect with beginners is to form a channel on either side of the weld throughout its length, these channels, shown in Fig. 11, being below the surface of the plate.

These channels not only weaken the whole length of the weld along each side but are the starting points of cracks, as is proved by bending tests on welds containing this defect.

With a little experience a welder soon reaches the stage in which he eliminates these irregularities and defects. We have proved by a wide range of tests that a welder who has not previously welded

backwards does not require much practice before he obtains excellent results.

(Characteristics of Well Executed Welds).—The welds are very regular in appearance and are not so wide as the welds done by the ordinary method. No channels which correspond to a thinning at the edges of the welds are present. There are no hollows or excess addition of metal on the surface. The underside

of the weld shows that the fusion has been complete throughout the thickness of the metal, in other words the penetration is complete and there is no starting point for cracks.

(Examination of Welds).—The following observations made during the examination of welds executed by this new method of welding, accompanied by the results of the first series of tests will conclude this first study. Some notes on the industrial results obtained over a period of eighteen months are also included.

It is incontestable, from the point of view of good penetration of the metal and the absence of the defect of adhesion, that welding backwards offers

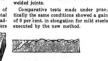

FIG. 11.—CHANNELS IN BADLY   FIG. 10.—BADLY EXECUTED
EXECUTED WELD.   WELD.

considerable advantages over other methods of executing welds and is capable of entirely eliminating these defects.

Etching tests on welds obtained by the new method show a perfect joining between the metal of the plate and the added metal. The welds show distinctly less oxide inclusions than those obtained by the ordinary methods, and are free from blowholes. In addition they possess ordinary hardness and the remaining mechanical properties are more regular.

Bending tests have given good results. It is possible to fold the weld without starting a crack which is a very good indication of excellent elongation properties and good penetration of the weld. The tensile strength of the weld is also greater than that of ordinary welds and im-

provement in the other mechanical properties is obtained.

A series of welds made by welding backwards, has been obtained from various works carrying out commercial oxy-acetylene welding and although we are not permitted to give the results obtained by testing, they are all favorable to the new method. Extremely favorable results have been obtained in Italian workshops.

From the point of view of instructing welders, there is no doubt that apprenticeship need only be brief as when the fundamental principles are well understood rapid progress is made. We have shown that after a very short period of instruction a beginner can produce welds free from the defect of adhesion, and the burning of the metal, with good penetration and good joining of the metal.

In the construction of projectiles, shells and bombs, by oxy-acetylene welding, the percentage of leaks, or rather sweating, in one workshop, when the welds were placed under the hydraulic tests was 15 to 16 previous to the introduction of the new method of welding. This figure was reduced to 4 per cent. and was practically eliminated at the point where leaks were common, namely at the junction of the circular and longitudinal welds, but it was found necessary to instruct workmen to give a just more attention to making the junction of the welds. In the case of aeroplane bombs the reduction in the percentage of leaks was of the same order after the new method had been applied. Hydraulic tests to destruction demonstrated the good quality of the welded joints.

Comparative tests made under practically the same conditions showed a gain of 6 per cent. in elongation for mild steels executed by the new method.

The following figures relative to the increase in the speed of welding, with a consequent economy in gas and labor, bears out the previous statement that the gain is in the neighborhood of 30 to 35 per cent. In repetition work an average welder executed 54 feet of ¾-inch plate in 10 hours, practically 5½ feet per hour, which can be considered as an excellent output, by the ordinary method of welding. The same welder, welding backwards, executed almost double this quantity, namely, 98 feet in 10 hours, almost 10 feet per hour.

In analogous work, but on ¼-inch plate, the output for numerous welders varied between 210 to 310 longitudinal cylindrical seams, each about 9 inches long, and in the welds were free from

Continued on Page 65

DEVELOPMENTS IN SHOP EQUIPMENT

DRILLING MACHINE

The Hoosier Drilling Machine Co., Goshen, Ind., have placed on the market their line of drilling machines, which are now being built in 16 in. and 20 in. sizes and in various styles. These comprise what is known as Nos. 1, 2, and 3 machines.

Fig. 1 shows a 20 in. stationary-head, back-geared machine, equipped with power feed and an automatic stop mechanism; and in Fig. 2 there is shown a No. 1, 16 in. machine with a plain drive. It will be seen that in one case the machine is provided with a round table and with a finished extension of the base in which T-slots are cut to provide for securing large pieces of work in position for drilling, while the other machine has a square table and a round unfinished base. Various combinations of these arrangements may be provided to suit the requirements of different shops.

On the power-feed machine shown in Fig. 1, a sufficient range of speeds and feeds is provided to cover average classes of work. There are eight speeds, four of which are obtained direct, and four through the back-gears; and three power feeds are obtainable in addition to the handwheel and lever feeds. The spindle is furnished with a ball thrust bearing, and it is counterbalanced by a weight in the column. An automatic stop in connection with a graduated spindle sleeve provides for accurate depth drilling and boring. Machines equipped with plain drive, as shown in Fig. 2, have the same general features of design as the power-feed, back-geared machines, but the construction has been simplified to meet the requirements of shops which do not need a drilling machine furnished with power feed and as wide a range of speed changes.

Fig. 3 shows a vise especially made by Hoosier Drilling Machine Co. for holding work on drilling and milling machines, although it is also adapted for use on shapers and planers. The jaws are faced with steel, and flanges with bolt slots are furnished on the body to enable the operator to securely anchor his vise to the table of the machine. A key slot milled the entire length of the body allows the vise to be brought into perfect alignment with the table of a milling machine or other tool on which it is used.

SEMI-AUTOMATIC TWO-SPINDLE POINTER

A new type of pointer has been brought out by the Kent Machine Company, Kent, Ohio, which will be known on the market as the Kent Semi-Automatic Two-Spindle Pointer. The machine, as shown in the accompanying illustrations, will accommodate bolts and rods up to and including ¾ in. in diameter and of any length from 1½ in. up.

The illustration of the interior of the machine shows it to be very simple in design and construction, but still a departure from the pointers at present on

VIEW OF POINTER WHEN CLOSED.

the market. The exterior view of the machine shows what great care was taken to entirely enclose all running parts so as to insure absolute safety to the operator, and yet make the machine easily accessible for adjustment, repair, or inspection. When necessary to adjust either the cutting tool, or the chuck jaws, only the front cover need be removed.

This two-spindle machine requires only one operator for bolts or rods of ordinary length, but it is so designed that rods of any length may be pointed. It is only necessary for the gripping chuck, where it is securely gripped, after which the pointing spindle advances and the pointing operation is accomplished. The spindle then withdraws, the jaws in the gripping chuck are opened and the bolt drops into a tote box or other receptacle below the machine. All of these operations except the placing of the bolt

FIG. 1—THE BACK GEARED, POWER-FEED MACHINE.

FIG. 2—THE PLAIN DRIVE TYPE OF DRILL

FIG. 3—STYLE OF VISE USED ON SUCH DRILLING MACHINES.

in the receiving chuck are entirely automatic.

The two-spindles, as shown, are actuated by the large cam and have alternate motion which makes it possible for one operator to feed both spindles. The forward motion of the spindle is spring actuated, so that no particular care need be exercised in placing the bolts or rods in the receiving chucks, except they must extend at least 1 inch beyond the face of the jaws to permit the pointing operation. The jaws of the gripping chuck are well protected, room being left only for the insertion of the bolt. Since all subsequent operations are entirely automatic one operator can easily feed both spindles with the cam shaft running at 25 r.p.m., making the total production of the machine 50 bolts or rods per minute.

As may be noted in the accompanying illustrations, the, pointing chucks are made from a solid piece of steel, which gives them great strength and rigidity. The cutters are securely held in milled slots by self-adjusting clamps, and by loosening the screw shown they may be easily adjusted or removed and replaced when necessary. The head of the chucks are provided with tool steel guide bushings for centralizing the work to be pointed. These guide bushings are removable for the different sizes of bolts or rods.

The gripping chuck consists of two slides made of tool steel with formed jaws at their end for gripping the work. One of these slides is stationary in the machine, the other being movable is opened by a cam mechanism. Like the forward action of the spindle, the gripping device is actuated by a spring which prevents injury to the machine should a larger bolt or rod than those being pointed be placed in the gripping chuck by the operator. The stationary jaws may be adjusted for the different sizes by a small set screw. The jaw actuated

by the cam and spring automatically adjusts itself. The jaws grip the work at four points which approximately centres it ready to enter the centralizing bushing in the pointing chuck.

There is plenty of chip space, and as bolts and chips are carried away in different directions, a pointed product, clean and ready for the next operation, is assured. The floor space required for the machine is 31 x 41 inches, and the distance from the floor to the centre of the spindle is 37½ inches. The total weight of the machine is 1,100 pounds.

New Things in Machine Tools

BRINELL TESTING MACHINE

The Brinell method of testing the hardness in metals is one which has been adopted in many countries, and various sizes of these machines have been placed on the market in order to conform to the different sizes of test pieces.

To meet the demand for a Brinell tester that will suit the requirements of the laboratory and toolroom, the Pittsburgh Instrument and Machine Co., 101 Water Street, Pittsburgh, Pa., has developed a style C machine as illustrated. There is a slight difference in the construction of this machine from the standard type, inasmuch as the oil reservoir is attached to the side of the supporting frame, and so arranged that the overflow of oil from the cylinder will always circulate back to the reservoir. The maximum distance between the 10-mm. steel ball and the press table of this machine is 3 in. The procedure of making tests is the same as that of their standard style A machine.

MARKING MACHINE

J. Merey, 2842 N. Maplewood Ave., Chicago, Ill., has placed on the market what is known as a Stampograph. This is, as can be noted from the illustrations, a rotary machine for lettering and numbering and can be used on all kinds of metals or substances that are in their soft state.

The impression is adjustable and can be perfectly spaced and aligned. The machine has an automatic feed, subject to the sizes of the characters there being a fixed feed provided for each size. The disks for the characters are made from the very finest tool steel and carefully hardened and tempered.

The feed is caused by means of cams and a multiple rack shaft; the impression is obtained by means of crankshaft connected with a handle and connecting rod, which is so constructed, that the characters should not break in case some part of the metal should resist, or from any other cause. The adjusting to obtain the impressions for various sizes of objects or work, can be gained by turning the ball handle in either direction and following the gauge affixed on the side until desired impression is obtained. The machine illustrated, as can be seen, is made for hand power. The machine as built for machine power weighs about 350 lbs. and requires space 2 x 2 feet, and can be run from a countershaft or motor driven. The handpower machine weighs about 200 lbs.

The power machines are only built on specification and the stock machines are fitted out with full alphabet, 1-16, 1-8, 3-16, 1-4 in.; numerals 1 to 0 in same sizes. Special multiple engraved dies can be furnished if desired. Both machines are outfitted with all engraved characters, and all necessary requirements, as cross table, vise and circular table with graduations to stamp in radius

SCREW PRESS

The Manhattan Machine & Tool Works, Grand Rapids. Mich., have placed on the market what is known as their four-post screw press, type BB. The illustration herewith gives readers an idea as to its general appearance, and following are other particulars regarding the same.

The massive construction is, of course, apparent, and all parts are properly balanced in so far as strength is concerned. Every convenience for facilitating work, and simplicity of operation, has been embodied in its design. Of course this four-post type is more adaptable to heavy work than the regular two-post style of press.

ONE CAN READILY SEE THE INTERIOR FROM THIS VIEW.

The MacLean Publishing Company
LIMITED
(ESTABLISHED 1887)

JOHN BAYNE MACLEAN, President.   H. T. HUNTER, Vice-President
H. V. TYRRELL, General Manager.

PUBLISHERS OF

CANADIAN MACHINERY
and MANUFACTURING NEWS ~

A weekly journal devoted to the machinery and manufacturing interests.

B. G. NEWTON, Manager.   A. R. KENNEDY, Managing Editor.

Associate Editors:
J. H. MOORE   T. H. FENNER   J. H. RODGERS (Montreal)

Office of Publication: 143-153 University Avenue, Toronto, Ontario.

VOL. XXIV.   TORONTO, JULY 29, 1920   No. 5

PRINCIPAL CONTENTS

Looking Over the Field

THE specialized shop is a good thing for the company, a good thing for the shell-buster type of "mechanic," but a mighty poor institution for the man who wants to develop.

That may be a queer way of stating the case, but there is good authority for the verdict. Only a few days ago I had a long chat with Mr. Robert Patterson, of Stratford, who was for twenty years or more master mechanic of the Grand Trunk shops at that city. Had his health remained good he would no doubt be to-day filling that or some other responsible railroad position. But he still retains a very keen interest in the mechanical field, and he can approach it from the standpoint of the apprentice, the mechanic, the employer, or the purchaser of equipment. More than once one young chap wanting advice takes a trip across to Robert Patterson's house, and he gets the best that the ex-railroader can give him from a wide experience in many shops and from a close connection with the mechanical world.

MR. PATTERSON was always a student. In fact he still is, and the chances are he will continue to be. Especially in his active days of shop work, he was a reader of books, technical papers, magazines, and anything that might contain something of interest. He went far afield from the railroad world to look for ideas, and he found them. "I remember one idea," remarked Mr. Patterson, "that we made good use of. We had difficulty in taking bolts from locomotive frames, and it was a matter of some three days to attend to this operation on an engine. I read in some of the technical papers published then, of a little hydraulic arrangement that was used for a job, and the idea struck me that it could be applied to our work. So we got it in the shop and figured it out, and it worked. Those bolts were driven out in about half a day, and it does not take much figuring to see what was saved in the way of time in the shop and getting the engine out quickly again. I suppose, in all, that idea was worth a good many thousands of dollars to the company."

MR. PATTERSON is not enthusiastic about the high wages paid to apprentices in railroad shops now, and he sees a chance in this tendency to either make or break the boy, depending largely on the in-fluences that are brought to bear on him. Of course it would be hard to convince a boy that high wages were not good for him, but it must be remembered that a boy in his apprenticeship age does not know nearly as much as he thinks he does. Mr. Patterson holds that the high wage to the apprentice is apt to make him independent of his home at a period in his life when he ought to be subject to a certain amount of home re-straint and correction. He has seen cases where boys, making more as apprentices than their fathers need to as journeymen, have picked up and left home at the least pretense, their high wage enabling them to have an independence that was very unhealthy for their after career. A certain amount of discipline is also necessary in apprenticeship, as no boy prizes highly a thing that comes easily to him and for which he has to make no sacrifice and very little effort.

"I REMEMBER very well," recalled Mr. Patterson in the course of conversation, "one time when the late Mr. Hays, then president of the road, was inquiring about certain makes of equipment in use in the shops. We were able to speak very highly of several them, and this was the reason: They stayed with us after they put the machine in the shop. To make that a little clearer, here is a case in point. We put in five tools of a cer-tain type, and they gave us good service. A few months after a representative of the firm came to the shop and asked if he might go out and see them in operation. So we went along, and after watching them he stated that there was an improvement out which would suit our work nicely, and that he would send the attachment along at once. This was done without any cost to us. That firm gave us a real service with everything we ever bought. It paid them and it paid us."

E. J. Dodd, writing from London, says that at the Canadian Exhibition, the MacLean business and technical newspapers and magazines, and "The Financial Post," were displayed at the stand of the Department of Over-seas Trade and that quite a number of copies were sold, mainly to persons interested in buying from Canada. One of these was a member of the staff of the Russian Bolsheviki, Trade Ambassador M. Krassin, who made special enquiries and asked if he might be allowed to follow up the reading of the papers by a call on Mac-Lean Newspapers' London office. It is difficult to know just what is going to happen in Russia, but one thing is certain, Russia cannot get on a normal producing basis without machinery plant and industrial brains. The Jews, who are controlling Russia, know this and there are bound to be developments favorable to Canada.

"Green Fields and Pastures New"—The Red Ped-dler at the Door of the Orient.

MARKET DEVELOPMENTS

Premium Material Keeps Them Going Now

Cars Sent Over from Canada to Handle a Shipment—Takes a Good Deal More Money to Handle the Same Tonnage—New York Market is Dull.

IT IS taking an increasingly large amount of money to carry on business in some sections of the steel and iron world, as well as in the machine-tool trade. Right now the steel warehouses of Canada face a position that gives them the option of sitting back and waiting until mill shipments come in, or going out and paying a high price for material they can procure at once. The firms that are carrying on under the latter conditions take on a greater risk with every purchase they make, for their only safe course is to get the material in and sell it at once before there is a chance of decline in prices. Many of these places have material on order, but not delivered, at 2.65 per pound, while they are paying 8 cents per pound for shipments which they can get inside of a week or two.

Deliveries at best are very poor and show no indication of being influenced by the talked-of improvement. A reduction was made during the week by makers of high-speed drills in Canada, amounting to 10 per cent. This equalizes prices to some extent, and also brings the Canadian price on level with the figures that are quoted by Old Country firms doing business here and which have been the means of losing quite a few orders for the Canadian trade.

There is nothing brisk in the machine-tool trade. Some of the houses are going as far as to quote spot delivery on a' large number of their lines, which is quite a change from the extended schedule that has been in force for many months. ·The New York market is dull, and it is expected that should this continue, some of the makers will be out looking for business, signifying a return to something near normal conditions.

Scrap markets are stagnant at Canadian points. Dealers are not anxious to buy, in fact in many lines they are out of the market.

CONSTRUCTION IN MONTREAL HELD UP BY UNCERTAIN DELIVERY OF STEEL

Special to CANADIAN MACHINERY.

MONTREAL, July 28—Transportation difficulties and car shortage would appear to retain the central position on which present industrial activity depends. It is doubtless true that shipments of material have shown some slight improvement but of such a meagre character that dealers and manufacturers are not yet enthusiastic regarding the better outlook. The promised advance in freight rates on American and Canadian railroads is a cloud on the horizon, that may eventually bring on another storm of general price boosting.

Steel conditions are showing gradual but very slow improvement and dealers here are reluctant to predict any pronounced betterment until the midsummer season is over. Passenger traffic is usually heavy during this period and this frequently interferes with the regular passage of railroad freight. Machine tool activity is seasonably light with supplies featuring the sales. Scrap movement is largely confined to local foundry requirements.

Freight Changes Will Affect Sales

"We are still waiting improvement in delivery, in order to meet the constant needs of our customers," said a steel dealer." The summer season is generally marked by·a falling off in production and likewise less buying, but this condition this year is aggravated by the fact that, in many instances, forced curtailment in operation of factories, is necessitated by the inadequate railway facilities, so that fuel and material is not being shipped regularly to the mills." Much construction work, both in the building and equipment line, has been greatly delayed or even deferred, on account of the uncertain delivery of supplies. Dealers, however, are generally hopeful that improvement will develop from now on, but will not commit themselves to any fixed time or period. The car situation has become so uncertain that many mills, while willing to accept business, will not guarantee delivery, only with the stipulation that it will be shipped at a later date. A factor that will require some attention shortly is what condition will apply on the shipment of material purchased at a specified figure but when delivered will carry the charges of increased freight rates. There appears to be only one logical solution to this problem, and that is the passing on of the additional charges, as usually happens. "We as dealers," remarked one, "are here endeavoring to fill the needs of our customers. If, in the interval between the placing of the'order and the delivery of the goods, unforseen charges are incurred we can only expect to have this taken on by the purchaser. We cannot carry on at a loss." Some cancellations are anticipated, but it is thought that these will be few, as nearly all contracts, at the present time, are made so elastic as to be easily adjusted to sudden changes in industrial conditions.

Tool Requirement Light

Apart from a slight flurry in railroad enquiry the demand for machine tools has been relatively quiet, but this is to be expected at this period of the year, owing to the general easing up of business in the holiday season. The movement of supplies continues almost normal, but prices on all commodities are maintained, or show advances. As soon as freight charge adjustment has been made dealers are anticipating proportionate increases on the cost of equipment.

Easier Tendency in Scrap

Although few dealers will commit themselves as to what effect a buying movement would have on price quotations, the impression gathered from different statements would imply that a more active market might tend to bring prices to a lower level, probably not sufficient to appear slumpy but enough to

ease the situation from the present state of tension and induce consumers to take a more active interest in the market. Present movement is exceedingly quiet and any visible signs of heavy buying is conspicuous by its absence. Local business is steady but comparatively light. "The foundries are taking considerable scrap but not greatly in excess of normal," said one dealer; "many of these places are kept very busy making up for strike losses." Wrought iron has been easier and dealers are asking around $20 per ton. Malleable scrap is also weaker and the price of $23 shows a recent decline of $2 per ton. Other steel lines are likewise easier but quotations are usually based on individual sales.

NEED MORE MONEY
TO CARRY ON NOW

Warehouses Either Have to Take on High Price Material or Wait Turn For Delivery

TORONTO.—The limit to which conditions in the machine tool, iron and steel trade can improve is largely fixed by the increase that can be shown in the ability of the railways to handle the business that is piled up and waiting for them.

Information coming to hand this week does not indicate that an improvement is in sight. For instance the Carnegie mills alone would require, according to their own estimate, 13,000 cars in excess of the supply granted them, to handle the finished and semi-finished material that is stacked up at their plant.

On the other hand some makers of machine tools are quite anxious to take on business, and New York advices indicate that if present conditions prevail for a couple of months, many of the makers will be out looking for orders, something that has not been the case for some time past. When we say "looking for orders" we do not mean that there has been any time when they would not consider business, but rather that their business was so far booked ahead that there was little use of them looking for business. It has been more a case of getting out the material than selling it.

There are some firms that even go so far as to state they are prepared to make spot delivery on nearly all their lines, no doubt depending on cancellations to some extent to make this promise possible.

A Testing Time

The steel market of to-day, that is, for the firm that wants to go ahead and turn over a good tonnage each month, is a greatly changed business from the well-defined business of the old steel merchant as we have known him in years gone past. If the steel warehouses depended right now on what they could get from business booked with the mills

POINTS IN WEEK'S
MARKETING NOTES

Canadian-made high speed drills have been reduced ten per cent. Prices are now listed plus ten to thirty.

New York machine tool market remains dull, and the prediction is made that in a short time the makers will be out looking for business.

British automobile makers have been making purchases in the American machine tool market.

Some warehouses having business booked at 2.65 with the Corporation, are paying 8c for material that can be secured on short notice.

The Carnegie mills need, in excess of their usual supply of cars, at least 13,000 freight cars to move the finished material that is piled at their mills.

Pittsburgh mills declare that if they do not get a better supply of cars, they will have to let up in their production of steel.

Several of the automobile plants are running on short time, reports from various centres asserting that cars are in stock now for the first time in many months.

some months past, they would be in for a quiet session. Their one big chance is to get out and buy what they can at what they have to pay for it, and then get busy and get out from under the obligation. And in many cases the obligation is increasing at such a rate each time as to cause a person to stop and think seriously.

Here is a case that shows what we mean: One Toronto firm right now has a good-sized tonnage of black sheets (28) on the way. The firm selling them is a high-price mill, and they made arrangements with one of the Canadian roads to get five cars across to handle the material which they claim will be back here in 20 days. The cars are marked and the promise is given to the railroads that they will not be used for any other purpose. These sheets cost eight cents at the mill. When all the extras, such as freight, exchange, etc., is added, the selling price, with the warehouse profit, will be about 12 cents per pound. Some $30,000 will be necessary to handle this one deal, and there is considerable more risk in this class of business than formerly, as many buyers state they are out until they can get a more reasonable price.

This same Toronto warehouse has a large tonnage placed on the books of the Steel Corporation, but they can get no

delivery. Hence they face this situation: (1) Sit back and wait for delivery from the Corporation at 2.65, or (2) go ahead and buy from the independents at eight cents, and they are choosing the latter course, although they make less money per dollar invested than they did in the old days when sheets were around two cents per pound.

Little Material In

Some plate and structurals have been arriving lately, but the amount is comparatively small. There would be more work done if more material were forthcoming. Firms building cars and railway equipment are finding it hard to get enough material together of a widely assorted character to permit them going ahead. The same thing is found in the automobile field. In fact the whole automobile industry is suffering, more or less, from this same cause. Some of the shops in this district are operating on a much reduced production schedule.

Reduction in Drills

A reduction has been made in the selling price of high-speed drills in Canada, at least as far as the Canadian-made article is concerned. The idea is to equalise prices. Large consumers are now quoted list plus ten up to 1-2-in., from 33-64 to one inch, plus 20 per cent. and over 1 inch plus 30 per cent. The smaller consumer will pay 10 per cent. additional to each of the above classes.

There is a lot of small business moving now in cutters, drills, taps, dies, etc., and dealers claim that if the larger concerns were buying, as they show in running to capacity, there would be a splendid trade moving for them.

Scrap Stays the Same

The scrap dealers are not happy. They are not busy. They are neither buying nor selling to any great extent, and if anything will put the damper on the scrap market it is lack of enquiry. Dealers plainly state that for the present they are out of the market in many lines, unless they can take on material at their own price, and be prepared to hold it for future improvement, which they are none too certain of realizing. The tightening of credits in many quarters is blamed for the dullness, but from whatever the cause, the dulness is in the scrap metal market, and it has been camping there for some weeks.

The American Engineering Standards Committee has requested the American Society of Mechanical Engineers and the Society of Automotive Engineers to act as joint sponsors in the matter, leaving the decision to the joint sponsors as to whether a new sectional committee shall be organized for the project, or whether the work should be done by a sub-committee of the Sectional Committee on Screw Threads, for which the same two societies are acting as joint sponsors.

BELIEVES THAT THE BUILDERS WILL SOON BE OUT LOOKING FOR ORDERS

Special to CANADIAN MACHINERY.

NEW YORK, July 29.—Not in many years has business in machine tools and allied equipment been so quiet. If the present inactivity in buying is prolonged for sixty days or so, some machine-tool manufacturers will need business, at least on some sizes of machines. While shipping conditions are by no means good, enough shipments are being made to catch up to some extent and machine-tool companies can now quote earlier deliveries than was possible two months ago. A few companies are able to make shipment of some machines from stock.

Much has been looked for from the railroads, but so far this prospective business has proved a disappointment, as the aggregate of orders is small by comparison with what the carriers are known to need. The Chicago, Rock Island & Pacific has distributed orders totalling $200,000, of which one large machine-tool company got about $150,000. The New York Central, the Pennsylvania and the Richmond, Fredericksburg & Potomac have each inquired for a few tools, but the large railroad

lists, which were expected to come out by this time, have not been issued. When the Interstate Commerce Commission announces an increase in freight rates, which will probably be within the next few weeks, the railroads will know better where they stand in the matter of finances, and it is quite possible that inquiries for shop equipment will follow that freight-rate announcement.

While domestic business is quiet, there is some export inquiry in which British automobile manufacturers are prominent. Several of the British companies have been buying in this country, and further purchases are in prospect. A. Harper Sons & Bean, Limited, of England, recently appointed W. J. Fuller, for many years manager of the American branch of Alfred Herbert, Limited, as American buyer, and it is reported that Mr. Fuller is now negotiating for considerable equipment of American make. A representative of the Sunbeam Motors Co., of England, has a representative in this country, and machine tools may be bought also for that company's plant.

shortages in the amount of work the men on payrolls will do, as the men will not put in full hours for any length of time. As long as cars are in shorter supply than mixing performance all the deficiency is attributed to "car shortage," for the labor shortage is not actually developed.

If a decided improvement in car supplies, for shipping steel products does not occur within 30 days, or perhaps 60 days at the outside, it will be necessary for the mills to decrease their production of steel, the production of pig iron being of course curtailed at the same time. Thus far the situation has been met, first by simply piling finished steel as made, and later, in the past few weeks, by curtailing finishing operations and piling part of the steel in unfinished or semi-finished form. This process could not be continued indefinitely, even if there were unlimited storage facilities for semi-finished steel such as billets and sheet bars, for the reason that eventually all the steel would have to be finished and the finishing departments can carry only a very moderate overload. A fair estimate seems to be that even the present accumulation of steel will at the very best require from three to six months' time for working off if the railroads get to the point of rendering the maximum service conceivable with their present physical equipment.

Picking Out Shipments

On account of the improbability of all the finished steel now in stock at mills, rolled to customers' specifications, being acceptable to the customers at the late date at which it will have to be shipped, the mills are now making an effort to pick out and ship the more perishable forms of steel, such as merchant bars that are cut to length, and other descriptions of steel at all special in character. This steel is picked out from stocks or from current rollings, as the case may be, so that as far as possible the steel left for later shipment will be more or less in stock sizes, either useful to the customers for whom rolled, or saleable in the open market.

Variations in Demand

Current market demand is excellent for merchant steel bars, particularly small sizes, and in fact almost everything under 2-inch. Demand for shapes and plates, on the other hand, is very light and has been so for several months. The testimony of the mills in this respect is confirmed by the June report of the Bridge Builders' and Structural Society, showing lettings of fabricated steel contracts this month to have been only 50 per cent. of the fitting shop capacity. June is sometimes a light month if this business has been done earlier, but the average for the six months ending with June is only 72 per cent. The fact is that investors are quite indisposed to go ahead with large construction jobs, on account of great delays prom-

MILLS CONTINUE TO PILE THEIR MATERIAL OWING TO CAR SHORTAGE

Special to CANADIAN MACHINERY.

PITTSBURGH, July 29. — A better freight movement is reported from the Chicago district, where it is said some of the accumulations of steel have gone from mill, in addition to current production. In the rest of the Central West, including Pittsburgh, shipping conditions are even worse than a week ago, and as production of steel continues substantially at the former rate this means that steel is being accumulated at mill at a more rapid rate than formerly. More of the steel is being left in unfinished form, for prudential reasons; and the actual production of finished steel is smaller than a month ago.

The blast furnaces and steel mills stood to gain some advantage in car supplies by the recent modification. At their request, Service Order No. 7, according preference to coal mines, was interpreted as not including as "coal cars" those flat bottom gondolas that are under 36 inches high, inside measurement, but it seems now that there are not many cars in this region that would be affected. Previous to the new interpretation most of the roads were regarding as coal cars any that were over 30 inches. The industry gets a definite backset in that the requirement that cars regarded as coal cars may be loaded only when the destination of the loaded material is toward coal mines

that should be furnished the cars. Previously this rule had been very largely evaded. As it is now, nearly all the coal cars that reach furnaces and mills have to go out empty, while formerly the shippers were taking advantage of their possession of the cars to load them out.

According to some claims the railroads are really handling a great deal of freight the insufficiency of movement being due to the particularly large volume of business offered. At any rate, the shippers at present have little hope of their condition being improved as long as any improvement in car movement redounds only to the benefit of the coal mines, under Order No. 7. However, the shippers have recently found hope that within a few weeks the coal mines will be supplied with all the cars they are able to load. The deficiencies in car supplies at coal mines, from the mine ratings, average about 50 per cent., and it is not thought that the car supplies are going to be doubled in the near future, or at any time, so as to provide 100 per cent., the expectation being that with a moderate further increase in car supplies a so-called "labor shortage" will be developed at the coal mines, say when car supplies get to 65 or 70 per cent. of ratings. These labor shortages are not shortages in the number of men, but

ised for such work and the high cost of labor and materials generally. The mill prices of the plain structural shapes are not particularly objected to, for as a matter of fact they are rather moderate compared with prices of some building materials and with the cost of labor on construction jobs.

All present indications suggest that with construction work involving the consumption of steel at a low ebb, with the operations of various manufacturing consumers materially curtailed by various causes, and with large stocks of steel accumulated at mills, the steel market is going to be an easy one, but the steel market is one of surprises, and present indications may possibly be found to be somewhat misleading.

Pig Iron

As to pig iron, the market such as it is shows rather an upward than a downward trend. Deliveries are very difficult to secure, and some furnaces are piling iron. The stiffness of the market may be due simply to the transportation condition, but a loosening in transportation may increase consumption so as to enable the stocks to be absorbed without weakening the market. The average consumer, at least in this district, has become very conservative and will buy no iron except what is absolutely needed in the near future. The buyers seem to expect the market to decline, but if they all overstay the market they may produce an advance when they come to buy. A fair guess seems to be that there will be simply hand-to-mouth buying of pig iron for months to come.

GUELPH MAY NOT HAVE TRUCK CO.

Some Difficulty Arose Regarding the Location—Subscribers Ask For Their Money Back

According to reports from Guelph, the Commerce Truck Co. is not likely to materialize. Several of those who subscribed for stock are now asking that their payments be refunded, and they have also notified their respective banking houses to stop payment on cheques they have given for future instalments on their stock allotment. An option was taken up on a tract of land in Guelph, for which the purchase price, $10,000, was to be paid in cash or stock, the owner of the land preferring to take cash in this instance. Where this cash payment is to come from is not quite clear.

Several Toronto parties were interested in selling the stock, the capital of the company being placed something like a million and a half, large parts of which were to go to the promoter and to the parent company in Detroit. Sales in Guelph, after the place was covered by a company of salesmen, amounted to about $47,000, most of the subscriptions being with the idea of securing a new industry. Some of the salesmen then tried out Kitchener, with better results, and the suggestion was made that the company be taken to the latter place. This, it is alleged, led to a controversy which caused the Detroit directors to withdraw, and the Canadian directors, left alone in the field, did not know exactly what to do.

Meanwhile, the expense of securing the charter has been incurred, also the obligation to pay $10,000 to the owner of the site in Guelph, while a number of salesmen will certainly want to be reimbursed for the time they took in selling stock to the people of Guelph and district.

One of the most notable improvements in electric furnaces for heating and melting metals has lately been perfected by a British firm. Many attempts have been made to produce a material which would get white-hot under the action of electric current and would yet remain unchanged so that it could be used again and again indefinitely. Long experience in furnace materials, combined with painstaking research, has enabled this company to produce a substance which can be moulded into any desired form, has a high resistance to electricity, and is for all practical purposes permanent. It is now being used in crucible form for melting brass, in tube form for the "heat treatment" of tools, and in the shape of a bath for other purposes.

SELECTED MARKET QUOTATIONS

Being a record of prices current on raw and finished material entering into the manufacture of mechanical and general engineering products.

PIG IRON

Grey forge, Pittsburgh	$42 40
Lake Superior, charcoal, Chicago	57 00
Standard low phos., Philadelphia	50 00
Bessemer, Pittsburgh	43 00
Basic, Valley furnace	42 90
Toronto price:—	
Silicon, 2.25% to 2.75%	52 00
No. 2 Foundry, 1.75 to 2.25%	50 00

IRON AND STEEL

Per lb. to Large Buyers	Cents
Iron bars, base, Toronto	$ 5 50
Steel bars, base, Toronto	5 50
Iron bars, base, Montreal	5 50
Steel bars, base, Montreal	5 50
Reinforcing bars, base	5 00
Steel hoops	7 00
Norway iron	11 00
Tire steel	5 75
Spring steel	10 00
Band steel, No. 10 gauge and 3-16 in. base	6 00
Chequered floor plate, 3-16 in.	8 40
Chequered floor plate, ¼ in.	8 00
Bessemer rails, heavy, at mill	....
Steel bars, Pittsburgh	3 00-4 00
Tank plates, Pittsburgh	4 00
Structural shapes, Pittsburgh	3 00
Steel hoops, Pittsburgh	3 50-3 75

F.O.B., Toronto Warehouse

Small shapes	4 25

F.O.B. Chicago Warehouse

Steel bars	3 62
Structural shapes	3 72
Plates	3 67 to 5 50
Small shapes under 3"	3 62

FREIGHT RATES

Per 100 Pounds.

Pittsburgh to Following Points

	C.L.	L.C.L.
Montreal	33	45
St. John, N.B.	41½	55
Halifax	49	64½
Toronto	27	39
Guelph	27	39
London	27	39
Windsor	27	39
Winnipeg	89½	135

METALS

	Gross. Montreal	Toronto
Lake copper	$25 00	$24 00
Electro copper	24 50	24 00
Castings, copper	24 00	24 00
Tin	62 00	65 00
Spelter	11 50	12 00
Lead	10 75	11 00
Antimony	13 00	14 00
Aluminum	34 00	36 00

Prices per 100 lbs.

PLATES

Plates, 3-16 in.	$ 7 25	$ 7 25
Plates, ¼ up	6 50	6 50

PIPE—WROUGHT

Price List No. 44—April, 1920.

STANDARD BUTTWELD S/C

	Steel Black	Steel Galv.	Gen. Wrot. Iron Black	Gen. Wrot. Iron Galv.
⅛ in.	$6 60	$ 8 50	....	....
¼ in.	5 18	7 26	$ 6 43	$ 7 56
⅜ in.	5 18	7 26	6 43	7 56
½ in.	6 84	8 42	7 27	8 84
¾ in.	8 45	10 58	9 03	11 16
1 in.	12 50	15 64	13 85	14 49
1¼ in.	16 91	21 16	18 06	22 81
1½ in.	20 21	25 80	21 59	26 68
2 in.	27 20	34 04	29 05	35 89
2½ in.	43 00	53 82	....	....
3 in.	56 23	70 38	....	....
3½ in.	71 30	88 32	....	....
4 in.	84 48	104 64	....	....

STANDARD LAPWELD S/C

	Steel Black	Steel Galv.	Gen. Wrot. Iron Black	Gen. Wrot. Iron Galv.
2 in.	$30 90	$37 74	$34 60	$41 44
2½ in.	45 34	56 16	51 19	62 01
3 in.	59 29	73 44	66 94	81 09
3½ in.	73 14	90 18	82 34	99 36
4 in.	86 66	106 82	97 56	117 72
4½ in.	0 93	1 23	1 24	1 49
5 in.	1 15	1 44	1 44	1 73
6 in.	1 49	1 86	1 87	2 25
7 in.	1 94	2 43	2 42	2 90
8-L in.	2 04	2 55	2 54	3 05
8 in.	2 35	2 94	2 92	3 51
9 in.	2 81	3 52	3 50	4 21
10-L in.	2 61	3 26	3 25	3 90
10 in.	3 36	4 20	4 18	5 03

Prices—Ontario, Quebec and Maritime Provinces

WROUGHT NIPPLES

4" and under, 60%.
4½" and larger, 50%.
4" and under, running thread, 30%.
Standard couplings, 4-in. and under, 30%.
Do., 4½-in. and larger, 10%.

OLD MATERIAL

Dealers' Average Buying Prices.

	Per 100 Pounds. Montreal	Toronto
Copper, light	$15 00	$14 00
Copper, crucible	18 00	18 00
Copper, heavy	18 00	18 00
Copper wire	18 00	18 00
No. 1 machine composition	16 00	17 00
New brass cuttings	11 00	11 75
Red brass cuttings	14 00	15 75
Yellow brass turnings	8 50	9 50
Light brass	7 00	7 00
Medium brass	8 00	7 75
Scrap zinc	6 50	6 00
Heavy lead	7 50	7 75
Tea lead	4 50	5 00
Aluminum	19 00	20 00
	Per Ton	Gross
Heavy melting steel	18 00	18 00
Boiler plate	15 50	15 00
Axles (wrought iron	20 00	20 00
Rails (scrap)	18 00	18 00
Malleable scrap	23 00	25 00
No. 1 machine cast iron	32 00	33 00
Pipe, wrought	12 00	12 00
Car wheel	26 00	26 00
Steel axles	22 00	20 00
Mach. shop turnings	11 00	11 00
Stove plate	26 50	25 00
Cast boring	12 00	12 00

BOLTS, NUTS AND SCREWS

	Per Cent.
Carriage bolts, 7-16 and up	+10
Carriage bolts, ⅜-in. and less	Net
Coach and lag screws	—15
Stove bolts	55
Wrought washers	—25
Elevator bolts	+10
Machine bolts, 7-16 and over	+10
Machine bolts, ⅜-in. and less	+10
Blank bolts	Net
Bolt ends	Net
Machine screws, fl. and rd. hd., steel	27½
Machine screws, o. and fil. hd., steel	+25
Machine screws, fl. and rd. hd., brass	net
Machine screws, o. and fil. hd., brass	net
Nuts, square, blank .... +25 add	$2 00
Nuts, square, tapped ....add	2 25
Nuts, hex., blank ....add	2 50
Nuts, hex., tapped ....add	3 00
Copper rivets and burrs, list less	15
Burrs only, list plus	25
Iron rivets and burrs	40 and 5
Boiler rivets, base ¾" and larger	$8 50
Structural rivets, as above	8 40
Wood screws, O. & R., bright	75
Wood screws, flat, bright	77½
Wood screws, flat, brass	55
Wood screws, O. & R., brass	55½
Wood screws, flat, bronze	50
Wood screws, O. & R., bronze	47½

MILLED PRODUCTS

(Prices on unbroken packages)

	Per Cent
Set screws .... —20%, 25 and	5
Sq. and hex. hd. cap screws	12½
Rd. and fil. hd. cap screws ..plus	25
Flat but. hd. cap screws ....plus	50
Fin. and semi-fin. nuts up to 1-in.	12½
Fin. and Semi-fin. nuts, over 1 in., up to 1½-in.	—5
Fin. and Semi-fin. nuts over 1½ in., up to 2-in.	+12½
Studs	+5
Taper pins	—12½
Coupling bolts	+40
Planer head bolts, without fillet, list	+45
Planer head bolts, with fillet, list plus 10 and	+55
Planer head bolt nuts, same as finished nuts.	
Planer bolt washers	net
Hollow set screws	+60
Collar screws ......list plus 20,	30
Thumb screws	40
Thumb nuts	75
Patch bolts ....add	+85
Cold pressed nuts to 1½ in...add	$1 00
Cold pressed nuts over 1½ in..add	2 00

BILLETS

	Per gross ton
Bessemer billets	$60 00
Open-hearth billets	60 00
O.H. sheet bars	76 00
Forging billets	56 00-75 00
Wire rods	52 00-70 00

Government prices.

F.O.B. Pittsburgh.

NAILS AND SPIKES

Wire nails, base	$6 10
Cut nails, base	7 00
Miscellaneous wire nails	50%

ROPE AND PACKINGS

Plumbers' oakum, per lb.	0 10¼
Packing, square braided	0 38
Packing, No. 1 Italian	0 44
Packing, No. 2 Italian	0 36
Pure Manila rope	0 35½
British Manila rope	0 28
New Zealand hemp	0 28

POLISHED DRILL ROD

Discount off list, Montreal and Toronto .... net

A GOOD BOON

"Geometric Tools are a good boon in our business. Always reliable and sure of turning out a class job; in fact, I don't know how we could secure a decent output without them."

This is what the Works Inspector of a London (England) shop says. He adds that they are making screw and small parts for air craft, and have quite a lot of Geometric Dies in constant use.

Because of the fact that Geometric threading tools are "a good boon" in the thread cutting business, the majority of screw machines and turret lathes are equipped with Geometrics.

THE GEOMETRIC TOOL CO.
NEW HAVEN, CONN.

Canadian Agents:
Williams & Wilson, Ltd., Montreal
The A. R. Williams Machinery Co., Ltd., Toronto,
Winnipeg and St. John, N.B.
Canadian Fairbanks-Morse Co., Ltd., Manitoba,
Saskatchewan, Alberta

For any thread, any size, any pitch,—there's a Geometric Collapsing Tap or Self-Opening Die Head. Be assured there is one to meet your particular threading need—whatever it may be. Ask us.

MISCELLANEOUS

Solder, strictly	$ 0 35
Solder, guaranteed	0 39
Soldering coppers, lb.	0 62½
White lead, pure, cwt.	20 35
Red dry lead, 100-lb. kegs, per cwt.	16 00
Gasoline, per gal., bulk	0 38
Pure turp., single bbls., gal.	3 15
Linseed oil, raw, single bbls.	2 37
Linseed oil, boiled, single bbls.	2 40
Wood alcohol, per gal.	4 00
Whiting, plain, per 100 lbs.	3 00

CARBON DRILLS AND REAMERS

S.S. drills, wire sizes	32½
Can. carbon cutters, plus	20
Standard drills, all sizes	32½
3-fluted drills, plus	10
Jobbers' and letter sizes	32½
Bit stock	40
Ratchet drills	15
S.S. drills for wood	40
Wood boring brace drills	25
Electricians' bits	30
Sockets	50
Sleeves	50
Taper pin reamers	25 off
Drills and countersinks	net
Bridge reamers	50
Centre reamers	10
Chucking reamers	net
Hand reamers	10
High speed drills, list plus 10 to	20
Can. high speed cutters, net to plus	10
American plus	40

COLD ROLLED STEEL

[At warehouse]

Rounds and squares	$7 base
Hexagons and flats	$7.75 base

IRON PIPE FITTINGS

	Black	Galv.
Class A	60	75
Class B	27	37
Class C	18	27

Cast iron fittings, 5%; malleable bushings, 22½%; cast bushings, 22½%; unions, 37½%; plugs, 20% off list.

SHEETS

	Montreal	Toronto
Sheets, black, No. 28	$8 50	$10 50
Sheets, Blue ann., No. 10	8 50	9 50
Canada plates, dull, 52 sheets	8 50	10 00
Can. plates, all bright	8 60	9 00
Apollo brand, 10¾ oz. galvanized	....	....
Queen's Head, 28 B.W.G.	11 00	....
Fleur-de-Lis, 28 B.W.G.	10 50	....
Gorbal's Best, No. 28	....	....
Colborne Crown, No. 28.	....	....
Premier, No. 28, U.S.	11 50	11 60
Premier, 10¾-oz.	11 50	12 00
Zinc sheets	16 50	20 00

PROOF COIL CHAIN

(Warehouse Price)

B

¼ in., $13.00; 5-16, $11.00; ⅜ in., $10.00; 7-16 in., $9.80; ½ in., $9.75; ⅝ in., $9.20; ¾ in., $9.30; ⅝ in., $9.50; 1 in., $9.10; Extra for B.B. Chain, $1.20; Extra for B.B.B. Chain, $1.80.

ELECTRIC WELD COIL CHAIN B.B.

⅛ in., $16.75; 3-16 in., $15.40; ¼ in., $13.00; 5-16 in., $11.00; ⅜ in., $10.00; 7-16 in., $9.80; ½ in., $9.75; ⅝ in., $9.50; ¾ in., $9.30.

Prices per 100 lbs.

FILES AND RASPS

	Per Cent.
Globe	50
Vulcan	50
P.H. and Imperial	50
Nicholson	32½
Black Diamond	27½
J. Barton Smith, Eagle	50
McClelland, Globe	50
Delta Files	20
Disston	40
Whitman & Barnes	50
Great Western-American	50
Kearney & Foot, Arcade	50

BOILER TUBES.

Size.	Seamless	Lapwelded
1 in.	$27 00	$.....
1¼ in.	29 50	.....
1½ in.	31 50	29 50
1¾ in.	31 50	30 00
2 in.	35 00	30 00
2¼ in.	35 00	29 00
2½ in.	42 00	37 00
3 in.	50 00	48 00
3¼ in.	.. ..	48 50
3½ in.	63 00	51 50
4 in.	85 00	65 50

Prices per 100 ft., Montreal and Toronto

OILS AND COMPOUNDS.

Castor oil, per lb.	
Royalite, per gal., bulk	27½
Palacine	30½
Machine oil, per gal.	54½
Black oil, per gal.	28
Cylinder oil, Capital	82½
Petroleum fuel oil, bbls., net	18

BELTING—No 1 OAK TANNED

Extra heavy, single and double		6½
Standard		6½
Cut leather lacing, No. 1		2 00
Leather in side	2 40	3 00

TAPES

Chesterman Metallic, 50 ft.	$2 00
Lufkin Metallic, 603, 50 ft.	2 00
Admiral Steel Tape, 50 ft.	2 75
Admiral Steel Tape, 100 ft.	4 45
Major Jun. Steel Tape, 50 ft.	3 50
Rival Steel Tape, 50 ft.	2 75
Rival Steel Tape, 100 ft.	4 45
Reliable Jun. Steel Tape, 50 ft.	3 50

PLATING SUPPLIES

Polishing wheels, felt	$4 60
Polishing wheels, bull-neck	2 00
Emery in kegs, Turkish	09
Pumice, ground	06
Emery glue	30
Tripoli composition	09
Crocus composition	12
Emery composition	11
Rouge, silver	60
Rouge, powder, nickel	45

Prices per lb.

ARTIFICIAL CORUNDUM

Grits, 6 to 70 inclusive	.08½
Grits, 80 and finer	.6

BRASS—Warehouse Price

Brass rods, base ½ in. to 1 in. rod	0 34
Brass sheets, 24 gauge and heavier, base	$0 42
Brass tubing, seamless	0 46
Copper tubing, seamless	0 48

WASTE

XXX Extra	.24	Atlas	.20
Peerless	.22½	X Empire	.19½
Grand	.22½	Ideal	.19
Superior	.22½	X Press	.17½
X L C R.	.21		

Colored

Lion	.17	Popular	.13
Standard	.15	Keen	.11
No. 1	.15		

Wool Packing

Arrow	.35	Anvil	.22
Axle	.28	Anchor	.17

Washed Wipers

Select White	.20	Dark colored	.09
Mixed colored	.10		

This list subject to trade discount for quantity.

RUBBER BELTING

Standard ... 10% Best grades... 15%

ANODES

Nickel	.55 to .60
Copper	.38 to .40
Tin	.70 to .70
Zinc	.16 to .17

Prices per lb.

COPPER PRODUCTS

	Montreal	Toronto
Bars, ½ to 2 in.	$42 50	$43 00
Copper wire, list plus 10.		
Plain sheets, 14 oz., 14x60 in.	46 00	44 00
Copper sheet, tinned, 14x60, 14 oz.	48 00	48 00
Copper sheet, planished, 16 oz. base	46 00	45 00
Braziers', in sheets, 6x4 base	45 00	44 00

LEAD SHEETS

	Montreal	Toronto
Sheets, 3 lbs. sq. ft.	$10 75	$14 50
Sheets, 3½ lbs. sq. ft.	10 50	14 00
Sheets, 4 to 6 lbs. sq. ft.	10 25	13 50

Cut sheets, ½c per lb. extra.
Cut sheets to size, 1c per lb. extra.

PLATING CHEMICALS

Acid, boracic	$ .23
Acid, hydrochloric	.04
Acid, nitric	.11
Acid, sulphuric	.04
Ammonia, aqua	.15
Ammonium, carbonate	.23
Ammonium, chloride	.22
Ammonium hydrosulphuret	.75
Ammonium sulphate	.30
Arsenic, white	.14
Copper, carbonate, annhy	.41
Copper, sulphate	.15
Cobalt, sulphate	.20
Iron perchloride	.62
Lead acetate	.30
Nickel ammonium sulphate	.20
Nickel carbonate	.32
Nickel sulphate	.22
Potassium sulphide (substitute)	.42
Silver chloride (per oz.)	1.30
Silver nitrate (per oz.)	1.25
Sodium bisulphate	.14
Sodium carbonate crystals	.06
Sodium cyanide, 127-130%	.38
Sodium hyposulphite per 100 lbs	8.00
Sodium phosphate	.15
Tin chloride	.80
Zinc chloride, C.P.	.30
Zinc sulphate	.10

Prices per lb. unless otherwise stated

Quality

THE
STEEL COMPANY
OF
CANADA
LIMITED
HAMILTON MONTREAL

Service

If interested tear out this page and place with letters to be answered.

INDUSTRIAL NEWS

NEW SHOPS, TENDERS AND CONTRACTS
PERSONAL AND TRADE NOTES

TRADE GOSSIP

Starting in Vancouver.— The Britannia Wire Rope Co., is getting in shape to operate at their plant on Granville Island, B.C. The machinery for most of the plant is now on the way from England. C. H. Gill is the British director, and Robert Gibson, managing director.

Ford Buys Mines.—Henry Ford, according to reports in Cincinnati, has purchased two coal mines in Kentucky, and is planning on joining his recently purchased railroad, the Detroit, Toledo & Ironton, with these mines to ensure coal for his factories.

Getting Them Organized.—The work of co-ordinating the various departments of the C.N.R. and G.T.R. has been concluded as far as the passenger business is concerned, and the freight departments will be unified by the end of the present month, according to the indications.

Ford Increasing.—The Ford Motor Co., is notifying the steel mills that it is planning a production of 4,000 cars per day, starting with its manufacturing year, Aug. 1, and the latter are asked to increase their allotments of steel accordingly. This plant has been on a daily basis of 3,000 to 3,500 cars recently.

Factory Taken Over.—C. W. Sherman, of the Dominion Foundries and Steel. Limited, Hamilton, stated that a new factory, affiliated with the local concern, started operations this week at Waterville, near Albany, New York. Mr. Sherman added that the new factory is a large one and affords employment to between 500 and 600 men.

Start in Victoria.—The Canadian Abrasive Co., Ltd., is starting operations at Victoria, having secured a loan of $22,000 from the Provincial Government Mr. J. L. Near, Major Martyn, Mr. Moore, of the Canadian Explosives, Ltd., Capt. D. S. Pullen, Col. Lorne Ross are among those interested in a company with a capital of $100,000. Plant and equipment cost $50,000.

Steel Prices Halted.—Prices received by the Saskatchewan Government in response to tenders for steel bridges are no higher than spring quotations, an indication to officials that a halt in steel prices has been reached. Present quotations are about 11 per cent. above those of spring, 1919. A contract was awarded to the Canadian Bridge Company of Walkerville, Ont., for four steel bridges at $9,363.40, for the four, f.o.b. shops.

Buy Steam Shovel.—At the session of the St. John, N.B., council, the need of a steam shovel for the preliminary work of the street paving, was the main topic for discussion. Representatives of the Kingston Steam Shovel Company were present, and it was ultimately decided to purchase a machine, with a trench digging

GEO. G. CUTTLE

George G. Cuttle has just purchased the entire interests of the Standard Machinery and Supplies, Limited, Montreal. The company is capitalized at $200,000, and has been successfully operated since 1916. . Mr. Cuttle became connected with the company in 1917 as manager of the metal department; in January, 1918, he was appointed general manager and six months later was elected vice-president. He first entered the iron and steel trade in 1904 with the Montreal Rolling Mills, now a part of the Steel Company of Canada, and worked through various departments until 1913, when he was appointed general purchasing agent, which position he held until joining his present company.

The Standard Machinery and Supplies are the Canadian representatives of the Waltham Grinding Wheel Company of Waltham, Mass.; J. L. Goodhue Company, Limited, Danville, Que.; and the Clinton Metallic Paint Company of Clinton, N.Y., in addition to carrying large stocks of tool steels, machine tools and other supplies.

ping attachment, so it may also be used in digging trenches for sewers and water pipes. The price is $7,000.

Coal Board Issues Order.—The Railway Board at Ottawa has issued an order which prohibits the exportation of coal from any Atlantic, St. Lawrence River, or Gulf port of Canada, excepting to the United States or Newfoundland. The order becomes effective on and after August 1st, and was issued after the consideration of information gained from the United States Inter Commerce Board.

New Rates Contract. — The Special Scale Committee of coal operators and miners sitting in District 18, Calgary, have completed their labors. The agreement is between the Western Canada Coal Operators' Association and District 18, of the United Mine Workers of America. The agreement provides for an all-round increase, and is in force till April next, and retroactive from March last.

Canal Work Handicapped.—Work on the Welland Ship Canal is held up on account of shortage of coal, shortage of Hydro power, and shortage of labor. While there is hope of overcoming the latter difficulty, there is not much hope for overcoming the first two. When work stopped last year there were 2,500 men working on it. There was little chance of the same number being employed this year.

Conference in Fall.—The second great Industrial Conference will be held in Canada some time in October or November. An appropriation was made by Parliament at last session to cover the expense of the conferences, and it is expected that the experience gathed at the last conference will result in making this one an even more representative gathering of labor, capital and the general public.

Conference at Welland.—The conference to be held at Niagara and Welland on August 2, 3 and 4, promises to be well attended. A large number of invitations have been accepted by business and industrial men in the United States and Canada. On August 2nd a power banquet will be given at the Clifton House at Niagara Falls, where Sir Adam Beck will deliver an address. An address will be given on August 3rd at Welland on the "Welland Ship Canal and Deep Waterways."

New Service Projected.—As a result of the deliberations of the West Indies Conference, a new steamship service will be inaugurated early in 1921, between Canada and the West Indies. The service will be carried on by two of the new steamers building for the Canadian Government, and will provide sailings from Halifax and St. John to Bermuda, Bahamas, British Honduras and return. The steamers will carry about 50,000 tons cargo and accommodate about 25 passengers.

Short of Material.—H. Frawley, manager of the Canadian Car and Foundry Company, at Fort William, stated that there are still over two hundred cars to be repaired before the old order is completed, and it was the shortage of material that prevented the plant from completing the order before now. Mr. Frawley also stated that the new order of the Canadian Pacific Railway's will be commenced by the end of the month. At the present time the plant is working on the sample car which, when completed, will be approved of by the officials of the Canadian Pacific Railways, and as soon as it has been approved by the officials the plant will be able to go right ahead with the order.

The Strike Over.—When Robb's Engineering Works at Amherst opened many of the moulders, whose six weeks' strike was terminated, reported for work. About half have already been taken on and the rest will be given their jobs in a few days. A committee from the moulders waited upon the management, expressing the men's appreciation of the sympathetic attitude of General Manager Newell and stated that they had no grievance and had desired only the higher rate. He in turn stated that there was no ill-feeling on the company's part. The agreement arrived at will continue in force until April of next year. After October 1st, the men now getting the 72-cent minimum are to receive 75 cents per hour.

Warns Coal Operators.—Speaking in Washington recently to the coal operators of the National Coal Association, Mr. J. D. A. Morrow, vice-president of the association, gave the members a solemn warning. Unless the coal operators of the country could meet the demands of the people for fuel, he told them, the industry was faced with a resumption of Government control, which would be very difficult to get rid of. The Northwest was short of 5,000,000 tons and a plan giving priority to coal for the Northwest had been agreed upon with the Inter State Commerce Commission. It was up to the coal operator to make this arrangement a success and fill the deficiency.

Production is Above Average.—That the coal shortage in the northwest of the U.S. and Western Canada is not due to any lack of coal itself, but of the means of distributing it, is shown by a statement compiled by W. B. Williams, vice-president of the Delaware & Hud-

son Railway Co. He states that the production of bituminous coal in the first six months of 1920 has exceeded the corresponding period of 1919 by 41,000,000 tons. The production for the six months of 1920 was 255,000,000 tons, and if this rate is kept up the production for the year will be some 4,000,000 tons above the average for the five years from 1914 to 1919 inclusive.

Ford-Smith Picnic.—The employes of the Ford-Smith Machine Co., Hamilton, held their annual picnic at Dundas Driving Park. About 300 of the employees and their friends left Hamilton in four special cars from the Terminal station at 11.15 a.m., arriving at the park about 12 o'clock, where lunch was prepared at tables tastefully laid out in the pavilion. Several noted features amongst the amusements kept the ball rolling and made every hour a happy one. A vaudeville competition was fruitful in bringing to light a galaxy of youthful artists. Dancing, recitations and music, in which the pipes figured, kept all in pleased surprise. Vaudeville was followed by community singing and many favorite songs, both humorous and sentimental, were heartily joined in by all. Tea was partaken of by all in the pavilion, after which a novel feature was presented by a Punch and Judy show, which was heartily enjoyed by old and young. The prizes for the various sporting events were presented by the genial head of the firm, after which the pavilion was cleared and dancing brought to a close a most enjoyable day.

Building Remodelled. — The Ruddy Building, 9 and 11 Wellington Street E., has been taken over on a long lease by the Burnside Realty Co., the president of which is Mr. E. C. Roelofson, of the Roelofson Machine Tool Co. This company on the first of July closed their Galt plant, the last of which has been disposed of to various industrial establishments in the district. The building in Galt was taken over by the McCaskey people for the manufacture of counter-check books. The Roelofson Machine Tool Co. will continue to act as distributors for the Potter & Johnson lines, and some other business may be handled as soon as arrangements are completed. The ground floor of the building is being taken over by the Roelofson Co., where they will carry in their showroom a full line of lathes, drill presses, shapers, etc. The facilities are excellent for shipping and receiving at the rear of the building. An elaborate suite of offices is being arranged on this floor for the company. The other three floors have been in the hands of the contractors for some weeks now, transforming them into offices, most of them built and finished to suit the requirements of the various firms leasing them. There are three floors above the main one being finished in this way, there being a total of some 20,000 feet of floor space in the building.

MARINE

Montreal.—The steamer "Alexandrian" sailed from Montreal bound for Dantzic, thus inaugurating a new service from Canada to Germany.

Vancouver.—The contract for the $3,000,000 dry dock to be built on Burrard Inlet by Coughlan & Sons, has now been signed, and it is understood that work will be started in sixty days from the date of signing.

London.—Lloyd's Register figures for the quarter ending June 30th, are the highest yet reached. The aggregate tonnage under construction is 3,576,000 tons, and this exceeds the United States tonnage under construction by 1,672,000 tons. No details of the figures are yet available.

Halifax.—A fleet of four trawlers and twenty-four drifters, a portion of the 100 or more of these vessels that were sold by the Canadian Government, sailed for Scotland recently. They will call at the Azores on their way across for coal. They will be employed in the fishing service round the British coasts.

Brockville.—The raising of the sunken steamer, Keystorm, sunk in the storm of 1913, near Sister Lighthouse, is getting near the final stages. Captain Leslie, of Kingston, who is engaged on the work, is confident that the feat can be accomplished despite the numerous failures that have taken place.

Alpena, Mich.—Two barges lost, and the steamer Charles Bradley seriously damaged, was the result of a recent storm on Lake Huron. The Bradley, which was towing two barges, the Mystic and the Blodgett, was forced to send out wireless calls for assistance, when her bow collided with her and smashed in her stern. The steamer Huron stood by the Bradley all night, and in the morning took off the crews of the barges, which were filling rapidly.

Welland.—The stern section of the steamer "Canadian Squatter" was launched from the yard of the British American Shipbuilding Company recently. After both sections have been launched, they will be towed to Montreal where they will be joined together by the Canadian Vickers Co. The "Canadian Squatter" will, when completed, be 320 feet long, 43 feet, 10 inches beam, and 25 feet deep. She will have a deadweight capacity of 4,350 tons, and a speed of 11 knots.

Montreal. — Considerable trouble is being experienced by masters of British vessels coming to Canadian and United States ports by the desertion of their crews. The men desert in order to ship in vessels under American and Canadian registry and to get the higher wages paid by them. In order to stop this epidemic of desertion, all cases will be prosecuted and the full allowable punishment will be sought for. In two days there were twenty desertions in one port alone.

Garlock.

MONTREAL

CONDITIONS IN IRON AND STEEL IN VARIOUS SECTIONS OF THE OLD COUNTRY

For the first time since the Armistice there has been a distinct falling off in the demand for tinplates, and new business has been slack during the past month. Makers in South Wales are, however, well supplied with orders, which will keep the works going until the end of the year. Only in November last standard plates could be obtained as low as 36s. per box. Then prices rose by leaps and bounds up to as high as 76s., but at the time of writing standard plates, 20 by 14, can be secured at from 72s. to 68s., according to position, with other sizes in proportion. Naturally, whilst abnormal figures prevailed makers were not slow to accept all the specifications offering, but a couple of months ago it became apparent that top prices had been reached, and that a downward tendency would soon set in. The chief indication of this change was evidenced by the decline in block tin. Prices for plates remain fairly steady, but the market is extremely dull, and it is possible that within a few weeks there will be a break, to the advantage of consumers. In the meantime the operatives are pressing demands for increased wages, which, if granted, will mean an advance of something like 300 per cent.

SCOTLAND

The first six months' trading in industrial Scotland has given bigger profits than at any time during the war years, and the outcome demonstrates the rapid manner in which Scotch firms have carried through the transition from war to peace conditions. Most of the plant is now running on original lines, and many new installations are helping to swell the ever-increasing output. However, apart from iron and steel, makers are not now having orders thrust upon them irrespective of cost and date of delivery. In point of fact, the trend is to hold out for reduced prices, which manufacturers are slow to consider so long as arrears of work are still heavy. Be that as it may, canvassing for orders is once more being resorted to. At the same time the pres-

sure of business in iron and steel is still very heavy. From a survey of the "key" industries the opinion is formed that the highest levels have now been reached, a belief which is strengthened by the greater eagerness of Americans to export to near-hand Continental ports, and even to British centres. Malleable-iron makers have still an abundance of work in sight, but sheets are now in weaker request, and speculators are now offering their holdings.

SHEFFIELD

The reaction which made its appearance in the iron and steel industry a couple of months ago has steadily developed, and the situation is now more unsettled than at any time since the signing of the Armistice. There are numerous reasons for the pessimism with which one meets almost everywhere. The last advance in prices was followed immediately by a cessation of new business. This circumstance had little effect upon the position at the works, for as a whole they are fully booked, and are certain of disposing of the whole of their output for the present year. This situation also suffered from the collapse of the Japanese market, as a large volume of orders from Japan was cancelled. The general unwillingness on the part of consumers to place new orders was due to a fear of a break in prices. These were at the peak, and people were naturally desirous of avoiding the heavy loss that might have been entailed if a reduction set in. Buying and selling have been on a hand-to-mouth basis for more than a month. There is at present something like a famine in the supply of foundry iron and castings. The foundries of the country would be unequal to the demands of engineers even if pig iron were available, as there are not enough moulders. These skilled men were in short supply before the disastrous strike, and it is estimated that 3,000 of them drifted into other occupations and have not since returned.

ares. It is planned to raise the boat with pontoons, tow it in a certain distance from its present position on its side and then beach it. The pontoons will then be rearranged so that the boat may be righted and brought to the surface.

The death occurred at the Moore Convalescent Home in Montreal, of Captain Elisha B. Smith, one of the oldest of the lake sailors. He had a record of service as captain of Lake Ontario, Bay of Quinte and St. Lawrence River steamers extending over a period of 53 years. Captain Smith was 88 years old at the time of his death, which followed three years suffering from acute rheumatism. Captain Smith only had one accident during his long career, and that was when his steamer was sunk in the Cedar Rapids. Happily no lives were lost, although Captain Smith was thrown from the bridge into the river when his vessel struck the rocks. Captain Smith was twice married and, is survived by his second wife, whose home is in Toronto. Captain Smith's body was taken to Picton for burial.

INCORPORATIONS

Notice is received of the incorporation of The Canadian Cinch Anchoring System, Ltd., for the purpose of the manufacture, sale, export and import of the "Cinch Anchored Systems" and allied products. Capital to consist of $20,000 divided into 200 shares of $100 each. Chief place of business to be at the City of Toronto.

Notice is received of the incorporation of The Newton-Dakin Construction Co., Ltd., to take over the concern known as The Loomis Dakin Construction Co. Ltd. The company to carry on the work of Engineers and Contractors. Authorized capital to consist of $250,000 divided into 2,500 shares of $100 each; the chief place of business to be in the City of Sherbrooke, P.Q.

PERSONALS

W. T. Colewell has been appointed superintendent of the Canada Car plant, exclusive of the Malleable Iron Works.

Mr. Walter Lambert, naval architect and marine surveyor of 14 Place Royale, Montreal, has undertaken the Canadian agency of Messers Cochran & Co., Ltd., of Annan, Scotland, makers of the well-known Cochran donkey boiler, oil or coal-fired. Mr. Lambert has also opened an office at 408 Bower Bldg., Vancouver, to look after his interests on the Pacific Coast.

F. W. Peters has been appointed general assistant of the C.P.R. in British Columbia, in succession to the late Richard Marpole. Mr. Coleman, vice-president in charge of the company's western lines, stated: "The company feels that it has in F. W. Peters, our senior officer at the coast, one in whom the public have the greatest confidence. His connection with British Columbia is

---

Montreal.—In connection with the legislation recently enacted by the Government to aid the shipbuilding industry, the Canadian Association of Shipbuilders and Engineers is preparing an active campaign to bring orders for ships from foreign countries to Canada. Captain Gerrard, the secretary of the association, says that they have secured the accession of several strong members, and that they feel now that they can make shipbuilding a permanent industry.

Montreal.—It is stated that J. W. Norcross will be the president of the executive committee of the new British Empire steel merger. In this position he will have full charge of all the undertak-

ings of the big corporation, and will at the same time serve as deputy chairman of the board of directors. As he will retain the presidency of the Canada Steamships Lines, which will be leased to the British Empire Steel Corporation, he will thus have full command of the shipbuilding and shipping activities of the corporation.

An Interesting Task. —Capt. W. B. Leslie of Kingston, who is engaged on the raising of the big steel freighter Keystorm, sunk in 1913 a short distance this side of Sister Lighthouse, at Brockville, expects that all will be ready for the raising of the vessel at the end of this week. He is confident that the feat can be accomplished after many fail-

He Knows

And Uses

MORROWS

Ingersoll   **Canada**

If interested tear out this page and place with letters to be answered.

not quite as long as that of Mr. Marpole at the time of his death, but he has spent almost 15 years here of the most interesting period of the history of the province, and has always placed himself freely at the disposal of the community for the promotion of any worthy cause."

CATALOGUES

The Buffalo Forge Co., Buffalo, N.Y., have issued a new catalogue No. 721, dealing with their dyehouse and bleachery ventilation equipment. This booklet describes the dyehouse apparatus in detail and gives various illustrations of actual installations.

WILL PURCHASE THE COMPANY NOW

Canada Foundries, and Forgings Likely to Take Over Mann Axe and Tool Co.

BROCKVILLE.—At the special general meeting of the shareholders of Canada Foundries and Forgings Company, held here, approval was given to the recommendation of the board of directors of the enterprise involving the acquisition of the Mann Axe and Tool Company, of St. Stephen, N.B., approximately 70 per cent. of the outstanding stock of the Brockville concern reported at the meeting either in person or by proxy.

If certain negotiations now being conducted with the municipality of St. Stephen are carried to a satisfactory conclusion, it is stated, the Mann plant at that town will be rebuilt without delay. The business will be conducted under the name of the Mann Axe Company, in which Canada Forgings will own a controlling interest.

The Forgings directors also held their quarterly meeting, following which the usual preferred dividend of 1¾ per cent. and the common one of 3 per cent. were declared for the current three months. The earnings of the company, it was stated, were well in excess of the requirements in this respect, the process of working the industry into a sound postwar basis being satisfactorily accomplished.

DOCTOR AND LAWYER DIFFER

A man in a Western town was hurt in a railroad accident, and after being confined to his home for several weeks he appeared on the street walking with the aid of crutches.

"Hello, old fellow," greeted an acquaintance, rushing up to shake his hand. "I am certainly glad to see you around again."

"I see you are hanging fast to your crutches," observed the acquaintance. "Can't you do without them?"

"My doctor says I can," answered the injured party, "but my lawyer says I can't."

COAL IN NORTH?

James Price of Cobalt has an idea that coal can be found in the district between New Liskeard and Englehart, and he has suggested to the town council that the municipality should engage in the search for it at the ratepayers' expense. Mr. Price thinks the mineral would be found at from 100 to 300 feet, and wants a test bore sunk, but the council negatived the idea. A. A. Cole, mining engineer of the T. & N. O., says there is no coal in the region indicated, as the geological formation is too old. There is a slight possibility of oil there, he says.

BLOWPIPE MANIPULATION

Continued from Page 122

defects such as pinholes, etc. These results were duly checked by the Secretary of the French Acetylene Assocation.

In conclusion, the process of welding backwards should appeal to all firms who are out for economy in the execution and improvement in the quality of their welds. It is only necessary to emphasize that common sense is necessary in applying the methods on the lines which have been outlined in these articles.

The United States Bureau of Mines makes the following recommendations for the prevention of spontaneous combustion of stored bituminous coal: — Piles not to be over 12 ft. deep, and no part of the interior to be over 10 ft. from the surface. Store only screened lump coal—if possible. Keep out dust as much as possible, and to do this avoid handling. Have lump and fine evenly distributed. Do not let lumps roll to the bottom and form air passages. Re-handle the screenings after two months, if possible. Store away from any sources of even moderate heat, and well away from the main buildings of the plant; never against a frame building. Allow six weeks' seasoning after mining before putting into storage piles. Avoid alternate wetting and drying. Avoid admission of air to the interior of the pile through interstices around timbers, irregular brickwork or a porous bottom such as coarse cinders. If wet coal is received, dump in small piles around the edges, where air can get to it freely to carry away moisture, and where other coal will not be packed on top of it.

DOMINION CHUCKS

STEEL OR CAST-IRON BODY BUILT FOR HEAVY DUTY

All Screws Are Reversible

SCREWS are made of the best grade steel. Both ends are broached and are heat treated after machining. They are reversible, so that either end may be used, are large enough in diameter to stand the torsional strains applied by operator when setting up his work. They are made to give the best service—and may be depended upon to stand up under the hardest usage.

DOMINION STEEL PRODUCTS CO. LIMITED

Engineers ⋅ Manufacturers

BRANTFORD, CANADA

Classified Opportunities

Next Export Number
August 5th.
Reserve Space At Once!

CANADIAN MACHINERY
AND
MANUFACTURING NEWS

Vol. XXIV. No. 6   August 5, 1920

Measuring in *Thousandths* Forty Years Ago

Measuring to a Thousandth of an Inch 40 Years Ago Was Some Task, But the Machine as Described Did Its Duty Nobly—The Progress of Measuring Methods is Also Discussed.

By J. H. MOORE

IN these days of interchangeability and unit production we often lose sight of the fact that it is not so very long ago since such conditions came into being. In olden days we have no record of even the foot and inch. Everything was measured by the cubit, and fraction thereof, and for the benefit of those not familiar with this term, we might say that a cubit is supposed to be the length of the forearm from the elbow to the extremity of the middle finger.

This view illustrates how the machine was used.

Evidently this system of measurement caused some controversy between the different countries, for each country adopted their own idea of a cubit. In England we find that 18", or 45.72 cm., was considered a cubit, while the ancient Egyptians decided 20.61", or 52.35 cm., equalled the same factor. The ancient Romans said 17.4", or 44.20 cm., was the correct measurement, while the ancient Greeks claimed 18.25", or 46.35 cm., was the one and only real cubit, and so it went on, each nation working to their own standard.

Leaving this confusing period behind and coming to the early eighties, that is about 1880, we find conditions very much improved, as far as standards are concerned, but not so greatly advanced in the working to those standards. Feet and inches, with their various fractions were, of course, in use, but anything less than a 1-64 was very seldom bothered with. When you worked to a 1-64 it was considered fine work. Interchangeability was practically unknown, the idea being to make each succeeding machine as near as possible to the last one.

Let it be understood that we are casting no reflection on the class of mechanic existent at that time, for machinists in those days were workmen of the highest order. They could run every machine in the shop, not merely one or two, as is sometimes the case to-day. Considering the equipment they worked with, they accomplished wonders. What we desire to point out is this, that accuracy and interchangeability did not seem to be considered as necessities.

To speak of a millionth of an inch in those days would have been an act of folly, and no doubt your friends would have persuaded the doctor to pay you a little informal call shortly after your flamboyant statement. Even a thousandth of an inch was closer than they could measure, micrometers not being existent at that time, and so conditions went on until one memorable day an important step was made in the right direction.

In the well-known town of Dundas, Ont., there existed a machine tool concern who were far from content with the measuring methods at their command. This was the firm now known as the John Bertram & Sons Co., Ltd., and their present

spirit of striving after simplicity, quality, and accuracy seems to have been existent even at this early date, for in a pioneering frame of mind they started to make a measuring machine for themselves.

Making such a machine was a real task in those days, for where was the standard to come from? Sixty-fourths were not good enough for them; thousandths was what they were after, but how to obtain this accuracy, that was the real puzzle. Of course, to us in these enlightened days it appears a simple problem, but remember conditions as they existed at that time.

Mr. John Bertram and Mr. Charles G. Draeseke, who later became chief draughtsman and designer for the firm, started on their ticklish proposition. It was decided to use a 12″ scale as a basis to work from; in fact, it was the only basis they could work from, and a special scale, as accurate as possible to secure, was purchased. This scale was over ⅛″ thick, and in our vocabulary, "Was some scale.".

The Start of the Machine

A piece of steel was secured and turned down to 1¼″ diameter. It was next decided that they would cut a perfect 8-pitch thread on the piece just turned. Sounds simple nowadays, but in those days an accurate lead screw on a lathe was unknown in this plant, and the change gear system was a scream compared to present-day practice. By figuring closely, however, and by cutting special gears for the change studs on the lathe, they proceeded to mark with a very sharp-pointed tool the path which the proposed thread would take. After experimenting for some time, they found the desired combination, and using the 12″ scale as a guide, they cut the thread, studying the result through a magnifying glass. Eureka! the scale was placed against the 12 inches of thread, the glass was run over the distance carefully. The thread was perfect—not an error existed. Needless to mention, joy reigned in this plant on the completion of the task, the screw being the most important part of the proposed machine.

The next problem consisted of laying out and marking what they termed the dividing disc, and in order that readers can get a good idea of the general appearance of the machine, we suggest they refer to either of the views shown, all of which depict the disc clearly.

This portion is divided in the following manner. On the outside rim there are 125 divisions, each division representing a thousandth. The method of arriving at this conclusion was very simple. They made the lead of the screw 8-pitch in order that 8 times 125 would make 1,000. In other words, jn eight turns of the disc

the screw would move one inch, at the same time 1,000 divisions would pass the guide scale shown at C in Fig. 2.

The side portion of the disc is divided into 128 divisions, which means that they actually get a decimal equivalent by using the side markings. In those days all gauges were made by 1-16ths, therefore, 64 of the 128 divisions equalled 1-16; 32 divisions, 1-32; 16 divisions, 1-64, and so on. Everything seems to have been thought of, for they even placed the 7″ scale, shown at C, Fig. 2, to act as a guide and reminder. This scale shows to better advantage in Fig. 3.

Further Details of Machine

As can be understood, some arrangement had to be made for adjustment. This was accomplished by means of the screw D. This screw has a fine thread, about 16 to the inch, and has a lock-nut to hold everything securely when once set. The disc B, which we have previously commented on, was marked in an old gear-shaper, which by the way, was used up to a few years ago. A shows the thread which caused the biggest task of all.

To give readers some idea of the machine's general proportions, we append the following approximate dimensions. The bed measures 17″ x 6″, and is made from cast iron. On top of the bed is placed a 3″ x ¾″ deep slot. In this slot the two top portions slide, each having a tongue to suit. These parts spoken of are fastened to the bed by means of washers and cap screws underneath, slots being placed in bed to allow for adjustment. The disc is 6″ diameter, and is marked as already explained. The two top portions are made entirely from the best quality bronze, and the nut on the main measuring head E is split and provided with shim. The screw illustrated in this nut takes up any adjustment necessary, and the nut itself is 4″ in length. The finer adjusting nut F is 3″ long, and the scale C is marked from 0 to 7″ in eighths of an inch. As already explained, this scale is merely a reminder and a guide.

These dimensions, briefly, bring out the main points in the machine, but what we would like to emphasize is the soundness of the fundamental principles involved. Remember that no one in the plant had ever seen a micrometer, yet, in a way, they adopted the very same principle. Some time after the machine's completion one of the first Brown & Sharpe micrometers was brought into their plant. Great rivalry was the natural result, and the owner of the micrometer claimed he had the only fine standard, and that the machine as illustrated was only a fad. This, of course, started the fireworks, and a test was made, with the result that both the machine

FIG. 1—GENERAL VIEW OF MACHINE BRINGING OUT THE POINTS OF INTEREST.

and micrometer were exactly similar. Messrs. Bertram & Draeske were happy men that memorable day.

The motive that prompted the making of this machine in the first place was, as we have already explained, the desire to obtain some standard to which they could work with perfect safety. On completing the machine, all their gauges, ring, plug, and otherwise, were made to the machine, feeling sure they were correct in every way.

This was the initial step to bigger and better things, and the awakening of the spirit of progressiveness for which this firm is famed. They have been pioneers in many movements and developments, and by that spirit have reached an enviable position in the machine tool trade that needs none of our championing. This same earnest effort after simplicity, accuracy, and efficiency is still existent, and the simplicity, yet effectiveness of this old-time measuring machine is a good example

FIG. 3—ANOTHER VIEW OF THE MACHINE.

of the same qualities in their present line. We do not say this in an effort to lavish praise, but merely as a statement of a fact which is recognized by all interested in the machine tool business.

Present Conditions

Coming to present conditions we find that things have changed since the measuring machine described was conceived. A thousandth of an inch is an exceedingly coarse measurement in these days of rapid production. The system of measurement has likewise changed.

The Pratt & Whitney Co., Dundas, which is handled under the same management as the Bertram concern, have developed a system of gauging that is well worthy of description, not only as a comparison to the old-time method already described, but for its own particular merits.

It is hardly necessary to ask readers if they know what a gauge made by this concern looks like, for they are to be found in practically every machine shop in some form or other. Have you ever realized that to make these gauges absolutely accurate it is imperative that some standard be adopted whereby accuracy can be assured? This is where the benefit of the modern measuring machine comes in, and no doubt every reader is well aware of the appearance of a modern P. & W. measuring machine.

Every gauge, tap, die, etc., manufactured by this concern goes to one of their measuring machines for checking purposes. The problem of building such a machine was a very different one from the other illustrated at Figs. 1 to 3. Briefly, the four principal problems confronted in the designing and building of the machine were as follows. First, the extreme difficulty of making a perfectly straight bed which would not be affected by temperature or torsional strains. Second, the production of a dividing screw of a degree of accuracy far ahead of anything that had hitherto been attempted. Third, the absolute necessity of automatically governing the measuring pressure of the instrument, and fourth, the accurate re-locating of the sliding head on the various positions on the bed at a known distance from the stationary head.

These problems were all the more difficult insomuch that the machine was being designed for general use and not for use by experts in their own plant alone. The ultimate object was to develop a commercial instrument which could be sold at a reasonable figure, or in other

words, manufacture a machine that would allow all manufacturers the means whereby standards could be originated and duplicated to within the narrowest limits possible by human skill. When one considers this viewpoint, one naturally sees where added thought had to be given to the varied uses to which such a machine would be put.

To speak of the success of this instrument is like telling the story of some past war. Both alike are history. The achievements are innumerable, and there is hardly a large manufacturing establishment that cannot boast of a P. & W. measuring machine.

Method of Operating

The writer had the pleasure of viewing various tests on this style machine, and a description of the same would not be amiss. As can be understood, all kinds of material can be measured. The bed itself is on a three-point bearing, and along the rear of bed is placed small silver bosses one inch apart.

On each of these bosses are placed cross-hair lines, the intersection of which is the dead centre of the boss. These marks are invisible to the naked eye and can only be seen through the microscope provided with the machine, and noticed clearly in the illustration. How the marks are placed on these bosses is a trade secret which, of course, we do not discuss. Various rests, attachments, etc., are used, depending on the nature of the test, but we will suppose in this case that we are going to measure a piece for its length. You first set the stationary head in a known and convenient position. Then, having fastened this head in place by means of the handle shown in photo, you proceed with the actual measuring.

If one looks closely at the stationary head shown to the left, in Fig. 4, they will see a small lever or pin supported by a spring tension. It is this feature that guarantees the governing of the measuring pressure. Should one press too hard on the micrometer dial of the sliding head this pin will fall out entirely, warning the operator that he has not the correct pressure, while on the pressure being perfect the pin will drop to a vertical position only, and not out entirely.

We will suppose that we have placed the piece to be measured on suitable rests, and in position between the measuring spindles. The proper pressure has been exerted and the pressure pin has dropped to a vertical position.

All one need do is read the dial for their measurement, which will be correct within a limit of 1-100,000 of an inch.

In the testing of the pitch of a tap depth of thread, angle, etc., a special attachment is used which detects errors in such a manner that the operator of machine cannot possibly miss the inaccuracy, if any. There are other innumerable attachments and devices for the machine that one cannot very well explain, and, in fact, must be viewed in use to be appreciated; but enough has been said to show the thought which has been placed into the manufacture of this instrument.

Although the machines are standard at 62 degrees Fahrenheit, it is not necessary to use them at such initial temperature as variations due to temperature will affect both the work and machine practically alike.

When the machine is used for scientific research, however, the initial temperature should be adhered to. In fact, wherever possible the temperature of 62 degrees should be kept to as near as possible. As a general rule concerns using these machines recognize their merit and value, and place them either in a special box and stand, or in a room for measuring purposes alone. The latter condition is the best of all, and it was in such a room that the writer viewed the machine at the Dundas plant. As a still further guarantee that only the best of products will leave these two plants, they have installed an up-to-date laboratory, equipped complete in every detail. It is in this department that all raw material is tested for its quality before being used. The analysis is very exacting, and no doubtful material is ever used. This is still another step in the right direction. Not only do they intend making their product to the correct size, but of the best procurable material.

Referring once more to the modern measuring machine just described, it would seem that we have reached the finest form of measuring possible. Without question good use has been made of the past 40 years in the development of an efficient and accurate system of measurement, but who knows what another similar period may bring forth.

This extract from Mr. Charles Draeseke's letter of July 6th, 1920, to Sir Alex Bertram, refers in an intimate way to the developments and tests in the earlier days of the firm;

The "Darling, Brown & Sharpe" scale that you alluded to was, if I remember right, an 18″ one, bought as you say about 1878 or 1879. What fixes the date in my mind is that it was in 1878 that Sir John got into power with the N. P., and things immediately began to boom. The rule was graduated in 12ths, 14ths, and 20ths, so as to be used for gear measuring and it was by it that your father and I discovered the error in your leadscrews. On second thoughts now it may have been a 24″ scale because I distinctly remember as though it were but yesterday, that you could distinctly see the error in 18 inches. I was turning up a lot of No. 4 drill spindles at the time, and before reducing the small portion to the required diameter I just took a smooth cut over that part, filed it a little and then ran a 4 PI. mark over it with a very fine pointed tool and when the rule was applied you could distinctly see the error in 18″, think it was some-thing like 1-64″, but cannot be sure of this. However, whatever it was, we geared up the lather to correct it, but cannot remember the change-gears used now, but they produced a thread with no visible error. By use of these gears I cut the screw and nut that are on that measuring machine you refer to. I then made two sets of cylindrical gauges for the shop, and as far as we were able to discover by applying them to any American made

machines they were correct, at least for all commercial purposes. The date of the measuring machine would be I should say, 1879. Shortly after that you bought a lead-screw from P. & W. Co., or Bements, don't remember which, for that long 20-ft. lathe that W. Fechnay ran, where most of our leadscrews were cut. You also put a corrected screw on Mike Cahill's lathe downstairs.

Note.—W. Fechnay and Mike Cahill mentioned in Mr. Draeseke's letter are still working in the Bertram plant.

DISPOSAL OF OLD COMPRESSED-GAS CYLINDERS

Some time ago a rather unusual incident occurred in a manufacturing plant, says The Traveler's Standard, showing the necessity of handling with great care cylinders or tanks used as containers for compressed gases. In the plant referred to, large quantities of steel and iron scrap of the most miscellaneous character are melted in an open-hearth furnace and the metal thus obtained forms the basis for the manufactured products of the plant. At one time a compressed-gas cylinder was received with other scrap, and upon being thrown into the furnace it exploded and blew the entire top off the furnace. This experience was remembered, and when a similar gas cylinder was received on another occasion, it was taken off the charging buggy and thrown on the charging floor. Several hours later one of the employes, observing the valve on the cylinder was still in place, secured a wrench and opened the valve. The cylinder had been filled with chlorine, and enough of the gas was released to cause serious discomfort to three men, and to make them unfit for work for several days. Not long after this occurrence still another gas cylinder was found in the scrap, and examination revealed the fact that it contained something like 25 pounds of anhydrous ammonia. This also might have caused trouble if it had been put into the furnace or left lying about the plant. As a result of these experiences, an order was issued to the effect that all cylinders thereafter received should be taken to the "skull-cracker" and broken up, before charging them into the furnace. There are at least two lessons to be learned from these experiences: (1) Compressed-gas cylinders should not be exposed to high temperatures unless it is known, positively, that they are entirely empty (and if the valve is still in place and closed, and there are no easily seen openings in a tank, it is safer to assume that it is not entirely empty); (2) all possible precautions should be used when opening gas cylinders, even if they are supposed to be empty—do it out-of-doors, be sure there are no fires or open flames anywhere in the vicinity, and have the work done by some person who understands the dangerous properties of gases. A warning of particular importance to the users of compressed gases is this: Be sure each cylinder is empty before discarding it, and if any gas remains in the cylinder, see that the cylinder is put in a safe place and marked in some way so that it cannot possibly be thrown on the scrap pile through error. Instead of scrapping a gas cylinder, return it to the gas manufacturer. If it is in good repair it has a money value and can be re-filled and sent out again and if it is no longer serviceable, the manufacturer will know how to dispose of it safely. If a cylinder is received in a leaky condition, leave it out in the open air and protect it so that inquisitive persons cannot be harmed by the escaping gas; and if the gas remaining in the cylinder cannot be used safely, notify the gas manufacturer and ask for instructions regarding it.

What is the Purpose of a Heat-Treating Process?

Any Such Process is Supposed to Impart Hardness, Toughness, Tensile Strength, Grain Structure and So On, as the Case May Be, to the Material Being Heat Treated

STEEL treatment is demanding and receiving more attention lately than at any other time in our industrial life. This being a fact we felt that readers would be glad to peruse a description of what is known as the Hump method of steel treatment, but before proceeding with the same we might add that this matter is information gleaned from the literature issued by the concern handling this method of heat treatment.

We are well aware that the purpose of heat treatment is to impart certain qualities of hardness, toughness, tensile strength, grain structure, etc., to the material. These qualities should be obtained and accomplished without change in shape or size of material, without scaling or decarbonization, without waste of material by poor treatment, and with the least possible expense of labor and supervision. Such is, of course, an ideal condition, yet that is what the firm handling this hump method claim to have succeeded in obtaining. They state the following claims regarding their method of heat treatment:

Our system gives unmistakable indications when the charge has reached the critical or transformation point, as a reference point for quenching. It insures uniform temperature of all parts of the charge. Distortion due to unequal expansion is minimized and rejections are practically eliminated. It gives accurate control of and information regarding the rate of heating and permits of a maximum rate of heating without injury to the charge. Pyrometer inaccuracy and incorrect information as to transformation point are eliminated as causes of in-

work is protected against contact with fresh supplies of oxygen or contaminating gases, avoiding scaling or decarbonization.

The electric furnaces are long lived and have low upkeep charges. There is

THIS VIEW SHOWS FIVE ELECTRIC FURNACES WITH RECORDERS FOR THE HEAT TREATMENT OF CARBON STEEL TOOLS.

less fire hazard than with fuel-burning furnaces and the many uncertainties of fuel furnaces are eliminated. Working conditions in the heat-treating room are improved, due to a lower temperature and the absence of smoke and noise from burners. The furnaces for hardening and drawing are clean in operation and can be installed in comparatively small space in the machine shop, where the materials can be heat-treated in the direct course of manufacture, thus reducing charges for space and handling. All heat-treating operations can be

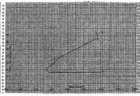

FIG. 1—CHART SHOWING THE HUMP CLEARLY.

correct treatment. A clear, explicit record of the rate of heating and of quenching point for each charge is made and preserved, and can be referred to in connection with subsequent physical tests. During the heating period the

brought under the complete and certain control of one man, who is guided by the unequivocal indications of instruments. High quality, uniform and regular production is assured. The same results are obtained day after day.

At Fig. 1, we depict a chart produced by a Leeds & Northrup Pyrometer in the heat treatment of a piece of steel by the process under discussion. But let us proceed with the description of the system as told by the firm themselves.

The Hump Method

In the hump method of heat treatment, which is covered by U. S. Patent No. 1,-188,128, the outward manifestation of the changes in internal structure which take place when steel is heated past the so-called critical or transformation point is employed to indicate when the work should be withdrawn from the furnace. In using this method the temperature of the furnace, and therefore of the work, is raised at a uniform rate until the transformation point of the steel is reached. At this time there will be a marked decrease in the rate of temperature rise. This change in the rate of rise is made plainly visible to the operator by an autographic recorder connected to a thermocouple placed close to or in contact with the work. The effect is clearly shown by a bend or hump in the curve as at C. See Fig. 1.

This hump corresponds to a pause in temperature rise or decrease in rate of heating of the steel, which occurs in spite of the fact that heat is being transferred to the work during this interval as rapidly as before or after. It is explained by metallurgists as being due to the dissolving of cementite, or carbide of iron, in the pure iron, or ferrite, and to other chemical and physical changes depending upon the composition of the steel. By microscopical and chemical means, it is known that the physical and chemical structure of steel after quenching is profoundly influenced by the relative position of the quenching temperature with respect to the temperature at which the arrest occurs. The

hardness, strength, ductility and toughness are all definitely influenced.

As a guide in hardening, the temperature pause, or decalescence point, as it is called, is much more reliable than is the furnace temperature as indicated by a pyrometer. The arrest in temperature signalled by the hump or bend in the line drawn by the pyrometer pen shows that the internal changes so momentous as

that surprising differences in temperature persist even after prolonged "soaking." To insure that the work shall actually reach the desired temperature within a reasonable time the furnace is often held at a higher temperature,

accuracy of the temperature measurement, and growth in grain size due to too long immersion may occur. Lead or salt baths give no control whatever of the date of heating. Hardening baths also have many practical disadvantages, such

FIG. 3—SOME MILLING CUTTERS AND TAP BLADES HARDENED BY THIS METHOD.

FIG. 4—THIS DIE WAS BROACHED BY THE PUNCH, THEN HARDENED AFTER. IT FITTED PERFECTLY.

affecting the qualities of the finished steel are actually taking place, and once the proper interval to be allowed after the hump before quenching has been determined, there is no uncertainty about the results of hardening.

If, on the other hand, the work be controlled with reference to temperature readings only, there are several possible sources of error, namely:

(a) The thermocouple may be at a temperature different from that of the work, due either to insufficient time having elapsed for the work to assume the furnace temperature or to inequalities in temperature between different parts of the furnace. Experiments which have been made with Leeds & Northrup commutating recording pyrometers connected to read alternately upon two or more thermocouples in a single furnace show

which is afterwards reduced. The "soaking" and possible overheating of the work may, and often do, result in injurious growth in grain size in the metal being treated.

(b) In attempting to hold the furnace at a constant temperature, the temperature may fluctuate, and even though it may subsequently be reduced to the proper temperature, the work may nevertheless have been overheated and injured. In this method the quench is made as soon as the work is at the right distance beyond the critical point and the work is given no opportunity to reach an excessive temperature.

(c) The temperature which is assumed as the critical temperature of the steel may not be correct. On the other hand, in using this method, the time at which each lot of material passes through the critical point is definitely located. The lead pot and the fused salt bath methods of heating work for hardening give greater assurance than does the ordinary furnace that the work shall reach a uniform temperature, but there still remain uncertainties as to the actual transformation temperature, and as to

as expense, dirt, necessity of subsequently cleaning the work, space occupied, etc.

(d) The pyrometer used for measuring furnace temperature may be incorrect.

The user of this method need not concern himself about the absolute accuracy of his pyrometer, nor bother with independent transformation point determinations. It is not at all necessary that the temperature indicated by the thermocouple should be the correct temperature of the work, for so long as the recorder connected to the thermocouple shows clearly the pause in temperature rise, the moment at which transformation occurs is definitely known. Having learned by trial just how many minutes or how many degrees should elapse after the beginning or end of the transformation before the work is removed from the furnace, he is upon sure ground and can repeat results.

Uniform standardized conditions and a controllable rate of heating the work are essential to this method of heat treatment. A small furnace for heat-treating tools, dies, etc., is shown at Fig. 3. The heating element consists of a vertical cylindrical resistor, surrounded by insu-

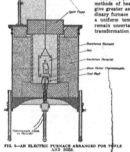

FIG. 5—AN ELECTRIC FURNACE ARRANGED FOR TOOLS AND DIES.

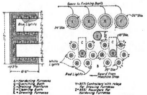

FIG. 6—ARRANGEMENT OF FURNACES AND RECORDERS FOR HEAT TREATMENT OF AUTOMOBILE GEARS.

lating material in a sheet-iron jacket. The resistor rests upon a refractory block which is supported by a cast-iron bottom plate. The heating chamber is closed at the top by a refractory cover, while a cast-iron top-plate confines the loose insulating material filling the space between resistor and jacket. An iron-constant thermocouple of bare No. 8 gauge wire projects upward from the centre of the bottom refractory block.

By means of a small wire attached to a tool support on the top plate of the furnace the work to be treated can be suspended in close proximity to, or touching, the end of the thermocouple. In production furnaces other methods of supporting the work are used. For example, in the furnace shown at Fig. 6, designed for the heat treatment of automobile transmission gears, the work is placed upon holders before insertion in the furnace. Covers placed on the furnace completely close in the heating chamber, preventing renewal of the atmosphere, and the work is thus protected against oxidation and scaling.

time. This is desirable in order to avoid stresses and distortion that would follow from unequal expansion or contraction if different parts of the work passed through the transformation point at different times.

The arrival of the work at the transformation point C causes an abrupt change in the rate of heating. Due to the suddenly increased capacity of the steel to store heat, the temperature stops rising or proceeds much more slowly than before, although the rate of supply of heat energy has not been changed. However, once the transformation is completed, as at D, the temperature again rises rapidly. The pause is plainly shown by the hump in the curve.

true whether or not the temperature represented at the point C on the chart is correct and whether or not it is the actual temperature of the steel at that moment. The important fact is that the chart tells the attendant when the steel is going through the transformation, from which he may know that quenching after a certain interval will secure the desired physical qualities. Furthermore, the chart remains as a record of just how each individual lot of steel was treated and can be referred to in connection with properties developed in physical tests of that steel.

It is found that the rate of temperature increase has a marked influence upon the properties exhibited by the steel

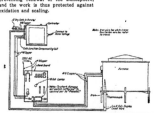

FIG. 1—WIRING DIAGRAM FOR L. & N. ELECTRIC DRAWING FURNACE AND RECORDING CONTROLLER.   FIG. 6—ELECTRIC FURNACE ARRANGED FOR HEAT TREATMENT OF TRANSMISSION GEARS.

Regular Procedure

At the moment when the work is introduced into the furnace the temperature of the latter is, say, 1400° F., but the heat storage capacity of the furnace walls being small compared with that of the charge, the temperatures of both thermocouple and furnace walls drop rapidly for a few hundred degrees, the current through the heating element or resistor being shut off during this time. The result can be seen in the chart at Fig. 1 The temperature drops rapidly from 140° F. at A to about 870° F., and then rises slowly to B, where it is stationary, the thermocouple, the furnace walls, and all parts, small and large, of the charge having reached approximately the same temperature. The switch is then closed, the input being so regulated that the temperature rises at the desired rate. The fact that the work and furnaces start from the same temperature at B, far below the critical point, coupled with a proper arrangement of the heating element with respect to the charge, insures that all parts of the work will go through the critical point at the same

Before quenching, it is necessary to heat the work to a certain distance above or for a certain time after this point, the time or distance depending upon the mass and shape of the steel, the quenching medium employed, and the qualities desired. The exact further heating to be allowed after the completion of the transformation point is learned by experience or trial, but once known, all uncertainty as to the result is practically eliminated.

In using this method, the metal is quenched without hesitation as soon as the pen has gone the prescribed distance past the hump which indicates the transformation point. Errors due to inaccuracy of pyrometers, non-uniform temperature in the furnace, failure of the work to reach the furnace temperature, or incorrect information regarding the transformation temperature, are avoided, and the steel is not injured by overheating or by holding it at a high temperature for too long a time. Each piece of work carried the same distance beyond the reference point C or D will show the same internal structure. This is

quenching. The resistance furnace is admirably adapted for controlling the rate of temperature rise, since the rate of energy input is easily regulated by reference to an ammeter supplied as part of the furnace equipment. The Leeds & Northrup potentiometer pyrometer, also forming part of the equipment, is likewise peculiarly suited for carrying out this method of heat treatment, as it is sensitive to small changes in thermocouple e.m.f. and exhibits changes in rate of temperature rise upon a magnified scale.

Results Achieved

This method of heat treatment has been found in practical work on a commercial scale to confer the following benefits:

The microscopic structure and physical properties of the hardened steel can be controlled with accuracy and certainty. Results can be studied and conditions, as shown by the recording pyrometer chart, repeated, or if desired, modified, by changing either the rate of heating or the location of the quenching point with relation to the transformation point,

or both. There is a great reduction in volume changes, distortion, and breakage as compared with the ordinary methods of hardening. The work is free from scale and decarbonization. The cleanliness of operation with electrical furnaces, and their small weight and bulk, make it practicable to locate the heat-treating department in the machine shop in the direct course of manufacture. The rehandling involved in taking the work to a separate heat-treating department and bringing it back again is obviated.

As an example of the use of the electric furnace in hardening tools, the manner in which punches and dies are produced in our own shop, where this method has been used exclusively for the last six years, will be described. Some of these punches and dies are shown in the accompanying illustrations. The punch is first made in the ordinary way by machining and hand finishing. It is easy to secure accuracy in the making of a punch, as all measuring, gauging, etc., is done on the outside. The die is made to approximately the correct size, but

calipers or micrometers can be used, it is not practicable to make the die exactly the same shape and size as the punch, as any hand work is necessarily more or less irregular.

The fact that dies hardened by this method fit exactly with the punches by which they were broached shows that there has been no volume change or distortion during the process of hardening. It has been found, however, that the previous history of the material, the rate of heating and the interval which is allowed to elapse between the transformation point and the quenching point have certain relations to volume changes

done in gas furnaces, followed by an oil quench. Upon their return to the machine shop the gears were wire brushed to remove adhering lead.

The electric furnace equipment replacing the lead pots and gas furnace is located in the machine shop itself. There are six electric hardening furnaces located on three sides about a quenching tank, with three electric annealing furnaces on the remaining side, also a drain grid and a cleansing bath between the quenching tank and the annealing furnaces. The tops of all furnaces and tanks are flush with an elevated platform. The work is brought on trucks to the edge of the

FIG. 5—THIS DRAWING FURNACE IS USED FOR DRAWING TRANSMISSION GEARS. TEMPERATURE IS CONTROLLED WITHIN 5 DEGREES FAHRENHEIT.

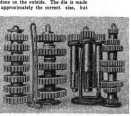

FIG. 9—GEARS ASSEMBLED ON HOLDERS READY FOR THE FURNACE.

slightly smaller. The punch, having been hardened, with little or no change in dimensions, is then used as a broach to cut the die. It is forced into the die a short distance, raising a burr, which is removed by filing, the operation being repeated until the cut is finished. The die is then relieved, but no further work is done on the cutting edge. After hardening the die by the hump method, its size and shape are found to be so exact with respect to the punch that subsequent work, as stoning or grinding, is not required.

Dies which have been broached by the punch and then hardened by the hump method are said to be superior, for purposes where close correspondence of punch and die are necessary, to hand-made dies hardened in the ordinary manner. Hand-made dies must be corrected by hand work after hardening, and such hand work can be guided only by reference to a templet. It is practically impossible to make the templet exactly the same shape as the punch, and it is also difficult to make the die fit the templet exactly. Even on work where internal

which are of great importance in the hardening of such objects as dies, milling cutters, gears, etc.

Production operations are greatly simplified by the use of electric heat-treating furnaces. As an example, refer to Fig. 5. Formerly in this plant automobile transmission gears were heated for hardening in lead pots located in the basement beneath the machine shop. After the gears had been machined, they were removed on trucks to an elevator, lowered to the heat-treating room, an uncomfortable place because of heat and fumes, and wheeled to the lead pots. The gears were placed in the lead pots in batches, and when they had come up to temperature were lifted out one at a time and dropped into the oil quenching bath. The time of heating in the lead pot was thus not the same for all pieces, and the temperature also varied with the location in the pot, possibly resulting in non-uniform hardness and grain structure. The gears were also sometimes injured by dropping upon one another in the quenching tank. The drawing was

platform, where a boy arranges the gears upon holders, which are easily picked up by the operator and upon which the gears remain until they have passed in turn through hardening furnace, quenching tank, cleansing bath and drawing furnace. Corresponding to each hardening furnace, there is a curve-drawing recorder and for each drawing furnace an automatic temperature recording controller, all mounted upon a wall near by, where also are the furnace rheostats. An assistant foreman on duty at this point directs the operations according to the indications of the curve-drawing recorders connected to the hardening furnaces. The automatic controllers of the drawing furnaces require no attention.

Leeds & Northrup drawing furnaces are installed with Leeds & Northrup recording controllers for regulating the temperature automatically. The recording controller is identical in principle and construction with the pyrometer already described, with the addition of switches, which are opened or closed when the temperature

(Continued on page 102)

Small Firms Can Enter the Export Field

Firms Should Be Prepared to Stay Behind Their Goods and Have Shipments Right Up to Standard of Samples—A Valuable Field to Exploit for Future Business.

By J. H. RODGERS, Montreal.

THAT there is good business to be secured for Canadian firms in the export field is the opinion of G. G. Hodges, vice-president and managing director of the Industrial Export Co. of Canada. He discussed various phases of the matter with "Machinery," and has some good advice to offer to shippers and to those who have not yet tried the foreign field.

"For some time past," stated Mr. Hodges, "there has been a great deal of discussion regarding the value of export business, and in spite of the evident interest in that phase of our commercial enterprise, very little action, except on the part of large corporations has been taken with a view to stimulating our business overseas.

"A close analysis of the situation leads to the belief that the fundamental cause for this lackadaisical attitude lies in the nature of the average business organization. Our industries in Canada are not made up of large units, but represent a multiplicity of smaller organizations, each of which is controlled by one man or a small directorate, and the output in each individual case is not very large. Domestic business prior to the war, served to keep these organizations busy for the best part of the year, and even though the plants were not running to capacity, the results achieved were sufficient to make it possible for the owners to continue operating probably with the same policy year in and year out. The war stimulated industry, causing in many cases an increased plant capacity, which is to-day in excess of the requirements of the domestic market, and the taste for export business acquired during the war has developed among many manufacturers a desire to develop the foreign field. Unfortunately, the average manufacturer does not know how to go about getting this business, and is either unwilling to admit his lack of knowledge, or is dubious about his ability to hold his own outside of the domestic field."

One Way of Getting Business

"In the Old Country, as in the United States in lesser degree, it is customary for manufacturers to place their samples in the hands of firms whose business it is to solicit and handle the business secured from foreign countries. Unfortunately Canada has not many such intermediaries, yet there are a few such organizations equipped to take out of the manufacturer's hands all responsibility relative to the handling of his foreign business. It remains for the manufacturer to confide in such an organization, and to place in its hands his samples and complete data, so that his lines may be brought to the attention of the organization's representatives, and in turn to the attention of the various firms in the foreign field.

"He must not be too impatient, however,—he must realize that unlike the domestic market, a letter or a representative cannot get in touch with a possible buyer in the course of a few days. Months may pass before the prospective buyer in the foreign field can be approached and his business secured. Samples may go out at great expense on the part of the selling organization, and be lost or returned without business, but sooner or later, with the manufacturer's co-operation, and with a reasonable amount of patience, the effort put forth by the selling organization will bear fruit, and once initial business is in, he can fairly well depend on repeat business assuming larger proportions as time goes by."

Chance for the Small Firm

"It matters not if the manufacturer is in a small country town, he can just as well handle the export business if his lines are in the hands of a reputable export firm, because it is the duty and desire of this firm to undertake not only the selling and the financing of the shipments, but also to look after the shipping, seeing that the merchandise gets on board the ship on its way to the customer.

"There is no need for the smaller manufacturer to keep out of the export field when he requires the business so much. It is true that without the aid of an export selling organization he might feel the expense and trouble involved too great to permit him to handle export business, but if the Canadian manufacturer, large or small, desires an outlet for his products, two things must be borne in mind: First, that the export business can be just as valuable as the domestic business, and the time is coming when the manufacturer who is going to make a success of the domestic market must be able to compete by bringing down his overhead with the export trade.

"Secondly, he must produce goods which are up to sample in every instance. Possibly it would be just as well to add a third injunction. If he places his line in the hands of an export house, it is essential that he give that firm his hearty support. He little realizes, possibly, when he does hand his samples and prices to that house, just what expense and trouble it will go to in order to place his line before the respective buyers, whether they are in England, South Africa or China.

"If only Canadian manufacturers would understand that foreign buyers desire Canadian merchandise, and that now is the time to go after the business, we need not fear a slackening of business with which to keep the wheels turning."

—Bally in Montreal "Star."

Would it be Wise to Adopt the Metric System?

If You Have Any Doubt in the Matter Read This Article Which States Sufficient Vital Facts to Send the Metric System Where It Belongs, Namely, Not in Our Country

THIS last few months has witnessed one of the biggest attempts of a lifetime to foist on the American public the metric system of measurement. We need not go into all the details relative to this movement for that is now history. What we do wish, however, to point out is that the movement has met with the fate it deserves, namely, received what it dreaded most, "Honest Publicity."

By honest publicity we mean statements of facts as they really exist. Various periodicals on Uncle Sam's side of the line went after the proposed scheme in great style showing kinks in the armor in practically every joint. They pointed out why the adoption of such a system would be nothing short of national suicide, and now that these facts have been compiled in definite form we feel it would make ideal reading for our readers who should be likewise interested. The committee on industrial development of the Cleveland chamber of commerce presented to that body the following information regarding the metric system, and why they believed its compulsory adoption should be opposed. They state as follows:

The metric system was invented by James Watt, the inventor of the steam engine, in 1773, and contemplated in its original form a decimal system of measurement not only for weights, distances and volume, but also for time as related to the hours of the day, which were to be ten; the days of the week, which were to be ten; and the months of the Year, which were to be ten. It also included a decimally divided circle and compass.

The measurements for time, a decimally divided circle and compass apparently were not practical, as, so far as your committee has been able to learn, they have never been used.

The measurements of distance, volume and weight have been adopted or legalized in many countries. France adopted by a compulsory law, this system in 1793. In 1812, under Napoleon, the law was repealed and the French adopted, by a compulsory law, this system called "Systeme Usuelle." In 1837, the metric system was again made compulsory in France. Despite the operation of this compulsory law for more than eighty years, it is stated that the old units are freely used in many industries in France; for example, the aune and denier are still used for measuring silk.

Many South American countries have adopted the metric system by compulsory and permissory laws. However, a survey conducted by the American Institute of Weights and Measures conclusively proves that the use of this system is not universal in these countries. Argentina adopted the system in 1863. Yet an examination of the results of a questionnaire shows the use of many units unknown in the metric system. For example, land is sold by the square vara. In marine measurements the kilometer, meter, pie, ton, mile, knot and cubic foot appear to be used about equally.

Brazil adopted the metric system in 1862. The same condition is found there, with such terms as gallao, arroba, alqueire, etc., appearing frequently.

The adoption of the metric system in Chile in 1858 has apparently not altered the use of the old system in that country. The units libra, quintal, pie, vara and inch are found to be used in many industries.

The same condition prevails in Colombia, which adopted the system in 1853. In fact, this is the condition prevailing in practically all of the South American countries.

With slight differences, the units other than the metric units in use in Latin American countries are remarkably similar to the English system.

The condition prevailing in South America is interesting in that, at the time of the adoption of the metric system by the several countries, little or no physical disadvantage, in the form of deep-rooted manufacturing practice, existed. Apparently the reluctance of the peoples of these countries to change is attributable to the greater convenience of their old system.

It is stated on good authority that 50% of all machine tools manufactured in the world are made in the United States. It is further stated that 82% of the total business of this character originates in the United States and in Great Britain, both of which countries use the English system. These machine tools are the basis from which all manufactured products spring, and it is not only possible to do, but manufacturing in the metric system is being done constantly with machines and machine tools in which the great majority of the parts are built to the English inch measurements. These machines, in most cases, are standardized on the basis of frames, bases and parts of a similar nature, it makes no difference whether the machine is to be used for manufacturing products to English inch measurements or metric measurements. In the majority of cases the only changes which are necessary are in such moving parts as lead screws and some change gears and parts of a similar nature. Where a company is doing a considerable export business, these metric parts have also become standardized, so that they present no more difficulty in their manufacture than do the corresponding parts, which are built to English measurements.

Proponents of the metric system urge as their main argument the advantage that would result in export trade from its adoption. To this argument your committee takes exception.

Many things manufactured in the United States are sold in tremendous quantities in foreign countries without a particle of hindrance by virtue of the system of measurement by which they were made. Foreign automobiles sell in the United States without a thought being given as to whether or not they are made to the metric or to the English system, and it is a well known fact that the American automobile made to the English measurement system has invaded every known country in the world. Examples of the ready sale abroad of articles made by the English system could be multiplied without end.

The American Institute of Weights and Measures recently conducted what might well be called a census of metric use in the United States. The response to their questionnaire is particularly illuminating when applied to a consideration of the foreign trade of many American manufacturers. For example, out of the replies received from automobile manufacturers, it was found that none were equipping their cars for foreign trade exclusively with tires and rims in metric sizes. In fact, the companies which perhaps do the largest automobile export business, such as Dodge Brothers, Ford, Maxwell, etc., ship their cars equipped exclusively with tires in American sizes.

The Paige-Detroit Motor Company states that occasionally a request for metric spark plugs is received. They further state that such requests are few and far between, because of the predominance of American spark plugs, and that in eighteen months they have not shipped a single car so equipped.

Dodge Brothers, who ship more than a million dollars' worth of automobiles a year to foreign countries, equip all of their cars with standard English thread spark plugs. Instances of this kind could be multiplied practically without end.

The Cleveland Twist Drill Company, which has been in export trade for a great many years, reports that in its particular line ninety per cent. of the

shipments to France, Sweden, Italy and Spain are made in metric measurements; to other so-called metric countries, roughly, fifty per cent. But the great bulk of its exports go to countries using the English system—Canada, Australia, South Africa and Great Britain. The Cleveland Twist Drill Company is absolutely opposed to making the metric system of weights and measures compulsory.

The metric system was made legal in the United States in 1866, and is open for the use of anyone desiring to adopt it. Contracts drawn in this system are legal and binding.

It is now proposed to introduce into Congress a bill which, if passed, will make the use of the metric system compulsory in the United States. In the opinion of your committee, the enactment of such a law would cause inestimable loss to the manufacturing interests of the country, and would create confusion and chaos from which the country would probably not recover for years.

An idea of the chaos that would be created in everyday affairs may be gained from glancing over the following list of changes that would have to be made.

In domestic life:

Grocers' scales all require new poise weights, all notched balance beams scrapped and new ones provided, with new sliding weights.

Peck and bushel measures discarded. Liter, larger than a quart, new containers required.

Hectoliter, equal to 2.8 bushels, not a practical unit.

Prices on all commodities to be read. Justed to new units.

In culinary matters:

All recipes to be readjusted to kilo, grammes and liters; cook books to be rewritten; general confusion in kitchen operations.

New milk bottles.

In other household affairs:

Gas meters to be replaced by new system of units of volume, or readings of meters taken in one system and converted into the other, to avoid scrapping meters in use.

Water meters in same category as gas meters.

Tape measures and yard sticks to be discarded.

In shopping:

Counter measuring machines to be reconstructed, yards to meters.

Dry goods to be folded at cotton and woollen mills in meter folds instead of yard folds, requiring change of machinery.

Photographic plates in common sizes to be known by awkward combinations of figures. An 8 by 10 plate becomes 203 by 254 millimeters.

Quires and reams to be displaced by decimal multiples, requiring changes at manufacturing plants.

All containers and cartons to be modified in sizes and shapes to be adapted to new unit sizes.

Shirts, collars and cuffs to be known by strange names of sizes. A 16 inch collar becomes a 406 millimeter collar. A 187 millimeter hat is worn instead of 7⅜ inches.

In building materials and construction:

Abandon board measure and substitute square decimeters, centares, or ares.

Doors familiar to all builders as 2 ft. 6 in. by 8 ft. 8 in. become 762 by 2082 millimeters.

An ordinary brick is 81 by 101 by 203 millimeters. Sizes of sash also are converted into strange units. Weights of tin, copper, zinc, lead sheets and plates placed before builders in unknown units, in awkward combinations.

Molds and pallets in brickyards to be changed to new units or inconvenient numbers used to represent sizes.

Earth excavation on basis of cubic meters, representing about 1.3 cu. yd. Designation of shovels and dippers in excavating machinery to be in fractions of cubic meters instead of definite yards.

All architects' drawings to be in new units, involving a most perplexing conversion of current building material units or made up on a system that will involve changes in all woodworking machinery to meet metric units.

In railroad affairs:

Change in position and renumbering of, say 100,000 mile posts, an incidental and minor affair relative to other changes.

Standard gauge of track becomes known as 1435 millimeters.

Changes in time-table mileages.

Books of rules rewritten and speeds given in new units instead of miles per hour. Slow boards all repainted with new speeds thereon.

Speed recorders scrapped or remodeled. Employees instructed in new methods of estimating speeds.

Dimensions of all wheels, axles, standard parts of car construction changed into new units.

Railroad track and warehouse scales to be reconstructed into metric units. Capacity of cars to be in new units.

In public land surveys:

Lands in many states surveyed and staked out in townships, sections, quarter sections and eighth sections, in none of which divisions is there an easy conversion into metric units.

In reconveyance of lands, present deeds giving metes and bounds in English units would require expensive and elaborate efforts to put dimensions and areas in metric units, which if correctly made would still be unintelligible to most people.

As great as the confusion in the few respects above mentioned and as expensive as the process would be in all the usual affairs of life, these few enumerated examples pale into insignificance beside the cost involved in making the conversion in the manufacturing industries. The cost of the war is but a fraction of that which would confront the general industries of the country. Virtual scrapping of all small tools and fixtures would be faced by manufacturers were such an act of Congress passed.

A colossal fortune in changes of lead screws and screw cutting gears in lathes alone is involved.

In all machinery the changes become of such a staggering nature their mere contemplation is most depressing. Days would be required simply to state the number of changes involved in the industries.

The ravages of war have fixed definite limits of territory. The ravages of the metric system changes would be universal in domestic affairs.

Compulsory legislation in this matter would plunge the nation into economic disorder, wiping out values in billions of dollars.

A dual system is confusing beyond description. Two systems do not admit of being in common use. A gradual change, piecemeal, is impracticable. A sudden sweeping change would throw the entire country into disorder.

In the chemical laboratory the metric system is used. It is there because it has advantages in chemical work. It was adopted because of its advantages there.

The metric system would be adopted by the industries if it possessed advantages for them.

The rewriting of all text books, engineers' tables, the changes in architects' drawings, in mechanical drawings, the education of the personnel of machine

shops and other industries in the use of a new system of units are matters of greater magnitude than any yet undertaken by this country.

Interchangeable manufacture is strictly an American invention. Indeed, it is referred to in Europe as the "American method." Standardization is for the purpose of interchangeability. American industries have spent time and money standardizing for the purpose of economical manufacture on an interchangeable basis and to-day outstrip in this respect European methods so far that there is no comparison.

Your committee is not able to believe that the adoption of the metric system will promote this sort of standardization and interchangeability. It is our opinion that it will not only retard this work, but that it will destroy all that has already been done, and that much time and much money would have to be expended in making the change.

The metric system does not provide for the purpose of the manufacturer convenient units of measure. American manufacturers are accustomed to making micrometer measurements. In gear cutting, for example, there is not a single pitch in the metric system that will fit the United States system now so universally used. An example of this is the 12-pitch gear which is used by thousands in this country. The closest pitch to this in the metric system is module 2. These gears will not run together because the thickness of the metric tooth measured along the pitch circle is .124", while that of the 12-pitch tooth is .131" thick. In order to run these gears together it would be necessary either to change the 2 module to 1.889 module, or to change our standard to 2 module. It would mean that millions of gears on hand would no longer be interchangeable, that millions of dollars would have to be expended for changing over and rebuilding gear cutters, and that the necessary gauges for checking the product would become obsolete, except for the purpose of repairing old gearing.

A common automobile cylinder dimension is that of the 3¾" bore. It is a standard size and is turned out by one concern alone in quantities of 12,000 per day. This standard is as well established that pistons and rings can be secured to fit it in every city in the United States. The metric translation of this size is 95.25 MM. Would our manufacturers be willing to adopt, or would other countries be willing to adopt this size of 95.25 MM., or would it not be found necessary to adopt a 95 MM. bore? If 95 MM. should become the standard, it is not difficult to estimate the effect on the manufacturer who would be forced to scrap his tools, fixtures, jigs, etc.

The United States makes more than one-half of the screw products of the world. Our system is the most interchangeable of any of the systems which are in use at the present time. The bolts and nuts made by one manufactur-

er are readily interchangeable with the corresponding sizes made by another manufacturer. It has even been found that our standard 1″-8 thread per inch bolts and nuts can be used with 1″-8 thread bolts and nuts of the English Whitworth system.

In the metric system there is not a single diameter of bolt or pitch which will fit those now being extensively made in the United States. The pitches in the English system are expressed in terms of a certain number of threads per inch of length; while in the metric system they are measured from a given point on one thread to a corresponding point on the next thread, and under this system all fine threads become an awkward fraction of a millimeter.

It was estimated at the time of the signing of the armistice that the gauges, checks, etc., made for war munitions alone cost the government $30,000,000. At least half of these will be of no use to the War Department or to anyone else if a compulsory metric law is passed. This gives an inkling of the tremendous cost of making the change proposed.

The English system, for the purpose of manufacturing, is as susceptible to decimal division as is the metric system. Beginning with the inch it is customary to halve for each sub-division up to a convenient fraction, 1-32 or 1-64, beyond which micrometers and other instruments are adjusted to hundredths, thousandths and ten-thousandths, all of which units are practicable and usable.

In the metric system the first decimal subdivision of the meter is the decimeter, a unit 3.937″ long, of no utility and rarely used. The next unit is the centimeter, a unit too large for good work and too small for use in distance measurements. The next is the millimeter, .03937″, the most widely used unit because of the adaptability of its size value. This unit is about as fine as can be used on a steel scale. For tools and fine, accurate mechanisms a millimeter is too large, and so the next sub-division is the 1-10 millimeter, which is .0039″. The great bulk of good work in machine tool, automobile, tool working and other industries requires units between 1-10 and 1-100 millimeter. The one is too coarse and the other too fine. The result is the halving and quartering of millimeters to get usable units approximating 1-100, 1-1000 or 2-1000 of an inch. Thus, it will be seen that the advantage claimed for the metric system of providing a decimal system fail by virtue of this forced use of fractions.

A great majority of manufacturers are on record as opposing the adoption of the metric system under one of the three heads:

(1) It offers no advantage over the present system of inch measurements.

(2) It is a very expensive procedure to introduce.

(3) It will produce great complications during the period of transition.

Your committee, after consideration of its investigation and of the facts that have come to its attention, earnestly urges the Chamber of Commerce to go on record as being absolutely opposed to the compulsory adoption of the metric system of weights and measures in this country, and strongly recommends that the Chamber of Commerce direct communications to the members of the Committee on Coinage, Weights and Measures of the Congress of the United States, and to all of the representatives of Ohio in the Congress of the United States, opposing a legislation which will in our opinion be so disastrous to all of us.

SAFEGUARDING MACHINERY AT ITS SOURCE

SPEAKING on this subject, Mr. H. A. Schultz of the U.S. Steel Corporation, remarked as follows:

"As chairman of this committee I might say that as we have been so recently appointed we have hardly had time to get into action, and this report can contain little more than a statement of present conditions and a promise for the future. Some three years ago your council appointed a committee using the same title as that of the present. The first committee working on this subject was very energetic but, as it was pioneering, it was rather difficult to point to results. Probably the most important feature of the work of the first committee was in paving the way with the Machine Tool Builders' Association, and with the American Society of Mechanical Engineers. That this pioneer work was in a measure successful and also as a reflection of the trend of things, we have but to turn to the advertising pages of our technical journals, where page after page show machines pictured with hazardous points guarded. It is true that the guarding is not always adequate, and also that the manufacturer sometimes fails to guard all of the hazardous points, but on the other hand think of the advertising pages of 10 or 15 years ago! In those days the gears, power transmission, etc., were devoid of any covering.

It is perfectly obvious that the best time to safeguard the machine is while it is being designed. Then only can we hope to incorporate the features essential to safety and obtain the kind of a job that is pleasing to the eye and satisfactory to the critical inspection of the engineer inspector and the plant manager. Too often, unfortunately, the guards that are installed after the machine is in its working position in the shop are frail, unsightly and inadequate.

Safety specifications should be based upon practical recognized safety standards, and should be drawn up in a form so that they may be included and used as a whole, or in part, for any contract for construction work, or for purchase and installation of machinery and equipment. A set of safety specifications should be attached to, or embodied in,

general specifications when originally submitted to contractors or manufacturers for bids. In this manner the contractor or manufacturer will be fully advised as to the safety requirements and these features will be included and properly taken care of during the process of construction.

Where the size of job, or the character of the equipment involved warrants, it has been found very desirable to go over the safety specifications in detail with the contractor before the work is started and during the process of designing, and then inspect the equipment in the contractor's shop before shipment is made. By so doing a mutual understanding between the contractor and the purchaser may be reached regarding the safety features, and unnecessary expense incident to additions in the field may be obviated.

The work of this committee will necessarily depend very much on the degree of co-operation it secures from the members of the section and probably the greatest help, in addition to suggestions, would be secured when purchasing departments demand that the machinery ordered be furnished with all necessary safeguards and that such guards must comply with state laws, insurance companies requirements, and the requirements of the safety department of the purchaser.

If we begin with our own companies we may feel reasonably certain that later such demands will be made by the purchasing departments of all companies —in fact, many of them have already begun, as you know. Much good can also be obtained if we will do missionary work with our own companies, that is, those of us who are connected with the companies having machinery for sale. Surely we can take steps to insure the proper safeguarding of machinery which is for sale by our own employers. Then, too, we can do missionary work with manufacturers from whom our company make purchases by pointing out the hazardous features of the machines they manufacture, and suggesting the necessary safeguards. Always, of course, we should emphasize the fact that manufacturers of machines that are adequately guarded have an advantage over their competitors whose machines are not guarded because the sales engineer can truthfully state that the machine, as delivered, will comply with the requirements of the safe practice inspection department of the State Industrial Commission, and also the requirements of the various insurance companies.

Let us all think, for a moment, of all the machines that are now on the market that could, with but slight effort and small expense, be redesigned to include proper protection for workers operating the machine. When we think of the large number of such machines we are inspired by the very size of the job ahead of us, and the opportunity that it gives for everyone to do at least one bit of definite constructive work for accident prevention.

Have We Any Definite Data on Brake Wheels?

The Author States His Knowledge Regarding the Design of Brake Wheels, But Regrets That More Definite Laboratory Tests Have Not Been Made

By JOHN S. WATTS.

THERE are many features in the designing of brake wheels, upon which we should have more definite information, and I would like to see some of the College experimental laboratories make some tests to determine the facts governing these points.

The first point to be decided upon in designing a brake band, is the co-efficient of friction, of the materials, of which the brake wheel and hand are to be made. These co-efficients are given in the various handbooks for all the usual materials, with one important exception, namely, asbestos, which is now largely used and has proved its usefulness for the work.

The co-efficient of friction for asbestos linings on cast iron wheels is 0.3, and is practically constant at all speeds and pressures. This is a smaller co-efficient than that of some of the woods used, but much higher pressures can be imposed on the asbestos than on wood, as it will not char or burn from the heat generated, therefore for heavy or long continued braking action the asbestos is much the superior of wood brake blocks.

In connection with the co-efficient of friction, it will be noted that the co-efficients for the various woods on cast iron is only three-fifths of the co-efficients for the same woods on wrought iron. It would seem, therefore, that the most effective arrangement would be, to bolt the wood blocks to the brake wheel, instead of to the brake band, and so take advantage of the higher co-efficient. This arrangement, though, may be dangerous if not carefully looked after, as the brake band, being then the part that will take the wear from wear, before it is detected, if not inspected at regular intervals.

I know of one case where this change was made, and has worked very satisfactorily, but the change was made to overcome trouble from the heat generated. The brake wheel was twelve feet diameter by twelve inches face, and was used on a drop table to lower freight cars from a higher track to a lower one. The heat generated, when the blocks rubbed on the wheel, was so great, that the rim of the wheel became heated to the point where the expansion caused the spokes to crack. This trouble was entirely overcome when the blocks were bolted to the wheel, as the wood did not then transmit the heat to the wheel.

The next point that arises is the required diameter and width of the brake wheel, which involves consideration of the area of the braking surface, required to absorb and dissipate the energy due to the falling load. If the brake is to be operated frequently or for long

periods of time, the ability of the brake wheel to radiate the heat generated will probably be the deciding factor. If the brake is only used intermittently and for short periods of time, the safe bearing pressure on the brake lining will be the point to be decided upon.

Taking up the question of the area required to dissipate the heat, we have only the following rather meagre data.

By one technical writer, we are informed that, on a friction disc brake of the type used on electric cranes, the number of thermal units conducted away per square inch of bearing surface per minute, may be considered to be from 4 to 7, in a current of cold air; and from .75 to 1 in a cool place; in both cases the quantities stated are for intermittent service.

This gives a range of from .75 to 7 thermal units per square inch, under the same class of service, and is rather too wide a range to be considered useful information without any details as to the causes for such a great variation. Converting these thermal units into foot pounds per minute, we have .75 x 778, to 7 778 ᐸ from 583 to 5,446 foot pounds per square inch per minute.

In another place we find it stated that from 200 to 250 foot pounds of energy can be dissipated when using wood blocks on a cast iron wheel, this much lower amount being on account of the danger of burning the wood. It is also stated that car brakes, iron on iron, under favorable cooling conditions, often absorb as much as 10,000 to 15,000 foot pounds of energy per square inch. The time is not stated, but is presumably one minute.

Collating this information we find that we have the following data:

For wood blocks, 200 to 250 foot pounds per square inch per minute;

For fibre on iron, 600 to 5,500 foot pounds per square inch per minute;

For iron on iron, up to 15,000 foot pounds per square inch per minute.

Unfortunately, we do not have information specifying just what condition imposes these limits. We should have tests made to determine the foot pounds of energy that can be dissipated by each of the various materials used for brake linings, up to the point where charring commences, when running under load for varying periods of time, with varying periods for cooling, the idea being to consider the brake lining alone, and determine the limits for the lining material separately from the wheel, as far as possible. For iron or asbestos linings it is obvious that the heat generated does not affect the question, so far as the brake block or lining itself is concerned, within practicable limits.

For any kind of brake lining, there is another limiting feature, that is, the safe

bearing pressure per square inch, to avoid mechanical failure or disintegration of the material. For wood this bearing pressure must be kept below the amount which would cause the wood to splinter. Tests should be made to determine this pressure for the various woods and different thicknesses, as the allowable pressure will vary, probably with the thickness. For fibre and asbestos, the bearing pressure may be a maximum of 200 pounds per square inch, provided that the material is riveted to the brake band with rivets spaced on centres close enough to keep the tensile strain, imposed on the fibre by the dragging effect of the friction, within safe limits. There seems to be no information available as to the tensile strength of the various kinds of fibre used, but for asbestos brake linings the ultimate tensile strength is about 4,000 pounds per square inch, and a working stress of 400 pounds per square inch is safe. For iron on iron the bearing pressure must be kept below the point where rapid abrasion would commence, which is at a pressure of about 600 pounds per square inch.

Taking up now the question of the ability of the brake wheel itself, to dissipate the heat generated, or at least enough of it that the temperature of the wheel will not rise to a dangerous point. The heat radiating power, in thermal units, per square foot, per hour, per degree of difference in temperature, for cast iron, is given as .648; which is equal to 504 foot pounds, or say, .06 feet pounds per square inch per minute. As practically all this radiation must be from the rim of the wheel and we cannot count as radiating surface that surface on which the band is rubbing, we will have on an average about three times as much radiating surface as we have surface on the brake band. We may, therefore, expect to radiate 1.8 foot pounds per square foot, per minute, while the brake is on, and if this is one-half of the total time, we will dissipate 3.6 foot pounds per square inch per minute per degree of rise of temperature.

The limit beyond which the temperature of the wheel should not rise, depends upon the material used in the brake blocks, and for wood should not exceed 300° Fahrenheit, and for any material, to prevent breaking the wheel by unequal expansion, should not exceed 400° Fahrenheit.

Taking the atmospheric temperature at 70°, we have, then, an allowable increase of temperature of 230° for wood or fibre, and 330° for asbestos or iron, which gives us 3.6 x 230 = 828; and 3.6 x 330 = 1,188 foot pounds of energy per square inch of braking surface, per minute, which can be radiated under

the conditions assumed, that is, the radiating surface being three times the braking surface, the brake operating during one-half of the time, and the increase of temperature as given above.

It must be noted that if the brake is on for any considerable length of time, it would be possible that the temperature might rise beyond the danger point before the period of rest occurred. On the other hand, a certain amount of the energy is absorbed in raising the wheel to this temperature, and this may be taken into account, when the wheel is relatively heavy, and the machine is operated for shifts of a few hours, with time between shifts to cool off.

If necessary the radiating surface may be considerably increased by the casting on of ribs, preferably of such a shape as to cause a current of air to pass over the radiating surface. If not otherwise a detriment, the capacity of the wheel can be increased by making the rim heavier, and so taking more heat to raise the temperature to the danger point.

I hope that we will have some authority make the tests outlined above, with an average size and design of wheels, using all the various brake linings, and so determine what the limits are in respect to the power which can be absorbed, due to the temperature rise with a number of trials at various proportions of loaded and rest periods.

To summarize, the data required is as follows:

The maximum bearing pressure for each lining.

The maximum temperature for each lining.

The maximum temperature for the average brake wheel.

The radiating capacity per square inch of brake surface of an average brake wheel, when revolving, and when stationary.

Plan Boards as an Aid to Administration

This Article, Which Appeared in the "Organizer," Describes a Very Interesting Form of Plan Board That Enabled the Management to Keep in Touch With the Various Operations.

By HARTLAND SEYMOUR.

IT IS generally admitted that the plan board is one of the most interesting and most helpful aids to administration and organization which scientific management has brought us. It is essentially a graphic index—is, in fact, a most complete and thorough follow-up system.

The type of plan or control board varies, of course, with the nature of the work. One of the most complete plan boards ever seen by the author was constructed and operated, briefly, as follows. The board itself covered one side of the room—that is, the production office. The height from the floor to the top of the board was 5 feet. The board itself was 3 feet 6 inches wide, so that the bottom was 2 feet 6 inches from the floor. Thus both top and bottom were easily reached from a standing position.

Across the top of the board, from end to end, ran a division which recorded the complete assemblies. In this case the product was a motor-vehicle chassis. The series number and the number of the chassis to be manufactured was drawn up on a strip of stiff card and pinned along the top of the board as shown in the first figure. Thus, in this case, an order had been issued for the manufacture of 150 of the standard chassis; this series was given the letter "L" by the order department. Accordingly a long strip of Bristol board with the series letter and the number thereon was pinned up across the top of the board.

Below this ran two parallel lines the length of the board and graduated up to 200. The full particulars of the current order were drawn up and pinned in the space on the left as shown, while the scheduled date of completion was pinned upon the extreme right. Then a cardboard outline of a chassis, colored red, was pinned up at the figure 150. A corresponding cardboard outline, colored blue, was then adjusted day by day so that the progress of completed assemblies could be watched.

The progress on each section of the chassis and each individual component was recorded as follows. The board was divided by five vertical lines into six equal sections. A line running across the top of each section was graduated to 200. Then strips of paper, each equal in length to the width of each section on the board, were entered up with full particulars of each component, name, number, and the number per set, and so forth, as in the second illustration.

These strips were then ruled across to correspond with the graduations on the board, so that when pinned up the graduation marks would run the entire length of the board. Horizontal lines were drawn on each strip as shown and colored red. The operation was then as follows:

The strips having been placed in proper sequence on the board, the progress on each component was recorded by painting over with black the red line on the strip. To take an instance. A screw on a certain part of the chassis will do. Well, this screw belongs to a "section" of the whole assembly, one only of which is required. So that of Section A, let us call it, 150 are required to make up the 150 chassis, with the corresponding number of the screw taken as an example. This screw will have a number, say A/35, so that it can be readily identified. Then the particulars of this screw are entered on the strip, as shown, and its progress is recorded as described.

Each section of the total assembly has a strip of its own. We will suppose there are ten such sections. Then the strip bearing Section 1 is pinned at the start, and the strips, bearing the components going up to make that section, follow it. Then comes Section 2, followed by its components, and so on right down to Section 10, with its parts. Each section is, then, a small assembly of its own, and as so many components are completed which go to the assembly of any one section, this fact is recorded on the assembly strip.

Another Example

We have considered the case of a screw, one of which was required per Section A. Suppose we take another case. If we look under Section B we find a small bolt B/78. Now, three of these small

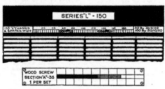

ILLUSTRATING THE IDEA, NOT ONLY OF THE BOARD, BUT THE STYLE OF INDIVIDUAL STRIPS USED.

bolts are required for each section. That fact is recorded on the strip for that bolt B/78, in the words "Number per Set —3." So that, altogether, 450 of the bolt B/78 are required for the whole 150 chassis.

What happens in this case is that each graduation represents three, so that when six of B/78 are completed the red line is blacked over as far as the second graduation, and so on right up to the 150 mark, which will then record the manufacture of 450 of these bolts.

This board was kept up to date by a reliable girl clerk. It may be added that

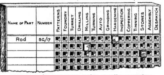

ANOTHER STYLE PLAN BOARD WHICH DESERVES ATTENTION.

she was responsible for taking care of the memoranda bearing the information from which the plan board was made up. These slips were bound in books of 100; they were filled in and detached by the progress clerk or chaser responsible for keeping in touch with that particular department. By having a definite routine of watching these components, complete information was available day by day.

The plan board, needless to say, was extremely valuable to everybody concerned with production, and well worth the initial trouble and cost expended on it.

Another form of plan board, illustrated above, as used in another works, is operated as follows. It might be mentioned here that the product is also a motor-car chassis with a great number of parts.

The plan board is built round three walls of the office, the top being 5 feet 6 inches from the floor and the bottom 2 feet 6 inches, so that the board measures 3 feet in depth. It is divided into fifteen equal parts by vertical wooden ribs. Each section of this board represents a complete sectional assembly of the chassis, and the name of the section is printed across the top. Running down the left-hand side of each section of the plan board are, in the first column, name of the component and, in the second, the part number of this piece. The remainder of the board is divided up into fifteen small pigeon-holes, each an inch square. Over each pigeon-hole is the name of a department, the order of these being the order in which the pads are carried by the progress chasers in their tour through the shops. At the

end of the day, when the chaser makes his report, he fills in one of these pads. His procedure is as follows: If the rod, part SC/17, was in operation in the drilling department, say, he puts a red mark in the space opposite "Rod, SC/17," and under the heading "Drilling Department." If, on the other hand, this part has just been sent in to the drilling department, perhaps for working on the next morning, he will put an ordinary black-ink mark instead of the red.

All these reports have to be in the planning office before nine o'clock the next morning, and from these the board

blue, are kept in a box. A clerk in the planning office, on receiving the reports, takes up the first and notes that the rod, part No. SC/17, is in operation in the drilling department, so he takes a red cube out of the box and fits it into the pigeon-hole in the plan board corresponding to the division on the sheet. This is done for every component every day.

The making-up of the board from these reports takes two girl clerks from a quarter of an hour to twenty minutes, at the outside, every morning. The board furnishes a very valuable guide to the work going on in the shops, and from it the production manager can see just where each part is.

COLOR SCHEMES FOR PLANT PIPING

By H. L. Wilkinson

The importance of some definite and unmistakable means of readily distinguishing between the various pipe and conduit lines and systems in the power plant has long been recognized by both designing and operating engineers. Particularly in large power plants in this question of vital importance, and it should be stated that in some of the well-ordered large power plants reasonably satisfactory systems have been installed. Also there has been some effort made to standardize this practice, a committee appointed by the American Society of Mechanical Engineers to investigate the question having reported upon the preferable distinguishing colours for the various divisions of piping systems in the usual power plant, as states the author at a recent meeting of the American Society of Heating and Ventilating Engineers.

While the importance of standardization in this detail of our power plants has been recognized, too little attention has been given to the matter of late, and the writer believes that its importance should not be overlooked. It is, therefore, suggested that the members of this society consider the following plan which is designed to apply to all piping systems.

It will be noted that the recommendation involves, in addition to piping lines, colour schemes for machinery, motors, hand-rails, dadoes, waste pails, elevator cages, etc., thus extending the usefulness of this distinguishing system to all parts of the industrial plant. It has been a revelation to the writer to find the cordial reception which this proposed colour scheme has received in many industries and concerns of national prominence, and the result is that new industries and concerns are taking it up continually. The application of this standardization might well be extended to the designating by distinguishing colours the tools used in various departments of a plant. This would bring about several benefits. In the first place, it would enable managers to hold each department of a plant responsible for its own tools. Then, because of the improved appearance of the tools, it would create more interest in their care on the part of mechanics. Last, but not least, it would surely be an insurance against tools being carried out of factories.

While the scheme designates definite colours for certain pipes, it is, of course, unnecessary to apply any one particular colour on any one pipe; the arrangement being flexible, so that an engineer may determine for himself his own design. In some cases, where there are as many as fifteen to twenty-five various pipe lines, it is necessary to use combination colours on some pipes, painting the straight pipe one colour and the joints, valves, elbows, etc., the alternating colour. Of course, some exceptions will be found where an established custom has been in effect for some time, as in the case of sprinkler-pipes and all fire lines, where vermilion is the logical colour.

The advantage of this standardization will become readily apparent from the case of a certain plant where there was a stoppage or break in one of the pipe lines. The workmen carefully followed the line to the side wall where the trouble apparently existed, and opened up the wall for a considerable area, only to be confronted with everything but the line they sought. Of course the expense of ripping out the wall was nothing compared to the inconvenience and loss sustained, due to the suspension of operation until this repair was made. If this pipe line had been printed with a distinguishing colour and a small arrow placed on the wall where the pipe entered, indicating the direction the pipe took after entering the wall, all this difficulty would have been obviated.

WHAT OUR READERS THINK AND DO

LIGHT AIR HOISTS
By A Draftsman

The use of the light air hoist is already well known, but it is particularly adaptable in plants where there is a regular supply of compressed air at a pressure of about 70 lbs. per sq. inch.

The beauty of this type of hoist is that it can be hung over any spot where frequent lifting or lowering has to be accomplished, such as a dip tank. It may also be suspended from an overhead trolley if desired, the trolley running on a track in the usual manner. It may even be invented and modified to work a light lifting elevator of short lift. In fact the uses of such a hoist is very varied, and its uses prove the value of knowing how such a hoist is made.

Briefly its essential parts are a two-way valve, a strong cylinder with a top flange, cap and eyebolt, lower flange, cover or stuffing box, a cast iron piston equipped with cup leather, and a piston-rod with crosspiece and V bolt for attaching to hook.

The accompanying illustration show the main features of such a hoist from 2½ in. to 8 in. in diameter. In this

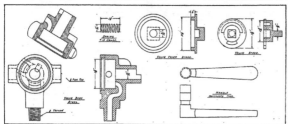

FIG. 1—DETAILS OF THE AIR VALVE.

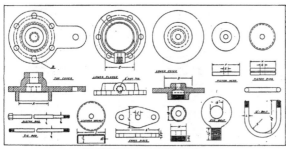

FIG. 2—VARIOUS OTHER DETAILS OF IMPORTANCE.

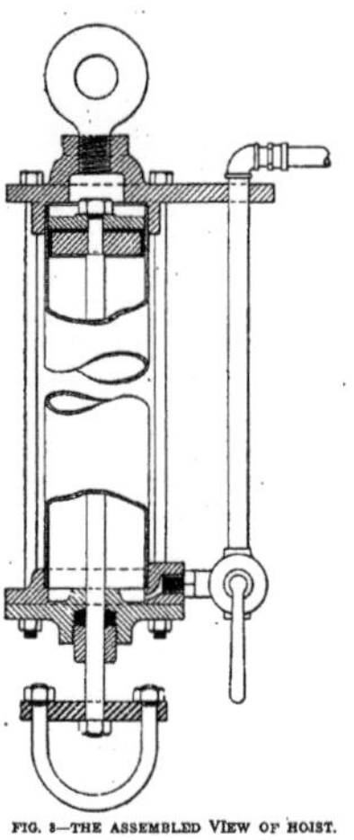

FIG. 3—THE ASSEMBLED VIEW OF HOIST.

type the cylinder is of commercial brass tube, ⅛ in. in thickness, to which the end flanges are held by longitudinal tie-rods. The two-way brass valve is common to all these sizes and its details can be noted clearly. For convenience of operations a light rod is often attached to the end of the valve lever.

At Fig. 1 is shown the details of air valve, while Fig. 2 illustrates the other necessary parts. The different parts are all named so that readers should have no difficulty in picking out the details. Fig. 3 shows the assembled view of the hoist. Fig. 4 gives the table of various dimensions from 2½ in. to 8 in. sizes.

THREADING AND BORING FIXTURE

By G. Barrett

The accompanying sketch illustrates a fixture that has been adapted to a lathe for the boring and threading of small headers. The boring bar B is held in a special holder, the base of which is provided with a slide that fits an ordinary tool slide that is secured to the face plate of the lathe. The bracket C is fitted to the saddle of the machine so that the front face is parallel to the face plate and at right angle to the travel of the carriage. Suitable holes are drilled in this bracket

SIZE OF CYL.	DIMENSION																	
	A	B	C	D	E	F	G	H	I	J	K	L	M	N	O	P	Q	R
2½	2¾	½	2¾	2½	½	2 13/32	½	2⅛	2½	⅝	2	½	⅝	1⅛	11/16	⅝	½	¼
3	3¼	½	3¼	3	½	2 15/16	½	2⅝	3	⅝	2	½	1	1⅛	11/16	⅝	½	¼
4	4¼	5/16	4¼	4	⅝	3 11/16	⅝	3⅝	4	¾	1 11/16	⅝	1¼	1⅛	11/16	¾	⅝	5/16
5	5¼	5/16	5¼	5	¾	4 11/16	¾	4⅝	5	1	1 15/16	⅝	1¼	2	11/16	1	¾	5/16
6	6¼	11/16	6¼	6	¾	5 11/16	¾	5⅝	6	1	1 15/16	⅝	1¼	2	11/16	1	¾	11/16
7	7¼	11/16	7¼	7	1	6 11/16	1	6⅝	7	1¼	1⅜	¾	1¼	2⅜	11/16	1¼	1	11/16
8	8¼	11/16	8¼	8	1	7 11/16	1	7⅝	8	1¼	1⅜	¾	1¼	2⅝	11/16	1¼	1	11/16

FIG. 4—TABLE SHOWING VARIOUS DIMENSIONS.

at three different points for bolting the flange D in the desired positions for locating the openings E, F and G in alignment with the spindle of the lathe. Fig. 2 is an end view of the fixture. The bore is finished to size by the adjustment of the tool by the handwheel shown. The cutting of the thread is accomplished in the usual manner, gearing the lathe for the desired pitch and manipulating the tool by means of the handle on the tool slide.

NON-DESTRUCTIVE METHOD OF BELT JOINING

An unusual problem in the handling of hot cement clinkers with a temperature of 200° and over, was recently solved at the plant of the Standard Portland Cement Company, Leeds, Ala., in an interesting and cost-reducing manner. The method decided upon for moving the clinker was a rubber conveyor belt, but the clinker could not be cooled sufficiently in the process previous to conveying, to prevent scorching of the belt and its rapid destruction. The answer to this problem was found by running the belt at an incline of 12 degrees, so that the lower pulley dipped into a trough of water, thus carrying a film of cold water upon the belt, on to which the hot clinker from the loading hopper was deposited.

At this point a new problem was encountered; namely, how to join the belt so that the belt's full strength would be retained, and in a way which would withstand the extremes of temperature, the wear on the pulleys and the abrasion of the clinker. For this purpose Crescent belt fasteners were adopted, because they brought the belt ends tightly together in a snug joint, which made the belt practically endless on the pulley side, so there was no opportunity for clinker ash to get into the joint and abrade the belt ends, and also because in this method of joining, no metal came in contact with the pulleys to cause wear, and a permanent joint was thus assured. Moreover, the exceptional strength of the heads of the rivets and the formation plates prevented destruction of belt joint through abrasion by the clinker.

In six months of operation, this conveyor has carried 61,000 tons of clinker, and the Standard Portland Cement Company credits the saving of $300.00 in belt cost alone to this conveyor. The belt used was Goodyear Hy-temp, which is made particularly to withstand temperatures up to 200°, and is adapted for work on conveying jobs in mines, coking plants and cement factories where heat resistance and ability to withstand hard wear are prime requisites.

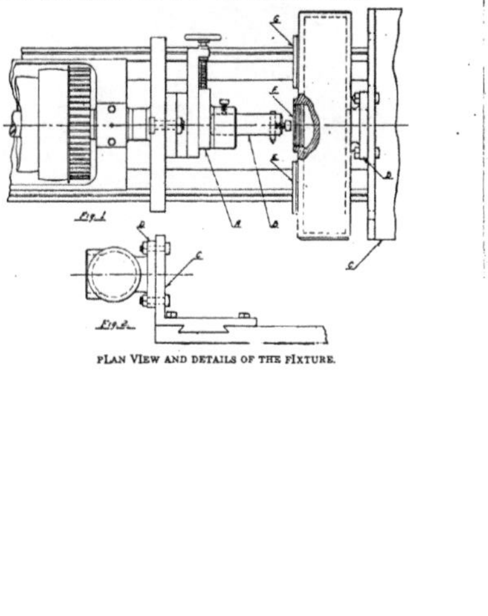

PLAN VIEW AND DETAILS OF THE FIXTURE.

DEVELOPMENTS IN SHOP EQUIPMENT

HIGH POWER LATHE

The Joseph T. Ryerson & Sons, Chicago, Ill., have placed on the market their line of high power selective head engine lathes as illustrated. These machines can be used for quantity production as well as general machine shop work, and embody the most advanced design, high class workmanship and finish. Speeds and feeds are also well looked after.

A prominent feature is the method of driving. With hardly a change these lathes can be adapted to constant speed, single pulley drive, with or without spindle reversing attachment, or direct reversing motor drive. In this latter arrangement the motor is bolted to the bed and the armature shaft directly connected to the main driving shaft, doing away with belts, tension idlers, and chain drives with their consequent trouble and inefficiency.

All controls have been centralized on the apron. From here the operator may start, stop and reverse the spindle instantly, engage, reverse, or trip the feed, as well as traverse the carriage and cross slide.

The 12 spindle speeds are changed by means of two levers on the headstock, the one controlling the various sets of change gears and the other engaging the back gears. The action of both levers

Headstock

The headstock is cast integral with the bed. This construction is more rigid as the two mutually reinforce each other and permit the use of a larger drive gearing. The bed also serves as a container for oil, into which part of the gears are placed, providing a perfect oil splash at all times.

The spindle is a .60 carbon alloy steel forging of exceptionally large diameter. This not only reduces all vibration to a minimum, but also increases the bearing surface considerably, and provides for a very large hole through spindle. A thrust ball-bearing which takes up all the heavy working strains is mounted on the spindle and an adjustable thrust bearing eliminates all longitudinal play. The phosphor bronze bearings are of the split conical type, adjustable for wear and accurately scraped to fit.

The belt driven machines have the pulley mounted on ball bearings which are carried on a stud, relieving the shaft of all bending strains. The reversing mechanism, which is of the planetary type, has one of its pinions extended sufficiently to receive the power directly from the pulley. The reversing attachment is omitted in the motor driven lathes and, instead, a reversible motor is mounted on the bed, applying the power directly to the primary shaft. A

tapered keys and wedges controlled by a lever mounted in a shift plate which indicates the various spindle speeds. This secondary shaft also carries the driving pinion and the back gear engaging device, which is operated by friction bands and a lever conveniently located on the headstock. The back gears, as well as all change gears, are permanently in mesh and the various changes can be performed with great rapidity and without danger of engaging more than one set of gears at a time. The power is transmitted to the spindle through a phosphor bronze driving pinion and an exceptionally large herringbone gear.

The spindle drives a large spur gear connected to the feed reversing mechanism, which is operated from the apron, permitting the operator to engage, trip, or reverse the feed screw without changing his position. The 28 feeds are obtained by two sets of geared cones controlled by dive keys, which in turn are actuated by a hand wheel and lever on the headstock. Provision is also made for transposing and compounding gears, permitting the cutting of threads of almost any desired pitch.

Apron and Carriage

The apron and carriage are of exceptionally heavy construction and unusual

GENERAL VIEW OF THE LATHE.

is instantaneous and all changes may be made while the lathe is in operation. The hand wheel visible on the headstock can be turned to any desired feed marked on its circumference, automatically selecting the feed indicated.

cone of six gears is connected to this shaft, the extension of which carries the back gears.

The change gears on the secondary shaft are of the internal friction band type and are operated by means of

design. The longitudinal and cross feeds are effected by independent sets of worm gears which are engaged by large friction cones. The positive feed consists of a large split nut made of phosphor bronze, and a very simple device pre-

vents the engagement of friction and positive feeds simultaneously. One lever on the apron controls the starting, stopping and reversing of the spindle, and one lever controls the engaging, tripping and reversing of the feeds. On the bottom a lug with micrometer adjustment is attached to automatically trip either the friction or the positive feed at any point along the carriage travel. As the apron is of box section, all shafts are supported at both ends, insuring perfect alignment and durability. The rack feed pinion is heat treated and arranged to be withdrawn from the rack when chasing.

The carriage has a very large surface bearing on the ways and is guided by V's of unusual vertical depth. The cross slide traverse is regulated by hand and power. The tool post is mounted on a swivel base equipped with angular hand feed. Means are provided to adjust all sliding surfaces for wear and all parts are protected from dirt and chips. All gears throughout are made from .45 carbon steel with the exception of the worm gears in the apron, which are of phosphor bronze. All bearings, with the exception of the S. K. F. ball bearings, are renewable. All gears and bearings in the headstock and apron are automatically splash oiled.

MARKING MACHINE

J. Merey, 2842 N. Maplewood Ave., Chicago, Ill., has placed on the market what is known as a Stampograph. This is, as can be noted from the illustrations, a rotary machine for lettering and numbering and can be used on all kinds of metals or substances that are in their soft state.

The impression is adjustable and can be perfectly spaced and aligned. The

SIDE VIEW OF MARKING MACHINE.

machine has an automatic feed, subject to the sizes of the characters there being a fixed feed provided for each size. The disks for the characters are made from the very finest tool steel and carefully hardened and tempered.

The feed is caused by means of cams and a multiple rack shaft; the impression is obtained by means of crankshaft connected with a handle and connecting rod, which is so constructed, that the characters should not break in case some part of the metal should resist, or from any other cause. The adjusting to obtain the impressions for various sizes of objects or work, can be gained by turning the ball handle in either direction and following the gauge affixed on the side until desired impression is obtained. The machine illustrated, as can be seen, is made for hand power. The machine as built for machine power weighs about 350 lbs. and requires space 2 x 2 feet, and can be run from a countershaft or motor driven. The handpower machine weighs about 200 lbs.

The power machines are only built on specification and the stock machines are fitted out with full alphabet, 1-16, 1-8, 3-16, 1-4 in.; numerals 1 to 0 in same sizes. Special multiple engraved dies can be furnished if desired. Both machines are outfitted with all engraved

FRONT VIEW OF THE MACHINE.

characters, and all necessary requirements, as cross table, vise and circular table with graduations to stamp in radius

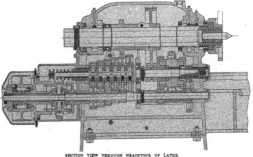

SECTION VIEW THROUGH HEADSTOCK OF LATHE.

The MacLean Publishing Company
LIMITED
(ESTABLISHED 1887)

JOHN BAYNE MacLEAN, President. H. T. HUNTER, Vice-President
H. V. TYRRELL, General Manager.

PUBLISHERS OF

CANADIAN MACHINERY
AND MANUFACTURING NEWS

A weekly journal deVoted to the machinery and manufacturing interests.

B. G. NEWTON, Manager.   A. R. KENNEDY, Managing Editor.

Associate Editors:
J. H. MOORE   T. H. FENNER   J. R. RODGERS (Montreal)
Office of Publication: 143-153 UniVersity Avenue, Toronto, Ontario.

VOL. XXIV.   TORONTO, AUGUST 5, 1920   No. 6

PRINCIPAL CONTENTS

Editorial Versatility

THE following question and answer in the Hamilton "Spectator" illustrate the fact that you can't stump an editor. The query is, "To settle a dispute, would you be so kind as to allow me to ask in your 'Knotty Points' column why the funnels on almost all steamers are painted with colors red, white, and black? Thanking you in advance, I remain, yours truly, Reader."

To which the busy editor replies: "The colors are most conspicuous."

While not disputing the fact that such a combination of colors is conspicuous, it would seem to us that the editor's lack of knowledge of the real reason was only equalled by his willingness to oblige the reader. This is, as every seafaring or shipping man knows, to identify the company the steamer belongs to. In the old days of the sailing ship, every line had their own house flag, and the hulls of the ships were painted with a particular color scheme. For instance, one large English sailing ship company made a rule of painting their ships white, from truck to water line, blocks, yards, masts and every fixture conforming to this color. They were known as the white ships. An Italian line of sailing ships was known by its fluted masts, the masts themselves being white, with the flutes picked out in light blue. Another line would have a white hull, with painted ports. With the introduction of the steamer, the funnel lent itself to distinguishing marks, and these were made as near as possible the same colors as the house flag, though of course this was not always the case.

For instance, the red funnel and black top, in conjunction with a square red flag carrying a golden lion and globe, is the symbol of the Cunard Line, while a cream-colored funnel with black top and a red swallow-tail flag with a white star on it signifies the White Star Line. The Allan Line and the Dominion Line both used a red funnel with a black top and a white band, the distinction being in the position of the band. In one case it was close up against the black, while in the other there was a band of red between the white and the black. Every line of any note has its own particular marks.

The reason the Hamilton reader concluded that red, white and black was the favorite color of the seas, is probably because a red funnel, with a white band and black top is the distinguishing mark of the Canada Steamship Lines, and in the thriving ocean port of Hamilton these are probably the most frequent visitors. We imagine our editorial friend, from his familiarity with nautical matters, must have just returned from reporting the America Cup races. While on the subject of funnels, we would like to ask him if a ship with two funnels goes faster than a vessel with one, or vice versa?

The Old System of Barter

An interesting story comes of Samuel M. Vauclain, president of the Baldwin Locomotive Works. While financiers talked of credits and exchange and the chances of opening trade with Europe, Mr. Vauclain went abroad and came back with orders that filled his plant. And he got the pay? He is taking it out in iron ore, grain, and oil. He probably finds that way of doing business easier than having to explain to the Rumanians, Serbians and Pole why United States take a bite out of their money when they want to spent it in the States. The armistice was not much more than signed when Vauclain was in Europe with a proposition to swap engines for oil, grain, ore, etc.

Where Do You Fit In?

Take 100 men at the age of 25. By 35, five of these have died; 10 are wealthy; 10 are well-to-do; 40 live on their earnings; 35 show no improvement. At 45 sixteen have died; one is wealthy; three are well-to-do; 65 live on their earnings; 15 are no longer self-supporting. At 55, 20 have died, one is wealthy; three are well-to-do; 46 live on their earnings, and 30 are not self-supporting. At 55, 36 have died, 1 is wealthy, 4 are well-to-do; 5 live on their earnings; 54 are not self-supporting. At 75 years, 63 have died; 1 is wealthy; 2 are well-to-do, and 34 are dependent. The estates of these men look like this: One leaves wealth, two leave comfort, 15 leave from $2,000 to $10,000, while 82 leave nothing.

What Strikes Cost

It is hard to realise just how much wealth and days of work are shot to pieces during the course of a strike. We have come to look upon strikes as part and parcel of our industrial life, while it is time that we started to regard many of the strikers as thieves and robbers that go about putting their hands in the pockets of the workers, taking out many times what they ever replace. For instance, strikes in Pennsylvania during 1916-17-18-19 meant 10,179,243 days' work lost, and wages lost amounting to $28,664,871. These figures do not take in coal or steel strikes.

A Canadian manufacturer has written to Canadian Machinery about the advisability of placing orders now for certain tonnages of iron and steel. If he can get his business taken on at Steel Corporation prices, then it is well to book, but if the purchasing is being done from the premium mills, buy only what is needed. It is not likely that prices will fall below the quotations of March 21, 1919, as many of the high cost mills claim they cannot produce at them and break even.

MARKET DEVELOPMENTS

Holiday Atmosphere in the Market Now

Business Reported Rather Quiet, But Improvement Is Looked For —Pittsburgh Still Piling Material and Making More Semi-Finished Than Finished—No Price Reductions in Machine-Tools

Although the steel mills are turning out a decreased tonnage, they are still producing more than the railroads are taking away from them, so piling is still going on. The Canadian yards, although much in need of this material, report no improvement in the matter of deliveries. The Canadian steel mills are hard pressed for their own coal and coke supplies, and on this account are not able to do as much as they otherwise would in the way of relieving the situation locally. Some of the Canadian mills have labor troubles to contend with, which brings their production figures down to a very low level.

The business that is coming up in the machine-tool market is not in large volume. Dealers are able to show improvement in the matter of securing deliveries from the makers of several lines. There have been some cancellations, and it is often possible to fill an order in this way much ahead of the promised schedule. But all this brings the makers of machine tools nearer the place where they will once more be out actively soliciting business. For several years it has been more a matter of trying to get the orders filled and satisfy the customers for having to wait so long. Any suggestion that this turn of the market will mean lower prices to the buyer must be taken guardedly. One well known maker has this week sent notices to the trade covering this point by announcing that there will not be any reductions during the year 1920 so far as his line is concerned, the idea of this being to reassure buyers who might be holding off in the hope of better price treatment a little later on.

There is something not far removed from stagnation settling down on the scrap metal trade. Certain it is dealers are not anxious to get out and take on tonnages to deliver on contracts accounts for about all the business that is moving.

PITTSBURGH MILLS ARE NOT YET SHIPPING AS MUCH AS THEY MAKE

PITTSBURGH, Aug. 5. — Taking the blast furnaces and steel works as a whole, the shipments are somewhat improved, and thus the rail transportation situation as relates to the distribution of the iron and steel industry's products appears at least to have rounded the turn. The worst condition fell towards the end of April, and for some time afterwards there was improvement, but late in June conditions started growing worse again, not because the total volume of rail transportation did not continue to increase, but because cars were diverted from general movement to the movement of coal. The original coal mine preference order was dated June 21, and was run for 30 days, while a 30-day extension carries it to August 19. The blast furnaces and steel works were deprived of a considerable number of gondola cars, but with increasing transportation all around the handicap is being overcome. The time is now probably near at hand when the coal mines will be supplied with all the cars they can load, whereupon any further increase in the volume of rail transportation will bring benefits to the iron and steel and other industries.

The point has not yet been reached at which all the furnaces and mills are shipping as much tonnage as they make, but the additions to stock are at a smaller rate, and at some plants, though thus far at only a few, the shipments exceeded the production. The Chicago district had been doing very well in this respect, having cleaned up the larger part of its accumulation of steel at mill, when last week the coal mine strike in Illinois and Indiana effected a large curtailment in iron and steel production in the district, about half the blast furnaces having to bank. There was an immediate effect upon the rolling of steel at the Gary plant, on account of the mills depending for power upon blast furnace gas.

The policy has become still more general among steel mills of holding steel, when it cannot be shipped, in the semi-finished form, whereby the production of finished steel is less, proportionately, than the production of crude steel, and shipments are approximately equal to the finished steel output. At some mills special orders rolled in the past are being picked out for shipment, while the mills are rolling stock sizes, the object being to get into consumers' hands such finished steel as might be desirable for use at some later time, leaving the stocks at mill in more mobile form.

Light Demand Continues

The market demand for steel has been light for some time past, and prospects now are that this light demand will continue for some months to come, general business and financial conditions being distinctly unpropitious for bringing out a full demand for steel, particularly when productive capacity is 40 to 50 per cent greater than before the war. There is railroad buying in prospect, but no large volume compared with the heavy purchases the railroads used to make, as in 1905 and 1906, in 1909 and 1910 and in 1912. With railroad buying extremely light in the recent past, even a fair volume of buying in the near future would not strike anything like the average that used to be common, and either in the next year or the next ten years the steel industry cannot expect the railroads to take nearly as large a percentage of the total steel output as they have in the past, although in point of tonnage the purchases may be fairly large.

In the matter of construction work generally the outlook for steel consumption is not particularly favorable, for even though the cost of steel can be shown

to be relatively moderate, the cost of other building materials, and the cost of labor, are very high, while there are vexatious delays in all construction work. Then, again, conditions as to financing new enterprises are rather unfavorable.

Equalization of Prices

It is a more common view than formerly that steel prices in general will decline to the level maintained steadfastly by the Steel Corporation since the Industrial Board arrangement of March 21, 1919. That there should be two or three markets is unprecedented, and is illogical, except for brief periods when there may be delivery premiums. That the corporation will not advance its prices was made clear long ago. Only recently Judge Gary issued a statement on business conditions in general, in which he took occasion to criticize in strong language the disposition of so many sellers of various commodities to exact the highest price obtainable and thus make inordinate profits. As to earnings the Steel Corporation's report for the second quarter of the year furnishes further testimony to the fact that the Industrial Board prices are quite profitable, for the earnings were far in excess of dividend requirements even though the shipments were much below capacity.

The second quarter earnings of the Steel Corporation, after payment of interest on subsidiary company bonds, but before depreciation allowances, sinking funds, etc., were $43,155,705, against $42,089,019 in the first quarter, and the surplus after dividends was $13,776,833.

Pig Iron Uncertain

The pig iron price outlook is particularly uncertain, because the market has no basis. Prices in the different districts are made from week to week by small transactions, while in steel products there is the stable basis afforded by the Steel Corporation' having adhered to the Industrial Board prices for over 16 months. The pig iron markets have advanced far above the Industrial Board schedule and thus are entirely out of line with steel prices. The markets as made of late have been made by demand for prompt iron, but generally of buyers having no inclination to buy for forward deliveries. Thus in the past fortnight the valley market has advanced $1 a ton on Bessemer, basic and foundry, on transactions of only a few thousand tons in each grade, and these small tonnages can hardly be considered adequate to advance the whole market for the future, when hundreds of thousands of tons will have to be traded in. What consumers are buying now is deliveries rather than pig iron. The only market that now exists is the prompt market, it being quotable, with the recent advances, at $47 for Bessemer and $46 for foundry or basic, f.o.b. valley furnaces, freight to Pittsburgh being $1.40.

Machine tool houses for the most part are having a fairly quiet session, and the chances are that it will remain this way until the holiday season is over. The arrival of more material would also help in the sale of equipment.

—

The supplies business is confined to small orders. Some railroad business is still pending, it not having been allotted yet.

—

One American house has sent out a circular letter announcing that there will be no reduction in its lines this year. The idea is to give confidence to the market, so that buyers will not hold back in the hope of a better price in a few months.

—

Deliveries of plate, sheets, bars, tubes, etc., are very poor. In some yards it is weeks since there has been a shipment in some of these lines. No improvement can be noticed in the matter of getting mill deliveries.

—

Several firms are paying a good deal of attention to collections now, and although they have their business in good shape, admit that they have more trouble in getting their money now than they did two or three months ago.

SELLING IS ONCE MORE A REAL THING

Machine Tool Dealers Report That Business is Quieter than for Some Months

TORONTO.—The week has been rather dull in all the iron, steel and machine tool lines. The holiday at the first of the week tended to emphasize this trend. One thing is becoming more noticeable, and that is the selling work that is being caried on by several of the leading houses. It has been said a good many times that the war experience, when every person was looking for the man who could secure equipment for him, had spoiled a lot of salesmen, and undoubtedly there was a certain amount of truth in the contention. It looks as though this spoiling period had been pretty well done away with. The buyer in many cases is not so keen as he was some month ago. He knows there is not likely to be the same demand for his finished product, and he also knows, if he is in the steel or iron line, that it is not going to be an easy matter for him to get the output. All this means that it is going to be more of a buyer's market than has been the case for some years now. The sellers

have been in command for a long time. The change may bring a lower price in some instances, although mounting production costs, which are the common property of all the makers, will keep the market from sagging very seriously. Several cases have been heard of this week where quite attractive propositions were put up to prospective purchasers in order to get them to come into the market now and keep the firm's booking up to capacity.

Travellers for machine tool houses report that in a good many centres in this district the trade is none too brisk. Several factories are seriously embarrassed by not being able to secure supplies, while lack of power is holding back operations in other centres.

Shipping Still Bad

The steel warehouses are not getting much better service than for some weeks past. It seems that deliveries secured are very often the result of serious effort on the part of the purchasers to get the stuff moving toward the Canadian border from the American mills. The Canadian mills turning out bars, etc., are not getting a large output now, and some of them are handicapped by having labor troubles on their hands—all this at a time when there is practically mountains of business to be secured.

Warehouses are having hard work to keep their monthly volume of business up to months when plenty of material was available. These figures are staying up remarkably well, but it must be admitted that in a good many cases it is the increased selling prices that fill up the blanks rather than the large tonnage that is being handled.

Watching Collections

"For the past few weeks," stated one warehouse to-day, "we have been paying special attention to getting our collections in shape, and the results have been very gratifying. There are some firms though, that we do business with who find their collections harder now than they were a short time ago, and on this account it is necessary for us to keep our payments well up. It is a very easy matter to get quite a large bill on your books. I think that is a point that a good many of your readers would do well to think about. It came as a surprise to us to see how quickly accounts grow. The present prices in the steel market are so much higher than they were for warehouse goods in past years that almost unconsciously these amounts grow to very large proportions."

The scrap metal dealers declare that there are days when they would be better off were their yards and offices closed up. Making all due allowance for that statement, there is little business moving at present, although increased activity in some of the foundries may make a call for small lots. But the big volume of trading is not being done, nor do the yards see where it is going to come from. The steel mills seem to be pretty well

supplied, and they are running short at best on account of not getting their usual supplies of coal and coke. Prices have not been changed, but it is only fair to say that the quotations may not always be paid this week. Yards at times will take on an offering, but it is generally found that when this is done, the deal has been put through at a price named by the yard.

NO REDUCTION IN PRICE THIS YEAR

American Firm Sends Word to the Trade In Regard to Several Lines

Black & Decker Mfg. Co., of Baltimore, are sending the following letter to the trade:—

"There will be no reduction in prices this year and starting January 1, 1921, we will protect you against loss through price reductions for 60 days after the purchase of goods.

We have recently had a number of inquiries from jobbers, asking if any reduction in price was contemplated in the near future. This has suggested to us, in line with our policy of co-operation with our jobbers, that we can work more closely with them in this respect.

'There is no price reduction of any kind possible in our line at present and you may accept this letter as guaranteeing you against any decline in the price of Black & Decker, portable electric drills, electric valve grinder and electric air compressors between now and December 31, 1920.

Furthermore, we take this opportunity of advising you that after that time you will be protected against loss through price reduction for a period of 60 days after the purchase of our products. This does not mean if conditions make it possible for us to get a lower price on our products that we will give you 60 days' notice and defer the reduction for that length of time. If a reduction is possible it will be made promptly so as to give the ultimate purchaser the benefit and we will absorb any loss which would otherwise be caused the jobber who bought within 60 days before the change in price."

WILL RE-OPEN THE SHIPYARDS

Shutdown of Dominion Plant is Only a Temporary Affair

George McKellar, operating superintendent of the Dominion Shipbuilding Company, Toronto, which laid off a large number of its men on Friday and Saturday, instructing them to return on Tuesday for their wages, said that the yards would be in operation again within seven days. This was the only information which Mr. McKellar could give regarding the temporary shut-down of the plant, which has employed more than 1,500 man.

"Any further information will have to be got from the firm's solicitors, and they

BUYERS AND SELLERS OF IRON AND STEEL

Some Philadelphia iron makers declare it will be years before they get an adequate supply of cars.

Boston reports that pig iron users there are not at all keen about contracting for 1921 deliveries.

A thousand tons of Bessemer iron was sold to a Pittsburgh firm this week at $47 per ton.

Two furnaces in the Buffalo district have been forced to bank because they could not get coke.

Cleveland reports that practically no orders are being received for either foundry or malleable for next year.

Prices of coke continue to advance, and spot furnace is quoted in Pittsburgh at $18 to $19, ovens. Railway troubles and the desire of furnace interests to be covered causes the high price.

Some of the independents have withdrawn from the bar iron market as they are too heavily booked for fourth and first quarters.

Railroads are coming into U. S. steel market for big tonnage, but find it hard to place orders for definite delivery or to get shipment of material already bought.

There is an increasing export demand for steel plate. Some independents are asking 3.75 for export, being for 3-16 in.

On account of the outlaw miners' strike in Illinois, mills are underpowered at the Gary plant of the American Sheet and Tin Plate Co., and complete suspension is predicted.

Scrap dealers in the States are able to reload about one car in five of those received. They are having trouble in keeping their contracts covered.

Coal operators of Alabama are making a strenuous fight for the open shop. For several weeks strikes have been in progress at a number of domestic coal mines, and production has been seriously affected. The operators have started a systematic campaign for the open shop, and public sympathy is sought.

Good sized premiums are being paid for prompt delivery of sheets in the New York market. Blue annealed are bringing 6 to 7.50c in Pittsburgh.

cannot be reached. They are in New York in connection with the matter. I understand that there will be a meeting of the directors of the company to-day, after which there might be a statement," said Mr. McKellar.

That the firm would resume operations within a week was the statement made by officials of the Shipbuilders' Union who sought some explanation of the layoff when the men left the yards on Friday and Saturday.

READY TO START WORK AFTER FIRE

St. Thomas Plant Badly Scorched, But Preparations Made to Get to Work

St. Thomas was visited by one of the worst fires in recent history, early Thursday morning of last week, when the large two-storey brick structure used as a pattern storage department, pattern room and store room by the Canada Iron Foundries, Limited, was completely de-

stroyed, as well as a large portion of the roof of the casting foundry and the material sheds and shipping room of the Dominion Brake Shoe Company. The total damage was roughly estimated by Stanley Gilbert, manager of the Canada Iron Foundry plant, at $100,000, partly covered by insurance. It is probable that this figure may be exceeded when a complete inventory of the contents of the buildings is available. Thousands of dollars worth of valuable patterns and material was destroyed.

Although the loss of virtually all their patterns, save those that were being used in the casting room, will seriously inconvenience the plant, Manager Gilbert stated that it was his policy to carry on the foundry as usual. In fact, even while the buildings were still burning, Mr. Gilbert had the Hydro-Electric workmen on the job repairing the power wires in order to enable the casting department to operate. The majority of the men in that department reported for duty at seven o'clock, and were soon engaged at their regular duties.

Sizes Up the Real Canadian Market

British Trade Commissioner in Ontario Points Out Chances for Trade Here—Japan Merchants Bidding for a Share of the Trade— Advises that More Advertising Should Be Done.

THE following are extracts from the reports on the trade of Canada and of Ontario by F. W. Field, British Trade Commissioner in Ontario, just issued as a British Government Blue Book:

Prominent Part For Canada

A review of the year which has elapsed since the signing of the armistice, suggests that economic, industrial and commercial conditions are sound, and that given more stable political and social conditions throughout the world than those obtaining at the beginning of 1920, the Dominion should achieve a degree of development far surpassing that achieved hitherto. Canada has gained a larger sense of nationhood through the war, and this feeling will become more pronounced as the time passes and conceivably find expression in the economic sphere. Her achievements in finances, in production of every kind, as well as those of her expeditionary forces, have engendered feelings of the kind mentioned. In common with her allies, she is confronted with grave problems, but is dealing with them with characteristic energy and hopefulness. With her vast resources she has reason to hope that in the difficult times ahead, she will create sufficient wealth to liquidate her heavy burden of debt and to play a more prominent part in world economy.

Market For British Manufactures

As a market for British manufactures Ontario presents an attractive field, more especially in the lines in which United Kingdom manufacturers have excelled for so long. The development of Canadian manufacturing, of which such a large proportion is carried on in Ontario, leads one to the conclusion that the greater part of British trade in this market will in future years be confined to the principal groups of manufactures which constituted the business here of United Kingdom firms before the war.

Japanese Activities

The activities of the Japanese continued during the past year and every effort is being made to obtain business in the Canadian market. A large range of samples of goods sold here have been obtained by the representatives of Japanese manufacturers and agreements made to duplicate them at a much lower price than quoted by their competitors.

A number of factors are operating against the acquisition by Japan of a large volume of trade in this area, including such matters as terms and methods of doing business, deliveries, quality, etc., but it would be a mistake on the part of United Kingdom manufacturers to ignore the growing strength of this competition. There are those who believe that this was done in some instances in past years when Germany commenced to gather information and samples in various markets, and to obtain a foothold in trades in which severe competition had not generally been anticipated.

Agencies in United States

It is still a matter for the consideration of certain United Kingdom firms as to whether their agency arrangements for Ontario, and Canada generally, should be treated as part of their United States agency arrangements. At least one British firm with branch works in the United States has now established a branch in Ontario. The tendency of United States firms with trade connections in Canada is to establish separate units in Canada, either branch works, offices, or agents, to deal with their Canadian trade.

Machinery Sales and Service

There appears to be an opportunity for increased sales of machinery of various kinds in this market, including air compressors, hoisting machinery, mining equipment, textile machinery, and various other lines in which United Kingdom firms have proved specialists. Three of the factors operating against greater business are the lack of sufficient local representation and active salesmanship, the failure of United Kingdom manufacturers to keep in constant touch with those who have purchased their machines, and the inability to effect repairs and supply spare parts at short notice. While the business in any one line may not be large enough to justify the establishment of a proper local organization, which should remove the difficulties noted, it might prove feasible for several manufacturers of non-competing machinery to have joint representation here, with a technical man to look after repairs; a good sales organization, which would adopt the American "follow up" system, as it is termed; and a stock of the spare parts which are most likely to be needed. Unless some such system is adopted, it is to be feared that the imports of United Kingdom machinery to this market will become continually smaller.

Advertising

There is room for much more advertising by the majority of United Kingdom firms doing business here. The local agent among others should be consulted as to the best methods of expending the advertising appropriation.

From War to Peace

The transition of industrial works from a war to a peace basis is being effected gradually and with varying results. Foundries and machine shops in some instances appear to have had difficulty in securing sufficient business to occupy their capacity since the armistice. Many works were enlarged to cope with the demand for munitions, and the increased capacity has not always been utilized, partly on account of labor troubles, dull market conditions in certain lines, and the sale of second-hand machinery previously used in munition shops. Labor troubles, market conditions, and the organization necessary for the establishment of export trade have retarded the manufacturing progress of industrial firms. On the whole, however, few adverse effects were apparent during the year as a result of post-war conditions.

A note of the actual position of a few typical industries in Ontario, which devoted their energies to war work, will be of interest:—

(a) Firm of founders, crane makers and machinists:—Sufficient business has not been obtained to occupy full capacity. The firm is now in a position to handle the lines previously made and one or two new ones, including machine tools.

(b) Firm making drop forgings and drop forged tools. The capacity of the works was doubled during the war. The company is operating to full capacity and is making several new lines.

(c) Firm making marine and heavy forgings. The works were enlarged by about 500 per cent. during the war, the product being mostly forged steel shell cases. The shell plant has been scrapped and is standing idle. The production of marine forgings and locomotive parts is being continued, but owing to a decline in the demand for these lines, the works are operating under capacity.

(d) Firm making small machinery, etc. The entire works were devoted to war work. Its plans for the future, after the armistice, were laid aside on account of the unrest in the ranks of the metal workers, causing a disorganization which in November, 1919, was still apparent.

(e) Firm making locomotives. This firm continued its regular lines during the war, making also shell forgings and finished shells. The forging and machining plant was sold, the buildings remodelled as blacksmith and tank shops for the company's regular work.

(f) Firm making motor cars. During the war, the company's works were applied to the manufacture of aeroplane

engines, 200 men being employed. The company is now making motor cars, its plans calling for the production of 15,000 cars during the twelve months ending October, 1920; 1,000 men are employed.

(g) Firm making shrapnel tubes during the war. The company is now making bushings for motor cars and is doing a domestic and export business to the United States.

(h) Firm making shells during the war. The company is now making motors for phonographs and lighting fixtures.

(i) Firm making shells during the war. The company is now engaged in the manufacture of agricultural tractors and is seeking export, as well as domestic business.

(j) Firm making machinery and shells during the war. The company's works were purchased by a United States firm for the manufacture of tungsten products.

(k) Machine and stamping company. After the completion of munition contracts, this concern commenced the manufacture of several lines, one department being devoted exclusively to the making of gears and transmission cases for motor cars. In the near future, beveled gears and differential cases for a similar purpose will be made. A heat treating plant to harden and heat gears has been established. The company is also making phonograph motors and gramophone cabinet hardware, such as hinges, lid supports, needle cups, etc. Other departments are making spark plugs for motor cars, and bicycle chains. In addition, the company is doing a commercial business in small stampings and automatic screw machine products. All these lines have been taken up and developed since the armistice.

Production Costs

The production costs in Canadian factories are generally reputed to be higher than those in the United States, although there are notable exceptions. In an Ontario factory making certain small brass goods, the cost is estimated at from 30 to 40 per cent. more than in the parent works in the United States.

The progress of manufacturing in Canada has been exemplified by the number of lines now made here which were previously imported, and the number of lines already made here, but the manufacture of which has been commenced by additional firms.

Among the new lines added by Ontario manufacturing firms in 1919 were the following: Standard structural shapes up to and including 15-in. beams and channels; car axles, glove silk fabric, rough woollen cloths, worsted yarn spun by French process, pumps, high speed planer (in experimental stage), mine forgings and drop forged tools, such as wrenches, screw drivers, etc., axes, hammers, carpenters' tools, school seats, lawn mowers, bushings for motor cars, pulp and paper machinery, water-proof paper, spark plugs, bicycle chains,

gramophone and piano hardware, safety razor parts, motor car axles, motor car transmission gears, differentials and radiators, heavy chains, knitted silk cloth for glove manufacturers, chamoisette cloth, collapsible tubes, steel wire rope, soda ash, forging bars, oxygen for industrial purposes, and heavy steel castings made by electric furnaces.

Several new lines of knitted fabric, such as jersey cloth and fabric, were produced during the year by an Ontario woollen mill. They are particularly adapted for ladies' dresses, suits and skirts in the jersey cloth and for men's light overcoats and golf suits in the fabric.

An Ontario company, which had confined its production to Kraft, pulp and Kraft building paper, commenced the manufacture of Kraft wrapping paper early in 1919.

United States Branch Factories

The past year was notable for the number of United States companies making inquiries in this area with regard to the proposed establishment of branch works.

The Industrial Commission of an Ontario city, in close proximity to the United States, was in correspondence as to this matter in November, 1919, with over a hundred United States firms. In many cases a decision was made to erect such works, and construction was commenced.

The character of these enterprises varies from small assembling plants to large factories employing several thousand hands. This action has been taken for the following reasons among others:

(a) To handle the Canadian market at close range.

(b) To grow, as a Canadian industry, with the expanding market and increasing purchasing power.

(c) To enjoy, as a Canadian industry, the benefits of any special trade arrangements made for the units of the British Empire.

(d) To handle the export orders, received by the parent company in the United States, for shipment to the United Kingdom and other points in the Empire.

In addition, several United States concerns with small works in Ontario, have been seeking sites for the establishment of larger factories.

United Kingdom Factories in Ontario

The inquiries of United Kingdom manufacturers as to the establishment of branch works in Ontario were more numerous in 1919 than ever before. Works were established for the production of elastic hosiery and abdominal belts, felts for paper-making machines, and silk labels.

Alterations and additions were commenced on large branch works of a Welsh tinplate mill, which acquired one of the national manufacturing, works at Toronto of the Imperial Munitions Board.

Representatives of many United Kingdom manufacturers visited Ontario to investigate conditions for local factories. While several branch works of United Kingdom firms may be erected in 1920, the number of such plants will probably never equal that of United States firms with branches here.

Assistance to Ontario Firms

The British Trade Commissioner at Toronto has been of assistance to Canadian importers and manufacturers in various ways. Lists of United Kingdom manufacturers, importers and exporters have been furnished locally, introductions to the Department in London handed to business men proceeding overseas, inquiries of Ontario concerns for quotations of United Kingdom firms forwarded and information as to the establishment of Canadian branch works in the United Kingdom. Ontario firms have made use of the office as a general source of information as to British trade matters. A number of letters received from His Majesty's Consular officers in various parts of the world have been transmitted to local firms making inquiries abroad. Acknowledging the receipt of such a letter, a Toronto firm wrote:—"The information supplied is of considerable assistance to us and it will avoid us incurring expenses in that territory which would eventually prove unprofitable."

On account of war conditions, more direct trading has developed during the past few years. Large Canadian imports of tea, for example, have come to Vancouver from the tea plantations rather than through London. Canadian buyers have established direct relations with hide and skin suppliers in Australia, the market previously being centered in London. While the United Kingdom, for many reasons, will continue to be an important centre for re-export trade, the changes which have occurred in the position during the war will undoubtedly leave their mark.

Acknowledgments

It is gratifying to be able to record the continued assistance and support which have been given throughout Ontario to the work of the British Trade Commissioner's office at Toronto. It is not possible to mention, in the space available here, the names of all those who have helped, but among others this expression of thanks applies to the Dominion and Ontario Government authorities, other public bodies, Canadian manufacturers and their associations and officials, the Boards of Trade and their secretaries, industrial commissioners, bankers, the railway companies, many local trade organizations, including the Toronto branch of the Canadian Association of British Manufacturers and the daily press and the trade journals. Their assistance has been greatly appreciated by this office and has given a much more substantial value to the work than otherwise would have been the case.

SELECTED MARKET QUOTATIONS

Being a record of prices current on raw and finished material entering into the manufacture of mechanical and general engineering products.

PIG IRON

Grey forge, Pittsburgh	$42 40
Lake Superior, charcoal, Chicago	57 00
Standard low phos., Philadelphia	50 00
Bessemer, Pittsburgh	43 00
Basic, Valley furnace	42 90

Toronto price:—
Silicon, 2.25% to 2.75%	52 00
No. 2 Foundry, 1.75 to 2.25%	50 00

IRON AND STEEL

Per lb. to Large Buyers — Cents
Iron bars, base, Toronto	$ 5 50
Steel bars, base, Toronto	5 50
Iron bars, base, Montreal	5 50
Steel bars, base, Montreal	5 50
Reinforcing bars, base	5 00
Steel hoops	7 00
Norway iron	11 00
Tire steel	6 75
Spring steel	10 00

Band steel, No. 10 gauge and 3-16 in. base	6 00
Chequered floor plate, 3-16 in.	8 40
Chequered floor plate, ¼ in.	8 00
Bessemer rails, heavy, at mill	....
Steel bars, Pittsburgh	3 60-4 00
Tank plates, Pittsburgh	4 00
Structural shapes, Pittsburgh	3 00
Steel hoops, Pittsburgh	3 50-3 75

F.O.B., Toronto Warehouse
Small shapes	4 25

F.O.B. Chicago Warehouse
Steel bars	3 62
Structural shapes	3 72
Plates	3 67 to 5 50
Small shapes under 3"	3 62

FREIGHT RATES

Pittsburgh to Following Points — Per 100 Pounds.
Montreal	33 45
St. John, N.B.	41½ 55
Halifax	49 64½
Toronto	27 39
Guelph	27 39
London	27 39
Windsor	27 39
Winnipeg	89½ 135

METALS

Prices per 100 lbs.
	Gross. Montreal	Toronto
Lake copper	$25 00	$24 00
Electro copper	24 50	24 00
Castings, copper	24 00	24 00
Tin	62 00	65 00
Spelter	11 50	12 00
Lead	10 75	11 00
Antimony	13 00	14 00
Aluminum	34 00	36 00

PLATES

Plates, 3-16 in.	$ 7 25	$ 7 25
Plates, ¼ up	6 50	6 50

PIPE—WROUGHT
Price List No. 44—April, 1926.
STANDARD BUTTWELD S/C

	Steel Black	Galv.	Gen. Wrot. Iron Black	Galv.
¼ in.	$6 60	$8 60		
⅜ in.	6 12	7 24	$ 6 48	$ 7 66
½ in.	6 12	7 26	6 48	7 66
¾ in.	6 84	8 42	7 27	8 94
1 in.	9 48	10 68	9 95	11 16
1 in.	13 50	16 44	13 50	16 49

	Steel	Gen. Wrot. Iron
1¼ in.	18 91 21 16	18 06 22 81
1½ in.	19 21 25 69	21 59 26 48
2 in.	27 20 34 04	29 05 35 89
2½ in.	42 00 52 52	....
3 in.	56 55 70 36	....
3½ in.	71 50 88 32	....
4 in.	84 48 104 54	....

STANDARD LAPWELD S/C

	Steel Black	G r.	Black	Galv.
2 in.	$30 90	$37 74	$34 60	$41 44
2½ in.	46 84	56 16	51 19	62 01
3 in.	59 29	73 44	64 94	81 09
3½ in.	73 14	90 36	83 54	99 36
4 in.	86 65	106 25	97 36	117 70
4½ in.	9 98	1 23	1 24	1 49
5 in.	1 15	1 44	1 46	1 78
6 in.	1 49	1 86	1 97	2 26
7 in.	1 94	2 42	2 42	2 90
8-L in.	2 04	2 56	2 54	3 05
8 in.	2 35	2 94	2 92	3 51
9 in.	2 81	3 52	3 50	4 21
10-L in.	2 61	3 25	3 25	3 90
10 in.	3 68	1 18	1 18	5 68

Prices—Ontario, Quebec and Maritime Provinces

WROUGHT NIPPLES

4" and under, 60%.
4½" and larger, 50%.
4" and under, running thread, 30%.
Standard couplings, 4-in. and under, 30%.
Do., 4½-in. and larger, 10%.

OLD MATERIAL

Dealers' Average Buying Prices.
	Per 100 Pounds. Montreal	Toronto
Copper, light	$15 00	$14 00
Copper, crucible	18 00	18 00
Copper, heavy	18 00	18 00
Copper wire	18 00	18 00
No. 1 machine composition	16 00	17 00
New brass cuttings	11 00	11 75
Red brass cuttings	14 00	15 75
Yellow brass turnings	8 50	9 50
Light brass	7 00	7 00
Medium brass	8 00	7 75
Scrap zinc	6 50	6 00
Heavy lead	7 50	7 75
Tea lead	4 50	5 00
Aluminum	19 00	20 00

	Per Ton Gross
Heavy melting steel	18 00 18 00
Boiler plate	15 50 15 00
Axles (wrought iron)	20 00 20 00
Rails (scrap)	18 00 18 00
Malleable scrap	23 00 25 00
No. 1 machine cast iron	32 00 33 00
Pipe, wrought	12 00 12 00
Car wheel	26 00 26 00
Steel axles	22 00 20 00
Mach. shop turnings	11 00 11 00
Stove plate	26 50 25 00
Cast boring	12 00 12 00

BOLTS, NUTS AND SCREWS

	Per Cent
Carriage bolts, 7-16 and up	+10
Carriage bolts, ⅜-in. and less	Net
Coach and lag screws	—15
Stove bolts	55
Wrought washers	+10
Elevator bolts	+10
Machine bolts, 7-16 and over	+10
Machine bolts, ⅜-in. and less	+10
Blank bolts	Net
Bolt ends	Net
Machine screws, fl. and rd. hd., steel	27½

Machine screws, o. and fil. hd., steel	+25
Machine screws, fl. and rd. hd., brass	net
Machine screws, o. and fil. hd., brass	net
Nuts, square, blank	+25 add $2 00
Nuts, square, tapped	add 2 25
Nuts, hex., blank	add 2 50
Nuts, hex., tapped	add 3 00
Copper rivets and burrs, list less	15
Burrs only, list plus	25
Iron rivets and burrs	40 and 5
Boiler rivets, base ⅝" and larger	38 50
Structural rivets, as above	8 40
Wood screws, O. & R., bright	75
Wood screws, flat, bright	77½
Wood screws, flat, brass	55
Wood screws, O. & R., brass	55½
Wood screws, flat, bronze	50
Wood screws, O. & R., bronze	47½

MILLED PRODUCTS

(Prices on unbroken packages)
	Per Cent
Set screws	—20¢, 25 and 5
Sq. and hex. hd. cap screws	12½
Rd. and fil. hd. cap screws	plus 25
Flat but. hd. cap screws	plus 50
Fin. and semi-fin. nuts up to ⅝-in.	12½
Fin. and Semi-fin. nuts over 1 in., up to 1⅛-in.	—5
Fin. and Semi-fin. nuts over 1⅛ in., up to 2-in.	+12½
Studs	+5
Taper pins	—12½
Coupling bolts	+40
Planer head bolts, without fillet, list	+45
Planer head bolts, with fillet, list plus 10 and	+55
Planer head bolt nuts, same as finished nuts.	
Planer bolt washers	net
Hollow set screws	+50
Collar screws	list plus 20, 30
Thumb screws	40
Thumb nuts	75
Patch bolts	add +55
Cold pressed nuts to 1⅜ in.	add $1 00
Cold pressed nuts over 1⅜ in.	add 2 00

BILLETS

	Per gross ton
Bessemer billets	$50 00
Open-hearth billets	60 00
O.H. sheet bars	76 00
Forging billets	56 00-75 00
Wire rods	52 00-70 00

Government prices.
F.O.B. Pittsburgh.

NAILS AND SPIKES

Wire nails, base	$6 50
Cut nails, base	7 00
Miscellaneous wire nails	50¢.

ROPE AND PACKINGS

	Per lb.
Plumbers' oakum, per lb.	0 10¾
Packing, square braided	0 38
Packing, No. 1 Italian	0 44
Packing, No. 2 Italian	0 36
Pure Manila rope	0 35½
British Manila rope	0 28
New Zealand hemp	0 28

POLISHED DRILL ROD

Discount off list, Montreal and Toronto ..... net

MISCELLANEOUS

Solder, strictly	$ 0 35
Solder, guaranteed	0 39
Soldering coppers, lb.	0 62½
White lead, pure, cwt.	20 35
Red dry lead, 100-lb. kegs, per cwt.	16 00
Gasoline, per gal., bulk	0 38
Pure turp., single bbls., gal.	3 15
Linseed oil, raw, single bbls.	2 37
Linseed oil, boiled, single bbls.	2 40
Wood alcohol, per gal.	4 00
Whiting, plain, per 100 lbs.	3 00

CARBON DRILLS AND REAMERS

S.S. drills, wire sizes	32½
Can. carbon cutters, plus	20
Standard drills, all sizes	32½
3-fluted drills, plus	10
Jobbers' and letter sizes	32½
Bit stock	40
Ratchet drills	15
S.S. drills for wood	40
Wood boring brace drills	25
Electricians' bits	30
Sockets	50
Sleeves	50
Taper pin reamers	25 off
Drills and countersinks	net
Bridge reamers	50
Centre reamers	10
Chucking reamers	net
Hand reamers	10
High speed drills, list plus 10 to 30	
Can. high speed cutters, net to plus 10	
American	plus 40

COLD ROLLED STEEL

[At warehouse]

Rounds and squares	$7 base
Hexagons and flats	$7.75 base

IRON PIPE FITTINGS

	Black	Galv.
Class A	60	75
Class B	27	37
Class C	18	27

Cast iron fittings, 5%; malleable bushings, 22½%; cast bushings, 22½%; unions, 37½%; plugs, 20% off list.

SHEETS

	Montreal	Toronto
Sheets, black, No. 28	$8 50	$10 50
Sheets, Blue ann., No. 10	8 50	9 50
Canada plates, dull, 52 sheets	8 50	10 00
Can. plates, all bright	8 60	9 00
Apollo brand, 10¾ oz. galvanized	....	...
Queen's Head, 28 B.W.G.	11 00	...
Fleur-de-Lis, 28 B.W.G.	10 50	...
Gorbal's Best, No. 28	....	...
Colborne Crown, No. 28	....	....
Premier, No. 28, U.S.	11 50	11 60
Premier, 10¾-oz.	11 50	12 00
Zinc sheets	16 50	20 00

PROOF COIL CHAIN

(Warehouse Price)

B

¼ in., $13.00; 5-16, $11.00; ⅜ in., $10.00; 7-16 in., $9.80; ½ in., $9.75; ⅝ in., $9.20; ¾ in., $9.30; ⅝ in., $9.50; 1 in., $9.10; Extra for B.B. Chain, $1.20; Extra for B.B.B. Chain, $1.80.

ELECTRIC WELD COIL CHAIN B.B.

⅛ in., $16.75; 3-16 in., $15.40; ¼ in., $13.00; 5-16 in., $11.00; ⅜ in., $10.00; 7-16 in., $9.80; ½ in., $9.75; ⅝ in., $9.50; ¾ in., $9.30.

Prices per 100 lbs.

FILES AND RASPS

	Per Cent.
Globe	50
Vulcan	50
P.H. and Imperial	50
Nicholson	32½
Black Diamond	27½
J. Barton Smith, Eagle	50
McClelland, Globe	50
Delta Files	20
Disston	40
Whitman & Barnes	50
Great Western-American	50
Kearney & Foot, Arcade	50

BOILER TUBES.

Size.	Seamless	Lapwelded
1 in.	$27 00	$.....
1¼ in.	29 50	.....
1½ in.	31 50	29 50
1¾ in.	31 50	30 00
2 in.	35 00	30 00
2¼ in.	35 00	29 00
2½ in.	42 00	37 00
3 in.	50 00	48 00
3¼ in.	.. ..	48 50
3½ in.	63 00	51 50
4 in.	85 00	65 50

Prices per 100 ft., Montreal and Toronto

OILS AND COMPOUNDS.

Castor oil, per lb.	
Royalite, per gal., bulk	27½
Palacine	30½
Machine oil, per gal.	54½
Black oil, per gal.	28
Cylinder oil, Capital	82½
Petroleum fuel oil, bbls., net	18

BELTING—No 1 OAK TANNED

Extra heavy, single and double		6½
Standard		6½
Cut leather lacing, No. 1		2 00
Leather in side	2 40	3 00

TAPES

Chesterman Metallic, 50 ft.	$2 00
Lufkin Metallic, 603, 50 ft.	2 00
Admiral Steel Tape, 50 ft.	2 75
Admiral Steel Tape, 100 ft.	4 45
Major Jun. Steel Tape, 50 ft.	3 50
Rival Steel Tape, 50 ft.	2 75
Rival Steel Tape, 100 ft.	4 45
Reliable Jun. Steel Tape, 50 ft.	3 50

PLATING SUPPLIES

Polishing wheels, felt	$4 60
Polishing wheels, bull-neck	2 00
Emery in kegs, Turkish	09
Pumice, ground	06
Emery glue	30
Tripoli composition	09
Crocus composition	12
Emery composition	11
Rouge, silver	60
Rouge, powder, nickel	45

Prices per lb.

ARTIFICIAL CORUNDUM

Grits, 6 to 70 inclusive	.08½
Grits, 80 and finer	.8

BRASS—Warehouse Price

Brass rods, base ½ in. to 1 in. rod	0 34
Brass sheets, 24 gauge and heavier, base	$0 42
Brass tubing, seamless	0 46
Copper tubing, seamless	0 48

WASTE

XXX Extra	.24	Atlas	.20
Peerless	.22½	X Empire	.19½
Grand	.22½	Ideal	.19
Superior	.22½	X Press	.17½
X L C R	.21		

Colored

Lion	.17	Popular	.13
Standard	.15	Keen	.11
No. 1	.15		

Wool Packing

Arrow	.35	Anvil	.22
Axle	.28	Anchor	.17

Washed Wipers

Select White	.20	Dark colored	.09
Mixed colored	.10		

This list subject to trade discount for quantity.

RUBBER BELTING

Standard ... 10% Best grades... 15%

ANODES

Nickel	.55 to .60
Copper	.38 to .40
Tin	.70 to .70
Zinc	.16 to .17

Prices per lb.

COPPER PRODUCTS

	Montreal	Toronto
Bars, ½ to 2 in.	$42 50	$43 00
Copper wire, list plus 10.		
Plain sheets, 14 oz., 14x60 in.	46 00	44 00
Copper sheet, tinned, 14x60, 14 oz.	48 00	48 00
Copper sheet, planished, 16 oz. base	46 00	45 00
Braziers', in sheets, 6 x 4 base	45 00	44 00

LEAD SHEETS

	Montreal	Toronto
Sheets, 3 lbs. sq. ft.	$10 75	$14 50
Sheets, 3½ lbs. sq. ft.	10 50	14 00
Sheets, 4 to 6 lbs. sq. ft.	10 25	13 50

Cut sheets, ½c per lb. extra.
Cut sheets to size, 1c per lb. extra.

PLATING CHEMICALS

Acid, boracic	$ .23
Acid, hydrochloric	.04
Acid, nitric	.11
Acid, sulphuric	.04
Ammonia, aqua	.15
Ammonium, carbonate	.23
Ammonium, chloride	.22
Ammonium hydrosulphuret	.75
Ammonium sulphate	.30
Arsenic, white	.14
Copper, carbonate, annhy.	.41
Copper, sulphate	.15
Cobalt, sulphate	.20
Iron perchloride	.62
Lead acetate	.30
Nickel ammonium sulphate	.20
Nickel carbonate	.32
Nickel sulphate	.22
Potassium sulphide (substitute)	.42
Silver chloride (per oz.)	1.30
Silver nitrate (per oz.)	1.25
Sodium bisulphate	.14
Sodium carbonate crystals	.06
Sodium cyanide, 127-130%	.38
Sodium hyposulphite per 100 lbs	8.00
Sodium phosphate	.15
Tin chloride	.80
Zinc chloride, C.P.	.30
Zinc sulphate	.10

Prices per lb. unless otherwise stated

HAVE SECURED NEW RULING ON SALES TAX ON RETURNABLE PACKAGES

The Canadian Manufacturers' Association Has Sent Out the Following Notice:—

The sales tax regulations applying to returnable packages, in its original form, created unnecessary labor and expense to 'manufacturers who ship goods to their customers in returnable packages. Under the regulation, manufacturers were compelled to collect and make monthly returns of a sales tax on the total invoice price including charge for the container, with the understanding that when such containers were returned, credit for the amount of the sales tax paid thereon might be taken.

The Association arranged for a deputation of manufacturers to visit Ottawa to lay before the Government the unnecessary labor and expense caused by this troublesome system of dealing with the tax.

After hearing and admitting the justice of the Association's case, the following ruling was approved and issued by the Department of Customs and Inland Revenue, viz:

Re Federal Sales Tax on Returnable Packages

"That manufacturers dealing in goods shipped in returnable packages may make to the Government, not later than the end of March in each year, an annual sales tax return as to such packages, instead of monthly returns; and the sales tax so payable to the Government by the manufacturer shall be paid on the difference between the amount charged for the returnable containers shipped during the year and the equivalent amount rebated for containers returned during the same period; and it shall be optional with the manufacturer to charge the sales tax on the value of returnable containers in invoices to the purchaser subject to credit when returned, or to pay such sales tax himself."

Yours faithfully,
(Sgd.) J. R. K. Bristol,
Manager Tariff Department.

A NEW FILLING METAL

There are frequent calls for a filling material to repair cracks and small holes in very light castings, which cannot well be welded. Various methods have been used for this purpose, but the latest is in the nature of a metal having a very low fusion point, and a comparatively high tensile strength. The new metal is known by the name "Metaloid." The composition of the metal is a trade secret, but in appearance it is about the color of solder, but is remarkably light in weight, being of less specific gravity than aluminum. In making a repair, the casting is heated to a temperature of 200 degrees Fahrenheit, and the bar of "Metaloid" is rubbed over the part to be repaired. It adheres closely to the casting, and makes a strong and satisfactory repair. The "Metaloid" requires no heating, and it also provides a guide to the correct temperature of the casting. If the latter is too hot, the "Metaloid" will give off a blue flame when applied to it. If too cold, the "Metaloid" will not run at all, while at the correct temperature it flows quite readily with no sign of burning. Holes in castings have been filled, and the casting afterwards subjected to 300 lbs. per square inch water pressure, the Metaloid showing no signs of leakage. The new metal has been largely used in the repairing of crank cases and radiators of automobiles, and should be of great use in the general engineering field. At the present time it is being tested out by the engineering department of a large Canadian sulphite mill with a view to determine its acid resisting qualities.

"Metaloid" is handled in Canada by Messrs. Kennedy & Kennedy, Tyrell Building, King Street E., Toronto.

HE HAS TRIED OUT THE WESTERN COAL

Orillia Man Has Some Shipped Down and Seems Satisfied With the Results Secured

Orillia.—W. W. McRain has this week received by express a small shipment of coal from his brother's mine near Edmonton. It is firm and bright, burns without making much smoke, does not dirty the pots, and leaves very little ash. It is announced that negotiations are now on foot with the railways to carry this coal at a rate that will enable it to be placed on the Ontario market at a reasonable price. The coal sells for $5 a ton at the mine, and with a $7 freight rate, could be retailed in Ontario for about $15 a ton. As many of the mines in Alberta are working only to about 25 per cent. of their capacity during the summer months, it would appear that they should be able to give material assistance in relieving the situation and preventing a coal famine. There is a mistaken idea that this coal, which is classified as semi-bituminous, will deteriorate if stored for any length of time. It can be kept for any length of time if placed under cover, in cellars or sheds. People will watch with much interest developments in the direction of getting a supply of coal for Ontario from the Western Provinces.

MOTOR-DRIVEN HAMMER

C. C. Bradley, Inc., of Syracuse, N.Y., are now making a motor-driven helve hammer, which is controlled in the usual manner by means of a foot treadle. The blow is regulated by the pressure of the loose running belt against the idler pulley.

PAPER DEARNESS
Effect on the Press

In view of the suggested further advance in wages and the very serious situation which must arise in that event, the British Weekly Newspaper and Periodical Proprietors' Association has issued a memorandum regarding the high cost of production in which it is pointed out that since 1914 proprietors have had to meet a continued succession of increases in cost of production and distribution; increases which, since the armistice, have been intensified. The conditions under which they are at present laboring include:

1.—High price of paper, blocks, contributions, and illustrations.
2.—Higher salaries and wages.
3.—Altered conditions as to working hours, overtime, holidays, and so forth.
4.—Higher railway and other transport rates.

Before the war newsprint (paper for newspapers) cost 2c per lb., whereas it is now 12½c per lb. and over; and there have already been eleven separate advances in printing charges, which have thereby increased by 120 per cent. to 200 per cent.

For a 32-page weekly paper, with a circulation of 150,000, this means an increase in the cost of printing alone of between $25,000 and $30,000 per annum.

A publication using 20 tons of paper per week was spending in the purchase of paper before the war $880 per week—just under $50,000 per annum. The present cost of the supply for one week's issue for a paper of the same size at 12½c per lb. is $5,780, an advance of $4,850 per week, or $252,200 per annum.

Attempts have been made to cope with these increases by reducing the size of publications, raising the sale price, and by considerably increasing advertisement rates. These efforts have, however, been quite inadequate to meet the increased charges, and many of the publications are, in fact, now running at a substantial loss.

These conditions necessarily affect a very large number of smaller firms to a much greater extent than they do the big corporations, and it must be realized that in the aggregate these smaller firms employ far more labor than the larger concerns, so that when, as is inevitable if conditions do not improve, many of the smaller papers drop out as their reserves are exhausted, unemployment must ensue.

The Trans-Pacific tells of a most peculiar strike. It comments as follows:

About 4,000 employees of the great Tata mills in India went on strike, demanding a special bonus because of an announcement that Sir Doradji Tata, head of the concern, had been presented with a son. It appears that the announcement was incorrect, but the workers refused to believe it, and decided to continue the strike and hold out for the bonus, son or no son.

INDUSTRIAL NEWS

NEW SHOPS, TENDERS AND CONTRACTS
PERSONAL AND TRADE NOTES

TRADE GOSSIP

Want Equipment.—A western firm is in the market for spring making machinery. In the list of requirements the following equipment is mentioned—rolls, for rolling automobile springs, shears for cutting same, drill presses, etc. We shall be glad to supply the name and address to firms in a position to supply this equipment if they will communicate with the Editor.

May Close Shops.—The Gananoque manufacturers have replied to the recent letter from Gananoque Lodge, No. 4, of the Amalgamated Association of Iron, Steel and Tin Workers of North America, requesting that they meet the executive committee of the Union for the consideration of the question of an 8-hour day. In their reply they state that if the men persist in their demand the factories of the town will all have to close down.

Better in Hamilton.—Except for the steel mills in Hamilton the coal shortage is causing no great worry there. The majority of the plants are able to get along during the summer without coal, using power to run the plants, says a Hamilton report. And as for the steel plant, officials there state that at present there is no probability of closing through lack of coal. The other plants, using the little they do, report that they are quite

prepared for a shortage, and, unless the shortage happens in the winter, will not be seriously affected by it.

Brantford Needs Power.—There is considerable worry on the part of Brantford officials owing to the inability to secure Hydro power. It is pointed out that several new industries have been landed for Brantford, and are nearly ready for production, but the Hydro Commission will accept no more new customers, and power cannot be got. Ald. F. C. Harp, chairman of the Fire and Light Committee, declared that a special meeting of the City Council would be called shortly, and a fight would be initiated to secure a supply of power even if other interests had to suffer for a time.

Making Extensions.—The L. & P. Mfg. Co., Niagara Falls, Ont. are extending their enterprise by the addition to their foundry of a machine shop completely equipped for the production of standard and special machinery. An outstanding feature of their production will be the Blystone line of concrete machinery and foundry equipment, on which they have placing on the market, and have ready for delivery a metal plower, 48 in. x 48 in. x 14. equipped for direct drive or direct motor connection. In addition to these they propose manufacturing a line of

small tools for garage and machine shop use.

To Meet in St. Thomas.—St. Thomas delegates to the Pere Marquette system federation convention held at Grand Rapids, Mich., have returned to St. Thomas, bringing with them the news that the next convention will be held in St. Thomas, the second week in July of next year. The delegates to these conventions number about 40 and represent 2,000 men employed in the allied crafts, including the metal trades and the carmen on the entire P. M. system. Working conditions, safety plans and all movements for the betterment of the men and the advancement of the company's interests are discussed at these gatherings. The delegates from St. Thomas were: H. Dukes, blacksmiths; George McKenzie, electricians; John King, boiler makers; A. Whalls and A. MacDougall, machinists, and Charles Spitler, carmen.

That Collingwood Plant.—The Collingwod Steel plant has caused a lot of worry to Collingwood town council and to many of the citizens as well. It was lately reported that the plant was to be scrapped and on Monday last when it was found that a couple of freight cars had been put on the siding at the plant, it looked as if this were so. The town immediately stepped in and attached the plant for taxes, $64,000, is the amount alleged to be due the town. The seizure was made and thus the movement of the machinery and equipment was staved off for a few days at least. The mayor went to Toronto to consult with the owners, and the result of the interview is not yet known. The plant is a big one and until a few months ago it was owned by Wm. Kennedy and Sons, of Owen Sound. The plant was in commission during the war and turned out a very large amount of steel for the making of munitions.

COMPANIES MUST GIVE ALL THE INFORMATION ASKED BY THE BOARD

Peterboro.—Judge Gunn. who presided as chairman of the Conciliation Board to inquire into the differences between the C.G.E. and their striking employees, is serving in the same capacity on a board nominated by the Department of Labor to adjust differences between Ottawa employers and their carpenters.

It appears that several of the mill-owners have failed to appear and give evidence required by the Conciliation Board, and in this connection Judge Gunn has made an important ruling that they can be compelled to do so. And His Honor gives these mill owners in question till to-day to appear before the Board; otherwise, he declared, they and every member of their staffs are liable to be subpoenaed and forced to give the required information.

"We are a tribunal appointed by the

Minister of Labor to investigate this dispute—and there seems to be a lot to investigate. We have the power to get the information," said Judge Gunn. He further stated: "All well-regulated bodies submit to such a board by sending representatives. If those interested do not do so in this case we will subpoena the whole staffs and members of the firms if they do not come at our invitation."

Full Investigation

"This investigation," continued His Honor, "is a public one, in the interests of the public, at the request of our city council; and we want full information. Four hundred and fifty workmen, who have been in the service of these companies for years, walking the streets, is not a refreshing state of affairs in a community of this size."

PERSONALS

Mr. Darrow, who has been connected with the Wilt Twist Drill Co., for some years, has been appointed assistant superintendent of the plant at Walkerville, in charge of production. Mr. Darrow, during the past few months has spent considerable time in Toronto territory and several other Canadian districts, making a study of plant problems, particularly as they had to do with drills, reamers, cutters, etc.

" Ill have no more guessing--Make Morrow Drills standard throughout the Plant --then we **Know** *we're right."*

JOHN MORROW
Screw & Nut Co.
Limited
INGERSOLL   CANADA

If interested, tear out this page and keep with letters to be answered.

CRUCIBLE STEEL CO. LOCATE IN WELLAND

Atlas Crucible Steel Co. of Dunkirk, N.Y. Secure Interest In Canadian Business

Welland.—Mr. T. J. Dillon, president of the Dillon Crucible Alloys of this city, announced that the Atlas Crucible Steel Company of Dunkirk, N. Y., has acquired a substantial interest in the Welland business, which will now be known as the Canadian Atlas Crucible Steel Company.

The Canadian Atlas Crucible Steel Company, Limited, has an authorized capital of 10,000 shares of 8 per cent. cumulative, preferred stock, par value $100, and 10,000 shares common stock, no par value.

Arthur H. Hunter, president of the American company, will become chairman of the Board of Directors, and T. J. Dillon will remain as an officer active in the management of the business.

The Atlas Crucible Steel Company, through Mr. Hunter, considered all the different commercial centres of the Dominion, and decided on Welland as the logical place to make a connection, owing to its close proximity to the sources of supply of raw materials and its unsurpassed facilities for the distribution of its finished products, both by rail and by water.

INCREASE OUTPUT IN NOVA SCOTIA

Montreal.—R. M. Wolvin, president of the Dominion Steel Corporation, who has returned to Montreal after a two weeks' inspection tour of the company's extensive coal and steel operations in Nova Scotia, says that the output of coal in the Nova Scotia mines is to be considerably increased. To secure this the sum of five and a half million dollars is to be expended in further develops and extending the properties of the company in that Province. This increased output, he believes, will do a great deal to meet the urgent needs of the country for coal.

Owing to prevailing high costs for material and labor, construction work at present will be much more expensive, but Mr. Wolvin stated that he believed that coal consumers of the country prefer a moderate and gradual increase in prices rather than face a serious coal shortage later, with its attendant abnormally higher prices.

Compared with current prices of coal at the mines both in England and the United States, Nova Scotia coal is selling at $1.00 to $2.00 lower per ton f.o.b. cars at the point of shipment, and consumers in Eastern Canada have been regularly supplied at the lower prices. While the manufacturer of Central and Western Canada had been paying from $15 to $18 per ton for coal, his competitor in Eastern Canada has been fully supplied at an average delivered price of $9 to $10 per ton, he stated.

WHAT IS THE PURPOSE OF A HEAT-TREATING PROCESS?

Continued from page 136

perature, as measured by a thermocouple within the furnace, exceeds certain predetermined limits. These contacts are operated with certainty by the small motor which adjusts the potentiometer and operates the recording pen. No current flows through the galvanometer needle, which serves only as a mechanical trigger to guide the adjustment of the potentiometer. The currents controlled by the switches are sufficient for the operation of a standard commercial relay panel, as illustrated on page 24 and shown in the diagram of connections opposite.

The automatic control mechanism holds the temperature of the furnace within a range of + 5° F. A chart supplies a continuous record of the temperature and therefore serves as a clock upon the correct operation of the automatic devices. The variations in temperature of the work are necessarily less than those of the furnace, as the energy is supplied to the heating element of the latter intermittently and diffuses gradually to the work. Once the work has been brought up to the drawing temperature, the current is on for short periods only, as for 30 seconds at intervals of 30 minutes approximately.

Due to the thorough insulation of the furnace and the fact that a continuous stream of air has not to be heated, the amount of energy consumed is very small, the current being on for perhaps less than two per cent. of the whole time. The furnace walls have a high heat-storage capacity which prevents rapid or extreme fluctuations of temperature. The winding of the resistor element is so proportioned that the temperature is uniform throughout the furnace.

For quantity production the work is usually placed upon holders for convenience in handling and for uniformity in the distribution of the charge in the furnaces. Overhead trolleys or telphers with hand or electric hoists can be arranged for lifting these holders in and out of the furnaces and baths. The various illustrations accompanying this article will show clearly other points of interest to this system of heat-treatment.

The latent power of the waterfalls in Iceland is estimated at 1,000 million horse-power. There is now a prospect, according to the "Economic Review," that these large resources may be utilized in the near future. The Icelandic Waterfall Commission is at present considering the granting of a concession to a Danish-Norwegian company for this purpose. It is intended to use the power for the production of nitrogen, and for an extensive scheme of electrification. Among other schemes is one for the construction of a 200-kilom. electric railway, running from Reykjavik, east and south through the agricultural district, which, if carried out, would be Iceland's first railway.

CANADIAN MACHINERY

AND
MANUFACTURING NEWS

Vol. XXIV. No. 7

August 12, 1920

Design and Application of Keyway Broaches

The Use of Keyseating Machines, and Their Broaches, Is Not as Well Known as It Should Be—The Data Contained in This Article Are Both Valuable and Authoritative

By G. Laidler

KEYWAY broaches are commonly of three types: one consists of a series of toothed blades drawn through the hub of the work and held in relative position by means of a locating centre; a second has a blade of rectangular section secured in its own stock or carrier; and a third type has an integral blade and stock, and is termed a solid broach.

None of these methods of broaching keyways, it will be understood, can produce the highest degree of accuracy, but for commercial work on large quantities of similar pieces, keyway broaching is often quite good enough.

Without discussing the comparative merits of the various methods mention-ed above, the writer proposes to describe the application of the second method, used by his firm to broach tapered single keyways in cast-iron hubs on horizontal keyseating machines of the Knowles and Lapointe make.

In these machines an adapter or arbor has a tapered end A, which fits into a socket on the machine. The outer end B of this arbor is parallel and is a sliding fit in the hub C to be keyseated. Through the arbor there is a hole D, which is concentric with the tapered portion, but eccentric with the parallel part. This hole has a slot E for the passage of the broach blade which "breaks through" the parallel end of the arbor as shown. The stock, F, of the broach, with its inserted blade G, or saw, slides in this slotted hole and is secured to the puller of the machine by a removable key which engages a slot H in the enlarged end.

The saw is of uniform height throughout. It has thirty-three teeth at ⅝-inch pitch, and is long enough to cut the deepest keyway at one stroke without unduly loading its teeth. By virtue of the inclination of the slot, in which it is held to the axis of the carrier, each tooth removes one thirty-third of the total depth of the keyway.

In operation the arbor is set so that the saw is on the underside to enable chips to fall clear; the work is placed on the parallel end, the broach is inserted

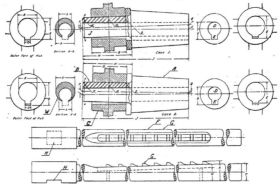

THESE VIEWS ILLUSTRATE DIFFERENT CONDITIONS, AND HOW THE DIFFERENT PARTS ARE PROPORTIONED.

and keyed up, and on starting the machine the saw is slowly pulled through. The machine stops automatically at the end of its stroke; the broach is disconnected and the work removed.

If the outer end of the arbor had its centre line coincident with that of the tapered portion, it is easy to see that the saw, in passing through, would produce a parallel keyway, that is, without any taper, and the calculations for the keyseating equipment for a given piece would be quite simple. But since most keyways are tapered and this taper is effected by setting the axis of the outer end of the arbor at an angle to that of the hole and tapered part, some careful figuring is necessary.

For turning this tilted end of the arbor a special offsetting fixture, Fig. 2, is used. It comprises a circular base, A, with ways planed across it at an angle B, corresponding to the required taper of the keyway, to receive a slide C, having a split boss D into which the tapered end of the adapter is clamped. This slide can be adjusted crosswise, and the amount of offset so regulated. For different tapers of keyway, the same slide is used in conjunction with bases planed to correspond to the required tapers.

In determining a standard method of calculation for several new sets of keyseating equipment, a number of arbors and broaches which had served for years was examined. Each particular piece to be keyseated requires its own arbor of the proper length, diameter and tilt, also its own broach with its inclined saw for the proper width and taper. It is evident that if the concentric hole D in the arbor be too large, the crescent-shaped section W of the hub end will be too weak, and conversely, if the hole be too small the broach will not be stiff enough to resist the bending moment at its outer end, J, as its saw enters the hub.

Taking the bore of the hub as a basic dimension, the mean ratio of broach diameter to arbor diameter was found on the existing sets to be as .68 to unity, and the mean ratio of thickness of the crescent at the centre of the hub to the arbor diameter as .28 to unity. These proportions were adopted in the table of formulae that was devised.

It will be observed that if the hub part of the arbor is tilted downwards at the outer end, the deep end of the keyway will be towards the machine, and the arbor will then be thickest there, where the greatest strength is required. This is the normal condition, termed Case I in the tables. A thick washer covers the strengthening fillet on the arbor and presents a true surface against which the preferably machined face of the hub is set. When the contour of an offset rim on the work prevents this condition by fouling the machine, the piece must be reversed and the deep end of the keyway is then outwards, giving a weaker pin, termed Case II.

As to the formulae, which explain themselves: commencing with the known data of the hub and the degree of taper required on the keyway, we decide to which case the work belongs, and then calculate the dimensions of the arbor and saw and tabulate them in the order given. The arbor is allowed to project ⅝ inch beyond the hub in order to support the outer end of the broach under its bending stress. Dimensions N and O are used as a check in turning the tilted portion of the arbor, whose centre line is offset R and Q at the ends. These latter dimensions will not be used if the offset fixture is applied.

In shop application the process of dimensioning equipment for a given hub and keyway is much simpler than the description would convey. The toolmaker is given the dimension sheet and the diagram of arbor and saw, the formulae being for office use only, and from these he reads off the figures to which he is to work. The results are uniform as the method of calculation eliminates errors that are liable to creep into unsystematic manufacture, for example when the toolmaker is expected to make an equipment "like the last one," which means in other word, "out of his head."

Dimension	FORMULAE FOR HORIZONTAL KEYSEATING EQUIPMENT	Standard Examples
	DESIGNATION OF VALUE	
A	Bore of Hub	.1.000 / .875
B	Actual Length of Hub on Sample.	2¼ / 2½
C	Width of Saw or Keyway	.375 / .375
D	(+) At Deep End.	1.137 / 1.096
E	(+) At Shallow End.	
F	Taper of Key per Inch.	
G	Case No. Deep End of Key Inwards = 1, Outwards = 2.	1 / 1
H	B + 1	2¼ / 2½
J	B + 2½	7¼ / 3¼
K	A − .005°	1.177 / 1.035
L	.68 K	.336 / .595
M	.68 K	.304 / .441
N		
O		
P		.123 / .147
Q		.243 / .266
R		1.181 / 1.091
T	L − .66	.534 / .574
V		.421 / .327
W		.313 / .330
R	W = −.651 + B = 0	1.143 / 1.143

TABLE SHOWING FORMULAE FOR THE VARIOUS PARTS.

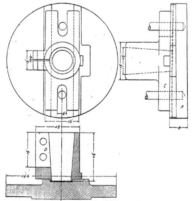

FIG. 2.—VIEW OF SPECIAL OFFSETTING FIXTURE.

Careful Tool Layout Means Added Production

You Can Offset the Benefit of Any First Class Machine Tool by Poor Tool Layout—A Combination of a High Class Tool, Together With Intelligent Tool Layout, is Essential

By J. H. Moore

THE average machinist is always on the qui-vive for something which will make him a still better workman. A medium of information that seldom fails is that of a photograph. Show him a photo of some new machine, a tool set up, or in fact anything in the machine tool line and he will remember it clearer and longer by this means than by mere written description.

Proceeding on this assumption we illustrate various views of set ups as taken on a Reed Prentice Automatic Lathe. The first two deal with the machining of pistons, and referring to Fig. 1 we note that this shows the rough turning operation.

The piston being manufactured is 3¾ in. diameter and the operations performed are rough turning, grooving, facing, and bevelling, all in the one set up. The photo is taken just as the cut is completed.

The outside of the piston is a rough casting when it comes to this machine, but the open end has been previously faced, and the inside bored where it fits on the spindle of lathe. The wrist pin hole is also finished, while the spindle is made with a draw in collet arrangement.

The piston is placed on the spindle nose, the draw in handle is pulled forward, and a false wrist pin is inserted through the piston with another draw in arrangement. The spindle draw in handle is next pulled back clamping the piston tightly to the nose. This brings us to the point of starting the lathe.

The first tools to cut are the roughing tools in the shaped tool block A shown in photo at Fig. 1. Directly following the roughing tool is a semi-finishing tool which is held in the block B.

When these tools have completed their cut the carriage comes to a positive stop. This occurs just at the point where the three ring grooving tools shown at C are in correct position. These tools are run in by hand and cut the grooves almost to size.

While these various operations have been progressing the back arm tools, that is, the bevelling and facing tools, have been forced into the work through the

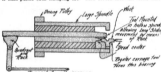

FIG. 4—ROUGH SKETCH SHOWING METHOD OF TURNING ONE SIDE OF JOINT.

medium of an inclined cam mounted on the back of the carriage. This cam transmits motion to the back arm tools, which are mounted on the arms supported by the large round bar at rear of machine. It might be well to add that no centres are used on this operation. We are able to give the actual production figure on the work, the time being 1 minute and 35 seconds floor to floor.

Finish Turning

The finish turning operation, Fig. 2, deserves close study owing to the ingenuity displayed. The piston is mounted on the spindle as previously explained. The carriage on this machine moves from left to right, and not from right to left as it did in Fig. 1. We explain this at this point in order to avoid confusion of direction.

The front tools now begin their cut. When the tool shown to the left starts work on the skirt of the piston, the two tools to the right start in on the ring grooves. These tools turn the head of the piston slightly smaller than the tool turning the skirt. This is conceded to be good piston practice.

While these turning operations have been proceeding the back arm tools have been fed into the work. As the photograph shows these consist of the ring grooving and relief grooving tools. The other remaining back tools are the bevelling tool, and the one used for facing. The closed end of the piston. As these can be clearly noted there is no need for comment.

A good view is given of the inclined cam which is attached to the carriage, and the method of performing the entire operation is apparent by a study of the photo. The time used up in this operation is 2 minutes 15 seconds.

FIG. 1—THE ROUGH TURNING OF THE PISTON.

The Third Example

The third and final example is shown at Fig. 3. This view illustrates a Reed Prentice lathe with some special features. The machine is set up to finish. The two bearing surfaces of a Ford universal joint at one operation. We, of course, know only too well the shape of this piece from driving our favorite fliver.

The very shape of the piece, four of which are shown in the photograph, some finished and some unfinished, causes one to wonder how on earth to drive the piece, yet perform the double operation. The task seems impossible, yet here is how it is accomplished.

A very large cast-iron spindle is bored hollow, and a chuck or grip if you prefer to call it such, is provided in this spindle as shown in photo. Inside is mounted a long tool post or guide, extending completely through the spindle. This guide has a rack cut at its extremity. A lever quadrant engages with this rack, and the lever in turn is connected to the ordinary carriage of the lathe.

This means that the work is held in two dead centres as shown on rough sketch at Fig. 4, and the method of transmitting motion to the tool on the guided slide is self apparent. By using this novel method of tooling the production figure has reached 940 pieces per day, a record to be proud of.

From these few examples, one can readily agree that it pays not only to

FIG. 3.—FINISHING THE TWO BEARING SURFACES OF UNIVERSAL JOINT AT ONE OPERATION.

study your equipment, but also the tooling you use and adopt the same.

HINTS TO USERS OF DIAMOND TOOLS

In handling these tools, says "Machine Tool Review," users must bear in mind that the diamond is a natural product and not a manufactured article. Whilst they will withstand considerable heavy pressure, a small jar or blow will often fracture the stone, rendering it unfit for further use.

Large wheels require large size stones, as the vibration is so great that small stones would soon split under the great pressure, or else work out of their holders.

Wheels should be frequently trued so, as to preserve an even or smooth surface. There is a limit to the amount of shock a rough diamond will stand from an uneven surface, and if a wheel is found to be badly out of truth, an ordinary type of Huntington or similar dresser should be used to bring the wheel to something like its proper condition before using the diamond tool.

Above all things, only take light cuts with the diamond, say about .002 at a time, as anything heavier than this will tend to burst the stone, owing to the great heat generated. Feed the diamond gradually to the wheel, don't bring it up with a bump, and take great care to see that the diamond is cutting properly and not merely rubbing away.

A diamond is similar in general effect to a steel cutting tool, and by the same rule its cutting edge should be watched to see that it is kept in proper condition. When it becomes worn or shows signs of having a flat surface, it should be immediately re-set to bring a new cutting edge into operation. Stones are frequently shattered owing to continued use after they should be re-set, as when the cutting edge is worn away to a flat surface it is similar to using a dull tool or drill. Also should a diamond show signs of fracture it should be examined and re-mounted immediately, as this may frequently be the means of saving a diamond from splitting into small pieces.

The greatest care should be taken to avoid grinding away the metal immediately surrounding the stone, otherwise the stone will soon work out and get lost. If possible, use an abundant flow of water on the diamond when in operation, as this will keep it cool, avoid strains and give it a much longer life.

FIG. 3.—THE FINISH TURNING OPERATION ON PISTON.

Can We Avoid Grinding Wheel Accidents?

The Author, Who is an Authority on the Subject, Gives Some Excellent Suggestions for the Care and Use of Grinding Wheels in General

By A. Rousseau*

TO INSURE safety in the use of grinding wheels the same fundamental principles apply as in the case of all other industrial operations—the prevention of accidents and protection of operators by mechanical means, and the education of the workmen to observe the proper precautions.

The operation of all high-speed machinery is hazardous, and the grinding wheel is no exception to this rule. The structure of a grinding wheel is fragile and under certain conditions can easily be broken. Under ordinary working conditions the periphery travels at a speed of approximately a mile a minute, so that if a breakage does occur the results are likely to be serious. Special precautions, therefore, must be taken; first, to prevent wheel breakages; and second, to guard the operators against injury in case breakage does occur.

When a grinding wheel accident occurs the cause can always be traced to one of the following sources: the wheel, the machine, the mounting, or the operation. By making all the conditions of each of these sources ideal, grinding wheel accidents could be prevented. In other words, if a perfect wheel is properly mounted on a satisfactory machine, and is operated under ideal conditions, the chance is very remote that the wheel will break while in operation. I will attempt to describe these ideal conditions by practical suggestions, and if all grinding wheel users would follow these suggestions, grinding wheel accidents would be cut down to a minimum.

However, there enter into the practical application so many uncontrollable factors, the most important of which is the human factor, that, even though every effort is made to secure these ideal conditions, it is necessary to provide protection devices of some sort. There are three standard forms of protection devices, which will be described later. One of these forms, or its equivalent, is considered absolutely necessary and should be provided, except in those cases such as internal grinding, where the work itself forms ample protection.

General Description

The modern grinding wheel consists essentially of a large number of abrasive or cutting particles held together by the bond, which in vitrified wheels is of about the same structure as porcelain or glass. Therefore, a wheel must be treated with respect and should not be thrown about as if it were a chunk of steel. Many users of wheels do not recognize the fact that grinding wheels

*Paper read before Ontario Safety League. Author is connected with the Norton Co., Chippawa.

are fragile, and as a result many wheels, especially thin ones, are broken by careless handling.

Manufacturer's Duty

All reputable grinding wheel manufacturers take every possible precaution during the process of manufacture to produce only such wheels as are free from defects. Wheels are inspected between operations and any which are found to be defective are rejected and broken up.

After the final manufacturing operation, every wheel, larger than 5-in. diameter, is subjected to a mechanical test of sufficient severity to rupture a wheel which contains any inherent weakness. In this test the wheels are mounted on a spindle and revolved at a speed much higher than the recommended speed. The testing speed is sufficiently high to provide a factor of safety of at least 2½, in most cases considerably higher.

Each testing machine is equipped with a tachometer connected directly to the spindle, which enables the operator to tell the exact speed at which the wheels are revolved.

A complete record of all wheels tested is kept for reference. Daily affidavits as to the accuracy of these records are sworn to before they are permanently filed. The manufacturing and shipping check must also contain a record of the test, otherwise the inspection department will not pass the wheels.

Immediately after testing the wheels go to the inspection department where they are given the final inspection before being packed. Every wheel is given a careful and thorough examination by experienced men with eyes trained to detect imperfections of all kinds. The inspectors also check the size, shape, grain, grade, quantity, and special features. The utmost care is also used in packing. Beyond this, however, the manufacturer cannot go. It becomes the duty of each user of grinding wheels to determine if wheels have been damaged after leaving the manufacturer, and further to see that they are kept in good condition before and during use.

Customer's Duty

Immediately upon receipt the wheels should be examined to make sure that they have not been damaged in transit or otherwise. Each wheel should be held free and clear, then tapped lightly with the handle of a screwdriver or some other such implement. If the wheel is in good condition it will give forth a clear ring. If it does not give a clear ring, it is a fairly good indication that the wheel is cracked.

Damp wheels and those in which the

spaces between the grains are filled with sawdust, also rubber and shellac-bonded wheels, do not give a clear ring when tapped in this manner. Such conditions should be taken into consideration when applying the test.

Care should also be exercised in the storage of wheels. They should be stored in dry places of even temperature. Shellac and rubber-bonded wheels, ½ in. or less in thickness, should be laid flat on a straight surface to prevent warpage. All other wheels should be supported on edge in racks.

Before mounting, all wheels should again be carefully inspected to make sure that they are in good condition and that they have not been damaged by being moved about or roughly treated while in storage. This is very important, inasmuch as a wheel may sometimes be cracked by merely falling over from an upright position. The same "ring" test should be applied before mounting.

The selection of the proper wheel for the work required of it is sometimes an important factor in considering wheel breakages. A wheel which is too hard or too fine may generate sufficient heat to cause uneven expansion, thereby breaking the wheel.

Moreover, a wheel of weak structure or one of insufficient cross section might be broken by applying heavy work.

The Grinding Machine

The design and condition of the grinding machine are very important factors in the prevention of grinding wheel accidents. Improperly designed machines, and machines that are in poor condition, have been the cause of many wheel breakages.

Machines should be sufficiently heavy and rigid to prevent vibration and should be securely mounted on substantial floors, benches, foundations, or other structures. Bearing boxes must be of proper length to provide ample bearing surface, thereby preventing heating and rapid wearing. An automatic method of lubrication should also be provided.

The ends of the spindles should be so threaded that the nuts on both ends will tend to tighten as the spindle revolves. A simple rule to determine the direction of the thread is as follows:

To remove the nuts they should both be turned in the direction that the spindle revolves while the wheel is in operation.

If threaded in the wrong direction, the nuts would tend to loosen as the wheels revolve and serious consequence might result. The spindle should be of sufficient length to provide bearing for the entire length of the nut when the wheel and flanges are in place. The spindle end, nut and flanges should be protect-

ed. (Properly designed hoods will take care of this requirement.)

Grinding machines should be located in well-lighted and ventilated rooms, and, in addition to the devices designed especially for grinding wheels, the usual safety rules for belt guarding and fast-moving machinery should be heeded.

Mounting

Wheels used for grinding on the periphery or on the sides near the periphery, are usually mounted on spindles, arbors, collets, or centres, and are held in place by flanges or collars.

Great care should be taken in mounting wheels. The hole in the wheel should be slightly larger than the diameter of the spindle or arbor on which it is to be mounted, so that it will not be necessary to exert any force to place the wheel in position. Experience has shown that the hole should be approximately 0.005 in. larger than the arbor.

Wheels of this type should never be mounted without flanges. The flanges should be made in accordance with the dimensions given in the Safety Code. The inner flange should always be either keyed, shrunk, or screwed to the spindle. The outer flange should be made to run perfectly true with the spindle. The outer flange should always be of exactly the same diameter as the inner flange. The hole in the outer flange should be made an easy sliding fit on the spindle or arbor. All flanges should be relieved or recessed at the centre to a depth of at least 1-16 in., in order that the flanges may bear at the outer edge only. All flanges must be true and in balance.

Compressible washers of blotting paper, rubber or leather should always be used between the wheels and the flanges. These washers would take care of any slight unevenness in the surface of the wheel, thereby insuring a uniform bearing around the entire periphery of the flange.

Before mounting the wheel, care should of any slight unevenness in the surface of wheel and the flanges are clean and free from foreign particles

When tightening the nuts, care should be taken to tighten them only enough to hold the wheel firmly. Excessive tightening may damage the wheel.

Other forms of mounting, such as cementing or sulphuring the wheel to iron backs and centres, or the clamping of segments to a metal ring, are only recommended for special purposes and no definite specifications can be given. Advice on doubtful cases of this kind should be obtained from experts before installing.

Mounting wheels with tapered holes on tapered arbors demands great care, for any forcing necessary to get the wheel in position would tend to crack the wheels.

Operation—Speeds

In the operation of a grinding wheel, speed is the first factor to be considered. Overspeeding is dangerous, as the higher the speed at which a wheel is running the less external force is required to break it. In fact, if speeded high enough the wheel would be broken from centrifugal force alone, without the aid of any outside force. For this reason, it is extremely important to keep the speeds correct.

Although the most efficient working speed for a wheel is usually a safe operating speed, conditions are frequently found where wheels are operated at speeds which are dangerous. An excessively high speed is allowed either because of the ignorance of the danger or through lack of proper supervision.

As a wheel wears down, the surface speed decreases, if the number of revolutions per minute is not changed. In order to keep the surface speed constant, various methods are used. The most common of these is to provide the grinding machines and countershaft with cone pulleys by means of which the revolutions per minute of the wheel can be changed.

Another method is to have a set of machines, each running at different speeds, and by changing the wheel from the slower to the faster machines, the surface speed can be kept approximately the same.

These methods are both satisfactory, but both require careful supervision to guard against accident. In the case of a machine which is provided with cone pulleys, the belts will usually be set for the highest speed when a wheel is removed. Unless a competent man is in charge of this work there will be a very good chance of the operator placing a new, full-sized wheel on the machine and starting it up without shifting the belts to the slow speed. Unless some sort of a locking device is provided there is also a chance of an operator increasing the speed before the wheel has been worn down to the proper diameter.

In the case of a set of machines running at various speeds, there is a chance of an operator placing a wheel of large diameter on one of the machines intended for a smaller wheel. This can be guarded against by the installation of some sort of a stop which would limit the size of wheel which could be mounted on each machine. In case machines are provided with protection hoods the opening in this hood can be made just large enough to permit the mounting of a wheel of the proper size. If protection hoods are not used, the same end can be accomplished by fastening a pin to the base of the machine in such a position that it will not permit the mounting of an oversized wheel.

For rough grinding on bench, floor, swing frame, and other machines, a speed of 5,000 peripheral feet per minute is recommended as the standard operating speed for vitrified and silicate wheels, except those known as cup and cylinder wheels. For these wheels a speed of 4,500 peripheral feet per minute is recommended. Speeds exceeding these figures should be used only upon the recommendation of the wheel manufacturer.

For some cases of precision grinding an operating speed of 7,000 peripheral feet per minute is sometimes recom-

mended. Precision machines are usually built rigid and the wheels are kept true and in balance. They are also usually provided with good hoods. For these reasons the higher speeds are not considered as dangerous on this type of machine.

The actual speed at which the machine spindles are run should be frequently tested with a speed indicator, or tachometer. If a spindle is driven by a variable-speed motor, the speed control for the motor should be enclosed in a locked case, or some device should be used, which would prevent anyone but a man assigned to that duty from changing the speeds. The maximum size of a wheel which should be used with certain operating speeds should be indicated in a conspicuous manner on a card posted near each machine.

Work Rest

The proper adjustment of the work is also extremely important as many grinding wheel breakages have been caused by work being caught between the wheel and the rest. Although wheels do not always break from this source, many injuries to the workman's hands have been caused in such a manner. The work rest should be kept adjusted close to the wheel at all times, the space between wheel and rest being not more than ¼-in. They should be of rigid construction and should always be securely clamped after each adjustment. Instructions should be issued, so that the rests will not be adjusted while the wheel is in motion.

Applying Work

When starting up a wheel it is well for the operator to stand to one side until the wheel has reached full speed and has been allowed to run at that speed for about a minute. If it stands up satisfactorily for that length of time it is usually safe to apply the work.

Work should never be forced against a cold wheel, but should be applied gradually, giving the wheel an opportunity to warm evenly and thereby eliminating possible breakage. This applies to starting work in the morning in cold grinding rooms, and to the use of new wheels which have been stored in cold places.

On such operations where work is applied continuously to the wheel, care should be taken that the heat generated will not be so great as to break the wheel. If a large amount of heat is generated, the condition can sometimes be overcome by a change in the grade of the wheel, or by applying a steady stream of water at the point of contact.

Truth and Balance

It is important that wheels be kept running true and in balance. The work of keeping wheels true should be assigned to certain men who have been given special instructions. An inexperienced man who attempts to true or dress a wheel frequently makes it worse than it was before. Wheels should be tested for balance occasionally, and if the out-of-balance conditions cannot be corrected

by truing or dressing, the wheel should be removed from the machine.

Wheels used for wet grinding should not be allowed to stand partly immersed in water. The water absorbed may throw the wheel dangerously out of balance.

Side Grinding

The question of grinding on the sides of a straight wheel has been much discussed. In some places this is absolutely forbidden. There are some classes of work, however, where it is necessary at times to use the side of the wheel for grinding. It is difficult to set a fixed role as to when this practice is permissible, as each case must be considered individually. It should be remembered that thin wheels are not as strong as thicker ones; also that coarse wheels and those of softer grades can be more easily broken than the finer and harder wheels.

If these things are taken into consideration, together with the nature of the work to be done, any mechanic of good judgment should be able to tell whether or not a certain operation is dangerous. We recommend, however, that for work where it is necessary to continually use the straight side of a wheel, or where it is necessary to exert considerable pressure on the side, cup wheels be used. The continued practice of grinding on the sides of a wheel intended for peripheral grinding will soon wear the wheel out of shape, thereby making it run out of balance, and sometimes weakening it beyond a safe point.

Eye Protection

In some cases of grinding there is considerable danger of eye injury. Where this danger exists the operators should be provided with goggles which should preferably be the property of the operator. They should be of such construction that the lens will be firmly held in the frame in case of breakage. In order to give the best protection from fine dust flying around in the air the frame should be made to fit snugly around the eyes, following the contour of the face. The edge of the frame which fits against the face should be provided with a cushion. At the Norton Company plant it has been found that the type which is held in place by means of an elastic head band is more satisfactory than those with the metal temple bows.

A piece of plate-glass held in a metal frame is sometimes fastened to the machine just above the point on the wheel where the work is done. For certain classes of work these shields have proved quite satisfactory. · Another device which has been used with more or less success consists of a leather flap or spare brush which is attached to the hood and adjusted so as to interrupt the particles and dust.

Types of Protection Devices

As the time allowed for the presentation of this paper is limited a detailed description of the various forms of protection devices cannot be given here. We can merely give a brief outline of the principles of each.

There are three approved forms of protection for use in connection with grinding wheels—protection hoods, protection flanges and protection chucks.

Protection hoods consist mainly of an enclosure for the grinding wheel which retain all of the parts of a wheel which might break in operation.

Protection flanges are designed to be used with tapered wheels or wheels of special shape, their function being to hold together the parts of a broken wheel.

Protection chucks are designed for use with cylinder or ring wheels. The jaws of the chuck, by being clamped around the periphery of the wheel, would prevent the broken pieces from being thrown out by centrifugal force.

These are listed in the order of their relative efficiency and importance. It will be noticed that protection hoods head the list. This arrangement was decided on after very exhaustive tests, which proved conclusively that properly designed hoods are more efficient than protection or so-called "safety" flanges. An endeavor should therefore be made to provide all wheels with properly designed hoods.

However, there are some cases where hoods cannot be used and in such cases protection flanges are considered the next best. Of these the tapered flanges used with tapered wheels are the most common and most efficient. Other types, such as hub and ring flanges or dovetail flanges may be used in special cases, but care should be used in their design.

Complete specifications for the construction of those devices will be found in the Safety Code for the Use, Care and Protection of Abrasive Wheels, prepared by the Grinding Wheel Manufacturers of the U.S. and Canada, and approved by the National Machine Tool Builders' Association. Copies of this may be obtained on application to practically any manufacturer of grinding wheels.

Summary

These remarks can be briefly summarized as follows: "To insure safety in the use of grinding wheels seek first to prevent wheel breakages by the observance of common-sense rules. And secondly provide all grinding machines with one of the standard forms of protection devices to' protect operators and others in case something goes wrong."

Mr. H. V. Hutton, Verity Plow Co., Ltd., Brantford, in speaking of this paper remarked that in his opinion Mr. Rousseau had given what he believed to be the last word on grinding wheels. As a representative of one of the big industries of our country, large users of grinding wheels, we have met with most of their problems, and this paper comes as a fine thing to us. It strikes me that a copy of same would be of untold value if placed in the hands of every foreman responsible for the care and maintenance of said grinding wheels: I personally will see that the Verity Plow foremen are so supplied.

I believe we should have a closer inspection of grinding wheels from both inside our plant and from safety and

Government inspectors who visit our plant from time to time. I personally cannot remember that an outside inspector ever checked up the speed of our wheels. To prove the necessity for this —only recently we found the foreman's son in one of our well-managed departments operating a new wheel on the high speed pulley running close up to 7,000 surface feet per minute. Mr. Rousseau's report teaches us that we must have a higher respect for this grinding wheel. It has perhaps caused more accidents than from any other source in our factory. He warns us that it is fragile and tells us that it must be well cared for before putting into service and well protected and guarded when in service.

To get the most out of this paper we should discuss its different phases. A few questions have been prepared and we hope you have others to put before the meeting. With Mr. Rousseau here to give us expert advice, we should avail ourselves of this opportunity. While the greatest thing in life is eternal life, most of us want to live our allotted time here. Then let us think and plan safety for the other fellow.

I should like to ask Mr. Rousseau whether a properly guarded grinding wheel ever caused a fatal accident.

Mr. Rousseau: No such case has ever been recorded, while many fatalities have been prevented by the use of approved guards and flanges.

I would recommend the hood as being the best means of safeguarding the wheels, as they will keep the pieces in the machine. A certain small portion of the hood is left open for the man to insert his work, and a small portion of the wheel might fly off and hit the man in the head, but it would not be fatal. I consider that the use of protective flanges is not enough for the large wheels as a heavy object might swing against it, causing a large piece to fly off and hit the man in a fatal spot. The flange diameter should be one-half that of the wheel.

If the wheels are properly equipped with the protective hood, the flange should be at least one-third the diameter, and if you do not use the hood the flanges' should be much larger.

The latter varies with the size of the wheel and is based on the exposure between the flanges. For a 24-in. wheel the minimum is an 18-in. flange, on 30-in. wheel minimum 24-in. flange; 12-in. wheel the minimum is a 6-in. flange. The larger the exposure between the flanges the larger the pieces that can be broken off.

Mr. Lennard: Do you have any trouble in getting the men to keep the guards on, as so many of them say it gets in the way of the work?

Mr. Rousseau: We have very little trouble in getting them to keep the guards in place; the reason is that a properly designed hood is part of the machine and belongs there. The outside hoods are sometimes left off, but they are more in the way when left off than when in place. I believe that a lot of this trouble is caused by people using the wrong size flanges, and then some-

times use substitutes instead of restoring them when they are worn out. One man using a 12-in. wheel had a 6-in. flange on one side, and en the other where they were worn out had put on crooked washers. The foremen do not always think of these things. If you have a 24-in. wheel and 18-in. flange and the wheel becomes worn down, you can use 18-in. wheel and 12-in. flanges. When it is worn down below this point it is time to put on a new wheel. Men will sometimes put on a new 25-in. wheel and use the same flanges that were on the 18-in. wheel.

Mr. Hutton: Is the manufacturer ever justified in making a special wheel the use of which is not safe?

Mr. Rousseau: We get orders for all sorts of wheels of all shapes and sizes. We have in fact received orders for wheels a diameter of hole larger than the diameter of the whole. It is possible that the man who has ordered does not know this. We take it up with him and try to point out to him the best way to have the wheel designed. We get very good responses from the men in most cases, but if the customer does insist we will make the wheel for him, we being free from every responsibility.

Mr. Lennard: What is the best plan to get the men to wear goggles?

It is hard to get the men to use the goggles when going to the machine shop where the tool grinder is kept, the men will not put on their goggles for a few minutes. In this case we have found that the small glass shield on the machine is of value.

Mr. Hutton:—One of our men made the suggestion that a movable shield could be used and carried forward as needed and then thrown back if in the way of certain work.

Mr. Maclachlan: As far as grinding wheels are concerned I think that the greatest trouble is in looking after the grinding wheel that is a utility wheel and used for everything. Guards will sometimes make it useless for certain types of work. Referring to goggles; I have found one method very good, and that is to make the men realize the strength of the glass in the goggles. The men are afraid of getting the glass in their eyes.

Mr. Rousseau: We have found that it is very effective to make a test of the strength of the glasses, and for this purpose use a hammer. The men are usually very much impressed by this. In regard to the first item you mention about the hood on the general utility wheel. This is one of the things that we run up against quite often. You will find it will interfere with the operation of the wheel, and if you cannot think of any other way, a special hood should be designed. Grinding wheel accidents, when they do occur, are very serious.

Mr. Maclachlan: Many manufacturers and employees buy cheap goggles, and these are often broken, causing the pieces to fly in the men's eyes. Goggles should only be purchased from a reputable dealer.

Mr. Kuechenmeister: In regard to covering emery wheels, glass shields on the machines, we find, do away with the argument for goggles. We provide the men with the best goggles money can buy, and if we find that the lens does not suit the man's eyes he is sent to an optician who will grind a lens to suit his eyes, and these are put in the regular frame. We consider that the goggles are an essential part of the man's tools and we are trying to put them on the same basis.

Mr. Rousseau: I do not claim that glass shields are as good as the goggles, because nothing can take their place, but they are a help.

Mr. Falk: The men do not want to wear the goggles because of the discomfort. If the rim of the goggles is bound they are more comfortable and the men will wear them.

Mr. Albert: We have a large frame outside the office, and all the goggles that are broken are attached to the frame with information as to how they were broken. The men are shown the goggles and shown the records, and this persuades them to wear the goggles.

Mr. Wanzer: I have found that the ordinary goggles do hurt my eyes, all I have tried them on. If they are bent to the shape of the head it helps.

Mr. Kuechenmeister: The goggles can indeed be made to fit the eyes and to suit the eyes of the men, and this work is done by our first-aid man. His eyes are tested if he complains and a special lens is ground.

Mr. Wanzer: Are these the large size goggles?

Mr. Kuechenmeister: The regular 1¼ inch goggles.

We find that 50 per cent. of the accidents, including all cases in the doctor's hands, are eye cases. We are trying to get the men to wear their goggles all the time, and also to get them to buy them as they own them then.

Mr. Rousseau: In regard to this corrected lens. Is it made of regular lens stock?

Mr. Kuechenmeister: These are made of the regular stock and we haven't found a case where the pieces went into the man's eyes.

Mr. Rousseau: We had a case where the man's eye was cut by a piece from his goggles. We have made a sample pair from special lens, which we are going to submit to goggle manufacturers to see whether or not a practical goggle can be made from these.

Mr. Falk: We have made investigations and find that the Julius King Co., of New York, grind a special lens, and this is believed to be the best on the market. It is a single lens and the glass will stand a very heavy blow from a hammer and will not shatter.

Mr. Maclachlan: As all double lenses are cemented together with gelatine, this cannot be used in hot surroundings or in summer, as the gelatine will melt and obstruct the sight. You can get a very strong lens that will not shatter.

NEW PUBLICATIONS

The Canadian Engineering Standards Association have informed us that a stock of the most important publications of the British Engineering Standards Association are now available for distribution. Of course a nominal price must be charged for these publications. Below is a description of some of these books.

Requests for copies should be addressed: Secretary Canadian Standards Association, Room 112 West Block, Ottawa, Ont.

3-1903. Report on the Influence of Gauge Length and Section of Test Bar on the Percentage of Elongation, by Professor W. C. Unwin.—This report deals with the variation of percentage of elongation with different gauge lengths and sections of test bar. The tests cover a wide range of conditions, and no such complete information as to variation of elongation with the form of test bar has been previously available. 25 cents net.

C. L. 2552. Interim Report on British Standard Sizes of Single Row Ball Journal Bearings for Automobile.—Gives inch sizes for light and heavy type bearings for shafts from ¾ to 4½ inches diameter, and metric sizes for light, medium. and heavy type bearings for shafts from 10 mm. to 110 mm. diameter. 25 cents net.

6-1904. Properties of British Standard Sections for Structural Steel.—This important and complete report is likely to undergo revision in the near future, as an endeavor is being made to obtain a larger measure of agreement between British and American standard sections.

53-1912. British Standard Specification for Cold-Drawn Weldless Steel Tubes for Locomotive Boilers. (Revised August, 1913.)—Covers material, manufacture, physical and chemical properties, tests and inspection. 25 cents net.

61-1913. British Standard Specification for Copper Tubes and Their Screw Threads. (Primarily for domestic and similar work.)—Deals with three classes of tube for low, medium, and high pressure, and covers composition, tests, and standard sizes, with specification for suitable screw threads, giving complete dimensions. 25 cents net.

51-1913. British Standard Specification for Wrought Iron for Use in Railway Rolling Stock. "Best Yorkshire" and Grades A, B, and C. (Revised August, 1913.)—Covers physical properties, tests, manufacture, and inspection for puddled wrought iron plates, bars, angles and other sections, also for billets and blooms for certain grades of material. 25 cents net.

15-1911. British Standard Specifications for Structural Steel for Bridges, Etc. and General Building Construction. Etc., and General Building Construction.—Covers manufacture, physical properties, tests, and inspection methods. 25 cents net.

24-1911. British Standard Specifications for Railway Rolling Stock Material. Part V.—Copper Plates, Rods Tubes, and Brass Tubes.—Covers analysis, tests, and inspection methods. 25 cents net.

The Prevention of Water Waste on Railroads

Water Waste on Railroads is a Very Serious Item, as Can be Gleaned From a Reading of This Paper Presented Before the American Waterworks Association Convention at Montreal

By C. R. Knowles*

IN presenting this paper on the prevention of water waste on railroads, I would like to emphasize the fact that I do not appear before you merely as a consumer of water pointing out the savings to be made only by reducing the quantities of water purchased, but also as a producer of water with problems and troubles very like your own. The sources of waste that may be enumerated and discussed at this time are as common to the supply furnished by a railway company's pumps as to the supply purchased from a city waterworks plant.

The duties of a city waterworks manager and the duties of a superintendent of railway water service are along parallel lines, namely, the economic production of water adequate in quantity and satisfactory in quality. There is this difference, however, the manager of the private or municipal waterwokrs is in constant touch with his plant or plants, and has direct supervision of their operation, while the plants on a railway system may be scattered over half a continent and are subject to the varying conditions peculiar to the territory in which they may be located. Consequently some of the problems encountered in the prevention of water waste on railroads may be novel to the city waterworks man.

The writer has been conducting a water waste campaign on the Illinois Central system for the past five years, endeavoring to impress on officers and employees the value of water and the importance of water waste prevention. It is very gratifying to be able to report that this campaign has resulted in a material reduction in the waste and unnecessary use of water.

The total consumption of water on the Illinois Central system for the past five years, divided between water pumped by company forces and water obtained from an outside supply, is shown in the following table:

	Company Plants Gal.	Outside Supply Gal.	Total Gal.
1915..	15,300,000,000	2,986,000,000	18,286,000,000
1916..	15,100,000,000	2,844,000,000	17,984,000,000
1917..	14,000,000,000	2,754,000,000	16,754,000,000
1918..	14,140,147,000	2,771,674,000	16,911,821,000
1919..	12,967,260,000	2,655,740,000	15,622,000,000

It will be noted that the consumption in 1919 was 2,664,000,000 gallons less than in 1915, while the average reduction for the five-year period is 1,468,619,750 gallons. During the above-mentioned period there was an increase of over 20 per cent. in tonnage handled, which would indicate that a still greater reduction was made in the waste of water than is shown by the above table. The reduction in the waste of water was accomplished by frequent water waste

*Superintendent Water Service, Illinois Central Ry.

surveys at all points of the system, these water waste surveys varying from an investigation of a single hydrant at an outlying station requiring only ten or fifteen minutes' time, to an investigation of the water supply at large terminals, sometimes requiring several days. As an example of conditions found in these water surveys a few instances may well be cited.

At a large Southern terminal a request had been made for authority for an expenditure of approximately $20,000 for new pumps and pipe lines, the request being based upon the assumption that the old pumps and pipe lines were too small to furnish sufficient water. A water waste survey disclosed the fact that 40 per cent. of the water pumped was being wasted. When these conditions were corrected there was no difficulty in providing all the water required without any expenditure for additional equipment.

Water Consumption

The consumption of water at a large office building used for general railway purposes had increased to approximately 8,000,000 gallons per month, and as this consumption appeared excessive a water waste survey was made, with the result that the consumption was cut to approximately 2,000,000 gallons per month, a decrease of 75 per cent. The conditions at this point were due to general waste of water through almost every water facility in the building. For example, the controlling valves on the boiler feed water heater were not operating properly, allowing unlimited quantities of water to pass through the heater to the bilge tanks, where it was being pumped by an electric bilge pump to the sewer, this waste causing a threefold loss:

1. The cost of furnishing the water.

2. The loss of the coal required to heat the water, which at the time the investigation was made was estimated to be the equivalent of 250 B.h.p.

3. An additional loss of electric current required to pump the water from the basement level to the city sewer, the bilge pump operating every 50 seconds at the time of the investigation.

The urinal tanks, eighteen in number, were found to be flushing at intervals of from 50 seconds to 1 minute and 10 seconds. There tanks were of the three-gallon flush type, probably 2,000,000 gallons of water being used through these urinals. The controlling valves to the house tanks were in bad order, and the overflow from these tanks to the sewer was practically constant. In addition to these large wastes there were a number of minor wastes of hot and cold water in the restaurant and other places in the building.

At a large engine terminal in the Middle West the automatic valves controlling the water supplied to a large water boiler washing system were found inoperative, the consumption of water through this system amounting to 300,-000 gallons per day. Upon repairing and adjusting the valves the consumption decreased immediately and the daily consumption at the present time is approximately 60,000 gallons, a saving of 80 per cent. of the water formerly used.

At another point the consumption was decreased nearly 3,000,000 gallons per month by adjusting and repairing automatic valves controlling water supplied to a boiler feed water heater and boiler washing system. In this particular instance the saving in heat applied to the water wasted was in excess of the cost of the water.

Numberless other instances could be cited where material economies in the use of water has been effected through these water waste surveys, those quoted above being merely given as examples. The favorable result of our efforts towards prevention of water waste have not been easily obtained as it is extremely difficult to convince the average railroad employee that he should exercise care in the use of water, as he cannot understand why one should worry about water with innumerable lakes and rivers on and adjoining the right of way. He cannot appreciate the fact that costly pumping stations, reservoirs, water-softening plants, storage tanks and pipe lines, are necessary to deliver the water to the point of use.

I have been asked why the city waterworks manager should be interested in water waste prevention where the water was supplied through meter and the cost of such waste billed against the consumer.

Prevent Waste

In the first place it is to the interests of all public officers to prevent waste in any form, as their interests are not limited to the individual but to the community as a whole, and they recognize it as their duty to lead in campaigns for the elimination of all waste in every form, as the creation of waste benefits no one and adds nothing to the wealth of the country, institution or the individual. The city waterworks manager, realizing his duty to the public good, in his efforts to eliminate waste, does not stop to question whether the waste goes to make up a part of his revenue or not. While there may be some few instances where a waste of water would increase the revenue of the company without materially affecting the operation of the plant, unlicensed waste on the part of one consumer jeopardizes the supply to

others, and in justice to all waste cannot be tolerated whether the waster pays for it or not. The water company cannot afford to encourage waste, even by metered consumers, on account of the example set to those who are not metered and have no interest in keeping down the consumption. Excessive waste such as might occur with a large consumer as a railway company causes a great fluctuation in the demand. While a million a day more or less would make but little difference with a plant pumping a hundred million gallons or so per day, it would create a serious condition with many smaller plants, and doubtless in many places the correction of waste, leakage and unnecessary consumption would eliminate the necessity of expenditure for additional pumping equipment and distributing systems, which is a serious matter under present material and labor conditions. This was appreciated by the committee on war burdens water works in the States, in their report, in which they stated: "Pressure will doubtless be brought to bear to force communities to husband their water supply by reducing waste, leakage and even unnecessary consumption in order to curtail unnecessary investment in plants thus made necessary." While the above statement had direct reference to war conditions, in a large measure they are just as true of conditions to-day.

The heavy migration to the cities in recent years has increased the urban population and the demand for water for domestic purposes to such an extent that many waterworks are facing heavy expenditures for additional capacity at a time when it is extremely expensive and difficult to make such extensions. The reduction of waste will undoubtedly postpone these extensions until conditions return more nearly to normal.

A writer in "Engineering and Contracting" calls attention to the enormous waste of water by one of the largest cities in the world and goes to show that the prevention of waste through installing meters at a cost of $13,000,000 would save an investment of $94,000,000 for additional equipment, not to mention a saving of $69,000,000 in coal, wages, and repair.

As the equipment and appliances used in modern works have increased in cost 50 to 300 per cent, there is no question that many waterworks plants are overworking their power houses to supply water to be wasted, and that many requests for appropriations are the direct result of waste beyond the control of the waterworks manager.

The annual consumption of water by the U. S. railroads is estimated at 900,-000,000,000 gallons per year, 225,000,-000,000 gallons of which is purchased from private or municipal waterworks plants, and undoubtedly represents no inconsiderable portion of the water pumped by these plants.

Many cities supply water at sliding rates, giving the large long hour user the benefit of low rates. As the railways are such large consumers, the water which they waste is nearly all furnished at the lower rates, which rates yield a comparatively small net profit to the waterworks company. If this water was made available for distribution to a large number of small consumers, the net revenue would be increased materially, thus benefiting both the railway and the water company.

It has been estimated that 20 per cent. of the water used on the railroads of the country is wasted. If this estimate is correct we have 180,000,000,000 gallons of water pumped per year for no purpose other than to increase the expense of railroad operation and burden the pumping plants with an additional load. Using the figures of 23 pounds of coal for each 1,000 gallons of water pumped, the waste requires the consumption of 2,250,000 tons of coal, or more than 6,000 tons per day, and no doubt the part that this waste pays in the expense for additions to power and pumping equipment makes the coal bill look insignificant by comparison.

What Do You Know Regarding Monel Metal?

The Story of Its Discovery, Its Varied Adaptability, and How It Should be Used is Stated Herein—Castings, Hot Rolled Rods, and Sheets Are Discussed in Turn

By J. H. Moore

HOW many readers can truthfully stand up and say, "Oh, yes, we know what monel metal is and what it can accomplish." While we are not of a gambling tendency we could safely wager that a good percentage might be tempted to ask the question, "What is monel metal?"

Believing that they will be interested in the uses and working of the same, we append the following:

This metal has only been known since 1905, yet is already being successfully used by industrial plants for a great variety of purposes. The metal, which is a natural alloy produced from the distinctive nickel-copper sulphide ores of the Sudbury district, Ontario, has a tensile strength equal to or exceeding that of mild steel. Its non-corrosive, acid-resisting, heat-resisting and ductile qualities have opened up for it a wide field of possibilities.

At first its resistance to the corrosive action of sea-water seemed to be the big feature of its commercial value, and the propellors of many a vessel are made from this metal. It was discovered later that it possesses another distinctive and commercially valuable peculiarity, namely, that it resisted the weakening effect of high temperatures on the strength of metals much more successfully than some other commoner metals. This fact led to the use of monel in fittings for cast steel valves, the blades of steam turbine impellers, and so on. Its varied adaptability is such that every progressive shop should not only get to know its uses, but also its working, for it can be readily machined, forged, soldered and welded. Owing to its unusual toughness and hardness certain precautions must be observed, so that proven and approved methods should be followed to assure low production costs and satisfactory results.

The metal cuts like no other metal. In respect to the power requirements, it is not unlike steel, owing to its high tensile strength and its resistance to shear, but so far as the cleaving action of the cutting tool is concerned, it is more like brass or the other more ductile metals. The metal is removed in ribbons, rather than in chip form, necessitating—on account of its toughness—the employment of high-speed-steel cutting tools with keen edge and decided rake.

The better grades of high-speed steel only should be employed for cutting tools, and care should be exercised to follow carefully the directions for treating the steels furnished by their makers.

As a general rule monel metal can be machined dry, though, of course, cutting lubricants and cooling solutions may be employed or may even be necessary for fine work.

It can be machined effectively at a wide range of cutting speeds—from a slow speed of 8 ft. or 10 ft. per min. with a heavy cut and feed to as high as 250 ft. per min. with a light cut and feed, provided ample power is available. As a rule, on general work a speed of 50 ft. or 60 ft. per min., with a ¼ in. cut and a 1-32 in. feed, will be found to be effectively satisfactory, but if a high finish is desired the depth of the cut should be decreased and a higher cutting speed employed, care being taken to keep the tool sharp. The metal is susceptible to a high and lasting polish, and when ground and buffed has an appearance resembling that of pure nickel, but with a shade difference in color that adds to its attractiveness for novelties, plumbing fixtures, automobile trim, yacht fittings, etc., the permanency of the polish making it also of particular value for instrument mechanisms

and fabrication. Unusual attention has been given to polishing operations, and standardized processes for the effective finishing of the various forms of metal stock have been established.

Castings

(1) Use a solid stone, of which there are several grades and makes. The Norton Company's grade "Q," grain No. 20; and Carborundum Company's grade "G," grain No. 16, are very satisfactory. (2) A rag, wood, or canvas wheel coated with No. 40 emery. (3) A rag, wood, or canvas wheel coated with No. 120 emery. (4) A rag, wood, or canvas wheel coated with No. 120 emery and finished with an ordinary buff, using buffing compound.

Hot-Rolled Rods

(1) A rag, wood, or canvas wheel coated with No. 90 emery. (2) A rag, wood, or canvas wheel coated with No. 120 emery. (3) A rag, wood, or canvas wheel coated with No. 120 emery and finished with an ordinary buff, using buffing compound.

Sheets

(1) A rag, wood, or canvas wheel coated with No. 90 emery. (2) A rag, wood, or canvas wheel coated with No. 120 emery. (3) A rag, wood, or canvas wheel coated with No. 120 emery and finished with an ordinary buff, using buffing compound.

This metal can be forged as readily as can iron or steel if certain but simple necessary precautions are taken. A low-sulphur fuel should be employed for the heating, oil or gas preferably, and the flame conditions should be neither strongly oxidizing nor strongly reducing. The metal should be heated to at least 900 deg. C. (1652 deg. Fahr.), but to not more than 1100 deg. C. (2012 deg. Fahr.), and to secure a forging of maximum strength, the forging should be finished at a temperature of from 500 deg. to 600 deg C. (932 deg. to 1112 deg. Fahr.).

Owing to the low heat conductivity of the metal, care should be exercised to have the bars or ingots uniformly heated throughout and to have the heat extend well past the working section. This is best realized by turning the ingot or bar quite frequently while in the furnace.

Castings, rods, and sheets of the metal can be welded with the aid of a monel-metal welding rod, and by either electric or oxy-acetylene processes.

The chief precaution necessary is to avoid overheating the parts, and this can be guarded against by limiting the welding temperature to approximately that recommended for forging operations. The strength of the weld can be materially increased by finishing it at a reduced temperature—one in the neighborhood of 500 deg. C. (932 deg. F.) or at a color slightly fainter than a dull blood red, the color of the work at the commencement of the weld being a yellowish white.

The effective annealing of the metal entails precautions against the formation of excessive oxide and anneals of somewhat longer duration than required for most other metals, but except for these safeguards, rods, sheets, and wire can be rendered soft and ductile without difficulty. Monel-metal castings are not annealed, but are used as cast. A reducing anneal should be employed, embedding the metal between layers of charcoal in tight boxes in order to prevent the formation of oxide so far as possible.

The increased duration of the anneal is due to the low heat conductivity and great heat-storing capacity of the metal, necessitating prolonging somewhat the period at which the metal should be maintained at the uniform annealing temperature. Compensating in some measure for the increased duration of heat, however, is the fact that no particular care is required to guard against either a too rapid heating of the metal or a reduction in temperature so rapid as to produce chilling. The low heat conductivity of the metal automatically assures an orderly molecular construction, conforming readily to the necessarily gradual heat variations.

The metal has, of course, to be thoroughly and uniformly heated. In

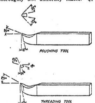

ILLUSTRATING THE CORRECT ANGLE FOR GRINDING THE TOOLS.

the case of rods the temperature should be between 800 deg. and 1000 deg. C. (1452 deg. to 1832 deg. F.), and in the case of sheets any temperature above 875 deg. C. (1607 deg. F.) that may be safely maintained without danger of burning the sheets. As a rule the higher the annealing temperature is carried, the softer does the metal become; but when 1,000 deg. C. is exceeded, in the case of rods, little or no additional softening is realized, and the danger of overheating becomes a factor to contend with, while in the case of monel-metal plates and sheets the advisable limiting temperature should not be more than 50 deg. higher, or 1100 deg. C. at the most.

The metal may be pickled, but the practice is not to be recommended if the metal can be softened as well by annealing, on account of the greater expense of the pickling operation and the fact that in pickling an objectionable oxide scale is formed and has to be removed. However, if pickling has to be resorted to, the bath should be made of a 12 deg. Baume solution of ferric sulphate in water and kept at a temperature of between 38 deg. to 60 deg. C. (100 deg. and 140 deg. F.), maintaining at all times sufficient ferric sulphate in solution to prevent precipitation of copper. A large excess of ferric sulphate is only objectionable in that it adds to the consumption of pickle, although at the same time somewhat decreasing the time required for the pickling.

A satisfactory pickling bath may also be made of a solution of 1.1 specific gravity hydrochloric acid at boiling temperature, provided the oxide formed is not so excessive as to be objectionable and a hindrance in the work.

The formation of nickel and copper oxides produces the objectionable scale that adheres to the metal and puts copper in the bath. The metal like iron has the property of cementing the copper, for which reason the adequate supply of ferric sulphate in the bath is essential. Even with a plentiful supply of ferric sulphate in the bath all danger of cementation is not removed, for should continued contact occur between the monel metal and some iron pin, rod, or other iron part, copper cementation will result; so care must be taken to guard against the contact of the monel metal, while in the bath, with any object constructed of iron—more than passing contact, in any event.

While the pickling takes place, part of the oxide scale that is formed dissolves, and the balance remains adhering to the monel metal, so that progress of the pickling is observed and measured by the tenacity with which the scale adheres to the metal. At intervals, therefore, a piece of the metal should be removed from the bath and the progress of pickling noted by washing it under a jet of water or by rubbing it with a piece of waste. As soon as the scale becomes loose enough to be easily removed—that is, when the metal can be washed clean by an ordinary water jet or by light rubbing—the pickling is completed, and the metal should be taken from the bath and washed clean, or the loosely adhering scale brushed or rubbed off.

From the foregoing, it will be seen that while the working of this metal presents no great difficulty, a certain skill is necessary in the handling, therefore, there are certain points to watch. Keep the tools sharp and ground to proper shape, and supply ample power for them to accomplish their work. As far as the other operations, welding, forging, etc., are concerned, follow the precautions stated in this article and you will find the task just as simple as the working of any other metal.

Arthur Jackson, formerly of the Gould-Shapley & Muir Company, Brantford, Ont., also the Jones & Lamson Company, Springfield, Vt., and for the past five years selling and demonstrating Gridley Automatics in Great Britain, has been appointed Potter & Johnston representative with the Yamtake Company, of Tokyo, who are Japanese agents for the Potter & Johnston Machine Co., Pawtucket, R. I.

WELDING
AND CUTTING

Lessons on the A.B.C. of Good Welding

Dealing With the Welding of Automobile Frames—Compensation for Service, Causes of Failure, Etc., Etc.

By W. B. Perdue

FRACTURES of frames are most likely to occur at or near some connecting cross member, and often in positions where it is inadvisable or impossible to insert or use any additional strengthening material, or where the use of such material might entail an expensive labor bill and keep the car out of service for several days.

Exhaustive tests have been made to determine the efficiency of welded work, the best of which may be mentioned as that of frames welded several seasons past, and which have since been in continuous service, some of them under the most trying conditions over rural and mountain roads.

More impressive tests have been made by testing such welding frames under powerful presses to determine their endurance under bending strains sufficient to buckle the original material. Crushing and tensile tests of welded section have also been made, in all of which the results of well executed welds have proved most gratifying, many instances being on record where the strength of the repair was equal to or greater than that of the frame itself.

Compensation for Services

Granting that the welder is qualified to do work of this character; one of the first questions that arise is "What should the car owner pay?" Shall he be charged the customary price for labor and material, or shall an additional charge be made?

At first thought it seems an extortion to charge an owner a fancy price for a repair simply because few shops are willing to undertake it and because without such repair his car is useless. He must either pay what is asked, purchase an entire new frame which may require weeks to deliver, or else lose his original investment in the car.

For welding repairs which can be made by the ordinary "metal melter"

*Welding Instructor, Heald Engineering School, San Francisco, Cal.

the shop is justified in charging for the services of the journeyman mechanic. When it is necessary to send out for an expert in any particular line the shop is justified in charging for his services plus a small percentage of profit. If the expert who is able to make this repair is available as one of the regular mechanics of the shop, it can not be denied that while engaged in work of this nature the shop is entitled to compensation for the services rendered just as though an expert had been called in for the purpose.

Few self-trained men are able to do dependable work of this nature.

In most cases those who can do this work have spent some time and a certain amount of hard earned money to secure instruction at the hands of other experts. The cost of such instruction represents an investment upon which they are entitled to returns.

Furthermore, the individual or shop selling service for less than that charged by others for the same grade of work eventually loses the respect of customers. The average man does not care to be considered as an object of charity. He expects to pay a reasonable sum for services rendered, and often forms the opinion that the fellow who works for little has a reason for doing so. To charge less than the other fellow is to admit the superiority of the man or the methods he employs. Price cutting is business suicide — the most effective means of advertising your competitor.

Causes of Failure

1. Improper Regulation of the Flame.

2. Working the Weld in the Wrong Direction.

Chiefest of all causes of failure should be mentioned the improper adjustment of the welding flame. Not one welder in a dozen knows how to secure the true "neutral flame" that is absolutely essential to the efficient weld. The freedom of a flame from being carbon-

ized or oxidized is not necessarily indicated by the clear outline of the cone. This clear outline merely indicates that all the particles of oxygen and carbon coming in contact with each other are burned. An excess of oxygen passed through the flame will drive before it a corresponding quality of acetylene. therefore it follows that:

The welding flame may be at one and the same time both oxidizing and carbonizing.

The consumption of equal quantities of oxygen and acetylene means nothing. The use of excess oxygen, commonly called a waste of oxygen, means a corresponding but slightly lessened waste of acetylene.

If a reducing flame is produced by an excess of acetylene the weld will be hard and brittle. The same result may obtain from the use of a tip too small or a torch of inadequate capacity to perform the work with the required rapidity. A bubble, slightly reddish in color, just beneath the surface of the melted metal is an indication of an excess of acetylene, and absorption of carbon by the metal. Frothing, foaming and excessive discharges of sparks are indications of excess of oxygen.

It is gratifying to state that with reasonable care in the regulation of the flame the ill effects of excesses of either or both of these gases may to some extent be neutralized or overcome by the experienced operator. Some suggestions follow:

(a) Commence vertical welds at the bottom and work upward. See Figs. 29 and 30, July 1st issue. This prevents the deposit in the weld of any excess of gases that may be present, or of the products of combustion in the outer envelope of the flame.

(b) Keep the torch in motion from side to side, making sure that each ripple extends the full width of the weld. To hold the torch "dead," that is without motion, will cause the deposit within the weld of the products of combustion, and with an inexperienced operator will cause the metal to become locally overheated and to spill out of the weld.

(c) To carry inadequate acetylene pressures is to effect improper combustion. The pressure given in tables furnished by torch manufacturers apply to pressures actually delivered to the tip.

From two to five pounds excess pressure should be carried at the regulator to allow for the lowering of pressure while passing through the hose to the torch.

(d) With equal pressure and injector types of torches the pressure at the regulator should be sufficient to cause the flame to break away from the tip of the torch with the torch valve two half turns open. With torches which mix the gases by the stirring process all that is necessary is to turn on enough acetylene to secure perfect combustion. When this point is reached smoke will cease to be given off from the end of the flame.

(e) The use of high oxygen pressures is unnecessary and fatal to good welding. The flame can not be forced back into the acetylene line if the oxygen pressure is slightly lower than that of the acetylene. Aside from the production of superior welds the use of low oxygen pressures is decidedly economical.

Lack of Penetration

This is as often the fault of experienced welders as of beginners. To secure proper penetration it is necessary:

(f) To secure the proper bevel. Experience has proved that a bevel of at least 45 degrees is absolutely essential to secure proper fusion of both faces of the parts to be joined. Attempts to melt out the bevel with the torch are as a rule unsuccessful. The bevel may be cut out with the cutting torch provided all traces of slag are afterward removed with a chisel or wire brush.

(g) As in boiler welding it is essential to allow an opening of one-eighth inch, NO MORE, NO LESS, between the beveled edges. This permits the flame to penetrate the weld properly without causing the bottom of the bevel to break down and spill out on the reverse side.

Plastic Welding

Many welders of the old school will say "Your ripple weld does look a little prettier, but my plastic weld is just as good." The plastic weld is defective because of:

(h) Interposition of oxide between the wide fills thereof through which sufficient heat can not be transmitted to cause proper fusion with the adjoining metal.

(i) In the plastic weld the added metal is not manipulated, and is therefore in a "cast" condition possessing very little strength.

Proper manipulation of the weld does for it what forging or rolling did for the original ingot. In addition to this it also helps to bring any impurities to the surface and float them out of the weld. The addition of the very small amount of metal necessary to make the ripple permits it to be thoroughly fused with the adjoining bottom and sides of the weld. The welder who holds a "dead rod" can not hope to produce a weld of very great strength. The grain of such a weld will be extremely coarse.

(j) Heating and hammering are not recommended in such cases. This practice may have its uses, but the automobile frame should not be so treated.

Pre-Heating

(k) Such pre-heating as may be necessary in welding automobile frames should be done with the welding torch. The use of other torches for pre-heating may unnecessarily anneal the parts to be welded causing loss of strength. In addition there may be deposited in the weld certain impurities that may cause the weld to become defective.

Improper Fusion

The average welder is over-anxious to "get the work going" by the addition of the welding rod. The appearance of a film of oxide on the surface of the heated metal is often mistaken for the fusion of the metal itself. This may be due to (l) the use of a torch in which the gases are not properly stirred together, (m) the use of an excess of oxygen, improper motion imparted to the torch—that is a successive lifting and lowering of the flame—causing the oxygen of the surrounding air to produce a destructive effect upon the heated surfaces; or to many other causes.

(n) Holding the torch at an improper angle, (o) a slovenly, jerky motion or otherwise improper motion imparted to the torch—generally the result of wrist instead of free arm movement, (p) the use of a welding tip of a size unsuited to the work, the too rapid or improper addition of the welding rod, (q) or its addition before the bottom and sides of the weld are in proper fusion are common faults. To overcome these faults it may be necessary to have some other welder study and observe your use of the torch. This in the end is the safest and least expensive plan, since the loss from the failure of even a single weld may amount to a far greater loss than would be thus expended. Those faults which do not detract from the appearance of the weld are extremely serious, since that which has a perfectly sound appearance may have a line of weakness throughout its entire structure.

(r) Another example of improper fusion is the tendency of some to cover too little area with the torch, and to "slick" away the fused metal from the edges of the weld by burrowing into them with the flame. The danger sign in this case is the rapid flowing away of the metal from an edge that presents a reddish appearance and distinctly shows the grain of the metal. (s) An excessive oxygen pressure may cause the same trouble.

Clean Surfaces

Clean surfaces are absolutely necessary for making good welds. (t) If scale, rust or grease is permitted to remain on the surface to be welded, a slag is formed that will make a streaked and seamy weld. All welded surfaces and those adjoining must be kept free from foreign substances. (u) All traces of brass, zinc, babbitt, solder, etc., must be removed from the welding zone.

SOME VALUABLE DON'TS

The Hart-Parr Co., Charles City, O., who issue a factory paper called Hart-Parrtners, have in their issue some hints, or rather don'ts, for the dictator, which are well worth repeating. Here they are, just as the willing but sorely tried stenog believes they should be said:

Don't turn your back to the stenographer. She likes to hear you talk.

Don't chew gum, a cigar, pipe, your mustache or correspondence. It isn't polite, and looks bad.

Don't keep your chair revolving and complaining. Sit still—it quiets our nerves.

Don't get up and take your morning exercise while dictating. It annoys us.

Don't get peeved if you drop half a sentence down your shirt front and the stenographer has to get you to dig it up again.

Don't use a barrage of tobacco smoke to put your stenographer out of business. If you must smoke, buy her a gas mask.

Don't be afraid to drop your voice at the end of a sentence, your stenographer won't run away with it.

Don't chin with the boys all day, then dictate at 5.29 and expect your stenographer to have it out by 5.30. Be reasonable.

Don't try to exceed the speed limit when dictating; none of us are court reporters, and anyway look what has been happening to the speeders around town lately — SAFETY FIRST.

Don't expect your stenographer to laugh hilariously at some of your old last year's jokes. Her sense of humor has its limits.

Don't wear out the batteries on the buzzer if your stenographer doesn't appear on the instant. She might be powdering her nose getting ready to vamp you when she comes in, and such a delicate operation should not be interrupted.

DEVELOPMENTS IN SHOP EQUIPMENT

AUTOMATIC MILLING MACHINE

A N automatic milling machine intended for the manufacture of duplicate parts in large quantities is a recent product of the Brown & Sharpe Mfg. Co., of Providence, R. I. It is essentially a manufacturing machine and is known as the No. 21 Automatic Milling Machine, see Fig. 1. The machine has structural characteristics common to the other styles of plain milling machines of the column and knee type made by this concern. However, in the application of the automatic control to that of a plain milling machine many new and important features were developed.

Automatic Controls

Embraced in the design of the machine is the automatic control of the spindle and table. By means of the adjustable dogs on the front of the table, the control of the spindle and table is entirely automatic. These movements include a variable feed, constant fast travel, and a stop for the table; start and stop, and

right and left hand rotation for the spindle. The table and spindle may be operated independently of each other and these movements may or may not be intermittent in either or both directions and may take place one or more times. The spindle may be stopped upon the return travel of the table, thus eliminating the possibilities of marring the work, and the spindle reverse allows of the use of two sets of cutters, with teeth facing in opposite directions so that one set may be in operation for one direction of table travel and the other set for the opposite direction of table travel. A constant fast travel and a slow variable feed in both directions is automatically controlled by the table dogs. The machine can be set and the dogs will operate the table independently of the spindle.

Adjustable Table Dogs

As the various automatic operations of the machine are performed through the medium of the adjustable table dogs it will perhaps be of interest to note a description of these dogs.

There are four different style dogs necessary to operate all the automatic movements of the machine. However, for all ordinary milling operations two or three of the styles are usually sufficient.

Referring to Fig. 2. A long dog "A" used at "A" or "B," controls the reversing of the table. This same dog also stops the table if it is so desired and the table stop lever is set.

Dog "C" controls the constant fast travel, and dogs "D1" and "D2" control the variable slow feed, it being possible to set these dogs to operate in either direction. The variable slow feed dogs are trip dogs and are made changeable to operate as shown at "D1" when the direction of the table travel is to the right, and as shown at "D2" when the direction of the table travel is to the left.

The table always moves at its constant fast travel when reversed, and

when the machine is set for reversing the spindle, the spindle is reversed when the table is reversed. When the machine is set for stopping the spindle, the spindle stops when the table is reversed and starts with the fine variable table feed. No extra tripping dogs are required for either reversing or stopping.

Continuous milling operations may be performed by employing two "A" dogs and two "D" dogs, and for intermittent

FIG. 1—GENERAL VIEW OF THE MACHINE

FIG. 2—TYPICAL EXAMPLE ILLUSTRATING THE AUTOMATIC CONTROL BY THE TABLE DOGS.

milling operations, dogs "A" and "B" and dogs "C" and "D" are employed, the number of pairs of "C" and "D" dogs depending upon the number of pieces of work on the table.

An interesting example of the application of these dogs is illustrated in Fig. 3. Two fixtures are set in a staggered position and at opposite ends of the table and two sets of cutters are employed. The dogs set in their proper position, the table travels at a constant fast travel of 210" per minute to a cutting position immediately before the subject to be milled when the cutting feed is automatically engaged. Upon the completion of this cutting operation, a dog reverses the table and spindle simultaneously and the table travelling at a fast constant travel brings the fixture at the opposite end of the table under the other set of cutters. This cycle of operations is then continuously

FIG. 5—DRIVE SHAFT MOUNTING, FRICTION CLUTCHES AND SPIRAL GEARS.

repeated, the operator but loading and unloading the fixtures.

Hand Control

Although the automatic control of the spindle and table is by means of the table dogs, the same results may be attained by hand, employing the two controlling levers located on the front of the saddle.

Occasionally the loading time of a piece exceeds the cutting time and the table is set to stop for the safety of the operator. Under these circumstances the machine is semi-automatic in operation, and the hand control levers are employed in place of the dogs.

The manipulation of these controlling levers is extremely simple and the ease and rapidity with which they may be operated, is in some cases faster than when it is fully automatically operated.

By means of the hand control levers the machine may be operated as a plain milling machine, accomplishing within its capacity, all that of an ordinary plain milling machine.

The constant speed type of drive permits the machine to be driven by belt, direct from the main shaft to the single driving pulley. This pulley runs at a constant speed and is mounted upon a

sleeve of the main driving shaft. Contained within the pulley is a friction clutch for the starting and stopping of the machine and an efficient brake for the instantaneous stopping of the machine. This brake also affords a spindle lock when removing an arbor or cutter. The friction clutch is operated by a lever located on either side of the column at the front of the machine.

Mounted upon the drive shaft, see Fig. 5, are the friction clutches for the starting, stopping and reversing of the spindle automatically. This mechanism mounted upon the drive shaft relieves the spindle of all unnecessary weight. Power is transmitted from this shaft to the spindle through a series of spiral bevel gears which furnishes a smooth and powerful drive.

Another feature of the constant speed type of drive is the complete separation of the spindle speeds and the table feeds, permitting any combination of the two within the capacity of the machine. Variations of the spindle speeds are obtained through change gears, giving sixteen changes of speeds in geometrical progression from 28 to 695 r.p.m. in either direction. The table feeds are positive and are entirely independent of the spindle speeds. There are 12 changes, ranging from 1.37" to 18.38" per minute. This provides a range of 0.002" to 0.026" per revolution of the spindle for small mills, and 0.026" to 0.656" per revolution of the spindle for large mills. Both sets of change gears are contained within heavy cast iron cases and are made readily accessible by covers, upon which are cast tables of the proper gears for the various spindle speeds and table feeds. This is clearly illustrated at Fig. 6.

Motor Drive

This type of constant speed drive milling machine is well adapted to the application of a motor. The motor is placed at the rear of the machine where it is completely out of the way and does not increase the floor space occupied. The motor is mounted on a heavy bracket that is firmly bolted to pads provided on the base of the machine. A belt transmits the power from the motor to the single driving pulley and a cast

iron guard protects these parts from dust and grit and the operator from injury.

Taper Nose Spindle

The front end of the spindle is tapered, hardened and ground and has a recess to receive a cutter driver and clutch on arbors and collets. Arbors and collets are provided with clutches and have a threaded hole in the end of the

FIG. 6—VIEW OF SPINDLE SPEED GEAR CHANGE CASE.

shank, the clutch fitting into the recess in the end of the spindle and the arbor or collet being drawn into place and held securely by the drawing-in bolt which passes through the centre of the spindle. The threaded end of drawing-in bolt enters the end of the shank of the arbor or collet and is tightened with a wrench from the back of the machine.

Operating Mechanisms

The reverse gearing and cams that are actuated by the table dogs are assembled as a unit in an oil tight case and operates the reverse and stop movements of the table and spindle. This unit of mechanism is automatically lubricated and is protected by a safety friction coupling set to slip at a predetermined load thus guarding against possible damage.

The other unit of mechanism that responds to the action of the table dogs is that which controls the constant fast travel of the table and the variable table feeds. This unit is provided through-

Continued on page 62

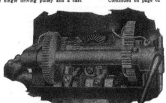

BACK GEARING WHICH IS THROWN IN AND OUT OF ENGAGEMENT BY A SINGLE LEVER.

The MacLean Publishing Company
LIMITED
(ESTABLISHED 1887)
JOHN BAYNE MACLEAN, President. H. T. HUNTER, Vice-President
H. V. TYRRELL, General Manager.
PUBLISHERS OF

CANADIAN MACHINERY
ᴿᵉ MANUFACTURING NEWS ᴿᵉ

A weekly journal devoted to the machinery and manufacturing interests.
B. G. NEWTON, Manager.   A. R. KENNEDY, Managing Editor.

Associate Editors:
J. H. MOORE   T. H. FENNER   J. H. RODGERS (Montreal)
Office of Publication: 143-153 University Avenue, Toronto, Ontario.

VOL. XXIV.   TORONTO, AUGUST 12, 1920   No. 7

Getting Back to Normal

ONE of the most encouraging signs that have been seen of late is contained in the replies of 23 manufacturers to a questionnaire recently sent out by the Merchants' Association of New York. These replies show that slowly but steadily the output of labor is reaching a higher percentage of its possible maximum, and that, man for man, the worker is working more steadily, and producing more during his working time, than he has been doing for some time past. If this tendency persists, and a higher plane of efficiency ensues and remains, then a long step has been made towards the reduction in the cost of living. It is a fact that there has been, pretty well throughout the world, a tendency on the part of the average worker to hold back, so as to make the work in hand last as long as possible, and thereby, in his opinion, preserve the era of plentiful employment and high wages. This tendency was perhaps not so much in evidence on this continent as it was in other countries, Great Britain especially suffering from this viewpoint of labor. Over there they had a regular name for it, thus giving it the dignity of a policy. It was known as ca' canny, which is a Scotch colloquialism for "take it easy," and for which the French use the expression "laissez faire."

Conscientiously carried out, this policy results in bringing the production costs of every article it is applied to up to an ever increasing figure, a figure which represents far more than the real value of the finished article. Brought to bear over a wide range of articles, it naturally affects the pockets of the men themselves who are responsible for it, and forces them to demand a correspondingly higher rate of pay. So the thing goes on. The surest way to obtain better living conditions and a reasonable wage, which will allow the worker to obtain the comforts of life, is for him to work while he is at it, without skimping and holding anything in reserve. An article produced by maximum efficiency, at lowest cost, will be in greater demand than when the same article is produced at a price which puts it out of the reach of many who would otherwise use it. The workman can receive a higher wage, the manufacturer can realize his profit, and a far larger demand is created. High wages in themselves are not an objectionable feature to any manufacturer. Where he loses money is in training men to supply a constant turnover, and in having machines standing idle or operating at low efficiency through the unwillingness of a man to give his best when he is receiving the wages he has demanded. If this condition is passing away, there is much to hope for. It would seem that the common sense of the workers was at last asserting itself, to the detriment of those who have so long been trying to lead him along what looked like the flowery paths of ease. This return to a normal viewpoint will be no doubt viewed with alarm by some of these erstwhile friends of the people.

The Niagara Industrial Conference

THE progressive people of the Niagara Peninsula have been demonstrating to a gathering of manufacturers from Great Britain, the United States and Canada the potentialities of the district for industrial purposes. They point with pride to the cheap power that will be available, although it is evident that it will eventually be only cheap comparatively. The chief engineer speaks of power that will cost $1,000 per horse-power to develop. That does not seem to be a very attractive figure, but, perhaps, spread over a sufficiently large industrial population, the cost to the individual might not be prohibitive. The most interesting part of the movement is that which deals with bringing the sea into direct communication with the Great Lakes, by making it possible for ocean-going steamers to navigate up the St. Lawrence and on to Port Arthur. This is an ambitious scheme, but we are told that the Federal Government has been conducting preliminary surveys for the last three years, looking to the development of the waterpowers along the route, as well as deepening and widening the canals. Before this can become an accomplished fact, many, many millions of dollars must be spent, and while it is possible that the development of the waterpowers would eventually prove an adequate return, it is doubtful whether the navigational end of the project would ever be justified. The present lake freighter, which has been evolved from many years of experience for a special purpose, can carry freight at a rate cheaper than any ocean freighter. When the great grain crop has to be removed, these vessels are available in large numbers to bring the grain down to the elevators, where it can be shipped into the ocean-going steamer for transportation overseas. Their special facilities for loading make them capable of a much quicker turn round than is possible with any ocean boat. The extra cost of the long inland navigation for a large ocean-going steamer, with her big crew, heavy insurance fees, pilotage dues, etc., would make the inland voyage a very expensive proposition. It must also be remembered that for navigational purposes the whole project would be useless for the larger part of the year. A waterway from the sea that is blocked off from the sea by ice for six or seven months is a questionable asset. It should be the subject of long and careful enquiry from every possible angle before the country is committed to the outlay of millions and millions of dollars, which might better be spent in developing the industries already here. Our national canal might easily be as expensive, or more expensive, than our national railways, and they are surely luxury enough for a young country like this. Festina lente is a good motto sometimes.

Making a Rebel Out of a Good Workman

How the Campaign Is Being Carried On in Various Industrial Centres—Holds Out Inducements That Example of Russia May Be Followed With Success in This Country

THERE is little doubt in the mind of the average man but that there is some secret and sinister influence at work, which is calculated to intensify and fan the smouldering embers of discontent in the mind of the man who earns his living by the exercise of trade or by his labor. The evidence of this is the epidemic of strikes and threats of strikes over grievances which sometimes exist only in the imagination of the men making them. What is the germ that, lodging in the mind of the usually reasonable and open minded man, turns him into an element of a mob, which is frankly and avowedly out to get everything it possibly can for itself, regardless of the effect of its action on any other section of the community, or even of the ultimate effect on the industry which provides the fields for their labor? The average member of a trade union is a good solid citizen, very often possessing a stake in the country in the shape of his home, and with children that he is educating with a view to giving them a better start in life than he has had. It is hardly this type of man one would expect to find developing into a revolutionary. He wants to get adequate return for his labor, and a fair share of the comforts of life for himself and his family.

Of his own volition, he would attain this object in a constitutional manner. The mounting costs of food, house rent and clothing have made it necessary for a man to earn higher wages, but on the other hand, the demanding of exorbitant wages can only have the effect of still further increasing the cost of commodities. When these demands are coupled with demands for the shortening of the working day, involving a lessening of production, the struggle begins to look like a war of extermination, involving both sides in irretrievable disaster, and with them all classes of the population who are not directly concerned with the dispute at all. There is strong evidence that this feeling of industrial unrest is being deliberately fostered and encouraged by a group whose object is to produce chaos and revolution of the type that has been scouring Russia for so many months. The means chosen are the distribution of literature, carefully graduated and dispensed in doses almost homeopathic, but continued steadily till the ideas suggested begin to take hold of the least susceptible of those treated. The goal to which the worker's mind is directed in this propaganda is the eventual destruction of all capital,

the abolishment of the wage system, and the dedicating of industry for the benefit of the worker only. This is the familiar Soviet doctrine, although any reference to Communist or Soviet ideas is carefully avoided until the way has been paved.

The first approach to the workman is by means of a pamphlet, which contains the preamble of the Industrial Workers of the World. The following ideas are expressed in this sheet:—"The working class and the employing class have nothing in common. Between these two classes a struggle must go on until the Workers of the World organize as a class to take possession of the earth and the machinery of production, and abolish the wage system. Instead of the conservative motto, 'A fair day's work for a fair day's pay,' we must inscribe on our banner the revolutionary watchword, 'Abolition of the wage system.'"

Then Starts the Work

This is the introductory step. The consideration it receives will naturally depend on the temperament of the recipient. A certain proportion of the men will consign the thing to the gutter after one glance, others will give it a certain amount of thought and then reject it, but there will be a few who will think it over and let the idea seep into their minds. The first pamphlet is then followed up after a short period with this: "Value of Industrial Action." "We cannot successfully meet the well organized employers with the antiquated trade union methods. United industrial action under which an injury to one means an injury to all. All power to the rank and file." This is calculated to open his mind to the possibilities of what is called in England direct action. In order that his mind may be kept in the right direction, he receives the third week another dose of the propaganda, which says:—"Revolt, Secession, and Re-organization are in progress. Do you desire to see forces of the workmen scattered or do you wish to see the One Grand Union for all the workers? There is one enemy to the working class—the Industrial Barons, the exploiters. There is room for only one industrial union. Act together. Unite together. We will stand together and fight together for the day of industrial freedom." Here the idea is frankly and plainly set forth that the only enemy of the worker is the man that employs him, or as the propagandist prefers, exploits him. By standing together the worker is told that he will hasten the day of industrial freedom,

whatever that is. The next pamphlet suggests that although wages have been increased, they are really being reduced, owing to the low purchasing power of money. It further states that the standard of living is being reduced by the fact that wages are not keeping pace with the soaring prices. The same pamphlet states that the remedy for all this is an organization with which to overthrow capital, and the capitalist wage system. Any strike or lockout in any department of any industry should be the signal for a complete cessation of every branch of industry. At the same time that he receives the pamphlet he receives a chart showing the organization of the Workers' International Industrial Union. The head body of this is known as the International Bureau of the Workers' International Industrial Union. It is divided into the American, Australian, British, etc., administration. The American branch of this is subdivided into various departments, the lowest of these being the local industrial union to which the worker belongs. After a pamphlet has been distributed showing that every class of worker is striking, and incidentally "showing the bosses what they can do when they stick together," while at the same time it attacks the American Federation of Labor, a further step along the revolutionary road is taken. It suggests that the union leaders from Gompers down have been playing their own game by keeping the various unions separate, and that the workers should form shop committees and workers' councils for agitating and organizing. "Establish industrial unions in the different industries and join them into one big revolutionary union of all the workers to fight the bosses and their Government."

Beginning to Unfold

Here is what the propaganda has been gradually leading up to. The worker is told that his union is working with the bosses and the government, and that he has to fight the government, and that therefore he must be well organized and strong enough to carry on that fight. The ugly head of revolution is beginning to be plainly seen through the camouflage of words.

Just previous to the first of May literature was distributed urging the workers to stop work on that date. They were invited to show labor's triumph over capital by a general cessation of work and a celebration of the International Labor Day. The final pamphlet shows clearly the source from which the whole poisonous doctrine is emanating.

Continued on page 63

MARKET DEVELOPMENTS

General Dullness Pervades All Lines

Railroad Rates May Bring More Activity—Steel Mills Are Not Finishing Up More Than They Can Ship—Pig and Scrap Iron Markets Show a Better Tendency

THERE is but little stirring in the local markets, and while there have been so far no changes in prices, this is, perhaps, due to the fact that there is so little business being done. In some lines of machine tools deliveries are better, but in most conditions are the same as for some time past. There has been a slight reduction in the production of pig iron and in finished steel in the United States mills. A number of mills are keeping their product in a semi-finished state, and finishing just enough to keep the cars available filled. Transportation facilities are said to be improving lately.

The dullness of the markets in the metal business is exciting no comment in the United States. It is pointed out that the two midsummer months are always dull, and that the first two weeks of August are the dullest of a slack period.

The settlement of the railroad rate case will result in a better outlook for the steel trade, not necessarily for the amount of business to be got from the roads, but because the roads will be put into better physical condition. They will thus be able to keep the wheels of industry in general going more smoothly, which will in turn react to the benefit of the steel industry.

In connection with this increase in the rates on U.S. roads, it is felt that it will be very soon followed by increases on Canadian roads. Then, again, a betterment of transportation facilities will help the production of plate, and this in turn will mean more cars being built. Some authorities think that it will be fairly late next year before the car situation is normal.

BETTER BUSINESS IS LOOKED FOR AFTER HOLIDAYS

Special to CANADIAN MACHINERY.

Montreal, Quebec, Aug. 9. — The opinion is widespread that industrial activity will show a gradual betterment from now on, due partly to the nearing close of the holiday season and partly to the fact that the railroads have been given increased inducement in the matter of rates on freight transportation. The car shortage, however, is an actuality and not as some people suppose a lever being used to obtain higher rates. A dealer here, who has studied the conditions from every angle, intimates that it will be well into next year before a normal condition of car supply will again be evident. The trade here is looking forward to renewed activity early in the fall.

Advances in Steel Prices Likely

The granting of increased freight rates to the railroads has given a new aspect to the industrial situation, but one that will not materially aid in removing the tension that is only too pronounced at the present time. There is a prospect, however, that the shipment of goods will show improvement, but the scarcity of cars is still of such a character that immediate betterment can hardly be expected. The congested condition at the steel mills has resulted in considerable curtailment in production. A dealer here, buying regularly from a large American mill, stated that this plant has been practically idle for several weeks, owing to the lack of yard space to store the finished product. Many mills are continuing to take orders, but on the basis that these will be booked under the no cancellation clause, that the material will be run off when convenient, and that the finished product will be stored at the expense of the purchaser, and finally delivered when transportation facilities are favorable. This is not conducive to heavy buying and present placements are generally made only for immediate requirements and delivery on this basis in usually on a pretty stiff premium. Dealers here are showing no anxiety reagrding warehouse supplies, as it would be undesirable to carry heavy surplus stock when same would have to be obtained at premium prices. Another factor that engages the attention of buyers, just now, is the possible effect increased haulage rates will have on prices of different commodities. It is believed that the added charges will be handed on to the consumers, so that early quotations are likely to show advances, particularly on lines that are brought in from the States. It is thought that it will only be a question of a few weeks when traffic rate revision will be adopted on the Canadian roads. The production of cars is dependent on plate supply and it is hoped that early betterment in transportation will aid in solving this problem. Railroads constructing cars are in a better position to secure material as they can usually furnish the necessary cars to the producing mills. Nominal prices prevail here, but dealers are anticipating an early upward revision.

Better Trade Looked For

Machine tool houses have experienced a quiet week and look for little improvement in the immediate future. Little inquiry has come through for new equipment, and with used tools being scarce, buying on this class of machinery has recently shifted to the States, where it is stated, considerable supplies are available. The difficulty is in getting delivery. The inadequate transportation of late has seriously interfered with many lines of activity, and industries in different sections of the country have had to curtail or suspend operations. There is an optimistic feeling that the railroads will, in the near future, find themselves in a position where freer movement of freight

will feature general traffic, and consequently bring about increased interest and renewed confidence in industrial trade circles, with better buying on the part of consumers, even though prices are expected to advance.

Quiet But Strong

The listless condition of the scrap market is a continued feature of dealers' activities, or rather the lack of them. Buying is confined to small quantities for foundry purposes and chiefly of a local nature. The general restriction of trade, due to continued car shortage, has curtailed buying on every side, and scrap movement has been reduced to a minimum. A slight upward revision has been made in some lines. Brass turnings are up 1-2 cent per lb. Wrought iron axles, at $25, are higher by $5 per ton. Steel axles at the same nominal figure show an advance of $3 per ton. Rails are up, the present price being $23 per ton. Malleable scrap is quoted at $26.00, an advance of $3 per ton.

The Local Market Remains Quiet

Practically All Branches of Metal Trades Experience Dullness

TORONTO.—Writing a market report under present conditions is like making a chronicle of events that never happened. The dullness characterizing last week's markets has been carried over into this, but in even greater degree. There are various theories mentioned to account for this, and one can take the one that appeals most to his fancy, or make a brand new one if one so desires. Among dealers in the machine tool trade there are several different opinions. One gentleman gave it as his opinion that everybody was holding back in the anticipation of another war developing, and that this was influencing the banks also. And the opinion was that business generally was marking time until the crop was harvested, and the wheat exchanged for money and credits to keep the country going. A third ascribed it to the fact that it was always a dull season during July and August, and so many people being away on holidays, caused decisions to be deferred to a later date. It is quite probable that while none of these reasons alone are sufficient to account for the stagnation, they are all contributory in some degree. Everybody will hope very sincerely that the first opinion is altogether without foundation. Steel warehouses are in the same position as the machine tool dealers, and of course the local situation has been affected by the closing down of the Dominion Shipbuilding Co. and Willys-Overland factory. Speaking to one local dealer on the subject of deliveries, we heard that in some lines these were better. In the majority of machine tool lines, however, deliveries were still requiring periods of from three weeks to five months. In the small tool business, such as drills, reamers, taps, milling cutters, etc., there has been a fair demand for small lots, over a wide range of customers, but those that buy in big quantities have

not been in the market. In the old material market about the same state exists as in the machinery lines. There is no change in prices as yet, but there is a possibility of a decline, with the market in its weak state.

MILLS ARE NOW SHIPPING MATERIAL AS IT IS FINISHED

Special to CANADIAN MACHINERY.

PITTSBURGH, August 12. — Production of pig iron by the merchant blast furnaces was 2.9 per cent. lower rate in July than in June, while production by the steel works furnaces decreased 2.3 per cent., making a decrease in pig iron production as a whole of 2.5 per cent. This is a smaller decrease than usually occurs in July, the high humidity in the midsummer period causing a decrease in the output of furnaces individually. As the production of a furnace decreases from this cause, the consumption of coke per ton of pig iron made increases, and the probability is that coke consumption was at a higher rate in July than in June. Thus the slightest decline in pig iron production could hardly be ascribed directly to coke shortage, although, it is a fact that there remains a coke shortage. If there were more coke, some furnaces that are banked or cold would resume operations.

Statistics of steel production in July are not available at this writing, but the presumption is that there was a decrease from the June rate of a very few per cent. The decrease in production of finished steel was greater, since not a few steel works have been accumulating semi-finished steel, fearing to put all their steel output in the finished form because the particular material rolled might not be acceptable to customers when shipped. The common

rule among mills now is to put into finished form only the amount of steel that can be shipped with such cars as are available.

Transportation Improving

There has been a decided improvement in rail transportation conditions as affecting the blast furnaces and steel mills. The producers are unwilling to admit all of the improvement, fearing the admission would bring demands from customers for heavier deliveries. In fact, the producers have experienced convincing customers that transportation conditions were as bad as they were. A difficulty not infrequently experienced is that of a given customer feeling that on account of his supposedly special condition he should receive special treatment.

The majority of mills now have an even break between their production of finished steel and their shipments, and in some cases there have been slight reductions in stock of finished steel. On the whole, however, the steel that has accumulated in the past four months or so on account of transportation difficulties has still to be moved, a quantity somewhere in the neighborhood of 2,500,-000 tons.

Unfinished Steel Easier

There are reports current of middlemen offering sheet bars at $68 a ton, and possibly even at $65, while the market for some time previous had been quotable at $70 to $75, and some important tonnages had brought the full $75. The mills themselves are indisposed to admit declines, and seek the highest prices possible, even when they have steel for early rolling. Among buyers of slabs, billets and sheet bars, there is a disposition to hold off as far as possible, in expectation of lower prices, and thus those who come into the market are only those who must have early deliveries. The market is in the curious position of being supported by inability to ship all the steel produced, when at the same time there are some large accumulations of unfinished steel awaiting cars for shipment.

Dull Markets

The pig iron and steel markets are extremely dull, but the dullness produces no particular comment, and is not necessarily an unfavorable feature since the markets are always dull in the two midsummer months, and the first fortnight of August is usually the dullest of the whole period.

The light demand, in point of tonnage,

that has existed for several weeks has not weakened the merchant furnace position, as they are encouraged by the fact that consumers in need of early deliveries have difficulty in making any purchases at all, and every now and then the market advances a trifle. Thus a couple of sales of basic iron of 2,000 tons each have just put the market up 50 cents a ton. The market now stands at $46 for foundry, $46.50 for basic and $47 for Bessemer, f.o.b. valley furnaces, freight to Pittsburgh being $1.40, but market for a 40 per cent. advance about August 26.

Plates continue to show a very easy tone, and a fairly attractive specification for early delivery can be placed at 3.25c. The decline in the independent market is attributed partly to the fact that the Steel Corporation's deliveries are now heavier in proportion to the total demand, all the Steel Corporation plates being at 2.65c.

Sheets, on the other hand, are stronger. The mills are so well filled for deliveries in the next few months that it is difficult to buy at all, and thus the occasional offerings go at still fancier prices than two or three weeks ago, a price of 8c on No. 28 black being not uncommon.

The Railroad Rate Settlement

The iron and steel trade feels that the settlement of the railroad rate case, by which the railroads are authorized to advance freight and passenger rates by amounts practically equal to all they asked, will have a favorable influence upon business, by removing another of the uncertainties. As to actual orders, these will not be large in proportion to the present capacity of the steel industry. There has been much misapprehension as to the ratio between railroad requirements and the tonnage capacity of the steel industry. The industry's capacity, in finished rolled steel, is about 40,000,000 gross tons a year, while if the steam roads take 2,000,000 tons of rails in a year for replacements and extensions of track they are doing quite well. The building even of 100,000 freight cars would require less than 2,000,000 tons of steel. However, there will, at any rate, be much more railroad business in steel in the next twelve-month than there has been in the past couple of years, or since the Government ordered a considerable number of cars which it had great difficulty in inducing the railroads to accept, many roads claiming either that they did not need the cars or that the price was too high, or both. The iron and steel industry, however, is taking a different interest in the railroads now than it used to do. Then it wanted orders. Now it wants the railroads to function, giving good service to industry in general, for then the demand for steel for general purposes will be increased.

PIG IRON TRADE

There is a better tone to the pig iron market in the United States. In the Philadelphia district furnaces are all sold up, and some of them oversold, the lowest price for iron now being $46.75 to $47 base furnace. No. 2 iron, which was sold recently at $49, is now being sold at $51. Chicago reports show a somewhat smaller inquiry recently, although some buyers are showing interest in iron for first half of 1921. Some iron has been sold for this delivery at $46 Chicago furnace. Malleable has sold recently at $48.25 furnace. A good demand for steelmaking grades has developed in the Pittsburgh centre, and basic iron has advanced about 50 cents a ton as a result. Furnaces are well sold up and in some cases are only now delivering iron contracted for first half of this year. There is a good demand in New England, although total sales are not quite up to the usual average. New York reports a considerable interest being shown in the 1921 market, for first half delivery. One large machinery manufacturer has an inquiry for 10,000 tons for first quarter delivery, mostly foundry iron. Furnaces are not generally anxious to quote on this business, costs being an uncertain feature, especially in view of the railroad increases just granted. The market is generally strong, with prices tending upward.

U.S. SCRAP METAL

Reports from various United States centres show that on the whole the market is on the upward trend. While demand in one or two places is weak, the general tendency is towards a stronger position. Thus, while both Chicago and Boston report the market as dull, New York is holding steady with an upward movement in heavy melting steel. This grade is now $20 to $21 f.o.b. New York. Philadelphia reports a stronger market, 5,000 tons steel being sold recently at $23.50 delivered to an Eastern consumer. A large amount of steel scrap is being shipped to England and Scotland on recent contracts, which amount to about 160,000 tons, one third of which has been shipped. Buffalo has good enquiries for practically every grade of scrap material, and prices are firm. It is anticipated that the new railroad rates will make an increase of from $1.20 to $1.50 per ton on all grades. Advances of from 50 cents to $2 per ton have been made in the Pittsburgh district, where heavy melting steel is now quoted at from $26.50 to $27 per ton. There is a heavy demand for export trade, and large shipments have been made to Italy recently. From Birmingham come reports of a better tone and strengthening prices.

OVERSEAS TRADE ENQUIRIES
The following opening for goods of Canadian manufacture are reported by the Overseas Trade Commissioner: A London, Eng., firm which handles large quantities of constructional steel, mild steel bars, rods, plates and other sections would be pleased to consider offers from Canadian manufacturers.—Reference No. 1367.

A leading firm in British Guiana desires to stock for the hardware department, brass foundry, plumbers and cabinet ware.—1368.

One of the leading firms in British Guiana desires to hear from Canadian manufacturers or suppliers of tin ware, enamelled ware, etc. Manufacturers at or near seaboard preferred, in order to save railway freight.—1369.

An important firm in Rome desires to hear from Canadian exporters of coal, metal, mineral oils and technical articles.—1370.

A Buenos Ayres firm of hardware importers is interested in receiving quotations for flat headed woodscrews, all sizes. These screws were formerly made in a local facory, which can no longer supply the demand. Samples can be seen at the Commercial Intelligence Branch department of Trade and Commerce.—1371.

A firm of Commission agents in Rome would like to hear from Canadian exporters.—1376.

A large firm in Milan, Italy, is desirous of purchasing electrical material in Canada.—1377.

A Natal, South Africa, firm specializing in agricultural machinery, asks for quotations on discs, the exact diameter being 24 and 2-8th inches by 5-32, the concavity is 3 7-8 inches and the weight about 21 lbs. This firm will order in large quantities if the price is satisfactory. Any Canadian firm enquiring can figure on an order of not less than 2,000 discs. Sample can be seen at the Exhibits and Publicity Bureau, Department of Trade and Commerce.—1382.

A firm in British Guiana, with large capital for the cultivation of cassava on a large scale, desires to be put in communication with suppliers or preferably manufacturers of engines for motor machinery, agricultural machinery, tractors and ploughs.—1383.

The hardware branch of one of the leading firms in British Guiana would like to import Canadian tin and steel travelling trunks.—1386.

An important London, Eng., firm are open to purchase lacrosse sticks, and invite quotations from Canadian manufacturers.—1380.

A retired younger member of a firm in Cape Town, with several years' experience in the paint brush trade, desires Canadian agency in paint brushes. Further details, including the names and addresses of the firms mentioned here, may be obtained from the Department of Trade and Commerce, or the office of Canadian Machinery, 143 to 153 University Avenue, Toronto.—1394.

If interested tear 'out this page and place with letters to be answered.

THE NIAGARA PENINSULA INDUSTRIAL CONFERENCE

The three days' conference of the Niagara District Industrial Association was a busy time for those attending, and the industrial possibilities of the district and Ontario generally were well canvassed. Following a visit to the falls on the first day, the guests of the Association were tendered a banquet, at which Senator G. D. Robertson, Minister of Labor, W. M. German, M. r. for Welland, and H. G. Acres, Chief Engineer of the Chippewa Creek development were present. Speaking after the banquet Mr. German traced the development of the power from Niagara from 1893 to the present day, and predicted the time when every drop of water going over the falls would be utilised in power production. Mr. German issued a friendly challenge to the American visitors to attempt a commercial conquest of Ontario by developing their industries in the province, and at the same time cement Anglo-Saxon relations. Mr. H. G. Acres gave some figures relative to the power developments. He said in part, "There is now 250,000 h.p. being developed, and in 1922 there will be 450,000 horse power needed for Ontario requirements. The final scheme includes power development at Niagara, in the Trent Valley and the interprovincial power resources of the Ottawa river." He also said that there was a further 80,000 h.p. possible of development from the drop between the Chippewa canal inlet and the Niagara river. He also said the time would come when this power would be cheap at $1,000 per horse power. Senator Robertson spoke on the necessity of conserving the coal supply from the United States, which was needed by Central Canada.

Transportation

The second day of the conference was devoted to transportation questions. Marine, railway and highway angles of the subject were discussed by men from each division. The slogan of "Bring the ocean to your doors" was very much in evidence. In his after dinner address Senator Robertson spoke on highway development in the peninsula. After stating that the Queenston-Hamilton road would be proceeded with immediately, and the Welland-Port Colborne road pushed to completion, he said he was in favor of building a higher bridge over the Niagara river to relieve the present traffic congestion at the frontier. Speaking of the National Railways, he said that service would be the basis of competition, with the C. P. R. as a standard. He said that the Federal Government had for more than three years been surveying the St. Lawrence canals with a view to power development and deepening the waterway from Lake Ontario to the sea. Several speakers dilated on the netessity of having deep water communication from the St. Lawrence to the head of the lakes, which they said was needed by both the United States and Canada. The third day was devoted to visiting various industrial plants in the Niagara peninsula, and Port Weller and the Welland Ship Canal works were also shown to the visitors. Thorold and Port Dalhousie were included in the itinerary. The industrial possibilities were fully canvassed during the three days' proceedings, and the Niagara Peninsula given a good deal of valuable publicity.

CLEANING AND STERILIZING WELL WITH STEAM

After other methods failed, steam generated by a locomotive, was used successfully for cleaning and sterilizing a well located on the line of a railroad.

A steam ejector was screwed on the lower end of a pipe, then the pipe was lowered into the well casing and held with ropes so that its lower end came within about 4 in. from the bottom of the well. A flexible hose fed the steam supply from the boiler to the ejector. As the steam was permited to flow through the apparatus discharging the water, workmen stirred the bottom of the well with poles. Thereby the sand and mud which had accumulated was caused to rise and, together with considerable water, was ejected through the discharge pipe. The water was heated almost to a boiling temperature. After the discharge had been operating for three hours, the water which flowed out became clear and clean, indicating that the job had been completed.

A previous attempt made to clean the well with steam from a pile driver was unsuccessful because of the relatively low steam pressure which its boiler developed.

CATALOGUES

N. A. Strand & Co., Chicago, and for whom R. E. T. Pringle, Ltd., Toronto, Ontario, are Canadian distributors, have issued a very interesting catalogue. It is known as catalogue No. 20, and deals with various types of equipment fitted with flexible shafts of both the wire core and link type. A copy of this book will be forwarded upon request to anyone interested.

The Buffalo Forge Co., Buffalo, N.Y., have issued their catalogue No. 421, dealing with their line of Niagara Conoidal Fans. Details of design and tables of performances are given, and, as the foreword of the catalogue announces, the tables will enable beginners and architects to make fan selections to meet any demands met in ordinary heating and ventilating practice. A copy of this book will be forwarded upon request.

The United Alloy Steel Corporation, Canton, Ohio, have published a very interesting book dealing with Electric Furnace Alloy and Special Carbon Steels. A comprehensive talk is given on the production and used use of the electric furnace steels. The various steels and their uses are described in detail. The concluding portion of the book is devoted to various tables of interest. A copy of this book can be obtained upon request.

The Dominion Steel Products Co., Brantford, Ont., have issued a very handsome catalogue dealing with their plant, its past accomplishments and its present capacity for service. They state that the purpose of the book is to trace briefly the development of their plants, and the meals that have made them so widely known.

The book contains 61 pages, and is on beautiful stock, the illustrations being in colors. The plant, the personnel, the consulting staff, their form of service, and other important items are discussed.

We understand that any executive interested in this volume can obtain the same upon request.

Catalogue No. 11 of the Independent Pneumatic Tool Co., Chicago and New York, is now off the press. It contains over 70 pages of real live information regarding their lines of drills, grinders, chisels, etc., etc., and a copy of the same can be had upon request.

The Orton & Steinbrenner Co., Huntingtou, Ind., have issued a very attractive catalogue, dealing with their line of material handling machinery. This includes locomotive cranes, gantry and santilever cranes, drag line excavators, clam shell buckets, coal crushers, orange peel buckets, bridge and pillar cranes and elevating and conveying machinery.

There are some 50 pages to the book and it is profusely illustrated.

The Clipper Belt Lacer Co., Grand Rapids, Mich., have issued a booklet dealing with their line of lacing. The story of the idea as first thought of is given, together with its development. Examples of lacing problems are also shown.

The B. F. Sturtevant Co., Hyde Park, Boston, U.S.A., have issued a new catalogue No. 264, dealing with their line of electrical apparatus. The book is very well prepared and covers various types of motors, generators, generating sets, propeller fans, air heaters, and other electrical apparatus for special applications. Varied and interesting data relative to the various lines is also given. A copy of this catalogue can be obtained upon request.

Before the war Canada's principal source of supply of plate glass was Great Britain, with Belgium second, the imports from the two countries in 1914 being 2,-307,670 and 1,976,563 square feet respectively, whereas during the fiscal year 1919 Canada obtained but 152,558 square feet from Great Britain and none from Belgium. At present over 99 per cent. comes from the United States, the quantity in 1919 being 1,736,805 square feet, whereas, in 1914, only 299,042 square feet came from that source.

Sharp Chasers
Cut Clean Threads

Accurate, uniform threads result only from dies which are maintained in the highest state of cutting efficiency. This means that chasers must be kept sharp, and ground uniformly.

Even if just touched up from time to time, the chasers respond splendidly, with clean threads. And with this machine — the Geometric Chaser Grinder—the matter of keeping threading tools up to scratch is a comparatively simple matter.

Various makes of chasers can be ground on this adaptable machine. The two wheels permit the easy grinding of both milled and tapped chasers. In addition, the plain wheel lends itself readily to various kinds of tool grinding.

Uniform grinding of a set of chasers is purely a mechanical matter through the use of adjustments which can be accurately set to govern the grinding of an entire set of chasers.

The Catalogue describing this machine is a mine of information on chaser grinding. Write for it.

THE GEOMETRIC TOOL COMPANY
NEW HAVEN   CONNECTICUT

Canadian Agents: Williams & Wilson, Ltd., Montreal; The A. R. Williams Machinery Co., Ltd., Toronto, Winnipeg, and St. John, N.B.; The Canadian Fairbanks-Morse Co., Ltd., Manitoba, Saskatchewan, Alberta.

If interested tear out this page and place with letters to be answered.

SELECTED MARKET QUOTATIONS

Being a record of prices current on raw and finished material entering into the manufacture of mechanical and general engineering products.

PIG IRON

Grey forge, Pittsburgh	$42 40
Lake Superior, charcoal, Chicago	57 00
Standard low phos., Philadelphia	50 00
Bessemer, Pittsburgh	43 00
Basic, Valley furnace	42 90
Toronto price:—	
Silicon, 2.25% to 2.75%	52 00
No. 2 Foundry, 1.75 to 2.25%	50 00

IRON AND STEEL

Per lb. to Large Buyers	Cents
Iron bars, base, Toronto	$ 5 50
Steel bars, base, Toronto	5 50
Iron bars, base, Montreal	5 50
Steel bars, base, Montreal	5 50
Reinforcing bars, base	5 00
Steel hoops	7 00
Norway iron	11 00
Tire steel	5 75
Spring steel	10 00
Band steel, No. 10 gauge and 3-16 in. base	6 00
Chequered floor plate, 3-16 in.	8 40
Chequered floor plate, ¼ in.	8 00
Bessemer rails, heavy, at mill	....
Steel bars, Pittsburgh	3 00-4 00
Tank plates, Pittsburgh	4 00
Structural shapes, Pittsburgh	3 00
Steel hoops, Pittsburgh	3 50-3 75
F.O.B., Toronto Warehouse	
Small shapes	4 25
F.O.B. Chicago Warehouse	
Steel bars	3 62
Structural shapes	3 72
Plates	3 67 to 5 50
Small shapes under 3"	3 62

FREIGHT RATES

Per 100 Pounds.

Pittsburgh to Following Points

	C.L.	L.C.L.
Montreal	33	45
St. John, N.B.	41½	55
Halifax	49	64½
Toronto	27	39
Guelph	27	39
London	27	39
Windsor	27	39
Winnipeg	89½	135

METALS

	Gross. Montreal	Toronto
Lake copper	$25 00	$24 00
Electro copper	24 50	24 00
Castings, copper	24 00	24 00
Tin	62 00	65 00
Spelter	11 50	12 00
Lead	10 75	11 00
Antimony	13 00	14 00
Aluminum	34 00	36 00

Prices per 100 lbs.

PLATES

Plates, 3-16 in.	$ 7 25	$ 7 25
Plates, ¼ up	6 50	6 50

PIPE—WROUGHT

Price List No. 44—April, 1920.

STANDARD BUTTWELD S/C

	Steel Black	Steel Galv.	Gen. Wrot. Iron Black	Gen. Wrot. Iron Galv.
⅛ in.	$6 60	$ 8 50	....	....
¼ in.	5 13	7 26	$ 5 43	$ 7 56
⅜ in.	5 13	7 26	5 43	7 56
½ in.	6 84	8 42	7 27	8 84
¾ in.	8 45	10 58	9 03	11 16
1 in.	12 50	15 64	13 35	16 49
1¼ in.	16 91	21 16	18 06	22 31
1½ in.	20 21	25 30	21 69	26 68
2 in.	27 20	34 04	29 05	35 89
2½ in.	43 00	53 82	....	....
3 in.	56 23	70 38	....	....
3½ in.	71 30	88 32	....	....
4 in.	84 48	104 64	....	....

STANDARD LAPWELD S/C

	Steel Black	Steel Galv.	Gen. Wrot. Iron Black	Gen. Wrot. Iron Galv.
2 in.	$30 90	$37 74	$34 60	$41 44
2½ in.	45 54	66 16	51 19	62 01
3 in.	59 29	73 44	66 94	81 09
3½ in.	73 14	90 16	82 84	99 86
4 in.	86 66	106 82	97 56	117 72
4½ in.	0 98	1 23	1 24	1 49
5 in.	1 15	1 44	1 44	1 73
6½ in.	1 49	1 86	1 87	2 25
7 in.	1 94	2 43	2 42	2 90
8-L in.	2 04	2 65	2 54	3 05
8 in.	2 35	2 94	2 92	3 51
9 in.	2 81	3 52	3 50	4 21
10-L in.	2 61	3 26	3 25	3 90
10 in.	3 36	4 20	4 18	5 03

Prices—Ontario, Quebec and Maritime Provinces

WROUGHT NIPPLES

4" and under, 60%.

4½" and larger, 50%.

4" and under, running thread, 30%.

Standard couplings, 4-in. and under, 30%.

Do., 4½-in. and larger, 10%.

OLD MATERIAL

Dealers' Average Buying Prices.

	Per 100 Pounds. Montreal	Toronto
Copper, light	$15 00	$14 00
Copper, crucible	18 00	18 00
Copper, heavy	18 00	18 00
Copper wire	18 00	18 00
No. 1 machine composition	17 00	17 00
New brass cuttings	11 00	11 75
Red brass turnings	14 50	15 75
Yellow brass turninbs	9 00	9 50
Light brass	7 00	7 00
Medium brass	8 00	7 75
Scrap zinc	6 50	6 00
Heavy lead	7 50	7 75
Tea lead	4 50	5 00
Aluminum	19 00	20 00
	Per Ton	Gross
Heavy melting steel	18 00	18 00
Boiler plate	15 50	15 00
Axles (wrought iron)	25 00	20 00
Rails (scrap	23 00	18.00
Malleable scrap	26 00	25 00
No. 1 machine east iron.	32 00	33 00
Pipe, wrought	12 00	12 00
Car wheel	33 00	33 00
Steel axles	25 00	20 00
Mach. shop turnings	11 00	11 00
Stove plate	26 50	25 00
Cast boring	14 00	14 00

BOLTS, NUTS AND SCREWS

	Per Cent.
Carriage bolts, 7-16 and up	+10
Carriage bolts, ⅜-in. and less	Net
Coach and lag screws	—15
Stove bolts	55
Wrought washers	—25
Elevator bolts	+10
Machine bolts, 7-16 and over	+10
Machine bolts, ⅜-in. and less	+10
Blank bolts	Net
Bolt ends	Net
Machine screws, fl. and rd. hd., steel	27½
Machine screws, o. and fil. hd., steel	+25
Machine screws, fl. and rd. hd., brass	net
Machine screws, o. and fil. hd., brass	net
Nuts, square, blank .... +25 add	$2 00
Nuts, square, tapped ....add	2 25
Nuts, hex., blank ....add	2 50
Nuts, hex., tapped ....add	3 00
Copper rivets and burrs, list less	15
Burrs only, list plus	25
Iron rivets and burrs	40 and 5
Boiler rivets, base ¾" and larger	$8 50
Structural rivets, as above	8 40
Wood screws, O. & R., bright	75
Wood screws, flat, bright	77½
Wood screws, flat, brass	55
Wood screws, O. & R., brass	55½
Wood screws, flat, bronze	50
Wood screws, O. & R., bronze	47½

MILLED PRODUCTS

(Prices on unbroken packages)

	Per Cent
Set screws .... —20%, 25 and	5
Sq. and hex. hd. cap screws	12½
Rd. and fil. hd. cap screws ..plus	25
Flat but. hd. cap screws ....plus	50
Fin. and semi-fin. nuts up to 1-in.	12½
Fin. and Semi-fin. nuts, over 1 in., up to 1½-in.	—5
Fin. and Semi-fin. nuts o.er 1½ in., up to 2-in.	+12½
Studs	+5
Taper pins	—12½
Coupling bolts	+40
Planer head bolts, without fillet, list	+45
Planer head bolts, with fillet, list plus 10 and	+55
Planer head bolt nuts, same as finished nuts.	
Planer bolt washers	net
Hollow set screws	+60
Collar screws ......list plus 20,	30
Thumb screws	40
Thumb nuts	75
Patch bolts ....add	+85
Cold pressed nuts to 1½ in...add	$1 00
Cold pressed nuts over 1½ in..add	2 00

BILLETS

	Per gross ton
Bessemer billets	$60 00
Open-hearth billets	60 00
O.H. sheet bars	76 00
Forging billets	56 00-75 00
Wire rods	52 00-70 00

Government prices.

F.O.B. Pittsburgh.

NAILS AND SPIKES

Wire nails, base	$6 10
Cut nails, base	7 00
Miscellaneous wire nails	50%

ROPE AND PACKINGS

Plumbers' oakum, per lb.	0 10¼
Packing, square braided	0 38
Packing, No. 1 Italian	0 44
Packing, No. 2 Italian	0 36
Pure Manila rope	0 35½
British Manila rope	0 28
New Zealand hemp	0 28

POLISHED DRILL ROD

Discount off list, Montreal and Toronto .... net

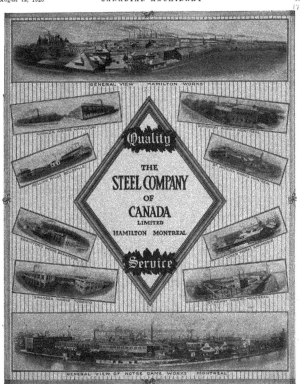

These Plants and the Organization that directs and operates them,
stand back of
and Guarantee Quality and Service at All Times

MISCELLANEOUS

Solder, strictly	$ 0 35
Solder, guaranteed	0 39
Soldering coppers, lb.	0 62½
White lead, pure, cwt.	20 35
Red dry lead, 100-lb. kegs, per cwt.	16 00
Gasoline, per gal., bulk	0 38
Pure turp., single bbls., gal.	3 15
Linseed oil, raw, single bbls.	2 37
Linseed oil, boiled, single bbls.	2 40
Wood alcohol, per gal.	4 00
Whiting, plain, per 100 lbs.	3 00

CARBON DRILLS AND REAMERS

S.S. drills, wire sizes	32½
Can. carbon cutters, plus	20
Standard drills, all sizes	32½
3-fluted drills, plus	10
Jobbers' and letter sizes	32½
Bit stock	40
Ratchet drills	15
S.S. drills for wood	40
Wood boring brace drills	25
Electricians' bits	30
Sockets	50
Sleeves	50
Taper pin reamers	25 off
Drills and countersinks	net
Bridge reamers	50
Centre reamers	10
Chucking reamers	net
Hand reamers	10
High speed drills, list plus	10 to 20
Can. high speed cutters, net to plus	10
American ... plus	40

COLD ROLLED STEEL

[At warehouse]

Rounds and squares	$7 base
Hexagons and flats	$7.75 base

IRON PIPE FITTINGS

	Black	Galv.
Class A	60	75
Class B	27	37
Class C	18	27

Cast iron fittings, 5%; malleable bushings, 22½%; cast bushings, 22½%; unions, 37½%; plugs, 20% off list.

SHEETS

	Montreal	Toronto
Sheets, black, No. 28	$8 50	$10 50
Sheets, Blue ann., No. 10	8 50	9 50
Canada plates, dull, 52 sheets	8 50	10 00
Can. plates, all bright	8 60	9 00
Apollo brand, 10¾ oz. galvanized	....	....
Queen's Head, 28 B.W.G.	11 00	....
Fleur-de-Lis, 28 B.W.G.	10 50	....
Gorbal's Best, No. 28	....	....
Colborne Crown, No. 28	....	....
Premier, No. 28, U.S.	11 50	11 60
Premier, 10¾-oz.	11 50	12 00
Zinc sheets	16 50	20 00

PROOF COIL CHAIN

(Warehouse Price)

B

¼ in., $13.00; 5-16, $11.00; ⅜ in., $10.00; 7-16 in., $9.80; ½ in., $9.75; ⅝ in., $9.20; ¾ in., $9.30; ⅞ in., $9.50; 1 in., $9.10; Extra for B.B. Chain, $1.20; Extra for B.B.B. Chain, $1.80.

ELECTRIC WELD COIL CHAIN B.B.

⅛ in., $16.75; 3-16 in., $15.40; ¼ in., $13.00; 5-16 in., $11.00; ⅜ in., $10.00; 7-16 in., $9.80; ½ in., $9.75; ⅝ in., $9.50; ¾ in., $9.30.

Prices per 100 lbs.

FILES AND RASPS

	Per Cent.
Globe	50
Vulcan	50
P.H. and Imperial	50
Nicholson	32½
Black Diamond	27½
J. Barton Smith, Eagle	50
McClelland, Globe	50
Delta Files	20
Disston	40
Whitman & Barnes	50
Great Western-American	50
Kearney & Foot, Arcade	50

BOILER TUBES.

Size.	Seamless	Lapwelded
1 in.	$27 00	$.....
1¼ in.	29 50	.....
1½ in.	31 50	29 50
1¾ in.	31 50	30 00
2 in.	35 00	30 00
2¼ in.	35 00	29 00
2½ in.	42 00	37 00
3 in.	50 00	48 00
3¼ in.	.. ..	48 50
3½ in.	63 00	51 50
4 in.	85 00	65 50

Prices per 100 ft., Montreal and Toronto

OILS AND COMPOUNDS.

Castor oil, per lb.	
Royalite, per gal., bulk	27½
Palacine	30½
Machine oil, per gal.	54½
Black oil, per gal.	28
Cylinder oil, Capital	82½
Petroleum fuel oil, bbls., net	18

BELTING—No 1 OAK TANNED

Extra heavy, single and double		6½
Standard		6¼
Cut leather lacing, No. 1		2 00
Leather in side	2 40	3 00

TAPES

Chesterman Metallic, 50 ft.	$2 00
Lufkin Metallic, 603, 50 ft.	2 00
Admiral Steel Tape, 50 ft.	2 75
Admiral Steel Tape, 100 ft.	4 45
Major Jun. Steel Tape, 50 ft.	3 50
Rival Steel Tape, 50 ft.	2 75
Rival Steel Tape, 100 ft.	4 45
Reliable Jun. Steel Tape, 50 ft.	3 50

PLATING SUPPLIES

Polishing wheels, felt	$4 60
Polishing wheels, bull-neck	2 00
Emery in kegs, Turkish	09
Pumice, ground	06
Emery glue	30
Tripoli composition	09
Crocus composition	12
Emery composition	11
Rouge, silver	60
Rouge, powder, nickel	45

Prices per lb.

ARTIFICIAL CORUNDUM

Grits, 6 to 70 inclusive	.08½
Grits, 80 and finer	.6

BRASS—Warehouse Price

Brass rods, base ½ in. to 1 in. rod	0 34
Brass sheets, 24 gauge and heavier, base	$0 42
Brass tubing, seamless	0 46
Copper tubing, seamless	0 48

WASTE

XXX Extra	.24	Atlas	.20
Peerless	.22½	X Empire	.19½
Grand	.22½	Ideal	.19
Superior	.22½	X Press	.17½
X L C R	.21		

Colored

Lion	.17	Popular	.13
Standard	.15	Keen	.11
No. 1	.15		

Wool Packing

Arrow	.35	Anvil	.22
Axle	.28	Anchor	.17

Washed Wipers

Select White	.20	Dark colored	.09
Mixed colored	.10		

This list subject to trade discount for quantity.

RUBBER BELTING

Standard ... 10% Best grades... 15%

ANODES

Nickel	.55 to .60
Copper	.38 to .40
Tin	.70 to .70
Zinc	.16 to .17

Prices per lb.

COPPER PRODUCTS

	Montreal	Toronto
Bars, ½ to 2 in.	$42 50	$43 00
Copper wire, list plus 10		
Plain sheets, 14 oz., 14x60 in.	46 00	44 00
Copper sheet, tinned, 14x60, 14 oz.	48 00	48 00
Copper sheet, planished, 16 oz. base	46 00	45 00
Braziers', in sheets, 6 x 4 base	45 00	44 00

LEAD SHEETS

	Montreal	Toronto
Sheets, 3 lbs. sq. ft.	$10 75	$14 50
Sheets, 3½ lbs. sq. ft.	10 50	14 00
Sheets, 4 to 6 lbs. sq. ft.	10 25	13 50

Cut sheets, ½c per lb. extra.
Cut sheets to size, 1c per lb. extra.

PLATING CHEMICALS

Acid, boracic	$ .28
Acid, hydrochloric	.04
Acid, nitric	.11
Acid, sulphuric	.04
Ammonia, aqua	.15
Ammonium, carbonate	.23
Ammonium, chloride	.22
Ammonium hydrosulphuret	.75
Ammonium sulphate	.30
Arsenic, white	.14
Copper, carbonate, annhy	.41
Copper, sulphate	.15
Cobalt, sulphate	.20
Iron perchloride	.62
Lead acetate	.30
Nickel ammonium sulphate	.20
Nickel carbonate	.32
Nickel sulphate	.22
Potassium sulphide (substitute)	.42
Silver chloride (per oz.)	1.30
Silver nitrate (per oz.)	1.25
Sodium bisulphate	.14
Sodium carbonate crystals	.06
Sodium cyanide, 127-130%	.38
Sodium hyposulphite per 100 lbs	8.00
Sodium phosphate	.15
Tin chloride	.80
Zinc chloride, C.P.	.30
Zinc sulphate	.10

Prices per lb. unless otherwise stated

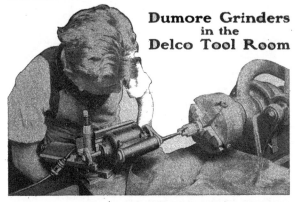

Dumore Grinders in the Delco Tool Room

BUSY, EFFICIENT, ADAPTABLE TOOLS

THE Dumore Grinders at the Dayton Electric Laboratory Company (Dayton, Ohio) are continually busy and the Delco tool makers surely like them. The job you see chucked in the heavy lathe is one of a lot of fine counterbores used in the production of distributor cams on Delco ignition sets. Fine limits are fixed on this job and the Dumore as an internal grinding machine has amply proven its dependability for the accurate grinding of duplicate diameters.

This is only one of the many uses made

of Dumore Grinders. For all work where speed and production costs are to be considered—where accuracy is of prime importance—and also for the multitude of odd jobs that arise every day — Dumore Grinders are indispensable

Dynamically balanced, no end play, no vibration. Equipped with S. K. F. and Norma ball bearings. Well built, adaptable, efficient—Dumore Grinders cut costs throughout the shop.

WISCONSIN ELECTRIC COMPANY
2928 Sixteenth Street   RACINE, WISCONSIN

If interested tear out this page and place with letters to be answered.

INDUSTRIAL NEWS
NEW SHOPS, TENDERS AND CONTRACTS
PERSONAL AND TRADE NOTES

TRADE GOSSIP

Opened New Depot—The L'air Liquide Society, in addition to opening new stations recently at Sydney, N. S., and Windsor, Ont., have also opened a depot at St. John's, Nfld., for supplying customers there with oxygen.

To Give Bond.—After his appointment as assignee for the Dominion Shipbuilding Company Mr. Osler Wade said there would be no statement of assets and liabilities for another three weeks. Mr. Wade gave a security bond of $200,000 in a security company. It is hoped that the wages due the employees will be paid in a few days.

Have Bought Factory.—The Dominion Cone Company have sold their factory at 14 Morrow Avenue to Tarbox Bros., specialty manufacturers. The building is of two storey brick, with a basement, the total floor space being 12,500 square feet. The frontage is 40 feet. The deal was closed through Robins, Ltd.

Guelph Moulders Strike.—Several firms in Guelph have been affected by a strike of the moulders employed in their foundries. The firms are: Crowe's Iron Works, White Sewing Machine Company, Griffin Foundry Company, and the Callender Foundry Company. The increase asked for, it is understood, is one of 25 per cent.

Brantford City Engineer.—Mr. Frank Adams, city engineer of Chatham, has been appointed city engineer of Brantford, at a salary of $3,500 for the present year, and $4,000 for next year. Mr. Adams is an old Brantford boy, and well acquainted with the locality, and its special problems. He will take up his duties on September 15th next, succeeding the late Major Harry Jones.

Want Cheap Rates.—The Calgary Board of Trade passed a resolution recently demanding that a western sitting of the Railway Commission be held. The matters they wish taken up at this meeting would be consideration of the railway's application for increased rates, with a suggestion that certain classes of freight such as coal be given special consideration.

Buys Plant in Montreal.—The plant of the Canada Stove & Foundry Company, Ltd., Montreal, has been purchased by the Gurney Foundry Co. The purchase was made to relieve the congestion at the Toronto plant, which did not lend itself to expansion. The head office of the company will be in Toronto still, and the West Toronto plant will be enlarged to take over certain lines now made in the King street factory.

Canada Car Defers Action.—It was decided at a recent meeting of the board of the Canada Car & Foundry Company to defer action on the payment of the 22 3-4 per cent. deferred dividend until a further meeting of the board on Sept. 2. A proposal to issue ten-year 7 per cent. income bonds in lieu of cash was favorably received, and it is likely that something of this nature may be the plan adopted at the forthcoming meeting. The majority of the board favor the early liquidation of the arrears.

Prepayment of Freight.—The situation caused by the exchange-rate in connection with shipping goods across the line is under discussion by the Interstate Commerce Commission and the Railway Commissioners of Canada. The railways in the States in order not to lose by the exchange situation demanded some time ago that the American portion of the haul be prepaid. The shippers protested and the Interstate Commerce Commission made an order restraining the roads from the practice. The roads ignored this order, and the whole question is now being gone into by the two governing bodies.

Extensive alterations and additions are being made to the plant of the National Farming Machinery Company of Montmagny, Quebec. The company has been making preparations to manufacture all lines of farming machinery and to meet the requirements for large production it has been necessary to carry out a plan of considerable expansion. One of the features of the new developments is the erection of a commercial rolling mill for the production of the grade of material used in the construction of farm machinery. This mill, which it is expected will be in operation next year, will be equipped with a 20-inch, a 12-inch and an 8-inch mill. The company are fabricating all the structural steel for the new buildings and also constructing the units that will be used in the mill. They likewise anticipate operating the plant with power developed from a dam which they are constructing several miles up the river. Plans have also been prepared for the erection of a stove plate foundry and a malleable iron foundry, but it is not expected that building operations on these two latter buildings will be commenced until next year.

MARINE

May Start Work Soon.—It was stated by the assignee of the Dominion Ship Building Company that work might soon be started again on the unfinished ships. This work would be carried on by the creditors, who are the owners of the ships.

The Rapids King, the river boat of the Canada Steamship Line which has been located at the Victoria Pier for some weeks doing the service of a summer hotel, has again been placed in service, as it is felt that the hotels of the city are now able to take care of the tourists coming to Montreal.

The Canada Vickers Company again celebrated the launching of another Government steamer when the hull of the "Canadian Conqueror" was placed in the water recently. This is the ninth vessel that this company has launched for the C. G. M. M. The keel for another Canadian Government boat was ready to be placed as soon as the Conqueror passed down the ways. Three vessels are still under construction for the Government.

Motor Ship in Montreal.—The 8,750 ton deadweight motor ship "Oregon" has arrived in the St. Lawrence and will load at Montreal for River Plate ports. She was built at Copenhagen in 1916, and is owned by the Scandinavian American line. She has Diesel engines of 2,600 h.p. and a speed of 10½ knots. She is commanded by Capt. Julius Lissner, and carries a crew of 35 men all told. The auxiliary machinery is electrically driven. Her usual port is New York, but labor troubles made her visit Montreal this voyage.

The Danish vessel Oregon, the first motor driven ship to come up the St Lawrence, arrived in the Port of Montreal last week. This vessel is a freighter of about 3,000 tons, with a deadweight capacity of 8,750 tons. This ship was built at Copenhagen and is fitted with two powerful Diesel motors of 2,600 h.p., capable of making 10½ knots. It was stated that the vessel consumes about 9¼ tons of oil per 24 hrs., while an oil burning ship of the same dimensions and

CANADIAN MACHINERY Volume XXIV.

making the same speed would require
about 35 tons of oil. Economy of space
is a feature, with less firemen and more
engineers.

PERSONALS

T. E. M. Murphy, who for many years
has been in business for himself, is now
associated with the Gates Engineering
Company, 348 St. James St., Montreal.

Herbert Johnson, of the Armstrong
Whitworth Company of Canada, has been
appointed general superintendent of the
plant at Longueuil. Lawrence Russell,
the general manager, will.be in charge of
the sales.

The Canadian Pacific Railway Company announce the apopintment of Mr.
F. S. Rosseter to be acting Superintendent, Chapleau Division, with headquarters at Chapleau, vice Mr. W. R.
Boucher, who has received a month's
leave of absence.

John W. Gates, formerly president and
general manager of the Gates Refractories, Limited, has resigned his position
with this firm and has opened an office at
348 St. James Street, Montreal, where he
will carry on business under the name of
the Gates Engineering Company, Limited. The new organization will be under
the direct supervision of Mr. Gates, and
will specialize in boiler settings, special
furnace brick work, and general contracting for power plant furnace installations.
Mr. Gates has had a life-long experience
in this line of work, and is the patentee
of the Gates boiler furnace blocks. He
has supervised some of the largest
boiler setting installations in Canada.

ACCIDENT PREVENTION

Accidents and accident prevention in
machine building are discussed in Bulletin
No. 256 published by the United States
Bureau of Labour Statistics. The booklet
gives the result of an investigation to
ascertain the frequency and severity of
accidents in the machine-building industry, and to study and analyse these accidents in such a way as to supply the information necessary for effective safety
work, To this end the report seeks, as far
as possible, to locate the accident hazards in particular departments and occupations, to discover the reason and
causes for the occurrence of accidents,
and to point out some of the more successful methods for their prevention.
The investigation was limited to certain
selected plants representing the more
important classes of product, and covered in all 194 plants. Accident rates
are analysed for the industry as a whole,
by character of products, and by departments. Inability to speak English as related to accidents, safety organization,
direct safeguarding methods in machine
building, and machine design as a factor
of safety, are among the subjects discussed.

Many proposals have been made for
utilizing the huge amount of "scrap"
left over from the war. One device recently patented was for a pump made
from shell cases. Still more recent is
the invention by a British engineer of
a simple arrangement for converting
shells into a hydraulic jack. The body
is formed by a 4.5 high explosive shell
and the ram by an 18-pounder shrapnel shell.

AUTOMATIC MILLING MACHINE
Continued from Page 167.

out with ball bearings and is also automatically lubricated, being contained
within an oil-tight case.

Automatic Lubrication

The automatic lubrication of all rotating parts within the frame of the
machine is another important feature of
the machine. Filtered oil is pumped to
a reservoir cast in the top of the frame
and by means of pipes and a gravity system oil is constantly distributed to
the various bearings. For manufacturing purposes an abundant supply of
cutter lubricant is pumped from a large
tank located within the base of the machine.

Realizing the importance of a rigid
construction in a high productive ma-

chine of this type, the designers have
embraced features adapted to these conditions. The column, knee, and table,
being provided with internal bracing and
reinforcing ribs; it can support work to
its full capacity. The arbor is amply
supported by means of an adjustable
overhanging arm and an arbor yoke employed at any intermediate point near
the cutter, furnishes added support for
heavy duty work. The wearing surfaces of the table and bearings throughout the machine are of such proportions
as to provide for the severe services that
a strictly manufacturing machine is subjected. Means of compensation for wear
is provided throughout the machine. The
conveniently located arrangement of
levers, etc., insures a flexibility and concentration of control that tends to develop a high degree of efficiency in the
operation of the machine.

Briefly here are the main specifications. Capacity: Longitudinal feed, 22"; transverse adjustment, 4½"; Vertical adjustment, 14½"; longitudinal feed, automatic. Spindle: No. 10 taper hole; hole through spindle, 2¹/₃₂" diameter. Drive: Diameter of pulley, 13"; width of belt, 3"; 5 H.P. required; constant speed, 300 R.P.M.; 16 changes of spindle speeds, 28 to 495 R.P.M. in either direction. Arbor Support: Diameter of hole in bushing, 1 13/16"; centre of spindle to under side of arms, 5½"; greatest distance, end of spindle to centre in arbor yoke, without arm braces, 16". Greatest distance, end of spindle to arbor bushing in arm braces, 10½"; greatest distance, face of column to arm braces, 14". Table: Working surface, 54" x 12"; overall dimensions of table, 48" x 11"; 3 T-slots, ⅝" wide. Feeds: Positive, independent of spindle speeds; 18 changes from 1.37" to 18.88" per minute; fast travel between cuts, 210" per minute. Cutter Lubrication: Tank cast in base. Pump furnished inside of frame; capacity of tank, 7 gallons. Vise: Flanged. Capacity: 6¾" wide, 1 9/16" deep, opens 5½". Floor Space: At right angles to spindle, 67"; parallel to spindle, 59½". Weights: Net, about 3,555 lbs.; ready for shipment, about 3,775 lbs. Dimensions for shipment, 62" x 42" x 66".

MAKING A REBEL OUT OF A GOOD WORKMAN

Continued from page 169

This has been carefully disguised in the preceding pamphlets, but it is evidently thought that by the time a man has read all the others, he is ripe for this one, and in this way they are pretty sure that the men who read this final pamphlet will not turn away from the propaganda on account of its source. It reads as follows:

"The sun of communism is rising in the east."

"From Russia its invigorating rays are awakening the Proletariat of the world.

"The revolutionary advance guard of the Proletariat calls upon the workers everywhere to break the bondage of economic and political slavery and demonstrate for the cause of real freedom.

"Finding the capitalist governments in a conspiracy to crush the workers' Republic of Russia, it becomes our task and duty to direct our demonstration on May first against the murderous conspiracy.

"All power to the workers.

"All industry to the workers.

"Long live the social revolution.

"COMMUNIST LABOR PARTY."

Now,—How to Meet It

This is the kind of stuff that is being secretly disseminated among the workmen. It requires a man of strong mind not to be affected eventually by the constant reiteration of these suggestions. The flaunting of their new made wealth by many who daily demonstrate their unfitness for it is a constant irritation to the men who are living close to the line that divides ease and comfort from a daily struggle. These men are in a state of mind that is easily fanned to a deeper discontent by just such means as we have been discussing. This propaganda must be met by a campaign of education, which will enlighten the average worker on the ordinary workings of the laws of economics. Where the one influence appeals to his passions, the other should appeal to his reason. He is a man of intelligence, and often of very high intelligence, and can distinguish right from wrong as quickly as any other member of the community. However, if he is continually having the wrong held up to him in an attractive garb, and nothing shown him from the other side, it is only to be expected that he will sometimes be led in the wrong direction. The whole world has suffered untold loss and misery since the end of the war, through the attempts to put into practice the visionary ideas of the Communists. If the doctrine is allowed to spread on this continent without any efforts being made to combat it, we are liable to see upheavals and disturbances which will make all those we have previously experienced pale in compari-

DOMINION CHUCKS

STEEL OR CAST-IRON BODY BUILT FOR HEAVY DUTY

All Screws Are Reversible

SCREWS are made of the best grade steel. Both ends are broached and are heat treated after machining. They are reversible, so that either end may be used, are large enough in diameter to stand the torsional strains applied by operator when setting up his work. They are made to give the best service—and may be depended upon to stand up under the hardest usage.

DOMINION STEEL PRODUCTS CO. LIMITED
Engineers · Manufacturers
BRANTFORD, CANADA

Classified Opportunities

HELP WANTED

SALESMAN FOR MACHINE TOOLS SUP-
plies. Must be a high grade man of engineering knowledge, and good address; for such there
are excellent prospects. Address giving full particulars in confidence. Jas. Buckley Co., St.
Nicholas Bldg., Montreal.   (clfm)

CANADIAN AGENTS WANTED

CANADIAN AGENT WANTED — AMERICAN
firm whose line (high grade factory specialty)
is being manufactured in Canada, wishes to get
in touch with live man who will act as Provincial
or Dominion representative. Box 679, Canadian
Machinery.

FOR HIGH-GRADE AMERICAN-MADE LATHE
—14 to 26. In replying state territory you can
cover and cover thoroughly. Box 682, Canadian
Machinery.   (c22m)

WANTED

WANTED TO PURCHASE BY U.S. FIRM,
small shop equipped with automatic screw
machines. Box 699, Canadian Machinery.

PATTERNS

TORONTO PATTERN WORKS, 65 JARVIS
Street, Toronto. Patterns in wood and metal
for all kinds of machinery.   (clfm)

BRANTFORD PATTERN WORKS ARE PRE-
pared to make up patterns of any kind—including marine works—to sketches, blue prints or
sample castings. Prompt, efficient service. Bell
'Phone 651; Machine 'Phone 753. Brantford Pattern Works, 49 George St., Brantford, Ont. (clfm)

RATES FOR CLASSIFIED ADVERTISING

Rates (payable in advance): Two
cents per word. Count five
words when box number is required. Each figure counts as
one word. Minimum order $1.00.
Display rates on application.

USED MACHINES FOR SALE

5—3 x 16 Jones & Lamson Geared Flat Turret Lathes.

3—3 x 24 Jones & Lamson Geared Flat Turret Lathes.

1—18" x 8' Mueller Lathe.

1—36" Boring Mill.

2—20" Excelsior Drills.

CHARLES P. ARCHIBALD & CO.
Machinery & Supplies
285 BEAVER HALL HILL, MONTREAL

FOR SALE

CLEARANCE LINE—GENUINE "BLOCOMBE"
Drills are offered at a sacrifice, sizes "C," "E,"
"F," and "G." Wholesalers and dealers will find
it of interest to communicate their requirements
to Box No. 692, Canadian Machinery.   (clfm)

FOR SALE—HARDWARE MANUFACTURERS
—the patent rights, or Canadian manufacturing
rights of letters patent No. 261,327, relating to
self-contained soldering iron. This tool generates its own fuel and would be a boon to mechanics and motorists. Fred Mayor, 96 O'Leary Ave.,
Toronto.

MACHINERY FOR SALE—MOTOR, 66 CYCLE,
220 Amp., 90 H.P., 220 Volt, 3 Phase Speed, no
load, 1015 R.P.Ms. 14" Diameter Drive Pulley, 17"
face, complete with starter and base. Apply
Midland Woodworkers, Limited, Midland, Ont.
  (clfm)

AGENCIES WANTED

QUALIFIED ELECTRICAL AND MECHANI-
cal sales engineer. Technical graduate having
travelled and made connections in South America,
France, Belgium, India, China, would like to
represent Canadian manufacturing and export
house. Box 694, Canadian Machinery.

MACHINE WORK WANTED

MACHINE WORK WANTED FOR LATHES,
shapers, milling machine and plans, etc.
Hourly or contract basis. Prompt delivery. W. H.
Sumbling Machinery Co., Toronto.   (clfm)

CANADIAN MACHINERY
AND
MANUFACTURING NEWS

Vol. XXIV. No. 8   August 19, 1920

172

The Construction of Spindle Ends for Millers

The Practice of Cutter and Arbor Driving on Milling Machines Offers in Itself an Interesting Study—Herein Are Contained Methods as Adopted by Various Concerns

By FRED HORNER

THE construction of spindle ends to suit them for driving the various classes of cutters, arbors, and shank ends, offers an interesting study. Practice is more varied here than it is in the case of drilling machine spindles. The majority of the latter are confined to the standard Morse taper hole, with a slot to receive the tang of the drill, or the chuck arbor, as the case may be. Only in recent years has any modification been effected, and that is in the direction of substituting a clutch drive for the usual tang fitting, the nose of the spindle being mortised out to embrace flats milled on the drill shoulder, thus providing the strongest form of drive. In one or two instances also a special type of key drive is adopted, the key engaging in a groove milled in the tapered shank.

The common tang drive is also employed for light milling cutters, though the Brown & Sharpe taper is used instead of the Morse. The Morse is chiefly brought into requisition when collets are made to

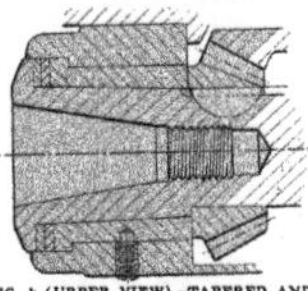

FIG. 1 (UPPER VIEW)—TAPERED AND SCREWED TANG FOR SMALL MILLS.

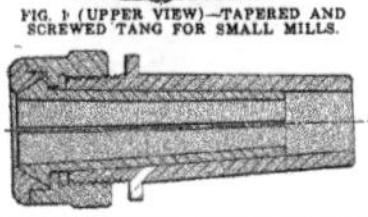

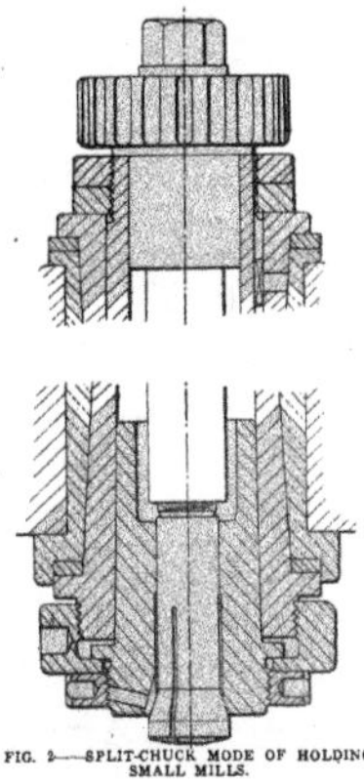

FIG. 2—SPLIT-CHUCK MODE OF HOLDING SMALL MILLS.

FIG. 2 (LOWER VIEW)—SPLIT COLLET FORM OF CHUCK FOR SMALL TAPER MILLS.

fit spindles, and receive various tools as desired, comprising drills, etc., in which event it is more convenient to have a Morse taper collet. On the class of cuts for which the end mills are used, there is sometimes a difficulty that arises from the tendency of spiral teeth to draw into the work, and pull the shank out of the hole, as well as giving rise to chatter. When this is found to happen, a mill should be chosen with handing opposite to the direction of rotation—this is, a left-hand spiral for a right-hand cutter, and a right-hand spiral for a left-hand

cutter. A convenient device for small mills, especially those employed in attachments for vertical and angular cutting, is the screwed shank, the cutter being made with a steep taper to match that in the spindle nose, Fig. 1, and being loosened when removed has to be effected by a wrench on flats. For convenience of making, a good many small mills have parallel shanks, especially those for experimental purposes, or odd ones quickly turned out to special shapes. When held in an ordinary taper hole spindle, a collet is generally fitted, having spring or divided jaws, to grip on the shank. But regular small machines, such as the bench millers, and some of those for light work, with swivelling or universal heads, are constructed with spring chucks.

Thus the Van Norman machine takes a spring chuck in the manner illustrated by Fig 2, tightened with a knob at the rear end. In instances where taper shanks are driven without a tang, the class of collet fitting shown in Fig. 3, may be adopted with advantage, since it affords the means of firmly gripping the shank, and of readily releasing it with-

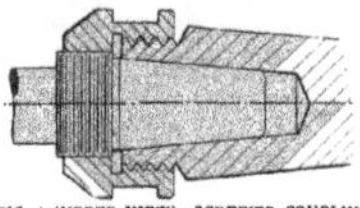

FIG. 4 (UPPER VIEW)—SCREWED COUPLING TO LOCK SHANK IN SPINDLE.

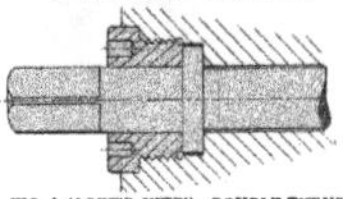

FIG. 4 (LOWER VIEW)—DOUBLE-THRUST DRAW-BOLT.

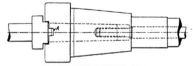

FIG. 5—MODE OF DRIVING ARBOR WITH CLUTCH.

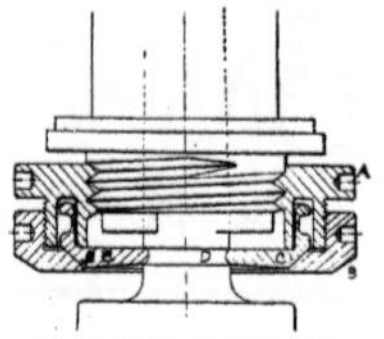

FIG. 10—BECKER DRAW-IN DEVICE.

out hard driving on the tail. A similar principle is met with in the threaded coupling device for larger cutters, the spindle nose having a coarse thread, Fig. 4, on which a flatted nut is run, and forces the taper shank into the spindle.

The Shoulder Clutch

Where neither of the foregoing methods of rotation is sufficiently powerful or is not otherwise convenient, the shoulder clutch transmission comes in. This gives a very strong drive, there is no tang to twist off, and the trouble of releasing is only that of disengaging the fit of the taper shank. The spindle nose

taking its thrust in each direction, for instance in the way represented in Fig. 6. The lay-out of a long spindle and arbor on a plano-miller is illustrated in Fig. 7, the clutch drive, draw-bolt, and the spindle drive being visible. The front head, A, has a lateral adjustment to bring the cutters to the desired position, hence the spindle, B, is splined to pass through the driving gear sleeve, C.

The trouble of reaching to the end of the spindle to get at the draw-bolt is not of much moment in horizontal-spindle machines, though sometimes a collar run on the threaded nose is used for retaining, Fig. 8. Vertical spindles are rather a nuisance, from the necessity of getting up to the height at the top, and

The Becker collar fitting, Fig. 10, consists of a chuck body, A, a cap-collar, B, and a collar cut in two so as to form half jaws, CC. The cap-collar serves first to adjust these half jaws to catch in the groove on the neck of the shank, D, following which the chuck A, is screwed up on the coarse threads of the nose, forcing the shank tightly into the spindle taper. An ingenious device is used on the range of vertical-spindle machines manufactured by Kendall & Gent, Ltd., of Manchester (England). A cotter is utilised in conjunction with a screwed

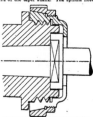

FIG. 8—RETAINING COLLAR FOR ARBOR.

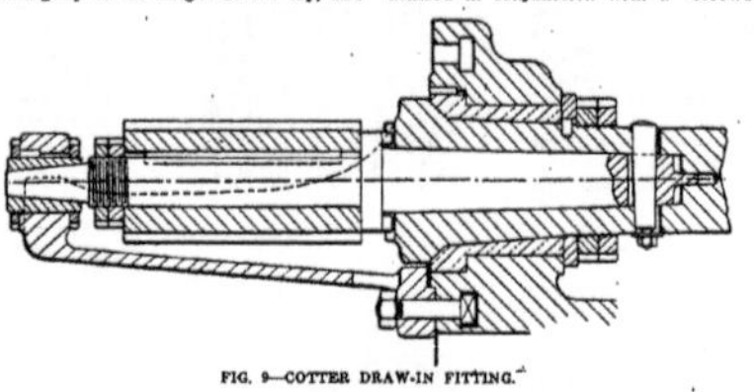

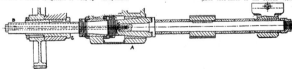

FIG. 9—COTTER DRAW-IN FITTING.

is slotted across, such as seen in Fig. 5, and the collar on the arbor or shank clutches therewith, being kept up to position by the draw-bolt passed through the spindle.

A slot is seen cut out at A in this example, for the purpose of driving in a cotter to separate the fit. But more usually the draw-bolt is arranged to furnish a means of ejection as well, by

a good many designs of fastenings are in use with the object of enabling the locking to be performed at the bottom. The methods comprise chiefly the application of screwed caps, and of cotters, examples of each being here shown. The cotter fastening necessitates a free space above the bearing, as in Fig. 9, and an upper cotter way is cut for the purpose of extraction. The same mode of fitting is also met with in some horizontal spindles.

collar, in the manner depicted in Figs. 11 and 12. The first view shows an external view of the spindle end, with a full size arbor held in position, and the second view a section through a similar fastening, but having a smaller size arbor contained, involving the addition of an adapter or collet. The threaded collar A, has a central space bored out to allow of revolution around the cotter, B, which passes through the slot in the shank (and the collet in the second view).

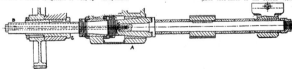

FIG. 7—SPINDLE ARRANGEMENTS OF PLANO-MILLER.

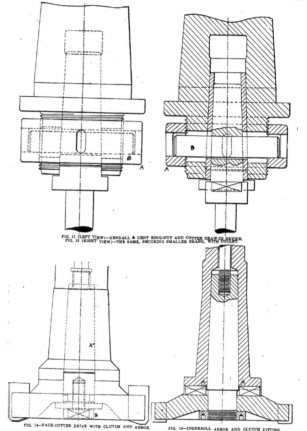

FIG. 11 (LEFT VIEW)—KENDALL & GENT RING-NUT AND COTTER DRAW-IN DEVICE.
FIG. 13 (RIGHT VIEW)—THE SAME, SECURING SMALLER SHANK, WITH COLLET.

FIG. 14—FACE-CUTTER DRIVE WITH CLUTCH AND ARBOR.   FIG. 16—INGERSOLL ARBOR AND CLUTCH FITTING.

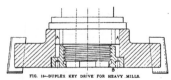

FIG. 19—DUPLEX KEY DRIVE FOR HEAVY MILLS.

Six slots are cut in the diameter of A, to enable it to be tightened or loosened with a hooked wrench, so forcing the cotter upwards or downwards and tightening or extracting the shank. The rotational effort is transmitted by the slotted nose of the spindle, clutching on to the flats located below the cotter hole, and in the case of the collet fitting the collet also has its nose mortised out to embrace the flatted shoulder of the arbor.

The problem of driving the larger cutters has given rise to much controversial practice, and to working difficulties through the sticking or freezing on of cutters after hard duty. The blind following of lathe methods in this case has perhaps been responsible for much of the trouble. It has always been the custom to attach lathe face-plates and chucks by a screwed fit on to the spindle, and the same idea was pursued by many manufacturers as regards face milling cutters. The two conditions are, however, not identical, since the face-plate or chuck affords more convenient means of release from its much larger relative diameter, and the fact that a bar can be applied to gain greater leverage. The cutter is by comparison smaller, and not easy to loosen, or to apply a bar on, without causing damage.

How Force Should be Conveyed

Designers have come to see that the tightening of a threaded fit ought not to be the means of producing rotation, but that the force should be conveyed in a positive manner, leaving the screwed element merely the duty of maintaining the cutter in place endwise. This modification in principle makes a vast difference in convenience of usage. Three main ideas are in use at present, the slotted spindle, the spindle with keys on the diameter, and the interior or box clutch. In any event, the aim should be

to reduce the overhang as much as practicable. The large cutter heads, such as those on rotary milling machines, are in a different category, being permanent fixtures on the spindles, and consequently being fastened with keys, or with bolts passing through a flanged spindle. But

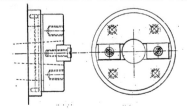

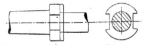

FIG. 16 (UPPER VIEW)—CYLINDRICAL NOSE WITH DRIVING KEYS.
FIG. 17 (LOWER VIEW)—ARBOR SLOTTED TO FIT FIG. 16.

where removal at intervals is necessary, the key or clutch method is preferable. A key drive is illustrated in Fig. 13, the keys standing out from the spindle body, and engaging in the grooves in the cutter hub. Retention is effected by the flange nut B. Fig. 14 gives an example

of a clutch drive, with a separate arbor included for the purpose of centering the cutter. This arbor, A, is screwed gently into place up in the spindle, by a wrench fitting on the flats at the end of the parallel part, and the cutter is slid on, its clutch teeth intermeshing with the mortise across the spindle nose, and the retaining washer, and screw B, set in position.

The Ingersoll drive, see Fig. 15, comprises an arbor pulled in by draw bolt, and used to centre and retain the mill.

The advantage of centering by the larger diameter of the exterior of the spindle nose is obvious, and several types of drives are made, thus combining the steadiness of fit of the screwed nose without its disadvantages. The matching of the cutter hub occurs through either a parallel contact or a tapered

one. The first-named is represented by Fig. 16 (J. Parkinson & Son, Shipley, England), the outside is finished to limit, and the bore of the cutter or head makes a close sliding fit thereon, the drawing up and retention being performed by screws tapped into the hole

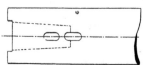

FIG. 21—BORING-BAR NOSE WITH SLOTTED FACE.

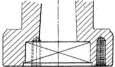

FIG. 18—BOX CLUTCH DRIVE.

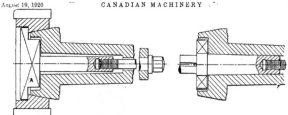

FIG. 19—BROWN & SHARPE TAPER NOSE WITH BOX CLUTCH.

in the spindle face. Arbors on which cutters are strung receive their drive from slots cut through shoulder, Fig. 17.

A similar form of nose is incorporated in the Cincinnati millers, and also in the Milwaukee machines, with the difference in the latter that face cutters are centred by an arbor instead of on the outside of the nose, and that. a loose supplementary collar. is screwed against the nose for driving arbors, this collar having slots on its outer face to catch the prongs or dogs sticking out on each side of the arbor.

Box Clutch Type

The strongest construction of nose is that embodying the internal or box clutch, a design which also possesses the merit of simplicity, and freedom from projecting parts, with minimum risk of injury. The centring of the hub, when a face cutter is mounted. takes place either upon a parallel exterior, or a tapered one, and the drawing up into place 's accomplished with screws in the face, or—more rapidly and conveniently—by a draw-bolt. Fig 18 gives the choice of either way, large

cutters being located by the spindle body, and secured with screws passed into the three tapped holes in the face, and arbors receiving their locking effort through a draw-bolt.

In the Brown & Sharpe construction (see Fig. 19) a draw-bolt serves for holding in in all cases. There is a driver, A, with round stem and flatted head, which latter engages in the box opening of the spindle, and also in the slots of the cutter head, which is drawn on to the tapered spindle nose by the action of this driver and the bolt. The arbors, or the collets are formed with prongs projecting to catch against the flats of the box clutch. Messrs. Alfred Herbert Ltd., of Coventry, (England), have also discarded the threaded nose principle, on account of the difficulties and inconveniences attendant upon its use, and now build their millers with the style drawn in Fig. 20.

There is a draw-bolt to pull the head up to location on the nose, and the drive in the specimen illustrated is taken through a centre flatted to fit the clutch, and tapped to take the draw-bolt. For removal, a pair of screws (one of which appears in the side elevation) is slip-

ped into position and screwed up against the spindle face. An incidental but important point in connection with these clutch drives is that the cutters can be turned with equal facility in either direction, thus enabling left-hand cutters to be mounted without special preparation, a fact which is not the case with screwed noses.

The principles of centering and driving are not very dissimilar in regard to boring machines. Two points of difference as to spindle ends may be specified however, one is the necessity for coupling up a long boring bar to the spindle, and of propelling the latter along as the feed goes on. This is so in certain designs of machines. The other special feature concerns some classes of boring machines, which have to be employed for end facing operations, and the nose mechanism combines provision for driving bars, or for dispensing with this feature and traversing a tool across the end of the work. The mode of coupling up the bar ends is usually by the familiar taper and cotter fastening, one cotter way for the attachment, and another farther along for ejection.

(To be continued)

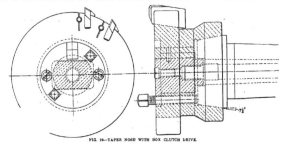

FIG. 20—TAPER NOSE WITH BOX CLUTCH DRIVE.

Automatic and Semi-Automatic Furnaces*

The Continuous Heat Treatment of Metals is a Very Important Study—Typical Furnace Installations for Various Materials Are Shown and the Advantages of Automatic Heating Are Discussed

COMPILED BY J. H. M.

QUANTITY production of quality product involves consideration of many factors frequently overlooked in those manufacturing operations which include the heat-treatment of metals.

It is generally assumed, for instance, that the uniform heat-treatment of large quantities of steel shapes of similar size requires merely a uniformly heated product, and that a uniformly heated product requires nothing more than a uniformly heated furnace. Nevertheless, variations in metallurgical quality of the finished product are likely to exist even though the product has been uniformly heated, and variations in the uniformity of heating are likely to occur even with an indicated uniform temperature in the furnace chamber. The real test is the degree of uniformity of the finished product, which is determined not only by the manner in which the heat is applied to each individual piece, but also by the manner in which it is cooled.

Uniform heating prepares the metal for uniform heat-treatment, but it is the final set or adjustment of the structure that actually determines the uniformity of the heat-treatment and the quality of the product.

To heat the charge uniformly it is necessary that each piece be subjected to the heat in the same manner, at the same temperature and for the same length of time, which requires some-

FIG. 2—Charging end of hardening furnace. Material is laid on slow-moving conveyor chain and delivered into heating chamber. Each piece is kept separate from the others while passing the heating zone.

thing more than a uniformly heated furnace.

If a large quantity of material is piled in a furnace chamber, it is certain that the outside pieces will be heated to, and finally cooled from, a higher temperature than the pieces at the centre or at the bottom of the mass. When a large, cold charge is introduced into a hot chamber at the same time, there is a natural drop in chamber temperature, which is recovered only as the charge absorbs heat. The temperature variation is in proportion to mass;

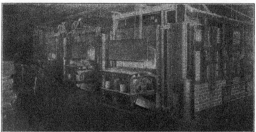

FIG. 1—Charging ends of automatic furnaces for heat treatment of crankshafts, etc.

FIG. 3.—Discharging ends of hardening furnaces. After reaching the final temperature, the pieces are removed through an opening in the side of the furnace, and quenched in a tank partly buried in the floor. A conveyor in this tank removes the material from the oil and delivers it to charging ends of tempering furnaces.

the time element in proportion to mass, surface and temperature; and the variations in proportion to the number of charges, all other conditions being equal.

Similarly, if a large number of uniformly heated pieces are plunged into a quenching bath at the same time, there is a tendency for the bath to become irregularly heated and the outside of the mass to cool at a lower temperature and in less time than the centre. Thus, the good effect of uniform heating is partly negatived by subsequent unequal cooling, and to the extent of such disparity the product is inferior in quality.

With such mass methods of heating and cooling it is unreasonable to suppose that each piece receives the same treatment. It is immaterial what the furnace pyrometer reading or the temperature of the quenching bath may be, for, unless all pieces are both heated and cooled exactly alike, the indication of uniform temperature in the furnace or quenching bath serves merely as evidence, not as proof, of a final uniformly heat-treated product.

The ideal heating condition is approached when the heat is applied to each piece in the same manner, for the same time and at the same temperature; and is most accurately, conveniently and economically accomplished when the furnace operates with a continuous input and output of material, a continuous input of fuel and discharge of spent gases, and a gradual heating of the material to the final and determinative temperature.

The principle of treating each piece individually when heating, applies equally to cooling, whether it be quenching for hardening, rapid air cooling for normalizing, or slow, prolonged cooling as for annealing.

The automatic furnace most easily meets these conditions by insuring the gradual heating of each piece in the same manner, to the same temperature, for the same length of time. It also has the advantage of increasing output in proportion to labor and fuel, as one man with an automatic furnace may do more work, at less cost, than two or three men with a furnace operating on the batch principle. Further than this, the automatic furnace requires less floor space for a given output. With these advantages the automatic type of furnace is entitled to consideration whenever the manufacturing conditions are in harmony with this method of heating.

It should be borne in mind, however, that the automatic type of furnace has its limitations as well as its field of usefulness. It performs to best advantage when the supply of material is regular and the heating operation continuous; when the size and shape of the pieces to be heated are fairly uniform; when the temperatures to be maintained are not unduly high, and when there is a sufficient quantity of material to keep the furnace going at full capacity throughout the day.

Its field of usefulness is narrowed when the supply of material is irregular and the operation intermittent; when there is a wide variation in the size and shape of the pieces to be heated; or by other conditions that do not make the operation as a whole strictly continuous and in harmony with the factory product.

Just as an automatic machine must be set up differently to accommodate pieces of different size and shape, and is unsuited to the handling of small batches of work, just so is the usefulness of the automatic furnace confined to the heat-treatment of particular kinds and quantities of material.

It is generally more difficult to adapt an automatic furnace to a wide range in shape and size of pieces to be heated than, for instance, to adapt a turret lathe or automatic screw machine to the varying demand for product within their limitations. With the furnace, a marked variation in the shape and size of pieces generally necessitates a difference in construction of conveyor, also provision for chang-

FIG. 4.—Automatic furnace for heat treatment of coil springs, etc.

FIG. 5.—General layout of automatic furnaces for continuously heating, hardening and tempering rod stock.

FIG. 6—Charging ends of tempering furnaces. In this case the pieces are removed from the conveyor of the quenching tank and charged into the tempering furnaces.

ing its speed to allow for any variation in heating time required by the difference in mass and surface of the pieces and variation of fuel input and discharge of spent gases. There is a very great difference between the operation of mechanism when subjected to heat and when cold. The necessity for difference in design to meet the conditions of expansion, contraction and comparative weakness of material when heated must be considered.

With the automatic type of furnace, the advantages from the standpoint of quality, denoting good heating, should be as evident as the advantages from the standpoint of production.

The necessity of producing a quality product is more important than the necessity of producing a product in large quantities. Unless this relationship is properly considered in the design and operation of an automatic furnace, the results fall short of what is actually required. This important point is frequently overlooked in the consideration of such furnaces. It is not only necessary to provide the mechanical features incident to the continuous movement of material through the furnace, the combustion features necessary to transform the energy of the fuel into hot gases, but to properly deliver and apply the heat in these gases to as much of the surface of the individual pieces as possible. Of equal importance is the conservation of the heat in the gases to as much of the surface of the individual piece as possible. Of equal importance is the conservation of the heat in the gases after they have been in contact with the material, in order that the fuel used for the operation may be held to a minimum.

The metallurgist will recognize the advantages of a method of heating by which the material is gradually brought up to the final temperature utilization of heat in spent gases to preheat material; the exposure of each piece to the heat in the same manner, to the same temperature, for the same length of time; the uniformity not only of the individual pieces but of the sections thereof as compared with the corresponding sections of other pieces; in short, the advantages that naturally follow what is virtually a method of continuously preheating, heating, cooling and handling each piece the same as all others.

The production manager will appreciate the advantages of increased production per unit of floor space; the reduction in manual labor per unit of production; the uniformity of product, resulting in decreased scrap, following less dependence upon the human element; the fuel economy that naturally follows the direct transfer of the heat in the spent gases to the incoming cold material; the improvement in working conditions as a result of the lesser amount of heat, in proportion to output, thrown off into the room by the automatic furnace; in short, the saving in the cost of the finished product due to economy in labor, time, fuel, floor space, decreased rejections, etc.

Typical Installations

The illustrations herein cover some of the applications of this method of heating and cooling which is coming more and more into favor. Owing to the great variety in working conditions it is impracticable to fully standardize the design or sizes of such furnaces, as the individual requirements warrant consideration of each case on its merits, in order that the heating cycle may be properly adapted to the given manufacturing conditions. However, the illustrations afford an idea of the manner in which typical heating problems are handled and of the opportunity of effecting further improvement in industrial heating operations.

A continuous furnace for the heat-treatment of automobile crankshafts, camshafts and the like is illustrated at Fig. 1.

The shafts are carried through the furnace on a slow-moving conveyor driven by a motor. The spent gases are not only employed to preheat the incoming material, but also to preheat the air for combustion.

The furnaces illustrated at Figs. 2 and 3, are well-suited for heat-treatment operations, such as the annealing, normalizing, hardening, tempering, etc., of such material as crankshafts, axles, shafts, springs, tubes, bars, rods, cylinders, etc. They are often arranged in series so that two or more operations such as annealing, hardening or tempering, may be conducted continuously and in sequence. Frequently, by simply adjusting the temperature or heating time, one furnace may be used for hardening, tempering, annealing or normalizing.

A hardening furnace of the automatic type, arranged in series with a quenching tank and tempering furnace, is illustrated by Fig. 2. The material to be heated is laid on slow-moving conveyor chain at the charging end and delivered into the furnace chamber. Special insulated shields and arrangement of blast protect the operator from the heat.

The pieces are brought up to heat by being slowly passed through the furnace, from a low temperature at the charging end until they reach the maximum temperature at the discharging end, Fig. 3. They are then removed

FIG. 7—Rotary hardening furnace for automatically heating and quenching small steel parts.

from a working opening at the end or at the side of the furnace, as the size and shape of the material requires, and quenched in a tank partly buried in the floor. A conveyor in the quenching tank removes the material and delivers it to the charging end of the tempering furnace chamber. After the pieces have been brought slowly to the tempering heat they are discharged on the apron at the discharging end of the tempering furnace.

In the passage through the chamber of each furnace the pieces are kept separate and suspended in the heat, with practically point contact on the hot conveyor chains.

Coil Springs, Etc.

The furnace illustrated by Fig. 4 is adapted to the continuous tempering of such material as coil springs and similarly formed products that may be conveniently handled on the type of conveyor shown.

The conveyor is provided with a pocket for each piece, to afford the advantage of individual heating, with no piece in contact with another. The construction of the conveyor is adapted to the size and shape of the material to be heated. The size of the chamber and arrangement of working openings may be adapted to the quantity to be handled and the related shop conditions.

Light rod stock may be continuously heated for forming, hardening and tempering with the type of furnace illustrated by Fig. 5, arranged in series as shown. The rods in the foreground, cut to length, are placed on the conveyor and automatically pass through the heating furnace on the right, from which they are removed through an opening in the side wall near the opposite end of the heating chamber. They are then passed through the shaping machine, thence to the oil hardening bath (visible between the first two columns), from which they are automatically conveyed into and through the tempering furnace (on the left of the second column). As discharged from the tempering furnace they are inspected and adjusted, if necessary, and made ready for final fitting.

The progress of the rods is continuous from start to finish. The heating and cooling of each piece individually is clean, rapid and systematic. Pyrometers are provided to record the temperature of each heating operation.

The type of furnace and arrangement of units, while advantageous from the standpoint of quality and cost of heated product, should be considered only when the manufacturing conditions are such as require the continuous production of large quantities of material.

Tubes, Billets, Etc.

For the continuous heat-treatment of comparatively short material, such as tubes, billets, etc., which do not necessitate the use of a chain conveyor, there may be employed, either individually or in series, the type of furnace illustrated at Fig. 6.

At the charging ends of some hardening furnaces, the material is placed on a table and delivered into the furnaces by an automatic motor-driven mechanism (enclosed beneath the table). The charging mechanism is employed to deliver the material continuously into the furnace and to fix the position, movement and time of exposure of each piece. The automatic movement begins immediately after the piece is placed on the table and is regulated by a controller convenient to the hand of the operator. To obtain the advantages of individual heating and cooling, the charging mechanism is so arranged that but one or two pieces are charged and discharged at the same time. This eliminates the disadvantages in heating and cooling common to any method that would involve the charging or discharging of a batch at one time. The charging mechanism is so arranged that the pieces are charged in proper rotation of heating and cooling.

At the discharging ends of other hardening furnaces, the pieces are plunged into the quenching bath one or two at a time at regular intervals, so that there is no opportunity for bunching of the material or irregular heating of the bath itself. It is but a matter of a second or two from the final heating in the chamber to the cooling in the bath and as there is practically no exposure to the atmosphere the steel is quite clean and free from scale.

The quenching tanks in front of other hardening furnaces are each provided with a conveyor and means for circulating the quenching fluid. The initial movement and sequence of the pieces as they are plunged into the bath is continued by the movement of the convey-

or, which with the circulation of the fluid insures a uniform cooling of each individual piece.

Covers are provided to protect the bath, and protected footwalks facilitate passage between the tanks.

The conveyors in some quenching tanks carry the pieces out of the bath direct to the charging ends of the tempering furnaces, as shown at Fig. 6, where the pieces are removed from the conveyor to the charging table of the tempering furnace, passing through this furnace in the same manner as through the hardening furnace.

At the discharging ends of the tempering furnace, the pieces are continually discharged, one or two at a time, in the order as charged. They may be removed by hand, as shown, delivered to a conveyor to facilitate the uniform cooling of each piece individually and the transfer of the pieces to the point of the next operation.

When the product is to be annealed only, the procedure is the same as that described for the tempering operation.

Advantages of Automatic Heating

When the product is heat-treated in this manner, with ordinary care and attention on the part of the operators the chances for error are minimized and the percentage of rejections greatly reduced. The individual pieces have been charged automatically and gradually brought up to the final temperature; the time of exposure has been determined identically for all pieces; the quenching has been performed rapidly with practically no fall of temperature between the final heating and cooling operations and without material exposure to the air. The delivery of each heated piece separately into the quenching bath and the sustained movement on the conveyor through the bath in the order of their

FIG. 9—Charging end of rotary furnace. This furnace is used for annealing pressed steel parts.

discharge from the furnace, together with the circulation of the quenching liquid itself, insures individual and uniform cooling.

Such individual and uniform heating and cooling results in a high-grade product, serves to decrease the cost of labor and fuel for the heat-treatment itself, the cost of subsequent machining operations, and the loss of product, rejected as the result of poor heat-treatment.

It is frequently desired to anneal nonferrous metal in a manner that will leave it clean or bright or both so as not to require pickling at the end of the heating operation. To accomplish this result with material in the form of wire and miscellaneous punched, drawn, spun, stamped or rolled parts, you proceed as follows:

The material to be heated is placed on the conveying chain at one end of the furnace, travels downward through a water seal and thence upward into a closed muffle. In this muffle the metal is gradually brought up to the required temperature as it progresses to the discharge end of the furnace, at which point it is carried downward through another water seal and is delivered clean and bright at the discharge end of the conveyor.

The determination of furnace size and the details of chamber arrangement and conveyor construction are dependent almost entirely upon the manufacturing and plant conditions. The furnace may be built with single or twin muffles.

This type of furnace, particularly in the larger sizes, should be considered only when the composition of the metal permits, when the size and shape of the material to be annealed are more or less uniform and the quantity sufficient to permit the maintenance of heat both day and night. Expansion and contraction, following the irregular heating and cooling, and the tendency of operators to rush the firing operations when starting, would unduly shorten the life of the muffle.

Rotary Furnaces

For the heat treatment of comparatively small pieces, the size and shape of which will permit of continuous stream movement, the internally-fired rotary type of furnace, with helical lining, may be employed for annealing, normalizing, hardening, tempering, bluing and miscellaneous heating operations on ferrous or non-ferrous metal products, such as bolts, nuts, rivets, balls, saw teeth, buttons, cups, shells, etc.

Recent modifications of this type of furnace, which has been used extensively for years, widen its field of usefulness to a point heretofore considered impracticable with this method of heating and handling.

While a great variety in methods is employed for conveying material to or from the furnace, details of which are determined largely by the size, shape and quantity of material to be heated and the general manufacturing conditions, the principle of heating and movement through the furnace remain substantially unchanged.

The material to be heated is charged into the charging drum in bulk, and a definite quantity is taken from the drum into the furnace at each revolution. The material is gradually and automatically wormed through the furnace, at a fixed speed, to the discharge end, where it reaches the desired temperature and is automatically discharged.

It is not only brought up to the desired temperature from the cold state gradually so as to afford time for the heat to penetrate its mass without overheating its surface parts, but each individual piece is constantly exposed to the heat in the chamber and to the ever-changing surface of the evenly heated spiral runway, being finally discharged at the time it reaches the desired temperature and degree of saturation.

The spiral runway permits a definite amount of material to pass from the charging drum into the furnace. Each charge is kept separate from the others, and each individual piece is continually turned over and moved forward in the spiral runway.

The horizontal position of the cylinder and the spiral runway are essential features of the furnace. Together with the variable speed regulator they determine the length of travel and time of exposure through the furnace, which is the same for each individual piece. This would be impossible in an inclined cylindrical furnace with a smooth lining and without provision to regulate the movement of the pieces. In such a furnace some of the material is bound to run ahead of the rest, causing a difference in time of exposure, resulting either in underheating or overheating. The horizontal position also permits of convenient charging and discharging and simpler driving mechanism.

Liquid or gaseous fuel, the control of which is practically the only manual part of the operation, is introduced into a combustion chamber at the discharge and meet the incoming cold material, which absorbs the heat in the gases, leaving their temperature comparatively low at the outlet of the charging end.

The temperature is continuously indicated, while the furnace is revolving, by pyrometer equipment adapted to the rotary motion of the furnace.

The lining, which is adapted to the temperature requirements and the nature of the material being handled may be refractory material or metal or a combination of both. In either case the principle of the spiral runway to move the material is retained.

The speed may be varied, according to the shape, size and quantity of material being heated, by means of a variable speed driving mechanism. When a conveyor is employed to deliver material to or remove it from the furnace, the speed of the conveyor may be synchronized with or maintained independently of the speed of the furnace.

When it is desired to quench immediately after heating, the furnace is generally provided with a quenching tank and conveyor, as shown at Figs. 7 and 8. With such construction, the material, in comparatively small lots, is discharged from the furnace into the quenching tank, being caught on the conveyor and discharged over the end of the tank.

Annealing Pressed Steel Parts

The successful adaptation of this

FIG. 19—Discharging end of automatic furnace for wire heating.

type of furnace, in the larger sizes, to the annealing of pressed steel parts in large quantities has indicated the possibilities of improving the quality and decreasing the cost in this line of manufacture. In the operation shown at Fig. 9, the heavy drawn steel cups are charged into the furnace hoppers directly from a conveyor carrying the cups from the press room. The cups are wormed through the furnace to the discharging end, where they are automatically discharged through a large covered chute to protect them from the atmosphere mtil they reach the quenching bath, from which they are automatically carried up and dumped into mill trucks.

By this method of protecting the material in its passage from the furnace to the quenching bath, it has been found in practice that the pickling operations usually employed in connection with low carbon-pressed steel parts have been very materially reduced and in some cases entirely abandoned.

The furnace illustrated at Fig. 9 has a capacity ranging from 5,000 to 10,000 pounds per hour.

In an installation of this kind, when the volume of work handled is sufficient, a furnace may be employed for handling the output of each press, thus keeping the press and annealing operations in step and maintaining a steady and continuous output.

The heat-treatment of large quantities of small and medium sized drop forgings, ranging up to 8 or 10 pounds each, is successfully accomplished with these rotary furnaces arranged in series for normalising, hardening and drawing. The sequence of operations and method of handling material from one furnace to another are similar to those already described. The advantages of this method from the standpoint of uniformity and cost per unit of output, are in marked contrast to the uncertain functioning of heating and handling equipment generally employed in this line

For drying small metal parts after washing operations, a modification of the rotary furnace may be employed.

In this method, the material, charged into the hopper, is conveyed automatically into the perforated revolving drainage drum, in which the greater part of the moisture is removed, the liquid discharging into a drip pan. The material is then conveyed through the dryer, in which it is thoroughly dried, finally passing out through a chute.

This type of dryer may be employed in the handling of chemical products which require a very thorough drying, for which operations the drainage drum is omitted.

The method of applying heat to the material is practically the same as that employed in the rotary heat-treatment furnaces previously described. The charge is divided into a number of small lots, each separated from the others and continually agitated, with a consequent uniform exposure of all surfaces to the heat, assuring a very thorough drying process.

Special Automatic Furnaces

The great variety in manufacturing processes and plant conditions frequently calls for automatic furnaces of novel design adapted to the individual requirements. One of these, shown by Fig. 10, is employed to heat the end or a portion of the body of hollow cylinders without overheating the base, in order to facilitate the subsequent crimping, tapering or other forming operation. Such operations are delicate and require very thorough annealing so that the structure of the metal may permit of forming without fracture or uneven surface.

This method of heating illustrates a departure from long-established heating practice, resulting in a very material improvement in quality and production cost. The practice for years on this operation had been to revolve the individual piece and play gas flames against the section to be annealed. Fairly satisfactory results could be obtained by this method provided the metal were a good conductor of heat and the walls of the cylinder comparatively thin, but more especially if the operator were careful not to prolong the time of exposure. As a low temperature is usually required in the metal and the temperature of the flame was nearer to the melting point than to the annealing point, there was great danger of overheating, heating unevenly or underheating, which danger was lessened only by the skill of the operator.

Another method was to suspend the pieces in a heating chamber or bath, the temperature of which approximated the temperature desired in the metal. While this method was better than the other from the standpoint of uniform heating, both were open to the added objection that the output was limited and the cost high.

To eliminate these disadvantages on large quantities of material, the operation is made continuous with the type of furnace illustrated at Fig. 10, which is built with single or twin conveyors to suit conditions. The type of conveyor used on this furnace makes possible the continuous employment of practically the entire length of the conveyor for handling material. In practice the operators are kept busy in loading and unloading the furnace, the output per man is increased hundreds of per cent. over the previous practice, and the loss due to improperly heated product is practically nil.

Another furnace of this type, designed to heat the ends of tubes for bending, is illustrated by Fig. 11.

With this continuous furnace the tubes are placed in holders forming part of a conveyor. Provision is made at one end to automatically fix the length of tube exposed to the heat, and at the other end to automatically strip the tubes from the conveyor. Means are provided to cool by water such parts of the conveyor as are exposed to the heat and to localize the heating of the tubes as desired. The tubes are brought up to heat gradually, and each is subjected to the same temperature, for the same length of time and in the same manner. The results in practice have shown a marked decrease in rejections due to improperly heated material, an increase in output due to the automatic method of charging, heating and discharging, and decreased fuel consumption.

When the manufacturing conditions prohibit the use of a strictly automatic type of furnace, it is frequently possible to modify the method of handling without materially departing from the principle of continuous operation. Whenever a furnace is required to meet variable manufacturing conditions, it is preferable to consider a type of furnace that is practically continuous through not strictly automatic in operation.

(To be continued)

FIG. 11—Discharging end of automatic tube heating furnace. The tubes are placed in holders which form a part of the conveyor. The length of tube exposed to heat is automatically fixed.

Steel Treaters Will Meet at Philadelphia

It Will Pay You to Attend This Convention—Herein is Given a List of the Papers to be Presented—There Will Also be a Huge Live Exhibit in Operation

By J. H. M.

TO say that a knowledge of Heat Treatment in these days of keen competition is essential is like mentioning that Columbus discovered America, both facts are already well-known. To those interested in the progress of the heat-treating branch of the industry, we can offer a valuable suggestion. We have already stated in a previous issue that the American Steel Treaters' Society are holding a huge convention at Philadelphia, Sept. 14 to 18 inclusive. We are now able to present the list of talent who will issue, through the medium of papers presented, the very cream of the progress and development in this most essential branch of industry. We advise you to read this list. It will surprise you. The talent is the best in America, and the information which will be issued in this way will alone be worth going to Philadelphia for. You do not require to be a member of this society to enjoy the privilege of attending this convention. Everyone is welcome. To the manufacturer we would say "Go yourself," and to the heat-treaters and those interested we would say "Tell your employer you wish to attend. He will more than likely pay your expenses as the information you will bring back will more than repay him." Let Canada be well represented at this convention, for in addition to the papers as listed below there will be a huge assortment of live exhibits similar to what we already explained in a previous issue of Canadian Machinery.

Following are the papers to be presented at the convention of the American Steel Treaters' Society and the Steel Treating Research Society, Commercial Museum, Philadelphia, Pa: J. D. Andrews, district manager, Brown Instrument Co., New York City. Pyrometers and their Application to Steel Treating. Arthur N. Armitage, Mesta Machine Co., Pittsburgh, Pa. Pyrometers from the Standpoint of the User. T. F. Baily, President, The Electric Furnace Co., Alliance, Ohio. Relative Economy of Electric, Oil and Gas-Fired Furnaces. Guy P. Bible, salesman, H. T. Potts & Co., Philadelphia, Pa. Relation of the Steel Salesman to the Development of Heat Treatment. A. Bensel, 1st Vice-President, Driver-Harris Co., Harrison, N. J. High Temperature - Resisting Alloys for Carbonizing. Eric E. Bilgart, President Pyromagnetic Instrument Co., Chicago, Ill., The Relation of Time, Temperature, Mass and Surface to the Hardening of Carbon steel.

James Brakes, Jr., Electrical Tester, Commonwealth Edison Co., Chicago,

Ill., Relative Economy of Electric, Oil, Gas and Coal-fired Furnaces. J. A. Brown, W. S. Rockwell Co., New York City, Factors Governing Production of Heated Product. N. E. Brown, Superintendent, U.S. Electrical Mfg. Co., Los Angeles, California. A Quenching Tank for Tool and Die Work. Peter Chambers, Owner, Angelus Steel Treating Co., Redondo, California. Forging Temperature of High and Low Carbon and Alloy Steels. E. F. Collins, Consulting Engineer Industrial Heating Devices, General Electric Co. Schenectady, N.Y., Relative Thermal Economy of Electric and Fuel Fired Furnaces, and its Influence on Process Costs and Plant Efficiency. J. D. Cutter, Climax Molybdenum Co., New York City, topic announced later. A. H. D'Arcambal, chief metallurgist, Pratt & Whitney Co., Hartford, Conn. Working Test of High Speed Steel, Illustrated. Ed. F. Davis, Celite Products Co., New York City. The Function of Insulation in the Heat Treating of Steel. Illustrated. A. K. Drury, H. A. Drury Co., Ltd., Montreal, Canada. The Heat Treatment of High Tensile Steels for Automobiles, etc. Wm. A. Ehlers, Industrial Furnace Engineer, American Gas Association, New York City. Fuels and Their Heat Utilization in Furnaces. Illustrated. Louis A. Elner, Assistant Metallurgist, H. H. Franklin Co., Syracuse, N.Y. Proper Heat Treatment for Carburized Steel Parts made from Plain Carbon Steels of Different Carbon Content, also from the Various Alloy Steels. The Advantages of the Double Quench, and the Direct Quench from the Carburiser. G. W. Fransheim, Supplee-Biddle Hardware Co., Philadelphia, Pa. Heat Resisting Metals. H. J. French, Bureau of Standards, Washington, D. C. The Heat Treatment of a High Chromium Steel. C. U. Geesey, Chemist and Metallurgist, The Garford Motor Truck Co., Lima, Ohio. The Functions of a Chemist and Metallurgist. R. L. Gilman, Experimental Heat Treater, Standard Steel and Bearing Co., New Haven, Conn. Carburizing. Hardening and Tempering High Carbon Alloy Steels in One Hundred Thirty Minutes. F. A. Hall, American Metallurgical Corporation, Philadelphia, Pa. Topic not announced. H. G. Hall. The Bristol Co., Chicago, Ill. The Essentials of Modern Pyrometry. H. H. Harris. Quigley Furnace Spec. Co., New York City. Containers Used in Heat Treatment Processes. Illustrated. O. E. Harder, Associate Professor of Metallurgy, University of Minnesota. Topic not announced. Dr. J. Culver Hartzell,

Dalton Adding Machine Co., Norwood, Ohio. No. 1—The Relation of the Electric Furnace to the Fabrication of Steels with Special Reference to the Chemical and Physical Changes Produced, and the Relative Economy of the Electric Furnace and the Gas, Oil and Coal-fired Furnaces. No. 2 — The Relation of Chemistry to the Fabrication of Steels, and the Effects of Small Percentages of Zirconium, Cobalt, etc., upon the Physical Properties of Steels. C. A. Isaux, Foreman, Avery Co., Peoria, Ill. Carburizing Temperature Best Adapted to Steels of Various Carbon Contents. O. C. Hedin, Treasurer, Gopher Machine and Tool Works, Inc., Minneapolis, Minn. Progress in Small Shop Hardening Methods. H. E. Hemstreei, General Foreman Spring Department, Sheldon Axle & Spring Co., Wilkesbarre, Pa. Heat Treatment of Axle Forgings for Heavy Trucks. R. L. Herrick, Editor, Raw Material, New York City. Molybdenum Steel. V. E. Hillman, Metallurgist, Crompton and Knowles Loom Works, Worcester, Mass. The Efficiency of Various Quenching Mediums and Their Applications. I. T. Hook, General Motors Co., Detroit, Mich. Factors Limiting the Strength of Materials. Carl Christian Jensen, Draftsman, Mahr Manufacturing Co., Minneapolis, Minn. The Current in the Thermo-couple. Charles Morris Johnson, Director of Research Department, Crucible Steel Co., Pittsburgh, Pa. Properties and Microstructure of Heat Treated Flame and Acid Resisting Steel. Illustrated.

G. W. Keller, Sales Manager, Brown Instrument Co., Philadelphia, Pa. - Injecting a Little Salesmanship into your Jobs as Metallurgists. R. B. Kerr, Foreman Heat Treating Department, John Deere Harvester Works, Moline, Ill. Hardening Die Blocks and Carburizing Thin Stock. W. O. Kellogg, General Combustion Co., New York City. Economy in the Use of Fuel Oil. Fred Viall Larkin, Head of Department of Mechanical Engineering, High University, South Bethlehem, Pa. Lesson learned from the Manufacture of Munitions and Ordnance in the Heat Treatments of Steels used in Peaceful Pursuits.

H. O. Loebell, Henry L. Doherty & Co., New York City. Oxidation of Steel During the Period of Heating. W. G. Lottes, Steel Expert, International Harvestor Co., Chicago, Ill. Various Quenching Mediums and Their Application. W. H. Lyman, General Superintendent, Warner Gear Co., Muncie, Ind. No. 1 —Relative Economy of Electric, Oil, Gas and Coal-fired Furnaces. No. 2—The

Role of the Metallurgical Laboratory in Relation to the Inspection Department. T. A. Lynch, Research Engineer, Westinghouse Electric and Manufacturing Co., East Pittsburgh, Pa. The Manufacture and Treatment of High Grade Helical Springs.

A. F. MacFarland, Metallurgical Engineer, Vanadium Alloys Steel Co., Latrobe, Pa. Topic not announced. W. H. Marble, Manager, American Stainless Steel Co., Pittsburgh, Pa. Stainless Steel. G. S. McFarland, Metallurgist, Jeffrey Manufacturing Co., Columbus, Ohio. A Research in Case-Carburizing. Illustrated. F. J. McIntyre, in Charge of Pyrometers, Hayes Wheel Co., Jackson, Mich. Practical Facts on Pyrometry. Marshall Medwedeff, Metallurgical Engineer, Samson Tractor Co., Janesville, Wis. The Role of the Metallurgical Laboratory in Relation to the Inspection Department. J. B. Morey, Foreman of Annealing in Wire Mill, Atlas Crucible Steel Co., Dunkirk, N.Y. The Efficiency of Various Quenching Mediums with Their Practice and Application.

W. F. Newhouse, Saranac Machine Co., Benton Harbor, Mich. Hardening Small Tool Steel Parts by the Molten Lead Process. T. Y. Olsen, Tinius Olsen Testing Machine Co., Philadelphia, Pa. Most Recent Developments in Testing Machines and Equipment for Metallurgical Use. Earl W. Pierce, Metallurgist, Maxwell Motor Co., Newcastle, Ind. The Role of the Metallurgical Laboratory in Relation to the Inspection Department and Machine Shop. Geo. Porteous, Supervisor of Forge and Heat Treatment, Minnesota Steel Co., Duluth, Minn. The Treatment of Open Hearth Steel of Suitable Analysis for Crucible Tool Steel, in the Manufacture of Hand and Pneumatic Chisels, Shear Blades and Kindred Tools. Chas. Ring, Chief Chemist, Ohio Steel Foundry Co., Springfield, Ohio. The Annealing and Heat Treatment of Miscellaneous Steel Castings.

Henry B. Smith, Metallurgist, and F. J. Olcott, Foreman Heat Treating and Forging, New Britain Machine Co., New Britain, Conn. The Forging and Heat Treating of Chrome Nickel Steel Spindles. E. P. Stenger, Metallurgical Engineer, Thompson and Black, Detroit, Mich. The Effect of Heat Treatment on Fatigue Strength. Illustrated. A. F. Strubing, Managing Editor, Railway Mechanical Engineer, New York City. Heat Treated Parts in Locomotive Service. O. E. Szekely, President, O. E. Szekely Co., Moline, Ill. Topic not announced. A. Tabachnick, Chicago, Ill. Effect of Repeated Heat Treatment Compared to Normal Heat Treatment. G. H. Trout, The Smith Gas Engineering Co., Dayton, Ohio. Producer Gas for Heat Treating Furnaces.

Henry Voltman, W. S. Rockwell Co., New York City. Factors Governing the Selection of Furnaces for the Heat Treatment Operation. D. A. Wallace, Waterloo, Iowa. Topic not announced. W. R. Ward, General Foreman, Tool

Treatment Department, Bethlehem Steel Co., Bethlehem, Pa. A Practical Aid to the Treatment of Steels. W. H. Wiegand, Vice-President, E. J. Codd Co., Baltimore, Md. Chain and Tubular Furnace Doors. Illustrated. J. A. Wilson, President and General Manager, The Brooklyn Steel Treating Corp., Brooklyn, N.Y. The Relation of the Commercial Heat Treating Plant to the Ultimate Consumer and the work of the Society in the Interests of Both. Henry Traphagen, Metallurgist, Toledo Steel Casting Co., Toledo, Ohio. Selection of High Speed Steels for Tools.

Major A. E. Bellis, Metallurgist, United States Arsenal, Springfield, Mass. Hardening, Quenching and Tempering High Speed Steel. H. O. Loudenbeck, Engineer of Materials, Union Switch & Signal Co., Swissvale, Pa. Topic unannounced. Mr. G. R. Brophy and Miss S. B. Leiter, Metallurgists, Research Laboratories, General Electric Co., Schenectady, N.Y. The True Action of Cyanide in Case Hardening Steel. Illustrated. R. A. Kunits, Chief Engineer, Advance Furnace & Engineering Co., Springfield, Mass. Topic not announced. H. P. MacDonald, Vice-President, Snead & Company, Jersey City, N.J. Electrical Heat Treatment of Steel. Illustrated by moving pictures. J. de Boves, President, Apex Steel Corporation, New York City. Heat Treating the Order.

C. B. Peck, Associate Editor, and J. C. Marsh, Metallurgist, Railway Mechanical Engineer, 720 Transportation Bldg., Chicago, Ill. "The Field for Heat Treating Locomotive Forgings." J. L. Thorne, Metallurgist, United States High Speed Steel & Tool Corp., 489 Fifth Ave. New York City. Heat Treatment of High Speed Steel. Robert B. Pottinger, Manager, R. B. Pottinger, Racine, Wis. Subject: "Forging, Welding, Tempering Light Steel." Theo. G. Selleck, Alfred O. Blaich Co., Chicago. Subject: "Case-hardening." Illustrated.

To give the program in detail would use more space than we can afford, but from start to finish this convention should be, to use an old familiar term, "A real humdinger." The ladies are also invited and special arrangements have been made to entertain all those attending, so the fair sex will no doubt be there in strength. Remember the dates, Sept. 14 to 18. Make it a point to be there.

INTERESTING DRILL TALE

An incident which has come to our attention, showing where electrical devices are helping to solve the skilled labor shortage, is worth repeating. In this particular case, a Cleveland manufacturer had plans for erecting a messannine floor to be supported by twenty-eight iron columns. In order to anchor these columns securely at the base, it was necessary to cut square openings through the floor, 13 inches x 18 inches. It was at this point that the difficulty was encountered. The flooring consisted of ¾ inch matched hardwood laid

on two-inch planks, both laid in a heavy concrete base. Sawing was impossible. By drilling a hole at each of the four corners and then chipping with a hammer and chisel, two men required four and two-sevenths hours to cut out the first opening. This was too slow, and then someone thought of drilling the entire outline of the desired opening with an electric drill.

By this latter process, the remaining twenty-seven holes were cut out with an average time of exactly one-third that of the former method. Had the first plan been followed through it would have required twelve ten-hour days for two men to complete the job. With the assistance of the drill, the entire work was finished in four ten-hour days or a saving of the time of the two men for eight full days. Figures were obtained of the combined wages of the two men who did the work, and it was found that eighty hours had been saved at the rate of $1.35 per hour, which equalled a total saving of $108. The electric drill used was a "Van Dorn," Code-D400 type, and its cost was practically paid in the saving in four days' operation.

The use of the electric drill for general millwright purposes in drilling or cutting openings in wood, iron and steel, etc., is something deserving of close attention.

USEFUL PUBLICATIONS

The Canadian Engineering Standards Association have informed us that a stock of the most important publications of the British Engineering Standards Association are now available for distribution. Of course a nominal price must be charged for these publications. Below is a description of some of these books.

Requests for copies should be addressed: Secretary Canadian Standards Association, Room 112 West Block, Ottawa, Ont.

24-1911. British Standard Specifications for Railway Rolling Stock Material. Part I. Locomotive Carriage and Wagon Axles. (Revised December, 1911.)—Alternative specifications (with and without limits on sulphur and phosphorus) are given for locomotive axles (crank and straight) and for carriage and wagon axles. 25 cents net.

24-1911. British Standard Specifications for Railway Rolling Stock Material. Part II.—Locomotive, Carriage, and Wagon Tires. (Revised December, 1911.)—Specifications are given for locomotive tires (with chemical analysis) and for carriage and wagon tires (both with and without analysis). 25 cents net.

24-1911. British Standard Specifications for Railway Rolling Stock Material. Part III.—Laminated, Volute and Helical Springs and Steel for Laminated Springs. (Revised December, 1911.)—Gives alternative specifications (with and without chemical analysis as to carbon sulphur and phosphorus) for each of the above. 25 cents net.

The Lubrication of Soft Metal Bearings

Friction, Flooded Lubrication, Bearing Design and Other Points of General Interest Are Discussed in This Article, Which Appeared in "Mechanical World"

FRICTION is the name given to the force which opposes motion and is, therefore, ever present between the journal and the bearing. It is found in all manner of mechanical devices and, strangely enough, is one of the most valuable and at the same time most destructive forces. Without friction, brakes would lose their value, and nuts would never be used on bolts. Trains would of necessity run on tracks provided with gear teeth, and we could not walk as we do now, but would be compelled to find other means of locomotion. Friction, however, is not desirable in bearings. Although much experimental work has been done on this subject, the laws of friction are as yet but little understood.

The surfaces of all materials which appear smooth are in fact made up of microscopic hills and valleys. When two surfaces in contact are moved relatively to each other, the clashing of the points creates a force which opposes motion. Wear results from this action, and the energy expended is converted into heat.

Fluids, as well as solids, show friction, and this has been described as the force encountered in rolling the particles of the fluid against one another. The laws of friction in fluids and solids are quite different, and these have been summarised as follows: For solids, dry or slightly lubricated, frictional resistance is proportional to the load; it is independent of the extent of the rubbing surfaces; except at very low speeds it decreases as the velocity increases.

In liquids the frictional resistance is independent of the load; it is directly dependent on the extent of the rubbing surfaces; and increases as the velocity increases.

The function of the lubricant in bearings is to separate the surfaces by a film so that metallic contact does not occur. If such a separation does take place the friction resulting will follow the laws for fluids. It has been well established by Tower that under conditions of perfect lubrication the journal is actually fluid-borne, and in this case the laws of fluid friction may be applied. He showed that when a bearing is plentifully supplied with lubricant the friction depends very little on the load or the character of the surfaces, but is dependent on the extent of the surfaces, the velocity and the character of the lubricant.

Tower's experiments were made with the load and bearing above a journal, the lower part of which was immersed in a bath of oil. He found that the journal carried the oil between the surfaces

and formed a film between them. One of the most interesting points of his experiments was noted quite accidentally. In the course of his work he had occasion to drill an oil hole at the top of the bearing and found that the oil flowed freely from it. He attached a pressure-gauge at this point and determined that a pressure of 200lb. per square inch was developed, although his load was only 100lb. per square inch of projected area. Later experiments showed that the pressure of the film at the top was greatly in excess of that at the sides and that it was greater on the discharge side than on the entering side. The thickness of the film has been determined as between 0.0013in. and 0.0029in.

However, in most applications such ideal conditions are not reached, and usually on starting the surfaces are in contact and subject to the laws of friction for solids. Lubrication is often interrupted or imperfect, by reason of improper distribution, and friction does not follow exactly the laws either of solids or of liquids, but is intermediate between them. This is the type of intermediate friction encountered in bearings with which the present paper deals, and it is necessary an indefinite quantity depending on all of the named variables. It will be seen that the matter of the character and supply of lubricant, as well as the nature of the surfaces, will be important factors in determining the friction and wear of the surfaces.

Since lubrication is so vital in the matter of friction and wear, prime consideration should be given to it in bearing design. Every effort should be made to create a film, although it is not always practicable nor desirable to provide bath flood, or forced lubrication. Various methods for supplying the lubricant are in use. Drop-feed lubrication, which is the simplest form, requires only a hole in the bearing through which the oil is introduced. Unfortunately, this hole is often placed at the point of greatest pressure, so that no opportunity is allowed for the establishment of a film. Introduction at the point of minimum pressure would probably reduce both wear and friction. Saturated-pad lubrication is employed in some cases, the most common example of which is the railroad car bearing. The bearing covers only the upper third of the journal, and waste, saturated with oil, is pressed against it from below.

Ring or chain lubrication is used on many line-shaft bearings and on bearings of electrical equipment. Chains or rings are provided of a diameter

considerably larger than the journal and resting on it, and these run in grooves in the bearing and through a reservoir of oil. Good results have been obtained by this method, and it is claimed by some authorities that conditions closely approaching perfect lubrication are reached.

Flooded Lubrication

Flooded lubrication consists of pumping the oil or carrying it by gravity in large volume to the bearing and delivering it at practically no pressure. Perfect films are often obtained, and the added advantage of dissipating the heat of friction brings it into use with large high-duty bearings. Forced lubrication is used in a limited number of cases. Oil is pumped to the points of maximum pressure and a perfect film is maintained. The pressure of delivery at the bearing must, therefore, be above the pressure of the film, and ranges from 15lb. per sq. in. to 600lb. per sq. in.

Grease lubrication is applied principally to heavy, slow-moving machinery. Considerable friction is encountered from the lubricant itself, but under heavy pressures the "body" of the grease prevents abrasion by the tenacity with which it clings to the respective surfaces and separates them.

Oil grooves are resorted to in many bearings in an endeavor to secure a film. However, when the film is once formed, the grooves are a distinct hindrance to its maintenance. Grooves should, in general, not lead into the region of maximum pressures, as in this case they may actually lead the oil away from, instead of towards, the place where it is most needed. Grooves should preferably be placed in the region of minimum pressures, and should run parallel to the axis of the shaft. Care should be taken to round the edges of the grooves to minimise the danger of injuring the film.

In bearings subjected to heavy loads the oil or grease may be entirely squeezed from between the surfaces when motion ceases. Grooves to the pressure points will provide convenient reservoirs of grease for starting and thus prevent abrasion, and this is the only case where such grooves should be countenanced. Errors in locating grooves may be avoided, to a great extent, by keeping in mind the desirability of securing films.

Another factor in securing proper lubrication is clearance between the bearing and the journal. Where the bearing covers only a portion of the journal, the latter can be made smaller in diameter, thus providing clearance at the

minimum pressure sides. This is often further increased by planing away additional metal from these sides. The amount of clearance desirable will vary with the velocity of the journal and the nature of the lubricant, but in general it can be said that too much clearance will decrease the opportunity for the formation of a film. The error, however, is often made on the other side—that is, too little clearance is provided. It should be remembered that the bearing is often rigidly held, so that with a temperature rise expansion of both bearing and journal tend to decrease the space between them.

Clearance should be provided between bearing and container whenever possible to allow free expansion. Without this, expansion of the back of the bearing may cause pinching off of the lubricant at the sides and what are apparently perfectly fitted bearings, when cool, may be in fact very badly fitted when they become warm.

Dissipation of heat from the bearing is a matter which is often overlooked. The heat of friction is usually carried away by radiation, but in some cases cooling is accomplished by currents of air, oil, or water. Water cooling is often employed, but in some cases this is not practicable and the bearing is called upon to run at high temperatures.

Bearing Design

Bearing design is sometimes checked up by the product of pressure, in pounds per square inch of projected area and velocity in feet per minute. Various values have been assigned, ranging from 24,000 to 1,720,000. One manufacturer of heavy machinery limits this value to 60,000 for ordinary lubrication, while 1,100,000 seems to be good practice for locomotive main crankpins.

As will be seen from some of the precautions in design, the gearing that has the best lubrication will last longest, other things being equal. Grit and dirt will often start scoring, and it may be of interest, in passing, to note that this remedy is sometimes used in curing hot boxes. Bearings are occasionally so tightly fitted that little lubricant can enter between the surfaces. Minute oil grooves may then be secured by introducing a small quantity of powdered emery, which makes circumferential scratches on both the journal and bearing surface. Care should then be used in cleaning the emery from the lubricant, as abrasion to a serious extent may be caused. Clean bearings, well lubricated and kept in alignment, should give little trouble when properly designed.

Shipyard Not Closing.—There is no truth in the rumors that the Davie Ship Building plant at Levis, P. Q., was about to shut down. The rumor also included the subsidiaries of the plant, and is equally unfounded in regard to them, according to Mr. A. A. Wright, vice-president of the company.

RUBBER VULCANISING

The story of the rubber industry is very entertaining, as told in Mechanical World. They state as follows:

The rise of the rubber industry, which was founded in Manchester by Macintosh in 1825, to its present position and magnitude, may be directly attributed to the discovery by Goodyear, in 1839, of the process known as vulcanisation. Raw rubber, which Macintosh employed in the earliest days as a proofing for fabrics, is by no means a suitable material, from the manufacturer's point of view, for the production of a finished article, in spite of the fact that it possesses a number of unique properties. Amongst other drawbacks to its use are the facts that it is strongly adhesive, and that it is sensitive to changes of temperature, becoming soft and tacky at a summer heat and stiff and hard in the winter. These defects alone would have sufficed to limit very seriously its uses in industry. In fact, it is safe to say that but for Goodyear's investigations and subsequent discovery, the industry would have been one of minor importance—if, indeed, it had continued to exist at all.

Goodyear found that by incorporating rubber with sulphur and heating for a suitable period to a temperature of about 140 degrees C., rubber becomes profoundly modified and acquires new properties which render it eminently suitable for the manufacture of a host of useful articles. It loses its adhesiveness, becomes resistant to temperature changes, and exhibits greatly enhanced strength, elasticity and durability. A year or two later, Hancock, in England, independently discovered the process, his method consisting in immersing sheets or formed particles made from raw rubber in a bath of molten sulphur at a temperature of 135 degrees to 140 degrees C. In principle the two methods are the same.

In 1846 Parkes showed that rubber could be superficially vulcanised by immersing it in a cold dilute solution of sulphur chloride in carbon bisulphide. This method is adopted to the vulcanisation of thin sheets or films of rubber only.

From these early days up to the present time all manufactured rubber goods have been vulcanised by one or other of these processes, that of Goodyear finding by far the widest application.

An entirely new process for affecting vulcanisation has recently become available as the result of the discovery, by Mr. S. J. Peachey, Lecturer in Chemistry at the Manchester College of Technology, that by exposing rubber alternately to the action of two gases—viz., sulphur dioxide and hydrogen sulphide—it becomes rapidly and completely vulcanised, even at the ordinary temperature. The process appears to be of fundamental importance for the following reasons:—

1. It is a true sulphur vulcanisation

(as distinct from the sulphur-chloride treatment) and yields a product entirely comparable with that obtained by the Goodyear process.

2. It eliminates the use of heat, and to a great extent the use of mechanical pressure.

3. It employs two gases, both of which can be produced on a large scale at a very cheap rate.

4. It is rapid in action.

5. It enables the manufacturer to employ organic filling agents, such as leather waste, sawdust, shoddy waste, and the like, which cannot be used in connection with the hot process. In this manner a number of cheap and highly durable materials may be fabricated from the various wastes, and employed as floor and wall coverings, for boot and shoe manufacture, and for upholstery work.

6. Coal-tar dyestuffs, and even natural dyes, which, with a few exceptions, are destroyed by the "hot cure," can be introduced into mixings to be cured by the new process with the production of delicate tints and shades hitherto unobtainable.

The process has the advantage of extreme simplicity, and its translation from the laboratory to the works should prove a very simple matter.

Already numerous samples of daintily coloured floor-coverings, fancy leathers, and felts, suitable for hat making, have been produced in the laboratory, and these have an appearance and a finish which could hardly be improved upon. Several pairs of boots have been soled with leather reformed from waste by the new process, and the practical test of several months' hard wear has shown that the reformed leather is even more durable than the real article.

The process can be extended to the vulcanisation of rubber in solution. If a solution of rubber in benzole or naphtha be saturated or partly saturated with hydrogen-sulphide and mixed with a solution of sulphur-dioxide in the same solvent, the liquid sets in a few moments to a stiff jelly, and on eliminating the solvent by evaporation a fully vulcanised rubber is obtained. The use of the mixed solutions for producing perfectly vulcanised seams and joints has proved highly satisfactory in practice, and inner tubes repaired in a few moments by the new process have an excellent life. Further, by the aid of the solution process, reformed leather soles and heels may be attached to boots without the aid of stitching or nailing, and indeed a boot may be produced from the reformed leather without a single stitch being necessary.

The above-mentioned applications of the process have been fully worked out in the laboratory, but they do not exhaust a fraction of its possibilities — and new technical effects in many other directions are continually presenting themselves.

The Peachey process is fully protected all over the world.

WHAT OUR READERS THINK AND DO

TAPER TABLE

By J. H. M.

The accompanying table of standard tapers should be of decided interest to every reader as it represents the latest table issued by the Brown & Sharpe Co.,

of Providence, R. I. This firm has made certain changes which they have found advisable, and we would strongly suggest that readers cut the table out for further reference. Should you so desire, you could write the B. & S. people for further copies of the chart.

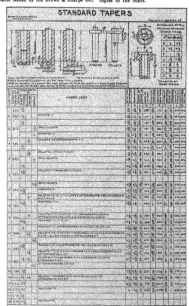

THIS TABLE IS WELL WORTH PRESERVING.

WOOD STEADY RESTS

By Donald A. Hampson.

Shops doing general repair work often feel the need of a larger steady rest for the lathe or lathes, as the case may be. This is noticeably true of those shops at long distances from the larger cities—shops that are usually none too well equipped and that endeavor to serve the wants of a community which may range from automobile work to power plant and mining machinery repairs.

The pocket book of the owner cannot stand the price of a big lathe—one that may stand idle eleven months in the year. And then anyway, the capacity of steady rests as regularly supplied is from one-quarter to one-third the largest diameter that may be turned in the machine. Drums, pipe threading, transmission machinery parts, special castings—these are a few of the more common parts that come to the lathe for some job that requires a steady rest much bigger than the one on the floor.

The photo shows the one best way out of these jobs. This is a drum for a variable speed drive—26″ at the big end, both ends to be bored out, and that in a 28″ lathe.

Two inch maple blocks form the rest. The lower one is cut out to the radius of the drum the small end of the drum with two notches at the bottom to line it with the ways of the lathe. The top piece is cut to two radii—one for the small end and one for the big end when the piece is reversed. In the foreground may be seen a bar 1″ x 2″ steel secured to the lathe by heavy C clamps; into this are tapped rods that draw down on a strap passing over the top wooden piece. The whole forms a secure and satisfactory arrangement and one that enabled two of these drums to be refitted in quick time.

Such blocks are put away to be ready

VIEW OF THE STEADY RESTS.

for some future job or re-cut for a different one if there is no possibility of needing again for the original purpose. But few shops have pipe centres for anything over 6" size, and when a piece comes along to be cut off or made up into a "Dutchman" they have to turn the work down. One of these wooden rests will do the trick most acceptably.

COLLARS FOR MILLING MACHINE ARBORS

By G. BLAKE

Collars for milling machine arbors are in constant demand in the machine shop, and it will be readily admitted that the majority of good first-rate shops, although stocked in tools and equipment, generally suffer from a lack of suitable collars. In most cases they are worn and badly knocked about, this being the result of the collars having been made from mild steel and left soft. How often does one notice this state of affairs and make up their mind to have proper sets made up hardened and ground.

With this object in view, the writer got out the following tables of sizes and standardized them in the works where he is engaged as mechanical superintendent. The collars were made from ordinary black mild steel, which will caseharden quite well for the purpose. The cost from a manufacturing point of view is not heavy, as they can be readily made on a capstan or turret lathe. The turret and capstan lathe is looked upon by some solely as a production machine,

but it is surprising to find that it is a very valuable asset for the toolroom, for roughing out work, for turners, and a boy can run the capstan, thus lightening the heavy costs of the toolroom. Of course such a condition can only apply where there are a number of centre lathes in the toolroom and sufficient work to justify the inclusion of a capstan lathe.

Returning to the subject of milling collars, the tables were got out as will be seen for 3 different diameters of arbors, 1 in., 1¼ in., and 1½ in., they being the most common sizes in general use.

Other special widths of collars can be included to suit the particular work engaged in, but the sizes given cover a wide range for milling work. The idea of making the collars on the capstan is that once set up they can be produced very cheaply in batches, whereas in many instances they are made in the toolroom by a skilled man, thus sending up the cost.

THE "JENTIL" PROCESS

The efficiency of the "Jentil" process for cleaning, derusting and rustproofing, is a development of the munitions making for the British Government. The "Jentil" process produces a non-magnetic oxide which thoroughly protects and at the same time cleans off rust, paint, enamel, lacquer, grease, etc., leaving a pleasant ebony finish. Rust, no matter how deep, may be removed, leaving one homogeneous metal. No distortion of the ferrous articles treated takes place. It is claimed that there is no

measurable difference in the size of any article after treatment. The process renders every article black in appearance, a rusty or dull surface will finish a dull black after treatment, and a burnished surface prior to treating will finish a burnished black.

The time required in cleaning, derusting and rustproofing will vary from fifteen to forty-five minutes, according to the size of the article treated, and at a very low cost. During the process the articles never come into contact with any acids. In cases where it is desired to lacquer, japan, enamel, etc., it is claimed that the "Jentil" finish provides an excellent primary or base, which does not discolor or otherwise affect the subsequent enamelling. No previous treatment is necessary when putting on a "Jentil" finish. Articles going through the process do not require to pass through a bath of pickling acid; rough iron castings can be treated without sand-blasting, or other form of cleaning. One advantage of this process is that assembled articles may be treated with equal satisfaction.

The uses for this treatment are so varied that it would be almost impossible to mention any particular industry to which it would especially apply, as it may be economically adopted to almost every commercial and industrial requirement. The "Jentil" process is being placed on the Canadian market by the Jentil Cleaning, Derusting & Rust-protecting Co., Limited, 137 McGill St., Montreal.

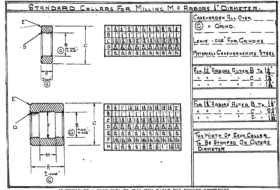

STANDARD COLLARS FOR MILLING M⁄C ARBORS 1" DIAMETER.

IT WOULD BE A GOOD IDEA TO CLIP THIS TABLE FOR FUTURE REFERENCE.

DEVELOPMENTS IN SHOP EQUIPMENT

WIRE NAIL MACHINES

The Joseph T. Ryerson & Son, Chicago, Ill., have placed on the market what is known as their line of Ryerson-Glader wire nail machines.

These machines are built in six sizes, similar to the one shown in Fig. 1, and range from the No. 00 to the No. 4 size. The production capacities of the six sizes range from 500 nails per minute for the No. 00 to 175 nails for the No. 4.

All important adjustments are within easy reach of the operator and can be made while the machine is running. All parts subject to strain or wear are of steel.

The wire is fed automatically from a bundle or coil between straightening rolls and then into the dies. While the wire is gripped in the dies the head is formed with one blow of the heading hammer, and as the latter recedes the straightener carriage moves forward the length of stock required for a nail. The wire is again gripped and the point cutting dies both make the point and clip it off, leaving enough stock outside the dies to form the next nail head. In advancing, the heading hammer operates an ejector which removes the finished nail. The top view, Fig. 2, shows the operating parts of the machine.

The wire is straightened by the straightener rolls moving back over the wire while it is gripped in the dies. The crankshaft is counterbalanced to compensate for the thrust of the header cross-head. The equipment furnished consists of one set of gripping dies for the larger size of wire to be used in the machine, together with two pairs of point cutters.

ELECTRIC-WELDING OUTFIT

The illustration shows a reproduction of a photograph of a single-operator electric-welding equipment manufactured by the Westinghouse Electric & Manufacturing Company, and is claimed to be exceptionally efficient because the generator operates at arc voltage and no resistance is used in circuit with the arc. The necessity for providing automatic moving devices such as relays and solenoid control resistors is eliminated. There are 21 steps provided, from 50 to

GENERAL VIEW OF THE OUTFIT

225 amperes, giving a current regulation of less than 9 amperes per step, making it much easier for a welder to do vertical or overhead work necessary in railroad shops.

For portable service, the motor generator set with the control panel is mounted on a fabricated steel truck,

equipped with roller bearing wheels, and is easily hauled about the shop or yards by one man. The suitable plugs and receptacles for three-phase, three-wire or two-phase four-wire, direct or alternating motors allow the set to be quickly and conveniently connected to the supply circuit at any desired point. Only one plug is desired for the motor but the number of receptacles required depends upon the number of points at which it is desired to do the welding work. These plugs and receptacles are supplied on special orders.

The panel is made up of a single section upon which are mounted indicating instruments, protective apparatus and switches for adjusting the welding current. If additional circuits are required it is necessary to use separate outlet panels.

Each outlet panel consists of slate panel on which is mounted a circuit breaker, line switch and necessary single pole resistance adjusting switches for regulating the heat at the arc by varying the amount of current supplied. These outlet panels may be supplied for either metal or graphic electrode welding or both and may be operated simultaneously and independently of each other, the only limit being the capacity of the generator. A metal or graphic electrode holder is supplied with each welding panel, according to the work for which the panel is designed. The holders are of light weight, are well balanced and are designed so that they may be used continuously without overheating.

GENERAL VIEW OF THE WIRE NAIL MACHINE.

PLAN VIEW OF THE SAME MACHINE.

CONTINUOUS MILLING MACHINE

The multiple spindle continuous vertical milling machine which is here illustrated and decribed is now being built by the Betts Machine Company, Rochester, N.Y. The machine as shown is provided with three spindles, but it can be furnished with additional spindles to meet any requirements of work where such a design is considered desirable.

It is intended for heavy production milling on duplicate parts and is of very simple and rigid construction. The spindles are steel forgings and they are driven through long splines and spur gears. Each of the spindle carries a milling cutter and may be adjusted vertically by hand. This, with a four spindle machine may be fitted with two roughing cutters and two finishing cutters, thereby completing the operation in one cycle. The power for driving this machine is furnished by either a pulley or an individual electric motor at the top of the machine.

The table has a flat bearing on the bed, and a split tapered bushing provides for taking up wear in the bearing on which the table revolves about the column. The table is driven through a large internal gear, all bearings are bronze bushed and all gears run in a bath of oil. Four rates of continuous feed are obtained through sliding steel gears. None of the gears are ever in mesh except those that are actually transmitting power, so that there is no unnecessary wear.

The holding fixtures are carried on the table that revolves continuously, and no time is lost in chucking the work as the

GENERAL VIEW OF THE MILLER.

pieces are changed while the fixtures are passing from one cutter to the next. It is claimed that one of these machines will take care of as much work as can

be conveniently handled by two men, and depending on the nature of the work it has a production capacity three times as great as a single spindle machine.

NEW THINGS IN MACHINE TOOLS

HEAVY DUTY MOTOR

The Westinghouse Electric and Manufacturing Company are placing on the market a special "H K" motor for providing maximum efficiency where heavy starting torque is required, such as that of crane and hoist service. This motor has been designed to meet the demands where the load consists of a series of starts, stops and reversals. A notable feature in its construction is its compactness, giving small overall dimensions, light weight and great mechanical strength. A blower is placed in the rotor which effectively ventilates both armature and field windings. Little energy is required to start the motor owing to the small diameter of the armature. To utilize to advantage the ventilating feature of the "H K" motors they are rated on the one-quarter hour basis.

GEAR HOBBING MACHINE

The Cincinnati Gear Cutting Machine Company has recently completed and placed in service a new type of gear hobbing machine. The capacity of the machine has a maximum for gears of 17¼ inches diameter and 12-inch face. Spur and helical gears may be cut, the latter either right or left and up to 4 degrees of angularity. The work spindle is horizontal and is rigidly supported by two long bronze-bushed bearings. The thrust bearings are of alternate steel and bronze washers. The outer bracket that supports the work arbor is of interesting design, being made in two portions, only one of which is released and backed off when removing the arbor. The second section is provided with a special attachment so that accurate return of the work arbor is assured after reloading. The hob spindle is driven by helical gearing to provide smooth running, and is adapted for eight changes of speed. Automatic knockout is provided for stopping the machine at the finish of the cut. The machine may be entirely controlled from the operator's position in front. The net weight of the machine is about 5,600 pounds.

LATHE ATTACHMENT.

A very useful milling attachment for lathes has recently been developed by the Hinckley Machine Works, of Hinckley, Ill. It is adapted for placing on the tool slide of the lathe and its construction permits of doing a wide range of work speedily and economically.

BAR AND PIPE FORMER

The Wallace Supplies Manufacturing Company, of Chicago, Ill., have recently produced a machine for the bending of cold bars to any desired shape, and also various sizes of pipe. The machine is power driven and adjustable stops are provided to obtain any degree of bend required. When angles, channels and similar shapes are worked forms and follower bars are necessary. The machine stops automatically when the bend is competed. Special provision is made for the prevention of distortion when bending.

LIFTING AND TIERING TRUCK

The Automatic Transportation Company, of Buffalo, N.Y., are now making a lifting and tiering truck, the duty of which is to pick up and elevate loads to any desired height up to six feet, such as the loading of material on to box cars, waggons, platforms, etc. The entire mechanism is operated electrically, and the machine has a capacity of about two tons. The platform is carried by two strong uprights and is raised or lowered by means of a large single pitch screw revolving in a heavy bronze nut. The speed of the lift is about four feet per minute. Automatic limit stops are provided to prevent overrun of the platform at either extreme position.

TURRET TOOL HOLDERS

The Lovejoy Tool Company, Inc., of Springfield, Vt., are introducing to the trade a new line of turret tool holders. The turrets are made of hardened steel and are interchangeable with any base, a feature that permits of using a number of turrets with the one attachment, so that the tools for one setting do not have to be removed from the holder, thus giving quick and accurate changes. One movement of the binding lever releases the turret, accurately indexes it to the next tool position and again rigidly clamps it to the base. These turrets can be used on lathes having a centre distance above the tool block as small as 1 3-8 inches.

PNEUMATIC RIVETER

The Baird Pneumatic Tool Company, of Kansas City, Mo., are placing on the market a specially constructed pneumatic riveter. The machine is supported on a stand and is intended for use in riveting traction plates on the rims of pneumatic tires, but with slight modifications, can easily be adapted to a great variety of work.

The MacLean Publishing Company
LIMITED
(ESTABLISHED 1887)

JOHN BAYNE MACLEAN, President. H. T. HUNTER, Vice-President
H. V. TYRRELL, General Manager.

PUBLISHERS OF

CANADIAN MACHINERY
AND MANUFACTURING NEWS

A weekly journal devoted to the machinery and manufacturing interests.
B. G. NEWTON, Manager.   A. R. KENNEDY, Managing Editor.

Associate Editors:
J. H. MOORE   T. H. FENNER   J. H. RODGERS (Montreal)
Office of Publication: 143-153 University Avenue, Toronto, Ontario.

VOL. XXIV.   TORONTO, AUGUST 19, 1920   NO. 8

PRINCIPAL CONTENTS

The Council of Action

THE creating of what is known as a Council of Action by the leading trades unions in Great Britain, with the avowed object of dictating to the Government its policy on certain questions of national importance, creates, as Mr. Lloyd George plainly states, "the most formidable challenge ever given to democracy." There can be no doubt that the people of Great Britain generally are war weary, and not labor alone. The greatest beneficiaries of the war were the trades unionists, and some of the capitalists, though not by any means all. It is the great middle classes who suffered the most and got nothing at all out of it who would probably be the most unwilling to see their country embarked on another period of war. Their voice, however, is not heard above the clamor made by the representatives of organized labor, which, after all, is by no means a majority of the community, but, on the other hand, a relatively small portion. Organized labor in Great Britain has, since the war, been frankly out to get everything it could possibly extort, and has not been afraid, in this endeavor, to attempt the complete paralysis of the country. The great transportation strike should have taught their representatives that when they go in opposition to the general public they are compratively powerless, and it is quite possible that in making their threats of a general strike in the event of Britain taking action in Poland, they did so with their tongues in their cheeks. As a matter of fact, the British Prime Minister says that the aims of the Government and the aims of the Council of Action are practically the same, but, he plainly and positively states that were they not, "dictation would be resisted by all means in the power of the Government." Trades unions can serve an excellent purpose, and have been instrumental in making life a more pleasant thing for the trades unionist. When they start to assume the role of the Government it is time to call a halt. While every right-minded person wishes to see labor, both organized and unorganized, get all the good out of life that it is possible to obtain, no sane person wishes to see any minority of the people holding a club over the rest. Lloyd George is to be congratulated on his uncompromising stand against such coercion.

The Mechanic of the Future

WITH no attempt at disparagement towards the present-day mechanic, we can point with pardonable pride to the old-time machinist whose locks may be quite few, but his experiences many. Talk to him a short time and he will unfold to you tales of accomplishments under the greatest of difficulties. "We had to be all-round men in those days," he will say, with a wag of his head, which brings us up to the point of weighing in our mind the calibre of present-day mechanics.

While still a trained workman, he is a specialist rather than an all-round man. He may be a lathe hand, milling machine operator, bench hand, etc., but in very few cases a man that can tackle intelligently any machine in the shop. Why is this so? For various reasons. It may be that he has never served a full apprenticeship, or he may have been an easy-going chap, content to be an adept at one particular duty. Then, again, he may never have had a proper chance to become an all-round workman. Many plants, unfortunately, encourage their mechanics to become specialists, and while this may be an easy way to obtain maximum production, what of the mechanic of the future?

A young chap need not serve a full apprenticeship these days to make real money. All he need do is serve a year or so, then shoot off to some automobile manufacturing plant, take over a mechanic's (?) job, and make a mechanic's wages. It is hard to credit that young fellows 19 years of age can make as much, and sometimes more, than a first-class tradesman by such a course, yet we have proof that this is so.

Do such men ever become real mechanics? To our mind, NO! They are merely handy men at a particular duty. Except we get together and make an apprentice ship in the machine tool trade attractive for the young man of to-day, the time is drawing near when the all-round machinist will be a thing of the past.

The Engineer's Opportunity

THE organization of the Federated Engineering Societies at their recent conference at Washington, D.C., is certainly a step in the right direction.

About 75 branches of the engineering profession were represented, these including Civil, Mechanical, Mining, Electrical, Sanitary, Heating, Marine, Agricultural, Radio, Refrigerating, Fire Protection, Industrial, Automotive, Military Engineers, Architects and Geologists, etc.

The need of such a body has been felt for years, but the problem was to successfully affiliate the various interests. The action taken at Washington has removed the only remaining nigger in the woodpile.

The object of this newly formed body is both ambitions and essential. They desire to raise the standing of engineers generally, and to teach the public how much they really depend upon the engineering profession. Any society, club or organization, the chief object of which is the advancement of the knowledge and practice of engineering and the application of applied sciences, and which is not organised for commercial purposes, is eligible for membership.

The success of this society, as we see it from our side of the line, depends entirely upon the engineers themselves. With proper support this body should grow to be a powerful influence and will form the official mouthpiece, so to speak, for engineers all over the country.

To our mind the forming of this society should be greeted with great enthusiasm by engineers in general and they should rally to its support. An opportunity such as this affords is not to be treated lightly. We will watch with interest the development of the organization, and once more emphasise the necessity of every engineer becoming a live member and actual booster.

MARKET DEVELOPMENTS

Pig Iron and Coke the Leading Features

There Have Been Advances in Most Grades of Pig Iron, Steel Products Not Showing Much Strength—Coke Situation Likely to Become Easier with Improvement in Car Supply.

IN the local market conditions are very much the same as they were last week. There have been no changes in prices of machine tools, or in stocks handled by the steel warehousemen. The same causes are operating, viz., summer vacations, tight money, and crop expectations. It is probable that if any change of prices occurs it will be in an upward direction, rather than downward. In the United States there has been a general advance on all grades of pig iron, while steel products have shown a tendency to decline. The coke situation which has been very acute, through car shortage, now is expected to show an improvement, and this in turn may affect the pig iron. At the present time, makers of pig iron are very loath to take on business for 1921, as they have no idea of what costs are going to be. There has been considerable curtailment of production among automobile manufacturers, though some of the principal firms are still working at full capacity. The falling off in these trades is expected to be made up by large buying by the railroads once the new freight rates come into force. In the case of the Illinois Central Railroad, which has an enquiry out for over fifty machine tools, it is thought that the purchase of these was decided on before the freight rate increase. One effect of the falling off in the automobile trades, and the cancellations occasioned thereby, has been a speeding up of delivery in the machine tool market. The old material market has shown a strong upward tendency in practically all grades, heavy melting steel being particularly strong. This, of course, refers to the U. S. market, the local market being quiet.

PIG IRON AND COKE ARE THE INTERESTING MATERIALS AT PRESENT

Special to CANADIAN MACHINERY

PITTSBURGH, August 19.—In the general iron and steel market situation pig iron and coke occupy the centre of the stage. Steel products are quiet and merely show, on the whole, a slight declining tendency. Pig iron on the other hand has lately assumed a very aggressive attitude, with some furnacemen predicting much higher prices, and some sales being made at actual advances. It is true that other furnaces deprecate advances and state they will be the last to quote higher prices. The advances in pig iron, when made, have been based upon scarcity rather than upon volume of demand, while the common excuse of furnacemen has been the high price of Connellsville coke. For some time this excuse held good, or at any rate coke stayed at an altogether fancy price. Now, however, coke seems to be lined up for a decline. At the moment of this writing no actual decline is found in the market but in some quarters it is held that everything is set for a decline within a couple of days or so.

A tense situation is accordingly produced, with pig iron having assumed an advancing attitude, partly on account of coke, and coke now threatening to decline and remove this supporting factor.

In some quarters it is thought that the furnacemen who of late have been actively engaged in endeavoring to push the market upwards were doing so because they felt that while the pig iron market could not stay up for any length of time there was a combination of circumstances in which they could push the market up temporarily and make something out of the movement. If the market was about to decline some day it might as well decline from a higher as from a lower level. Some furnacemen do not agree with this reasoning and hold instead that if the market were held steady it would last longer than if it rode for a fall.

Prices in Detail

For some time past foundry iron has been quotable at $46, valley. For a fortnight or more predictions have been common that $50 iron was going to be seen shortly, and two merchant interests at least announced that their price was $50 and they would not sell until they got that price. One of the two has now succeeded, making a sale at $51, furnace, equivalent, considering the point of delivery, to about $50.50, valley. The quantity was 500 tons, for delivery over the remainder of the year, and the buyer a New England foundry interest.

It chances that the brand sold is a very well established one and the buyer might have paid the price asked on account of desiring that particular brand for the mix, so that the one transaction while interesting does not mark up the market as a whole.

In basic pig iron the curious situation is presented that over a week ago a sale of a good sized tonnage was reported at $48.50, valley, when the recognized market had been $46.50. It is alleged that the reputed buyer denied having made any purchase at any price. The market could hardly advance $2 without being active at the same time, yet since that report no sales at any price have been reported. In a normal market, if the sale reported made the market, other transactions should have followed by this time, and thus a very curious situation is presented. In some quarters "the market" on basic iron is regarded as $48.50, while in other quarters the market is still called $46.50.

In the case of Bessemer iron, which had been $47, valley, there have been some odd lots sold at this price in the past few days, while on the other hand one sale is reported at $50. The trade is decidedly puzzled as to where prices really do stand.

Improved Coke Situation

Monday, August 2, saw a great improvement in car supplies in the Connellsville coke region. Monday is always a particularly good day, but the improved placement continued all week,

and the next week Monday and Tuesday saw practically 100 per cent. placement, the supplies during the remainder of the week being relatively good also. The Connellsville Courier reported coke production in the week ended August 7 at 194,140 tons. This was 21,270 tons gain over production reported in the week previous, and about 5,000 tons more than the best production that had been reported since the middle of April, when the outlaw rail strikes suddenly reduced the production greatly. In the week following, not yet reported upon, it is believed there was a further increase.

The improved car placement seems to rest upon a solid foundation. It is not a case of a few more cars getting into the service, as there were plenty of cars as it was, the difficulty being that they moved altogether too slowly. One furnace after another was reporting that it had two or two and one-half times as much coke en route as was normally necessary in order that daily receipts should average up to the requirements, and of course those cars were placed at coke works less than half as often as normally. Now, however, the furnaces report much better coke reports to-day that coke has been getting through from ovens in three days, when not long ago many cars took a couple of weeks. By reason of heavier shipments and also the quicker movement some furnaces have received more coke in the past few days than they are able to unload.

The outlook accordingly is that there will be a sharp break in the spot coke market, which in the past few weeks has been at the record high level in the history of the Connellsville coke trade. Up to this writing no definite decline has occurred in the market, but in most quarters the decline is momentarily expected. Generally speaking furnace coke is quotable at $17.50 to $18.50, though rarely has $18 been shaded. Foundry coke has been showing a wide range according to quality, or lack of quality. Ordinary 72-hour coke, which is produced in large quantities when car supplies make the running time so irregular, has been available at $19, but most foundries would not consider the $19 coke as really of foundry quality. For $19.50 some coke has been obtainable that would pass muster with perhaps the majority of furnaces. For selected brands higher prices have been paid, and after coke had passed through one or two brokers some foundries remote from the region have paid as high as $21 per ton at ovens.

Steel Market Quiet

The steel market continues quiet, even quieter than it was in July. It is waiting for some definite lead. Several sales of sheet bars have gone through at $70, against $75 previously paid, and $70 might possibly be shaded. Six steel interests are counted up as being willing to sell at not over $70. Finished steel products are hardly changed. While the market has been very dull the unfilled tonnage statement of the Steel Corporation indicated that the corporation sold 150,000 or 200,000 tons more in July than in June, this of course being due to the independents having so much higher prices than the corporation.

LOCAL MARKET STILL ON VACATION

TORONTO.—Looking for market news in the Toronto machinery market is like looking in a dark hole for a black cat that isn't there. Most of the men in this business seem to be out of town for vacations, and those that are not might as well be. There is practically nothing moving, although there is a certain amount of enquiry. There is little real buying sentiment behind this, and it is more in the nature of a feeling out of the market than anything else. Naturally, there has been no change in price lists since last week, although it is by no means certain that there may not be some slight increases in the near future. Deliveries in some lines are better, and in the case of milling ma-

chines, these can now be supplied in some instances from stock. Steel warehousemen state that they are keeping going in a quiet way, and have nothing to complain about. considering the general conditions. Deliveries are if anything better than last week, though the change is very slight. There is little likelihood of any change in prices. There is no doubt that the first two weeks in August are sustaining their reputation of being an off season. The prospects of the Toronto ball team seem to be a livelier topic than the prospects of the machinery market just now. When we get through the holiday season, and the exhibition, and get the harvest gathered, perhaps we shall settle down to business.

THE RAILROADS ARE DOING SOME BUYING NOW AND MORE EXPECTED

Special to CANADIAN MACHINERY

NEW YORK.—While business in machine tools has been steadily declining in volume during the last two or three months, the increase in freight rates granted to the railroads is expected to have a salutary effect on railroad buying of shop equipment within the near future. The announcement of the decision of the Interstate Commerce Commission on the freight rate advance is too recent to have brought any new railroad business into the market as a direct result of that action, but the fact that the carriers will soon have funds with which to make necessary purchases is a cause for optimism among those makers of and dealers in tools which are used for railroad shop work. The Illinois Central issued a list at Chicago calling for more than 50 tools, but it is probable that this inquiry had been fully decided upon before the news of the rate increase was made public. Several of the eastern roads are known to have fairly large lists prepared and quotations will be asked for as soon as the directing heads of the roads can persuade their boards to make appropriations. The fact that higher rates have been granted will probably make it easier for some of the roads to market new securities with which to buy equipment.

With the exception of the General Electric Co., there is none of the Eastern industrial companies doing any buying worthy of note. The General Electric Co. has been a consistent buyer for months, its latest inquiry calling for about 20 tools for its Schenectady plant. An eastern company affiliated with the automobile industry has purchased 25 internal grinding machines for railroad work.

Delivery periods on machine tools are becoming shortened, due to the falling off in new business and also to the fact that there have been some cancellations, particularly in the Detroit district, where automobile production has dropped off in certain plants. It is reported that the Cadillac, Buick, Ford, Dodge and Hupmobile plants are working practically full, but in other factories production has been cut from 25 to 50 per cent. and some workmen have been laid off. Cancellations of machine tool or-

POINTS IN WEEK'S MARKETING NOTES

The Illinois Central Railway has an enquiry for over fifty machine tools.

Toronto steel warehousemen find business fairly good for the time of year.

Delivery periods on machine tools are getting better.

Pig iron reaches the $50 mark in the U. S. market, while coke is expected to ease off.

The scrap iron market has assumed a firm tone, and several advances are reported from U. S. points.

If interested tear out this page and place with letters to be be answered.

BUYERS AND SELLERS OF IRON AND STEEL

Pittsburgh sheet mills are curtailing production on account of the car shortage. It is estimated that production has now fallen to about 40 per cent. below capacity.

For the first seven months of 1920, the Ford Motor Company turned out 900,000 cars. This was 10,000 short of their schedule.

Deliveries on plates by Pittsburgh mills are improving. One result of this is a slightly easier market, but the price ranges generally from 3.25 cents to 3.50 cents.

In order to relieve the shortage of cars, it is proposed to move a much larger amount of grain over the Great Lakes. This in turn will cut into the ore shipments, which are already far behind.

The Dominion Steel Corporation recently shipped its first consignment of plates for foreign order from its new plate mill at Sydney, Cape Breton.

Increased car supply, and conservative buying, is expected to make the market price of coke in the U. S. drop considerably in the near future. A drop of from $1 to $3 is spoken of in some quarters.

The putting into force of the freight increases in the United States is expected to provide an impetus to the heavy machine tool trade. One road has already come into the market for 53 tools of various kinds.

European production, coupled with the increase in freight rates, is leading some of the steel exporters in the United States to fear the coming of competition from that quarter.

Although the United States production of July was greater than that of June, the daily production was less, counting the one day extra in July. The figures for the month were 3,060,626 tons in July, as against 3,046,623 tons in June.

---

ders in the east have not been numerous, but some companies have been affected by the partial slump in automobile production, notably the Simms Magneto Co., East Orange, N.J., which has cancelled a considerable number of machines. The Empire Cream Separator Co., Bloomfield, N.J., which had extensive plans for increasing its output of cow milking machines, has also cancelled a large part of the shop equipment it recently bought.

Machine tool sales offices in New York are receiving few inquiries and are largely marking time.

U.S. SCRAP METAL

The scrap iron market is showing generally stronger according to advices from the various U.S. points. Heavy melting steel has been sought and $28.50 to $29 offered for it. Buying by English mills is expected to take place shortly, there having been several enquiries noted. Rerolling rails are in in good demand, and very hard to get. New York reports an active and upward moving market, especially in steel, grades. No. 1 heavy melting steel is quoted at $21 to $22 f.o.b. New York, while stove plate has advanced to $25, an increase of 50 cents. Shipments are very difficult to make, the railroad situation still being very bad. This latter fact has been the cause of a quiet market in Boston, though there has been some heavy melting steel sold for export at $22 at Boston dock. No. 1 machinery scrap is bringing $43 delivered. With the exception of heavy melting steel, the Pittsburgh market is easier, but this grade is quoted at $27 to $27.50 delivered, while a special grade of this material has been sold at something over

$23. There is a strong market at Buffalo for all grades. The demand is mostly for prompt shipments, the object being to get the material before the new freight rates go into effect. Some dealers fear that after that date the demand will fall off. Chicago reports a firm market with not much actual business, but a general advance in quotations. There is not much moving, as cars are hard to obtain. Railroad offerings amount to about 13,000 tons. There is an improving enquiry reported from Cincinnati, but there is a good accumulation of stocks there which is moving very slowly. St. Louis is supporting what is practically a dealer's market, consumers staying out. Prices remain steady, and there is not much material available.

Have you had a meal of seaweed yet?— Before long this saying may be very common, for a scientist in the Government service in California has recently developed a process for pickling the kelp of the Pacific Coast. Los Angeles stores are already offering it for sale. Who knows, but here we may have at our hands, the solution to the H. C. of L.

When you speak of boycotting anything, do you know where the statement first came from? The average person never bothers where it comes from so long as the meaning is clear, but for the novelty of the thing, we will state how the expression came into being. A Captain Boycott was an agent for an estate in Ireland, and the tenants, dissatisfied with his management asked the landlord to remove him. On declining to do so, the tenants and their friends refused to work for Boycott, and that's how the word began.

PIG IRON TRADE

Pig iron has reached the $50 mark for foundry No. 2 grade, and basic has also advanced. Philadelphia reports a sale of No. 2 plain for spot shipment at $51, while No. 2x is $50 eastern Pennsylvania furnace. The railroad situation is still very difficult, and iron is hard to get. Basic is worth anywhere from $45 to $50 eastern Pennsylvania furnace. The Chicago market is quiet, but at the same time shows no signs of weakness. On the contrary whatever tendency to change is apparent is in the direction of a raise in prices. While there is likely to be less demand from the automobile trade, there will be more than enough coming from the railroads to make up for this. $1.75 to $2.25 silicon for 1921 delivery has been sold at $46, furnace. There is not much interest taken as a whole in next year's delivery sales. The outstanding purchase in the New York market was the Worthington Company's which closed for nearly 10,000 tons for its first half supply. There are new enquiries for first half aggregating 5,000 tons in addition to this. From Pittsburgh the reports also show advances, basic having gone up $2 a ton and foundry iron from $1 to $3 for No. 2 foundry. Basic is quoted at $48.50 valley, while No. 2 foundry is quoted at $50 valley. Some sales have been made at the lower figure of $47. Other grades are strong but no sales have been made to establish a price. Cleveland also reports a tendency to higher prices. Foundry iron has been sold at $47 and $48, while some is being held for $50. The advance of prices by the last remaining Virginia furnace that was in the market has stopped the buying of this iron in the New England district. There have been some sales of Alabama iron in this district, the price for No. 2 x delivered being $54. There is no Pennsylvania iron offering. There is a good demand coupled with a scarcity of iron in Buffalo, last half and first quarter being the deliveries favored. Iron for last half delivery sold for $49. There is an enquiry for 10,000 tons for export basic iron. No. 2 x for first quarter delivery has been sold for $50, but there is not very much being offered for 1921 as yet. St. Louis reports light sales with prices about the same as last week. 1921 enquiries are in evidence, without receiving much attention from makers. Cincinnati iron is quoted at $42 for remainder of the year and first quarter by some makers, while others ask $45 for the same. Birmingham market is strong, but shipping difficulties are great and a large tonnage has accumulated. Production is being kept up. There are some enquiries for next year's iron.

AN UPSET GRINDING OPERATION

By Donald A. Hampson.

The term "upset" is used in the heading because the operation to be described upsets all our common theories as to grinding, more than that, long practice has proven that the process is correct for this particular job, which leads us to repeat, "When theory and practice do not agree, it is time to change the theory." It is a special case only, however, where "slow" grinding has resulted in the right kind of production.

Our drawing shows a piece of grey-iron that is one of a score used as part of a friction clutch made in large quantities. The surface marked A must be

PIECE TO BE GROUND

machined so that it will hold cigarette papers at all four corners—this is an accurate working surface in the clutch assembly and it, is also used as a holding surface for milling the notch B, which likewise must be within very close limits.

All the accepted methods of milling and of grinding were tried out and though they turned out blocks that were acceptable, they were only made so by going through a roughing and finishing operation, which proved too laborious for good manufacturing. The casting was too light to stand up under any kind of milling cut and not show a change of size where the roughing cut struck the cross ribs;—grinding was tried with discs or ring wheels but unless each piece was individually clamped for wet grinding, it could not be turned out flat—with the dry disc grinding, the heat of cutting and the inevitable pass-off from the disc would leave one or

SHOWING METHOD OF SET-UP

more low corners or sides on the work. Just how the slow grinding process was evolved is a matter of conjecture. The photo shows the set-up on a lathe that was discarded for conventional

work. The emery wheel that does the business runs at 27 r.p.m.

This wheel is held between large flanges on a mandrel that is supported on the lathe centres. Eighteen inches in diameter, in general of cup shape but with a working "edge" five inches wide, this wheel is good for a year's steady grinding. During that time, it will grind 200 pieces an hour, of required accuracy, on the average, and a laborer who gets the hang of things has doubled that production.

As will be seen, a jack shaft is mounted in wooden bearings at the rear of the headstock, this shaft carrying a crank that drives the cross slide with a reciprocating motion through the medium of a hickory pitman. On the cross slide there is an angle plate with guides between which the castings are dropped. To secure power enough, the lathe is run with the gears "in," which also gives a slow speed that grinds without heat. From 1-32″ to 1-64″ is removed from each piece.

The operator sits at the front of the lathe. The carriage is restricted by stops to about a half inch of travel, which is limited at the grinding end by an adjustment for sizing purposes. With the carriage moved back, one or two castings are dropped in the space in the angle plate after which the carriage is moved to bring the work against the emery wheel. A treadle and rod is provided so that the operator's weight may be utilised for pressure in grinding, but the stop prevents overdoing and grinding castings too small. A "kicker" ejects the finished pieces, which come out cool and uniformly flat. About once in two weeks of steady running, it becomes necessary to dress the wheel, this being done with an ordinary Huntington dresser guided by a bar of steel held parallel to the travel of the cross slide.

In production and accuracy, as well as cost, this unique arrangement has proven all that could be wished. The concern has several times invited specialists in the grinding and milling machine field to submit proposals for tools that would offer greater production, but after trying out sample lots, the answer has come back each time, "We cannot better your present production, quality considered."

OXYGEN PLANTS TO BE BUILT

The sensational development of Canadian industries in pre-war days was a weak and toddling growth compared to the mighty strides the Dominion is making industrially to-day. From 1905 to 1915 the capital invested in Canadian factories increased more than 135 per cent. and the value of factory products advanced roughly from $700,000,000 to $1,400,000,000. It would be rash to attempt an estimate of Canada's industrial growth since 1915, but it is not a guess to assert that the Dominion has forged ahead industrially as never before in its history.

An important recent announcement

bearing on Canada's future industrial development is that of the Dominion Oxygen Company's promised erection of a chain of great oxygen plants for separating oxygen from the atmosphere. The first of these plants will be situated in five of the Dominion's industrial centres, and service stations will be established throughout Canada to supply oxygen users wherever there is a demand for the gas, either on or off the railroad.

The first of these plants has already been completed at Toronto, and it is said to be the largest oxygen plant ever built in Canada. It will supply the Canadian market through an extensive warehousing system, pending the erection of the aditional plants. Sites for these additional plants have been secured at Montreal, Winnipeg and other points, so that the work of construction may be prosecuted as rapidly as possible.

ANOTHER FIRM TO BUILD DIESELS

The Worthington Pump and Machinery Corporation have issued a bulletin announcing their entry into the marine Diesel engine construction field. The bulletin reviews the marine situation and the case for the Diesel internal combustion engine, and quotes instances of actual working to demonstrate the reliability of this prime mover. The announcement of the building of a 2,400 I.H.P. six cylinder marine Diesel by the company follows. This engine was built at the company's Snow Works, Buffalo, and the cylinders are 29 inches bore by 46 inches stroke, and the crank shaft makes 120 revolutions per minute, the engine being of the 4 cycle type. The engine is built to the Worthington Company's own designs, but the leading features of successful foreign builders have been incorporated. The company have made plans to enable them to produce these engines in large numbers as the demand increases.

One of the most conspicuous forms of British engineering activity since the war lies in the growth of locomotive production. Every country in the world wants locomotives, and the extension of many locomotive shops in both England and Scotland, is designed to ensure that the demand will, as in the past, be met with.

It is so far to the good that most of these shops were retained on locomotive work during the war. One firm alone supplied 132 locomotives to British railways, 380 to the French railways, and 300 for other overseas countries during the war. This was in addition to producing many narrow gauge locomotives, hundreds of "Tanks" and hundreds of thousands of shells, numerous aeroplanes, torpedo-tubes, gun carriages, and other "munitions." The total value of the company's output during the war was sixteen million pounds.

CANADIAN MACHINERY

AND
MANUFACTURING NEWS

Vol. XXIV. No. 9   August 26, 1920.

Some Interesting Examples of Tool Layout

It Pays to be Thoroughly Conversant with Tool Layout Practice. This Article is a Continuation of Former Material Dealing with Different Style Machines, and the Tools Used.

By J. H. MOORE

IN preceding articles we have discussed the advisability of being thoroughly conversant with the different styles of machine tools on the market, and their method of tooling. We have already pointed out that a first-class machine tool cannot do its best work if equipped with poorly designed tools, and to aid readers in this regard we have shown various styles of machine tools, together with good examples of tool layout. In the article to follow we will illustrate tool-layouts on what are known as Potter & Johnston manufacturing automatic chucking and turning machines. This series of talks is well worth preserving for future use.

For the benefit of those not familiar with this type of machine we append the following. As can be noted from Fig. 1 the machine is equipped with a turret head supported by the arm A. On this turret are four stations upon which tools can be bolted, and as is the usual practice, every new machine

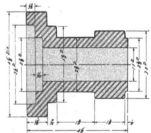

FIG. 2.—A DETAIL DRAWING OF THE PART BEING MACHINED AT FIG. 1.

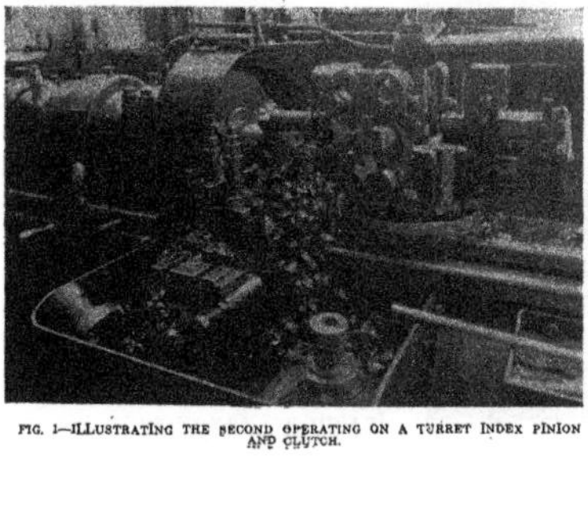

goes out equipped with an outfit of turning tool holders, etc., etc. What we are particularly interested in, however, is the manner in which this portion of the machine is operated.

When in operation, the turret is advanced, revolved, and returned automatically. These movements are performed at as high a speed as practical, thus making comparatively little lost time in the operation of the cutting tools. When the turret slide has advanced to the cutting point, the feeding motion is automatically thrown in.

Of course a regular cross slide forms part of this type machine, and a portion of the slide can be noted at B, see Fig. 1. On the cross slide are used the regular front and rear tools. They can be arranged and adjusted to work at the same time as the turret tools, are cutting, or separately, depending upon the nature of the work. In some cases special blocks are fitted to the cross slide, these carrying blades, cutters, etc., but again this depends en-

FIG. 1.—ILLUSTRATING THE SECOND OPERATING ON A TURRET INDEX PINION AND CLUTCH.

tirely upon the class of work being
manufactured.

Relative to speeds. It might be well
to state that the spindles of such machines have three automatic changes of
speed, and with the spindle change
gears provided, one can procure the
proper speeds for handling any work
within its capacity.

The feed question is also very simple,
the feed being transmitted from the
same gear shaft that drives the spindle.
This ties, so to speak, both speed and
feed together. Two automatic changes
of feed are provided, known as ordinary
and auxiliary, each having an independent set of feed change gears. The
ordinary feed is used for turning; boring, etc., while the auxiliary feed is
adapted for turning and boring operations where the character of the work
is such that a coarse feed will suffice.
This latter feed is also suitable for ream-

FIG. 3—SECOND OPERATION ON FRICTION HEAD.

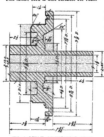

FIG. 4—DETAIL OF THE HEAD BEING
MACHINED AT FIG. 3.

ing and threading. The gearing in this
style machine is so arranged that quite
a variety of different pitches of threads

can be cut, using automatic collapsing
taps, or automatic opening die heads.
Either feed can be engaged at any
time automatically, and they are entirely independent of the constant high
speed for the idle movements as already
explained.

An automatic back facing bar is
generally a part of such machines. This
bar, which is placed through the spindle,
is provided with a taper hole to carry
drills, cutters, facing tools, etc. These
can be carried either singly or in combination. By using such an attachment
a variety of parts such as gears, pulleys, etc., which would otherwise require
two chuckings, can be finished at one
setting; the back facing tools machining the inner end of the hub, while the
turret and cross slide tools are turning
the periphery, facing down both edges
of the rim, the outer hub, and boring
the hole. It is quite a common occurrence
to see all three parts moving in harmony, and as many as 15 cutting tools
working in simultaneous operation. This
once more proves what we have already
stated, namely, use your machine to its

maximum cutting power at all times.

On such machines as described the
chuck is supplied with a pilot bush
which supports the turning and bor-

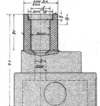

FIG. 6—DETAIL OF THE PUMP GEAR CASE.

ing tools during the cutting operations.
This feature of course makes for extreme rigidity. In some cases a two
jawed chuck is used, while in others the
three jawed type is adopted. Both
styles are of the universal scroll design.

For small work, an automatic lever
chuck is generally used. This allows
the placing and removing of work with
one hand. Another advantage is that
it is unnecessary to stop the spindle to
put the piece in the chuck or remove it
when the cut is completed. Air chucks
are also used when deemed advisable.
In other words, depending upon the
nature of the work, the machine is
equipped accordingly.

The automatic action of the turret
and cross slide is secured by means of
cams, these covering practically every
class of work coming within their range.
If working on parts of very special
character, special cams are installed, but
for general work this is unnecessary.

From the foregoing it is evident that

FIG. 5—MACHINING A PUMP GEAR CASE.

all cutting operations being automatic, the operator is engaged only when placing the work in the machine, or removing the finished product. This means that an attendant does not require to confine his attention to one machine, but can handle a battery of two, to six, depending on the class of work being handled.

Having placed before readers the main points in connection with this style of machine, let us proceed to examples of work performed.

At Fig. 1 is illustrated the second operation on what is known as a turret index pinion and clutch. The drawing of this piece is depicted at Fig. 2, and the heavy lines denote the portion machined during the operation.

The material is machine steel, and the piece is held on a roll grip arbor. As will be noted, very few special tools are employed. The piece is rough and finish turned, faced and formed, all tools being supported by pilot bars. This is the feature that we mentioned which ensured rigidity. One of the finished parts can be seen resting on the tray of machine. Do not merely glance at these set ups, study them, for they teach a lesson.

The next photograph, Fig. 3, depicts the second operation on a Friction Head. The machined surfaces are marked with heavy lines, as at Fig. 4, and in this case the material is cast iron.

The work is gripped on its large diameter on the regular chuck, and practically nothing is required in the special tool line. The piece is rough and finish turned, faced and grooved and the tools are supported by a revolving pilot in the finished hole. Note the revolving pilots, and in fact the complete tool layout. This should be easy, after a study of Fig. 4.

At Fig. 5, we see a job of rather a different nature. This view depicts the machining of a Potter and Johnston gear case, and referring to Fig. 6, we note the machined surfaces from the heavy lines. The material is cast iron, and

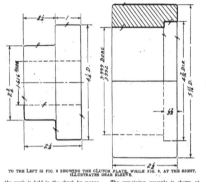

TO THE LEFT IS FIG. 8 SHOWING THE CLUTCH PLATE, WHILE FIG. 9, AT THE RIGHT, ILLUSTRATES GEAR SLEEVE.

the work is held in the chuck by means of jaws shaped to suit the case. The work is rough and finish turned, faced, counterbored, reamed, and the threads cut. The position depicted at Fig. 5 is that of the threading operation, an automatic die head being used. Finished parts can be noticed resting on the machine tray.

Fig. 7 depicts the second operation on what is known as a quick return clutch plate. The machined surfaces are denoted at Fig. 8 by the heavy lines as usual. The material in this case is machine steel, and the piece is held on a roll grip arbor. The tools used are all standard, and the piece is rough and finish turned, and faced. All tools are supported by pilot bars. Two finished pieces can be noted towards the front of the photograph.

The remaining example is shown at Fig. 9. This depicts the first operation on a spindle gear sleeve tooth section, and the machined surfaces are marked at Fig. 10 by means of heavy lines. The material is machine steel, and the work is cut from stock. The drill used for the hole is a flat twist drill, and this drilling position is illustrated at Fig. 9. The piece is rough and finish turned, bored, faced, and counterbored, all tools being supported by pilot bars. Finished parts can be noticed in tray of machine.

From the examples shown readers can form at least a fair idea of the style of work performed by these machines, but of course it is impossible to depict even a small percentage of the variety of work manufactured on such machines.

Continued on Page 209.

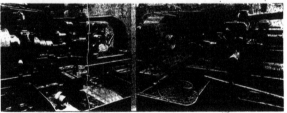

FIG. 7—SECOND OPERATION ON QUICK RETURN PLATE.　　FIG. 9—FIRST OPERATION ON SPINDLE GEAR SLEEVE TOOTH SECTION

Automatic and Semi-Automatic Furnaces*

The Conclusion of Article from Last Issue, Dealing with the Subject of Automatic and Semi-Automatic Heating — Further
Examples of Furnace Installations Are Given.

Compiled by J. H. M.

IN heating large dies it is desirable
that they be thoroughly saturated at
a temperature slightly below final
temperature before the formed corners or
edges on the face reach the final temperature. As the life of a die is no
greater than the life of its corners or
edges, it naturally follows that every
possible effort should be made to prevent overheating these corners or edges
or exposing them to the final temperature any longer than is absolutely necessary. The corners and edges of the
die are the first to heat and the first to
cool, and it is at these sections that the
ruinous effects of unequal contraction
are manifested. These corners and
edges heat and cool more rapidly because the surface in proportion to the
mass of these sections is much greater
than that of the body of the die, which
naturally reveals a difference in time
of heating and cooling. Obviously, it is
essential to consider the difference in
time required to reach a given temperature in heating or cooling, due to the
difference in surface and mass of the

*Copyright by W. S. Rockwell Co., New York.

FIG. 13—Discharging end of semi-automatic furnaces for heavy dies. The quenching tank
and tempering plate are shown in the foreground.

sections and embodying the outline of
the piece and the body of the die itself.
Improper heating or cooling, resulting
from a disregard of these factors, is responsible for a great deal of the limited
die service with consequent loss of production, material and labor. Such loss
greatly offsets any saving in time or
fuel by heating the die quickly without

due regard to all the factors involved.
The type of furnace illustrated at
Fig. 12 affords a practically continuous
method of heating and cooling, with
every facility for bringing the dies up
to the final heat through a slow, soaking process. The dies to be heated are
placed upon carriages and slowly conveyed from one end of the furnace to

FIG. 14—View of charging ends of straight-line billet heating furnaces. The billets are charged in one end and passed straight
through and out the other end.

FIG. 12—Single chamber semi-automatic die-heating furnaces, each arranged for two rows of dies. Note the carriages for dies in the foreground.

the other. They are withdrawn one at a time, quenched, and tempered on the small plate furnace as shown in Fig. 13.

This semi-automatic principle may be employed in heating heavy pieces, such as bars, billets, ingots, axles, shafts, etc., for miscellaneous heating operations preparatory to forging, pressing, rolling or extruding. The construction of the floor, arrangement of chambers and working openings, and the method of delivering material to, through and from the furnace, vary with the nature of the operation, the size, shape and number of pieces to be handled, and the local manufacturing conditions.

As a rule the furnaces for this class of work are divided into two types: the "L" and the "straight-line." In the "L" type the pieces are charged into one end, pushed through to the opposite end and taken out of an opening located in one or both of the side walls. Furnace is often arranged with openings on both sides to serve a machine located on each side of the furnace.

This furnace may be built in single-chamber form, with openings on one or both sides or in twin-chamber form. Provision may be made to handle two rows of material in a single chamber. On account of the advantages of this type of furnace in the application of heat and decreased area of exposure in working openings, it should be given preference over the "straight-line" type whenever the manufacturing conditions will so permit.

In the "straight-line" type the material is charged in one end of the furnace and pushed or rolled straight through and out at the opposite end.

This type is to be preferred whenever the size or weight of the individual

piece heated renders difficult its removal through a side door, or when it is desirable to discharge direct to a machine or quenching tank located substantially in line with the furnace chamber. It has the disadvantage of greater area of exposure incident to the wider working opening, which is apt to react somewhat on the uniformity of the material, the comfort of the operators and the furnace structure.

At Fig. 14, the pieces to be heated are placed upon an elevating platform to raise the charge to the hearth level. The material is then pushed through the

chamber and delivered direct to machines located at the opposite end of the furnace.

Selection of Furnace

Determination of the type and size of furnace, number and position of working openings, and number of chambers, is governed entirely by the individual manufacturing conditions. The controlling factors are the quantity, shape and size of material to be heated; temperature; position of furnace with reference to adjacent machinery; floor space; facilities for moving material to, through and from the furnace; fuel to be used; location of stacks, vents, etc. These factors differ in each plant, make a study of each individual case necessary, in order that the type and size of furnace selected may be adapted to the manufacturing conditions.

SOME SPEED

An average speed of a little over 1.2 miles per hour was made in a recent flight from Rome to Turin by Lieut. Brack-Papa in an A. R. F. biplane constructed by the Fiat Company and equipped with a 750-horsepower twelve cylinder Fiat engine. The machine started from the Mirafiori aerodrome, in the suburbs of Turin, at 11.45 a.m. on April 5, carrying two passengers in addition to the pilot, and landed on the Centocelle ground, just outside Rome, at exactly 2 p.m. The intervening distance of 388 miles was thus covered in two hours fifteen minutes. A few days previously the same machine was officially timed over the Mirafiori aerodrome and found to give an average speed of 161.5 m.p.h. with four passengers and a pilot. In this flight the effects of wind were, of course, eliminated, and the speed attained, we understand, constitutes a world's record.

Discharging ends of semi-automatic furnaces for heavy dies. Note the hoist and quenching tank.

The Construction of Spindle Ends for Millers

Further Examples of Spindle Ends Are Shown, Each Style Being Fully Discussed in Detail—The Article in Its Complete State is Well Worth Filing for Future Reference.

By FRED HORNER

THE addition of a clutch transmission for heavy bars, Fig. 21, relieves the cotter of the driving strain. Fig. 22 gives plan and section of a special formation of nose, which serves two functions. In this drawing it is represented as driving a boring-bar, locked by cotter, and ejected when necessary by a key driven down the aperture A. For securing and driving a snout boring head the parallel outside of the nose comes into use; a longer cotter is put through the hub of the head, drawing it up so that clutch teeth on the hub mesh with teeth B on the nose and furnish a powerful drive, see Fig. 23. For the sake of safety to the operator, a brass guard collar is slipped over the clutch teeth when these are not in use, the guard being seen in Fig. 22.

For Heavy Work

That type of boring machine which revolves a large tee-slotted face-plate in addition to the central bar that slides through it often has to make use of a special bearing or socket casting, bolted to the face-plate, and receiving a milling head or facing head, the mass of which or the stress on which is too great to be taken on the end of the bar. There is manufactured in England a highly useful class of machine, in which the spindle nose carries a powerful spur driving gear, on which is bolted a circular head having a transverse slideway,

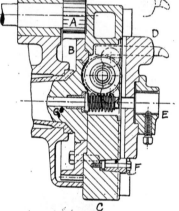

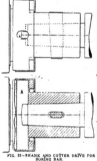

FIG. 21—SHANK AND COTTER DRIVE FOR BORING BAR.

FIG. 24—TOOL-HOLDER SLIDE OF FEARN-RICHARDS MACHINE, WITH BORING-BAR SOCKET IN PLACE.

through which moves a slide. This latter receives a facing tool, so that by the geared mechanism actuated from a shaft passing through the spindle facing cuts can be taken either to or from the centre. For driving the ends of bars, and the shanks of drills, reamers, milling cutters, tapping attachments, etc., a socket is bolted on to this slide, and is set central by means of a dead stop.

The sectional appearance of the mechanism is shown in Fig. 24 (George Richards & Co. Ltd., Manchester, England), the various parts being driving pinion A, spur gear B, head C, slide D, socket E, and stop F. The last named item is put in or removed by turning the shouldered screw which passes through the slot in its body. The traversing of the slide is brought about through the turning of the central shaft G, revolving the worm gear indicated,

and thence through a spur gear, the two pinions shown dotted in mesh with the rack, on the back of the slide. The two pinions afford a very steady motion, and remain in mesh with the rack for ordinary diameters; on an extra

FIG. 23—SNOUT HEAD COTTERED OUTSIDE SPINDLE NOSE.

long traverse, shown by the dotted lines, the outer pinion transmits the feed. This may be had in either direction, for instance, a flange may be run across with

body flanged to bolt on to the facing slide direct, Fig. 27.

For operations demanding the utilization of a snout head, that is, blind

Fig. 28, is bolted on. The cast steel socket portion receives a sliding toolholder, clamped with two hollow screws, thereby arranging for the required degree of projection. An offset holder is also made, with the hole in the socket set to one side, and the holder cranked out to enable it to reach to larger diameters than practicable with the other fitting.

Precautions to be Observed

All the skill in design and manufacture of these various methods of fitting and driving is set at nought unless certain precautions are observed in finishing the details, and in subsequent use. The presence of keen edges on any parts of the shanks, or arbors, or clutches, cannot be allowed, because these are so easily burred, and thereby produce small lumps and roughness on the corners and adjacent surfaces of the metal. This fault prevents accurate mating of the parts, and deflects the shanks, arbors, or cutters out of true running. Even when the amount of this is very slight, it may happen that the want of close fitting may lead to a little slackness, tending to vibration and chatter, which would not occur if the faces met solidly all over.

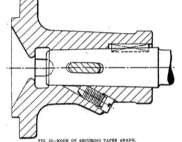

FIG. 25—MODE OF SECURING TAPER SHANK.

a coarse feed on the inward motion, and a finer one on the reverse to the edge again.

For driving ordinary boring bars the socket illustrated is employed, there being a key in the bar to receive the drive. For holding the shank of a tapping attachment a similar kind of socket, but longer, is bolted on. To carry the taper shanks of milling cutters, the shape of socket, Fig. 25, occupies the central position. A rectangular key provides a positive drive, and a cotter holds the tool shank back. When a collet or sleeve has to be set in the socket, the set-screw seen disposed at an angle is applied against a notch on the sleeve, in the manner visible in Fig. 26, a screw also in the sleeve holding in the cutter shank. Larger cutters, if they have shanks, are treated similarly to Fig. 25, but the heavier ones have a cast

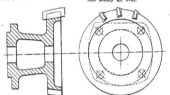

FIG. 27—FACE CUTTER WITH FLANGED BASE.

holes where it is not possible to pass a bar right through, or interiors that have to be faced across on the ends, or interior flanges, the attachment

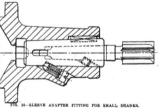

FIG. 26—SLEEVE ADAPTER FITTING FOR SMALL SHANKS.

The practice of hardening, of course, reduces the chances of bruising and burring, but does not absolve the manufacturer from the necessity of adequate chamfering. In operation, much care is required in mounting and dismounting the elements. All dirt and grit should be carefully cleaned off before assembling, the hammer should be used very sparingly—preferably not at all—and in the case of boring-bars or other details where cotters are employed, and blows cannot be avoided. they should be as few as possible, and firm and well-directed. Carelessly struck blows, or those delivered needlessly hard, are apt, in course of time, to produce opening of a bar or spindle, changing its form subtly, and causing a degree of eccentricity of running, due to the warping out of longitudinal truth. Another error that is sometimes committed is that of not taking sufficient care when altering the taper of spindle noses or other fittings, as

when one desires to change from a B. & S. to Morse, or vice-versa, or to any special taper. The finishing out should be done with a carefully-rigged up internal grinding attachment, taking very slight cuts at the finish, so as to avoid inaccuracies due to the spring of the grinding spindle, or its bearings or holder.

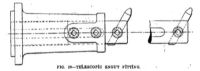

FIG. 29—TELESCOPIC SNOUT FITTING.

On What Does Production Primarily Depend?

According to the Author, Who is Secretary and Chief Engineer of the National Safety Council, Production Depends Upon Two Factors, Namely Machines and Men—Read His Views.

By SIDNEY J. WILLIAMS

PRODUCTION depends primarily on two factors—machines and men. Production also requires materials but these in turn are produced by other machines and men—using "machines" broadly to include all equipment. Production also requires capital to provide the machines and the men, and management to direct them. But machines and men are the immediate producers. They and they only can increase or decrease production. The things that cause them to do the one or the other are the things that we must study.

The basis of modern production, with respect to both machines and men, is regularity-standardization. The Swiss watchmaker a hundred years ago built his watch from the ground up, fitting each part to the parts already completed, as a carpenter builds a house. He never made two watches exactly alike. I believe the Swiss watchmaker still follows somewhat the same method. But we in America make our dollar watches (which now sell for two dollars) and our two dollar alarm clocks (which now sell for four) with dies and jigs and fixtures; we stamp out the parts on power presses or machine them in automatic screw machines, so that part number 106 is always fits part number 105 exactly, although the men who make the two parts may never have seen each other and may not speak the same language. Whether in the eyes of an artist all the alarm clocks in America are worth a single beautiful Swiss watch, is a question that I will leave to philosophers. We as a nation are committed to the principle of mass production; and while we may deplore the vanishing of the old craftsman, few will seriously suggest a return to the former order.

Motion Study Expert

But we do not stop with standardizing machine operations. We standardize also the operations of men. The motion study expert tells us that practically every man wastes a large percentage of his time and effort in even the simplest operation—that a dozen skilled mechanics will do the same thing in a dozen different ways, all of them wrong. So, the motion study expert determines the one best way to do a job, and in one case after another we find that by following his method we not only increase production but we decrease fatigue. While so called "efficiency engineering", like other pioneer movements, has suffered from the pretensions of quacks, there can be no serious question that modern industry demands the elimination of lost motion and the standardization, so far as is practicable, of every job in the plant.

If this is true, as it is true, I do not believe that any one will question the next proposition that I want to make—that anything which interferes with this absolute regularity of operation is inefficient and uneconomical, and that it is one of the chief functions of the engineer and the executive to bust out these disturbing influences and eliminate them.

Production Interference

The things which thus interfere with regular production are many and varied. Some of them are external to the plant itself—such as a war, or a nationwide strike, or a railroad tieup. The prevention of these is largely a function of government, and people will not long tolerate conditions that permit such national inefficiencies.

Then there are catastrophes within the plant; a strike; a disabling fire; a break-down in the power plant. Every one knows that it is the function of management to prevent such occurrences and that a management which does not in general prevent them cannot be permanently successful.

Then there is a third group of apparently minor disturbances. A laborer pushing a truck strikes an uneven place in the floor and a casting falls off the top of the load. The casting is heavy and he goes for help to put it on again. Meanwhile the lathe operator is waiting for the casting. The blockade of the passageway stops another truck coming up with material for another operator, who must also wait. The total loss of time may not be more than four minutes for each of five men—twenty minutes in all or, say, twenty-five cents worth of time.

What does the foreman do when he finds that the lathe operator is waiting for material? If he is a foreman of the old school, he goes back and bawls out the "blankety blank wop" for running his truck into a hole in the floor. The man, thus admonished, is henceforth more careful—that is, he is slower in his movements. He takes pains to avoid the holes in the floor, the posts in dark passageways, the other various sundry obstructions which sprinkle his pathway. A little later the foreman is surprised and grieved to find that he must put on another man to help the truckers because they cannot keep up. He discourses feelingly and eloquently with the Assistant Superintendent on the total depravity of laborers in general and of his laborers in particular. Am I exaggerating? Not very much. How many of us carefully walk around a hole in the floor, or a slippery place, ten times a day, because it would take a little mental and physical energy to fix it up?

The Up-to-Date Spirit

Of course, the foreman who is really on to his job, has the floor fixed at once. If he is unusually intelligent, he also looks around for other things which interfere with efficient trucking. He may find that the lighting in the passageway is poor, and recommend to the Superintendent that it be improved. He may find that the truck itself can be slightly changed so as to make it less likely that anything will fall off. The up-to-date foreman realizes that when his laborers each lose twenty-five cents worth of time a day it is up to him, not to bawl them out for it, but to find out what is wrong and correct it. And it is up to the Superintendent and higher executive officers to supply what the foreman lacks in this regard. Once labor was the

cheapest part of the cost—but the dollar-a-day man is gone.

In machine operations it is even more obvious that regularity is the essence of modern production. I dare say that every punch press foreman or superintendent in the country has as his idea of heaven a place where—if punch presses are tolerated at all—not one of them ever misses a single stroke. Unfortunately this kind of heaven, like other kinds, is seldom met with on earth. I know of large and successful companies where punch press production is from 25 to 50% less than perfect —that is, where only half or three-quarters of the strokes are productive, simply because proper means are not provided for placing the material in the press and taking it out again. The important point is that the greater part of this loss is positively preventable through the use of simple and inexpensive automatic or semi-automatic devices for placing and removing the material. With comparatively few exceptions, even in the most efficient plants, we have not had time to study these little things—the castings falling off the truck, the material sticking to the punch press, causing in each individual case a loss of time reckoned only in seconds or minutes, but causing in the aggregate a loss running undoubtedly into many millions of hours and many millions of dollars. In the past, so vast were our resources that we could overlook these "small" losses, just as the city of Chicago pumps millions of gallons of water that are distributed through leaky pipes. But now the time has come, with our diminishing man power, diminishing natural resources, and prospective sharp competition from abroad, when we must meter and stop these leaks if we are to maintain our volume of production. I have tried thus far to make the point that modern production is built absolutely on standardized regularity of operation and that anything that interferes with such regularity interferes with production. It is putting the same thought into other words to say that it is our constant effect to find the one best way of doing things and then always to do them in that way; and that anything which happens unexpectedly, whether it is a coal strike or a casting falling off the truck, is bound to interfere with our regular program and therefore to cut down our production. Therefore it is interesting to observe that the Standard Dictionary under the word "accident" gives as its first definition this: "Anything happening unexpectedly." In a second definition the dictionary recognizes, of course, the common usage of the word "accident" as meaning an occurrence in which some one is hurt. But in the broader meaning it is plain that, from the standpoint of industry, a coal strike is unexpected and is therefore an accident; a breakdown of the power plant is unexpected and is therefore an accident; the sticking of material

in a punch press is unexpected—not contemplated or desired by the designer of the machine—and is therefore an accident, whether any one happens to be injured or not. In short, "accident" in the broad meaning given it by the dictionary is exactly synonomous with all the disturbing things which interfere with production—it is the exact opposite of production efficiency. It stands precisely for all those things which we as engineers and executives must fight unceasingly. Now I come at last to the point where safety enters into the proposition. Some of the "accidents" which I have mentioned—some of the things which happen unexpectedly—result in personal injury. From the standpoint of the man interested only in production, the accidents which happen to injure some one are no more and no less important than those which do not. It is true that from other standpoints the accidents causing personal injury are much more important. To the Legal Department, they mean the payment of compensation. To the Employment Department, they mean the necessity of providing another man. To any man with a spark of human feeling in his heart they mean pain and suffering to a fellow human being, perhaps poverty and distress to his dependents. But leaving these considerations out of account, considering ourselves as absolutely non-human engineers or executives, with an abundant supply of skilled labor and no compensation laws to worry about, even then we must recognize that at the very least the accident which injures some one is no less important than the accident which does not injure any one, and that all accidents—all these things happening unexpectedly—interfere with the regularity of our operation, and cut down our production. No engineer or executive living can afford to say, "I am not interested in accident prevention," unless he is willing to say, "I am not interested in efficiency" —because accidents and efficiency are absolutely incompatible; they cannot exist in the same plant; the plant where things are permitted to happen unexpectedly is an inefficient plant. Every accident is an indication that there is something wrong with men, methods, equipment, or material.

Danger Signals

And if the engineer or executive must then be interested in all accidents, whether they cause injury or not, those which do cause injury have for him a peculiar importance, because they stand out conspicuously and serve as danger signals to warn him of the inefficiency that is undermining his output. This is not theory but fact. The improvements in punch press operation, which I have already mentioned, were suggested and carried out—by whom? By the production department or by some imported efficiency engineer? No—by the safety man! He found that men were losing fingers in these punch presses because

they had to reach in to the press to place or remove material. He changed the method of operation by introducing automatic or semi-automatic feeds and kickouts. He did this primarily to prevent the loss of fingers. Having done it, he found that he had increased production from 25 to 100%. The foremen were astonished. They need not have been astonished. Reflection would have shown them that the movement of the operator in reaching into the press, while it occasionally caused the loss of a finger, always caused the loss of time, and that an arrangement which would save the occasional finger would also save a fraction of a second every time the machine was operated. To quote another instance of a more general nature, the manager of one of the largest paper mills included in the membership of the National Safety Council once said to me: "Before we had a safety committee, every little while we would have to shut down our machine because of a belt breaking or something of that sort. Now our safety inspection catches these things before they happen and we have no more shut-downs. Our safety work has more than paid for itself through the increase in production, aside from cutting down our compensation costs."

You may say—"these plants were not run efficiently. If they had been, it wouldn't have taken a safety man to show them how to feed punch presses or keep belts in repair." I freely grant this. Yet the fact remains that both of the plants which I have mentioned were and are in the very front rank of their industries, they were well managed, they were making money. The management simply, like every other management that I know of, was so busy with development and with the outstanding problems which it had to face that it could not watch every single detail.

In another instance, even more noteworthy than those I have mentioned, the Chief Engineer of a large company conceived an entirely new method of accomplishing an important process in the industry—a method wholly mechanical, to replace one which required constant attention by skilled men. It happened that these men were exposed to injurious dust. When the new plan was presented, it was opposed by some of the experienced factory executives who declared on general principles their firm conviction that it would not work. The President of the company said "Even if the new method does not increase production as we hope it will, it will at least remove this danger to the health of our men. Therefore, we will try it out." It was tried out. It worked. It cut down the number of attendants required from 23 to 3 and removed all hazards to those that remained. It increased production so tremendously that it will undoubtedly revolutionize the entire industry. This

(Continued on page 205)

Plant of the Dominion Engineering Works Ltd.

By J. H. RODGERS

IT is surprising but interesting to note that in this comparatively short space of thirty years the export of pulp and paper from Canada has shown an approximate increase of one million per cent. In figures this might be represented as an increase in value of from $122 in 1890 to the astounding figures of $122,000,000 in 1919. This remarkable growth has been due, very largely, to the vast natural pulp wood resources and the unlimited supply of available water power, together with the initiative of the progressive business men engaged in the development of this industry.

There is, however, one feature of these activities that, until very recently, has received very little attention, and that is in the construction of the machinery essential to the problem of paper-making. In the past the bulk of this equipment has been imported, and this particularly applies to the better grade of high-speed newsprint and writing-paper making machines.

The unprecedented and rapid expansion of the pulp and paper industry in Canada, during recent years, has necessitated an abnormal demand for paper making equipment, and the opportunity for establishing plant for Canadian-made machinery was readily recognized by many of the leading business men directly associated with this great development. The result of investigation and negotiations was the incorporation of the Dominion Engineering Works, Limited, at Lachine, Quebec. This company has acquired the modern plant of the old St. Lawrence Bridge Company at Rockfield, Que., a short distance from the works of the Dominion Bridge Company, the executive of the latter being chiefly responsible for the inauguration of the new enterprise.

The plant is particularly well adapted for this new line of work, as it is provided with every modern facility for rapid and economic handling of the complexity of parts required in the construction of this type of machinery. The main building has a length of 660 feet, and is 220 feet in width. There are two erecting bays, one with a span of 85 feet and one of 75 feet, both having a length of 220 feet. There is also an additional bay 440 feet by 60 feet, that contains the boiler and power plant, tool room, smith shop, flask department, works offices, etc. The original building was provided with two 35-ton cranes and four 40-ton cranes. These cranes may be operated in pairs when desired, giving a combined capacity of 75 tons. A new 75-ton crane will shortly replace one of the smaller ones now in use. Several assembly and jib cranes are likewise provided for general erection work. The erecting department will be divided into two sections, one for paper making machinery and the other for hydraulic equipment.

An interesting feature in connection with the installation of the machine tools is the grouping of the machines on the unit principle. The overhead shafting and motors are carried on specially designed framework supported from the floor, and this arrangement relieves the roof trusses from unnecessary weight, and likewise provides free movement for the cranes overhead.

A 72" DRYER BEING TURNED ON A LARGE LATHE.

VIEW SHOWING FARRELL GRINDER GRINDING A GRANITE ROLL.

This concern at the present time are engaged in the construction of two 166-inch high-speed newsprint machines for the Laurentide Paper Company. The installation will be of the most modern design. The drier rolls will be sixteen feet in length and 72 inches outside diameter and each roll in a finished condition will weigh approximately 16 tons, and in operation will revolve at a surface speed of 1,000 feet per minute, or about 56 rev. per min. Very close accuracy is required in the machining and assembling of these units, as any slight variation in balance will be detrimental to the smooth running of the machine, and consequently the quality of the product. .

The inner and outer walls of the cylinder must be perfectly concentric . to assure uniform expansion and contraction by the action of the steam drying medium. These rolls, ready for assembly in the frames, are balanced to the fraction of a pound. The making of paper is comparatively a simple process, the fundamental principle being the passing of a film of wood pulp over a series of dryer drums, heated internally with steam, and thence through a number of smaller rolls covered with rubber, felt, or made of polished metal. This train of drums and reels is required to operate at high speed, and for this reason it is imperative that the mechanism and parts comprising the entire machine be fitted with extreme accuracy.

The company are likewise building a 148-inch Harper tissue-paper machine for the Interlake Tissue Mills, and prior to the present incorporation the Dominion Engineering and Machinery Company, then a subsidiary of the Dominion Bridge Co., has already constructed several paper making units, including a 48-inch Fourdrinier writing-paper machine for the Howard Smith Company, of Beauharnois, Que.,

In addition to the paper making machinery activities they are manufacturing hydraulic equipment from the original designs of the I. P. Morris Company of Philadelphia, having secured the Canadian rights for the manufacture of this firm's equipment. They are at pres-

ent constructing two 20,000 h.p. waterwheels for the Laurentide Company, these being duplicates of the units now operating at the latter's plant.

The character of the work engaged in will necessitate additional capacity to the existing building now used for a foundry, which has a length of 181 feet. and a width of 145 feet. This will be enlarged to 300 feet by 145 feet, and will be equipped with an 84-inch cupola in place of the two now in use, which are 37-inch and 54-inch in diameter. At present the foundry is provided with two small casting pits, but the new extension will be equipped with a casting pit 120 feet by 30 feet, and 8 feet deep. This is to facilitate the making of the castings used in the large water turbines. The foundry will be equipped with every modern requirement for effective and economic service. A large special sandblast room has been constructed in the main foundry for cleaning the large castings. Large and small core-ovens are installed and likewise a cinder mill for reclaiming metal and coke from the cinders. Two overhead electric cranes are provided in the foundry, one of 70-ton and the other of 15-ton capacity.

PRODUCTION FACTORS

(Continued from page 203)

two-fold result was not a coincidence. It was a natural result of the fact that the old process was wasteful in both materials and time. The injurious dust that should have gone through the manufacturing process was blown out into the faces of the men. The new method saved both the dust and the men.

The man who did this remarkable piece of work was not a "safety engineer." He was an engineer who believed in safety. As a man, he believed in safety as a human necessity. As an engineer, he believed in safety as a thing inseparable bound up with engineering efficiency.

On the human value of safety, I leave each of you to judge for himself. I know how I feel and I think I know

how you feel. But that is not my topic to-night. But on safety as an engineering job: I know that the engineer whom I have mentioned, and the safety man in the punch press shop, and the paper mill manager, are right—I know that safety and engineering efficiency are inseparable. You cannot have one without the other. If you are safety men and think you have to put on a guard that interferes with production, you may be sure that your remedy is only a temporary one and that it must give way eventually to an improved machine or method which will be safe without being inefficient. And the engineer or executive who thinks that accident prevention is not in his department—that he will leave that to the safety man or the insurance company or the state inspector —is missing something—something big. He will find out some day that every accident in his plant or on the machine which he designed is a danger signal for him—a symptom of time and dollars wasted as well as lives and limbs. The rough shod methods of industry a generation ago, which left a trail of human wreckage, were not only inhuman—they were inefficient. They and the man who stood for them are gone on the scrap heap and will never return.

THE INTERCRYSTALLINE BRITTLENESS OF LEAD

According to an abstract in the "Journal of the Franklin Institute" of a U.S.A. Standard Bureau Note, sheet lead sometimes assumes a very brittle granular form during service, due to corrosion. An explanation which has been offered by previous investigators for this change, in properties is that it is due to an allotropic transformation, the product resulting from the change being analogous to the well-known "grey tin." Contact with an electrolyte, particularly a weak acid solution of a lead salt, has been claimed to be the agency by which the transformation is brought about. Metallographic examination of the granular "allotropic" lead shows that each grain has the characteristic properties . of the ordinary form of lead. The intercrystalline cohesion of the grains for one another, however, has been so weakened that the material has a granular appearance. The rate at which the intercrystalline brittleness is brought about is proportional to the amount of impurities and to the concentration of acid in the solution in which the lead is placed. Practically all the impurities which are found in lead are lodged in between the grain. The preferential attack by the corroding agent for these impurities and perhaps also for the "amorphous intercrystalline cement," accounts for the brittleness produced.

VIEW SHOWING A PORTION OF THE FOUNDRY.

WELDING
AND CUTTING

Lessons on the A.B.C. of Good Welding

This Article Deals with Side, Overhead, and Pressure Work—Considerable Interest is Being Taken in These Practical Lessons.

By W. B. PERDUE *

LET us review carefully the position of the rod and torch shown in Fig. 3 of the April 22nd issue. Note that the angle between the torch and the weld is from 60 to 70 degrees, and that between the rod and the bevel is about 30 degrees. If we maintain these same angles with respect to our work and also the same relative positions with regard to our arms, the torch, and rod, we will experience no difficulty in executing a weld in any imaginable position.

Causes of Failure

There are three principal causes for the failure of the welds described in this chapter, viz: 1.—Lack of proper penetration and insufficient fusion of the parts welded. 2.—Unsteadiness of motion, or lack of training of the muscles of the arms in the positions in which these welds are required to be executed. This is clearly indicated by the swishing

* In charge of welding instruction, Heald's Engineering School, San Francisco, Cal.

noises of the flame and the abundance of sparks which indicates oxidation of the metal within the weld. 3.—Lack of attention to minute details necessary to produce a perfect weld, resulting in a weld of rough "plastered" appearance.

Side Welds

Many of the large containers and pressure cylinders now being built by the acetylene process require the execution of welds along the sides of same. To the unversed operator of the torch these welds are almost impossible of execution. Figure 32 shows a phantom photograph of a weld of this type, looking through the plate at the operator. Note that the alternate layers of added metal extend from the outer or ripple surface to the bottom of the bevel, and slope at about 45 degrees with relation to the direction of the weld.

Working either backward or forward the position of rod and torch when at the outer or ripple surface is as shown in the

posed photograph, Fig. 32. At the bottom of the bevel the torch is lifted so that the downward inclination is about equal to the upward inclination here shown. The hand holding the rod does not come below the level of the weld.

With this method of operation the tendency to form adhesions or to slop the metal from the weld is greatly minimized. The necessity for special preparation of the parts with a special bevel is also obviated, since the bevel used in ordinary work serves as well as any other.

Overhead Welds

Overhead welding is not so difficult as some may at first imagine. There is a tendency on the part of beginners to hurry the work and to add metal from the rod without first bringing the sides to proper fusion, and adhesion of the parts which are not thus securely bonded together. This difficulty mastered, and proper position secured, the student will encounter no other serious difficulties and should soon become proficient in this work.

Overhead welds may be worked either forward or backward, the same movement being used as is employed in welding on a horizontal surface. Figure 32 is a photograph posed to show the

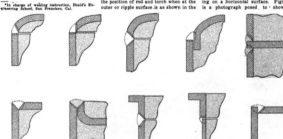

Reading from left to right and starting at top figure, we have Figs. 34 to 43. These show preparation of parts for pressure welding.

proper position of the operator in close quarters lying prone on his back. A fireproof cover stuffed with fibre asbestos will be a great comfort if used as a pillow. Long gauntlets should be worn to prevent sparks from entering the sleeves of the operator; and the use of a face shield instead of goggles is recommended.

Although it is almost impossible to dislodge any of the metal from an overhead weld, there might at times be some danger from solder or spelter which might be in or above the parts being welded. At any rate nothing is gained by placing one's face, arms or body directly beneath a large body of metal in fusion.

An imaginary line of welding has been indicated on the posed photograph shown in Figure 32. This position gives freedom of motion and prevents such sparks as may be dislodged from the vicinity of the weld from alighting on the person or clothing of the operator. The work may be carried on from a point a few inches to the left of the point directly over the top of the operator's head to a point almost two feet distant, when it will be found necessary to change positions.

When working inside a boiler or other confined space, the best of protective clothing should be worn and the operator divested of all other garments save his underwear. Otherwise the combination of a spark on the clothing and a leak of oxygen might produce disastrous if not fatal results to the operator. The helper should be cautioned never to leave the operator while working in any confined space.

Regulators should be in good repair, so that there will be no leak from them when the pressure regulating key is removed. To remove this key by turning it to the left, first from the oxygen and then from the acetylene regulator should be the first duty of the helper in case of any accident such as the bursting of a hose. The operator who fails to thoroughly instruct his helper in these very necessary precautions is taking an unnecessary chance. Be careful. Take no chances.

Containers and Pressure Cylinders

In the construction of containers it is not always possible or even advisable to separate the edges or otherwise make allowance for contraction in those welds which extend around the circumference of the parts to be joined. In such work two welders often work together. The welds are commenced at opposite sides and worked as side welds, thus producing less strain than where the weld is worked all the way round by one operator.

The welding of bottoms, either dished or flat, to cylinders caused the cylindrical part or shell to expand and become larger than the bottom or head in which the effects of expansion are not so great. This produces a deformation due to the bulging of the shell where the weld is commenced and worked all the way round without interruption. Therefore, in order to prevent distortion and the consequent internal strain of the parts being joined it is necessary even while tacking to work alternately on opposite sides, by this means the effect of expansion on one side is counterbalanced by that produced immediately afterward on the other.

In the construction of containers when either a bottom or heads in both ends are required, the method of putting in and their preparation for welding must be governed by the purpose for which the container is intended.

Pressure Containers

Containers which are to withstand considerable pressure should have the heads dished and flanged in accordance with boiler-maker's standard specifications. Furthermore, they should be so prepared that the line of welding is one of tension

(Continued on page 209)

FIG. 32.—Posed position, facing operator, showing rod and torch at the finish of ripple in Side Welding, working backward. In Forward Welding the same position is assumed at the commencement of the ripple.

FIG. 33.—View from directly above operator, lying prone on the floor, showing position for execution of overhead weld. Note that the line of welding is even with the top of operator's head and some distance to the left.

Learning the Machinist Trade is not all Sunshine

So Says John Conley, Whose Former Efforts We Have Previously Recorded—Breaking Drills, Chasing Centre Punches, and Figuring If His Square is Square, Are John's Latest Troubles.

By J. DAVIES

HITHERTO John has simply been making holes that did not need to be located very exactly, but now he is given a job that requires that the holes must be in a certain place. It has been marked off at the marking-off table as is nearly always the case, and the centre is a very small one. The drill is not running very true and he tries in vain to strike the centre. Not being successful, he decides to make the drill run true and hits it with a hammer, but this procedure breaks the drill. Off he goes to the foreman to get an order for another drill. "What's the matter with the one you had?" asked the foreman, and John meekly replied, "I broke it." "How did you break it?" "Trying to straighten it," says John. "Haven't you learned yet that you can't straighten a hardened twist drill with a hammer?" "Yes, I learnt that about five minutes ago," sighed the would-be mechanic.

"All right," commented the foreman kindly, but firmly, "but listen here. Don't forget your experience. Everyone makes mistakes. It is those who do not make the same mistake twice that succeed; come and I'll show you how to do the job."

The foreman cleans out some little bits of chips that have got inside of the chuck, moves the adjusting screws a little and makes the drill run true, then enlarges the centre punch holes with a centre punch to give the drill a chance to strike the centre. Turning, he said, "You will notice that when you start to drill a hole, the drill will not always cut true to the circle marked off, but will cut to one side. In such a case cut a light groove with a round-nosed chisel, on the side toward which the drill is to be drawn, in other words on the wide side. Don't wait until your drill is cutting nearly the full size before you draw it over or you may not be able to draw it over. Draw it over as soon as you detect the error, then cut out your chisel marks with the drill and repeat the operation until your drill is cutting true to the lines."

"Do you think you will be able to manage that job now?" he asked suddenly.

"Yes, I think so," remarked John.

"All right, let me know when you have them done," and away the foreman went.

About five minutes later John could be seen crawling on his hands and knees under the bench after the centre punch that flew out of his hands. After several minutes' fruitless search Chas. Bailey lifted his foot off it and kicked it towards him, at the same time advising

him to glue it to his hand so that he could hold it. In spite of this advice John finished the job in a fairly creditable manner and lost no time in informing the foreman of the fact.

"I'll next give you a somewhat difficult job," said the foreman, "but if you will listen to the instructions carefully you should be able to manage it.

"This is a little repair job. I want to drill a hole through the collar as shown on this sketch to suit a hole in the round shaft. You will notice that owing to someone's bad workmanship the hole in the shaft does not go through the centre.

"Place the round bar in the vee block, then put the drill through the hole, bolting the vee block securely and fastening the table in such a position that the drill is free to move up and down in the hole. After this is done replace the bar with the collar to be drilled, leaving the vee block in the same place. Providing you take the proper care the holes in the collar will coincide with the holes in the bar. If the hole is to be drilled in the bar, set your drill from the collar in the same way. You can use this method for many other jobs which would be very difficult to mark off. Last but not least insert the round bar in the collar and drill through both collar and bar.

"Just by way of variance suppose we were merely attempting to drill through the centre of a round bar. In this case all you would have to do would be to simply bolt a vee block down to the table and set your drill to the centre of the vee, secure your work to the block and drill it through."

John was so impressed with the usefulness of the vee block that he told his father of his experiences in the few minutes' chat that they usually had before

bedtime. He also informed him that he intended to add a pair to his own personal tools.

"Oh, that reminds me," said his father. 'I intended to ask you if you knew if your square was square, and if your level was level.

"Yes, I know for a positive fact, I took them over to Mr. Glenn, who has charge of the tool room and asked him to show me how to test them. Here is what he did. First he took the combination square, put in the 12 in. blade, taking care that there was no dirt or grit in the slot where the blade goes in. He then placed the base of the stock against the planed edge of a surface plate or marking-off table, then made a fine line close up against the blade, and reversed the square. He next drew another line parallel with the first one, leaving just as small a space between the lines as the eye could detect. He told me you could draw one line exactly on the top of the other if you desire. If the two lines thus drawn run exactly parallel, your square is correct, but if the lines do not run exactly parallel then the error in your square is equal to half the error shown by the lines. Now take out the blade and turn it upside down to test the other edge, or you may try it with micrometers to see if it is parallel, which amounts to the same thing.

"To test your center finder, you practically repeat the same operations, except that instead of the surface plate, or marking-off table, you must get something that is perfectly round.

"The end of a short shaft turned and faced in the lathe, or the end of a plug gauge will serve very well, the diameter being such as will allow the sides of the

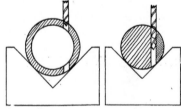

THIS SKETCH FOLLOWED JOHN'S INSTRUCTIONS, AS TO THE METHOD OF DRILLING.

unshine

centre finder to bear somewhere about the middle."

I He also told me how to test other tools, here is what he said:

"To test a spirit level place it upon a surface plate and pack up the plate, not the level, until the level shows perfectly level, then draw a pencil mark around the edge of the base of the level, reverse the level, placing the base inside the pencil mark, and if the level still shows true, it is true. If you are testing an adjustable level, adjust it until it fulfills the conditions as stated. If your level is not adjustable then you will have to scrape, file, or machine the base of the level to correct any error there may be in it.

To Test Your Protractor

"Since there are 360 degrees in every circle, if two lines forming a square or right angle are placed inside a circle, they will divide that circle into exactly four-equal parts, and as 360 divided by four is equal to 90, the angle of the square will be 90 degrees. Therefore set your blade at 90 deg. and test exactly as you did for your square; if the base of the stock does not form a perfect square with the edge of the blade, it follows that every other angle will be out an equal amount.

"He said he couldn't test my straight edge except by reference to the shop straight edge. To test a straight edge properly, he mentioned we would have to have three, and if any one of these fit the other two they would be perfect.

"I don't see how he makes that out. If his straight edge and my straight edge fit each other, I don't see why they wouldn't both be right, do you?"

"Why yes, let me explain. Suppose one straight edge was hollow or concave, and the other was round or convex, it would be possible for them to fit each other perfectly and yet neither of them would be straight by any means. Now suppose we had a third one which was straight. When the first two were tried out on the straight one it would show the hollow on one straight edge and the round on the other. You can see that it would be impossible to make three pieces fit each other perfectly unless each piece was perfectly flat. Here are some other facts worth knowing.

"A pair of angle plates are among the most useful tools for fastening work to the drill. If they are both alike and properly designed, they form quite a number of very useful combinations. Bolted face to face, they can be made to form any angle; standing on end they form vees for drilling round stock end. ways; side by side they form parallels for drilling pipes, etc.

"An overhanging arm is often useful for drilling wheels and pulleys that are too large to go on the table.

"In drilling, tapping, and clearance holes, as in bearings and caps for same, drill the clearance holes in the cap first, then use the cap as a guide or jig for the tapping hole, putting the clearance drill into the tapping hole until it begins to cut full size, which will prevent the tap

raising a burr; then change the drill to tapping size and drill the tapping holes.

"In some cases where two holes are to be drilled in line with each other at opposite sides of a casting, it is impossible to drill right through at one setting. In such a case a centre is fixed either in a casting bolted to the base of the machine, or else fitted to the hole in the centre of the table. The centre punch mark for one hole is placed in the fixed centre, and the other is brought directly under the drill. Great care must be taken to have the fixed centre directly in line under the centre of the drill spindle. Drill one end, then turn the job over and repeat the operation. When turning the job end for end, if the hole drilled is too large to sit on the conical part of the centre, it must be temporarily bushed with a concentric bush—that is, one that is turned and bored true. Fasten the job securely before you start to drill, the method of doing this depending entirely upon the nature of the job and the tools available. Now off to bed, son. I've told you enough for one night. I feel you may develop into an honest-to-goodness mechanic yet. One thing I like is you are a good listener, and good listeners are bound to learn."

A, B, C OF WELDING

(Continued from page 207)

and not of bending. This precaution is taken chiefly because of the fact that few operators secure complete and proper penetration of the parts welded, with the result that the bottom of the weld contains a line of weakness that under continual bending strains will extend deeper until it eventually produces a fracture.

The forms illustrated in Figures 34 and 35 are, therefore, subject to this criticism. Where pressures are not too great, and the welder is sure of his ability to secure complete penetration and has a torch which stirs the gases into such complete and perfect mixture as to produce no change in the metal, these welds may be used, provided proper manipulation is given thereto to eliminate impurities and produce proper refinement of the grain of the weld.

The form illustrated in Figure 36 is ideal, but costs a great deal more to prepare than the others. That given in Fig. 37 is permissible only for riveting. It should not be used where welding is to be employed.

Figure 38 illustrates the common method of preparation for welding a partition in a cylindrical container. Where this method of welding is used two welders should be employed, one working immediately behind the other. This prevents the strain which would ensue from a second heating of the metal in the immediate vicinity of the first weld and also utilizes the heat of the forward weld to speed up that of the one following. The faster of the two workmen should carry the front weld.

Storage Tanks

Where the container is to be used for storage purposes only and is not to be subjected to high pressures the head may be cut the inside diameter of the shell and tacked in position as shown in Figure 39, where both shell and head are beveled. Figure 43 shows the same arrangement with the head beveled and the shell left square. This beveling may be done more rapidly and cheaply with the cutting torch than with the circular shear, even where such circular or bevel shear forms part of the shop equipment. This form of welding is preferable where the workmen are competent and can be depended upon to secure proper penetration without adhesions or burning the metal in the bottom of the fill.

Figure 40 shows the head for a storage tank chamfered, flanged, and backed into the shell until the edges are even. It is then tacked on opposite sides in several places and welded in position. Welders can work in pairs, on opposite sides, to great advantage on this work.

Angle Iron Rings or Tops

In welding angle iron tops to cylinders either as a reinforcement for the top to keep the container in proper shape or to secure thereto a bolted head, it is necessary that both the angle iron and the shell shall be flanged where the thickness of the metal exceeds ¼ inch. The proper preparation is shown in Figure 41. The same method of preparation is used where two cylindrical shells are to be welded end to end as is often necessary in the manufacture of long containers which can not be rolled from a single sheet of metal.

The method for welding a heavy ring to a light shell is shown in Figure 42. It is important that the greater force of the flame shall play upon the thicker section and that complete penetration shall be effected as shown by the dotted outline of the weld.

SOME INTERESTING EXAMPLES

(Continued from page 197.)

There is hardly a line of industry where they are not used, but our main idea is to familiarize readers with the different classes of tools and their uses.

All the operations depicted were performed on what are known as the 6A type of automatic, and for a later article we will present still another style of machine tool and its work.

Large Mill in B. C.—The construction of the B. C. Spruce Mills, Ltd., Wattsbury, B. C., 18 miles west of Cranbrook, B.C., is well under way. It is the first all steel and concrete mill to be built in B. C. The flume is ten miles in length and it will be the largest mill in the interior and have an output of 250,000 ft. every 16 hours. It is estimated there is enough timber for 12 years' cutting. B. F. Wilson, Warsau, Wis., is the manager.

DEVELOPMENTS IN SHOP EQUIPMENT

PORTABLE SLOTTING MACHINE

The accompanying illustrations depict the latest type of portable slotting machine recently placed upon the market by the Newton Machine Tool Works, Inc. of Philadelphia. U. S. A. This machine has a maximum stroke of 76″—cross feed to the tool slide of 40″ and in-and-out feed to the tool slide of 6″. There is a vertical adjustment of the tool slide of 24″ and a cross feed of the upright on the sub-base by motor of 84′. Saddle is counterbalanced with counterweights inside of the upright running in guides. The feed to upright is by separate motor through spiral and bevel gears and stationary screw. These feeds are automatically intermittent through friction cloth.

The upright has power rapid traverse in either direction through friction clutch controlled by hand lever. The saddle is operated by screw with one piece bronze nut, and is supplied with taper shoes to compensate for wear. The saddle traverse is controlled by means of a coarse pitch large diameter screw which is fitted with roller thrust bearings both top and bottom to insure operation in tension. The drive to the screw is by reversible motor through spur and bevel gears, and the driving

screw is carried in tension bearing with thrust collars running in oil and is supplied with a safety clutch which is disengaged by the saddle at top or bottom to prevent jamming.

Operating dogs for the reverse are mounted on the side of machine with a latch so that they can be easily changed while the machine is in operation. These operating dogs have no connection with the feed.

The tool apron has swivel for relief and is arranged to swivel through a full circle for the making of angular cuts. The tool slide intead of having bolted straps on the side for the square gibs is integral with the slide itself and bronze taper shoes are used for taking up the play.

All gears on this machine are fully enclosed. A reversing planer type motor of 15 h.p. is recommended to drive the machine. A 7½ h.p. motor is recommended for the feed to the upright.

FOUR-POST SCREW PRESS

The Manhattan Machine & Tool Works, Grand Rapids, Michigan, U.S.A., has placed on the market a four-post screw press which is illustrated herewith. This press is of very massive

GENERAL VIEW OF THE SCREW PRESS.

THESE VIEWS SHOW BOTH SIDES OF THE SLOTTER.

construction, and possesses both rigidity and strength.

It will be noted that a ratchet action is secured by means of a ratchet disc, and a lever, which is of sufficient length for all purposes within the capacity of the machine. This ratchet can be reversed if desired. It can also be disengaged so as to leave the hand wheel free for rapid adjustment of the ram.

The two post presses manufactured by this concern are for the regular run of tool room work, etc., but this four post construction is especially suited to the heavier class of work.

Every convenience for facilitating quick work and simplicity of operation have been embodied in the design, and all parts are balanced as to proper strength, so that there is no danger of breaking or failing just when you want it most. Following are the principal specifications:

Capacity, 80 tons; size of bed, 20 x 36 inches; space between standards, 26 in.; floor space, 31 x 41 in.; height over all, 77 in.; height to top of bed, 24 in.; diameter of standards, 3 in.; maximum height from bed to ram, 22 in.; thickness of ram, 14 in.; thickness of bed, 3¾ in.; diameter of screw, 3 in.; diameter of hand screw, 23 in.; diameter of punch shank, 1¼ to 3 in.; net weight, 2,270 lbs.; shipping weight, 2,425 lbs.; boxed for export, 2,600 lbs.; boxed for export, 58 cu. ft.

NEW THINGS IN MACHINE TOOLS

STAYBOLT CUTTER

The Baird Pneumatic Tool Co., Kansas City, have placed on the market the staybolt cutter as shown in the illustration. It is said that this cutter will clip off staybolts up to 1½ in. diameter at the rate of 1,200 per hour.

The machine is composed of a 15 in. air cylinder, the piston head of which connects directly through toggle movement with a pair of lever arms, into which are fastened, but removable, a pair of cutter knives of sufficient strength to clip the stay bolts.

The machine is operated at a working pressure of 100 lbs. air pressure. It delivers 88 tons pressure on the knives and the blades are so designed that when placed against the crown sheet the stay bolt is cut off just the proper distance to allow for heading over.

The machine can be operated on modern rail stay, or marine boilers, and is capable of being held at the gravity point at any angle, being perfectly balanced. This feature allows it to be easily handled on either straight side or radial stay work. Different sized blades are used depending upon the nature and length of bolt to be cut.

The machine is so designed that it can be operated by one man, and all control valves are placed convenient to the operator. Apart from use in staybolt cutting, these machines can be used for the reclamation of scrap. Following are its principal specifications: Diameter of cylinder, 15 in.; travel toggle, 3 in.; maximum capacity between cutter knives, 1¼ in.; air pressure required, 100 pounds; maximum tonnage on cutters, 88 tons; net weight complete, 430 pounds.

AIR OR STEAM ENGINE

A small engine that may be operated either by steam or air for driving portable mechanism, such as boring bars and drilling machines, has been developed by the H. B. Underwood Corporation, of Philadelphia, Pa. The working parts are inclosed in an oil-tight case which forms part of the engine base. The engine operates at a speed of 250 r.p.m. on a pressure of from 70 to 125 lbs. per square inch. These engines are made in two sizes, 3 and 5 H.P. Weight about 270 lbs.

MULTIPLE MICROMETER

The A. T. Brush Tool Company of Erie, Pa., are now making a micrometer that can be used for measuring from 0 to 2 inches, without any attachments. The spindle has a screw of 220 threads per inch while a separate screw with 40 threads per inch is used for traversing the thimble. It is stated that the readings are quite easy and accurate. Provision is made for taking up the wear of the anvil and the spindle.

CYLINDER REAMERS

The Wetmore Reamer, Company of Milwaukee, Wis., is placing on the market a special set of reamer tools, designed for the production of round, straight and thoroughly smooth holes in small cylinders. A set consists of three reamers, one for roughing, one for semi-finishing and the third for finishing. Suitable arbors are provided for all makes of machines. The finishing reamer is of the floating type. All cutter blades are made of high-speed steel and accurate adjustment is provided to 0.001 inch.

IMPROVED HELVE HAMMER

An improved design of helve hammer has been brought out by Long and Allstatter, of Hamilton, Ohio. It is of the rubber-cushioned type. The treadle is arranged to prevent lost motion between it and the belt-tightener and is very sensitive in its action. Increased area of brake surface assures quick and accurate stoppage of the helve hammer when in an "up" position.

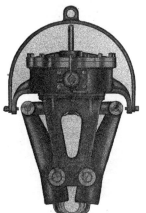

GENERAL VIEW OF THE STAYBOLT CUTTER.

The MacLean Publishing Company
LIMITED
(ESTABLISHED 1887)

JOHN BAYNE MacLEAN, President. H. T. HUNTER, Vice-President
H. V. TYRRELL, General Manager.

PUBLISHERS OF

CANADIAN MACHINERy
and MANUFACTURING NEWS

A weekly journal devoted to the machinery and manufacturing interests.

B. G. NEWTON, Manager.   A. R. KENNEDY, Managing Editor.

Associate Editors:
J. H. MOORE   T. H. FENNER   J. H. RODGERS (Montreal)
Office of Publication: 143-153 University Avenue, Toronto, Ontario.

VOL. XXIV.   TORONTO, AUGUST 26, 1920   NO. 9

PRINCIPAL CONTENTS

THE WEEK IN PASSING

CANADIAN manufacturers do well to keep going as they have done, considering the trouble they have met in securing supplies. Tube mills find it difficult to operate to capacity, or anything like it, because they cannot secure skelp; our steel mills are doing their best on a very indifferent supply of coal, but they are away behind in what they should be shipping; makers of boilers and kindred lines are hampered, because they cannot get a plentiful supply of standard sizes of tubes; stove makers and electrical workers want more steel sheets, and many construction concerns are hard put to keep up their supply of rods for reinforcing. The industrial community of the Dominion is making a remarkable showing considering the handicaps and hurdles that are being placed in front of them.

THERE are now over six hundred American factories with branches in Canada, and indications are that the number will increase. The tariff is the first consideration that has to be met if U. S. firms are to do business in Canada. The war added a 7½ per cent. tax to that, and the exchange rate since then, at times as high as 17½ per cent., has made the question one of such seriousness as to compel action. U. S. firms coming to Canada do so to handle the Canadian market at close range, and also to gather the advantage of trade relationships that may exist between the various parts of the British Empire. It might mean in the long run that Canada would purchase less direct from United States by the opening of so many U. S. branches in this country, and in this way improve the exchange situation, but it will take some time for this to come to pass.

WORD comes from the Advisory Committee of Railway Executives of New York that the following are the aims in view: (1) an average daily minimum movement of freight cars of not less than 30 miles per day; (2) an average loading of 30 tons per car; reduction of bad orders cars to a maximum of four per cent.; reduction of the number of locomotives now unfit for service, and more effective efforts to bring about the return of cars to the owner roads. Now that is all very well, but the great bulk of it rests with the roads themselves. Many of the cars that come into the yards in this and other cities, especially from the other side of the line, would sag in the middle were they to get thirty tons piled into them. When Canadian business men have to pay charges from U. S. points in New York funds, and pay the increase of some forty per cent. in rates, they will have an item that will make even the strongest sit up and blink.

EVERY week or so some very well meaning financial critic rises and assassinates the premium mills in the steel and allied industries. And after this has been done the premium mills keep right on operating just the same. There seems to be no reconciling some of the prices that prevail in the steel market now. For instance, why should No. 10 gauge steel sheets sell at 9.50, and No. 12 at 11.25? Simply because the mills that have them for sale ask that difference. There is some improvement in the railroad situation. It may be a little hard to get some interests to admit that such is the case, but yards report cars coming from Pittsburgh in a week and less, a thing that was out of the question a few weeks ago. With the continuance of this movement—and the increased freight rates will make cars valuable—Canadian manufacturers should look for improved service in the matter of getting their much-wanted supplies from U. S. mills, and with a betterment of service the premium mills will dwindle, or at least their high prices will.

A SPLENDID summer home is pointed out in Northern Ontario, and the remark generally goes with it: "He made his money out of the war." Before the war he was just an ordinary business man, making a decent sort of a living. The war came and, to use the phrase that best fits the case, he cashed in on the thing. Cashed in while others gave their lives, and their limbs, while others sacrificed sons, husbands and brothers, homes and everything that seems worth while. And apparently there was nothing wrong in what this worthy citizen did. He wound up his business and paid all his debts, and pocketed the hold-over. And all this while others were winding up their little estates and leaving to return no more. It's a mighty poor compliment to pay to a man to hang on him the fact that he made his money out of the war. And it's a mighty poor compliment, to pay to the Government of any country to say that it allows one section of its manhood to fight and die, while another lives and waxes fat and wealthy on their agony.

ENVOYS EXTRAORDINARY

—From London "Punch"
Prime Minister (To Bolshevist Delegates:) "Happy to see you, gentlemen. But would you mind going round by the tradesmen's entrance, just for the look of the thing?"

MARKET DEVELOPMENTS

Establishing Pig Iron at $50 Per Ton

Quite a Number of Sales Made at that Figure—Large Concern Recently Opening in Canada is Out Asking for Cancellations of Reductions in Price.

CHARGES are being added to business in this country. The hope that the prepayment of freight in New York funds was to be done away with seems to be doomed to failure, according to the interpretation that many houses have of the ruling, and now there is the added charge of through freights from U. S. points. These things tend to take up any sag that may be developing in favor of the consumer.

There is some improvement in the matter of railroad delivery, as yards are reporting that they receive cars from Pittsburgh in from five days to a week now. But there is still a severe shortage in some lines, such as sheets and boiler tubes. Canadian manufacturers are embarrassed more or less in several lines, because they cannot get the sizes they require, and industrial concerns all over are after tubes for repair to plants. Canadian mills are giving little relief to the steel market just now for the reason that they are operating on very indifferent and insufficient coal supplies.

One of the largest automobile concerns that has started work in Canada is asking that there shall be cancellations of a large number of its orders, or failing this, that reductions in price be made. The automobile and allied industries are having a quiet season, and the dealers in machine tools are looking for some other industry to take their place in the purchasing of equipment. The railroads have failed to buy up to expectations, in fact some of their more recent requisitions have been cancelled.

Premium prices are still quoted on a large number of lines in the steel warehouses.

The scrap metal market is uninteresting for either buyers or sellers. There is very little material being offered, and the yards are not going out of their way to take on any more tonnage. The one thing that would help would be the operating of the big mills, such as steel plants, to full capacity, and this cannot be done until the supply of coal improves.

PIG IRON IN PITTSBURGH IS GETTING FIXED AT $50 TON

Special to CANADIAN MACHINERY

PITTSBURGH, August 26.—Pig iron and coke still occupy the centre of the stage. Pig iron was going to advance because coke was so high. Now coke has declined slightly, with prospects of further declines, while pig iron has advanced sharply, with no proof that it will not advance indefinitely beyond the present level.

Car placements in the Connellsville coke region have continued good, this being due of course to the cars moving less slowly and thus making round trips more frequently. The total number of cars in the coke service rarely varies much. The increase in production, due entirely to better car supplies, is shown by the Connellsville Courier's reports of coke production in the Connellsville and Lower Connellsville region, which for three successive weeks have been as follows: 172,870 tons, 194,140 tons and 219,440 tons. Thus the last week reported upon shows an increase of 25,300 tons over the preceding week and an increase of 46,570 tons over the week before that, an increase of 27 per cent. in two weeks. The immediate influence upon the market has been small because the increase

has been chiefly in the production by the furnace ovens, rather than in production by the merchant ovens, but it is going to come to the same thing in the long run, as if the steel interests make more coke in their own ovens they will seek less in the open market. In the past week spot furnace coke has been rather soft at $17.50 per net ton, f.o.b. ovens, while formerly and for many weeks it was $18.00 and $18.50, with a strong leaning towards the higher figure. "Off" coke has sold down to $16 or less.

Advancing Pig Iron

A curious caper was cut by valley foundry iron, for after being quotable at $46, valley, for several weeks, suddenly there were sales at $50. The sales were not heavy in point of tonnage, only a few thousand tons altogether, but they were in numerous lots, and in a few cases the deliveries ran to the end of the year, so that there is no question about the market being established at $50. The jump was a quick one, but from a long range view foundry iron has not had a particularly rapid advance, since three months ago it was

$45, and six months ago it was $40. Bessemer and basic grades are probably going to be established at $50, but this far not enough sales are on record to establish the fact. The last well established quotations were $47, valley, for Bessemer, and $46.50 for basic. There was a large sale of basic reported at $48.50, but the reported buyer has always required a limit of .03 per cent. on sulphur, and that is not standard iron. A furnaceman would be throwing money away if he could not get an extra for such sulphur in these times, hence the sale did not make a general market for basic.

An interesting commentary on the pig iron market situation is that when foundry iron was $4 some of the furnaces were discussing whether the advance in freight rates on the raw materials, coke, ore and limestone, ought not to be passed on to the pig iron buyer. For valley furnaces the advances were computed to amount to between $1.50 and $1.75 per ton of pig iron. Now the furnaces have passed $4.00 to the consumer, and the consumer must pay the advanced freight on the pig iron as well. Even to a Pittsburgh consumer the advance on pig iron freight is about 60 cents, as all rates have gone up this week 40 per cent. and the old valley-Pittsburgh rate was $1.40. On New England buyers the 40 per cent. advance

will fall much more heavily. Thus th. pig iron buyer has to pay ms increasu.. freight on pig iron, the furnaceman's increased freight on the raw materials assembled, and an advance in the price besides.

The pig iron advance is made the more remarkable because pig iron was already high relative to steel, yet steel shows a slight declining tendency and pig iron advances. In last report sales of sheet bars. were recorded at $70, Pittsburgh, a decline in the market of $5. It has since developed that some of the sales were equal to $1 or $2 below $70, and there is now a possibility that the next sales will be at $65, as some of the mills are anxious to sell sheet bars and the demand has slackened off. Several sales of billets are reported from the east on the basis of $60, Pittsburgh, a price that was very difficult to do two or three weeks ago.

Market Anomalies

The general market, of course, is up-side down. If coke really cost all the furnaces $17.50 ovens, $50 would be al-together too low for pig iron, for at 2,200 pounds coke consumption the cost of coke delivered to valley furnaces, with the new freight, would be about $22 per ton of pig iron made. With $50 pig iron billets at $50 would be alto-gether too low. With $60 billets bars and shapes would be too low, for they can be bought from large independent mills at little if any above $60 per net ton. If one started from the coke he could argue that prices ought to advance all along the line, but when one came to bars and shapes at $60 or there-abouts, by the independents, he would en-counter the fact that the Steel Corpora-tion, which produces between 40 and 45 per cent. of all the steel, has prices of $47 per net ton on bars and $49 on shapes. The corporation has had those prices since the Industrial Board adjust-ment of March 31, 1919. It is nearly a year since the independents began to expect that the corporation would ad-vance its prices, but the corporation has not done so.

Transportation Improved

The steel interests as a whole com-plain almost as much as formerly about transportation conditions, but the fact is that almost every week there is a de-cided improvement. The complaint is due in most cases to the improvement not being as rapid as was expected. Several weeks ago the turning point oc-curred in the matter of accumulating steel at mill, the rate of accumulating decreasing, and in the past week or two the condition has been of the mills ship-ping as much steel as they were making. Some have shipped less and have added to their accumulation, while others have shipped more, thus moving some of the steel they had previously accumulated.

Ever since steel started to accumulate at mills the question has been discussed what effect there would be upon the market by the steel eventually being shipped, in addition to current produc-tion. There is a little weakening in

BUYERS AND SELLERS OF STEEL

It is a moot point whether the raising of freight rates will not be balanced by the slowing down of some lines of industry.

The coke market is getting easier, but not to any marked extent. The increase in production and the better car supply are having their effect.

Germany has recently purchased 14,000 tons of ship plate and sections in the and it is probable that an additional 30,000 tons will be placed.

Pig iron prices continue to advance, foundry and bessemer grades being now firmly settled at $50 for this year's delivery, valley furnace. There is much caution on the part of buyers for next year iron.

There is a better market for plates, enquiry being about 5 per cent. greater than during last week. The price for sheared tank plates remains at 3.25 to 3.50, U. S. and it is probable that an additional 30,000 tons will be placed.

United States railroad car builders are expected to come into the market shortly with large plate orders.

Demand for warehouse material is active, and in some grades of black and galvanized sheets an advance has been noted. ..

Some automobile firms have asked sheet makers to hold up shipments for a short period, but generally there is a continued demand for delivery of sheets on order, and a good enquiry for new orders.

With the improvement in the car situation, the old material market has also taken a turn for the better, and advances on various grades run from $1 to $4 a ton.

Germany is reported to be offering barbed wire for export to South America at a price about 1½ cents lower than can be done by U. S. makers.

Mill accumulations of steel are being moved now, and the car situation looks very much better.

Germany has cut prices of pig iron, and prices of all iron and steel products, the prices to remain in effect till October 31st subject to withdrawal. No. I foundry is quoted at a price equivalent to about $37.65. ·

France has removed all import restrictions, and the only metal products now excluded are portable firearms, empty cartridges, and projectiles of all kinds.

---

prices, but it is hardly discernible. One reason for the market's remaining rather strong probably is that the ship-ments are far from uniform to different customers, since on account of embar-goes and other influences the mills can ship much more freely to some customers than to others, and thus there are left many steel consumers who are still very short of steel.

QUIET PERIOD IN MACHINE TOOLS

DEMAND IS NOT BRISK — SOME CONCERNS ASK FOR CANCEL-LATION OR REDUCTION

TORONTO. — There are numerous cases of cars of material reach-ing Toronto from the Pittsburgh district in from six days to a week. This is one indication, and a fairly reliable one, that in spite of all other evidence otherwise, there is an improvement in the ship-ment situation. It may take some time for this to be generally felt, and of course there are always those who are quick to hint that the U. S. railroads are likely to find a large number of cars now that the rates have been advanced.

Information which was printed some days ago to the effect that the prepay-ment of freight charges from United States points in New York funds had been discontinued seems to be true in a very limited degree. Advices reaching the Canadian offices of some of the larger U.S. concerns state they have found this is not the case, excepting on a few south-ern lines. The roads are still insisting that freight be prepaid in American funds, and the business world still feels

that an injustice is being done. One of these advices contains the following mention to the rates: ". . . the suspen-sion only applies to tariffs which were filed by certain south-western lines, but does not affect the territories in which our mills are located (which is mostly in Pittsburgh) and there will therefore be no change from the present practice of prepaying such shipments."

It seems that the Canadian lines are not entirely agreed on the question, and it has been suggested that they must come to some understanding before any-thing will be done.

The increased freight charges which are due on U.S. roads in a few days now, will no doubt help to keep the price of material up. The trouble with many of these rates is that by the time they get passed on to the consumer they have grown out of all proportion and relation to the original increase.

The Machinery Market

The automobile business can be said to be out of the place it has occupied in the machinery market for some time past as the chief buyer of tools and equipment. For some time past it has been common knowledge that the mar-ket has been looking for a new buyer to come in on a large scale and take up the slack left when the automotive shops fell off in their purchases. It was ex-pected that the railroads both in Canada and United States would come into the market in a large way, but in Canada the buying of the roads has not been very extensive, in fact some of the re-quisitions have been cancelled in the last few weeks.

Some of the industrial concerns are

not inclined to view with alarm the halt that has taken place in the making of automobiles. They claim that the auto industry has spoiled the labor market in every centre in which it has located, on account of the high wages paid to operators of automatic and semi-automatic machines.

One of the largest automobile concerns to come into Canada during recent months has sent out a large number of letters respecting various orders it has placed with Canadian houses, the drift of them all being that they be allowed to cancel the orders or that a substantial reduction be made in the purchase price. The chances are that if anything takes place it will be the allowance of cancellation, as it is almost out of the question for a dealer to make a reduction, as it would simply mean that more money would have to be made elsewhere to maintain the balance of profit, a thing which is most necessary in these times.

Several firms are waiting now until after Toronto fair to see the turn affairs may take there, relying on information secured at that gathering to judge their course by.

An Erratic Market

The state of the steel market is shown by the fact that No. 10 gauge is selling at 9.50, while for No. 12 the figure is 11.25. There is no apparent sense or reason between the two figures, and in the ordinary course of events they cannot be reconciled at all. The heavy sheet mills roll the 10 gauge, but beyond that the buyer had to get out and do business with the premium mills. These latter are taking full advantage of the present peculiar market. They have some material, and they are turning out more, and their books are not overburdened with orders. Hence they are selling to a trade that is willing to pay the price if it can get quick action in the matter of delivery. Canadian dealers are paying from eight to nine cent base for sheets from a number of these mills. There is considerable scheming going on too in order to get this material into this country on time. Some of the dealers are able now and then to get Canadian railway cars to send to United States mills, on the understanding that they must get them back promptly. In this way some very good tonnages have been brought across the line. The delivery of bars has been very unsatisfactory of late, and Canadian mills, owing to the trouble in securing shipment of coal, have not been able to do very much to help out in relieving the situation.

Some shipments of tubes have been secured by Toronto yards, but the amount coming in is very small compared to the orders booked up against the arrival of the supplies. Prices in United States warehouses have been increased about ten per cent. during the past week, but this has not been added here yet. The tubes that are coming in are fairly well assorted from two to four inch. Industrial concerns in many points want tubes for repairing purposes, and the shortage in supplies is hurting operations in several of the large plants

in Canada where they specialize in building boilers. The big call is for 4 inch 18, and for 3½ inch. 16. A salesman going around announcing that he had a good supply of these would be a welcome caller at any port just now. The demand for plate appears to have fallen off as the business in this line is quite normal.

The scrap metal market is a lifeless affair now, and there seems to be little chance of it recovering until some of the

larger mills are busy again to the point where they are eating up the material. There is a limited amount of trading going on now, but the "at a price" element is always in evidence. As one of the dealers stated to-day, "You can always sell at a price, but you can't always buy at a price." Yards are not keen to load up with much material, as they seem uncertain in regard to their views of the future trend of events.

NEW YORK THINKS THAT PRICE
IS DECIDING FACTOR IN MARKET
Special to CANADIAN MACHINERY

NEW YORK, Aug. 26. — There is a sharp division of opinion in the machinery trade as to the extent to which the present business readjustment will go. Some seem to feel that the readjustment must go much further before there is any marked improvement in demand for machine tools and allied equipment, while others believe that business is suffering only a temporary reaction, and that inquiries and orders will begin to flow in more freely in the early fall, or at latest after the results of the coming Presidential election are known.

It is admitted in some quarters that the peak of prices has about been reached in all products made from iron and steel; it is contended that a readjustment of values must come about before there is any marked improvement in buying. At present there is no suggestion of lower prices on machine tools. As long as manufacturers have a few months' business on their books they are not likely to jeopardize such business by reducing prices, but such price reductions may come when builders of tools are actually "hungry" for orders. Some will undoubtedly be in this position before the first of the year if the present inactivity is prolonged. Some sellers

say that if there are to be lower prices on machine tools there must also be a readjustment downward of wages paid to labor. Raw materials are high, but this cost comes secondary to that of labor costs.

The dullness which has prevailed in machine tool markets for some weeks continues. In Chicago there has been a little activity among the railroads, and there are prospects that nearly all of the carriers will come into the market for shop equipment as soon as the full effect of the new freight and passenger rates has been felt. At present the purchasing departments of the roads do not know just how much of such equipment will be bought because no funds have been appropriated by boards of directors. Should railroad buying develop in fairly large volume this fall it would be a great boon to the machine-tool industry. There are some who predict that there will not be much railroad buying before spring.

The Federal Signal Co., Albany, N.Y., which manufactures railway signal systems, is in the market for about a dozen tools, but otherwise inquiry in this market is at a minimum. Some dealers and factory representatives report conditions as dull as they have ever known them to be.

TIRE MAKERS ARE
LAYING MEN OFF

Production on Hand Enough to Warrant Curtailing at Toronto Factory

The night shift of the Goodyear Tire and Rubber Company's plant at New Toronto, consisting of about 600 men, is being laid off. President C. H. Carlisle confirmed this report and stated that, while the plant had always worked to capacity, it had been deemed advisable to cut down the production. The "lay-off" will apply only to the New Toronto factory.

President Carlisle ascribed the necessity of the slowing down to overproduction. "We have not had trouble with the men over wages," he said, and paid tribute to the loyalty and co-operation of the employees.

The first of the New Year was set by Mr. Carlisle as the possible date when the men would be taken on again. "We do not anticipate laying off any more

men," he said. This will leave between 2,400 and 2,500 men still on the payroll of the firm.

There was the possibility of keeping all the men on the pay-roll, but they would be working only four days a week, and it was decided it would be better to let them out in order that they could get continuous work in other places.

Speaking of over-production, Mr. Carlisle said that the wet months of June and August had retarded the demand for automobiles and the mileage was greatly reduced. The lack of steel, caused by the steel strike, the coal miners' strike and the switchmen's strike had all had their bearing on the automobile industry.

Open Canadian Branch.—Clemson Bros., makers of Star back saw blades have opened a branch warehouse and office in Canada, the location being Room 304 Imperial Building, Hughson & Main Sts., Hamilton. F. J. Macdonell is the Canadian manager.

Decide To Arbitrate and Stay at Work

British Steel Masters and Employees Reach Agreement—Joint Committee Will Consider All Matters that May Come in Dispute— Three Main Points in the New Understanding.

LONDON.—The details of an agreement recently arrived at between the iron and steel employers and their skilled craftsmen and helpers engaged in the repair and maintenance of blast furnace plants, coke ovens and steel rolling mills, which have recently been made known, must gladden the hearts of all who take a friendly interest in the relations between capital and labor. The agreement is signed by, and on behalf of, the Steel Ingot Makers' Association, the North of England Iron and Steel Manufacturers' Association, the Cleveland Iron Masters' Association on the one hand, and the Amalgamated Society of Engineers, the Electrical Trades Union, and the United Machine Workers' Association on the other—whose representatives are to be congratulated upon accomplishing an exceedingly useful piece of work, that must remove anxiety and assure peaceful and uninterrupted running of machinery for many months to come.

The working arrangements of the men concerned, engineers, electricians, plumbers, blacksmiths, roll turners and a number of other trades, have always been a source of anxiety to the management, as they invariably are in all industries forced to keep their plant running day and night, inasmuch as owing to the rates for overtime, it is uneconomical to have more mechanics on the night shift than is absolutely necessary to effect minor adjustments. But there is always the possibility of a breakdown, necessitating the employment of more men than are available, thereby causing delay in the restarting of the productive machinery, not to mention annoyance on the part of the steel-workers themselves, who are paid by results.

The Main Points

The three main points that distinguish the agreement concern, in the order of importance: (1) Arbitration and a joint committee to consider matters in dispute; (2) recognition that day workmen are interested in tonnage output; (3) payment for abnormal hours (overtime) in plainly stated terms. Again the principal and outstanding feature of (1) is that there is to be no cessation of work pending a settlement of the matter in dispute, either by the joint committee or by arbitration. And the agreement cannot be terminated unless three calendar months' notice in writing has been given.

In other respects the procedure of the joint committee follows pretty closely that of the "Whitley scheme," particulars and details of which are familiar. The committee is to consist of not more than eight employers' representatives and eight workmen's representatives,

who will adjudicate upon any question submitted to them from any works owned or controlled by the association mentioned. As each establishment will have its own joint committee and shops committees, the above will provide an additional court for the hearing of a grievance, strengthened by the circumstance that the majority of its members will not be directly concerned.

Arbitration Proposals

Even in the event of this committee being unable to agree, however, the matter in dispute is to be submitted to arbitration. If, after all this procedure has been tried, the results are unavailing in averting a strike, it cannot be said that the "walk-out" savored of the lightning variety, so beloved of the syndicalist. As regards (2), the details of the tonnage bonus scheme have not been worked out, but it is significant that employers are beginning to realize that a mechanic—an engineer's fitter, for instance—is, and ought to be interested in output.

It will be remembered that the committee advised Winston Churchill to advance the wages of engineers, who, by virtue of their occupation, particularly tool and gauge makers, could not be employed on piece work or other system of payment by results. The necessity for some such increase was manifest by the fact that these men were taken from the productive shops into the tool room—from mass production payment by results and high earnings—on to a day-rate system and comparatively low earnings wholely and solely because of their skill.

In other words the more highly skilled suffered very considerably because they were better craftsmen than their colleagues. How the special increase of 12½ per cent. first granted to the tool makers eventually covered the whole round of industrial activity and was paid to every man for whom it was originally intended relatively in the same position, is a byword, and a subject of laughter even to-day.

Three-Shift System

Regarding the third point in the agreement, an arrangement for establishing a three-shift system appears to have been worked out in an eminently simple way without the complications commonly associated with maintenance men, such as bare time for so many hours, time and a quarter for so many, and time and a half afterwards, and so forth. A man engaged on the first shift, namely 6 a.m. to 2 p.m., will be paid 8 hours; the second shift, from 2 p.m. to 10 p.m., will be paid 10 hours, and from 10 p.m. to 6 a.m., 12 hours. On Saturday the first-shift will be from 6 a.m. to 1 p.m., for which 8 hours will be paid. An in-

terval of 50 minutes for each shift will break the monotony and give opportunity for a meal. None will remain permanently on the same shift; every man will take his turn in doing the three-shift, alternate weeks.

The advantage of this arrangement lies in the opportunity that is given to the management to apportion the requisite number of maintenance men in each shift, and the desire engendered in each group of men to keep their plant running as long and as free from delay as their colleagues, whom they relieve. Coupled with the fact that the worker output, indifference is changed to zeal and a desire to get things going, and to preparations and intelligent anticipation of things needed in cases of emergency.

STEEL EARNINGS VERY GOOD SO FAR

First Four Months of the Year Are in Advance of the Same Period Last Year

MONTREAL.—Among the constructive events of the day is the information that the Dominion Steel Corporation's earnings for the first four months of the year are running well ahead of those for the same period last year, in fact, running very much ahead.

At a meeting of the directors the financial report for the four months mentioned was discussed, and pleasure expressed at the improvement. As a result of the meeting, plans for improving the status of the corporation and the opening of two new coal mines were authorized.

In reference to the opening of the mines, it is stated that the demand for coal is so urgent that every effort of the company to increase the output will be extended to take advantage of this demand. That is the immediate reason for rushing the opening of the two new mines.

The financial expenditures necessary to this and to carry out other improvements will be paid for out of the improved current earnings, and will not necessitate any raising of money.

"Our current earnings are good enough," said the president, "to pay for all improvements; etc., as we go along."

Coal Arriving.—Railway officials claim that a record was established at the Niagara frontier on Sunday when nine hundred cars of coal were cleared for destinations in Ontario. Five hundred of these cars passed through Hamilton and the remainder went via Burlington Junction to Toronto and points

The Planer With the Second-Belt Drive

A smooth, easy reverse—which permits
higher cutting and return speeds yet prolongs the life of the entire machine—that is
the outstanding feature of this distinctive
Whitcomb-Blaisdell Second-Belt Drive.

**Our Planer Book gives the details of design
and construction. Write for it.**

**24 x 24 x 6 one head and 26 x 26 x 8 two heads
in stock.**

26 x 26 inch
Whitcomb-
Blaisdell Planer
Widened Pattern

THE A. R. WILLIAMS MACHINERY CO., LIMITED

Halifax, St. John,
Montreal, Winnipeg, Vancouver   "If It's Machinery---Write "Williams""   64 Front St.
TORONTO

SELECTED MARKET QUOTATIONS

Being a record of prices current on raw and finished material entering into the manufacture of mechanical and general engineering products.

PIG IRON

Grey forge, Pittsburgh	$42 40
Lake Superior, charcoal, Chicago	57 00
Standard low phos., Philadelphia	50 00
Bessemer, Pittsburgh	43 00
Basic, Valley furnace	42 90
Toronto price:—	
Silicon, 2.25% to 2.75%	52 00
No. 2 Foundry, 1.75 to 2.25%	50 00

IRON AND STEEL

Per lb. to Large Buyers	Cents
Iron bars, base, Toronto	$ 5 50
Steel bars, base, Toronto	5 50
Iron bars, base, Montreal	5 50
Steel bars, base, Montreal	5 50
Reinforcing bars, base	5 00
Steel hoops	7 00
Norway iron	11 00
Tire steel	5 75
Spring steel	10 00
Band steel, No. 10 gauge and 3-16 in. base	6 00
Chequered floor plate, 3-16 in.	8 40
Chequered floor plate, ¼ in.	8 00
Bessemer rails, heavy, at mill	....
Steel bars, Pittsburgh	3 00-4 00
Tank plates, Pittsburgh	4 00
Structural shapes, Pittsburgh	3 00
Steel hoops, Pittsburgh	3 50-3 75
F.O.B., Toronto Warehouse	
Small shapes	4 25
F.O.B. Chicago Warehouse	
Steel bars	3 62
Structural shapes	3 72
Plates	3 67 to 5 50
Small shapes under 3"	3 62

FREIGHT RATES

Per 100 Pounds.

Pittsburgh to Following Points

	C.L.	L.C.L.
Montreal	33	45
St. John, N.B.	41½	55
Halifax	49	64½
Toronto	27	39
Guelph	27	39
London	27	39
Windsor	27	39
Winnipeg	89½	135

METALS

	Gross. Montreal	Toronto
Lake copper	$25 00	$24 00
Electro copper	24 50	24 00
Castings, copper	24 00	24 00
Tin	62 00	65 00
Spelter	11 50	12 00
Lead	10 75	11 00
Antimony	13 00	14 00
Aluminum	34 00	36 00

Prices per 100 lbs.

PLATES

Plates, 3-16 in.	$ 7 25	$ 7 25
Plates, ¼ up	6 50	6 50

PIPE—WROUGHT

Price List No. 44—April, 1926.

STANDARD BUTTWELD S/C

	Steel Black	Steel Galv.	Gen. Wrot. Iron Black	Gen. Wrot. Iron Galv.
⅛ in.	$6 60	$ 8 50	....	....
¼ in.	5 13	7 26	$ 5 45	$ 7 56
⅜ in.	5 13	7 26	5 45	7 56
½ in.	6 84	8 42	7 27	8 84
¾ in.	8 45	10 58	9 08	11 16
1 in.	12 50	15 44	13 55	16 49
1¼ in.	16 91	21 16	18 06	22 31
1½ in.	20 21	25 30	21 59	26 68
2 in.	27 20	34 04	29 05	35 89
2½ in.	43 00	53 82	....	....
3 in.	56 23	70 38	....	....
3½ in.	71 30	88 32	....	....
4 in.	84 48	104 64	....	....

STANDARD LAPWELD S/C

	Steel Black	Steel Galv.	Gen. Wrot. Iron Black	Gen. Wrot. Iron Galv.
2 in.	$30 90	$37 74	$34 60	$41 44
2½ in.	45 34	56 16	51 19	62 01
3 in.	59 29	73 44	66 94	81 09
3½ in.	73 14	90 16	82 84	99 36
4 in.	86 66	106 82	97 56	117 72
4½ in.	0 98	1 23	1 24	1 49
5 in.	1 15	1 44	1 44	1 73
6½ in.	1 49	1 86	1 87	2 25
7 in.	1 94	2 43	2 42	2 90
8-L in.	2 04	2 55	2 54	3 05
8 in.	2 35	2 94	2 92	3 51
9 in.	2 81	3 52	3 50	4 21
10-L in.	2 61	3 26	3 25	3 90
10 in.	3 36	4 20	4 18	5 03

Prices—Ontario, Quebec and Maritime Provinces

WROUGHT NIPPLES

4" and under, 60%.

4½" and larger, 50%.

4" and under, running thread, 30%.

Standard couplings, 4-in. and under, 30%.

Do., 4½-in. and larger, 10%.

OLD MATERIAL

Dealers' Average Buying Prices.

	Per 100 Pounds. Montreal	Toronto
Copper, light	$15 00	$14 00
Copper, crucible	18 00	18 00
Copper, heavy	18 00	18 00
Copper wire	18 00	18 00
No. 1 machine composition	17 00	17 00
New brass cuttings	11 00	11 75
Red brass turnings	14 50	15 75
Yellow brass turninbs	9 00	9 50
Light brass	7 00	7 00
Medium brass	8 00	7 75
Scrap zinc	6 50	6 00
Heavy lead	7 50	7 75
Tea lead	4 50	5 00
Aluminum	19 00	20 00
	Per Ton	Gross
Heavy melting steel	18 00	18 00
Boiler plate	15 50	15 00
Axles (wrought iron)	25 00	20 00
Rails (scrap	23 00	18 00
Malleable scrap	26 00	25 00
No. 1 machine east iroh.	32 00	33 00
Pipe, wrought	12 00	12 00
Car wheel	33 00	33 00
Steel axles	25 00	20 00
Mach. shop turnings	11 00	11 00
Stove plate	26 50	25 00
Cast boring	14 00	14 00

BOLTS, NUTS AND SCREWS

	Per Cent.
Carriage bolts, 7-16 and up	+10
Carriage bolts, ⅜-in. and less	Net
Coach and lag screws	—15
Stove bolts	55
Wrought washers	—25
Elevator bolts	+10
Machine bolts, 7-16 and over	+10
Machine bolts, ⅜-in. and less	+10
Blank bolts	Net
Bolt ends	Net
Machine screws, fl. and rd. hd., steel	27½
Machine screws, o. and fil. hd., steel	+25
Machine screws, fl. and rd. hd., brass	net
Machine screws, o. and fil. hd., brass	net
Nuts, square, blank .... +25 add	$2 00
Nuts, square, tapped ... add	2 25
Nuts, hex., blank ... add	2 50
Nuts, hex., tapped ... add	3 00
Copper rivets and burrs, list less	15
Burrs only, list plus	25
Iron rivets and burrs	40 and 5
Boiler rivets, base ¾" and larger	$8 50
Structural rivets, as above	8 40
Wood screws, O. & R., bright	75
Wood screws, flat, bright	77½
Wood screws, flat, brass	55
Wood screws, O. & R., brass	55½
Wood screws, flat, bronze	50
Wood screws, O. & R., bronzé	47½

MILLED PRODUCTS

(Prices on unbroken packages)

	Per Cent.
Set screws .... —20%, 25 and	5
Sq. and hex. hd. cap screws	12½
Rd. and fil. hd. cap screws ..plus	25
Flat but. hd. cap screws ....plus	50
Fin. and semi-fin. nuts up to 1-in.	12½
Fin. and Semi-fin. nuts, over 1 in., up to 1½-in.	—5
Finp and Semi-fin. nuts o.er 1½ in., up to 2-in.	+12½
Studs	+5
Taper pins	—12½
Coupling bolts	+40
Planer head bolts, without fillet, list	+45
Planer head bolts, with fillet, list plus 10 and	+55
Planer head bolt nuts, same as finished nuts.	
Planer bolt washers	net
Hollow set screws	+60
Collar screws ......list plus 20,	30
Thumb screws	40
Thumb nuts	75
Patch bolts ....add	+85
Cold pressed nuts to 1½ in...add	$1 00
Cold pressed nuts over 1½ in..add	2 00

BILLETS

	Per gross ton
Bessemer billets	$60 00
Open-hearth billets	60 00
O.H. sheet bars	76 00
Forging billets	56 00-75 00
Wire rods	52 00-70 00

Government prices.

F.O.B. Pittsburgh.

NAILS AND SPIKES

Wire nails, base	$6 10
Cut nails, base	7 00
Miscellaneous wire nails	50%

ROPE AND PACKINGS

Plumbers' oakum, per lb.	0 10¼
Packing, square braided	0 38
Packing, No. 1 Italian	0 44
Packing, No. 2 Italian	0 36
Pure Manila rope	0 35½
British Manila rope	0 28
New Zealand hemp	0 28

POLISHED DRILL ROD

Discount off list, Montreal and Toronto	net

Geometric Solid Adjustable Die Heads

When one uses the word "ONLY," he needs to be sure of what he is talking about. With confidence we use the word in saying that a Geometric Solid Adjustable Die Head, fitted with a set of milled dies, is the only solid die head that will back off without stripping the thread.

Geometric Solid Adjustable Die Heads may be fitted with either a releasing shank, plain shank, or with special shank for the Gridley Automatics. The releasing shank permits the head to disengage from the shank upon completion of the thread. The plain shank fits the head for use on the turret of a lathe or on a live spindle, such as a drill press.

Apart from the fact that it is not equipped with the self-opening feature or the roughing and finishing attachment, the Geometric Solid Adjustable Die Head is the same in principle and construction and will do equally as accurate work as the other styles of Geometric Die Heads.

**Whatever the requirements, there
is a type of Geometric Die Head
best adapted to the work.**

THE GEOMETRIC TOOL COMPANY
NEW HAVEN   CONNECTICUT

Canadian Agents:

Canadian Fairbanks-Morse Co., Ltd., Manitoba, Saskatchewan, Alberta.
Williams & Wilson, Ltd., Montreal. The A. R. Williams Machinery Co., Ltd., Toronto, Winnipeg;
St. John, N.B.; Halifax, N.S.

If interested tear out this page and place with letters to be answered.

MISCELLANEOUS

Solder, strictly	$0 35
Solder, guaranteed	0 39
Soldering coppers, lb.	0 62½
White lead, pure, cwt.	20 35
Red dry lead, 100-lb. kegs, per cwt.	16 00
Gasoline, per gal., bulk	0 38
Pure turp., single bbls., gal.	3 15
Linseed oil, raw, single bbls.	2 37
Linseed oil, boiled, single bbls.	2.40
Wood alcohol, per gal.	4 00
Whiting, plain, per 100 lbs.	3 00

CARBON DRILLS AND REAMERS

S.S. drills, wire sizes	32½
Can. carbon cutters, plus	20
Standard drills, all sizes	32½
3-fluted drills, plus	10
Jobbers' and letter sizes	32½
Bit stock	40
Ratchet drills	15
S.S. drills for wood	40
Wood boring brace drills	25
Electricians' bits	30
Sockets	50
Sleeves	50
Taper pin reamers	25 off
Drills and countersinks	net
Bridge reamers	50
Centre reamers	10
Chucking reamers	net
Hand reamers	10
High speed drills, list plus	10 to 30
Can. high speed cutters, net to plus	10
American plus	40

COLD ROLLED STEEL

[At warehouse]

Rounds and squares	$7 base
Hexagons and flats	$7.75 base

IRON PIPE FITTINGS

	Black	Galv.
Class A	60	75
Class B	27	37
Class C	18	27

Cast iron fittings, 5%; malleable bushings, 22½%; cast bushings, 22½%; unions, 37½%; plugs, 20% off list.

SHEETS

	Montreal	Toronto
Sheets, black, No. 28	$8 50	$10 50
Sheets, Blue ann., No. 10	8 50	9 50
Canada plates, dull, 52 sheets	8 50	10 00
Can. plates, all bright	8 60	9 00
Apollo brand, 10¾ oz. galvanized	....	....
Queen's Head, 28 B.W.G.	11 00	....
Fleur-de-Lis, 28 B.W.G.	10 50	....
Gorbal's Best, No. 28	....	....
Colborne Crown, No. 28	....	....
Premier, No. 28, U.S.	11 50	11 60
Premier, 10¾-oz.	11 50	12 00
Zinc sheets	16 50	20 00

PROOF COIL CHAIN

(Warehouse Price)

B

¼ in., $13.00; 5-16, $11.00; ⅜ in., $10.00; 7-16 in., $9.80; ½ in., $9.75; ⅝ in., $9.20; ¾ in., $9.30; ⅞ in., $9.50; 1 in., $9.10; Extra for B.B. Chain, $1.20; Extra for B.B.B. Chain, $1.80.

ELECTRIC WELD COIL CHAIN B.B.

⅛ in., $16.75; 3-16 in., $15.40; ¼ in., $13.00; 5-16 in., $11.00; ⅜ in., $10.00; 7-16 in., $9.80; ½ in., $9.75; ⅝ in., $9.50; ¾ in., $9.30.

Prices per 100 lbs.

FILES AND RASPS

	Per Cent.
Globe	50
Vulcan	50
P.H. and Imperial	50
Nicholson	32½
Black Diamond	27½
J. Barton Smith, Eagle	50
McClelland, Globe	50
Delta Files	20
Disston	40
Whitman & Barnes	50
Great Western-American	50
Kearney & Foot, Arcade	50

BOILER TUBES.

Size.	Seamless	Lapwelded
1 in.	$27 00	$.....
1¼ in.	29 50	.....
1½ in.	31 50	29 50
1¾ in.	31 50	30 00
2 in.	35 00	30 00
2¼ in.	35 00	29 00
2½ in.	42 00	37 00
3 in.	50 00	48 00
3¼ in.	.. ..	48 50
3½ in.	63 00	51 50
4 in.	85 00	65 50

Prices per 100 ft., Montreal and Toronto

OILS AND COMPOUNDS.

Castor oil, per lb.	
Royalite, per gal., bulk	27½
Palacine	30½
Machine oil, per gal.	54½
Black oil, per gal.	28
Cylinder oil, Capital	82½
Petroleum fuel oil, bbls., net	18

BELTING—No 1 OAK TANNED

Extra heavy, single and double		6½
Standard		6½
Cut leather lacing, No. 1		2 00
Leather in side	2 40	3 00

TAPES

Chesterman Metallic, 50 ft.	$2 00
Lufkin Metallic, 603, 50 ft.	2 00
Admiral Steel Tape, 50 ft.	2 75
Admiral Steel Tape, 100 ft.	4 45
Major Jun. Steel Tape, 50 ft.	3 50
Rival Steel Tape, 50 ft.	2 75
Rival Steel Tape, 100 ft.	4 45
Reliable Jun. Steel Tape, 50 ft.	3 50

PLATING SUPPLIES

Polishing wheels, felt	$4 60
Polishing wheels, bull-neck	2.00
Emery in kegs, Turkish	09
Pumice, ground	06
Emery glue	30
Tripoli composition	09
Crocus composition	12
Emery composition	11
Rouge, silver	60
Rouge, powder, nickel	45

Prices per lb.

ARTIFICIAL CORUNDUM

Grits, 6 to 70 inclusive	.08½
Grits, 80 and finer	.6

BRASS—Warehouse Price

Brass rods, base ½ in. to 1 in. rod	0 34
Brass sheets, 24 gauge and heavier, base	$0 42
Brass tubing, seamless	0 46
Copper tubing, seamless	0 48

WASTE

XXX Extra	24	Atlas	20
Peerless	22½	X Empire	19½
Grand	22½	Ideal	19
Superior	22½	X Press	17½
X L C R	21		

Colored

Lion	17	Popular	13
Standard	15	Keen	11
No. 1	15		

Wool Packing

Arrow	35	Anvil	22
Axle	28	Anchor	17

Washed Wipers

Select White	.20	Dark colored	.09
Mixed colored	.10		

This list subject to trade discount for quantity.

RUBBER BELTING

Standard ... 10% Best grades ... 15%

ANODES

Nickel	.55 to .60
Copper	.38 to .40
Tin	.70 to .70
Zinc	.16 to .17

Prices per lb.

COPPER PRODUCTS

	Montreal	Toronto
Bars, ½ to 2 in.	$42 50	$43 00
Copper wire, list plus 10		
Plain sheets, 14 oz., 14x60 in.	46 00	44 00
Copper sheet, tinned, 14x60, 14 oz.	48 00	48 00
Copper sheet, planished, 16 oz. base	46 00	45 00
Braziers', in sheets, 6 x 4 base	45 00	44 00

LEAD SHEETS

	Montreal	Toronto
Sheets, 3 lbs. sq. ft.	$10 75	$14 50
Sheets, 3½ lbs. sq. ft.	10 50	14 00
Sheets, 4 to 6 lbs. sq. ft.	10 25	13 50

Cut sheets, ½c per lb. extra.
Cut sheets to size, 1c per lb. extra.

PLATING CHEMICALS

Acid, boracic	$ .23
Acid, hydrochloric	.04
Acid, nitric	.11
Acid, sulphuric	.04
Ammonia, aqua	.15
Ammonium, carbonate	.23
Ammonium, chloride	.22
Ammonium hydrosulphuret	.75
Ammonium sulphate	.30
Arsenic, white	.14
Copper, carbonate, annhy	.41
Copper, sulphate	.15
Cobalt, sulphate	.20
Iron perchloride	.62
Lead acetate	.30
Nickel ammonium sulphate	.20
Nickel carbonate	.32
Nickel sulphate	.22
Potassium sulphide (substitute)	.42
Silver chloride (per oz.)	1.30
Silver nitrate (per oz.)	1.25
Sodium bisulphate	.14
Sodium carbonate crystals	.06
Sodium cyanide, 127-130%	.38
Sodium hyposulphite per 100 lbs	8.00
Sodium phosphate	.15
Tin chloride	.80
Zinc chloride, C.P.	.30
Zinc sulphate	.10

Prices per lb. unless otherwise stated

Quality

CONFIDENCE

THE GREATNESS OF BRITAIN IS FOUNDED ON HER INTEGRITY.

HER STRENGTH SHE DERIVES FROM WORLD COMMERCE—HER ENDURANCE FROM CONFIDENCE.

THE WAR TEMPORARILY RETARDED HER CONSTRUCTIVE EFFORTS, BUT THE POWER OF THE CONFIDENCE OF THE PEOPLES OF THE WORLD WILL RESTORE TO HER IN MORE BRILLIANT FORM THE LEADERSHIP SHE HAS JUSTLY EARNED AND PROUDLY HELD.

CONFIDENCE IS THE FOUNDATION OF OUR POSITION IN THE WORLD OF IRON AND STEEL— OUR PROGRESS DEPENDS ON IT.

WE HAVE BEEN SUCCESSFUL IN THE PAST— THE FUTURE IS BEFORE US. TO ADVANCE, WE MUST CONTINUE TO MERIT THE CONFIDENCE OF THE BUYERS OF STEEL AND IRON PRODUCTS OF EVERY DESCRIPTION IN CANADA, AND WE ARE DETERMINED TO ADVANCE.

THE
STEEL COMPANY
OF
CANADA
LIMITED
HAMILTON MONTREAL

Service

The Rational Use of the Cutting Blow Pipe

By Marcel*

THE technique of welding is better known than the use of the cutting blow pipe, and if this last named is often found in welding stations with the welding blow pipe, it is surely more often put to a worse use.

It can be asserted that the bad results obtained in the cutting of iron and steel come because the cutting blow pipe is an instrument that the exact regulating control of which is delicate.

If it is true that in the use of the welding blow pipe it is always possible to know at first sight the regulating of the flame and the aspect of the fusion bath, it is also true that very good practitioners of welding do not know sufficiently the practice of cutting.

A well-regulated cutting blow pipe for a certain kind of work is that in which the oxygen pressure, the exhaust aperture of the cutting jet and the heating flame are in proportion to the thickness of the piece to be cut.

We could cite many examples in which cutting with a blow pipe shows an enormous waste of oxygen, and very poor work. One of the most striking and the very latest one which we have been able to witness is the following: In a shop near Paris, a firm was demolishing a large quantity of car frames which the Germans had set on fire in a northern station, during their retreat. The frames were delivered in pieces easily transportable but of some other purpose. The work consisted, in fact, of cutting with the cutting blow pipe some U-shaped beams and flat iron pieces, the thickness of which ranged from 20 to 25 mm, as the larger pieces were left intact with the frames.

The blow pipes employed for this work were of the oxy-acetylene type with central jet, fed with acetylene by movable generators of the contact type. We are not exaggerating when we state that eight times less oxygen could have been used for this work if the operators had had in their possession to appreciate a good execution of their work. The cutting operation was done in the following manner: The blow pipe after being lit, the oxygen used for cutting escaped without the flame having been regulated, which during the whole operation showed an excess of acetylene, then the operator, forming the fusion of a part of the iron to be cut, was afterwards swinging the blow pipe from left to right of the sectioning line, obtaining thus a fusion of the metal, increased by the oxygen. This operation, being rather from this fact, very slow, gave a cut from edge to edge large enough

*From the March (1920) issue of Revue de la Soudure Autogène (Paris).

to allow the thickness of the hand to pass through!

If we add to that that the aperture of the cutting nozzle had been chosen without discernment, too strong, and that a six-kilos pressure was in current use, without mentioning the many leaks at the fittings and joints of the manometers and blow pipes, one can imagine the great saving which could have been realized in such a shop if it had been equipped with four or five cutting stations during several weeks.

It seems to us that it is necessary to recall the principal conditions and the rational and economical use of the cutting blow pipe.

The first thing most commonly found is the too frequent and almost general use of a too high oxygen pressure. One finds too often also inexperienced workmen who compare cutting to a physical phenomenon and attach much importance to the "force of the jet." One on the contrary must be impressed with the fact that the cutting of iron and steel is only a destruction of the metal through a chemical phenomenon, the oxygen pressure helping, only to bring the gaseous stream over the whole thickness to be cut.

An operator who is not well experienced in oxygen pressure to be used is naturally tempted to employ high pressure, for the manufacturers of valves deliver to their customers, especially for cutting purposes, apparatus equipped with manometers graduated up to 25 and 30 kilos.

In comparison with the graduations which are made for the valves for welding purposes, the operator is tempted to use, through simple deduction, pressures from 5 to 6 kilos for cuts not any thicker than from 10 to 20 mm, and from 10 to 15 kilos for work for which a pressure half as high would be sufficient.

Is the result better and the cutting done quicker?

It is certain that a slight increase of oxygen pressure may give a quicker cutting, but the expense in gas is also increased. There is thus no real economy by using a higher pressure.

By way of compensation, experience shows that the use of oxygen at exaggerated pressure rather slackens the speed in cutting through the cooling off caused by the expansion of unused gas and of the air forming around the jet. Thus, one must be very careful to use just the pressure necessary for a normal outlet of the cutting jet. These pressures can be settled very approximately as follows, when sheet iron and ordinary iron for construction purposes

and under normal conditions are concerned:

Thickness in Mm.—	Pressures in Kilos.
Up to 12 mm.	1 to 1.500
From 12 to 30 mm.	1.500 to 2
From 30 to 50 mm.	1.700 to 2.500
From 50 to 100 mm.	2.250 to 2.500

It is to be observed that the oxygen pressures for cutting leave a certain margin for their use on different thicknesses. Thus these pressures in certain proportions are the function of the aperture of the cutting nozzle.

Cuts of the same value may be obtained, for example, by using a pressure of 1 kilo 500 with a cutting nozzle of 10/10th and a pressure of 1 kilo with 12/10th; the table below shows the diameter of nozzles which can be practically used.

For cutting blow pipes with separate jets one can easily go as low as 10 mm, especially if the separate feeding of heating oxygen and cutting oxygen is realized. In the most current case in which the blow pipe is only fed by one stream of oxygen and especially in cases of central jet type more subject to back flame through obstruction of the nozzle, the pressure of 1 kilo 500 is practically necessary to obtain a good heating flame.

Another important cause to figure out the cost of the work is the section allowed for the flowing of the cutting oxygen. One must remember that the sections of circular apertures are in themselves like the quadrate of their diameters and that theoretically a cutting nozzle of 20/10th, must give way to four times more oxygen than a nozzle of 10/10th under the same pressure conditions, supposing that the waste of gas is equal in both cases, which is not absolutely exact. It is easily seen thus that the mere mistake of choosing a cutting nozzle of 2 mm instead of 1 mm will immediately give a quadruple expense of oxygen. It would be already doubled in the case of using a nozzle having an aperture of 15/10th instead of 10/10th.

In the work shops, the thicknesses to be cut are generally from 8 to 30 or 40 mm, when sheet-iron, in large plates is concerned, and up to 80 to 100 mm, for bar sections, profiles or round pieces.

This table gives the diameters of the nozzles:

Thickness in Mm.—	Apertures of Nozzles
Eight and below	7/10th
From 8 to 30	10 to 12/10th
From 30 to 40	12 to 15/10th
From 40 to 75	15 to 17/10th
From 75 to 120	19 to 20/10th

For cuts on thicknesses below 15 mm, there exists smaller types of cutting blow pipe with central jet, having series of nozzles the apertures of which

are very small and begin from 6/10th for sheet iron to 5 mm.

We also call the attention of the reader to the importance of the purity of the oxygen to the cutting mechanism. There are also in this case chances of saving in the shops, if the purity of the gas is controlled and if nothing but the purest of gas is used in the cutting stations.

The effect of impure oxygen is the slackening of speed in the cutting and this is so clearly noticeable that it must attract the attention of the practitioners.

As per the experiments of the Laboratory of the Old Union of Welding, (Union de la Soudure Autogene), the speed in cutting, under equal conditions, is reduced to 30 per cent. or a quantity of azote of 6 per cent. for a speed obtained with oxygen at 99 per cent. At 91.5 per cent. this speed is reduced to 50 per cent!

The influence of the purity of oxygen is well known by heavy thickness specialists who know that below .95 or .96 per cent. of the purity of oxygen they cannot make a cut.

Besides, the use of impure oxygen is one of the causes which often oblige the practitioner to use a higher cutting pressure, leading to an enormous waste of gas without appreciable improvement.

The power of the heating flame must itself be in proportion to the thickness to be cut. If it is too weak, it slackens the speed in cutting and favors the loss of the oxidation point; if too strong, it provokes the fusion of the cutting parts.

A thing which is not observed enough in the cutting done in shops is the space which must exist between the sheet and the nozzle. This space must be the smallest possible, so that the oxygen jet coming from the nozzle produces a narrow cut. Any exaggerated space between the nozzle and the sheet tends to increase the width of the cut on account of the divergence of the gaseous stream which is produced when coming out of the apparatus. This divergence is attenuated by an inside profile in the cutting nozzle especially studied for that purpose.

The oxy-acetylenic and oxhydric blow pipes which separate jet permit of getting closer to the sheet iron than the central jet blow pipes, the heating flame of which, though very short, leads to a certain separation of the nozzle from the surface to be cut. The construction of this kind of blow pipe forces one to place on a same level the extremities of the heating nozzle and of the cutting nozzle.

An improvement on the results of the cutting·blow pipes is obtained by previously heating the cutting oxygen. Many work shops where extensive cutting is done would reduce their expenses if they would adapt to their blow pipes or on one of the cutting oxygen pipes a device of some kind, properly studied,

allowing of increasing the temperature of the combustible gas.

This suggestion is the result of trials made at the laboratory of the Old Welding Union, a few years before the war, and that we intend to resume for the finding of devices suitable for the industry.

It seems thus that an improvement on cutting blow pipes could be obtained, and this simply by the strict application of the most elementary principles of technic and practice on this subject. Great saving could·be made in shops where cutting is done without care, precisely in those places where cutting with the blow pipe is considered as too expensive.

The cutting blow pipe is really a marvelous instrument which can be used with advantage and economy, not only for tearing down or ordinary cutting of steel parts, but also in new undertakings and many very interesting cases in metallic construction.

On this subject the Americans have gone ahead of us and they have created special implements and tools which allow them, with an excellent technic and practical use, to perform all sorts of cuts and on all thicknesses just as fine as if they were made with mechanical tools, much quicker and much more economically.

BRITISH REVERSING GEAR FOR MOTOR BOATS

An ingenious magnetic reversing gear for motor boats has been introduced by a leading British firm of engineers. The shaft carries a magnet which, when excited, attracts a flat steel disc behind it and causes it to revolve, giving a direct drive to the propeller. The reverse magnet is of a similar type and, when excited, draws the disc to the right and holds it stationary. The engine drive is then transmitted through planet wheels so as to drive the propellor in the reverse direction. These magnetic clutches and the engine throttle valves are operated by the one lever. When the lever is in the central position the throttle is closed and neither of the magnets is excited; therefore the engine runs idly at a low speed. Movement of the lever in one direction or the other excites the ahead magnet or the astern magnet as the case may be, opening the throttle and speeding up the engine at the same time.

NEW BOOKLET

The Hammond Steel Co., Inc., Syracuse, N.Y., have issued an interesting booklet dealing with their line of status, known as the oil hardening, non-shrinkable tool steel. The booklet touches on the uses of their line, and hints as to hardening and drawing temper of the same. Illustrations are given, showing examples of status steel used with surprising results, and one punching die depicted has 390 holes in it. There are other items of interest which make this booklet well worth having.

SECOND-HAND MACHINES DISAPPEARING

In speaking of the business now done in used equipment one dealer here stated that the Canadian market is gradually being depleted of standard equipment. There is still a large quantity of shell machinery stored in different places but little general use can be made of what is left of such equipment. What would seem a peculiar feature is the large volume of second-hand machinery, especially lathes, that is at present stored in American warehouses. A recent picture taken in a Chicago warehouse, and now in the hands of a dealer here, shows many hundreds of lathes and shapers, piled up from eight to ten deep, just leaving room enough for the overhead crane to pass. On some of this used equipment it is possible to get delivery in Canada at figures considerably lower than similar tools made in Canadian plants. However, one difficulty is in delivery, and this is a factor that is entirely out of the hands of the dealer or the buyers, as they must await the receipt of cars and the convenience of the railways regarding shipments.

Change Announced.—The company of G. & J. Weir (Canada), Ltd., has been incorporated to take over the business of the Clyde Engineering Co., and will manufacture their complete line of marine auxiliary machinery and power plant auxiliaries for land installations. For this purpose land has been secured and work has been commenced for foundry and machine shop in Montreal. This extension has been found necessary as a result of the increasing demand for G. & J. goods.

Locating in Peterboro.—The "Laundryette," a copper domestic washing machine which dries the clothes by centrifugal force as well as washing them and is almost a complete laundry, will be manufactured in Canada beginning November 1 by the Laundryette Manufacturing Company at Peterboro, Ontario. This company was organized in Cleveland, Ohio, five years ago and has so far been successful. It will incorporate in Canada. It has leased the warehouse of Rishers', Limited, wholesale grocers, in Peterboro, with the option to purchase. At the start it expects to employ 25 hands with the probability of increasing this number to 50 within a short time.

Material Wanted.—This paper has an inquiry from an American concern asking where the following material may be secured in Canada in less than car lots: Steel channels, angles and plate cut to size, cast iron doors weighing about 1,100 to 1,200 pounds each, steel rods, bolts, nuts, etc., 1-inch diameter steel rods, turn-buckles for 1-inch screws. The address of the firm can be had by addressing the editor of Canadian Machinery.

CANADIAN MACHINERY
AND
MANUFACTURING NEWS

Vol. XXIV. No. 10   September 2, 1920.

Do You Realize the Value of Your Scrap?

IF IT were good practice some years ago for firms to pay attention to the saving of scrap around the plant, the high prices of material make it ever so much more necessary to pay attention to this matter now.

When the war was at its height the War Industries Board of United States classified the various industries that were claiming attention, throwing out some that had been looked upon as quite essential, but holding that those engaged in such callings as scrap metal sorting and shipping were working at essential callings. On this ground exemption from military service was granted to those that applied for it.

There is just a danger that in a good many of our plants we are too much engaged in production, in turning out the material and in devising ways and means to cut down time on operations. This may be carried on to such an extent that a real waste takes place of odds and ends that could, with a reasonable amount of thought, be turned to good account.

There is an old saying that if a thing is kept around for seven years some use will be found for it. That is about as far removed from the modern idea of scrap reclamation as anything could well be imagined. The idea is not to keep material lying around in hope of finding some use for it later on, but rather to make use of it as soon as possible.

The big railway yards have made a keen study of this business. It is safe to say that the roads are saving hundreds of thousands of dol-lars every month by a strict attention to the scrap heap.

A certain well-known maker of pickles has his varieties numbering now up to 57, but the railway scrap record shows that the classification has gone to almost double that number, showing where some 99 bins are arranged to look after the various parts of engines and cars.

Station trucks and wheelbarrows are made out of old material. Old bars make the most of the truck, spot welding doing the holding together. The wheels and the frame for them are the only things made especially for the purpose. The wheelbarrows come to life from the old pieces of material, and the wheels for these are nothing more or less than the old hand brakes which the brakeman was wont in his strong and daring way to wind around to stop the train before air brakes became as common or as efficient as they are at present.

There is nothing whatever in the idea of reclaiming scrap to make it appear necessary for the shop to keep on indefinitely with the old material, even after it has lost its use, or to keep on patching up instead of buying new. On the contrary it works the other way. The wise shop wants to realize as soon as possible on its scrap —wants to get it out of the way and turn it into ready money.

The article on the following pages dealing with the system in vogue in the Angus shops of the C.P.R. at Montreal gives a good idea of what it really means to turn the scrap heap into an asset. Railway companies have gone a long distance in this work as the story will show.

Making an Asset Out of the Scrap Heap

by J.H Rodgers

IT is only in recent years that any concentrated effort has been made, in industrial plants, to establish an organized department for the reclamation of scrap. The old material problem has, too frequently, been isolated from the regular routine of manufacturing plants, invariably considered as being outside the pale of constructive industial enterprise. Large organizations, however, are now recognizing this so-called scrap problem as one that has a vital bearing upon the economic operation of the company's business.

When the Canadian Pacific Railway Company constructed their Angus shops at Montreal, the reclamation of scrap received more than the ordinary amount of attention, as it was officially recognized as being a very essential factor of railroad operation, and one that would require to be placed under careful supervision in order to effect a saving.

In the operation of an extensive railway system the possibility of material loss, in the way of scrap and discarded materials, might easily run into many hundreds of thousands of dollars, unless some systematic effort is made to prevent, or at least, minimize the leakage. The multitudinous ways in which scrap may be created in ordinary railroad service makes it an imperative necessity to exercise the greatest care in preventing any of the discarded material in getting away from the company, before it has been determined that such equipment is beyond the stage of useful service. Not only is it necessary to collect these scrapped parts from every available source, but it is equally essential to transfer the material to some reclaim terminal and general stores department, so that serviceable parts and reclaimed equipment may be listed and stocked for further construction purposes, or distributed to different points on the system to effect needed repairs.

FIG. 1—PARTIAL VIEW OF THE SCRAP YARDS AT ANGUS C.P.R. SHOPS.

FIG. 3—PNEUMATIC PRESS FOR STRAIGHTENING BRAKE BEAMS.

Facilities are provided at every termi-nal on the C. P. R. system for the accu-mulation of old material, and this is loaded, periodically, on suitable cars and shipped to the main terminal at Mont-real. The greater volume of the material is transported during the summer months when haulage is most favorable, the scrap accumulating at the different ter-minals throughout the winter. A part of the Western portion of the Angus pro-perty is utilized as a yard for scrap; approximately 10 acres are devoted to this purpose. A partial view of this sec-tion is shown in Fig. 1, the cars in the center being those in which the scrap is received. From these cars the material is taken and sorted, and placed in the various bins on the platform, according to their different classifications. Mat-erial that is to be reclaimed is kept sep-

arate from the rest, and subsequently goes through the necessary process of re-pairing or reconstruction for further ser-vice.

The line drawing Fig. 2 shows the gen-eral arrangement of the sorting dock and reclaim departments. The cars for load-ing the sorted scrap are located on the opposite side of the platform and paral-lel to the track on which the incoming cars are located. The arrangement of the several buildings and the mechanical equipment has been so planned as to facilitate the handling of the different materials from the time they are received until they are again in a repaired con-dition and placed in the stores.

In order to better acquaint the reader with the general layout, the summary be-low has ben given, the equipment and lo-

cation being indicated in the sketch, Fig. 2.

1. Paper Baling Press.
2. Scale.
3. Machine for cutting rubber hose.
4. Workmen's time clock.
5. Reclaim dock office.
6. Storage bins for old wrenches.
7. Switch board.
8. Two dynamos.
9. Emery wheel.
10. 10, 15, 18. Benches.
11. Electric welding screens.
12. Six spindle tapping machine.
14. Small tapping drill.
15. Bins for keeping bolts.
16. Machines for punching washers.
17. Pipe threading machine.
19. Bolt threading machines.
20. Bolt shears.
21. Power hammer.
22. Heaters.
23. Small drill.
24. Furnace for couplers.
25. Hydraulic riveting machine for couplers.
26. Shears.
27, 28, 29. Furnaces.
30. Pneumatic riveting machine.
31. Pneumatic press.
32. Furnace.
33. Bench.
34. Forge for spring work.
35. Anvil.
36. Lead melting furnace.
37. Seal oil tanks.
38. Rolling mill furnace.
39. Re-roll mill for bar iron.
40. Straightening rack.

Owing to the numerous sources from which the material is collected it is quite obvious that the incoming cars will con-tain a jumbled mass of old material. Examination of the contents is made as it is unloaded, so that little difficulty is experienced in classifying the material and determining whether it has to pass through the reclaim process or to be transferred to the bins set apart for that particular class of scrap. When it is only a question of sorting the scrap for trans-shipment to the large rolling mills, foundries, and other manufacturing plants, the work entailed is simply un-

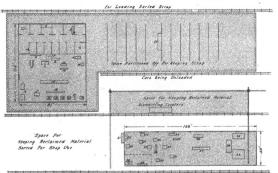

For Loading Sorted Scrap

Space Partitioned Off For Keeping Scrap

Cars Being Unloaded

Space For Keeping Reclaimed Material. Dismantling Couplers

—— 168' ——

Space For Keeping Reclaimed Material Sorted For Shop Use

FIG. 2—GENERAL LAYOUT OF SCRAP DOCK AND RECLAIM DEPARTMENTS.

loading, sorting, and subsequent reloading. In many instances, however, the discarded equipment is a combination of two or more metals or materials, so that for profitable disposal of such parts it is necessary to separate them in order to place each in its proper scrap classification.

Some of the more important phases of railroad reclaim may be outlined briefly as follows:—

Brake Beams

Causes for the scrapping of brake beams are so numerous and varied that discarded equipment of this character will arrive at the scrap dock in every degree of twisted form. Many of those delivered to the yard at Angus are in good condition but have had to be replaced by heavier beams, to meet the latest M.C.B. specifications. These discarded beams are generally of the 5 in. type, this size being now obsolete for brake purposes. However, the various fittings are nearly always in good condition and can be used in a like capacity on the heavier beams. The rivet heads are severed by the use of oxyacetylene, the rivet knocked out, and the brackets

FIG. 4—ASSEMBLING BRAKE BEAMS.

secured to the lower frame. Each pair of cylinders is controlled independently of the other. A beam that has just

been straightened is shown resting on the cylinders.

For reassembling the brake beams a

large pneumatic riveter is embedded in the ground; this is shown in operation in Fig. 4.

Drawbars and Couplers

A feature of the reclaim department is that of the handling of the couplers and drawbar equipment. The continual bumping and straining to which couplers are subjected provides an excellent opportunity of developing defects, and these must receive immediate attention to guarantee against mishap or accident. Dismantling is the first operation, and like the brake beams, the rivets holding the shanks to the drawbar are cut away with the oxyacetylene torch. If inspection shows that the coupler has been sprung it is placed in a furnace (24), and when heated sufficiently is placed in the hydraulic press (25) and straightened to its original shape. Slight cracks often develop in the couplers, due to internal stresses in the steel, or from some severe strain in service. These are carefully examined and chalk marked, and afterwards electric welded. Worn knuckles and corners are built up with new metal,

FIG. 3—ASSEMBLING DRAWBARS.

and rod connections utilized for the reconstruction of serviceable equipment. The small 5-in. beams are seldom scrapped as various uses have been found for them in construction work in the shops and roundhouses.

The majority of these obsolete brake beams are used for roof corlines for box cars, two of the 1-in. beams being welded together to obtain the desired length.

The greater number of the larger brake beams arriving for overhauling and repairing are usually bent out of the original form, and these must be straightened before they can be used. The beam is first dismantled, as stated above, and when entirely stripped, is placed in the furnace (29) and heated, after which it is taken to the home-made pneumatic press, shown in Fig. 3. This press is provided with four air-brake cylinders, two set for a vertical and two for a horizontal pressure, a thrust block of the desired shape being rigidly

FIG. 6—REASSEMBLING TRUCK SPRING SETS.

FIG. 7—PUNCHING WASHERS OUT OF SCRAP PLATE.

so that the coupler, after repairing, is
equally as serviceable, if not more so,
than the new coupler, as initial strains
have been largely eliminated. After
welding the coupler is again assembled,
interfering portions being ground off
by means of a portable pneumatic grinder. New or repaired shank straps are
provided, and these are riveted to the
drawbar in the hydraulic press shown in
Fig. 5, both rivets being inserted and
headed at the one operation.

Truck Springs

In every car arriving at the scrap dock
large numbers of coil springs are received. Aside from exhibiting exposure to
the weather many of these springs are
in perfect order, but require sorting out
according to their different classifications. A truck unit generally consists
of four heavy coil springs bound together
with top and bottom plates, depressions
being provided at each of the four corners so as to retain the springs in their
proper position, the whole being firmly

secured by a central bolt. The heavier
type is made of eight springs, four small
ones centered in the outer springs. Every
spring is examined and tested with a
gauge to determine if its original length
has altered. When below size the springs
are placed in the forge (34) and heated
to a dull red, taken out and opened to the
desired length, after which they are reheated and tempered in oil. The assembling operation is shown in Fig. 6. Leaf
springs are generally turned over to the
blacksmith shop, where they are overhauled and placed again into service.

Washers

The extensive use of washers for every
branch of railway service makes the
question of supply an ever-prominent
one. Facilities are provided at the Angus
shops for utilising all the light scrap
plate available for conversion into wash-

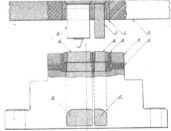

FIG. 8—DIE FOR MAKING WASHERS.

ers. The plates, when necessary, are cut
into convenient size, but little attempt is
made to flatten it, as this is accomplished
at the time the washer is punched out.
This operation is shown in Fig. 7. The
Mason press in the background was especially made for washer work, and the
die used is of the combination type, the
central blank going through the die,
while the washer is carried up on the up-
stroke and ejected on to a plate that
swings below the punch automatically,
this being done by means of a twisted
rod operating in a nut secured to the
upper portion of the frame, to the left of
the operator.

The punch press in the foreground is
especially interesting, for the reason that
it, in itself, is a product of scrap reclaim. It is an old Bertram shear adapted for washer service. The die in the
machine is of duplex design, where the
outside and inside dies are distinctly
separate but contained in the one holder,
and both operations performed simultaneously and a completed washer obtained

FIG. 9—RE-ROLLING MILL FOR SHORT LENGTH ROUND BARS.

FIG. 10—SLITTING REINFORCED RUBBER HOSE.

at each stroke of the press. A sectional sketch of the die construction is shown in Fig. 8. The die bed A is designed to give extreme rigidity and is counterbored at the top to receive the die holder B, which is bored and counterbored to retain the dies D and E, these being secured in position by headless screws (not shown); the die holder is held by screws C. The cored holes F and G are located in an inclined position, the lower opening being at the front of the die. The lower surface is filed smooth to aid the exit of the slug and the washer, each of which is guided into a separate receptacle. The punch plate H is bored part way and slotted to receive the punch holder I, which is shaped off on two sides to fit the plate, this providing an easy means of interchangeability and insuring accurate setting. The large punch K is provided with a pilot that enters the hole

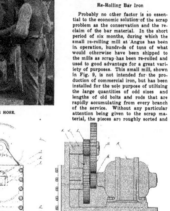

FIG. 11—MACHINE FOR SHEARING REINFORCED HOSE.

punched by J, so that the inside and the outside of the washer are concentric. Various sets of dies for the different sizes of washers may be fitted in the one die bed and punch plate.

Re-Rolling Bar Iron

Probably no other factor is so essential to the economic solution of the scrap problem as the conservation and the reclaim of the bar material. In the short period of six months, during which the small re-rolling mill at Angus has been in operation, hundreds of tons of what would otherwise have been shipped to the mills as scrap has been re-rolled and used to good advantage for a great variety of purposes. This small mill, shown in Fig. 9, is not intended for the production of commercial iron, but has been installed for the sole purpose of utilizing the large quantities of odd sizes and lengths of old bolts and rods that are rapidly accumulating from every branch of the service. Without any particular attention being given to the scrap material, the pieces are roughly sorted and

FIG. 12—OVERHAULING SCOOPS AND SHOVELS.

piled in readiness for being put in the furnace (39). After passing through the rolls a sufficient number of times to reduce the rod to the desired size, the finished bar is placed on the straightening rack (40). When cool the stock is weighed and then transferred to the stores warehouse. The great advantage of this local installation is that rush orders coming in from any part of the system for small quantities of certain sizes can be filled almost immediately, even though none is in stock at the time. All that is necessary is to charge the furnace with available scrap which will give the required size. The convenience of the re-rolling mill has been a boon to the system, when at different times it would have been difficult, if not impossible, to obtain quick delivery of the material from the market.

Reinforced Water Hose

To make the reclamation of reinforced water hose a profitable enterprise it is

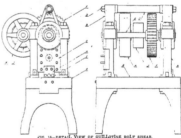

FIG. 14—SORTING DEPARTMENT FOR FREIGHT CAR BOLTS AND NUTS.

necessary to separate the fabric from the coil of steel wire contained in the hose. To accomplish this the C. P. R. have designed a special motor driven shear for slitting the hose lengthwise. A view of the machine in operation is shown in Fig. 10. A line drawing of the device is shown in Fig. 11. The motor A, is secured to the base B, which is bolted by means of a suitable bracket to the bed of an old lathe. A slide in the front of the frame carries the shear plate C, which is operated by means of the crank shaft D. The crank of the shaft is fitted to the hole in the oscillating block E, operating in the slot F, cut in the shear plate C. The drive pinion G, secured to the motor shaft, meshes with the large gear H, keyed to the extension of the pinion I, which, in turn, drives the gear L, geared to the crank shaft D. The shear blade M, is fastened to the movable plate while the stationary shear blade N is secured to the small lug O, which is integral with the frame. The old hose is pushed over the horn of the movable shear plate and the end grabbed by the small claw clamp fitted to the carriage of the machine. This carriage is forced backwards, drawing the hose through the shear, which operates at a speed of about 100 strokes per minute. When ゜‥‥ the hose is removed from the shear and the wire, now in rings, can easily be taken out.

Scoops and Shovels

An average of 500 scoops and shovels pass through the reclaim department every month. These are gathered from every conceivable source, from the Atlantic to the Pacific, and eventually find their way to Angus, where they are repaired and again placed in the stores ready for distribution to the service for which they are best suited. The bulk of these only require to have the ragged edge of the scoop trimmed off, while others may need the scoop or the handle replaced. The available supply is always ample to obtain the necessary repair parts without recourse to new material. Fig. 12 shows a section of the shop devoted to shovel repair. The pile of wheelbarrow bodies in the middle background has been made from light steel plate scrap.

Trucks and Barrows

The heading of this article is a typical example of what may be made from scrap material. To the left is shown an all-metal truck constructed without the use of rivets, the electric process being used to weld the various sections. The side frames, cross pieces and axle are all made of old piping, the feet and front made of plate and bar material. The handles are covered with pieces of air hose placed on the pipe when the latter is heated, so that the rubber will stick more firmly. The wheels are the only parts specially made for the truck. The wheel barrow is constructed entirely of scrap, the main frame and axle of pipe; the feet and axle brackets of light bar iron; the bodies of light plate and the wheels are made from old hand-operated brake wheels.

Freight Car Bolt Reclaim

A detail that might well be considered as coming under the general heading of scrap reclaim, although it is not included in the records of this department, is the gathering and sorting of the bolts out of the freight cars that are in for repairs. When the cars are brought into the shop the stripping operations are carried on without any attempt to keep

track of the numerous bolts with which the different parts of the cars are held together. The stripping crew simply throw these on the floor and the bolts and nuts are collected afterwards and taken to the local sorting room shown in Fig. 14. Here the bolts are examined, and if in good condition are placed in the small wall compartments shown at the left. If threads are defective, the thread is cut off and the bolt re-threaded and stored according to its new length. A bolt threading machine and a double-ended guillotine shear are located at the opposite end of the room. The shear has been specially designed for this particular duty, and was constructed in the Angus shop. An elevation sketch of the shear is shown in Fig. 15. The bed A carried the two housings B,B, and a spacing brace C is used to tie the upper ends together. The eccentric shaft E carries the gear F, that is driven by a pinion on the main drive shaft, which is provided with a fly-wheel K. The ram G carries the upper half of the guillotine shear H, while the lower portion I is secured to the bed of the machine.

Monkey Wrenches

The ordinary monkey wrench is a small item when considered individually, but on a system like the C. P. R. the aggregate consumption of this indispensable piece of mechanical equipment runs into tens of thousands, and the usual depreciation of this small tool is generally quite rapid. To maintain an efficient supply a section of the reclaim department is devoted to the reconstruction of monkey wrenches. Discarded wrenches, no matter for what cause, are gathered at local scrap terminals and finally find their destination at the Angus scrap dock, where they are separated from the other material and the different parts sorted according to their size and make. As hundreds of these are received every month it is seldom a difficult problem to obtain sufficient good parts to reconstruct a large number of highly-

FIG. 15—DETAIL VIEW OF GUILLOTINE BOLT SHEAR.

serviceable wrenches that may be placed in the stores for further distribution.

No small part of the duties performed by the scrap reclaim division is the conservation of bolts and nuts. In the majority of cases bolts are discarded because the nut has frozen on the thread, or the thread has been stripped. Seldom will it be found that the thread in the nut is seriously affected, and nuts can usually be reclaimed by simply re-tapping them. The sequence of bolt reclaim is to sort out to the different lengths and sizes, remove the nuts, straighten the bolt, cut off the defective portion, and re-thread, finally placing them in bins for specific uses.

Reclaiming of Tinsmith Shop Material

It would doubtless be a very difficult undertaking to dispose of manufactured products made from scrap sheet metal, but where the distribution of the finished articles is made to various branches of the one service, the question is not one of competition but rather of an economic character. The C. P. R. through the reclaim department are able to use enormous quantities of discarded sheet metal for many lines of shop and train equipment. All steel, copper, brass and tin sheeting, that comes to the scrap dock, is inspected, and if found satisfactory, is handed over to the sheet metal shop, where it is made into various articles.

Amongst the articles that are made from scrap material in the tinsmith shop are the following:

Window deflectors, hopper deflectors, smoke jacks, tee pipes, stove pipes, hopper chute pipes, ventilators (all classes), garbage cans, ash pans, funnels, drip covers, soiled cup holders, switch lamp discs, van lamp (casings), card cases, engineers' tool boxes, flat boxes, valve caps, hopper pans, water runs.

Fire guards, pyrene extinguisher holders, pipe guards, foot rests, filler block covers, drip shields.

Brine tank ventilators, soiled towel holder covers, sanitary cup holders, vestibule curtain boxes, false tops, cab lamp shades, conduit covers.

Washers, van lamp brackets, pipe clips, vestibule diaphragm shields, electric lamp shades, valve boxes.

Coal chutes, coal boxes, gear guards, rag holders, broom holders, heater reducers, ventilator extensions.

All the above articles are regular products of the department that are made in large quantities. A great many other uses are found for the reclaimed material which it is, of course, impossible to enumerate.

Miscellaneous Items

In railroad service the range for the adaption of the old piping is almost unlimited. Large quantities are constantly being used for the construction of wheelbarrows, warehouse trucks, guard rails, farm gates, ladders, etc. Old boiler tubes are utilized for making engine pilots, and other new uses are being discovered for this class of material. Car and locomotive axles are turned into various forgings for locomotive and car service. Old chains are likewise utilized for a great variety of purposes.

Worn out tires are usually sold to the large rolling mills, but many of them are taken to the forge shop where they are made into suitable tool holders for the larger machines. It might be mentioned here that all heavy machine cutting tools have the high-speed tip welded on to the shank. All castings that arrive in the scrap laden cars are examined and if found in good condition are recorded and placed in stock. Many parts not seriously damaged are repaired by means of welding and then turned in to the stores. All broken scrap is arranged according to its classification and subsequently sold. The bulk of the machinery cast iron scrap is absorbed in the local foundry. All old material having solder in its makeup is specially treated in a furnace, where the solder is melted off and cast into ingots. This also applies to lead and babbitt. These metals in the shape of ingots are turned over to the stores department. Every conceivable form of scrap is given attention and careful study is made to profitably dispose of it. Wire ropo, netting, lead dross, zinc, glass, waste, etc., are sorted and packed ready

for shipment, and in this way the best possible market is always available.

An interesting detail in connection with the handling of reclaimed equipment is the method of transferring the finished work to the various departments or stores. A large Maple Leaf motor truck is used for this purpose and it never leaves the boundaries of the company's property, being devoted exclusively to inter-department transfer of material to and from the reclaim section. This unit is shown in Fig. 16 being loaded with rebuilt couplers in readiness for taking to the freight car stores. A jib crane, fitted with an air-operated lift, is provided for handling the finished couplers.

NEW USES FOR COMPRESSED AIR

It is not generally known that compressed air is used in various ways in the tobacco industry, yet such is the case, and the following description, which appeared in "Compressed Air Magazine" should be of special interest. The author is chief engineer of the R. J. Reynolds Tobacco Co., Winston-Salem, N.C., and he comments as follows:

"The regular tools such as the rock drills for concrete work, and the machine shop equipment for drilling and driving rivets, chippings, etc., and also a few hoists using compressed air are in use but do not require any further mention as they are of standard design. Compressed air is used for blowing motors and machinery in general, to remove the dust, and in places in automatic machines where the fine tobacco droppings have a tendency to gum and clog the machine in a few hours; a fine jet of air was placed to remove this dust, the trouble was removed, and the machine could run along its full time.

"The same principle is used in another process where it assists in drying and also blows away any dust, while a quick-drying paste is setting. This has very extensive application in plants and has made the installation of several large compressors necessary.

"In making the 'Prince Albert' can tops they are blown clear off the die by air pressure, and just to show that it's an ill wind that won't work both ways, sheets of tin used in making the cans are picked up by means of un-compressed air, or a vacuum.

"A battery of boilers is so placed that they are below any drainage or storm sewer line. When these boilers are blown down a few inches under pressure, no complications result, but when they are cooled down naturally, as they should be, and so have no steam pressure on them to blow out their water up to the sewer, a compressed air connection empties them very successfully.

"The air is of great use and assistance in spraying liquids for conditioning dry tobaccos. The sprayer acts on the principle of the ejector and will pick up the liquid by suction for several feet rise.

"With a pressure-reducing valve in series, there is much use made of the compressed air, for the purpose of low pressure control of heating valves, humidity apparatus and similar equipment."

FIG. 16—MAPLE LEAF TRUCK USED FOR HANDLING THE RECLAIMED SCRAP MATERIAL.

Cutting of Steel Plates, Castings and Billets

Rapid Cutting of Metals by Radiograph and Oxy-Acetylene Torch Are Described—This Matter Should Interest Steel Mills, Boiler Shops, Shipyards, Locomotive Shops, and Machine Plants.

By J. H. MOORE

TO the uninitiated, the accompanying illustrations will cause considerable surprise, and believing an article dealing with the details and uses of such a machine would interest our readers, we have prepared the following:

The machine to be discussed is known as a radiograph and is manufactured by the Davis-Bournonville Co., Jersey City, N. J. It is used for varied purposes as the photos will clearly show. To convey an adequate idea of the power adaptability of such a tool is impossible by illustrations alone. Seeing the machine in actual operation is a very different matter. Were one to go through some of the large shipbuilding plants and watch these machines cutting plates, angles, billets, heavy forgings, etc., etc., they could then realize its wonderful power.

Briefly the machine is of portable nature, weighing about 50 pounds and designed for cutting all thicknesses of steel from ¼ inch to 20 inches, or more. The speed of cutting varies with the thicknesses, but not directly. For instance, the cutting of billets 10 inches thick is accomplished at a rate of about five inches per minute and approximately the same rate is employed on thicker sections up to 12 or 15 inches. A billet 18 inches square was cut off at the League Island Navy Yard in 6½ minutes, and about the same rate would be used for cutting billets 20 inches thick. The reason for this is that higher oxygen pressures are employed and the combustion rate is maintained throughout the kerf of a heavy section at nearly, if not quite, the same rate for all thicknesses from 16 to 24 inches.

We do not know what the ultimate limit of thick cutting with this machine

FIG. 1—FRONT VIEW OF MACHINE SET UP TO CUT TO STRAIGHT LINES.

really is. It might be placed at 24 inches, but development now in progress will probably make it feasible to cut 30 inches, if not 36 inches. A reason for this indefinite limitation is an indefinite demand. The construction of large battleships and fast cruisers now in progress requires the use of forgings much heavier than ever used before, and

this method of cutting is widely employed for such class of work.

The machine is used to cut either straight or curved lines, using a grooved track to guide for straight cutting, and a

FIG. 2—REAR VIEW OF THE MACHINE SET UP TO CUT TO STRAIGHT LINES.

FIG. 3—THE MACHINE SET UP TO CUT CIRCULAR PATHS, USING SCREW POINT AS A POINT.

CUTTING DISCS 11 INCHES DIAMETER AND 9 INCHES THICK.

FIG. 5.—A CLOSE UP VIEW OF DISCS SHOWN AT FIG. 4.

FIG. 6.—CUTTING STEEL PLATE IN NEW YORK SHIPBUILDING YARD.

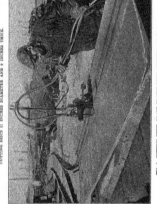

FIG. 3.—TRIMMING HEAVY 24" "T" BEAM. USING THE MACHINE FOR THIS PURPOSE HAS CUT THE FORMER COST IN SIX.

radius bar and center point for circular cutting. It is supported on three wheels, two of which are tractor wheels and the third a caster wheel. It carries a machine cutting torch and is fed along at a uniform rate by an electric motor working through a train of gears and driving the tractor wheels. A change gear feed box provides for various rates of speed suitable for cutting ¼-inch up to 20 inches thickness or more.

It might be well to add that the machine is fully covered by patents, both as to design and basic principle. Referring to Figs. 1, 2 and 3, we see the outfit set up for cutting to both straight and curved lines.

Cutting Straight Lines

Should readers ever use one of these machines, they should keep the following in mind. When a steel plate is to cut, it should be supported on blocks or horses a few inches from the floor. The supports should be so placed that the torch flame will not strike them in its path and the machine will not fall when the cut is finished. In some cases, it may be advisable to mark the the line to be cut with chalk, and prick punch it at intervals of a few inches, but generally this is unnecessary for straight cutting. The track supporting the machine should be laid on the plate parallel to the line to be cut and set six to eight inches away. When the position of the rack has been determined, a chalk line may be drawn along the side so that in case of being shifted accidentally it can be replaced to the original position.

The machine is now placed on the grooved track and connected to the motor with a flexible electric cord to the nearest lamp socket or source of 110 or 220-volt current, according to the winding of the motor.

The machine operates in either direction, depending on the setting of the

FIG. 9—AN EXAMPLE OF CIRCULAR CUTTING. THE PLATE IS 2½" THICK AND WAS CUT AT A SPEED OF 6" PER MINUTE.

reverse gear knob. The torch should be adjusted horizontally and vertically on the cross arm so that the tip is directly over the line to be cut and with the end about ¼ inch above the surface of the plate. The vertical adjustment depends

on the style of tip and thickness of cutting. The acetylene and oxygen cylinders are placed alongside in convenient positions where the pressure gauges can be seen. The cylinder valves are next cracked and attached to the regulators. Connect the hose to the torch, red hose to acetylene and black hose to oxygen, taking care to blow out first to free the hose of dust.

Open the gas cylinder valves carefully, turning on the oxygen valve first. The regulator adjusting screw should be in the "off" position. Open the valve in the torch, adjust the acetylene working pressure with the regulator screw, and light the torch with the sparklighter. Then open the oxygen valve, and adjust the oxygen working pressure.

Next, adjust the preheating flame with the oxygen needle valve, making it neutral, or slightly oxidizing. Open the cutting oxygen valve when making this adjustment. Cutting should begin at the edge of the plate wherever feasible but if the pattern to be cut is some distance from the edge, the plate may be perforated with the flame. Preheat the plate to a high red or white hot temperature and then turn the cutting oxygen valve handle. Opening the cutting oxygen valve operates the switch turning the current on to the motor. The motor starts to feed the machine along at a rate that will depend on the setting of the gears in the feed box. The feed rate varies with the metal thickness, size and style of tip and oxygen pressure, and it should be determined before starting to cut the material.

Should the flame cease to cut, it may indicate that the travel rate

FIG. 7—CIRCULAR CUTTING. (Note the true shape and thickness of finished pieces.)

FIG. 10—AN 108,000-POUND STEEL CASTING. (Note the 56" main riser, also four smaller 9" ones.)

is too fast, in which case move the machine back to the termination of the cut, and start cutting· again at a slower speed. On the other hand, too slow a speed may cause the cut to stop, and then the remedy is to increase the feed rate. The act of turning on the cutting oxygen starts the motor into operation, therefore do not turn on the cutting oxygen until the preheating flame has raised the metal to the burning temperature.

If the length of cut requires more than·one section of grooved track, two may be used, which will serve for any length of cut, the first section being taken up when the machine has passed over it, and laid down in front of the one in use. In case of cutting that requires the machine to travel off the plate, it will be necessary to provide support for the track by laying another support·alongside at the same level. To bevel a plate when cutting, adjust the torch sideways by loosening the swivel and clamping the torch at the desired angle. Notched, stepped or offset cuts can be made by stopping the feed motion by disconnecting one of the motor feed wires and feeding the torch sideways by hand the required distance. and then starting the feed again for straight-away cutting. Any combination of right-angle cuts can be made in this manner.

When starting a cut at the edge of a billet having a sloping side, the torch should be inclined parallel to the slope. When the cut has been started, the torch should be

gradually brought to an upright position for straight-away cutting.

To Cut Curved Lines

Curved line cutting is done without using the grooved track, the machine being guided by a radius rod set in a punch mark at the center of the desired circle or arc. Lay off the required arc with dividers or a trammel bar, and prick punch the center of the arc to be cut. Set up the machine with the center

point in the punch-marked centre, then adjust for the required radius.

The·third wheel of the radiagraph (not shown in the illustration) is a caster that is locked with a set-screw for straight ·cutting. The set-screw should be loosened for circular cutting so that the caster will be free to swivel. Set the counterweight on the radius rod beyond the centre point.and clamp it. The purpose of the counterweight is to hold the centre point down in the punched centre. If trouble.is caused by the centre point slipping out of .the centre, move the counterweight further out on the radius rod, and thus bring more pressure to bear.

Turn on the gases as directed · for straight cutting and start the same as for straight lines. The radiagraph will automatically follow the curved line, swinging about the centre point. The machine is traversed by the outer tractor wheel. The rate of torch travel will vary with the setting of the cross arm and the position of the adjustable leg on the radius rod. It may be necessary to do some experimenting before the proper setting of the feed box gears can be determined. Remember that the torch, when outside of the tractor wheel, will move at a faster rate than on straight cutting, and it will generally be necessary to provide for slower feed rate, or to increase the oxygen pressure.

A circle having a radius greater than about 81 inches requires special radius rods made to order. Circles of very short radius are cut by setting the center point close to the frame and

FIG. 11—THE RESULT OF A LITTLE OVER AN HOUR'S WORK. (Note the machine on top of larger riser to right.)

swinging the cross arm around the vertical post until the tip is on the inside of the radiagraph frame with the center point. This is done by unclamping the cross arm, lifting it up off the post, turning around over the frame and then slipping the socket down on the post and clamping. If the center point cannot be moved in to the desired position because of interference with the tractor wheel, remove it by loosening the clamp, holding it to the radius rod. The center point clamp can then be set close to the frame. This will usually give the short radius of action required.

It is possible to place the torch within about two inches of the center point but it is rarely feasible to cut circles of much less than six inches radius or twelve inches diameter. A faster feed rate must be set on the feed box than for straight cutting as the torch is traveling at a slower rate.

Cutting Combination Straight and Curved Lines

The path to be cut should be laid out and marked with the prick punch. The centers of the arcs should be punched and circled with chalk so that they can be readily located as the work proceeds. Lay out the work to make the straight cut first. Clamp the bracket on the radius bar at the correct position for cutting the arc of required radius, and screw the center point up so that it runs about 1-16 inch above the plate. The grooved track is not used when cutting to straight and curved lines on steel plate. The straight line cut is first made and as the center point reaches the punch mark it is quickly screwed down, cutting the point in the plate. The machine then swings around the center, cutting to an arc without stopping and interrupting the cut. To cut to a straight line from the termination of an arc, unscrew the center point

when the end of the arc is reached, which will permit the tractor wheel to rest on the plate, and then the machine will carry the torch forward in a straight line.

Having commented on the machine and how it should be used, let us discuss the illustrations shown. At Fig. 4, we see the machine cutting steel discs, 31 inches diameter, 9 inches thick, at the rate of five lineal inches per minute. This means that a cut is completed in 20 minutes. Some of the discs shown were 9 inches thick, others 5 inches, and from 25 to 36 inches diameter. At 1 is illustrated a D-B portable acetylene generator. At 2, some manifold oxygen tanks, and the radiagraph itself is shown at 3. A close-up of some of these discs is shown at Fig. 5.

Cutting steel plate with the oxyacetylene flame in a New York shipbuilding yard is shown at Fig. 6. The machine can be clearly noted, and these plates were cut at speeds varying from 18 inches to 2 inches per minute on plate from ¼ inch to 20 inches thick.

Fig. 7 illustrates some circular cutting in the New York shipbuilding yards. The true and finished cut is self-apparent in this view, and the thickness is also considerable.

At Fig. 8 is illustrated the work of trimming heavy 24 inch. I beams, while Fig. 9 shows a good example of a circular cut in steel plate. Fig. 10 depicts a 108,000 pound steel casting with a riser 36 inches in diameter, weighing about 30,000 pounds. Two of the four smaller 8-inch risers can be noted clearly.

The result of cutting by the method we are discussing is shown at Fig. 11. The main riser was cut off in a little over an hour with one of these machines, and an oxy-hydrogen torch. The complete outfit weighed less than 50 pounds. The four smaller risers were cut off

with a D-B hand-cutting torch. This casting was for the steel cylinder of a 7,500-ton forging press and represented a pouring of 140,000 pounds of metal. It is amusing to note the size of the machine in comparison with the work accomplished. Note the machine on the top of the riser.

From the foregoing information, and the illustrations accompanying the same, one cannot help but admit that we are progressing in the art of overcoming the bugbear of cutting heavy material.

NEW CATALOGUE

The Wilt Twist Drill Company of Canada, Limited, Walkerville, Ont., have just recently published a very elaborate catalogue of drills, reamers, and cutters, which not only lists the various types of tools, but gives a description of the different styles and types and various suggestions on their proper uses. This catalogue is now ready for distribution and can be had by writing to the jobber or direct.

Ship Plates Overseas.—The Dominion Steel Corporations are shipping about 1,000 tons of steel ship plates to Australia and New Zealand by the steamer "Otira."

CUTTING OUT CIRCULAR PLATES. A CLOSE UP OF ONE OF THESE PLATES IS SHOWN AT FIG. 9.

Grinding Wheels and Their Uses in Industry

Abrasives Used, the Process of Manufacture, and Other Important Points Are Spoken of in This Article, Which Was Presented Before the Engineers' Society of Western Pennsylvania.

By W. T. MONTAGUE

GRINDING is one of the most ancient of arts. Prehistoric man shaped instruments of stone, and later of metal, by rubbing them on rocks which possessed abrasive qualities. It is not a matter of record when the idea of cutting out a circular block of stone, mounting it on a spindle, and revolving it by hand was first thought of.

Sandstones were originally used in the industries where grinding operations were performed. although the applicability of emery was generally recognised by the Greeks of the early ages, who found it on the island of Naxos. The extensive use of emery in competition with the sandstone was limited until around the year 1870, when a method was invented for binding the grains together with a suitable medium and forming the wheels into necessary shapes for use in grinding.

Improvements in the methods of manufacture of grinding wheels naturally included a betterment of the abrasive material, with the result that the artificial abrasive was developed to overcome the imperfections of the natural emery, and to make available an abrasive in sufficient quantity to meet the ever-increasing needs of industry.

Abrasives

In general there are two types of abrasive—aluminous, and silicon carbide, the former consisting essentially of aluminium oxide, and the latter of a chemical union of the elements carbon and silicon.

Aluminous abrasives occur in nature as minerals in the form of emery and corundum. Aluminous abrasives are also manufactured by electric-furnace methods, and sold under the trade names, "alundum," "aloxite," "borolon," etc.

Silicon carbide abrasives do not occur in nature but are manufactured in the electric furnace and sold under such trade names as "crystolon," "carborundum," "carbolon."

The Norton brand of aluminous abrasive—alundum—is made from the natural mineral, bauxite, containing as high a percentage of aluminium oxide as it is possible to obtain. The ore is carefully analyzed and a mixture so made that the product of the furnace operation is fully controlled. The mixture is fused in an electric furnace of the arc type, and during fusion the material is purified and changed from soft bauxite into hard crystals of aluminium oxide.

The Norton brand of silicon carbide abrasive—crystolon—is manufactured by heating pure silica-sand and coke together in a special resistance type of electric furnace. The material is not fused, but a chemical reaction results from the high temperature employed, with resulting crystals of abrasive.

Sizing the Abrasive

The alundum and crystolon abrasives are received at the Norton grinding-wheel plant in irregular pieces about six inches in diameter. This material is passed through a series of jaw crushers, rolls, washers, etc., and is finally sized by being passed through standard-mesh screens. The standard-grain sizes begin at 8-mesh and continue through 200. The flour which is left is designated as 200F, and is further refined and classified by hydraulic means into such sizes as F, 2F, 3F, XF, etc. The number giving the size of the grain indicates approximately the number of holes to the linear inch, in the screen through which the grain will just pass. For instance, a 30-grain will just pass through a screen having 30 holes to the linear inch, or 900 holes to the square inch.

After passing over the sizing screen, the abrasive grain is stored in tanks, ready to be sent out as a polishing material or to be used in the manufacture of grinding wheels.

Manufacturing Processes

Grinding wheels are manufactured by four processes—vitrified, silicate, elastic and rubber.

By far the larger proportion of grinding wheels is manufactured by the vitrified process. In this process, the abrasive grain is mixed with the proper amounts of clay and water until the mixture has the consistency of a mud, which can be easily poured into moulds. After pouring into moulds, the wheels are taken to drying-rooms and left until thoroughly dry. They are then shaved on special shaving machines to the approximate shapes and dimensions required, and placed in dry storage, ready to go into the kilns for vitrification. Vitrification in the kilns takes place at about the melting point of steel and the length of time required for heating, the length of time held at high heat, and the cooling period are very important. Kilns are of the type used in the pottery industry, and are fired by a series of hard-coal fires uniformly spaced around the base of the kilns. In the larger type of kilns, it is approximately three weeks from the time the kiln is charged until it is drawn. This time is absolutely necessary, and regardless of the emergency nature of any order, every wheel burned in the large kilns must remain there for the full time. After the wheels have been burned, they are sent to machines where they are shaped to exact size by means of hard metal cutters.

Silicate wheels, as the name implies, are made by using a bonding material composed of silicate of soda. These wheels are made by tamping into iron moulds, and they are baked at a comparatively low temperature. By this process all wheels thirty or more inches are manufactured, 60-in. in diameter being the maximum size that can be made.

Elastic wheels are made by using a bond having quite a degree of elasticity. The bond is of an organic nature and is composed mostly of shellac. These wheels are also tamped into iron moulds and are baked at a comparatively low temperature.

Rubber wheels are made by mixing abrasive with rubber, and later vulcanising the resultant product.

Silicate wheels are used largely in the cutlery industry in general to replace sandstone; and for all wheels over 30 inches in diameter, which cannot be manufactured commercially by the vitrified process. Elastic wheels are used where very thin wheels are required for cutting off stock; also for finishing grinding of chilled iron rolls, and for other work where a fine finish is desired. Rubber wheels are used on the same class of work as elastic wheels, except that they have a somewhat harder action, and are, therefore, used only where grades harder than those made by the electric process are required.

Abrasive Action

The abrasive action of an aluminous abrasive is dependent upon the amount of crystalline aluminous oxide present, and also upon the temper or brittleness of the abrasive.

The best grade of emery comes from Turkey. It contains up to about 65 per cent. of corundum, which is the form in which the cutting element occurs in emery. Emery mined in America has as low as 10 per cent. corundum. The chief impurity of emery is magnetic iron oxide. Alundum abrasives contain more than 92 per cent. of this cutting element and a special alundum known as No. 38 contains more than 98 per cent. aluminium oxide. Practically no magnetic iron oxide is present in these abrasives.

Artificial aluminous abrasives are more efficient than emery, because (1) they contain a much higher percentage of

cutting element; (2) being free from impurities, they are capable of variation in toughness to suit the work to be done; (3) they are made to a definite standard of composition and temper.

The grinding wheel to meet present day requirements must be a scientifically developed cutting tool. Its action when at work is similar to that of the steel milling-cutter. The face of the wheel, are millions of cutting teeth at work every minute, and although these teeth are not as long or as strong as the teeth of the steel cutter, and cannot cut as deep, they are capable of working at a much greater speed. Each little cutting-tool, which in substance is a grain of abrasive material, cuts off a chip at each revolution. The chips resemble, in shape and character, the chips cut off by the milling-cutter.

Uses in Industry

The uses of grinding wheels in industry to-day are many and varied. The great refinement attained in the case of the gasoline motor used for automobiles could not have been reached without the use of the grinding wheel and artificial abrasives. Likewise, the grinding wheel plays an important part in the manufacture of tractors, motor-trucks, gas engines used for farm purposes, etc.

The ball and roller-bearing industry, which has grown up alongside the automobile industry, is likewise absolutely dependent upon artificial abrasives for the refinement and accuracy of its product. Being composed of hardened alloy steels, the only way that these could be brought within the required limits of accuracy and finish was by means of such aluminous abrasives as alundum. Likewise, the requirements of this industry are so great that existing supplies of natural abrasives, even though they were of proper standards of purity, would not begin to meet the demand.

The phonograph, the typewriter, the adding machine, the cash register and other apparatus of similar nature, could not be made at economically to-day if it were not for the grinding wheel industry and artificial abrasives. The agricultural implement industry uses grinding wheels and abrasive grain in large quantities for the manufacture of harvesting and threshing machinery, ploughs, planters, etc. Even such activities as the textile industry require grinding wheels for sharpening cards, snagging castings, and maintaining tools, cutters, and dies used extensively in keeping up its equipment.

The leather and shoe industry uses grinding wheels for the buffing of hides and for the sharpening of leather cutting and shaving knives.

The steel mills use alundum grinding wheels for grinding out seams of high-speed steel billets preparatory to rolling into bar stock. The steel foundries use alundum wheels for snagging steel castings. Crystolon grinding wheels are used in foundries for snagging cast-iron

castings, and for cleaning castings of brass, bronze, and aluminium.

The railroad industry has extensive use for wheels composed of artificial abrasives. Such parts as locomotive piston-rods and valves must be ground on cylindrical grinding machines; guide-bars must be surface ground with alundum grinding wheels; steel car-wheel treads and flanges sometimes are ground with alundum wheels, and manganese-steel frogs and switches have to be surfaced and fitted with grinding wheels composed of aluminous abrasive.

The optical industry uses aluminous abrasive wheels for lens-grinding, and aluminous abrasive grain for roughing-out lens blanks prior to polishing.

The cut-glass industry employs the artificial grinding wheel to a very large extent in cutting the intricate designs that go to make up the beauty of this ware.

The marble industry employs silicon-carbide abrasives in thin wheels for sawing marble into slabs, and in thick wheels for surfacing or moulding the marble into various shapes and designs.

The final polish on marble slabs used in interior building operations is obtained by means of abrasive blocks composed of very fine grit silicon-carbide or alundum abrasive, followed by putty powder.

Selection of Wheels

The main points to consider in the selection of grinding wheels are as follows:

Material.
  High tensile strength. (Aluminous abrasive.)
  Low tensile strength. (Carbide of Silicon abrasive.)
Operation.
  Cylindrical
  Surfacing
  Internal
  Sharpening
  Off-hand grinding
    Bench stands
    Floor stands
    Swing frames
    Portable
      Electric
      Pneumatic
      Flexible shaft
Wheel speed
Work speed
Contact
Condition and type of grinding machine.
Personal factor

Whenever a grinding job is presented to you, the first thing to think of is the nature of the material to be ground—whether it is hard or soft, etc. If it falls under the general head of a high tensile-strength material—such as all steels and down as far as the hard grades of bronzes—then an alundum wheel of some kind should be used. If, on the other hand, the material falls in the class of low-tensile-strength materials—such as cast iron, chilled iron, brass, soft bronzes, aluminium, and copper—then you should use crystolon wheels.

Operation

The next thing to consider is the nature of the operation to be performed by grinding—that is, whether cylindrical, surface internal, sharpening, or off-hand grinding is demanded. There are so

many small points to be considered in connection with each operation that it is impossible to go into this in detail.

Wheel Speed

This should be considered, but with any given class of grinding it is more or less fixed. We recommend the following speeds for different classes of grinding:

Application	Speed in Feet per Minute
Cylindrical grinding	5500 to 6500
Snagging and general off-hand grinding on bench and floor stands	5000 to 6000
Surface grinding	4000 to 5000
Knife grinding	3500 to 4000
Hemming cylinders	2100 to 2400
Wet tool grinders	4000 to 5000
Vertical surface grinding machines	4000 to 4500
Elastic and rubber cut-off wheels	9000 to 12000

If the speeds deviate very much from these, and it is impossible to change them to suit our recommendations, then this must be taken into account in your recommendations. Speeds higher than those recommended call for slightly softer grades to offset the harder cutting action; and speeds lower than those recommended call for slightly harder grades than would be ordinarily supplied.

It is impossible to tell a customer the exact speed at which his work should be done on any given grinding job. It is largely a matter of experiment. The work speed should be suited to the wheel in use and the nature of the material to be ground. On the Norton cylindrical grinder a speed of from 60 to 80 surface feet per minute is often used for roughing and from 30 to 40 surface feet per minute for finishing. On most types of precision grinding machines it is customary to rough-grind at a higher surface speed of work than on finish grinding.

Contact affects grade selection. Broad contact calls for softer grades, and narrow contact for harder grades, as the case may be. This is especially true in snagging and off-hand grinding. Where wheels are used for grinding the burr left by welding, or for grinding sharp fins from castings, extremely hard grades, such as S, T, and U, are called for. In cylindrical grinding the contact varies with the diameter of the wheel and the work, increasing with larger work or with a larger wheel, and thus making a softer grade of wheel desirable.

Condition of Grinding Machine

This is something which you would really have to observe personally in order to understand how it would affect grinding-wheel selection. If the spindle is loose and cannot be put in good condition, it means an irregular motion and in this case harder wheels must be used than would ordinarily be recommended. This is in order to overcome the tendency to pound the wheel face to pieces. Light, flimsy machines, and machines improperly secured to the foundation, also call for harder grades than would ordinarily be used. Machines are frequently placed in the middle of

Continued on Page 233

John Conley Secures Some Further Advice

On This Occasion His Dad Takes a Hand in the Matter, Telling John Where He Made His Mistakes—Drilling Out a Cap Screw, Drilling Holes Through Thin Sheet Steel and Other Jobs

By J. DAVIES

"DAD, I think the boss picked the most miserable jobs he could for me to-day. The first one was putting an inch hole through a piece of sheet steel about 1-32 of an inch thick. I placed the sheet steel on a piece of wood. The drill wasn't exactly running true, and when it broke through the plate into the wood it wobbled somewhat, with the result that the hole wasn't quite round. Mr. Smith had the nerve to ask me if I had put the hole in with a dumdum bullet.

"As if that wasn't enough, he gave me a bracket with a broken half inch cap screw in it and told me to drill out and replace it with another screw. Of course I drilled it out with a 13-32 drill, the standard tapping size for half inch, then I put the tap in and managed to twist out the remainder of the broken screw. To my surprise I found I had drilled all the thread out on one side, so I next had to drill a 17-32 hole and tap that out ⅝ in.

"Next, I had to put a number of holes in a piece of cast steel and ream these out with a No. 6 Morse taper reamer. The foreman told me to be careful not to break the reamer. I didn't know how it was but the boys seemed to be awfully interested in that job. They asked me if it was the same one, was I coming or going, where was I piling them, hadn't I better put some salt on it, and so on. I had an awful job drilling the holes, but when it came to reaming I made no headway at all, and for once I was glad to see the foreman coming down the shop. I asked him what the trouble was, and after critically examining the reamer he said 'The reamer is worn on the cutting edge. It looks as if the steel you are reaming has not been properly annealed. I will have it softened again.'

"Surely my troubles are over for to-day, I decided, and I proceeded to centre my next job, a cast iron piece. At my first blow I broke a piece out of it, and Dad, I don't mind saying, I'm feeling pretty blue to-night and don't care if I never see the machine shop again."

Some Real Advice

Robert Conley carefully adjusted his spectacles and then said: "John, here is a little rhyme I learnt some time ago"—
"It's easy to be pleasant when life flows along like a song, but the man worth while is the man with a smile, when everything goes dead wrong."

"I consider you have spent a very profitable day if you will remember not to make the same mistakes twice. You know, nothing happens without a cause. Now let's look for the cause.

"The irregular-shaped hole in the sheet steel was caused by the point of the drill having no support when it broke through the sheet. The remedy is to secure a piece of scrap iron, bolt your sheet down to it solidly, and drill through into the scrap iron until your drill has gone through the sheet steel, when you will have a true round hole.

"Drilling the cap screw the full tapping size was just lack of thought. It is practically impossible to drill a broken screw in a casting the full tapping size without drilling away part of the threads in the casting. You should drill a thirty-second or a sixteenth less. You can then drive in a square drift, and in most cases you can twist out the remainder of the broken screw.

"Learn to use your head as well as your hands, my boy. By the way, do you know the tapping sizes of standard nuts and bolts?"

"No, father, I do not."

"You ought to learn them off by heart up to an inch anyway, also tapping sizes for pipe taps; it will save you a lot of trouble. Just dot them down and learn them at your leisure."

SIZES OF TAPPING DRILLS—U.S. STANDARD.

Dia. of Tap	No. of Thds.	Tapping Size
1/4	20	13/64
5/16	18	1/4
3/8	16	5/16
7/16	14	23/64
1/2	13	13/32
5/8	11	17/32
3/4	10	41/64
7/8	9	3/4
1	8	7/8

PIPE TAPS AND DRILLS

Pipe	No. Threads	Drill
1/8	27	11/32
1/4	18	15/32
3/8	18	19/32
1/2	14	23/32
3/4	14	16/16
1	11½	1 5/16

"Did you get those sizes from a text book?" "I did get a list of sizes from a text book, but I found I had to revise them to suit the circumstances. The list I have just given you is not to be followed exactly in all cases. You have to use your own judgment according to the depth of the tapped hole, the material, and purpose of the job. The most fruitful source of broken taps is making the tapping hole too small.

"Regarding the taper holes you tried to ream, if your reamer refuses to cut at first, don't try to force it, but stop your machine and look for the trouble at once. Your reamer may be soft, or the cutting edge worn away, or it may not have proper clearance. Cast steel is pretty tough stuff to ream, and in a deep hole there is great danger of breaking the reamer. In cutting hard steel, turpentine is often used as a cutting compound, and it helps considerably. A standard Morse taper is 1 in 20, so that if a 1-32 larger drill is put in for every ⅝ in. depth it will make it much

easier for the reamer and leave a safety margin for the reamer to take out."

"How about speeds and feeds?"

"The speed of the drill and the amount of feed per revolution can only be determined by the man on the job, depending as it does on so many conditions, i.e., the hardness of the material to be drilled, the nature of the job, the quality of the drill, the depth of hole, etc. It is impossible to state a rule to cover all cases, therefore, the correct cutting speed must be determined by good judgment. It is safe to start drilling with carbon drills at 30 ft. per minute for machinery steel, using a feed of about .004 inch per revolution for small drills up to about ⅜ in. and a higher feed up to about .010 for larger drills. These are only trial speeds from which the correct speed can be obtained to suit the job."

"There was one other job that gave me a lot of trouble, Dad. It was drilling a hole square to the base in a block of iron that had an inclined surface. I was rather afraid I would not be able to draw it back to the right place."

"A pilot drill would have been useful in that case."

"What in the world is a pilot drill?"

"A pilot drill is a very useful tool. It is made in various forms, the chief characteristic being that of its having a round pilot turned on the end to suit the standard sized drills. Some pilot drills are made with four cutting edges, and some with only two. They are employed for counter-sinking counter-boring, and for enlarging small holes where accuracy is desired. Perhaps their most useful feature is the drilling of a hole on an inclined plane, the pilot preventing the drill from running to one side. A small hole is first drilled the size of the pilot."

He Wants to Know More

"Now, Dad, I want to know something about boring bars, how, when, and where to use them."

"All right, John, it is a pleasure to tell you anything I know. When a hole is too large to be finished with a drill, it is either drilled first or the hole is cored out in the foundry. It is then finished with a boring bar so that boring may be said to be the finishing of a hole that has been previously made. This is done by a bar carrying one or more cutters, but the practical method of operation is much the same in all cases.

"Use a boring bar as large in diameter as the cored or drilled hole will permit and as short as is consistent with the depth of hole to be bored, leaving a good length of bearing in the guiding or supporting bush when the cutter commences to cut. See that the tang at the end of the boring bar fits the drill spindle, and

in large bars it is well to have a small hole right through the end of the drill spindle and the taper shank of the boring bar, so that a pin driven in this hole supports the weight of the bar.

"A boring bar is much more likely to come loose in the drill spindle than a drill, so that some such device is necessary. Some people use a set-screw, but this is unsatisfactory and dangerous unless a safety set-screw is used. Be sure that the bar is a good fit in the table bushing, an easy, free fit, yet not too loose. This is very important as it is impossible to do a good job if the end of the bar is allowed to wobble. In setting up the job, see that the bar will enter the table bushing without being deflected to one side. Set your table first, that is, if the job will permit you doing so, with the bar free to move round and up and down in the bush. It might not be possible to set up some jobs in this way because the bar could not be raised high enough to pass over the job underneath. In these circumstances place the job into the table bushing and swing the whole arrangement under the drill spindle together. All the holes to be bored ought to be marked off carefully. The centre of the bar must be in the centre of the finished hole, independent of the centre of the original or cored hole. In setting up the job with the hermaphrodites, or jenny-legs, or odd legs, as they are variously called, put the pointed leg in the punch, mark on the circumference of the circle to be finished, and let the other leg just touch the bar. Move the job around until each punch mark is an equal distance from the bar. After taking one cut through and the hole cleaned up, which means that the cutter is cutting all around, test with the hermaphrodites again, only this time measure from the punch marks so that measurements from all the punch marks in the finished circle to the edge of the hole will be equal.

"It is not good practice to finish-bore a cored hole with one cut, as the cutter is nearly always sure to cut more to one side than the other. This causes the bar to spring and chatter, and the hole will not be round. When the hole is uneven to begin with, always finish with a light cut.

"In some jobs the centre of the table cannot be used as a support for the bar. In such cases improvise a bearing by bolting a piece of cast iron bored out to suit the bar, to the bedplate or baseplate of the drill, or perhaps to an angle plate. Again, the job may not permit of any guide or support, such as boring a blind hole. Finish with a reamer if possible; if you haven't one, the only thing you can do is to finish with light cuts. When boring to a standard size the finishing cutter should be notched out to fit the bar and turned up in the lathe. Having very little clearance in the finishing cutter makes a smooth hole, too much clearance develops chattering. But enough just now."

GRINDING WHEELS AND THEIR USES

(Continued from page 231)

a wooden floor which vibrates badly, and in this case harder wheels must be used for a machine on a firm, solid foundation.

This is extremely important in the operation of grinding wheels, frequently influencing the results obtained as much at 100 per cent. We mean by this that different men working on the same kind of machines and on the same work in the same shop will get one result (say, 15 hours' life), whereas other men under exactly the same conditions might get 30 hours. This is based on records obtained, and not on impressions, and explains why the same wheels will work differently in different shops.

MISSOURI SYSTEM OF CLASSIFYING MACHINISTS

By John Homewood.

THAT the all round machinist and the employers of such should be protected from the influx of make believes produced by the late war is very evident. Half made mechanics are a cancer in the body of the industrial world and some means should be adopted whereby they would be put in the proper place. Some method must be adopted that will handle this new class. When a man applies for a job and claims this and that, some way should be arranged whereby his claims should be put to the test before he enters into employ. To have a man fill out an application blank with his family history included in the other information is not sufficient to protect the employers of all round machinists.

It is a moral detriment to the good mechanic to have one of these bluffers working alongside him that claims that he is an all round mechanic when in fact he is an all round prevaricator. There is a class of men that get their experience chain fashion; they get it link by link until they have enough to pass along as all round men. During this period they adopt fradulent methods in order to get by. I do not believe in taking the bread from any one's mouth but I like to see him get it by honest means. It is not an uncommon thing for a man that has never been in a shop before to get a job as a machine operator and after working for a short period start welding his chain. He craftily obtains enough information to get him into another shop; with the assistance of a inexhaustive amount of nerve, and a few tools in his pocket, he furthers his efforts until the foreman catches on to him. The damage he does to tools and machinery is very great and the loss he causes is shouldered by the trained man as well as the employer.

Some will say it is the foreman's duty to detect these critters. The foremen have their hands full in attending to other duties besides attending to the details of detecting a poor workman. A man does not have to go very far in detecting a bluffer in some classes of work. It is a feeling among good men that these should be reported to the foremen. Who is going to do it unless he is particularly assigned to that task? There is no mechanic that is not in sympathy with a movement to suppress the present evil, if some method was adopted to handle it. A man may be well worth two dollars an hour, but does he get it? No, he has to support the loss caused by the offending one. When we purchase an article we naturally expect to get something near what the clerk claimed it was. When a firm hires a man who terms himself an all round mechanic, it isn't in a position to see what he is until the damage is done. He has robbed the firm of some "experience" that he should have obtained by other methods at a lower rate of pay.

The essential thing in a machine shop is co-operation. Can a worker be censured for failing to assist a bluffer in a small detail? A certain chap claiming to be a first-class toolmaker was given some hollow mills to make out of high speed steel. He got along all right although very slowly. The time came for him to harden the mills and the fact that he never looked into a hardening furnace before resulted in something that resembled slag drops. He carried his nerve along with him right up to the last minute by taking them up to the foreman, and, laying them on his desk, asked, "What next?" The boss picked them up and told him that he didn't order fried oysters but hollow mills.

The above experience and loss was shouldered by the firm. Another chap worked with a pattern maker's shrink rule for several days before he was detected. His reply to a query was "Oh, I just allow a little."

The writer was the guilty one that hired this latter bird, and was too busy on other matters to detect him. I had another chap apply for a position and decided from his talk that he was worth taking a chance on. He told me that his tools were on the way and he was assisted by the loan of a few. I took particular interest in this chap and took notice as to how he "carried" himself around the lathe. He carried himself around it all right; I thought he would get dizzy; he finally landed in front of the carriage and I honestly believe he thought the thing was charged with electricity as he seemed afraid to touch it. I began to get conservative after that and spent a little more time with a new comer. Can you blame one for yelling for the Missouri system of hiring labor, where they will be put to a practical test under a man that thoroughly understands his business.

I am sure that any foreman would welcome the idea of having a man put through the grill and be tested and classified before being thrust upon him. There is another phase in this matter that will not be out of place to mention and that is that a very good man can come into a shop and be improperly classified owing to the fault of the machinery that has been permitted to get into a very bad state.

All Roads Lead to Machinery Hall at Exhibition

At Least They Do for the Man with a Mechanical Turn of Mind— Exhibits of Interest Are Shown, and a Visit to This Hall Should Be Considered a Duty.

By J. H. M.

THE exhibition authorities picked on a real motto when they used the slogan "Work and Prosper." There is good reasoning to the cry and if one is interested in the mechanical field, a visit to the machinery hall will show the latest developments in the machine tool and other kindred lines. Space is once more at a premium in spite of the new space charges, and as usual one hears on all sides the comments relative to the inadequate provision made for the machinery industry. One exhibitor tells of the floor breaking twice while he brought in a rather heavy press and he mentioned that he would have to take it out by another exit where ne knew the floor to be stronger.

As the machine tool industry is a well established one in Canada, there is no reason why an up-to-date building could not be erected, and we feel sure that if this were done and on a larger scale than the present edifice, the response by the various manufacturers would repay the outlay. Here's hoping that when another year rolls around we will be able to talk in glowing terms of the new machinery hall, with all its latest improvements. Many concerns who have the machinery in place, but not under the belt, would have them in actual operation were the facilities up to date.

Let us take a tour through the build-ing and see what we discover. Suppose we decide to start at the east entrance and go down the right hand side, clean around the outside, then take the middle section last, but not least. No attempt is made to give the exhibits in their order of importance, but rather as we viewed them when we passed by.

Exhibits

The Acme Steel Goods Co., Chicago, is the first to greet the eye. This concern is showing a complete line of box straps and corrugated fasteners. Acme shook fasteners are the proper name, and these with wire are said to tie up any bundle in record time, and present a very neat appearance. It is claimed that such fasteners are easily tied, quickly applied and cannot come loose. Wire pullers, wire cutters and hammer combined are also shown.

The Wilkens Automatic Regulator Co., Ltd., Toronto, are next in order. All types of fuel and draft regulators, steam pump regulators and reducing valves are shown. A system of automatic control of boiler feed water and a thermostat regulator are included in this exhibit.

The Climax Baler Co., of Hamilton, are showing a complete line of their balers; that is the hand power machines. These machines are used wherever waste paper or other materials require being made up into small bales.

The Canadian Hauck Burner Co., Ltd., Port Hope, Ont., have a very complete showing of thawing outfits, preheating outfits, blow torches, portable oil burners, torches and furnaces. There are also some portable heating outfits, furnace burners of both low and high pressure and lead melting outfits. The latter is said to melt 200 lbs. of lead in less than 15 minutes. Rivet forges, brazing forges and crucible melting furnaces are also shown. The crucible furnace is said to melt 120 lbs. of brass in 45 minutes. We also noted in this exhibit plumbers' furnaces, kerosene furnaces and oil burners used for welding and preheating. A concrete heater next claimed our attention. This heater is used on concrete mixers in the colder weather and heats the sand, etc., in the mixer, in this way eliminating the necessity of thawing out lumps in the frozen sand. A smoothing iron as used on asphalt work is also shown heated by the Hauck burner principle. This latter idea is of course particularly for road making. The representatives in charge are W. W. Yates, R. W. Ware and A. M. Thomson.

The Elliott Machine Co., Ltd., Belleville, Ont., have a line of portable woodworking machinery, including scroll

Just a sample of the crowd at the Exhibition.

saws, floor and drum sanders, band saws, planers, etc.

The Dominion Steel Products Co., Ltd., Brantford, Ont., come next, with both a very neat and interesting exhibit. First, and perhaps most important of all, comes their showing of a 75 K. W. 425 R.P.M. high compression oil engine. This engine is of the Diesel type and is claimed to be the first Diesel engine completely designed and built in

unit is self-automatic and privides both electric light and power to the farm. In connection with this latter exhibit we might mention the showing of the Ivey storage batteries. These batteries are used in connection with the lighting outfits and to show how small a space they occupy in relation to those of other makes the demonstrator has placed a set of similar strength, but of another make, directly alongside. The

full line of dies, taps, reamers, milling cutters, etc., are shown and as the Pratt & Whitney line is already well known we need only add, see this exhibit before you leave the grounds.

L'Air Liquide Society have various representatives, these being F. Staddon, manager; L. Sykes, Mr. Burns, Mr. Davis, Mr. Cohen, Mr. Richards, and Mr. Hougten. They show a complete line of oxy-acetylene apparatus,

A general view of the Manufacturers' and Liberal Arts Building.

Canada. Mr. W. W. Brown, the designer, gave the writer a complete description of the engine and while one cannot go into a full talk on the matter in this small space, we can say that the exhibit is well worth a visit. Each cylinder is identical in assembly, and interchangeability is an accomplished fact. The writer saw the arrangements made and if desired one can disassemble any unit without disturbing its neighbor. Even the governor and other smaller parts are arranged on the same principle. The engine is very compact and is completely self-contained. They are said to be fully efficient under both steady and variable loads, this being of considerable importance to any heavy duty engine. The electric generator is, as stated before, part of the complete outfit, and the crane can pick up the complete self-contained unit and place it into position wherever desired. Full details are given in the literature, which is given to those interested. W. W. Brown, H. J. Jordon and E. Hopkins are representatives.

In addition to the engine, the concern are showing a complete line of cast iron and steel chucks of the design which has already been explained in our columns. These chucks are very neat and their strength is already well known. Various tire moulds are also exhibited, one in particular being engraved, or profiled at the outside diameter for the tread. This work is very neat indeed.

A tire applying press is also shown, this press being used for the applying of tires on trucks, etc.; 200 tons is the capacity of the press in question.

This concern has also an outside exhibit of their Dominion light plant. The

latter set takes approximately four times as much space as the former.

Another feature is said to be that the Ivey batteries come packed ready for setting up, that is, after the boxes are opened the batteries are lifted out on a specially made tray, the box set upside down and the tray placed on top of the box. Other features are that this style of battery has three regulator bulbs, which show when the same is 25, 50 and 75 per cent. discharged. The cases are transparent and made from pyrolin, this material being both non-inflammable and indestructible. The set shown are of 160 ampere hours and 32 volts. G. R. Archdeacon is in charge and mentioned he would be pleased to give all information possible.

The Henry Engineering Co., Toronto, come next. This concern is showing a full line of Pierce renewable fuses, an economy tiering machine, a full line of Jewel electrical instruments and the American steam jet ash conveyor. They also are exhibiting the diamond soot blowers for boilers. The ash conveyor is especially interesting and briefly here is the idea. The fireman shovels his ashes down a hole convenient to the ash pit. The ashes drop into a pipe, are sucked up into a hopper and from there can be unloaded at will. Several installations have been put in recently, the Robert Simpson Co. being one of the concerns to use this method. Jas. A. Rumzay is the gentleman in charge of this exhibit.

The Pratt & Whitney Co., of Canada, Ltd., have a very interesting exhibit at booth 69A in the industrial building. Mr. Ferguson is in charge and will be pleased to receive anyone interested. A

consisting of gas cylinders of various sizes, torches of various kinds and complete outfits ready for work. A cylinder has been cut in half and those not familiar with the interior construction and filling of such a tank should view this sectional cut tank. A thermos of the real honest to goodness liquid air is shown in its liquid state. It might be well to add that liquid air in its liquid state must be at least 300 degrees below sero. A drop of this on the hand would freeze, yet seem to actually burn, so low is the temperature. This concern is also showing a new form of tank for acetylene.

The Clemens Electrical Corporation, Ltd., Hamilton, Ont., have an interesting and varied exhibit, consisting of refillable fuses, soldering irons and a new type of toaster. As the soldering iron is of most interest to machinery readers we will discuss that feature. These irons are of different models and heat in from 30 seconds up depending on their size. The copper point is brought in contact with two carbons by releasing a spring. When the contact is made the iron immediately starts to heat. They are designed especially for tinning, running of seams and general soldering work. They are shown in various sizes from the 150 to 800 watt sizes. A two handle portable soldering outfit is also shown and this works on the same principle. Mr. R. C. Guest is representative at this booth.

The Carter Welding Co., of Toronto, Ltd., are next in order and they are exhibiting a very complete line of Rego welding and cutting apparatus. These torches and other parts are well displayed, and as the Carter Co. are al-

ready well known we need not suggest a visit, for you will no doubt go anyway. It will be worth while, as Mr. Carter or Mr. Sorely will be pleased to have you call.

The Dunlop Rubber Co., Toronto, have a very interesting display of rubber belting, fire hose, separate, and on the fire waggon. They also show an installation under power of a Gibraltar conveyor. This is a large rubber belt to which cast iron buckets are fastened. Fire extinguishers and fire hose in general are shown, also a few examples of garden and industrial hose. Mr. Richards is the gentleman in charge.

The Canadian Calcium Carbide Co. are showing the different sizes of their product as used for various purposes and by the different countries. Quite a display is given and it is well worth while. Mr. A. McMillan and H. E. Mussett are in charge.

The, Independent Pneumatic Tool Co., Chicago, have a very complete exhibit of their Thor air and electrically driven tools. These, of course, include their chipping, caulking, and flue beading hammers, riveting hammers, air drills, wood borers, air grinders, staybolt drills add electric drills and grinders of all kinds. This concern is featuring its new electric screw driver, which with certain attachments will not only place screws in by electricity, but will also remove nuts by the same means. Mr. McCrae, and Mr. W. H. Rosevear, the latter the Canadian manager, are in charge and will be pleased to see all comers. As the exhibit is of considerable importance, no doubt the crowd will gather there for at least a look.

The Goldie & McCullough Co., Galt, have a very complete exhibit, including examples of their well known lines and they have even got a splendid example of one of their safes on exhibition. We could not get in touch with their representative in time for this matter going to press, but their showing is certainly well worth a visit.

Peckover, Ltd., are displaying a complete line of steel, including metal lath,

plates, triangular mesh, and reinforcing sheets, bars, structural sheets, boiler tubes and tool sheets are also included. They are showing the Feralum and Vulcanum slip proof tread, these treads being made of iron with a heavy layer of alundum grit embedded in the surface at the time of casting. R. C. Peckover, J. G. Near, R. C. Rice, G. Jolly, A. Jacks, H. Raine, and W. B. Rubidge are all on hand at this exhibit. Incidentally the twelve tons of steel which they have on exhibit were sold 24 hours after being in place.

The Dominion Machine & Tool Co., Ltd., are makers of jigs, dies and special machines and have a goodly assortment of these displayed. They are also featuring their sherardizing process, this being a rust proof process for all kinds of materials.

The Cleveland Pneumatic Tool Co., of Canada, Ltd., are displaying a very complete line of pneumatic tools. These include chipping hammers and riveters, grinders, sand rammers, caulking and beading hammers, breast drills, rock drills, manifolds and all types of air hose couplings. Four piston air drills, ball bearing type, are shown, also compound geared air drills. Corner drills are included and the new Cleveland pocket in head rotator is also displayed and in fact everything necessary in this line is shown, even to the plain and armored hose. A very fleat display of valves are in a case, some 30 kinds being shown, these including the 1, 2, 3, and 4 way valves, Y's, etc., etc. C. D. Garner, Canadian manager, A. W. Hughes, Toronto office, A. M. Whittle, Montreal office, and G. H. Hall from Cleveland, are in attendance to seekers after knowledge.

The Canadian Ice Machine Co. have an excellent exhibit of refrigerating supplies and materials; fittings, ammonia, calcium chloride and other supplies being featured. This firm are the manufacturers of the York refrigerating systems and compressors, installations being made from ¼ ton up in the direct expansions, absorption or CO_2 sys-

tems, the latter being popular and of considerable adaptability for hospital and marine work. Mr. Haine, C. Lennox, and P. G. Hender are in charge of this booth.

The Goodyear Rubber Co. have a very full display of all transmission belting for mechanical purposes. Packing of all kinds is shown, this including steam, water, etc., etc. Corrugated matting, hose for air, steam, water, gasoline, etc., are included in this very complete exhibit. F. Fox, Mr. Dodge, Mr. Crosby and Mr. Woods are looking after this display.

Prest-O-Lite of Canada, Ltd., with Mr. A. M. Gimble, A. V. Tuthille, and F. T. Northwood in charge, are showing a full line of welding torches and general welding apparatus, including preheaters. They are also displaying the different types of filler rods used in welding. In fact anything relating to acetylene can be found at this exhibit.

The Davis-Bournonville Co., of Toronto, have a very fine exhibit of portable welding and cutting plants. They have a generator of 300 pounds capacity, also a portable one, mounted on a neat truck of 50 pounds capacity. They have a full line of welding torches and are featuring their new improved non flash back torch. The writer had this torch explained in detail and it is well worth looking into. They have also an oxygraph in operation cutting metals of all thicknesses up to 18 in. and the writer received a demonstration which showed this machine up to good advantage. It cuts on the patograph principle, being ratioed at 2 to 1. 18 in. thick material can be cut at a speed of 3 in. per minute, while ¼ in. material can be cut at 18 in. per minute. The writer found M. W. Gibbs an interesting speaker and he mentioned he would be only too pleased to meet readers of Canadian Machinery who were interested in any of his lines. J. F. Crowley and H. Willson are also at this exhibit.

D. K. McLaren, Ltd., have a very interesting display of single and double beltings and supplies for the textile trade in general. The belts include waterproof leather and balata belting and a very interesting demonstration is given in this connection, namely the running of two belts which pass through a pan of water while turning over pulleys. Other lines shown are split pulleys, belt dressing, belt fasteners and other accessories. W. S. Hamilton is in charge, with Craig Hamilton assisting, and they state they will be only too pleased to welcome visitors.

Jones & Moore Electric Co., Ltd., Toronto, are showing a neat line of motors from 1-6 h.p. to 25 h.p. A shoe repair shop outfit in actual demonstration is also shown.

Dominion Belting Co. are showing a full line of conveyor and transmission belting. This is the well known Maple

A busy scene around the grand stand.

Leaf cotton stitch belting. Belt dressing, fasteners, etc., are included in this exhibit, and E. Kinrade is representative.

E. Pullman Wipers & Waste Co., Ltd., Toronto, are displaying a full line of their wiper cloths, klenser kloths and mop cloths. They are especially featuring on the latter, because as it comes to the customer in very large pieces, it leaves them to cut it up at their own discretion. They are also very soft and absorbent, two essential features. Mr. F. Kribs and G. A. Hardie are in charge.

The W. R. Sexton Boiler Setting Co. are showing their line of Elco special fire brick, their style of nonpareil dumping grate bars, also boiler settings.

The Wayne Forge & Mach. Co., Toronto, who are agents for the Laclede Christy Clay Products Co., are showing a good varied line of heat treating furnaces for manufactured and natural gas. They are also showing different style fire brick for the same.

The Geo. W. Cole Co., Ltd., Toronto, are showing a very complete exhibit of their well known lines. A boiler feeder, condenser trap, return trap, receiver, various swing check valves, angle check valves and pressure reducing valves are shown. This is a very neat and complete exhibit and deserves attention from anyone interested in such a line. G. E. Cole, vice-president, W. H. Hughes, J. E. Farrell and D. Johnston are at this booth and will be pleased to see all callers.

Baines & David, Ltd., Toronto, are showing all kinds of iron and steels, sheets, bars, steelcrete, etc., etc. They are also displaying the universal stair tread, which is said to be absolutely non-slipping.

The Centre Portion

Coming down the centre portion we have the P. B. Yates Machine Co., Hamilton, Canada. This concern has a very complete exhibit including a C, moulder. This is a new machine on the market and has several features, among which is the entire open side and top and bottom heads. These heads are of the slip-on principle. The machines speed up to 100 feet per minute.

A type G2 chain feed straight edge jointing and ripping saw is also displayed and this machine has various commendable features of interest. They show a No. 431 endless bed sander, which is especially adapted to the sanding of narrow and short stock. A type S2 edge sander with tilting table is also displayed and this machine is so arranged that two men can work at it at one time. A very interesting display of cutter heads is shown, these including the square and round types, also special matching heads, jointer and ship lap heads. One has to see this exhibit to appreciate it, and literature covering all the Yates lines can

be had for the asking. Mr. C. H. Dankert and J. T. Donnolly are in charge. By all means stop and meet these men.

The Canadian Milk Products, Ltd., Toronto, are showing the Elyria glass enamelled equipment. This is used in dairies, ice cream plants, and in all chemical plants where a vessel of acid-resisting qualities is desired.

The United Shoe Machine Co. have a full line of machines used in the shoe repair industry. They have a 22 ft. outfit, including stitcher, a 8¼ ft. outfit, and a smaller 6 ft. outfit. All these are self-contained. Other accessories used in this business are shown. F. Webster and Mr. Hanson are in charge.

Brown and Boggs, Ltd., Hamilton, with W. J. Graham in charge, have a very fine exhibit of presses and other tools. They show three inclinable bed presses, from 1½ in. to 4 in. strokes, also another solid bed punch press of 1½ in. stroke. These presses are of the improved model, instead of cast iron as formerly. They have also a foot squaring shear, a steel brake for bending metals up to 10 gauge, and a full line of tinsmiths' groovers, rolls, etc. The presses are all quiet power.

Bond Engineering Works, Toronto.—This concern has a very full line of double sure power transmitting appliances, including shaft hangers, Spiro compression couplings, universal adjustment friction clutches, Grundy flexible insulated couplings, all loose ball bearing countershafts, and loose ball bearing pulleys. The majority of these lines are covered by patent. Pulleys, and in fact all transmission equipment, are included in this exhibit, which is very complete. W. G. Grant, the exhibit manager, is very anxious to meet anyone interested in such lines and issues a warm invitation. Joseph Walton is also at this stall.

The Dodge Co., Toronto.—The Dodge pulleys, both wood and steel, are shown in practically all sizes, and their line of bearing boxes, solid and split pulleys, couplings, friction clutches, friction couplings, etc., are shown. We could not get the name of the exhibit manager, but all are welcome, we feel sure.

Williams Tool Corporation of Canada, Ltd., Brantford, Ont.—This company, formerly the J. H. Hall Co., have a very complete line of pipe threading machinery. They show five machines from the ¼ in. to 2 in. size up to the 2½ in. to 8 in. size. Of course they make larger machines, in fact they have made what is claimed to be the largest pipe machine in Canada. This was of 18 in. capacity, which went to Crane, Ltd. Photographs of the machine are shown, also views of the recent patented die head, which has 16 streams of lubricant flowing on the die during the work.

Copies of their service bulletins are also shown, and the exhibit is certainly well worth a visit. E. L. Williams and L. S. Hall are waiting to welcome you.

The Canadian Fairbanks-Morse Co., Ltd., have a very complete exhibit, and we had better enumerate it bit by bit. There is a 21 in. Gisholt turret lathe, a No. 2 Brown and Sharpe surface grinder, a 14 in. Star screw-cutting lathe, and a 14 in. Walcott lathe of quick change engine design. There is also a No. 1A Brown and Sharpe Universal miller, a Francis Reed 7 in. sensitive drill, and a No. 1 Norton universal tool and cutter.

They have a complete line of valves and steam goods, also SKF ball-bearing hangers, Hyatt roller-bearing hangers, ring oil bearings, some C. F. M. gasoline motors, and Gould pumps, also quite an assortment of Norton wheels and grain. They have some Yale and Towne blocks of various sizes and a complete line of Brown and Sharpe small tools. R. E. Turner, P. F. Cleal, and C. R. Gall are in charge.

In addition to this exhibit they have an outside display of their type Z farm engines, one of these being suspended from a spring to show the small amount of vibration. Taken in all, you cannot afford to miss these two exhibits.

Jones and Shipman Ltd. have a very complete exhibit of their line of J. and S. small tools. These include their tool holders, centre drills, small chucks, vises, etc. They are also showing a 10 in. ball bearing sensitive drill with a gravity feed. This drill is going to be used to demonstrate actual tests on their centre drills. A 14 in. drill is also displayed.

A very cleverly designed machine is that of a 24 in. by 12 in. universal grinding machine. This machine has a specially arranged head that allows both external and internal work to be done. The change is accomplished by swinging the work head 180 degrees. Another feature of importance is the provision made on the driving spindle for added speed if desired for small work. A 5 in. pulley is on the drive spindle but this is so arranged that it can be taken apart, and another pulley half its size put on in its place. Sectional construction allows the following out of this idea, but its advantages are easily understood.

The grinder spindle can run at 60,000 r.p.m. if necessary, being equipped with ball bearings, and they intend running it to the fastest speed possible with the facilities at their command, which will be approximately 50,000 r.p.m. This type of machine is used for all classes of cylindrical, parallel, external, internal and taper grinding. The workhead has eight speeds, and there are many other features which we cannot go into in detail in this small space. An 18 in. x 8 in. universal grinder, a 20 in. and 22 in. power feed drill are also shown. T. E. Hewitt is in charge of this display.

The A. R. Williams Machine Tool Co., Ltd. This concern have a very complete exhibit as follows. The Normack Tool Co. have a Mr. Lovell as their representative, and he is demonstrating a very clever die and templet-making machine. They also show a 12 inch Hendey lathe, a 12 inch shaper made by the Rhode Mfg. Co., Hartford, Conn., a Le Blonde heavy duty miller, a Dalton 6 in. lathe, a No. 19 Le Blonde heavy duty lathe, a No. 2 Cochrane Bly filing machine, a ¾ in. Geometric Tool Co. thread machine, a Barnes 20 in. drill, a Beach combination saw, planer, drill and countersink combined, a Preston ball-bearing shaper and a No. 1 Lapointe broaching machine.

They also show two Williams double-end emery grinders, a Williams 14 in. sensitive drill, a New Way engine, some electric pumping outfits, some marine engines of both two and four cylinder types, a gasoline driven pump, an Evinrude bilge pump, an Evinrude boat motor, a splendid display of Starrett tools, and a full line of steel pulleys, belting, portable benches, stools and jacks. Mr. Cronk, from A. R. Williams, is in charge.

The Twentieth Century grade bars and a patented pick are also shown at this exhibit, Mr. Andrews being in charge of the former, and Mr. Harrington in charge of the latter.

Owing to our not being able to get in touch with Mr. Garlock Jr., we left the exhibit to be spoken of next to the last, for we wished to do it justice.

The Garlock-Walker Machinery Co., Ltd., have spread themselves to provide material of interest. They have various machines in operation, and the best plan would be to go over, item by item, the machines being shown. They have a Millholland turret lathe, a Cataract precision tool-room lathe complete with a unique and compact self-contained electrical drive, a Steptoe 24 in. single pulley drive shaper, a Briggs manufacturing miller, a Grant Manufacturing Co. noiseless riveter, an American No. 25 edging and ripping machine, and an American No. 3 double cut-off saw, which has a very neat and efficient adjustment for spacing of saw to table.

In addition to the above they show an American No. 9 knife grinder for both thick and thin knives, and they have a full line of Cataract precision lathes, grinders, drills, millers, both horizontal and vertical as manufactured by the Hardinge Bros., Inc. These are all fixed up on an attractive stand.

They have also the McClean hole-drilling tool in actual demonstration. This tool drills practically any shaped hole, square, hexagon, etc. You should see it for yourself. A Globe universal tool-room grinder is also shown, and a Racine No. 10 hack saw equipped with swivel vise. Some motor-driven disc and spindle sanders as used in pattern shops are displayed.

This exhibit is very well arranged, and they have various men in charge. A list follows: Wm. Garlock Jr., A. B. Walker, J. Albert Brown, W. F. Parsons, W. A. Marshall and L. McCormack, of Hardinge Bros., Inc. Either of these gentlemen will be pleased to see visitors, and a welcome is assured. Visits to these booths should be made if readers wish to keep in touch with the latest things in machine-tool developments.

In a report of the Department of Overseas Trade it is stated with regard to the iron ore deposits in Brazil that it is doubtful whether any country possesses a greater quantity of deposits of this mineral. Minas Geraes alone is said to contain more than 3,000,000 tons of mineral, and the metal in the ore is stated to average 50 per cent. A great future is also claimed for the petroleum deposits in Brazil. It is claimed that Brazilian petroleum is purely of animal derivation, and contains therefore more calories and a much less quantity of sulphur than other petroleum. The oil "area" has been estimated at 200,000 kiloms., and the deposits are said to be all close to the sea coast.

Some research work on high chromium steel, which has recently been carried out by the United States Bureau of Standards to determine the influence of various kinds of heat treatment, indicates that the maximum hardness is obtained by quenching at about 1066 deg. Cent., but that quenching from 955 deg. Cent. gives the best combination of strength and ductility. Steel quenched at temperatures above 1010 deg. Cent. shows a very low elongation and reduction of area. Such brittleness can be decreased by short-time tempering up to about 427 deg. Cent., while tempering above this heat rapidly decreases the strength and hardness. All quenching was done in oil. The material studied had the composition: C, 0.29 per cent.; Mn, 0.38 per cent.; Si, 0.70 per cent.; Cr, 13.2 per cent.

Yellow brass castings can be colored golden bronze by cleaning and spraying or brushing with a very weak muriatic acid pickle. They are next sprayed or brushed with a solution made by using 4 ounces of sulphide of potash, 2 ounces of thialdine crystals, and 4 ounces of chloride of ammonium per gallon. By heating the castings any shade of color, from golden yellow to a dark brown, can be produced.

Another view of the Exhibition grounds near the main entrance.

DEVELOPMENTS IN SHOP EQUIPMENT

ELECTRIC RIVETING MACHINE

Many cases arise in practice in which it would be desirable to employ power riveting, but in which either the amount of work to be done, or the isolated situation of the work, is such that it becomes very questionable if the installation of an hydraulic or pneumatic plant is justifiable. Most large works have now either one, or both, of these forms of power available, but many small ones have neither, and in the large works it is unusual to find the distributing mains, for either pressure water or air, carried over the whole ground. These and other circumstances make the possibility of riveting by means of electric power a matter of considerable importance in many cases and add interest to the electric riveting machine which we illustrate herewith. There are but few shops nowadays in which a supply of electric power is not available, and the ease with which mains can be run to a new position should, in many cases, make an electric riveter an attractive proposition. Even where alternative forms of power are available, the supply mains cannot, as a rule, be extended with the facility with which a

cable can be run. In exposed situations the cable also, of course, has advantage over an hydraulic main in eliminating the possibility of trouble owing to freezing.

The machine which we illustrate has been patented by Mr. Edward Adamson and is manufactured by the Mada Engineering Co., Ltd., of 12 Bevington Hill, Liverpool. The machine is known as the "Remca." Its general appearance is well shown in Figs. 1 to 3, from which it will be seen that it consists of a bow-frame with the operating mechanism carried on one arm. The machine may be used either vertically or horizontally, and may be suspended, as shown in Figs. 1 and 2, or fixed to a stand as shown in Fig. 3. The operating mechanism consists of a long-stroke solenoid designed so that full advantage is taken of the fact that increasing power is developed as the core approaches the end of its stroke. This gives a type of action particularly suitable for riveting, the final pressure being the maximum pressure. As will be seen from the figures, the whole arrangement builds up into a neat and compact tool, which, for portable work in particular, should be very convenient owing to the

power being carried by quite a flexible connection.

The electrical connections of the riveter are of a simple nature. An earth wire is incorporated in the cable so that the frame of the machine may be earthed. The tool is, of course, operated by alternately closing and opening the switch and, if necessary, more than one operating point may be arranged for by fitting more than one switch. When a rivet is closed the pressure may, of course, be maintained while the rivet is partially cooling, simply by leaving the switch closed. Trials have shown that a ⅝-in. iron rivet may be closed by one of these machines with 16 amperes at 230 volts, the time from closing the switch to opening it being 3 secs. The riveters are made to deal with rivets from ¼ in. to 1¼ in. diam. and with various depths of gap.

NEW BORING MILL

The Storm Mfg. Co., 6th Ave. and 4th St., S., Minneapolis, Minn., have placed a new boring mill, as shown in accompanying illustration, upon the machinery market.

THESE THREE VIEWS GIVE A GOOD IDEA OF THE GENERAL APPEARANCE OF THE RIVETER.

This machine is especially adapted for the boring or reboring of motor cylinders and other parts such as large gears, heavy bushings, tractor wheels, and large holes in castings which are of unusual shape, size, or outside dimensions. Referring to the illustration, it will be seen that the machine consists essentially of the following parts: A main frame or casting of the heavy pedestal type which is provided with flat table on which the cylinders or other work can be mounted; a boring head which is mounted on a vertical boring-bar supported within the base casting; a de-

GENERAL VIEW OF THE BORING MILL.

vice for clamping work to the table, consisting of two upright adjustable bars 40 inches long, a heavy arch, and a handwheel and clamping screw.

The boring-bar is hollow, and is 48 inches long by 2 9-16 inches in diameter. The heat-treatment to which this bar is subjected gives it great strength and stiffness, and it is claimed that it will withstand a thirty-ton load applied at the centre of a twenty-inch length of the bar without springing. Four cutter-heads of the regular Storm multiple-cutter adjustable type are furnished as regular equipment, giving a range for boring holes from 2⅝ to 7⅞ inches in diameter by 20 inches deep. However, special equipment can be furnished for reboring cylinders up to 12 inches in diameter. The cutter-heads are so designed that they can be quickly adjusted by simply turning the adjuster until the desired diameter is obtained. A centreing cone is furnished for each head so that the work can be quickly centred. When centreing a cylinder, it is placed on a table directly above the boring-bar with the head and centreing cone in position. The bar is then raised by means of a handwheel, until the cone fits into the bore of the cylinder and raises it slightly from the table or bars. This action

draws the bar directly over the cutter-head. The work is then clamped in place, and the boring-bar lowered to remove the cone.

A variable boring-bar speed is obtained through a countershaft and three-step cone pulleys, and positive feed is obtained through cut gears and heavy central steel screws. An automatic return and stop is a feature in-

corporated in the design of this machine, which is a means of saving considerable time, as the operator can leave the machine after the operation has been started without danger of spoiling the work. This machine, which occupies a floor space of 30 by 36 inches, is 44 inches high, has a table or upper face of 24 by 30 inches, and a shipping weight of approximately 10,000 pounds.

NEW THINGS IN MACHINE TOOLS

STRAIGHTENING MACHINE

The W. J. Pine Machine Co., of Kenosha, Wis., have placed on the market a machine which they believe to be a new departure in the machine tool line. Every user of cold-rolled steel knows the difficulties they have to contend with in its use. It is full of kinks and twists, and the usual method employed is to hammer the convex side of a kink to drive it in, but this only draws the grain on the long side longer, and if the kink is reduced at all there must be additional kinks of lesser degree. It is impracticable under present methods to use cold-rolled flat or square stock, where one edge or side is to be broken the full length owing to tension of all surfaces which, when broken, causes the stock to bow. Cold-rolled flat stock could be used to advantage for a great number of purposes where it is now impractical, if a practical means for straightening were at hand.

With this machine devices are at hand for doing this work in the only way practical, that is by pressure on the long side of a kink which contracts the long grain and expands the short grain.

The upper bar is a testing table on which stock is laid to find the kinks; it is then placed under the pressure screw, which is convenient for the purpose. When used to take the twist out of stock, one end is clamped on the parallel under yoke vise, and the other end in the twisting head vise.

The machine provides a convenient means for doing innumerable operations that come up in machine shops in general, such as offsetting connection rods and bars, bending irregular shapes in forms or dies, where pressure is used. Four sizes are being built which apply mainly to the length of the testing table.

GAS BURNER

The Berg Burner Company of Brooklyn, N. Y., are now manufacturing a new design of burner for the burning of oxy-hydrogen gas, automatically produced from oil and water. Superheated steam is generated by the burner and combined with the oil flow as it leaves the nozzle. The hydrogen of the steam unites with the hydrogen and the carbon of the oil while the oxygen set free in a superheated state is said to create perfect combustion when the torch is ignited. The

burner will operate in any position and is designed to consume the cheapest distillates or refined oils but not gasoline or similar products.

GRINDING MACHINE

The Bryant Chucking Grinder Company of Springfield, Vt., have recently developed a machine for the grinding of holes and exterior surface. The machines are made in two types, one with a single spindle for the operation on holes and the other with a double spindle for the grinding of both holes and outer surfaces. The work spindles run in adjustable bronze bearings. Two speeds are provided for both work spindle and the traverse slide. A taper attachment provided for the grinding of work up to an angle of 30 degrees. The wheel spindle is mounted on ball bearings of the inclosed type. The machine may be belt or motor driven.

FRICTION CLUTCH

The Link-Belt Company, of Chicago, are now making a new design of "Twy-cone" friction clutch that embodies the following features:—One-point adjustment, perfect balance, and complete inclosure of all moving parts. The friction cones are lined with thermoid, and the clutch may be operated when running at full speed.

INVERTED DRILLING MACHINE

A drilling machine for advantageous drilling of deep holes in cast iron, has been produced by the National Automatic Tool Company of Richmond, Ind. The work is held in a special fixture that is fitted to the under side of the table that feeds downwards. The work is securely locked in position by means of two large handwheels. The drill spindles are amply protected from the falling chips.

POWER BENCH PRESS

The La Salle Machine Works of Chicago, are placing on the market a line of bench press designed for economic rapid and accurate production of such parts manufactured by jewellers, typewriter and adding machine plants, novelty work, etc. The press is provided with a hardened tool steel clutch and an automatic safety device which disengages the clutch at each revolution of the press. The machine is of the open back type and may be used on the bench or fitted with legs if desired.

The MacLean Publishing Company
LIMITED
(ESTABLISHED 1887)

JOHN BAYNE MACLEAN, President. H. T. HUNTER, Vice-President
H. V. TYRRELL, General Manager.

PUBLISHERS OF

CANADIAN MACHINERY
↣ MANUFACTURING NEWS ↢

A weekly journal devoted to the machinery and manufacturing interests.

B. G. NEWTON, Manager.   A. R. KENNEDY, Managing Editor.

Associate Editors:

J. H. MOORE   T. H. FENNER   J. H. RODGERS (Montreal)
Office of Publication: 143-153 University Avenue, Toronto, Ontario.

VOL. XXIV.   TORONTO, SEPTEMBER 2, 1920   NO. 10

PRINCIPAL CONTENTS

Keeping Up High Wages

ONE hears many yarns now about firms that have laid off a large number of their men, and taken them on at a date a short time after at a reduced rate of wages. There have been cases where this has been done, but in many of these there are peculiar circumstances, and the trouble is that the bald statement gets circulated without including any of the extenuating circumstances.

There is no desire on the part of Canadian manufacturers to batter down the rate of wages, irrespective of the cost of living.

One thing that is going to sustain the present wage schedule longer than anything else is a fair and reasonable performance on the part of the men on the payroll.

A high wage is a good investment if it is being earned. The theory of being successful with a lot of poorly paid people was exploded a long time ago.

Employees generally have a lot to do with determining whether the present rates of wages shall continue or be reduced to make operations possible on a plane of reduced value.

The Pessimist Is Here

DID you ever meet the man who knows for certain that the bottom has dropped out of business? He gets you in a corner and pulls a paper out of his pocket, showing where some thousands of men have been laid off by a tire-making plant, where some other concern has laid off the night shift, and he has well pencil-marked all references to money being tight. There is always a danger of his not properly understanding what tight money means, failing entirely to realize that such a turn is a good business tonic for those houses that, if left to themselves, would never get their accounts in proper shape.

As a matter of fact it requires no ability to circulate pessimistic news; it is the easiest thing imaginable to pass it along, and any blatherskite can appear apparently well informed if he has rolled together a half dozen instances of this kind.

Take all the unfavorable circumstances, and add them together. Alongside them put the favorable ones, including the large orders still on the books of many of the industries, the splendid crops, the increasing efficiency of labor in many places, then weigh your results and see what you have.

It is much better, as far as possible, to do your own industrial survey work, rather than depend on others who have not had your point of view, and whose interests may sag while yours would prosper.

The Steel Corporation Prices

THE Steel Corporation continues to adhere to its selling schedule as far as United States orders are concerned. For some time past there has been a spread of at least $20 per ton on an average between Corporation prices and the figures obtained by the premium mills. Some of the independents have secured as high as $90 for sheet iron, while the Corporation price was $42; galvanized sheets have been delivered by the Corporation at 5.70, against 9.50 charged by the independents. In 1914 the average price at the mills for galvanized sheets, No. 28 gauge, was 2.87.

The Steel Corporation, it is figured, in the last year, has delivered to the domestic trade in United States some ten million tons, and although price-tendencies have been toward the sky, the Corporation has consistently stayed with the old schedule of March 21, 1919, as far as domestic trade is concerned.

Figuring, for sake of comparison, that their price was $15 a ton lower than the figures secured by premium operators, this selling of ten million tons at the lower figure has meant $150,000,000 that they could have collected had they joined the procession of those that are out for every last dollar.

Watching the Scrap Heap

IN another section of this week's edition of Canadian Machinery is an article showing in considerable detail the system used by the Canadian Pacific in their Angus shops for the utilizing of all forms of scrap material.

There has been great progress in this work in recent years, and the railroads all over the country have worked out various ideas that have resulted in the saving of hundreds of thousands of dollars.

Scrap material is valuable. The metal itself is not impaired, and it has a much larger value than might be indicated by a glance at prices some firms are willing to pay for this class of material.

The old idea of a scrap heap at the back door, with a periodical visit from some of the dealers with their wagons, is a thing of the past with the big shops. Material is of use only when it is employed, and nothing is worse than having a lot of trash that has outlived its usefulness.

The experience of the C.P.R. at Angus ought to contain ideas that will be valuable to Canadian shop executives in many lines.

PROSPERITY!
—Alley in the Memphis "Commercial Appeal."

MARKET DEVELOPMENTS

May Be Upward Revision in Steel Prices

Added Freight Charges Bring $2 to $3 Per Ton—Has Pig Iron Been Boosted to the Point Where the Trade Will Not Buy It?— New York Machine Tool Market is Dull.

INCREASED freight rates, and the fact that freight charges from points of origin in the steel and iron trade are paid all the way in New York funds, are some of the causes that seem likely to bring in their wake an upward revision in prices of steel and iron stocks, having special reference to the warehousing trade. This will probably amount to between $2 and $3 per ton all around.

There is no let-up in the price tension applied by the premium mills. They know they can get the price for their material if they can make delivery, and prevailing prices show that the ability to make delivery is being turned into cash at a generous rate. There is no limit to the demand for certain lines in the steel trade, such as sheets and boiler tubes. Some carlot sales of the former have been made during the week, the buyer not even asking for a figure, being concerned entirely about having the material landed at his plant.

Machine tool dealers report that in some cases they have an improvement in the inquiries that are coming in.

No large business is in the tool market, but a number of additions to existing plants are being made in the way of new tools. With reports of better returns from labor in the plants, dealers believe that several projects that have been hung up for some time may be gone on with. The New York machine tool market is dull, and rumors from there have it that if this condition lasts much longer some of the makers may be in a position where they will have more thought to revising prices, which are still double and more what they were in pre-war days.

The Canadian steel mills are making better deliveries now in several lines, but they are still having trouble in getting anything like the capacity production of their plants.

The scrap metal trade remains where it has been for some months, stagnant. The big buyers are not in the market, and yards are not particularly keen to take on anything for speculation, unless they can do so "at a price."

LOOK FOR UPWARD REVISION IN MONTREAL TO TAKE UP FREIGHT COSTS

Special to CANADIAN MACHINERY

MONTREAL, Sept. 2.—The upward revision in freight rates will be the basis for readjustment of steel prices and other commodities, and it is expected that quotations on many lines will be advanced proportionately to the increased costs of additional freight charges. The new rates now applying on iron and steel lines from Pittsburgh to Montreal are 55½ cents per hundred on car lots, and 69½ cents per hundred on less than car lots. The rates to other principal cities are given in the selected market quotations.

Steel Prices Firmer

Inadequate railroad facilities continue to influence the volume of business that steel industries are able to engage in. The congested condition that features some of the American producing mills is a controlling factor of the output, and of late it has been impossible to maintain capacity operations owing to the lack of space to store the finished material. It is often the case that a

department of a mill will require to close down until some space is cleared by shipments from the yard. The vital factor is the need of more cars and the ability of the railroads to take care of the shipments, both coming in and going out. For the next few weeks transportation companies will be busy in handling the crops and little relief is looked for in steel shipments until the early fall. There is a decided falling off in the sheet demand and it appears that the automobile industry is preparing for a quieter period. Dealers here were anticipating and preparing for an easier market, but latest reports are to the effect that conditions are firming and prices strengthening. It is believed that the changes in freight rates will mean advances in other directions and this tendency is already showing itself in some lines. Light gauge sheets are hard to obtain and likewise blue annealed stock. Canada plates, dull, are now quoted at $13, and bright at $14, but the latter is virtually impossible to get.

Queen's Head, 28 gage, is quoted at $13.50 and a price of $13.00 prevails for Fleur-de-Lis.

Quiet Business in Ingots

The seasonable quiet spell is increasingly emphasized by the additional decrease in the general demand for ingot metals and finished sheets. The consuming requirements at the present time are below normal and it is quite evident that the call for sheets is less marked than has been the case for many months. Dealers here are inclined to think that automobile factories are curtailing production, and this condition may be still more evident as time passes. The movement of ingot metals has fallen off slightly, but the demand for sheets of all kinds is well in keeping with the available supply, the latter being controlled by the producing mills which are considerably handicapped owing to the difficulty experienced in getting raw materials to the mills or in shipping the product after it is rolled. With the exception of aluminum, which is up 2 cents to 36 cents per pound, all ingot metals are slightly easier. Electro copper is quoted at 24 cents, castings at 23½ cents, tin at 56 cents, spelter at 10¾ cents, lead at 9¼ cents, and antimony

at 9 cents. Copper sheets show an ad-, vance on that previously quoted, these advances ranging from $3 per hundred on 14 ounce plain sheets, to $10 on 16 ounce planished sheets, the current quotation being $49 for plain sheets, 14 ounce 14 x 60, and $56 for 16 ounce planished.

Dullness in Scrap

The lack of activity in the scrap market is largely due to the prevailing conditions on the railways, especially in the States, where the scarcity of cars makes it difficult to get adequate supplies through for the normal operation of the mills and other metal industries. Local business is quiet but sufficient is passing to maintain a market interest. Of the business now passing through the hands of the dealers the volume seems to favor the steel lines. Large individual transactions are seldom recorded. Price changes this week are confined to heavy melting steel, which is up $2 to $20 per ton. Wrought iron axles have advanced $5, the quotation being $30 per ton. It is anticipated that further revisions will become effective when the freight rate is increased.

HE BOUGHT THE WHOLE EXHIBIT

Steel Prices May Be Revised Upward as Far as the Warehouses are Concerned

TORONTO.—Some of the dealers in the machine tool market, while quite willing to admit that there has been a falling off in the volume of business passing in recent weeks, claim that they find a better inquiry now than for some weeks. "I am certain," stated one dealer to Machinery this morning, "that as soon as employers feel they have the backing of labor to a greater extent than they have experienced during the past months, they will be going ahead with extensions. In our business there has been a good demand in the last ten days, and sales are resulting. There are no very large orders, but they are coming from places that are buying because they need the equipment, and in many cases they want to get delivery right away." This same dealer also is of the opinion that right now we are in the reconstruction period about which so much was heard after the war.

The Small Tools

Small tools are being sold to many concerns that are not large buyers. Several of the dealers seem to be paying more attention to the smaller buyer now than they did heretofore, claiming that their business is apt to fluctuate less if they have the accounts of a number of small buyers. Some of the large users of small tools are operating at a very small proportion of their capacity at present, although in some cases there are signs of a betterment, and shops are taking on a few men, whereas they have previously been laying them off.

Reports come from a number of jobbers that they are having more difficulty

---

Pig iron appears to have advanced to the point where buying ceases, unless it is for actual and pressing needs.

Pittsburgh market reports that sheet bars in some cases are selling at $67.50. A month ago the market was firm at $75. This is not reflected by any corresponding reduction in sheets.

Some of the independent mills state that instead of coming down to Corporation prices, they are considering making an increase.

During August most of the steel mills, through improved shipping facilities, were able to reduce their stocks, while the piling carried on in July really added to the stock in hand.

New York reports that although the machine tool market there is very quiet, there is no talk of the makers cutting prices. The continuance of cancellations might have something to do with the attitude.

Tight money is stated to be one of the largest considerations in the New York market.

Some Canadian dealers report it harder to make collections than formerly, and for that reason they are paying special attention to that end of their business.

The scrap metal market shows no tendency to improve. Dealers are afraid to stock beyond commitments unless they can buy "at a price."

Some of the furniture factories report that they are able to secure a much better result from labor now.

Large tonnages of sheets from some of the premium mills are being brought into Canada, one warehouse getting a dozen cars at the start of the week. One day's mail brought in orders for over 100 tons.

Premium Material Arriving

Most of the sheets that are reaching the city now seem to be from the pre-

in keeping their books free of held-over debts than for some time. As soon as money starts to get tight it is felt in the whole business circle. For this reason dealers in general, and they seem to be well-advised in the course, are paying more attention than usual to cleaning up odds and ends that have been hanging fire on their books for some time.

---

mium mills. The fact that Pittsburgh is quoting a slightly lower figure on sheet bars does not mean that there is any reduction in the finished material. One yard brought in about a dozen cars of black sheets and blue mixed at the first of the week. On the first day of the week there arrived orders for over 100 tons of sheets, so there is no difficulty in getting rid of the stuff even at the high price needed to cover premium material.

Sold the Exhibit

Here is an instance that shows the way material is grabbed up in certain corners of the steel market. One Toronto concern wanted to make a showing at Toronto fair this year, so they decided to truck out a good sized pile of sheets and make a real showing. So they took some seven tons and piled them. That was on the opening day of the fair, and the exhibit had not been arranged when a buyer loomed up.

"What's that?" he inquired.

"Sheets," was the answer.

"Sell me that pile when the fair is over?"

"Sure. That's our business."

"And another car lot as well?"

"Yes, another car on quick delivery."

The promise was made of a confirming order the next day, which arrived in due order. Price was never mentioned in the whole conversation, nor in the closing of the deal.

There is a nice tonnage of boiler tubes on the way to some of the Toronto yards, and they are of the much wanted sizes, viz., 3½-inch by 16 feet. One shipment that we heard of had about 1,000 of these in it. Orders have not been accepted in many of the yards for tubes for some time past, and there will be no trouble in placing all the tubes that come into this country for some time.

A Higher Steel Market

That there is likely to be an adjustment of prices in the iron and steel warehouses is the word being passed around in the trade this week. When asked about it, one of the warehouses put the case this way: "We look to see an advance that will average about $2 or $3 per ton. There is a 40 per cent increase in the freight rates on goods coming from the States, and these charges have to be paid all the way in New York funds. It should be remembered also that the great bulk of our business is brought in from U. S. points, where the freight rates have increased. The change has not gone into effect yet, but the chances are that the revision will be made around the first of the month."

Canadian mills are beginning to give much better service in the matter of bar mill material, and a fair tonnage is coming from this source.

Scrap Market is Unchanged

There is no improvement or change for the better in the scrap metal situation. The one chance for business to be resumed seems to be on a lower plane. "People have become accustomed

to high prices for their material and it is hard to get away from that feeling." stated one of the large dealers this week. "In our case the yard is empty and the warehouse is full." This means of course, that the more expensive metals are stocked away from the weather, and such lines as can be thrown in a heap are pretty well cleaned out. There is a lot of scrap to be had, with the possible exception of cast iron and stove plate, the offerings in these continuing to be very small. Some of the dealers put forth the claim that when the reds and yellows are moving, the whites are dead. What buying is going on now is mostly "at a price" where a holder wants to liquidate his stock.

NO THOUGHT YET OF LETTING PRICE DOWN

Business Will Have to be Much Worse Before Any Such Move is Made

Special to CANADIAN MACHINERY.

NEW YORK, Sept. 2.—Despite the general dullness of the markets, there has been more inquiry in the past week in this market than has developed in some weeks. The General Electric Co. is again in the forefront, having issued inquiries for its various plants calling for more than 100 tools. About 80 tools are wanted for its Western Massachusetts plant; 17 hand and automatic screw machines are inquired for for the Bridgeport, Conn., plant, and a list for Schnectady, N.Y., calls for 25 tools. These new inquiries are somewhat surprising to the trade, which had been led to expect that the General Electric Co. would considerably curtail its expenditures for machine tool equipment. For some months the General Electric Co. has been the most active industrial buyer in the East. The new plants it has acquired at Bridgeport, Conn., Baltimore, Md., Rochester, N.Y., and Fort Wayne, Ind., have all needed new equipment, and the orders placed have aggregated many hundreds of thousands of dollars.

Another indication that the railroads are expected to buy equipment of various kinds is given by the inquiry of the Safety Car Heating & Lighting Co., Jersey City, N.J., for about a dozen tools. This inquiry follows one for a similar number issued by the Federal Signal Co., Albany, N.Y., which manufactures railroad signal systems. There has been some railroad buying at Chicago during the past week or two. In the East little has been heard from the railroads, but the Baltimore & Ohio is preparing a large list, which the trade expects will be issued within a week or two.

On the whole, the machine tool business is not good. Sales during August have been fairly satisfactory in some instances, but some sellers report the poorest business in years. Prices are firm, however, and unless the volume of cancellations should grow considerably, most of the machine tool manufacturers

will have nothing serious to worry about for at least two or three months and in some cases longer. Railroad transportation difficulties are being ironed out, and the worst feature of the present situation is tight money.

PIG IRON MAY HAVE GONE TO A POINT WHERE BUYING WILL QUIT

Special to CANADIAN MACHINERY

PITTSBURGH, Sept. 2. — Pig iron seems to have been advanced to a standstill, as the market is now very quiet, with almost no inquiry at all, and buyers very generally are saying positively that they will buy no iron except what they absolutely must have for immediate requirements. The furnaces, however, do not seem to have expected any large buying movement and thus, perhaps, are not greatly disappointed. The offerings have been light and some furnace interests state that in the recent movement they scaled down the inquiries in many cases, offering only half the tonnage inquired for. Since last report prices for Bessemer and basic grades have been established. As noted a week ago, last sales had been at $47 for Bessemer, and $46.50 for basic, f.o.b. valley furnaces. Sales of both grades have since been made at $48.50, and although the lots sold were small the transactions are regarded as establishing new market prices. Foundry remains quotable at $50, to which it jumped recently from $46.

The furnaces may be entirely right in their contention that pig iron is scarce and that production is likely to fall short of requirements, but the factor of buyers having no confidence in the market will certainly count for. something, as that will move consumers to draw upon stocks and thus reduce the requirements. There is also the point that with many foundries activity has fallen off, new orders being scarce, and some of the high priced iron bought of late has been simply averaged with much lower priced iron previously bought, so that the cost of producing castings has been much less than it would be if nothing were used but iron bought at present prices.

Coke Steady

Connellsville coke has shown a remarkable indisposition to recede from its extremely fancy prices. There have been three adverse influences, a slight decline in coal prices, an improved car supply, resulting in Connellsville coke production increasing, and a greater disinclination on the part of furnaces to buy. For several weeks Connellsville furnace coke for spot shipment held at $18 to $18.50, while a week ago we reported the market at $17.50. In the past few days the market has been stronger, there being sales at $18 as well as at $17.50. One furnaceman reports buying coke at both prices in the same morning.

As to foundry coke, the quotable market remains at $19 to $19.50 for spot. The range that has obtained as the regular market for several weeks. There has been some change in quality, but that is all. At $19.50 one can get a fairly

good coke now, while three or four weeks ago $19.50 coke was rather indifferent, and $19 coke was difficult to distinguish from ordinary furnace grade.

Semi-finished Steel

Several sales of sheet bars have just been made at $67.50, Pittsburgh, though the total tonnage bought did not amount to much. Previously the market was about $70, some sales having been made at that, and others at a shade below. A month ago the market was $75. The remarkable thing is not that sheet bars have declined, but that they have declined so little, for the circumstances are such as to show that the mills are yielding in prices very grudgingly. The majority of the producers are not willing, but anxious to sell sheet bars, whereas recently it was a great favor to sell sheet bars at $80 or $90. It should be noted also that billets have been sold freely, at least in the east, on the basis of $60, Pittsburgh, and before the war billets and sheet bars ran at approximately the same price. There is no material difference in the cost of manufacture, and for producers anxious to sell sheet bars to hold out at $67.50 differential shows that they are in no very yielding mood. There is hardly any demand for sheet bars, as the sheet mills are fairly well covered, and consumers of sheets who would like to make conversion contracts, buying sheet bars and having them converted, cannot at this time find any mill willing to make the conversion, so there is no use buying the bars. Slabs are said to be held by mills at $65, but that is purely nominal as mills could not begin to pay the price. A few weeks ago a plate mill was in the market with a bid of $50 for slabs, but did not get any.

Finished Steel Holds Up

In face of a relatively light demand the independent mills hold up their prices, showing no disposition to recede to the level of the Steel Corporation prices, and many of them even tell customers that the prospects are of advancing prices. Few, if any, customers are convinced, there being too many indications that prices of commodities in general are in a receding mood, and that demand for steel is not going to be heavy until fundamental conditions improve. The mills evidently intend to squeeze the profit out of the present situation, one of scarcity, as they can, and to reduce prices would be to jeopardize the business now in books. There are cancellations as it is, and price declines would of course provoke more. As illustrating the temper of mills, recently a large independent in-

terest sold several lots of plates, aggregating no large tonnage, at 3.25c, and has since put its price up to 3.50c, probably to "protect" the 3.25c business, though it is evident that 3.50c cannot be secured. In sheets some very fancy prices are going, as there is scarcely anything offered. The leading interest sold its second tonnage months ago and on account of restricted operations it will require until some time in February to complete the business taken. The independents, as a rule, sold only for third quarter, but a recent count is said to show that the independents, on an average, are sold not only through November, but 41 per cent. into December, and thus it is exceptional for an independent to offer any sheets. Demand is light in point of tonnage, but is easily sufficient to support what market there is. Despite cancellations the mills are in strong position. Thus, while the automobile industry has cancelled a good bit of sheet business, one finds that the mills that for years have made a specialty of automobile finishes are very busy. The explanation appears to be that the automobile trade had gone to other mills to eke out supplies, taking almost anything down to common black, and when the trade found it needed less it simply cancelled the outside tonnage, so that at the reduced operation now promised for the automobile trade, say 60 to 70 per cent. of capacity, it will still keep its regular sources of sheet supply busy.

Transportation

Shipping conditions at the steel mills continue to improve. There is much more tonnage being moved than a month or two ago, but the distribution is still poor in that there are many embargoes. A mill may be shipping its output very comfortably, but at the same time some customers may be getting full supplies and others practically nothing. The mills as a whole reduced their stocks of steel slightly in August, while during July they were adding to stocks. In many cases mills are shipping considerable tonnages out of stock, but at the same time are adding nearly as much to stock, out of current production, the idea being to keep the steel relatively fresh.

PIG IRON TRADE

Market activity has dropped off to slight proportions during the past week, and in the Chicago district it was noticeable that the majority of sales were in few cases exceeding 500 tons. The market may be definitely fixed at $48.50 Valley. On the whole the market has been quieter than for some weeks past. Furnaces appear to be well sold up but there is a marked cautiousness amongst consumers owing to forecasts on the coke situation.

U.S. SCRAP METAL

The scrap market has shown decided strength during the past week and the recent heavy buying of the Carnegie Steel Company has left the market steady and heavy melting steel can be quoted firm at $29—$29.50. The rush by all interests to make delivery before the new freight rates go into force, coupled with the car shortage, threw the market into a rather chaotic condition.

MONTREAL NOTES

It is expected that the Montreal fire department will close the contracts early next month for the necessary additional equipment for fire-fighting purposes. The expenditure will be in the neighborhood of $150,000 and will include two gasoline-operated fire engines to replace steam units now in service.

The city commissioners of Montreal have decided that a new incinerator plant will be erected to replace the one recently destroyed by fire. It is not expected that construction operations will be started until next spring, but plans and specifications will be prepared shortly. The cost of the new plant will be approximately $200,000.

* * *

Promising developments are now under way for the erection of a modern match factory in this district. Sir Alexander Maguire and his brother, of the British firm of Maguire, Paterson and Palmer, were in the city last week in connection with the preliminary negotiations. It was stated that machinery was in readiness for shipment from England and that building operations will be proceeded with at once so that manufacturing could be commenced early next year.

* * *

Reuben Morris, of 889 Durocher St., Montreal, is the inventor of a new safety auto-bumper that can be operated as a fender for the prevention of serious accidents to pedestrians. A recent demonstration before city officials and prominent citizens proved the practicability of the device, which is the product of twelve years of experimental work. The appliance weighs about 35 pounds, is controlled by the steering gear, and may be locked and used as an ordinary bumper.

* * *

The Foster Motor Car and Manufacturing Co., Ltd., has recently been organized to manufacture the Foster motor car, a European type of car, of which over 10,000 are now in use in England. These machines will be fitted with the famous Herschell-Spillman motor. The company has acquired property in the east end of Montreal, and the head offices will be located at the corner of First Avenue and Ernest Street. Capt M. L. Fitgerald will be in charge of the Canadian organization, and the majority of the office and mechanical staff will be former members of the C. E. F.

SHOPS ARE BUSY IN UNITED KINGDOM

Iron and Steel Industries are Fully Occupied With Orders to Capacity of Raw Material

Cable advices have been received by F. W. Field, the British Government Trade Commissioner in Toronto, from the Department of Overseas Trade, London, England, that the United Kingdom iron and steel industries are fully occupied with home orders to the capacity of raw materials available. Manufacturers of tinplates, galvanized sheets, hardware, cutlery and pottery are in a little better position and can deal with an increased export business. Among the industries in the United Kingdom which can now handle larger business are machine tools, motor cars, except the higher grades, electric batteries and accumulators, drapery, clothing, hosiery and hats, and musical instruments. Indications in those industries point to a larger export trade in the near future. The same remarks apply to glass and glass ware.

United Kingdom manufacturers of machinery and engineering products are in a position to accept export orders.

MORE SHIPS TO TOUCH THE ORIENT

Negotiating With an Old Country Firm To Secure Passenger Vessels

Vancouver.—Speaking here at a joint luncheon of the Canadian Club and the Board of Trade, Hon. C. C. Ballantyne, Minister of Marine and Fisheries, referring to transportation and the growth of the Canadian mercantile marine, declared that Canada's ships were built at from $25 to $100 a ton lower than the American, and Canada's ships were much better built ships. Canada has 17 shipyards, he stated, and the Government received the lowest prices of all for shipbuilding from the city of Vancouver.

Sixteen Government ships had been built or were being built here, at an expense of $22,000,000, and all were to be left on the Pacific coast to care for the growing export trade, the Minister stated, and he predicted that in a very short time ships now connected with the Government railways would be plying to the Orient from Vancouver.

"While the Government has not sufficient money to build the passenger ships for this, negotiations are now under way with a large Old Country firm, and when completed the ships should be here not later than in a year's time, and perhaps sooner, the Minister said.

Toronto.—Two sections of the new floating dry dock which is being built in Montreal for the Russell Shipbuilding Co., of Toronto, have arrived here. The dock will consist of six sections and will cost about $80,000.

CANADIAN MACHINERY

AND
MANUFACTURING NEWS

Vol. XXIV. No. 11

September 9, 1920

Making 30,000 Nuts in 9 Hours on One Machine

Nut Tapping, Cold Nut Making, Bolt Heading, and Bolt Thread Rolling Are Discussed—Interesting Facts Are Also Given on How an Up-to-Date Bolt and Nut Department is Conducted

By J. H. MOORE

THE work entailed in conducting a large nut and bolt department is more than appears on the surface, and we feel that the description, dealing with such a department at the plant of the International Harvester Co., Ltd., Hamilton, Canada, will be of particular interest to our readers.

When one considers the numerous bolts, nuts, etc., used on the various styles of harvesting implements made by this concern, they can readily see that not only will the number be great, but the varieties many. We need not go into the clerical work, although this is a large item in itself, but will confine ourselves strictly to the routine and mechanical features involved. Every conceivable size, length and style of bolts and nuts are used. In some cases the threads are cut by means of the regular bolt cutter, while for other purposes they are rolled.

Before going into the actual thread rolling, let us dwell on other features of interest. Certain bolts are made by the cold heading process, others by the hot method. This latter course means that suitable furnaces are installed, and with all the foregoing, plus nut tappers in batteries, etc., etc., the department presents a very busy appearance indeed.

The storing facilities are of special interest, large bins arranged, tier upon tier, and plainly labeled as to their contents. All bolts and nuts upon completion reach this storing and shipping section, and are weighed on a ratio pan or counting scale.

Some of our readers may be familiar with this type of scale, but for the benefit of those who are not, we might explain that apart from the ratio pan, it presents a very similar appearance to any standard portable, or floor scale.

A special counting beam is installed, and it is this feature that allows the accurate and quick method of obtaining the total number of the product. For example, suppose you have a truck filled with bolts and you wish to know

FIG. 1—ILLUSTRATING THE MAKING OF COLD PRESSED NUTS.   FIG. 2—THIS SHOWS THE BOLT HEADER IN ACTION.

FIG. 3—NOTE THE DETAILS OF THIS THREAD ROLLING MACHINE.

the number contained in the truck. To secure the count you first set out the tare weight of the truck on one of the verifying beams, place the number of pieces in the ratio pan as required by the particular size of the weighing bar in use. This number may be two, ten, and so on depending upon the size of the scale, and how it is arranged.

Having placed the pieces in the ratio pan, you move it along the bar until the beam is in balance. Upon reaching this position the indicator above the pan will rest upon the figure corresponding with the exact number of pieces contained in the truck. This scheme does away with counting, thus overcoming the possibility of error. When weight is required in addition to the count, it is obtained in the same manner as on any other scale.

All materials are weighed in, and weighed out at this department, and entries made in the ledger accordingly. Absolute tab is kept on all stock, and every deduction from the same is entered in the ledger in red ink, thus clearly separating the entries from the parts deducted. By this method they can tell at a glance the stock on hand.

Nut Making Machines

Leaving the routine, and entering upon the actual manufacturing itself, let us refer to Fig. 1 which illustrates a machine busily engaged in turning out cold pressed nuts. Making nuts by the hot process method is also adopted at this plant, but the cold pressed machine is by far the most interesting.

The first thing that impresses one on viewing a cold nut presser is its solidity. It is a heavy massive affair and built for strength, as it might well be, owing to the work it must perform. Briefly, the procedure is as follows on the type of machine shown: The stock is fed in from the side by means of a series of rolls. These not only guide, but straighten the stock. From this point the stock is carried in by a grip feed which holds it securely by the aid of springs. This is previous to shearing off.

It next goes between two guides, the centre hole is punched, the nut is sheared, crowned, trimmed and pushed out, all these movements being automatic.

The completed nut, as far as the machine is concerned, drops out while the slugs are separated and drop out at another part of machine. The dies used for this class of work are made from the very best of steel, and we saw one set of trimming dies which had trimmed 330,000 nuts without regrinding and was still going strong. The punches are provided with pilots to prevent any spring, undue strain, etc.

The die box on one of these machines is especially interesting, it having all facilities for clamping the dies tightly, yet so arranged that adjustment can be made at a moment's notice. Cams play an important part in the operation of these machines as the various slides receive their action by such means. Some of these cams can be noted in the photograph. Note the box of nuts to the right of picture. This is where they are ejected from the machine, they come out at a great rate of speed. An average day's run is approximately 30,000.

The next illustration Fig. 2 depicts a bolt header in action, and in this case also, the stock is used in its cold condition. Once more weight plays an important part in obtaining rigidity, and all ways, slides, etc., are made as long and as wide as possible. In this example the ram of the machine has a double motion, the first gathering the

FIG. 4—THERE IS NO WASTE MOVEMENTS ON A MACHINE OF THIS NATURE.

stock, and the second finishing the head of bolt. As can be noted the stock is placed on reels and enters the machine through the two guide wheels, clearly noted in the picture. These rolls acts as straighteners as well as guides.

After entering the machine a predetermined distance, the blank is sheared off, the gathering die prepares the stock for head, and on the second motion of the ram the bolt is completed, brought to a chute, and ejected. Watching a bolt being headed on this type of machine is very interesting, the work being accomplished very rapidly. The use of cams is again resorted to in this class of machine.

Thread Rolling

Readers of course are well aware of the regular method of threading by means of revolving dies, but are they familiar with the thread rolling process? Believing there are many who are not acquainted with this operation we show Fig. 3 as a good example of such work being performed.

The bolts after heading (if thread is to be rolled) are taken to these machines and placed in the hopper A. In this hopper is a revolving dog or bolt catcher which not only leads the bolts into the chute B, but will allow them to enter in the correct manner. Down goes the bolts until a feed plunger engages with them one at a time, carries them through to the dies, rolls the thread, and ejects them.

A description of both the machine and the dies employed will be worth while. Rigidity is again an absolute necessity, and the parts are proportioned accordingly. Two dies are used, one being stationary, the other moving in a horizontal direction, and with a reciprocating motion. Depending upon the pitch of thread to be cut, the dies are made accordingly. Imagine one die 7 inches long, the other 6 inches, each about 1½ inches wide by 1 inch thick.

On the die is cut a thread of proper pitch, depth, etc., and this concern cuts their dies on both sides, so that when one side is worn, they merely reverse it and use the other. The writer saw one set of dies, used on one side only, which had rolled the thread on 800,000 bolts and was still going strong. These figures speak for the class of steel used in these dies.

The bolts, or blanks as they are called, are introduced at the proper moment by a slide, or starting tool as it is termed. This portion is so arranged that a bolt must enter the die straight. As soon as the work has passed through the dies, it drops from the machine into a waiting box. The dies are carefully adjusted to produce a perfectly formed thread, and the design is such that adjustment can be made very easily. To speed up the return of the moving die a quick return motion is provided.

Nut Tappers

Leaving the bolts and referring once more to the nuts we see at Fig. 5 a multi-spindle nut tapper busily engaged at work. For those not familiar with such machines we append the following:

The action is secured by means of cams in the rear of the machine, and as can be noted from the illustration, the tool shown is equipped with eight spindles. This means that at the rear, and near the top of each spindle, is a set of cams so located that each spindle completes its downward motion at a little different time from its neighbour. The one operator attends to the eight spindles, and in fact attends to two machines in some cases. The nuts being tapped in the photograph are for ⅝-inch bolts, and 10,000 nuts are completed by the operator every nine hours. A constant stream of lubricant flows on the taps at all times, and as can be noted these are of long stem design to allow numerous nuts being tapped before removal of the tap. Quick action holders are used in gripping the taps, so that it is the work of a moment to remove or insert the same.

The remaining illustration Fig. 5 shows another style of nut tapper, the work in this case being performed automatically. The nuts are placed in the hopper A, and by means of a revolving guide inside this hopper they find their way to the four chutes as shown.

On arriving at the end of these chutes they are directly above four automatically operated tapping spindles. To further illustrate this feature we have opened the door B to show these four spindles. On the nuts coming above the spindles the taps rise and enter them, complete their work, then an ejector pushes the nuts out of machine into a convenient box. By using this automatic feature a small battery of these machines can turn out 50,000 nuts per nine hour day, one operator attending to all machines.

Although we have by no means gone into details regarding this nut and bolt department we have at least given readers a fair idea of the methods involved, and the magnitude of the task of providing a large manufacturing plant with all necessary bolts, nuts, washers, etc.

FIG. 5.—THE OPERATOR OF SUCH NUT MACHINES AS ILLUSTRATED ATTENDS TO A BATTERY OF MACHINES.

Two Practical Ideas of Decided Interest

A Jig for Drilling of Boxes of Wheels Previous to Operation of Reaming is Described, Also a Discussion on the Application of the Inverted Tool for Use on Lathes.

TWO very interesting views of machine shop practice appeared in a recent issue of Mechanical World, and believing our readers will appreciate a reading of these experiences, we reprint them in their entirety. The first of the schemes is shown at Fig. 1, 2 and 3, and represents a drill jig for taper-pin holes. The author quotes as follows:

When drilling bosses of wheels, etc., previous to the operation of reaming, for the reception of taper pins, it is well to keep in mind several points. One method of procedure is as follows: First, drill through the boss — that is, at right angles to the axis of the bore—with a drill slightly smaller than the small end of the taper pin. With a drill slightly less in diameter than the large end of the taper pin open out the hole on one side of the boss—that is, by just drilling until the drill enters the bore. The shaft which is to go through the boss is pierced with a drill, the diameter of which is midway between the large and small ones with which the boss has been drilled.

On the two parts, the shaft and the component to be mounted thereon, reaching the assembler, he fits them together, and with the drilled holes in alignment proceeds to reamer out until the taper pin, which is to secure the two parts together, is a good drive fit. The taper pin should not fit too tightly, or the boss may be fractured when the pin is driven home, especially if the surrounding metal is thin.

Our intention is to consider a method of accomplishing the double drilling of the boss in an expeditious manner, and leaving the shaft to be dealt with by a separate jig, which could advantageously be one of the universal type employed extensively for various shaft work.

We will take, for a concrete example, a piece of work as illustrated in Fig. 1, which is a light grey-iron casting. The boss diameter is 1⅜ in., and has a 11-16 in. stud-hole bored through, the outer ends also being faced. Upon looking up a table of taper pins, the correct size for such a shaft and boss diameter we find to be: Taper pin, 0.219 in. large end; ¼ in. taper per foot. Use Nos. 13 (0.185 in.) and 4 (0.209 in.) drills.

In considering the class of jig most suitable for dealing with this job one or two factors had to be considered, as several methods are available, one being that previously described, first -drilling the small hole right through and afterwards opening it out with the larger drill. This, to attain accurate results, would necessitate slip-bushings, and as the boss through which the taper pin was to pass was so short it would be difficult to provide these to be of sensible size on account of there not being sufficient room, as will be seen on referring to the drawing. Again, opening out holes with twist drills, when only a small amount of metal is to be removed, imposes heavy strains upon the cutting lips, frequently causing breakages, especially if the feed is not absolutely rigid, as twist drills possess a decided tendency in the case under consideration to feed themselves rapidly forward on account of their helical construction forming through the hole a kind of thread, so reaching a point where jamming takes place, with the result that the drill is broken. Therefore, the method used was to drill either hole first, index round, 180 deg. and drill the other hole, then the work-holding cup containing the guide holes and the index holes within itself, as will be described later.

Fig. 2 shows the jig complete, the ribbed angle-plate A being an iron casting bored to receive the main spindle and spring plunger. The work B is shown mounted in readiness for drilling.

Fig. 3 is a section through the head of the jig, showing its detail construc-

tion. The ribbed angle-plate A is provided with suitable projecting bosses, bored and reamed to make the spindle C a good fit, not light, but so that it may be readily revolved by hand. On this spindle, by means of a taper pin, is attached a cup D, made from tool steel hardened throughout.

This cup serves quite a number of purposes. The work B is clamped to this, and it also contains the drill guide holes and serves as the index plate. At the rear end of the spindle a collar E is fixed which keeps the end-play out of the spindle, the boss of the cup running up against the casting on the front side, as seen in the illustration. The spring-plunger F, controlled by the knurled knob, engages the holes in the back of the cup, these being spaced 180 deg. apart and in line with the drill guide holes.

Clamping of Work

The clamping of the work is accomplished through the medium of a quick-acting device now to be described. The end of the spindle C is threaded, and upon this is screwed a circular piece G, slightly less in outside diameter than the bore of the work, to facilitate the removal of the latter over the former. This piece G contains, at its outer end, a slot into which the swing handle H is fitted. J is an ordinary U or slip washer. To load the work on the jig the slip washer, which should be chained to the jig to prevent its loss, is removed and the work passed over the swing-handle, up to its locating face in the cup D. The slip-washer is inserted and the swing-handle brought to the position shown in full lines and given a turn to clamp the work. After it is drilled the handle is turned back to release the work and the slip-washer removed.

By pulling off the work the swing-handle will be knocked into its horizon-

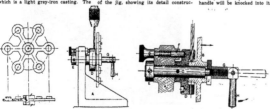

DETAILS OF THE PIECE AND OF THE JIG USED.

tal position, thus allowing the work to pass over it. It will be found to be very rapid in use, apart from the impracticability of using an ordinary nut on the spindle-end on account of the small bore of the work. The need for providing a loose spanner is also avoided.

The jig should be used on a two-spindle drilling machine, and guide strips should be bolted to the table between which the jig should be arranged to slide, the operations then being as follows: (1) Load and clamp work in jig; (2) drill either hole—say, small one first, for example; (3) slide jig along, index table to under next spindle; (4) pull out plunger, swing work through 180 deg. (when plunger will enter the opposite hole) and drill the other hole; (5) unload jig. Finish.

Load up again and with jig still under the same spindle, drill the large hole first (as this was the last to be drilled in the other piece), slide jig along, index round and drill small hole, etc.

It has been found that by a little care and forethought many parts can be designed to have the same boss diameter, etc., and shaft, and thus the jig may be used for quite a number of similar components. Also, by a little modification, the jig could be made to have interchangeable, work-holding cups, etc., used for various jobs.

An Inverted Tool

The second idea deals with the inverted tool position in turning operations, and here is what the author has to say:

The application of tools in lathes, semi-automatic and full-automatic machines, occurs either in the normal position, with the top face of the tool standing upwards, or in the reverse attitude. The latter setting becomes necessary for several reasons, the chief being the necessity for multiple cutting operations. Sufficient tooling capacity cannot always be obtained by the employment of what for convenience sake may be termed front-cutting tools. This applies not merely to cross-slides or double slide-rests, but also to certain turret tools, to boring, recessing, or threading tools, and to some special forms of swing arms utilised in the semi-automatic turning machines. The effects upon the working conditions are several in number, the principal being differences produced in the mode of holding the tools, questions of chatter or vibration, the access of lubricant, and the escape of cuttings. Some classes of operations are better accomplished with inverted tools, quite apart from any points of convenience in planning the tool layout. It is also easier in many instances to maintain the tools in efficient condition under difficult circumstances by the adoption of the inverted method for some of the processes.

By curbing one's natural wish to adopt bold methods, which appear at first to be the most economical and productive, a really greater output may be secured. Thus, instead of going the whole way when forming the contour of a certain shape, and putting the profile along the edge of a single tool, the operation might be divided up with advantage, a portion of the shape being tooled with a separate tool on the opposite side of the work.

An example of this may be specified in a contour which has one or more specially deep grooves; these are obviously more destructive and wearing to the tool edge than a normal depth of cut, and incorporating the complete shape in one tool is likely to give rise to trouble in keeping the cutting edges in good condition. Using separate tools as the rear for the grooving, these can perhaps be made of stronger shape to resist the special stresses, or at any rate be ground more often than is necessary with the main tool. A more drastic subdivision is sometimes beneficial, as when a row of ribs has to be turned, with interspersed valleys between.

A form tool made solidly is then frequently dispensed with, the valleys machined with a set of separate points — say, at the front,—and the ribs with another set lying at the rear. This mode of procedure offers an advantage in that lateral feed can be imparted to one set of tools without affecting the others, the consequent relief experienced from such side movement assists the cutting process, and is less distressing to the work when it is thin and liable to distortion.

Two other reasons for division of tasks may be instanced. One is that cuts proceeding at right angles (such as turning the diameter of a pulley, and facing the sides of each step) are readily done simultaneously in suitable machines, there being no interference of slides or tools. The other is that double tapers, converging from the ends to the middle, may be carried on simultaneously, the respective tools starting at their ends and travelling towards the middle of the piece, being controlled by their separate taper attachments. This general principle of multiple tool operation, which has been vastly extended by the several designs of automatic lathes, realizes the highest present possibilities in output, and differs from ordinary turret lathe practice in bringing, as a rule, all the tools into play simultaneously,

instead of batches of them successively. When longitudinal space allows, two or three articles can be chucked at one setting and finished with respective gangs of tools at front and back, a great increase of output per machine and tooling period.

A lesser advantage of subdividing operations lies in the disposal of chips, which may accumulate to an awkward degree and impede the working if all cutting is done from one side; whereas, a proportion of the tools being inverted, the chip problem is simplified, especially as the cuttings fall from the underside of inverted tools, instead of piling up on them.

In some types of automatics the inverted tools occur in multiple, there being a swing arm in addition to the ordinary cross-slide moved across in a linear direction. The swing arm comes down over the back toolpost at the appropriate time and cuts into the work with a forming, or a cut-off, or a chamfering tool. A special application of the path of movement of a swing arm is that utilised for successive roughing and finishing by superimposed tools. This interesting principle (from an example of tooling in the Fay lathe) is shown in end view at A, the operation done with the rear tools being the roughing and finishing of a pair of ball-bearing inner races, tooling the concave surfaces, and facing down one shoulder. The lower and roughing tool F is set to encounter the work first, and after an interval of travelling, when it has reached nearly to the finished depth, the upper finishing tool G comes into operation and smooths out the race, reaching the position indicated in the sketch.

The question of chatter, induced by looseness or vibration of slide parts, is affected by the inverted position of a tool under certain circumstances. The presumable opinion that a tool should work better when the pressure is in such a direction that it forces the slide faces down tightly together does not seem correct in regard to forming tools, particularly the broad ones. The principle which is found to work well with broad turning tools for finishing in the ordin-

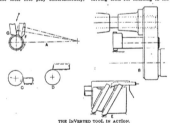

THE INVERTED TOOL IN ACTION.

ary lathe—i.e., that of the goose-neck or spring shape—is naturally good also for broad forming tools fed in radially in cross-slides.

The inverted position consequently furnishes this desirable action from the slight slackness of the slides and the upward spring of the holder. Instead of digging in, therefore, the forming tool is allowed a small degree of latitude to spring the edge away, and this movement, combined with faint give of the work or the spindle, results in a sweet cutting effect and a nice finish. It is not necessarily the placing of the tool at the back of the work which ensures this smooth cutting; the same would happen if the tool were inverted at the front and the spindle run backwards: in fact, this is frequently done when for some reason of convenience the form tools are mounted at the front of a cross-slide. If there should happen to be a roughing operation, the tool or tools for this are located at the outer side of the work, the completion of the shape and the smooth finishing being reserved for the inverted tool.

A Special Case

A special case which may be noted, where a smooth and springing action occurs, although the tool is not inverted and the work runs towards its top, is that of the combined cut-off and form. ing rests which are bolted to turret faces. These fittings carry tools in posts opposite to one another, and either is racked towards the centre by a long handle. Owing to the elasticity present in the body and derived from the turret and its slides, it is a matter of indifference whether a forming tool is used inverted or otherwise.

A class of forming holder—the vertical slide design—used largely in American practice has a bearing upon the question of finish. When a form tool of the radial type is fed in it has to stop at the required diameter, and the finish then existing on the work depends on the sweet cutting of the tool. But in regard to the vertically fed tool, which cuts tangentially, this can be fed past the work again after the first pass; at this second smoothing-out the elasticity of the parts asserts itself and sends the cutting edge a shade farther in towards the central axis of the work, thus proving a fine finishing cut without disturbing the adjustment.

On the automatic turning machines a commonly used kind of tool — the one which projects from a knee casting se. cured on the turret—comes under the classification of inverted, since the front face does not lie in a horizontal plane. It may be termed a semi-inverted tool, but is subject to rather peculiar in. fluences, from its unsupported nature, by comparison with the standard front and rear holders of the cross-slide. Ex. tending out in the air, as it were, and depending for its accuracy of movement upon the stiffness of the holder and of the turret elements, there is a risk of deviations from the proper path of movement, especially as the cutting edge is influenced by inequalities on the work surface. This disability in contrast to

the solid support of a traversing slide on the bed renders it imperative to provide suitable means of steadying, unless the cut is so light or the surface in such good condition that no trouble with springing may be anticipated. In the ordinary types of turret lathes, as well as in the automatic turning machines, a usual practice is to provide a central pilot bar, sliding through a bush in the chuck or the spindle, and often holding one or two boring cutters to serve a double function.

Such a central bar cannot always be employed—where the work has no opening, or one of too small diameter to be of any value. Then the overhead support is brought into use, this being really a more effective device than the central bar; the latter does not take care of the backward or the torsional spring of the holder away from the centre, whereas the overhead support B resists these deflections at the best possible place—that is, beyond the radius of action of the tool,—thereby confining the spring of the same between two binding points — the overhead support, and the stem of the holder which is gripped in the turret. The amount of such spring may be only but slight, given a stiff holder, but it is still to be reckoned with, and when tooling a rough casting or forging may interfere with the accuracy of travel of the tool, or at any rate produce chatter marks on the surface.

Inverted tools, as fitted to some classes of box tools, called the "over-shot" type, are subject to one fault in working that does not apply to the common horizontally set cutter C. A little downward wear of the bed exercises practically no influence upon the diameter turned, as the saddle reaches the lowered part. The effect upon an overshot tool is more marked, as revealed by a comparison of the dotted lines in C and D, the edge in the second instance moving in a direct radial line towards the work centre, and consequently modifying the cut diameter more quickly.

As regards the relative effectiveness of lubricant with ordinary and with inverted tools, there is usually a balance in favor of the latter, whether they are set at the back or at the top of the curve. The stream can be directed full upon the scene of cutting, it floods perfectly over the tool, and as the chips instantly fall away from the spot, there is no encumbrance to the coolant immediately and constantly reaching the surfaces of work and tool. The only difficulty that may be encountered with overhead tools arises when they are of a complex nature, and the box holding them occupies such an amount of space that the stream cannot gain access to the cutting edge with certainty because the shape of the box and its clamps and screws is such that the lubricant is deflected and largely wasted away from the location where it is wanted. Here some special provision may be desirable, such as illustrated at E for example, comprising pipe supply and grooves. Chasing tools are very often used in inverted fashion in turret lathe prac-

tice, the object being to run the chaser away from the headstock. In producing external threads this prevents the tool from running into the shoulder of the work through accidental over-running, and enables a full thread to be cut up as far as possible without risk of danger. As regards internal threads, the inversion avoids any chance of the nose of the chaser from striking against any interior projections—those on the work, or objects such as chuck jaws, collet ends, or pilot bushes. It is, of course, unnecessary to invert a tool for left-hand threads, the spindle being run as usual—forwards,—which causes the chaser to be traversed from left to right.

STERILISATION OF SOLUTIONS

Many ideas have been tried out relative to the sterilisation of cutting solutions, but the most successful one for the average industry, according to "Safety Engineering," is to have a set of drains running from the machine to a large pit consisting of two large tanks. The solution is run into one tank, which collects a great deal of the sediment, and then over to the other tank to be used again. Each week, and sometimes twice a week, the solution is taken out and the pit cleansed thoroughly. The solution is then returned to the pit, if it is not broken down, and is heated up to 148 deg. by steam pipes running through the pits, and maintained at that temperature for a half-hour. This process has been in successful operation at the plant of the Ford Motor Company. Examination with the microscope of the solution at the Ford plant showed from 10,000,000 to 15,000,000 bacteria per cubic centimetre in the infected solution, whereas but 400,000 to 600,000 bacteria per cubic centimetre appeared in the sterilised solution.

NEW PAMPHLETS

The Barber-Colman Co., Rockford, Ill., have issued a series of pamphlets dealing with their line of hobbing machines. The No. 2 to No. 12 size are included, and a detailed description of these machines, their action and principle, are given. Anyone interested in this line can procure a set of these pamphlets by mentioning Canadian Machinery.

AN INFORMATION CATALOGUE

The Williams Tool Corporation, Brantford, Ont., will issue, in about two months, a 100-page catalogue on pipe threading machinery. This will not only be fully descriptive of their entire line but will deal with the history of pipe threading machinery from days when threads were cut by means of hand tools down to the present time, when the operation has been reduced to a simple process by means of highly efficient equipment. In addition to this some pages will be devoted to the care of machinery, dies, gauges, etc., and to a discussion of best practice in the operation of pipe machinery.

The Williams Tool Corporation are making a display in the Machinery Hall at the National Exposition, Toronto.

Proper Railway Operation and Maintenance

This Refers to Operation and Maintenance Under a Divisional Organization—The Author Who Presented This Talk Before the Central Railway Club at Buffalo Knows His Subject Well.

By ALFRED PRICE, General Manager Eastern Lines, C.P.R.

IN the very early days of railroading on this continent there was no necessity for an elaborate official organization. One can imagine that the rules and methods of operation first adopted were somewhat similar to those now in effect on rural electric lines. The trains were then few in number; they cannot now be satisfactorily handled on single track lines. The locomotives weighed from 4 to 6 tons; the latest achievement weighs 427 tons. The passenger cars were simply stage coaches coupled together; they are now elegant palaces on wheels. The rails were short wooden beams, covered with strap iron, and after a short use failed mechanically under 6 ton locomotives; they are now from 30 to 41 ft. long, and are made of steel by the Bessemer, open hearth or other processes. In those good old days the trains were known as accommodation trains. The passengers were accommodating, for it is recorded that they habitually alighted from trains on steep ascending grades and climbed to the summits on foot; the modern passengers object to climb even into upper berths. But as traffic increased, trains multiplied, locomotives and cars were enlarged, and all railway facilities, equipment and appurtenances, which at the beginning were exceedingly crude, by a continuous series of improvements reached their present degree of excellence. Co-incident with this great physical change there was naturally and necessarily a development in methods of operation, varying from time to time as traffic increased and conditions altered.

As one surveys today the whole railway field on this continent, and considers the number of miles of railway under operation, the extent of the territory served, and the prodigious sums of money invested in property, material and equipment, it is almost impossible to realize that the first railroad in the United States was built only about 90 years ago. Since then not only have the achievements in railway building been stupendous, and the improvement in equipment and facilities marvelous, but the organization which has been developed to handle the immense traffic over an interlaced system of rails, with its complicated movements, has been amazing. In this development each railway company was at first a law unto itself and worked independently. All roads were not built to the same gauge, and the interchange of cars was thereby rendered impossible. The necessity of standardizing the gauge was therefore recognized. Every railway ran its trains according to the local time of the city in which its head offices were located, or on some other arbitrary time. The various railways had their own system of operating rules; and in giving signals by hand or lamp what was a "stop" signal on some roads was a "proceed" or "back-up" signal on others.

The need for reaching agreements on many matters and the desirability of standardizing methods brought railway officials together for an exchange of ideas and railway associations were the outgrowth of these meetings. Of the many railway associations in existence to-day the most important, although not the oldest, is the American Railroad Association, which was organized in 1872, its object being the discussion and recommendation of methods for the management of American railways. Probably the oldest organization of the kind is the Master Car Builders' Association, formed in 1867. Its objects are the advancement of knowledge concerning the construction, repair and service of railway cars, to bring about uniformity and interchangeability in their parts, and to adjust the mutual interests growing out of their interchange and repair. There are also important associations, representing all branches of railway work, including maintenance of way, car service, railway telegraph, railway signal, passenger traffic, freight traffic, accounting, baggage, stores and claims agents, which discuss and legislate upon the various

matters over which they respectively have jurisdiction. Through the recommendations and decisions of these associations, agreements have been reached on almost every known railway subject and almost every article used in connection with railway construction, maintenance and operation.

Upon one subject, however, no agreement has ever been arrived at. Both the divisional and the departmental organizations are in effect upon railways that are known to be efficiently and economically managed and neither system is without its champions. It is believed by some successful railway executives that the maintenance of way department should be under the direct supervision and sole control of men who are technically trained engineers, and that the track and bridge maintenance should be something entirely separate from the operation of the railway. Similarly it is their opinion that the mechanical department should be managed exclusively by men having had a thorough mechanical training and that there should be a well defined line of demarcation between it and the operating department.

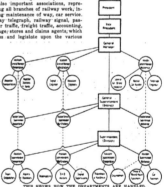

THIS SHOWS HOW THE DEPARTMENTS ARE HANDLED.

The theory is that, in this day of specialization, the right principle is to have experts in sole charge of the three important departments, operating, maintenance of way and mechanical, and that they should be handled as separate entities.

What is Organization?

Organization has been defined as "the systematic union of individuals in a body whose officers, agents and members work together for a common end." Those who favor the divisional organization believe that better results are obtainable by a fusion, under one head, of the three important departments referred to above and that by such an organization "the officers, agents and members" are much more likely to "work together for a common end" than if the departments are kept separate.

The departmental idea is carried a great deal higher up on some roads than on others. In some cases the general superintendent, besides being in charge of operation, controls all maintenance and mechanical work on his district, the officers of these departments reporting direct to him. In other cases the departments are kept entirely separate on districts, as well as on divisions, and the departmental officers report direct to the general manager. Again on other roads a great gulf is fixed between the working forces in the different departments, the general manager having no control of maintenance of way and mechanical matters. Instead the department officers have supreme authority over their respective departments and report direct to the vice-president in charge of operation and maintenance.

The absolute necessity of having highly trained experts supervise these two technical departments is fully appreciated, but this is quite possible under a system that will co-relate under the divisional superintendent all the forces of the operating, maintenance of way and mechanical departments, so as to make them complementary to one another. The plan recommended is clearly shown on the accompanying diagram. The solid lines indicate direct jurisdiction and control, whereas the dotted lines indicate a medium of communication between men engaged in the same department, so that the benefit of the technical knowledge and expert training of those at the top may be transmitted to those who are directly in charge of the work.

Instead of designating the departmental officers as "assistants," some would prefer to use the old familiar titles, such as superintendent of transportation, district master mechanic, division engineer, trainmaster, chief dispatcher, etc., but this is not material. By a reference to the diagram, it will be seen that the general manager, general superintendent and superintendent each has assisting him men who may be regarded as specialists in maintenance of way, transportation and mechanical work, respectively.

The assistant general manager (maintenance of way) prescribes standards in connection with track, bridges and buildings; allocates new rails supplied for, replacement; passes upon all plans submitted to the general manager by general superintendents for approval; criticizes maintenance of way expenditures, etc., etc. The assistant general manager (transportation) is responsible for the distribution of cars as between districts; the preparation of timetables, fixing the time of through trains at inter-district points; the issuance of instructions about preference and special passenger train movements; notices of embargoes; criticizes transportation expenses, etc., etc. The assistant general manager (mechanical) is responsible for the distribution of power as between districts; he prescribes locomotive shop, car shop and roundhouse practices; controls the movement of air brake inspection cars; supplies of dynamometer car and attendants for the making of tonnage rating tests; criticizes mechanical department expenses, etc., etc. All three officers report to the general manager and all instructions to district officers are addressed to the general superintendents over the signature of the general manager.

Their Respective Duties

The three assistant general superintendents bear pretty much the same relation to their general superintendent as the three assistant general managers bear to the general manager, communications and instructions being sent out over the signature of the general superintendent. They confer with officers of higher rank on work in which they are especially concerned, and guide those of lower rank so as to ensure the work being prosecuted in accordance with the prescribed standards and practices. The assistant superintendents report direct to their superintendent. They bear pretty much the same relation to him as corresponding officers of higher grades bear to the general manager and to the general superintendents. They are held directly responsible to him for the work of the men under them and have the advantage of the assistance and advice of the district officers in the same department.

The above gives in brief outline a general idea of a divisional organization, but does not show in any definite way the advantages to be derived from it. Time will not permit of more than a mention of a few of these. Such a system absolutely removes any departmental friction, and tends to promote harmony throughout all branches of the service. If passenger trains fail to maintain their schedules, the superintendent is not in a position to blame the engineering department for not keeping the track in proper condition for high speed trains, nor the mechanical department for not maintaining locomotives in condition to make time. He

is responsible for the condition of both the track and the power.

When there is an abnormal demand for service, locomotives turned out for traffic can, when necessary, be utilized to do odd jobs of maintenance of way work, such as loading cinders or ties en route and likewise locomotives, supplied for work train service, can often be utilized to handle revenue traffic before reaching working limits or beyond them.

When a division of railway is required to take care of some extraordinary rush of traffic, so that the facilities and power are taxed to their utmost, if the superintendent finds that a few extra men in the locomotive house, or at the ash pit, or coaling plant, would result in a quicker outturn of locomotives he is in a position to authorize their employment or to transfer them from some other class of work. When there is an accident—and no railway is immune from such unfortunate occurrences—it is very much better to have one man responsible for clearing the line, repairing the track, picking up the wreckage and resuming the running of trains, than to place the responsibility of clearing the line and picking up the wreckage upon the mechanical department, and for repairing the track upon the maintenance of way department, while the superintendent's forces stand aside waiting for the other departments to repair the damage and make the line passable for the resumption of traffic. In an investigation to determine upon whom to place responsibility for an accident, the superintendent can have no object in attempting to fix the blame, except where it belongs. Under a departmental organization, all departments interested are represented and every representative naturally desires to escape the necessity of admitting responsibility. This is unfortunate, but as human nature is constituted, it is inevitable.

A superintendent has a greater number of officers available for special emergencies. If it is suspected that men engaged in train service are becoming lax in the observance of any of the important general rules, or if it is considered advisable to check up any feature of track work, he is in a position to use all his assistants for checking or efficiency testing. There is an added advantage in that so many assistants obtain a general all-around knowledge of and experience in the operation and maintenance of a division, fitting them for greater responsibilities. Not the least important benefit to be derived from a divisional organization is the broader training which officers in the lower positions receive. A man occupying the position of general manager, or general superintendent, should not only know something theoretically, but a great deal practically, about the maintenance of track, buildings, bridges, signals, cars and locomotives—

(Continued on page 59)

WELDING AND CUTTING

Lessons on the A B C of Good Welding

A Digest of Text Book Information on Cast Iron, and Its Application to Correct Oxy-Acetylene Welding Practice.

By W. B. PERDUE,

Director Welding Dept. Healds Engineering School, San Francisco, Cal.

THE welding of cast iron is not as difficult to accomplish as the welding of wrought iron or steel, but offers a little more difficulty to the beginner because the metal flows so much more freely and some difficulty is experienced in controlling the molten metal. The two principal difficulties experienced by the beginner, however, are the overcoming of hard, brittle metal in the weld and the casting cracking either in the weld or adjacent to the weld, owing to internal strains set up by unequal contraction.

The melting point of cast iron is approximately 2,100 deg. Fahrenheit, while iron oxide has a melting point of about 2,400 deg. A bright red heat is sufficient to cause the combination of the oxygen of the air with the iron of the casting, thus forming iron oxide. It is not possible to melt this iron oxide and flow it from the weld, so it remains in the casting in the form of thin flakes or cakes. This not only prevents the alloying of the molten metal, but also combines with the free carbon. It is consequently conducive to the formation of white iron, therefore this oxide must be removed or destroyed.

The proper method of doing this is to puddle the molten metal with a filling rod, using the motion described in Chapter 1 of this series, causing the oxides to float to the top and out of the weld. In the same way blow-holes can be worked out which are formed by gases being pocketed in the molten metal.

Carbon exists in cast iron in different states. In what is called white iron, which is very hard, the carbon is combined with or dissolved in the iron. In the grey iron, which is soft and easy to work, most of the carbon is in a free state in the form of graphite. Since it is generally necessary to machine or file a weld in cast iron, it is indispensable that the line of weld be constituted of soft grey iron. Thus in welding cast iron always remember that too rapid cooling brings about a combination of carbon and iron, forming hard, brittle white iron, while slow cooling, or reheating after the weld is completed, keeps the carbon in a free state, resulting in a softer and more workable material.

Expansion and Contraction

Expansion and contraction should be treated with more importance in the welding of cast iron than in any other metal. Cast iron is absolutely lacking in elasticity and its tensile strength is very low. In preparing work for welding it is always necessary to take fullest precautions against the effects of expansion and contraction where the part to be welded is not free to expand and contract, such as cylindrical or box-shaped castings or any shaped casting where the expansion of the metal and subsequent contraction would be restrained due to the shape. Where the internal strain produced by contraction is greater than the tensile strength of the section to which it is confined, cracking or breaking will occur; but, where a strain not so great, still exists, internal strains are set up which may cause a break later when subjected to shock, and for this reason the importance of properly taking care of the forces of expansion and contraction will be apparent.

There are three general methods of coping with the forces of expansion and contraction. One method is to preheat the entire casting with some cheaper fuel agency than an oxy-acetylene flame, such as a gasoline, kerosene, or crude oil preheating torch or a charcoal fire, depending on the size of the casting and the importance of getting a very even heating. In some cases this is not necessary and oftentimes not possible. In

THIS ILLUSTRATES THE GATE VALVE SPOKEN OF.

such cases it is only necessary at the time of welding to heat simultaneously similar parts to a good red heat in order that the stiffness of the frame may be overcome and thus take care of the contraction.

As a good example of this take the case of a fly-wheel which has a cracked spoke. It will be found advantageous to heat adjacent spokes on both sides of the broken one. After the weld is completed these spokes will contract proportionately with the welded one. In some shops a different practice is followed. Instead of heating the adjacent spokes the spoke to be welded is covered with wet asbestos within a few inches of the weld and water is allowed to drip on the asbestos while the welding is going on. By this means very little of the spoke is heated up, with the result that the contraction at the completion of the work amounts to practically nothing. This method, however, is not in general favor.

Wherever complete preheating is possible it should be done. In the case of a large casting which has to be repaired in position, without dismantling, it is generally impossible to preheat the whole casting, but it is safer and more economical to build a coke fire or apply gas or oil preheating torches around the immediate vicinity of the break. While the welding operation is in progress, and also after its completion, it is often advisable to apply a blow torch or light a fire under some other part of the casting away from the weld to counteract the effects of expansion and contraction at the welded part.

If it is impossible to apply preheating, another method may be used. By the use of jackets, wedges or similar devices a casting may be sprung or bent out of shape and the portion of the part to be welded may be thus separated, and after the weld is executed and contraction sets in the jackets or wedges, etc., may be withdrawn. The return of the sprung parts to their original position will compensate the contracting strains. It is important, however, to be able to judge just how much to pry the edges apart in order that the contraction will bring them back to proper alignment. This depends upon the bulk of the metal being added and is usually not over a small fraction of an inch, say from 1-64 to 1-32 inch.

Filling Material

Owing to the fact that iron combines more readily with silicon than with carbon, a special welding rod for cast iron should be used containing a percentage of silicon. Thus the carbon in the iron being welded is prevented from combining with the iron and remains in a free or graphite state, insuring a soft grey iron weld. The silicon also tends to prevent oxidation. It will, therefore, be seen that the use of the proper filling rod will help in overcoming the hard spots previously mentioned as one of the difficulties met with by the beginner, especially when he is using low-priced cast iron rods, which really are not economical because they jeopardize the success of the weld. Manganese in the filling rod has exactly the opposite effect of silicon. At best it will be seen that in order to get soft, workable welds a special filling rod for cast iron must be used. This rod must contain a percentage of silicon, there must be no manganese present. The percentage of phosphorus must be very low, and there must be slow cooling after the welding is completed.

Use of Flux

As shown by the figures previously given, the melting point of ordinary cast iron is much lower than that of wrought iron or mild steel, and considerably lower than the melting point of iron oxide. As previously stated in this chapter, therefore, if any oxides are formed during the welding process they cannot be melted and blown away by the flame, as is the case where iron or steel is used. In the cases of wrought iron and steel the melting point is higher than that of the oxide. It will, therefore, be seen that when welding cast iron a proper flux is necessary to prevent the formation of oxides and to break down oxides when formed. It is only necessary to use flux sparingly, and in view of the small amount that is used it is unwise to endeavor to use cheap substitutes, such as borax, which will not give the proper results in the weld.

With the proper flux the metal flows readily and the resulting joint will be as strong as the original metal. The heated filling rod should be dipped into the flux and then plunged into the bevel and melted on the blowpipe frame, the inner cone of which is held about a quarter of an inch away from the metal. The molten metal is gradually produced, which must be puddled or stirred with the filling rod. This stirring helps to bring the oxides, slags and impurities to the surface. However, care should be used not to go over the work any more than absolutely necessary. In the case of a poor weld, it is better not to reweld the metal previously added.

After welding, the whole piece, if possible, should be reheated and allowed to cool slowly. Reheating as well as preheating is best performed in a furnace, in which the work should be allowed to cool, if possible. If not, cover the casting with hot sand or lime, or for smaller castings provide a bin of asbestos pulp. Avoid all drafts while cooling, as this will cause unequal strains in cooling, which may cause distortion or cracking.

27-Ton Gate Valve Saved by Welding

We do not wish to go on record with the statement that our Canadian readers can become proficient welders by merely reading these articles. To become proficient in the art of welding requires much practice, preferably under the tutorship of a competent instructor. We do wish to state, however, that there is nothing to prevent the intelligent welder who gives these articles careful study from duplicating the results here shown.

Mr. R. E. Travis, of 224 North Ninth Street, San Jose, Cal., who welded this gate valve through the bottom of toe (inside), and for a considerable distance on the rim, had never handled a torch when he entered as a welding student in January, 1919. After leaving school he did not seek the easy things, neither did he forget the careful study of such trade journals as might give him a wider vision of his calling.

When the industries of San Francisco were brought to a standstill by a disagreement between capital and labor, he sought employment with the Joshua Hendy Iron Works (which concern had previously considered the oxy-acetylene process an impractical experiment). The welding of this enormous gate valve, after several previous months of satisfactory service with that concern has demonstrated that the man who has the initiative, technical training, confidence and ability need not be hampered by local labor troubles, or by the prejudices created by incompetents.

The applications of this process are being steadily extended, probably more rapidly in the Canadian field than any other portion of the globe. Since the commencement of this series in the April 22nd issue upward of 40 inquiries have been received from Canadian manufacturers and operators. The author, whose address is given under the title of this article, will gladly continue to answer such inquiries as may be addressed to him, and will in the very near future boil down much of the information given into a chapter on "Kinks for the Welder."

VIEW FOR AEROPLANE PILOTS

A most ingenious invention is the "clear-view screen" which a British firm has produced for the benefit of navigators. Rain, spray, or snow renders it difficult or impossible to see through an ordinary port hole or other glass screen. The invention consists simply of a glass disc, which forms part of the screen and is rotated at so high a speed that any water or snow falling upon it is immediately spun off. The glass remains so clear that its rotation is extremely difficult to detect. This device has now been introduced on aeroplanes in order to assist the pilot when rain or snow is encountered. The disc is driven either by the struts, or by vanes fixed to the edges of the disc itself. When the aeroplane is travelling at 75 miles per hour the disc spins at 1,200 revolutions per minute. At anything over 800 revolutions the glass will keep clear in all weathers.

Notice is received of the incorporation of the "Jumbo Metal Works, Ltd." for the purpose of dealing in metal products and the manufacture and dealing in farm implements. Authorized capital $300,000 divided into 3,000 shares of $100 each. Chief place of business to be at the City of Winnipeg.

DEVELOPMENTS IN SHOP EQUIPMENT

GEAR HOBBING MACHINE

The Cincinnati Gear Cutting Machine Co., subsidiary of the Cincinnati Shaper Co., Cincinnati, Ohio, have placed on the market what is known as their 16-inch gear hobbing machine.

These machines are designed for strength, accuracy, speed, and convenience of operation. The machine is adapted for all spur gears within its range, but in cutting spiral gears, however, data must be furnished so that proper change gears can be supplied. The machine as supplied with standard parts is for spur gear cutting only. The gears for spirals can be supplied however wherever requested. The specifications given at the conclusion of this data will show the machines range and capacity. Both right and left hand spirals can be cut to approximately 45 degrees angularity.

The bed and housing are of box section and heavily ribbed throughout, while the work spindle is horizontal and is supported rigidly by two long bearings in the work saddle. Both bearings are bronzed bushed, and the front bearing is tapered for taking up wear. A No. 12 B and S taper is provided.

The work saddle is taper ribbed to long narrow slides in a manner to prevent any sagging when the clamping bolts are loosened. Elevating and lowering are accomplished by means of a crank handle, the movement being recorded by a graduated dial reading to .001-inch. The load is supported on ball bearing thrust collars.

Indexing Mechanism

The indexing mechanism consists of a double thread worm and a cast iron gear which are entirely closed and run in a bath of oil. Suitable adjustment is provided for taking up wear. The indexing is continuous and automatic. Variations are obtained through removable change gears.

The hob slide guide is exceptionally long with square ways taper gibbed. This long guide reduces binding action to a minimum, and the design and combination is such as to provide for the proper swiveling of the hob spindle. This can be adjusted approximately 50 degrees either side of zero, and is set by a vernier reading to five minutes.

An automatic stop for the entire machine at any point is provided for use on both belt and motor driven equipment. There is also a trip to stop the feed mechanism only.

The hob spindle is bronze bushed and adjustable for wear. Suitable means are provided for shifting to hob and hob spindle, so that the entire surface of the hob may be used before resharpening. This adjustment can be made without resetting the hob on its arbor. A cooling pump of large capacity is provided. The bed of the machine forms a reservoir for the oil or compound and strainers are provided for cleaning the returning oil after it has passed over the hob. The chips are retained in a separate chip pan. All pulleys and gears are properly guarded.

Changes in speeds and feeds are secured by means of removable change gears and sufficient range is provided to meet all average conditions. As already stated both belt and motor driven styles are made. Following are the principle specifications:

Rated capacity, diameter 16-ins.; rated capacity, width of face 12-ins.; capacity, actual diameter max. gear, 17 1-4-ins.; rated capacity, spur gears C.I., 3 D.P.; rated capacity, spur gear steel, 3 D.P.; rated capacity, spiral gears C.I., 3 D.P.; rated capacity, spiral gears steel, 3 1-2 D.P.; max. distance hob centre to spindle nose, 19 1-2-ins.; maximum diameter of hob, 4 1-2-ins.; diameter hob arbor, 1 1-4-ins.; taper hole in work spindle, B & S No., 12; driving pulley, size, 15-ins. x 3 1-4 ins.; speed of driving pulley, R.P.M., 400; number of changes of hob feeds, 26; hob feeds per revolution of work, .015-ins. to .250-ins.; number of changes of hob speeds, 8; hob speeds, R.P.M., 50 to 200; overall dimensions, covers open, 56-ins.x103-ins.; overall dimensions, covers closed, 52-ins.x82-ins.; net weight of machine, with countershaft, pounds, 5,600; domestic shipping weight, pounds, 5,900; shipping weight, boxed for export, pounds, 6,600; contents, boxed for export, cubic feet, 210. These weights do not include electrical equipment.

WHEEL TRUING MACHINE

The Precision Truing Machine & Tool Co., Cincinnati, Ohio, has recently placed on the market the grinding wheel truing machine shown in the accompanying illustration. It is claimed that the results obtained by the use of this truing machine are equal, and in some cases superior, to those obtained with a diamond. This machine can be applied to any make and style of grinder, and is

THESE TWO VIEWS GIVE A GOOD IDEA OF THE GENERAL APPEARANCE OF THE MACHINE.

operated by either alternating or direct current of 110 or 220 volts. The attaching bracket shown at the right is furnished with each machine, but additional brackets for special or unusual applications can be supplied. Three general-purpose abradant nibs are also furnished with each machine, and nibs for special purposes can be supplied to meet requirements. The general-purpose nibs are 1 inch in diameter by 1⅛ inches long and it is claimed that one of these nibs will keep an average grinding wheel in proper condition for 100 hours' continuous operation. The machine has ball bearings throughout, with provision for taking up wear.

BLUSH MULTIPLE MICROMETER

A. T. Blush Tool Co., 1145 West 11th St., Erie, Pa., have placed on the market the micrometer as illustrated. The purpose of the instrument shown is to eliminate the necessity of using two instruments in taking measurements ranging from 0-in. to 2-in. This is accomplished by the use of two threads of different pitches, namely, a 20 pitch for the measuring screw and a 40 pitch for the thimble carrier. Although the mechanism embodies two screws, the accuracy of the instrument depends only and entirely upon the 20 pitch thread as the 40 pitch thread only functions in keeping the thimble in proper alignment with the graduation on the hub.

The difference in the pitch of these threads makes it possible to obtain a 2 inch movement of the spindle with only a 1 inch movement of the thimble. The adjustment for wear on the anvil and the spindle is taken care of without moving either the anvil or the measuring nut which makes a very desirable feature as it is impossible to throw the anvil and the spindle out of parallel when making adjustments.

Due to the fact that the spindle is actuated with a 20 pitch thread this makes a very fast acting instrument as the thimble travels half as fast as the spindle. Another feature is the graduation, which is said to be much easier to read than the old style of graduation.

It is their intention to make these instruments in larger sizes having a range of 2 inches on each size, also a special instrument for very close work ranging from 0 inch to 1 1-2 and 1 1-2 to 3 inches.

TURRET TOOLPOST

The Lovejoy Tool Co., Inc., Springfield, Vt., has recently placed on the market a turret toolpost as shown in the accompanying illustration. The original Lovejoy positively locked cutter principle is used on all tool-holders which are held in this toolpost, and the turning and facing cutters are adjustable for height as they become worn, which permits their cutting edges to be kept in line with the lathe centre without sacrificing strength and rigidity.

By one movement of the binding lever, the operator can release and accurately index the turret to the next tool position, where it will be securely clamped in place by the completion of the single movement. The turret rings are approximately 4⅝ inches square, made of hardened steel, and are interchangeable on any base. This interchangeable feature permits the use of additional rings carrying a variety of tool combinations for various jobs, without disturbing the individual cutter adjustment, as well as a quick method of changing tools from outside to inside work.

The boring-bars are 1 inch in diameter and are free from projections, and they will cut to the bottom of a hole which is only slightly larger than the bar itself. However, special boring-

bars ¾ inch in diameter with bushings can be supplied if desired. The turning tools have a shank diameter of one inch, and are furnished with 19/32-inch high-speed steel cutters. The round shank permits of rotating the holder to produce any desired side clearance, and also provides for a small amount of end adjustment without disturbing the height of the cutting edge, which is a convenience when used on lathes equipped with diameter stops or feed-screws having direct-reading dials. This tool-post will interchange with any regular engine lathe toolpost without requiring any special fitting of the lathe.

OPEN-SIDE PLANER

The Universal Machine and Tool Company, of Canton, Ohio, are now making a 24-inch open-side planer which combines the advantages of the shaper and the planer. The supporting column is on the right hand side and operating controls are centralized. Variable stroke mechanism is used for driving the table, the speed of which is varied by means of shift gears, four speeds being available. Power feed is provided for the tool. Table has a surface of 17¼ by 46 3-4 inches, with a height under the rail of 36 inches. Maximum stroke of 27 inches. Weight about 4,500 pounds.

PLANER IMPROVEMENTS

The Cincinnati Planer Co. have incorporated a number of changes in the design of their planers which add to the serviceability of the machine. The top arch is now being made of the box pattern, which gives increased strength and bearing surface on the housings. The cross-rail is likewise of the box type and it is claimed that increased rigidity is a feature when the tools are cutting. An automatic stop has been added to the machine so that the clutch for elevating the cross-rail may be released at any desired point in the vertical range. Improvement has also been made in the holding device for the tool post harp.

THREE-JAW MACHINE VISE

A special vise that incorporates many advantageous features for tool-room and production work is now being manufactured by Manning, Maxwell and Moore, Inc., of New York. The three jaws can be swiveled independently and can be locked in any desired position, thus providing a wide range of positions particularly applicable for holding odd-shaped pieces. The jaws are fitted with loose grip plates, which tend to draw the work down on the body of the vise when pressure is applied by means of the screw. The body of the vise is made of steel alloy and the working parts are made of mild steel, case-hardened. The vices are made in five sizes ranging from 4 to 18 inches.

TOP VIEW SHOWS GRINDER, LOWER LEFT HAND VIEW THE MICROMETER, AND REMAINING VIEW THE TOOL POST.

The MacLean Publishing. Company
LIMITED
(ESTABLISHED 1887)
JOHN BAYNE MacLEAN, President. H. T. HUNTER, Vice-President
H. V. TYRRELL, General Manager.

PUBLISHERS OF

CANADIAN MACHINERY
~ MANUFACTURING NEWS ~

A weekly journal devoted to the machinery and manufacturing interests.
B. G. NEWTON, Manager.   A. R. KENNEDY, Managing Editor.

Associate Editors:
J. H. MOORE   T. H. FENNER   J. H. RODGERS (Montreal)
Office of Publication: 143-153 University Avenue, Toronto, Ontario.

VOL. XXIV. TORONTO, SEPTEMBER 9, 1920 NO. 11

PRINCIPAL CONTENTS

A Machinery Exhibit

IS there an opportunity for a machinery exhibit in Canada? This matter has been discussed by several of the dealers, and there seems to be a variety of opinion. The question came up in the first place owing to the absence of nearly all the Canadian makers, and many of the Canadian dealers, in the Machinery Hall at the Toronto Exhibition.

Opinion differs quite sharply as to the advisability of placing a large exhibit in Machinery Hall as it now stands at Toronto grounds. It is an out-of-the-way place on the grounds, arranged to accommodate things as they were many years ago. There is no overhead work to assist in the handling of heavy machinery, and placing equipment to advantage is a laborious and strenuous piece of work, as well as a costly one.

It would be necessary to broaden the scope of a machine exhibit in Canada to include machinery in general as well as machine tools. In fact it might well be made an exhibit of wood and metal working machinery, and other lines that are generally used in the industrial life of the Dominion. Then, too, there could be taken in all sorts of equipment and accessories.

It would be advisable to hold such an exhibit at a time separate and apart from the Toronto Exhibition. The educational side of the affair would have to be kept uppermost, and every effort should be made to impress upon the minds of exhibitors and those who would attend that the whole thing was being staged with a serious aim in view.

For instance, in the field of machine tools, manufacturers and their mechanical heads should make a study of their problems, and come to such an exhibition with the expectation of meeting experts in machine shop practice who could advise them on any points they wanted. In this way the exhibit could be made something much more than a mere show where people could come and walk past and go away again.

Unless the undertaking were approached with the idea of making it a clearing house of real information, and a place where a maker or dealer would be prepared to offer the advice of his most expert staff, little or nothing of a lasting nature could be accomplished.

The fact that it is harder year after year to interest Canadian machine tool houses in Machinery Hall at the Toronto Exhibition has prompted the above suggestions. It may be that there is an opening in this country now for a well-staged and well-conducted machinery exhibit at a time apart from other exhibitions and at a place where suitable buildings could be secured.

Looking for Ideas

FORBES MAGAZINE, in a recent issue, refers to the success of Thomas Edison, and remarks "One invention was the result of 5,000 experiments. When he was working on his improved electric lighting he sent men into remote corners of the world in search of some fibre or other material which could provide the electric bulbs with the right kind of filament. Such is his spirit of enterprise that he reads more trade publications probably than any other business leader."

Leaders in manufacturing life in this country are coming to the point where they are depending more and more on the trade and business papers for help in their particular line of business.

Only a few weeks ago there was recorded in these columns the case of the master mechanic of one of the leading Canadian railroads. In going through one of the mechanical papers he found an idea that he was able to apply to his work in the shop, and by so doing saved the railroad company hundreds of thousands of dollars in time of repair, and also enabled the rolling stock to be put out much sooner in order that it might be earning revenue instead of eating its head off in the repair shop.

There is something worth thinking about here. If you put your ideas in print in a mechanical or technical paper you are sure of an audience of interested readers, and it is almost certain that some one is looking for the idea you have. The amount that may be paid for the material is only a small part of the service. It may not be very much, but it may be as much as the paper can afford. The big idea is that you can, through the medium of the technical pages in such a paper, pass along an idea that may be the means of giving great assistance to some mechanic who is having troubles of his own. The passing on of these ideas does not impoverish the person having them, neither is he enriched by withholding them.

Trade Follows the Flag

THE saying which is used as a caption for this article is not always true. In some cases, and perhaps the majority, the flag follows trade, and it is in this manner that the British Empire has been developed. The originators of the Canadian Government merchant marine, however, saw that the possession of a fleet of modern commerce carriers, which could go to all parts of the world, could become a great drawing power for Canadian trade, and that in this case trade would follow the flag. The Canadian Government fleet, conceived in the first instance as a measure to tide the shipbuilding yards over a doubtful period, has expanded into a ship-owning concern that will have no less than sixty-six first class modern cargo carriers under its flag. This flag is the flag of the Dominion of Canada, and it does not require a very great imagination to see what an advertisement for Canada this fleet can be. Sailing from Canadian Atlantic and Pacific ports, these Canadian built and Canadian manned steamers will carry holds full of goods manufactured in Canada to every port on the shores of the seven seas. This fleet would justify its existence were it as a great advertising scheme only, but it has already justified itself as a commercial success. The Canadian Government Merchant Marine is a fleet in being, of which every Canadian can be justly proud.

MARKET DEVELOPMENTS

Steel Prices Are Not Weakening in the Least

Independents Even Talking Now of Increasing Prices, Rather than Coming Down to the Level of the Steel Corporation Figure— Scarcity Not Believed in Canadian Market.

JUST when the steel market had it all figured out that prices were due for a drop, especially in the premium mills, rumors come into the market that there is serious talk of an upward revision of the already high prices. And coupled with this is the statement that the United States Steel Corporation is considering bringing up some of its schedules, although there is nothing to give weight to this report. It is a fact, though, that the Steel Corporation will have an added bill of some $50,-000,000 per year for haulage of material and semi-finished shipments, and it is estimated that only one-quarter of this gets passed on to the consumer.

There are warehouses in this district paying around nine cents for sheets, while at the same time they have a very large tonnage on the books of the Corporation at around 2.65 or a little better. Delivery is the big item just now.

Buyers, or prospective buyers, of machine tools are not coming very rapidly into the market now. One explanation is that they may be staying out as long as possible, in the hope of a reduction being made in the price of the lines they wish to secure. There has been nothing coming from the makers that would confirm this attitude, but a continued refusal on the part of buyers might lead to some unexpected alterations in the policy of selling.

Scrap metal dealers report business as being very dull. Some of the yards are taking on material, but not at a high figure. Most of the dealers are not heavily stocked. In fact not since the war has there been anything like the stock of scrap metal, especially in iron and steel, that there was in pre-war days.

THE CONDITIONS PREVENT NORMAL VOLUME OF TRADE FOR MONTREAL

Special to CANADIAN MACHINERY

MONTREAL, Sept. 9.—The gradual betterment in railroad transportation is improving the industrial outlook but the relief is so slight that an early return to normal is not anticipated. The process of adjustment will require time, and, above all, close co-operation from all sources of production, as this is the vital element in maintaining equilibrium in trade conditions. The coal scarcity is still a factor here, but less acute. Shipments have been more regular, and plants are fairly well supplied for current needs. Operations at the Canadian Car shops are about normal, with nearly all men back at work. Canadian dealers are watching with considerable interest the developments of the coal situation in the States, and although a two-year contract has been entered into between the operators and the miners, the latter attached their signatures under protest that the wage scale was not satisfactory. Many collieries are idle as a result of men still out, and the possibilities of underproduction, at a time when the demand is increasing, has created a nervous tension among large consumers and manufacturers. The general impression, however, is that labor in general is

recognizing the essential need of uninterrupted output in all lines of activity, and that the present disturbances will be short-lived.

Price Revisions Expected

The higher freight rates are a factor now, and on present orders must be taken into account, but where goods are obtainable from stock, previous quotations are generally considered. Replacements, however, are a problem that adds to the difficulties of the dealer, and it is only a question of a week or so when all prices will be based on new freight charges. In commenting on the steel situation one dealer said: "We are now taking into consideration the increased cost of hauling materials and making such revisions as will be necessary in the readjustment of prices to meet this advance in rates. These revisions will be announced in a short time. In the meantime quotations from stock are made on the old basis, but invoices on material delivered from American points include the advanced freight charges." There is still little trouble in disposing of the sheets coming in, although the demand appears less marked at present. Tank

plates are available in better quantities but the demand has lessened slightly.

Fall Improvement Expected

The month of August was a comparatively quiet one for machinery dealers, and business turnover was below the average. Some report a fair month, but the balance has been possible more from the movement of small equipment and supplies than from the sales of machine tools. The feeling that labor is more settled is having an influence, and hopes are expressed that the fall season will see a partial return to greater activity. The requirements are apparent, but unsettled conditions and constantly rising prices have prevented a more healthy state of progressive development in many lines of industry. American tools may be slated for advances, but this will apply principally to heavy equipment, as small supplies will be affected but little by the increase in freight rates.

Quiet But Firm

The listless condition of the old material market provides little of interest, and dealers are waiting for the awakening when consumers start a buying campaign. At present the buying is on a small scale and only for immediate needs, largely for local operations. Price quotations are more or less nominal, as individual sales are generally transacted on their respective merits.

MARKET NEEDS MORE MATERIAL

Little Improvement Noted in Securing Steel From the Mills—Prices Stay High

TORONTO.—The holiday season and the Toronto Exhibition are two factors that are quite in evidence in the market this week. Dealers in several lines make it a point to come in contact with a number of outside men at the Fair, and after many of these consultations are in better shape to plan their business for the future.

The consensus of opinion seems to be that there is going to be a lot of buying done, not primarily in machine tools, but in lines that will call for machine tool equipment to turn out the material.

Some of the dealers are going as far as to suggest that the time is ripe for a machine tool exhibit in Canada, claiming that the present one at Toronto Fair is not representative of the industry as it exists in this country. There can be no question about that, and the suggestion of a Canadian machine tool exhibit is worth thinking about.

The Shortage Still Here

There may be times when it appears that there is some improvement in the supply of steel coming through, but when one hears of the way in which manufacturers and builders are taxing their ingenuity in order to keep going, these stories of improved conditions simply pass out.

The slowing down of the automobile industry does not seem to be affording relief to other users who want the material they might have been taking had they remained open at full blast. The auto industry uses about eight per cent. of all the steel turned out in United States. As one house figures this, suppose they do cut 25 per cent. in their production, that is only two per cent. of the production. Others seem certain that the fact that labor was planning a big strike in the automobile industry has much to do with the attitude of some of the larger plants on both sides of the line, particularly in United States.

Toronto yards are having a hard time securing sheets. Prices at which these are sold run around 12 cents per pound. One of the yards put the case this way to-day: "Many users may imagine they are paying a high price for sheets, and perhaps they are, but it does not follow that we are making any more money out of it. The real state is this, that at present prices we are making less on our money than we used to when iron bars were to be had for two cents a pound. We have to pay fancy figures in order to get shipments. While we have a very large tonnage on the books of some of the large rollers at quite a bit less than 3c per pound, we are right now buying heavily and paying 8 and 9 cents for quick delivery."

Tubes Very Scarce

There is no improvement in tubes. The replacement business is very urgent now

POINTS IN WEEK'S MARKETING NOTES

Pittsburgh claims that markets at this season of the year should be growing more active, instead of which they are becoming quieter and giving less evidence of sustained buying power.

Predictions are that there will be a prolonged deadlock in the pig iron market. Makers hold for $50 in the Pittsburgh market, while the trade is not prepared to buy at that figure.

The new freight rates in at United States will make a difference of $50,000,000 a year to the U. S. Steel Corporation on materials hauled to its various plants.

Independents are keeping up their high prices for steel and even talk of increasing them.

Metals may show an increase owing to freight changes of materials carried to the smelters.

Deliveries are not improving as much as desired in the steel market.

Toronto warehouses are paying as high as 9 cents for sheets for quick delivery, while at the same time they have a large tonnage on order with the Corporation around 2.65.

The New York machine tool market reports several firms holding off, waiting for lower prices, but the same advice states there is nothing to indicate a dropping market.

for supplies to get ready for the winter season, and there is no relief in sight. There are many public institutions, such as hospitals, asylums, etc., where it is impossible to shut down.

There are, in the steel market, occasional rumors that some of the steel mills may close for some time. There is, of course, nothing definite to hang the story on, but there seems to be a feeling that such a thing might happen if present conditions keep up much longer.

In the Toronto field there are some reports of price cutting in some of the contracts, even in fields where goods are scarce.

Contractors are calling for material for reinforcing in some quarters, and the supply of this is limited. Fortunately some of the warehouses were fairly well supplied some time back, but thing would be poor word they depending entirely on mill deliveries to provide the necessary material.

There is not such an insistent demand for plate as there was some months ago

when ship yards were busier. The vessel for the pen-stock of the Hydro work at Chippewa has not been placed yet, and there may be some difficulty in getting prompt delivery. There will be about 2,500 tons required.

The Scrap Metal Trade

There have been some instances in the market within the last few months where scrap iron has sold up to and beyond pig iron. But the reverse of that is the case now. It looked a week or so ago as if the furnace men had put iron—around $50 in Pittsburgh—on a plane where buyers would leave it alone. According to precedent scrap iron should follow, but it has not.

Yards are still taking on a certain amount of scrap, but there is a general "don't care" atmosphere in the whole trading, because dealers are not keen to buy when they don't see a higher market or a ready sale ahead. The yards of the country are generally pretty well depleted. In fact none of the yards are carrying anything like the stock they did before the war. In that period they were stripped of anything that looked like iron or steel, and their stocks have been short in many lines ever since.

Red and yellow metals are about the same. Copper is not moving. The needs of the consumers seem to be fairly well fixed, and there is no extra business demanding material. It is likely that copper will go up a cent a pound as a result of the new freight rates. Not that the new rate on the shipment of copper will amount to that, but there must be taken into consideration that supplies of coal and other materials are shipped to the smelters before the copper is made. Prices in the metal market remain unchanged.

HAVE VERY LARGE ORDERS ON HAND

Canada Car and Foundry Have Bookings That Run Into Many Millions

Montreal.—Action in respect to the liquidation of the 22% per cent. dividend arrears on Canadian Car & Foundry preferred stock was again deferred by the directors of the enterprise, according to a statement made at the conclusion of the board meeting here.

As the full board does not meet again until late in October, it is not anticipated that any steps will be taken prior to that time, when the matter will probably be discussed by the directors. It is understood, however, that there has been no change in the plan by means of which the arrears are to be liquidated, the majority of the board favoring the proposal to issue 7 per cent. income bonds in lieu of cash.

The prime factor in reaching such a decision, it is stated, was due to the fact that the Car enterprise has nearly $40,-000,000 in unfilled orders on its books, requiring a working capital of well upward of $15,000,000 to provide the necessary outlay for labor and materials.

INDEPENDENTS KEEP PRICES UP AND EVEN TALK OF MAKING INCREASES

Special to Canadian Machinery

PITTSBURGH, Sept. 9.—Producers, dealers and consumers in all branches of the general iron and steel market, including coke, pig iron, semi-finished steel and finished steel products, unite in declaring that there is a condition of quietness if not of absolute stagnation. Coming on the heels of the ending of the July-August mid-summer period, which is traditionally a dull one, the situation today is important and suggestive. The markets ought to be growing more active. Looking back to June, one can now see that the markets have grown more and more quiet as time passed, and the fact that what is normally a dull season of the year was passed through seems to have had nothing to do with it.

There is nothing in the technical position of the markets that would presage any increase in activity in the near future. Producers are moderately well sold up and consumers are moderately well covered. No large number of producers will run out of orders in the near future, requiring them to force sales, while buyers are unlikely to be forced to buy heavily for quite a while to come. The attitude of producers is that of "standing pat" on prices as there is nothing to be gained by trying to stimulate activity. The position of buyers is that they will buy only for absolutely known wants in the near future, and as they are already covered, in the main, the current purchases are very light.

Pig Iron

The pig iron markets are stagnant, particularly in the western Pennsylvania and valley district. Men are discussing now how it happened that foundry pig iron recently was shoved up $4 a ton to $50 valley. As there has been little buying at the new price the advance does not seem to have done the producers much good, and it may prove to have done them harm by straining the patience of consumers to the breaking point. If the furnaces had left the market at $46 they might have pulled the consumers along and gotten them to buy from time to time. One man who is practically neutral as between buyer and seller, but very well posted, remarks that the advance was simply a sort of notion, so many furnacemen having made up their minds it would look nice to have pig iron advance to the round figure of $50, and it is known that quite a number of bets were made as to whether or not pig iron would reach $50. Bets can be collected even if little pig iron can be sold. The same observer just referred to remarks that the southern furnacemen seem to have handled their market with more skill. They got the price of pig iron at Birmingham up to $42 at the end of last April, and they have held it there steadfastly,

though sometimes obtaining delivery premiums. When northern iron started advancing over again there was afforded a wider field for the sale of southern iron, of which the southern furnaces have been taking advantage. For instance, years ago the alignment of prices became such that Birmingham iron could rarely get into Pittsburgh, being shut out by valley iron, but even with the advance in freights Birmingham iron can beat valley iron for Pittsburgh delivery by over $2 a ton, the freight Birmingham to Pittsburgh being advanced from $5.70 to $7.60, making southern iron at $42 Birmingham $49.60 delivered Pittsburgh. Valley iron, on the other hand, at $50, furnace, is $51.90 delivered Pittsburgh. The Pittsburgh consumer may not buy southern iron, but he is deterred from buying valley iron unless he needs a small lot for very early delivery, in which case the price matters very little.

A fair conclusion is that there will be a prolonged deadlock in the pig iron market, with only very light transactions. A certain amount of iron has been sold and bought, and the question is whether the producers will first need orders or the consumers will first need iron. For a long time past production and consumption have balanced, while the present outlook is that production will increase and consumption will decrease. Unless fundamental conditions change, therefore, it would look as though sellers will have to come into the market first, and necessarily with cut prices, but this may not be, possibly, until several months have elapsed.

Steel Prices Unchanged

The gap between Steel Corporation and independent prices is not being narrowed, despite the dullness of the market. In the past week or two there have been little cases of price declines and also little cases of price advances, on the part of some independents. The point seems to be that when there is competitive business the price declines while when there is no business a mill may mark its prices up, as it loses nothing by doing so and makes safer from cancellation the business it already has on its books. It is about a year now since there first appeared a distinct difference between Steel Corporation and independent prices, the difference increasing until substantially the maximum was reached last February or March, since when there has been practically no change, taking the general average of the market. Expectations used to be entertained that the phenomenal and almost unbelievable condition of there being two markets would be terminated by a gradual sagging of the independent prices. The course of the market re-

cently indicates that nothing of the sort will occur. There will be no sagging, but rather the prices will be maintained to the last moment, the banners waving and the band playing. The independents still talk of there being further price advances, and have even revived talk of the Steel Corporation being about ready at last to advance its own prices, which have been maintained, without change since the Industrial Board adjustment of March 21, 1919. While there is no official statement, there is an intimation from New York, possibly inspired, to the effect that the corporation has no intention of advancing its prices. Now would be the time of all times for the corporation to advance its prices, for by the recent advance in freight rates its costs are increased by about $50,000,000 a year in increased freights it pays on materials assembled at its various plants and on shipments between plants. The corporation gets a part of this back, however, possibly a fourth or a third, in that it makes a greater profit on such material as it sells on the "Pittsburg basis" to point taking a lower freight from the producing mill to the point of delivery. The $50,000,000 a year is approximately equal to the rate at which the corporation accumulated surplus during the first half of this year, after paying charges and dividends. However, the corporation does not necessarily need to accumulate surplus, all the time, having a good surplus now, while its book-keeping has been particularly conservative, large sums having been written off lately, keeping the property account down.

NEW YORK TOOL MARKET IS DULL

Some Buyers Reported Staying Out in Expectation of Getting a Lower Price

Special to CANADIAN MACHINERY.

NEW YORK, September 9.—During the week Labor Day has cast its shadow over the machine tool business, which has been unusually quiet. Aside from the inquiries from the General Electric Co., reported last week, there has been virtually nothing in the market worthy of note. The General Electric Co. is inquiring also for a considerable number of tools for its Bridgeport, Conn., works, in addition to 84 wanted for its North Easton, Mass., plant, and 25 for Schenectady works.

One large manufacturer in the East, who had been on the point of buying a large list of tools, has decided to hold off, possibly for several months. The impression seems to be gaining that prices of machine tools will come down within a few months, but there is as yet nothing tangible upon which to base such an expectation other than the fact that current business is in small volume compared with what was done in the early part of the year.

Not only has domestic business in ma-

chine tools fallen off, but there has been a decided slump also in export buying, particularly in England. The recent increase in the profits taxes in England has considerably upset the manufacturers of that country, and they are not in a mood to take on new commitments.

LINE TO TAP THE
. FELDSPAR FIELDS

Revival of an Ontario Mining Industry As Result of Increased Demand

Kingston.—There is likely to be a branch line of railway run from the Canadian National Railway into the township of Bedford, starting probably from Westport. This line is designed to open up the development of feldspar deposits in the district that cannot now be reached owing to the absence of railway facilities. The feldspar in this district is regarded as one of the richest

in North America. The only two other important deposits are in North Carolina and Maine, but the deposits in the Kingston district are said to be engaging the attention of the manufacturers of porcelain ware, electrical insulation, Old Dutch and Sapolio, bathroom fixtures, fine tile and enamelware, including automobile enamel. The demand for feldspar as the basis of a great many manufactured ornamental articles is growing by leaps and bounds and the Frontenac deposits have increased enormously in value.

There is a likelihood, too, of the location of grinding mills at the pits so that there may be some control over the process of manufacture, thus utilizing local labor as far as possible. Up to the present time, however, the ore has been exported direct from the mines by vessels from Kingston and Cobourg to the United States, where it was refined, and some deposits have already been purchased outright by Americans.

OPERATIONS AT THE SAULT WERE ON LOWER BASIS—OUTLOOK BRIGHT

President Wilfrid H. Cunningham, of the Lake Superior Corporation, referred at length to conditions in the steel trade in his annual address to the directors. Regarding work at the Soo mills he said:

"Operations of the steel plant for the fiscal year," he says, "continued throughout the first quarter at below 50 per cent. capacity on account of scarcity of new orders, and subsequently, until April were seriously curtailed, notwithstanding heavy orders on hand booked in November and December. The winter was unusually severe, and operations were seriously hampered by heavy snowfalls and extremely cold weather. These conditions affected our own transportation facilities, and interfered with shipments by rendering it impossible at times for railways to supply sufficient cars. Sickness was very prevalent during the winter, at one time 20 per cent. of the operating force being absent from this cause. Not until the last quarter of the year was the plant able to

operate approximately to capacity."

But the outlook is now brighter, according to Mr. Cunningham. "Orders on hand," he goes on, "at the close of the year in steel and iron products, amounted to 268,000 gross tons, which will assure capacity operation until December, 1920. At the present time business is offering in large tonnage for delivery in the first six months of 1921, and booking will be made as soon as the advanced prices are definitely settled. The outlook is considered satisfactory as to new business throughout the coming year." Equipment was completed for rolling structural shapes up to 15 inch beams and channels, and chrome vanadium and nickel steel are now being made for the automobile industry.

"Contracts for new structural mills were placed," says Mr. Cunningham, "but have since been suspended, pending an improvement in the money market, when it is expected more favorable arrangements can be made for the financing."

TRYING TO CUT
OUT ACCIDENTS

So a Committee is to Go Over the Buildings in Course of Construction

An innovation in making provision for the safety of workmen engaged in building and construction work has been introduced in the erection of large extensions to the plant of the American Rolling Mill Company at Middletown, Ohio.

At two p.m. each Tuesday a committee composed of foremen, mechanics and laborers inspects the entire job from sewers to roof with the one purpose of seeing that proper methods are taken to safeguard the employees against accident. This committee makes a detailed report of each inspection to a representa-

tive safety committee, which considers and puts into effect the recommendations of the inspection committee.

The success of this program is being closely watched by various safety societies which are now encouraging day laborers as well as trained men to speed up their work. Experts in all types of building construction are agreed that the first marked reduction in building cost will come through increased production. The best features of this new safety plan are being copied by many large construction companies in the United States and Canada. The public is interested in the plan because indirectly it will lower rentals by reducing construction costs. This applies equally well to industrial construction, large building projects and homes.

MONTREAL NOTES

William P. Millar, an old resident of Morrisburg and Brockville, who has been travelling representative in China and Japan for the Canada Foundry and Forgings, died recently at his home in Virden, Man., at which place he had been residing for some years past.

Roy M. Wolvin, president of the Dominion Steel Corporation, left Montreal last week on a business trip to the steel and coal interests of the company at Sydney. While away he will attend the launching of the C. G. M. M. steamer, "Canadian Mariner," which has been constructed by the Halifax Shipbuilding Company; the ceremony of christening will be performed by Mrs. Wolvin.

At a recent meeting of the Montreal Board of Trade W. S. Leslie, chairman of the committee on municipal affairs, was appointed to confer with the Administration Commission in connection with the application of the Bell Telephone Company for increased rates and the manner of collecting them. On the question of increased facilities for communication with the South Shore, the Board appointed W. B. Ramsay, chairman of the harbor and navigation committee, and A. M. Irvine, as a committee to confer with the South Shore Board of Trade.

In connection with the forthcoming trans-continental air flight from Halifax to Vancouver, the Quebec Harbor Commissioners have been approached as to the using of Lampson's Cove as a seaplane harbor, which will serve as a permanent station for transcontinental flights, and will place Quebec on the allred air route. Capt. Leroyer, of the Canadian Air Board, has stated that one of these flights will take place towards the end of this month. The object of these flights is not only of military value but it is to demonstrate the practicability of the airplane for commercial purposes.

The Superheater Company, Limited, will manufacture the type of fire-tube superheaters now in general use in locomotives of all Canadian railways, and also fire tube superheaters for marine installations, of which there are now over two thousand in daily service. In the case of the water tube boilers in marine service, the company designs and manufactures the superheater to suit each individual case. This likewise applies to stationary installations, either watertube, fire-tube, or separately fired. The conditions are studied in each case with a view to supplying the most economical and effective installation. The company employs a form of return bend which is made by a forging process from the metal of the superheater pipes, making an exceptionally strong construction which is proof against high temperatures. This form of return bend has also been considerably sought after for the use in pipe coils for refrigerating purposes, condensers and other uses.

Extending Canada's Trade to Every Harbor

Development of Canadian Marine is Proceeding on a Profitable Basis—Ships Are Now Sailing from Canada to Many Points— Equipment and Service of a Very High Order.

ALTHOUGH Canada has come into the world view in maritime matters of recent years, many of her people, far removed from the sea board, are but dimly aware of her position in this respect. As a means of putting first hand information on the activities of the Government marine before these people, a party of press representatives from Toronto, Hamilton, Ottawa, London, Quebec, and Montreal were entertained to luncheon on board the "Canadian Victor," the latest acquisition to the fleet. The party travelled over Government-owned lines from their respective towns to Montreal, where the vessel was loading for Liverpool at Pier 12, Montreal. The Toronto and Hamilton representatives travelled via Napanee, Smiths Falls and Ottawa over the C.P.R., transferring to the Grand Trunk at Ottawa for the remainder of the journey to Montreal. The trip to Ottawa was made by daylight, and the scenic beauties of the route were much enjoyed by the newspaper men. The portion of the route through the Rideau chain of lakes is especially beautiful, and the excellence of the road bed contributed materially to the pleasure of the trip. The train at times made considerably over 60 miles an hour, and there was no evidence of the usual bumping and swinging which so often accompanies railroad travelling on the roads of this country. Montreal was reached about 11 p.m.

Luncheon on Board

Luncheon on board the "Canadian Victor" was set for 1.30 p.m. on Tuesday, August 21st, and at that hour a deeply interested company gathered around the table in the smoke room. Yes, the "Canadian Victor" has a smoke room in her accommodation, which is far in advance of that on any cargo steamer the writer has so far run across. The head of the table was taken by Captain Coffin, master of the vessel, while facing him at the other end was Mr. R. B. Teakle, manager of the Canadian Government Merchant Marine, who had on his right Mr. M. P. Fennell, secretary of the Montreal Harbor Commission and on his left Capt. S. E. Tedford, marine superintendent of the C.G.M.M. Besides the representatives of the press, there were present Captain Bourassa, Harbor Master of Montreal, Captain Bales, Port Warden of Montreal, Mr. H. Milburn, assistant manager of the C.G.M. M., Mr. H. J. Whiteside, publicity agent of the C.N.R., and Mr. Walter Thompson, publicity agent of the G.T.R. During the luncheon, the subdued clatter of the winches rapidly lowering the cargo into the holds formed a bass accompaniment to the hum of conversation and

served to remind one that this was no pleasure ship, but a busy commerce carrier, carrying Canada's goods and Canada's flag to the far confines of the Empire.

Mr. R. B. Teakle, in a post-prandial speech, gave a highly interesting and closeup view of the attainments of the C.G.M.M. from its inception to the present day, and the achievements that are aimed at for the future. Mr. Teakle's speech impressed one as that of a man who knew his subject thoroughly and was imbued with a very warm enthusiasm for the project he had under his charge. In fact, every official of the C.G.M.M. one meets seems to be infected with this same spirit.

Started After the War

Mr. Teakle spoke of the inception of the fleet, which was really started in

order to keep the shipyards, then concluding their Imperial Munitions Board contracts, employed for a further period. The world shortage of shipping induced the Government to increase their earlier orders, and then the vision of what such a fleet might mean to Canada decided them to bring the fleet up to its present dimensions. The "Canadian Victor" was the 38th ship to be delivered, and brought the total tonnage now in commission up to 203,000 tons. There were 28 more steamers to be delivered, which would make a total of 400,000 tons dead weight and a fleet of 66 steamers. A fleet of this size was a very respectable one for any company to own. The first vessel to sail was the "Canadian Voyageur," a product of the same firm that had built the "Canadian Victor," the Canadian-Vickers, Ltd. At the end of 1919, the total tonnage in commission was 93,900 tons. The fleet

when all are delivered will virtually girdle the earth, services being established to every part of the world. Close by the "Canadian Victor" the "Canadian Pioneer" was loading for Bombay, Calcutta and Kurrachee, while from the British Columbia coast the "Canadian Importer" and "Canadian Exporter" had inaugurated a service to Australia and New Zealand, Hong Kong, Shanghai and Japan, and the Dutch East Indies will have services and the Canadian vessels from the Pacific coast will meet in far Eastern ports with Canadian vessels from the Atlantic coast, thus forming virtually a girdle of the globe.

Training Canadian Officers

Not only were these vessels Canadian built, but they were 80 per cent. Canadian manned. In order to provide future officers from the youth of Canada

ONE OF THE CANADIAN MARINE VESSELS

every steamer carried two apprentices, selected from various towns in Canada. Mr. Teakle was one of the originators of the Navy League and through this organization many of the boys were supplied. These apprentices, after four years' service, would be eligible to sit for examination as junior officers, and then climb up the ladder in the usual way, until they attained the highest position attainable at sea. The "Canadian Victor" was the "first of the Government vessels to be fitted with a cold storage hold, and some of the future vessels would probably be equipped with oil fuel. Speaking of the organization, of the fleets, Mr. Teakle mentioned that they had as Australian manager, Mr. Geo. E. Bunting, a Canadian of long experience and training in the shipping business, while in London, England they were represented by Mr. Wm. Philips, also a Canadian. The ships

run in exactly the same manner as the large English companies ran their ships. In fact in some ways they were run better, and he instanced cases where Canadian Government ships were turned round in four or five days, while old country ships in the same ports took ten or fourteen. They handled in the port of Montreal as many as six or seven ships a week, and they came and went quietly with no fuss, but with full holds. So far the ships had been run at a handsome profit. It had been said, and there was some truth in it, that even if the vessels were run at a loss they would be of value to the country. However, that was not the view that he and his assistants took of it, and they were running them as a business organization, to make a profit for the owners, who were the people of Canada.

At the conclusion of his speech, Mr. Teakle was thanked very heartily, and congratulated on his clear exposition of the purpose of the C.G.M.M.

The "Canadian Victor"

Following Mr. Teakle's speech, the company were taken on a tour of inspection of the vessel, and the various features pointed out by the officers in charge. As before mentioned, the accommodation for all hands is very much above the average to be found in a cargo steamer, and there is in addition accommodation for eight passengers, should there be need at any time to carry someone from port to port. The men's accommodation under the poop is spacious and considerably better than used to be provided for third class passengers up till quite recently. The poop deck itself is long and spacious and makes an excellent promenade deck.

The facilities for handling cargo are very good, there being four winches at each hatch, besides one on the poop, and the anchor windlass on the fo'c's'le head. The steering engine is located in the poop, and is operated by telemotor gear from the wheel-house on the bridge. Hand gear is fitted in case of emergency and a powerful brake is fitted that can hold the rudder stationary in case of a breakdown of any part of the gear.

The vessel's leading dimensions are: 400 feet long by 52 feet beam and 32 feet deep, while her deadweight tonnage is 8,432. She has triple expansion engines, with the contraflo type condenser, the air pump and bilge pumps only being operated off the h.p. cylinder. A winch condenser is fitted for port use. The main feed pumps are independent, made by the well known G. & J. Weir, of Glasgow. There is also a Lamont pump. The circulating pump is of the centrifugal type. There are three boilers of the Scotch marine type, with three furnaces in each, operated under the Howden system of forced draught. Mr. McGregor, the chief engineer, had two vessels blown up under him during the

war, but looks none the worse for his unpleasant experiences. . . . . .

Canadian-made Goods

Going through the sheds afterwards under the guidance of Capt. Tedford, it was highly impressive to observe Canadian-made autos in huge crates consigned to India, and even Arabia. Those were part of the cargo of the "Canadian Pioneer," which was loading for these ports. At another shed was observed the "Canadian Observer," loading for Newfoundland and Gulf ports, while at another wharf the Canadian Government steamers have been the means of bringing 8,000 tons of Cuban sugar to Montreal, which otherwise would have been difficult to get. Captain Tedford becomes enthusiastic when speaking of the steamers of the fleet and claims that there is no line of ships that can show better operation than the C.G.M.M. He wants the funnel of the line to carry a maple leaf to distinguish them, and let everyone who sees them know that there is a Canadian ship. He rightly contends that there is no better advertisement for Canada than a Canadian ship in every port of the world.

A Little "Extra" Work

Despite the work of looking after a fleet, Captain Tedford is ready for any special duty that comes along. Quite as a matter of course he mentioned that he had lately returned from a little salvage operation on the coast of the St. Lawrence. The steamer J. E. McKee went

ashore on the rocks at the mouth of the Saguenay River in a thick fog, and at high tide. A vessel going ashore at this place under these conditions usually stays there, and most people thought that the McKee would. However, Captain Tedford thought she could be salved and proceeded to salve her. They left Quebec for the scene of the wreck in the afternoon and arrived at 2 o'clock the following afternoon, which was a Saturday. At 2 o'clock Sunday the vessel was afloat, after a portion of her cargo had been discharged. Her propeller was gone and her rudder badly damaged but she was towed into Quebec three days after the expedition started. This was a highly successful and expeditious salvage feat, for which Capt. Tedford came in for many congratulations.

Canada possesses an organization in her merchant fleet which is a distinct asset to the country. It is carrying her flag to countries which have hardly heard of her, and incidentally opening up trade routes for her manufacturers. The company, for it is a company, is to be congratulated on the men it has secured to organize and operate the fleet. The fleet is Canadian-built and 80 per cent. Canadian manned. May it soon be 100 per cent. Canadian and 100 per cent. successful.

Mr. H. J. Whiteside, of the Canadian National Rys., and Mr. Walter Thompson, of the Grand Trunk Ry., were responsible for the guidance and comfort of the party, and were tireless in their efforts for the well being of their guests.

HERE ARE FIRMS THAT ARE ASKING FOR SEVERAL LINES

Firms interested in the following lines are advised to write to the Intelligence Department, Trade & Commerce Dept., Ottawa, mentioning key numbers:

1584.—Electrical apparatus. A firm in Roumania desires to receive prices and quotation as well as other particulars from Canadian firms in a position to export electrical apparatus.

1585.—Industrial machinery and tools. —A Genoese firm would be glad to enter into negotiations with Canadian firms with the end in view of representing them in Italy.

1586.—Machinery. A house in Milan, Italy, wishes to purchase machinery in Canada.

1587.—Cranes, machinery, hardware, etc. A house in Turin, Italy, is willing to purchase the above in Canada.

1588.—Machinery. A firm in Roumania desires to receive quotations and other particulars from Canadian firms in a position to export all kinds of electrical machines.

1589.—A house in Milan, Italy, is anxious to represent Canadian houses dealing in machinery.

1590.—Agricultural and industrial machinery. A most important house in

Rome would be glad to hear from Canadian firms dealing in the above.

1593.—Agricultural and industrial machinery, pumps. An old established firm in Naples, dealing in agricultural and industrial machinery and pumps would welcome proposals from Canadian exporters.

1598.—Electric machinery and material. A Cape Town firm of electric engineers are prepared to take up the South African representation of Canadian electric machinery and materials.

1607.—Machinery, hardware, pumps, etc. A most important firm in Genoa is willing to consider offers from Canadian firms dealing in machinery, hardware, pumps, etc.

116.—Iron and steel goods, etc. A Cape Town commission agent is seeking agencies for all of the lines mentioned above.

1617.—Machine tools. A Cape Town commission agent is seeking agencies for all of the lines mentioned above.

1617.—Machine Tools.—A Cape Town representative is prepared to consider offers of agency for the Union in machine tools, gears, etc.

Everything in Woodworking & Metalworking Machinery

A Corner of our Warehouse
163 Dufferin St. Toronto

When You Want
MACHINE TOOLS
New or Used

Complete Equipment for Railroad Spring Shops, Flue Shop Equipment, Boiler Shop Tools or Machinery for any other purpose; High Speed Drills, Reamers, &c.; Pneumatic and Electric Drills and Hammers, Transmission Supplies, &c., &c.

And WOODWORKING MACHINERY of All Kinds
SEND US YOUR ENQUIRIES

Garlock-Walker Machinery Company, Limited
32 Front Street West
MONTREAL   TORONTO   WINNIPEG

If interested tear out this page and place with letters to be answered.

INDUSTRIAL NEWS

NEW SHOPS, TENDERS AND CONTRACTS
PERSONAL AND TRADE NOTES

Have Contract.—A contract for the mechanical equipment of the Hudson's Bay Company's store at Victoria, B.C., has been awarded to Greene and L'ster, plumbing and heating contractors, Winnipeg.

A Busy Shop.—Favorable reports are received regarding operations by the Ontario Steel Products, Limited. The plant at Gananoque, where shovels are made, is running at capacity, and business at the other factories is also reported to be very good.

Make Steel in West.—Rolling mills and blast furnaces for the manufacture of steel, will, it is expected, be established in this province as soon as the necessary market for the product of such an undertaking has been assured, according to a statement made by Major Martyn, Industrial Commissioner for the province, who, with James H. McVity and Nichol Thompson, of the Advisory Council of the British Columbia Department of Industries, has left for Chicago and San Francisco in connection with the matter.

Toronto.—The Canada Steamship Lines will build a new steamer for their Lewiston service, which will be in commission for next season. She will be the largest passenger boat in the fleet, and will be thoroughly modern in every respect. She will be propelled by geared turbines, and will have a speed of about 22 knots. She will have four decks, and a restaurant, dancing floor, moving picture theatre, and a children's playground are spoken of. Pressed steel fittings will be used entirely in the interior, reducing the fire risk to a minimum. Passenger accommodation will be provided for 4,000.

Customs Tariff Inquiry.—In preparation for the approaching revision of the Canadian Customs Tariff by Parliament, a committee of cabinet ministers, consisting of Sir Henry Drayton, Minister of Finance; Senator Robertson, Minister of Labor, and Hon. Tolmie, Minister of Agriculture. will begin an inquiry into the tariff at Winnipeg, on September 14 next. The itinerary as announced is as follows: Winnipeg, September 14, 15 and 16; Vancouver, September 20, and 21; Vancouver, September 24 and 25; Nelson, September 30 and October 1; Edmonton, October 6 and 7; Regina, October 11 and 12; Winnipeg, October 14; Sault Ste. Marie, October 18; Medicine Hat, September 17; Victoria, September 22 and 23; Vernon, September 27; Calgary, October 4 and 5; Saskatoon, October 8; Brandon, October 13; Fort William and Port Arthur, October 15; Windsor, October 21. The itinerary for the rest of Canada will be announced later.

Build Machine Shop.—The Gunnel Company of Canada, 2410 West Dundas Street, Toronto, received a permit from the city architect for the erection of a three-storey brick and reinforced concrete machine shop, which, when completed, will cost in the neighborhood of $100,000.

Want Yards Open.—Business Manager Herbert Wright, of the Toronto Union of Shipbuilders, is in Ottawa conferring with the Minister of Marine and Fisheries, urging that they proceed with the construction of the two ships on the ways of the Dominion Shipbuilding Company. These were started before the company was forced into liquidation. The third vessel, that was under way for a New Orleans firm, is being finished by the owners. Mr. Wright will urge upon the Government the fact that if the work does not go on many men will be out of work this winter.

Will Rebuild.—The first sod of the new factory of the Imperial Steel & Wire Co. at Collingwood was formally turned by Deputy Reeve William Williams, in the presence of members of the municipal council, the chamber of commerce, and other business men. Lt.-Col. J. A. Currie, president of the company, said the new buildings would be much larger than those destroyed by fire in May, 1919, while the plant would be so expanded as to give an output of about 200 tons of wire products per day. Contracts for the building, Lt.-Col. Currie also said. have, practically been awarded, local firms being the successful tenderers. Construction will be proceeded with immediately.

Change in Welland.—The plant of the Dillon Crucible Alloys, Limited, Welland, has been purchased by the Canadian Atlas Crucible Steel Co., Limited, and Canadian interests will be strongly represented in the company as well as on the board of directors. Atlas products, which include "L-XX" brand of high speed steel and other Atlas brands, carbon and special alloy tool steels, will now be made in Canada. The general sales office of the company will be located at 133 Eastern Ave., Toronto, Ont., with a warehouse in conjunction, and branch warehouses and sales offices at 326 Craig St. W., Montreal, also at Winnipeg, Man.

Toronto.—Mr. F. S. Wood, formerly Agent for the Canadian National Railways at Pittsburgh, is being transferred to Toronto to look after the interests of the Canadian Government Merchant Marine. Mr. Wood has been located at Pittsburgh for the past five years.

To Make Matches. — Sir Alexander Maguire and his brother Dr. D. P. Maguire, who have been in Montreal at the Ritz-Carlton Hotel in connection with the plans for the erection of a match factory plant to be operated by the British firm of Maguire, Paterson and Palmer, Limited, have left for New York, where they will embark on the Olympic for Southampton. They expect to return in about three weeks to complete all details of their plans. It was stated by Sir Alexander Maguire that machinery is already awaiting shipment on the other side, and they anticipate being able to commence manufacturing operations early in January.

INCORPORATIONS

Dominion Motor Castings, Ltd. Incorporators: Clifford Albert Ripley, manufacturer; and others, all of Detroit. Capital $250,000. Head office, Windsor.

Canadian Farm Power and Machinery Co., Ltd. Incorporators: Milton Ira Adolphe, farmer, Listowel, Ont.; and others. Capital $1,000,000, divided into 10,000 shares of $100 each, of which 5,000 shares shall be preference shares. Head office, Toronto.

Auto Supplies Co., Ltd. Incorporators: Henry Howard Shaver, barrister-at-law; Alexander Findlay Boyle, automobile builder; and others, all of Toronto. Capital $150,000,. divided into 1,500 shares of $100 each. Head office, Toronto.

London Concrete Machinery Co., Ltd. Incorporators: Henry Pocock and John Charles Doidge, manufacturers; and and others. all of London. Capital, $500,000, divided into 5,000 shares of $100 each. Chief place of business, London.

"With those results make *Morrow* standard throughout the entire plant."

You get more holes with less regrinding by using

MORROW DRILLS

Get them at your jobbers

JOHN MORROW SCREW & NUT CO., LIMITED
INGERSOLL, CANADA

MONTREAL	WINNIPEG	VANCOUVER
131 St. Paul Street	Confederation Life Bldg.	1290 Homer St.

7 Hop Exchange, Southwark St., London, England

If interested tear out this page and place with letters to be answered.

NEW ZEALAND IS A GOOD SPOT
FOR CANADIANS TO DO BUSINESS

F. C. Brookbanks, a prominent merchant of Auckland, New Zealand, who has been in Canada, believes there is a chance to develop very satisfactory trade relations between Canada and New Zealand.

"New Zealand trade is worth cultivating by Canadian manufacturers, said Mr. Brookbanks, "since we are purely a pastoral and agricultural country with little or no manufactures to speak of. During the war we did an extensive business with Japan, but the Eastern sources of commerce are no longer satisfactory. The grades are too low in quality and business methods are unsuitable to us in every way. Goods are far too often not sent to sample and as a result nobody wants their merchandise. Latterly, it has become impossible to judge colors from their dyes, which are especially bad in dark shades.

United States Busy

"The United States," said Mr. Brookbanks, "has fully realized the importance of trading with us and is already doing a tremendous business, having sent out its own representatives and appointed resident agents in the various large cities such as Auckland, Wellington, Dunedin and Christchurch. Furthermore, for the last two years certain American firms have been allocating a considerable portion of stock for New Zealand trade. In some cases these goods could have been disposed of in their own country, but in order to encourage and foster New Zealand commerce they have adopted this course.

"Now, in theory, the ideas of the American manufacturer are excellent, but in practice they are bad," stated Mr. Brookbanks, "and English goods are therefore much preferred by our stores. I am here to investigate personally trade conditions in Canada and to judge at first hand whether Canadian markets offer greater opportunities to the New Zealand importer than do the European or American.

"After a careful study carried out in all parts of the world I am convinced that Canada alone can satisfy our requirements, if she will but take the trouble to rectify and overcome the drawbacks existing at present so detrimental to the expansion of her foreign export trade."

When asked what criticism he had to make regarding the matter, Mr. Brookbanks replied: "I can readily give several instances of the difficulties I refer to, but my criticisms are meant in the kindest way as between cousins, so to speak.

"To begin with, we merchants invariably experience annoying delays in the execution of our Canadian orders. A delay in delivery is just as vital a loss to us as to Canadians themselves. When you consider that we are five months behind you as regards seasons, it ought to be the easiest thing in the world to accommodate us, since what is unseason-able in Canada is just the reverse in New Zealand. Again, when a manufacturer accepts an order from us, he must give us some idea of the date of delivery, otherwise we could not assume the risk at so great a distance. Another thing, some of your manufacturers have a bad habit of substituting without authority, and the result is invariably unsuitable. In every case, we should prefer to place a second choice, if this would be adhered to.

"Carelessness in shipping arrangements is another fruitful cause of annoyance to us. The omission, for example, of the statutory customs declaration to the effect that 'these goods were made in the British Empire' entails the payment of enormously increased duty, whereas, a little foresight anad extra trouble on the part of the exporter would save us a considerable amount of needless expense."

MARINE

Montreal.—The Canadian Government Merchant Marine has taken delivery of the Canadian Victor, built at the plant of the Canadian Vickers Co., Montreal. The vessel is of 8,400 tons, and will load for Liverpool. The line to the Far East will be opened in the immediate future by the departure of the Canadian Pioneer.

Washington.—The U.S. Shipping Board has authorized increased freight and passenger rates on the Great Lakes, to be made effective after one day's notice not later than January 1st next. The freight rates are enabled to be increased by 40 per cent. and the passenger rates by 20 per cent. It is expected that the vessel owners will put the new rates in force without delay.

Kingston.—The steamer Chamberlain, recently damaged by fire, has been sold by Captain Harry Martin to A. A. Laroque of the McNaughton line. Capt. Martin has also sold the barge "White Friant" to R. C. Weddell, Trenton, for the coal and stone trade.

Vancouver.—The Hon. C. C. Ballantyne, speaking at Vancouver, said that eventually many more millions would be spent on the harbour, in addition to the five millions which would be devoted to the Government pier.

Brockville.—The work of lightering the cargo of the stranded steamer Phelan at Iroquois Point is proceeding rapidly. The Donelly wrecking fleet has about completed this work, 32,000 bushels having been removed, and the work of pumping out the wet grain is now being proceeded with. The grain is being loaded into barges, the dry and wet being kept separate.

PROPER RAILWAY OPERATION
Continued from page 250

tives, besides being a competent transportation officer, and the best way to acquire this general knowledge is to be placed in a position to gain the practical experience. A superintendent given such an opportunity will naturally make a more capable general officer than one whose training is confined to one department.

It has always been stated that some railways are being efficiently and economically managed under a departmental organization. The question naturally arises, "Would still better results be produced under a divisional organization?"

Trying the Coal.—E. H. Oliver, chief engineer of the Alberta Government power plant, has been sent to Winnipeg to conduct a series of tests with Alberta coal for the purpose of demonstrating to consumers in that province that coal from Alberta will stand the shipping and the climate conditions and will burn well.

Vancouver.—The Canadian Pacific Railway coast-service will be augmented by a new steamer which will replace the "Princess Sophia" which was lost last year in the Alaskan straits. The contract has been let to the Wallace Shipyards of Vancouver, and the vessel will be 326 feet long with a speed of 17 knots. She will cost about $1,500,000.

If interested tear out this page and place with letters to be answered.

CANADIAN MACHINERY

AND
MANUFACTURING NEWS

Vol. XXIV. No. 12   September 16, 1920

Various Devices Which Have Proved of Value

The Fixtures as Explained Herein Have Proved Themselves to Be Both Time and Money Savers—It May Be Possible that You Can Adapt the Same Principle to Your Line.

By J. H. MOORE

WE will make no attempt in the following article to confine ourselves to any one type of fixture, but will illustrate and describe four distinctly different devices which we noticed on a short tour through the plant of the Willys-Overland Co., Ltd., at Weston, Ont.

Of course, everyone understands that this concern manufactures the Overland Four automobile, therefore the devices shown are used in the manufacture of the various parts going into the completed product.

The first of these fixtures, Fig. 1, is a simple, yet very efficient milling jig, used on a vertical spindle miller. The work being performed is the milling of the crank shaft bearing caps, and ten caps are held in the fixture. One station is, of course, used for loading, as is the general practice. Note the fact that this fixture has a plate on its side fastened by four button head screws. On the plate is stated the name of jig, what it is for, etc, etc.

The scheme of marking fixtures in this way is one that should be more generally carried out. The tool and fixture room attendant has no trouble in giving out the correct fixture, and when a great amount of somewhat similar fixtures are used, as in this particular plant, the idea becomes still more valuable. As the fixture shows up so clearly in the photograph we need go into no further details.

Still another milling fixture is shown

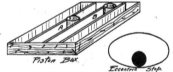

FIG. 4—ILLUSTRATING BOTH THE ECCENTRIC STOP FOR JIG IN FIG. 3, AND THE STYLE OF PISTON BOX AS USED IN TESTING.

at Fig. 2. In this case the work being performed is the milling of the bosses on what is known as the Front Spring Yoke. These yokes are made of forged steel, which later goes through a hole in the yoke, rests on this step, and is thus prevented from turning while tightening the nut. Twelve cutters are used on the arbor, and six yokes are placed in the fixture, three to each side.

The fixture is of the revolving type, being locked in position by a spring pier actuated from a convenient handle. While the machine is performing the work on one side of the fixture, the operator loads the other. The turned portion of the yoke rests in Vee blocks, and the clamps shown have a spring behind them in order to aid quick release. Note that good lubrication of cutters is provided for, as the lubricant tube stretches entirely across the gang of cutters. In 8¾ hours somewhat over 200 of these yokes are machined.

Boring Fixture

The third photo, Fig. 3, illustrates a fixture very much out of the ordinary. The portion shown is the rear axle housing, and the operation is that of boring out, on a Bullard mill, the hole in which one of the boring heads is shown.

It is absolutely essential that this hole be central with bosses on the inside of the partially assembled casting, for it is in the inside that the gears perform their duty. A little deviation from the

FIG. 1—A VERTICAL MILLER FIXTURE OF INTEREST.

FIG. 2—MILLING FRONT SPRING YOKES ON DOUBLE JIG.

correct measurement means that the gears will not meet properly, and readers know what such a condition leads to.

Picture to yourself two bosses in the interior, approximately about E and F. It is the boss E that we are particularly interested in. An eccentric oval shaped disc, as shown at Fig. 4 is fastened to the shaft on which the handle A is placed. In loading the jig, the end C is placed in the guide bush as shown, and the other end is placed on a similar shaped guide, attached to the starwheel B.

The handle A is now used to bring up into position the eccentric disc already spoken of. The wheel B, which has a threaded portion working in the boss of jig, is screwed up until the complete housing is tightened and set in position from the inside, boss E resting against the eccentric disc.

The adjustable screw D is merely a leveller and chatter stopper, and the clamp at right hand end of the fixture swings out of position when you wish to unload the work.

Assembling Stand

The stand shown at Fig. 5 is one used with great success in making any necessary repairs or adjustments to the engines after testing. It will be noted that the complete engine is swung on the moving member and that this portion of the stand can be locked at various angles. No matter what part of the engine you wish to get at, the use of this stand makes it easily accessible. The remaining details being so self-apparent, we will not dwell further on the matter.

Other Items of Interest

In our little tour through the plant we saw various other features of interest, one of which was the quantity of single spindle drills made into multiple spindle machines, through the use of multiple spindle drilling heads. In some cases two spindles were used, while in other cases all distances of centres and numbers of spindles up to as high as seven were used. On special duplicate work, such as is accomplished in this plant, the use of multiple heads is a splendid idea, for you secure at low cost, a double purpose machine.

We noticed a very novel tapping fixture for tapping the holes on the rear axle housing. This housing is shown at Fig. 3. The fixture was nothing but a flat plate on which run a series of balls. These balls, of course, ran in a circular race. Another plate, which was of correct shape to hold the housing, rested on these balls and floated, so to speak, with no appreciable effort. The operator simply brought down his spindle, the tapping chuck accomplished the work, and, as the tap entered the hole, the fixture floated itself central to the tap.

Another scheme we saw that was worthy of note was the method of segregating the pistons after testing. It might be well to add that all pistons are tested on a Prestwich fluid gauge, and must be correct to size. So important are they in this regard that they have arranged a box, shown at Fig. 4, to accommodate the different sizes encountered.

For example, suppose a batch of pistons measure 3.371" to 3.373". They are placed in section A, but should the pistons measure exactly 3.370" they are placed at B, and so on. Every piston is stamped as to its measurement, and there can be no mistake on this point.

From these few examples one can readily draw the conclusion that when you have a standard product to manufacture the answer to maximum production is efficient fixtures, and the Willys-Overland Co. have certainly paid good attention to this very essential detail.

FIG. 5—BORING FIXTURE FOR REAR AXLE HOUSING.

CARE OF PATTERNS

By H. Bentley

The treatment usually accorded to patterns is far from what it should be, considering the time and money that are expended on them. In recent years some of the larger engineering works have paid particular attention to this department, but with the exception of these the methods usually adopted are very haphazard, so says the author in "Mechanical World." Possibly the explanation for this state of affairs is to be found in the fact that patterns, like drawings, are considered a necessary evil, and only a means to an end, and so expenses in connection with same are cut down to a minimum. The time and money that are wasted, and the annoyance that is caused, due to the bad handling and misplacement of patterns, have caused many a manager to think furiously. With the large works, where an elaborate system is in operation and a man has control of the patterns both in and out of the stores, we need not deal. But, obviously, when such a system is in vogue successfully in a large works, one can reasonably expect a less elaborate system to work satisfactorily in the smaller shop. The works which allows its patterns to be huddled together, large and small, cannot realise the wastage of time caused by this lack of system. If once the care and storage of the patterns were allocated to some responsible person, the saving effected would many times outweigh the extra cost involved.

Where the foundry is integral with the engineering works the consequences are perhaps not so serious as in the case of general iron foundries. At the present time, when castings are so difficult to obtain and patterns are being duplicated and sent all over the country to try to keep up with the requirements, one can readily understand the irritation and delay which are caused by an incomplete or a dilapidated pattern being despatched. The proper provisions for storing the patterns systematically are, of course, just as essential in the foundry as in the works. The usual practice is to maintain patterns at the foundry that are in constant use, while those that are only used occasionally may be returned to the works. It appears rather strange that many firms with excellent stores for finished parts consign their patterns to any dark old corner in the works. Proper care and attention to the patterns mean better and quicker work in the foundry, for dilapidated patterns usually produce unsatisfactory castings.

An Essential

One of the first essentials to a satisfactory system is, of course, a proper record of all patterns, and for this purpose a record book should be kept, with full particulars as to loose bosses, coreboxes, etc. This book should be divided into sections for the different types of machines that are manufactured, and some system of numbering adopted which allows of easy reference by person, telephone, or letter. The writer prefers the method which adopts an initial letter for the type of machine—say, A for lathe, B for drilling machine, C for planing machine—and where various sizes of these machines are built a figure should precede the letter—say, 1A for 6 in. centre lathe, 2A for 7 in., 3A for 8 in. centre lathe, etc., and similarly with the other types of machines. The arrangement allows the type and size of machine to be quickly determined. Further, the separate patterns for each machine should have an individual number, which should follow the letter, and as far as possible the similar part in each machine should have the same number. For example, for 8 in. lathe parts the following system may be followed: Bed, A1, gap-piece, 3A2; large cabinet, 3A3; small cabinet, 3A4.

Brass figures can be obtained which can readily be fastened to the patterns, and each loose boss and corebox should have the numbers imprinted by metal letters and figures in an unmistakable manner. If proper supervision is made of the numbering, no difficulty should be found in allocating the proper pattern with bosses and coreboxes. The pattern-maker usually marks both loose pieces and coreboxes · in such a way that no mistake will occur in fixing either the loose pieces or arranging the cores.

In addition to the proper identifying of the patterns, a simple plan of storing is necessary unless considerable time is to be wasted sorting through a mass of patterns. The writer is familiar with stores where provision is made for both the patterns that are in regular use as well as those which have been discarded, due to improved design, but which have to be occasionally utilized on account of renewals and breakdowns. The arrangement in each case is identical, the stores being divided into the two sections—up-to-date and out-of-date patterns. These sections are subdivided for the various types of machines. Each subdivision has the patterns arranged on the general principle of large patterns on the floor, medium-sized patterns on shelving, and the smallest patterns in lockers. This arrangement allows ready access to any of the patterns, and between the two main sections sufficient floor space is provided for the fitting together and laying out of any patterns prior to their being sent to the foundry. Similarly, sufficient space is provided to allow for the withdrawal of any pattern without upsetting the others. Pattern numbers should not be transferred from a discarded pattern to a new one, or trouble will result. New patterns should have new numbers.

FIG. 8—THIS ADJUSTABLE STAND HAS PROVED ITSELF TO BE OF PARTICULAR VALUE.

Considerable Saving

The system is equally good for works or the foundry, and if the issue and care of the patterns are placed in the hands of a responsible person the consequent smooth working which is bound to follow will mean considerable saving in a year's time. If the pattern store is placed in the hands of a person whose only knowledge of them is that they are of wood, then failure will result. Immediate attention and overhaul of incoming and outgoing patterns are essential so that errors and omissions may be quickly rectified.

This arrangement is probably as simple as possible, and for this reason is to be recommended in preference to some of the more elaborate arrangements, which, while answering quite well in a large works, are too expensive to uphold in the smaller works. Printed sheets, with list of parts suitably arranged, facilitate the checking of patterns in and out.

Various Examples of Efficient Safeguarding

The Punch Press is Perhaps the Chief Offender as Far as Accidents Are Concerned, and This Article Describing Safety Devices Should Be of Particular Interest.

By A. L. KEAMS, Safety Engineer Simmons Company

PUNCH presses have always been one of my hobbies, because they are the most dangerous. They cut off more fingers than any other machine. The reason for this is that the operator is compelled to place the material into the die, and also to take it out with his hands. For that reason, my plan has always been to do this work in such a way that this will not be necessary, and I have found that you can push it or slide same into the die, and have the material taken out automatically.

You will have two things to contend with. First, the superintendent and the foreman, who are thinking of production, and second, the operator, who believes it interferes with his work. The only way to convince is to show them an easier method of doing the job. I always pick an operation which is very easy to change. You know, nothing hurts more than to find something that is easily changed, especially if you have been doing it the other way for a long time.

Take for example Fig. 1. This illustrates a piece of work that was manufactured exactly the reverse of the method depicted. That is, they placed the part with the round portion in first, which was very dangerous. By turning it around, and cutting part of the die away, I could then slide this piece in, as the round part was the widest and would not go through. It is kicked out automatically, and all the operator does is to set it on the slide and let go, without any danger to himself.

Another example is shown at Fig. 2. In this case, the work was accomplished with the portion shown standing up on the press with a straight piece on top of same. This was in the bottom of the press, and the die was in the crosshead. I simply took the part out of the bottom, and placed it in the top, and put the forming die in the bottom, attached the slide to same, and instead of having to place it on the pin as heretofore, we simply slid it in, and it is kicked out automatically. This was

a forming job, and is now done in a safe way.

At Fig. 3 is shown a forming job. We were unable to tilt the press in this particular instance, because it had such a short stroke, so could not kick out the finished piece. We made a slide, however, and the operator knocks out the piece with a stick.

Fig. 4 is doing a similar job as Fig. 3, except that the press had a longer stroke, and we were able to tilt it, kicking out the finished piece. Once more the operator merely places the piece on the slide, and has nothing more to think of.

Fig. 5 illustrates a job that allowed the material to pass through the die. The operator was compelled to set the piece into the die, but we built a slide for this, and now he has only to place the piece on the slide.

At Fig. 6 is shown a re-drawing job. A stripper was placed over the die, and the operator was compelled to place the piece in the die, underneath the stripper,

ILLUSTRATING FIGS. 1, 2, 3, AND 4—FIGS. 1 AND 2 ARE AT TOP. FIGS. 3 AND 4 AT THE BOTTOM.

FIGS. 5 AND 6—SEE TEXT FOR EXPLANATION.

and take it out in the same way. We removed the stripper, and placed it on the ram, but we found that when the piece stripped, it would fall directly back into the die; so some way had to be devised to get rid of it. We took a piece of spring steel, fastened it to the side of the press by a brace, and found when the piece stripped, it would strike this and bound off, but as the piece had to be out of the way when the ram descended, we placed an extra part and fastened it to the cross-head, as shown. This pushed it out of the way, clearing the ram. We next built a slide, sliding the piece into the die where it was kicked out automatically as shown in the photograph. We used scrap to try this scheme out, and its success was beyond our expectations.

Fig. 7 depicts a case where re-drawing work is being done in such a manner as to allow the piece to go through the die. The nearest the operator's hand can get within; the danger zone is about 21 inches in this particular example.

Fig. 8 is a very difficult proposition. This is a re-drawing job, but if you will note the bottoms of the cups are round. When we tried to push them into the dies, we found that they would tip. A great deal of comment was made on this operation, and many said it was a fool thing to try. After several different things were tried, a tray was made, the bottom hollowed out for the cups to fit on, and to overcome the tipping we placed an extra piece on one side, and put tension springs in between same. This had the desired effect of holding up the heads of the cups, and stopping the tipping. We placed a piece across the tray, close to the die; for when the cup left the tray it again had the desire to tip. This piece held them up so that they would fall properly into the die.

Fig. 9 depicts a trimming operation. Where, heretofore, a man had set the piece into the die, we now place it in a slide which takes it to an automatic pusher which in turn places it into the die. We do this work without first mak-

ing blue prints, it being all experimental, as it is impossible to tell the angle these slides must be to allow the piece to drop properly into the die. Each job is just a little different from the others, and must be worked out for itself. Quite often we merely take a piece of tin, keep bending same until the proper angle and shape have been found, then make a permanent one from this model.

We have learned with these simple devices that the spoiled work has decreased wonderfully, simply because the pieces get into the die in a more perfect way than when placed by the hand. We have instances of men working for several days without making a single piece of scrap.

Some there are who still claim production cannot be increased by press safeguarding, but I believe otherwise. I have tried to get my increased production out of the machine, and not out of the man. Heretofore, the man was doing 2-3 of the work, and the machine, 1-3. I make the machine do 2-3, and the

man, 1-3. The operator places his material on slides, instead of into the die, and in most cases he does not have to bother about the piece after it is finished; so his mind is only occupied with one thing.

Following are the figures of the increases of production on the different operations, as they were taken at the time when installed. I feel confident that these have increased since that time for the simple reason that the operators have become accustomed to the same. In some instances, we have new men put on these jobs who had never operated presses before.

The increased production on Fig. 1 was 55 per cent.; on Fig. 2, 37 per cent.; on Fig. 3, 25 per cent.; on Fig. 4, 50 per cent.; on Fig. 5, 100 per cent.; on Fig. 6, 150 per cent.; on Fig. 7, 50 per cent.; on Fig. 8, 66 per cent.; and on Fig. 9, 125 per cent. These figures may seem strange, take Fig. 6, which had the largest increase. In this instance, the increase was not larger, because the press did not run faster, because we got every stroke. The operator in the old way, if he had caught every other stroke, would have increased 100 per cent., but it showed that he was not able to do that, and hence the increase was 150 per cent.

We consider these experiments as our first aid work. Our real task will begin when automobile feeds are worked out. We have about 500 presses running; one department alone has about 3,000 different dies. This department lost 24 fingers last year. It has run for the months of April, May and June without a scratch, which goes to show what can be done when safety devices are given serious thought.

We have overcome the objections of the superintendents and foremen, because we have increased their production, and we have made the operators contented, as they can earn more money, and not work as hard as they did before. For that reason, we have the co-operation of all, and have instances in which

FIGS. 7 AND 8—TWO EXAMPLES OF RE-DRAWING WORK.

the operator refused to work without first having a slide placed on the machine, not only for safety's sake, but because he reasons he can earn more money by having the slide installed.

HANDY SUGGESTIONS

The Railway Association of Canada have issued their bulletin No. 12, which contains many valuable suggestions to would-be shippers. They start off as follows:

"A machine broke down from being run at too low a speed! For lack of moisture in the factory air, a cotton fabric failed to take smoothly its rubber surfacing—the air made the nappe stand out. In other words, your goods, or any man's goods or services require at least some consideration from the user if they are to do justice to themselves. All your pains go to waste if your customer or client neglects to use your product as it is intended to be used.

"And so with railway service to you. While the constant vigilance of the managements must keep up the performance of actual railway staffs, nothing but voluntary co-operation can maintain the 'shipping efficiency' of that other part of the transportation machine—the consignor and the consignee.

"Consequently, while attending as closely as possible to their own interior efficiency, the railways of Canada must appeal to the users of the exterior efficiency of the transport system. May we suggest, therefore, that you instruct your secretary to prepare for you a report somewhat along the following lines:

First—As to Containers

"What is our shipping room staff using to pack our goods? Is it really a cheap container? Or is its seeming cheapness offset by the loss of time, or temper, and possibly of customers by its frequent failure to protect our shipments? Or by the worry and fuss and loss of time in having to duplicate orders that should have gone through unbroken in the first place?

"Is our container pilfer-proof? Is it, for example, a carton sealed with strip-paper? Easily slit and resealed? Or is it protected with wire sacks less easily violated?

"Are our men using a crate where they could use plain sacking—as in the case of a certain cushion manufacturer, who cut the weight of a shipment in two by using burlaps? Or should we crate our goods on six sides where we now protect only four?

"Are we using cardboard where we should use crating, or a solid box?

Second—Marking

"How does our shipping room staff address our shipments? With good clean whole stencils? Or with daubs of a worn-out paint brush? Do the invoices show the name and address of our customer clearly and accurately—including the name of the country and province? Or is there a 'temperamental' scribbler making them out? Do the markers copy

them from the invoice just as clearly? Or with one eye on the invoice and the other on Bill Brown's imitation of Babe Ruth on the ballfield?

"Do they use tags that are easily torn off? Or is the marking placed where it can't come off and will be seen easily?

"Has your head shipper a copy of the rules and regulations of the Canadian Freight Association?

Third—Bills of Lading

"Who makes them out? Does he or she write 'a fair hand'? Even so, why are they not made out on the typewriter —most people can use a machine well enough for that purpose?

Fourth—Time of Shipment

"Does your shipping room work overtime? If so, is it necessary? Is there any reason why the work should be 'bunched' at the end of the day? Or could some of it be got rid of, for example, in the morning, when the freight sheds are not so rushed?

"We venture to suggest inquiry along these lines, and improvement wherever possible along the lines indicated. Your co-operation in using railway service in the best possible way will be of great assistance to the carriers, and, therefore, to all Canadian, industry." ·

FOR BETTER MAINTENANCE OF AIR COMPRESSORS

By Mark Purcell*

From a survey of the general situation on railroads in connection with train operation and when direct inquiry is made, the usual impression is that compressors are giving good service, although at the same time service records may show a large amount of wheel sliding and breaking in two of trains that, from first analysis, would not appear to have any relation to, or to any extent to be due to, the manner in which the compressors were operating, and yet careful, or even casual, investigation develops in many cases that

---
*General air brake inspector, Northern Pacific Railroad. Paper read at the annual meeting of the Air Brake Association, Chicago.

the compressor is largely responsible for the trouble. .

If a compressor runs slow, or for any reason raises the pressure slowly, there may not be enough accumulated after a brake application is .made, for slowing down or stopping, to make a proper release, and the engineer, being necessarily anxious to avoid delay, attempts to release and proceeds with the result that only some of the brakes are released and wheels are slid or drawbars pulled out; and, with passenger trains, passengers are annoyed and sometimes injured by rough handling of the train, and still the compressor has not actually stopped and refused to work.

From such conditions as those outlined in the foregoing, and those cases coming up occasionally where the compressor fails completely, we investigate further as to cause and remedy, and find that most of the low efficiency and some of the failures are due to lack of repairs, or to the quality of the repairs made being slightly, the standard required. In some cases it is an error in a single particular, but more often in a combination, such as a number of leaks, each of which may be too slight to cause trouble, yet when combined becoming serious. The above applies more particularly to compressors having steam and air cylinders compounded, on account of the balancing effect that may be caused by leakage.

It is not uncommon to find a compound compressor that is chronically inefficient, and be advised that one part after another has been repaired or replaced, until all or nearly all of the working parts have been gone over, and yet the operation is not improved. Some such cases are a matter of record, where a careful inspection showed everything in fair condition, except the high pressure air piston rings were 3-16 to 5-16-inch open, or rings in both high and low pressure air pistons 3-16 to 1-4-inch open when in the smallest part of cylinder, of rings in some of the pistons of the main valve 1-8 to 3-16-inch open (Continued on page 272)

FIG. 9—SHOWING HOW TRIMMING OPERATION WAS ACCOMPLISHED.

Thickness of Plates, and Riveting of Stacks

The Common Guyed Steel Stack is the One Under Discussion—
This Subject is a Bit of a Mystery to the Average Person; and
Should be of Particular Interest.

By JOHN S. WATTS

THIS article deals only with the common guyed steel stack, and not at all with the self-supporting stack, as the former has not been usually designed at all, but built by rule of thumb, with a consequent wide variation of practice in the thickness of plate used. I have seen all kinds of riveting from 2" pitch up to 12" pitch. The lack of any data published on the life of these stacks has led me to gather up the information collected over a number of years, to try and arrive at some more dependable method of specifying the scantlings required than has been in vogue in the past.

My reason for not including the design of self-supporting stacks is that this problem is pretty well understood, and is really a very different affair, and is therefore not touched upon in what follows.

The first point to be decided upon is the thickness of plate to be used, and while it is well known that a new stack of any thickness almost will stand the wind pressure, until quite badly corroded, no attempt has been made, so far as I know, to determine the thickness required to carry the wind pressure, and to resist corrosion over a pre-determined number of years.

In a guyed stack, the stresses are analagons to those in a beam uniformly loaded, and supported at each end, the load being the wind pressure, and the stack being supported by the foundation at the bottom end, and the guy ropes at the upper end.

D. K. Clark, in his book "Rules, Tables and Data," gives a formula for thin hollow tubes, as beams supported at both ends and loaded in the center, which agrees very closely with the results of experiments made by Sir William Fairbanks on tubes having dimensions similar to those of the stacks we have under consideration.

This formula is

$$W = \frac{3.14 \times D^3 \times T \times S}{l}$$

Where W = Load in pounds
D = diameter of tube, outside, in inches
T = thickness in inches.
S = Stress in pounds per square inch
l = length between supports in inches; in our case this is the length from the foundation to the guy band.

That portion of the stack, above the guy band, being a cantilever, uniformly loaded, will be equal to the other portion of the stack in strength, if made one-fourth of the length of the stack between the foundation and the guy band.

For very high stacks two or more sets of guy bands may be used, and, l is then tween the foundation and the guy band.

In the case of a stack, the load to be carried is that due to the wind pressure, and is uniformly distributed, and therefore the load that can be supported will be double that given in the above formula, which considers a concentrated load. We therefore change the formula to read

$$W = \frac{6.28 \times D^3 \times T \times S}{l}$$

It will make it more convenient to use the length, L (which is in inches), in feet, so making L = the length in feet between supports, we have

$$W = \frac{6.28 \times D^3 \times T \times S}{12 \times L}$$

The wind pressure is generally taken as 56 pounds per square foot as a maximum, and in a round stack, the effective area against which this pressure will act as half the diameter, multiplied by the height, both in feet. The load W will then be

$$W = \frac{D}{12} \times L \times 28$$

Substituting for W, in the beam formulae, we get

$$W = \frac{6.28 \times D^3 \times T \times S}{12 \times L}$$

$$\therefore \frac{D}{12} \times L \times 28 = \frac{6.28 \times D^3 \times T \times S}{12 \times L}$$

$$\therefore T = \frac{\frac{D}{12} \times L \times 28 \times L \times 12}{6.28 \times D^3 \times S}$$

$$= \frac{6.28 \times D \times S}{6.28 \times D \times S}$$

Taking the safe stress to be 16,000 pounds per square inch, and substituting in the above formulae, we have

$$T = \frac{L^2 \times 28}{6.28 \times D \times 16,000}$$

$$= \frac{3589 \times D}{}$$

To the thickness given by this formulae must be added a sufficient amount to cover the loss by corrosion, which may be expected to take place during the number of years which it is desired the stack should last.

Under average conditions, burning soft coal with natural draft in a fairly pure air, and the stack having received one shop coat of graphite paint and, another coat of paint just previous to erec-

tion, the plates will corrode away at a rate of about .01 in. in thickness per year. In very close proximity to the sea this rate will be increased to about .013 in. per year.

If the stack could be repainted once a year, the life of the stack would be about doubled, but the expense of erecting staging to do this painting, which must be done both inside and out to be of any benefit, is prohibitive.

It will, generally speaking, be found cheaper to make the plates of a thickness that will give a life of ten to twelve years as at the end of that period it is very probable that a larger stack will be required to take care of an increase in the boiler plant.

The actual thickness of plate to be employed, then, will be = T + (.01 in. × the number of years of life required).

Experience has shown that the plates do not corrode under the rivet heads, nor do the rivets themselves waste away, it is therefore not necessary to have the riveting any stronger than a solid plate would be of a thickness = T.

The actual thickness of plate used, being designated as P, the required efficiency of the riveted circumferential joint, to equal the strength of plate T, will be

$$E\,ff = \frac{T}{P}$$

The diameter of the rivet used must be such that the bearing surface of the plate, on the rivet, will be sufficient in area to develop the full shearing strength of the rivet. Taking the safe bearing stress as 24,000 pounds per square inch, and the safe shearing stress as 12,000 pounds per square inch, the diameter of the rivet being d, the safe load on one in single shear will be $d^2 \times .7854 \times 24,000$; and the safe load on one rivet in single shear will be $d^2 \times .7854 \times 12,000$.

$$\therefore d^2 \times .7854 \times 12,000 = d \times P \times 24,000$$

$$\therefore d = \frac{P \times 24,000}{.7854 \times 12,000}$$

$$= P \times 2.55$$

The required efficiency of the joint being that of the rivets in single shear, will be

$$E\,ff = \frac{d^2 \times .7854 \times 12,000}{Pitch \times P \times 16,000}$$

$$\therefore Pitch\ of\ rivets = \frac{d^2 \times .7854 \times 12,000}{P \times E\,ff \times 16,000}$$

$$= \frac{P \times E\,ff \times 16,000}{d^2 \times .7854 \times .75}$$

$$= \frac{P \times E\,ff}{}$$

(Continued on page 265)

Reducing the Cost of Locomotive Repairs [1]

The Subject of Cost is a Vital One in Any Business, But Especially so in a Railroad Shop—The Subject of Repairs, While an Unpleasant One, Must Be Met.

By S. W. MULLINIX

MATERIAL represents approximately 40 per cent. of the cost of repairing and maintaining locomotives, a large percentage of which is purchased either rough or finished, and owing to the high cost of labour and the absence of modern and special production machines in many railroad shops, there are many articles which can be purchased in a finished condition more cheaply than they can be manufactured by the railroads. There are also many articles and tools sent from the purchasing to the mechanical department to be tested out to see how they compare with home-made articles and tools, as well as to compare with similar articles from other manufacturers. These tests should be given careful consideration, as they very often decide a standard for the entire system, and sometimes what is considered a successful article or tool in one shop is complained of by another. For that reason, when an article is being tested it should be gone into thoroughly, having competent employes pass on it.

Were one to stop and figure, one would be appalled at the cost in dollars and cents of the unnecessary labour which is wasted on material such as hard castings, requiring twice as long to finish, or the many hours of labour that are wasted on some pieces of material that turn out to be full of blowholes or cracked. They may have been passed on by the test department, but no matter how efficient the inspectors are, they will not find all of the defects, especially in our rough material. Even though defective material is replaced by the manufacturers, they do not pay for transporting back and forth to the store department, and from there the handling to the shop and back, nor do they refund the amount spent for labour in the shop.

The mechanical department should work very closely with the store department on unsatisfactory material, calling attention promptly to any article that is not standing up or giving the service that should be expected, so that they can get in touch with the manufacturers for correction.

Locomotive Cost

During the war period the locomotive cost has also been greatly affected, due to shortage of material, making it necessary to substitute all classes of material. The mechanical department is to blame, to some extent, for this condi-

*Proceedings of the Western Railway Club.

tion, due to the changing of patterns and specifications, which at times may be a paying proposition, but it tends to discourage the store people from carrying any large amount of stock. The mechanical department is also responsible for the great number of parts which the store is compelled to carry in order to supply the various styles and types of equipment, such as air brakes, injectors, safety valves, and lubricator parts. It would be far cheaper to scrap the obsolete equipment, replacing with standard material, which will give better results and reduce the number of repair parts to be carried in stock. We must check over our material and equipment, simplifying stocks wherever possible, and we must not merely go over them once and then sit back and imagine that the job has been completed. It never will be as long as we operate railroads.

Another very expensive practice is the robbing of material from one engine for another. We not only pay for removing and replacing the part removed, but the chances are that it will cost more to apply the piece removed than it would cost to apply a new one. Each of these items traced down will disclose that if someone, whoever he may be, had been on the job it would not have been necessary. A large percentage of the material which we use does not give the service that should be expected, on account of extraneous conditions. For instance, a piston rod may be packed half-a-dozen times, never getting it tight on account of imperfect rod, or guides not being properly lined. Had the trouble been found out in the first place and corrected, it would have saved five sets of packing and the cost of applying. This applies to many items used on locomotives.

Material, where carried in stock by the store, should not be ordered before it is to be applied, especially small finished parts, as they will be thrown about and either broken or lost, necessitating ordering other pieces, which will double the cost. Shop people should be more familiar with the cost of different items of material which they use, as it is surprising to many good railroad men to know just what many small items really do cost, and probably they would think twice before throwing old parts away and putting on new ones.

Another item of material where the quality has a great deal to do with the cost is in connection with the painting of locomotives, for no matter how care-

fully or how scientifically it is applied, poor paint will not give good results, and will have to be removed much oftener than good paint. We are not merely painting our locomotives, especially our freight engines, with a view to beautifying them, but to preserve the material. An old coat of good paint is much better than a new coat of poor paint.

We have the opportunity to be far more efficient in the handling of material than with labour, as the human element does not enter into it to any great extent, and if we watch more carefully the things which we scrap there would not be so much material to reclaim, for the more lax we are in scrapping material the better showing the reclamation department can make.

Manufacturing Locomotive Parts

The use of forging machines in the blacksmith shop has worked wonders in the way of producing forgings which require very little or no finish, such as small side-rods, eccentric rods, motion work, spring hangers, and yokes, large hexagon nuts, and like material. Many large shops to-day are equipped with modern forging machines, which with the proper dies are a great help in reducing costs, as many of the parts now made require very little finish, where formerly they were finished all over. Blacksmiths have a great chance these days to do their work so as to eliminate finish. For example, in main and side-rods centre sections may be forged to size, requiring no finish except where the rods are channelled. Guide blocks can also be forged so as, to require no finish except where they fit on the cylinder heads and the face for the guides and yoke. Guides should be forged so as to eliminate as much finish as possible. Parts such as piston rods and valve stems should be cut to the proper length. Wedge bolts should be forged and ends squared, so that all that is necessary to complete them is to thread them in a bolt cutter. These were formerly made from bar stock on turret machines. Eccentric blade forks, ends, and jaws for other motion work should be made allowing finish only where they fit on to the links—outside contours may be ground off. Eccentric blades for outside valve gears should be forged, allowing finish only on the sides, on the back end and for the link fit on the front end. Large hexagon and castle nuts should be forged and holes punched allowing very little finish. There are many other items where finishing can be reduced or eliminated, and, naturally,

the less the number of operations necessary to produce a finished part, the less the cost will be. Dies for this work are very expensive and should be well taken care of, as they will pay for themselves many times over.

Castings

Patterns should be changed so as to reduce or eliminate finish where possible, and changed from brass to cast iron or steel or from cast iron to steel where parts are apt to break. If it is absolutely necessary to use brass, use it; if not, use either cast iron or steel, but no matter what is used no more finish should be called for than is absolutely necessary. We are using many castings in the rough to-day which a few years ago we thought necessary to finish all over, such as side-rod collars, other collars for motion work wedge-blocks, and collars for main rods, front and back ends. also for strap middle connections, and many other small castings. Finish may be reduced on parts such as cast-iron piston heads, allowing finish only on the outside for cylinder size, and the inside for the rod fit. Front cylinders and valve heads may be cast with finish for the joint only. Core slots in driving-box wedges for wedge bolts to size, and maintain one standard for all wedges. All grate, side, and centre bars and supports, ashpan and grate shaker-rigging castings should be cast ready to apply on engines with all holes cored and all castings the proper size and shape. The use of good metal patterns is a help in securing good castings of the proper size and shape, as they will not warp or twist out of shape, and once they are made right they will always be that way. Eliminate as much as possible finish on whistles, cab fixtures, blow-off cocks, boiler checks, cylinder cocks and valves, bells and yokes, relief valves, parts for piston and valve-stem, packing, oil cups, swab holders, grease cups and plugs. Many of these items are used in large quantities, and every little saving that can be made will help to reduce the cost.

Where it is necessary to finish castings or forgings, enough allowance for finish should be made, as it is much cheaper to remove ⅜-inch than to just scrape off the sand or scale.

Reclaiming Material

Wonderful progress has been made in the last few years in the way of reclaiming material. Although 'the results will not be so noticeable in the next few years, they will be as good at least, as a more careful study is being made of the things which might be reclaimed, as well as of the new places where reclaimed material can be used. At Silvis we are making many special machines, such as rattlers for cleaning nuts and other small material, cutting shears and machines for straightening iron, special cutters for cutting gaskets and washers from old leather and rubber. The electric and acetylene welders have also worked wonders in the way of reclaiming. Most of this material is reclaimed after reaching the scrap dock, and is handled in most cases by a special reclamation department.

There is another side to the reclamation question, one which is not so often heard of. This is the reclaiming of material such as broken parts on locomotives without removing them, or by removing, reclaiming, and replacing on the same engine they were removed from. In most cases this is done with either the electric or acetylene welders. There is very often a close margin between the cost of reclaiming some certain piece and purchasing and applying a new one. In these cases we should apply the new piece, as too often in figuring the amount which we save we colour the figures to make them look as attractive as possible, or to otherwise try to make a good showing, without taking into consideration the life or wearing qualities of the piece we reclaim against that of a new one.

We must guard against letting our enthusiasm get the better of our good judgment in the things which we reclaim. Unquestionably, there are parts being reclaimed to-day at a loss, both in material and in labour. Other cases will actually show a great saving. Where parts, such as piston rods, valve stems, and other like material have been reclaimed by building up for fits, they should be properly annealed before being turned to fit. Where rods or stems are worn near to the scrapping limit, or to the age limit, if we have one, it is not advisable to attempt to reclaim them. The practice of welding small broken cast-iron parts which, in the first place, do not require much finish, is not worth the chances one takes in getting a good weld. In many cases new parts will be just as cheap and much better.

Looking back through the years one can plainly note the progress made day by day and year by year—made possible both by more modern machines and tools and by improving the methods of doing work. Still progress has not kept pace with the modern locomotives, which are not only much larger, but are equipped with modern appliances, such as mechanical stokers, super-heaters, feed-water heaters, brick arches, and power reverse, all of which add to the cost of maintenance. The average cost of repairing engines has increased year by year, but by united effort it should be possible, by more efficient methods and practices, to stop its upward trend, which is necessary before we can hope to reduce it.

Thickness of plates and riveting of Steel Stacks.

CONTINUED FROM PAGE 263

To illustrate the mode of procedure we will take a concrete example, say a stack 40 in. diameter by 100 feet high, to last an estimated time of twelve years.

As explained above, the length of the stack from the foundation to the guy band should be four times the length of the stack above the guy band, the length L will therefore be four-fifths of 100 feet, that is, 80 feet. Substituting in the formulae we have

$$T = \frac{L^2}{3589 \times D}.$$

$$\therefore T = \frac{80^2}{3589 \times 40}$$

$$= .044 \text{ in.}$$

For corrosion during twelve years, we must add .01 in. $\times$ 12 $=$ .12 in., making a total of .044 in. $+$ .12 in. $=$ .164 in.

The nearest gauge to this is No. 8 B. W. G., which is .165 in.

The efficiency of the joint, which will be required to equal the strength of a plate .044 in. thick, will be

$$\frac{.044}{.164} = .268.$$

The diameter of the rivet will be

$$d = P \times 2.55$$
$$= .165 \times 2.55$$
$$= .4207 \text{ in.}$$

which will call for a $\frac{7}{16}$ in. diam. rivet hole.

The pitch will be

$$\text{Pitch} = \frac{d^2 \times .7854 \times .75}{P \times E \text{ ff}}$$

$$= \frac{.4375^2 \times .7854 \times .75}{.165 \times .268}$$

$$= 2.54 \text{ in., say } 2\frac{1}{2} \text{ in. pitch.}$$

While this riveting is only necessary at the points of maximum bending moments, it is probably as cheap to make the riveting uniform throughout as it would be to having wider pitches at the points of lower strain.

The rivets in the vertical seams have no other function to perform except that of holding the joint together without openings that would leak. air into the stack, and a pitch of sixteen times the thickness of the plate will answer the purpose well.

In a paper read before the American Chemical Society at St. Louis, Jerome Alexander recently announced the discovery of a new fuel fluid, greater in heat value than either coal or present fuel oils. He asserted that the fluid would prove valuable to navigation, permitting a wider cruising radius, and could be used for smoke screens. The new fuel utilises coal waste and cheap tars.

Evolution of the Electric Brass Furnace

Electric Crucible Furnaces and Overhead Resister Furnaces Are Discussed in This Interesting Paper which Was Presented Before the American Electrochemical.

By H. M. ST. JOHN

AN electric furnace classification is based, as a rule, on the difference of method utilised for the application of heat to the material under treatment. In this case the species are three: The induction or direct resistance furnace, in which heat is generated in the metal itself by virtue of its own resistance to the passage of an electric current.

The arc furnace, in which heat is generated between an electrode and the metal, or between independent electrodes, and transferred to the metal by conduction or radiation, direct or indirect.

The indirect resistance furnace, in which heat is transferred from an incandescent resistor to the metal by conduction or radiation, somewhat as in the arc furnace.

All of the electric furnaces so far proposed for melting brass naturally fall in one of these classes, or a combination of some two of them.

In the beginning the would-be inventor of an electric brass-melting furnace was obsessed by the idea that such a furnace should bear a close resemblance to the combustion furnaces then in use for that purpose. In the electric crucible furnace as in the fuel-fired crucible furnace, the heat must, in general, be generated at some point outside the crucible and transmitted to the metal through its walls. A favorite suggestion was to surround the crucible with an electric resistor of granular nature, which was heated to incandescence by the passage through it of a suitable electric current.

Most of these fundamental difficulties were obviated by using a resistor which surrounded the crucible but did not come in contact with it. In this way the heat generated in the resistor was first transmitted to the crucible by radiation, and, finally, by conduction through the walls of the crucible to the metal.

To the writer's knowledge, the only furnace of this type which was ever operated with any degree of technical success was equipped with wall resistors of thin carbon slabs, provided with means of variably adjusting the contact pressure between the slabs. The generation of heat in this furnace depended, not on the resistance of the carbon itself, but in the contact resistance between the slabs which could be varied at will in such a way as to provide an excellent means of controlling the current and voltage, and thus the rate of power input. Despite its good qualities, it was soon apparent that this furnace could never be commercially successful. Its maintenance cost, both for carbon electrical parts and refractories, was a serious handicap, but more serious still was the fact that its thermal efficiency was inherently and irremediably low.

The latest proposal to melt brass in an electric crucible is one which has recently been very thoroughly described and discussed, in which the metal within the crucible is heated by a high-tension, high-frequency induction or eddy current. The walls of the crucible itself are electrically conducting and serve to heat its contents during the period while the metal is still solid and in pieces not in good electrical contact with each other. Most of the disadvantages already described as characteristic of the crucible furnace have been overcome in this design. This new type has, however, so far been built in sizes more suitable for the laboratory than the foundry.

The low thermal efficiency of crucible furnaces heated from without naturally suggested the possibility of utilizing the walls of the crucible itself as a resistor. This principle was tried out quite thoroughly rather early in the development of electric furnaces for melting brass. In one type the crucible was built of a special mixture with suitable electrical conductivity, but no attempt was made to insulate the walls of the crucible from the metal which it contained. In another design this feature was taken care of by means of special insulating lining separating the conducting walls from the metal. It proved almost impossible, however, to maintain this insulating layer, and short circuits invariably resulted.

Overhead Resistor Furnaces

Not all, even of the early experimenters, limited their attention to the crucible furnace. The advantages of a larger furnace capacity and the elimination of crucibles were sufficiently obvious, and had already resulted in a considerable use of various types of direct-flame oil and gas furnaces. The earliest attempts to melt brass electrically in a furnace of this sort made use of an incandescent resistor supported above the bath and radiating heat directly to the metal. This construction, applied to an open-hearth furnace of small capacity, gave rapid melting and a fairly high thermal efficiency. The principal difficulties were two; the development of a resistor which would stand up continuously under the required conditions without an excessive maintenance cost, and the invention of some reliable method for supporting the resistor in the desired position over the bath. Neither of these basic problems has ever been adequately solved.

The difficulties which interfere with the use of an overhead resistor are partially avoided if the resistor is located above the bath, but at either side or surrounding the central portion of the melting chamber. One well-known type of furnace now in commercial use employs this principle, utilizing for the purpose a granular resistor contained in a nearly circular refractory trough. This trough is exposed to very severe usage, since the resistor temperature must be much above that of the molten metal. The roof also is at a temperature considerably in excess of that of the metal, and must be highly refractory. Another type of furnace which is in somewhat limited commercial use employs a combination of granular resistor and smothered arcs at either end of the melting chamber. In this type also most of the heat must first be radiated to the roof and then to the metal.

A high thermal efficiency was early recognized by investigators in this field as an absolutely essential qualification for the permanently successful electric brass-melting furnace. It semed obvious that if some practical method could be devised for generating heat in the metal itself, conditions most favourable for a high efficiency would be produced. The celebrated pinch-effect phenomenon is too well known and has been too often discussed to require definition here. The pinch-effect can be utilised to increase substantially the electrical resistance of molten brass through which a heavy electric current is flowing. This was done with considerable success in designing the first practical direct-resistance furnace for brass. It has recently been proposed to change the construction of this furnace in such a way as to eliminate the massive metallic electrodes.

The elimination of electrodes from the design of the direct-resistance furnace was evidently of the highest importance. This was done in a somewhat later type of furnace by constructing it as a vertical-ring induction furnace with the resistor channels joined at the bottom to form a complete circuit for the passage of electric current. The limitations of the vertical-ring induction furnace are due, first of all, to the fact that it is an induction furnace, and, as such, cannot be constructed in large sizes without introducing electrical disadvantages such as low power factor.

Furnaces of the Arc Type

The widespread success of the arc fur-

naces in the melting of steel was not overlooked by those more particularly interested in the brass industry. Many attempts were made to apply both direct and indirect arc furnaces directly to the melting of brass, without changing the design which had been found most suitable for steel melting. These attempts were pretty uniformly unsuccessful, because copper and its alloys—particularly the high-zinc alloys—suffered under the direct application of such a high temperature heat source as the electric arc.

A great deal of study was devoted to the discovery of some method which would make possible the utilisation of th good features of the electric arc for brass melting. The direct type of arc furnace was evidently out of the question. In the indirect arc furnace, overheating was less localised. It seemed probable that if the metal could be stirred with sufficient vigour, the entire bath could be maintained at a uniform temperature, and tendency toward local overheating could be entirely neutralised. It was found by experiment that rocking the furnace mechanically, at the rate of approximately two oscillations per minute, resulted in a degree of agitation ample to maintain complete uniformity of temperature throughout the molten bath, and that the obvious advantages of the arc furnace could be utilised in this way without the slightest injury to the metal, even in the case of alloys containing 40% or more of zinc.

The so-called rocking electric furnace resulted from this development and is in wide commercial use at the present time for melting all classes of copper alloys, as well as copper itself. As in the induction furnace, the vigorous stirring of the metal results in a uniformity of temperature and of composition throughout the alloy, a feature which is of particular importance in the melting of high-lead alloys.

The pronounced success of the rocking type of arc furnace has prompted many suggestions for modified designs, similar to it in principle but differing from it in details of construction. For example, it has been proposed to rotate the furnace body instead of merely oscillating it.

The evolution of the electric brass furnace has now proceeded to such a point that fundamental improvements in design are henceforth likely to be rather slow to materialise. There will, of course, be constant progress in the development of refinements in mechanical and electrical design, calculated to make the furnaces more reliable, more durable, and more nearly foolproof than they are at present.

This brings us to a brief consideration of what the brass-foundryman can expect and ought to realise from the use of electric furnaces.

The first and probably the most important point is the saving of metal—commonly wasted during the melting process—which electric melting makes possible. If the charge to be melted

consists of new metal or clean scrap, the net metallic loss during melting and pouring from the furnace should not exceed 1 per cent. For yellow brass and 0.5 per cent. with red brass. With clean yellow brass, containing 40 per cent. of zinc, losses as low as 0.75 to 0.85 per cent. have been experienced as an average for a considerable tonnage of metal melted. In melting a scrap charge containing a high percentage of non-metallic, such as oily borings, chips, grindings, etc., the net loss should not exceed 2 per cent.

It is also true that the furnace which melts without agitation is not particularly well suited for melting a charge which contains a high percentage of finely divided, dirty scrap, while these can be handled without difficulty in either induction or rocking arc furnaces. Brass melted in the electric furnace is practically free from metallic-oxide drosses and has no opportunity to pick up sulphur or other contamination from combustion gases.

The consumption of electricity energy under average conditions of 8 to 10-hour operation is as low as 240 kw.-hours per net ton for yellow brass, and 275 kw.-hours per ton for red brass, in the induction or arc furnaces. In 24-hour operation figures as low as 200 kw.-hours per ton or less have been obtained. Less efficient furnace types use from 400 to 500 kw.-hours per ton, depending upon conditions.

Flexibility, which in this case may be defined as the suitability of a furnace for radical changes in operating conditions or for an abrupt change in the composition of the alloy to be melted, is a marked characteristic of the resistance and rocking arc furnaces, which is notably lacking in the induction furnace.

The net melting cost, considering all factors which should properly be considered under this head, is naturally lower in those furnace types which melt the metal most rapidly and efficiently, since their use of electric energy is more economical and their higher rate of production reduces the fixed charges per ton of metal melted. In many cases the cost of electric melting is not more than half the cost of melting the same alloy in combustion furnaces. Even in the less efficient furnace type, electric melting is usually less costly than the older methods.

Electric brass melting can no longer properly be called "the coming thing." It has arrived in a most convincing fashion, as is evidenced by its rapid adoption by the larger and more progressive rolling mills, foundries, and manufacturing establishments which use brass in large quantities.

ADJUSTABLE TAPER GAGE

The Knauel Tool Works of Rock Island, Ill., have brought out an adjustable gage that may be quickly set to any desired taper and locked in position. These gages are only made in the open size but will cover a wide range of taper work.

GRAPHITE IN CANADA

A report on graphite, just published by the Mines Branch of the Department of Mines, contains a wealth of information on the subject of this interesting and important mineral. The report is written by Mr. H. S. Spence, mining engineer of the department, and treats of ore properties, occurrence, distribution, mining, and uses of graphite in a most comprehensive manner. Interesting information is given on the present status of the graphite industry in Canada, and on the outlook for the industry as it is likely to be affected by foreign competition.

The report points out that Canada possesses deposits of flake graphite superior in richness and quality of flake to any on the American continent. What is probably the largest and richest deposit of flake graphite known in the world occurs in Ontario, and is worked by the Black Donald Graphite Company of Calabogie. Difficulties of concentrating and refining the graphite, however, have long hampered operators and have mitigated against the establishment of a flourishing industry. Quite recently these difficulties have been overcome by the employment of the oil-flotation system of ore concentration, which yields far better results than were obtainable by the old methods, both in the richness in carbon of the concentrates made and in the amount of graphite recoverable from the ore treated. Several Canadian mills have now been equipped with the above flotation process and are producing refined graphite equal, if not superior, to the best graphite in the market.

Crucibles used in the melting of steel and alloys consume a large proportion of the graphite produced, and other important uses are in lubricants, paints, foundry facing, pencils, stove polishes, dry batteries, dynamo and motor commutator brushes, electrodes, and boiler scale preventives. In all, about 50 different uses of graphite are listed in the report, which, in addition, gives much interesting information on the methods of manufacture of a number of the more important graphite products. The report consists of about 200 pages, and is profusely illustrated with photographs, drawings, and maps. Copies may be obtained by application to the Director, Mines Branch, Department of Mines, Ottawa.

Montreal.—Another addition to the fleet of the Canadian Government merchant marine will be made about the middle of this month when the complete "Canadian Runner" will be turned over to the company by the Canadian Vickers, Ltd. Launched on May 8, 1920, from the yards of the Port Arthur Shipbuilding Company, the vessel is of 4,350 dead-weight tons, and had to be cut into two sections to come through the Lachine Canal. The bow section is already at Vickers and the stern is expected to arrive to-morrow. She should be ready to sail, probably for the United Kingdom, about the end of the third week in September.

Industry's One Law "Be Square With Your Men".

Lasting Success Only Attained by Understanding and Utilizing Human Element in Accordance with Principles of Justice— Scientific Achievements Heralded, While Greatest Forces Have Not Been Cultivated.

By ERNEST E. BELL

INTEGRITY of purpose coupled with ability for accomplishment, or the lack of these, I doubt not, influences the fixing of credit with everyone of you more often than does the financial statement. In taking a broad survey of industry generally and casting a balance of assets and liabilities for industrial progress, we find one of the most important, if not the most important factor, to be the human element. That industrial institution which under present conditions fails to realize this fact and grasp the fundamental principles upon which the human element can be utilized for industrial progress, regardless of its financial statement, is neglecting to cultivate one of the most important factors for future profit.

An appreciation of this fact, with intelligent inquiry into the attitude of management toward labor, and the attitude of labor toward that particular management, is becoming of more and more importance to the banker and those entrusted with the fixing of credits. It is to be hoped, however, that those responsible for extending financial support to industry will not satisfy themselves with inquiry into these relationships, but will inform themselves as to the fundamentals underlying right industrial relationships, so that they may intelligently awaken and guide effort in this direction.

It all resolves itself, in the final analysis, into a question of human relationships, and the question of human relationship is as old as the Garden of Eden. As between man and woman, it started when Eve asked Adam to partake of the apple. As between men, it was raised, and has been a question since Cain inquired, "Am I my brother's keeper?" As a factor in industry it has existed since the herdsmen of Lot had strife with the herdsmen of Abraham on the plains south of Bethel and since Pharaoh's taskmasters required the children of Israel to make bricks without straw.

As for you and me, our earlier teachings of love, sympathy and kindness were gained at our mother's knee, while our impressions as to right conduct and punishment for wrongdoing were received over our father's knee. Such thinking as we may have done since has been influenced very largely by these early impressions, and whether the teachings of our mother or the experien-

ces with our father in the woodshed have predominated, depends somewhat upon the kindness or justice with which these were administered.

Human relationships find expression through family ties, through friendships, through contacts which result from the performance of our daily tasks. Systems of government have been established in an effort to control and co-ordinate human action and protect human rights. Men differ in their opinions as to personal rights and as to forms of government. But the most rapid advances of civilization and industry are recorded under those forms of government which, within proper limits and with justice to all, have permitted of individual initiative and action. The principal effort of every human being is to find contentment and happiness, and here again men differ in their conception of what is required to produce contentment and happiness. But the majority of men have concluded that true · contentment and happiness are not to be found in the acquisition of material wealth at the expense of others nor in the wielding of political or industrial power for purely selfish ends.

It has been said that all men are born free and equal. That all men do come into this world with equal right, and should have equal opportunity for the pursuit of liberty and happiness, is unquestionably a right premise. But whatever their right we certainly cannot agree that man's opportunity is equal, and even if it were all men are not born with equal ability to succeed, nor with equal capacity for appreciation of the many and various purposes for which individuals may rightly strive. St. Paul tells us that some are born to prophesy, some to preach, some to teach. And the greatest Teacher of all ages tells us that there shall always be rich and poor, masters and servants, those who exercise authority and those who submit to authority. But this same Teacher gives justice, kindness and love as the controlling factors in all human relationship. He condemns with equal emphasis oppression by those in authority and disobedience on the part of those who should serve.

Improvement in forms of government undoubtedly is possible and betterment in industrial relationship should come with it, but the lessons which we draw from the wisdom and experience of the past lead to the inevitable conclusion that there must be an authority with

power to direct action; that individual capacity plussed by individual effort will make for individual advancement, and that all individuals are not equal in natural capacity nor is their willingness to make sacrifice and to wisely apply their effort. Some men, therefore, will outstrip their fellows and thus will acquire advantages and benefits, to which they are rightly entitled, but which they must use with the true spirit of consideration for the rights of others and not for the unjust exploitation of those less fortunate.

Regardless of forms of government men are contented and happy only when they have opportunity for individual expression in effort, when they are free to apply themselves, with proper reward, to the limit of their natural capacity. It is this that makes industry, and industry creates those things which minister to the physical necessities of humanity. Unless work is rewarded by a fair return for effort, and to be a fair return it must be sufficient to provide for the normal necessities of the individual and those dependent upon him, no form of government is sufficient to guarantee the happiness and contentment of the human race.

Industry, therefore, as it provides opportunity for individual effort and accomplishment, becomes of paramount importance in human relationship and to human happiness. Industry started when God found Adam clothed in a fig leaf, led him to the nearest exit from the Garden of Eden and told him to go out and earn his bread by the sweat of his face.

But man can never escape his responsibility for individual effort—work— and to the extent that we try to escape that individual responsibility will we find discontent, want and distress of every character. For the past four or five years most of the civilized world has devoted its energies to destruction instead of industrial construction—the result largely of mutual reaction from oppression. In the utilization of material forces men have learned to study and recognise natural laws. But in human relationships we seem to have overlooked the natural laws of cause and effect, of action and reaction. Where oppression is the cause the effect must eventually be the reaction which we now witness in Rusia, which tears at the very foundations of civilization by the roots, turns order into chaos and makes men reckless in their strife for the recogni-

*Address delivered before the Pittsburgh Association of Credit Men. Ernest E. Bell is Vice-president of the Hydraulic Steel Co., Cleveland.

tion of those things which they demand as natural endowment.

Less Thought Given Human Relations

Man seems to have acted with less intelligence in conducting human relationships than in the subjugation of material elements. He seems to have given less thought and less study to the natural laws which control humanity. Just as water can be utilized for power through recognition of the natural law of gravity, so can the power of the human race be developed and utilized only when we recognize the natural endowments of mankind and operate in accordance with them. In our relationships with family and friends we recognize a natural law of love, and through acts of kindness show appreciation of this natural attribute and produce a reaction of like character. When we step out of the family into industry we seem to assume immediately that this natural attribute of the human race fails to exist and apparently set ourselves to operate upon the theory that all men are selfish, exclusively selfish, and through operating upon this basis, produce only selfishness as a reaction.

Seek Cause for Dissatisfaction

Scientists, psychologists and economists have taken up the study of this problem in an effort to find the cause for dissatisfaction in the industrial relationship between individuals and classes. An article recently appeared in one of the national weeklies which gave some very interesting data in regard to the reaction of certain laborers to music. It said that some who gave no evidences of refinement, without education, who were seemingly capable of only the most ordinary tasks, showed marked appreciation for music of a high order, and left noon hour concerts with tears of emotion in their eyes, as they thanked those who had provided this pleasure and diversion for them.

Another experience is that of a steel company which is making intelligent effort to interpret right relationships between employer and employe, to gain the confidence of their men by, perhaps, unusual methods. This company has secured one of the best artists in the country to produce paintings of men and factory scenes, some of which are reproduced in their plant publication. This artist tells of a common laborer, a roustabout in the steel mill, who looked over his shoulder one noon hour as he tried to portray on canvas the molten metal flowing from an open-hearth furnace. The man said to the artist, "You paint God." In surprise the artist inquired what he meant, and the laborer went on to express in broken English his idea that God is in the hot metal, God is in the iron, God is in nature. And this artist has had a lot of experiences, which show that under any spark-burned flannel shirt beats a heart which is full of sentiment.

Sociologists have stories to tell of the reaction which comes from kindly deeds, performed not as acts of charity but of common justice, and in such manner as to recognize the self-respect of the individual, from which they conclude that regardless of employment, environment or the station of life in which the individual may be found, the same human impulses actuate all mankind. Science through continuing this effort eventually may make the unique and remarkable discovery that the human race is human; that there is in the hearts of men those things which God originally planted there, sympathy, kindness, love, and that these will react to sympathy, kindness and love. But when the psychologists, the sociologists and the scientists get through, if their science leads to right conclusions, they will go back to the teachings of the Carpenter of Nazareth and proclaim in unison that the only right basis for human relationship is to "do unto others as you would have others do unto you." And when men learn to treat with each other on that basis there will be no difficulty in the reconciliation of differences in industrial life.

But, you say, this is sentimentalism, and there is no room for sentiment in business; that the function of business is to acquire gain, and that its very character, therefore, makes necessary the elimination of those human attributes, in the full expression of which we find pleasure and satisfaction in the intimate relationships of family and friendship. Let us look at this problem, considering business from the most selfish and narrow standpoint, as being the art of acquiring gain in worldly goods. To this end we employ our every energy and seek out every element which can be utilized for the accomplishment of this purpose. The most important of these is the human element. From a purely selfish standpoint, therefore, it behooves us to study the potential power of human endeavor for industrial progress will be released.

Natural Laws Aid Progress

Man's ability to use material forces is dependent upon his ability to understand and recognize natural laws. We recognise the natural law of gravitation in damming up streams for the creation of power. Men of unusual ability, of particular training and knowledge, have spent decades in an effort to learn the natural laws through which the power of electricity can be utilised. But have we used the same intelligence in our study of the human element as we make effort to utilise its power for industrial progress? Conditions today in industry give positive answer. Industry in many parts of the world is paralysed, and in our own country is suffering to an extent from which it will take years to recover, for the very simple reason that we have not understood the hand-ling of men; have not applied ourselves with the same degree of intelligence to a study of human attributes and the proper treatment of human beings as has been used in the study of material forces for their utilization. Our very selfishness has reuited in a reaction of selfishness, and industry and individuals are now paying the price, not only in discontent and unrest, but in dollars.

If men react and become more constructively useful through our giving consideration to the human attributes of love and kindness, then from the most selfish standpoint of individual gain we can find no excuse for refraining from giving full expression to these sentiments in our industrial human relationships. Did the lords of Russia who practiced oppression during decades and centuries past act with far-sighted wisdom from a purely unselfish standpoint? No, and against them must the indictment be drawn for the present destruction, which is nothing more nor less than reaction from centuries of injustice and oppression. And we may well consider against whom our children and our children's children will draw the indictment, should our failure to properly conduct ourselves in human relationships in industry result in destructive reaction in years to come. Possibly this is sentimentalism, in that it treats with attributes of sentiment, but in the last analysis it is only commercial self-interest, because it makes for the utilization of the potential power of the human element in industrial progress and for protection against that destruction and loss which reaction from opposition to natural laws must and inevitably will bring about.

This does not necessarily mean that business is to become a love feast or a Sunday School picnic. "Business is business" and doubtless always will be. But it may be good business for us to pause long enough to think seriously of these matters from a purely business standpoint; to make an inventory of business generally; to face without prejudice the debit and credit columns of those factors which advance or retard constructive progress; to let our vision extend beyond our own profit and loss statement for this year in protection of right profits in years to come.

It may not be expedient to give expression to love and kindness in the same manner in business as we express these feelings in the family circle or with intimate friends. Men do not expect kindness to find expression in business through the giving of something for which an adequate market return is not required. In fact, they resent such an attitude as has been shown where so called welfare work has been attempted with kindness as the only compelling motive. But where kindness has resolved itself into sincere interest in the other fellow's well-being and is expressed in terms of justice and not charity, better relations have been and are being

established. And this is true because there is in every human being an instinct which demands the recognition of self-respect and this self-respecting individual says: "I don't need your kindness in the form of charity. Give me justice in the right spirit and I ask no more." And whether men realize the fact or not, few indeed, fail to appreciate and react to conduct which finds its impulse in recognition of the principles of the golden rule and in an effort to observe those principles.

This also is human instinct which demands the privilege of self-expression and it is this instinct which causes men to resent and resist ready-made plans which are handed down by well-meaning but misguided ownership; plans toward the forming of which labor has given no thought and in the carrying out of which it has no function. We are perfectly willing to have an architect draw the plans for our new home, but every member of the house likes to be consulted before the plank are finally approved. If each one of us, therefore, will consider the other fellow as having the same human attributes as we and accord him the same treatment to which we would naturally respond were our positions and conditions reversed, reactions from such conduct would soon adjust all differences. Men only want to be treated human, not humane.

The leader of a labor union came into my office some weeks since and asked in connection with some statements which had been made over the signature of our company, just what we meant by management having an equal responsibility to capital and to labor. In reply I asked him, "Just what do you mean by capital?" His statements indicated that he looked upon capital as that vague, intangible, unapproachable, unpreachable, inhuman power, which is labor's enemy, and the sole purpose of which is to exploit labor. With evidence of this mental attitude on his part, I inquired if a fair definition of capital would not be the conservation of energy, and used this illustration: "All men have been enjoined to work, to exert the energy, physical or mental, with which nature has endowed them. Let us turn two men loose in the wilderness. One applies his energy to clearing off a patch of ground, to building a cabin, to getting a garden started, so that he may have food for the winter, to make provision generally against future needs. When winter comes he has a warm cabin, plenty of firewood and plenty of food. He has built up a reserve to make him comfortable. The other, instead, spends his time in loafing, hunting and fishing a little when he is in the mood. This is all very well for the summer, but when winter comes, will he not find himself some cold morning knocking at the cabin door of the man who worked and conserved, asking for a handout and a place to warm his feet? The first man has simply created capital in

the form of those things which protect him against want. The second man has failed to do so.

The Lesson of Tony and Mike

"And to bring the illustration a little nearer to present day conditions: Tony and Mike started to work in a steel mill some years ago. They both started on the same day and at the same rate of pay as laborers in the yard. Tony was an energetic worker and applied himself to his task. When the day's work was done he used to go home to his family, conserving his energy for the next day's work. He always managed to save something out of his pay, and by taking care of his physical welfare, he made himself an increasingly valuable man. Within a few months he asked the foreman of the yard how he could fit himself for a better job. The management provided opportunity for Tony to get a better education by studying nights. In the course of time he became a unit boss. His increased pay enabled him to lay by a larger amount in the savings fund, and when the company gave an opportunity for employes to buy stock Tony had a cash down payment ready and got hold of a block of stock. In the course of time Tony became a foreman. He had become a capitalist through his investment in the company's stock, and when he was taken sick and died, it was with the thought that his wife and family were cared for through this stock which had been acquired by his thrift and conservation. Mike shirked his work from the first day. He just about held his job. His evenings, as well as his pay, were spent in the saloons. Physically, he became less capable every year. He still works as a roustabout in the yards, but, he is not nearly as good a man to-day as he was 10 years ago. And he hasn't saved a cent."

At this point I asked, "To whom does the management of the company owe the greater responsibility? To the capital created through Tony's energy, represented in the stock which he left for the protection of his family, or to Mike who failed to conserve his physical energies or the result in pay which they produced?" You can guess his answer.

So, when actual human relationships as to cause and effect are pointed out, and it is realized that capital, whether in large or small amounts, is simply the accumulation of conserved energy, which comes from somebody's application, hard work and ability to save, labor would, indeed, be unreasonable did it not recognize the responsibility which management has to those who have acquired capital in this manner. But labor is resentful at being oppressed in order that great accumulations of capital may be accomplished, and rightly so.

And thus the greatest differences which exercise humanity to-day, those existing between capital and labor, may be reasoned out and reconciled if we but approach this problem in the right spirit. If we will subject it to the same

analysis of cause and effect, of action and reaction, as we apply in the utilization of material forces, both capital and labor inevitably will come to see that difference in station and accomplishment are not differences which necessarily lead to controversial strife. The unreasonableness in men is very largely reaction from the unreasonable manner in which they have been treated, whether capitalist or laborer. Industry is suffering on account of our having forgotten or failed to recognize the humanity of the human race, on account of our having acted upon impressions which were gained in the woodshed entirely, and having overlooked those other more important attributes which were the basis of our mother's teachings.

And what are we going to do about it? The Almighty, for the preservation and development of civilisation, has in His wisdom implanted certain instincts in the human soul. They constitute the most potent force in the whole scheme of creation, and until these human impulses have been released, so that the human element in industry may extend itself naturally, labor is no more able to relieve its own unrest than the fever patient is to reduce his own temperature without the cause having been removed.

We can no more change the instincts with which the human race has from the beginning and will until the end of time come into the world than we can cause water to flow uphill. Man has learned to utilize material forces through studying natural laws and adjusting his operations to certain fundamental principles which nature has prescribed. Why, then, is it not good common sense, good business, for us to study the natural forces which control humanity and so adjust our thinking and our doing that industry may benefit in full measure from the power for progress which is to be found in the human element? We don't fight natural forces, but adjust ourselves to natural laws in the utilization of material elements. And until industry shall have learned that all human beings are fundamentally the same, that all characteristics which prompt human action and stimulate human endeavor, are fundamentally the same, differing only in degree, and shall appreciate this fact and recognize it in human relationships, industrial unrest will never be quieted and labor will never be satisfied regardless of the hours of toil, the rate of pay or the cost of living which are offered.

There is no single influence in industry to-day which more quickly changes the balance sheet or more vitally disturbs credits than misunderstandings and strife between those who apply their dollars and those who apply their skill and brawn to industrial progress. Inventories of raw materials accumulate and lie dormant instead of being fabricated into finished product and converted into the wherewithal with which to pay bills. And this may be true, not because of differences at that particular plant, but for causes which have their origin

in other plants, possibly in other industries, and not unlikely in other cities. Our process of industrial development has brought about an interweaving and an interdependence of interests which has wiped out industrial and geographical boundaries.

You pay a high price for flour in Pittsburgh, while wheat rots in the elevators of the great Northwest. Your warehouses are filled with finished product, ready for shipment, and your banks find it difficult to supply the capital with which to keep industry going. These specific instances are due very largely to differences which have arisen between individuals in the transportation industry. If transportation had not been interrupted, the wheat from the Northwest ere this would have reached the consuming market; dollars which have been sent from the East to carry this grain in elevators would be in circulation and your stores of finished product would have been converted into cash.

It is confidence that lubricates the wheels of industry; it is confidence upon which credit is based, but until confidence shall have found expression in right relationships between so-called and miscalled "capital" and "labor," it will be impossible to establish that uninterrupted flow of commodities upon which all credits depend and through which obligations are liquidated.

NOT SATISFIED
WITH OUTLOOK

Increase in Miners' Wages Has Resulted in Decrease in the Output

Roy M. Wolvin, president of the Dominion Steel Company, returned to Montreal from a trip to the Maritime Provinces, where he inspected the company's plants and conferred with the officers in charge of its steel and coal activities. "The steel works at Sydney are at present very prosperously employed and working to full capacity," Mr. Wolvin stated. "We have plenty of orders ahead, and, taken altogether, the outlook is good. Production at the coal mines, however, is disappointing, due in large measure to the fact that the miners are taking many summer holidays, while there is a feeling of restlessness abroad, while waiting for the report of the Royal Commission now investigating conditions in Nova Scotia."

Mr. Wolvin expressed the view that, owing to the conditions existing in the country at the present time, and with the cost of living on the decline, together with the prospect that there would be considerable unemployment during the winter, the commission could not with consistency recommend an increase in wages. The investigation of that body, he stated, would undoubtedly show that the increased wages granted during the past three years had generally resulted in a decrease in the output of coal at the mines.

WAR CONTRACTS
IN THE COURT NOW

Dominion Steel Was Ordered to Roll Rails and Stop Its Work On Munitions

Ottawa.—That the condition of Canadian railways was such in the early months of 1918 as to justify the Dominion Government taking steps that practically stopped the making of steel for shells by the Dominion Iron and Steel Company in order that rails necessary for renewals might be rolled was shown in the hearing of the case of the Dominion Iron and Steel Company versus The King, which opened in the Exchequer Court.

The point at issue was the price the company thought should be paid for 116,000 tons of rails rolled and delivered subsequent to April, 1918, by the steel company, the rails being furnished in consequence of an order in council passed by the Government under the authority of the War Measures Act enjoining the company to supply at least 100,000 tons in order that the railways might be put in a position to operate up to full capacity during the period of the war.

The cause of the difference of opinion as to the value of the rails, it developed during the hearing, was that when the Government order was issued the Dominion Iron and Steel Company had in hand large orders from the Imperial Munitions Board for shells, and the manufacture of these had to be practically discontinued, with the result that 99,000 tons were not delivered. On behalf of the company it was stated that for this steel the company would have received about $80 per ton, but cost of production would have exceeded the cost of making rails by from $3 to $5 per ton.

CAN'T GET RAILS
TO PUT DOWN TRACK

So Building of Many Branch Lines Has Been Held Up For the Present

Ottawa. — Difficulty in securing supplies of steel rails and also shortage of labor are assigned as causes of little progress this year in projected construc-

tion of branch railways. The Canadian National had planned to take up slightly-worn main line rails and use them for branch extensions, replacing them with heavy 85-pound rails; but the latter cannot be secured save with great difficulty and delay, and, as a result, branch line work, especially in the West, has been retarded.

There is likewise a shortage of railway labor due to the fact that Canadians do not take to that kind of work, while thousands of foreigners who used to do it went home during or after the war, and have stayed there. Others have gone to the cities. As the urban labor markets are becoming slack, it is hoped that in these quarters the necessary men may soon be secured for railway extension and maintenance.

BALDWIN'S READY
IN SHORT TIME

Ready to Go Ahead With Production of Material at Toronto Plant

"Baldwins, Limited, hope to have eight mills in operation within two months, all at Ashbridge's Bay," Sir George Wright, vice-president of the company, said as he was leaving Toronto for England. Sir George has been in Toronto for several weeks organizing the plant. "We have sufficient electric horse-power from Niagara to run the eight mills," said Sir George, "and we have been assured by Sir Adam Beck that within the next year we will be able to obtain sufficient power to run any extensions that we may care to make.

"We intend to put down sufficient mills to supply the whole of Canada with black plates, tin plates, galvanised sheet metal and black sheet.

"I have been very much impressed with the help that Mr. Cound, our general manager, has received from the people of Canada. The working people of Canada seem to take a great interest in the plant.

"Like everywhere else, it is true here that the only way in which a plant may be successful is by the help and support of the people who work with us."

Baldwins, Limited, have purchased several reducing furnaces at Collingwood, which they are moving to Toronto.

RAILWAY RATE INCREASE AT A GLANCE		
	Eastern Canada	Western Canada
Freight Rates—		
September 13 to December 31	40%	35%
January 1	35%	30%
Coal increased 10 to 20 cents ton. Fuel wood, 10%. No increase on milk, sand-gravel, crushed stone.		
Passenger Rates—		
September 13 to December 31	20%	20%
January 1 to July 1	10%	10%
July 1	Return to present rates	
Increases not to exceed 4 cents per mile. No increase on commutation fares.		
Special Services—		
Sleeping and parlor car rates		50% increase
Excess baggage		20% increase
Switching, demurrage, stopovers, weighing, etc., must be made basis of special application.		

Did You See Mechanical Drawings at the Ex.?

J. H. Moore.

IT may be possible that many visitors to the Exhibition at Toronto this year have missed the excellent display of mechanical drawings, owing to their peculiar and unsuitable location.

The majority of these drawings are the work of railroad shop apprentices, and the students are to be highly commended upon their work. A wide variety of subjects are covered, and the writer noticed that these included the following:

Drawings of valve motion, section of locomotive boiler, steam pump, model beam engine, air compressor, steam engine, locomotive cylinder details, locomotive inspirator, section of pneumatic drill, electrically driven pump, pneumatic hard grease press, steam hammer, pneumatic motor, and quick action triple valve.

Other subjects were hydraulic wheel press, hydraulic piston remover, control panel for electric welding and cutting, suburban type locomotive, duplex pump, flanging press, electro turbo generator, and a set of lever shears.

All drawings showed signs of promise, while some in particular were almost without fault. The chief features to the writer's mind were the neatness of arrangement, and careful workmanship throughout.

Quite a number of contestants won prizes for their efforts, and as it was impossible to secure the names of all those exhibiting, we will speak only of the prize winners. These prizes were divided into certain sections, so that when you read of various first prizes being awarded you will understand we speak of prizes for the different sections.

First Prizes

The lucky artists to land first prizes are as follows: W. Grandison, G. T. R. shops, Stratford, Ont., his showing being a splendid drawing of a duplex pump; E. T. Driver, of Toronto, won first prize by his drawing of a continuous current generator, while Arthur Sansfacon, of the C. P. R. Angus shops at Montreal, landed a first prize for his splendid drawing of some columns of the Corinthian order. Frank Brown, of Toronto, got first prize for his showing of the Gloucester Cathedral cloisters. The two latter drawings were, of course, architectural.

Second Prizes

The gentlemen who came close seconds and who captured the coveted blue tickets, were as follows: A. T. Dutand, G. T. R. shops, Montreal, got his prize for a view of a suburban tyne locomotive. Another G. T. R. Montreal, man to land a second prize, was L. A. Tuveral, a pair of lever shears being his exhibit.

The judges evidently felt some special prizes should be awarded, and C. Kyle, C. P. R. shops, Montreal, and E. Harris, G. T. R. shops, Montreal, were the fortunate ones. The former secured his prize for a showing of locomotive 2,300, while the latter displayed a neat flanging press. It might be well to add that C. Kyle also won another second prize in the architectural class by an exhibit of some columns of the Corinthian order. J. B. Helme, of Toronto, landed a second prize for a neat drawing showing the east entrance of the main building of the University of Toronto.

From the foregoing one can see that quite a number of prizes were given and well they deserved them. In some cases the drawings were colored with different shades of ink, making a rather pretty effect. In practically all drawings appearance was aimed at from the border lines in.

Help Them Along

It is very encouraging to see such a display of earnest effort by the apprentices in the railroad shops in particular. The executives who have the destinies of these young men in their hands are to be congratulated on the splendid showing of their boys, and if all apprentices in the machine tool line could do equally as well, it would tend to raise still further the present standing of the machine tool business.

We were rather disappointed on not seeing some efforts by the students of our various technical schools, but perhaps by next year we will have that pleasure. We feel sure there is plenty of hidden talent in our many technical and evening classes, and would suggest that the teachers of such classes encourage their students to prepare work for next year. It is also to be hoped that by the time the next exhibit is due, the place of location will have been changed to one of more suitable locality, such as the machinery hall.

MAINTENANCE OF AIR COMPRESSORS
Continued from page 262

and a poor fit in the grooves, and when these are properly repaired the compressors work all right.

What Examination Proved

In one case an 8 1-2-inch cross compound compressor developed pounding, would not supply sufficient air, and the engine had to be taken off an important train. Examination showed that the rings in both the high and low pressure pistons were 3-16 to 5-16-inch open at the ends and lacked about .012-inch of filling the ring grooves. New rings that would fit the cylinders, but were .012-inch smaller than the grooves, were applied to both the air pistons, and the compressors gave good service for four months, and probably longer, but records do not show.

Still another, 8¾-inch cross compound compressor pounded some, and ran so slow with 190 pounds steam pressure that it would not keep up 90 pounds standard brake pipe pressure on a regular passenger train of ten cars, against normal leakage. Examination showed:

(a) Main steam and air cylinders slightly worn. (b) Steam cylinder head parts in good condition. (c) Main steam and air pistons about 1-32-inch smaller in diameter than their cylinders. (d) Packing rings in steam and air cylinders came together at ends in smallest part of cylinders and were a little less than 3-32-inch open in largest part of cylinders. (e) Rings lacked .005 to .008-inch of filling grooves in high pressure steam cylinder. (f) Rings in low pressure steam cylinder lacked .012 to .015-inch of filling grooves. (g) Rings in high pressure air cylinder lacked .015 to .018-inch of filling grooves. (h) Rings in low pressure air cylinder lacked .012 to .015-inch of filling grooves. (i) Upper final discharge valve had a 3-16-inch lift, one upper intermediate valve had 7-32-inch lift and other air valves had only slightly more than standard lift.

New pistons that would fit the cylinders closely at the smallest part, and in which the rings fitted the grooves with just enough clearance so they would expand by their own tension when compressed into the grooves, were applied and no other repairs made, and the compressor given a severe running test in which its capacity and speed developed were fully up to standard.

Recommended standards to cover repairing and adjustment of the different parts of compressors have been extensively published and discussed ever since the cross compound compressor came into use, but general observation would indicate that there are not taken serious. ly and appear to be seldom considered in connection with compressor repairs and maintenance. On account of this, the general standard of service of these compressors is considerably below what may reasonably be expected and what may be obtained by uniformly raising the standard of quality of work done in maintaining them, which may be done at slightly increased cost, and by simply following standards such as those outlined in recommended practice by the Air Brake Association.

The Northwest Air Brake Club believes that it is time for a general awakening to the need of progress and improvement along these lines, and that the logical and most effective means of bringing about such improvement is through the influence of the Air Brake Association and the air brake clubs throughout the country. The number of engine failures chargeable to the air compressor alone indicate the need for improved compressor maintenance, and the other troubles that can readily be traced directly to a lowered compressor efficiency emphasize it to a high degree.

WHAT OUR READERS THINK AND DO

SHORT CUTS
By Fred Horner

The following hints, or short cuts, will no doubt be of interest to the readers. Some may know of them, then to others the schemes will be new. The first idea is shown at A. This really gives two suggestions for set-screw guards, to those who do not care much about hollow set-screws. In the first instance the idea is to thread the neck of the dog, and screw upon it a stout steel tube. The other notion also incorporates a renewal, when the thread in the dog has worn badly. Bore the neck out and thread afresh to a larger size. Then screw in a renewal bush which is extended into a protective tube. A socket wrench is used to operate the screws.

The next scheme is shown at B. This depicts a miniature bolt driver.

When small bolts or other flatted pieces have to be driven between centres, a useful driver may be made in the manner shown. A piece of stock is bored out to slide snugly over the centre, and is locked thereon by a grub-screw bear-

ing upon a flat. The nose of the fitting is slotted out suitably. to embrace the head, the sharp corners left from the slotting being preferably removed as shown.

The third suggestion is shown at C. In this case nuts are used to aid in the chucking of screws. In any case where small bolts or screws have to be chucked for thinning, trimming, chamfering or pointing operations, it is convenient to run a nut on the thread as illustrated, so that the jaws will have a bearing to centre, and grip the work properly. This avoids all damage to the thread, and risk of the piece tilting and pulling over to one side.

The next example is shown at D. For those who wish to make a simple fine adjustment for their surface-gauge, the sketch C is shown. The method is to interpose a U-piece of steel between the base and the pillar, and so get a slight up and down movement for the final fine setting by springing this loop together or letting it open, the control being through the screw. The lower

half of the loop is fixed by screwing it over a stud screwed into base.

At E is a hint of an entirely different nature. This illustrates a simple device for bringing cylindrical pieces axially under the drill, so that the resulting hole will pass right through the centre. It consists of a small block of metal milled with a vertical vee-groove, to lie against various diameters of drills, and a transverse groove at the bottom to rest on the curve of the work, this permitting the latter to slide into correct relation to the drill.

The last hint is shown at F. It depicts a device for cross drilling in the vise. Readers are doubtless familiar with the tip of gripping a bit of plate on top of a shaft or spindle in the machine vise, the plate having a central hole to guide the drill axially across the shaft centre. A modification, which extends the usefulness of the idea, is to cut flats, on the plate as seen in the illustration, giving for example three different spans, to suit the vise opening on corresponding diameters of shafts.

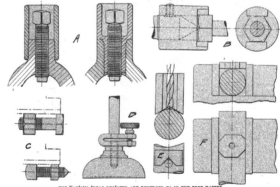

THE VARIOUS IDEAS DEPICTED ARE REFERRED TO IN THE TEXT MATTER.

DEVELOPMENTS IN SHOP EQUIPMENT

ACKLAND SCREW DRIVER

The screw driver as depicted is the product of the Ackland Specialty Co., Springfield, Mass. It is said that this driver will remove any rusted or stuck screw which could not be budged by an ordinary driver.

The principle involved is an old established law of mechanics. A few pounds of pressure at the end of the upper bar and on the rachet handle results in a much greater pressure at the slot of the screw. This feature of course eliminates the danger of the driver slipping and ruining the screw slot.

HOW THE SCREWDRIVER IS USED.

Automatic and machine manufacturers are said to be especially interested. The blades are made from special heat-treated and tempered steel. The post is attached to the blade by a case hardened socket and sliding on the post is a rachet which permits using the driver in narrow and restricted places. The cross bar is secured to the post by a ball joint permitting movement in all directions.

The chain is attached on the end of the cross bar and looped around some projecting lug of the machine. A glance at the photo illustrates clearly the method of using.

PROFILE MILLING MACHINE

The Automatic Machine Company, Bridgeport, Conn., have placed on the market what is known as their automatic multiple-spindle profile milling machine. It is said to be specially adaptable for automobile engine work, or in

other manufacturing where it is necessary to finish a large number of irregularly-shaped pieces.

A glance at the illustration will be worth while before going into further details. The spindle head can be made to accommodate any number of spindles, and their positions also can be arranged to suit the work. When a change is desired the spindle carrier is easily removed and a different one substituted, using the same spindles. These spindles are equipped with bronze thrust and taper bearings below, and straight bearings above, the central point where they are driven by spiral gears.

In operation, one cam moves the head or spindle carrier longitudinally and another moves the table or platen transversely. These are so designed and synchronised as to make the cutters follow the desired outline. After the cycle has been completed the feed trips out, while the cutters continue to revolve to clear themselves. The operator's duties are in this way confined to merely chucking and unchucking the work.

There is an adjustment for wear in cutter diameters in profiling any single design, and there is an additional adjustment on the traverse motion of the table

MILLER AS EQUIPPED WITH SIX HEADS

to aid in setting the jig correctly and rapidly. The table is also provided with adjustable taper gibs.

An independent vertical adjustment for position of cutter on each spindle is provided, and the spindle bearings can all be adjusted for wear. The complete spindle head is counterweighted and is raised or lowered by the handwheel shown in the illustration.

The material and workmanship is claimed to be of the very best, and ball bearings are used to take the thrust on the spindle driving shaft, and the vertical worm shaft. Sight feed oilers are placed on the head and principal bearings, and accurate timing of the cams and gears is ensured by thorough running tests before shipment. Following are the principal specifications:

Number of spindles—as required. Table, 24 in. by 24 in. Height, from top of table to bottom of spindle, when up, 22 in. Longitudinal-transverse motion —to suit work. Dimensions over all, 4 ft. 6 in. wide, 5 ft. deep, 6 ft. 9 in. high.

REPAIR JACK

The R. S. Whitney Manufacturing Co., Lewiston, Me., have placed on the market a safety repair jack which is illustrated herewith. This type of jack is designed to meet the demands of all classes of car, truck, and tractor repair work, and combines simplicity with strength.

All parts are properly proportioned, and a lever raises and lowers the screw, and is attached by means of a collar to the rotating nut. This, of course, increases the adaptability and flexibility of the jack. The rotating nut is placed at the extreme lower end of the screw in order to provide the greatest possible

VIEW OF THE JACK.

clearance. An extension supporting arm is also included, this portion being shown in the sketch. You can thus accomplish varied work with the one jack. The height closed is 19 inches, and it has a raising radius of 11 inches plus that of 7¼ inches, obtainable by using the extension arm, therefore the scope of usefulness and utility is practically unlimited.

FLUE AND PIPE BENDER

The Baird Pneumatic Tool Co., Kansas City, have placed on the market an arch flue and pipe bender as shown in the accompanying illustration. This machine is especially adapted for bending into difficult shapes locomotive or marine arch flues, also air pipe connections as used under the regular freight or passenger cars.

It is rigidly constructed to meet the hard usage of boiler shops, and will bend arch flues or pipe to any desired angle. It is not necessary to sand or heat the flues or pipes with this machine.

It is fitted with special blocks and forming rollers the exact contour of the outside diameter of 3 in. and 2½ in. boiler tubes. It is said that the round bend obtained by using this machine permits a flue-cleaning motor to pass through in the cleaning process. This is very often necessary in cleaning arch flues.

As can be noted from the illustration the complete machine includes cast iron table with planed, finished top. To this table is fastened the air cylinder supporting the piston directly connecting with the forming rollers and yoke, which in turn produce the radius bends in the pipe through the forming blocks and quadrant located at opposite ends of the table. This quadrant is fitted with a plug which can be placed in spaced stops to secure any radius within the capacity of the machine.

A patented four-way valve is fastened to the cylinder and within easy reach of the operator. Should more than the bend required be secured, by reversing

GENERAL VIEW OF THE BENDER.

the air valve the operator can take out the amount necessary. This notice should particularly interest railroad shops, marine plants, boiler shops, and shipyards. Following are the machine's principal specifications.

Diameter of air cylinder, 3 in.; stroke of piston, 26 in.; maximum travel of forming roller, 26 in.; air pressure required, 100 pounds per square inch; dimensions of forming table, 3 ft. by 5 ft.; bending blocks — one with capacities, pipe, ¾ in., 1¼ in., 1½ in., 2 in., 2½ in. and 3 in.

NEW THINGS IN MACHINE TOOLS

UTILITY SCREW PRESS

A screw press that has been designed for a wide range of work has been placed on the market by Carl Pletz and Sons, of Cincinnati, Ohio. The machine is adapted for the straightening of shafts, bending bars or shapes, pressing in bushings, etc. The lower end of the steel screw is fitted with a special pad that remains stationary when in contact with the work, the thrust being taken on a bronze washer running in oil. When using a four foot bar it is claimed that a pressure of about 20 tons is applied to the screw. Two sizes are made, with distances under the screw pad of 14 and 20 inches. The approximate weight is 500 and 600 pounds respectively.

MILLING ATTACHMENT

A milling attachment that has been adapted for use on shaping machines, has been placed on the market by the Tri-State Milling Machine Company of Memphis, Tenn. The spindle is worm driven and is supported by a rigid overhanging arm, and the housing is intended to be secured to the ram slide, the ram being moved back out of the way. The table of the attachment is fastened to the table of the shaper with special provision for latter adjustment.

MULTIPLE-SPINDLE MILLING MACHINE

A multiple-spindle vertical milling machine has been added to the lines of

the Betta Machine Company, of Rochester, New York. The machine is of rigid construction and is designed for heavy production work. The machine is made with three or more spindles as desired. The spindles are made from steel forgings driven through long splines and spur gears; spindles being adjusted vertically by hand. The machine may be belt or motor driven. Work holding fixtures are located on the continuous revolving table, the pieces being changed while the work is passing from one working position to the next. The table has a flat bearing on the bed and is driven through a large internal gear, all bearings being of bronze and all gears running in oil. It is claimed that one of these machines has a production capacity three times as great as a single spindle machine.

AUTOMATIC-STROKE BELT SANDER

A belt sander that automatically moves the sand belt across the surface being sanded has been brought out by the Mattison Machine Works, of Rockford, Ill. The machine is of rigid construction, the heavy casting carrying the power drive mechanism being mounted on two ample supports at either end. The shoe for applying the sand belt slides in a bar extending across the front of the machine, and is operated with a rapid reciprocating motion by a steel belt running over two flanged pulleys. Geared quadrants, which may be adjusted to suit any length of work, operate the pinions that drive the pulleys. All shoe connections are made of aluminum, and are counterbalanced. The work table is mounted on ball bearings.

FACE GRINDING MACHINE

The Diamond Machine Company of Providence, R.I., are now manufacturing an extra heavy duty face-grinding machine, that is intended for grinding operations on the heavier equipment that formerly has been accomplished on planing machines. The grinding wheel is made of the sectional type, a number of abrasive blocks secured in a chuck. Special attention has been given to the handling of the coolant, so that clean, cool liquid is returned to the circulating system. Duplicate control is provided so that the machine may be operated from either side. The machine is motor driven through a Morse silent chain.

The MacLean Publishing Company
LIMITED
(ESTABLISHED 1887)
JOHN BAYNE MACLEAN, President. H. T. HUNTER, Vice-President
H. V. TYRRELL, General Manager.

PUBLISHERS OF

CANADIAN MACHINERY
MANUFACTURING NEWS

A weekly journal devoted to the machinery and manufacturing interests.
B. G. NEWTON, Manager.   A. R. KENNEDY, Managing Editor.

Associate Editors:
J. H. MOORE   T. H. FENNER   J. H. RODGERS (Montreal)
Office of Publication: 143-153 University Avenue, Toronto, Ontario.

VOL. XXIV. TORONTO, SEPTEMBER 16, 1920 NO. 12

Paying the Real Increase

AN Ottawa despatch credits Hon. F. B. Carvell, head of the railway commission, with a statement to the effect that the increase in freight rates could not possibly amount to more than 15 cents per ton of coal coming to Ottawa. This figure came out because it was reported that certain coal dealers were to raise the price of coal 50 cents per ton on the head of the new freight schedule.

Right here is where the big complaint is constantly raised. Wholesalers and retailers alike complain that by the time increased freight charges are passed on to the customer they amount to three or four times the amount granted to the carrying interests.

The statement of Hon. Mr. Carvell that 15 cents is the increased cost of a ton of coal in Ottawa suggests that it might be well for the railway commissioners to have worked out a table showing the exact increases on a given list of staples in the principal points in Canada.

By so doing they would be giving the public the protection to which they are certainly entitled, and the absence of which is liable to leave openings for increased prices that have no foundation in fact or reason.

It is true that many papers are publishing figures showing how the rates affect certain lines, but these are not accepted as official, and they lack, for that reason, the weight an official ruling would have. For instance, were it handed out by Ottawa, in an official way, that the increased carrying charges on a ton of steel from Pittsburgh to Toronto, or Montreal, having regard to the added increase in the cost of moving the raw materials in the first place, amounted to one or two dollars, or whatever the figures actually were, the trade would accept that ruling and adjust schedules accordingly.

The trade generally, as far as lines covered by this paper are concerned, believe that if the freight service is improved forty per cent., along with the forty per cent. increase in rates, it will have been a fairly good investment. They know that industries are having a hard time in many cases because they cannot get the service they require from the roads, and they know that many firms have been put to a lot of expense in having to provide their own carrying service. On the other hand, if they put up the increase of forty per cent. and do not receive a better service it will simply be that much money passed over to the roads.

Mechanical Drawing Is Not Fancy Work

MECHANICAL drawing at Toronto Exhibition seems to appeal to the directors as worthy of a place in a building devoted to needle-work, etc.

The fancy work is placed first, and the room that is left at the top of the cases is devoted to a display of mechanical drawing. Then, a few pieces being left over mechanical drawing is a part, and they are hung upstairs to the accompaniment of more lace and photographs.

Mechanical drawing is a part, and a very important part, of the machinery field, and cannot be alienated from it.

It is not fancy work—neither is it a pastime. It is serious, constructive and essential work and the people who would be most interested in it, and who would profit most by it, are the men who visit Machinery Hall.

It might be possible also to introduce something in the way of designing, rather than in copy work, bringing perhaps from twenty young minds an idea or two that might be of the real worth-while type.

The Double Steel Market

RIGHTLY or wrongly, there is a growing feeling in the steel trade that there will be a levelling process soon that will do away with the two-headed market that has existed for some time. By a two-headed market we mean the selling price of the corporation as opposed to the selling price of the independents.

It may be that the corporation figures will come up a little, while it can be only a matter of time before some of the prices now secured by the independents will have to sag, although there are some of them brave enough even yet to talk about the prices increasing.

Right now Canadian warehouses are buying sheets in United States, for which they are paying nine cents a pound in New York funds. The average price for these covering some years is shown in the following table, prepared from the current market quotations at Pittsburgh:

1905	2.28	1913	2.20
1906	2.47	1914	1.89
1907	2.60	1915	1.93
1908	2.48	1916	3.06
1909	2.26	1917	6.29
1910	2.29	1918	4.98
1911	2.03	1919	4.43
1912	2.00		

An explanatory note in the Pittsburgh market records states that the prices in 1919 are subject to certain premiums for quick delivery.

Those who must have material are buying and paying for it, but the present price schedule does not encourage buying for stock, thus giving the volume of orders that the mills must sooner or later have to operate at capacity and at low cost.

The pith of such plants as the sunflower, rush or elder is advocated by the "Scientific American" as a good material for cleaning the lenses of scientific instruments.

An extension of the Temiskaming and Northern Ontario Railway from the present terminus at Cochrane to James Bay is shortly to be undertaken by the Ontario Government.

MARKET DEVELOPMENTS

A Readjustment of Steel Prices Likely

Report Gains That Corporation May Come Up While Independents Ease Off—Freight Rates Should Result in Better Service— Railroads Are in Market for Some Equipment.

WHERE and how the new railway rates will fall is one of the topics that the market discusses at length this week. In some cases it will make no difference in prices quoted, although the difference will come when the buyer pays for the haul from the present location of the material to his own premises. Much material in the steel trade has already felt the total cost of haulage increases because the U. S. freight rates cover the case. Considerable of the concern is that the purchaser shall pay only the increase granted to the roads.

Although there is nothing official to warrant the belief, the trade seems to feel that there is bound to be a move that will do away with the dual steel market that exists now, viz., the selling schedule of the corporation and of the independents, between which there is a difference so great that it cannot be reconciled. Most of the deliveries that are finding their way into this territory now are com-

ing from the independent mills, and as a result the prevailing prices are very high.

Machinery houses report in several cases that they have had a number of very interesting inquiries during the past few days, and that deliveries on orders that have been booked some time ago are coming through in a very satisfactory way.

The Government railroads are buying small tools now, particularly carbon and high speed drills. As a general thing though the trading that is being done is in smaller lots.

Several houses report that they find it necessary to pay more attention to their collections than usual, reporting some of these as being rather slow.

In hardly any section of the steel, iron or allied trades can it be said that there is any buying being done for stock now, prices and deliveries being very much opposed to any such policy.

MONTREAL TRADE NOT WORRIED OVER THE INCREASE IN FREIGHT

Special to CANADIAN MACHINERY.

MONTREAL, September 16. — Business and industrial conditions are developing a gradual improvement as the mid-summer holiday season comes to a close. Some factories that have been curtailing operations, due to a shortage of fuel, are more actively engaged, and in many lines the prospects for a busy fall season are quite marked. Following the announcement of the increase in freight rates the Railway Board have received many requests for deferred action, but while recognising the right to appeal, the Board authorises the increase to become effective at once. In the statement issued, the following clause is inserted: "If it should appear to be a case where the appeal should be granted and the decision modified or reversed, there could, if deemed proper, be provision for rebates of rates charged beyond those ultimately fixed."

Better Delivery Offsets Freight Increase

It is expected that the 40 per cent. increase in freight rates will ultimately mean that steel commodities will be advanced in price sufficiently to meet this additional charge. Where the steel is taken from warehouse stock the price

that has been effective is still quoted, but where material is shipped from the mill the added freight charges approximating 17 to 20 cents per hundred pounds, are included in the invoice. This condition will apply generally when local supplies are exhausted. The car situation in this district is quite normal but congestion in the vicinity of American mill point prevents the regular shipment of steel, and dealers and consumers in distant sections are frequently inconvenienced by the non-delivery of much needed material. The demand for steel lines has shown no marked improvement, but the movement is steady and of an encouraging character. The situation may be summed up in the statement of a dealer here when he said: "Users of steel are not buying any more than they need but are willing to pay for what they want." This would appear to indicate that buying is not restricted for urgent requirements, but that only immediate necessities are given consideration. The supply of sheets is still below the maximum trade requirements, but owing to the failing demand from the automobile industry the quantity of available sizes is

showing improvement, and it is not improbable that the market conditions may offset the increased cost of transportation, and as a result the quotations on some lines may not show the added costs of haulage.

Present Activity Slight

"The increased freight charges will have little material effect upon our business, as present operations are on a restricted scale owing to the continued dullness of the market. Movement is almost confined to local activities and little railroad transportation figures in its shipment, but where material is required to be placed on the railroads for distant haulage the customer naturally or automatically pays the extra charge." This statement shows the quiet state of the old material situation, and in some respects the dealers are disinterested in the present market. There is, however, a feeling that the railroads will make every effort to establish normal conditions before the winter actually sets in, and if this can be effected there is a large possibility that improved operation would follow in many lines of industry and open the way for a better demand for a greater range of scrap materials. Last week's quotation may be considered as nominal with an upward tendency.

WANT TO GET A BETTER SERVICE

Believe Rates on Railway Will be Worth While if it Results in Quicker Shipments

TORONTO.—The market is adjusting itself this week to the new freight charges. It was only a couple of weeks ago that the new rates from U.S. points were taken care of, and in doing this a revision of rates on account of Canadian changes has been made unnecessary, as much of the shipments come from points in United States.

While there is in some concerns an alarm at the principle of continually increasing costs, there is no alarm over the new rates as a general thing. Here are a couple of opinions, one of them from a leading warehouse, and another from the manager of one of the largest selling forces for machine tools and other lines.

"As far as the business of the steel yards is concerned, we believe the rates are all right. Much of the material we get is from Pittsburgh, and the rates we have announced, amounting to an increase of $2 per ton in most lines, cover all the increases that are necessary. One of the worst things that the manufacturers with whom we have business in Canada have to contend with is the fact that railway service has been very indifferent for many months past. There are plants that have had to resort to all sorts of schemes in order to give service to themselves that they should get from the railways. If the increase in rates will enable the roads to get their equipment whipped into better shape, then we believe it will be a good investment to pay the increased carrying charges."

The manager of the other business spoke along similar lines. "I can show you a good many lines where our increases have been over 40 per cent., which is the amount now allowed the roads. The whole thing hangs on the performance the roads are prepared to give for the increase granted them. If there is no improvement over what we have been experiencing, then we will look upon it as just that much needlessly added to the cost of carrying on business, which we have to pass on to the men who buy from us, but if we get a 40 per cent. improvement in the service from the roads, then good and well, we will have something for our money, and it will be a good investment."

Must Watch Closely

Others point out that it is necessary for dealers and manufacturers to watch very closely all the increased charges that have been put into effect, and see that they are not trying to absorb too many of them, and by so doing wearing down their margin of profit, past which it is not safe for them to go.

In this same connection, the market reports that more attention is being paid to collections as a general thing

POINTS IN WEEK'S MARKETING NOTES

August production rate of iron and steel was the largest since December, 1918, barring last February and March.

Pittsburgh claims a decided improvement in freight movement, saying that in many lines the business is nearly normal.

There are few new buyers in the steel market and business is dull as far as new bookings are concerned.

August shipments of steel are stated to be about 90 per cent. of the output.

The Government railroads are in the market for carbon and high speed carbon drills.

The steel trade feels that if higher freight rates can bring about a more efficient service it will be money well spent.

There is not the insistent demand for plate there was a few months ago, and dealers claim there is enough in the yards to care for all the business that is offering.

A shortage of cement is holding up work on the Welland canal. The contractors claim the large shipments going to United States are responsible for this.

Shipments of machine tools are very satisfactory now. Some firms think September will be one of the largest delivery months of the year.

Warehouse prices on bar iron and steel, plate, sheets, etc., were put up $2 per ton.

All soft steel bar makers are practically sold up for the remainder of the year.

Some rail tonnages of importance for 1921 delivery have been accepted by mills, but the price is left open till Jan. 1.

English makers of ferromanganese have cut their price to this country $20 per ton.

Lack of old rails is apparently all that is holding up large trading in hard bars, one interest reports having to turn down inquiries aggregating nearly 10,000 tons for this reason.

than for some time past. Some dealers report that they are having more trouble now in keeping their books well cleared up, and firms that before always paid promptly, are now taking advantage of every circumstance to extend their payments over as long a period as possible.

The Steel Market

The opinion is expressed, although there is nothing in fact to back it up, that the U. S. Steel Corporation prices will be raised and the prices of some of the premium mills will be lowered. "For a long time," stated one manager this morning, "there have been two separate and distinct markets, a high one and a low one, but unfortunately there has been only one seller, and that is the premium operator. We have had business booked with the Corporation for months, and we can't get an answer as to when it is coming through. But the continued existence of two steel markets is too absurd to last long."

There has been a fairly heavy shipment of tubes, and they are much needed. Most of them are 16 ft. 3¾ in., and many of the larger power plants are persistently in the market for these and the four inch tubes. A fairly large tonnage of two inch tubes is coming through, these being used more for smaller plants, for locomotives, etc.

There is nothing like the demand for plate that existed some months back, and of this line it can be said that there is not much trouble now in getting all that is required for ordinary purposes. The sheet users are all in trouble together, and no one seems to be any worse off than his neighbor. Sheets are coming through in a week to ten days, from some of the independents, the price for black 28 gauge being 12.10 per lb.

The Machinery Market

A fairly good business is being done in machine tools, some places having disposed of a good assortment of these during the past few weeks. There is an improvement in the number of inquiries coming into the market, and dealers are hopeful that there will be a decided improvement later on.

The railroads—National—have specifications out for a fairly good list of carbon and high speed drills for some of the eastern shops, but otherwise the buying is being done by smaller interests. The automobile interests are not taking on anything like the stock they were a few months ago, and no large buyer has come in to take up the sag.

The scrap metal business is having a short week, because nearly all the yards are closed owing to the Hebrew holiday season following their new year celebrations. Trade during the past few days has shown no improvement over what has been experienced. Few buyers are in the markets for material, and the yards are buying very closely in what they do take on.

SHIPMENTS IN AUGUST WERE 90 PER CENT. OF STEEL PRODUCTION

Special to CANADIAN MACHINERY.

PITTSBURGH, Sept. 16.—The iron and steel market continues stagnant in practically all branches. It is a case of the trade going ahead on its momentum, as producers are well filled with business and production is fully absorbed by deliveries against orders already in hand. Consumers desire to be obligated only to take such material as is required for business they have in hand, and as deliveries are better there is less demand for prompt lots to piece out with contracts.

Heavier Steel Production

The monthly report of the American Iron and Steel Association shows 3,000,-432 gross tons of steel ingots produced by the 30 companies that make monthly reports, against 2,802,819 tons reported for July. Computing on the basis of the number of working days in the month and year, and allowing for the unreported production, ingots were produced during August at the rate of about 42,700,000 tons per annum, showing a 7.0 per cent. gain over the rate in July. This is a large gain from one month to the next, and is particularly striking in view of the fact that August, being a midsummer month, would be subject to the same adverse weather conditions as July. Barring last February and March, the August production rate is the highest since December, 1918.

Shipments Heavier

In August there was a double gain in shipments, since the production was larger and the mills shipped more than their production, instead of shipping less as they had done in July and in the three months preceding. During those four months they had to stock some of their output, on account of car shortage. About the end of July the turn came, shipments and production being approximately equal. Since then shipping conditions have constantly improved. As to the total volume of freight movement, the last week reported upon, that for the week ended August 14, shows the largest number (962,352) loaded for that week in any year, the number just passing that in the record year, 1918. An authority estimates that by this time the total freight movement is not only ahead of that in the same period in any year, but is ahead of the best record made in any week in any year. As to steel shipments, they were naturally very heavy when production was at the rate indicated and shipments were in excess of production.

Of the finished steel products accumulated during the car shortage easily the major part has now been moved. There remains considerable tonnage of semi-finished steel, the steel having been put in that form as a convenience, with the idea that it could gradually be fed to finishing mills as supplementary to the current production of steel. The finishing departments can, of course, stand only a slight overload, and thus with the best shipping and market conditions it would be quite a while before all the semi-finished steel could be worked off.

Momentum of the Trade

It has been made clear to everyone that the pig iron and steel markets have gotten into a condition of dullness from which they will only be resuscitated by a marked improvement in general business conditions. There is nothing in the markets themselves that would lead to there being a fresh buying movement. The influence must come from without, as buyers are now to be supplied with heavier deliveries than they were getting, on an average, in the past five months, so that the requirements of buyers must increase before they take much interest in making fresh commitments.

In the circumstances, the only question is how long the industry will be able to keep busy on its present commitments, or in other words how long the momentum will carry the trade. At a juncture like this there is always a difference of opinion. Some observers point to "full order books" and compute how long the orders will last the producers. Others assert that order books count for nothing, that orders can be canceled and buyers order shipments suspended or curtailed. Experience gathered over many years, from various situations like the present, indicate that neither party is exactly right, but that the former come nearer being right. There are always cancellations at a time like this, and there have been this time, but the cancellations amount altogether to but a very few per cent. of the total amount of business on books. Quite possibly a great deal of business would be canceled if the market broke, but the producers always look out for this. When the time comes that they see little new business to be secured either at existing prices or even at deep cuts, while a great deal of business is to be squeezed out of existing contracts, they simply "stand pat" on their prices. To cut prices, to engage in competition, would be simply for the mills to swap orders with each other at lower prices all around.

Accordingly there is the experience of mills in general adhering stiffly to their previous quotations, with the exception in some cases of such small mills as have been securing the outside prices on account of being able to make

very early deliveries. When such mills reduce prices they are still above the general average of the independents. The Steel Corporation is in a class by itself, its prices being so much lower than the lowest of the independents.

Steel Corporation Tonnage

For the first time since May, 1919, the United States Steel Corporation has shown a decrease in its unfilled obligations, the last report showing 10,805,-038 tons of unfilled business at the close of August, this indicating a decrease of 313,430 tons in August, comparing with an increase of 139,651 tons in July. Thus there was a change of about 450,000 tons from one month to the next. About half of this was due to an increase in shipments and about half to a decrease in bookings, or rather in the net bookings, as there may have been some cancellations in August, offsetting a small part of the bookings. Shipments in July may be estimated at 75 per cent. of capacity, and August shipments at 90 per cent. As the August decrease in unfilled obligations was about 23 per cent. of capacity the net bookings in August were about 67 per cent. of capacity. The unfilled obligations left are equal to about nine months of production at say 90 per cent. Should the corporation book month by month only one-half as much as it shipped, this being a very low estimate, the orders, if properly distributed to the different departments, would theoretically carry the corporation for 18 months instead of nine months.

Pig Iron

Production of pig iron increased in August as compared with July and the increase in merchant output was particularly sharp, being 6.7 per cent. The August rate of output was barely a shade under the rate last March, and otherwise was easily the heaviest rate shown by the merchant furnaces in a year and a half. This puts somewhat of a new aspect on the price advance that most pig iron producing districts brought about late in August. It is notable, now, that the Chicago district furnaces refrained from advancing their prices. They are still on the basis of $46 for No. 2 foundry iron, while the valley and Cleveland furnaces are at $50, and the eastern Pennsylvania and Buffalo furnaces are still higher. More important still is the refusal of the southern producers to put up their prices. They are still at the $42, Birmingham, figure attained at the end of last April. The furnaces could undoubtedly have advanced prices above $42 if they had desired, but they considered it wise not to do so. There is now a deadlock in the pig iron market, and one should scarcely have difficulty in deciding which party is going to yield first. Nothing may occur, however, for several months, as the present engagements will last both parties quite a while.

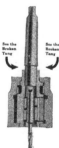

CANADIAN QUERY IN U.S. MARKET

Railway Specifications Are Being Look-Over—Business is Not Brisk

Special to CANADIAN MACHINERY.

NEW YORK, September 16—Several New York machinery dealers are quoting on the recent inquiries of the Grand Trunk Railway System (George W. Caye, purchasing agent, Montreal, Que.), which call for both metalworking and woodworking tools for various Grand Trunk shops in Canada. Among the machines called for are the following: Two driving wheel lathes, 90 in. swing; two gap lathes, Bertram or equivalent; one 60 in. duplex horizontal driving box boring machine; one universal horizontal boring machine, equivalent to No. 32 Lucas; two single frame steam hammers, 800 pound; engine lathe, 26 in x 12 ft.; crank shaper, 20 x 22 in.; Fox turret lathe; radius link-grinding machine; Gisholt horizontal axle-box boring machine; McCabe pneumatic flanging machine; 35-ton hydraulic spring-testing machine; six Bridgeport grinders; 42 in. car wheel lathe; triple head bolt cutter; four sliding head Sibley drill presses; 26 in. woodworking tools as follows: one double arbor universal saw bench, one 36 in. band saw, one 16 in. hand jointer, five swing cut-off saws, one combined radial car boring machine, four self-feeding rip saws, four vertical hollow chisel car mortisers and borers.

The American Brake Shoe & Foundry Co., New York, is asking for bids on about ten machines. Other inquiries are few. Eastern railroads are not doing any buying of consequence, but there is a fair amount of buying in Chicago, principally by the Chicago & Northwestern and the Santa Fe roads.

Machine tool business has suffered a severe set-back in the past six weeks or two months, and the effects of this are seen in the slowing down of manufacturing operations at some machine-tool plants and the catching up on deliveries, some machines now being available out of stock. Prices, however, show no apparent weakness; in fact there have been some advances recently. The outlook for the future is somewhat uncertain; there are some in the trade who do not look for any pronounced revival in business before spring, while others think that a renewed demand will develop much sooner, possibly after the presidential election in November.

The full-rigged three-masted sailing vessel, Grand Duchess Maria Nikolawna, now in the harbor of Montreal, presents a striking contrast to the modern present-day steel freighter. This ship has a registered tonnage well below 2,000, but carries masts of such a height that it was necessary to step her main-topgallant mast in order that she pass beneath the Quebec bridge. The vessel was originally built in 1873 by R. Steele & Co., of Greenock, Scotland, and on her recent voyage out from Newcastle-on-Tyne made the trip in thirty days.

SIGNS OF PROGRESS

Imperial Oil is to build a $250,000 distributing plant at Brandon.

Dominion Bank is building a three-story premises at Brantford.

Masonic Hall at Bridgeburg is planned to cost $20,000.

Imperial Steel and Wire Co. are to start on new buildings at Collingwood.

Fisher Body Corp. is to build a factory at Ford City to cost $175,000.

Dominion Oxygen Co., Ltd., may build a factory at Halifax.

Steel Co. of Canada, will build a storage warehouse on Harvey Lane.

Taylor and Campbell are building an electrical factory on Adelaide Street, London.

Bank of Nova Scotia is building at Moncton, N.B. Dominion Bridge Co. have contract for the steel.

Willard Automobile Storage Battery Co., Cleveland, plan half million plant in Toronto.

Canada Bread Co., Toronto, build brick and steel stables and factory in Toronto, at Danforth and Ladysmith.

The Engineering Sales and Service Co., 55 St. Francis Xavier Street, Montreal, have obtained the Canadian agency for Haughton's Patent Metallic Packing Company of London, Eng., manufacturers of high grade anti-friction packings.

The C. P. H. Gas Engine Company of Canada, Ltd., with offices in the Dandurand Building, Montreal, expect to build a plant for the manufacture of a new type of internal combustion engine which is particularly adapted for all classes of service. J. Van der Gbote is secretary treasurer and general manager.

SHOP SECRETS ARE PROTECTED

Justice Guerin has handed down, in the Superior Court, an important decision on the question of the ethical and conventional obligations between employees and employer in condemning Siegfrid Sanders to pay $100 damages and the costs of an action brought against him by the firm with whom he was formerly employed in this city. The company maintained that Sanders, in the course of his work, acquired in confidence knowledge and information respecting the making and methods of a special machine, and that after he left their services he revealed this information to his new employers. His former employers allege that they sustained a severe monetary loss.

The company further alleged that Sanders had been employed by them to devote his time exclusively to their manufacturing department, and they also alleged that he agreed not to "directly or indirectly" make known to any persons, either during the period of his employment or thereafter, any of the special processes or knowledge of special machinery which he might learn from the company during his employment.

YARDS HERE TO FINISH THE BOATS

Ottawa Makes That Promise and Winter Work at the Yards Will be Assured

Herbert Wright, business agent of Toronto Local 128, International Brotherhood of Boiler Makers and Iron Shipbuilders, whose members are out of employment through the failure of the Dominion Shipbuilding Company, announced, following his return from Ottawa, that ships which the company was building for the Government would be proceeded with. Osler Wade, assignee of the company, had refused to continue the contracts, since he feared that they would be built at a loss.

"I saw Senator Robertson, Minister of Labor, and he assured me that the Government would proceed with the ships this fall. I don't know exactly when the work will commence, but I believe that it will be when the present vessel is finished." said Mr. Wright. "There was nothing said during the interview regarding labor conditions, which were given as one cause which contributed to the failure of the company. He did not mention it, and I presume the same rate of wages will be paid."

To Build Tankers.—According to Mr. G. A. Reading, representative of Robt. Reford Co., steamship agents, the keel of a ten-thousand-ton oil tanker will, in a few weeks, be laid on the ways recently vacated by the "Canadian Mariner," at Halifax. If all goes well the Halifax shipyards will contract for the construction of several oil tankers of large tonnage. Mr. Reading added: "The recent launching of the Canadian Mariner marked the fulfillment of the dreams of shipbuilding officials in Nova Scotia. Built entirely of steel by Canadian workmen, it is the largest ship ever launched in the province." As this was the first ship to be built by the Halifax shipyards the launching was naturally viewed with some trepidation by the promoters, and it was with relief that they saw the big vessel slide easily from her stocks.

J. B. McAndrew, of St. Catharines, Ont., has been appointed by the Civil Service Commission, Ottawa, to the position of Structural Engineer on the Welland Ship Canal.

W. F. V. Atkinson, consulting engineer and formerly chief forester of the Spanish River Pulp and Paper Mills, Ltd., Sault Ste Marie, Ont., has moved to Toronto, where he will continue in private practice.

H. Dixon, formerly chief engineer of the Canadian Northern Railway, has been appointed chief engineer of both the C.N.R. and the G.T.P., following the amalgamation of the two lines. Mr. Dixon's headquarters will be in Winnipeg.

CANADIAN MACHINERY

AND

MANUFACTURING NEWS

Vol. XXIV. No. 13   September 23, 1920

Belting as a Means of Power Transmission

In a Previous Article We Dealt with the Chain as a Medium, But in This Present Case We Speak of the Uses of Belting and How to Care for It.

By J. H. MOORE.

READERS will no doubt recall an article dealing with chain drives, which appeared in a previous issue. In the present article we will take up the other extreme, namely that of transmission by means of belting. It is of course understood as before that we are not advocating any particular style of drive, but are merely presenting for readers' benefit the two sides of the case.

When we speak of a belt drive we can infer anything from the 1 in. single-ply belt up to the larger and heavier class of belting as used for extreme heavy service. We might not necessarily refer to leather belting alone, for there are many fabric belts of various natures being used at the present time. Each manufacturer has his own particular claims regarding the merits of his own product, and as we have no intention of entering into a controversy on this subject we suggest to readers, who desire comparison, that they study at all times the advertisements as appearing in this and other technical journals. Manufacturers in general are only too pleased to tell you about their wares in such advertisements, and a study of the same will prove worth while.

The point we wish to emphasize is not so much what each type of belt will accomplish but rather where belts can be used and what care they require.

Take a tour through any manufacturing plant to-day and what do you find? Line after line of shafting on which are pulleys almost innumerable. Over each pulley we notice a belt leading to still another pulley, this latter one being on either a countershaft or on the machine which is being driven. If the former case holds true, we see yet another belt leading from the countershaft to the machine, and so it goes on. Trace the source of power from the motor to the line shaft and we generally find that even the motor has a belt leading from its pulley to the driving pulley on the line shaft.

In other words it is a case of belts,

and then more belts. In some instances you will find the machine tools driven by individual motors, but even in such cases a belt is often used to transmit the motion from the line, or motor shaft itself, to the other portions of the machine. To state everywhere that belts can be used is an utter impossibility, for we would be sure to miss something or other, but generally speaking we have the manufacturing plant, the machine shop, the power house, agricultural machinery, and in fact every source of revolving motion can make use of belts of some kind.

The Belt to Use

To find the type of belt best adapted for a particular drive is not always an easy problem as many demands are made under various conditions. In some cases weight of material is the big requirement. For instance, belts used for compressors and hammers require good weight. Then there is the belt used

on a grinder. Such a belt should run absolutely true and have no side play.

So it goes on, each problem entirely different. No one type of belt is good for all conditions, so that it is up to the user to watch how his various belts stand the grind. When you find one that gives you real service, stick to that type and do not go around experimenting with some inferior belting which may cause a little less but cause unending trouble. The cost of a belt is only one item in the programme, and a very small item at that. Of course there are cases where you can make a change for the better, but be sure it is for the best. This may sound contradictory after saying do not experiment, but what we imply is this, never purchase a large quantity of untried belting at one time. If you hear of a belting that you believe would be better for your purpose, secure enough for one machine only. Run a trial run on this new belt, making it undergo a real test, studying the result.

FIG. 1—A 34" BELT ON A GENERATOR DRIVE.

FIG. 3—A NEAT BELTING ARRANGEMENT.　　　　　FIG. 4—A GOOD ARRANGEMENT OF BELTING.

You are then able to form a true opinion as to its merits.

Points to Watch

If testing a belt be sure you give it a fair trial. You can ruin a belt's chances in various ways. For instance, don't expect any Tom, Dick, or Harry to place your belts on the machines. Have a man for that purpose who knows enough to make his belt joints properly and so on. Metal fasteners put in crooked, holes too large, in the case of belt lacing, rivets put in any old way, if belt is riveted, etc. These are just a few things that one must not do if a belt is to be given a fair trial. Another point to watch is that of roof clearance.

In many cases the roof of factories consists of beams or joists, these having to be cut away slightly to give the belt clearance. The average worker seems to be afraid to weaken the roof or something, for he cuts barely enough to allow the belt to pass. The result is that if the load on the belt fluctuates at all the belt flops up and down, this action naturally decreasing its life to a considerable extent, especially at the joint. This point of proper clearance is one that should be carefully watched. Crowned pulleys are a great help to the smooth action of a belt, and luckily manufacturers in general are realising this point more than ever before.

Quantities of Belt Used

To illustrate clearly that this discussion is not one-sided, we will boldly state that very often more belt is used than is absolutely necessary. For example, heavy machinery should, whenever possible, be arranged near the main drive so as to economize both in the length of heavy shafting carried and in the length of belting used. Where machines do not run continuously these should be grouped together, as continual starting and stopping involves considerable wear and tear on any belting. We have viewed many a case where a machine has been planted down with no thought of ideal position, with the result that not only more belt was used than required, but loss of power resulted.

Another cause of wear and tear on belts is that of improperly aligned shafting. Poorly laid out shafting means wobbly pulleys, these in turn adding to the already existing strain on the belt.

Care for Your Belts

And now a word regarding the care of belting. The life of any belt depends upon the good judgment with which the belt is selected for its work and the care which it receives. Take for example a leather belt. It requires very little attention in service, but nevertheless it requires some. It should be kept clean, and if the ends are not lapped the lacing or whatever connection is used should be kept in order, so that the ends of the belt may be held perfectly square with each other.

Whenever possible a belt should be made endless on the pulley. A new endless belt will of course require to be tightened after being used a short time, but this is a small matter, and after this has been done the belt will run for a long time with no other attention except occasional cleaning and dressing. Should there be lacings or hooks which have injured the ends of a belt, new pieces may be cemented on them. If possible never have more than one joint to a belt. A wide belt is often cut in half, longitudinally, and used as two belts for smaller width pulleys. This is practical where one cannot procure the width they desire, but it is good practice to stick to standard widths of belt wherever possible.

Belt Dressing

To increase the pull of a leather belt the use of belt dressing is often resorted to. When judiciously applied this is a good idea, but where careless or thoughtless application is made, the result is detrimental and is very destructive to the belt. First, before placing belt dressing on any belt be sure of the belt dressing. Be positive that there is nothing injurious to the leather in the dressing. As we are all aware, the driving power of a belt depends upon its flexibility and adhesive properties. The object of a dressing therefore must be such as to retain these essential qualities. The adhesive action can be increased by careful application of a guaranteed dressing, but it takes a liquid or semi-liquid preparation to penetrate the

pores of a belt, thus giving it pliability. Never use a dressing to maintain the pull on a belt. As an aid it is ideal, but at the first opportunity take up the unnecessary slack.

Should you use too much dressing the pulleys will get all gummy, large patches of dressing sticking up on the periphery. This forms air pockets, so to speak, between the belt and these patches, thus decreasing instead of increasing the driving power.

Never allow such a condition to exist. If necessary knock off the belt, scrape both the pulley and the belt, and start afresh. In other words, if you must use dressing, use it wisely, not lavishly.

Examples of Belt Drives

Fig. 1 depicts an interesting belt drive at the Morristown Water and Electric Light Co., Morristown, Tenn. This fl. lustrates a 34 in. Heart double belt on a generator drive. Note the smooth action of the belt, and as this cut is from an untouched photograph it still more emphasizes the fact of the belt's smooth motion.

The second example, Fig. 2, illustrates a 34 in. Spartan double belt handling a heavy drive at the Railroad building in St. Paul, Minn. Note the idler pulley used and its method of adjustment. Both these photographs are reproduced through the courtesy of the Canadian Graton and Knight, Ltd., of Montreal, Canada.

It will be noted that two different style belts were used in these cases, which once more proves that the belt used depends entirely upon condition. In some cases a belt is specially tanned to resist exposure, steam, oil, chemical fumes, etc., and it is always a safe plan to state your conditions to a reliable belt manufacturer and secure his opinion as to the style of belt to use.

Fig. 3 is merely shown as an example of neat belting arrangement. Note how the lathes are set at a definite angle and the belt drive arranged according. Personally we prefer a straight drive to the twist drive depicted, but in this case shop conditions were such that the style illustrated was absolutely necessary, the saving in floor space alone being considerable.

The last example, Fig. 4, illustrates a good arrangement of belting. As the photograph is self-explanatory we need not dwell upon the matter.

To depict all the conditions of belt drives would take a volume, but enough has been shown for practical purposes.

In conclusion we would strongly emphasize the need of caution in the choice of any belt drive. Study it, weigh the conditions carefully, see if the load is steady or irregular, and above all, get to know the merits of the various styles of belts, both leather and fabric. It is only by being conversant with such conditions that one can expect belt satisfaction.

RECOVERING OIL FROM SHALE
By E. R. Hassrick

The next ten years will see a revolution in the use of raw coal, and oil shales will be used extensively in oil recovery, according to a statement by Dr. George H. Ashley, chief of the Topographic and Geological Survey of the Pennsylvania Department of Internal Affairs.

Believing that the increased consumption of oil and gasoline in this country will necessitate the finding of additional sources of supply, he has commenced a study of the possibilities of recovering oil from shales. The use of shale for distillation purposes, according to Dr. Ashley, is an extremely expensive proposition, and he believes that this method of procuring oil for the making of gasoline will hardly be brought into general use until the price of gasoline reaches 50 cents a gallon. At the present time, he says, there has been no decrease in the supply of oil and gasoline in the country, but rather a continuous increase in production. During the last year, however, the demand has exceeded the supply, and the country's reserve stock has been considerably reduced.

In addition, it has been necessary, on account of the exportation of gasoline from this country, to import it from other quarters. He also points out that, although there has been an increase in oil production in this country, the initial production of oil in new wells has been

far below the initial production of wells in years gone by. With increasing costs of machinery and operation and a smaller initial flow from new wells, he believes that the present high price of gasoline can, in a measure, be attributed to these items.

In discussing the possibility of securing oil from coal, Dr. Ashley says: "Today, except for metallurgical purposes, we burn coal in a raw state, wasting, when used for power, from 80 to 95 per cent. of the heat and energy in the coal, and all of the oil, gas, tar and other by-products, except as they are burned in the fire. "It may be safely predicted that the next ten years will see a revolution in our use of raw coal. Already the by-products coke oven is replacing the old bee-hive oven. The coke is made by the distillation of bituminous coal at a high temperature, around 1,200 degrees F., the by-products being used for heating and illuminating, tar and ammonia. If, however, bituminous coal is distilled at a moderately low temperature, below 750 degrees F., the quantity of gas given off is reduced and coke is replaced by a smokeless fuel, called in one process 'coalite,' having the same heating value as an equal weight of coal, and—oil.

Products Obtained

"A plant operating in England with English coal is reported as obtaining from one ton of coal three gallons of motor spirits suitable for every purpose for which gasoline is used, 16 gallons of oil similar to crude petroleum, which can be used for burning, light, lubricating or for distillation; 7,000 feet of gas, richer in heat and higher in candle-power than ordinary coal gas; 20 pounds of sulphite of ammonia for fertilizing and high explosives, and 1,400 pounds of smokeless fuel.

"Contrasting these figures with those showing the production from by-product coke ovens at high temperature in the Pittsburgh district, it should be noted that these ovens give only 1,100 pounds of coke and no oil, but give 300 pounds of tar and only six and one-half pounds of ammonia sulphite, three gallons of benzol and 7,000 feet of gas."

Pennsylvania is at this time mining about 180 million tons of bituminous coal a year. "If one hundred million tons of that were first passed through such a process as that mentioned," continued Dr. Ashley, "it would increase our oil supply by over one billion gallons, and, remembering that smoke is unburned coal, it may be questioned if the 'coalite' left over and the distilled gas would not supply nearly or quite as much heat as the one hundred million tons of coal."

A Danish company proposes to develop some iron ore deposits in the north-western peninsula of Iceland. These are said to contain from twenty to thirty million tons of ore. The material is said to be a brown oxide containing from 40 to 55 per cent. of Fe, and lies at a height of some 1,000 feet above the base of the basalt rocks.

FIG. 2—A 34" DOUBLE BELT HANDLING A HEAVY DRIVE.

Using Rock Drills in the Industrial Plant

The Modern Machine Shop and Industrial Plant Find Many an Awkward Job Which Can Be Easily Accomplished by Use of Such a Tool as Described Herein.

By F. A. McLEAN.

IN the installation of conduit for electrical wiring, gas, water, steam or air pipes, in the erection and re-erection of motors, generators, machine tools, crane runways, shafting, etc., in the fitting of steel door frames, window sashes, buffers, corner plates, in the modern machine shop or factory, the electrician, the plumber, the steamfitter, the millwright, the master mechanic and the mason, often find it necessary to drill holes of various diameters and depths in stone, brick or concrete walls, ceilings, floors, columns, stanchions or foundations.

The usual method of drilling these holes with a hand steel and a hammer is by no means satisfactory for several reasons. Hand drilling, whether it be in stone, brick or concrete, is a slow and monotonous job of which men soon tire. Then it is almost impossible to strike the steel with a uniform blow and the holes vary widely in diameter, often being larger than actually needed, are of irregular shape and require a lot of patching up with cement or mortar. This makes an unsightly job, examples of which may be seen by glancing at the chinks or dabs of mortar around pipes or bolts protruding from the walls and foundations of many manufacturing plants.

Where there are a number of holes to be drilled or a portion of the plant is forced to remain idle, pending the drilling of anchor bolt holes, the cutting or trimming of rock for a foundation or the installation of wiring or pipes to accommodate some new piece of equip-

ANOTHER USE TO WHICH THIS STYLE OF DRILL CAN BE PUT.

ment or change in the location of a motor, boring mill, lathe, drill press or pump, the cost of drilling the holes by hand is likely to be excessive and in some instances may be greater than the cost of the power drilling equipment necessary to do the work.

Hand rock drills of the standard Jackhame: type have been successfully used in a number of plants for the heavier forms of this kind of work for several years. A notable example of their value in this respect occurred during the rebuilding of the Thomas A. Edison plant

at East Orange, N.J. On this job it was found necessary to drill 9,500 holes in the concrete walls to take expansion bolts used for holding the steel window frames in place. Through the use of the Jackhamer the work was accomplished in about one-fifth of the time that it would have taken by the hand method.

With the exception, however, of such jobs as cutting large hard rocks for engine and machinery foundations, digging holes, boring frozen ground or tearing out old concrete or masonry foundations, most of the rock drilling requirements of the average factory or machine shop can be met by a somewhat smaller and less powerful drill than the regular Jackhamer, which is a little too heavy for light work and is not as easy to handle when drilling overhead as a smaller tool would be.

With this idea in mind, the Canadian Ingersoll-Rand Company has recently brought out a new drill of the Jackhamer type, known as the BAR-33, or Industrial Jackhamer, and also a simple attachment whereby the larger sizes of their "Little David" chipping hammers may be rapidly and easily converted into light hand drills or "plug drillers" as the small drills used for plug and feather work and other light drilling jobs by stone masons and monumental workers are usually called.

This Industrial Jackhamer weighs only 21¼ pounds, uses ⅞-inch hollow hexagon steel with bits of up to 1⅜-inch, consumes 34 to 55 cubic feet of air per minute at pressures of 80 to 100 pounds and is adapted to all forms of drilling

THESE THREE VIEWS SHOW VARIED USES FOR THIS TYPE OF DRILL.

where the rock is not excessively hard and the depth of the holes required is not over five feet. Its light weight 'and small over-all dimensions relieve the operator from unnecessary fatigue and enable him to drill a far larger number of holes in a given time than can be done by the best hand driller and to reach places that he could not get at with a larger and heavier machine.

In general design this new drill resembles the company's standard BCR-430 Jackhamer, the valve mechanism being similar. All ports are short and direct, the inlet being separate and distinct from the exhaust. The rotation is simple, sturdy, automatic and positive. A rifle bar and ratchet are located in the back of the cylinder and impart a rotary motion to the piston which in turn rotates the chuck holding the drill steel, on much the same principle as that made use of in the ordinary automatic screw driver. The handle, front head, through bolts and cylinder of these drills are drop forgings and other parts are made from special alloy steel. The photographs accompanying this article illustrate a few of the ways in which these drills are made use of in industrial plants.

In the small plant where tnere is only an occasional need for drilling holes in stone, concrete or brick, and only a few holes are required at any one time, 'he "Little David" plug drilling attachment will be found useful. When applied to a No. 3 or No. 4 "Little David" chipper, a simple plug driller is formed that will drill ¼-inch to ⅝-inch holes to a depth of six or eight inches at a rate of speed that far eclipses hand work, and when the required number of holes have been drilled, the tool is quickly converted into a chipper again ready to do the work usually done by these tools. In addition it may be used for toothing out bricks, enlarging drilled holes, dressing up concrete and stone, etc.

The plug drilling attachment consists of an alloy steel bolt constructed so that a portion of the air from the exhaust of the drill is deflected and carried through a short hose to a special blowing attachment which clears the cuttings from the hole, a rotating wrench and a bit of such size and shape as will be most suitable for the work required, complete the equipment. As shown in one of the accompanying photos, the tool is held in one hand and a semi-rotary motion imparted to the rotating wrench by the

THE EASE OF HANDLING IS SELF APPARENT.

free hand, thus ensuring the ultimate roundness of the finished hole.

It is perhaps needless to point out that while the largest sized bits that are recommended for use with the Industrial Jackhamer and the plug driller are respectively 1⅝-inch and ¾-inch, neither tool is limited to the production of holes of these diameters or under, since a number of holes can be drilled very close together and the intervening webs then broken out by a few sharp blows from a hammer, by chiseling or wedging or by holding the tool against them and turning on the air, allowing the vibration of the tool to break them loose. Large holes drilled or cut in this way are generally more regular in shape than those drilled with the hand steel and their size is subject to more accurate control.

In machine shops or manufacturing plants operating foundries, both tne Industrial Jackhamer and the chipper used to make the "plug driller" can be adopted to knock out cores and clean the sand

from all sorts of castings and will do as much work as two or three men could do by hand. When using the Jackhamer for this purpose it may sometimes be found desirable to remove the rotation pawls so that the "steel" or "bit" does not turn. The Jackhamer will handle steels up to five or six feet in length, permitting the removal of cores from engine and machine tool frames or other heavy castings, while the chipper is better suited to use on smaller and more intricate work, although in large plants where there is a considerable amount of core breaking to be done, better results are sometimes obtained by tools designed especially for this purpose.

BRITISH OIL ENGINE PIONEERS

The way in which a name, whether good, bad, or indifferent, sticks to a thing which has once received it is illustrated by the Diesel engine. In many vital respects the modern oil engine as used for propelling ships and for generating electricity is distinct from the type invented by the late Dr. Diesel, but people go on calling them "Diesels" as if the German inventor were the father of them all. There is a good deal in a name when applied to an engine in common use, because a wrong name not only gives a mistake notion of the machine but it does an injustice to the real inventor. This point came up at a recent meeting of the British Institution of Mechanical Engineers, when a paper on the progress of oil engines was read. It was there pointed out that the so-called Diesel engine was really the Ackroyd Stuart engine, as a British inventor of that name had, long before Diesel's day, worked out the principles on which large oil engines could be made to work satisfactorily. British engineers feel that it is the name of Ackroyd Stuart and not that of Diesel which should be immortalized in connection with this great branch of engineering.

GIANT FLOATING CRANE

A floating crane recently supplied from Great Britain to an Australian dockyard is one of the giant cranes of the world. It is capable of lifting a load of 150 tons and swinging it within a radius of 90 feet. In this British design the whole of the deadweight is carried on a large "roller path" fixed to the deck of the pontoon and not on the top of the steel tower which rises from the centre of the pontoon. Dangerous stresses are thus avoided, material is saved, and the life prolonged. The crane has been officially tested up to 200 tons with complete success, all the performances calculated for by the builders having been realized. No moving ballast is used in the pontoon, and at no time is the angle of tilt more than about four degrees. Two cranes of a similar type have been at work in the same dockyard for six years. A noteworthy feature of the crane is that, although designed for heavy loads, it handles light loads with ease and rapidity. All the machinery on board is electrically driven, power being obtained from a steam dynamo or from shore by means of a flexible cable.

DETAILS OF THE DRILL ITSELF.

Do You Know That Steel Melts from the Inside?

The Author, Who is the Discoverer of the Fact That Steel Melts from the Inside, is Allowed in This Article to Tell His Story in His Own Way.

By G. P. BLACKISTON.

NEW things continue to happen, new inventions continue to be found, and new discoveries are occasionally made despite the oft heard proverb "There is nothing new under the sun." Perhaps in no field more than in the steel field have greater, more far-reaching discoveries of scientific value been made in recent years. Only twenty years ago a man, whose name is found in the records of mechanical achievement, made the momentous discovery that steel containing certain alloys possesses an entirely new and distinct property, called red-hardness. It was about this time that F. W. Taylor stumbled, as it were, while engaged in a work of an entirely different nature, upon this remarkable fact, a fact of so much importance that the new steel, which became known as high-speed steel, was the impetus that was responsible for the revolutionizing of the then common manufacturing processes. For this new steel was able to remove metal, when shaped into cutting tools, at so fast a speed and feed and at a depth of cut so great, that truly the world was astonished when its properties were demonstrated.

And now comes a gentleman named G. P. Blackiston, and announces to the world that steel melts from the inside. Like Taylor's, Blackiston's discovery was an accident. We do not know what its value will be to the steel trade but will tell the story as it comes from the lips of Mr. Blackiston himself, whose desire is to give the information freely and let all who may profit from it.

Mr. Blackiston is a practical steel man, and made this discovery while at work in a crucible steel plant. As is well known in the manufacture of crucible steel, the mixture for melting consists of definite and carefully weighed amounts of certain grades of wrought iron, blister bar, alloys and scrap which

are packed in pots. These pots, when filled and covered, are placed in a furnace which contains a number of holes—six pots, three deep and two abreast, being placed in each hole. The holes are then covered and the pots are subjected to a heat of about 3,000 degrees F., for a period of from two to three hours. After the mixture is melted, the pots are allowed to stand in the furnace until all the gas in the solution has boiled out and the molten steel is then ready to be cast.

In the mixture used at the time that the discovery was made, numerous pieces of octagon bar steel such as is used for chisels, and rectangular bar steel, as used for lathe and planer tools, were found. After the pots had been packed, placed in the furnace, and the melting was well under way, one of the pots broke and was immediately removed from the furnace and set to one side.

Let us examine briefly what was revealed by an investigation into the mass of metal found in this broken crucible, which had been subjected to the high heat of 3,000 degrees F. for a short time. After the pot had cooled, it was observed that chunks of metal were present which seemed to have the original shape they possessed at the time they were inserted in the pot. By tapping them with a hammer, a hollow sound was noticeable and upon further investigation the pieces were indeed found to be hollow. They were carefully removed and are shown in the accompanying photographs, Figs. 1 and 2. It will be observed that the pieces removed from the pot are two hollow shells that have retained their original contour, except that in the process of melting the molten metal in the inside found a weak spot in the outer shell and passed out through the opening shown. The reasons for this are not so

obvious. As is well known, pure iron has a higher melting point than steel and the more carbon there is in the steel, the lower its melting temperature. The outer layer of these bars of tool steel received the heat first, but this heat was, no doubt, immediately transferred to the inner portion, as the heat continued to inner portion, as the heat continued to be added, it is believed that the carbon was burned out of the outer layer and that it became, in fact, a low-carbon steel approximating iron itself. The inner layer still retained its carbon and would, of course, melt at a lower temperature, which explains why it had entirely melted, while the outer layer was intact. This important discovery was, of course, due to the accidental breaking of the pot containing this certain mixture just at the right time; and the subsequent handling and investigation revealed this interesting phenomenon. We do not state where this discovery may lead to, but perhaps the principle may be applied to the manufacture of articles, now made in an entirely different manner, in a new and cheaper way.

MEASURING MACHINE

A microscopic measuring machine has been placed on the market by Alfred Herbert, Limited, of Coventry, England. The machine is mounted on a rigid box bed, and the table carries a pair of centers, one of which can be adjusted sidewise for accurate alignment of the work. The reading of the instrument is accomplished by the relative position of the two cross hairs that are fitted to the microscope, one of the hairs rotating with the outer tube and the other with the eye piece. The outer tube has a dial graduated in half-degrees and the eye piece has a vernier reading to one minute of arc. Accurate reading is obtained by the angular position of the two cross hairs. The machine is likewise fitted with a light projector.

HERE ARE THE PHOTOS ON WHICH THE AUTHOR BASES HIS CLAIMS.

Have You Been Reading About This Society?

By J. H. MOORE

CONSIDERABLE interest is being shown in the recent formation of the Federated American Engineering Societies in the United States, and for our readers' benefit we will tell briefly what has transpired.

On June 3 an organizing conference was held at Washington, with the result that the new society came into being. About 75 branches of the engineering profession were represented, and this alone tends to show the solidity of the movement. It has been apparent for many years that with the constant increase in number of engineering and allied technical societies, each carrying on its work independently of the others, that some form of comprehensive organization was desirable which could speak for these societies in matters of common concern. There were frequent occasions where united action by these professions was desirable, and as a result of a serious need to meet the conditions arising from the war, the Founder Societies created an engineering council—"to provide for the consideration of matters of common concern to engineers as well as those of public welfare in which the profession is interested in order that united action may be possible." This was affected through the agency of the United Engineering Society, of which Engineering Council was made a department. In February, 1919, the American Society for Testing Materials became the fifth member and the American Railway Engineering Association was admitted as the sixth member in April, 1920.

The Engineering Council held its first meeting in June, 1917. It has a chairman, two vice-chairmen, who with three other members elected by the council constitute an executive committee. Its officers and members of its 24 committees, membership on which is not limited to representatives on the council, total 125 engineers representing all parts of the United States.

The appointment by the four founder societies of committees on development was for the purpose of determining the functions and objects of these societies and of making recommendations as to the changes in their activities that were desirable both as to internal relations and as to their relations to other societies.

Joint Conference Committee

Confreres of the four societies met in August, 1919, and organized as the joint conference committee. The purpose of the latter was to determine in what manner these four societies could co-operate in non-technical or welfare work affecting the relations of the engineer to, and his service in public affairs. The committee presented a report to these societies in September, 1919, and recommended the formation of a comprehensive organization, the purpose of which should be "to further the public welfare wherever technical knowledge and training are involved, and to consider all matters of common concern to these professions."

This plan of procedure was approved by engineering council at its meeting in October, 1919; the joint meeting, held in New York in January, 1920, of the members of the governing boards of the four founder societies and of the American Society for Testing Materials, the members of Engineering Council, and the trustees of the United Engineering Society, unanimously requested the joint conference committee to call, without delay, a conference of representatives of the national, local, state and regional and engineering organizations of this country for the purpose of bringing into existence the comprehensive organization recommended by the joint conference committee.

The Organizing Conference

At the organizing conference held in Washington, June 3 and 4, 1920, in response to the call issued by the joint committee, representatives of about 75 engineering and allied technical organizations created the Federal American Engineering Societies and provided for the administration of its activities by the American Engineering Council and its executive board. The constitution and by-laws adopted were substantially those recommended by the joint conference committee. At its session of June 4, the following resolution was unanimously adopted:

Resolved: That it is the sense of this organizing conference that the joint conference committee should be entrusted with making provision for putting the conclusions of this conference of engineers into effect, and that engineering council be requested to carry on its work until the new organization has been established, and by all proper means to further the programme of the new organization. The conference further recommends to the contributing societies that they continue supplying the funds required by engineering council until its work is taken over by the new organization.

Thus the way has been opened for the American Engineering Council to succeed Engineering Council as soon as the former is ready to take over the work that the latter is now doing, which will probably be on January 1, 1921.

What of Engineering Council?

In view of this fact it will be of interest to review briefly what Engineering Council has accomplished in its three years of existence. Under the terms of the charter granted by the State of New York to the United Engineering Society, Engineering Council has defined the field of its activities as follows:

"Council may deal with any matter of general interest for which joint action of two or more of its member societies would have been appropriate, if council had not been established. Council may initiate and carry through projects of the general character defined in the by-laws, for which the necessary financial provision has been made.

"Council may take up, and in its discretion act upon, any matter of general interest referred to it by any member society or by any other society, national, state or local, or by any branch of government, or by any individual or group of individuals."

In a circular issued in December, 1919, outlining the "Aims and Work of the Council," in discussing what it has done, the following statement is made:

"Council's greatest contribution, possibly, although little known, because indirect and intangible, has been the development through its own discussions and those of the governing bodies of its member societies, of the problems involved in bringing together for united action the fragments of a profession so much broken by specialization as is the profession of engineering."

The working together of committees of representatives of the several member-societies of engineering council has unquestionably had a leavening influence for the betterment of the engineering profession, and the council has certainly value in initiating the joint work of these societies.

It was not until seven months after its organization that funds were available for carrying on the work. The secretary states: "For 1918, the appropriation for council averaged 40 cents, and for 1919, 60 cents per individual member of the member-societies." With this limited support we have accomplished a great deal of work.

Field of the Federated Societies

The questions are frequently asked: "What will be the field of activity of the Federated American Enginering Societies?" and "What does the organization intend to do?" The work of this organization will be a kaleidoscopic character, with few fixed lines of activity, its major work depending on conditions as they arise from time to time; it will take over and extend the work of Engineering Council.

What it intends to do is to use its power for the service of the community, state and nation in public affairs wherever engineering experience and technical (Continued on page 288.)

A Talk on the Art of Practical Casehardening

The Author of This Article is Well Able to Speak of the Subject, Being a Recognized Authority—He Once More Emphasizes the Importance of the Man as Well as the Method.

By THEODORE G. SELLECK

IN concluding this series of articles on the subject of the casehardening of steel, the writer wishes to emphasize the point he has tried to make clear throughout the article,—no matter what the method of operation employed, *the Man is More than the Method.*

We may give all the instructions for the operation of the process of carburization that are known, practically or theoretically, and with the most complete details of procedure, and yet such instruction cannot be so exact that the human element is not the largest factor in the successful operation of the process.

The most complete specifications for the carbonizing of steel yet published are those of the American Society of Automotive Engineers, and a close study of them reveals just how much is left to the judgment of the casehardener. These specifications have been compiled as the result of the experiences and experiments of many expert engineers and are, undoubtedly, the best guide we have for the casehardening of automotive parts.

As an example of what is meant by the assertion that even these specifications are in nowise complete and that much is left to the judgment of the operator, let us consider one of the specifications,—that for the treatment of ".20 per cent. Carbon Steel,"—which is used in casehardened parts more than any other type of steel. The specification follows:

"1· Carbonize at a temperature between 1600 degs. F. and 1750 degs. F."

"2· Cool slowly or quench."

"3· Reheat to 1,450 degs. F. to 1,500 degs. F. and quench."

The wide range of temperatures for carbonizing, the advantage of the direct quench or slow cooling, the time the parts are to be held at the carbonizing temperature, the nature of the carburizer and every practical detail is left to the judgment of the operator. It cannot be otherwise in the operation of our present system of carburizing.

There are many variables to be considered: Variations in the chemical constitution of the steel; even the specifications demanded by the S. A. E. allow a variation of ten points in the carbon content of the steel just mentioned and thirty points variation in the manganese content. This wide range of temperatures is to provide for those variations and also to meet the varying qualities that may be desired in the "case," the depth of penetration, etc.

The specifications mention that the

*Fro mthe Journal of American Steel Treaters' Society.

above treatment is sufficient for a certain class of work, "such as gears, steering-wheel pivot-pins, cam-rollers, push-rods, and many similar details of an automobile." But if we were to step out of the automobile industry and wished to use the process for something foreign to that line, the above instructions would not be particularly helpful unless the casehardener possessed sufficient knowledge and judgment to make a comparison between that line of work and the work he wished to treat. The simple treatment given above is used on a much wider range of work than the S. A. E. suggests it may be used for and, in many cases, with perfect satisfaction, but in more cases with losses and failures that should convince the operators (who fail with it) that they should change methods; but there are some who are so determined that double-treatment of case-hardened parts is not necessary; that they persist in the single-treatment when they have a constant, and sometimes heavy loss.

Double-treatment for the same steel is specified as follows:

"1· Carbonize at a temperature between 1600 degs. F. and 1750 degs. F."

2. Cool slowly in the carbonizing mixture.

3. Reheat to 1550 degs. F. to 1625 degs. F.

4. Quench.

5. Reheat to 1,400 degs. F. to 1,450 degs. F.

6. Quench.

7. Draw in hot oil at a temperature which may vary from 300 degs. F. to 400 degs. F., depending upon the degree of hardness desired."

In these operations, seven in number, we have the most complete and perfect method for the heat-treatment of casehardened parts; however, there are but three of the seven operations that are not left to the judgment of the operator, (2, 4 and 6). In the other four the operator must follow his judgment as to the proper carbonizing temperatures, quenching temperatures, and drawing temperatures; these are naturally the most important factors of the process and upon the proper choosing of them depends the success of the operation. Most operators know that the same piece, or the same quality, of steel carbonized at 1600 degs. F. for eight hours will have quite a different quality of "case" than it will possess if carbonized for the same period of time at a temperature of 1750 degs. F., and it is up to the operator to determine the exact carburizing temperature which will give him the quality and depth of case he desires. If he knows his steel and his carburizer, he may be able to

determine these points without making preliminary tests, but in most cases they can be determined only by experiment.

There are other types of steel that require the same treatment as that given above; in fact, i· · a sort of standard treatment that may be followed almost universally for steels of the same carbon content, the operator being the judge of the proper heat to apply within the wide range allowed. Steel castings are left out of the list of S. A. E. specifications, but the writer has followed the above single-quench method very satisfactorily on such parts. But if a casehardener is not experienced, he may have a world of trouble trying to follow the same method, because of the variations in the quality of the castings. The double-treatment will hardly ever fail, and the "drawing" may, in most operations, be omitted.

There are times when a case hardener is called upon to carbonize steel of higher and lower carbon content than that given above.

Steel carrying less than 10 per cent. carbon is not used for machined parts to any extent, or seldom for forgings; consequently, it is rarely used as a casehardening steel; there are some uses made of it, however, where it is advantageous to carbonize it. It may be carbonized at much higher temperatures without danger of injury to the "core" than steels of higher carbon content, and for that reason is sometimes used for the manufacture of agricultural steel, for such parts as cultivator-shovels, plow-shares, plow-moldboards, etc. This steel, when carbonized at a temperature between 1650 deg. F. and 1750 deg. F, will stand considerable working at high temperatures, such as forging, welding, and forming, without losing any appreciable amount of its carbon content, and at the end of the shop operations will, by the application of the double-heat treatment, develop a very tough "core" and a very hard surface. Where this steel can be used, single-heat treatment is sufficient, if the carburizing temperature has not been carried above 1650 deg. F. If the quality of the case is not of much importance, quenching direct from the pot will be the only treatment necessary. If a more refined case is desired, it will be better to let the parts cool down in the carburizer and reheat to 1450 deg. F. and quench.

It is sometimes necessary to carbonize steel with carbon content as high as .20 to .30 per cent. Such steel should never be carbonized except as an emergency, and the most painstaking care should be used in the heat-treatment,

which should always be the double-quench and draw method.

Alloy steels require only slightly different treatment than that given the plain carbon steels.

Points to be Watched

Nickel steel, containing 3¼ per cent. nickel and .10 to .20 per cent. carbon, will take practically the same treatments as the S. A. E. 1020 except that the second quench should be carried out at somewhat lower temperatures. In the use of this steel the treatment depends upon the character of the parts to be carbonized. This steel is used extensively in the manufacture of gears and these should be very carefully quenched, and drawn, if the highest quality is to be established in the finished gear. The second quenching temperature should be as low as can be used and get the necessary hardness of the case. Where the hardening temperature for plain carbon steel is given as 1,400 deg. F. to 1,450 deg. F. it is given as 1300 deg., so it is wise to do a little experimenting before adopting a temperature for the refinement of the case. Parts of irregular shapes and thickness should be drawn in oil at from 250 deg. to 500 deg. F., according to the degree of hardness required.

Nickel-chromium steels of the usual analyses require the same treatment as the plain nickel steels but must be more carefully watched during the carburizing operations as they are carburized more rapidly than plain nickel.

Chromium-vanadium steels are the best casehardening steels we have so far as the process of carburizing goes, although there are steels better adapted to certain parts.

The carbonizing temperatures for these steels are the same as those recommended for the plain carbon steels, but the quenching temperature for the refinement of the core should equal the carbonizing temperature. That is, if the steel is carbonized at 1650 deg., the first quench should be at that temperature, the metal being reheated after cooling down in the carburizer. For the refinement of the case a temperature between 1475 deg. and 1550 deg. is recommended. For the most particular class of work a "draw" at 250 deg. to 550 deg. is indicated.

These specifications, compiled and issued by an organization of highly technical men, seem unnecessarily indefinite, but as a matter of fact that are as precise and definite as can be laid down for the treatment of a metal that comes so far from being regular and constant in its chemical constitution; and until steel can be manufactured to exact analyses, and present variations of carbon and its other elements are eliminated, we shall have to have flexible rules for its treatment.

In following these, or any other, instructions, for the heat-treatment of steel, the operator should have as an aid a good pyrometer, and be always sure that it is in proper working condition.

The inexperienced steel-treater will find considerable difficulty in arriving at conclusions concerning the accuracy of his instrument unless he has some means at hand for checking it. Usually an extra thermo-couple is kept for that purpose and by it the regular service couples are checked at very frequent intervals, sometimes as often as every day. This is by far the best way and the simplest.

With the pyrometer, the steel-treater should equip himself with a "color chart," giving the temperature values of the various heat colors. By making comparisons between these colors and the temperatures indicated by the instrument he soon gains a knowledge of colors that will enable him to readily observe any important error in the reading of his instrument.

A color-blind steel-treater would seem to be about the last anomaly we would look for in the working of steel, but they exist altogether too commonly. Some are not entirely so and yet sufficiently to make the accurate working of steel a mere chance, while others cannot discern the difference between the primary colors, to say nothing of their various tints. This latter class rest entirely upon the pyrometer, of course, and their instrument goes wrong their work is sure to do so. For that reason, employers of steel-treaters should insist upon an examination for such optical defects before they allow a man to function in the treatment of steel parts that require positive and careful handling at high temperatures.

For inexperienced case-hardeners who do not know the value of the more common colors and tints as expressed in temperature, the following tables are given. These should be supplemented by color charts showing the actual colors. There are several of these that may be obtained from manufacturers of pyrometers and heat-treating equipment, without cost.

Degrees Centi-grade	Degrees Fahren-heit	Color
400	752	Red heat, visible in the dark.
474	885	Red heat, visible in the twilight.
525	975	Red heat, visible in the daylight.
581	1077	Red heat, visible in the sunlight.
700	1292	Dark red.
800	1472	Dull cherry-red.
900	1652	Cherry-red.
1000	1832	Bright cherry-red.
1100	2012	Orange-red.
1200	2192	Orange-yellow.
1300	2372	Yellow-white.
1400	2552	White welding heat.
1500	2732	Brilliant white.
1600	2912	Dazzling white.

It is conceded by the best authorities that the most skilled observers may vary as much as 100 degrees Fahr. in their estimation of relatively low temperatures by color, and beyond 2,000 degrees Fahr. it is practically impossible to make estimations with any certainty. So that the use of the color chart is limited, practically, to the use it may be in giving warning of the failure of the pyrometer, and the judging of the uniformity of the temperature throughout the heating chamber of carbonizing furnaces.

In reheating, for hardening case-hardened parts, the liquid bath offers the best means of heating the parts uniformly, because the temperature required may be easily maintained and none of the parts can ever get hotter than the bath.

The materials for such baths most generally used are molten lead, mixtures of barium and potassium chloride, barium chloride, cyanide of potassium, and other metallic salts. These various baths are adapted each to a certain range of temperatures. The lead bath for instance, may be used for temperatures between its fusing point and about 1,290 deg. F., above which it is not practical because of the volatilization of the metal at that temperature; lead is chiefly used for "drawing" operations, and as a quenching medium for "high speed" steel. For temperatures between 1,400 deg. and 1,650 deg. F. a mixture of 60 per cent. barium chloride and 40 per cent. chloride of potassium may be used, or, 50 per cent. calcium chloride and 50 per cent. sodium chloride; these chloride solutions have one very bad property however, and that is the rusting of the material treated in them. This may be overcome by washing the parts in a solution of soda after quenching in clear water. A better heating solution, or bath, is a mixture of soda-ash and the common commercial cyanide-chloride mixture containing about 75 per cent. of sodium-cyanide, in proportions of 60 per cent. of the former and 40 per cent. of the latter. This gives a solution containing about 30 per cent. of cyanide.

In the use of any of these solutions, or baths, the furnaces should always be provided with ventilating hoods for carrying off the fumes and gases from the operator. Even lead yields a very poisonous gas when subjected to a volatilizing temperature.

Preheating Necessary

In the use of these heating baths the work should always be preheated to about 800 degrees before immersing in the solution, otherwise the solution will be so chilled that the operation will be greatly slowed up. Some pot furnaces, specially designed for such work, have special pre-heating chambers attached that permit of doing that part of the work by the use of waste gases from the furnace, so there is no extra expense attached to the preheating.

Parts heated in solutions should be suspended on hooks or in baskets from the top of the pots, and should not be allowed to touch the bottom or sides of the pot.

The pyrometer thermo-couple should be protected by a pure nickel protecting tube and should extend well into the pot without touching the bottom. Care should be observed in the matter of overcrowding the pot as work can be done more rapidly if the volume of solution is much larger in proportion than the volume of metal being heated.

In applying the carburizing process to any class of work it should be always borne in mind that each particular form

of parts to be carburized constitutes an individual problem in the art, and should be so considered, and its particular method of treatment should be worked out, without any reference to the methods applied to any other part. In other words the treatment best adapted to the carburization of cam-shafts is not suitable for wrist-pins, and the treatment best adapted to ball-races will not answer for very light bushings, caps, etc., and any operator of the process who knows of but one method to follow, one carburizing temperature to use, and but one quenching temperature to apply, is in line for a lot of trouble and loss.

With all the information that he can obtain from others, either by word of mouth, ·demonstration, or through the medium of books, the case-hardener will find himself frequently confronted with problems for which he has no answer. So it becomes the operator of the process to be always studying his art, even to making problems and finding their solution.

The writer is frequently consulted by casehardeners concerning various questions relative to the successful carburization of certain parts with which they have difficulty. Some of the questions asked are so very simple that it would seem that a man of just ordinary intelligence would not need to ask them, while others are so deep that the only answer that can be given is the suggestion that the successful treatment of certain important automobile parts depends entirely upon the development of new methods of carburizing.

The carburizing process is of so much importance in the manufacture of certain lines of the common necessities of life, as we are living it these days, that it cannot be allowed to linger in its present very incomplete stage of development. It must keep pace with the other important branches of steel-treatment.

When we consider that, as it is commonly applied, the operation of the carburizing process involves the use of more than double the fuel that should be required; that the containers used to hold the work, on the average, weigh more than the metal parts they contain, which means added labor, increased furnace capacity, and increased wear and tear; that the important, perhaps we should say the all-important, quality of uniformity, is difficult to realize under present conditions; the necessity of bringing out a new process is apparent.

At present we have no adequate system of heating carburizing boxes. The practice of placing them in the semi-muffle furnace, usually so constructed that the uniform distribution of heat is a problem greater than the average operator can solve, generally results in some of the boxes getting vastly more heat than others, frequently to the utter destruction of some of them while others are not heated above the necessary carburizing temperature. Recently the writer was "on the job" when one of a charge of four boxes in the furnace was partially melted while the other

three were not at an excessive temperature, the cause in this case being the improper placing of the oil-burner from which a needle flame shot against the wall of the box. The fault, primarily, was with the manufacturer of the furnace, since the furnace was practically new and such conditions should not have been possible; secondarily the fault was with the operator, because he did not watch his equipment and know whether the furnace was operating properly or not. But the operator being "a green-horn," and this the first casehardening job he ever had, he could hardly be expected to know that cast iron boxes can be melted in a casehardening furnace. It was a case where the employer believed, as so many of them do, that "any Dub can caseharden steel," and this man fitted the appellation to a nicety.

Metal containers for the carburization of steel, as they are now used must eventually be discarded and new methods of applying heat be developed before the process can come into its own and reach its highest point of usefulness.

We must get away from the thought that experience, intelligence and carefulness are not necessary qualifications in the selection of men to operate the process, however, before any improvements will be of the least advantage to us. We may be able to eliminate the man to some extent, but there will never be a time when the proper operation of the carburizing process, as well as all other branches of steel-treating will not depend upon the direction of an intelligent mind for successful application, and we will always be compelled to admit that "THE MAN IS MORE THAN THE METHOD."

THE FEDERATED SOCIETIES

(Continued from page 285.)

nical knowledge are involved and to consider and act upon matters of common concern to the engineering and allied technical professions. In the conduct of many public matters which are essentially of an engineering nature, it is vital to the public welfare that engineers and allied technologists should lead.

There will be an increasing number of questions arising in which the opinion of these professions will be of fundamental value to the welfare of the nation, and it behooves these professions to so support the Federated American Engineering Societies that it may so function as to supply this great public need. It is only the exceptional individual of these professions who can fail to see that in increased activities in these broader and less selfish fields the standing of the profession will be greatly improved, and this of necessity must improve the position of the individual engineer.

It is the duty of the engineer to take up public service work. It is incumbent upon him as a citizen to "do his bit," and there is the added responsibility due to his special knowledge and experience which is primarily of importance in the execution of public work.

The Federated American Engineering Societies will speak for a group of citi-

zens who, by reason of special training and knowledge represent a high order of intelligence, and who individually and collectively would be derelict in their responsibility for the talents they possess if they did not apply them unselfishly for the common weal. In unity of action there will of necessity be strength and power. We shall look forward with interest to the progress of this new organization.

HIGH SPEED PUNCHING

Thoughts of the boxing ring are aroused by the title: "The Laws of High-Speed Punching," which belongs to a paper prepared by Captain Tresidder for the British War Office. It deals, however, with a more formidable kind of struggle, as it investigates what happens when a projectile pierces armor-plate. More than a military interest attaches to this question as Captain Tresidder has found that the laws which apply to a shell going through armor plate are the same as those concerned in the ordinary engineering process of punching holes. The extraordinary interest of this problem is suggested by the fact that a candle fired like a bullet will go right through a wooden board without changing shape. Another striking fact is that a 12 inch plate must exert 60 million horsepower in order to stop a 15 inch shell at a range of ten miles. This enormous power is exerted only for one· five-hundredth of a second, but in that brief space the plate does more work than a 33-horse-power engine can do in an hour. Many results of great scientific and engineering interest have been attained by the author of this paper, who has worked out an entirely new theory of what happens when a projectile drives a hole through solid steel.

AN IMPROVED HAND VISE

It is a well known fact that the simplest things are the last to undergo improvement. People take them so much for granted that they seldom think about bringing out an improvement upon them. A typical case is the ordinary hand vise, which is used in every workshop all over the globe. It is far from being a perfect instrument, because it works on a hinge and the gripping jaws therefore do not keep parallel as the vise is opened. In spite of this drawback it is only recently that a successful effort was made to turn out a better type. A British firm is making a patent vise in which the hinge is replaced by a right and left-handed screw. When this screw is turned by means of a small wheel the lower end of the vise opens to the same extent as the jaws, which thus remain parallel to each other. With this vise it is possible to hold round articles quite firmly.

It is interesting to know that the annual timber consumption in France is not more than 100 board-feet per capita, or less than one-third that used by Americans. Of the timber required, France imports 30 to 40 per cent.

WELDING AND CUTTING

Welding of a Corroded Boiler Shell*

This Tale Describes How a Corroded Lancashire Boiler Was Repaired by Means of a Portable Electric Welding Apparatus.

By A. K. DAWSON.

HAVING recently had the opportunity of witnessing repairs to a corroded colliery Lancashire boiler by the use of a portable electric-arc welding apparatus, it is the author's intention to describe the method of using this plant, and the work done by it.

The boiler in question was one of a group supplying steam at 100 pound per square inch to Curtis turbo-alternators and other engines. It was 30 feet long by 7 feet 6 inches diameter, and hand-fired. External corrosion had so seriously damaged the shell underneath, between the sludge-pipe mounting and the front end-plate, that the shell was holed through at one point alongside the angle-iron ring, the latter also being seriously wasted away. To remedy this evil, much time and expense would have been incurred in cutting out part of the corroded shell round about the blow-off hole away from the front end-plate, with a view to repair by means of a new cover plate. To secure the latter at a moment's notice would have been by no means easy at the present-day rate of supply of such commodities. Furthermore, it was necessary to renew the lower portion of the angle-iron ring, and a new cover patch would have been to cut, shape, dress, and drill out for the rivet holes in it, as well as in the boiler shell; also, a 7 5-16 in. hole to cut for the sludge-pipe connection. After fitting this plate in position by numerous rivets, it would have to be "caulked" or "fullered," and finished off neatly to leave no roughness.

To save time, and yet make a thoroughly reliable repair, it was decided to build up the corroded shell and angle-iron by means of Messrs. Tilling-Stevens' mobile arc-welding equipment, which is embodied in a petrol-electric-driven motor lorry.

A description in outline of the plant and the method of using it is as follows: A 40 h.p. petrol engine drives the generating dynamo, the current from which supplies an electric motor in connection with the back axle to propel the vehicle

*Abstract of paper read before the Associates' and Students' Section of the North of England Institute of Mining and Mechanical Engineers.

when travelling on the road. For welding purposes special switchgear and resistances cut off the current from the motor and convey the electric energy by a pair of flexible cables to the electric arc at the work. On the back portion of the lorry is a twin-cylinder vertical air compressor supplying from 30 to 60 cub. ft. of free air per minute, at a pressure of 100 lb. per sq. in., and driven by a twin-cylinder petrol motor. After standing the lorry in the nearest available position to the work (in this instance about 40 yards), the compressor was used to dress off the corroded surfaces of the steel plating by means of a pneumatic chisel, air being conveyed by a small flexible armored hose; this operation ensured the clean condition of the shell surface for welding, which was then commenced.

First Operation

The first operation was to secure a good electrical contact for the negative side of the arc with some part of the boiler, and this was attained by slightly welding the head of a small iron bolt temporarily on to the face of the manhole collar. The cable from the negative terminal of the switchboard was connected to this bolt, whilst the other cable from the positive terminal at the lorry was joined to a simple pair of steel tongs holding the electrode. These latter are made of Swedish charcoal iron, each measuring 12 inches in length, with a diameter of 5-32 inch, and are smeared with a flux before using. One striking property about them is the amount of bending and twisting they will tolerate without breaking. To commence welding, the end of one of these iron pencils is pressed into the tongs, the handle of which is electrically isolated with insulating tape: the operator places a hand-observation screen of red glass before his eyes, and makes contact with the electrode and the boiler shell. On withdrawing the pencil slightly an electric arc is produced of sufficient heat intensity to fuse the electrode end, the molten particles of which are carried over on to the negative side of the arc —i.e., the boiler shell, which is also rendered intensely hot at the point of contact. Great care is necessary to maintain the air gap between the electrode and the metal to be welded, the most suitable gap evidently being about ⅛ inch or 3-16 inch. If the pencil is allowed to touch the shell for even a second or two a complete circuit is thereby made, and as soon as the air gap ceases to exist, the arc, of course, fails, and the metals cool, with the result that the electrode becomes fixed to the boiler plate. It has then to be broken off and a fresh arc struck. Under such circumstances a heavy rush of current is caused by the absence of the air gap, consequently a steadying resistance and an electrical governor are installed in the circuit on the lorry to reduce as far as possible the current flutuations.

In doing the welding of this boiler, the switchboard at the lorry gave the following information: When no welding was being done, 50 volts, 100 amperes; when the continuous arc was being maintained during welding, 40 volts, 150/200 amperes; when the electrode was permitted to touch the boiler shell, making a closed circuit, 40 volts, 300 amperes.

When welding was done in the underneath side of the boiler shell, the electrode in fusing had the tendency to form molten globules, which were liable to fall off on to the bottom of the flue if the arc were allowed to play at the same point for any length of time. When this occurred the electrode was brought away from the shell, thus destroying the arc, and the soft globules of iron gently flattened against the plating with a hammer. Under average conditions it was possible to use 20 of these Swedish iron electrodes per hour, and for the building up of the boiler-plate in question it took, roughly, 300 of such pencils.

Time Involved

To give some idea of the time taken to complete the work, the welding apparatus arrived at the colliery about noon on one day; the cables, etc., were run out to the boiler, and the corroded steel plating was dressed with the pneumatic chisel in the afternoon, and half the welding done by 7 o'clock that night. By the following afternoon the welding was completed, and the work smoothed off with the compressed-air hammer, whilst the front lower edges of both furnace tubes were also built up where slightly damaged through frequent stoking and raking out of fires. One fitter with his assistant and a boy were with the equip-

Continued on page 63

My Experience as a Would-be Foundry Foreman

By F. H. BELL

ANYONE who has never had experience as a foundry foreman cannot possibly have any conception of what he has missed. In fact he has missed so much that he is in no position to appreciate how really thankful he should be.

However it must be admitted that the young man who has no further ambition than to remain dormant on a journeyman's job is terribly deficient in self-esteem. My own experiences along these lines have been so extremely interesting to me that I have been forced to laugh about them myself on frequent occasions, and for that reason have thought that perhaps while not posing as humorous or funny, but on the contrary being of a very prosaic, staid disposition, I might, nevertheless, at least amuse others by relating them.

To begin with I served my apprenticeship in a little jobbing foundry which was owned and operated by my father, who I might say was a molder with no mean qualifications. From this it may be inferred that I was a privileged character, which in reality I was—coming and going as I pleased.

I made various trips to other shops throughout Canada and the United States, but always fetching up at home when there was nowhere else to go. The experience gained through these spasmodic jaunts added to the experience in which I received at home doing such jobs as attending to the cupola and making cores and so on down the line, imbued me with a store of conceit which led, me to believe that there was not much about a foundry with which I was not familiar.

However there came a day when my preordained programme destined that I should say goodbye to my old home town and seek a livelihood elsewhere, knowing this time that I had to make good, as I no longer had a haven to fall back on. Then came the crucial test. With all my good opinion of myself, was I capable of holding a real job? I began to fear that I was not. I had different positions offered me—some in the States and some in Canada, but being a good Canadian I decided to remain among my own, and forthwith selected as my future abode the thriving industrial city of Brantford. Brantford, at that time, boasted of a population of about eighteen thousand, and according to the usual method of accounting—six to a family—there should have been about three thousand families. But, alas! It was not so. There were about two thousand families of six and about six thousand extra or spare fellows. These poor fellows used to spend their evenings standing around in bunches on the street corners, looking at each other until about ten o'clock, when they would turn in for the night, and by half past ten a cannon might be fired down most any street without doing bodily harm to anyone, because there would be no one on the street to harm.

Next to Moscow, Brantford had probably more churches in proportion to its population than any other city in the world, and the inhabitants were of the class that when the vote was taken on local option and a majority of three-fifths was necessary to make it become law, it was carried, all but eleven measly little votes, so it will be seen that the population of Brantford was made up of an exceptionally righteous class of citizens.

As every one acquainted with Brantford knows; it is a city of foundries and practically every boy works at the foundry business. Now here was a city with a superabundance of the highest type of manhood which would have made excellent catches for the fair sex, had there been any of the fair sex to catch them.

Up the shore of the noble Grand River about seven miles was another thriving little borough, by the name of Paris. Here everything was entirely the reverse. The main industry of the town was a knitting mill, which in fact was the largest institution of its kind in Canada. I am not prepared to give exact figures as to how many hands were employed at this establishment, but if they had eight hundred, one hundred would be old and experienced men who had been in the business long enough to be holding such positions as foremen and experts of one kind or another, and would in all probability be heads of families. The other seven hundred would be girls. These girls all had to be of unblemished character, as one bad one would be like a bad peach in a basket—likely to contaminate the whole basket of peaches. Just think of it, hundreds of A1 girls and hundreds of fellows such as we have just pictured and seven miles of country road dividing them. It could not be expected that such a state of affairs could long endure. A bicycle works was started in Brantford which relieved the situation to some extent, by providing the means of lessening the seven mile difficulty and in course of time a trolley line was put into operation which improved matters to a much greater extent. Still all of this was only make-shift and as something real had to be done it was decided that a foundry would have to be started in Paris. Accordingly the proprietor of the mills hit upon the idea of instituting a stock company, consisting of his funds and a few names for fillers and launching an enterprise known as the Paris Plow Company. A thoroughly modern plow shop was built and a plow maker of renown was imported along with a supply of operatives, and everything was in readiness to be put into full swing with one exception—they required a competent and efficient man to take charge of the foundry and with rare judgment they decided upon me.

I reported promptly at the office and had a lengthy interview with the manager, the result being that I was engaged at the handsome salary of three dollars per day with absolute authority over my department and no interference from anyone. I was shown through the foundry and learned that they had all the floors rammed up, but were afraid to take off a heat and they wanted me to come on at once.

Bright and early next morning I was on the job to put the cupola into commission and get the work poured off. The manager was there also, not in the office, but in the foundry. Of course he was not interfering with me or my work; he was just showing me how to proceed. Before putting up the bottom doors of the cupola he decided that we should try out the fan to see if it was properly speeded. It was set up in accordance with instructions from the makers, but he would take no chances. He had a pressure gauge on the wind pipe and he would have a certain stipulated pressure on that gauge or we would not attempt to run the cupola. By running the fan at the speed mentioned in the catalogue he did not seem to get any pressure on the gauge and he concluded that it was a lucky thing for him that he did not listen to my advice and make a try at it. In order to get off the heat it was decided that some of the machinery in the machine shop would have to be shut off so that we could borrow the pulleys until such time as we could have proper ones brought in. We were not able to make the alterations in time to pour off on that day, but we had a good early start for next day and I had the honor of lighting the fire for the first heat. Of course in the meantime we had put up the bottom doors and put on the coke and iron which made all the difference in the world with the pressure gauge when the blast was put on. Common sense should have dictated to the manager that it would, but common sense appeared to be the one thing which he lacked. However, we got the heat off all right. We had plenty of pressure on the gauge; in fact the gauge registered right up to the top and no telling how much further

Continued on page 62

DEVELOPMENTS IN SHOP EQUIPMENT

LUBRICATION SYSTEM

Modern experiments in cutting metal under a lubricant and coolant have led the Universal Boring Machine Co., Hudson, Mass., to provide their machines with equipment to pump and handle sufficient coolant.

As is well known the purpose of any coolant is to assist in arrying away heat from the place where the work is being done, thus keeping down the temperature of the cutting edge of the tool.

On the assumption that it is far better to use too much than too little coolant, they have equipped their machine to deliver twelve gallons a minute to the tool, if necessary. The pump used is claimed to never lose its prime. The system is said to be clog-proof, because anything that can get in can proceed through without injury to the pump, because there is no passage in the entire system smaller than the intake, which is ¾ in. The impellor pumps by centrifugal force, and any chips that can enter pass right through the system, thus avoiding the necessity of strainers.

The bed of machine has oil troughs on the front and back. These slope towards the centre. The front trough is on a higher level than the rear. The lubricant is carried by a 1¾ in. pipe from the centre of the front trough, through the bed to the centre of the rear trough and from there into the tank. The troughs project 4¾ in. and are 4 in. deep.

Between the ways on the top of the bed, the surface slopes toward the centre into a rectangular trough 16 in. long. From there the lubricant is carried into the tank. The table has oil pockets on each end and grooves in the T-slots, and delivers the lubricant toward the rear. From there it is carried by a flexible tube of ample diameter into the tank in the rear. The pump is attached to the motor bracket and is driven by belt from the motor shaft. It has a capacity of twelve gallons per minute. The lubricant is carried from the pump to the tool by piping and elbows with a flexible tube nozzle to facilitate the flow in any direction. The photograph shows clearly the

arrangement. We understand that the concern will equip their machines with this lubrication system on special order.

END VIEW OF THE FIXTURE.

BORING AND GRINDING MACHINE

The Sunderland Machine Shops, Omaha, have placed on the market what is known as the Fox cylinder and boring machine. The two views shown give a good idea of this machine's general appearance. It will fit in any lathe, 14 in. to 24 in. swing, and can be used on either open head or closed cylinder blocks.

In use, the cylinder is mounted in the fixture as shown in illustration, and is so adjusted by the fixture screws and the cross-slide screw that the centre of the bore is in line with the centre of the boring head.

The boring head is equipped with an outer spindle to which the boring tool can be attached. This spindle revolves at a slow rate of speed and is driven through a large heavy-duty enclosed gear bolted to its outer surface. An overhead belt transmits power to the pinion shaft, which is equipped with tight and loose pulleys. The inner or grinding spindle rotates inside the outer boring spindle and is supported by three sets of S K F double-row self-aligning ball bearings which give it the required bearing support, thus assuring a true and free-running shaft that will not throw out, whip, or vibrate. It is claimed that a degree of exactness of 0.00025 inch can be readily obtained with this machine if desired.

Among some of the advantages claimed for it are its low cost, the fact that it can be used for boring or grinding

SHOWING METHOD OF ATTACHING LUBRICANT SYSTEM.

with equally good results, that it can be mounted or removed from the lathe in ten minutes, that changing the work from one cylinder bore to another requires only a few seconds, and that its use does not interfere with chucking work in the lathe. The capacities are:

GENERAL VIEW OF THE MACHINE IN ACTION.

Swing of spindle, 22 inches; boring and grinding capacities, 2½ inches and larger; and length of spindle, 15½ inches. The equipment includes one main spindle and operating standard, one counter-shaft, one cross-slide table, two adjustable work-holding brackets, one cutter-head, two grinding wheels, one diamond wheel ,tracer, one truing-up indicator, and two wrenches. The machine equipment, boxed for shipment, weighs approximately 500 pounds.

KEY-SEATING MACHINE

A horizontal key-seating machine is a new line being put on the market by the Hercules Manufacturing Company, of Portland, Ore. The work-holding face plate is secured to one end of the machine and may be tilted to accommodate pieces with tapered keyways.. The machine cuts on the forward stroke and the cutting tool is relieved on the backward stroke by the action of wedge shaped blocks operating in the base of the bar rest. The machine is light and particularly well adapted for taking to heavy work, in place of bringing the work to the machine.

PLATE MILLING MACHINE .

A plate milling machine for milling the edges of boiler plates has been placed on the market by the Marshalltown Manufacturing Company, of Marshalltown, Iowa. The operating mechanism is carried on a sliding head that operates the full length of the machine, a 5 h.p. motor being provided for driving the spindle. Both a friction slip and a quick return are provided in the feed mechanism. Ball bearing thrust is fitted to the spindle. Rollers are provided on each side of the cutters for holding down the work. Machines are made in four sizes, having a range from 8 to 24 feet.

NEW THINGS IN MACHINE TOOLS

COMBINATION TOOLHOLDER

Maurice H. Derringer, of 3133 N. Eighth St., Philadelphia, Pa., has placed on the market a special toolholder for lathe use. One end of the bar is fitted with a yoke that may be swung to either side of the bar as desired. One side of the bar is adapted for the holding of square stock while the other is provided with a groove for the holding of round stock. The opposite end of the bar is fitted with a gooseneck that can be locked in position at any angle.

FALCON PIPE WRENCH

J. H. Williams and Company, of Brooklyn, N.Y., has added to its line a new Falcon wrench that has a wide range of application and is particularly adapted for piping installations. The jaw is of curved design with open ends, which makes it very suitable for operations on awkward fittings. The wrench is designed for one way operations. The jaw is symmetrical, so that when one side becomes dull it may be reversed, thus doubling its life, These wrenches are made in six sizes, with ranges from ½ inch to 12-inch pipe.

REVERSIBLE DRIVING CHUCK

The Gustin-Bacon Manufacturing Company, of Kansas City, Mo., are offering to the trade a reversible driving chuck, that will grip the stock when running in either direction, being adapted for use on such work where it is desired to reverse the direction of rotation. The jaws are of tool steel and it is claimed that the work is held centrally at all times. A large number of jaws are obtainable for different classes of work.

OIL AND WATER STRAINER

The Griscom-Russell Company of New York, are now making a strainer that is intended for use in the straining of lubricating, fuel or quenching oil, or in water supply lines. The body is of cast iron and the strainer is of perforated sheet steel, and when the strainer is used for oil it is lined with fine mesh wire. The strainer connections may be made to suit any desired purpose, and may be installed in units or in sets. When used in sets either one of the strainers may be cleaned without interrupting the service.

BENCH HORN PRESS

The Enterprise Machinery Company, of Chicago, Ill., have recently placed on the market a bench horn press known as the type "V." The press is intended for rapid assembly of small parts and for operation such as seaming and riveting on hollow work. The machine is of the single stop type with strokes of ⅞, 1¼, and 1⅜ inches. The ram has an adjustment of ½ inch. The press is 25 inches high and operates at 300 r.p.m.

DROP HAMMER

Long - and Allstatter Company, of Hamilton, Ohio, are now manufacturing a new type of drop hammer. Frame adjustment is provided and a positive automatic arrangement is used to prevent the frames from loosening. Increased distance is provided between the V-guides, giving a wider range to the dies used. The weight of the ram is approximately fifteen times that of the ram, and meanr are provided for recording the various heights at which the ram should be released for each set of dies, thus assuring duplicate work at different periods. The machine is made in eight sizes, ranging in size from 100 to 1,200 pounds falling weight.

LO-SWING LATHE

The Fitchburg Machine Company of Fitchburg, Mass., are now making a 4-inch Lo-swing lathe. It is of the constant speed type and both speeds and feeds are controlled by levers in the headstock. The two carriages, each carying three tools, operate on heavy V's so located that the carriages may be run past the tailstock. The feed gears in the carriages are driven by means of Oldham couples which eliminate the necessity for accurate alignment. Sliding gears are provided for the reverse feed. These lathes are made in three lengths, 60, 80 and 108 inches.

INSERTED CENTER POINTS

The Eastern Tube and Tool Company, of Brooklyn, N. Y., are introducing a new line of machine center points made of high speed steel for insertion in to the mild steel shank, which are hardened and ground, and furnished in all standard sizes and tapers.

HYDRAULIC TURRET LATHE

Albert Herbert, Limited, of Coventry, England, through their American offices, are placing on the market a special design of hydraulic turret lathe that was originated by an Italian, and intended for use by disabled men. The chuck turret and the cross slide are all operated by means of hydraulic cylinders conveniently situated. A minimum pressure of 75 lbs. is used for the operation of the machine.

Japan has another plant to its credit, that being a new one just completed for manufacturing thick steel plate. It is owned by the Imperial Steel Works at Yawata, Kyushu, and cost 4,000,000 yen.

The MacLean Publishing Company

LIMITED

· (ESTABLISHED 1887)

JOHN BAYNE MACLEAN, President.   H. T. HUNTER, Vice-President
H. V. TYRRELL, General Manager.

PUBLISHERS OF

CANADIAN MACHINERY
⚙ MANUFACTURING NEWS ⚙

A weekly journal devoted to the machinery and manufacturing interests.

B. G. NEWTON, Manager.   A. R. KENNEDY, Managing Editor.

Associate Editors:

J. H. MOORE   T. H. FENNER   J. H. RODGERS (Montreal)

Office of Publication: 143-165 University Avenue, Toronto, Ontario.

VOL. XXIV.   TORONTO, SEPTEMBER 23, 1920   NO. 13

Wanted—A Superintendent

THERE was a vacancy to be filled in an organization not many miles from Toronto. It was one of the worth-while jobs that crop up only on rare occasions.

The superintendent was leaving to go into business for himself. He had been with the firm for some years, was a good man, and his departure meant finding another good man to take his place.

There was a conference right straight off, and the president and general manager, for he held the dual position, wanted to know if any of the others present had in view the name of any other man who could be brought in to take over the work.

A number of names were suggested, but one by one they were discarded, and the only result of the meeting was to make it plain that the gathering was almost back where it started, with the possible exception of having gained the knowledge that it was no small task to find the man they wanted.

The vice-president of the company, who also held the position of secretary, at the first chance the following day, started on a new line.

He called the manager to his office and proposed the name of the foreman of the machine shop.

"But he's had no experience as superintendent," objected the manager.

"Neither had he any experience as foreman of a machine shop until he was put in that position here," was the answer.

The manager admitted that he had turned out to be a good foreman.

"Well, then," came back the vice-president, "if he made a good foreman, why not a good superintendent?"

The result was that the foreman of the machine shop, one of the company's own men, was taken in and made superintendent. He is just as good a superintendent as he was foreman.

The point is here:—Is your shop a place where you are growing foremen or superintendents, or are you dwarfing them so that they stay as machine operators and clock punchers?

How do you make a good foreman? By selecting a skilled mechanic with an honest personality, one with some leadership preferred. Very well—that is just a starting point. You must put responsibility on him—give him a chance to make good, and put that most important of all things on him, the seal of approval in the shape of increasing wages.

The average man wants to develop. It is the most natural thing in the world that he should.

If you find it necessary to start a gum-shoeing trip all over the country every time a worth-while position opens up in your organization, it's high time you turned the X-rays on the organization to find out the trouble.

Some shops grow foremen and superintendents, and some shops do not.

What kind of a crop are you growing?

Mr. Findley's Statement

THE address of Mr. Thomas Findley, of the Massey-Harris Co., before the Tariff Commission at Winnipeg, has been published in many Canadian papers as paid advertising.

It may be that it is necessary to do this in order to get his views before the public. It is a fact, though, that the same papers publish every day a tremendous amount of material that is not half as interesting nor one-quarter as instructive as the facts presented by Mr. Findley.

He reviewed the position of his company; he told of all the tinkering that had been done with the tariff on agricultural machinery; mentioned the loss the farmers were inflicting upon themselves by not housing their implements; showed that the farmer in Saskatchewan was buying his binder cheaper than the farmer south of him in North Dakota or Montana.

This paper cannot recall offhand where any man has so thoroughly laid before the public the business of his company.

It is just such frankness, and such adherence to facts that will stand up after investigation that will carry conviction.

Mr. Findley has done much to spike the guns of free trade shouters—or at least he has made it increasingly necessary for them to verify their statements before parading them before the public in print.

The Why of Time Clocks

THERE are plenty of men in shops who object to "punching the clock." Every time they shove in their ticket they do so with a grudge in every move, and there is often far more ginger put into a vigorous smash on the clock lever than in any other operation that takes place in the shop hours.

Whence come clocks and the demand for them?

Look back a bit and the thing is very plain. Men used to work long hours and the wages were low. If a man were a few minutes late, the firm lost very little. This few minutes would amount to only a few cents, and the loss could be made up or absorbed in some other way.

It is a different proposition now.

Men are working many hours a week less than they used to, and they are getting a much greater amount for the shorter work day.

This has become so marked that a manufacturer has to reckon in minutes where he used to speak and think and plan his work in hours. A man in the shop working a short day has less time in which to speed up for any time lost in the morning.

Labor unions have insisted upon the shorter day, and in self defence it becomes necessary for the employer to see to it that he gets full measure for his work-day.

MARKET
DEVELOPMENTS

Trading Quieter, But Prices Remain Firm

Many of the Machine Tool Centres Report Little Inquiry, But There is No Evidence Yet that Any of the Makers Are Thinking Seriously of a Price Reduction.

THERE is not a large volume of business moving in any of the machine tool markets. None of the Canadian centres report any great buying, and the New York market says that it is quieter there now than for months past, with very little hope of recovery. Some of the Canadian dealers are inclined to blame tight money for keeping some prospective buyers out of the money, and they claim they could dispose of a number of machines were their prospects able to finance. There is a fair number of orders as it is, but they are very much scattered. Dealers in this country do not complain of the stagnation that seems to have hit the New York market.

The supply of steel is a little more liberal, but the price that must be paid is liberal in proportion. The premium mills are making the great bulk of the deliveries that are reaching this market in the way of sheets, and they are still romping off with a high figure for their material. Warehousing interests are getting more or less nervous about going in and tackling large tonnages of this stuff,

fearing that the premium market might crack, and leave them with a shipment on their hands in a skidding market.

As far as machine tool dealers can interpret the situation, they cannot see any price reduction yet. Many of the firms they represent are still booked for months to come, and the orders are all of the non-cancellation variety. It is seldom that a request comes for cancellation, a healthy sign for Canadian industry.

A further revision of prices has been made in Toronto warehouse lists bringing bar iron and steel and bar mill shapes up to $5.75 per hundred pounds, a new high mark, and one not equalled in the memory of most of the steel men in Toronto.

The scrap metal market is still waiting for something to start it off on busy days again, but that something has not appeared, although some of the dealers claim that the only thing to restore the volume of trade is a lower level of prices.

NEW RATES SEEM TO BRING A QUICKER MOVEMENT OF THE CARS

Special to Canadian Machinery

MONTREAL, Que. Sept. 23.—Apparently, one good result that the increased freight rates have brought about has been the improvement that has materialised in the delivery of goods to consumers. This condition, while not pronounced, has nevertheless given encouragement to many buyers of material. A feature that has prevented a greater betterment in United States shipments of steels and other metal lines is the tendency of the railroads in Canada to keep their equipment on their own roads so as to take care of the heavy demand now being made to move the western crops. The car situation has been a sore point during the past year, but improvement has gradually been evident, and when the grain shipments have been taken care of the movement of ordinary freight will soon reach normal.

The steel situation has developed little of interest during the past week, and operations in this district are much the same as what they have been for the

past several months. In some respects the reports indicate that the activity is quieter in general and that the market is taking on a weaker tone. It was thought that the increased freight rates would be followed by a combined effort on the part of the dealers to adjust prices to the new conditions, so as to take care of the extra cost of transportation. This tendency, however, has been offset somewhat by the quieter tone of business, and although some dealers are tacking on the extra freight charges to quotations, others are allowing the old prices to stand and are absorbing the additional freight charges themselves. Consumers are in the market constantly, but buying sparingly, and trading is carried on in the hand-to-mouth style that has been more or less the order of things for the past eighteen months. Dealers report that no large developments are anticipated and this gives the impression that the fall activity will be of a conservative character. Another factor is the fact that mill production at present is above the existing demand, so that supplies

are gradually assuming the position when an easier market should obtain. The lesser requirements for some commodities, notably the automobile industry, has enabled the mills to devote more time to other needed lines, so that the trade may shortly see steels listed at lower figures. One naturally looks for increased costs when freight rates are advanced, but present conditions may show that the higher charges will provide the necessary means of more rapid transportation of material, with the result that the congestion at producing points will be relieved and raw materials delivered to the producer more regularly. Shipments to this district have shown gradual improvement of late.

Light Machine Demand

As a result of the curtailment that has evidently developed in the automobile industry the activity in machine tool lines has been somewhat disjointed, due to cancellations of equipment or extended delivery. This has reacted in many ways, and foundries and other metal trades have experienced a movement to a quieter period. This trend has come at a time when users of tools were contemplating the call for higher charges on equipment haulage. The de-

mand of late, however, has been easing off, so that dealers are willing to lower the quotations slightly to counteract the extra cost that would figure in the additional freight charges. While local dealers are reporting a quiet spell, the sales are a fair average for the year. Sales of heavy tools are light, but other machines and small supplies are of sufficient volume to maintain a balance of trade.

Scrap is Listless

There is little to report on the old material situation as the quiet character of the market has nothing of interest for the trade. Operators are almost entirely confined to local trading and in this respect the buying is solely for current needs and no covering is made for future requirements. Dealers are devoid of interest and transactions are carried on more on the nature of the individual sale than on market prices or conditions. Prices locally are only to be considered in a nominal way.

PREMIUM MILLS SUPPLY TRADE

Otherwise Some Lines Would Be Very Short in the Canadian Market

TORONTO. — The trade is becoming more or less accustomed to the new freight schedules that are now in effect on the Canadian roads, at least very little is ever heard about the matter in the markets. One matter that is quite conspicuous is the scarcity of ready money, and the difficulty with which existing lines of credit are given a little elasticity in order to meet certain situations.

Some of those who are in the machine tool business seem to be quite certain that the difficulty in getting credit and financial assistance is keeping not a few buyers out of the market, and that when matters ease off a little, and there is more ready money to be had, there will be more business in their line.

"We have had men come in here," stated the manager of one Toronto house this morning, "and tell us that they have certain extensions to their plant in mind, and asking what we can do for them in the way of extending payments. The banks apparently cannot see their way clear to give them the assistance or the time they desire, and they think they can get it from the machine tool dealers, but as a matter of fact we would soon be in for a lot of trouble were we to undertake to finance things on paper that the bank would not accept or deal in.' There will come a time when this business will come into the market, as we are certain these men want to make extensions, and they are not bluffing about it either."

No Price Crack Coming

Dealers generally are confident that prices as they stand now will be the prices at which machine tools will sell for some months to come. "As far as

Reports from Pittsburgh claim that the shop facilities have improved to such an extent that they are rapidly reducing accumulation of steel in the yards there.

There is very little demand in the United States steel market for structural shapes, plates, and the larger size of merchant steel bars. Prices on these are not reduced, the mills holding that such a course would not bring them any new business.

One report from the steel mill section is that some of the mills will be forced to curtail production in November.

The deliveries of sheets in the Toronto market are nearly all through the independent and premium mills.

Some machine tool dealers believe that the tightness of money at the present time is keeping a number of buyers out of the market.

Any suggestion in the reduction in the prices of machinery finds little encouragement in the markets. Some firms are still many weeks behind in their deliveries.

There was another change made in the price of steel and iron bars and bar mill shapes in the Toronto warehouses, the face price now being $5.75 per hundred, which is a record high price.

The stove markets are finding it difficult to secure enough of the proper gauge of sheets to keep their plants in operation.

we can see there is no chance of anything happening to the price situation, and we are very certain that there will be nothing approaching a price cutting business. If such a thing were to happen it would simply mean that we would have each other's business in a couple of months at reduced prices. Most of the bookings that are on the way through now have a non-cancellation clause attached to them, and in that way there is a very small chance of anything approaching trouble coming on the trade."

The makers will need to be a great deal more keen for business before reductions will be thought of. Some of them are showing a better delivery schedule, but in many lines there is still quite a few weeks between bookings and deliveries, and this sort of thing will

have to be worn down a lot more before price reduction will be a necessity.

The dealer who complains that he is not in on any big orders in the Canadian market just now need not feel hurt, as there is no big buying being done, but there is a good showing of smaller business, and it is fairly well distributed.

Small tools remain the same in price. Some of the railroad business has not been placed yet. Most of what is being purchased now is for the eastern shops.

The Steel Market

There has been a further revision of the Toronto market upward to take care of the freight rates and other costs. It now means that the base price for steel and iron bars and bar mill shapes is 5.75 per 100 pounds. This is the highest price that has been reached in the memory of some of the oldest dealers in the warehousing business here.

There is a fair arrival of sheets, but it is all from the premium mills, and on orders that were placed only a few days ago. These premium mills are out to cater to the hand-to-mouth trade. The chances are that they never get booked up over a week or so in advance, always holding themselves in readiness to turn out a rolling in very short order for a customer who is willing to pay the price for immediate delivery.

One well-known stove maker was making the rounds this morning looking for sheets. He was almost out of certain gauges, especially anything near 22 or 24, which is the size much in demand by this class of trade. He was mightily pleased to find what he wanted, although he had to pay the high price for it. With him it was simply a case of getting the material or shutting off that corner of his shop, which is particularly busy at this season.

Just how long the warehouses will keep on going into the premium market is problematical. There is always the danger of dealers waking up some fine morning to find that there is a crack in the premium market, and it is not a happy chance to have a good sized tonnage on the way when such a thing takes place. Those who watch the market closely believe that there will be a chance, but they cannot see it right away, and apparently they are willing to take a chance of getting away with premium goods for a time yet.

Boiler tubes are coming across in fairly good numbers, but the assortment is far from complete yet. There are some of the much-wanted sizes that are not to be had, particularly in the case of 3-inch, several of the houses being short of these. There is a big market in this country now for boiler tubes, and a very large tonnage could be disposed of were it possible to secure them.

There is not the volume of delivery coming from the Canadian mills that could be hoped for, as they are still working without adequate supplies of coal.

MILLS ARE REDUCING THE PILES STACKED AROUND THEIR PREMISES

Special to CANADIAN MACHINERY.

PITTSBURGH, Sept. 23.—In the main, the production of steel tends to increase, being helped by constantly improving rail traffic conditions, but here and there some of the smaller mills are reducing output or planning to reduce output, on account of their orders having fallen off. The larger mills, which sold farther ahead, are well supplied with orders.

While production has been increasing somewhat, shipments have been increasing still more, as the mills are able to move more of their accumulated steel. The Carnegie Steel Company is reported to have reduced its accumulation of steel by about one-third, from the top point, leaving about a third of a million tons, a trifle over half of which is in ingots and semi-finished steel.

Delivery Demands

In more than half the finished steel lines the deliveries are such that consumers are in easy position and are not pressing the mills for heavier shipments. In tin plates, wire nails and the smaller sizes of merchant steel bars there is still urgent request on the part of many buyers for heavier shipments against contracts, though there is relatively little new buying. In sheets there is considerable pressure from some consuming lines, but the deliveries on the whole are moderately good, being helped by the fact that the automobile industry has canceled some orders entirely, while on many others it has called for reduction or suspension in shipments. An extreme case is that of an automobile factory which ordered a mill to suspend shipments instantly, and raised an issue because the mill allowed a car to go forward that was already loaded and shipped—two cars with material that had already been made to the buyer's specifications. The mill was, of course, far within its rights in this matter, since its established policy that consents to hold up the filling of specifications. When the material has actually been produced it will always expects the buyer to take away.

Present Market Quiet

That at some animes and among all classes of steel some buying of certain lines this summer has been pursued. The situations have moved along to fill increased requirements, than in recent years, particularly in the fabricated products for consumptive requirements of some buyers following the decline in prices. Having contracted for the amount of material at most current prices and accepting deliveries than they expected, these buyers went into the open market and bought additional tonnages. Now both classes of steel are

coming, resulting in some cases in there being more than is needed, and in nearly all cases in a cessation of the prompt buying.

Stocks on Hand

The usual question at this time is being asked without getting an answer, how much steel the buyers have in stock. After a period of stringency in deliveries the common assumption is that buyers necessarily cannot have any stocks, yet in the past it has always developed that many of them had stocks. Buyers are in various positions, and usually the case is that the trade hears of those who are short of steel while it hears nothing about those that are well stocked. At the present time buyers are content to carry their stocks, those that have any, for while the railroad situation is greatly improved the better position may not last. Usually the peak of the load on the railroads comes in October, on account of the crop movements, and after that there are the usual dangers of winter weather slowing down the railroad movement. At this time, therefore, the buyer concludes that a stock is a good thing to have. His position may, however, change radically in the course of a month or two, when he gets near to inventory time, for he may then suddenly become desirous of having his inventory as small as possible.

Items in Light Demand

Demand is practically absent for structural shapes, plates and the larger sizes of merchant steel bars. Mills are maintaining prices on these partly because there is no business and reducing prices would not create business, partly because they can put their steel into forms that are in more request, and partly because they have contracts on books at full prices, which they desire to protect. Structural shapes are not offered at less than 3.10c. so far as can be ascertained, by any of the independents, and this is $1.13 a ton above the Steel Corporation price of 2.45c. Plates are now commonly quoted at 3.25c. as several independent mills will make this quotation. Some mills are asking more, but their prices are simply nominal. The mills seem to have made 3.25c a sort of dead line, below which they will not go.

Duration of Business Readjustment

That business generally is in process of readjustment is admitted on all hands. The only divergence of opinion is as to the extent of the readjustment and the length of time that will be required before markets turn more active again. Some observers think there is to be merely a shortlived lull in business activity, while in other quarters the view is that it will be more than a year hence before business is really on the up grade again. There is agreement on this

point, that lessened activity is needed to bring down commodity prices and the cost of labor, measuring the latter not necessarily in wage rates per hour or per day, but rather in the amount of work done for a given sum of money. Thus the question seems to resolve itself into one of how long the present momentum of business will carry things before there can be sufficient slackening to produce this readjustment. For the steel trade, the common prediction is that the industry in general, drawing inferences from the position of both producers and consumers, has enough steam up to run it not only through this year but into next year. According to this view the steel mills are going to run, with an occasional exception, at the best rate the physical conditions permit, at least, through this year and, probably into early spring. In a few quarters, however, this view is considered altogether too sanguine, and the idea is expressed that it should not be surprising if the mills should be forced to curtail production sharply in November. As to steel prices, practically the unanimous opinion is that they will stay up until after the mills have had a period of relative idleness. The mills will stubbornly resist price declines and they have made so much money that they can well afford to be independent.

Pig iron is stagnant, with prices not quotably changed.

NEW YORK TOOL TRADE VERY QUIET

Various Views Taken on What is Liable to Happen There in Near Future

NEW YORK, Sept. 23.—Some in the machine tool trade describe conditions as "the quietest in years." Both inquiries and orders are few and far between, and with some machine tool manufacturers orders in the past month or so have been entirely offset by cancellations. It is surprising how quickly the machine tool trade has altered from unexampled prosperity to uncertainty. Even those machine tool plants whose order books are in the best of condition of any have scarcely more than three or four months business ahead, and unless there should be a buying movement by the first of the year, at the latest, a majority of the plants of the United States will find themselves at that time with scarcely any business ahead. Some companies are even now caught up on orders on some lines and either are making tools for stock or have curtailed their manufacturing operations.

Despite the present uncertainty as to the future of business in the near three to six months, there are some who do not view the situation pessimistically, but on the contrary look for a resumption of buying on a fairly good scale within the next 30 or 60 days; at any rate after the presidential election. Many manufacturing plants are extremely busy, but others are running on part

time and quite a number have let men go. The industrial situation is spotty, but not alarming.

Except for an inquiry from the Standard Oil Co. of New Jersey for a few machines, there is scarcely an inquiry in the market worthy of note. There are occasional orders for single machines, but even this business in the aggregate is not large.

MONTREAL NOTES

The new factory of the L. E. Waterman Company at St. Lambert was officially opened last week, when a number of notable men from England, New Zealand, United States and Canada, were guests of the company. H. L. Symonds, chairman of the London, England. Chamber of Commerce, officiated at the opening.

—

Henry Berry, vice-president and general manager of the Canadian Asbestos Company, died last week, after an illness of several months. Mr. Berry had been associated with the asbestos industry nearly all his life and his own personal efforts in this direction had done much to widen the field of usefulness of this great Canadian mineral. Mr. Berry was 55 years of age; he leaves a widow and three sons.

—

The Grand Trunk management has appointed Mrs. E. H. Gaudion as welfare superintendent of the female clerical staff at their head offices in Montreal. Mrs. Gaudion will devote her entire time and special training to the duties of this position, the desire of the management being to make the environment of office life as healthy and happy as possible. The Welfare Superintendent will have charge of the lunch room and rest rooms and will also extend her help to the homes of girls in cases of sickness.

—

Negotiations have been finally concluded for the erection of a mammoth 1,000 room hotel in Montreal. Frank A. Dudley, president of the United Hotels Company of America, and a number of prominent commercial and industrial men of the city have formed the Mount Royal Hotel Company, Limited of Montreal, with a capital of 10 million dollars. The site secured for the new hotel is the block bordered by Peel, Burnside, Metcalfe and St. Catherine streets. It is expected that an early start will be made on the construction work. The officers of the newly organized company will be: Frank A. Dudley, president; Hon. W. J. Shaughnessy, first vice-president; and W. J. Cluff, second vice-president.

—

Authorization has been given for the increase of the capital stock of the Atlas Construction Company, Ltd., from the sum of $100,000 to $500,000; such increase to consist of 4,000 shares of $100 each.

PIG IRON TRADE

There is not much general buying in the pig iron market, but shipments from furnaces are being readily taken up. Where automobile plants are suspending delivery, other plants are taking the iron they refuse. The reports from Philadelphia show that furnaces are well booked up, and are offering but little material. Steel works which are not very well provided for are still clamoring for delivery, particularly of basic grade. This grade is now selling at $50 furnace. There has been no change in foundry grades.

Chicago producers are well booked ahead, but the market is by no means weak though inquiry in pig iron is slack. Requests for shipment are being made by the smelters, as few have accumulated iron beyond their present needs. There is no demand for basic iron, owing to buyers waiting for orders from the railroads before buying.

New business in Cleveland is exceedingly small, owing to a continued fall in the market. Furnace shipments are continuing, however, on a higher scale and with the exception of automobile foundries, consumers are buying. Prevailing prices were $46 to $47 for next year, $47 to $48 for this year.

Boston reports that requirements for New England consumers are covered for this year. All the principal railroads are now open. Castings are in demand and foundries busy.

The market for pig iron in New York has been good. Sales for scattered lots amounted to several thousand tons for delivery this year at established prices. New inquiry has been light. Consumers are able to procure iron more easily owing to improved transportation. Eastern Pennsylvania iron has been sold at prices ranging from $50 to $51, for 1.75 to 2.25 silicon, to $53 to $54 for 2.25 to 2.75 silicon. Some southern foundry iron is selling at $42, furnace, for delivery this year and next.

Pittsburgh market for pig iron remains dull. The prices are firm owing to the orders on hand.

Buffalo. — There appears to be very little business in the market here. Enough has gone through, however, to establish $50 as the prevailing market price for No. 2 plain foundry iron. Bookings for 1921 are very small, buyers apparently hoping for something more favorable in the way of price.

Reports are made that business is almost at a standstill at St. Louis. The western and north-western sections, however, are better — prices quoted were from $42 to $45 for 1.75 to 2.25 silicon, for prompt and future delivery.

Southern markets are inactive, but the market is considered strong. No. 2 foundry is quoted on a basis of $42.

U.S. SCRAP METAL

Pittsburgh. — The market for open-hearth grades of scrap has softened, owing to the fact that deliveries are better and also to the withdrawal of one of the large steel companies. Some of the smaller dealers who had tonnages which they were unable to move because of shortage of cars are now getting permits for immediate transportation.

Boston.—The market here is strong despite the heavy advance in freight rates, and dealers are able to place fair sized tonnages for shipment to the steel mills.

New York.—There is a fair amount of inquiry for most grades. Prices of iron and steel scrap are firm, and while a firm tone still prevails no further advances have been reported. Cast scrap continues in chief demand and is scarce.

Philadelphia.—An eastern steel maker bought several thousand tons of cast iron borings during the past week. Some borings have been sold as high as $25, the highest level for this grade in 30 years. This is attributed to the increased freight rates, and the decreased production on the part of Detroit automobile builders.

Buffalo.—Steel is in greater demand than other grades. Dealers can sell almost any tonnage of turnings and borings. Most of material going forward is on old contracts. New orders placed are for prompt delivery, there being little business for future delivery.

Cleveland.—The market for iron and steel scrap is downward. Quotations have been marked down sharply during the past week.

Cincinnati. — The iron and steel scrap market in this district is firm, with large consumers buying. Higher prices are quoted for some foundry grades.

Chicago. — Prices of iron and steel scrap have continued to decline, owing to consumers having no interest in the market and the movement of scrap being negligible.

St. Louis.—The scrap market is very uncertain. The underlying trend seems to be toward lower prices, but as yet dealers' lists do not reflect this.

Much Coal Arriving.—A Hamilton report says:—Unless something very unexpected happens, the much-talked-of coal famine this winter will not be realized, according to local railwaymen. For the last two months thousands of tons of coal have been rushed into this province, and the railways are still busy moving shipments of coal from the mines and returning empties with all possible speed. On an average of 400 cars of coal a day is cleared at Niagara Falls on the G.T.R. The T. H. & B. and C. P. R. at this point also move a large number of cars each day. The record established by the G. T. R. several weeks ago, when they moved 900 cars of coal from the border has not yet been broken.

SELECTED MARKET QUOTATIONS

Being a record of prices current on raw and finished material entering into the manufacture of mechanical and general engineering products.

PIG IRON

Grey forge, Pittsburgh	$42 40
Lake Superior, charcoal, Chicago	57 00
Standard low phos., Philadelphia	50 00
Bessemer, Pittsburgh	43 00
Basic, Valley furnace	42 90
Toronto price:—	
Silicon, 2.25% to 2.75%	52 00
No. 2 Foundry, 1.75 to 2.25%	50 00

IRON AND STEEL

Per lb. to Large Buyers	Cents
Iron bars, base, Toronto	$ 5 75
Steel bars, base, Toronto	5 75
Iron bars, base, Montreal	5 75
Steel bars, base, Montreal	5 75
Reinforcing bars, base	5 75
Steel hoops	7 00
Norway iron	11 00
Tire steel	5 75
Spring steel	10 00
Band steel, No. 10 gauge and 3-16 in. base	6 10
Chequered floor plate, 3-16 in.	8 50
Chequered floor plate, ¼ in.	8 10
Bessemer rails, heavy, at mill	....
Steel bars, Pittsburgh	3 00-4 00
Tank plates, Pittsburgh	4 00
Structural shapes, Pittsburgh	3 00
Steel hoops, Pittsburgh	3 50-3 75
F.O.B., Toronto Warehouse	
Small shapes	5 75
F.O.B. Chicago Warehouse	
Steel bars	3 62
Structural shapes	3 72
Plates	3 67 to 5 50
Small shapes under 3"	3 62

FREIGHT RATES

Per 100 Pounds.

Pittsburgh to Following Points

	C.L.	L.C.L.
Montreal	58½	73
St. John, N.B.	84½	106½
Halifax	86	108
Toronto	38	54
Guelph	38	54
London	38	54
Windsor	35	50½

METALS

	Gross. Montreal	Toronto
Lake copper	$25 00	$24 25
Electric copper	24 00	24 50
Castings, copper	23 50	24 00
Tin	56 00	58 00
Spelter	10 75	10 75
Lead	9 25	10 50
Antimony	9 00	12 00
Aluminum	36 00	37 00

Prices per 100 lbs.

PLATES

Plates, 3-16 in.	$ 7 25	$ 7 35
Plates, ¼ up	6 50	6 60

PIPE—WROUGHT

Standard Buttweld Pipe

Per 100 Ft.

	Steel Blk.	Galv.	Gen. Wrought Iron Blk.	Galv.
⅛	$ 6 50	$ 8 50	$.....	$.....
¼	5 31	7 41	5 91	8 01
⅜	5 31	7 41	5 91	8 01
½	7 10	8 63	7 95	9 48
¾	8 80	10 87	9 95	12 02
1	13 01	16 07	14 71	17 77
1¼	17 60	21 74	19 90	24 04
1½	21 04	25 99	23 79	28 74
2	28 31	34 97	32 01	38 67
2½	44 73	55 28	....	....
3	58 52	72 29	....	....
3½	74 08	90 62	....	....
4	87 75	107 37	....	....

Standard Lapweld Pipe

Per 100 Ft.

	Steel Blk.	Galv.	Gen. Wrought Iron Blk.	Galv.
2	$32 01	$ 38 67	$35 71	$42 37
2½	48 26	58 79	54 11	64 64
3	63 11	76 88	70 76	84 53
3½	75 90	92 46	85 10	101 66
4	89 93	107 55	100 83	120 45
4½	1 05	1 29	1 30	1 54
5	1 22	1 50	1 52	1 80
6	1 58	1 95	1 97	2 33
7	2 06	2 53	2 53	3 01
8L	2 14	2 66	2 66	3 16
8	2 49	3 07	3 07	3 64
9	2 98	3 67	3 67	4 36
10L	2 77	3 41	3 41	4 05
10	3 56	4 39	4 39	5 21

Prices—Ontario, Quebec and Maritime Provinces

WROUGHT NIPPLES

4" and under, 60%,

4½" and larger, 50%.

4" and under, running thread, 30%.

Standard couplings, 4-in. and under, 30%.

Do., 4½-in. and larger, 10%.

OLD MATERIAL

Dealers' Average Buying Prices.

	Per 100 Pounds. Montreal	Toronto
Copper, light	$15 00	$14 00
Copper, crucible	18 00	18 00
Copper, heavy	18 00	18 00
Copper wire	18 00	18 00
No. 1 machine composition	17 00	17 00
New brass cuttings	11 00	11 75
Red brass turnings	14 50	15 75
Yellow brass turninbs	9 00	9 50
Light brass	7 00	7 00
Medium brass	8 00	7 75
Scrap zinc	6 50	6 00
Heavy lead	7 50	7 75
Tea lead	4 50	5 00
Aluminum	19 00	20 00
	Per Ton	Gross
Heavy melting steel	20 00	18 00
Boiler plate	15 50	15 00
Axles (wrought iron)	30 00	20 00
Rails (scrap	23 00	18 00
Malleable scrap	26 00	25 00
No. 1 machine east iron	32 00	33 00
Pipe, wrought	12 00	12 00
Car wheel	33 00	33 00
Steel axles	25 00	20 00
Mach. shop turnings	11 00	11 00
Stove plate	26 50	25 00
Cast boring	14 00	14 00

BOLTS, NUTS AND SCREWS

	Per Cent.
Carriage bolts, 7-16 and up	+10
Carriage bolts, ⅜-in. and less	Net
Coach and lag screws	—15
Stove bolts	55
Wrought washers	—25
Elevator bolts	+10
Machine bolts, 7-16 and over	+10
Machine bolts, ⅜-in. and less	+10
Blank bolts	Net
Bolt ends	Net
Machine screws, fl. and rd. hd., steel	27½
Machine screws, o. and fil. hd., steel	+25
Machine screws, fl. and rd. hd., brass	net
Machine screws, o. and fil. hd., brass	net
Nuts, square, blank .... +25 add	$2 00
Nuts, square, tapped ... add	2 25
Nuts, hex., blank ... add	2 50
Nuts, hex., tapped ... add	3 00
Copper rivets and burrs, list less	15
Burrs only, list plus	25
Iron rivets and burrs	40 and 5
Boiler rivets, base ¾" and larger	$8 50
Structural rivets, as above	8 40
Wood screws, O. & R., bright	75
Wood screws, flat, bright	77½
Wood screws, flat, brass	55
Wood screws, O. & R., brass	55½
Wood screws, flat, bronze	50
Wood screws, O. & R., bronze	47½

MILLED PRODUCTS

(Prices on unbroken packages)

	Per Cent.
Set screws .... —20%, 25 and	5
Sq. and hex. hd. cap screws	12½
Rd. and fil. hd. cap screws ..plus	25
Flat but. hd. cap screws ....plus	50
Fin. and semi-fin. nuts up to 1-in.	12½
Fin. and Semi-fin. nuts, over 1 in., up to 1½-in.	—5
Fin. and Semi-fin. nuts over 1½ in., up to 2-in.	+12½
Studs	+5
Taper pins	—12½
Coupling bolts	+40
Planer head bolts, without fillet, list	+45
Planer head bolts, with fillet, list plus 10 and	+55
Planer head bolt nuts, same as finished nuts.	
Planer bolt washers	net
Hollow set screws	+60
Collar screws ......list plus 20,	30
Thumb screws	40
Thumb nuts	75
Patch bolts ....add	+85
Cold pressed nuts to 1½ in...add	$1 00
Cold pressed nuts over 1½ in..add	2 00

BILLETS

	Per gross ton
Bessemer billets	$60 00
Open-hearth billets	60 00
O.H. sheet bars	76 00
Forging billets	56 00-75 00
Wire rods	52 00-70 00

Government prices.

F.O.B. Pittsburgh.

NAILS AND SPIKES

Wire nails, base	$6 10
Cut nails, base	7 00
Miscellaneous wire nails	50%

ROPE AND PACKINGS

Plumbers' oakum, per lb.	0 10¼
Packing, square braided	0 38
Packing, No. 1 Italian	0 44
Packing, No. 2 Italian	0 36
Pure Manila rope	0 35½
British Manila rope	0 28
New Zealand hemp	0 28

POLISHED DRILL ROD

Discount off list, Montreal and Toronto .... net

MISCELLANEOUS

Solder, strictly	$ 0 35
Solder, guaranteed	0 39
Soldering coppers, lb.	0 62½
White lead, pure, cwt.	20 35
Red dry lead, 100-lb. kegs, per cwt.	16 00
Gasoline, per gal., bulk	0 38
Pure turp., single bbls., gal.	3 15
Linseed oil, raw, single bbls.	2 37
Linseed oil, boiled, single bbls.	2 40
Wood alcohol, per gal.	4 00
Whiting, plain, per 100 lbs.	3 00

CARBON DRILLS AND REAMERS

S.S. drills, wire sizes	32½
Can. carbon cutters, plus	20
Standard drills, all sizes	32½
3-fluted drills, plus	10
Jobbers' and letter sizes	32½
Bit stock	40
Ratchet drills	15
S.S. drills for wood	40
Wood boring brace drills	25
Electricians' bits	30
Sockets	50
Sleeves	50
Taper pin reamers	25 off
Drills and countersinks	net
Bridge reamers	50
Centre reamers	10
Chucking reamers	net
Hand reamers	10
High speed drills, list plus 10 to 30	
Can. high speed cutters, net to plus	10
American	plus 40

COLD ROLLED STEEL

[At warehouse]

Rounds and squares	$7 base
Hexagons and flats	$7.75 base

IRON PIPE FITTINGS

	Black	Galv.
Class A	70	85
Class B	30	40
Class C	20	30

Cast iron fittings, 5%; malleable bushings, 22½%; cast bushings, 22½%; unions, 37½%; plugs, 20% off list.

SHEETS

	Montreal	Toronto
Sheets, black, No. 28	$12 00	12 10
Sheets, Blue ann., No. 10	9 50	9 60
Canada plates, dull, 52 sheets	13 00	13 00
Can. plates, all bright	14 00	.....
Apollo brand, 10¾ oz. galvanized	....	....
Queen's Head, 28 B.W.G.	13 50	.....
Fleur-de-Lis, 28 B.W.G.	13 00	.....
Gorbal's Best, No. 28	....	....
Colborne Crown, No. 28	....	....
Premier, No. 28, U.S.	11 50	12 60
Premier, 10¾-oz.	11 50	13 00
Zinc sheets	16 50	20 00

PROOF COIL CHAIN

(Warehouse Price)

B

¼ in., $13.00; 5-16, $11.00; ⅜ in., $10.00; 7-16 in., $9.80; ½ in., $9.75; ⅝ in., $9.20; ¾ in., $9.30; ⅞ in., $9.50; 1 in., $9.10; Extra for B.B. Chain, $1.20; Extra for B.B.B. Chain, $1.80.

ELECTRIC WELD COIL CHAIN B.B.

⅛ in., $16.75; 3-16 in., $15.40; ¼ in., $13.00; 5-16 in., $11.00; ⅜ in., $10.00; 7-16 in., $9.80; ½ in., $9.75; ⅝ in., $9.50; ¾ in., $9.30.

Prices per 100 lbs.

FILES AND RASPS

	Per Cent.
Globe	50
Vulcan	50
P.H. and Imperial	50
Nicholson	32½
Black Diamond	27½
J. Barton Smith, Eagle	50
McClelland, Globe	50
Delta Files	20
Disston	40
Whitman & Barnes	50
Great Western-American	50
Kearney & Foot, Arcade	50

BOILER TUBES.

Size.	Seamless	Lapwelded
1 in.	$27 00	$.....
1¼ in.	29 50	.....
1½ in.	31 50	29 50
1¾ in.	31 50	30 00
2 in.	35 00	30 00
2¼ in.	35 00	29 00
2½ in.	42 00	37 00
3 in.	50 00	48 00
3¼ in.	.. ..	48 50
3½ in.	63 00	51 50
4 in.	85 00	65 50

Prices per 100 ft., Montreal and Toronto

OILS AND COMPOUNDS.

Castor oil, per lb.	
Royalite, per gal., bulk	27½
Palacine	30½
Machine oil, per gal.	54½
Black oil, per gal.	28
Cylinder oil, Capital	82½
Petroleum fuel oil, bbls., net	18

BELTING—No 1 OAK TANNED

Extra heavy, single and double		6½
Standard		6½
Cut leather lacing, No. 1		2 00
Leather in side	2 40	3 00

TAPES

Chesterman Metallic, 50 ft.	$2 00
Lufkin Metallic, 603, 50 ft.	2 00
Admiral Steel Tape, 50 ft.	2 75
Admiral Steel Tape, 100 ft.	4 45
Major Jun. Steel Tape, 50 ft.	3 50
Rival Steel Tape, 50 ft.	2 75
Rival Steel Tape, 100 ft.	4 45
Reliable Jun. Steel Tape, 50 ft.	3 50

PLATING SUPPLIES

Polishing wheels, felt	$4 60
Polishing wheels, bull-neck	2 00
Emery in kegs, Turkish	09
Pumice, ground	06
Emery glue	30
Tripoli composition	09
Crocus composition	12
Emery composition	11
Rouge, silver	60
Rouge, powder, nickel	45

Prices per lb.

ARTIFICIAL CORUNDUM

Grits, 6 to 70 inclusive	.08½
Grits, 80 and finer	.6

BRASS—Warehouse Price

Brass rods, base ½ in. to 1 in. rod	0 34
Brass sheets, 24 gauge and heavier, base	$0 42
Brass tubing, seamless	0 46
Copper tubing, seamless	0 48

WASTE

XXX Extra	.24	Atlas	.20
Peerless	.22½	X Empire	.19½
Grand	.22½	Ideal	.19
Superior	.22½	X Press	.17½
X L C R	.21		

Colored

Lion	.17	Popular	.13
Standard	.15	Keen	.11
No. 1	.15		

Wool Packing

Arrow	.35	Anvil	.22
Axle	.28	Anchor	.17

Washed Wipers

Select White	.20	Dark colored	.09
Mixed colored	.10		

This list subject to trade discount for quantity.

RUBBER BELTING

Standard ... 10% Best grades... 15%

ANODES

Nickel	.55	to	.60
Copper	.38	to	.40
Tin	.70	to	.70
Zinc	.16	to	.17

Prices per lb.

COPPER PRODUCTS

	Montreal	Toronto
Bars, ½ to 2 in.	$42 50	$43 00
Copper wire, list plus 10		
Plain sheets, 14 oz., 14x60 in.	49 00	48 00
Copper sheet, tinned, 14x60, 14 oz.	52 00	48 00
Copper sheet, planished, 16 oz. base	56 00	55 00
Braziers', in sheets, 6 x 4 base	48 00	46 00

LEAD SHEETS

	Montreal	Toronto
Sheets, 3 lbs. sq. ft.	$11 50	$14 50
Sheets, 3½ lbs. sq. ft.	11 25	14 00
Sheets, 4 to 6 lbs. sq. ft.	11 00	13 50

Cut sheets, ½c per lb. extra.

Cut sheets to size, 1c per lb. extra.

PLATING CHEMICALS

Acid, boracic	$ .23
Acid, hydrochloric	.04
Acid, nitric	.11
Acid, sulphuric	.04
Ammonia, aqua	.15
Ammonium, carbonate	.23
Ammonium, chloride	.22
Ammonium hydrosulphuret	.75
Ammonium sulphate	.30
Arsenic, white	.14
Copper, carbonate, annhy	.41
Copper, sulphate	.15
Cobalt, sulphate	.20
Iron perchloride	.62
Lead acetate	.30
Nickel ammonium sulphate	.20
Nickel carbonate	.32
Nickel sulphate	.22
Potassium sulphide (substitute)	.42
Silver chloride (per oz.)	1.30
Silver nitrate (per oz.)	1.25
Sodium bisulphate	.14
Sodium carbonate crystals	.06
Sodium cyanide, 127-130%	.38
Sodium hyposulphite per 100 lbs	8.00
Sodium phosphate	.15
Tin chloride	.80
Zinc chloride, C.P.	.30
Zinc sulphate	.10

Prices per lb. unless otherwise stated

MakeScrewCutting the Most Satisfactory Operation in Your Shop

Others have done so by equipping with

Geometric
Screw Thread Cutting Tools

Wherever Screw Machines and Turret Lathes annex a Geometric Die Head or Collapsing Tap,

N
O
I
T   C
C   O
U   Goes Up and   S Goes Down
D   T
O
R
P

Prove for Yourself

what a Geometric installation will do for you.

Tell us the specifications of your work, and the kind of screw machine used, and we will make you a proposition.

THE GEOMETRIC TOOL CO.
NEW HAVEN, CONN.

Canadian Agents:

Williams & Wilson, Ltd., Montreal
The A. R. Williams Machinery Co., Ltd., Toronto,
Winnipeg, St. John, N.B., and Halifax, N.S.

Canadian Fairbanks-Morse Co., Ltd., Manitoba, Saskatchewan, Alberta

If interested tear out this page and place with letters to be answered.

FREE TO YOU

Geometric
Die Heads

Instructions for
Their Care and Use

HAVE you one of these pamphlets on the Care and Use of Geometric Die Heads?

YOU WILL WANT THIS BOOKLET SOMETIME

Are you using Geometric Die Heads?

YOU WILL WANT IT NOW

If not using Geometric Die Heads now, you will want it later, when

YOU DO USE GEOMETRIC DIE HEADS

It is a small 11-page pamphlet, 6 in. x 4½ in., which tells you how to take care of your Die Heads.

WRITE US FOR YOUR FREE COPY

A post card request will bring the booklet to you, and placed in the hands of the man who uses the Die Heads, it will help.

CUT DOWN EXPENSE

Inspectors and Car Foremen in. Session

Large Gathering at Montreal Where Many Useful Matters Were Discussed—A Large Number of Firms Were Represented at the Gathering.

THE 20th Annual Convention of the Chief Interchange Inspectors and Car Foremen's Associations of America, which was held for three days last week at the Windsor Hotel in Montreal, was one of the most successful in the history of the organization. Over 300 delegates were present from all parts of the United States and Canada. The meeting was held in the large assembly hall of the hotel, and the rear portion was devoted to the mechanical appliances exhibit. In addition to the general business sessions several papers were presented followed by a considerable discussion. These papers included:—Lubrication of Freight and Passenger Equipment by M. J. O'Connor, special lubrication inspector of the N.Y.C.R. R.; Freight Claim Prevention, by T. A. Ward, chief clerk freight claim department of the N. Y. C. R. R.; Examination of Car Inspectors, by E. H. Hall, superintendent of car department of the Chicago Great Western R. R.; Best Means of Repairing Cars in Train Yard, by O. E. Sitterly, general car inspector of the Penn. R.R. The closing session was concluded with a considerable display of international amenities, when the majority of the members, being Americans, spoke of the freedom and hospitality they had enjoyed in Canada, and the Canadians replied with similar courtesies. The following officers were elected for the ensuing year:—President, Edward Pendleton, general car foreman, Chicago and Alton Ry., Peoria, Ill.; First Vice-President, A. Armstrong, chief joint car interchange inspector, Atlanta, Georgia; Second Vice-President, W. F. Westfall, special inspector of the New York Central Ry., Cleveland, Ohio; Secretary-Treasurer, W. R. Elliott, general car foreman, St. Louis, Mo., terminals. The following supply firms were represented:—American Brake and Shoe- and Foundry Company, Chicago; Steel Foundries, Chicago; Bettendorf Company, Bettendorf, Ia.; Boss Lock Nut Company, of Canada, Limited, Montreal; Bradford Draft· Gear Company, Chicago; Buckeye Steel Casting Company, Chicago; Camel Company, Chicago; Canada Grip Nut Company, Montreal; Canadian Car and Foundry Company, Montreal; Canadian Gold Car Heating Company, of New York; Canadian Johns-Manville Company, of Montreal; Canuck Supply Company, of Montreal; Chicago-Cleveland Car Roofing Company, of Chicago; Chicago Railway Equipment Company, of Chicago; Columbia Nut and Bolt Company, of Bridgeport, Conn.; Paul Dickinson Inc., of Chicago; Duff Manufacturing Company, of Pittsburg, Pa.; Galena-Signal Oil Company, of Canada, Montreal; Griffin

Wheel Company, of Chicago; Holden Company, of Montreal; Independent Pneumatic Tool Company, of Montreal; Imperial Appliance Company, of Chicago; Imperial Belting Company, of Chicago; Liberty Tool Company, of Baltimore; Mahr Manufacturing Co., of Minneapolis; McCord and Company, of Chicago; W. H. Miner, of Chicago; National Malleable Castings, of Cleveland, Ohio; A. O. Norton, of Boston; Okweld Railway Service Company, of Chicago; Pyrene Manufacturing Company, of Montreal; Railway Review, of Chicago; Robinson Connector Company, of New York; Union Draft Gear Company, of Chicago; Westinghouse Air Brake Company, of Wilmerding, Pa.

The following is a list of· those registered at the meeting of Chief Interchange Car Inspectors' and Car Foremen's Association:—

RAILROAD MEN

C. S. Adams, Asst. Genl. For., N.Y.C., Malt Haven, N.Y.

C. W. Broa, Gen. Car For., N.Y.C. & St. L. Ry., Chicago; A. Berg, Gen. Car For., N.Y.C. Lines West, Erie, Pa.; A. J. Baumbuth, Gen. Car For., N.Y.C.R.R., New York City; F. Paul Busch, For. Inspector, Buffalo Creek Rr., Buffalo, N.Y.; Valentine Balts, Chief Jt. Inspector, Wheeling, West Va.; John C. Burke, Car For., Mo. Pacific Ry., St. Louis; A. J. Boyd, Car For., N.Y.O. & W. Norwich, N.Y.; Jas. Bannon, Genl. Car. For., C.P.R., West Toronto, Ont.; Robt. Barnaby, For. Car Inspector, D.L. & W. Ry., Buffalo, N.Y.; B. F. Blanton, Asst. Gen. Car For., E. A. L., Jacksville, Fla.; W. B. Buckle, Gen. For., C.P.R., Angus Shops, Montreal; A. A. Burkhard, Gen. For., N.Y.C., West Albany, N.Y.; John A. Burns, Car Inspr., D. & H. Albany, N.Y.; J. H. Campbell, Car For., G.T. Ry., Niagara Falls, N.Y.; E. S. Campbell, Gen. Car For., Minn. Transfer Co. Ry., St. Paul, Minn.; T. S. Cheadle, Chief Car Inspector, B.F. & P., Richmond, Va.; Glenn E. Chesney, Car For., G.T. Ry., Island Pond, Vt.; James Coleman, Asst. to Gen. S.M.P., G.T. Montreal, Que.; D. P. Crillman, Gen. Car For., M.C.R.R., Detroit; M. Culligan, Gen. For., N.Y.C.R.R., Dunkirk, N.Y.; R. Copland, Asst. Shop For., N.Y.C.R.R., N.Y. City; M. J. Curley, Trav. Car Inspr., N.Y.C.R.R., N.Y. City; Carl Dierks, For. Car Insp., D. & H. Oneonta, N.Y.; C. E. Donoghue, A.D. Gen. Car For., N.Y.C., Oswego, N.Y.; F. A. Donahue, Car Inspector, C.N.O. of L. Cincinnati, O.; J. E. Donoghue, Car For. N.Y.C., Watertown, N.Y.; P. A. Dufour, Car For., M.C.R.R., Kalamazoo, Mich.; F. Dunstern, Ry. Mech. Engr., N.Y. City; B. Dennum, Gen. Car For., D.H.R.R., Wilkesbarre, Pa.; E. D. Dennum, Gen. Car For., N.Y.C.R.R., Clearfield, Pa.; I. M. Douglas, Car Repair Asst., B. & O. R.R., Keyser, W. Va.; R. J. Duval, Car For. N.Y.C., Lyons, N.Y.; C. F. Ennson, Gen. Car For., Hocking Valley, Columbus, O.; W. P. Elliott, Gen. Car For., Terminals, St. Louis, Mo.; Jos. Ettwein, Asst. For., N.Y.C.R.R., Erie, Pa.; R. D. Freeney, Gen. Car For., Seaboard Air Line, Hamlet, N.C.; John F. Finnerty, Car For., N.Y.C., Carthage, N.Y.; F. L. Fierbier, Gen. Car For., N.Y.C., Cleveland, O.; Allen Foster, Travelling M.C.R. Insp., N.Y.C., Cleveland, O.; F. J. Freesburg, Car Inspr., N.Y.C., Ohio, Pa.; C. W. Frey, Gen. Car For., M.C.R.R., Detroit, Mich.; W. A. Forbes, Gen. For., G.T., Belleville, Ont., Geo. P. Fox, Rond. House, N.Y.C.R.R., Albany, N.Y.; J. H. Ford, Gen. Car For., L. & N.R., Penn Argyll, Pa.; Joseph Forcke, Gen. Car For., L. E. & W., Lima, Ohio; George Fisher, M.C.R., G.T.R., London, Ont.; G. M. Fish, G.T. Car Dept., C. & M. Ry. Middleport, O.; J. Fletcher, G.T. Car Dept., Seaboard Air Line, Portsmouth, Va.; A. H. Farther, Shop Supt., N.Y.C.R.R., Buffalo, N.Y.; Gabriel Ferro, Chief Int. Inspr., F.E.C. Ry., Havana, Cuba. J. B. Gillum, Gen. Car For., Frisco R.R., St. Louis, Mo.; J. E. Gordon, Gen. Car For, N. Y. C. & St. L., Buffalo, N.Y.; J. J. Gainey, Gen.

For, C.R. & Inspr., Southern Ry., Ludlow, Ky.; J. F. Griffin, Div. Gen. For., N.Y.C., Utica, N.Y.; J. F. Gordon, Car. For., F.E.C., Key West, Fla.; L. A. Gearville, Car For., G.T., Richmond, Que.; J. M. Getsen, Asst. Chief Inspr., Niag. Front. Car Inspr.' Assn., Buffalo, N.Y.; C. J. Griswold, Dist. Gen. For., N.Y.C.R.R., Chicago, Ill. A. A. Helwig, Trav. Gen. For., C. & A R.R., Bloomington, Ill.; W. W. Halbert, O.I.I., All Lines, St. Louis, Mo.; Chas. Hilderbrand, Car For., B.R. & P., Buffalo, N.Y.; W. H. Hall, Chief Car Inspr., C. of N.J., Jersey City, N.J.; Eugene Head, Chief Clerk, Car Dept., Wabash Ry., Decatur, Ill.; F. J. Heatherton, Car Inspr., L. & N., Newport, Ky.; F. L. Hendricks, Chief It. Inspr., Wilkesbarre, Pa.; A. Heylster, A.M.C.B., N.Y.C. R.R., Chicago, Ill.; W. J. Hatch, Gen. Air Brake Inspr., C.P. Ry., Montreal; W. M. Herring, Ch. C. to Asst. to V.P., Southern Ry., Washington, D.C. Bert Johnson, Asst. Supt. Air Brakes, N.Y.C., Ashtabula, O.; Adolf Johnson, Gen. For., North Am. Car Co., Chicago; C. J. Justus, Spec. Inspr., N.Y.C.R.R., N.Y. City. John C. Kenja, Secretary, C.I.C.I. & C.F. Assn., Chicago, Ill.; J. H. Kinter, For. Car Inspr., Penn. Ry. Enola, Pa.; A. B. Kipp, Gen. Car Inspr., N.Y.O. & W. Middleton, N.Y.; Joseph Krans, Car For., M.C., Toledo, O.; T. B. Koeppke, Supt. Equipment, Indianana, Battg. Co., St. Louis, Mo. A. Lachange, Genl. Contract For, C.P.R., Montreal; Geo. Landry, Car For., Rutland, Rouse's Point, N.Y.; C. E. LaMontagne, Gen. Car Inspr., Rutland, Malone, N.Y.; H. A. Lightage, Gen. Car For., I.C.R.R., East St. Louis, Ill.; J. L. Latta, Car For., D. & H. Schenectady, N.Y.; S. Lindman, Gen. Car For., N.Y.C., Youngstown, O.; Geo. Lynch, Chief It. Car Inspr., Cleveland, O.; W. C. Lang, Gen. Car Inspr., P. & L.E., Corapolis, Pa.; W. H. Long, Dist. Car For., Chi. Natl. Rlys., Toronto, Ont.; Alex. Lawson, Chief Jt. For., N.Y.C.R.R., Buffalo, N.Y.; D. C. Laws, Car For., Virginian Ry., Norfolk, Va. F. P. McNally, Chicago Stad Car Co., Harvey, Ill.; J. N. McWood, Gen. Car For., G.T., Ottawa, Ont.; John McCormick, Gen. Car For., L.V.R.R., Buffalo, N.Y.; D. McCarthy, Car For., N.Y.C., Buffalo, N.Y.; J. McLeod, Chief M.C.B. Clerk, Wabash Ry., Decatur, Ill.; W. R. McMunn, Asst. to Supt. R.S., N.Y.G.R.R., N.Y. City; E. E. Mo-Munn, Car For., N.Y.C.R.R., Williamsport, Pa. A. Marshall, Car For., G.T., Portland, Me.; Edwin Moses, Gen. Car Inspr., N.Y.C., New York, N.Y.; L. W. Martin, Gen. Car For., Ten. Ry. Asst., St. Louis, Mo.; J. A. Masters, M.C.B. Billing Inspr., C.N.R., Port mouth, Va.; C. W. Madden, Chief Car Inspr., C. & O., R.R., Richmond, Va.; Chas. F. Masters, Chief It. Inspr., D. & H. R.R., Carbondale, Pa.; E. H. Mattingley, Jt. Genl. Car For., B. & O. and B. & O.C.T., Chicago, Ill.; A. J. Mitchener, Div. Gen. Car For., M.C.R.R., St. Thomas Ont.; W. G. Milburn, Chief Jt. Inspr., Pontoria, O.; Elli Minick, Gen. Car For., L.V.R.R., Wilkesbarre, Pa.; A. L. Miller, Asst. Gen. Car For., Wabash Ry., Moberly, Mo.; W. R. Morris, H. of R. Asst., A.B. & A. Ry., Atlanta, Ga.; Livingston Martin, C.C.R. Asst., Bureau, B.O.R.R., Baltimore, Md.; D. C. Messroie, M.C.B., G.T. Ry., Montreal. E. R. Norquist, Asst. H.C.I., G.T. Ry., Battl Creek, Mich.; L. H. Nettels, Ch.C.-G.S.R.R., N.Y. C.R.R., Buffalo, N.Y. P. J. O'Dea, Gen. Inspr. Mech. Dept., Erie R.R., Buffalo, N.Y.; T. J. O'Donnell, Chief Int. Inspr., Wabash Front. Car Inspr.' Assn., Buffalo, N.Y.; W. J. Owen, Chief It. Inspr., P. & P.U., Peoria, Ill.; M. J. O'Connor, Mech. Inspr., N.Y.C.R.R., N.Y.C. Malone, N.Y. E. W. Pellerz, Supt. Transp. & Labour, Domer Steel Co., Ins. Buffalo, N.Y.; T. E. Park, Jt. Car For., B. & H. and N.Y.O. & W. Sidney, N.Y.; B. F. Patram, Gen. Car For., Southern Ry., Richmond, Va.; H. C. Payne, Div. Car For., Erie Ry., Elmira, N.Y.; C. A. Park, For. Car Inspectors, N.Y.O. & W.R.R., Norwich, N.Y.; J. W. Puckett, Gen. For. Car Repairs, Southern Ry., Alexandria, Va.; Pearl Parker, Gen. For. Car Dept., N.Y.C. R.R., Kankakee, Ill.; W. H. Price, Gen. Car For., C.N.R., Halifax, N.S.; H. A. Palmer, Car For., G.T., Rouse's Point, N.Y.; Harry Pearce, Car For., G.T., Montreal; M. B. Perry, Ch. Car Rep. Asst., B.O., Zanesville, O.; F. T. Price, Ch. Gen. Asst., B.O., Zanesville, O.; F. T. Price, Ch. Gen. J. V. Richards, Chief Car Inspr., L. & H.R. Ry., Phillipsburg, N.J.; C. E. Roscoe, Ch. Clerk, Supt. R.S., N.Y.C.R.R., Cleveland, O.; C. G. Roberts, Chief M.C.B. Clerk, C. & O. R.R., Richmond, Va.; M. J. Ryan Gen. Car For., N.Y.C.R.R., Oswego, N.Y.; W. A. Rogers, Chief Jt. Inspr., P. & L.E.

(Continued on page 58.)

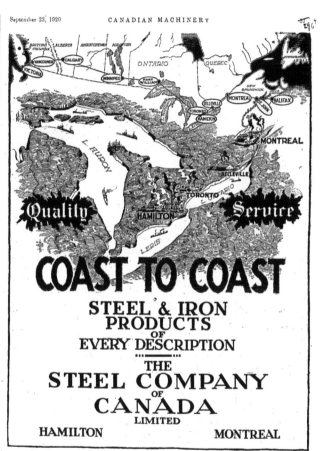

TRADE CAN BEST BE DEVELOPED APART FROM GOVERNMENT OFFICES

MR. LLOYD HARRIS, of Brantford, brought forward at the Congress of Chambers of Commerce of the British Empire one of the most plausible suggestions offered for the development of trade within the Empire, and the better understanding of the problems of all sections.

Mr. Harris' chief complaint is with the appointees of the various governments having to do with affairs of business, which, he claims, they do not understand, having had no practical training, and being bound around with conditions that keep them from developing.

After the signing of the armistice Mr. Harris was in London, England, for some months, as head of the Canadian Trade Mission there, and he had a splendid opportunity to see where the trouble was coming from in the development of more trade between the various parts of the British Empire. Mr. Harris cannot reconcile the idea of more trade and leaving the government trade organization as it is at present.

"The resources of the British Empire need co-ordination," stated Mr. Harris. "A government is not greater than public opinion, and there are times when it must be forced. A gathering such as this (the Chambers of Commerce)

should have men with the vision and the knowledge to develop Empire trade as it should be developed."

Mr. Harris goes further. He believes that the body to take care of this work should be known as an Empire Trade Council, to be composed of three representatives from each outlying part of the Empire, the British Government also to appoint members. This should form a permanent trade council in London. The appointments to that body should not be made for political purposes. Those receiving the appointments should not be paid members. The idea of Mr. Harris is that there are enough men in the country who would tackle such a proposition with the sole idea of being able in this way to render some service to the country, and to bring the outlying parts of the Empire closer together. The appointments would be made for a term of three years, with new men coming in at intervals before all the older members dropped out, in this way carrying out the continuity of the programme.

The speaker was particularly plain in making the meaning clear, that on no account should the carrying out of such a scheme be left in the hands of any government officials.

FIRMS WANT TO ESTABLISH TRADE RELATIONS WITH CANADA

INFORMATION regarding the following lines can be obtained by communicating with this paper and mentioning key numbers.

1830. Engineers' and Shipbuilders' Supplies.—Firm in Scotland desire to receive catalogues and quotations from Canadian firms who are able to export anything of interest to the trade of engineers', shipbuilders' and allied trades.

1831. Motor Trucks.—A firm of manufacturers' representatives of reliability desires to secure the agency for Canadian motor trucks—from three to five tons.

1832. Motor Cars.—A firm of commission merchants in Barbados would like to secure the agency for a Canadian motor car in Barbados, should be a five seater with all the latest improvements.

1852. Wire Rods. — A London firm wishes to hear from Canadian manufacturers of wire rods who can offer supplies for export.

1853. Nails. — A firm of travelling agents with branch offices in Jamaica, Demerara and Barbados desire to obtain a Canadian connection for nails and other small hardware.

1854. Steel Products. — Established trading corporation, with head office in Chicago, and branches throughout the world, ask to be placed in touch with Canadian manufacturers of iron and

steel products, particularly structural iron, steel plates, pipes and pipe fittings of all kinds.

1855. Wire Rods.—Large ironworks in Scotland would be glad to receive quotations from Canadian producers who are in a position to export. The immediate requirements of this company are filled, owing to the arrival of a recent shipment from Canada, but they are regularly in the market.

1856. A Buenos Aires firm of general importers is interested in receiving quotations, c.i.f. for black, galvanized, oval high tension and barbed wire.

1857. A Buenos Aires firm of general importers is interested in receiving quotations, c.i.f. for iron and steel bars, beams, galvanized sheets and black sheets.

1858. A Buenos Aires firm of general importers is interested in receiving quotations, c.i.f. for tin plate.

1869. Brake-Shoes.—A Newfoundland commission agent, long in a position of trust with an important firm, wishes to be placed in communication with Canadian manufacturers of brake-shoes for trucks and sleds.

1588.—Machinery. A firm in Roumania desires to receive quotations and other particulars from Canadian firms in a position to export all kinds of electrical machines.

Annual Meeting.—The annual meeting of the shareholders of Canadian Locomotive Co., Ltd., was held in the general offices of the company, at Kingston, when a by-law providing for the adding of two directors was passed and the following board of directors elected: Aemilius Jarvis, Robert Hobson, Warren Y. Soper, John L. Whiting, K.C., James Carruthers, M. J. Haney, F. G. Wallace. At a subsequent meeting of the board of directors the executive officers for the coming year were appointed:—F. G. Wallace, president; Aemilius Jarvis chairman of the board; J. H. Birkett, treasurer; William Harty, Jr., and William Casey, were elected to the board and Mr. Casey appointed a vice-president and Mr. Harty secretary. The financial showing of the company for the past year is most gratifying, regardless of the fact that the works were closed during the first three months of the company's financial year owing to strikes. The company at the present time is working to full capacity with sufficient orders on hand to run it for several months.

———

Mr. J. H. Davey, who has been connected with the Brown & Sharpe Mfg. Co., of Providence, R.I., for about 25 years, has taken up the work of representing that firm in Canada. Mr. Davey has been in Glasgow as the Brown & Sharpe representative previous to coming here. He is now working with the Canadian Fairbanks-Morse Co., the sales representative of the Brown & Sharpe lines, and all sales will be made through this company as heretofore.

INSPECTORS AND CAR FOREMEN
(Continued from page 56.)

R.R., Youngstown, O.: G. K. Reynolds, Can. For., N.Y.C.R.R., Corning, N.Y.; F. W. Roth, Car For., N.Y.C.R.R., Buffalo, N.Y.; J. V. Nottammer, For. Car Dept., N.Y.C.R.R., N.Y. City.
Wm. Shamp, Gen. Insp., C. B. & Q., Chicago, Ill.; P. F. Spangler, Trav. Car Instr., S. L. & St. F. Ry. Co., Springfield, Mo.; W. H. Sherman, Car For., G.T. Ry., Sarnia, Ont.; A. F. Schatler, Car Insp., G.T. Ry., Port Huron, Mich.; J. E. Severy, Chief Int. Insp., D.L. & W. Ry., Binghampton, N.Y.; F. A. Starr, Gen. Car Insp., C. & O. Ry., Covington, Ky.; E. S. Smith, M.C.R., Florida S. Coast Ry., St. Augustine, Fla.; C. C. Stone, Spec. Tray. Aud., M.C.R., Southern Ry., Richmond, Va.; W. J. Stoll, C.I.I., Toledo, O.; J. R. Sutton, Car Frt. Car Inspr., C.I. & W. Ry., Indianapolis, Ind.; C. F. Smith, Ch. C-D.M.C.B., N.Y.C.R.R., Chicago, Ill.; R. C. Saxler, Gen. Car For., N.Y.C.R.R., Elkhart, Ind.; Fred. Schuets, Gen. For., Live Poul. Tran. Co., R.P.D. No. 2, Forks, N.Y.; O. B. Setterly, Car For., P.R.R., Buffalo, N.Y.; W. P. Simpson, For., P.R.R., Erie, Pa.; H. J. Smith, Gen. Car Insp. D. L. & W., Scranton, Pa.; D. W. Stephens, Car For., T. & C., Fort Worth, Tex.; A. F. Sturbing, Mgr. Ed., Ry. Mech. Eng., New York, N.Y.
F. W. Trapnell, C.I.I., Kansas City, Mo.; F. A. Teed, Gen. Car For., B. & O. Ry., Washington, Ind.; W. F. Tidswell, Car For., M.C. Ry., Detroit, Mich.; Geo. Treacy, Gen. Car For., Ann Arbor R.R., Owosso, Mich.; Jas. B. Turner, Gen. Car For., Can. National, Moncton, N.B.
J. E. Vitcum, Ch. It. Inspr., All Columbus Rys., Columbus, O.
W. F. Westall, Special Inspector, N.Y.C.R.R., Cleveland, O.; C. W. Wolfe, Car Insp., Southern Ry., Richmond, Va.; B. S. Wright, Gen. For. Car Dept., N.Y.C. & St. L. Ry., Conneaut, O.; B. M. Waido, Chief Int. Insp., All Lines, Dallas, Tex.; J. H. Weale, For Car Dept., N.Y.C.R.R., N.Y. City; J. W. Wilson, Chief It. Inspr., N.Y.C. and P. & R., Newberry Jct., Pa.; G. B. Williamson, Car Repair Asst., B. & O. R.R., Baltimore, Md.
A. Ziebold, Gen. Car Inspr., Hocking Valley R.R., Columbus, O.; C. A. Zielke, C.I.I., G.T. Ry., Port Huron, Mich.

If interested tear out this page and place with letters to be answered.

INDUSTRIAL NEWS

NEW SHOPS, TENDERS AND CONTRACTS
PERSONAL AND TRADE NOTES

TRADE GOSSIP

New Catholic Union.—The National Catholic Union has established a new local in the plant of the Canadian Car & Foundry Company, Limited. According to an announcement made by Gerard Tremblay, General Secretary of the Montreal National Catholic Union, this new body is the result of the recent strike, which was followed by a shutdown of the company's plants. He stated that men, finding the company obdurate, appealed to Abbe Hebert, head of the National Union of Catholic Workers, to use his influence to bring about a termination of the trouble, with the ultimate result that the company reopened its plants. Abbe Hebert stated that the new union comprised several hundred men.

Secured Contracts.—The MacKinnon Steel Co., Ltd., Sherbrooke, Que., have contracts for the steel for Royal Bank of Canada, Bridgewater, N.S.; general contractor, Russel & McAuley, Halifax, N.S.; steel for News Boiler House Extension, Grinder Room, Sulphite Mill Boiler House, Brompton Pulp and Paper Co., East Angus, Que.; steel superstructure for bridge, 55 feet clear space, over River Pointe du Joui L'Assomption municipality at the site known as Point Jourdain, L'Assomption, Que.; general contractor, Alex. Venne, L'Assomption, Que.; steel for Wagon Shed for Imperial Oil Co., Halifax; general contractor, McDonald Construction Co., Halifax, N.S.; steel superstructure, 65 feet clear space bridge to be erected across outlet of Brome Lake, at Fulford, Que.; general contractor, C. J. Dryden Co., Ltd., Montreal, Que.; steel work for highway bridge, 60 feet long, in two spans, with 25 feet roadway and 8 feet sidewalk for City of Granby; steel for Royal Bank Lennoxville, Que.; general contractors, Newton Dakin Construction Co., Sherbrooke, Que.; steel for post office, Yarmouth, N.S.; Russel & MacAulay, general contractors, Halifax, N.S.; steel for store for L. R. Steel Co., Wellington St., Sherbrooke, Que.; architect Mr. Gregoire, Sherbrooke, Que.; general contractors, Sherman Construction Co., Sherbrooke, Que.; steel for Canadian Bank of Commerce, Wellington St., Sherbrooke, Que.; general contractors, Anglina Norcross, Ltd., Montreal, Que., steel for Tuberculosis Hospital, Halifax, N.S., general contractors, Rhodes, Curry, Limited, Halifax, N.S.

WILLIAM CASEY
JOINS THE BOARD

Vice-President and Manager of the Canadian Locomotive Plant Started There as Office Boy

William Casey was this week elected to the Board of Directors of the Canadian Locomotive Co., Kingston, and also appointed one of the vice-presidents.

WILLIAM CASEY

His promotion with this company has been a fine illustration of the possibilities of mechanical training and initiative.

He is a Kingston boy, being born there in October of 1887, so at present he is still a young man of 33. After completing his schooling he entered the employ of the Canadian Locomotive Co. as office boy. Leaving there he went to the machine shop where he served an apprenticeship, afterwards going to the drafting room. In 1907 came his first important promotion, when Mr. Casey was made machine shop foreman, remaining there until 1912 when he returned to the office to take charge of piece work and estimating. In 1914 he was made assistant to the vice-president and general manager (A. W. Wheatley) and following the latter's resignation early in 1916, Mr. Casey was appointed manager.

SIGNS OF PROGRESS

H. L. Hewson plans erection of clothweaving mill at cost of $50,000, at Amherst, N.S.

An extension of 126 by 85 feet, one storey high, will be made to the Kindel Bed Co., Stratford, Ont.

Buildings to the value of over one and three-quarter million dollars have been undertaken in Ottawa this year up to the end of August.

Building permits in Moose Jaw for the eight months of this year show an increase of almost a million dollars over the same period of 1919.

William Newlands and Son are architects for a first-class, modern hotel, costing $450,000 to be built at Kingston, Ont.

The Acme Sash and Door Co., Ltd., Winnipeg, have been awarded the mill work contract for a store costing $150,000 for Ramsay's, Ltd., Dauphin, Man.

Contracts have been awarded by the Technical School Commissioners, Hull, Que. The Dominion Bridge Co. secured the iron, and D. O. La Violette the stone work.

L. A. Amos, architect, 78 Crescent Street, Montreal, is preparing plans for a match factory to be erected in Quebec for Maguire, Paterson and Palmer, Ltd.

One and a half million dollars is to be expended on reconstructing the abattoir of the Pat Burns Co., Regina, Sask. The building will be of brick and concrete.

The value of building permits issued in Vancouver, B. C., the first eight months of the year is almost double the value of the permits issued during the same period of last year.

A motor truck industry, with head office in Indiana, will erect a Canadian plant at London, Ont. The incorporation of the Canadian company with a capitalization of approximately $5,000,000 is now being negotiated.

The general contract for erecting a hotel costing $1,000,000 for Benson-Hines Co., London Chambers of Commerce, has been awarded to Geo. A. Fuller Co., Ltd., 285 Beaver Hall Hill, Montreal.

Colthurst, Trace and Nichols, architects, of Windsor, are preparing plans for the construction of a huge cold storage plant, of reinforced concrete, three storeys in height. Windsor, Detroit, London and Toronto capital is behind the scheme, and construction is expected to start in the spring.

What the Manager Said :

"You'll have to get that order out by the 15th if you work the men overtime," said the manager to the machine shop superintendent. It's promised, and this firm has built up its business by keeping its promises. If your drills are not boring holes fast enough put a rush order through for some Morrow High Speed Forged Drills."

Morrow drills will bore more holes in a given time without regrinding than any other drills. They're harder, they're tougher, they're stronger. They save time, trouble and worry.

Test them. A competitive test in your own shop will prove it.

MORROW'S

INGERSOLL - CANADA

Distributors of PXH and Imperial Files

"They cut faster and wear longer."

H. W. Welsh, former manager in
Sherbrooke, Que., for the E. and T.
Fairbanks Co., Limited, and latterly
manager of the scale department of the
Canadian Fairbanks-Morse Co., Montreal, Que., has joined the organization
of the MacKinnon Steel Co., Limited,
Sherbrooke, as manager and has already assumed his duties.

WOULD BE FOREMAN

Continued from page 290

it would have gone if it had been of
greater capacity. We had all the tuyere
doors open and then had iron clean to
the top of the stack and coming down
like rain in the yard, but we got there
just the same. Next day the pulleys
were replaced as they were in the beginning. After we got the castings out
and found that they were good, the manager came to me and explained that he
was well satisfied, but that it would
now be necessary to get to work and
make some money out of the place, so
he proceeded to start operations on this
money-making plan of his. Not interfering with me or my work he pointed
at one corner of the shop and said: "We
will call this No. 1 floor,—Alex. you
take this." Next, he says: "This will be
No. 2,—Herbert you take this." And
so he numbered off each floor and put
a man on each of them until the men
were used up. After this he turned to
me and informed me that he had engaged a core-maker, who had not yet
arrived, but that in the meantime I was
to make what cores were required, after
which I was to mold some things which
he had laid out for me. On the fourth
day he came to me and told me that

he would be out of town for a
while but that I should mold
some plow shares for which he showed
me the patterns and gave me to
understand that they would have to be
cast with the face down. I suggested
that there would have to be some getting
ready done first, to which suggestion
he replied in a very stern voice that
there would be no more getting ready,
but that getting the work out would be
the only consideration. He then left me
and I have never seen him since.

I wended my way back to Brantford
that night and wrote to him advising
him that we had parted company and
that on pay day I would draw on them
at the bank for my money. To this I
received a rather waspy reply from the
manager who seemed to be put out at
my not liking his methods. True to
my promise I drew on them at the bank
but my draft was neither paid nor returned for over a week, when I learned
that the directors had held a meeting
and accepted my resignation and also
asked the manager for his, so that they
could both be accepted at the same time.
This was done and we were both paid
off—two birds killed with one stone—
leaving two vacancies to be filled. Thus
ended my first experience as a foundry
foreman.

WELDING BOILER SHELLS

Continued from page 289

ment, and carried out everything in connection with the welding, etc.

The extent of the new metal forming the weld between the sludge-hole and the front end-plate was approximately 21 inches in length circumferentially round the shell; 3 inches in width and ½ inch in thickness (average), the hole through the plate being completely filled up. The total cost of repairing the boiler by means of electric welding was £54. Although this may appear at first sight a large amount for a small repair yet much time was saved by adopting this means of making up the original wasted plate instead of fitting a new part altogether. Further, with regard to the tensile strength of an electrically-welded joint, this may be as much as 95 per cent. of that of the original unwelded metal, the average strength being frequently from 80 to 90 per cent.

Owen Sound.—Work will be started almost immediately upon the new dock on the east side of the Owen Sound harbor. It was authorized about a year ago, and an appropriation of $89,700 has been made for it. It will be of concrete, and will be nearly 1,000 feet long. The contract has been let to a Peterborough firm, and work should start at once. When completed all the water front owned by the Government will have concrete dockage.

Classified Opportunities

HELP WANTED

SCREW MACHINE FOREMAN WANTED FOR Cleveland and Brown & Sharpe Automatics. Give full particulars of age, experience and salary wanted. Location, small city near Toronto. Box 705, Canadian Machinery. (c18m)

WANTED—MACHINE SHOP FOREMAN FOR small shop manufacturing pumps, windmills and cylinders. Must be a man of thorough experience. Give full details first letter. Box No. 709, Canadian Machinery. (c14m)

POSITION WANTED

MECHANICAL ENGINEER WITH BROAD DE-signing and construction experience desires re-sponsible position as assistant engineer or machine designer. First-class references available. Further particulars. Box 708, Canadian Machinery. (c14m)

RATES FOR CLASSIFIED ADVERTISING

Rates (payable in advance): Two cents per word. Count five words when box number is re-quired. Each figure counts as one word. Minimum order $1.00. Display rates on application.

CANADIAN AGENTS WANTED

AGENTS WANTED. — REPRESENTATIVE IN Eastern Canada for well-known American firm manufacturing oxy-acetylene welding appara-tus, etc. Box 705F, Canadian Machinery.

FOR HIGH-GRADE AMERICAN-MADE LATHE —14 to 26. In replying state territory you can cover and cover thoroughly. Box 682, Canadian Machinery. (c23m)

AMERICAN FIRM MANUFACTURING PAT-ented process case hardening specialty wants the services of party capable of building up sales in Canada. Box 706F, Canadian Machinery.

CANADIAN REPRESENTATIVE WANTED by U.S. manufacturer of well known line of high speed and alloy steels. Correspondence in-vited and will be treated as confidential. Box 707F, Canadian Machinery.

CANADIAN AGENT WANTED — AMERICAN firm whose line (high grade factory specialty) is being manufactured in Canada, wishes to get in touch with live man who will act as Provincial or Dominion representative. Box 670, Canadian Machinery.

FOR SALE

14 in. Bertram Hand Feed and Perfect, Lever Feed, Rising Head, Bench Type Drilling Machine.
No. 1½ Brown-Boggs Hand Lever Punch.
40 in. x 8 ft. Back-geared, Horizontal Bor-ing Mills.

Apply Box 704, Canadian Machinery.

PATTERNS

TORONTO PATTERN WORKS, 65 JARVIS Street, Toronto. Patterns in wood and metal for all kinds of machinery. (etfm)

BRANTFORD PATTERN WORKS ARE PRE-pared to make up patterns of any kind—in-cluding marine works—to sketches, blue prints or sample castings. Prompt, efficient service. Bell 'Phone 681; Machine 'Phone 733. Brantford Pat-tern Works, 49 George St., Brantford, Ont. (etfm)

MACHINE WORK WANTED

MACHINE WORK WANTED FOR LATHES, shapers, milling machine and planer, etc. Hourly or contract basis. Prompt delivery. W. H. Sumbling Machinery Co., Toronto. (etfm)

Used Machinery Bargains

3—5 x 36 Jones & Lamson Flat Head Turret Lathes, geared feed.
2—2 x 24 Jones & Lamson Flat Head Turret Lathes, geared feed.
2—16" x 8' Cincinnati Lathe.
1—21" x 8' LeBlond Lathe.
1—24" x 10' Milwaukee Lathe.
1—36" x 10' C.M.C. Lathe.
1—54" Boring Mill.
1—Sellers Tool Grinder.

CHARLES P. ARCHIBALD & CO.

MACHINERY and SUPPLIES

265 Beaver Hall Hill   Montreal

International
Railway, Marine and Industrial Supplies

Large Stocks on Hand for Prompt Shipment

Emery wheels of all sizes or for any use.	Rope, Manila and Hemp.	Gaskets of all kinds and sizes.
Files.	Hose of all kinds and sizes.	Full line of Ship's Hardware.
Railway Jacks and Screw Jacks.	Blow Torches.	Boiler Grates.
Boiler Compound.	Winches.	Full line of Engine-room Equipment.
Flue Cleaners.	Pneumatic tools.	Taps, Dies, Drills and Reamers.
Electrical equipment to meet your requirements.	Brass Castings, finished and unfinished.	Anti-Corrosive and Anti-Fouling Paints.
Ship's Telegraphs.	Warehouse Trucks.	Metallic Packing of all kinds.
Compasses.	Bearing Metals of every de-scription, finished and unfin-ished.	

Wire or Write Your Requirements.

International Machinery and Supply Co., Limited
421 St. James Street, Montreal

If what you need is not advertised, consult our Buyers' Directory and write advertisers listed under proper heading.

CANADIAN MACHINERY
AND
MANUFACTURING NEWS

Vol. XXIV. No. 14

September 30, 1920

Blind Workers
Prove *Efficient*

Cleveland Shops Have Opened Their Plants to the Blind and Sixty-nine Operations Are Successfully Accomplished by These Sightless Workers

By EVA B. PALMER*

"WE have in our employ a totally blind man who has been with us since Sept. 23, 1913, assembling compensator switches. This work is being done on piece work, and the fact that he is making his rate is proof positive that he is very efficient at this work. It is, I believe, a very severe test for a blind man. The fact that his work is perfect and that his speed is high enough so that he makes out on a piece work rate, which is the same as the rate for a man with all his faculties, is pretty good proof of the fact that he has been able to master a difficult job very satisfactorily. He also is a universal favorite, and I believe, is respected and loved more than any other man in the factory."

The above statement, signed by J. F. Lincoln, vice-president of the Lincoln Electric Co., is an indication of the value placed upon blind workmen by many Cleveland employers. This city was among the first in the country to see the possibilities in industry for blind artisans, and a start was made in February, 1913, when the Society for the Blind placed one man in the Lake Erie

*Executive Secretary Cleveland Society for the Blind.

Bolt & Nut Co. He did hand nutting, and his pay ranged from 80c to $1.25 a day. He remained two years and left of his own accord, having given entire satisfaction.

The movement progressed slowly until war-time conditions created a shortage of labor, and since then many openings have been secured. To-day there are 81 blind men and women working in 40 different factories, and each placement makes the next one easier. The employer finds himself supplied with a dependable worker, for the blind man realizes that chances for him are not numerous, while the man without sight declares that he would far rather have a *man's* job than a blind man's job. The time-worn trades of chair caning, broom making and carpet weaving are still useful, but to the man who asks no odds of the world because of his affliction a regular man-size job is preferred.

The electrical field has so far revealed more practicable operations than any other, but the automobile industry, with its necessary machine shops, will undoubtedly take the lead in time, especially since machine operations have been proven feasible for blind workmen.

Nutting bolts is still being done in several shops, and W. B. Alexander, general superintendent of the National Screw & Tack Co., has this to say: "We have employed as many as ten blind men at a time, and now have very close to that number. They are employed in nutting bolts by machinery and hand, and we are well pleased with their work and production. I might also state that they are very prompt and lose very little time. During the winter months we had five of our blind employees form an orchestra, composed of piano, drum, violin and two saxophones, for the benefit of our office help. It was very enjoyable, and furnished good music."

Regarding Accidents

Three questions are nearly always raised when the matter of employing blind labor is brought to a concern for the first time. One is the danger of accident. A large percentage of Cleveland factories carry State insurance, but many insuring with private companies are interested to experiment with blind workers. In 1919 the following modification was made in the Ohio law: "No agreement by an employee to waive his

rights to compensation under this act shall be valid, except that an employee who is blind may waive the compensation that may become due him for injury or disability in cases where such injury or disability may be directly caused by or due to his blindness. The Industrial Commission of Ohio may adopt and enforce rules governing the employment of such persons and the inspection of their places of employment."

It is interesting to note that only two concerns employing blind workers have availed themselves of this opportunity to escape paying compensation for possible injury. The blind man is easily the most careful in the entire plant, as loss of sight has doubled his caution, and as a result there have been no accidents so far. As employers become educated to consider a man according to what he has left rather than by what he has lost, people with handicaps will not find it so difficult to obtain work.

The second query raised is the one as to production. The person having the responsibility of fitting the blind man into his job must not only be familiar with labor conditions, but must know blindness, its splendid possibilities as well as its painful limitations. When the blind man's production is up to normal it is because most careful selection has been made, first of the job for the man and then of the man for the job. A general error of the uninformed public is to consider that blind people have traits in common, owing to a common

handicap. When the employer realizes that their abilities are as varied as possible, and when he has been shown that lack of sight is no hindrance on certain operations, he is willing to try one worker after another until the right person is found. Much of the success of the Cleveland work is due to the skill and personality of the young demonstrator who is seen at work in one of the accompanying illustrations. He went blind in January, 1919, and had no factory experience whatever before losing his sight. He is thus able to meet every objection on the part of employers, as becoming blind and goes everywhere alone. He tried out 150 operations in 31 Cleveland factories during a period of three months, and proved nearly all of them practicable. In many instances he exceeded the average speed for sighted workers. This is not exceptional, as in several cases blind men are maintaining a higher speed rate than seeing people on the same jobs. In one large plant a blind man arranging carbon brushes on trays does almost twice as much as any one else, and is paid accordingly. In the same factory a blind man does all the work formerly performed by two sighted girls with perfect vision.

In each instance the blind worker conforms to the rules of the shop. If piece work is the rule, he receives the same rate as the other workers, while if day work is the rule he is paid in

proportion to his production and receives his bonus if such is given to others in his shop.

E. G. Greene, superintendent of the Ferry Cap & Set Screw Co., made the following statement which shows the production of the blind men in his shop:

"In regard to the two blind men employed at our factory I will say that they have proved very satisfactory in every way. They are running drill presses and produce as much work as those who can see."

What About Transportation?

The third question raised is as to transportation. "But how in the world will he get here?" asks the skeptical employer. Here again the skill of the placement agent must be shown. Every effort is made to place the workers in shops accessible from their homes, and while most of the blind men and some of the women go to and from work alone, in some instances this is impracticable. There are several ways of handling this difficulty. Sometimes another member of the family is placed in the same shop, or search is made throughout the factory for an employee who lives near the blind person's home and can act as guide. In several cases the entire family has been moved near the factory.

The placement of blind women in Cleveland factories was undertaken at their own earnest request. Good news travels fast and the girls learned with

DRILLING OIL HOLES FOR SPRING BOLTS ON SINGLE SPINDLE DRILL PRESSES AT THE FERRY CAP & SET SCREW COMPANY.

STACKING COMMUTATORS AT PLANT OF RELIANCE ELECTRIC ENGINEERING CO.

BLIND DEMONSTRATOR OPERATING ELECTRIC SPOT WELDER FOR THE BROWN SPRING OILER CO.

astonishment of the high wages being earned by the blind men. Some of the operations now being done by women are: Counting by weight on balanced scales, taping small coils, assembling small parts of electrical apparatus and of sewing machines, running single spindle drill presses on small work, foiling mints, packing candy, stacking commutators and small laminations, setting up cartons and wrapping butter blocks.

The list given at the end of this article shows the large variety of operations which have been found entirely practicable for both blind men and women. Contrary to the expectation of many employers, the impression made by the blind worker on his fellow employees is good and one superintendent stated that the standard of promptness and also of neatness had been raised throughout the entire department as "Gus" was always on time, clean shaven and smiling. Gus is now earning $40 a week.

A good friend of the Society for the Blind, in its efforts to provide every employable blind person with a well-paid job, is C. L. Collens, president of the Reliance Electric & Engineering Co. He has promoted the undertaking among Cleveland electrical concerns and writes the following letter. "I am very glad to comment on the work of the blind employee in our plant. He is employed on the operation of stacking commutator segments, and we can speak very favorably of his accuracy and speed in this

work. One noticeable feature is the steadiness with which he applies himself to his work, as there is no tendency to be distracted or interested by other work going on around him in the plant." The blind have long been an object of charity—and injustice. The new movement demands that the community do for the blind what it does for the seeing: judge each individual according to his merits and give him a fair chance to make good.

Operations Successfully Performed by Cleveland Blind

Arranging carbon brushes on trays.
Assembling:
  Ball-bearing cups for Ford cars.
  Chimneys for oil stoves.
  Compensator switches.
  Control levers.
  Controller slates.
  Door bell transformers.
  Drive flanges and shafts.
  Generators.
  Grease cups.
  Ground wire telephone clamps.
  Hot air boxes.
  Junction boxes.
  Kelly handle bars.
  Locks.
  Pittman rods.
  Plug, nut and bolt in oiler.
  Radiator parts.
  Steering gear.
  Tension studs and shuttles for sewing machines.

Tools.
Trolleys for electric cranes.
Vacuum cleaner parts.
Vibrator spark coils.
Vacuum gas tanks.
Wick raisers for oil stoves.
Wire rope clamps.
Bench Work:
  Cleaning castings.
  Cleaning fire brick.
  Clamping wicks in metal rings.
  Counting by weight.
  Creasing boxes.
  Cutting and skinning cables.
Etching electric light bulbs.
Foiling mints.
Inspecting separators in storage battery cells.
Making hand-woven hats.
Nutting bolts by hand and machine.
Operating:
  Broaching machine.
  Centering and counter-sinking machine.
  Drill press.
  Grinding machine.
  Lathe for facing brake hubs.
  Milling machine, power and hand.
  Nut facing machine.
  Polishing machine.
  Porter-Cable lathe.
  Punch press.
  Single-spindle and multiple-spindle drill presses.
  Spot welding machine.
  Stamping machine.
  Tapping machine.
(Continued on page 306.)

Honor Memory of the Late Hugh McCulloch, Sr.

Nurses' Home Opened at Galt—As a Memorial to Founder of One of the City's Best Industries—Building Complete and Furnished Throughout is Presented to the City.

AS a memorial to the late Hugh McCulloch, Sr., one of the founders of the Goldie-McCulloch industry in Galt, a splendid nurses' home has been opened and presented to the city by Mr. R. O. McCulloch, one of the donors. The opening of the home, which is complete in every detail as to furnishings and equipment, was a notable occasion, being presided over by His Honor Lieutenant-Governor Clarke. Fitting reference was made to the career of the late Hugh McCulloch, and Mr. R. O. McCulloch in making the presentation gave some interesting facts in connection with the McCulloch family, and also a history of the steps leading up to the decision made by himself and Mrs. Shearson to erect a nurses' home, Mrs. Hugh McCulloch joining in the undertaking by furnishing the building throughout.

"My revered father," said Mr. R. O. McCulloch, "the late Hugh McCulloch, came to Canada from Scotland in 1850, arriving in Galt on August 24th of that year. With the exception of a few months spent in the neighboring village of Ayr, from which place he returned to Galt in 1851, his whole life until his death in 1910 was spent in this community. From 1851 until 1859 he was in the employ of James Crombie. In that year the firm of Goldie & McCulloch came into existence, when the late John Goldie and my father purchased their employer's business. In 1891 the present company was organized, and my

father became its first president, which position he occupied until his death in 1910.

"My late brother, my sister, Mrs. Charles A. Shearson, and I desired to erect a suitable memorial in affectionate remembrance of my father, and many plans and ideas were suggested and discussed, but no definite conclusion had been arrived at when the war broke out. Further consideration of the matter was, of course, postponed, and then my brother became very ill, and after a long illness died in 1917. When the war ended in 1918 the question was again taken up, and my sister-in-law wished to join with us in memory of her late husband. Different plans were canvassed, but none satisfied us until the needs of the hospital were brought to our attention. The appeal of physical suffering makes the strongest appeal to our common humanity. Hospitals were established to alleviate suffering and heal the sick; and trained nurses are the most necessary part of the equipment of a modern hospital. We learned that the accommodation for the nurses was entirely inadequate, and that consequently they were compelled to live in part of the main hospital. Everywhere there was expansion; increasing numbers of patients and more nurses required, but no accommodation for them. A residence or home for nurses appealed to us as the greatest need in the community, and we felt it would commend itself greatly to my father and brother, as they were both much interested in the hospital."

A Worthy Example

His Honor Lieut.-Governor Clarke, in opening the home, said in part:

"Mr. McCulloch was in more than one sense a real pioneer. He left his native Scotland before the old sailing vessels had been superseded by the steamer, and his passage was a long and rough one, not fewer than seventeen passengers dying during the voyage, which took about seven weeks to complete. He came, like many another pioneer in those days, specially well equipped for a successful career in Canada, which was then at the very beginning of its industrial history. He settled in Galt as far back as 1851, and with the late Mr. John Goldie gradually and surely developed the business associated with their names and known, not only throughout Canada and the United States, but in the Old Country as well. We are sometimes apt to forget the underlying principles on which national life is founded, but we have only to think a little—to look a little below the surface—in order to realise that without a well-developed, healthy and expanding industrial inter-

est, no country can long stand, nor can the amenities of civilized life be enjoyed by its people. Therefore, it may be well to-day to think of the immense debt we owe to the contribution which the little parish of Sorn, Ayrshire, Scotland, made to Canada, when in 1850 it sent one of its sons in the person of Hugh McCulloch, a well-educated, robust and enterprising young man, then only twenty-four years of age, to this country. He very soon found his opportunity and in conjunction with Mr. Goldie, as I have said, organized the business which shares his name. The expansion of that business meant much to Galt, and beyond Galt it meant much to the Dominion, for the products of its industry not only furnished a needy market with the goods required but gave high-class, skilled labor a continually widening field. I mention this side of Mr. McCulloch's service to the country because I believe that a career such as his in this respect ought to be held in the highest esteem. An upright, honorable man, conducting business with enterprise and conscientious integrity is one of the greatest possible assets any country can have and certainly Mr. McCulloch was such a man. . . .

"One word more, I think, ought to

MR. R. O. McCULLOCH,

One of the donors of the Nurses' Home, erected to the memory of his father, who made the formal presentation of the building to the Galt Hospital Trust.

The Late HUGH McCULLOCH, Sr.,

To whose memory the new Nurses' Home was erected. He located in Galt in 1851, and in 1859, with the late John Goldie, purchased the Crombie foundry, establishing the firm of Goldie & McCulloch Co. The late Mr. McCulloch passed away on August 24, 1910.

The beautiful Hugh McCulloch Memorial Nurses' Home, officially opened by His Honor, Mr. Lionel Clarke, Lieutenant-Governor of Ontario, and formally presented to the Hospital Trust by Mr. R. O. McCulloch, one of the donors. This view of the building is taken from the east side and shows some of the stalwart pines that surround it.—Photos courtesy Galt "Reporter."

be said and it is this: In erecting this home by Mrs. Shearson and Mr. McCulloch and in the furnishing of it so admirably by Mrs. Hugh McCulloch, this family has set an example to others in Galt and in other places, which would be for the public welfare were it followed, and I am sure that not only would the public welfare be advanced, but the givers themselves would have much joy in their gifts. The privilege of giving—of doing good—is one of which we can all, more or less, avail ourselves in our several ways, but those who exemplify their benevolence in such a way as we witness here to-day are specially fortunate and are to be envied in having afforded to them an opportunity to gratify the heartfelt desires with which they are so worthily inspired."

PLAY SAFE

During the week of October 10th to 16th, the Ontario Safety League will ask the public in Ontario to play safe. It will be known as Safety Week, and during this time an intensive drive will be made against accidents.

Similar large scale campaigns have been held in other cities and have been a notable success. The "Safety Week" held in St. Louis in the autumn of 1918 is the most conspicuous example, as fatal accidents in the corresponding week in 1917 were 24 and in "Safety Week" there was one fatality. Cleveland cut its fatalities in its "Week" last year from 15 in the same week of the previous year to 6 in "Safety Week." Pittsburgh cut its fatalities from 28 to 16. These facts are vital and experience has shown that the good effects of the safety propaganda have been continued month by month ever since. Similar intensive drives against accidents have been a success in industry and on the railroads, and the League looks forward with confidence to success.

During Safety Week, the various committees plan to teach accident and fire prevention to men, women and children in a spectacular educational campaign in which various forms of publicity will be used. The public will be reached by means of posters, bulletins, bill boards, street-car cards, letters, advertisements, meetings and safety shows and given a liberal education in Safety. The Safety Week idea has been approved by big industrial managers, by boards of trade, chambers of commerce, Canadian Manufacturers' Association, Toronto District Labor Council, Ontario Motor League, fire departments, police departments, boards of education, Empire Club, Rotary Club, Kiwanis Clubs, Canadian Clubs, and others.

Subcommittees of the League will cover such subjects as : Publicity, bulletins, motion pictures, schools for industrial safety, statistics, boy scouts, women, traffic, schools and colleges, etc., etc. The various days of the week have been specially designated—Safety Sunday, with special sermons in the churches; Monday is careful day for motorists; Tuesday will be careful day for street car men; Wednesday is set aside for school safety; Thursday will be at home day for safety; Friday is to be called careful day for pedestrians; and the slogan for Saturday will be "Safety for All, All for Safety."

To protect aluminum and aluminum alloys from corrosion, a German inventor has tried the experiment of browning the metal electrolytically. The aluminum is suspended in an electrolyte consisting of a sulphur compound of molybdenum, and zinc is used for the anode. The cell is maintained at a temperature of 60 deg. to 65 deg. Cent. The aluminum is soon covered with a dark brown coating. The metal may be bent or rolled without cracking the coating. A piece of aluminum thus coated is said to have been immersed in a salt solution for two months without showing the slightest trace of corrosion.

The Painting of Agricultural Implements

Painting Methods Differ Considerably Depending Upon the Nature of the Product—The Method of Mixing Paint is Described, Also Other Items Which Will Prove Valuable For Future Reference

By J. H. MOORE

IT is hardly necessary to state that practically all agricultural implements manufactured are noted for their bright and cheering appearance when leaving the factory.

True, they lead a varied life, on reaching their destination, for some farmers take care of their machines, while others allow them to lay out in some field, rain or shine, winter or summer, with the result that in a short time the cheerful look disappears to be superseded by a dilapidated appearance which is far from comforting.

This condition is, of course, up to the farmer himself. and yet, we could not commence with this article without giving our personal comment on the fact that if farmers could only see the pains spent in preparing and painting the machines they abuse, they would be more careful in future.

The Procedure Adopted

Taking it for granted that the various pieces have reached the manufacturing stage where the painting operation comes in natural sequence, we will first consider how the paint is prepared.

The various colors, of which there are quite a few, come to the factory in powder form, or in other words, in dry color form. Depending on the color to be mixed, the painter, or paint mill operator, as he is termed, takes the dry color and

FIG. 2—DIPPING McCORMACK MOWER FRAMES.

FIG. 1—NOTE THE TROLLEY AND HOOK ARRANGEMENT.

places it in an upper can above the grinding mill. He also places the oil which is to be mixed with the dry color in a adjoining pan. Should readers not be familiar with the appearance of a paint mill, we will ask them to consider in their mind's eye, two grindstones revolving slowly in a horizontal position.

Imagine these stones grooved with grooves about 1-8-in. wide, and semicircular in form, each groove leading to the centre of the stones. The size of the stones depends entirely on the capacity of the paint mill.

Directly above the stones is a hopper in which is an agitator somewhat after the shape of a ship's propeller. This should give a fairly good mind picture, so let us proceed to the actual mixing of the paint.

The dry colors, and oils, are now allowed to run into the hopper, where the agitator mixes the two quite thoroughly. The paint, which is now of a thick creamy nature, passes through the grooves of the stones, into a pan placed for the purpose. The paint is thinned out at later stages to suit requirements. Briefly, this explains the method of preparing the different colors. Usually there are 12 grooves to each stone, but again

(Continued on page 306.)

Calculations and Analysis of Riveted Joints

This Article Instructs You How to Design a Riveted Joint of Any Required Efficiency—The Calculations Are by no Means Complex and Are Easily Followed.

By T. H. FENNER, Editor Power House

THE efficiency of a riveted point is expressed always as a percentage of the strength of a solid plate. When holes are drilled or punched out of a plate to allow of riveting, the plate is weakened by the amount of metal cut out. If we have a plate of one square inch section, and remove .25 square inches all drilling, the section remaining is only .75, and the resistance of the plate to stress will be only 75 per cent. of its original resistance. This would represent what is known as per cent. of plate strength to solid plate. This is only one factor. The second factor is the proportion that the strength of the rivets bears to the strength of the solid plate. This will depend on the diameter and number of rivets in the unit section of plate considered, and is also expressed as a percentage. We shall consider the means of arriving at these efficiencies shortly.

There are several styles of riveted joints, and two distinct classes. The lap type of joint is so called because it is made by lapping the edges of the plate over each other and riveting them together. The second type is known as a butt joint and is made by bringing together the edges of the plates to be joined and placing a broad plate over them, riveting all three plates together. In many cases two cover plates are used. These two types of joints are subject to various modifications. Thus a lap joint may be single, double, or treble riveted and the rivets may be placed in regular rows or in staggered rows, known as chain or zig-zag riveting. The butt strap joint amy be made with one or two straps, with equal or unequal straps, with double, treble or quadruple riveting. The type of joint selected will depend upon the pressure to be carried, cost of manufacture and rules of the province in which the boiler is to work. The cut here illustrates the type of joints mentioned above.

To Find Efficiency of Joint

The efficiency of a riveted joint varies with the diameter of rivets, pitch of rivets, number of rivets in a pitch, and the thickness of the plate. In considering riveted joint efficiency, the unit of plate area considered is always contained within a pitch or between the centre lines of two adjacent rivets. Where there is more than one row of rivets the pitch of the rivets in the outer row is taken. We will explain each type as we come to them. The most simple is of course the single riveted lap joint. Suppose we have a plate of 5-16 inch in thickness, and with. ⅝ rivets pitch 2

inches. We have to consider the strength of the joint as compared to the solid plate. We will consider the plate to be of 60,000 lbs. tensile strength. We shall therefore have a section, before any drilling is done, of 2 x 5-16=⅝ square inches. This section will have a resistance of 60,000 x ⅝=37,500 lbs. Now, after drilling, the amount of metal removed will be in diameter just equal to the diameter of one rivet because in the section we are considering there will be just half a hole on each side. A glance at the sketch which shows the pitch of line passing through the centre of rivets will make this clear. Therefore the amount of metal removed will be equal to the diameter of one rivet multiplied by the thickness of the plate, or

$$\frac{5}{16} \ x \ \frac{5}{8} = \frac{25}{128}.$$ This amount subtracted from the original area will give the section remaining, and if this remainder be divided by the original section we will have a fraction representing the proportion of original section remaining. To express this as a percentage we multiply by 100. Now to follow out the procedure step by step.

Let P= pitch
d= diameter of rivet.
t= thickness of plate.

We have a plate section of pitch P, and thickness t, the area being Pt. The rivet area will be equal to dt, so that when the holes are drilled we shall have Pt—dt = P—d the t on each side cancelling out. The proportion of plate section remaining is $\dfrac{P-d}{P}$ and the percentage of plate to solid plate is $\dfrac{P-d}{P} \times 100.$ Substituting figures, we get

$$Pt = 2 \times \frac{5}{16}]$$

$$dt = \frac{5}{8} \times \frac{5}{16}]$$

$$Pt - dt = 2 \times \frac{5}{16} - \frac{5}{8} \times \frac{5}{16}$$

$$P - d = 2 - \frac{5}{8}$$

$$\frac{P - d}{P} = \frac{2 - \frac{5}{8}}{2}$$

$$\frac{P - d}{P} \times 100 = \frac{2 - \frac{5}{8}}{2} \times 100$$

$$= \frac{2 - .625}{2} \times 100$$

$$= \frac{1.375}{2} \times 100$$

$$= 68.75$$

So that the proportion of plate strength to solid plate is 68.75 per cent. This formula is always used, and may be set down

$$\frac{P - d}{P} \times 100 = . \quad \text{per cent. of plate}$$

strength to solid plate.
In the Ontario regulations the plate efficiency is symbolized by the letters Kt, so the formula is

$$\frac{P - d}{P} \times 100 \ Kt$$

The next point to consider is the efficiency of the rivets in the joint. This is the proportion that the area of the rivet in the pitch limit bears to the unit section of plate. As rivets are not as strong in resisting shearing stresses as they are in resisting tensile stresses, an allowance must be made for this difference. The Ontario regulations use the constant C, and this is expressed

Shearing strength of rivets
$$\frac{\text{Shearing strength of rivets}}{\text{Tensile strength of plate}}$$

C may be taken as
.85 for iron rivets in iron plates.
.70 for steel rivets in steel plates.
.65 for iron rivets in steel plates.

In the joint we are considering there is but one rivet in the pitch, and as there are but two plates which the rivets pass through they are in single shear. We will then have an area of $d \times d \times$ .7854, divided by the area of the unit section, which is P×t. This is expressed in the formula:

$$\frac{a \times n}{P \times t} \times C \times 100 = \text{per cent. of rivets'}$$

strength to solid plate.
So we have

$$\frac{5 \times 5 \times .7854 \times 1 \times 16 \times C}{8 \times 8 \times 2 \times 5} \times 100 = Ks$$

This worked out becomes, if we take C as .85,

$$\frac{5 \times .7854 \times 85}{8} = 41.72 \text{ per cent.}$$

If we neglected the constant C we would get about 40 per cent. This lowest percentage is the one used in calculating the working pressure of the boiler.

In the previous joint considered the plate strength was 68.75 while the rivet strength was only 41.72.

A single riveted lap joint should be about 50 per cent. strength, so that

there is evidently some factor that requires altering. As the plate strength is higher than the rivet strength, it is observed that the factor that is wrong is the pitch. This is arrived at by considering that the plate is of minimum thickness, and the rivet cannot be decreased in diameter. As the pitch is a divisor in the formula for finding rivet strength, it is plain that by decreasing this factor the rivet strength percentage will be brought up. How much must we alter it to get the required figure? We can find that out very easily.

The formula for rivet strength we have seen is:

$$\frac{a \times n}{p \times t} \times C \times 100 = \% \text{ of rivet strength}$$

Now, let us insert the quantities we know. We have 50 per cent. for rivet strength, the other factors, excepting p remaining the same, and it is p we want to find. So we have

$$\frac{5 \times 5 \times .7854 \times (1 \times .85 \times 100 \times 16}{8 \times 8 \times p \times 5} = 50$$

Then, P is the unknown quantity. We will work this out. First by cancellation we simplify the expression and get

$$\frac{5 \times 5 \times .7854 \times .85}{p} = 50$$

Transferring we get

$$50p = 5 \times 5 \times .7854 \times .85 \times 5$$
$$p = 5 \times 5 \times .7854 \times .85 \times 5$$
$$\frac{50}{p} = 1.66$$

The nearest fraction to this is 1¾. Let us see how this works out on the plate strength. We have

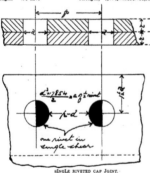

SINGLE RIVETED LAP JOINT.

$$\frac{p - d}{p} \times 100 = \% \text{ of plate strength}$$

Substituting figures we get

$$\frac{1¾ - ¾}{1¾} \times 100 = \%$$

This is simpler expressed in decimals, thus

$$\frac{1.625 - .625}{1.625} \times 100 = 61.53\%$$

This is about right, and we have a better balanced joint, with 61.5% plate and 50% rivet strength.

Analysis of Riveted Joints

In considering riveted joints, the various ways in which they may fail must be considered. The joint we are dealing with, being the most simple kind, has the least possible ways of failing. It can fail through the shearing of the rivets, and as there is only one rivet in the pitch limit the resistance of the joint to failing in that way will be the shearing strength of one rivet. It may fail by the plate tearing between the rivet holes, or it may fail by the plate crushing in front of the rivets.

First of all consider the strength of the solid plate. We have a pitch of 1.625 inches, with a thickness of 5-16 of an inch, and a tensile strength of 60,000 lbs. per square inch. The strength of the section will then be

$$p \times t \times 60,000$$
$$= 1.625 \times .3125 \times 60,000$$
$$= 30468.75 \text{ lbs.}$$

Then the strength of one rivet in single steel will be if tensile strength of rivet stub is 51,000 (.85 of rivet strength) ¾ × ¾ × .7854 × 51,000.

This worked out is 15,300 and the percentage to the solid plate is

$$\frac{15,300 \times 100}{30,468.75} = 50.2\%$$

The resistance offered by the plate to tearing between the rivet holes will be

$$p - d \times 80,000$$
or $(1.625 - .625) \times .3125 \times 60,000$
$1 \times .3125 \times 60,000 = 18,750$
and the percentage to the solid section will be:

$$\frac{18,750 \times 100}{30468.75} = 61.53\%$$

This is in exact agreement with the previous result.

We must now consider the plate crushing in front of the rivet holes. We will have a section of plate equal to the diameter of one rivet and the thickness of the plate. The crushing strength of good boiler plate is about 90,000 lbs., so that we have ¾ × 5-16 × 90,000 = 17571

This gives

$$\frac{17571 \times 100}{30468.75} = 54.3\%$$

which is higher than the rivet strength percentage and may therefore be neglected.

We may say that a joint of this kind is practically never used in the longitudinal seams of a steam boiler, as its low percentage of strength would penalize every other part of the structure. However, they are sometimes met with on steam domes, where the small diameter in comparison to weight of plate elbows allows a low percentage joint to be used. In tank work, where the thickness of plate is much more than is necessary to resist the pressure, a single riveted lap joint is often used. For purely steam boiler work it is hardly to be considered. The next type of joint, and one very widely used, is the double riveted lap joint.

The Double Riveted Lap Joint

This joint has two rows of rivets, and two rivets in a pitch. It is usually designed for a strength of about 70 per cent. of the solid plate rivet strength and plate strength being approximately equal. The same formulæ are used in making the calculations, but a new factor is a formula for determining the distance between the rows of rivets. The Ontario regulations prohibit chain riveting, all rows of rivets having to be staggered, or zig-zag.

Let us take a concrete example and work it out, which is the best way of explaining it. Let us assume we want a boiler built of ¾ inch plate and having a double riveted lap joint of about 2 per cent. strength. We want to get the other dimensions. The diameter of rivet may be taken as ¾ inch, or double the thickness of the plate. Very often in practice the rivet hole will be 13-16 in this size of plate, and of course the rivet size after driving

will be the same as the diameter of the hole. Using the larger diameter in the calculations makes a slightly higher percentage of rivet strength and a slightly lower plate strength. However for our purpose we will consider the diameter as ¾. Now we want to find the pitch that will give the requisite strength with this diameter of rivet. We come back to the formula.

$$\frac{p-d}{p}\times100=\%\text{ of plate strength}$$

Substituting we get

$$\frac{p-d}{p}=.72$$

$$p-d=.72\ p$$
$$d=.28\ p$$

Then if d=.75 we get

$$.28\ p : p :: .75 : x$$

which gives

$$\frac{.75\ p}{.28\ p}=2.679\text{ inches}$$

as the pitch we require. Knowing the diameter of rivet and thickness of plate, and number of rivets in a pitch, we could have found the pitch from the formula for rivet strength, as we did in the case of the single riveted joint. We will in this case, see how the pitch we have found works out for rivet strength. We have

$$\frac{a\times n}{p\times t}\times C\times100=\%\text{ rivet strength}$$

Substituting figures we have, using .70 as the constant C in this case, and 2 11-16 as the pitch

$$\frac{3\times3\times.7854\times2\times70\times16\times8\times100}{4\times43\times3\times100}=\%$$

This worked out gives a percentage of 61.37, which is too low. We wish to make our joint about even, though if the plate strength is a little higher than the rivet strength it is all to the good, as the plate strength may deteriorate in time, through the corrosion or thinning of the plate, which will not likely occur with the rivet.

We can use still another method if we wish, and that is to equate the two strengths to each other, and thus arrive at a pitch that will satisfy both.

This means that if the strength is to be equal, the two formulas must be equal to each other. Let us state this in the form of an equation. We have

$$\frac{p-d}{p}\times100=\%\text{ and }\frac{a\times n\times C}{p\times t}\times100=\%$$

If these % are equal the other sides must be equal, so we have

$$\frac{p-d}{p}=\frac{a\times n\times C}{p\times t}$$

$$p-d=\frac{p\times t}{a\times n\times C\times p}$$

$$p=\frac{p\times t}{a\times n\times C}+d$$
$$t$$

Here we have arrived at an expression for pitch which should satisfy us. Let us substitute figures and see

$$p=\frac{3\times3\times.7854\times2\times8\times70}{4\times4\times3\times100}+\%$$

$$p=2.39,\text{ say } 2.4$$

so we get 2.4 as a pitch. Let us try this again for our plate and rivet strength. First for plate percentage:

$$\frac{2.4-.75}{2.4}\times100=68.75$$

Then for rivet strength

$$\frac{3\times3\times.7854\times2\times5\times8\times70\times100}{4\times4\times12\times3\times100}=68.72$$

Here we have got a perfectly evenly balanced joint, and though it is some 4% lower than we at first determined on, that is not a very serious matter.

Let us see now how this joint stands comparison in its resistance to failure at different points. It may fail by rupturing between two rivet holes in the outer row and shearing one rivet. It may fail by shearing two rivets, or by tearing between the rivets on the inner row. It can also fail by crushing in front of the rivets in the outer row, and shearing the inner rivet, or by shearing the rivets in the outer row and crushing in front of the rivets in the second row. Let us take these in order. First the unit section is p×t×60,000

$$2.4\times.375\times60000=54,000$$

If the plate tears between two rivets on the outer row, and shears the rivet in the inner row it will have to over-

come a resistance of

$$60000\ (p-d)\ t+a\times.7\times60000$$

That is, the plate has diameter of one rivet, plus the area of one rivet in single shear, whose resistance to shearing is .7 that of the plate tensile strength. Substituting figures we have

$$(2.4-.75)\times.375\times60000+.75\times.7854$$
$$\times42000$$

which gives 74.250 or more than the solid plate. If it fails by shearing two rivets we have

$$a\times2\times42000$$
$$=.75\times.75\times.7854\times2\times42000$$
$$=37102\text{ lbs.})$$
$$\text{which is } \frac{37102\times100}{54000}=88.7\%$$
of the solid

If it tears between two rivets on the inner row, we shall have again

$$\frac{(p-d)\times t\times60000}{(2.4-.75)\times.375\times60000}=68.7\%$$
$$\frac{}{54000}$$

If it crushes the plate in front of the rivets in the outer row and shears one rivet in the inner row we have then

$$(d\times t\times90000)+(a\times42000)$$
$$(.75\times.375\times90000)+(.75^2\times.7854$$
$$42000)=43872.3$$
$$438723$$
$$\text{and } \frac{}{54000}\times100=81.2\%$$

which is much higher than either the plate or rivet %. The last two cases are exactly the same as each other so we see that 68.7% is the strength of the joint. The minimum distance between

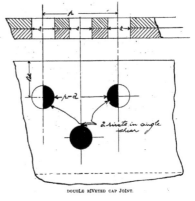

DOUBLE RIVETED CAP JOINT.

rows of rivets is found from the formula.

$$\frac{(11\frac{1}{2}+4\ d\ \ (p+4\ d)=\text{minimum}}{10}\ \text{distance}$$

The treble riveted lap joint shown in the sketch is not very much used, as there is not much advantage in employing it. The percentage obtainable is not much greater than with the double riveted joint, and the cost of manufacture is increased. The calculations are made exactly in the same way as for a double riveted joint, excepting that there are three rivets in single shear instead of two. It may fail by the plate breaking between the outer rivets and shearing two rivets. It may crush the plate in front of one rivet and shear two rivets. It may shear three rivets. The plate may break between the rivets in the centre row and shear one rivet in the inner row. It may fail between the rivets in the inner row, and this of course is the weakest part, so that p—d

$\frac{}{p}$ —is the governing factor.

The calculations would be as follows:
1st case — (p—d) t×60000.+a×2× 42000.
2nd case—d×t×90000.+a×2×42000.
3rd case—a×3×42000.
4th case — (p×d)t×60000+a×42000. 42000.
5th case—(p—d) t×60000.
In the next article we will take up butt joints.

STEEL TREATERS
J. H. Moore

The writer had the pleasure of attending the recent convention of Steel Treaters in Philadelphia and brings back pleasant memories of the event. The advance notices predicted were lived up to in every respect, and the various exhibits were both neatly and efficiently displayed. The floor space was arranged in ideal style, and those attending had a splendid opportunity to view each exhibit without confusion. To speak of any particular exhibit is unnecessary, as we printed in a previous issue the firms who were to be represented. Enough to say that every concern showed good taste in their display and ideas of considerable educational value were to be found at the various booths.

The meetings in the lecture room were of vital importance, over seventy papers being presented. Some of these were presented by the authors themselves, others by title. We intend issuing through our columns some of these papers, so that readers will get the next best thing to an actual visit. A glance over our list previously printed will show the subjects covered in these papers and will emphasize the treat in store.

The amalgamation of the Steel Treaters Research Society and the American Steel Treaters Society took place on the T—day morning, Sept. 14th, and the new body is now known as the American Society for Steel Treating. The joining of these two bodies makes the society even a more important one than before. To go into all the details is impossible at this time, but as soon as the official proceedings are issued, we will give more particulars.

PAINTING IMPLEMENTS
(Continued from page 302.)

that depends on the stone, and the paint being milled.

The various methods of painting are many, varying from the hand to the dip method. In some cases the nature of the pieces necessitate the painting by hand, in order to protect some other portion, for in many cases the parts are painted after partial, or even complete assembly.

Special trucks are used where painting by hand is necessary, these trucks being so constructed that the workman can get at all parts of the piece without any trouble. The number of coats given the piece depends entirely on its duty, and exposure to weather when in use. As a general rule the pieces are varnished in addition to being painted. Wherever possible the work is handled by air hoists, this speeding up the operation considerably. All work is carefully routed, for a delay in any one portion would more than likely hold back subsequent operations.

A good method of handling their product is shown at Fig. 1. In this case the hooks are of individual nature running on a single trolley line as shown. The work being performed is that of varnishing mower frames, and the manner of suspension is self-apparent. By this means the workmen can easily reach all parts to be varnished without undue effort.

The method shown at Fig. 2 is known as the dipping process, and the work being dipped is one of the McCormack Mower Frames. This portion gets two coats of paint and one of varnish, and as can be noted an air hoist is used for handling purposes. The paint in these dip tanks is kept in constant motion, some by screw agitators and others by the use of air, thus ensuring a thoroughly mixed paint.

Ideal ventilation is provided throughout the entire paint building, and every convenience for the workmen has been provided.

BLIND PROVE EFFICIENT
(Continued from page 293.)

Warner & Swasey hand screw machine.
Warner & Swasey turret lathe.
Packing:
    Heel plates.
    Tools.
Packing and sorting metal stampings.
Setting up cartons.
Splitting mica for condensers.
Stacking:
    Laminations.
    Commutators.
Straightening wire brushes.
Taping:
    Field coils.
    High-speed magneto coils.
    Open armature coils.
    Shunt coils.
    Stator coils.
Wrapping and packing:
    Butter.
    Candy.

Galvanizing, What it is and How it is Done

Describing and Illustrating the Process as It is Practised in One of Canada's Leading Industries, together with a Short Story of Its Origin.

By F. H. BELL, Editor Canadian Foundryman

THE subject of galvanizing is one which will be read with interest by many foundrymen, and being a comparatively simple process, and in many respects a part of the foundry business, I will endeavor to describe in a brief way how it is accomplished.

There are two processes by which galvanizing is performed. The original method, and the one from which the art derived its name, is that of an electro-galvanic battery operating on a bath in which zinc anodes are placed along with the articles to be galvanised, and is in many respects similar to electro-plating of any kind. The battery which has since been succeeded by the electric dynamo was originally the work of Aloysius Galvani, an Italian electrician, who was born at Bologna, Italy, in 1737, and died in his native town in 1798. Galvani did not discover the art of galvanizing, but one, Febroni, observing that the zinc in batteries became oxidised in contact with the acidulated water, and adhered to copper or whatever metal it came in contact with, adopted the name of galvanise. Electro-galvanising is still practised on some classes of work, but it is not on this process that we will dwell on this occasion.

While the word "galvanise" rightfully belongs to the genus "electrique," it has become so associated with the process of zinc plating that its original meaning has been forgotten. As a matter of fact it never rightfully belonged to the original process for the reason that Galvani's discoveries and inventions never reached the stage where the zinc pole was used, but his successes gave his contemporary, Alessandro Volta, a foundation upon which to build, and which led up to the discovery of the zinc and copper positive and negative poles, known as the "Voltaic Pile," but usually referred to as the "Galvanic Battery."

The process of galvanizing which I will attempt to describe on this occasion is that of immersing the article to be treated in a vessel of molten zinc. While this may appear to be simply a mechanical process it is in reality a chemical process, although perhaps unknown to the mechanic who performs the operation. In chemistry there are two elements to be considered, one the metal, lic and the other the acid. Different acids act differently in contact with different metals, but we will not go into the chemical aspect of the subject, suffice it to say that the acid and the heat generated by the electric current caused the two metals to adhere to each other in the former process, while the acid and the heat generated by fire cause

the two to adhere to each other in the process about to be described.

In the illustration will be seen a typical galvanizing plant, being that of the Goold, Shapley and Muir Co., Limited, Brantford, Ont. This company manufactures a line of windmills, pumps and sundry other specialties which require galvanizing, and incidentally do an enormous amount of the custom galvanizing which is done throughout Canada. At the extreme right of the illustration and behind the workman is a large wooden vat, a portion of which is shown. This vat is approximately 20 feet long, six feet wide and two feet deep, and contains diluted sulphuric acid. It is connected up to steam pipes for the purpose of heating the acid, not by radiation, but by the direct action of the steam being forced into the acid. The vat, as we have seen, is of wood. The acid would dissolve an iron one, but strange as it may seem, does not affect the wood.

Next to this and in front of the workman is another vat containing muriatic acid.

To the front of the picture and behind the pile of pipes will be seen the end of another vat containing water.

Beyond this will be seen one of the kettles in which the molten zinc is contained, and at the rear left corner of the room will be seen another of the kettles.

At the rear, to the right, will be seen the open doors of the hot air ovens. Thus we have the equipment of a modern galvanizing plant.

The kettles are of electrically welded steel, one and three-quarter inches in thickness on the sides, slightly thicker on the bottom, and flanged at the top. These kettles are built into furnaces with fire doors the entire length. Soft

coke is the fuel used and the metal is kept melted continuously. One of these kettles is long and shallow, being in dimensions 21 feet long, 22 inches wide and 30 inches deep, and with a capacity of approximately 20 tons of melted metal. The other kettle is 5½ feet by 8 feet by 22 inches, and holds approximately 16 tons of metal.

By having two kettles of such different dimensions it is quite possible to galvanize almost any piece which could be offered.

When material is brought to be galvanized, if it is a casting with sand burned into it, the sand should be removed by means of an emery block or file, and if necessary, by the application of hydrofluoric acid, which will dissolve the silica even though burned into glass. A plant where much of this would require to be done would have a vat of hydrofluoric acid, but where the bulk of the work to be done is on sheet metal this is not necessary.

For the regular run of work such as is done at this establishment, the articles to be galvanized are put into the sulphuric acid, where they may remain all night if the acid is not boiled. In the day-time steam is turned into it, keeping the acid in a boiling state. This hastens the action and several lots may be done in one day. When the article shows that all scale has been loosened, it is removed from the acid in a slimy state (from the action of the acid) and is taken to the water vat and washed. From this it is taken to the muriatic acid vat and allowed to remain in this for a short time, when every vestige of the first acid will be removed and the article will be perfectly clean metal with a thin film of chloride of iron on the surface. Muriatic acid is a combination

Continued on page 310

VIEW OF GALVANIZING DEPARTMENT, GOULD, SHAPLEY, & MUIR CO., LTD., BRANTFORD.
(Photo taken by star-light.)

Some Interesting Examples of Gauge Making

Making Gauges is at Any Time a Ticklish Proposition, as Accuracy is the Main Consideration—Herein is Described, from Practical Engineer, How Various Gauges Were Completed.

By A. G. ROBSON, A.M.I.Mech.E.

AN internal cylindrical gauge is referred to in Fig. 1 as a plug gauge. After the steel is cut from a bar it should be annealed and then turned in the lathe, the diameter being made 0.010 in. larger than the finished size of the plug. The surface should be fairly smooth without any irregularities. The plug is made "dead" hard by uniformly heating to a very bright red and suddenly cooling in water; it will then be found to have increased in diameter by 0.002 in. to 0.003 in. A grinding arrangement is now secured in the slide rest of a high-speed lathe, while the plug is placed between the centres. In Fig. 1 the plug is indicated by the letter P, and the emery wheel by EW. H is the holder, FP the fixed pulley, and B the bearings. An additional countershaft is required for driving the emery wheel. The arrows indicate that the directions of rotation of plug and emery wheel are opposite. The bearings must be very carefully fitted and should be examined from time to time. The grinding is done by moving the hand slide rest whilst the plug and wheel rotate, and should be continued until the plug is .001 in. larger than the finished size.

Lapping

The excess of .001 in. on the plug is removed by lapping off, a process which will remove any unevenness which can only be detected by applying the micrometer, and at the same time will produce a smooth, shiny surface. The lap consists of a lead ring (Fig. 2), which is bored to fit just over the plug, but with a portion removed as shown. The lathe being started, the plug rotates whilst the lead ring is held stationary by hand, and oil and fine emery powder are placed in the slot in the ring. The ring will exert a pressure upon the plug and the emery powder and oil will be sucked in around it. The powder tends to become fixed in the lead surface, and consequently the lapping is more effectual. The plug should be tested by the micrometer to see that it is parallel.

A plug of lead is used for lapping the hole, and about .001 in. would be allowed for this operation after grinding. (It is advisable in this case to allow slightly more than .010 in. for grinding. In the case of the internal gauge above described any increase on the diameter due to hardening is an advantage, but any increase on the diameter of an external gauge will result in there being less to grind out of the hole.) The lead plug would be turned to fit the hole and the smaller diameter in Fig. 3 would be secured in the lathe chuck. The lathe is started, and whilst the lead plug revolves the ring gauge is held over the plug to prevent rotation. The ring is gently moved to and fro to secure even lapping. Oil and fine emery powder are placed on the plug, and as the lapping proceeds the powder tends to "pin in" the lead surface. If a large number of holes re-

quire lapping provision must be made to compensate for the wear on the lead plug. It is therefore sawn in four places and a conical hole made in the end. As the plug wears a slight tapping of a steel pin shown in Fig. 3 will spring the lead and increase the diameter.

Plate Gauges

There are many varieties of plate gauges. A very common one is shown in Fig. 4. If a number of these were required, cast steel blanks would first be cut from ⅛ in. plate. After being roughly milled the faces and sides would be cleaned up and the particulars of the gauge stamped upon the face. The letters H and L refer to the high and low sizes respectively. The next operation consists of making "working checks" (working check gauges), which will appear as shown in Figs. 5 and 6. The making of working checks is of great importance. The pieces of plate are filed larger than their sizes, after which they are case-hardened by ferro-cyanide of potassium being placed on the faces and edges when heated, and they are then plunged into water. By washing the gauge in water the powder will easily be removed. The checks should now be lapped. A piece of copper about ⅛ in. or 3-16 in. diameter has two flat surfaces filed on one end. Fine emery powder and oil are then mixed together and the copper lap is placed in the mixture and rubbed on

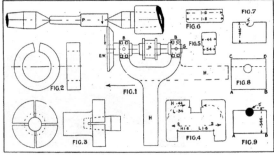

FIG.7
FIG.6
FIG.5
FIG.1
FIG.2
FIG.8
FIG.3
FIG.4
FIG.9

THIS VIEW INCLUDES FIGS. 1 TO 9.

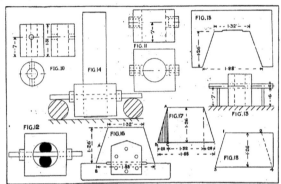

FIG. 10 TO FIG. 18 INCLUSIVE.

the finished faces of the checks. This gives a bright surface and a remarkably good finish. The checks should be occasionally tested by the micrometer during the process of lapping, and a small straight-edge applied to the faces.

Working Limits

Working limits are usually allowed on gauges, so the checks must also be made within these limits. The high measurement is given as 1.8 in. If 0.0003 in. were allowed on this high size then the size of the check should be the mean of this added to 1.8 in.; that is, 1.80015 in., or one and a half ten-thousandths larger than 1.8. The check would also be correct if it were made 1.8003 in., but then it would have to fit tightly in the gauge, as this measurement is the high limit to the high size of the gauge.

In the case of the low measurement, 1.6 in., assuming the same allowances were permitted, this part of the check should be the mean of 0.003 in. deducted from 1.6 in.; that is, 1.59985 in., which is one and a half ten-thousandths less than 1.6 in.

Fitting of Checks

The gauges should now have the checks fitted to them. The first step is to obtain a true surface on one side of the openings shown in Fig. 4 and then work the other faces to suit the checks. It is well to make the checks tight-fitting at this stage prior to case-hardening. Gauges will behave in a very peculiar manner during the process of case-hardening, but it is found that by heating them to the least temperature sufficient to fuse the case-hardening powder there is very little change in size after the gauge is immersed in water. Overheating will prevent the gauge from contracting to its original size and it remains larger than the check gauge. The amount may be as much as 0.002 in. When this is the case and the gauge has to be closed in to fit the check it is hammered in the direction of arrows, 1, Fig. 4. Care must be taken that the high measurements of the gauge are not increased also. Should the gap in a gauge be too small and require enlarging the metal face should be hammered lightly in the direction of arrows 2, Fig. 4, commencing from the centre. Having now made the gauges nearly to fit the checks, the surfaces of the gauges must be lapped by the copper lap until the checks fit easily by hand. Care should be taken to prevent the checks becoming affected by the heat from the hand or by holding them too near the light. Gauges may be made a "proper" fit to a check gauge in the warm atmosphere of the workshop at night, but in the low temperature of the following morning they will be found to have altered. In working to the ten-thousandth part of an inch due regard must be paid to the temperature of the room, and to assist in this direction micrometers should be mounted on wood or metal stands.

External Semi-Circular Gauge

In Fig. 7 an external semi-circular plate gauge is shown. The important parts of this gauge are the size of the hole and the semi-circle. In Fig. 8 a piece of ¼ in. plate A B C D is shown. The face A B C D should first be trued up by a grinding machine. This face should now be placed upon the face of a magnetic chuck, secured to the table of the grinding machine, whilst the opposite face is ground. The position of the hole in the plate is not very important. In Fig. 8 the hole is drilled as shown. As the finished size of the semi-circle is 0.2 in., then the hole in Fig. 8 should be about 0.003 in. smaller than the diameter, 0.4 in. The hole is enlarged so as to be 0.001 in. less than 0.4 in. This 0.001 in. is removed by lapping the hole. The size of the hole is finally determined by a lapped plug tightly fitting in. The plate is then sawn through along the dotted line.

It will be observed that this line is above the centre of the hole. Now the surface A B of the plate is made true by applying a straight-edge to it and removing the high places by an oil stone. Having completed this lower surface, the plug is then inserted as shown in Fig. 9. The micrometer is then applied to ascertain the measurement from this surface to the top of the plug. In Fig. 9 this is shown to be 0.546 in. In order to have a semi-circle in the plate the distance from the centre of the plug to the base will be 0.546 in.—0.1 in. (the radius of the plug) = 0.446 in. The upper face is then filed and finally lapped until the measurement, 0.446 in., is obtained as shown in Fig. 7.

Cross-Hole Gauge

A cross-hole gauge is seen in Fig. 10. It consists of a piece of circular cast steel with a central hole. Two other holes (one on each side) are at right angles to the central hole and also in the same plane. The central hole would

first be drilled and lapped to the correct size; the gauges would then be turned on the outside by fitting a mandrel in the hole. A drill block is then made as shown in Fig. 11. It is a rectangular piece of steel, bored out to suit the outside diameter of the gauges. Two holes are drilled and these are then bushed. In order to test whether these holes are at right angles a rod resting on two micrometer is passed through the bushes (see Fig. 14). Another rod, with a hole through it to permit the first rod passing through, is placed in the block as shown. A square tried each side of the rod will show whether the holes are at right angles. To test the alignment of the holes a round pin has a tapered face filed on it, and by trying this pin as shown in Fig. 12 and measuring the distance it projects through the other side for each position the holes are checked for being in line with each other.

After the cross holes are drilled they should be finished by a rimer, which can also be used for rectifying the holes should they not be in the same plane or at right angles to each other. To finally determine the distance 0.7 in. two "slips" should be made. If the cross hole is 0.2 in. then the distance 0.7 in. is determined as shown in Fig. 13. A tight-fitting 0.2 in. pin is placed through the holes, and the two slips, which should be made 0.6 in., must just touch the pin easily.

A type of gauge introducing a little calculation is seen in Fig. 15. It is a simple plane surface gauge and the small undercut on the inner face is only clearance. It is at once apparent that the check will be to the shape and sizes shown in Fig. 16. It consists of two pieces A and B held together on a back plate by small screws. The piece B requires its upper face to be lapped true. Difficulty with the check will be experienced in making the face A, as the micrometer cannot be used to determine the measurement 1.32 in. The easiest and most reliable means will be to find what the angle A B C in Fig. 17 measures. If this angle is accurately determined the measurement 1.88 could be measured by the micrometer, and then the 1.32 in. should agree when the height is 1.26 in. Looking at the shaded right-angled triangle A B C the base B C is found to be 0.28 in.

Now the tangent of angle A B C $= \dfrac{A C}{B C}$

$= \dfrac{1.26}{0.28} = 4.5.$

Looking at a table of tangents of angles, it will be seen that 4.5 is between the tangent of 77 deg. and 78 deg. The tangents are as follows:

Tan 78 deg. = 4.7046
Tan 77 deg. = 4.3315
∴ Difference for 1 deg. = 0.3731

Now the angle between 77 deg. and 78 deg. must be found. Let this angle be called 77.x deg.

Then Tan 77.x deg. = 4.5
and Tan 77 deg. = 4.3315

∴ Difference for 0.x deg. = .1685
but 0.3731 is difference for 1 deg.

∴ 0.1685 is difference for $\dfrac{0.1685}{0.3731} = 0.45$

∴ 77.x deg. = 77.45 deg.,

which is the angle A B C. The standard protractor in the workshop should be carefully set to this number of degrees. A piece of metal is next obtained and the two opposite faces rubbed down and lapped until they are 1.26 in. apart. The angle on one end, as shown in Fig. 18, should now be obtained by filing, rubbing down, and lapping. When the angle 1 3 4 agrees with the protractor the distance 3 4 should be made a few thousandths of an inch greater than 1.88 in.; then the angle 2 4 3 is obtained by the protractor. The whole face 2 4 should now be rubbed down and lapped until the distance 3 4 is 1.88 in. Care must be taken in checking the angle whilst this is being done, as unevenly distributed pressure upon the plate will cause it to alter. It will facilitate adjustment, and ensure the two faces of the check coming together if each face is undercut as shown in Fig. 16. The back plate is now drilled, and after the holes are marked off upon the two parts of the check these may be drilled. The check must be light-tight to the gauges, and whether this is so can be observed by holding the check and gauge upon a piece of glass up to the light.

GALVANIZING
(Continued from page 307.)

of chlorine and hydrogen, sometimes referred to as hydrochloric acid. This acid not only cleans the surface of the iron but eats into it, and the chlorine combines with the iron, forming the coating referred to. The article is taken in this condition and placed in the oven and dried with this coating on it. When thoroughly dry and preferably when still warm, it is immersed in the melted zinc. When dipping anything long like a pipe or even a flat plate it should be put into the zinc very carefully, one end first, and should be taken from the tank with equal care.

When the melted zinc comes in contact with the chloride of iron it becomes impregnated and the zinc is actually alloyed with the iron through the influence of the chlorine rather than adhering to the surface. The acid, while assisting in the work of securing the zinc to the iron, also attacks the melted zinc as well as the article being galvanized and also to a certain extent the tank in which the metal is melted. Oxidation also takes place and the effect of these various influences is to form a sort of dross or crystallized zinc, and, unlike zinc oxide or iron oxide which would float on the surface, this dross settles to the bottom, and if allowed to remain there would soon cause the ruination of the kettle. To avoid this lead is introduced, and the lead being heavier than the zinc or dross settles down to the bottom and as the dross settles it remains between the zinc

and the lead, and is scooped out with a perforated ladle and pounded into iron molds, from which it is taken and shipped back to the refinery where it is again converted into good zinc.

As we have said a certain amount of oxidation takes place and that some of the oxide was in the dross which settled to the bottom, but as a matter of fact the percentage of oxygen in that material would be small. The bulk of the oxidized zinc gather on the top and is skimmed off. Some of this will be seen in the pile in front of the tank. This would be much more in evidence were it not for the use of flux in the metal. The flux removes the oxygen from the metal and at the same time forms a covering on the melted metal which to some extent prevents it from further oxidation. The zinc is only exposed by skimming when a piece is to be dipped, and only at such a portion as is necessary to allow the dipping to take place. The flux used consists of sal-ammoniac and sawdust, and in some cases glycerine and sawdust. The pile of skimmings seen in the illustration contains a certain amount of good zinc which can be recovered.

As might be imagined, coats are not required in this department. The fellow with the coat on is only a visitor. The gentleman next to him is Mr. Wm. Lusby, who for the last 20 years has been superintendent of this department and knows the galvanizing business from A to Z.

This galvanizing room is but one department of the Goold, Shapley and Muir Co.'s works, and the windmills, etc., which require to be galvanized form only part of the output of this plant. Gas, gasoline and oil engines, stationary and portable, are among their products. They also manufacture a line of pumps, from the smallest kitchen pump to the largest of mining pumps. But the windmill which they have manufactured in unlimited numbers and which was their earliest venture, has gained such a hold on the public mind that the plant is commonly known as "The Windmill."

The company, in their various departments, employ some 300 hands, and were among the first to undertake the art of galvanizing in Canada.

In conclusion I would say that while galvanizing is a comparatively simple art, it represents quite an outlay. The electrically welded steel kettles constitute a big item, and the 30 odd tons of zinc at 18c. per pound, which is the prevailing price, is no small item. This being kept in a molten state year in and year out would consume a considerable quantity of fuel. But such is business.

RAIL BENDING PRESS

The Hydraulic Press Mfg. Co., Mount Gilead, Ohio, have placed on the market a hydraulic rail-bending press which has a capacity of 35 tons. It is of the horizontal type, and mounted on wheels so that it can be readily be moved into the position where it is to be used.

Temperatures for Heat Treating Carbon Steels

It is Well Known the Lowest Temperature Which Will Give the Best Results is the Best Temperature, But What is the Lowest Temperature?—This Question Has Puzzled Many.

By F. L. MOISTER

THE lowest temperature which will give the desired results is the best temperature. This leaves open the question—which is the lowest temperature.

Heat treatment begins with the ingot. The size of the ingot grain is in direct ratio to the rapidity in freezing. Rapid freezing means small grain, slow freezing means larger grain. In practice the ingot is stripped as soon as the walls have solidified enough to permit handling without danger of bleeding and immediately charged into the soaking pit. If the pit temperature is right the heat equalizes and freezing is normal. If too hot, as sometimes occurs, freezing is delayed. The writer is of the opinion that the wide variation in physical properties of steel of same melt and same product is due more to variations in size of ingot grain than to any other cause.

High temperatures and delayed cooling in rolling or forging mean large grain. Low temperature and normal cooling mean fine grain.

Some authorities claim that the lowest temperatures for annealing and hardening are always above recalescence. This is true for annealing, but even in that operation the lowest temperature above recalescence is best, for low annealing temperature with longer time is infinitely better than higher temperature. The low temperature annealed steel will be finer grained and superior in ductility and elasticity.

Consider forgings, both open and die; few forge shops pay much attention to the forging temperature; what they aim at is a soaking heat, the steel highly plastic, so hammer work is reduced to a minimum. The finishing work is often at such high temperatures that the grain grows in cooling and frequently the hot forgings are piled and cooling delayed. Delayed or interrupted cooling from high finishing temperatures induces grain growth.

A glaring instance of this occurred in a shop using their own make of smith's hand tools. A 4 in. flatter broke off at the base before it had been in use one hour. The fractured surface was very coarse grained and showed pronounced evidence of what might be termed "Acute Detail Fracture." The smiths, to a man, called it burned steel. The writer held it to be a case of slow cooling from a high finishing temperature.

The flatter was forged from a 2 in. x 2 in. bar of flatter steel, carbon .82, the bar was cut into lengths, handle hole punched, tapered to enter die, reheated a bright orange and flashed in the die with three blows of a trip hammer. They were piled while very hot and cooling delayed. Such tools were issued to the shop without further treatment.

The shank of the broken tool was quartered longitudinally and two diagonal pieces broken transversely by pressure. The fractures were very coarse crystalline, of pronounced fiery lustre, and there were no signs of bending before occurred.

The other two pieces were heated to 1,450 deg. F, held for some time to allow for lag, and cooled in the air. They were broken in the same manner. The fractures were fine grained, no fiery lustre, and the pieces had bent considerably before rupture occurred. This test proved the writer's contention, that the steel had been injured, not by burning, but by delaying cooling from a high finishing temperature.

Two flatters were taken from the same lot, heated 1,450 deg. F, cooled in the air. Then heated to 1,400 deg. F, and quenched in water. One was drawn to 900 deg. F, and the other to 850 deg. F, and both put on same work as the broken one. The first was in constant use for 26 days, had mushroomed to 3⅛ in. across the hammer end with but one slight crack showing. The second was in use for nine weeks, had curled down at hammer end, no chips had broken from it and it was good for many more weeks' service. An excellent tool.

In forging a cold chisel of carbon .44, the smith will heat the steel cherry red, forge it into shape with light blows and quench in water when the color is a dull red. It will be hard as a diamond but brittle, and drawing back to a straw color will make it tough and hard. A good tool. Had he quenched in water at cherry red the steel would be full of shrinkage cracks and the tool useless. He selected the lowest temperatures for his purposes.

The effect of quenching in water from too high temperatures can be seen by etching the face of most any make of machinists' hammers. The etched surface will show a network of fine shrinkage cracks which cause the steel to spall off in use. Some years ago the writer investigated such hammers which were causing injury to workmen by chips flying from them. Several makes were selected for the test and all but one make, when etched, showed the network of shrinkage cracks. Under an intensified work test—striking the face of an anvil with a full arm blow—some hammers spalled at less than 20 blows, and none showing cracks stood 50 blows. The ones that did not show cracks stood over 100 blows without spalling.

In heat treating locomotive forgings to a specification which called for:

Carbon	.40	.55
Manganese	.40	.70

Silicon	.10	.30
Phosphorus		.05 Max.
Sulphur		.05 "
Chromium		.30 "
Nickel		.40 "

55000 lb. Elastic Limit in forgings up to 4 in. dia. or 2 in. wall,
90000 lb. Tensile Strength in forgings up to 4 in. dia. or 2 in. wall,
50000 lb. Elastic Limit in forgings over 4 in. dia. or 3½ in. wall,
85000 lb. Tensile Strength in forgings over 4 in. dia. or 3½ in. wall.
20.5 per cent. min., elongation in 2 in.

F, hold for heat to be equalised, and allow for lag, cool in air, with forgings separated so as not to delay cooling, put the steel in excellent condition for heat treating. The forgings were made by several smiths, observation showed widely varying finishing temperatures, and therefore the annealing temperature had to be held long enough to correct the grain in the hottest finished ones.

Heat slowly to 1450 deg. F, hold at that point till thoroughly heated, and quench in freely circulating water. Draw 950 deg. F. to 1000 deg. F. if 55000 lbs. Elastic Limit is required, 1000 deg. F. to 1050 deg. F. if 50000 lbs. Elastic Limit is required.

Temperature varying for carbon content and section of forging.

Tests made from such forgings gave perfect results. All forgings were accepted. The elongation and reductions were in excess of requirements and the fractures were fine, dense gray lipped.

Forgings quenched in oil from higher temperatures, when tested, did not have as fine grain; high elasticity or ductility. The fractures were not as regular and did not have the clear silky appearance of the water quenched.

A fairly wide range of quenching temperatures is possible, if tensile qualities only are desired, and the steel can be made reasonably uniform by varying the draw. But if tensile and shock resisting qualities are desired, the quench should be,as low as possible so the grain will be fine and the draw low. The steel will have equal ductility, greater elasticity and thus superior shock resisting qualities.

Grain change above recalescence is not instantaneous. Time is as important as temperature. There is a definite grain for a definite temperature. Assuming a 1700 deg. F. grain steel; it will take more time to refine such a grain down to a 1450 deg. F. grain than to 1600 deg. F. grain.

This brings us back to the original premise: The lowest possible temperature is the best temperature.

In all cases the grain mentioned is the MACROGRAIN.

WHAT OUR READERS THINK AND DO

SOME HANDY HINTS
By Fred Horner

The first scheme is that depicted at A. The illustration shows the use of a dog on an arbor, and this idea is very useful under certain conditions, such as where the work is too delicate to stand hard forcing, or the bore is a trifle too large to give a good driving fit on the arbor. Provided the work has arms or holes through which the tail of a dog may be passed, the latter can be gripped on to the mandrel (with strips of brass or copper for protection) and applied to carry the work round as depicted.

Spot Marking

The second hint is shown at B. This illustrates the marking of spots on opposite sides of round work.

There are many ways of getting locations at exact opposite sides of a shaft. The marking may be done with vee blocks and surface-gauge, or in the lathe, or with one of the special double centering punches. As an emergency dodge, in the absence of all these facilities, it is possible to secure indications by means of a toolmaker's clamp, gripping it lightly upon the shaft and moving it through, with a slight motion sufficient to make bright spots that may be centerpopped. Chalk or whiting can be applied to the shaft first if preferred. The same method may be pursued with a slide gauge, or even a pair of ordinary calipers if passed over the shaft carefully.

The Third Idea

The last idea, shown at C, is a dodge for setting and maintaining a thin wash-er square on an ordinary arbor. A toolmaker's clamp is gripped on the arbor behind the washer, as seen in the face view, so that it can be noted whether the washer is square across. The clamp backs up and sustains the washer against the pressure of the facing tool. This hint is necessary, because the bore of a thin washer is too short to ensure locating the work square upon the arbor.

THE MACHINING ALLOWANCE
By G. Blake

In these days of mass production and the consequent cutting down of all unnecessary expense, some firms are inclined to err on the allowance for machining on castings, stampings, forgings, etc. In many cases there is not sufficient metal left for machining up, with the result that the saving in weight and cost of metal is very small compared with the cost and upkeep of tools. For instance, the cost of tool grinding is far higher than it should be on account of the scanty amount of metal taken off in the machining. As the tool is only taking off the black skin it very soon puts a dull edge on the best high-speed steel tools. When the allowance is increased, i.e., more metal left on to be machined, the tool stands up better, requires less grinding, and lasts longer, and above all makes a better finish.

Of course where there is a pickling and sand blasting plant the allowance can be cut down somewhat, but the difference is so small that it is of no account in the final cost of the job. The pickling and sand blasting plant saves trouble in the way of tools, and it is a matter of wonder to the writer why so few firms are in possession of such an outfit.

Another point brought before the writer's notice some time ago in the motor trade was the number of small forgings, castings, and hot stampings having cored holes which might well have been left solid as it cost more to open the holes out to size than if they had been solid. This was on account of extra drill and cutter grinding involved.

Another point is the normalizing and annealing of all castings, forgings, stampings, etc. Much can be done in this way to overcome machining costs. Of course this applies more to the small firms as most of the large concerns have their own foundry, forges and drop-stamping department. With small firms it is very essential to insist that all castings, forgings, and stampings supplied from outside makes be thoroughly annealed. In some cases a Brinell test is misleading as very often hard spots are encountered in the machining.

KEEPING THE PLANER BUSY
By Donald H. Hampson.

Machine shops that have planers over thirty-inch frequently find that it is hard to keep them busy more than a quarter of the time. These shops know the facilities of the other shops in the region and they know the requirements of the manufacturing plants in the same territory from this knowledge, they can see no way to keep the planer busier than it is. To stand idle three-fourths of the time is to have

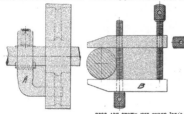

HERE ARE SHOWN THE THREE IDEAS

so much capital idle — not to speak of the room the planer takes up. An idle planer invariably becomes the target of all the odds and ends conceivable, so that when a job does come along it takes a machinist and the floor boy about an hour to dig away the debris.

The printing industry offers work for idle planers in all well populated districts. Printers do not go around hunting machinists to do planer work — the machinists have got to solicit the work, in fact, they have got to call the printers' attention to the need of having the work done. Once the need is shown, the printer will consent quickly enough. There was a time when printers were the most forlorn and run down business people on the list—and this accounts for the presence of vast quantities of second-hand machinery in their plants that needs repairs. But thanks to the Standard Cost System and the Ben Franklin Clubs, the printing industry is on a well paying basis to-day and printers' credit is good at the banks. When it comes to figuring costs on a job, printers are ten years in advance of most machine shops—their "Taylor-White" systems go right down to the two-man establishments and show them everything exactly to a penny.

Returning to planer work. Job presses, also called "platen presses," have two cast iron plates flat on one side and ribbed on the other like a surface plate. Between these two flat surfaces the printing is done. In time the surfaces get worn so that to get an even impression on the "low" spots, the pressman must stop and paper up by trial amounts to turn out good work. The amount of time he spends in a week on such tinkering would pay for the re-planing of the two castings.

The same is true of cylinder presses. Oftentimes the re-planing of the bed of a cylinder press, including a cut over the ways, will make an old press do work almost as good as a new one. These castings may be as large as 40 in. x 60 in. and weigh a thousand pounds, whereas the job press castings are found as small as 10 in. x 12 in., and of only 50 pounds weight.

Paper cutting machines also depend upon the planer work for the accuracy of the product they turn out. The table of the machine is a flat casting with light ribs beneath; it gets some wear and the seasoning of time coupled with standing on uneven floors frequently get them an eighth of an inch out. There is a slot about an inch wide and the full length which guides what is known as the "back gauge" (a part corresponding to the gauge on a sheet metal squaring shears.) This slot gets badly worn; in that condition it is safe to say that it costs the average printer a dollar a week for spoiled work and time spent in dickering with it. Several other parts subject to wear are planer parts.

Then the proof presses come in for a share of attention. These are of two types—one in which a cylinder travels (rolls) along a track at a fixed distance

from a planed flat cast iron plate and the other, the old-fashioned Washington hand press, which has two plates similar to a job press and which are caused to approach each other by hand operation, giving the impression in that way. Both types of proof presses are used in some cases for actual printing of commercial work as well as for proving (testing) cuts, advertising matter, set up pages, etc. The proof press must be more accurate than the power presses or their test of the work is void. The worn or bruised plates require a light cut over them to become as good as new.

The above mentioned are the most common examples of planer work to be secured in printing offices. Others might be mentioned according to the extent of the equipment. In many cases there is other work that must be done along with the planer cutting, in others the newly planed surface with a little adjusting will suffice. It is doubtful if any printer would refuse to have such repairs made after his attention had been called to their need and to the shop's facilities for handling the work.

A visit to the shop of any manufacturer of presses or folders or cutters or other printers' equipment will show at once the pre-eminence of the planer as a machine tool on such work, and if any shop with idle planers is within reasonable distance of one of these plants, they have only to qualify to obtain plenty of extra work, for with planers at $8,000 and up, they are scarce even with the shops that have ample capital.

SIMPLE PIN CUTTER
By Tyke

Having received an order for one thousand small pins ⅝ of an inch long and .068 inch in diameter, to be made of drill rod, the question as to how these were to be cut off accurately, quickly, and with reasonably square ends, was gone into. After a little consideration the following method was adopted, which gave every satisfaction as to accuracy, and proved to be exceedingly cheap and rapid. The accompanying sketch is virtually self-explanatory, XX being the jaws of the shaper vise, B a tool steel block drilled a few thousandths larger than the drill rod used, and hardened at the drilled end only. C is a machine

steel block with a step ¾ x ⅝ inch milled at one end. D is a tool steel tool grooved as shown in order to eliminate producing a flat on one side of the wire when shearing. This tool was given the merest suggestion of clearance and suitably hardened. The entire device was now ready to use; C and B were firmly held in the shaper vise, as shown, and the tool properly located in the caliper box, barely touching the tool steel block S. A length of drill rod was inserted in the hole at Y and advanced untiil it touched the opposite face of the piece C. The stroke was set at 3 inches and the speed about 60 per minute. The whole thousand pins were cut off in twenty minutes and the tool was good for several thousand more. This simple method more than paid for itself, and with slight modifications could be made adjustable, such as a screw stop in C and different size holes in B, thus taking care of larger or smaller pins or varying lengths.

HINTS ON USING NEW TOOLS
By G. Blake

To my mind these are precautions which should be observed in using new tools such as drills, reamers, broaches, etc. Many tools are broken through the operator's ignorance and neglect. A new drill, reamer or broach has been drawn from the tool store and smashed in a few seconds, especially in cold weather. Most tools give best results when slightly warmed.

Say in the case of a new drill, the operator will fix it in the socket, start up the machine, and put the feed on right away. The result is that the sudden shock on the new drill causes the drill to break. When using new drills over ½ inch diameter, reamers or broaches, the writer recommends that they be warmed by dipping in hot oil. In the case of drills, you should use a light feed for the first few holes.

The same applies to broaches, which, if treated in the above manner, will amply repay the trouble and time spent. Of course this is not necessary in very hot weather, but it is essential in cold climates and during frosts. It has been observed during very severe frosts that machine tools, etc, are very easily broken.

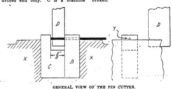

GENERAL VIEW OF THE PIN CUTTER.

DEVELOPMENTS IN SHOP EQUIPMENT

COAL OIL LAMPS

The Standard Meter Co., Ltd., 10 Morrow Ave., Toronto, Canada, have placed on the market a very interesting line of lamps which are described herewith. At Fig. 1 is shown what is known as the Monitor high power coal oil brazing lamp. This lamp is used for brazing

FIG. 1—SHOWING BRAZING LAMP.

and work of similar nature on a large, scale and can be used in any position. It is said to be both portable and powerful. They are made in 2, 4, and 5 pint sizes.

The next type of lamp shown at Fig. 2 is known as the Monitor high power coal oil blow lamp. This style of lamp is both wind and rain proof and is used extensively by millrights, plumbers,

FIG. 2—SHOWING BLOW LAMP.

electricians and hot water engineers. This lamp is made in 2-3 pint, 1 pint, and 2 pint sizes.

Another type of lamp carried by this

concern is known as the Petropark high power inverted incandescent coal oil lamp. The latter style lamp is used as portable light for country railroad stations, farm buildings, fair grounds, large workshops, wharves and other innumerable places where portable lights are desired.

TOOL STANDS, RACKS AND STOOLS

The old days when a manager would discharge a man for sitting down to his work have passed. It has been proved that for certain classes of work the workman can become more effective by sitting, and the same holds good insomuch that men are no longer allowed

VIEW OF THE STOOL AND THE TOOL STAND.

to waste time running all over the shop to find tools. Nowadays tool stands, drawers and benches are always handy to the various machines, or at least such is the case in the real up-to-the-minute plants.

The National Engineering Co., Ltd., Sarnia, Canada, are firm believers in all the above statements and to follow out their belief have placed on the market a varied line of time savers. These include tool stands of various nature and sizes, some being supplied with drawers

and some without. They also market an all steel storage rack which can be built in any number of sections and combinations for use against a wall, or placed back to back. Such racks are especially useful in warehouses, or in the cut off department.

In addition to the above mentioned lines they make a shop stool of novel design. This stool is depicted in the illustration standing alongside one of the tool stands as made by this concern. The stool is so simple, yet efficient, that it requires no explanation except that it is made in three heights, 24-ins, 27-ins. and 30-ins.

The A. R. Williams Machinery Co., Toronto, are handling this line and will

be pleased to give all information regarding the same upon application.

METAL CUTTING MACHINE

The Yoder Co., Cleveland, Ohio, have issued a machine for use in cutting sheet metal into strips suitable for various purposes. It is claimed that this equipment provides a rapid means of performing the operation and that the work is done in a way that insures parallelism of the edges. Machines of this type

are made in several sizes to accommodate various widths and thicknesses of metal.

TURRET CHUCK

The Scully-Jones & Co., 647 Railway Exchange Building, Chicago, Ill., have placed on the market what is known as a "Wear-ever" turret chuck. This chuck is of the same general construction as other chucks of this company's manufacture, but is designed to hold square-end tap shanks or other tools with shanks of the same form. Various sizes of chucks are made to fit different diameters of tap shanks, and standard bushings to be furnished to allow a chuck to be used in large turret holes.

SHEET TINNING MACHINE

The Aetna Foundry & Machine Co., Warren, Ohio, have placed on the market a machine designed for use in tinning heavy sheets used in the manufacture of milk-can bodies. There are three pairs of rolls, the lower pair acting as feed-rolls, the second pair, which run in a bath of tin, acting as spreaders, and the upper pair of rolls, which run in an oil bath, removing the excess tin, and leaving the sheets with the desired coating and finish.

AIR OPERATED CHUCKS

The Logansport Machine Co., Logansport, Inc., have issued a new line of air chucks with a duplex control mechanism which allows them to be used on double spindle turret lathes built by the Jones & Lamson Machine Co., of Springfield, Vt. Frank G. Payson Co., 9 S. Clinton St., Chicago, Ill., is the selling agent.

PORTABLE PROJECTOR

The Graphoscope Development Co., 50 E. 42nd St., New York City, have issued a moving picture apparatus especially designed for the use of salesmen. This complete unit, ready for operation, is packed in a case of about the same size as an ordinary suitcase. It is equipped with a motor drive and is of simple construction. The machine uses standard films and will project a picture 16 feet square at a distance of 35 feet.

U. S. SUB-PRESS

The U.S. Tool Company, of Newark, N.J., are now manufacturing sub-presses for various classes of stamped work. Stock sizes with platens of 2 by 3 inches and 6 x 8 inches are carried, and special sizes for particular work may be obtained.

STANDARDIZATION OF MATERIALS

For the first time in Australia, systematic and comprehensive steps are being taken for the standardization of various materials used in industry. This question was taken up some time ago, quotes the "Commonwealth Engineer," by the Institute of Science and Industry, and the first Australian standard specification for the Commonwealth, relating to steel, used for structural purposes such as angles, tees, beams and channels, has now been published. This is the first of a number of Australian standard specifications which will be issued by the institute, which has already brought into line the views of both manufacturers and users as to uniform standards for not only structural steel, but also of railway rails and fishplates and tramway rails and fishplates. Standard specifications for the two latter will be published by the institute shortly.

AIR PLANER

The Turbine Air Tool Co., 710 Huron Road, Cleveland, Ohio, have placed on the market a line of air planers which are adapted for such uses as shaving telegraph poles, telephone poles, etc. They are also used for railroad tie making, pile peeling, removing bark from logs for foundation building, and general dock, wharf, and dam building. There are of course many other uses for such a type of tool but let us go into a description of the same.

As can be noted light weight has been considered. The device has only thirteen primary parts, and has no reciprocal parts. The only wearing portions are the ball bearings and the only parts requiring replacement are the knives and ball bearings. The work of making these replacements is exceedingly simple.

The power used is compressed air, and the speed is from nothing to 12,000 r.p.m. at the knife. The model P

FIG. 1.—SHOWING MODEL P PLANER.

shown at Fig. 1 should be operated from an air compressor with 100 pounds pressure, giving 45 cubic feet of free air per minute.

The cost of maintenance is said to be

FIG. 2.—ILLUSTRATING MODEL D.

negligible and the machine is easily kept in working order. The model D shown at Fig. 2 is similar to Fig. 1, with the

exception that it is designed primarily for work on flat surfaces.

AIR OR STEAM ENGINE

The H. B. Underwood & Co., 1015, 1025 Hamilton Street, Philadelphia, Pa., have placed on the market what are

GENERAL VIEW OF THE ENGINE.

known as their 3 and 5 h.p. steam or air engines. These engines are built in the two sizes above mentioned, and the illustration herewith gives one a good idea of their general appearance. They are used and designed for driving portable tools such as boring bars, valve seat rotary planers, drilling machines, etc., and are constructed to use either compressed air or steam.

All working parts are enclosed in an oil tight case, which is formed in the base of the engine, making it dust proof, and reducing wear to a minimum. The cylinder of the 3 h.p. engine is 3½-inches diameter and of the 5 h.p., 4½-inches diameter, both engines having 3 1-2-inch stroke. The design of the piston valve used is said to reduce friction and wear to a minimum.

The crankshaft is a solid forging, finished 1½-inches diameter. The engine is well balanced, uses 70 to 125 pound pressure, and runs at 250 r.p.m. This engine should particularly interest railroad shops.

The floor space required is 12-in.x 21-inches, and the height from base to extreme top of governor 46 inches. The net weight is 270 pounds.

The MacLean Publishing Company
LIMITED
(ESTABLISHED 1887)
JOHN BAYNE MACLEAN, President. H. T. HUNTER, Vice-President
H. V. TYRRELL, General Manager.

PUBLISHERS OF

CANADIAN MACHINERY
MANUFACTURING NEWS

A weekly journal devoted to the machinery and manufacturing interests.

B. G. NEWTON, Manager.   A. R. KENNEDY, Managing Editor.

Associate Editors:
J. H. MOORE   T. H. FENNER   J. H. RODGERS (Montreal)
Office of Publication: 143-153 University Avenue, Toronto, Ontario.

VOL. XXIV.   TORONTO, SEPTEMBER 30, 1920   NO. 14

Steel Market Still High

THE premium market may crack, but the steel market in general is not on the skids.

The premium market is a peculiar sort of a creation. If you desire you can refer to the operators in it as a lot of gougers, but if you take the trouble to sit back and think it over a bit, you may find that the frenzied bidding that was done by some lines of industry for available material had much to do with the creation and the upkeep of the premium market.

The automobile business has done more than any other line to build up a premium market in sheets. There were stove makers who, months ago, parted with their stocks of sheets, letting car builders have them at prices that were larger than the profits that would have been made by putting these sheets into stoves.

When any man has a group of buyers around him bidding for something he has for sale, the chances are he to one that the price is going to go up. The buyers made the premium market in the first place, and the mills are showing no tendency to let the profitable thing die on their hands as long as they can pulmotor an ounce of steam into its system.

The Canadian trade stands to benefit by a better supply of steel coming from the mills of United States. Stocks are low, as buying has been done on a very limited scale.

Buyers may get some relief shortly. It is coming to them that any reduction shall be passed on to them at once, for most of them have been buying from the premium mills and paying the long price.

Heat Treating of Steel

THE convention and exhibition of Steel Treaters held during the week of Sept. 14 to 18 at Philadelphia marks an important step in metallurgical progress. It is only a matter of two years since this organization came into being, yet in that short time they have obtained a membership in the neighborhood of 4,000, and local chapters are now to be found in practically all cities interested in the treatment of steel.

The need of development in the steel treating industry has been felt for some time, and the society is doing a

splendid work in this regard. The demand for quality steel is now a common one, and steel treatment is well known to bring out the best qualities in the steel.

The heat treatment of steel is no longer considered a fad. It is a recognized science. The eye is not sufficient these days to judge or guess at the correct heat. Up-to-the-minute apparatus is employed, such as optical pyrometers and testing equipment of all kinds. The age of scientific steel treatment is here to stay.

Toronto has so far the honor of being the only Canadian Chapter, but in the next few years we will, no doubt, see other Canadian branches arising.

The Strike of Iron Workers

THE argument has been put forward many times that union workmen are out for the benefit of all the workers.

The strength of concerted action is supposed to help the little man who otherwise would have a hard time going up against some of the large combinations of capital and talent.

Just now iron workers in Toronto are striking for recognition of their union and $1 per hour as their rate of wages. They have left several large jobs, and are already tieing up the work.

The secretary of the Building Trades Council has this remark:—

"The serious feature of it is that the roofs of many structures will not be on by the time winter sets in if the strike is allowed to continue. No inside work could be done during the winter."

That is plain enough. The strike of the steel workers for an increased wage and recognition of their own union may mean that tradesmen in half a dozen other lines shall be kept from securing employment on these buildings during the winter months.

This is a point where unions have failed to show the consideration for others that one might reasonably expect.

Do Boilers in Sinking Ships Explode?

DID you ever read about the boilers in a sinking ship exploding? The average man has neither been ship-wrecked, nor has he been gazing on the boilers of a sinking ship to see just what does happen. And further, most people who happen to have gone down, lingering in the vicinity of the boiler in a sinking ship, have not returned to make out a report on what really did happen.

Mr. T. H. Fenner, editor of Power House, and a close student of such matters, claims the old theory is all wrong. In a recent issue of his paper he says:—

"In a newspaper report of a recent disaster on the Great Lakes, the statement is made that as the vessel was sinking, the boiler exploded. This is not an unusual statement, but we venture to say it has little but the reporter's imagination to back it up. Is there any good reason why a boiler should explode on a sinking ship? The thoughtless individual usually jumps to the hard-worked explanation of the sudden contraction and lets it go at that. What, then, does this sudden contraction perform? Does the shell contract so much that the pressure is increased beyond the resisting strength of the metal? Most assuredly not. The sudden contraction theory does not hold water. The fact is that the boilers do not explode when the ship is sinking. If they did, it would be of little use trying to salvage sunken steamers, as they would be damaged to such an extent that salvage would be impossible. Of all the steamers sunk by mine and torpedo during the war, were there any cases of boiler explosions unless the boiler shell was actually ruptured by a piece of shell, or other projectile? As a matter of fact, the boilers on steamers sunk and afterwards salvaged are usually found intact and fit for use again. This idea of the boilers on a sinking ship exploding is one that has been fostered by non-technical writers in the daily press, and accepted without question by men who should know better, and who would recognize the error of the statement if they gave the matter a little thought."

Price Tendencies Much Discussed This Week

WHEN the announcement was made from Detroit a few days ago that the Ford Company had cut deeply into the price of cars, the question at once came up: What will the makers of cars in Canada do to meet this?

It creates a very real situation. For some months past the automobile industry in Canada has been doing the great bulk of the buying in the machine tool market. They have been speeding up production, and there has been a market for their cars. There has been a lot of money in the country, and makers were able to pass on increasing costs to the consumer.

To-day the makers of motor cars are not speeding up production. They are making what cars they can sell in the present market. Were they to do otherwise they would be ruined. All that would be necessary to accomplish this would be to go ahead and fill up warehouses and agency branches with high cost cars, and then sell them in a declining market. One automobile maker, discussing the matter with Canadian Machinery this week, stated:

"The move of Henry Ford certainly puts it up to us in such a way that there is no use trying to dodge the issue. We will have to face it. I have no doubt but that our salesmen in the farthest corners of the country were asked that question the first morning the Ford cut was announced: What is your firm going to do about it? You can take it for granted that the makers of cars did not put up prices this year because they wanted to. The buying public were expecting a move in the opposite direction and there is nothing that would have been better for the automobile business. But what happened? We were in a very high market for everything that went into a car, and when we announced an increase in price we did the very same thing that other makers did, acted because our cost sheets pointed the way we must take. It was either a case of lower the cost of the car or get more for it. When this condition came about there was only one thing to do, viz., keep production well in hand with immediate sales.

"Now then, for the lowering of costs in Canada. That is your question, and my reply would be off-hand that if selling prices go down then everything must crack: the steel market, the wages, the dividends, and general commodities. There will undoubtedly be an announcement made in this matter soon, but the bolt—and it is a real bolt—came so suddenly that the situation cannot be figured out in a day or so."

The Implement Manufacturers

Another of the fields that is already feeling the demand for a lower price is the plow shops. As a general thing these places are working a year in advance. That is, they are buying now for what they will make in the last part of 1921 and the first of 1922. Several of these firms have warehouses pretty well filled up with high-cost product, and they are anxious naturally to speed up their sales in order to get the stock out before a lower price becomes recognized by the trade. The implement business recognizes that the rupture that has taken place in the price equilibrium of the automobile market is bound to have an effect on the whole metal working field, especially since Ford has cut into the price of tractors as well as touring cars.

Welcomes the Move

Here is another idea of the president of a firm doing particularly fine work in the mechanical field:

"The move that is taking place now and which was started by Henry Ford, is right. It should have been made a long time ago, and I hope it will spread all through our industrial life. I am not afraid of it in the least, for fundamentally we are sound. The trouble is we have been joking ourselves too long. There are actually people who believe that all that is necessary to have good times is to keep on printing paper money. I sat in at a conference of manufacturers and bankers two days after the signing of the armistice, and we discussed this very thing for hours. At that time I wanted a readjustment and I have never seen anything since to make me change my views. For two years now we have been fooling ourselves. We have been like the squirrel in the cage that chases his tail. It has been higher prices, higher wages, higher prices, higher wages until we have got to the point where there is no stability in anything. We are paying out a lot of money to men and when they go out in the markets to buy they find they have not so very much after all.

"So what have we? On one hand we have the workers and the wage-earners calling for more money, we have the men selling raw materials to us calling for more money, we have the men who are going to build a factory addition calling for more money, and every person who does anything is calling for more money, and in most cases getting it. On the other hand, we have the bankers, who are trying their best to save the credit of the country, working in the opposite direction, trying to keep down credits, and make the amount of business conform to the amount of money we had in circulation. There are too many fine adjustments in connection with economics and finance that we have overlooked, or if we have not overlooked them we have not had the courage to face them. As a result of all this we have the man in business having to get in touch with about two and a half times as much money as he did in 1914 to finance a given volume of trade and to take care of the decreased output of the men in the shop. The thing cannot go on as it is.

"When the war was over we should have done generally what Henry Ford has now set out to do. People were expecting something to happen then. We all knew we had been living on war tension. Men knew they were making war wages, and they expected that they would drop with the close of the war. Men who had put money into war ventures made big returns, and they never expected them to continue after the war. Commodity prices never had a good reason for staying up except for the artificial bolstering that has been put under them. I contended right after the war was the time to take our bumps, and I have never seen anything since then that has caused me to change my mind. I could get the backing of only two men at the conference I mentioned before to such a course. The rest were afraid of that thing they call unrest, and feared the consequences of an effort to let things take their own natural course in the process of readjustment. We were expecting something to happen right after the war, and right after the war was the time when this something should have happened that they are trying to bring about to-day.

"Now it is here, at least to my mind there is nothing more certain than that prices are going to crack. When we get some of the big leaders setting the pace in that direction the movement is on, no matter if it is in United States, because the markets here directly reflect any influence that is set in motion there. As I said before, our firm does not fear the bust, the crack, the explosion, or whatever else you care to call it. It will simply mean that we can get more money, that the money we get will finance more business, and we will be nearer our proper position having regard to the gold standard, and the value of our money in the foreign markets."

MARKET
DEVELOPMENTS

Prices Not Tumbling in Steel or Tools

Business Being Placed at To-day's Prices for Six Months to Come
—New York Reports That Machine Market There Does Not Look
for Improvement Before First of the Year.

TALK of price cutting in the iron, steel and machine tool market finds no reflection in fact. It is safe to presume that the premium steel market is due for a setback, and this may be brought about by the attitude of buyers themselves rather than by any desire on the part of the mills to have cheaper steel. Buyers, by their persistence in bidding for material, have had much to do with keeping the premium mills busy. This is particularly true of the sheet mills.

A pretty good indication of what dealers think of the chances for lower prices can be secured from the attitude of the warehouses. One firm this week, in the face of price lowering talk, placed an order for tonnage that will take care of his business for the next six months; this was in bar iron, which is to-day selling at 5.75 per pound,

a record price for this district. This dealer cannot see easier steel prices for some time, and he is willing to back his opinion with his own money.

Machine tool dealers have nothing on which they can announce a lower price. Makers speak of increasing transportation costs of labor increases, and they see no relief in all the things they put into their machines. Some of the schedules are still some weeks behind. The buyers of tools figure that there is more real selling going on now than for months past and from past experience they believe lower prices may come. Dealers say nothing to encourage this view.

The scrap metal market is very dull. The large buyers are not in the market yet, and until they come trading is going to be confined to small lots.

MACHINE TOOL BUSINESS BELOW
NORMAL IN MONTREAL MARKET

Special to CANADIAN MACHINERY.

MONTREAL, Que., Sept. 27—Business appears to be reaching another stage in its history when uncertainty will be the ruling factor, and this will result in a further policy of conservatism in the buying attitude of consumers. This feature has been more or less pronounced during the past year, but the belief that the present is an opportune time for a determined start to be made in peace time adjustment seems to be taking hold of many of those directly associated with industrial enterprises. This trend of thought has been developed in seclusion and will likely be gradually impressed in a more concrete form from now on.

Steel Market Unsteady

While there is no visible evidence to substantiate the statement, the early trend of prices in many steel commodities will probably be in a downward direction. Dealers here are not looking forward to any drastic reduction in prices, but there are very few but what believe that before the close of another month the conditions will warrant a decline on existing quotations. Observations by dealers' representatives, who have been in the States recently, state that the mill congestion is much less marked at present, owing to the better

facilities of the railroads and the improved movement of freight of every description. This condition has enabled producers to operate to increased advantage and the greater output will, no doubt, be felt in trade circles in the near future. The sheet situation is far from normal, but the lessening activity of the motor industry has created an increased supply of sheets that has materially assisted in bringing relief in other directions, notably to stove and car manufacturers. Consumers, apparently, are better pleased with the way material is coming to hand, and dealers are being freed from the necessity of urging heavier shipments from the mills. This may be due to several causes, among which may be mentioned a visible decline of general industrial activity, and the fact that builders and manufacturers are assuming a waiting attitude in the expectation that prices are coming down. Such a situation may develop, but there is little to support a belief that any such movement will affect, to any great extent, the present prevailing prices. It is true that production is on the increase, but this is primarily due to the more regular operation of the mills, the cost of producing being generally maintained by the high character of wages.

Machine Tools Quieter

Despite the occasional announcement that machine tool business is good. the general impression gathered from reports from dealers is to the effect that present activity, while showing slight improvement over that of the midsummer months, is, nevertheless, below the normal for this period of the year. No one branch of industry seems to be responsible for this condition as orders are widely distributed and general in character. The aggregate business, however, is reasonably well maintained by the steady supply business that is practically unaffected by the noticeable trend to industrial curtailment. Even here, however, the buying is of a conservative nature and it would not be surprising to see lighter buying in the near future.

Metals Unsteady

The metal market has not been featured by any outstanding development but has been subject to slight fluctuations in price quotations. Coppers have been moving normally with prices well maintained, conditions tending to weaker quotations. Tin has shown a little unsteadiness owing to the uncertain industrial conditions in England, but apart from disturbing the even tone of the local market, the English situation has influenced prices here but slightly. Present quotations are on the nominal basis of 54 cents per lb. Spelter is steady at 10¾ cents, while lead is easier at 9

cents per lb. Antimony is in good demand at 9¼ cents per lb.

Scrap Continues Quiet

"We are here for the sake of appearances. Trade is very quiet and the buying policy of the consumer is not a stimulant to greater business. The market is not strong, neither is it weak, but it is very quiet. Price is little inducement in making sales. Material is bought only when it is needed." This statement outlines the old material situation and is a good indication of the dull state of the situation generally. Foundry iron and steel scraps, as they form the heaviest movers.

CAN SEE NO CRACK
IN PRICES HERE

Some Steel Dealers Are Buying at To-day's Prices For Six Months to Come

TORONTO.—The talk of price cutting is something that one hears much of in the markets to-day, but signs to show where it has actually visited the steel, iron and machinery market are lacking. It might be expected that steel warehouses that carry on a merchant business would be first to respond to price reductions, were any seriously in sight. No evidence of such a move can be found.

Machinery dealers claim that they have heard nothing from their principals, and until they do there is absolutely no chance of them making any reductions, as they for the most part simply sell the machines at a fixed rate of profit.

Some of the dealers in the city have sold a lot of machine tools during the past week. There are no large orders in the market, but a number of smaller deals have been put through. The trade does not expect much from the automobile industry just now, realising that they are not likely to go ahead with much rapidity in face of the recent price slashing in certain makes of cars.

"Our deliveries are coming through very well," said one dealer, "but we are still some weeks behind in many lines. I can't see where prices are going to crack, although I will admit there is more or less of a demand for such a move on the part of the buyers. I do not see any reduction yet in iron, steel or wages, nor yet in transportation costs. The material the makers are working out now is all high price stuff, and I cannot see, unless something very unexpected happens, why or how we can have cheaper machine tools for some time to come."

Other dealers hold views along this line. They all admit that it would be a nice thing to be authorized to announce a 10 or 20 per cent. reduction, but they are all equally emphatic in stating that they do not see how it can come about yet.

POINTS IN WEEK'S
MARKETING NOTES

Some of the independent steel producers in the Pittsburgh district admit that they expect their prices to fall.

It is claimed that most of the U.S. automobiles that have been placed on the market at high prices have been manufactured from steel bought at the March 21, 1919, War Board schedule, and that the prices paid for the steel did not warrant the prices asked for cars.

There is very little buying being done in the United States steel markets. Deliveries have been quite rapid in the last few weeks and the amount of steel piled at the mills has been rapidly decreasing.

The talk of price cutting in the steel, iron or machine tool markets finds no reflection in actual reduction. Dealers in Toronto say they have no word from their principals in regard to price changes.

One Toronto warehousing firm this week placed orders at present prices for their supply of bar iron for the next six months, showing that they do not anticipate any reduction in price during that period.

The scrap metal market in Canada is a still and uninteresting affair. Larger buyers such as the steel mills are not in the market and little improvement is looked for until there is something definite done in the way of stabilizing the prices.

The Steel Market

Here is the opinion of one of the leading steel merchants in Toronto, and it covers the case pretty well:

"We look to see some of the premium prices taken off first, and at the same time it is likely that prices of the Steel Corporation will go up. Of course I have no knowledge of the affairs of the Steel Corporation, but I have been through the steel country lately and that seems to be the generally accepted opinion there. Now we are not looking for a drop in prices of steel for some months. We may be wrong, but I have gone pretty carefully over the situation, looking to such things as the supply of coal and iron ore, and the closing in of the shipping situation on the lakes. These are all factors that have a great deal to do with the output for the coming six months or a year for that matter. I cannot see anything to make me believe that the prices of steel are going to come down."

Backing His Beliefs

"Have you enough confidence in that prediction to buy for some months to come?" we asked.

"I have, and only this week we placed orders for a large tonnage of bars from a Canadian mill, covering deliveries for the next six months. The price paid is the price right now, and it means that we will still have to sell bar iron and steel at 5.75c. per pound, which is the highest figure that we have ever experienced in a good many years of business in this city. We are willing to take a chance on our guess being right. If the market cracks, as people say, we will have an interesting time of it, but the only thing that can crack it is an over-production, and do you see it coming? What is more, it does not pay us to pay too much attention to every little price flurry. We are dealing in the most requisite thing possible, steel, and it will not be possible in years to get steel at anything like the prices that were prevailing before the war."

There is a fair tonnage of premium mills sheets in some of the warehouses. These were taken in at a high price, but there is no intention on the part of the warehouses to reduce the price. "Were we to start that we would at once defeat our own object, and what is more we are not going to trim the price. We paid a lot of money for them and we must sell on the basis of cost. The market is empty of these lines and we have no fear about getting rid of them."

A Weak Market

The scrap metal market is decidedly weak. The interests relied on to keep up the buying in large quantities are not in the market, nor do they show any inclination to come in. Copper is not moving in any degree. "Until prices get down to something definite we do not look for an improvement," is the way one dealer sized up the situation.

NO IMPROVEMENT
LOOKED FOR IN 1920

New York Machine Tool Market Is Quiet and Nothing Is in Sight to Revive It

Special to CANADIAN MACHINERY.

NEW YORK, Sept. 30—There has been no change in the machine-tool situation. There is very little inquiry and orders are few. Except for a fair amount of railroad buying at Chicago, the demand for machine shop equipment over the entire United States is quieter than it has been since 1913-14. Some plants are letting men go and are reducing operations.

It is apparent that there will be no general resumption of machine-tool buying until industrial conditions are on a more stable basis. The deflation of prices, which has been in progress for some months in certain lines, and which within the past week has hit the automobile industry, must progress to a

point where buyers will have more confidence.

The announcement of the Ford Motor Co. of a reduction of automobile prices to the pre-war basis is the most far-reaching single item in the present industrial situation. This reduction has been followed by similar cuts by the Franklin and Crow-Elkhart companies and undoubtedly others will follow suit soon. It was the activity in the automobile trade which brought about general business prosperity in 1919, and it was the slump in automobile demand, caused largely by tight money, which has brought about the present inactive conditions. This is particularly true with relation to machine tools. When a considerable number of automobile man-

ufacturers have reduced their prices the way may be paved for a resumption of automobile buying, followed by the same business boom which resulted from the automobile buying fever last year.

Many automobile manufacturers had plans in mind for further expansion, and a few new plants were in course of construction, when the slump hit the automobile trade. Wholesale cancellations of machine tools, which followed this slump, have hit the machine-tool industry hard. It is quite possible that resumption of automobile activity, together with increasing railroad buying, will restore prosperous conditions to the tool industry next year. Few expect better business conditions before January 1.

count of car shortage. Since about the beginning of August they have been moving their stocks, and the highest estimate is that one-third of the stocks remain, while the lowest estimate is that practically nothing remains of the stocks. Thus the stocks have been moved at a much greater rate than that by which they accumulated, and the production itself, also moved, has been greater than formerly. Obviously the tonnages of steel received by jobbers and manufacturing consumers have been much greater in the past few weeks than in July and preceding months. This change would in itself be amply sufficient to account for the stagnant market.

It is the common view, however, that there is a decrease in consumption of steel, with further decreases in prospect, and thus no one seems to expect the market to recover. The common view is that prices will have to go through a course of liquidation, so that the only questions are when the readjustment will occur and how far prices will drop.

PREMIUM MILLS MAY COME DOWN BUT NO GREAT REDUCTION IN SIGHT

Special to CANADIAN MACHINERY.

PITTSBURGH, September 30.—Independent steel producers make considerably less effort than formerly to conceal the fact that they expect steel prices to decline. A common prediction now is that prices will not go below the Steel Corporation level. As no one had predicted that they would the statement is not particularly important, except as a definite admission of an abandonment of the hopes independents had entertained from time to time that the Steel Corporation would advance its prices. The main question now is one of time, when it will be that the independent steel prices will drop. Thus far they have not declined to any really noticeable extent, for the decrease in premiums for particularly prompt shipment that has been in progress at intervals for three months or more does not represent a general market decline. Naturally, in keeping with traditions, the aim of independent mills is to hold their prices up as long as they have much contract business on their books, as to reduce prices would be to encourage cancellations and suspensions of shipment.

The Automobile Declines

Up to date, Ford, Franklin and Crow-Elkhart have announced reductions in the prices of their cars, from 15 to 30 per cent. In some quarters these reductions, doubtless to be followed by other automobile makers, are regarded as containing a challenge to the steel industry to reduce its prices. The steel industry, however, shows some indisposition to admit that any challenge of this sort is necessarily implied. They point out that during the past nine months, or say since the first of the year, the automobile makers have received steel at a very wide range of prices. In the aggregate, by far the major part was at the Industrial Board prices, and a relatively small part was at very fancy prices. It is asserted that automobile prices of late have been in relation with the highest priced steel,

constituting the smallest tonnage, rather than in relation with the lowest priced steel, constituting by far the bulk of the tonnage. The fanciest prices were made by the automobile industry, seeking additional tonnages to supplement deliveries due on contract, and practically disappeared when that particular trade stopped buying.

In view of the fact that the sharp price advances in steel products early this year were due in great part to the eager and extravagant bidding of the automobile industry, and were recognized at the time as being caused chiefly by that industry, when other consumers who wanted steel were unable or unwilling to bid fancy prices, it might be considered natural enough that now the automobile industry should be able to exert an influence in the opposite direction, and produce low prices as it formerly produced high prices. One difficulty, however, is that the automobile industry does not really take any large part of the steel output, and another is that the trade has been so sudden in its recent cancellations of orders and instructions to suspend shipments that its contracts cannot well be set at a high valuation, excepting, of course, one or more large makers who are good in living up to contracts.

Heavy Deliveries of Steel

The great increase in steel deliveries that has occurred in the past few weeks does not seem to be rated as high as it deserves to be in constituting a market factor. The stagnation in new buying is properly attributable chiefly to this increase in deliveries. It should be noted, first, that the production of steel in August was seven per cent. more than the production in July, while the production August 1 to date has been well in excess of output in the same length of time prior to August 1. Next it must be noted that up to about August 1, and beginning in April, the steel mills were stocking steel, on ac-

As to decreases in consumption, there is the automobile industry, which is consuming less than one-half as much steel as it did at the top point, and then there is the showing made by the fabricating shops, the report of the Bridge Builders' and Structural Society showing bookings of fabricated steel contracts in August equal to 40 per cent. of the fabricating capacity, against 50 per cent. for July and an average of 72 per cent. for the first six months of the year. Again, there is the case of the railroads, which are buying but little now, and are hardly likely to be in the market again in the near future. In the first eight months of the year about 65,000 freight cars were ordered by the railroads and industrial concerns, but there is no likelihood that the last four months of the year will add much more business.

Steel Prices

The independent plate market is quotable at about 3.25c. There are rumors of lower prices being done, but these lack positive confirmation. Merchant steel bars can generally be bought at 3c by buyers with regular mill connections, and mills that are quoting higher prices, up to 3.50c or more are probably getting little, if any, business. Shapes are 3.10c, but with scarcely any buying. Black and galvanized sheets show a declining tendency, but are very difficult to obtain for early delivery at any price. A fair average of the independent sheet market is 7c to 7.50c for black, 8.75c for galvanized, 5.25c for 5.50c for the heavier gauges of blue annealed, say 12 and heavier, with 5.75c base for 16 gauge.

The Steel Corporation prices, which are going to be more and more in evidence, are: Bars, 2.35c; shapes, 2.45c; plates, 2.65; blue annealed sheets, 3.55c; black sheets, 4.35c; galvanized sheets,

5.70c. These apply only to the domestic market.

Pig Iron

The local pig iron market is quotable unchanged simply because nothing has occurred, and the "market" must be given at the levels at which business was last done, $50 for foundry and malleable, and $48.50 for Bessemer and basic, f.o.b. valley furnaces. Lower prices might be quoted even now if there were any inquiry, but there is none. Pittsburgh foundries do not even buy southern iron, which can be had at $2.36 less, delivered Pittsburgh, than valley iron.

PIG IRON TRADE

The pig iron market continues firm. Orders on producers books equal five months' output of furnaces. Shipments are moving freely. Resale iron is more plentiful on the market at Philadelphia owing to better railroad conditions. Enough basic has been sold to establish the price at $50 eastern Pennsylvania, or around $52.50 delivered. More than 1,000 tons of malleable was bought under $52, which appears exceptional as malleable has been strong and scarce.

The market is very quiet in Chicago. Cancellation of automobile castings has affected the melters in Michigan and Indiana, who are booked up with automobile business.

Sales of pig iron at Boston have been small, mostly limited to prompt shipment. The price for 1.75 to 2.25 per cent. silicon iron is quoted at $51; $52.25 for 2.25 to 2.75 silicon, and $55 for 2.24 to 3.25 silicon. One Tennessee furnace was closed down during the week on account of lack of coke.

New York reports shipments of pig iron have increased owing to improved railroad conditions. In this connection, Buffalo foundry iron at $48. Buffalo for the 1.75 to 2.25 silicon grade. Some Central Pennsylvania foundry has been sold for prompt shipment during the past week at $50 furnace for 2.25 to 2.75 silicon.

The market at Pittsburgh is extremely dull, consumers showing practically no interest. Furnaces are well booked for the remainder of the year. Better car supply and larger deliveries of fuel have increased the pig iron output somewhat and the Carnegie Steel Co. is operating 41 of its 59 furnaces, which is an increase of one stack over last week.

Basic is quoted at $50 on the Buffalo market. All the sales last week were for last half delivery.

The tone of the pig iron market at St. Louis is less favorable than recently. In the past few weeks a considerable tonnage has accumulated in the yards of the southern furnaces. Prices have not changed excepting that several makers who had been quoting $45 for No. 2 southern (1.75 to 2.25 silicon) have gone to $42, there being plenty of iron offered at that figure.

U.S. SCRAP METAL

CHICAGO reports that the decline in quotations on iron and steel scrap continues steadily on the market. Consumers are avoiding purchases as long as possible to reduce their reserves to the lowest point and avoid losses by further decline.

On the whole the market in New York is quiet. There has been a slightly improved demand and as a result an advance of 50 cents in turnings, clean cast borings, and wrought pipe of special dimensions.

Considerable scrap has been sold at Philadelphia following the purchase by one firm of 5,000 tons at $27 delivered about ten days ago. There is a scarcity of rerolling rails.

The American Steel & Wire Co., Worcester, Mass., has come into the market again at Boston, for heavy melting steel. As high as $25 at shipping point is being offered for sizable lots of heavy melting steel.

The market at Pittsburgh is very quiet, with no inquiries or new business. Heavy melting and No. 2 railroad wrought is somewhat firmer.

The demand for all kinds of scrap is heavy for delivery in from one to two months, at Buffalo. There is a strong demand for machine shop turnings and cast iron borings, as well as heavy melting steel. Railroad conditions are much better.

The recent buying movement in the Detroit market has subsided. Much scrap is available, but the higher prices developed by the movement continue to prevail.

Prices of scrap on iron and steel have declined from 50 cents to 2.50 per ton during the past week at St. Louis. All dealers want to sell at once. Melters are taking advantage of this state of affairs.

HYDRAULIC NEVER TO REPLACE COAL

Interesting Figures Brought Out by Authority in United States Field

Dr. Charles P. Steinmetz, chief consulting engineer of the General Electric Company, New York, warns against a too general belief that the power of water may ever completely replace the use of coal and oil for power purposes.

The maximum possible hydraulic energy of the nation, he estimates at the tremendous figure of 370,000,000 horse-power, which is little more than the total energy which we now produce from coal, and is about equal to the present total energy consumption of the country, including all forms of energy.

"This is rather startling," said Dr. Steinmetz. "It means that the idea that when coal once begins to fail we may use the water power of the country as the source of energy, is, and must, remain a dream, because if all our potential waterpowers were developed and every raindrop used, it would not supply our present energy demand.

"Thus hydraulic power may and should supplement that of coal, but can never entirely replace it as a source of energy. This probably is the strongest argument for efforts to increase the efficiency of our methods of burning coal.

"Obviously, of course, we can never hope to develop more than a part of the maximum potential energy of the water, and I merely cite these figures to show that our water power resources, if developed, to theoretical perfection, would still fall short of meeting the energy demands of the country to-day."

The total present developed water power in the United States exceeds 7,000,000 horsepower with 19,000,000 still undeveloped, and potential water power capable of development amounting to 31,000,000 horsepower.

The present use of water power saves the country from upward of 33,000,000 tons of coal annually.

BRANTFORD NEEDS MORE POWER NOW

But the Chances of Getting the Supply Seem to Be Very Limited

The shortage of coal, the scarcity of wood in the civic yards there were augmented when the Brantford city council was informed by the local Hydro-Electric Commission that this city was now carrying a load of 200 more horsepower than it was getting, and that the situation would get worse instead of better with the approach of winter, with lessened supply and greater demand. The only ray of hope was shed when the Gas Committee reported that with new fields opening up and available because of increased prices, the Dominion Natural Gas Company, would be in a position to give assurance that a greater supply of better quality gas would be available for Brantford this winter.

Sell Forging Plant—It is understood from a reliable source that the offer received by Canada Forgings for the Delaney plant at Buffalo will put the company in a strong position financially. A meeting of the shareholders has been called for Sept. 28th at the head office in Brockville to deal with the offer, which is said to be made by a syndicate. The operations of the company are stated to be running along smoothly and satisfactorily, and it was asserted by one of the directors that James Smart Manufacturing Co., one of the three component parts of the company, is alone earning enough to pay the common dividend.

CANADIAN MACHINERY
AND
MANUFACTURING NEWS

Vol. XXIV. No. 15

October 7, 1920

Looking After a Big Family of Freight Cars

A Great Amount of System and Supervision Necessary to Avoid Endless Complications—How the C.P.R. Look After the Handling of Reports in the Car Record Office.

IN the early days of railroading, car records were not much in vogue for the reason that cars usually were confined to service on the rails of the owner. Shipments moving from one road to another were transferred at the junction point to cars of the receiving road. This condition obtained for many years, but with the expansion of industrial activities came through rates and through routes, and these aided by that all important factor—competition—gradually compelled all roads to permit their equipment to be loaded to points on other roads, with the understanding that the car and the contents would move through to destination without breaking bulk.

As cars commenced to move from one road to another, always subject to possible delays and diversions, it became necessary for car owners to keep a close tab on the movements of both home and foreign equipment. The same conditions made it necessary for the roads to

get together and formulate rules and regulations to cover the interchange of equipment and to protect the accounting of car hire.

Prior to July 1, 1902, the mileage system of remuneration for the use of freight cars was in vogue, but on this date the Per Diem Rules became effective, and the latter are now in general use on all Canadian and American roads. Under the old system the owner of the car seldom received full return from the foreign roads for the use of the cars, but with Per Diem the road receiving a car must account for it to the car owner at the current per diem rate from the date of its receipt up to the date it is delivered to the owner or to another road, or in the case of a destroyed car, up to the date it is reported as a "destroyed" car to the owner, in accordance with the per diem rules.

Give all Information

Generally speaking, car records are maintained at all stations and contain everything there is to know in connection with the handling of the car contents, including the charges assessed or collected while in the terminal. Apart

from the accounting features these records show the following information: Initials and number of car, kind of car, date and time of arrival, train reference, point of shipment and kind of commodity if a loaded car, date and time of notification to consignee of arrival of his shipment, date and time of placement, date and time of release, date and time car is forwarded, train reference, and, if car is loaded, kind of commodity and destination.

At stations where a great number of cars are handled there is also maintained a daily on-hand report, usually referred to as a yard check. This report shows the initials and number of the car, kind of car, on what track located and whether loaded or empty, and if held, for what purpose.

At stations where cars are interchanged with other roads, an interchange report is maintained which shows the initials and number of the car, kind of car, name of road to which car is delivered, date and time of interchange, name of billing station, name of destination station and kind of contents.

The movements of all cars on any part of the road and on foreign roads are re-

*Compiled by J. H. Rodgers, from Canadian Pacific Car Service office records, and the paper, "Car Records and Their Relation to Transportation and Car Accounting," read before the September meeting of the Canadian Railway Club, by J. D. Altimas, Asst. Gen. Supt. Car Service, C.P.R.

A GENERAL VIEW OF THE ACCOUNTING OFFICE. CAR SERVICE DEPARTMENT, C.P.R.

corded in the books of the Car Accountant, whose office is usually at headquarters. The basis of these records is the agents' interchange report, the conductors' freight or mixed train report and the foreign roads' junction report.

The interchange report now in use on all roads is authorized by the American Railroad Association and is known as the reciprocal form of report. The agent of the delivery road prepares the report in quadruplicate showing information for all cars delivered, and passes the completed report to the agent of the receiving road who checks the information shown thereon, and if he finds it correct, signs all four sheets, certifying to the receipt of the cars. He retains one copy of the report for his station record, sends one copy to his car accountant, and returns two copies to the agent of the delivering road. The latter keeps one copy for his station record and forwards one copy to his car accountant.

Under American Railroad Association rules where a railroad delivers a foreign car to a connection, not the owner of the car, he is obligated to advise the car owner the name of the road the car was delivered to and the date of the movement and to show whether the car was loaded or empty. This is called the junction report. Roads which use the cut up system preserve the interchange slips for foreign cars delivered

and send them to the car owners as the junction report. This saves the delivering road considerable work and gives the car owner a first hand record, avoiding all errors due to transcribing.

The Interchange Report

Practically all of the larger roads now use the cut up or self transcribing system for conductors' train report and the agents' interchange report. Under this system the conductors' train report consists of one wide form giving all the information required, including an extra column for the date of the movement, also one narrow form, which is a duplicate of the wide form in so far as it refers to the initials and number of the car, number of station car is taken at and left at and date of movement. This narrow form has small holes punched in the margin between the lines to the left of the "initials" column for the use of the sorting clerks. These two forms are made at the one time by the use of carbon paper. In making entries on the narrow form it is not permitted to use ditto marks, as each slip, being separated from the rest, must show the complete information covering the car referred to.

The agents' interchange report consists of four wide forms giving all the information required, also two narrow forms which are duplicates of the wide

forms in so far as they refer to the initials and numbers of the cars, name of road delivered by and name of road delivered to, name of station at which delivery is effected and date of delivery. The narrow forms have small holes between the lines to the left of the initials column for use of the sorting clerks. The narrow forms are of different colors—one pink for the use of the delivering road and one canary for the use of the receiving road. The pink indicates a delivered record, the canary a received record. The original and five copies of this report are made in one operation by the use of carbon paper. The information called for is shown on the section of the form, Fig. 2. For reasons before mentioned "ditto" marks must not be used on the narrow portion. Those forms, when completed by the agents and conductors are forwarded promptly to the Car Accountant's office where they are entered into the car record books.

The handling of these reports in the car record office is a systematic and interesting operation. As soon as the mail is opened the reports are passed from the mail desk to the examining clerks who see that all information called for by the form is properly entered, after which the narrow forms are separated from the wide forms. The wide forms of the conductors' train reports

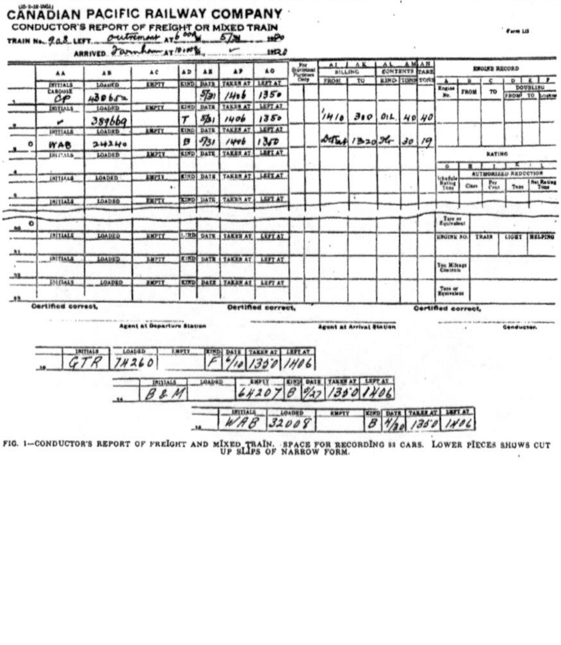

CANADIAN PACIFIC RAILWAY COMPANY

FIG. 1—CONDUCTOR'S REPORT OF FREIGHT AND MIXED TRAIN. SPACE FOR RECORDING 33 CARS. LOWER PIECES SHOWS CUT UP SLIPS OF NARROW FORM.

are immediately turned over to the Statistical Department for their use, while the wide forms of the interchange reports are filed for ready reference. The narrow forms are then sorted and arranged according to the number of the car entries on the form. The forms are then piled into packages having about 200 or 300 in each package, the forms showing the smallest number of car entries being placed on top. This is done to enable the cutter to dispose of the blank slips as they come from the machine and thus eliminate the unnecessary handling of blank slips by the sorting clerks. Interchange reports are given preferred handling.

Working by Weight

When the narrow forms reach the cutter they are put through the machine, and as the slips leave the machine the cutter separates the blank slips from the record slips, throwing the blanks aside, and placing the record slips into small boxes enclosed on three sides only, these boxes having their tare weight (equivalent number of record slips) stencilled on them so as to facilitate the weighing. These tare weights are checked occasionally to see that there is no variation. The cutting is accomplished by a machine of the advance lever type and when the reports are of uniform size with spacing and holes also uniform there is little chance for mutilation of the slips. The reports are printed on paper of uniform weight and when cut up weigh approximately 272 per ounce. They are weighed after cutting into slips and also at the close of the day, so that each sorting and entry clerk is credited with the correct number of slips worked. An ordinary druggist's scale is used and gives every satisfaction.

Sorting the Slips

The sorting table is usually a high desk, with its top surface on an incline,

and fitted with two rows of spindles one inch apart, one row being about seven inches behind the other. The spindles are 4 3-4 inches high with a shoulder or collar about 3-4 of an inch from the bottom. The part below the collar fits into holes bored in the top of the desk. One or more spindles are alloted for each record and sufficient spindles to accommodate all records are looked after by one sorting clerk. The spindles are a little less in diameter than the holes in the slips so that the sorters can slip them on the spindles very rapidly. Lists are prepared which show what records are alloted to each spindle, and these are located in a suitable position on the sorting table. When sorting the slips are held in the right hand and sorted with the left, otherwise the car initials would be covered and the handling would be much slower. Sorting clerks generally handle about 20 to 30 slips per minute. The original sorting is done to suit individual requirements. Usually the first sorting is done in book form and the second in page order. The latter is not always economical.

The junction records when received on the standard junction report form are transcribed by typewriter to the cut-up form and are passed through the machine in the same manner as other forms and are then passed on to the sorters. Junction records received on the cut-up forms are at once turned over to the sorters for handling. At regular periods during the day the record clerks obtain the slips from the sorting desk and proceed to enter them in the books. By this method a record clerk does not have to handle any records except those that are to be entered in his book. The record books are of the loose leaf type and usually about 18 inches long by 17 inches wide. The home car record sheets are ruled about as follows: See Fig. 3.

One column about one inch wide for car number, then 31 columns 3-8 of an

inch wide for the days of the month and one column about one inch wide at the extreme right for mileage as per diem purposes. Each page holds fifty cars and a book usually includes 5,000 cars in numerical order. The home car record usually is made up to take care of two months' records in order to avoid the necessary transferring of the records from one book to another.

The foreign car record sheets are ruled as follows: See Fig. 4. One column about 3-4 of an inch wide for the last account, another of the same width for initials, another for the car number, 1-4 inch columns for, kind of car, per diem days earned, checking, for each day of the month, and three or four 3-8 of an inch wide for mileage purposes. Each page holds fifty cars and each book has space for about 6,000 cars of which from 1,500 to 1,800 are actually in service at any one time. These books are made up to show cars in alphabetical road order and in numerical order and are indexed for ready reference.

Watching Interchanges

When a clerk is ready to enter records he first finds the space alloted to the car and then enters in the block which represents the date of the movement, the number of the station at which the car was taken and the number of the station at which the car was left. Usually the number of the station at which the car is taken already appears in the block and it is only necessary to enter the number of the station at which the car is left and a check mark drawn after the number of the station at which the car is left indicates that the record is complete. When a car moves empty the pencil line drawn above the number of the station at which the car is left shows the record is complete.

In the case of interchange records, a receipt from the connecting road is entered by showing the symbol representing the road in the upper left hand cor-

THE PICTURE TO THE LEFT SHOWING THE FOUR SORTING CLERKS, ILLUSTRATES THE FIRST SORTING AFTER THE FORMS ARE CUT INTO SLIPS. TO THE RIGHT—SORTING THE SLIPS INTO PAGE ORDER.

ner of the book immediately followed by the number of the station. To indicate a delivery to a connecting road the symbol of the road is entered in the lower right hand corner of the block preceded by the station number. In practically all cases the number of the station appears in the record and it is only necessary to enter the symbol. Home record clerks average about 300 entries per hour, whereas foreign record clerks average about 225 entries per hour.

· Roads which do not use the cut-up form of reports simply take the interchange reports, conductors' train reports and the junction reports from the mail desk, place them in suitable packages for handling and pass them around to the several record clerks who enter the movements direct from the reports to the books. Under this method each clerk must examine every car shown in the report in order to pick out the records which are to be entered in his book. This takes considerable time and labor.

With a record of every movement made by cars on his own rails, with the interchange record of cars delivered to and received from connections and with the junction record of his cars furnished by foreign roads, a car owner has a complete record of every one of his cars during each month of each year. This enables him to keep a close check on the handling of his cars by foreign roads and to check the earnings of his cars under any and all circumstances. He is also in a position to know what other owners' cars are doing on his rails.

Car records properly maintained and kept up-to-date are of great benefit and serve many purposes of the Transportation Department in addition to being the basis of the accounting for car hire. The following are some of the uses to which car records are constantly put:—

· To furnish records to the public in connection with shipments in which they are interested.

To furnish records to the Transportation Department to assist in tracing freight urgently required at destination or which may have been unduly delayed in transit or which may be required to be diverted in transit.

FIG. 2—AGENTS' INTERCHANGE REPORT. SPACE RECORD FOR 22 CARS. CUT UP SLIPS SHOWN AT BOTTOM.

To furnish location of special class equipment, such as refrigerator, tank, potato, Eastman heaters, palace horse, special grain cars, etc., to enable the Car Service Department to keep them in the special service to which they are allotted, or to move them to points where they are required, also to see that such cars are used to full advantage.

To furnish car movements, loaded and empty, to the Transportation Department to assist in tracing delays and placing responsibility and to the Claims Department to assist in locating overs and shorts and disposing of claims.

To check the handling of cars on foreign roads and thus overcome unnecessary delays and diversions.

To furnish weekly or monthly statements of the distribution of all classes of equipment by districts or divisions.

To furnish records of car loads transhipped in transit.

To furnish mileage, loaded and empty, made by special class cars or cars of any class.

To check the record of foreign equipment to see they are not unduly delayed or used contrary to Car Service Rules.

To check repair bills as to location.

A record clerk works on the average about six hours a day in entering records and about two hours a day in furnishing locations, movements and other information.

The car record is the basis of the car hire accounting system and the car records of foreign equipment handled are usually turned over to the Accounting Department about the tenth day following the close of the month. The first duty of the clerks on accounts is to ascertain what records are incomplete, i. e., what cars are short on interchange receipt or delivery and to check back against the interchange reports to see if such records were skipped by the record clerks. The second operation is to extend the number of days each car was on the rails, and where there is an incomplete record, to list the initials and the numbers of the cars so as to aid the correction staff in completing the records.

Where a missing delivery is involved per diem is usually allowed up to the date the car is last reported moving. In the case of a missing receipt per diem is allowed from the date car first moved. When the record has been completed, the balance of the per diem, if any due, is allowed in a subsequent month's report.

After the per diem days have been extended, a per diem report is prepared for each road, showing the car number and the number of days earned and same is forwarded to the car owner, within forty days from the last day of the month in which the per diem was earned. On the receipt of the per diem report by the car owner he enters in the record opposite each car the number of days per diem allowed by each road, and after all the per diem reports have been carried, if his records indicate a shortage, he has the privilege of making claim against the road which, according to his records, has short-paid the per diem. Such claims to be valid must be handled in accordance with the rules.

In addition to preparing reports for per diem earned currently, reports are also prepared to cover errors and omissions in per diem reports of previous months, to cover switching reclaims, to cover rule 14 reclaims, namely, cars held by one road on account of inability

FIG. 1—HOME CAR RECORD SHEET. SPACE ON EITHER PAGE, OR SIDE OF THE SHEET, FOR TEN MONTHS' MOVEMENT OF ANY OR ALL CARS IN THE SERIES.

of another road to receive and to cover reclaims due to special conditions.

Under Per Diem Rules, per diem must be paid by a road using a car to the owner for each and every day car is in service on that road, but where a road handles a car in terminal switching service for another road it is entitled to an arbitrary reclaim from the carrier road of an agreed number of days not to exceed five for each car handled in such service. Per Diem Rule, 15 also provides that where a road holds cars on account of another road to receive them, the holding road is entitled to reclaim from the road on whose account the cars were held the full amount of per diem involved, always provided that due notice is given in accordance with the rules.

The Per Diem Rule Agreement, to which practically all American and Canadian roads subscribe, is promulgated by the American Railroad Association and lays down rules to govern every phase of per diem accounting. These rules in their original form were adopted and made effective on July 1, 1902. Since that date there have been changes from time to time in the rate to take care of the increased cost of owning and operating a car. At the present time the rate is 90 cents per day. The main objection to the rules in their original form was that no penalty was provided for the non-payment of per diem earned

CANADIAN PACIFIC RAILWAY COMPANY.
SUMMARY OF PER DIEM AND PASSENGER CAR MILEAGE.

W. PHELAN,
Car Accountant

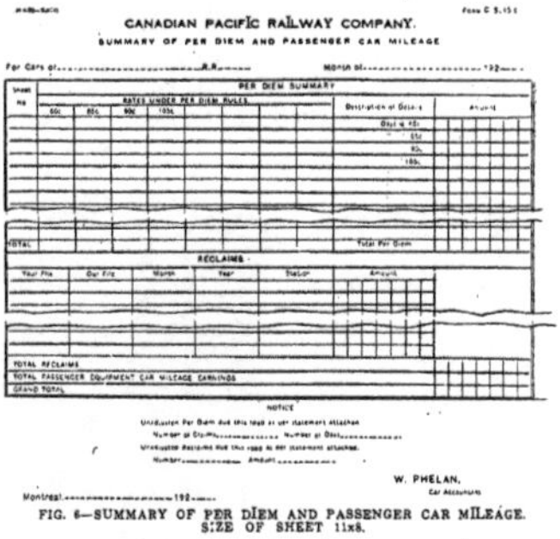

FIG. 6.—SUMMARY OF PER DIEM AND PASSENGER CAR MILEAGE.
SIZE OF SHEET 11x8.

While an improvement this ruling did not effect the end desired, and revised rules became effective on the 1st of March, 1920. The next few months will show how these new regulations will work out.

The Per Diem Rules have done more

system a car owner knows exactly what his cars earn and methods are provided for accurate accounting, whereas, under the mileage plan the car owner had no means whatever of checking the earnings of his equipment on foreign roads and had virtually to accept what-ever was allowed. In the old mileage days many cases of deliberate dishonesty in accounting were developed and the conditions helped a great deal in the agitation that was carried on in favor of the adoption of the per diem system.

Car records to be of value must be properly maintained and kept up-to-date. This can only be accomplished by the hearty co-operation of all concerned in the preparation and handling of the reports and the records. In the Car Accountant's office surprise checks are used to a great extent and with good results to ascertain if the correct and complete information is being recorded in the books. With this system the clerks soon realize that they must do their work properly or make way for others who will. The secret of the whole transaction is system and supervision.

FIG. 4.—FOREIGN CAR RECORD SHEET. SPACE EITHER SIDE FOR MONTH'S
MOVEMENT OF FOREIGN CAR WHEN ON TRACKS OF C.P.R.

and as a result the car owner was put to considerable expense to collect what rightly belonged to him. On July 1, 1913, a penalty of 5 cents per car per day was made effective when per diem was not paid to the car owner within six months from the last day of the month in which the per diem was earned.

than anything else to force the railroads to keep proper car records. While it may be true that the cost of accounting for car hire under per diem rules is much greater than under the mileage plan, there is little question that the extra saving more than balances the additional cost. Under the per diem car

CANADIAN PACIFIC RAILWAY COMPANY.   PER DIEM REPORT.
Cars at........W.ABASH........R......   Month of........MAY........1920....

W. PHELAN,
Car Accountant

FIG. 5.—PER DIEM REPORT. SIZE OF SHEET, 11x8.

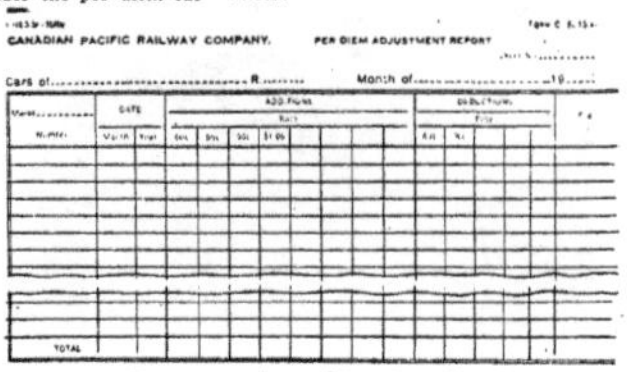

CANADIAN PACIFIC RAILWAY COMPANY.   PER DIEM ADJUSTMENT REPORT.
Cars of........................R........   Month of........................19......

FIG. 7.—REPORT FOR PER DIEM ADJUSTMENT. SIZE OF SHEET, 11x8.

Nine Examples of Jig and Fixture Practice

Various Principles Are Discussed Such as the Snapper, Universal Chuck, Turnover and Slide Stop Motion. Each Jig Depicted Has Proved to Be the Best for Its Own Particular Work.

By J. H. MOORE

TAKING it for granted that everyone realizes the advantages of efficient jigs and fixtures, we will proceed to describe seven interesting examples of fixtures which have proved real producers for .The International Harvester Co.. Hamilton, Canada.

The first of these depicted at Fig. 1 is what is known as a snapper jig. It is so-called because it snaps into place quickly and holds the work very firmly. Referring to the illustration we see that A is a solid cast iron base, while B shows one of three jaws very similar to those used on an ordinary lathe chuck. These jaws are adjustable and come in at the same rate of speed towards the centre, receiving their motion from the action of handle shown to right of the letter A. They, of course, slide in the Tee shaped slides C. This feature means that the work E, which is a portion of a manure spreader, is always centred before the top portion of jig comes down on it.

Having centred the work, the handle F is released which brings down the cup shaped brush D on top of the boss on the work E. The pin H acts as a guide, and underneath this pin at each side of jig is fastened a strong spring. The action is apparent. The spring pulls the top down with considerable pressure, the lever G being part of the movement. This type of jig is widely used at the plant we speak of and can handle a large variety and range of work. The operation in this particular case is the drilling of a 1 1-4-in. hole through the centre of the work E. The casting as shown is approximately 16-in. diameter, but the jaws of jig can

FIG. 2—THIS STYLE OF JIG PROVES THAT IMPROVEMENT IS ALWAYS POSSIBLE.

come in to take a much smaller piece if necessary.

An Improved Type

The jig as described at Fig. 1 was the original snapper type, and hundreds of the same have been made with splendid success. There is a saying that no matter how good a thing is, it can always be improved upon, and this holds good in the present case as Fig. 2 will illustrate.

The fixtures shown in this view are really improved offshoots of the snapper jig, but are termed the universal chuck type, owing to their having an ordinary universal chuck as the holding unit.

It will be noted that a universal chuck is placed directly in the centre of the

jig base, and that two bosses are placed on each side of the aforementioned base. In these bosses are fastened two guide pins. The swinging member A has a cup shaped bush in the centre boss C. The quick action clamp D grips the other guide rod, and by referring to E you can see the clamp released. In action you loosen a purled knot a quarter turn which brings the clamp to the position shown at E allowing you to swing the top members of jig out of the way. B illustrates the work in one case, and the other view shows a still different type of gear being drilled.

The great advantage of this type of jig over the one depicted at Fig. 1 is the fact that by changing the bushings at C to suit the nature of the work, you can use the fixture for an almost endless variety. Of course it is only adapted to parts of round shape, parts which can be gripped in the chuck. Since installing the improved design, the former shown at Fig. 1 have taken a back seat, although for certain classes of work they are still ideal.

Reaming Jigs

The next two examples are shown at Fig. 3. Both jigs are used on the same piece, and we will take them up in order starting with the one shown to the left. The work being performed is that of reaming from the cored hole a 2 3-8-in. hole. Note the action of the clamp A. On inserting the piece in the jig, the shaper wedge B is driven in, this in turn raising the back end of clamp A, and pressing the point of the same down firmly on to the work. The swivel screw C is next brought into action, this guaranteeing that the work goes up into correct position.

To the right is shown the reaming of

FIG. 1—THIS STYLE JIG IS KNOWN AS A SNAPPER JIG.

FIG. 3—TWO EXAMPLES OF REAMING JIGS. NOTE THE METHODS OF HOLDING.

the second hole, 1 5-8 in. diameter. The work is gauged from the former hole reamed by means of the pin D, and is held in position by both the small screw shown at top of jig, and the taper wedge E. The view at F. illustrates clearly the details of the piece for which these jigs were made.

Turnover Jigs

Fig. 4 illustrates two other jigs known by the term turnover jigs. The fixture to the left is used to drill two holes in the piece A at the points shown by cross marks in chalk. The method of obtaining these two positions is very simple. Note the slot C at lower portion of the fixture. It is cut at the correct radius, and of sufficient length to allow the correct distance of travel. Two bushings, which are hidden from view by the work in the jig, guide the drills to their proper positions. In action the operator merely places the work on the circular shaped portion of jig, fastens it by the taper key B, drills one hole at one extremity of the slot, and one at the other. The pin which is inserted in the slot acts as a positive stop. This type fixture is easily made and is very efficient.

The second jig, that to the right, is used to drill the tongue for a whiffletree shown at D. Note the two chalk cross marks. This denotes where the holes are drilled. There are four holes drilled, two on either side of the casting.

The method of procedure is as follows: The work is placed in position and held by the taper key shown. The two holes are drilled on one side of the casting, then the moveable portion of the jig is pulled out of a slot in the stationary

member, and revolved 180 degrees, then pushed back into locking position once more. The collar F prevents the moveable portion being pulled out too far. The remaining two holes are now drilled, and of course they cannot be anything else than opposite the other two, owing to the safe method of locating.

This turnover principle is used to great advantage in the plant we speak of, hundreds of various style turnover fixtures being used.

Three Varied Jigs

The last photo, Fig. 5, illustrates three distinct types of jigs. Starting from the left we find as follows: The fixture

to the left is used to drill the two top holes on the part A which is a portion of a fertilizer. These holes are on a radius, and here is how they are drilled. The work is placed on the moveable portion B, over the centre rod, then fastened in position by means of the taper key shown. This portion B slides on a guide slot about 4-ins wide, and of the correct radius to accommodate the holes. Two bushings are placed in correct position at top of piece B. First one hole is drilled at extremity of slot partially shown in stationary member, then the other is drilled at opposite position. To allow for adjustment, an adjustable screw with lock nut is placed at either end of the radius shaped slot, and any adjustment necessary can be accomplished very easily.

The jig shown at the centre is simply a pin jig to drill the pin D as shown, but the ejector used is worthy of note. The work is placed in Vee block E and drilled through a bushing in the regular

FIG. 4—TWO JIGS THAT EMPLOY THE TURNOVER PRINCIPLE

manner. On being completed, the operator merely touches with the palm of his hand the collar at end of F and the spring accomplishes the rest. It shoots the pin out of the jig, and of course springs back into gauging position as soon as the hand leaves the collar F. The output on this particular piece has greatly increased since installing this style of ejector.

The last fixture to be discussed is used to drill the four holes in piece G. Four bushings H are placed in correct location and work fits into jig up against stop pin shown, and held in position by screw with handle.

This portion of the jig moves in a radius similar to fixture shown on the left, but in a horizontal position. A radius shaped slot in the base of jig limits the travel.

From the foregoing examples of jig practice readers will no doubt get ideas for the adoption of similar principles to their work, for after all it is the fundamental principle underlying the making of any jig that really counts.

FIG. 5—THREE JIGS, EACH WITH ITS OWN PRINCIPLE.

The Heat Treatment of Locomotive Forgings

The Article Deals Not Only With Locomotive Forgings, But Similar Parts as Well—The Author Presented This Matter Before the Buffalo Chapter of American Society for Steel Treating.

By LAWFORD H. FRY

THIS paper presents some notes on the theory and practice of the heat-treatment of steel forgings with particular attention to larger forgings such as locomotive driving axles and similar parts. In dealing with the theory my idea is to give only a general outline of the metallurgical theories of hardening and to compare these theories with what we see every day in practical work in the treating plant.

The term "heat treatment" is a somewhat loose one. It may properly be used to describe many processes, and some critics assert that it has been used to cover a multitude of sins. In the present paper the term is used to cover quenching from a temperature above the critical range, followed by a drawing at a temperature below the critical range.

Theory of Heat Treatment

The practice of heat treatment as defined above is based on the fact that the quench produces great hardness and brittleness, while the draw reduces the hardness and restores ductility, in proportion to the temperature reached.

It is not too much to say that our present civilization—based as it is on engineering—is dependent on the simple fact that steel hardens on being quenched. In spite, however, of the great importance of the process, and of the amount of research work that has been done on it, there is still room for doubt as to the exact nature of the reactions which produce hardening, and some of the most eminent metallurgists hold divergent opinions on the subject.

We know well enough how the physical properties of the steel are affected by quenching and the microscope has carefully followed and recorded the visible changes in the interior structure of the steel. It is true, however, that the scientific explanation of the hardness produced by quenching still remains no more than a theory, and as yet the most eminent metallurgists are not unanimous in their choice of a theory. The changes in structure as revealed by the microscope can be summed up in a few words. As it comes from the forge the steel has a coarse irregular structure with comparatively large grains of pearlite inclosed in a network of ferrite. On being heated up through the critical range the carbon, to which the pearlite structure is due, dissolves in the surrounding iron so that at temperatures above the critical points the steel becomes a uniform homogeneous solid solution of carbon

*Standard Steel Works Co., Burnham, Pa.

in iron. By rapid quenching this structure can be preserved into the cold state, and the resulting steel will be extremely hard. If slow cooling is allowed, the carbon comes out of solution as the critical range is passed, pearlite and ferrite re-form and the steel reverts to its former condition except that the grain is refined. The steel is then as soft as or softer than before. An intermediate rate of cooling produces an intermediate structure and an intermediate hardness.

Theories of Hardening

To explain these possible changes in hardness many theories have been offered. Those backed by the best known metallurgists may be collected into three groups and designated:

 (1) Allotropic iron theory.
 (2) Carbonist theory.
 (3) Strain hardening theory.

Osmond in developing his theory of the first group assumed that iron occurred in three forms, or allotropic states, through which it passed with a change in temperature. He concluded that the change from one form to another required time, so that a rapid cooling prevented the iron being transformed to its usual low temperature form, and trapped some of the intermediate form which was hard and brittle at ordinary temperatures.

Arnold attributes hardening to a direct action of the carbon in the steel. He supposes the existence of a carbide of iron called hardanite, which is very hard, and which by being retained on quenching imparts hardness to the steel. It has even been suggested by other exponents of the carbon hardening theory that the sudden compression of the metal on quenching transforms the carbon into minute diamonds which produce the hardness.

In the more modern theories, hardening is held to be produced by the strains set up on quenching. We are all familiar with the fact that if a piece of steel be strained beyond its elastic limit it will be hardened. This has been explained by Beilby by his theory of amorphous vitreous cement. He showed that when the surface of a metal is polished, the pressure used transforms the outer layer from a crystalline structure into a glassy amorphous film which is very much harder than the crystalline material. Recent experiments with balls for ball bearings have shown that the polished skin on the balls adds 30 per cent. to their strength. This hard amorphous material accounts for the increase in strength and hardness produced by cold working. When a piece

of steel is stretched or compressed beyond its elastic limit the crystals throughout the metal slip over each other, at each slippage surface a film of hard amorphous vitreous material is produced. The films are too thin to be visible even under the microscope, but are numerous enough to harden the metal throughout. Beilby further noted that the amorphous material reverted to the crystalline condition on being heated. As Carpenter has pointed out the change from the crystalline to the amorphous condition can be effected by work but not by heat, while the change in the reverse direction from amorphous to crystalline can be produced by heat only and not by work.

Beilby's theory having been found to fit all the conditions of cold hardening, was proposed in 1914 by three different groups of scientists as an explanation of the phenomena of hardening by quenching. The form in which the theory was proposed by Edwards and Carpenter appeals to the writer as giving the simplest explanation of the observed facts. An account of this theory is given in the paper on the Hardening of Steel by Carpenter, read before the Royal Institution, March 7, 1919, and reprinted in Engineering, March 21, 1919. It is pointed out there that during the quenching of the steel severe internal strains are set up—

 (1) By the shrinking of the outer skin;
 (2) By the suppression of the critical change.

These strains develop intercrystalline films of hard amorphous material just as do the strains produced in cold working. The heat in the steel would tend to recrystallise the films, but if the quenching is sufficiently rapid the temperature at which the films are stable is reached before they have time to revert and the metal is hardened. The greater hardness with high carbon is accounted for by the greater strains set up in the suppression of the critical change.

To the practical steel treater it is interesting to know the hardness at which he is aiming thus linked up with the quenching strains which are the source of the difficulties and dangers of the process. If the strain hardening theory is accepted, the process of heat treatment is seen as the art of producing and controlling strains. The first step in the study of the practical application of heat treatment is an examination of the strains produced by quenching.

As mentioned above, Carpenter point-

ed out that the strains producing hard-cuing might be produced in two ways:

(1) By the shrinking of the outer skin.

(2) By the suppression of the critical change.

I do not know that any attempt has been made to separate the effect of the two classes of strains, or to compare the part played by each. A general consideration of the matter leads to the belief that the strains from skin shrinkage are of secondary importance in producing hardness, but have a primary part in producing troubles. It is to be noted that the strains from skin shrinkage distribute themselves unequally through the pieces, being greatest at the surface and decreasing toward the center. The strains from suppression of the critical change on the other hand have their origin in the interior of each grain and can, therefore, be uniformly distributed throughout the piece. Further these intragrain strains can be produced without distortion of the piece as a whole. It follows, therefore, that the ideal condition in quenching would be reached if it were possible to produce the critical change strains without the skin shrinking strains. Unfortunately this is, from the nature of things, impossible. When the piece is plunged into the bath the skin is thus put into tension while the interior is in compression. The more rapid the cooling of the skin the greater are the stresses set up, and in the case of a large cylindrical forging such as a locomotive driving axle it is possible by too rapid a quench to split the piece longitudinally. As the piece cools a second danger develops. The outer skin cools and contracts while the core remains hot and expanded. To accommodate itself to the core without cracking the skin must stretch. In this stretched condition it becomes rigid, while the core still has a considerable amount of contraction to carry out. Hence in the final stage of cooling the core is pulling away from the skin and if this final cooling is too rapid the core may be ruptured.

Let us now examine the action of the quenching medium on the hot steel. Each particle that is effective in quenching comes in contact with the hot piece, moves along its surface taking up heat for a time, and then is swept away by the convection or other currents in the medium. The amount of heat taken from the piece in a given time will depend on the number of particles brought into contact in that time and on the amount of heat each particle can take up. Hence a light oil quenches quicker than a heavy oil because being more fluid the convection currents bring more particles per minute into contact. Water quenches more rapidly than light oil both because it is more fluid and because each particle can take up more heat. It also follows that a given medium will quench more rapidly if the convection currents are supplemented by a forced circulation. This is illustrated by the figures given below. They are obtained from two similar locomotive axles made from the same steel, quenched from the same temperature in the same medium, a solution of one part cutting compound to three parts water and drawn at the same temperature. For axle A the quenching medium was circulated only by its natural convection, while for axle B vigorous circulation of the medium was obtained by stirring with an air jet. The time of cooling through the interval 1450 degs. F. to 700 degs. F. was decreased from 16 to 9 minutes by this stirring. The physical results were:

Axle	Circulation	Time to Cool from 1450° F. to 700° F. Min.	Elastic Limit Lb. Sq. In.	Tensile Strength Lb. Sq. In.	Elon. in 2" Per Cent.	Red. of Area Per Cent.
A	Natural	16	49,500	95,000	29.5	45.0
B	Forced	9	69,000	103,000	21.0	42.0

Comparative figures for the quenching speed of different media are given in Table I. These are obtained from experiments made with locomotive axles, and while they are perhaps not universally applicable, they serve as a basis for illustrating the difference in cooling rates, and the differences in stresses set up by quenching in the different media. To see the effect on the stresses produced consider a section of the piece quenched having the face A A A A in contact with the medium, while the face B B B B is towards the hot interior of the piece. The cooled face A will contract while the face B is kept hot and expanded. This strains the material. The strain which we can measure most conveniently by the stress it sets up will depend on the difference in temperature between the faces A and B, or, more strictly speaking, on the temperature gradient between the two faces. The faster the heat is taken away from the face A the greater the temperature gradient and the greater the stress. The tables show the calculated figures for temperature gradient and stress for the various rates of cooling.

The effect of the size of the piece must now be considered. The time required to cool a piece of steel through the critical range depends on the rate at which heat is taken out per unit of surface and also on the amount of metal back of each unit of surface from which heat has to be extracted. For example, a bar 1 inch in diameter has 0.25 cu. in. of metal for each square inch of surface, while a 10 inch bar has 2.5 cu. in. of metal per square inch of surface, that is, 10 times as much metal to be cooled through each square inch of cooling area. Therefore, cooled in the same medium the 10 inch bar would take something like ten times as long to cool through the critical range as would the 1 in. bar. This effect of size is clearly shown in the difference in physical properties between large and small parts when both are given the same treatment. Table 2 gives the results of an experiment to illustrate this point. From two locomotive axles test pieces were cut with a hollow drill before treatment. These pieces about 1¼ in. in diameter by 5 in. long were put back into the hole from which they were cut and the axles heated for quenching. Just before quenching the pieces were taken out of their holes and quenched independently of the axles, but at the same time and in the same medium. They were then put back into place and the axles drawn back. The physical properties of the test pieces and axles are shown in the table. It will be seen that the test pieces are considerably harder than the axles. As a further example of the effect of size Tables 3 and 4 have been compiled showing the physical properties of large and small axles quenched in oil and of small test pieces quenched respectively in air, oil and water. The property most affected is the elastic limit which is about 50,000 pounds per square inch for the 12 in. locomotive axles and 57,000 to 59,000 pounds per square inch for the five and six inch trolley axles. Test piece No. 6, which was 1¼ in. in diameter, quenched in the same oil as the axles, gave 93,500 pounds per square inch elastic limit. The steel in all cases is a carbon steel with about 0.50 per cent. carbon.

Column 4 gives the approximate time taken to cool from 1450 deg. F. to 850 deg. F. The elastic limit (hardness) in-

TABLE 1.

Quenching Speeds, Temperature Gradients and Stress Due to Temperature Gradient with Various Quenching Media.

1	2	3	4
Quenching Medium.	Rate of Cooling in B.t.u. per sq. ft. of Surface Per Hour.	Temperature Gradient in Deg. F. Per Inch of Thickness.	Stress Due to Temperature Gradient in Lb. Per. Sq. Inch for Each Inch of Thickness
Air	8,000 B.t.u.	22° F.	4,000 lb. sq. in.
Heavy Oil	20,000	55	16,000
Light Oil	30,000	68	13,000
Water	80,000	176	50,000

TABLE 2

Effect of Size of Piece Quenched on Physical Properties

1 Axle No.	2 Size of Piece Quenched	3 Elastic Limit Lb. per sq. in.	Tensile Strength Lb. per sq. in.	Elongation in 2" %	Reduction of Area %
1	Axle	86,000	124,000	15.9	27.5
1	Test Piece	75,000	124,500	17.9	43.5
2	Axle	45,500	89,000	25.5	46.5
2	Test Piece	56,500	93,000	20.0	55.0

TABLE 3
Size and Physical Properties of Various Forgings

| 1 | 2 | 3 | 4 | 5 | 6 | 7 | 8 | 9 | 10 | 11 |
| | | Diameter | | Appr. | | | Elastic | Tensile | Elong. | Red. |
Line	Material	O.D.	Bore	Wgt.	Qnch.	Draw	Limit	Strength	in 2"	Area
1	Loco. Axles	12"	3"	2000	1450 Oil	1050	51,500	97,000	22.0	52.0
2	Loco. Axles	12"	3"	2000	1450 Oil	1050	50,000	98,000	24.0	48.5
3	Troller Axle	6"		840	1450 Oil	1050	59,000	107,500	24.0	51.0
4	Trolley Axle	5½"		630	1450 Oil	1050	57,000	90,000	25.0	49.6
5	Test Piece	1½"		1.6	None	1450	77,000	124,000	16.5	39.0
6	Test Piece	1½"		1.6	1450 Oil	1200	93,500	128,000	19.5	61.0
7	Test Piece	1½"		1.6	1450 Water	1200	112,000	125,000	19.5	55.5

TABLE 4
Quenching Speeds and Physical Properties

Line No.	Material	Treatment	Approx. Time to Cool from 1450° F. to 500° F. Min.	Elastic Limit Lb. sq. in.	Elastic Ratio %	Tensile Elongation Factor
1	Loco. Axle	Quench Oil	30	51,500	53	2,140,000
2	Loco. Axle	Quench Oil	20	50,000	51	2,310,000
3	Trolley Axle	Quench Oil	15	59,000	55	2,340,000
4	Trolley Axle	Quench Oil	15	57,000	63	2,250,000
5	Test Piece	Cool in Air	6.5	77,000	62	2,040,000
6	Test Piece	Quench Oil	1.8	93,500	73	2,500,000
7	Test Piece	Quench Water	1.0	112,000	90	2,450,000

creases steadily as the time of cooling is reduced. Test piece No. 5, which was cooled in air, coooled more rapidly than the quenched axles and is consequently harder. Micro-photographs of the three test pieces quenched respectively in air, oil and water show a fineness of grain in proportion to the rapidity of cooling, but this is an accompaniment of the hardness and not its cause.

Coming now to the application of the principles we have considered, let me present some notes on the practice of the works with which I am connected. Before the war our heat treating experiences covered locomotive forgings.

With the entrance of this country into the war we were requested by the Government with all the force of a command to undertake the manufacture of nickel steel forgings for the 155 m/m guns and howitzers. Of these forgings the largest piece is the gun tube which is about 20 feet long with a bore 5½ in. at the breach and 5½ at the muzzle, the weight being about 3,900 pounds.

Under the advice of the Ordnance Department, we put in vertical oil fired furnaces and experience led us to choose water as a quenching medium.

In gun manufacture the tests are taken two from each end of each forging transversely, that is to say, the axis of the test piece is taken across the direction of forgings. As a consequence, if the steel is fibrous, the fibre runs across the test specimen and the latter fails to give proper elongation. This trouble cannot be eliminated by heat-treatment, as it is inherent in the steel. I think it is safe to say that with a properly operated treating plant, at least ninety per cent. of the trouble encountered has its source not in the treating plant, but in the open hearth where the steel is made.

In the gun treatment the greatest chance of purely treatment trouble lies in the time of immersion in the water.

It has been pointed out that if the final cooling of a quenched piece is carried out too rapidly quenching cracks may be produced. The nickel steel gun forgings with their varying sections proved to be very susceptible to this trouble. Finally we developed the practice of varying the quenching time to suit the varying sections of the piece. For example the gun tube would be completely immersed and quenched for two minutes, then the thinner muzzle section withdrawn and the heavier breech section given another minute in the water. As a rough approximate figure it was found proper to give an immersion of one minute for each inch thickness of wall in the forging.

In returning to a peace basis of manufacture, the equipment and experience obtained in gun making is available for the manufacture of locomotive forgings. The first question that arises concerns the relative merit of oil quenching as formerly practiced and of water quenching as used for guns. In giving preference to water we follow the lead of the Pennsylvania Railroad, who have been pioneers in the introduction of heat-treated locomotive parts. Of the two media, water by reason of its drastic quenching power takes first place in the production of desired physical properties in the steel. Any argument in favor of oil is based on the danger of quenching cracks if water is used. The answer to this is to have a carefully organised shop and to watch closely the time that the material is left in the water.

In conclusion I sum up the requirements in successful heat treatment as follows:

1. Select sound, well-made steel.
2. Heat uniformly to the proper quenching temperature.
3. Quench rapidly but withdraw the bath at a sufficiently high temperature.
4. Draw uniformly and accurately.

5. Submit to a shock proof test before putting into service.

By correct practice along these lines the steel will give from 40 to 60 per cent. better service than the same steel untreated.

(Acknowledgment to Mr. H. P. Tiemann is due for assistance obtained from his book, "Iron & Steel.")

TRAINING THE WORKERS

The General Electric Co. has established in connection with its educational system a business training course for college graduates without technical education.

According to the company, this course is aimed to give training in the principles of higher accounting, a knowledge of which is of great importance to those who hope to become executives; to explain the essential elements of business law made necessary by Governmental supervision of corporations, the tax laws, and other complexities of our modern economic life; and lastly, to apply this general knowledge concretely to the business of the company, which is necessarily intricate owing to the size of the organization, the volume of its sales and the wide range of articles manufactured.

The training course consists of actual employment during the business day, in one of the departments where the student will become familiar with the practical work and the departmental functions. The class work will engage the best efforts of the student for eight or ten hours each week outside the class periods, which are held two evenings a week from 5.30 to 7.30 p.m. The course is divided into semesters and is in progress during the usual months of the college year.

In a paper entitled: "Some Aspects of Electrolysis of Pipes," presented to the American Waterworks' Association, Professor G. Alleman points out that damage only occurs where the leakage current either enters or leaves the pipe. It can be easily calculated that one ampere flowing continuously from an iron pipe for a period of one year is equivalent to 20 pounds of iron or 74 pounds of lead, and it must be remembered the current which leaves the pipe at one point may return at another place if the electrical conditions are favorable. In consequence, and as supported by experience, the effect of the current may correspond to any multiple of these two equivalents. Iron pipes are not corroded when electrically negative, and the opposite condition applies to lead pipes, but it is also possible under certain conditions for lead pipes to be corroded when electrically negative.

The British American Shipbuilding Co. has closed its doors and the activities of one of Welland's foremost industries are ended. The loss of this splendid industry is a serious one, and hopes are entertained that the plant, which is for sale, will be purchased.

A Real Home for Out-of-Town Apprentices

Realizing That Many of Their Apprentices May Be Out of Town Boys, This Concern Has Set Aside a Special Home for Their Use—Read the Particulars, It is Well Worth While.

By J. M. HARRISON

NO doubt many an old timer will remember the period when machinist apprentices worked for the handsome sum of 75 cents weekly. The writer himself did not come under such a classification, but received the small fortune of $3 for his initial year at the business. Contrast the two former conditions with those of the present day. An apprentice receives a better salary, works under better conditions, and in every respect has a real chance to advance. In some plants, including the one we will mention later, apprentice classes are formed, and in these classes are taught the different branches of mathematics. Some concerns hold these lessons right in their plant, while others send the boys to some nearby technical school, but in either case the principle is the same.

The chief aim is to advance the apprentice as speedily as possible, allowing him to obtain both the practical and technical knowledge. In the old days if an apprentice made up his mind to reach more than the usual standard, it meant burning the midnight oil, and plunging into whatever textbooks he could find on the subject. These unfortunately were sometimes not of the highest order.

The young chap learning his trade at the present time has very little excuse to offer if he does not advance in his profession, for even should his concern not hold apprentice classes as already described, there is always the modern technical school evening classes which

are open at a very low cost. Competent instructors are there to see that the boys following up their various professions get a thorough technical training, and it depends on the interest taken by the student what the final result will be.

The well-known firm of Brown and Sharpe Manufacturing Co., Providence, R.I., have always been noted for what they have done for their apprentices, but they have added still another feature of interest, namely, a home for out of town apprentices.

Realizing that many of their apprentices may be out of town boys who are taking the training in their apprentice department, they have set aside the home as shown in the photograph for such lads. It is located in a good, residential section of the city, and is only a short distance from the works.

On the ground floor are the housekeeper's apartments, two sleeping rooms, a living room and a closed in sun porch. On the upper floors are good sized halls, sleeping rooms, toilets and a large wash room containing lavatories and shower baths. Hot water is always available from an automatic heating system, and there is steam heat in every room.

Each boy has an extra width spring, cot with mattress, pillows and bedding of the best quality, a chiffonier, an armed rocker and rug, and each room contains a study table and chair. There is no gas in the boys' apartments, all lighting being by electricity. Bath and hand

towels are furnished and these, together with the bedding, are laundered by their own electrical equipment in the basement of the house.

The living room is furnished with an aim toward comfort and homelike environment, and a cabinet Victrola adds to the enjoyment of the evening and week-end hours of the boys. A supply of magazines is always on hand, and through the extension service of the Providence Public Library the book shelves are kept filled with good, wholesome reading matter. The closed in sun porch, opening through French windows from the living room, makes an attractive place for lounging and games.

The house is under the supervision of the apprentice department, but is directly conducted by a responsible housekeeper and an assistant. There are no set rules, the boys simply being asked on coming into the house to do there as they would in their own homes. The policy is to maintain a thoroughly homelike atmosphere and to avoid the institutional tendencies often evident in such places.

No meals are served, but there are boarding places nearby where board may be obtained at the current rate. Room rent is not taken out of the pay, but each boy pays the housekeeper a fair amount weekly, this charge including the towel service.

Application for a room is made with the application for an apprenticeship by sending a note with the application or by writing on the margin of the appli-

APPRENTICE HOME OF BROWN & SHARPE—EXTERIOR AND LIVING ROOM.

cation blank. If there are no openings when the apprentice begins his course his name is placed on the waiting list and he is helped in finding a satisfactory room to use until there is a vacancy in the dormitory. Some of these temporary rooms are with families of men long employed by the company.

Our boys like to live at "Apprentice House." It is clean and quiet, and affords opportunities for good companionship that may well be compared with the valuable associations of a college fraternity house.

A scheme such as this is to be highly commended, and it goes to show how much this concern has the interest of their apprentices at heart. No wonder the boys like to live at such a house, for under such careful management it is always kept clean and quiet, and affords opportunities for good companionship which may well be compared with the system adopted by various colleges.

HANDY INFORMATION

In the September copy of "Threads" there is some information which is well worth reprinting. The matter we refer to is used in the "He wants to know" section, and here are the questions asked, and the answers given.

The first deals with information on the tolerance of diameters of screws before threading, the answer being as follows:

The best practice is to have the stock turned to correct diameter before running the thread chasers on. When the stock is much oversize, it puts the die head in a very bad position, acting both as a milling tool and a threading tool. This one point causes as much screw threading trouble and breakage of chaser teeth as any other one cause.

A customer has been successful in threading ⅝-inch—16 malleable iron lugs with a geometric threading machine, where the thread was cut directly on the casting, the pieces being 1-32-inch oversize, rough, and the threading operation reduced the diameter to about 1-64-inch under size, because the pieces were galvanized after threading. The chasers threaded 1,000 pieces between grinds. This customer referred to their work as "a strenuous job."

The next question was this "Are high speed steel chasers preferable to carbon?"

High speed steel chasers are used when speed is the chief consideration. They are not recommended where an especially fine finished thread is required. On a thread for what might be called a commercial finish, and where quantity rather than quality is desired, high speed steel has the advantage. High speed steel thread chasers are not recommended for cutting brass, for they have a tendency to stick to the work.

Fine finish and highest speed do not go together, and where threads of fine finish are required carbon steel chasers must be used. All grades of carbon steel are of much finer grain than high speed steel.

The next item is an answer rather than a query, and is presented in letter form as follows:

In the July, 1920, issue of your joyful little publication is stated the fact that you know of no satisfactory method of measuring internal threads except with standard plug gauges. Undoubtedly this is the most generally adopted method for production work, although there is a formula and method which is given in Communication B 523 of the Gage section of the Bureau of Standards. The formula is as follows:

$$E = M - \frac{\cot . a}{2 n} + G (1 + \mathrm{cosec}. a)$$

in which E = pitch diam.; M = measurement between balls; n = No. of threads per inch; G = diam. of balls; a = half of included angle of thread.

Instead of wires as used on plugs or external threads, balls of the diameter best suited to the size of the thread are used and the measurement between these made with gage blocks in combination with a slide parallel in order to secure the correct adjustment and pressure.

I think that this is the formula which the party desires, but is practical only in exceptional cases on ring thread gauges.

Next on the list comes this query: "Why should two chasers do the cutting while the others of the set do none?" Here is the answer:

Uneven chamfer is most likely the cause of this unequal cutting. Prove the chamfer even from the cam slot. The grinding should be done from the keyway and not from bottom or top of chaser. The cam may be out of true, and cause the condition; or, possibly, the distance from the cam slot to the cutting face of the chasers is not true.

Lastly we have this one: "A 1¼-inch collapsing tap does not always close up, and often tears the threads."

It is probable that the threads, when tearing, clog the chasers and hold the tap stiffly, so as to interfere with its collapsing properly. Our suggestion is to try grinding the chasers with a slight hook, which will, no doubt, stop the tearing of the threads. The cam spring in the tap may have become weakened, and this would account for the tap not closing. A weakened cam spring could readily be replaced by a new one. The beauty of presenting these excerpts is that the very problems under discussion may help you solve a similar difficulty.

A movement is under way at Windsor to maintain a floating hotel to relieve the lack of hotel accommodation in the border cities. President Bourke, of the Great Lakes Transportation Co., declares his willingness to moor one of the company's vessels where it would be easily accessible to travellers.

ONE OF THE SLEEPING ROOMS AND WASH ROOM IN BROWN & SHARPE APPRENTICE HOUSE.

Gears From a Purchaser's Standpoint

By D. G. STANBROUGH

IN a paper read before the American Gear Manufacturers' Assoc. convention at Detroit, the author deals with questions relating to personnel and organization, labor market, plant buildings, and equipment, in so far as these affect the gear-manufacturing industry. He next dealt with the fact that satisfactory gears from a purchaser's standpoint can only be produced as a result of conformity to good practice along the following lines: (1) Design; (2) materials; (3) forgings and castings; (4) heat-treatment; (5) machining; (6) hardening, and (7) inspection.

There are certain fundamental considerations of design, he said, that can be counted upon to produce good gears, although a partially satisfactory product may be turned out if all of these fundamentals are not strictly adhered to. The good features of gears are uniform sections and a tooth form which will meet all the requirements as to strength. The design must also take into consideration the practice of the plant in which the gears are to be manufactured. Take, for instance, the familiar cluster gear used in automobiles. This cluster can be made from the blank with integral gears, or can be made with three integral gears and a fourth gear riveted or fastened to a flange. In the first case, a certain method of cutting has to be employed, and there will be unquestionably a considerable loss from warpage in heat-treatment. In the other case, the loss from warpage is reduced, but, on the other hand, considerable grinding is introduced, and a nice fitting job is necessary. The decision as to the design should naturally rest with the practice of the shop. Good gears can be produced by either method, and the one to be adopted depends more upon which is in vogue than upon any technical consideration as to the results that can be obtained with either method.

Method of Mounting

Another big thing which needs careful attention in designing gears, if satisfactory results are to be obtained, is the method of mounting. The best cut gear that can be made will not give satisfaction unless properly mounted. Many a noisy gear would not be noisy if it were run on quiet bearings. The mounting of gears should be such as to take care of the stress incident to the pressure angles. The shafts should be stiff, the mounting rigid, and the bearings as close to the centre of the load application as it is possible to make them. In designing a gearcase, strains should receive careful consideration. The manufacturer cannot be expected to turn out a finished product which conforms to the customer's specifications unless the product be given at least fair conditions

under which to operate. Much time and energy can be saved by giving more consideration to the design of gear mounting before the manufacturing work is undertaken. Too little attention has been paid to the mounting of gears by designers in the past. The manufacturers can do the purchaser a favor by pointing out the deficiency in design before undertaking contracts.

Materials for Gears

In selecting the materials from which gears should be manufactured, the conditions under which the gears are to be operated should be considered. First, those subjected to practically no stress and very little wear should be made of bronze, cast iron, untreated bar stock, and aluminum. Second, if the service will produce great wear, but not severe stresses, the material should be "straight" carbon steel, and should be carburized. Third, if the gears are subjected to extreme wear and considerable stresses, they should be made from forgings of alloy steel, and should be carburized. Fourth, where severe impact and other stresses are involved with a normal rate of year, high-grade alloy steel forgings should be used, oil-treated and not carburized. Fifth, where the gears are to be subjected to extreme wear and extreme stresses, they should be made from self-hardening steels which have been forged.

Castings and Forgings

No serious difficulty is encountered when the gear blanks are cast, although it is important to keep the sections uniform in order to guard against shrinkage cracks, especially in the case of gear blanks made from steel castings. With cast iron it is important, in order to obtain long life, that the combined carbon be high, and that the metal be cast so that the blank will be chilled, thereby retaining a sufficient quantity of the carbon in solution. The blanks should not be so hard that they cannot be machined without annealing, the most satisfactory results being obtained with a scleroscope hardness of about 35; the combined carbon content should be between 0.30 and 0.50 per cent. No particular difficulty is experienced with bronze and aluminum castings.

The forging of gear blanks represent a considerable problem. The temperatures at both the beginning and end of the operation should be held within certain well-defined limits, according to the forging temperatures of the steels being used. High temperatures produce burnt gears, which sometimes can be used only after an expensive corrective treatment process has been employed. A low finishing temperature results in cold strains which, when relieved in heat-treatment,

reveal cracks, seams, and fissures. Another phase of the forging of gears is in regard to the type of forging machines employed. Generally speaking, the flow lines of the metal in a forging should be at right angles to the forces applied to the gear teeth while in operation. It follows, then, that a gear that is flat while gears having bosses, and those in which length is the predominant feature, can best be produced by a forging machine. The size of the hammer has an important bearing on the subject; a large gear cannot be satisfactorily forged under a light hammer, as the time required to close the dies permits the gear to cool and results in too low a finishing temperature.

Preliminary Heat-Treatment

Practically all gears that have been forged should be given a preliminary heat-treatment before machining, particularly those gears that will be subjected to great stress and that must be practically free from distortion after hardening. On mild steel this initial heat-treatment consists simply of annealing or normalizing. Normalizing the steel eliminates whatever stresses have been set up in the forgings and renders the gears suitable for machining. On the alloy and high-carbon steels the normalizing treatment is not sufficient. Gear blanks made of these steels are first quenched from a suitable temperature and then drawn sufficiently to enable them to be machined.

Machining of the Blanks

In manufacturing gear blanks it is important that the surfaces be held parallel and that they be machined true with the holes in the blanks. In the author's experience, broaching will produce a straighter hole than reaming. Gears that have splined holes should be machined from the splines, both in the blanking and cutting operations. It is also essential that the pitch line of the gear be concentric with the hole.

Probably the simplest method employed for locating the gears when finish-grinding the bores and faces is the one in which the work is located from the periphery. Another method employed is that in which the gear is located from the root diameter; and still another method—and the most difficult of the three—is that by which pins carried in the chuck contact with the gear teeth at the pitch line. Locating a gear from the root diameter is the most practical method. When this practice is followed it is necessary to finish-cut the root diameter. This method also enables the locating pins in the chuck to be made stationary for each size of gear, which is

(Continued on page 339)

WELDING
AND CUTTING

Possible Economies in the Application
of the Oxy-Acetylene Process

THE following article, by the secretary of the French Acetylene Association, on possible economies in the application of the oxy-acetylene process, is both timely and of great interest.

The cost of blowpipe welding has increased to such an extent compared with other methods of making joints that there is a risk of its remarkable progress being stopped. An investigation to determine what methods should be adopted to avoid this reveals the fact that economies of all kinds can be realized in the application of the process. With the present day methods economy in one way or another will give a total saving of 20, 30, 40 and even 50 per cent.

In the first place consider the raw materials which at the present time are so costly. How many firms call for a test of their supplies of oxygen and carbide? The majority of users of the process assume that there is only one quality of carbide, and consequently the question of whether the carbide is good, bad, or indifferent never crosses their m'n1. Suppose, for example, carbide is bought yielding 200 litres of acetylene per kilogramme of carbide instead of 300 litres, unless a repayment claim is made based on the analysis, the cost to the buyer is increased by 'one-third, to say nothing of the impurities in the gas, which accompany inferior carbide, or the additional manipulation necessary.

Regarding Oxygen

With regard to oxygen, in France it is guaranteed to be 95 per cent. pure, but very few firms test to verify this degree of purity. Suppose, for example, it is not more than 90 per cent. pure, a fact which would not be discovered when using a welding blowpipe, the buyer's costs are increased by at least 12 per cent.; of this figure 5 per cent. is loss in purity of the gas and at least seven per cent. due to the reduction in the speed of welding. The welds produced are also distinctly inferior.

The utilization of carbide and oxygen calls for economy. Certain acetylene generating plants will, through excess production and bad design, discharge into the atmosphere one-quarter and even one-third of the gas developed. With

others the generation is accompanied by intense polymerisation, as is shown by the colour and nauseous odour of the residue; the loss in this case may be from 20 to 30 per cent. Again, there are avoidable losses in charging and cleaning, using wrong sizes of carbide, and wastage of small carbide. There may be small leakages in the apparatus, at the joints and valves, which in the aggregate amounts to considerable waste and represents appreciable loss.

In the case of oxygen, how many firms check the cylinder contents or possess an accurate pressure gage for this purpose? Suppose, for example, the pressure is 100 or 110 atmospheres instead of 120, there would be a loss of 10 or 15 feet of oxygen per cylinder, which, at the present price of oxygen, might mean considerable loss. There may be a leakage at an oxygen valve, or through a defective regulator, or through porous flexible hose, or through a defective union or connector, consequently it is not difficult to use almost 200 cub'c feet of gas where 100 cubic feet should be used. There is also considerable loss in the return of empty oxygen cylinders. Frequently, owing to lack of supervision, carelessness or negligence on the part of the welder, cylinders are returned containing gas at a pressure of 80 pounds per square inch and upwards. It is not uncommon to find full cylinders mixed with empty ones.

The Blowpipe

Consider next the blowpipe. This offers a very large field for economy. It is well known that the consumption of oxygen and acetylene should approach equal volumes and that the average for well-known blowpipes will work out about 1.2 to 1.3 cubic feet of oxygen for each cubic foot of acetylene. But if consumption tests are made in any workshop it will be found that the figure is nearer 1.8. In other words, there is a waste of half a cubic foot of oxygen per cubic foot of acetylene used; thus, with a blowpipe consuming 20 cubic feet of acetylene per hour there is a waste of 10 cubic feet of oxygen per hour, in addition, the speed of welding is reduced and the welds inferior.

The rates of the consumption of oxygen to that of acetylene does not entire-

ly depend upon the type of blowpipe used, it also depends upon the proper maintenance and the working conditions of the blowpipe, and, in a very large measure, upon the pressure at which the oxygen is used, a pressure which is invariably too high.

With regard to the actual welding operation, there is the case where a blowpipe of insufficient power for the work in hand is used, as a result of which the welding speed is too slow, even for the expenditure of gases. If the blowpipe is too powerful it is necessary to raise the flame in order to avoid holes; in this way heat is wasted and, of course, oxygen and acetylene in addition to the loss of time.

In one workshop we found for a given class of work that 12 feet per hour was the speed of welding, whilst in another works doing exactly the same work the number was 18 feet, and the welds were good and carefully executed. This result, which shows an economy in gases and labor of 33 per cent. is simply due to the better preparations and better execution of the welds.

Welding Backward

It has been shown that the new method of executing welds known as welding backwards and which can be applied to varying from ¼ to ⅜ inch in thickness, will give an economy in labor and gases of at least 25 per cent. How many firms have studied this important development and applied this method?

There are also important economies to be shown when well designed and appropriate jigs are used, when the edges are well prepared, and when the comfort of the welders is studied. The blowpipes which burn in the air whilst a welder turns the work, searches for a welding wire or, perhaps, rolls a cigarette, are wasting unnecessarily the principal materials which, as we know, are very dear.

The objection might be raised that it is an easy matter to describe methods of economising but in industrial application the proposals become more or less theoretical. Our answer to that is that the figures and methods of economising spoken of have been obtained and applied in the welding firm of Girel & Co. This firm employs 500 welders and consumes about 150 tons of carbide per day and 110,000 to 130,000 cubic feet of oxygen per day. The carbide and oxygen are analyzed on their arrival and the contents

of the cylinders verified on receipt and in returning them; the yield of the generating plant, tests or leakages, the overhaul of blowpipes and regulators, are made a regular part of the work. Records as to the oxygen used daily by each welder are kept, so that it is an easy matter to determine the consumption of acetylene and thus to fix the cost per foot of welding, and then to constantly endeavor to lower the cost by a better utilization of the flame, using better devices for assisting the welding operator, etc. As a part of excellent collaboration on the part of the several foremen, the foreman welders, and the welders themselves, and the excellence of the materials used and the general organization, we have obtained results which, in these days, would be considered remarkable. As an example, in manufacturing certain articles which other workshops were also engaged upon, the cost of oxygen and acetylene was only one-half of that of other important firms. The production for the same amount of labor was double that of other firms, and finally, after Government inspection of the work, there were no rejections.

This is only one example where great economy has followed the paying of attention to the points indicated. It is certainly time that all firms engaged in welding operations paid more attention to the economical side of the process if only from the point of view of the quality of the work.

AN UNUSUAL WELD

A repair job of interest to all engineers was recently carried out by means of the electric arc welding process to the main engine cylinder of the steamer Chippewa.

Our readers will be familiar with the Chippewa's old-fashioned beam engine or "walking beam," as it is sometimes called. Needless to say the engine only has one cylinder, so that when a vertical crack some 40 inches long developed in the cylinder wall during the busiest part of the season, something had to be done about it. The cylinder is 5 ft. 6 in. in diameter by 12 ft. long, and carries a working pressure of 60 lbs. per sq. in.

After consultation with the Lincoln Electric Co. as to the possibility of successfully welding the crack, it was decided to proceed with repairs as quickly as possible, by means of a method which has been very successful in repairing large iron castings subject to heavy working stresses.

Holes were drilled down each side of the crack, and steel studs were screwed in and cut off about 3-4 in. from the cast iron. By the time the last stud was in, the Lincoln Electric Co's. motor truck carrying the portable electric welding outfit was on the dock within 20 feet of the Chippewa's engine room. The cables were connected up and the welder went to work to weld those studs and the cast iron of the cylinder into a

SHOWING CRACK EXTENDING 40 in. DOWN FROM TOP OF CYLINDER, STUDDED READY FOR WELDING.

FINISHED WELD RESEMBLING STEEL PLASTER ABOUT 1 IN. THICK, APPLIED TO OUTSIDE OF CYLINDER WALL FOR FULL LENGTH OF CRACK.

Note:—It was found that this method of welding had the effect of drawing the edges together and closing up the crack as the weld was made, from the bottom up.

solid mass of metal. Twenty-four hours later the job was complete and shortly afterward the Chippewa steamed out into the lake for a trial trip.

The weld was found to be quite tight and appeared to be a perfect job. The Superintendent Engineer deemed it advisable, however, to make assurance doubly sure by re-inforcing the weld with two steel bands placed right around the cylinder.

Electric arc welding has been used extensively of late along the water-front in Toronto for boiler and hull repairs but this is one of the first local instances that has come to our notice of the use of the process for a major repair to a large and important iron casting.

There was recently put into service in America, at the Government League Island Navy Yard, Philadelphia, a crane which is described in the General Electric Review (U.S.A.) as the largest hammer head type fitting-out crane which has yet been built. The crane is approximately 250 feet high overall and can handle a 350-long-ton load at 115 feet radius, and a 50-long-ton load at a 190 feet radius. The maximum lift of

the main hook from the deck of the pier on which the crane is located is 145 feet, but the drum holds sufficient cable so that the hook can be lowered 25 feet below the deck of the pier, making a total lift for the main hook of 170 feet. The 50-ton hook has a maximum lift of 180 feet. The total length of the swinging jib is 300 feet, of which 100 feet is on one side of the centre line and contains the machinery house and counterweight, while the other 200 feet is on the opposite side of the centre line and contains the runways for the trolley carriages.

HEAT TREATMENT OF STEEL

In a paper recently read before the Detroit Convention of the American Gear Manufacturers' Association, Mr. G. W. Yale advocated the use of the electric furnace for the heat treatment of steel. A lead bath or an ordinary coal or gas fuel furnace has, he observes, a very great heat capacity, and is accordingly but little cooled down when the material to be heated is placed in it. The consequence is that small parts heat up more quickly than large ones and the result is often disastrous. He considers for example, the case of a "chunk" of steel 4 in. in diam., with a spur 1-4 in. thick which is put into a big furnace or lead bath at a temperature of 1,500 deg. F. The bath or furnace is chilled but slightly, and the spur being small in volume heats up rapidly and gives a large grain structure, whilst the heavier portion becomes heated more slowly.

Steel expands when heated, both when the temperature is below and when it is above the transformation point, but at the actual transformation point it contracts, and hence if different portions of a mass of steel under treatment pass the transformation point at different times, distortions are likely to occur, and there may be cracking and failures.

If the same piece of steel be heated in an electric furnace of small heat capacity, the furnace is chilled when the steel is placed in it and the temperature can then be raised slowly, in which case both light and heavy portions of the article under treatment pass through the transformation point at the same time.

Although the cost of electric heating per B.Th.U. developed, is much higher than in direct heating by coal, gas or oil, the cost per B.Th.U. usefully employed is not very different, since, with electric heating, the heat is generated just where required, whilst with other systems of heating 60 per cent. to 80 per cent. of the total heat passes off up the chimney. The labor costs are also markedly different since with direct heating the total cost of the labor includes that expended in cleaning off scale, straightening warped pieces, or grinding operations, which are said to be unnecessary when the heat treatment is conducted along the lines which Mr. Yale advocates. Taking the case of such articles as automobile gears, it is stated that two men can, with the electrical method, turn out 2,000 lb. and 2,400 lb. of treated metal per day.

DEVELOPMENTS IN
SHOP EQUIPMENT

CONTINUOUS MILLING MACHINE

THE machine about to be described
is the product of the Newton
Machine Tool Works, Inc., Philadelphia, Pa., and is known as the first
model continuous milling machine. Figs.
1 and 2 will give readers a good idea of
its general appearance without work or
cutters, and starting from this point
we will discuss a few of this machine's
features as spoken of by the makers.

The column and base are a one piece
casting in order that all strains resulting from bolted joints may be obviated.
The table is circular in form, with special
provision for mounting the various forms
and types of holding devices as is required for the rapid machining of a
variety of parts. It will be observed that
this table is adjustable upon the base
in order to provide for the proper positioning of the jigs carrying the work in
relation to the cutters, in order that the
work may be placed upon the smallest
possible radius, and utilize the least possible diameter of cutter.

It will be realized the advantages of

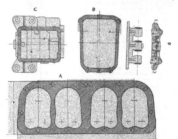

FIG. 4—SKETCHES OF WORK DESCRIBED IN TEXT.

FIGS. 1 AND 2—SHOWING THE MACHINE FROM BOTH SIDES.

FIG. 5—THIRTY PIECES PER HOUR ARE COMPLETED AT THIS OPERATION.　　　　FIG. 6—MACHINING GEAR CASE COVERS. NOTE FIXTURES USED.

positioning, from the viewpoint that the production is largely controlled by the number of pieces removed to one revolution of the table; hence, it is essential that these be placed with the least possible space between them, whether this space be in the form of an open block, or whether it be in a wedge shaped form, due to the pieces being laid out with too much angularity in relation to each other, and this can only be overcome by having the table adjustable as described. These tables may be 24-inch, 36-inch or 48-inch diameter on their working surface, in addition to which they are surrounded by a pan for draining lubricant when working on steel.

The adjustment to the table provides for the machining of pieces with a flat surface, or machining pieces of a forked character, in which a gang of cutters are used. The rotative movement is controlled by a fixed feed, so that there is no possibility of the operator increasing or decreasing the production of the machine, as any holding device passing the loading station will immediately at-

tract the attention of the man in charge of the section, and in addition to this, as the functions of the machine require so many stations per hour to pass the loading station, that number of pieces must be loaded and unloaded or a clear explanation made as to the loss. There is, however, a provision in the feed mechanism by which this rate of feed can be increased or decreased by those in authority.

The spindle head is equipped with two spindles, each of which is individually adjustable for setting the cutters to gauges. The left-hand spindle is used for the roughing operation, and the right-hand spindle for finishing. The distance between centres of these spindles varies from 22½ inches on the largest machine, to 13 inches on the smallest machine, so that very long pieces can be machined with the roughing operation complete before the finishing cutter starts on the same piece. Demonstration has illustrated that this is perhaps the most accurate method of machining parts in quantities, because

any wear or any variation in the machining occurring, due to inequality of castings, is absorbed in the finishing operation, and owing to the small amount of work to be performed by the finishing cutter, accuracy of both finish and dimension are maintained with infrequent grindings of both cutters required, due to this condition. Secondly, with the ability to pick up and obtain accuracy with the finishing spindle, it permits of the running of much higher cutting speeds and feeds than otherwise practical.

The makers state as follows: We have frequently taken pieces 6 inches x 20 inches in length out of the jig and laid them on a surface plate to find that they will hold six papers, and also that it will be impossible to at any point enter a thickness gauge of .0015, indicating that we not only secure a high finish but we do not have any warpage of the casting, or any digging in of the cutter on the leading side. The speeds of the spindles are controlled in the same manner as the feed described above, and a

FIG. 7—FIRST OPERATION ON GEAR BOX REFERRED TO AT FIG. 6.

FIG. 9—A BETTER EXAMPLE OF JIG CONSTRUCTION.   FIG. 10—MACHINING BRACKET AS SHOWN AT "D." SEE FIG. 4.

large element of the success of this work lies in the design of the worms and worm wheels used in the driving of the spindles these worms and wheels being made over a method peculiarly our own, and as the result of very exhaustive experiments carried on in the year 1914.

The housing, or head, carrying the spindles is adjustable on the column, and this again is a necessary element of this type of machine, in order that the holding device may set as flat or low to the table as the casting itself will permit, and the head adjusted downward provides for the minimum extension of overhang to the cutter and spindle; whereas, if this head were a rigid frame having a fixed distance above the table, it would be necessary to build a jig up into the air in order to bring the work to the cutters, instead of bringing the cutters to the work as is done in this design.

One of the big elements entering into the high rates of feeds and speeds that these machines maintain is the manner in which the work presents itself to the cutter. As the work is revolving it

will, as you will realize, present a constantly changing angle to the cutter; thereby providing the benefits of the shearing action and overcoming the flat, heavy cutting which is obtained by operations maintained on a straight line.

The machine is made with all driving gears hardened, enclosed and running in oil, with the result that the combined elements of a heavy, rugged frame, exceedingly large bearing surfaces, large diameter spindles, a well balanced drive, providing a surplus of power, means the complete elimination of vibration, and some of these machines have been in operation for several years without having required any replacements or any stoppage other than for changing holding fixtures or cutters.

Fig. 3 illustrates an automobile cylinder head chucked from the rough casting by the gauge appearing in the photograph. In other words, the casting is placed in the holding device, the gauge set in position and the casting then adjusted to the gauge which locks the casting from the combustion chamber, in order that a uniform area of combustion

is obtained. The production of these pieces is 30 per hour. The sketch Fig. 4 at A, gives the outline dimensions of these castings. Fig. 5 shows the same casting, and is intended only to illustrate the statement made relative to the angularity of the pieces and the closeness with which they are chucked one to the other. In this instance, owing to the water connection appearing at one end of the casting, it would not be practical, even though the circle were increased, to reduce the space appearing between each casting. Were the casting free of this water connection, then a larger number of pieces on a larger circle would increase the production, because the idle time, as indicated by the gap between the castings, would be omitted.

Fig. 6 illustrates a casting which is a gear case cover or lever stand of an automobile. Observe how closely these pieces come to each other, so that there is no loss of time in the cut leaving and starting from one piece to another. The sketch B, see Fig. 4, shows an outline of this piece, production of which

FIG. 3—THE GEAR BOX AT ITS SECOND OPERATION.

is 120 per hour. Both these illustrations will clearly show the relationship between the size of the cutter and the spindle, so that some comparative idea may be formed of the ruggedness of the entire apparatus.

Fig. 7 shows method of chucking gear box referred to above for the first operation, and Fig. 8 illustrates method of chucking the same gear box for the second operation. The production of these boxes runs from 100 to 125 per hour.

Fig. 9 shows a better construction of jig for a similar piece, which is made of malleable iron, the production of which is 100 per hour, this piece being show at C, in Fig. 4. The photograph Fig. 10 shows method of facing bosses on a bracket. This bracket is shown at D, in Fig. 4, in which you will observe

there are six faces to be machined, three of which are on one line, and three on another line, at an angle to the first, so that two operations are required to complete this piece. At a feed of only 4 inches per minute, a production of 30 pieces per hour is provided. Much greater production could be obtained if required. It will be readily understood that it would be quite impossible for one operator to do anything like this amount of work by spot-facing the bearings in place of the process described. These various views are used as typical illustrations of two different classes of work applicable to this type of machine, but they are also used for a variety of other operations, which are too numerous to describe in an article such as the present one.

FROM A PURCHASER'S STANDPOINT

(Continued from page 333.)

an advantage. Although the method of locating from the pitch line has given satisfactory results, it is probably just a little less accurate owing to the fact that warpage produced in heat-treatment will have a greater effect at the pitch line than upon the root or at the top of the teeth. Locating the gear from the periphery is open to the objection that greater accuracy is required in the primary operations in order to get satisfactory results. Further difficulties to locating from the periphery arise from the fact that burrs are often thrown up, which, if not removed, will cause errors.

Final Hardening of Gears

All gears made from carburized stock, and, as a rule, all oil-treated gears, are finish-machined before final heat-treatment. Although considerable difficulty may be experienced in the final heat-treating process, the "hump" method developed by Leeds and Northrup Company, Philadelphia, Pa., is one of the most satisfactory of the various methods which may be employed. Distortion in gears is dependent to a large degree upon heating. The range of heating, together with the establishment of constant quenching temperatures, can be relied upon to solve problems in gear distortion which formerly were believed to be uncontrollable. To a large measure the solution of these temperature-control problems is due to the development of the electric furnace.

Methods of Testing Gears

After the purchaser has satisfied himself that the various manufacturing conditions required to produce a satisfactory gear have been complied with, attention should be directed to the final inspection process. The hardness is of much importance, but it should be remembered that tests such as the scleroscope furnishes are comparative only, and that standards of hardness to suit the service to which the gears will be subjected should be established. The limits of these standards should be such that the gear will withstand a reasonable amount of wear, but it should not be so high as to make it brittle. The work of inspection, then, so far as hardness is concerned, is to determine whether the gears are within the limits allowed.

After the hardness test, each piece should receive a rigid visual inspection under good lighting conditions for the purpose of discovering cracks and seams which may develop during the final hardening operation. With reference to inspection of tooth form, a rolling fixture in which the gears are mounted at the correct centre distance and rolled with a master gear is recommended. A hand-rolling fixture will give better results than a power type. The gears should be tested for backlash and for conformity to the dimensions specified by the established practice of the manufacturer.

NEW THINGS IN MACHINE TOOLS

OT-STEEL BENCH LEGS

The Angle Steel Stool Company, of Otsego, Mich., are placing on the market a special line of steel angle bench legs made in various heights and adapted for many purposes. The legs are flat on top and are of pleasing design and where extra width is required the sections may be placed back to back. All legs are finished with a baked-on coat of green enamel.

STATIONARY HEAD DRILLING MACHINE

A heavy-type drilling machine with a stationary head has been developed by the Sibley Machine Company, of South Bend, Ind., which is particularly adapted for production work. Apart from the fixed head the machine is very similar to the sliding head type of the same company. Back gears are of the sliding type and are operated by a lever. Feed change gears run in oil and all gears on the machine are completely enclosed with suitable guards. The machine may be furnished with special equipment if desired. The floor space is approximately 23 by 63 inches.

20-INCH DRILLING MACHINE

The 20-inch drilling machine of the Silver Manufacturing Machine, of Salem, Ohio, has recently been re-designed, the features of reconstruction including a new style of frame and an increased capacity beneath the drill spindle, the maximum distance being now 28 1-4 inches to the base plate. The spindle is fitted with S. K. F. ball thrust bearings. The machine may be obtained with either hand or power feed and can be supplied for both belt and motor drive as desired. The machine is also built in gangs of two, three or four spindles.

HERCULES PRESS-BROACH

The 15-ton arbor press and vertical broaching machine formerly made by the Hercules Machinery Company has re-

cently been redesigned and it is claimed that the number of operations capable of being performed by the new machine has been increased. The operating ram has a stroke of 18 inches and may be driven either by belt or individual motor. The machine is fitted with a knee that has a vertical adjustment of 14 inches, and this knee may be removed and replaced by fixtures or conveyors, if desired. A safety pin is provided for over-load of the press. The travel of the ram is governed by means of an automatic trip.

AUTOMATIC ACETYLENE GENERATOR

An automatic acetylene generator that is primarily intended for use in oxy-acetylene welding, cutting and lead-burning, has been developed by the Imperial Brass Manufacturing Company, of Chicago, Ill. The apparatus is mounted on a truck, making a portable unit. The operation of the generator is automatically started when the gas begins flowing to the torch. The pressure is also under automatic control, and by means of a blow-off is maintained always below 15 pounds. The gas is passed through a water seal as it passed from the generator. Several sizes of the apparatus are manufactured.

NEW TYPE OF GUARD

The Modern Tool Company, of Erie, Pa., are now equipping all their machines with a new type of guard that has the advantage of affording visibility of all the movable parts, at the same time adequately protecting the workmen from coming in contact with any of the mechanism. The guards are of the phantom type and are made of expanded metal and secured to the frame of the machine by means of angle irons. Hinged portions are provided to facilitate the oiling and cleaning of the different parts of the machine protected.

The MacLean Publishing Company
'   LIMITED
          (ESTABLISHED 1887)
JOHN BAYNE MACLEAN, President. H. T. HUNTER, Vice-President
        H. V. TYRRELL, General Manager.

          PUBLISHERS OF

CANADIAN MACHINERY
✒ MANUFACTURING NEWS ✒

A weekly Journal devoted to the machinery and manufacturing interests.

B. G. NEWTON, Manager.   ·   A. R. KENNEDY, Managing Editor.

          Associate Editors:
J. H. MOORE   T. H. FENNER   J. H. RODGERS (Montreal)
Office of Publication: 148-153 University Avenue, Toronto, Ontario.

VOL. XXIV.   TORONTO, OCTOBER 7, 1920   NO. 15

PRINCIPAL CONTENTS

British and American Trade

THE extent to which United States producers of steel, finished and semi-finished, have captured the Canadian market can not be appreciated without a study of the figures.

If one cares to go back a few years to 1883 it will be found that the United Kingdom and United States were doing about an equal share of the business of selling steel goods to Canada, there being not more than a couple of hundred thousand dollars' difference in the figures.

But see what has happened in recent years. Taking the figures for the period ending March 31 in each case the years 1913 to 1920 show this:—

Year	Steel Imports From U.K.	Steel Imports From U.S.
1913	$10,394,276.00	$106,471,913.00
1914	10,132,543.00	85,729,001.00
1915	7,402,894.00	55,462,477.00
1916	3,950,000.00	71,459,771.00
1917	4,683,103.00	119,754,365.00
1918	4,081,681.00	154,113,633.00
1919	5,912,788.00	154,426,648.00
1920	5,167,100.00	144,669,402.00

How did this come about? The Canadian Association of British Manufacturers, in a little book published for the purpose of increasing British trade in Canada, points out that there was a great opportunity in 1896 following the coming of preferential tariff treatments in Canada for the British to expand their business here. The article in question points out that "United Kingdom manufacturers never had a better opportunity to capture the now vast trade of this Dominion than that which presented itself in the years immediately following. But what was the result? Most pitiable! Not only did they not hold their established proportion of trade, but by a want of confidence in Canadian integrity, by neglect, by a pursuance of the lines of least resistance, by failure to adequately and substantially support accredited representatives, actually drove the business into the mills of their shrewd and more aggressive competitor."

The result, according to the same source, can be summed up as follows: In 30 years British trade with Canada in steel goods increased 50 per cent., while that of United States increased 1,250 per cent.

American firms have learned the value of publicity in dealing with the Canadian market. Their goods are advertised, and they conform to the wants of this country. British firms have a good opportunity now to study the Canadian market, our ways of buying, the conditions and acceptances on which we finance, and many other things, an observance of which would make a great deal of difference in the trade return figures of months and years to come.

Could You Fill This?

HERE is an advertisement that appeared in a Toronto paper this week. It occupied two inches and carried a heading sufficiently conspicuous to make it noticeable:—

We have an opening for a man at present as Assistant Superintendent of a large plant in Canada. Desire Canadian, and a technical education would be beneficial, but not absolutely necessary. If you feel you have made good, and your record will stand investigation, it may pay you to get in touch with us. Should have general knowledge of manufacture of mechanical lines. We are looking for a real good man of proven ability, who could take full charge when necessary. No others considered. State experience, age, present connection and approximate salary desired. Negotiations confidential.

Now, these openings keep cropping up now and then. Perhaps not very often, but a glance at the papers shows quite a number in the course of a month.

The position may be offered to-day, or six months from now. That depends on a good many things.

If it were put up to YOU to-day, what would YOU do? Could you walk into the office, make application and state your case, knowing that you would be able to fill the position?

Or would you have to do, as many others, watch that chance go to someone else, and be content to stand back and comment that some chaps had all the luck, and that none of the breaks came your way.

The day of study has not passed, neither has the prospect of dividends for the man who burns the midnight oil.

The openings go to the men who are ready to take them. They seldom wait months or years for a man to get ready.

If you are prepared it may take a long time for the big chance to come, but if you are not prepared, you may take it for granted the big chance will never come.

Guaranteeing the Price

THE practice of cutting rates in piece-work shops is coming to be one of the things that will soon be known only as a memory. It was done in many places. One job comes up from years past of a plow factory where the rate had been set at "so much per" on the polishing operation. Polishing plows is not easy work, at least it was a husky man's calling at that time. When the boys in the polishing room had dusted off enough shares and consumed enough dust to chalk up around $2.25 per day, the new price schedule was posted on the wall, and it became apparent that $1.85 was the limit.

It was another way of punishing a man for working hard.

Under that system the wise piece worker got a job and knew when he was at the limit. He jogged along, killed time and at the end of the day had an output that could be cashed in for $2, or thereabouts. It was encouraging laziness on the part of the men.

All over the country we have shops now where the prices are fixed after a careful study—they are guaranteed for a given period or until such time as there may be an alteration in the operation itself. Production men claim they are getting better results this way than from any other system they have tried.

Must Specialize to Get Mass Production

"THE day of the all-around mechanic is done for with the possible exception of some of the jobbing establishments, and I believe that the change is the best thing for the men themselves and the only possible thing for the shops. The general training still has its place and should not be neglected, but it should only serve as the good groundwork for the development of our specialists."

That, in brief, is the idea of Mr. R. W. Gifford, superintendent of the Massey-Harris Co. plant in Toronto, and what is more, Mr. Gifford can bring out arguments to convince a man that his line of reasoning is right. He is willing to admit, though, that the thorough training is desirable, and that it may be necessary for the shops themselves to bring about a training school of their own in order to keep up the supply of men who shall have a good groundwork from which to draw the specialists and maintenance men of the future.

Where the Move Started

"The specialized shop started with the automobile business," stated Mr. Gifford, in discussing this matter with Canadian Machinery, "and from that it has spread to every other concern that feels the necessity for mass production. There is no other way in which a business can be maintained and face the competition of the world. For instance, I suppose if you were to go into our shops here, and we gave you enough iron or steel, hammers, a fire and anvils or places to work, you could, in time, by sufficient hammering, turn out a plow by yourself, but we couldn't sell it, as there would have been too much cost attached to it, and the time consumed in using all the equipment would also amount to a very great deal.

Nor does Mr. Gifford believe that specializing is a bad thing for the men. "The man who has the all-round training, if he has work that calls for it, can find full scope for every bit of it, but if his occupation calls for another line of specialized work, he may find his previous training a handicap. That may seem a queer statement," continued Mr. Gifford, "but it is a fact that if a plant where production counts for much is looking for a foreman or an executive or a production man, they have to look to these specialized shops now. A shop is no use, neither is its immense equipment, unless that shop is getting production. Now that is the standpoint from which we must start to talk about the specialist from either the angle of the management or of the men. In districts where there are no specialized shops, the manufacturers find that they are handicapped because the men in that district do not understand how to go about it to get mass production."

Even in the Tool Room

Even the sacred precincts of the tool room have been invaded, and in that sphere, where the all-round man was supposed to revel in his wide application of the ways and devices of long training, to-day we have the specialist.

"In the tool room of the Massey-Harris plant," continued Mr. Gifford, "we have some 75 men. We doubled the staff there some time ago, and it may be that we will make large additions to the force there later on. It pays to do it. In that force of tool makers they are nearly all specialists now, and it is hard to see how it could be done otherwise. There is the man working on dies, some on those that are to work on cold metal, others on those that will come in contact with hot metal. Others specialize on jigs and fixtures, some on grinding. What is the result? We get men who we believe are the best to be had on these special jobs, and from these men it is only natural to hold that we get the best work that can be secured."

In Other Lines as Well

That everything is being specialized, and that machinist trade can be no exception, is the view of the superintendent. "Suppose there were something wrong with your eye, to whom would you go to have it fixed, to an ordinary doctor or to a specialist? The specialist, of course, and which one of these men would get the most money for his services? Once more the specialist. When a man goes to college he gets a good training. He may take degrees. But after all that is over, what happens? Does he hang up a sign saying that he has gone through all these subjects and is prepared to give advice on any of them. Not a bit of it—he specializes. He goes in for engineering, for construction, or for any one of many lines. And does it pay us to specialize? Not long ago we finished a die worth over $3,000, but it will save us $15,000. We turned out one die for making seats for agricultural machinery, and out of these seats we punch enough washers to pay for the thing."

How About the Men?

"But where do the men come in on this specializing?" was asked.

"Guarantee the prices paid in a shop and the men will benefit by the specializing and by the system that will get mass production," replied Mr. Gifford, without any hesitation. "There are plenty of shops where the men have been afraid to let out because they knew that once they reached a certain limit the price on the job would be adjusted so they would have to work harder to get up to their former mark. The result of this policy—a mistaken one, I believe—has been that the full production possibilities of many departments have never been let out. Guarantee the prices, and the same plant will turn out a great deal more work, and it will be better for all concerned."

Mr. Gifford has had a good opportunity to study this point in a number of plants. He can draw on numerous cases to prove his points. "Here is one case that will show the advantage of specializing. One man is in the erecting department, and he has a certain part to bring together. There is a nice little knack in the thing, and he has been at that a long time and can turn out a great deal more work than any other man. He is making more money than a good many tool makers. An all-round mechanic might not make much more than his salt on that job, but the ability of that man to do that little job quickly and thoroughly is a mighty good thing for the man and a good thing for the shop. I can think also of another man on a hammer. He is a specialist on that particular operation. Formerly in that shop they were afraid to pull out because they had a pretty good idea that once they got past a certain limit the price would be cut. The promise was given that the price would stand and the man let loose. There are seasons when he makes four times as much as he used to, and the firm is willing that he should go ahead and make as much more. He specializes on that work and he knows the rate will not be altered. The result is production, and I know of no surer way to get it."

In the Foundry, Too

Mr. Gifford holds that it is possible to specialize all the way through, and it is being done in the foundry. "In our shop out of 150 molders there are probably not more than six or seven general purpose men. The rest are bench, floor men, squeezer, match plate, etc. The use of molding machines tends to this method, and it is the best for all concerned. We do not want a man to work on a dozen things. We want him to do one thing well, and the better he can do it the better will our product be and the more money he will earn. There is one department where shops have not yet got operators in sufficient numbers who are thoroughly adept, and that is in the automatics, especially in operators for automatic screw machines. I do not believe that, as a general thing, the output of this class of machine is much more than fifty per cent. of what the machines are capable of doing. Firms making these machines would find it to their advantage to train operators."

(Continued on page 345)

MARKET DEVELOPMENTS

Steel and Tool Prices Have Not Given Way Yet

Some Few Concessions Being Made, But as a General Thing Conditions Remain Unchanged—Demand for the Delivery of Steel on Order is Just as Insistent as it Was Six Months Ago.

SOME price revisions on machine tools have come out during the week. One maker has announced a schedule that follows the lead of the automobile industry, while the Canadian end of one British firm has given out a revision which makes a few changes, but leaves the list a little higher than previously. New York reports that buyers there feel that the machine tool market belongs to them, and they are reported to be getting some odd concessions that were not available a few months ago. That is about the extent of the price changes, and when taken against the entire field, it is hard to see where much progress has been made in the matter of price reduction.

Men who have covered the field in Canada for steel interests say, that with the possible exception of the automobile business, there is no let-up whatever in the demand for steel, and buyers are just as insistent as ever in wanting to get their orders filled and delivered at their plants. New business is not coming in rapidly to the steel men,

this probably being accounted for by the fact that it is not yet possible to get rapid deliveries, although there is a great improvement in this direction.

The departments that specialize in small tools claim that there is more competition going on now than for some months past, and the trade is being thoroughly canvassed in this district to get any business that may be going. Prices, though, remain firm and there is no complaint of any price concessions beyond those generally recognized by the trade.

The scrap metal market shows no change at all. Dealers in most cases state plainly that they can be counted for the present as being out of the market unless there is something offered at quite a price concession. The big users have not returned to the point yet where they will buy, and until that takes place there is going to be continued stagnation.

BUYERS WAIT IN HOPE THAT THE PRICES MAY BE MORE FAVORABLE

Special to Canadian Machinery

MONTREAL, Oct. 7.—Trading conditions are more or less affected by the movement that is becoming general throughout the States and Canada towards an adjustment to lower price levels on many commodities, both industrial and domestic. Business, in consequence, has reflected the conservative policy of consumers, many of whom are adopting a waiting attitude and holding off buying until a decline is announced. It has been stated that increased European production of steel goods has reacted on the Canadian mills and export from here has been on the downward grade.

Lower Prices Anticipated

The trade is undoubtedly facing an early revision of steel prices, but it is difficult to get anyone to commit himself as to the time such adjustment will take place. It is becoming increasingly clear that the demand for finished products is gradually falling off, and with the mills and foundries in a better position to increase production, the result of such conditions can only mean that any change in prices would be in a downward direction. Shipments of American steel

are better in volume, and owing to past demand and existing requirements, the greater portion of this material is going direct to the purchaser, but current demand is not adequate to absorb all the steel that is being produced, and with dealers reluctant to stock at present prices, the movement of steels can only continue by removing the barrier of excessive high costs by adjusting the normal level of steel prices to a point where consumers will come into the market with more confidence and assurance of a stabilized condition. It is thought that a concerted effort to lower the cost of steel materials would have a good effect on labor, and remove some of the factors that at present are disturbing many branches of the industry, pave the way for better co-operation between industrial classes, and enable the manufacturing industries to engage in many activities that have been suspended or curtailed owing to the impossible conditions that have intervened.

Metals Quiet But Steady

The movement of metals is of a quiet character but the volume of sales is relatively light, due largely to the conservative attitude of buyers in taking on

only sufficient for immediate requirements. The prediction seems to favor lower prices but the existing market is comparatively firm. Lake copper is the only metal showing a change, a decline of one cent, leaving the present quotation at 24 cents per pound.

Supplies Moving Steadily

Coupled with the announcement that business is light, machine tool dealers are living in expectation of lower prices on many of their lines. Many are unable to figure out just how a cut of any magnitude could be reasonably made but admit that this action may take place at any time. The inquiry for equipment is quite light and below normal for this time of the year but the movement of supplies continues in such volume as to indicate that industrial activity is well maintained though easing up in some directions. The better shipments on tools and small equipment reflect a lighter demand, but the cost of production and transportation are still so high that any early drop in prices must be based on trade requirements rather than on manufacturing costs.

Foundry Iron Higher

Apart from steel and cast iron scrap the old material market shows little signs of activity and the general movement of all metals is below normal; as a matter of fact, present business is

duller than it has been for many months. Export of scrap is practically nil and current trading is largely of a local character. The possible downward trend of prices on manufactured products has apparently affected trading, and in consequence the buying of materials has been held up to some extent. The curtailment of production in some industries has likewise been felt in scrap circles and demand is proportionately lighter. The quotations obtaining on all lines are subject to adjustment to suit individual sales but those listed are a fair average of those prevailing in this district. The general situation shows firmness with machine cast iron quoted at $36 per ton, and stove plate at $28, these prices representing an advance of $4 and $1.50 per ton over those of last week.

NOT A REQUEST
FOR CANCELLATION

That Is Report of One Machine Tool House — Demand Keeps Up For Steel

TORONTO.—Another week has passed since the talk of price cutting had its start, and still it is hard to find any sign of its striking the machine tool, iron or steel market. Some of the dealers are authorized by their principals to guarantee their prices for some months to come, and in nearly all cases buyers are protected, and they would get the benefit of any reduction that was made.

"We have had no requests for cancellation from any of the business that we have on our books," was the report of one firm to-day. The dealers would be the first to get wind of any such movement in this country, and it would be forwarded from them to the makers. Much of the machinery is sold under the non-cancellation understanding, but the fact that there has been no request for cancellation in any case looks as though there were no doubters in the manufacturing world here. As a matter of fact, as explained by one house to-day, most of the business booked is of a very reliable sort. Very few of the orders are in large lots now, and generally represent changes or additions to existing equipment, and they are not to be upset by any temporary setbacks.

In discussing the matter of prices, the resident representative of a large Old Country firm stated to Machinery that prices had been sent out just a few days ago from the home office. In some cases there was a reduction of a few pounds, and in others an increase. On the whole the prices seemed to have gone up a little. He did not seem to think there was any chance of British houses at present sending in machine tools at a lower figure than they were doing now. "What may take place in the future I don't know, and would not care to predict. There may be a number of influences at work that we know

Pittsburgh reports that not very much progress has been made toward low prices for steel, though it admits that the general sentiment has become much stronger on this subject.

The cancellations that have come, through the automobile makers to the steel mills, released a tonnage that has been taken up by other industries.

Information from Pittsburgh for steel at fairly quick delivery is quoted this week at 3 cents for bars, 3.10 cents for shapes, and 3.25 cents for plates. Some predictions are that these prices will yield, but there is no evidence of it yet.

There is very little attention paid to lower prices talk in the machine tool market, although there are reports of one maker having issued a new price list.

One of the largest machine tool houses in this district reports this week that so far they have not had a single appeal for cancellation of any business on their books.

One Old Country house doing business in Canada has forwarded new lists on machine tools. Some of these show increases, others are down a little, but as a general thing the whole list is said to be a little upward.

Some of the small tool dealers report that there is more competition at the present time in the Toronto field than there has been for a good many months past.

The scrap metal market is waiting for a turn that will bring some business its way. Some of the largest yards state frankly that for the present at least they are out of the market. The large users are not buying, and dealers show no tendency to take on material for stock.

nothing about, and these may have much to do with the prices. For instance, were buying to fall off, or were there many cancellations, the makers might find that they were accumulating stocks, and money is not the easiest thing in the world to secure for getting out finished material right now. It might be that they would have to get rid of some of their stock in order to secure money. In that case it would influence prices. Of course that is entirely sup-

positions, but I have seen things work out that way before."

The same dealer, when asked regarding the attitude of the buyer in Canada, stated that he did not believe that buyers were holding off in the hope of better prices. A new machine tool, especially of the labor saving variety, would pay its way from the start, and did not depend on the cheese-paring policy in matter of prices.

Hard After Business

Dealers selling small tools report that there is more real business hunting going along now than at any time for months past. "People are getting out to sell at a faster pace than they have been doing for a long time. Many of the stores here have been putting men on small tools and nothing else, and as a result we have to get out more." There is a lot of small business turning up, although there are few large orders coming into the market now. The National Railways have a specification out for a list of carbon drills, but otherwise the business is made up largely of small orders. There is no change in any of the prices.

Little Improvement

In some cases deliveries of steel are not so marked in improvement as they were a week or so ago. A representative of one firm has just returned from going over a good deal of the territory in this district. Speaking of the attitude of the buyers he said, "There is no let-up in the demand for steel. Apart from some of the automobile industries, and other lines that have contracts from them there is the same insistent demand for steel that there has been for months past. I can name a dozen places where I called where my reception was the same. It would run like this: 'Glad to see you, come on in,' and in a very short time it would be around to this, 'You remember that order we had booked with you some time ago? Wish you would take that up and give it your personal attention to see if there is nothing that could be done to get that shipment to us.' Of course there were some variations from that request, but that was the feeling all over. There is certainly no falling off in the demand for steel, apart from the one line to which I referred to at the outset."

The Scrap Metal Market

"We are practically out of the market for the present," is the way one of the large yards stated its position this week. "We have been expecting the turn to come for some time, and believe that a day one way or the other is going to make the difference. If the market remains weak now for a few days longer, or if it breaks, I believe that there will be business resumed, but it will be on a lower plane than we have been used to."

The unsettled conditions following the price cutting move seems to have some thing to do with the scrap market. Many of the users are continuing to draw on reserves, believing that they

will be able to make replacements at a lower figure. It all tends to keep the buyers out of the market and the dealers do not feel as though they would be justified in going ahead and buying for stock.

TOOL MARKET IS THE BUYER'S NOW

New York Looks For a Quiet Period For the Remainder of the Year

NEW YORK, Oct. 7. — The recent trend of price developments points to rather quiet conditions in the machine-tool trade over the remainder of the year at least. This is a buyer's market, and it is but natural that buyers should be considering the possibility of lower prices on all of the products they buy, including machine tools. As yet there are no indications of a downward movement in machinery prices, but concessions are being made by sellers that were not necessary a few months ago. For example, some sellers, at the request of buyers, are guaranteeing their prices for a period of three to six months.

While the outlook for the remainder of the year in the machine-tool trade is for very light buying there is a deal of optimism as to next year's prospects. The railroads are expected to come into the market eventually, and some sellers have fixed the beginning of substantial railroad buying as about January 1. Railroads are known to be working on budgets for 1921, and if the financial situation warrants it is believed likely that they will buy quite heavily of shop equipment, which they are known to need very badly. Some of the railroad shops are antiquated, and a considerable part of the shop equipment of some of them is in a bad state of disrepair. It would seem as if buying of machine tools for the railroad shop cannot much longer be deferred.

At present, however, there is practically no railroad buying in the East, though a number of roads have made fairly good purchases at Chicago.

BRITISH PRICES ON THE WAY DOWN?

Galvanized Sheets Show a Definite Rrop of $7 Per Ton For Export

British prices show a downward tendency, particularly for export. A definite drop of $7 per ton is noted in galvanized sheets, which are now about 6c a pound at to-day's rate of exchange.

A wide divergence in wages is indicated in puddling. The current British puddling rate, expressed in American money, is $8.25 per ton, while the last settlement at Youngstown, Ohio, established a rate of $18.02 for September and October.

AUTO CANCELLATION NOT GOING TO BATTER DOWN STEEL SCHEDULE

(Special to Canadian Machinery)

PITTSBURGH, Oct. 7.—In itself, the iron and steel market shows very little by way of new developments and not much progress has been made, visibly, towards lower prices. General sentiment, however, has become much stronger on the subject of substantial price declines in automobile prices have had a great deal to do with molding public opinion as to prices generally. It seems that the public takes particular interest in whatever occurs in the automobile field. In last report it was noted that Ford, Franklin and Crow-Elkhart, had reduced prices. To the list may now be added Locomobile, Studebaker, Mercer and Overland. One may well conclude that a general and substantial decline in automobile prices is in progress, no matter what technical denials individual automobile makers may give to the press. The reductions are not sporadic, the names on the list of reductions thus far being typical of altogether too many classes of machines. Locomobile, Franklin, Studebaker and Overland each fall in a different class.

Steel and Automobile Prices

The steel industry is quick to refuse to admit that steel price declines can be connected with automobile price declines. It is commonly admitted in the steel trade that steel price declines must occur, the common view being that independent prices will recede to the Steel Corporation level, whether in a few weeks or a few months, and it is also the common view that the Steel Corporation prices will not be reduced. Some independents may have to cut under Steel Corporation prices to regain trade lost by what some buyers have not hesitated to call profiteering practices. The steel trade wants it to be understood, however, that the automobile trade itself cannot force down steel prices. It is admitted that the automobile trade was in large part responsible for the particularly fancy prices for prompt deliveries of steel that obtained early in the year, but it is asserted that the trade's power to put prices down is by no means measured by its influence in putting prices up earlier in the year, when automobile and part makers bid for prompt deliveries when the mills had already sold their product.

As to possible declines in Steel Corporation prices an impressive comparison is made, showing that from the Steel Corporation's standpoint automobile makers can have no argument that since they have reduced their prices the Steel Corporation should reduce its prices. The comparison takes an automobile that has been selling at between $2,000 and $3,000. The reduction in price on this movement might be $150, or $300, or even $400. Less than one gross ton of rolled steel, in the form

in which the average steel mill sells steel, is represented in that automobile. The Steel Corporation's rated capacity is 16,200,000 tons a year, of products weighed in the form in which sold. A fair production can be taken at not under 15,000,000 tons. The common shares outstanding amount to about $508,000,000, so that $15,000,000 would represent $1 a ton on a year's output and would also represent a three per cent. dividend. The corporation certainly would not reduce prices to the automobile trade and keep them up to the railroads, to investors putting up bridges, buildings, etc., and to other buyers, hence a horizontal reduction in all prices is the only thing that could be considered. A reduction, therefore, that would be equal to a three per cent. common stock dividend would amount to less than $1 in the mill cost of the steel entering into a $2,000 or $3,000 automobile. In other words, the steel mills cannot help the automobile trade to recoup when it reduces prices of automobiles.

Another item that dissociates the automobile price reductions from the steel market as it is made from day to day and week to week is that the automobile trade is not in position to offer orders of any consequence at this time. Many of the automobile makers have contracts running to the end of this year, with the prospect that on account of reduced consumption deliveries will run into the new year. General Motors is commonly understood to have a contract running through next June, covering considerable parts at least of the requirements of its various subsidiaries. If there is such a contract it is with the Steel Corporation and at the Industrial Board prices, since the corporation has sold to domestic trade at no other prices since the schedule became effective March 21, 1919, and an automobile manufacturer can hardly complain about Industrial Board prices.

Two Steel Markets

There used to be three steel markets, the Steel Corporation set of prices, a schedule of prices, not entirely definite, at which certain large independents were selling to regular trade for extended delivery, and a wide range of prices representing prompt deliveries, charged chiefly by smaller mills that kept their order books relatively clear. The last named market has now practically disappeared. One can, as a rule, secure rather prompt deliveries at the prices formerly ruling only for somewhat extended or delayed deliveries. That market is substantially unchanged. It is represented by 3c for bars, 3.10c for shapes and 3.25c for plates. In time these prices will doubless yield, but there is no distinct evidence that they have yielded thus far.

Last week the American Pig Iron As-

sociation, composed of merchant furnacemen from a majority of the producing districts, held its monthly meeting at the William Penn Hotel, Pittsburgh. The prevailing, practically the unanimous, sentiment at the meeting, derived from 'information obtained from customers, was that a general industrial depression is impending, to exhibit its full force in the first quarter of the new year, with greatly reduced consumption and production and materially altered pig iron prices. The atmosphere of that meeting was as different as can be imagined from that of the meeting two months earlier, when several furnacemen were loudly predicting "pig iron is going to $50." It did go to that figure, in the case of foundry iron valley, Cleveland, Buffalo and Eastern Pennsylvania, but apparently the advance occurred chiefly because the furnacemen set out to get it. The furnacemen did not, however, get the orders. That was another matter. Consumers stopped buying. The trade has if that foundry iron at valley and Cleveland furnaces has now declined $3, to $47, furnace, but the decline rests upon even flimsier foundation than the preceding advance, as there was simply a small sale by a middle interest. No one doubts, however, that a buyer with an attractive inquiry could buy foundry iron at $47. The buyers do not want to pay that price and do not know what price they would pay. They are in position to wait until conditions become clearer.

PROSPECTS GOOD
IN THE EAST NOW

Steel Mills Run at Capacity, and Output of Coal is More Satisfactory

Montreal.—Roy M. Wolvin, president of the Dominion Steel Corporation, left for Cape Breton to inspect the Sydney steel plant and the Cape Breton coal fields. From there Mr. Wolvin will proceed to Newfoundland, where he will make a tour of inspection of the corporation's mines at Wabana.

Reports from Sydney, Mr. Wolvin stated prior to his departure, were most encouraging, the steel mills there working to capacity, with the production at the coal mines showing a gratifying increase over previous levels, the output for September being particularly satisfying, although definite figures for the month had not yet been received at the Montreal office.

Mr. Wolvin's visit to the Wabana mines is in connection with the resumption of the export trade in iron ore, which during the war period and since the armistice, was reduced to negligible proportions. Mr. Wolvin states that it was hoped that the advent of 1921 would see a profitable trade in ore again established, the new overseas connections formed through the British Empire Steel Corporation being an important factor in the market for the ore.

PIG IRON TRADE

The iron market is developing soft spots. Valley foundry iron sold this year at $47 base. Basic is being offered at a lower rate by steel makers than the regular sellers. The prices are uncertain as there are few sales. Resale is the main feature of the pig iron market in Philadelphia. Very little interest is shown in next year's iron.

The market at Chicago is strong 1.75 to 2.25 silicon sold at $46 furnace. No iron is being piled, delivery being equal to production.

It is reported in Cleveland that basic could be had at $46.50 valley. The demand from stove foundries is unusually heavy, especially from the hot air furnace men.

The pig iron situation in New England has resolved into a struggle between consumers and furnaces to see which can hold out the longer.

The foundrymen almost without exception believe that the prices are going to drop, and refuse to buy beyond their immediate requirements. Eastern furnaces feel there is much uncertainty as to the future market. Requests have been received urging deliveries. New business has been small on the New York market during the past week. Prices remain unchanged for next year. Basic is wanted to some extent.

Owing to the feeling that prices are about to decline, consumers of pig iron are staying out of the Pittsburgh market. The foundry market shows a sagging owing to a sale last week of 700 tons of No. 2 foundry iron (1.75 to 2.25 silicon) to an Ohio consumer for last quarter delivery at $47 valley furnace.

No apprehension has been caused among Southern pig iron producers by the lull in the market. $42 for No. 2 foundry (1.75 to 2.25 silicon) is still the base. A few new inquiries are coming in but sales are very limited.

U.S. SCRAP METAL

Prices in the U.S. scrap market are declining slowly. Dealers are doing some buying, but consumers have reserves and are waiting for lower levels, refusing to be interested. On the Chicago market the consumers feel they can safely work on their reserves and expect to buy at a lower rate when forced into the market later.

The market in steel scrap shows weakness in New York, and though reductions have been made, demand is dull.

Dealers in the scrap market in Philadelphia attribute the present lull in the scrap market to the influence of price reductions in automobiles made by the Ford Motor Co. Shipments on contracts continue heavy.

Prices in all grades have declined in Boston. Dealers are unable to place tonnage except by making heavy concessions.

Practically no inquiries are being made for iron and steel scrap in Pittsburgh. Consumers are comfortably supplied. Prices are unchanged.

Many Buffalo dealers are anxious to obtain material to cover orders taken in the last month, but are holding back hoping for a break in the market. Lower prices are regarded as inevitable.

On the Cincinnati market, iron and steel scrap shows a lagging tendency, but some good orders for future delivery have been made and enquiries continue. Heavy shipments of recent purchases are being made.

BALDWIN PLANT
STARTS ROLLING

The Tin Plate Mill Had Its First Trial Run a Few Days Ago

Baldwin's new $2,000,000 plant at Ashbridge's Bay has commenced rolling tinplate. Mrs. H. A. Cooch of Toronto at the opening threw the switch and the huge 1,500 horsepower electric motor set in motion the tremendous 160-ton main rope drive wheel, the largest flywheel yet used in Canada. The battery of eight mills commenced to grind. Then, cheered by hundreds of watching workmen, Toronto society women fed plates of white hot steel into the mills and rolled them again and again.

Mrs. A. M. Russel, wife of the president of the company, rolled the first sheet in No. 3 finishing mill. Other women who demonstrated their ability as steel workers were: Mrs. Roy Miller, Mrs. Fred Miller, Mrs. E. L. Cousins, Mrs. H. A. Cooch and Miss Church. Controller Maguire rolled the first plate through the roughing mills.

MUST SPECIALIZE

(Continued from page 341)

The whole shop organization is rapidly changing, and there should be a change in the training of men for the mechanical field to keep pace with the times. Mr. Gifford does not believe that the man who wants to be a success should think any the less of the fundamental training he needs. "Don't think that I am opposed to having the well-trained man. Every shop should see to it that an apprentice gets a thorough grasp of the fundamentals, and shops are not paying enough attention to this matter. We need these men, need them for maintenance, need them all the time to develop into branches that are almost passing out for want of well-trained men. Just one class need be mentioned, viz., the die-maker, and the real die-maker is becoming almost an extinct species."

Mr. Gifford has no apologies to make for the specialized shop. He firmly believes that it is the only way to secure results, and he is convinced that it is the best thing for the men and the best thing for the large industries—in fact he can see no other way in which they can be carried along successfully.

SELECTED MARKET QUOTATIONS

Being a record of prices current on raw and finished material entering into the manufacture of mechanical and general engineering products.

PIG IRON

Grey forge, Pittsburgh	$42 40
Lake Superior, charcoal, Chicago	57 00
Standard low phos., Philadelphia	50 00
Bessemer, Pittsburgh	43 00
Basic, Valley furnace	42 90
Toronto price:—	
Silicon, 2.25% to 2.75%	52 00
No. 2 Foundry, 1.75 to 2.25%	50 00

IRON AND STEEL

Per lb. to Large Buyers	Cents
Iron bars, base, Toronto	$ 5 75
Steel bars, base, Toronto	5 75
Iron bars, base, Montreal	5 75
Steel bars, base, Montreal	5 75
Reinforcing bars, base	5 75
Steel hoops	7 00
Norway iron	11 00
Tire steel	5 75
Spring steel	10 00
Band steel, No. 10 gauge and 3-16 in. base	6 10
Chequered floor plate, 3-16 in.	8 50
Chequered floor plate, ¼ in.	8 10
Bessemer rails, heavy, at mill	....
Steel bars, Pittsburgh	3 00-4 00
Tank plates, Pittsburgh	4 00
Structural shapes, Pittsburgh	3 00
Steel hoops, Pittsburgh	3 50-3 75
F.O.B., Toronto Warehouse	
Small shapes	5 75
F.O.B. Chicago Warehouse	
Steel bars	3 62
Structural shapes	3 72
Plates	3 67 to 5 50
Small shapes under 3"	3 62

FREIGHT RATES

Per 100 Pounds.

Pittsburgh to Following Points

	C.L.	L.C.L.
Montreal	58½	73
St. John, N.B.	84½	106½
Halifax	86	108
Toronto	38	54
Guelph	38	54
London	38	54
Windsor	35	50½

METALS

	Gross. Montreal	Toronto
Lake copper	$24 00	$23 50
Electric copper	24 00	24 00
Castings, copper	23 50	24 00
Tin	54 00	55 00
Spelter	9 00	10 50
Lead	9 50	9 50
Antimony	9 50	12 00
Aluminum	36 00	37 00

Prices per 100 lbs.

PLATES

Plates, 3-16 in.	$ 7 25	$ 7 35
Plates, ¼ up	6 50	6 60

PIPE—WROUGHT

Standard Buttweld Pipe

Per 100 Ft.

	Steel Blk	Steel Galv.	Gen. Wrought Iron Blk.	Gen. Wrought Iron Galv.
⅛	$ 6 50	$ 8 50	$....	$....
¼	5 81	7 41	5 91	8 01
⅜	5 81	7 41	5 91	8 01
½	7 10	8 63	7 95	9 18
¾	8 80	10 87	9 95	12 02
1	13 01	16 07	14 71	17 77
1¼	17 60	21 74	19 90	24 04
1½	21 04	25 99	23 79	28 74
2	28 31	34 97	32 01	38 67
2½	44 73	55 28	....	....
3	58 52	72 29	....	....
3½	74 08	90 62	....	....
4	87 75	107 37	....	....

Standard Lapweld Pipe

Per 100 Ft.

	Steel Blk.	Steel Galv.	Gen. Wrought Iron Blk.	Gen. Wrought Iron Galv.
2	$32 61	$ 38 67	$35 71	$42 37
2½	48 26	58 79	54 11	64 64
3	63 11	76 88	70 76	84 53
3½	75 90	92 46	85 10	101 66
4	89 93	107 55	100 83	120 45
4½	1 05	1 29	1 30	1 54
5	1 22	1 50	1 52	1 80
6	1 58	1 95	1 97	2 33
7	2 06	2 53	2 53	3 01
8L	2 16	2 66	2 66	3 16
8	2 49	3 07	3 07	3 64
9	2 98	3 67	3 67	4 36
10L	2 77	3 41	3 41	4 06
10	3 56	4 39	4 39	5 21

Prices—Ontario, Quebec and Maritime Provinces

WROUGHT NIPPLES

4" and under, 60%.

4½" and larger, 50%.

4" and under, running thread, 30%.

Standard couplings, 4-in. and under, 30%.

Do., 4½-in. and larger, 10%.

OLD MATERIAL

Dealers' Average Buying Prices.

	Per 100 Pounds. Montreal	Toronto
Copper, light	$15 00	$14 00
Copper, crucible	18 00	18 00
Copper, heavy	18 00	18 00
Copper wire	18 00	18 00
No. 1 machine composition	17 00	17 00
New brass cuttings	11 00	11 75
Red brass turnings	14 50	15 75
Yellow brass turninbs	9 00	9 50
Light brass	7 00	7 00
Medium brass	8 00	7 75
Scrap zinc	6 50	6 00
Heavy lead	7 50	7 75
Tea lead	4 50	5 00
Aluminum	19 00	20 00
	Per Ton	Gross
Heavy melting steel	20 00	18 00
Boiler plate	15 50	15 00
Axles (wrought iron)	30 00	20 00
Rails (scrap	23 00	18 00
Malleable scrap	26 00	25 00
No. 1 machine cast iron	36 00	33 00
Pipe, wrought	12 00	12 00
Car wheel	33 00	33 00
Steel axles	25 00	20 00
Mach. shop turnings	11 00	11 00
Stove plate	28 00	25 00
Cast boring	14 00	14 00

BOLTS, NUTS AND SCREWS

	Per Cent.
Carriage bolts, 7-16 and up	+10
Carriage bolts, ⅜-in. and less	Net
Coach and lag screws	—15
Stove bolts	55
Wrought washers	—25
Elevator bolts	+10
Machine bolts, 7-16 and over	+10
Machine bolts, ⅜-in. and less	+10
Blank bolts	Net
Bolt ends	Net
Machine screws, fl. and rd. hd., steel	27½
Machine screws, o. and fil. hd., steel	+25
Machine screws, fl. and rd. hd., brass	net
Machine screws, o. and fil. hd., brass	net
Nuts, square, blank .... +25 add	$2 00
Nuts, square, tapped ....add	2 25
Nuts, hex., blank ....add	2 50
Nuts, hex., tapped ....add	3 00
Copper rivets and burrs, list less	15
Burrs only, list plus	25
Iron rivets and burrs	40 and 5
Boiler rivets, base ¾" and larger	$8 50
Structural rivets, as above	8 40
Wood screws, O. & R., bright	75
Wood screws, flat, bright	77½
Wood screws, flat, brass	55
Wood screws, O. & R., brass	55½
Wood screws, flat, bronze	50
Wood screws, O. & R., bronze	47½

MILLED PRODUCTS

(Prices on unbroken packages)

	Per Cent
Set screws .... —20%, 25 and	5
Sq. and hex. hd. cap screws	12½
Rd. and fil. hd. cap screws ..plus	25
Flat but. hd. cap screws ....plus	50
Fin. and semi-fin. nuts up to 1-in.	12½
Fin. and Semi-fin. nuts, over 1 in., up to 1½-in.	—5
Finp and Semi-fin. nuts o.er 1½ in., up to 2-in.	+12½
Studs	+5
Taper pins	—12½
Coupling bolts	+40
Planer head bolts, without fillet, list	+45
Planer head bolts, with fillet, list plus 10 and	+55
Planer head bolt nuts, same as finished nuts.	
Planer bolt washers	net
Hollow set screws	+60
Collar screws ......list plus 20,	30
Thumb screws	40
Thumb nuts	75
Patch bolts ....add	+85
Cold pressed nuts to 1½ in...add	$1 00
Cold pressed nuts over 1½ in..add	2 00

BILLETS

	Per gross ton
Bessemer billets	$60 00
Open-hearth billets	60 00
O.H. sheet bars	76 00
Forging billets	56 00-75 00
Wire rods	52 00-70 00

Government prices.

F.O.B. Pittsburgh.

NAILS AND SPIKES

Wire nails, base	$6 10
Cut nails, base	7 00
Miscellaneous wire nails	50%

ROPE AND PACKINGS

Plumbers' oakum, per lb.	0 10¼
Packing, square braided	0 38
Packing, No. 1 Italian	0 44
Packing, No. 2 Italian	0 36
Pure Manila rope	0 35½
British Manila rope	0 28
New Zealand hemp	0 28

POLISHED DRILL ROD

Discount off list, Montreal and Toronto	net

CANADIAN MACHINERY

AND
MANUFACTURING NEWS

Vol. XXIV. No. 16

October 14, 1920

Producing Spindle Bracket for Upright Drill

The Machining of Such a Piece Must Be Followed Out With Great Care—Various Tools and Fixtures Used Are Shown and Described in Detail.

By FRANK H. MAYOH

MOST every mechanic is familiar with the spindle bracket of the upright drill, but perhaps few consider when sliding this spindle bracket up or down the dovetail slide of the machine what care must be exercised in machining this piece to obtain interchangeability of product consistently day in and day out on an economical basis. It is the purpose of this article to describe various tools used to this end and present in some detail the methods which have proved successful for accomplishing this as adopted by one drill press manufacturer.

This spindle bracket is shown by Fig. 1, the finished surfaces being indicated in the usual manner. It is made of a good grade of cast iron and on account of the deep slot cut in the dovetailed end must be allowed to season well during the operation between roughing and finish

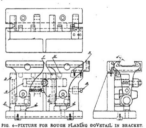

FIG. 4.—FIXTURE FOR ROUGH PLANING DOVETAIL IN BRACKET.

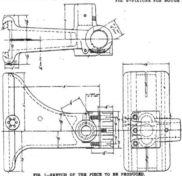

FIG. 1.—SKETCH OF THE PIECE TO BE PRODUCED.

planing of the dovetail. In connection with the machining of this a list of operations are furnished, and these are shown in table form at Fig. 2, the first operation being to snag the casting, removing all fins and rough spots. The second operation noted is to rough plane the dovetail and the fixture illustrated by Fig. 4 is used for that purpose.

Referring to Fig. 4 this fixture is one of ten made to hold in all 20 pieces, two parts being contained in each fixture, and these fixtures being set in two rows on the planer table. The brackets are held in pairs, using a heavy clamp A, and set screws B and C, which hold each part respectively against lugs on the casting at D, E, F, G and H, thereby providing means for taking the thrust off the cut, and spots for squaring the work. To set the work, pins are employed at J, thus all brackets are located definitely from the boss, and as the fixtures are made to match, all the parts in a row are planed alike. On one fixture, at the start of each row, a bracket K is fastened by screws and dowels, and to this bracket is attached a steel plate cut out to match the dovetail of the work except that it is ⅛ inch large all round. By means of this, and 5-32 feeler gauges interposed

between the cutting tool and the gauge, the dovetail is planed accurately to size, allowing 1-32 to be removed by the finishing operation.

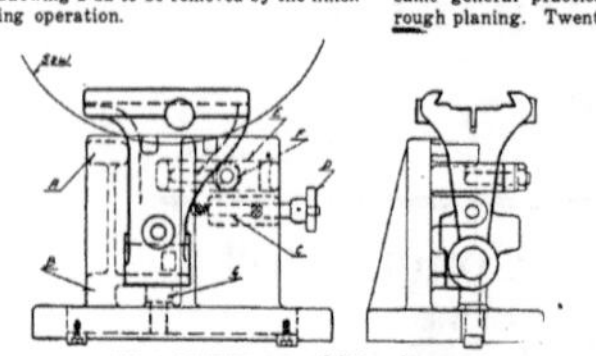

FIG. 6—FIXTURE FOR MILLING SLOT.

To illustrate the method of arranging these fixtures on the planer tabel a diagram, Fig. 5, has been drawn, and it will be noticed that between each double line are two numbers, 1 and 2, 15 and 16. These represent that two parts are held on each fixture, and the fixtures are set end for end in two rows, while at the right hand, as indicated, are the gauges previously mentioned for setting the planer tools.

For milling the slot between the dovetail, a fixture Fig. 6, constructed along the same general lines as those used for planing is provided, which, together with a large circular saw fed vertically into the work, cuts the slot across both ends at the same time. This fixture holds one part only. The work is held against lugs A and B by screw C, operated by hand knob D, while clamp E, tightened by the nut F, secures the work in place, which is also set on pin G. The fixture is held to the table of the machine by two tee bolts in the usual manner.

After cutting this slot it is found in practice that the casting springs out of shape, consequently they are set aside a short time to allow the castings to season before performing the finishing operation of planing, milling and drilling.

Taking up the work again to finish, plane the dovetail, which is operation four on the list; recourse is had to the same general practice as was used for rough planing. Twenty parts, set in two rows, with two in each fixture same as shown by Fig. 4 and 5 is the method employed, except that ⅛ inch feeler gauges instead of 5-32 are employed to set the planer tool, thereby ensuring that the dovetail on all parts will be substantially alike and finished correct to size, which size is tested to gauge, this gauge, together with the feeler gauges being shown by Fig. 7.

A drill jig of more than usual interest is shown by Fig. 8. This jig is for drilling and reaming the 1⅛ inch hole through which the drill press spindle slides, and this must line up with the dovetailed end of the bracket. Referring to the drawing, the work is pushed over the steel seat A until it rests on pin B, in which position it is clamped to the seat by means of the two clamps C and screw D. At the other end, two spring pins, E and F, are released against the boss, and these are clamped by screws G and H through the medium of beveled pins in the usual manner, while screw J and hand knob are used to bind the boss against the spring pins. Spring pin K is now released until it is pushed up and bears against the work, in which position it is clamped by screw M. The work is now securely held and we are prepared to do the drilling. This is accomplished as follows: cover N, hinged on pin P, is swung down into the position shown, and is locked there by nut and washer Q on hinge bolt R. This hinged cover contains fixed and slip bushings suitable for guiding the drill (which is of the three lip core drill type) and reamer size bushing, all of which together insures the production of good work in duplicate.

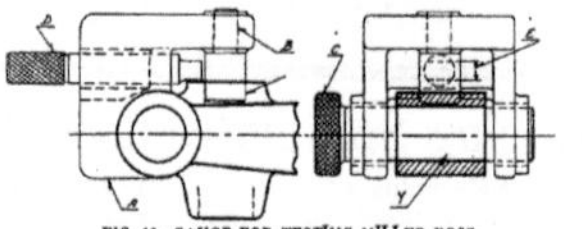

FIG. 12—GAUGE FOR TESTING MILLED BOSS.

To continue the drilling as called for in operation six on the list, the jig, Fig. 9, is employed. This has some of the characteristics of the jig shown in Fig. 8. inasmuch as part of the clamping is similar. The same type of dovetail slide A is used, over which the work is slipped while clamps B and C are used to tighten the work on this slide. Before tightening the clamps the work is also slid over stud D which fits into the hole previously machined. In this instance the work rests against the shoulder E on the stud D for locating it in the vertical plane, and thus supported, a wrench is used for tightening nut F, this securely holding the work in position, as all holes are crossways, and the thrust of drilling comes against the solid dovetail seat, or the stud D.

At this time the long hole G in the work is drilled from both sides, while the cross hole at the opposite end of the work is machined on the two diameters as necessary.

Following the drilling of hole G the bosses on each side of the work are faced. Solid steel bushings H and J are provided for guiding the spot facing tools, which are of standard make, while

Op. No.	Operation.	Tools Used.	Tool No.	Op. No.	Operation.	Tools Used	Tool No.
1	Snag Casting			8.	Drill Small Holes except #28 holes.	Drill Jig Stock drills	S.T.-1323
2	Rough Plane Dovetail	Fixtures (Hold 20)	S.T.-1312				
		with Set Blocks.	S.T.-1314	9.	Drill Six #28 Holes.	Drill Jig. Stock drills.	S.T.-1324
3.	Mill Slot.	Fixture - Slotting Saw.	S.T.-1313				
4.	Finish Plane Dovetail.	Fixtures (Hold 20)	S.T.-1312	10.	Tap All Holes.		
		With Set Blocks	S.T.-1315	11.	Mill Slot In Spindle Hole.	Fixture	S.T.-1325
		Gauge	S.T.-1316				
5.	Drill & Ream 7/8" Hole	Drill Jig	S.T.-1317	12.	Burr		
		1 11/32 Core Drill		13.	Scrape Dovetail	Surface Plate.	S.T.-1326
		1⅜ Adjustable Reamer		14.	Assembly.		
		Plug Gauge	S.T.-1320				
6	Drill 1⅛ Hole, Spotface 1¼ Bosses, Drill & Ream. Pinion Hole	Drill Jig - Slip Bushing	S.T.-1318				
		1⅛ drill 1⅛ Spotfacer					
		Core Drill	S.T.-1319				
		Reamer	S.T.-1328				
7	Mill 1½ Finished Boss.	Fixture	S.T.-1321				
		Gauge	S.T.-1322				

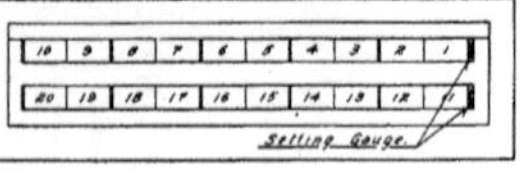

FIG. 2 SHOWS LIST OF OPERATIONS, WHILE FIG. 5 (SEE LOWER RIGHT HAND CORNER) ILLUSTRATES THE ARRANGEMENT OF FIXTURES, SHOWN BY FIG. 4 ON PLANER TABLE.

slip bushing K guides the 33-64 drill, this being transferred first from one side, then to the other side of the jig. Bent pins L are used to engage with pin M

bushing Q, Fig. 9. Following this, the reamer shown is employed for finishing both of the holes true to size. This reamer likewise passes through both

inch and a half hole which was machined at operation five, this being a wringing fit in the hole, as the same must be kept very accurate as regards size, and have a good smooth finish.

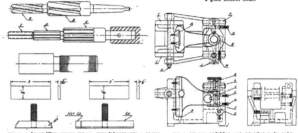

UPPER HALF OF VIEW SHOWS FIG. 10. THIS ILLUSTRATES A GROUP OF TOOLS USED WITH DRILL JIGS. LOWER PORTION OF VIEW SHOWS FIG. 7, WHICH ARE TESTING TOOLS FOR DOVETAILED SEAT.

FIG. 8—JIG FOR DRILLING AND REAMING MAIN HOLE.

in the bushing to prevent the same from turning, this being a method quite commonly employed in jig construction. At the other end of the jig are bushings N, P and Q, which are used for guiding the first two tools, which are illustrated in Fig. 10, it being understood that this hole is cored, as the bracket comes from the foundry, therefore the first tool used is the drill A, Fig. 10, which is a three-lipped core-drill of two diameters, the smallest diameter being small enough to pass through the cored size of the large hole.

This means that both the small and large drill may be made to cut at the same time simultaneously, opening up both holes, the drill at this time being guided on diameter B, Fig. 10, in slipped

holes, and the pilot end C, Fig 10, enters bushing N, Fig. 9, while reamer D finishes the small hole, and reamer E finishes the large hole. The diameter E, Fig. 10, is also guided in suitable slip bushing at Q, Fig. 9, which would be a good running fit for the reamer. Still referring to Fig. 9, the bushing P is hardened and ground, and is driven into the cast iron body of the jig so as to provide a lasting hole construction into which the slip bushings can be pushed for guiding the core drill and reamer.

A tapered socket is used in the drilling machine for holding the drill at this operation, while a universal drive connection is adapted for driving the reamer. The plug gauge shown in the lower portion of this illustration is for testing the

Now that all the important holes are machined, advantage of this fact can be taken during the remaining operations. It will be noticed in the following operations that the clamping methods are somewhat simple. Referring to Fig. 11 this is a milling operation which is performed on a vertical milling machine, and is noted as operation 7 on the list Fig. 2. In Fig 2 the bracket is slipped over the dovetailed seat A, and stud B. A long bolt C was placed through the hole where shown before putting the work in the fixture. This bolt was used to tighten the work, as by placing the wrench on nut D the work may be drawn together tight on the dovetail in the usual manner. The pin E is provided as a holding means to prevent the sutd

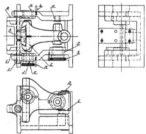

FIG. 9—JIG FOR DRILLING AND REAMING LARGE CROSS HOLES.

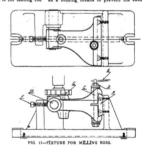

FIG. 11—FIXTURE FOR MILLING BOSS.

from turning while tightening the nut. At the opposite end of the fixture where the milling is done, a square head-screw F has been provided, and by tightening this against the boss on the work the same is securely held against the cutting stress, due to the feeding across of the large surface mill for finishing boss G.

After this milling operation it is necessary to gauge the distance from this boss to the large hole B at the end of the bracket. This is done by making a special limit gauge which is shown in Fig. 12. It consists of a body A, made of cast iron, in which is placed the stud B. The illustration also shows a knurled stud C. This is made a slip fit, and in use the gauge with the knurled stud removed is slipped over the work in the manner shown, the stud B entering hole X in the work. The knurled stud C is then pushed through the hole Y and into the holes of the gauge in the manner illustrated.

In this position the gauge is located relative to both holes previously reamed in the jigs. Another feature of this gauge is the knurled pin D. This is made of tool steel and is a snug sliding fit in the bushing. The pin is flattened on both sides at E, and with the gauge in place it is pushed over the boss on the work in the manner shown. If the flat side B' will go over, and the other flat A' will not go over the boss on the work, the same is milled correct, as the distance from the centre of the pin D is made three thousandths different, thereby providing a maximum and minimum to insure accurate work.

The remaining holes to be drilled are of a general routine nature requiring only such accuracy as to provide a reasonable degree of interchangeability common to machine tool construction. With this end in View the jig shown by Fig. 13, in unison with the base, has been provided for machining the six holes which are done at this time. This is referred

to as operation eight on the list, Fig. 2. In connection with this jig, Fig. 13, it is to be noted that when drilling the series of four holes in the direction of

the arrow, the base provided is not used as the work rests on the table of the drill press upon the surface A. The jig proper, B, contains a plug C, which is slipped into the hole of the work, while the surface on the jig D comes against the finished boss, being held thereto by the square head-screw E.

With the work setting on surface A, three holes F, and one hole M, are drilled. In machining the two screw holes, the drill used is tapping size for the screw, ignoring for the time being the counter-bore and tap size. In the hole between these two holes, which is for a headless adjusting screw, the drill used is first body sized to the desired depth, followed by tapping size the remaining distance. The bracket with the jig in place is now slipped over stud G in the base H, while the two holes J and K are drilled. Two pins L keep the work in its proper location on the base.

At the same time as the first three holes mentioned were drilled, hole M was also machined, using bushing noted by the same letter to guide the drill. Removing bracket entirely from the jig and placing the same on the table of the drill press, resting it on surface A, the two screw holes are drilled out body size and counter-bored, it being unnecessary to guide these tools as they will follow the holes previously drilled in the jig. All tools used for this purpose are stock tools, such as drills and counter-bores.

The simple jig shown at Fig. 14, consisting of two parts only, contains all the elements necessary for drilling the six small holes round the thirty-three sixty-fourths hole. This jig has a stud A, which fits into the hole of the work, and the jig itself B is made of steel. It is slipped over the boss and the wide surface X squares it in the work. The dovetail on the jig in the bottom of the work. This feature

in connection with the pin A locates the jig radially with sufficient degree of accuracy. By guiding the drill through the six small holes C this operation is

completed in what is obviously a very simple manner. No. 28 drills are used for this purpose, and the work sets on the table of the drilling machine with blocks to keep it approximately level.

Operation 10 calls for tapping all holes. As a means of adjusting the bracket to a snug sliding fit on the drill press spindle the large hole has been slotted. This comes diagonally across the screw holes, and the last tool for this work is the milling fixture shown.

Referring to Fig. 15 the base A contains a stud B. This is driven in the boss on the base, and is securely held therein by the pin C. A long collar D with a hexagon nut completes the fixture. The collar D has a groove milled in it to allow the cutter to run into it, while the boss on the fixture also has a groove cut across it for the same purpose. This is why the stud C has a head on it with a nut to hold the base together when stud is in place.

In operation the bracket is slipped over the stud and is clamped in the manner shown by hexagon nut. One of the finished bosses resting on lug E locates the bracket radially on the stud. The fixture is clamped to the table of a horizontal milling machine in the customary manner, using two bolts, one at each end of the fixture, and a tongue in the T slot in the table to line the same up.

Referring once more to Fig. 2, the twelfth operation is burring or removing of sharp corners from the bracket. Operation 13 calls for scraping the dovetail. Operation 14, which is a perfunctory one, calls for assembling. For scraping the dovetail seat a special surface plate is required, but no drawing of this has been shown as it merely has a seat cut into it so that the dovetail can slide over this seat with about ¼ inch clearance, so that the bracket may be tested for angularity.

It will be noticed in following this sequence of operations that nothing has

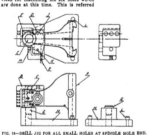

FIG. 13—DRILL JIG FOR ALL SMALL HOLES AT SPINDLE HOLE END.

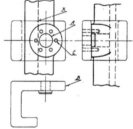

FIG. 14—DRILL JIG FOR SIX SMALL HOLES.

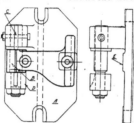

FIG. 16—FIXTURE USED FOR MILLING SLOT.

been spared which will facilitate the necessary degree of accuracy, although the tools provided are by no means elaborate, and as the production is in hundred lots it does not warrant wide expenditure.

ALCOHOL MOTOR FUEL RESEARCH

Vigorous efforts are being made by British men of science and engineers to discover all the possibilities of alcohol as a motor fuel. The increasing scarcity of petrol, and the prospects that the supplies from all visible resources will gradually shrink until they disappear in twenty-five years or so, lend a special urgency to the problem of finding a cheap and plentiful substitute. Realizing how serious the situation was, the Royal Automobile Club and the Commercial Motor Users' Association, both of Great Britain, united in paying the cost of an ambitious programme of research at the University of Manchester, into the value of alcohol as fuel. At the request of the Alcohol Motor Fuel Committee, appointed by the British Government to survey the whole question, the programme was put in the hands of Professor Dixon, who recently gave the results of his investigations in a lecture delivered in London. A scientific study was made by Professor Dixon and his assistants of the behaviour of alcohol and of mixtures of alcohol and benzole after ignition. The main object was to find out the conditions which would give the highest efficiency with existing internal combustion engines when run on alcohol or alcohol mixtures. Information of the greatest value to designers has been collected. Investigations in this field have extended to the value of potatoes, beet, maize, and other sources of power alcohol. It is quite possible that many tropical and subtropical plants which have hitherto been regarded as a nuisance or as of little value will prove to have a high yield of the substances out of which, by fermentation, alcohol may be produced in large quantities at a low price. The research now being conducted may therefore be the means of transforming the economic prospects of more than one country.

COAL RESEARCH

The economical use of coal is a problem which concerns engineers in all parts of the world. It is especially interesting in connection with the "carbonisation" of coal at temperatures in order to obtain smokeless fuel, gas, oil and numerous valuable by-products. This problem involves a close scientific study of exactly what happens to different kinds of coal at various temperatures. Mr. Sinkinson of the Imperial College of Science, London, has hit upon a most ingenious method of discovering at what temperature coal begins to change into coke. He makes use of the fact that coke conducts electricity, while coal does not. By measuring the electrical resistance of a sample of coal as it is gradually heated he has fixed the temperatures at which various coals enter into the critical stage at which they give off their valuable products.

PEACE TYPE AEROPLANE

Most of the aeroplanes hitherto used for commercial purposes have been converted from models designed for bombing during the war. British designers have realized that better results could be obtained by designing the whole aeroplane afresh for its particular work of carrying passengers and goods. One of the first of the true Peace type of aeroplane has lately been put in service on the London to Paris and London to Amsterdam routes. This machine is able to carry eight passengers (in addition to the pilot) and up to 500 pounds of luggage. Alternatively, for goods transport only, it can carry a load of 2,240 pounds. The speed at 5,000 feet up is 121 miles per hour; calculations show that this aeroplane can convey a load almost as cheaply as the average cost of transport by road. The seats are comfortably upholstered and large windows are provided to give the passengers a good view of the country over which they are travelling.

ELECTRIC WELDING RIVET HEATING

Electric welding has already been applied in Great Britain to the building up of the cutting edge of high speed tools which have become worn or chipped. A later development of electric welding in this connection is provided by a British machine which welds a tip of hard cutting steel on to the shank of the tool. The process takes only a second or two and the weld, which is made under pressure, is the strongest part of the completed tool. This device should appeal to users of machine tools in parts of the world where facilities for repairing tools are not ample.

Another British invention which should appeal to the same class of engineer is a new electric rivet heater. Electricity has been used for this purpose before, but always by way of direct heating in an electrically-heated element. In the new machine the rivet itself forms part of the electrical resistance which generates heat. The tip of the rivet is placed on the machine, completing the circuit. In a second or two it heats up to riveting temperature, the process being so rapid that no scale forms on the metal. As many as two hundred and fifty rivets (half-inch) can be heated in an hour by this machine; and the current consumption is so low that one unit of electricity can heat about 5 pounds of rivets.

NEW ENGINEERING STANDARDS

The work of the British Engineering Standards Association is still proceeding steadily. At the request of British makers of small tools and of machine tools the Association recently undertook an inquiry into the standardization of milling cutters and reamers. After a very exhaustive study the association issued a report setting out the standards decided upon. Another interesting report from the same body deals with standard specifications for electric cooking ranges. The two sizes standardised are suitable for five persons and ten persons respectively. It may be added that when, quite recently, engineers in the United States decided to arrange for standardization on a systematic plan they adopted the model built up by the British Engineering Standards Association.

Report on Coarse and Fine Threads—Part 1

This Report Should Be of Interest to Everyone, for the Advance- ment Up to Date Will Reduce the Variety of Screw Threads in General Use.

IT might be well to say that the National Screw Thread Commis- sion was created by an act of Congress in the United States on July 18, 1918, for the purpose of ascertain- ing and establishing standards for screw threads for use of the various branches of the Federal Government and for the use of manufacturers. The commission was to exist for a period of six months. Before the expiration of this time an extension was granted until March 21, 1920. On March 1, 1920, a further ex- tension of two years was granted. The commission is composed of two repre- sentatives of the army, two representa- tives of the navy, four from engineering societies and the Director of the Bureau of Standards, Washington, D.C., who is chairman. The progress report repre- sents the greater part of the work which it was expected would be accomplished by the commission and has been submit- ted for approval to the Secretaries of War, Navy and Commerce. The law provides that when the report is thus approved it is binding upon the depart- ments in question and must be used by other Federal departments whenever possible. The following is a digest of the report so far as it concerns the fine and coarse thread series recommended.

The aim of the commission in estab- lishing thread systems for adoption and general use was to eliminate all unne- cessary sizes and in addition to utilize so far as possible present predominat- ing sizes. While from certain stand- points it would have been desirable to make simplifications in the thread sys- tems and to establish more thoroughly consistent standards, it is believed that any radical change at the present time would be out of place and interfere with manufacturing conditions, and would in- volve great economic loss.

The testimony given at the various hearings held by the commission is very consistent in favoring the maintenance of the present coarse and fine thread series, the coarse thread series being the present United States standard threads, supplemented in the series below one- quarter inch by the standard establish- ed by the American Society of Mechani- cal Engineers (A. S. M. E.), and the fine thread series being substantially standards that have been found neces- sary, consisting of sizes taken from the standards of the Society of Automotive Engineers (S. A. E.) and the fine thread series of the American Society of Me- chanical Engineers (A. S. M. E.). The recommendation of these standards will tend toward their more universal use, and will constitute that important gain which is effected by standardization, with a minimum handicap.

National Coarse Thread Series

There is specified in Table I. a thread series which will be known as the Na- tional Coarse Thread Series. This series contains certain sizes known previously as the United States standard threads and also certain sizes known as the A. S. M. E. machine-screw threads. There are included in the National coarse thread series only the sizes which are essential. The National coarse threads are recommended for general use in en- gineering work, in machine construction where conditions are favorable to the use of bolts, screws and other threaded components where quick and easy as- sembly of the parts is desired, and for all work where conditions do not re- quire the use of fine pitch thread.

National Fine Thread Series

The threads specified in Table II. will be known as the National Fine Thread Series. This series contains certain sizes known previously as the S. A. E. threads, and, also, certain sizes known

as the A. S. M. E. machine-screw sizes. There are included in the National fine thread series only the sizes which are essential. The National fine threads are recommended for general use in automotive and aircraft work, for use where the design requires both strength and reduction in weight, and where special conditions require a fine thread, such as, for instance, on large sizes where sufficient force cannot be secured to set properly a screw or bolt of coarse pitch by exerting on an ordinary wrench the strength of a man.

The notation used throughout the re- port is shown in Fig. 1. The standard form of thread profile known previously as the U.S. standard, or Sellers profile, shall be used as shown by Fig. 2, which represents the form of thread for mini- mum nut and maximum screw. No al- lowance between nut and screw is shown in Fig. 2, this condition existing in class II., medium fit, where both the mini- mum nut and the maximum screw are basic.

TABLE I. NATIONAL COARSE THREAD SERIES

Identification		Basic Diameters				Thread Data	
1	2	3 D	4 E	5 K	6	7	8 D
Numbered and Fractional Sizes	Number of Threads per Inch	Major Diameter	Pitch Diameter	Minor Diameter	Metric Equivalent of Major Diam.	Pitch	Depth of Thread
		Inches	Inches	Inches	Mm.	Inches	Inches
No. 1	64	0.073	0.0629	0.0527	1.854	0.0156250	0.0101
2	56	0.086	0.0744	0.0628	2.184	0.0178571	0.0116
3	48	0.099	0.0855	0.0719	2.515	0.0208333	0.0133
4	40	0.112	0.0958	0.0795	2.845	0.0250000	0.0162
5	40	0.125	0.1088	0.0925	3.175	0.0250000	0.0162
6	32	0.138	0.1177	0.0974	3.505	0.0312500	0.0203
8	32	0.164	0.1437	0.1234	4.166	0.0312500	0.0203
10	24	0.190	0.1629	0.1359	4.826	0.0416667	0.0271
12	24	0.216	0.1889	0.1619	5.486	0.0416667	0.0271
1/4	20	0.2500	0.2175	0.1850	6.350	0.0500000	0.0325
5/16	18	0.3125	0.2764	0.2403	7.938	0.0555556	0.0361
3/8	16	0.3750	0.3344	0.2938	9.525	0.0625000	0.0406
7/16	14	0.4375	0.3911	0.3447	11.113	0.0714286	0.0464
1/2	13	0.5000	0.4500	0.4001	12.700	0.0769231	0.0500
9/16	12	0.5625	0.5084	0.4542	14.288	0.0833333	0.0541
5/8	11	0.6250	0.5660	0.5069	15.875	0.0909091	0.0590
3/4	10	0.7500	0.6850	0.6201	19.050	0.1000000	0.0650
7/8	9	0.8750	0.8028	0.7307	22.225	0.1111111	0.0722
1	8	1.0000	0.9188	0.8376	25.400	0.1250000	0.0812
1 1/8	7	1.1250	1.0322	0.9394	28.575	0.1428571	0.0928
1 1/4	7	1.2500	1.1572	1.0644	31.750	0.1428571	0.0928
1 1/2	6	1.5000	1.3917	1.2835	38.100	0.1666667	0.1083
1 3/4	5	1.7500	1.6201	1.4902	44.450	0.2000000	0.1299
2	4 1/2	2.0000	1.8557	1.7113	50.800	0.2222222	0.1443
2 1/4	4 1/2	2.2500	2.1057	1.9613	57.150	0.2222222	0.1443
2 1/2	4	2.5000	2.3376	2.1752	63.500	0.2500000	0.1624
2 3/4	4	2.7500	2.5876	2.4252	69.850	0.2500000	0.1624
3	4	3.0000	2.8376	2.6752	76.200	0.2500000	0.1624

Specifications

A clearance shall be provided at the ·minor diameter of the nut by removing the thread form at the crest by an amount equal to 1-6 to ¼ of the basic thread depth. A clearance at the major diameter of the nut shall be provided by decreasing the depth of the truncation triangle by an amount equal to 1-3 to 2-3 of its theoretical value.

The following general specifications will apply to all classes of fits hereinafter specified. In order to conform to the general ideas of standardization the pitch diameter of the minimum threaded hole or nut should correspond to the basic size, the errors due to workmanship being permitted above the basic size. The maximum length of engagement for screw threads manufactured in accordance with any of the classes of fit specified herein shall not exceed the quantity as determined in the following formula:

$$L = (1.5) D.$$

Where L = Length of Engagement;
D = Basic Major Diameter of Thread.

The specifications established for the various classes of it are applicable to the National coarse threads and the National fine threads.

Classification and Tolerances

The National coarse and fine threads shall be manufactured in accordance with the following "classification and tolerances."

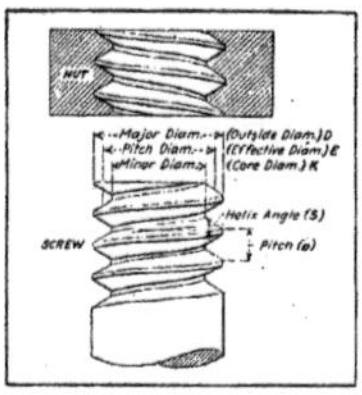

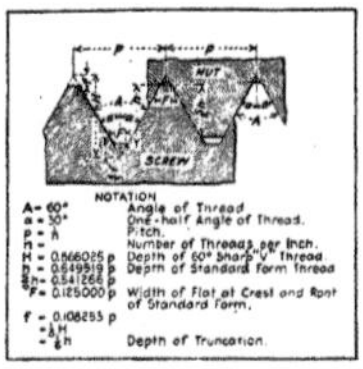

Class I., Loose Fit

The loose fit class of screw threads is intended to cover the manufacture of strictly interchangeable threaded parts where the work is produced in two or more manufacturing plants. In this class will be included threads for artillery ammunition and rough commercial work, such as stove bolts, carriage bolts, and other threaded work of a similar nature, where quick and easy assembly is necessary and a certain amount of shake or play is not objectionable.

The pitch diameter of the minimum nut of a given diameter and pitch will correspond to the basic pitch diameter as specified in the tables of thread systems given herein, which is computed from the basic major diameter of the thread to be manufactured. The dimensions of the maximum screw of a given pitch and diameter will be below the basic dimensions as specified in the tables of thread systems given herein, which are computed from the basic major diameter of the threads to be manufactured, by the amount of the allowance given in Table III. The tolerance on the nut will be plus; to be applied from the basic size to above basic size. The tolerance on the screw will be minus; to be applied from the maximum screw dimension to below the maximum screw dimension. The allowance provided between the size of the minimum nut, which is basic, and the size of the maximum screw for a screw thread of a given pitch will be as specified in Table III. The tolerance allowed on a screw

TABLE II. NATIONAL FINE THREAD SERIES

Identification		Basic Diameters			Thread Data			
Numbered and Fractional Sizes 1	**Number of Threads per Inch** 2	**Major Diameter** D 4	**Pitch Diameter** E 5	**Minor Diameter** 5	**Metric Equivalent of Major Diam.** Min. 6	**Pitch** p 7	**Depth of Thread** h 8	
		Inches	Inches	Inches	Mm.	Inches	Inches	
No. 0	80	0.0600	0.0519	0.0438	1.524	0.0125000	0.00812	
1	72	0.0730	0.0640	0.0550	1.854	0.0138889	0.00902	
2	64	0.0860	0.0759	0.0657	2.184	0.0156250	0.01014	
3	56	0.0990	0.0874	0.0758	2.515	0.0178571	0.01160	
4	48	0.1120	0.0985	0.0849	2.845	0.0208333	0.01353	
5	44	0.1250	0.1102	0.0955	3.175	0.0227273	0.01476	
6	40	0.1380	0.1218	0.1055	3.506	0.0250000	0.01624	
8	36	0.1640	0.1460	0.1279	4.166	0.0277778	0.01804	
10	32	0.1900	0.1697	0.1494	4.826	0.0312500	0.02030	
12	28	0.2160	0.1928	0.1696	5.486	0.0357143	0.02319	
¼	28	0.2500	0.2268	0.2036	6.350	0.0357143	0.02319	
5/16	24	0.3125	0.2854	0.2584	7.938	0.0416667	0.02706	
⅜	24	0.3750	0.3479	0.3209	9.525	0.0416667	0.02706	
7/16	20	0.4375	0.4050	0.3725	11.113	0.0500000	0.03248	
½	20	0.5000	0.4675	0.4350	12.700	0.0500000	0.03248	
9/16	18	0.5625	0.5264	0.4903	14.288	0.0555556	0.03608	
⅝	18	0.6250	0.5889	0.5528	15.875	0.0555556	0.03608	
¾	16	0.7500	0.7094	0.6688	19.050	0.0625000	0.04060	
⅞	14	0.8750	0.8286	0.7822	22.225	0.0714286	0.04640	
1	14	1.0000	0.9536	0.9072	25.400	0.0714286	0.04640	
1⅛	12	1.1250	1.0709	1.0167	28.575	0.0833333	0.05413	
1¼	12	1.2500	1.1959	1.1417	31.750	0.0833333	0.05413	
1⅜	12	1.3750	1.3209	1.2667	34.925	0.0833333	0.05413	
1½	12	1.5000	1.4459	1.3912	38.100	0.0833333	0.05413	
1¾	12	1.7500	1.6959	1.6417	44.450	0.0833333	0.05413	
2	12	2.0000	1.9459	1.8917	50.800	0.0833333	0.05413	
2¼	12	2.2500	2.1959	2.1417	57.150	0.0833333	0.05413	
2½	12	2.5000	2.4459	2.3917	63.500	0.0833333	0.05413	
2¾	12	2.7500	2.6959	2.6417	69.850	0.0833333	0.05413	
3	10	3.0000	2.9350	2.8701	76.200	0.1000000	0.06495	

TABLE III. CLASS I—LOOSE FIT. ALLOWANCES AND TOLERANCES, SCREWS, NUTS AND GAGES

	1	2	3	4	5	6	7
					Master Gage Tolerance		
No. Threads per Inch	**Allowance**	**Extreme or Drawing Finish Dia. Tolerance**	**Diameter**	**Lead**	**Thread Angle**	**Net Pitch Dia. Tolerance**	
	Inches	Inches	Inches	Inches	Degrees	Inches	
80	0.0007	0.0024	0.0002	±0.0002	±0° 30'	0.0020	
72	0.0007	0.0025	0.0002	±0.0002	±0° 30'	0.0021	
64	0.0007	0.0026	0.0002	±0.0002	±0° 30'	0.0022	
56	0.0008	0.0028	0.0002	±0.0002	±0° 30'	0.0024	
48	0.0009	0.0031	0.0002	±0.0002	±0° 30'	0.0027	
44	0.0009	0.0032	0.0002	±0.0002	±0° 30'	0.0038	
40	0.0010	0.0034	0.0002	±0.0002	±0° 20'	0.0030	
36	0.0010	0.0036	0.0002	±0.0002	±0° 20'	0.0032	
32	0.0011	0.0038	0.0002	±0.0002	±0° 20'	0.0034	
28	0.0012	0.0043	0.0003	±0.0002	±0° 15'	0.0037	
24	0.0013	0.0046	0.0003	±0.0002	±0° 15'	0.0040	
20	0.0015	0.0051	0.0003	±0.0003	±0° 15'	0.0045	
18	0.0016	0.0057	0.0004	±0.0003	±0° 10'	0.0049	
16	0.0018	0.0063	0.0004	±0.0003	±0° 10'	0.0055	
14	0.0021	0.0070	0.0004	±0.0003	±0° 10'	0.0062	
13	0.0022	0.0074	0.0004	±0.0003	±0° 10'	0.0066	
12	0.0024	0.0070	0.0005	±0.0003	±0° 10'	0.0071	
11	0.0026	0.0085	0.0005	±0.0004	±0° 10'	0.0077	
10	0.0028	0.0092	0.0005	±0.0004	±0° 5'	0.0084	
9	0.0031	0.0100	0.0006	±0.0004	±0° 5'	0.0092	
7	0.0034	0.0111	0.0006	±0.0004	±0° 5'	0.0103	
6	0.0039	0.0124	0.0006	±0.0004	±0° 5'	0.0116	
5	0.0044	0.0145	0.0006	±0.0005	±0° 5'	0.0135	
4½	0.0052	0.0169	0.000s	±0.0005	±0° 5'	0.0157	
5	0.0057	0.0184		±0.0005	±0° 5'	0.0172	
4	0.0064	0.0204	0.0006	±0.0005	±0° 5'	0.0192	

* Allowable variation in lead between any two threads not farther apart than the length of engagement

or nut of a given pitch will be as specified in Table III.

Class II., Medium Fit

The medium fit class, Subdivision A, regular, is intended to apply to interchangeable manufacture where the threaded members are to assemble nearly, or entirely, with the fingers and where a moderate amount of shake or play between the assembled threaded members is not objectionable. This class will include the great bulk of fastening screws for instruments, small arms and other ordnance material, such as gun carriages, aerial bomb dropping devices and interchangeable accessories mounted on guns; also machine screws, cap screws, and screws for sewing machines, typewriters and other work of a similar nature.

The pitch diameter of the minimum nut of a given diameter and pitch will correspond to the basic pitch diameter as specified in tables of thread systems given herein, which is computed from the basic major diameter of the thread to be manufactured. The major diameter and pitch diameter of the maximum screw of a given pitch and diameter will correspond to the basic dimensions as specified in tables of thread systems given herein, which are computed from the basic major diameter of the thread to be manufactured. The tolerance on the nut will be plus; to be applied from the basic size to above basic size. The tolerance on the screw will be -minus; to be applied from the basic size to below basic size. The allowance between the size of the maximum screw and the minimum nut will be zero for all pitches and all diameters. The tolerance for a screw or nut of a given pitch will be as specified in Table IV.

The medium fit class, Subdivision B, special, is intended to apply especially to the higher grade of automobile screw thread work. It is the same in every particular as Class II-A, medium fit (regular), except that the tolerances are smaller. The tolerance for a screw or nut of a given pitch will be as specified in Table V.

Class III., Close Fit

The close fit class of screw is intended for threaded work of the finest commercial quality, where the thread has practically no back lash, and for light screw driver fits. In the manufacture of screw thread products belonging in this class it will be necessary to use precision tools, selected master gauges, and many other refinements. This quality of work should, therefore, be used only in cases where requirements of the mechanism being produced are exacting, or where special conditions require screws having a precision fit. In order to secure the fit desired, it may be necessary in some cases to select the parts when the product is being assembled.

The pitch diameter of the minimum nut of a given diameter and pitch will correspond to the basic pitch diameter as specified in tables of thread systems given herein, which is computed from the basic major diameter of the thread to be manufactured. The major diameter and pitch diameter of the maximum screw of a given diameter and pitch will be above the basic dimensions as specified in tables to thread systems given herein, which are computed from the basic major diameter of the thread to be manufactured, by the amount of the allowance (interference) specified in Table VI. The tolerance on the nut will be plus; to be applied from the basic size to above basic size. The tolerance on the screw will be minus; to be applied from the basic size to below the maximum screw dimensions. The allowance (interference) provided between the size of the minimum nut, which is basic, and the size of the maximum screw, which is above basic, will be as specified in Table VI. The tolerance for a screw or nut of a given pitch will be as specified in Table VI.

Class IV., Wrench Fit

The wrench fit class of screw threads is intended to cover the manufacture of threaded parts ¾-inch in diameter or larger which are to be set or assembled permanently with a wrench. Inasmuch as for wrench fits the material is an important factor in determining the fit between the threaded members there are provided herein two subdivisions for this class of work, namely, subdivision "A" and subdivision "B". These two subdivisions differ mainly in the amount of the allowance (interference) values provided for different pitches.

Subdivision "A" of Class IV., wrench fit, provides for the production of interchangeable wrench fit screws or studs used in light sections with moderate

---

TABLE IV. CLASS II-A—MEDIUM FIT (REGULAR). ALLOWANCES AND TOLERANCES. SCREWS, NUTS AND GAGES

No. Threads per Inch	Allowance	Extreme or Drew. Net Pitch Dia. Tolerances	Diameter	Lead	Half Angle	Net Pitch Dia. Tolerances
	Inches	Inches	Inches	Inches	Degrees	Inches
80	0.0000	0.0017	0.0002	±0.0002	±0.0002	±0° 30'	0.0013
72	0.0000	0.0018	0.0002	±0.0002	±0.0002	±0° 30'	0.0014
64	0.0000	0.0019	0.0002	±0.0002	±0.0002	±0° 30'	0.0015
56	0.0000	0.0020	0.0002	±0.0002	±0.0002	±0° 30'	0.0016
48	0.0000	0.0022	0.0002	±0.0002	±0.0002	±0° 30'	0.0018
44	0.0000	0.0023	0.0002	±0.0002	±0.0002	±0° 30'	0.0019
40	0.0000	0.0024	0.0002	±0.0002	±0.0002	±0° 20'	0.0020
36	0.0000	0.0025	0.0002	±0.0002	±0.0002	±0° 20'	0.0021
32	0.0000	0.0027	0.0002	±0.0002	±0.0002	±0° 20'	0.0023
28	0.0000	0.0031	0.0003	±0.0002	±0.0002	±0° 15'	0.0025
24	0.0000	0.0033	0.0003	±0.0003	±0.0002	±0° 15'	0.0027
20	0.0000	0.0036	0.0003	±0.0003	±0.0003	±0° 15'	0.0030
18	0.0000	0.0041	0.0003	±0.0003	±0.0003	±0° 10'	0.0033
16	0.0000	0.0045	0.0004	±0.0003	±0.0003	±0° 10'	0.0037
14	0.0000	0.0049	0.0004	±0.0004	±0.0003	±0° 10'	0.0041
13	0.0000	0.0052	0.0004	±0.0004	±0.0003	±0° 10'	0.0044
12	0.0000	0.0054	0.0004	±0.0004	±0.0004	±0° 10'	0.0048
11	0.0000	0.0059	0.0005	±0.0004	±0.0004	±0° 5'	0.0051
10	0.0000	0.0064	0.0004	±0.0004	±0.0004	±0° 5'	0.0056
9	0.0000	0.0070	0.0004	±0.0004	±0.0004	±0° 5'	0.0062
8	0.0000	0.0076	0.0004	±0.0004	±0.0004	±0° 5'	0.0068
7	0.0000	0.0085	0.0004	±0.0004	±0.0005	±0° 5'	0.0077
6	0.0000	0.0101	0.0005	±0.0005	±0.0005	±0° 5'	0.0089
5	0.0000	0.0115	0.0006	±0.0005	±0.0005	±0° 5'	0.0104
4½	0.0000	0.0127	0.0006	±0.0005	±0.0005	±0° 5'	0.0115
4	0.0000	0.0140	0.0006	±0.0006	±0.0005	±0° 5'	0.0128

* Allowable Variation in lead between any two threads not farther apart than the length of engagement.

TABLE V. CLASS II-B—MEDIUM FIT (SPECIAL). ALLOWANCES AND TOLERANCES. SCREWS, NUTS AND GAGES

No. Threads per Inch	Allowance	Extreme or Drew. Net Pitch Dia. Tolerances	Diameter	Lead	Half Angle	Net Pitch Dia. Tolerances
	Inches	Inches	Inches	Inches	Degrees	Inches
80	0.0000	0.0013	0.0002	±0.0002	±0.0002	±0° 30'	0.0009
72	0.0000	0.0013	0.0002	±0.0002	±0.0002	±0° 30'	0.0009
64	0.0000	0.0014	0.0002	±0.0002	±0.0002	±0° 30'	0.0010
56	0.0000	0.0015	0.0002	±0.0002	±0.0002	±0° 30'	0.0011
48	0.0000	0.0016	0.0002	±0.0002	±0.0002	±0° 30'	0.0012
44	0.0000	0.0016	0.0002	±0.0002	±0.0002	±0° 30'	0.0012
40	0.0000	0.0017	0.0002	±0.0002	±0.0002	±0° 20'	0.0013
36	0.0000	0.0018	0.0002	±0.0002	±0.0002	±0° 20'	0.0014
32	0.0000	0.0019	0.0002	±0.0002	±0.0002	±0° 20'	0.0015
28	0.0000	0.0022	0.0003	±0.0002	±0.0002	±0° 15'	0.0016
24	0.0000	0.0024	0.0003	±0.0002	±0.0002	±0° 15'	0.0018
20	0.0000	0.0026	0.0003	±0.0003	±0.0003	±0° 15'	0.0020
18	0.0000	0.0030	0.0003	±0.0003	±0.0003	±0° 10'	0.0022
16	0.0000	0.0032	0.0004	±0.0003	±0.0003	±0° 10'	0.0024
14	0.0000	0.0036	0.0004	±0.0004	±0.0003	±0° 10'	0.0028
13	0.0000	0.0037	0.0004	±0.0004	±0.0003	±0° 10'	0.0029
12	0.0000	0.0040	0.0004	±0.0004	±0.0003	±0° 10'	0.0032
11	0.0000	0.0042	0.0004	±0.0004	±0.0003	±0° 5'	0.0034
10	0.0000	0.0045	0.0004	±0.0004	±0.0004	±0° 5'	0.0037
9	0.0000	0.0049	0.0004	±0.0004	±0.0004	±0° 5'	0.0041
8	0.0000	0.0054	0.0004	±0.0004	±0.0004	±0° 5'	0.0046
7	0.0000	0.0059	0.0004	±0.0004	±0.0004	±0° 5'	0.0051
6	0.0000	0.0071	0.0005	±0.0004	±0.0005	±0° 5'	0.0059
5	0.0000	0.0082	0.0005	±0.0005	±0.0005	±0° 5'	0.0073
4½	0.0000	0.0089	0.0005	±0.0005	±0.0005	±0° 5'	0.0079
4	0.0000	0.0095	0.0006	±0.0005	±0.0005	±0° 5'	0.0085

* Allowable variation in lead between any two threads not farther apart than the length of engagement.

stresses, such as for aircraft and auto-mobile engine work.

Subdivision "B" of Class IV., wrench fit, provides for the production of inter-changeable wrench fit screws or studs used in heavy sections with heavy stresses, such as for steam engine and heavy hydraulic work.

The pitch diameter of the minimum nut of a given diameter and pitch for threads belonging to either subdivision "A" or subdivision "B," will correspond to the basic pitch diameter as specified in tables of thread systems given here-in, which is computed from the basic major diameter of the thread to be manufactured. The major diameter and pitch diameter of the maximum screw of a given diameter and pitch for threads belonging in either subdivision "A" or subdivision "B" will be above the basic dimensions as specified in tables of thread systems given herein, which are computed from the basic major diameter of the thread to be manufactured, by the amount of the al-lowance (interference) provided. The tolerance on the nut will be plus; to be applied from the basic size to above basic size. The tolerance on the screw will be minus, to be applied from the maximum screw dimensions to below maximum screw dimensions. At the present time the commission does not have sufficient information or data to include in its tentative report values for tolerances and allowances for wrench fits. It is hoped, however, that suffi-cient information resulting from investi-gation and research will enable the com-mission to decide at an early date the allowance and tolerance values for the two classes of wrench fits included herein, which will be applicable to the various materials, and which will meet the requirements found in manufacture of machines or product requiring wrench fits.

Tolerances

There are specified herein for use in connection with the various fits estab-lished, three different sets of tolerances, as given in Tables III, IV, V and VI.

The tolerances as hereinafter specified represent the extreme variations allow-ed on the work.

The tolerance limits established rep-resent, in reality, the sizes of the "Go" and "Not Go" master gauges. Errors in lead and angle which occur on the threaded work can be offset by a suit-able alteration of the pitch diameter of the work. If the "Go" gauge passes the threaded work interchangeability is se-cured and the thread profile may differ from that of the "Go" gauge in either pitch diameter, lead or angle. The "Not Go" gauge checks pitch diameter only, and thus insures that the pitch diameter is such that the fit will not be too loose.

The tolerances established for Class I, loose fit, and Class II, medium fit, permit the use of commercial taps now obtainable from various manufacturers. For Class III, close fit, in which it is desired to produce a hole close to the basic size, it is recommended that a selected tap be used.

The pitch diameter tolerances provid-ed for a screw of a given class of fit will be the same as the pitch diameter tolerances provided for a nut correspond-ing to the same class of fit. The allow-able tolerances on the major diameter of screws of a given classification will be twice the tolerance values allowed on the pitch diameters of screws of the same class.

The minimum minor diameter of a screw of a given pitch will be such as to result in a basic flat (⅛ x p) at the root when the pitch diameter of the screw is at its minimum value. Note: When the maximum screw is basic the minimum minor diameter of the screw will be below the basic minor diameter by the amount of the specified pitch diameter tolerance.)

The maximum minor diameter may be such as results from the use of a worn or rounded threading tool when the pitch diameter is at its maximum value. In no case, however, should the form of the screw as results from tool wear be such as to cause the screw to be rejected on the maximum minor diameter by a "Go" ringe gauge the minor diameter of which is equal to the minimum minor diameter of the nut.

The maximum major diameter of the nut of a given pitch will be such as to result in a flat 1-3 of the basic flat (1-24 x p) when the pitch diameter of the nut is at its maximum value. (Note: When the minimum nut is basic the maximum major diameter will be above the basic major diameter by the amount of specified pitch diameter tolerance plus 2-9 of the basic thread depth.) The nominal minimum major diameter of a nut will be above the basic major dia-meter by an amount equal to 1-9 of the basic thread depth plus the neutral space. This results in a clearance which is provided to facilitate manufacture by permitting a slight rounding or wear at the crest of the tap.

In no case, however, should the mini-mum major diameter of the nut as re-sults from a worn tap or cutting tool be such as to cause the nut to be rejected on the minimum major diameter by a "Go" plug gauge made to the standard form at the crest.

The tolerances on minor diameter of a nut of a given pitch will be 1-6 of the basic thread depth, regardless of the class of fit being produced. In Fig. 3 there are shown the various re-lations previously specified for tolerances on both the screw and the nut.

The specifications establishing the various sets of tolerances for the dif-ferent classes of fit specified herein will apply to the manufacture of national coarse threads, national fine threads, and wherever applicable to the production of all special threads.

Where tolerances are desired for a special thread and the pitch is not list-ed in the tables given, the tolerance values should be chosen corresponding to the number of threads per inch near-est to that of the special thread being

TABLE VI. CLASS III—CLOSE FIT. ALLOWANCES AND TOLERANCES. SCREWS, NUTS AND GAGES

No. Threads per Inch	Interference or Negative Allowances	Screws or Diameter and Lead Tolerances	Diameter	Lead	Half Angle	Net Truth Flat Tolerances
	Inches	Inches	Inches	Inches	Degrees	Inches
80	0.0001	0.0006	0.00010	± 0.00010	± 15' 00"	0.0004
72	0.0001	0.0007	0.00010	± 0.00010	± 15' 00"	0.0005
64	0.0001	0.0007	0.00010	± 0.00010	± 15' 00"	0.0005
56	0.0002	0.0007	0.00010	± 0.00010	± 15' 00"	0.0005
48	0.0002	0.0008	0.00010	± 0.00010	± 15' 00"	0.0006
44	0.0002	0.0008	0.00010	± 0.00010	± 15' 00"	0.0006
40	0.0002	0.0009	0.00010	± 0.00010	± 10' 00"	0.0007
36	0.0002	0.0009	0.00010	± 0.00010	± 10' 00"	0.0007
32	0.0002	0.0010	0.00010	± 0.00010	± 10' 00"	0.0008
28	0.0002	0.0011	0.00015	± 0.00010	± 7' 30"	0.0008
24	0.0003	0.0012	0.00015	± 0.00010	± 7' 30"	0.0009
20	0.0003	0.0013	0.00015	± 0.00010	± 7' 30"	0.0010
18	0.0003	0.0015	0.00020	± 0.00015	± 5' 00"	0.0011
16	0.0004	0.0016	0.00020	± 0.00015	± 5' 00"	0.0012
14	0.0004	0.0018	0.00020	± 0.00015	± 5' 00"	0.0014
13	0.0004	0.0019	0.00020	± 0.00015	± 5' 00"	0.0015
12	0.0005	0.0020	0.00020	± 0.00015	± 5' 00"	0.0016
11	0.0005	0.0021	0.00020	± 0.00015	± 5' 00"	0.0017
10	0.0006	0.0023	0.00020	± 0.00020	± 2' 30"	0.0019
9	0.0006	0.0024	0.00020	± 0.00020	± 2' 30"	0.0020
8	0.0007	0.0027	0.00020	± 0.00020	± 2' 30"	0.0023
7	0.0008	0.0030	0.00025	± 0.00020	± 2' 30"	0.0026
6	0.0009	0.0036	0.00030	± 0.00025	± 2' 30"	0.0030
5	0.0010	0.0041	0.00030	± 0.00025	± 2' 30"	0.0035
4½	0.0011	0.0044	0.00050	± 0.00025	± 2' 30"	0.0038
4	0.0013	0.0048	0.00050	± 0.00025	± 2' 30"	0.0042

*Allowable variation in lead between any two threads not farther apart than the length of engagement.

produced. Where the number of threads per inch is midway between two of the pitches listed, the tolerance corresponding to the coarser pitch should be used.

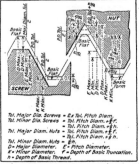

Tol. Major Dia. Screws = Ex.Tol. Pitch Diam.
Tol. Minor Dia. Screws = Tol. Pitch Diam. + ⅜ f.
                   = Tol. Pitch Diam. + ⅜ h.
Tol. Major Diam. Nuts = Tol. Pitch Diam. + ⅝ f.
                    = Tol. Pitch Diam. + ⅝ h.
Tol. Minor Diam. Nuts = ⅝ h.
D = Major Diameter.   E = Pitch Diameter.
K = Minor Diameter.   F = Depth of Basic Truncation.
h = Depth of Basic Thread.

FIG. 3—Relation between pitch diameter tolerances and tolerances on major and minor diameters. (Drawing shows one side of thread only and therefore spaces indicate half tolerances or tolerances on radii.)

For instance, the tolerance on a screw having 11½ threads per inch would correspond to the tolerances specified for a screw of 11 threads per inch.

With reference to the classification of screw thread fits attention is called to the fact that the minimum threaded hole or nut corresponds to the basic size; that is, the pitch diameter of the minimum nut is basic for all classes of fit. This condition permits the use of taps, which, when new, are oversize, and which are discarded when the whole cut is at the basic size. In order to secure the desired fit the screw size is varied; the maximum screw corresponds to the basic size for the medium fit class, is slightly above basic size for close fit class, considerably above the basic size in the wrench fit class, and below the basic size for the loose fit class.

The tolerances specified in column 7 of Tables III, IV, V and VI are the net tolerances, which are in no way reduced by permissible manufacturing tolerances provided for master gauges. These master gauge tolerances are provided for by being added to the net tolerances. Thus the net working tolerances increased by the master gauge increment or equivalent diametrical space required to provide for the master gauge tolerances. The limits established for the extreme tolerances should in no case be exceeded. The application of gauge tolerances in relation to tolerances allowed on the work can be best understood by considering that the extreme tolerances represent the absolute limits over which variations of the work must not pass. The manu-

facturing tolerances required for master gauges are then deducted from the extreme working tolerances, producing the figures specified as net tolerances. Further reduction of the extreme tolerances is caused by the manufacturing tolerances required for the inspection gauges and working gauges.

It is essential that the proportion of the tolerance used by the workmen producing the work at the machine be well within the net tolerance limits. The net tolerance limits as established by the master gauges may be considered as the largest circle of the target, the space

representing the width of the line establishing the largest circle. The marksman always aims to hit the bull's-eye. Any mark inside of the largest circle or

occupied by the master gauge tolerances cutting the circle scores. Any mark outside of the largest circle does not score. The same is true in producing work—the careful manufacturer will aim to produce work which is in the centre of tolerance limits. The bull's-eye in this case, which is the working tolerance used at the machine, will be considerably less than the net tolerance and the result will be that a very large percentage of the work will be accepted, and spoiled or rejected work will be reduced to practically nothing. If the net tolerance limits are used as working limits at the machine there will be a larger percentage of rejections due to differences in gauges and wear of both tools and gauges. The application of this principle is illustrated in Fig. 4, which is a diagram showing the relative position of master gauge, inspection gauge and working gauge tolerances with reference to the net tolerance allowed on the work.

Extreme Limits

The extreme limits as shown by the lines at A and a in Fig. 4 represent the absolute limits within which all variations of the work must be kept, including permissible variations provided for manufacturing tolerances on master gauges. The manufacturer of the product should not be concerned with the extreme tolerances but should work within the net tolerance limits. The extreme tolerance limits are included for the manufacturer or inspector of master gauges, and in no case should master gauges be approved which are outside of the dimensions established by these extreme limits.

The lines at b and B represent the net working tolerance limits within which all manufactured product must come.

The regions AB and ab represent the space required to provide for the "Go" and "Not go" master gauge tolerances respectively.

Master gauges provide physical stand-

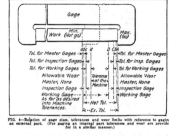

Tol. for Master Gages
Tol. for Inspection Gages
Tol. for Working Gages
Allowable Wear Master, None
Inspection Gage
Working Gage—As far as desired Into Machine Tolerances.

Tol. for Master Gages
Tol. for Insp. Gages
Tol. for Working Gages
Allowable Wear Master, None
Inspection Gage
Working Gage
Net Tol.
Ex. Tol.

FIG. 4—Relation of gage sizes, tolerances and wear limits with reference to gaging an external part. (For gaging an internal part tolerances and wear are provided for in a similar manner.)

ards representing the limits placed on the work. The master gauge tolerances are placed within the extreme tolerance limits. However, the manufacturer re-

ceives the full benefit of the specified net tolerance. So far as the manufacturer is concerned he should in no case permit variations in the work produced to extend beyond the limits established by his master gauges.

The regions *BC* and *be* represent the space required to provide for the "Go" and "Not go" inspection gauge tolerances respectively. The inspection gauge tolerances are placed inside the net tolerance limits.

The regions *DC* and *dc* represent the space required to provide for the "Go" and "Not Go" working gauge tolerances respectively. These working gauge tolerances are placed within the net tolerance limits. This insures that any work accepted by the working gauge will be accepted by the inspection gauge, and that work accepted by both working gauge and inspection gauge will be within the net tolerance limits.

The "Go" master gauge is not to be used on the product. It serves as a standard for comparative measurements or as a check for verifying the inspection or working gauge. It also serves as a standard representing the wear limit for the inspection or working gauge. The "Go" master gauge is, therefore, not subject to wear.

The "Go" inspection gauge may wear until it reaches the size represented by the master gauge. As shown in Fig. 4 the wear provided for the inspection gauge is that which takes place within its own tolerance region. However, a definite allowance for wear may be provided for the "Go" inspection gauge in addition to its tolerance region if desired.

The "Go" working gauge wears within its own tolerance limits and through the inspection gauge tolerance region, and continues to properly accept work until worn to the dimension established by the "Go" master gauge. It is good practice to transfer the "Go" working gauge to use as an inspection gauge when it is worn so that its dimension corresponds to that of the inspection gauge.

The "Not go" master gauge is not to be used on the product. It serves as a standard for comparative measurements or as a check to verify the inspection or working gauge. It is, therefore, not subject to wear.

The "Not go" inspection gauge wears within its own tolerance region and into the tolerance region established for the "Not go" working gauge. It is good practice to transfer the "Not go" inspection gauge to use as a working gauge when it is worn so that its dimension corresponds to that of the "Not go" working gauge.

The "Not go" working gauge wears within its own tolerance region into the working tolerance. It is purely an economic question as to when the "Not go" working gauge should be discarded, due to wear, inasmuch as continued use reduces the working tolerance, the result of which must be balanced against the cost of a new gauge.

(To be continued)

MOTOR INDUSTRY TAXED TO THE LIMIT, FINDS GOING HARD

SPEAKING to a newspaper reporter in Oshawa a few days ago, the manager of General Motors of Canada, Ltd., expressed the hope that conditions would improve, but admitted that business at present was well below normal. The present slackness in the motor industry has been caused by many conditions. The chief reason for the dull conditions lies in the fact that the motor industry has been hard hit by the imposition of heavy excise taxes. Under the old taxation prior to May 1, which the automobile industry and accessory trades considered very high, the General Motors of Canada, Ltd., have paid to the Dominion Government approximately $2,500,000 in special taxes. Since May 1 these taxes have been very much increased, and in the case of the highest priced McLaughlin cars have been doubled. It is the impression of the General Motors executive that the motor industry is unduly taxed. Taxes are now so high that prices must be correspondingly increased, which has caused demand to slow up.

Tax of $780 on One Car

On the lowest priced automobile made by the McLaughlin Motor Car Co., the tax amounts to $238 per car. The highest priced car is taxed $780. The special taxes are levied when the car is sold to the dealer and do not include customs duty and business profits' taxes. It has been the contention among manufacturers that they are being taxed out of proportion to other staple industries. The tax is 240 per cent. over and above that imposed by the United States Government upon the American buyers of cars.

General unsettled trade conditions and adverse rate of exchange, together with unsatisfactory conditions in Europe, where money is scarce and borrowing power at its lowest ebb, and the heavy excise and other taxes imposed by the Government, have for some time past had their effect on the operations of the General Motors of Canada, Ltd., in Oshawa, with the result that it has been expedient to decrease production for the time being.

No Easy Task

"Nobody knew what would happen as things were so uncertain." The company was doing its utmost to improve its export trade, but with the buying powers of Europe greatly curtailed as a result of the war and its after effects, and with the present adverse rate of exchange, the task was no easy one.

With business in the unsettled condition it is at present, it has been found necessary to decrease the staffs in many of the departments of the General Motors of Canada, Ltd. Where men have to be temporarily laid off, care is being taken to discriminate between men who have families to support and those who are single. Men who own their own homes and live in Oshawa permanently are also given consideration. It might be necessary to resort to half-time in order to keep as many men employed as possible until conditions have righted themselves.

IMPROVEMENT IN ROAD MAKING

The spraying of roads with tar has now become almost a standard means of rendering them waterproof and free from dust, but the process has certain difficulties attached to it. One is the difficulty of keeping traffic, both vehicular and pedestrian, off the road until the tar has been covered by chippings. Several ways have been devised of carrying out both operations at the same time, and perhaps the most effective is that recently adopted by a British road engineer. He uses a 5-ton tractor to haul a 320-gallon tar tank followed by a 5-ton tipping waggon filled with chippings. At the back of the wagon is a V-shaped trough through which the chippings fall upon the freshly tarred surface. To prevent the choking of this trough it is traversed by a shaft with projecting spikes, and this shaft is rotated by a chain geared to the rear axle of the wagon. This simple equipment is a great labour saver and it spreads the chippings much more evenly than can be done by hand.

SELLING GAS SCIENTIFICALLY

British gas undertakings are leading the way in the sale of gas on a scientific basis. Henceforward, instead of selling gas at so much per thousand cubic feet, they will sell it according to its heating value. A new unit, christened the "therm," will be used for this purpose. A therm is equivalent to one hundred thousand British Thermal units. Consumers will now pay something like eleven pence per therm. The pressure of the gas will be carefully regulated under penalties.

The discovery of a new alloy, with important industrial possibilities, is announced by the Academy of Sciences, Paris. It is to be called Elinvar, an alloy which is extensively used by makers of scientific instruments in France. The new alloy, it is claimed, is an improvement on Invar, and is peculiarly valuable by reason of its uniform elasticity.

The erection of the Deep Creek viaduct on the Pacific and Great Eastern Railway of British Columbia has been commenced, and is noteworthy, as the structure will be one of the highest of its type in America. The base of the rail will be over 300 feet above the pedestals.

A Course of Instruction for the Graded Certificate

We Will in This Course Go Through the Mathematical and Theoretical Work to Cover Every Grade of Certificate Issued by the Ontario Board

By T. H. FENNER

WE considered in the last article the various forms of lap joints, and now come in logical sequence to the butt strap type. These are double riveted, triple riveted, and quadruple riveted, the triple riveted being most in use. They are called butt strap joints because the edges of the plates to be joined are butted together and are secured by straps or plates of metal, which cover the joint, inside and outside, and are riveted through straps and plate to make a solid joint. This type of joint makes for a high percentage of strength, and at the same time preserves the truly cylindrical shape of the boiler. The rivets, being in both double and single shear, thereby gain in strength. A rivet is said to be in double shear when it holds together three plates, and therefore before it can be sheared has to be cut through in two places instead of one. The shearing strength of such a rivet is usually allowed to be 1¾ times the shearing strength of a rivet in single shear.

In the double riveted double strap butt joints the straps are usually made equal in width on the inside and outside of the joint and the full number of rivets is used in each row. In the treble riveted double strap butt joint the outside strap is only made wide enough to take in the two inside rows, and every alternate rivet is left out of the outside row. In the quadruple riveted double strap butt joint the straps are also unequal, and besides every alternate rivet being left out of the third row, two rivets are left out of the outside row. In some cases the strap is cut away on the outside edge to conform to the rivet line of the third row, thus securing a better caulking edge. This is not universal practice, but is a feature of some well-designed joints. The writer knows of at least one joint of this type with an efficiency of 92½ per cent.

When designing a joint of the butt strap variety, the same procedure is gone through as in the others. We have also to consider the thickness of the straps, and the Ontario regulations give a definite rule for determining this dimension. The thickness of straps where the full number of rivets in each row is used and double straps is found from the formula

$$T_1 = \frac{5}{8} T$$

where T_1 is the thickness of strap and T thickness of shell plate. For double straps when straps are unequal in width and every alternate rivet omitted in outer row

$$T_1 = \frac{5T \times (P - d)}{8 \times (P - 2d)}$$

Suppose we consider first a double strap double riveted joint, with equal straps and full number of rivets in each row. In a joint of this kind all rivets are in double shear, and therefore we must reckon for the number of rivets in a pitch 2 × 1.75 = 3½. This will of course have the effect of increasing the rivet strength, and in order to equalize

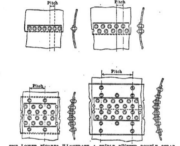

it we must increase the pitch, which brings up the plate strength.

Suppose we wish for a plate strength of about 80 per cent., let us see what results we can obtain. We have $\dfrac{p - d}{p}$ × 100 = 80, and if we assume a ⅝-inch diameter rivet we get

$$\frac{p - d}{p} \times 100 = 80$$

$$p - d = .8p$$

$$p - .p = .8p = .2p$$

Then if d = .75, the pitch will be

$$\frac{.75}{.2} = 3.75$$

With a pitch of 3.75 inches, diameter .75, and plate strength efficiency of 80 per cent., what will be the rivet strength efficiency? Let us work it out, taking C as .7:

$$\frac{a \times n \times c}{p \times t} = \%$$

$$\frac{3 \times 3 \times .7854 \times 7 \times 4 \times 8 \times 70}{4 \times 4 \times 2 \times 15 \times 3 \times 100} = 76.9$$

say 77 per cent. which is within 3 per cent. of the plate strength, and may be taken as a pretty evenly balanced joint. We could increase the strength of the rivet section by decreasing the pitch, but this would also decrease the plate strength, and as we have allowed a low constant for the rivets, we may feel safe in leaving it as it is.

We must now consider the joint from the point of view of possible failure, or as we say, analyse it. In a butt joint the rivets on one side of the centre line only are considered, as conditions on each side of the centre line are exactly the same. Taking the plate from the centre line it may crush in front of a rivet in the inner row and shear one rivet in double shear in the outer row. It may shear two rivets in double shear, or it may tear between the rivets of the inner row and shear one rivet in the outer row. It may tear between the rivets on the outer row, and this is usually the weakest point.

Suppose it crushes in front of one rivet and shears one rivet. We have

$$d \times t \times 90{,}000 + a \times 1.75 \times 42{,}000$$

⅝ × ⅝ × 90,000 × ⅞ × ⅞ × .7854 × 7 / 4 × 42,000 = 57,726

The unit section is 3.75 × .375 × 60,000 = 84,000 lbs. Therefore the percentage

$$\frac{57{,}726}{84{,}000} = 68.7 \text{ per cent.}$$

in this case is = 68.7 per cent. which is lower than either percentage already obtained. If we look into this a moment we will see why. Comparing

THE LOWER FIGURES ILLUSTRATE A TRIPLE RIVETED DOUBLE STRAP BUTT JOINT, AND A QUADRUPLE RIVETED DOUBLE STRAP BUTT JOINT, BOTH WITH UNEQUAL STRAPS.

this joint with the double riveted lap ye, see that although we have the advantage of the rivets being in double shear, for this particular case of crushing the plate and shearing one rivet, the greater pitch of the butt joint offsets this advantage. It is obvious that to bring this part of the joint up to a closer strength proportion, we must increase the rivet area. This will decrease the percentage of plate strength, but we shall have to make a compromise. By making the rivets 13-16 dia. instead of ¾ in. we shall get a closer relation.

$$\text{First} \quad \frac{3.75 - .8125}{3.75} \times 100 = 78.3\% \text{ plate}$$

strength,

$$\text{and} \quad \frac{13 \times 13 \times .7854 \times 7 \times 4 \times 8 \times 70}{16 \times 16 \times 2 \times 15 \times 3 \times 100} = 90.3 \text{ rivet}$$

strength.

Now to consider again crushing in front of one rivet and shearing one rivet in double shear.

13/16 x ⅞ x 90,000 +13/16 x 13/16x.7854
× 1.75 x 42,000
    = 27,414 + 38,106
    = 65,520 lbs.

As the unit section is 84,000 lbs. we get

$$\frac{65,520}{84,000} \times 100 = 78\%$$

In the case of shearing two rivets we will have

13/16 x 13/16 x .7854 x 1.75 x 2x42,000 = 76,212 lbs., which gives a percentage of 90.

For tearing between one row and shearing one rivet we have

(p—d)t x 60,000 + a x 1.75 x 42,000
66,090 + 38,106

which gives over 100 per cent.

Tearing between the rivets of the outer row gives (p—d)t x 60,000 = 66,090, and this, of course, gives 78 per cent.

We have therefore a joint not very much higher in strength than a good lap joint, and considerably more expensive to manufacture. This is the reason that comparatively few of these joints are met with.

The Triple Riveted Butt Joint

Coming to the triple riveted butt strap joint, we have reached probably the most widely used joint in high class boiler manufacture. The efficiency of this joint is high, usually running about 86 per cent. It is not unusual to find the rivet strength in these joints considerably higher than the plate strength, and this is accounted for by the large number of rivets in double shear. The weakest part of the joint is the plate section at the outside row of rivets, and if this is high enough the rest will be pretty near right. The high efficiency is due to the large pitch in the outer row of rivets where each alternate rivet has been omitted. In a joint of this type there are five rivets in a pitch, and of these four are in double shear, thus making 4 × 1.75 + 1 = 8 rivets practically, when considering rivet strength. The methods of considering the joint are

exactly the same, except that the analysis is more complicated. After the dimensions are determined on, we must also check up and see that the diagonal pitch does not exceed the limit set by the Ontario regulations. We will take for purpose of example a double strap triple riveted butt joint, with unequal straps, and every alternate rivet omitted from the outer row, and an efficiency of say 86 per cent. The plate thickness is 7-16 inches.

The first step is to obtain the pitch for the outer row of rivets. With 7-16 inch thickness of plate, we can assume a 15-16 inch rivet, Then we have

$$\frac{p - d}{p} \times 100 = 86$$

p—d=.86p
d = .14p

And if d = .9375 then p = $\frac{.9375}{.14}$ which

gives 6.69, or nearly 6¾. We can take this pitch as a basis to work on.

For rivet strength we shall have then

$$\frac{.69 \times 8 \times 4 \times 16 \times 70}{27 \times 7 \times 100} \times 100 = 130\%$$

Which is ample, being stronger than the solid section.

In how many ways can this joint fail? As in the last case, take from the edge of the plate and work outwards. The first case in which failure may occur is by crushing the plate in front of the rivets in the inner row, and shearing two rivets in double shear, and one rivet in single shear.

The unit section is

p t 60,000=6.75 x .4375 x 60,000=177,000
15/16. x 7/16 x 2x90,000+.69x4.5x42,000
    = 204,210

This is sufficiently high to allow it to be neglected.

In the next case the plate may tear between the rivets of the inner row, and shear two rivets in double shear, and one in single shear. This would be

(p—2d)t x 60,000+ax1.75x2+a x 42,000
    = (6.75—1.875) x .4375 x 60,000+.69 x
    4.5 x 42,000
    = 258,378

which is also above the unit strength. The joint may fail in the third instance by tearing at the second line of rivets and shearing one rivet in single shear. (p—2d) t x 60,000+a x 42,000 = 156,948 and the percentage to unit strength is 88.6, which is higher than the plate strength at outer row.

If the joint fails through shearing all the rivets it will have to shear four rivets in double shear and one in single shear.

4a x 1.75 + a x 42,000 = 231,840 which is much higher than the unit strength. The last case to consider is tearing at the outer row, which is (p—d)t x 60,000 = 15,258 and the percentage is 86.2, which may be considered the efficiency of the joint.

The minimum diagonal pitch is found for the inner and middle row from the

$$\text{formula, Diag. pitch} = \frac{3P + 4d}{10}$$

and for the outer and middle row = D.P

$$= \frac{3}{10} P + d.$$

The minimum distance between middle and outer rows is obtained from the formula

$$\sqrt{\left(\frac{11}{20} P + d\right)\left(\frac{1}{20} P + d\right)}$$

and for the inner and middle rows from

$$\sqrt{\frac{(11 P + 8d)\ (P + 8d)}{20}}$$

We will not enter into the solution of these formulae immediately, but will use them when designing the actual joint for the boiler we are considering.

The Quadruple Riveted Joint

This is a further development of the butt joint, and is used where a very high joint efficiency is required. The rivet section in this joint is very large, there being eight rivets in double shear, and three in single shear, giving a total of 17 rivet areas in the joint. The outside strap covers only the two inside rows of rivets, and the third row of rivets leaves out every alternate rivet, and the outer row has twice the pitch of the third row. The efficiency of the joint will depend on the efficiency at the outer row, which will be as usual P—d. This pitch must never exceed 10 inches in any case. With a 90 per cent. joint the pitch of the outside row would be 9¾ inches, that of the next row 4 3-16, and the inside row 2 3-32 inches. The joint may fail by crushing the plate in front of four rivets in double shear and three in single shear. It may tear between the rivets of the inside row, and shear four rivets in double, and three in single shear. It may tear across the second row of rivets and shear three rivets. It may tear at the third row and shear one rivet, or it may tear at the outer row. Shearing all rivets is hardly likely. The reason for fixing a minimum diagonal pitch is to preclude failure by tearing across a diagonal pitch line. The least value of the joint is the plate percentage at the outside row, and this, as we know, is 90 per cent.

MONEY FROM "WASTE" COAL

Close to the pit-head of many coal mines there lie huge "tips" of small coal. When coal of good quality was cheap, these tips were looked upon as waste because the heating value of the stuff was so small that no one would pay enough to cover the carriage. The huge increase in the cost of coal has, however, turned attention to the economic possibilities of these heaps of low-grade fuel. A large company has been formed to acquire the heaps and to transform the coal dust into briquettes of a form and composition suitable for domestic and industrial purposes.

John Conley is No Longer a Drill Press Operator

The Shaper Now Receives His Attention, and Incidentally He Learns a Few Pointers About the Action of the Stroke of This Machine.—He Also Tells of an Interesting Jig.

By J. DAVIES

READERS will no doubt be glad to hear that John has been promoted. He is no longer a lowly drill press hand, but answers to the name of shaper operator. The foreman, who in spite of his seeming brusqueness has John's interest at heart, is giving him some advice and as we creep up behind him, so to speak, let us hear what he has to say.

"Now, John," he commences, "don't be afraid to ask questions. If you have any doubt about how to tackle a job ask how it was done before. Many a time you will have to exercise your own ingenuity and improvise methods of your own.

"You know it is not the method that counts so much as the result, but they must both be taken into consideration. In some cases the end does not justify the means. For instance, a shell mechanic came in the other day and was given a job on a large drill press. He put the job on the table and got a drill from the stock room which happened to be a long one and would not go over the job without lowering the table. He evidently did not know that the table could be lowered, so he took the drill to the blacksmith and asked him to cut 2 inches off it. That was a method we didn't quite approve of so we asked him to kindly introduce that idea into some other shop.

"The shaper, or the fitter's friend, as it has sometimes been called, and the planer are so nearly akin that most of the methods and principles are applicable to both. Much of the same work can be done on either shaper or planer, the same methods of securing the work and the same kind of tools being used, but the shaper is best adapted for light

work and short cuts, or where a cut of a definite length is required. For example, cutting a key seat in the end of a shaft, or cutting up to a square shoulder. Most shaper work is held in a vise bolted to the table, the vise being usually graduated so that it can be set to any angle.

"The first consideration in planer or shaper work is to decide if the work can be finished in one setting. Never move or reset a job if it can be avoided. Now here is a blanking die with a hexagon hole through it. I want it shaped so as to leave a margin of steel around it about ½ inch wide. This is to reduce the work of subsequent grindings and it can be done at one setting by swinging the vise around to suit the different angles. Be sure not to move the tool up or down when taking the finishing cut."

With a little friendly advice from his neighbor John got along with this job fairly good until the overhead belt came off. He secured the belt stick to put it on again, but unfortunately got on the wrong side of the pulley. John was conscious that several of the men were looking on smiling, so determined to show them that he could do the job correctly. He struggled, pushed and got red in the face, but finally jammed the belt stick between the shaft and the pulley, breaking the belt. This necessitated a trip to the belt repair man, who on repairing the belt put it on for John, who was watching intently to see how it was accomplished. John learnt something on seeing him go to the other side of the pulley, give the belt one little flip, and it was on.

A little later the foreman came down

the shop with a short piece of round stock in his hand. Stopping opposite the shaping machine, "John," he said, "I see you have just about finished that job. Here is a short piece of round stock. I want you to shape a square on the end of it to suit a wrench. Make it about 1 in. long and as big as the stock will allow."

John carefully measured the size of the bar and found it was just one inch. He had found out at the evening school that a square drawn inside a circle was .7071 times the diameter of the circle, so he reasoned to himself that if the finished size was .7071, and it was an inch to begin with, he would have to take off .293 or .1465 from each side. As the piece was short he stood it up on end in the vise, using a pair of vee blocks to hold it straight. He cut .146 from one side, then cut the other side until it measured .707. He was just about to loosen up the vise, turn the piece around to do the other two sides, when he remembered how he had finished the last job by moving the vise around, and setting it by the graduations marked on the base. He moved the vise 90 degrees, repeated the operation and completed a very creditable job which he was proud to show to the foreman. He was rather disappointed when the foreman said:

"The job is all right, but it took you four times longer than it should. I told you it was a square for the end of a wrench, so you didn't need to monkey around with micrometers measuring half thousandths, also it wasn't necessary to cut so much wind at the end of each stroke. You could have shortened the stroke and speeded up the machine.

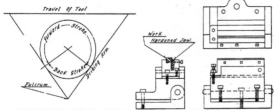

SKETCH ILLUSTRATING ACTION OF STROKE OF SHAPER, ALSO THE JIG SPOKEN OF IN THE TEXT.

The amount of time required depends upon the degree of accuracy, and the degree of accuracy depends upon the job. I suppose I should have told you it didn't matter to a thirty-second. That old saw about making everything as perfect as you can is all nonsense. The requirements of every job must be considered separately and your own common sense will tell you the nature of the finish that the work calls for.".

When John reached home that evening his father noticed that he had a liberal supply of cotton bandage on his hand.

"Hello, John, what is the matter with your hand?" "I was shifting the belt from one speed to another while it was running pretty fast, and there was a piece of wire sticking out where the joint was and it made a nasty gash in my hand. I'll bet you I don't get caught that way again.

"By the way, father, I had quite an argument in the shop today about the return stroke of the shaper. I agreed that the length of the stroke makes no difference in the time required for each stroke, but I can't see how the return stroke is faster than the forward one as my friend claimed it was. Here is my argument. The slide that carries the tool is driven by a disc that has a slot in it, and the length of stroke is governed by the position of the driving pin in the slot. Now that disc is revolving at a uniform speed, one half a revolution drives the slide forward, and the other half drives it back. Since each half would require the same time, the backward and forward motion must take the same time. What do you think about it, father?"

"Your conclusion would be correct, John, if your statement were true, but unfortunately for your argument your statement is not true. The forward stroke takes more than half a revolution and the backward stroke less, according to the length of the rocker arm. Here is a sketch that shows it very clearly. You will notice that the shorter the rocker arm, the faster the back stroke, also that the stroke is not uniform, being fastest at the middle and slowest at the ends.

(See sketch on previous page)

"I see how it works. I shall have to take it all back what I said.

"I am glad you investigate things for yourself and know for sure the how and the why. It is surprising what a lot of erroneous ideas pass muster even with good mechanics. Well, is there anything I can help you with before we go to bed?"

"Yes, there is one other job, but it seems so simple I am ashamed to ask the boss to show me how to do it."

"Don't be afraid to ask for information. That's what the boss is there for and I know he would be glad to show you anything on any job and get you started right. What is the job?"

"To shape two blocks of steel for bolt dies perfectly square, and to a given length. The limit of error allowed being .005."

"What seems to be the trouble?"

"Well they are just cut off the rough stock, and after shaping one side, I can't get them to lie flat on the bottom of the vise no matter how I try to hammer them down. When I tighten up the vise they seem to raise up on one side or the other, the result being they are not square."

"I think I can give you an idea that will help you. In the first place, see that the sides of the fixed jaw are perfectly square. Then take your block and machine one side. Now place the side you have just machined against the solid jaw of your vise, and take a piece of angle iron for a packing piece and place between the work and the other jaw in such a manner that the pressure will be on 3 points. This will cause the side you have just finished to bear all over the fixed jaw and will be perfectly square with a cut taken off the top.

"Test the work as you go along and if you should find you want, say, .005 in. more off one end than the other, let your tool just touch the high spot and make a .005 feeler pass between your work and the tool at the low end. Setting a job with a tool is more accurate than with a lever or surface gauge, or in fact any other tool."

"Why so, father? Suppose I got an indicator and set the job .005 higher one end than the other, wouldn't that be just as good and save me putting the machine backwards and forwards?"

"It might, and it might not, but if you want a really accurate job I wouldn't trust it, because your shaper table may not be in the same plane as your tool and your indicator wouldn't show it. You could put your indicator in the tool post and any error in the machine would then automatically correct itself."

"I see," said John, "I used a jig today, father, to shape up some dovetail taper keys. These keys were to hold some inserted forming tools in a holder and were to be shaped to a given angle on the side, and ¼ inch taper to the foot lengthways. They were to be all alike, or interchangeable, as the boss said. I thought I would have an awful job holding them, but that little jig did the trick all right."

"Is that so—what was the jig like and how did it work?"

"Well, Dad, here is a little sketch I made of it. You simply grip the base of the jig in the jaws of the vise and adjust the screws to give the right taper both sideways and lengthways. I left them a little large for fitting, but I noticed they didn't get any fitting. The tool maker that was doing the job simply placed the base of the same jig on the magnetic table of the universal grinder and ground them to fit. He said he could do it that way better and quicker than he could file them and I believe him."

"Yes, John, you will find that jigs and fixtures are absolutely necessary for accurate repetition work and if you follow the practice of making a sketch of the essential features of jigs and fixtures as you come in contact with them you will soon have a very valuable collection, but look here, it's eleven o'clock, so off to bed. We will continue this talk on some future occasion."

CASE CARBONIZING

The Driver Harris Co., Harrison, N.J., have issued a very attractive book entitled Case Carbonizing. It contains one hundred and twelve pages 5 inches by 7 1-4 inches, and has 31 illustrations and several useful tables.

This is one of the most compact little books yet issued for the use of those interested in case-hardening or carbonizing. While primarily gotten up to call attention to the use of nichrome boxes and pots, it is not a mere advertising booklet, as the information is of general use to all those having to do with carbonizing or heat treatment of low-carbon steels.

It starts out with a chapter on case-carbonizing, divided into sections on the definition of case-hardening, its history, mechanics, oil tempering vs. case-hardening, requirements for case-hardening, quality of steel used, effect of temperature and time, pack-hardening. Chapter II, Cyanide-Hardening, gives details of the practice and formulas used for cyanide work and outlines its limitations. Chapter III, Gas-Hardening, covers the theory and use of various gases for case-hardening. Chapter IV, Lead-Hardening, tells of the use of the lead bath in connection with case-hardened work. The temperature range of the lead bath is given also. Chapter V, Carbonizing Containers, mentions and discusses the various kinds of containers and their merits or demerits. Chapter VI, Nichrome Commercial and Technical Data, gives the strength of richrome, its acid-resisting properties and other features. Chapter VII, Cast Nichrome Containers, takes up cast nichrome for cyanide hardening, cast nichrome for pyrometer protection tubes, dipping baskets, additional uses of nichrome castings. Chapter VIII, Commercial Methods of Using Cast Nichrome, describes its uses for treating automobile starting and lighting equipment, studs, set-screws, small bolts, nuts, etc., ring gears for automobile differentials, roller bearings. The appendix is filled with tables of stock patterns and special container sizes. The latter is especially valuable for the man wanting containers to suit his particular requirements.

Discover Coal?—A report from Shelburne, Ont., says that on a farm near there a farmer claims to have discovered coal while drilling for water!

DEVELOPMENTS IN SHOP EQUIPMENT

CHUCKING MACHINE

The McDonough Manufacturing Co., Eau Claire, Wis., have placed on the market what is known as the Daniels automatic multiple spindle chucking machine. The illustration herewith not only shows the machine itself, but points out clearly various parts of decided interest.

The machine is designed to obtain the most efficient means of machining parts in large quantities, with the use of semi-skilled labor to operate the machine after it is set up and timed for the various operations. It has five tool carrying spindles and a blank station. The turret upon which the spindles are mounted, and whose axis is concentric with that of the table, does not revolve, but through its automatically controlled vertical movement, it feeds the tools to the work.

The table has six chucking positions and revolves step by step, bringing each chucking position successively into alignment with each spindle. The tools, mounted on the spindles, perform their different operations and a finished piece is removed when it is at the blank station and an unmachined piece takes its place. The time of machining a piece is the time of the longest operation, plus the time of one index.

Detailed Description

The spindles have a rapid approach to and a rapid reverse from the work. The spindle carrier is operated by means of a quadruple screw, operating through a bronze nut in the centre column.

The screw is revolved by means of a worm gear and worm and the worm by means of three friction clutches—one for rapid approach, one for the feed and one for a rapid reverse. The indexing of the table is also controlled by a friction clutch. By obtaining these movements through friction clutches, the machine is protected, for should a tool break and jam in the work, the feed pressure goes up and causes the feed friction to slip, and should the movement of the rapid approach or reverse or indexing mechanism of the table become jammed in any way, the friction clutches controlling these operations will slip, thus preventing damage.

The stopping, starting and tripping of the machine is controlled by two levers so placed that they are always within easy reach of the operator from the chucking position.

The Table

The table is 36 inches in diameter and six 10-inch chucks or workholders can be mounted on it. These chucks can be made larger, if necessary. The travel of the spindle carrier is 13 inches, it taking four inches of this vertical movement to withdraw the locating pin and to control the indexing of the table. The maximum distance between the spindle carrier and the table is 21 inches.

The machine will handle work from three to six inches in diameter and perform the following operations: cup turning, boring, facing, reaming, tapping with collapsible taps, and drilling multiple offset holes by attaching a multiple spindle auxiliary head to any of the spindles. The table can be set to index one, two, or three spaces.

The speeds are varied by means of change gears. One set controls the speed of all the spindles, which can be regulated to give a spindle speed from 30 to 300 revolutions per minute, all depending upon the work to be done. Each spindle has an individual set of change gears, which will vary the speeds of the spindles from one to three and one-half, so that each spindle can be given the most efficient speed for the particular operation performed by it. Change gears vary the feed from ⅝ of an inch to 4⅝ inches per minute.

The spindles are equipped with taper bushings that can be adjusted to compensate for any wear, and each spindle has an individual adjustment on the spindle carrier of nine inches. All the shafts and the feed screw are made of heat-treated chrome nickel steel. All of the gears are made of steel or bronze with the small important gears made of nickel steel and carbonized. All bearings are bushed with bronze bearings.

The machine can be used with countershaft, or direct motor drive, and uses a 10 h.p. constant speed motor. The net weight is 10,000 pounds, and the total projected floor space is 38 inches wide by 77 inches long.

COMBINATION MELTING FURNACE

The Canadian Hanch Burner Co., Port Hope, Ont., have improved their combination melting furnaces in many ways, and for readers' benefit who may not know the particulars of this style of furnace, we append the following information. Their uses are varied, and they are adapted wherever lead and other soft metals have to be melted.

The illustration shows clearly the idea. It is actually two outfits in one,

GENERAL VIEW OF CHUCKING MACHINE ILLUSTRATING POINTS OF INTEREST.

a melting furnace, and a portable oil burner. The burner can be instantly detached if desired and used for a wide 'variety of purposes, such as melting babbitt out of bearings, heating bearings for rebabbitting, expanding to shrink fits, melting lead out of pipes, fitting joints, preheating for welding, etc., etc.

GENERAL VIEW OF THE OUTFIT.

The furnace illustrated is the smallest size with a 125 pound pot. This melts 100 pounds of lead in 14 minutes. The burner consumes 3 pints of kerosene per hour. With the larger furnaces 200 pounds of lead can be melted in 15 minutes, and 450 pounds in 20 minutes.

FLUE CLEANING RATTLER

The Baird Pneumatic Tool Co., Kansas City, U.S.A., have placed on the market a flue cleaning rattler and unloader as shown in accompanying sketch. This machine is a friction and not a drop rattler. It cleans by friction, first removing the scale, then using this scale in powdered form as a scouring medium.

It is capable of handling, at one rattling, a full set of boiler tubes averaging 250 to 300 flues, cleaning them throughout their entire length inside and out, in from one to six hours, depending upon the hardness of the scale. Any length flues up to 24 feet can be cleaned, the adjustment for length being obtained by a moveable piston head in the drum. A six-inch play permits of a slight lateral motion, but not enough to

allow the flues to become tangled and destroyed.

No dust is allowed to escape into the shop as the scale is gradually worked to either end of the rattler, and from there into the dust rooms provided for its accommodation.

The cleaned flues are unloaded by being pushed from the rattler by compressed air out through an end door

into a push car, ready to be safe ended and installed.

Power is applied to rattler by either direct motor drive. The machine is adapted for all classes of fire tube boiler flues, handling large diameter super-heater flues as efficiently as the smaller sizes.

NEW THINGS IN MACHINE TOOLS

GRINDING AND BUFFING MACHINE

The Van Dorn Electric Tool Company, of Cleveland, Ohio, are now manufacturing a 1-h.p. heavy duty electric grinding and buffing machine adapted for floor, bench or aerial work. Motor ventilation is provided by means of a fan drawing air through screened 'openings at the rear of the frame. Interchangeable assembly provided for changing over from d.c. to a.c. at the minimum of expense. The speed on d.c. is 2,000 r.p.m. The shaft is mounted on ball bearings enclosed in dust-proof housings. . The frame construction provides for work of long length being held parallel to the axis of the grinder. The floor type is fitted with foot-operated switch.

CHAMFERING MACHINE.

The Grant Manufacturing and Machine Company, of Bridgeport, Conn., has recently produced an automatic double-spindle chamfering machine that is intended for chamfering simultaneously both ends of automobile piston pins, and on pins 5-8-inch in diameter and 3 inches long, it is claimed that a speed of 40 pins per minute can be maintained. Pins of the proper length are placed in a hopper from which they roll down a slight incline to the feeding mechanism,

where they are taken, one at a time, and placed in a clamp which retains them while the cut is being made. Other uses may be made of the machine, such as drilling, facing or counterboring opposite ends of small work. Square or other shape work may be handled equally as well as round pins.

COUNTER-SINKING MACHINE

The Langelier Manufacturing Company, of Arlington, R.I., are now making a special line of opposed-spindle counter-sinking machine for centering or drilling both ends of small pins at the same time. The spindle at either end runs in phosphor-bronze bearings and operates at a speed of 2,000 r.p.m. The pulleys for the spindle drive are run on ball bearings and are located on stationary sleeves on the drill heads, thus eliminating any bending tendency on the spindles. Spindles are simultaneously operated by means of a hand lever. Adjustable stops are provided for feed control. The jig-heads, for holding the work, are actuated by a locking cam, and may be operated by hand or foot lever for releasing the work. The coolant is controlled automatically so as to flow only when the tools are cutting.

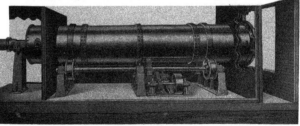

FLUE CLEANING RATTLER AS INSTALLED INSIDE A PLANT WITH DUST ROOMS ON THE END.

The MacLean Publishing Company
LIMITED
(ESTABLISHED 1887)
JOHN BAYNE MACLEAN, President. H. T. HUNTER, Vice-President
H. V. TYRRELL, General Manager.

PUBLISHERS OF

CANADIAN MACHINERY
~ MANUFACTURING NEWS ~

A weekly journal devoted to the machinery and manufacturing interests.

B. G. NEWTON, Manager.   A. R. KENNEDY, Managing Editor.

Associate Editors:
J. H. MOORE   T. H. FENNER   J. H. RODGERS (Montreal)
Office of Publication: 143-153 University Avenue, Toronto, Ontario.

VOL. XXIV.   TORONTO, OCTOBER 14, 1920   No. 16

PRINCIPAL CONTENTS

Market Tendencies

CANADIAN buyers are leaving the premium steel mills, and prices of these rollers will likely sag in accordance. It is reported that sheets from some of the corporation mills are arriving freely now, and in some cases there are applications for cancellations.

There is nothing to indicate that anything approaching a crash or even a sudden drop is coming in the steel market. There is a lot of business on the books of the mills, especially the corporation's.

It is much better that prices should step down little by little than that there should be a- sudden drop. Steel prices, apart from the schedule of the premium mills, have not been high.

We' are approaching a stage in the markets where it may be possible to once more do business on the basis of value, taking as a definition of value the following:—

"The price arrived at between the least urgent buyer and the least urgent seller."

For years now in many lines of steel, iron and machine tools there have been only a few months when there were any "least urgent buyers." The market has been full of buyers, and urgency has been one of their outstanding, rather than absent qualities.

When we get' to the stage, and we are now very close to it, where two or three men present the claims of an article of equal merit, the matter of price is bound to enter into the sale to a greater extent than it has since 1914. There have been years when it can truly be said that price was one of the least considerations. It should never be allowed to dominate the situation, but in the natural course of events it will become a greater factor from now on than it has been for many years and months past.

The manager of a large Canadian concern, after a prolonged visit to Western Canada, and points in United States, has this to say:—"You can look long and hard before you will find a better country than Canada. We are in better shape than we realise. There is a lot of good, legitimate business moving, in spite of tight credits, and the bad spots exist mostly in the brain cells of some of the croakers."

"I am afraid the high wages the apprentices get now has lowered the desire for more mechanical knowledge."

THIS statement was made by one of the apprentice instructors on a Canadian railroad. It was put down in writing as the firm belief of the man who had made the remark.

It cannot be said that he is talking at random, because the statement was made in all seriousness. He comes in contact with the apprentices daily, and has a good opportunity to see the development of any tendency that may become apparent in the training of apprentices.

The high pay for apprentices is part of the system wished on us by the war shops. These same practices spoiled good shop training.

The superintendent of one of the largest Canadian car shops told this paper that the high rate of pay for apprentices had not improved the work of the apprentices; had not brought in lads with any better education; had not increased their desire to go on and become high-class men.

Instead, it had worked the very opposite in many instances. Apprentices in their second and third years came to the point where they were making enough money to take care of themselves. They were independent of home influence and parental discipline just at a time when many a boy needs to be led to the shed and told exactly "where he gets off at."

The added money was not bringing any better work. It may have contributed to meeting the added cost of keeping the apprentice boys; apart from that it was sheer waste.

Industry is changing, and specialists are in demand. Firms want mass production. They desire speed and excellence in each minute operation. In the majority of these training is not necessary. A few weeks' practice fills the bill. The money is there without the skill that comes of long training. The finer points of design may be looked after by those who care to bother; mathematics and calculations can be attended to by others who foolishly follow this branch. But why bother about these matters? Here's a little job that needs no training in particular, Why, in a few weeks or so, one can make as much money as the man who learns his trade and gets a thorough mechanical training.

Quite so, and some fine morning you come down to find that the little job is not needed any more. There is no call for it.

What then? You are in the class of the thousands of unskilled, "looking for a job," and thankful to get one at any price.

Now the proposition is here, and it is as plain as the nose on your face:—

Shops that are specializing in order to get mass production are bedevilling the mechanical field in the broad sense, unless they face their full responsibility in the matter. They must build up the ranks of trained and skilled men to as great an extent as they are making that fine training and skill unnecessary by their shop practice. Unless they do this they are going to find, in a few years—and it may be very few—that they will have a famine of tool makers, maintenance men, and designers.

The situation is real—it is serious, and it is one that cannot be sacrificed to the desire of any branch of industry to secure great and record-breaking production at the expense of a well trained, thoroughly educated and capable mechanical population.

Grinding at Speed of 100,000 Revolutions Per Minute

THE present-day tendency towards ever increasing speed has been often deplored by the more staid and conservative elements of the community. Where the endeavor to procure speed has only been actuated by a desire for new sensations, this disapproval is, perhaps, well merited. All speeding-up processes are not developed with such meretricious objects, and in the field of industry speed is a desirable thing, both in general production and in the details which make rapid production possible. From this necessity has arisen the various high speed machine tools, and high speed cutting steel, with which we have all become familiar in recent times.

To what extent this tendency has developed was forcibly brought home to us by an account of a high speed grinding spindle attachment recently exhibited by a well-known British firm of tool makers. Grinding is one of the operations in metal working where speed is of great value, and especially on small internal work, such as lapping gauges, internal tapers, bushings for precision tools, etc. The wheels employed are of necessity very small, and this permits of the attainment of high speeds. The higher the speed of the abrasive surface, the less tendency will there be for the surface of the metal being dragged, and thus roughened, a condition strictly to be guarded against.

Practice in grinding had reached a point where these wheels had been revolved at 75,000 revolutions per minute, which is a very respectable rate for any spindle to revolve. The spindle we are alluding to exceeds this speed by 33 1-3 per cent. That is, it is guaranteed to be capable of running safely at 100,000 revolutions, and has been run at a speed of 104,000 revolutions per minute.

This is getting to a point where the mind fails to grasp the significance of the figures involved at the first glance. Try to picture a small spindle and the wheel mounted on it which makes a complete revolution in just .0006 of a second. If the spindle on which the wheel is mounted is one-half inch in diameter, its circumference would cover a distance equal to 13,090 feet, or almost 2½ miles per minute. A locomotive with a 5 feet diameter driving wheel would have to turn that wheel 833 times a minute to attain the same distance in the same time. The spindle, of course, will, on the smaller sizes of wheels, be of less diameter than one-half inch, because it is very obvious that with a wheel of ½ inch or ⅝ inch diameter, one could not well have a ⅓ inch spindle.

The wheels for this speed must be of good construction, for they will have to resist the force tending to tear them apart, due to the speed. Of course, they will be of no great weight, probably averaging 2 to 4 ounces. Suppose the wheel weighs only 2 ounces, and is of one inch diameter, the force tending to make it fly off at a tangent, or the centrifugal force would be equal to about 18,000 lbs.

These figures seem hard to realize, and they serve to show what has been reached in engineering practice when such speeds can be employed with perfect safety. There can be no doubts as to the material employed, or the manufacturing methods used, when speeds of one hundred thousand revolutions per minute can be guaranteed.

Making It Harder to Do Business With Canadian Firms

THE Canadian Association of British Manufacturers and their representatives are doing their best to bring to the attention of the British houses any incidents or practices that retard the development of business in Canada.

The terms insisted upon by some of the British firms are found to be particularly objectionable, and are set forth in a little booklet called "Sales Craft," issued by the Toronto branch of the association:

"A Canadian manufacturer of steel and iron products recently had occasion to replenish his stock of certain materials used in various articles manufactured in his mills, and as there are only two sources of supply, Great Britain and France, and the producers few in number, the business was offered to a United Kingdom manufacturer. A natural preference in the matter of trading, national bias in conjunction with Imperial sentiment, prompted this loyalty to British enterprise. The offer, as many another prospective Canadian purchaser has been forced to testify, was not met in kind, the intending customer being informed that the order could only be undertaken on the basis of "cash with order" or by the instituting of a "letter of credit" with a London banking house to the order of the U. K. manufacturer. Such stipulations were, and are, quite out of the question, as no responsible Canadian business house will consent to do business under those conditions nor submit to the dictation of trading on terms of that nature. The result of the matter was the booking of the order by a French manufacturer on open account at sixty days dating. Hence, for ten years past this business, though offered repeatedly to a British manufacturer and more than once to the same house, has been executed in France. There is no question as to financial standing of the Canadian firm, it is quite good and annually improving—the firm buying largely in the United States, but nothing in England. In a recent conversation, the head of the business expressed an opinion, that is in no sense isolated, that United Kingdom manufacturers and business houses can expect no great extension of trade in this country until they accredit to Canadian business men a measure of integrity beyond that of a band of robbers. In this one instance, orders amounting to practically $50,000.00 have been lost to British mills merely on a question of terms of sale."

The same publication goes on to point out that the time extended to Canadian customers by the American manufacturers of iron and steel products for payment of accounts differs somewhat in various mills and also in some States, but the most usual and generally accepted terms are 2 per cent. for cash in 30 days or net in 60 days. In some instances a cash discount of 2½ per cent. is conceded and others call for cash in 20 days, less 3 per cent. With some manufacturers of machinery the terms are quite elastic and large buyers of good credit can obtain very favorable concessions. The method of collecting bills by sight or demand draft against documents or other similar methods are practically unknown to American shippers, who are inclined to figure that if an account is a doubtful risk on regular credit, it would be a precarious venture on any, the object being to make the terms of payment the least possible obstruction in the field of distribution.

A visitor from South Africa tells us that during the war they paid as high as $490 per ton for steel plate. Which leads one to think that if there were much business in plate carried on there wouldn't be much left of the famous Gold Coast in that country.

Here is the opinion of another Canadian manufacturer whose product needs a lot of money before being put on the market, there is no trouble selling, but there is trouble in financing just now. He says: "Credit is the thing that is bothering us now. There is plenty of business. In fact we could run our place twenty-four hours a day were we able to get the material and the help from our bank. We look to see the move toward price reduction continue. In our lines falling prices of material do not worry us as we tender for work, and then cover ourselves for everything we need.

MARKET
DEVELOPMENTS

Canadians Leaving the Premium Steel Markets

Some High Priced Stuff Still in the Warehouses Here—Cancellations Asked on Some Lines of Sheets, But in Other Lines There is Still a Very Keen Demand for Many Lines of Steel

THE market in steel is running pretty true to expectations, and one development of the past week has been that Canadian buyers are deserting the premium mills. Although there is still a demand for the products of these mills the chance of financing any considerable tonnage at a high figure is too speculative to be inviting. Steel sheets are showing a better delivery, but as a general thing the car situation, judged by the recent performance, has been worse instead of better. Some reports have it that applications are being made to the mills for cancellation on sheets, but it is doubtful if there will be any concession in this line. Cancellations are the one thing that the whole business fabric will likely fight hard to prevent. Once they start they are too far-reaching in their influence, and the feeling is that the time to

stop them is at the start, and this can probably be done, as nearly all business has been booked in the last six months or a year on the no-cancellation understanding.

There is no large buying in the machine tool market, although several of the houses say they have done a good business in the last month, and are finding a number of good inquiries coming into the market. Buyers anticipate some announcement in regard to price adjustment, and while this has the tendency of slowing business generally, it does not cancel the orders, but merely postpones their coming into the market. So far there has been no general move on the part of the machine tool builders to make concessions, although stock must be accumulating fairly rapidly at some of the plants now.

MONTREAL PRICES REMAIN VERY FIRM IN NEARLY EVERY LINE

Special to CANADIAN MACHINERY.

MONTREAL, Que., Oct. 14.—Contrary to the expectations that prices would develop a tendency to lower levels, the market in general retains a firmness that apparently indicates that the time has not yet arrived for the cutting into the prevailing costs of materials. While it is understood that the stocks of consumers and dealers are not very large the supplies in the hands of producers and the orders on the books are of such a character that producers and likewise many of the jobbers would not care to see any sudden decline in present quotations. While easier conditions are a possibility, the strength of the present situation appears ample to sustain the belief that no drastic reduction will take place for some time. Some dealers are looking forward for a definite revision about the New Year, but any outward sign of early change is noticeably absent. The standpat attitude is in the face of the universal report that activities in many directions are slackening.

Firmness Likely to Continue

"Buyers of steel and other metal lines are not looking for supplies beyond their present requirements, but aside from a few isolated cases there is little to show that consumers are holding off buying in the hope that prices are coming down. The trade as a whole realize the uncer-

tainty of the early future but at the same time recognize the conditions existing in the steel industry and kindred activities and while appreciating the advantages of lower priced commodities cannot very well see how any excessive decline can be successfully carried out at present or in the immediate future. Of course we are living in times when overnight changes are not only possible but frequently foregone conclusions, and it is this knowledge that prevents any reliable prediction of the trend of the market. We look for a fairly steady market for some little time." This statement by a dealer here would indicate that, under the atmosphere of uncertainty, there exists a general belief that conditions as now obtaining in the industrial field would not permit of any drastic lowering of price quotations on any of the regular lines of steel materials. The new demand for steels is undoubtedly lighter, but requirements of customers are still ample to absorb the bulk of the material coming through from the mills. The transportation problem is much better and delivery is becoming quite normal. This condition will accomplish much in bringing about a more reliable basis of working operations and aid in removing the uncertainty that prevails in all branches of activity. There is still a marked scarcity of cold rolled steel and some sizes are impossible to secure.

Plates are more plentiful and sheet supply is showing gradual improvement.

Supply Movement Good

"The bulk of the machine tool movement is in the way of the smaller equipment, and while sales are not heavy the volume of business is quite seasonable, but probably below normal for this time of the year," remarked a dealer this morning. "It requires considerable more effort to sell equipment today, not that the buyer does not need the machinery, but rather because many are content to worry along with what they have in the past better prices will obtain in a few months." The dealer added, however, that apart from the acquisition of machine tools the buying of supplies continues in fair volume.

Non-ferrous Metals Lower

"The situation continues very quiet and movement of scrap is exceedingly light. Steels and irons are very dull and weakness has developed in non-ferrous lines." In addition to this statement the dealer intimated that foundry operations had fallen off during the week but expected a renewal of activity for the coming week. The best business of the present is done in cast iron scrap and stove plate. Coppers are lower by one cent, light being quoted at 14 cents and heavy and crucible at 17 cents per lb. Brass is down a half, light bringing 6½ cents and medium 7¼ cents per lb. Scrap zinc on a similar decline is quoted at 6 cents per lb.

NO REAL SLUMP
IN TOOL MARKET

Firms Report That Signs Point to Much Better Booking in the Near Future

TORONTO.—There are some reports of cancellations this week in the steel business, but the chances are that the firms making application will not be allowed to cancel as the business for the most part has been placed under a non-cancellation clause. There are some suggestions now on the part of sellers that some industries have during the past six months or so resorted to pyramiding of orders in the hope of getting a fair amount through quickly, that is they would come on the company's books for two or three times what they actually needed in the hope of getting delivery of a portion of the order, which would be about their real wants. There is nothing to substantiate such a charge, but it is a report in the market. ·

The machine tool market believes that the turning point in the market situation is here, or very close at hand. None of the firms report very large orders in the district, but some of them tell of a good month's business closed for September, now that the figures have been added up. In one case a dealer had an enquiry for a good sized string of equipment, that looked more like an order from a war shop than anything we have seen for some days.

But the important thing seems to be prices in the machine tool market just now. On the face of it there seems to be no good reason why prices should come down, and in one isolated case one maker has actually announced an increase in one line. But schedules for delivery are getting much quicker, and some tools can be ordered out of stock now for the first time in a good many months, or years, for that matter.

There is an expectation on the part of buyers that there will be a reduction in price. Dealers admit that this feeling is holding back quite a bit of buying that is not of the urgent sort. The belief is that by holding off for a while longer there will be a better price than by coming into the tool market now.

One dealer sized it up this way: "Some of the newer concerns that sprang up with the war will likely be the first to make some cut in their prices. They will have to do it sooner or later because they cannot get the backing to go on and pile up for stock, so they must either cut or curtail . operations, and when they start to cut down operations they will find that their manufacturing costs will grow higher instead of getting less. I do not expect reductions for some time from the older firms in the machine tool business. We have nothing from our principals that would lead us to believe that reductions are coming, neither do I think that a cut in prices now would result in selling many more tools. We have no applications for cancellations on business that is on our

POINTS IN WEEK'S
MARKETING NOTES

The steel buyers in Canada are leaving the premium mills as fast as they can. there being too much risk attached to financing any more high priced material.

—

Boiler tubes and cold rolled shafting are the black spots in the steel market just now, and there is a big unfilled demand for both.

—

Machine tool houses report that inquiries are better, and some nice lists are coming into the market. The amount of business being closed is not large yet, but the prospects for a good season's trade are bright.

—

The scrap metal market is not recovering. Dealers are not anxious to buy, and those wanting to sell fear to do so as concessions are about the only way to sell.

—

Montreal reports that dealers there do not expect a revision of prices in either steel or tools before the first of the year.

—

· The New York machine tool market continues very quiet, and purchasers realize that the market is theirs more than it has been for many months.

—

Some applications are made for cancellation of orders for steel, but there is not much chance of these being granted except under very exceptional circumstances.

books, neither is there any request for reductions in prices on this business."

The sale of motors in this district is being more or less interfered with by reason of the scarcity of power. There is a fair demand for motors of the 60-cycle type, because power of that frequency is more available, but for motors of the 25-cycle type the sale is not so good.

Many of the firms handling both wood and metal working machinery report a very good movement in wood working machinery, and say that the business is running well ahead of the figures for the same period last season.

The Steel Market

Some reports are heard of attempts to stop some large deliveries in the steel market, this extending even to sheets, which have been a much sought after line for months past. This does not mean that the market is full of sheets, but it does mean that some lines find now that they have bought well beyond their requirements. Should this business is taken on the understanding that a contract is a contract, and the chances are that the deliveries will be made and the

buyers will have to deal with the financing as best they can. There should not be much difficulty in disposing of sheets yet, as there is a big demand, according to reports, and most of the available material is at a high figure from the premium mills.

The change in the steel market is coming surely, if slowly. As has been predicted in these columns, and in our Pittsburgh letter for some months, the premium market is the first place to sag. Canadian warehouses that have been drawing heavily on the premium rollers for their supplies have simply ceased to buy from them. These warehouses have something to sell, and for the present they are not taking on any more material. There is bound to be an effect on the independent schedules, and these mills will have to keep their plants going, and the one way they can do this is by coming down on their prices. The prospects seem to be that it will be only a short time before their figures will be down to the level of the Steel Corporation.

Cold rolled shafting is a scarce article. There are plenty of places where a good tonnage can be taken on if some mill can be located that will handle the business. Several machinery houses that make it a business to instal a plant complete, are in a bad way to get shafting, and the chances are that they will find little improvement for some time yet.

Boiler tubes are not much better, in fact they are about the one black spot on the map. There is not the finishing capacity for them that there is for sheets, for instance, so it is impossible for the plants in operation to catch up with the business that is offering and urgently required. "If anything the tube situation seems to be getting worse," was the cheering comment of one dealer this morning. "Our advice is that there will be little improvement, and in the meantime there are a large number of plants that are urgently in need of tubes."

SAYS COAL PRICES
NOT JUSTIFIED

Values Should Come Down in a Short Time Is the Claim Now Made

. Washington.—Reductions in coal prices may be expected soon, Charles L. Couch, president of the American Wholesale Coal Association, announced.

Mr. Couch, following a meeting of the executive committee of the organization at which the entire coal situation was canvassed, declared that it is the opinion of the committee that a great change in the whole coal market situation is imminent. He suggested to buyers of bituminous coal who have a month's supply in storage to confine their purchases to immediate needs.

Prices which have prevailed for bituminous coal, Mr. Couch declared, are not justified by the present satisfactory production.

HERE ARE THE HIGH SPOTS AND
THE LOW ONES IN N.Y. MARKET
Special to CANADIAN MACHINERY.

NEW YORK, Oct. 14.—The machine-tool situation is made up of somewhat contradictory elements, which are both favorable and unfavorable. It must be admitted that at the moment the unfavorable factors are much to the forefront. Among these factors are:

(1) A falling off in business during the past few months to an extent that has placed some machine-tool manufacturers in a position where they have no orders ahead and are making tools for stock. This has resulted in curtailment of operations, which in a few instances has caused reduction of the working days from six to four days a week.

(2) The expectation of buyers is that machine-tool manufacturers will lower their selling prices. While most prices are holding firm, there has been a reduction of 15 per cent. in one make of internal grinding machines, and in a great many instances buyers have insisted that sellers guarantee their prices when sales are made.

(3) The credit situation is still an important factor in the situation. Not only are collections slow on old accounts, but manufacturers generally have little liquid capital with which to finance the purchase of new shop equipment.

The above factors, together with a degree of uncertainty as to the course of business in the next few months, are mainly responsible for the decided slump in machine-tool buying. In spite of present conditions, discouraging as they are in many respects, there is a considerable degree of optimism. This optimism is based largely on the following premises:

(1) That machine-tool prices cannot decline very far, due to the continued high costs, which to a certain extent will remain high because of the fact, generally recognized, that labor costs will not come down much, if any, in the immediate future.

(2) The fact that only a few weeks remain before the Presidential election, which is nearly always a disturbing influence in business.

(3) The fact that the railroads of the country must eventually buy very heavily. Their shop equipment is in deplorable condition, and the mechanical departments are committed to extensive purchases as soon as the finances are available; when this time will come is somewhat in doubt, but it is expected that by the first of the year or shortly thereafter, railroad purchasing will be begun on a fairly large scale.

(4) The fact that, despite the lack of actual orders, there are many pending inquiries, which represent actual needs;

buying against these inquiries will result, it is expected, when the money situation has improved, and when a confidence in the price situation has been restored.

In addition to the above may be mentioned the fact that it is generally believed that the automobile industry, by reducing prices on cars, has gotten past the worst phase of its depression, and that sales of automobiles should show an increase during the next six months that will bring automobile manufacturers back into the market as buyers in the spring, at the latest, possibly before.

Nowhere is there any denial of the fact that business was due for a slump, and it is generally believed that the present depression will be a good thing. A readjustment had to come, but it is contended that the period of reaction will be relatively brief, and that next year should be an exceedingly prosperous one.

This is the way the machine-tool trade sees the situation. The significant thing is that, although buying of tools is at the lowest point in many years, there should be much general confidence in the future; that is the future beyond the next few months.

There has been a little business in the last week, but nothing of importance. An inquiry for about 125 machines comes from the Brazil Railway, but actual purchases may be deferred for two or three months.

PIG IRON TRADE

Pig iron prices are very uncertain. Buyers are waiting and the absence of direct sales by producers leaves market level in doubt. The market at Philadelphia is very quiet.

Consumers and producers on the Chicago market are waiting for developments and it has become a game to see which can hold out the longest before making a definite move. Practically no buying for 1921 delivery is being done. While the basic market has weakened to some extent its level is still a matter of speculation, but it is understood that some furnaces will do business at $46 or less, but they want offers of definite business before making a quotation.

Little activity is shown on the Boston market excepting unusually good delivery. The feeling is that the present is not the time to buy for the first half. Foundries in general are operating at full capacity and have sufficient on hand for the remainder of the year.

Demand is very limited in the New York district and all the iron sold was for prompt and nearby shipment.

Owing to the Buffalo market being disturbed by reports of sales at considerably under quotations, furnace men are watching the situation with very uncertain minds. Very little selling is being done.

There is very little activity in the pig iron market in the Birmingham district.

STEEL CORPORATION PRICES ARE NOT LIKELY TO YIELD—INDEPENDENTS WILL
Special to CANADIAN MACHINERY.

PITTSBURGH, Oct. 14.—While the iron and steel market is in process of readjustment not much change occurs from week to week. The various stages of readjustment are likely to be passed through in much the same manner that has been exhibited at various times in the past when a buying movement was ended and a period of slack operation and declining prices had to intervene before business was ready to go ahead again. There is one difference of some consequence, however, that being that the present re-adjustment in prices and decline in industrial activity is a general one in all lines of business, and may even be less marked in steel than in other commodities, whereas in the past the steel market has undergone wider swings than commodities in general. For instance, after the panic of October, 1907, the iron and steel industry was very much depressed for about a year and a half, whereas business in general decreased much less. This time the iron and steel industry may get off more easily than other industries, for everyone is now talking about price declines and heavy declines, in all sorts of commodities.

The fact that steel prices have not declined in any important way thus far, except for the disappearance of the particularly high prices, involving premiums for very early shipment, does not, of course, mean that the steel industry will get through this period without any price

re-adjustment. The market is simply slow in responding to the disappearance of aggressive buying, this being for the reason that the mills, with occasional exceptions, are very well supplied with contracts, and to reduce prices would be to jeopardize the contracts. By price cutting the mills would lose much more in contract business, through cancellations and postponements, than they would gain by securing new orders.

It still remains the prospect, according to the judgment of nearly everyone in the trade, that the Steel Corporation's prices will not yield, the re-adjustment being confined to the independents. It is considered certain that the independent prices will recede to the Steel Corporation levels, and a common view is that if they go as far as that they will go somewhat farther, the independents cutting under the Steel Corporation prices, in order to buy their way into the market again, for patrons of independents who have had to pay higher prices for months than patrons of the Steel Corporation are not likely to be anxious to place orders with independents at Steel Corporation prices when they have a wide latitude of choice as to where they will place their orders.

There is even a little talk of the Steel Corporation advancing its own prices a trifle, for the double reason that the corporation suffers a loss of several tens of millions of dollars a year by reason of the freight rate advances last Au-

The Planer With the Second-Belt Drive

A smooth, easy reverse—which permits higher cutting and return speeds yet prolongs the life of the entire machine—that is the outstanding feature of this distinctive Whitcomb-Blaisdell Second-Belt Drive.

Our Planer Book gives the details of design and construction. Write for it.

24 x 24 x 6 one head and 26 x 26 x 8 two heads in stock.

26 x 32-inch
Whitcomb-
Blaisdell Planer
Widened Pattern

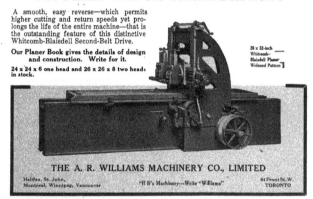

THE A. R. WILLIAMS MACHINERY CO., LIMITED

Halifax, St. John,
Montreal, Winnipeg, Vancouver

"If It's Machinery---Write "Williams""

54 Front St. W.
TORONTO

If interested tear out this page and place with letters to be answered.

gust, and may feel that it should recoup, while the independents might be more ready to adopt Steel Corporation prices if a little concession were made to the independents by way of advances of say $3 or $4 a ton. The independent decline would in that case be much greater than the Steel Corporation advance, but the thing would look somewhat like a compromise, and would be less distasteful to the independents, most of whom have been preaching more or less constantly for a year or so that the Steel Corporation would be forced to advance its prices.

Production

Production of steel ingots in the United States in 1919 is officially reported at 33,694,795 gross tons, against the record of 43,619,200 tons made in 1917. Capacity lies somewhere between 50,-000,000 and 55,000,000 tons, given that manufacturing conditions, so that operations in 1919 barely averaged two-thirds of capacity. For this defection lack of orders was chiefly responsible. The strike that began September 22 restricted output considerably, but it is doubtful whether if there had been no strike, the industry would have operated at capacity, though capacity operations either late last year or early this year.

An interesting point in the statistics is that the Steel Corporation produced 51.0 per cent. of the steel ingot output in 1919, against a proportion of .45.5 per cent. in 1918. The lower percentage is representative of the Steel Corporation's relative capacity, rather than the higher percentage. In the fore part of 1919 the Steel Corporation had much less restriction of output than the independents, due to the fact that the Steel Corporation late in the period of hostilities forecast the future properly, and quietly booked business at prices attractive to the buyer, on the understanding that shipment would be made some time after the war. During the period of Government regulation such orders would fall in Class D, the steel that could be shipped after all priorities and preferences had been taken care of, and of course practically no Class D was shipped as long as the war requirements lasted. The independents

either failed to consider the matter at all, or thought they would be able to secure higher prices when the war restrictions were removed. The strike was probably another cause for the Steel Corporation making more than its normal proportion of the 1919 steel output, through the corporation being slightly less affected by the strike than the independents, even though the strike leaders acted as though they were directing the strike particularly against the independents.

Other production items for 1919 were: Steel castings, 976,437 tons; rolled iron, 1,059,451 tons; rolled steel, 24,042,000 tons.

The report of steel ingot production in September shows that the same rate obtained as in August, about 42,700,000 tons a year. There had been rather a sharp increase from July to August, and the failure of September to show a further increase was probably due to some of the mills being under less pressure from their customers for deliveries, since manufacturing conditions were improved and some mills, well enough supplied with orders, increased their production. The cases of mill operations being restricted by the state of order books are still exceptional, and it remains the prediction that this month will show an increase in tonnage output. This month will probably be the turning point in output, with declining production afterward until the general industrial readjustment is completed and business starts growing more active again. Whether the readjustment will require three months or six months or a longer period is thus far only a matter of conjecture.

Pig Iron

The pig iron market has been so extremely quiet that new price levels have not been clearly developed. Everyone recognizes that the prices to which the market advanced in August, $48.50, valley, for Bessemer and basic, and $50 for foundry and malleable, and which were maintained fairly well through September, are booked to yield to much lower prices, but it requires actual transactions to show where values really stand.

It is reported that a middle interest bought 3,000 tons of basic iron from a valley furnace at $45, selling 1,000 tons to a valley consumer at $46, and selling the other 2,000 tons in the east. Other buyers, however, expect to do much lower prices, possibly as low as $40. As noted in last report, foundry iron became quotable at $47,. valley, instead of $50, on a small transaction, but nobody else seems willing to pay as much as $47.

U.S. SCRAP METAL

The decline continues in practically all grades of iron and steel scrap at Chicago. Much of this is due to the efforts to get under cover material reaching them from outside sources for which they have no means for disposal.

On the Boston market buying is limited almost entirely to a few dealers who have old orders to fill. One of the largest interests is refusing to buy. There is no demand for heavy melting steel.

Prices continue to fall in the local iron and steel market at New York. Over half of them have dropped from 50 cents to $1.50. Dealers say that with the exception of car wheels, which are never very plentiful, they are able to obtain supplies without difficulty.

The unsettled condition of the automobile industry following reduction in prices is largely the cause of the dull scrap market at Detroit. The auto factories are the chief consumers of iron and steel in Detroit and the conditions in the automobile trade are expected to be reflected in the scrap iron market for some time.

The easier tone lately in the markets generally has resulted in depressing the scrap trade in Philadelphia. Eastern consumers practically refuse to buy. No activity is expected on the Pittsburgh market until after the presidential election. Consumers are well covered and with few exceptions are getting good deliveries. Inquiries are limited.

The Southern market is reported as weak. In some cases melters are asking that shipments on contracts be held up.

THE STEEL WORLD IS WAITING TO HEAR FROM JUDGE GARY

New York.—Wall Street looks forward to Judge Gary's speech before the American Iron and Steel Institute on October 22 at the Hotel Commodore for a cue as to the true situation in the steel industry. Judge Gary reiterated yesterday his declaration to give no hint of his views before making that speech. Charles M. Schwab, at his country home in Loretto, Pa., declared that the steel industry was in a good, healthy state. "The conditions which make necessary a reduction in prices in certain lines and with certain companies will also tend to lower costs and increased production and will make for stability and normal operations," he declared. "The railroads of the country in normal times and under normal conditions consume about one-third of all the steel produced. Railroads during the past five years have ordered nothing like their usual amount and this deficiency of past years will have to be made up, which, added to their usual consumption, will keep the steel business steady and normal. The same conditions apply to building requirements and other lines of trade."

NEW CATALOGUE

The Armstrong Bros. Tool Co., Chicago, Ill., have issued a new catalogue, B 20. This catalogue embodies and makes effective various changes in list prices, and it is suggested by the concern that all previous lists be destroyed.

The book contains 143 pages of real live information regarding their varied line of tool holders, boring tools, lathe tool sets, grinding holders, rachet drills, C clamps, ordinary clamps, lathe dogs, wrenches, etc., etc.

For the convenience of the automotive trade they have segregated a special section devoted to drop-forged wrenches for automotive use. A copy of this book may be had upon request.

CANADIAN MACHINERY

AND
MANUFACTURING NEWS

Vol. XXIV. No. 17   October 21, 1920

Handling Two Iron Discs on Production Basis

The Advantage of Combining Machining Operations is Self Apparent in This Material—The Same Principle Could Well Be Adapted to Other Lines, and This Article May Start You Thinking

By F. SCRIBER

AN example of how two cast iron discs are handled as one, which has the advantage of combining the machining operations, is the subject of this article and Fig. 1 illustrates the disc, the drawing at the left being a view of finished disc. It will be noticed that the disc is beveled and has a flange cast on it. At the right of Fig. 1 is shown how two of these discs are cast together, while the description which follows will make obvious the manner of handling the same on a production basis.

Having the two discs cast together the first operation consists of gripping the work on the periphery of the two flanges, and in a three-jawed chuck. The hole is now machined and both hubs faced. The next operation consists of turning the two flanges, while at the third operation facilities are provided for machining both beveled surfaces and facing of flanges. The last operation of importance is milling the three elongated slots in the beveled face, and this is followed by cutting the discs apart and facing them true.

Referring back to the first operation, the tools used are illustrated in Fig. 2. A represents the chuck jaws in which the work is gripped, this operation being performed in a turret lathe. In the first turret, face tools as shown are used, these

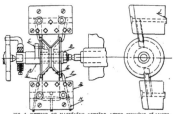

FIG. 3—METHOD OF MACHINING OUTSIDE AFTER TURNING FLANGES.

being a short drill B for spot drilling the hole, and a cutter C for facing the boss. Both tools are held by suitable screws in the holder D, this in turn being held in

the turret while the bushing E is used to securely grip the drill. Another feature of this set-up is the back facing cutter F which is operated through the

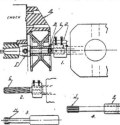

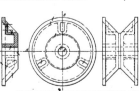

FIG. 1—SHOWING APPEARANCE OF DISC WHEN FINISHED, ALSO HOW THEY ARE MACHINED TOGETHER.

FIG. 2—FIRST OPERATION FOR MACHINING BEVELED DISCS.

spindle of the machine for facing the inner hub at the same time as the spot drilling and facing at the opposite end are being performed.

In the second turret, is a longer three-lipped drill G, which is of course used for opening out the hole, it being apparent from this that hole is cast rough in the discs. In the third turret the boring bar and cutter H are held, this cutter being of the flat type and it is used for enlarging the hole after drill has been passed through, thereby leaving a fairly accurate and true hole which is afterwards reamed by the machine reamer J, shown below, which is held in the fourth turret. This reamer is carried in the bushing, K, which allows it to float slightly as is common for this class of work.

Before proceeding with the next operation of importance it is necessary to

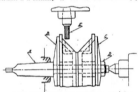

FIG. 4—METHOD OF MILLING IN VERTICAL MACHINE.

turn the two flanges on the discs, and this is done by driving a plain arbor in the work and machining in the customary manner between centres. When machining the beveled surfaces on these two combined discs it is necessary that each piece be located on the arbor in exactly the same relation to the cross-slide of the lathe as its predecessor, thereby eliminating the necessity of moving the cross-slide of the machine before making each cut. This is nicely taken care of with a sufficient degree of accuracy by providing a special arbor A. The work is held on the arbor against a shoulder by the nut and washer B, while a lug on the arbor carries a pin C which comes against another lug on the inside of the disc and acts as a driver for the same against the thrust of the cutting tools.

The machine used at this time is a lathe with a heavy cross-slide which carries tools on the front and rear side of the work. It is obvious that the arbor is held between centres and is driven by a dog in the usual manner. The block D on the front of the cross-slide contains a flat cutter E for rough facing the two beveled surfaces and the inner sides of the flanges. This cutter is notched to break apart the broad chips which would otherwise form on account of the wide cut, while cutters F and G face the outer sides of both flanges.

These three cutters are held at a five

degree angle in the cross-slide block by square head screws and are adjusted by means of the headless screws H. On the rear side of the cross-slide is a similar block, this likewise carrying three tools, and is used for finishing the same surfaces. The cutter which machines the beveled surfaces at this time has no notches cut in it. These blocks are held to the cross-slide by means of tee bolts, while tongue J is used for squaring the same on the cross-slide. The manner of using is to feed the front block in until the cross-slide comes against the atop which governs the roughing size, and then reverse the cross-slide and feed the rear tools in until the other stop is reached, thereby controlling the finished size.

Having completed the boring and turning of the discs the three elongated slots are next taken care of by placing discs on an arbor A Fig. 4. Two large collars B and C are clamped against the outer sides of the work by means of a nut D. The arbor, which has a tapered shank on it, is placed in the dividing head of the milling machine and the tailstock centre of the arbor supports the outer end of the arbor. Using an end mill E in a vertical milling machine, and traversing the table longitudinally back and forth with the centre of work under the centre of the cutter, one of the elongated slots is milled, and by indexing the same the three slots are consecutively completed.

This completes the essentials and all that is now necessary is to cut the two discs apart. This is done by placing the work on an arbor, when by using a cutting off tool they are separated. After the discs have been thus cut apart they are replaced on another arbor and a light facing cut is taken across the ends to bring them to correct length.

STANDARDISATION OF TEST PIECES

A report has been issued by the British Engineers Standard Association dealing with the standardisation of test pieces for the notched bar test, now so widely coming into use. The subject of the report has engaged the attention of the association for some considerable time past, and the committee, which has been under the chairmanship of Dr. W.

Cawthorne Unwin, has had the benefit of much experimental and research work carried out at the National Physical Laboratory, with funds provided by the Department of Scientific and Industrial Research. The Aeronautical Inspection Department of the Air Ministry has also co-operated, as well as a number of manufacturers, including Sir Robert Hadfield, Bart.

The report commences with a short description of the usual methods adopted for fracturing the test piece, useful notes and data being given of the principal machines used in these tests, viz., Charpy, Izod, Fremont, Amsler and Guillery. The actual form of notch is dealt with, and this is followed by the dimensions recommended for the British standard test pieces and for subsidiary standard test pieces of smaller cross section for use when the former cannot conveniently be obtained. The report concludes with a valuable series of plates illustrating to scale the various test pieces dealt with, the direction and point of impact of the pendulum or falling weight of the testing machine and grips for holding the test pieces. The report can be obtained from the secretary of the B.E.S.A., 28 Victoria Street, S.W.1. Price 30c., post free.

Cutting Production.—Owing to the unsettled conditions of the market, the officials of the Electric Steel and Metals Co., Ltd., of Welland, have deemed it advisable to curtail the operations of their plant for the time being, and on Saturday most of the employees were laid off. Mr. G. C. Mackenzie, the general manager, stated that it was not definitely decided to close the plant altogether, and that the heads of departments, foremen, etc., had been retained in the hope that more favorable conditions will prevail. The Electric Steel & Metals have been in operation in Welland for the past seven years, and during that time have been a valuable asset to the community. They opened up during the war and operated entirely on munition work, until almost the close of hostilities, when they turned their attention to other classes of foundry production.

Mining Tools for Greece.—There is a chance for Canada to sell mining tools and machinery in Greece, according to a statement by Trade Commissioner Clark after a visit to that market. Mr. Clarke sums up the situation as follows: Greek mining for its more progressive and extended exploitation requires principally (a) capital; (b) modern machinery and plant installations; (c) more economical methods in mining operations, and (d) more convenient means and greater facilities of transport. It may be expected that further developments will occur in this industry, whose productive capital already is estimated at at least 69,000,000 drachmas, and Canadian manufacturers of mining tools and machinery should note the existing industry and watch carefully whatever additional progress takes place.

Machining Friction Gear Blank   r riction Co.

Tool Layouts Are Always Interesting, an   , Are Shown
Various Examples of Work Performed Ac   to the Practice
of the Foster Machine Co., Elk   .nd.

By J. H. MOORE

WE have in previous issues shown various tool layouts of interest, and in the present article will illustrate some typical examples of work being performed on Foster turret lathes. These machines are made in Elkhart, Ind., and readers will no doubt remember a description of the machines themselves, which we gave some time ago in our new equipment section.

The examples shown herein deal with the machining of different styles of castings, also the production of parts from the bar stock. The tooling equipment used, as well as a detailed description of the various steps in the operation, are stated and dimensioned drawings of the parts are also shown.

It will be noted that certain reference numbers have been used in the text, and the same numbers apply to both the drawings of the parts and their layout. The surfaces finished on the lathes are indicated by heavy lines, thus making them stand out from the other portion of the view.

In most instances the time of machining is given, this including the time consumed in chucking the work prior to the machining, and in removing it from lathe after completion of operation.

Finishing Friction Gear Blank

Fig. 1 illustrates a cast iron friction gear blank that is finished all over in three operations, the work being performed on a No. 2B universal turret lathe. A layout of the tooling used in the first of these operations is shown at Fig. 2, and the operation is completed in 10.1 minutes, the various steps being taken in the following sequence.

The work is mounted in a three-jawed Barker chuck, which is equipped with a set of corrugated and hardened false jaws that grip the work along the outside of the rough rim. The hole A is rough bored by the cutter mounted in the boring bar, and which is held in the multiple turning head in position 1 of the turret, while the surface B is rough turned by tool K, and surface C is rough

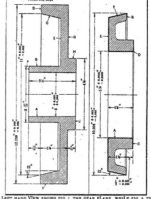

LEFT HAND VIEW SHOWS FIG. 1, THE GEAR BLANK, WHILE FIG. 9, THE RIGHT HAND VIEW, ILLUSTRATES FRICTION CONE.

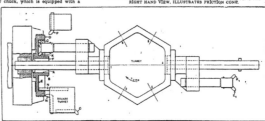

FIG. 2—TOOLING EQUIPMENT USED IN FIRST OPERATION ON GEAR BLANK.

spindle of th
inner hub r
drilling r
are be'
In
lir

LEFT HAND SHOWS FIG. 3, THE FIRST OPERATION ON THE GEAR BLANK. FIG. 4, TO THE RIGHT, IS A CLOSE UP OF THE SAME OPERATION PARTIALLY COMPLETED.

straight bored by the tool L. These tools are held in a special tool holder which is also held in the same fixture mounted at position 1 of the turret.

The end of the boring bar used in the boring of hole A enters a grinding bushing in the chuck before the cut is taken, so that concentricity of the several surfaces is assured. At the same time surfaces D, E and F are rough faced by the tools M, N, and O, respectively, these being held in the square turret shown.

The hole A is again bored by the cutter held in the boring bar which is mounted at position 4 of the multiple turning head. The end of this bushing also extends into the guiding bushing in the chuck while this step of the operation is being performed. The surface B is finish turned by tool K1, and surface C is finish straight bored by tool L1, these tools being held in a special holder that is also mounted at position 4 of the turret. At the same time surface F is finish faced by tool P, which is held in the rear tool holder shown.

The gear blank is now removed from the chuck, and the machine as used is shown at Figs. 3, 4 and 5. The arrangement of the tools in these illustrations differs slightly from that shown at Fig.

2 in which the tool used in finish facing surface F is mounted in a rear tool holder, while the other views show the tool mounted in the square turret. Figs. 3 and 4 show the front of the machine, while Fig. 5 shows the machine looking from the top. All these illustrations depict clearly the tooling equipment as employed.

The equipment used in the second operation is shown at Fig. 6. The operation in this case is completed in 9.1 minutes, the sequence of the steps being as follows: The work is first secured in a three-jawed Barker chuck equipped with a set of jaws that hold the work by bearing on the previously finished surface C.' The surfaces H and I are next rough turned by tools Q and R. These tools are mounted in the multiple turning at position 1 of the turret. This fixture is provided with a pilot that extends into a guiding bushing in the chuck while the cuts are being taken, thus ensuring that the surfaces machined will be concentric with hole A.

At the same time surface G is rough faced by tool S, which is mounted in the square turret. The surfaces H and I are next finish turned respectively by tools Q1 and R1, these being held

in a special tool holder in the multiple turning head, mounted in position 3 of the turret. The hole A is finish bored by the cutter held in boring bar T, which is mounted in the same turning head. At the same time surface J is rough faced by tool S, and finish faced by tool U. Both these tools are held in square turret. The hole A is now reamed by the expansion reamer held in position 5 of the turret. After this the work is removed from the lathe.

The photograph Fig. 8 was taken at the end of the third step of the series, and the tooling used is shown at Fig. 7. This operation is performed in 2.4 minutes, and the successive steps are as follows:

Third Operation

The work is secured in an air operated draw back adapter, which is inserted in the hole A. The surface C is rough taper bored by tool V, which is held in the square turret, this tool being used in conjunction with the taper attachment on rear of machine. Surface C is finish taper bored by a special tool held in the multiple turning head in position 1 of the turret. A pilot on the operating rod of the adapter enters a guiding hole in the multiple 'u.r,.

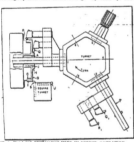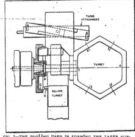

FIG. 6.—TOOLING EQUIPMENT USED IN SECOND OPERATION ON GEAR BLANK.   FIG. 7.—THE TOOLING USED IN FORMING THE TAPER SURFACE ON THE INSIDE OF THE GEAR RIM.

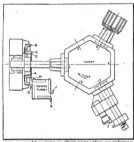

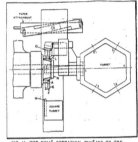

FIG. 10—TOOLING USED IN FIRST OPERATION ON FRICTION CONE.

FIG. 11—THE FINAL OPERATION. TOOLING ON THE FRICTION CONE.

ing head as the head is advanced for this step, thus locating the tool so that surface C is bored concentric with hole A. After this the finished work is removed from the machine. These three operations are completed in a total time of 21.6 minutes.

Machining Friction Cones

The friction cone, used in connection with the gear blank just described, is illustrated at Fig. 9. This part is made of cast iron, and two chuckings are required in finishing it, both of which are performed on the No. 2B machines, in the fast time of 9.7 minutes. Fig. 10 shows a layout of the tooling used in the first operation, this being completed in 5.3 minutes. The steps taken are as follows:

The work is placed in a three-jawed Barker wrenchless chuck equipped with corrugated and hardened false jaws that grip the work along the rough sur-

face of the rim. Hole A is now rough bored by the cutter held in the boring bar mounted at position 1 of the turret, this being provided with a pilot that extends into a guiding bushing in the chuck while the cut is being taken.

At the same time surface B is rough faced by tool G, which is mounted in the square turret, and surface C is rough faced by the tool H, which is

(Continued on page 61)

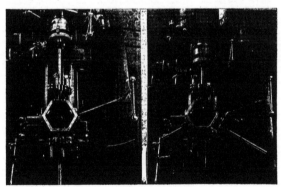

FIG. 5—PLAN VIEW OF MACHINE WORKING ON FIRST OPERATION ON GEAR BLANK.

FIG. 8—THE SECOND OPERATION ON THE GEAR BLANK.

Report on Coarse and Fine Threads—Part II.

This Portion Concludes the Article from Last Issue, and Specifies the Gauges in Detail—It Also Supplies Tables of Tolerances for Them, and Includes Specifications for Screw Thread Productions.

THE general subject of gauging screws is too extensive to be fully covered in this report. Reference is made, however, to bulletins published by the Bureau of Standards covering various inspection methods, including the standard ring and plug gauges, and the optical projection method of gauge inspection; also, to an article in the "Journal" of American Society of Mechanical Engineers of February, 1919, with reference to the use of the projection lantern for gauging work.

Inasmuch as the threaded plug and ring limit gauge is the most universally used scheme of gauging, and one that has been brought to the highest state of refinement, there is set forth herein what is considered the best practice used in the production and use of such gauges. It is understood, however, that it is not the intention of this commission to confine manufacturers to any particular method of gauging, as that would tend to hinder progress.

It has been the practice of many manufacturers, including Government shops, to work with "Go" gauges only and to depend upon the judgment of a good workman to keep within proper limits by the amount of "shake" or difference between the work and the gauge. With a highly skilled and trained force working on but one kind of work and also referring the working gauge to but one master gauge, a fair degree of interchangeability can be maintained under these conditions.

In the recent military preparations the Government required munitions in such vast quantities and in such a short period of time that this method of insuring interchangeability failed, and a method of gauging had to be established which did not rely entirely upon the skill and judgment of the workmen or inspectors. One reason for the necessity of a complete gauging system was that it was not possible to obtain a sufficient number of skilled workmen or inspectors. Furthermore, one master gauge could not be used all over the country and consequently discrepancies in measurement between different shops had to be guarded against by the use of properly tested standards and by approved methods of measuring.

It is believed that the experience gained by manufacturers producing war material will result in a much more extensive use of gauges than was ever thought practicable during pre-war times. The gauge specifications which are given herein cover the manufacture, use and application of a system of gauging which has been thoroughly demonstrated in the execution of war contracts as being adequate for the production in large quantities of strictly interchange-

able screw-thread product. It is not the intention of this report to limit manufacturers to any particular methods of test in checking either the manufactured product or in measuring gauges, for the reason that any specification which would tend to limit the development of new and improved methods of measuring would be very undesirable. However, when the ordinary forms of thread gauges are used, the specifications given herein will apply.

Gauges and Methods of Test

The specifications for gauges given herein are built upon the following fundamental assumptions:

Approved limit master gauges do not reduce the net working tolerance. Permissible errors in angle of thread specified for "Go" gauges tend to reduce the net working tolerance, while similar permissible errors on the "Not Go" gauge tend to increase the net working tolerance. These two factors, therefore, balance each other. Permissible lead errors specified for the "Go" gauge reduce the net working tolerance, while permissible lead errors on the "Not Go" gauge tend to increase the net working tolerance. In order to realize the full net working tolerance, the permissible diametrical variation specified for both "Go" and "Not Go" gauges (gauge increment) is placed outside of the net tolerance limits. The extreme tolerance equals the net tolerance plus gauge increment. The "Go" gauge should check simultaneously all elements of the thread (all diameters, lead, angle, etc.). The "Not Go" gauge should check separately the elements of the thread.

The following general specifications refer in particular to gauging systems which have been found satisfactory by the Army and Navy for the production of interchangeable parts as specified under the subject of "Classification and Tolerances." These specifications are included for the use of manufacturers where definite information is lacking. They are not to be considered mandatory.

Thread gauges may be included in one of four classes, namely: standard master gauges, limit master gauges, inspection gauges, and working gauges.

The standard master gauge is a threaded plug representing as exactly as possible all physical dimensions of the nominal or basic size of the threaded component. In order that the standard master gauge be authentic, the deviations of this gauge from the exact standard should be ascertained by the National Bureau of Standards and the gauge should be used with knowledge of these deviations or corrections.

Limit master gauges are for reference only. They represent the extreme upper

and lower tolerance limits allowed on the dimensions of the part being produced. They are often of the same design as inspection gauges. In many cases, however, the design of the master gauge is that of a check which can be used to verify the inspection or working gauge.

Inspection gauges are for the use of the purchaser in accepting the product. They are generally of the same design as the working gauges and the dimensions are such that they represent nearly the net tolerance limits on the parts being produced. Inasmuch as a certain amount of wear must be provided on an inspection gauge, it can not represent the net tolerance limit until it is worn to master gauge size.

Working gauges are those used by a manufacturer to check the parts produced as they are machined. It is recommended that the working gauges be made to represent limits considerably inside of the net limits in order that sufficient wear will be provided for the working gauges, and in order that the product accepted by the working gauges will be accepted by the inspection gauges.

The following list enumerates the inspection and working gauges required for producing strictly interchangeable screws as specified for national coarse threads, national fine threads, or other straight threads.

(a) A maximum or "Go" ring thread gauge, preferably adjustable, having the required pitch diameter and minor diameter, cleared to facilitate grinding and lapping.

(b) A minimum or "Not Go" ring thread gauge, preferably adjustable, to check only the pitch diameter of the threaded work.

(c) A maximum or "Go" plain ring gauge to check the major diameter of the threaded work.

(d) A minimum or "Not Go" snap gauge to check the major diameter of the threaded work.

The following list enumerates the inspection and working gauges required for producing strictly interchangeable nuts, as specified for national coarse threads, national fine threads, or other straight threads.

(a) A minimum or "Go" thread plug gauge of the required pitch diameter and major diameter. The minor diameter of the thread plug gauge may be cleared to facilitate grinding and lapping.

(b) A maximum or "Not Go" thread plug gauge to check only the pitch diameter of the threaded work.

(c) A "Go" plain plug gauge to check the minor diameter of the threaded work.

(d) A "Not Go" plain plug gauge to

check the minor diameter of the thread-ed work.

The following list enumerates the limit master gauges required for the verification of the working or inspection gauges as previously listed for verifying the screw.

(a) A set plug or check for the maximum "Go" thread ring gauge, having the same dimensions as the largest permissible screw.

(b) A set plug or check for the minimum or "Not Go" thread ring gauge having the same dimensions as the smallest permissible screw.

(c) A maximum plain plug for checking the minor diameter of both the "Go" and "Not Go" inspection thread ring gauge.

The following list enumerates the limit master gauges required for the verification of the working or inspection gauges as previously listed for verifying the nut.

(a) A minimum or "Go" threaded plug to be used as a reference for comparative measurements, corresponding to the basic dimension, or standard master gauge.

(b) A maximum or "Not Go" threaded plug to be used as a reference for comparative measurements, corresponding to the largest permissible threaded hole.

(c) A minimum plain ring gauge to check the major diameter of the "Go" and "Not Go" master threaded plug unless suitable measuring facilities are available for this purpose.

Gauges may be made of a good grade of machinery steel pack-hardened, or of straight carbon steel of not less than 1 per cent. carbon; or preferably of an oil hardening steel of approximately 1.10 per cent. carbon and 1.40 per cent. chromium.

Handles should be made of a good grade of machinery steel plainly marked to identify the gauge.

The following specifications will be helpful in the design and construction of gauges used for producing threaded work.

All plain plug gauges should be single-ended. Plain plug gauges of 2 in. and less in diameter should be made with a plug inserted in the handle and fastened thereto by means of a pin. Plain plug gauges of more than 2 in. in diameter should have the gauging blank so made as to be reversible. This can be accomplished by having a finished hole in the gauge blank fitting a shouldered projection on the end of the handle, the gauge blank being held on with a nut.

The "Go" plain plug gauge should be noticeably longer than the "Not Go" plain plug gauge, or some distinguishing feature in the design of the handle should be used to serve as a ready means of identification, such as a chamfer on the handle of the "Go" gauge.

Both the "Go" and "Not Go" gauges should have their outside diameters knurled if made circular.

The "Go" gauge should have a decided chamfer in order to provide a ready means of identification for distinguishing the "Go" from the "Not Go" gauge.

Snap gauges may be either adjustable or non-adjustable. It is recommended that all snap gauges up to and including ⅛ in. be of the built-up type. For larger snap gauges, forged blanks, flat plate stock or other suitable construction may be used.

Sufficient clearance beyond the mouth of the gauge should be provided to permit the gauging of cylindrical work.

Snap gauges for measuring lengths and diameters may have one gauging dimension only, or may have a maximum and minimum gauging dimension, both on one end, or maximum and minimum gauging dimensions on opposite ends of the gauge. When the maximum and minimum gauging dimensions

TABLE VII. MANUFACTURING TOLERANCES ON PLAIN GAGES					
Manufacturing Tolerance Allowed on Work	Allowable Tolerance for Master Gage		Allowable Tolerance for Inspection Gage	Suggested Tolerance for Working Gage	
	Minimum Gage	Maximum Gage	Minimum Gage	Maximum Minimum Gage	Maximum Gage
0.002	+0.0000 +0.0001	−0.0000 −0.0001	+0.0001 +0.0001	−0.0001 +0.0005	+0.0005 −0.0005 −0.0005
0.002 to 0.004	+0.0000 +0.0002	−0.0000 −0.0002	+0.0002 +0.0004	−0.0002 −0.0004	+0.0005 +0.0007 −0.0004 −0.0007
0.004 to 0.006	+0.0000 +0.0003	−0.0000 −0.0003	+0.0004 +0.0007	−0.0004 −0.0007	+0.0007 +0.0011 −0.0007 −0.0011
0.006 to 0.010	+0.0000 +0.0005	−0.0000 −0.0004	+0.0005 +0.0015	−0.0006 −0.0015	+0.0010 +0.0015 −0.0010 −0.0015
0.010 to 0.020	+0.0000 +0.0005	−0.0000 −0.0005	+0.0010 +0.0015	−0.0010 −0.0015	+0.0015 +0.0021 −0.0015 −0.0021
0.020 to 0.050	+0.0000 +0.0006	−0.0000 −0.0006	+0.0020 +0.0026	−0.0020 −0.0026	+0.0026 +0.0033 −0.0026 −0.0033

TABLE VIII. TOLERANCES ON MASTER THREAD GAGES FOR LOOSE FIT AND MEDIUM FIT WORK				
(This applies to both Standard and Limit Master Gages)				
Number of Threads per Inch	Allowable Variation in Lead Between Any Two Threads Not Farther apart than Length of Engagement	Allowable Variation in One-half Angle of Thread	Allowable Tolerance on Diameter of Minimum Thread Gages	Allowable Tolerance on Diameter of Maximum Thread Gages
4 to 6	±0.0005	±0° 5′	+0.0000 +0.0004	−0.0000 −0.0004
7 to 10	±0.0004	±0° 5′	+0.0000 +0.0004	−0.0000 −0.0004
11 to 18	±0.0003	±0° 10′	+0.0000 +0.0004	−0.0000 −0.0004
20 to 28	±0.0002	±0° 15′	+0.0000 +0.0003	−0.0000 −0.0003
30 to 40	±0.0002	±0° 20′	+0.0000 +0.0002	−0.0000 −0.0002
44 to 80	±0.0002	±0° 30′	+0.0000 +0.0002	−0.0000 −0.0002

TABLE X. SUGGESTED MANUFACTURING TOLERANCES FOR WORKING GAGES FOR LOOSE FIT AND MEDIUM FIT WORK				
Number of Threads per Inch	Allowable Variation in Lead Between Any Two Threads Not Farther apart than Length of Engagement	Allowable Variation in One-half Angle of Thread	Allowable Tolerance on Diameter of Minimum Thread Gages	Allowable Tolerance on Diameter of Maximum Thread Gages
4 to 6	±0.0006	±0° 5′	+0.0010 +0.0020	−0.0010 −0.0020
7 to 10	±0.0005	±0° 10′	+0.0010 +0.0020	−0.0010 −0.0020
11 to 18	±0.0004	±0° 15′	+0.0008 +0.0015	−0.0008 −0.0015
20 to 28	±0.0003	±0° 20′	+0.0006 +0.0012	−0.0006 −0.0012
30 to 40	±0.0002	±0° 30′	+0.0005 +0.0010	−0.0005 −0.0010
44 to 80	±0.0002	±0° 45′	+0.0004 +0.0008	−0.0004 −0.0008

TABLE XI. MASTER GAGE TOLERANCES FOR CLASS III. CLOSE FIT WORK				
(This applies to both Standard and Limit Master Gages)				
Number of Threads per Inch	Allowable Variation in Lead Between Any Two Threads Not Farther apart than Length of Engagement	Allowable Variation in One-half Angle of Thread	Allowable Tolerance on Diameter of Minimum Thread Gages	Allowable Tolerance on Diameter of Maximum Thread Gages
4 to 6	±0.00025	± 2′ 30′	+0.00000 +0.00010	−0.00000 −0.00010
7 to 10	±0.00020	± 2′ 30′	+0.00000 +0.00010	−0.00000 −0.00010
11 to 18	±0.00015	± 5′ 00′	+0.00000 +0.00020	−0.00000 −0.00020
20 to 28	±0.00010	± 7′ 30′	+0.00000 +0.00015	−0.00000 −0.00015
30 to 40	±0.00010	± 10′ 00′	+0.00000 +0.00010	−0.00000 −0.00010
44 to 80	±0.00010	± 15′ 00′	+0.00000 +0.00010	−0.00000 −0.00010

TABLE IX. SUGGESTED MANUFACTURING TOLERANCES FOR INSPECTION GAGES FOR LOOSE FIT AND MEDIUM FIT WORK

Number of Threads per Inch	Allowable Variation in Lead Between Any Two Threads Not Farther Apart than Length of Engagement	Allowable Variation in One-half Angle of Thread Gages	Allowable Tolerances on Diameters of Minimum Thread Gages	Allowable Tolerances on Diameters of Maximum Thread Gages
4 to 6	±0.0006	±0° 5′	+0.0006 +0.0015	−0.0006 −0.0015
7 to 10	±0.0005	±0° 10′	+0.0004 +0.0010	−0.0004 −0.0010
11 to 18	±0.0004	±0° 15′	+0.0004 +0.0008	−0.0004 −0.0008
20 to 28	±0.0003	±0° 20′	+0.0003 +0.0006	−0.0003 −0.0006
30 to 40	±0.0002	±0° 30′	+0.0002 +0.0005	−0.0002 −0.0005
44 to 80	±0.0002	±0° 45′	+0.0002 +0.0004	−0.0002 −0.0004

are placed on opposite ends of the gauge, the maximum or "Go" end of the the corners of the gauge on the "Go" end decidedly chamfered.

All plug thread gauges should be single-ended. Thread plug gauges 2 in. and less in diameter should be made with a plug inserted in a handle and fastened thereto by means of a pin.

Plug gauges of more than 2 in. in diameter, unless otherwise specified, should have the gauging blank so made as to be reversible. This can be accomplished by having the finished hole in the gauge blank fitting a shouldered projection on the end of the handle, the gauge blank being keyed and held with a nut.

"Not Go" thread plug gauges should be noticeably shorter than the "Go" thread plug gauges, in order to provide a ready means of identification, or the handle of the "Go" gauge should be chamfered.

End threads on plug thread gauges should not be chamfered, but the first handle of the "Go" gauge should be flattened to avoid a feather edge.

Inspection and working thread plug gauges should be provided with dirt grooves which extend into the gauge for a depth of from one to four threads.

The length of thread parallel to the axis of the gauge should, for all standard "Go" thread plug gauges, be at least as much as the quantity expressed in the following formula:

Where

$L = (1.5) D$

L = Length of thread
D = Basic major diameter of thread.

For threaded work of shorter length of engagement than (1.5) D, the length of thread on the "Go" gauge may be correspondingly shorter.

"Not Go" thread gauge for pitch diameter only: All "Not Go" thread plug gauges should be made to check the pitch diameter only. This necessitates removal of the crest of the thread so that the dimension of the major diameter is never greater than that specified for the "Go" gauge, and also removing the portion of the thread at the root of the standard thread form.

All ring thread gauges should be made adjustable.

The "Go" gauge should be distinguished from the "Not Go" gauge by having a decided chamfer and both gauges are to have their outside diameter knurled if made circular.

The end threads on ring thread gauges should not be chamfered but the first half turn of the end thread should be flattened to avoid a feather edge.

The length of thread parallel to the axis of the gauge should, for all standard "Go" ring thread gauges, be at least as great as the quantity determined in the formula perviously given. For threaded work of shorter length of engagement than (1.5) D, the length of thread on the "Go" gauge may be correspondingly shorter.

"Not Go" ring thread gauges should be made to check the pitch diameter only. This necessitates removal of the crest of the thread so that the dimension of the minor diameter is never less than that specified for the maximum or "Go" gauge and also removing the portion of the thread at the root of the standard form. There are specified herein for use in the production of national coarse threads and national fine threads, and for other straight threads, several tables of gauge manufacturing tolerances.

Table VII will be found practicable for all plain plug, ring and snap gauges used in connection with a measurement of screw thread diameters. In addition to the master gauge tolerances, suggested tolerances for inspection and working gauges are also given in Table VII.

Table VIII will be found practicable for both standard and limit master thread gauges for thread work designed in accordance with the manufacturing tolerances for Class I, loose fit and Class II, medium fit, made to Tables III, IV and V.

Table IX contains suggested manufacturing tolerances for inspection thread gauges with a small allowance for wear for use in quantity production of Class I, loose fit, and Class II, medium fit thread work, made to Tables III, IV and V.

Table X contains suggested manufacturing tolerances for working thread gauges with a small allowance for wear for use in quantity production of Class I, loose fit, and Class II, medium fit thread work, made to Tables III, IV and V.

Table XI contains the tolerances suggested for both standard and limit master thread gauges for work designed in accordance with manufacturing tolerances for Class III, close fit thread work, made to Table VI. As the component tolerances for this class are relatively small, it is believed that the working gauges will be required to be held within the gauge tolerances shown in Table XI.

For plain plug gauges, plain ring gauges, and plain snap gauges required for measuring diameters of screw thread work, the gauge tolerances specified in Tables VIII, IX, X and XI should be used. Attention is called to the fact that the tolerances on thread diameters vary in accordance with the number of threads per inch on the screw or nut being manufactured. In manufacturing a plain plug, ring or snap gauge, in the absence of information as to the number of threads per inch of the screw to be made, or for gauge dimensions other than thread diameters, the tolerances for plain gauges given in Table VII may be used.

The tolerances on lead are specified as an allowable variation between any two threads not farther apart than the length of thread engagement as determined by the following formula:

Where

$L = (1.5) D$

L = Length of thread engagement
D = Basic major diameter of thread.

The tolerances on angle of thread as specified herein for the various pitches are tolerances on one-half of the included angle. This insures that the bisector of the included angle will be perpendicular to the axis of the thread within proper limits. The equivalent deviation from the true thread form caused by such irregularities as convex or concave sides of thread, rounded crests, or slight projections on the thread form, should not exceed the tolerances allowable on angle of thread.

The tolerances given for thread diameters in Tables VIII, IX, X and XI are applied in such a manner that the tolerances permitted on the inspection and working gauges occupy part of the extreme tolerance. This insures that all work passed by the gauges will be within the tolerance limit specified on the part drawing as represented by the limit master gauges. The tolerances given also per-

mit the classification and selection of gages so that if a gage is not suitable for a master gage it may be classified and used as an inspection or working gage provided that the errors do not pass outside of the net tolerance limits. The application of the tolerances on diameters of thread gages is exactly the same as explained herein for plain gages. For an example of this application see Table XII.

Typical Specifications

Material: The material used shall be cold-drawn bessemer steel automatic screw stock.

Composition:

Carbon, 0.08 to 0.16 per cent.
Manganese, 0.50 to 0.80 per cent.
Phosphorus, 0.09 to 0.13 per cent.
Sulphur, 0.075 to 0.13 per cent.

Method of manufacture: Bolts and nuts may be either rolled, milled, or machine cut, so long as they meet the specifications herein provided. Bolts and nuts to be left soft.

Workmanship: All bolts and nuts must be of good workmanship and free from all defects which may affect their serviceability.

Finish: All bolts and nuts to be semi-finished; that is, the bodies to be machined, under side of head and nut faced, upper face of head and nut to be chamfered at an angle of 30 deg., leaving a circle equal in diameter to the width of the nut.

Form of Thread: The form of thread shall be the "national form," as specified herein, and formerly known as the United States standard or Sellers thread.

Thread Series: The pitches and diameters shall be as specified in Table I, herein, and known as the National coarse thread series.

Class of Fit: Class II-A, medium fit (regular).

Dimensions: Nominal size: ½ inch. Number of threads per inch: 13. Length under head: 3±0.05 inches. Minimum length of usable thread: 1 inch. Diameters: (Specified in Table XI of complete report).

Tolerances and Allowances: See Table IV.

Nuts: Form: Hexagonal. Thickness: ½±0.01 inch. Short diameter (across flats): ⅞±0.01 inch.

Heads: Form: Hexagonal. Thickness: 7-16±0.01 inch. Short diameter (across flats): ⅞±0.01 inch.

Gages: The gages used shall be such as to insure that the product falls within the tolerance as specified herein for Class II, medium fit (regular):

The following gages are suggested and will be used by the purchaser:

For the Screw:

A maximum or "Go" ring thread gage.

A minimum or "Not-Go" ring thread gage to check only the pitch diameter of the thread.

A maximum or "Go" plain ring to

check the major diameter of the thread.

A minimum or "Not Go" snap gage to check the major diameter of the thread.

For the Nut:

A minimum "Go" thread plug gage.

A maximum or "Not Go" thread plug gage to check only the pitch diameter.

A "Go" plain plug gage to check the minor diameter.

A "Not Go" plain plug gage to check the minor diameter of the thread.

Inspection and Test: Screws and nuts shall be inspected and tested as follows:

At least three bolts and nuts shall be taken at random from each lot of 100, or fraction thereof, and carefully tested. If the errors in dimensions of the screws or nuts tested exceed the tolerance specified for this class, the lot represented by these samples shall be rejected.

Delivery: Unless otherwise specified the assembled bolts and nuts are to be delivered in substantial wooden containers, properly marked, and each containing 100 pounds.

The commission, in formulating this progress report, has acted largely in the capacity of a judiciary, basing its decisions upon evidence received from authorities on screw-thread subjects and upon the conclusions drawn by other organizations having to do with standardization of screw threads. In addition, the various subjects dealt with have been considered with a knowledge of present manufacturing conditions and with anticipation of further development in the production of screw-thread products. Above all, it is the intention of the commission to facilitate and promote progress in manufacture.

In the time provided the commission has devoted its attention to the standardization of only those threads, sizes, types and systems which are of paramount importance by reason of their extensive use and utility. There remains much to be accomplished along the lines of standardization of special but important threads, and of maintaining progress in our standardization work in keeping with the developments of manufacturing conditions.

According to an estimate of the mining bureau of the Department of Agriculture and Commerce at Tokyo, based mainly on coal veins over 2 ft. thick, and generally lying not more than 2,000 feet below drainage level, the available coal deposits of Japan amount to about 822,000,000 metric tons of 2,204 pounds. In addition, it is judged that approximately 2,940,000,000 metric tons, which lie at greater depths, can be mined if highly efficient methods are used. It is also surmised that the concealed and undetermined deposits of coal may possibly amount to 5,030,000,000 metric tons.

Change in Name.—The Machine & Stamping Co., Ltd., 1209 King Street West, Toronto, Ont., changed its name on October 1 to the Russell Gear & Machine Co., Ltd., without changing the management.

TABLE XII. TOLERANCES AND LIMITS ON MAXIMUM AND MINIMUM MASTER, INSPECTION AND WORKING GAGES FOR SHAFT AND HOLE

	Work to be gaged:		
Hole,	1.250 + 0.004	1.254	max.
	− 0.000	1.250	min.
Shaft,	1.248 + 0.000	1.248	max.
	− 0.004	1.244	min.
Gages for hole:			

Maximum Gages

	Dimension of Part	Gage Tolerances	Gage Limits
Limit master gage	1.254	−0.0000 −0.0003	1.2540 1.2537
Inspection gage	1.254	−0.0004 −0.0007	1.2536 1.2533
Working gage	1.254	−0.0008 −0.0013	1.2532 1.2527

Minimum Gages

Limit master gage	1.250	+0.0000 +0.0003	1.2500 1.2503
Inspection gage	1.250	+0.0004 +0.0007	1.2504 1.2507
Working gage	1.250	+0.0008 +0.0013	1.2508 1.2513

Gages for shaft:

Maximum Gages

Limit master gage	1.248	−0.0000 −0.0003	1.2480 1.2477
Inspection gage	1.248	−0.0004 −0.0007	1.2476 1.2473
Working gage	1.248	−0.0008 −0.0013	1.2472 1.2467

Minimum Gages

Limit master gage	1.244	+0.0000 +0.0003	1.2440 1.2443
Inspection gage	1.244	+0.0004 +0.0007	1.2444 1.2447
Working gage	1.244	+0.0008 +0.0013	1.2448 1.2453

By comparison of the above figures, it will be seen that it is not possible for the master, inspection or working gage dimensions to overlap.

WELDING AND CUTTING

Lessons on the A. B. C. of Good Welding
A Knowledge of the Torch You Are Using is Absolutely Essential if You Feel You May Require to Make Some Repair to It.

By W. B. PERDUE *

THE ordinary individual, who has no ambition other than to work in a shop under the direction of a competent foreman, need not worry about the construction of his torch. In fact, the less tampering he does with it the better he will get along. Simple as they may seem, torches are instruments of precision. In the best regulated welding shops they are treated as such; their repair is left to experts; those in readiness for operation are ranged in good order; kept free from dirt and grime; in good preservation and working order, as all tools of precision deserve to be treated.

The operator who feels that he may at some time be required to make emergency or other repairs should see to it that he has expert knowledge and the thorough understanding that will enable him to get directly at the seat of trouble and remedy it without delay.

Necessary Principles

There are certain fundamental principles involved in the construction and use of any tool. The expert is the person whose knowledge of these principles enables him, by examination, to locate the source of trouble, and to determine what corrective measures may be required to produce or restore normal working conditions.

We will now discuss at length the principles upon which successful operation of oxy-acetylene equipment depends, giving credit where due to the originators and inventors of new ideas and for quotations which follow to Messrs. Granjon & Rosenberg of Paris, the pioneer oxy-acetylene experts.

1. "The chief difficulty to be overcome is to obtain the necessary stability of flame; (a) The velocity of propagation (the speed with which the flame will burn) of the flame is about 330 feet per second; (b) It is therefore necessary, in order to avoid the striking back of

*Mr. Perdue is in charge of welding instruction at Heald's Engineering School, San Francisco. He wishes to state that owing to sickness he has been unable to answer queries, but is again able to do so.

the flame, that the velocity of the mixture at the exit should be of the same value, or at least such that it constantly prevents the return of the flame to the interior."

The theoretical calculations of the author proved the velocity of propagation given above to be about one-third more than it would be in a stirred and stabilised mixture. It then occurred to him that in maintaining this excessive and unnecessary speed the temperature of the flame was actually diminished, to the extent that a great deal less effective heat energy was developed than resulted from the complete combustion within the flame of exactly proper proportions of gases travelling at much lower velocities.

No matter how well constructed are the regulators which govern the flow of gases to the torch, it is necessary to employ some means within the torch to prevent the passage through the flame of alternate excesses of oxygen and acetylene.

Changes in Flame

These changes may be so slight as to cause no visible change of the flame, but careful examination of the weld will reveal adequate proof of the fluctuations which take place. The amount of gas wasted is considerably more than might be imagined, since the passage through the flame of even a slight excess of either gas very materially lowers its temperature and efficiency. To avoid flashing of the torch it is necessary that the velocity of the flow be greater than the propagation of the flame at the time when it is intermittently "lean"; that is, at the time when the amount of acetylene is lowest and an excess of oxygen predominates.

It is a very significant fact that in our large power plants where the steam

CUTTING AN INGOT FOR STEAMSHIP CRANKSHAFT.

pressure seldom varies more than a pound in 250 no well-informed engineer would consider the installation of an engine without a governor. Here economy is the principal factor. With gas pressures of say ten pounds the fluctuation of a mere fraction of an ounce would be proportionately equal. Here the soundness of the weld and the possibility of endangering life and property through its failure must be considered.

The Modern Welder

Is it any wonder then that the modern, progressive welder demands to see the stabilizing chamber in a torch before buying, and that he is favored with more work than he can handle, while his competitor with antiquated equipment blames hard luck for the failures of his welds and his lack of patronage.

For his success in developing the stirring and stabilizing chamber in torches the author is indebted in a measure to Remoud J. Tobin, of Richmond, Calif., the inventor of the Tobin Patent Adjustable Welding Head, which was used with such great success in torpedo boat construction during the recent war.

"The intimate mixture of gases should be perfectly accomplished before their escape from the blowpipe. . . . In some torches there is an energetic mixing of the two gases, but the contact is not made molecule to molecule, and stream lines of oxygen or acetylene can escape at the nozzle without being mixed."

Inasmuch as oxygen can be used without limits of pressure, oxy-acetylene torches are divided according to pressure of acetylene required into three principal classes:

(1) Equal pressure type — for high pressure acetylene.

(2) Positive pressure type — for medium pressure acetylene.

(3) Injector type — for low pressure acetylene.

Low pressure acetylene generators can be used only with injector type torches. Medium pressure generators have a range up to twelve pounds and can be used with the smaller tips furnished with equal pressure torches or with tips of any size on positive pressure torches.

Change of Torches

Inasmuch as all torches designed by the author can be changed to any of the foregoing types by substitution of another mixer, and as he has nothing to gain by extolling the merits of any particular type, he feels competent to discuss the various types from a standpoint of merit. The patents issued to most torch producing concerns limit them to one particular type of construction. It is extremely interesting to observe salesmen inviting the attention of prospective customers to the "low pressures" with which their torches operate, or to the "extremely small orifices" which are supposed to result in economical operation.

Now any reasoning person knows that with a small orifice the greater the required pressure to force therethrough a fixed or definite amount of gas and vice versa. It is also well known that it requires a certain definite amount of gas to generate a certain definite amount of heat, therefore, the principal points of merit in all classes of gas burning devices are those outlined in requirements 1 and 2.

"Blowpipes for high-pressure acetylene were first to be constructed. They were designed for using dissolved acetylene, and are practically used only for that purpose now."

It is a lamentable fact that. many makers of this type of equipment advertise that their torches will give satisfaction with generator pressures. Conditions over which the manufacturer has no control, such as the length and size of hose, piping, and connectors, are such that the use of printed pressure tables is quite misleading; therefore, the author invariably advises operators to set the regulator at that pressure which causes the acetylene flame to cease giving off smoke before turning on the oxygen. If desired, these pressures may be adjusted at the regulators for the larger tips and necessary adjustments for smaller sizes made by means of the valves on the torch.

Any intelligent workman, familiar with torch construction, can produce an equal pressure torch. Where means are not provided for stirring and stabilizing the flow of gases such torches are producers of inferior welds, are expensive in operation, and should not be used for professional welding work.

This idea is best illustrated by comparing the effect produced by a clear stream flowing at the same velocity into a muddy river. Miles below their junction can be discerned the clear water still flowing independently on.

Now let us imagine the speed of the clear stream to be increased. This produces an eddying effect which tends in some measure to commingle it with the other water. It is possible, however, to increase this speed to a point where the tendency to mix with the muddy water is lessened. Thus we learn from facts known since childhood, that extremely low pressures of acetylene, which require correspondingly high pressures of oxygen, as well as those pressures of acetylene which are equal to the oxygen, are not, without special stirring devices, conducive to efficiency.

Effects of Streams

Next let us imagine the effect of passing these streams through a conical expansion (diamond shaped) chamber such as is used in many torches. The tendency to commingle will be slightly increased, but hardly enough to be noticeable. These tests, made with colored gases and liquids under every possible variation of pressures and conditions, and the use of specially sensitized plates to determine the efficiency of the mixtures produced, have proved the conical chamber mixing theory to be impractical.

Lastly let us imagine these streams driven through a broken conduit of shape similar to the stabilizing chamber of a modern torch. Here they are completely commingled. Furthermore, little if any convulsive effect is imparted to the stream flowing out and away from such chamber. Thus is secured the mixture and stability required to produce perfect combustion of gaseous mixtures.

In low pressure injector type torches the acetylene, under slight pressure, enters a suction chamber from whence it is aspirated by the suction of a stream of oxygen under higher pressure. This is accomplished by means of a device known as the Giffard injector, which operates by intermittently creating a partial vacuum, drawing in the acetylene, expelling it, and repeating the process.

Torches of the positive pressure type, when equipped with stabilizing chamber of proper construction, combine the advantages of the other two types, and can be used for any and all varieties of work.

3. "The torch must be so constructed that the delivery of the proper proportions of gases to the nozzle will not be affected by its becoming heated."

Here an examination of the torch prior to purchase to determine what means, if any, have been taken to insulate the mixing device and the stirred gases from the heat is the only precaution that can be recommended. The construction of tip and mixing device of one piece of metal certainly simplifies the construction of the torch, and enables it to be sold more cheaply. It often happens that the heating of the tip of such torches causes a leakage of oxygen into the acetylene chamber, producing a condition which invariably causes a tendency to flash.

As an emergency measure, recommended only when better equipment cannot be procured, the acetylene pressure may be increased, and the oxygen orifice drilled or reamed to such size as requires a slightly lower oxygen pressure than that of the acetylene.

Weight of Importance

The weight of a torch is a matter of some importance. The good mechanic likes a tool which is light and "feels good in the hand." It goes without saying that more and better work will be done with torches which "fit the hand" and are light and easy to handle than with those . which are built without thought of convenience in handling.

Metal to metal seats on tips and other movable parts should be kept clean and free from grit. Tips should be kept clean both inside and out, since any accumulation on the outside of the tip causes it to become heated more quickly than a clean tip. It is still more important that the orifice of the nozzle be kept free from incrustations and obstructions. This must not be done by means of wires or broaches inserted from the flame end, or by rubbing the face of the tip on the bricks of the welding table. In old fashioned non-stabilizing injector type torches any widening of the nozzle causes a derangement of the torch. Such types may be cleaned only by means of

(Continued on page 61)

DEVELOPMENTS IN SHOP EQUIPMENT

NEW OPEN-SIDE PLANER

The L. & P. Manufacturing Co., Niagara Falls, Ont., have placed on the market both open-side planers and the double column planer shown in illustration. The beds of these machines are of closely ribbed type fully enclosed between the ways. The fourteen-foot machines have three oil wells provided to each of the vees. The worm bearings are cast solid in the bed with reservoirs to maintain oil above the centre of the drive. Outside housing for thrust bearing and reduction gears are cast solid with the bed.

The housings of the double column planer as shown in illustration are of heavy box pattern, provided with clamping screws for holding in working position to both housings.

The planers are of double deck box type, having ports in sides for removal of chips, stake holes being drilled through top deck only. This prevents chips dropping into driving mechanism. Three tee slots are provided and four rows of 1-inch stake holes spaced twelve inches apart each way.

The columns are of heavy box pattern and are well ribbed and securely fastened to the bed.

Four tool heads are furnished on the two column planer, two on the open-side planer. Each of these heads are of standard heavy swivel type provided with parallel gibs with set screw take-ups. The feed screws are fitted with micrometer collars.

The parallel drive is obtained by forged steel worm and cast iron machine cut rack, the worm lying at such an angle that thrust is parallel to the ways.

All shaft bearings are split bronze lined boxes, excepting those supporting the worm, the latter being solid. The drive thrust bearings are readily adjustable, and consist of a number of steel, bronze and fibre disks, having radial oil grooves across their faces. They run in an oil well located outside the bed. Reversing thrust bearings are similar but are located inside the worm oil well.

Worm bearings and oil reservoir are piped to filler connection outside of bed, and thus platen need not be removed for refilling. The balance of drive bearings, being outside of bed, are accessible at all times.

The motor is connected by flexible coupling to driving pinion shaft on left hand side behind planer close to bed.

The reduction gears consist of one pair of wide face spur gears fully enclosed, arranged to run in oil, the drive pinion being provided with an outboard bearing.

Following are the specifications for the double column machine. Planer width 48 inches, width between housings 50 inches, length of table over all 15 feet 6 inches, length of bed over all 23 feet 4 inches. The length of down feed of head or cross rail is 9 inches, and the minimum distance centre of tool blocks is 19 inches. The width over gear belt is 8 feet 1½ inches, and the height over gears 8 feet 10 inches.

The other specifications herewith on the open-side type are self-explanatory, and these machines are made in both the belt and motor types.

Planer width 48 inches, length of table over all 15 feet 6 inches, length of bed over all 23 feet 4 inches, length of down feed of head on cross rail 9 inches, floor space length, including table travel 32 feet, planer height 48 inches, maximum height under rail 59 inches, width of table 44 inches, horse-power required 15 to 25. The drive motor used is a 25 horse-power reversible variable speed

planer motor and runs from 250 to 100 r.p.m. The cross rail motor is a 2 horse-power constant speed motor running at 1,150 r.p.m.

ROTATABLE COAL HOPPER

Firemen on big locomotives find promise of lightened work in the rotatable coal hopper that has been invented for engine tenders. This invention is so planned that the coal is kept handy at the firing deck, making it an easy matter for the firemen to reach the coal without shoveling from the rear of the tender or using power apparatus. This special hopper is in the form of a great segmental tub, or drum, having a diameter that is the approximate width of the tender. This drum is inclined toward the firing deck; it is so mounted on a ball-bearing centre plate as to turn readily. In its outer wall are openings, one for each segment, through which the coal falls by gravity. As soon as the coal is emptied from the segment, the brake that controls the drum is released. Naturally the greater weight above the centre of the drum makes it rotate, bringing the next loaded segment into position.

GENERAL VIEW OF THE DOUBLE COLUMN MOTOR DRIVEN PLANER.

NEW ARBOR PRESS

The National Engineering Co., Ltd., Sarnia, Canada, have placed on the market a new arbor press, an illustration of which is shown.

Instead of using the old "goose-neck," a "three-point" construction with a direct strain is adopted. The machine, neatly

GENERAL VIEW OF ARBOR PRESS.

finished, is of light weight and occupies the small floor space of 20 inches.

Some of the features of superiority in this three-ton press are as follows: It has a 36-inch range, which makes it a very useful tool in garages for removing gears, etc. It also has an adjustable table with nine openings on the dial, and the ratchet wherein the strain is placed on all the teeth instead of one tooth. The machine has a capacity of 16-inch diameter. The A. R. Williams Machinery Co. are selling agents for the "NECO" line.

DRILL PRESS

The Sibley Machine Co., South Bend, Ind., U.S.A., have placed on the market a new stationary head drilling machine with positive geared feed, back gears, etc. It is modelled very much after the style of their sliding drill head, and the illustration accompanying this article gives one a good idea of its general construction.

One of the new and exclusive features of this machine is the fitting of all drive shaft bearings with die cast split

bushings of a high quality anti-friction metal. These bearings are interchangeable and insure long life and easy replacement. Drive pulleys and cones carry wide belts, while with the generous bevel and back gear ratios abundant power is delivered to the spindle. Long range of the head and spindle on column permits the use of large jigs. Back gears are of the sliding type, fully enclosed and operated by a single lever. Spindle is provided with ball thrust bearing.

Another feature is the new geared feed with four changes and a neutral position. Feed changes are obtained by moving a knob conveniently located in the centre of the hand wheel in connection with a sliding spline, four feeds being provided for each spindle speed. The initial drive of the feed shaft is from the top cone shaft through spur gears and two-to-one spirals, the latter bronze and steel respectively. This arrangement is very compact. All gears in the lower feed box run in an oil bath and are completely enclosed. The steel worm which meshes with the extra large worm gear also runs in an oil pocket. The adjustable automatic stop of new design is quick and positive in action. Hand feed and quick return of spindle is through the medium of a three-lever spider, at right of feed box.

Special attention has been given to the oiling of the machine; all bearings are equipped with oil cups, grooves and channels, insuring proper distribution to the vital points. The loose pulley is self oiling and will run several months without attention. All feed gears in the feed case are completely enclosed and running in oil. Particular care is given to the finish of this tool, spindle, sleeve and all shafts being ground. All machines are thoroughly tested before shipment.

GENERAL VIEW OF GEARED HEAD DRILL PRESS.

Positive geared tapping attachment, geared or belted motor drive, round or square table with tee slots and oil pocket, oil pump outfit complete and quarterturn countershaft are furnished as special equipment.

INSERTED TOOTH SAW

An inserted tooth metal saw is being placed on the market by the Simonds Manufacturing Company, of Fitchburg, Mass. This saw, which is designated as No. 000, was especially designed to meet the requirements for a saw to cut structural iron, I beams, channels and stock with thin walls that could not be cut as successfully with saws of somewhat similar design, because the pitch or spacing of the teeth was not fine enough to keep two teeth in the cut at all times in this work.

This new saw through its arrangement and closer spacing of teeth provides for smoother running and eliminates chatter which is often noted where the teeth are placed further apart in the plate. It is a fast cutting saw and is fitted with as many teeth as it is possible to securely fasten into the plate in the Simonds proven method of holding, and at the same time retain the perfect alignment of teeth and strength of plate necessary to stand up and cut true. These saws are being made from 10 inches in diameter with kerf of one-quarter inch and maximum of 40 teeth to 50 inches in diameter, 15-32 inch or 7-16 inch kerf with 210 teeth as the maximum number.

TABLE FOR FACE GRINDING

The Badger Tool Company, of Beloit, Wis., are placing on the market a face-grinding table to be used in connection with their regular line of single-spindle disk and cylinder-grinding machines. The overall dimensions of the table are 14 by 73 inches and the longitudinal travel is 32 inches. Hand-wheel feed provided for a 4-inch adjustment toward the grinding wheel; greater movement may be obtained by resetting the main saddle on the bed. All table and grinding wheel equipments are interchangeable.

BRITISH INNOVATION IN WEAVING

It is very rarely that anything really new in weaving machinery makes its appearance. Nevertheless a British firm has just introduced a new type of loom which involves many radical departures from the traditional type. From the practical standpoint its main feature is that it can be tended by a young girl, who can be taught the necessary operations in a few hours. The new loom has a wide range of products and already artificial silk, sponge cloth, and even blankets have been woven on it.

The MacLean Publishing Company
LIMITED
(ESTABLISHED 1887)

JOHN BAYNE MACLEAN, President. H. T. HUNTER, Vice-President
H. V. TYRRELL, General Manager.

PUBLISHERS OF

CANADIAN MACHINERY
MANUFACTURING NEWS

A weekly journal devoted to the machinery and manufacturing interests.

B. G. NEWTON, Manager.   A. R. KENNEDY, Managing Editor.

Associate Editors:

J. H. MOORE   T. H. FENNER   J. H. RODGERS (Montreal)
Office of Publication: 143-153 University Avenue, Toronto, Ontario.

VOL. XXIV.   TORONTO, OCTOBER 21, 1920   No. 17

The Cost of Strikes

THE Bureau of Labor Statistics at Washington has concluded that it is no use trying to compute reliable figures on the cost of strikes. All data pertaining to the matter, where an attempt has been made to reduce it to money, is of no value.

The bureau refers to the steel strike as showing the difficulties of gaining accurate information as to costs. Because of the difficulty of ascertaining the number of persons actually on strike and the different statements made by the two sides, it is held to be impossible to get figures of any value.

The innocent third party, the public, the bureau says, seldom suffers a money loss, its losses being mostly in the matter of inconvenience.

In Britain, right now, there is a concrete case of what it means to the third party. If the miners persist in their strike it will mean in a short time that hundreds of thous. ands of workers, who have no quarrel with their em. ployers, and who cannot afford to do so, will be thrown out of work by reason of the absence of coal to keep their places of employment in operation.

The Matter of Trades

ONE finds more attention given in the daily papers than even to the matter of trades and apprentices. In a recent issue of the Woodstock Sentinel-Review there was the following:—

"In the good old days there were plenty of barrels at 25 cents each. But in those good old days there were cooper shops in almost every town and village. Cooper shops are disappearing. Is anybody learning the cooper trade these days? The old-time shoemaker has practically disappeared. Where he survives at all it is as a cobbler. The old-time tailor is disappearing. Many of those who still keep up the struggle for existence earn more of their living from mending than from making. Tinsmiths and plumbers tell you it is almost impossible to get boys to learn these trades. Boys will not serve the apprenticeship necessary to learn a trade when they can go into a factory, and, in a few weeks, learn to run a machine sufficiently well to earn a man's wages."

The New Liskeard "Speaker" discussed the matter from another viewpoint, saying in part:—

"But the war has brought about a still greater change for the boy and young man, because he can get such good pay at almost any kind of work, will not 'lose the time' to learn the trade. We expect it would be next to impossible for any of the manufacturers to get apprentices to-day, unless such apprentices are paid next to journeymen's wages. These boys and young men only think of the present. So far as they are concerned, the future may look out for itself. If we are right in our view of this matter, soon there will be no master mechanics. So called 'handy men' will take their places. Who can foretell what changes will take place during the next ten years?"

There is a condition of affairs here that Canadian Machinery has, on many occasions, tried to place before its readers.

The years of discipline that were spent in learning a trade were not wasted. The small wages of those earlier days enforced thrift often to the point of actual want, but it left an appreciation of the worth of a dollar that is painfully absent in many a young man to-day.

It will be a sorry day for the future of this or any other country if we pass up the learning of a trade as waste time, forgetting the stalwarts in all lines who have come up through the ranks, and the foundations for whose greatness were laid in the stringent and memorable apprenticeship days.

THERE are an increasingly large number of people in the business world who are believing that Royal Meeker, commissioner of the United States Department of Labor, said something very true when he stated: "It will take a long time to deflate the world's inflated currencies or to inflate the world's deflated supply of goods." In another way, we are in for a period of high prices for some time yet. It is not until one gets down to facts that he realizes just where prices have hopped in the last few years, say since 1914. Unfortunately, we have not yet developed in this country the art of keeping statistics in such available form as they have them in United States, but conditions are much the same, as in many cases we get our supplies from that country. In the matter of lumber, the average of increases since 1914 shows: Yellow pine, 340 per cent.; Douglas fir, 527 per cent.; spruce, 165 per cent.; hemlock, 240 per cent.; and oak, 444 per cent. In the structural side of building the following materials show: Structural steel, 126 per cent.; steel bars (used for reinforcing), 135 per cent.; wire nails, 188 per cent.; steel rivets, 16c per cent.; and electrolytic copper, 41 per cent., now, but it has shown increases as high as 119 per cent. Cement shows 77 per cent., common brick, 350 per cent.; crushed stone, 95 per cent.; sand, 150 per cent.; and gravel, 122 per cent.

WHAT France intends to do with the money derived from the sale of goods which she purchased from the United States at the close of the war is a question which is now puzzling that country. On the termination of the war the French Government bought from the United States stock or "surplus war material" which cost the United States approximately $1,500,000,000 for $400,-000,000. This stock consisted of both moveable and stationary material ranging from equipment that had never been used to some almost worn-out, and included everything needed in the prosecution of modern warfare. It was supposed that France would use this material in her own reconstruction work, but several sales from this stock at home and abroad since the armistice, the most recent being to Harris Bros., a U.S. firm, have raised the question as to whether France intends to apply the proceeds from these on her loans from the United States amounting to $2,997,477,800. Those in touch with the situation believe that France may have applied the proceeds on her indebtedness resulting from her popular loan of 1917. It looks as though the Frenchman was highly efficient in financial matters to buy at bargain prices and sell in a higher market and receive payment in premium money.

The Secret of Henry Ford's Outstanding Success

Was forced to develop his factory to supply the demand he created.

THE editor of the New York "Sun" shatters accepted theories and thereby shocks his readers by a habit he has of doing his own thinking, of leading, not following the public, and making his own analysis of conditions. When the press and public of the United States were enthusiastically and thoughtlessly happy over the enormous business coming to them for steel and other material for a war-devastated world, that paper burst the bubble by showing that there was more steel in an ordinary New York skyscraper than would rebuild the whole damaged area in Belgium and France.

If we were to ask the why of Henry Ford's supersuccess, there would be a hundred per cent. reply that it was entirely due to his ability as a manufacturer.

The "Sun" editor, applying the analytical test, finds that the great, unthinking public is again wrong. Instead he shows that Mr. Ford is primarily a great salesman or merchant, that he was forced to develop his wonderfully efficient factory to supply the demand made by himself as a merchant. The "Sun" says:

Henry Ford Points the Way

Henry Ford is both a great manufacturer and a great merchant. He could never have become the great producer he is if he had not been a great merchant. It is because Ford is the outstanding genius of the automobile world as a merchant that he has become the outstanding genius of the automobile world as a producer.

The Ford automobile is primarily what it has come to be because Ford has greater genius than any of his competitors in the distribution of his car. With the price right and his selling organization right the production of his car became a matter of ways and means of keeping up with the demand. And as the output increased the cost of production decreased, and as this decrease in cost was deducted by the great merchant from the selling price the demand automatically increased.

In his early days Henry Ford was a mechanic at the bench. It was as a mechanic that he dreamed out the now world-famous Ford car. He had never been a merchant, had never fancied that he was endowed with the qualities of a merchant, though it was as a merchant that he was and is nothing short of a genius of the first order.

And now he has given us fresh evidence of his vision as a matchless distributor of the products of his mammoth workshops in the drastic slashing of prices to pre-war levels. In this stroke Henry Ford has pointed the way for all other producers, all other merchants. The country is headed for lower prices, pre-war prices or prices thereabout.

No man, no body of men, no association of interests working in collusion or co-operation, however gigantic, can hold up these unnatural and exorbitant war prices. What Henry Ford has done in his far-reaching vision other men must do or they cannot survive.

A review of the successes and failures in Canadian manufacturing confirms the "Sun." Walter H. Cottingham, probably our outstanding present-day success, developed and acquired the largest paint and allied factories in the world to fill the demands created by him as a merchant.

Senator Nichols was induced to take over and direct the destinies of the General Electric Co. and allied interests because as a merchant he had created his own market for electrical products in Canada.

Sir George Foster, our Minister of Trade and a member of the Imperial Trade Commission, expressed this view in 1914. He said the commission had found in Australia that German trade was growing steadily because the Germans were better merchants. Their superior salesmanship was overcoming the handicap of the superior quality of British goods and the strong national sentiment that extends throughout the Empire in favor of the British manufacturers.

Canadian manufacturers are in a less fortunate position than they of the Mother Country. The latter have well organised plants, trained workmen and established connections. The majority of men at the head of Canadian industries have worked up from apprentice boys. They have had to build new organizations and train workers. It has been a mighty strenuous career. Neither the thousands of employees nor the country as a whole appreciate what we owe these pioneering executives. But their thoughts have been on production and they love their factories and their employees. They have neglected their accounting and merchandising departments. The "Sun's" editorial suggests a new thought to many, and a very timely one. We must give more attention to selling our products at home and abroad. In this way only can most of us hope to keep our workers and plants running full time.

But salesmanship means more than taking orders. The buyer must be able to pay. The products must be what he wants, and what he orders; and our service must be prompt, reliable and satisfactory in every way. That it has not been is shown by the persistent complaints of our late Overseas Commissioner Lloyd Harris and many of our local Trade Commissioners. But it can be made satisfactory.

BROWNED ALUMINUM

To protect aluminum and aluminum alloys from corrosion, L. von Grotthuss has tried the experiment of browning the metal electrolytically. The aluminum is suspended in an electrolyte consisting of a sulphur compound of molybdenum and zinc is used for the anode. The cell is maintained at a temperature of 60 deg. to 65 deg. C. The aluminum is soon covered with a dark brown coating. The metal may be bent or rolled without cracking the coating. A piece of aluminum thus coated was immersed in a salt solution for two months without showing the slightest traces of corrosion.—Abstracted from Metal and Iron.

SUPPOSE HE DID!

MARKET DEVELOPMENTS

Which Way Will Steel Market Move Now?

Dealers Claim That It Has Become a Case of Seeing Which Party Can Wait the Longest—No Change Likely in the Prices of Small Tools in Canadian Market.

PITTSBURGH openly admits, that the steel market there has developed more than usual flatness, and that there is a falling off in consumption. While this may be so, yet it is not reflected to any great extent in Canada. In fact some of the merchants in the steel and iron business claim that the situation has become stronger during the last week. A few days ago, they claim, there was a feeling that the bottom might fall out of the market, but it has become fairly well established since then that there is not going to be any sharp reduction in prices.

Some of the material that has been turned loose by the cancellations of the automobile plants is being taken up by other concerns, and there are cases where considerable relief has been afforded by this action.

Dealers in small tools do not look for any change in the prices of their products, claiming that right now they

are selling high speed goods at from fifty to seventy-five per cent. below the schedule that was prevalent during the war. There have been reductions made in several lines of small tools during the last few months, and there seems to be a feeling that present prices will stick for some time.

It is too early to predict just what effect the British coal strike is going to have on British firms exporting steel and tools to this country, but some dealers hold that the effect will be felt in a very short time, as there is not much stock held in this country.

The scrap metal market is very dull just now. Dealers do not want to, take on any more tonnage than they can dispose of almost at once in the present market, so the most of the business is being done in small lots. Prices quoted are more or less nominal, and cannot be guaranteed. Some of the trade look for a lower price, and when this comes they believe trade will increase.

CANCELLATIONS CAN BE TAKEN UP TO MAKE SPOT DELIVERIES NOW

Special to CANADIAN MACHINERY.

MONTREAL, Que., Oct. 21.—There is still a tense atmosphere surrounding business transactions and buyers continue to show a conservative attitude when looking for material. Consumers are purchasing only for urgent requirements and the placing of orders for material and equipment is only done when the goods must be obtained to carry on immediately. While the situation in England is not expected to have a direct bearing on Canadian conditions, the tension contingent on a possible strike there has been felt in this district. Should this trouble be of long duration it would first affect the coal situation here, as efforts would probably be made to bunker trans-Atlantic vessels at Canadian ports.

Under the surface of apparent firmness the steel market is undoubtedly developing a nervous tendency which will ultimately be manifest in a slowing down of demand and a consequent effort to meet the situation by lower prices. Already some dealers have been asked regarding the possibility of cancelling a portion of orders placed for material, but little definite action has been taken to

advise the mills to this effect, as the bulk of orders now on the books are not subject to any cancellation. In some instances it may transpire that dealers will endeavor to dispose of some of this stock by redistribution to customers requiring spot supplies. The better service given by the railroads has facilitated the transportation of raw and finished materials, so that delayed deliveries of steel and other commodities is not so serious a factor in market conditions.

Outward indications of price reductions on machine tools are not pronounced, but the reluctance of the dealer to assume a definite attitude of optimism might imply that weakness may develop at any time. "The general demand for tools of all kinds is less in evidence and enquiry for next year's business is falling off considerably. This, we feel, is only temporary, and we are looking forward to a renewal of activity with the turn of the New Year." This statement by a dealer here gives the impression of a quiet period for the remainder of the year. The old material market continues in a lifeless mood and trading is very light, particularly in the non-ferrous

lines. Machinery cast retains the centre of interest in the volume of business done.

WAITING GAME IN STEEL TRADE

Can Dealers Hold Out Longer Than the Rollers in Price Adjustment?

TORONTO.—Although the week was broken by the holiday, business is by no means that of a small week in many lines. Machine tool dealers believe that the understanding is becoming general that price reductions are not going to materialize for some time. They also hold that price reductions would not have brought any more business into the market. The importation of British machine tools may be interfered with if the strike of the coal miners is not settled. Reports are that there is not much coal in any of the plants, and the steel makers are beginning to cut down already. The blast furnaces in the Middlesborough district are already damping down, and thousands of furnace men and steel workers are idle already. The Tees-side district provides a third of the whole British output of pig iron.

The steel market is just now 'at the

point where it is waiting to see what is going to take place. One of the merchants sized up the situation to-day as follows:

"We are through with the independent mills for the present and it means that we are in a waiting game, and it is a question of who can stand it the longer, the mills or us. I believe that in the past week there has been a settling of the market. Just a few days ago everything seemed more or less up in the air, and there was a feeling that prices were likely to be shot to pieces all at once— that the premium mills would at once hop down to the level of the corporation prices, and there might even be a sagging of these, but there has been a change in the last few days. The trade has come to realize that the bottom has not dropped out of the market, nor is it going to. There is still a great demand for many lines of steel, and the cancellations that are coming in are nearly all from one or two lines of industry, and their tonnage can quickly be taken up for spot deliveries if it is decided to allow them to cancel." It was hinted in some quarters that the dealers here would be willing to pay up to 6.50 base for black sheets from the independents for quick delivery. The corporation price for Canada is 5.55.

There is a better supply of cold rolled shafting than for some weeks past owing to cancellations by the automobile trade, which formerly had been a large buyer of this class of material. There are many shops all over the country, and many firms which make a business of furnishing complete equipment, that will be glad to take on any cold rolled they can find.

The boiler tube shipments are reported as better than for weeks past, but it will be some time, even with the very best of shipments, before the deliveries will begin to even up with the demand.

The Small Tool Market

There is a nice lot of business moving in this market, at least some of the dealers are quite well satisfied with what they are getting. They admit that there is no large buying going on now, but for all that there is a good turnover and the stock is moving out well.

"Do we expect any changes in small tools? No, we do not," was the opinion given by one of the dealers this week. "It was only a few months ago that reductions were made in high-speed drills and cutters. This was done to enable us to meet outside competition, but the reduction has remained, and I think you will find that right now the prices of high-speed drills, cutters, etc., is about from 50 to 75 per cent. below the prices that were being charged in war times. You can look a long time in the machine-tool field before you will find anything that has done that. We believe that

present prices are going to stay for some time unless something very much out of the ordinary takes place."

The scrap metal market shows little or no change this week. Prices that are quoted elsewhere in this paper may or may not be paid by the dealers, depending on the need they experience for purchasing. "In some lines we do not want to take on material at all," is the way one dealer stated the case to-day. "In others if we take anything at all we want only lots that we can handle in the present market. It looks to us very much as though the market is due to come to lower levels."

NOT SO DULL IN
YEARS AS NOW

That is What Report Says of the Tool Trade in New York This Week

NEW YORK, Oct. 21, 1920.—The machine-tool industry is marking time. Buyers show very little interest, and the efforts of machine-tool manufacturers and their representatives to "stir up" orders are generally unavailing. The market has not been quite so dull in years. As has been previously reported, the machine-tool industry is centering its hopes for future business largely on railroad purchases. But it must be admitted that the Eastern roads

as yet have shown no definite indication as to when their inquiries will come out.

With the exception of a reduction of 15 per cent. in one line of internal grinders, there has been no move on the part of machine-tool manufacturers generally to make price reductions. While reductions have been considered by some companies, their ultimate decision has been that lower prices would not at this time stimulate any substantial amount of buying.

The price guarantee is becoming more general, and this seems to meet the

view of those buyers who have actual need for tools immediately. Some guarantees run for as long as six months, which means that the buyer will be given a rebate equal to any reduction which might be made within that period. In most cases this is a one-sided agreement; that is the buyers decline to agree to pay an increase in price should there be any. Legally, of course, a contract is not a contract unless it works both ways, but sellers in some instances seem willing to make the one-sided arrangement if by so doing they can land an order.

STEEL MARKET HAS VERY FEW URGENT BUYERS AND IS WEAKER

Special to CANADIAN MACHINERY.

PITTSBURGH, October 21.—The iron and steel markets continue stagnant, as is naturally to be expected. When dull times have prevailed for some time there is necessarily some business being done, but in the early part of a generally dull period there is particularly little business because buyers are getting deliveries on previous purchases. Conditions to-day show that buyers have done what they usually do in an active period when material appears scarce, they have overbought. It looks as though consumption had greatly decreased, because the same buyers who a couple of months and more ago were importuning producers for heavier shipments are now acting as if they were receiving too much material. Consumption, however, has decreased very little. The main change is a change in the mental attitude of buyers. Such changes frequently occur almost overnight.

Pig Iron Slumping

Whatever may have been true of the advances in pig iron prior to the sharp advancing movement of last August, there was no difficulty in seeing last August that the advances made in that month were ill-timed. The volume of buying was very small, much less than is usually required to send prices upwards, and it looked very much as if the advances were based upon little more than a whim of the furnacemen to get pig iron up to the interesting objective of $50. The course of affairs now in the pig iron market reflects the artificial character of the latest of the advances, for pig iron instead of declining is slumping. A difficulty is that inquiry is so light that sometimes the prices at which producers would really be willing to sell are not disclosed. Thus Bessemer iron is still quotable nominally at $48.50, valley, the price to which it advanced in August, although it is certain that furnaces would sell at less. It requires a precise quotation, or an actual sale, to develop the price at which furnaces really would sell.

In the case of basic iron there are clear cut developments. Basic had also advanced in August to $48.50, valley, and

during September the market remained, nominally at least, at that figure. Since the beginning of this month it has been clear that some furnaces would be glad to sell at less. A couple of weeks ago a middle interest was understood to have bought 2,000 tons at $45, valley, then selling 1,000 tons to a consumer at $46. Since then there has been further weakening. It is thoroughly established that several sellers have quoted $42, valley, without effecting sales. In the case of

one consumer at least, the reason they did not sell was that the consumer got the iron he needed, 1,500 tons, from a middleman at $40, valley, and the middleman had offered 3,500 tons, so that it looks as though there is 2,000 tons ready for some one to buy at $40. According to common practice in quoting the pig iron market, transactions in which a middle interest figures are not regarded as setting market prices, hence the quotable market to-day on basic iron is $42, valley, a price at which iron has been offered, with no takers, representing a decline of $6.50. It was early last February, on the advancing market, that basic got up to $42, and thus the market has declined as much in about three weeks as it had advanced in six or seven months. Usually the market declines at about the same rate as that at which it had previously advanced, and the more rapid decline this time confirms views entertained previously by many in the trade that the advance was partly artificial, and not well based.

As to foundry iron, it is in the same class, marketwise, as Bessemer, not enough business having been done to show where values really stand. The trade calls the foundry iron market nominal at $47, valley, or $3 under the price attained in the August advance. Doubtless on an attractive inquiry furnaces would quote considerably less than $47.

Attitude of Steel Producers

In past times when the steel market became quiet the mills would hold together in the matter of prices, keeping up their quotations until they had delivered most of their contract tonnage. There was no advantage in reducing prices to effect a few fresh sales if thereby cancellation of existing contracts would be invited. At those times, however, the independents and the Steel Corporation were together. Now the independents have to play their game alone, and it is a question whether they will hold together. Some steel makers profess confidence that they will do so, and that production will be allowed to decline and decline again before mills reduce their quotations to the Steel Corporation level, which common opinion has it is the ultimate objective. Not a few observers expect prices to continue to weaken as they have been doing in the past few weeks, and if they continue the pace set thus far it will be only a few weeks, or two or three months at most, until the independents, with few exceptions are down to the Steel Corporation level. Some observers even profess doubt whether, even if the independents adhered to their present quotations, the demands on their contracts would be sufficient to enable them to keep above a 60 per cent. rate of operation in the winter.

Steel Prices

There has been particular weakening in strip steel. While the automobile industry does not consume any large part of the steel output as a whole, say 4 or 5 per cent., averaging the year as a whole,

---

POINTS IN WEEK'S MARKETING NOTES

Pittsburgh reports that the iron and steel markets continue stagnant, and the belief is expressed that consumption has greatly decreased.

Some observers claim that if the independents maintain their prices they will not have enough booking to operate at 60 per cent. capacity in the winter.

The top of the plate market in Pittsburgh is now 3.25, while as low as 3c has been done in consideration of an attractive tonnage.

Some of the independents have sold black sheets as low as 6.50.

Cold rolled shafting is more plentiful owing to cancellations that have been made by automobile concerns.

Dealers in small tools point out that they expect no reductions as they are selling high-speed now from 50 to 75 per cent. under war figures.

The scrap metal market is weakening toward lower prices. Dealers do not want to accept large deliveries at present prices, and only small lots are moving.

OUR FRIENDS

THE GREAT POPULARITY OF GEOMETRIC SCREW CUTTING TOOLS HAS BROUGHT THEM TO US

Thousands of Screw Machine users have come to know The Geometric Tool Company through the perfect work performed by the Geometric Automatic Adjustable Die Heads and Collapsing Taps.

Connecticut Section of the A.S.M.E., Guests at the Works of The Geometric Tool Company.

A demonstration of the efficiency of Geometric Tools was made. You can make a like demonstration right in your own shop. Let screw thread cutting be the most satisfactory operation there.

THE GEOMETRIC TOOL COMPANY
NEW HAVEN   CONNECTICUT

Canadian Agents:

Williams & Wilson. Ltd., Montreal.   The A. R. Williams Machinery Co., Ltd., Toronto, Winnipeg, St. John, N.B., Halifax, N.S.
Canadian Fairbanks-Morse Co., Ltd., Manitoba, Saskatchewan, Alberta.

If interested tear out this page and place with letters to be answered.

they do use a great deal of strip steel and thus their light operation at this time is reflected in the strip market. Last August, when independent prices were well established at 5.50c for hot rolled and 8.50c for cold rolled, there were predictions that the market would go up half a cent, but instead of doing so it has been declining, and it is probable that one could buy now at about 5c for hot rolled and 8c for cold rolled. Cold finished steel bars are easier, there being

little difficulty in buying at 4.25c, while a few weeks ago sales at 4.25c were rare. In plates 3.25c is now the top of the market instead of the bottom, single carloads being available at that figure, while attractive inquiries bring out quotations down to 3.10c, and 3c has been done in at least one case. In structural shapes there is conflicting testimony. The independent price had been 3.10c, against the Steel Corporation's 2.45c. Some interests claim 3.10c is being adhered to, while

others report transactions at 3c and under, down possibly to 2.70c.

The sheet market is easier. Independent quotations generally are 6.75c to 7.25c on common black, but it is said that one mill has sold at 6.50c. It is known positively that a mill "married" contracts by selling to a customer at 6.50c when the customer wanted to cancel a contract placed some time ago at 7.50c. Galvanized 'run from 8.25c to 9.75c and blue annealed from 5c to 5.25c.

The Week's Events in Montreal Industry

Fraser S. Keith, general secretary of the Engineering Institute of Canada, officiated last week at the organization of a new branch of the Institute at Moncton, N.B. Officers of the branch are W. A. Duff, chairman; J. D. McBeath, vice-chairman, and M. J. Murphy, secretary treasurer.

Further airing of the Ross rifle controversy is promised, following a recent address by Canon Scott of Quebec, senior chaplain of the First Canadian Division during the war, before a large gathering at Calgary last week. He is asking for an investigation and a definite finding so as to enlighten the people of the country regarding the facts of the entire case.

A portion of the plant of the Davie Shipbuilding Company at Levis has been acquired by the lumber firm of Knox Bros. of Montreal. The buildings will be equipped with machinery for resawing and dressing lumber, and it is expected that the finished output will approximate twenty carloads of lumber per day. Yard storage capacity will be about 15,000,000 feet of lumber, and the drying shed will accommodate one million feet. It is stated that operations will be commenced at once.

At the October meeting of the Canadian Railways Club the subject for discussion was a paper on "Patents of Invention," by William McFeat, Patent Attorney of Montreal. In his address Mr. McFeat gave a brief but clear description of the early history of patent laws, the importance of invention to community and country, the fundamental principles of obtaining and maintaining Canadian patents, and likewise how to realise on patents granted. Thirty new members were added to the roll at this meeting.

Following the recent ruling of the board of railway commissioners, the Province of Quebec has taken steps to regulate the supply and distribution of fuel by appointing J. W. Lewis as fuel administrator. Mr. Lewis held this position when the situation necessitated control last year. While coal distribu-

tion will be left almost entirely in the hands of the trade, the office has been recreated to keep a rigid check on tonnage movement and to be prepared to deal with complaints should the occasion arise.

In order to provide themselves with adequate supplies of silicate bricks for the maintenance of their blast furnaces, the Dominion Iron and Steel Company are taking steps to develop a large tract of land rich in silica, which the company acquired at Neils Mountain, near North Bay. The road bed for the necessary hauling of the material from the property to the adjoining railroad is being prepared by slag brought from Sydney, N.S. The silicate brick plant is now being constructed at Sydney and manufacturing is expected to commence in the near future, as soon as sufficient material has been shipped from the clay beds up north.

The business firm of Forrest, Jackson and Forrest has recently been organised and will henceforth carry on business under the name of the Sheffield Engineering Supplies, Ltd. In addition to the Dormer high-speed twist drills now carried in stock, the firm will carry a complete line of Dormer high-speed steel milling cutters and reamers. These lines are all made by the Sheffield Twist Drill & Steel Company. Other lines carried by the firm will be high grade hack saw blades and files. Capt. H. H. Pinch has been added to the organization and will have the position of vice-president. E. Jackson will retain the position of president, and E. B. Forrest will be secretary-treasurer.

Emile Belanger, 26 years of age, an employee of the Consolidated Rubber Company of Montreal, was, last week, awarded $17,500 for the loss of two arms, which were amputated following an accident in April last year. It was proven that the machine was inadequately protected with safety appliances, that these appliances had been recommended by workmen, and that delay in installing the necessary protection was the cause of the accident. "In these circumstances," Justice Surveyor said, "there is reason to declare that there was in-

excusable fault on the part of the company defendant, and therefore it ought to be condemned in damages as though the accident in question were governed by the common law." Plaintiff sued for $25,000 damages under the Workmen's Compensation Act.

On the afternoon of October 14 an interesting event took place at the C.P.R. Angus shops at Montreal, it being the annual presentation of prizes to apprentices for the work accomplished during the past year. The prize winners in order of merit were: W. Morrison, machine apprentice; A. E. Watson, machine apprentice; C. Girdwood, tinsmith apprentice; T. Brown, machine apprentice; R. Thomas, machine apprentice; L. Kinlough, carpenter apprentice, and E. Hardy, carpenter apprentice. The prizes were in the form of several volumes of reference books particularly selected for the various branches of the work in which the different boys are engaged. In addition to awards for local work, prizes won at the Canadian National Exhibition, by apprentices A. Sanfacon and L. Leboue, for architectural drawing, and a special prize to T. Brown for mechanical drawing, were also presented. The presentation was made by Mr. John Burns, works manager, and several works officials gave short addresses on their past experiences and the progress that has been made during the past fifteen or twenty years.

FIRMS WANT TO ESTABLISH TRADE RELATIONS WITH CANADA

Information regarding the following lines can be obtained by communicating with this paper and mentioning key numbers.

2067. Wire.—An established firm of manufacturers' representatives in Barbados wish to secure a Canadian agency for barbed and mesh wire.

2073. Agate Bearings.—A firm at Utrecht, Holland, manufacturing weighing machines would like to obtain the addresses of Canadian manufacturers of agate bearings.

2076. Wire Nails.—A firm of manufacturers' representatives in Barbados wish to secure a Canadian agency for wire nails.

If interested tear out this page and place with letters to be answered.

Sheffield Makes, Thinks and Talks Steel

Mr. Swallow, of Ingersoll, Has Had a Chance to See the Development of Changing Industrial Life in That Centre—Men There Are Experts in All Matters Pertaining to Steel

MR. H. SWALLOW, manager of the Ingersoll File Co., returned a week or so ago from a brief trip to Sheffield. In reply to a query from Canadian Machinery regarding conditions in the steel world of Sheffield—for it's all steel there—Mr. Swallow writes as follows:

To understand the conditions in Sheffield to-day it is necessary to step back a generation or so.

Originally it was a city of small units, each unit self-contained and making articles from raw material to finished product. Several of these units would occupy one building whence grew the factory which might contain up to a hundred such units each doing all the operations on a similar article. Eventually for financial and other considerations they grouped together under one executive and one firm name, but maintained their working identity. An order for five dozen tools would be received and would be handed to one of the groups where it would be completed. When orders grew in magnitude overlapping was inevitable. The system evolved craftsmen of wonderful ability but it also evolved weird and wasteful methods of handling and routing materials.

The war taught what could be done in mass production. The urgency of patriotism was behind that. Then came the armistice, and found a world utterly unstocked of all the products Sheffield loves to make—and can. But operating costs kept rising, the demands of labor increased, until the English steel man found himself in the position of his Yankee competitor years ago who was faced with the greater individual production, and the lower labor costs of Sheffield.

The demand, therefore, was for speed of production, smoothness of operation, and modern factory methods. Sheffield gasped at the prospect, grunted its disapproval of the departure from grandad's ways of doing business, but went to work to solve these problems. The modern buildings erected for shell production were utilised, expert factory organizers were called in, deals were made with the unions, and everybody rode on a wave of prosperity such as they never dreamed of. Labor kicked, and is still kicking and therein is the fly in the ointment. Production has slackened in the past few months because of the unions' resentment of labor-saving innovations.

If Sheffield can solve that problem she will again be the steel centre of the world. She has the right materials, the best methods, and the finest steel craftsmen on earth, steel men with generations of steel workers behind them who think, talk and live little else but steel and steel products. Any Canadian tool-maker who could command the same class of help would cut a swath a mile wide through the American continent.

The failure of the much-heralded coal strike in England was due to two causes: The union was not financially ready for it, and the public was, as a whole, hostile. No strike can hope to succeed in the face of popular disapproval, however "red" its leaders may be. An armistice has been called, but it may be broken at any moment. The next big strike in England will be a show down. It may culminate in a lockout by the employers who have already adopted that plan with the electricians. After a settlement (not another compromise) look out for Sheffield.

The executives who run the industries there are young men, educated by practical steel workers, and schooled in the amazing production of war years, live wires like Harry Brearley, the inventor of stainless steel, who are always a lap ahead in the production of special steels. The writer met one man much under 40, who is managing director of five large enterprises, and manages them profitably and well.

Recently, gigantic mergers of some of the oldest and wealthiest firms in Sheffield have taken place. This will further consolidate the varied interests, and has undoubtedly been caused by the understanding arrived at by the French and American steel makers who have eyes on the fat market of the European continent.

It is to be seen that Sheffield is gradually in her thorough English manner assimilating all that she can profitably use of new ideas and methods. She has always traded in the far corners of the earth, and when she gets her house in order, other nations will have to set their pace by hers.

Already her cutlery and small tools are appearing in quantity in this and other countries. Her steel makers are catching up and making fairly quick delivery. She is backed by immense wealth and can wait.

CLAIMS OF CANADIAN SHOPS FOR U.S. SHELL WORK ARE SETTLED

THE work of settling claims that Canadian munitions plants had against the U.S. Ordnance Department has been completed, and the last of these has been paid. The matter of disposal of the property of the Motor Trucks at Brantford is still in court, but a decision is expected very shortly.

One of the officials of the U.S. Ordnance Department, speaking to Canadian Machinery of the disposal of the property and the work connected with the making of the settlements, stated: "I do not think any of the men who have been on the settlement board would care to go through the experience again. In the rush of getting results when they were very badly needed there was not the time nor the machinery necessary to get a complete record of everything, and some of the munition makers themselves had to carry a good many of their accounts and memos in their pocket. When it came to making a settlement of course we looked for documents, for vouchers, for written orders, etc., and in a good many cases they were not in existence.

"When we came to dispose of the tools that were the property of the Ordnance Department, we met the machine tool builders of United States, and agreed then to do everything in our power to keep from upsetting the market for new tools, which would have resulted had we gone ahead and dumped our stock on the market. Single purpose machines were scrapped, and we had to see that they were sold as scrap. We found there were parties in both countries who planned to secure the machines at scrap prices and rebuild them into some sort of mongrel lathes and put them out at a fair price as a cheap machine. Our stand was that they had no business to do this, and that if they bought a machine as scrap and at scrap prices, it was up to us to see that they got scrap and nothing else. We did not believe that they had any more right to profiteer in this matter than we would have, and so, when a machine, generally a single purpose affair, was set aside to go at scrap prices, we generally undertook to see that it was well struck with a drop hammer before it had a chance to leave our hands."

The U.S. Ordnance Department is undertaking the publication of a magazine at Washington, the material for which comes from men who are or were engaged in the Ordnance work during the war. The object is to preserve and place on record as far as possible material in connection with ordnance activities, and also to review the work that was carried on in the shell shops of Canada where United States contracts were placed. A number of articles that appeared in Canadian Machinery dealing with various operations in the shell shops will be drawn on for the furnished information for the Washington magazine.

Quality

CONFIDENCE

THE GREATNESS OF BRITAIN IS FOUNDED ON HER INTEGRITY.

HER STRENGTH SHE DERIVES FROM WORLD COMMERCE—HER ENDURANCE FROM CONFIDENCE.

THE WAR TEMPORARILY RETARDED HER CONSTRUCTIVE EFFORTS, BUT THE POWER OF THE CONFIDENCE OF THE PEOPLES OF THE WORLD WILL RESTORE TO HER IN MORE BRILLIANT FORM THE LEADERSHIP SHE HAS JUSTLY EARNED AND PROUDLY HELD.

CONFIDENCE IS THE FOUNDATION OF OUR POSITION IN THE WORLD OF IRON AND STEEL—OUR PROGRESS DEPENDS ON IT.

WE HAVE BEEN SUCCESSFUL IN THE PAST—THE FUTURE IS BEFORE US. TO ADVANCE, WE MUST CONTINUE TO MERIT THE CONFIDENCE OF THE BUYERS OF STEEL AND IRON PRODUCTS OF EVERY DESCRIPTION IN CANADA, AND WE ARE DETERMINED TO ADVANCE.

THE
STEEL COMPANY
OF
CANADA
LIMITED
HAMILTON MONTREAL

Service

If interested tear out this page and place with letters to be answered.

Quick Drop in Steel Prices Not Desirable

Mr. F. H. Whitton, of the Steel Co. of Canada, Traces Conditions That Have Been Met in the Market Since the Signing of the Armistice—Costs Are Still High

SPEAKING before the Canadian Wholesale Hardware Association on the iron and steel industry, Mr. F. H. Whitton, of the Steel Company of Canada, made it clear that his company was not looking for a sharp decline in prices of steel or iron. He believed that such a course would be dangerous, and not in the best interests of our industrial life.

"In January of 1919," said Mr. Whitton, "we believed that the period of readjustment had arrived. We made reductions of $4 to $6 a ton, business was depressed, some wanted to cancel orders, and there was a great curtailment of business up to June, 1919. But that was a false sign because the demands of peace time requirements immediately caused a rebound in business and you all know the pressure which has taken place from that time to the present. Up until that time labor was very scarce; it was an era of wage upon wage increase, decreased production, general dislocation of business and restlessness among the men. The public were extravagant and there was a great indulgence in luxuries. Speculation has increased during the past eighteen months. But at last a halt has been called. Our war debts have to be paid, our Government has called upon us to be economical and is taking our money to pay interest and principal. On behalf of itself the Government has had to curtail expenditures. Banks have wisely curtailed credit. The object of our banks is to maintain the interests of the country and their action indicates that the time has arrived when the first steps should be taken to adjust business towards normal conditions. There is nothing to be feared if we show a few jolts but the lessening of industrial activity will not be sufficiently severe to be alarming and those who fear a depression will likely be wrong.

"There is much work in arrears, much construction work to be done. A large demand is assured to supply our own people and the whole world. A downward trend in prices is taking place and the market will find its normal level, but what is its normal level? No one knows. When the public take all the goods produced at high prices it is useless to argue for lower prices. Speaking of hardware, we must not allow our products to be classified in our minds or in the minds of our customers, in the same list as the more speculative commodities such as coal, sugar, cotton, silks, etc.

"Regarding general conditions there is no question about them whatever. In regard to the crops this year, we have a most bountiful harvest and the very immensity of it will probably bring to normal levels the price of food stuffs. Those lines which require employing the least labor will come down first and to the greatest extent. Iron and steel premium prices will probably disappear, but prices will not come down as rapidly as other lines and the new normal level of steel will probably be a surprise to most of us.

"Iron and steel in its many finished forms passes through so many processes and probably employs a larger percentage than any other general commodity. Going back to the fundamental articles, we take ore. Ore in the ground is worth 10 to 20 cents a ton; coal in the ground is worth 20 to 30 cents a ton. These materials amount, with other alloys, to 15 per cent, while 85 per cent of the value represents labor costs. You know the position of labor to-day. We may consider wages high but there is no rush downwards. . There will be a recession in labor but only as conditions warrant. So we see that articles with the least labor will be affected the least.

Comparison of Prices

"I have compared prices of materials of a year ago with the present. These figures include 10 per cent. United States exchange in Canadian currency and exchange is figured on just that portion bought from the United States. The prices are for the third quarter compared with 1919 and are as follows:

Ore is 31 per cent. higher.
Coal 109 per cent. higher.
Coke 118 per cent. higher.
Fluxing stone 22 per cent. higher.
Alloys 49 per cent. higher.
Scrap 45 per cent. to 80 per cent. according to grade.
Fire brick 52 per cent. higher.
Ingot molds 57 per cent. higher.
Repair parts (iron) 42 per cent. higher.
Repair parts (steel) 20 per cent. higher.
Oils 22 per cent. higher.
Engine oil 90 per cent. higher.
Black oils 145 per cent. higher.

"Wages are the most important item," said Mr. Whitton, stating that all those present knew the increase in this line. He, however, advised that no undue pressure be brought to bear to bring down wages too suddenly.

Mr. Whitton pointed out that the above figures have greater significance when it is borne in mind that there are four or five tons of material in one ton of steel. "There is a great shortage of iron and steel," said Mr. Whitton, "it being essential in so many different requirements such as machinery and production of other goods, in farming, mining, etc.

The demand in Europe is also large and their production has been decreased."

Discrepancy in Steel Prices Causes Misapprehension

"Speaking of prices, there is a very great variation in the price of iron and steel between the largest corporation in the United States and the other makers and no one attempts to explain. I have no doubt that there are good and sufficient reasons for that position. Everyone concerned thinks that in the iron and steel business. Of course we know they have advantages over any other iron and steel producer, when you take into account their costs, properties, ownership of railroads and boats, their large production. But there is some motive other than this because their exports are at much higher prices than their domestic prices. There is that discrepancy and this is apt to cause misapprehension as to values in the minds of buyers.

"In the New York Times of October 5th Judge Gary, speaking as to the effect of the advance in freight rates on prices, said: 'I hope it will not cause the Corporation to advance prices; however, I am not certain as yet what the effect will be and prices may be affected to some extent. Rails should be slightly advanced.'

"This is interpreted to mean that the Corporation is at least more likely to advance their prices than to recede," said Mr. Whitton. Continuing: "Take the case of the independents. Their costs of manufacture are very high and the difference in prices between the corporation and the independents does not mean that this is a margin of profit which the independents enjoy over the corporation. We must not expect too rapid or severe cuts in steel prices. There are only a few people who can wave the magic wand and reduce prices, but too rapid reductions would cause unemployment and the lessening of purchasing power. Only steady production will bring values down.

"Labor conditions are improving in efficiency. Men respect their jobs more than they did, they are willing to give you a little more labor in return for the prices that you are paying them. We have probably not reached the pinnacle by long odds of the efficiency of labor. Wages are at the top notch and the men and the unions will fight reductions. On the other hand is the tendency to-day for the open shop. That is the only basis on which we should work.

Immigration

"Immigration is large and increasing. A great number of foreigners went back home after the armistice and when they

get money again they will be back. There are many who do not want to come back and there are a number who will not be allowed to come back. But this immigration which is coming in at the present time is going to bring us a supply of labor and more efficient labor than we have had for some time. We have been very short of common labor. British subjects do not like common laboring any more. At the same time you must recognize that the unions will make strenuous efforts to restrict immigration as much as possible through politics and other ways.

Must Keep More Trade at Home

"Another thing we can do in helping on the downward trend. Canadians have got to buy more Canadian goods and curtail purchases in other countries. Figures for August show that the United States bought from us $47,000,000 worth of goods, an increase of $3,000,000 over the previous year. In the same month we bought from the United States goods to the value of $85,000,000 which was $22,-000,000 increase. We had an adverse trade balance in that month of $38,000,-000.

"We must look after ourselves first as they do and which we must respect them for. This will apply to all of us in the coming year. One point you are apt to forget is that when we pay over that $38,000,000 we have to send along another $3,800,000 to pay the exchange for one month on the increase alone without the exchange which had to be paid on the total value of goods purchased. In 1920 so far we have bought from the United States goods to the value of $445,000,000 as against $385,000,000 last year, or an increase of $110,000,000 up to the end of August.

"Steel is scarce in the markets of the world. Railroads use one-third of the production of the mills and during the past five years purchases by the railways have been much more than normal. The deficit has to be made good. Construction needs of the present and future are very large. This is not a time for buyers to lose confidence. We knew how to go up, we knew how to meet our difficulties. We know how to come down and we have the ability to do it safely. All we have to do is keep our heads level, to instil in our men the idea that we have all still got to work, that all this theory that living is possible without work is wrong, and if we so order ourselves there is nothing in the way of disaster in any shape or form ahead of us."

H. W. Hutchinson, of Winnipeg, who has been elected a director of the Cockshutt Plow Co., Brantford, Ont., is well known in Western business circles but has also been associated with a number of Eastern Canadian industrial and financial enterprises. He is vice-president of the Sawyer-Massey Ltd., Hamilton, a director of the Dominion Bank, and president of the Consolidated Mortgage Co., Winnipeg Trust & Mortgage Co., Mortgage Investors Co., and other companies.

UNSETTLED IN OLD COUNTRY DISTRICTS

Too Many Strikes and Rumors of Trouble to Permit of Rapid Recovery

The Toronto office of the Department of Overseas Trade, speaking of machinery and engineering in the home country, reports:

The demand on home account has, for most descriptions, continued quiet, and manufacturers are turning their attention increasingly to export matters. Some sections are, it is true, well employed, notably in the case of manufacturers of textile machinery, who are stated to have sufficient orders on hand to keep them busy for some years to come; but, as a whole, these trades are suffering from something approaching a slump in the volume of orders coming forward.

The labor situation does not tend to improve matters, for, with the trouble in the electrical engineering trades still unsettled, and a lock-out having been declared, with its consequent adverse effect upon practically all branches of the engineering trades, manufacturers are acting with considerable caution.

On the other hand, the threatened strike has been averted, the claim for an additional 6d. per hour having been conceded; but the general labor outlook, especially in view of the latest developments in the coal situation, cannot be said to be favorable to improved conditions in these industries.

Signs of Progress

Oakville, Ont., is to have a new fire hall, to be built at a cost of $15,000.

Architect Carrothers, of the Board of Education, London, Ont., is preparing plans for a new Central High School.

Announcement has been made that $10,000 is to be spent in repairing and renovating the Prince George Hotel, Toronto.

Building operations are under way on the new factory to be built by the Canada Ingot Iron Co., Guelph, Ont. The cost is estimated at $16,000.

A new high school is being planned by the High School Board at Burlington, Ont. It is expected to be ready by the new year at a cost of $60,000.

The C.P.R. is having a survey made for a branch line from Harriston to Listowel, which will likely run through Palmerston with a passenger depot at Palmerston.

The Imperial Oil Co. are planning to have all the buildings on their property at Brandon, Man., removed this fall in order that an early start may be made in the spring on their $300,000 plant.

It has been announced by Maj. C. McLaurin, superintendent of the Government hydroplane station at Jericho, that an aerial substation will shortly be built at Kamloops, B.C.

A new Chinese Y.M.C.A. is to be erected in Toronto at a cost of $100,000, under the leadership of Edward Gung, a Chinese

medical student at the University. The intention is to tear down the present quarters on Chestnut and build on that site in the spring.

Manitoba Medical College is to have a two-storey brick addition at a cost of $172,000. The contract has been awarded to the Sutherland Construction Co.

Engineering

A new Collegiate Institute costing $300,000 is planned by the Board of Education at St. Catharines.

Thos. Kennear & Co., 49 Front Street, E., Toronto, are planning the erection of a warehouse at Galt, Ont.

A report is in circulation that the Canada Steamship Line intend building a new terminal at Hamilton.

The construction of a telephone system from Hamilton to London to operate trains is planned by the G.T.R.

The city of Kingston is planning to build a modern 110 room hotel at a cost of $450,000 if the money can be raised by subscription.

Horwood & White, architects, Toronto Street, Toronto, Ont., are preparing final plans for dept. store estimated to cost $1,000,000 for Hudson Bay Co.

The erection of a library estimated to cost $25,000 at Lethbridge, Alta., is contemplated by Carnegie Library Corp., James Bertram, Secretary, New York.

The property at 157 Sherbourne St., Toronto, has been purchased by Burgers, Ltd., and will be converted into a candy factory.

The contract for the construction of new stables costing $45,000 for the Royal Canadian Mounted Police at Point Grey, B.C., has been awarded to R. Moncrieff.

The Mackinnon Steel Co., Ltd., Sherbrooke, Quebec, has received an order for steel for the coal trestle and steam pipe bridge from the Brompton Pulp & Paper Co., Ltd., East Angus, Quebec.

Lieut.-Col. A. G. Peuchen is erecting a four story building costing $150,000 at 191 King W., Toronto, which will be leased to the Ford Co., at a rental of $17,000 for their head office and salesroom.

A wire mill is to be added to the present plant of the Steel Company of Canada, Hamilton, at a cost of $250,-000. It will be the largest of its kind in Canada, and will provide employment for 300 men when completed.

When the engineer, Mr. Achison, revises the plans for the new Deaf and Dumb Institute at Portage La Prairie, new tenders will be called as the tenders offered previously were for a minimum of $1,300,000, while the committee have but $1,000,000 to spend.

The Dominion Envelope and Carton Co., Toronto, have completed a transaction by which they take possession of seven acres of the commission's property at the Eastern Gap, Toronto. Work will commence in the spring upon a building 300x500 feet and one story high.

"Through for the Day"   *A bunch of happy workers who have been putting thought and quality into Morrow Drills for YOU.*

Specify
Morrow Drills

Try Your Jobber First

IT'S worth your while to see that you get the make of drill that gives the best service. The drill of long life —say *Morrow's* and mean it.

The John Morrow Screw & Nut Co., Limited
INGERSOLL, CANADA

Montreal	Winnipeg	Vancouver
131 St. Paul St.	Confederation Life Building	1290 Homer St.

7 Hop Exchange, Southwark St., London, England

If interested tear out this page and place with letters to be answered.

TRADE GOSSIP

More Coal Coming.—It was stated by an authority this week that coal would probably be more plentiful next month with the closing of navigation on the upper lakes. At present the mines are sending the largest portion of their output to Western Canada while they can take advantage of cheaper freight rates by water. With the closing of navigation in the near future, Toronto and central Ontario district will come in for its share of attention.

Germany Gets Order.—Canada has lost the order for 2,400 steel railway cars which the Belgian Government, through the Canadian Trade Commission, sought to place with manufacturers in this country. Instead of "made in Canada" cars, the Belgians will ride in coaches of the "made in Germany" variety when the order is filled. Advices to the Trade Commission are to the effect that the Belgians have placed the order with Germans and will pay for the railway material with foodstuffs which are to be supplied to Germany.

Plate Mill Record.—The new plate record of the Dominion Iron and Steel Company, at Sydney, broke all previous records for production when 260 tons of plates were rolled in one day. The regular daily production of the mill is about 200 tons a day. About two months ago the 240-ton mark was reached. Last week the mill ran without a hitch of any kind, resulting in the establishment of the new 260-ton record. The greater part of the plates which are being run at the present time are being shipped to Halifax, Montreal and other Canadian cities. As yet no further foreign orders have been received.

Locating in Toronto.—West Toronto's building operations include another large factory, which will employ a large staff. Grinnell and Company, of Canada, situated near Bloor street on Dundas street, having decided to manufacture in Canada are putting up large machine shops and foundry adjoining their present office building. The work is now well in hand, and the buildings, which will stand upon about two and a half acres of land, are to be of reinforced concrete, and will extend from Dundas street right back to the railway tracks. The machine shops will be 180 by 52 feet, and the foundation 210 by 100.

Rather Buy Than Build.—Although no statement is being made in advance of the declaration of the successful bids, the Canadian Pacific Ocean Services have put in tenders for some of the German passenger vessels which are being disposed of by the British Government. It is probable that at least one first-class and one second-class boat may be secured. The company prefers to buy rather than to build, because of the high cost of construction to-day, which Sir

Thomas Fisher recently commented upon very strongly. The Empress of Canada recently launched on the Clyde, cost, roughly, £1,700,000, as compared with £550,000 for which it could have been constructed before the war.

Who Did This?—Trailers purchased by the Ontario Power Commission for service on Windsor Street Railway System cannot be used until extensive alterations have been made. According to an employe of the Commission, it has been found impossible to keep the cars on the rails, because the wheels are one inch narrower than those in use. It will cost several thousand dollars to make the necessary changes.

Ask Injunction.—Following the sale of the Cramp Steel Company's plant in Collingwood to the Baldwin's Canadian Steel Corporation Limited, the municipality made a seizure of the plant when it was found that it was the intention of the purchasers to move the equipment to Toronto. The reason for the seizure was the alleged arrears of taxes, which, with the compound interest to date, reaches almost $200,000. The original owners claim that there was an agreement with the municipality whereby the arrears in taxes would not be collected, but this was cancelled following the sale. The Baldwin Corporation claims that before the sale was completed they secured a statement from the town treasurer that there were no arrears of taxes, and also obtained a receipt for the 1919 taxes. Following the seizure an injunction was asked by the company restraining the town officials from seizing the plant, but no answer has been filed in defence.

A B C OF WELDING

(Continued from page 61)

valves which prevent change in the preheating flames when cutting oxygen valve is thrown open. The uses for a good cutting torch in the modern repair and welding shop are so varied that they need not be mentioned. With a little practice it is possible to cut out shapes almost as smoothly as can be done with any machine tool, bevel steel parts for welding or cut out a rivet in hurry up time.

The accompanying illustration shows the author amusing himself cutting off the end of a heavy steel ingot preparatory to its being forged into a crank shaft for one of the steamers which helped us win the recent war.

In the plants of the large steel companies many small shapes are cut with the torch from plate purchased as scrap in old boilers; steel shafting is cut to convenient lengths and salvaged. Many things are done to prevent the final loss of the dollars which the small town dealers in such articles let slip through their fingers.

TURRET LATHE PRACTICE
(Continued from page 373)

mounted in the same turret.' The hole A is next finish bored by the cutter held in the piloted boring bar mounted in the multiple turning head at position 3 of the turret.

The chamfer D is produced by tool I, this tool being mounted in the same fixture. The surface B is finish faced by tool J, which is held in the square turret, and surface C is finish faced 2½ tool K, which is also held in the same turret. The hole A is now reamed by the expansion reamer in floating holder in position 5 of the turret. After this the work is removed from the chuck.

Second Operation

The second operation is completed in 4.4 minutes, and the tooling is shown at 11, and the various steps are as follows: The work is placed on an automatic drawback adapter that expands to suit hole A, which was finished in previous step. The surface E is rough faced by tool L, which is mounted on the rear tool holder. The surface F is rough taper turned by tool M, which is mounted in the square turret, this tool being used in conjunction with the taper attachment.

The surface F is now finish taper turned by a special tool O, which is mounted in the multiple turning head at position 1 of the turret. The turning head is provided with a pilot that enters a hole in the spindle, thus ensuring that surface F is machined concentric with hole A. The work is next removed from the machine.

These various layouts are of particular value, insomuch that in themselves they teach a valuable lesson in careful layout. To machinists and tool designers in general, we can recommend a study of these layouts, as it will be time well spent. In a later article we will describe other tooling equipments as used on turret lathes for machining automobile differential- gear cases, cast steel chuck bodies, bar stock parts and other interesting portions.

A B C OF WELDING
(Continued from page 379)

a soft brass wire. Tips for torches of the more modern stabilizing types may be cleaned by means of a steel broach inserted from the rear end.

Nothing equals pure drawn red copper for the construction of welding tips, and makeshifts for this metal should not be tolerated. There is a tendency on the part of some manufacturers to substitute "special alloy," "nickel copper," or other metals which cost less to drill and manufacture.

A popping sound emitted from the tip of a cutting torch when the preheating gases are lighted indicates a leak caused by defective seating.

Cutting Torches

Torches for rapid and efficient cutting must be built with specially constructed (Continued on page 60)

Classified Opportunities

CANADIAN MACHINERY
AND
MANUFACTURING NEWS

Vol. XXIV. No. 18   October 28, 1920

Some Interesting Examples of Bulldozer Work

The Uses of a Bulldozer Are Explained, and Four Typical Examples of Work Performed Are Shown—Mechanics Should Become More Familiar with the Possibilities of Such Machines.

By J. H. MOORE

BEFORE going into a discussion of the uses of a bulldozer, and examples of work performed on the same, we will imagine that there are readers who have never watched such a machine in operation. This may appear, to those who are familiar with a bulldozer, like explaining the A, B, C's of the alphabet, but one must remember that many a good mechanic has never entered a forging department, therefore has had no chance to watch the bulldozer perform. It is for the benefit of such readers that we append the following. Study Fig. 1, not with an idea of looking at the die, but at the machine itself. This view illustrates the main essential working parts of a Williams and White bulldozer. To be brief, this style of machine is used for all types of bending of both hot and cold metals. It consists of a heavy cast iron, or semi-steel bed, with a block or bolster at one end. This block is stationary. The other end has a moving head which slides back

and forth on the bed. This head is clearly shown on the photograph. In operation, heavy dies are clamped against the stationary bolster, and on the moving head. These dies are so made that when the moving head comes up to its extreme stroke the form left between the two dies is exactly the desired shape.

The cold bending of course is limited to partly light stock, while very heavy stock is first heated by furnaces in close proximity to the machine. The material is placed on or in the stationary die, when the moving head is back as far as it will go. The operator then pushes his clutch lever, as shown on the photograph, the clutch is released, and along comes the moving head, or ram as it is termed, bending the material as it comes.

This description will no doubt be sufficient to allow readers to grasp the object of such machines. It might be well to add that they are built of extremely

FIG. 2—A SPREADING OPERATION. THE STATIONARY PORTION OF DIE ONLY IS SHOWN.

strong design and can withstand very hard usage.

For Forge Shops

The feature as spoken of above makes them ideal for forge shop work, and at the plant of the International Harvester Co., Hamilton, Canada, where these photographs were taken, a great number are in use for all styles of bending. A typical example is that shown at Fig. 1. The part being bent, or formed, as some express it, is a portion of a harvester. The piece in its finished condition is shown resting on top of the stationary bolster B. The work, before being formed, is illustrated at C. It will be noticed that in this case the work is placed flat on the stationary die, and by close inspection of the photograph the path which the metal will take to fit the die can be followed. A is the moving die. The stock in this case is about 2 inches wide and ½ inch thick. It is first heated, then placed on the die. Near the end of the stock, closest to the line of vision on this photo, is a hole. Advantage is taken of this fact, and the hole is slipped over a gauge pin to prevent the stock moving. This is at the

FIG. 1—GENERAL VIEW OF A BULLDOZER, WITH EXAMPLE OF WORK PERFORMED.

FIG. 3—A SQUEEZING OPERATION PERFORMED ON BULLDOZER.

end C. The operator pushes the handle, along comes the ram, and a perfect bend is the result. Twelve hundred of these parts are bent in one day.

Spreading Operation

At Fig. 2 is shown a very different style of bulldozer work. A represents a guide for the moving die. We do not show this moving portion as it would conflict with the rest of the view, but its shape can be understood from the nature of the work performed. C represents the part before bending, or spreading, while B shows the finished work.

This piece is made from the U-shaped stock as shown, and the work is gauged from the slot. This is an operation which would be very awkward to perform but for the bulldozer, and when we mention that 300 are completed every nine hours, it speaks well for the efficiency of the operation.

A continuation of the idea is shown at Fig. 3. This illustrates not a spreading but a squeezing operation. Small trun-

nion supports fit in between these parts, a pile of which can be noticed at right of the photo. After leaving the operation depicted at Fig. 2, the parts are drilled, then passed on to the bulldozer at Fig. 3. Here the operator picks up a trunnion, which has two small pinelike projections on it, and slips these into the two holes at top of piece as shown at A. He now places them in the stationary die, and sets the moving rams in motion. Readers may imagine that the rectangular-shaped nose shown on photo is really squeezing on the piece B. This is not the case, as the nose is really a holding down member and passes directly over the work B, while the die portion of the moving ram comes up and squeezes the work directly to the point where the trunnion pins come through. This might really be termed an assembling operation as the work is all cold. Incidentally it has proved much faster than the old hand assembled method.

Typical Bulldozer Die

The remaining illustration, Fig. 4, is

shown more with an idea of presenting the appearance of a complete bulldozer die. C represents the moving die, while the two rams A are attached to this die for guiding and holding down purposes. This system is often adopted to hold the stock securely to the stationary die while being bent. The two portions B are merely housings for the rollers under which the rams A pass. On close inspection of this die a guide pin can be seen toward the left-hand side of the left ram A. D is the stationary die. Another gauge pin can be noted between the two rams A, and somewhat nearer the right hand side.

There are of course more intricate dies made which give a side, and even an angle, motion of separate dies, but for the purpose of this, the first article on bulldozer work we have touched only on the dies of simpler design. At some later date we will present more intricate problems in buldozer practice.

MOLYBDENUM IN STEEL MAKING

A prominent American metallurgist states that ferromolybdenum is added in a fixed condition. It is supposed to give the steel properties similar to those of tungsten steel, but only one-third to one-half as much molybdenum is necessary; that where regular, high-speed steel contains 18 per cent. tungsten, 6 to 9 per cent. of molybdenum may be substituted. However, it gives these properties only when the addition is properly made and proper heat treatment follows. The regulation of these factors caused so much trouble and expense that, in the United States, the manufacture of molybdenum high-speed tool steels has been practically discontinued for several years. It is used for this purpose in other countries, however, to a considerable extent. At the present time it is mainly employed in tool steel as an auxiliary rather than as a major constituent. Various reasons have been assigned for the discontinuance of the use of molybdenum.

FIG. 4—A GOOD EXAMPLE OF A TYPICAL BULLDOZER DIE.

How One Shop Maintained its Margin of Profit

It is Not Often We Get a Chance to Print Such Experiences—
Contract Work is at All Times a Peculiar Proposition—How
Certain Work Was Completed is Told Herein

By DONALD A. HAMPSON

CONTRACT work is like job work in one respect—the margin of profit is so small that it keeps the engineering and production heads constantly on the alert to turn defeat into victory. The way one shop maintained its small margin of profit is told herein and the several instances may present ideas which others can use in their business.

For years it had been a struggle to get castings when we wanted them, and as we would like to have them. The congestion in the grey iron foundry business was so bad that we were obliged to make anything go that we could, even if we had to spend money to re-make the parts after they left the foundry. One great trouble with foundries in general is the lack of attention a casting gets after it is poured. The place is pretty hot by the time pouring starts and the men want to get home as soon as that part of the work is completed which takes any skill. This is where a big mistake is made. Hundreds of dollars worth of spoiled castings could be saved if an intelligent, skilled man were left behind to knock apart the moulds, but the general practice is to have Tony or Mike, the watchman, knock out the castings as soon as he comes on the job, and this he proceeds to do in a heroic manner that leaves all castings in the same state—uncovered, and supported in a still red state, by whatever part may happen to rest on the floor, regardless of how the casting will warp and sag in cooling.

We were building a number of machines that had a "track" that was planed to accurate dimensions. It was the hardest piece to machine we had, because of its length and the inside facing of rows of pads sunk below the general surface. This latter was made worse if the castings were warped, so that the pads cut almost through at one place while barely cleaning up at others. Cautioning the foundry did no good. They were supplied with a straight pattern and requested to be careful about the cooling, but all to no avail, and we continued getting castings that would not clean up.

To make these go we heated them red at the sections of bend, weighted them in various ways, and let them cool off. By so doing we were able to get very straight castings and turn our work out with no delay. A casting so weighted is shown at Fig. 1, which illustrates one "symmetrically crooked"—merely having dropped at the ends.

A Talk on Threading

Shops that have to do accurate threading on screw machines or threading machines have been troubled in maintaining a constant lead and in getting a good thread at the end where the die starts. Workmen and executives are apt to think their results and their tools are not as good as those used in other lines and to wonder just how these favored people turn out this perfect work. For instance, they ask, "How do tap makers get such good threads clear to the points of their taps?" The answer is simple. "By selection." Tap blanks are all made up as plug or bottom, taps and threaded as such. Then an inspector goes over the run and picks out those of poor lead and with poor threads at the start; the good ones are fluted and relieved as plug taps, the poorer ones are cut into taper taps. It is commercially impossible to maintain the production of threaded parts to the highest standard and some such scheme as mentioned saves money without sacrificing needed accuracy. Our shop was making an automotive part on contract, including a screw that was exceedingly hard to keep in shape right up to the end. Our equipment was of the very best. The way we did the work was to set the automatic to make an extra ⅛-inch of length and by another operation to cut this part off

afterward, it being the only part that had threads that were not acceptable.

Ball Bearings

Our milling department was up to the latest wrinkles in steels and feeds and speeds. Contract work has to be forced where there is clear going. This forcing wears the shoulders of the feed screws and the seat that takes the thrust. It means constant re-facing and collaring. We decided to settle that job once and for all. Each machine, as it was taken out of service for any repairs whatever, was disassembled by removing table, screw and saddle bearing of the screw. Enough material was cut away to permit the insertion of a standard ball thrust bearing of the three ring type, costing around a dollar apiece. Putting this bearing between the shoulder and the cast iron bearing placed all the thrust load on hardened parts. It settled our problem then and there, at very little expense. The first of the millers so changed has now been in daily service for eight years and there is not yet enough wear to make repairs advisable.

Rapid Facing

A customer decided on a change on a dozen presses we were making, substitut-

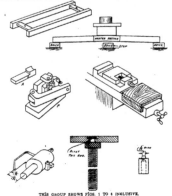

THIS GROUP SHOWS FIGS. 1 TO 4 INCLUSIVE.

ing a solid 3-inch square steel bar for the cast head piece that was intended to go between the uprights. We got the steel and sawed the bars off, 24 of them in all. As they had to be individually fitted and everyone knows what an uncertain job vertical planed cuts are, we cast our eyes on milling machine equipments that would do the work better and quicker. The photo shows what we found and how it was done. A knee with tongue to fit the table slot was used for both a stop and to align the bars. A single clamp sufficed to hold the work in place. And the inserted blade end mills that we had made to screw directly on the spindle nose for various surfacing jobs were just the proper tool for this work. With plenty of power, we left the

the wisdom of this. This continual centring wears the ways in one short section (for cuts are rarely of capacity length) and it results in uneven wear and abuse of the table. Workmen have a propensity for dropping hammers, wrenches, and miscellaneous junk on the exposed section of the table between the job set up and the front end—they never put them down carefully. In consequence, that section looks decidedly second-hand in less than a month, and no good end has been accomplished from this malpractice. We found it almost impossible to break the men of this habit so, as it is more convenient to keep tools near by, we made up the protection boards for each machine and the men took to them at once. By making the

when located by the spindle, fitting in the top and extending up to the taper and tang end, which is driven by the machine spindle. The lower end of the work spindle carries the roller, which "rolls" over the screw end. Production was increased four-fold and the rejects became practically nil. By giving the head a spin with a finger, the boy knows when it still needs more, which he applies by another pull on the feed lever and another trial up to the point where the head will not spin but will push.

Trepanning Tool

After we had built and delivered several cold heading machines to the wire company's new plant, they wanted a change made that involved cutting two

REFERENCE TO THE TEXT WILL EXPLAIN THESE PHOTOGRAPHS.

sawed pieces long, avoiding any loss through a crooked cut, and let the cutter "walk right through the end," as the chips will show.

The piece shown in Fig. 2 is one of a pair used as a split nut in some ten thousand patented devices we contracted to make. It is made of cold drawn steel ¾-inch square, milled to shape. The half round cut was made with a formed radius cutter, doing the work in a hand milling machine with lever feed, milling one piece at a time. Though production was acceptable, the finish was not, the tendency being to crowd the work past the arbor, springing the latter and leaving the cut shallow and rough. This resulted in slower tapping and broken taps.

A better scheme was worked out as shown at B, Fig. 2. Two pieces at a time are held in a drilling jig, upright, spaced properly and clamped by the hinge. Then a three lip twist drill is passed down through the bushing and cuts the concave in both pieces at once. Greater production and real uniformity resulted at once from this improved fixture. In general, it might be said that outside of manufacturing shops, the advantages of three or four lip drills and the possibilities of drilling vs. other methods are rather overlooked.

Keeping the Miller New

The best practice in milling and planing is to set up jobs at different points on the table—not to set each and every one right in the centre—but nine out of every ten workmen cannot be taught

sides a tight fit on the table, the boards hold themselves in position in spite of the jar of cutting. If no other good were accomplished, the feeling that the machines were not abused would compensate for the little effort expended. Fig. 3 shows this kink.

The drawing Fig. 4 illustrates an assembly of a screw set into a head piece, and the teat end riveted over upon a washer in the counterbore, the combination forming a screw that could be turned when desired, but held without an excess of side play. We had been riveting over the teat by hand and we knew all along that we were not getting good work—fully half of the screws were so tight that they had to go through a long loosening process or else they were so loose they had to be sent back, with risk of incurring trouble No. 1 the second time.

The simple fixture shown by the photo is the way in which we turned out a product that "looked as if it was made somewhere." The head piece may be seen under the arch of the fixture, and the lower end of a screw appears below the clamp. This clamp is a bar of steel bent at right angles, the short end having a jaw that holds the screw in a channel, and the long end forming the handle which the boy grasps in connection with the stationary bar, which is nearby; pressure sufficient to keep in place is thus afforded, while as soon as the hand is taken away the work may be removed. The arch shaped main casting is fastened to the drill press table

three inch holes through the side frames of each, a distance of two inches. The machines had been set up but as yet there was no power in the building and so we had to resort to hand methods— and in these days hand methods have to be arranged for the easiest possible working.

First we put ¾-inch holes through in the centre, using a common ten-inch ratchet. Meanwhile we were making the trepanning tool shown at Fig. 5, which was to be piloted by the ¾-inch hole and was to cut away a chunk three inches in diameter. And this it did quite nicely, using the same ratchet with a pipe extension on the handle. Our experience in cutting tank and boiler plates led us to adopt this method as the best and in preference to a series of smaller holes to be finished with a larger tool.

Spotting Bearings

Go into any garage and watch a man scraping a bearing. His can of Prussian blue is in constant use. In fact, the garage machinist would not know how to fit a bearing if he did not have this material. But a little Venetian red or vermilion powder mixed with oil will do as well. The old school of machinists who had to go miles from the nearest habitation on a repair job never had any Prussian blue—they used a little red lead or red paint or lamp black—anything they could find after they got there—and they did good work with it too. Really, a marking material is not essential ex-

cept on the finest class of work for the bearing spots always show up a visible black or polish that is sufficient. And lamp black is as good a marking material for fine work as any compound that can be bought, and much cheaper.

Concrete Cams

We had a four inch Cleveland automatic and often set it up for as few as a hundred pieces. On several occasions we required feed cams that we did not have and to avoid the expense and delay of these large pieces, we made concrete cams. These served quite nicely. After the machine had been set up in other particulars we would lay out the feed cam in chalk and lead pencil directly on the drum, which was something like three feet in diameter. By turning over the machine and checking the movements we assured ourselves as to the accuracy of our work. Then a strip of 1-16-in. x ¾-in. steel would be bent to follow this contour and wired in place, standing on edge of course. A sheet of wrapping paper would be cut and stuck to the face of the drum to cover the screw holes. Two or three cam screws were greased and inserted to the usual depth, their function being to anchor the cam-to-be. Then using sections of pasteboard for the exterior of the "form," cement was poured (or rather trowelled) on; the procedure being to put the cement on top, cover with pasteboard at the right thickness, fasten the latter securely, and turn the drum until a further portion was on top. The work was left overnight to harden. The resulting cam was solid and secure and had a smooth steel roller face. In point of expense it was highly satisfactory.

Something About Burrs

It is impossible to wholly avoid the process of removing burrs, but in so far as drilled and tapped holes are concerned, two simple means are at hand to avoid this work. Many reamed holes have set-screw or oil holes intersecting them or are intersected by unimportant holes. These unimportant holes should be drilled after the others but before the main holes are reamed, and the tap holes should be tapped before any reaming is done. Then when the reamer is put through, the finished hole is smooth and does not have to be hand scraped or again reamed for the swell that the tap would leave.

The practice of countersinking all holes should be followed; it is a saving of time always, general impressions to the contrary notwithstanding. For ordinary holes a touch is sufficient to remove the burr for tapped holes, they should be countersunk up to or slightly over the outside size; the saving in time and in files will make up for the extra operation. Apprentices trained in the Brown and Sharpe plant are taught early to countersink all holes.

During the war we had a contract that included hundreds of thousands of parts shown by Fig. 5 which were made on the automatics, heat treated, and assembled.

The vital part is the little extension, or teat, which had to be hard to resist constant blows without any change of shape, but not so hard as to break or shatter. We had assumed that case-hardened machinery steel would answer the requirements, but tests on the first few thousands proved to the contrary. The few that were strong enough to stand the racket were so hard that they cracked off at the shoulder; the rest, though hard by the file test, battered up in a short time because of the soft interior. The parts had been heated for about 24 hours.

We knew that we could get results by using tool steel, hardening, and then drawing to the proper temper, but this latter material was too expensive both in first cost and in treatment throughout the cutting and tempering. And the proportion of "lost" parts was sure to be greater.

After some trial, we arranged the work satisfactorily. The same machine steel parts were used but they were put through a continuous run of 48 hours in the furnace before quenching. They came out glass hard, of course, but in addition the depth of case went clear through the teat. And the adjacent face of the shoulder was deeply hardened, its state being such that there was no strain at the line where it joined the teat. The parts were hard enough on the surface and far enough under the surface to give the desired backing. They had only to be "let down" to take out the "snap" and become ideal for the work.

The carbonizing transformed the entire end into tool steel at a lesser cost and was then tempered as tool steel would have been. In not a few cases this same process may be followed to advantage; it is reported that a famous file company was short on a certain "shape" and bought three tons of mild steel at 2 cents a pound, which they proceeded to bring up to their standard in a similar manner, and no complaint was ever received, nor did anyone suspect what they were getting.

LUBRICATION AND THE GERM PROCESS

The following is a synopsis of a paper on "Lubrication and the Germ Process," by Henry M. Wells and James E. Southcombe, M.Sc., read at the British Association meeting:

The authors draw attention that the modern view of lubrication on the physico-chemical side is the existence of a residual valency or affinity between the oil and the solid surfaces of the bearing and journal. The problem has, therefore, to be considered as a kind of intimate relationship between the oil and the metal. This is quite a different view to that held by investigators in the past, who thought that the property of "oiliness" so strongly manifested by fatty oils as compared with mineral oils, depended in some particular physical characteristic of the oil itself. The modern point of view has

been elaborated recently by a number of students of the subject.

The authors have discovered the reason why the fatty or fixed oils possess superior friction-reducing properties to that of mineral oils. They have shown that this is due essentially to the presence in fatty oils of minute quantities of free fatty acids, which are absent in the majority of mineral oils. In consequence of this discovery they have added small quantities (about 1 per cent.) of fatty mineral or organic acids, which are easily accessible and relatively cheap to hydrocarbon oils. The products so obtained possess remarkably low frictional co-efficients. As an example, the addition of 2 per cent. of common fatty acid to a mineral hydro-carbon oil reduces the co-efficient of friction shown by this oil on a friction testing machine from 0.0084 to 0.0052—a diminution of 26 per cent.

By suitably choosing the type of fatty acid to be added to mineral oil, the behaviour of oils in the presence of water and other contaminating substances can be modified at will. It is also shown that the risk of metallic corrosion is even less with these oils than the usual compounded oils, because the latter frequently develops in working notable quantities of free acid.

The principle of making mineral lubricating oils possessing increased frictional properties by adding to mineral oils suitably chosen fatty acids in relatively minute amounts has been patented throughout the world, and has been somewhat fancifully called the "germ process," because the fatty acid is the germ of the idea.

The value of the process lies in that oils possessing friction-reducing properties much superior to mineral oils can be prepared at relatively little extra cost to mineral oils. Secondly, by this means an immense quantity of fatty oil suitable for the manufacture of margarine and other foodstuffs is liberated for the country's needs and substituted by a much smaller amount of commercial fatty acids, which are quite unsuitable for food and are more or less in the nature of a by-product.

Finally, it may enable the lubricating oil manufacturer to produce the highest class of friction-reducing oils from comparatively thin — therefore cheaper — mineral oils, and incidentally this may have the utmost significance to-day, when we are striving to foster and develop the mineral oil production of our own country and the Empire.

ERRATA

In our August 12th issue we published an article entitled "Can We Avoid Grinding Wheel Accidents?" In the third paragraph from the bottom of the last column on page 159 the term "exposure between the flanges" should read "outside the flanges," and not the word "between." The wording used was rather confusing so that we call readers' attention to this matter.

Manufacture of Helical Springs for Heavy Duty

In This Age of Mechanical Design, New Problems Arise Continually in the Use of Springs Demanding the Highest Possible Quality—Herein is Shown an Example of Spring for Extreme Heavy Duty.

By T. D. LYNCH, Research Engineer Westinghouse Electric & Mfg. Co.

SPRINGS of great variety of design and quality have been manufactured from flats, rods, ribbon and wire during the ages past. Some of them have been heat treated before forming, others after forming and still others have had no heat treatment, the high elastic properties being given to the latter by the work done in cold rolling or drawing the raw stock to its finished size.

Early spring making was for the use of buggies and spring waggons, then for clocks, watches, cars, locomotives, etc., and still later for automobiles. To the automobile manufacturer belongs the credit for the development of the superior quality of springs now used in the construction of automobiles. This progress, however, has not been without concentration of purpose relative to both material and treatment and each step has inscribed its mark of success or failure.

In this age of mechanical design, when every ounce of strength is exacted from each member, new problems arise continually in the use of springs demanding the highest quality possible to produce. To meet this growing demand for quality, every phase of spring manufacture must be co-ordinated and carefully processed at the furnace, at the rolling mill and at the spring making plant.

The steel maker has offered his carbon spring steel made in large open hearth furnaces, which has its field of application but is not strong enough for the most severe conditions that must be met at this time. Chrome vanadium steel, silico-manganese steel and chrome molybdenum steel have much higher elastic properties and are capable of withstanding proportionally greater stresses. Therefore, these and other alloy steels must be considered in connection with this high class work.

Springs for Heavy Duty

Within recent months one of the writer's duties has been to assist in the procurement of helical springs for a unique and unusually severe duty. The accompanying photograph shows the general problem involved, which consists in the transmission of the driving power of a motor to the wheels of a locomotive through a set of cushion springs. The arrangement is such as to permit free vertical and lateral wheel play. In this application the stresses are tension, compression, torsion and shear, separately or in combination, and these stresses are augmented from time

to time by shock, producing a condition that makes necessary a material of high elastic strength, and, at the same time, a large degree of toughness.

The first step in the study of this problem was to investigate the general practice of spring makers—to find that the general process in vogue for the manufacture of springs from bars larger than 3-8-inch diameter is to obtain the bars from the steel maker, produced in large open hearth furnaces, rolled in a commercial manner, representing tonnage production. Bars thus made are heated to a temperature varying more or less from a bright to a dull red at different parts of their length, coiled into springs and quenched from this irregular coiling temperature. It is needless to say that springs made by such a process cannot be depended upon in severe service.

A few spring makers have realised the need for greater care in making and heat treating springs for heavy duty and a receptive attitude towards better spring making is now being taken by them. Spring steel, and especially alloy spring steel, cannot be too carefully made, nor can it be too carefully coiled and treated in order to produce a strictly reliable finished product. Therefore, let us assume that the steel is carefully made and rolled into rods by the processes known to the steel makers as their best, coiled into springs at a proper and uniform temperature and

finally treated by steel treaters under their best practice. Only such manufacture and treatment can be expected to give us the very best results. This hypothesis being granted, let us propose a process for consideration, covering the manufacture of a silico-manganese steel spring made from 1-inch rounds, including the making and rolling of the steel, the coiling, treating, testing and inspection of the spring, in order that we may procure the reliable quality above indicated.

There are many alloy spring steels on the market, any one of which would require similar care in their making and fabrication, but for this article we have chosen a silico-manganese steel. The question now arises how should a spring of the highest quality be processed, including the chemical composition, manufacture of the steel, rolling of the billets and bars, heating and coiling of the spring, the heat treating, inspection and general care required throughout the different steps in order to produce the most reliable product. The following process has been worked out and is submitted for criticism and suggestions from those who may have similar problems to solve.

Chemical Analysis

The steel shall be what is known as silico-manganese steel of the following chemical analysis, both on ladle test and check test from finished rolled

HEAVY DUTY SPRINGS USED ON THE WHEELS OF A LOCOMOTIVE.

bars: Carbon .50-.60, manganese .60-.80, phosphorus not over .04, sulphur not over .04, silicon 1.50-2.20.

The steel shall be made by the crucible or electric furnace process and by a company that has been known to have produced high quality spring steel of the class specified. The ingots shall be not less than 9 inches square at large end and 8 inches square at smaller end, so poured and the molds so coated as to give a smooth surface to the ingot. Each ingot, when cold, shall be carefully inspected and any blemishes shall be chipped or ground out, leaving a surface that will be free from laps or seams after rolling.

Rolling of Rounds

The ingots shall be slowly and carefully heated to approximately 1,100 degrees Centigrade and rolled or forged to squares of approximately 3 inches by 3 inches and sheared into suitable billet lengths for final rolling. Excessive reduction will not be permitted. Sufficient discard shall be made so that no signs of piping or segregation can be found and a very careful inspection made, especially of the top cut.

The billets shall be allowed to become cold and a very careful inspection made for surface defects and any slight blemishes ground out, leaving a smooth, even surface, no ragged corners or slivers being permitted. The billets shall then be heated to approximately 1,100 degrees Centigrade and rolled to finish size, great care being taken to avoid excessive reduction at any one pass.

The bars shall be sheared to length and carefully inspected for piping, segregation and shall be true to section. Each bar shall be straight, free from surface cracks, scratches, seams, folds and indentations and shall be true to section. The diameter of the bar shall not vary more than two and one-half per cent. from that specified.

All bars shall be tied in bundles and a metal tag securely attached to each bundle. The tag shall have stamped on it the requisition number, heat number, size of rod and manufacturer's identification mark. When bundles are opened great care must be exercised so as not to nick or in any way injure the bars. A nick or scratch in a bar, however small, can not be permitted.

Coiling of Springs

The bars shall be heated slowly to a uniform temperature of approximately 925 degrees C. (1,700 deg. F.) and immediately coiled over a mandrel preheated to at least 100 degrees C. The mandrel shall not be water cooled nor shall any water be allowed to touch the spring while hot.

Notching for length shall be done at a dull red heat and in such a manner as not to cut, scratch or otherwise injure the surface at any other point on the spring. The springs after notching shall be allowed to cool slowly and uniformly in such a manner as to prevent local chilling which may cause surface stresses or cracks to form.

The springs shall be slowly and uniformly preheated to approximately 700 degrees C. (1,290 degr. F.), transferred to a furnace held at a quenching temperature of 900 degrees C. (1,650 deg. F.) and uniformly heated as near as possible to this temperature and quenched in oil. The quenched spring shall be drawn in a salt bath at approximately 455 degrees C. (850 deg. F.) to relieve quenching stresses. The drawn springs shall be cleaned from the adhering salt by a hot soda wash, followed by an oil or lime dip to protect them from corrosion.

Physical Properties and Tests

The properties specified below shall be determined in the order given. The spring shall not be rapped or otherwise disturbed during the test.

Solid Height.—The solid height is the perpendicular distance between the plates of testing machine when the spring is compressed solid with a test load of at least 125 per cent. that necessary to bring all the coils in contact. The solid height shall not vary more than 1.5 per cent. from that specified.

Free Height.—The free height is the height of the spring when the load specified in paragraph (a) has been released, and is determined by placing a straight-edge across the top of the spring and measuring the perpendicular distance from the plate on which the spring stands to the straight edge, at the approximate centre of the spring. The free height shall not vary more than 1.5 per cent. from that specified.

Loaded Height.—The loaded height is the distance between the plates of the testing machine when the specified working load is applied. The loaded height shall not vary more than 1.5 per cent. over nor more than .75 per cent. under that specified.

Permanent Set.—The permanent set is the difference, if any, between the free height and the height (measured at the same point and in a similar manner) after the spring has been compressed solid three times in rapid succession with the test load already specified. The permanent set shall not exceed .4 per cent. of the free height.

Hardness Tests.—The Brinell hardness number shall not be less than 375 nor more than 450. This test shall be made on the coupon resulting from notching to length and broken off after spring has been heat treated.

Structure.—The grain structure of the finished spring should be troostitic or troosto-sorbitic.

Dimensions

The springs shall be straight, with uniform pitch, and shall remain straight throughout the working range. The diameter of single coils of the spring shall conform to the tolerance given on the drawing. The concentricity of the coils of the spring shall be determined by means of a plug-gauge and ring-gauge, extending the full length of the spring, "D" is specified outside dimensions:—

	Length of Spring	
	10" or less	Over 10"
Outside diameter of ring gauge	d minus .063	d minus .094
Inside diameter of ring gauge	D plus .063	D plus .094

where "d" is specified inside diameter of spring. "D" is specified outside diameter of spring.

When the drawing specifies a spring with ends squared and the spring is placed with either end on a flat plate, its axis shall not make an angle of more than 5 degrees with a perpendicular to the plate.

Springs varying from the required finished dimensions may be reheated and reset but all such reset springs shall be completely re-treated—that is quenched and drawn again.

Tests and inspection shall be made at the end.

Packing and Marking

All springs shall be plainly marked with serial numbers and manufacturer's identification letter at one end of the finished spring and on each coupon resulting from notching to length. The purchaser shall assign the serial numbers to be used with each order. Each size of spring shall be packed separately in such a manner as to protect the springs from injury during shipment.

Each package shall be plainly marked as follows:—Requisition number, P. D. Spec. No., number of springs, style number of spring (or drawing and item number if there is no style number), name of manufacturer.

For example:—Req. C-5,100; P. D. Spec. No. ; 6 springs; style No. 591,500; Blank Company.

Tests and inspection shall be made by the purchaser at the place of manufacture prior to shipment. The inspector representing the purchaser shall have free entry at all times while the contract of the purchaser is being performed to all parts of the manufacturer's works which concern the manufacture of the springs ordered. The manufacturer shall afford the inspector, free of cost, all reasonable facilities to satisfy him that the springs are being furnished in accordance with this specification.

This paper has included more than the heat treatment of springs and is intended to emphasize the great importance of a better knowledge of complete spring manufacture and bring together on common terms the designing engineer, the steel maker and the spring maker.

A full discussion of this article is desired, hoping that the combined experience of the members of this organization or others may bring out additional data that will assist in establishing a complete and efficient process for the manufacture of truly reliable springs for heavy duty. It is further hoped that this paper has shown the need of co-ordinating the talent and experience of the men at the steel mill, rolling mill and spring coiling plant as well as the heat treating and testing of the finished product.

Conley Continues Along the Road of Experience

John's Trials Are Varied—He Performs a Bit of Dentistry, Spoils a Job, Then Finds Out Why He Spoiled It—He Hears of the Peculiarities of Cast Iron, Etc., Etc.

By J. DAVIES

AFTER John had been in the shop for six months his father said jokingly to him, "I suppose you have learnt all there is to learn in the shop by this time." To which John replied, "Not much. I have just been in the shop long enough to find out that I don't know anything."

"If you have found that out in six months you are making pretty good progress. It takes some men six years to find it out and some never do. What have you been doing to-day?"

"I have been doing a bit of dentistry to-day. Some teeth were broken out of a lathe gear and I had to put some fresh ones in."

"How did you manage it?"

"Mr. Smith had the fragments of the broken teeth chipped out, and the spaces filled up with the oxy-acetylene welder, then acting under instructions I bolted the gear against the angle plate and marked around two of the teeth on the angle plate. I next ground a tool to fit one of the spaces I had previously placed at the top. I now put the tool in the tool post, fitted it correctly into the space, and tightened it up firmly, setting the indicator for depth at zero.

"I next loosened up the job and turned it round until the broken part came under the tool, and two of the good teeth coincided with the marks I had put on the angle plate. All I had to do after this was to feed the tool down to the zero mark, which indicated the depth, and repeat the operation as many times as there were teeth to be replaced."

"That sounds pretty good, but how did you know that you were moving your job round in a true circle?"

"Oh, I forgot to mention that the boss had the surplus metal turned off in the lathe, and had a piece of flat iron bent at right angles bolted against the angle plate for the gear to rest on. I thought I did that job pretty slick but the next job, oh, boy!"

"What happened to the next job?"

"Well there is no use mincing matters, I completely spoiled it, and I don't hardly know yet how I did it."

"That sounds interesting. What kind of a job was it?"

"It was a cast iron block. I had to put a 60 deg. dovetail on each side, and make it to fit a female gauge. This was the first dovetail that I had ever done. The first thing I did was to put the head of the shaper round until the 60 deg. mark coincided with the zero, then I tried to cut the dovetail but I found the tool rubbing hard on the back stroke and when I tried the 60 deg. thread gauge on it it wouldn't fit within a mile. I moved the head to a different angle, and by the time I made the

work fit the gauge, I had it too small. I must confess I was completely up a tree, and now the job is resting on the scrap pile waiting for a cheap trip to the cupola. What was the trouble, dad?"

"The first difficulty, my boy, was in misunderstanding the graduations on your machine. The machine was at zero when your cross and vertical slides were at right angles to each other. Nearly every shaper and planer that I have seen is marked this way. It is apt to be very confusing to a beginner, consequently when you wanted an angle of 60 deg. with the horizontal, you had to move your head 30 deg. because you were already set at 90 deg.

"Your graduation would now read 30 deg. which does not mean that you would cut a 30 deg. angle, but that you have moved your head 30 deg. from the vertical, which would be 60 deg. from the horizontal. It doesn't follow that after your machine is set to the correct angle that you are going to cut that angle. It is quite possible to cut a right angle and have your machine set for a 60 deg. angle."

"How do you make that out?"

"The shape of your tool has a good deal to do with it. Suppose you had your shaper head set at 60 deg. and the cutting edge of your tool was vertical to your work, you would make a right angle because in that case you couldn't make your tool cut any other than its own shape. In every case of dovetails or undercuts you must be sure that your tool is a smaller angle than the one you desire to cut.

"It will help you to understand angles if you get a common two-foot rule, place one 12-inch length on top of the other and open it out to form a square. Remember that you start at 0 and keep increasing your angle to 90 deg., when your rule will form a square. If you continue in the same direction until you have straightened out your rule you have increased your angle from 90 deg. to 180 deg.

"To avoid the tool dragging on the back stroke, it is necessary to move your clapper box so that your tool will lift away from the work. Before starting a cut always lift up your tool by hand to see that it lifts away from your work. If you were cutting a slot for the head of a bolt you would not be able to move your clapper box. You would have the choice of two alternatives, either lengthen your stroke so that your tool could lift clear through the slot and lifted up at one end and put down at the other, or the tool and clapper box could be fastened rigidly so that they could not move at all.

"I once saw a fellow with an ingenious device for this kind of work. He had a common door hinge clamped to the back of the tool in such a way that it would bend on the forward stroke and when the tool got past the end of the slot it would fall down and on the return stroke lift the tool clear of the slot."

"Thanks, very much, for your explanation, dad. I see the idea clearly, no wonder I spoiled the job."

A short time after John had this conversation with his father, he was promoted to the planing machine, and as the shaper and planer are so near akin we will omit any further reference to the shaper and follow him to the planer.

When he was running the shaper he had often looked with envious eyes at the planer hand riding on the end of the table with his legs dangling down. Now that John is running the planer he is keen to do the same thing himself the first chance he gets.

It so happened that the first job he had was one with a fairly long stroke. John perched himself up on the end of the table and started to kid the other fellows and do all in his power to make them jealous of his soft snap as he termed it, but unfortunately for John the dog on the side of the table which operates the reverse slipped, and before John could grab the wrench and tighten it up, the table had run off the rack to the delight of those around. John naturally came in for some chaffing, and he had hard work to organize a crow bar gang to put the table back in its place, which he was anxious to do before the boss came around.

He was planing a surface plate, and he had received instructions to just rough it out, then take it off the machine, put it on one side, and go on with something else. He didn't understand why he was not to finish it when he had it set up, so he decided he would ask the boss the reason. So when Mr. Smith came around, John, pointing to the casting, asked him why he had had that casting taken off the machine before it was finished.

Mr. Smith replied: "It is like this, John. Cast iron is subject to more or less change of shape, particularly after the entire surface, or skin, has been removed, by turning or planing. One method of avoiding these changes in the finished part is to allow the castings to "season" or stand for several weeks or even months after taking the roughing cuts, before finishing them. These changes are caused by the internal stresses in the casting, which are

gradually readjusted and neutralized by seasoning.

"Sometimes in order to avoid a long seasoning period, castings that must retain their shape as much as possible are subjected to some kind of heat treatment. A method which has proved successful, as applied to the cast iron beds of measuring machines, is in the nature of an annealing process. After rough machining castings, they are placed in a furnace and heated to a dull red heat and allowed to cool very gradually, then machine finished and scraped.

"Another peculiar feature of cast iron parts which are subject to repeated heating and cooling is that they grow or increase in size. This growth is so pronounced that annealing ovens eight feet in length have been known to grow to eight feet six inches."

"There is another very important thing I want to call your attention to. Few machinists realize how easily a casting, even of very large size, can be sprung by careless bolting down to the table. Before finishing a casting that needs to be machined very accurately, it should be roughed all over as there is a certain amount of internal stress in every casting, and any uneven places at the bottom should be packed up, so as to give a good solid bearing surface before being finally bolted down. Otherwise the work will be bolted down, and when the bolts are released it will go back to its original shape, with the result that the planed surface will not be true.

"Where there is a large surface to be machined, like the surface plate you have just taken off, it could have been roughed out with a gang tool holder. Tool holders are commonly used in most planer and shaper work. They are an advantage, not only from the great saving in tool steel, but from the facility with which the small inserted tool can be removed and ground up to shape. One special form of tool holder, peculiar to planers, is for holding three, or four inserted roughing tools. These are called gang tools. It has been found that the resistance of a cut varies as the square of the thickness of the chip, and that a rougher feed can be taken if the cut is divided up among three or four tools than would be possible if only one tool was used.

"Cast iron should be finished with a broad flat tool unless the metal is very hard and causes chattering, when the breadth of the tool will have to be somewhat modified. In roughing out cast iron it is sometimes advisable to take a cut with a chisel along the edge, where the tool leaves the work, to prevent it breaking away and leaving ragged edge, but, look, it is 11 o'clock, and time for bed, so off you go, so we will continue this discussion later."

ates less light; its emissivity is always only a fractional part of that of the former and always less than one.

In consequence of its lesser light-emitting power the temperature of the metal, when measured optically, appears to be less than it really is and the difference between the true and the "apparent" temperature is greater the smaller is the emissivity of the metal. This true temperature may, however, be calculated from the apparent or measured temperature by using the following equation:

$$\frac{1}{\text{true temperature} + 459}$$

$$\frac{1}{\text{apparent temperature} + 459} = .0000579 \log E$$

where E is the emissivity and the temperature are in Fahrenheit degrees.

Tests at the U.S. Bureau of Standards have shown that the emissivity of both nickel and Monel metal in the molten states is .38 and consequently almost equal to that of iron or steel. In the solid state both nickel and Monel become covered with a thin layer of oxide when heated in air to incandescent temperature; the emissivity of solid oxides of nickel and Monel are .89 and .70 respectively.

Optical Reflecting Power of Nickel and Monel Metal

SOME recent tests of the U.S. Bureau of Standards are of interest in connection with the use of nickel and Monel metal for the construction of mirrors and of reflectors for automobile and locomotive headlights, for searchlights and other similar equipment. The tests show that polished surfaces of both metals reflect a high percentage of light, and greater than most other metals as is indicated in the following table containing the results of reflectivity tests.

and it is in this form that most people are familiar with them. Much interest is being manifested, however, in solid nickel and Monel metal reflector construction, i.e., in reflectors constructed of formed nickel or Monel metal.

Optical Emissivity of Nickel and of Monel Metal

In order to determine the temperature of an incandescent mass of any metal in the open with an optical pyrometer such as that of the Leeds and North-

Metal	Authority	Per Cent. of White Light Reflected
Nickel	Bureau of Standards	62
Monel Metal	Bureau of Standards	62
Silver	Smithsonian Physical Tables	92
Gold	Smithsonian Physical Tables	65
Speculum Metal	Smithsonian Physical Tables	67
Platinum	Smithsonian Physical Tables	61
Steel	Smithsonian Physical Tables	55
Copper	Smithsonian Physical Tables	55
Nickel Bronze	Rubens and Hagen	49
Tungsten	Coblentz	50

Due to the fact also that nickel and Monel metal are not corrodible metals a polished surface of either will retain its polish and reflecting power and will not become dulled and less efficient upon exposure to air and moisture as surfaces of other metals will.

These valuable properties of nickel are most commonly utilized at present in the form of nickel-plated reflectors rup Company, or the Scimatco, it is necessary to know the radiating or light emitting power of the metal; namely, the emissivity or the ratio of the light emitted by the metal to that emitted by a perfect black body at the same temperature. The metal always appears cooler or darker than a black body, such as carbon or graphite, at the same temperature for it emits or radi-

REGO CUTTING APPARATUS

We have received a very fine catalogue dealing with Rego welding and cutting apparatus. The Carter Welding Co., Toronto, Canada, are sole sales representatives for this line, and to those not familiar with this apparatus we suggest a perusal of the following:

The Rego principle of gas mixture, whereby the acetylene enters the mixing chamber at a slightly higher pressure than the oxygen, is said to positively insure the elimination of the flashback under all operating conditions. It is claimed that it materially reduces the oxygen pressure, thereby preventing any excess and wasteful oxygen, which is detrimental to good welding. Economy of operation, safety and perfection of flame, are other features claimed for this line.

The catalogue shows some actual photographs of tests for flashbacks, and it is clearly shown that there no flashback occurs. The book deals with the carrying-cases, trucks and various welding apparatus. There are also shown some welding and battery outfits, some lead burning and light welding apparatus, and a decarbonizing outfit.

Another item is a description of the Rego torch, with complete data on the same. The various models are shown, and specifications given. The latter part of the book describes tripod manifolds, wall manifolds, and the Rego diaphragm regulator, together with detailed description. Other styles of regulator are shown, also adapters, hose connections, wrenches, etc., etc.

Calculating Dimensions on 100-h.p. H.R.T. Boiler

The Shell and Joint Portions Are Discussed in This Present Article, and in Later Issues We Will Go Still Further Into This Matter—It is a Good Thing to Know the Whys and Wherefores of Boiler Design.

By T. H. FENNER, Editor Power House

IN our October 14th issue we published an article by the above author dealing with the design of joints of the butt strap type. Believing that readers are interested in these calculations, we continue in the present article to discuss the application of the previous joints discussed to the design of an actual boiler.

Assuming that the boiler is of 100 h.p., the leading dimensions being 72-in. diameter, by 18-ft. long, and a working pressure of 125 lbs. per square inch, we have the following:

The first thing necessary to find is the thickness of plate to carry the required pressure, and this is accomplished as follows: It is a well known fact that the bursting pressure of a sphere is found from the formulae below.

$$f = \frac{pd}{2t}$$

When we wish to apply this to an actual boiler, we have some variations to introduce, these being the least percentages of longitudinal joint, and the factor of safety. Neglecting these for a moment, we will show how to find any one of the factors from the others. For instance, to find the pressure the formula would be

$$p = \frac{f2t}{d}$$

To find the thickness we would have

$$t = \frac{pd}{2f}$$

While to find diameter the formula becomes

$$d = \frac{2tf}{p}$$

The formula to find the working pressure of a boiler is

$$WP = \frac{S \times \% \times 2t}{d \times F}$$

where S = tensile strength of material, % = least percentage of longitudinal joints, t = thickness of plate in inches, d = diameter in inches, and F = factor of safety.

We have already set the required working pressure, and as we are to get a boiler of good workmanship and material we can assume 4.5 as a factor of safety. We will use a triple-riveted double-strap butt joint, as a efficiency of 86% for the longitudinal seams, so that all that remains is to find the required thickness of shell. We must, therefore, substitute the known figures in the formula, and thus find the unknown. We have then

WP = 125,
d = 72,
S = 60000,
F = 4.5,
% = 86.

Putting in the figures, we get

$$\frac{60000 \times .86 \times 2t}{72 \times 4.5} = 125.$$

It is easier to use all fractions instead of decimals, so

$$\frac{60000 \times 86 \times 2t \times 2}{72 \times 100 \times 9} = 125.$$

By cancelling out we get

$$\frac{50 \times 86 \times 2t}{8 \times 9} = 125$$

$$2t = \frac{125 \times 27}{50 \times 86}$$

$$t = \frac{125 \times 27}{100 \times 86}$$

$$t = .3924.$$

The nearest fraction to this is 25-64, but as the plates are not made in variations closer than 1-32, we will call it 3-8 inch and see if that will give the desired result. If not, we will increase it 1-16. Then

$$\frac{60000 \times 86 \times 2 \times 3}{72 \times 100 \times 9 \times 4} = WP.$$

Cancelling

$$\frac{50 \times 86}{9 \times 4} = 119.$$

This gives only 119 lbs., so we must take 7-16 as the size of plate, and try again.

$$\frac{60000 \times 86 \times 2 \times 7}{72 \times 100 \times 9 \times 8} = WP.$$

Cancelling

$$\frac{25 \times 43 \times 7}{6 \times 9} = 139.$$

This gives us 14 lbs. more than we require, but as plates in this country are not rolled to dimensions of thirty-seconds of an inch, we must use the 7-16 plate. This is a good fault, as it allows a margin against corrosion, and makes practically no difference to the heat conductivity. It also allows a bigger factor of safety. This will now be, assuming the pressure is carried at the figure first set

$$\frac{60000 \times 86 \times 7}{72 \times 100 \times F \times 8} = 125.$$

Then

$$F = \frac{60000 \times 86 \times 7}{125 \times 72 \times 100 \times 8}$$

Cancelling

$$F = \frac{86 \times 7}{5 \times 24}$$

$$F = 5.$$

Designing the Joint

We have now got the diameter, length and thickness of plate of the boiler, and now we will determine the particulars of the joint.

First, as we have seen in the articles on riveted joints, we will determine the pitch of rivets in the outer row. Then

$$\frac{P - D}{P} = .86$$

$$P - D = .86P$$

$$D = .14P.$$

Now we have 7-16 plate, we will have a 15-16 inch rivet, and if the diameter is .14 of the pitch, the pitch will be

$$\frac{.9375}{.14} = 6.69 \text{ or say } 6\,11\text{-}16.$$

Let us see how close this will come to the required percentage.

$$\frac{6.6875 - .9375}{6.6875} = 85.98\%.$$

This is sufficiently close for all purposes. We will now try the .rivet strength, using the pitch found. We

shall have four rivets in double shear and one in single shear, making eight rivet areas in all. Then by formula, using .7 as the constant for steel rivets in steel plates

$$\frac{a \times n \times C}{D \times t} = \%.$$

Then

$$\frac{15 \times 15 \times .7854 \times 8 \times 70 \times 16 \times 16}{16 \times 16 \times 107 \times 7 \times 100} = \%.$$

Cancelling

$$\frac{9 \times .7854 \times 20}{107} = 132\%.$$

This gives a much higher percentage of rivet strength than plate strength. By reducing the rivet diameter to ⅝ inch we get a higher plate strength, and a lower rivet strength, thus

$$\frac{7 \times 7 \times .7854 \times 8 \times 70 \times 16 \times 16}{8 \times 8 \times 107 \times 7 \times 100} = 111.3\% \text{ rivets}$$

and $\dfrac{6\ 11\text{-}16 - \frac{7}{8}}{6\ 11\text{-}16} = 86.9\%$ for plate.

This is more in proportion, and the smaller size rivets will reduce the cost of manufacture.

We will now see if we are within the limits of the maximum pitch as allowed by the Ontario regulations. The maximum pitch is found by using the formula—

$$PM = (C \times T) + 1\frac{5}{8}$$

where PM = maximum pitch,
C = Constant,
T = Thickness of plate in inches.

The constant C varies with the type of joint as follows:

No. of Rivets in one pitch	Constant for Lap Joints	Constant for Double butt Strap Joints
1	1.31	1.75
2	2.62	3.5
3	3.47	4.35
4	4.14	5.52
5	....	6.00

As there are five rivets in a pitch, we must use the constant 6. Then

$$PM = (6 \times .4375) + 1.625$$
$$= 2.6250 + 1.625$$
$$= 4.25.$$

Although the regulations omit to state this, in a joint with every alternative rivet omitted in the outside row, this formula should be multiplied by 2, and in this case the maximum pitch for the outside rivets is therefore 8½ inches, and we are well inside this dimension. The combined efficiency of plate and rivets must be found from the formula given in the regulations:

$$\frac{(A \times C) + (P - 2d)T}{P \times T}$$

This, as can be readily seen, is the shearing strength of one rivet, multi-

plied by the constant, and added to the net section of the plate between the outside row of rivets, divided by the unit section. Substituting figures we get

$$\left\{ \left[\frac{7 \times 7 \times .7854 \times 70}{8 \times 8 \times 100} \right] + \left[6 - \frac{11}{16} - 1\frac{7}{8} \right] \times \frac{7}{16} \right\} =$$

$$6 - \frac{11}{16} \times \frac{7}{16}$$

Combined Efficiency.

Reducing to decimals it becomes

$$\frac{.42 + 2.16}{2.92} = \frac{2.58}{2.92}$$

$$= 88.3\%,$$

which is higher than the plate strength as determined.

We must now determine the distance between the rows of rivets, and the exact allowable arc definitely decided by the regulations. Of course one could arrive at this distance by calculating the section of metal to be left to secure the necessary strength, but for the purpose of those passing examinations it will be better to work out the formulae set down by the Ontario board. The minimum distance between the outer and middle rows is found by the following expression, where V = the minimum distance:

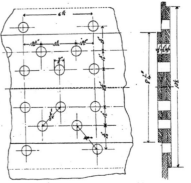

TREBLE RIVETED DOUBLE UNEQUAL STRAP JOINT.

$$V = \sqrt{[(11\text{-}20P + d)(1\text{-}20P + d)]}$$

In order to make these clear we will show the various steps in full. Substituting figures we get

$$V = \sqrt{[(11\text{-}20 \times 6.6875 + .875)}$$
$$(1\text{-}20 \times 6.6875 + .875)]$$

$$= \sqrt{[(.55 \times 6.6875 + .875)}$$
$$(.05 \times 6.6875 + .875)]$$

$$= \sqrt{[(3.6781 + .875)(.3343 + .875)]}$$
$$= \sqrt{(4.5531)(1.2093)}$$
$$= \sqrt{5.4793}$$
$$= 2.33$$

Therefore 2.33 inches is the minimum distance between outer and middle rows. For the distance between inner and middle rows we have

$$\sqrt{[(11P + 8d)(P + 8d)] + 20}$$

Substituting figures

$$\sqrt{[(11 \times 6.6875 + 8 \times .875)}$$
$$(6.6875 + 8 \times .875)] + 20$$

$$= \sqrt{[(73.5625 + 7)(6.6875 + 7)] + 20}$$
$$= \sqrt{[80.5625 \times 13.6875 + 20]}$$
$$= \sqrt{[1102.69 + 20]}$$
$$= 33.2 + 20 = 1.66.$$

Then 1.66 inches is the minimum pitch between the inner and middle row. For the minimum diagonal pitch between inner and middle rows the formula is

$$P = (3P + 4d) + 10$$
$$= (3 \times 6.6875 + 4 \times .875) + 10$$
$$= (20.0625 + 3.5) + 10$$
$$= 23.5625 + 10 = 2.35.$$

(Continued on page 68)

Relation of Fuel and Furnace to Heat Treatment

Technical Appellations Are Avoided as Much as Possible—Principal Fuels Used Are Discussed, Also Furnaces and Burner Nozzles—This Article Should Be Read by All Interested in Heat Treatment.

By R. P. MAYNARD *

IN dealing with this subject it is not the purpose of the writer to avoid in so far as is possible any technical appellations, and to refrain, moreover, from reference to any commercially advertised appliances, such as furnaces and burners of which there are legion. My vocation brings me in intimate touch with the heat treatment of steel in all its many phases and in shops utilizing appliances from the most modern to the crudest makeshift.. In the course of this constant contact with the steel treater, I find that he is invariably well trained in the handling of both high and low carbon steel, and possesses almost infallible knowledge of the requisite textures, grain and lustre of the metal in all its varying degrees of carbon content, and that, while he is aware that heating is the most important of all the operations to which steel must be subjected, still I find that in the majority of cases he is grossly ignorant of that which makes the heat possible—namely, fuel application.

Besides the steel, there are three other prime elements involved in its treatment, which are (1) furnace; (2) fuel, and (3) operator. The latter is often like the average automobilist, whose knowledge of his car extends only to the hood and footboard. It is my intention to convey henceforth a brief discussion of combustion and its application, augmented by a few criticisms of what is hoped will be of a constructive nature.

The principal fuels used in heat treating are bituminous and anthracite coals, coke, fuel oil, natural gas and gas manufactured from solid or liquid fuels. These fuels consist of combinations of carbon, hydrocarbon and hydrogen, which, when combined with oxygen at certain temperatures, unite and form a combustible mixture. The source of oxygen is, of course, the air, and the more intimately it is mixed with a given fuel, the more efficient the combustion derived.

Fuels Used

Solid fuels are gradually being displaced for heat treating by liquid and gaseous fuels, with the latter, gas, predominating, principally because of its cleanliness and convenience. Of all the available fuels used commercially, gas is the only one which the steel treater does not have to handle, store and distribute, and is, moreover, unique in that it is not paid for until after being used.

Most gas companies maintain a corps of trained industrial fuel engineers, whose business it is to assist the steel treater in his heating problems, and it is partly because the oil companies do

* From Journal of Steel Treaters' Society.

not offer this engineering service that this splendid fuel is not in more general use for the heat treatment of steel. Properly burned, the use of oil affords a marked economy, especially when used in connection with the larger furnaces, for in the use of oil the heat treater is simply operating miniature gas plants, as the liquid hydro-carbons are expanded into a gas and united with oxygen to produce heat. To burn a given quantity of oil requires the same amount of air as is required to burn the equivalent heat value of gas. The luminosity of the flame is due to the carbon content of the fuel, which insures a soft reducing furnace atmosphere, thus eliminating the possibility of scale.

Must be Thoroughly Screened :

To properly burn a liquid fuel requires that the oil be thoroughly screened, as no oil burning system can burn oil that is not clean. The liquid must then be as finely divided as possible, and while in this atomized state must be mingled with the proper amount of air to support combustion. The liquid hydro-carbons can be finely divided by several methods. Air at a high pressure may be made to travel at cross-current with the oil stream, and the oil caused to spray by the impact of the compressed air. Steam may also be used in the same manner, with the added advantage that it imparts a certain amount of hydrogen gas which unites with the oil. However, latter day practice is treading away from compressed air, and the prevailing custom shows that the average shop is using air at only six to 12 oz. pressure with oil at from 20 to 50 lbs., and with gas at from two to four oz.

One of the principal sources of grief which the heat treater has to contend with is the blower problem. The two prevailing methods are the central air installation as embraces the distribution of air through pipe lines, and the unit blower application, where a small fan is connected to each furnace. Both processes have their advantages, but it has been conclusively proven that no fuel burning equipment has as low power factor as the small unit blower, which enables not only individual control, simplicity and quietness of operation, but which has frequently saved its first cost from the difference, as against the central system, in operating power during a reasonably short period. No matter how small any installation of a central system, where additional equipment is at all probable, the subsequent additions to furnaces usually exceed the original provision for air, whereas the unit system easily provides for this.

Just a word about furnaces and burner nozzles. There are more than a dozen large builders of reliable heat treating furnaces in the United States, and their products are, in the main, uniform. While many of these manufacturers have made progressive strides in furnace design and construction, still there appears to be in most furnaces—especially in the semi-muffle type—a conservative standardization. Whether this is due to lack of imaginative ingenuity, or simply a frank admission that the present type of furnace has reached its perfection, I do not pretend to advance. From time to time a builder will improve a door lifting device, or add sturdier legs, or heavier refractory linings, but I will venture to predict that the coming year will see some radical departures in furnace construction.

Already I notice a tendency to decrease the number of burner nozzles on heat treating furnaces. It has ever been the theory of the furnace builder that inasmuch as heat must be applied with uniformity and not in spots only; numerous burner tips must be located along the side of the furnace. May I not draw a comparison that the days of the bucket brigade are over, and that with two healthy streams of water a modern fire department can do more than with 1,000 water buckets? Any furnace fired with 20 nozzles can be heated just as efficiently with two, or at most, with four, when properly spaced and of adequate capacity. It is like comparing 20 little water falls to the concentrated power of Niagara.

Let me also advise against the use of the "pepper-box" or "honeycomb" type of nozzle now in general use. These nozzles can easily withstand high temperatures in a large combustion box, but are apt to fuse and oxidize in a small furnace. They should never be used in a high speed furnace. Also I find that in most furnaces the nozzle tips are set well back in the channel-port of the refractory lining, and if the nozzle tip is of the pepper-box variety, the restricted channel only shortens its usefulness. My experience has convinced me that a nozzle should be of a single orifice, set flush with the inside wall of the furnace lining and cemented in so that no metal is exposed to the heat.

A thing of paramount importance in heat treating is the furnace atmosphere, which is influenced directly by the proportions of fuel and air, as well as the intimacy of their mixture. The result of incorrect proportions of fuel and air and imperfect mixture thereof is often the cause of injury to the steel, and will even cause carbonizing boxes to scale ex-

cessively and become prematurely us.-less.

Furnace atmosphere also has its bearing on the heating of high-speed steel. In shops where the high-speed furnaces are kept constantly at a working temperature, there is little chance of damaging the most delicate high speed steel. But in the hardening room, where a high-speed furnace is used intermittently there is danger because the operator will employ a highly oxidizing blast to bring a cold furnace up to heat quickly, and then neglect to change the character of the flame when entering the steel.

This calls to mind an incident which happened recently, which might well be called the "Parable of the Small Gas Meter." A steel treater had a carbonizing oven and cyanide furnace, the combined gas consumption of both taxing the capacity of the single gas meter, but he was not aware of it. He later installed a high speed furnace, and in making the burner installation of this appliance I suggested that he had better notify the gas company to install a larger meter. Evidently he thought a larger meter would register more gas in proportion than the present meter, thus increasing his unit fuel consumption on his present basis of production. At any rate, the small meter remained, and when all three furnaces were in operation he could not keep from "pitting" the high speed steel because of insufficient gas and excess air. The moral of this is that if you use gas in your heat treating room, always consult the gas company in making changes or additions to your equipment. They are genuinely anxious to sell you service—not merely fuel.

Another thing, never use a furnace that is too big for the work required. It is poor economy. I know of a shop in Chicago where they bought a big semi-muffle furnace, which plays as many roles as the bass-drummer in an "Uncle Tom's Cabin" troupe. They use it for forging, annealing and case hardening, and I might also add that in the latter practice this same firm also seal their carbonized boxes with Portland cement instead of fire-clay.

There are constantly little occurrences in the hardening room which are perplexing, especially to the inexperienced steel treater. I have often heard a heat treater assail the gas·company for poor gas service, when all in the world the matter was that the salt mixture in his potassium chloride (clean heat) bath was ready for the embalmer, or else perhaps he was quenching his steel in oil or water which was far too hot. On the other hand, I have seen the heat treater blamed when gears which he had hardened and sent on into another department had subsequently been ground with the wrong sort of abrasive wheel.

The Question is "Will It Work?

A Claim That the Introduction of Chemical and Metallurgical Science Has Not Proved to Be a Panacea for All Industrial Evils.

By HENRY TRAPHAGEN

THE introduction of chemical and metallurgical science in our industries has not proved to be the panacea for all industrial evils that the layman imagined it to be. Profound scientific learning or skill in ·chemical technique will not always solve our industrial problems. After all is said and done, it appears to me that the one big outstanding dominant question in the industries must be, "Will it work?"

Is the material suitable for the purpose for which it is sold. Is the material capable of weathering abuse in use. Is it uniform; is it the finished product of skilful craftsmanship or a flimsy makeshift designed to meet an arbitrary formula? These are some of the questions that the conscientious manufacturer must ask himself if he would know the real worth of his product.

The consumer is little concerned with how a product is made; it is of little moment to him what the manufacturer puts into the material. But he is vastly interested in how much service he can get out of it. Again, it is the same old question: "Will it work?"

It requires more than a knowledge of metallurgy to run a steel mill. The most expert chemist on earth cannot make a good drill if he is not a mechanic, nor can he devise tests that will show up the practical defects in such a drill if he is not acquainted with the mechanical details of such tools. The most carefully computed mixture that ever graced a charge slip will not make a good heat if the cupola ·tender does not know how to patch his lining or if the chemist does not know how to recognize physically defective coke. All the chemical analyses in Christendom will not make a porous casting sound. The best ingot that was ever poured will not produce a trustworthy rail if the roller tears and pinches the billet in rolling, and so it goes. No product is any better than its weakest point. There is that ever present chain of operations, each link separate, yet indissolubly joined with the other. If one link is faulty the chain is useless. Truly no product is any better than the man who. makes it. And this is doubly true in· the various metal industries.

Science is Wonderful

Metallurgical science is a wonderful thing. Wizards of the past and present are the creditors of the world, for the debt can never be fully paid. Still it must be remembered that metallurgy and chemistry are not cure-alls. The scientist cannot make a workman out of an idiot; nor can he evolve that mystic formula that can wipe out and supersede craftsmanship.

Things of metal have always been and always will be products of personal skill and experience. It would be well for us to pause, think, and think deeply and earnestly, before we embrace any mechanical contraption that purports to be a substitute for human intelligence. The light of science has shown the way through dark and devious paths. The flaming torch of the pioneer has given way to the searchlight of organised science. Yet that all-powerful light is in imminent danger amid the disconcerted flashes of a million lesser ill-directed lights.

Weeds grow thicker and faster than flowers. The broad-minded, practical, experienced scientist is one thing. The arbitrary, shallow, inexperienced chief chemist or metallurgist, found in many plants, is quite another. How often do we see perfectly good steel rejected because the phosphorus content may be one thousandth of one per cent. above an arbitrary limit. And on the other hand, how often do we witness the curious procedure of accepting a carload of defective castings on the evidence of a doctored and carefully nursed test-bar.

Producer Must Analyze

The chemist and metallurgist in the plant of a large producer functions in quite a different manner than the fellow in the shop of a consumer. The producer must analyze in order that the product may be sorted. He must attempt to cater to the wants of many and varied customers. Hence, the number of laboratories that are to be found in our large producing mills. Keep·in mind that a chemical analysis is a sorting process and nothing else. It is no guarantee of workmanship, it is no protection against physical defects, and the fact that metal may be of proper composition will not prevent a sloppy, ignorant workman from ruining good material.

The gentleman who buys must have a laboratory. He must imitate the big fellow. He must analyse a package of drillings from every shipment; he must copy impossible specifications from handbooks and force them upon the producer. He must rave about accuracy in composition, and at the same time let the question of workmanship go to pot. He sees fit to allow a youngster fresh from college who cannot drive a· nail with six hammers dictate to the producer, whose product is backed up by years of experience.

Now what has been the result of all this? Naturally, the customer must be pleased. The producers have carefully

Continued on page 69

WHAT OUR READERS THINK AND DO

HERE'S A PROBLEM

Small jobbing shops frequently are called upon to do work that taxes the ingenuity of the man in charge, and when small quantities of certain pieces

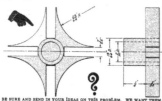

BE SURE AND SEND IN YOUR IDEAS ON THIS PROBLEM. WE WANT THEM.

are required to be produced, the time allowed does not permit of expensive jigs and fixtures by which the work could be carried out successfully. On the other hand the piece may be so intricate or necessitate such close accuracy that great care must be exercised in deciding on the process of manufacture so as to avoid loss through the possibility of errors arising from the method adopted.

Such a job is shown in the accompanying sketch. These pieces represent a Geneva movement that forms part of the mechanism for moving picture machines. Unless the different sections of this piece are absolutely uniform and to dimensions they are not suitable for the duty they are required to perform. A short time ago a small shop in Montreal secured a trial order for about one hundred of these, with the prospect of obtaining a considerably larger order if the work proved satisfactory. The trial quantity was completed, but quite a number of pieces failed to pass the customer's inspection. At present the superintendent of the shop is devising other methods of doing the work, and he is desirous of learning of ways and means from some of the readers of Canadian Machinery as to the best method of accomplishing the job.

The piece is made of cold rolled steel and the dimensions and uniformity must be very accurate. Production would approximate 5,000 yearly.

Readers are requested to send in their ideas as to the best way of manufacturing this Geneva movement, and an estimate of the cost of production per 100. We shall, of course, pay for all acceptable material as usual.

BORING JIG FOR THREE-PIECE BRASSES
By J. H. Houldsworth

The accompanying drawing illustrates the assembly of a box jig used for boring out the two holes in the three-piece brasses for the cutter levers of the tiffin nail machines. This is a marked improvement over the old method of holding the brasses in a chuck, as it insures greater accuracy, both in centre distance and alignment of the holes. In the fixture shown the jig A is secured

to the special face plate B by a bolt on either side, and midway in the centre line of these bolts is the swivel bolt E, held in position by the nut D. Ample space is provided for operating the nut by fitting the auxiliary face plate with four bearing lugs. The jig face plate is secured to the lathe face plate by three equally spaced bolts. The three brasses F, F, F have semi-circular grooves on the sides, and are firmly bolted together by means of the round strap bolt G and the nuts H, H. The swinging clamp I is provided with two small strips for clamping the brasses in the fixture. The four screws J are used to line up the brasses before boring. The swivel bolt E is located the same distance from the centre of the two holes, so that when one is bored the clamping bolts are removed and the stop pin K drawn back, to permit of turning the fixture through 180 degrees, when the stop pin engages with the small hole L, thus bringing the small hole in the correct position for boring. When these brasses are worked in quantities this jig is highly efficient and economical.

Want Locomotive.—The Department of Railways and Canals, Ottawa, is asking for tenders on a new or second hand switching locomotive: One (1) only 0-4-0 switching engine, 50 or 60 tons, saddle tank type preferred, but would accept engine with tender if former not obtainable. Quote price for delivery at Government siding, Mervitton, Ont., freight, duty and all other charges paid.

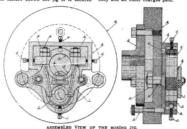

ASSEMBLED VIEW OF THE BORING JIG.

MAKING EFFICIENT BEVELS

By W. S. Standiford

ONE of the most important tools in frequent use by machinists and other metal workers for testing the accuracy of angle work as it progresses is the bevel gauge. This instrument, as bought in the hardware stores, has two drawbacks which are as follows: The blade is made too wide so that when a machinist, etc., has to measure an angle on the edge of a narrow slot which he is turning in a lathe, the tool is altogether useless, as it cannot be used at all in such a case. Where a very shallow slot, having sufficient width to allow insertion of bevel gauge blade, is being turned, it will be found that the end of the opening slot in blade does not allow the tightening screw to approach the blade point, which is essential in order to measure angles on shallow depth work.

Most bevel gauges have the end of slot placed about $1\frac{1}{8}$ inches distant from the blade's extreme point, thus preventing the use of the tool for narrow slots and shallow slot work. Of course, the end of the hole can be filed down toward the point and the blade ground down to narrow dimensions on an emery wheel so that it could be used, which was done by the writer and enabled him to finish a piece of emergency work on a breakdown job on which, as is usual in cases requiring rapidity of action, not much time can be lost; the gauge thus ground down did not last many years as the blade, having been made out of too soft a steel, wore out rapidly. Since then, the writer has made the gauge depicted in Fig. 1 out of high carbon steel, and it still shows no appreciable wear.

To make one like it proceed as follows: Get from a hardware store a small threaded square shouldered bolt of sufficient length to suit thicknesses of steel for blades, and $\frac{1}{8}$ or 5-32-inch diameter. If a bolt with a square shoulder is unobtainable, it will either be necessary to make one or else rivet a piece of steel placed lengthwise on one piece so as to prevent the bolt from turning when the wing-nut is tightened up. Select a bolt with a wing-nut on it instead of one having a round thumb screw, as the former allows of greater tightening pressure.

If a bolt of $\frac{1}{8}$ or 5-32-inch is not procurable, send away to any model maker's supply house. It is essential to use as thin a bolt as possible in order to make use of as narrow a blade as is practicable, which allows of working in small slots. Bevel gauges can be made any desired size, but the one illustrated is just the right length in ease of handling and for general all-round work. Make the slotted blade out of $\frac{1}{8}$-inch thick steel, the slot to be made first and of sufficient width so as to let the fastening bolt slide easily along it. The blade should be made either $\frac{1}{2}$ or $\frac{5}{8}$ inch wide, these sizes making a tool that can be used in various chamfering work and other difficult places where the ordinary article would be entirely useless. Five inches long is a good length to make the movable blade and also the horizontal ones. The latter is to be constructed out of two pieces of $\frac{1}{8}$-inch thick steel, both pieces should be made the same size, the width being one inch and having its sides filed, planed or ground to a straight edge. Make a piece of steel of the same thickness as the movable blade, and as wide as the stationary ones. Fasten this piece on one end and between the two blades by rivets. This steel washer allows the movable blade to slide in a slot and also lets it go to one end of the stationary pieces. When the wing-nut is tightened up, the side pieces are sprung together and caused to grip the central angle blade tightly, the result being that it will stay in any position and withstand a fair amount of hard knocks before it alters its setting. In use this type of device proves to be very handy as it can be adjusted for chamfering work on both deep and very narrow slots. The flat edge, or side, on the movable blade makes a very handy depth gauge, thus two tools are combined in one, either of which can be used separately when the other is not set.

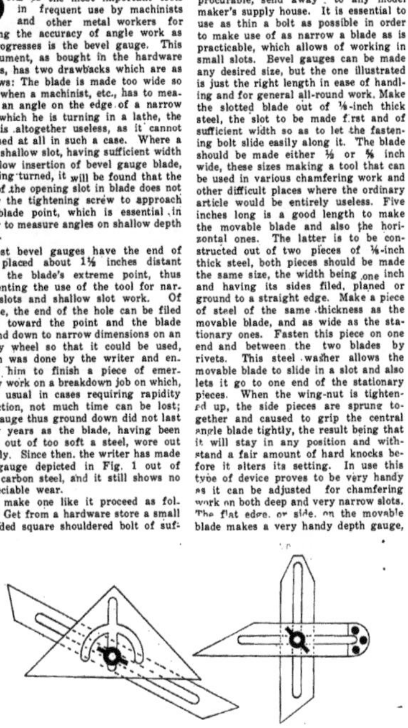

FIGS. 1 AND 2—SHOWING TYPES OF GAUGES DESCRIBED IN TEXT.

Another novel and very effective form of bevel gauge is shown in Fig. 2. This tool is made out of high carbon tool steel plate annealed; $\frac{1}{8}$-inch thickness is a good size. It is cut out in triangular shape, the shoulder on the fastening bolt sliding in the curved slots. When the writer first made this gauge the slots were curved in circular form as shown in the photograph of sketch, but if I ever make another one, the slots will be extended out to the extreme corners of the triangle, as I believe that it would be much better, although experience shows that the tool is very handy as it is. The beam, like the three-cornered piece, is made out of $\frac{1}{8}$-inch high carbon steel plate, as this thickness makes a very compact and long wearing tool. No description of it is necessary, as the illustration shows the construction plainly, it only being essential that the edges of the square shouldered holding bolt should be rounded slightly with a file, so as to allow it to slide freely in the curved parts of slot. This gauge is shown with its beam set at an angle to the long side of the triangular section, but as will readily be seen, it can be used with any one of its angles, thus allowing of chamfering work in and around and close to metal shoulders on metal being turned in the lathe or for shaper or planer work. Metal shoulders being in the way generally interfere with the setting and working of the ordinary type of bevel gauge. The sharp point of the triangle will allow it to be set and used in chamfering the edges of a very narrow and shallow slot.

In conclusion, any mechanic who constructs either type of instrument, will find them to give good satisfaction on all classes of work, and if they are made out of high carbon steel, long wearing tools will be the result of the construction.

PURIFYING OIL BY CENTRIFUGAL ACTION

For many years there has been in use in most of the large engineering works of this country a centrifugal separating machine which removes the oil from iron and steel turnings or borings, and thus permits it to be used over again. The idea which underlies this machine has now been carried further and applied to the removal of dust and dirt which unavoidably collect in oil used in the lubricating systems of marine and other large engines. The new machine is very similar in form and operates on the same principle as the now well-known centrifugal cream separator. The foul oil is fed continuously into the machine through a strainer which catches any gross particles, precisely as milk is fed into the cream separator. The purified oil escapes continuously, while the dust and dirt collect on the inner surface of the separating drum. Provision is made by which the attendant can instantly tell when the machine is choked with the deposited impurities.

For many years an employee of the Steel & Radiation Co., Limited, Toronto, H. M. Alexander relinquishes his position with the company to become secretary-treasurer of the Toronto Refining & Manufacturing Co. and the Mona Oil Co.

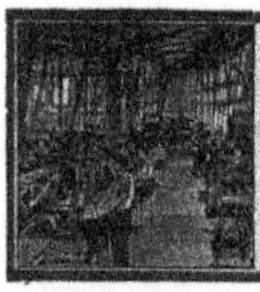

DEVELOPMENTS IN SHOP EQUIPMENT

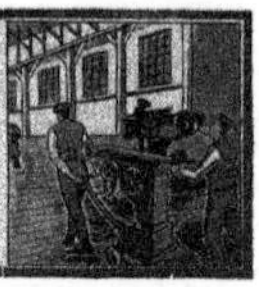

AUTOMATIC TURRET LATHE

THE illustration depicts a turret lathe recently put out by Alfred Herbert, Ltd., Coventry, England, with offices at 1-3 Jarvis Street, Toronto, Canada. This machine has a maximum swing over the bed of 18½ inches, and is known as the No. 5 model. All operations are automatic, except chucking, and the machine stops automatically at the finishing of a piece. There are a number of interesting features to the lathe, such as the longitudinal adjustment of the headstock, the independent operation of the front and rear cross-slides, and the low position of the spindle, which makes setting up an easier task.

Constant-speed drive by a 4 inch belt is employed, the 10 inch driving pulley running on ball bearings. The drive can be engaged for starting the machine by means of a friction clutch operated by a hand lever. The headstock is adjustable along the bed for a length of 6 inches and its mechanism is completely enclosed, the gears running in oil.

Spindle Speeds

There are five spindle speeds that are changed automatically. It is claimed that the changing is performed instantaneously and silently and that it can take place while the tools are cutting, if necessary. Change gears are provided to give six principal or substantive speeds, by which the range of the automatically-operated speeds is changed. Since these five automatically changed speeds can be used with each substantive speed, a total number of 30 steps is provided in a range from 14 to 411 r.p.m., the maximum gear ratio being 28.5 to 1. If desired, change gears can be furnished to give additional speeds.

The hollow spindle is 5⅛ inches in diameter at the front neck and has a flange on the nose 10 inches in diameter. It has a bore of 3⅝ inches, runs in white metal bearings and is provided with a ball thrust bearing. It can be stopped automatically at any instant and then re-started automatically in time for the next cut. This action is valuable, particularly to enable the withdrawing of a tool at the end of a cut without leaving a helical mark on the work.

The turret has four faces, machined square with the spindle. The tool holes are 2⅛ inchs in diameter. The turret is mounted on a slide having a working stroke of 13 inches. It is indexed at the extreme back position and then clamped automatically.

The cam controlling the action of the turret consists of a drum running in oil and driven directly by a worm. It is provided with two grooves, one for moving the turret slide and the other for indexing the turret. Since the latter action occurs only when the turret is in its extreme back position, all of the working stroke is available for actual machining operations. The turret-actuating groove is cut like a screw-thread and returns upon itself, so that the turret moves with a smooth and steady motion. It is claimed that no irregularity or jumping occurs, even when taking heavy cuts.

For each complete forward and return movement of the turret the cam makes three complete revolutions. Rotation of the turret to present a different tool to the work requires but 1.1 second, and the time for a complete cycle of the machine at high speed is only 48 seconds. The turret is 11¼ inches in diameter, and the centres of the tool holes are 4⅛ inches above the slide. The distance between the face of the turret and the face of the spindle varies from 15 inch as a minimum to 34 inches as a maximum.

Cross-Slides

The front and back cross-slides are independent of each other in their action, and can be set to work either at the same time or separately, as required. For example, roughing and finishing facing cuts can be taken simultaneously. The cross-slide cams can be adjusted to operate at any part of the cycle of operations. Each cross-slide is provided with a double toolholder and a stop witk screw adjustment for regulating the diameter turned.

The centre of the spindle is 4½ inches above the top of the front slide and 3½ inches above the back slide. The slides have adjustments along the bed of 7½ inches independently of each other. The horizontal distance from the spindle to cross-slides can be varied from nothing to 11½ inches, power feed being available for a stroke of 4½ inches. The maximum diameter admitted between the cross-slides is 16 inches. Ample room is provided under the tools for the chips to get away, the chute for this purpose being plainly visible in the illustration.

Ranges of Feed

The "self-selecting" feed motion provides for both the turret and the cross-slides, ranges of 7 feeds, any one of which can be brought into action instantly at any part of the cycle, even during a cut, if required. The feeds are driven from the headstocks, but the quick or idle motions are driven from the constant-speed driving pulley, so that the idle motion always takes place at the same rate of speed. It is claimed that the range of feeds is sufficiently broad so that it is never necessary to change the feed cams. The feed of the turret slide varies from 1 to 144 revolutions of the spindle per inch of feed, and that of the cross-slides from 40 to 366 revolutions per inch.

The bed is a box casting. It is built low, the spindle being only 40½ inches above the floor, so that the machine can be tooled up and attended more conveniently. The tray extends the full length of the bed and provides ample room for chips.

A pump with fittings, including splash guards, is supplied at extra cost. Oil supply through the turret can be furnished for use with hollow tools for carrying lubricant to the cutting edges.

A back-facing attachment can be furnished if desired. It is claimed that its

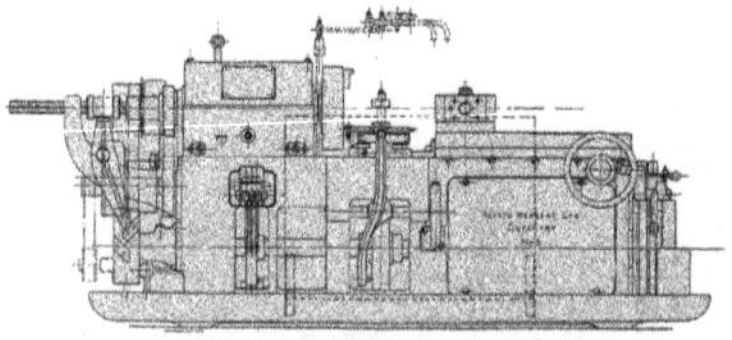

LINE DRAWING OF THE NO. 5 TURRET LATHE.

use often enables a piece to be completely machined at one setting instead of two.

The 15 inch Coventry chuck is suitable for use with this machine, although special chucking fixtures can be furnished for work that cannot be handled by it. A standard tool outfit can be supplied and the maker recommends its use for general work. The outfit includes a centering tool, facing tools for the cross-slides, boring bars and combination boring, turning, and facing tools that are provided with an overhead support.

The machine is intended to be driven by a 7 h.p. motor. The floor space required is 6 x 12 ft., the approximate net weight is 8,500 lbs., and the size when boxed for shipment is 340 cu. ft.

COMBINATION TOOLHOLDER

The toolholder as illustrated has been placed on the market by Maurice H.

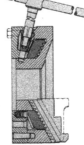

GENERAL VIEW OF THE HOLDER.

Derringer, 3133 N. Eighth Street, Philadelphia, Pa. It is intended for medium sized work on lathes from 12-in. to 20-in. swing. On the end shown at the left is a yoke which can be swung to either side of the bar. One side of the bar is adapted for holding square stock up to 5-16-in. and the other side, formed as a V-block, will hold round stock from ¼ to 9-16 in. in diameter. This groove is useful for holding indicator stems as well as boring or internal-threading tools. When the holder is placed in position in a lathe, the screw in the lathe toolpost can be tightened on the end of the tool, thus holding it rigidly. The yoke may be easily removed by taking out the taper pin which holds it. The other end of the holder is provided with a goose neck which can be locked in position at any angle. This neck will hold 5-16-in. tool stock, being especially adapted to thread cutting.

IMPROVED SPIRAL CHUCKS

The Charles Taylor, Limited, of Birmingham, England, have placed on the market their patent improved spiral chucks. These chucks are being handled in Canada by Messrs. Williams & Wilson, 84 Inspector St., Montreal, and they are exclusive agents for the same.

These chucks, of which we show illustrations and line drawings, embody many interesting and important features of design that add greatly to their efficiency and durability under continual and exacting service, and are the recent outcome of 35 years of manufacture and study of self centring chuck requirements.

The construction of the chuck is such that the movement of the jaws is not at right angles to the lathe spindle, but inclines inwards at an angle of about 30 degrees. This design permits of a thrust upon the spiral rack and more directly back of the jaws, thus minimizing the tendency to tilt that is said to be always present in chuck jaws where the movement of the jaws is square with chuck axis. The V shape of the rack permits of a finer pitch, giving greater gripping power and the inclined position provides for a better distribution of the pressure, as the bulk of this is transmitted through the solid

PHOTOGRAPHS OF CHUCK.

surfaces of all moving parts are hardened and ground. This adds very greatly to the life of the chuck. Special provision is made for taking up any slight wear of the moving parts. Every part of these chucks is made to limit gauges and is interchangeable and every chuck is guaranteed to grip true to within .002 measured 3 inches from face of jaws. They are claimed to be the tightest gripping self-centring chucks made.

The strength of the parts of this improved chuck permits of a length of key handle approximately twice that of ordinary chucks, giving corresponding greater grip and providing this extra grip by leverage that loses nothing in extra friction. In addition to the usual 3-step and bar jaws, "soft" blank jaws are supplied. These blank jaws are hardened on the teeth to withstand wear, but are left soft on front face and grip, so they can be machined to hold any special or odd shaped article. They are of high class case-hardening steel and after machining to hold the special article can be case-hardened, if necessary.

metal of the jaws and square with the axial line of the rack.

The pinion for operating the spiral rack is non-projecting and the bearing

LINE DRAWING OF CHUCK.

NATIONAL STANDARDIZING BODY ORGANIZED IN AUSTRIA

There has been formed in Austria a national engineering standardizing body. It is called the "Normenausschuss der Osterreichischen Industrie" and is organized under the auspices of the "Hauptverband der Industrie Deutsch-osterreichs." The secretary is Dr. Jaro Tomaides.

This is the tenth national standardizing body to be formed, the others being in Belgium, Canada, France, Germany, Great Britain, Holland, Sweden, Switzerland and the United States. With the exception of the British Engineering Standards Association, whose work dates from 1901, they have all been formed since beginning of the European war.

The MacLean Publishing Company
LIMITED
(ESTABLISHED 1887)
JOHN BAYNE MACLEAN, President. H. T. HUNTER, Vice-President
H. V. TYRRELL, General Manager.
PUBLISHERS OF

CANADIAN MACHINERY
MANUFACTURING NEWS

A weekly journal devoted to the machinery and manufacturing interests.
B. G. NEWTON, Manager.   A. R. KENNEDY, Managing Editor.

Associate Editors:
J. H. MOORE   T. H. FENNER   J. H. RODGERS (Montreal)
Office of Publication: 143-153 University Avenue, Toronto, Ontario.

VOL. XXIV.   TORONTO, OCTOBER 28, 1920   No. 18

PRINCIPAL CONTENTS

Keep Buyers in the Market

PRICE deflation is a matter that many papers do not care to say much about, as it is a question that can cause a great deal of misunderstanding. There are always those willing and ready to read into an article things and ideas that were never intended.

But the question of prices is so tied up with every phase of business that it must be considered. Men are talking about the process, and it is more marked now than it has been for many months. It was expected when the armistice was signed that there would be a rapid drop in prices. In fact many people were looking for a crack right then and there, but it was postponed, and from the way things are shaping now it would appear that the postponement had been all for the best.

The danger in this situation lies in the fact that we may have a rapidly falling market, a thing that is dangerous and undesirable. Were this to come to pass we would find buyers leaving the market entirely, or if they remained it would be only to supply their immediate and urgent wants, and there is nothing more miserable than that sort of trading.

Canadian Machinery has discussed this matter with many men in the iron, steel and machine tool field during the past weeks, and when one finds Judge Gary making it the centre of his annual address to the Iron and Steel Institute there is little room to doubt that it is the real question of the hour.

The desirable solution is to keep the buyers in the market. By this we are not urging any buyer to go in hard now. His own common sense, and his knowledge of his own field, must, in the last analysis, determine the size or value of his purchases.

The business world, in many cases, was quick to respond to the touch of higher costs with higher selling prices. Some of them undoubtedly went the limit in this and worked on the principle of piling on all the traffic would bear.

When the tide is running in the opposite direction these same interests, if they are to serve their field well, should be just as keen to make adjustments in the other direction.

If there is a determined all-round resistance to price declines, if there are no mutual concessions, no tempering of contracts to meet changed conditions in any line, then we are facing a serious situation.

What will result? The buyer will be forced to keep out because he will be afraid of being ruined if he buys at high prices, with the prospect ahead of him of selling at a lower level.

If, on the other hand, there is a spirit of reasonable business action, of concession wherever possible, of a desire to consider the proposition and the problem of the other man, then it may be possible to get the country cleared out of the high-cost materials in all lines, much of which is still held in various centres. If the whole loss were shouldered on any one section it would mean ruin in many cases.

The manager of one of our large establishments puts the case well as follows: "Let us be reasonable, and not afraid; we are going through the period of deflation that we thought was coming two years ago, and until we get through it we will never be down to a solid basis where we can go ahead and do constructive work again. Let us guard against doing anything that will bring us face to face with unemployment. Let us keep the buyers in the market even if we make very little money—this is the market even if we make very little money—this is not a season of large profits, and we must realise it. This situation will work out well if we keep our heads, use common sense and attend to business. I am not afraid of it."

Nothing But Wages

IT seems difficult to realize at times the length to which workers have gone in making demands, each of which was like a millstone tied to the neck of industry. In Italy, for instance, when the shops refused a 50 per cent. wage increase, unions determined on the following programme in order to coerce the employers, which is given in the words of the manifesto issued by the labor leaders:—

1. Production must be reduced to a minimum.
2. Work must be simulated, thus obviating any absurd pretext on the part of the management to resort to dismissal or lockout.
3. No files or other tool should be used unless they be in perfect order and fully suitable to the job on hand.
4. Every repair to machinery must take the longest possible time.
5. Nobody, for any reason, should attend to a different job from the one on hand, as repairs, oiling, etc.

Finally it was stated that if employers dismissed the workmen they should take no notice of the dismissal, but continue to enter the plant. Should the management resort to forcible ejection the men were instructed by their union leaders to proceed to violence. Should a lockout be enforced it was declared that the works should be re-entered at all costs, even by breaking in.

No doubt you will say that was an extreme case. Quite so, but these extreme cases are simply the logical outcome of any system that looks upon successful negotiation of wages and working conditions as securing a jug-handled document, with the handle in their own grasp.

Employers to-day are willing in many cases to meet their men on grounds quite different to those of some years past—if they only get the chance to do it.

There is a new school of employers, and they are, for the most part, vastly different to the old autocratic type.

It is a serious thing for a man himself when the matter of hours and wages get so impressed upon his mind that he speaks, thinks and acts only in this narrowed sphere.

The employer who will not recognize initiative, effort, and industry may exist now, but if he does he represents a fast-dwindling minority.

The sense of accomplishment is greater than the recompense of wages, and the realization of work well done cannot be measured in terms of dollars and cents.

Judge Gary Sanely Optimistic About Future

(By a Staff Representative)

NEW YORK.—The address of Judge Gary to the members of the Iron and Steel Institute is looked forward to each year as a statement that will give the estimate the steel world of United States has of business conditions, and prospects of their betterment or otherwise.

Judge Gary spoke plainly on this matter. There was no effort to mince matters. When the speaker was through reading his address there was a feeling that he had indulged to too great an extent in a description of affairs in Europe, but when, in the lobbies afterward, steel men started to analyse the statement, they found that he had put his finger on the real situation, viz., that it was possible to resist price deflation to the point of keeping buyers out of the market, or, on the other hand, make the concessions necessary to keep the buyers in the market.

Just What Judge Gary Said

Speaking of the matter of prices, Judge Gary said:—

"In certain lines of the iron and steel industry there have of late been some decreases in the volume of new business and also voluntary reductions in selling prices. I consider this decidedly healthful. All, or nearly all, of us have for months been unable to supply the demands of our customers as to quantities or deliveries and our prices, considered as a whole, have resulted in profits. As a matter of course some adjustments will need to be made. The average of the general scale ought to be reduced equitably and relatively. Without referring to individual cases or lines of general business, I believe in many instances prices have been outrageously high. This observation applies more especially to middlemen, so called, and to smaller departments of industry. It also includes employes in certain trades; but it does not pertain under present conditions to the masses of workmen.

"The present tendency is toward a lower, more reasonable and fairer relative basis. The whole community desires and strives for this. The difficulty is found in the fact that every individual is perfectly willing that all others shall make reductions—the larger the better. As there was more or less a scramble for higher and still higher prices when they were advancing, there will be just as much selfishness in the enforced use of brakes when there is a tendency toward decreasing prices. Now, a general public, including particularly those who are neither sellers nor buyers to a large extent, will in one way or another bring about a fair and reasonable readjustment of prices. The law of supply and demand will be the principal factor.

"What shall be our attitude 'in these circumstances'? I answer:—We must evidence the same disposition which was displayed before the Industrial Board in March, 1919, when our steel committee co-operated with the Government's representatives in the endeavor to secure a general, equitable, orderly and methodical reduction in the prices of all commodities and services. Let us be reasonable and just, reducing our prices if and when other reductions and costs permit, and then with level heads, clear minds and honest convictions, stand solid as against panic or lack of confidence in the industrial situation. Let us strive to be right. If we are right we can be determined and courageous. Let us as individuals consider the interest of all others. Our business is basic. It is perhaps one of the most important. We may, we must, exert an influence for stability throughout the business world at a time when readjustments are, more than usually, liable to provoke disturbed conditions."

Agreed That He Was Right

The iron and steel industry of United States attend the Institute meeting. The corporation heads and the independents are always there. It is the big chance the technical and operating heads have to meet the big executives, and also it is the big chance the former have of making the latter sit still and listen. When Judge Gary finishes, the meeting goes on with the discussion of highly technical matters. The steel barons may understand some of the discussion, but the chances are that much of it goes over their heads. But they do not bolt. They occupy the front seats and take their medicine in spite of the fact that practically all the steel men of the country are waiting around to meet them—for, as a matter of fact, there is a general scatteration of the members as soon as Judge Gary is through with his address.

One hears many exchanges of market news—for instance, that a fairly large tonnage of pig iron has sold at $40—that premium mills are doing in some cases as low as six cents on black sheets for Canada, with fair delivery, against 5.35 of the corporation, with no promise of quick action in the matter of shipment.

The rumors that the corporation might advance prices to cover the $50,000,000 which has been added to its freight bills, did not materialize in fact. If there had been any intention of making such a move, or had there been any necessity for it, three months ago was the time when it could have been put across. It would have a hard time trying to gain popularity now—in fact it could not be reconciled with the trend of Judge Gary's remarks, which were all pointed and tuned toward lower levels as soon as possible.

Of blue ruin talk there was none at all, in fact, the head of the Steel Corporation is noted for pruning very severely anything that might be twisted or misconstrued to read like bread-line literature. All through the meeting there was a marked tendency to be frank and honest in discussing affairs and tendencies, and the impression was created that sane optimism and a wholesome confidence in the future could handle successfully the deflation tendencies which are now upon us.

Among important steel makers present besides Judge Gary were Charles M. Schwab, chairman of the Bethlehem Steel Corporation, and Eugene G. Grace, president of the Bethlehem Corporation; John A. Topping, chairman of the Republic Iron and Steel Company; James A. Campbell, president of the Youngstown Sheet and Tube Company; E. A. S. Clark, president of the Consolidated Steel Corporation; Willis L. King, vice-president of the Jones & Laughlin Company; R. P. Lamont, president of the American Steel Foundries Company; Clarence Howard, president of the Commonwealth Steel Company, and W. H. Brevoort, president of the Republic Steel Company.

The Descent.
—Leecoq in the Portland (Ore.) "Telegram."

MARKET DEVELOPMENTS

Volume of Business is Not Very Large Now

Pittsburgh Reports That Buyers Are Not Anxious to Take on Large Tonnages Yet—Machine Tool Market Finds Many Buyers Waiting to See What is Likely to Happen.

THE market is discussing prices more this week than for some time past. The annual address of Judge Gary at the meeting of the Iron and Steel Institute has no doubt helped to direct thought in that channel. The feeling is expressed in many places that the real question of the hour is keeping prices attractive enough to hold the buyers in the market. The opinion is freely expressed that by a resistance of the tendency toward lower levels the buyers may be forced out of the market.

Pittsburgh reports that adjustments are being made in some existing contracts, although they are booked on the non-cancellation basis. Business there is fairly dull, and some of the mills are turning out more tonnage than is being placed on their books. New York also reports that the machine tool market in that centre is quiet. There is a feeling expressed in several quarters that the first of the year will see a revival of buying in many lines. Supplies of steel are coming to hand fairly well, but

there are lines yet where a large unsatisfied business still waits for deliveries. Sheets are scarce and wanted. Tubes come in, but not sufficient, and the prices are firm. Compared with pre-war days, it may be stated that four-inch tube—which is a standard size—to-day sells for 65 cents per foot, against 20c, which may be called a fair pre-war figure.

Machine tool houses are not doing a large business, but there is a scattered selling that is making figures fairly representative, when compared with previous months this year. The small tool trade has the same experience. A large number of small orders is the general report, no very large buying being done.

Scrap metal dealers are waiting for a turn in affairs, as their business is very much unsettled. Even were it to become established on a lower basis they might be able to go ahead, but it does not seem possible to reach that lower level with the knowledge that it will provide a real working basis.

THE IRON AND STEEL MARKETS REMAIN DULL AND LIFELESS

Special to CANADIAN MACHINERY.

PITTSBURGH, Oct. 28.—Few parallels can be found in history for the present dullness of the iron and steel markets. There is no new buying of any consequence in pig iron, none at all in semi-finished steel and scarcely any in finished steel. Pig iron and semi-finished steel contracts are, by trade custom, firm and not subject to cancellation, but there is latitude, particularly in pig iron, as to delivery. Sometimes, as a few months ago, the furnaces are behind in deliveries. Now many consumers are instructing furnaces to defer shipment or reduce the rate of shipment. In finished steel products, while there was a considerable volume of cancelling a few weeks ago there is further cancelling even now, together with many requests for postponement of shipment. If a correct statement could be prepared it would probably show that the independents as a whole are losing more by cancellations and postponements than they are gaining by booking new business, while, of course their order books are thinning each day by the amount that they ship.

Sometimes when orders fall off the

iron and steel producers get into a pessimistic frame of mind and look on the dark side of things. That is not the case at this juncture. Those who held their prices down and those who sought the highest price obtainable, during the scarcity, realised very well that the situation could not last and that a stable basis with reasonable profits would have to be established some time, fair prices resulting in a healthy and full demand eventually. That is the prospect the producers are now looking forward to. There will have to be lower prices and there will have to be a period of light operation, but prices are not expected to become really unprofitable and the period of light operation is not expected to last long.

Producers Variously Situated

The steel market is a very disorderly one, and it is uncertain in what manner prices will decline and how long the decline will require. This complexity results from the producers being situated so variously. In times past, when demand stopped, the mills, or at least the large

mills, would find themselves all in about the same position as to the volume of business on books. Then it was a clear case of holding prices until as much tonnage as possible had been squeezed out of the contracts. Eventually a point would be reached when there was more business to be obtained by reducing prices than by holding prices and shipping on contract.

Now, however, the mills are very variously situated. The Steel Corporation is booked for an average of about nine months. Some of the large independents are booked for three or four months, provided the market stays up moderately well. Other large independents have much less business on books. Among the small mills a few have a good volume of business but some have practically nothing. The state of order books to-day depends upon the price policy pursued this year. The mills that sought the highest prices had to keep their order books clear so as to be able to make early deliveries, for it was early deliveries that brought the highest prices, and at the same time they were quite unable to sell ahead. These mills have made a great deal of money and can well afford to sell at very low prices, even below cost, for the purpose of buying their way into the trade again. Some,

however, seem to prefer idleness. Instances are cropping out of a mill that was formerly one of the very highest priced sellers being now the lowest priced seller among the independents.

Steel Bars Cut

The Pittsburgh Crucible Steel Company suddenly began offering merchant steel bars in rounds, squares and flats, at 2.35c, which is the Steel Corporation price. Other independents have been quoting 3c to 3.25c, and while they learned within a day or two of the cut, they assert they will not meet the cut or even reduce prices, preferring to sit back and let the company involved fill up. In the past few weeks bars have been considered one of the strongest, if not the strongest, item in the whole list of steel products.

In plates the majority of the mills are naming 3.25c as "the market" and then cutting this figure, on individual inquiries, down to 3c as the limit. There are exceptions, however. Two western mills have quoted 2.65c and 2.75c respectively, and a sales representative of one of the mills is reported to have canvassed his trade offering to go $1 a ton below any quotation received elsewhere. While it is still uncertain whether steel prices will experience most of their decline in the next few weeks or will hold, for, say, two or three months and then begin their real decline, predictions are already beginning to appear as to how long the general readjustment will require, the prediction being that next April will see the turning point, with prices at their minimum and with buying increasing.

There is much discussion as to the attitude of the United States Steel Corporation, which has maintained up to date the Industrial Board prices of March 21, 1919, except for two or three little advances made last August. There have been continued rumors for two months past that the corporation would advance its prices slightly, by an average of $3 or $4 a ton, based on the increasing costs caused by the freight rate advances of August 26. This would give the independents some encouragement to drop to the corporation price level. There is, however, the factor that since the war-time control there has been no realignment of prices, one price with another, on the basis of production cost and relative demand, and a horizontal advance in the Steel Corporation prices would hardly produce a fair and equitable price structure. Very few predict, however, that the independents will reduce their prices just to the Steel Corporation level, whatever that may be, the view being that when the independents get down to the corporation level they will have to cut under the corporation prices in order to get business.

Pig Iron and Coke

Two or three small lots of basic pig

The independent steel mills are probably losing more by cancellations and postponements than they are booking in new business.

One Pittsburgh company has cut the price of merchant steel bars in rounds, squares, and flats to 2.35.

Quite a tonnage of basic pig iron has been disposed of at $40, which is quite a reduction from the high figure that was being exacted in August.

Machine tool business is fair in this district, but buying is scattered and not in large lots.

Boiler tubes are arriving, but there is still a large unfilled demand. They are still selling at 65c for a 4 in. tube against a price of 20 cents prior to the war.

New York machine tool market continues very dull, with few large buyers in sight.

A number of U. S. machine tool makers have been approached with a view to lowering their prices, but they are standing firm against such a move.

---

iron have been sold by furnace interests at $40, valley, marking a further decline in the market of $2 a ton, a total of $8.50 from the top price, reached late in August. In Bessemer iron there is recorded a decline of $1.50, to $47, valley, as a furnace interest has formally offered iron at that figure, though, as a matter of fact, a single carload represents the only sale actually reported. Foundry iron remains quotable at $47, valley. An inquiry for a round tonnage would no doubt develop a lower price, but the inquiry must first appear.

Pig iron cannot find its true level as long as coke is at an altogether fictitious level. There has been a spectacular decline in coke in the past week, but there is still much distance to travel. In approximately a week Connellsville furnace coke for spot shipment has dropped from $17 to $14. Foundry coke lags in the decline, being quotable at about $17, but in time it will respond to the furnace coke market. A price of $8.50, or half the recent $17 price, may be seen within a fortnight or a month, but as even that would be nearly six times the $1.50 price that prevailed in May, 1915, no one can say where the coke market will stop.

KEEP BUYERS IN THE MARKETS NOW

That Should be the Aim of the Dealers in all Lines as Far as Possible

TORONTO.—The machine tool, steel and iron trade has seen much better days than it is experiencing in this district just now, but the dealers are not inclined to complain. Some of them thumb over their records and find that June was just as bad as October is likely to turn out to be.

As has been the case for some time, there are no large buyers in the market. Rumors of fairly large lists having been heard, but buying has not started with the movement that is likely to come a little later on.

Prices are still holding firm in nearly all lines of tools, although dealers are not saying much about this matter. Makers have given little indication of any intention on their part of cutting their selling prices.

Firms that sell to farmers report a good business. There is an increasing volume of trade in this way, as farmers are more and more equipping their premises with lighting sets and other forms of modern equipment.

Clearing the Stock

Steel merchants are having a good business, but there is not the pressure on them there was a month or so ago. Many local dealers are not buying from the independents now. Neither are they getting deliveries from the corporation. Bookings placed with the latter a year ago in some cases are still unfilled. "We are going to get better figures from the premium mills now than before," stated one dealer, "and I would not be surprised to see them doing better than six cents, although I do not know of an actual case where less than 6.50 has been paid. When one considers that the Corporation price to Canada is 5.35, with no quick delivery, it is much better for us to get a tonnage from the mills at 6.50 with delivery. We don't mind paying 6.50 for black sheets now, as costs are still high, but when we had to put up as high as 8c a pound to bring sheets over here, it was too much and too dangerous to last long."

Making Concessions

The head of one of the merchandising firms in the steel trade stated to Machinery this morning: "There must be now a gradual bringing down of prices in all lines, and in order that this may be done there will have to be concessions made, and we are following this policy now. I believe it should be generally done, not in the way of price slashing, but in order that the price list may be kept interesting enough so that the buyers will not go out of the market and stay out. If we can work along this way for a time, we will arrive at a basis where business can go on without any

trouble. In this way it will be possible to avoid a period of unemployment, and it is very desirable for the present that we have as little unemployment as possible."

Have Lists Out Now

Quite a number of the manufacturers have lists out now of material on hand of which they wish to dispose. "It all goes to show," stated one dealer this morning as he looked over several of the lists, "that it is easy to get people to buy when the markets are high and rising, and when it is difficult to secure material. Many firms have gone ahead and bought wherever they could because they were afraid of having no supplies at all to fall back on. Now they are not so busy and stock is piling up on them." Dealers as a general thing are not much interested in the lists that are being sent out by the manufacturers, as in many cases they, the dealers, have all the material in stock and on the way that they can take care of in the present market."

Tubes are arriving in fair quantities now, but they show no sign of coming down in price. A fair comparison in price shows that in 1914 a four-inch tube was selling at about 20c per foot, while now the quotation is 65 cents. There is, the mills claim, difficulty in securing skelp for the making of tubes, and the feeling is that this will be one of the last lines to come down.

The demand for reinforcing is not very keen now, as the season is pretty well over.

One dealer made it plain to-day that he thought buyers would make a mistake if they stayed out of the market now in the hope of buying much cheaper later on. He did not counsel placing a heavy order now, as he was of the opinion that price changes might be made, but his point was that if buying ceases it will react all the way through and bring about unemployment, whereas, if all the different factors in trade adopt a reasonable attitude it will be possible to gradually get the high cost material out of the trade entirely without an overwhelming loss having to be borne by any particular person or firm.

Some of the offices seem to be getting a very fair share of business in small tools, although in no case is anything heard of large lists having been sent out. There is a rumor in the market that there may be some readjustment of prices, having special reference to high speed drills. There is nothing official in regard to the matter, but it comes from a very reliable source.

The scrap market has no interesting spots. The trade was in hopes that the recent adjustment of prices, which brought prices down, would react on scrap and cause dealing to get under way again at a lower level. This has not happened at a lower, higher, or the same level and business drags.

BRITISH TOOL MAKERS IN NO RUSH TO BRING DOWN THEIR PRICES

THAT British machine tool makers are in no hurry to lower prices, but that competition in weeks to come would decide their course to some extent, is the impression one gains from discussing the matter with Mr. P. V. Vernon, a director of the firm of Alfred Herbert, of Coventry, England.

"The condition of labor is not the most settled thing with us," remarked Mr. Vernon. "In fact we often are in the position of wondering what is coming next. Not many months ago there was a ruinous strike of the moulders, and just now the miners have quit work. Then machinists were out, and other engineering trades are talking strike. It all makes manufacturing conditions rather unsettled. With material we are not having much trouble obtaining it all at home, but the price we are paying is still high."

The system of taxation that is in

TOOL BUILDERS ASKED TO REDUCE

But So Far Their Attitude is Based on Relation to Cost Prices

Special to CANADIAN MACHINERY.

NEW YORK, Oct. 28.—Aside from inquiries from the Reading Steel Casting Co., Reading, Pa., for 25 or 30 tools for an addition to its plant, there is little activity in Eastern machine tool markets. Business certainly shows no signs as yet of revival, and sellers do not look for any change in the situation until after Presidential election, and some go so far as to predict that present conditions will continue until the first of the year or later.

There are some scattered orders for single tools, but a majority of sellers in this market are doing little or nothing in the aggregate, as compared with what might be considered a normal business.

Some of the automobile companies have been conducting a sort of organized campaign to induce manufacturers in all lines to bring down prices. Some machine tool manufacturers, thus approached, have signified that they have no present intention of reducing prices; that their selling prices are based on production costs, plus a reasonable profit, and that they do not intend to change prices until costs can be reduced. There are some evidences of labor costs coming down, through the willingness of men out of jobs to work for less than they have been getting in the past year or two. Material costs, however, are not lower, as most manufacturers of machine tools are still getting high priced pig iron, castings and steel on contracts made some time ago.

vogue seems to have the tendency of keeping prices up. Manufacturers are allowed to make the same profit as they did in the two years previous to the war. Anything over and above that comes under the heading of excess profit, and of that the Government takes 60 per cent. and leaves the manufacturer with 40 per cent.

The competition in the machine tool business of the Old Country makers is pretty well divided between their contemporary firms at home and the U.S. builders. "Business has changed since the signing of the armistice," stated Mr. Vernon. "When the war was on our product was almost 100 per cent. for home consumption, it being all used in the turning out of munitions. Shortly after there came a change until we found that our business was one-eighth foreign and the rest home. When I left the figures showed that one-third of the business was foreign and the rest for home use. There has been a falling off in many lines, but it is common all over the world. I have been for some weeks in the machine tool and manufacturing centres of United States and I know that many of the big shops there are not booked very far ahead with business, neither are the big shops, such as the auto works, buying anything in the way of equipment just now."

British firms are not paying much attention to German trade yet. The German mark is so low in value that the Germans themselves cannot afford to buy from British firms, and the impression is gathered that as a rule the British firms are not particular about opening up trading with the German people.

Mr. Vernon is a firm believer in British houses wishing to open up for business in Canada making a personal study of the field here. "The ways of doing business are quite different in various countries," is his contention, "and once a person has been in Canada he understands far better the problem of the office or the representative here. Our British firms have been advised many times to do this very thing, to make a visit to the field and meet the people and find out their ways of doing business, and I am more firmly convinced than ever that it is a splendid thing for a British manufacturer to do."

The British makers of machine tools of the better class are not at all afraid of not being able to meet the competition of the rest of the makers. For a time the Japanese were putting a cheap tool on the market. They took one of the Alfred Herbert machines and called it—or rather the imitation machine they made from it—the "Alf," and put it out to the trade. "That move did not hurt us at all," remarked Mr. Vernon, "in fact the imitation and the use of part

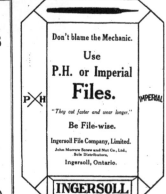

of our name never injured our business. As soon as we were able to put out tools after the war the Japanese imitation did not provide any real competition."

This is Mr. Vernon's first visit to Canada. With Mrs. Vernon he has been spending some weeks in United States before coming to this country. Together with Mr. Blair, the Canadian representative of the Alfred Herbert Co., he is making a study of the Canadian field with the idea of further extending their business in this country.

WON'T TAKE GOODS— EXCHANGE BLAMED

South American Firms Take Action That May Cause Great Loss

Buenos Ayres.—At a meeting here of representatives of American commission houses it developed that the total value of American goods rejected at this port owing to the falling prices and high exchange on the American dollar approximates $12,000,000.

Other large consignments of American goods are now en route to Buenos Ayres, and leading American firms are fearing a crisis unless the Federal Reserve administration relieves them from operation of the re-discount rules.

Most of the rejections were noted with regard to textiles, foodstuffs, machinery and steel.

CONDITIONS O.K. IN SCOTIA FIELD

Agreement Has Been Reached in Regard to Taking Iron Ore From Mines

Montreal.—Roy M. Wolvin, president of the Dominion Steel Corporation, returned after a two weeks' trip to the company's properties in Nova Scotia and Wabana. The steel industry, he said, was in a satisfactory condition, and he had been able to arrive at an agreement with the Newfoundland Government whereby the export of iron ore would be carried out under conditions favorable to the corporation and along the lines desired by those who had conceived the British Empire Steel Corporation.

A Worthy Booklet.—The reasons why American manufacturers are establishing plants in Canada at the rate of one a week are summarised in a booklet just issued by the Union Bank of Canada, entitled "A Canadian Plant—Why?" In the face of the growing strength of the "Made-in-Canada" movement, and the extension of the British Preferential Tariff, Canada, the booklet claims, is the logical location for a plant with which to handle British and Canadian trade.

F. H. WHITTON DIES IN HAMILTON

General Manager of Steel Co. of Canada Failed to Rally from Operation

Hamilton. — F. H. Whitton, general manager of the Steel Company of Canada, died at St. Joseph's Hospital, Hamilton. Mr. Whitton failed to rally from a serious operation performed on October 21.

In the death of Mr. Whitton, Hamilton loses one of the leading figures in its commercial life. Deceased was born in Northamptonshire, England, 61 years ago, and spent his early youth in that country. He attended college at Cam-

F. H. WHITTON.

bridge, but when 18 years of age came to Canada and entered the employ of the Grand Trunk Railway at Montreal. He later removed to the United States, but returned to Canada to assume the management of the Ontario Tack Company, which position he held till 1907. When the Steel Company of Canada was formed Mr. Whitton was appointed assistant general manager and in 1916 became general manager.

Mr. Whitton was prominent in social and religious life, as well as in the commercial world, being a devout Roman Catholic and a member of many clubs. He attended St. Joseph's Church. His wife and one son, Corbett F. Whitton, assistant secretary of the Steel Company of Canada, survive.

Building Addition.—A four-storey addition is being made to the plant of the Fisher Body Company, of Canada, Ltd., at Walkerville, Ontario, and will be completed about the end of the year. The new building is of heavy mill construction and will have 75,000 square feet of

floor space. It will be used for painting and trimming closed automobile bodies. When it is in operation the capacity of the closed body department will be increased about 80 per cent. The estimated cost of the addition is $275,000.

INCORPORATIONS

Incorporation has been granted to the British Radio Corporation of America to make, buy, sell, lease, etc., electric appliances of every description, including wireless, telephone apparatus, etc. Head office at Montreal, and capital stock $10,000.

The Hamilton-Bothwell Oil Co. has been incorporated at Bothwell with a capital of $300,000, to operate and deal in gas and oil lands, etc.

Ira Sessenwein Ltd. has been incorporated to carry on the business of scrap dealers in metals, rubber, etc. Place of business to be at Montreal, and capital placed at $30,000.

Incorporation is granted to "All-British Signal Company," with a capital of £80,000, and head office in Toronto. The company is to make or sell submarine signalling made by John Gardner.

Incorporation is granted to the Overseas Export Co. of Canada, to import, export, etc., all kinds of merchandise, machinery, chemicals, etc. The head office is to be at Montreal, and the capital is $100,000.

Dunwin Motors has been incorporated to buy, sell, trade in and make motors, autos, etc. The place of business is at Montreal and the capital is placed at $100,000.

During the war the shortage of copper induced the Germans to experiment with all kinds of substitutes, and it was found that iron acted satisfactorily for spacing pieces in steam turbines, but not for vanes. Nickel-plated or enamelled steel also failed to answer for this purpose. More promising results are expected, according to "Technik und Wirtschaft," from nitrogenized iron, prepared by a process in which the heater vanes are exposed to a current of gaseous ammonia, the nitrogen of which combines with the iron, while the hydrogen reduces any superficial oxide.

Coming to Canada?—A London, Eng., despatch says: "A big increase in emigration to Canada may be one result of the strike of British coal miners. Reports received from Morriston, near Swansea, Wales, state that many of the steel workers there who have been made idle by the strike have decided to emigrate to Canada at an early date.

Kingston. — The largest locomotive ever turned out of the Canadian Locomotive Works left the plant to go into commission on the G. T. R. It is of special design for hauling heavy freight trains over difficult grades, and will be employed as a transfer engine between Toronto and York.

5-16 inch Size
Capacity:
3-32" to 5-16"

4 1-2 inch Size
Capacity; 3 1-2" to 4 1-2"

The Screw Machine Man Says:

"Everybody recommends Geometric Die
Heads. Better Threads in Quicker
Time is the reason"

It doesn't matter how strenuous the job is,
a Geometric is built to handle it easily.

Make Screw Thread-Cutting the Most Satisfactory Operation in Your Shop.

————Write————

THE GEOMETRIC TOOL COMPANY
NEW HAVEN   CONNECTICUT

Canadian Agents:

Williams & Wilson, Ltd., Montreal
The A. R. Williams Machinery Co., Ltd., Toronto, Winnipeg, St. John, N.B., Halifax, N.S.
Canadian Fairbanks.Morse Co., Ltd., Manitoba, Saskatchewan, Alberta

If interested tear out this page and place with letters to be answered.

SELECTED MARKET QUOTATIONS

Being a record of prices current on raw and finished material entering into the manufacture of mechanical and general engineering products.

PIG IRON

Grey forge, Pittsburgh	$42 40
Lake Superior, charcoal, Chicago	57 00
Standard low phos., Philadelphia	50 00
Bessemer, Pittsburgh	43 00
Basic, Valley furnace	42 90
Toronto price:—	
Silicon, 2.25% to 2.75%	52 00
No. 2 Foundry, 1.75 to 2.25%	50 00

IRON AND STEEL

Per lb. to Large Buyers	Cents
Iron bars, base, Toronto	$ 5 75
Steel bars, base, Toronto	5 75
Iron bars, base, Montreal	5 75
Steel bars, base, Montreal	5 75
Reinforcing bars, base	5 75
Steel hoops	7 00
Norway iron	11 00
Tire steel	5 75
Spring steel	10 00
Band steel, No. 10 gauge and 3-16 in. base	6 10
Chequered floor plate, 3-16 in.	8 50
Chequered floor plate, ¼ in.	8 10
Bessemer rails, heavy, at mill	....
Steel bars, Pittsburgh	3 00-4 00
Tank plates, Pittsburgh	4 00
Structural shapes, Pittsburgh	3 00
Steel hoops, Pittsburgh	3 50-3 75
F.O.B., Toronto Warehouse	
Small shapes	5 75
F.O.B. Chicago Warehouse	
Steel bars	3 62
Structural shapes	3 72
Plates	3 67 to 5 50
Small shapes under 3"	3 62

FREIGHT RATES

Per 100 Pounds.

Pittsburgh to Following Points

	C.L.	L.C.L.
Montreal	58½	73
St. John, N.B.	84½	106½
Halifax	86	108
Toronto	38	54
Guelph	38	54
London	38	54
Windsor	35	50½

METALS

	Gross: Montreal	Toronto
Lake copper	$24 00	$23 50
Electric copper	24 00	24 00
Castings, copper	23 50	24 00
Tin	54 00	55 00
Spelter	9 00	10 50
Lead	9 50	9 50
Antimony	9 50	12 00
Aluminum	36 00	37 00

Prices per 100 lbs.

PLATES

Plates, 3-16 in.	$ 7 25	$ 7 35
Plates, ¼ up	6 50	6 60

PIPE—WROUGHT

Standard Buttweld Pipe

Per 100 Ft.

	Steel Blk.	Steel Galv.	Gen. Wrought Iron Blk.	Gen. Wrought Iron Galv.
⅛	$ 6 50	$ 8 50	$.....	$.....
¼	5 31	7 41	5 91	8 01
⅜	5 31	7 41	5 91	8 01
½	7 10	8 63	7 95	9 48
¾	8 80	10 87	9 95	12 02
1	13 01	16 07	14 71	17 77
1¼	17 60	21 74	19 90	24 04
1½	21 04	23 99	23 79	28 74
2	28 31	34 97	32 01	38 67
2½	44 73	55 28	....	....
3	58 52	72 29	....	....
3½	74 06	90 62	....	....
4	87 75	107 37	....	....

Standard Lapweld Pipe

Per 100 Ft.

	Steel Blk.	Steel Galv.	Gen. Wrought Iron Blk.	Gen. Wrought Iron Galv.
2	$32 01	$ 38 67	$36 71	$42 37
2½	48 26	58 79	54 11	64 64
3	63 11	76 88	70 76	84 53
3½	75 90	92 46	85 10	101 66
4	89 98	107 55	100 83	120 45
4½	1 05	1 29	1 30	1 54
5	1 22	1 50	1 52	1 80
6	1 58	1 95	1 97	2 33
7	2 06	2 53	2 53	3 01
8L	2 16	2 66	2 66	3 16
8	2 49	3 07	3 07	3 64
9	2 98	3 67	3 67	4 36
10L	2 77	3 41	3 41	4 06
10	3 56	4 39	4 39	5 21

Prices—Ontario, Quebec and Maritime Provinces

WROUGHT NIPPLES

4" and under, 60%.

4½" and larger, 50%.

4" and under, running thread, 30%.

Standard couplings, 4-in. and under, 30%.

Do., 4½-in. and larger, 10%.

OLD MATERIAL

Dealers' Average Buying Prices.

Per 100 Pounds.	Montreal	Toronto
Copper, light	$14 00	$14 00
Copper, crucible	17 00	18 00
Copper, heavy	17 00	18 00
Copper wire	17 00	18 00
No. 1 machine composition	17 00	17 00
New brass cuttings	11 00	11 75
Red brass turnings	14 50	15 75
Yellow brass turnings	9 00	9 50
Light brass	6 50	7 00
Medium brass	7 50	7 75
Scrap zinc	6 00	6 00
Heavy lead	7 50	7 75
Tea lead	4 50	5 00
Aluminum	19 00	20 00
	Per Ton	**Gross**
Heavy melting steel	20 00	18 00
Boiler plate	15 50	15 00
Axles (wrought iron)	30 00	20 00
Rails (scrap	23 00	18 00
Malleable scrap	26 00	25 00
No. 1 machine cast iron	36 00	33 00
Pipe, wrought	12 00	12 00
Car wheel	33 00	33 00
Steel axles	25 00	20 00
Mach. shop turnings	11 00	11 00
Stove plate	28 00	25 00
Cast boring	14 00	14 00

BOLTS, NUTS AND SCREWS

	Per Cent.
Carriage bolts, 7-16 and up	+10
Carriage bolts, ⅜-in. and less	Net
Coach and lag screws	—15
Stove bolts	55
Wrought washers	—25
Elevator bolts	+10
Machine bolts, 7-16 and over	+10
Machine bolts, ⅜-in. and less	+10
Blank bolts	Net
Bolt ends	Net
Machine screws, fl. and rd. hd., steel	27½
Machine screws, o. and fil. hd., steel	+25
Machine screws, fl. and rd. hd., brass	net
Machine screws, o. and fil. hd., brass	net
Nuts, square, blank .... +25 add	$2 00
Nuts, square, tapped ....add	2 25
Nuts, hex., blank ....add	2 50
Nuts, hex., tapped ....add	3 00
Copper rivets and burrs, list less	15
Burrs only, list plus	25
Iron rivets and burrs	40 and 5
Boiler rivets, base ¾" and larger	$8 50
Structural rivets, as above	8 40
Wood screws, O. & R., bright	75
Wood screws, flat, bright	77½
Wood screws, flat, brass	55
Wood screws, O. & R., brass	55½
Wood screws, flat, bronze	50
Wood screws, O. & R., bronze	47½

MILLED PRODUCTS

(Prices on unbroken packages)

	Per Cent
Set screws .... —20%, 25 and	5
Sq. and hex. hd. cap screws	12½
Rd. and fil. hd. cap screws ..plus	25
Flat but. hd. cap screws ....plus	50
Fin. and semi-fin. nuts up to 1-in.	12½
Fin. and Semi-fin. nuts, over 1 in., up to 1½-in.	—5
Fin. and Semi-fin. nuts over 1½ in., up to 2-in.	+12½
Studs	+5
Taper pins	—12½
Coupling bolts	+40
Planer head bolts, without fillet, list	+45
Planer head bolts, with fillet, list plus 10 and	+55
Planer head bolt nuts, same as finished nuts.	
Planer bolt washers	net
Hollow set screws	+60
Collar screws ....list plus 20,	30
Thumb screws	40
Thumb nuts	75
Patch bolts ....add	+85
Cold pressed nuts to 1½ in. ..add	$1 00
Cold pressed nuts over 1½ in. ..add	2 00

BILLETS

	Per gross ton
Bessemer billets	$60 00
Open-hearth billets	60 00
O.H. sheet bars	76 00
Forging billets	56 00-75 00
Wire rods	52 00-70 00

Government prices.

F.O.B. Pittsburgh.

NAILS AND SPIKES

Wire nails, base	$6 10
Cut nails, base	7 00
Miscellaneous wire nails	50%

ROPE AND PACKINGS

Plumbers' oakum, per lb.	0 10¼
Packing, square braided	0 38
Packing, No. 1 Italian	0 44
Packing, No. 2 Italian	0 36
Pure Manila rope	0 35½
British Manila rope	0 28
New Zealand hemp	0 28

POLISHED DRILL ROD

Discount off list, Montreal and Toronto ..... net

MISCELLANEOUS

Solder, strictly	$ $ 29½
Solder, guaranteed	0 31½
Soldering coppers, lb.	0 62½
White lead, pure, cwt.	20 35
Red dry lead, 100-lb. kegs, per cwt.	16 00
Gasoline, per gal., bulk	0 42
Pure turp., single bbls., gal.	3 15
Linseed oil, raw, single bbls.	2 37
Linseed oil, boiled, single bbls.	2 40
Wood alcohol, per gal.	4 00
Whiting, plain, per 100 lbs.	3 00

CARBON DRILLS AND REAMERS

S.S. drills, wire sizes	32½
Can. carbon cutters, plus	20
Standard drills, all sizes	32½
3-fluted drills, plus	10
Jobbers' and letter sizes	32½
Bit stock	40
Ratchet drills	15
S.S. drills for wood	40
Wood boring brace drills	25
Electricians' bits	30
Sockets	50
Sleeves	50
Taper pin reamers	25 off
Drills and countersinks	net
Bridge reamers	50
Centre reamers	10
Chucking reamers	net
Hand reamers	10
High speed drills, list plus	10 to 30
Can. high speed cutters, net to plus	10
American plus	40

COLD ROLLED STEEL

[At warehouse]

Rounds and squares	$7 base
Hexagons and flats	$7.75 base

IRON PIPE FITTINGS

	Black	Galv.
Class A	70	85
Class B	30	40
Class C	20	30

Cast iron fittings, 5%; malleable bushings, 22½%; cast bushings, 22½%; unions, 37½%; plugs, 20% off list.

SHEETS

	Montreal	Toronto
Sheets, black, No. 28	$12 00	$10 50
Sheets, Blue ann., No. 10	9 50	9 00
Canada plates, dull, 52 sheets	13 00	13 00
Can. plates, all bright	14 00	.....
Apollo brand, 10¾ oz. galvanized	....	....
Queen's Head, 28 B.W.G.	13 50	.....
Fleur-de-Lis, 28 B.W.G.	13 00	.....
Gorbal's Best, No. 28	....	....
Colborne Crown, No. 28.	....	....
Premier, No. 28, U.S.	11 50	12 00
Premier, 10¾-oz.	11 50	12 40
Zinc sheets	16 50	20 00

PROOF COIL CHAIN

(Warehouse Price)

B

¼ in., $13.00; 5-16, $11.00; ⅜ in., $10.00; 7-16 in., $9.80; ½ in., $9.75; ⅝ in., $9.20; ¾ in., $9.30; ⅞ in., $9.50; 1 in., $9.10; Extra for B.B. Chain, $1.20; Extra for B.B.B. Chain, $1.80.

ELECTRIC WELD COIL CHAIN B.B.

⅛ in., $16.75; 3-16 in., $15.40; ¼ in., $13.00; 5-16 in., $11.00; ⅜ in., $10.00; 7-16 in., $9.80; ½ in., $9.75; ⅝ in., $9.50; ¾ in., $9.30.

Prices per 100 lbs.

FILES AND RASPS

	Per Cent.
Globe	50
Vulcan	50
P.H. and Imperial	50
Nicholson	32½
Black Diamond	27½
J. Barton Smith, Eagle	50
McClelland, Globe	50
Delta Files	20
Disston	40
Whitman & Barnes	50
Great Western-American	50
Kearney & Foot, Arcade	50

BOILER TUBES.

Size.	Seamless	Lapwelded
1 in.	$27 00	$.....
1¼ in.	29 50	.....
1½ in.	31 50	29 50
1¾ in.	31 50	30 00
2 in.	35 00	30 00
2¼ in.	35 00	29 00
2½ in.	42 00	37 00
3 in.	50 00	48 00
3¼ in.	.. ..	48 50
3½ in.	63 00	51 50
4 in.	85 00	65 50

Prices per 100 ft., Montreal and Toronto

OILS AND COMPOUNDS.

Castor oil, per lb.	
Royalite, per gal., bulk	29½
Palacine	32½
Machine oil, per gal.	65
Black oil, per gal.	34
Cylinder oil, Capital	1.01
Petroleum fuel oil, bbls., net	19

BELTING—No. 1 OAK TANNED

Extra heavy, single and double		6½
Standard		6½
Cut leather lacing, No. 1		2 00
Leather in side	2 40	3 00

TAPES

Chesterman Metallic, 50 ft.	$2 00
Lufkin Metallic, 603, 50 ft.	2 00
Admiral Steel Tape, 50 ft.	2 75
Admiral Steel Tape, 100 ft.	4 45
Major Jun. Steel Tape, 50 ft.	3 50
Rival Steel Tape, 50 ft.	2 75
Rival Steel Tape, 100 ft.	4 45
Reliable Jun. Steel Tape, 50 ft.	3 50

PLATING SUPPLIES

Polishing wheels, felt	$4 60
Polishing wheels, bull-neck	2 00
Emery in kegs, Turkish	09
Pumice, ground	06
Emery glue	30
Tripoli composition	09
Crocus composition	12
Emery composition	11
Rouge, silver	60
Rouge, powder, nickel	45

Prices per lb.

ARTIFICIAL CORUNDUM

Grits, 6 to 70 inclusive	.08½
Grits, 80 and finer	.6

BRASS—Warehouse Price

Brass rods, base ½ in. to 1 in. rod	0 34
Brass sheets, 24 gauge and heavier, base	$0 42
Brass tubing, seamless	0 46
Copper tubing, seamless	0 48

WASTE

XXX Extra	.24	Atlas	.20
Peerless	.22½	X Empire	.19½
Grand	.22½	Ideal	.19
Superior	.22½	X Press	.17½
X L C R	.21		

Colored

Lion	.17	Popular	.13
Standard	.15	Keen	.11
No. 1	.15		

Wool Packing

Arrow	.35	Anvil	.22
Axle	.28	Anchor	.17

Washed Wipers

Select White	.20	Dark colored	.09
Mixed colored	.10		

This list subject to trade discount for quantity.

RUBBER BELTING

Standard ... 10% Best grades... 15%

ANODES

Nickel	.55 to .60
Copper	.38 to .40
Tin	.70 to .70
Zinc	.16 to .17

Prices per lb.

COPPER PRODUCTS

	Montreal	Toronto
Bars, ½ to 2 in.	$42 50	$48 00
Copper wire, list plus 10..		
Plain sheets, 14 oz., 14x60 in.	49 00	48 00
Copper sheet, tinned, 14x60, 14 oz.	52 00	48 00
Copper sheet, planished, 16 oz. base	56 00	55 00
Braziers', in sheets, 6x4 base	48 00	46 00

LEAD SHEETS

	Montreal	Toronto
Sheets, 3 lbs. sq. ft.	$11 50	$14 50
Sheets, 3½ lbs. sq. ft.	11 25	14 00
Sheets, 4 to 6 lbs. sq. ft.	11 00	13 50

Cut sheets, ½c per lb. extra.
Cut sheets to size, 1c per lb. extra.

PLATING CHEMICALS

Acid, boracic	$ .23
Acid, hydrochloric	.04
Acid, nitric	.11
Acid, sulphuric	.04
Ammonia, aqua	.15
Ammonium, carbonate	.23
Ammonium, chloride	.22
Ammonium hydrosulphuret	.75
Ammonium sulphate	.30
Arsenic, white	.14
Copper, carbonate, annhy	.41
Copper, sulphate	.15
Cobalt, sulphate	.20
Iron perchloride	.62
Lead acetate	.30
Nickel ammonium sulphate	.20
Nickel carbonate	.82
Nickel sulphate	.22
Potassium sulphide (substitute)	.42
Silver chloride (per oz.)	1.30
Silver nitrate (per oz.)	1.25
Sodium bisulphate	.14
Sodium carbonate crystals	.06
Sodium cyanide, 127-130%	.38
Sodium hyposulphite per 100 lbs	8.00
Sodium phosphate	.15
Tin chloride	.80
Zinc chloride, C.P.	.30
Zinc sulphate	.10

Prices per lb. unless otherwise stated

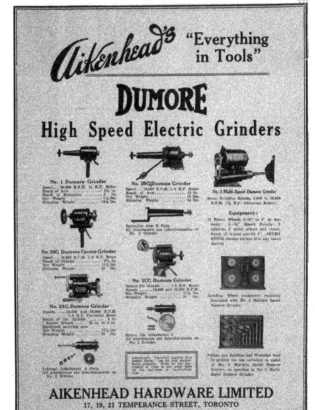

INDUSTRIAL NEWS

NEW SHOPS, TENDERS AND CONTRACTS
PERSONAL AND TRADE NOTES

LOOKING FOR BETTER TRADE
EARLY IN THE COMING YEAR

RE. ERENCE to business conditions, p esent and prospective, is contained in the following opinions:

"In connection with the present and future possibilities of galvanized sheets," said Mr. A. T. Enlow, president of the Dominion Sheet Metal Corporation, Ltd., Hamilton, Ont., "we might summarize the situation as follows: present supply, none; immediate prospective supply, fair; next spring, usual rush.

"It is evident therefore that the wise t⁻⁻er will order before January 1 and as prices are declining somewhat and may sag a little lower he will lose nothing by so doing.

"There will not be any radical decline, in fac: the decline already made is possibly the limit in this swing.

"Underlying basic business is good, especially in Canada, and with the election in the U.S. over on November 2 and the Republican candidate elected, as now seems very probable, business there will also be stimulated and Canada will get the usual reflex glow."

Prices Are Firm

"our present request for our opinion cf the business outlook for the immediate future is rather difficult to answer, b:: we will begin by saying that we are optim:stic," said Mr. L. H. Laythe, sales m.n.r.er Butterfield & Co., Rock Island, C-ne.

"We can see nothing in the present situation to cause undue alarm. That w have reached the crest of the long-extend high price movement and are

r⁻⁻ beginning the descent on the other side cannot be questioned.

"It was all expected, however, by clear thinking business men, and they have only been fooled in that it took longer to reach the peak than was anticipated.

"The law of supply and demand is inexorable, and was bound to bring about a readjustment of values sooner or later, and as long as the condition thus brought about is looked upon sanely and all steps and actions in business ventures governing and carried out accordingly, we can see no cause for worry.

"From our own standpoint as manufacturers of small tools, we can see little possibility of any change in prices for the present at least. Radical increases in freight rates, coupled with the fact that raw material in the shape of tool steel and malleable castings shows no reduction, make it impossible for us to reduce our prices now.

"Our advice to jobbers and large users is to buy what they need to keep their stocks well sorted, but not more than can be turned over in a reasonable length of time.

Progressive Swing Next February

"Finally it is our opinion that with the presidential election over in the United States, the crops moved in the Canadian West, a highly satisfactory holiday season at Christmas time, and the usual January first inventories out of the way, that about the first of February of the new year will see things on the progressive swing again, and in a much more healthy condition than at any time since the war began."

NATIONAL CASH REGISTER TO
ADOPT PROFIT-SHARING PLAN

The profit-sharing plan adopted by the National Cash Register Company, Ltd., was explained to the employes recently by Frederick B. Patterson, vice-president of the company. The arrangement i: characterized by an extreme liberal'ty towards the employes, calling for an equal division of profits between the company and those employes who are not members.

The profits of the company are to be

determined by an outside firm of accountants. From the total profits will be deducted an amount equal to six per cent. on the company's investment, not including the value of patents and "goodwill." To offset this six per cent., the employees will receive the highest wages paid for their class of work.

The remaining profits will then be divided equally, the fifty per cent. which goes to the employees being paid in cash,

as near as possible to the half-yearly accounting dates. The fifty per cent. which goes to the company may be kept in the business for new buildings, machinery, extensions, etc. The employes receive their share unconditionally.

For the operation of the plan the employes are divided into five groups. The first, executives, will receive 12 per cent. of the total profits; the second, foremen, 5 per cent.; the third, job foremen, 8 per cent.; and the fourth, all who have been employed over one month and are not in the first three classes, 25 per cent. Only the fifth group, those who have been in the company's employ less than thirty days, do not share. Each man's share is based on the salary or wages received at present.

Officials of the company expect that the gain in loyalty and production and the decrease of labor turnover resulting from this scheme will make for an increased fall's business. The policy of the company has always been to consider the good of the workers and it has been justified by their success.

Better Times Ahead.—In connection with the annual meeting of the Russell Motor Car Co., Lloyd Harris, president of the Willys-Overland Co., said in referring to the automobile plant on Weston Road, Toronto: "This company has completed the conversion of its plant from aeroplane engine work to automobile work, and during the first six months of the calendar year enjoyed an active demand for its product. Since July 1, in common with other motor companies, it has experienced a contraction of sales and its operations are being carried on at the present time on a reduced basis. It is expected that there will be a renewed demand after the first of the year when the returns from the crops have had time to get into circulation."

Order Engines.—The Canadian Locomotive Company has received an order for four locomotives for the Timiskaming Line, Timiskaming & Northern Ontario Railroad. It is stated that this order will make about 28 engines yet to be completed. At the present time the men in the works are busy on ten locomotives which are being built for the Grand Trunk. It is understood, with the present orders on hand, it will mean that the company will have lots of work until some time after the first of February.

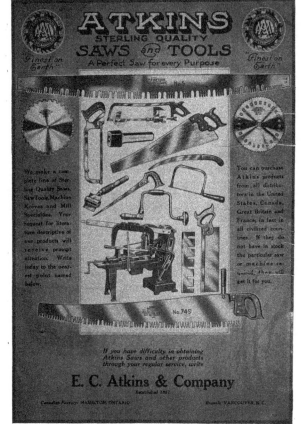

TRADE GOSSIP

Operating at Midland. — Industrial prospects are exceedingly bright for Midland, where there is plenty of hard coal for every home at $18 a ton. The Midland Iron and Steel Company's plant resumes operations this week after a shut down of several months. The plant closed down early in the spring in order to re-line the furnace, but the shortage of coke caused a further cessation.

Erecting Big Plant.—Announcement is made of the sale in Toronto of five acres of land at the corner of Hanson Street and Coxwell Avenue, where there is at present a Grand Trunk Railway siding. Plans and specifications have been prepared for the erection of a modern daylight plant of the best brick and mill type construction. The cost of the building will be in the neighborhood of half a million dollars. The main building will be 500 feet long and 80 feet wide. Upon completion three companies will commence operations in the buildings: Crouse-Hinds Company of Canada; Harvey-Hubbell, and the Hubbell-Mack Machine Screw Company. The new plant is not expected to commence operations until November 1, 1921.

Brantford Firm Changes.—The Brantford Machine & Tool Co., Ltd., has just been incorporated to continue and add to the former business of the Brantford Machine & Tool Co. Mr. Wm. J. Ham, a son of Mr. Jos. Ham, of Ham Brothers, has taken an interest in the concern and will take an active part therein. In addition to their old line of machine and machine tool work, they intend extending and manufacturing a line of garage equipment, such as bench presses, boring tools and small air compressors, and expect to find quite a ready market for same.

Canadian-Detroit Twist Drill.—The Detroit Twist Drill Company has announced the establishment of a manufacturing plant in Walkerville, Ontario, known as the Canadian-Detroit Twist Drill Company. The Detroit Company decided several months ago that the growth of Canadian industry warranted the construction of a plant across the border, and the new company was accordingly incorporated. High-speed drills and reamers will be manufactured and a complete stock will be built up as soon as possible. The nucleus of the Walkerville organization was formed from the personnel of the Detroit plant. About thirty men will be employed at the start. The officers and executives of the new company are as follows: President, Muir B. Snow; vice-president, Lewis H. Jones; treasurer, H. H. Sanger; secretary, P. C. Hill; manager, R. A. B. Goodman; superintendent, Geo. Johnston. Manufacturing operations have already begun, and the first drills are expected to be completed by November 15.

REPORT HAS IT THAT OIL IS TO BE FOUND IN HASTINGS COUNTY

BELLEVILLE.—Startling events of the past few days uncover the possibility of this section of Ontario becoming an important oil producing centre. In Tyendinaga Township, Hastings County, discoveries have been made which are very promising, and one of the largest companies in the world interested in development of oil properties is already in possession of options on over one thousand acres of land in that township. A farmer named Roach, living about three miles from Lonsdale, had a well which he couldn't use on account of the strong oil flavor, so he decided to drill another one. When down forty-two feet on the new well the rush of gas compelled suspension of operations, and there seemed more gas and oil than water. The farmer wanted water badly, but the recurrence of oil indications set him thinking. A relative from Pennsylvania, by profession an oil driller, had visited the farm a year or so ago, and noticing indications of oil, expressed great surprise that the oil scouts who cover the world looking for new fields had overlooked this locality. Mr. Roach came to Belleville and told Mr. John Elliott, manager of the Standard Bank, his story. Mr. Elliott at once communicated with his son-in-law, Mr. Reeves, travelling representative for the General Oil Fields, Limited, of London, England, who is spending a vacation in Belleville, having recently returned from Mexico. What Mr. Reeves saw convinced him that the prospects for an oil field here were excellent; in fact he states that indications are better than anything he saw in Mexico, one of the greatest oil countries in the world.

Mr. Reeves thinks so well of the property that he secured an option on over one thousand acres of land, including the Roach farm. If oil is discovered in quantity the farmers will be paid a good rental besides royalties, which means sudden fortune for a number of farmers in the vicinity of Marysville and Lonsdale.

Geologists have been sent for by Mr. Reeves and a thorough examination of the property will be made at once; if the property proves up as Mr. Reeves thinks it will, he thinks that his company will spend at least a million dollars in development within the next year.

SAY HELP IS NEEDED TO MAKE USE OF THE LOW-GRADE IRON ORES

FORT WILLIAM.—Presentations dealing with aid to the iron industries of this district in Ontario and a lengthy disposition of the doctrines of free trade occupied the time of the tariff commission here.

Sir Henry Drayton, chairman, Hon. G. D. Robertson and Hon. S. F. Tolmie were in attendance and the first witness was J. E. Marks of Port Arthur, who presented the case for the imposition of a bounty of at least 50 cents per ton upon iron ore mined in Ontario. He urged that the former bounties given for iron-smelted from Canadian ores failed in their effect because they went to the smelter and not to the miner. He adduced figures in an effort to show that there were millions of tons of ore available close to railway and navigation facilities which required aid in order to be merchantable. He urged that the Dominion of Canada ought to cause the Canadian iron to be mined and utilised in Canadian furnaces.

The chairman asked whether he did not think that, with an apparent surplus of $100,000 the iron industry could not be trusted to take care of development without aid.

Mr. Marks replied that the raw material of the iron manufacturers was largely the half-finished product from the States.

To a remark by the chairman that the British Columbia bonus had apparently not been effective in building up iron industries, Mr. Marks said that the smelters got all the bonus and not the miner.

The chairman asked concerning the deleterious elements in the local iron ores and was told that the ores were free of all injurious elements except sulphur and this was disposed of by roasting, which was successfully done with the ore from the Atikokan range.

Dr. R. J. Manion, M.P., said that he had not intended to speak but he had advocated some form of aid to the iron industry on the floor of the House, and one reason that he would approve of the bounty was that if it was not earned by production the bounty would not be paid. He believed that Canada ought to produce many of the steel products which she now imports.

Owners of Iron Lands

J. J. O'Connor spoke as representative of owners of iron ore lands, and said that the time was coming soon when the lower grades of ore would be needed, and it was high time that Canada attempted to make use of her low grade ores. He pointed out that one American firm had spent $750,000 in experimenting with magnetic substances on the ranges in Minnesota and now began to install a three million dollar plant to utilise low grade iron ore.

TO PROVIDE EDUCATION FOR THE OPERATING ENGINEERS IN ALBERTA.

A DELEGATION from the steam Engineers in conference in Edmonton, consisting of E. E. Howard, Edmonton, F. J. Dobson, Edmonton, and W. G. Singer, Red Cliff, in the absence of the Minister met Mr. J. T. Ross, the Deputy Minister of Education, and D. A. Campbell, Provincial Director of Technical Education at the Department of Education, and pointed out the great need of providing instruction to the thousands of men in the province who have charge of boilers and engines in manufacturing and industrial establishments, in large business houses where elevators are used, in mining centres, in lumber mills, and in power plants.

It was pointed out that a technical knowledge of steam engineering, in addition to practical experience, is essential to the man in charge of such pivotal positions throughout the province. All are required to hold provincial certificates, and during the past two years the following numbers were issued by the Steam Boilers Branch of the Public Works Department, to men operating various types of portable and stationary boilers and engines.

	1918	1919
First class	6	4
Second class	26	18
Third class	135	142
Traction final	163	101
Firemans' final	55	45
Provisional	447	345
Renewals (Provisional)	177	209
	1,009	864

At the present time there are over two thousand five hundred portable and traction boilers in the province available for use, and the stationary boilers number seven hundred and forty, located in three hundred and thirty-six plants. All the men in charge of these boilers must be holders of provincial certificates obtained as the result of examinations, covering a wide range of practical scientific and technical knowledge.

The deputation was informed that the present policy of the department of education would assist the Steam Engineers in a manner similar to that in operation for the miners.

It was suggested that even classes for steam engineers might be organized at Edmonton, Calgary, Medicine Hat, Lethbridge, Red Cliff, Drumheller, and in the Crow's Nest Pass. These classes would be under the control of the local school boards. Provincial grants are already available, and will be paid to boards which conduct classes in these subjects.

The deputation was told that the department was aware that a real need existed and had announced steam engineering as a type of work which fell properly within the scope of the new provincial institute of technology and art at Calgary. The Steam Engineering department of this institute might undertake instruction by correspondence, to steam engineers located in all points in the province. The whole matter will be given further consideration upon the return from the east of the Hon. George P. Smith, Minister of Education.

The delegation left with the intention of taking advantage of the present provisions which enable school boards to supply this type of education in evening classes to adults.

The city of Lethbridge already has such a class organized.

BUSINESS OPPORTUNITIES

By communicating with this paper and mentioning key numbers, information can be obtained by firms interested in the following lines:

2035. MILD STEEL BOLTS AND NUTS.—Importers in Scotland are desirous of getting into touch with Canadian manufacturers of these in small sizes from ¼ in. thick up to ½ in.

2036. MACHINE TOOLS.—Engineering apparatus and tools, instruments for controlling heat treatment of metals, industrial machinery metals, special steels, electrical heating appartus, mica, etc. A most important limited company at Milan with very good business connections in Italy and abroad wishes to do business with Canada in the foregoing.

2045. A Trinidad commission agency wish to make connection with Canadian exporters of agricultural and artisans' tools.

2033. IRON AND STEEL BARS. — Firm in Scotland is open to receive quotations for the following: each five tons; iron or steel bars:
1½ x 1½ x ¼-ins. Angle x 20 to 21-ft.
2-in. x 2-in. x ¼-in. Angle x 20 to 21-ft.
3-'n. x 2-in. x ¼-in. Angle x 18 to 20-ft.

2034. GALVANIZED SOLID WIRE AND BARBED WIRE.—Firm in Scotland is open to receive quotations for the following: 3 tons No. 6 galvanised solid wire; 3 tons No. 7 galvanized solid wire; 2 tons No. 8 galvanized solid wire; 4 tons galvanised barbed wire, 4 points fi/nch. apart.

Starts British Branch.—The Simonds Canada Saw Company, Limited, of Montreal, Quebec, have just secured a charter and established a new company in Great Britain, to be known as Simonds Saws, Limited. An office and shop has been opened at 53 Bayham, Camden Town, London, N.W. The shop will provide for maintenance and repairs of saws, machine knives, and other edge tools produced by the Simonds Company. Guy A. Eaves, formerly connected with the Fitchburg, Mass, plant of the company, has assumed the duties as office manager, and Leon E. Wilbur, who has covered the Great Britain territory since his discharge from the army in France, will be associated with him. The Simonds Manufacturing Company now maintains plants in Fitchburg, Mass., Chicago, Ill., Montreal, Canada, and a steel mill at Lockport N.Y.

STAINLESS OR RUSTLESS STEEL

Stainless or rustless steel consists essentially of an alloy of iron and chromium, containing usually 0.1-1 per cent. carbon and 12-14 per cent. chromium. It is produced in crucibles or electric furnaces, owing to the tendency of the chromium to oxidize at the melting point. The metal is cast in the usual manner into ingot molds, and the ingots are forged or rolled into bars or sheets. The material can be forged fairly readily into various forms if heated to a bright orange temperature; and if after forging the metal is allowed to cool in air, it will be found to possess good cutting qualities. Quenching in water enhances the hardness to a considerable degree, especially if the steel contains more than 0.4 per cent. carbon, but oil quenching gives the best results. This steel resists formation of scale during forging to a great extent, and is therefore adapted to high-temperature uses, such as engine valves, distilling apparatus, etc. When ground and polished the material resists tarnishing to a remarkable degree. It is slowly attacked by dilute or strong sulphuric and hydrochloric acids, practically unaffected by nitric acid, and is unaffected by nearly all the fruit acids and strong vinegar.

Automobile engineers in all parts of the world will await with keen interest the receipt of a report recently completed by the research committee of the Institution of Automobile Engineers of Great Britain. The report deals with the physical properties of the ten automobile steels standardized according to the recommendations of the British Engineering Standards Association. Colored charts will be given showing the result of tests of tensile strength, hardness, and other qualities as affected by heat treatment. Automobile construction is becoming more and more scientific every day, and it is evident that British designers intend to keep in the forefront of the advance which depends upon a minute knowledge of the materials employed. Only a limited number of reports will be printed and early application is therefore advised.

In advising an engineer as to the best means of treating an old oak staircase which had been attacked by woodworm, "The Builder" says that the most effective means of treating oak work when attacked with woodworm is to give it several coats of paraffin, well soaked into the wood. This should be done by a careful workman, making quite sure that every hole has been well soaked with the paraffin; if treated with paraffin it does not damage the oak, but has a tendency to improve the color.

CALCULATING DIMENSIONS ON 100 H.P. H. R. T. BOILER

Continued from page 395

And for outer and middle rows

$3\text{-}10P+d$

$=3 \times 6.68875 + .875$

$=2.00625 + .875$

$=2.881.$

Of course these dimensions are all the minimum allowable, and if we exceed them it will not matter.

For the thickness of the metal forming the straps we have

$$\frac{5 \times T \times (P-d)}{8 \times (P-2d)} = \text{Thickness.}$$

Then

$$\frac{5 \times 7(6\ 11\text{-}16 - \frac{7}{8})}{8 \times 16(6\ 11\text{-}16 - 7\text{-}4)}$$

$$\frac{5 \times 7 \times 93 \times 16}{8 \times 16 \times 16 \times 79}$$

$=.321$ inches.

This is practically 5-16, but it would be just as well to use ⅜ plate, and thus have a margin over size to allow for wasting, and to make a better caulking edge.

The dimensions for the joint will therefore be:

Pitch outside row, 6 11-16 inches.
Pitch inner rows, 3 11-32 inches.
Diameter of rivets, ⅞ inch.
Thickness of straps, ⅜ inch.
Minimum distance between outer and middle rows, 2.33 inches.
Minimum distance between inner and middle rows, 1.66 inches.
Minimum diagonal pitch between inner and middle rows, 2.35 inches.
Minimum diagonal pitch between outer and middle rows, 2.88 inches.
The width of straps will be the sum of the distance between the three rows of rivets, plus three diameters, multiplied by two.
The accompanying dimensioned sketch shows all the particulars.

Company Reorganized.—The reorganization of the Mami Axle Co., Ltd., St. Stephens, N.B., has just taken place. This is one of the oldest companies on the coast, and has been operating for about 25 years. The capital has been increased from $50,000 to $200,000, and it has been decided that they will rebuild the forge plant which was burned down last July. This building will be a modern forge plant with the purpose of making axes. It will be 60 ft. x 160 ft., and of steel and brick construction. A full line of equipment will be installed. The following are the new board of directors: Chas. E. Henstis, president; James Arnold, Brockville, vice-president, and J. A. Briggs, secretary-treasurer. The other gentlemen are: Hon. George P. Graham, Brockville; T. J. Dillon, and B. J. McCormack, of Welland.

WILL IT WORK
Continued from page 397

fostered this chemical delusion and have educated the public to buy chemical analyses and test bar evidence; the public wanted it, and got it. The whole thing may be summed up in the following words: production, and still more production; composition, accurate to the dot, but quality be hanged. Craftsmanship and skill have given way to arbitrary formulae and the pseudo scientists are happy.

If the consumer will be guided by the question: "will it work?" he will devise simple, practical, sensible tests. He will employ practical, experienced inspectors. He will seek to test material for its suitability to the work at hand, and not merely for its composition.

In the series of articles to follow I will attempt to lay before our readers conditions in the metal world as they are and not how some people think they ought to be.

THE
NATIONAL AUTOMATIC TOOL CO.
Richmond, Indiana, U.S.A.

Classified Opportunities

CANADIAN MACHINERY

AND
MANUFACTURING NEWS

Vol. XXIV. No. 19   November 4, 1920

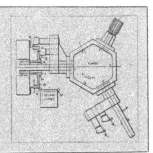

A Great Variety
of Work Done *on*
the Turret Lathe

*Tool layouts for a Gear Case,
friction pulley, semi-steel casting, steel spindle, and other
parts are shown. These layouts should be filed for future
reference.*

IN our previous article dealing with the practice of the Foster Machine Co., Elkhart, Ind., we showed the finishing of gear blanks, friction cores, etc., and to those who did not see it we suggest their looking it up. In the present article, which is the last one, we will go further into the question of tool layout and describe tooling equipment as used for machining automobile, differential gear cases, semi-steel castings, chuck bodies, friction pulleys, and bar stock parts.

Differential Gear Case

The case shown at Fig. 1 is machined on a No. 1B universal turret lathe, the machined portion being shown by heavy lines. This portion is made of malleable iron, and is machined in 6.4 minutes. A machine equipped for this operation is illustrated at Figs. 3 and 4. A good idea of the set-up can be obtained from these views, Fig. 4 showing a close up view of both the work and tools. This photo-

FIGS. 3 AND 4—THE VIEW TO LEFT SHOWS TOOLS REQUIRED IN FINISHING DIFFERENTIAL GEAR CASE. VIEW TO RIGHT IS A CLOSE UP OF FIG. 3 FROM THE REAR.

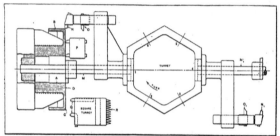

FIG. 5—SHOWING TOOLING USED IN FIRST OPERATION ON CHUCK BODY.

graph was taken from the rear of the machine.

The various steps of the operation follow in sequence. The casting is placed in a three-jaw air chuck provided with special jaws that grip the work along the periphery of the flange. The chuck is also provided with a guide bushing to suit the pilots of the various tools mounted in the turret. The holes A (see Fig. 1) are rough-bored, and surface B is turned by the tools held in double cutter piloted boring bar, this bar being mounted in a turning slide tool in the turret. The surface C is rough-turned by a tool held in the square turret, and the surface D is rough-faced by another tool in the same turret.

The hole A is next finish-bored and surface E is faced by a double cutter piloted boring bar held in a turning slide tool mounted in the turret. The surface F is rough-faced by a tool mounted in a piloted bar which is held in the turret, and the surface F is finish-faced by a tool mounted in another piloted bar, which is also held in the turret.

Holes A are now reamed by an expansion reamer mounted in the turret. The chamfer G is produced, and corner H is formed by tools mounted in a piloted bar held in a sliding tool in the tur-

ret. The casting is now removed from the machine as this completes the operations on this case.

Machining Cast Steel Chuck Bodies

The chuck body, illustrated at Fig. 2, and which is made of cast steel, is completed in two operations, the machined portions being marked in heavy lines. The time of machining is 1 hour 21.8 minutes, and the piece is completed on a No. 3B universal turret lathe. The layout of the tooling equipment as used in the first operation is shown at Fig. 5. This operation requires 42.2 minutes and the sequence of steps are as follows:

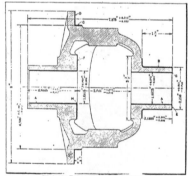

FIG. 1.— DETAIL OF DIFFERENTIAL GEAR CASE.

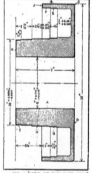

FIG. 2.—VIEW OF THE CAST STEEL CHUCK BODY.

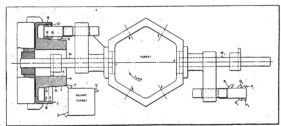

FIG. 4—THE FINAL OPERATION ON THE CHUCK BODY.

The work is mounted on a chuck equipped with three corrugated and hardened jaws that grip it on the rough surface of the hub. One half the length of hole A is rough-bored by the cutter held in piloted boring bar M. The surface B is rough-turned by tool N, and corner C is rough-turned by tool O. These tools and bars are held in the multiple turning head at position 1 of the turret.

The surface D is now rough-faced by the tools in post P on the rear of the cross-slide. The portion of hole A rough-bored in the previous step is now finish-bored by the cutter held in boring-bar M; surface B is finish-turned by tool

N, and the corner C is finished-formed by tool O. These tools and bar are mounted in the multiple turning head at position 4 of the turret.

At the same time the surface D is finish-faced by tools Q, which are held in the square turret. Concentric grooves are cut on surface D by tools R, which are also mounted in the square turret. The casting is now removed from the chuck, the operation being completed.

The second operation on the chuck body is completed in 39.6 minutes. A layout of the tooling is shown at Fig. 6, and the steps are as follows. The work

is mounted in a three-jaw chuck, which grips it on the previously finished surface B. The half length of hole A that is still unfinished is rough-bored by the cutter held in bar M. The surface E is rough-bored by tool N; surface F is rough-bored by tool O, surface G is rough-turned by tool P, surface H is rough-turned by tool Q, and the surface I is rough-turned by tool R. All these tools and the bar M are mounted in the multiple turning head in position 1 of the turret.

At the same time surface K and J are rough-faced by tool S, this tool being held in the square turret. The portion of hole A just previously rough-

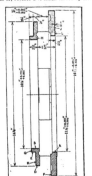

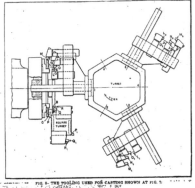

FIG. 5—THIS SHOWS THE SEMI-STEEL CASTING MENTIONED IN TEXT.   FIG. 6—THE TOOLING USED FOR CASTING SHOWN AT FIG. 5.

bored is now finish-bored by cutter held in bar M.

The surface E is finish-bored by tool N, surface F is finished-bored by tool O, surface G is finish-turned by tool P, surface H is finish turned by tool Q, and surface I is finish-turned by tool R. All the above tools and bar are held in the multiple turning head at position 4 of the turret.

At the same time, surfaces K and J are finish-faced by tool T, which is held in the square turret. The threads of surface I are now cut by tool U, mounted in square turret, and which is used in conjunction with the screw-cutting attachment. This completes the operation and the work is removed from the chuck.

Machining Semi-Steel Castings

At Fig. 7 is shown a semi-steel casting which is machined in two operations on No. 3B universal turret lathes. Both operations are completed in 16.5 minutes. The tool equipment layout is shown at Fig. 8. Following are the steps taken:

The work is placed in a three-jawed chuck equipped with corrugated and hardened false jaws which hold the work by bearing against hole A. The hole B is rough-bored by tool L, the surface C is rough-turned by tool M, the surface D is rough-turned by tool N, and the surface E is rough-turned by tool O. All these tools are mounted in the piloted turning head at position 1 of the turret.

At the same time surface F is rough-faced by tool P, surface K is rough-faced by tool Q, and surface G is rough-faced by tool R. These tools are held in the square turret.

The hole B is again bored by tool L, and surfaces C, D, and E are again turned by tools M, N, and O, respectively, all of these tools being mounted in the piloted multiple turning head in position 3 of the turret. At the same time surface F is finish-faced by tool P, surface K is finish-faced by tool Q, and surface G is finish-faced by tool R,

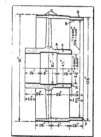

FIG. 10—SHOWING THE FRICTION PULLEY.

these being tools mounted in the square turret.

The hole B is now finish-bored by tool L, and surfaces C, D and E are finish-turned by tools M, N, and O respectively, all of which are mounted in the turning head in position 5 of the turret. The casting is now removed from the chuck.

The remaining tool layout is shown at Fig. 9. This operation is completed in 6.6 minutes, and the sequence of steps are as follows: The work is mounted in a three-jaw chuck that grips it along the previously-finished surface C. Hole A is rough-bored by tool S, and surface H is rough-faced by tool T, these tools being mounted in the piloted multiple turning head held at position 1 of the turret. The surfaces J and I are mounted on a tool block on the rear of the cross-slide. Hole A is now finish-bored by tool S, and surface H is finish-faced by tool T, these tools being mounted on

the piloted multiple turning head at position 4 of the turret. The work can be noted by the dot-and-dash line. This completes the operation and work is removed from the chuck.

Machining Friction Pulleys

The cast iron friction pulley depicted at Fig. 10 is machined in one operation on a No. 2B universal turret lathe. A machine provided with the tooling used is shown at Fig 11. This illustration shows the rear of the machine, and the photograph was taken after the pulley mounted on the lathe had been partially machined. The series of steps in this operation is taken in the following order: The pulley is placed in a three-jaw Barker chuck, equipped with special jaws that bear against the inside of the pulley rim. This chuck is also provided with a guiding bushing to suit the pilots on the various tools mounted in the turret.

Hole A and surface B, Fig. 10, are rough-bored, and surfaces C, D, and E are rough-turned by tools mounted in a multiple turning head which is held in the turret, while surfaces F and G are rough-faced by tools mounted in the square turret. Hole A and surface B are finish-bored and surface D is finish-bored by tools mounted in a multiple turning head held in the turret, while crown C is finish-turned by a tool held in the square turret and a tool mounted on a tool-holder on the rear of the cross-slide, these tools being operated in conjunction with a taper attachment on the rear of the lathe. The relief in hole A is bored by a cutter held in a boring-bar which is mounted in a sliding tool attached to the turret. Hole A and surface B are reamed by a special combination expansion reamer mounted in the turret. The finished pulley is now removed from the lathe.

Producing a Part From Bar Stock

Fig. 15 shows the drawing of a part machined from bar stock in one operation on a No. 1B universal turret lathe.

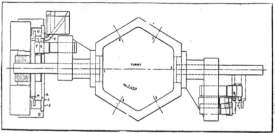

FIG. 9—THE SECOND OPERATION ON THE PIECE SHOWN AT FIG. 7.

A machine engaged in producing one of these pieces is shown in Fig. 12. This illustration shows quite clearly the method employed in the manufacture of this part. The steps in this operation are taken as follows: The stock is fed to a stop on the turret and then held by means of an automatic chuck. Hole A, Fig. 15, is drilled by a drill mounted in a flanged tool-holder held in the turret. Hole A is chamfered as shown, by means of a taper reamer mounted in a flanged tool-holder held in the turret. Surface B is rough-turned by a tool held in the square turret. A revolving centre which is mounted in the turret is placed in hole A during this and the succeeding operation so as to support the work adequately during these operations.

All the remaining unfinished surfaces except surface C (which is not finished, the part being made from cold-rolled steel) and the surface produced when the piece is cut from the stock are finish-formed by means of a special forming tool that is held in the rear tool-holder. The completed piece is cut off from the stock by a tool held in the square turret.

Machining a Long Steel Spindle

The steel spindle illustrated in Fig. 13 is machined in two chuckings on No. 2-B universal turret lathes. It will be noted that this piece is over four feet long, while the greatest diameter is only 2¼ inches. Both operations on this spindle are performed in 35.7 minutes. A layout of the tooling equipment used

in the first operation is shown in Fig. 14, this operation being performed in 19.9 minutes. There are no reference letters shown on the various surfaces of the work in Fig. 14, and so Fig. 13 must be referred to in following the method of manufacturing this part. The various steps in the operation are taken in the following order:

The stock is fed up to the stop on the turret corner between Sides 1 and 6, which permits somewhat more than one-half of the length of the piece to project beyond the chuck. Surfaces A and B are rough turned by the tool held in the single cutter turner in Side 1 of the turret, and surfaces C, D, and E are rough turned by tool T, which is held in the square turret.

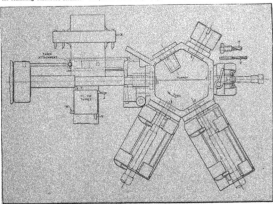

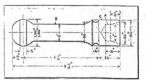

FIG. 13—PIECE FORMED FROM BAR STOCK IN ONE OPERATION.

Surfaces A and D are finish-turned and surface C is finish-straight-turned by the tools held in the multiple cutter turner in side 2 of the turret. The work is indicated in this turning fixture by dot and-dash lines. At the same time, surface F is turned by tool T. Surfaces B and E are finish-turned by the tools held in the multiple cutter turner mounted in side 3 of the turret. The work is also indicated in this fixture by dot-and-dash lines. A centre is drilled in end G by the roller-rest centering tool in side 4 of the turret. Then the centreing tool is replaced by drill H and hole H is drilled after which drill H is replaced by the tap drill V and hole I is drilled in preparation for a tapping operation which is performed in this end by hand. The centering tool is then replaced in the fixture. While the centering holes are being drilled surface C is rough tapper turned by tool W mounted in the square turret which is used in connection with the taper attachment. While end G of the shaft is still supported by the centering tool in position 4 of the turret, the necking tools mounted in fixture X on the rear of cross-slide are first located, then the four grooves between the various surfaces on the right hand end of spindle are turned by these tools.

The threads on surface E are now cut by the self-opening die head mounted in position 5 of the turret. The stock is fed up to the special stop in position 6 of the turret, which permits slightly more than the total length of the spindle to project beyond the chuck.

While the stock is supported in the stop just mentioned, the shaft is chamfered at end O by tool Y, which is held in the square turret, so that the single cutter turner which turns surface I in the succeeding operation can be properly started.

While the piece is still supported in

the stop in position 6 of the turret, the spindle is cut off from the stock by tool Z, which is also held in the square turret. The projecting end of the stock is chamfered at the same time by tool Z so that the single cutter turner used to turn this end of the succeeding spindle can be correctly started.

The second operation on the spindle consists of machining the left hand end, and is performed in 15.8 minutes. The work is held in a chuck equipped with a collet bushing that grips the work around the previously finished surface F. The tooling equipment used in this operation is identical with that shown at Fig. 14, because surfaces J, K, L, M, and N, and holes P and Q, are of the same dimensions as surfaces A, B, C, D, and E, and holes H and I, which were machined by means of this tooling equipment. The only difference between the two operations is the machining of surface F, which is performed in the first operation. For this reason a detailed description of the steps in the second operation is not necessary. This concludes the present article and readers will do well to preserve these various layouts for future reference.

IDENTIFYING THEIR PRODUCT

The Skinner Chuck Company, Inc., of New Britain, Conn., have recently joined the ranks of American manufacturers who identify their products with a trade mark. The new trade mark of this company, while very simple, is really unique. It shows an alligator in the form of the letter "S," superimposed on a solid black circular background, with the words "Skinner Chucks" drawn around the outside. The alligator was chosen after long deliberation as the symbolic figure best suited to the exploitation of their idea, because both the

alligator and their chucks have strong jaws and long life.

The trade mark will be stamped on all their chucks in the future to indicate that the same was made and is guaranteed by the company. A special advertising campaign extending over two months in prominent metal-working publications and other publications of interest to chuck users is the vehicle by which the company will make the trade mark known to the present and potential buyers of their products.

A NEW RADIO STATION

A new wireless station at Siasconset, Nantucket Island, Mass., has been opened by the International Radio Telegraph Company. The site is a historic one, from a wireless standpoint, for temporary stations have been located here ever since the beginnings of wireless, but permanent, commercial service is now established at this point.

This new station, which is the fifth to be opened by the company, is now the nearest one on the Atlantic coast for vessels coming from Europe. Its working radius is more than 250 miles by day and 1,000 miles by night. It has direct cable connections with Wood's Hole, and is thus in touch with the entire country.

Current for operating this plant is obtained from a storage battery which is charged when necessary by means of a generator driven by a gasoline engine. The transmitter is of the spark-gap type of 2 kw. capacity, and is equipped for operating on either a quenched spark gap or a rotary synchronous spark gap. The frequency of the radio motor generator set is 500 cycles, with a sparking rate of 1,000 per second. The current in the antennae circuit for 2 kw. input is 15 amperes. The receiver is adapted for either damped or undamped waves over a range of wave lengths of from 300 to 3,000 meters.

The radio call for this station is WSC. Land line charges will be computed from Wood's Hole.

DUMPING BODY FOR TRUCK

The Karry-Lode Truck Company, of Long Island, City of New York, are now manufacturing an all-steel dumping body for their electric trucks. These have a capacity of 40 cubic feet, dump over the end of the truck, and are especially adapted for coal handling.

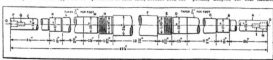

FIG. 15—A LONG STEEL SPINDLE PRODUCED FROM BAR STOCK IN TWO CHUCKINGS.

Portable Boats of Early Railroad Practice

Did You Know That the Transportation of freight and passengers Between Philadelphia and Pittsburgh Was Successfully Conducted in CANAL BOATS? —It Was—But Many Years Ago.

By J. SNOWDEN BELL

IT WILL be a matter of surprise, not merely to the general reader, but as well to most persons familiar with the operation of railroads, to know that many years ago the transportation of freight and passengers between the cities of Philadelphia and Pittsburgh was largely and successfully conducted in canal boats, which traversed the entire distance between these terminals partly by railroad and partly by canal. There are but few persons now living who, like the writer, have seen these boats in service, and the data as to their origin and development, and their rail-

ment by which, in their earlier service, they were carried over the railroad sections of the route.

The Portable Boats

The transfer of boats from one level of a canal to another, upon carriages running on inclined planes, instead of floating them through locks, of course involved, broadly speaking, their movement both on land and in water. This was a well known practice long prior to the operation of the Pennsylvania public works system, and is even said to have been practiced by the ancients,

of boats over the railroad tracks of the Pennsylvania public works—which, however, was not on a commercial scale— or by the use of sectional boats, is recorded in Day's "Historical Collections of the State of Pennsylvania," Philadelphia, 1843, as quoted from a prior publication, which is not named, as follows:

"In October, 1834, this portage (the Allegheny Portage Railroad), was actually the means of connecting the waters of Eastern Pennsylvania with those of Mississippi; and as the circumstance is peculiarly interesting we here place

FIG. 1—A REDUUCED REPRODUCTION OF THE ELGAR PATENT CANAL BOAT.

road transportation equipment, are so meagre and so scattered that a record of it, while merely a matter of history, may be found also to be of sufficient interest to warrant its presentation.

What was known as the main line of the Public Works of the State of Pennsylvania, extending from Philadelphia to Pittsburgh, and aggregating 394.6 miles, was constructed by the state and completed in March, 1834. It was for a number of years—that is, until June, 1857, when it was sold to the Pennsylvania Railroad Company—operated by the state, under the control of a board known as the Canal Commissioners. It comprised a line of double track railroad from Philadelphia to Columbia, 81.6 miles; a canal from Columbia to Hollidaysburg, 172 miles; the Allegheny Portage Railroad, crossing the Allegheny Mountains and extending from Hollidaysburg to Johnstown, 36.6 miles, and a canal from Johnstown to Pittsburgh, 104 miles. Horses were the first motive power on both these lines of railroad, but were soon superseded by locomotives, and mules were always used for haulage on the canals. At noted in Wood's "Practical Treatise on Railroads," 2nd edition, 1832: "This railroad is therefore the first which was undertaken in any part of the world by a government."

The boats used in this system of combined land and water transportation, which were generally termed "portable" boats, will be first considered, and be followed by general notes of the equip-

who did not appear to have known of canal locks. An illustration appears in Stevenson's "Sketch of Civil Engineering in North America," 1838' of a boat and "boat car" used on the Morris Canal of New Jersey, in which the author states was "the only canal in America in which the boats are moved from different levels by means of inclined planes instead of locks." This prior practice, however, has obviously no bearing derogatory to the merit and novelty of the Pennsylvania portable boats.

The initial step in the transportation

it on record. Jesse Chrisman, from the Lackawanna, a tributary of the north branch of the Susquehanna, loaded his boat, named "Hit or Miss," with his wife, children, beds and family accommodations, with pigeons and other livestock, and started for Illinois. At Hollidaysburg, where he expected to sell his boat, it was suggested by John Dougherty, of the Reliance Transportation Line, that the whole concern could be safely hoisted over the mountain and set afloat again in the canal. Mr. Dougherty prepared a railroad car calculated to bear

A SECTIONAL BOAT ASCENDING ONE OF THE INCLINED PLANES ON THE ALLEGHENY PORTAGE RAILROAD.

FIG. 2—THE SECTIONAL BOAT PATHFINDER REFERRED REFERRED TO IN TEXT.

the novel burden. The boat was taken from its proper element and placed on wheels, and under the superintendence of Major C. Williams (who, be it remembered, was the first man who ran a boat over the Allegheny Mountains) the boat and cargo at noon on the same day began their progress over the rugged Allegheny. All this was done without disturbing the family arrangements of cooking, sleeping, etc. They rested at night on the top of the mountain, like Noah's Ark on Ararat, and descended next morning into the valley of the Mississippi, and sailed for St. Louis." The author of the "Historical Collections," referring to the conditions of operation when he wrote, adds:

"The trip of a boat over the mountain is now no novel sight, except that instead of going over whole, they are so constructed as to be separated into three or four parts on reaching the railroad. After thus mounting the cars piecemeal, with their loads of emigrants, baggage and freight on board, they wend their way over the mountains, and resuming their proper element at Johnstown, they unite their parts again and glide on to the waters of the great West."

The earliest record of the idea of constructing a canal boat in separate

port, dated February 7th, 1827. In this report, Mr. White says:

"I made a partial examination of the country over which the railway must pass, and from the general appearance I think the ground is favorably situated, considering the formidable barrier interposed between the eastern and western wtaers. A good turnpike road would probably answer all the purposes of transportation for several years, and a part of the bed could be occupied by the railway whenever the business should require its construction. I would suggest the idea of making the canal boats in three or four pieces, to be divided transversely, and transported over the portage without changing the cargo."

The credit of the invention of means for operating on a commercial scale, transportation of freight and passengers, without unloading and reloading or transferring, over a route comprising both canals and railroads, is due to John Elgar, a civil engineer of Baltimore, Md. After having devised and patented other appliances relating to railroads, a patent of the United States was granted to him on November 7th, 1835, for what he designated as "certain improvements in the art of, and in the

thirteen months than the date when the small boat "Hit or Miss" was carried over the Allegheny Portage Railroad, as before noted, and while that railroad transfer from one canal to another may, perhaps, have suggested Elgar's invention, that invention is none the less meritorious, by reason of its inauguration of a commercially valuable practice on an extended scale.

Fig. 1, which is a reduced reproduction of the drawing of the Elgar patent, is apparently the earliest representation of a sectional canal boat anywhere recorded, and the principle of the Elgar invention, upon which the subsequent practice on the main line of the Pennsylvania public works was based, is fully and clearly stated in the specification of the patent, as follows:

"The object which I have in view, in the first instance, is to prevent the necessity of removing the goods from the vehicle within which they are first loaded by constructing cases which serve on railroads as car bodies, and on canals as boats. This I effect by making such vehicles, or car bodies, of sheet iron, in the manner of iron tanks, riveting them up watertight in the same way. The dimensions of these bodies must be determined by that of the canal locks

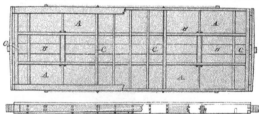

FIGS. 3, 4 AND 5—A PLAN VIEW OF TRUCK, A SIDE VIEW AND AN END VIEW.

sections for transportation over a portage between two lines of canal, appears in a report on a survey of the Juniata route, made by Canvass White, a civil engineer, to the Canal Commissioners of Pennsylvania, which is printed on page 83 et seq. of the commissioner's re-

apparatus for the conveyance or transportation of goods, on a line where canals and railroads form alternate links in the chain of communication, as for example, on the great Pennsylvania line from Philadelphia to Pittsburgh."

While this patent is later by about

through which they are to pass when used as boats. If, for example, the lock will admit a boat of fourteen feet in width, and eighty in length, the bodies may be made seven feet wide and twenty feet long so that eight bodies, two abreast, and four in length, may pass at the same time. I intend sometimes to make the bodies wholly of sheet iron, but they may be made of that material to a height of about three feet only, with an additional height, say of three feet, made of wood. The bodies, when made of this length, are to be carried upon eight-wheeled cars. If four-wheeled cars are preferred the bodies must be made of a length suitable thereto, and a greater number of them will then, of course, be connected together when in the water.

"As these bodies are, by their combination, to form canal boats, the requisite number of them are to be so formed at one end as to constitute a well-shaped bow, and the same number are to be so shaped as to constitute a stern; the other ends are to be made square so that when connected by proper fastenings they will be in one continuous inflexible line, to the length of the lock through which they are to pass."

On February 24, 1843, patent No. 2973 was granted to John Dougherty, of Hollidaysburg, Pa., who was doubtless the constructor of the car on which the "Hit or Miss" was carried over the Allegheny Portage Railroad, in October, 1834, as before noted. The specification states the invention to be "certain improvements in the apparatus for the transportation of goods on canals and railroads; said improvements consisting

in a more perfectly carrying out of the method of transportation for which Letters Patent of the United States were granted to John Elgar, dated on the 7th day of November, 1835, and of which I am the assignee for the State of Pennsylvania."

Fig. 8 of the Dougherty patent is stated in the specification to be a side view of a four-section boat, and Fig. 9, a top view of a double series of sections, bolted or keyed together, but the sheet containing these views cannot be found in the Patent Office, having unfortunately been lost or accidently destroyed. The general features of the design may, however, be understood from the following exerpt from the specification:

"The boats, or boxes, may be made of sheet metal, or of wood; or their lower portions may be of metal, and their upper of wood; the invention not being in any way dependent upon the kind of material employed. I connect these boats, or boxes, together, when they are used on canals, in such manner as that there shall be two sections in width, and three, or more in length. In this respect my plan of connecting the sections is not the same with that adopted by Mr. John Elgar, who proposed to connect them in a continuous line, and in such manner as that they should possess a certain degree of flexibility at the places where they were joined to each other; but when so joined, they have not been found to operate well, as they cannot be kept with their sides and bottoms coincident, but vary laterally, as well as upward and downward, from which cause they are liable to be injured by snags or rocks, and have their

motion retarded by the water. A still more frequent difficulty resulting from the original mode of connecting them has arisen from the want of a free passage of the towing lines from end to end of the boat, all of which objections I obviate by attaching to the fore end of each section, which is to have a rear section joined to it, a plate of iron, six or eight inches more or less in width, and of such length as that it shall extend entirely across the under part of the section, from side to side, and sufficiently high on each side to confine the two parts, or sections, in place.

Such plates are to be bent so as to conform to the curvature of the bottom, are to be fastened to one of the segments by bolts, or otherwise, and to project over and form a ledge, say two-thirds more or less of their width, so that the rear section may be received and rest upon it. The sections are then to be firmly secured end to end by loops and keys, or in some analogous mode, until the intended length is obtained, and two such series of sections are to be secured by bolts, bars, or clasps, side by side, and are thus to constitute a combined boat, of the ordinary width of a canal boat, and in length adapted to the locks through which they are to pass."

An extended search has failed to develop any drawing illustrating the structural details of the sectional boats as actually built and operated, and the recollection of the few now living who have seen them is so much impaired by the long lapse of years as not to be fully reliable. Their characteristic features may, however, be sufficiently understood by reference to Fig. 2, which represents

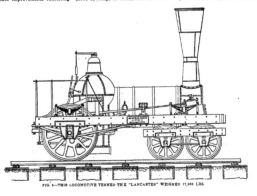

FIG. 8—THIS LOCOMOTIVE TERMED THE "LANCASTER" WEIGHED 17,000 LBS.

the sectional boat "Pathfinder," a model of which formed part of the exhibit of the Pennsylvania Railroad Co., at the Chicago Exposition, in connection with the following description, which is given by Capt. H. A. Walters, of Lewistown, Pa., who, in 1849, started as a driver on the Pennsylvania Canal.

"These, (the sectional boats) were 82 feet in length, 13 feet in width, and in depth 12 feet, and were divided into four sections, each 20½ feet long; the boats were round on the bottom and not flat. The sections were fastened together by irons about half way down the side—the iron projected out from the one section into a V-shaped iron on the other section, then a T-iron fitted down through both of these irons and locked them together. One section was placed upon one railroad truck which was a little bit longer than the section—say about 23 to 24 feet, and had four wheels. . . . The trucks were round in the bottom to fit the boat's sections."

Capt. Walters' statement of the width of the boats is manifestly erroneous, as they could not, for the reason hereafter stated. have been in excess of about 7 feet 9 inches in width. This is also indicated in the first illustration, which is reproduced from an illustration in the Philadelphia Commercial Museum, and represents the stern section, and an intermediate section of a boat, loaded on the so-called "trucks," ascending an inclined plane on the Allegheny Portage Railroad, to be delivered to a locomotive at the summit.

The Railroad Hauling Equipment
The Transporting "Trucks"

In transporting the sectional boats over the railroad divisions of the main line of the Public Works of Pennsylvania, each section was loaded upon an eight-wheeled car, known as a "truck," although it was fitted with two four-wheeled trucks, one at each end, and the trucks were coupled up in a train which was hauled over the railroad divisions by a locomotive. On arriving at the canal terminal of a division, the trucks were lowered into the canal, and the sections of the boats were coupled together and hauled through the canal to the next railroad division by horses. The writer has, when a boy, seen sections of these boats, on their trucks, being loaded in a forwarding house or private freight station at Eighth and Market Streets, Philadelphia, from which they were hauled by horses to the Willow Street Railroad, over which the trucks were hauled a few miles by a locomotive to the inclined plane at Belmont, on the Schuylkill River, and thence by a locomotive to the canal at Columbia, Pa.

The only illustration of these flat cars or so-called "trucks," which has been developed, or which it is probable is now in existence, is that which appears in the drawing of the Dougherty Patent before referred to, three of the views of which are here reproduced, as Fig. 3, a plan view of the car body; Fig. 4, a side view, and Fig. 5, an end view. The four-wheeled trucks, not involving any departure from the ordinary practice at their date, are not here shown.

The description of the car, given in the specification of the patent, is as follows:

"In constructing the cradle it is necessary to limit its width to about 8 feet 9 inches; the passing along the road forbidding that this width should be exceeded; it is necessary, also. that the cars should be capable of running upon curves of fifty or sixty feet radius. The truck wheels which I use are about 2 feet 9 inches in diameter, and instead of being kept to the level of the upper sides of the side pieces of the truck frame, as formerly, they are allowed to rise about a foot above them."

The drawing of the patent does not show side stakes, or any other means for holding the boat section in position on the cradle, and if the stakes were used as would appear to be the case, the width of the boat section would probably not be greater than about 7 feet 9 inches in view of the limitation of clearance before mentioned.

The specification further says that: "The longitudinal timber, B, B, may in this case be a foot in depth, and the width of the space, A, A, must be such as, but need not be greater than will admit of the trucks adapting themselves to the curves of 50 or 60 feet radius, as above named. The trucks work on centre bolts, a, a, as usual. The cross timbers, C, C, of the cradle, are on their upper sides, adapted to the form of the bottom of the boat."

The Early Locomotive Power

The Canal Commissioners, under whose control, as before stated, the connected system of railroads and canals was operated, were authorized by an Act of the Pennsylvania Legislature, approved April 15, 1834. to use locomotive engines as motive power, which theretofore had been horses. In pursuance of this authorization they ordered a number of locomotives from the works of M. W. Baldwin, of Philadelphia, the first of which, the "Lancaster," was put in service on the Philadelphia & Columbia R. R., June 28, 1834, and the second, the "Columbia," September 10, 1834.

The "Lancaster," which is shown in Fig. 6, weighed 17,000 pounds, and had cylinders 9 by 16, and driving wheels 54 inches in diameter. The "Columbia" was of the same weight and dimensions. It is shown by the records that the

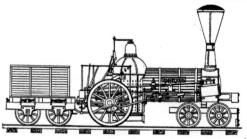

FIG. 7—THE BRANDYWINE. THIS STYLE HAS OUTSIDE CONNECTIONS INSTEAD OF THE HALF CRANK DRIVING AXLE

"Lancaster" hauled a train of nineteen burden cars over the heaviest grades between Philadelphia and Columbia, which was characterised at that time by the officers of the road as an "unprecedented performance." The weight of these locomotives was in excess of that which was estimated by the Canal Commissioners to be within the capacity of the rails of the Philadelphia & Columbia Railroad, as, in their report of December 7, 1833, they say: "The Wiggan (English) rail, weighing forty-one and a fourth pounds per yard, has been adopted for both tracks of the sixty miles now in progress. It is calculated for carrying locomotive engines weighing six tons." The weight of the Baldwin locomotives was not, however, found to be objectionable, and the reports of their performance were, in all particulars, satisfactory.

In his report to the board of Canal Commissioners of November 7, 1834, Mr. Edward F. Gay, principal engineer, referring to the engines "Lancaster" and "Columbia," says:

"Indeed, these engines are justly considered superior and beautiful specimens of mechanism, and reflect great credit on the ingenious mechanic (M. W. Baldwin, Esq., of Philadelphia), who constructed them. They are each supported on six wheels, which is found to be the only arrangement that will enable a locomotive engine to overcome the severe curves connected with the high grades upon this road without injury to the engine or railway."

In the same report these engines are further referred to as follows:

"As all the engines preparing for the road are designed to be of the same class, the following statement of the capacity of the 'Lancaster' may be applied to the others. Weight of engine, 8 tons; capable of drawing 36 tons, exclusive of cars—say 56 tons gross. Amount taken at each load limited to 30 tons, or about 48 tons gross. Running time between the inclined planes (77 miles) with the above load, 8 hours, including stoppages.

Expenses of the Trip

20 bushels coke at 20c $4.00

1½ cords wood at $4	6.00
Engineer and attendants	4.00
Oil	.60
Total	$14.60"

Eight more locomotives, all of the same general design as the "Lancaster" and "Columbia," were built by Mr. Baldwin for the State up to the close of the year 1835, these being the "Philadelphia," November 26, 1834; "Pennsylvania," January 3, 1835; "Delaware," February 7, 1835; "Susquehanna," March 12, 1835; "Schuylkill," April 1, 1835; "Kentucky," July 14, 1835; "Juniatta," September 5, 1835; and "Brandywine," October 21, 1835.

As shown in Fig. 7, the "Brandywine" differed from the first engines in having outside connections instead of the "half crank" driving axle, and an iron frame instead of a wooden one. The report of Mr. Gay, the principal engineer, rendered November 7, 1834, includes an estimate for "Eighteen locomotive engines and tenders complete, at $6,300 each, $113,400," and in his report of October 30, 1835, which covers all the ten before noted Baldwin locomotives, he says:

"The engines upon this road have generally performed their trips with great regularity; and it affords me pleasure to add that the American engines, delivered within the present year, are capable of doing more work than was estimated in my last report; the most of them, in their ordinary trips, draw a gross load of 75 tons. The engine 'Schuylkill' has drawn over the road a gross load of 100 tons, and others have drawn, over the highest grade, from 80 to 90 tons gross. When the curves and grades upon the road are taken into consideration, it is believed that the performance of these engines will be found equal to any in America."

The second locomotive that was built by Mr. Baldwin, the "E. L. Miller," which was the immediate predecessor of the "Lancaster," was equipped with a form of traction increaser by which a portion of the weight of the tender could be transferred to the driving wheels of the locomotive, thereby increasing their tractive power. This device was brought out by E. L. Miller, president of the South Carolina Railroad, for which road the engine bearing his name was constructed, and a patent for it was granted to him, June 19, 1834. As the Miller device was applied on locomotives built for the Canal Commissioners, a description of it may be found of interest in this connection.

Fig. 8 is reproduced from the drawing of the Miller patent, the specification of which specifies the invention as consisting in "using the tender, or car, next to the engine, for the purpose of adding weight to the driving wheels of the engine at such times only as a greater adhesion is required than the weight would give, which it would be practicable to carry as a fixed weight on those wheels without injury to the road."

The specification describes the construction and manner of operation of the traction increaser in the following terms: "The mode which I have used, and found to answer perfectly in practice, is simply to connect the car, or tender, next the engine, to the engine by a strong iron bar, or lever, one end of which is bolted to the under side of a cross timber in the frame of the car, and which lever extends under the frame of the tender to the end of the frame of the engine and into the iron frame, together with the drawing bolt, secures it to the engine.

"Transversely to this lever, I attach to the end of the tender next the engine, two levers, so that their fulcra shall be six or eight inches on each side of the main lever, or drawing bar. These levers have a jaw, or pivot, five or six inches in length, directly over the main lever, and should be about 4½ feet in length.

"When the increased adhesion is wanted, the engineer has only to place his foot upon the ends of these levers and press them into a hook, or groove, for that purpose, on the corner post of the tender; and a portion of the weight of the car, or tender, next the engine, is thus thrown upon the driving wheels of the engine; and when the increased adhesion is no longer wanted, this weight is detached by simply loosening the ends of the levers."

In his report of October 30, 1835, Mr.
(Continued on page 100)

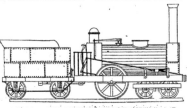

FIG. 8—A DRAWING OF THE MILLER PATENT.

TER

"The Tools
You Buy
Again"

Milling Cutters

Drills and Reamers

Taps and Dies

R FIELD

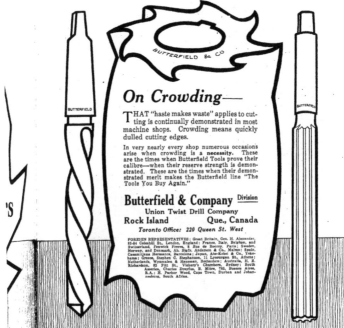

BUTTERFIELD & CO

On Crowding—

THAT "haste makes waste" applies to cutting is continually demonstrated in most machine shops. Crowding means quickly dulled cutting edges.

In very nearly every shop numerous occasions arise when crowding is a **necessity.** These are the times when Butterfield Tools prove their calibre—when their reserve strength is demonstrated. These are the times when their demonstrated merit makes the Butterfield line "The Tools You Buy Again."

Butterfield & Company Division

Union Twist Drill Company

Rock Island   **Que., Canada**

Toronto Office: 220 Queen St. West

FOREIGN REPRESENTATIVES: Great Britain, Geo. H. Alexander, 83-84 Coleshill St., London, England; France, Italy, Belgium, and Switzerland, Fenwick Freres, 8 Rue de Recroy, Paris; Sweden, Norway, and Denmark, Ab. Sigfr. Anderson & Co., Malmo; Spain, Casamitjana Hermanos, Barcelona; Japan, Abe-Kobei & Co., Yokohama; Greece, Stephen C. Stephatzos, 11 Lycourgan St., Athens; Netherlands, Wynmalen & Hausman, Rotterdam; Australia, H. R. Richardson, 92 Pitt St., Vickery's Chambers, Sydney; South America, Charles Dreyfus, B. Mitre, 783, Buenos Aires, R.A.; H. Parker Wood, Cape Town, Durban and Johannesburg, South Africa.

If interested tear out this page and place with letters to be answered.

WELDING AND CUTTING

Facts Regarding the Oxy-Acetylene Process

The Examples of Work Shown with Data
Pertaining Thereto, Have Been Compiled by
Author from Information Applied by L'Air
Liquide Society

By J. H. MOORE

THE fundamental principles of oxy-acetylene welding are easily understood, and when one has already a rudimentary knowledge of the physical and chemical properties of metals, it does not take long to become proficient in the work after some experience has been acquired in the use of the necessary apparatus. The two gases, oxygen and acetylene, are well known and need no explanation. Oxygen is compressed and supplied in steel cylinders. Acetylene is supplied in dissolved state in cylinders first filled with a porous material soaked with acetone, which removes any risk of explosion, or may be obtained from a suitable generator.

Acetylene, when burnt in equal proportions with oxygen produces a neutral flame (that is, neither oxidizing nor carbonizing), of an estimated temperature of over 7,000 degrees Fahrenheit. All metals under the direct influence of this intensely hot flame melt rapidly. Suitable additional metal is added to fill up the groove between the two pieces and results when allowed to cool in a perfectly homogeneous joint of practically equal strength to the metal adjacent. The gases pass through regulators specially designed to reduce the pressure of gas emanating from the cylinders to a suitable working pressure, afterwards passing through special flexible tubing into the body of the blowpipe in which they mix in predetermined proportions.

Perfect results, however, can only be obtained by the use of perfect apparatus. In selecting apparatus it is essential that only the best is purchased. The question of cost therefore should not be the only consideration to enter into when the question of purchase is considered. So called "cheap" (or inferior) apparatus is not only detrimental to good work but represents an element of risk to the operator, and a few dollars presumably saved on the first cost would be more than lost the first time of using it; besides being wasteful in the use

of gases, with the likelihood of discredit to the worker or his employer arising in the future.

On the other hand, when perfect apparatus is used the results obtainable are only limited to the skill of the operator. The cutting of wrought iron and steel is based on the well-known fact that when iron is heated to a high temperature in the presence of pure oxygen it rapidly burns, and forms oxide of iron, which melts at a lower temperature than the metal itself. The apparatus for oxy-acetylene cutting is similar to that used for oxy-acetylene welding except that the blowpipe or cutting torch is so constructed that a stream of pure oxygen acts on the heated metal and oxidizes it. The resulting oxide liquefies

almost simultaneously and is blown away by the action of the blowpipe. This process of oxidation and liquefaction is rapidly repeated until the metal is severed by a clean and narrow cut, either through its entire depth or such part of it as may be required.

Cutting by Oxy-Acetylene Process

Cutting all thicknesses of wrought iron and steel by the oxy-acetylene process has now become a common method. Not only is the cutting by this process rapid, but it is also very much cheaper in consequence of the enormous saving of time and labor. It is practically as easy for the operator to cut a circle or any other design as it is for him to cut straight across, which is all that can be done, of course, by sawing.

There is hardly any limit to the thickness of steel, etc., which can be cut by this process if proper equipment is used, and the many machines that are on the market designed for the cutting of heavy thicknesses of steel, working more or less automatically, and the success they have attained, bear testimony to the essentialness of such a process.

For ordinary purposes, where the metal

FIG. 1—A GOOD EXAMPLE OF HOW THIS STYLE OF WELDING CAN BE USED TO ADVANTAGE

FIG. 1—THIS ENGINE CYLINDER WAS REPAIRED IN 48 HOURS.

FIG. 3—A NAIL MACHINE THAT ANSWERED TO THE ROLL CALL IN 48 AFTER DISABLEMENT.

is only a few inches in thickness, the work is very easy and requires no mechanical assistance; in fact, a man can become an expert operator after very little training.

The question is sometimes raised as to whether cutting by the oxy-acetylene flame has any harmful effect on the material, and the answer is emphatically, No. Mr. P. F. Willis, who is an impartial authority on the subject, makes the following allusion to this subject in his book "Oxy-Acetylene Welding and Cutting." "Is there any detrimental change in material when cut with oxygen?"

No. On high carbon steels there is a slight softening of the metal for a distance of about ⅛ inch from the cut, due to the annealing effect of the heat. Numerous tests have been made and all bear out the above assertion. The results of one such test may prove interesting. A steel plate ⅝-inch thick and 14⅝-inch long was used.

	Test No. 1	Test No. 2
	Original Stock	
	Before Cut	After Cut
Tensile	47,620 lbs...	50,110 lbs.
Elastic Limit	31,640 lbs....	29,980 lbs.
Elongation	33%...........	33%
Reduction of Area...	35%........	42%

It is an advantage, when cutting, to have the blowpipe perfectly steady, with the tip at a distance of about ¼-inch from the surface of the metal, for ordinary thicknesses.

Cutting discs out of steel plate is a comparatively simple process, and requires very little skill on the part of the operator. Some concerns supply a circular cutting guide which enables the operator to cut a perfect circle of almost any diameter with no more effort than is required to control the blowpipe. Cutting heads off rivets when repairing boiler and structural steel without injury to the plate, is also a particular advantage of this process. By cutting off the head, it only requires a tap of the hammer to knock out the rivet. In certain instances, this is the only practical method of taking out rivets, especially in places where they

are inaccessible to a hammer and cold set.

For cutting out discs and intricate designs in iron and steel this method is especially valuable. Cutting up scrap and dismantling steel structural work is also an important field for the oxy-acetylene process, and is the solution to many problems of a similar nature, such as clearing up wreckage after a fire, train disasters, etc.

Its Efficiency and Economy

The time was, and not long ago, that when an important part of machinery broke there was no alternative but to replace it by a new part if procurable, or patch it up by some crude method which usually could be depended upon to give but temporary service only. Sooner or later the time would come when a new part had to be obtained or the machine discarded and condemned

FIG. 4—A FEW HOURS MADE THIS FACE PLATE AS GOOD AS NEW.

FIG. 5—A FEW HOURS WAS ALL THAT WAS NECESSARY TO PUT THIS GEAR IN WORKING CONDITION.

FIG. 6—THE CYLINDER BEFORE WELDING.   FIG. 7—HOW IT LOOKED AFTER BEING WELDED.

altogether, and replaced by a new one.

Probably at that time, when such machinery was comparatively cheap to buy, this might have in the end been considered economy as one could naturally get more service from a new machine than an old one more or less out of order, but with the advent of the oxyacetylene process of welding these methods came to be looked upon as unnecessarily wasteful, and at the present time would be regarded as positively inefficient.

With the speeding up of all industries directly or indirectly concerned in the production of war material, machinery and equipment became taxed to the utmost so that when an important part or a casting became broken oxy-acetylene welding was looked upon as a highly practical and efficient method of restoring it to use again—not only from the standpoint of economy, but because it might be welded in a few hours and made practically equal to new, whereas to procure a new part would take very often several weeks, and in the case of foreign made machinery, would be altogether unobtainable.

The more the price of machinery has risen and the greater the scarcity of metal of all kinds, the more valuable the oxy-acetylene process has come to be regarded until it is now looked upon as an indispensable necessity in all industries, and for all purposes where the joining of metals is involved. There is practically no inherent weakness in the welded joint when the work is properly performed, especially on cast iron and steel. The tensile strength of a weld on cast iron is equal to or greater than the metal itself. In the case of steel, the welded part is about 90 per cent. to 95 per cent. of the strength of the metal.

Even on brass and copper, upon which welding is not attempted as often as it should be, the results are better than can be obtained by riveting or any other method of joining which are consequently rapidly being superseded by welding.

It will therefore be seen that this process holds out many advantages, especially to the manufacturer of metal furniture and utensils, cylinders, drums, containers, etc., where a joint of neat appearance and absolutely leak-proof is necessary; and where the article has to be subjected to great pressure. After smoothing off the welded part it is quite

impossible to detect the joint so that it is for all practical purposes seamless.

In certain cases such as the manufacture of bakers' dough vats, where it is necessary that the article, when in use, has to be absolutely clean in every respect, the fact that there are no seams or apertures which would make it difficult for the receptacle to be kept quite clean and free from deleterious matter, the practically seamless article made by oxyacetylene welding has much to commend it even when strength is not of great importance; furthermore, more often than not it can be made cheaper by this than any other method.

There is hardly an industry of any kind but has its own use for this process in some way, either for repairing or manufacturing, or for some other purpose incidental to either of these two. To recount them all would be quite impossible, but the following are a few of the uses to which it is most commonly applied, and will be sufficient to give a general idea as to the many other similar uses to which it could be profitably adapted.

Railroads are saving thousands of dollars yearly and keeping their rolling stock in constant service by welding broken locomotive cylinders, drive wheels, welding cracks in fire boxes,

FIG. 8—AUTOMOBILE CYLINDER BEFORE WELDING.

mud rings, flues in flue sheets to eliminate leakage, broken forgings of all kinds, cracked or defective steel castings, broken flanges on cylinders, frames, crossheads, quadrants, reverse levers, etc., etc.

Shipyards, by repairing broken cylinders, broken propeller blades, damaged engine parts, boilers, davits, stern posts, cutting manholes and portholes in steel plates, cutting up wreckage, etc. Steel and brass foundries by building up defective castings, filling up blowholes, welding cracks and otherwise saving

parts which, by being somewhat defective, would be useless for the purpose required; cutting off steel gates, risers and runners, welding cracks in bronze bells, etc. Pulp and paper mills by welding broken gears, machinery parts, paper cores, etc.

Machine shops are reclaiming parts that have flaws in them, either by accident or other fault, and saving the amount of work spent on them; welding broken machinery parts, etc. Big savings are also being effected in various industries by welding high speed steel to low carbon shanks and using up every part of a tool after it has become too small for ordinary use. This admits of the whole of the high speed steel being utilized, which, owing to its high cost, is an important consideration.

Garages also have many and important uses for this process for welding broken crank cases, cylinders and water jackets, lengthening and straightening chassis, welding broken crank shafts, gear shift rods, fenders, gasoline tanks, radiators and many other broken or damaged parts but which it would take too long to enumerate. So manifold and varied are the uses for the process that it has almost come to be regarded as an axiom that, whenever two pieces of metal are to be joined, the best and most economical method is to weld them.

Believing readers will be interested in actual illustrations of work performed we show throughout this article some examples of good judgment in welding. The first of these, Fig. 1, shows a part of a handkerchief-making machine, which, together with about 100 small parts, was broken in transit. As this machine was made in France, an

FIG. 5—THE CYLINDER AFTER WELDING.

attempt to secure a new casting would have caused untold delay. The dimensions of all the slots shown were in milli-metres, which made it still harder to even contemplate making a new pattern. The expense of replacing, not to speak of delay, would have been considerable.

The problem, however, was easily solved. The casting was welded by the oxyacetylene method, and restored to use again within a few hours. It was practically equal to new, and the cost of

(Continued on page 100)

DEVELOPMENTS IN SHOP EQUIPMENT

UNIVERSAL GRINDER

The Jones & Shipman, Ltd., Leicester, England, have placed on the Canadian market the universal grinder shown in the illustration. It is known as their model 273 and is used for both internal and cylindrical grinding. There are several features of interest which are herein described.

The columns are high grade castings of rectangular section designed to afford maximum rigidity. The wide spread bases have three-point bearing contacts proportioned to eliminate vibration. The body carries the reducing gear and change feed for the sliding table, together with the clutches for the automatic change feed. The automatic reverse is enclosed in a box on the front. The table is pivoted on a secondary slide provided with a hardened central spigot, and is adjustable for taper grinding by means of a hand-wheel and screw. The setting is registered on a scale reading the taper per foot and the angle in degrees. A T-slot is planed the full length of the table to hold any special work, in addition to the angular slot carrying the work-head and tailstock; the workheads are pulled up by this angular slot against the master straight-edge in front of the table, ensuring perfect alignment.

The table slide is always covered and guards are provided to exclude grit and water. As it always travels over the full length of the slide-ways of the bed, such slight wear as ultimately occurs is evenly distributed and accuracy is maintained over the whole length. The traverse is automatic with eight variations of speed, lowest 15 inch, highest 109 inch per minute. The wheel head is of massive construction, ensuring a perfectly steady running wheel. The wheel spindle—of special case-hardened steel—is finished to a high degree of accuracy. The bearings are of phosphor bronze, lubricated with wick feed, with double ball thrust for taking up all end play, and exceptionally well protected against dust. It is bolted on to the swivel table on the cross slide, and can be securely locked in any position. It also carries an adjustable driving bracket for the drive to the internal spindle. The head carrying the internal spindle is a separate fixture mounted on the same swivel table, and is of suitable dimensions to carry super high-speed spindles. The design throughout is carefully pro-

portioned to eliminate all vibration.

The cross slide carries the wheel-head and also the internal grinding head, and is arranged to swivel in a complete circle enabling either the external or internal wheel to be brought into use without disturbing the work; its movements are fully graduated. The cross feed to the wheel head is automatic and variable, and is rendered very sensitive by an arrangement of counter-weighting the cross slide. It is fitted with accurate and reliable trip mechanism for sizing the work. Automatic sizing of work is attained by setting the adjustable knock-out to the required diameter, and engaging the automatic cross feed, which is instantly stopped when the wheel has advanced the predetermined distance.

The work head is fitted to a graduated swivel base on the table. It carries a double pulley to give fast and slow speeds for a large variety of grinding,

and is arranged to carry a face plate or chuck, and may also be locked for dead-centre grinding. The tail stock or back centre is of massive design, lever operated, the compression of the spring being so arranged that not only is it adjustable for light and heavy work, but can be set for any given position, thus preventing the centre from pressing into the work more than the desired amount. Its adjustable spring tension accommodates the expansion of the work when grinding. The water service is incorporated in the machine; the water taken is easily accessible for cleaning purposes. A high grade geared pump forces a heavy volume of coolant through flexible armored hose to the adjustable nozzle attached to the wheel guard, the return being through inclined water ways and strainer to a removable filter tray, thence over a weir to a sump and tank.

Lubrication. A simple and efficient

GENERAL VIEW OF THE GRINDER.

system throughout enables every working surface (which is not already self-oiling) to be readily supplied with oil with a minimum of trouble. Following are the chief dimensions: Capacity between centres, 12 in. dia., 24 in. long; table traverse, 24 in.; table taper adjustment, 30 deg. included; height of work head centres, 6½ in.; working surface of table, 43½ in. by 6 in.; cross traverse of wheel slide, 6½ in.

Number of work speeds, 8; number of wheel speeds, 3; size of disc wheel, 10 in. by 1 or 1½ in.; countershaft driving speed, 500 r.p.m.; horsepower, 4½; floor space, 96 in. by 62 in.

FILTRATION AND STERILIZATION

To the man who is not vitally concerned the matter of machine tool lubrication may not appear of much importance. On the contrary, it is of proven importance, otherwise there would not be the demand for such a large number of different cutting oil liquids and compounds.

It was not so long ago when the ability to increase production by liquids applied to cutting tools was recognized and the initial steps taken along that line. Water was the first liquid tried, used as a stream feed, or provision made so it would be allowed to trickle upon the tool and work. A little experience in this way proved that a mixture of soda and water was more satisfactory all around, as the tendency to rust and corrode was largely overcome.

The evolution of machine shop practice, requiring high speed tools, carried with it efforts for extreme accuracy in machining and the element of increased production. With the speeding up of tools it was readily recognized that, in order to keep down operating costs and produce a product of the highest quality, it would be necessary to use a cutting fluid which had not only the cooling properties necessary to absorb excess heat, but lubricating qualities as well, to increase accuracy, decrease the power consumed, and at the same time speed up production.

With the production of a cutting oil which would serve the double purpose came the increased cost and this item must, of necessity, be kept at the lowest possible figure, otherwise the product could not be marketed. It was, therefore, apparent that an equipment which would filter and sterilize the liquid would eliminate the necessity of throwing the used oil away. This method would reduce cutting oil expense and still promote high speed operation.

The S. F. Bowser Company, Limited, oil storage and filtration engineers, Toronto, are out to meet this need and have designed a filtering and sterilizing system which will soon be placed on the market. The illustration gives a good idea of its appearance and construction. These embody the tried and true principles of screening, precipitation, filtration and sterilization, all combined in a commercial and practical manner. This class of filtration is not a venture or experiment, and the company have had sys-

tems, essentially the same, in service for several years, which are giving the very best results.

An installation of this nature will naturally increase the life of the cutting tools, decrease power consumption, permit higher operating speed, increase accuracy, give better lubrication and save throwing away oil which has been used. By this conservation a better grade of oil can be used—one which will not foam, rust or corrode the work or tools. While the filtering process, which is automatic, is under way, the oil is also sterilized, thus eliminating obnoxious odors and preventing infection, thereby preserving the health of the employees.

The systems are so designed that they can be adapted to conditions prevailing and arranged to accomplish the work required. From the machines and chip separators the oil, either by gravity or special provision, can be delivered to the filter and sterilizer which automatically removes foreign matter and sterilizes the fluid. After going through a series of compartments, screens, filtering devices, etc., the liquid is delivered to the filter tank, which acts as a temporary storage. From this tank the oil can be returned to the different machines the same as new oil and used again.

NEW THINGS IN MACHINE TOOLS

ADJUSTABLE WRENCH

The Allen-Diffenbaugh Wrench & Tool Co., Baraboo, Wis., have marketed an adjustable wrench which automatically grips a nut when the handle is pulled. It is made in several sizes from drop-forged steel and has jaws that are machined and hardened.

UNIVERSAL CLAMPS

The Burvin Co., Inc, 87 Warren Street, New York, are marketing tools known as the Force universal clamps. These are made by the Black Rock Mfg. Co., Bridgeport, Conn. in several sizes adapted to hold irregular work on milling machine tables.

PLANING MACHINE

The Powell Machine Co., 245 Stafford Street, Worcester, Mass., are marketing a 48-inch x 17 foot belt-driven planing machine of the high-speed type. The feature of this machine is the accelerating drive which overcomes many of the

difficulties so far met with in designing planers of the heavy type.

OIL COOLER

The Griscom-Russell Co., 90 West Street, New York, have placed on the market a device known as their multi-whirl cooler. The device is designed for cooling the oil used in the lubrication of turbine bearings and reduction gears, or the quenching oil used in the heat treatment of steel. It is claimed that by means of this device the oil can be maintained at the proper viscosity for efficient results.

STARTER FOR MOTORS

The General Electric Co., Schenectady, N.Y., have issued an automatic starter for use with squirrel cage induction motors driving lineshafts, pumps, compressors, and similar devices. The starter is operated by a push button or by a float switch, pressure governor, or some similar automatic accessory.

VIEW SHOWING THE FILTRATION AND STERILIZATION SYSTEM.

The MacLean Publishing Company
LIMITED
(ESTABLISHED 1887)
JOHN BAYNE MACLEAN, President.   H. T. HUNTER, Vice-President
H. V. TYRRELL, General Manager.

PUBLISHERS OF

CANADIAN MACHINERY
MANUFACTURING NEWS

A weekly journal devoted to the machinery and manufacturing interests.
B. G. NEWTON, Manager.   A. R. KENNEDY, Managing Editor.
Associate Editors:
J. H. MOORE   T. H. FENNER   J. H. RODGERS (Montreal)
Office of Publication: 143-163 University Avenue, Toronto, Ontario.

VOL. XXIV.   TORONTO, NOVEMBER 4, 1920   No. 19

These Changing Times

THE business community in Canada can well heed the remarks of many of the best thinkers on this continent who are trying to bring home the fact that there is a real and vital responsibility resting upon it right now in maintaining a substantial activity in operations during the period of price adjustment.

A policy of moderation and sympathy is needed. The forcing of any man or of any company to the wall should be avoided if there is any honorable way out of such a move.

Manufacturers of either raw, semi-finished or finished material should do their utmost to prevent unemployment, and in this they should have the co-operation of every employee.

The man in the shop should recognize that the management has a real problem on its hands now. Financing is not easy—it is hard, especially for firms that have not been in existence long. There are managers and presidents who are drawing hardly anything at all in the way of salaries at present. The business is simply taking it all, and their optimism for the future and their determination to participate in the development of months to come keep these men in a state of suspended, if not actual poverty.

Occasionally one hears a remark to the effect that "the worm has turned, and we—the management—are going to have our turn now. The men in the shop have run this place for the last four years and we are going to let them know now what a pinch feels like." This attitude is not general, but it exists in some places. Its application to shop management now would be the veriest nonsense.

Nursing a grudge is wasted energy—it will not turn out one extra casting in a year—it will not get any better results in the tool room or the machine shop—neither will it make any other department any more capable or efficient.

There should be a reasonable policy to bring about as much employment as possible. Unemployment is not desirable at any time—it strikes too quickly and too surely at the homes, at the wives and children.

The Canadian trade has so far kept its head fairly well in the matter of cancellations. It is highly desirable that this attitude should continue. An avalanche of cancellations, were they allowed, would start a circle of trouble so vicious that one could hardly measure the element of danger it contained.

All things considered, there is a fair volume of business passing in Canada. It is not to be expected that much expansion will take place until prices and values find new working levels.

The business community in Canada is going through these changing conditions successfully.

Common sense, confidence in the future, an honest day's work and a fair amount of actual and honorable sweat will put us across safely and give us a start on an era of development such as we have not witnessed heretofore.

Another Harmless Boiler Explodes

ONCE again has a boiler, used on a farm for the purely agricultural purpose of threshing proved that it is not exempt from the physical laws that govern other steam-containing, or pressure-containing vessels, though it is free from legislation. It proved this elemental fact by the very impressive means of blowing up, fortunately without hurting anybody. The men in the immediate vicinity had a very lucky escape, and those who were further away from the firing line were also fortunate, as some pieces weighing several hundred pounds were reported to have landed several fields away.

It shows surely a lack of consideration, or a sense of the fitness of things, on the part of these boilers to blow up when several successive governments have decided in their wisdom that these boilers cannot blow up, and are perfectly harmless domestic pets. The process of reasoning which decided that because a boiler is on a farm it is going to act differently under given circumstances than a boiler in an industrial establishment is not very clear to the ordinary mortal. Furthermore, the boiler on the farm is supposed to be possessed of sufficient intelligence to enable it to perform its functions without the aid of any skilled help. It is very likely that the real reason of the exemption of boilers on the farm from the Steam Boiler Act was on account of the farmers all possessing votes, and the fact that a farmer suddenly called upon to pay inspection fees, and probably higher wages to skilled help, would have emitted a howl that would have been heard uncomfortably in Queen's Park.

Now that we have a Government composed largely of farmers it is hardly likely that they will pass legislation that will cause them to be under this necessity. We must therefore submit with what degree of patience we can to the spectacle of occasional boiler explosions on farms, with or without loss of life, until someone who is not afraid of adverse votes gets a measure put through that will include farm boilers.

Our own opinion is that if there is any place on earth where a steam boiler should be subject to rigid inspection, it is on a farm, because anyone who has noticed the way many a farmer treats his farming implements can form a fair estimate of the way he will look after such a thing as a steam boiler.

We have pointed out from time to time in these columns the potentiality of danger that lies in any vessel containing a quantity of water under high steam pressure, and the fact that it was of relatively small size did not eliminate the danger. If one were to take the kitchen kettle and carefully seal the spout and tie the lid on after filling it with water, and then place it over the flame of the gas heater or the coal fire, it would probably blow the kitchen to pieces. The smallest boiler is much larger than the biggest kettle. Why should it be exempt?

Machinist Should be Given a Variety of Work

By T. DALY, Works Engineer International Harvester Co., Hamilton.

THE status of the all-around machinist has always been an interesting subject and readers no doubt read with interest the views of R. W. Gifford in our October 7 issue. T. Daly, works engineer for the International Harvester Co., Hamilton, Canada, has his own ideas on this subject, and here they are:

To my mind the all-around machinist is not passing out completely, but I do believe that the modern machine shop methods do not produce the same proportion of all round men as formerly. The tendency of to-day is to develop mechanics who are, more or less, specialists. The volume of modern industry and keener commercial competition have quickened the effort for increased production. Production is influenced by three important factors: the division of labor, the co-operation of labor, and the use of machinery.

Primitive man lived in a state of sturdy independence. He made his own weapons of defence, hewed his own log canoe and ground his own corn. When civilization came and community life developed, men chose definite trades. The blacksmith or armourer made implements and weapons, the carpenter worked in wood, and the mason built in stone. This division of labor was thus a natural outcome of the growth of communities. To my mind the advantages of divided labor are as follows:

1. Increased skill in the workman; as for example, in the tempering of tools by a toolsmith. 2. Saving of time; as in operating a crane, the slinger securing the load and the driver controlling the hoist. 3. The tendency for the workman, when encouraged, to improve the machine or method, if his mind is concentrated on his work. Many improvements have thus originated. 4. Greater efficiency of the workman by being kept to the work he can do best; it would be wasteful, for instance, to keep a good pattern maker working on rough carpentry.

Against this it may be contended that: 1. The worker's pride in his craft is often destroyed by its subdivision into simple and monotonous processes, and he may feel no responsibility for the product as a whole; hence much inferior workmanship. 2. Excessive monotony tends to make the workman a mere machine, and does not develop his intelligence. 3. If his task is changed, his usefulness is often lost. 4. The function of industry being to provide the material means for a good social life, the undue subordination of human beings to any mechanical system should be resisted. By the co-operation of labor the combined action of numbers can accomplish what one man alone cannot do; as in the erection of a steel bridge.

On the use of tools and machinery, all industry is dependent, for the rudest agriculture could not be carried on without a spade, nor traveling at 60 miles an hour on the railway without an engine. During the war the necessity of the times caused a further subdivision of labor. Unskilled help was brought into the munition industry, given a brief training on special machines, thus producing numbers of specialists who are just "machine hands." Had sufficient trained men been available this work would have been more in their hands, but under the circumstances the real machinist had the better class of work, and the making of safer and quicker tools and machines for use by unskilled operators. This influx of unskilled labor into the mechanical field makes the machinist imagine his services to be underrated.

Will It Lower Standard

If asked the question, Will the passing of the all-round machinist lower the standard of his calling? I would say this, that the man who has served an old-time apprenticeship will regretfully nod his head. He recollects how in the old days he used to mark off his own work, do his own simple forging, or chalk on a board how he wanted the blacksmith to make it. He would do his own shaping, turning and drilling and finally assemble the work at the bench. To the individual this is much more interesting than one particular piece of work, for it has the spice of variety about it, but, of course, it cannot be termed efficient as to production. So it has yielded to the methods brought about by better machinery and preciser measuring tools. The old-time all-round man, however, may now be kept to one machine or process, and thereby be prevented from realising the full value of his experience. On the other hand, the machinist who specializes becomes more rapid and accurate and probably drifts into some phase of work at the lathe or milling machine or at the bench, where he finds ample scope for his manual skill, his eye and his brain.

There is, and always will be, a place for all-round machinists; in the small shop, where there is not work enough to keep a man for each kind of machine; in the repair shop of larger works, where he may have to do anything from facing a slide valve to hanging a countershaft; and on experimental work—in fact, wherever there is frequent variety. For the position of foreman, a man with a broad experience is generally required, and here the all-round man will serve better than the specialist, provided that he can accept responsibility. It is also generally agreed that the bench hand has a wider outlook than the man at the machine, for it falls to his lot to interpret the drawings, sublet the machine work, and assemble the parts.

On the whole, therefore, it does not appear to me that the standard of the machinist trade is being lowered, as the matter tends to adjust itself. Relatively fewer all-round machinists are being produced, but proportionately fewer are required, and the standard of the work of the average machinist is continually being raised by the demand for more accuracy in machine manufacture.

Can We Improve Present Conditions

Can we improve present conditions or raise the standard of the machinist or toolmaker? is a question which could be asked. To this I would say, that to remedy the present tendency to over-specialize, the machinist should be given as much variety of work as is consistent with good production, to maintain his interest. Much can be done by employers to cure the tendency and prevent too-early specialization, by having a well-regulated system of apprenticeship. There is no doubt that love of the vocation in which a man is engaged breeds both efficiency and contentment of mind. Hence boys of good education who are interested in things mechanical should be encouraged to serve a definite term in learning the machinist trade. They should follow a schedule of changes from one machine to another, and have a good spell at the vise.

Technical training should not be undervalued, and the youth who acquires some knowledge of mathematics, mechanics and drawing in his own time will value it more than if it is given gratis by his employers. The wages during apprenticeship should have sufficient margin above the cost of board and lodging, so that the apprentice will not be tempted to drop his prospects of becoming a skilled mechanic with chances of promotion, for the higher initial pay of the machine hand on piece work.

By the end of his time the apprentice will have discovered his preference for some machine or class of work; or, if his inclination for the technical side is strong enough, he may push his way into the realms of machine design. In any case his varied apprenticeship will give him a good foundation for a successful career in a skilled and useful calling.

MARKET DEVELOPMENTS

Premium Steel Mills Let Down the Prices

Some Business Going Their Way for Spot Delivery, But the Price is Better Than for Some Months—Cancellations Are Asked for in Some Lines, But Are Not Generally Being Granted.

THERE are signs that the deflation of prices is settling more seriously on the iron and steel markets than for some weeks past. Not that there has been any sudden drop, but there is an eagerness to sell now that has been absent for some time past. The premium mills in United States are still selling steel in Canada, but they are doing this at a much more favorable price than formerly.

Cancellations are asked for in some lines, but about the only chance of having a cancellation recognized is when shipment has been delayed so long that it has become unseasonable. In that case consideration is often given. Otherwise requests for cancellation are receiving little sympathy. The trade holds that, were they allowed, it would leave a lot of unsaleable material on hand. The policy of no cancellation is being adhered to in the iron, steel and tool field in almost every case. One moderation is that some of the firms will defer shipment to anything within reasonable time, if by so doing they are going to

benefit their customer. In many cases firms report that they have no requests for cancellations.

The fact that the National Machine Tool Builders' Association is meeting about a month sooner than scheduled is seized upon by many critics to indicate that there will likely be some announcement following such a gathering, although the association does not exist for any such purpose. Some of the builders have little business left on their books. In Canada machine tool builders are still fairly well employed, especially those engaged on heavy work.

The scrap metal market in Montreal was cut down this week on all lines, but even at the new figures there is nothing more moving. The new metal market is easier in many lines, and reductions in prices have been made at selling points.

Reports on collections vary. Some firms tell of trouble in getting their money, even from large and substantial companies, while in other cases the financiers will tell you that they have no trouble at all in this matter.

MONTREAL SCRAP METAL PRICES ARE CUT; BUSINESS STILL DULL

Special to CANADIAN MACHINERY.

MONTREAL, Nov. 4.—There is little doubt that the present easing of prices on many commodities is the commencement of that period of readjustment that the trade, and the public generally, has been looking forward to ever since the signing of the armistice. At different times during the past two years there have been developments that gave promise of a return to something like pre-war conditions, but the abnormal requirements for all classes of materials, which were excessive during the war period, continued throughout the succeeding year and showed little abatement until a few months ago. Trade conditions are now taking a turn where a lowering of prices becomes almost a necessity, but many circumstances combine in making it almost an impossibility to take drastic action in reducing quotations. The first real evidence of price cutting was reported recently when one of the mills in the States offered plates at $1 per ton less than they could be purchased elsewhere. When mills are

openly taking an attitude of this nature it apparently foretells an early adjustment of prices. At present there is more in keeping to existing demand. Before any marked reductions can be made in steel prices it will be necessary for all concerned to co-operate in bringing about a fuller knowledge of what is required in removing the obstacles that at present prevent any marked reductions in the metal industry. There is a tendency on the part of labor to take a more reasonable view of conditions and production per man has shown an increase during the past few months.

Dealers Adjusting Prices

Steel mills have reached the point where actual competition is more than a by-word. Supply and demand are becoming real factors, with supply surpassing the trade requirements, the logical consequence being a marked trending to lower levels. Whether these reductions will be heavy or light is, at present, an unknown quantity, but the so-

lution apparently hinges on the attitude of labor as to the question of wages. That labor is anxious to see a reduction in the cost of living is a well understood fact, but without their co-operation it is virtually impossible to establish a permanent basis of readjustment, as the present cost of production prohibits any excessive declines in the cost of producing nearly every necessity of industrial and domestic life. The pendulum of trade is swinging with considerably more uniformity as the supply of materials, which has been more or less restricted during the past year or so, is fast becoming normal. The sheet situation, since the collapse of the motor industry, has rapidly adjusted itself, and at the present time the supply of sheets, with the exception of brights, is almost adequate to trade requirements. Black sheets, 28-gauge, have declined, and local quotations are on a basis of $10.50 per hundred pounds. Plates are quite plentiful with delivery good; warehouse stocks may be had at $6.50 for 3-16 inch, and $6 for ¼ inch up. Tank plates, Pittsburgh, are now quoted at $3.50 and dealers here are revising prices to conform to the new American base. Boiler tubes are still scarce and prices are firm.

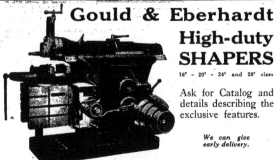

Conditions in the general industrial field are reflected in the metal market and price declines are in order. The quiet character of the demand has become additionally emphasized, and the present trading is quite listless. The local trading has decreased despite the lower price levels, and this is due to the lighter demand together with greater available supply. While dealers are still of an optimistic turn of mind there is a feeling that the remainder of the present year will see still further readjustment of conditions that will tend toward temporary trade depression. Quotations on copper have declined about 3 cents per pound. Lake is quoted at 21 cents, electro at 20½ cents, and castings at 20 cents per pound. Tin is now selling at 50 cents, a decline in a couple of weeks of 4 cents per pound; demand is lighter. Spelter is firm at 9 cents with a tendency to weakness. Lead is weak and requirements light; dealers are quoting 8¼ cents per pound. Antimony is down 1 cent to 8½ cents per pound. Aluminum is steady at 36 cents per pound.

Prices Decline on Light Demand

Aside from light local activity on the part of iron foundries, the old material situation has little of an interesting character and trading is exceeding light, almost to the point of being stagnant. The price decline has been little inducement for the consumer to come into the market as his requirements do not call for large quantities of scrap. The movement of old materials is dependent on the activity maintained in every class of industry, and with this activity showing marked signs of falling off the requirements for scrap are relatively reduced. Dealers here anticipate a quiet period until the turn of the year, and do not look for any heavy buying until well on toward the spring. All lines of scrap have been marked down by local dealers. Coppers are now quoted at 12 cents for light, 15 cents for heavy and wire. Brasses are down from 2 to 3 cents per pound. Price reductions are also general in steel and iron scraps, these declines ranging from 50 cents per ton on boiler plates to $4 on cast iron borings. Intermediate revisions are listed in the selected market quotations.

MAKE IT HARD TO CANCEL ORDERS

Firms Believe That There Should be An Adherence to Contract Wherever Possible

TORONTO.—There are certain lines of business reporting a good turnover now, while others claim that business with them is quiet. No large sales are reported, although the wood-working end has turned in some fairly good orders; There can be no denying the fact that buyers of machine tools are asking more and more what is going to

---

POINTS IN WEEK'S MARKETING NOTES

Montreal scrap metal market has been marked down all along the line, but no more business resulted, and trading there, as in Toronto, is very dull.

—

Some of the dealers are still buying from the premium mills, but the prices paid are much more reasonable; for instance, black sheets being secured for about 6.50 against as high as 7 and 8 a short time ago.

—

The National Association of Machine Tool Builders have advanced the date of their meeting almost a month, holding it early in November. Some announcement regarding prices and selling policies is expected to follow this session.

—

Some of the dealers of machine tools in Canada are still well employed, and contrary to reports in the market they have not laid off men, neither have they decreased wages.

—

Small tool trade is quiet, there being few orders of any size going. Reports indicate that there may be some change in prices, especially of high speed within a few days.

—

The production of iron and steel is going down, and for the month just closed was not over 99 per cent. of the totals done in August and September.

—

Current prices for pig iron this week at Pittsburgh are: Basic, $40; foundry and Bessemer, $45. These prices show a decline, but even at these levels selling is very limited.

—

Warehouses report a good volume of business in most cases, while one goes so far as to state that the month just closed, October, was as good as any during the year.

---

be the attitude of the builders in regard to prices. The fact that the annual fall convention of the National Machine Tool Builders' Association has been advanced from the first week in December to the first week in November gives strength to the theory that there is an anxiety to get together and discuss matters of policy, especially having a bearing on the schedule of prices. It is only natural to expect that in such an atmosphere buying is going to be rather dull.

Need New Equipment

"Constructive selling in the Canadian field is hard work." is the opinion of one well-known manager this morning. "Our

---

men have worked out several problems showing where they can give far better results than are now being secured by older methods, but it seems very hard to make much progress in this way. The Canadian manufacturer has a liking for getting the last final kick out of every old machine there is in the plant, no matter whether it pays him to do it or not. Many of the plants here will have to mend their ways if they are going to stand the pace with some of the newer plants that are being put in here now."

Another machine tool man, who has just returned from a rather extended survey of the field, claims that many buyers—or firms that should be buyers—are waiting and putting off coming in. to the market just as long as they can. They think they can secure something better in the way of terms or conditions or prices, and for that reason they are not buying. The business they have to offer is good, and it is not going to go out of existence, but it is not helping in the meantime to keep up the volume of trade.

About Cancellations

Some of the warehouses are receiving requests for cancellations, especially in business that has been on their books for some months. The policy of some of the firms is quite well marked on this point, and one that seems to take in quite a number of others was set forth as follows: "We do not accept cancellations but we will make postponements in shipments, and do anything within reason to make it easier for a man to take on his obligations and settle for them. If we were to start accepting cancellations we would be lost in a very short time. It would mean that we could go ahead and turn out material, finance it, and then, when ready to ship, we would find that our market had gone and the stuff was left on our hands. There is nothing worse for business in general than that we should make it easy to start a procession of cancellations. One cancellation is sometimes enough to start a whole lot of others in operation, and for that reason we do all in our power to keep the first cancellation in the ring from being made."

Here is the opinion of another man in the steel merchandising business: "We have some requests for cancellations but they are neither numerous nor serious, nor do we expect them to become so. Most of them are on material that has been on our books for some time past. If it is possible for us to make arrangements with our principals to cancel we will do so, but otherwise we insist that the business stand. Some of the material that cancellations are asked for has a seasonable value that is apt to be spoiled by our inability to secure it soon enough. In such a case we are willing to give every consideration to the request. Of course we try to get it held over for the next season if possible. Just a few days ago we had a request for a cancellation on a big piece that had been

Geometric Solid Adjustable Die Heads

When one uses the word "ONLY," he needs to be sure of what he is talking about. With confidence we use the word in saying that a Geometric Solid Adjustable Die Head, fitted with a set of milled dies, is the only solid die head that will back off without stripping the thread.

Geometric Solid Adjustable Die Heads may be fitted with either a releasing shank, plain shank, or with special shank for the Gridley Automatics. The releasing shank permits the head to disengage from the shank upon completion of the thread. The plain shank fits the head for use on the turret of a lathe or on a live spindle, such as a drill press.

Apart from the fact that it is not equipped with the self-opening feature or the roughing and finishing attachment, the Geometric Solid Adjustable Die Head is the same in principle and construction and will do equally as accurate work as the other styles of Geometric Die Heads.

> Whatever the requirements, there
> is a type of Geometric Die Head
> best adapted to the work.

THE GEOMETRIC TOOL COMPANY
NEW HAVEN   CONNECTICUT

Canadian Agents:

Canadian Fairbanks-Morse Co., Ltd., Manitoba, Saskatchewan, Alberta.
Williams & Wilson, Ltd., Montreal.   The A. R. Williams Machinery Co., Ltd., Toronto, Winnipeg;
St. John, N.B.; Halifax, N.S.

If interested tear out this page and place with letters to be answered.

specially forged, and which had cost the forging plant a lot of money. There was only one thing to do in such a case, and that was to insist that the firm ordering the work should meet its full obligation, and I think that standard is generally accepted."

Most of the warehouse men report that they are doing a satisfactory business. One of them told Canadian Machinery that the month just closed (October) would be one of the very best in the whole year. There is still a good demand for sheets, and some of the dealers state that they are still taking on material from the premium mills, but the premium mills have come down in their price to some extent. Some of them, it is said, are doing as well as 6.25 on black sheets for the Canadian trade, and on blue annealed one quotation gave 5.90.

In tubes there is still a big demand and enough bookings to take care of all the shipments that come in for weeks yet.

Some peculiar reports get in circulation in the machine tool market. This paper was told in all seriousness that one well known maker had laid off a number of men, and was generally slack. Within an hour or so a representative of this paper met one of the officials of this company on the street and put the question to him. The answer was that they had not laid off a man and that the factory was comfortably busy.

The Small Tool Business

Buyers are not taking on very much just now in the way of small tools. They are buying just what they need. In some cases one will hear of a fairly large user buying a dozen or so of a standard and much-used size of drill cutter, or tap, while in ordinary times the buying would be done by the gross: There are rumors in the market this week of some readjustment in the prices of certain lines of high-speed tools, and it is likely that an announcement will be made shortly in regard to this.

Several lists are in the market from firms that find they have on hand more stock than they can make up profitably. Buying was done at high speed in times when material was short, and there may have been more or less pyramiding of orders. Now the jobbers are being appealed to to place some of this overstock. As a general thing the jobbers just now have all they can handle, and are not likely to take on anything unless there is a very attractive price clause attached, and this is just the thing the manufacturers are trying to avoid. In many cases they know they cannot make their material up and get their money out of it in the ordinary way.

Dealers in small tools state that the general stock is getting fairly low in the country and that outside of a few industries that bought in excess of their requirements there is a lot of business that will undoubtedly come into the market before many months.

ORDERS ARE NOT COMING IN
AND THE BUSINESS IS NEEDED

Special to CANADIAN MACHINERY.

PITTSBURGH, Nov. 4. — There has been a further decline in steel production, the rate now being not over 90 per cent. of the average rate in August and September. The Steel Corporation is operating its various plants as well as car supplies and fuel supplies permit, and a few of the large independents are doing almost as well, but some large independents and the great majority of the smaller mills have begun decreasing output in the past week or two, or have carried farther a decrease already begun. In a few instances mills are operating at only 50 or 60 per cent of capacity.

With many of the mills that have decreased production the cause is simply that the mills have largely run out of orders. These are mills that sought the highest possible prices, and necessarily had to keep their order books relatively clear so as to be able to make the deliveries that brought the highest prices. When the market turned dull these mills could not fill their order books by reducing prices because sufficient business was not available at any price.

With the other mills the case is one of cancellations and postponements of delivery. The mills seemed to be sold up for several months, but cancellations and postponements have made quite a hole in the weekly requirements of customers. Mills in this position are indisposed to cut their prices, as they still have a large volume of business ahead, though not for as heavy deliveries week by week as they could make by operating full. It would be disadvantageous for such mills to cut prices, since that would jeopardize the contract business on books and would not bring enough new business in to make up.

Steel Price Changes

The independent reported a week ago as having put out a price of 2.35c on merchant steel bars, equal to the Steel Corporation price, has since withdrawn the price, and is now quoting 3.00c, without having booked a great deal of business at the cut. This illustrates how little business can be gotten by the mills that are maintaining prices. Quotations commonly run from 3.00c to 3.25c, these prices being adhered to for the sake of protecting contract business on books.

The plate mills still call 3.25c the general market on plates, but admit that it is practically impossible to obtain more than 3.00c. In actual competition this price is shaded, the lowest priced transaction lately reported being one involving 3,000 tons at 2.85c. Western mills, however, have been quoting a lower Pittsburgh basis than this, but would not ship eastward, as that would be against their freight.

Black sheets are now available from several independents at 6.50c, if the specification is at all desirable. On galvanized 8.75c has been done, but this is considered exceptionally low, 9.00c being a more common quotation. Blue annealed can be had on the basis of 5.00c, except perhaps for gauges 14 to 16.

The American Sheet and Tin Plate Company will probably open its order books before the end of the month for first half contracts with manufacturing consumers and first quarter with jobbers. On account of restricted operation the company will carry considerable business into the new year that, was booked for the present half year, probably about two months of work in sheets and three months in tin plate. Customers' allotments for the half year will be correspondingly restricted, making the tonnage carried over plus the new contract equal a half year's estimated requirements. As is customary, the independents are expecting, or at least hoping, that the Steel Corporation will advance its prices on sheets and tin plates, but there is no indication whatever that any change will be made from the prices adhered to since the Industrial Board adjustment of March 21, 1919. these prices being: Blue annealed sheets, 3.55c; black sheets, 3.35c; galvanized sheets, 5.70c; tin plate, $7 per base box.

Coke and Pig Iron

Connellsville coke for spot shipment has continued its decline in the past week. Last week's report recorded a decline in spot furnace coke from $17 to $14, while at this writing the market is down to $10. Foundry coke, which had declined only slightly, has been declining more rapidly and is now down to $13. It lags behind furnace coke in the decline but will no doubt eventually show the natural differential above furnace coke, which is hardly more than $1 a ton, if as much. The decline in spot coke is due to so many furnaces having banked or blown out, or deciding to do so in the very near future, whereby shipments on requirement contracts had to be curtailed, leaving more coke to be offered in the spot market.

There is much speculation now as to where coke will stop in its decline, but there is very little foundation for making predictions. While some observers seem to think that because coke sold at $18 to $19 last August it cannot decline more than so much below that level, there is no definite percentage named as permissible under market conditions. On the other hand, coke sold at $1.50 in May, 1915, and at $3.60 in April, 1919, just after the Industrial Board adjustment in pig iron and steel prices. One theory is that coal is going to find a level at about $4, and that this will put

"Everything in Tools"

Get all your tools from Aikenhead's, "Canada's Leading Supply House." Your requirements will be promptly and carefully filled.

If you have never ordered from Aikenhead's, send your next order to them. You will find the quality, price and service right!

Full particulars of these or any other tools upon request.

Wahlstrom
Automatic
Tapping Attachment

The Wahlstrom Automatic Tapping Attachment is ten times faster than hand tapping and its simple to and fro motion frees the lands of chips, thus preventing jamming and tap breakage.

The Wahlstrom is constructed of hardened and ground parts and has nothing whatever that can get out of order. It is manufactured in two sizes. No. 1 takes taps of all sizes from 5-40 to ½-inch standard. No. 2 takes taps from ¼ to ¾-inch.

Wahlstrom Chuck
Grips the Entire Shank. Never Slips

Because the Wahlstrom Chuck grips the entire shank you may use numbers 1, 2 and 3 M. T. Shank Tools, whether they have a tang or not. The cam action is such that the resistance of the tool governs the gripping power of the jaws. Slippage is impossible.

Automatic and instantaneous in action, the Wahlstrom Chuck makes every minute and every movement profitable. The operator can make any change of tools in two seconds without stopping the spindle, simply by grasping the loose shell of the chuck.

Wahlstroms are made in two styles—one for straight shank tools, the other for M. T. shank tools—and they are both automatic. BUY THEM AT AIKENHEAD'S.

Aikenhead Hardware, Limited
17 Temperance Street, Toronto, Ont.

If interested tear out this page and place with letters to be answered.

coke at about $7, on the basis of its requiring one and a half tons of coal to make a ton of coke, with $1 for the conversion.

While everyone agrees that pig iron is going to decline until it finds a stable level, the market only makes an occasional step towards the requisite readjustment, inquiries being so infrequent that furnaces have little temptation to reduce their quotations. Basic remains quotable at the reduced price of $40, valley, reported a week ago. Foundry and Bessemer have each declined $2 in the week, to $45, valley. The decline in foundry quotations was precipitated by an inquiry for 200 tons for delivery at Pittsburgh. Several valley furnaces tried to get the order by quoting $45, furnace, or less, but a lake front furnace is understood to have gotten the business by making a still lower price. Bessemer declined to $45 without any transactions being reported, the case being simply that a producing interest offered iron at the $45 level.

One valley furnaceman predicts that

pig iron will recede to the Industrial Board schedule of prices, evidently on the basis that in steel the Steel Corporation has remained on the Industrial Board schedule and the independent prices are universally expected to come down to the Steel Corporation level. Industrial Board prices were $25.75 for basic, $26.75 for foundry, and $27.95 for Bessemer, these prices being, generally speaking, f.a.b. furnace. No other producer admits that pig iron will get as low as that, the common prediction being that pig iron will drop to between $30 and $35. This probably means $30. There is good ground for argument that decline in steel prices to the Industrial Board schedule does not prove that pig iron should decline correspondingly, for the Industrial Board schedule, reflecting the philosophy that obtained in fixing prices during the war-time control, put a rather larger percentage spread between pig iron and semi-finished steel, and between semi-finished steel and finished steel, than has obtained under open market or competitive conditions.

MACHINE TOOL BUILDERS ARE PUTTING THEIR MEETING NEARER

Special to CANADIAN MACHINERY.

NEW YORK, Nov. 4.—The annual fall convention of the National Machine Tool Builders' Association has been brought forward from the first week of December to November 11 and 12. This change of dates is presumably for the purpose of giving the machine tool manufacturers an earlier opportunity to consider the present business slump so that a plan of action may be suggested that will best meet the situation.

Three questions of importance face the machine tool trade: First, the business outlook for next year; second, the question of reductions in selling price, and third, the possibility of a revision downward of labor costs.

As for the first, it may be said that the trade as a whole is very optimistic regarding conditions in 1921, but the outlook for the remainder of this year is none too good, especially as it now appears quite certain that little or no railroad buying will be done before January 1. There is a fair number of inquiries for tools from industrial plants, but prospective buyers seem in no mood to close business. Individual opinions in the trade are that price reductions at this time will not stimulate much buying. However, the pressure that has been brought to bear by some buyers, particularly automobile companies, for lower prices has resulted in serious consideration on the part of many machine tool builders as to whether prices can be brought down. Material costs are coming down, but these reductions have not as yet been reflected to any extent in machine tool manufacture. Likewise, some plants are bringing down labor

costs by insisting upon greater individual efficiency among workmen, but this reduction is offset to some extent by the fact that few plants are working full time.

While the National Machine Tool Builders' Association does not, and cannot, consider the question of selling prices officially, it is quite probable that the gathering of many machine tool manufacturers in New York this month will result in informal conferences on the subject, and the members of the association will go home with some definite idea as to what should be done to meet the present situation.

WARN USERS WHEN JUICE RUNS LOW

St. Catharines Having Trouble in Securing Power for its Industries

St. Catharines.—The local Hydro power crisis, precipitated when on account of the cold spell customers threw on their electric heaters despite repeated warnings, and burned out half of the transformers, is somewhat easier, due to co-operation on the part of consumers, which has been to an extent forced.

For two days some of the smaller factories have been cut off and an effort will be made to have the street lights reduced to half and illuminated signs eliminated.

Manager Yates, explaining how the load has been reduced thus far, said: "If at any time a reduction is necessary the signal will be flashed to all the users,

as was done last night by successively pulling out the switches feeding the different sections of the city. That section which shows the least response to the signalled request will be the one allowed to suffer."

AMERICAN CO. IS NOW INTERESTED

And Reorganization of the Beaver Motor Truck Co. Has Been Announced

The announcement that the American Axle Company, capitalized at $6,000,000, had become heavily interested in the Beaver Motor Truck Company, of Hamilton, and in future would supervise the operation of the local factory, has been made. The American Axle Company operates axle and truck plants in a number of American cities and is rated as the largest exclusive manufacturer of worm drive axles in the world. The American company has become heavily interested in the Beaver Motor Truck Company and will supervise the operation of the local factory.

Under the scheme of reorganization, H. R. Williams, who organized the original Beaver Truck Company, becomes president and general manager, with E. W. Macavay, president of the American Axle Company as vice-president. Captain Harry Trenaman has been appointed treasurer, while William Mulvaney, E. D. J. Stares, W. H. Dawson, of the T. Eaton Co., F. C. Wolf and P. J. Schultz, of the American Axle Company, are the new directors. The reorganization of the local companay is expected to prepare the way for some big things. Plans are now under way to turn out 3,000 trucks during the next twelve months, while arrangements are being made to place a new model on the market.

MORE CAPACITY FOR WIRE MILL

Steel of Canada Making Important Additions to Existing Plant

When the buildings now in the course of erection are completed, the capacity of the wire mill of the Steel Company of Canada at Hamilton will be doubled. The plans call for duplicating the capacity of the pickling room, annealing department, and also of the coppered wire department. The latter will then be capable of turning out 200 tons of finished wire per day, which will make it the largest wire mill in Canada. In addition to these extensions there is also being erected a large storage shed, steel construction throughout, equipped with travelling crane for handling raw material in the most economical manner.

INDUSTRIAL NEWS

NEW SHOPS, TENDERS AND CONTRACTS
PERSONAL AND TRADE NOTES

TRADE GOSSIP

Equipment Wanted.—T. McAvity & Sons, Ltd., St. John, N.B., are in the market for a few No. 6 turret screw machines.

Locating in St. Thomas. — The Canadian Edison Phonographs, Ltd., have taken over the buildings in St. Thomas, formerly occupied by St. Thomas Cabinet, Ltd., and before that by Thomas Bros. The Edison Company took over the buildings, much of the equipment in the woodworking shops, and get six acres of land along with it. They have also placed orders for considerable new equipment in the line of woodworking machinery. The deal for the sale of the plant to the Edison Company was put through by Mr. William Garlock, of the Garlock-Walker Co., Toronto.

George A. Marshall, of Messrs. Marshall, Son & Bunney, 39 Richmond St. E., Toronto, has gone to England on the SS. Metagama, where he will spend a month or more on business matters.

Will Finish Ships.—Arrangements are being made by the Department of Marine with the liquidator of the Dominion Shipbuilding Co., of Toronto, for the completion of the two steamships which that corporation had under construction for the Government of Canada. The vessels, which were being built for the fleet being operated by the Canadian Government Mercantile Marine, are to be of 3,500 tons each, and about 75 per cent. of the work upon them had been done when the company went into liquidation. The Department of Marine will either undertake to do the remainder of the work or have it done through the liquidator.

May Earn Less.—In commenting on the business prospects, President T. J. Dillon, of Canada Foundries & Forgings, Welland, Ont., says: "In order to realize on the full possibilities of the plants this company found on its hands at the end of the war, it was considered advisable to appropriate a fairly large amount for additional equipment for rounding out their different units to lower the cost of manufacture of regular lines and add a few new ones. Indications point strongly to gradually lowering on steel commodities, and lessened dividends, which will result in declining earnings. Although the common dividend declared was somewhat less than those previous, it makes 10 per cent. for the year 1920."

J. F. MACKAY WITH MORROW COMPANY

Formerly Business Manager of Toronto Globe and Latterly With The Willys-Overland Co.

J. F. MacKay, formerly business manager of The Globe and latterly secretary and treasurer of the Willys-Overland, Ltd., has resigned his office with the latter company to become vice-president and general manager of the

J. F. MACKAY.

John Morrow Screw & Nut Co., Ltd., of Ingersoll. He will also occupy the position of vice-president of the allied industry, the Ingersoll File Co., Ltd.

This and other important changes are consequent upon the retirement of J. Anderson Coulter from the presidency of the two companies. Mr. Coulter has been actively connected with the progress of the industry for more than thirty years. The directors announce the election of Col. F. H. Deacon, of Toronto, as president; he has been vice-president of the companies. The other appointments are: Assistant general manager and treasurer, H. P. Stoneman; sales manager, F. N. Norton; secretary, Lt.-Col. H. L. Edmonds.

As to his successor in the important positions he has held with Willys-Overland, Ltd., no announcement has been made, but it is understood that the directors will ratify an appointment within the next week.

POWER CAUSES MORE TROUBLE

Industrial Centres Have Hard Time in Keeping Their Plants in Operation

BRANTFORD.—The Hydro Commission here has been unofficially notified that a 20% reduction in Brantford's Hydro allotment will be made shortly. A conference of manufacturers and the local Hydro Commission will be arranged to take steps to prepare for the cut. The situation here is already serious and the city council committee in charge of street lighting is prepared to curtail this to provide more power for domestic users.

Windsor's Shortage

WINDSOR.—Deeming it necessary to bar use of Niagara power for signs and window lighting, the Windsor Hydro-Electric Commission will ask the city council to pass a by-law giving effect to a resolution passed at a meeting of the commission. O. M. Perry, manager of the local commission, said every effort will be made to discourage the use of electric heaters. Co-operation is necessary also on the part of business men. Manager Perry said, in order that street lighting may be provided in new residential sections.

New Incorporations. — Incorporation has been granted to the following: United Shoe Repair Shops, Ltd., Montreal, $50,000; Canadian Feature and Production Co., Ltd., Winnipeg, $1,000,000; The Bee Starch Co., Ltd., Montreal, $100,000; Canadian Harvester, Ltd., Toronto, $40,000; Canadian Skirmisher, Ltd., Toronto, $40,000; Canadian Rover, Ltd., Toronto, $40,000; Pacific-Atlantic Construction Co. Ltd., Vancouver, $5,000,-000; The Davies Studios, Ltd., Montreal, $40,000; The Live Fish Co., Ltd., Quebec, $50,000; Cosmo Cotton Company, Ltd., Yarmouth, N.S., $600,000; Copeland Hotel Co., Ltd., Pembroke, Ont., $100,000; Sheffield Engineering Supplies, Ltd., Montreal, $50,000; Canadian Highlander Ltd., Toronto, $40,000; Canadian Libbey-Owens Sheet Glass Co., Ltd., Toronto, $1,680,000; Canadian Queen Manufactoring Co., Ltd., Vancouver, $100,000; Canadian Challenger, Ltd., Toronto, $40,-000; Canadian Logger, Ltd., Toronto, $40,000; Commercial Investigators, Ltd., Montreal, $50,000.

The Week's Events in Montreal Industry

H. B. Smith, formerly president of the Northern Navigation Company, has been appointed managing director of shipyards for the Canada Steamship Lines.

Thos. Robb, secretary of the Shipping Federation, has sailed for Europe, where he will represent Canadian shipping interests at the meeting of the League of Nations to be held shortly at Geneva.

D. M. Brown, formerly office manager of the Montreal branch of the Holden Company, has been appointed manager of the Ontario division, with headquarters at Toronto. C. M. Gray will succeed Mr. Brown as office manager at Montreal.

The Montreal Chambre de Commerce has been considering the advantages that would follow the building of a tunnel under the Champ de Mars and Jacques Cartier Square, for the purpose of relieving the congestion that frequently occurs in this vicinity on market days and when traffic to and from the harbor front is excessive. No definite action has been taken but the project has been laid before the Administrative Commission.

At a recent meeting of the Montreal Branch of the Engineering Institute of Canada the entire evening was devoted to moving pictures, both of an instructive and entertaining character. One of the reels shown was in connection with the work of McGill University. Every branch of the work was covered by the film, including educational, athletic and military features. Between the showing of the pictures brief addresses were given by J. L. Busfield and the secretary, F. S. Keith.

Following a conference held between the Administrative Commission and an influential delegation of business men and citizens of Montreal, headed by Lord Shaughnessy, a deputation will go to Ottawa to confer with the Government on the question of constructing a second bridge across the St. Lawrence in the vicinity of the harbor of Montreal, to connect the Island with the south shore. It is felt by a large number that this additional means of communication between the two points is an urgent necessity and that the next few years will demand some relief from the present conditions.

At a recent meeting of the Montreal branch of the Rotary Club, F. E. Rejall gave an interesting address on high-speed steel, touching on the electric and crucible processes of manufacture. While intimating that considerable progress had been made with the electric furnace in this connection, he favored the crucible method, as the product thus obtained was more homogeneous and much better quality although much more expensive to manufacture than the other, owing to the ability of the electric furnace to take a charge of six tons and over, while an average crucible would contain from 60 to 80 lbs.

On Saturday last the Canadian Vickers Company successfully launched the Canadian Commander, the eleventh vessel constructed by this company for the Canadian Government Merchant Marine. The ceremony of christening was performed by Mrs. Ballantyne, wife of the Minister of Marine and Fisheries. This new addition to the National fleet is a steel vessel of 8,350 deadweight tons, 413 feet long, 53 feet beam, and 31 feet deep. The vessel was built under the supervision of C. F. M. Duguid, of the Department of Marine, and W. J. Alderson, representing Lloyd's Register of Shipping. The boat is equipped with triple expansion engines and Scotch boilers fitted with Howden's forced draft system.

The Montreal Board of Trade at a recent meeting had before it a communication from the Jamaican Boards of Trade in which those bodies had passed resolutions endorsing the movement to increase trade relations between Canada and the British West Indies. The letter pointed out a number of impediments to trade which would require adjustment before further action would be taken. The chief obstacle is the present so-called "16 Dutch" standard and color test on Jamaican sugar coming into Canada. The West Indies Boards of Trade are also in favor of a mutual preferential tariff, which, they claim, would not only contribute to a large degree in the extension of trade between the two countries.

The Canadian Vickers Company of Montreal has recently received word from the aviation branch of Vickers Limited of England that the parent company has been awarded first prize of £10,000 as the winner of the amphibian class of aeroplanes. The "Viking" machine was designed by Mr. R. K. Pearson, who also designed the famous and successful "Vickers Vimy." The Viking is fitted with a 450 h.p. Napier engine, is capable of carrying five passengers at a speed of 120 miles per hour, and can be landed on water and driven ashore without any special slipway. It is worthy of note that this is the fourth big prize won by Vickers machines in air or air and water competitions. These include the prize of £10,000 won by the late Sir John Alcock in his flight across the Atlantic, and a like prize won by Sir Ross Smith in his London to Australia flight.

It might be well to mention that extensive experimental work has been constantly carried on by the Vickers Company since the armistice in the development of air machines and that the expenditure in this connection far exceeds the amount of the prize just awarded.

BUSINESS PAPER EDITORS' MEETING

Interesting Sessions in New York Followed by Election of Officers

Clay C. Cooper, Mill Supplies, Chicago, was elected president of the National Conference of Business Paper Editors for the ensuing year at a meeting held in connection with the annual deliberations of the Associated Business Papers, Inc., at the Hotel Astor, New York City, Oct. 20 and 21. The other officers chosen were: Vice-president, C. J. Stark, The Iron Trade Review, Cleveland; secretary-treasurer, R. Dawson Hall, Coal Age, New York. The executive committee is composed of A. I. Findley, Iron Age, New York; C. T. Hawson, Railway Maintenance Engineer, Chicago; A. R. Kennedy, Canadian Machinery, Toronto; James H. Stone, Shoe Retailer, Boston; H. L. Parmalee, Metallurgical and Chemical Engineering, New York; and Arthur L. Rice, Power Plant Engineering, Chicago.

FIRMS LOOKING FOR CONNECTIONS

MAY BE SOME ITEMS HERE YOUR FIRM CAN SUPPLY THEM WITH

By communicating with this paper and mentioning key numbers, information can be obtained by firms interested in the following lines:

2118. Tree-Felling Machinery. — A London company have an enquiry for tree-felling machines and would be glad to hear from Canadian manufacturers with catalogues.

2119. Hardware, etc.—A commission agent in Barbados, in good standing and representing a number of English firms, is desirous of obtaining a Canadian connection for general hardware, including such lines as valves, stop-cocks, etc., for machinery, also nails, locks, etc.

2229. Iron and Iron Products.—A firm in Jassy, Roumania, is desirous to receive quotations c.i.f. Roumanian port as well as other particulars from Canadian firms in a position to export iron and iron products, tools, utensils, stoves, etc.

2230. Steel Baling Wire.—An Ormskirk firm asks to be placed in touch with exporters of Canadian grain and steel baling wire used for pressing hay and straw by steam.

MORROW

F ORGED drills are built on scientific principles with a view to MAXIMUM PRODUCTION. They are made from hammered bars of high speed steel (not less than 18% tungsten content).

The drills are forged at plant No. 3 and are hot TWISTED (the grain of the steel runs with the twist). They withstand the torsional strain under most exacting duty and give more holes with less re-grinding.

Order from
Your Jobber
No Reliable Jobber will
Substitute

YOU SEND AN ORDER NOW.   ORDER BY THE NAME "MORROW"

The John Morrow Screw & Nut Co., Limited
INGERSOLL, CANADA

7 Hop Exchange, Southwark St., London, England.

| Montreal | Winnipeg | Vancouver |
| 131 St. Paul St. | Confederation Life Building | 1290 Homer St. |

Set and Cap Screws, Engine Studs, Nuts, Files and Rasps.

PORTABLE BOATS OF EARLY RAIL-ROAD PRACTICE

(Continued from page 417)

Gay notes that the number of engines in service had increased to 17, ten of which were manufactured by M. W. Baldwin; five by Robert Stephenson, of England; one by Coleman Sellers & Sons; and one by Long & Norris. Commenting upon the engines, he says:

"The two latter have been but recently put upon the road, and their capacities are not yet fully tested; they are, however, believed to be excellent engines. The engines from Mr. Baldwin have all been tested and found to be of the first class. The five engines imported from England are not as efficient as those manufactured in this country; the workmanship of them is good, but many important parts of the machines are too light to enable them to encounter (with a heavy load) the high grades and severe curves on this railway, in consequence of which frequent repairs are required upon them."

(To be continued)

FACTS REGARDING THE OXY-ACETYLINE PROCESS

(Continued from page 422)

repair was comparatively low. This casting was broken through the middle and by looking closely the welded section may be seen.

Fig. 2 illustrates a gas engine cylinder weighing 1,500 lbs. that got cracked by frost. Again the outlay to. replace would have been considerable, and the cost an item to be seriously considered, but welding proved the solution. In 48 hours the engine was dismantled, the bed welded and the engine re-erected ready for service again.

Consider the case of Fig. 3. This was a snail machine bed that broke through the middle and looked like a serious matter for a while. In 48 hours, however, the machine was again fit for service and the repair made at very little cost.

The next example, Fig. 4, depicts the face plate of a large wheel lathe, which was used for machining locomotive drive wheels. It became broken by accident and represented a loss of some hundreds

of dollars. Four hours was all that was necessary to have it again equal to new.

When a large flywheel of the size shown at Fig. 5 and weighing 3,500 lbs. is incapacitated from service, a serious loss must inevitably follow unless it is possible to repair it again quickly. This was done by oxy-acetylene welding as before, and the saving in time and money effected may be easily imagined and further comment is hardly necessary.

The next two illustrations, Figs. 6 and 7, show a Hart-Parry gasoline engine cylinder that weighed 325 lbs. It has a ten-inch bore. The views show it before and after welding, and the work in its entirety only took about three hours.

The remaining two illustrations, Figs. 8 and 9, dep'ct an automobile cylinder before and after welding. For reclaiming such parts this process is ideal, and has saved many a dollar. There are, of course, hundreds, yes, thousands, of cases that could be quoted and illustrated showing how this system has proved of decided value, but enough has been illustrated to start the ball of thought arolling. We appreciate the courtesy of L'Air Liquide Society, who very kindly loaned us these illustrations, as examples of work performed. Oxy-acetylene welding is only in its infancy, and the chief thing to do is to have confidence in it, then follow up that confidence by actual use.

The Seneca Falls Mfg. Co., Inc., have recently acquired the business of the O. R. Adams Manufacturing Company, Inc., of Rochester, N.Y., which was organised about two years ago to develop and manufacture a new type of production lathe called the "Adams Short Cut." It is a compact, sturdy, geared head machine, capable of taking a cut said to be equal to the average 16-inch engine lathe, occupying less than one-half the floor space with a corresponding reduction in investment.

The Seneca Falls Manufacturing Company, Inc., are planning to consolidate the business of both companies at Seneca Falls, N.Y., where additional facilities will be provided to increase the manufacture of both the "Star" and the "Short Cut" lathes.

Classified Opportunities

Makin

CANADIAN MACHINERY

AND
MANUFACTURING NEWS

VOL. XXIV. No. 20   November 11, 1920

Making One Thousand Milling Machine Saddles

Suppose You Had to Produce One Thousand Saddles Every Year, Like the One Shown—Would You Follow the Method Suggested by the Author?—If Not, Let Us Hear How You Would Do Them

By JOHN T. WATTS

A WELL-KNOWN authority once remarked that there is no method of manufacture existent that cannot be improved upon. Commencing on this assumption, Canadian Machinery decided to give readers something entirely original by requesting certain contributors to state how they personally would produce a well known article at present being manufactured. Various parts were sent out for opinions, and the present article, dealing with the saddle of a milling machine, is both instructive and interesting. It is, of course, understood that the suggestions to follow are the author's and not ours. We take an absolutely neutral stand, merely presenting the author's views. Other material of like nature will follow in later issues, and by thus interchanging ideas mutual benefit will result. We especially wish comments and criticism regarding this series of opinions, and suggest that readers let us have their viewpoint on the present article and its productive suggestions. Here is what the author has to say:

We are to assume that the saddle shown in Fig. 1 has been laid before us to determine and design the tools, Jigs, etc., and the best methods of production which will give the lowest cost of manufacture and an interchangeable product for an estimated output of one thousand parts per year.

As interchangeability is the essential point, it will be necessary to make a jig for every operation, even if the output is not increased or the cost decreased by the use of the jig.

The correct procedure for us as the

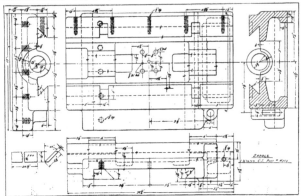

FIG. 1.—THIS SHOWS IN DETAIL THE SADDLE UNDER DISCUSSION.

production department is to first review the design in a general way with the idea of suggesting modifications in the shape of the part which would enable us to use standard tools or jigs which are already available and to recommend any changes which would assist us to increase production and decrease cost.

Before returning the drawing with the suggested changes to the designing department, the feasibility of carrying out the suggestions should be carefully checked and if possible a sketch made showing our idea of how the change could be made without interfering with the correct functioning of the part in question, or fouling some other part of the mechanism.

The making of this sketch will also serve to assure ourselves that we are not going to make a change that will introduce still worse troubles in some other part.

Taking the saddle before us, we decide that the most desirable way to machine the vees on the top would be by using a form milling cutter, but it looks as though the central boss would interfere, and laying out this part to scale to see just where we stand, as in Fig. 2, we find it impossible to get a milling cutter through with the boss as shown.

As the only alternative method is to plane this vee on a planer or shaper,

which is slower and more expensive and more difficult to attain equal accuracy, we are justified in spending some time in an effort to eliminate this central boss.

We will suggest, then, that the design be changed to one of the following, giving the alternatives in the order of choice, that is, from a production standpoint we would prefer the first suggestion first and so on.

Suggestion No. 1. In the absence of drawings of the other parts, we assume that these bosses are only to carry the screw spindle, and possibly this function would be fulfilled as well if only one-half a bearing was used as indicated in Fig. 3.

Suggestion No. 2. In the event that it is deemed essential to have a full bearing, this could be better made a separate casting, bolted in place as shown in Fig. 4.

If either of these suggestions are adopted, the vees can be machined by milling, having one cutter for roughing out and another for finishing, both taking the whole surface at once and being adjustable for wear. The accuracy, being then dependable entirely on the cutter, can be maintained easily and the operation needs only semi-skilled labor.

While suggestion No. 2 may be criticized as involving the machining of more

surface than the original design, the total cost would be less, due to the greater ease with which the various operations can be performed and the fact that equal accuracy can be obtained with cheaper labor.

Suggestion No. 3. Continuing our criticism, we find that the vees are shown with sharp corners, which are a practical impossibility because the tool which produces these corners will soon wear off at the point and will then produce the shape shown in Fig. 5 and make it impossible to get a decent working fit at this point, as the fit will only be on the point and after a few days' use the slide will be a sloppy fit. Further, the edge at (a) is easily damaged, for that reason being very poor design. The vee should be changed to the shape shown in Fig. 6, for production reasons and to give a better product.

The dimension (b) in Fig. 6 can be about 1-16 of the height (c).

Experience has shown that the angle of the vees should be 55 degrees instead of 45 degrees as called for in Fig. 1, and 55 degrees is nearly standard practice, it will be easier to get milling cutters to the 55 degree angle, and therefore it is recommended that the design be changed to give this angle.

Suggestion No. 5. The drawing does not state to what depth the tap drill for

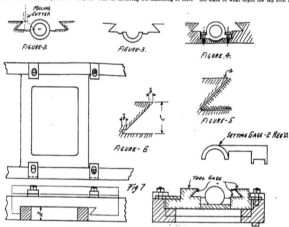

MILLING CUTTER

FIGURE-2

FIGURE-3.

FIGURE-4.

FIGURE-5

FIGURE-6

FIGURE-5

SETTING GAGE - 2 REQ'D.

Fig 7

TOOL GAGE

FOLLOW THE VARIOUS SUGGESTIONS CLOSELY. FIGS: 2 TO 7 INCLUSIVE

the ¾-inch tapped holes is to penetrate. This is a common omission on the part of some draftsmen and gives considerable trouble in the shop as, if the hole is not tapped deeply enough, the cap screws will bottom in the hole before taking a grip on the part they are to hold, and necessitate dismantling the machine to redrill and retap the holes. Or, as sometimes happens and is still worse, the error is not discovered until damage has been incurred to the machine.

To cover the ground thoroughly we will assume that the above suggestions have been rejected and that we are required to follow the drawing as shown in Fig. 1.

The Start

Starting with the casting as delivered to the machine shop, snagged, cleaned and wire brushed, the first point to be considered is the possibility of the casting warping from shrinkage strains during or after machining. In this particular piece, its shape is not one that would lead us to expect any trouble in this direction, provided that the casting has been allowed to season for a few weeks before work is started.

While experience is the only guide to a decision as to whether a casting is likely to warp or not, it may be stated that a casting whose length and breadth are approximately equal and the shape of which is nearly alike above and below the centre line, is not as a general rule subject to unequal cooling strains which would cause warping.

Nevertheless, the possibility of trouble in this respect should be borne in mind, and the first batch of parts run through should be carefully watched to determine the actual effect of the shrinkage strains. If the casting shows a tendency to warp, we can first try a longer period of seasoning, up to three months, which is usually sufficient to cure any warping in castings of a fairly uniform shape, and is the most economical remedy, as well

as being one that tends to a better product, inasmuch as it gives assurance of a shape that will not alter after the machine is shipped out from the factory.

If the trouble still continues, we must rough machine both top and bottom surfaces and then leave the piece to season for a while before taking the finish cuts. This involves a double handling and twice setting up in the machine, and therefore, for economy, should not be used until all other means possible have been tried, even to a redesign of the part to give a more uniform shape.

When a piece is not too large the time required to season can often be very materially shortened by running the castings in the tumbling mill for a few hours. The hammering which the parts receive in the tumbling mill will frequently relieve the internal strains and so remove the cause of the warping.

Before commencing actual production, we will take one of the castings and mark it off on the surface plate, to show the location on the castings of all the finished surfaces, using the centre of the bosses for the 2⅜-inch hole as the starting point. This, because it is important that when finished the 2⅜-inch hole be central in the boss, while a little more or less metal at the other places than the drawing calls for would not be of so much importance. The lines on this casting give us the heights to the working surfaces of our jigs, with the assurance that in working to these heights we will not be taking so much or so little off one surface as to prevent cleaning up on some other surface.

The laying off of this casting also serves as a check on the pattern as to the amount of stock left for machining, and on the foundry as to the casting being true to pattern.

The upper surface seems to be the logical point at which to start the machining, and we must design a jig to hold the castings on the planer, making it to hold as many pieces as the length of the

planer table will allow. The general idea of such a jig is indicated in figure 7.

As will be noted the sketch shows only one section of the jig, the idea being to simply repeat this section to fill the length of the planer table, but casting them all in one piece. The jig is made with a key on its lower surface to fit in one of the slots of the planer table as indicated, thus lining up the jig. Lugs for clamping the jig to the table should be provided, but are not shown for the reason that they would complicate the drawing unnecessarily.

The modus operandi is very simple as it is only necessary to drop the saddle into the jig, pressing the projecting part of the saddle against the cross-piece of the jig, to take the thrust of the cut, then slip the two setting gages in place and tighten up the clamps. The setting gage fits the boss on the saddle and the side of the jig and so lines up the centre line of the boss with the jig. The clamps are made with an oval hole, so that it is only necessary to slack up the clamp nuts a turn, when the clamps can be slid back and the saddle removed and the next placed.

At one end of the jig is cast a part for the tool gage, which is machined to the exact shape of the vees to be machined in this jig, but a certain fixed distance larger all over. For clarity we will suppose this distance to be 1-32-in. The tool gage is then machined so that if the planer tool in its travel is always at a distance of 1-32-in. from the surface of the tool gage it will cut the vees to the desired sizes.

The height "d" in figure 7 should be arranged to suit the finish lines on the sample casting laid off on the surface plate.

To machine, first finish the top flat surface, being careful to start the cut from the centre and feed out, to avoid the possibility of breaking off the weak corner, at the top, see "b" figure 6. Next rough out the vees, allowing not over

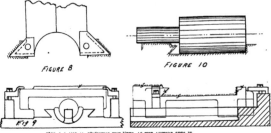

FIGURE 8　　　　FIGURE 10

Fig 9

FIGS. 8, 9 AND 10—FINISHING THE VEES, AS THE AUTHOR SEES IT

1-64-in. for the finish cut. To finish
the vees a form tool should be made,
about as shown in figure 8. The holder
would preferably be split to admit of
being expanded by liners between the
two halves to compensate for the wear
of the tools. At first sight it might
seem simpler to overcome the effect of
the wear of the tool by traversing the
tool sideways, but, while this is
possible, much more accurate results
will be obtained if the tool is cutting on
both sides simultaneously, as the cut
on one side counteracts the tendency of
the tool to spring away from the cut on
the opposite side.

The next operation will be to machine
the lower slide, and a similar jig to that
just described will be required, the work
being done in a planer precisely as in
the last operation. This jig, however,
will have ways machined to fit the vees
just finished in the last operation, as
shown in figure 9, the saddle being pushed onto these machined ways until the
central boss hits against the stop provided. The clamps are then tightened
up, being similar to those already described for the flat jig.

A still better method, if the machine
is available, would be to use a multi-
spindle planer type miller, the jig would
be the same as shown in figure 9, and
the surfaces would be machined, first
by a milling cutter in the horizontal
spindle as indicated in figure 10, and
then completed at the same setting by
another milling cutter in the vertical
spindle. This second cutter is illustrated in figure 11. Doing the work in
this way accuracy is easily attained
by the use of stops fitted to the feed
mechanism.

Boring the 2 3-8-in. Hole

The 2 3-8-in. hole is to be bored next
and we must design a jig to do this job
in a drill press. The jig must of necessity be made to fit the upper vees, because these were set by the jig to be
central with the boss, when being planed.
The boring jig is shown in figure 12, the
boring tool indicated being of the adjustable multiple type, which can be made
or purchased on the market.

The upper guide being hinged can be
swung out of the way, and a saddle
placed on the jig. A loose bushing is
used in the top guide while the first
cutter or drill is passing through the
top hole, when the top hole is drilled.
This bush is removed and the reaming of
the top hole completed. The boring bar
is then traversed down and the machining of the bottom hole completed. The
bottom guide also is fitted with a loose
bush which is pushed out by the drill,
the drill then fits the guide and holds
the bar central until the bottom hole is
reamed.

The next operation is to face the bosses, which, the top guide being swung
out of the way, can be easily accomplished with a standard spot facing tool substituted for the boring bar in the drill

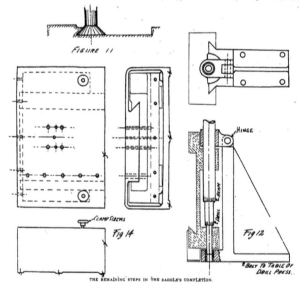

FIGURE 11

Fig 14

CLAMP SCREWS

HINGE

DRILL BUSH

DRILL BUSH

Fig 12

BOLT TO TABLE OF DRILL PRESS.

THE REMAINING STEPS IN THE SADDLE'S COMPLETION.

press chuck. The correct location of these faces can be got by marks or stops on the drill press spindle.

The final operation is to drill the small holes, and for those we chose a box type of jig, which will permit drilling all the holes with the one jig. This jig is shown in figure 14. and a study of the drawing will show that the saddle is slid into the jig and tightened by two clamp screws, the casting is located by a part of the jig which fits the lower vees. After drilling the holes on one side, the jig is turned over ninety degrees and the remainder of the holes drilled.

Having laid out the jigs it remains to determine how many machines it will be necessary to have to get the desired output, as it may be that we will have to have more than one of each kind of jig.

Taking up each operation in the order in which they are executed, we have first the planing of the upper vees. The area of the surface to be machined is 24 13-16-in.x15-16-in. or 374 square inches.

Allowing that we can plane 160 square inches per hour, on this type of work, the operation will require 2.4 hours per piece. Bear in mind that in estimating the time required to execute these operations, we should not base our estimate on record outputs made on similar parts, but on the average actual output that we can safely figure on over a whole year allowing for all the accidents, stoppages and other delays that are sure to occur during the course of a year.

Taking 300 working days per year and eight hours per day, gives us 2,400 working hours per year, or exactly the required output of 1,000 per year at the rate of 2.4 hours each. As there should be no difficulty in accomplishing

the operation in this time, we may allow that one jig will be sufficient.

The second operation, planing the lower vees, which have a surface to be finished of 17-in. x 24-in., or 408 square inches, will require, at the rate of 150 square inches per hour, 2.5 hours. This is 2,500 hours for the yearly output, and will require two jigs to make sure of getting the output.

The third operation, boring the 2 3-8-inch holes and facing off the bosses will consume about four hours, or a total of 4,000 hours per year. This indicates that we require to make up two jigs.

The drilling and tapping of the small holes should be accomplished at the rate of approximately four per hour, and the one jig will be ample for this operation.

Generally the management will ask for an estimate of the cost of the jigs before issuing instructions to proceed with their manufacture, and the simplest and most accurate way of arriving at this cost is to have a record of the costs of all jigs that are made, with their weight in an accessible form. From this record we take the cost of jig most similar to the one to be estimated on and taking its cost per pound, we can, by calculating the weight of the proposed new jig, and multiplying this weight by the cost per pound arrive at a very close approximation to the actual cost.

Taking up the four jigs, designed for this saddle, we estimate the costs to be as follows: No. 1, $660.00; No. 2, $750.00; No. 3, $200.00; No. 4, $200.00; total $1,810.00

This is a cost of $1.81 for each saddle, assuming that one year will be the life of the jig. Of course its useful life should be much longer than this, but changes in design, or in demand for the product may occur and a year is long

enough to calculate on to be on the safe side.

The saving effected by the use of the jigs will, at the lowest calculation, be the saving made by using semi-skilled labor, instead of the highly skilled labor which would be necessary if the jigs were not available.

Assuming that skilled labor would execute the work in the same time without jigs that we have allowed for the semi-skilled labor with the jigs, that is a total time of 9.15 hours per saddle, the saving per piece will be the difference between the rates for that length of time of the two classes of men. Taking this difference at 30 cents per hour, the saving will be 30x9.15, or $2.75 per saddle, to be credited against the cost of $1.81 per saddle for the jigs.

Evidently then, even without crediting the jigs with the value of the assurance of interchangeability of their product, they will pay for themselves in less than one year.

———

Using New Metals.—A very significant hint was thrown out by the chairman of the Institute of Metals at a recent convention of that body. He pointed out that while the internal combustion engines used to drive merchant ships contained only 3 per cent. of metals other than iron and steel, the driving equipment of the latest British submarines (oil engines, steam turbines, and electrical machinery combined) showed a proportion as high as 37 per cent. Therefore the marine engineer would find in naval practice suggestions for increasing power and reducing weight by using some of the wonderfully strong yet light alloys which British metallurgists had developed.

The Engines Used in Early Railroad Practice

Continuing Our Story of Last Week on Transportation We See Herein a Few More Examples of the Early Type—In a Later Issue the Modern Locomotive Will Be Discussed

By J. SNOWDEN BELL

MR. A. MEHAFFEY, superintendent of motive power of the Allegheny Portage Railroad division of the line, to which the British engines had probably been transferred, goes much further than Mr. Gay in reporting unfavorably on them. In a report of November 1, 1836, he says: "Two of them, viz., the 'John Bull' and 'Red Rover,' both British engines, have recently been sold, and it would have been a saving to the Commonwealth had they been given away for nothing the first day they were placed on the track."

Data on English Engines

This statement savors so strongly of prejudice as not to be worthy of credence, and the temperate and reasonable criticism of Mr. Gay correctly indicates the reason for the failure of the British engines to give satisfactory results in service. As they were not equipped with trucks, it is obvious that they were not well adapted to traverse the short curves of the line, as also their com-

The five British engines, which are stated, in Mr. Mehaffey's report, as being the "Albion," "Atlantic," "John Bull," "Fire Fly" and "Red Rover," were built by Robert Stephenson. of Darlington, England, and were all four-wheeled machines, having one pair of driving wheels and one pair of carrying wheels, which were journaled in pedestals rigid with the main frame, and consequently had no capacity of relative radial or lateral motion. Messrs. Robert Stephenson & Co., Ltd., of Darlington, England, have furnished a drawing entitled "Steam Engine No. 54" (builder's number) and a more complete blue print, "Working Drawing of Nos. 110, 112, 113 Locomotives, March 24th, 1835; Nos. 129, 139, March. 4th, 1836," together with tables of dimensions of "Locomotives supplied to the United States Railways," and the writer has also received from an English correspondent a copy of a list headed "Locomotives built by Robt. Stephenson & Co. for the U. S. A. between 1831-1836," purporting to be

wheels were of the same diameter, and in the other two the driving wheels were of greater diameter than the carrying wheels. The engines of these two wheel arrangements are not, however, indicated by name in the lists.

Fig. 9 is a reduced reproduction of the builder's blue print showing the engines having driving and carrying wheels of different diameters, the dimensions of one of the engines being given as follows: Light weight, 8 tons, 11 cwt.; cylinders, 12 x 8 inches; driving wheels, 5 feet diameter; carrying wheels, 3 feet 6 inches; boiler, 3 feet 4 inches diameter, 8 feet long; firebox, 2 feet 5 inches x 3 feet 2¾ inches x 3 feet 5¼ inches; height of chimney above rail, 14 feet 6 inches. These dimensions are not, however, entirely in accord with those appearing in the list from Wishaw's book, in which the cylinders are stated to be 10 x 16 inches. The cylinders of the engines having driving and carrying wheels of equal diameters (4 feet 6 inches) were 10 x 16 inches.

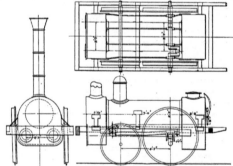

FIG. 9—A REDUCED REPRODUCTION OF THE BUILDER'S BLUE PRINT. SEE TEXT   FOR FURTHER DATA

paratively light construction, noted by Mr. Gay, impaired their hauling capacity. It does not, however. by any means, necessarily follow that they were either worthless or could not have been used, with reasonable advantage. in lighter service.

made up from Wishaw's Railways of Great Britain and Ireland, 1st Ed., 1840, Tables 27 and 28, Sec. 2.

From the above data it would appear that in three of the five locomotives built for the Philadelphia & Columbia R. R., the driving and carrying

The principal engineer's report of October 30, 1835, states, as before noted, that among the seventeen locomotives then in service there was one that was built by Coleman Sellers & Sons, and Mr. Mehaffey's report of November 1, 1836, specifies two, the "America" and

"Sampson," as built by C. Sellers & Sons, and first run September 1, 1836. The only data that have been obtained regarding these engines is that contained in a letter from Mr. Charles Sellers to his brother, Mr. George Escol Sellers, dated July 23, 1884, and a rather crude wood-cut of a locomotive appearing in an advertisement of Coleman Sellers & Sons, Cardington Iron Works, which is here reproduced as Fig. 10.

The following excerpt from Mr. Sellers' letter contains all that he says which is descriptive of the locomotive:

"Our first locomotive was put on the railroad in 1835. I ran it for one week before we asked the commissioners to take a trial trip to Lancaster and back. It was outside connection, centre-bearing on the truck and iron frame. I recollect great objection was made to the iron frame, as it would not give to the inequalities of the road; all of the English locomotives, and I think two of Baldwin's on the Pennsylvania Railroad, were wooden frames.

"They (the commissioners) said that we had to pay Baldwin $500 for a patent attachment to throw part of the weight of the tender on the driving wheels, which we did not use, but that we had put on a patent attachment of our own to throw part of the weight of the forward end of the engine on the drivers, which they thought was much better, as it was self-acting. The drawing for the engine most have been made in 1835 (I think earlier, because we had to get the boilers made in New York, and very poorly made at that)."

The traction increaser referred to in this letter was doubtless that of the patent of C. & G. E. Sellers, May 22, 1835. The principle of the invention is stated to be: "So coupling or connecting the cars containing the load to be drawn to the body of the locomotive engine, as that the load by its action upon a lever or standard shall tend to raise the fore end of the locomotive in any desired degree, and thus to loosen the pressure upon the fore, and transfer the same to the hind wheels." The appliance is very crudely shown and described, and it is not at all clear how the tender or train is to be connected to it, but, if operative, it must have been, as stated in the letter of Mr. Sellers, "self-acting."

The early motive power of the state system also included a locomotive built by Long & Norris, the "William Penn," stated in Mr. Mehaffey's report as having been first run October 14, 1835, and a number of locomotives of the same type, built by William Norris, among which were the "George Washington," "Robert Morris," "Benjamin Franklin," and "Washington County Farmer," which are stated, in the same report, to have been put in service at different dates in the year 1836.

All the locomotives built for the state by William Norris and his successors up to the year 1840, or thereabouts, were of the 4-2-0 type, and were, in all essen-

tial particulars, similar to that shown in Fig. 11, which represents one of a lot of seventeen constructed by him for the Birmingham & Gloucester Railway, England. They did not materially vary in dimensions from the "William Penn," the weight of which was 23,560 pounds, of which 14,600 pounds were on the driving wheels; cylinders, 10 x 20 inches; driving wheels, 4 feet diameter; boiler, about 14 feet long and 36 inches diameter; grate area, about 9 square feet.

Respective Merits

The merits of the respective locations of the driving axle in the Baldwin and the Norris engines, i.e., behind the fire-box in the former and in front of it in the latter, were often the subject of discussion during the period that the 4-2-0 type continued to be practically the only one in service. Those who favored the Baldwin arrangement correctly maintained that it gave the engine a longer wheel base and thereby rendered it steadier at high speeds, while the advocates of the Norris design, with equal correctness, pointed out that it increased the adhesion and tractive power of the engine. The advent of the 4-4-0 type rendered the question an academic one, and meanwhile, on the lines of the state system, the Baldwin engines were generally preferred for passenger service and the Norris engines for freight. Both classes undoubtedly gave as efficient service as was possible with their limited capacity, and the reports of the performance of the Norris engines were uniformly favorable. Indeed, the report of J. W. Patton, superintendent of the Allegheny Portage Railroad, October 31, 1838, went so far as to make the statement (which would seem to be of doubtful correctness), that "of the locomotives now on this road those made by Mr. William Norris, of Philadelphia, are much superior, doing double the amount of work, with half the quantity of oil and fuel, and not requiring half as much repairs."

All the motive power thus far referred to was constructed in Philadel-

phia and first put in service on the Philadelphia & Columbia Railroad division of the line, although some of the locomotives were, from time to time, transferred to the Allegheny Portage Railroad and their places supplied by new ones. For some reason, which is not apparent, the canal commissioners placed their orders for locomotives intended for service on the Portage road in other cities, and on the opening of the road for traffic, in the season of 1835, three locomotives were ready for service on the long level between inclined planes 1 and 2, these being the "Boston," "Delaware," and "Allegheny." The following matter relating to these three engines is taken from the extremely interesting "History of the Pennsylvania Railroad," by William Bender Wilson, Philadelphia, 1899 (Vol. I., pp. 121-123):

"The 'Boston' was the first locomotive to do service on the Allegheny Portage Railroad. It was built by the Mill Dam Foundry Company, of Boston, Mass., and delivered at Johnstown just before the close of navigation in 1834. It was put in condition during the winter and sent to Pittsburgh to be used as a pattern. It was returned to Johnstown, March 28, 1835. Without water or fuel it weighed 8½ tons. Its cost, exclusive of tender, on the wharf at Boston, was $6,996.75. The cost of transportation to the railroad amounted to $223.25. It was put into regular service May 10th, and until November 1, 1835, made its regular trips, covering 52 miles daily, with the exception of 2½ days, when it was laid off for repairs, which cost $17. Engineer Welch, in reporting upon its services in the time mentioned, said of it: 'This engine during the greatest part of the season, in connection with its other work, has hauled the passenger cars in both directions each day. This detained it; otherwise it might have made three trips a day for the greater part of the time. It performed the labor every day of eighteen horses, and it might do easily one-third more if it were not necessary to reserve it for the

FIG. 10—THIS TYPE OF ENGINE WAS PUT ON THE ROAD ABOUT 1835

transportation of passengers. The daily expenses of running it are $7.12½, exclusive of repairs.' Its cylinders were 8 inches in diameter, with a 18-inch stroke, whilst its driving wheels were a small pair, 4 feet in diameter, with wooden felloes and spokes. The wheels were tired with iron and were flangeless. During the season of 1835 it was in service 174 days, averaging 52 miles a day-distance, and 10 miles per hour speed. Its steam pressure was 125 pounds to the square inch.

A Different Story

"The 'Delaware' and 'Allegheny' were not so satisfactory, and were a source of expense and vexation during the season. They were built by Edward A. G. Young, of New Castle, Del.; reached Hollidaysburg, April 15, 1835, and were sent to Johnstown where the parts were fitted together and the necessary alterations made in an ordinary blacksmith shop, there being no machine shop in operation at the time. Their contract price was $5,500 each, and it cost $158 additional per locomotive to transport them from Philadelphia to Hollidaysburg. Better results were expected of them than from the 'Boston' because the boilers were larger and would generate more steam. The machinery was arranged differently from that of most other engines built upon the same general principles. It was apparently more simple, but less substantial. The builder had had several years' experience in the use of locomotive engines, and it was expected that the deviations made by him from the general plan, and from the engine designated in the contract as the model according to which he was to build for the Portage Railroad, would be an improvement inasmuch as they were to be put up and tried upon the railroad by persons furnished by the builders and approved of by the engineers before they were finally paid for.

"The 'Delaware,' after running for four days, broke its crank axle, and had to remain idle until the 1st of September before it was repaired by the contractor. The 'Allegheny,' after considerable refitting, was accepted. It ran about two weeks when its crank axle broke, rendering it useless for the balance of the year. These three locomotives performed all the service they did for the year on the 13-mile level. The 'Pittsburgh,' built upon the plan of the 'Boston,' was constructed by McClurg, Wade & Co., at Pittsburgh, at a cost of $4,500, and was delivered on the road on September 3, 1835."

On page 127 of the same volume there is given the following description of the performance of a locomotive which had been ordered for the Philadelphia & Columbia Railroad, and was tested on its way thereto on the Portage road:

"It was during this year (1836) that a question of what power should be used on the Hollidaysburg level, that had been agitated for some time, was

settled. As the steepest grade on that level was fifty-two feet to the mile there was a great diversity of opinion as to the ability of a locomotive engine to work on the level. The authorities had contracted March 24, 1836, with McClurg, Wade & Co., of Pittsburgh, for the construction of a locomotive named the 'Backwoodsman,' for use on the Columbia & Philadelphia Railroad, and as that machine was ready for delivery, the board of Canal Commissioners ordered that it be delayed en route to be experimented with on the level.

"Arriving there in the latter part of September, it was worked under the charge of Messrs. Bridges and Whitney for several days, and proved that locomotives could be used with ease and economy there. At the first trial it arrived at the Hollidaysburg scales from the foot of plane 10 in eleven minutes, hauling eight heavy bloom cars. Its next trip, with thirteen heavily laden cars, occupied twelve minutes."

The report of J. Snodgrass, superintendent of motive power of the Allegheny Portage Railroad, who took charge February 15, 1839, states that there were then seventeen locomotives on that road, the largest portion of which had been used on the Columbia road previous to 1835 . These included nine built by William Norris; one, the 'Boston," by R. M. Houton; three, the "Allegheny," "Delaware" and "Comet." by E. A. G. Young; and four, the "Backwoodsman," "Mountaineer," "Conemaugh" and "Pittsburgh." by McClurg, Wade & Co., of Pittsburgh, Pa.

Development of the inefficient and wasteful management, and demoralizing results of improper exertion of political influences, which, in the view of the writer, are characteristic of government ownership, began almost upon the inception of the operation of the main line of the Public Works of Pennsylvania thereunder. A recital of details would be foreign to the purpose, and beyond the permissible compass of the present paper, but an instance is presented in the explosion of the boiler of the engine "Bush Hill," on the Allegheny Portage Railroad. April 23, 1847, a report of which was made June 10, 1847, to the Committee of Science and Arts of the Franklin Institute, Philadelphia, by Mr. Edward Miller, a civil engineer of the Pennsylvania Railroad. This report is published in the "Journal of the Franklin Institute," whole No. Vol. XLIV, 1847 (pp. 69-71), and being believed to be specially interesting, it is here reproduced in full, attention being particularly called to the italics, which are those of Mr. Miller, and to his concluding paragraph:

Details of the Accident

"John C. Cresson, Esq.,
"Chairman of Committee on Science and the Arts, Franklin Institute.
"Dear Sir:—Upon the day I passed over the Allegheny Portage Railroad,

April 23rd, 1847, one of the locomotive engines attached to a freight train exploded, killing the engine driver and injuring severely two other persons. I could not, at the time, obtain a correct statement of the facts, as business required me to proceed the same night to Pittsburgh. Being at Johnstown on the 4th inst., I had an opportunity of examining the wreck and enquiring into the facts connected with the explosion, which I now send you, to be laid before your committee, if you consider then of sufficient importance

"The locomotive 'Bush Hill' was built by Messrs. Norris, and put on the Columbia Railroad April 17th, 1837. In the fall of the same year it was transferred to the Portage, where it has since been in constant use. It was a six-wheeled engine, weighing 12 tons, being one of the heaviest on the road, used the adhesion of two wheels, and was considered one of the best machines on the Portage. It was generally used as a freight engine, but had been running on the 13-mile level for some days prior to the morning of the accident, with the passenger train.

"On that day the steam was raised by the fireman before breakfast, with the expectation that the 'Bush Hill' would have to take the passenger train; but the regular passenger engine, which had been undergoing repairs in the Johnstown shops, arrived and took its place. But for this change the loss of life might have been much greater.

"The fireman, John B. Davis, states that he tried the gauge-cocks before going to his breakfast, and found the water above the upper cock. He also tried them after breakfast, before the engine started, and found a full head of water.

"The engineer of the 'Bush Hill,' James Patterson, had gone down to Johnstown the night before to ask that another person might be sent to run his engine, as he wished to attend a funeral. James Barron was accordingly sent up with him by the passenger line in the morning to Incline Plane No. 1; but, as the funeral was to take place at the half-way house, it was concluded that Patterson should run the engine that far, Barron accompanying him. The 'Bush Hill' was attached to a freight train, and after proceeding about one-quarter of a mile exploded, killing Patterson and wounding Barron and Davis badly.

"The only part of the boiler which gave way was the forward flue-sheet, the upper part of which was torn from the flanch by which it was riveted to the boiler, as far down as the upper row of copper flues, which was also torn loose. The remaining flues, and the whole cylindrical part of the boiler, together with the dome and fire-box, sustained no injury from the explosion. The rent followed strictly the angle of the flanch without starting a rivet. The rush of steam forward threw the engine entirely over, and it fell backward, bending the platform against the dome, and

crushing the tender and one of the cars. The working gear was much bent and broken.

"The flue-sheet was made of two plates riveted together; the lower one, through which the flues passed, being of three-eighths of an inch iron, and the upper, one-quarter inch. The upper plate was strengthened originally by two stay-bars, both of which were broken off, their fractures showing that they had given way long before. No signs of want of water, nor of overheating, appeared in the flues, fire-box, or any part of the boiler.

"The angle of the flanch of the flue-sheet where the rent occurred was probably injured originally in the bending, as a very bad flaw extends not only around the fractured part, but also around the portion below the flues, which was not injured by the explosion. I have sent two portions of the flanch by Mr. Power to the Institute, from which this will be apparent.

"Mr. James Bowstad, the foreman of the Johnston shops, says *he noticed last winter that the flue-sheets of several of the locomotives, including the 'Bush Hill,' were sprung,* and he believes that in all these cases the stay-bars are broken. The other engines which are in this condition are in daily use on the road, and they are unable to repair them, because there are no spare engines to supply their place. He considered 'Bush Hill' in good order in all other respects. He says that he examined it the day before the accident in order to ascertain whether a new patch which had been put on the dome leaked. The steam was then high, but it did not leak of any consequence. James Barron states that when he and Patterson

got on the engine the steam was escaping rapidly from this patch, and he thought it too high, for *it was very blue.* He did not, however, at the time, consider it dangerous. Neither Barron nor Davis knew anything about the con-

dition of the safety-valve, and they have no recollection of hearing the steam blowing off at the valve.

"Mr. Power, the superintendent of the road, believes that the valve was screwed down by somebody while the fireman was at breakfast, during which time the engine stood on the track, fired up and waiting for the train, with nobody to look after it. This is, however, mere surmise; the injury to the valve preventing any conclusions as to its condition. The safety-valve was two inches in diameter, and the scale of the spring-balance would not indicate a greater pressure than 130 pounds to the square inch if screwed down as far as it would go.

"In conclusion I will remark that the cause of this accident is more manifest than in any explosion I have had occasion to investigate; the evident flaw at the flanch of the flue-sheet, and the fracture of the stay-bars rendering an explosion almost inevitable, if any accident or carelessness should produce an unusual head of steam.

"What can be said of the policy which compels the use of locomotives in such a condition as this? for there is good reason to believe that several others are in the same dangerous state. The number of these machines on the Portage is entirely insufficient to convey the traffic upon it with economy and safety; all of them have been many years in use, and many of them are engines which had been used on the Columbia road until antiquated before being transferred to the Portage.

"Very respectfully,
"Edward Miller, C.E.
"Pittsburgh, Pa., June 10th, 1847."

A resolution of the board of Canal

Commissioners, dated June 2nd, 1838, directed that two engines be fitted to the use of anthracite coal as soon as practicable, pursuant to which a series of experiments was commenced, which were stated by A. Mehaffey, superintendent,

Philadelphia & Columbia Railroad, in his report of that year, to have "produced the most gratifying result." The plan first tested was "to attach a fan to the front of the boiler of one of the locomotives, and thus create a draft. This, together with some trifling alterations in the engines, promised, in theory, to answer a good end, but when brought into practice was found insufficient to keep up the fire at all times and under all circumstances, as the fan could be put in requisition only when the locomotive itself was in motion." This having failed, the master mechanic, Mr. Brandt, suggested placing a fan of a different construction, immediately in front of the ash pan, and driving it by a separate engine. An application of this arrangement was accordingly made, the engine having a cylinder of a 4-inch bore and 8-inch stroke and being placed on the left side of the boiler.

The performance of the locomotive thus equipped was, according to the report, very satisfactory, stress being laid upon the fact that the blower was available whether the locomotive was standing or moving. The report makes the following statement as to the performance of the locomotive, the name or builder of which is not, however, mentioned:

"The quantity of coal consumed from one plane to the other (distance 77 miles) with a train of twenty or twenty-five loaded cars, is about one and a half tons, at four and a half dollars per ton, retail price. A train of the same weight would require two cords of oak wood, the average price for which, cut and split ready for the engine, is four and a half dollars a cord, thus showing a gain in favor of the coal, thirty-three

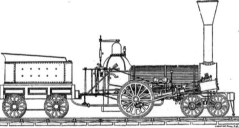

FIG. 11—THE TYPE BUILT BY WILLIAM NORRIS AND SUCCESSORS

per cent. Besides, coal enough can be loaded on the tank at Columbia, to carry the train to Philadelphia, without the loss of time and cost of labor, consequent upon stopping and loading wood about every twelve miles.

"The locomotive above described is now performing her regular work from day to day. Another on the same plan will be ready in the course of ten days." The Canal Commissioners' report of October 31, 1839, contains that of James Cannon, then superintendent of motive power, in which he mentions the purchase of three locomotives, built by Henry R. Campbell, of Philadelphia, costing, with tenders, $7,500 each, of which he says that they "are machines of the very highest order. They weigh thirteen tons and combine great strength and power with beautiful finish. Their performance has fully equalled all my anticipations." He also refers favorably to two locomotives built by D. H. Dotterer & Co., of Reading, and to the purchase of an engine with vertical boiler, which had been placed on the road nearly a year before by Ross Winans, a similar engine being daily expected. He then refers to the experiment made by "the late superintendent" (Mr. Mehaffey), in the burning of anthracite coal, as to which he says that it "did not succeed and was abandoned before the undersigned took charge of the road." He then describes the plan in use under his supervision as follows:

"The plan now in use is a very small rotary attached to the bottom of the boiler which is driven by a small quantity of steam taken from the dome. The fan is enclosed in an iron casing from which the air is conducted through a funnel to a perfectly tight iron chest, which encloses the whole of the ash pan. The air being forced into the chest, it is constantly working itself through the fire by the power of its own pressure and thereby keeps up a constant blaze, and an amount of heat equal to all its purposes. This fixture was first tried upon a new engine called the 'James Clarke,' one of Mr. Baldwin's first-class machines. It has been making its regular trips for about six weeks, and as yet has given no sign of a failure. It has drawn very heavy loads at every trip and has done its work with more apparent ease than when wood was used upon it as a fuel. As a test, the coal was weighed one trip. The engine drew twenty-two loaded cars, or one hundred and twenty-one tons, and consumed one ton and a half of coal, and had it been of the purest quality and well selected, the trip could have been made with a ton and a quarter, costing about seven dollars and twenty-five cents, while the same amount of freight, drawn by the same engine, would have required two cords of wood, which, including cording, cutting and splitting, costs about eleven dollars. This coal is from the Pottsville mines, and it is the only anthracite which we have yet found to answer the purpose. I may also state that there is other coal of a similar quality, as convenient or more so, that will probably answer as well upon trial.

"There are now five engines upon the road which are propelled by steam, generated with anthracite exclusively, and it is believed that by the first of January all the heavy engines on the road will be ready to use it.

"Numerous experiments have also been made in the use of bituminous coal. This, when it can be had of a pure quality free from sulphur and in masses large enough to prevent its falling through the grates, has been found to answer admirably by a very simple alteration of the grate bars. Four engines are now using it exclusively. Two are employed in drawing the day lines of passenger cars, and two on the night lines of passenger and freight cars."

In the annual report of the commissioners, January 15, 1841, Mr. Cameron says that the engine contracted for with Ross Winans is similar in principle to the previous engine, but entirely different in its proportions, and adds: "It is intended exclusively for the transportation of heavy trains of burthen cars. It will haul double the ordinary train, but, owing to its great weight, must be run very slowly over the road."

Prior to the above experiments on the Columbia Railroad, anthracite coal is stated to have been burned successfully on the Baltimore & Ohio Railroad, in the "grasshopper" and "crab" engines of Ross Winans, with vertical boilers, the fire-boxes of which did not have a greater area than those of the engines of the Columbia road. In the Winans engines, the fan was operated by a rotary engine worked by the exhaust steam so that they were subject to the objection, noted in Mr. Mehaffey's report, of being inoperative when the engine was standing, although this does not appear to have impaired her performance. In view of the successful results in burning anthracite coal, reported by the two superintendents of motive power of the Columbia road, it is remarkable that the use of this fuel was not continued, but this does not appear to have been done.

The substitution of a motor-operated fan for an exhaust blast, as a means for creating the necessary draft in a locomotive fire-box, which had been practised prior to the year 1834, has latterly been the subject of consideration by engineers of acknowledged ability in the design of locomotives, and a number of designs for the application of the principle have been proposed. So far as the writer has been able to ascertain, none of these has gone into actual service, but, the principle appears to be a correct one, and results of practical value may reasonably be expected from its development.

The haulage of the sectional boats constituted a considerable portion of the service of all the locomotives which have been referred to, and while such haulage also formed part of the service of the larger and more powerful locomotives which superseded them, it was not continued for a very long period in the later engines. The consideration of the original and following early motive power which has been herein presented, would, therefore, seem to be sufficient in connection with the general subject sectional boats.

The importance and value of inland navigation upon canals and canalized or deepened rivers has for a long time been fully recognized, and large expenditures, both governmental and private, have been devoted to its development. Economical considerations have determined that the dimensions of vessels most desirably adaptable for service in such navigation, should be such that they would not be capable of haulage over railroad track. It is, consequently, while not impossible, altogether improbable that the system ferring freight without breaking bulk over a line of alternate links of railroad and canal, will be reproduced in the future. Nevertheless, it is not too much to say that Canvass White's conception, nearly a century ago, of such a system, rises to the level of engineering genius, and the practical development of that conception into successful operation by his successors, the leading features of which, so far as information was obtainable, the writer has endeavored to present, evidence those who participated in it to have been sufficiently able and ingenious mechanics, and energetic operators, to merit that a record of their achievements should not be omitted from the pages of historical technical literature.

NEW CATALOGUE

The Saint Louis Machine Tool Co., St. Louis, Mo., have issued a new catalogue dealing with their line of machine tools. This catalogue is known as No. 16. Their varied line of grinding, polishing and tapping machines are shown by many illustrations and full data are given relative to each style of machine. It is also stated that safety collars are provided when called for. Roughly speaking some 25 pages are used to display this line.

NEW CATALOGUE

The Graton & Knight Mfg. Co., Worcester, Mass., have issued a very interesting book dealing with their lines of standardized leather belting for power transmission. This bulletin is numbered 101, and is done up in very attractive style. The book consists of almost 100 pages and is well illustrated throughout.

After the various lines have been discussed and their qualifications stated, the remainder of the book is filled with real live information for anyone using belting. Data of considerable value is given, in fact the latter portion of the book is a small text book in itself.

W. M. David, of the firm of Baines & David, Toronto, together with Mrs. David, is on an extended trip to Wales.

WHAT OUR READERS THINK AND DO

ANSWER TO PROBLEM

On page 398 in our October 28th issue, we published an article dealing with a problem in machining a Geneva movement for using in a moving picture ma-

HERE IS ONE SOLUTION TO THE GENEVA MOVEMENT PROBLEM

chine. This problem seems to have aroused considerable interest, for we have so far had several replies to the query.

The first of these is published below.

The Fellows Gear Shaper Co., Springfield, Vermont, on reading of the problem, decided it was up to them to state their opinion on the matter. and as is well known, they are well able to speak on manufacturing problems. To our mind, they state, this problem should be solved as follows:

We presume that the difficulty encountered by the manufacturer in producing this Geneva stop mechanism is due to the curves indicated by the 29-64 radius and the location of the slots. It is almost impossible to mill four curves and have them all accurate, even though the cutter itself is accurate. This is due to the great difficulty of accurate indexing.

In our experience we have had a lot of irregular shapes to produce on the gear shaper, and a suggested method for handling this would be as follows: A cutter could be made which would accurately reproduce all four curves and at the same time cut part way into the slots, thus locating them. Then it would be a simple matter to finish the slots to depth with a milling cutter. The sketch accompanying this suggestion is self-explanatory as it clearly shows the outline of the proposed cutter.

PUNCH PRESS ARRANGEMENT
J. W. Rodgers

A handy arrangement for catching the finished product of a punch press is shown in the accompanying sketch. This

particular device was operating on a washer press where the washer was blanked out in a duplex die and the fin-

ished washer carried up in the punch, from which it was ejected as the ram reached its top position. The two brackets A, A were secured to the stationary portion of the press in such a position as to allow free movement for the operating shaft B. This shaft is of square section and twisted or milled on the upper half of the length for operating in the nut C, the latter having a lug at one side for fastening to the bracket D, which is secured to the ram E. To the lower end of the rod B the bracket F is secured, and this bracket carries the pan that is swung under the punch as it comes to its top position. The swinging piece may be adjusted to the desired height by means of the collar, G, the lower portion of this collar being turned to fit the bore of the bracket A.

TOOL ROOM ALIGNING V-BLOCK
By Robert Mawson.

A fixture that is practically necessary to any department where accurate tools are made is some type of aligning device. Before elements such as shafts,

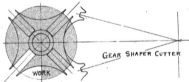

THIS SHOWS THE PUNCH PRESS ARRANGEMENT

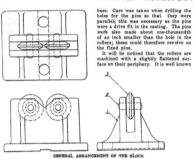

studs or pins can be machined with shoulders or keyways, it is absolutely necessary to know that these parts are parallel. This is more so when such pieces fit the entire length or operate in bearings or bushings. If the least bend or misalignment were in evidence, it is obvious that the piece could not turn on its axis.

Of course a shaft could be tested in a lathe, but this would entail centring, which is an expense; and further, the process of testing would mean the use of a machine tool. The lathe is prevented from performing some other machining operation, therefore such procedure entails a double expense, unnecessary centring, and use of the lathe. To obviate this the tool shown in the illustration was designed and has been used successfully not only in our tool room, but also in the machine shop. The base (1) is made of cast iron, and is 6-in. long, 4-in. wide, and was designed with such generous proportions so that the tool would rest squarely under any desired condition.

The vertical walls of the base are provided with ribs so that the castings will not distort when in service. Two hardened tool steel rollers (3) 1½-in. in diameter are held by means of steel pins (2) in the vertical walls of the tool

base. Care was taken when drilling the holes for the pins so that they were parallel; this was necessary as the pins were a drive fit in the casting. The pins were also made about one-thousandth of an inch smaller than the hole in the rollers; these could therefore revolve on the fixed pins.

It will be noticed that the rollers are machined with a slightly flattened surface on their periphery. It is well known

GENERAL ARRANGEMENT OF THE BLOCK

that a perfect knife edge is the ideal condition for an aligning block roller, but when such a tool is employed for heavy or frequent use, it would prove impracticable, as the hardened sharp edge would soon be damaged. The roller as designed, therefore, provides us with a tool of lasting service and proves accurate enough for any purpose demanded in the tool room or machine shop.

By referring to the illustration it will be seen that two slots have been machined in the base, one at right angles to the other. These slots are employed to accommodate tongues of the proper width to suit slots in a milling machine table or any other machine tool. When the device is used in this manner, two of them are placed at the proper distance apart, to suit the element being tested. The V-blocks may then be fastened to the table, holes being provided in the base to accommodate such mediums. It is then a simple operation to test the part resting on the rollers, using an ordinary dial or other type of indicator. Revolving the piece around will show by means of the indicator if it is out of alignment and how much the error is. The element, when necessary, is then

straightened in the usual manner and further tests made until the desired results are obtained.

REPAIRING OF BOX CHUCK WRENCHES
By Harry Moore

No attempt is made in this article to give instructions on the manufacture of chuck wrenches, but merely some instructive notes to the mechanic who may, have one to repair occasionally.

After a chuck wrench has been used for some time it will, as a general rule, have worked itself bell mouthed, and the moment you try to tighten with it will slip off the chuck square. A good way to make a permanent repair in such a case is to turn off the upturned edge to a slight taper as shown in Fig. 1.

The corners of the square are next slotted down with a hand saw to their full depth. A piece of machine steel is now bored out to fit the taper on wrench, and hammered on until the square is its original size once more. The ring is then faced off and turned, completing the job as shown in Fig. 2.

Hardened chuck wrenches do not give the same trouble with regard to wear, but as they are generally made of thinner stock, a piece will sometimes break entirely away. These wrenches are invariably a fairly close fit in the circular hole in chuck, therefore, a slightly different procedure is necessary in making a repair. The wrench is first of all annealed, and if the broken piece can be found it is sweated into place. The wrench is then turned as shown in Fig. 3. A machine steel ring is made several thousandths smaller, then shrunk on, the completed repair being shown in Fig. 4.

If the broken piece is not at hand the wrench is filed, or ground, and a piece made to fit as at Fig. 5, when the procedure is as before. The making of a chuck wrench is a comparatively simple matter if tackled in the proper manner. Very many mechanics fail to make a good job for two reasons: First, they leave too much stock for the punch to remove; second, they endeavor to force the punch in order to remove a lot of stock at once. The result in the first case is that the punches are broken, and in the second case, jagged, uneven holes are procured.

A good method for laying out a wrench

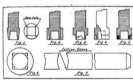

THE DETAILS OF THE REPAIRING OF BOX WRENCHES　　　　BROKEN CAST IRON ROLL—HOW THE REPAIR WAS MADE

preparatory to drilling, is shown in Fig. 6. The four small holes are drilled first, then the large centre hole (which should be slightly smaller than the width of square) is drilled to the same depth. This method leaves very little stock for the punch to remove. This punch is made as shown in Fig. 7.

It will be noticed there are only two cutting edges on the punch. By making it this way a proper rake can be ground on the end which is impossible in the case of a punch with four cutting edges. It is placed on the lines at one side, driven home, then placed at right angles to complete the hole. It will be found that such a punch requires only a few light taps and leaves a clean, square hole.

MECHANICAL REPAIR ON A BROKEN CAST IRON ROLL
By W. S. Standiford.

In every mechanic's life emergencies will arise that call for rapid thinking and original work on his part in order to keep the wheels of industry running smoothly, as machinery breakdowns will happen in all factories and mills at the most unexpected times. In this article the writer shows how he made an emergency repair some years ago, on a broken roll, that lasted until the roll was worn out. When a bar of iron was going into No. 2 pass, a sliver on it jerked the tongs from the hands of roller, the tongs being jammed between the side of bar and the collar separating numbers 2 and 3 grooves; the result being that an eight-inch long section of collar was broken out. Owing to the conditions of the country, welding supplies were unobtainable, a new cast iron roll could not be gotten in less than three weeks, it taking four more days to make a new roll out of it.

As the company was anxious to get their rolled bars out on time, the writer was delegated to see if a mechanical repair could be made, provided it were possible to make it withstand the enormous side pressure exerted by the iron during rolling. A patch the size of the broken section and screwed to the roll was first thought of and decided to be impracticable, as both joints being near one another, would cause the patch to spring and shear off some of the metal on the bar.

I finally decided to turn the cast iron collar off of the roll altogether and replace it with a wrought iron one, the joints of the two halves being a long distance apart would enable the collar to withstand the spring when a bar was entered in the pass. In order to make the collar as rigid as possible under the circumstances, a slot having a slight taper and 2½ inches deep was cut into the roll, the width being 2¼ inches. After the slot was turned, a wooden templet was made, the length of the two halves were ascertained by the "cut-and-try" method, until when placed in the roll groove opposite one another, the wooden collar made a very snug fit, at the joints and bottom of groove. From these templets a wrought iron collar was

forged by a blacksmith, put in lathe and turned with a taper matching the slot; the tapered section being a few thousandths of an inch larger so as to make a force-fit in the slot. One half of the wrought iron collar was forced into the slot as far as it would go, by the careful use of hammers, a heavy iron ring was then put over the half collar and four jacks, one opposite each joint, and the others, being spaced equal distances apart. Part of the ring on the opposite side of the jacks was then filled with iron blocks. The jacks were then screwed up, care being taken to get equal pressure so as not to spring the half-collar, the pressure exerted by the jacks being continued until the collar section touched the bottom of slot. Collar was drilled and tapped to take three-fourths inch thick machine screws that extended one inch into the roll body. The other half of the collar was next fitted, a good job resulting. The roll was then put into the lathe and the collar turned to the same shape and size as the others. When the time arrived to use these rolls in the mill, the writer's heart was, for a few minutes' time, in his throat (metaphorically speaking), but the new collar withstood the heavy pressure admirably, a slight mark on the iron bars produced by the joints being removed in the finishing pass of the rolls. The net result of this repair job was: The company got their contract out on time.

Over $300 was saved on a new roll, and the writer received an unsolicited raise in wages. No sketch is shown of the roll as readers are no doubt familiar with a set of rolls, and how they look. Should you not be conversant with the same, look up the issue of Canadian Machinery, July 15th, page 70. The sketch of the collar is self-explanatory.

See Page 440

TWO USEFUL FACING TOOLS
By "Draftsman"

The facing tools described are types that have many applications in the machine shop. They are intended for use on vertical drilling machines, but may be similarly used on horizontal boring machines. Fig. 1 shows a tool for facing the rim around a previously machined hole. The steel body A fits into the socket of the drill spindle and is secured by a set-screw bearing on a flat B. The centre of the outer end has a hole of two diameters, the inner one being threaded. Into this fits a tool steel pilot which is centred by the bevelled shoulder C. A slot through the pilot and across the end of the body carries the cutter D, whose projecting faces are backed off right and left. The outside of the body is threaded for a hexagon nut which holds the cutter by pressure on an intervening washer E. Two longi-

Continued on Page 443

DETAILS OF BOTH FACING TOOLS

DEVELOPMENTS IN SHOP EQUIPMENT

BRASS FURNACE OF NOVEL DESIGN

Metallurgical and chemical engineers for a considerable period of time have been endeavoring to solve the question of melting brass and other non-ferrous metals in the electric furnace in the most efficient and economic manner, realizing as they have that there was no question of higher production when melting electrically so long as certain metallurgical features of the process could be properly overcome. As a result of considerable research work, single-phase and even two-phase electric brass furnaces have been designed and put on the market, some of which seemed to solve the question of melting these metals electrically. It has, nevertheless, been a desire of those interested that a three-phase unit might be developed which would do the melting more efficiently and economically, and it will be interesting to engineers in the metallurgical and electrical field to know that such a furnace has been marketed.

The Volta Manufacturing Company, Limited, Welland, Ont., have had their engineering staff at work for some months past developing a three-phase electric brass furnace along the lines of what they considered would be the most efficient piece of apparatus to meet the requirements of the brass and non-ferrous metal industries, and have now put on the market a three-phase electric furnace which they claim is superior to anything heretofore developed for this class of work.

As will be seen by the illustrations of the furnace, the shell is of cylindrical design and is operated in an upright position—similar to the electric steel furnaces. The furnace is lined with refractory brick and the roof is also made of refractory material. It is provided with a cast iron base equipped with a motor-driven set of gears and rollers, which gives to the furnace a natural gyratory movement.

One of the strong features is that it is a three-phase unit, which means a balanced load on the lines of the power company and consequently a better power rate for the operator. It is quite generally known that many power companies having a limited amount of power on their lines, do not desire and will not permit the operation of a single-phase unit on their lines except at a very high rate. On account of having three electrodes it is not necessary to maintain the same degree of perfect alignment of electrodes as is necessary with the single phase unit in order to maintain a high degree of efficiency and inasmuch as the arcing takes place between the three electrodes the arcing is distributed over a much wider area than in the single phase two-electrode furnace, consequently better melting facilities are provided.

As shown in the photographs, the electrodes enter the furnace through the side of the shell at points 120 deg. apart. Graphite electrodes are used and are supported on heavy cast iron arms which are provided with guides to take the electrode holder. The construction of these arms provides a substantially insulated sleeve around electrode, which is kept cool by means of a water-cooled collar. The latter is in the form of a ball point, and by this arrangement the position of the electrodes can be changed relative to each other, at the same time making provision so that they can be lowered or raised in respect to the metal bath.

The electrode holders move in the guides provided in the cast iron arms. This movement is obtained by means of heavy screws actuated either by hand or through a simple system of motor-driven gearing. The charge, as will be noted, enters the furnace below the electrodes, and it is not necessary to withdraw the same from the furnace at any time, either during the charging or pouring period. This eliminates any chance of electrode breakage due to the heavy scrap falling on them. It gives a longer life to electrodes and also eliminates the loss of time and the expense ordinarily incurred in removing and replacing the broken ones.

The roof of furnace is independent of the shell. This permits of the relative height of the roof to the bath being designed in such a manner as to insure the best melting conditions as well as the least possible fusing of the refractories. A vent hole with a light cover is provided in the roof of furnace which allows operating under tight conditions with perfect safety. Should gases, formed from the use of oily or dirty material in the charge, reach a certain pressure, they are allowed to escape through this vent in the roof.

In actual practice it has been found that the heat from the arc is sprayed down and over the metal in such a manner as to heat the whole surface of the metal evenly. The effect of this is very important as it reduces the volatiliza-

NOTE THE BASE AND ROLLERS WHICH GIVE THE FURNACE THE GYRATORY MOVEMENT

GENERAL VIEW OF THE FURNACE. NOTE THE MOTOR AT SIDE

TWO USEFUL FANCY TOOLS
Continued from Page 441

tudinal grooves F in the pilot gather any grit that may pass.

When in use the cutter removes metal until the pilot bears on the top of the pin G, on which the work is mounted. The tool may be reground many times and its cutting edges will always be the same distance from the end of the pilot, thereby uniformly limiting the cut.

In Fig. 2 is shown a lighter type of tool for facing black hexagon nuts in the simple adjustable jig. Here the body fits into a tapered socket and the nut is knurled. The pilot hole is of one inch diameter, and the threaded part of the body has two through slots for the separate cutters. The latter are held, at the lower end, by a collar with two grubscrews at right angles, and this also pinches the pilot. At the upper end each cutter is retained by a bevel engaging that of the thick washer.

Such a tool has quite a long life before the cutters need to be renewed.

tion losses in the metal being melted to a minimum, and the linings, not being subjected to any high temperatures, last much longer. The motion given to the furnace by means of the combination of gears and rollers, has proved itself to be a very efficient and natural mixing one by means of which superheating of the metal is practically eliminated.

This furnace has been subjected to some very severe tests and is said to have proved itself an efficient and economical melter, very simple of construction and easily handled. When running on yellow brass, metal has been produced with a current consumption as low as 225 k.w.h. per ton of metal produced, and when running on aluminumbronze the current consumption averaged about 300 k.w.h. These figures were not taken at a time when the furnace was operating continuously and it is only reasonable to expect that the power consumption will be considerably reduced when the furnace is thoroughly heated and is being operated continuously. At the present time the manufacturers are producing these furnaces in the following standard sizes: 500 lbs., 1,000 lbs., and 2,000 lbs., having a transformer capacity of 125 kw., 175 kw. and 300 kw. respectively.

EVER-HOT SOLDERING IRON

The Peterson-Plummer Manufacturing Company are now making an everhot soldering iron which is light in weight and a convenient size for every requirement. The gasoline is contained in a reservoir in the handle which is made of seamless brass tubing, the pump being located at the outer end. The flow of gasoline is through cotton wick contained in a rustproof iron pipe between the handle and the burner. The burner is so designed that it may be used effectively in any kind of weather or high winds. The iron is fitted with one large and one small point.

NEW THINGS IN MACHINE TOOLS

MULTI-SPEED GRINDING MACHINE

A grinding machine that is very suitable for both production or tool-room work has been placed on the market by the Wisconsin Electric Company, of Racine, Wis. The device is provided with a set of interchangeable spindles and pulleys, so that a wide range of speeds is obtainable. Dust-proof ball bearings are fitted to the spindle. The device can be swiveled to any desired angle for every class of work.

TOOL GRINDING MACHINE

A motor-driven ball-bearing tool-grinding machine, fitted with a quick make and break oil switch, is now being manufactured by the Ransom Manufacturing Company, of Oshkosh, Wis. A starting pedal is placed at either side of the base of the machine, and when the foot pressure is released the machine is automatically stopped. Self-aligning S K F ball bearings are fitted to the spindle. The machine is operated by means of a 3-4 h.p. General Electric motor.

QUICK-ACTING MACHINE VISE

The Nelson Tool & Machine Company of Bloomfield, N.J., has recently brought out a new machine vise, and it is claimed that no hammer is needed in bedding the work on the vise bottom or to the parallels. The operating screw is set at an angle so that when pressure is applied the sliding jaw shows no tendency to lift upwards. The vise is made in two sizes, six inches wide by two deep and opening of 5 3-4 inches, and four inches wide by 1 1-2 inches deep, with opening of four inches.

IMPROVED ABRASIVE DISK

An improved type of abrasive disk for use in disk grinding has been placed on the market by the Gardner Machine Company, of Beloit, Wis. The construction of these disks increases the efficiency in grinding operations. The thickness is double that of the ordinary glue-bond disk, the bond being a special cement that powders away during the grinding operations. The corrugated surface causes the abrasive grains to assume a better position, always presenting new cutting points to the work.

DRILL RACK

The Peerless Machine Co., Wis., have placed on the market what is known as a gauging drill rack. This rack provides a place for drills, and each drill that is in the rack has been accurately gauged and must be in the correct place. This rack provides a practical gauge for all sizes of drills and reamers ordinarily used in any shop, large or small. In highly organized shops these racks are often installed in the tool crib where one rack may be used exclusively as a gauge while another series of racks are used to take care of the different tools.

In small shops these racks are sometimes attached to the drill press, and a complete set of drills maintained at the machine tool. This rack is claimed to meet a need which has long been apparent in all shops, namely, on account of tools slipping in the chucks, the numbers indicating the drill sizes are soon d=st=oyed or badly blurred, and a practical gauge is desired to eliminate errors and save the time lost in securing a micrometer or gauge.

The MacLean Publishing Company
LIMITED
(ESTABLISHED 1887)

JOHN BAYNE MACLEAN, President.   H. T. HUNTER, Vice-President
H. V. TYRRELL, General Manager.

PUBLISHERS OF

CANADIAN MACHINERY
MANUFACTURING NEWS

A weekly Journal devoted to the machinery and manufacturing interests.

B. G. NEWTON, Manager.   A. R. KENNEDY, Managing Editor.

Associate Editors:
J. H. MOORE   T. H. FENNER   J. H. RODGERS (Montreal)
Office of Publication: 143-153 University Avenue, Toronto, Ontario.

VOL. XXIV.   TORONTO, NOVEMBER 11, 1920   No. 20

PRINCIPAL CONTENTS

Quoting New York Funds

SOME weeks ago this paper, in an interview with J. Vernon McKenzie, formerly trade commissioner at Glasgow, referred to the fact that some Canadian exporters were asking for payment in New York funds. He explained that there was marked resentment to this practice on the part of many of the Old Country firms, the latter holding that Montreal quotations should be sufficient to cover the situation.

An Ottawa despatch this week refers to the same practice, complaint being made that trade commissioners are finding decided objection to this system, which they regard as trying to pinch off a little extra from the buyer.

One section of the Ottawa despatch points out that Canadian Trade Commissioners in every part of the British Empire state that general indignation is expressed by importers that Canadian firms almost invariably collect for goods exported in United States funds. Latterly the complaint has been that Canadian steamship companies also collect for freight on such goods in United States funds. One Trade Commissioner reports a British merchant as saying, "You keep asking us to give a preference to Canadian producers when buying because it is a British country. Why should I do that when Canadian exporters always ask us to pay in United States funds? Are we to believe that Canada is only a commercial adjunct to the United States? It looks so to us when such things happen."

Much of the material that some Canadian manufacturers use is bought in the U. S. market, and paid for in New York funds, but no doubt provision is made for this in fixing the selling price. Unless there is some very good explanation to be offered on behalf of the Canadian manufacturers adopting this practice, it will be charged that they are taking undue advantage of a passing situation, and in so doing injuring their reputation to a very serious degree, all of which will be hard to overcome in the future when goods are plentiful and orders badly needed.

Investing in New Machines

A STORY comes from Hamilton telling of the demonstration given to the civic authorities of a new mechanical excavator. The machine cost $10,000 or so, and, naturally, the civic officials wondered if they were going to get that much real value for their money.

After the machine was seen in operation it was estimated that forty men would be required to keep up with it, and one of the controllers remarked, "People criticized the spending of this much money on a piece of machinery, but this proves to me that it was a wise investment."

That same situation comes up a good many times in a good many places. Firms look at the amount of money that they will have to pay for a new machine, and they keep on looking at that sum until it grows into a foothill or a mountain.

They do not look on past that obstacle to see what is on the other side in the way of greater production and better work. They have an old machine, and by some strange process of mental juggling they think every extra day's work they take out of the carcass of that old thing they are just that much ahead.

The amount that can or should be spent on a machine can be safely judged by what that machine will do. If it can produce sufficiently in excess of an existing tool to pay its way and then some, why not let it?

There are cases where an old machine is humping along, causing the man who runs it considerable trouble and loss of time, putting the shop "up against it" if that man happens to be off for the day—and yet they imagine they are getting easy money by not having to go into the machine tool market and replace the old tool.

And once they get into the market and buy the new machine, put it in line and see it perform, watch the production records and compare them with the best limpings of the old tool—well, as the civic officials in Hamilton with their excavator, they are convinced that it was a good investment and a money maker right from the start.

You can save money in more certain ways than by hanging on to your old equipment.

Sessions of the tariff commission convince some critics there is no connecting link between farmers and manufacturers. Is that so? How about pig iron?

—Knott in the Dallas "News."
The Optimist.

Poor Insulation or a High Tension Spark

There are certain mechanical laws that must be conformed with in any line of activity. When some or any of these are violated, there is going to be trouble. A recent case occurred at Buffalo, where a number of men were cruelly burned, most of them fatally, when new apparatus in a generating station went wrong. We asked T. H. Fenner, the editor of Power House, to discuss this matter, and in the following he has done so.

THE generation of high potential electric current and the transmission of it over many miles of country is attended by comparatively few accidents in proportion to the amount of electrical power development. The chief cause of fatalities through high voltage electric lines is by the coming into contact with them of linemen engaged in their usual work of running new lines, and repairing and locating faults in existing ones.

Accidents in the generating stations themselves are happily few in number, and fatalities even less, and with the constantly better efforts towards safety that are being put forth, the number should steadily decrease. It is therefore with something of a sensation of horror that we read of the destruction of no less than eleven men, almost instantly, through an accident, which, although it may have happened before, does not recur to our memory.

The accident in question was the explosion by some fault in the insulation, presumably, of the oil used for the cooling of a transformer in the Tonawanda power house, Buffalo. The change of electric current, which is generated at high pressure, to current at a much lower pressure is accomplished by means of transformers which consist of two separate coils of wire wound on iron cores. The high potential coil will consist of a large number of turns of small wire and the low potential side of a less number of turns of a larger wire. The effect of the high potential current being passed through the high potential coil is to induce a current in the low potential coil of the same periodicity, and the relation between them is such that if the high side A consists of 50 turns, and the voltage is, say 100 volts, then if the low side consists of 25 turns the voltage in that coil would be reduced to 50. This, of course, applies to any voltage, and any relation of the number of turns. There is little loss between the two sides, by which we mean that the product of volts and amperes in the low side will be almost the same as that of the volts and amperes in the high side.

What loss there is appears in the form of heat, and in order to carry off this heat, it is usual to immerse the whole apparatus in a bath of oil. This oil in turn has very often coils running through it which carry a circulating stream of cooling water, and thus the heat is effectually dissipated.

It was the bath of the oil that exploded and thrust forth a sheet of flame which licked out the lives of eleven men in an instant. The cause of the accident is stated to have been a short circuit. The apparatus was under test at the time, and it is probable that the oil bath had been raised to a dangerously high temperature without it being noticed.

A piece of defective insulation and a high tension spark would easily supply the little extra heat necessary to cause the flash. No doubt the enquiry that will follow will throw light on the exact cause of this sad accident, and make the recurrence of it extremely unlikely in the future.

A Buyers' Market Here

HERE is the opinion of a man whose business it is to watch the markets. T. F. Bentley, writing in a recent issue of "The Purchasing Agent," says:—

"There is a definite downward trend in all commodities and in all countries. Prices are not declining because of the election campaign, or because it is harder now to get funds for speculative purposes than it was a year ago, but because business is gradually getting on a healthy basis, that is, we are coming again into a "buyer's market," which will be better for general business. A buyer's market means that the buyer may require definite information as to the price he must pay, not only for immediate needs, but for a reasonable time to come. It means that the buyer may demand reasonable assurance that the goods he agrees to take and pay for will actually be delivered and will not be diverted to some other point that may offer a trifle more payment. Finally, it means that the buyer may insist, within reasonable limits, that specifications of quality be carried out."

What's the use of having your trousers creased, boots shined, clean shave, clean collar and latest tie, fine line of talk about the weather, politics, reduced prices, Bolshevik, League of Nations, if you can't succeed in getting the customer's signature on the dotted line?

What the Business Papers Say

"Marine Engineering."—This is a plea for greater frankness on the part of engineers and manufacturers, as a matter of policy for their own advantage. There is a very proper idea that a man may well keep his private business to himself and that corporations may well practice a like reserve. But when a man wants to sell something it at once becomes a matter of business to the purchaser. There are, of course, instances in which the purchaser may well use his purchase without too much curiosity; the average man had best wind his watch regularly and think no more about it till he wants to know the time. But usually a machine needs intelligent care, and the more the user knows about it the better service he will get from it.

The Changing Foundry Aspect

"The Foundry."—During the past two weeks the shift and readjustment in foundry conditions has been most rapid. For a short time previously a decrease in demand in most lines had been noted, and in some directions this even had taken the form of cancellations of existing contracts for castings. To-day the cycle has swung entirely around until with few exceptions a buyers' market for castings governs, and some shops are taking on work foreign to that which they have followed for the past two years in order to keep their forces engaged.

What Determines Selling Price?

"Iron Age."—"In the textile trade and in others in which the ultimate consumer buys over the counter of a store, it is true that the declines came first in the raw materials and that wholesale markets underwent drastic readjustment. But the reductions that have been made in retail prices, most of them moderate thus far, were not the result of a decrease in the actual cost of the goods sold at retail. The dealer's stocks had been bought at high prices and were sold at reductions, going in some cases below cost, because they could be replaced on the dealer's shelves at prices considerably below what he had paid for them. Cost did not determine selling price."

Boilers Keep on Exploding

"Power," (New York).—Certain states that have adopted the A. S. M. E. Boiler Code have exempted oil-country and agriculture boilers with the exception of periodic inspections. Undoubtedly in some cases this has been an expediency resulting from political influence and is most unfortunate, especially when it is considered that many of these boilers are of lap-seam construction and are operated by very unskilled labor. Many of these boilers do explode and the loss of life is appreciable.

MARKET DEVELOPMENTS

Price Talk the Market Topic for This Week

Some Reductions Have Gone Through, But the Most of the Firms Have Made No Move Yet—The Scrap Market in Canada Has Been Pruned, But Business is Dull at the Lower Levels.

PRICES and price talk engage the attention of the market this week even more than before. There is a sincere desire on the part of a large section of the trade to get down to a workable basis as soon as possible, but there are very real difficulties in the way, chiefest among these being that costs do not show much tendency toward lower levels.

One maker-of machine tools in Canada stated this week that there was very little chance of them, announcing a lower price under present conditions, as there was nothing in their cost of manufacturing that was showing a lessening. However this may be, the fact remains that a number of American makers of machine tools have announced reductions to their representatives in Canada, and they are now working on the new schedule. New York is to be the centre of interest this week, and it is generally believed that following the convention of the National Machine Tool Association there will be some announcement, concerted or otherwise, dealing with the price feature. Cuts already made range from 15 to 20 per cent.

Small tools have been reduced in Canada by some ten per cent.

The steel market itself is quieter, but there is still a good deal of business to be secured. Manufacturers are not buying in large quantities, nor have they for some years past. The reasons for this are two in number: (1) they could not get the material if they were able to pay the price, and (2) the price has been too high in most cases to get them away from the hand-to-mouth class. A slight reduction is announced this week in warehouse prices of bar iron, now selling at 5.50 per pound. Most of the bar iron trade is supplied from the Canadian mills.

The scrap metal market went through a severe pruning at the hands of the dealers this week. They cut fast and deep into the lists they had been working on for buying purposes. Whether this decides the matter remains to be seen. A dealer can name his price, but if the seller will not accept it there is still room for an argument. Just now there are only three or four lines that are wanted, and these are scarce, and mostly in the iron and steel lines. Copper and brass scrap are drugs on the market, and business is stagnant.

BUYERS IN MONTREAL MARKET WAITING FOR SETTLED PRICES

Special to CANADIAN MACHINERY.

MONTREAL, Que., Nov. 11.—It is seldom now that we hear of mills complaining of a shortage of raw materials for the steady maintenance of plant operations. There are cases still reported where delivery to the mills has been interfered with, but this situation is overshadowed by the more serious one of the declining number of orders being placed on the books, and to the fact that the supply is now equal to, if not greater than the demand. This condition evidently is quite general throughout the steel industry, and is likewise more or less true with all lines of activity at the present time. The interest in the market appears to be influenced by the price adjusting that has featured sales activities on many commodities. The downward trend that marks the movement of materials of nearly every description has acted as a limit gauge by which the majority of buyers are now measuring their requirements. Current purchases are based on urgent needs for immediate consumption, and with declines a factor of weekly or daily occurrence, the place in the sun will be occupied by the seller rather than the buyer. However, this policy of conservatism may be carried to the point where industrial depression would loom up as a serious factor, as present producing costs will hardly permit of continual cutting of material prices, unless something is done to enable the manufacturers to make these reductions without serious loss to themselves. Curtailment in mill and blast furnace operations has been among the recent developments in American steel centers, and in consequence many men have been temporarily unemployed. This action on the part of producers has been found necessary, as it would be inadvisable to continue manufacturing materials at existing high costs when the demand for the product is showing a marked falling off. The movement in this district is steady in character but the volume indicates that requirements are becoming less. Price quotations given last week still hold, but further declines are not improbable.

Prices Unsteady; Demand Light

Weakness and quiet demand continue to feature the metal situation and, with few exceptions, the demand is below normal. The movement of galvanized sheets has shown slight improvement and general distribution is much better. Ingot metals are moving steadily but volume is light, evidently indicating that purchases are made with the idea of meeting the needs of the present rather than the acquisition of material for early future requirements. The prices on all ingot metals are at present unchanged but a further shading is not improbable.

Retrenching Attitude Evident

Dealers here are interested in the possible action to be taken at the meeting now being held in the States by the

machine tool builders. It is expected that a decision for price revision will be arrived at during the meeting, and the trade will be advised of such action in the very near future. There are many indications of the trade requiring equipment, but the attitude of the buyer today is one of retrenchment in face of lowering prices, and every effort is made to extend the time when the placing of orders for new equipment will become imperative. It is nevertheless true that these placements will ultimately be made, but where a few short weeks may mean a substantial saving in tool cost, the user is evidently content to weather the present storm of uncertainty and inconvenience in order to effect the saving that would result by the reduced cost of the new tools. On the other hand, the slackening of trade, that already shows on the horizon, may spread to many quarters and minimize the necessity of the manufacturer to purchase the equipment which, apparently, a few short weeks ago, was so urgently needed.

Low Prices No Inducement

The old material situation still dwells in a listless atmosphere. as the buying at the present time is exceptionally quiet. What trading is carried on is largely that of local movement in supplying foundries with regular requirements. Even here it is reported that demands are less insistent than formerly. The industrial situation generally offers little inducement for the absorption of scrap for future consumption, and with the uncertainty that continues to feature trade outlook, the prospect for even normal buying during the remainder of this year is not encouraging. Some dealers go so far as to say that trading will continue dull until the spring. The decline in price quotations noted last week is still effective and further shading is probable, in face of the quiet demand.

SOME REDUCTIONS HAVE BEEN MADE

Several Lines are Working Now Toward a Lowered Selling Schedule

TORONTO.—In most lines of the machine tool, iron and steel trade this week business would plead guilty to being very quiet. There are variations from this both ways, some firms claiming that their business is good, going ahead of the same month last year all the way along, while others, on the opposite side, report trading as very quiet and with no large buyers in the market.

It cannot be denied that there are a lot of machine tool quotations out just now, and if they could be converted into real business there would be nothing the matter. The turning of these quotations into real business is not being done rapidly. There can be no doubt that many buyers are anticipating something in the way of an announcement regarding prices or better conditions. Whether this would result in more business being placed at once it is hard to say. Canadian Machinery asked one of the dealers today if such would be the case, and he replied: "That is something I would not undertake to answer. There are some cases where it would not make any difference were I able to go out with a reduction of 30 per cent. On other lines it would make a difference. The standard lines are the ones that are looked to to set the pace in this matter of price reduction. Some of the newer companies sent out notices some weeks ago that they would allow us to deduct as much as 20 per cent. from their April lists. but this did not make any marked difference in sales. There are other cases, though, where it will make a difference. One of the largest makers of boring mills has announced a reduction, and it amounts to quite an item, especially when the quotation in the first place runs around ten to twelve thousand dollars. The reduction is around 15 per cent., and the company, in its report that it is prepared to do this, warns the buying public that they have gone the limit in doing this, and that prices will stand for a long time to come without going any lower."

Questions at other of the machine tool houses brought out the information that no reduction in prices had been authoriz-

ed, although there is a general feeling that something more definite in this can be expected following the meeting of the National Machine Tool Builders' Association at New York.

Discussing the matter of prices with an official of one of the machine tool companies in Canada a few days ago. Canadian Machinery brought up the matter of prices and inquired whether there was any intention on the part of the firm to move toward a lower schedule. The answer was: "Not as long as present costs keep up. The machines we have on hand were made of high .cost material, and were we to cut the price without any cut in the price of the material in them we would lose money. I don't see how Canadian makers can cut prices unless there are reductions in materials and wages." '

There is a growing feeling in the market that the matter of prices will be dealt with in a definite way shortly. The absence of any announcement does not help matters, and the dealers ,would be strengthened were something done one way or another.

The prices of high-speed steel tools have been reduced 10 per cent. and on some other lines there have been reductions made in carbons. The list on high-speed drills, for instance, is divided now as follows: ⅛ inch and under, net, 33-64 to 1 inch, plus 10 per cent.; over 1 inch, plus 20 per cent. The list formerly ran from list plus 10 to plus 30. The new prices came into effect on the 8th of the month, bearing out the intimation in these columns of last week that an adjustment was being made in prices of small tools.

"Will you sell any more at the new prices ?"

"That remains to be seen." was the reply of one dealer. "I do not think the price has much to do with it, although the buyer will very often insist that it has. Just now business is small, but there are a fairly large number of the small pieces. We are looking for renewed buying before very long."

Steel Market is Easier

There are few Canadian orders on the books of the premium mills in United States, about the last of this material coming over during the week. Many of those who find it necessary to import are promised consideration very shortly by the Steel Corporation, which is beginning to catch up with some of its bookings. There are certain lines that will still come from the premium mills. For instance, sheets, although Canadian rolled can now be available in better supply, will be brought in from the premium mills, but the price is better than prevailing for some time, the chances being that buying can be done now at slightly over 6 cents per pound, a real improvement on the figures .that were prevalent some few months ago.

There has been a reduction of $5 per ton in the warehouse price of bar iron in Toronto. This will, of course, also

---

POINTS IN WEEK'S MARKETING NOTES

American builders of machine tools are getting better labor results than for months past.

—

Some reductions in machine tool prices are out now, varying from 15 to 20 per cent.

—

National Machine Tool Builders' Association meets in New York on November 11 and 12.

—

U.S. railroads are expected to come into the market with fairly good lists early in the year. •

—

Scrap metal prices, with the exception of a few "whites," have been severely pruned. Very little business is offering.

—

Our Pittsburgh correspondent claims that bv the first of the year the independent steel mills will not be running over 60 per cent. ·

—

Pig iron prices have broken quite sharply, and a large tonnage would probably set á new low market figure.

—

High speed small tools are reduced 10 per cent. by Canadian makers. The new list became effective November 1st.

—

Bar iron is selling at 5.50c per pound in Toronto warehouses.

take in bar steel as well. The price is now 5.50 cents per pound, still a mark that stands as very high compared to the usual run of prices for this commodity. Bar iron is coming quite freely from the Canadian mills now, and they are able to look after the great bulk of the business offering, with the exception of the large size rounds and squares.

One Toronto warehouse had a voluntary reduction from one of the premium mills during the last week.

"Were you surprised or insulted ?"

"Neither," came the reply. "We are used to anything happening now." Lest there be any doubt, it might be added that the warehouse accepted the reduction.

Delivery of tubes is only fair, prices are firm, and there is a lot of business yet to be attended to.

"Firms are not buying anything that is not absolutely necessary yet," explained one of the dealers to Canadian Machinery to-day. "In no 'case is a firm taking on stock material. From the 15th of November each year until the end of the holiday season we never look for a great amount of business. It is between seasons, and a few off weeks need occasion no alarm."

Scrap Prices are Cut

Scrap dealers are out with the axe this week, and they have cut deep into the price schedules that have been standing for some time. Heavy melting, wrought axles, scrap rails, machinery cast iron, car wheels, steel axles and stove plate are the items that escaped notice when the killing was on. Otherwise the list is pretty well battered down. Here is the way in which the largest yard sized up the situation: ·

"Conditions are, if possible, worse than last week, and there is no guarantee that prices are staple yet. The Steel Co., we hear, is interested in heavy melting steel but at a lowered price, perhaps down three or four dollars from last week. For the first time in a year cast iron shows signs of weakness, and prices are not strong. Steel turnings are also weaker, delivered consuming districts; there is no market for borings and turnings here, although some may be placed across the line. Agricultural and railroad malleable is not in demand at all. Scrap pipe is a drug on the market and cannot be sold. Car wheels and grate bars are in fair demand. Stove plate is weakening; and rerolling rails are a drug now.

"In the non-ferrous lines heavy copper and wire shows a further decline, and dealers are now offering for No. 1 copper 12 to 12½ cents; light copper is quoted at 10½ cents; No. 1 composition scrap, which is the firmest of all non-ferrous metals, is quoted at 12 cents, the same as No. 1 copper wire. There is no market for heavy yellow brass, and the quotation of 7½ cents is nominal.

"I believe," he said in conclusion, "that there are many dealers in the .scrap metal business who are not making their salt at present. Business is poor and there is nothing in which to speculate with a fair chance of profit.

TO GUARANTEE THE COMPANY'S BONDS

Town of Paris to Lend Aid to the MacFarlane Engineering Firm

A proposition has been made to the Paris, Ont., Council by the MacFarlane Engineering Company for the municipality to guarantee bonds for the firm to the amount of $40,000. A committee appointed to investigate made a highly favorable report to a special meeting of the Council as regards the security offered. There were also present some thirty representative members of the Board of Trade, of whom a number spoke in favor of the adoption of the offer. On request of Mayor Patterson, a unanimous vote was passed to support the Council in the movement. A by-law has been prepared and will be submitted to the ratepayers early in December. The firm has already some 75 or 80 men employed, and it is confidently expected that within a few months after the passing of the by-law 200 men will be on the pay roll.

5-16 inch Size
Capacity:
3-32" to 5-16"

4 1-2 inch Size
Capacity: 2 1-2" to 4 1-2"

The Screw Machine Man Says:

"Everybody recommends Geometric Die
Heads. Better Threads in Quicker
Time is the reason."

It doesn't matter how strenuous the job is,
a Geometric is built to handle it easily.

Make Screw Thread-Cutting the Most Satisfactory Operation in Your Shop.

—Write—

THE GEOMETRIC TOOL COMPANY
NEW HAVEN   CONNECTICUT

Canadian Agents:

Williams & Wilson, Ltd., Montreal
The A. R. Williams Machinery Co., Ltd., Toronto, Winnipeg, St. John, N.B., Halifax, N.S.
Canadian Fairbanks-Morse Co., Ltd., Manitoba, Saskatchewan, Alberta

If interested tear out this page and place with letters to be answered.

TOOL MAKERS IN U.S. ARE GETTING BETTER RESULTS FROM WORKMEN

Special to CANADIAN MACHINERY.

NEW YORK, Nov. 11.—One factor which has been somewhat disturbing to business—the presidential election—is now out of the way, and the result is such as to inspire a greater degree of confidence in the early recovery of trade from the depression in which it now finds itself.

However, there are other things to be adjusted before renewed buying on a fairly large scale may be expected. One of the most important is the deflation of prices, which has scarcely touched the machine-tool industry as yet, either as regards costs of manufacture or selling prices.

In a measure, the manufacturers of machine tools have brought about slightly lower labor costs, this being due not so much to reduction in basic wage rates as in greater individual efficiency among workmen. For example, there are many plants which have discharged 10 or 20 per cent. or more of their employees which find themselves getting as much production from the force remaining as they were obtaining with the entire working force. In other words, the men who have jobs are working harder to retain them. This productivity is of even greater importance, from the standpoint of the manufacturer, than any reduction in basic wage rates, which probably none too high if the men will turn out the work.

There have been a few reductions in selling prices by machine-tool builders, notably one of 15 per cent. put into effect November 1st by a large New Eng-

land maker of turret machinery. Other reductions are known to be under consideration, but action will be deferred until after the meeting of the National Machine Tool Builders' Association in New York on November 11th and 12th. It is regarded in the trade here as practically certain that many tool manufacturers will go home from this meeting prepared to make substantial cuts in prices, say from 10 to 20 per cent. It is contended, however, that this will be about the limit, as machine tool prices have not been abnormally high.

Immediate business prospects are slightly better in some respects. The Erie Railroad has put out inquiries for six or eight tools, and more inquiries are expected from the same road. One or two other Eastern roads are also on the point of submitting lists to the trade, and this, it is hoped, is the beginning of railroad buying in the East. There has been a fair amount of railroad buying at Chicago in the last few months, but the Eastern roads have done practically nothing since the lines were returned to private ownership. The new freight and passenger rates have been in effect for more than two months and railroad earnings are showing a fair increase; hence it is expected that some of these earnings will soon be expended for rolling stock and shop equipment. In pre-war years it was customary for the railroads to issue their lists of tool requirements about January 1st and do their heaviest buying in February. There is some reason to expect that the same will be true in 1921.

however, is doubtless well under 37,000,000 tons.

In the matter of steel production the curtailment depends on the state of order books. As the Steel Corporation is very well filled with business its output has not decreased, and has possibly increased a little. Those of the independents that pursued a relatively conservative course in the matter of prices are moderately well filled, but with cancellations and suspensions are unable to run full, and thus a number have reduced output only by say 10 or 15 per cent. Others, with lean order books, which they cannot replenish to any great extent by cutting prices since there is little business obtainable at any price, have had to curtail more, and probably quite a number of small independents are running at 60 per cent. or less. The necessity for curtailment strikes one mill after another. At the present time many mills are counting up how many weeks their order books will carry them at 50, 60, or 70 per cent. By the end of the year the independents, on an average, will hardly be operating at more than say 60 per cent., and perhaps the average will be under 50 per cent., while even without any improvement in business the Steel Corporation ought to be able to maintain its present rate of output until April or later.

Steel Prices

On bars, shapes and plates the common quotation among independents is now 3 cents. In instances an independent is asking more, but merely as a nominal price to protect contracts on books. On bars there is little, if any shading below 3 cents. On shapes a concession of a dollar or two a ton would probably be obtained without difficulty. In the case of plates there is much irregularity, but it requires an attractive inquiry to bring out a cut price.

In sheets, particularly black sheets, the market seems to be tumbling rather than declining. The market on black sheets had been going down at the rate of say a quarter cent a week until a few days ago 6.25 cents became the common quotation for mills that were really seeking business. There are various reports of prices being done under this figure. It is believed that several mills would quote 6 cents very promptly if they saw chances of booking an order that would put them in a comfortable position for a week or two, say a tonnage running into four figures. It is reported, but not confirmed, that a valley mill sold 2,000 tons at 5.75 cents. Another report is that a mill made a first quarter contract at 5.50 cents, but this is doubted. In one well-informed place the prediction is made that first quarter business with independents will be done, when the time comes, at about a cent per pound above the Steel Corporation price, which has been 4.35 cents, and is likely to remain there, and that when second quarter business is done it will be at about the

PITTSBURGH NOT QUITE SURE WHERE PRICE LEVELS MAY REST

Special to CANADIAN MACHINERY.

PITTSBURGH, November 11.—There is no material change in the iron and steel situation as to the volume of demand appearing in the open market, the demand being extremely light. The amount of stocks consumers had when the market turned soft cannot be estimated closely. It may have been considerably greater than buyers represented, when they were importuning mills and furnaces for heavier shipments, but whatever the stocks are the buyers wish to reduce them, and they desire to have as small stocks as possible on the inventory date. The quadrennial election last week has had no appreciable effect on the general situation.

The changes that have occurred in the past week have been in connection with prices and the volume of production, both showing more or less declining tendency. The decline in prices is spotty and irregular, affecting some products greatly, others scarcely at all. The de-

cline in production, in the aggregate, is steady, though it is not steady as regards any individual producers.

Pig Iron Production

Since furnaces began blowing out, early in October, about two score stacks in various parts of the country have gone out, the greater part being late in October and early this month, while there were proportionately somewhat more steel works furnaces than merchant furnaces that went out. The blowing out occurring so largely late in October, the month's output was not greatly affected, and with the operating furnaces in receipt of better supplies of coke the month's output as a whole showed an increase over the September rate. In fact the October rate was the best shown for any month this year excepting last March, being about 39,000,000 tons a year, while the rate last March was about 40,000,000 tons. The rate now,

Steel Corporation level. The corporation will open its order books next month for the first half, but in both sheets and tin plates it will have to carry over about three months of work, so that the new contracts will be in reality for second quarter. This will make no material difference, of course, since the contracts will be with regular customers, who will be getting deliveries in the first quarter on old contracts instead of new contracts.

Pig Iron Breaks

The condition continues in the pig iron market that each inquiry of any size brings out further price declines. The pig iron sellers, who consider themselves salesmen when iron is scarce and they can get higher prices simply for the asking, are now exhibiting the form of salesmanship that consists in quoting a prospective buyer a price and then explaining that the price is largely nominal and they want to hear from the buyer again when he is ready to close. Hitherto the foundry iron market has been considered $45 valley, after the declines from the $50 price reached in August, and forge would naturally be say $1 less. In the past week an inquirer for forge was quoted all the way from $44 to $41, while an inquirer for No. 2 foundry was quoted from $45 down to $41. No purchase was made in either case, the prospective buyers concluding to allow the sellers more time in which to make up their minds positively. On an inquiry for 2,000 tons of basic for the east $38.50 valley was done, or $1.50 below the market as previously established. Bessemer has been offered at $42, or $3 off, finding takers only to the extent of two carloads. An inquiry now pending for 150 tons will bring out a further cut. We quote the market at the moment as follows: Bessemer, $42; basic, $38.50; foundry, $41, f.o.b. valley furnaces, freight to Pittsburgh being $1.95.

Connellsville furnace coke for spot shipment is down $1.50 in the week to $8.50 per net ton at ovens. In the two weeks preceding there has been a decline of $7. Foundry coke is now quotable at $10.50 to $11, though in remote districts, where the buyers have not learned the full size of the slump, $12 has been paid in some instances in the past few days. No one knows where coke and pig iron will land in the final adjustment, and various predictions are made, and can be defended to an extent, but in quarters where there is the best information and the greatest disposition to form an unbiased opinion, the idea is that furnace coke will land at somewhere in the neighborhood of $7, with foundry coke say a dollar per ton higher, and that pig iron will find its level at not far either way from $30 furnace.

It has been decided to call for tenders for the building of the new Lottridge Street School, Hamilton. The approximate cost is to be $300,000.

MR. IRVING TO TORONTO OFFICE

Change Announced in Selling End of the C.M.C. Business in This District

Mr. W. J. Irving has taken over the management of the Toronto sales branch of the Canada Machinery Corporation. He has been with the C.M.C. ever since there was a C.M.C, and he has covered the ground pretty well from Fort William to Sydney. There may be some metal working shops, and there may be an occasional wood working shop that does not know him at sight, but if so the

W. J. IRVING

plant is probably a very new one, and if they stay in business any length of time W. J. will be on their calling list.

Mr. Irving learned his trade as a pattern maker with John Ballantyne at Hespeler, after which he took up the work of a draftsman, finally spending a good deal of his time as a designer. Several of the lines made by the C.M.C. are Mr. Irving's own ideas.

It was almost ten years ago that he went on the road for the firm, and although he has done several other things, selling machine tools has been his chief business.

One well-known machine-tool salesman, speaking to the writer about Mr. Irving, remarked, "I have been up against him several times, and my complaint about him is that I had to be around too early in the morning and too late in the evening to beat him out. I bet dollars to doughnuts that he has taken orders in the early hours of the morning, and in the late hours of the day. I have known him to sign orders on a stump in the field, and I have seen him getting the necessary signature in the train was pulling in at the station."

For the present at least, Mr. Irving will keep his house at Galt, going there

for the week-ends. Mr. Irving, like a good many of the rest of us, believes that there are too many people trying to squeeze into Toronto to live comfortably. However, his activities will be taken up entirely now with his work in connection with the Toronto office.

David King, who has had charge of the Toronto office for many years, is retiring from active work to enjoy a well-earned holiday. Mr. King has passed the three score and ten mark and has had a wealth of experience in the selling end of the machinery business.

J. H. QUIRT DIES IN CALIFORNIA

Was One of the Best Known Machinery Salesmen in This and the Northern Districts

The machine tool trade, and many of the older firms in the machine shop and mill business, will regret to hear of the death of J. H. Quirt in Los Angeles, California. He was born in Ogdensburg, 79 years ago, and at the age of 13 with his parents moved to Arthur, Ont. Later he was apprenticed to the blacksmith trade, working at that for some years. Some time after a partnership was formed by Mr. Quirt with James Shand, now superintendent of the Dodge Co., Toronto. This business took in both the machine shop and molding branches. Following this, Mr. Quirt became connected with the A. R. Williams Machinery Co. and spent nearly all his twenty years with them on the road and attending to collections. Speaking of him this week, Mr. T. A. Hollinrake, president of A. R. Williams, said: "He was one of the best known and best liked men in the machinery business. I doubt if there ever was a man who knew the north country better than he." On account of his health he retired from business about seven years ago, moving, with Mrs. Quirt, to Los Angeles, where he resided until his death. He is survived by his wife, in California, and five daughters: Mrs. James Shand of 96 Uxbridge Street, Toronto; Mrs. Walter Findlay, 143 Cottingham Street, Toronto; Mrs. J. Black, Denver, Colorado; Mrs. F. W. Ritchie, Los Angeles, California, and Mrs. W. D. Young, Covina, California.

An experiment in paving is being carried out on the Victoria Bridge across the St. Lawrence at Montreal, and if the method of paving is successful, the entire length of the bridge will be completed next year. Two hundred feet of mastic flooring are being laid, with the object of determining how well it will withstand the weather during the winter. One advantage claimed for the material is that it is fireproof, and will prevent such disasters as the starting of fires caused by the overheating of gasoline and oil from passing automobiles.

CANADIAN MACHINERY

AND
MANUFACTURING NEWS

VOL. XXIV. No. 21   November 18, 1920

Using Turret Lathes in Railroad Shops

The well - managed railroad machine shop has found the value of turret lathes. Their wide range of standard tooling, together with their ease of setting up, makes them a very profitable proposition. Many shops have found it an advantage to concentrate on the making of standardized parts in one shop.

By J. H. MOORE

THE real mechanic of to-day is the man who refuses to sit back and say, "Oh, our present method is good enough." The man who gets ahead is the one that is always on a hunt for a still better method of producing his work. For some time past we have been presenting through our columns different styles of work accomplished on varied classes of machine tools. Not one of these articles has been published with an idea of comparison in mind, but rather with a desire to place before readers how the various firms are producing their products. They themselves can form their own conclusions. By carefully studying these various tool layouts valuable information can be procured which will stand you in good stead on some later occasion when similar problems present themselves.

There is an old saying that it pays to keep posted on all conditions from the foundry to the finished piece, and if more mechanics followed out this advice, production in many lines would soar in consequence. You cannot learn too much about your work, and it is only

by studying it carefully that you can master it.

Anyone will, I feel sure, concede the point that the study of motions is a very essential one. There was a time when such an idea was ridiculed, but that time has passed. Henry Ford would never have achieved his present success had not lost motion been reduced to a minimum. Every movement counts if you desire maximum production. Granting that such is the case, how much more so is this true if lost motion is caused through lack of judgment in tool arrangement.

You could take the finest machine tool existent, and mar its possibilities by poor arrangement of tooling. To obtain the best from any machine you must first study your product. Think carefully on what you consider the most economical method of producing such a product, then, having decided on the operations you wish to perform, cast your eyes around and discover the type of machine you require. Manufacturers are rather lucky in these modern days, inasmuch as the makers of various

machine tools will, upon request, study their product for them, giving them drawings showing what they believe the best method of producing the same. They even go so far in some cases as to guarantee a certain production, quote both price of machine and tools, or in other words practically solve the production problem for them.

Such service was impossible in the old days, and when a manufacturer can in this way secure such valuable help, it makes his task a much simpler one. All he need do is study the various plans submitted, modify his first ideas if necessary, then choose the equipment he thinks best suited for his own particular requirements.

We will suppose, however, that all this has transpired, that the machines are installed, and the work is being completed as per the submitted plan. Does this mean that the plan is above improvement? Unfortunately such a belief does exist in some plants, but luckily in the majority of shops they are always on the lookout for a shorter method. Improvements to the arrange-

FIG. 1—TURNING A SHAFT. NOTE THE HEAVY CUTS TAKEN BY BOTH TURRETS SIMULTANEOUSLY.

FIG. 4—A PROSSER BEING MADE FROM THE BAR.

FIG. 3—MAKING RIVET SETS ON THE TURRET LATHE.

FIG. 2—FINISHING A GEAR BLANK ON THE TURRET LATHE.

ment of the tooling, and sometimes to the design of the tools, add still a better production record. It is this spirit of eager striving after maximum results that is to be commended. We can well afford to have more of this spirit existent.

One of the finest methods of learning is by a study of the other fellow's method, not in a spirit of critical comparison, but actually in a spirit of seeking for further knowledge. This is why we have been presenting the previous articles dealing with tool layout work, and the present material will be followed up later with an account of other methods of producing work on turret lathes. The practice to follow will be that of some examples of work performed on Warner & Swasey turret lathes, in railroad shops. Although these examples are of work in railroad shops, there are many good points to be obtained from them for use in other classes of manufacture.

Examples of Work Performed

The classes of work going through a good sized railroad shop almost defy description, so that we will be able to give but a few of the great many. At Fig. 1 is seen a typical example of shaft turning on a No. 3A machine. Note the heavy cuts being taken by both turrets simultaneously. As this is a straight turning proposition we will pass on to a further example. Fig. 2 represents a gear blank, being finished on a 2A machine. The tooling used is clearly shown and this photograph is shown more to illustrate the variety of work passing

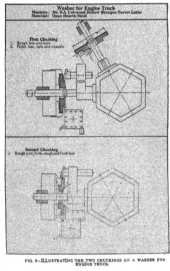

FIG. 3—ILLUSTRATING THE TWO CHUCKINGS ON A WASHER FOR ENGINE TRUCK.

FIG. 4—A PISTON VALVE FOLLOWER COMPLETED IN ONE CHUCKING.

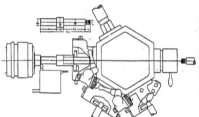

FIG. 7—THIS VIEW SHOWS LAYOUT FOR PINION.

In the second position (see view on the other page) it is turned 1-in. diameter for the distance of 1¼-in. length, as shown. The third station forms end, the fourth threads the 1-in. diameter portion, and at the fifth position the stock is cut off.

Fig. 8 depicts two chuckings of a washer for an engine truck. The material is open hearth steel, and the sequence of operations follows: At first chucking the first station rough faces and bores out the work, while the second position is used to finish face, turn and chamfer.

The second chucking rough turns, forms, and rough and finish faces. As the dimensions of the piece are given on the sketch, readers can readily form their other proportions.

Two chuckings are also necessary for the centre ring for piston valve shown at Fig. 9. The machine used is a No. 3 universal hollow hexagon lathe, and the material is cast iron. At the first chucking, the initial station is used for rough turning and facing the work. The second station and the last in this

through a railroad shop than for any other reason.

In some railroad shops universal hollow hexagon turret lathes are used to make rivet sets, and one such example is shown at Fig. 3. The work and tools can be clearly seen and require no comment. A prosser being made from the bar is shown at Fig. 4.

The photograph making up the heading of this article depicts a No. 4 universal screw machine busily engaged turning a stud. Both turrets are used simultaneously, and the nature of the work being accomplished can be seen from the pile of studs on the stand alongside of the machine.

Piston Valve Follower

The next example, Fig. 6, illustrates the turning of a piston valve follower in one chucking. To still further discuss this layout in detail, let us refer to the lower part of Fig. 7. Here we see the details of tooling and sequence of operations. The machine used is a No. 3A universal hollow hexagon turret lathe, and the follower is made from cast iron.

At the first station, the work is rough turned to 12¼-in. diameter, bored 2¼-in., and the piece faced on its 3⅝ diameter surface. At the same time the square turret rough faces the 12¼-in. and 13 7-16-in. diameter, and also forms the radius on the 13 7-16-in. diameter.

At the second position the work is finish turned to 12¼-in. diameter, the 2¼-in. hole is completed, and the 12¼-in. and 3⅝-in. diameters finish faced. The third station finish faces the 13 7-16-in. diameter portion. The fourth, and last, position used, reams out the 2¼-in. hole.

Further Examples

Let us next consider the first portion of Fig. 7. In this case we have a 2¼-in. cold rolled steel pinion, the dimensions of which are plainly seen on sketch. Five stations are used in this case, the sequence of operations being as follows: The stock is fed to length and the 1½-in.

diameter portion turned for 6½-in. in length. It is now formed ½-in. to the 1⅜-in. diameter, and turned to 1⅝-in. with the square turret.

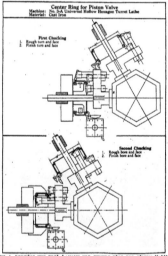

Center Ring for Piston Valve
Machine: No. 3-A Universal Hollow Hexagon Turret Lathe
Material: Cast Iron

First Chucking
1. Rough turn and face
2. Finish turn and face

Second Chucking
1. Rough bore and face
2. Finish bore and face

FIG. 9—SHOWING THE TOOL LAYOUT FOR CENTRE RING FOR PISTON VALVE.

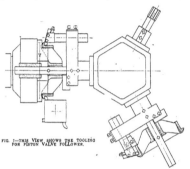

FIG. 9—THIS VIEW SHOWS THE TOOLING FOR PISTON VALVE FOLLOWER.

first chucking, finish turns and faces the work.

The second chucking is clearly noted on the sketch. First the work is rough bored and faced at the portion now turned toward the tools, and in the second position the finish boring and facing is accomplished.

Fig. 10 presents another tool layout of interest. In this case the work is made of open hearth steel and is known as a friction joint for a locomotive trailer truck equalizer. At the first chucking the work is spot centred. The second position rough turns and faces, while the third finish turns and faces the work, aided by the square turret.

The second chucking is shown at bottom portion of sketch. First the work is rough turned and chamfered, then, at position No. 2 it is finish turned. The dimensions of outside diameter of the piece being given, we need state no further as to its other proportions.

The last example, that of Fig. 11, depicts the operations on a certain type of bushing. In this case the machine used is a No. 2 universal hollow hexagon turret lathe, and the bushing is made from a forging.

At the first chucking the operations are as follows. First the six-inch diameter hole is rough bored, and the work is rough faced and turned on its eight-inch diameter. Next the hole is finish bored to six inches and finish faced and turned to its eight-inches diameter.

At the second chucking, the work is, of course, reversed, and the first station rough turns the 8¼-in. diameter, also rough faces. The second position, incidentally the last, finish turns the 8¼-in. diameter, and finish faces and forms the radius in the 6-in. hole.

As stated previously, we are only able to portray a few of the problems pos-

sible of solution on this type of lathe. We have in our files, gear blank layouts, friction clutch core work, pulley work, valve bodies, pinion sleeves, pump bodies, differential cases, separator cores, flywheels, and other innumerable examples of work performed. Enough has been shown, however, to illustrate our intentions, and as a concluding remark we might add that, even after you have installed any type of machine, remember to study your tool layout. Keep the layout up to date, and above all do not neglect your tools. Keep them sharp, in good condition, and free from dirt and grease as much as possible.

TINNING BLACK IRON SHEETS AND CASTINGS

By W. S. Standiford

In repairing of automobile radiators and for general use in machine shops in emergency work where it is desired to give black iron sheets and objects made out of bar iron or malleable castings a coating of tin, it will be found that by using the process given below any iron or steel object can be given a thick coat-

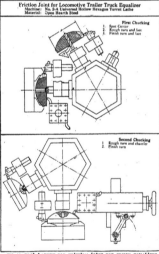

Friction Joint for Locomotive Trailer Truck Equalizer
Machine: No. 2-A Universal Hollow Hexagon Turret Lathe
Material: Open Hearth Steel

First Chucking
1. Spot Center
2. Rough turn and face
3. Finish turn and face

Second Chucking
1. Rough turn and chamfer
2. Finish turn

FIG. 10—TOOL LAYOUT FOR FRICTION JOINT FOR TRUCK EQUALIZER.

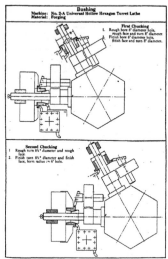

Bushing
Machine: No. 2-A Universal Hollow Hexagon Turret Lathe
Material: Forging

First Chucking
1. Rough bore 4″ diameter hole, rough face and turn 8″ diameter.
   Finish bore 4″ diameter hole, finish face and turn 8″ diameter.

Second Chucking
1. Rough turn 8¼″ diameter and rough face.
2. Finish turn 8¼″ diameter and finish face, form radius :- 6″ hole.

FIG. 11—TOOLING NECESSARY TO MACHINE BUSHING.

ing of tin which will protect it from rusting in exposed places for a long time, as a good deposit of metal can be obtained by this method. All the equipment needed is two crocks of suitable size to hold the raw and cut commercial muriatic acids and also an iron ladle to hold the melted tin. The cut acid should be first made by putting small clippings of sheet zinc into the raw muriatic acid until it will dissolve no more. For use it should be diluted with twice the amount of water. In some cases, it is better to use the cut acid without dilution as some muriatic acids seem stronger than others and work quicker.

All objects to be tinned should have the scale and grease removed from their surfaces. This is best done by wiping them with a clean piece of waste and immersing in the raw acid until all scale has disappeared, and the iron turns white. The time of pickling depends upon the strength of the acid and amount of scale. If much pickling is done, the acid will work slower, but it can be used until it is pretty weak and then thrown

away and fresh material used. This process allows of coating iron sheets and other articles with not only solder, but with block tin and also lead, if preferred.

After the objects have had their scale removed by immersion in the raw acid, put them into the cut acid for a few moments and shake the surplus off before dipping into the molten tin or solder. This insertion should be done carefully in order to prevent splashing of the hot metal. If the coating is not satisfactory on the first trial, dip article again into the cut acid and put into the hot metal.

The chemical action of the cut acid is this, that the muriatic acid unites with iron or steel, and in so doing leaves a thin deposit of zinc film which facilitates the adherence of the molten metal to iron or steel objects, causing it to stick better than if the ordinary raw acid were used alone. Many shops accumulate small pieces of solder and these can be used for work of this kind, so that the expense is trivial. Even the dross that is skimmed from a ladle of hot solder can

be used for this purpose, if there is enough of it, as there is always a certain amount of pure metal in it to do the work.

RUST REMOVER

A new preparation known as "Meno" rust remover and cleanser has recently made its appearance on the market. The inventor of the preparation is a chemist of wide experience, who, realizing that the present methods of removing rust, corrosion, etc., from machines, engines, tools, parts and all metal surfaces requires considerable time, and high labor costs, commenced the task of devising a better and more economical cleansing method. Here is what he claims for it. This preparation is a scientific combination and blending of certain chemical ingredients, which in combination produce an electro-chemical action that rapidly loosens and dissolves rust, corrosion, grease, oil, dirt, carbon, paint or any other foreign substance that is adhering to the metal—irrespective of its age or hardness—and its action automatically ceases when contact between the cleanser and the metal is established. This is as far as it will go, for it will not injure or mar the surface of the metal itself in any way. There are two methods of using the preparation as follows: (1) Apply it to the machine or part with a brush and allow it to remain for a short time, then brush or rub it off and it leaves the metal bright and clean. (2) Mix the preparation in a vat, tank or container with water, then attach the machine or parts to wire or chain so that they will hang in the solution. No further attention is required since the process of cleaning goes on while the parts are immersed. It is stated that the preparation is absolutely safe in every way and that it will not burn or explode. Another point is that it will not cause corrosion or rust to form, in fact, it protects the metal and makes it exempt from corrosive or disintegrating action for a long period after it has been treated by this preparation, and there need be no fear of injury to the most delicate part no matter what metal it is composed of. The preparation is said to be an economical one to use as the same solution may be used many times over, and does not deteriorate or lose its cleansing power. Peter A. Frasse & Co., of 417 Canal Street, New York, are the sole distributors, and are now establishing agencies in various parts of the country.

SCREWDRIVER ATTACHMENTS

Several types of screwdriver attachments for use on air-operated and electric-operated tools, are now being made by the Independent Pneumatic Tool Company of Chicago. The attachments are made in three sizes to fit the different tools manufactured by the company.

Invention is Important Factor to Community

We All Have, More or Less, That Desire to Invent Something, But Even After the Invention is Completed, Certain Procedure Must Be Followed to Ensure Protection.

By W. McFEAT, Patent Attorney, Montreal.

SOME interesting points relating to patents were brought to light in a recent paper presented by the above author at the Canadian Railroad Club in Montreal, and these facts are of importance enough to warrant their reappearance. After all is said and done, we are all inventors at heart, and for this reason the data will be all the more welcome.

I will first tell, says Mr. McFeat, of whence come our laws by virtue of which these patents are granted. I will also try to give my idea of what an important thing for the community invention is. And I will tell upon what conditions Canadian patents of invention may be obtained and maintained. Also how to realize on an invention.

As early as Edward III, in the year 1477, special acts were passed giving relief against monopolies in the form of Letters Patent, under which great restraint of trade was practised for the benefit of the patentee and the Crown. Our old friend, Sir William Blackstone, in his commentaries, first made in a course of lectures at Oxford in 1753, states briefly the condition of affairs before the reign of King James the First of England (1603). Blackstone informs us that during these early years monopolies were granted by the king for the sole buying and selling, making, working or using, of all things whatsoever, and by these Letters Patent manufacturers were restrained from doing that which they had done before. These monopolies had been increased enormously during the reign of Queen Elizabeth. The penalties provided in these early days consisted of heavy fines, with or without the pillory, and for repeating the offense the loss of one ear. We can imagine some of these early infringers sitting in the stocks for days with their ankles locked in the boards and their hands tied behind their backs, where they were made to suffer all manner of indignities by the loafers in the neighborhood. The practice was at the time to raise large sums of money for the use of the Crown and the Government by means of these patents without the slightest advantage to the general public. Amongst the monopolies granted by Letters Patent was the patent granted to the cloth makers of Ipswich, and the patent for playing cards. In neither of which cases was there any pretence of invention.

Up to this time patents had been granted for inventions, besides many more granted, as I have said, for carrying on certain trades or producing various articles or, in fact, importing them from other countries. The monopolies granted under the Tudor Sovereigns be-

came so monstrously oppressive that the Parliament of England prevailed upon James the First in 1621 to consent to a bill which is the celebrated Statute of Monopolies. This Statute of Monopolies of England is the foundation of all modern patent laws. This original law provides for the grant of a monopoly as a reward not only for the benefit that the invention confers upon the public but also for the ingenuity of the first inventor. Consequently if he who first obtained the patent was not the man whose ingenuity first discovered the invention and if proof existed that it could have been taken from a book or other publication, then, although he may have been the first to bring it to the attention of the public and thereby benefit the public by it, he would not be entitled to a monopoly because he was not the first and consequently should not have reward for ingenuity. This was the belief of the progressive branch of the manufacturers of the early part of the seventeenth century as laid down by

---

Modern patent laws of different countries vary considerably. Each country aims to obtain the best results for the community. Read how our Canadian law is enacted.

---

this initial patent law, and as I have just said, it has formed the basis of all modern patent laws.

In order to obtain the monopoly granted by Letters Patent of invention it is necessary for a man to rise above the commonplace in whatever art he may be engaged, whether the building of cars, the construction and laying down of track work, the erection of bridges or the manufacture of steel and other metal parts and textile and wooden things, each contributing to the railway and its operation. He must perceive the obstacles to increased or improved production, or greater comfort for the public or the worker, propound the problem the obstacle offers and conceive a remedy. In this way he rises above the commonplace and by seeking a monopoly in the form of a patent upon the device or process conceived he brings himself in a measure before the public and if he is diligent in promoting the result of his ingenuity he will attract the notice of others working in the same art or those who are financially behind this work. From my years of activity in connection with patents of invention I have come to look upon invention as the dramatic, as compared to the commonplace, in the development of the arts. It is this which to my mind directs

that attention to the fortunate one today whose activity makes the first step forward in any art, which brings to him the reward intended by that old law, defined in the Statute of Monopolies back in the early part of the seventeenth century. The dramatic is present in almost all cases where attention is attracted. Take for instance the destruction of lives and buildings in the recent explosion in New York. Our attention was directed to the death of these men and women and the destruction of parts of the buildings, and we all think and talk about it; while just as many lives, just as valuable, are lost, and just as much damage to property takes place each week in the ordinary course of events by street accidents and fires in New York, but they are the commonplace; whereas the explosion was the dramatic. It was the same in the bombing of the City of London by German flying machines when lives were destroyed but not in any greater number than are destroyed in almost an equally short time by street accidents. I wish to impress upon you who are engaged in production that by making a first step forward you are entitled to the notice of those upon whom you depend for advances, and this because of the dramatic nature of the development as compared with the commonplace work.

I have endeavored to bring to you an appreciation of the field that is open particularly to those whose work is the production of that which is necessary to maintain and develop the means of rapid transit.

Patent Laws Vary

The modern patent laws of different countries differ very considerably in their application of the broad principles I have stated. Each country makes its laws with regard to patents of invention, to obtain the best results for the community. In some countries limited publication of the invention there or in fact anywhere in the world makes the invention public property if this publication has not been preceded by the application for a patent. Other countries require that the invention be manufactured to a reasonable extent in that country in proportion to its manufacture abroad. Other countries ignore these requirements and as in the case of the United States, for instance, permit manufacture and sale of the invention in that country for two years before the patent is applied for. In some countries importation of the patented invention is allowed and the patentee may stop all others from importing while he, under his monopoly, may import and thus supply the demand of those countries. These

foreign laws, or the majority of them, had been in existence for many years before the present Canadian patent law came into existence and apparently our law makers adopted in a single law what appeared to be the features of these other laws which are good for Canada. Being Canadian producers we are, of course, interested principally in the Canadian law. From my point of view it has been necessary to emphasize the importance of looking for and overcoming the obstacles to progress in the arts in order that you may appreciate what a patent of invention is; and what I have said regarding the stepping forward in the art is universal, and consequently is essential to the application for letters patent of invention of Canada.

Assuming that a worker in one of the arts has perceived his obstacle and conceived his means for overcoming it and making his forward step, and wishes to obtain the reward to which he is entitled, he is, according to Canadian patent law, permitted to manufacture and sell the invention for one year, wherever he may be, even in the Fiji Islands. If he makes his application for a patent in Canada within that year and is the first to have the ingenuity to make the forward step he is entitled to his patent, provided he can truthfully make oath that his invention has not been in public use or on sale with his consent or allowance for more than one year. He must make this oath in his application. An examination is made in the Patent Office at Ottawa in order to ascertain, if possible, whether he is the first to make the invention and if nothing to the contrary is found the patent is granted. During the time between the filing of the application and the grant of the patent and during the year before the application is filed the inventor may have the invention manufactured wherever he pleases and may import it into Canada without jeopardizing his rights; and he may also continue to import the invention for one year after the patent has been granted. There is a provision in the Canadian Act for an extension of this year but the Commissioner of Patents must be satisfied that the further importation after the first year will be for the benefit of the community and that it will lead to manufacture in Canada, otherwise he will not grant any extension. With regard to the manufacture of the invention in Canada—as the patent issues it is subject to a provision of the Patent Act which requires that manufacture of the invention be commenced in Canada within two years after the date of grant of the patent and after such commencement the manufacture must be continuously carried on, even if it has been commenced before the application is filed. Failing commencement to manufacture, and continuously manufacture, the patent is forfeited and all rights and privileges thereby granted cease. The law is very strict in this regard and the

Privy Council has stated that there is no room for interpretation, its reading being perfectly clear. There is no doubt about this as the section of the law covering manufacture reads thus:

"Such patent and all the rights and privileges thereby granted shall cease and determine and the patent shall be null and void at the end of two years from the date thereof unless the patentee or his legal representative within that period or an authorised extension thereof commence and after such commencement thereafter continuously carry on in Canada the construction or manufacture of the invention patented, in such a manner that any person desiring to use it may obtain it or cause it to be made for him at a reasonable price, at some manufactory or establishment for making or constructing it in Canada."

Before 1903 it was the common belief amongst manufacturers who took advantage of patents of invention to improve their business that this law could be interpreted to mean that manufacture need not be carried on unless there was a bona fide demand; but in 1903 this section of the law was tried out and our Supreme Court decided that the law was clear and that there was no room for interpretation, this decision being upheld by the Privy Council. The result of this decision was that the owners of a large majority of Canadian patents found that their monopolies had been forfeited, and they got after our law makers and made them pass what is known as the Remedial Act, which revived all the patents forfeited, upon the condition that manufacture should forthwith be commenced. But a difficulty presented itself. How could the inventor of a grain elevator or a railway bridge or a round house commence to manufacture his invention and continuously carry on the manufacture or erection if the Government or the railways could not be made to see sufficient advantage to give an order? This was taken care of in the Remedial Act by providing for relief from the restrictions of the law regarding manufacture, and provision was made for an order to be granted by the Commission relieving such patent from the restrictions of the law and making the owner responsible for the manufacture of the invention in Canada so that the reasonable requirements of the public in reference to the invention would be satisfied. This may be done by granting a license, and any one interested in the art to which the invention appertains may go before the Commissioner, and upon proving to him that the owner of the patent either by neglect or wilfully monopolizing the invention has prevented him from obtaining it, may obtain an order from the Commissioner on the owner of the patent compelling him to grant a license to the sufferer. The owner of the patent, however, may be heard and if he proves that the alleged sufferer is not in the true sense of the word "of

the public," but, on the other hand, a competitor of the owner of the patent in the supply of the invention to meet the requirements of the public, then it may be expected that the Commissioner will deny the petition for the license and the alleged sufferer will be compelled to send his customers to the owner of the patent. If, however, it appears to the Commissioner that the owner of the patent is exercising his monopoly in the restraint of trade or is otherwise not satisfying the reasonable requirement of the public in reference to the invention, then he will order the owner of the patent to grant the license; and if the license so ordered is not granted within three months, then the patent will be forfeited. I may draw attention here to the fact that the spirit of Canadian patent law is that manufacture for Canada's need must be carried on in Canada. This is further demonstrated in another condition upon which the Canadian patent is granted, namely: the limit of time set for importation of the invention. According to law if the owner of a patent or his licensee or an assignee of a share in the patent, imports the patented invention or any machine or car or locomotive or frog members, switchstand or anything else containing the invention after the expiration of one year from the grant of the patent, then the patent shall be void as to the interest of the person importing it. For instance, supposing John Jones has a patent on a detail structural part for the reinforcement of the roof of a box car and he grants a license to a car dealer who has the cars built in the United States and equipped with the patented structural detail. These cars are imported by him into Canada say one year and three months after the issue of the Canadian patent and sold to a railway company. In this case the patent is forfeited because of this importation, but it is only forfeited as to the interest of this dealer and consequently his license is lost; and if John Jones has had nothing whatever to do with and no knowledge of this importation, then his patent is not affected. On the other hand, if John Jones has knowledge of this transaction of his licensee, which may be proved by the obtaining of license fees for the cars sold, then, he has connived and because of this connivance, not having prevented the licensee from importing, John Jones also forfeits his rights and the patent itself becomes null and void and the invention is thereby free to be used by the general public. There is an erroneous belief that the Canadian patent law does not apply to railway inventions and that cars and locomotives and railway stock generally may be built or equipped in the United States according to or with inventions patented in Canada and imported into Canada without jeopardizing the Canadian patent rights. In fact, I have heard it stated positively by lawyers in the United States that the Canadian Patent Act makes an excep-

tion of railway cars. They base their contention on a section of the Canadian Act which makes an exception of inventions in any foreign ship or vessel touching our shores. The law, however, is not any more open to interpretation than the law governing manufacture, because its wording is simple and very clear. This interesting law reads as follows:

"No patent shall extend to prevent the use of any invention in any foreign ship or vessel, if such invention is not so used for the manufacture of any goods to be vended within or exported from Canada."

There is nothing equivocal in these words. It is clear that the law relates to any foreign ship or vessel and it does not require a legal luminary to make our learned judges understand what is meant by the words "foreign ship or vessel." Anyone having a rudimentary knowledge of the English language knows that a foreign ship or vessel cannot be a railroad car on rails and I have no doubt that when the time comes for a court of competent jurisdiction to decide what this law means that there will be no hesitation in deciding that it has reference to ships or vessels or in other words steam ships or sailing vessels coming from foreign ports to the ports of Canada, and that it is limited strictly to these ships or vessels and cannot be extended in its scope to include railway rolling stock. There is considerable controversy about this, but our friends across the border who persist in ignoring our laws governing manufacture and importation cannot have a conviction of the justice of their cause, when they argue that the railway rolling stock is a foreign ship or vessel, otherwise they would take the steps available to obtain a decision on the question by our courts.

Interesting Phase

Another phase of the Canadian patent law which I should bring to your attention is the fact, not before mentioned, that an applicant must swear that his invention has not been in public use or on sale for more than one year with his knowledge or consent, before his application for Canadian patent. When the invention is conceived and developed in Canada it is not likely that the inventor will sacrifice his rights. The case is different with the inventor who conceives and develops his invention in the United States, because if he takes advantage of the United States law and puts the invention on the market for almost two years before he applies for his United States patent, then he will have forfeited his Canadian rights even before he applies for his United States patent. In many cases this is done inadvertently, the United States inventor being ignorant of the Canadian law, and after his United States patent application has been allowed and he finds his invention is a money-maker in the United States he turns his eyes to Canada and seeks advice from his attorney regarding Canadian patent rights. The

grant of a Canadian patent is also subject to the filing of the application within one year from the date of the issue of the first foreign patent for the same invention. It has happened in cases which have come before me for investigation that the United States inventor has not filed his Canadian application until sometimes five or six years after the first public use or offer for sale of the invention in the United States. This comes about in this way, he makes his forward step in the art and in order to ascertain if the idea is practical and if practical whether there is any money in it he gets some made and places them on the market, or if it is a comparatively large plant interests a railway or other large corporation by offering the invention by means of blue prints. A bridge, of course, could not be built and then offered for sale, so the inventor would bring his drawings to the bridge department of the railroad and submit his idea. Whether the bridge is built at that time or not the inventor must file his application for a Canadian patent within twelve months from the date he offers it for sale. If he does not offer it for sale but constructs the invention in private and then puts it in use where it may be seen by the public, a new headlight for instance, he must file his application for a Canadian patent within twelve months from the day of first public use. Our friend, however, usually has no knowledge of this condition of grant of the Canadian patent and may have been told, as has too frequently happened to my knowledge, that the Canadian patent may be obtained by filing an application within twelve months after the issue of the corresponding United States patent, With this idea in the United States inventor's mind he proceeds with his United States work and after having satisfied himself that the idea is practical and a moneymaker, he files application for United States patent. It is a well known fact that an application for United States patent takes all the way from six months to six years and longer to get through the Patent Office. Assuming that this inventor's patent issues in the United States two years from the date of its filing and it is filed one year and eleven months after his first offer for sale or public use after the issue of his United States patent and within a year he files his application for a Canadian patent, ignoring the statement in the petition that the invention has not been in public use or on sale with his consent or allowance for more than one year previous to his application for a patent in Canada, then his application is void as is also his patent if granted. Sometimes these inventors have been advised that what is meant by this statement is public use or offer for sale in Canada only, and that public use or offer for sale outside of Canada does not affect in any way the Canadian rights. This is not so. The Canadian application

must be filed within one year from whichever is the earliest date, the offer for sale, public use or issue of the foreign patent. For this reason a large proportion of the inventions which have been conceived and developed in the United States and form the subject of Canadian patents have been dedicated to the public of Canada before the filing of the application for the Canadian patent and consequently all these Canadian papers are not worth the paper on which they are written.

Contrary to common belief, invention is a simple process of the mind and it may be interesting for me to give some examples:

It was found that in casting metal rotary motion would result in throwing out the dross. This was due to an understood law of nature, an operation of centrifugal force, the advantage being that, for instance, rollers with more perfect surfaces could be cast. This was known and it was the practice to stir or otherwise cause the metal to rotate it being an understood law of nature that rotary motion would so result, as an operation of centrifugal force. The point I wish to convey is that those skilled in the art of casting metal knew that by imparting a rotary motion to the molten metal, the dross would be caused to separate; and the problem before all molders was to produce the motion more conveniently and uniformly than by stirring the liquid metal, as soup would be stirred by a spoon in a pot. The commonplace workers in the art employed any kind of implement suitable for the purpose and overlooked the inconvenience and want of uniformity, but a man named James Harley, a workman in a foundry in Pittsburgh, conceived the idea in 1834 that he could make the metal revolve by injecting it into a mould tangentially. He obtained a patent for it and the patent was upheld. Another interesting simple invention was made by Asa Whitney, of Philadelphia, in 1848. He discovered that hardness once given to iron will not be destroyed or seriously impaired by the immediate reheating of the iron and its subsequent very slow cooling. What he discovered was simply a law of nature and as such could not be patented, but he went further and worked out a process consisting in taking the wheels from the moulds very soon after their rims were chilled and in putting them immediately into a chamber which had previously been heated about as hot as the heat of the wheels when put into the chamber and thereupon raising the temperature of all parts of the interior of the chamber or furnace and its contents to an equally high point, and finally in causing all parts of the wheel to cool with equal slowness. Mr. Whitney obtained a patent for his process and his patent was upheld. In each of these cases the invention consisted in simply applying a well known law of nature to a certain

art and the simplicity resides in the fact that once the obstacle is seen and the problem is lined up there is usually very little difficulty for the trained mind to conceive an idea by which the obstacle may be overcome and a step forward made in the art.

Another Example

Let us take as an example an invention which has been produced in Montreal and assume that the inventor, contrary to the usual practice of inventors in the Province of Quebec, wishes to obtain a reward for his invention. He makes his application for a patent within one year from first public use, and if he has not already commenced to use his invention he introduces it to the trade. Of course, if he has a backer he may be relieved of the necessity of introducing the invention to the trade by his personal effort, but the case I am taking as an example is one where the inventor has no backing. I have known cases where the inventor has succeeded by first having his invention manufactured, then obtaining a price from the manufacturer, after which he approaches the jobber, the retail dealer or the consumer, turns over any order he obtains to the manufacturer to be filled at a price stipulated in a contract, his sale price being, of course, in advance of this. A demand may be created this way, having for its purpose to induce the manufacturer, jobber, or retailer to become interested. There are many other ways of course to realise on inventions.

There is a great advantage to employers from patents upon inventions developed by their employees. This is so whether the patents are held by the employer or employee, and the advantage is due to the mere fact of the patent being obtained which usually proves a bar to a competitor obtaining a patent by which he may hold up his employer.

With regard to contracts many of our prominent concerns make contracts purporting to be based upon patents of invention but when investigated and analysed prove to be a guarantee of profit to some outsider which could not be exacted except for the contract itself. Business troubles of this nature can usually be foreseen and avoided by a careful investigation of the antecedents of the patents involved before the contract is signed.

In conclusion I would say that the individual granted a monopoly on an invention by the Patent Laws of Canada is afforded a protection which is not equalled by the patent laws of any other country, and he is encouraged to explain his invention to those engaged in the same kind of work. His patent of invention is a property of which he cannot be despoiled by any Act of Parliament either Federal or Provincial. His right is not excelled in point of dignity by any other property right, and equals the right of authors in their copyrighted works. Furthermore the benefits he confers are greater than those he receives, for he confers upon his community a means of either lessening toil or of increasing comfort. It is also an indestructible heritage of posterity. In return he receives a limited and conditional monopoly from the Dominion of Canada the grant of which costs neither the Government nor the people anything. And his reward shall be all the more sure because it is for creating things unknown before, the creation of which is a step forward in the commercial or scientific life of this country.

HE WANTED A LITTLE TIME
TO GATHER MONEY AND GOT IT

IN these days of financing, men are doing things at times because they have to, and not because they want to. Here is a story related to Canadian Machinery to-day as illustrating the effect of tight money on some lines:

"A friend of mine," stated our informant, "had a shipment of parts coming in amounting to about $25,000, and he had secured these on the understanding that it was to be practically a cash deal. The bargain was made some time ago, when it was a great deal easier to finance than it is now. He was in need of the parts to finish up work that he had under way in his factory, but to tell the truth he did not have the $25,000, and there was not one chance in a thousand of the bank helping him to get it, or any portion of the amount. Without the parts the stuff he had in the process of manufacture would be no use at all. So there he was.

He Started Something

"Well, this friend of mine is somewhat of the plunger type, and it is a hard thing to get him stopped entirely. So he wired at once to cancel the order and stop shipment of all goods ordered by him. He knew quite well that the parts made up would be of no value to the firm that had made them for resale, and the chances were they would protest against any such thing as a cancellation, although there was no mention made of such a thing in the sale. The expected happened, and it was not many hours before a wire came back to the effect that the order would be cancelled. My friend was highly pleased to hear this because he wanted the parts. So there started a letter-writing contest that lasted for some days that ran on into weeks. In the meantime he was busy digging up money. His proposition is all right and safe enough, but that does not seem to make much difference just now when it comes to raising cash. At the end of three weeks he had $18,000 in the bank against the shipment, and in order to save further legal trouble the man who was making the parts agreed to ship on payment of $15,000 and give him thirty days for the balance. Well, that was exactly what he wanted—nothing more than a little time, and he got it and made good use of it.

"'I claim,' continued our informant, "that my friend's nerve was a good thing for all concerned. It was a good thing for the man who had the goods ready to ship to him, and who was moved to the point of legal action by his threat of cancellation. It was a good thing for the men in those two plants that he did. Just what he did. It saved the situation, and I know of a good many men who, if they faced such a problem, would simply be licked. It all goes to show that a certain amount of daring—and shall I say nerve as well?—can get people out. Nine-tenths of all trouble as well as into it."

Now on Canadian Market.—The Magic Leather Treatment Co., Detroit, have opened a warehouse in Windsor, Ont., whence they will cover the requirements of the Canadian market, thus eliminating details of exchange and customs duty. This product will thus be placed on the Canadian market for the first time. It is a purely neatsfoot oil proposition. A high viscosity is secured through a scientific chemical process, so that when the treatment (it is not a dressing) is stuffed into the belt it becomes a good substitute for rubber, making the belt impervious to all shop or climatic conditions. It is alike applicable to leather, canvas, and manila rope and, it is claimed, increases the life of the belt four times. It is especially useful for belts in bakeries, dairies, confectionery and other food manufactures, where odorless, sanitary conditions are necessary. Mr. J. Horrocks will have charge of the Windsor office and is the chief selling agent for Canada.

NEW DISAPPEARING WINDOW SHUTTER

Continued from Page 457

by the bar Q resting on the bevelled portion of the frame.

Fig. 1 is a skeleton view of the window, with shutter lowered and before closing. Fig. 2 is a detail of shutter completely closed and slats K locked by lateral movement of the vertical strips M. Fig. 4 illustrates the mechanism for moving the strips to and from the shutter slats. Two Views of the shutter and the elevating supporting bar are shown in Fig. 5. A partial section of the upper chamber is illustrated in Fig. 6, and a general cross section of one side is shown in Fig. 7. Negotiations are now progressing for the formation of a company to manufacture this appliance.

New Design of Disappearing Window Shutter

By J. H. RODGERS

THE accompanying drawings illustrate the general construction features of a new design of window shutter intended for use on office buildings, factories and private residences. This device has been invented by Charles A. Bull, of Montreal. It is claimed that the shutter can be constructed to conform to existing windows, or as is primarily intended, may be incorporated into a window casing, thus forming part of

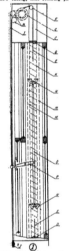

the whole. When not in use the shutter is elevated by means of cables and located in a chamber attached to the upper framework of the window. When in a closed position the shutters are held firmly closed by the pressure of the side strips bearing against either side and at each end of the shutter.

In the design shown the construction calls for the placing of the shutter within the space between the movable windows and the storm sash. The design may be modified to suit any particular style of casing. Provision has likewise been made for adequate fire protection in the automatic closing of the shutter, consequently eliminating the creation of draft for feeding the flames.

In the sketch shown A represents the sheet steel frame of the window. B is the upper portion in which is the chamber C for retaining the shutters when same are not in use. The lower or elevating bar D is supported by a cable E —one on either side. This cable passes over the pulley F, and is wound upon the spool G. On the same shaft that carries the winding spool is a small drive pulley around which passes the control cable or link chain I, this cable passing over the guide pulleys H and thence in a downward direction in front of the window frame to the pulley J. This pulley is fitted to an adjustable bracket for regulating the tension of the operating cable. The shutter slats K are suspended an equal distance apart and the end trunnions are slightly off-centre so that they automatically tilt into an inclined position when lowered from the upper retaining chamber.

When completely lowered the piece D rests upon the sill and the vertical locking strips M are pressed against either side by the movement of the link mechanism N, operated by the connecting rod O, which is moved in a vertical direction by means of the lever P, the outer end of this lever protruding through the casing of the window frame and within easy reach of a person in the room.

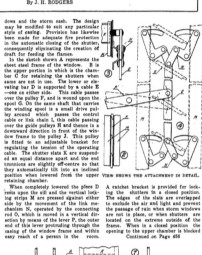

VIEW SHOWS THE ATTACHMENT IN DETAIL.

A ratchet bracket is provided for locking the shutters in a closed position. The edges of the slats are overlapped to exclude the air and light and prevent the passage of rain when storm windows are not in place, or when shutters are located on the extreme outside of the frame. When in a closed position the opening to the upper chamber is blocked

Continued on Page 456

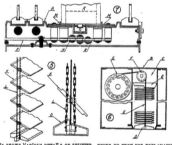

THIS SHOWS VARIOUS DETAILS OF SHUTTER. REFER TO TEXT FOR EXPLANATION.

WELDING
AND CUTTING

Lessons of the A.B.C. of Good Welding

This Material Deals With the Ideals of the Welding School

By W. B. PERDUE

SEVERAL years ago the prospective purchaser of a welding torch hung out a shingle with the stereotyped sign "Welding and Brazing" upon it. Many times the shingle was displayed before the torch arrived, so that several jobs might be waiting long before the new welder figured out just where to connect the hose and just how he could connect up all the tips which came with the outfit at one and the same time.

To him the torch and its seemingly intricate regulators were a thing of mystery—he had never seen one before—and you can wager your last nickel that when he once succeeded in getting the regulator pressures adjusted so as to get some definite action that he never moved them. There they stayed until the terrific impact of the full cylinder pressure pounding down on the seat each time the tank was opened rendered the seats useless. Then, of course, the company who sold them came in for a "good hauling over the coals."

Recently, however, much of the injury to the good name of the welding industry has been retrieved by progressive and up-to-the-minute welding concerns who have brought the merits of welding before the public so forcibly that the trade schools throughout the country are devoting more interest to this trade than to many others.

Scientific Interest Justified

When he catalogues the various subjects in which the welder must be versed, or of which at least a working knowledge is required, the college professor or the dean of the engineering school is well justified in giving the art of welding a place in the sunlight. It is gratifying to see this awakening of scientific interest and the great amount of research work

AUTHOR'S NOTE.—The increasing number of inquiries from high schools and trade schools which teach welding, as well as large concerns who are unable to secure competent help and must therefore train their own men, necessitates a more detailed reply than can be given in an ordinary letter. Hence this article. The regular routine of instruction will be resumed next issue. The principles given are not all original with the author, but are such as have been found most successful in welding schools throughout the entire country.

that is at present being conducted, and more especially the demand by many of the large users of welding for engineers with sufficient knowledge of the process to determine by visual inspection of a weld whether or not it is sound and has been executed by a skilled operator.

A thorough study of the requirements necessary to secure complete and proper penetration coupled with the ability to properly execute the work oneself will enable the inspector to pass upon welding more rapidly than upon the work of any other skilled mechanic. The fact that the author in addition to his other duties inspects more than one thousand pressure containers per month, and in the event of defective work not only replaces the container but bears all expenses that have been entailed by the purchaser, should convince the skeptic that this work can be done with great rapidity and that the percentage of defects that can escape the trained eye is almost negligible.

The decline of the apprenticeship system of recent years and the demand for skilled workers during the war led many inexperienced laborers to represent themselves as skilled mechanics and to secure employment as machinists, carpenters, toolmakers and what not. The scarcity of labor at that period enabled many of these to hold positions for a few days and thereby to secure some valuable schooling at the hands of the foreman who was compelled to "turn orders out." In many machine shops there were mechanics detailed to set and grind tools for "operatives" who knew nothing more than "stop the machine" if something goes wrong."

After drifting from one shop to another, and devoting his leisure to study in the public library it often happened that the "operative" learned enough to enable him to hold down a permanent job. The field of the modern trade school is to save the employer, the hundreds of dollars' worth of scrapped material resulting from the work of inexperienced operatives, or in other words, to form a link between capital and labor.

Welders Require Skillful Training

To the uninitiated the art of welding

appears simple. The beginner or the helper who has watched his welder for many months is very prone to remark that he has noticed how clumsy others have been in their work and that he is confident that he can master the entire process in a very few days. Others quite as firmly believe that they can master the art by home study and self instruction. We do not wish to go on record as making the statement that such procedure is impossible, but we do state that the cost of any materials required to master the art under such conditions would by far exceed the cost of tuition and expenses in a school of recognized merit where the help of one or more expert instructors is at all times available.

The average student is inclined to be over-anxious to get to the point where torch manipulation and practice is no longer necessary. The man who cannot execute a bronze weld with a ripple of absolute regularity and cannot continue to move his torch hand without interruption while his left carries the welding rod after more flux needs more practice. Welding looks simple, but appearances often deceive. It can be learned in a short time by the average man who is readily discouraged. After the third or fourth unsuccessful attempt to execute a neat ripple he is very likely to give up.

This is fortunate rather than unfortunate. Men who are so readily put aside would not become an ornament to the profession. They would not have the patience to carefully study what others are doing or attempting to do, and to keep themselves abreast the times through the medium of their trade journals. On this particular subject a well known authority on training welders states: "It is essential that he (the student) should familiarize himself with the physical and chemical laws embraced in the welding and cutting business." Such knowledge will instil great respect for the powers that he controls and afford the necessary resourcefulness to overcome difficulties that may be encountered. The more knowledge he has at his command the greater will be his respect for his trade and his desire to assist in its upbuilding.

An ignorant workman becomes a mere machine, able to turn out sound welds only under such conditions as have been taught him by others. The calloused villain of the movie screen who is made to

exclaim, "—— them, they are nothing more to me than machines," may have been nearer to the truth than we like to imagine.' Yet we have the consolation that it depends entirely upon the mechanic himself whether he does the work and receives the pay of a "machine" or uses his thinking powers to lift himself into the class of the thinkers and doers.

The more highly trained and intelligent workman may be no better welder on common every-day routine and repetition work, but his broader knowledge of the principles on which his trade is founded will make him more resourceful and consequently better able to cope with new problems and to handle high-class work that may be a little out of the ordinary. Hence, the welding school must not stop with the manual training of the welder and such lectures and lessons as will enable him to gain a general notion of the physical and chemical laws which govern his work, but there should be a decided effort to train each individual student to think for himself.

Honesty Above All Things Else

The first thought to a student of any trade school should be the necessity of absolute honesty both with himself and his employer. He is old enough to realize that any neglect to properly assimilate the lessons given him for home study will but delay his progress and that it is up to himself to devote enough time to the home study work outlined, to enable him to keep ahead of schedule, and to be prepared at each session for the work to be done that particular day.

He should be taught principles of welding so thoroughly that he will be able to explain and illustrate to the instructor just what is required before attempting any new work. With a thorough understanding of what he desires to accomplish he is able to draw a mental picture of the work to be accomplished and from this picture to develop the ideal with practice. He must learn to test his own welds, discover defects in his own and the work of others by visual inspection; in other words, he must eventually become his own critic and get into that state of mind that will not permit him to turn out anything that may be unsound or of slovenly appearance.

Short Class Periods Essential

As previously stated, the periods of instruction should be brief. This not only materially lessens the cost of gases used (the one big item of expense in training welders) but gives the student a change from physical exercise with the torch to mental exercise devoted to home study, and (at such periods as the student can command) to muscular training with a dummy torch under a light in his own home. (This dummy can be constructed from an old broom handle or other short stick with a nail driven through near the end to resemble the angle of the tip. The shadow of this point is made to describe the motion to be imparted to the tip of the torch.)

It has been found by common experience that any new art can be more quickly imparted to the adult mind by means of a short lesson given daily than by means of continuous instruction. Some of the best operatives of the Standard Oil Company (records of Richmond, California, plant) are those who were trained in the company's own shop by attendance at their war-time night school three evenings per week. The same also holds good in training adults to operate motor vehicles. For example, the owner of a new automobile can be taught driving and considerable of the care of the machine in ten daily periods of one hour each. To try to impart the same instruction in one ten-hour session would be rank folly and no auto school in the country would make the attempt.

The operation of any mechanical device requires not brain work as is commonly assumed, but the training of the subconscious mind which is directly connected with the spinal column. Short lessons repeated at frequent intervals have been found to develop the subconscious mind most rapidly—in fact, more rapidly than when the lesson is drawn out to the point where fatigue causes the student to become nervous and to lose poise and self-control.

No operative is competent to operate a torch or other device requiring constant attention until his subconscious mind is so developed as to relieve the brain of all detail. If the driver of an auto must brain-think which way to turn the steering wheel when he turns a corner he is liable to make a mistake and come to grief. The untrained driver who fears a collision with a tree a hundred feet in front of his car will invariably effect such a collision. But when his subconscious instinct has been trained to operate the car any necessary movement of the wheel or gears is made before the brain has time to think.

In the training of welders it is necessary and essential to encourage home practice to cause the various details of operation to be passed on to the subconscious mind as rapidly as possible. "Whistle a waltz and move the torch back and forth across the weld." Whistling relieves the mind of worry and enables the instinct to grasp the details of operation more quickly. If the operator must brain-think to move the torch he cannot give proper attention to other details essential to the making of a sound weld. In fact, one can concentrate the brain on but one idea at any one time—the subconscious mind can carry out a dozen details simultaneously. Therefore, the welder must have the training that leaves him entirely free to watch the welding, and that training is possible only when he has completed a wise course of instruction that has corrected his faults, imparted sound principles, given him a correct perspective of his work and taught him to think for himself.

PRODUCTION METHODS FOR WORKING LEAD

When casting ingots of lead or other metals of a similar nature, where rapid handling is necessary, the following method is advised. In the molten metal a centrifugal pump is situated. The method of driving is by a vertical shaft, which in turn is driven by a set of bevel gears driven from the horizontal shaft. From the pump a pipe leads out to a revolving fixture. The control of the lead is taken care of by the valve, which necessarily has to be kept hot by a flame from a burner so that the metal will not chill and stick the valve. The valve must be constructed of cast iron and the pump must have iron bearings, the shaft through which must be a "sloppy" fit.

Another instance is where one desires to cast small parts in quantities and the weight of them must conform to certain limits. This can be accomplished by the use of a quantity valve which I have shown in the sketch? This valve sits immersed in the molten metal and must also be constructed of a non-adhering material—cast iron. The stem of the valve has a chamber in it that holds just the proper amount of metal. It must be a taper-seated valve of course, and held in place by lock nuts at bottom. These can be of steel. The valve, as shown in sketch, is in the discharging position, having been turned 180 degrees from the loading position. In order that the valve will properly discharge, it is necessary that no vacuum be caused to retard the discharge of the metal. A vent pipe A is necessary to accomplish this, but admittance of air will cause the formation of oxidized refuse in the chamber, which will cause a variation in weight of the mold.

If the intake of this vent pipe is situated where it will draw in the flame from the burner beneath the tank of metal, no oxidization will take place and the weight will be uniform. The discharge B leads to a fixture of a revolving type where the molds are situated.

AIR TRIP FOR POWER PRESSES

A pneumatic device for the tripping of power presses and other machinery where difficulty is often experienced in reaching the starting treadle when handling the work has been brought out by the Lovejoy Tool Works of Springfield, Vt. The appliance consists of a cylinder and valve, and is secured to the frame of the press in such a position that the connection at the top of the piston may be bolted to the trip lever of the press. When the air is turned on the piston is raised and the trip mechanism is released. The piston and valve are brought back to the idle position by means of two springs. The valve may be operated from any position about the machine by the use of a cord that is held in the hands of the operator.

DEVELOPMENTS IN SHOP EQUIPMENT

RING TABLE

READERS will no doubt remember the article we had in our October 7th issue dealing with a Newton continuous miller. This present material deals with another type of miller produced by this firm and which has certain features well worth going into.

Figs. 1 and 2 illustrate this machine. The view shown at Fig. 1 depicts the roughing spindles, while Fig. 2 illustrates the finishing spindle. It will be observed that the same fundamentals of construction referred to in previous machine form a part of this one.

The base is circular in form, providing a central taper column. The table casting is fitted to the central taper column of the base, and in addition is provided with an annular bearing close to the periphery of the table. The cross rail and the central upright are made in one piece so as to reduce even on this large machine the number of bolted connections. The table is 84 inches in diameter and the depth from the annular bearing to the top of the table is 12 inches. The least diameter of the taper fit between base column and table is 36 inches. Table is provided with a finished hub 42 inches in diameter to assist in locating jigs on the table. The central column is bolted and keyed to the base and the cross rail is fitted in the front with one housing containing two spindles for the roughing cut. On the back of the cross rail and at a distance of 42 inches is a similar housing carrying a single spindle for the finishing cut.

Roughing and Finishing

You will observe that both the bousings for the roughing and finishing spindles can be positioned on the cross rail so that where machine is used for a variety of work the spindles can be positioned to the most economical location of the jigs upon the table so that the jigs can be made with the least angularity and the least loss of space between them. The outer end of the cross rail is supported by a column which is bolted and dowelled to the base as well as the cross rail.

Drive is on the top of the machine, motor gearing to a jackshaft, and at the extreme outer end of the cross rail will be noticed a box which is used to transmit the motion from the jackshaft to the roughing and finishing spindles. This permits of a provision for varying the rotative speed of the spindles independently of each other, and whilst the speed is predetermined and fixed, this provision permits of changing the speeds when the grade of material or size of cutters is changed.

The rotative movement of the table is controlled by a fixed feed which is pre-determined, but provision is supplied by which this rate of rotation can be changed to suit any change in the grade of material. There is not, however, any possibility of the operator increasing or decreasing the production of the machine as the rotative speed has been pre-determined, which means that a given number of stations per hour must pass the loading station, hence that number of pieces must be machined. The table itself is rotated by a herringbone gear 81 inches in diameter. Each of the spindles is provided with an individual adjustment for setting the cutters to gauges.

Different sizes of housings providing varying centres between the roughing spindles are used, depending upon the dimensions of the work. Generally, however, these centres are either 12 or 14 inches. Both spindles are rotated inward or clockwise on the left-hand spindle and counter clockwise on the right-hand spindle. With the distance of 42 inches from the centre of the roughing cutters to the centre of the finishing cutters, it

FIG. 1.   FIG. 2.

FIG. 3.    FIG. 4.

is quite clear that the roughing operation has been performed on a given casting before the finishing operation commences, hence the finishing cutter is
relieved of any influence on the part of
the roughing operation and due to the
slight cut taken by the finishing cutter,
any inaccuracy resulting from either
dull roughing cutters or in equality of
castings is picked up by the finishing
cutter, insuring accuracy of both finish
and dimensions. Correspondingly, by
the division of the work, the number of
grindings of cutters is reduced. Many
instances are found wherein a high degree of finish is not required, but by
taking the finishing and the correspondingly faster feeds, increased production
is obtained.

The principle of roughing and finishing cuts permits of operating at much
higher cutting speeds and table feeds
than is practical by any other process of
surfacing operations. All bearings, except the spindle bearings, are oiled by
cascade method of lubrication, which is
pumped from a reservoir in the outer upright to the box on the top of the machine from which point it is distributed.

For illustration purpose we show various photographs of work performed.
Fig. 3 depicts the roughing operation
on the base of a cylinder block, the dimensions of which are shown at A in
Fig. 4. From the roughing operation depicted the finishing operation can be
readily understood. On this work a feed
of 20 inches per minute on the maximum
radius of the work was used.

Fig. 5 depicts the method of jigging
casting shown at B, Fig. 4. It will be
noted that this jig is of the open front
type, this permitting quick loading and
unloading from a roller conveyor. The
time for jigging the piece shown at
Fig. 5 is less than one minute. These
castings have been machined in two
minutes, but the usual practice runs from
2½ to 3 minutes each, depending upon
the requirements and supply.

Some of the features claimed for these

machines are ruggedness in all operating parts, and an excessive amount of
weight in view of the fact that the parts
themselves are quite incapable of absorbing any of the vibration set up by
all the cutting actions, hence the machine is designed to absorb such vibration. The character of work usually
presented to these machines does not require a great range of adjustment, hence

with the standard model this is taken
up, first, by the adjustment in the spindle
heads, and second by variation in the
height of the jigs themselves.

All gears are enclosed and run in oil
and all essential gears are hardened.
Provision is made for stopping the table
for setting the cutters by a conveniently
placed lever. The bearings are all sealed
to prevent the escape of lubricating oil.

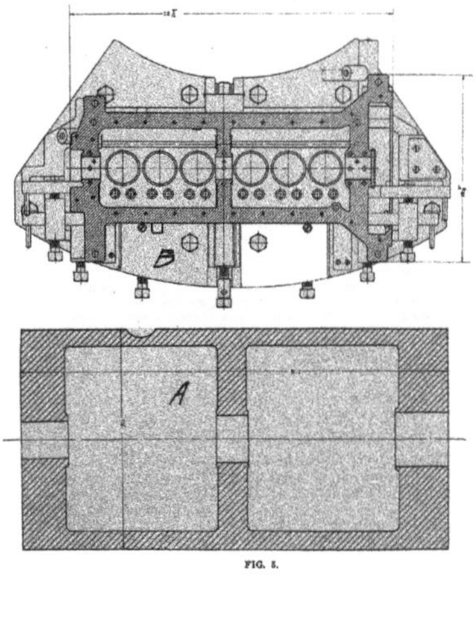

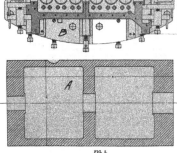

FIG. 5.

The MacLean Publishing Company
LIMITED
(ESTABLISHED 1887)
JOHN BAYNE MACLEAN, President.   H. T. HUNTER, Vice-President
H. V. TYRRELL, General Manager.
PUBLISHERS OF

CANADIAN MACHINERY
and MANUFACTURING NEWS ✐

A weekly journal devoted to the machinery and manufacturing interests.
B. G. NEWTON, Manager.   A. R. KENNEDY, Managing Editor.
Associate Editors:
F. H. MOORE   T. H. FENNER   J. H. RODGERS (Montreal)
Office of Publication: 143-153 University Avenue, Toronto, Ontario.

VOL. XXIV.   TORONTO, NOVEMBER 18, 1920   No. 21

PRINCIPAL CONTENTS

The Bogey of Cheap Tools

CARE should be taken lest the talk of cheaper machine tools goes farther down the actual conditions of the case will permit developments in this direction to measure up to what the buyers expect.

There are greater considerations than the bringing down of the prices of machine tools.

There have been cases where buyers have purchased cheap machine tools. They have gone on the assumption that a lathe is a lathe and a miller is a miller. At the start they saved money on the purchase, but that was the only place where there was any saving.

A buyer in the average market gets what he pays for. If he clamors for a cheap outfit, he can get his wants attended to. And when he has got this outfit housed in his premises he has exactly what he paid for—a cheap outfit.

There are greater considerations now than that there should be a slash in the price of machine tools, and the chiefest of these is that the last word in quality and stamina shall be built into the equipment that is going to be used in the industrial areas of Canada.

It would be much better for the buyer of machine tools to pay the price that will allow the maker of them to put in them the best work, brains, steel and iron that can be secured. And having paid that price he will be within his rights in insisting that when he buys he shall get the last word in shop equipment.

There is room for improvement in the quality of the work turned out in many Canadian plants. Men who have been abroad and seen Canadian products in competition with those of other lands have recognized that in some cases there was need for greater exactitude in machine shop practice.

There are makers of machine tools in Canada who have never received enough for their product to allow them to go ahead and develop along proper lines. There has been, in some cases, not enough margin on the sale to permit of experimental and engineering expenditures.

It will pay handsomely to invest a sum for mechanical equipment that will include possibilities for betterment and expansion to the maker of those tools.

Prices cut to the bone mean production at a figure rather than at a standard. A machine tool is not like a pound of prunes or a bag of sugar. Its sale does not and should not depend on the shading of the price to the last fraction.

Don't clamor for cheap tools. Much rather pay the price and insist on the quality being as high as the price.

Adequate Boiler Inspection

IN an article published in the Toronto Star Weekly the Hon. W. R. Rollo has something to say on the subject of making labor safe. In connection with this, he states, there is a demand for the inspection of every boiler in the province, in addition to the insurance inspection. Mr. Rollo intimates that this demand among others will probably be met.

While we are entirely of the opinion that every boiler should be inspected, we can see no reason why a boiler owner who wishes to carry insurance should be penalized by two inspections. If he carries no insurance, then the Government should undertake the inspection in the interest of the public and at the owner's expense. Where the owner carries insurance, an inspection is made by the insurance company's inspector, who is generally, at least, as capable as the Government inspector. Getting a boiler ready for inspection is a matter entailing time and expense, and whereas an insurance company usually make their inspection fit the owner's convenience, this will hardly enter the Government official's mind. In many cases the insurance company make their inspection at a time when the factory is closed down, and the boiler opened up for cleaning in the ordinary way.

The Province of Quebec makes every insurance inspector carry a Government certificate of competency as a boiler inspector, and holds regular examinations for candidates for these positions. Having given this certificate, the inspector has all the status of a Government official as far as his reports and inspections go. Most large and many small concerns carry insurance for the protection it gives them. To compel them to submit to two inspections would be inflicting an unnecessary hardship.

The Business Outlook

THE average man in any line of business is always on the lookout for indications of how trade is moving. This is true now in the machine tool trade in Canada and United States.

Here is a report from a large American firm that is worth noting:—

"With the election over, the general tone of business in the last week or ten days has been better than at any time in the last three months. Collections, of course, are still very bad, but incoming inquiries and actual orders from all over the country indicate that the American people have every confidence in the next administration."

And in the Canadian field? Here's the opinion of one manager who handles a lot of business in the machine tool and equipment field:—

"We are doing less business now than last month, but we are not complaining. I believe we are right in the midst of the period of price adjustment, about which we have heard so much since the end of the war. When I see a plant laying off men I might feel very much discouraged were it not for the fact that at these same premises they are building large extensions for next year's trade."

Just Taking a Chance On That Extra Trip

THE month of November is not very far advanced before we hear of the loss of a Great Lakes freighter. Fortunately, in this first case, it was unaccompanied by loss of life, although the crew went through great hardships before they were rescued. In the same storm several vessels had narrow escapes, and one barge foundered, the crew being rescued.

As the end of the season of lake navigation draws near every voyage undertaken is a gamble with the god of storms. The lake freighter, evolved after many years' experience as the ideal vessel for the transportation of bulk cargo, has been so evolved without any thought being given to her seagoing qualities.

Frankly, she is expected to keep out of the way of bad weather, being utterly unfitted to contend with it. Yet every season a certain number of these vessels are lost, too often with all hands, or a large portion of the crew.

How does this happen? Is it that the captains of these vessels, knowing fairly well what weather to expect, and well aware of the capabilities of their craft, are so imbued with zeal for the owners that they start voyages in which every element is against them, in the hope of winning through?

Or is it the owner that drives them out, anxious to get the last cent that can be made from his property?

In some cases the captain may be the owner, or part owner of the vessel, but in that case he would not be inclined to take risks. The Great Lakes are well equipped with weather observation stations, and meteorological science has reached a point where the observed conditions, compared with similar conditions over a number of years, show the probabilities for at least twenty-four hours ahead with reasonable accuracy. Harbors are fairly frequent, and, for the most part, well buoyed and well lighted. Where then is the weakness? The weakness, undoubtedly, lies in the vessels themselves.

Pitted against the terrific storms that blow across the inland seas the structure collapses. The heart of the fabric, the power plant, is not capable of holding the body up to face the seas in the most efficient manner. The mass of steel, hundreds of feet in length, becomes, instead of a ship, a mere hollow girder floating on the waves. It is either driven ashore, where it is held as in a vice while it is battered to pieces by nature's hammer, the waves, or before ever it reaches shore the hammer has done its work. Whichever way it happens, the job is very completely finished, and, too often, a toll of human life is taken.

Is there a remedy? The only one that would be of any use is one which no government would dare to take. That would be to set an arbitrary date for the close of navigation, and is out of the range of practical ideas. So long as there are men willing to take the risk, so long will we open our morning papers to read of another wreck on the Great Lakes.

The Star Weekly, in its financial section says:—"According to Iron Age, the present average price of eight leading iron and steel products is $7.75 a ton, whereas a week ago the average was $80.10." Now, then, who is going to rise to say that steel prices have not been cut, slashed, pruned, trimmed, skimmed, thinned, decreased, flattened out, deflated, reduced, battered down, and all else?

Pay Promptly If You Can

FINANCING is not an easy matter just now. It is a strange situation for a firm to find that it can secure plenty of business, but cannot finance them. There are in the steel and metal industries of Canada where that is the state to-day.

Collections are slower also. Firms that do not need to do so, in many cases, are taking advantage of every means of long distance financing known to them. They issue notes and keep postponing payment as long as possible.

The serious thing is that when any number of firms start this sort of business it is not long before we find the whole business community financing on long terms instead of prompt acceptances.

It is one sure and certain way to start a vicious circle that is extremely difficult to break or wind up at a desirable time or point.

It is necessary that every leniency should be extended to worthy institutions to-day, and it is not desirable that any of them be put to the wall.

On the other hand, it is equally necessary that, under the guise of this treatment, no financially able firm shall grab the opportunity of joining the list of paper and note financiers when they are able to pay their way.

What the Business Papers Say

GET STARTED ON THE RIGHT ROAD

Mechanical World, Manchester.—We know full well the difficulties manufacturers are laboring under, and there is no quick road to the deflation of prices. At the same time a start must be made, and it is not putting it too high to say that manufacturers as a body can more easily afford to bear some initial losses due to the effect of a fall in prices on existing stocks. Behind them, of course, often come the importers, who stand to be hit most. Good profits, however, have been made in the past, and if a long period of serious unemployment is to be avoided, heads had better be put together to find a way out.

THE INFLUENCE OF WAR ON MACHINERY

Iron Trade Review.—There can be no question that designers of machinery are profiting from the experiences of the war. Many specially designed tools created several years ago under stress are now being redesigned and improved. The most noticeable feature of this remodeling process is the tendency to adapt old types to semi-automatic operation. This is accomplished on boring mills, milling machines and similar tools by providing rotating tables with from four to eight chucks, so that the work may pass from one set of cutters to another for the various operations. For duplicate work, these machines greatly increase the output per man and in some instances it is possible for one operator to handle more than one machine.

OBSOLETE TOOLS ARE A NUISANCE

Machinery (New York).—Light cuts and slow speeds greatly decrease production, and the inaccurate work from many obsolete tools increases the time required for assembling machined parts. These old machines are also a source of discouragement to the men who must use them—a factor not to be overlooked. It is well known that locomotives are kept in repair shops longer than is necessary owing to the use of obsolete tools, and a careful study of conditions in representative repair shops shows that the loss from this source is even greater than is generally realized.

HIGHER WAGES NOW ARE DANGEROUS

Canadian Engineer.—Contractors and supply dealers are of the opinion that a reduction in wages, rather than an increase, can be expected. They say that living costs are slowly but surely reaching a lower level, and that propaganda for wage increases in the face of existing conditions in the construction industries is having a serious effect on general business. A vast amount of construction throughout Canada has been postponed in the expectation that prices of materials and labur would not remain at war-time levels very much longer, and propaganda for further increases is causing the total abandonment of many of these projects.

Motors and Their Correct Application

The Characteristics of Motors in Terms of Their Speed and Torque in Relation to Different Kinds of Loads is Discussed by the Author

PAUL AND LINCOLN

AN interesting paper bearing on the above title was read before the Toronto Electric Club by Mr. Paul Lincoln of the Lincoln Electric Co., recently.

Mr. Lincoln prefaced his remarks by some reference to the historical side of power development, and quoted largely from a book called "The New Epoch", by George S. Morison. Mr. Lincoln especially drew attention to a quotation occurring in that book, as follows: "Wherever needed, we can now produce practically unlimited power." Mr. Lincoln used this quotation as a text for his talk, which here follows:

The particular point that I wish to make is that the significant thing that is bringing about Morison's "New Epoch" is not entirely the "manufacture" of power, but largely the ability to distribute this power to "wherever needed," and that latter ability is the distinct contribution of the electrical profession. It is the problem of distribution that has been solved most completely by the alternating current polyphase system. The ease, efficiency and safety with which alternating currents can be changed from one voltage to another by means of the static transformer has given that system so great an advantage that it has no competitors, in so far as the distribution problems are concerned. In point of adaptability to driving the great multiplicity of loads that motors are applied to, the alternating current motor lacks some of the flexibility of the older direct current motor.

Before the advent of the alternating current polyphase motor the direct current had a considerable vogue, and in some applications its greatest flexibility has caused its retention in service in spite of the increasing prevalence of the alternating current polyphase distributing systems.

The contest that has been waged between A. C. and D. C. during the last 25 years has gone in favor of A. C. not because the A. C. motor is more rugged, but solely due to the superiority of the A. C. for distribution. It is the ability to put power "wherever needed," in the words of Morison, that has won the battle for the alternating current. The problem of applying motors to the diversified work that they have to do is in general more easily solved by the D. C. motor than the A. C. However, by careful analysis of the work the motor has to do as well as a complete knowledge of the characteristics of the A. C. polyphase motor; the A. C. motor has been successfully applied to every

class of work that motors are called on to do.

The ease of distribution of A. C. polyphase currents is bringing about a complete revolution in the method of operating our modern factories. Time was when the water wheel or the steam engine was a source of power, that it was customary to equip a factory with a single prime mover, and then distribute this power throughout the plant by means of shafts and belts. During this period, shafting and belting took so firm a hold that it is exceedingly difficult now to think of a factory installation without thinking of a room full of shafting and belts.

The "wherever needed" is not a single point in the factory, but the individual machines that are operated therein. To carry Morison's dream to its logical conclusion, we must never equip, not a factory, but the individual machines in that factory. The electric motor is successfully taking the place of the steam engine in factory drive, but its job is only half done when we substitute a motor for the former engine and leave the belts and shafting. The job is completely done only when the idea of motor drive is carried to the ultimate of the

"wherever needed" that Mr. Morison emphasizes.

Now the problem of applying motors to the ultimate machine wherein power is used involves first an analysis of the power characteristics of these machines, and second a complete knowledge of the nature of the motor to be applied and of the modifications of design that are possible thereon. The most important characteristic of a load, at least, when considered in connection with a driving motor, is its speed torque character. The most usual demand in this respect is that the speed shall be independent of the torque or practically so. When factories were driven by a single prime mover, either water or steam, this constant speed characteristic was the one demanded and when electric motors began to supplant other prime movers, it was natural that this constant speed characteristic should be the one most in demand. Both the A. C. and the D. C. motor were able to meet this requirement easily.

A second requirement often demanded in a load is the ability to adjust the motor speed to a new value and when so adjusted, still to have its speed independent of its load. This is a require-

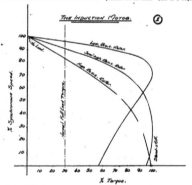

The Induction Motor. ①

% Synchronous Speed

% Torque.

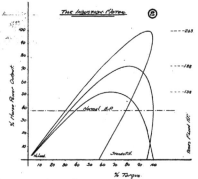

LINCOLN ELECTRIC CO.

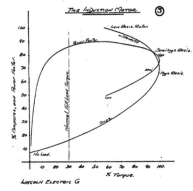

LINCOLN ELECTRIC G

ment that can be met by the D. C. motor with relative ease, but is impossible of fulfillment by the A. C. motor. It is this requirement that has done more to keep the D. C. motor in service in competition with the A. C. than any other one requirement. Machine tools is the particular duty where this requirement is most in demand. The absence of this ability to give constant speed characteristics over an adjustable range of speed is the A. C. motor's weakest point. Methods of securing this ability with the A. C. motor undoubtedly will be forthcoming in time; but whether they can be secured at a reasonable cost and, what is most important, without a sacrifice of the ruggedness and reliability of the induction motor, is a still unanswered question.

A third load requirement frequently called for is one where torque forms the important element rather than speed. Such loads are typified by the elevator and the hoist. In applying motors to such loads, constancy of speed with carrying torques, is of no consequence. The type of motor to meet such requirements successfully is quite different from the first two where constancy of speed is essential. The operation of railway cars and trains is another example of load where torque plays a much more important part than speed.

The above three classes of load are only a part of those that are met in practice. Loads may be classified not only in regard to their speed-torque characteristics, but also in regard to their power-time character or the "intermittancy" of the load. Consider for

instance the power-time character of a punch-press or a forging press. During a large part of the time of operating a machine of this kind, the only duty of the motor is simply to turn the machine over idle. At intervals, more or of the work, power is expended on the

work for a short time, after which there is another period of comparative rest. In a punch-press for instance, the time during which power is actually being expended in shearing the metal is exceedingly small, but during that small instant the rate at which power is used may be exceedingly large. In such applications, it is customary to use a fly-wheel which stores sufficient energy to do the active work of shearing the metal and then rely on the motor to restore the energy to the fly-wheel during the comparative long interval between successive strokes. The characteristics of a motor for this work obviously differ materially from that to operate a lathe for instance, where the work is done at a uniform rate. Or take the case of an elevator motor; the rate of doing work here varies greatly between the starting period, when it is necessary to accelerate the car and its load, and the running period, when it is necessary simply to lift the load. Also, the load on the motor varies greatly depending on whether the car with its maximum load is being lifted or lowered. The motor obviously must be of different size and character to care for the maximum conditions that are imposed upon it, and the problem of the proper rating to give the motor that does the work properly is not an easy or a simple one.

As indicated in the foregoing, the problem of applying a motor to its load is first—a thorough understanding of the nature of the work to be done, and second—a thorough understanding of motor characteristics and the modifica-

Continued on Page 64.

MARKET DEVELOPMENTS

A Quiet Season in Tools, Iron and Steel

Trading is in Small Quantity, But There is Generally a Dull Season at This Time of Year—Steel and Iron Prices Stay Up Much Longer Than Figures in the Non-Ferrous Lists.

BUSINESS is quiet in the machine tool, iron and steel market. But in saying that one should take into consideration that at any time this is a quiet season in these lines. Little purchasing is ever done in the steel market during the last half of November and December, and it is hard to say how much this accounts for the present state. Prices are holding firm, though, and if there is any tendency in the U. S. steel market to let prices sag seriously it is not being reflected here. In some lines the demand is good and deliveries not in keeping with the immediate requirements.

It is doubtful if a decrease in prices would bring much new business into the market just now. As a rule, machine tools are not sold because the price is attractive, but because they are wanted to do a certain work, buyers preferring to pay the high price and insist on the quality being equally high. Some makers have announced reductions, but the movement has not become general. It had

been expected that the meeting of tool builders in New York would be followed by something approaching a general announcement regarding policy and prices for the ensuing year, but so far there has been nothing forthcoming apart from the announcement of the president that costs were still very high, and would hardly permit of reductions. Tool builders in Canada are in the same position, and it is very doubtful if they can make reductions that would assist sales.

Buying in the small tool division is small, and the territory is being combed for business more thoroughly than any time since the beginning of the war.

The scrap metal market is becoming known for the business it is not doing, rather than for any volume of trading that is taking place. Prices have been cut to the bone, but the market remains the same, listless and inactive. The big buyers, even at low levels, are not in the market.

CLOSE OF NAVIGATION IS FELT IN BUSINESS AT MONTREAL NOW

Special to CANADIAN MACHINERY.

MONTREAL, Nov. 18.—The near approach of the close of navigation is having its effect on trade in this district. It is expected that for a short period at least the railroads in this locality will be additionally active in hauling freight to the seaport of St. John, N.B., which port will carry on the bulk of the winter trade to and from overseas. Reports seem to indicate that business is gradually developing a quieter trend, particularly in manufacturing lines. Business in machine tools has shown a slight falling off, especially in orders on new equipment. Supply demand is steady. Some dealers report fair business while others infer that turnover is light. The aggregate business is possibly lighter than for some time, but dealers are not discouraged as they realize the present dullness is quite seasonable, owing to the nearness of the year's close.

Lower Levels Anticipated

There is a marked tendency in all

quarters to liquidate supplies on hand before further placement of orders. When buying is necessary a conservative policy seems to rule, and in consequence the market in every direction has taken on a dullness that promises to continue for the remainder of the year, with possibly a slight fluctuation both in price and volume of business. The consumption of materials for industrial enterprise is gradually lessening, and with mills and furnaces endeavoring to maintain operations, rather than reduce their staffs, the available supply of many commodities is in the ascendency, and with very few exceptions the cry of "shortage of materials" is seldom heard. Premium business is seldom reported now. Dealers are not anxious to acquire large warehouse stocks, but are trying to keep sufficient on hand to meet their customers' needs as they arise. While the factor of price is, no doubt, an influence in keeping the consumer from buying ahead, the chief difficulty is the

uncertainty that features the industrial situation at the present time. Should weakness develop to the point of heavy price cutting the element of labor would necessarily have to be taken into serious consideration, as existing conditions in this field prohibit profitable operations unless relative reductions are effected in the cost of production. Steel quotations are generally firm but the factor of price is of minor importance in face of the poor demand. Price revision on the market basis doubtless is pending, but dealers are reluctant to take definite steps in this direction until the turn of the year. It is possible that actual orders would bring out better figures than those quoted, but with trade on the ebb, dealers are satisfied to let the nominal prices stand.

Additional Weakness Develops

Absence of activity in the old material market has added to the dullness that has been more or less marked in this field for some time. Despite the declining trend of prices on almost every line of scrap, the lack of interest on the part of consumers keeps the situation in a state of listlessness. Conditions are not conducive to heavy buying of scrap and

users are content to acquire scrap sufficient to meet their present needs, leaving the future to look after itself. There seems little hope that business will show improvement during the next two months, and the outlook for the early period of 1921 is shrouded in uncertainty. Further reductions have been made by dealers here and quotations reflect the inactivity of the general situation. The range of decline in non-ferrous lines is from ¼ cent on heavy lead to 2½ cents on heavy copper. Crucible copper is now quoted at 13¾ cents and light at 10¼ cents per lb. Almost the entire list has been affected. Iron and steel lines are again weaker, the declines ranging from $1 a ton on machine shop turnings to $5 per ton on boiler plate and wrought iron axles. Malleable scrap at $23 is quite firm. Volume of scrap movement is exceptionally light.

BUSINESS QUIET IN SOME LINES

But Even In Normal Times There Is Generally a Lull at This Season of the Year

TORONTO.—The market is quiet now in several departments. It is very easy to go wrong in estimating how much of this is seasonable and how much is not. Many lines of business became accustomed to going through war years with no off-seasons, so that they have come to expect the same thing is going

to hold good in the markets for all time to come. There are lines, several of them in fact, in the iron, steel and machine tool fields that have always been, more or less flat at this time of the year, and this season, being about the first when trade is on the way back to more normal conditions, is apt to be looked upon as exceptional because it does not measure up to war years or to the unexpected conditions that followed the war.

Trade in machine tools is quiet. One may advance many explanations for this, but it does not alter the fact. Many firms have some very interesting quotations out now, and they expect to convert some of these into real business. The matter of price is discussed openly, and there is a very marked difference of opinion as to whether a reduction would assist in the sale of any more machines under present conditions. The experience of several dealers can be summed up in the opinion of one who stated, "Nearly every sale that we are making now is made because a firm realizes the need for a new tool in some department. They see where there can be a saving made by the introduction of the new equipment."

"I am open to admit that our business is not brisk now," stated one firm to Machinery this week, "but what of it? One has only to look around a little to see ahead and anticipate what is coming. I can take you to a plant

where a number of men have been laid off. On the face of the thing that is not very hopeful, is it? But go on around to the other side of the same premises, and you will see that they are busy putting up a great extension to their existing place. Does that really look as though they see the bottom dropping out of things around here? No, it means that they are getting ready for a lot of business they hope to do next year."

Selling Small Tools

Firms handling Canadian high speed are working on the new list, announced in last week's issue of Machinery. This price schedule puts them on an equal footing with the British firms, which have heretofore been able to sell under the Canadian quotations. Business is not brisk, although there are a number of small orders to handle. One thing is certain, and that is that the shops are being combed for business now as they have not been for some time, and the dealers are putting on more steam than usual. In fact, there is a revival in selling in all lines of machine tools, including the small tools.

There may be a tendency on the part of some dealers to put certain parts of their stock up for clearance at this time of the year. Many of them end their fiscal year in the near future, and in order to make the year's business as good as possible, an extra selling cam-

paign may be inaugurated to move as much stuff as possible, even in this dull season.

Referring to the recent reduction in the price of high speed, one of the dealers remarked, "In this case we simply have to take our loss and go ahead. When the prices of goods in stock go up we stand to gain, and when they go down we have to take our losses. It can't be a case of 'we win' all the time."

The Steel Market

There has been some improvement in deliveries during the week, and some shipments are coming to light that have been on the books of the U. S. mills for almost a year.

Prices are remaining firm, and there is nothing in the local markets to reflect a dropping away from the selling schedules that have been in use for some time. It would be nothing out of the ordinary were business to fall off a little at this season, as it has been the experience of steel merchants in normal times that from now on to the first of the year business has never been brisk. Allowance should be made for this in considering the volume of trade that is passing now. There is a good demand for certain lines. The larger sizes of tubes seem to be fairly well supplied for this season. Many plants have decided to go ahead with existing equipment for the season as they cannot close off now. This affects the larger sized tubes, but there is still a demand for the 2-inch variety for the smaller plants, and it may be some time before the demand for these is satisfied.

Scrap Still Dull

"Nothing better in sight," was the comment in the scrap metal market this week. Trade is very weak, and the tonnage that is moving is limited to contract commitments. Dealers find that buyers are not coming into the market even at lowered prices.

In this connection it seems necessary to again call attention to the fact that the prices quoted on another page of this paper are Dealers' Buying Prices. That is, Toronto dealers will pay those prices for material. We have had several requests wanting to know where scrap material could be secured at the prices mentioned. It is almost impossible to get together a fair resale list, as selling is done subject to a good many conditions. If firms want to buy

---

POINTS IN WEEK'S MARKETING NOTES

. Cancellations are coming in, although they are not at as great a rate as formerly.

Lead is now 56% above the 1913 price; zinc 20% above; copper 6% below; tin 15% below; pig iron is 172% above, and finished steel products in general 110% above.

Independent steel mills are maintaining prices in nearly every case, believing that reduction would not bring them any new business.

Pittsburgh reports state that independent steel mills are freeing bars, shapes and plates at 3c, and that 6c is a common quotation for black sheets 28 gauge. The National Machine Tool Builders met in New York, and although the quotation of prices was discussed at some length the general feeling seemed to be that when the costs of production were so high at the present time there was small chance for reduction to be made in price.

Firms that are selling lines that are used largely by farmers report business as being very active and sales are very satisfactory.

Steel merchants in this district are doing very fair business, although at this season of the year, even in normal times, there is always a falling off in the volume of trading.

The new price list which is being used by the Canadian high speed tool trades, now meets the price that has been used by the British firms for some time. Heretofore the old country high speed has been able to undersell the Canadian production.

---

direct from those having scrap metal to sell, we would advise their use of the advertising columns of this paper, stating just what they wanted to secure, and they would probably find out then how much they could get.

---

of manufacturing machine tools are little if any lower. Mr. Newton expressed the conviction that price reductions at this time would avail nothing in stimulating renewed buying activity. This sentiment was echoed generally by the manufacturers attending the meeting, and the impression was gained that many, if not all of them, will cling to their present price schedules for the present, at least.

Cancellations were also discussed. The fact that some machine-tool manufacturers have had cancellations aggregating several hundred thousands of dollars made this subject of unusual importance. The machine-tool builders were at a loss to suggest any way by which cancellations may be minimized in a time like the present, except by a process of educating the buyer that he must feel a larger share of responsibility in placing an order. "A reawakening of the business conscience is needed," according to one speaker.

While there was no concealment of the fact that machine tool business in the United States has at present hit what is possibly the worst slump in its history, there was no pronounced feeling of gloom at the convention. Rather, the machine-tool manufacturers look for a resumption of business activity within a few months, by spring at the latest, and possibly sooner, and most of them are understood to be financially able to weather the storm.

In the local tool market there is very little change. Some sellers seem to detect a slightly better tone, as evidenced by a few more inquiries, but it is certain that if there is any improvement it is not general. Some tool builders are doing little or nothing here. There are expectations of railroad buying, but as before reported these prospects are somewhat nebulous as yet.

The General Electric Co. has issued a new list of about 20 tools, including 10 lathes, for a new plant which it is reported may be built in or near Newark, N. J. As the General Electric Co. has other pending schedules, on which no action has been taken, the above-mentioned list has not occasioned as much interest as might otherwise be expected.

Would Sell Its Plant—The plant constructed by the Brantford Industrial Realty Co. for the A. C. Spark Plug Company in the Holmedale district of Brantford, has been placed on the market as a result of uncertain conditions in the automobile industry. An officer of the realty company stated there was a possibility that the Spark Plug Company would start operating next spring, but any offers for the purchase of the plant will be entertained.

Increased Output—Alberta's output of coal for 1920 has increased 30% in excess of last year's. A total output of over 6,500,000 tons is expected by the end of the year.

MACHINE TOOL BUILDERS HAVE NOTHING ON PRICE REDUCTION

Special to CANADIAN MACHINERY

NEW YORK, Nov. 18.—The National Machine Tool Builders' Association held in New York last week its nineteenth annual convention. Quite naturally the present business situation was the principal topic of discussion both in and out of the convention hall.

President Albert E. Newton, in his opening address, pointed out the problem which the average machine-tool manufacturer faces, namely, the expectation on the part of the buyer that prices will go lower and the fact that costs

STEEL AND IRON STAY UP MUCH
LONGER THAN MANY OTHER LINES

Special to CANADIAN MACHINERY.

P ITTSBURGH, November 18.—Open market demand for steel products continues extremely light. There is a fair, though slowly dwindling, demand for steel products against contracts. Some cancellations are still coming in, though not at as great a rate as formerly. and there are some postponements. Thus far at least the Steel Corporation has not been affected, but all the independents have had to restrict production more or less, even those that had order books apparently well filled, and this decline in independent steel production promises to continue until there is a fresh buying movement to replenish order books. The prospect of such a change is more indistinct than it appeared two or three weeks ago, for the general business situation is worse than it seemed to be recently.

Iron and Steel Exceptional

Indeed, a proper appraisal of the prospects requires the fact to be emphasized that the iron and steel industry is exceptional among industries as to its present position. No matter what the prospects may be, the iron and steel producers are making money now, and they have been making a great deal of money up to date. Practically all other industries have been in bad shape for weeks or months. The automobile, tire, textile and various other industries have come almost to a standstill, and with large stocks of manufactured goods which they find it difficult to liquidate. In the non-ferrous metals there have been very heavy losses. Lead is 50 per cent. over its average price in 1913, zinc is 20 per cent. above, copper is 6 per cent. below and tin is 15 per cent. below. Iron and steel prices make a sharp contrast with these metal prices, for pig iron, even with the recent declines, is 172 per cent. higher. The finished steel excess over 1913 would be greater still were not the Steel Corporation prices given some weight in determining the general average at present.

So much for price comparisons. As to demand, it is true there is very little demand for pig iron and steel products in the open market, but the production is being taken on old contracts, and the production is off only a few per cent., hardly much more than 10 per cent. from the high point, and it is fully one-fourth, perhaps one-third, greater than production in 1913. As the production is being taken. and as both producers and consumers are endeavoring to decrease stocks rather than increase them, the rate of consumption is evidently large, whereas for many commodities there is almost no demand at all.

The natural deduction to be made from this remarkable divergence is that the iron and steel industry is simply running on its momentum. It cannot expect to continue prosperous when other industries are not, but rather, when it will certainly take considerable time for business in general to revive from the very poor situation now existing, one should conclude that in iron and steel "the worst is yet to come."

Steel Production

The monthly report of the American Iron and Steel Institute indicates that production of steel ingots in the United States during October was at an average rate of about 42,950,000 gross tons a year. That is one-half of one per cent. heavier rate than the rate in September or August, the highest rate since last March, and the highest rate since December, 1918, barring February and March of this year. Yet during October it was well known that many producers were decreasing output. It seems strange that the total should have increased at all. The explanation is that the Steel Corporation, being well fitted with business, endeavors to run full, and experienced somewhat better operating conditions in October, while as to the independents their decreases were progressive during the month and their average for the month was well ahead of their present rate of output. The November and December figures of production will probably show sharp decreases from the October rate.

Prices Slow to Yield

If any independent steel producer is selling any description of steel at the Steel Corporation (Industrial Board) price the case is altogether exceptional. As a matter of fact the independent prices are very slow indeed to yield to what is considered the inevitable, a decline in the independent market to the Steel Corporation level. As an augury for the future that can hardly be considered as altogether favorable. The mills have no agreement to maintain prices, nor are they receiving anything like enough business to support the market. It seems clear that the mills that are maintaining prices are doing so because there is so little temptation for them to reduce prices, no business of consequence being offered. There is a lack of competition not because mills refuse to compete but because the subject matter for competition is lacking.

On bars, shapes and plates the common quotation of independents is 3.00c. No doubt there is shading in some cases but there does not appear to be much. Sheets, however, show a constantly declining tendency, 6.00c being now done easily on No. 28 black. For several weeks the black sheet market has declined at an average rate of about a

quarter cent a pound per week. Blue annealed in the heavier gauges could probably be had at about 4.00c. though for 14 to 16 gauge the price might be nearer 5.00c. basis. Galvanized are about 1.50c above black.

Easily the strongest item in the finished steel list is tubular goods. There is heavy demand for shipments on contract, customers desiring mills to make heavier shipments than they are able to make. While the independents generally have a basis price, $7 a net ton above the Steel Corporation (54 per cent. as against 57½ per cent.) some orders even now are being taken at delivery premiums ranging up to $10 a ton, or $17 altogether above the Steel Corporation price.

Pig Iron

Since last report foundry pig iron has been offered by producers at $40, valley, or $1 under the reduced price of a week ago. Bessemer and basic are steady but extremely quiet. The market now stands at $40 for foundry, $42 for Bessemer and $38.50 for basic, f.o.b. valley furnaces. these prices representing declines from the top point, reached last August, of $10 on basic and foundry and $6.50 on Bessemer. Prices have slid off very easily, with practically no resistance, but furnaces are taking a new stand. They feel that enough has been done by way of reductions that were almost voluntary in character, and assert that if consumers want still lower prices. they will have to offer orders that would make it worth while to cut prices. As there is no likelihood of consumers getting in the mood in the near future to buy to any extent, no matter what the prices, the furnaces feel they might as well save some of their ammunition. Very recently it was a common guess that pig iron on this decline would find its level at not under $30 nor over $35, but it looks now as if the readjustment will prove more drastic than a decline only to $30.

Serious Car Shortage.—Another serious car shortage is said to threaten Drumheller's mines. There are only enough cars to keep seven mines working part time, and there are no more cars in sight, although the demand for coal from the rest of the prairies which are not stocked up with coal is increasing. One thousand men are idle and two thousand will be idle when the daily output of 7,000 tons will be cut off.

Annual Meeting.—One change was made in the officers of the Cockshutt Plow Company at the annual meeting, E. A. Mott being added. The directors are: Col. H. Cockshutt, George Wedlake, G. K. Wedlake, Sir Augustus Nanton, E. A. Mott, Sir Lomer Gouin, H. W. Hutchinson, James Adams, E. Perry. The reports presented were reported as very satisfactory to the shareholders.

SELECTED MARKET QUOTATIONS

Being a record of prices current on raw and finished material entering into the manufacture of mechanical and general engineering products.

PIG IRON

Grey forge, Pittsburgh	$42 40
Lake Superior, charcoal, Chicago	57 00
Standard low phos., Philadelphia	50 00
Bessemer, Pittsburgh	43 00
Basic, Valley furnace	42 90
Toronto price:—	
Silicon, 2.25% to 2.75%	48 00
No. 2 Foundry, 1.75 to 2.25%	46 00

IRON AND STEEL

Per lb. to Large Buyers	Cents
Iron bars, base, Toronto	$ 5 50
Steel bars, base, Toronto	5 50
Iron bars, base, Montreal	5 75
Steel bars, base, Montreal	5 75
Reinforcing bars, base	5 50
Steel hoops	7 00
Norway iron	11 00
Tire steel	5 75
Spring steel	10 00
Band steel, No. 10 gauge and 3-16 in. base	6 00
Chequered floor plate, 3-16 in.	8 50
Chequered floor plate, ¼ in.	8 00
Bessemer rails, heavy, at mill	....
Steel bars, Pittsburgh	3 00-4 00
Tank plates, Pittsburgh	3 50
Structural shapes, Pittsburgh	3 00
Steel hoops, Pittsburgh	3 50-3 75
F.O.B., Toronto Warehouse	
Small shapes	5 50
F.O.B. Chicago Warehouse	
Steel bars	3 62
Structural shapes	3 72
Plates	3 67 to 5 50
Small shapes under 3"	3 62

FREIGHT RATES

Per 100 Pounds.

Pittsburgh to Following Points

	C.L.	L.C.L.
Montreal	58½	73
St. John, N.B.	84½	106½
Halifax	86	108
Toronto	38	54
Guelph	38	54
London	38	54
Windsor	35.	50½

METALS

	Gross. Montreal	Toronto
Lake copper	$21 00	$21 50
Electric copper	20 50	21 00
Castings, copper	20 00	20 00
Tin	52 00	52 00
Spelter	9 00	10 00
Lead	8 25	9 00
Antimony	8 50	10 00
Aluminum	36 00	35 00

Prices per 100 lbs.

PLATES

Plates, 3-16 in.	$ 7 25	$ 7 35
Plates, ¼ up	6 50	6 60

PIPE—WROUGHT

Standard Buttweld Pipe

Per 100 Ft.

	Steel Blk.	Galv.	Gen. Wrought Iron Blk.	Galv.
⅛	$ 6 50	$ 8 50	$.....	$.....
¼	5 31	7 41	5 91	8 01
⅜	5 31	7 41	5 91	8 01
½	7 10	8 63	7 96	9 48
¾	8 80	10 87	9 95	12 02
1	13 01	16 07	14 71	17 77
1¼	17 60	21 74	19 90	24 04
1½	21 04	25 99	23 79	28 74
2	28 31	34 97	32 01	38 67
2½	44 73	55 28	....	....
3	58 52	72 29	....	....
3½	74 06	90 62	....	....
4	87 75	107 37	....	....

Standard Lapweld Pipe

Per 100 Ft.

	Steel Blk.	Galv.	Gen. Wrought Iron Blk.	Galv.
2	$32 01	$ 38 67	$35 71	$42 37
2½	48 28	58 79	54 11	64 64
3	63 11	76 88	70 76	84 53
3½	75 90	92 46	85 10	101 66
4	89 93	107 55	100 83	120 45
4½	1 05	1 29	1 30	1 54
5	1 22	1 50	1 52	1 80
6	1 58	1 95	1 97	2 33
7	2 06	2 53	2 58	3 01
8L	2 16	2 66	2 66	3 16
8	2 49	3 07	3 07	3 64
9	2 98	3 67	3 67	4 36
10L	2 77	3 41	3 41	4 05
10	3 56	4 39	4 39	5 21

Prices—Ontario, Quebec and Maritime Provinces

WROUGHT NIPPLES

4" and under, 60%.

4½" and larger, 50%.

4" and under, running thread, 30%.

Standard couplings, 4-in. and under, 30%.

Do., 4½-in. and larger, 10%.

OLD MATERIAL

Dealers' Average Buying Prices.

	Per 100 Pounds. Montreal	Toronto
Copper, light	$10 50	$10 50
Copper, crucible	13 00	12 00
Copper, heavy	12 50	12 00
Copper wire	12 50	12 00
No. 1 machine composition	13 00	12 00
New brass cuttings	7 00	9 00
Red brass turnings	10 00	10 00
Yellow brass turnings	7 00	7 50
Light brass	5 00	5 00
Medium brass	6 50	6 00
Scrap zinc	5 00	5 50
Heavy lead	5 25	5 60
Tea lead	2 50	3 00
Aluminum	16 00	16 00

	Per Ton	Gross
Boiler plate	13 00	12 00
Malleable scrap	20 00	23 00
Axles (wrought iron)	25 00	20 00
Rails (scrap)	18 00	18 00
Malleable scrap	23 00	25 00
No. 1 machine cast iron	32 00	33 00
Pipe, wrought	9 00	10 00
Car wheel	30 00	33 00
Steel axles	20 00	20 00
Mach. shop turnings	8 00	9 00
Stove plate	25 00	25 00
Cast boring	8 00	12 00

BOLTS, NUTS AND SCREWS

	Per Cent.
Carriage bolts, 7-16 and up	+10
Carriage bolts, ⅜-in. and less	Net
Coach and lag screws	—15
Stove bolts	55
Wrought washers	—25
Elevator bolts	+10
Machine bolts, 7-16 and over	+10
Machine bolts, ⅜-in. and less	+10
Blank bolts	Net
Bolt ends	Net
Machine screws, fl. and rd. hd., steel	27½
Machine screws, o. and fil. hd., steel	+25
Machine screws, fl. and rd. hd., brass	net
Machine screws, o. and fil. hd., brass	net
Nuts, square, blank .... +25 add	$2 00
Nuts, square, tapped ... add	2 25
Nuts, hex., blank ... add	2 50
Nuts, hex., tapped ... add	3 00
Copper rivets and burrs, list less	15
Burrs only, list plus	25
Iron rivets and burrs	40 and 5
Boiler rivets, base ¾" and larger	$8 50
Structural rivets, as above	8 40
Wood screws, O. & R., bright	75
Wood screws, flat, bright	77½
Wood screws, flat, brass	55
Wood screws, O. & R., brass	55½
Wood screws, flat, bronze	50
Wood screws, O. & R., bronze	47½

MILLED PRODUCTS

(Prices on unbroken packages)

	Per Cent
Set screws .... —20%, 25 and	5
Sq. and hex. hd. cap screws	12½
Rd. and fil. hd. cap screws ..plus	25
Flat but. hd. cap screws ....plus	50
Fin. and semi-fin. nuts up to 1-in.	12½
Fin. and Semi-fin. nuts, over 1 in., up to 1½-in.	—5
Fin. and Semi-fin. nuts over 1½ in., up to 2-in.	+12½
Studs	+5
Taper pins	—12½
Coupling bolts	+40
Planer head bolts, without fillet, list	+45
Planer head bolts, with fillet, list plus 10 and	+55
Planer head bolt nuts, same as finished nuts.	
Planer bolt washers	net
Hollow set screws	+60
Collar screws ......list plus 20,	30
Thumb screws	40
Thumb nuts	75
Patch bolts ......add	+85
Cold pressed nuts to 1½ in...add	$1 00
Cold pressed nuts over 1½ in..add	2 00

BILLETS

	Per gross ton
Bessemer billets	$60 00
Open-hearth billets	60 00
O.H. sheet bars	76 00
Forging billets	56 00-75 00
Wire rods	52 00-70 00

Government prices.

F.O.B. Pittsburgh.

NAILS AND SPIKES

Wire nails, base	$6 10
Cut nails, base	7 00
Miscellaneous wire nails	50%

ROPE AND PACKINGS

Plumbers' oakum, per lb.	0 10¼
Packing, square braided	0 38
Packing, No. 1 Italian	0 44
Packing, No. 2 Italian	0 36
Pure Manila rope	0 35½
British Manila rope	0 28
New Zealand hemp	0 28

POLISHED DRILL ROD

Discount off list, Montreal and Toronto .... net

MISCELLANEOUS

Solder, strictly	$ $ 29½
Solder, guaranteed	0 31½
Soldering coppers, lb.	0 62½
White lead, pure, cwt.	20 35
Red dry lead, 100-lb. kegs, per cwt.	16 00
Gasoline, per gal., bulk	0 42
Pure turp., single bbls., gal.	3 15
Linseed oil, raw, single bbls.	2 37
Linseed oil, boiled, single bbls.	2 40
Wood alcohol, per gal.	4 00
Whiting, plain, per 100 lbs.	3 00

CARBON DRILLS AND REAMERS

S.S. drills, wire size	35
Can. carbon cutters, plus	10
Standard drills, all sizes	35
3-fluted drills, plus	10
Jobbers' and letter sizes	35
Bit stock	40
Ratchet drills	10
S.S. drills for wood	40
Wood boring brace drills	25
Electricians' bits	30
Sockets	50
Sleeves	50
Taper pin reamers	.25 off
Drills and countersinks	net
Bridge reamers, carbon	50
Centre reamers	5
Chucking reamers	net
Hand reamers	10
High speed drills, list net to plus	20
Can. high speed cutters, net to plus	10
American plus	40

COLD ROLLED STEEL

[At warehouse]

Rounds and squares	$7 base
Hexagons and flats	$7.75 base

IRON PIPE FITTINGS

	Black	Galv.
Class A	70	85
Class B	30	40
Class C	20	30

Cast iron fittings, 5%; malleable bushings, 22½%; cast bushings, 22½%; unions, 37½%; plugs, 20% off list.

SHEETS

	Montreal	Toronto
Sheets, black, No. 28	$10 50	$10 50
Sheets, blue ann., No. 10	9 00	9 00
Canada plates, dull, 52 sheets	13 00	13 00
Can. plates, all bright	14 00	.....
Apollo brand, 10¾ oz. galvanized	....	....
Queen's Head, 28 B.W.G.	13 50	.....
Fleur-de-Lis, 28 B.W.G.	13 00	.....
Gorbal's Best, No. 28	....	....
Colborne Crown, No. 28.	....	....
Premier, No. 28, U.S.	11 50	12 00
Premier, 10¾-oz.	11 50	12 40
Zinc sheets	16 50	20 00

PROOF COIL CHAIN

(Warehouse Price)

B

¼ in., $13.00; 5-16, $11.00; ⅜ in., $10.00; 7-16 in., $9.80; ½ in., $9.75; ⅝ in., $9.20; ¾ in., $9.30; ⅞ in., $9.50; 1 in., $9.10; Extra for B.B. Chain, $1.20; Extra for B.B.B. Chain, $1.80.

ELECTRIC WELD COIL CHAIN B.B.

⅛ in., $16.75; 3-16 in., $15.40; ¼ in., $13.00; 5-16 in., $11.00; ⅜ in., $10.00; 7-16 in., $9.80; ½ in., $9.75; ⅝ in., $9.50; ¾ in., $9.30.

Prices per 100 lbs.

FILES AND RASPS

	Per Cent.
Globe	50
Vulcan	50
P.H. and Imperial	50
Nicholson	32½
Black Diamond	27½
J. Barton Smith, Eagle	50
McClelland, Globe	50
Delta Files	20
Disston	40
Whitman & Barnes	50
Great Western-American	50
Kearney & Foot, Arcade	50

BOILER TUBES.

Size.	Seamless	Lapwelded
1 in.	$27 00	$.....
1¼ in.	29 50	.....
1½ in.	31 50	29 50
1¾ in.	31 50	30 00
2 in.	35 00	30 00
2¼ in.	35 00	29 00
2½ in.	42 00	37 00
3 in.	50 00	48 00
3¼ in.	.. ..	48 50
3½ in.	63 00	51 50
4 in.	85 00	65 50

Prices per 100 ft., Montreal and Toronto

OILS AND COMPOUNDS.

Castor oil, per lb.	
Royalite, per gal., bulk	29½
Palacine	32½
Machine oil, per gal.	65
Black oil, per gal.	34
Cylinder oil, Capital	1.01
Petroleum fuel oil, bbls., net	19

BELTING—No 1 OAK TANNED

Extra heavy, single and double		6½
Standard		6½
Cut leather lacing, No. 1		2 00
Leather in side	2 40	3 00

TAPES

Chesterman Metallic, 50 ft.	$2 00
Lufkin Metallic, 603, 50 ft.	2 00
Admiral Steel Tape, 50 ft.	2 75
Admiral Steel Tape, 100 ft.	4 45
Major Jun. Steel Tape, 50 ft.	3 50
Rival Steel Tape, 50 ft.	2 75
Rival Steel Tape, 100 ft.	4 45
Reliable Jun. Steel Tape, 50 ft.	3 50

PLATING SUPPLIES

Polishing wheels, felt	$4 50
Polishing wheels, bull-neck	2 00
Emery in kegs, Turkish	8¾
Pumice, ground	06
Emery glue	30
Tripoli composition	9½
Crocus composition	12
Emery composition	11
Rouge, silver	64
Rouge, powder, nickel	38

Prices per lb.

ARTIFICIAL CORUNDUM

Grits, 6 to 70 inclusive	.08½
Grits, 80 and finer	.6

BRASS—Warehouse Price

Brass rods, base ½ in. to 1 in. rod	0 34
Brass sheets, 24 gauge and heavier, base	$0 42
Brass tubing, seamless	0 46
Copper tubing, seamless	0 48

WASTE

XXX Extra	.24	Atlas	.20
Peerless	.22½	X Empire	.19½
Grand	.22½	Ideal	.19
Superior	.22½	X Press	.17½
X L C R	.21		

Colored

Lion	.17	Popular	.13
Standard	.15	Keen	.11
No. 1	.15		

Wool Packing

Arrow	.35	Anvil	.22
Axle	.28	Anchor	.17

Washed Wipers

Select White	.20	Dark colored	.09
Mixed colored	.10		

This list subject to trade discount for quantity.

RUBBER BELTING

Standard ... 10% Best grades... 15%

ANODES

Nickel	.55 to .60
Copper	.38 to .40
Tin	.70 to .70
Zinc	.16 to .17

Prices per lb.

COPPER PRODUCTS

	Montreal	Toronto
Bars, ½ to 2 in.	$37 50	$37 00
Copper wire, list plus 10		
Plain sheets, 14 oz., 14x60 in.	44 00	44 00
Copper sheet, tinned, 14x60, 14 oz.	47 00	46 00
Copper sheet, planished, 16 oz. base	51 00	50 00
Braziers', in sheets, 6 x 4 base	43 00	42 00

LEAD SHEETS

	Montreal	Toronto
Sheets, 3 lbs. sq. ft.	$11 50	$14 50
Sheets, 3½ lbs. sq. ft.	11 25	14 00
Sheets, 4 to 6 lbs. sq. ft.	11 00	13 50

Cut sheets, ½c per lb. extra.
Cut sheets to size, 1c per lb. extra.

PLATING CHEMICALS

Acid, boracic	$ .23
Acid, hydrochloric	.04¾
Acid, nitric	.11
Acid, sulphuric	.04½
Ammonia, aqua	.15¾
Ammonium, carbonate	.23
Ammonium, chloride	.22
Ammonium hydrosulphuret	.75
Ammonium sulphate	.30
Arsenic, white	.16
Copper, carbonate, annhy	.41
Copper, sulphate	.13
Cobalt, sulphate	.20
Iron perchloride	.62
Lead acetate	.30
Nickel ammonium sulphate	.20
Nickel carbonate	.32
Nickel sulphate	.20
Potassium sulphide (substitute)	.40
Silver Chloride (per oz.)	1.15
Silver nitrate (per oz.)	1.10
Sodium bisulphate	.13
Sodium carbonate crystals	.04
Sodium cyanide, 127-130%	.39
Sodium hyposulphite per 100 lbs	9.00
Sodium phosphate	.15
Tin chloride	.30
Zinc chloride, C.P.	.30
Zinc sulphate	.08

Prices per lb. unless otherwise stated

INDUSTRIAL NEWS
NEW SHOPS, TENDERS AND CONTRACTS
PERSONAL AND TRADE NOTES

Building in Galt.—The local commission of the Hydro have decided to erect a fine modern public utilities building at the corner of Dickson and Wellington streets, Galt. They expect to proceed with the work during the coming year.

New Coking Plant.—It is reported that the United Gas and Fuel Co. intend commencing the erection of a three-million-dollar coke oven gas plant in the northeast end of the city of Hamilton.

New Library.—Plans are under way for the beginning of the new Carnegie library at Lethbridge, with the promise of $25,000 made by the Carnegie Foundation for this purpose. It will consist of a daylight basement and one storey above it.

Call for Tenders.—Plans are nearing completion and it is expected that tenders will be called for during the winter for the addition to the Ritz Carlton Hotel, Montreal. The addition will add 150 rooms, will be nine or ten storeys in height, and cost about $750,000.

May Build Hotel.—An unofficial report has spread about that tentative plans are entertained by the Canadian National Railways to erect a hotel in Quebec along the lines of the Chateau Laurier, Ottawa.

Steel Contract.—The Hydro Electric Power Commission of Ontario has awarded the Dominion Bridge Co., Ltd., the contract for five large penstocks for its Queens plant. The contract calls for early erection, as part of the plant must be in operation by September next year. Over 2,000 tons of steel are involved in the contract.

New Molding Shop.—A new molding shop is being started at Kingston, Ontario, by Messrs. Kelly and Driver. It has been stated that the ratepayers will be asked to give the company a free site on Montreal St., at a cost of $2,000, and exemption from taxes for ten years.

Large Building.—Although the actual construction work will not start till next spring, preliminary surveys are being made of the grounds along the Trent Road on which the new Albert College is to be erected at Belleville. Chapman, Oxley & Bishop are the architects, and Mr. Oxley is the engineer. It is expected that an outlay of a million dollars will be necessary.

GIFT TO ASSIST HOSPITAL WORK

H. W. Petrie Remembers the Hospital In the City Where He First Started Business

H. W. Petrie, president of the H. W. Petrie Co., Ltd., Toronto, has remembered his old city, Brantford, where he first started in business. The remembrance has taken the form of the setting aside for a term of five years of $25,000 Victory bonds, the interest to be paid half yearly to the hospital. In his letter to C. H. Waterous, president of the Brantford General Hospital, Mr.

H. W. PETRIE

Petrie pointed out that his "first business venture took place in Brantford, of which for many years I was a citizen, and, despite my removal some thirty years ago to Toronto, the interests and developments of the place have always been very dear to me. In this regard I shall deem it a privilege to do something on behalf of your splendid public institution."

C. H. Waterous, in accepting the gift, stated: "Your citizenship here is still recalled with pleasure and profit by many old friends, who take pride in the success which has so worthily crowned your business enterprise and integrity, and who deeply appreciate that in the days of your well-earned prosperity you have decided to remember in so tangible a manner the place and the people associated with your early life and commercial efforts."

The Brantford "Expositor," comment-

ing editorially on this, remarks: "The hospital board has not received much encouragement of this kind in its labors, and the generous action of Mr. Petrie is consequently greatly appreciated by them. It is hoped also that it will prove an example to others."

DRAWN STEEL PLANT IS BUSY

Hamilton.—Speaking of the steel industry in general, S. J. Waddie, president of the Canadian Drawn Steel Company, stated that its immediate future was not likely to be affected to anything like the same degree as other lines of industry might be.

As for the company he is associated with, Mr. Waddie said there were sufficient orders on hand to insure uninterrupted operations for a long time to come. Retrenchment was something his company had had no occasion to consider so far. All employees were at work and the plant was operating at full capacity. The greater part of its production was for the domestic market.

"We could get much more work to do if we could only get the raw material," observed Mr. Waddie, "but we experience the greatest difficulty in this connection, with the result that our export market has been crippled somewhat."

Use Northern Rock.—The Silica Granite Products Company, Ltd., is a company formed under Ontario charter, having headquarters in Hamilton, Ont., president, A. R. C. Smith, secretary-treasurer, A. Caddie. The company's property is situated in the township of Henwood, New Ontario, and consists of 160 acres, whereon is a very extensive formation of granite rock. This rock, when quarried and crushed to various sizes is a base for cleansers of various kinds, such as dry and wet hand cleaners, household cleaners, etc. A medium size of the rock is found to be exceptionally good for foundry use in sand blasting of castings. The rock is used for monumental work, is similar in color and hardness to Scotch granite. There are in all, at present, twenty-one known uses for this product, which assures it an extensive market. Installation of machinery is almost completed, and the company expect to ship about December 1, 1920.

The Week's Events in Montreal Industry

Industrial incorporations for the past week include the Forged Steel Specialties Company of Montreal, $500,000, and the Canadian Piston Ring. Company. of Montreal, capitalized at $75,000.

Hodgson Freck, Ltd., of 369 St. James Street, Montreal, foundry, mill and metal merchants, have secured the Eastern Canadian agency for Robert Drury of Canada, Ltd., of Toronto, manufacturers of hardwoods and veneers.

At the November meeting of the Canadian Railway Club, in Montreal, a paper on "How to heat railway buildings economically" was read by R. H. Black, engineer of power plant construction of the Grand Trunk Railway Company. A large gathering of the members was present and the discussion following the reading of the paper indicated the great interest that was taken in this subject.

Malcolm Lemieux has recently purchased large interests in Motor Car Distributors, Ltd., and has been elected a director and appointed general manager

of the entire concern. He has secured district agencies for all of Eastern Canada for Peerless and Chandler cars, and Beaver Hall, the new four-storey spacions building on Beaver Hall Hill, Montreal, has been acquired for offices and warerooms. Mr. Lemieux has spent his entire lifetime in the automobile business.

The Armstrong Whitworth Company of Canada are entering on a new line of activity in the shape of marketing an English make of automobile manufactured by the Armstrong Siddeley Company of Coventry, England. Every part of this car, even to the tires, is made by the Armstrong Siddeley Company. This car is of exceptionally high grade quality and is designed especially for the convenience of the owner-driver, simplicity of operation being an important feature. In design it is entirely original, and embodies experience gained during the past four years. This is particularly true of the six-cylinder engine, which is that known as the Deasy aircraft engine. These cars will sell at $8,500 and $11,000.

MACHINE TOOLS CANNOT COME BACK TO LOW PRE-WAR LEVELS

"HOW about prices? Well, for the life of me I can't see where there is any great cut coming, and I don't think it is well that we should do anything to lead the buyers to believe that such will be the case for the present at least."

That, is the way the manager of a firm of machine tool dealers sized up the situation for Canadian Machinery.

"For a long time," he continued, "there have been makers of machine tools that have never made much money. They have never had enough margin to allow them to go ahead and develop as they should develop. They have never been able to afford a research department or do the experimental work they should. The machinist, too, for years, was a low paid man when compared to the outside trades, in many of which there was not nearly the skill required that the machinist needs in his work. I can look back over my own experience, and can remember the many prices for building a lathe. Very often the completed machine would represent the cost of actual production, and the usual prof. it, but there would be nothing allowed for the number of castings that were poured and broken up as being not good enough.

"Now, here is a line that will show how prices compare with the old days. As a rule 1914 is taken because it was

in that year the war started, and everything dates from before or since the war. But figures I am giving are for 1910 and 1911. In 1910 a chucking and turning machine cost $1,200, and the present price is $2,010. Here is another style of the same machine, a little larger. In 1911 the price was $1,500 against a quotation today of $2,405. Then here is the 15-inch shaper that in June of 1911 would have sold for $390. and today is listed at $875. A 24-inch shaper that in 1911 brought $595 is today selling at $1,195. That is, as far as I can learn, about the rate of increase that has been made by the standard builders of machines. It is not out of keeping with the way in which everything has increased. Pig iron and steel have increased more than that. Not only has the price of the tool increased, but one must not forget that the standard and quality of the output has increased. We are making a better machine today than we were before. There are more attachments on it for good work. There may be reductions made, and I do not hold that they would not be desirable, but it would be a mistake to imagine that business is going to come back to the same poor level it had before the war. I can't see where it is desirable from any point of view, or where it would be to the lasting advantage of any person."

THE YEAR'S WORK AT SCOTIA MINES

Increase in Production Found in Comparing With Last Year's Figures

If increase in production is a dependable criterion, the earnings of the Nova Scotia Steel & Coal Company for the 1920 year will show a big gain over last year. Due to adverse market conditions the 1919 production fell off sharply as compared with the two previous years. That there has been an equally decisive recovery is indicated by the following figures (in tons):

	Nine months to	
	Sept. 30.	
	1920	1919
Coal	456,312	387,151
Pig iron	63,632	26,697
Ingots	95,354	38,306
Blooms and billets	82,202	36,124
Merchant mills	44,859	27,864

Nova Scotia Steel last year on greatly curtailed output earned $5.73 a share for the $15,000,000 common stock, as against $5 paid in dividends. On this small number of common shares (150,-000) the company this year is undoubtedly making a much more satisfactory showing, thus confirming the conviction of leading shareholders that their property is one of the most valuable participants in the pending British Empire Steel merger.

The city of St. Johns, Nfld., is to have a new 200-room hotel, costing $1,000,000, to be operated by the United Hotels Co. The capital invested is to be local capital.

A new garage is to be erected by the Consolidated Motors, Ltd., at Winnipeg, at a cost of $80,000.

MOTORS AND CORRECT APPLI-
CATION

Continued from Page 465

tions that may be made in the motor
by control of its design. Both of these
points should be thoroughly understood
and analyzed, before a proper motor application can be made.

Figure 1 herewith, shows the speed-
torque characteristics of a standard
polyphase induction motor. The upper
one of the three curves shown is the
speed-torque curve of a typical standard induction motor. In these curves,
synchronous speed is taken as 100%,
and the maximum torque that the motor will develop is also taken as 100%.
The full-load torque—that is, the torque
at which it will operate continuously
without overheating—is only about 30%
of the maximum torque. It will be noted
that the full-load torque occurs when
the speed has dropped to about 96% of
the synchronous speed; that is, the slip
is about 4%. If over-loads are applied,
the speed continues to drop and the
torque to rise, until the torque has risen
to a value of a little more than three
times that at full-load, and the speed
has dropped to about 75% of the synchronous. Then occurs the so-called
"pull out" point; that is, an attempt to
pull still more load with torque causes
the motor rapidly to come to a standstill because beyond that point the
torque decreases rapidly, with decrease
in speed. This is accompanied by a
rapid increase in current as is shown
by Figure 3. This "standard motor"
is provided with a low resistance motor
and is the type of motor that is usually referred to as "standard."

If instead of a low resistance rotor,
one is provided with approximately double the resistance of the "standard," the
middle curve of Fig. 1 will apply. The
speed-torque curve of the motor will be
distinctly modified by this change in
resistance. The slip for a given torque
is much greater than in the "standard"
motor. Instead of 4% slip with full-
load torque, we now have nearly 10%.
The "pull out" instead of occurring at
about 75% synchronous speed, now occurs at about 30% synchronous speed.
Also, the torque at standstill is nearly
a maximum, and the current taken at
standstill, as shown in Fig. 3, is considerably less than with the standard
motor. In comparing these two curves,
it is quite obvious that for work that
requires large motor torques at speeds
lying between approximately one-third
and two-thirds of the synchronous speed,
the motor with the middle curve is better adapted than the "standard motor."
Also, for "fly-wheel" jobs—that is,
where a fly-wheel does the work, and it
is the duty of the motor to bring the
fly-wheel up to speed again at its leisure, the middle curve is obviously the
better, since it allows the fly-wheel to
give up more than double the energy without bringing the motor to a given load,
than in the case of a "standard" motor.
For hoist work also, where heavy torques

are necessary at start and the loss of a
small fraction of speed under load is
immaterial, experience has demonstrated that this type of curve is highly desirable.

If the resistance of the rotor is still
further increased; that is, approximately doubled again, the lower curve of Fig.
1 applies. In this case it will be seen
that the slip is still further increased,
being about 18% at full-load torque, and
that at standstill the torque is still increasing; that is, the "pull-out" point is
at some negative speed. This type of
motor is applicable to cases where the
maximum torques are required at standstill and at speeds up to 40% or 50% of
synchronous. The elevator is a typical
example of this duty. In this duty, the
maximum torque is required at standstill since the friction of rest must be
overcome at this point and, as is well
known, the friction of rest in the typical elevator engine is several times that
after the first few revolutions. The
overcoming of this friction of rest and
the acceleration of the car from standstill requires the maximum torque at
low speeds in this application, and experience as well as theory has shown
that this curve is best adapted for elevator work.

By using a Slip Ring motor the rotor
resistance may be changed at will, and
a motor of this type may be changed
at will to meet any of the speed-torque
characteristics shown in Fig. 1. In securing this advantage, however, we have
been obliged to use slip-rings on the
rotor, thereby increasing the cost and
reducing the reliability and ruggedness
of the machine, but more important, we
have very considerably reduced the maximum torque which the motor will produce. These disadvantages make it highly desirable to stick to the Squirrel Cage
construction, if it is in any way possible, to fit any single speed-torque characteristic shown in Fig. 1 to the work
to be performed.

Fig. 2 shows how the horsepower output of the three motors shown in Fig.
1 varies with torque. In considering
these curves, we should bear in mind
that the main part of the work done
by the low resistance and semi-high resistance rotors are preferably close to
the synchronous speeds, while the design of the high resistance rotor is such
that it can do its best work at speeds
below 50% of synchronous.

Fig. 3 shows how the amperes and
power factor of these motors cary with
torque. For a given torque, the amperes and power factor of any of the
three motors is exactly the same.

However, the maximum amperes of
the high resistance rotor machine at
standstill is only about two-thirds of
that of the low resistance rotor. The
much higher torque of the high resistance rotor machine gives us at standstill a torque per ampere about three
times as great in the high resistance
rotor machine as in the low resistance
rotor.

CANADIAN MACHINERY
AND
MANUFACTURING NEWS

VOL. XXIV. No. 22   November 25, 1920

Here Is One Method of Making Back Gear Parts

Can You Suggest a Better Plan of Producing These Parts? If So, We Want to Hear About It—The Author Merely States His Views Upon the Problem.

By G. BARRETT

RELATIVE to the manufacture of back gear parts these pieces could be made in various ways, using screw machines, Jones and Lamson bar turrets, semi-automatics, disc grinders, etc. However, we will take a Warner and Swasey for our lathe work, as it is particularly adapted to the work, and by the use of specialized labor would give very good results at the least cost of production.

In machining the lever, the first operation would be the facing of the two sides of the large boss. These pieces would be clamped on the indexing fixture, Fig. 1, the boss being supported by the V-block A, shown in the sectional sketch. This support is to take the strain of the cut of the straddle milling cutters B, the clamps C and D holding the lever firmly in position.

The bolts E work in T-slots and are used to clamp the upper revolving part of the fixture after indexing, thus keeping the fixture rigid while milling the boss. The advantage of this method is that one piece can be placed in position while work is proceeding on the other.

The approximate output from this machine would be 16 per hour.

Second Operation

The second operation would consist of drilling and reaming the 3-4 inch hole in the boss which is accomplished in the jig, Fig. 2. The lever is held flat on the boss by the knurled bush A, which is just sufficiently tightened to bring the lower face square on the block B; then by means of the taper drift C, the lever is held up to the V-block D. The outer portion of the lever

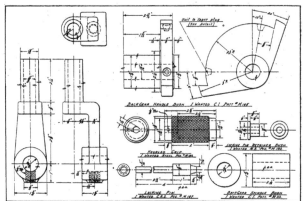

HERE WE SEE THE VARIOUS DETAILS. STUDY THEM CAREFULLY.

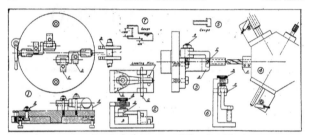

THIS SHOWS ARRANGEMENT NO. 1, AND INCLUDES FIGS. 1 TO 6.

is centred by the two locating pins shown. The average production from this operation should be about 22 per hour.

The third operation would consist of boring and turning for the knurled grip. This would be performed on a turret lathe and held in position by the pin A, Fig. 3. The fixture used is a special casting and is held to the face plate by suitable cap screws. After being perfectly aligned the casting could be doweled, so that the same could be removed and replaced at a future date for quick and accurate setting. The fixture proper is slotted at B, leaving about 3-16 inch clearance for the offset in lever. This clearance is to provide for adjustment of irregular castings.

This detail is taken care of by two set screws, one on either side of the fixture. The first operation on the shaft portion is spotting the end and then drilling, using a straight shank drill, this being done to allow of close setting when using the facing tool A

in Fig. 4. The gauge for this operation is shown in Fig. 5. We next rough turn the outside, leaving about 1-16 inch for finish. This roughing is done by means of a box tool, using Stellite tool bits. The next face of the turret would carry the reamer for finishing the hole to size. The next detail is the finish turning of the outside. The time required for these operations would provide for an output of about 100 per day of nine hours.

The final operation on this lever would be that of drilling the 1-2 inch tap hole in the boss. The jig for this is shown in Fig. 6. The drill bush acts as a clamp for holding the work while drilling. Two of these jigs are used, one for drilling and one for tapping. Production from these jigs would be about 300 per day.

Knurled Sleeve

The knurled sleeve is made from cold rolled stock of the required diameter, the bars being held in the collet chuck A, Fig. 1 in arrangement

2. This piece has eight operations, comprising centering, drilling, squaring bottom of hole, reaming, turning clearance for knurl, knurling the body and drilling the end and pin holes. The lathe setup is shown in Fig. 1. The 5-16 inch hole is drilled in the jig Fig. 2, the machined end resting on the plug A and the upper end held by the drill bush B. The lathe would produce about 50 per day and the drilling would account for approximately 250 per day.

Work on the locking pin is shown in Fig. 3. This pin would require three operations on the 5-16-inch shank. Box tools are used for the turning of the stem, three of them being used for taking off different amounts of stock. The tapered end of the pin is formed by a special tool fitted to the rear of the cross slide. The piece is then parted off by the front tool C. The production would be about 200 per day.

The small retainer bush is made from 15-16-inch cold rolled steel, using the collet chuck on the lathe spindle A in

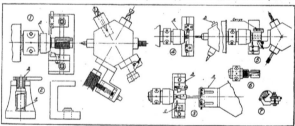

ARRANGEMENT NO. 2, SHOWING WORK ON THE KNURLED SLEEVE.

Fig. 4. The stock is fed through the spindle to the stop B, and then drilled to the required depth. The tool on the rear of the cross slide is then used to turn the small diameter, after which the piece is cut off by the parting tool in the front holder. About 400 per day should be made on this operation.

The spindle bushes are made of cast iron, cast in lengths of about 2 feet, roughed out to within 1-16-inch of the finished size, and then placed in the turret lathe where they are drilled, reamed and finish turned and cut to the required length, as shown at Fig. 5. The possibility of a burr on the inner end would require the pieces to be placed on a separate arbor and the end faced off, as shown at Fig. 6. The first operation would give a production of about 150 per day and the facing would produce about 90 per hour.

Machining the Sector

To machine the sector the piece would be gripped by the hub in the two jaw chuck Fig. 1, arrangement No. 3. The sector is drilled, reamed and faced, using gauge B. The facing tool C is made from bar stock with inserted blades and pilot. About 200 per day should be got from this operation. The next operation is drilling and reaming for stop pins. This is done in a box jig Fig. 2, the sector being centered from the hole already bored and held against locating pin A by the pointed set screw B. This pin is used to bring the lugs on the sector to the proper position from the centre of the stop pin holes.

After tightening the set screw the sector is held by the taper drift C. The drill bushes are made a slip fit so as to be renewed while reaming holes. For this operation we use a two spindle drill, one spindle carrying a drill and the other the taper reamer. Output about 300 per day. The final operation on the sector is shown in Fig. 3. This is the turning of the hub. This is done on a lathe fitted with a hardened steel arbor A, inserted and ground to fit the hole in the hub. The special face plate B is made to accommodate two set screws C and D, which are located in the desired position to screw into the stop pin holes. These set screws need very little tightening in order to hold the work. The facing tool E has a hardened steel pin inserted and set to the required distance from the cutters. The output from this operation would be about 100 per day.

In the making of all these parts it is essential that every operation be accurately performed to avoid trouble in succeeding operations.

In my estimation the cost of these jigs and fixtures would be around $400, and this figure would mean a considerable saving in the first 1,000 plates.

A safety device for punch presses, developed at the works of the Ford Motor Company, makes it impossible for the workman to get his hands in the way of the plunger. To operate a press thus equipped the workman must press two buttons about a foot apart, so that it is necessary for him to use both hands. A two-man press so equipped will not trip until four buttons are pressed at one time; a three-man press will not trip until six buttons are pressed simultaneously. According to a director of the company, the device has not only practically eliminated punch press accidents, but has increased production on an average of 10 per cent, owing to eliminating the fatigue resulting from the operation of punch presses by a foot trip. All punch press operators are required to use tongs for inserting metal and taking it out of the presses.

INTERESTING CATALOGUE

The Canadian Blower and Forge Co., Ltd., Kitchener, Ont., have issued a new catalogue, No. 19C, dealing with their varied line. This includes forges, blowers, exhausters, fans, drills, punches, shears, binding machines, tire setters.

Combination wood working machines, fan system apparatus for heating, ventilating, drying and mechanical draft, air washers, humidifiers, dehumidifiers, pumps, etc., etc., are also shown. This book contains 186 pages of real, live information to those interested in such lines.

NEW CATALOGUE

The Machinery Co. of America, who are distributors for Baldwin, Tuthill & Bolton, Covel Mfg. Co., and Hanchett Saw Works, have issued a very nice catalogue dealing with the lines mentioned above.

These include saws, automatic band sharpeners, saw sharpeners and gummers, roll stretchers, saw stretchers, lap grinders, brazing clamps, saw brazing forges and shearing and cross cutting machines.

Other lines shown are filing clamps, stands for band saws, patch machines, band wheel grinders, segment grinders, band saw swages, swages of all kinds, swage shapers, swage hammers, gauges of all natures, knife grinders, two-wheel grinders for high speed steel knives, balancing machines, and other innumerable lines, 224 pages in all being shown and well illustrated throughout.

The American Mining Congress economist states that gold production has declined from $101,000,000 in 1915 to $58,500,000 in 1919, and probably will be $40,000,000 in 1920.

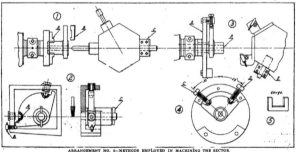

ARRANGEMENT NO. 3—METHODS EMPLOYED IN MACHINING THE SECTOR.

Do You Safeguard Your Existing Equipment?

Accidents Can Be Reduced to a Minimum by Guarding Moving Parts of Your Machine Tools—This Applies Especially to Existing Equipment, Which May Have Been Manufactured Before Guards Were Considered Essential.

By J. H. MOORE.

THE splendid results accomplished by safety engineers within the past few years have shattered, beyond shadow of a doubt, the old fallacy that accidents occurring while at work were necessarily incidental to industry. No longer do we believe that a certain percentage of casualties is inevitable.

Not only has it been shown that accidents are preventable, but that, apart from being the humane thing to do, the very prevention has been profitable to the firm concerned. Take, for example, the power press. This type of machine was responsible, year after year, for the maiming of thousands of employees, and the position of press operator was an exceeding hazardous one. But conditions have changed since the installation of efficient guards, and at the present time a tour through any up to date press room is a pleasure. Should an operator lose a finger these days the cause can usually be traced to rank carelessness on his part.

These same conditions exist in many other lines of industry, and machines in general are better guarded than ever before. Mr. H. A. Schulf, of the U. S. Steel Corporation, speaking upon the subject of safeguarding machinery at its sources, remarked as follows: "We have but to turn to the advertising pages of our technical journals, where page after page shows machines pictured with hazardous points guarded. True, the

guarding is not always adequate, some parts being exposed that shouldn't be, but we cannot afford to be too critical. Think of the advertising pages of 10 or 15 years ago. In those days the gears, power transmission, etc., were devoid of any covering."

It is obvious that the best time to safeguard a machine is while it is being designed, for we can then incorporate the features essential to safety, at the same time producing a design that is pleasing to the eye. This, however, is impossible in the case of machinery built 10 years ago, but that fact need not deter one from installing guards on these early models. True, they may not look quite as neat, yet neatness is a secondary consideration alongside of safety. The chief point to watch is this: Make the guards as complete as possible, and of good stout construction. A frail guard never was a success.

Safeguarding Existing Equipment

To illustrate what can be accomplished in safeguarding existing equipment we have selected four views from the plant of the International Harvester Co. at Hamilton, Canada. This concern has made a very careful study of the safety question, and have at each of their plants various safety experts who make it their chief business to design and instal adequate guards in all departments.

It is, of course, impossible to show

even a small percentage of these guards, but we have contented ourselves with the four views depicted herein. Take for example Fig. 1. This photograph shows a portion of one of the aisles running through their woodshop. Note the guards at every machine, each being of a standard style. This is a feature well worthy of consideration, because by so doing they reduce the expense in making. In some cases the guards are made in standard sections, then bolted together as required. The aisle space in this photograph was not specially prepared for the occasion, but represents how the aisle is kept clean at all times. There are no chips, shavings, or dust to be seen, and the reason is self-evident. Every machine is equipped with dust and shaving removers, the material being sucked up the pipes, going from there into a huge hopper.

Fig. 2 illustrates a close up view of one of these guards, and its construction can be clearly noted. The framework is 1¼-in. angle iron, while the lattice work is usually ⅛-in. to 3-16-in. thick, and ¾-in. to 1-in. wide. Note that certain portions of these guards are hinged, so that they swing out to allow repairs to the machine if necessary. A castor is placed at the bottom of the legs in some instances, as for example at A and B in the present photograph. The method of holding these sections in place is the same as used on any gate, and this part can be noted at C.

FIG. 1—NOTE THE CLEAR AISLE SPACE AND WELL PROTECTED MACHINES.

FIG. 4—ONE STYLE OF EMERY WHEEL GUARD AS ADOPTED AT THIS PLANT.

FIG. 2—CLOSE-UP VIEW OF ONE OF THE GUARDS.

The third example, Fig. 3, depicts a safety device installed on a rip saw. This particular work is of the following nature: The board and its proportions can be clearly seen. In the centre of the board there is a slot, but only for a certain length. This means that the saw has to be fed up vertically into the board after it has been placed in proper location to the saw. Before the installation of this guard, accidents were frequent, but by placing the device on the machine no further trouble has occurred. The complete attachment cost but a trifle, yet its results are satisfactory in every way. While on the subject of expense, it might be well to state that it is not always the most expensive guard that is the best.

The chief aim in the design of any guard should be simplicity. It stands to reason that the simpler it is the less expensive it will be, and this is not the only advantage. Give a workman an intricate, elaborate safety device, and he will distrust it. He will swear that it holds back production, and will have every possible excuse why he should not use it.

On the other hand instal a simple, yet efficient guard, explain the reasons for its installation, then watch the difference. It pays to take the workmen into your confidence. This statement may sound unreasonable to some, yet to prove its truth we quote the following experience of a large institution that decided to go into the matter of safeguarding all their presses. Their accident toll was very heavy, and they were, to say the least, worried about it.

Believing that what they did with their equipment was entirely their own business they proceeded to design and instal guards on all their machines. Imagine their surprise to find that the employees practically refused to use them, contending they would hinder production, thus reducing their wages. The development of this attitude was entirely unlooked for, and things were at a merry pass until the manager struck on the happy idea of getting the employees together for a heart-to-heart talk.

This was done, and at the talk he explained why the safeguards had been installed. Figures were quoted to show the extent of the accidents and the dire need of reform. "The idea is not a selfish one," he said, "but one that is for the workers' particular benefit. It cost real money, and lots of it, to put these guards in place, and our chief desire is to lower the accident hazard. Production will not decline, but will actually advance by the addition of these guards if you give them a fair trial. The very fact of their being there gives you confidence to work, knowing the punch cannot catch you unawares."

To make a long story short, the employees, approaching the matter in an entirely different spirit, found that what the manager had said was true, increased their production, and even went so far as to offer suggestions for further guards and improvement. Thus by taking the workers into their confidence this concern reduced their accidents from 120 to nil in one year.

Saw Guard

To proceed, however, to a description of Fig. 3. A is merely the gauge to which the work is held and there is another stop, not shown on the photograph, to locate work in opposite direction. B is simply a box inverted to protect the worker from the rising saw. The workman places the board to both gauges, presses his foot on lever shown, and up comes the saw.

An elaborate fixture on this work would have been superfluous, but by keeping to simple bounds the task is accomplished at practically no expense worth speaking of.

The last example, Fig. 4, depicts one of the styles of emery wheel guards adopted at this plant. They are of the dust-removing type, being connected up to the exhaust system in the usual manner. The guard is entirely home-made, and while not very classy to look at, or very elaborate in design, it accomplishes its purpose.

We leave the article at this point, not because the subject has been exhausted, but in the belief that sufficient has been shown and said to start the germ of thought working. Think to yourself, "Is there any machine in our plant that needs guarding?" Then, if there is such a machine, get it protected before it slips your mind. Accidents always remind us of this need, but it is better to lock the stable door before the horse departs.

FIG. 3—A NOVEL AND SIMPLE FORM OF SAW GUARD.

Hardening Screw Gauges With Least Distortion

How to Harden Screw Gauges With the Least Distortion Along the Pitch Line—That Was the Problem Confronting the Above College, and Herein is Explained How the Work Was Accomplished

By WILFRED J. LINCHAM, Goldsmith's College, London.

THE first law of hardening (known for many years) asserts that sudden quenching after heating retains the steel in the condition that prevailed at the quenching temperature.

The second law of hardening (of much more recent discovery) states that quenching, while crossing an arrest or recalescence point, produces the least amount of distortion, and it is one of the purposes of this paper to endeavor to prove the truth of this law. A second reason for the paper is to offer the author's results to other screw-gauge manufacturers for their immediate assistance.

The second law had already been used by other investigators and practically applied with material success, but the author was unaware of the exact researches that have led to the adoption of fixed temperatures. The Wild-Barfield process he believed to be the first practical example, the principle of which is to use the change in magnetic properties of wrought iron or steel when passing from the β to the α state to indicate the temperature at which to quench. Reference to the Roberts-Austen diagram, Fig. 1, shows this to occur along the Ar2 line, temperature 765 to 770 deg. C. (1,409 to 1,418 deg. F.).

The Horstmann Brothers arrived at a similar result by quenching at 700 deg. C. (1,292 deg. F.), which is the temperature of the eutectoidal line Ar1. The screw gauges they manufactured were produced, they said; with no alteration in pitch, but a slight swelling always occurred in diameter, which was lapped down afterward. They kindly communicated their method to the author, who thereupon decided to make his own investigations, the results of which are now set forth.

The author's first-sought ideal was a quenching temperature which would produce no distortion whatever, but it will be seen that this has not been achieved, nor does it seem likely that anything more than a minimum distortion can be arrived at. The Ar1 line for the use of the second law appeared to have manifest advantages over that at Ar2 because of its constancy, of temperature over a large range of carbon percentages; and it seems reasonable to expect that steels of any carbon content will act in the same manner when cooling across this line, if all are free from impurities.

Before the war the manufacture of screw gauges was confined to the use of high-carbon steel, say, from 0.85 to 1.2 per cent. carbon, which was usually left unhardened, on account of the difficulties caused by distortion on quenching. The making of hardened screw gauges had been practised to a very limited extent before 1914, the correction after harden-

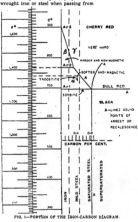

FIG. 1—PORTION OF THE IRON-CARBON DIAGRAM.

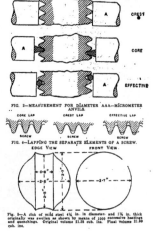

FIG. 3—MEASUREMENT FOR DIAMETER AAA—MICROMETER ANVILS.

FIG. 4—LAPPING THE SEPARATE ELEMENTS OF A SCREW.

Fig. 5—A disk of mild steel 4¼ in. in diameter and 1½ in. thick originally was swollen as shown by means of 1000 successive heatings and quenchings. Original volume 21.25 cub. ins. Final volume 21.60 cub. ins.

ing being obtained by the use of laps of cast-iron or boxwood. · The method taken generally has now been revived and extended, and is the only one followed at the present time, so a short description will be useful in order that the author's intentions may be understood.

Assuming that the gage can be hardened with so little change that the pitch tolerance, plus or minus, is not exceeded, which it will be shown can be done, the process of making a gage involves, first, the accurate screwing of the blank, and then hardening the screwed piece.

The first was carried out in a lathe accurately corrected in pitch by a cam bar. · The screwing tool, known as a form tool, was of disk shape and supplied with chasing teeth on its circumference. Two of these "form chasers" were used, both being quite accurate in pitch and of perfect sectional contour for the thread to be cut. One was used to rough out the gage screw to about half or two-thirds the thread depth; and the other, very carefully adjusted in position, to finish the thread correct on core, crest and effective diameters and full form. Constant measurement was, of course, necessary, but as the lathe slides were provided with micrometer disks, rapid and almost automatic manufacture was soon secured. The intention of the author at this stage of the work was to just clean up the gage with a full-form cast-iron lap, accurately made from a tap cut in the before-mentioned lathe. This method of lapping proved a failure, the abrasive clinging to the soft gage and causing "scoring," so he fell back· on the cutting tool itself for the accuracy demanded, assisted by the use of a very

mild steel of 0.14 per cent. carbon for the blanks that cut with a, polish. The object of the screwing operation was to produce a gage that was perfect in all dimensions while yet in the soft state.

The next step was to harden. The hardening of screw gages, and, indeed, some others, was performed in a cyanide furnace, made by John Wright, of Birmingham, which consisted of a rectangular cast-iron bath, covered and ventilated to remove the fumes, filled with sodium cyanide, and heated by coal gas and air, the latter being fed at a pressure of three pounds per square inch. The cyanide, which melted at about 600 deg. C. (1,112 deg. F.) and boiled at or near 800 deg. C. (1,472 deg. F.), was maintained at 750 to 770 deg. C. (1,382 to 1,418 deg. F.) when in use, the temperature being indicated by a thermo-couple pyrometer. The gages (of mild steel) were placed on trays suspended clear of the bottom of the bath and soaked from ten minutes to one hour, according to size, or until they had acquired the requisite depth of case.

The Cyanide Bath

If a greater depth of case than 0.005 in. be required, packing the gages with barium carbonate and bone black in closed boxes and soaking in a muffle furnace at about 850 deg. C. (1,562 deg. F.) must be resorted to, but the final heating should be completed in the cyanide bath.

It is possible to allow the cyanide bath to cool to the quenching temperature (supposing that to be exactly known) after the requisite soaking of the gage, but the process absorbs much time and

increases expense, and is not to be advised for continuous working. Also the temperature may be ascertained by the color of the gage, which may be taken as a very dull red, but this is not advisable. The author therefore installed, in addition to the cyanide bath, a Brayshaw salt-bath furnace, using "pyromelt," and kept that furnace at the constant quenching temperature. This furnace was called the "store furnace," for when a gage had had its full time in the cyanide it was stored here at quenching temperature until its heat had evened up; and it was afterward quenched to the store furnace and of ample capacity, with inflow and outflow cocks.

The screw gage was next measured for pitch and diameters. The diameter measuring machine has already been illustrated in the Proceedings, 1917, page 54, and Fig. 4, page 55, and the mode of its use is merely suggested in Fig. 2 (page 5), which shows the micrometer anvils at the moment of measurement.

The Pitch-Measuring Machine

The pitch-measuring machine, Fig. 3, was designed by the National Physical Laboratory and constructed at Goldsmith's College. The drawing is somewhat diagrammatic in order to show the principle clearly. The gage is supported between centres on the main or fixed casting. A movable carriage supports an indicator on the left and a micrometer on the right, being always pressed leftward by means of a weight. The indicator carries a stylo, which engages with the screw thread, and a pointer in connection therewith, whose reading is

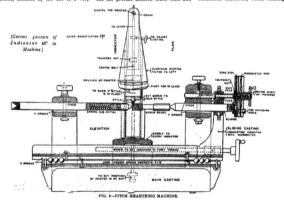

FIG. 3—PITCH MEASURING MACHINE.

multiplied so as to be in the region of one-millionth of an inch of sensitiveness. The micrometer abuts against the main casting and can be read to 0.000001 of an inch and very rapid charting of screw errors can be obtained.

After hardening, it was found that the screw pitch had not distorted beyond the tolerance, plus or minus, if precautions had been taken such as suggested later. The diameters, however, had increased, at least the core and crest were larger, but the effective surface was often unchanged. The operation of lapping followed, three laps, Fig. 4, being used; but the enlargement being only about 0.0001 to 0.0003 in. in diameter, the amount of lapping was very little. Naturally the operation was of a delicate character, and was the particular part of the whole process that most held up delivery; many gages were spoiled in its exercise; but it is believed that the whole hardening process will yet be so perfected that only a full-form cleaning first, will be necessary.

Gages Examined by Optical Lanterns

Finally each gage was examined by an optical lantern magnification of exactly 50 to one, as already described in previous papers; and, of course, the pitch and the diameters at all three elements of section were thoroughly tested during the stages of manufacture and just before delivery, by a competent staff

of examiners. It will be said that if only a screw gage could be made to travel through its various manufacturing steps in such a simple and well-oiled manner, the difficulties of screw gage making would be completely overcome. It is one intention of this paper to show that the difficulties of manufacture are being overcome by the adoption of definite scientific method, and there are strong and practically certain hopes of doing away with lapping altogether.

The crucial point of the manufacture is the hardening. Lathes have been specially built that will produce accurate pitch, and mild steel is being used that can be finished in those lathes with a polish. It will be shown that a quenching temperature can be found that only lengthens the pitch by 0.0002 in. over 0.7 in. of length, which is within the pitch tolerance, 0.0003 in., on such a length for inspection gages of correct effective diameter.

Two remarkable examples of distortion are shown. The first, Fig. 5, was, before treatment, a cheese-shaped disk of 4¼ in. diameter by 1½ in. thickness, of common mild steel. It was heated in a coke fire and quenched in water, the operation being repeated about a thousand times. The continued distortion has swelled the disk into an approximate spherical form, but that the volume is unchanged was ascertained by displacement measurement.

The second specimen, Fig. 6, was at first a rectangular disk, and here again the spherical form is attempted, the treatment being the same as in the first example.

These two specimens came to the author as a piece of exceptional good fortune, being presented by F. A. Thompson. Their history is: A garage boy required buckets of hot water at meal times and kept three of these pieces hot on coke fires, using which he pleased to heat the water by immersion, and continuing his unintentional experiments over a period of about a year.

The author commenced his experiments in April, 1918, by providing seven specimens for each experiment, the material being J and L steel of 0.14 per cent. carbon. Each specimen was ground and carefully surfaced by stoning the ends, and the dimensions were about 9-16 in. in diameter by 0.7 in. long. Previous experiments upon pieces of full gage shape showed the wisdom of using simple cylinders for the temperature tests so as to obtain symmetrical distortions.

One form of distortion resulting from hardening a cylindrical piece of steel was found to be a swelling or bulging at the ends, but the length, measured at the edges, remained very little changed indeed.

The experiments were begun without any clear notion of the form the dis-

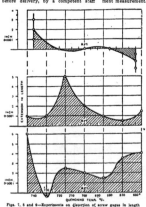

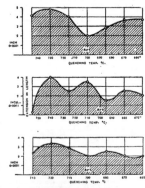

Figs. 7, 8 and 9—Experiments on distortion of screw gages in length only; viz.: On pitch line. J and L mild steel 0.14 per cent. carbon. Cased in sodium cyanide for 30 min. at 750 deg. C. Specimens 9-16 in. diameter by 0.7 in. long. Fig. 7.—Average change of length (measured over the end surfaces). Fig. 8.—Distortions all extensions. Fig. 9.—Curve which seemed to indicate no fixed law.

Figs. 10, 11 and 12—Experiments on distortion of screw gages in length only: Viz.: On pitch line. J and L mild steel 0.14 per cent. carbon. Cased in sodium cyanide for 30 min. at 750 deg. C. Specimens 9-16 in. in diameter by 0.7 in. long. Fig. 10.—Curve which indicates that 700 deg. C. (1,292 deg. F.) is the best hardening temperature. Fig. 11.—Wavy curve with minimum at 795 deg. C. (1,263 deg. F.). Fig. 12.—Minimum at 700 deg. C., but also at three other temperatures.

tortious were likely to assume, and axial measurements were at first made, which proved to be wrong for the purpose required. Also the ends of the cylinders were not parallel to less than 0.0001 in., and the average dimension was taken in the hard and in the soft states respectively. When, however, the circumference was divided into three or four parts and marked, and the change of length taken at the marked positions only and afterward averaged, quite consistent results were obtained, the figures fairly representing pitch line distortions on a screw gage.

Figs. 7 to 12 show, for six sets of experiments, changes of length in units of 0.0001 in. plotted on a base of quenching temperature.

Fig. 7 (April 5, 1918) is a record of the average change of length measured over the end surfaces, and contains the axial bulging. It was on this set that the bulge was first noticed, and from this time onward measurements were made on the edges only. The temperature of 700 deg. C. (1,292 deg. F.), however, is indicated as that of no distortion.

Fig. 8 (April 24, 1918). The distortions are all extensions, and 670 deg. C. (1,238 deg. F.) indicates least distortion. Several screws were hardened at this figure with success.

Fig. 9 (April 27, 1918). This curve was distressing. It seemed to show that there was no fixed law and that nothing would result from the labor. The figure 685 deg. C. (1,265 deg. F.) was taken as best, for the drop at 730 deg. C. (1,346 deg. F.) might not be trustworthy.

Fig. 10 (May 4, 1918). A good curve showing 700 deg. C. (1,292 deg. F.) to be decidedly the best temperature. Two specimens were quenched at 670 deg. C. (1,238 deg. F.) and are shown to fairly agree.

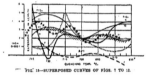

Fig. 11 (May 9, 1918). Another wavy curve, with minimum at 695 deg. C. (1,283 deg. F.).

Fig. 12 (May 11, 1918). There is here a minimum at 700 deg. C. (1,292 deg. F.), but also at three other temperatures.

The problem, which now appeared hopeless, suddenly cleared itself. It occurred to the author to plot all the curves on one sheet by superposition, as in Fig. 13, and this operation was at once fruitful in showing that the distortion, or, as it may now be called, extension, had definite laws and was not the result of mere chance. Disregarding the first curve shown dotted, all cross at exactly one point, namely, 700 deg. C. (1,292 deg. F.), the recalescent point Ar1 of the equilibrium diagram. The extension is, unfortunately, not obliterated, but it has a fixed and definite value of 0.0002 in., and is within the tolerance limits. It indicates a condition of stability, and if 700 deg. C. (1,292 deg. F.) be made the quenching temperature, this extension can be allowed for when cutting the screw.

Other parts of the curves show nothing but instability. The use of the position 700 deg. C. (1,292 deg. F.) therefore is calculated to assist correct manufacture, while all the other temperatures are unreliable. On that account, an at-

tempt to find a stable temperature of maximum extension, for use when needed, was practically unsuccessful; for gages quenched at 715 deg. C. (1,319 deg. F.) did not always extend well. A very important point is that the temperature of stability being so very limited.

In Fig. 14 the distortion scale has been magnified and plotted for a range of 15 deg. ∓ of 700 deg. C. Maximum instability occurs at 715 deg. C. as well as maximum distortion, and in all cases temperatures below 700 deg. C. are safer than those above. A fair and safe latitude for quenching lies between 690 deg. C. (1,274 deg. F.) and 700 deg. C. (1,292 deg. F.).

The author found no difficulty in making sure of 700 deg. C. with the use of a platinum-rhodium thermo-couple, a thermometer within a galvanometer-case giving warning of any change of temperature of the cold junction. Upon retesting the thermo-couple after about six months' use it was found by the makers to be quite accurate.

It now appears that the second law of hardening proposed at the commencement of this paper is proved so far as the line at Ar1 is concerned, and could no doubt be proved for other arrest points.

In the early stages of this research

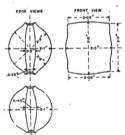

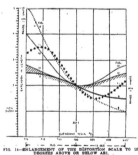

measurements of distortion on diameter were made, which always showed an increase of about 0.0001 to 0.0003 in., but the purposes of the research being the measurement of length distortion in screw gages, and time being pressing, diameter measurements were not proceeded with.

Before obtaining the pyrometers, which were not in use before April 5, some experiments were made on a color scale extending through

Bright red
Gray just appearing (due to freezing of the cyanide)
Full gray
Dull red
Black just appearing
Mid black
Full black

the specimen being held in a half light. The results pointed to a best temperature between dull red and black just appearing. Some experimenters have further perfected the color test, but the author considered the pyrometer method undoubtedly superior, and with his system of two furnaces could be practised by a somewhat unskilled operator.

Further experiments are needed on ring gages, which have laws of their own as regards diametric distortion, and the author had hoped to make experiments on other gage shapes.

It has not been exactly discovered what the first stages of distortion assume, for in the boy's experiments the diameters are all decreased,'though it is certain they increased at first. It was believed that a barrel shape was first produced, coupled with end swelling, but careful test has shown a festoon shape on the rim, and it must be always remembered that nature insists on keeping the volume constant.

Instead, however, of waiting till all these facts could be obtained, it appeared wiser to make known the results that had already been acquired, in order that screw gages might be produced more easily and accurately by every manufacturer in a time of need.

The fact of the emulsive constitution of steel at Ar1 led the author to believe that gages quenched on this line would not be thoroughly hard. He was glad to say that, on the contrary, he had found them in every case to be glass hard. It must not be forgotten that the process of hardening by casing in cyanide produces a high-carbon steel at the surface, while the interior remains in the mild state, and this may have something to do with it.

The most important of the conclusions arrived at in this paper are:

(a) The second law of hardening is proved.

(b) A temperature of minimum distortion that can be relied upon is 700 deg. C. (1,292 deg. F.).

(c) No variation is allowable above 700 deg. C.

(d) An allowance of 10 deg. below 700 deg. C. is practically permissible.

THE PETROLEUM SITUATION IN THE CAUCASUS

The scarcity of the petroleum oils and the rising prices of the same in this country give a special interest to a report that has been made on the petroleum situation in the Caucasus by M. Aldebert as published in Le Geine Civil.

Communication with the oil regions of the Caucasus, which was becoming more and more difficult since 1914, was cut 'off entirely in 1917 because of Bolshevism and the Turco-German occupation. At the present time we are still very poorly informed on the subject. Widely different and contradictory reports are in circulation, and the report of M. Aldebert sets out to classify the situation as it existed on January 1, 1920.

Almost the whole of the Russian oil production comes from the Caucasus. In 1916 this amounted to about one-sixth of the total production of the world, and ranked Russia as next to the United States as a' producer of mineral oil.

Although the Caucasus forms a continuous oil-bearing basin, for indications are found upon both the northern and southern slopes of the mountains, M. Aldebert divides the territory into three sections where the principal developments' are concentrated:

1. The Maikopp section, which yields about 1.2 per cent. of the Russian output.

2. The Grosny section, which yields 17 per cent.

3. The Baku section, yielding 78 per cent.

In the Maikopp section, which is divided into three groups, there are 138 wells, of which only 38 are worked. These have a daily yield of about 67,000 gallons. The principal group at Maikopp is connected with the refineries at Ekaterinodar by an 8-inch pipe line about 87 miles long. The other two groups are connected by pipe lines to the tanks on the Ekaterinodar and Nova-Rossisk railway. The production of the Maikopp section rose from 503,000 gallons in 1908 to 47,140,000 gallons in 1912, but fell to 10,309,000 gallons in 1916, while in 1919 the output had again risen to 15,400,000 gallons.

These operations have suffered very little from the war, being situated in the centre of the provinces of Conban and Tereck, where the population is Cossack and, from the very outset of the revolution, have resisted the Bolshevik movement and have made up the nucleus of Denikine's army.

At Grosny the section is divided into two parts, that of the Old and New Rayon, located at a distance of about 9 and 3 miles, respectively, from the city. The operations at Old Rayon have not been destroyed, either by the revolution or the uprising of the populace, but at New Rayon the operations have been almost totally destroyed.

In November, 1917, seven flowing wells were fired and burned until February, 1919. Some of these wells had a yield of more than 283,000 gallons a day. An immense layer of lava produced by the burning oil covers the slopes of the hills

to a depth of more than a foot. This lava consists of a petroleum coke and is actually used as a fuel under boilers. The wells which were not burned were destroyed with the axe and dynamite.

At New Rayon, out of 122 wells which existed before the revolution, only 5 are now being worked; 7 are being repaired, while the others are idle or have been abandoned.

The operations of the Grosny basin were begun in 1896, and grew continuously, except for a slight fluctuation during the first revolution of 1906, and reached a total yield, in 1917 of 550,300,-000 gallons; but it would have reached 786,000,000 gallons had transportation facilities been sufficient.

In addition to tank cars there were 8-inch pipe lines leading from the district to the Caspian and Black Seas which, before the war, had a daily capacity of 10,310,000 gallons, while to-day, after repairing the pipe line destroyed in 1917 and the two pumping stations out of four, the capacity is only 471,000 gallons a day.

At Baku the production of the Akcheron peninsula, which was 258,000,000 gallons in 1882, steadily increased until it reached its maximum of 3,457,000,000 gallons in 1901. The revolution of 1905 cut the production of this section down to 2,106,000,000 gallons, but as matters became settled it rose to 2,467,000,000 gallons in 1916, although the scarcity of repair parts, machinery and material began to make itself felt in consequence of the war and the closing of the Dardanelles. The year 1918 brought in anarchy, and industry was completely paralyzed.

The year 1918, with a production of 630,000,000 gallons, saw the nationalization of the petroleum industry, followed by a Turkish occupation after several days of bombardment with its horrors and massacres. The economic disorganization is complete; there are now neither new operations nor exportations, and output is reduced to a minimum, so that we have to go back to 1889 to parallel it.

At the end of 1918 English and Russian troops occupied Baku anew. It had suffered from bombardment, but the operations were intact.

In 1919 a Tartar government was established, which later became the Republic of Azerbeidjan and was actually recognized by France. When order was established in August, 1919, the English evacuated Baku and, contrary to the forecasts, there were no massacres and no disorder.

But oil production was reduced to a minimum because there were almost no exportations, Astrakan being in the hands of the Bolsheviki and Astrakan, before the war, handled about 75 per cent. of the Baku exportations.

In order to span this crisis the government advanced money to the operators in order to enable them to carry forward new operations, but as the tanks are full the situation is a difficult one. The ex-
(Continued on page 64)

WHAT OUR READERS
THINK AND DO

DOUBLE FACING TOOL
By F. SCRIBER

To reduce the time required for facing discs which were held in a chuck in a lathe, the tool holder shown in Fig. 1 was used. The fundamental feature of this is that one tool (say for instance A) is set for rough facing, while tool B is set so it will face to the finished size. Having these two tools A and B set in the block C they are clamped securely thereto by the square headed screws shown. The necessary top and side clearances are ground on these tools, so they need not be set on an angle. The other units are provided to bring the tools successively into position for doing the facing, in other words the block complete with tools is mounted on a cross-slide of a lathe with tool A set for roughing size. The tool is then fed across the work in the usual manner thereby completing the roughing out. After this nut D is loosened by the operator, which allows block C to be lifted above tongue E by means of spring F. The block C is then turned by hand quarter way around, bringing tool B into position now occupied by tool A, and the finishing cut is taken across the work after tightening nut D, which compresses the spring and holds block C tight in position.

The various details which are used in this connection have been shown below

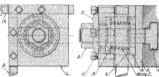

FIG. 1—THE ASSEMBLED TOOL READY FOR USE.

the assembly in Fig. 1 and in Fig. 2. These consist of a centre stud which is used in the centre of the block and holds the various members in place, a locating block H upon which the spring rests and above which the block is centralized by locating sleeve I and a clamping nut J used inside the swivel block C, the base block K is shown in Fig. 2, upper right-hand corner. Over this is slipped the swivel block C with locating sleeve I and block H between them.

When assembling this tool, locating block H is pinned to base K, tool-carrying block C has locating sleeve I pinned to it, clamping-nut J is screwed over centre stud G with a spacing collar M, which is not detailed between it and locating block H, thereby holding base K

securely to the lathe cross-slide and locating it square by means of the tee section of the bolt in the cross-slide tee slot. Tool-carrying block C with locating sleeve I pinned to it is next put in position with spring in place, following which washer N is put over the centre stud. The nut is put in place and the arrangement is used as previously described. This type of block could also be arranged for turning by suitably locating the cutting tools.

It will be obvious from this description that loosening the hexagon-nut D about one revolution allows the swivel block to lift the tool-carrying block clear of the tongue on the locating block thus allowing the swivel block to be turned the necessary amount.

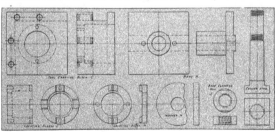

FIG. 2—ILLUSTRATING THE VARIOUS DETAILS OF THE TOOLS.

MACHINING ROD BRASSES

By Clyde D. Thorburn

For a considerable period we have been up against it for a quick method of machining the strap fit on main rod and side rod brasses. At one time our small crank planer worked about sixteen hours per day and could keep up with amount required, but with an ever-increasing demand for these brasses it was deemed necessary that we should endeavor to find some quicker means of production.

We used a long angle plate on planer and set up a row of brasses against it, but with the extra setting required we did not gain anything on cost of machining. In these days of high rates of pay it becomes more and more necessary for us to cut down cost of production, and as is well known, suitable jigs not only mean a reduced cost but uniformity of pieces and more per day.

Jigs as shown were designed specially for this purpose, and we expect to gain these favorable results. The jig is designed to suit a Bateman high-speed planer and to hold 10 brasses (5 on each side, two heads working). The body A is a double-angle plate, planed on base B to suit table of planer, and on both outside and inside faces. In the outside faces 5 holes 3 in. x 1 in. deep are bored at suitable centres (according to the size of our largest brasses) for cones C to fit into, and further holes tapped 1½ in. diam. to take studs D, which hold the step cones in position. Directly above these stud holes five more holes are drilled to take the sleeves E containing the locating pins F and springs G. Step cones are made of cast iron, and sizes of steps correspond with all our common classes of brasses.

Across the face of each cone four centre lines are cut to enable jobs to be set quickly and without the aid of surface gauge. In these cones four ¾ in. holes are drilled right on dead quarters to take end of locating pin. This is for brasses having four square sides. For brasses having one or more sides taper, other holes would have to be drilled in cones to give that correct taper.

Brasses are faced on both sides and bored out to standard size. If bore is not required to be standard or probably left rough, these brasses would only be

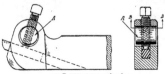

DETAIL VIEW OF THE TOOL HOLDER.

bored ¼ in. deep, or sufficient to fit on cone. Before brasses are bored they have previously been sweated together. In setting brasses on jigs, planer hand must see that joint of same is right on centre line marked across the face of cones. He applies the big washer H (which has a slot extending from centre hole to outside diameter to enable quick application and removal) and tightens up, only having one nut on each brass to tighten. An ordinary brass roughing tool will be used for planing between flanges, and a square nose-finishing tool to plane down inside of each flange. Probably a square nose tool would do the complete operations if the amount of stock is not too great. The first side of brass machined, all that is required is to slacken the nuts on inside of jig, pull back the locating pin and revolve cone until locating pin falls into next position. A small jack I is used (made of a nut and set-screw) under front end of brass to prevent a possible slipping of brass on cone. This jig practically eliminates the setting, previously a part of the operation that accounted for a lot of time.

During 1919, France, England and Sweden were the principal buyers of cash registers, which, together with parts, amounted to $4,375,526.

MODIFIED TOOL HOLDER

By Tyke

Time in a small shop is of vital importance, but no matter how careful one is small tools are very apt to get—well not exactly lost, but very badly misplaced. This is especially true when any device is made up of two or more loose pieces. These are liable to be dislodged by vibrations or shocks. Such an occurrence happened the other day and the part lost was the cam out of a tool holder.

Tool holders when desired are wanted at once, and mean lost money. The holder and wrench were located in their customary places, but the cam had vanished. After considerable time had been spent in the usual shop hunt for the lost parts, it was decided to improvise. When looking at the main body of the holder and seeing only a hole instead of the familiar cam it looked almost hopeless, but after a few minutes' cogitating a very simple, practical and efficient method was decided upon.

A piece of 3-4-in. cold rolled steel about 7-8-in. long was cut off, drilled and tapped for a standard 3-8 set screw. The upper portion of the tool holder was drilled 13-22-in. and the job was completed. The cold rolled piece "A" was placed in the hole and the set screw inserted and screwed home. It will be clearly seen that as soon as the set screw bore down on the tool bit the piece "A" rose in direction of arrows "B" until its progress was arrested by the upper part of the hole, after which the tool bit proper received the full pressure exerted by the set screw.

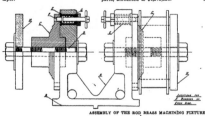

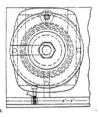

ASSEMBLY OF THE ROD BRASS MACHINING FIXTURE.

Would You Like to Receive One of These Scales FREE?

You can do so very easily, at the same time increase your knowledge and powers of perception.

The scale is 6 in. long and is made from finest quality steel. One side is marked in 32nds, the other side in 64ths. A table of decimal equivalents is also stamped on one side, and a table of tap drill sizes on the reverse side. This scale is well worth securing.

Rules of the Contest

The answers to this novel contest continue to pour in, and for the benefit of those who do not understand its value and scope let us explain. Every week we publish a list of puzzling queries similar to those below. It is up to the reader to read the query, turn to the advertising pages, and find the firms we refer to. These puzzling references are interesting to follow, and will quicken your powers of perception. It costs nothing, you get acquainted with the different lines, and have an opportunity to win a prize worth while.

The Lucky Prize Winner for Nov. 4th Issue List is GEORGE PILGRIM, Box 407, Dundas, Ont. He Had All Answers Correct.

INTERESTING FACTS TO BE FOUND IN THIS WEEK'S ADVERTISING SECTION

1. Something that wins friends even under adverse conditions.
2. A product that is shipped in four ways.
3. A machine that tells the cold blooded truth.
4. How to assure accuracy.
5. A machine that will produce two distinct classes of work at the same time.
6. An interesting operation economically performed.
7. How to save time, labor and money.
8. A product that is said to be breaking production records.
9. Something that can be cast quickly and at slight cost.
10. A product that can be made to meet individual requirements.
11. A book is often dull, but somewhere in this issue we read of another article that gets dull, and like the book, should be laid aside. What is it?
12. Something about a machine that when it is in place is as firm as the spindle itself.

Make a Start and Send in Answers Promptly. Closing Date for This List is Dec. 9, 1920

DEVELOPMENTS IN SHOP EQUIPMENT

CAM-GRINDING ATTACHMENT

In designing cams of various shapes for automobile crankshafts, the tolerance has gradually been reduced so that grinding machines are called upon to produce many difficult shapes to meet the requirements of accuracy. In order to obtain a silent cam, clearances have been reduced and at the opening and closing points in some cases there is a slight rise in a number of degrees; for example. 0.005 in. in from 20 to 25 degrees: To provide for these requirements a new attachment for use in a cam-grinding machine has been developed by the Landis Tool Co., Waynesboro, Pa.

The accompanying photographs show the attachment mounted on the Landis 10-inch plain self-contained machine. The camshaft A is driven by dog B, which in turn is driven by driver C, mounted on spindle E and is adjustable to facilitate in setting camshaft in proper relation to master cam. On this spindle is worm wheel F, the spindle being worm driven throughout. On spindle E is master cam G, which is made from solid steel and mounted on spindle by means of taper and nut. In contact with the master cam is roller H, which is moved from cam to cam by means of lever I. The relative position of the roller is indicated by pointer J. By means of a cam on shaft K the roller and master are disengaged by a half turn of lever I and the roller shifted to the next cam by completing the turn. Thus, the roller and master cam are disengaged and roller shifted to proper position by a single movement. On spindle is also placed brake L, its object being to compensate for any lost motion. which would be reflected on cam as the pressure is reversed when passing over the nose.

The outstanding features of the new underslung cam grinding attachment as explained by the manufacturer are as follows: The swing bracket is tubular in the cross section and is well supported directly under the master cam and work centres close to the machine bed. This construction balances the swinging bracket, eliminating all strain. The work being carried directly over the fulcrum centre reduces the vertical movement and grinding contact from changing over the periphery of the wheel, thus cutting down to a minimum the difference in contour of cams caused by the changing diameter of the grinding wheel. The

DETAIL VIEW OF THE CAM ATTACHMENT.

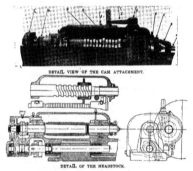

DETAIL OF THE HEADSTOCK.

GENERAL VIEW OF THE GRINDER.

master cam being mounted directly on the headstock spindle, substantially connected and in line with the work, eliminates lost motion and errors. The master cam is kept in contact with the master cam roller by means of compression springs, the tension being adjustable to suit different conditions.

It is pointed out that accurate work is secured when the grinding wheel is of a diameter equal to that of the roller used in contact with the model cam when generating the master. This condition is caused by the contact between the work and grinding wheel travelling above and below the wheel centre. The more the grinding contact varies from the wheel centre, the more pronounced the variation in the contour of the cams ground with different sized wheels. Due to the work being swung an equal distance from a point directly over the fulcrum centre, the movement of the work is practically in a horizontal plane, reducing to a minimum the travel of the grinding contact above and below the wheel centre.

GEARED HEAD TURRET LATHES

The Warner & Swasey Co., Cleveland, Ohio, have placed on the market what is known as their No. 4 and No. 6 geared head construction turret lathes. These lathes have a geared-head construction which is said to deliver four times the power transmitted by the ordinary geared friction head of the same size machine, and twice the power of the double friction back-geared type. Even with this increased power, less effort is required to move the controlling levers; consequently the machine is easier to operate.

The front view of the No. 6 turret lathe with a geared head is illustrated in Fig. 1. The first model of this type built has been thoroughly tested under every condition that might be encountered in service, and six machines of each size have been installed in outside shops for similar experimenting. The results obtained from all these machines have been very satisfactory.

Fig. 2 shows a view of the geared head with the cover removed, from which it

DETAIL VIEW OF GEARED HEAD.

will be seen that the steel gears are of a coarse pitch and have wide faces. The gears run in a bath of oil which also lubricates the bearings. Two sets of gears are mounted on the front shaft, while a third set and the reverse friction clutch are mounted on the back shaft. Twelve spindle speeds and a reverse rotation are obtained. Another advantage claimed for machines of the geared-head construction is their adaptability to the various types of motor drive. The head may be geared directly to the motor or driven from it by means of a chain or belt.

The No. 6 turret lathe can be equipped with the standard carriage used on Warner & Swasey turret lathes, one of which is shown on the machine illustrated, or with a heavy-duty carriage that is specially valuable when taking heavy cuts or performing heavy forming operations. A range of six power cross-feeds is provided on this heavy-duty carriage for facing, forming, and cutting off work at various speeds. The No. 4 turret lathe is equipped with a standard carriage. The maximum capacities of this turret lathe are as follows: Diameter of bar stock, 1¾ inches; length turned, 10 inches; swing over bed, 16 inches; and swing over cross-slide, 7

inches. The maximum capacities of the No. 6 turret lathe are as follows: Diameter of bar stock, 2¼ inches; length turned, 12 inches; swing over bed, 20¾ inches; and swing over cross-slide, 9¼ inches.

AUTOMATIC MILLING MACHINE

The Potter & Johnston Machine Co., Pawtuckett, R.I., have placed on the market what is known as their No. 1M automatic horizontal milling machine. This machine has been designed in response to a demand for a smaller automatic milling machine built upon similar lines to their present 2-M automatic milling machine.

In designing this machine, two tables have been used, making it a duplex milling machine, a feature which will prove valuable in a great many ways, this arrangement making it possible to handle two distinct classes of work upon the same machine without the operator moving away from the operating position. On the automatic milling machine the operator is loading up the fixture on one side of the table, while the cutter is milling work mounted on the other side, and with the exception of ten seconds required for indexing the table turret and advancing the table automatically to the cutting point there is no lost time, thereby insuring a greatly increased output. As soon as the operator has loaded the fixture on one table he performs a similar operation on the other table, keeping both sides of the machine loaded with work. There are no overhanging arms or projections to interfere with the operator while replacing work in the vise or fixture and all controlling levers are convenient to the operator. There is no waste of power as the machine is practically continuously milling.

The table base carrying the work tables is at a fixed height, has walls of ample thickness, well braced, and runs to the floor, giving a rigid construction and solid foundation. In this base is the

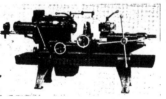

GENERAL VIEW OF THE TURRET LATHE.

mechanism for revolving the table turret and feeding.

The tables are mounted upon circular saddles or turntables, and the rotation or indexing of these is of the turret principle. Each table has two tee slots, and a wall which runs entirely around the table serves the double purpose of providing a channel for lubricant as well as supporting the vises or fixtures, which are greater than the width of the table, the top of this wall being finished off level with the top of the table. One vise or fixture is carried on each side of the table.

The feeds for each table, which are independent of each other, are accomplished by .trains of gears. Sixteen changes of feed are obtainable in practically geometric progression, ranging from 13-16 to 17⅜ inches per minute. With each set of change gears two variations of feed are obtained by changing the hand lever at side of machine to either slow or fast position. The feeds for each table are also independent from the spindle speeds.

Provision is made for operating the tables by hand, also revolving the turret tables when setting up. Automatic stops are provided which are adjustable to suit the different working positions at which tables may be set. When indexing, the tables come to their working positions rapidly and automatically, thus avoiding

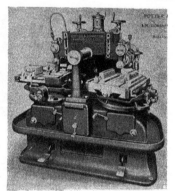

THIS VIEW SHOWS BOTH TABLES CLEARLY.

GENERAL VIEW OF THE MILLER.

any hand motion to bring the work up to the cutters. As soon as indexing is completed, the table feed is automatically thrown in. An independent, fast table feed is provided which is a time saver when straddle milling, groove milling, or space milling is being performed, as it brings the work clear of the cutters before the table revolving motion is thrown in.

The revolving and indexing is accomplished by foot pressure on lever at front of machine (each table having its own lever) and the revolving mechanism brings the table turret to the exact position before the lock bolt takes its seating. Ten seconds only are required to revolve and index the table turret, and this is the only period when the cutters are not cutting metal. The design permits of all automatic mechanism being operated by hand.

The spindles are made of crucible steel forging, ground, and running in bearings of phosphor bronze. Compensation of wear is provided in spindle bearings; the front spindle bearing is tapered and the rear spindle bearing straight. The spindle nose is clutched for driving an arbor; the arbor is held in place by a draw-back rod through the spindle, and the spindle is tapered at the front end to receive arbors. The spindle is carried on a vertical adjustable slide, which is in turn carried in a horizontal adjustable sleeve, which is responsible for the traverse of spindle. The adjustment for both spindles is independent. Graduated dials facilitate the settings.

The spindle drive is actuated by means (Continued on Page 64.)

The MacLean Publishing Company

LIMITED
(ESTABLISHED 1887)

JOHN BAYNE MACLEAN, President.   H. T. HUNTER, Vice-President
H. V. TYRRELL, General Manager.

PUBLISHERS OF

CANADIAN MACHINERY
~ MANUFACTURING NEWS ~

A weekly journal devoted to the machinery and manufacturing interests.
B. G. NEWTON, Manager.   A. R. KENNEDY, Managing Editor.

Associate Editors:
J. H. MOORE.   T. H. FENNER.   J. H. RODGERS (Montreal)
Office of Publication: 143-153 University Avenue, Toronto, Ontario.

VOL. XXIV.   TORONTO, NOVEMBER 25, 1920   No. 22

PRINCIPAL CONTENTS

The Price of Machine Tools

THE automobile situation must give an effective answer to the idea that cut prices will bring about a worth-while buying movement.

When the first cuts were announced, there was some buying in almost immediate response to the move, but, upon analysis, it was found to be almost entirely dealer trade.

When these orders were filled the movement was largely stopped, as there was no buying on the part of the ultimate consumer.

And that is where the business in cars stands to-day. The cut has apparently been made in vain.

Machine tool men will bear this out. Reductions have been made in some cases, but is there evidence that it has brought in a corresponding amount of new business?

The feeling at the National Machine Tool Builders' Association was against reduction, claiming that no good would result. Against that was the minority report, expressed at the Dealers' Convention, "Prices must be made lower on machine tools, for the buying expects it. I think that a general reduction is almost essential.'

The trouble with all conventions is that they bring together men with the same mind, leaving aside for the time being the men to whom they look for the bread and butter of the business, viz., the buying public.

The buying public can be conveniently ignored at conventions, but they must be reckoned with when the dealers get out to put into actual operation the extra steam generated at conventions.

The dear general public may demand lower prices in machine tools, but the thing the dealers and makers would like to know is—would they come along and buy simply on the strength of a price reduction?

The Joy of Accomplishment

MR. J. H. BAIRD, founder of the Baird Machine Co., of Bridgeport, Conn., has died at the ripe old age of 93 years, and his passing will be regretted by lovers of machines and inventions.

Mr. Baird will not live as a great business man, although he was wise enough to surround himself with good business executives. He will be remembered and loved for his own words, "Well, I don't care if I did lose money, I made the dern thing work." That is the frame of mind that extracts pure, unadulterated pleasure out of accomplishment. He would take more out of gaining his point in mechanics than out of making a million in cash.

It was this feeling that made Mr. Baird the success he was as an inventive genius. Were he alive to-day he would probably object to the term "genius," but he would have to admit that at least it was the reward of patience. He made pins. He did away with the idea that ?it takes ten men to make a pin." He made pins and made machines that put them in the papers—and after that he made another machine that made safety pins.

Many of his inventions bore out the fact that an inventor is often a poor business man. For instance, years ago Mr. Baird made a machine that fastened the hoops in hoop skirts. This he sold for $30, whereas the man who purchased it made a resale for $50,000, which was almost a fabulous sum in those days.

And there is nothing on record to show the regret of the inventor. His name will live as the inventor of that machine, whereas the man who made the $50,000 out of selling it has been forgotten long ago.

The inventor put something real and lasting into the world—the man who made himself rich over a short handling did not.

Mr. Baird was a relic of the old school, where accomplishment was counted greater than reward, where, in fact, accomplishment was its own reward.

And after all is said and done, it's a pretty fair record to leave in these money-mad days:—"Well, I don't care if I did lose money—I made the dern thing work."

A profit-sharing scheme is a popular thing when there are profits. A plan that will hold interest in the lean days is more to the point.

German engineers are working to secure a substitute for coal, the idea being to draw electricity direct from the atmosphere. The idea is well within the range of possibility that in years to come our present expensive power plants will be scrapped, and mechanics and engineers of years to come will regard them as we do the tread mills of past generations.

What the Business Papers Say

Paying on Basis of Results

Engineering, London.—"When we turn to the question of the acceptance of payment by results by the unions as a whole we are again obviously at once confronted with difficulties. In this case, however, we think the opposition would come more from certain unions officially than from the individual work people. This should make the matter more easy to deal with. Various systems of payment by results are, of course, in operation in many trades, and in general are certainly favored by the people who work under them. In other trades the system is forbidden, but we know of more than one case in which such a system is being worked to in certain shops with the full concurrence, and even at the desire, of the work-people in defiance of union rules. A general acceptance of payment by results is not going to be granted off-hand, but the bringing about of such an acceptance is a sufficient possibility to make the matter well worth bringing forward in a definite way."

Where Prices Really Stand

British Trade Review.—Common talk on this question of price-reductions has been often misleading. To a large extent recent downward movements are due to abstinence from buying, and it is quite clear that this holding off on the part of purchasers cannot go on indefinitely. Since present quotations in some cases represent the minimum of profit, or even a loss, to the seller, it is clear that the position is to some extent artificial, and it would seem to follow that when buying is resumed prices will harden in the classes referred to. The world is still hungry for goods, and when activity is resumed in the markets it is more than probable that buying will be conducted on a large scale. Under the circumstances the question arises whether the present would not be a suitable time for buyers to consider the advisability of recommencing operations.

MARKET DEVELOPMENTS

The Markets Afford an Interesting Study

Many Sides Are Shown This Week—U.S. Steel Prices May Come Down Below Corporation Levels—New York the Scene of Gatherings of Dealers and Makers, But No Price Announcement Results.

THE conventions of both makers and dealers in machine tools have come and gone, but so far they have not been followed by any announcement regarding reduced selling prices. At both of these meetings the point was made a good many times that present costs are very high and not out of line with the selling prices. Cancellations were also spoken of, and it was found that conditions in United States in this particular are probably worse than in the Canadian field, as the number of cancellations here has been comparatively small in all lines of machine tools.

Iron and steel warehouses are doing a fair trade, and prices remain about the same. This week sheets dropped back a little, as there are quite a number of re-sales in the market, coming from manufacturers who find that they are rather heavily stocked with high-cost material and prefer to get rid of some with the idea of stocking up later on at a better figure. There is a good demand for most lines of sheets though, and the surplus stock is not going to influence the market very much one way or the other.

Boiler tubes are in demand, many of the boats being tied up now ready for the season's repairs.

The automobile shops are not in any better condition than they have been for some time back. The reductions announced in the price of cars were taken up by dealers who placed orders for next season. When these were taken care of the buying ceased, and orders are badly needed in nearly every car shop in the country.

The scrap metal market can stand almost any phrase that one cares to apply to it this week, and the more strenuous the phrase the more pleasing to the dealers. Some of the yards must be having a hard time in getting ends meet with the amount of trade that is in sight, and there is nothing in particular that would encourage the belief that things are going to get better. The steel mills are not buying, neither are the foundries, while offerings for non-ferrous metals do not exist.

MATERIAL COMES VERY QUICKLY FROM THE STEEL MILLS NOW

Special to CANADIAN MACHINERY.

MONTREAL, Que., Nov. 25. — The steel market today reflects the conditions that exist in almost every phase of industrial life. Doubtless, the seasonable quietness is responsible for much of the dullness, but other factors are influencing the apparent depression that is becoming more pronounced every day. The tendency to lower prices on steel commodities is more or less marked in every direction, but nothing drastic has yet been done in the price-cutting movement. Operations at the mills are becoming more restricted and production is fast becoming a problem. It is easy to get the output but the mills are reluctant to produce steel far beyond the immediate requirements of the trade, and these demands are below normal, and at the same time very irregular.

The cry of delayed delivery on many classes of steel is now a thing of the past, and if heard at all, is a feature of transportation rather than production. This might be exemplified in the statement made here by a dealer. "Just recently," he remarked, "we mailed an order to our American mill for some material that had to be rolled to special dimensions. At the beginning of the following week we were requested, by the consumer, to make some slight changes in the sizes previously stated. We immediately wired to the mill, only to be advised that the material had been rolled and was ready for shipment. This, while being a specific case, is nevertheless true of the general market." In commenting on the decline of price quotations the dealers could not say how far this would extend but could not see where any great reduction could be made in the face of existing cost of labor and materials. Raw materials, of course, were becoming more reasonable in price, but production costs per ton showed little decrease. There is, however, a feeling that general adjustment of wages may follow the movement in this direction by individual firms. Canadian mills are showing the trend of the times, as reports indicate a decline in the volume of business on the books. One local dealer stated that the general quotation in Pittsburgh for tank plates was $3, and that even lower prices had been reported. The price here, from warehouse, was on the base of $6 for ¼-inch and up, and $6.50 for 3-16 inch plates. Demand in this regard shows a marked falling off. While some mention has been made of an easier situation in boiler tubes, the needs of the trade for this commodity is still ample to maintain the present strength of the prices quoted. The easier market in sheets shows the effect of the inactivity in the auto industry, and the prices reflect the volume of the supply over that of the demand. Quotations here are on a basis of $9.25 for No. 28 gauge black and $7.75 for blue annealed No. 10.

Inquiries Still Come In

"It is only the man who is definitely persuaded that he needs a tool who is in the market at the present time." This statement seems to reflect the condition generally prevailing in machine tool circles. Some dealers report a good week, but the sales in the aggregate are below normal. While admitting a falling off in heavy machine tools sales, nearly every dealer agrees that the trade is in need of equipment and that the slackness now evident is more than a seasonable one. The consuming trade in ma-

The Planer With the Second-Belt Drive

A smooth, easy reverse—which permits higher cutting and return speeds yet prolongs the life of the entire machine—that is the outstanding feature of this distinctive Whitcomb-Blaisdell Second-Belt Drive..

Our Planer Book gives the details of design and construction. Write for it.

24 x 24 x 6 one head and 26 x 26 x 8 two heads in stock.

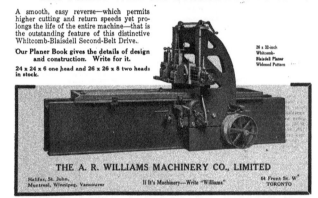

26 x 32-inch
Whitcomb-
Blaisdell Planer
Widened Pattern

THE A. R. WILLIAMS MACHINERY CO., LIMITED

Halifax, St. John,
Montreal, Winnipeg, Vancouver

If It's Machinery—Write "Williams"

64 Front St. W
TORONTO

---

chine tools is no different than in other lines where people are influenced by the lowering trend of prices. When the market was in the ascendant the cry of replacement values held the attention of all, particularly of the consumer, and this is one factor of present trading that is responsible for much of the business let-up. One is encouraged, however, at the optimistic tone of some dealers' remarks, and the turn of the year is expected to open the way for a renewal of more active trading. In speaking to Canadian Machinery one dealer stated that during the past month he had received many inquiries for different lines of equipment, and while realizing that the great majority of these were nothing but price feelers, he felt it his duty to answer such inquiries, but in most cases nothing further developed.

Coppers and Tin Down

Metals, like other commodities, are absorbing some of the dullness that is prevalent throughout the country. Sales are quieter than for many months, partly due to seasonable decline and partly to the general retrenching policy of many lines of industry. Few deny the normal requirements of the trade; but until the country as a whole becomes a free buyer the metal situation will experience a quiet period. Coppers have again declined locally to the extent of 1 cent per pound. Lake is quoted at 20 cents, electrolytic at 19½ cents and castings at 19

cents per pound. Tin is affected by the decline on the English market and shows a drop of 4 cents per pound. The quotation here is 48 cents per pound.

Dullness Very Pronounced

"Still in the same position," remarked a dealer in old materials. "Waiting for something to turn up and convert the dullness into cash. There is practically no market and prices are little guide to the situation. Quotations are necessary, I suppose, and those given are as good as any other." When asked as to what prices would actually be paid, the dealer stated that it would all depend on the nature of the sale, and with sales of a negative character, the prices were of little consequence. The scrap movement here is very light and irregular.

NOT MUCH CHANGE IN THIS MARKET

Prices For the Most Part Are Holding Firm—Scrap Market Remains Inactive

TORONTO.—There are several lines of business in this district that could, just now, report trading almost the same from week to week. There are a few firms with high spots, there are a few with the very low spots, while most of them are in between. For the most part dealers recognize that they are going through a period where their busi-

ness acumen is going to be taxed, but they have looked forward to this period coming for a long time back. Generally they are prepared to stand it, although in some cases credit and finance are receiving a great deal more attention than ever before.

The local machine-tool dealers do not see much in the proceedings of the National Machine Tool Builders' Association to justify the belief that there will be a cut in the prices soon. One of the salesmen from a large American firm, who spent some time in the district this week, stated that their price reduction had been made several weeks ago and they were selling now about 50 per cent. of their capacity. He was certain that, as far as his company was concerned, it would not be possible to make any more price reductions for many months to come.

The matter of price is also referred to at some length in another section of the markets, in reports from the convention at New York where these matters were freely discussed.

There are a number of inquiries coming into the market here, but the majority of them are for single machines, and many of these inquiries are not yet turned into real sales. Machines on order are coming in rapidly as a general thing, and the delivery schedules are getting pretty well worked out.

Small Tools Market

There is a real honest-to-goodness

selling going on in the small tool market. There are no large orders. The purchasing agents in this district are getting so used to seeing a number of the salesmen from the small tool houses or branches that they must be at the point where they can call most of them by their first name.

The National Railways are inquiring for quotations on an assortment of high-speed drills,' but the business has not been placed yet. The prices, recently revised, are standing and are likely to for some time.

The Steel Market

There is still a fair amount of business in the steel market. In some cases there are lines that are moving well. There has been a marking down in the warehouse prices of black sheets and of blue annealed. The base price of 28's is now given as $9.50, and blue annealed at $8.50; galvanized 28 are still $12, and 10% are $12.40, although the galvanized prices are weak at this point, and a reduction is quite possible.

There is considerable reselling going on in sheets now. Manufacturers find they have lots on hand that are not needed just now, and they figure it would be good business to clear out their high-priced stock if they have the chance to do it. Against this there is the fact of a good demand for sheets, and the chances are that this surplus stock will be taken up shortly, and when this is done the market will have to face the ordinary conditions. It is possible to secure sheets at present on fairly short notice. For the most part the available stock is from the independent mills and fairly high in price. Corporation deliveries are expected by several of the dealers as they have been on the books for as long as twelve months. In view of the fact that they are likely to be getting this material they are not booking with the independants.

Tubes are in fairly good assortments in several of the local warehouses, and they are moving out well. The demand at this time of the year is from lake boats that are now tied up for the season and are undergoing necessary repairs. This makes a call for the larger sizes, 3¼ and 4 inch, while the demand for tubes for industrial plants has not fallen off to any degree, as might be expected owing to the lateness of the season.

Bar iron is selling in fair quantities. This is hardly the season when one looks for a big movement in this.

Structural men figure that there is a lot of tonnage ahead of them in connection with a number of the larger buildings that are planned for this city and the surrounding district.

Pig iron prices vary according to tonnage taken and other terms of contract. A man who knows the situation pretty well sized up the situation this week as follows: No. 1 foundry iron, of 2.50 to 2.75 per cent. silicon is selling around

Pittsburgh says that production of steel is decreasing rapidly and that many of the independent mills are now not operating at 60 per cent. of their capacity.

—

It is rumored that the decrease in prices of steel, finished and semi-finished, may go below the Steel Corporation schedule.

—

Merchant blast furnaces have in many cases gone out of business. For a good many months users have been buying in very small quantities, and it has caused no surprise to find that some of the stacks are out of orders.

—

Machine tool dealers do not find much in the reports of the National Tool Builders' Association to cause them to believe that there will be a reduction in prices.

—

The National Railways are inquiring about high-speed drills, but so far no business has been placed.

—

There is a fair amount of business in the steel warehouses, especially in sheets and boiler tubes. the latter being called for at this time of year by boats that are now laid off for season's repairs.

—

Blast furnace men in Canada report that purchases of pig iron are being made in very small quantities.

—

The scrap metal market shows no signs of improvement in any line. Steel men are not taking on anything at present and the foundries are not doing much better. Even the present very low price in scrap metal may have to give way to still lower levels before business is done on a larger scale.

—

$59 delivered in Toronto now, while No. 2 iron or 1.75 to 2.25 per cent. silicon would be about a $1 a ton under that figure.

Several U. S. brokers who have quite ample tonnages at their disposal have been going through Canada offering attractive figures on resale iron. Apart from brokers there are a couple of U.S. firms making a special effort to break into the Canadian market. Of course there is the duty of $2.75 per ton, exchange at current rates, which are a little higher than usual just now, and the carrying charges.

There is very little pig iron being purchased by the Canadian trade at the minute. A number of inquiries that come into the market are for purposes

of estimating on contemplated work, and for the purpose of comparison.

A well-known maker of iron stated to Canadian Machinery that Canadian furnace costs were as high to-day as they were several months ago, and there were not many indications of reductions being in sight. However, he admitted that if it were found necessary there might be a readjustment in their selling schedule, although he still held that the law of costs and selling price did not warrant such a move.

The scrap market has been referred to as dull, dead, inactive, stagnant and everything else that gives the idea of no business being done.

Well, one might take any one of these new phrases out and apply it with perfect safety to the scrap metal trade. The export market is not lending any strength to the situation, as the outside markets are apparently facing much the same situation as the markets here. The steel consumers, if they are in the market, are offering prices that the dealers would not consider. Neither are the foundries doing anything better, nor has wrought iron scrap been asked for. Buyers of non-ferrous have not done much better in the way of taking anything out of the scrap heap. Small lots have been wanted, otherwise there is no movement. Some of the dealers are of the opinion that the list, quoted elsewhere, which is low enough, should be regarded as purely nominal, and it may be battered down even more, although no dealer had the heart this week to knock any of the figures down lower.

BIG STEEL PLANT SHUTS OFF QUICK

In Response to Demands That Claims of Railroad Men Be Met At Once

Sydney.—When 125 railroad employees of the Dominion Steel Corporation served an ultimatum on the management demanding a settlement of their wage differences before five o'clock the same day the company replied by closing down the various mills at noon and ordering the banking of the blast furnaces. This action automatically threw 4,000 men out of employment, pending the settlement of the dispute with the railroaders. It is estimated that it will require at least a week to get the furnaces in operation again, and it is believed that the big plant will be tied up for that period at least. The company, in the meantime, it is understood, is making application to Ottawa for redress under the provisions of the Industrial Disputes Act. It has been stated that lack of orders has been threatening the operations of steel plants for some weeks past. For the past two years the railroaders employed within the steel plant have been contending for the scale awarded all other railroaders in Canada following the McAdoo award in the United States.

THE WORLD CALLS

NEVER BEFORE, AS NOW, HAS THE WHOLE WORLD CALLED

"PRODUCTION"

Are you doing any Screw Thread Cutting?

How are you doing it?

Do you know that Geometric Thread Cutting Tools have effected an enormous saving of time and costs in many thousand shops?

The first step is to investigate. Tell us the job you have to do, and hear what we can do for you.

THE FINAL SHIPMENT OF THE DAY—*THE FREIGHT SHIPMENT*

Express   --   Parcel Post   --   Auto Truck   --   Freight

These are the four channels by which Geometric Screw Thread Cutting Tools and Machines are shipped.

There is no place where more business is done to the square foot than in our Shipping Department, because of the demand for Production Tools—Automatic Die Heads and Collapsing Taps—everything for more and better screw threads.

Make Screw Thread Cutting the Most Satisfactory Operation in Your Shop.

——Write——

THE GEOMETRIC TOOL COMPANY
NEW HAVEN   CONNECTICUT

Canadian Agents:

Williams & Wilson, Ltd., Montreal. The A. R. Williams Machinery Co., Ltd., Toronto, Winnipeg, St. John, N.B., Halifax, N.S.

Canadian Fairbanks-Morse Co., Ltd., Manitoba, Saskatchewan, Alberta.

If interested tear out this page and place with letters to be answered.

PRICES NOT LIKELY TO STOP AT STEEL CORPORATION LIMIT

Special to CANADIAN MACHINERY.

PITTSBURGH, November 24.—The pig iron and steel markets continue stagnant as regards new buying, while there is a further decline in specifications against contracts, and a fresh wave of cancellations and postponements. Production, accordingly, is decreasing, and at a more rapid rate than formerly.

There is, in fact, a very sudden ending to the spell of steel scarcity, with heavy pressure upon mills for deliveries. It is by no means remarkable that the change should be sudden, the remarkable thing being that the pressure continued so long after various other industries had become quiet. The end of the period of feverish activity and the decline in inflation was foreseen long ago, even early in the year, when not a few observers were predicting a panic by August, and the Federal Reserve Board undertook to bring about a gradual deflation so as to prevent a smash. The steel industry maintained its activity longer than might have been expected, and naturally when the end came it came suddenly.

Prospective Prices

In the past two or three weeks the opinion has been spreading rapidly that before this readjustment in the iron and steel industry is completed there will have been declines in the Steel Corporation prices. Until recently, the common almost universal prediction was that the steel market would find its level, in the readjustment, by the independent prices dropping to the Steel Corporation level and staying there, the whole market being thus equalized and stabilized, with a fresh buying movement beginning. Many observers predicted that the operation would be completed by March or April. It is strange that the prediction gained such popularity, for it was essentially an unsound one. There was a time when, if the independents had voluntarily reduced their prices to the Steel Corporation or Industrial Board level, buyers would have had confidence in the prices thus made, but the opportunity is past. The order books of some independents are almost exhausted, yet they continue to quote their higher prices. Buyers see that the law of supply and demand is being disregarded by producers and are indisposed to have confidence in any prices quoted. When independents reduce their prices it will be only because they are forced to do so and when the Steel Corporation level is reached the independents naturally will not stop there, as there is no business waiting to be placed at that particular level.

The common prediction now, therefore, is that the independents will eventually go below the Steel Corporation prices. It is seen that the readjustment in business and financial conditions generally, in the direction of prewar prices, is going to go farther than was expected a few months ago would be the case. Production costs, therefore, will come down, and with a decline in costs selling prices for steel below the present Steel Corporation prices are to be expected.

Chairman Gary of the Steel Corporation late last week made a carefully worded public statement. The substance was that the corporation has had good grounds for advancing its prices above the Industrial Board schedule of March 21, 1919, for several reasons, including in particular the wage advances and the freight rate advances, but has refrained from doing so, and that it has been decided to continue present prices "unless and until it becomes necessary and proper to make changes to meet altered conditions." The iron and steel trade has experienced no difficulty in concluding that the qualification means that prices eventually will probably be reduced. As prices have not been advanced when costs increased, and costs are certain to decrease rather than increase in future, no other conclusion is logical. It is inferred that the corporation will have little disposition to reduce its prices in the near future on account of such shading of its prices as the independents will probably be indulging in within a short time, say within two months at the most. The corporation will then still be moderately well supplied with orders while the independents will have practically nothing to dispute the right of the independents to get some business, even though the independents have made such large profits of late that they could afford to be idle a half year or longer and still have greater profits in the aggregate than the Steel Corporation.

Production

Many of the independents, representing perhaps half the total independent capacity, are operating at 60 per cent. or less, this representing a sudden decrease. One large independent has been making scarcely any steel at all, while operating its finishing departments in a moderate way, working up accumulations of semi-finished steel. Very few independents are operating within 90 per cent. of their rate in August or September. The Steel Corporation's operations are well maintained, and with a possible exception here and there the corporation plants are producing larger tonnages than at any time since last March. However, it looks as though even the corporation would have to decrease operations soon at some of its plants, though probably only to a small extent.

As to the merchant blast furnaces, they have almost run out of business, and this is only what was to be expected, since for more than six months consumers have been buying very sparingly, only for early deliveries and then in as small tonnages as possible. There is no demand in the market now, irrespective of price, and as furnaces would not think of piling iron when costs are on the eve of a great readjustment the only course is to blow out as soon as orders run out. While a furnace will not make up stocks for future sale, it will before blowing out produce enough iron to take care of deliveries that its customers are certain to take in the near future.

Iron and Coke Prices

Since last report market quotations on pig iron are down $2 a ton as to Bessemer and $1 on both foundry and basic. The declines occurred on extremely limited transactions, and the next business that develops will in all probability see further declines. The market just now is quotable at $40 for Bessemer, $39 for foundry and $37.50 for basic, f. o. b. valley furnaces, freight to Pittsburgh being $1.96.

Before buyers of pig iron and steel products take hold again with any freedom prices will be much lower, and the buyer will have much to say as to price. Profits will have to be telescoped. The pig iron buyer will not take hold in pig iron on the basis that the furnaceman is making no profit if he is paying a big profit to the coke maker. The coke maker must do his share. Thus far the coke maker has not seen the light. Connellsville furnace coke for spot shipment is $8 to $8.50, when $5 would afford a very good profit. As to Lake Superior iron ore, prices for which for the 1921 season will hardly be set before April or May of next year, little if any reduction is to be expected, as ore prices have been moderate right along. There may be a slight concession by the ore producers just to show their willingness to co-operate, and if there is any decline in the lake rate, which in the past season was $1.20, including the unloading charge at Lake Erie docks, that will be handed on to the ore buyer.

———

Street Railway Steel.—A number of conferences have been taking place during the past week between Toronto civic representatives and engineers and officials of the United States Steel Corporation. The whole matter rests with the proposed changes and extensions to the street railway system in Toronto, to be undertaken after the city takes over the road. The city is anxious to secure information as to the best way to make the change in track and possibly gauge, and at the same time keep the service in operation, and the experience of the U. S. Steel Corporation engineers was desired in this connection.

SELECTED MARKET QUOTATIONS

Being a record of prices current on raw and finished material entering into the manufacture of mechanical and general engineering products.

PIG IRON

Grey forge, Pittsburgh	$39 96
Lake Superior, charcoal, Chicago.	53 50
Standard low phos., Philadelphia.	44 79
Bessemer, Pittsburgh	41 96
Basic, Valley furnace	37 50

Toronto price:—
Silicon, 2.25% to 2.75%	57 50
No. 2 Foundry, 1.75 to 2.25%	56 00

IRON AND STEEL

Per lb. to Large Buyers	Cents
Iron bars, base, Toronto	$ 5 50
Steel bars, base, Toronto	5 50
Iron bars, base, Montreal	5 75
Steel bars, base, Montreal	5 75
Reinforcing bars, base	5 50
Steel hoops	7 00
Norway iron	11 00
Tire steel	5 75
Spring steel	10 00

Band steel, No. 10 gauge and 3-16
in. base	6 00
Chequered floor plate, 3-16 in.	8 50
Chequered floor plate, ¼ in.	8 00
Bessemer rails, heavy, at mill	.....
Steel bars, Pittsburgh	3 00-4 00
Tank plates, Pittsburgh	3 50
Structural shapes, Pittsburgh	3 00
Steel hoops, Pittsburgh	3 50-3 75

F.O.B., Toronto Warehouse
Small shapes	5 50

F.O.B. Chicago Warehouse
Steel bars	3 62
Structural shapes	3 72
Plates	3 67 to 5 50
Small shapes under 3"	3 62

FREIGHT RATES

Pittsburgh to Following Points	Per 100 Pounds.	
	C.L.	L.C.L.
Montreal	58½	73
St. John, N.B.	84½	106¼
Halifax	86	108
Toronto	38	54
Guelph	38	54
London	38	54
Windsor	35	50½

METALS

	Gross.	
	Montreal	Toronto
Lake copper	$20 00	$20 50
Electric copper	19 50	20 50
Castings, copper	19 00	20 50
Tin	45 00	50 00
Spelter	9 00	10 00
Lead	8 25	9 00
Antimony	8 50	9 00
Aluminum	36 00	35 00

Prices per 100 lbs.

PLATES

Plates, 3-16 in.	$ 6 50	$ 7 00
Plates, ¼ up	6 00	6 60

PIPE—WROUGHT
Standard Buttweld Pipe

	Per 100 Ft.			
	Steel		Gen. Wrought Iron	
	Blk.	Galv.	Blk.	Galv.
⅛	$ 4 66	$ 5 49	$ ...	$ ...
¼	4 21	7 41	5 91	8 91
⅜	4 91	7 41	5 91	8 91
½	7 15	8 63	7 15	9 92
¾	8 85	10 87	9 95	12 92
1	12 51	16 07	14 71	17 77

1¼	17 40	21 74	19 90	24 94
1½	21 04	25 99	23 79	29 74
2	28 31	34 97	32 61	35 67
2½	44 73	55 28	...	...
3	56 52	72 39	...	...
3½	74 06	90 62	...	...
4	87 75	107 87	...	...

Standard Lapweld Pipe

	Steel		Gen. Wrought Iron	
	Blk.	Galv.	Blk.	Galv.
2	$32 61	$ 38 67	$35 71	$42 37
2½	48 24	58 79	54 11	64 64
3	63 11	76 88	70 76	84 68
3½	76 90	93 46	85 10	101 66
4	89 92	107 55	100 85	120 66
5	1 06	1 29	1 36	1 64
6	1 22	1 50	1 62	1 99
7	1 55	1 95	1 97	2 55
8	1 96	2 50	2 52	3 01
8L	2 16	2 66	2 66	3 16
9	2 49	3 07	3 07	3 64
10	2 96	3 67	3 67	4 36
10L	3 27	3 41	3 41	4 06
10	3 40	4 29	4 39	5 21

Prices—Ontario, Quebec and Maritime Provinces

WROUGHT NIPPLES

4" and under, 60%.
4½" and larger, 50%.
4" and under, running thread, 30%.
Standard couplings, 4-in. and under, 30%.
Do., 4½-in. and larger, 10%.

OLD MATERIAL

Dealers' Average Buying Prices.

	Per 100 Pounds.	
	Montreal	Toronto
Copper, light	$10 50	$10 50
Copper, crucible	13 00	12 00
Copper, heavy	12 50	12 00
Copper wire	12 50	12 00
No. 1 machine composition	13 00	12 00
New brass cuttings	7 00	7 00
Red brass turnings	10 00	10 00
Yellow brass turnings	7 00	7 50
Light brass	5 00	5 00
Medium brass	6 50	6 00
Scrap zinc	5 00	5 50
Heavy lead	5 25	5 60
Tea lead	2 50	3 00
Aluminum	16 00	16 00

	Per Ton	Gross
Boiler plate	13 00	12 00
Malleable scrap	20 00	23 00
Axles (wrought iron)	25 00	20 00
Rails (scrap)	18 00	18 00
Malleable scrap	23 00	25 00
No. 1 machine cast iron	22 00	33 00
Pipe, wrought	9 00	10 00
Car wheel	30 00	33 00
Steel axles	20 00	20 00
Mach. shop turnings	8 00	9 00
Stove plate	25 00	25 00
Cast boring	8 00	12 00

BOLTS, NUTS AND SCREWS

	Per Cent.
Carriage bolts, 7/16 and up	+10
Carriage bolts, ⅜-in. and less	Net
Coach and lag screws	—15
Stove bolts	55
Wrought washers	—25
Elevator bolts	+10
Machine bolts, 7/16 and over	+10
Machine bolts, ⅜-in. and less	+10
Blank ends	Net
Bolt ends	Net
Machine screws, fl. and rd. hd., steel	27½

Machine screws, o. and fil. hd., steel	+25
Machine screws, fl. and rd. hd., brass	net
Machine screws, o. and fil. hd., brass	net
Nuts, square, blank	+25 add $2 00
Nuts, square, tapped	add 2 25
Nuts, hex., blank	add 2 50
Nuts, hex., tapped	add 3 00
Copper rivets and burrs, list less	15
Burrs only, list plus	25
Iron rivets and burrs	+40 and 5
Boiler rivets, base ⅜" and larger	$8 50
Structural rivets, as above	8 40
Wood screws, O. & R., bright	75
Wood screws, flat, bright	77½
Wood screws, flat, brass	55
Wood screws, O. & R., brass	55½
Wood screws, flat, bronze	50
Wood screws, O. & R., bronze	47½

MILLED PRODUCTS

(Prices on unbroken packages)

	Per Cent
Set screws	—20 ex, 25 and 5
Sq. and hex. hd. cap screws	12½
Rd. and fil. hd. cap screws	plus 25
Flat but. hd. cap screws	plus 50
Fin. and semi-fin. nuts up to 1-in.	12½
Fin. and Semi-fin. nuts, over 1 in., up to 1½-in.	—5
Fin. and Semi-fin. nuts 1 or 1¼ in., up to 2-in.	+12½
Studs	+5
Taper pins	—12½
Coupling bolts	+40
Planer head bolts, without fillet, list	+45
Planer head bolts, with fillet, list plus 10 and	+55
Planer head bolt nuts, same as finished nuts.	
Planer bolt washers	net
Hollow set screws	+60
Collar screws	list plus 20, 30
Thumb screws	40
Thumb nuts	75
Patch bolts	add +85
Cold pressed nuts to 1½ in.	add $1 00
Cold pressed nuts over 1½ in.	add 2 00

BILLETS

	Per gross ton
Bessemer billets	$60 00
Open-hearth billets	60 00
O.H. sheet bars	76 00
Forging billets	55 00-75 00
Wire rods	52 00-70 00

Government prices.

F.O.B. Pittsburgh.

NAILS AND SPIKES

Wire nails, base	$6 15
Cut nails, base	7 00
Miscellaneous wire nails	50c.

ROPE AND PACKINGS

Plumbers' oakum, per lb.	0 10½
Packing, square braided	0 38
Packing, No. 1 Italian	0 44
Packing, No. 2 Italian	0 36
Pure Manila rope	0 35½
British Manila rope	0 28
New Zealand hemp	0 28

POLISHED DRILL ROD

Discount off list, Montreal and Toronto ... net

MISCELLANEOUS

Solder, strictly	$ $ 29½
Solder, guaranteed	0 31½
Soldering coppers, lb.	0 62½
White lead, pure, cwt.	20 35
Red dry lead, 100-lb. kegs, per cwt.	16 00
Gasoline, per gal., bulk	0 42
Pure turp., single bbls., gal.	3 15
Linseed oil, raw, single bbls.	2 37
Linseed oil, boiled, single bbls..	2 40
Wood alcohol, per gal.	4 00
Whiting, plain, per 100 lbs.	3 00

CARBON DRILLS AND REAMERS

S.S. drills, wire size	35
Can. carbon cutters, plus	10
Standard drills, all sizes	35
3-fluted drills, plus	10
Jobbers' and letter sizes	35
Bit stock	40
Ratchet drills	10
S.S. drills for wood	40
Wood boring brace drills	25
Electricians' bits	30
Sockets	50
Sleeves	50
Taper pin reamers	25 off
Drills and countersinks	net
Bridge reamers, carbon	50
Centre reamers	5
Chucking reamers	net
Hand reamers	10
High speed drills, list net to plus	20
Can. high speed cutters, net to plus	10
American ... plus	40

COLD ROLLED STEEL

[At warehouse]

Rounds and squares	$7 base
Hexagons and flats	$7.75 base

IRON PIPE FITTINGS

	Black	Galv.
Class A	70	85
Class B	30	40
Class C	20	30

Cast iron fittings, 5%; malleable bushings, 22½%; cast bushings, 22½%; unions, 37½%; plugs, 20% off list.

SHEETS

	Montreal	Toronto
Sheets, black, No. 28	$ 9 25	$ 9 50
Sheets, blue ann., No. 10	7 75	8 50
Canada plates, dull, 52 sheets	13 00	13 00
Can. plates, all bright..	14 00	.....
Apollo brand, 10¾ oz. galvanized	....	....
Queen's Head, 28 B.W.G.	13 50	.....
Fleur-de-Lis, 28 B.W.G.	13 00	.....
Gorbal's Best, No. 28...	....	....
Colborne Crown, No. 28.	....	....
Premier, No. 28, U.S. ..	11 50	12 00
Premier, 10¾-oz.	11 50	12 40
Zinc sheets	16 50	20 00

PROOF COIL CHAIN

(Warehouse Price)

B

¼ in., $13.00; 5-16, $11.00; ⅜ in., $10.00; 7-16 in., $9.80; ½ in., $9.75; ⅝ in., $9.20; ¾ in., $9.30; ⅞ in., $9.50; 1 in., $9.10; Extra for B.B. Chain, $1.20; Extra for B.B.B. Chain, $1.80.

ELECTRIC WELD COIL CHAIN B.B.

⅛ in., $16.75; 3-16 in., $15.40; ¼ in., $13.00; 5-16 in., $11.00; ⅜ in., $10.00; 7-16 in., $9.80; ½ in., $9.75; ⅝ in., $9.50; ¾ in., $9.30.

Prices per 100 lbs.

FILES AND RASPS

	Per Cent.
Globe	50
Vulcan	50
P.H. and Imperial	50
Nicholson	32½
Black Diamond	27½
J. Barton Smith, Eagle	50
McClelland, Globe	50
Delta Files	20
Disston	40
Whitman & Barnes	50
Great Western-American	50
Kearney & Foot, Arcade	50

BOILER TUBES.

Size.	Seamless	Lapwelded
1 in.	$27 00	$.....
1¼ in.	29 50	.....
1½ in.	31 50	29 50
1¾ in.	31 50	30 00
2 in.	35 00	30 00
2¼ in.	35 00	29 00
2½ in.	42 00	37 00
3 in.	50 00	48 00
3¼ in.	.. ..	48 50
3½ in.	63 00	51 50
4 in.	85 00	65 50

Prices per 100 ft., Montreal and Toronto

OILS AND COMPOUNDS.

Castor oil. per lb.	
Royalite, per gal., bulk	29½
Palacine	32½
Machine oil, per gal.	65
Black oil, per gal.	34
Cylinder oil, Capital	1.01
Petroleum fuel oil, bbls., net	19

BELTING—No 1 OAK TANNED

Extra heavy, single and double		6¼
Standard		6½
Cut leather lacing, No. 1		2 00
Leather in side	2 40	3 00

TAPES

Chesterman Metallic, 50 ft.	$2 00
Lufkin Metallic, 603, 50 ft.	2 00
Admiral Steel Tape, 50 ft.	2 75
Admiral Steel Tape, 100 ft.	4 45
Major Jun. Steel Tape, 50 ft.	3 50
Rival Steel Tape, 50 ft.	2 75
Rival Steel Tape, 100 ft.	4 45
Reliable Jun. Steel Tape, 50 ft...	3 50

PLATING SUPPLIES

Polishing wheels, felt	$4 50
Polishing wheels, bull-neck	2 00
Emery in kegs, Turkish	8¾
Pumice, ground	06
Emery glue	30
Tripoli composition	9½
Crocus composition	12
Emery composition	11
Rouge, silver	64
Rouge, powder, nickel	38

Prices per lb.

ARTIFICIAL CORUNDUM

Grits, 6 to 70 inclusive	.08½
Grits, 80 and finer	.6

BRASS—Warehouse Price

Brass rods, base ½ in. to 1 in. rod	0 34
Brass sheets, 24 gauge and heavier, base	$0 42
Brass tubing, seamless	0 46
Copper tubing, seamless	0 48

WASTE

XXX Extra	.24	Atlas	.20
Peerless	.22½	X Empire	.19½
Grand	.22½	Ideal	.19
Superior	.22½	X Press	.17½
X L C R	.21		

Colored

Lion	.17	Popular	.13
Standard	.15	Keen	.11
No. 1	.15		

Wool Packing

Arrow	.35	Anvil	.22
Axle	.28	Anchor	.17

Washed Wipers

Select White	.20	Dark colored	.09
Mixed colored	.10		

This list subject to trade discount for quantity.

RUBBER BELTING

Standard ... 10% Best grades... 15%

ANODES

Nickel	.55 to	.60
Copper	.38 to	.40
Tin	.70 to	.70
Zinc	.16 to	.17

Prices per lb.

COPPER PRODUCTS

	Montreal	Toronto
Bars, ½ to 2 in.	$37 50	$37 00
Copper wire, list plus 10..		
Plain sheets, 14 oz., 14x60 in.	44 00	44 00
Copper sheet, tinned, 14x60, 14 oz.	47 00	46 00
Copper sheet, planished, 16 oz. base	51 00	50 00
Braziers', in sheets, 6 x 4 base	43 00	42 00

LEAD SHEETS

	Montreal	Toronto
Sheets, 3 lbs. sq. ft.	$11 50	$14 50
Sheets, 3½ lbs. sq. ft. ..	11 25	14 00
Sheets, 4 to 6 lbs. sq. ft..	11 00	13 50

Cut sheets, ½c per lb. extra.
Cut sheets to size, 1c per lb. extra.

PLATING CHEMICALS

Acid, boracic	$ .23
Acid, hydrochloric	.04¾
Acid, nitric	.11
Acid, sulphuric	.04½
Ammonia, aqua	.15¾
Ammonium, carbonate	.23
Ammonium, chloride	.22
Ammonium hydrosulphuret	.75
Ammonium sulphate	.30
Arsenic, white	.16
Copper, carbonate, annhy	.41
Copper, sulphate	.13
Cobalt, sulphate	.20
Iron perchloride	.62
Lead acetate	.30
Nickel ammonium sulphate	.20
Nickel carbonate	.32
Nickel sulphate	.20
Potassium sulphide (substitute)	.40
Silver Chloride (per oz.)	1.15
Silver nitrate (per oz.)	1.10
Sodium bisulphate	.13
Sodium carbonate crystals	.04
Sodium cyanide, 127-130%	.39
Sodium hyposulphite per 100 lbs	9.00
Sodium phosphate	.15
Tin chloride	.30
Zinc chloride, C.P.	.30
Zinc sulphate	.08

Prices, per lb. unless otherwise stated

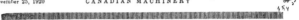

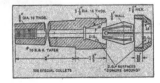

Dealers Talk of Prices and Cancellations

Meeting Held in New York, Where These Matters Were Much Discussed—One Dealer Alone Speaks in Favor of a Reduction—How They Have Come Out on Cancellations.

REPORTS of the National Supply and Machinery Dealers' Association convention indicate that price recession and cancellations formed the chief topics of discussion, and that in the case of prices the sentiment was not at all favorable to urging for reductions. The opposite view to this was expressed by a Milwaukee dealer, who stated: "Prices must be made lower on machine tools, for the buying public expects it. I think that a general reduction is almost essential."

J. D. Doan, of the American Tool Works, Cincinnati, a manufacturer, who had at one time been a dealer, gave reasons which he said prevented a general reduction in the price of machine tools. He claimed that the wages of tool builders' labor had not been inflated as in the case of a certain stove manufacturer, whose molders had been getting from $15 to $30 a day. Therefore, wages could not be reduced. It costs about $40 a ton to make pig iron, he said, and iron is selling at that price now. So iron cannot be reduced. Moreover, thirty furnaces blew out in October, curtailing production. Neither is there great hope for a reduction of steel.

"There must be an honest-to-goodness reason for reduction of machine tools before any can take place." he continued. "I have had several instances called to my attention where business has not been placed because of the expectation of lower prices. A list would be issued; some dealer would quote a reduction on one tool, then the entire list would be held up, the purchasing agent expecting a general reduction. Our duty as an association is to stabilize confidence on the part of the buyers. One can't compare the machine tool industry with a more speculative one like the dry goods trade.

"If the cast iron entering one of our 18-in. lathes should be reduced 25 per cent., we could reduce the price of our finished machine not more than 3 per cent."

W. A. Viall, Brown & Sharpe Mfg. Co., Providence, R.I., agreed with the majority that prices could not be reduced at this time. He said that to-day there are no heavy stocks in the hands of the manufacturers. Their policy will be a receding one, tapering off production and placing in stock a few representative sizes. Nobody will take steps to increase plant capacities. Dealers can look forward to a period of quiet. The speaker had talked with several machinery men, some of whom placed the expected revival in April, others in July.

Probable Date of Revival

In the discussions, some named February as the month when the revival of business would set in. No one present admitted that he would be willing to reduce prices in order to liquidate stocks of tools on hand. In the matter of cancellations, all agreed that no two requests for cancellation could be handled exactly the same. The particular circumstances of the contract and the date of delivery of the tool would enter into the question. It was generally agreed that where delivery of machinery is delayed long after the specified time, the buyer may have legitimate reason for canceling. Many dealers had received such letters as this from buyers: "If you ship the tool, we shall be unable to pay for it."

One dealer narrated how the clause in his contract, stating: "No cancellations allowed without our consent," had been altered in recent contracts by striking out the last three words. A dealer suggested that buyers canceling be required to pay a certain percentage of the cost of the tool for the privilege, thus to compensate the dealer for his costs in making the sale.

TARIFFS DISCUSSED BY IRON AND STEEL MEN AT MONTREAL

Various phases of the iron and steel industry were brought out at the session of the tariff commission in Montreal. Some extracts follow:

W. W. Butler, on behalf of the iron and steel interests, upheld the need for a tariff to protect these industries. Steel and iron imports, he said, amounted for last year to $189,897,602, or 18 per cent. of the total imports. The bulk of pulp and paper machinery was imported, although the Canadian industry was now in a position to supply this material.

The Canadian industry was doubly hampered by having to import much of its raw material, and by the competition of a highly organized, well protected iron and steel industry in the United States.

The lead Great Britain once had in this industry had steadily decreased from the time free trade gave the protected German and Belgian iron and steel works an advantage in competition.

Adverse Exchange Rate

The purchase in the United States of goods which could be manufactured in Canada kept the exchange rate in an adverse position and unnecessarily increased the cost to Canadian consumers of other goods which are not manufactured here. At best the home market was none too large, and a long free list gave American companies an advantage in the Canadian market in many lines.

An enquiry by Sir Henry elicited the information that only the Algoma Mills were rolling structural steel, and they had just started. Apart from this, only small structural parts were made in Canada. Some lines which had been commenced lately by Canadian mills had had to be abandoned on account of United States competition. The American firms preferred their home market in time of shortage, and dumped their excess on the Canadian market, in slack periods. The anti-dumping law, although good in principle, was not an absolute protection against dumping.

T. B. Howard, president of the Phoenix Iron Works, stated that the output of his industry had gone down very seriously since the war. This was an essential Canadian industry, and worthy of protection.

Sir Henry suggested that part of the Canadian structural people's troubles resulted from the fact that they had to buy structural steel from firms who later competed with them structurally. Would it not be to their advantage to have structural steel manufactured here?

Mr. Howard agreed that it would. If the Algoma company developed in this direction it would be of great use to them.

The industry had attracted to itself $26,000,000, represented by expensively equipped shops. It provided a field of employment for many university graduates and scientifically trained men. Its mechanics were highly skilled and big wages were paid to its construction workers. The United States firms had the advantage in climatic conditions and larger contracts and could therefore easily invade the Canadian market.

James Percy McNaughton, who for nearly twenty years has been general sales agent for the Dominion Iron and Steel Company at Montreal, passed away this week after a brief illness. Before associating himself with the steel company, Mr. McNaughton was in partnership in a brokerage business under the firm name of Lamplough and McNaughton. He was born in Ottawa 53 years ago. He leaves a widow and two sons.

It Costs Money To Build High-Class Tools

President Newton Reviews the Outlook at New York Gathering—Necessary to Have Best Workmen to Keep Up the Standard—Cutting Prices Does Not Help Matters

ADDRESSING the nineteenth annual convention of the National Machine Tool Builders' Association at the Hôtel Astor, New York, President Albert E. Newton, after reviewing the business situation, said that "any reduction in prices on machine tools will not stimulate demand, and is more likely to work to the contrary."

"For the past few months business in general has been on a down grade," stated Mr. Newton, "attaining greater velocity as the time went on, so that now, even if the brakes were set hard, we could not expect to stop the down ward trend at once. Therefore, let us face the situation as it is and realize that we are face to face with a real depression, and that the orders for machine tools are few and far between.

"Let us ask what are we going to do about it? Can we by any act of ours better the situation? The answer is 'no.' Most of us know from experience that the demand, or lack of demand, for machine tools is entirely beyond the control of the machine tool builders. We also know that any reduction in prices on machine tools will not stimulate the demand, and is more likely to work to the contrary.

"We should realize that the present condition of our industry requires us to be calm and courageous, think clearly and deliberately, and deal in facts and not theory.

"During the past five years the machine tool industry, as a whole, has not increased its prices as much as the costs have increased, and today our margin of profit is not on as a high a plane as it should be. Bear in mind that we must continue to improve our product, and that experimental and development work is a necessary part of the cost of machine tools. Also, that to produce a high-grade machine we must have high-grade workmen, and to secure and keep high-grade workmen, we must grant comparatively high wages.

"There is only one cause that I know of that should either increase or decrease the price of a machine tool, and that is cost, and I submit that our costs are today higher than ever before. It has been stated, and I believe it to be true, that the recent increase in freight rates will add 5 per cent. to this cost.

"We know that decreased production means higher overhead cost, so that even if our raw materials may cost us less in the near future (and I am not so sure that this will happen), this possible saving will be more than offset by the increased overhead cost due to reduced output.

"The depression with which we are now contending is no new experience for us, and is not a surprise, as we have been expecting it for some time; in 1913-'14 we experienced a situation in the machine tool industry very similar, if not worse than that which we are now passing through; that experience taught us the real value of our association, a value that has never, and can never, be too highly rated.

"I believe that most of our members fully realize that the values which accrue to them from the fruits of our association are proportionate to the time and mutual support given by them, and that our collective future as members of the National Machine Tool Builders' Association is dependent upon the unselfish and friendly co-operation of each individual.

Cancellations of Orders

"There is another important subject which always appears in times such as we have with us at present; that is, the cancellation of orders.

"It certainly would seem reasonable that an industry such as ours, which of necessity must purchase in advance its materials and supplies, sufficient to last for months, should be protected from indiscriminate cancellation of the orders on which it has based its plans.

"I hope that we may some time solve

this condition and adopt some method that will raise the order for a machine tool to the dignity of a real contract, rather than a mere option that may be given up at will by the purchaser.

"It is an old saying, 'If you want a short war, prepare for a long one.' Let us attack this depression as though we thought it would last for some time; let us conserve our energies (meaning cash) so that whatever comes we shall be able to meet conditions in a businesslike manner. Then we shall be ready to resume our full productive capacity when the demand for our product comes back to us, as it surely will in the not very distant future."

At the election of officers of the National Machine Tool Builders' Association, August H. Tuechter, of the Cincinnati-Bickford Tool Co., Cincinnati, was elected president, succeeding Albert E. Newton. Mr. Newton continues as a member of the executive committee. Other officers elected were: First vice-president, E. J. Kearney, Kearney & Trecker Co., Milwaukee, Wis.; second vice-president, C. Wood Walter, Cincinnati Milling Machine Co., Cincinnati; secretary, Carl E. Deitz, Norton Co., Worcester, Mass.; treasurer, Winslow Blanchard, Blanchard Machine Co., Cambridge, Mass.

Winnipeg Bridge.—The Winnipeg City Council are to be recommended by the civic committee on works to advertise for tenders for building the Maryland Bridge. Plans and specifications have been prepared and tenders must be made by December 14. It is specified that the bridge be completed ready for traffic by November 1, 1921, and that the entire structure, entailing railings and other features, must be completed by July 15, 1922. The final cost will approximate $300,000, and a penalty clause of $50 a day is to be included in any contract made. Traffic will be maintained over the bridge by temporary arrangements.

Plants Are Busy.—A despatch from Hamilton stated that one of the busiest spots there is at the National Steel Car Co. This concern has 844 men on its payroll, exclusive of the office and sales departments. While not a few Hamilton manufacturing plants have begun to retrench to a moderate degree, this large concern has found it necessary to put on a night shift, with the result that the important feature of its output is from 20 to 24 box cars per day for the government railways. These cars have a capacity of 40 tons. They are built to carry grain. Every afternoon the pre-

ceding day's output is hauled from the company's yards, and thence to Western Canada as fast as they can be taken there, to assist in expediting the movement of grain to the head of the lakes. In addition to the order for box cars for the government's railways, which was reported to be for 1,500, a large order has also been secured from the C. P. R. for flat cars. The C. P. R. has also placed its order for several all-steel sleepers. With the exception of the wheels, these sleepers, like all other cars built by the company, will be constructed completely at its local plant, and when shunted to its yards will be ready for service.

Thomas Findley, who for some years has held the dual office of president and general manager of the Massey-Harris Company, Limited, has resigned from the latter position, and in his place the board has appointed Thomas Bradshaw general manager. Mr. Bradshaw will be assisted by C. L. Wisner as assistant general manager in charge of sales, and by George Valentine as assistant general manager in charge of manufacturing. Mr. Findley, whose health has much improved of late, as president will continue to act as the company's chief executive.

Trade Gossip

Howard S. Dick has been appointed city engineer of Kingston. He graduated from Queens in 1913, and was overseas with the Canadian Engineers.

Enlarged Foundry — The Canadian Hanson & Van Winkle Co., Toronto, have completed a large addition to their foundry, for the handling of gray iron castings, especially soft machine castings for the machine trade.

Have More Power.—The Hydro power cut in Kitchener has been reduced to 10 per cent. as the result of a number of industries agreeing to use electrical power at night. These factories have arranged to take a flat load over 24 hours rather than a heavy day load of 10 hours.

The United States furnished the entire supply of candles used in Cuba in 1918, which amounted to $1,492,037.

AUTOMATIC MILLING MACHINE

(Continued from page 482)

of helical gears and the main spindle gear is of herring-bone type, thereby eliminating tooth chatter. The overhanging arm is made in rigid form in the centre of machine, carrying a support extending to each arbor. Provision is made in these supports for a vertical adjustment. The cutter arbor is supported at its outer bearing in a phosphor bronze bushing.

The oiling system has been given special attention, all bearings being oiled from reservoirs. The spindle bearings are lubricated by sight feed drip oilers. In addition to the above there are three reservoirs with piping leading to other bearings of the machine. Reservoirs are provided with covers to prevent chips and dirt from getting into the same. Lubrication for the cutters is supplied by a geared pump. The vise is of a new design, its construction being very compact, without any loss of capacity. The driving pulley is 14 inches diameter and is driven from a countershaft, a 3½ inch belt being used.

PETROLEUM SITUATION
(Continued from Page 476.)

portation of oil from Baku is greatly reduced because of the lack of railroad transportation and the difficulty of passing through Georgia. The estimated production for 1919 is 1,257,000,000 gallons.

The average production of wells, which was 4,400 gallons per 24 hours in 1904, was 1,445 gallons in 1917, 1,165 gallons in 1918 and 1,730 gallons in 1919. Since the war the drilling of new wells has shown a constant decrease, falling from an average of 37,720 feet per month in 1913 to 13,120 feet at the end of 1917 and 1,400 feet at the end of 1918. For 1919 the average is still lower.

Meanwhile living expenses are very high, materials scarce and high priced; though the government is making every effort to relieve the situation, which is one that affects the industries of the whole world and especially those of the United States by the unprecedented demand the low output of the Russian oilfields causes to be placed upon those of the United States.

High Priced Coal.—Notwithstanding a general decrease in the prices of foodstuffs, Windsor dealers increased the price of anthracite coal to $26 a ton, approximately $10 more than the price being charged in Detroit. The difference in exchange and a shortage of coal are responsible for the step, it was explained.

Classified Opportunities

MACHINE WORK WANTED

MACHINE WORK WANTED FOR Lathes, shapers, milling machine and planer, etc. Hourly or contract basis. Prompt delivery. W. H. Sumbling Machinery Co., Toronto. (e1m)

PLANT WANTED

ENGINEERING WORKS WANTED—Preferably in Ontario, to produce grading, excavating, road-building machinery. Advertiser will provide designs, patterns, etc., and is prepared to make contract immediately with suitably equipped firm. Replies should be in detail necessary to determine suitability of plant, location and the extent and nature of present manufacturing operations. Principals only deal with. Box 710, Canadian Machinery. (e1m)

HELP WANTED

WANTED — ESTIMATING AND DESIGNING engineer for transmission towers, fabricated steel telephone poles, fire escapes and steel stairs. Give full information in first letter. Box 734, Canadian Machinery.

WE WANT TO INTEREST SUCCESSFUL TRAvellers, at present covering Northern and Eastern Ontario with mechanical supplies, in a proposition offering unlimited possibilities capable of development into a main line. Box 735, Canadian Machinery. (c22m)

MACHINERY WANTED

WANTED—ONE 10 H.P. GENERAL ELECTRIC or Westinghouse motor, 60 cycle, 3 phase, 440 volts, 1200 R.P.M. The Frost & Wood Co., Ltd., Smith's Falls, Ont. (c24m)

Machinery Wanted

Advertisements under this heading will be inserted once free of charge. Readers are invited to use this department when in need of equipment. Replies may be sent to box number, care of Canadian Machinery, if desired.

For Sale

One 52-inch John Bertram & Sons Co. Boring Mill, late pattern, with two heads on the cross rail, with independent jaw chuck built into the table, arranged for motor drive using ten horse power motor, without friction.

In use at the present time on very fine class of work.

Price $6,450.

Box No. 729, Canadian Machinery. (C25M).

FOR SALE

ONE 5 H.P. JONES & MOORE, 1440 R.P.M. motor, and one 10 H.P. Canadian General Electric Company's 1500 R.P.M. motor. Both for 25 cycle, 550 volt, 3 phase current; 10 H.P. motor equipped with standard oil-immersed starter with no voltage release attachment. Both in good condition and may be had at a reasonable figure. Mathews Gravity Carrier Company, Limited, Port Hope, Ontario. (c22m)

FOR SALE—ONE BURY AIR COMPRESSOR, 12 x 14, Class M. 330 cub. ft. per minute. belt driven, only slightly used, $750; one air compressor tank 42 in. x 9 ft., man-hole on top, two 4½ in. side openings and one 2 in. side opening, $100. Taylor-Forbes Co., Ltd., Guelph, Ont. (c22m)

BELTING—AT LESS THAN USUAL PRICES. All sizes up to 24 inches in width. 8-inch, 4-ply new rubber at 30 cents; 4-inch, 4-ply at 36 cents; 5-in. 4-ply at 45c; and 6-in. 4-ply at 50c.; 4½-in., 4-ply, new stitched canvas at 30 cents; new, solid, tooth circular cordwood saws, 30-in., at $9.00, and 26-in. at $8.50. N. Smith, 138 York St., Toronto.

ONE 24-IN. DIA. x 48-IN. STROKE SIMPLE horizontal side crank Corliss steam engine, with 20-ft. x 25-in. flywheel and outboard bearings, made by Wood, Bolton, Eng.; one Northey jet condenser, 10-in. steam inlet, 8-in. water inlet, and 5-in. outlet; one London Machine Tool Co., 24-in. x 24-in. metal planer, single head, 6-ft. table, power cross feed, hand vertical feed and countershaft. Apply The Nova Scotia Underwear Co., Ltd., Windsor, N.S. (c24m)

PATENTS FOR SALE

R. M. WHITE, TITULAR TO CANADIAN Patent No. 202132, dated July 20th, 1920, relating to stair-climbing hand truck, wishes to dispose of his full rights and privileges in Canada. For further particulars apply to Adrian A. St. Laurent, I.E., Patent Broker, 75 Stewart St., Ottawa, Ont. (c22m)

HENRY R. ADAMS, TITULAR TO CANADIAN Patent No. 204644, dated September 21, 1920, relating to a gramophone reproduction device, wishes to dispose of his full rights and privileges in Canada. For further particulars apply to Adrian A. St. Laurent, I.E., Patent Broker, 75 Stewart St., Ottawa, Ont. (c22m)

CASTINGS

Medium and light, grey iron castings for manufacturing purposes. We carry bushings, solid and cored, from 1 in. to 6 in., or larger.

TORONTO FOUNDRY CO.
1864 Davenport Road, Toronto. (c31m)

FOR SALE
USED MACHINERY
IN A 1 CONDITION

1—10" Hendey Shaper.
1—16" x 6" Mueller Lathe.
1—16" x 5" Cincinnati Lathe.
1—20" x 10" U.M.C. Lathe.
1—20" x 10" Milwaukee Lathe.
1—24" x 24" Engine Lathe.
3—6 x 24 2 ½ in. Turret Lathes.
We are offering the above machines at very attractive prices. Write us for quotations on your requirements.

CHARLES P. ARCHIBALD & CO.
Machinery and Supplies
205 BEAVER HALL HILL, MONTREAL

PATTERNS

TORONTO PATTERN WORKS, 42 JARVIS Street, Toronto. Patterns in wood and metal for all kinds of machinery. (c1m)

BRANTFORD PATTERN WORKS ARE PRE-pared to make up patterns of any kind—including marine works—to sketches, blue prints or sample castings. Prompt, efficient service. Bell 'Phone 451; Machine 'Phone 735. Brantford Pattern Works, 49 George St., Brantford, Ont. (c1m)

AGENCY OPPORTUNITIES

WE ARE FREQUENTLY ASKED TO SECURE representatives for manufacturers in Canada, Great Britain and the United States. From time to time we shall publish particulars regarding agencies available and correspondence is invited. If the information has not been previously supplied please give particulars regarding experience, sales organization, territory, etc.

AMERICAN MANUFACTURER OF SANITARY drinking fountains who has made arrangements to have his line manufactured in Canada wishes to get in touch with a man who can sell the appliance in industrial plants of all kinds. Box 730A, Canadian Machinery.

CANADIAN AGENTS WANTED

FOR HIGH-GRADE AMERICAN-MADE LATHE —14 to 26. In replying state territory you can cover and cover thoroughly. Box 692, Canadian Machinery.

CANADIAN REPRESENTATIVE WANTED by U.S. manufacturer of well known line of high speed and alloy steels. Correspondence invited and will be treated as confidential. Box 707F, Canadian Machinery.

CANADIAN MACHINERY
AND
MANUFACTURING NEWS

VOL. XXIV. No. 23

December 2, 1920

Drop Forging

A Short History of Forging is Given, Together With Various Examples of Drop Forging Practice. These Examples Include Axles and Crank Shafts.

By J. H. MOORE

FORGING is at all times an interesting study, so before describing the new portion of the plant of the Canada Forge Co., let us delve for a short while into the art of forging itself.

The art of forging iron and steel by power hammers was introduced to America about the year 1850. This discovery proved so successful that research work was immediately commenced, with the result that forging has now become a scientifically developed art, and one of our important mechanical industries.

There are two commonly known types of forgings, namely, drop forgings and flat hammer forgings. The former type are so called because they are forged in dies under vertical power hammers, operated by efficient hammermen. while the latter class are made under either powerful steam driven hammers or hydraulic presses, operated by ham-

mersmiths. Flat hammer forgings are usually of quaint and unusual design, and, of course, as can be understood, no dies are used. This is why the operators are called hammersmiths. and they must be smiths in every sense of the word to produce good work. This latter class of work is indeed an art, because forgings of unusual shape are being continually required in one piece from the very smallest up to forty tons in weight. It is easily seen that highly capable men only can be employed on such work.

The hammers used vary. Some are capable of striking each blow equal to from one-quarter of a ton to 60 tons. The hydraulic presses can in turn exert a pressure from one-quarter of a ton to 12,000 tons, if necessary. Take for example some ship and locomotive parts. These are so heavy that the hammersmith and his crew require power rotat-

ing devices and electrically driven cranes to handle the material to and from the furnaces, until the desired shape and dimensions are completed. The hammersmith is the man who must use his skill and judgment as to whether a light or heavy blow is required. While at the plant (later to be described), the writer saw them working on some huge billets 36 inches in diameter, and handled by rotary device, and it was a treat to watch the work proceeding. In this case a hydraulic press was used of 6,000 tons capacity.

Steels and Their Treatment

Wrought iron was at one time considered the best material for forgings on account of its ductility, and the great amount of vibration it could withstand before crystallization would set in. There are still uses for it, and in some cases it is claimed to be superior to

FIG. 1—THE BILLETS BEING HEATED.   FIG. 2—WORK COMING FROM THE ANNEALING OVENS.

FIG. 4—DIES USED TO PRODUCE THE FRONT AXLE OF AUTOMOBILE.

other metals. Such cases are hull forgings for ships, stern frames, posts, rudders, etc. Modern engineering practice, however, with its demands for greatly increased loads and hard service, requires, for certain parts of heavy machinery, forgings which will have two or three times the strength and rigidity of iron, and will at the same time combine with these qualities the factor of ductility. These qualities are to be found in what is generally known as open hearth steel forgings.

The steel must, of course, be of proper chemical analysis for the purpose required. In the case of small forgings, rolled steel billets are used and worked to shape under steam hammers, while the larger work from steel cast ingots is generally done under hydraulic presses.

It is a well known fact that the working of the hot steel not only provides a means of shaping it to form, but at the same time refines the grain structure, and tends to increase its density and dynamic strength. The steel is first heated to what is known as the critical range of temperature, and in order to acquire the best results the finishing temperature in working should be just within this critical range, in order to increase the elastic limit and the period of endurance in service of the forging. This range of course depends upon the nature of the steel.

During the process of working, and the alternate cooling, internal stresses and strains are created, which can only be neutralised and made homogeneous through heat treatment in special furnaces regulated usually by electric pyrometers. The importance of heat treatment is to make the steel possess certain physical properties best suited to the purposes intended, and this is accomplished by increasing its ductility, elastic limit, and tensile strength.

To obtain proper heat treatment, you require, what is obvious, namely, good material, proper equipment, and a knowledge of correct procedure. Experience is a great thing in the art of heat treatment. This plant does its heat treating in car type overfired furnaces. This style of furnace removes all forging strains and stresses, thus insuring the greatest degree of safety by still further refining the grain and creating a higher degree of toughness in the finished forging. All the forgings at the plant we speak of are thoroughly tested

and inspected before being permitted to leave the factory.

The foregoing information has been mostly on their flat hammer work, and before leaving the subject, let us look at Figs. 1 and 2. In the first view we see two billets being heated in the furnace, while in the second picture we notice some of the work which has just been annealed. The size of these forgings is self apparent, and a good method of comparison is to look at the size of the crane chain links, then study the size of the work. It might be well to add that this concern make practically all their own furnaces, and use burners of their own design.

Before the material is used for forging purposes, it undergoes a severe test in the chemical and testing departments. An analysis is taken of all material entering the plant, and every particular is known and indexed on file. Usually the chief constituents are marked on the end of the billet as it rests in the yard. For example, the writer saw one particular billet which was marked Chrome 1.37, Nickel 3.02, 22" x 22" x 7'-2". 11900 lbs. This information gives the smith all necessary data for his purpose.

Drop Forging

And now, having touched on the flat hammer work, let us consider some examples of drop forging practice. The first example depicted is the heading. This shows a separator stem made for

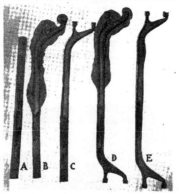

FIG. 5—THE VARIOUS STEPS NECESSARY TO COMPLETE A FRONT AXLE FOR AUTOMOBILE.

the Massey-Harris Co., and its method of manufacture is very interesting.

The work is first heated in the furnace, and this is done in a furnace of double construction of their own make. It is equipped with four burners of their own design, and by means of these burners the furnace can be run 10 hours on 24 gallons of crude oil. This is a splendid record, and speaks well for the burner construction. The billets are heated from 2200 to 2300 degrees Fahrenheit, and the hammer used to drop them to shape is of 6000 lbs. capacity. It might be well to explain that the billet is 3" square and 13" in length, thus allowing for the making of two stems.

We will suppose the billet has been heated and is ready for the hammer. The following steps are now followed out. The stock is first edged down near the middle of billet on the portion of die marked A. It is next placed over the center impression of bottom die B and the blows struck. Usually 6 to 7 blows complete the form. After this, it goes to the trimming press and is trimmed.

This operation is worthy of special mention as a hydraulic press of 150 ton capacity is used. This press was employed during the war for shell work. They simply switched it over to peace time duty, and as they had it all piped up and the water there, it costs practically nothing for maintenance. Of course such a press could not be used in the average case, but the fact of its being in use makes it worthy of mention.

The fash is next cut off at end of billet, and the hammer makes a tong hold on the remaining stock. It now goes back into furnace and after reheating, the operation is repeated. Twelve billets are carried at one heat of furnace, or in other words, 24 forgings. The production of this piece is 300 stems per 9 hour day, and in the making of the die some 1800 lbs. of steel are used.

The next photograph, Fig. 4, depicts the die used to forge the front axle of the McLaughlin Buick Car. The billet in this case is 2⅛" diameter, and 41¾" long in its rough state. To further explain the procedure in forging of the axle we have prepared Fig. 5, which illustrates the various steps in its progress.

The stock is first placed into what is termed the roller (see A) to gather balls, or a greater amount of stock in certain sections, and reduce the amount in others, so as to distribute it properly to place in the binder, or second portion of die (see B). The purpose of the second form is to bend the stock up in proper shape so as to fit the third impression C as closely as possible. It is now placed over final impression C, and hammered to shape. Usually six to seven blows are sufficient. This view shows the roughing set of dies only, and as will be noticed, takes care of only one-half of the axle. This is the method of making these axles one-half at a time.

After half of the axle is rough finished, it goes to the finishing die, this consisting of but one impression. Five or six blows bring the forging to correct form, and the axle must be correct as to bosses, pad and web of beam. From here it goes to the trimming die to remove fash, or surplus material. It

is then returned to the hammer and restruck, as sometimes the trimming die will bend the forging. The other end of axle now goes through similar operations.

After leaving the hammer the axle next arrives at the bulldozer. It is practically impossible to forge a piece of such nature to accurate length, so that this is where the bulldozer comes in nicely. The forging is made a shade heavy on the beam, and in this last operation the bulldozer stretches it to exact length, and makes sure all pads are correct. In action, here is what happens.

The axle is placed in a form of correct shape, up comes the moving head, and a spreader strikes the center portion of axle, thus spreading it to accurate length. The axle is gripped tight on each end by clamps shaped to suit the web. The spreader has two rollers on it which engage with tool steel tapered guides. These spread, and in turn spread the axle. While this occurs two rams attached to the die heads flatten the spring pads. This operation eliminates any machine work on these pads. The machine can turn out 450 axles every 9 hours, and the production on the hammers is as follows: 400 ends blocked and 400 ends finished every 9 hours.

The forgings used in the axle making are blocked in one hammer, taken over to the furnace of second hammer, then heated there. This saves a great amount of heat, and the two hammers, namely blocking and finishing, work in tandem. It is estimated that 75% of

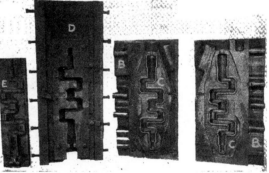

FIG. 4.—THIS SHOWS BOTH-DROP FORGING AND TRIMMING DIES USED ON AN AUTOMOBILE CRANK SHAFT.

FIG. 7—STEPS IN THE MAKING OF AN AUTOMOBILE CRANKSHAFT.

heat is saved in this way. For example, when the first operation is completed, the forging is still a cherry red. This being so, it does not take long to heat it up to the proper forging heat for finishing operation.

The bending impression B (see Fig. 4) is not made with any idea of forming to finished shape, but the idea is rather to have sufficient stock at the correct place to form up nicely on the impression C. The dies which we have just been discussing required two tons of steel to make them, and cost in the neighborhood of $1,500. Referring to Fig. 5, we note that A represents the rough stock, B one-half of axle rough forged, C one-half finished forged and trimmed, D the other half rough forged, and E the completed axle. These steps, together with previous description, will no doubt be perfectly clear.

Crank Shaft Die

The next example, Fig. 6, depicts the dies used to produce the crank shaft for the Chevrolet Baby Grand. In this case they use stock 3½" diameter, and draw it down to 2½" diameter on both ends. This is accomplished on breakdown, or swedger, shown at extreme side of each die block, see Fig. 6. This portion of die prepares the stock for the finishing impression. The stock as it appears after the swedging operation is shown at A, in Fig. 7.

Let us again refer to Fig. 6. The stock is next placed in the bender at B, when it is formed up to shape and rough outline of impression. It is next placed over the finish impression C, and

struck sufficient blows to bring to required size. From this operation it goes to the trimming press and fash removed. These trimming dies are shown at D and E and they are so designed that forgings will drop through freely. After trimming, the piece is returned to hammer and restruck to make sure there is no twist in the crank.

From here it goes to an upsetting machine, which has a pressure of 200 tons. The stock is inserted in the lower die of

the machine, and a vertical ram with upper die attached grips the crank firmly. The horizontal ram now comes forward, and a heading tool enters the die and upsets the flange shown at C in Fig. 7. B in this same view illustrates how the crank appears before upsetting. This flange is completed with one stroke of the ram. In the forging operation 180 cranks are completed in 9 hours, and 350 are upset in a similar period.

Designing Dies

Before proceeding with further examples of dies, it might be well to explain that this concern prepares all their own dies, even to supplying the steel for the same. They are, in other words, a self-contained organization. They have a die sinking department, a tool room, a machine shop, etc. These various units are all in Welland and under the heading of the Canada Foundries and Forgings, Ltd.

Suppose, for example, that a new piece comes into the plant, which has to be drop forged. The following vital points must be observed. First, the blue print, or sample, must be studied and measured in order to decide the size and width of block to be used. As a general rule, on a heavy hammer such as the 6000 lb. we have been speaking of, a die block is never made less than 12" deep. This, of course, is in order that the block will be able to resist the strain exerted on it. The width must also be sufficient to withstand the strain, apart from the planning of proper width to accommodate the main impression, and other operations such as bending, edging, rolling, etc.; etc. The rolling, by the way, is really the gathering of the stock in a certain place, as for example in Fig. 4, at A. Another general rule is to leave at least 3" of each side of impressions.

Having decided upon these points, a

FIG. 8—SINKING A CRANK SHAFT DIE.

FIG. 9—CONNECTING ROD DIE FOR GAS ENGINE CONNECTING ROD.

die. The average bearing is 1.9985" diameter, the collar at the end of crank is 5" diameter, and the section is 1⅜" thick plus 2½" wide with a 7 degree flare.

Another Example

The next example, Fig. 9, illustrates the dies used to forge a 1½ h. p. gas engine connecting rod for the Kerr & Goodwin Co., Brantford. The steel for this job is approximately 1¾" square in the rough and it goes through the following steps in sequence. The center of stock is drawn down at corner of die, then is placed in edger. This rolls the stock to proper shape and size. It is now placed in finish impression and blows struck. It is next trimmed, from there going back to the press to be straightened. It now goes to the punch, and has the eye punched out, thus completing the rod, as far as they are concerned. The rod is 11⅝" center

the equipment. The action of such a furnace is so self-evident that we make no special comment on the same except to say it is of their own construction.

The machine shop connected with this portion of the plant accomplishes the rough and finish machine work if desired, and is well equipped for both large and small work. In addition to the chemical and test departments, a fine first aid room is installed, and in this room is even the equipment necessary for operations should they be required. The workers in charge of the department are ex-army medical men, who thoroughly understand their business. While this new portion of plant was primarily intended for making forgings for automobile work, it is well equipped to handle any work of a forging nature, and we are led to understand that still further additions will be made as soon as working conditions permit.

Compressed air and a new species of sound box are declared to be the secrets of the invention. The air is supplied by means of an electric motor, and when it passes through a small valve controlled by the phonograph needle, it expands and creates a burst of sound that in the case of vocal selections, at least, surpasses by many times the volume of the original voice. It is said that to a person standing about 30 yards from the instrument the illusion of a brass band playing at about an equal distance is almost complete. The most entertaining results, however, are derived, when a vocal phonograph record is placed on the stentorphone. At the time of the public test, when Sir Harry Lauder's "We Parted on the Shore" was played, it sounded like the voice of the mightiest giant that ever stalked through the pages of a fairy tale, according to the reports.

Mr. Gaydon thinks that the machine can be used as a drill sergeant, giving the words of command before playing the marching accompaniment. In battle it might be used to drown out the shrapnel crescendo and scare the enemy to death, is another militant suggestion. It has also occurred to the inventor that possibly it can be used to deliver the speeches of political candidates in a way that will baffle the obstructive tactics of the loudest-voiced and most persistent heckler.

The Federation of British Industries has devised a unique scheme for advertising home manufactures in all parts of the world. A new steamship, to be called the Federation, is to be constructed by some of the leading firms of the country as an exhibit of British marine engineering. The cargo of the ship will consist of various articles of British manufacture. Representatives of the firms interested will make the voyage for the purpose of showing their goods and interviewing foreign buyers.

Some Interesting Machine Shop Problems

Various Problems on Machining Bearings Are Solved, Also the Fitting of Brass Lines for Marine Propeller Shafts—A Simple Centre Jig is Described, and Other Ideas.

By JOHN S. WATTS

THE notes that follow are on a variety of topics, connected the points made were common knowledge and practice with the old time all-around machinist, my experience with the younger mechanics leads me to believe that they are unfamiliar with much of the useful knowledge acquired by their predecessors, when a man usually followed a job through to actual operation, and so gained experience which is not generally available to the mechanic of to-day.

The lack of technical schools and journals in the years past had a tendency to make the mechanic of those days a man who studied things out for himself, and acquired a habit of analytical study which enabled him to arrive at the correct solution of any problem which he encountered. On the other hand, I find that knowledge gained from the experience of others does not become a part of oneself, and is not so thoroughly grasped as that which has been acquired by the use of one's own intellect.

While it would be absurd to condemn the training of apprentices, it unfortunately does tend to weaken their self-reliance, and the mechanic of to-day, will, 99 times out of a 100, refuse to tackle a job which he has not been taught to do. This is a fault which he should fight against, and is in the last analysis simply a form of mental laziness, as the well trained and educated mechanic of these days should, with his wider knowledge of the laws of mechanics, be more able to handle the unusual job than his predecessors. The only advantage the older men had was that they were naturally used to taking on any job that came along, having never had any assistance from teachers or books.

A case in point came to my attention a short while ago in a shop finishing large marine propeller shafts. The brass liners are about 12 inches diameter, and anywhere from four to six feet long, and after boring out are heated and shrunk on to the shaft. These people were losing the liners, sometimes three or four consecutively, on the same shaft, by reason of the liners developing a crack circumferentially when cool; this cracking they attributed to the material, as they thought a crack caused by a shrinkage strain would be a longitudinal one.

Now in the shop where I served my apprenticeship we were turning out propellor shafts continuously, and I never saw one crack, and this although they were made of common yellow brass, nearly all scrap, while the ones mentioned above which did crack were the 88-10-

2 mixture cast of all virgin metal, and subject to analysis.

Investigation showed that both the shaft and the liner were rough turned, and no attempt had been made to get a smooth surface where the liner fitted. Considering that at the temperature at which these liners are shrunk on the shrinkage longitudinally will be around one-eighth inch per foot, or ¾-inch on a six foot liner, it is clear that either the liner must slip along the shaft as it cools, or crack circumferentially as actually happened. Having been used to shrink fits whose length was comparatively short, it did not occur to them that the liners had to shrink in length as well as in diameter, and the solution of their trouble was obviously to turn both shaft and liner smooth enough to allow the liner to slip along the shaft when contracting.

Another Case

Before mass production reached its present dimensions, it was common practice for the machinist to devise his own jigs, and generally to decide as to whether it was worth while to fix up a jig or not. The practice now seems to be that if no jig is provided the machinist makes no attempt to improvise one. For instance, even the smallest shop will at times get a quantity of wheels, flanges or similar circular castings to bore out, and to true up each individual piece in a chuck is a waste of time, even if only six are required. By taking three or four short pieces of bar and after bolting them to the face plate, bore them out to the diameter of the casting, as shown in Fig. 1, we have a jig which will centre the castings accurately enough for ordinary work, at a trifling expense. The casting is then clamped to the face plate and the time spent in trueing up is eliminated.

For boring out plain solid bearings,

when the quantity is too small to warrant the expense of a special jig the best method is to bolt an angle plate to the face plate, at a distance below the centre of the lathe equal to the height from the base of the bearing to the centre of its bore. Also drill holes in the angle plate to match the bolt holes in the base of the bearing, and we are then assured that all the bearings so machined will be alike as to centres.

While on the subject of bearings it should be noted that solid bearings for hand operated, or slow moving gearing, such as chute hoists and the like, where the bearings are bolted to the wood work of the building or structure, should be bored out one-sixteenth of an inch larger than the shaft, to allow for mis-alignment, due to the possible warping of the wood and settlement of the structure. To the man who has not had the operating of such mechanism this may seem like sloppy work, but actual experience dictates this clearance as essential for really satisfactory performance.

A point often overlooked in the machining of gears, bearings, etc., which are called for in the drawing to be faced, on one side of the boss only, is that it is not correct to merely machine off enough to clean up the surface. It is necessary that the distance from the centre line of the gear, or of the bearing, to the machined face, must be as shown on the drawing, or the erecting crew will have trouble, and this dimension should be worked to in machining.

When a parallel or taper reamer, of a size not available, is wanted in a hurry, one can be made up very quickly by turning up a piece of flat tool steel, and fastening to it hard wood blocks as shown in Fig. 2. The shank can be made square or tapered to suit the driving medium and a reamer so made will do

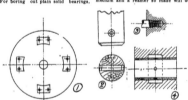

a surprising amount of work of a high quality before being worn out.

Some Other Ideas

Cases arise occasionally when a stud has to carry its load in shear, and if screwed in in the usual way the area to resist shear is only that at the bottom of the thread, and it is further weakened by the sharpness of the thread, which tends to make it easy to shear off. This can be improved upon by counter-boring the hole about ½-in., as shown in Fig. 3, to the neat diameter of the stud. The stud is then screwed in until the end of its threaded part is at least ¼-inch under the surface, the result being that the solid part of the stud now takes shearing strain.

For heavy pressures or for places where the joint would be hard to get at, if a leak should start, a sure job must be had, and the safe way is to spot face the casting under the nuts and bolt heads, thus getting a greater tensile strain on the bolt with an equal pull on the wrench, due to the lesser friction under the nut, with a smooth surface. The nut and bolt head under these conditions will also give a better support to the flange, against springing under the pressure. When heavy bolts are used, ¾-in. diameter or over, additional safety can be gained by slightly heating the bolts, so that they will shrink after being tightened up by hand.

It would seem that a shaft coupling, shrunk or pressed on a shaft would remain true, after driving the key, but experience shows that the driving of the key will often throw the coupling out of line slightly, and it is good practice to true up the coupling after it has been keyed to the shaft. Cases of hot bearings, in line shafts, otherwise mysterious, may often be traced to the coupling being forced out of true when the key was driven.

Loose pulleys, ring oiling bearings and such, where the inside is cored out to form an oil reservoir. should have the rough surfaces carefully painted over to prevent any of the moulding sand adhering to the surface, afterwards becoming loosened and getting mixed with the oil. A good idea in loose pulleys and rollers is to drill about six holes in the bush

and plug them with felt, this will keep the oil from running out of the reservoir, and provide a continuous feed of clean oil to the bearing; the general idea is illustrated in Fig. 4.

OSCILLATING BROACHING FIXTURE
By "Draughtsman."

THE operation is to broach a piece of drawn brass tubing 2 inches long, 1 inch inside diameter, and 3-32 inch thick, and to form a small fluting running its whole length, as in Fig. 1.

The piece is subsequently punched for an oil hole and then pressed into the end of a wooden connecting rod, thereby forming a bush which cannot turn because of the fluting, which also acts as an oil groove. The operation is accomplished on a press of 3 inch stroke, and the fixture is designed so that one piece is broached and another ejected per stroke.

The upper holder B carries a broach C and a pusher D. The broach, shown in detail, has an inserted blade E held by two pins. The enlarged head butts against the counterbore shoulder, and is prevented from turning by a "half-and-half" pin. The pusher is simply a pin shouldered at its upper end and held by a set-screw. Its body is slightly less in diameter than the brass tube, and its lower end is shouldered and reduced to a short centering taper.

The bolster F has a recess in which rests the hardened die block G. By means of a lever H and handle it can be rotated through 180 degrees, its movement being limited by stop pins J.

The dieblock is retained by a steel ring K. The stationary stripper plate L has two bushings, the first one M, under the broach is fluted to suit the work, and resists upward thrust, while N, the one under the pusher, has a plain hole and resists downward thrust. The front part of the stripper is reduced in thickness to give ample room for the operator's fingers. The broach hole in the bolster carries a heavier bushing O to take the thrust of broaching, while a clearance hole P under the pusher communicates with the hole in the bed of the press.

Operation

With the ram up and lever over, say to the right, loosely insert bush No. 1 in the hole under the pusher. Trip the press and the pusher will force the bush into the die-plate.

With the ram up, swing the lever to the left and insert bush No. 2 under the pusher. Bush No. 1 is now under the broach, which never leaves the stripper. Trip the press and bush No. 1 will be broached while bush No. 2 is pushed into the die plate.

With the ram up again, swing the lever back to the right and insert bush No. 3. Trip the press and bush No. 3 will enter the die plate causing bush No. 1 to be ejected below, while bush No. 2 will be broached. And so on.

An option has been obtained by a New York organization from the School Board, on a site on the north side of Bloor Street, at the head of Church Street, Toronto, and a ten-story private hotel will be built there. The cost is estimated at between $4,000,000 and $4,500,000.

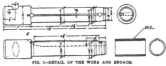

FIG. 1.—DETAIL OF THE WORK AND BROACH.

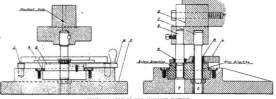

SECTIONAL VIEW OF THE COMPLETE FIXTURE.

Development of 8 Driving Wheel Locomotive

By J. SNOWDEN BELL.*

LOCOMOTIVES having eight driving wheels, the latest designs of which exemplify the most generally approved practice in heavy freight and passenger road service, the Mallet and Decapod types being adapted to, and ordinarily used only under exceptional conditions, were not, as a class, the outgrowth of increasing requirements in the operation of railroads, but were introduced at a very early period in their history, dating from some time prior to the year 1825. A review of the progressive steps of improvement that have brought them to their present perfected state will present features which have passed out of general recollection, and may, from a historical standpoint, be found of interest.

Wylam Colliery Railroad, 1825

The first eight driving wheel locomotives of which a record has been found, are those which are described and illustrated in Wood's "Practical Treatise on Railroads," London, 1825, as being then used on the Wylam Colliery Railroad, in England, and are shown in Fig. 1, which is reproduced from Plate VI of the Treatise. The boiler is of the return tube type, and, with the two vertical cylinders, is supported on a wooden frame, which is carried on two independent truck frames, these being fastened to the main frame "by bolts in the middle, which allow a lateral motion in the frames, to conform with the occurrence of curvatures in the line of the road."

Two pairs of driving wheels are fitted in each of the trucks. The piston rods are connected to beams, supported at one end on the main frame, from which connecting rods extend to cranks on the ends of a transverse shaft, located at about the middle of the length of the locomotive, which shaft carries a spur gear, engaging a similar gear on a shaft below it. The latter shaft is geared to the adjacent driving axle of each truck, and these to the other axles, intermediate gears being interposed to ef-

*From Record No. 189 of the Baldwin Locomotive Works.

fect the rotation of all axles in the same direction, the locomotive thus having no less than eight spur gears for transmitting power from the pistons to the four driving axles.

The dimensions of the locomotives are not given, but from the scale accompanying Plate VI the cylinders appear to be about 8½ inches bore, for 24 inches stroke of pistons; the driving wheels 43 inches diameter; and the boiler 42 inches diameter and 9 feet 9 inches long. These locomotives probably weighed in the neighborhood of 10 tons, which was an extremely heavy weight at their date, and Nicholas Wood, the author of the Treatise, gives the reason for their distribution of weight as follows: (Pp. 155, 156.)

"The railroad on which these engines travel being too weak to support the weight divided on four wheels, recourse was obliged to be made to eight wheels; and this rendered the use of so many

cog wheels necessary, to obtain the adhesion of the whole weight of the engine.

"This circumstance adds much to the complication of the engine, which, of course, increases the friction by the multiplicity of cog wheels and other moving parts."

Ross Winans, of Baltimore, Md., who will be remembered as one of the most prominent and successful locomotive builders of his time, obtained, on July 23, 1843, a patent of the United States, No. 3201, for "a new and improved manner of constructing locomotive engines with eight driving wheels," in the specification of which he states that:

"I do not pretend, therefore, to have originated the idea of constructing a locomotive engine with eight driving wheels, but limit my claim to invention to having so brought together, and combined, various known elementary parts of such engines as to have produced one

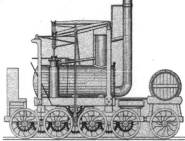

FIG. 1.—AN EARLY TYPE OF EIGHT DRIVING WHEEL LOCOMOTIVE.

which is substantially new, attaining the desired end in a manner more simple and effective than on any of the plans heretofore devised."

As shown in Fig. 2, which is reproduced from the drawing of the Winans' patent, the locomotive of that patent had a vertical boiler and horizontal cylinders, the pistons of which rotated a transverse spur wheel shaft at one end of the engine, which was geared to the nearest driving axle, with which the other three driving axles were coupled by end cranks and side rods. Two spur gears only were used, instead of eight as in the Wylam engines.

Three locomotives of this design were built to the order of Mr. Winans, by M. W. Baldwin, of Philadelphia, for the Western Railroad of Massachusetts, and were completed in April, May and June, 1842. No drawings of these engines are in existence, and it is believed that they were the only locomotives of this type that were built with vertical boilers.

Ross Winans, 1844

The "Report on the Use of Anthracite Coal in Locomotive Engines on the Reading Railroad," made by Geo. W. Whistler, Jr., April 20, 1849, states (p. 23) that "In October of that year, 1844, Ross Winans produced the first successful coal burning engine with a horizontal boiler," and the report further states that 12 of these engines were placed on the Baltimore & Ohio Railroad, between October, 1844, and December, 1846, and that a substantially similar engine was built by the company in May, 1847.

These engines, which were the first eight driving wheel engines on the Baltimore & Ohio Railroad, were known as the "mud diggers," doubtless by reason

FIG. 3—THIS TYPE OF ENGINE WAS KNOWN AS A MUD DIGGER.

of their action on the then comparatively light superstructure of the road.

As shown in Fig. 3, which is reproduced from a photograph of engine No. 37, made at Mount Clare shops in 1863, the main connecting rods were, as in the three engines of the Western Railroad of Massachusetts before referred to, coupled to cranks on a shaft extending across the frames, in the rear of the boiler, and geared, by spur wheels, to the back driving axle. The four driving axles carried end cranks which were coupled by side rods.

The 12 Winans' engines of the Baltimore & Ohio Railroad, mentioned in Mr. Whistler's report, were of the above construction, and were the "Hercules," No. 33, October, 1844; "Gladiator," No. 34, November, 1844; "Buffalo," No. 35, November, 1844; "Baltimore," No. 36, December, 1844; "Cumberland," No. 37, July, 1845; "Elephant," No. 38, July,

1845; "Opequan," No. 40, July, 1846; "Elk," No. 41, August, 1846; "Catoctin," No. 42, October 1846; "Youghiogheny," No. 43, November, 1846; "Tuscarora," No. 45, December, 1846, and "Allegheny," No. 46, December, 1846. The engine built by the company was the "Mount Clare," No. 49, May, 1847.

These engines weighed 23.5 tons. and had 17 x 24-inch cylinders and 33-inch driving wheels. Some of them were in yard service as late as 1865, and probably for some years afterwards. A number of their cylinders and main connections were used as parts of stationary engines in the shops of the road after their road service was terminated.

M. W. Baldwin, 1846

The first locomotives, of what is now known as the 0-8-0 type, in which the objectionable feature of gearing was eliminated, by coupling the main con-

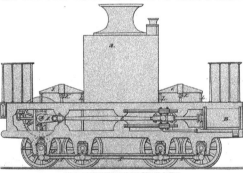

FIG. 2—A REPRODUCTION OF THE WINANS' PATENT.

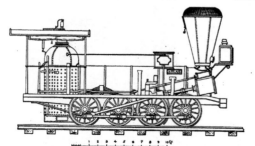

FIG. 4—SHOWING INCLINED CYLINDERS AND DROP HOOK VALVE GEAR.

necting rods directly to the crank pins of the main driving axle, were built by M. W. Baldwin, seventeen of them being constructed in a single order, for the Philadelphia & Reading Railroad, in 1846. Two of these engines, the "Atlas" and the "Hercules" (Fig. 4), which were used in pushing service on the Falls grade, at Philadelphia, weighed 27 tons, and their cylinders were 17¼ x 18 inches and driving wheels 42 inches in diameter. The other engines of the order weighed 22.4 tons, and their cylinders were 15½ x 20 inches, and driving wheels 46 inches in diameter.

The characteristic feature of all of these engines was the very ingenious appliance known as the Baldwin "flexible beam truck," in which the boxes of the first and second axles were fitted in beams, which were articulated independently, to the main frames on opposite sides of the locomotive, by spherical joints. The driving boxes were turned cylindrically and fitted in correspondingly bored pedestals, and the brasses of the front coupling rods were cylindrical and fitted on corresponding pins. The wheels of the third axle had flat or "blind" tires. The engines were thus enabled to pass curves of very short radius. The flexible beam truck is the subject of Mr. Baldwin's patent No. 2759, of August 25, 1842, and the principle of its operation is well and clearly stated by Zerah Colburn, in his treatise "Locomotive Engineering," 1871, p. 85.

All the engines of the first order had inclined cylinders and drop hook valve gear, as shown in Fig. 4. The 15 lighter engines were also fitted with independent cut-off valves, but single main slide valves were applied on the "Atlas" and "Hercules." The fireboxes of the latter engines were small, having only about 8.45 square feet of area, and the engines burned wood fuel. The total heating surface was about 1153 square feet,

and, on the basis of the evaporative value of the tubes and firebox, the boiler power appears to be in excess of the cylinder requirements. The tractive power of the engines may be estimated at about 10,850 pounds, and it is probable that no engines of greater power had, up to their date, been built. They did efficient service for many years, and a number of them were rebuilt by the railroad company, with wide fireboxes for anthracite fuel, and link motion valve gear.

The general design of the Reading engines continued to be the practice of the Baldwin works, in the 0-8-0 type, until that type was superseded by the addition of a leading truck. In some instances, as in wood burning engines, the firebox was located between the third and fourth axles, instead of behind the rear axle, and the final modification of the design, which was made some time prior to 1854, provided an increase of

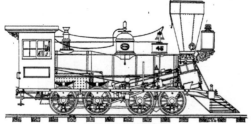

FIG. 5—THIS TYPE WAS BUILT IN 1857.

grate area and firebox surface, by setting the firebox on top of the rear axle. A number of locomotives of this construction were built for various railroads, one of which, No. 45 of the Pennsylvania R.R., built in 1857 is shown in Fig. 5. This engine had 18 x 20-inch cylinders, 42-inch driving wheels, and weighed 63,700 pounds.

Ross Winans, "Camel," 1848

As has been stated by Zerah Colburn, whose eminence as an authority on locomotive engineering is beyond question, the so-called "camel" engine of Ross Winans was, "as a whole, the most peculiar engine in use in the United States. In every detail of construction this engine is alike peculiar, and in the strongest possible contrast with the proportions, arrangement and workmanship of the standard American engine." Its design is quite clearly shown in Fig. 6, which is reproduced from an advertising lithograph issued by Mr. Winans.

The distinctive features of novelty of the camel engines were the following: 1. The use of eight driving wheels, set closely between horizontal cylinders and a long overhung firebox, the width of which was equal to, or greater than, the distance over frames. 2. A firebox having a downwardly and rearwardly inclined top, which was exclusively a water casing, there being no steam room in it. 3. A dome and engineer's cab, placed on the top of the boiler, close to the forward end. 4. An upper chute or pair of chutes, for feeding coal through the top of the firebox. 5. A firebox having no water space on its rear side, which was closed by doors, so as to expose its entire area when desired. 6. The abandonment of crown sheet stay bars, and the substitution of stay bolts connecting the crown sheet with the outer shell. 7. The use of a half stroke cam, as a means for effecting cut off. All these engines were substantially of the same construction, except as to the firebox, of which there were three classes, the short, the medium and the long.

The first engine of the class, the "Camel," No. 55, was placed on the Baltimore & Ohio Railroad in June, 1848, and 112 in all were comprised in the company's equipment. A number were also built for the Philadelphia & Reading, Pennsylvania, Northern Central, Delaware, Lackawanna & Western, and other railroads. Some of them were rebuilt and modified, and many remained in yard service for a long time after more modern equipment took their place in road work.

Ross Winans, "Centipede," 4-8-0 Type, 1856

The first locomotive in which eight driving wheels were combined with a four-wheeled leading truck, and what is now known as the "Mastodon" or 4-8-0 type, was the "Centipede," which was built by Ross Winans in 1856, and sold to the Baltimore & Ohio Railroad in 1863. This engine is of special interest, not merely as being the first of its type, but also by reason of numerous novel and peculiar features of structure which it embodied, and it is greatly to be regretted that no description or drawings of it appear to have been preserved, and, therefore, that it can be described only from the recollection, naturally dimmed by the lapse of years, of the few now living who, like the writer, have seen the engine.

The boiler was of the same form, and about the same size, as those of the camel engines; the cylinders were 22 inches bore and 22 inches stroke; and the driving wheels 43 inches in diameter. The weight was probably not far from 29 tons. The engine had independent cut-off valves, and the cab was located on the front bumper platform, the throttle rod passing through the stack. The bearings of the truck axles extend ed nearly the entire length between the wheels, and there were other peculiar details which cannot now be remembered.

Comparatively few locomotives of the 4-8-0 type have been built for American railways. Probably the most extensive user of this type is the Norfolk & Western, which in the years 1906 and 1907, placed in service 100 4-8-0 type engines built by The Baldwin Locomotive Works. These were followed, in 1910, by 50 heavier engines of the same type. One of the latter is illustrated by Fig. 7.

These locomotives, as originally built, used saturated steam, and were equipped with Walschaerts valve motion and piston valves. They were of the following dimensions:—weight of engine, 261,100 pounds; weight on driving wheels, 215,- 200 pounds; cylinders, 24 x 30 inches; driving wheels, 56 inches diameter; boiler, 80 inches diameter; heating surface, 4041 square feet; grate area, 44.7 square feet; boiler pressure, 200 pounds; factor of adhesion, 4.07; maximum tractive power, 52,500 pounds.

Baltimore & Ohio R.R. Co., 1865

With a view to the avoidance of certain structural objections which obtained in the camel engines, as well as to incorporate additional features of improvement, Thatcher Perkins, then master of machinery of the Baltimore & Ohio Railroad, designed a class of locomotives of the 0-8-0 type, which were variously termed the "Perkins' eight

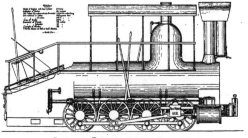

FIG. 6—THE ROSS WINANS TYPE, TERMED THE MOST PECULIAR ENGINE EVER USED IN THE UNITED STATES.

FIG. 9—THIS STYLE ENGINE WAS BUILT IN 1917.

wheel connected engines," the "green backs," and the 'Jerseys." These engines are shown in Fig. 8, and 24 of them were built in 1865, two by the company, two by Reaney, Son & Archbold, of Chester, Pa., and 20 by the Grant Locomotive Works, of Paterson, N.J.

The Perkins 0-8-0 engines were similar to the "camels" which they were intended to supersede, only in the particulars of having overhung fireboxes, of the full width over frames, and 43-inch driving wheels, and differed therefrom in the following particulars. Their boilers were larger, being of the straight type, of ⅝-inch iron, 48 and 48¾ inches diameter at alternate rings. Their cylinders were bolted to an intermediate bed or saddle, on which a round smoke box was supported as in present practice, and were originally bored out to 19½ inches, but were designed to be rebored to 20 inches, being 1¼ inches thick.

Their frames, which were of the bar type, were very substantial, weighing about one ton apiece, and having top rails finished to 3¾ x 6 inches. The link motion was of the Gooch type, and

a radius of 55 inches was obtained for the links by connections to and from a supplemental rocker shaft. The links were of cast iron, of box form, and weighed 434 pounds each. The draw bar extended through the ash pan, and its front end was coupled to a pin fitted in cross braces bolted to the bottom frame rails. This arrangement, by which the boiler was entirely relieved from strains of draft and buffing, was applied by Mr. Perkins at a much earlier date, and it has, within a few years past, been brought out as a new and original design.

The principal particulars of these engines were as follows: Weight, 65,000 pounds (empty); cylinders, 19½ x 22 inches; boiler, larger diameter, 48¾ inches; thickness, ⅝ inch; pressure, 100 to 110 pounds; 115 flues, No. 12 W. G. 2½ inches outside diameter and 15 feet long; inside firebox, length, 66 inches; width, 42 inches; depth, 57 inches; grate area,, 19.25 square feet; flue-sheet (copper), ⅝ inch; back sheet (copper), ⅝ inch; side sheets (copper for 36½ inches from bottom 3-8 inch; exhaust nozzle (single), 4½ inches. Weight of tender,

23,000 pounds; water capacity, 2140 gallons.

Yard Engines of 0-8-0 Type

The earlier engines of this type were usually relegated to maintenance of way and yard work, when, as they became old, they were replaced by heavier power in road service; but within the last few years engines of this type, of weight and power greatly exceeding the earlier road engines, have been designed for, and advantageously operated in, yard service, where comparatively heavy loads were required to be moved.

Fig. 9 shows an engine of this class, built in June, 1917, by The Baldwin Locomotive Works, for the Duluth, Missabe and Northern Ry., which embodies the latest and most improved practice in the 0-8-0 type, and is also very much larger and more powerful than the average road engine of that type which preceded it.

The weight of this locomotive is 216,-000 pounds; cylinders, 24 x 28 inches; driving wheels, 51 inches diameter; heating surface, 2690 square feet; superheater surface, 621 square feet; grate area, 48.2 square feet. The boiler pres-

FIG. 8—SHOWING THE PERKINS EIGHT-WHEEL CONNECTED ENGINE.

FIG. 11—REFERENCE TO THE TEXT WILL GIVE DETAILS REGARDING THIS P.R.R. TYPE.

sure is 185 pounds; maximum tractive effort, 49,700 pounds, and factor of adhesion, 4.35.

J. P. Laird, 2-8-0 Type, 1864

The next, and most generally adopted, step in the design of the eight driving wheel locomotive, consisted in the application of a two-wheeled leading truck, thereby providing the 2-8-0 (usually termed the "Consolidation") type. This was originated by John P. Laird, then master of machinery of the Pennsylvania R.R., who, in 1864, rebuilt the Baldwin engine No. 98, which was originally of similar construction to engine No. 45 (shown in Fig. 5) and among other changes; put in a two-wheeled leading truck.

As shown in Fig. 10, the rebuild also included an increase of the driving wheel base, and the addition of side tanks and a rear coal bunker; also the fitting of all the driving axles on pedestals rigid with the main frame. The rebuilt engine 98 was used in pusher service on the Allegheny Mountain grade, and in the latter part of 1864 or early part of 1865, another Baldwin engine, No. 97, which was originally similar to No. 45, was rebuilt on the same lines as No. 98.

A. Mitchell, "Consolidation," 1866

In July, 1866, the Baldwin works built to the order of Alexander Mitchell, master mechanic of the Mahanoy division of the Lehigh Valley R.R., the 2-8-0 type locomotive "Consolidation," which has been usually believed to be the first of that type; and prior to the adoption of the present system of classification by wheel arrangement, the type was known by the name of this engine. It was however, as above stated, originated two years previously by John P. Laird, his first locomotive of that type (P.R.R. No. 98) differing from the "Consolidation" in being a tank engine for pusher service, while the latter was a road engine with tender, as shown in Fig. 11.

The weight of the "Consolidation" was 90,000 pounds, of which 80,000 pounds was on the driving wheels; cylinders, 20 x 24 inches; and driving wheels, 48 inches diameter. The design soon became practically the standard for heavy freight service, and up to the present time continues to be generally approved practice, under ordinary conditions.

A NEW BRITISH UNDER-WATER PUMP

In salvage operations, in well-making, and in many kinds of dock and harbour engineering great advantage is gained by using a pump which can work under water. Several types of submersible electric pumps have been designed by British engineers, and the latest type has some special features of interest. The usual practice is to allow the water free access to the stationary part of the motor, thick rubber being used to insulate the electric circuits. In the new invention this part of the motor is cased in a special steel alloy, the casing being filled with oil. This arrangement makes the motor very compact and of high efficiency. By using suitable materials for the casing and pump parts, and a special incorrodible steel alloy in the rotating part of the motor, the pump can be adapted for handling weak acids and other corrosive liquids.

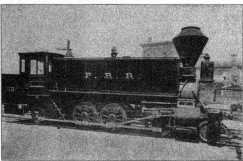

FIG. 10—NOTE THE INCREASED DRIVING WHEEL BASE, AND ADDITION OF OTHER PARTS

Heat Losses in Operation of Electric Furnaces

The Chief Causes of Heat Losses in Electric Furnace Operation Are Described—Various Experiments Were Made to Eliminate These Losses and the Results of Same Are Stated Herein.

By F. HODSON, President Electric Furnace Construction Co.

SOME very interesting data were recently presented before the American Electrical Society, by M. R. Wolfe and V. de Wysocki on the heat losses through electrodes on a 3-electrode 6-ton furnace. The figures given to follow will no doubt surprise many electric furnace users as to the amount and cost of power they are losing through the furnace roof. After a close study of this particular problem for the last four years, the writer is only surprised that the losses recorded were not larger. If the amount of heat in the shape of flame that escapes from most furnace roofs were taken fully into account, the loss would be found to be nearer 30 than 18.7 per cent. The method employed by the authors of the data we referred to above, whilst giving actual heat losses from the electrodes, only gives heat absorbed by the water cooler—a great deal more than this is usually lost.

In considering these heat losses, which perhaps would be better described as through the furnace roof rather than through the electrodes, it might be interesting to consider how the conditions governing the loss of heat and power can be improved.

The Causes

Broadly speaking, the causes are: (1) Heating up of electrodes, due to faulty electrode joint or defective electrode; (2) Heating up of electrode due to imperfect contact or badly designed electrode holders; (3) Heating up of electrode, due to electrode pencilling off at point of arc; (4) Heating up of electrode due to oxygen from the air combining with the carbon of electrode and burning off.

Special care should always be taken in jointing up electrodes—troubles due to this become increasingly worse as the melt proceeds. Good quality electrodes can now be purchased, and it pays to pay even a little higher price for proved electrodes. The design of electrode holder should be carefully considered and changed if undue losses occur. With proper contact, little or no heating should occur here; on our own furnaces we have not found it necessary to install water-cooled holders, looking upon these as an admission of failure to get proper electrical contact.

Pencilling of Electrodes

This is probably the indirect cause of more heat losses than any other feature. It is accentuated in cases where too large a space is left between the electrodes and the cooling ring, or where the furnace doors are not kept properly closed at the

bottom or where, as in some cases, the melter has to unduly open his doors to inspect or rabble his molten charge. The air strikes the white-hot electrode and burns away carbon, the hole in the roof acting as a flue draft, and the electrode pencils off, so that the area at the bottom, where the arc is, is often only one-fifth of what the furnace builders have designed to carry the power behind the furnace. The current continues to flow through this restricted area of electrode and its resistance to the excessive current over the smaller area results in the whole electrode heating up instead of going into the arc. This also causes trouble with the joints, with the holders, and on all the outside electrical connections. Sometimes large pieces or flakes of electrodes break off, due to the increased resistance.

FIG. 1—VIEW OF FURNACE IN OPERATION.

With the above facts in mind, our metallurgists have been working for the last five years to try and prevent this pencilling and heating up of electrodes. It was realised early in the investigation that if the flame could be prevented from blazing out at the top that one of the chief causes, viz., pencilling, would be largely overcome. Experiments have been made with over a dozen different types of water cooler and also attempts were made to insert an inert gas under the furnace roof.

As a result of these experiments, Mr. F. W. Brooke, vice-president of our company, has developed a type of electrode cooling ring or electrode economizer, now fitted as standard on all Greaves-Etchells furnaces, that really does prevent this flame and consequent pencil-

(Continued on page 95)

Would You Like to Receive One of These Scales FREE?

You can easily do so and at the same time add to your store of knowledge by entering our weekly contest of which details are given below.

The scale is 6 in. long and is made from finest quality steel. One side is marked in 32nds, the other side in 64ths. A table of decimal equivalents is also stamped on one side, and a table of tap drill sizes on the reverse side. This scale is well worth securing.

What You Have to Do

We publish every week a number of interesting facts or statements selected from the advertising pages for that week. The selections for this issue are given below. Read these through, then turn to the advertising section and see if you can pick out the advertisements to which they refer. The work is interesting, it will train your powers of perception and of memory, it costs you nothing, it will make you better acquainted with the various lines of machinery and tools in the market, and with perseverance you are bound to win one of these useful scales as a prize.

Contest for December 2nd Issue

Contestants are required to write us stating to which advertisements in the advertising section in this number the following references are made:

1. Where to get valuable advice.
2. How to figure the net value of a reamer.
3. Something that can be used with confidence.
4. A material that is said to be ideal for use in the construction of heavy duty machinery.
5. A machine said to be quick and easy of control.
6. Something said to have enormous gripping power.
7. Something that will take larger work and turn out larger production.
8. How to secure something that is erasure proof.
9. How to perform long drilling operations.
10. Something that is said to have a powerful positive drive
11. A product that has been manufactured constantly for 60 years.
12. Something said to be highly endorsed.

Closing Date for This Contest is December 23rd

WELDING
AND CUTTING

The Welding of Wrought Iron and Steel

The Equipment and Materials Necessary, the Preparation of the Weld, the Use of the Carbon Electrode and Other Points Are Described

By H. L. UNLAND

THE welding of wrought iron and steel in simple sections by the electric arc welding process presents no serious difficulties. Reasonable care on the part of the operator in keeping the weld clean and in the preparation of the weld will, with ordinary skill in welding, result in a successful weld.

The subject of welding may be divided into three steps:

1. Equipment and materials.
2. Preparation of weld.
3. Welding.

Equipment and Materials

In addition to the equipment and auxiliary apparatus, special jobs render it desirable to have on hand other miscellaneous pieces of equipment.

Odd pieces of carbon block or of copper are of much assistance as dams in holding the molten metal in place. In cases where the weld must be smooth on one side, a piece of copper or carbon is held against the weld and metal filled against it. Iron or steel can be used if care is taken not to weld to it.

In filling a hole the bottom is often closed by holding a plate of carbon or copper against it until sufficient metal is filled in to hold.

Care should be taken to flow the molten metal against the guide pieces and not to allow the arc to play directly on them, otherwise the weld will probably be contaminated by this material, or else the guide piece may be welded solid and cannot be easily removed.

A steel wire scratch brush is used to remove light scale and rust, before commencing to weld if necessary, and also at intervals during the welding, usually when changing electrodes.

For small work the positive lead may be bolted to an iron plate forming the top of a work bench. The work may be set on this bench, the contact being sufficient to carry the current. In many cases a vise mounted on the table will be found desirable. If the work is too large for the table it may be set beside the table and a bar laid across to it.

This will provide sufficient current-carrying capacity, provided scale and rust do not entirely prevent contact. The rails in a round house, if bonded, are usually connected to the positive lead, and any car on these tracks may be welded by running only the cable leading to the electrode, the return cable being unnecessary since the current will be carried back through the rails.

A convenient terminal for the positive cable consists of a copper hook of proper size to which the cable is bolted. The terminal may be laid on the work or

---

The author of this material occupies a position in the Power and Mining Engineering Department of the General Electric Co., and is well able to speak on the subject. He emphasizes the importance of keeping the work clean, and describes various methods of cleaning the same.

---

hooked on a projecting part. It is seldom necessary to actually clamp the return lead to the work unless the metal is thickly covered with scale or dirt, which acts as insulation, in which case it is easier to chip or brush off a clean place for the contact than to use a clamp.

If welding is to be done in a room where other employees are doing other work, screens should be provided around the welding operator. They should be high enough to prevent the light striking a large part of the ceiling since the flicker of this light would probably affect the other workmen. The effect, while probably not injurious, would be irritating. White walls and ceiling should be avoided in a welding room.

Gas burners for preheating, and fire brick, sand, or sheet asbestos for covering, are useful, especially for cast-iron work, which in many cases should be preheated uniformly to a red heat and welded while at this temperature. A receptacle of water is desirable, in which the electrode holder can be cooled when

it becomes too hot after continued use.

Some operators feel that gloves are necessary to protect the hands from the arc. In many cases, however, operators find gloves to be in the way, especially when working with the metallic electrode. If desired, however, any leather or cloth glove will give sufficient protection to the skin of the hands which is much less sensitive than the skin on other parts of the body. The arms, neck, and face should, however, be covered, since exposure of these parts will probably result in burns similar to sunburn, which, while painful, are not serious.

The Flux

It is the experience of a great majority of welders that flux of any kind is unnecessary in welding, and further, that it is a source of danger in that there is liability of contaminating the weld. If the work is kept clean by brushing at frequent intervals and ordinary care is taken in the operation of the arc a good weld can be made without flux and if these attentions are lacking, flux will not make a good weld.

Preparation of Weld

Metal that is clean is much more likely to make a good, strong weld. Scale, rust, grease, soot, and foreign matter will contaminate the weld and such inclusions necessarily weaken it or else make it hard. Impurities may also make the metal porous and spongy, due to liberation of gas. Pieces of foreign matter may prevent the molten metal filling all parts of the weld and cause cavities.

Various methods for cleaning are in use—pickling for small parts, washing with gasoline or lye, boiling with lye, sand blasting, chiseling, scratch brush, ing etc., the method depending on the local conditions.

Preparatory to welding locomotive tubes to the sheets, it is sometimes advantageous to send the locomotive out on a run to burn off the grease and then clean off the oxide and soot by sand blast. Another method is to heat the boiler to normal by steam pressure and then clean by sand blasting or scratch brushing.

In welding heavy sections where it is necessary to deposit several layers of metal, the surface of the preceding layer should always be cleaned before starting the next.

For long seams the edges should be ⅛ inch apart at the one where the weld is started, and at the far end the space should be ¼ inch plus 1½ per cent. of

the length. This takes care of the expansion of the metal in the sheet and also of the contraction of the metal in the weld as it cools.

Another method of reducing expansion is to put in short sections at intervals, welding in one layer at a time, starting at the centre and working alternately toward either end. Then put one layer in the open sections, and continue in the same way till the weld is completed. The welded section of any layer should not match those in the layer below or above, the joints being broken as in laying brick work.

The welding of complicated shapes such as flywheels and some castings may require preheating at certain points to produce initial expansion which will be overcome as the weld cools. In some cases the entire piece must be preheated and in some cases after welding the whole piece must be annealed. This is sometimes done by heating the piece uniformly and then covering it with sand-asbestos, etc., and allowing it to cool slowly.

In welding cracks in plates, forgings, or castings the crack should be chiseled out to get a good bevel entirely through the plate with ⅛ to 3-16 in. clear opening on the back or to the bottom of the crack in castings or forgings. In boiler work ½ inch holes are sometimes drilled well beyond the ends of the crack and the crack chiseled, beveled, and welded.

Welding With the Metallic Electrode

The arc should be kept short, not over ⅛ inch in length.

The current should not be greater than those indicated in the table for the electrode diameter. Excessive current causes burnt or porous metal to be deposited.

The arc should be kept constant in length to insure uniformity in the metal deposited.

In welding a seam the electrode should be moved in a zig-zag or circular path advancing along the seam. The metal will adhere only to the surface of the work actually played on by the arc so care must be used to bring the arc in contact with the whole surface to be welded.

Be sure the electrode is connected to the negative terminal. If the polarity is reversed the arc will be more difficult to maintain and the deposited metal will not be as good as it should be.

In starting the arc the electrode should be just touched to the work and withdrawn immediately to the required distance. If the electrode is held too long in contact it will weld to the work, causing some delay in freeing it and starting over.

In welding be sure that the arc plays over the entire surface of the joint. The metal of the work is fused by the direct impact of the arc and if molten metal merely runs ahead of the arc over the solid metal of the work, it will not result in a weld.

The metallic electrode used is generally from 14 to 18 inches long. It may be gripped in the holder, either at one end or in the middle as required by the skill of the operator or the nature of the work.

The operation of welding overhead is the same as in normal welding. The difficulty largely lies in holding the electrode steady in the cramped position usually required. If the arc length is kept constantly short the metal will be successfully deposited, and practice is required to accomplish this. The appearance of an overhead weld is sometimes marred by drops of metal projecting or by uneven thickness of the deposited metal, but this can be overcome by proper manipulation of the electrode. A rest for the arm will sometimes assist the operator to hold the electrode steady.

Use of the Carbon Electrode

The holder should grip the electrode from 4 to 5 inches from the end. The electrode should for ordinary work be tapered to a blunt point at the working end to keep the arc from wandering over the end of the electrode. As the electrode burns away with use, the holder is moved back along the electrode to keep length of working carbon constant. The burning away of the electrode will tend to keep the taper approximately constant.

The arc is struck in the same manner as with the metallic electrode, but a longer arc should be used, from 1 to 1½ inches is the average. The arc should not be too short when welding or depositing metal as there is danger of depositing carbon in the weld with the probability of a hard weld resulting. In cutting or melting off metal, the arc should be kept short, about ½ inch being an average length.

To cut (for which purpose the carbon electrode must be used) the arc is operated like a gas torch. It is held in one place long enough to fuse the metal and allow it to run off. For thin plates laid flat a hole is melted through by holding the arc on one spot, then the electrode is slowly advanced along the desired line, the molten metal dropping out below. For thick pieces, such as shafts, castings, etc., it is desirable to start at the top on one edge and work down, allowing the molten metal to run down through the cut. It is often necessary to follow the molten metal down with the arc to keep it melted until it runs off.

The width of the cut will depend on the size of the electrode used and on the skill of the operator in keeping to a straight line. The cut will be slightly wider than the diameter of the electrode in order to allow the arc to be played on the bottom of the cut, and it will be wider for thick sections than for thin ones. The edges of the cut will not be smooth, due to the masses of the molten metal not running away and also to the fact that the arc will tend to jump from one point to another and cause an uneven cut.

To deposit metal with the carbon electrode, the arc is struck as above, but is not held in one place long enough to melt through. A pool of molten metal is established and a melting rod of metal is fed into the arc and melted down on the work. It should all be heated thoroughly to insure complete union before more metal is added.

Since heavier current can be used with the carbon electrode than with the metallic, faster work can be done in depositing metal. The quality of the weld is not quite so good, however, as when the metallic electrode is used. However, for filling holes in castings, building up worn spots, etc., the carbon weld is satisfactory and should be used.

Due to the high temperature and large amounts of heat liberated when using the carbon electrode, the electrode holder is liable to become very hot and under some conditions to melt away at the end. When the holder begins to get hot it should be plunged in a receptacle of water kept conveniently near the operator.

Cutting speed with the carbon electrode for various values of currents strengths and thicknesses of material will be approximately in accordance with the curves shown in Fig.

Cast Iron Welding

This metal, due to its properties, is unsatisfactory for welding by any method. Its low strength and brittleness causes it to break from expansion and contraction strains unless precautions are taken, and even then a successful weld cannot be assured.

Pieces of simple cross section and heavy pieces present much less difficulty than complicated shapes, but due to the undependable nature of cast iron care should be used in all cases.

The experience and skill of the operator are large factors in determining whether or not a given weld will be successful.

By experimenting, a number of operators have learned to weld copper to copper, copper to brass and steel, and bronze to bronze, as well as Monel metal, high-speed steel and Stellite. The special uses are rather limited in their application and the methods vary, so it is recommended that each operator experiment along the lines suggested by his experience.

Steel Plant in B. C.—Incorporation of the Coast Range Steel, Limited, with a capital of $15,000,000 and controlled by British capital, was announced at Vancouver. It is the intention of the company to create an iron and steel plant with blast furnaces on the lower mainland of British Columbia. Plans of the company provide for the ultimate expenditure of $50,000,000 in the establishment of a great iron and steel industry, and it is the intention to start active construction within six months. Engineers have been in the province for two months making a survey of the situation. Vancouver men named directors are H. J. Landahl, Fred T. Congdon, J. D. Kearns, John Steta and Major Montague Moore. The provincial government has approved the project and will pay a bounty per ton on pig iron production.

DEVELOPMENTS IN
SHOP EQUIPMENT

AUTOMATIC CHUCKING AND TURN-
ING MACHINE

A new automatic chucking and turning
machine, termed the 8B, has been developed by the Potter & Johnston Machine Company, Pawtucket, R.I. This
machine is built with a geared automatic
change speed head, and the spindle-
driving mechanism is all contained in
the headstock, which is of unit construction. The machines are heavy, powerful, and accurate, well suited to the
manufacturing of multiple parts. The
drive is by a single pulley transmitting
20 h.p.

Four combinations of six automatic
variations of speed are available, giving
twenty-four spindle speeds in geometric
progression from 6 r.p.m. to 92.5 r.p.m.
Any one of these combinations may be
instantly obtained by levers conveniently located on the headstock. The gearing for driving the spindle is self-contained within the headstock, all gears
running in oil, which is pumped through
all bearings. A gear on the spindle takes
care of the higher spindle speeds, while
the lower ones are taken care of by a
gear fastened to the chuck or faceplate.

The feed gearing is driven from the
spindle. There are seven combinations
of three automatic variations of feed,
making a total of 21 feeds from .005 in.
to .125 in. per revolution of the spindle.
By changing one train of gears conveniently located on the outside of the
machine the range of feeds may be
doubled. The hand changes for the
feeds are obtained by means of tumbler

levers, conveniently located at the front
of the machine. The ratio of the second
feed to the first feed is constant. The
third feed is independent of the first two
and any desired combination may be ob-

REAR VIEW OF THE MACHINE.

tained. The changes obtainable on the
machine give the proper lead for cutting
from twenty to four threads per inch,
using automatic collapsing taps or automatic opening die heads. Any one of
the three feeds can be obtained automatically at any time. The feeds are independent of the high constant speed for
idle movements of the turret slide while
withdrawing, revolving and advancing
the tools to the point of cutting.

The cross slide is provided with front
and rear blocks and two pairs of tool
posts are furnished with each machine.

The front and rear cross slide operate
independently by screw feed, having a
feed of 10 in. and can be arranged to feed
into the work at any predetermined
time, and at any desired relation, one to

the other. The turret slide is of rugged
construction and travels on ways so designed that all wear will be even and will
not affect the accuracy of the machine.
The turret slide has a 28-in. feed and no
allowance needs to be made for revolving as the turret revolves at the extreme
end of its travel. It has 13-in. longitudinal adjustment by means of a hand
crank and is securely clamped in any desired position by three bolts, besides being located by the adjusting screw.

The turret has four stations upon
which tools may be mounted, and with
each machine an outfit of turret-turning
tool holders, stems and cutters is furnished. Turrets with five or six faces
may be furnished if desired. The turret
is revolved by power through an intermittent pinion and gear and is so designed as to give an easy stop and start,
the turret being stopped when the lock
bolt engages, thus removing any shock
from the lock bolt. It is clamped into
position by a powerful binder working
on the largest diameter of the turret
seat. Levers are conveniently placed to
release the binder and lock the bolt so
that the turret can be revolved by hand.
Both cross and turret slides are adjusted in relation to each other by con-
veniently-placed clutches.

All operation of the speed clutches,
feed and quick return clutches is accomplished by a patented method operated
by dogs located on the dog wheel or

FRONT VIEW OF THE MACHINE.

drum. This method gives instantaneous movement to the clutches and enables the feed, speed and quick return to begin at exactly the same place each time. An oil pump and piping and oil arrangement through the turret are furnished on machines handling material requiring a lubricant. A 24-inch three-jaw geared scroll chuck regularly accompanies the machine and is furnished with standard set of jaws and wrench. The chuck is provided with pilot bushings to receive pilot bars for supporting the tools during the cutting operation.

A swing of 35 inches is possible over the machine bed and 24 inches over the cross-slide. The travel of the cross-slide (front and rear separate) is 10 inches. A 15 h.p. motor is required to drive the machine, when motor driven.

NEW REINFORCED BEARING

The Lang Manufacturing Co., Ltd., Guelph, Canada, have placed on the market a bearing of most unusual design. This bearing is known as the Ridsdale reinforced bearing, and is of course covered by patents. It is a well known fact that the efficiency of your machinery depends to a great extent upon the bearings used, and the makers of this new bearing state that their patented bearing was first used in a case where a certain style bearing had given trouble for years.

Briefly, the idea is this: A spring or spiral made of a certain alloy is placed in the bearing cap or housing. Babbitt is next poured into the bearing, in the usual manner, but in this case the babbitt is merely a holder of the reinforcements. These are shown at Figs. 1 and 2. At Fig. 2 is depicted the reinforcements in place, and held securely by the babbitt.

The big claim for this bearing is that there are no hard spots in such construction, also that even the cheapest grade of babbitt can be used with good results. The wear comes on the reinforcement and not on the babbitt itself. In fact in some cases the babbitt, if of a proper grade, can act as a lubricant. The makers also claim that it can be used in any bearing including those where the speed and pressure is extreme.

They state that this bearing has a tensile strength of 36,000 pounds per square inch against a much lower figure for babbitt only. They also claim that its resistance to crushing strain is much

greater. Its resistance to heat is also a large factor. Instances have been cited where this type of bearing has run for some considerable time without a drop of oil.

Summarized, they claim as follows: This type bearing requires less attention and can use a cheap grade of babbitt as a binder. It requires practically no rebabbitting and will stand under endurance tests to prove that it will not heat up and seize.

PATENT DIAL GAUGE

The Albert Herbert Ltd. 1-3 Jarvis Street, Toronto, have supplied us with the details of a patent "Capstan" dial

VIEW OF THE GAUGE.

gauge which they have placed on the market. This gauge is shown in the accompanying illustration and the makers claim for it the following:

The gauge is said to fill the requirements for an accurate, speedy and economical method of testing the accuracy of production. It is used for alignment tests in machine construction,

for testing finished parts, for testing concentricity of finished surfaces, and for setting up lathe, planer, milling, or other machine work.

The Capstan head screw, which is fitted to the spindle, is threaded 100 to the inch and is used in setting the hand of the gauge to zero on the test piece. The zero is thus kept always in one position —the top of the dial gauge. This is a great improvement as with most gauges it is necessary to turn the entire dial round for this purpose. This feature eliminates the possibility of the operator, by accident or out of curiosity, altering the adjustment, as the Capstan head screw must be deliberately turned by means of a tommy pin.

The push pin is conveniently placed and is used to lift the spindle or plunger above the part to be tested and allow it to gently descend on the work at one constant pressure—thus eliminating the personal touch. No downward pressure on the work can be given by the operator.

An adjusting screw is used for setting the tolerance to be allowed on the piece tested by simply rotating the knurled head of the screw. For example, after adjusting the screw to bring the hand to the zero mark on the sample or test piece on which a tolerance of say .002 inch high and low is allowed. The screw is rotated to bring the hand two divisions of the dial on the low side of zero. Thus set, any work passed under the gauge would read between the tolerance and two divisions on the left or high side of the zero mark. As the spindle thus has only a movement of .004 the speed in which work is tested is limited only by the operator. The push pin in such an instance is not used.

The dial, as can be noted, is mounted on a base. Various sizes and styles of bases can be provided if desired, and if required metric dials can be procured.

SIMPLE WAGON TIPPER

Many devices, some operated by hand and some by power, have been invented to "tip" a wagon so as to empty it of its contents. One of the most interesting of these devices, worked by hydraulic power, has recently been demonstrated by a leading British firm. The engine of the wagon drives two hydraulic rams, which raise the body and clear it of its load within half a minute. The whole operation is carried out by moving a single lever, this action being performed by the driver without moving from his seat.

THESE VIEWS SHOW NOT ONLY THE BEARINGS THEMSELVES, BUT THE REINFORCEMENTS.

The MacLean Publishing Company
LIMITED
(ESTABLISHED 1887)
JOHN BAYNE MACLEAN, President.   H. T. HUNTER, Vice-President
H. V. TYRRELL, General Manager.
PUBLISHERS OF

CANADIAN MACHINERY
~ MANUFACTURING NEWS ~

A weekly journal devoted to the machinery and manufacturing interests.
B. G. NEWTON, Manager.   A. R. KENNEDY, Managing Editor.
Associate Editors:
J. H. MOORE   T. H. FENNER   J. H. RODGERS (Montreal)
Office of Publication: 143-153 University Avenue, Toronto, Ontario.

VOL. XXIV.   TORONTO, DECEMBER 2, 1920   No. 23

Mechanical Accuracy

TAKE a hair from your head (the average is about 2½ thousandths of an inch in thickness) and if you could split that hair into ten strands of uniform dimensions, just one of those strands would give a fair conception of the closeness to a mean standard prescribed in more than 300 operations on our car.

"Our work calls for more than 5,000 operations in which the deviation from a mean standard is not permitted to exceed the one one-thousandth of an inch; more than 1,200 in which it is not permitted to exceed a half one one-thousandth, and more than 300 in which it is not permitted to exceed a quarter of a thousandth."

These statements appeared recently in a wide advertising campaign carried out in various national magazines, and were prepared to appeal to mechanical men in general.

Can you imagine such technical copy reaching the columns of popular magazines, even ten years ago?

It is only another proof that this is becoming more and more a mechanical age and that things are changing for the better.

Why are the limits held so close? Simply to obtain accuracy and an interchangeable product. We owe a great deal to the automotive industry in this respect, for they have been largely responsible for our seeking after interchangeability.

Working to fine limits became common practice in Canada during the days of the shell shop. We hear much of the shell-buster-funny chap no doubt. But we don't hear so much about many mechanics who had never measured anything much closer than a sixteenth.

Memory recalls one chap who used to carry a brass rule in his overalls pocket. He could work to sixteenths all right. The question of shells was put up to him and after a study of the required standards he explained that it was out of the question—it couldn't be done in his shop.

What happened? Simply that in a few months he was working in thousandths of inches and turning out a good shell.

Fine limits indicates high class machine shop and tool room practice, as well as excellence of design. Canadian shops should never drop from these standards of mechanical superiority.

Oil Versus Coal

REPORTS from Vancouver state that on account of the oil shortage, oil-burning equipment in boats and locomotives is being taken out and replaced by coal burners.

There is something in this move that may cause a number of very careful students to make one of those "I-told-you-so" remarks. For some time the view has been held that oil-burning equipment was being boomed out of all proportion—that it could never live up to all the advance notices that were being put up for it.

There is a shortage of oil, so marked in fact that it is not possible to secure the 70,000 barrels daily the Canadian Pacific requires.

There must be fuel for the automobiles, joy-riders and others, no matter if there is no fuel for locomotive and steamships.

Many a firm that has spent its good money in perfecting coal-burning apparatus could hardly be blamed for getting a little hot under the collar by the booming that has accompanied every move made by the oil burners.

The oil burning type has many advantages; it is easier to operate; it gives almost a maximum heat right at the start.

But when there is no fuel oil to be had, or when it is in very limited quantities, what then?

The oil burning equipment is desirable on railroads that run through wooded sections, for the foot-hill country of British Columbia, Oregon and other coast territory can bear gaunt testimony of the way in which the old belching locomotive used to strew its live coals on dry material along the right-of-way.

It is an expensive proposition changing over from oil to coal burning equipment. For a locomotive the cost might be from $10,000 to $12,000, while a fair-sized steamer of the ocean-going sort could hardly be altered for less than $50,000 or $60,000.

The coal burner will be with us for some days, until the part of wisdom is to use every possible means devised and improved for getting as much energy as possible out of every ton of coal burned.

The oil burner will be here, too. There is plenty of room for them both.

One Optimist in Business

THE remarks uttered by Lord Leverhulme at the recent opening of the extension of the Toronto plant come as a tonic at this particular time. So many men are preaching hard times, bad outlook, country going to the dogs, etc., that to hear a man who refuses to believe anything of the kind is refreshing. When this man is a practical, successful business man, who can back up his ideas with his money, and is doing so, his opinions are entitled to a hearing. Lord Leverhulme thinks that it would be a backward step to lower wages at the present time, and he is undoubtedly right in the main. To lower wages at the same time that the cost of living is about to be lowered brings the whole situation back to where it was when costs and wages were at their highest. It comes back to the old story, as the speaker emphasized strongly, that maximum production will support itself, by creating a demand, and giving a high wage to those engaged in industry. Of course maximum production depends entirely on the attitude of labor, and in this connection it is noted by some observers that in some shops where men have been laid off, the production of those remaining shows a very favorable increase. The orgy of extravagance, high living, and a false standard generally is passing away, and it is a good thing for us that it is. That there will inevitably be some distress before times are normal again is also, unfortunately, true. If employers generally are imbued with some of the spirit of Lord Leverhulme, and employees willing to live up to their part of the contract, there will be the minimum of distress, and the speediest readjustment.

A Piece of Good Advice

WHEN Mr. Lloyd Harris told the business men of Canada not to depend too much on the Government, he spoke words of wisdom. We do not mean by this-the present Government, the past Government, or the future Government in particular, but Government in general as an institution. A government is after all only a collection of ordinary men, who have attained prominence and position sometimes through merit, and sometimes through the fact that having no particular ability, they can be useful to the stronger men in the Government that have. These several men, whatever their abilities, upon attaining place, have their whole outlook on affairs tainted with the view of the office holder, whose primary object is to retain office, and do what he can for the country next. It is obvious that it is vain to look to such an institution for leadership in any vital trade or business development. In fact, any country which depends on its Government to advance its business interests is usually disappointed. It is the enterprise and daring of the individual trader that extends trade, and although a Government can carry out the wishes of the business community to a greater or less extent, it is very rarely that they suggest anything. Generally speaking, Government intervention in matters of business is usually attended with loss and inconvenience to those in the business, as has been abundantly shown in Great Britain by the Government control of shipping. Probably the only exception of note was Germany before the war, but it must be remembered that there they had no representative Government in the sense that we have. Therefore, they had men in the various departments of the Government, who were specialists, and not politicians. The Government holds about the same position in relation to the country that a chief office clerk or office manager does to a big corporation, and should not try to overstep that mark.

Strength of Materials

OUR knowledge of materials, particularly metals, has increased to a great extent within the past few years, but according to Professor F. C. Jenkins, president of the Engineering Section of the British Association, we are, as yet, only on the threshold of that complete knowledge which is essential before we can really tell the cause of a failure, or provide with certainty against accidents.

He states as follows: surely an engineer should be able to say whether a cylinder is safe without dropping it from the roof or rolling it down the front doorstep to see if it breaks. The time has come when the fundamental data on which the engineering theories of the strength and suitability of materials are based require thorough overhauling and revision.

When one considers the matter carefully, the professor is not far from wrong. Judging from some of the formulae we have seen, it would be a much better plan to drop the part under discussion and see if it would break, rather than try to figure out the theoretical chances of its proving too weak.

This condition should not exist, and if sufficient research work is made in this particular field, there is no reason why the formulae can not be simplified to a great extent. We should not have to depend on actual physical tests, but should be able to figure with certainty the strength of our various materials.

Cases exist where theoretical tests alone would not suffice. For example, suppose some structure is being built on which life depends. In such a case both theoretical and physical tests are made. Every boiler must undergo a water test before being fit for use, the various members of a bridge must be tested for their physical properties, and so on.

These precautions are essential with all due respect to theory, because defects can, and do, occur in process of manufacture. Theory does not allow for such defects, therefore they must be checked up by physical tests.

A great need exists, however, of our present formulae being simplified. Here is a chance for the heads of our colleges. Let them concentrate on the task, for by so doing they will be rendering a real service to the machine tool industry.

Can We Harness Radium

IT is a well-known fact that matter is composed of infinitely small particles, called atoms, but have we any conception of the internal structure of an atom? Professor McLellan, F.R.S.O.B.E., in speaking on this subject quotes as follows: "The difficulty of investigating atomic structure might be realised from the fact that if the atoms contained in a litre box of bird-shot were enlarged to the size of the shot, they would cover the entire surface of the earth to a depth of eight inches."

The discovery of electrical elements had greatly facilitated atomic research—the positive element being the nucleus of the atom around which varying numbers of negative electrical elements called electrons perpetually revolve. The study of radium emotions, however, had been the most important source of the accepted atomic theory.

Radium had been formed at an early stage of cosmic evolution from the condensation of highly energized elements at a high temperature. Now that it had cooled to the temperature of the earth it was in a position to de-energize itself. This energy was dissipated in the form of streams of electrons and alpha rays, the latter being gistotopic helium and traveling at a rate of 20,000 miles per second. It would be the problem of the future to harness this enormous energy and put it to economic use in the form of heat, light and electricity.

Although no immediate relief to the power situation can be expected, it is good to know that a time may come when radium energy will come under our control and allow itself to be harnessed.

KEEP GOING!   "Motor World."

MARKET
DEVELOPMENTS

The Steel Market Holds Interest This Week

Several Local Steel Men Left at the Week-End for Steel Centres of United States to Find Out Just How Matters Stand—Reports From Several Centres.

DEVELOPMENTS in the steel market are the outstanding feature of the week in machine, iron and steel circles. Some of those who watch the market closely believe there are symptons of a runaway market if those at the head of affairs do not hold the thing in check. The first answer of the corporation to the independents was the announcement—hardly made officially yet—that Canadian business would now be regarded as domestic trade by the United States rollers. For some time past Canadian orders have been, regarded as export trade, and carried a higher price as a result. It means that there will have to be considerable readjustment in order to protect the business already on the books of the corporation, and some of it has been waiting there for a year now. Where a definite price has been given on a contract it is not considered necessary to make allowance, but where material is bought for warehousing, or where the sale of the product depends on current quotations, there will probably be a transfer of the business from the old figure to the new. At least that is the expectation in the Canadian market.

There are a number of inquiries in the market for machine tools and kindred equipment, but the trouble lies in the fact that much of it cannot be closed. In some cases it is hard for firms to finance the purchase now, as they are anxious to keep all available cash for working capital. On the other hand there is a marked tendency on the part of many buyers to rather glory in the fact that this is a buyer's market, and the firm that wants the business must come along and bid lustily for it.

The scrap metal market remains just where it was last week, in the dumps. Dealers will not guarantee to pay the prices they authorize, and there is almost nothing going on. Buyers are out of the market, and dealers are also in many lines. Prices are still downward, but no dealer seems to have the heart to batter down the present lowly list. The figures, though, are only nominal.

SUSPENSION IS ONLY TEMPORARY
PROSPECTS BRIGHT FOR NEW YEAR

Special to CANADIAN MACHINERY.

MONTREAL, Que., Dec. 2.—The keener competition that features the trading in all lines of commodities foretells a further adjustment in selling prices. Mills are realizing that they must share of the losses incidental to a declining market, and the present policy seems to favor a disposal of surplus stocks before heavy production can be considered on the changing basis. There is, however, an effort of conservatism in taking the downward steps to avoid a fall that might have serious after effects, should the dormant demand, now supposed to exist, suddenly develop to an active reality.

Despite the fact that present demand is below normal, the feeling still exists that industrial and manufacturing requirements are only held in temporary suspension and that early in the year may see a return to a more healthy condition in respect to future bookings. Speaking of the relative cost of production, one dealer said: "It should be borne in mind that during the war exports of all steel and metals from the States were excessive, and that production was carried on in an extensive manner. For many months after the war this same condition obtained, but during the past year the rehabilitation of European countries resulted in re-establishing steel activities in Europe, and consequently the necessity of importing fell off, so that mills here and in the States had to look to their home requirements to absorb the output from the mills.

"This condition became more pronounced with the general falling off of trade so that curtailment in manufacturing was the logical outcome. Owing to many circumstances this lessening of activity meant a proportionate increase cost per ton in the making of the steel, as in many instances the overhead costs remained the same on the smaller tonnage output. The adjustment now going on is the first real move in this direction since the war ended, and as we must have it, the sooner the better." The steel demand in this district is rather quiet and will likely remain so during the month of December. Dealers, however, are not discouraged as reports and inquiries are of a character to indicate that placement of good business is only awaiting a more settled condition of trade.

The plate situation has developed no new features of interest other than 'a lighter demand. Shipyards are not so busy and material coming in for this purpose shows a marked falling off. The sheet situation is easier and the supplies are more readily obtained; this applies more particularly to black sheets as the call for galvanized is still good. Queen head and fleur-de-lis are down 50 cents per hundred, the respective quotations being $13 and $12.50 per hundred.

Further Declines in Metals

The further reductions that have been made in the price of ingot metals indicates the effort that is being carried out in the universal adjustment to a more reasonable working basis. This is not literally true, as production costs, to-day are in excess of what they should be in comparison to what metals are selling for on the current market. What is actually taking place is the liquidation of accumulated stocks, and this will probably require some time, especially as consumption just now is below normal.

The demands from Europe since the war have rapidly declined and no corresponding outlet has been found for American metals, the consequence being that refiners here have had to adjust their output to existing needs, which, at present, is in excess of trade requirements, and in face of the high production costs the stocking of metals has no inducements for the refiners. Ingot metals are marked down from ½ cent to 1 cent per pound. In copper products, bars are quoted at $35; plain sheets at $40; copper sheets, tinned, at $43, and planished copper sheets at $47. Lead sheets are also easier, the price ranging from $10 to $10.50 per hundred pounds. Brass rods, sheets and tubing are likewise lower, the quotations being 30, 38 and 42 cents respectively. Copper tubing, seamless, is now 44 cents per pound.

No Market in Scrap

"The prices you are quoting," said a dealer in old materials, "are just as good as any other, as trading at present is not influenced by the price. We offered 13 cents just this week for a quantity of copper wire, but the sale was not made, and we are just as well pleased. It would have meant carrying it in stock, and this we are not anxious to do." The listless character of the old material market is not conducive to enthusiasm on the part of dealers, and the apparent disinterest reflects the trade depression. Some little movement is always in evidence but the volume is very light.

Dealers look for a quiet period for the remainder of the year, and do not anticipate excessive activity for some months, although many are hopeful of a gradual betterment after the holiday season. Prices quoted might be shaded a little, but lacking an actual sale, the list may be considered as that generally prevailing throughout this district.

NEW STEEL PRICES SOON EFFECTIVE

Trade in Canada Should Get Advantage of These in a Short Time

TORONTO.—The new prices that are coming to the steel market formed the chief topic of conversation this week. In fact so interesting was this matter that the week-end saw several of the local steel merchants hurrying off to New York and other steel centres to see just where they stood, and what the chances were for stability, many of them seeming to think that there were present all the makings of a runaway steel market which might develop before many days were over.

The machine tool market is quiet. At this season of the year there is generally a period when little business is in the market, but the facts are that right now it is more than seasonably dull, although there are a number of very good inquiries in the market. The feeling is

POINTS IN WEEK'S MARKETING NOTES

Cleveland does not deny that there is a slump, but feels sure it will be on the mend by the New Year, and normal by April.

—

Tendency in the machine tool business will be to cater to a diversified custom in future instead of specializing on one line.

—

Steel Corporation prices now are the ruling factor in the markets, most of the independent mills having adopted them.

—

The price of sheets has declined, but the lower rates have not had the effect of stimulating buyers.

—

Bessemer and basic are still declining but the decline is not so rapid as has been the case lately.

—

The local machine tool market has quite a few good enquiries out, but business is rather uncertain.

—

One independent mill is offering the Canadian trade black sheets at 5.25 base, this being the lowest figure given here yet. It cuts under the rate that has heretofore been given by the Steel Corporation.

—

U.S. reports indicate that if the new steel market revives buying of steel, there will be better business in sight for machine tool trade.

—

Pittsburgh claims that the new prices on steel are not bringing in any new business. Buyers look for more cuts and are waiting.

—

The local scrap market will answer to almost any reference that is made to it, provided it is uncomplimentary and indicative of stagnation and inaction.

—

Dealers having lines that are sold to farmers report a very active trade. For instance, engines are selling rapidly.

—

Firms that are having the hardest time now are those that specialized almost entirely in making parts for automobiles. They narrowed their field in that way to such an extent that when the auto business fell off they were almost done for.

—

Prices published this week may be changed on short notice, so allowance should be made for that.

growing that a frank statement on the part of the machine tool builders in regard to selling prices and policies would do something to clear the air, and allow the work of placing business to proceed. Right now the price change is a sort of checkerboard affair, as some of the makers have announced reductions, while others have not, and their agents in this part of the country seem to be quite satisfied that they will not name any lower selling costs.

There is generally a desire on the part of firms to make as much of a clearing-out at this time of year as possible, in order to keep up the total figures for the year, especially if their fiscal year ends with the calendar year, and this is the practice with the majority of firms. No doubt there is some of this being done, but not to the extent that might be imagined. For instance, one report tells of the sale of a new radial drill, never taken out of the crate, that went out at almost half the regular price. Other sales of this sort are also heard of, but after all they are only incidents in the market, and do not indicate at all that the general policy of selling is moving in this direction.

Price or Uncertainty?

"We have a number of very attractive inquiries, but we can't sew up the business," was the remark of two different offices this morning.

"Why? In some cases they find it a little hard to finance the business, wanting to hold on to all the available cash they can. We are looking for something a little easier in this line before very long.

"Then again there are buyers who seem bound to ride around on the strength of the fact that this is more of a buyer's market than it has been for some time past. It is not an unusual proposition now for a man to have the specifications for a machine up his sleeve, and go around as though saying, 'Here's what I want, and the man who can come across with the most attractive looking proposition is going to get the business.'"

Then, again, there are reports of where firms and salesmen cannot come to an agreement regarding the price. There is one case where a nice order is waiting for a certain line of equipment, but there is a strong difference of opinion over the figure, and so far it has not been settled.

The Steel Market

A stock list coming to one of the local steel warehouses this morning conveyed the information that one of the independents offered sheets at 5.25 for immediate shipment. Formerly, when prices looked as though they would have to come down, the independents reduced business on their books from 8c to 6.25, largely with the idea of protecting it, but that is as far as they had come previously, while the new price

they are quoting means that they are under the Steel Corporation price as it has been quoted here. The Corporation price to Canadian users has been all along 5.35.

The United States Steel Corporation is now prepared to regard Canadian business as domestic rather than as export business. This statement held good from the first of the week on bars and structural material and a similar statement is likely to come out at any moment governing the selling of sheets and other lines. This is a concession that has been looked for for some time back, and one which the trade in this country formerly enjoyed.

"How will existing business be attended to in view of the cut that has been made?" Canadian Machinery put that question to the Canadian manager of the U. S. Steel Products. It is something that the trade looks for at once when a new price is announced. This is the plan, as we understand it:

When a customer has figured on a definite contract, and has placed his order for the steel at a figure that will allow him his legitimate profit, he is not entitled to a reduction to meet the new market, for there is no necessity of him doing so in case the contract is still standing. On the other hand, if a firm or a warehousing concern has placed an order that they will have to resell, they will be protected on the new price. There may be a few variations from that programme, but as we understand it, that plan is substantially correct.

Some of those who are in the market all the time profess to believe that there is the makings of a real runaway market in existence if a certain amount of care is not exercised. Some dealers believe that the Corporation will now find it necessary to protect their own bookings.

The Scrap Market

Nothing is changed in the scrap metal situation, and it remains in about as dead a condition as could be imagined. Dealers will not state whether or not they will pay even the rather lowly price list which we publish in another column. Although the figures were furnished by the dealers themselves, they will not guarantee for the present to pay them unless they have some place where they can place the material.

Pittsburgh can best be described as waiting.

One War-Time Lesson

Firms that are in good shape are going carefully in order to keep their credit sweet. They are curtailing expenses and making more investigations before placing advertising contracts. And right here is a peculiar development. A short time back advertising was looked upon by some of the firms as a good place to dump some of their surplus in order to keep it from the Government. New equipment was often put in on the same basis. But this advertising done by many firms on this rather loose basis has been the means of convincing the firms that advertising is a real, positive force, and these newer firms are looking upon it as a real factor in their business development. They now look upon advertising as an investment and not as a place to hide profits. There is a stated appropriation now in firms that just used to "spend money."

Travellers are working their territory harder. They have instructions not to make long jumps. Sales managers in some cases have taken over the work of the advertising department.

Some sales managers are working half crews all the time, and others, in order to hold their organization, work crews half time. This latter seems the more favored idea, but it cannot always be done because they must work them long enough to give the men enough in their week's wages to live on.

The Matter of Wages

Labor is shifting in many cases. Machinists are working in several plants for 65 and 75 cents per hour where $1 was the rule before. For the present at least that fine old idea of "not liking your job" has ceased to be. It's a case of "have you got one?" The best men are being retained, and figures show that these men are working harder so there is a consequent improvement in production.

Another market being watched closely is steel, and it is mostly a waiting game to see when it is going to hit bottom.

This is Interesting

Speaking of association gatherings, one of the large builders of tools tells me that they all go to the annual convention—just finished a week or so ago in New York. Well, when they are there it's a case of scratching each others' backs and making statements about there being no cut in their prices. Then on the way back from New York many of them get out their lists and put the axe into them.

But after all—and this is from a visit to a number of centres, Cleveland, Pittsburgh, Erie, Canton, etc., there is a feeling of splendid optimism. Business has been running mad in some directions, and there is a determination to get down to a more secure foundation, and once that is done it will be a case of full speed ahead.

NARROWED THE FIELD BY NOT CATERING TO DIVERSIFIED TRADE

Special to CANADIAN MACHINERY.

CLEVELAND, December 2.—The firms or cities that have been taken up largely with supplying machinery, material or parts to the automobile trade have learned that it is not wise to put all the eggs in one basket. The dependence on one line has meant the neglect of developing others that might have been good to them now. They find that through trying to reach a very high efficiency in this line they have narrowed their field.

The best thinkers around this district, men who are heavily interested in the machine tool, iron, steel and metal industries, make no bones about the matter. They say they are caught in a slump and that it is a real slump, but in the next breath they tell you that it came quick and hard, and is going to move away rapidly. If one were to make a summary of all these opinions, a conservative answer would be that the start back will be around the first of the year, and that by April the new movement will be well under way.

These firms that were so geared up that they sank or swam with the automobile trade are quietly saying "never again," and are now preparing to cater to a much more diversified field. Erie, they tell us, is the third city in the world in the diversity of her manufactures, and the tone of business there right now is absolutely the best to be found in this whole district.

CHANGE FROM OIL TO COAL BURNER

Scarcity of Fuel Oil Makes it Necessary For C. P. R. to Alter Equipment

Vancouver. — The Canadian Pacific Railroad Company, on account of the acute oil shortage, is changing all its equipment, including steamships, to coal-burners.

It is estimated that the change will cost close to $60,000 for each steamer and $12,000 for each locomotive. The first coal-burning engine used by the road in the last five years left the station here last night with a transcontinental train.

H. W. Brodie, general agent of the company in the West, said that the oil shortage was due to the fact that the Standard and Union Oil Companies could no longer afford to supply the road with 70,000 barrels daily. He had heard, he also said, that the United States Navy Department was about to take over the Union Oil Company for its own use, and that the Standard was too busy supplying fuel for automobiles and transportation in the United States to provide for Canada.

ALL ON CORPORATION PRICES—
BUT WILL THAT SCHEDULE LAST ?

Special to CANADIAN MACHINERY.

PITTSBURGH, December 2. — Late last week the remaining independent steel makers came down to the Steel Corporation prices on bars, shapes and plates. More than a month ago an independent at St. Louis quoted the corporation price on plates. Later a Chicago independent came down on bars and over a week ago some Cleveland mills were ' quoting the corporation prices. However, on Wednesday of last week a canvass showed that five or six large independents were quoting 3.00c on all three products, the corporation prices being 2.35c for bars, 2.45c for shapes and 2.65c for plates. Thursday was a holiday and Friday morning an independent in Youngstown and an independent in Pittsburgh began quoting the Steel Corporation prices, the others immediately following, as it requires only a few hours for information of this sort to get around the trade. The established custom in the trade is that when a buyer is quoted a cut price by a certain mill he does not place an order at the cut price, but immediately tells all the other mills about it.

The reductions represented a case of the mills practically running out of orders at higher prices. The larger mills were not expected to run out nearly so soon. A couple of months ago several were claiming to be filled with. business at least to January 1, if not longer, but it seems enough tonnage was cancelled to make a difference of several weeks. There is some tonnage left on books, but not a great deal. In general, the remaining tonnage will be adjusted in price to meet the new market.

Return to Industrial Board

The Industrial. Board prices, representing the second reduction in steel prices from the war control· schedule, became effective March 21, 1919. The Steel Corporation has consistently adhered to those prices. The first thing the independents did was to · cut the prices, practically all the independents selling at cut prices during April and May, 1919. Afterwards, as they filled up, they began to look to the 'Steel Corporation to advance prices. This the corporation would not do. Late last year, or about a year ago, the independents began advancing prices on their own account, and at the beginning of this year practically all the independents were at advanced prices, a notable exception being the tin plate mills. When the time came to make contracts for the second half of this year the independents demanded higher prices on most of the business, but not on all. Within the past two or three weeks the independents returned to the Steel Corporation price on tin plate. Now the action is followed by the independents on bars,

shapes and plates. The remaining important products are sheets, pipe and wire. On these the independents are still above the corporation. · The reason the independents are still above the corporation on these products is that they still have a considerable tonnage of business on books, but of late this has been greatly diminished by cancellations, and it may not be long until these products fall in line.

Sheets Decline

Sheets were so extremely high in the independent market that they could decline a long time before getting to the Steel Corporation. or Industrial Board level. For a couple of months or more the black sheet market declined at the rate of about a quarter cent a pound per week, but in the past fortnight there has been a decline of one cent. One can readily buy now from several independents at 5.00c for 28 gauge and on an attractive order even this price would probably be shaded $2 or $3 a ton. Galvanized sheets can be had at 6.50c. At the beginning of last week one mill was quoting 6.70c, others being at 7.00c or higher, but at the close of the week there were several· sellers at '6.50c. In blue annealed the common quotation is 4.50c, but this basis could probably be shaded a trifle on the lighter gauges, while it could be shaded considerably on the 'heavier gauges, those that can be produced by light plate mills.

No Buying

Reduced prices have not brought the independents any additional business, nor was it expected that business would be created by this means. Buyers are bent upon having the smallest possible in-·ventories and commitments January 1. The business of jobbers and manufacturing consumers has slowed down very greatly and naturally the buyers will not purchase from the mills except as they have occasion to sell or consume the material.

Manufactured Products

The market for cold finished steel bars of late has been 4.00c, with some producers still quoting 4.25c. The market was 3.60c when the makers of cold finished steel bars were getting their bars at 2.35c, but prices were advanced afterwards even by makers who were getting their bars from the 'Steel Corporation. The question now is whether the makers will reduce their prices at once below 4.00c. Thus far they show no inclination to do so. Standard railroad spikes remain at 4.00c, but this price is too high by comparison with bars at 2.35c. Before the war, when costs and profits were less, the spike makers used

to consider a spread of $9 a net ton between bars and spikes as a fair one. '

Mill Operations

Independents as a whole are probably operating at an average rate of less than 50 per cent., while the Steel Corporation continues to operate as well as at any time in recent months, being very well supplied with business. A few independent mills are closed entirely, but such closings are not permanent, being for the purpose of accumulating specifications. For some time to come, and until demand develops, independents are likely to close for two or three weeks at a time and then run for a week or two at a fair rate, this being much more economical than to maintain a low rate of output continuously. The Steel Corporation is likely to run for months at nearly if not quite its usual rate.

Pig Iron

Bessemer and basic are down $2.50 in the week and foundry $2, the market being now quotable at $37 for foundry, $37.50 for Bessemer and $35 for basic, f. o. b. valley furnaces, freight to Pittsburgh being $1.96. The declines in Bessemer and basic occurred by way of there being offerings, rather than quotations against actual inquiries. In foundry there were one or two small sales to bring out the lower quotation. The bottom, of course, is not yet reached, but as declines of $13 in foundry, $11 in Bessemer and $13.50 in basic have occurred from the top points reached last August it is expected that further declines will be at a slower rate. Furnaces admit that prices will decline to cost, but there is a question where the cost line lies. No close figuring can be done as long as the coke market has not settled. · For the past fortnight the Connellsville coke operators have maintained spot furnace coke at $8 to $8.50 and furnaces consider such prices ridiculous, being in doubt whether they would pay even $6.

Semi-finished Steel

The first sale of any semi-finished steel at the Steel ·Corporation price occurred a few days ago, 'a sheet mill in the Pittsburgh district buying 2,000 tons a month of sheet bars for December, January and February at $47 Pittsburgh, or $48.40 delivered. The identity of 'the seller is not disclosed and the independents do not yet admit a willingness to sell at the Steel Corporation prices.

According to tests carried out by the United States Forests Products Laboratory on fences of charred and untreated posts of various species, charring is of little value in protecting the butts of fence posts and telephone poles from decay. The charred posts proved in these tests to be even less durable than the untreated ones.

SELECTED MARKET QUOTATIONS

Being a record of prices current on raw and finished material entering into the manufacture of mechanical and general engineering products.

PIG IRON

Grey forge, Pittsburgh	$39 96
Lake Superior, charcoal, Chicago	53 50
Standard low phos., Philadelphia	44 79
Bessemer, Pittsburgh	41 96
Basic, Valley furnace	37 50
Toronto price:—	
Silicon, 2.25% to 2.75%	57 50
No. 2 Foundry, 1.75 to 2.25%	56 00

IRON AND STEEL

Per lb. to Large Buyers	Cents
Iron bars, base, Toronto	$ 5 50
Steel bars, base, Toronto	5 50
Iron bars, base, Montreal	5 75
Steel bars, base, Montreal	5 75
Reinforcing bars, base	5 50
Steel hoops	7 00
Norway iron	11 00
Tire steel	5 75
Spring steel	10 00
Band steel, No. 10 gauge and 3-16 in. base	6 00
Chequered floor plate, 3-16 in.	8 50
Chequered floor plate, ¼ in.	8 00
Bessemer rails, heavy, at mill	....
Steel bars, Pittsburgh	3 00-4 00
Tank plates, Pittsburgh	3 50
Structural shapes, Pittsburgh	3 00
Steel hoops, Pittsburgh	3 50-3 75
F.O.B., Toronto Warehouse	
Small shapes	5 50
F.O.B. Chicago Warehouse	
Steel bars	3 62
Structural shapes	3 72
Plates	3 67 to 5 50
Small shapes under 3"	3 62

FREIGHT RATES

Per 100 Pounds.

Pittsburgh to Following Points

	C.L.	L.C.L.
Montreal	58½	73
St. John, N.B.	84½	106½
Halifax	86	108
Toronto	38	54
Guelph	38	54
London	38	54
Windsor	35	50½

METALS

	Gross. Montreal	Toronto
Lake copper	$19 50	$20 50
Electric copper	19 00	20 50
Castings, copper	18 25	20 50
Tin	45 00	50 00
Spelter	8 75	10 00
Lead	7 75	9 00
Antimony	8 25	9 00
Aluminum	35 00	35 00

Prices per 100 lbs.

PLATES

Plates, 3-16 in.	$ 6 50	$ 7 00
Plates, ¼ up	6 00	6 60

PIPE—WROUGHT

Standard Buttweld Pipe
Per 100 Ft.

	Steel Blk.	Galv.	Gen. Wrought Iron Blk.	Galv.
⅛	$ 6 50	$ 8 50	$....	$....
¼	5 81	7 41	6 91	8 01
⅜	5 81	7 41	6 91	8 01
½	7 16	8 63	7 95	9 43
¾	8 80	10 87	9 95	12 02
1	13 01	16 07	14 71	17 77
1¼	17 60	21 74	19 90	24 04
1½	21 04	25 99	23 79	28 74
2	28 31	34 97	32 01	38 67
2½	44 73	55 28	....	....
3	58 52	72 29	....	....
3½	74 06	90 62	....	....
4	87 76	107 37	....	....

Standard Lapweld Pipe
Per 100 Ft.

	Steel Blk.	Galv.	Gen. Wrought Iron Blk.	Galv.
2	$32 01	$ 38 67	$36 71	$42 87
2½	48 26	58 79	54 11	64 64
3	63 11	76 86	70 76	84 53
3½	75 90	92 46	85 10	101 66
4	89 93	107 55	100 85	120 45
4½	1 06	1 29	1 30	1 54
5	1 22	1 50	1 52	1 80
6	1 58	1 95	1 97	2 33
7	2 06	2 53	2 53	3 01
8L	2 16	2 66	2 66	3 16
8	2 49	3 07	3 07	3 64
9	2 98	3 67	3 67	4 36
10L	2 77	3 41	3 41	4 06
10	3 56	4 39	4 39	5 21

Prices—Ontario, Quebec and Maritime Provinces

WROUGHT NIPPLES

4" and under, 60%.
4½" and larger, 50%.
4" and under, running thread, 30%.
Standard couplings, 4-in. and under, 30%.
Do., 4½-in. and larger, 10%.

OLD MATERIAL

Dealers' Average Buying Prices.

Per 100 Pounds.	Montreal	Toronto
Copper, light	$10 50	$10 50
Copper, crucible	13 00	12 00
Copper, heavy	12 50	12 00
Copper wire	12 50	12 00
No. 1 machine composition	13 00	12 00
New brass cuttings	7 00	9 00
Red brass turnings	10 00	10 00
Yellow brass turnings	7 00	7 50
Light brass	5 00	5 00
Medium brass	6 50	6 00
Scrap zinc	5 00	5 50
Heavy lead	5 25	5 60
Tea lead	2 50	3 00
Aluminum	16 00	16 00

	Per Ton	Gross
Boiler plate	13 00	12 00
Malleable scrap	20 00	23 00
Axles (wrought iron)	25 00	20 00
Rails (scrap)	18 00	18 00
Malleable scrap	23 00	25 00
No. 1 machine cast iron	32 00	33 00
Pipe, wrought	9 00	10 00
Car wheel	30 00	33 00
Steel axles	20 00	20 00
Mach. shop turnings	8 00	9 00
Stove plate	25 00	25 00
Cast boring	8 00	12 00

BOLTS, NUTS AND SCREWS

	Per Cent.
Carriage bolts, 7-16 and up	+10
Carriage bolts, ⅜-in. and less	Net
Coach and lag screws	—15
Stove bolts	55
Wrought washers	—25
Elevator bolts	+10
Machine bolts, 7-16 and over	+10
Machine bolts, ⅜-in. and less	+10
Blank bolts	Net
Bolt ends	Net
Machine screws, fl. and rd. hd., steel	27½
Machine screws, o. and fil. hd., steel	+25
Machine screws, fl. and rd. hd., brass	net
Machine screws, o. and fil. hd., brass	net
Nuts, square, blank .... +25 add	$2 00
Nuts, square, tapped ....add	2 25
Nuts, hex., blank ....add	2 50
Nuts, hex., tapped ....add	3 00
Copper rivets and burrs, list less	15
Burrs only, list plus	25
Iron rivets and burrs	40 and 5
Boiler rivets, base ¾" and larger	$8 50
Structural rivets, as above	8 40
Wood screws, O. & R., bright	75
Wood screws, flat, bright	77½
Wood screws, flat, brass	55
Wood screws, O. & R., brass	55½
Wood screws, flat, bronze	50
Wood screws, O. & R., bronze	47½

MILLED PRODUCTS

(Prices on unbroken packages)

	Per Cent.
Set screws .... —20%, 25 and	5
Sq. and hex. hd. cap screws	12½
Rd. and fil. hd. cap screws ..plus	25
Flat but. hd. cap screws ....plus	50
Fin. and semi-fin. nuts up to 1-in.	12½
Fin. and Semi-fin. nuts, over 1 in., up to 1½-in.	—5
Fin. and Semi-fin. nuts over 1½ in., up to 2-in.	+12½
Studs	+5
Taper pins	—12½
Coupling bolts	+40
Planer head bolts, without fillet, list	+45
Planer head bolts, with fillet, list plus 10 and	+55
Planer head bolt nuts, same as finished nuts.	
Planer bolt washers	net
Hollow set screws	+60
Collar screws ......list plus 20,	30
Thumb screws	40
Thumb nuts	75
Patch bolts ....add	+85
Cold pressed nuts to 1½ in...add	$1 00
Cold pressed nuts over 1½ in..add	2 00

BILLETS

	Per gross ton
Bessemer billets	$60 00
Open-hearth billets	60 00
O.H. sheet bars	76 00
Forging billets	56 00-75 00
Wire rods	52 00-70 00

Government prices.

F.O.B. Pittsburgh.

NAILS AND SPIKES

Wire nails, base	$6 10
Cut nails, base	7 00
Miscellaneous wire nails	50%

ROPE AND PACKINGS

Plumbers' oakum, per lb.	0 10¼
Packing, square braided	0 38
Packing, No. 1 Italian	0 44
Packing, No. 2 Italian	0 36
Pure Manila rope	0 35½
British Manila rope	0 28
New Zealand hemp	0 28

POLISHED DRILL ROD

Discount off list, Montreal and Toronto .... net

559½

HEAT LOSSES IN OPERATION OF ELECTRIC FURNACES

(Continued from page 498)

ling of electrodes and other troubles caused thereby. These economizers have been fitted to a number of furnaces and have stood up remarkably well. Some have been in constant operation for six months and are as good now as the day installed. The invention is covered by patents, but can be supplied for all types of furnace. A full description of the device was presented at the recent meeting of the Association of Iron and Steel Electrical Engineers, by Mr. E. T. Moore, chairman of the electric furnace committee, but a brief description will probably interest.

The Brooke economizer depends for its action on the well known scientific fact that hot gases under pressure, if suddenly allowed to expand, quickly lose their high temperature. The gases generated in an electric furnace are only combustible at high temperatures and in the presence of oxygen. All other cooling rings simply content themselves with cooling these burning gases. In the Brooke economizer the gases never do ignite, the roof ports and the electrode are kept reasonably cool and the gases leaving the furnace are at too low a temperature to ignite.

The gases pass first of all in between the electrode and the port hole roof and

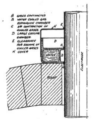

SECTION VIEW OF FURNACE
ROOF AND ELECTRODE ECONOMIZER
FIG. 3—DETAIL OF COSTRUCTION.

A GASES CONTRACTED
B WATER COOLED GAS
    EXPANSION CHAMBER
C AIR OBSTRUCTION OF
    COOLED GASES.
D LARGE COOLING
    CHAMBER
E CLEARANCE
    FOR ESCAPE OF
    COOLED GASES
H COVER

ELECTRODE

then through the small clearance marked A, Fig. 2, into the cooling ring and into a relatively larger chamber B, which causes the expansion of the gases and thereby giving up a large amount of sensible heat in the gases. This heat is absorbed by the water in the cooling ring. From chamber B they are once more contracted through the small clearance C and then passed to a very large chamber D for a second and chief expansion. The design differs slightly

with amorphous or graphite electrodes, but the effect is that no flame is seen from the top of the furnace and the electrodes do not pencil off.

The attached photograph, Fig. 1, shows a 3-ton Greaves-Etchells furnace at the Dodge Steel Co., Tacony, Pa., and illustrates the actual size of the cooling ring. It will be noticed that although the electrodes are withdrawn and the furnace is pouring there is no flame or reduction of diameter. Even when carbon is thrown on the slag no flame appears. On a recently installed four-electrode furnace the monthly figures were 12 pounds of graphite electrodes per ton. The average over six months was 12½ pounds, with a power consumption of 550 kw. hours per ton of steel.

Tests are now being made on a furnace fitted with the special type cooling ring, but it can be seen with the naked eye that the temperature of electrodes is far less, and therefore considerable saving of heat and power does take place.

The Labor Temple, Vancouver, B.C., is to be remodelled for a technical school.

Excavation has commenced on the site of the new Gendron factory, north of Danforth Avenue, Toronto. It is expected that the factory will be completed about the end of February and will employ about 500 men at the start.

MISCELLANEOUS

Solder, strictly	$ 0 28½
Solder, guaranteed	0 30½
Soldering coppers, lb.	0 62½
White lead, pure, cwt.	20 35
Red dry lead, 100-lb. kegs, per cwt.	15 00
Gasoline, per gal., bulk	0 42
Pure turp., single bbls., gal.	3 15
Linseed oil, raw, single bbls.	2 37
Linseed oil, boiled, single bbls.	2 40
Wood alcohol, per gal.	4 00
Whiting, plain, per 100 lbs.	3 00

CARBON DRILLS AND REAMERS

S.S. drills, wire size	35
Can. carbon cutters, plus	10
Standard drills, all sizes	35
3-fluted drills, plus	10
Jobbers' and letter sizes	35
Bit stock	40
Ratchet drills	10
S.S. drills for wood	40
Wood boring brace drills	25
Electricians' bits	30
Sockets	50
Sleeves	50
Taper pin reamers	25 off
Drills and countersinks	net
Bridge reamers, carbon	50
Centre reamers	5
Chucking reamers	net
Hand reamers	10
High speed drills, list net to plus	20
Can. high speed cutters, net to plus	10
American	plus 40

COLD ROLLED STEEL

[At warehouse]

Rounds and squares	$7 base
Hexagons and flats	$7.75 base

IRON PIPE FITTINGS

	Black	Galv.
Class A	70	85
Class B	30	40
Class C	20	30

*Cast iron fittings, 5%; malleable bushings, 22½%; cast bushings, 22½%; unions, 37½%; plugs, 20% off list.

SHEETS

	Montreal	Toronto
Sheets, black, No. 28	$ 9 25	$ 9 50
Sheets, blue ann., No. 10	7 75	8 50
Canada plates, dull, 52 sheets	13 00	13 00
Can. plates, all bright	14 00	.....
Apollo brand, 10¾ oz. galvanized	....	....
Queen's Head, 28 B.W.G.	13 00	.....
Fleur-de-Lis, 28 B.W.G.	12 50	.....
Gorbal's Best, No. 28	....	....
Colborne Crown, No. 28	....	....
Premier, No. 28, U.S.	11 50	12 00
Premier, 10¾-oz.	11 50	12 40
Zinc sheets	16 50	20 00

PROOF COIL CHAIN

(Warehouse Price)

B

¼ in., $13.00; 5-16, $11.00; ⅜ in., $10.00; 7-16 in., $9.80; ½ in., $9.75; ⅝ in., $9.20; ¾ in., $9.30; ⅞ in., $9.50; 1 in., $9.10; Extra for B.B. Chain, $1.20; Extra for B.B.B. Chain, $1.80.

ELECTRIC WELD COIL CHAIN B.B.

⅛ in., $16.75; 3-16 in., $15.40; ¼ in., $13.00; 5-16 in., $11.00; ⅜ in., $10.00; 7-16 in., $9.80; ½ in., $9.75; ⅝ in., $9.50; ¾ in., $9.30.

Prices per 100 lbs.

FILES AND RASPS

	Per Cent.
Globe	50
Vulcan	50
P.H. and Imperial	50
Nicholson	32½
Black Diamond	27½
J. Barton Smith, Eagle	50
McClelland, Globe	50
Delta Files	20
Disston	40
Whitman & Barnes	50
Great Western-American	50
Kearney & Foot, Arcade	50

BOILER TUBES.

Size.	Seamless	Lapwelded
1 in.	$27 00	$.....
1¼ in.	29 50	.....
1½ in.	31 50	29 50
1¾ in.	31 50	30 00
2 in.	35 00	30 00
2¼ in.	35 00	29 00
2½ in.	42 00	37 00
3 in.	50 00	48 00
3¼ in.	.. ..	48 50
3½ in.	63 00	51 50
4 in.	85 00	65 50

Prices per 100 ft., Montreal and Toronto

OILS AND COMPOUNDS.

Castor oil, per lb.	
Royalite, per gal., bulk	29½
Palacine	32½
Machine oil, per gal.	65
Black oil, per gal.	34
Cylinder oil, Capital	1.01
Petroleum fuel oil, bbls., net	19

BELTING—No 1 OAK TANNED

Extra heavy, single and double	6¼
Standard	6¼
Cut leather lacing, No. 1	2 00
Leather in side	2 40 3 00

TAPES

Chesterman Metallic, 50 ft.	$2 00
Lufkin Metallic, 603, 50 ft.	2 00
Admiral Steel Tape, 50 ft.	2 75
Admiral Steel Tape, 100 ft.	4 45
Major Jun. Steel Tape, 50 ft.	3 50
Rival Steel Tape, 50 ft.	2 75
Rival Steel Tape, 100 ft.	4 45
Reliable Jun. Steel Tape, 50 ft.	3 50

PLATING SUPPLIES

Polishing wheels, felt	$4 50
Polishing wheels, bull-neck	2 00
Emery in kegs, Turkish	8¾
Pumice, ground	06
Emery glue	30
Tripoli composition	9½
Crocus composition	12
Emery composition	11
Rouge, silver	64
Rouge, powder, nickel	38

Prices per lb.

ARTIFICIAL CORUNDUM

Grits, 6 to 70 inclusive	.08½
Grits, 80 and finer	.6

BRASS—Warehouse Price

Brass rods, base ½ in. to 1 in. rod	0 30
Brass sheets, 24 gauge and heavier, base	0 38
Brass tubing, seamless	0 42
Copper tubing, seamless	0 44

WASTE

XXX Extra	.24	Atlas	.20
Peerless	.22½	X Empire	.19½
Grand	.22½	Ideal	.19
Superior	.22½	X Press	.17½
X L C R	.21		

Colored

Lion	.17	Popular	.13
Standard	.15	Keen	.11
No. 1	.15		

Wool Packing

Arrow	.35	Anvil	.22
Axle	.28	Anchor	.17

Washed Wipers

Select White	.20	Dark colored	.09
Mixed colored	.10		

This list subject to trade discount for quantity.

RUBBER BELTING

Standard ... 10% Best grades... 15%

ANODES

Nickel	.55 to .60
Copper	.38 to .40
Tin	.70 to .70
Zinc	.16 to .17

Prices per lb.

COPPER PRODUCTS

	Montreal	Toronto
Bars, ½ to 2 in.	$35 00	$37 00
Copper wire, list plus 10		
Plain sheets, 14 oz., 14x60 in.	40 00	44 00
Copper sheet, tinned, 14x60, 14 oz.	43 00	46 00
Copper sheet, planished, 16 oz. base	47 00	50 00
Braziers', in sheets, 6 x 4 base	39 00	42 00

LEAD SHEETS

	Montreal	Toronto
Sheets, 3 lbs. sq. ft.	$10 50	$14 50
Sheets, 3½ lbs. sq. ft.	10 25	14 00
Sheets, 4 to 6 lbs. sq. ft.	10 00	13 50

Cut sheets, ½c per lb. extra.
Cut sheets to size, 1c per lb. extra.

PLATING CHEMICALS

Acid, boracic	$ .23
Acid, hydrochloric	.04¾
Acid, nitric	.11
Acid, sulphuric	.04½
Ammonia, aqua	.15¾
Ammonium, carbonate	.23
Ammonium, chloride	.22
Ammonium hydrosulphuret	.75
Ammonium sulphate	.30
Arsenic, white	.16
Copper, carbonate, annhy	.41
Copper, sulphate	.13
Cobalt, sulphate	.20
Iron perchloride	.62
Lead acetate	.30
Nickel ammonium sulphate	.20
Nickel carbonate	.32
Nickel sulphate	.20
Potassium sulphide (substitute)	.40
Silver Chloride (per oz.)	1.15
Silver nitrate (per oz.)	1.10
Sodium bisulphate	.13
Sodium carbonate crystals	.04
Sodium cyanide, 127-130%	.39
Sodium hyposulphite per 100 lbs	9.00
Sodium phosphate	.15
Tin chloride	.30
Zinc chloride, C.P.	.30
Zinc sulphate	.08

Prices per lb. unless otherwise stated

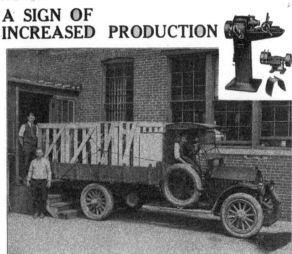

NEW YORK LOOKS FOR MOVE
TO BETTER BUYING NEXT YEAR

Special to CANADIAN MACHINERY.

NEW YORK, Dec. 2.—The development of the past week of outstanding importance is the announcement by leading independent manufacturers of steel products that they have reduced their prices on plates, shapes and bars and some other products to those which have been in effect since March 21, 1919, on all orders taken by the United States Steel Corporation. While the major products have come down, certain other products such as rails, wire and nails, sheets, etc., are still higher, as quoted by independent mills, than the prices quoted by Steel Corporation subsidiaries.

The reduction in steel prices has a direct relation to machine-tool trade, as any revival of business in iron and steel lines must of necessity be preceded by an adjustment of prices on steel that will bring buyers back into the market with a greater degree of confidence that the bottom had been about reached.

The machine-tool trade sees ample evidences of a willingness on the part of manufacturers to buy shop equipment, but the general uncertainty has caused them to hold aloof. Even actual and pressing needs for machine tools have been put off until after January 1 in numerous instances. The readjustment of steel prices by independent mills is an important step toward the expected revival of business next year. If the railroads, for example, buy steel in fairly large quantities they will surely buy machine tools, as many of their shops are greatly in need of complete rehabilitation. Several large railroad lists have been held up by Eastern roads until after January 1, but there is a strong expectation that they will be issued early in the new year.

Current business in machine tools is very light. Some machine-tool builders are not booking enough business to pay expenses. Dealers have not done more than 10 to 15 per cent. of a normal business during November. There is no expectation of any buying in December, but the turn of the year is being looked for with interest and a greater feeling of optimism than has prevailed in many weeks.

well in excess of our own needs. We mention this to bring out the fact that while it is to be hoped that industry will, in the course of a few months, be again fairly active, the element of competition will be very prominent and the margin of profit correspondingly small.

HAVE NO MORE
ORDERS ON HAND

Director of Vickers Ltd., Says Not Much Future for Shipbuilding

Montreal.—The "Canadian Leader," 3,340 net tonnage, the twelfth ship built for the Canadian Government Merchant Marine, and the sixth this season, by the Canadian Vickers Ltd., was successfully launched from the yards of the company. At the subsequent reception A. R. Gillham, managing director of Vickers Ltd., stated that his firm had launched more than 50,000 tons of shipping during the current year.

He painted a gloomy picture of the conditions that now confronted his company and the Canadian shipbuilding industry in general. More than 40 per cent. of Canadian shipyards had sung their swan song, he said, and, unless Government aid were forthcoming the same fate confronted the Vickers company. "The sorry fact is," said Mr. Gillham, "that we have no orders for next year. We have two boats now on the stocks which we hope to launch next April, and if there are no further orders by that time the yards will be forced to close."

KEEP AS MANY
ON AS POSSIBLE

C.M.A. at Conference Discuss Ways of Dealing With the Present Situation

At a joint meeting of the Ontario Division executive and Toronto branch executive of the C.M.A., very full consideration was given to the problem of unemployment throughout the Province of Ontario. The executive viewed with apprehension the present condition of unemployment generally, but they had in mind that there is always a certain amount of unemployment at this season of the year, which is accentuated by the fact that a number of unemployed, who do not belong to this city, locate in Toronto in search of work, thus aggravating the local labor situation.

The executive strongly recommended to all manufacturers to do everything possible to prevent the laying off of too many men where it is at all avoidable, and it was agreed when a reduction of working time is necessary to carry out the reduction by closing for whole days rather than for parts of days, so as to minimize factory disorganization and inconvenience to the workers and thereby keep as many men employed as possible.

GOODS WILL MOVE IF PRICE IS RIGHT
WAR PRICES AND WAR PROFITS ARE PAST

THE final chapters in the industrial liquidation are now being written. The retailers are at last falling into line, and where they have shown a disposition to meet their customers' views as to prices, they have demonstrated that there is an enormous latent purchasing power. There is no doubt of the ability to move goods into the hands of final consumers, provided the price is right. It should not take long to clean up present stocks once the retailers, generally, adopt the policy that the leaders have already started. Then will follow the replacement of stocks which, thanks to heavy reductions in price by manufacturers, can be done to good advantage. Of course, all this means that war prices and war profits have gone for good. All must disabuse themselves of the idea that these can be retained at any stage. But the point is, first, that there is now promise that the period of industrial recession will be comparatively short-lived and secondly, that even what used to be considered normal profits would, in the large majority of cases, amply support present stock prices. In this way Hayden, Stone & Co. review, the market situation in their weekly letter.

Even the most stubborn of the obstacles to a return to healthy active conditions is beginning to give way, they continue. We allude to the question of wages. During the war period, these were largely paid out of capital. Of course, this could not long continue without complete disaster. Henceforth, they will be paid according to the ability of industries to meet them. It will be, in the end, more satisfactory to be assured of a certain wage, even if less in dollars and cents, that will have a purchasing power equal to a normally higher but actually far more uncertain remuneration.

There is, however, one rather disturbing thought; our prosperity in the last five years has been largely based on Europe's demand for our goods, at abnormal prices. This demand will, henceforth, not only be radically diminished, but will be replaced by the ability of Europe to compete in our own markets. Even after severe markdowns, our prices will be very attractive to European producers, especially on present exchange rates. Europe still owes us enormous debts. These cannot be paid in money; they must be met with goods. Whether it is better to let these goods in at such prices as to jeopardize our own manufacturers, or to protect them against this influx, even if to do so it may be necessary to forego payment of this debt, is a deep economic problem, which it is hardly the function of this letter to take up, but the mere fact that such a problem exists emphasizes that world competition is being rapidly restored and that some outlet must be found for our own goods, as our productive capacity is probably

WAR TIME PRACTICE WILL NOT HELP IN THE PRESENT MARKETS

RIGHT now is the time when a man must show whether or not he has any qualities as a manager, salesman or financial adviser.

That is the opinion of one well-known Toronto manager, speaking to Canadian Machinery this week. Here is the way he figures it out:

"During the war we got a poor idea of what business amounted to. There was nothing in the way of real serious selling done. Now there may be some variation from that statement, but as a rule it will hold true. The shell business was not sold. It was taken away, and there was only one necessary thing to do, viz., produce good shells. Any lessons of the war were for those in the production department, and I think we are a better people industrially then we were before on account of the experience we gained in that way.

"Many firms that came into being since the war have a real situation to face now. In many cases they have not been in business long enough to have seen such seasons coming and developing as we are facing now. There is nothing unprecedented in the developments that we are face to face with today. Many of us have gone through the same thing before, and have managed to come out in fair shape recently, although sometimes more or less ruffled up.

"You may know there is a tremendous difference between producing, with the government taking the product away, and financing much of the work, and getting out on your own hook to produce, finance and sell a certain given line. That is where we have got to do real hard work now, and as I remarked before, right now gives a man a good chance to show what kind of managerial stuff he has in his system. Problems of financing and selling are, to my notion, the real tests of good business training and capacity. Give me the man who can adjust finances and keep the sales on the move and you can take over the rest of the organization."

are to continue to fill the dinner pails of the workers they must have protection for their produce.

STEEL PRICES AT MARCH '19 LEVEL

Largest of the Independents Announce New Prices at Which They Will Sell

Pittsburgh.—Return of steel prices to the base established by the United States Industrial Board, March 21, 1919, developed here on Saturday, when the Jones and Laughlin Steel Company, the largest of the independent interests, adopted new selling rates on certain finished steel products.

This reduction, the first to be made by independents in the Pittsburgh district, will not affect the wages, it was understood, of the approximately 25,000 persons in the company's employ.

The new prices, which are effective immediately, are: Bars, $2.35; structural sheets, $2.45; plates, $2.65, base Pittsburgh. Adjustments will be made on the prices for wire and cold-rolled steel products.

Independent steel companies of the Youngstown district, employing about 30,000 workers, announced to-day they will meet the reduction of prices announced to-day by the Jones and Laughlin Company of Pittsburgh, the cut to be effective at once.

Heads of the three largest independent corporations here declared a readjustment in the cost of both coal and labor will have to follow. They said that wherever labor costs with the independents are higher than the United States Steel Corporation, the wages will have to be reduced.

GALT FIRM GETS THIS BUSINESS

The Exchange Rate This Time Worked in Favor of the Canadian Firm

Irritation over United States exchange bold-up has been the means of securing for Messrs. Goldie and McCullough Ltd., of Toronto and Galt, a large order for boilers and heavy machinery for the new municipal lighting plant at Murwillumbah, Australia. Several American machinery manufacturers tendered for the business. The order is an important one. Mr. A. McArthur, of the Empire Engineering Supply Company, Sydney, Australia, at present at the Walker House, had the satisfaction of booking this order, and is expected to throw other important trade to Canadian manufacturers.

Mr. McArthur is now in consultation with officials of the Canadian Manufacturers' Association, urging that organization to inspire greater activity among its members to take advantage of present opportunities and lay the foundation for a large volume of the Australian trade. The Antipodean Commonwealth now has a tariff of 40 per cent. against all importations foreign, with a preference in favor of Great Britain and such of her Dominions as are on the same plan as Australia with respect to white labor conditions. Mr. McArthur says that if Canadian manufacturers are prepared to quote reasonable prices in many lines of articles, Australia is in the market with a large volume of orders to be filled.

PROTECTION FOR THE STEEL MILLS

Case Presented by the Steel Co. of Canada at Hamilton Sessions

Some of the bigger interests from the financial standpoint which sent representatives to claim a continuance of tariff protection at the Hamilton session of the Tariff Commission were the Steel Corporation of Canada, the iron and steel industry of Hamilton, Dominion Foundries and Steel Limited, the flat glass industries, the Spirella Company of Canada, all the above being located in Hamilton, and the Canadian Yale and Towne Co., Ltd., of St. Catharines. The statement of the Steel Corporation of Canada as presented by Mr. Robert Robson, president, was an exhaustive one, and it followed out the attitude adopted by the large corporations in the same business in Nova Scotia and at Sault Ste. Marie. As Mr. Hobson put it, the tremendous advantages of the self-contained industries across the line make it particularly difficult for the Canadian plants to compete with them, and he had plenty of statistics to show that the steel industry in Canada is subject to keenest competition from United States producing units owing to greater capacity, shorter freight haulage, and convenience to markets. The general statements from the different iron and steel interests were consistent ones of growth, with increasing payrolls and larger output, but their contentions were also consistent in the claim that if they

MANY THINGS TO BE CONSIDERED

Mr. Hollinrake Says C.M.A. Urges Members to Keep On As Many as Possible

Discussing matters of unemployment, T. A. Hollinrake, a president of the Toronto branch of the Canadian Manufacturers' Association, reminds the public that anxiety is not confined entirely to those who are out of work.

"The attitude of the C.M.A.," says he, "has been favorable to Labor. We have a committee on industrial relations which is keeping in close touch with the unemployment situation.

"At the present time it is the C.M.A.'s policy to ask all members of the association to keep as many employed as possible, even if they might easily be able to dispense with them, feeling that at this time it is a necessary and wise action.

"The power situation has been heartbreaking to employers. That condition might very well have aggravated un-

employment. But employers as a whole bore the brunt of the power shortage and kept on their men, even though idle.

"Unemployment depends on the state of business. And manufacturers are experiencing no little uneasiness because of the tariff question, luxury and business profits taxes. Manufacturers are hampered in many ways just now, and the result is not beneficial as regards unemployment."

PIG IRON TRADE

The iron market is at present the quietest in years, and prices are being determined by small transactions on the Philadelphia market. Some small sales were made during the past week by Eastern Pennsylvania makers, up to $44 for No. 2 plain (1.75 to 2.25 silicon), al. though the usual prices are from $41 to $43 furnace. Malleable is offered at $46, while basic, in absence of business, is nominal, but could be secured from eastern makers on a basis of $35 valley or $41.16 del.

Pig iron dealers in Chicago who have been familiar with the market for years say that the present situation is the quietest ever experienced. There is not only practically no inquiry, but it seems impossible to interest any one in taking either spot iron or future tonnages. The policy that is seemingly followed with furnaces is to go out of blast rather than continue production under present conditions. Three furnaces have blown out in the South Chicago district. Northern iron is nominally quoted at $46 and southern $42.

Basic iron is being offered by a valley merchant furnace in the Pittsburgh district at $35 and Bessemer and malleable at $37.50 a ton as compared with prices of a week ago. No. 2 foundry (1.25 to 1.75 silicon) is quoted at the same price as last week—$39.

Pig iron sales in the Cleveland district are mostly of a resale character. Furnace prices are difficult to determine because the producers are either not selling at all, or in very small quantities. Carload sales were made by producers during the week at $42, base Cleveland furnace, and $41, northern Ohio furnace.

The volume of resale iron on the St. Louis market is increasing. It has been stated by men long in trade that they have never seen such a rapid change from a period of seeming scarcity to one of large surplus. Agencies quote $42 for No. 2 (1.75 to a 2.25 silicon). This price however does not mean anything as it is not being realized on sales.

Production is being curtailed sharply as producers claim that costs have not declined materially, even despite the heavy cuts in coke prices and they are unwilling to pile iron at present levels.

Despite the fact that foundries have further curtailed production and trading is almost at a standstill, there is a more cheerful undertone in the Boston market. Dealers who had given up hope of selling first half iron before the first of the year now seem to think that some buying will take place before the middle of December. A Connecticut consumer bought 300 tons of resale Virginia iron at $47.26 del. for No. 2X 2.25 to 2.75 silicon, which is approximately $40 furnace base.

On the Buffalo market there is very little furnace iron moving and transactions in resale have diminished considerably. The selling price in resale is varied. One lot of 500 tons sold at $43, while another small lot sold at $40. One furnace has quoted $44 for 1.75 to 2.25 silicon foundry, but no sales are noted at this price.

Transactions on the New York market are principally in resale iron and these sales usually consist of not more than one or two cars each. Buffalo foundry has been resold during the past week at prices ranging from $38 to $42 Buffalo for 1.75 to 2.25 silicon grade.

In the Cincinnati district one furnace which has been out of blast has been relined and is again working.

The disposition seems to be not to produce iron at the present high costs. The Jenifer furnace has been blown out, which further reduces the output. A little tonnage is being offered by an independent furnace company for $38 for No. 2 foundry, but the majority of furnaces are holding the price at $42.

Scrap Metal Market

Sharp drops are noted in many districts in the scrap metal market, but the general opinion is that it has nearly reached the bottom and that after another dollar or two decrease it will settle to a definite level. Ten thousand tons of heavy melting steel, recently sold at $19.50, delivered Chicago. Consumers are taking practically nothing and quotations are largely nominal. Railroad offerings are moderate.

After a remarkable record throughout the summer, continuing after all the other grades had softened, No. 1 machinery scrap at last has dropped in the Boston market. During the week the situation changed from a small supply to a surplus which accounts for the drop of $6 and $8. The fallen market seems to have brought some new business. Several hundred tons were bought at $33, del.

Prices in iron and steel scrap have declined from $1 to $3 in practically all grades. Rerolling rails city wrought and stove plate have dropped from $2.50 to $3 a ton.

Heavy melting steel has declined and is now holding at $14 to $15 f. o. b. New York. Clean cast borings and machine shop turnings remain unchanged, as these materials are not accumulating as fast as most of the others.

On the Pittsburgh market the actual quotations are largely nominal and unchanged from those of a week ago. Small lots are available at from $3 to $4 a ton lower, but these represent surplus quantities that consumers have on hand and are trying to dispose of, apparently with very little success.

On the Buffalo market the price of scrap is still declining, with few sales being made. Many cancellations of old orders have taken place and some dealers are overloaded with stock as they failed to gauge the market correctly and kept on buying. Most grades show a decrease in price over the past week and heavy melting steel is now quoted at $19 to $20.

Sharp declines have taken place in the steel scrap market this week at Cleveland, owing to an absolute dearth of business.

The general impression is that the market is still far from the bottom and as a result cancellations are numerous. Heavy melting steel is down from $21.50 and $22 to $19 and $19.25, and other grades are similarly cut.

The market continues extremely dull at Cincinnati. Surpluses have been accumulated in some yards and this is largely responsible for the price slump.

On the St. Louis market re-rolling rails have been marked down from $22.50 to $20.50, compared with a price of $40 on September 12. Reductions, however, have failed to stimulate buying. Consumers are more interested in cancelling orders than acquiring new tonnages.

PERSONAL

W. H. Fairchild, city engineer of Galt since March, 1916, has handed his resignation to City Clerk McArney, to take effect December 31. When interviewed officially he gave no reasons for his action and said he had no plans for the future. During Mr. Fairchild's term of office, Galt launched its permanent paving programme and he has supervised construction of all the pavements laid to date. Citizens generally have been well satisfied with his work. It is understood an effort will be made to retain him.

On behalf of the office staff of the John Morrow Screw and Nut Company, of Ingersoll, the presentation of a handsome electric reading lamp to Mr. George Duncan, of Hamilton, of the company's sales force, took place at the office. Mr. Duncan, who formerly resided there, has been in the employ of the firm about fifteen years. He has severed his connection with the company to accept a position at Buffalo, being associated with J. Anderson Coulter, who recently purchased the plant of the Curtis Screw Company of that city.

CANADIAN MACHINERY

AND MANUFACTURING NEWS

VOL. XXIV. No. 24   December 9, 1920

Efficient Automatic Machine Tool Records

An Automatic Machine Operator Should Be Provided With Suitable Charts to Enable Him to Change His Sets-ups Quickly. —Skeleton Set-ups on Turrett Chucking Lathe Are Explained.

By F. SCRIBER

TO enable automatic machinery operators to work on something better than a hit or miss plan in changing from one job to another, it is becoming more and more the practice to provide the operators with suitable charts or descriptive matter to enable them to quickly change their machines from one set-up to another, that is to say, that they can change from one class of work to a different class of work by following a sequence of set-up instructions and setting up the machine by scale measurements and also to arrange the proper feeds which will enable them to start cutting on the work set up with some degree of surety that the work being done will be very near to what is wanted and that the machine will be operating at practically maximum capacity.

An analysis of a turret chucking machine set-up which shows in some detail how this can be accomplished is the subject of this article. It is of course apparent that definite types of machines require somewhat different analysis even though they do the same class of work.

Referring to Fig. 1, we have a plan skeleton view of the actual set-up members on a turret chucking lathe; this

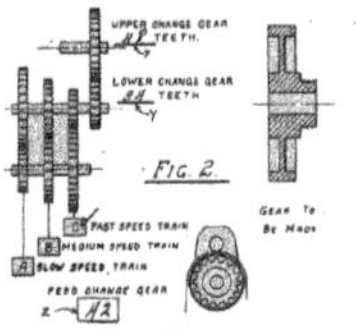

FIG. 2—SHOWING DIAGRAM OF CHANGE GEARS.

shows a chuck A, the turret has four sides or faces B and the cross-slide carries a rear block for tools C, and a front tool block D; various dimension lines are shown in this view with dimension and indicating arrows referring thereto. The purpose of this is that with a given piece of work gripped in the chuck an operator of this machine can set his cross-slide from the chuck to a figure say 1¾ inches, ad given, and can then locate his first turret face to the dimension 17 inches given. Likewise he can set the

blocks on his cross-slide to the dimension 8 and 29 inches, which is shown, and then these various units will be in the correct cutting relation to the chuck for the work in hand.

The first part of this tool-setting description refers to locating the headstock carrying the chuck with relation to the turret as in this particular type of machine the headstock is adjustable

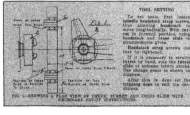

FIG. 3—ILLUSTRATING HOW THE TRIPPING DOGS CAN BE SET AT DEFINITE DISTANCES.

lengthwise. Reference is also made here to change gears in the following terms: "Use change gears as shown in diagram," and a diagram of these change gears mentioned is shown in Fig. 2, where it will be noted that the number of teeth in the various change gears is supplied in places provided for that purpose X and Y.

The train of gears to be used is indicated by a check mark, while the feed change gear Z also has to be determined for the particular job in hand. To the right of Fig. 2 is shown a special gear having a thread on it which we will assume is to be machined at this time, but before proceeding with the machining of this gear it is first necessary to set the tripping dogs for controlling the movement of the turret and the cross-slide,

TOOL SETTING

To set tools, first loosen spindle headstock strap screws, thus allowing headstock to move longitudinally. With turret in forward position, bring headstock and cross slide to measurements given.

Headstock strap screws can then be tightened.

If it is necessary to revolve turret by hand, stop the turret slide at extreme return stroke. Use change gears as shown on diagram.

After this is done set the tripping dogs to suit the conditions.

FIG. 1—SHOWING A PLAN VIEW OF CHUCK, TURRETT AND CROSS SLIDE WITH NECESSARY SET-UP INSTRUCTIONS.

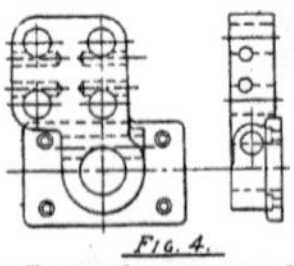

FIG. 4.—ILLUSTRATING TYPE OF HOLDER DESCRIBED IN TEXT.

this being referred to in the lower portion of the tool-setting memorandum in Fig. 1.

To set these tripping dogs another diagram is provided, this diagram illustrates how, by starting at a point A the dogs may be set definite distances apart around a circle as indicated by the small circles S and F, to trip the various turrets and cross-slide operating member at the proper time. Having set the machine up as called for in these diagrams and shown by the first three illustrations we are now ready to consider the accurate setting up of the tools for machining this special threaded gear, but before proceeding with this it is perhaps better to note that a tool holder can be attached to each turret face which has in addition to the centre hole three or four other tool carrying holes, an illustration of this holder is shown in Fig. 4.

Fig. 5 shows a blank diagram suitable for use with this holder where it will be noticed that on the centre line A at the right are the words No. 1 turret face, and below this at the centre on line B, No. 2 turret face; there is a note at the top of this diagram as follows: "These numbers refer to the holes in standard tool-holder on turret," these being numbered 1, 2, 3, 4, and the purpose of these will be obvious by referring to Fig. 6.

This diagram is made for a machine having five turret faces. By referring to the gear-drawing, Fig. 2, and referring to the successive set-ups in Fig. 6, it will be noted by following the reference letters that on the first turret face A, Fig. 6, we have a spotting drill. At B we have a drill with facing cutter and

two turning tools located in the holes 1 and 3 of the holder; at C we have a special boring tool in the centre hole and two turning tools in holes 1 and 3 for finish-turning the work. At D we have a reamer which is used for finishing the hole and at E is provided a threading die for threading the hub of gear, all of these tools are set up to the dimensions given from the various turret faces of the machine and are quite accurate, so they will locate the tools in the proper cutting position for finishing the gear, although it is usually necessary to take some trial cuts to obtain accurate work.

It will be apparent by following out this sequence of operations that a great saving of time will be had with much less setting up time and with less spoiled work than would be the case if the operator had to work by cutting and trying out each tool as he proceeded with his set-ups. While this plan is by no means new it is not so broadly used but what a good many variations of this general plan will readily suggest themselves.

BRASS ITCH

One of the most interesting of the occupational diseases, and one about which but little has been known as yet, is the result of brass-poisoning. The matter is quite satisfactorily treated by Dr. R. P. Albaugh, in a recent issue of Modern Medicine.

The symptoms are exhibited for several hours after exposure to the fumes from molten brass or zinc. They start with a dry, parched thirst, an irritating and unproductive, cough, a feeling of constriction in the chest, lassitude and anorexia—which latter is simply loss of appetite—often followed by nausea and vomiting; a headache sometimes develops and chilly sensations are noticeable within one to four hours. The chills rapidly verge into a distinct rigor which lasts from one and one-half to two or three hours. The application of warm clothing or external heat seems not to diminish the rigor of a chill. Muscular cramps and sharp pains in the joints usually accompany the chill. The symptoms end rather abruptly, almost by crisis, and are followed at once by

a profuse perspiration. The patient will probably fall into a deep sleep following the stage of relaxation and without any apparent ill effects.

Brass poisoning does not occur from handling of the metal or the zinc alloys which go into its composition, but is limited to those exposed to the inhalation of the whitish smoke and sublimation products from the molten metal.

There is no specific treatment for brass poisoning, "brass itch" and "brass chills." Zinc is supposed to be responsible for the bad effects. The affection was formerly confused with malaria. Some workmen find relief in drinking hot milk to which pepper is added. A good purge also seems beneficial. Dr. Albaugh advises the prevention of brass poisoning by better hygienic arrangements in foundries and smelters, elimination of careless habits of workmen and large, roomy quarters.

SAFETY IN THE MACHINE SHOP

In connection with its efforts to reduce the number and severity of industrial accidents, the Travelers' Insurance Company, Hartford, Conn., has recently added another book to its accident-prevention series. This new publication is entitled "Safety in the Machine Shop." It contains 188 pages, six by nine inches, and is bound in paper covers. It is a splendidly arranged book dealing with the various phases of the safety problem as encountered in machine-shop practice.

The book is divided into fourteen sections, the first of which sketches the development of the machine shop from the time such an establishment consisted of the proprietor and one or two assistants, up to the present-day shop employing thousands of persons. In the remaining sections the safety problems that arise in the modern shop are discussed in detail and careful consideration is given to cranes, lathes, automatic machines, grinding apparatus, and other machine-shop equipment. In addition to the discussion of safeguarding and safe-operating methods, subjects indirectly related to accident prevention are considered. One section treats of "Infection From Cutting Oils" and another deals with the subject of proper illumination and its bearing on the accident prevention problem.

These and other special features will undoubtedly be of interest to machine-shop proprietors and to other persons concerned in accident-prevention work. This book will be sent without charge to anyone making request for it to the supply department of this company.

Western Branch.—The Gordon MacLay building at Brandon, Man., has been purchased by the Massey-Harris Company, Ltd., at a purchase price of $130,-000. A large distributing plant will be established there in the spring.

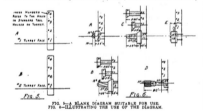

FIG. 5.—A BLANK DIAGRAM SUITABLE FOR USE.
FIG. 6.—ILLUSTRATING THE USE OF THE DIAGRAM.

Do Machinists Correct the Mistakes of Others?

Drive Problems Are Often Misunderstood—Electricians Are Not Machinists; Therefore Should Stick to Their Own Trade—The Repair Man is the One Who Suffers.

By DONALD H. HAMPSON

IT IS human to make mistakes; machinists make as many as any other class of workmen, while those machinists in the repair business are called upon to correct more of the mistakes of other workmen than men in any other calling. We learn more from mistakes than we do from direct instruction, probably because the thing is visualized right before us in all its glaring error. A number of mistakes innocently made by electricians will be cited.

The Gordon Coal Co. had a big elevator plant with a vertical bucket line taking coal from the track pocket and delivering it on a horizontal chute above, whence it was conducted to the proper bin for that size of coal. This machinery had been run for ten years by a 20 h.p. gas engine and had given satisfaction except for an apparent loss of power during the last couple of years. Naturally it was concluded that the engine needed a little attention—just cleaning and tuning up, for the work of unloading three cars of coal a week had been no tax upon the wearing qualities of any part. The people who had sold the engine were very indifferent in the work they did when called upon, with the result that the Gordon Co. in their inexperience came to the conclusion that the outfit was worn out.

At this juncture a couple of telephone electricians who were starting in the wiring business for themselves heard of this opportunity and induced the coal company to instal very nicely. They placed the motor pulley in the same plane as the old engine pulley, driving with the same belt to a four-foot pulley overhead. Apparently, the item of speeds did not occur to the electricians or, if it did, they concluded that the reduction from twelve inch to the nine inch pulley on the motor was quite sufficient, not taking into consideration the fact that their motor ran at more than three times the speed of the gas engine.

When they finally got started, the belt slipped so that it would run off the motor pulley. By shortening it till it was as taut as a violin string, they made it stay on and the bucket chain began to move. It looked fine and the coal people remarked that their elevator system had more pep in it than it had shown for years. There was no coal being moved just them—due to one of the numerous schemes that are enacted to jack up the price—and the house could do nothing else than pay the bill for what looked like a good job well done.

Some weeks later when a car did get through, it was found that the conveyor could not be started with coal in the track pit. After shovelling this out and dumping in a limited amount from the car at a time, the load was successfully distributed. The electricians had been called in and they concluded that the speed was too high so they had the motor pulley reduced an inch and a half and took another piece out of the belt. The next car was unloaded in about the same manner, but not until the belt had been again shortened, and in trying to unload the car in less than five hours the pit was allowed to fill—just as in

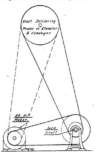

HERE WE SEE THE CHANGES MADE TO THE DRIVE.

the old days with the gas engine. It was stove coal this time and how that chain and buckets did rasp through the pile! Six buckets ripped off and several links bent—bent so badly that they jumped the sprockets.

Machinists repaired the damage. Asked to report on the causes, they stated that the speed was entirely too high. When taking material from the side of a pile or when the material is dumped into a horizontal chute, almost any speed within reason may be employed, but when buckets must be forced in and through a compact mass at the bottom of a vertical run, the speed must be slow for commercial types of elevators to stand up at all. In this case a countershaft was interposed that reduced the

surface speed from over 150 to 50 feet per minute—which is a safe maximum for such work. Better to have a mechanism which can do its work without injury and be always ready to work, better to fill each bucket every time than minutes saved while the outfit is running to be followed by a protracted season of tinkering. And this is true of many things besides coal elevators.

In this case, the machinists changed the motor pulley to a size larger than the original—both to get greater adhesive surface and to lessen the bending of the belt as it passed around it, the latter being two-ply. The drawing shows the changes made. It shows also how a means was provided of taking up on the belt without cutting it every time, which had to be done when the drive was overhead. The changes proved all that could be wished. If the electricians had given thought to speed or had consulted anyone who understood this and the best practice with belts, they would have saved themselves and their customer much expense and protected the reputation they were trying to build. And this information could have been obtained gratis; workmen in one line are only too happy to help others who occasionally encounter problems in that line as a sort of side issue to their own particular field of work.

The Double Belt Fallacy

Departing for a moment from the original purpose of the article, a point in connection with the instance just recorded applies with equal force to other belt problems, i.e., the use of double belts. The motor drive named was through a double belt to the first pulley and that double belt had to form itself around that small motor pulley running at a high rate of speed.

The writer is firmly convinced that a double belt should never be used on any pulley under 18" and better not under 24", for the reason that on smaller pulleys the normal surface speeds at the inside and outside of the belt where it passes over the pulleys differ so greatly; it may be that there is a difference of 10 per cent. If the two plys were separated—not cemented together—and could be made to run that way, the difference in speed could be plainly noted, but they are not, and so the belt must stretch and slip and buckle as it curves around the pulley. What this does may be imagined. Very soon the cement bond is broken here and there and we have

two belts, somewhat stuck together, but neither of them pulling half the load as they go flip-flopping around before final separation.

To show what double belts will do on large pulleys, a certain plant put in three 8" double leather belts nineteen years ago as a means of transmitting 16 horse-power from the engine room to the machine shop. The pulleys on the shafts are all 48" and travel at a surface speed of 1800 f.p.m. To-day those same belts are transmitting 30 to 35 h.p.—they have been treated three times, the stretch taken out twice and there is no sign of a separation of plys. Contrast this with a drive the writer was asked to improve recently. The "electric light gang" had installed a motor in place of a steam engine to drive a blower in the city gas plant of the same company, the installation being a 10 h.p. motor with 8" pulley driving to a 6" pulley on the blower through a double belt. It had been a new belt, first class, and it had slipped so well that the friction had heated the inner ply, loosening it and forming it in puckering waves all around the inside. A double belt had been used because of the generally known fact that "they transmit more power and stretch less."

The writer admits to having been sold on the "more power" point of the double belt, and learned by bitter experience the falsity of it. A grinder with 6" pulleys was so equipped, and, as the belt was the best that money could buy, he concluded that the purchase of a new one every six months was a necessary expense that could not be avoided. It was not until a "drummer" (as they called them then) came along with one of the early woven cloth belts that we learned our mistake. This sample belt, at half the cost of the leather one, lasted five years—simply for the reason that it was soft enough to hug the pulley throughout the entire arc of contact and hug it so closely that it did not slip. The moral is: if you have a hard drive and the pulleys cannot be enlarged, use wider faced pulleys and a single belt in preference to narrow pulleys and a double belt. To make the point clearer by going to an extreme, just imagine how much power a pulley would deliver to a belt made of sole leather.

Returning to the subject of electricians, the boys had secured a job to motor equip the machinery in a food product plant, including the can and bottle washing machines. Somewhere they had picked up a number of 140-tooth gears which they were putting on the can washers, driving the gears through 9-tooth pinions on the motors. They went to a local shop to get the 9-tooth pinions made. The shop protested, giving their reasons as the weak teeth and the extra noise as well as a 9-tooth gear being less than standard; but they finally made the gears on a "no responsibility" basis. And in a few short months the pinions were done for, being

replaced by worm drives of similar reduction. In this particular case a standard, based on theory and practice, was disregarded as a matter of little consequence—but men in every craft are apt to hold lightly the restrictions of another, so the electricians cannot be too severely censured.

The Roe Brothers firm had the only coal business in the village of Florida —progressive men who early established a coal elevating plant run by a steam engine, changing later to a gasoline engine, and finally to an electric motor as these prime movers were developed. Their motor drive was an entirely different arrangement from that of the elevator previously mentioned; in this case the motor was installed on 4 x 6 beams up near the roof, driving through two gears a length of shaft from which the bucket lines took their power. One gear, a 10-inch steel pinion of two and a half face, was on the motor shaft—it meshed

MORE POWER

When you have a hard drive and your pulleys cannot be enlarged, use wider faced pulleys and a single belt in preference to narrow pulleys and a double belt.

Be sure that when you install gears, the surface speed is not too high, and above all see that the centre distance are correct. If the atmosphere is dirty and gritty, decide to lubricate the gears.

Another point to remember is this: Many a drive has been improved through the substitution of a chain in place of the gears. These are a few of the lessons contained herein.

with a gear on the shaft which latter was supported by hangers bolted to the same 4 x 6's. The motor was of ample size for the work. Its speed was 1200 r.p.m.

Well, any mechanic can picture in his mind what happened to that soft steel pinion delivering 15 h.p. at 1200 revolutions. The teeth were strong enough from a theoretical consideration but there was not width enough to distribute the load over an amount of surface to insure long life. The pinion wore out in three months, and not three months of steady running either. Mr. Roe brought his worn pinion in to us to make a duplicate just as quick as we could hustle it out aided by telegraph and express in getting a blank.

About three months later he came in again on a similar errand. This time our curiosity was aroused and we ob-

tained the features of his engineering layout. The worm gears showed that the centre distances were too far apart to have the teeth mesh proper depth and our questioning brought out the point that the shafts spread apart when the load was applied, though they were close enough together under the friction load only. We guessed (aloud) that the beams sprung and that the hangers on the shaft were not rigid enough. Our suggestion was to move the motor farther away, substitute sprockets for the gears, and use a silent chain drive. The suggestion met with favor, but, of course, could not be carried out in a hurry, so we made another pinion, this to be carefully installed and the beams strengthened and the gears well lubricated.

A month later our Mr. Roe 'phoned to order the chain and sprockets at once because the new pinion was wearing and the gear also was in bad shape. Luckily the new parts came before the gears gave out altogether, and after being duly installed they ran for years and years without giving any trouble, and are running yet with entire satisfaction.

It was a case of the electricians not knowing the ill effects of high surface speed on soft gears, too narrow to wear under the most favorable conditions and in the gritty atmosphere of the place subjected to unusually rapid abrasion. Then gears at this speed should run constantly lubricated and they should be fully in mesh to insure perfect rolling contact. The substitution of the chain was the employment of a device especially designed for high speeds, long life at those speeds, and with an entire disregard for minor changes in the centres.

NEW LINE

The National Electro Products, Ltd., Toronto, who are at present manufacturers of electrolytic oxygen and hydrogen by an electrolytic process, have added still another line, namely that of Purox equipment. This line includes a complete range of oxy-acetylene and oxy-hydrogen welding and cutting torches, decarbonizing and lead burning outfits, gauges, regulators, hose, goggles, welding rods, fluxes, etc. This is a new line in Canada but has been used in the United States for the past five or six years. The first time this line was shown in Canada was at the Toronto Exhibition, and they drew favorable comment at this time.

GIVE CREDIT

In the November 4 issue of Canadian Machinery there appeared an article on Turret Lathe Practice as adopted at the plant of the Foster Machine Co., Elkhart, Ind. In this there were drawings and portions of the reading matter which should have been credited to "Machinery," New York. We regret that at the time of publication there was omission of these credit lines to our American contemporary.

The Manufacture of High Speed Tool Steel

The Crucible and Electric Furnace Methods of Producing Steel
—The Base Materials Must Be Carefully Selected to Obtain a
Satisfactory Product.

THE manufacture of iron and steel is always a very interesting subject, but the story of tool steel is even more fascinating. At least such proved the case at a recent meeting of the Toronto chapter of the American Society for Steel Treating, when Mr. Cremp, metallurgist for the Canadian Atlas Crucible Steel Co., Ltd., Welland, Ont., gave an illustrated talk on "Modern Methods of Tool Steel Manufacture."

The speaker was greeted by a very large number of men interested in steel treatment in its various phases, and embellished his talk with four reels of moving pictures, the first of these showing various views of the American plant, how they weighed carefully the various ingredients, placing the material in the crucibles, the heating of same, the pouring of the metal, etc.

The second reel illustrated the electric furnace method of producing tool steel, these views showing both three-ton and six-ton furnaces in operation. Various tests were shown, also the grinding of all surface defects before hammering.

The third reel was devoted to rolling work, also depicted annealing and straightening of stock. The last reel continued on this subject and illustrated various cold drawing operations, and other interesting procedure in tool steel manufacture.

The talk was so cram full of real up-to-the-minute data that the writer decided to get the details from the speaker, so that readers of Canadian Machinery who could not personally attend might get as near as possible the full benefit of the lecture. This he was able to do, so from now on let us imagine we are sitting back in our chair, listening to Mr. Cremp.

In considering the manufacture of tool steel it is very important that the process be not compared with the Bessemer or open hearth method. Those readers who are somewhat familiar with these various methods will know the great amount of hand labor necessary for a small production of tool steel. This is due to the fact that in tool steel manufacture the foremost thought has to be, or at least should be, quality rather than quantity.

It might be as well to define what I mean by tool steels, as the term is very broad. By the statement I mean all steels which are capable of being made into tools, and as such, and after suitable treatment, are capable of doing work upon other steels, metals or substances. This will include the simple carbon tool steels and all alloy tool steels. By simple carbon tool steels is meant steels which contain no elements other than silicon, manganese, phosphorus and sulphur and whose physical properties depend upon the carbon content only; the carbon content carying from 0.05% to 2.20%, depending on the use, but for most purposes between .75% and 1.35%. By alloy tool steels is meant steels which contain all of the above elements and in addition varying percentages of one or more of the following elements: tungsten, chromium, vanadium, cobalt, molybdenum, etc., in sufficient quantities to affect the physical properties of the steel.

Probably the chief requisite in any tool steel is uniformity in quality, both chemically and physically, for one must have both chemical and physical uniformity in a tool steel to obtain best results. Either may be individually perfect in a steel, yet that steel will fail when made into tools; and quite naturally it usually is the physical or structural uniformity which is overlooked. How often have all of us noted the case of two steels being identical in chemical analysis, yet one would make excellent tools while the other would fail miserably when made into the same tools? It is due to this absolute necessity of uniformity that tool steel has been limited to certain methods of manufacture for best results. In order to

A NEAT INSTALLATION OF ELECTRIC FURNACES USED BY THIS CONCERN.

THIS ILLUSTRATION DEPICTS A HEAVIER HAMMER JOB.

obtain this uniformity the method of manufacture must be capable of accurate control and for this reason the crucible and electric furnaces have met with best results and greatest approval.

As in other manufacturing processes, what are termed its raw materials are really the finished product of some other industry. So in the manufacture of tool steel few virgin materials are used, but rather the finished products of some other process. Thus are used various kinds of scrap steel, the products of the crucible, electric and open hearth steel processes; wash metal, the product of a special open hearth process; pig iron, the product of the blast furnace; melting or muck bar, the product of the puddling furnace; ferro-tungsten, ferrochrome, ferro-vanadium, ferro-manganese, ferro-silicon, etc., all finished products of still other processes.

Selection of Base Materials

The selection of the raw or base materials is very important and only the very best are suitable for tool steel manufacture. This is especially true in the crucible process, where little or no refining can be done and the quality, that is quality as far as analysis is concerned, depends entirely upon the base materials used. As a result, all these materials are purchased on analysis, with special attention to their sulphur and phosphorus content, which two elements, as you know, are the greatest enemies in tool steel. It is, of course, true that a wider latitude of materials can be used in the electric furnace, but this should not be presumed upon. Too many have made the mistake of expecting to get first quality product from their electric furnace by using junk to begin with. This is not possible in the

electric any more than in any other process of making steel, although, as above stated, the possibilities of refining are greater in the electric furnace than in any other. It is not a wise or an economical policy, for what is saved in the purchase of materials is more than offset by the extra time required to refine those materials and consequently decrease the production from the furnace.

We now arrive at the steel compounding and melting operation and will discuss the crucible process first. The various materials which constitute the base of the mix for the charge are carefully weighed out into trays. This consists of muck bar, various scraps consisting of crop-ends from ingots, billets and bars, as well as other solids of known analysis. Just so much of each different one can be used, depending upon its analysis and the result desired. The importance of care and accuracy in this step needs no comment, for, as before stated, upon the charge depends the chemical content of the product. In a separate room, technically known as the "medicine" or "dope" room, the alloys and other costly elements which largely determine the particular properties of the finished product are very carefully and accurately weighed out into metal pails. These are transported to the melting floor and there charged into the crucibles with the base materials previously described, the entire comprising what is called the "mix."

The crucibles used are manufactured from either clay or clay and plumbago, holding anywhere from 50 lbs. to 100 lbs. of material. A cap is now placed on the pots to exclude any furnace gases and is ready to be lowered into the furnace. In charging a furnace a person

known as the "puller-out" (for he pulls out as well as sets-in) grasps the pot in a pair of tongs and lowers it into the hole of the furnace. The construction of the furnace is such that six pots can be placed in each hole, the total capacity of the furnace depending on the number of holes. The various sized furnaces are usually spoken of in terms of pot capacity as a twelve, twenty-four or thirty-pot furnace. These furnaces are of the regenerative type, having two sets of checker chambers underneath. The object of regenerators or checker chambers is to preheat the incoming air and gas, thus building up the temperature in the melting hole. Either producer or natural gas can be used.

In melting down the charge the direction of the incoming gas and air is reversed at regular intervals. By this reversal of direction the exhaust gases are made to flow first through one set of checkers and then through the other, giving up their heat to the brick in passing, which in turn is reabsorbed by the incoming gases. The length of time required to melt the heat varies with the nature of the mix and the elements contained therein.

Two and one-half to three and one-half hours are usually required to melt the charge thoroughly, and it is then "killed," which requires about thirty minutes. This "killing the melt" is a term which applies to that portion of the process allowed to bring the steel to the proper casting condition and to permit any gases in the molten steel to escape. This, as well as the entire melting operation, is in charge of the "melter," who must be highly trained, skillful and reliable. The exact time for pulling a crucible heat is very important and definite and must be constantly watched, as a heat having too little fire will probably contain particles of undissolved alloys, and if allowed to soak too long after melted, or if carried to too high a temperature, the pots will be badly attacked and the steel will absorb more or less carbon.

The Puller-Out

At a word from the melter, the "pullers-out" proceed to pull the white hot pots containing molten steel from the furnace. These men must necessarily be of exceptional physique, experienced and steady nerve. They prepare themselves for their work by removing all of their regular clothing and putting on special clothing, wrapping their limbs in burlap and donning "clogs," which are very heavy shoes shod with wood and iron. In order to remove a pot from the furnace it is necessary for the "puller-out" to step directly over the hole, reach down into the furnace, whose temperature is approximately 3000 deg. Fahrenheit) and grasp the pot. This is no easy task, when it is remembered they contain close to 100 pounds of metal besides their own weight; in all,

probably 120 to 140 pounds. Special tongs are used for the purpose.

As the pots are pulled from the furnace they are delivered to the melter, who in turn empties them. It is at this point that different manufacturers of crucible steel differ. Some proceed to pour the contents of each pot individually into molds whether casting small or large ingots, while others first teem all the individual crucibles into a ladle and then proceed to cast their ingots from this ladle. In my estimation there is little comparison between the two methods as far as obtaining uniformity of product is concerned. In the first method you are apt to have as many different analyses as you have pots where single pot ingots are cast, and where large ingots are cast the casting must necessarily be more or less intermittent. Outside of a few improvements in construction and handling, the entire process of crucible melting has changed but little since its first introduction by Huntsman in 1740.

Electric Furnace Method

The electric melting process is spoken of as being acid or basic, depending upon the nature of the lining in the furnace. If the furnace is operated under acid conditions, the lining is of silica brick and sand; if under basic conditions, it is lined with magnesite brick and magnesite or dolomite. Whether it is operated acid or basic depends upon the result desired. The reduction of sulphur and phosphorus and especially the latter is more easily accomplished by the basic process. As these are most objectionable to tool steel, most electric tool steel furnaces are operated on the basic principle.

In this case, instead of weighing out the raw material in 90 or 100 pound lots, the entire furnace capacity is weighed up as a charge. This charge,

be what it may, ingot crop-ends, punchings, boiler plate, flashings, tubing, etc., is charged by hand into the furnace. Upon the completion of this operation a certain amount of lime and slag-making material is charged into the furnace, which is the chemically active portion of the bath and upon it depends the amount of refining that is accomplished. The entire charge is now ready to be melted down and the current is applied. The current, or "juice," as it is commonly known to the men, is usually about 100 to 110 volts, the amperage consumed depending considerably on the type of furnace and the operator using it. The usual consumption per ton of liquid steel is about 500 to 650 k.w. when starting from a cold charge. The melting down operation, that is, bringing the entire charge to a molten condition, requires from 1½ to 2½ hours.

At this point a test or preliminary sample of the bath is taken and sent to the laboratory for analysis. While awaiting the report, which takes from twenty minutes to one hour, depending on the type of steel, it is generally necessary to pull a slag. During the melting down and before the temperature of the bath becomes too high, certain impurities in the charge, and especially phosphorus, are taken up by the slag and held as long as the slag is basic. As the temperature increases and the old slag remains in contact with the bath, it has a tendency to become acid or at least less basic, so it is usually advisable to remove this slag and put on a new one.

The period which has just elapsed in the electric melting process is technically known as the oxidation period, during which the phosphorus is removed. The second period is known as the reduction period. A strongly reducing atmosphere is maintained throughout this period as well as a strongly basic slag, the object

being the elimination of sulphur. The length of time necessary depends upon the degree of desulphurization required. It may even be necessary to pull one or more slags. It is also necessary, during this period (in order to hasten chemical action) to maintain a higher temperature than during the first period.

Bringing to Proper Analysis

As soon as the laboratory reports the results of the preliminary sample, the melter proceeds to make whatever additions are necessary to bring the liquid steel up to proper analysis. Here is a distinct advantage of the electric furnace in making steel; absolute control of the temperature and analysis at all times. Even if some error has been made in the original calculations or weighing up, upon receipt of the preliminary report this can be corrected. As soon as the additions have become thoroughly melted and uniformly distributed throughout the bath by frequent and sufficient rabbling, the molten metal is "killed" or deoxidized as in the crucible process. The correct pouring temperature as well as the condition of the metal is determined by pouring small test ingots or blocks. At a signal from the melter the entire furnace is tilted and its contents poured into a heated ladle. From the ladle the steel is cast into molds as in the crucible process.

The liquid steel, made by either process, is now cast into molds, which are usually square or hexagon in shape, these molds being made of cast iron. Small molds such as are used for tool steel work, say 3½ in. square, up to 12 in. or 14 in. square are usually split with a solid bottom, and when in use are fastened together with rings and wedges. The object of using split molds is the greater ease in removing the cast ingots and also greater ease in smoking the molds, which operation is neces-

GRINDING OUT THE DEFECTS. EVERY FLAW MUST COME OUT OF THE BILLETS BEFORE THEY ARE PASSED.

sary to keep the liquid steel from sticking to the sides when casting. Considerable care must be exercised on the part of the molder in preparing and setting up the molds to see that they are properly aligned and plumb. If they are not exactly perpendicular the metal, when flowing from the ladle into the mold will stick on the sides or splash, producing rough and bad ingots.

The ladles are usually of such capacity as to hold the entire heat, and consist of a metal shell which is lined with fire-brick. In the bottom of the ladle is a hole into which a clay or graphite nozzle is fitted, the size of which varies with the size of the ingots being cast. This nozzle, through which the liquid steel flows, is opened and closed by means of a rod on the bottom. The stopper, as the entire rod and stopper is called, is actuated by means of a handle hinged on the outside of the ladle.

transported to the annealing department. Not all ingots are annealed before working, but high-speed and certain other alloy steel ingots should be. After the ingots are loaded into the furnace they are slowly brought up to the annealing heat and then allowed to soak at this temperature. The operation is not so much to obtain a soft ingot, although an annealed ingot will work softer than one which has not been annealed, as it is to remove the cooling strains of the ingot and to allow the internal structure to rearrange itself. After they have soaked a sufficient length of time the firing of the furnace is stopped, and the furnace and ingots are allowed to slowly cool. When they reach a temperature below a black the ingots are pulled out and passed on to the first inspection and grinding department. The entire annealing operation requires about twenty-four hours.

element determined. A complete report of every heat of ingots is made, and if the content of each element does not come within certain specifications the heat is scrapped immediately. The specifications for different grades, of course, vary with the use for which they are intended, but in each case must be rigidly adhered to. It is also important and absolutely necessary in order to avoid error that no material should be applied on orders until a complete analysis of the heat is at hand. It is the frequent case in many mills that certain ingots are applied on orders on expected or hoped-for analysis, but this is poor policy as too often your hopes are wide of the mark.

The analysis being approved, the annealed and ground ingots are now ready for the cogging hammers and are transported to that department where they are charged into the reheating hammer furnaces. The firing is started, and the bringing up to forging temperature is carefully and slowly accomplished. With each hammer is the hammer crew, consisting of hammer, man and helper, hammer-driver and lever men, in each case named after the duty performed. The hammerman is in charge of the gang and is responsible for the work done, frequently working on a contract tonnage basis and hiring his own men. He must be a man of experience and know exactly how high to heat each different grade of steel, within what limits of temperature each grade of steel will forge without disaster, and just how much reduction they will stand. The size of the ingots must be properly proportioned to the different sized hammers, or correct results cannot be obtained. Overheating and bursting are two of the greatest difficulties encountered in this department.

The average hammers used in tool steel mills range in size from 1,000 pounds to 12,000 pounds and are steam driven. Steam and hydraulic presses have been experimented with on tool steel to a considerable extent, but in general have not met with best success, the action being rather more drastic than most tool steels will stand and especially high alloy tool steels. When the ingot has reached the correct temperature, which is measured by a pyrometer, it is taken out of the furnace, placed under the hammer and reduced to what is known as a billet. A certain amount of the pipe end of the ingot is cropped off at this stage, depending upon the soundness of the ingot. This is known as the crop end and will amount to 10 to 20 per cent. of the ingot weight. Another appreciable yet frequently overlooked loss starts to assert itself with the heating operations on the ingots and billets. This is the loss due to scaling. Few would believe that this loss alone amounts to 10 to 20 per cent. of the weight of the ingot. Neither of these are total losses by any means as both are kept separate according to the grade

AN EXAMPLE OF LIGHT HAMMER WORK.

Obtaining perfect cast ingots is of the utmost importance to the tool-steel maker, and is his first great source of difficulty. I mean by perfect ingots, ingots which are free from pipe, aggregation, blow holes, fins, rough surfaces, and a hundred other objectionable imperfections which render the ingot wholly or partially useless. It is a well known fact that in order to obtain good finished material, it is absolutely necessary to start with good ingots. Piping is one of the greatest, and certainly the most noticeable difficulty encountered in the ingot. Pipe is the term which is applied to designate the shrink cavity which forms in the ingot as it solidifies.

Annealing the Ingots

As soon as the ingots have cooled sufficiently so that when tipped over they do not bleed, the rings are knocked down and the ingots stripped. They are then

The inspectors carefully go over each individual ingot and mark with crayon any seams, cracks, or surface imperfections which might develop into larger imperfections in the billet. The grinder operator now takes the ingots and painstakingly grinds out all places marked by inspector. This is an expensive, and to many, an unnecessary operation, but experience has shown that with certain types of steel, too much care and caution cannot be exercised, as surface defects never grow smaller but always work larger in subsequent operations.

Chemical Analysis

It is at this point in the progress of the ingot that drillings are taken from every heat of ingots and sent to the laboratory for complete chemical analysis. The laboratory must be in charge of an experienced and reliable chemist whose only duty is to report accurately every

of steel and returned to the melting department where they are remelted and start their journey over again. The amount of reduction varies with the result desired, but usually runs about 65 to 80 per cent. in area. This entire operation is called cogging. In fact, every operation of reduction, whether on the hammers or rolling mill which does not produce the finished size is called a cogging or recogging operation.

The cogging or forging operations have more for their object than the mere reduction to the size and shape. The steel as it exists in the ingot is in a very unrefined and coarse crystalline condition, with the carbon content and other elements more or less segregated according to their solidification temperatures. The forging operation tends to uniformly distribute the various carbide segregations by breaking up the large, weak crystals, of formation and producing a refined, uniform structure throughout. In order to accomplish this objective it is necessary that sufficient work be put upon the ingot and at a proper temperature.

The billets are once more taken to the inspection and grinding department, and if of a type of steel which makes it advisable, they are annealed. The inspecting and grinding of the billets is carried on in exactly the same manner as described for ingots. There are certain types of steel which can be chipped with air chisels. On this material it is easier to chip the imperfections out than it is to grind them.

Rolling

The ground and inspected billets are now delivered to either the hammers or the rolling mills as the case may be for recogging or for finishing. The recogging would be carried out just as previously described in the cogging operation. If the work is such that it can be done on the mill, it is advisable to do so as their work is much quicker and more accurate. On a mill the reduction and working is brought about by passing the billet back and forth through grooves in rolls. Each succeeding groove is slightly smaller than the one previous, and in this way the billet continually decreases in cross section and increases in length. The shape and succession of the different grooves for efficient rolling is a unique science in itself and must be laid out so that the material is not burst, seamed, lapped, or scratched in any way. A mill is always designated according to the diameter of its rolls as a 10 in. mill, a 14 in. mill, these meaning 10 mill having 10 and 14 in. rolls respectively. Most mills are built with rolls three high and driven in such a manner that the material is fed through one way between the bottom two rolls and back through the top rolls. The length of a train of rolls is spoken of in so many stands as five, six, or seven stands. The different sets of rolls which make up the stands are known as roughing rolls, strand, leader and finishing rolls, according to the shape of the grooves in them and the duty performed. The mill forms a very efficient and rapid means of working steel and has many advantages over a hammer but must be in the hands of a competent roller to produce the results desired.

Great care must be exercised in setting up the mill to get everything in alignment and have the work pass through smoothly. Not only rounds, but squares, hexagons, octagons, flats and many remarkable shapes can be rolled on a mill, one of the principal advantages being that once the mill is set up and the size and shape found to be correct, any amount of material can be put through, and it will all be exactly alike. The greatest difficulty encountered is the time lost in changing rolls to meet the different sizes and shapes. Consequently it is advisable to segregate all the rounds, squares, and flats and roll them at once while the mill is set up. This partly accounts for the reason why sometimes when an order is placed with a mill the delivery seems unreasonably long, while at other times most extraordinarily short, all depending on whether the order arrived just after or just previous to a rolling of that size and shape.

The ingot has now arrived in its finished shape and size and is known as a hot rolled or hammered bar, but is still in a condition not suitable for most uses. The bar is still in a raw condition, uniform and quite hard; structurally it is in a sorbito-pearlitic condition. In order to put it into a usable condition it must be annealed; accordingly, the raw hot-rolled bars next go to the annealing room. Here they are packed into tubes with some material such as charcoal and ashes, lime, sand, etc., and the ends of the tubes sealed with fire-clay. Some manufacturers do not pack their material for annealing but its use seems very desirable and beneficial. First, it tends to a more uniform condition throughout the entire bar, and third, it causes the cooling down of the bars from their annealing temperature to the slower, all of which are essential to good annealing.

The sealed tubes and their contents are now rolled into the furnaces, which are then sand sealed and the heat applied. The annealing temperatures are all carefully predetermined and controlled by pyrometers. The particular annealing temperature in each case depends upon the type of steel being annealed. The time of heating, time held at temperature and cooling must all be correctly worked out and varied according to the size and kind of bars. The usual annealing operation requires about twenty-four hours, and it is usually considered best practice to allow the tubes

ROLLING OUT THE BARS.　THIS, OF COURSE, IS ON SMALLER SIZED STOCK.

and material to cool down below a black before removing from the furnace. The annealing operation not only softens the steel but it also puts it in a uniform and stable condition.

Final Inspection

The annealed bars are now removed from the tubes and go into the final inspection department. Here each individual bar is carefully gone over for size and surface conditions, a portion from each end is broken off so that the condition of the grain can be examined. From this examination can be learned whether or not the annealing has been correct, whether the bars have been decarburized in the mill or in the annealing, and whether the bars have any minute pipes or other internal imperfections. The man doing this inspection work becomes highly skilled and can pick out small pipes and imperfections which the ordinary person could not see. Here also several ends from every different lot of bars from each heat are taken and tested for hardness. By far the best method in my opinion for making this test is the Brinell, the principle of which is well known. Certain limits of hardness are established for the different grades and tempers of steel, and unless the tests come within these limits the material is rejected and reannealed. After the bars have passed this final inspection they are sawed, straightened, stamped, and bundled ready for shipment.

Due to the many different grades of steel, different shapes, and the thousands of different sizes, a very accurate yet simple system of keeping track of the material from the ingot to the finished bar must be worked out. Not infrequently a billet or two of one kind of steel gets into a lot of billets of the same size of another steel and probably will not be caught until the final inspection and sometimes not here if the only difference is in the temper, both being the same grade. Consequently, the entire process requires constant vigilance and attention in order to produce quality material, which every tool steel manufacturer should desire. By far the greatest part of the task of maintaining the quality of the product falls to the metallurgist as most everyone else is primarily interested in the production and are very apt to overlook many vital matters to accomplish that end.

He usually has at his disposal a well-equipped physical or metallurgical laboratory and capable assistants. I wish to particularly emphasize the vital importance and usefulness of the work which is capable of being done in this department. In the past it has been frequently the case that the physical laboratory has been more of a show place, or at best an experimental department, but in the present day demand for uniformity and quality in tool steel this department becomes a prime factor. One of his most useful friends in this work

is the microscope. By the proper application of this apparatus he is able to control the internal structure of his product and is able to tell when things are not going right, and by making frequent observations of the ingots, billets and finished product he is able to keep the entire material lined up on a good basis.

FORM TOOL MAKING
By J. Homewood

WHEN rapid production of intricate formed faces on brass products is desired, at the same time maintaining uniformity of shape and size, the following adapter and attachments for an engine lathe are recommended. The material produced from a turret lathe will be rapidly made, and with form tools such as I describe, will enable inexperienced operators to produce accurate work in rapid time.

The form tool shown in Fig. 1 will obviously have to be held in an offset tool holder on the turret lathe; one that can be adjusted for center distance to suit the size of work to be produced. Inasmuch as the product in this particular case is of brass, the form tool will be made out of carbon steel, preferably a nonshrinking and nonchanging steel such as Ketos, or any other steel on the market with the necessary properties. If it is impossible to secure steel of this class, extra caution should be exercised in hardening owing to the shape of the form tool, as the contractions taking place in a tool of this type when same is hardened will very often result in a crack or a total break.

The writer has had considerable experience with this type of form tool and after considerable grief in the way of broken tools finally resorted to the following method of hardening, where ordinary carbon tool steel was used,

which proved successful in almost every instance. Heat the tool up to as low a heat as it will harden under, then dip it in warm salt water until the vibrations can no longer be felt in the tongs; after which put the tool in oil.

The form tool is first turned up in the lathe and faced square on the working end. A hole is drilled in the center, as shown so as to provide clearance for the cutting tool. One-quarter of the tool is cut away in the miller so as to provide room for the master tool to back out of the way. After the blank is milled out, it is put into the cam chuck adapter and secured from turning by a set screw. The cam is screwed on the spindle of the lathe, and this cam operates the carriage back and forth by means of the roller (b) which rides on a shouldered stud (c). This stud is fastened in the center of a strap that spans the lathe carriage and is fastened thereto by cap screws.

The compound rest is set parallel with the ways so that it can be used for feeding longitudinally. A lathe on which the rack and pinion can be disengage is most suitable for this class of work so that no drag will be put upon the carriage. It is sometimes found advisable to have a tension of some kind taking the slack out of the carriage and roller engaging with the cam. The rise of the cam is enough to give cutting clearance on the forming tool.

A committee is now discussing ways and means for the proposed new county hospital to be built at Simcoe, Ont.

The erection of a fourteen-storey building at the corner of Bay and Temperance Streets, Toronto, is being planned by the York County Athletic Club. The plans are well under way, and the estimated cost is $2,811,000.

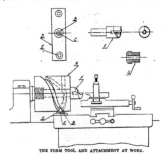

THE FORM TOOL AND ATTACHMENT AT WORK.

Would You Like to Receive One of These Scales FREE?

You can easily do so and at the same time add to your store of knowledge by entering our weekly contest of which details are given below.

The scale is 6 in. long and is made from finest quality steel. One side is marked in 32nds, the other side in 64ths. A table of decimal equivalents is also stamped on one side, and a table of tap drill sizes on the reverse side. This scale is well worth securing.

What You Have to Do

We publish every week a number of interesting facts or statements selected from the advertising pages for that week. The selections for this issue are given below. Read these through, then turn to the advertising section and see if you can pick out the advertisements to which they refer. The work is interesting, it will train your powers of perception and of memory, it costs you nothing, it will make you better acquainted with the various lines of machinery and tools in the market, and with perseverance you are bound to win one of these useful scales as a prize.

The Lucky Prize Winner for Nov. 11th Issue List is FRANK WALSH, 93 S. Bentruck St., Sydney, N.S. He Had All Answers Correct.

CONTEST FOR DEC. 9TH ISSUE

Contestants are required to write us, stating to which advertisements we refer to in this number.

1. Something that has four heads. Oh, no—it's not a man. What is it?
2. Something that has been accumulating for over sixty years.
3. How to secure prompt shipment of a certain product which is made in all standard sizes.
4. How to overcome the present power shortage.
5. Something that is at your beck and call.
6. How to keep track of every heat treating operation.
7. How to produce exact work in record time.
8. A company that produces a product in Canada different from anyone else.
9. Something that your lips never touch.
10. A product that will withstand heavy pressure without increasing its temperature.
11. A machine that can cut pieces up to 120 inches diameter.
12. How to secure the reward of greater production.

These are Correct Answers for List from Nov. 11th Issue:

1. H. W. Petrie Co.
2. The National Acme Co.
3. L. S. Starrett Co.
4. Bilton Machine Tool Co., and Fisher Motor Co.
5. A. B. Jardine Co.
6. Union Mfg. Co.
7. Clipper Belt Lacer Co.
8. R. McDougall Co., Ltd.
9. Holden Co., Ltd.
10. Norton Co. of Canada.
11. Canadian Machinery advertisement.
12. Canadian Driver-Harris Co.
13. Brown & Sharpe Mfg. Co.
14. Clemson Bros., Inc.

Closing Date for This Contest is December 30th

WHAT OUR READERS THINK AND DO

John Conley Gets Some Planer Experience

Planing Thin Work, Curves, Locomotive Links, Irregular Shapes, and Spirals, are a Few of the Problems.

By J. DAVIES

JOHN has found out, that to "pack all your troubles in your old kit bag" is not the way to get them solved, not in a machine shop anyway. He prefers to share them with the other fellow, so that when he had a thin plate to plane all over on both sides, and found that it was too big for the vise and he couldn't put plates on the top because he had to plane the entire surface, he simply made tracks for Tom Clark, an old planer hand, and asked him for advice.

Tom came over to John's machine and examined the job critically and then said, "I should advise you to hold it down with toe dogs and use the Johnny blocks to get the size." "Quit your fooling, what's a toe dog and a Johnny block?" "In a job like the one you have, there is only two ways to fasten it down that I know of. If the plate is thick enough and holes in the sides of the plate are no detriment to the job, we generally drill holes and use what we call finger clamps, some call them toe clamps."

When it is undesirable to drill holes in the work, toe dogs are usually resorted to. This equipment consists of pieces of wrought iron or steel made to fit the slots of the table, with a hole drilled and tapped at a slight angle to receive a set screw. Pieces of steel of a suitable length are ground, one end with a ball point to fit the cupped end of the set screws, the other with either a chisel or centre punch end to press against the work. Care must be taken that the angle of the toe dog is not too great—the angle of the tapped hole is usually 8 or 10 deg., just enough to insure that the dog will press the work hard down to the table. When using toe dogs, stop pins must be used to prevent the work slipping on the table."

"I understand you now, but what about the Johnny blocks?"

"Johnny blocks is a shop term for Johanson precision blocks. They are simply blocks of steel, hardened and ground very accurate and can be built up to any size. They are very useful to get your sizes on the planer."

"How do you use them?"

"How thick is this plate to be finished?"

"Three-quarters of an inch."

"Well, simply get a block that is ¾-inch thick, place it on the table and set your tool to just touch it or to pinch the thinnest piece of paper you can get, or if you have a job with a number of sizes get the blocks to correspond to the different heights and set your tool to the different heights and set your tool very useful and accurate.

"I will have to go though, for my cut is nearly over, still there are one or two other things you want to look out for. When you lower your cross rail, lower it as close to your work as you can and if you have an important job on, test if it is parallel to your table with the indicator. Put your tool in straight so that if it should happen to move it will swing away from your work and not dig into it."

"Thanks very much, perhaps I may be able to do something for you some day."

"Oh, that's nothing, if you are interested come round to my place some evening and I will tell you of some of the unusual planer jobs I have seen."

"Righto, how would next Monday night suit you?" "It suits me all right." "All right, I'll be right there."

Some Interesting Tales

On the following Monday evening John duly presented himself at the home of Tom Clark and found him deeply engrossed in the pages of Canadian Machinery. Although Tom was a regular old timer he kept himself informed of all the latest developments of the trade and could relate some interesting experiences. When they were comfortably settled down John said, "Now, Tom, what do you consider your most interesting planer experience?"

"I don't know, but probably planing some taper radial sections for a swing bridge." "Do you mean to tell me you can plane a radius or a curve? I thought a planer was for straight work only."

"Sure thing, you can do all kinds of things on a planer if you have the attachments. When I was working at the Canada Foundry they planed all the links for the Canadian Northern locomotive that they were building."

"How did they do it?"

"In planing links such as are used for locomotives, the tool is stationary except for the downward feed, and the

 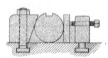

SOME LESSONS ON PLANING CAN BE LEARNED FROM THESE SKETCHES.

work is made to travel through a curved path. This is accomplished in several ways. The most satisfactory is where the work is bolted to. an extra table held between strips, which hold it from moving lengthwise on the table, but allow it to travel transversely across the table. Attached to this extra table is a bar with a hole in it the required radius. Through this hole passes a pivot pin, which is fastened to some secure support. If it is a very large radius which requires but a very slight cross motion of the extra table and necessarily a very long arm, the support for this pin would probably be on a special concrete foundation of its own. If the radius is a very small one and the pivot pin must be close up to the planer table, it can be very conveniently supported by a bracket bolted to the side of the planer.

"As the table moves backwards and forwards, the extra plate to which the work is fastened rotates about the centre of the pivot pin, and in so doing slides backwards and forwards across the table. Where frequent variations in radius occur, the link may be slotted and the pivot pin adjusted to different radii, as shown in sketch."

"Another radius planing job is to plane a casting to sit on a boiler, where the curve would have to be the same radius as the boiler. This is a little different from the link planing attachment. In the link planing attachment the work was made to swing to a given radius. In this attachment the tool is made to swing in a given arc. If the construction of the planer is such that it is not convenient to bolt your attachments to the housing direct, a bracket A is bolted across the housing B, and carries an arm C.

"One end of the arm C swings on a pin attached to A and the other end is pivoted to the tool slide. The feed screw is removed, which allows it to slide up and down freely. After the work and the tool are properly set the planing is done by simply feeding the saddle along the cross rail. The horizontal movement causes the tool point to follow a radius equal to the length of arm. C, which is made to suit the required radius. This type of attachment can only be used when it is a true radius that is required."

"Does that mean that it is only possible to plane to a true curve?" "Oh, no, not by any means. By making a template the shape required, an irregular form, convex and concave, can be produced. The template is bolted to the cross rail by any convenient attachment, this, of course, would depend upon the construction of the machine. "The template is made with a slot in it and a roller that is attached to the tool slide is made to fit this slot when the saddle is fed horisontally along the cross rail, the tool slide moves up and down and conforms to the shape of the template, thus producing the shape of the template on the work."

A Planer's Varied Uses

"Isn't it surprising the variety of work that can be done on a planer?" "I should say so, why in a case of emergency I have seen them cut gears on a planer." "How were they set up?"

"The gears were put on a mandril and a bracket containing centres was bolted to the planer table, and a gear, a small gear would do if it had the same number of teeth, put on the same mandril. A spring finger was made to fit the space in the index gear and bolted to the table so that the number of teeth could be correctly spaced." "Wouldn't it be handier to do all those kind of jobs on the miller?"

"Generally speaking, yes, but occasionally the planer is the best. I remember a case in point, the Elliott paper box factory wanted some corrugated rolls made, such as they use for making corrugated paper. The planer was the only machine long enough to do the job so the superintendent fitted up a milling cutter with a suitable bracket to the cross rail and devised a special slow forward feed with a worm and worm wheel arrangement. This could be thrown out of gear with a lever and the table run back at its ordinary speed at the end of every tool. The roll was about 64 inches long.

"Another interesting planing attachment is used at Grey's in Toronto. They not only make corrugated rolls for flour mills there, but special corrugations too. In cutting spirals, a clamp is fastened to the work with the end extending across the table. The end of this clamp travels up and down an inclined plane that is fastened to the housing of the planer and is kept in contact with it by a suitable weight. As the table travels backwards and forwards, on account of the inclined position of the piece bolted to the housing, it forces the work to make a partial revolution with each stroke of the planer. By varying the angle of this bar the amount ·of travel on the work can be regulated. The steeper the angle formed, the steeper will be the pitch of the spiral. The planing is done with a tool formed to the desired shape. The right number of spirals can be obtained by a ratchet wheel arrangement which can be made to move any number of teeth each stroke of the table.

"Another very useful attachment for a planer is a pair of vee blocks, and by the way, the planing of a pair of vee blocks involves a principle that may be very useful to you. First plane a tongue on the base of the two blocks to fit one of the slots in the table, fasten the two blocks in position, set over the head to 45 deg. and take a cut off. Then without moving the angle of the head or position of the tool, turn the blocks round and take a cut off the other side. This ensures that they will be both alike, equal angles, and will be in line at any part of the table. · Round stock is also held by means of a special planer strip. This strip has a tongue on the base to keep it in a line and it has an angular face so that the pressure on the shaft forces it down upon the table, but enough for one night."

In a communication by the Director of Mines, U.S.A., to the "Journal" of the Franklin Institute, it is recommended that carbon tetrachloride fire extinguishers should not be used on fires in confined spaces by persons who are not protected from the gases arising or who are unable to escape from the immediate vicinity of the fire at once after applying the ·extinguisher. The following gases, it is .stated, are generated when carbon tetrachloride liquids are vaporized: Phosgene, one of the most poisonous gases used in warfare, chlorine and hydrogen chloride. The vapor of carbon tetrachloride itself is a dangerous anaesthetic. It is only in confined spaces, however, that dangerous concentrations of these gases are produced.

EFFICIENT BAND BRAKE
By G. Wilson

The accompanying sketch illustrates a band brake that can be used for a great variety of purposes. The principle of the brake can be readily seen. The main casting A carries the pawls D and is free to move on the shaft. When hoisting, the racket wheel B is revolving only.

To release the weight, W has to be lifted; C is merely a cover plate fastened to the main casting by small screws.

THIS SHOWS THE DETAIL OF THE BRAKE.

DEVELOPMENTS IN
SHOP EQUIPMENT

POWER PRESS BROACH

The Hercules Machinery Co., Detroit, Mich., have placed on the market a power press broach which is illustrated herein.

The working elements consist of a ram to which motion is transmitted by means of gearing which is controlled by a single lever through a system of friction clutches. The ram is a high grade steel rack, cut 4-5 pitch, and ground perfectly true with removable hardened steel head, and key-seated the full length, insuring perfect alignment. All gears are cut stub tooth, 20 degree pressure angle. This press broach is exceedingly simple, sturdy, powerful and convenient, and does by power a large variety of operations requiring pressure, such as forcing arbors, bushings, shafts or pins in or out of holes, seats in valves, linings into pumps, etc., straightening or bending, forming, sizing holes with hardened plugs, testing castings, sealing valves for testing, expanding babbitt bearings, etc.

It is an efficient tool for all broaching operations, and is designed for rapid quantity production, and can be conveniently placed for different operations on any assembling conveyor line. The knee can be removed and replaced with other fixtures that might be required for different draw or push broaching or other operations. The three speeds of the Hercules power press broach are all handled with the movement of one lever. By the upward movement of the lever the "quick-return" speed is obtained. By the inward movement of the lever the "fast down" speed is operated, while the downward movement of the lever gives the slow or pressure-down speed. The hand wheel is for use only when setting up some operation requiring close adjustment. An automatic throw-out also forms a part of this outfit, and this throw-out has the same effect as the automatic stop on a screw machine and permits the operator to set and govern the travel of the ram. The knee of this press is carefully fitted to perfectly machined ways with a shear pin support, making it impossible to overload and thereby avoiding all breakage. Not only is the shear pin a safety factor, but in the event of extreme overload the clutches would slip before b eakage could occur.

BENCH DRILLING STAND

The Black & Decker Mfg. Co., Baltimore, Md., has placed on the market a bench drilling stand as illustrated. It can take from ⅜-inch to ⅝-inch portable electric drills of this company's design, and they can be quickly detached when desired.

The bracket carrying the drill can be raised or lowered on vertical column and is secured in any desired position by means of split collar and clamping screw. Drill may be swung clear of base, making it possible to use this stand for such work as applying ring gears to automobile axles, drilling in the ends of shafts, and other work too high to be drilled on the bench. Both vertical and horizontal adjustment are secured by means of the clamping screw.

Extra long feed lever gives feed ratio of 6 to 1. 100 pounds pressure applied to handle feeds drill under 600 pounds pressure. In the base are six tapped holes to accommodate half-inch studs, used to clamp work in place. One stud with nut and clamp is supplied with stand. This bench drilling stand is said

to be exceptionally rigid, all parts being of unusually generous size. For instance, the vertical column is a solid steel shaft 1 7-16 inches in diameter. Base is provided with four holes for fastening stand to bench. Four ⅜-inch lag screws are supplied.

Following are some interesting dimensions: Height (bottom of base to top of vertical column) 30 inches, vertical adjustment of drill 12 inches, drilling radius (distance from centre of drill bit to circumference of vertical column), 7 inches, horizontal adjustment of drill 360 degrees, feed (vertical travel of drill when operated by feed lever) 4 inches.

INCLINABLE PRESS

The Toledo Machine & Tool Co., Toledo, Ohio, have placed on the market a straight column press that possesses several unique features. Reference to the photograph will show that this press may be used in either an upright or inclined position by simply turning the hand wheel. It is also entirely self-contained, being driven by a direct connected electric motor. Attention is called to the

VIEW TO LEFT DEPICTS POWER PRESS BROACH. THE ILLUSTRATION TO RIGHT SHOWS BENCH DRILLING STAND.

GENERAL VIEW OF THE PRESS.

motor bracket which is so arranged that it always keeps the motor in a vertical or upright position no matter whether the press is upright or inclined to any angle within its limits. The pinion is always in mesh.

The distance between the uprights is somewhat greater than is ordinarily furnished on a press of this size. The slide is of liberal proportions and is carefully guided in substantial gibs, to insure the greatest possible accuracy.

The makers state that this machine can be furnished with or without gearing and with or without the bar knockout in slide and the direct connected lower lift-out, as well as with or without the motor bracket. It is also built with the frame arched out from front to back and with an opening 14 inches wide in each housing to permit of feeding the stock from right to left or vice versa. Following are its principal specifications:

Weight as shown, 6,500 lbs. Bed, 24 x 23 inches. Bottom of ram, 21 x 19½ inches. Opening in bed, 12 x 12 inches. Distance between housings, 24 inches. Stroke, 2 inches. Distance from bed to ram, stroke down and adjustment up, 13 inches. Thickness of bolster, 3 inches.

ELECTRIC GENERATING APPARATUS

The Dodge Manufacturing Co., Mishewaka, Ind., have placed on the market an electric generating apparatus that is of rather interesting design. The photograph herewith shows clearly its general construction, and this outfit consists of complete power units for electric generating. Tests have been made on these units in strict accordance with the rules adopted by the American Institute of Electrical Engineers.

The generating units are made up either direct connected or belted in all standard voltages using Dodge heavy oil engines as motive power. The sizes manufactured at present range from 12½ h.p. to and including 75 h.p. The wide range of sizes in these outfits makes the fact evident that the power units are equally adaptable to either marine or stationary power plant service.

Oil Engine Used as Prime Mover

The heavy oil engine used is said to be sufficient guarantee of low cost operation, which of course results in a great saving in the cost of current generation. The engine is sold under a standard guarantee which specifies a consumption of fuel oil of .5 lb. per B.H.P. hour, using fuel as low as .28 Baume, containing not less than 18,500 B. t.u. per pound. It is said that every engine turned out must use less than the amount guaranteed on actual brake test; otherwise such guarantee could not be made with any degree of safety.

The engine illustrated has a brake output of 50 h.p. and is direct connected to a 30 K.W. 220 Volt D.C. generator, running at a normal speed of 425 R.P.M. It would therefore use 10,000 gallons of fuel oil per year, if operated 10 hours a day under full load for 300 days. This fuel costs on the average of 5 cents a gallon and would mean a total of $500 for annual fuel cost. This reduces the fuel-cost of current to $.005 per K.W. hour.

Construction of Generator

The generator used has been developed especially for this particular service by the Engberg Electrical & Mechanical Works at St. Joseph, Michigan. The armature is of the iron clad ventilated type, with laminated core, being built of electrical sheet steel, thoroughly japanned before assembling. The drum and core are provided with air ducts, permitting a thorough circulation of air through same. Nothing but the very best double-covered magnet wire of the highest conduc-

tivity is used, and the insulation is thoroughly waterproof. A very desirable feature in connection with the armature is that it has been so designed as to eliminate a coupling on the armature shaft, the crankshaft coupling is direct connected to the armature drum, by such construction that the engine drives through the armature and not on the armature shaft. This method of construction is very substantial and makes it very easy to remove the armature—in fact this can be done without disturbing the engine in any manner whatever. Close regulation is assured by having the flywheel of the engine located close up to the armature and on the same end of the crankshaft.

The commutator is made of copper bars, insulated with the best grade of mica plate and is very heavy, thus insuring years of continuous operation without need of renewal. It is built up on a separate sleeve and bolted to the armature drum, so that the shaft can be removed without disturbing the windings. The bars are collected in a steel chuck, especially designed for this particular purpose, which is provided with large steel screws. The chuck is heated and allowed to expand, at which time screws are tightened to the very extreme. When the chuck cools and shrinks, the commutator is drawn into a perfect and lasting position.

Every part of both the engine and generator is of the highest possible grade in regard to material, workmanship and design. All bearings are extremely large and accurately fitted, thus assuring smooth and constant operation with comparatively little attention.

Both engine and generator are carefully tested and inspected at frequent intervals during the course of construction. Both are mounted on a heavy cast iron sub-base and dowelled in place, after which they are subjected to a final operating test under load.

GENERAL VIEW OF THE GENERATING APPARATUS.

The MacLean Publishing Company
LIMITED
(ESTABLISHED 1887)

JOHN BAYNE MACLEAN, President. H. T. HUNTER, Vice-President
H. V. TYRRELL, General Manager.

PUBLISHERS OF

CANADIAN MACHINERY
~ MANUFACTURING NEWS ~

A weekly journal devoted to the machinery and manufacturing interests.
B. G. NEWTON, Manager.   A. R. KENNEDY, Managing Editor.

Associate Editors:
J. H. MOORE   T. H. FENNER   J. H. RODGERS (Montreal)
Office of Publication: 143-153 University Avenue, Toronto, Ontario.

VOL. XXIV.   TORONTO, DECEMBER 9, 1920   No. 24

PRINCIPAL CONTENTS

The British Empire Steel Corporation

THE British Empire Steel Corporation has not yet become an accomplished fact. The proceedings were started months ago by which several Canadian companies were to be brought together, and in future operated in line with others in the Old Country.

There have been violent changes in the stock market value of many of the holdings that were to have been included in the merger in the first place, and it has been possible several times to go into the market and buy these securities at a less figure than that at which they were to be included in the Corporation.

It is probable that there will have to be considerable alteration before the British Empire Steel Corporation is successfully handled by the underwriters. There has been a lack of information in several lines that has hindered the work and shaken the confidence of Canadian investors who were identified with the plan in the first place.

Several of the leading companies in the plan of reorganization have become so far committed to the proposal that it is unlikely that anything approaching the disruption of the plans will take place, but it seems not only desirable but necessary that there should be a new basis worked out before the plan can be successfully consummated.

Looking to the future, the iron ore deposits, the coal beds, and the limestone of the eastern section of the Dominion should form a rare combination in the carrying out of industrial dreams of promoters and workers. The situation is unique, right on the seaboard and with ready access to the markets of the world.

If, with all that, they cannot in time turn out steel in competition with the steel companies of the world, there must be something wrong that does not appear on the surface.

Exchange Rate Again

THE exchange rate is again at a point where it is causing serious thinking, during the past few days it having touched 15 per cent. on several occasions. It is not an uncommon thing to hear of buyers asking for the longest possible period of time in which to settle for shipments from United States, not because they have not the means to make payment, but because they hope for a better turn in the exchange situation, and the gaining of a few points will mean quite an item to them.

Others, again, have been fooled by doing this thing. They have held off to the last, hoping for a more favorable rate, only to find that instead of coming down the rate of exchange has been going up, and that each day delayed has cost them more money instead of saving them a few dollars as they had hoped.

It is no doubt desirable that we should sell more to United States and buy less from the people there. That, in theory, is very good, but the fact is that there are many things, especially in the world of iron, steel and machine tools that we must secure there or at other points outside this country.

YOU may stay out of the market so long that prices will have touched bottom and started on the way up again before you act. Staying out of the market is a doubtful way of making a success of anything.

* * *

LESS than a year ago a man with a pile of steel sheets in his yard was sitting on them surrounded by a shotgun and an axe for fear some one buyer would insist on taking the lot. Now said buyer can come in unmolested and take the heap.

What the Business Papers Say

Time for Them to Get Together

"Syren and Shipping."—The question is, as has been rightly remarked by Mr. Launcelot E. Smith, the chairman and managing director of Smith's Dock Company, not one of asking the men to accept low wages. If they will look after production, wages will look after themselves. If, he says, the British working man and the British employer will get together they can beat the world, but the men are hampered by their unions and the employers by their federations. When an employer of Mr. Smith's character talks in this strain, our opinion is confirmed that it is up to both sections to tackle this grave problem in earnest, and at once.

Says Men Value Their Jobs Now

"Iron Trade Review."—Workmen have begun to value their jobs again and it is well that they may for their own good. Trade depressions are great stimulants to individual efficiency. Under such conditions only the really efficient may be reasonably sure of their jobs, this as workmen, or their ability to compete, as manufacturers. The working out of the law of practical economics now is to be seen on every side. In the end it means restored prosperity to the machine tool and other industries.

About the Power of the Future

"Power."—It is no longer a problem where to find a market for oil, but where to find the oil to supply the market. What are we going to do regarding our oil supply? Shall we wait until it is exhausted, as is the supply of natural gas, when it will be too late to act, or shall we join with the rest of the world in drawing up a world policy to safeguard the future?

Criticizing English Motor Cars

"Engineering."—It is, of course, a purely commercial question whether or not this is a wise policy, and it may be that there is room for all these expensive machines, and that more profit may be made in selling a few large cars than a greater number of small ones. Even this seems doubtful, for the number of people who are at the present time prepared to spend very large sums on motors is probably not great, and the moderate-priced machine has improved so much of late years that there is not such a very great gain in having an extremely expensive one.

A Difference in Methods

THE "Railway Age" gives statistics of the operation of American and British railroads during the first six months of 1920. The article was prepared by the Bureau of Railway Economics and the figures regarding the British railroads are the first of the kind ever available.

The average freight rate per ton per mile on the British railroads, which are still under Government control, was, in the first six months of the year, 26.3 mills, exclusive of charges for the collection and delivery of freight to and from the railroad stations. The average rate on American railroads in the same period was 9.7 mills, or about one-third as great.

The principal reason why American railroads can haul freight at an average rate only one-third as great as that of the British railroads is that they handle it in so much larger units—that is, in so much larger carloads and train loads. The average number of tons carried in each train by the British railroads was only 180, while in the United States it was 710.

The average number of cars in each freight train in England was 35, which is about the same as the average number of cars in an American freight train, but the average load carried in each car in England was only 6 tons, while in the United States it was about 28 tons.

It was, of course, this great difference in the average number of tons carried in each car which was chiefly responsible for the much larger number of tons carried in each train in the United States.

On Bein' the Boss

IN years when things was movin' fast and cash was stalking 'round, when clinkin' dimes and nickels was an ordinary sound—I used to think in them there days before we felt a pinch, that bein' the boss around the place was one outstandin' cinch.

Why, say, we never worried much, things seemed to hump along, with everything a-breakin' right and nothing ever goin' wrong.

His golf friends came in after lunch to coax him from the scene, to go and whack the little pill and clod-hop on the green. He gave his orders right and left, by heck, it was a cinch, this job of bossin' on the works before we felt the pinch.

And I looked forward to the day when they would pussyfoot to me, and say, "young man, this job of boss is henceforth handed unto thee."

Ah, then I would sit down in ease, a rift of wisdom on my brow, I'd keep an auto and a farm, a bacon hog, likewise a cow. My feet I'd plant upon the desk, my wealth would prosper inch by inch—oh, bossin' the whole shootin'-match would be to me a pleasant cinch.

But since things aint so fresh just now as they was 'bout three months ago, I'm not so certain of the fun of bein' the boss around the show. Some funny things turn up these days between the mornin' and the night, and I hear them tellin' 'round- the plant that money's gettin' powerful tight.

The boss aint quittin' early now, he's settin' tight around the place, they're stickin' grey hairs in his head and plantin' wrinkles in his face. There's papers piled upon the desk, enough to fill a good-sized cart, and he's figurin' this here thing and that, and squintin' hard upon a chart. He's sittin' there long after dark, and says we aint a-goin to bust, but he's puttin' on the pressure there and changin' off from may to must.

I use to think that bossin' job was one grand, easy, fat affair, but I aint a-hankerin' for them now to lead me forward to his chair.—Ark.

The Retroactive Award

THE manner in which retroactive wage awards work out in some cases makes a hardship that is very difficult for the company against which the award is made.

For instance, such an award was made in the case of the Nova Scotia Steel & Coal Co., in the case of the miners, and by the award the increase in the cost of producing a ton of coal was about $1.25. This does not mean that the miner got that amount extra for digging a ton of coal, but that the combined increases paid to the several men who handle the coal from the time it is taken from the working place in the mines until it is loaded and ready for shipment amounted to the $1.25 per ton on the output of the mine.

The wage award was made retroactive over a period of two months or so. The company, in some of its contracts covering the sale of coal, have a clause which makes the customer liable for this increase, and in such cases it is possible to recover the increase for these sales, but a great deal of coal is sold in the open market without any such clause.

When a manufacturer buys coal he uses it for his steam plant, it goes into the manufacture of some line, and is passed out to the trade in this way at a fixed price. The manufacturer has no chance to recover on the increase he has to pay, for his goods have been sold.

The coal company used at that time about sixty per cent. of its coal for its blast furnaces and open hearths, and the coal product was being sold to the customers of steel at current prices, with no string tied to the sale by which the buyer might find himself liable to an extra tax some two months after to pay for the increased cost of coal used in making the steel. In this case the coal company had no come-back.

An award that is made retroactive as a general thing leaves the company no chance to adjust its selling prices to meet the changed conditions, and is apt to impose a burden on its financial strength that it may not be able to bear without having to curtail seriously in some other direction.

THE scrap iron market is variously described by dealers as stagnant, dead, inactive, listless, lifeless, rotten, uninteresting, etc. Now, if any of our dear readers have a phrase that will nicely sum up these opinions, will he kindly shoot?

"I Wonder if he Has Really Reformed?"

MARKET DEVELOPMENTS

Buying Should Be Better in the New Year

Tool Buyers in New York Claim They Will Not Come Into the Market Until January—The Premium Market Has Simply Ceased to Exist—Exchange Once More a Serious Matter Here.

ONE might very well imagine that matters were dull indeed in the machine tool markets, and in the steel trade generally, were it not for the fact that occasionally a dealer comes along who has landed a very respectable order, and that at his regular price, without any special consideration. In the steel market it is to be expected that trade will be small in volume between now and the first week in January. The trade has always found it that way, and this year is no exception. Boiler tubes are much wanted just now, and that is the one line that moves best at this season. As soon as the boats tie up for the season - work is started on repairing.

The exchange situation is again causing some serious thinking on the part of firms having shipments coming from United States points. They are in many cases ask-ing for the longest possible time in which to take up the payment of their bills, hoping that the rate will turn a little in their favor. Some of them have been fooled by the recent rise, and their delay has cost them money instead of saving any for them.

The premium mills are getting very little business from the Canadian territory just now. In fact the premium market has ceased to exist. It may be some little time before the buyers here will get the full advantage of cheaper steel and iron.

The scrap metal market does not improve at all. Buy-ers are remaining out of the market. Dealers are not buying anything to speak of unless they have some con-tracts that have to be filled.

MONTREAL LOOKS FOR BETTER TRADE SHORTLY AFTER FIRST OF THE YEAR

Special to CANADIAN MACHINERY.

MONTREAL, Dec. 9.—Business at the present time seems to be regulated more by financial conditions than by the general trade requirements. There is little doubt that trade in general would be more brisk if the difficulty of financ-ing was less pronounced. The problem of economic adjustment is one that re-quires considerable initiative, and cour-age on the part of the manufacturers, as under existing conditions it necessitates carrying on, either at a loss or fore-going all profits during the period of clearing out old stocks and replenishing at the new basis. Even this policy is, enshrouded in great uncertainty, as the declining prices offer little guidance as to the ultimate stopping point. This in-fluences the trade to such an extent that buying all along the line is curtailed or restricted to the point that stocks, which normally would be disposed of in a rea-sonably short time, are carried over, with the result that further buying is postponed, the consequence being a tem-porary suspension of activity in many lines of industry.

Local Price Revision Expected

The revision of American steel prices to lower levels has not resulted in any pronounced activity on the part of con-sumers. It will probably be some little time before mills are turning out steel at the new base, but the movement of price reduction appears to be general and it is now a question whether the lower quo-tations will open the channels for re-newed business. "We are now in a pe-riod when it is usually expected that business will be slack, and at the present time this dullness is emphasized, particu-larly when compared with that of the past several years, when the dull pre-winter months differed very little from other periods of the year. However, when the present is compared with corre-sponding months in pre-war days, there is very little to show where a striking difference can be drawn." This state-ment by a dealer here indicates that gen-eral business is quiet, but that too much stress is placed on the relative falling off between the activity created by abnormal world conditions and those which should obtain had the commercial trail been blazed in the ordinary peaceful manner.

"It is logical to suppose," continued the dealer, "that a transient period must be passed through before a sound work-ing base can again be established, but this will slowly right itself, and probably before spring we will see a healthy re-vival of industrial enterprise, but it is only after the tide begins to rise that full confidence will return and open the way to renewed activity. We are anticipating a revision of prices, but it will probably be the first of the year before these changes will be announced. It will be some time before American prices will affect this market, and in the meantime we will continue to quote present prices on warehouse material."

Little Demand for Machine Tools

General lessening of activity through-out the district has reacted to some ex-tent on machine tool dealers, as reports are to the effect that business is more than seasonably quiet. In some respects it may be stated as exceptionally dull. Many of the factories in this lo-cality have made reduction to their operating staffs, and while nothing definite can be learned as to the duration of curtailment, it is not thought likely that any marked effort will be made to replace or install new equipment until business has again resumed some of its former activity. The problem of selling machinery and supplies is be-coming real work, as users are not as easily persuaded to the advisability of renewing tools. This attitude on the part of the consuming trade will likely mean a period of several months when buying of equipment will be restricted to the most urgent needs. Manufactur-ers realize that early adjustment is im-perative, and this condition, while tem-porarily holding trade in suspension,

must obtain during the transition period. Dealers are far from being pessimistic on the future outlook, and many are confidently making plans for the coming year, in the hope and belief that before the spring the reversal of the tide will take place and the stability of the market be established. "We are a little reluctant to revert to former prices, but it is probably the best move after all, as such action is an essential factor in solving the problem of supply and demand. At the present time the trade is kept from buying by the weakening tendency of all commodities, a condition that is not conducive to active buying. This remark from a dealer apparently explains the prevailing dullness in many quarters.

Regular But Light Movement

The metal situation is keeping step with other lines of activity in gradually relinquishing the high hold on prices that have been the order of the day for many months. These lower quotations are the result of trade adjustment as much as to the falling off in demand. There is still reason to believe that solid ground has not yet been reached and further sinking will likely be reported. Local quotations show a further decline this week. Tin and aluminum are down one cent, the quotations being 44 cents and 34 cents respectively. Other ingot metals are down from ¼ to ½ cent per pound; electro copper is now 18½ cents, spelter 8½ cents, lead 7½ cents, and antimony 8 cents per pound. Movement of metals is light but quite regular.

Interval of Inactivity

Old materials are in a position where the market is conspicuous by the entire absence of any active interest. Dealers are unanimous in stating that there is "nothing doing." This applies generally, with the exception of such local trading as is necessary to carry on the regular business of the foundries in the district. Even this activity has been below normal for some time. Dealers are not particularly interested in prices, as the small demand offers little opportunity of reselling the material, and therefore they are not desirous of accumulating any excessive stock in face of the weakening market. The prices quoted are a fair average of what dealers here are paying but actual transactions would probably be carried on at figures slightly lower.

THE EXCHANGE IS BOTHERING THEM

Buyers Like Long Terms on U.S. Goods to Take Advantage of Any Drop

TORONTO.—One might imagine that the machine tool trade in this district were very quiet were it not for the fact that now and then on the way around the machine tool houses good mixed orders are heard of. There is bus-

POINTS IN WEEK'S MARKETING NOTES

Large buyers have advised the machine tool trade that there will be no buying until the first of the year.

New York reports indicate that makers of machine tools are in no hurry to announce any reduction in price, believing that such a step would not bring any new business into the market.

Some of the American sheet mills are opening their books for new business now, but the Canadian sales office of the Corporation has not yet taken on any new business.

There is a rumor on the street to the effect that the price of sheets in Canada from American mills will be the same as the price at which they are sold in the United States, but there is no official confirmation of this report as yet.

The United States Steel Products announce that Canadian business will now be regarded as domestic by their mills. This comes back to the same position that they occupied some months ago, before the increase was put on Canadian orders.

The average independent steel mill in United States is not operating at much over 50 per cent. of its capacity at the present time, while the Corporation is running at 85 to 90 per cent.

Makers of pig iron have very little business on hand, and a number of the furnaces in the United States have blown out during the past week.

On account of the high exchange rate, many buyers who have shipments coming in from the United States are asking for as much time as possible in the hope that the exchange may be a little more favorable to them.

With the prospect of securing more power, dealers are looking forward to an improvement in the sale of motors, both for group and individual drives.

iness moving, although it is not as good as it was some months ago, and those who study matters carefully believe that it is going to be better shortly after the break of the new year.

One of the Toronto men speaking about the matter this morning said, his

principals had announced no change in their selling prices, and if any were to be made at the first of the year the chances were that he would be advised of it by now, or have some tip that it was coming, although the policy of some houses is to send out their changes immediately. Some of them believe that prices of tools that are made up and in stock are such that they cannot cut them much, and that if there is to be a new price they will have to secure labor and material at a lower price.

The prospects of a better supply of power in this district will be reflected in the sale of motors very shortly. There is undoubtedly a marked preference on the part of many manufacturers for motor drive for group or individual machines, but it has not been possible to secure power enough, so the sale of motors of the voltage sent on the Hydro line has been much curtailed.

In the Steel Market

Few changes are announced in the steel market this week. There seems to be some little doubt as to what extent the new prices apply to Canadian trade. The Carnegie mills, for instance, announce that in their products, such as structurals, bar mill material, plates, etc., they will now regard Canadian business as domestic. This ruling applied from the first day of December, and all orders that were on the books before that time are not put down at the new price. This ruling is satisfactory to the trade, as it had been expected that the old price would stand where a definite figure had been named on a contract.

The American Sheet and Tin Plate books have been opened for business in the States, but so far their Canadian office is not taking on business, although it is expected that this will be the case here in a few days. Neither is there any official confirmation of the rumor that they had decided to regard Canadian business as domestic trade, although on the other hand it would cause no surprise were this to be confirmed.

Some of the warehouses are not ready to take on all the shipments that are being made in their direction. One of the dealers stated his case as follows: We have been month after month waiting for delivery of our orders, particularly in sheets. Much of this was seasonable business, and now that stuff is coming through faster and the demand is lower, there is a danger of them swamping us with material. This does not mean that we are going to cancel, for we want to place the material in stock, but it does mean that we are not going to allow the mills to take advantage of the situation now that deliveries are more in their favor. As long as they spread their deliveries over a period, well and good, but there is not going to be any smothering done if we can prevent it."

The Matter of Prices

Prices remain much the same in the local market. Galvanized sheets are not by any means plentiful, the mills

having been turning out only a limited tonnage.

Black sheets are not a plentiful article, and this condition is not likely to improve, as the mills are not turning out many of these. The independents are largely down to a low rolling rate per day. They have very little business on their books and are reducing staffs in keeping with the amount of business that is offering. In some lines the advantage still rests with the independents, because they are in a position to give almost immediate shipment out of stock, whereas there are many lines on which it is still, impossible to book business with the Corporation.

Tubes are in good demand now, and a brisk business is being done. Of course there is nothing unusual in this, as it has been the same year after year. As soon as a ship ties up for the season work is started on repairing, and for the next two months the tube business ought to be a pretty fair trade.

Bar iron is selling well, but perhaps the most active line in this section is band iron, the stovemakers having requirements in this line at present.

The Matter of Exchange

Now that the exchange on Canadian funds in New York has gone up over the 15 per cent. mark, there is a tendency on the part of many buyers to urge as long consideration as possible, in order that they may have a chance to get something better a little later on. Some of them have recently been fooled on this, because exchange, instead of crawling down has shown a marked tendency in the other direction. Some of the firms selling steel in Canada give fairly long terms of payment, and at any time in that period the buyer can take advantage of any drop in the exchange rate and make his payment against the account. Especially now with the rate high again, buyers are watching every day to see if there is any advantage in their favor in the changing quotations.

black sheets, 28 gauge, 4.35c; galvanized sheets, 28 gauge, 5.70c.

End of Independent Market

The famous and spectacular independent market for steel products has thus come to a mute and inglorious end, excepting in the case of pipe. No one, perhaps could have imagined the independent market giving out so quietly. It can hardly be said to have done so active and energetic a thing as to die. It simply faded away. There was no important business offered as an incentive. The independent mills that reduced their prices first did not get any business to speak of by the operation, nor has buying been stimulated to any appreciable extent. Such wants as the country has seem to be supplied almost wholly by the Steel Corporation which so far as can be ascertained is operating as well as at any time, say at between 85 and 90 per cent. It must be that a large number of buyers have been on the corporation books, making purchases from independents simply as the corporation's allotments to them were insufficient. With respect to much of the country's business, then, the independents were simply getting the overflow.

Independent Operations

The different independents are operating at various rates, and there is also much starting and stopping from week to week, so that a close estimate of the general average rate of operation by the independents is difficult. The common view seems to be that the independents are not operating at an average of over 50 per cent. In slack times it is a common practice for a mill to close for a time, to allow orders and specifications to accumulate for a moderately full run, and it is considered probable that many mills will seize upon the last fortnight of the year as a good time in which to close. Thus the year may end with the independents' almost altogether closed. As for the Steel Corporation, it looks as though it would be able to maintain a moderately full rate of operation for months even if there is no improvement in the general situation.

Manufactured Steel

Declines in the hot rolled products are naturally working their way through to the manufactured products. Cold finished steel bars have come down from 4c to 3.60c. Bolts and nuts have been reduced various amounts, averaging about 20 per cent. Rivets have not been formally reduced. For some time there has been shading of the old official price of 4.75c to 4.50c and 4.35c, but a reduction in keeping with the declines in hot rolled steel is expected to carry the market down at any rate to 4c. Chain is still nominally at 7.25c.

Hot and cold rolled strips have been reduced a cent a pound from the old official price and a half cent from the market actually existing in the past two or three weeks, the new prices being 4c

THE PREMIUM MARKET DID NOT DIE— IT SIMPLY WENT AND FADED AWAY

Special to CANADIAN MACHINERY.

PITTSBURGH, December 9.—Two more important steel mill products, wire and sheets, have come down to the Steel Corporation or Industrial Board level, and only one large line, tubular goods, remains. The readjustment in pipe may not come for quite a while, for pipe continues in excellent demand, with heavy requirements against contracts and some new buying of no small consequence in the open market. Pipe stands quite alone among finished steel products.

In the wire product readjustment one large independent interest took the lead, reducing its prices to 3.25c for plain wire and $3.25 for nails, these being the Steel Corporation prices. The $3.25 on nails is an Industrial Board price. The 3.25c on plain wire is an advanced price, the Industrial Board price having been 3.00c, the American Steel & Wire Company having advanced its price on wire last August by $5, leaving nails unchanged. While it may seem strange to have wire and nails on the same price basis, the differential has never reflected the actual cost of manufacture. For many years it was 10 cents, and for a short period long ago it was only five cents. The 25-cent spread in the Industrial Board schedule was a relic of the war time control.

The other independent wire interests showed no alacrity in following the lead of the first independent. This indifference is to be attributed to the fact that there is scarcely any demand in the open market, hence little temptation to engage in competition, while perhaps it was thought that some of the contract tonnage remaining on books could be saved. One independent, for instance, offered to adjust 4.25c nail contracts to 4.00c. Another reduced its nail price from 4.25c to 3.75c instead of to 3.25c, although it did make the full reduction, to 3.25c, on plain wire.

The Decline in Sheets

At the time of last report the · sheet market, on a really desirable order, was practically down to a level just $10 a ton or a half cent a pound above the Steel Corporation prices, although no mill was openly quoting altogether so low. For a few days the position of several mills was that while they were quoting $10 a ton above the corporation prices they were willing to do the corporation prices if they were offered a good sized order, say 1,000 or 2,000 tons. The difficulty was that no such business appeared in the market. At this writing it is clear that at least one mill would quote the Steel Corporation prices even on an ordinary order, not waiting for an impossibly large order. Hence the independent sheet market is now quotable at the Steel Corporation level.·

Late last week the American Sheet & Tin Plate Company opened its order books for the new sales period, nominally the second half of the year in the case of regular manufacturing consumers, but practically for the second quarter, since about three months of work has to be carried over from this year's contracts, the holdover being due to the company not having been able to operate at capacity on account of steel and fuel shortage and car shortage, at various times over the past six months and more. The prices are of course as formerly: Tin plate, $77 per base box for 100-lb; blue annealed sheets, 10·gauge, 3.55c;

If interested tear out this page and place with letters to be answered.

for hot rolled and 7c for cold rolled. The producers did not expect to get any new business of consequence by the reduction, and state the reduction was made chiefly for the purpose of adjusting contracts on books at higher prices, so that customers would be free to file additional specifications.

Foundry pig iron is still quotable nominally at $37, valley, and perhaps this is so chiefly because there has been no inquiry to tempt producers to make lower quotations. On quotations made voluntarily by furnaces rather than in response to important inquiry Bessemer is down $2.50 to $35, valley, and basic $2 to $33, valley. It is possible that these prices will hold for several weeks. No great decline, at any rate, is to be expected until there is enough business developed to promote active competition. As matters stand the furnaces are rapidly running out of business and have nothing to do but blow out. Of the 13 merchant furnaces in the Mahoning and Shenango valleys seven are out and the six still in blast are referred to as being "on the ragged edge." Perry at Erie is out.

BUYERS ARE OUT OF THE MARKET
Special to CANADIAN MACHINERY.

May Come In at First of the Year if any Inducements are Held out to Them

NEW YORK, . December 9.—(Special to Canadian Machinery).—Most of the large prospective buyers of machine tools, which includes the railroads and companies such as the General Electric Co., have advised machine-tool dealers and sales representatives that there will be no buying until after January 1. They do not even promise for a certainty that there will be buying at that time, but the intimation is given that if other conditions, such as credit and prices, are satisfactory they may do some buying early in the new year. The General Electric Co. has plans for continued expansion, but these plans have been held in abeyance for many months. Quotations have been received by this company in the past several months on scores of tools, which have not been bought, and if orders for the entire number inquired for should be issued early next year it would have a very beneficial effect in stimulating the market.

November was an exceedingly dull month, and there is even less buying so far in December. With the inventory period so close at hand, it is natural to expect that even urgent requirements will be put off until the new year.

Little has been heard lately regarding prices. There are so few buyers in the market that the subject is not brought up with the same frequency and insistence that prevailed a month or so ago. Machine-tool builders are principally concerned in getting their present orders shipped. There is a good deal of talk about an adjustment of wages for labor before the machine-tool industry starts out on a new price basis.

MANY CHANGES TO BE MADE IN BRITISH EMPIRE STEEL CORPORATION

A REPORT in a Montreal paper says that the British Empire Steel Corporation is to be dropped entirely. The paper goes on to say:

"The probability of this being the fate of the venture has been a subject of much discussion in financial circles for weeks past, but in circles close to merger interests no admission in the connection was available.

"It is now generally admitted both in Montreal and London circles close to the company that it will be impossible to carry out the financial arrangement necessary to the completion of the contracts with the various units of the merger, and that in consequence it will be impossible to proceed.

"On the other hand it is generally understood that an alternative scheme will be an arrangement to bring Dominion Steel and Nova Scotia Steel together, but all the other units are to be dropped.

"The proposition was one of the most ambitious ever undertaken in connection with Canadian industries, and the need of letting the matter drop will be heard with genuine regret by many Canadians who believed that as a constructive scheme it had many fine and attractive qualities, particularly with respect to the support it promised to bring from the Mother Country.

"The programme had been carried through up to the point where all that practically remained to be done was to bring out the issue of $25,000,000 of 8 per cent. preferred stock of the corporation, and the money market in London has been most unreceptive for a long time, and the issue was put off from time to time in the hope that it would improve."

This report is not in keeping with that received by Canadian Machinery this week from one of the officials connected with the deal from the very start. He admitted that there were obstacles in the way which had prevented the underwriters from finishing their work, but those connected with the deal had committed their companies to it to such an extent that it would be impossible to call the deal off. It seems that the details will have to be worked out on a new basis as the figures used in the first instance have been found to be unsatisfactory. The official giving information to this paper did not know for certain when the deal would go through, but intimated that for the present there was nothing being done.

Equipped at Cost of a Million, Will Cater to the Needs of Western Canada

The first open-hearth reverberatory furnace in Manitoba recently commenced operations at Selkirk, about ten miles northeast of Winnipeg.

The plant, which represents an investment of about $1,000,000, is known as the Manitoba Rolling Mills, and its equipment is said to be of the most modern type, consisting principally of one 15-ton furnace, three 15-ton ladles, crane, dumps, molds, and other machinery. Electricity is supplied by the Winnipeg hydro-electric stations.

It is expected that the surrounding territory will furnish an ample quantity of steel scrap.

The fuel used by the new rolling mill is Canadian bituminous coal, crushed into 1-inch cubes, elevated into a hopper, passed through a dryer, and from there into a pulverizer.

Orders have been received which will engage the output for several months, it is reported, but shortage of labor, high freight rates, and the scarcity and high cost of raw materials imported from the United States are said to have hindered capacity production. It is expected that the output of this plant will supply in a measure the demands of the market in Western Canada and that additional units will be equipped as fast as business warrants.

MALLEABLE IRON MEN STATE CASE

A Basic Industry, and Its Worth Proved By Experiences During the Great War

Presenting a memorandum on behalf of the various Canadian manufacturers of malleable iron, F. L. Storey addressed the Tariff Commission, Mr. Storey pointed out that the American industries of a similar nature, with their immensely larger market, operate under a condition of mass production which is non-existent and unattainable with the Canadian foundry in sparsely-populated Canada. If the tariff were removed these United States plants could reduce the selling field of Canadian manufacturers by their competition. Another handicap against the Canadian foundry is that the laid down costs of raw materials are greater than those of the U.S. foundries. This extra cost arises from the duty which the Canadian manufacturer has to pay on pig iron, coal and fire brick and the extra freight to Canadian foundry points.

He submitted these figures in respect to the Canadian malleable iron industry: Capital employed, $2,605,000; wage roll (yearly), $3,410,000; output, $6,816,000; employees, 2,634. "This," he argued, "is a key industry, and the war has taught each country the necessity for its protecting and preserving its key industries."

CANADIAN MACHINERY

AND
MANUFACTURING NEWS

VOL. XXIV. No. 25

December 16, 1920

535.

Six Examples of Simple Chucking Devices

Chucks for Machining Small Washers, Facing Rough Nuts, Machining the Ends of Short Threaded Parts, Drilling Small Holes at Right Angles to Main Axis, and Other Parts

By F. SCRIBER

DRAWINGS of a number of chucking devices are here shown. These were originally designed for lathe use, but slight modifications would make the same adaptable for turret or screw machine service. The chuck shown in Fig. 1 is used for machining small washers on the face. It consists of a spring holder A, bored to take the washer and split in one place only. The piece B is a sliding fit on the body D, and is threaded to engage with the knurled tightening nut C. This nut is grooved in the shoulder at the rear. A hole is drilled in line with this groove, so that when the nut is in position on the arbor, the locating pins may be driven into place, thus allowing the nut to turn freely without any lateral movement. Work of any desired diameter, within the range of the chuck, may be done by making suitable spring holders. This particular type of chuck has a comparatively long life as it does not rely on screw threads for maintaining its concentricity.

The piece shown in Fig. 2 is more in the nature of an arbor than a chuck. It was designed to be used in facing a large number of black or rough nuts square with the thread. In order to do this it was essential that the nuts should tighten up on the thread and not against the opposite shoulder. It will be noted that the arbor proper has a slightly tapering thread on which the split collar is located. The nuts are threaded on the collar by hand and the collar is turned by means of a pin wrench, until the nut is firmly held. When machining the end of short threaded pieces, threaded studs, etc., it is often difficult to get them off after the machining operation. The chuck shown in Fig. 3 was designed to overcome this trouble.

The work when screwed into the chuck, bears against the pin A (which is a sliding fit in the hole) and in turn, this piece presses against the pin B, which is inserted at right angles and with a nice turning fit, threaded on end and provided with a square on the other end for using a box wrench. The pin B has

a portion in the centre filed in the form of an eccentric so that the work can be easily released by simply turning the pin after the machining has been performed. The small thread on the end of the pin B is simply to retain it in position. In Fig. 4 is shown a simple but serviceable chuck for use when drilling or boring small holes in round stock at right angles to the main axis. A chuck of this kind will take care of a great variety of work, such as the enlarging of holes or the turning of odd shapes in previously drilled holes.

When a large number of short round pieces require to be centred the chuck shown in Fig. 5 will be found to be very handy. The chuck is bored to take the desired diameter of stock and is slotted on one side to enclose the lever A. Before slotting, a hole is drilled at the forward end to take the small brass pin B. This pin is provided with a slot in which the short end of the lever operates; this slot must not be cut too wide as there is a possibility of the pin falling out when work is removed, or interfere with the

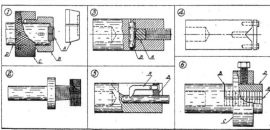

THIS GROUP SHOWS THE SIX IDEAS; EACH OF WHICH THOUGH SIMPLE IS VALUABLE.

work when placing in the chuck. When pressure is applied to the outer end the inner end presses against the tail of the lever, thus forcing the pin B down upon the work. The advantage of this chuck is that the machine need not be stopped to change the pieces of work, as the work can readily be removed when pressure is released.

The chuck shown in Fig. 6 is used when drilling or reaming a large number of washers, several of these being placed in the chuck at the one time. When making the chuck the centre hole is first drilled and bored to the desired size and afterwards split as shown, so that the top half may be removed to facilitate loading and unloading. The ring C was bored to fit closely and drilled and tapped for the tightening screw. When loading the clamping ring is released and slipped back out of the way. In a chuck of this description the operator has a clear view of the work as the pieces are placed in the holder. All of these chucks can be made from discarded material about the shop and frequently eliminates the necessity of securing castings for the making of impromptu equipment.

Development of 8 Driving Wheel Locomotive
By J. SNOWDEN BELL

IN the December 2nd issue we touched upon certain phases in the development of the eight driving wheel locomotive. This article is the conclusion of the story. Readers have no doubt followed this development with interest, and it might be well to add that we secured the data from record No. 89 of the Baldwin Locomotive Works. Let Mr. Bell continue with his story.

A comparatively recent locomotive of the type, engine No. 19 of the Lake Superior & Ishpeming Ry. (Fig. 12) indicates the great increase in size and power which has been made in this type since its origin in 1864, and development in the "Consolidation" of 1866. This engine was built in April, 1916, and its dimensions are as follows: Weight, 268,-000 pounds; on driving wheels, 238,000 pounds; cylinders, 26 x 30 inches; driving wheels, 57 inches diameter; boiler, 88

inches diameter; heating surface, 3643 square feet; (superheater, 844 square feet); grate area, 58.7 square feet; boiler pressure, 186 pounds; factor of adhesion, 5.25; maximum tractive power, 56,000 pounds.

"Mikado," 2-8-2 Type, 1897

The next succeeding step in the development of the eight driving wheel locomotive was the addition of a two-wheeled trailing truck, this wheel arrangement characterizing what is known as the "Mikado" or 2-8-2 type, which was first applied by The Baldwin Locomotive Works in 40 engines built by them for the Nippon Railway of Japan, in March, 1897. The advantage of this type, in affording a substantial increase in boiler capacity, was immediately and generally recognized, and it has practically superseded the 2-8-0 type in heavy

freight service. Many 2-8-0 engines have been rebuilt in conformity with the 2-8-2 type.

Fig. 13 represents a high power 2-8-2 type locomotive which is one of a lot built by the Baldwin Locomotive Works for the Chicago, Burlington and Quincy R.R., in June, 1917. This locomotive weighs 314,700 pounds; weight on driving wheels, 239,200 pounds; cylinders, 28 x 32 inches; driving wheels, 64 inches diameter; boiler, 88¼ inches diameter; heating surface 4,465 square feet; (superheater, 1031 square feet; grate area, 78 square feet; boiler pressure, 180 pounds; factor of adhesion, 3.98; maximum tractive power, 60,000 pounds.

"Mountain," 4-8-2 Type, 1911

The latest design and probable limit of capacity of the eight driving wheel locomotive appears in the "Mountain" or

FIG. 12.—THIS ENGINE WAS BUILT IN 1916, AND IS USED ON LAKE SUPERIOR AND ISHPEMING RAILROAD.

FIG. 12—A HIGH-POWER 2-8-2 TYPE LOCOMOTIVE BUILT FO R THE CHICAGO, BURLINGTON AND QUINCY RAILROAD.

4-8-2 type, in which a four-wheeled leading truck is substituted for the two-wheeled truck of the 2-8-0 and 2-8-2 types and which was originated by the American Locomotive Co., in locomotives built by them for the Chesapeake & Ohio Ry., in 1911, for passenger service, in which they have hauled trains of 645 tons over grades of 1.82 per cent, at a speed of 26 miles per hour.

The heading of this article illustrates one of a lot of locomotives of the type, built for the Southern Ry. by the Baldwin Locomotive Works in 1917. The principal dimensions are as follows: weight, total, 314,-800 pounds; weight on leading truck, 53,800 pounds; weight on driving wheels, 209,800 pounds; weight on trailing truck, 51,200 pounds; cylinders 27 x 28 inches; driving wheels, 69 inches in diameter; boiler, 76½ inches diameter; heating surface, 3,568 square feet; (superheater, 942 square feet); grate area, 66.7 square fet; boiler pressure 190 pounds; factor of adhesion, 4.38; maximum tractive power, 47,800 pounds.

Except as to the greater safety of a

four-wheeled leading truck at comparatively high speeds, and the capacity of a slightly longer boiler, the 4-8-2 type presents no substantial advantage over the 2-8-2, while on the other hand, it is subject to the objection of reducing the proportion of weight available for adhesion. Thus, the 4-8-2 type for the Southern Ry. carries 66½ per cent. of its total weight on driving wheels, while in the 2-8-2 type for the Burlington, illustrated in Fig. 13, the proportion is 76.0 per cent. The speed of passenger trains on mountain grades is not so high as to render a two-wheeled leading truck unsafe, and the possible increase of boiler capacity in the 4-8-2 type is not great.

The exceedingly great increase, shown by the foregoing review, in the size and power of the eight driving wheel locomotive over the Baldwin direct coupled engines of 1846, which may properly be considered as the earliest instances of the designs that were adapted to continued practical railroad service, has, of course, been accompanied, and rendered economically available by the numerous

improvements in structural features and accessories which have been embodied in locomotives of the various other designs that have from time to time been introduced. The advantage of the design, as practically perfected in the improved types in which it has been applied, is unquestionable, and other than for exceptional conditions of service it will doubtless continue to be the preferred one for freight train work.

AN EXPLANATION

Through a regrettable error in, our Advertising Copy Department engravings were taken from an article on Die-Castings, which showed examples of work done by the Franklin Die-Casting Corp., and the cuts were used in the advertisements for the Fisher Motor Co. The Franklin Die-Casting Corp. have written us stating that they are the originators of the process described in the article, and that the Fisher Motor Co. are in no way connected with the Franklin Die-Casting Corp. either by license or otherwise. We have pleasure in making this explanation and correction.

AN 8 COUPLED, FLEXIBLE BEAM TRUCK LOCOMOTIVE BUILT FOR SABANILLA AND MAROTO RAILWAY, CUBA, IN 1866. THE WEIGHT IS ABOUT 43,000 POUNDS.

Canadian Firms Discuss Their Production

Some Firms Find That the War Shop Still Exerts an Influence—
Many Forces Have Been at Work in Recent Years to Break Up
the Organizations in This Country

TO what extent has production fallen off in Canadian factories during the past three or four years? This paper put the question to a number of manufacturers, holding that if such were the case a remedy should be found, and if, on the other hand, production had not dropped off, it was not right that the Canadian workmen should be credited with having anything to do with such a condition. Some of the information was secured a few weeks ago and conditions may have changed somewhat since, but there is, in the main, a very fair analysis of the whole situation.

One Kitchener shop says: "Unfortunately we are not in a position to give you much information regarding this point, as our line of manufacture is such that we have no accurate means of determining just what the production per man is, as compared with years past, and are not able to give you particulars regarding this point.

"The situation with us, however, has been such that the housing accommodations in our city were totally inadequate and we have been obliged to employ a great many young unmarried men. These we have found would not work steady, owing apparently to the high wages which they were receiving. This has caused us a great deal of trouble and has reduced our production very materially.

"With the change, however, in labor conditions we find a better class of help available and are gradually replacing men of this type. We regret that we cannot give you any more information, but shall be interested in the results which you obtain from various firms from whom you are asking for information along this line."

Goes Over the Whole Situation

One Canadian manufacturer discussed the matter at considerable length. He has been a careful observer of passing events, and looking back over a period of several years, says:

"You have raised a large question. Almost any manufacturer will tell you offhand that production per man has fallen off sadly in the last year or so, but it is hard to lay one's finger on the exact reason for this. General Booth, on his return to England after a world tour, recently stated that what the worker wanted, in his opinion, was not a bigger wage, but the opportunity at all times to do less work. From my own observation, I am in agreement with the General.

"Plants have changed so in the past six years that we have literally no standards to guide us in an impartial judgment of what constitutes a real day's work. Before the war there was always some record which had been set up at one time or another by a highly skilled operator; this passed into shop history, but became a standard by which performance and pay were unconsciously regulated.

"Then came the shell business, with the consequent breaking up of organizations by enlistment and the bigger pay offered in the new shell shops. We have all been obliged therefore, to enlist any help we could get hold of to keep the plants running at all. The men who were known as steady workers before the war were absorbed immediately and gladly, but the waverers before became the shirkers after. This, mark you, was not entirely their own fault. They had lived for five years under abnormal conditions, having no responsibility, just doing what they were told to do, and living from day to day. They had no fear of not being clothed, or of having their meals at some time or other.

"When the trade boom of June, 1919, came along, factories were just beginning to settle down with a fairly complete organisation. The motor trade in particular, and allied industries also, were in dire need of all sorts of help

and commenced to offer inducements to get meh into their shops. This again broke up the working forces of all factories. Here in Western Ontario we felt that we were merely feeding the factories of Detroit. The labor turnover was out of all proportion to the force employed, and most factories were left with the mass of their workers half trained, and consequently, production per worker was again cut. Here again the question of standards enters. With an almost new force, there can be of necessity no actual standards. It is a matter of laborious training to secure any.

"It is unfair to compare the production of semi-skilled men with that of old workers in any given industry. However automatic machinery may be, there still enters the personal element and aptitude to any particular job. But in our own particular industry, we are only getting a little more than half the production we were before the war on the skilled and semi-skilled jobs. We introduced a bonus scheme, but the necessary proportion of labor to output was never attained.

"Perhaps the experience of this factory has been unusual, as all our help has had to be trained from the raw, so to speak, but the results have been as noted above and production has been far below that of pre-war days. I believe the men and women we employ are doing their best, as far as they know, but for the reasons briefly outlined above it is not a very good best. I do not believe that the majority of our helpers are wilfully trying to hold up production, but the era of higher wages has unsettled them to the extent that they are not anxious to learn thoroughly any particular job.

"With regard to coming into close contact with men, in normal times this is not difficult in small towns, as the outside interests of Quebec employees and employees move along the same lines. In lodges and churches, sports and clubs, they meet on the same ground and have very often been brought up in the same school. But latterly we have had a large percentage of "floaters" with whom we have nothing in common. These men do not enter into the social activities of a town at all. An employer has therefore no means of judging these men except by their performance in the shop, and for some months can get no 'line' on them."

Labor Is Inefficient

A manufacturer in the steel trade in one of the industrial centres of Quebec tells his experience as follows:

"Our experience is and has been for some time that labor is very inefficient; production per man has fallen from 30% to 40%, and is about on that basis now as compared with pre-war conditions. We have seen it stated that in some plants production per man was increasing, but this has not been our experience. We can see no improvement, but possibly it may reach us later. Conditions with us are probably about the same as they are with other manufacturers. Labor is inefficient, careless and independent. Spoilage is much greater than normal because of carelessness of labor.

"We have made use of no particular method except to at all times talk with our men on the desirability of increasing output and becoming more efficient. We do not pay bonuses, but advance our men regularly as they improve. There is no reason that we know of why production should be more difficult, in fact everything tends to the contrary, as we are always on the lookout for new machinery and new tools, and our plant is probably the best equipped plant in Canada for the manufacture of the line of goods that we market. Shop conditions could not be improved; living conditions are above the average. We

are inclined to think that with the gradual decline in the cost of living, and the fact that a great many plants are reducing their forces, in a short time there will be a surplus of help and that they will be more efficient."

Piece-Workers the Same

A St. John, N.B., firm, discussing the matter, has had the following experience: "In our finishing shops we could not say that we have had occasion to find any decrease in production per man, as all our work of this nature, to a large extent, is handled on piece work; the same also applies to our moulding shop. However, with our laborers we have noted a remarkable decrease, but at the present time, indications seem to point for an improvement in this class of work."

Supply and Demand

This firm is in a good industrial district in Ontario and the experience would no doubt be that of many others: "In a plant the size of ours we have not, the figures to prove the actual falling off in production that may have taken place in the last number of years. In parts of our plant where a premium is paid for increased production the falling off is not so noticeable, but in the parts of our plant where we have not been able to do this we get considerably less production per man than we did before the war.

"The situation arose, of course, out of the very great need for men during the war and the lack of a sufficient supply. We think that the condition in our plant is a little better now than it was a few years ago, owing possibly to the fact that business is not so brisk and we may need to lay off help before many months have passed. They seem to sense things such as this and naturally we get a little better effort on their part.

"The whole question, to our minds, simmers down to a question of supply and demand and we think chances are that within the next six months all concerns in the country can expect much more efficiency than they have been favored with before."

Piece-Work the Best

A Nova Scotia firm believes that piece-work is the best plan to keep production up, and gives its experience as follows, this being an engine and boiler industry: "We would advise that in our opinion there has been a decided slowing down of the production per man as compared with pre-war conditions. We have found that this can be remedied and almost completely overcome by introducing piece-work and offering the men an inducement for speeding up. We are introducing piece-work now in almost every department, and the result has been most surprising. In some cases we have actually reduced our costs to one-third of that previously prevailing by this means."

Compares 1918 and 1920

This firm is in the foundry business in a Quebec centre and apparently has taken some trouble to keep in touch with past performance on the part of its force. Its experience is as follows:

"In reply to your letter we wish to say that a few days ago we took up the question of production of our plant for the month of August in the years of 1918 and 1920. This showed a falling off of nearly 20% and this indicates that there is considerable truth in the statement which seems to be general that labor is not producing to the same extent as prior to the war. The production in a foundry producing miscellaneous castings is naturally affected by the class of work handled in the periods under observation, and as we are now making more miscellaneous castings than in 1918, we think that this is one reason for the falling off in production. We have no special methods to secure better effort on the part of the workmen, but we arrange as far as possible to pay them on a piece-work basis, and find that production on a piece-work basis will vary from 10% to 50% over day work. Generally speaking, we are free from labor troubles. Ours is an open shop and whilst we pay union wages and better in some cases, we do not deal with the men collectively.

"International organizers or agitators have endeavored at different times to stir up trouble, but the majority of our men are satisfied with the working conditions and will not listen to them. We cannot truthfully say that there is any tendency on the part of our workmen to hinder production. We may have some employees who hinder production by laziness, but not by intent."

Still Feels War Influence

The war shop is still felt in many places, and it has not been eliminated as a disturbing element where fine work is necessary. This firm operates its Canadian plant in Quebec. The general superintendent discussed the matter of production at some length:

"It is pretty hard for me to give you any figures regarding this subject. I do know that the production was increased in all departments of this plant, but whether it has also been successful from a monetary standpoint is a question that I am not in a position to answer. We feel that the production in any plant can fall off just the same as the quality of supervision will fall off, and not come up to standard, and when this latter condition exists, the plant will fall into a listless condition and if not wakened up by those in direct touch with the figures, will gradually fall into a state where it is almost impossible to again reach the old standing.

"My reason for the loss in production, in a great many plants is that the plants are now full of war-time mechanics and other workmen who served their apprenticeship and became master workmen in one operation. The regular men of a great many plants were influenced to change from one plant to another, to sell their labor to the highest bidder so that the old-time mechanics were scattered to the four winds, some of them never returning to the plant in which they worked before the war. When the different plants returned to their standard productions, it was a case of a few men who were left out of the original organization, having to break in a great many, who I think are only apprentices. We are not entirely free from this condition and it requires that we keep an active watch on all operations, feeds, speeds and other methods pertaining to rapid production, or we would have to agree that we are losing ground. On the contrary, however, there are three conditions: first, standing still; second, going back to failure, and the third, going on to success. And I for one am unwilling to admit anything in connection with our plant excepting the last condition, because I believe if I had to I would be admitting that this company was working on the down and out course."

A Good Word for the Men

An Ontario firm, making several lines in the iron, steel and metal industries, believes that some care should be taken about statements made regarding the way in which men abuse shop privileges. This firm holds that in 85 per cent. of cases the men are trying to be square.

"It is a most difficult matter for us to answer with any degree of accuracy. We have made very considerable changes during the past five or six months which makes it a very difficult matter to make positive statements along the lines your letter asks information upon. My own judgment is that in any factory there are a certain number of men, say 15%, who are out to get as high a wage as possible, for as small amount of work as they can get along with, giving in return, but the other 85% are good, decent chaps who want to do the fair thing and give a return for the money paid them.

"We believe this applies to our own staff here and we naturally do not want to be advertised as giving out the information that production per man has been universally less during the past years. Hours of labor have been considerably shortened, while wages have been increased, and the cost of production to manufacturers has, without doubt, risen on a given amount of the company's product. We have consequently been compelled to assess our customers a higher price for this reason, and also, for the reason that the same principle has made the cost of our materials higher or at least has been one factor in doing so."

Rope as a Means of Power Transmission

English Practice Regarding Rope Drives—The Three Versus Four, or More, Strand Ropes—Spiral Elongation, the Life of Rope, Rope Guards, and Effect of Moisture.

By J. MELVILLE ALISON

THE subject of ropes for power transmission is one that has received close attention in England. In fact engineers in that country have given closer technical and practical consideration to the subject than we have perhaps on this side of the water. For this reason we append the following article which was presented before the Engineering Institute of Canada with an object in view, viz., to create interest.

Features of English Practice

The outstanding features of English practice may be briefly stated as follows:

(1) Cotton has entirely supplanted any other material for transmission work in England—manila is of the dim and distant past.

(2) The ropemaker is consulted as to plant lay-outs speeds, sheave diameters, centres, grooving, etc., and his opinion is deferred to.

(3) Guarantees of maintenance are required and given in almost all installations.

since the contest lies mainly between the three and four strand, it will suffice to demonstrate the advantage held by the three-strand over the four or any other construction.

Examining the section of a three-strand rope, Fig. 1, we obtain an equilateral triangle by connecting the centres of each strand, demonstrating the triangulation of the strains, a principle adopted by all engineers in erecting angular constructions, such as bridges, etc. In a rope of this construction we have a trinity of elastic spirals, so supple that they will bend to and fro without disturbing their formation. With the four-strand rope, adopting the same method of connecting the strand centres, we obtain a square and an illustration of the parallelogram of forces from the four centres. Again, the four-strand rope cannot be constructed without a supporting core, and since this is indispensable to its construction, its collapse must mean the dislocation of the whole structure, and as the core represents only about one-fortieth of the area,

It will, therefore, be acceptable that the medium engaging the greatest number of turns should prove to be the most resilient and more capable of disposing of the shocks and stresses set up, for instance, in rolling mill work, than any other. Again, as affecting the bending capacity of the rope, another important difference is observable—only two strands appear between every turn of a three-strand, while three occupy the spaces between every turn of a four-strand.

Longevity of Cotton Ropes

As to the life of cotton ropes, much, of course, depends upon their size and the conditions under which they have to work. All things being equal, durability may be gauged by sectional area, and the most economical diameters range from 1½ to 1¾ inches, more of the latter being used in England than any other size. For rolling mill work, however, it so often becomes a matter of sheave widths that 2 inch ropes are generally used. A remarkable case of longevity may be mentioned of twenty-four cotton

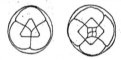

FIG. 1—ILLUSTRATING THE SECTION OF ROPES.

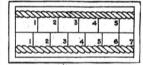

FIG. 2—SHOWING SECTIONS OF SPIRAL ELONGATION.

(4) The manila and steel rope manufacturers employ cotton ropes for transmitting their power.

(5) The efficiency and longevity of cotton ropes have commanded the employment of this material in the manufacture of transmission ropes.

In discussing the question of cotton or manila for transmission ropes, one must not lay too much stress upon initial cost, for although cotton is dearer than manila, its superior resilience, grip and groove impact add so greatly to driving force that up to one-third more horsepower may be transmitted. Further, its prolonged life being reckoned at approximately five to one will reduce apparently excessive first charges to an insignificant item.

Three versus Four (or More) Strand Ropes

Rope of three, four, or even seven strands is sometimes employed, but

it may be reasonably assumed that the superior force exerted by the alternate contraction and extension of the spiral as the rope passes over the sheaves, must tend toward the breaking up of this core, the period of endurance being limited by its elasticity and the tension at which it is laid. This is what actually takes place.

Spiral Elongation

This will probably be the more readily grasped by referring to Fig. 2, which shows a three and four-strand rope in parallel, with the turns of one strand indicated by the figured lines. Thus the turns of the three-strand gradually gain upon the four in the proportion of one in five, or, in other words, six turns of the three-strand occupy the space of five turns on the same thickness. The strand spiral of a seven-strand rope is nearer the straight line, being almost double the length of a three-strand rope.

ropes, 1¾ inches in diameter, employed to transmit 820 horsepower at a velocity of 4,396 feet direct from engine flywheel, 28 feet in diameter, in a Lancashire cotton mill. These were fixed in October, 1879, and are still running in 24-hour a day service, a period of over forty years. Another set has been working twenty-eight years on an average of 20 hours per day, and appear little the worse for wear,

Such cases of longevity lead to the conclusion that fatigue of material due to constant activity does not readily manifest itself in well-make cotton ropes. Their quick recovery from driving strain is suggested as the cause, since they pull down on the working side and bulge out to their normal diameter immediately upon passing to the idle or slack side of the drive.

English and American Systems

There are advocates of both systems

on this side of the Atlantic. We in England are almost exclusively under the English multiple system, and have little experience of the American continuous method. Our education, however, in this latter has been considerably advanced during the last six to ten years on this side, and it has been amply demonstrated that good cotton rope on the continuous system will pay for itself.

In England we are satisfied that the multiple rope system is the better in all-round practice. Immunity from stoppage is the most important factor, since it is seldom that more than one rope gives way at a time. The offender may be laid aside to await a favorable opportunity for replacement. An admission made to me by one of the leading authorities of the American system, was that while all his experiments had been made with manila rope, he had the impression he could get still better results with a good cotton rope.

A drive has been erected for the Armstrong-Whitworth Company of Canada, Ltd., at Longueuil, Que., in which the sheaves are 108 inches and 72 inches in a rolling mill plant, with only 15 foot centres and grooves for twelve 1¾ inch ropes, which allows for only 7 feet 6 inches between rims. This drive was fitted under guarantee of maintenance from the ropemakers, who originally were consulted on the scheme.

Over Roping

Over roping certainly retards efficiency. The fault sometimes manifests itself by the slack changing with the pulling side, and at other times by the ropes travelling across the grooves or even leaving the pulleys altogether. When such troubles arise it is best to remove one or several ropes if need be without interfering with the splicing. Pass them over both pulleys and tie them well back to avoid revolving shafts. They can then be readily put back to work as the others give out. We have known instances where half the original number were taken off before satisfactory driving was obtained. This state of things frequently arises, even under fairly regular loads, when actual requirements come below the calculated power.

Over or Under Driving

While it is, and has been, generally accepted that the slack side of the drive should come on top, this position is not always obtainable, nor is it under all circumstances desirable. With an erratic or fluctuating load such as obtains in rolling mill practice, we recommend the slack to come on the bottom. The variable load of the mill tends to set up considerable oscillation in the ropes, and if the slack is above, may cause the ropes to ride or even jump off the sheaves. On the other hand, if the slack is below, any tendency of the rope

to wander from its appointed track held very largely in check by the pull on the tight side, also gravity and the weight of rope will control the oscillations.

Rope Guards

If, however, ropes persist in wandering from their appointed track, we recommend the erection of a rope guard, in comb fashion, which should be placed approximately 10 per cent. of the centres away from the driven sheave. In rolling mill practice in England this is very freely adopted.

Sheave Diameters

Sheave diameters have a very important bearing upon the question, and while in England the minimum standard is taken as approximately thirty times the diameter of rope, in America forty to fifty times is accepted.

This, however, is largely a question of speeds to sheave size, as, for instance, a ½ inch rope will run comfortably over a 4 inch sheave at 1,000 revolutions per minute, a 1 inch rope over a 22 inch sheave at 3,000 revolutions per minute, and it is only when the speed rises to 5,000 feet per minute that we meet the thirty times rule for 1 inch rope.

Here, again, cotton scores over manila, since its softness readily admits of superior bending properties, while the native hardness of manila and almost entire absence of resilience compel a higher minimum sheave diameter to which ropes of this material can be bent over.

Centrifugal Force

Groove impact also acts in opposition to centrifugal force, which has proved so potent a factor in belt driving, that whatever is done by way of compounding with narrower belts, adhesive dressings, or squeezing out the inevitable air-cushion, it will assert its retarding influence at a velocity fixed roughly at about 3,000 feet per minute, when the power calculated on the basis of width must be discounted in proportion to the increase in speed.

As effecting the rope drive, however, many tables are compiled whereby a steady decrease of transmitted power over 4,800 feet per minute is shown, until we find ropes at 7,000 feet capable of transmitting only half the load of the same rope at 4,800 feet.

Effect of Moisture

Cotton ropes are also less susceptible to atmospheric changes than manila, for the reason that moisture more readily evaporates through the fibres, thus rendering them practically immune from internal mildew which so often is apparent in a discarded manila rope.

Inter-Stranded Cotton Driving Rope

Allowing the claims in favor of the three-strand rope, without further ad-

vocacy of the general principle involved, it may be logically contended that if such a rope possesses constant equality throughout, if the yarns are equal in counts, number, tensile strength and tension, each one following its appointed track without deviation, and are all of the best procurable quality, then we have arrived at a point bordering on perfection in the making of driving ropes, and their manufacture and application has been a close study for over fifty years.

This rope is made expressly for power transmission from the finest selected American yarns. Each of the three strands is made up of a succession of sheaths or layers of yarns, which may be peeled off until the centre thread is reached. The rope is machine built, under hydraulic tension, so that a perfect and combined tension is secured to each thread, from the moment these threads are led to the machine. The crinkling effect to be found more or less in all rope made by the ordinary method is thereby eliminated.

As previously mentioned, the triangulation of strains is secured in three-strand ropes, and in this particular rope is maintained to the individual threads. In effect this construction means that with the rope at work each thread takes its relative proportion of duty, and again in wearing only the external layer will show abrasion, leaving the next layer intact. This feature is of significant value, since there are many instances where, say, a 1¾ inch rope, after years of work, is reduced in diameter, but has still sufficient life left in it to enable it to be removed to a drive where ropes of about 1½ inch diameter may be in use.

There is no limit to length in constructing this rope other than may be controlled by convenience in transportation.

Splicing

All that has been said relative to the making of cotton driving ropes and their application to transmission of power would be of little value without some reference to the question of splicing.

This is so important to the well-being of the system in general that rather than have installations spoiled by indifferent workmanship expert splicing mechanics, properly trained to this duty, are sent out from England to all parts of the world. We have experimented with different devices of metal and other couplings, with a view to dispensing with splicing, but nothing has yet been found which, in England, would be considered against the long splice. This is usually calculated at about eighty-two times the diameter of the rope.

Can a Twist Drill Live up to its Reputation

It Can, and Will, if You Give It a Chance—Make Sure the Angle of Cutting Lip is Correct—See That the Clearance and Centre Angles are Right, Also Study the Data on What Will Happen if You Are Not Careful Enough

By H. WILLS

WHEN considering the precautions taken in the various operations in the making of a twist drill from the first laboratory test of the steel bars to the final inspection, it is a pretty safe conjecture that if a drill gives trouble some of the conditions surrounding its use should be investigated.

Twist drills will stand probably more strain in proportion to their size than almost any other tool, and a very large percentage of drill troubles could be eliminated were proper attention given to grinding the points. The form of the drill point controls the rate of production, accuracy of the hole, frequency

MEASURING LENGTH AND ANGLE OF CUTTING LIP.

of necessary grinding and the very life of the drill.

If the illustrations and instructions in this article are followed religiously, a uniform and satisfactory result will follow to an extent hardly appreciated by the average user of drills. In order to simplify the instructions contained herein we suggest that users of twist drills be supplied with a drill grinding gauge chart of similar nature to that shown.

Twist drills must be properly ground and run at a suitable speed and feed in order to do their work efficiently. With the aid of such a tool a skilled workman can attain the best results, this of course increasing the drilling production.

The tool as shown combines a gauge ground to an angle of 118 degrees and an accurate 4-inch scale graduated in

32nds and 64ths on one side and 16ths on the other. This gauge gives the proper angle and length of lip to the drills when they are being ground. The scale also contains a table of speeds and feeds

FIG. 1.—SEE THAT BOTH CUTTING LIPS ARE INCLINED AT THE SAME ANGLE.

for drilling steel and cast iron. Opposite each ¼ inch mark is a number showing the proper speed at which to run a drill of corresponding diameter.

Length and Angle of Cutting Lip

Make sure that both cutting lips are inclined at the same angle with the axis of the drill and of equal length. The point angle of 59 degrees has been universally adopted as best suited for average conditions, and the result is shown at Fig. 1. To properly grind a drill, proceed as follows:

Correct Clearance and Centre Angles

In this case the drill point must have the proper clearance or contour of surface back of the cutting edges and such clearance must be identical on both sides. Approximately a 12 degree clearance

FIG. 2 AND 3.—WATCH THE DULL POINT.

angle, Fig. 2, combined with the centre angle of 130 degrees, which will give a constantly increasing clearance towards

the centre, Fig. 3, has proven best for average conditions.

Incorrectly-Ground Drill Points

Some of the undesirable conditions resulting from drill points improperly ground follow:

When both lips are not ground at the same angle with radius, here is what happens. If both lips are not ground at the same angle with the axis, Fig. 4, one lip will fail to counteract the tendency of the other to spring away from the cut; consequently one lip will do more work than the other, which will result in its becoming dull more rapidly than if both lips were cutting equally, and it will be subjected to an abnormal torsional strain.

When Cutting Lips are Different Lengths

When the cutting lips of a drill have the same point angle, but are of differ-

FIG. 4 AND 5.—POINTS TO WATCH.

ent lengths, Fig. 5, the point of the drill will be "off centre" or eccentric. As a result the hole will be oversize to an extent equal to double the amount of this eccentricity.

FIG. 6.—WATCH THIS DANGER.

If the drill point is ground both with lips at different angles, and of different lengths, Fig. 6, there will be a combination of the undesirable results already described.

Insufficient Clearance at the Point or Centre

Fig. 7 shows a side and end view of a drill with the proper angle of point (59 degs.), and the proper angle of clearance at the periphery (12 degs.), but with

insufficient clearance at the point or centre.

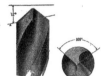

FIG. 7.—THIS IS A COMMON TROUBLE.

Fig. 8 shows a drill with insufficient clearance both at the periphery and at the centre. The line A-B-C is at an angle of 12 degrees, but there is no clearance immediately back of the cutting edges B-C, and the excess of clearance at the heel A-B is of no benefit.

Have You Experienced This?

Fig. 9 shows what is liable to happen to a drill ground with insufficient clearance, especially if an attempt is made to obtain maximum production.

Fig. 10 shows a drill with a clearance angle of about 12 degrees but it does not have the proper contour back of the cutting edges. This manner of grinding leaves the cutting edges thin and weak, causing them to crumble away under heavy feeds.

Estimating Approximate Centre Angle

This gauge can also be used to approximate the centre angle. Although the included angle of the gauge is only 118 degrees, a centre angle of 130 degrees is recommended. The use of this gauge in the manner shown at Fig. 11 enables a very close estimate to be made.

Most twist drills are made with a gradual increase in the thickness of the web or centre of the drill toward the shank. As the drill becomes shorter

FIGS. 8 AND 9.—TWO COMMON TROUBLES.

and the web thicker, greater force is required to drive it. To overcome this, it is good practice to thin the web by grinding away the excess thickness, re-

ducing it to its original dimensions. This grinding must not extend too far up the flute of the drill and care must be exercised that the cutting lips are not injured and that the same amount is ground out of each groove. Fig. 12 shows a drill with the web properly thinned. In Fig. 13 the grinding is excessive, leaving the web entirely too thin and liable to crumble. When this happens a split drill is practically inevitable.

Twist drills are made with a slight taper from point to shank so that the largest diameter is always across the corners of the cutting lips. This prevents the drills from binding in the work, when they are sharp. If the outer corners are allowed to become badly worn, the drills will bind and cannot perform satisfactorily. Whenever the outer corners of the cutting lips show wear, the drills should be re-ground and every particle of worn surface removed, or the drill will continue to bind and very quickly be damaged beyond repair.

In grinding high speed drills, care should be taken not to overheat them,

FIG. 10.—WATCH THIS DANGER.

and when heated they should never be plunged into cold water. Doing so is likely to cause small surface cracks, which reduce the efficiency of the drill and may result in serious damage to it. Forcing the grinding on a wet grinder may also bring about the same condition.

If the suggestions for grinding drill points contained herein are followed and drills are run at the proper speeds and feeds, satisfactory results are practically assured. It is, however, hardly possible to do this grinding as accurately by hand as by using a good twist drill grinding machine, of which there are several on the market.

Broken or damaged tangs of drills are generally the result of an imperfect fit of the drill shank in its socket, which may be caused by a "worn out socket," "dirt or chips accumulating in the socket," or "bruises on the shank of the drill." In either case the driving power of the taper is reduced or destroyed, resulting in an abnormal strain being put upon the tang.

A drill of either carbon or high speed steel that can be filed is not necessarily

FIG. 11.—ESTIMATING THE APPROXIMATE CENTER ANGLE.

too soft for service; in fact, if drills were tempered so that a good file would make no impression, they would be entirely too hard for general use. If you have any doubt regarding the temper, try the drills in actual service.

Speeds and Feeds

There are so many conditions affecting drilling operations that it is extremely difficult to establish "hard and fast" rules for speeds and feeds. The tables given can be safely followed when drilling in commercial materials. Experience will enable the operator to determine what changes, if any, can be made from them. Assuming the drill is properly ground, when the corners of the cutting lips wear away rapidly it is an indication that the speed is too great. If the cutting edges roughen or break out in minute particles it indicates that the feed is too great.

In conclusion, let me say a word of caution regarding the use of very small drills. It is seldom that these are run at more than a fraction of the speed necessary to obtain the best results, and

FIGS. 12 AND 13.—CORRECT AND INCORRECT THICKNESS OF WEB.

excessive breakage is inevitable. These are delicate tools; be sure they run true and that the cutting edges are kept sharp. A fine grade emery stone is best suited for this purpose.

Would You Like to Receive One of These Scales FREE?

You can easily do so and at the same time add to your store of knowledge by entering our weekly contest of which details are given below.

The scale is 6 in. long and is made from finest quality steel. One side is marked in 32nds, the other side in 64ths. A table of decimal equivalents is also stamped on one side, and a table of tap drill sizes on the reverse side. This scale is well worth securing.

What You Have to Do

We publish every week a number of interesting facts or statements selected from the advertising pages for that week. The selections for this issue are given below. Read these through, then turn to the advertising section and see if you can pick out the advertisements to which they refer. The work is interesting, it will train your powers of perception and of memory, it costs you nothing, it will make you better acquainted with the various lines of machinery and tools in the market, and with perseverance you are bound to win one of these useful scales as a prize.

CONTEST FOR DEC. 16TH ISSUE.

Contestants are required to write us, stating to which advertisements we refer to in this number.

1—Something that can be procured with or without a certain attachment.
2—What to use if accuracy is of the utmost importance.
3—A machine that uses a herringbone gear drive to the spindle.
4—How to show a vast improvement.
5—Something said to be of unusual purity.
6—How to keep a shop clean and at the same time reduce insurance rates.
7—Something said to be more economical than coal.
8—A product that claims to have winning features.
9—A product that uses no electricity, and gives you very accurate results.
10—An important factor in every contract job that can be successfully mastered.
11—Something that is said to be safe, economical, and efficient.
12—A product that possesses strength enough to stand up under hard usage.

Closing Date for This Contest is January 6th

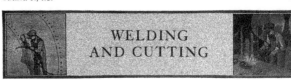

WELDING
AND CUTTING

The Welding of Locomotive Fireboxes

Full Welded Fireboxes, Semi-Welded Fireboxes, Semi-Welded Half Side Sheets, and Full-Welded Half Side Sheets Are Taken Up in This, the First Part, of Complete Article on Above Subject.

Compiled by J. H. M.

IT IS not often that one is able to state in concise form the experience of two experts, but in the article to follow, dealing with the welding of locomotive boilers, the data contained therein have been gathered by Messrs. Geo. L. Walker and R. T. Peabody, both of the air reduction engineering service department of L'Air Liquide Society. This matter was placed before those interested in book form by L'Air Liquide Society, and through their courtesy the author is now able to compile this material in the form best suited for our purpose.

The oxy-acetylene process has stepped very much to the fore in late years, and has proven of great value in the repairing of fire boxes and flue sheets, principally because the joints are perfectly

Failures have been recorded in the past but these have been traced to incorrect methods employed. This has been largely owing to inexperience on the part of the workman, or even utter lack of knowledge of the conditions. No one can make a success of welding until they become acquainted with the principles involved, choosing suitable apparatus, supplies, and so on.

General Instructions

The information to follow is given chiefly to assist the practical welder who might be or is working on locomotive boilers, and who is ambitious enough to read about what others are doing in that line. The methods advocated constitute the standard practices adopted by many railroad shops on this continent, and if

welding, to ensure success it is necessary to observe five important things: First—The cleanliness of the sheets at welding seam. Second—That sheets are bevelled at a proper angle. Third—That sheets are in correct position. Fourth—That provision is made for expansion and contraction, and fifth—that you have suitable welding rods as filling material.

Before cutting a new patch, carefully inspect part to be welded and cut out all defective places, then cut only where there is a good foundation for the weld. When cutting out patches, carefully avoid cutting through stay bolts if possible. Do not under any circumstances cut out sheets or patches through old welds. For side sheets or three-quarter door sheets, extend new sheets at least one

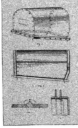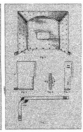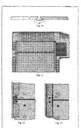

FIGS. 1 TO 13, INCLUSIVE.—TEXT DEALS FULLY WITH THESE EXAMPLES.

water-tight, while the speed with which the work can be accomplished adds still more to its value.

carefully observed will ensure satisfactory results.

In the preparation of the work for

stay bolt higher; for crown sheet, one stay bolt lower; for patches, extend weld at least one row of stay bolts lar-

ger; for flue sheets, extend weld to one row of stay bolts lower.

When cutting out sheets or patches with cutting torch, cut them straight and bevel afterwards—preferably with a chisel. Clean surfaces are essential to good welding. Scale, rust, oil or grease must not be allowed to remain on the part to be welded, otherwise satisfactory welds are impossible. The most satisfactory angle for bevelling plates is an angle of 45 deg., making between the two plates an angle of 90 deg., or a right angle. All sheets and patches must fit with an opening of from ⅛ in. to 3-16 in. between patch and sheet all around—never less than ⅛ in. and never more than 3-16 in.

The welding of cracks in side sheets, door sheets, fire doors, crown sheets, bottom and top of flue sheets should not be attempted, or only resorted to in cases of extreme emergency, to be corrected at the very first opportunity. Welding cracks in barrel of boiler, welding over stay bolts and welding stay bolts is not allowed under any circumstances.

Welding of Full-Welded Fireboxes

A full-welded firebox is one which has all of its joints welded instead of riveted. The welded joint, to the investigator's opinion, is stronger, more economical, will not leak, and has more durable expansion and contraction properties than the riveted seam. The welding should be performed as follows:

First place the firebox upon the mudring A B A B, Fig. 1, and bolt it fast. Next, prepare to weld the outside, or water side, of say the flanged flue sheet to the firebox, as shown at Fig. 2, by cutting a bevel of about 45 degrees all the way along its edge from mudring to mudring. Now bevel the matched edge of the firebox in like manner. The best practice is to make the flange of the flue and door sheets so as to include a row of radial staybolts. See Fig. 4.

After this set the firebox so that the weld joint remains open at its bottom about 3-16 in., Fig. 3, allow for expansion, to clinch the welding metal and insure a perfect weld through the vee. Start the joint by welding it for about an inch just above the mudring, say at 1, Fig. 1, using 3-16 in. or ¼ in. Swedish iron welding rod, and a good welding blowpipe with suitable tip. From a point about 10 in. up from the mudring weld downward from 2 to 1. Then, continning the method, weld from 3 to 2, from 4 to 3, and so on. This practice will prevent the spreading of the joint from unequal expansion. Add enough welding rod to raise the surface of the weld about ⅛ in. above the original metal. If practicable, a second welder should follow the outside welder and smooth-flow the joint on the inside while it is red hot. This will be certain to make the joint tight.

Working from the inside, weld the firebox to the mudring at the corners A and B, Fig. 1, then weld on the door sheet the same as the flue sheet.

Welding Semi-Welded Fireboxes

A semi-welded firebox is one on which welding is substituted for riveting along the lower vertical joints of the flue and door sheet with the side sheets. This practice is followed where it is desired to avoid riveted joints in the fire zone, and some shops also weld the bottom horizontal joint of the firebox and the mudring for about 12 inches both ways from each corner. The chief advantage of this welding is the elimination of leakage. The work should be done as follows:

First bolt the firebox to the mudring and rivet the flue and door sheets to the side sheets down to within about 36 inches from the mudring. Omit the rivet holes along the corner joints to be welded from 1 to A, see Fig. 5. Also leave out temporarily the rivets next to the top of the joint at 1, Fig. 5.

PREPARING WORK

To secure a successful weld, you must first prepare the work properly. There are five important things to observe, viz., the cleanliness of the sheets at the welding seam, that the sheets are bevelled at proper angle, that they are in correct position, that provision is made for expansion and contraction, and that you have suitable welding rods for filling material.

Next prepare the side sheets as shown in Figs. 6 and 9, but cutting into the flange about ⅝ in. so that it can be set down flush. Butt the joints and bevel both edges at about 45 degrees on the outside of the firebox. Leave a 3-16 in. opening at the bottom of the vee as shown. Some welders have used a lap weld like that shown at Fig. 7, but it does not make as strong and tight a joint. The butt weld as shown at Fig. 8 is the best for such work.

The next step is to tack-weld the flange at 1, see Fig. 5. Use a good welding blow-pipe for this purpose, making sure you have a suitable tip, and a good grade of welding rod.

Starting on the outside at about 10 inches below 1, weld from 2 to 1. Then from 3 to 2, and 4 to 3. Complete the joint by welding on the inside from A to 4. Weld the other three corners in like manner. If a second welder is available the metal should be flowed on the inside while being made.

Welding of Semi-Welded Half Side Sheets

Semi-welded half side sheets are those which are welded along the horizontal joints to the crown sheets, and are riveted to the mudring at the bottom and to the door and flue sheets at the ends. The welding of the long seams should proceed as follows where the door and flue sheets have not yet been set.

First proceed to bevel the edge of the half side sheets and the corresponding edge of the crown sheet as shown at Fig. 10. Next screw in the second row of staybolts and bolt the sheet to the mudring before starting to weld. To avoid delay, all of the staybolts, except the rows adjacent to the weld may be screwed in and set, and the side sheets riveted to the mudring before the welding is done.

On this being completed, weld about 2 inches of the joint at 1, Fig. 11, then, beginning at 2, about 10 inches from 1, weld back to 1, from 3 to 2, and so on to the end of the joint. Use a good welding blowpipe, with suitable tip, and as before, make sure a good grade of welding rod is used. The door and flue sheets should now be riveted in.

If the side sheets are to be put in after the door and flue sheets have been set, the flanges of the door and flue sheet should be heated and raised as shown at Fig. 12. This will allow the part of the joint under the flange to be bevelled.

When the new door and flue sheets are put on, the adjacent rivets A, see Fig. 12, should be left out until the welding is completed.

In Fig. 13 is shown the practice of welding the end of the long seam when it is desired to change an old riveted joint to a welded one. The vertical weld should be carried to the second rivet each way.

Full-Welded Half Side Sheet

A full-welded half side sheet has no riveted joints at the top or sides, all three edges being welded. The work should be performed as follows: Cut out the flange of the door or flue sheet next to the joint of the crown and half side sheets, as shown at A and D, Fig. 14. Then bevel the flange, including rivet hole C, and set it flush with the half side sheet. Next bevel the crown sheet, including the rivet hole B. Now set the side sheet so that a ¼ in. space is left at its top and sides, and then bevel those edges. If this sheet is not riveted to the mudring, bolt it fast. Screw in and set all staybolts except those next to the welds.

Now weld the horizontal seam for about 1 in. at 1, as shown at Fig. 15. Then from point 2, about 10 in. from 1, weld back to 1, and from 3 to 2, continuing until the joint is completed. Next start about 10 inches down the vertical joint of, say, the door sheet at 10L and weld up to 1, see Fig. 15. Then from

10 inches below 10L and 11L weld back to 10L, from 12L to 11L, and so on to within a few inches of the mudring.

Weld the flue sheet joint the same way as the door sheet, from 10R to 9, see Fig. 15, from 11R to 10R down nearly to the mudring. Weld the bottom of the joints to the mudring at 15L and 15R last. When the flanged joints of the door and flue sheets are in good condition, but the rest of the plate needs renewing, a good job may be performed by setting in the half side sheet, as shown in Fig. 16. It should be tacked at 1. Then weld from 2 to 1, 3 to 2, and

section xxxA to cool with the weld at 1. This provides for uniform shrinkage. Weld hole 2, Fig. 18, in the same way, and then weld the crack from 2 to 1. Preheat the section xxxB when welding the last hole of that line. The weld should be reinforced, as shown in Fig. 24.

In some cases where it is not practicable to replace a badly cracked side sheet by a new one, or put on a patch, the welding may be done as follows: Referring to Fig. 17, preheat the top of the left row of holes at xxxa. Then weld in hole c of this row. Preheat the area

providing over 50 per cent. of the world's supply. Canada produces about 25 per cent., and the United States and other countries the remainder.

In Canada, mica occurs pretty generally. The most productive areas are situated along the lower St. Lawrence below Quebec, north of the Ottawa near Mattawa, and in the townships of Burgess in Leeds county, Lanark in Lanark county, and Loughborough in Frontenac county, also in a few areas in British Columbia. The production of 1919 was valued at $273,305.

Mica mining is attended with many difficulties. For successful exploitation it is essential that the miners be experienced in the mining of this material, and be familiar with the special conditions and problems it presents. Many good mica deposits have been abandoned on account of the lack of experience of the operators.

The general run of mine mica is of a small size. A very small percentage produces sheets of 4 x 6-inch surface, while fully 50 per cent. will cut to 1 x 3 inch sheets only. Fortunately, a process of cementing the small sheets enables the building up of larger surfaces. This product is known as "micanite" or "mica board" and is mostly used in the electrical industry for insulation. Mica is largely used in the manufacture of boiler and steam pipe covering, its insulating properties, exceeding by far that of any other known substance. Comparative tests have demonstrated that the loss of heat from bare pipes has been reduced by 90 per cent. when the pipes were enclosed in mica covering.

Owing to its resistance to shock, mica is used for spectacles or goggles worn by workmen in industries where flying chips or sparks endanger the eyes, and in observing processes of melting and fusing in furnaces. The small pieces of mica, formerly wasted, are now used for various purposes. When ground fine in oil, mica forms a valuable lubricant, especially for shafting or journal boxes on locomotives or railway cars. Ground mica, when mixed with a flux, is also used in giving to wallpaper and other substances a silvery effect.

So many uses are being found for mica that what was formerly an industry with a very large proportion of waste is now one in which the material is almost completely utilized.

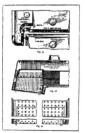 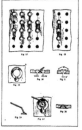

FIGS. 14 TO 24.—FURTHER EXAMPLES.   SEE TEXT FOR DESCRIPTION.

so on, following the practice described above.

Welding Cracks in Side Sheets

Cracks in side sheets between staybolt holes are not easily welded, because of the difficulty of making proper allowances for expansion of the metal. In fact, if a side sheet has many cracks like those shown in Fig. 17 it would be better to put in a new sheet rather than resort to welding. Much welding of side sheets is being done, however, and the following method may be successfully used when the staybolt holes are to be closed:

To weld a crack like that at Fig. 18, bevel the sides of the hole and crack, as shown in Fig. 19. Prepare for each hole a soft steel disc about 3-16 in. thick and of say ¼ in. smaller diameter than the hole. Bevel the edges of the disc, as shown in Figs. 20 and 21. Melt a piece of welding rod to the disc to serve as a holder, hot-bending the rod, as shown in Fig. 22.

Next place the disc in the hole and melt down the upper edge a, see Fig. 23, to tack the disc in position. Preheat the metal between the two holes at xxxA, Fig. 18, and keep it red hot. Melt off the rod and complete the welding of the disc into the hole. Now allow the hot

xxx down to d, as well as that at xB. Weld the crack at the bottom of the hole e of the first row, then the staybolt hole e, the crack at the top of this hole, and finally weld the joint around the head of the staybolt d, which is assumed to be in good condition. Reheat the area xxxa to relieve the shrinkage strains, and let the job cool down before proceeding with the other rows.

Weld the last row on the right next by preheating at xxxa, filling the holes at d and e, and then the crack between them. Reheat the lower part of the row at xxx, and allow the whole job to cool uniformly. Now weld the second row from the left in the usual way, and allow the sheet to cool again. Weld the third row from the left last, which is done in this case because it is the longest continuous crack.

MICA AND ITS USES

Mica is one of the most useful minerals, the production and distribution of which is little known. Of the many varieties, only three are of commercial importance, and of these but two are available in any quantity—the muscovite, or white mica, and the phlogopite, an amber mica. The latter is the most important of Canadian micas.

India is the largest producer of mica,

W. Stuart Cooper, a commercial traveller with a wide knowledge of trade conditions in the West Indies, is now in Canada for the purpose of further developing trade between the two countries. Mr. Cooper intimated that trade conditions in the islands are steadily improving, and emphasized the fact that an increasing desire for Canadian trade is evident. Mr. Cooper has established his headquarters in Montreal, with offices at 401-403 Sun Life Building.

DEVELOPMENTS IN
SHOP EQUIPMENT

RADIAL DRILL.

The Niles-Bement-Pond Co., with general offices at 111 Broadway, New York, have placed on the market a new line of radial drills, which are termed right line radial drills.

These machines embody many radical changes, and the drive has been greatly simplified so that a higher percentage of the horsepower is delivered to the spindle. The design also gives increased rigidity, greater convenience of operation, and wider adaptability.

The feature that is perhaps the most decided departure and that adds to the rigidity of machine and simplicity of the drive is the patented double column. The design of this column is clearly noted from the illustration accompanying the article.

The column is a single casting formed of two box section members cast integral at the top and bottom, the arm saddle being mounted between them. This construction permits the simplest and most direct drive from the motor to the spindle. The motor is mounted on the back of the arm saddle and drives the spindle through a single horizontal shaft running between the column members, two-thirds of the usual driving gears and shafts being eliminated.

Column of Beam Section

Since the column is not of the stationary type but rotates with the arm, it has been possible to place the metal to the best advantage in the form of a patented beam section which is much stiffer than a circular section of the same weight. Furthermore, the bending stresses in the column are always in the direction for which its section is designed. In order to give additional support to the column, the trunnion or circular part extends through to the bottom of the base.

Another advantage of this type of column is that it permits the use of V-type tracks at the front and back for guiding the arm saddle. When clamped by means of a wedge action on these tracks, the arm and column form a rigid unit. The column can be instantly clamped rigidly to the pedestal by means of a motor operated device. Its simplicity and convenience are shown in the illustrations. The operator need not leave his working position at the drill, as the clamp is engaged and disengaged by simply throwing a switch located on

the drill head. A lever for clamping by hand is also provided.

The clamping mechanism consists of a hinged conical ring acting on the column trunnion. When the ring is contracted by hand or power, it pulls the column flange down firmly to the pedestal.

The electric clamp is operated by a small motor through worm-wheel and nut. The mechanism is self-adjusting, wear being taken up automatically. As this clamp is operated by electricity instead of air, it uses the same source of power as the driving motor. Therefore, when installing these machines, no consideration need be given to a compressed air supply.

The arm is especially easy to swing, the entire weight of the arm and column being supported on a ball bearing at the bottom of the column, in addition to which any side thrust is taken on two roller bearings. When the clamping mechanism is relieved, special springs lift the column a few thousandths of an inch so that the column flange clears the pedestal. Friction at this joint is thus eliminated and the arm and column turn entirely on the ball and roller bearings.

The arm is of a new patented section and is especially designed to give greater resistance to the drilling pressure.

The most important features are an upper narrow guide for the saddle and the lower bearing set in a plane back of the front surface. This construction brings the driving shaft closer to the spindle and gives a greater depth from the front to the back of the arm; therefore, for a given amount of metal, the arm is much stiffer than arms of the usual designs.

The arm is raised and lowered by power from the driving motor operating through the stationary elevating screw and a revolving nut in the arm saddle. The mechanism is engaged by throwing the clutch lever located on the driving gear box and is started and stopped by the controller handle on the drill head. The elevating and clamping mechanisms are interlocking so that the elevating mechanism cannot be engaged until the clamp is relieved nor can the arm be clamped until the elevating mechanism is disengaged.

Another valuable feature on these machines is the automatic stop to prevent accident or damage to the machine, should the operator carelessly run the arm to the limit of its travel at either the top or bottom. This device also stops the arm in case the arm or spindle strikes an obstruction in lowering.

GENERAL VIEW OF THE DRILL.

The elevating screw is hung at the top of the column on a friction ring. When the arm reaches the top or bottom of its travel, the revolving nut comes in contact' with a pin on the screw causing the screw to turn and thus bringing the arm to a stop. When the spindle or arm meets an obstruction in lowering the elevating screw is lifted and turns freely, thus stopping the arm.

Spindle Drive

The most important advantage of the double column construction is that it makes possible a drive so simple and direct that there are only four gears and one double-faced pinion from the motor to the spindle. The power is transmitted direct to the spindle by a horizontal shaft running through the column and is applied at the lower end of the spindle, as close as possible to the drill.

The gears for direct drive consist of two spur gears at the motor end and two bevel gears and an intermediate double-faced pinion in the drill head. The back gears are located in the gear case next to the motor and run in oil.

The motor controller is located on the head and the spindle is started, stopped, reversed or varied in speed by a convenient lever.

The spindle counterweight is geared directly to the spindle and is supported at its centre of gravity to eliminate friction and binding on the guides.

The range of the feeds and speeds is sufficient to take care of any boring ordinarily required of a radial drill, in addition to drilling and tapping. There are eight positive geared feeds, which are changed by means of a disc, graduated to show the feed for each position. The feed change gears are entirely enclosed and run in oil. They are so constructed that they can be taken out of the head as a unit.

UNIVERSAL SHAPER

The Universal Machine & Tool Co., Canton, Ohio, have placed on the market what is known as their 24-inch Universal Shaper. This machine combines various features well worth describing.

The machine is claimed to fill the gap between the shaper and planer, and is said to be able to handle work of awkward shape. The ram stroke has been replaced by a speedy table stroke, and the natural lifting tendency of the ram is replaced by the action of the universal table, giving as high as 40 strokes per minute.

Four speeds are controlled by two levers on the operator's side of the machine, these giving speeds of 7, 20, 30 and 40 strokes per minute. All operating levers are placed on the operator's side, and with the open side feature it is possible to clamp long work to the table without obstructing the operator. The machine is of rugged design and is said to be capable of taking an extremely heavy cut. It is single belt driven, has power feed to head and saddle, and all shafts run in ball bearings.

GENERAL VIEW OF THE SHAPER.

SCREW DRIVING MACHINE

The Reynolds Machine Co., Massillon, Ohio, have developed a machine that sets and drives small screws in place, thus speeding up light assembly work. Some of the applications to which this machine is suited are in driving electrical binding screws, switch box screws, washer head screws, etc.

GENERAL VIEW OF THE SCREW DRIVING MACHINE.

This concern has hitherto built machines which automatically feed and drive screws of ordinary proportions, but this new machine handles screws having a length less than the diameter of the head.

In operation, screws are thrown at random into the magazine at the right of the machine. This magazine automatically arranges them in single file with heads upward and delivers them in this position into the inclined track shown leading from the magazine. Opposite, and closing the lower end of the track is a finger recessed to receive one screw and hold it in line with the spindle. The spindle, which carries a screw driver bit, is brought down by the foot lever, the bit engages slot in screw head and finger and screw are carried downward until the point of screw engages work, when the finger is automatically withdrawn to allow the screw to be driven home.

The spindle is adjustable frictionally driven so that screw driver bit stops when screws are driven as tight as desired, or screws can be merely started, or driven to a uniform depth. Table is twelve inches in diameter and is adjustable on column to take up to fifteen inches high and has screw for close adjustment. Foot lever is pivoted well above the pedal to give an easy swinging position, and can be adjusted both for position and resistance to suit convenience of the operator. Operation of machine with foot lever leaves both hands of operator free to handle or assemble work. Spindle may be run several hundred R.P.M., setting screws practically instantly, and the machine is driven with tight and loose pulleys as shown, or with individual motor.

The MacLean Publishing Company
LIMITED
(ESTABLISHED 1887)
JOHN BAYNE MACLEAN, President.   H. T. HUNTER, Vice-President
H. V. TYRRELL, General Manager.
PUBLISHERS OF

CANADIAN MACHINERY

⟶ MANUFACTURING NEWS ⟵

A weekly journal devoted to the machinery and manufacturing interests.
B. G. NEWTON, Manager.   A. R. KENNEDY, Managing Editor,
Associate Editors:
J. H. MOORE   T. H. FENNER   J. H. RODGERS (Montreal)
Office of Publication: 143-153 University Avenue, Toronto, Ontario.

VOL. XXIV.   TORONTO, DECEMBER 16, 1920   No. 25

PRINCIPAL CONTENTS

A Good Time To Buy Tools

RIGHT now is a good time to buy in the machine tool market. That may be a peculiar remark, but it is a fact.

You can get service now that you could not possibly get six months ago, and which you may not be able to get six months from now.

Many of the larger makers announce that they will protect buyers in the matter of price decline, so there is nothing to be afraid of in this respect.

Buyers should take advantage of the fact that it is a buyers' market now. They accuse the sellers of taking advantage of the situation when the tables turn and the market becomes one that is dominated by the sellers. Why, then, would it not be good business for a buyer, when a buyers' market turns up, to go in and make the very best bargain he can? He certainly can talk with greater freedom now than he could a year ago. The situation is here and it is all in his favor.

It may be that we are facing another situation like the one that followed the close of the war. Every person stopped still for a few weeks. There was no particular reason for it, but it happened just the same.

Right now we are in a similar place. People are stopping. There is no good reason for it. We have not had a crop failure—we have not had a panic, neither are we facing another war.

It all seems to be because our dollar is growing bigger when we come to spend it. It is worth more now than it was three months ago when we take it out into the markets and exchange it for the things we need.

If you are a prospective buyer right now should be a good time to go in and talk to the machine tool dealers about your requirements. You will find that they have time to talk to you, and give you good attention. You may find that you can buy to advantage. It is worth try-

ing. We are not going to stay where we are for very long and the first thing you know it will be turning around again to a sellers' market and you will have to take what you can get and stand in line for some months to secure it.

Picking War Plants To Pieces

RUNNING stories of how United States wasted money in the war is getting to be a favorite subject with a good many papers.

For instance, investigation shows that the Government let 111 contracts to the amount of $478,828,334 for the construction of artillery of all calibres, guns, howitzers, gun carriages, limbers and recuperators. Of this immense programme of expenditure there reached the American troops and were actually used in combat 39 75-millimeter anti-aircraft mount trucks, 48 4.7-inch guns of the 1906 model, 48 4.7-inch gun carriages of the same model, 24 8-inch howitzers and 24 8-inch gun carriages.

Then for the turning out of explosives the expenditure ran into the millions, while the explosives ran into nothing at all. In fact, the amount of explosives and gas from the United States war shops that ever reached the front was very, very small.

It might be possible to go ahead and deliberate upon this waste and extravagance at great length, but look at the other side. Take a few of the cases right around here. It might be possible to go up to the splendid Leaside plant in the latter days of the munitions manufacturing when they were getting tuned up to turn out the 12-inch shell, and say "Look at all this plant here, fine machine tool equipment, great shops and layout. All this and no production."

Again it might be possible to go to Motor Trucks at Brantford, when a large building for this purpose was put up at an abnormally high cost, and look at that place. There it was, hundreds of thousands of dollars' worth of equipment being rushed in there and put in position, work going on at feverish haste in order to get that one thing most necessary—production. Did they get any? No, they did not. As far as we know that great plant never turned out a cent's worth of shells.

Well, it would be a very simple matter to go to that plant and say, "Look at that for waste. A great steel structure in the first place. All rigged up with the best of equipment, all heated, power transmission machinery there. A big organization and yet not a cent's worth of output to be put over against all that expenditure." And that charge would be perfectly true. And yet it would be just as unfair as it is true.

No doubt much of the dust that is being kicked up in United States could be traced to the same source.

In this country firms were not sorry that they lost their production. Any such feelings were swallowed up in the joy at the war coming to a close.

War is a waster, and the waste is not confined to operations at the front, nor to the districts in which battles are fought. War plants were largely wasted afterward. They were not put up for commercial pursuits, neither were they erected with the idea of competing with any other plant afterward. They were rushed up for a purpose, regardless of expense, which was in many cases very high when compared to days when things were done on a competitive basis.

It is hardly a fair thing to put a peace-time analysis to things that were done in the feverish haste that characterized much of the war work.

Chas. M. Schwab, Optimist, States His Views

CHARLES M. SCHWAB, speaking in New York a few days ago, gave some plain talk to those who are afraid of the present situation. To quote Mr. Schwab's own words in this matter:—

"But I want to go on record here as saying that nothing could be healthier for American business than the very condition through which we are now passing. It had to come. I only wish it had come sooner. The severer the storm is now, the quicker it will be over and the sooner we can emerge into clear weather and shape our course upon the sea of prosperity.

"Business in the United States ever since the war started had been, until very recently, upon a false basis. The disposition of many manufacturers had been to say not: 'Let me see your costs sheet,' but 'Let me see your statement of profits.'

"Now the true test of success in business is not profits, but economy. Profits may be the result of good fortune, of a fleeting period of inflation, of temporary conditions of any kind, but a business structure which is built simply upon the profit sheet of the moment is built upon the sands. The only business foundations which are sure and steady are erected upon the rock of economy.

"The result of this experience through which we have passed is that our costs have been inflated and we have had in America both our business institutions and American individuals generally indulging in wasteful and extravagant methods.

"The supreme virtue of the existing situation is that it is compelling every business man in America, in fact, every individual in America, to examine thoroughly his costs of doing business and his costs of living. The result of it all is to force business and to force individuals to start to economize and to save.

"I hear men say that the greatest need of the hour is the liquidation of labor. In saying this many have in mind the high wages now being paid to labor and industry, and mean that labor must adjust itself to a new standard of living.

"Now the laboring man is primarily interested not in the amount of money paid to him, but what his money will buy. If the cost of living comes down there is no question that our American laboring man can maintain his present standard of living even though his wages as stated in dollars amount to a lesser sum than before.

* * *

"In the steel industry one-half the total cost of making steel is the cost of labor. You can, therefore, see how important efficiency and labor costs are in all items of manufacture.

"The laboring man is entitled to his full share for the contribution he makes to the value of an article. The laboring man should be taken into the fullest confidence of his employers. He should be so sure of getting his fair share of the wealth he produces that he will work with zeal and enthusiasm.

* * *

"The world during the past few years has been a spendthrift and a waster. It has spent its life and its blood in prodigal living. It has spent a large part of its liquid wealth and has now come to a period where it can go forward only by energy and effort of the most vigorous kind.

"When, therefore, our business men thoroughly study and cope with these problems of economy and make up their minds to go ahead and to produce to the utmost on a smaller margin of profit than they have been accustomed to realize in the years of inflation, then our course will have been set toward a haven of safety and progress.

* * *

"At the moment, our export trade hangs by a slender thread. The exchanges have been moving altogether too much in our favor. We are rapidly getting into the position where the whole world owes us money and yet cannot pay it. We have recently seen a little cloud on the horizon out in North Dakota. Certain farmers owed the bankers money. The value of the produce which secured the loans went down below the value of the principal. The bankers called the loans and the farmers could not pay.

Appointed General Manager

C. R. BURT, M.E., has been appointed general manager of Willys-Overland Co., Weston Road, Toronto. Mr. Burt is no stranger to this country, having been connected with the Russell Motor Co. for some years. He is a thorough mechanic, as well as a capable plant executive. Starting as a boy, he was an apprentice with Brown & Sharpe, Providence, R.I., where he remained for twelve years, being sent by that company in 1900 to Europe to study conditions and practice there.

C. R. BURT.

A few years later, he became connected with the Barber-Colman Co., at Rockford, Ill., makers of gear cutters, etc. While engaged at this plant Mr. Burt designed and perfected the Barber-Colman gear hobbing machine.

It was in 1913 that Mr. Burt came to Canada as factory manager of the Russell Motor Co., and the following year he was appointed assistant general manager and given a place on the board of directors. In this year he made another trip to Europe for the Russell Co., to look into conditions of trade and motor car practice in general.

Shortly after this the munitions contracts started to come to this country and Mr. Burt had much to do with the successful operation of the big munitions plant put up and operated by the Russell Motor Company.

This process has resulted in closing something like twenty small banks in North Dakota during the past few weeks.

"The whole world owes the United States money and the whole world is hungry for the things the United States can supply. Not quite a year ago, one of our great financiers returned from Europe and suggested that we create a revolving fund of $500,000,000 to finance the absolute requirements of Europe and start the countries in distress on their way to self support. Perhaps he did not state the correct sum, but for lack of our ability to sell to Europe through having sufficient credit available, the value of the farm products of this country, which Europe would like to buy, has decreased during the past few months not hundreds of millions, but billions of dollars.

"If the business of the United States is to go forward as it must go forward, our people must take a world view. We must think internationally. We must trust in the good faith and in the productive power of Europe, sending to them our raw materials and goods to enable them to resume productivity, and accepting in payment therefor securities representing their productive activities."

"Iron Age":—To sum up, certain readjustments of past periods of depression that required time for completion are not needed now, while the readjustments that are needed to-day are theoretically unending, cannot be pursued to their end because they have none, and hence will not be pursued indefinitely. We may be able to resume work, therefore, and on an intelligent and safe basis, in a shorter time than many new seem to think is probable.

MARKET DEVELOPMENTS

Some Bright Spots Seen In The Dull Market

Dealers Report Some Good Orders, but Markets are Mostly Quiet With Little Trading—Independents May Move to Cut In on the Corporation Prices if They Can Get Business

A FEW nice orders were secured in the machine tool market during the week and these provide the bright spots. Otherwise trade was not up to the mark. Several of the principal makers have notified their agents that they will protect buyers against price declines for the coming six months. Buyers do not seem to be taking advantage of the fact that it is a buyers' market now, and they could probably come in with their requirements to much better advantage than at any time during the past three or four years.

Local warehouses have announced new prices on some lines. Bar iron and bar steel are selling at 5c base. Sheets and plates were also put on a lower level. There is a fairly large stock of sheets in store, quite enough to take care of the present requirements of the trade. Manufacturers still send out lists of surplus material on hand, and the peculiar situation is found at times of the dealer trying to sell the manufacturer the very materials that the manufacturer would like to see the former taking off his hands for resale.

One of the largest dealers in scrap metals notified some 3,000 firms during the week that for the present they were entirely out of the market as far as taking on anything in the line of brass, copper, zinc, lead, spelter, etc., is concerned. In fact they are out of the whole non-ferrous line, and in iron and steel they are buying only against contracts. This is the first time in the history of their business that they have sent out notices. They intend to keep out until something definite is established as a working base in the metal market. During the past month they have had to write off some hundreds of thousands by way of depreciation on stock in their yards and warehouses.

BRIGHT SPOTS SHOW UP IN THE WEEK'S BUSINESS IN MONTREAL

Special to CANADIAN MACHINERY.

MONTREAL, Dec. 16.—The varying reports of fairly good trade and dullness from dealers and agents, makes it difficult to state definitely, what the actual market conditions are at the present time. Intermingled with the usual seasonably quiet, one is confronted now and then with the information that activity is even above normal. One steel and heavy hardware firm here, during the past week, has had exceptionally good business of late, and reports that the present period compares very favorably with corresponding weeks of previous years, barring the abnormal demand during the years of the war. However, business in general is comparatively quiet, but in the opinion of many is more the result of bank regulations than that of actual trade requirements. Business today is temporarily diverted to retail activity, and for the next two or three weeks will overshadow other lines of endeavor, but with the turn of the year the prospects appear good for a gradual return of general industrial and commercial enterprise.

Waiting For The New Year

No outstanding features have developed to alter the general quiet condition of the steel market. There is little doubt that American mills are doing their best to induce buyers to come into the market owing to the depleted condition of the mills' order books, but the recent declines have, as yet, failed in doing this to any great extent. The existing demand is not such as would call for any keen competition, for if it was so, it is thought that prices would show a more rapid decline. Local trade shows no change from that of last week, the movement being light but fairly regular. It is anticipated that price quotations here will be revised about the beginning of the year, when it is expected that activity will show a gradual, though perhaps slow return to that of several months ago. Plate movement has eased up a little, and likewise sheets, although galvanized lines are none too plentiful.

Anticipating Good Business

While few dealers are denying the slackness of trade and the hard efforts that are necessarily required in obtaining sufficient orders to keep things going to the extent of "just paying the way," the majority of the tool dealers in this district have a cheerful note to sound when speaking of next year's business. There may be nothing of a tangible nature to warrant this attitude, apart from the fact that the needs of industry, in the way of increased equipment or replacements, while apparently adequate for a curtailment period through which we are now passing, are nevertheless insufficient to maintain even normal operations, should trade suddenly return to former activity. Dealers report fairly good business in the lighter supplies but admit a slight falling off from previous weeks. Some little business has been done in used equipment but sales of new machinery, particularly of the heavy variety, are few and far between.

No Interest In Scrap

Dealers are almost indifferent to the present situation, fully believing that

nothing of a special character will develop this month to change the market from the listlessness through which it is now passing. Price, apparently, is not the obstacle that must be surmounted, as any reduction in scrap quotations fails to bring other than casual inquiries. As long as retrenchments are the order of the day, with industrial production being curtailed in many quarters, the consumption of scrap is reduced to a minimum, and consumers will not buy more than they actually need to carry them over the immediate present. At the same time dealers are not anxious to accumulate yard supplies, so that the result is one where inactivity is the dominating feature. Some of the steel scraps have been reduced. Boiler plate, heavy melting steel, malleable scrap, and stove plate, show a decline of $2 per ton. Wrought iron pipe has declined from $9 to $8.50 per ton.

SOME GOOD SALES ARE BEING MADE

But in the Main the Machine Tool, Iron and Steel Market Continues Quiet

TORONTO.—It would be wrong to get the idea that there is no business in the machine tool market. There is a certain amount of selling, and the tools being sold are good sized ones, involving in several cases a pretty fair sale price. Dealers say that buyers are not asking much about the prices. Several of the larger firms have gone this far—they are protecting their buyers in the matter of price for the ensuing six months—that is, if in that period the price is put to a lower figure than that paid by the purchaser, he is entitled to a rebate of the difference.

One of the dealers seen this morning has just returned from a trip to several of the centres, and he is of the opinion that there is not going to be a return to pre-war prices, but that the values will find their new levels fairly well above the old prices. "There were firms in the machine tool business that never made enough out of it to grow as they should. I can think of several that have always turned out a good product, and almost at any time they have been ready to throw up their hands and quit because they never made enough to see them through rough spots. Not only that, but there is a pretty well fixed belief that the cutting of prices would not bring any more purchasers into the market. We are quite satisfied to go ahead and sell on the present lists, feeling that we are getting just as much business as though we had a new and considerably revised schedule."

Although nothing definite has been announced it is felt that the automobile firms will be getting under way, probably on a small basis at first, around the beginning of the year, and they are look-

Steel ingot production in United States for November was 15 per cent. less than in any of the corresponding three months.

—

Independent steel mills are shortage of business, several of them operating at the present time at not more than 60 to 70 per cent. It is generally conceded that it does not pay to run a mill much under 70 per cent. of its capacity.

—

Pittsburg believes that the next market development will be cutting of Corporation prices by some of the Independents. The big question is whether such practice would bring them any business.

—

The production and sale of pig iron is in smaller quantities than for some time past. Foundry iron is quoted at several places at $33 and $35.

—

Several machine tool firms are announcing that they are willing to protect buyers against price reduction for a period of six months.

—

Some very nice orders have been secured during the week by machine tool firms, although on the whole business is still quiet.

—

Several lines of steel were marked down in warehouses this week. Bar iron and steel are now selling at 5c base. Black sheets are quoted from warehouses at 9c.

—

There is still a good demand for boiler tubes and there is a complete range of these in stock at most of the steel merchandising warehouses.

—

The largest scrap metal dealer in this district has sent out word to some 3,000 customers that for the present his firm does not intend to buy any non-ferrous metals. They will not take these lines at any price, while dealing in iron and steel is still at a very low ebb.

—

Some of the scrap metal dealers have found it necessary to write off some hundred of thousands of dollars in the last few weeks, owing to the depreciation of stocks in their yards and sheds.

ed upon as likely to have some pretty fair inquiries to attend to. Taken as a whole, the machine tool business is quiet, but there are bright spots, and the trade in general is not in a mood where it is complaining.

Steel Market Easier

Several prices were marked down at local warehouses this morning. Steel and iron bars are now selling at $5 base. Steel plate is quoted in most cases at 6c. Sheets are to be had for 9 cents for 28 and 7.50 for No. 10. Galvanized is selling around 11.50, and is none too secure at that figure.

Very few of the warehouses are taking in sheets now. In some cases they claim to be well enough stocked to run along now for three months. Fortunately prices are mixed enough to enable most of them to keep pace with the decline in prices. There are still some sheets in stocks that were taken in at high prices, and these, put in with shipments of lower priced goods, makes the present price. There are still a number of manufacturers who have a certain amount of surplus material which they are trying to unload. In fact there have been cases where a dealer was trying to sell the same line to a manufacturer that the latter was thinking of trying to sell to the dealer. There is not sufficient quantity of this in the market to make any serious difference in prices. The market seems to be pretty well filled with sheets for present requirements.

Tubes are fairly active, and there is a good assortment in several of the warehouses. Bar mill products are not moving very rapidly.

Preparations are being made in some of the warehouses for getting in shape for renewed buying after the first of the year. Some of the dealers believe that we are passing through a period much the same as that which came shortly after the war, when there was a lull in trade for a short time, to be followed by a period of keen buying. Were such a situation to develop it would result in stiffened prices, as there are no large reserves.

Out of the Market.

There is nothing in the scrap metal trade to indicate that dealers are looking for a revival before long. One of the largest yards stated to Canadian Machinery: "We are sending out notices to over 3,000 firms that we deal with stating that until further notice we will not take on any non-ferrous metals under any consideration. We cannot see our way clear to keep in the market at any price until something happens to settle the schedules for dealing and give us some reason for believing that we are down to the bottom in the matter of prices. It looked for a short time at the start of the week that the market was going to gain strength, but just at the moment when this looked possible both New York and London went weak,

and the thing was off. This is the first time possibly in the history of this firm when we have absolutely and positively left the market in any particular lines. We are out on the non-ferrous lines until something definite is established either up or down."

"And how about the iron and steel end of it?" was asked.

"Little better, but not much. Foundries are not buying, and the steel plants are not taking on a pound. We are not cutting the prices any lower, but it would be well for you to make it very plain to your readers that the prices quoted for old material are purely nominal. We would not undertake to pay any of them unless we had some place where we were going to place the material at once."

It is understood that some of the yards have had to write off very large amounts in the past few weeks, running into a good many hundreds of thousands.

NEW YORK MARKET IS STILL VERY DULL

And a Number of Used Machines Intended for Export Are Offered At a Price

Special to CANADIAN MACHINERY.

NEW YORK, Dec. 16. — Business in machine tools is exceedingly quiet and will doubtless remain in this condition until after the first of the year. Some of the railroads and other large prospective buyers indicate that their purchases will be made after January 1, but just how early in the new year they will act is not known. It is believed in some quarters that the volume of buying which may be done by the railroads in 1921 is exaggerated. Officials of the roads point out that a good many of the contemplated purchases must be financed out of current surplus profits, and with a decline in freight movement it is not apparent that there will be any large surplus profits in the next few months even with the present high freight rates. Moreover, if money is borrowed by the railroads for such large purchases as locomotives and cars a lower money rate will be necessary. There has been a slight easing in the money rates, but not enough as yet to encourage any large borrowing.

There are varying opinions as to when a revival of machine-tool buying may be expected. Some in the trade put off the time until spring.

In the next two weeks a large number of machine-tool plants will shut down for periods ranging from a few days to two weeks. This will afford an opportunity for annual inventories and for repairs to shop equipment. In practically all plants working forces and the number of working hours have been reduced. Some companies are virtually without orders and are making a few tools for stock. However, the amount of manufacturing for stock that will be done under present

high costs is apt to be exceedingly limited.

Buying of used tools is more active than the market for new machines. A considerable number of new machines is being offered for sale at concessions in prices, these machines in most instances have been bought for export. Owing to cancellations from abroad the exporters are obliged to offer the tools for sale here.

CONDITIONS IN THE BUFFALO MARKET

Steel and Iron Trade Is Down to the Level of New Quotations

Buffalo. — A further curtailment of production by blast furnaces and steel plants in this district amounting to possibly 50 per cent. of the previous scale of production is announced.

One large mill has suspended steel operations altogether.

Three out of four blast furnaces operated in connection with this interest are banked. Another large steel-making in-

terest is operating about 50 to 60 per cent. Four out of nine blast furnaces are down and the open-hearth equipment has been partially closed. Another steel-making plant is operating about 75 per cent. This plant is operating its two blast furnaces. Two other furnaces are operating from 75 to 100 per cent.

Local mills have not published a lowered schedule conforming with present Steel Corporation and Jones & Laughlin prices, but their price policy is practically one of conformance. They state that future orders will be figured on individually, and special prices made. This policy works out to what is tantamount to the corporation schedule. One maker here has taken several orders for bars, plate and structural. The price on the bars is 2.35; shapes, 2.45, and plates, 2.65.

Pig iron sales just now are light with resale continuing to make the market. About 1,000 tons in all have been sold in the past week. This iron is being sold at various prices. Some of it is sold as low as $37; some at $38 and $40 for the base grade at foundry.

CORPORATION MILLS ARE STILL OPERATING CLOSE TO CAPACITY

Special to Canadian Machinery.

PITTSBURGH, Dec. 16. — Production of steel ingots in the United States was about 15 per cent. less in November than in either of the three months preceding, which had shown substantially the same rate. The Steel Corporation's production rate was fully maintained, while the independents had a decrease of about 28 per cent. It is possible that the Steel Corporation's output increased somewhat, and if so the independent output decreased by more than 28 per cent. The estimate of independent production is based upon the November report of the American Iron and Steel Institute. These monthly reports show the output of 30 companies, producers having 15 or 16 per cent. of the capacity not making reports. As the non-reporting producers are independents and as there is a great difference how between Steel Corporation and independent operations the report must be analysed and an allowance made for this fact. The tonnage actually reported shows a decrease of 13 per cent. for November.

The decrease in independent steel production has been continuous since the latter part of October and thus the present rate is far below the average of November. The common prediction is that by the end of this month the independents will be very largely idle, many plants closed and others operating at only 50 to 70 per cent.

The Steel Corporation's output is, if anything, increasing. The corporation has blown in several additional blast furnaces in the past few weeks, not because it did not want to operate them previously, but because it has only lately been

able to secure enough coke, partly because there had been a shortage of cars in the Connellsville region and partly because the by-product plants, at the blast furnaces, could not get sufficient coal.

Steel Corporation's Orders

The unfilled obligations of the Steel Corporation at the end of November are reported at 9,021,481 tons, showing a decrease during the month of 815,371 tons, against decreases of 537,952 tons in October, 430,234 tons in September, and 313,480 tons in August. Previously, for 14 months, there had been increases.

While the November decrease in unfilled tonnage seems large, the showing is really a very favorable one to the Steel Corporation in view of general conditions. The decrease represents about 61 per cent. of capacity, but shipments were probably about 89 per cent. of capacity, and thus the net bookings were about 28 per cent. of capacity, net bookings being the difference between total bookings and the cancellations, for doubtless even the corporation had a little by way of cancellations. The average independent, on the other hand, had scarcely any bookings, while most of them had heavy cancellations, and undoubtedly many independents would show for November a loss in unfilled tonnage actually greater than the shipments.

The 9,021,481 tons of unfilled obligations at the end of November is equal to the corporation's output for over seven months at 90 per cent. of rated capacity. Of course the business would not keep the corporation operating at that rate for

seven months, since some of the business is strung out over longer periods. The corporation is, however, in a very comfortable position. The tables are turned, as compared with the time when the independents, on account of their much higher prices, were making much larger profits per ton than the corporation.

Market Stagnant.

The semi-finished and finished steel markets remain absolutely stagnant. It is probably no exaggeration to say that never has the steel market been so dull. That does not mean that consumption is at an unprecedently low ebb, by any means. The consumption is being supplied chiefly by the Steel Corporation on its old orders. The corporation has about 44 per cent. of the capacity and is shipping at the rate of about 90 per cent. of capacity. The steel shipped is doubtless all going into consumption, and there is still some production by independents.

During the period of steel scarcity there was much reference to the Steel Corporation making sales and shipments only to "regular customers." A quite erroneous view of the situation would be formed if it were assumed that the buyers of steel are divided simply into two classes, customers of the corporation and customers of the independents. A great many buyers patronize both the corporation and one or more independents. In such cases the corporation allotted steel to the customer and the customer covered his remaining requirements outside. Now, in many cases, the amount the consumer gets from the Steel Corporation is sufficient, because the requirements have decreased.

This situation must be taken into account in weighing the predictions made in most quarters that there is likely to be a mild revival of buying just after January 1. The prediction is based on the perfectly correct premise that steel consumption never ceases, that even in the poorest times in the past there has been consumption equal to 50 to 60 per cent. of capacity. If the Steel Corporation operates full that takes care of a very large part of the "hard times demand." Then it is to be considered that the country's steel capacity has increased about 50 per cent. since 1914 and it is not yet proved that the normal consumptive demand is increased in that same proportion.

Steel Prices

As noted in previous reports, tin plates, bars, shapes, plates, wire products and sheets have come to show an independent market at the Steel Corporation level. The independent market that began about a year ago is ended. Pipe, however, remains an exception. The independents are still on their 54 per cent. list, the Steel Corporation having maintained the 57½ per cent. list of the Industrial Board. The difference is about $7 a net ton. However, some of the independents were formerly obtaining large delivery premiums, beyond their regular prices.

The common view is that the next market development will be shading of the Steel Corporation's prices by some, perhaps many, of the independents. Certainly there would be a disposition in that direction, if the operation seemed to be worth while. It is all a question of whether a mill could secure enough business by cutting prices to make an economical operation. Some of the independents used to claim that their costs were so high that they could not afford to sell at the Steel Corporation prices, but of course they included overhead in their cost. If the mill is idle the overhead is lost anyhow. On the other hand, it is not feasible to operate a mill at a very low rate, that entailing a greater rate than closing entirely. Thus, unless a mill can get enough business to operate at 60 or 70 per cent. for at least a week or two at a time it is unlikely to try to get business at all.

Pig Iron

Furnaces continue to go out of blast. There is no inquiry for pig iron and furnaces are making no effort to force sales. The foundry iron quotation has dropped from $37 to $35, valley, Bessemer and basic remaining quotable at $35 and $33 respectively.

Germany and Belgium Are After Business

Making a Strong Bid to Win Back the Connection With South Africa—Many Lines in Which Canadian Makers Might be Interested for Their Export Business are Mentioned Here.

GERMANY and Belgium are both actively engaged in their efforts for the South African trade in iron and steel, including bar iron, iron wire and tools. In barbed wire, 12 gauge, 100 pounds per gross, barbs six inches apart, the present offers from Germany are 35s 3d c.i.f. Cape Town.

In such tools as pincers, pipe wrenches, punches and other items similar to those of which German samples are shown at the Department of Trade and Commerce, Exhibits and Publicity Bureau, in Ottawa, shipments are coming in fairly freely from Germany. Occasion has been taken to see several invoices of shipments on open indents, the items on which were taken from the old 1913-14 catalogues. The several buyers interviewed point out that although open orders were placed it was only the articles which German manufacturers knew would be competitive that were shipped; in other words, no advantage was taken of the open order.

The present prices of the German tools referred to are an increase over their 1914 prices of 175 to 225 per cent.

Provided that the quality is satisfactory and the prices and terms competitive, a good business could be worked up in Australia in such lines as twist drills, reamers and milling cutters, grinding wheels, drop forged spanners, taps, dies and screwing tackle, files, vises, forges, drills, metal thread screws, and power transmission appliances; also electric house supply meters.

Australian requirements in the twist drill line are confined to carbon tools. There is a big demand for high speed tools. American prices are usually found in excess of British prices. United States manufacturers do the largest business, though Canadian houses get a portion of the trade. Business in grinding wheels is almost entirely confined to United States makers. A large and growing business is being done and the opportunity is good to participate.

The United States houses are strong competitors in the drop forged spanner business. There is an exceedingly good opening in this line but care should be taken that the right articles are supplied.

For many years traders looked to North America for their supply of files, but many merchants have turned to British supplies. There is no reason, however, why Canadian manufacturers should not obtain a share of the business. At present the largest line of taps, dies, and screwing tackle comes from the United States, which has a strong hold on the market. Canadian manufacturers, however, who will supply good tools, which are interchangeable, and of good quality,

at competitive prices are assured of good business.

There is a ready sale for vises, forges and blacksmiths' drills, and Canadian manufacturers should give their attention to Australian markets.

Metal thread screws, bolts, etc., are largely used by stovemakers and motor car body builders, as well as in general engineering and standard Whitworth threads are used.

A large number of pressed steel pulleys, hangers, compression couplings, etc., are imported, but Australia makes her own wood pulleys.

There is a large demand for electric house supply meters, as great difficulty has been experienced in obtaining these for some time past.

To obtain business in any of these lines it is necessary to appoint a representative. English and United States houses are well represented, and it is almost useless for Canadian manufacturers to try and obtain business otherwise.

Developing Steel Plant

Since the Broken Hill Proprietary Co. opened its steel works at New Castle, near Sydney, Australia, in 1915, 637,213 tons of pig iron has been produced, while the output of the steel plant has been 503,223 tons. The demand for their products is so great that it has been found necessary to add to the present extensive premises. It is expected that the proposed new plant will cost £3,500,-000. The extent of the company's output has enabled a number of new and important industries to spring up, the latest being for the establishment of a wire netting industry which is expected to start operations next year. When the new furnaces are running it is expected there will be an output of 450,000 tons.

The importation of agricultural machines into Czecho-Slovakia is at present prohibited, but licenses for limited quantities are granted to co-operative societies and other importers for the purchase from abroad of machines which are needed and cannot be produced in Czecho-Slovakia. The rate of exchange is a great obstacle, so that if importations were unrestricted the turnover would not be large as the price to the farmer would come high.

Another Market Here

Bohemia, Moravia and Silesia before the war offered a market for certain lines of machines and industrial equipment, such as spinning machines and other textile machines, paper, machinery, etc. There is a considerable manufacture of machinery in Czech-Slovakia and the present policy of the government permits only the purchase of machines from

abroad which cannot possibly be obtained in the country. Import licenses for textile machinery from Great Britain has been refused, but certain special machinery have been admitted. When the industries of the country are on a more normal and independent basis it may be expected that the trade restrictions will be modified.

Grain harvesting machines are manufactured to some extent in Czecho-Slovakia, but not so as to preclude the necessity for importation. The United States and Canada formerly supplied the bulk of mowers, reapers and binders, and importers are again looking to these countries for supplies. The low exchange value of the crown and import restrictions hinders business at present, but on the other hand, there is a lack of these implements and also a lack of agricultural labor and importations from abroad will soon be necessary. Canadian firms should now prepare the ground for future business by getting in touch with importers and co-operative agricultural societies.

Tractors and motor ploughs are another line of agricultural machinery which offers openings for future trade from abroad, home production does not cover the demand.

Prague is the most important centre for the trade in agricultural machinery and implements, while as a distributing point for Slovakia the town of Bratislava (Pressburg) also promises to be important in the future.

SCRAP METAL MARKET

The prices on the scrap metal market still continue to lower. The consumers continue out of the market.

The outstanding feature of the situation in scrap iron on the Chicago market is the increasing tonnage being offered by railroads which may be partly due to the wild weather, which makes it possible to gather and sort this material at a season when the cold and snow usually prevents it.

Sharp reductions are the outstanding feature in cast scrap and iron car wheels on the New York market. Iron car wheels have been sold as low as $24 f. o. b. New York and heavy machinery cast at $25 and $26 respectively. This is a reduction of $5 to $6 and in the other grades of $3 and $4 respectively.

The business on the Philadelphia market is confined mostly to forced sales and trading among dealers. The continued absence of buying is bringing further drops in scrap prices. Heavy melting steel has been bought by dealers as as low as $15.50 delivered.

CANADIAN MACHINERY
AND
MANUFACTURING NEWS

VOL. XXIV. No. 26

December 23, 1920

Planning Tooling and Operations for Lever

The Lever Under Discussion Is For Use on a Standard Milling Machine—Five Operations Are Entailed, the Only Equipment Used Being a Lathe, Milling Machine and Drill Press.

By E. N. DAVEY

AN important feature in designing the tools and operations for this piece, to my mind, is the equipment of the shop. For my purpose let us suppose it is to be made with an ordinary engine lathe, milling machine, and drill. If a turret lathe and multiple spindle drill are to be used the same equipment can be modified to suit, and would naturally cut the price of production a little. I think this piece could be best handled in five operations, namely.

1.—Lathe operation on the large end drill, ream, counterbore, turn outside and face. 2.—Face both sides of the boss on the small end with a pair of milling cutters. 3.—Face the back of the large end on the drill. 4.—Drill the half-inch hole in the small end. 5.—Mill the teeth.

The first operation would require, first, a small fixture to bolt on the faceplate of the lathe for quick setting up. Fig. 1 shows this fixture. It is bolted to the faceplate by two ⅝ bolts, and prevented from sliding by two dowel pins. The piece is placed in the casting as shown.

The casting is made to form, and bored out in position a little larger than the boss, enough to give a working clearance, and held in position by three set screws and a plate. To remove the piece, the top set screw only is loosened, and the plate is allowed to drop down. The outside turning is completed while the drill is being fed through. After drilling and reaming, the counterboring and facing is accomplished with a tool which has a pilot to fit the hole. Fig. 2 shows construction of this tool.

It is a piece of tool steel turned the required shape with a one-inch pilot on the end. Four slots are milled longitudinally to take four pieces of high speed tool steel, and on the outside a containing ring is turned to fit. These tools can be adjusted and are held in position by keys. This tool has the great advantage that in case of breakage it can be pushed further out until the high speed steel is practically worn away. This is the major operation and would occupy about ten minutes.

For the second operation we will take an angle plate and screw into it a one-inch hardened steel pin, threaded at the end for a ⅝ nut. This nut is to be made so that the casting will pass over it without having to remove the nut. A slotted washer is inserted between the

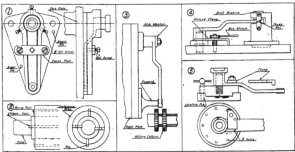

FIGS. 1 TO 5—ILLUSTRATING HOW THE AUTHOR WOULD COMPLETE THE LEVER.

nut and the casting to enable us to tighten up the work.

Some sort of support will have to be provided for the small end to bear on, and the direction of rotation of the milling cutter arranged to thrust the work downward. The angle plate must be bolted to the milling table, so that the distance between the faces after the first piece is completed will come automatically correct. The time for this should be about two minutes.

The third operation, namely, facing the boss, and the fourth operation, drilling the half-inch hole, could be done on the same jig. If a round 12-inch blank flange could be picked up somewhere it would serve to build this jig upon. If not, a casting would have to be made. Drill and tap the base for 1-inch cast steel pin. Make the pin about one and a half inches long. This would leave room for the pilot on the facing tool. Swing the small end on to a boss or raised piece, the right height, and up against a stop, when the piece would then come directly under a hardened drill bushing properly located. It could be held in place by some locking lever device such as at Fig. 4. Time for the two operations, three minutes.

The final operation is milling the nine teeth. The very fact that there is an odd number of teeth allows us to run right across the face with the milling cutter and cut the side of one tooth and the opposite side of the other, and takes nine cuts. If we have any horizontal indexing attachment, the problem resolves itself into one of quick and accurate setting up. If there is no suitable fixture belonging to the milling machine, then a simple holding and spacing jig could easily be made by taking two plates.

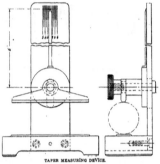

TAPER MEASURING DEVICE.

Make one rotate round the other, hold the work on locating pins and drill nine holes in for spacing. See Fig. 5. Time, about six minutes. Total time, approximately twenty minutes.

TAPER MEASURING DEVICE
By John Homewood

The occasion often arises when it is desired to know how much taper there is on a certain piece of work. Very frequently this is required when a piece is wanted for replacement. Many methods for ascertaining this taper are laborious and subject to mistakes in calculations. The sketch herewith illustrates a simple bench gauge that should be a valuable asset to any shop using tapers. The details of construction, and its action, are self evident, and need little explanation. The upright that extends from the rear of the base carries the adjustable swivel bar and the indicating finger, that shows at a glance the amount of taper that is on the work. The working faces of the gauge are hardened and ground and pointer should be set to zero by means of a cylindrical standard.

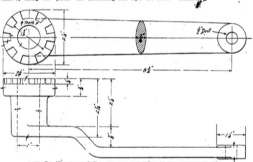

A DETAILED VIEW OF THE LEVER. WHAT DO YOU THINK OF THE AUTHOR'S PLAN?

Practical Pointers on Rolling Mill Operation

English and American Methods of Rolling—Getting Large Output in Rolling Thin Plates—How to Avoid Waste—Universal and Plain Mills—How to Obtain the Best Results

By W. S. STANDIFORD

AMONG the vast diversity of metal products turned out by the various rolling mills, the rolling of ship plates, automobile sheets and strip steel occupies a front rank position. In view of the above, readers will undoubtedly be interested in reading about roll designs, difficulties encountered in rolling, and other factors entering in the production of the above metal sections.

The term "sheets" is generally applied to flat pieces of metal such as iron, steel or the non-ferrous varieties which are less than one-fourth of an inch thick. From one-fourth inch and upwards, the metal is called "plates," tin plates, being an exception to this, the term being applied to very thin sheets of iron and steel which are coated with tin, or a mixture of lead and tin. Plates are rolled in various lengths, their dimensions being limited to the railroad transportation facilities, such as the sizes of

an inch thick, are produced in this way. The waste of plates rolled on a universal mill will average about 15 per cent. less than those rolled on the average plate mill. The length and thickness of plates which a mill can turn out is independent of the dimensions of the rolls, but the width is determined by the length of their barrels. Thus a 150-inch mill means that the mill will roll a plate 150 inches wide. Plates can be rolled from ingots or slabs, both giving very good results in the finished product. They can be rolled in either two high, or three high mills, the English using the two-high system. In most cases, it is the usual English practice to have a separate set of roughing rolls connected up by means of boxes and spindles to the finishing ones, the rolls being driven by a reversing engine. The reversing engine is not an economical piece of machinery to run, but it seems

of plate-making. These advantages are generally considered by English manufacturers to be more than an offset for the extra amount of coal consumed.

On an English plate mill, the same crew does the roughing and finishing. The crew works the cut slab until the roughing rolls reduce it to a suitable thickness; it then is transferred to the finishing rolls. Two crews cannot be used on this style of mill for the reason that the jerking and uneven action of the roughing rolls interferes with the smoothness of the surface finish of the plates and also with the uniformity of gauge thickness, the plates being rendered with great variations in thickness, hence the advisability of doing the roughing by itself. In the English system the rolls are worked dry, no water being used upon them. This necessitates turning the rolls slightly concave in the middle in order to allow for ex-

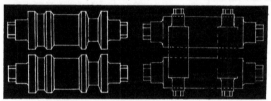

FIG. 1—LEFT HAND VIEW. A TWO-HIGH SET OF BLOOMING MILL ROLLS.
FIG. 2—RIGHT HAND VIEW. ILLUSTRATING HOW THE ROLLS ARE PLACED IN A UNIVERSAL MILL

cars. There are two styles of plates sold on the market, viz.: "Sheared" and universal plates. Sheared plates are wide ones as rolled in the mill, they being cut to certain sizes in widths as desired by the consumer. In all plate mills there is more or less waste of metal due to the plates having imperfect edges. The economical working of any plate mill depends largely upon the percentage of the metal which has to be sheared from it to remove the imperfect edges.

Avoiding Shear

To avoid having to shear any metal whatever from the sides of long and narrow plates, they are most generally rolled in a "universal" mill. Plates as high as one hundred feet long by six feet wide, and one-half an inch thick down to one foot wide, and one-fourth

to be well liked and largely used by English plate mill manufacturers.

The New Plan

The English plate manufacturers formerly used to roll the hot ingots directly into plates without their being reheated, but of late years they have adopted the plan of rolling the large ingots into slabs of four inches or more in thickness, then cutting the slabs while hot into the desired lengths and stored away until required. When needed, they are put into a heating furnace to raise their temperature for the final rolling. More fuel is consumed by the latter method; but, on the other hand, fewer defective plates are made, and taken in general, there is much less strain on the machinery than is the case when the ingot is rolled in one heat, which is the case with the American system

pansion as the metal expands in the middle more than it does upon the ends of the rolls. When a mill of this type is first started up on a Monday morning, it is customary to roll narrow plates at first so as to get the rolls gradually heated and expanded to a level surface. The general practice in America is to finish sheared plates from ingots or slabs—without re-heating them—on one stand of rolls, the three-high system being used and running water being employed to cool them and prevent over-heating and undue expansion.

It may be thought that the use of running water on the plates would cool them very quickly and interfere with their reduction in thickness. The water does not cool these plates as much as might be reasonably expected, because the heat of the plate suffices to maintain the water in a spheroidal state, thus pre-

venting any close contact between the two.

Water Has Beneficial Effect

Water on iron and steel plates during rolling has a beneficial effect, as it loosens and causes the scale (which contains impurities) to break away from the metal. Iron plates while being rolled contain more scale than the steel ones. This is due to the fact that iron is more impure and porous than steel and it contains lots of cinder and slag, which are worked out by the action of the rolls. Scale on plate, whether of iron or steel, when being rolled, is a nuisance, especially near the finishing stage, because if it is not swept or blown off by compressed air it will invariably mark the finished product. This is due to the fact that the pieces of scale laying on top of the plate will be rolled sufficiently into the surface as to make a slight depression on it. When a set of rolls are used for both roughing and finishing purposes, the wear on the same is greater than is the case where a separate set of rolls are need for roughing down the ingot, as the drafts used are far heavier in the roughing down process. The result of this is that the rolls have to be dressed at more frequent intervals when they are used for both roughing and finishing.

Rolls in Tandem

The American system of using running water to cool plate rolls obviates the necessity of turning then with a concave surface as is used in the English system. When thin plates are to be rolled, the largest output is secured by using a separate stand of finishing rolls, and placing the roughing set in tandem. In this case each set is driven by its own engine, or an electric motor as the case may be in that particular mill. All sets are also in line with each other and connected by tables. By this method, the finishing rolls can be run at as high a speed as is practicable; which not only gives a better surface to the thin plates, but allows a greater output to be secured. The work of the roughing rolls can be so distributed as to synchronize with the time cycles of the roughing and finishing mills. One of the most well liked and generally used types of plate mills is the "Lauth" mill. This is a three-high mill with one stand of rolls. A unique feature of this mill is the manner in which the middle roll is used to support the top and bottom rolls when they are in action. The top and bottom rolls of the mill are 34 inches in diameter and are both of driven variety, while the middle one, which is 24 inches in diameter, runs loose and revolves by friction with the plates being rolled. It is raised or lowered by a hydraulic cylinder to enable the plate to enter the rolls between the middle and top, and the middle and bottom rolls. As the middle roll is ten inches smaller in diameter than the top and bottom ones, it is pressed against the surface of the top roll when the hot steel is being rolled, the thickness of metal in the top roll preventing the middle one from boiling.

Another important point is due to the fact that as the middle roll is smaller in diameter, it has a sharper periphery or contact of surface than both the top and bottom rolls, the result being that the middle one bites deeper into the metal being rolled, thus it has a more rapid reducing action than the larger diameter roll whose periphery is flatter and does not elongate the metal to the extent of the smaller one. In practice, the Lauth mill has been found to be very efficient and it gives good results, being used extensively. This type of mill is made to use cast steel slabs, which are tapered slightly so as to permit of their removal from the ingot mold in which they are cast. As there is no means of edging the slab, the top roll not lifting sufficiently for the purpose, it would naturally roll out a plate that would be wider at one end than

---

VALUABLE ADVICE

The author of this article is a practical rolling mill man himself, and knows whereof he speaks. His information is brief and to the point. An article of this nature is well worth filing away for future reference, as it contains many valuable pointers in rolling mill operation. This is only the first part of the article. It will be concluded in our Jan. 6th issue.—Editor.

---

at the other, did not the roller take advantage of this fact, that in. rolling metal the piece rolled is extended in length, but not appreciably in width.

Bringing Slab to Width

One way to bring the slab to an even width would be to insert it between the rolls, having its long side parallel with their axes. But the main drawback to the method exists in the fact that the slab is too narrow and not wide enough to allow of its being carried sideways on the ordinary roller tables that are supplied with mills of this kind. As it is absolutely necessary, in order to prevent a waste of metal in the finished sheets, that the slab should have an even width at the start of rolling, it is inserted diagonally in the rolls with one corner facing upwards. This reduces it to a diamond shape, it is then turned over by the manipulator and entered on the opposite diagonal corner, with the result that the second reduction has again squared the slab, the process spreading it in both directions. If necessary, this operation is repeated

several times by the roller. It all depends upon the amount of carbon that the steel contains, thus affecting the drafts. A slab rolled by this method has a uniform width. The rolling is now resumed in the ordinary manner until the plate has been reduced to the desired gauge. We now come to the design most generally used, namely, the two-high set of blooming mill rolls to reduce ingots for making into slabs and billets. These are illustrated in Fig. 1.

This set of rolls is shown in sketch as having one extra wide slabbing pass in the centre, and two different sizes of billet grooves. As a matter of fact, five or more passes could be cut into the rolls, depending upon the size of the billets desired and also the width of slabs produced. It will be noticed in the sketch that the corners at bottoms of passes are left sharp. This was done in making the sketch to save as much work as possible, the idea being to show the appearance of this type of rolls only. In actual practice, the roll turner always puts rounded fillets at the corners of passes in this style of rolls. This is done so that the corners on the ingot will be worked as much as possible and have a fibrous nature.

Should Be Turned Over Frequently

Rolls of this design are used all over the States and Canada, also in other countries. They are used for reducing the large ingots, the latter being worked in the slabbing pass as they come from the furnace. The ingot is turned over on its side, each one alternately rolled until, if billets are desired, the metal is sufficiently reduced in area to allow it to enter one of the billet passes. It soon gets reduced to a long rectangular shape in these grooves. In order to have the edges of the finished material as free from cracks as possible, it should be worked as rapidly as can be done while it is at a high heat and be frequently turned over so as to facilitate the metal getting into a fibrous condition, which is the object desired by all iron and steel manufacturers.

The slabbing pass in these blooming rolls are put in level by the roll turner, which is the case in rolls used in some mills. In other cases, however, the pass is a convex one; the highest part being in the middle. This design makes an excellent working pass, which does not strain the machinery as much as does a groove having a straight bottom. As the ingots in all blooming mills are turned over at a right angle to the previous position which they occupied, in order that all of their four sides should be thoroughly compressed; the two concave sides of the ingot when rolled at a right angle to the axes of the rolls, straighten up when their top and bottom are being compressed in the roll pass.

If the ingot is being reduced in a slabbing pass, the metal, owing to the heavy drafts used by the roller, bulges

out at the joints of the rolls. When the bulged or uneven side is turned over and faces the top and bottom roll slabbing pass, it lies over at an angle out of a perpendicular line to the roll axes and not at a right angle on the feed rollers. Should the draft be reasonably heavy, the ingot is apt to turn over in the groove altogether while being reduced. The result is that the ingot is mashed out over the collars of the roll pass next to the one in which the rolling is done, and if the engine is very powerful the roll is either broken or badly bent so that it cannot be used again. On the other hand, if the ingot has its concave sides resting upon the feed rollers, the position of the ingot will be at an exact right angle to the roll axes and also feed rollers, the ingot resting on its corners only, its centres not touching the feed rollers at all.

Gives Fine Results

The advantage of this is that the concave surface of the metal allows the ingot to be easily gripped and quickly dragged into the pass, the merits of curved passes for this class of rolls being of the utmost importance as regards the rapid reduction of steel and it gives fine results under actual working conditions in every class of mill using this style of roughing grooves.

Slabbing and plate mill rolls depend mostly for their drafts on the amount of up and down play which their housings and mechanism permit, it being in the roller's power to use either light or heavy drafts as is considered necessary. It is the usual practice to reduce the steel quickly as possible while rolling. Many rollers in their eagerness to get their day's work done as soon as possible abuse the rolls and machinery by reducing the metal at the utmost possible limit of drafts, which rolls and machinery will stand, thus throwing a continual strain upon the apparatus. The danger of ruining rolls by twisting necks off of wobblers as well as damaging the housings and engines is one that a roller will take risks upon when he is at work, as it seems to be a peculiarity of human nature for men to be careless with other persons' property. By watching the workings of the rolls from a position where the roller cannot see him, a works manager can easily find out for himself if his roller is handling his machinery roughly. When extra heavy drafts are being taken, a heavy cloud of black dust will arise from the rolls while they are in action. This is an infallible test, the black dust coming from the pass surface which is being ground off by the hot ingot metal, which has reached its utmost limit of resistance. The machinery will also be heard to vibrate heavily and make more than the usual amount of noise when extra heavy drafts are being used.

A Waste of Money

Rolls roughly handled require an extra amount of metal taken off by the roll turner when dressing, in order to bring the passes into shape again. Thus there is another waste of money which is totally unnecessary. In working of blooming mill rolls, it is much better to operate them at their maximum safe rate of reduction, which they will stand. A steady, even pace, with no interruptions caused by breakdowns, is much better than a spasmodic effort being made to break tonnage records with its evils of straining the machinery, that sooner or later causes a shut-down ahead of its time for repairs. Excessive reductions also produce slabs that are thicker in their centres than at their edges. To give an idea of the amount of wear that a set of normally used blooming rolls undergoes while they are rolling all kinds of ingots comprising both soft and hard steels, the following measurements were taken from a set of worn rolls that were sent to the roll lathe to be dressed. These rolls contained four billet and one slabbing pass; the sides of the billet grooves being worn out 3-16 of an inch wider than a templet which is used by the roll turned to keep them the same width. It will be observed in the sketch of blooming mill rolls that the grooves have tapering sides. There is a twofold object in putting them in with tapering collars, namely that by having grooves with slanting sides, the hot metal touches a bottom point of the groove before it grips the wider part at top, the result being that there is no shearing of metal from ingot at the top part, which would be the case if the groove sides were straight, the sheared metal from ingot coming off in the form of narrow splinters.

Sides Parallel With Roll Axes

Should they be put in the rolls straight—that is, having their sides parallel with the roll axes, and they wore out 3-16 inch at their sides, it would be impossible to make them fit templets no matter how much metal was taken off of the rolls in turning. To make these worn roll grooves again fit their templets, it was necessary for the writer to take 11-16 inch from the collars, or tops of the rolls. A corresponding amount was removed from their bottoms so as to keep the depths the same. In regard to the total amount of time consumed in doing this work on both rolls, it required 4½ days, the tool used having a width of 3½ inches.

The roll lathe was driven by a 15 h.p. direct current electric motor, the latter having more power than, was really necessary. The writer while doing the turning desired to know the exact amount of power that was needed, and had the chief electrician discover the amount necessary to operate the lathe on both roughing and finishing cuts. With the above width of tool, it took 7 h.p. to sustain a steady roughing cut and 3¼ h.p. for the finishing one. Each roll was 26 inches in diameter and weighed over 8,000 lbs.

A type of mill also much used in this country is called the "universal" mill. It is adopted extensively for plate-making and jobbing work. This mill consists of a pair of ordinary plain cylindrical rolls mounted in the usual way, and at the back of these a shorter pair of similar rolls, which are mounted with their axes in a vertical position; so that they will compress the edges of the bar at the same time that the horizontal rolls compress it on its flat sides. This arrangement obviates the necessity of the ingots having to be turned over on their sides so as to work the edges that face the joints between the collars of rolls, which is the case with slabbing rolls.

The vertical rolls are, as a general rule, made smaller in diameter than the horizontal ones, and they are driven from their upper ends by means of mitre-wheel gearing which operate other bevel gears keyed onto a shaft crossing the top of the housings. On the end of the shaft is a spur wheel working into the upper pinion in the pinion housings, or with a spur wheel keyed to it. Originally, the vertical rolls were driven from their bottom ends. but scale from the steel being rolled kept falling into the gearing and caused the teeth to wear out rapidly so that they are now driven from their top ends. Fig. 2 illustrates how the rolls are placed in a "universal" mill, no gearing being shown in the sketch.

Width or Thickness

It will be seen from this combination of four rolls that any width and thickness of flat plates within the compass of the mill can be produced with equal facility. Universal mills generally make plates up to 6½ feet wide. Of course, there can be wider plates made by this method of rolling if the rolls are made sufficiently wide and thick in proportion to the work that they have to do. Plates made on such mills usually have their edges straightened by pressing them while hot on the floor between two bars, which are squeezed against the plates laterally. If the piece of metal rolled in a universal mill is not entered by the roller perfectly square, or if it bears much harder on one vertical roll than on the other—it will be curved in a horizontal direction; and the same thing will occur if the plate is unevenly heated; the hotter side elongating more than the colder one, and once it is bent it is not an easy matter to straighten it. The effects of uneven heating in the working of the steel in the rolls, shows in the finished product, it being of vastly greater importance that the material shall be uniformly heated, even if it does take more pounds of fuel to do it. Unevenly heated and crooked plates not only give trouble in turning. them over and handling them on the castors, but their

thickness varies in different parts, which is due to the many hot and cold spots on the plate. When the outside of a plate is cold and its interior hot, this condition causes the rolls to spring and the housings to stretch appreciably when rolls reach the cold spot. It will readily be seen from the foregoing that it is a very difficult matter in the case of long plates to keep the gauge thickness uniform, so that if one end of a long plate is much colder than the other, it will be found to be thicker than the other.

When very thin plates are rolled this is a constant source of trouble, for in the manufacture of thin plates and long narrow strips, they cool so rapidly that their last ends are always colder and thicker than the first. In the universal mill the speed of the surface of the vertical rolls located at the back of the housings must be greater than that of the horizontal ones through which the bar is passed, the amount depending upon the extent of the draft which the roller uses at each pass. Any deficiency of the surface speed of the vertical rolls will prevent them from passing the metal contained in the bar as fast as they received it from the horizontal ones, and it would—if thin—accumulate in a mass of crumpled and bent steel between the rolls and be spoiled.

To avoid this, the vertical rolls should always be run somewhat faster than is required theoretically, which is a matter depending upon the skill and judgment of the roller if good results are to be obtained. As it is impossible at times to guard against a workman using too much draft at a pass, its speed must in practice be very much in excess of the correct amount. On the other hand, if the speed is too great this involves much slipping of the edges of the bar against the surfaces of the vertical rolls, which rapidly wears them out and absorbs much power uselessly, causing a considerable "backlash" which is very destructive to the gearing.

Avoid Backlash

Backlash on rolls takes the form of a succession of slips and jerks. Even if the surface speed of the two sets of rolls is correctly proportioned when they are new, the relation between their diameters is soon destroyed as they are turned down too correct for wear; as vertical rolls do not wear evenly with the horizontal ones.

It is the usual practice in most mills using steam power to operate their rolls to drive the horizontal ones by a reversing engine, while the vertical rolls are driven by a second separate set, but of less power. Taken all in all, the universal mill is not an easy one to work, much skill being needed on the part of the roll adjuster to keep the same work. ing smoothly. Constant vigilance must be exercised on his part as the bearings

and roll surfaces keep wearing and need looking after. The advent of the automobile has created a demand for sheets of the highest grade and quality, a demand which is exceedingly costly and difficult on the manufacturer's part to meet, as in the rolling of these sheets there are always some that contain slight imperfections such as surface blemishes, buckles, roll marks, furnace scale, pitting caused by the rolls wearing down near the end of the chill depth, and also dirt dropping on the metal from the point of the workman's tongs, which are allowed to rest upon the floor while metal is getting heated.

Sheets having slight imperfections are termed "seconds," and in order to make any money at all on this class of sheets the manufacturer generally insists that the automobile plant absorb a small proportion of these seconds, a price reduction being made on defective ones. Automobile sheets are mainly divided into three classifications. Number one is called body stock. The second is termed splash-guard, hood, door, apron and flat fender stock. The third, crown fender, radiator casing and cowl stock. There are further subdivisions of these classifications that consist in the steel analysis and also the hardness of temper which is imparted to the sheets by cold rolling. The bars from which automobile sheets are rolled are treated differently from those used to make ordinary ones, such bars being pickled in acid so as to remove scale and other dirt. There are also many other operations gone through, such as repeated picklings and annealings that the ordinary brands of sheets used for other purposes do not undergo.

Each mill that manufactures sheets for the automotive industry has its own method of getting results, but they all resolve themselves into rolling and producing the highest grades of sheets that are possible. Cold rolling figures largely in the obtaining of a high finish on sheets, but it also has the drawback of hardening the steel to such an extent between successive rollings, which are necessary to reduce the metal to the desired gauge, that frequent annealings between each rolling are required, otherwise the metal would become very brittle and crack.

Even the softest quality of steel is hardened or stiffened by means of hammering, drawing, or cold-rolling. Cold rolling increases the elastic limit and ultimate strength but also decreases the ductility. Body steel is divided into the following varieties or classifications: bright finish, and velvet or dull finish. These are again divided into standard stamping or drawing metal, deep drawing metal, and extra deep drawing quality. The bright finish is a glossy shining surface and it is intended for japanning or enameling. The velvet surface sheets have a grain that is more or less open, which is intended

to hold the paint, these sheets being made for hand coating. The painting operations are priming, rough stuffing and an air-drying enamel that needs no baking. Such sheets are admirably suitable for work of that character. The deep drawing quality is also used for parts that have shallow depth. Back and side panels usually are made from this quality of steel. The extra deep drawing quality of metal is produced for very difficult drawing operations such as seat backs, wheel housing, and other parts. The seconds from this class of sheets can be used by glazing over the imperfections with putty. Door, apron, splash-guard and fender-stock are made from a grade of sheets that need to have a very high finish as it has to be enamelled or japanned.

This grade of sheets has a closer grained surface than the body stock, but it will not stand much bending and it should be used for work that has no very sharp angles and that is comparatively flat. Deep drawing and extra deep drawing steel ought to be ordered for work requiring deep bending. After sheets for automobile bodies have been rolled they are next run through a series of small rollers to straighten them. This process removes any small buckles or strains that were caused by uneven annealing or other causes.

Making Sheets Straight

This method of straightening does not leave them absolutely flat. To make them level it is necessary to stretch them in a hydraulic operated device called a "stretcher leveler," which makes them absolutely straight and flat. It is not necessary to use stretcher-levelled sheets if they are to be drawn through dies, as the drawing operation will usually absorb any small buckle in the metal, providing the sheet is not badly or deeply buckled. As the sheets made for automobile manufacturers are liable to rust while they are in a warehouse, they should be oiled and crated ready for shipment.

Most automobile sheets are made in Nos. 19, 20, 21 and 22 United States standard gauge, and should be ordered in that gauge and not in thousandths of an inch. The writer believes that the time is coming when some of the alloyed steels such as vanadium, chromium, etc., will be made into sheets for motor cars as it certainly would have great advantages in regard to lightness and strength. If the new stainless and rustproof steel could be used in sheet form for automobiles it would give good satisfaction, as once the enamel peels off of the ordinary sheets used in our autos it doesn't take long for rust to form on the metal's surface, the result being that when a car has to be re-enamelled it takes a lot of work to make a smooth job on the rust spots. Using rustproof sheets would save a lot of time and money in this respect.

To Be Continued

Increasing Production With Pneumatic Tools

The Selection, Operation, Maintenance, and Repair of Pneumatic Tools Are Important Points. Different Methods of Driving Rivets, Method of Starting Hammer, Etc., Etc.

THE need of accelerated production in the shipbuilding and engineering industries has resulted in a greatly increased use of the portable pneumatic tool (which has now been developed to a very high degree of efficiency). Pneumatic tools with their percussive action are peculiarly adapted for riveting, chipping, caulking, etc., and are extensively used in shipyards, on bridge and constructional work, while in foundries and steel works their value for fettling castings and removing seams from billets is fully recognized.

The selection, operation, maintenance and repair of pneumatic tools are subjects which demand careful consideration. The hammer selected should be powerful and rapid in action and should operate with a minimum of vibration. The design should be well balanced and the form of the tools such that they are convenient to hold and guide without undue fatigue. The driving of rivets with a pneumatic hammer is a strenuous job and calls for a man of strength and stamina seasoned by experience.

The riveting hammer may weigh anything from 16 to 25 pounds and will strike from 600 to 1,000 blows per minute. The length of stroke varies, for different hammers between the extremes of 5 and 9 inches. For general purposes, however, it may be assumed that a hammer of 6 inch stroke will answer most needs in a satisfactory manner. Vibration is unavoidable in a pneumatic hammer, but in modern designs this trouble has been reduced to a minimum. Excessive vibration greatly increases the difficulty of holding and guiding the tool, and, of course, occasions the operator undue fatigue.

There are two methods of driving rivets at present in use on shipwork, first, the "two man hand gang" method, and, secondly, the one man "pneumatic machine" method. Driving rivets by hand requires natural aptitude, training, and experience with the result that, generally speaking, the hand gang riveter is probably more adept than the machine riveter and this is undoubtedly the reason why hand riveting has in the past been considered superior to machine work.

One authority has said, "Give a good machine riveter who understands his business a good pneumatic hammer, proper air pressure, well heated rivets, and he will do more work than the hand gang, with the results equally good."

In the actual driving of rivets the results depend largely upon the skill of the man operating the tool. Given average endurance, however, almost any man of reasonable intelligence can quickly learn to perform the operation with success. The points to be observed in riveting, if good tight joints are required, are as follows:

The holes must be practically in line. The plates must be well bolted together. The rivet heads must be well up against the plates. The rivet must be hot enough to expand and fill the hole for its entire length when driven. The holding up device must keep the rivet firmly in position. The riveting hammer must have sufficient power and speed to form the rivet before it becomes too cold to fill the hole.

A rivet which is loose in its hole may be caulked up to pass test, but will not carry its estimated proportion of the load under stress, and a large percentage of such rivets will lower the efficiency of a joint to the danger point.

POINTS TO REMEMBER

Pneumatic tools with their percussive action are adapted for riveting, chipping, caulking, etc. They are used in shipyards, bridge and construction work, while in foundries and steel works they can fettle castings, and remove seams and defects from billets.

Be careful in selecting the tool to get one suited for your work. After you get it, take care of it. A tool cannot do good work if it is not cared for properly. The most trifling faults are liable to throw pneumatic tools into disorder, causing them to stick and kick.

When rivets fit easily in the holes, and the latter are clean cut and sound on the head side, excellent work may be performed by the use of a simple pneumatic "holding-on" machine. In repair work, especially where old rivets have been cut away, these conditions do not hold good, and it will be found that the holes are ragged on the head sides, preventing the rivets from being pushed home. In these circumstances recourse may be made to one of the hammer or jam holding-on machines in which the thrust cylinder is combined with a hammer which may be operated to drive up the rivet and then shut off without interfering with the thrust. In button head riveting, the head assists the operator to hold the hammer in place, but when flush riveting is required, such as is used on the outside of a ship, the hammer must be guided and controlled without this assistance.

The appliances known as shell rigs are frequently resorted to in these circumstances for supporting the hammer when on the outer shell of a vessel. In driving button headed or cone headed rivets, the machine should be sufficiently powerful to form perfect heads without rocking to work down the edges. The hammer should be started off lightly until the rivet has settled somewhat into the hole to prevent bending to one side.

The following are actual times taken to drive button headed rivets with an AW No. 6 riveting hammer: ¾ inch diameter rivets, 3 to 4 seconds; 1 inch diameter rivets, 5 to 6 seconds; 1¼ inches diameter rivets, 12 to 14 seconds; 1½ inches diameter rivets, 17 to 20 seconds.

The necessary equipment for driving flush rivets includes a pneumatic chipping hammer and several chisels for cutting away the surplus material. The difficulty of driving flush rivets, is the necessity of getting the rivet's head to a point in order that the surplus metal may be chipped off while it is still hot, and soft. The operation is complicated by the difficulty of holding and guiding a flat set on a flat surface.

The duty of the holder is to ensure that the rivets are placed in holes hot and that the heads are well up to the work conditions, which are essential if good tight work is to be done. For general use with air hammers the "pneumatic holder up" is coming into increasing favor. The construction of these tools is very simple. A solid piston of about 3¼ inch diameter works in a shoot cylinder, and is operated by compressed air at the same pressure as that used for the riveter's. One end of the piston carries a snap or die adapted to fit the head of the rivet, while the other end of the machine is extended by screwing on suitable lengths of tubing to reach some solid backing. Upon opening the valve, the piston moves outwards and the rivet is firmly pressed into position.

As previously remarked, there are other forms of pneumatic holders in which percussive action is available in addition to thrust, which are of value where the holes are tight or the plate surface unlevel.

The maintenance of pneumatic tools is an important subject which is not always given the care and attention it deserves. The pneumatic tool store should be equipped with a plentiful supply of spare parts and such accessories and conveniences as experience has shown to be necessary, while repair men should be competent to remedy all minor

defects with a minimum of delay. Only a small percentage of repair men ever learn exactly why or how pneumatic tools operate, but they can all learn to keep them in their original condition as delivered from the factory.

Even this is not such a simple matter as it would appear, but it may be accomplished with a little care and observation. The most trifling faults are liable to throw pneumatic tools into disorder and cause them to stick and kick or start badly. As bad tools invariably lead to bad work, it is important to impress upon workmen the necessity of at once returning defective hammers to the stores. When a hammer kicks and sticks or is otherwise out of order, a riveter may frequently put up with it rather than risk the delay involved in journeying to the store for another tool, and as a result work is done which later may have to be caulked or cut away.

When machines are returned for cleaning and adjustment, the repairer should carefully examine surfaces which fit together to form air-tight joints, remove any burrs with a fine flat file, take away sharp corners, and slightly countersink pin holes. He should try, on a heavy block of wood arranged for this purpose, one of the hammers which is known to be in first-class working condition. With the working characteristics of this machine as a standard, he will readily learn how fast a machine should run, how much vibration is to be expected under normal conditions, and how rapidly the steel should cut into the wooden block when the machine is developing its full power.

With these standards in mind the repairer can test doubtful machines and judge by comparison whether they are right or wrong. A good repairer will become so familiar with the machines under his care that he will be able not only to state positively if they are in good or indifferent order but to readily diagnose faults. When a machine has plenty of power, but is slow, kicks, or refuses to start promptly, it will be generally found that the piston or the valve is not working freely. When in good order the piston should slide the entire length of the cylinder when tilted by reason of its own weight. If it fails to do this an examination will usually reveal high places on the pistons which are fouling the cylinder walls. These should be carefully reduced by rubbing with fine emery cloth. Rough spots on the cylinder walls should also be looked for and treated in a similar manner.

The valve should be carefully examined for burrs or high places which would prevent it from shutting closely on its seating and it should be noted that it is sliding freely through the whole length of its travel. A badly seated valve permits an escape of air which, although apparently insignificant, will be sufficient to cause a hammer perfect in all other respects to work badly. If the valve is working too easily it will open and close the ports and so reserve

the flow of air before the piston has had time to follow its action, and care should therefore be exercised in easing a tight valve to ensure that the operation is not carried too far.

It is assumed that the repairer will clean out all drilled parts and pin holes, being careful not to overlook the smaller ones. It frequently happens that a mysterious loss of power may be traced to a leakage occurring between faces which should form airtight joints, this being owing to the edges becoming burred by careless handling. Such burrs should be removed by means of a smooth file and the faces tested for inequalities by the use of a little blacking. Occasionally the failure of a hammer to start promptly may be traced to leakage at the throttle valve situated in the handle.

Such a leak, by filling the ports and passages of the hammer with live air, forces the piston to take up a position in the middle of the cylinder covering the exhaust ports, with the result that when full air pressure is applied the piston remains stationary in this position. Usually a tilting of the hammer combined with a slight tap is sufficient to overcome the equilibrium of the piston and start the hammer, but a machine which persistently exhibits this fault should be fitted with a new throttle valve without delay.

It may be taken as granted that the only way to ensure that pneumatic tools are properly cared for and maintained in good order is to insist upon their return to the stores at the end of each day's work. The repairer will then run each machine for a few moments and lay aside for further investigation any which in his opinion exhibit signs of disorder.

Those passed as satisfactory should be arranged on racks and lowered bodily into a tank of mineralised oil for the night, thus ensuring that no internal corrosion can take place as a result of the condensation of moisture from the air supply. It will often be found that a simple soaking in a tank of mineralised oil for a period of fifteen minutes will remedy a defect and obviate the necessity of dismantling a tool for repair.

Many of the foregoing remarks apply with equal force to other types of pneumatic tools as well as to riveting and chipping hammers, and the question of maintenance and repair should receive the most careful attention if good work and high efficiency is to be obtained from an installation.

LARGE STEEL PLANT

A huge electrically driven steel plant called the Winshu works, situated at Yawata, Japan, have had their equipment sent from the United States, the estimated cost being $3,700,000. The initial plant will include three 50-ton basic open-hearth furnaces, a 20-unit producer plant, an 84 in. plate mill, and a 24 in. structural mill. A motor-driven 35 in.

blooming mill is to be installed later. The three basic open-hearth furnaces will be served by twelve Smythe gas producers. The 84 in. three-high plate mill, which is nearing completion, is of the Lauth type, and will be driven by a 2000 horsepower motor through herring-bone reducing gears and pinions; the motor running at 420 revolutions per minute makes the mill speed 56 revolutions per minute. This company has its own blast furnaces, coal and iron ore mines, and firebrick plant in China, so that it will be the most completely self-contained organization in the Far East.

USEFUL INFORMATION

Where should I use Babbit, and when should I use bronze? If you have ever asked yourself that question, read what Brass World has to say on the subject. Babbitts as a class show high plasticity combined with low Brinnell hardness and compression strength, and these properties limit their application. Bronzes are therefore used where higher physical properties are desired to resist pressure or impact or to provide longer service. Whereas in the case of the babbitts, the softer element is the major constituent, so in the bronzes it is the minor constituent. Bearing bronzes are in general, copper-tin matrices filled with lead. The term bronze applies strictly to copper-tin mixture as differentiated from brass—a copper-zinc mixture. These terms are frequently loosely used, and it is quite common to speak of bearings as brasses. On the other hand, manganese bronze, Tobin bronze and a multitude of other so-called bronzes are in reality brasses, to which possibly one or two per cent. tin has been added, and aluminum bronze is usually a mixture of copper and aluminum, sometimes with a small quantity of iron.

NEW USE FOR AEROPLANE

A new use for the aeroplane has been discovered in Canada in carrying out surveys of mosquito-breeding areas. In undertaking such surveys, says Dr. C. Gordon Hewitt, the Dominion entomologist, one is often confronted with the difficulty of mapping out the swamp areas and other breeding places quickly with any degree of accuracy. By means of an aeroplane photographic surveys can readily be made. In forest work also it may be possible to use such a machine for making surveys of timber that is being killed or has already been destroyed by forest insects.

A process is said to have been developed in England for extracting alcohol and its derivatives from coke-oevn gas on a commercial scale. If applied to the whole of the coal carbonised in Britain, the process is estimated to yield 50,000,000 gallons of motor spirit per annum.

Lubrication and Its Importance to Industry

By A. H. NOYES

THIS is truly the age of machinery. The mechanical inventions of the past fifty years have been marvelous. Watching certain modern machines, with their giant arms, manipulating with absolute ease seemingly impossible loads, and watching still another type of modern machine performing operations delicate enough for a woman's fingers, it is almost impossible to refrain from looking for the god, sometimes the demon, who must be hiding behind the machine and working the levers.

Did you ever stop to think that this intricate and often delicate machinery depends absolutely upon one thing, a thing that can not only stop every current and cog and wheel but damage the whole apparatus beyond repair? Does one person in a thousand realize that this tiny thing, which nevertheless has the life of the whole tremendous mass of machinery in its power, is a drop of oil? What irrevocable damage is caused by two metal surfaces sliding, rolling or grinding over each other, if there is no lubricating substance between.

Ever since machinery has been in existence, lubrication has been its companion. In fact, it existed even before there were machines, when it was simply a case of reducing the power necessary to push one surface over another. People did not know the word friction, did not even know what lubricant meant, but they did know that a little bit of fat or grease made the sliding a lot easier. Even today, in the island of Madeira, the natives' ox-carts have runners instead of wheels, and to assist the patient animal, the driver runs ahead with the "grease rag," with which he smears the cobble stones of the roads. From years of greasing they look like polished ebony. It is a far call from a wooden sled sliding on cobble stones to two highly polished metal surfaces gliding over each other in a delicate machine, to the powerful piston of a great ship sliding back and forth in the cylinder pushing the whole weight of the ship before it, or the wonderful roller and ball bearings which do so much to reduce friction, and consequently the cost of motive power today. So it is a far call from the grease rag of the Madeira native to the marvelous automatic force feed lubrication of the present day's genius. If the yelling, gesturing native become a bit too tired of running back and forth with his "grease rag" it is only a little more sweat and effort from the patient ox that is at stake, no one hurt, no damage done, only a little more time wasted, and time is a drug on the market with the natives.

But when two polished metal surfaces slide over each other, unoiled, they have a fatal tendency to stick together. Then what? Bearings destroyed, cars delayed, ships at the mercy of wind and wave, time which is worth millions lost, contracts forfeited, and incalculable damage done. Lubrication, then, becomes the sine qua non of industry. This seemingly unimportant thing becomes of prime importance and must be considered from two viewpoints. The first of these is the selection of the lubricant, and the second is the continuous application of this lubricant to the bearings in the amount required.

Only a child would attempt to lubricate the highly polished bearings of an

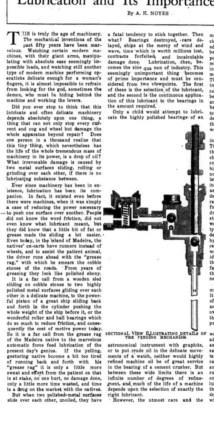

SECTIONAL VIEW ILLUSTRATING DETAILS OF
THE FEEDING MECHANISM.

astronomical instrument with graphite, or to put crude oil in the delicate movements of a watch, neither would highly refined machine oil be of great service in the bearing of a cement crusher. But between these wide limits there is an infinite number of degrees of refinement, and much of the life of a machine depends upon the selection of exactly the right lubricant.

However, the utmost care and the most expert knowledge might be exercised in selecting just the lubricant suited for a machine, and little use would it be if it never reached the bearings. A poor lubricant that reaches the spot is much better than a perfect one that stays in the oil tank. This is where the element of human frailty formerly entered. The man with the oil can was often more important than the president of the company, his mistakes would sometimes be more costly, and both the president and the oiler were liable to error. Men have such excellent "forgetteries." Add to that characteristic the universal tendency to carelessness and to "slighting the job," and you have in humanity a woefully deficient oiling machine.

Some accurate and reliable lubricating instrument had, therefore, to be devised. There were, however, certain advantages that the human oiler did have: the machine which should replace him must incorporate all of these advantages to a higher degree, and, in addition, it must not get out of order. So came into existence the force feed lubricator. For instance, the lubricator must start work as soon as the machine starts and must deliver oil in proportion to its needs. A ratchet driven pump was found to best fulfill these two conditions. The motor driven pump of course does not vary its stream of oil and the hydrostatic pump was no more successful. In the latter type the steam condenses in the condensing chamber and forces the oil drop by drop to the bearings. This is successful enough when the machine is working slowly, but when it works rapidly, so much live steam is passing into the chamber that it does not condense fast enough and therefore does not feed the required amount of oil just when the need is greatest.

Hence the ratchet driven pump, to my mind, has almost supplanted other types. It derives its power from a moving part of the machine and consequently stops and starts and accelerates or retards its own speed exactly with the machine. It therefore regulates its delivery in direct proportion to the needs of the machine. Some of the pumps are fitted with worm gears. These are especially adapted to high speed steam units such as direct connected light engines and sawmill steam feeds of the reversible rotary type. But the "man with the oil can" was not affected by heat or cold, and if working in flying dust he might rub his inflamed eyes, but he still kept on oiling. The best modern force feed lubricators have heaters incorporated in them and they are more oblivious to cold than any man. They do not solidify in winter, therefore, and they do not over-

feed in summer. The only way to avoid dust clogging was to place the oil in a solid box free from the working parts. This principle, as soon as discovered, was applied in the most advanced type of pumps. In this way it was made absolutely impossible for flying dust and grit to get into the oil and be carried to the bearings.

THE LUBRICATOR IN SECTION.

The working parts should be as simple as possible and they must be easily reached and repaired. One model has a one-piece valve which eliminates practically all troubles, for it is obviously impossible for a solid casting to get out of order. All the working parts of this model are on the outside of the oil reservoir, which probably accounts for the absence of repair bills for continuous service.

The sight feed is one of the latest and most important devices added to the force feed lubricators. The sight feed leaves the action of the lubricator visible so that an interruption in the flow of the oil is immediately detected. By its means, also, the engineer can see exactly the quantity of oil going to the various bearings.

The lubricators also have several chambers in the oil reservoirs, as many as required, in fact, and can therefore deliver different oils to different bearings. This is necessary to the life of any machine. Pumps of all sizes are in use, from the very small one-feed pump to the one holding several gallons of oil and delivering by forty feeds. The larger pumps or, as they may be called, pump systems are designed for the location they are to occupy. The lubrication of an entire plant is thus controlled from a central point. The economic advantage of such a system is unquestionable. To such an extent is this recognised that practically all the large power plants and all marine engines are equipped with automatic force feed pumps.

Although accuracy and reliability of action were probably the chief considerations in the mind of the originator of these pumps, yet they have been found to effect a very great economy in oil, as well as in man power. Some of the pumps are credited with saving from 50 per cent. to 80 per cent. of oil—a not insignificant matter in these days; and the cost of the wage which would have been paid to an oiler doing the work of its forty feeds.

Recent years may have produced more marvelous and more intricate machines than the force feed lubricator, but none to my mind that have demonstrated a greater usefulness or a more perfect adaptation to the needs they were designed to serve.

HOW TO KILL ANY SOCIETY

We often hear how we can help things along, but it remains for the Michigan Architect and Engineer to tell us how we can kill any society or association. There are ten ways to kill an association, they say, and here they are.

1. Don't come to the meetings.
2. But if you do come, come late.
3. If the weather doesn't suit you, don't think of coming.
4. If you don't attend a meeting, find fault with the work of the officers and other members.
5. Never accept an office, as it is easier to criticize than to do things.
6. Nevertheless, get sore if you are not appointed on a committee, but if you are, do not attend the committee meetings.
7. If asked by the chairman to give your opinion regarding some important matter, tell him you have nothing to say. After the meeting tell everyone how things ought to be done.
8. Do nothing more than is absolutely necessary, but when other members

roll up their sleeves and willingly, unselfishly, use their ability to help matters along, howl that the association is run by a clique.
9. Hold back your dues as long as possible, or don't pay at all.
10. Don't bother about getting new members. "Let George do it!"

TAPPING SOFT METALS

In tapping soft metals such as aluminum, there is quite often some difficulty in starting the thread, as the tap will at first either ream out the hole in a partial manner or strip the thread after it is started.

Both these dangers can be avoided, and perhaps the simplest method of all is to make an accurately threaded guide block about the same depth as an ordinary nut.

By placing this guide block in correct position over the hole to be tapped, no trouble will be experienced as the tap must then travel through the soft metal at a predetermined rate. Of course the guide block must be clamped to the work. It is essential that the block be fastened, and another point to be watched is the use of a good lubricant.

Whenever dealing with soft metals, do not rush the job. Humor it, for owing to the weakness of the metal it is likely to break away if too much force is used. Even copper can be stripped if two much force is adopted. If cutting threads on the lathe the danger is not so great, but even there it is necessary to use the best of care. Above all see that the cutting tools are sharp. If you don't the results will be far from perfect.

Some recent tests to determine the proper amount of air required for special thermit welding gasoline and compressed air preheaters showed that 25 pounds per square inch seems to be a practical minimum. At this pressure, a single burner preheater will require approximately 25 cubic feet of free air per minute and a double burner preheater approximately 50 cubic feet. For very large welds, where the walls of the molds are thick and the preheater gates longer than usual, a pressure of 40 pounds per square inch would be advisable. This would require approximately 35 cubic feet of free air per minute for a single burner and 70 cubic feet of free air per minute for a double burner.

A coating of magnesia cement on the timbers of mines is stated to be an economical and efficient assurance against fire, especially in the arid regions where timber becomes highly inflammable and is difficult to replace. It is elastic, adheres firmly to almost any surface, and is stable and durable.

WHAT OUR READERS THINK AND DO

The Design of Hinges and Hinge Pins

A Door Can Be Prevented From Sagging by First of All Determining "Are the Hinge Pins of Sufficient Strength"

By J. S. WATTS

WE have all seen machines of a neat and pleasing appearance spoiled by the doors being loose on their hinges, and hanging at a decided slope when opened, thus giving a worn out, dejected look to an otherwise good looking machine. It would seem that a few minutes of the designer's time, spent in making sure that the hinge pins load, without rapid wear, would be well repaid in the much better appearance of one with one long pin, for comparison. It will be noticed that the two pin arrangement allows the door to sag about twice as much as the one pin design, with the same slack. Moreover the two pin design tends to wear more rapidly than the other.

In the placing of the hinge bars, there are two ways in which these can be placed, as shown in Figures 2 and 3, respectively. If made as in Figure 2, it will be necessary, either to grind or machine the face of the hinge bar to get the door in place, or to leave sufficient clearance to allow for the roughness of the casting. This leaves a gap at "a" which is unsightly and becomes more so with wear. Making the hinges

as shown in Figure 3 obviates this difficulty, and the fit becomes better with wear.

To determine the size of hinge pin, etc., is purely a matter of calculating the bearing surface required, as the pin will be of ample strength in other directions if the bearing surface is sufficient.

Referring to Figure 3, the weight acting through its centre of gravity has a leverage "b," while the hinge has a leverage equal to "c." The load "P," on the hinge pins, is therefore

$$P = \frac{W \times b}{c}$$

and if we allow a maximum bearing pressure of seventy pounds per square inch we can expect the bearing to resist wear for a reasonable number of years. The diameter of the pin and the length of the bearing should be about equal to each other.

It is obvious that the greater we can make the length between the hinge bars "c," the lower will be the pressure on the pin, and the less the effect of the wear on the sagging of the door.

BROACHING ON THE DRILL PRESS
By Harry Moore

In shops where a limited amount of broaching is done the drill press is very often the broaching machine, and it is the purpose of this article to give some ideas on drill press broaching that may perhaps save many hours of experimenting. Keyways, both single and multiple, form common broaching operations, and therefore will be of interest to many readers. It is always good policy to drill out as much as possible of the keyway to be broached, and this can generally be done if the hole is not too long. In making a jig to drill out the semi-circular grooves, that will afterwards be broached, many mechanics make the bushings as shown at 1, relying on the bushing guiding the drill supported by grooves in the jig itself. This answers very well if only a few pieces are to be made, but the grooves wear out in time, become deeper and out of line, with the result that when the broach follows, it has too much metal to remove, and consequently, the work is not satisfactory.

A better method is shown in Fig. 2, where the bushing is extended to the bottom of the hole. It is a simple matter to replace the bushings when worn. The jig should be laid off and drilled first, then turned to the diameter of the hole in the work; the bushings are made solid and ground away after hardening.

Holding the broaches in drill press spindle is important for obtaining good

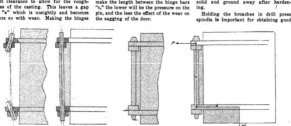

FIGS. 1 TO 3—ILLUSTRATING THE PRINCIPLES INVOLVED.

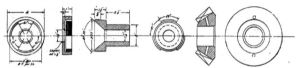

THE VARIOUS DETAILS OF THE CHUCK FOR THE BEVEL PINION.

results. The holder shown in Fig. 3 locates the broach (which is slotted at the top) and holds it lightly in place until it has passed through the work, when the withdrawal of the spindle will release it. The holder is provided with a small pin, behind which is a rubber pad or small spring, which keeps the pin pressed into a groove, thus preventing it dropping out when not in use. Broaching a long hole in the drill press can best be accomplished by using a bar with inserted blades, similar to a boring bar.

Since three or four of these blades will be needed it is essential to have some method of quickly changing them after each cut. The broach bar shown in Fig. 4 is designed with this object in view and answers the purpose admirably, since the only requirement is to keep the cutters central.

Fig. 5 shows the same bar adapted to cut four keyways. It will be noted that the bottom cutter locates the top one, and that by withdrawing the plunger the cutter can be removed, then the top one is free to slide out, providing a quick change; both cutters are slotted at the bottom as shown. The plate screwed to the bottom of the bar is slotted to allow the pin to pass through, so that when the plunger is withdrawn it is given a slight turn, which keeps it out of the way while setting or removing the cutters.

It would take too much space to go into details regarding width of teeth, amount of clearance, etc., necessary on a push broach, but a good general rule is—do not try to remove too much stock at the one time, the longer the hole the less metal per tooth, also leave as much chip clearance as possible, for if these get crammed with cuttings before the broach is through, one of two things will happen—either the broach will break, or the path it has traveled since the cramming period will resemble a miniature trench more than a keyway.

The location of the work on the drill press will, of course, depend on the shape of it; simple jigs will nearly always accomplish the end desired. To do quick work, it is better to have both work and broach located, making sure also that spindle is securely locked. Do not allow the broach to fall through the center hole and crash into a soap box on the floor. A better method is to place a block of wood under the table, so that the broach will have only a slight drop after passing through the work. In this way the broach can be replaced in the spindle without sprawling under the table. This may seem obvious to many, but it is remarkable how many instances one comes across in the shape of otherwise well managed concerns, where, with a little thought and effort, a heart or back-breaking job could be transformed into a pleasure.

CHUCK FOR BEVEL PINIONS
By G. Blake

We had a large number of small bevel pinions to grind out in the bore, and as will be seen from the sketch they were not suitable for holding in a 3 or 4 jaw chuck, so the writer got out the special chuck as shown in the accompanying sketch.

The plain part was gripped in the ordinary 3 jaw chuck. The special feature of the idea was the quick handling and loading, which was accomplished by employing the interrupted thread method as is used on the breech block of a gun.

A small groove cut on the cap, and another on the body of the chuck, served as a guide for the operator, so as to enable him to put the cap on in the right place for the thread to engage. Much time was saved in handling the job in this manner and the machine used was a Heald internal grinder.

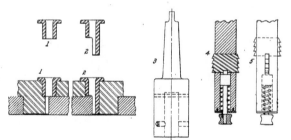

DETAIL VIEWS OF THE FIXTURE. REFER TO TEXT MATTER.

DEVELOPMENTS IN SHOP EQUIPMENT

OPTICAL MEASURING EQUIPMENT

The Van Keuren Co., 362 Cambridge Street, Boston, Mass., have sent us some interesting data regarding their optical measuring equipment and reference gauges. This optical equipment for the mechanical measurement of gauges by means of light waves is a new development, and is of worthy importance.

Take for example Fig. 1. This view depicts the optical measuring equipment, and this consists of a set of optical glass flats, and a source of monochromatic light, that is, light of one wave length or color.

An optical flat differs from a lens in that it has no magnifying power, but its utility consists solely in that it is a very accurate transparent flat test surface. Working flats 2 inches diameter by ⅝ inch thick are supplied, each having one surface accurately flat within 5 millionths of an inch. Only one of these is required for flatness tests, and the other two suffice for length measurements or comparisons.

One master flat, 2½ inches in diameter by ⅝ inches thickness, with one surface accurately flat, within 2½ millionths of an inch, serves to complete the set of optical flats. Thus the two working flats and the master, can be proven absolutely by testing them against each other, employing the familiar principle that three surfaces fitting together in all combina-

tions establish or originate flat surfaces. The master flat, in addition to its value for reference purposes, can be used on larger work than the working flats.

Monochromatic Light

The monochromatic light is a most practical arrangement specially designed for every day use. It consists of a conveniently constructed, finished oak box, provided with a tungsten filament lamp and a special selenium diffusing glass. This selenium diffusing glass cuts out all of the wave lengths of the violet, blue, green, yellow and orange light; and transmits only a very definite red wave length, resulting in an equivalent measuring unit of 8 dark interference bands to the ten thousandths of an inch or 12.5 millionths per band. Daylight, which results in merging colored bands with an average equivalent measuring unit of approximately 10 millionths per band is quite satisfactory for flatness tests. The monochromatic light is preferable for flatness tests and is particularly recommended for length measurements. It can be attached immediately to any alternating or direct current lamp socket.

The following rules must be observed when testing for flats:

1. To test a surface by interference bands it must be sufficiently well finished to reflect light. It should be clean and

dry; and nicks or feather edges should be removed. The optical flat is then placed on the surface and with a slight pressure manipulated to bring the point of contact to the place desired and the bands a convenient distance apart (usually about ¼ of an inch).

2. Straight bands indicate a flat surface. Curved bands indicate a curved surface. 3. In testing flatness with daylight, judge the straightness of the bands with reference to the distance between bands of like colors, and use an approximate measuring unit of 10 millionths per band. With the Van Keuren monochromatic light use the dark bands and an equivalent measuring unit of 12.5 millionths per band. 4. When the bands curve one-half the distance between their centre lines the surface is out of flat ¼ unit. Other errors are in proportion to the amount of curvature. 5. When the centre of curvature of the bands is on the same side of the bands as the point of contact the surface being tested is convex—when on the opposite side it is concave. 6. The pronounced light spot indicates the point of contact. 7. When comparing two gauges for length as illustrated, they are of the same thickness or length when the bands on both gauges match each other. 8. When a series of straight bands occurs between two flat surfaces, there is al-

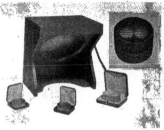

FIG. 1—THE OPTICAL MEASURING EQUIPMENT.   FIG. 2—A GROUP OF REFERENCE GAUGES.

ways a wedge of air between them. The bands run at right angles to the direction of the wedge. In comparing two gauges the shorter gauge will have its point of contact lower down on the wedge than the longer gauge, and the number of bands from the point of contact on the shorter gauge to the point of contact on the longer gauge indicates the number of units difference in length.

9. When the bands on two surfaces are parallel, the surfaces themselves are parallel. 10. When a flat gauge is properly wrung on a glass flat it is in contact within about 3 millionths of an inch, and no hands or colors occur.

Reference Gauges

The reference gauges shown on the other illustration are known as set R16. They are specially adapted to departmental and small shop requirements, and consist of five gauges, namely, 1-16 inch, ⅜ inch, ¼ inch, ½ inch, and 1 inch. Using these in various combinations makes available an accurate standard for every 1-16 of an inch. This set is said to be of particular value because it includes the great majority of stock sizes of drills, reamers, plugs, balls, tools, and drawing dimensions.

The flat faces establishing the standard of length are practically true surfaces, both for flatness and parallelism. The maximum variations of one or two millionths of an inch can be detected only with the aid of light waves, the most remarkable and modern measuring facility contributed by science.

A certificate showing the degree of perfection and the actual length measurements in millionths of an inch, compared by means of light waves, with authentic standards certified by the Bureau of Standards, Washington, D.C., is furnished with each set of reference gauges. Each gauge is held within a guaranteed accuracy of 10 millionths of an inch.

The contact face provided on the ⅞ inch diameter reference gauge has an area of .6 square inches. This is from 20 to 40 per cent. larger than usually provided on gauge blocks. A very important fact, however, is that experience has shown that these ⅞ inch diameter gauge blocks wear surprisingly uniformly flat and parallel. This is due to the fact that owing to their circular shape they are always wrung together in a different direction.

Expansion and Contraction

A 1-inch gauge block will expand 0.0000069 inch with a rise of temperature of one degree Fahrenheit. An equal length of iron or steel will expand from 0.0000058 inch to 0.0000072 inch depending upon the character of the material.

Thus, a tool or part made in accordance with these gauges, when the temperature is 80 degrees or 50 degrees, will contract or expand as the case may be almost the exact amount to bring the part being made correct at 68 degrees Fahrenheit, the standard temperature of measurement.

This automatic temperature control is of fundamental importance, for without it great difficulties would result.

TWO SPINDLE DRILL

The Black & Decker Mfg. Co., of Towson Heights, Baltimore, Md., has recently designed a two-spindle drill, specifically adapted for drilling holes necessary for the Murphy top fastener. This fastener is used for attaching automobile side curtains and when the base of the fastener is put on the automobile body, two holes ⅞ inch apart are necessary. This new machine drills these two holes the correct distance apart in the same operation.

The handle of the drill is like an automatic pistol and a pull on the trigger sets the two drill bits in motion. A sec-

GENERAL VIEW OF TWO-SPINDLE DRILL.

ond pull stops the drill and in this way the operator can control the tool without changing the position of either hand. The motor and the gear construction necessary for driving the two spindles is simple but rugged. The gears themselves are of the finest grade steel, heat treated. The two spindles have specially designed chucks to hold the two ⅛ inch drill bits.

The tool is exceptionally light and by holding it in the natural way with the forefinger on the trigger the operator is able to drill two holes at the same time ⅛ inch in diameter and ⅞ inch apart.

COMBINATION DRILL AND REAMER

A tool that has been specially designed for use on jig work has been brought out by the Fastfeed Drill and Tool Company of New York City. This is a combination drill and reamer made of high-speed steel. The drill size is made about .010 to .015 inch below that of the reamer which is the desired size of the hole. The tool has four blades and they may be sharpened on the front edge as in the ordinary reamer. It is claimed that a high rate of production can be maintained with these tools. These tools can be obtained with shanks to fit any existing equipment.

SURFACE GRINDER

The Perfect Machine Co., Galt, Ont., have placed on the market a new surface grinder. This machine has a vertical travel of 8 inches, and is equipped with a table 18 inches by 8 inches, having a micrometer adjustment. The machine will take a 12-inch by 2-inch emery wheel, and is suitable for all kinds of surface grinding work.

They are also manufacturing a new 20-inch light type drill press that is equipped with both gear and power feed.

STRAIGHT-COLUMN INCLINABLE PRESS

Several interesting features are included in the recently designed straight side press brought out by the Toledo Machine and Tool Company of Toledo, Ohio. The press is direct motor driven and the motor pinion is always in mesh no matter what the inclination of the press. A special elevating screw is provided at the rear of the press for angular adjustment. The press may be furnished with or without gearing or other auxiliary equipment, and the frame may be designed to permit side feeding of the stock.

SELF-CONTAINED GRINDING MACHINE

The U. S. Tool Co., Cincinnati, Ohio, has recently placed on the market the self-contained tool-grinding machine shown in the illustration, ease of starting and total enclosure of the motor and control apparatus being features claimed by the maker.

The machine is equipped with a 5 h.p. shunt-wound, adjustable-speed Westinghouse motor running at from 1,120 to 1,600 r.p.m. on 230-volt direct current. The motor is started and stopped by a push button that controls a type C Westinghouse starter equipped with a speed-adjusting rheostat and located in the base of the machine. The push button is mounted on the top of the motor so that

GENERAL VIEW OF THE GRINDER.

the workmen will not start the motor by jabbing the button with the material which they intend to grind, since it is easier to operate it by hand.

The MacLean Publishing Company
LIMITED
(ESTABLISHED 1887)

JOHN BAYNE MACLEAN, President. H. T. HUNTER, Vice-President
H. V. TYRRELL, General Manager.

PUBLISHERS OF

CANADIAN MACHINERY
~ MANUFACTURING NEWS ~

A weekly journal devoted to the machinery and manufacturing interests.
B. G. NEWTON, Manager. A. R. KENNEDY, Managing Editor.
Associate Editors:
J. H. MOORE T. H. FENNER J. M. RODGERS (Montreal)
Office of Publication: 143-153 University Avenue, Toronto, Ontario.

VOL. XXIV. TORONTO, DECEMBER 23, 1920 No. 26

A Christmas Message

CANADIAN MACHINERY extends to its readers its best wishes for a Merry Christmas.

"Just another year, another link,
To bind our friendship closer."

In this money-mad age we are apt to forget at times the real motive behind Christmas giving. It is a fact that a great section of the world has been so busy with grabbing and getting that it has had little time to pay much attention to the development of those finer traits that find expression in giving, and in considering the happiness and comfort of others.

It may clash with the teachings of certain radicals, both in the ranks of employers and employees, that it is a more wonderful thing to give than it is to receive. Despite this, that great old truth has stood for centuries and it will be here long after all the radicals, money-grabbers, agitators and the get-rich-quick group have passed on and been forgotten.

Christmas, 1920, finds the world rid of nearly all its war—finds nations, after months of skirmishing, still seriously considering the best way to weed out the chances of future wars through a combination of their resources and energies.

Things have not gone to the dogs, though there have been enough adverse currents to force them in that direction. The general bulk of business in this good land is still sweet and wholesome and showing signs that it is ready to sprout and grow into bigger and greater things in 1921.

To the advertisers in Canadian Machinery—we have appreciated your confidence during the year in our ability to carry your message to the industries of Canada and to distant parts of the Empire. We have tried to merit this appreciation by producing a paper that is sold to the readers on its merits, without premiums and without any inducement other than its educational and business value to the subscriber.

And so may your Christmas be a happy event, filled with kind deeds and inspiring thoughts, crowned with love and enriched with the sense and satisfaction of duty well performed and days profitably invested.

The Luxury Tax Off

THE removal of the Luxury Tax should help the machine tool business and many kindred fields indirectly. There has been so much tinkering with business in this country that the wonder is it has survived at all. Buying and selling, losing and making, have been twisted and bent out of form entirely, until one is forced to the conclusion that the more Government officials keep their hands off business in general the better for all concerned, especially for sound business itself.

The Luxury Tax did not directly touch the iron, steel and machine tool business, but it did touch the automobile business, and that is coming pretty close to the former, and it did touch many other lines, the stopping of which backed up on the tool and steel business in short order.

The machinery, iron, steel and metal working industries in this country need all the assistance they can get right now. They are buying from an admittedly high market and they are taking chances on selling into one equally as good as far as recovering their investment is concerned.

The manufacturer who is going ahead and doing his best to provide as much employment as possible is performing a far greater service than the class who are tub-thumping for tariff removals and decrying the importance of the manufacturer to the community.

Is This a Fair Practice?

OTTAWA reports indicate that complaints are still coming in that "freights on Canadian goods shipped from Canadian ports to Britain on Canadian lines are demanded in American instead of Canadian dollars. The Canadian Government Merchant Marine," continues the despatch, "has written to the Commissioner of Commerce that they have always made their freight collections in Canadian dollars. The complaints from Great Britain continue in regard to the practice of other lines demanding American dollars for Canadian service."

The report that the Canadian service never asks for payment in New York funds may be quite correct as far as present practice is concerned, but there are cases on record where it has been demanded, and bills of lading to this effect have been forwarded to the Department at Ottawa. Trade commissioners in various parts of the Old Land have complained several times about this practice.

Canadians are hurting their connections abroad by trying to nip in on American exchange when they have no right to do it.

If it is a Canadian product, shipped on a Canadian line to a country that is in no way connected with United States, it is hard to dig up a good excuse for adding a line to the effect that payment must be made in New York funds.

It is a short-sighted policy, and one that looks long and hard at a little present gain, rather than on developing and building for the future a trade that is going to continue to come back with its repeat orders in days when the business may be badly wanted.

MARKET DEVELOPMENTS

All the Markets Have Had A Quiet Week

Prices, Say Machine-Tool Builders, Are Not Going to Come Down as Costs are Still Up—Many Independent Mills Are Now Down for Lack of Orders—Carbon Drills are Marked Down.

DEALERS state that this time of the year should be rather dull even in normal times, and the machine tool trade will admit that these predictions are being fulfilled in this district. Nor is there any isolation in regard to the condition, New York reporting conditions much the same as here. The New York market is getting the worst of it from the standpoint of the dealers as a number of tools are being put out there for resale at figures well below lists. Yet the makers of machinery do not seem to be in the mood to be stampeded, as only this week several of them sent letters to their representatives in Canada that they were not making any changes in their selling prices. Their costs of production were not easing off, and that was the basis on which they were working, hence they could make no better selling prices than those already authorized. That, in brief, was the drift of the announcements.

Steel warehouses are quieter than they have been for some weeks past, although some of them claim a fairly good volume of trade. This may be accounted for in part by the fact that manufacturers prefer to draw from warehouse stocks rather than make up a mill tonnage themselves. So the warehouse trade cannot always be looked upon as a reliable barometer of business conditions.

Boiler tubes head the list of good sellers during this season. Most of the lake fleets are tied up now, and work starts at once on going over the equipment. Stocks are in good shape to supply all sorts of needs for this work, much better, in fact, than they have been for many months past.

Carbon drills were brought down by Canadian makers during the past week. They have been selling at 35 per cent. off the list, while the new price is 40 and 5 off.

The scrap market does not improve, nor do dealers see where any improvement is coming from. They refuse to buy except to fill contracts, stating that for the present they are out of the market.

MONTREAL GETTING IN SHAPE FOR THE DEVELOPMENTS OF 1921

Special to CANADIAN MACHINERY.

MONTREAL, December 23. — While here and there one hears the remark that business is good, the general impression is that quietness prevails, but of a character that would imply "the lull before the storm." The magnitude of the aftermath may be of small proportions, but the opinion is general that early in the New Year we will experience a partial return to more active participation on all sides in the development of industry and commercial activity. The C. P. R. management are taking advantage of the slack period and are closing the Angus shops for about two weeks, for the purpose of generally overhauling the plant and equipment. This is only one of the many instances where factory operations are being temporarily suspended to take inventory and make necessary repairs. In no small degree this has been reflected in the decline that has been noted in other lines of the metal industry.

Steel Movement Very Quiet

The steel situation continues to move along in the listless manner that has marked its movement for several weeks past, with such sufficient business passing to maintain the interest of dealers and producers, the consumers invariably being a passive spectator as to iron and steel activity. This does not mean that buyers are completely out of the market, but that their needs are gauged by the general requirements, and these have been materially reduced during recent weeks owing to the conservative policy that has been adopted on all sides. "The volume of business at the present time," remarked a dealer here, "is not of an encouraging character, but when conditions are taken into consideration, it is not surprising that this falling off in demand has developed. We are able now to get shipments from the mills almost as soon as the order is turned in, and, as a matter of fact, material previously on order is coming through in such quantities that our warehouse is beginning to take on a well stocked appearance. Some of this has been laid down at fairly high figures, while later quantities are coming in at more reasonable prices, so that in making sales we are able to equalize on material on hand. We are expecting some slight adjustment on some lines, as black sheets and plates, but believe that prices in general will hold comparatively firm."

Stock Taking Affecting Business

The fact that many industrial plants reserve the closing month of the year for their periodical stock taking has reacted on the business carried on, not alone in the plants directly affected, but also in other lines of activity that depend on co-operative enterprise for equalization of trade. Some shops throughout this district are taking this opportunity to readjust their operating staffs, and take a more than usual interest in the inventory of shop and equipment. "During the past week," stated one machinery dealer, "our representatives have been brought in touch with the stock taking phase of industry, where plants are practically closed to the outside world, as regards the purchase of material other than that required for absolute necessities. In some instances we are told that closing down is contemplated, for varying periods, for the purpose of taking stock and general readjustment of

working conditions. This attitude is the feature of present conditions and for the next few weeks will probably be the chief factor in industrial activity, or rather the lack of it." Asked as to the outlook for 1921, the dealer stated that he had every hope of a good year, slow at first but accelerating in the early spring. "With the gradual resumption of activity, which we anticipate early in the year, we may see a readjustment of prices, based on the new order of things as developed out of revised production costs, chief of which will be labor."

Buying at Low Ebb

The situation in old materials is still one where the interest of the dealer is hard to arouse. The consuming trade is almost nil and dealers are reluctant to take on any additional supplies in the hope of making a later sale. Present business is largely of the hand-to-mouth variety, where a direct transfer is made from seller to the buyer, through the dealer. Some light business is carried on regularly in meeting the needs of local foundries, but nothing of a special character has featured the scrap market for some time. "We are more than ordinarily quiet for this time of the year," said a dealer, "and we lay this to the readjustment that is taking place in many of the industrial plants throughout the country. Until this phase of industry has become more settled we are not looking for a renewal of active buying of scrap." Asked as to the possible duration of the slack period, the dealer would not commit himself, but the impression gathered was that before February there would likely be a gradual return to more active operations. In the absence of any marked movement of old materials dealers are satisfied to let quotations stand, as prices can only be based on individual sales. However, those quoted are the nominal average for the current week.

TRADE IS QUIET IN THIS DISTRICT

Makers Say They Are Not Changing Their Prices—Carbon Drills Are On a New Schedule

TORONTO.—If the removal of the luxury tax affects the machine tool business it will be in a reflex way, and will not be apparent for some time yet. The machine tool field is essentially in the class of essentials, and has not felt the tax as directly as many other lines. At the same time the removing of anything that will change the mental attitude of the buyer toward coming into the market to make purchases is going to work out to the good of trade in general.

Makers have not budged recently in the matter of prices. Such announcements as were going to be made came on some weeks ago, and since then the

The machine tool market in New York continues to be remarkably dull, and a number of makers are closing down their plants for the holiday season.

Many new machine tools are being offered by the United States manufacturers at a reduction from list prices. They have bought more equipment than they see use for now and are anxious to get rid of it even at a loss.

The price of carbon drills has been reduced in Canada. The old price was 35 per cent. off the list, while now the selling price is 40 and 5 off.

Some machinery dealers in this district have received word from their principals during the past week that they have no intention of reducing prices, as their costs are still too high to permit of any such action.

The steel warehouses are still doing a fair amount of business, as many manufacturers prefer to draw from the stocks in these places, for actual wants rather than to make up a mill tonnage under present conditions.

Boiler tubes are the bright spot in the steel market just now, as many of the lake fleets are tying up for the season and will immediately undergo repairs.

The scrap metal market is exactly where it was a week ago. Dealers in this district are refusing to buy except against contracts.

Pittsburgh reports state that a number of the smaller independent mills are closed entirely, and that the operating average of them all is not much over 35 per cent. of capacity.

Some reports have it that one of the next developments will see the independents cutting the Corporation prices.

There is very little business being done in pig iron at any of the furnaces, and reports from blast furnace centers indicate that some of the stacks are blowing out.

other makers have done nothing at all toward revising the selling schedules on which they are working. This morning one of the dealers showed this paper a letter which he had received from one

of his principals, in which it was stated plainly that they had made no reduction in price, and were not contemplating making any reduction. The letter went on to point out that labor was not down yet, and there was no lessening of the cost of production or of carrying costs. This firm held that it would be a serious thing for the company to go out and make a cut in the price now. The dealer who received the letter said he was not particularly anxious to have any reduction made at this time. He was quite satisfied that were he to start out today with a reduced list on his lines he could not use it to advantage. That feeling seems to be general, although one of the Toronto men in the machine tool business remarked this morning that it had often been the case, in his experience, when there was considerable talk about there being no drop in prices, the change came.

Hard to Sell Here?

"It is a hard market in which to sell machine tools on the basis of performance," was the complaint of one dealer this morning. "I may be wrong in the firms I work with, but in many cases I find that after going thoroughly into the working details with a man who is interested in a machine tool, and showing him where he will be money ahead every day by putting in the new equipment, he will buy a machine for less money. Many of the firms talk about going in for export trade, but I believe that they will have to manufacture to better advantage before they ever make much of a success in this line. It is one of the most discouraging things a machine tool salesman can run into, viz., trying to sell on a good sound basis, only to find that you lose ground many times by following this method."

A Quiet Month in Warehouses

Warehouse trade in steel business is quiet this month. It is in almost any year, and dealers are not prepared to state what percentage of their lessened trade can be blamed to seasonable conditions. It used to be said that steel was a good market barometer, but there are some exceptions, and the warehouse business is one of them, in that it may be very good while industry in general may be very quiet. Right now there are quite a number of firms in the country that are not buying for stock but drawing on the warehouses instead. This brings quite a bit of trading to the warehouse, but it does not mean that any more goods are being turned out, because the manufacturer in many cases would rather buy in small lots from the warehouse than make up a tonnage good enough to be forwarded to the mills in the ordinary way. Thus mills and factories may be operating at a very small percentage of their capacity and the warehouses may be quite busy supplying business that does not come to them at all when these same customers

are busy enough to warrant them placing mill tonnages for building up their own stocks.

Canadian warehousemen, it is claimed, cannot make as much on their turnover as the warehouses in United States. By the time the steel is laid down here they are up against a much higher cost than the U. S. warehouse and cannot recover proportionately as large an amount from the consumer.

Yards Not Busy Now

Apparently many of the warehouses are finding difficulty in keeping their men busy in the yards and shipping departments. There is not the volume of business moving there was a few months ago, although up to date there have been very few cases where men have been let out. A number of the dealers mentioned this and the chances are that some action may be taken toward making a reduction in hours or days of employment until there is some improvement.

Tubes are the brightest spot in the business now. Fleets are tying up for the winter at a good many of the lake ports, and that replacement and repair business is starting to come in nicely now. There is a good assortment of tubes on hand, much better than several months ago, when it was a difficult matter to keep all the wanted sizes in stock.

Carbon Drills Down a Little

A change was made in the selling price of carbon drills, the new price becoming effective the 20th of the month. These have been selling at 35% off the list for some time now, but the new price is 40 and 5 off. It works out about like this. Take a one-inch drill which is listed at $3.50, with 35 off would mean a price of $2.28 to the buyer. Now it will mean with 40 and 5 off, a price of $2 to the user. All the Canadian makers have adopted the new price.

There is a fair amount of "pick-up" business going now, but the average order is not very large. The National Railways have a requisition out now for some small tools.

No Improvement Here

The scrap metal market is just where it was last week. There is only contract dealing going on, and that is in very limited volume. The yards are still very much disinclined to take on any material and they are assuming the same attitude as formerly, viz., that they are absolutely out of the market, and they want it clearly understood that any prices they are responsible for quoting in print are purely nominal.

THE INDEPENDENTS MAY CUT THE PRICE IN EFFORT TO GET ORDERS

Special to CANADIAN MACHINERY.

PITTSBURG, December 23.—There has been a still more impressive decrease in the operations of the independent steel interests. One large independent is reported to be operating only three of its twelve blast furnaces, with steel making departments slowed down almost as much. Not a few of the smaller mills are closed entirely. The operating rate for the independents as a whole probably does not average over 35 per cent. and the common prediction is that by the first week in January the independents will be down to practically nothing.

This condition does not indicate that steel consumption has ceased by any means, for the United States Steel Corporation, instead of making less steel than formerly, is making more. The corporation is operating 91 per cent. of its steel ingot capacity, against 88 per cent. two or three weeks ago. It is not on account of increased pressure upon it for deliveries that the corporation has lately increased its production, but because it can now do so. For more than a year the corporation has been endeavoring to produce all the steel possible, and lately, with the loosening up of the trade situation, it has better supplies of coal and coke whereby it has been enabled to blow in some blast furnaces that had been idle perforce.

When the Steel Corporation operates at 91 per cent. of capacity, which is approximately 23,000,000 gross tons a year, it is producing at the rate of 21,-000,000 tons a year. During the first ten months of this year production by the industry as a whole averaged a rate of 42,000,000 tons a year, there being only relatively minor variations, in individual months, from this rate. Thus the Steel Corporation is now producing one-half as much steel as was formerly produced by the whole industry. Beyond question the consumption of steel has greatly decreased, and thus there is left little business for the independents. It is not a case of the Steel Corporation taking business away from the independents, but one of the corporation having previously booked the business, when the independents were charging high prices and buyers would patronize them only for early deliveries. Many consumers were buying both from the corporation and from independents and now, with reduced requirements, they can get along with their Steel Corporation deliveries.

There is an interesting aspect to this alignment. For years there has been a question whether or not the Steel Corporation is a "trust." In the suit of the Government, the Supreme Court decided in favor of the corporation, but by a narrow margin, and the matter was left in such shape that if the corporation should alter its policy there would be ground for re-opening the case. It has been quite a common view that the Steel Corporation should keep its proportion of the total capacity under 50 per cent. The corporation is on the safe side in that, for at the outside, its capacity, measured by steel ingots, is only 44 per cent. of the total. Obviously there can be no harm in the corporation making more steel than its "proportion" if it does so by letting the public have the steel at lower prices than the independents ask. Last year the corporation made 51 per cent. of the steel, instead of 44 per cent., this year there will be something of the same sort, and if present conditions continue any length of time the corporation will show a very high percentage of next year's production. Nobody could complain about that unless it would be the independents, and they would hardly be able to find a point of law on which to hang such a complaint.

Strip Steel Lower

The mills have made additional reductions in strip steel, following those of a fortnight ago. At that time the nominal prices were 5c for hot rolled and 8c for cold rolled, but the open market as shown by a few sales was half a cent less. The reduction was to 4c and 7c, respectively. The present reduction is to 3.30c for hot rolled and 6.25c for cold rolled. Strip steel had scored much heavier advances than other rolled steel products, except sheets,

the reason no doubt being that the automobile industry played such a large part, and the automobile factories were quite reckless earlier in the year in bidding for steel. The new prices, however, are really in keeping with the general steel schedule. The 3.30c price for hot rolled strip steel is simply the standard price of 3.05c for hoops, plus an extra or advance of 25 cents per 100 pounds for the quality, including accuracy, surface finish and stamping quality. Under the war control there were two extras, one of 25 cents per 100 pounds, over the hoop and band base price of 3.50c for "stamping quality," and one of 50 cents for "deep stamping quality." Now no distinction is made, all hoops being regarded as "deep stamping," and the material is simply called hot rolled strip at 3.30c instead of 3.05c, base, plus 25c extra for stamping quality.

Steel Prices Steady

At various times recently the independent market on tin plate, bars, shapes, plates, wire products and sheets declined to the Steel Corporation level. Pipe did not decline, and has not declined to date, the independent list being $7 to $10 a ton above the Steel Corporation prices, these being on a list with a basing discount of 57½ per cent. As the independent pipe mills have been receiving lately a few cancellations and a great many postponements or suspensions they will probably have to drop within a short time.

In the commodities in which the independent market has declined to the Steel Corporation level, no important shading has as yet appeared, although it is a common expectation that in due course some of the independents will cut the corporation prices. There is, of course, some quoting by western mills that technically represent price cutting, but the common view of this is that it is simply an ignoring of the "Pittsburg base," mills naming delivered prices that represent less than the regular Pittsburgh price, plus rail freight from Pittsburgh to destination, but a higher net price f. o. b. the mill than the Pittsburgh price. Particularly with the advanced freight rates in effect since August 26 there is a natural disposition, when orders are scarce, for the removed mills to ignore the Pittsburgh base.

The pig iron market remains stagnant. Nominal prices remain: Bessemer, 36c; basic, $33; foundry, $35, f.o.b. valley furnaces, freight to Pittsburgh being $1.96. There is no incentive to furnaces to reduce these prices as there is no inquiry, hence no opportunity for competition. Next month will probably see some declines, possibly to $30 for basic and foundry and $32 for Bessemer, but there may not be enough furnaces in blast for any actual reductions to occur. Only a third or a fourth of the valley merchant furnaces are now in blast, and these may not stay in.

While furnaces in general have been indisposed to contract for Connellsville coke for the half year, feeling that they do not know whether they are going to run and that the coke operators are not yet down to their lowest terms, a fair amount of business has gone through in the past week or two, involving perhaps 75,000 tons a month, but of course on a "requirement" basis, so that if the furnaces do not run they are under no obligations. All the business was closed on a ratio of 5 to 1 against basic pig iron at valley furnaces. If basic iron, now $33, is $30, something not at all improbable, the coke will be invoiced at $6 month by month until there is a change in pig iron. The contracts generally have a minimum below which coke cannot go, irrespective of pig iron, and some at least provide for a stiffer ratio in case pig iron advances above a certain point.

Spot coke is $5.50 to $6 for furnace, and $7 to $7.50 for foundry, per net ton at ovens, Connellsville region, though some particularly well known brands of foundry coke command more. Three important producers of foundry coke are quoting on first half contracts $10, $9 and $8.50, respectively, but none of them is booking any business of consequence.

NEW YORK HAS A VERY QUIET WEEK

And Tool Makers are not Likely to Operate Their Plants at Full Capacity

Special to CANADIAN MACHINERY.

NEW YORK, December 23.—The year is drawing to a close with no improvement in the machine-tool business, but eyes are eagerly turned toward the new year, when at least a modicum of betterment is expected. Some in the trade are not sanguine that there will be any marked increase in sales for several months, but it does not seem possible that conditions as dull as those which have prevailed during November and December can continue for long. December sales of some concerns in the trade scarcely provide for postage stamps.

New machine-tools are appearing on the market in increasing numbers and must be absorbed before there is substantial buying from the manufacturers. Many of the tools are being offered at sacrifice prices. A large company has placed on sale about 125 standard machines, many of which have never been removed from shipping crates, and at prices about 25 per cent. below those today quoted by the makers. Some companies overbought and are now anxious to dispose of their surplus equipment, even at a loss.

Most machine-tool plants will operate at a low rate over the holiday period, and some will shut down entirely for a week or more.

The **BERTRAM** Page
MACHINE TOOLS

84-inch Vertical Boring
and Turning Mill

*Constant
Speed
Motor drive
through
speed box*

The John Bertram & Sons Co., Limited
DUNDAS.　　ONTARIO.　　CANADA

MONTREAL	TORONTO	VANCOUVER	WINDSOR	WINNIPEG	HALIFAX
723 Drummond Bldg.	1002 C.P.R. Bldg.	609 Bank of Ottawa Bldg.	Davis Bldg.	1205 McArthur Bldg.	Roy Bldg.

Canadian Machinery and Manufacturing News, December 30th, 1920. Vol. 24, No. 27. Published weekly at 143-153 University Avenue, Toronto, Canada. Subscription price in Canada, $4; United States, $4.50. Entered as second-class matter at the Post Office Department, Ottawa, Canada. Entered as second-class matter July 1, 1912, at the Post Office Department at Buffalo under the Act of March 3rd, 1879.

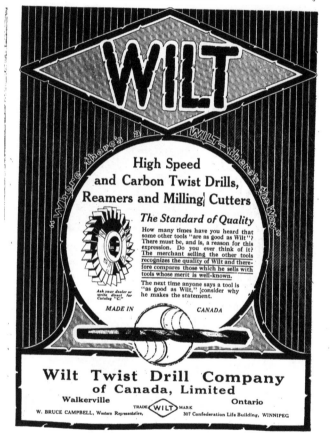

ALFRED

Horizontal and Vertical Milling
Machines, manufactured in a
range of sizes.

The Coventry Self-
Opening Die Head,
made in seven sizes—
5-16" to 3".

Combination Turret Lathes, made
in five sizes—13" to 28" swing.

THE machine tools included in the equipment of a modern engineering works may be roughly divided into two classes —main and auxiliary. Whilst the character of the former will vary with the nature of the work, the latter are, generally speaking, the same throughout the entire industry.

Thus the main plant of an automobile factory will consist largely of full and semi automatic turret lathes, milling machines, gear cutting machines and grinding machines, which are used only to a comparatively small extent in a marine engine shop.

The auxiliary plant, however, will be much the same in both establishments; cutting-off machines, tool, cutter and twist drill grinding machines, universal milling machines, tool-room lathes and universal grinders being necessary in some degree to all metal working undertakings.

For the last 25 years both main and auxiliary plant have been supplied by this firm for every kind of engineering work produced by machine tools.

HEAD WORKS:

COVENTRY, ENGLAND

HERBERT

, A   LIMITED

The Coventry Concentric Chuck, made in six sizes, 9" to 28".

THE manufacturing energies have been principally devoted to the development of a line of labour-saving machine tools used for repetition work. Turret lathes, capstan lathes, auto lathes and full automatic screw machines have always been a specialty. These machines are offered fully equipped with all the special tools and fixtures necessary for producing any specified work and at the same time guarantees of production can be given; these form a reliable basis on which labour cost of the work can be estimated.

A variety of both auxiliary and general purpose machines, comprising universal tool and cutter grinders, twist drill grinders, horizontal and vertical milling machines, ball-bearing drills, precision screw lathes and measuring machines are also produced.

In addition, sole selling agencies are held for the products of many prominent British and American firms. A comprehensive range of both machine tools and small tools can thus be offered which adequately fills the requirements of the modern machine shop.

Ball - Bearing Drilling Machines, made with one to four spindles.

Branches and Associate Companies at :
Toronto, New York, Paris, Lyons, Lille, Brussels, Milan, Calcutta, Bombay, Yokohama, Sydney, Etc.

Hexagon Turret Lathes for bar work, made in three sizes for bars up to 1⅝", 2¾" and 3" diameter.

ENGLAND

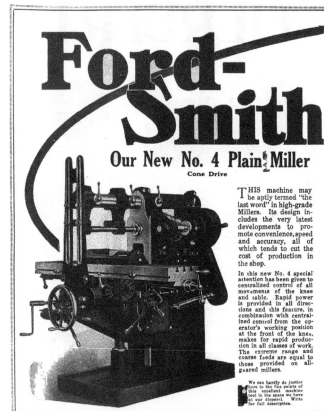

SPECIAL, alloy and regular analysis castings produced from electric and open hearth furnace steel. Single castings up to 50 tons. All heats subject to careful chemical and physical tests.

ION FOUNDRI
HAMIL

WILLIAMS

PIPE MACHINES

Used Throughout Canada and British Empire

THE Williams Tool Corporation of Canada, Limited, has taken over the business of John H. Hall & Sons, Limited, Brantford, Ont., and will continue to make the Pipe Machines that made the name of "Hall" favorably known throughout Canada and the British Empire. The new company will maintain the "Hall" standard of **quality** and **service**, and will combine the features of the "Hall" and "Williams" products with the aim of keeping them the best on the market.

A Few Outstanding Features

1—Single Pulley Drive, through.
2—Dodge Friction Clutch, on machine.
3—Clutch Lever at operator's left hand.
4—Rigid Bearing Bracket on drive shaft.
5—Gear Box Drive away from operator.
6—Individual Adjustment for each bearing.
7—Large Spindle Bearings are ring oiling.
8—Compact Control Levers on operator's side.
9—Large Die Cabinet and Tool Tray.
10—Substantial Oil Trough around top of ways.
11—Specially constructed Reservoir and Filterer.
12—Low-down Sliding Head.
13—Rotary Geared Pump reversible for left-hand thread.
14—Specially designed Carriage to drain off oil and cuttings.
15—All Gears amply protected.

WILLIAMS TOOL CORPORATION
of Canada, Limited - Brantford, Ontario
Successors to
JOHN H. HALL & SONS, LIMITED

If what you need is not advertised, consult our Buyers' Directory and write advertisers listed under proper heading.

If interested tear out this page and place with letters to be answered.

TRAHERN PUMPS are Standard

equipment on numerous makes of Machine Tools. Here are three illustrations showing typical installations.

Trahern Coolant Pumps are Rotary Geared. They will pump water, oil or compound; throw heavy streams free from pulsation; operate against 100 lbs. pressure; and reverse with machines. Speed 200 to 500 R.P.M., making the pumps very long lived. Will not lose prime when properly installed.

Let us send a list of firms using Trahern Pumps as standard equipment on their machines.

Prices low, delivery from stock.

TRAHERN PUMP DIVISION
GEO. D. ROPER CORORATION
ROCKFORD, ILL., U.S.A.

Automatic

ROELOFSON MAC
LIMITED
9 Wellington Street East,

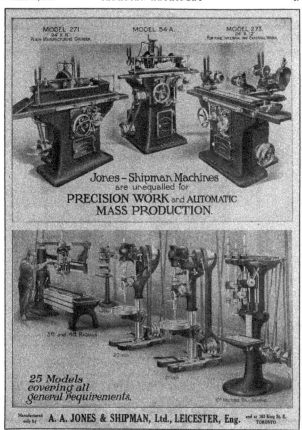

SPEED!

The "Landis" Threads 2,000 of These Nipples Per Day

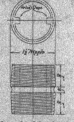

Landis Machine Company, Inc.
Waynesboro, Pa.

THERE is an interesting job in the line of threading nipples at the plant of the Fowler & Wolf Manufacturing Company, Norristown, Pa. The sketch shows what these nipples are and the photograph how they are threaded. One end receives a right hand and the other a left hand thread. Every one of these threaded nipples must have good, clean, full threads, because they are used on radiator work, and faulty threads would not be steam tight.

A double spindle Landis Bolt Threader is used, one side doing the right hand threading and the other side the left hand threading. Output is 2,000 nipples per day—4,000 threads.

The Landis bolt cutter has been in continuous service for the past six years, and not one cent has been spent for repairs.

Ask for the catalogue and more details of Landis Machines and Landis Die Heads.

For Sale by **The Canadian Fairbanks-Morse Company Limited**

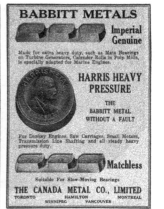

When "SHEPARDS" blanket the shop—

The Shepard line includes Electric Hoists and Monorail Cranes in capacities of ½ to 30 tons; Traveling Cranes, 1 to 50 tons; and Electric Cargo Handling Equipment for docks and ships. Our engineers, competent and willing to assist, will serve you without obligating you.

it results in unlimited hoisting service for each department.

The Shepard Electric Crane and Hoist pictured, make possible the transferring of a load to any part of the bay. The crane covers the length. The hoist operating on the crane covers the width.

At the will of the operator a piece can be lifted and conveyed to any desired point.

Invest in "Shepard" equipment—it pays big dividends by hoisting and conveying **more** in less time, at less cost, with a fraction of the labor otherwise required.

Shepard Electric Crane & Hoist Co.
426 Schuyler Avenue   -   -   Montour Falls, N.Y.

Canadian Representative:
RUDEL-BELNAP, Montreal

#111-S

Equip Efficient Tools with Efficient Drives

HIGHLY efficient tools are purchased to perform certain operations at a lower cost.

Effective as those machines may be, their maximum efficiency is largely dependent upon the form of drive used.

Link-Belt Silent Chain Drives deliver 98.2% of the power to the driven machine. There's no slippage of belts—no waiting for belt repairs—no inferiority of product due to ununiform application of power.

Every revolution of the motor shaft registers a like revolution of the driven shaft. Speeds are uniform. Greater speed reductions are possible. A better product results.

For Link-Belt Silent Chain Drives are "flexible as a belt—positive as a gear—more efficient than either". Write for our 40-page Machine Tool Drive Book No. 512. Shows many applications and gives full details.

CANADIAN LINK-BELT COMPANY
LIMITED

TORONTO   Wellington and Peter Streets
MONTREAL   10 St. Michaels Lane

LINK-BELT
SILENT CHAIN
FOR POWER TRANSMISSION

If interested tear out this page and place with letters to be answered.

"STERLING"

The man who knows says this brand of Hack Saw Blade is good—he believes there are none better. The band blades in all pitches are the best that can be made; also for power machines; they cut true, cut fast, and last long. They give real value used in any first-class machine.

This blade, combined with a "STERLING" No. 3A Improved Power Hack Saw, has no superior—it is a real winner.

"STERLING"
High Speed Power Saw

Improved Power Hack Saw
Machine No. 3A.
Solid but Adjustable Frame.
Automatic Lift on Return
Stroke.
Plunger Pump.

Manufactured by

DIAMOND SAW & STAMPING WORKS

357 SEVENTH STREET   BUFFALO, N.Y., U.S.A.

HOLES for 1921
Will Be Drilled Faster and Truer
with
CANADIAN-DETROIT
TWIST DRILLS

CUT down the cost per hole by specifying them on your next order.

We make milled drills of all types and sizes.

We also manufacture the famous DD rolled drills, in all sizes from ½ inch to 1½ inches, specially designed for drilling drop forgings and hard alloy steel castings. The Double-D combines the strength of the forged drill with the accuracy and uniformity of the milled type. The rolling process, which we control exclusively, gives greater density to the steel, without introducing the rough inaccuracies of hammering. Try DD drills on the toughest operation in your plant and watch them "hog in."

We invite comparative tests.

CANADIAN-DETROIT TWIST
DRILL COMPANY, LIMITED
WALKERVILLE, ONTARIO

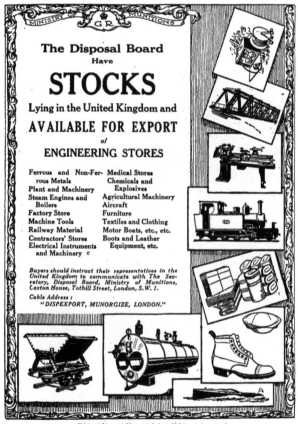

CRANK GRINDING MACHINE

Grinding Machines
Plain
Universal
Internal

SPECIALIZED TYPES

Crank Shaft
Cam Shaft
Ball Race
Piston
Roll
Gap
4-A Special
for Automotive Repair Work

Grinding Machines and Boring Machines have created a reputation for the "making good" qualities expected of machine tools. They have been especially designed and built for efficient and dependable service, and no effort has been spared to make them as perfect as possible.

See a LANDIS in operation and you will need no further proof to convince you that LANDIS machines are essential to rapid and accurate production.

**Illustrated Booklet
Sent on Request.**

UNIVERSAL GRINDING MACHINE

LANDIS TOOL CO., Waynesboro, Pa.
NEW YORK OFFICE: 50 CHURCH ST.

A
Big Job

Boring 44 holes, 1 13-16″ in diameter in a 14 ft. section of a rolling mill table is a big job. That's what this **Landis 6 ft. Boring Mill is doing.** The photograph, which was taken in the plant of the Dominion Steel Products Co., Brantford, gives a good idea of the type of work this machine performs.

Horizontal
Boring
Machines

If interested tear out this page and place with letters to be answered.

Looking
Ahead

FORESIGHT is the guiding hand of every really great enterprise. This ability of looking ahead—of accurately foreseeing trade demands and conditions—has enabled Garlock-Walker to meet these demands with products exactly suited for modern requirements.

We are now on the threshold of a New Year. A year of greater accomplishments, larger undertakings! Every individual in our entire organization will be striving with but one policy, every unit of our extensive facilities will be working with but one aim —to still better the already high standard of Garlock-Walker service.

Garlock-Walker Machinery
LIMITED

32 Front St. West, Toronto
Montreal and Winnipeg

GARLOCK WALKER
HIGH GRADE MACHINE TOOLS

If interested tear out this page and place with letters to be answered.

EVERYTHING IN METAL-WORKING

Cataract Bench Miller with vertical head is valuable for optical and small tool and die manufacturing.

Domes Four-spindle Bench Type Drills, with the latest Domes improvements, are capable of a large volume of high quality work.

Shapers of various sizes, with single pulley or cone drive, both new and used. In stock for immediate shipment.

Bath Universal Grinders will do your cylindrical, internal, surface, disc and tool and cutter grinding. Write us about your grinder requirements.

There is a Troy tool built especially suitable for your work.

Heavy Duty and Gap Lathes up to 60" swing and of different makes, in A-1 condition, are offered at attractive prices.

Many Silver, Champion, Rockwell and other back geared drills for your selection.

We have a good selection of manufacturing Engine Lathes, Turret Lathes, Screw and Automatic Machines in our warehouse.

Engine Lathes with complete equipment, with quick-change gears and double back gears, new and used, all sizes.

GARLOCK-WALKER

32 Front Street West, TORONTO

If what you need is not advertised, consult our Buyers' Directory and write advertisers listed under proper heading.

Interior View of Garlock-Walker Warehouse

20,000 Sq. Ft. of Floor Space Containing Every Variety of New and Used Machinery

YOU see merely a corner of our warehouse in the above illustration, but it will serve to give you an idea of the variety and extent of our line of machinery. Practically everything is found here for Woodworking and Metal-working purposes — lathes of all kinds, high speed drills, millers, planers, power hammers, electric and pneumatic tools — tools, in fact, of every description for shops large and small.

There are many used machine tools included in our exhibit, a number of them just as serviceable as new tools. It will pay you to look them over, for if they suit your purpose a considerable saving will result.

Remember this: Whether you are to personally inspect our stock or not you can place confidence in our desire and ability to furnish you with tools that will give absolutely good service. We aim to always supply our customers with machines best suited for their requirements. Success in this respect accounts for our continued and ever-growing patronage.

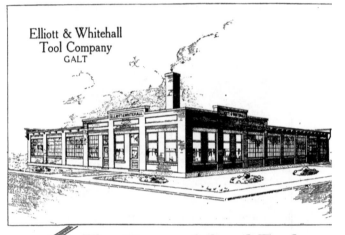

Bro *Sheet*

No. 15—Power Press

No. 10—Bench Press

No. 140—Double Crank Trimming Press

1,000 lb. Board Lift Drop Hammer

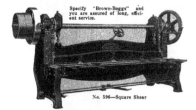

Bradley Hammers
f o r   p r o d u c t i o n —
25% and even greater has been the reported production increase in many instances.

Not only that, but Bradley Hammers stand up five, ten, twenty, thirty and even forty years, giving *continuous* service under the most severe conditions. They always assure *greater production*, more speed and *absolute accuracy*—because Bradley Hammers never fall down. For over 47 years Bradley Hammers have pounded out efficiency in hundreds of forge shops in America.

Made up in four styles—Bradley Cushioned Helve Hammer; Bradley Upright Strap Hammer; Bradley Upright Helve Hammer; Bradley Compact Hammer.

Let us convince you as we have convinced others.

C. C. BRADLEY & SON, Inc.
SYRACUSE   **N.Y., U.S.A.**

FOREIGN AGENTS

England—Burt & Hiscox, Ltd., London.
France—Fonderie Ferro & Co., Paris, Lyon, Rouen.
Alsace-Lorraine—Fenwick Freres & Co. Luxemburg.
Italy—Fenwick Freres & Co., Turin.
Norway, Sweden, Denmark—Otto A. Berntzd,
Copenhagen, Denmark.

Belgium—Fenwick Freres & Co., Liege.
Brazil—Fenwick Freres & Co., Rio de Janeiro.
Switzerland—Fenwick Freres & Co., Zurich.
Portugal—Fenwick Freres & Co., Lisbon.
China—Anderson, Mayer & Co., Shanghai.
Spain—Fenwick Freres & Co., Barcelona.

If what you need is not advertised, consult our Buyers' Directory and write advertisers listed under proper heading.

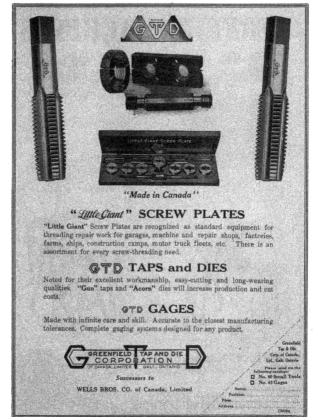

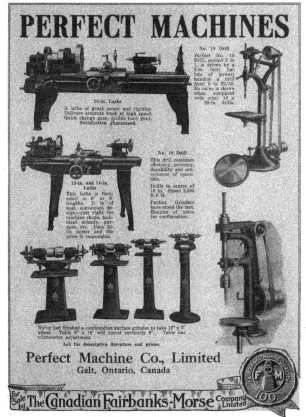

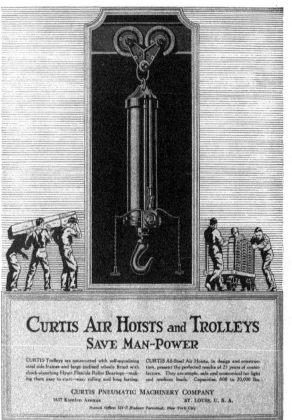

CURTIS AIR HOISTS and TROLLEYS
SAVE MAN-POWER

CURTIS Trolleys are constructed with self-equalizing steel side frames and large inclined wheels fitted with shock-absorbing Hyatt Flexible Roller Bearings—making them easy to start—easy rolling and long lasting.

CURTIS All-Steel Air Hoists, in design and construction, present the perfected results of 25 years of manufacture. They are simple, safe and economical for light and medium loads. Capacities, 800 to 20,000 lbs.

CURTIS PNEUMATIC MACHINERY COMPANY
1637 Kienlen Avenue   ST. LOUIS, U. S. A.

Branch Office: 511-5 Hudson Terminal, New York City

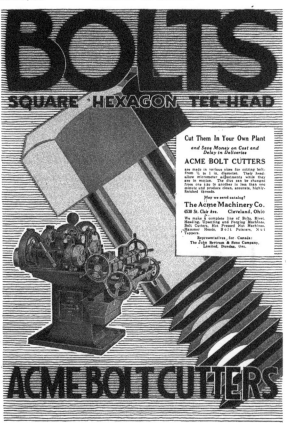

If the Pyramids were built today—

PIECE by piece, with years and men no object, clumsy hand-power slowly built the ancient Pyramids.

But, today, when men and minutes count big, man-power stands little chance in a tug-o'-war with the swift lifting power of Little Giant Pneumatic Geared Hoists.

They lift from one to ten tons without a hitch, jerk or jolt. A Little Giant load never slips because its automatic air-brake always grips. It holds what it hoists. A protecting limit-stop prevents jam-ups from over-hoisting.

Hoist pneumatically! Reduced labor costs and increased production prevail wherever Little Giants are used. These rugged all-steel hoists put *lifting* on a *production* basis.

Capacities of from 1 to 10 tons, in stationary, trolley, and geared trolley types. Ask for instructive Pneumatic Tool Bulletins.

Sales Representatives

The Holden Company, Limited
354-356 St. James Street, Montreal, Canada

Sales and Service Branches: TORONTO, 301 Adelaide Street, West WINNIPEG, 150 Princess Street VANCOUVER, 81 Pender Street
Canadian Factory: Canadian Pneumatic Tool Company, Montreal

COLUMBIAN
Sledge-Tested VISES

The Hollow Jaw IS Stronger

EVERY mechanic knows that a hollow tube has greater bending strength than a solid bar of the same weight of material. By actual test a tube 5 feet long is about three times as strong as a solid bar of the same weight and length.

Columbian Sledge-Tested Malleable Iron Vises are built according to this widely used principle of mechanics. The hollow jaw construction makes them much stronger than solid vises of the same weight. This construction, as applied to vises, is patented and can be used by no other manufacturer.

Columbian Sledge-Tested Vises are the only vises made of malleable iron. This feature alone makes them twice as strong as cast iron vises. The jaws will not chip, crack or break.

Removable steel jaw faces, cold-rolled steel handle with forged ball ends that never come off, are additional advantages that make Columbian Sledge-Tested Vises standard equipment in many of the largest machine shops in the world. They are a distinct aid to large-scale production.

Columbian Hardware Division
of The Consolidated Iron-Steel Mfg. Co.
World's largest makers of vises and anvils
CLEVELAND

If interested tear out this page and place with letters to be answered.

The City of *St. Johns* Quebec

It Will Pay You to Bring Your Plant to

the City of

St. Johns
QUEBEC

LET US BE YOUR TOOL ROOM

BROWN ENGINEERING SERVICE means up-to-the-minute equipment and highly skilled workmen, which enables us to build high quality and long service into all the special machinery and tools we make.

We have one of the best equipped plants in Canada with superior quality machine tools of every description, including a complete line of millers, grinding machinery, Johansson blocks, etc.

Send us your blue prints for estimates when you require any kind of special machinery, gauges, punches, jigs, dies, precision tools, etc.

Brown Engineering Corporation
LIMITED, TORONTO
415-419 King St., W.   Tel. Adel.425

MADE IN CANADA

Whitney Portable Hand Metal Punches

Allow Work to be Done Anywhere

These Whitney Metal Punches are not necessarily bench tools. Being convenient in size, they make powerful Hand Punches, allowing work to be done anywhere, with or without a vise.

Made of the best material—Drop Forged parts. Punches and dies quickly changed. No tools needed to dissemble; no bolts to remove, as in other makes.

W. A. WHITNEY MFG. CO.
715 PARK AVE., ROCKFORD, ILL.
Prices and full particulars upon request.

No. 2 Punch.
Length, 23 inches. Weight, 14 pounds. Capacity, 5-16-inch hole through ⅜-inch iron, or equivalent. Punches and dies in 13 sizes—⅛ to ¼-inch by 1-32-inch.

No. 1 Punch
Especially built for heavier capacity. Length, 34 inches. Weight, 22 pounds, well distributed, to nicely balance the tool. Capacity, ⅜-inch hole through ⅜-inch iron. Always specify SIZES of Punches and Dies desired when ordering.

Channel Iron Punch
A companion to No. 2 Punch. Every part of the two tools interchangeable. Length, 23 inches. Weight, 16½ pounds. Capacity, ⅜-inch hole through ¼-inch iron, or equivalent.

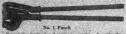
Canadian Agents:—Canadian Fairbanks-Morse Co.; A. R. Williams Machinery Co.; Williams & Wilson; McClary Mfg. Co.; H. W. Petrie, Ltd.

If interested tear out this page and place with letters to be answered.

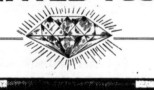

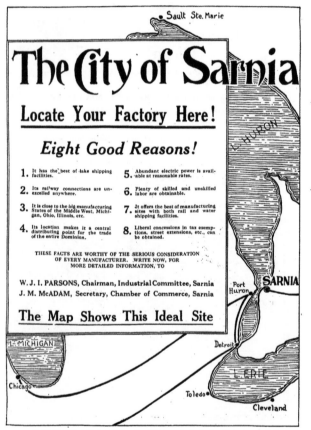

The City of Sarnia

Locate Your Factory Here!

Eight Good Reasons!

1. It has the best of lake shipping facilities.

2. Its railway connections are unexcelled anywhere.

3. It is close to the big manufacturing States of the Middle West, Michigan, Ohio, Illinois, etc.

4. Its location makes it a central distributing point for the trade of the entire Dominion.

5. Abundant electric power is available at reasonable rates.

6. Plenty of skilled and unskilled labor are obtainable.

7. It offers the best of manufacturing sites with both rail and water shipping facilities.

8. Liberal concessions in tax exemptions, street extensions, etc., can be obtained.

THESE FACTS ARE WORTHY OF THE SERIOUS CONSIDERATION OF EVERY MANUFACTURER. WRITE NOW, FOR MORE DETAILED INFORMATION, TO

W. J. I. PARSONS, Chairman, Industrial Committee, Sarnia

J. M. McADAM, Secretary, Chamber of Commerce, Sarnia

The Map Shows This Ideal Site

"Stubby"

A New Type Bridge Reamer

"Stubby" is about 25 per cent. shorter than the regular length bridge reamer, and the taper of the flutes is considerably sharper.

This results in a stronger, cheaper reamer which will materially lower reaming costs.

A "Stubby" reamer in the chuck of an air or electric drill is more easily handled—awkwardness of finding the hole and getting to work is done away with. "Stubby" Reamers will ream out a hole in little more than half the time required by the old-type bridge reamer.

All "Imco" Tools are good tools — built to remove more metal in less time; to be more accurate; to last longer between grindings; to have longer life!

Catalog of "Imco" 3 and 4-lip drills, reamers and milling cutters upon request.

Ingersoll Machine & Tool
Company, Limited
INGERSOLL   -   ONTARIO

Toronto Branch: 80 Bay St., Phone Adelaide 7227
Chas. A. Strelinger Co., Ltd., Windsor, Ont.

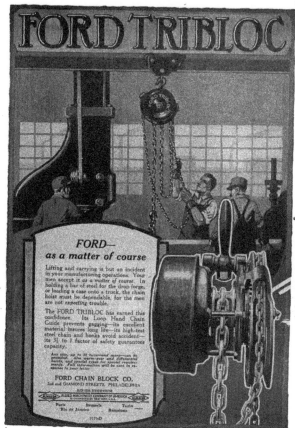

STEEL & IRON
PRODUCTS
OF EVERY DESCRIPTION

HAMILTON PIG IRON
Basic　　Malleable　　Foundry

STEEL AND IRON BARS

OPEN HEARTH STEEL SHEETS

RAILROAD TRACK MATERIAL
Angle Bars　Track Bolts　Tie Plates　Tie Rods　Spikes

SCREWS	WIRE	NAILS
Steel	Steel and Brass, Copper and	Wire
Brass	Bronze, Heavy and Fine,	Cut
Bronze	Bright, Annealed, Coppered,	Boat
Wood and	Galvanized, Tinned, Stranded,	Horseshoe
Machine	Steel and Copper Cable,	Shoe Nails
Screws	Barb, Woven Clothes Line	Tacks

POLE LINE HARDWARE
Pole Steps　　Cross Arm Braces　　Guy Clamps　　Guy Rods

WROUGHT PIPE	FORGINGS	LEAD PRODUCTS
Black Pipe	Car Axles	Lead Pipe
Galvanized Pipe	Shape and Drop Forgings	White Lead
Nipples	Carriage and Automobile	Shot
Couplings	Hardware	Putty

BLOOMS
BILLETS
PLOW BEAMS
ANGLES
CHANNELS

THE
STEEL COMPANY
OF
CANADA
LIMITED
HAMILTON　　　　MONTREAL

WIRE RODS
HORSE SHOES
FENCING
RIVETS
BURRS

If interested tear out this page and place with letters to be answered.

"ECONOMY" 22-inch Swing, Double Back Geared Quick Change Engine Lathe

Taper Attachment

ON the left is illustrated the taper attachment as furnished for our 12", 14" and 18" lathes. Its design is simple and consists of very few parts and its construction is extremely sturdy. No special bed is required; as all carriages are drilled, this attachment may be applied at any time. It is permanently fastened to the back of the carriage, making it instantly available at any position on the bed.

The regular compound rest and cross slide are used. The extra extension required is furnished with attachment and is interchangeable with short one supplied with regular lathe equipment.

& Drill Company
Illinois, U.S.A.

Builders of Engine Lathes

Power is Expensive—SAVE IT

But not by reducing production.

In fact, by the use of ball bearings, you can largely increase production and consume no more power than formerly, for in the average plant friction destroys at least 25 per cent. of all the power produced or bought.

High grade ball bearings, such as S K F and Hess-Bright, are almost frictionless. Not only is less power required, but belting life is lengthened, wear is practically eliminated, bearing renewals are postponed almost indefinitely, frequent tying up of equipment is avoided, oil costs are lower, and hot boxes are unknown.

Ball bearings, by conserving the energy formerly wasted, benefit the whole power line from the machines back to the boilers.

Canadian SKF Company Limited
Montreal   Toronto

A Service that contemplates more than the sale of a bearing.

Every S K F Industries service-salesman is a plant engineer equipped to consult on your bearing problems. S K F Service made available through these men is built on a laboratory-backed ideal that seeks to increase Canadian output by lifting the friction load from industrial machinery.

Manufacturers are invited to avail themselves of this freely offered co-operation.

Hess-Bright Ball Bearings
S K F Ball Bearings
Atlas Steel Balls
Gronkvist Chucks
Transmission Hangers

The SKF Research Laboratories where friction problems are solved

If what you need is not advertised, consult our Buyers' Directory and write advertisers listed under proper heading.

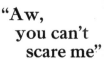

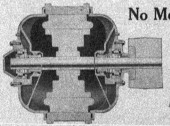

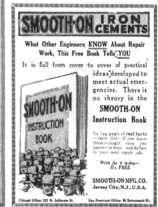

HEALD

STYLE No. 60 CYLINDER GRINDER

The Heald Cylinder Grinders are especially designed for internal grinding of holes in work of such shape that it is inconvenient or impossible to rotate in the usual manner when machining.

These machines are especially valuable for grinding airplane, marine or automobile engine cylinders. The ordinary method of smooth boring and reaming results in inaccuracies due to hard and soft spots in the casting, as well as springing of the thin cylinder wall. In grinding, the wheel revolves at a high speed in a true circular path, getting hard and soft spots alike, leaving an absolutely round hole.

Machine Style No. 60, as shown above, has a large vertical adjustment of 3½ inches, obtained by movement of the knee casting clamped to the bed. This machine is particularly suited to garage work or any shop where a variety of different sized work requires a large vertical adjustment.

Railroad shops are now using them to grind bearings in valve motion parts and piston rods, valve rods or brake parts, such as triple valves, etc.

Machine No. 65 is our latest style, built for precision and production work. Notice the massiveness, giving rigidity,—centralized control of levers, giving convenience of operation, and that it is self-contained.

STYLE No. 65 CYLINDER GRINDER

 THE HEALD MACHINE CO.

BRANCH OFFICES: New York, 699 Singer Bldg., Philadelphia, 1308 Commonwealth Bldg.; Chicago, 24 South Jefferson St.; Detroit, 401 Marquette Bldg.; Cincinnati, 911 Provident Bank Bldg.; Cleveland, 721 Engineers Bldg.; WESTERN AGENTS—Eccles & Smith Co., Los Angeles, San Francisco, Portland, and Seattle; Salt Lake Hardware Co., Utah, and Idaho; Hendrie & Bolthoff Mfg. & Supply Co., Colorado, and Wyoming.

INTERNAL GRINDERS

Heald Internal Grinders are built to handle work which can be easily revolved, such as bushings, cutters, ball races, hardened gears, gauges, cones, etc.

Each wheelhead is made up in a separate unit, instantly interchangeable. Twelve sizes of heads enable these machines to cover a wide range of work, which, together with the swivelling of the workhead, makes this machine indispensable in tool-rooms for grinding straight and tapered holes.

STYLE No. 85 INTERNAL GRINDER.

Style No. 85, as shown above, and whose capacity is 10″ swing by 2″ long, was built for small, short work. The construction is very simple, and a large number of mechanical features have made it exceptionally well suited for rapid production. It works best on holes smaller than ¾ dia. x 1½ long.

Style No. 70 handles work 15″ dia. x 11″ long. It has both automatic and hand feed for the table. Several new features have been lately added, including friction clutch workhead for starting and stopping the work, longer base to enable the operator to run the wheelhead back when plugging. Also the guard over the wheel has been re-designed so that it entirely covers the wheel.

Let us send you catalogues, together with literature showing these machines working in various factories.

STYLE No. 70 INTERNAL GRINDER

51 New Bond St., Worcester, Mass., U.S.A.

FOREIGN AGENTS—Alfred Herbert Ltd., England; Societe Anonyme Alfred Herbert, France, Switzerland; Societa Anonima Italiana Alfred Herbert, Italy; Horne Co., Ltd., Japan; Wilh. Sonesson & Co., Sweden, Denmark, and Norway; Henri Benedictus, Brussels, Belgium; American Machinery Corporation and Sindicato de Maquinaria Americana, Bilbao, Spain.

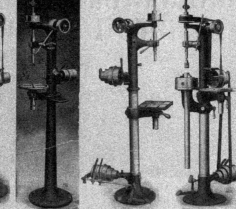
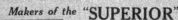

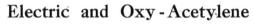

If what you need is not advertised, consult our Buyers' Directory and write advertisers listed under proper heading.

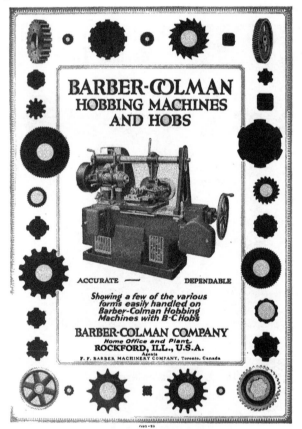

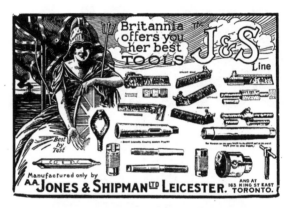

TRADE MARK REGISTERED

STRENGTH—GOODNESS—QUALITY
Performance Facts of Durabla Gasket Material

Tight joints are made and maintained by DURABLA under
the following severe services and conditions:

Steam:	275-300 lbs. pressure; 275°-300° superheat.
Oil:	Pressure still joint re-made over 160 times.
Gasoline:	Cylinder heads of gasoline engines and pumps.
Acids:	All acids that can be carried in pipe lines.
Water:	Equally satisfactory with temperature of 40° or 210° F., at any pressure.
Ammonia:	Both absorption and compression systems of refrigeration.
Air:	Air lines and cylinder heads of compressors—all type.

One standard material for all gasket work—Sold on a Service Basis

DURABLA is sold in sheet or gasket form and the trade
mark is stamped all over the material.

Manufactured by
DURABLA MANUFACTURING CO, NEW YORK

Did You Get Your Copy?

It's worth dollars because it contains the very information which you yourself would gather if you had the time to visit modern heat-treating plants throughout the country. But it costs you nothing, only the coupon, with your name, address, etc.

The cost of compiling and printing has been so great that distribution is limited to those directly interested in heat-treating work or responsible for results. The book is offered gratis in an effort to gauge more fully the value of "Canadian Machinery" as an advertising medium in the heat-treating field.

Please be sure, therefore, to use the coupon.

CANADIAN DRIVER-HARRIS CO.

WALKERVILLE, ONT.
CANADA

What "CANADIAN MACHINERY" Says:

This is one of the most compact little books yet issued for the use of those interested in case-hardening or carbonizing. It is not a mere advertising booklet, but contains information of general use to all those having to do with carbonizing or heat treatment of low carbon steel.

One hundred twelve 5 x 7¼ pages, 21 illustrations and several tables.

Coupon

C.M

Driver-Harris Co., Harrison, N. J.

Please send me (without obligations) a copy of the new book on CASE CARBONIZING. The nature of our heat treating is........................

No. of Furnaces Operated....................................

Name...

Title..

Company...

Address...

If interested tear out this page and place with letters to be answered.

If what you need is not advertised, consult our Buyers' Directory and write advertisers listed under proper heading.

WRIGHT HOIST

THE
HOIST
INDUSTRY
DEPENDS
ON

LATHES

or BAR JOBS

In view of present manufacturing conditions it is invariably performance that counts, and Steinle qualities given above are guarantees of satisfactory performance on your own work.

"If you want a recommendation of our Steinle you can get it from anyone in our organization who has come in contact with its production capabilities," said the general foreman of a manufacturing plant of creamery and dairy machinery and supplies who has had the opportunity of watching its performance since the installation at their plant in 1912.

Similar equipment will do equally well in your own plant.

Why not send us your blue prints if your requirements call for the installation of equipment of this character? We shall be glad to furnish you with production and tooling recommendations, or to have a qualified representative call and take up the question of Steinle methods as applied to your problems

MADISON, WISCONSIN, U.S.A.

ORIGINATORS OF THE SIDE CARRIAGE TURRET LATHE

If interested tear out this page and pin or glue with letters to be answered.

If what you need is not advertised, consult our Buyers' Directory and write advertisers listed under proper heading.

MULTI - DRILLERS 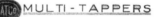 MULTI - TAPPERS

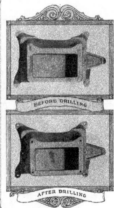

BEFORE DRILLING

AFTER DRILLING

Natco Multi-Driller helps make wash day easier

The Maytag Co., manufacturers of one of the best Washing Machines, use the Natco Multi-Driller to increase production on these labor saving machines.

In the piece shown here, which is the aluminum base for the engine, they drill three holes of different sizes to a depth of 5⅛ in. They formerly used a single spindle drill on this job and got a production of 40 per hour. Now, using the Natco Multi-Driller, they get a production of 80 per hour.

And yet with this 50% increase in production they have reduced their drill breakage 20% and state the ? ow it is practically zero.; But this

result is to be expected when you consider the strictly NATCO Patented Feature which makes possible an Independent Change of Speed for each Individual Spindle. Those not in use are thrown into neutral, saving wear and power.

The NATCO is the only Multi-Driller on which you can drill large and small holes, perform different operations, such as drilling, reaming, spot facing, etc., simultaneously and correct working speed. The independent changes of speed are made easily and quickly while the machine is running.

NATCOS are made in sixteen (16) different sizes, and it does not matter how large or how small the work is, or whether you have one hole or many holes to drill, there is a NATCO that will suit your requirements better and more economically and with greater production than any other Drilling Machine.

We also design and build jigs and fixtures for use in connection with NATCO Multi - Drillers and Multi-Tappers.

THE NATIONAL AUTOMATIC TOOL CO.
Richmond, Indiana, U.S.A.
Largest exclusive manufacturers of Multi-Drillers and Multi-Tappers.

If what you need is not advertised, consult our Buyers' Directory and write advertisers listed under proper heading.

Showing battery of NATCO Multi-Drillers in the plant of
Harper Sons & Bean of England, doing drilling and tapping.

The above letter tells the "Hole" story.

THE NATIONAL AUTOMATIC TOOL CO.
RICHMOND, INDIANA, U.S.A.
Largest exclusive manufacturers of Multi-Drillers and Multi-Tappers

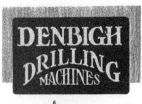

PERRIN
400-Ton Hydraulic Wheel Press

This Made-in-Canada Press is built to operate direct from Accumulator at 1500 pounds per square inch pressure.

Perrin for Reliability

WILLIAM R. PERRIN, LIMITED
TORONTO

GENUINE EMERY

Sizes 180, 160, 140, 120, 120, 100, 90, 80, 70, 60, 54, 46, 40, 36, 30, 24, 20, 18, 16, 14, 12, 10, 9, 8, hole.

EMERY FLOUR AND WASHED FLOUR

		In
Emery		
Glass	Paper	Sheets, Rolls,
Flint	and	Discs, Bands,
Garnet	Cloth	Strips and
Corundum		Tapes, &c.
Carborundum		

JOHN OAKEY & SONS
LIMITED
WELLINGTON MILLS
LONDON, S.E.1, ENGLAND

AGENTS:
F. Manley, 345 Carry St.
Winnipeg, Man.
Sankey & Mason
850 Beatty St., Vancouver

ROCKFORD
—the best for varied work

General manufacturing plants prefer the **Rockford Miller** because they require a machine which they can depend on to do any work within its range to the complete satisfaction of the inspection department.

Simplicity of operation and **absolute rigidity under every** condition are distinctively **Rockford features.** They have made this the very best machine for general manufacturing.

Full details on request

Rockford Milling Machine Co.
Rockford, Illinois

Agents:

Rudel-Belnap Machinery Co.
Toronto

A. A. Jones & Shipman, Ltd., Leicester, England, Great Britain; Louis Besse, Paris, France and Belgium; Societe Anonima Adler & Co., Milano, Italy, Italy and Colonies; Rylander & Asplund, Stockholm, Sweden; Scandinavia; Casemijians Helmanos, Barcelona, Spain, Spain and Portugal; Seisat Engineering Co., Malbourne, Australia; M. Meil Engineering Co., Petrograd, Russia.

If interested tear out this page and place with letters to be answered.

Kitchen and Table Cutlery

Produced Entirely on "BLISS" Presses

These photographs, taken in the plant of H. C. Hart Manufacturing Company, Unionville, Connecticut, illustrate the method of producing low-priced Kitchen Cutlery in large quantities.

"BLISS" Presses are used for all operations excepting grinding and polishing. The first operation is cutting blanks. Fig. 1 shows blank for handle; Fig. 4, blank for knife blade, and Fig. 5, blank for fork.

The second step is panelling the handles, Fig. 3; the cork filler for the handles is blanked, Fig. 2, and the two successive operations in cutting the fork tines, Figs. 6 and 7.

The final steps are bending, assembling, and closing the handles, illustrated in Figs. 9, 10, 11 and 12.

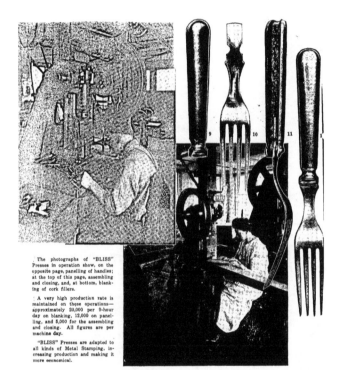

The photographs of "BLISS" Presses in operation show, on the opposite page, panelling of handles; at the top of this page, assembling and closing, and, at bottom, blanking of cork fillers.

A very high production rate is maintained on these operations—approximately 20,000 per 9-hour day on blanking, 12,000 on panelling, and 5,000 for the assembling and closing. All figures are per machine day.

"BLISS" Presses are adapted to all kinds of Metal Stamping, increasing production and making it more economical.

E. W. Bliss Company

Main Offices: BROOKLYN, N.Y., U.S.A.
American Factories: BROOKLYN, N.Y., and HASTINGS, MICH.

————————SALES OFFICES————————
CHICAGO, People's Gas Bldg.   DETROIT, Dime Bank Bldg.   CLEVELAND, Union Bank Bldg.
CINCINNATI, Union Trust Bldg.   BUFFALO, Marine Bank Bldg.   ST. LOUIS, Boatmen's Bank Bldg.   PITTSBURGH, Keenan Bldg.
————————FOREIGN SALES OFFICES and FACTORIES————————
LONDON, ENGLAND, Pocock Street, Blackfriars Road, S.E.   PARIS, FRANCE, 100 Boulevard Victor-Hugo, St. Ouen

No. 140

If what you need is not advertised, consult our Buyers' Directory and write advertisers listed under proper heading.

4,000 Ton Steam Hydraulic Forging Press for Marine and Other Heavy Forgings.

The product of a modern steam-hydraulic press is denser and more homogeneous than can be obtained with the steam hammer, due to the effect of the blow of the latter not penetrating to the centre, in contradistinction to the uniform kneading effect of the press; while the amount of work that can be done by the latter at one time with little variation in temperature, strongly tends towards a better product. The greater uniformity and reliability of steam hydraulic forgings make their use imperative wherever high-class products are required.

We Will Be Glad to Answer Your Enquiries.

NOVA SCOTIA STEEL & COAL COMPANY

General Sales Offices :   **LIMITED**   Head Office :
WINDSOR HOTEL, MONTREAL   NEW GLASGOW, N. S.

the Driller
sold by shopmen

You can talk on an equal footing with a Hoefer Sales Engineer. For he's been all through the very same mill that you are grinding—he's struggled with the problems of planning his production the quickest, best and most economical way, and the harder problem of making performance match promise, and he fully appreciates the importance of selecting equipment that will prove the most suitable to handle a particular set of conditions. So he knows just what you're up against, is qualified to be of real assistance to you and is always ready to apply his own shop experience to help you over a rough spot or two.

We think it a mighty fine combination—Hoefer Drillers and Hoefer Sales Engineers. Drop a line today to the nearest sales office and let a Hoefer engineer prove to your own satisfaction how helpful Hoefer Service and Hoefer Drillers can be to you.

Send for the Hoefer Catalog

HOEFER MFG. CO., FREEPORT, ILL.

If what you need is not advertised, consult our Buyers' Directory and write advertisers listed under proper heading.

Selecting Wheels for "Off-Hand" Grinding

Part only of a treatise on "Grinding Wheel Selection," by Carl F. Diets, General Sales Manager of Norton Company.

OFF-HAND grinding includes all sorts of rough grinding on castings or other parts where precision is not required, but where the quick removal of metal is the primary object. It also includes certain kinds of tool grinding which depend on the skill of the operator, but is usually confined

Grinding with a swing-frame machine.

work unless they are firmly held; hence, harder grades must be selected.

The flexible shaft machine is not much used, but finds application where the parts to be ground are clumsy and have corners or recesses that can only be reached by having a wheel so mounted that it can be presented to the work by the operator at will. Bath tubs, sinks, certain railway castings, and the

Off-hand grinding on a floor stand.

to the rougher tools for lathes and planers. There are three types of machines employed. Floor and bench stands, swing frame, and flexible shaft machines. Obviously the floor stand, being generally the most rigid of these types, may for certain operations be fitted with comparatively soft wheels. For various kinds of tool grinding or light general work, grades M to P can well be employed, while for heavy castings such as journal boxes and the like, or very sharp work, grades as hard as U are necessary.

Swing frame machines are subject to much vibration, due to their light construction and also to the manner of application. In grinding with swing frame machines the wheels are generally subject to much abuse, being brought down upon the casting with much force. Vibrations transmitted as a rapidly intermittent series of shocks at the wheel face can readily be seen to exercise a destructive influence on the structure. Grain particles are split off before they have performed their measure of

Flexible shaft equipped with Norton wheel.

like are ground in this manner. Obviously with little to support and only the hand of the operator to guide or steady it, the wheel has to overcome the shocks resulting from the many irregularities of surface ground and therefore needs to be of hard grade.

NORTON COMPANY OF CANADA
HAMILTON, ONTARIO

Electric Furnace Plants:
Niagara Falls, N.Y.
Chippawa, Ontario, Canadian

Canadian Agents:
The Canadian Fairbanks-Morse Co., Ltd., Montreal, Toronto, Ottawa, St. John, N.B.; Winnipeg, Calgary, Saskatoon, Vancouver, Victoria.
F. H. Andrews & Son, Quebec, Que.
Simonds Canada Saw Co., Vancouver, B.C.

If interested tear out this page and place with letters to be answered.

CANADIAN MACHINERY
AND
MANUFACTURING NEWS

VOL. XXIV. No. 27   December 30, 1920

1921 Finds Canadian Business Sound

THE words deflation of values and readjustment are so threadbare from overwork that they have lost their power to convey the meaning they should express.

Our best business men saw readjustment coming as soon as the war was over. Steps were taken then to fight it off—to continue our fake prosperity—and these were successful to a much greater degree than the most ardent exponents of that system ever thought possible.

Nothing cracked following the war.

We continued to imagine that our dollar was one hundred cents, and not merely a medium of exchange, depending for its value on the selling price of commodities.

The deluge of war work brought with it habits of waste—an unbalanced organization with all stress on production and none at all on selling—a government ready to take the product and assist in financing its production led to a neglect of scientific market surveys and study.

Men and women from all trades and callings went to war shops and earned more money than they ever thought possible. They acquired habits of extravagance in living and lavishness in spending that read like new ventures of the prodigal son.

In plain, honest words—we were not ready to settle down after the war, and the volume of business that has passed since then has too often lacked the sincerity that springs from intrinsic value and the stability that can be secured only from sound basic conditions.

Readjustment is bound to hurt. It would be against all business and physical laws were it otherwise.

It means passing from an atmosphere of imaginary prosperity down to an era of common sense.

It means that earnings must be less, where there was war inflation—it means that competition is again taking its old place in the business world, and that sales will more and more rest, not upon emergency or undue haste, but upon mechanical excellence and administrative sincerity.

Labor efficiency is increasing, but the lamentable, grotesque feature is that the improvement comes when the demand for production has fallen off.

There has been progress made in killing the strike germ.

Workmen seem slow to realize what a costly weapon the strike has been to labor. It has not been a boom, but a boomerang.

The year 1921 will require patience and consideration from employers and employees. It would be a wonderful thing were the labor agitator and the autocratic employer to link arms and depart from the industrial district. Their absence would be of more assistance than their presence and interference.

Fundamentally Canadian business is sound, and becoming schooled to changed conditions. Financial restrictions have caused firms to cut off the frills and clean their houses, and they have put them in shape for healthy growth and certain expansion.

Scores of splendid American firms have decided on their location in Canada, and from here will cater to the great markets of the Empire.

The development of the Canadian West is going on on a more sane basis than ever before. The people there realize that their future rests in working the soil instead of the people. This means more implements. More implements means more machine tools, iron, steel and equipment.

The year 1921 is before you, and it comes a day at a time. If you think there's a greater or a better land than Canada, or a place where you will be better treated, make haste and locate—if you believe that this great Canada is your best chance, then help make it what you want it to be.

Tracing the *Development* of the Year

Canadian Cities Have Prospered, and, in Spite of High Costs, Extensions Have Been Made—Many U. S. Firms Getting Ready To Open Up New Branches in Canada During the Year 1921.

CALGARY HAS HAD A BUSY YEAR

Reports Say That Business Has Been Good In All Lines—Work Still On in the Oil Fields in That District

By J. H. HANNA, Secretary Calgary Board of Trade

CALGARY.—During the past year Calgary has enjoyed increasing prosperity and this is shown in the bank clearings, which for the first ten months of the year were $348,549,415, as compared with $209,839,671 for 1919, an increase of $78,709,744. The live stock industry is one of the important branches of Calgary's trade and during the past year the Alberta Stock Yards Company have spent $400,000 in improvements and additions to the yards and for new offices. They have also secured an additional sixty acres of land in order to increase their facilities, and provide for the rapidly growing business.

Notwithstanding the comparative high cost of building, considerable work has been done and the building permits for the first ten months of the year show a total of $2,532,200, an increase of $411,200 over the previous year. At the present time two large and modern picture theatres are under construction and will doubtless be completed in the course of the next couple of months.

In order to keep pace with the increasing population, it has been necessary for the School Board to build five new schools, and even that extra accommodation has not been quite sufficient to meet the need.

Calgary has also benefited by the increased output of coal in this part of the Province, most of the leading coal mining companies having their head offices in this city. In the Drumheller district particularly, there has been a greatly increased output of coal, due to the new markets obtained in Manitoba and Saskatchewan. There is every prospect that this trade will develop, in view of the tests which have been made and which have proved satisfactory. This will mean a greatly increased output for the Alberta mines in future.

Calgary's principal industries, including packing house products, flour, breakfast foods, soap, saddlery, and other commodities, also the large shops of the C. P. R., have been kept busy during the year, although latterly, owing to the readjustment in prices which is now going on, business is not quite as active as in the earlier part of the year.

Development work is still going on in the oil fields adjacent to the city, and operators are hopeful of finding further oil deposits in the near future.

ST. JOHN HAS MORE POWER NOW

Existing Plants Have Been Much Enlarged, And Shipping Facilities Are Being Improved for Ocean Ports

By R. E. ARMSTRONG, Secretary St. John Board of Trade

ST. JOHN, N.B.—Marked progress has been made along one important line of industrial activity on this locality during the past year, viz., the dry dock and port development at Courtenay Bay, East St. John. The chief purposes of this development are to provide St. John with additional port facilities, supply it with one of the largest dry docks in the world, and at the same time furnish the Canadian National Railways with adequate terminals. The C. N. R. has had for many years track con-

nection with the western side of Courtenay Bay and recently the eastern side has been connected with the main line of the C. N. R. by means of a spur track. The inner basin of the bay is being dredged out to conform to the depth of the main harbor which lies westward of the peninsula upon which the city proper stands.

Adjacent to Courtenay Bay, further industrial promotion is about to be developed, including the establishment of a mill for the manufacture of fibre cases and also a pulp mill. These industrial works will be established by a corporation, of which the principals are the Clarke Bros. firm of Bear River, N.S.

The pulp mill of the Nashwaak Pulp and Paper Company, located alongside the reversing falls, just above the main harbor, has been enlarging its capacity lately and has arranged for an added water supply from the city to cover its new development programme.

Immediately adjacent to this mill the Canadian Pacific Railway is engaged in the construction of a steel bridge across the Falls. The piers for the bridge are well advanced and the placing of the metal work will be begun within the next few months.

The lime industry of St. John has made excellent progress this year. The plant of the Provincial Lime Company, which was destroyed by fire some months ago, has been rebuilt and enlarged, and adjacent to it the Provincial Government has an agricultural lime industry at work.

The lumber industry, insofar as the overseas trade is concerned, has been quiet this year, chiefly because of the unsettled conditions in the Mother Country. Large quantities of lumber are piled up here, awaiting an improvement in the market conditions. A somewhat better state of affairs prevails in the United States lumber market. The shipments of lumber to the United States during the first nine months of 1920 amounted to $3,737,992, as compared with $1,506,897, for the same period of 1919. The entire lumber export to the United States last year from this district was valued at $2,504,798. Almost all the lumber mills are now closed down for the season. It is not expected that there will be a large lumber cut this winter.

There have been a few changes in the metal industries. The Portland Rolling Mills plant on the Strait Shore has lately changed hands, and the company now operating it is known as the New Brunswick Rolling Mills. A number of lines of iron and steel products are being manufactured by this Company.

The foundry on Courtenay Street, known as the Smith Foundry, has been acquired by the firm of Harrigan and Howard, who are operating it as the Courtenay Iron and Brass Works.

The McAvity foundry plant on the Marsh Road is not yet operating to capacity, but the city plants of this energetic firm are very busy.

The Maritime Nail plant, the Pender Nail Works and the Canada Nail and Wire Company have been fairly busy throughout the year.

There has been a merger of the edge tool works formerly operated by Wilfred Campbell & Son and Josiah Fowler Co., Ltd. The new corporation operates under the firm title of Campbell & Fowler.

The brush and broom factory of T. S. Simms Co., Ltd., has been operating to full capacity, necessitating some important additions to the plant. The Simms plant is one of the best in Canada.. The factory of the Canada Brush Company has also had a good season's work and

some improvements are being outlined for the coming season.

The year has been a quiet one among the fish industries, due to scarcity of fish and disturbed market conditions. The sardine plant of the Booth Fisheries Company, of Canada, has not opened at all this year. The Connora Bros. sardine factory had a fairly good season, but with most of the fish plants the season has been a far from satisfactory one.

One of the largest of the city's industrial plants—the Atlantic Sugar Refinery—has been one of the greatest sufferers from the sudden drop in sugar prices. This factory had been doing a capacity business until a few months ago and had found it necessary to expend a quarter of a million dollars or more in additions and improvements to its storage facilities. The suspension of its business a few weeks ago has not only forced hundreds of men out of employment, but it has also greatly added to the storage troubles of the Company. In addition to its own warehouses being filled with raw sugar, several city warehouses on the water front are crowded with the raw article. It is expected that the refinery will resume operations about the first of January.

The port industrial workers have not been overworked this season and, unless an improvement in the overseas markets occurs very quickly, the amount of freight passing through the port will not be as large as usual. The shipments of lumber and general cargo threaten to be light this winter. The shipping merchants are optimistic and are hoping for better things by the first of the year.

Since last winter an additional grain conveyor, in connection with the C.P.R. grain elevator, has been placed on one of the west side piers.

One of the most encouraging features of the industrial outlook in this vicinity is the development of Hydro-Electric power, which has just been begun under provincial auspices on the Musquash River, 12 miles from St. John. From this one source about 8,000 horsepower will be developed for industrial and general use in the city. With cheaper power added to the export possibilities and distributing advantages of St. John, the industrial outlook is most encouraging.

In closing, a word or two concerning city industrial development. This year, for the first time since the close of the war, the city has been indulging in an extended programme of street paving, killing two birds with one stone—beautifying the city and at the same time furnishing employment for returned soldiers and others out of work. The programme will be revived in the spring. It is intended during this winter to indulge in some road extensions at the farther end of the city, adjacent to the Kennebecasis river, where the city owns a beautiful natural park that was bequeathed to it by a former federal member, Colonel Tucker. The city also proposes to do a lot of stone crushing this winter to help out next season's street programme.

THREE RIVERS HAS NEW PLANTS

Big Building Programme is Being Finished— An Active Board of Trade is Working on Several Schemes There Now

By WILLIAM TROTHIER, Assistant Field Secretary Three Rivers Board of Trade

THREE RIVERS, QUEBEC.—Three Rivers, in the Province of Quebec has, this year, more than maintained its rapid growth of the last ten years.

Industrially it has taken a long step forward, with the coming of a new plant, the Three Rivers Pulp and Paper Co. The capitalization of this latter company is set at $4,400,000. It has contracted for a plant costing a million and a half, which, when completed, will employ approximately 5,000 men and have a capacity of 200 tons daily.

Three plants, the Canada Iron Foundry, the Wayagamack Paper Co. and the St. Maurice Paper Co. have opened club-houses for the workmen, showing a healthy development of the spirit of brotherhood between capital and labor.

Few strikes have disturbed the industry field. Wages are moderate, the workers satisfied.

The problems that affect the city, such as a good water supply, housing, a new station and new transportation facilities to meet the growing demands of business, are all on the coming year's program of the reorganized Board of Trade, which starts the new year with a revenue from memberships exceeding anything it has had in its history.

Plans ready for the building of the million and a half dollar plant of the Three Rivers Pulp and Paper Co., capital, $4,400,000. Capacity, 200 tons per day; employs 5,000 hands.

Plans ready for a million dollar addition to the Wabasso Cotton Co., making increase of $200,000 in payroll.

Tidewater Ship Co. notes big increase in business.

Mechanical Engineering Co. buys 365,000 feet of land for expansion purposes.

St. Maurice Lumber Co. completing new mill for paper manufacture. Will be ready to manufacture pulp in the spring. Will probably require 5,000 hands in new venture. This company has also spent $50,000 in the construction of homes for workmen.

The Canadian Consolidated Rubber Co. has opened a branch in this city.

A steel concern is negotiating with the city for a plant in Three Rivers.

Other developments indirectly affecting the industrial progress of the city is the building of a bridge at Batiscan, which will greatly facilitate transportation here, and the opening of industrial classes for workmen.

NEW INDUSTRIES IN WOODSTOCK

And a Very Large Number of the Men in Factories Own Their Own Homes— Look for a Good Year

By HARRY SYKES, Secretary Woodstock Board of Trade

WOODSTOCK, ONT.—Woodstock has experienced a busy year in almost all lines of industry. A review of the labor situation in Woodstock reveals the pleasing fact that there is little unemployment in the city at the present time. While some of the local factories have reduced their working hours during this month, none of them have closed down, and only a small proportion have reduced their staff to any extent. A large programme of city improvements is under way, which is helping materially to solve the labor problem and the volume of work thus made available appears to be nearly equal to the demand.

Twenty-three principal industries represent a total investment of $4,430,000. The output of these industries during the past twelve months was $8,418,000. The annual payroll amounts to $1,797,000. It is a matter of congratulation to the factory employees that 534 own their own homes, which are valued at $720,000.

During the past year a Canadian Branch of the De Long Hook and Eye Company was established here. Also the Kennedy Car Liner and Bag Company purchased the William Stone Company factory. The newly organized Karn Piano Company commenced operations in the Karn Factory. A few days ago arrangements were completed with the American Ironing Machine Company, Chicago, to establish their Canadian branch in Woodstock. They anticipate operating their plant here about March 1st, and will make Woodstock their head office for Canada.

The heads of industries are quite optimistic as to the future, and are looking forward to the factories being busy as usual in a short time.

ST. THOMAS' INDUSTRIAL ERA

Plan to Buy Tract of Land for Firms That are
Likely to Come—Several New Ones
Came in 1920

By J. L. LODGE, Secretary St. Thomas Board of Trade

ST. THOMAS, ONT.—The Thos. A. Edison, Inc., have
purchased the plant and buildings of the St. Thomas
Cabinets Limited, comprising about one hundred thousand feet floor space, to which they will add two large
wings next year, and this company, which will be known
as the Canadian Edison Phonographs, Limited, will begin operations about the first of the year.

Hutch Bros., of Canada, manufacturers of wooden vehicle toys, and the Universal wallpaper hanger, and the
Provincial Machine and Supply Co., manufacturers of
phonograph hardware, were incorporated about the first
of the year, and are now both under production.

The E. T. Wright Co., manufacturers of "Just Wright"
shoes have disposed of their plant and business to the
Talbot Shoe Co., Limited.

The Canada Iron Foundries have spent $150,000.00 in
extending their plant this year, and anticipate further
extensions in 1921.

Early in January the citizens will vote on a by-law
providing for the purchase of 75 acres in the east end of
the city for the purpose of an industrial area. The proposed area is bounded by the Michigan Central, Canadian
Pacific, Wabash, and Grand Trunk Railways.

KITCHENER CONTINUES TO GROW

There Are Now 123 Industries With a Payroll
Of $11,000,000—Large Extensions
Made to Factories

By E. J. PAYSON, Secretary Kitchener Board of Trade

KITCHENER, ONT.—Kitchener continued during 1920
to experience a steady expansion and prosperity of
established industries. Population increased by 1,975,
and in the city proper there are now 23,027 people. As-
sessable property increased by $3,265,344, and now stands
at $18,053,251. There are 123 industrial establishments
in the city, giving profitable and steady employment to
some 8,000 employees, who receive about $11,000,000 in
wages. The aggregate output of manufactured goods
last year was approximately $40,000,000.

Among the larger additions to industrial plants is
that of the Lang Tanning Company, Limited, who are
spending about $200,000 on buildings and equipment. The
Kaufman Rubber Company, Limited, has under erection
a five story addition to its big factory, which addition
would make a substantial industry of itself. The Dominion Tire Company, Limited, has erected building and
equipment the present year which, when completed, will
have cost nearly half a million dollars, and their daily
output of automobile tires and allied products has reached mammoth proportions.

The Ames-Holden McCready System completed this
year the automobile tire factory begun late last year, and
will shortly have reached the quantity production stage
in that line. This firm has built and equipped this year
a factory for the production of felt footwear, which is
running to capacity. They also have begun to manufacture rubber footwear in premises in their tire factory not
yet occupied by that industry. This trio of industries
under the A.H.M. control form a big and valuable addition to Kitchener's manufacturing interests.

Salts & Potash, Limited, have doubled the space and
capacity of their plant and are approximating the dimensions of a large and important industry. Their products
consist of salts, potash, chlorides and other chemical products.

To cope with the increased demand for electrical power the Light Commissioner has erected a modern equipped

sub-station in the industrial centre of the city. Transportation has been improved by the construction by the
C.P.R. of a special freight line entering the city at the east
end as a continuation of the Grand River Railway, which
is the electrically operated branch of the C.P.R. between
here and Port Dover, on Lake Erie.

The addition to Victoria school was completed and occupied this year and at present a big extension is being
erected to the King Edward school, and a wholly new
school is being built in the South Ward.

This year has also seen the building and equipment of
a Nurses' Home for Victoria Hospital, built, furnished
and presented to the hospital by the late Jacob Kaufman
and family.

BRANDON GETS BIG CONCERNS

Gaining in Importance As a Distributing
Centre—They Look for Many More
In Coming Years

By H. L. CRAWFORD, Brandon Board of Trade

BRANDON, MAN.—Brandon's importance as a distributing centre is being recognized more and more each
year. The number of large concerns locating here established a record during 1920, that has not been equalled
since the palmiest days before the Great War.

Some of the really large concerns that have now located here are the Massey-Harris Company, Limited, and
the Imperial Oil Company. The Massey-Harris people
bought the Gordon Mackay building for $130,000, and
will establish a large distributing plant for Western
Canada here.

The Imperial Oil Company has commenced operations
to erect a $350,000 plant here, and their offices are al-
ready established in the city. The L. R. Steel Company
has recently made a purchase of a large business in this
city, and will open up January 1st. It is also stated that
the Grain Growers Limited will place a large plant here
for distribution purposes in the West. Several other
smaller concerns have located here during the year, and
the prospects for a large number of manufacturing plants
coming here in 1921 are very bright.

CHATHAM SECURED TWO IN 1920

Industries Have Been Well Employed and
Plans Are Being Made for Full Operations in Short Time

By M. M. MAXWELL, Manager Chatham Chamber of Commerce

CHATHAM, ONT.—During the year 1920, Chatham has
located two industries, one of which moved here from
another town, and the other, a new Canadian corporation,
in which American interests are largely concerned. The
first company is making knit goods, while the latter is
putting on the market a new furnace, modelled after a
very popular and wide selling furnace in United States.
Locally, conditions have been very good during the year,
and the majority of concerns have extended their businesses considerably.

The Hayes Wheel Co. are extending their rear axle
assembly department, as well as their truck wheel section; the Canadian Fertilizer Co. have added quite an extensive addition to their building, making commercial
fertilizer; the Gray Dort Motors Ltd., have added a large
plant, in which they have installed extensive and modern
machinery for making some of the parts which originally were made in other plants, particularly the pressed
steel parts; the Libby, McNeil & Libby Co., canners and
packers, have added to their buildings during the year,
while the Canadian Des Moines Steel Co., making structural steel and tanks, have a considerable addition under
way; the Chatham Malleable and Steel Co. is constructing a building which practically doubles their capacity

and installing their late, modern machinery for the manufacture of automobile accessories.

Other industries in the city are doing some small extension work, while in one or two cases considerable expansion is now planned, but actual construction work has not been undertaken and will not be for a few weeks or possibly six months.

The city of Chatham has not experienced as much of a commercial and industrial slump as has been experienced in many cities, because of the varied nature of the industries located here. There are very few idle men at the present time, even when other cities are experiencing considerable difficulty in this particular connection. Two plants in the city have been running small force for some weeks, but both are now making preparations for reopening their plant to full capacity by the first of the year at the latest.

BIG PLANTS PUT UP IN OSHAWA

General Motors Added to Their Plant—Several Large Buildings in the Course of Construction

By J. A. McGIBBON, Secretary Oshawa Board of Trade

OSHAWA, ONT.—During the past year Oshawa has made wonderful steps in industrial development. General Motors have added considerably to their plant, which now includes the Oldsmobile Motor Co., in addition to the McLaughlin and Chevrolet plants and the Samson Tractor. Pedlar People have moved into their spacious new plant, one of the finest in the province. W. J. Trich & Co. are rebuilding their planing mill and woodworking factory.

The Moffatt Motor Sales have erected a fine new showroom for the display of their Oldsmobile cars. The new Theatre Paramount has opened a fine, up-to-date structure seating 1,100 people. L. R. Steel & Co. will move into their new store in a few days. Ward .& Dewland, dry goods merchants, are moving into a fine new store erected by Bradley Brothers. A 65-acre park was given to the town this year by General Motors Ltd.

WINNIPEG IS GROWING RAPIDLY

Building of Factories and Shops Calls for Expenditure of Many Millions—More Power to be Available

By J. M. DAVIDSON, Winnipeg Board of Trade

WINNIPEG.—The year 1920 has seen substantial progress in the city of Winnipeg. Building figures for the year will run close to eight and a half million dollars. Not only is this the largest total since 1915, but it equals the combined figures of 1917, 1918 and 1919. Winnipeg's building totals for the first nine months of 1920 were equal to the combined totals of twenty other Canadian cities. Three large motion picture theatres have been erected in Winnipeg during the year, and operations have just been commenced on a six story addition to the Opympia Hotel.

Bank clearings have shown a remarkable increase during the year. For a single week ending November 10th, clearings reached the enormous total of $110,806,-325.00. This is not only the highest weekly figure ever reached in the Winnipeg clearing house, but it actually exceeded the clearings of the City of Toronto for two weeks previous.

The total capitalization of new companies established in the Province of Manitoba has already exceeded one hundred million dolars. Some of the larger industries established in Winnipeg during the year are: The Canadian Community Cut Glass Co.; Red River Pulp & Paper Co.; Western Film Mfg. Co.; Orange Crush Bottling Co., soft drinks; Tractioneer's Ltd., farm implements; Western Suit Case Mfg. Co.; Canada Fibre Products Co., hemp

products; Winnipeg Motor Cars Ltd.,'automobile manufacturers; Picardy Candy Shops, Ltd.; Crown Broom Mfg. Co.; Martin's Limited, clothing manufacturers; Campbell Mfg. Co., clothing manufacturers.

The Winnipeg River Power Company, a subsidiary of the Winnipeg Electric Railway Co., has worked during the year on the first unit of a $10,000,000 power plant on the Winnipeg river.

U. S. PLANTS IN STRATFORD

Two Have Started and Already Have Secured Production—Employment For Many More Hands is Provided

By A. W. DEACON, Secretary Stratford Chamber of Commerce

STRATFORD.—In the past year the Canadian Edison Appliance Co., Ltd., and the Gerlach-Barklow Co., Ltd., both American plants, opened up Canadian branches in this city. The former manufacture all sorts of electrical appliances and is employing approximately three hundred hands. The Gerlach-Barklow Co., Ltd., manufacture high-class art calendars.

The Kindel Bed Co., Ltd., which opened its Canadian branch in Stratford about two years ago, have, during the past few months, largely increased their plant. The Grosch Felt Shoe Co. have commenced operations during the past year, manufacturing felt shoes and are now employing approximately seventy-five hands.

NEW PLATE MILL AT SYDNEY

Development of the Steel Mill Has Been Carried On—Making Several More of Its Own Products Now

By F. C. KIMBER, Secretary Sydney Board of Trade

SYDNEY, N.S.—In February, 1920, the new 110″ plate rolling mill of the Dominion Steel Co. was placed in operation. This new department involved an expenditure of approximately $5,000,000 and has a capacity of approximately 12,000 gross tons monthly of sheared plates up to 98″ in width. Since the beginning of operations it has been shipping its entire production to the Canadian Government, with which the company has a contract for from 50,000 to 75,000 tons of plates per year for five years. In the early part of the year the first two units were placed in operation in the company's new power house, which has been erected at a cost of $1,250,000 and which, when completed early next year, will have a maximum capacity of approximately 10,000 W.W. This additional power will permit the electrification of some of the producing departments that have heretofore been using steam, and will enable the company to produce its electric power requirements at the lowest possible cost per K.W.H.

About two years ago the company completed an installation of 120 latest type Koppers by-product coke ovens with all auxiliary apparatus. This installation proved entirely satisfactory. A contract was let for another two batteries of thirty ovens each, which it is expected will be in operation about March 31st, 1921.

During June, 1920, the new electric drive for the rail mill was placed in operation and it has since given entire satisfaction. This replaces the former expensive drive in this department and the additional electric power required is made possible by the erection of the new power house above referred to.

Prior to the end of the year the company should have in operation its new plant for the production of silica brick. This will have a capacity of about 15,000 9″ straights daily. After experimental stages have passed it should enable the company to discontinue purchasing a large part of its silica brick requirements from Pittsburgh district. The silica rock required in this new department is to be brought from a quarry situated only fourteen miles from the steel works.

During September, 1920, the company reopened an old

plant which had been discontinued some years ago by other parties and commenced the production of cement from blast furnace slag. The resulting product has been found to be quite satisfactory for the company's purposes and this department is now operating at the rate of about three hundred barrels per day.

It is reported that the Cross Fertilizer Co., Ltd., which manufactures fertilizer from the Dominion Steel Co.'s slag, and which has built up a very successful business in this very essential product, is contemplating an extension of its plant and producing capacity in the near future.

BORDER CITIES GET U. S. PLANTS

In Negotiation With Large Number of These— Many Will Come Over as Soon as Building is Reasonable

By F. MACLURE SCLANDERS, Border Chamber of Commerce, Windsor

BORDER CITIES—Ford, Walkerville, Windsor, Sandwich, Ojibway. During the past year thirty-one new industries have located at the Border Cities, so that they now have 206 of such, large and small. During 1919, forty-four new manufacturing concerns came in. Practically all of such industrial development emanates from the United States.

At present, we are negotiating with no fewer than seventy-six American concerns, over a third of which have already intimated their decision to come here whenever conditions are more stable. In addition, we are working upon other eighty-six prospects which are shaping favorably in most cases, but are still in the preliminary stages and not yet in the "live" class which embraces the seventy-six first mentioned.

In view of the foregoing, we have obvious reason for encouragement. As a matter of fact, the truly extraordinary development characteristic of these Border Cities during the past few years is still largely unrecognized or uncomprehended in centres further east. This may be because our development mostly has its source in the United States. At any rate, we at this southernmost corner of Canada are forging ahead in a sense that is literally phenomenal. Ours is not mere development. It is transformation. Within the lifetime of a child now at kindergarten we have simply leaped from the town cate, gory into the truly metropolitan. This statement is at once conservative and perfectly safe—for, its truth may be demonstrated by the evidence of one's own eyes. It is visible on every hand; in street after street of new homes; in new factories and business premises of every description, and in a prosperous population which now approaches the 60,000 mark! Indeed, one is vividly reminded of the Western boom period of 1910-12, and of the wondrous strides then made by certain more progressive prairie centres. However, there is one fundamental difference: our progress and expansion are built upon the most solid of all foundations—the location and successful operation of new industries.

It may interest to remark that the growth of our industrial life owed practically nothing to munition or other war work, very little of which was conducted here. Few war orders came our direction. Consequently, we pursued the peaceful tenor of our productive ways; and in the end it is probable that we gained rather than lost, even although there might have been some disappointment at the time. Others got the war orders; and while they were working thereon, we were expanding the production of our normal commodities.

As to the situation and outlook, we are conscious of no logical reason for gloom. Rather do we welcome the present lull as the one circumstance absolutely required to restore general conditions to the more nearly normal and thereby stimulate a safer and more substantial business than ever.

In our opinion, prices and production costs are grotesquely hysterical. They are idiotic. They are also im-

possible. They have killed commerce. At last, people have torn the war bandage from their eyes. No war haze clouds their vision. Things appear at their true value. Buyers are disillusioned. The $35 suit that may now be purchased for $120 appeals as a stupid impertinence. The $35 home which its owner is willing to rent at $80 no longer attracts; the $6,000 which offers at from $17,000 to $20,000 still further accentuates the crass absurdity of the greediest, craziest period in the world's history.

Fortunately, however, the crisis is over; the fever diminishes. The candle of high costs is now brightly burning at both ends. Material costs have already indicated a marked decrease while a remarkable increase characterizes the production of the individual worker. We have now actually entered upon the first real period of readjustment. It took us quite a time to wrestle with the purely preliminary stages thereof. Now we are about to make real progress.

As it is, however, American and other manufacturers flatly state that they won't build factories at the prices of to-day which, in the comparatively near to-morrow, must be much lower. Some of them will come in immediately, provided they can be suitably accommodated in rented, ready-made premises.

The outrageous prices of to-day constitute an almost insuperable obstacle against the further, fullest industrial development of this country. But, whenever such revert to the more nearly normal, all the pent up progress of months will be let loose upon us. Business will be better than ever, in our opinion; because business will then be safer.

We do not look for a violent depression: but rather a, swift and comparatively easy reversion to a truer basis. Of course, even such a process involves loss for many. Nevertheless, past years of good business, wonderful business, should place intelligent business men in a position to take their goats along with their·sheep.

Upon the whole, we were never more optimistic; and, even while waiting for more normal times, we are doing wonderfully.

BIG THINGS AT PORT ARTHUR

Pulp and Paper Industry is Becoming an Important One—Plenty of Power Available For Use There

By M. FRANCIS, Secretary-Treasurer Port Arthur Board of Trade

RECENT expansions industrially are the inclusion of manufacturing pulp and paper and mining machinery, in the operations of the Port Arthur Shipbuilding Company, Ltd., which will mean an increase of approximately 600 skilled employees, and erecting an open hearth furnace, also the building of the Kaministiquia Pulp & Paper Co., Ltd.'s new pulp mill. It is understood that the local branch of the Provincial Paper Mills, Ltd., is about ready to increase its output to 100 tons pulp and 50 newsprint, per day. Plans for two other pulp and paper mills are about completed.

During the year the Port Arthur Wagon Works has become the property of Jno. Stirrett & Sons, who are expanding into a general woodworking business.

While many manufacturing centres have been hampered by lack of Hydro-Electric energy, Port Arthur is in the fortunate position of having supplies adequate to take care of considerable industrial expansion. The development in the Nipigon river at 72,000 h.p. places Port Arthur in a very enviable position.

The city has set aside for incoming industries 1,000 acres of favorably situated land served by both trans-continental railways and harbor.

Port Arthur is also extremely fortunate in being surrounded with a country that offers unlimited recreational and vocational facilities. These attract each summer many thousands of tourists, 32,000 in 1920, and this is rapidly becoming one of the community's chief industries.

ADDITIONS TO THE GALT SHOPS

C. P. R. Spending Large Amount For New
Service There—New Companies Enter
The Shoe Manufacturing Field

By H. J. FOSTER, Secretary Galt Board of Trade

GALT.—The industrial development of Galt during 1920
has, on the whole, considering the almost prohibitive
building costs, been very satisfactory. Our machine
shops, until recent weeks, have been very busy, and a few
additions have been made, the R. McDougall Company
leading in this respect.

Two new shoe factories, Scroggins Brothers, and the
Yale Manufacturing Company, Limited, have been added
to the colony, while the Getty & Scott Company, Limited,
spread out into two annexes, now having Factories No. 2
and 3.

Among the new firms is that of McCaskey Systems,
Limited, in a new building bought by them before removing from Toronto to Galt. In transportation field the outlook is good. At the present time the C. P. R. is engaged
in spending its appropriation of $1,000,000 in the city.
A new spur line is being built to the heart of the city, the
project involving the erection of a downtown station for
the Grand River Railway and the Lake Erie & Northern,
the company's electric feeders, stretching from Waterloo
and Kitchener to Brantford and Port Dover.

GUELPH HAS EIGHT NEW ONES

And Several of the Plants That Are in Operation There are Going to Extend Their
Premises in 1921

By H. WESTABY, Secretary Guelph Chamber of Commerce

GUELPH, ONT.—The City of Guelph has made steady
progress during 1920. Eight new industries have
established themselves in the city, and are as follows:—
Moncrief Furnace & Mfg. Co., the Canadian branch of
the Henry Miller Foundry Co., of Cleveland, Ohio.
The Erin Casket Works.
The Guelph Brass Works.
The Federal Electric Washing Machine Co., being the
Canadian branch of the Chicago firm.
The Regent Textile Co., from Montreal.
The Sherer-Gillett Co., from Chicago.
The Shinn Mfg. Co., from Niles, Michigan.
The National Standard Co., from Niles, Michigan.
The Moncrief Furnace & Mfg. Co. are to erect a mod-
ern foundry building on their property acquired on the
York Road. The Erin Casket Works, the Regent Textile
Co., the Federal Electric Washing Machine Co. and the
Sherer-Gillett Co. have all purchased existing factory
buildings; which in nearly every case they have enlarged
and improved.

The industries of the city at the present time number
129, and among the improvements made are:—

An addition, three storeys in height, 53 ft. x 90 ft., to
the Shinn Mfg. Co.

The Canada Ingot Iron & Culvert Co. are building an
entirely new factory building between George and Clarence streets. The plan calls for two buildings, 48 ft. x
290 ft. and 120 ft. x 132 ft.

The International Malleable Iron Co. have completed
and are now occupying an addition to their plant of 240 ft.
x 260 ft., two storeys in height, and in addition to this
they have just acquired 12¼ acres additional property,
on which to erect a drop forge foundry building and other
additions.

The Gilson Mfg. Co. have just completed an addition
to their premises 80 ft. long by 120 ft. wide, three storeys
in height.

The Louden Machinery Co. are now occupying an addition to their foundry 50 ft. x 106 ft.

The Northern Rubber Co. completed their new building this year. The building is 80 ft. x 200 ft., 4 storeys

in height, and is said to be one of the most complete fac-
tory-buildings in Ontario.

The Partridge Rubber Co. are occupying an addition
just completed 150 ft. x 120 ft. wide, four storeys in
height.

In addition to the above the Taylor-Forbes Co. are
planning to erect a new factory building that, it is said,
will cost $280,000. The Guelph Carpet & Spinning Mills
have announced their intention to erect an entirely new
building in 1921. The Sherer-Gillett Co., of Chicago, have
purchased ground on North St. and have announced their
intention to erect an additional building on this property
next year.

The population of the city has shown an increase for
1920 of 980 people, and the assessment has increased during the year by $1,162,000. In this is included new houses
that are valued at $410,498, which are now in various
stages of completion. The prospects for the city for 1921
are very bright and apparently there is no let-up as far as
can be seen in any of the local factories, all of which are
running on full time with plenty of work available for
everybody.

OWEN SOUND HAS NEW PLANTS

Several Now Operating and Another Expects
To be Turning Out Electric Washers
Early in the Coming Year

By GEO. MENZIES, Secretary Owen Sound Board of Trade

OWEN SOUND.—Owen Sound is growing. During the
past year several new industries have located here,
viz.:
The Clinton Knitting Co.
The Circle Bar Knitting Co.
The Grey Mattress Co. and
Taylor Bros. Jam Factory.

The Clinton Knitting Company have already found
their premises too small and are enlarging their factory
to double capacity.

The Circle Bar Knitting Co. occupy a large three story
building and are contemplating building an addition.

Another new industry is just about ready to commence
manufacturing, viz.: The Slade Manufacturing Co., manufacturers of electric washing machines. This firm is now
installing machinery and expect to commence operation
about the 1st of January, 1921.

WIRE PLANT IS REBUILDING

Collingwood Industry is Being Reorganized
And Expects to be Operating in
The Near Future

By W. A. COPELAND, Secretary Collingwood Chamber of Commerce

COLLINGWOOD.—The largest development of importance in the industrial life of Collingwood during the
present year is the fact that the plant of the Imperial
Steel and Wire Company, Limited, destroyed by fire two
years ago, is being rebuilt and is rapidly approaching
completion, and expects to be again in active operation
early in the coming year.

BELLEVILLE HAS SEVEN IN ALL

Claim to Have a Very Good Labor Market
And Populous Surrounding Country—
Prospects of More Factories Coming

By J. BONE, President Belleville Chamber of Commerce

BELLEVILLE.—During the past year seven new industries have located in Belleville, one of them being a
branch of an American industry, namely, the Teco Company, makers of pancake flour.

Other industries are: H. A. Wood Manufacturing Co.,
Limited, makers of automobile valves; Toronto Hat Co.,

Limited, hats and caps; Natural Tread Shoe Co., Limited; A. S. Richardson Co., Ltd., manufacturers of wax figures and shop fixtures.

The Elliott Woodworking Machinery Co., Limited, along with the four last named industries, moved here from Toronto where they found much improved conditions in labor and power. Another new industry is the Judge-Jones Milling Co., Limited, who have built an elevator with storage capacity of about 40,000 bushels.

There is also the prospect of our securing a new glass industry and some others, all of them being attracted by the very desirable labor conditions and the populous surrounding country from which this city draws labor by motor bus, etc.

MORE U. S. PLANTS ARE COMING

Hamilton in Negotiation* With a Long List, Some of Which Will Come as Soon as More Power is Available

By C. W. KIRKPATRICK, Commissioner of Industries and Publicity, Hamilton

HAMILTON—Despite the handicap of electric power shortage, from which practically all of the Ontario municipalities have been suffering for the past two or three years, Hamilton during 1919 enjoyed another period of unusual industrial development and prosperity. Lack of a sufficient power supply unquestionably kept a number of large manufacturers, particularly Americans, from locating, but inasmuch as the conditions existing in Hamilton were typical of power conditions existing in other desirable Canadian manufacturing centres, the majority of the industries did not decide to locate elsewhere, but wisely deferred action until such time as conditions right themselves and the situation becomes normal. Present indications are that with the coming of spring, or about April 1st, power problems, at least so far as industries are concerned, will become a thing of the past, and as a result increased industrial activity may be looked for during 1921. A number of American manufacturers, with whom the local industrial department had negotiations during 1920, intimated their intention and desire to resume these negotiations just as soon as the Hydro Department was in a position to discuss all-the-year-around power contracts and there are, therefore, good reasons for expecting that 1921 will be a banner one in industrial activity.

But, to get back to 1920, Hamilton has no reason to complain of its industrial growth during this year of more or less abnormal conditions. In fact, with the exception of 1919, it was probably the best year the city has enjoyed industrially since 1911. A total of 25 new industries was added to the city's already long list, and while the majority of them were comparatively small, being capitalized at from $25,000 to $100,000, many of them are branches of large American concerns that give promise of speedy development, providing that tariff and other conditions continue favorable to Canadian industrial growth. Among the branches of American concerns to select Hamilton as their Canadian headquarters were:

Peterson Core Oil Co., Chicago, Ill.
Canadian Nathan Co., New York, N.Y.
Don-O-Lac Co., Rochester, N.Y.
Metiskin Co., Rochester, N. Y.
Moto-Meter Co., Long Island City, N.J.
Canadian Libbey-Owens Sheet Glass Co., Charleston, W. Va.

American capital is largely interested in others of the new industries, although they are not actually Canadian branches of existing American companies.

The largest new industry to locate in Hamilton during the year was the Canadian Libbey-Owens Company, manufacturers of sheet glass under the Libbey-Owens patent. The Canadian company is capitalized at $5,100,000, and its Hamilton plant, which is now under construction, will, when completed, represent an expenditure of upwards of $2,000,000. Although building operations are being rush-

ed, the contractors estimate that it will take nearly a year to complete the buildings, but when completed they will be the last word in glass factory engineering and construction.

The city's industrial building permits for the year total considerably over $2,000,000 and this figure does not represent more than one-half the actual value of the buildings erected.

In addition to the new industries secured, there was, during the year, marked development among the existing industries. The Dominion Foundries and Steel Company recently completed the erection of new buildings and installation of one of the finest plate mills on the continent, at a cost of a million and a half dollars; the Canadian Cottons, Ltd., has, nearing completion, a seven story spinning mill, which represents an outlay of close to a million dollars; the Steel Company of Canada has under construction a wire products mill, which represents an expenditure of about half a million dollars, while the Mercury Mills, Chipman-Holton Company and other textile industries made very considerable extensions during the past twelve months. The progress of the city can best be gauged by the assessment returns completed in October 1st, last, which show an increase for the year in the assessment of upwards of $30,000,000 and in the population of 6,623.

Hamilton's industrial growth is the more remarkable in view of the fact that the city does not offer any special concessions in the way of bonuses, tax exemptions, fixed assessments or land grants to incoming industries. It has found from experience that the really desirable industries do not seek special favors but are willing, when manufacturing and distributing conditions are favorable, to become part of the community on equitable terms, bearing their fair share of the tax and other burdens and enjoying to the full the privileges of the municipality offers. One or two industries were lost to the city during the year by reason of the evasion of Anti-Bonusing Act by some of the smaller municipalities, but, on the whole, Hamilton has no reason to regret its stand against the purchasing of new industries at the expense of the smaller taxpayer. That its policy meets with the approval of the manufacturers themselves is shown by the fact that the biggest boosters for Hamilton in the securing of new industries are manufacturers already located here.

SHERBROOKE HAS HAD MANY

Many Industries Went There in 1920, and Conditions Are Reported as Being Very Favorable There Now

By J. H. BRODEAU, Secretary-Treasurer Sherbrooke Board of Trade

SHERBROOKE, QUE.—The following manufacturing companies with details in regard to their production and number of employees may be of interest to your publication:—

Julius Kayser & Co., Ltd., manufacturers of high-grade silk gloves, hosiery and underwear; number of employees 500; cost of plant $1,000,000.

The Superheater Co., Ltd., manufacturers of superheaters for locomotives and marine and stationary boilers; number of employees 150; cost of plant $200,000.

Canadian Sturdy Chain Co., Ltd., manufacturers of shirts, collars and cuffs; number of employees 150; cost of plant $50,000.

Regal Tire & Rubber Co., Ltd., manufacturers of automobile tires; number of employees 100; cost of plant $300,000.

Cluett, Peabody & Co., subsidiary, manufacturers of shirts, collars and cuffs; number of employees 150; cost of plant $80,000.

The Canadian Connecticut Cotton Mills, manufacturers of tire cotton fabric; number of employees 1,000; cost of

extension when completed will represent an investment close to $5,000,000.

Presure Proof Rings Co., Ltd., manufacturers of pressure proof rings; number of employees 25; cost of plant $25,000.

English & Scotch Woollen Co., manufacturers of men's clothing; number of employees 100; cost of plant $25,000.

Also Office Requirements, Limited, Goupil, Limited, J. H. Bryant, Limited.

One or two of our industries are feeling the effect of the depression to a certain extent, but the great majority have orders on hand which will tide them over to the revival·of business. As a whole, will say that the industrial situation in Sherbrooke is in a prosperous condition.

A technical school will be built at the cost of several hundred thousand dollars. Municipal Hydro-Electric power development at Two Miles Falls, at an expenditure of over $500,600, will provide 5,000 additional horse-power for new industries. Sherbrooke has over 60 industries and its manufacturing products total over $50,000,000 per year. There are ample housing facilities to accommodate all the labor required to work all the plants now standing. In addition the city is now spending over $600,000 on a model city suburb of workingmen's dwellings and the program calls for 200 houses of modern design. The labor situation was never better, all kinds of labor is available, including male and female, both skilled and semi-skilled; the working man is happy and contented in this section of the country, no foreigners or Bolshevists in this locality. Our labor population understands that capital is absolutely necessary and capital knows that labor is equally needed, consequently with this reasoning, strikes are foreign to our people.

MORE WAREHOUSES ERECTED

Regina Grows as a Distributing Centre—More Building in View For the Coming Year —Many Homes Going·Up

By CHAS. A. COOKE, Secretary-Manager Regina Board of Trade

REGINA—Despite adverse conditions created by the enhanced cost of material and labor, the City of Regina has experienced a large amount of industrial progress. Commodious warehouses have been erected by Wood-Vallance Company, Fairbanks-Morse Company, Goodyear Tire Company and the Saskatchewan Co-operative Creameries. The Leader Publishing Company have also added to their holdings a large storehouse.

Other constructions during the year embrace the new Presbyterian Carmichael Church built at a cost of $40,000; power house at the Parliament Buildings; four-storey business block of the Regina Trading Company; new Capitol Theatre, and upwards of 200 residences.

Projects for next year include a distributing house for Canadian Swift Company and a college to be erected by the Lutheran body.

SASKATOON'S RAILROAD SHOPS

Building Figures of the Year Show Very Marked Increases Over Previous Year— Big Car Shops are Contemplated

By GERALD GRAHAM, Commissioner Saskatoon Board of Trade

SASKATOON—The outstanding features of the 1920 building program in Saskatoon have been the beginning of construction of new C.N.R. shops and round house, the new Provincial Normal School and the new Physics building of the University of Saskatchewan. As part of the C.N. Railway program a new bridge is also to be erected on the side of the existing one. With the establishment of the mid-Western headquarters at Saskatoon it may reasonably be expected that further building will take place this year, also solve the completion of extensive new freight office and general superintendent head· quarters and extensions of the express companies prem· ises now under way. The sum of $800,000 has been set aside· for immediate work here, but confirmation of the figure is not obtainable. The new Normal School is well under way and the site, overlooking the whole·city, is exceptionally advantageous. The building is to be in brick and Bedford stone in the Gothic style, similar to that which the University of Saskatchewan has in all its buildings. The building will cost approximately $750,000. The Physics building at the university will, of course, conform to the style of others on the campus and the building permit shows a minimum expenditure of $403,000. Some of the other larger permits during the latter part of last year and during·the past summer are as follows:—

Description of Building	Minimum.
Coca-Cola Company	20,000.00
Franklin Garage	12,000.00
Chevrolet Garage	15,000.00
Y.M.C.A. Hut	20,000.00
Roy Garage	15,000.00
Auto Garage	20,000.00
Barries' Limited store	20,000.00
Addition to C.P.R. station	20,000.00
Riddell Carriage & Motor Works	20,000.00
Green Court apartment house	20,000.00
Imperial Bank alterations	15,000.00
Drycleaning establishment	20,000.00
Quaker Oats Co., Ltd., additions	86,000.00
Physics building at university	403,000.00
Mayfair school	184,000.00
C.N.R. roundhouse	45,000.00
New skating rink	25,000.00

For the year ending September 29th, 1919, the total of building permits was $1,359,390 and for the year ending September 20th, 1920, the total is $5,160,105, thus indicating that a considerable building program of a public character has been under way. One of our most pressing needs is a larger hospital which has been considered for some years and which will be undertaken as soon as the city has funds. The matter of a public library is in the same category.

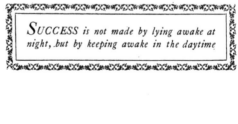

Turning Out Small Motors in Quantity

Intricate die work enters largely into the making of motors, and a peculiar yet splendid system is adopted at this plant.

By J. H. Moore

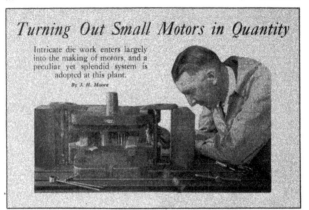

SPLENDID working conditions, ideal lighting, novel and accurate die work, these are some of the features found at the Canadian plant of the Robbins and Myers Co., Ltd., Brantford, Ontario. The data to follow will not only cover a brief description of the plant itself, but will include details of operations performed, items of general interest, and a detailed account of how this concern prepare their product.

Before entering the plant, one is struck with its neat exterior appearance. The building is of one story construction. The main office occupies the centre portion of the building at front, while the women enter at a right hand entrance, the men at a similar left hand door.

Beautiful flower beds, and a lawn of deep green give the building a regular country aspect, but on entering the plant one is really not sure whether he is actually inside a building or not. The lighting is perfect. A good high roof, well windowed and ventilated, throws ample light down, while the sides and end of the factory are also well provided with similar windows. Any plant arranged in the manner of the one we are speaking of, certainly deserves favorable comment.

The flooring is of wood blocks tarred in, and, as is well known, this type of floor saves to a considerable extent the workers' feet, the placing of machinery has also received close attention, the various departments being well laid out, and the machine tools segregated to the best advantage. Ample aisle space is allowed, in this way avoiding any confusion or congestion of trucking. Quite a number of operations performed at this plant are accomplished in a sitting position, for which splendid benches and stools are installed. Even the tool room operators are blessed with stools to sit on while their work is being completed. Tool stands are placed alongside each machine tool, and while the initial expense may be greater, the benefits derived soon pay for these stands.

Production Aids

For transporting material in the different stages of completion throughout the plant, trucks, and special stands are used wherever possible. It pays to make these, as it speeds up the work and keeps it so separated that it is easier handled.

Air lines are placed handy to all machine tools, and are used extensively to clean the tools of dirt, cuttings, etc. There is no need to go into detail regarding their raw stock and cutting off department, except to say that it is modern in every respect, and well arranged. As this is not intended to be a mere descriptive article, let us leave the generalities, and tackle the various departments one by one.

The Press Department

This section of the plant is set out by itself and the presses are arranged in a long line. Brown and Bogg presses are used, also presses made by the Bliss Co. The work performed is piercing, blanking, forming, and drawing, but as we intend showing these dies further on in the present article we need not dwell on this point at the present time.

Automatic Screw Department

This section is very well equipped and includes some large and small motor driven Cleveland and National automatics. There are also some No. 4 Warner and Swasey small lathes. The machines are arranged in ideal form. Screw work of all kinds is done in this department, also the turning of short shafts, pins, etc., etc.

Machine Department

In this section they have a full complement of floor grinders, buffers, disc grinders, Jones & Lamson lathes, Natco, and Leland Gifford drills. To give readers some idea of the work going through this department we have taken the operations performed on the washing machine motor body and head. This is of course only one type of motor, but will give readers an idea of the work entailed.

The Jones & Lamsons used are mostly of the double head type, and following are the sequence of the operations. The castings first go to the grinding end of the department and rough portions taken off. Now comes the turning of the head, which entails two operations, the turning of the inside, and outside fit.

The bodies are next bored out and faced. The photograph shown at Fig. 1 illustrates this operation, and it might be well to state that the hole is approximately 6⅝" diameter, and is finished 3¼" deep. The first station faces off the work by means of two ordinary tools, the second position has roughing mill cutters, and a roughing reamer for the centre hole, while the third position carries a boring tool for the centre hole. The last station holds the finishing mill cutters, also the finishing reamer, which is of the ball floating type.

Drill press operations on both the heads and bodies come next in order, but there is hardly any need to enter this in detail. Sufficient to say there are eleven operations.

After these operations have been performed, the parts go to the paint department, where they are sprayed with two coats of enamel. They are next placed in a Hoskins electric oven, which holds two racks, or approximately 600 to 700 parts at a time. They are baked in this oven for about three hours.

The operations on the Hoover suction sweeper motor while not exactly similar, entail the use of like equipment, so that we will pass on the winding department, together with the line of all machines used in conjunction with this department. These machines include Heald, Norton, and Landis grinders, Porter cable motor and belt driven lathes, some small Dalton bench lathes, some Mulliner lathes, some Rockford small lathes, a high speed riveter made by the High Speed Hammer Co., some Leland Gifford drill presses, and other tools of like nature.

FIG. 1.—BORING AND FACING THE BODIES OF WASHING MACHINE MOTORS.

Winding Department

The work entailed in the making of a motor for a suction sweeper is more than one would imagine, and while we cannot, in this space, tackle all parts and operations in detail, we give a few to illustrate the varied work. Take for example the shaft shown at H, see Fig. 2. This part goes through the following operations. First, the two shoulders are nicked by means of a special double tool on the lathe, then the second and third steps are turned to size. The wick hole is now drilled, and this operation is worthy of special mention. The hole is 3/16" diameter, and is drilled through the centre of the shaft until a depth of 4" is reached. The drilling of such a small hole to a depth of 4" is a rather ticklish proposition, but they have overcome the difficulty very nicely.

A circular tank has been placed on the drill press table containing a good supply of soap water. In the middle of this tank, and in a vertical position, is the drill. It is held stationary of course. The work to be drilled is held in a quick grip chuck in the drill spindle, and revolves at a good rate of speed. The operator feeds down his spindle, relieving the work as often as deemed advisable. The method as described has resulted in greater production with less breakage of drills, insomuch that the work being drilled in a tank of soap water cannot clog, thus saving both time and drills.

The next operations in sequence are as follows: An oil hole is drilled, the slots for the two keys shown are milled, the first, second, third and fourth steps are ground, then the discs are assembled

FIG. 2.—THE VARIOUS STEPS ARE SHOWN IN THIS PHOTOGRAPH. FOLLOW THE TEXT TO GET THE BENEFIT OF THIS LAYOUT.

FIG. 3—WINDING AN ARMATURE.

er, H a completed shaft with a key partially inserted. I depicts the armature ready for inserting the leads, J the armature ready for the commutator, K the armature ready for soldering the commutator, L represents the shaft in the rough, while M depicts the steel thrust washer.

The view opens one's eyes to the number of parts entering into the makeup of a complete armature, and the photograph of course shows but one type of armature made at this plant.

Assembling

We will now suppose we are going to assemble an armature as shown completed at A Fig. 2. The shaft is first banded with gummed paper, then the slots are insulated with fibre sheets forced into a sort of a U shape as shown. Previous to this, of course, one of the keys has been driven into position and the discs placed on the shaft. First comes a fibre disc next to the key, then come the metal discs, finishing up with another fibre disc, and the second key.

After the insulation has been placed in the slots, the armature is wound as shown in detail view at Fig. 3. These machines have dials on them somewhat similar to a milling machine disc plate. A series of holes go around the outside, and there is a brass pointer which can be set at any position. The holes are numbered from one, up. Depending upon the armature to be wound, the wheel or dial must turn so many holes past the pointer, which, by the

on the shaft. Following this, the discs are straightened in the core, then sent to the winding department. Fig. 3 illustrates one of these winding machines at work.

In order that readers may get the greatest benefit from the description of these various operations we have photographed at Fig. 2 the shaft, and the

parts going to make up a completed armature. We have lettered these parts so that the description can be easily followed. For example, A shows the complete armature, B illustrates one of the sections, while C shows a commutator complete, D depicts one of the wedge shaft keys, E the metal sheet disc, and F the fibre disc. G shows a fibre wash-

FIG. 5—HERE IS A GOOD EXAMPLE OF A DOUBLE ARMATURE DISC DIE.

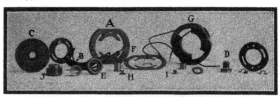

FIG. 4—SOME OTHER DETAILS OF INTEREST, DEALING WITH MISCELLANEOUS PARTS.

way, never moves. To watch the girls judge the correct number of turns, which may run into so many revolutions, plus so many holes is a treat. Like everything else, they get so used to it that the action becomes second nature.

After winding, the fibres are bent down, then the work is tested for opens

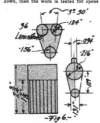

FIG. 6—DETAILS OF SLOTS, AND DEPTH OF PUNCH MILLING.

and grounds, after which it is thoroughly inspected.

The commutator is next pressed on, and the leads inserted. The soldering of the leads comes next. Inspection of the soldering follows, also greasing of the shaft. The commutator is now rough turned, and the shell turned to size to suit a little protection cap that goes over it. It is now tested for grounds, opens, reverses and shorts.

After these operations, the work is dipped in insulating varnish, and put in an electric oven, where it is baked for 10 hours. The thrust washers are next pressed on, and the shaft straightened, as it is imperative that the shaft run absolutely true. The first washer is now faced, and the distance over all between the face of this thrust collar, and commutator shell is procured. The commutator is next finish turned and sanded with No. 00 sandpaper.

The core is now ground, and the front end of shaft is finish turned to size. An oil hole is next drilled in the commutator shell, and is so drilled, that a continuous stream of oil reaches both the top and bottom bearings. The work is now balanced, and the balance is made up by means of small wedge-shaped pieces of brass. The fibre washer is next pressed on the shaft, after which it is faced. A small slot is milled at the end of the shaft, when the now

completed armature is sent to the test department. Considerable information of value can be gleaned by following these steps of manufacture with the photograph showing group at Fig. 2.

Field Core and Other Parts.

To follow the next operations, let us refer to group shown at Fig. 4. A, represents a field core, and of course these are first punched out in the press department. On reaching the assembling

FIG. 7—ILLUSTRATING THE METHOD OF FINISH FORMING SLOTS IN ARMATURE DIE.

FIG. 5—SHOWING TWO OF THE FILING MACHINES AT WORK.

after which they are dipped in, an air drying black varnish. On coming out of the dip they are allowed to dry in the air. There are, of course, other innumerable little details such as turning of brush rings, bearings and so on, but sufficient has been given to illustrate the systematic method of following up this class of work.

To make a washing machine motor is an entirely different proposition as the design is different in many respects, but we will not enter into this matter, as some of the work is very much a duplicate of that already described. The other parts shown in group at Fig. 4 are as follows: F illustrates a single disc of which the completed field core is made, and C shows the same in an assembled state with the coil attached. H and I illustrate the copper and mica sections going to make up the commutator shown at J. The other portions need not b lettered as they are merely parts going into the make-up of a motor. The brush rings can be clearly noted and need no special comment.

stage they are put together in the following manner. An assembling fixture is used which gauges the correct number of discs to make up the core. Two rivets hold the core together.

They are next ground, and as there is a flat spot on each side of these cores, they are placed on a mandrel, and are staggered so that the grinding wheel is always removing stock. The grinding of the outside diameter is the next operation in sequence. It might be well to mention that in the assembling of these cores this concern has employed a blind man as one of their assemblers, and in spite of his handicap he can turn out as good work as the others. To watch him perform is a sight in itself, and were one unaware of his affliction, they would hardly believe such to be the case, so dexterous are his movements.

The cores being ground, the coils are next given a slight curve in an air press. This coil is noted at B in the photograph. After being formed they are inserted in the core, and the terminals are next soldered to the coils.

Making the Commutator

The punchings, both copper and mica, are made in the press department, and on arriving at the assembling department are placed in circular form as follows: A copper section, a mica section, and so on. The sections are pressed into a steel ring as shown at C. The sleeve is now insulated, as shown at D, after which sleeve and collar are assembled. The commutator is next locked by means of a wire and pressed out of the ring. This is shown at E.

They are now inspected to see that copper and mica are perfect, after which they are covered with bakelite and baked. A rough cut is next taken off the outside diameter of commutator,

then the rear shell is faced to length. The holes are next reamed to size, and one end of the hole is chamfered. The sleeve is now milled with a small slot at its end. They are next tested for shorts and grounds, after which they are ready to assemble on the armature.

Field Coils

The field coils are first wound to the specified number of turns, and taped with what is known as empire cloth. This cloth is really a sort of linen, so coated that it takes 2,000 volts to jump through it. This provides a perfect insulation. They are next taped with muslin tape, the leads are soldered,

The Die Department

It is hardly necessary to call reader's attention to the fact that work of the nature we have just described entails considerable punch press work. This work is also of a very exacting character so that the data to follow on the die department will in itself be of real value to the mechanic. It is not often one runs into a similar line of die work and for that reason we have gone into the matter very carefully, giving figures of value and interest to the mechanic in general.

Die work at any time is a fascinating subject, but the dies about to be de-

FIG. 9—A NEW AND USED DIE BLOCK. THE NEW ONE IS 1¼" THICK, THE USED BLOCK ONLY ⅝" THICK.

scribed are of more than passing interest. Their armature disk dies are the chief feature, but before speaking of the die, it might be well to explain that these disks are made from a grade of sheet iron approximately 0.025" thick, termed electric iron. This material is used because of its low resistance to an electric current, or to use the proper expression, the material is used because of its permeability. Though very soft, it is difficult to punch, the wear on the dies being considerable. In the average plant, a die of the nature shown at Fig. 5 would be made altogether different from the method employed by this concern. The photograph depicts a double armature disk die, the outside punching being for a larger motor than the inside one. To give readers some idea of the dimensions of the slots in

blanking die A is made from a solid piece of steel in which the required number of slots are machined. The die fits into a slide in the die shoe B, and is held as shown. C illustrates a portion of stock punched, D the punch block itself, E some assembled disks of the large size, F as a single disk after centre portion has been blanked out, and G two of the centre disks. It will be noted that the first portion of the die punches out both the slots for large and small size disk, and to give readers some idea of the task let us refer to Fig. 6 together with the following description:

There are 36 slots in the outer die, these being .156" at the small end, and considerably wider at the other. Two .124" holes are drilled 3 degrees and 30 minutes apart, then slots are filed to a finish. The inside disk has 18 slo..

concern at the present time. Ketos steel is also used with considerable success.

The die block is first cleaned up all over to ensure getting below the decarbonized surface. They now heat the block to either a blue or a yellow color. This is done for marking off purposes, as they do not use the regular style of wiping the block with soapstone. They claim such an artificial surface wears off before the die is completed and is not suited for their purpose. Having colored the block, they proceed to mark off the die, but again deviate from all regular practice. No templets are used, the die being laid off direct from the drawing. This statement may not sound much to the layman, but to an experienced tool maker it speaks of much needed accuracy and confidence in one's laying out powers.

FIG. 11—ARMATURE DISK DIE FOR 1¼WW MOTOR.

these dies we have drawn the sketches shown at Fig. 6.

It has been considered bad practice by some concerns to make such dies or punches solid, because the thin strips of metal remaining are likely to warp in hardening and break. Another argument made is, that if made solid, repairing becomes more difficult.

These objections have been swept to one side as far as the firm we speak of are concerned, for not only do they make their dies solid, but their punches as well. This procedure not only calls for good tool work, but mighty particular care in the hardening operations.

Construction of Die

It would seem that all regular practice has been tossed aside in the preparing of these dies, for the die shown is not even of the sub-press type, as usually followed out, but is of the plain blanking type of construction. The

the slot in this case having a tit as shown. This tit is .094" diameter. The large portion of slot is .216", the small part .131". The distance between the holes drilled is .129". These dimensions if studied will show the care that is necessary in completing such a die correctly. The punch is another ticklish proposition; as can be noted the slots go down 1⅜" deep, and of course the punch must suit the die.

Let us suppose we are watching the making of a die similar to that shown in the photograph. The die shoe is first machined to size on the shaper, and as this is ordinary tool work, we need not dwell on the matter. The shoe being completed it is passed over to the man making the die.

The steel used in the making of the die block is always of a guaranteed oil hardening, non-shrinkable brand, and it might be well to state that Vasco and Deward steels are being used by this

The block is now ready for working out the elongated slots which form the projections on the armature disks, and the first step is to drill a series of equidistantly spaced holes entirely through the block. This work is done on the milling machine, the dividing head being used for spacing of holes exact. A bushing type of drill chuck is used in the miller spindle, and this chuck carries a drill of the correct diameter. The internal ring of holes is usually drilled first, then the size of drill is changed when the exterior holes are next drilled. Sometimes the procedure is reversed. Following this, a series of holes are drilled so that they just break into each other. After these holes have been completed the intervening webs are removed by shifting on the large disk shown at Fig. 5. 296 holes were drilled before the webs would be drifted out.

The next step is to finish form the slots, this being performed as shown

FIG. 12—DETAIL OF A FIELD CORE DIE.

nt Fig. 7. The work is done on a No. 2A Brown & Sharpe miller with a slotting attachment placed on the spindle head. The die is mounted on a tapered arbor, which is held in the dividing head of the miller, while slotting tools of the correct shape are fastened into the head of slotting attachment. The dividing head is held perpendicular, and the slotting attachment is also set perfectly square, this producing a straight hole. The clearance is produced in the filing operation which will be described later.

The sides of the slots are machined first, after which the two arcs are completed. The outer arc has a slot as already described cut into it, and for this operation a tool of suitable shape is used to cut the slot. As before, the indexing attachment of the dividing head is used for spacing of the holes. Very light cuts are taken during their operation, as were one to attempt a cut of even 1-64", distortion and general springing of the tool would result. Patience is a virtue on this work, care, rather than speed, being the chief consideration.

Finishing Operations

When these slots have been machined to their approximate size, the final operation consists of filing them to exact size on small filing machines as shown at Fig. 8. These machines are made by the Oliver Instrument Co., and consist of a table that can be set to any desired angle, and a file comes up the centre of the table of such a nature as to conform to shape of slot desired. The machine is driven by an individual motor, which by the way is one of their own make, and in addition to filing the slots to their correct shape, the machine also produces the correct clearance angle.

Here is another point where this concern follows a peculiar trail. They only allow .003 inch a side on these dies, the block being 1¼" thick. This means that the table of the filing machine is tilted over approximately 7½ minutes. Imagine a die 1 1/8" thick with only .003 clearance aside, and still this allowance has proved a decided success, and allows them to actually grind their dies down to actually one-eighth of an inch thick.

Sounds like a rather wild statement, but look at the photograph Fig. 9 of a thick and thin die, and be convinced. The die illustrated was made from Ketos steel.

The die block is now hardened, being heated up to 1375°F. to 1650°F., depending upon the nature of the steel and the shape and size of die. It is now drawn all the way from 300F. to 500F. For hardening, a Hoskins electric furnace, made by Hiram Walker & Sons, Metal Products, Ltd., Walkerville, is used. This is known as a high speed tool furnace and can be heated up to 2500F, if necessary. The heating is done gradually and uniformly until about 1100 F. has been reached before the hardening point heat is raised. The block is dipped into linseed oil and left to cool, after which the temper is drawn very slowly to relieve the strain.

The hardening of such a die as shown calls for considerable care, as the slightest amount of shrinkage or warpage would render it unfit for use. Being of the solid type it would be practically impossible to correct any distortion in

hardening. It takes from five weeks to two months to make a complete die as shown, so that the loss would be considerable. It might be said, however, that although following out such a practice this concern have never yet lost a die in the hardening operation.

Making the Punch

The die block being now completed, let us consider how the punch is made. In all cases this concern make their dies first, then fit their punches to the hardened completed block.

The punch is made from a solid, round bar of stock, drilled, turned and recessed in the lathe. It is next slotted in the miller. The punch is driven on a tapered arbor, which is held in the dividing head of the milling machine, and a slotting saw of the required thickness is used to form the individual members of the punch. Depending on the size of the punch, two to five cuts are taken. The position of the cutter is always changed on the last cut. On some types of armature punches, where the slots between the different punch members vary in width at the inner and outer extremities, slotting saws of different thicknesses are used in order to reduce the amount of machining. When this is necessary, the saw is not inserted as already stated, but is raised slightly and the second saw is also raised. The remaining web is machined out with the slotting attachment.

In the case of a large punch, where five cuts are taken, the semi-circular, and the tit form are milled after the third or fourth cut. The change of cutters is made with a definite idea in mind, namely to use the form cutters before the members are split right through. This avoids chatter and distortion. The reason five cuts are necessary is self-apparent, namely, the punch must not

FIG. 13—DETAIL OF A FAN PUNCHING DIE.

be sprung in any way and a heavy cut would be sure to do this. The operations spoken of so far have been accomplished with the punch in a horizontal position, but in order to perform the finishing work, it must be turned vertically and the slotting attachment with proper shaped tools used. After this operation is completed a slight touch up with the file is all that is necessary.

The punch is now sheared into the die block for a distance of about 1/8" to ¼", when any excess stock is again removed. This shearing operation is performed on a regular style arbor press, and they are very careful to shear a very little stock at a time in order to

found that the punch and die go together perfectly.

Believing it will be valuable data for tool makers in general, we append still further information on the die we have been speaking about. The clearance on the shoes to allow the punchings to drop out ranges from 1/64" to 1/32" a side, depending upon conditions, and every operation is set up on the machine by means of dial indicators. Every set up receives like care, for it is imperative that die be absolutely true. In order to secure an accurate measurement of the diameter of punches when milling the arcs, they use two round pieces of drill rod, letting them rest between two members of the punch.

turn it. The adding of this small slot has overcome this difficulty.

As a rule the shoe is made about 2 11/16" deep, the die blocks 1⅛ to 1¼" deep and the stripper plate ⅜" to ⅝" deep and the guide pins from 1¼" to 1⅜" diameter. Wherever it is thought wise a pilot punch is used. For example, see the punch at Fig. 11.

The die Fig. 11 is used to punch the armature disk for their 17 W. W. motor, used on washing machines, and is certainly worthy of attention. As can be noted, there are 21 small punches used, each with a very small tit at the exterior. There is also the three web punch, and the centre punch, which has a tit on either side of it. As before,

FIG. 14—ILLUSTRATING TWO DIFFERENT DIES. DETAILS ARE EXPLAINED IN THE TEXT.

place no undue strain on the finished die block.

The Guide Pins

As soon as the punch has been fitted correctly, it is assembled to the punch holder, and placed in the die for about 1/8 of an inch. The shoe with die in place, also the punch holder with punch in place is taken to the vertical miller, where the guide pin holes are drilled as shown in heading of this article. Note that parallel strips and paper keep the die and punch in perfect position while the holes for guide pin bushings are being bored. A Marvin & Casler offset chuck is used, which has a micrometer dial that makes it easy to set the boring tool to correct size of hole. This operation is known as the final drilling of guide pin holes, and after the work has been taken from the machine, the guide bushings are ground to fit the holes just bored. The bushings are next pressed in place, the guide pins assembled, when it will be

The micrometer measures the distance from these drill rods, then the distance equal to their diameter is deducted from the measurement found.

It will be noted by referring to the further pictures of various dies used, that one style of gauge is universally adopted. This is a very simple gauge and can be seen at F at Fig. 11. It consists of a steel finger, connected with a small spring to the stripper. On the punch block is a pin which hits the end of this finger as the ram descends, thus lifting the other end of the finger where gauge pin is inserted. This, of course, allows the stock to press through.

Before milling the punches, a small notch of about ⅛" wide, and of similar depth, is placed at the back of the punch. This notch keeps the punch from turning while being milled. They did this because they found that even though the punch was forced on a tapered mandrel, the vibration had a tendency to

the die is made solid, and entails considerable accurate work. This portion is of course covered by the stripper A, but as it practically takes the same form, readers get the idea quite clearly.

The blanking punch B is made solid, and it will be noted that a small pilot is used to centre the blank correctly. The 21 punches already spoken of are made separately, but the three web punch is made from a solid piece. The centre punch is made separate, so that in reality, there are actually 23 punches used at this portion of the die. C represents an assembled core, D the disks used in the centre of the core, while E depicts the disk for which the die shown is made. Note that the only difference in these pieces is the absence of the tits in the disk D. F depicts the gauge, which has already been explained.

Field Core Die

The die depicted at Fig. 12 needs little comment as it is self-explanatory.
Continued on Page 592

Making Parts for Universal Plate Mill

Operations in the making of universal rolling mill equipment—The machining of rolls, and boring work on sections—Machine work on certain parts of rubber mill equipment.

By J. H. Moore

N EW tendencies and developments in Canadian industry are always interesting. This is especially true of the variety of work at present proceeding in the plant of the Dominion Steel Products Company, Limited, Brantford, Ontario. To discuss all the classes of work being produced is not possible in this space, but we shall take up the two chief propositions which they are handling.

To explain the first proposition, we must in fancy travel to Hamilton, Ontario, there entering the plant of the Dominion Foundries and Steel, Ltd.

Here we see this latter concern installing a huge Universal plate mill, and it was this very installation that created the demand for tables for the aforementioned mill.

To make tables for a plate mill sounds like an easy proposition, but is it? . Ask the company at Brantford, and see what they say. The number of parts to tables of this nature is surprising, and to form some conception of the general appearance of one of these tables, let us refer to Figure 1. As will be noticed, the table consists of a series of rollers, driven by motors conveniently placed. Depending upon the length of the table, one, two, or four motors provide the driving power. Each roller has a

bearing at both ends, and, as can be seen, a bevel gear is keyed to one end of each roller. This in turn meshes with another bevel gear on the power shaft. As a rule the power shaft is coupled up at various place, depending upon the nature and length of the tables. This of course means that all motors driving the power shaft must revolve at the same rate of speed, and to accomplish this the motors are arranged in synchronism. While we have shown the bevel gears in full view at Fig. 1, these are of course covered in actual operation.

Without going into any details, let us run briefly over the various tables used, and the procedure in the making of a plate. The stock is of course taken from the yard by means of a crane, and placed in the furnace pusher. From this point the billets are pushed into the furnace, and on reaching required temperature come out on the discharge pusher. . The stock next reaches the approach table, and, rolling up on this table, it soon reaches the front mill table.

It now passes through one mill, landing out in rolled shape on the back mill table. Continuing its journey, it passes from the back table to the runout table, and from there to a plate leveler. Next comes the hot bed, and in some cases more than one hot bed is used. The procedure from here on varies considerably, depending upon the nature of the plant, but sufficient to say there is still one more table, namely the back shear table.

We have quoted this routine of operations, not with any special plant in mind, but rather to bring before readers a general conception of the procedure.

The Sections

The construction of these tables is somewhat similar, being made up of various sections, bolted and keyed to-

FIG. 1—GENERAL APPEARANCE OF ONE OF THE TABLES.

gether. Take for example Fig. 2. Here we see six of the sections being machined on a Pond planer. The bottom of these sections is planed on four spots only, the rest being bedded in concrete when installed in place. The piece is then turned, and the top planed all over. Next the sides are planed as shown in the illustration, when other minor operations follow. As the photograph is self explanatory, we will pass on to a description of the main mill table section shown at Fig. 3.

This table section is 18 ft. 4½ in. long, 4 ft. 3 in. wide, and 2 ft. 1 in. deep. It has 13 roll bearings, each of 6½ in. bore. The first operation is performed on the planer. The bottom is planed on four spots, and the rest of bottom is bedded in concrete in actual use. Next, the work is turned, and the top planed all over. The side bearings are now machined to width and depth, there being five of these bearings, as shown at A. The work is next set up on a Landis boring mill, and the ends milled as shown at Fig. 4. This is a simple operation and needs no special comment.

The section is now set up as shown at Fig. 3 and the cap holes drilled. There are 44 of these holes to be drilled, the majority of them being 13-16 in. diameter, and they are drilled 10½ in. deep. The work is of course moved along as the job proceeds. The bearings are also finished at this boring mill and a peculiar practice is adopted in the boring of these bearings. The bottom half, that is the portion shown in sketch, is bored out to 3 1-16 in. radius, or equal to 6⅛ in. diameter, this being to accommodate a brass liner which has a bore equal to 5½ in. diameter.

The liner however is of half circle section, and goes to the bottom only. The cap has no liner, and is merely bored out equal to 5½ in. diameter. The reason no liner is placed in the cap is as follows: All the wear is downward on these bearings, therefore there is no need of a liner where it would never wear out. The method of supporting the work shown at Fig. 3. is

FIG. 1.—DRILLING THE CAP HOLES ON ONE OF THE TABLE SECTIONS.

readily seen, so that no comment is necessary.

The next photograph Fig. 5 illustrates a few of the rolls going to make up one of the tables. The turning work on the rolls alone is a considerable item, and the few figures to follow will be of particular interest. The approach table, as made for the mill we speak of, has some forty rollers, ten table beams, seventy bevel gears of thirty-three teeth, 1½ D.P., and other numerous parts of like importance. The front mill table has twenty rollers, and the rear mill table contains a like number; 42 mitre gears go on each table, and all drive gears are keyed to shaft with two keys and coupling gears are held in a similar manner, but the roller gears have only one key. The rear mill table is driven with two 80 H.P. motors, and is also the front mill table.

Run out table has 58 rollers, and it might be well to add that these rollers are 14 in. diameter, 5 ft. in length of rolling surface, and 7 ft. 9 in. overall. The borings are 4½ in. and 5½ in. in some cases. Steel castings of course enter largely into the construction of these parts, although the caps for roller

bearings are of cast iron. The motor drive gears on some of the tables are 29½ in. outside diameter, 87 teeth, 5 in. face, and 3 D.P. These gears are cast steel.

The run out table has two 50 h.p. motors to drive it, and the hot bed table has a similar equipment. The hot bed table has a grand total of 61 rollers. The reason these hot beds are made so long is in order to allow the stock to cool, much as this may sound contradictory. In other words, the hot bed is a resting place for the stock while it is cooling.

The keyways used to hold the bevel gears to the drive shaft are 1¼ in. wide, and two are used. The shaft for approach table is 5 in. diameter at the largest part, and is over 12 ft. long. As stated previously, all sections are aligned and keyed together, and this must be done accurately. Every bolt hole is spot faced on these tables and all bearings are specified on the drawings as follows: The shaft must have a perfect bearing over its full length.

The charging furnace pusher itself is an intricate piece of work. Most of the mechanism is underground, but roughly,

FIG. 2.—SIX SECTIONS BEING PLANED ON A LARGE PLANER.

FIG. 4—MILLING THE ENDS OF A SECTION ON A LANDIS BORING MILL.

here is what occurs. A horizontal skid motion is developed through a lever and sliding arm, this pulling the billets over to the furnace. Two 20 h.p. motors drive the arrangement and there are

it can be easily understood that no sequence of operations can be given in an article of this kind owing to varied parts being made, but enough to say that other parts almost innumerable en-

table being more than 145 ft. long. There is over 550 ft. of table space up to the shear delivery table alone. This is a new industry for this section of the country, and necessitates both a need of accuracy and a keen knowledge of requirements. The mill for which these tables are being made is of the Universal type, and can produce billets, blooms, sheets, bars, narrow universal plates, etc., etc.

Rubber Mill Work.

Leaving the work on plate mill at this point, let us next consider the variety of work being accomplished by this concern for rubber working plants throughout the country. A sample of such work forms the title piece of the present article.

This view depicts the machining of a large ram for a special vulcanizing outfit. The work is 22 in. in diameter, 17 in.-5 in. long, and has a screw hole in one end that is threaded for a 4 in. taper plug. The other end has three ribs cast into it, spaced 60 degrees apart. This work must not only be machined but must be polished and free from scratches. There are also four 1 in. holes drilled at the webbed end of roll, and these holes are drilled to a depth

FIG. 5—A FEW OF THE ROLLS GOING TO MAKE UP A TABLE.

over 14 gears in the mechanism. There are various 118T—2D.F. herringbone gears used in the discharge pusher, which is, of course, the reverse idea of the furnace pusher, insomuch as the discharge pusher takes the billets away from, and not to, the furnace.

ter into the makeup of such tables. These include roller plates, lock washers, cap bolts, bracket bolts, nuts, keys, thrust collars, oil rings, gear guards, separator bolts, bearing caps, etc., etc.

The work entailed in the preparation of tables is surprising, one particular

FIG. 6—SECTIONAL VIEW OF 48" VULCANIZER. TEXT EXPLAINS THE METHOD OF OPERATING.

LEFT HAND VIEW FIG. 7—THIS SHOWS A MOTOR BED PLATE BEING MACHINED.
RIGHT HAND VIEW FIG. 8—AN INTERESTING ASSORTMENT OF TIRE MOULDS.

of 21¾ in. The work on the latter operation is performed on a Landis boring mill. There is approximately 1¼ in. to 1½ in. to come off the diameter of these rams, in order to get to the perfect metal. The work is accomplished on a 33 in. Le Blonde lathe and takes approximately 50 hours to complete.

Let us next consider Fig. 6. Here we see a sectional view of a 48 in. vulcanizer that this concern recently completed, and a short description of same would not be amiss. The tires are placed in the chamber shown, top lid is closed, and pressure put on the ram.

The ram is 22 in. diameter, and has a hydraulic pressure of 1500 lbs. per sq. inch. The steam pressure used is 40 to 50 lbs. to the square inch. The stroke is 12 ft. 1 in. and the top lid is arranged with a special breech block lock for quick handling. The machine work necessary on a vulcanizer of this nature is easily seen by reference to the sketch.

The planer job shown at Fig. 7 is one of rather decent size. It is a motor bed plate for a 400 H.P. rubber mill line, and the base of the piece is approximately 7 ft. square, the height being almost 3 ft. As will be noted there

are four heads to the planer, and use is made of all four whenever possible. In the picture shown only two are being used. The bottom is first planed, then it is turned over, and top is completed. The piece is next turned round and the edges planed. This bed plate is made of cast iron, and the job is completed in approximately 30 hours.

Another interesting photograph is that at Fig. 8, this view showing an assortment of tire moulds. This concern makes practically any type of mould, but the style shown is a collapsible core, for straight side tires. Note that the

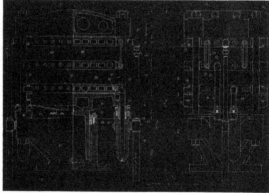

FIG. 9—A 75-TON 36" x 36" HEATER PRESS RECENTLY COMPLETED.

FIG. 10—MACHINING A BEAM ROLL FOR WIRE ROLLING MILL.

ring with bolts holds the two sections together. The machine work on one of these moulds is self apparent. The majority of the operations are completed on vertical boring mills, and all work is held to templates. This concern also does considerable tire mould engraving, and the engraving machine used is of unique design. Any style of tread can be arranged for, and engraved into the mould. Some very interesting treads were shown the writer, and the information was given him that there was not any impression they could not do on these machines.

The next illustration, Fig. 9, depicts a 75 ton 36 in. x 36 in. heater press. This has two six inch openings, and the steam platens are so designed as to be interchangeable. This means that only one spare platen is required in place of the regular three. The steam chests are also interchangeable, and this press has a water pressure of 1,000 lbs. per square inch, and a steam pressure of 40 lbs. per square inch. The steam of course is used to keep the platens up to the correct heat for vulcanizing. They are used for various kinds of rubber goods, in addition to bead molds for tires, and this short description will be sufficient for illustrative purposes. In conclusion we will generally summarize the varied class of work going on at this plant.

Fig. 10 depicts the machining of a beam roll for wire rollin¬ mill. These rolls are 24 in. in diameter, 11 ft. 8 7-16 in. overall, and have 66 grooves and 15 ribs in their finished state. The grooves are 1 11-16 in. wide and the ribs 5-16 in. wide. The very fact of these ribs being so narrow necessitates great care to prevent the breaking of the same during the process of machining. The grooves are 1¼ in. deep, so that considerable machine work is necessary. Following is the method adopted in the machining of these rolls. The first operation consists of taking a rough-

ing cut across the complete roll when it is now removed from the lathes for the purpose of balancing. It is placed on straight edges for the purpose, and is balanced by means of sticking clay on the light parts. After sufficient clay is placed on the necessary portions of roll it is weighed and the amount of offset computed to balance roll when turned. When this is done, the center is drawn over, the work placed in the lathe once more, and turned to within ¼ in. of its finished diameter.

A square nosed tool is next used to rough out the 66 grooves, these being cut down 1 3-16 in. deep. The steady rest is placed in its correct position and the ends bored out. First, the hole is bored for bearing spindle at the carriage end, after which work is reversed and used with steady rest at end as before.

The remaining end of work is now bored out, and the regular bearings driven into place. After the first side has been bored, the regular center at headstock is knocked out, and replaced by a bell center of suitable shape to accommodate hole previously bored. The bearings being driven in, the work is completed on centers as shown in the photograph, and all grooves are carefully finished. This work is also accomplished on a 33 in. Le Blonde lathe.

TURNING OUT SMALL MOTORS
Continued from Page 587

The work performed is the punching and blanking of field cores, and one of the completed disks is shown at A. The punching, taken from the centre, is shown at B, while C depicts an assembled field core. Both the die blocks and punches are made solid in this case, but of course the four small punches are separate as can be noted. The same style gauge is used, and the procedure of making the die being the same as

already described, we will pass on to Fig. 13.

Fan Punching Die

In the make-up of the Hoover suction sweeper, there is a fan used to keep the motor cool. The die used to punch out this fan is shown at Fig. 13. This is an intricate piece of punching, and each of the web-shaped punches A are made separate. The centre punch in this group is of course also a separate punch. The blanking punch B is solid, as is the die block itself. A completed punching is shown at C. Following up this fan, let us refer to Fig. 14.

To the left of this picture is depicted the die used to form the blank made at Fig. 13. A represents the punch, with pilot attached, B the forming die. At C is shown the flat blank, D depicts the formed fan, while F shows this fan mounted in its hub. The punch for forming is made solid, but the pilot punch is separate. As before, the die block is solid. The forming portions of the die operate from a spring stripper underneath the lower bolster plate. The action can of course be understood. The blank is placed on the die, down comes the ram and the stock must naturally form the punch and die, being formed to the shape shown.

The die to the right at Fig. 13 shows the equipment used to complete the punch. Then each arm of a separate punch. One of these arms can be noted at F. The peculiar shaped main punch measures approximately 2 5/32″ across, so that this measurement will allow readers to form a good idea of the proportions of the rest of the die. The smallest punches on this die are only .050″ diameter, and the others are 3/32″ diameter. There are also 2 oval punches as can be noted. The work on this piece is rather delicate, and mechanics can appreciate the fun in making punches .050″ diameter, not to speak of the tempering of the same.

Having touched on some of the important dies as used in this class of work, let us in conclusion consider the life of such dies. Take for example the die shown at Fig. 5. Here we have a double armature die of intricate design, yet 12,000 to 16,000 punchings can be made between grinds. The life of the die is clearly shown at Fig. 9 and a very conservative estimate could be placed at around the one million and a quarter mark. Some dies are ground from seventy to eighty times, or in other words, reduced from their original 1⅛″ thickness to ⅛″ thickness. The solid punches have also been found to stand up in like proportion. Such figures tell more than appears on the surface, and although the small clearance of .003 per side is not universal practice, it is the very fact of their using this clearance that makes for the long life of the die. Were the clearance greater the die could not be ground as far without affecting the character of the work. The experience stated herein are definite and authentic, and deserve careful consideration.

Taking the *Dust* Out of *Industry*

Rewarding Honest Effort Is the Basis on Which Success Can Alone Be Secured—How One Firm Is Dealing With Matters of Employment, Good Conditions and Fair Pay

WHEN a person mentions bonus systems to encourage production there comes to the mind the history of the old piece-rate work, and there are not a few people who make a face when that is mentioned. The old piece-work shop was a queer proposition in some ways, and even yet it is hard to figure out how some of the men who operated them convinced themselves that it was good policy to set a piece-work rate, and then, when a worker got to the point of efficiency where he could pile up a decent pay envelope, cut the rate and start him all over again.

Well, a real bonus system is a different thing; at least it is in the shop of Robbins & Myers at Brantford. This firm has been making small motors at the parent plant in Springfield, Ohio, for a long time, and in March of 1920 they were ready to start operations in the Canadian field, following out here many of the ideas and plans that had worked successfully in the other place.

Robbins & Myers believe in bonuses. They believe, in paying for service rendered. They believe in their employees making good wages, and they believe in keeping their hands as long as possible. But at the back of all this is the one thing that makes the rest possible, viz., when they set a price on a piece of work, they will not cut that figure because an operator or a mechanic can make good money. The only thing that would lead to the change in a price is when the operation is changed, or where it has been found in actual practice that the price on a new piece of work is at fault.

Mr. Roy Booth, the manager of the Canadian plant, makes this point very clear: "Were we to cut a price that we have set, our whole system of bonuses would go to pieces, and the employees would lose confidence in us." That, in brief, is the way he regards the bonus plan.

"In the old plan of day work, and often in the old days of piece-work, people did not know what their product cost them," continued Mr. Booth. "You know what used to happen in the piece-work. An employee would study every movement and cut out every detail of waste in the operation, and by hard work he would become so efficient that his wages would grow, sometimes over that of the foreman, or many of the people in the lesser executive positions. Instead of patting that man on the back and telling him to go ahead, there used to come the cut to a figure that would make it a killing job to make a decent wage. The idea was wrong, all wrong, and the wonder is that even yet we have some shops that follow this practice."

Pay for Production

The Robbins & Myers plan of pricing work is reasonable. There is nothing that points toward the sweat shop tactics. For instance, if 60 units is a fair amount of work for an hour, and if an operator, after becoming skilled, can put 80 or 100 or 120 units of work into the same period, is there any good reason why the reward should not increase in proportion? None that we can see, and the Robbins & Myers people figure it out the same way.

In the making of these time standards, care is taken to be reasonable. The operator who does the work when the time is being taken on the piece knows that he is being timed. Then the same operation is timed again, and allowance made for the delays that invariably come. The standard set is aimed at being as near reasonable as possible, and if there is any work where it can be shown that the rate is not fair it is reconsidered.

When, for any reason, that the operator cannot control, production is held up, the company pays for the time on the basis of the 60 units.

There are Exceptions

Just a few weeks ago some new stock was taken in. In this case it was largely for grinding and drilling. The stock that had been in use previously was particularly good and there was a uniformity in the steel that made it very easy on the tools. When the new stock was put in a change was found. It was a different thing. The steel was hard, and a person working there could hear the tool grinding and cutting as on a piece of flint. There was need to take the tools to the grinder more frequently, and there was quite a difference in the speed

THE CANADIAN PLANT OF THE ROBBINS & MYERS CO., AT BRANTFORD. ALTHOUGH THE BUILDING HAS BEEN COMPLETED ONLY A SHORT TIME, SPLENDID PROGRESS HAS BEEN MADE IN BEAUTIFYING THE GROUNDS AT THE FRONT. FLOWERS AND SHRUBS MAKING A PLEASING SHOWING.

with which the operation could be done. That was adjusted just as soon as the operator and the foreman found that there was a change in the steel from the quality used when the work was timed in the first place. A new timing was made, and it will remain there until the harder stock is run out.

The Inspection Department

Well, if the rate is good, and there's plenty of work, why not steam ahead and get rich? Simply because there is an inspection department on the job to see that quality survives along with quantity. There are the floor inspectors and the bench inspectors. The floor inspector has to see that work is being made according to drawing, and that end comes under his notice. He is paid on the output of the department—that is, his bonus, rests on that. But of course neither he nor the workers in the department are paid for work that is rejected. Then the bench inspector has a peculiar undertaking in that he is paid more for detecting a poor piece than he is for passing a good one.

"Well, hold on," I broke in here, "will that not mean that a bench inspector would be in a poor way financially were the work of the whole department brought to such a high standard that there would be very few rejections?"

"It might work out that way were we to let it go at that," replied Mr. Booth, "but we look after the bench inspector in other ways, and he is not allowed to go poor because there are few rejections. It has been our experience that the system brings the operators, the floor inspector and the bench inspector together to get the best results for all, for it is in completed and accepted work that the payment is made. There is very little work thrown out now, and very seldom is there any dispute over the matter. I am certain that I have not been called in to settle any such matters since the establishment was opened in Brantford."

The Securing of Help

Do they have trouble in securing help at the Robbins & Myers plant? They do not, and neither did they when help was scarce some months ago. The situation was not quite similar then to what it is to-day, when men and girls are looking for work. At the start of the year they were in need of help. They were not known in this country, and people who might be looking for employment were not "sold" on their shop. They are now. But to get back to the start, Mr. Booth practically tapped a new labor market in a new way. He wanted girls to work principally in the winding department, and he advertised. But he did not stick in a little two or three line ad. stating that they wanted a number of hands. He went about it in another way. He stated in his advertising matter that his firm would be ready to commence work in a short time, and that girls desiring to secure employment should call and receive an employment application form, which they would be required to fill in, and on this they would need the signature of two reputable business men certifying to the standing of the applicant for employment. Did the girls in the community resent this? Not a bit of it. They came along with their application forms, and they secured employment, and, what's more, they liked the place, and stayed with it. In the little strip photo published elsewhere in this article are some of the first girls who came with the Robbins & Myers plant. Some of them have left since the photo was taken, but they had very good reasons for so doing, and their leaving only goes to show that no matter how good your system of employment and treatment, it is liable to be broken the break here and the girls are now keeping house in Brantford.

Largely Skilled Help

Now don't get the idea that girls form the majority of the help employed at the Robbins & Myers plant. They have a highly organized plant for the manufacture of small motors, and in it have a large number of highly trained men, and it is through all the plant that the system of bonus and other forms of reward is in operation. The firm operates an open shop, but there are union men working in it.

Mr. Booth, as far as one can judge, can see no difference between the office and shop, and the chances

THE CAFETERIA IS SELF-SUPPORTING, BUT NOT A MONEY MAKER. THE REST ROOM IS A QUIET, COMFORTABLE SPOT. MODESTLY FURNISHED.

are that this doctrine is pretty well understood around the whole premises, and after all his way of figuring it out is right. Here. it is:

"Every operation that goes on in this plant, whether it is in the office or the shop, or on a truck, is necessary. That is the point from which we start to work, and we try to keep before every employee the fact that all are needed and all are necessary, and that one department cannot get along without the other. We are succeeding in doing this, and there is not a regulation that applies to the shop that is not in force in the office."

I remember being in Mr. Booth's office some months ago when one of the employees came in and told him that a number of them, from either the office or some corner of the shop were having a corn roast that evening and would like the use of the cafeteria for a dance after the corn roast.

"First rate," responded Mr. Booth, "by all means have your corn roast, and by all means have your dance after it."

The employee thanked him, and was just about to withdraw, when Mr. Booth started in again: "But remember this, that when you come here and open the cafeteria for a dance in the evening, it is open to all the employees of this company and not to any particular few. If others who are not invited to the corn roast happen to be along here during the evening and see the cafeteria lit up and ready for a dance, it must be understood that they are just as welcome as the party holding the corn roast."

The corn roast was held and the dance was held, and, in the words of the country correspondent, "a good time was had by all."

It Pays do Do These Things

The Robbins & Myers Co. believe in time clocks—in fact they are all clock punchers. There may be some who escape, but I have not heard of it. And the strange thing about it is that they are keen to punch the clock, four times a day. It pays them to do it; not that they are penalized for not doing it, but rather they are rewarded if they attend to this promptly.

There's a bonus for the punchers, and the good punchers can punch out a five per cent. bonus for a week's good performance. So if a girl makes $20 a week she can add a dollar to it by simply punching the clock promptly morning, noon, and night. If a mechanic makes $30 he can add $1.50 per week in this very simple manner.

"And what per cent. of the employees are bonus punchers?" was asked.

"Oh, at least 95 per cent.," was the answer. "We make no allowance on this, and there are no excuses accepted. I remember not long ago that one of the officials here stayed until 6.30 in the evening talking to a customer from out-of-town. When he went out he forgot to punch the clock and it spoilt his week. But we have to make a hard-and-fast rule in this matter, because once it becomes known that there is a single exception made to it, your whole system is off and done for. The bonus for a perfect week is paid on the total earnings for the week, including any production bonus that may have been earned."

And the company admits that it pays to take this positive measure of rewarding good attendance rather than by taking the negative view and penalizing for slow arrival, and the result is that the employees are all on good terms with the time clock.

Then There's a Service Bonus

The Robbins & Myers people are anxious to do away with their labor turnover as far as possible. In their line of work, as in many others, the longer a person is with them the more valuable they become, and the more the company desires that they should continue. In or-

A Word About Mr. Booth

ROY BOOTH.

Roy Booth, works manager of the Canadian plant of Robbins & Myers, Ltd., is not a theorist alone. He knows what a fair day's work is, because at the age of 15 he started as an apprentice with the Dayton Brass Co., where he served his time, upon completion of which he went as a bench mechanic to the American Tin Can Co. Entering the plant of the Dayton Computing Scales Co., he worked up to the position of assistant superintendent, and in 1908, when the branch was started on Vincent Square, Vauxhall Bridge Road, he was placed in charge. In 1910 he came back to the Dayton plant, remaining with this same company until 1917, when he entered the employ of the Robbins & Myers Co., at Springfield, Ohio, first as chief inspector, then as assistant electrical superintendent. In September of 1910, the company informed Mr. Booth that he was to be the man to head their peaceful invasion of this good land to the north. Work was started on the Brantford plant in October of 1919, and in nine months the plant was complete, machinery placed and running, a staff trained, and production of 150 motors per day was secured.

Mr. Booth is making a pretty fair Canadian. He can't swallow his accent yet and it continues to come to the top of his vernacular. Rotarians and other good chaps at Brantford have taken kindly to him—they like to have him about and vote him a good citizen. His wife holds much the same views and two young Booths claim the privilege of calling him Dad. He likes Brantford and Canadians in general, and it looks as though he were planted up here for keeps.

der to get this they use a service bonus. It's not a complicated piece of business, and works out something like this:

When an employee has been with the firm for six months his bonus starts, and at that time it is 1¼ per cent. of all the money he has earned in the previous three months. This keeps on until at the end of five years, the increases coming every six months, the employee finds that ten per cent. of all earnings in each previous six months are handed to him in a lump sum. This bonus includes ten per cent. on everything, even to the five per cent. bonus that is paid for prompt attendance and punching of the clock.

"Well, does it pay to go that strong?" was asked of Mr. Booth.

"There's not even room for an argument about it," was the answer. "It has almost eliminated our labor turnover. It holds our employees here, no matter how hard other firms may try to get them. We know how our wages compare with others paid in various shops here, and when there is, after five years, a ten per cent. bonus every three months, do you think any man or girl

is going to pass that up - without a little more than passing thought?"

Come to think of it, the thing works out to a nice little item. For instance, suppose you were working for the Robbins & Myers Co., and had been there for five years, and were receiving the ten per cent. bonus. Your wages are, supposing again, $40 a week—not an unlikely figure—that would be for thirteen weeks, $520. Forget the clock bonus for the time being, and think only of the $40 a week. Well, around Christmas, or around anything else for that matter, it would not be a bad sensation to have an extra cheque coming across for $52, and it would be a still more pleasant sensation to have this happening every three months.

This service bonus applies to every person in the establishment, office, tool room, winding department, shipping, trucking, etc.

Some Ideas of His Own

Mr. Booth, the Canadian manager, has some ideas of his own in regard to the handling of a staff, and I remember quite well that on one occasion he discussed this matter, although he was not speaking for publication at that time. But the idea is a good one and worth repeating:

"I do not like to see a man leave here, and I dislike still more to see a man or a girl discharged. We try to find a place where they can make good. I make it a point to see every person myself who is leaving our employ. There are cases where men leave machine shop and foundries because they are not getting a square deal from the foreman. I have run across just such cases; not here, but in other works. A workman may be 'in wrong' with the foreman, and it is a very easy matter for the foreman to give him all the lean jobs if he is on a bonus basis, and make it almost impossible for him to get ahead and make money. If a man leaves this establishment for a fault that is not his own, I think we are entitled to find out just where the trouble is, and if it is ours we want to set it right, and if it is his we want, if possible, to set him right. A great deal can be accomplished with a little good sense and reason, and the more you deal with men the more you should know about them. There are often cases where a man will not take his complaint over the head of the foreman or the superintendent; he will simply pass in his slip and fill out a quit card rather than make a show for it. He leaves, and gets work elsewhere, but he is very apt to entertain a poor opinion of our works, and this can be very easily spread. When this is going on to any great extent you cannot keep the community 'sold' on your shop being a good place in which to secure employment."

Every employee who takes employment with Robbins & Myers is insured with no cost to the individual under their group insurance system. It starts at $500, and increases $100 each year until a maximum of $1,000 is reached. If, through illness, the employee has to leave work, the insurance continues, and will be carried, if necessary, for a term of twenty years, provided of course that the insured person does not accept employment elsewhere in the meantime, in which case the insurance ceases. This insurance feature is a popular one with the employees, as they are glad to get the added pro-

tection and are apt to think seriously before leaving the employ of a shop extending them this amount of free protection.

Some months back now it was decided to put in a cost-plan cafeteria. The building was constructed with this in view, and the cafeteria is over the office. The floors are of hardwood, highly polished and waxed with a view to using it for dancing purposes. The tables and chairs are so made that the entire floor can be cleared out.

"Make much out of the cafeteria?" we asked.

"Sure," laughed Mr. Booth. "For the first three months we were $7.93 ahead of the game, and I have just received the figures for the last month which show a profit of $1.96, and you know that could be very easily wiped out if one were to let the knife slip a couple of times in slicing up the ham at present figures. We post a statement every month so that all the employees can see it, showing what we have in stock, what we have paid out in wages and for provisions, and against this what we get in the way of receipts. This has worked out very satisfactorily, and we are able to keep the confidence of all in the prices that we charge for the service."

Then there is the first Saturday in each month. It may be known locally as Fox Trot Saturday or Barn Dance Saturday, but whatever the name, on the first Saturday evening of each month the company opens the cafeteria in the evening for the employees to have a dance. They can bring along their friends, apparently the more the merrier. Where does the company come in? They do everything but dance. They provide the dance hall, the music, the cake and the coffee, and it is not on record yet that they have squealed about the number of friends that any employee leads out. The office staff and the shop mix freely at these gatherings, in fact there is no distinction one way or the other.

Sports are encouraged, and the Robbins & Myers team won the soft ball championship of the city in 1920. There is a clean, wholesome air about the whole place, and the evident desire of the company is that its employees shall have a good living and have a fairly decent time while they are doing it.

Judging a Wheel

We often hear the query what is a hard wheel and what is a soft grinding wheel. These terms are rather confusing and are not really comparative terms.

A wheel may be too hard for one job and too soft for another. If a wheel is too hard for one job, its bond insists on holding onto the abrasive particles after they have lost their sharpness or cutting power. Such a wheel is very inefficient and will glaze badly. If the wheel is too soft the abrasive particles will break away from their setting in the bond before they have lost their cutting power. Such a wheel wears away very rapidly, but on the other hand, it cuts very well. The ideal wheel is the wheel with a bond that permits the abrasive particles to drop off and expose new particles just at the point where the old ones become dull. With the wide variety of grits, grades and bonds manufactured, it is possible to purchase a wheel with just the right characteristics for most any job.

THESE ARE THE FIRST GIRLS COMING TO THE ROBBINS & MYERS PLANT WHEN OPERATIONS WERE STARTED. EXCEPT FOR MARRIAGES THEY ARE STILL WITH THE COMPANY.

Works Council After *Two Years'* Trial

A LMOST two years ago the International Harvester Co., Hamilton, started the Industrial Council plan in their establishment, doing so on a favorable vote of the employees. Their evidence in the matter is valuable now because it has the weight of real experience behind it. Canadian Machinery found officials of the company very frank in discussing the system, and anxious to give other manufacturers the benefit of the experience they have had in this direction.

M ANUFACTURERS are trying to find some plan that will put the "US" into indUStry. The men in the shops are ready to meet this idea half way. No man likes to feel that he is working in a shop, and yet is not part of it. There is something in the make-up of the average person that calls for a "say" in the conduct of anything that he is connected with. When one can satisfy that feeling, and link up with it a force that will actually make a man a better and more competent workman, then the chances are that you have something worth while.

Now that's just about what the Industrial Council is doing. It is not a matter of theory any longer. The average manufacturer or employer is not much impressed with theories now. He has seen too many of them flivver —but he is interested in anything that will work when it is put up against shop conditions for six days a week. The Industrial Council started as a theory, and it is now established as a fact. The International Harvester Co., at Hamilton, have been operating under the Industrial Council plan for almost two years, and they are willing to continue it. Nor do they hide their light under a bushel. As you enter the premises at Hamilton you are informed, by means of a large notice, that this establishment is operated under the Industrial Council Board, and when you reach the large offices upstairs there is another indication in the form of a commodious board room, devoted to the affairs of the Industrial Council.

The International Harvester Company believe that this system has gone past the experimental stage. They are willing to admit that it is a success, and are satisfied that conditions are better all round. It is almost two years since the plan was inaugurated. The secretary of the Works Council devotes his entire time to its working, and he has plenty to keep him busy, for the work grows in new directions every month, and the idea of "doing things" seems to be breaking out in new places all the time.

Here's What It Does

If you had a system in your shop that would make a mechanic a better mechanic, or that would cause a moulder to be a better moulder, don't you think you would be satisfied that you were working along the right lines? Well, there's no revolution going on at the Harvester plant, but it is a fact that the work that is being done through the Industrial Council plan, is doing this very thing. For instance: Some dispute arose in the foundry about the rejection of castings, and the matter was taken up at once by the elected representative of the men in that department. But it never got to the Industrial Council, nor is it recorded that there was even a special session about the matter. How was it done? Simply this: The men in the shop looked into the matter, and they found that the work that this particular man had been turning out had not been up to the quality required by the firm. They went right to him and showed him how he could improve his work a great deal, and make more money, by so doing. What happened? Why only one thing could happen. That moulder began right there and then to do better work, and that was the end of the complaint and of the whole matter.

The Works Council, if I get the thing right, does not aim to have a truck load of complaints and squabbles to deal with when it comes to its monthly meeting. Rather is it considered a good meeting when there are none of these matters to come before the board, and where the whole time can be taken up in planning new things and better methods.

Of Course There Are Wages

Canadian Machinery discussed the working of the Council with the Works Council secretary at some length, and in response to a query about the most-discussed thing dealt with, the answer came in the form of an admission that wages crowded pretty well to the top of the list.

"And can you come to an agreement about wages?" was asked, for it is a fact that hardly any man ever thinks he is regarded financially high enough by the firm for which he works.

"When we get together on this ticklish matter of wages," answered the secretary, "we find that the men are not unreasonable when they have all the facts placed before them. Of course we try to have as many real figures as we can when we discuss these matters, and it is our desire to bring our statistics as near correct as possible when we come to discuss the cost of living in Hamilton and else-

where, for that is what most of the demands for wages are based upon."

The secretary admitted that they did not accept the Ottawa figures, nor yet did they rely on the figures gathered by their branches in United States. "All these are good enough, but when our men are living in Hamilton, and keeping house here and buying their stuff here, this is the place where we must base our figures if they are going to have a real weight with the men, and going to be fair for all parties concerned."

And so the Harvester people took a very real way of finding out what it cost to keep house in Hamilton. They took a family budget, not one that a bachelor might think necessary, but a real one, worked out in the council from figures from real Hamilton homes, and they had the purchases made in the stores where their men dealt. The idea was to get as near as possible to real facts, so that when a readjustment of wages was made it would represent the real burden that was going to be placed on these wages when they were spent.

It took some little time to get this family budget brought as near correct as we wanted it, but in the end we found that the figures as worked out by the company, and the figures prepared by the men were just 88 cents a week apart for a family of five, so it was not a very serious matter to get over that difference."

Well, Not Yet

"Any cases yet that the Council have not been able to settle?" "Not yet," was the answer. "The principal thing in settling cases is to make sure that they go through the proper channels in the shop, especially to the foremen, before they are ever allowed to come before the Council. We do not want them brought in here for discussion or settlement. The Harvester Industrial Council makes provision for disputes being discussed in the shop and never brought in here at all, and we encourage that. Some of the men's representatives in the shops are very capable in the role of negotiators. We do not claim that there will not be questions coming up that we cannot handle, but we do know that more and more there is being created a get-together feeling in the works, and men who were hostile at the start, or luke-warm at least, are being won over to the proposition."

"Any particular instance of this, or just a feeling in general?" "Yes," went on the secretary, "here's a particular case. One of the foremen was very skeptical about the whole thing, and he was not in sympathy with the move. It may be that he thought matters would be carried over his head by the shop representative of the men. Well, we asked him to come in one day and attend a meeting. He did, and heard all the discussions that went on, and had an opportunity to see for himself what there was in it for the men and for the company. When the session

was over he told me he was sold on the idea. He had not understood how it would work out before. Since then he has been one of its best friends, and does all he can in the shop to see that the Council has fair treatment there.

Feel They Are In It

"Let me put it this way," I suggested. "Are there any results now, that you did not look for when the plan was first put into operation?"

"One big one," was the secretary's opinion, "and that is the men have the feeling that they now have a say in matters around here that would not be possible in any other way. That in itself is worth a very great deal and something many manufacturers are very anxious to secure. There is nothing a man values more than the feeling that he is really part of the works, and has a voice in the direction of the concern."

Nor does the work stop here. There is a mile or so more things than simply attending to the conditions in the shop. Did you ever get a new job in a big shop in a strange city? Not a particularly desirable sensation sometimes. You wondered who the chap next to you was, and perhaps didn't like to ask him. You didn't know where things were, and you felt generally strange for a time. Well, how would you like to have some one come along the first morning and give you a hand shake, and make you feel at home—tell you the run of the place, and everything you wanted to know—make you acquainted with the man on the next machine, tell you where the cafeteria is, and give you an idea of the run of the day, and what was going on around the place. Not bad, eh? Or if you were laid up for a day or so, you used to wonder if your little old job would be waiting there for you when you got back, or if you'd have to start in again all over. Much better to have some one from the shop call at the house, and see how you are getting along, see if you are getting proper attention, see if there is anything they can do for you. Why, of course. Well, that's the sort of work the Council encourages, and when a body of men get headed in that direction it's hard to say just when or where they will stop.

The Cost Plan Cafeteria

The Harvester cafeteria is a busy and popular place at noon hour. Realizing the important bearing a hot meal has on the workers' health and efficiency, Works Council petitioned the company for space and equipment to establish a cafeteria, which they did, setting aside a large floor on one of the existing buildings, and installing a complete modern outfit, then the project was turned over to Works Council to operate for the benefit of employes. There superintendent and janitor are all served alike with a substantial plain hot meal, as well as cold service at cost, and

These Views Tell of Happy Days at Harvester Picnics

The Plowites' Band is an Ever Popular Organization

wherever it is found possible to reduce prices it is done at once.

The Harvester Company in all its plants has a benefit society to which both the employe and the company contributes. It costs the company something like $50,000 per year for this work in its various branches. A man is entitled to 50 per cent. of his wages when sick, and this will be paid for 52 weeks. In case of death the heirs receive the equivalent of one year's salary of the deceased.

They also have in operation a Pension Plan, the expense of which is borne entirely by the company, and which is purely a voluntary expression of the company's desire and purpose to stand by the men who have stood by it, not as pensioners, but as the company's old guard, who have well earned this reward for long and loyal service and are justly entitled to all that the company does to make their lives happy and more comfortable.

Employes are pensionable at 65 years after 20 years' service, at 60 after 25 years' service, and at 55 after 30 years' service. Women are pensionable at 50 after 20 years' service. Special consideration is given to employes whose age and service approximate this requirement, and who become totally and permanently incapacitated.

There there are other directions in which the Council find room to work. The first aid class is an efficient body, although fortunately their services are not called upon very often, as the works at Hamilton have been fairly free of serious accidents for some time. This may be partly due to the fact that suggestions are welcomed at all times regarding safety appliances, and improved methods for doing any operation. Directly across from the main office is the hospital for the works, where a doctor and nurse are in attendance.

The same committee that looks after the cafeteria, health, etc., have charge of such affairs as gardening and the beautification of the grounds of the homes where the employes live, and they hold to the belief that a man is a better workman if his surroundings out of the shop are bright, pleasant and congenial. During the last season there were some 275 garden plots provided by the company, in addition to a contest between all the Harvester plants for the best kept gardens at the homes of the employes, two Hamilton gardeners winning the third and fourth places. Other committees look after publicity for anything and everything that goes on in connection with the work of the Council around the plant, in connection with social gatherings, sports, picnics, etc. Very useful work is done by a committee that looks after the sick of the shop. A list of these is furnished to the committee, and a point is made at once of calling on them. The spirit of the thing is growing around the Harvester plant at Hamilton, and it does not take much imagination to see where the Council plant is going to grow into a good sized tree, well worth the space it occupies in the organization, and fully justifying all the watering, pruning and enriching it has received.

Real Suggestion Brought Out

But what is there of real help in the meetings of the Council. You must remember that the company pays the men for the time they spend attending these sessions, pays expenses if they find it necessary to make a trip, pays for the time matters are being discussed outside of the Council chamber. What comes in return? Well, a casting came back from away out in the west one day. It was broken, and the matter came before the Industrial Council. It wasn't simply replaced and the broken one put on the scrap heap at the foundry. The men responsible for the making of these parts had a chance to talk the thing over. One man believed that there might have been

Sports of all sorts are encouraged by the Harvester council, and here is seen one of the ladder stunts, in which a record was made that still stands.

some trouble in loading explaining how it could be done to avoid undue packing stress. Another was of the opinion that from the nature of the work, it should have been made from malleable in the first place, and gave reasons for stating that. · Another had a suggestion to make about the design which might lend added strength to the part. Every suggestion that was made was practical, and worth money to the firm. Could all these suggestions have been secured and recorded in any other way? Not that we know of in so short a time · and under such practical circumstances.

Again, the committee that has charge of health, safety and kindred work around the plant, bring in their report. Do they deal in glaring generalities, or do they have something real to talk about. Judge for yourself from this:

These recommendations were made to the superintendent, and the report given carries with it the action that has been taken:

The crane in malleable foundry be repaired, thoroughly overhauled, and put in good running order as soon as possible.—Ordered done and nearly completed.

Concrete floor in malleable sorting room be repaired. —Ordered done and completed.

Platform outside of malleable foundry be repaired.— Ordered done and completed.

Installation of air hoist over tumbling barrels in malleable foundry be completed.—Ordered and being done.·

Hole in platform at dock be repaired.—Ordered and done.

Exhauster hoods for emery grinders in forge be pushed along.—Being moulded.

Crane for die rack in press room be completed and installed.—Ordered and drawings being prepared.

All telephone transmitters be cleaned and disinfected. —Ordered done.

Those instances are given to show that the men themselves, under this plan, are largely responsible for the conditions under which they work, and if there is anything that is not right, or that makes it an undesirable place in which to work, the remedy is in their own hands. There is nothing academic about the discussions. They are real and pertinent. If good they are acted upon, if not, they are considered and passed up.

How the Shop is Divided

The workings of the plan for selecting the members of the Industrial Council are similar to those published previously in this paper. The method of dividing up a shop

for the purpose of giving proportional representation to each department is interesting. In the Harvester plant it is divided as follows:

Division 1.—Cold Drawn, Iron House, Steamfitters, Press Room, Forge Shop, Bolt and Nut:—277.

Division 2.—Grey Iron Foundry, Grey Iron Mill, Grey Iron Core.—284.

Division 3.—Malleable Foundry, Malleable Mill and Anneal, Malleable Core.—254.

Division 4.—Metal Pattern, Wood Pattern, Malleable Finish, Tool, Machine Repair, Fire and Watch, Electrical, Engineers, Experimental, Planning, Drafting.—273.

Division 5.—Wood, Canvas, Lumber, Traffic, Truck and Storage.—263.

Division 6.—Box, Erecting No. 1, Mower, Machine, Sheet Metal.—267.

Division 7.—Shipping, Packing, Repairs, Stock Room, Office, Production, Cafeteria, Yard.—251.

World Wireless Service

An extensive wireless service is contemplated between the French Colonies and the Mother Country. The scheme, according to Wireless World, is divided into four sections. The first section, which aims at providing communication between ships at sea and coastal stations, is already working satisfactorily, but further extensions will be made. The second section will be used to maintain regular communication with the mountainous regions, which, owing to the weight of snow invariably bringing down wires, are out of communication during the winter. The coastal system will ensure communication between France and islands round the coast; it is hoped also to link up Algiers with Paris. The third, or Franco-European system, is already started, wireless communication having been established since May with Hungary, and it is hoped shortly to open up communication with Belgrade also. The fourth section covers the linking up of the French Colonies, including Saigon, Noumea and Tahiti, with France. Stations with a transmitting range of 7,500 miles will be erected at Paris, Saigon and Tahiti. Less powerful stations of about 4,300 miles range will be constructed at Jiboutil, Antananarivo, Noumean and Martinique. Smaller stations will link up the French Possessions in Africa, connecting Martinique with French Guinea and Saigon with Pondicherry. The Ministry of War has already begun the construction of stations in the French Congo, Anatanarivo and Saigon.

THE LOOK-UP MAN

If you are away from sickness or any other cause, they don't just wait for you to come back. The look-up man is sure to call and see what's wrong and how you are being attended to.

The hospital at the Harvester plant is in a separate building and the services of a doctor and nurse are always available. Through the efforts of the council, accidents are being reduced in the plant.

MANUFACTURE OF WIRE ROPE

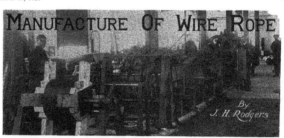

By
J. H. Rodgers

THE first requisite in the manufacture of wire rope is the selection and blending of the different iron ores. The different processes through which the metal passes, and the hammering and drawing into rods require great experience, and give to it peculiar properties that are essential for the finished product. The story to follow tells how the Anglo-Canadian Wire Rope Co., Limited, Rockfield, Que., manufacture their line of rope, this varying from the smallest wire rope up to cables 1¾-inch diameter.

The care necessary regarding the selection and blending of the iron ores is equally needed in the annealing and hardening processes, during which the wires are drawn through dies to the required gauge. It is very important that the operation of drawing be performed at a uniform speed through the entire length of a hank of wire, otherwise there is a great possibility that the strength of the wire will vary at different points. After drawing the wire that is used for certain duties is subjected to special processes of galvanizing in order to make it proof against atmospheric and other influences. Experiments have shown that when the first wire in a rope breaks, it may be assumed that the rope has passed through one-half of its life, and as no one knowingly works a rope until it breaks entirely, then the breakage of a few wires is a sign that a rope should be carefully watched and replaced by a new one at the earliest opportunity.

Various Types of Ropes

The flexibility of a wire rope depends upon the number of wires of which it is formed; consequently the use to which a rope is to be put will partly determine the number of wires used in its construction. In some cases nearly 400 individual wires are employed in making one rope. Fig. 1B shows in section some of the types of construction, the particulars of which are as follows:—

1.—Laid rope of 6 strands of 7 wires each. This is the class of rope most frequently used for hauling ropes, where the size of the drum and sheaves will permit; it is also the make of rope in general use for standing rigging, and is such as is required by Lloyd's regulations.

2.—Hauling rope made of 6 strands, each being of 7 wires covering 7 smaller ones.

3.—Hauling rope made of 6 strands, each strand made of 8 wires covering 7 smaller ones.

4.—Hauling rope made of 6 strands, each strand made of 10 wires covering 7 smaller ones.

5.—Formed rope made of 6 strands of 19 wires each. In larger sizes this make of rope is used for standing rigging on vessels. In smaller sizes it is sometimes used for running rigging, and it is the usual make of rope for travel warps.

6.—Flexible steel rope made of 6 strands each of 12 wires with hemp heart and hemp centre in each strand. This is the usual make of flexible steel wire rope 4½ inches in circumference and smaller, used for hawsers, running lifts, hoists, etc.

FIG. 1.—GENERAL VIEW OF MACHINE SECTION.

FIG. 2.—HIGH SPEED STRANDING MACHINE. DRAWING OFF GEARS IN FOREGROUND.

7.—Extra flexible steel wire rope made of 6 strands of 24 wires each.

8.—Special extra flexible steel wire rope made of 6 strands of 37 wires to the strand.

FIG. 3.—SMALL TRIPLE STRANDING MACHINE.

9.—This is the make of rope usually adopted for large ropes—say over 10 inches in circumference—largely used for slipways and salvage purposes.

Testing the Wire

At the plant of the Anglo-Canadian Wire Rope Co. every bale of wire is subjected to rigid test before it is allowed to be worked up into rope. Before the wire is reeled on to the bobbins a short piece is cut from each hank and given a physical test to determine its torsional and tensile strength. The test for torsion or twist is made in a small hand machine similar to that shown in Fig. 1A. The bed of this machine (which is located on a bench) is graduated into inches, so that the length of the section of wire under test will be indicated by the setting of the front end of the tailstock A. The spindle of the tailstock is so fitted that it will not revolve, but has a short endwise movement to accommodate the shortening of the wire when twisting. The two springs B are provided to keep the spindle in the extreme open position. The hollow shaft allows the wire to be run through from the rear so that it may be secured in the two chucks C and D. The fixed headstock E is fitted with a revolving spindle, operated by gearing F and the crank G. To the middle of the spindle the worm H is secured, which meshes with the worm wheel I, fastened to the vertical shaft that carries the indicating dial J, each graduation representing one turn of the spindle and likewise the wire.

Making Strands on "Snake" Machine

Fig. 1 shows a general view of the machine section of the plant, with two "snake" stranding machines in the foreground. These two stranders working conjointly are capable of turning out the standard 19 wire strands at the approximate speed of 300 feet per minute, the barrel of the machine having a maximum speed of 500 revolutions per minute. Every machine is operated by individual motor drive and the controller is located at the drawing off end of the machine where the operator stands. Fig. 2 shows the finishing end of one of the "snake" machines, with the gearing that is necessary for drawing off the completed strand. The speed with which the strand is pulled through the dies, in relation to the speed at which the machine is revolving, determines the length of lay. This factor of lay, or helix, is the fundamental essential in successful rope making. This is especially true when strands are composed of more than seven wires. It is very important

that "crossing" of the wires in a strand be avoided, as this will cause excessive stress and eventually result in a weakened or broken wire.

The "snake" machines are of the most modern type for making the strands, as they are capable of being operated at high speeds, accelerate rapidly, and are quickly stopped when the occasion arises; suitable braking facilities being provided and always under the control of the operator when standing in a working position. These machines have no central shaft, the revolving section being in the form of a hollow drum, open at intervals to permit of inserting and removing the bobbins, and threading the wires through the various rollers to the forward end, where they all converge toward the centre wire, around which they are twisted and thence pulled through the die by the action of the drawing-off gear. The revolving portion of these machines is made of cast iron, and built up in sections, rigidly secured together to form one continuous hollow drum. Within this drum, and uniformly located, are the suspended cradles which support the bobbins.

The wire from the bobbins is passed through a number of small pulleys or rollers supported in brackets secured to the inner surface of the drum. Plenty of space is allowed in the groove of these rollers so that the movement of the drum permits the wire to swing about the centre, but without any perceptible twist until it reaches the front head. Each bobbin is controlled by a brake which acts as a tensioning device, so that equal strain can be applied to each, allowing the wires to unwind uniformly. If the tension is insufficient there is a possibility of the wire buckling, and if too great there is danger of the wire breaking.

While the run of wire is generally of such a character as to permit of uninterrupted operations for long periods, it sometimes happens that a wire will break when the strand is being made. This necessitates immediate stopping of the machine. If the wire breaks before the twist is made, it simply means that the ends are prepared and placed in a small portable electric welding machine and joined together, and afterwards ground to eliminate any irregularities on the surface. This does not appreciably affect the strength of the rope. The tension is again adjusted on the bobbin and work is continued.

The drawing off is performed by passing the finished strand around a large pulley, the circumference of which is of a specific length. The shaft that carries this pulley is provided with a worm that drives a worm gear and

FIG. 4.—EIGHTEEN BOBBIN SECTION OF LARGE TRIPLE MACHINE.

vertical shaft, to which, at the upper end, a graduated dial is secured. This dial indicates the number of feet of strand that has been drawn off. After making several turns of the winding wheel the strand is reeled on to the bobbins. These are likewise power driven, geared in unison with the winding equipment. The bobbin shaft,

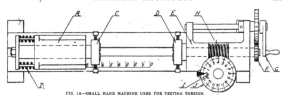

FIG. 1A—SMALL HAND MACHINE USED FOR TESTING TORSION.

however, is driven by means of a friction clutch, to provide for the varying diameters as the wire on the bobbin is increased in volume. Oscillating forks are fitted on the lead side to assure a uniform tracking of the strand. When strands of 19 wires are being made on these machines, the inner section of six wires over one is made on the seven bobbin machine, and this strand is covered by 12 wires in the twelve bobbin machine.

FIG. 2—BOBBIN CONTROL MECHANISM ON CLOSING MACHINE.

Making Combination Strands

A small triple stranding machine is illustrated in Fig. 3. This machine is capable of making strands of every construction for ropes up to 3-4-inch in diameter. The complete strand can be made in these machines at one operation, each head taking care of one gallery of the strand. To roughly show the various details in the

making of a strand of 37 wires, the sketch Fig. 3A is given. This is a skeleton diagram of a triple machine, the small head at the lead end taking care of the first section (2) of seven wires (which may, if desired, be used as a strand itself); the second head winds twelve wires about the first seven, giving a strand as shown at (3). The third and last gallery of eighteen wires is wound on by the third or largest head, the result being a strand of 37 wires as shown at (4). It will be noted that the wires here shown are relatively much larger than the machine parts shown, but the purpose of the sketch is to illustrate the general arrangement of the wires as they are located in the different heads, and the process of construction from the single wire to the completed strand.

It should likewise be noted that the centre shaft (which carries the wheels G, for supporting the bobbin cradles and shafts) is not included in the drawing. It should be understood that the central shaft in each head is of the hollow type for the free passage of the largest wire or strand passing through. It may also be stated that one or more heads may be disconnected, or operated in opposite directions, so as to obtain any desired strand construction.

In these triple machines the bobbin that carries the centre or core wire (1) is placed on a reel apart from the machine proper and drawn through the central shaft by means of the drawing off gear. The six bobbins D, for the first gallery (2), are located in the primary head, the bobbins being kept in a vertical position by the action of the eccentric ring A and cranks B, one of these cranks being fitted to the end of each bobbin shaft. This detail will be explained later. The forward bearing of the bobbin cradles is hollow to permit the passage of the wire. After passing through these holes, the wires converge towards the centre and pass through the lay plate F, the combined forward movement of the wires, and the revolving motion of the heads, giving the proper lay de-

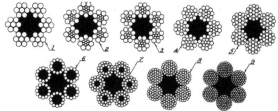

FIG 1B—SECTIONAL VIEWS SHOWING VARIED CONSTRUCTION OF WIRE ROPE.

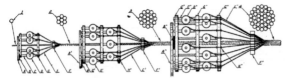

FIG. 3A.—SKELETON DIAGRAM OF TRIPLE MACHINE.

sired for that particular strand. Immediately after the twisting operation, the strand, or that specific section of the strand, is drawn through a die that assists in equalising the lay of the wires. The different heads, with the exception of the varying size, are quite similar; the larger ones have the bobbins divided into two sections to enable the required number of bobbins to be located in the frame without interference. In this case, the wires from the rear division extend across the forward space and through a bush inserted in the forward containing ring G. The bobbin shafts C on the double heads extend back to the rear ring for connecting up with the eccentric ring A. An eighteen bobbin section of a large triple stranding machine is shown in Fig. 4. This machine has a capacity for making strands of 37 wires of any construction for rope from 1-2-inch to 1 3-4-inches in diameter, in one operation. The machine complete weighs about 21 tons.

Making the Rope

The title cut illustrates the closing machine used for the last operation in the manufacture of wire rope. This particular machine has a capacity for rope up to 1 3-4-ins. in diameter. With the possible exception of cables required for special duty, every wire rope is constructed with a hemp heart, around which the six built up strands are located. This hemp rope is passed through the hollow shaft of the machine in much the same way as the core wire on the triple strander. Before passing to the machine, however, the rope is drawn through a small tank of special lubricating grease which is kept warm by means of an electric heater. The bobbins containing the strands are mounted the same as in the triple stranding

machines, and, as a matter of fact, the different machines in a wire rope plant vary only in detail and not in principle.

A detail view of the bobbin control is shown in Fig. 5 and a line sketch of this mechanism is illustrated in the line sketch Fig. 5A. The general design, which is typical of all machines, calls for the use of a separate ring to control the movement of the bobbin shafts. It is essential that premature twist be avoided when closing the rope, and therefore it is imperative that the bobbins be kept in a vertical position while they revolve about the axis of the machine. The central shaft A is supported at either end of the bobbin frame, but owing to the construction it is impossible to provide intermediate bearings; adequate support is provided by running the large wheels B on the rollers C, one of these being located at either side of the machine and secured to the base plate D. The roller brackets are adjustable for centre distance by means of the tie rod E. The bobbins F are carried in cradles G, which in turn are supported in the containing wheels B. The shafts of the forward cradles extend back to the rear for crank connection. These cranks are keyed to the respective shafts so that the cradle bearings for the bobbins are horizontal when the cranks are vertical. The cranks K are all of the same centre distance and when connected to the eccentric control ring H, bring the centre of the latter a like distance below the axis of the machine. This ring H is free to revolve about its own centre and is kept in vertical alignment by means of the two guide rollers I, which are located on studs passing through a cross bar J of the machine frame. It is quite obvious that, no matter in what position the machine may be, the cranks, and likewise the bobbins, will always be vertical.

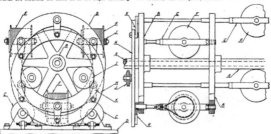

FIG. 5A—DETAILS OF MECHANISM OF BOBBIN CONTROL.

Treat *Machines* Better than the *Men?*

Well, Perhaps They Didn't Mean It To Work Out That Way, But the Foundry of To-day Filled With Molding Machines is Brighter Than the Old One Filled With Men

By F. H. BELL

Editor "Canadian Foundryman"

A TYPICAL MODERN FOUNDRY BUILDING. ABUNDANCE OF LIGHT, HEAT AND VENTILATION.

COULD there be anything more interesting or fascinating than the moulding and founding of metal? To see the agile artisan plying his tiny tools and his dexterous fingers in the skilful preparation of the moulds, and to see this same artisan and his nimble fingers converted as though by magic into a giant of herculean strength with brawny arms conveying the ponderous ladles of molten metal, boiling and seething from the melting furnace, to be poured into the moulds which he had just completed a few short moments before? To see the honest sweat streaming from his brow as he tears the moulds asunder, and exposes to view the red hot castings, true to every detail, and to note his beaming countenance as he gazes with well-deserved pride upon his handiwork?

Truly, it is a magnificent sight.

Yes, the programme of the moulder is wonderful, but for those of us who have spent our days within the gloomy precincts of the moulding shop as a means of keeping soul and body together, much of the grandeur and magnificence has been overlooked, or perhaps the fascination and the interesting features in connection with the moulders' noble art did not present themselves to us with the same degree of prominence as did some of the other more notable characteristics of this most necessary institution—the foundation of all mechanical achievements—the foundry.

The Surroundings Were Dingy

But they were there just the same, hidden away behind a sand heap or buried beneath a pile of debris, and while our eyes were bleared with core smoke and our brains befogged with foul gas and dust, we allowed the fascination of the surroundings to go unnoticed.

It is all right for the uninitiated foundryman who has received his foundry experience at the office desk to say that we take the wrong attitude, which perhaps we do, but under conditions as they existed there was no other attitude which a person could take.

But with all its shortcomings, the foundry, with its various features, is one of the most interesting of callings and one which bids fair to hold a more prominent place in the manufacturing world than it has done since the days of Tubal Cain.

Improvements Have Come

If the moulder, who perforce sees nothing but the dismal features, cannot become interested enough to suggest improvements, it behooves the employer to do so, but up to a comparatively recent date moulders have been so plentiful and apprentices so eager that no thought of change was ever considered, and if Tubal Cain could have come back to earth forty years ago and could have visited the average foundry he would have identified most of the equipment as his invention and would no doubt have been able to point out improvements which he had devised and which had since been forgotten.

With modern improvements in other lines of occupation and with increased educational advantages, the youth of the last few decades has not taken kindly to foundry work, with the result that moulders were becoming extinct and the trade a lost art. With the shortage of workmen came the idea of replacing them with machines, but little attempt was made to improve the conditions so that machine operatives could be retained and as a result common laborers drew the line at machine operating amid such surroundings as were met with in the foundry. This resulted in the employment of foreigners who, so far, have done most of the machine moulding in this country.

Must House the Machine

But now comes the difficulty which touches the tender spots. It has been discovered that the foundry is no place in which to install a machine, or in other words, it does not pay to keep a foundry in such a condition if machines are to be used. It has been found that no matter what precautions have been taken by the manufacturers of foundry equipment, the dust and dirt of the foundry is too much for a machine to endure, notwithstanding the fact

that human beings were expected to endure it through all the hundreds of years since foundry work began.

As a result of this latest discovery, an organization has been started which embraces practically all the manufacturers of foundry equipment and is known as the Foundry Equipment Manufacturers' Association. They have begun a determined campaign to induce foundrymen to make their foundries fit places for the equipment, and this campaign is already having its effect and to-day no foundryman would think of building a foundry without engaging an industrial engineer to design and superintend its construction. The result of this move has been most satisfactory and now we may look forward to a new era in foundry work. Like war, which with all its horrors, brings out inventions which would probably never have been thought of in the ordinary course of events, so the foundryman, with his careless indifference, has been the means of bringing out improvements which would not have been considered necessary had it been possible to continue in the old way.

Some Practices Contrasted

As a comparison between the methods of my boyhood and those of to-day, I will give a few examples which should convince the reader that the foundry is no longer lagging behind. In those days the apprentice was put through the entire foundry process, which included attending to cupola. This part did not appeal to him, but it was the proper thing to do and should still be the rule. He cleaned castings, helped cut over all the sand heaps in the shop, made cores, etc.

To begin with, the castings had to be removed from the sand by hand. To-day they are placed upon the table, suspended from the trolley car.

Then the water was thrown on from a pail—now it is put on with the sprinkling system.

After being wet down two men took opposite sides and with shovels smashed the lumps and cut the sand up in a heap. Now the sand-cutting machine straddles over the floor and in moving the length of the room leaves a finished sand heap in its wake.

In the core room the sand was prepared by getting a barrow load of sharp sand, a pail full of moulding sand and a measure full of flour. These were all pitched back and forward until considered mixed, after which the mass was run through a No. 8 sieve. It was then spread out

THE MODERN METHOD OF CUTTING UP THE SAND HEAP.

and punched full of holes with the shovel, and wet down in a similar manner to the sand heap and cut over several times to be sure it was mixed, when it would this time be run through a No. 4 riddle. Now the sand is mixed and tempered in a power-driven sand mixer.

Formerly the cores for every purpose were made in halves so that there would be a flat surface against the core plate. The halves would be dried and afterwards

pasted together and dried again and black washed. Now, cores are jolted into the core boxes, or blown in by compressed air, and for straight round cores or even cores which are not round, a machine similar in many respects to a sausage stuffer is used and the core made perfectly

THE REGULAR WAY OF POURING COMPARED WITH THE METHOD WHICH IS COMING INTO VOGUE.

true, of any length required and with the vent hole the entire length.

More Laborious Work

In charging the cupola the only means of getting the coke and iron on to the platform was to carry it up the stairs and dump it on the floor, after which it was rehandled and put in the cupola. The cleaning away from under the cupola was equally as interesting. The coke, iron, cinders and sand had to be separated by hand, by picking out the coke and putting it in a basket. Bits of iron were put in a tote box and the cinders which did not appear to carry much iron were wheeled to the dump. The balance was put in the cinder mill along with the siftings from the sand, and from the gang-way a few chunks of pig iron put in with them, and the whole outfit tumbled until nothing was left but the iron. The cinder mill was, of course, well supplied with openings to allow the ground cinders to escape and that which was not heavy enough to fall to the floor did the next best and went all over the shop in the form of dust.

But It Is Different Now

To-day there is no part of the plant better taken care of than the cupola. Wet cinder mills grind up the cinders and separate the coke and iron and are dustless. Sand sifters of various types are provided for sifting the sand which will form the bottom, while magnetic separators recover the last speck of iron from the gangway and hoists to suit every condition are used to deliver the material to the charging floor.

The scratch room of to-day instead of being the dust-laden department of former days is as free from dust as any part of the plant. Instead of putting the castings in the old-fashioned tumbling mills, similar to the one back of the cupola, the tumbling is done in tight mills which are connected to exhaust fans which carry all the dust outside of the building. The grinding wheels are likewise, connected to the same system. In addition to these there are new devices which were not known many years ago and which are not only labor-savers, but are reasonably dust-proof. Instead of the hand scratch brush and the

chisel and hammer, the modern foundry is equipped with revolving scratch brushes which are either operated on a machine similar to the grinding machine or carried about on a flexible shaft or on a portable grinding apparatus. Pneumatic chisels and sand blast apparatus of innumerable types do the work cheaper, better and healthier than formerly was the case.

The Modern Way of Pouring

But the part which is of vital interest to the employer and employee alike is the progress which has been made in methods of making and pouring the moulds. Up to a very recent date this was done entirely by hand and the moulder had to keep up a continuous motion every minute in the day or else he did not do a full day's work. To-day machinery does the bulk of it and the moulder's part is not the slavish job it used to be. As we have shown, the sand heap is prepared by machinery and no matter what class of mould is to be made there is a machine which can be used to advantage on it. In pit work the grab bucket cleans the pit out and the automatic sand sifter riddles the sand; the sand mixer prepares the facing sand; the pneumatic rammer does the ramming and the power crane does the lifting. In doing floor work the jolter does the ramming and if the work is such as will warrant the outlay, the jolting machine is built to turn the mould over. If the work is large the electric crane is used for lifting, but for medium work the pneumatic hoist is used to exceptionally good advantage and in many places the differential pulley block such as is built now can be used where no other device is as satisfactory. For small work such as has been considered as bench moulding, the squeezer has been perfected to such an extent that it will do most any job which was formerly done on the bench, and for work of a more difficult class the jolter, squeezer, stripper and pattern drawer offer a sufficient variety from which to choose the required combination to do practically any job.

Oven for the Job Shop

For jobbing work where the pattern-drawing device cannot be used to advantage the air hoist has many advantages. The air hoist is also used advantageously for lifting copes, handling cores, etc. For heavy lifting, cranes have been used for many years, but they were hand-operated. To-day almost any foundry will have electric cranes traveling the entire length of the shop and in addition have numerous small pneumatic cranes which can be lifted from place to place by the electric crane and connected to the compressed air pipe by means of a hose. In addition to this a modern foundry is equipped with trolley track connecting the cupola spout with the different floors, so that the moulder has no carrying of metal excepting on his floor and this is being dispensed with now to a considerable extent by the use of pouring systems which connect up with the main track. These pouring devices are so perfected that the ladle can be raised to any height and the mould poured to as good advantage as could be done by hand and with greater ease.

All told, the foundry of to-day is on a par with any other business, with the possible exception of the atmos-

CUTS CORES TO EXACT LENGTH AND TAPERS THE END.

pheric conditions and these as we have already pointed out are being rapidly brought into line in order that the equipment will give the maximum of efficiency and the men who had been overlooked are automatically absorbing the benefits which were primarily intended for the machines.

By employing competent industrial engineers who have had actual experience in the foundry to superintend the construction as well as the heating and ventilating of the new foundries, it has been found quite possible to have good, clean air which does not stick to the oil on the machinery and which does not stick in the lungs and nostrils of the workmen.

SIFTS THE FACING OR CORE SAND AND MIXES IT IN ONE OPERATION

HEATS AND VENTILATES THE FOUNDRY AT THE SAME TIME.

Every Man Cannot be an Executive

Some men prefer their regular shop work to any style of office position. This tale, while fanciful, illustrates that dissatisfaction is, after all, largely imaginary.

WHEN are you coming home, Jack?" My wife was phoning me at the office, and in my tired condition, her voice sounded as if it were a thousand miles away.

"I'm sorry," I replied, " but I will be at least a few hours yet. I have some important appointments, and it may be ten or eleven o'clock before I get through. Eat your supper alone. I'll be home as soon as possible."

"I'll bet you're tired," came back my wife's voice. "You win," I replied, wearily, hanging up the phone.

Reviewing the situation, I could not help feeling both disappointed and disgusted. Here was I, working till all hours of the night in an endeavor to pacify, and perhaps satisfy, a plant filled to the door, so to speak, with disgruntled workers. Truly the position of manager in a manufacturing plant was no sinecure, much as some would have you believe it so.

For the past few months things had gone far from smoothly. Threats of strikes had been numerous, and the cry had always been the same. "MORE MONEY, THAT IS WHAT WE WANT." Never had there been a word about production. In fact, had I stated that production had gone down instead of up, the workers would have keenly resented my critical attitude, and perhaps have struck, just to show they were of an independent nature.

And yet, had I not the right to speak on that very point? viz., that if wages go up, production should soar accordingly. Suppose, I pointed out, that profits were lower—but there again was a nasty snag.

Profits—why the workers hated the very word. They felt that the firm was making millions, when truth to tell we were just working on a safe margin. True, I was making a comfortable salary, but why shouldn't I? Was I not working for it just the same as the men in the shop? To the men in the factory, I was the fellow who held down the soft snap, who wore a clean collar and a silk shirt every day, and rolled down in my limousine around 11 a.m.

The truth of the matter is, I never wear a silk shirt, and can afford nothing larger than a Ford. I am at the office every morning not later than 8.30 a.m., but you know the old saying, "Give a dog a bad name."

In thinking these things over, I must have dozed off,

for suddenly I awoke to find a tall dignified stranger confronting me.

"Pardon me," I said, "but I don't seem to know you."

"You do—and you don't," said the unexpected visitor. "You used to know me, but lately you've neglected me terribly."

His manner of speech puzzled me and I must have looked it, for he continued, "If agreeable to you, I want to help you out of your present quandary. Don't bother explaining, for I know all about your trouble. You wonder who I am, and how I know of it? Well,—my name is Courage. all men need me but seldom use me. Lately, they have been using my brother, who is known as Let-well-enough-alone. Candidly, if you don't mind my saying so, that is your trouble so far."

I was taken aback and must have looked it. Finally I stammered, "Well, what to your mind is the solution to the present trouble?" My visitor seemingly paid no attention to my last remark for he replied, "Why did you stay at the office to-night?—don't bother answering, for I know already. You have made engagements with various men in your plant who are dissatisfied. You have also arranged for them to come at different times, a rather clever idea in its way,—but why are you meeting them in this secret manner? I'll tell you."

The method of my visitor's asking and answering his own questions was rather disconcerting, but wait, he was continuing: "You expect, by meeting your men one by one to pacify them, thus avoiding the latest of the threatened strikes. Am I right?"

I had to admit he was, but ventured to ask how he guessed.

"I never guess, I know," came back the dignified reply. "Look here, Jack (why, he even knew my first name), you've got to do something definite, or a strike is assured. Why not take me as a side partner? I want no wages or profits, and my cousin Good Judgment will come at the same figure as myself, namely, consideration. Between the three of us we can see this matter through. What do you say?" I nodded my head, and the moment I did so, in walked Good Judgment.

He recognized me as an old friend and the three of us bent our heads closely together, discussing the situation to our hearts' content. We laid a plan, that had better not be

told for the present. Enough to say that it was none too soon, for there was a sudden knock on the door, telling me that the first of my expected visitors had arrived.

"I don't want these men to see you," I said to my two companions.

"They won't," replied Courage laughingly. To them we are invisible. To you, we are ever present, but remember this, we are of no use to you except you use us. You must ask us to help. Good Judgment will help you out of many a hole, and personally I'll give you the courage to see it through, but remember, you, and you alone, can command us.

With these warnings ringing in my ear, I shouted "Come in," and the door opened to admit Waeburn, the head of my automatic. department.

The Plan Works

Waeburn seemed ill at ease. "What is it you want?" he commenced. I felt Courage right beside me so started in with my heavy artillery.

"Look here," I replied, "you asked me to-day for an increase of $600 per year, and threatened to call your men out, if you didn't get what you asked. Why?"

This certainly was not the sort of conversation I had intended having at the start, but Good Judgment whispered in my ear that it was for the best, so I kept on that track.

"Well," replied Waeburn, "I want more money because I've recently got married and I find the going rather rough."

"Is that the only reason I should give you $600 more every year?" I asked.

"No," came back Waeburn, "but I've done my work for the past five years as best I know how."

"That is your opinion," I shot back.

This rapid fire reply staggered Waeburn, for he spluttered out that he didn't see why I should suggest he was not doing his duty.

"Look here," I continued, "let us talk this thing out in a candid manner. You listen to me first, hear my viewpoint, then see if you can still honestly ask for the extra $600. First of all, I admit you've worked for us for five years, two as foreman. In that time you have had a good chance to study your men, the machines, the operations performed on them and so on. In other words you should know your department pretty thoroughly by now."

"I do," he replied.

"Very well," I countered, "we will see if you do. Take operation XXX on that small spindle you are doing out in the shop at the present time; do you think we are doing that work in the best way possible?"

Waeburn looked amazed at me. "Why, we've done that job the same way for the past five years," he said.

"Sure," I replied, "that's the trouble. How do we know that's the best way. Have you ever tried any other method?"

The look on his face was answer enough, so I followed up my advantage. "Do you know that I had a young fel-

low up from your department to-day who suggested what looks like a better plan of doing that work? Of course I told him such a suggestion should come through you, but his reply was that he had tried to get your ear before, but his reception was far from pleasant."

"You mean young Smith, don't you?" sneered Waeburn.

"Exactly," I replied.

"Oh! he makes me sick, always some new idea. If he had his way we would reorganize the whole plant."

"Well, he's going to get his chance," I replied. "I'm placing him in charge of your department and starting to-morrow morning."

. This statement was wholly unpremeditated on my part. I had never even thought of it before, but Good Judgment seemed to suggest it and Courage carried it through for me. Waeburn's face was a study.

"Where do I get off at?" he asked.

"You come into my position," I replied. You could have knocked him down with a feather. "Remember telling me yesterday I had a cinch of a job?" I asked. "You said you wished you could get such a position. You would show us what you would do. Well, you've got your chance now. I've been working too hard lately and I'm going to take a well earned rest."

"But look here," said Waeburn, hastily, "I was only fooling. I couldn't run your job. I don't know the first thing about financial matters and I don't know ———oh, what's the use, Mr. Blank, my place is in the shop."

"That's where you are mistaken," I replied, "but we will consider my first suggestion. In place of giving you full charge you'll work under my supervision. Tp start with, you get no more money than you are making at the present time, but if you make good, leave the salary matter to me and it will be very much in advance of the $600 you are asking for.'

I looked at my watch, saw that my next visitor was about due, so decided to close the interview.

"Well, then, that's settled. Report to me at 8.30 in the morning," and I showed Waeburn out in a much more dazed condition than when he came in. In spite of the novelty of the move, I felt it was a good one, in fact Good Judgment told me it was and as for Courage, I felt quite proud of him.

My workers imagined that I wore a silk shirt, and rolled down in my limousine about 11 o'clock every morning. Instead of that I often work from 8.30 a.m. until 11 p.m.

The Second Visitor

Suddenly, in walked Grogan, my foundry foreman. It was like Grogan; to walk in unannounced. He never knocked at my office door, and though I had often felt like telling him about it, I never had Courage with me before.

"Look here, Grogan," I said, Courage backing me up, "Before you come into my office knock at the door first, then wait to see if it is convenient for you to come in."

Grogan's face was a study. Never had he heard me speak like that. Somewhat abashed, he said: "Why, I never knocked before."

"All the more reason you should start now,' I replied. With this remark, I considered that I had said enough and started on a new tack.

"You were raising Cain with me the other day," I commenced, "regarding the poor supplies you were getting for the foundry."

"You said something," he growled. "Why the purchasing department you have is a joke. There isn't a practical man in the lot. I could buy better than that bunch with my eyes shut."

"Do you think so?" I ventured.

"I'm sure of it," he replied.

"Very well, then, consider yourself purchasing agent starting to-morrow morning," I said, and Grogan's expression was something beyond description.

"Do you mean that?" he asked.

"Certainly." I answered.

"What about the foundry?" "Bell, your assistant, will run that," I replied. "Now look here. You've said a great deal about the need of a practical man in charge of the purchasing department. You've been with the firm since it started. As far as I know you started in the shipping department, did you not?"

A nod of the head was his only answer.

"Since then you have been through every department. From previous experience you should make an ideal man for the purchasing end of this business, so be prepared to dig in at 8.30 to-morrow morning. If you make good you get even more of an advance than you asked me for last week." With this remark I ushered him out in much the same condition as Waeburn, my previous visitor.

An Easy One

The next visitor was easy to deal with. He was a new recruit in the drafting department, named Jones, who could see no reason why he should not be chief. As for the salary he was making,—well it was disgraceful, to use his own expression. In a previous talk with his chief I had decided to discharge this youngster, but on the chief's recommendation, I had agreed to let the matter slide. "There is good material there if we can develop it," said the chief. Personally I felt he would never discover it, as this fellow was one of the never satisfied type, but my rule is, give every man a chance to make good, so on he stayed.

The talk with him was brief and to the point. "You want to be chief, don't you?" I said. ,

"Yes," replied Jones.

"Very well, you are," I replied.

This remark took the wind out of his sails.

"You mean the present chief is discharged?"

"No, promoted," I answered. He is now chief engineer, but you are chief draftsman.

"What salary do I get out of this?" he asked. In a way I admired his nerve, but replied that his salary would remain as it was until he proved himself.

"In that case, I won't take the job."

This was a new one on me, and I had to call on Good Judgment for advice. "Take him up on it," was the advice given me, and I did.

"All right, how much do you want?" I said. With a smile at thus winning his point, the young upstart, if I can call him that, stated a figure equal to the salary of the regular chief. "You see," he explained, "I know what he gets." .

I merely nodded, signified my acquiescence and told him to start in the morning. Somewhat cocky at my ready acceptance, he strutted out of the office with a "Good night, see you in the morning."

Courage smiled as he departed. "It took all my patience to go through with that," he said. "Instead of tolerating him, I felt like throwing him out." "Still," interrupted Good Judgment, "I believe it's best for the department. There are quite a few in that department who think they ought to be chief, and this will be a lesson to them." My next task was one of diplomacy. I called up the suddenly deposed chief, explained my scheme told him to let matters go only so far to the bad and wished him a cheery good-night. When all was said and done, I felt better than I had earlier in the evening.

This being my last visitor I closed the desk, and turning to my two friends, said:

"Where will I see you in the morning?" The reply was what I expected.

"We're going home with you," they replied, "and what's more, we're going to stick with you until the end, always providing of course, that you want us."

"I never needed two such friends as much as I need them now," I answered earnestly, and with that we all departed:

The Morning After

The morning following these sudden changes was a hummer. The men in the shop, who were thus suddenly promoted to be foremen of their various departments, had nothing on the three who were raised to positions of assistant manager, purchasing agent and chief draftsman respectively. I was on the job as usual to tell the new seekers after fame what they were expected to do. Rumors floated from department to department, and the one most generally conceded to be correct, was that I had suddenly gone crazy. Some of these wild guesses came to my ears, but I merely smiled, sat tight, kept right alongside of Courage and Good Judgment who told me to see all and say nothing.

The climax came when the following notice was placed on the bulletin board of every department throughout the plant: "Three men have lately been promoted to executive positions. This has been done with a definite purpose, viz., to help these men advance. If capable, they will be retained in their present capacity. The firm wish their men to be satisfied, and, if after three months, others desire an executive position, let them make their application to me personally."

If rumor was rife before, it was greater after the posting of this notice. The men could not understand it. Their dissatisfaction, which after all was only imaginary,

He departed with a cheery "Good night—see you in the morning."

turned to wonderment. What on earth was the G. M. trying to put over them? Wonder changed to worry. Were the newly appointed executives capable of running the plant, and so it went on, until like all nine day wonders, interest waned and the shop went on very much as before.

Three Months Later

It was three months later that I had an uninvited visitor at my home. It was Waeburn, and a very changed Waeburn at that. After some humming and hawing, he blurted out the object of his visit.

"I'm a failure," he commenced, and then, out poured a tale of woe, that was already known to me through another source.

"I guess I'm not such a world-beater after all," continued Waeburn. "As an automatic department foreman I'm passable, but as assistant manager, well—I've made a mess of things, but thanks be I'm man enough to admit it. I've changed my mind a great deal since our last interview. Managing a firm is not all sunshine. I've tried hard. I've spent hours every night attempting to handle the job successfully, but I give up, it's no use, so, if agreeable to you, I'll leave a week from to-day."

"What seems to be your chief trouble," I asked.

"Frankly," replied Waeburn, "I realize now that one requires more than bluff to be come a manager. I fell down on the job simply because I lacked training, and in falling down, I also see why I was not satisfying you in my last position. As assistant manager I was entirely out of my field. Instead of liking the job, as I imagined I would, I much prefer work along my own line, namely, automatic machinery. I see plainly where I did not make the most of my opportunities in the shop. My policy was Leave-well-enough-alone, but I assure you, the lesson taught me has fallen on good soil. I tell you all this, to show that your giving me a chance to make good is appreciated. Having thought the matter over I decided to hand in my resignation."

"Where do you expect to go from me," I asked.

"I don't know," he replied.

"Would you consider taking over your old department?" At my question Waeburn stiffened. "Would that be fair to Smith, the present foreman?" he asked. I laughed. "Smith was in to me about an hour ago," I said, "and told me the job was too much for him. He is rather young yet and lacks experience, but Waeburn, you and he would make a great combination. What do you say? Will you accept?" It did not take him long to see my viewpoint, with the result that a notice was posted in the shop announcing his return the following week. On returning to the shop, he was honest enough to tell his men frankly why he had returned and that he had not made a success of the new position.

News of this character travels fast, and it was just a few days after Waeburn's return that I received a note from Jones, the recently appointed chief draftsman. His letter was written from his home explaining why he had not appeared at work the previous day. The note commenced in this strain:—

"When I took over the position of chief I was a hairbrained ass. Of course I didn't see it that way at the time, in fact I felt rather proud of myself, that at last I had convinced you of my worth. I felt it was a shame the way the firm had held me back. I was thoroughly convinced that my chief had a soft snap and that my knowledge was greater than his.

"Since taking over the position things have gone far from well. I pity my late chief when he once more takes over my position. I say this because, as far as I am concerned, I'm through. Unlike Waeburn, I'm not man enough to come and tell you personally, and I cannot go back to my old job again. By the time this letter reaches you I'll be on my way home. My folks often wanted me to go to college, but I like the crowd of goodfellows better. However, I've learned my lesson. I see that Education is the thing, and I'm going to get it. I intend taking a short College course for a year or two in the hope that by that time I will really be capable of holding a position such as you gave me an opportunity at. Where I fell down on the job was lack of experience. My theory was only half matured and my practical knowledge was nil. Incidentally, I intend to work in the summer months for some machine shop to build up on the practical end of the business. As I said before I've learned my lesson. From now on personal effort is what I will rely on."

I could not help but remember the chief's words as I read this letter. Truly the good in this fellow had come to the surface.

As for Grogan his story is different. He had luckily made the most of his time, and knew the requirements of the plant like a book. He not only held his position but made a wonderful success of it, saving the firm thousands of dollars through his insight and experience. A notice was posted on the bulletin board to the effect that Grogan was to be retained in his present capacity, and the reasons for his success were given. "Grogan has made good," read the notice, "because he has made the most of his time in the shop. He has saved the firm considerable money since his appointment. This he has done because he knows our requirements. For some time past we have heard rumors of unfair treatment to our employes. Any such employe who has a reasonable complaint can see the manager any evening at his home, and will be given a chance to prove himself. Above all the firm wish their men to be satisfied."

After the posting of this notice I waited breathlessly, yea, almost anxiously, for the result. Would a stream of dissatisfied workers call at my home, or would only a few take advantage of this opportunity to unload their troubles? I needed Good Judgment and Courage very much, but they both assured me the posting of the notice was a wise move.

So far, no one has called. Evidently my workers are more satisfied than formerly. Six months have passed since the posting of the notice, and production has soared wonderfully in that time. The results coming from the automatic department are especially good. As I expected, Waeburn and Smith make a great combination, and what's more, they work together. I take no unfair advantage of any man in the plant. Every one who makes a suggestion is thanked for it, and if adopted is rewarded accordingly. Courage and Good Judgment still stick with me, in fact I told them the other night they must never leave me, and they laughingly told me, that depended entirely upon myself. A rather fine answer when you think it over.

* * *

Has the above story impressed you? True, it is imaginary, but after all what is labor unrest? Is it not also imaginary trouble. Men are always aiming at the position they think they would like, not to make better the job they have. Let us get more of the spirit to make good on the job we've got, and the higher position will come; when we ourselves are ready for it.

Systematic Control of Locomotive Repairs

THE FINISHING TOUCHES TO THE ENGINE, JUST PREVIOUS TO ITS LEAVING THE SHOP.

By J. H. Rodgers

THE wide and varied detail work that is required in the repairing of locomotives would seem to present, particularly to the layman, an engineering problem where considerable difficulty might be experienced in obtaining a satisfactory solution by adopting any systematic method of pre-arranged classification of repairs on each and every locomotive as it is taken into the shop.

Three score years and ten are set down as the allotted age of man, but his own weaknesses, and the uncertainties of life, provide many opportunities for relentless fate to place him under the care of a physician or in the hospital, for varied periods, to undergo light or heavy repair, or general rebuilding. Locomotives are not unlike human beings in this respect. The periodical "life" of a locomotive is estimated on a mileage basis, this depreciating as life is extended, and while many engines may, and frequently do, cover the scheduled distance before being shopped a great number must necessarily be taken out of service at irregular intervals and placed in the shops to undergo "hospital" treatment, in order to fit them for the service they are called upon to perform.

Engines may be taken off the road for one, or more, of a multitude of causes, and it is this varied combination of "infirmities" that has, until recently, made it so perplexing to draw up effective repair schedules. The principles of production control have been universally practised for many years by large manufacturing industries, but apart from localized effort in some particular department no general organized movement has been made by railway companies to prove its adaptability to general repair activities.

In the case of new equipment units, where the work, from the rail up, is that of straight manufacture, the problem

PROGRESS SHEET – WEEK ENDING DEC. 4th. 1920.								
Div.	Eng.	M	T	FB	In Shop	Out Shop		
A	3927	4	1-1		11-1	29		
NB	2663	4	1-1		11-5	29		
A	3825	4	1-1		11-17	29		
O	2551	4	1-1		11-9	30		
Q	3356	5A				30		
O	750	3	1-1	4	10-4	30		
A	3864	4	1-1		11-2	12-1		
O	1099	3	1-1	1	10-25	1		

FIG. 4—A PORTION OF A PROGRESS SHEET.

ANGUS SHOP REPAIR ENGINE OUTPUT.									
DATE	DAY	Gang 1 Clarke	Gang 2 Larocque	Gang 3 Price	Gang 4 Coates	Gang 5 Sly	Gang 6 Baker	Gang 7 Hogan	Gang 8 McWaters
	Mon.								
	Tues.								
	Wed.								
	Thurs.								
	Fri.								
	Sat.								
	Mon.								
	Tues.								
	Wed.								
	Thurs.								

FIG. 1—THIS FORM IS USED FOR REPAIR ENGINE OUTPUT.

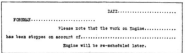

DATE.........................

FOREMAN..........................

Please note that the work on Engine.............

has been stopped on account of......................................

Engine will be re-scheduled later.

FIG. 2—THIS FORM IS USED WHEN AN ENGINE HAS BEEN SET BACK.

does not present any serious obstacles, but on repair work—and this constitutes the bulk of any railroad shop activity—the inauguration and subsequent successful operation of an efficient production control system, is, apparently, encompassed with such complicacy that considerable courage and executive ability are fundamentally essentiol, on the part of those in charge, before the work can be carried on with any degree of permanen success.

The Canadian Pacific Railway is probably the pioneer railroad company to recognize the economic possibilities from a well-organized production department, not alone in the construction of new locomotives, freight and passenger cars, but also in the general repair and maintenance of all motive power and rolling stock passing through the shop. When the system was initiated at Angus shops, nearly three years ago, it was not started on an experimental basis. The executive heads, believing that systematic methods from a production standpoint could be successfully applied to railroad repair work as well as other lines of industry, placed the control of the organization work in the hands of capable and experienced officials, to "establish the department on a firm foundation so that future building would improve, and not detract from, the strength of the structure.

The estimated repairs to any locomotive is based on the road report and the initial inspection made by the shops inspector and the general shop foreman. It might be stated that road requirements, to a large extent, determine the particular engines that are to be taken into the shops from time to time. Under normal conditions engines would take their turn on a mileage basis in being shopped, but urgent demands are often made for extra output of a certain class of passenger or freight engine which must be met irrespective of any regular order.

A VIEW SHOWING A PORTION OF A BOILER UNDERGOING REPAIRS.

After the preliminary examination the engine repairs are classified as 4A, 3B, 1C, etc. At Angus these classifications are indicated as follows: No. 1, new boiler; No. 2, new firebox; No. 3, any new firebox sheet; No. 4, all tubes; No. 5, 2-inch tubes; No. 6, specific repairs. With the exception of No. 6, all repairs include the replacement, or re-turning of the tyres, and general inspection and necessary repairing of all engine parts. Suffixes are used in conjunction with repair classification to indicate work not otherwise specified. These suffixes are: A, denotes wreck; B, denotes initial application of stoker; C, denotes initial application of superheater; D, denotes initial application of valve gear; E, denotes conversion from compound to simple, or vice versa.

For instance, if an engine, after preliminary inspection, is found to require a new set of tubes throughout, and to have a stoker applied, she will be classified as a No. 4B repair, and this symbol will be placed on all charts and correspondence in connection with the work on that particular engine; 2C would indicate that repairs called for a new firebox and the initial application of a superheater.

DAYS IN SHOP	CHARGE HAND / DATE IN SHOP / OPERATION	Eng. No.		
1	Engine In Shop " Off Wheels			
2	Hydro Test Gen. Stripping Complete			
3	Material Delivered Headers Out			
4	Tubes Out C & V Bushes Insp.			
5	Cyl. Patched			
8	Mountings In C & V Bushes Bored			
9	Binders Up Shoes & Wedge Laid Off Widths O K			
10	Tubes In Springs & Gear O K Guide Bars Cross Heads			
11	Spring Gear Up Boiler Tested Guides & XHd. Lined Motion & Vs. O K Wheels O K			
12	Engine Wheeled Headers In Motion Hung			
13	Supr. Pipes In " Tested Main & Sde Rods Fet. Inspirator O K			
14	Valves Set St. & Exh. Pipes In Ash Pan O K			
15	Pistons In Brake Gear O K Brass Work O K			
17	Eng. Trucked Side Rods On Eng. On Trial			
18	Eng. Fin. & Del. Date			

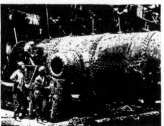

FIG. 3—A SUMMARY FORM OF AN 18-DAY SCHEDULE.

The form shown in Fig. 1 is that for the repair engine output. This specifies

the gang and the foreman in whose charge the engine has been placed, and the date the engine is to be delivered. These forms are made out for a period of thirty-one days but are revised weekly to conform to changes necessary in the schedule. These alterations, when necessary, are made after the engine has been a couple of days in the shop. The reason for this is due to the fact that the preliminary inspection does not always disclose the exact nature of the total repairs required. This can only be determined after the hydrostatic test has been made, which is carried out after the motion work has been dismantled and the driving wheels and truck have been removed. This pressure test, together with a closer physical examination, either confirms the first inspection, or brings to light additional defects, the latter invariably being confined to boiler conditions.

In the preliminary inspection the engine may have been classed as a No. 4 repair, but when subjected to the cold water test, may develop further firebox or boiler trouble, calling for the renewal of a firebox or throat sheet, or perhaps an entire new firebox. Should such conditions arise, it becomes necessary to re-classify the engine as a No. 3 or No. 2 repair; in which case readjustment of schedule becomes imperative, so as to co-ordinate the work of the different departments to which the various parts of the engine have been distributed.

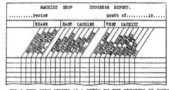

FIG. 5—THIS FORM SERVES AS A CHECK ON THE PROGRESS OF WORK PASSING THROUGH THE MACHINE SHOPS.

Under ordinary conditions locomotive repairs at Angus are based on an eighteen-day schedule. Cases are on record where engines have been overhauled and delivered in fourteen days, but the general time ranges from sixteen to thirty days, depending on the special work required. In every case, the first four days of dismantling and the last seven days of erection, are practically identical on all engines, these periods having become fixed items in all repair schedules, The hydro-test on the boiler is the main factor in determining the "in-between-period," as all other repair requirements can usually be classified before the engine enters the shop.

However, once the patient is taken into the hospital and assigned a bed, the true

nature of his illness and the character of the operation necessary can only be definitely decided after professional diagnosis and X-ray examination.

When changes are made in repair schedule, due to the developments of additional defects, it necessitates prompt action on the part of the schedule department in notifying all divisions of the shop that this particular engine has been "set back"; that is, her delivery date has been extended to conform to the new conditions. If this means that a delay of several days will be necessary, work on detail repairs on that engine must be suspended and the effort directed to work that is required. It is useless to provide clothes for a sick man until he is well enough to put them on. It might

TUNING UP THE LOCOMOTIVE BEFORE GOING ON A RUNNING TEST.

be said that it is good policy to have the garments ready. This would be true if it was not for the fact that another man was being kept in bed a day or two longer because his clothes were not at hand. Idle locomotives are items of expense on a railroad company's cost sheets. In order, therefore, to maintain the monthly output from the shop, it is necessary to so arrange the incoming engines that ample provision may be made for extended repairs to one previously taken in.

When an engine has been "set back" on schedule another one is advanced, and the different departments are advised as to the changes effected. The small form shown in Fig. 2 is used for this purpose.

On the ordinary repair, detail work on certain engines is carried on simultaneously in all departments during the period of dismantling and the erection date for each particular piece of equipment. A summary of an eighteen-day schedule is shown in Fig. 3, this representing a No. 4 repair. The first four days' work, which is the same for all engines, comprises: First day—taking the engine in the shop, hoisting the engine and removing the wheels; second day—cold water test and general stripping operations; third day—material and

Date.	Crossheads.	Wheels.	Pistons.	Eng. Trucks.	Idler Whls.

EAST MACHINE SHOP.

DELIVERY DATES OF MATERIAL TO ERECTING SHOP.

Week Ending Jan. 10th. 1920.

Date.	Crossheads.	Wheels.	Pistons.	Eng. Trucks.	Idler Whls.
Still Due	5805 2115	2518		1877 2518	
1-5	3936	5805W. D. 2115	5805 2115	5805	5805
6	790 6113	3936 2508	5014	2115	
7	3388 5755	5805 377	3936 2508	3936 2508	2508
8	5806	790	377	377	
13	3040	746	3388 5755	5806. 746	5806

#3 Repairs.

Engine	3419 711 2538 5754 3043	Due Out " " " " "	1-6-20 1-7-20 1-8-20 1-9-20 1-13-20

FIG. 2—THIS FORM SHOWS DELIVERY DATES OF MATERIAL TO ERECTING SHOP.

THIS IS THE MODERN METHOD OF HOISTING LOCOMOTIVES.

parts delivered to different departments and the headers taken out; fourth day—

tubes taken out and cylinder and valve bushes inspected. For the next seven days repair work is proceeding in each and every department so that on the eleventh day—or in the case of more extensive repairs, on the seventh day prior to scheduled date of engine delivery—work is started on reassembly and erection, which on the form here shown is being performed on the dates included between the eleventh and the eighteenth day, the latter corresponding to the date on which the engine is again placed in commission.

In order that the entire program of repairs may proceed with uniform regularity and insure delivery to the erection shop on or before the date specified, progress sheets are kept constantly up to date, so that a glance at the day's report will show the actual conditions as they exist. If any particular department has "fallen down" on the schedule, the schedule man loses no time in making an investigation, with the result that the cause of the delay is located, reported, and recorded, and immediate steps taken to remedy the trouble. A portion of one of these progress sheets (for the week ending Dec. 4) is shown in Fig. 4. Every shop foreman, and every other department head, is given one of these sheets, so that he is constantly advised of changes in the schedule and the delivery date of each engine. This enables him to so regulate his work that his own shop schedule will not fall behind, unless it be for causes beyond his control; in which case the schedule man has knowledge of it before it is a day old, and every effort is made to adjust the irregularity. The responsibility of this adjustment is one of the routine features of the production department.

The form shown in Fig. 5 serves as a check on the progress of the work pass-

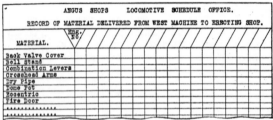

ANGUS SHOPS   LOCOMOTIVE SCHEDULE OFFICE.

RECORD OF MATERIAL DELIVERED FROM WEST MACHINE TO ERECTING SHOP.

MATERIAL.	Eng. No.										
Back Valve Cover											
Bell Stand											
Combination Levers											
Crosshead Arms											
Dry Pipe											
Dome Pot											
Eccentric											
Fire Door											

FIG. 6—A RECORD OF MATERIAL DELIVERED FROM THE MACHINE SHOP TO ERECTING SHOP.

ing through the machine shops. Provision is made for thirty-two engines and check marks are made in the small squares when machine work on that specific part has been completed. These progress sheets are collected daily by the schedule man and the progress recorded on the reference charts kept in the office of the supervisor of production. A section of the form used for recording the date of delivery of finished material from the West machine shop to the erecting shop is shown in Fig. 6. The material list contains 38 separate items, these being component parts of the equipment shown in the third division of the machine shop progress report. In conjunction with the two last mentioned reports, loose-leaf, letter size, typewritten reports are compiled for the current week, showing the delivery dates to the erecting shop, of the various units for all engines on that week's schedule. Included in the report is the equipment still due and also what is required for the first two days of the following week; this latter serves as a guide when making up the report fot the next period. Appended to this report is a list of the engines due to be com-

pleted that week. This form is shown in Fig. 7.

When new cylinders are required in the specified repairs to an engine, the production of these is so scheduled. that they will be cast, machined, and delivered to the erecting shop at the proper time for reassembly. The charge hand has no personal worries as to where he will get the work, does not require to visit the machine shop to see if the cylinder is ready, nor has he to take the responsibility of getting the cylinder to the erecting shop after the work is done. Every detail, apart from the actual duties of the men, is taken care of by the production department.

The form used for grey iron foundry work on cylinders is shown in Fig. 8. This gives the requisition number, pattern number, engine or shop order number, and the various operations necessary in the making and cleaning of the casting. This schedule provides for delivery of the rough cylinder to the machine shop, eight working days after the order has been given to the foundry foreman. If it should so happen that the shaking out takes place on a Saturday, the cool-

ing would take place over the week end, and in this way one full day would be saved.

In every branch of repair work carried on at Angus, the object has been to so co-ordinate the tasks of the various departments that a spirit of harmony pervades the entire organization. As the inauguration of the present system proceeded it became more and more evident to all foremen and supervisors, that because certain of their duties (which might be classified as clerical duties) were transferred to the production department they suffered no loss of prestige or position, and the real effect was to enable the foremen to spend more time in their shops in supervising the men, getting out the required work and doing it right.

The production department acts in the capacity of a clearing house for shops' information on output, and it obviously follows that the routing of work and the tracing of material to the shops can be better done by specialists than by foremen whose time can be spent to better advantage in the shops.

ANGUS SHOP   SCHEDULE DEPARTMENT.

GREY IRON- FOUNDRY   CYLINDER SCHEDULE.

REQN. No.	PATTn. No.	Eng. or S/O No.	Core I. Cast	Cores Made	Ram-Up Box	Pour Cstg.	Shake Out	Cool Cstg.	Clean Cstg.	Del. to Shop

FIG. 8—THE FORM USED FOR GREY IRON FOUNDRY WORK ON CYLINDERS.

A Score Board to *Prevent* Accidents

How the Dodge Mfg. Co. Have Secured the Co-operation of Their Men in an Effort to Keep Down the Number of Accidents —Falling Off in the Hours Lost Through Injuries to Workmen

By HARRY BOTSFORD, Dodge Mfg. Co.

IT has been eight years since the Dodge Manufacturing Company of Mishawaka, Indiana, first started its accident score board. During this period we have answered thousands of letters from firms all over the United States asking for methods of operation, explanations and comparisons. Believing that our method is unique, interesting and worth-while, we are glad to give other manufacturers our experience with our sign board.

In the first place when the sign board was erected and planned there was no thought of the pioneering work that the Dodge Manufacturing Company was doing in welfare work. Our sole idea was to develop some means of cutting down preventable accidents in every department in our plant. W. L. Chandler, our Supervisor of Insurance, decided the best plan to do this would probably be to make use of a competitive spirit in workmen and foremen. Today, after eight years of operation, the score board is as much a pride to the worker-personnel as is a record-breaking pulley or transmission equipment. One represents an achievement in human conservation and the other marks a mile stone in mechanical progress—both of which are vitally essential.

The size of the board as we erected it just inside the main gate of our No. 1 plant was 17 feet x 24 feet, which gave plenty of room for large lettering. The mechanism for scoring points is as simple and fair as possible. The starting point is 1000, both for year and for month; each division is penalized according to its accidents; each day's absence bears a percentage charge in proportion to the total number of "men-days" per month per division.

How Factory Is Divided

Our departments were divided into 26 divisions, according to the various degrees of natural hazard of variation in numbers of men employed. The degree of hazard is disregarded in our business which covers the same general subject throughout the plant; the differentiation being considered as equalized in the choice or selection of men with reference to their ability and fitness for their respective class of work.

As to the variation in the sizes and groups of workers, we set this by establishing a differential charge per man per day for time off—which is computed by reducing each

This is the score-board used in the Dodge plant. It is not an experiment, but a plan that has been in operation for many months, and the firm is so thoroughly convinced it is all right that they are anxious to pass it along to others, recommending it as workable and profitable.

division to men-days for each month; and using a multiplier of 10 to raise the figures to a more workable and understandable basis.

This system of scoring was finally arrived at after considerable experimental work and the present system seems to be fair in every respect and we have had no occasion to change it during the past few years in spite of the fact that during this time almost every known condition has cropped up.

In the fourth column of this score board will be noticed the figures which represent deductions for absence in that division.

Perhaps we have been too tolerant of the minor accidents which do not necessarily entail the loss of an appreciable length of time, as we do not penalize for any loss of time on the day of the accident. It is possible, however, by this means to secure a prompt report of all accidents, however small, so that we may be sure of proper attendance, and avoid, as far as possible, such suffering as may be otherwise charged to secrecy on the part of either men or division superintendents.

The Reward Comes Each Year

At the end of each year the employees of the divisions scoring 1000 receive two days extra pay, or such part of that amount as their time and employment bears to a full year. If none score 1000, then the highest ranking division receive the two days pay. General foremen of any division under them earning these premiums also participate on the same basis, but may earn but one prize if other divisions under them make a perfect score.

Our original plan was to distribute $25 in cash each month to all foremen of divisions winning perfect scores, but due to the relative importance and the efforts of the foremen with a widely varying number of men to deal with, we were, in all fairness, obliged to change this arrangement so that one-half of each prize is paid on a flat basis and one-half distributed according to the number of men overseen; thus a foreman in charge of 50 men will get a proportionately larger premium than the one in charge of 10 men.

It will be noted that the cash prize is small, and to some it may be even considered as trivial, but it was our intention not to make the men strive only to secure the cash prize as much as to awaken in them a healthy competitive spirit. In this we have succeeded beyond our fondest hopes for the competition is keen and constant.

Every member of each division is keenly alive to the fact that responsibility for keeping his division with a perfect score rests squarely on him—consequently, he gauges his work and actions to conform with all the precepts of safety-first. This interest is stimulated and fostered by making up the yearly basis out of the monthly average. The great thought is then concentrated on the yearly contest, and the discouragement of any unfavorable monthly showing is avoided because any other division may have a penalty sufficient in some months throughout the year to equalize these unfavorable periodical conditions.

Competition Is Often Keen

So keen does the competition become that more than once considerable tact and diplomacy had to be used in some divisions where the carelessness of an individual caused a disabling accident.

As a means of preventing accidents the score board has rendered excellent service for us. It is no uncommon occurrence to have 10 divisions out of 26 showing perfect scores. Included in some of our perfect scores is our Steel Foundry where we make castings weighing as high as 50 tons. This work is usually classed as hazardous.

Some divisions have perfect scores year after year

and there is hardly a single division that has not shown marked decrease in accidents during the past eight years.

In addition to the sign board we also maintain a safety first organization among our workers that deserves much credit for our extremely low accident experience in our many departments. We unquestionably recommend our plan to any manufacturer, and know any organization will find a similar plan to be well worth while.

Metal That Lubricates Itself

Now comes an inventor who says he has a metal which will run hot and which will not "seize", because he has found a way of combining graphite with it, so comments Popular Science Monthly. This is, probably the first time that any one has succeeded in combining a non-metal with a metal. An alloy of aluminum and graphite is said to give better results than the usual phosphor-bronze-bearing alloy at the same speed and pressure.

Substitute for Copper

During the war Germany was hard pressed for copper electric conductors, and Von Bett, a physicist, suggested that sodium would be an acceptable substitute.

The conductivity of sodium is about equal to that of zinc and about one-third that of copper, but since the specific gravity of sodium is small, it is a better conductor, weight for weight, than any metal.

On the other hand, sodium has a very low melting point and cannot be drawn into wire. Moreover, it cannot be exposed to the air.

The experiment was made of filling thin-walled iron tubes, painted as a protection against rust, with fluid sodium, and it was successful.

Whether or not such conductors as these can be introduced commercially depends upon the possibility of drawing the tubes cheaply. Sodium is now made cheaply from common table salt by electrolysis.

The exportation of Portuguese salt before the war was chiefly to the Netherlands, France, Newfoundland, Norway, Sweden, and Ireland, says Engineer. After the war conditions changed; the Netherlands, Sweden, and Norway now buy mineral salt from Germany, and France makes its purchases from England. According to the American Consul-General at Lisbon, the annual production of salt in Portugal for export is 150,000 to 200,000 tons. The stocks of 1917 and 1918 are still unsold. In Norway the importation of salt will be eliminated soon, owing to the building of large electrical sea water evaporating plants for obtaining salt direct from the sea. The representatives of the Portuguese salt industry have called the attention of the Government to their difficulties.

In a paper recently read before the Institution of Automobile Engineers, Mr. B. H. Thomas gave some particulars concerning the repair of worn motor parts by means of electrically deposited iron. He stated that the bath that gave the most satisfactory results, was composed of 75 grammes of ferrous ammonium sulphate to 1 litre of water, with a current density of 1 amp. per 30 square centimetres. The work is suspended in the vat from a standard fixed to the bench and in order to ensure a uniform deposit the anode is moved about the work by means of a small motor. The best anode proved to be that made up of 16 S.W.G. Swedish iron wire wound into the form of a cylinder and thoroughly annealed. The temperature of the solution should be kept above 68 deg. Fah., and it is desirable to decant and settle it every three days or so to remove oxides of foreign matter, which cause roughness in the deposited metal

Looking Over the *Shipping* Situation

By T. H. FENNER, Editor Marine Engineering of Canada

OOKING back over the past twelve months' activities in the maritime world they cannot be said to have been without interest, in fact in some ways they have been spectacular, but have gradually been slowing down as the year wore on, until now shipping has entered the doldrums. How long it will tack about there before catching the first whisper of the wind of prosperity, depends upon a number of factors, the chief and most uncertain one being the time when the world will settle down to work and produce.

The expected slump in freight rates has materialised, and as the writer has pointed out in previous articles, the more inexperienced among the operators of shipping have felt the pinch the most. At the same time, passenger rates have kept up their previous high level, and have in many cases advanced, and that they will keep this advance up, is most probable. The conditions governing in the freight world are altogether reversed in the case of the passenger vessels. In the shipbuilding trade, work is also falling off, after the gigantic efforts put forth by various nations during the past two years, and in many cases plants which cost millions to put down are practically abandoned, and of no use to anyone at the present time. In Canada, the year has been marked by the failure of some of the yards, and the prospects for the future are at the moment, none too bright. The position of shipping all over the world is that of passing through a period of depression, and its effect is shown in falling freights, cancellation of building contracts, and in the laying up of vessels to wait for better times. The passing of the legislation known as the Jones Act in the

United States was attended with much excitement, but inasmuch as its provisions have not yet been put into force, and are probably destined to be shelved, matters are being left in status quo for the meantime. The United States Shipping Board has at the moment probably enough troubles on its hands to suit even its fire-eating chairman, Admiral Benson, who has been reappointed. France has found the possession of a Government owned fleet not an unmixed blessing and is considering ways and means of getting rid of it to private owners. The blighting hand of the Government is being gradually moved from British shipping, but the Ministry of Shipping still survives, much to the disgust of the shipowning fraternity, who are anxious to be unfettered in their efforts to rehabilitate themselves. If the events of the past twelve months have shown one thing more clearly than another, it is that a Government is incapable of carrying on a commercial adventure such as operating ships or railways. The good showing of the C.G.M.M. up to the present has been largely due to the fact that practical shipping men were appointed to the executive positions, and the Government left the management of the fleet in their hands. The management have shown their strong commonsense by linking up the C.G.M.M. with some of the strongest and most powerful lines in the old country, by which arrangement they are sure of the best facilities in the various ports abroad and in the U. K., and are also assured of valuable feeders for cargo at outside points where a new line would have great difficulty in getting a look in. Such is the general state of affairs at this time of writing, and it will be profitable and instructive to look a little more closely into details.

The World's Tonnage

We have heard a great deal of late about the tonnage of the world being greater than in the pre-war days of 1914. This is true, but it does not take into account the natural increase that would have taken place in normal times, nor the fact that a lot of tonnage is very old, and in some cases almost obsolete. There has been a very large amount of reconditioning carried out during the past year, many vessels which had no overhaul during the strenuous days of the war having been taken out of commission and re-fitted completely. Another feature is that for a long period following the armistice the efficiency of the shipping in use was very low.

This was due to congestion in the various world ports, necessitating long waits for unloading, and loading again. The congestion was caused in a great measure by restrictive methods pursued by labor, and by the natural difficulty in clearing warehouses and storage sheds from the accumulation of merchandise that had piled up during the period of Government control. The breakdown of land transportation during the same period contributed also to the general tie-up.

This condition is being cleared up now, but is still bad in spots. The result of all this was that many ships were actually in their loading and discharging ports for a greater period than they were actually on the voyage between ports. This meant that there were too few ships to carry on the business, and the costs being so high, freights went up by leaps and bounds to cover these extra costs.

Owing to poor fuel, and general rundowness, ships took longer to cover the same mileage than in normal times, and

all these factors contributed to make the efficiency of the merchant fleet very low. Another point is that a large percentage of the increase in the world's tonnage is represented by wooden vessels, which are mostly tied up at the present day, and represent a dead loss to their owners. The United States Shipping Board possesses about $200,000,000 worth of these vessels, which they can neither operate, sell, or even give away.

What the Figures Show

Comparing the figures for 1914 with those of the present day we see some points of interest. The totals are taken from Lloyd's Register of Shipping, which is the world authority on the matter. Vessels of less than 100 gross tons are not included, neither are warships. In 1914 the world possessed a total of 45,404,000 gross tons. The total in June, 1920, was 53,905,000, or a total increase of 8,501,-000 tons. The relative position of the great shipowning nations has materially altered. For instance, in 1914 the United Kingdom had 18,892,000 tons, while the United States had only 4,287,000, including all her Great Lake craft. The percentage of the world's tonnage owned in the United Kingdom was 43.9, while that of the U. S. was 4.7. At the present day the percentage of the United Kingdom has decreased to 35.1, while the United States has increased to 24. The increase in the other countries has practically all taken place since the closing of the armistice, during which period the increase in the world's shipping has amounted to over 6,000,000 tons. By carefully estimating the net increase in the tonnage of the various nations, and eliminating all wooden and iron ton-

nage, the compilers of the Registry arrive at the conclusion that the world is still 3½ million tons to the bad as far as steel steamers are concerned. This effectually disposes of the claim that there is too much tonnage afloat.

Great Britain Still Building

The outstanding feature of the shipbuilding situation has been the increase in the number of vessels now under construction in the United Kingdom, and the decrease of those under construction in the United States. The building programme of the U.S. is now about finished. The same thing applies to the construction in Canada.

The following figures for the quarters ending March 31, June 30th and September 30th will give a good idea of the progress of events. There is no doubt that the immense amount of reconstruction work undertaken in the United Kingdom, as well as the dilatory attitude of labor, has kept production of new tonnage down, but even at that a very steady increase is noted.

In the quarter ending March 31st, the amount of new construction under weigh in the United Kingdom was 3,394,425 tons. This was made up as follows 814 steel steamers, 7 ferro-concrete, and 4 wood and composite steamers. There were also 35 steel sailing vessels, 4 ferro-concrete vessels and one wood and composite ship. In the United States at the same date there were 450 steel steamers, 29 wooden steamers, 10 steel sailing vessels, and 46 wooden sailing vessels, or a total of 535 vessels of all kinds, with a total tonnage of 2,573,298 tons. In Canada at the same date there were 50

steel steamers, or a total tonnage of 157,000 tons under construction.

Many Large Carriers Built

A notable feature of the steamers under construction in the United Kingdom was that there were 62 steamers of over 10,000 tons included. These were made up of 10 between 10,000 and 12,000 tons, 29 between 12,000 and 15,000 tons, 20 between 15,000 and 20,000 tons, and three between 20 and 25,000 tons. These vessels were mostly of the purely cargo, or the combined cargo and passenger type, very few companies building the purely passenger liner class of vessel. In the following quarter, that ending June 30th, the figures altered to the steamers were 888, ferro-concrete 3, and wood or composite 4. The sailing vessels were 41, ferro-concrete 3, and wood or composite 2. The tonnage represented was 3,388,000. In the United States the steamers had fallen to 366 of steel, 18 of wood, while the sailing vessels were 15 steel and 16 wood. The total tonnage represented was 2,105,956. In Canada the tonnage under construction was 58 steel steamers, and 12 wooden steamers, and 14 wooden sailing ships. The total tonnage was just over 200,000. In Great Britain the steamers over 10,000 had gone up to 63, nine of them being between ten and twelve thousand, thirty-one between 12,000 and 15,000, the remainder being the same as in the previous quarter. The peak of Canadian building for the year had been reached in this quarter, as the next tonnage of the vessels on the stocks and being completed.

THIS VESSEL MARKED A GREAT DEVEVLOPMENT IN SHIPBUILDING. SHE IS ENTIRELY ELECTRIC WELDED AND FITTED WITH AN OPPOSED PISTON INTERNAL COMBUSTION ENGINE DESIGNED BY CAMMEL, LAIRD & CO.

The figures for the quarter ending December will of course show this still more, but they are not yet available.

What Other Periods Show

The quarter ending September 30th shows a still higher figure for Great Britain, and a further drop in the U.S. Thus the steel steamers in the United Kingdom numbered 912, ferro-concrete 5, and wood or composite 4. Their total tonnage was 3,720,000 while sailing vessel tonnage made another 11,000.

United States figures are 260 steel steamers, with a total tonnage of 1,734,-000, wooden steamers being 12 of 18,000 tons and ten each of wooden and steel sailing vessels of 18,000 tons. There is thus a considerable decrease in the number and tonnage of all vessels building in the United States. In Canada the number of steel steamers had decreased to 46, with a tonnage of 158,000, three wooden steamers with a tonnage of only 1,057, and 15 wooden sailing vessels of a tonnage amounting to 11,000. Thus the figures show the gradual rise of British shipbuilding and the falling off of United States shipbuilding. The cause of the decrease, both in the U. S. and Canada is of course the completion of the Government building programmes. The United States now has a fleet of sea going merchant vessels operated partly by the Shipping Board and in part by private firms who have been allocated tonnage by the board. A large number of the vessels which were under private management have been handed back to the Shipping Board since the freight market fell in, and the Board has laid up a large number of their vessels which they have been unable to fix.

The United States Merchant Fleet

Probably the events of most interest in the year's operation of the now merchant marine in the United States, has been the passing of the Jones Act and the investigation of the Shipping Board. The Jones Act is the name given to the legislation introduced by Senator Jones, which had as its avowed object the bolstering up of the new merchant fleet against competition by the nations which were doing the most of the carrying trade. Chiefly, it was directed against Great Britain. The untoward events of the last five years having resulted in Great Britain losing a great portion of her shipping in the service, and for the benefit of the rest of the world, and the United States having become possessed, at enormous expense, of a more or less efficient fleet, it was felt by the Senator and those associated with him, that something had to be done to ensure its ability to live. Legislation passed in former years by Senator La Follette has resulted in making the ships of the U. S. the most expensive in the world to operate. This was on account of the extraordinary accommodation and living standard set for the U. S. seaman, and when this was coupled to the necessity of paying overtime for every little job that was done out of the regular routine, running ships at a profit even during the era of high freights was a problem.

It was not an uncommon thing for a Shipping Board vessel to arrive in a United States port with the bill for overtime to the crew totalling more than the actual wages, which themselves are very high.

Preference for U. S. Shipping

The method employed and backed by Admiral Benson, chairman of the Shipping Board, was to pass legislation by which all goods imported or exported to or from the U. S. in American ships should receive a preferential rate on the railroads, and also a preferential tariff. This meant that treaties in force with some twenty-nine other nations would have to be abrogated before this legislation could be put into operation. President Wilson, who even if he is an idealist, has a larger view than some of the U.S. senators, refused to give his assent to the bill, and it has hung fire ever since. In reply to this Admiral Benson issued a statement that the protest was engineered by alien enemies of the United States, who wished harm to the new fleet, and threatened the port of Seattle with drastic punishment if their protests did not cease.

The claim of the Western seaport was that the bill, if put into effect, would ruin their trade. However, Admiral Benson has been made Chairman of the reconstituted Board, and with the coming into office of a new administration, pledged to further the interests of the U. S. mercantile marine against all comers, there will likely be some interesting developments in 1921. It is noteworthy that the British shipping press hail with satisfaction the reappointment of the bellicose Admiral, as if they feel that no worse harm could come to the Shipping Board fleet than having him in control.

What the Inquiries Showed

The investigation into the activities of the United States Shipping Board was undertaken by two former officers of the Board, and they have so far found that the Board committed everything but murder. Such things as incompetency, ignorance, graft, etc., are charged in connection with all its activities, and if half the charges are true, one can only sympathize with the great American nation. If half the blunders and mistakes that are charged could be committed by any body of presumably big business men in that land where efficiency is supposed to roost on the same perch as the eagle, how ever could any ordinary body of men be expected to succeed? The British press, which is naturally highly interested, is of the opinion that a lot of the charges are due to pre-election vigor.

They take the view that when a totally inexperienced body like the U. S. Government take on a job like the building and operating of a huge merchant fleet, there were bound to be lots of mistakes, and the charges of graft they dismiss as due to the zeal of the investigators.

Competition Has Cut Rates

The vessels of the fleet have been put on various routes during the year, and have been in some cases responsible for the lowering of rates at a sudden clip. For instance, during the coal strike in England, when a new business in export coal sprang up suddenly from the Virginia ports, the board put such a number of ships on the trade that freight rates dropped rapidly. They also insisted on seeing the charter of any ships other than their own, American or foreign, so that they were in a position to know just what everybody else was doing, which of course gave them a considerable advantage. The deal by which Mr. Harriman has arranged a service between Hamburg and New York in connection with the old Hamburg-American line was consummated under the supervision of and with the consent of the Shipping Board. This gives the ships of the American corporation the advantages of Hamburg-American organization in Hamburg, but it also, in the opinion of a large number of people, gives the Hamburg-American line the entering wedge which they are looking for, in resuming active competition with the vessels of their erstwhile enemy. It is rather a peculiar thing that with the country still technically at war with the U. S., business arrangements can be made between German and American citizens, with the active help of a department of the American Government.

British Shipping

British shipping has been fighting its way back to its old time trades against a combination of handicaps which might well discourage any industry. The ship-owners have been put to all kinds of inconvenience and expense by Government departments, notably the Ministry of Shipping, which having been created during the war, is doing its best to perpetuate a useless life. The Government has promised that its activities will shortly be discontinued, but it still drags on, and affords a proof if such were needed, of how hard it is to make men give up the easy hours and emoluments of a Government position, even when there is no possible excuse for their retaining them. The large liner companies have been reconditioning their ships, and getting them back on the routes, but it is notable that no companies are going in for the large fast passenger liner. Costs of building and operation have increased to such a point that the liner de luxe is a commercial impossibility.

Change in "Style" of Vessels

The type of vessel in favor now is the moderate size and moderate speed vessel that can give good accommodation to a fair number of passengers while carrying a large cargo. The shipowning business has been very badly hit by the low production of coal, and the embargoes placed on the export of coal to foreign countries, and the export of bunker coal. It is of course well known that the bulk of the steamers leaving England in the past have had cargoes of coal, which they took to all parts of the world, bringing back the produce of those countries. The coal formed outward freight and also created credits to pay for the goods brought back to the United Kingdom from other countries. The attitude of the coal miners in restraining production of coal, and the vexatious delays and restrictions placed on steamers by the Government department which has control of coal, has resulted in vessels having to leave England in ballast, to make long voyages in many cases, in order to get a return cargo. This resulted in high freight rates, which made the price of all goods brought back still dearer to the people in England.

During the coal strike, coal was actually imported to Britain, but worse still, the market in France and the Mediterreanen which had always been supplied by Britain was opened to her competitors.

Building Costs are Away Up

Costs of shipbuilding have gone to such a pitch that many firms have paid large sums to cancel the contracts for ships they had under construction, rather than continue with them and have to pay an exhorbitant price. With falling freights, it is impossible to run the ships at a profit when the capital cost is abnormally high. In this connection some very straight words have recently been delivered to the workmen in the British yards. They have been told plainly that they are hanging back and not delivering an hour's work for an hour's pay, and that if this continues they will soon be without a job of any kind. A very serious side of the high cost, and not only the high cost, but the altogether unnecessary length of time taken is that it has resulted in a large amount of repair work being sent to the Continent that should have been done in the United Kingdom. A number of contracts for ships received from Norwegian and other foreign shipowners have been cancelled, but this is probably not so much a matter of high costs, as the slump in freights, which make it appear a bad time to acquire new tonnage. At a recent launch of a Cunard steamer, the Chairman of the company said that the vessel was six months late in delivery, and had cost in excess of her estimated cost a matter of nearly $3,000,000. This sort of thing is naturally a big handicap. There is one thing that is to be said as far as the British shipowners are con-

cerned, they have been through hard times before, and know just how to face them, and where to squeeze a living for a ship if it can be done.

If labor will assume anything like a reasonable attitude, especially the men actually sailing the ships, Great Britain will pull through the present depression in better shape than most of her competitors, and be on hand when the good period comes round again. There is no doubt that if the world in general were functioning properly, there would be ample work for all the ships at present afloat, and for a good many more. The elimination of a huge country like Russia from the trading field of the world has had a disastrous effect on many other countries, and the low rate of foreign exchange has also been a handicap on trading. British shipowners generally ask for nothing but a relief from the

ONE OF THE LATEST C.P.O.S. EMPRESSES—THE "EMPRESS OF CANADA."

irritating restrictions of incompetent and stupid Government officials, and then they feel they can meet competition as they have hitherto done, and beat it.

Canada's Shipping and Ship-building

The past has been a busy year for Canada, both in her shipbuilding yards and on the high seas. Besides the ships that have been built for the Canadian Government Merchant Marine there have been a few built for export, and there is no doubt that there could be more export orders secured if there was any method of overcoming the exchange rate handicap. It has been rumored on several occasions that large orders have been secured for vessels for export, but these have not materialized. A suggestion was made during the year that the Government should finance the building of ships for foreign order, taking a mortgage on the vessel as security. This was finally agreed to by the Government, but up to the present no orders have

materialized under this arrangement. Shipbuilders claim that the proportion of the cost of construction that would be advanced by the Government should be about 50% instead of the 25% they suggest. At the present time building is active at Halifax, Levis, Three Rivers, and Port Arthur and Collingwood. Some of the Pacific coast yards are also building vessels.

At to the outlook for the future we quote the opinion of one of the leading men in the industry, which should be of interest as coming from a practical shipbuilder, who knows whereof he speaks. The gentleman in question is Mr. C. M. Gillam, managing director of Canadian Vickers Ltd., who speaks as follows:

"During the past year the supply of materials, especially steel, was considerably handicapped by the labor troubles in the States in the railroad and mining industries, but this condition has much improved in the past few months. The tendency now as regards prices is downwards in practically all commodities. The supply of materials is distinctly good and promises to improve.

"There is an ample supply of labor and its efficiency has shown a steady improvement, although the wages which gradually increased during the past years to meet the increased cost of living, still remain at a high figure. In this connection it is confidently hoped that as the cost of living improves the saner elements in labor will be able to guide their less informed brothers, realizing that a reduction in labor costs is essential if business is to be obtained in competition with European countries.

"In the shipbuilding industry there is a serious falling off in the number of employes, due to lack of orders for 1921. The cause of this is the adverse exchange rate and the fact that the Canadian Government shipbuilding programme has come to an end. There is no market in the

United States for Canadian built ships due to the financial inducements offered by the United States Government to United States owners to build their vessels in the States. The consequence is that the only market for Canadian shipbuilding at present is Europe and, as stated above, the exchange rate is sufficiently high to prevent business.

"With reduction in prices of materials and increasing efficiency of labor, prices asked for in Canada for ships are steadily coming down, and it is confidently predicted that, with the further reduction, both in wages and materials, anticipated in the near future, such prices will be so reduced as to enable Canada, apart from the exchange question, to compete to advantage in the open markets of the world. This is the only feature which justifies the hope that, as exchange becomes normal, there will continue to be a limited shipbuilding industry in Canada which can operate on a commercially sound basis.

"It should be noted in this connection that the cost of construction of ships in Canada is not materially higher than the present cost in England when the exchange situation is left out of account, although it must not be forgotten that there will undoubtedly be a tendency towards cheaper production in England when labor conditions there become more settled, although there is no likelihood of wages in Great Britain ever being reduced to pre-war rates.

"It is worthy of note that, roughly speaking, 50% of the business of Canadian Vickers, Limited, in shipbuilding has been for export, and the performances of ships built by them for foreign countries has more than justified the confidence in the ability of Canadian workmen and Canadian shipbuilders to turn out ships in all respects equal to those of any other country.

"The outlook for 1921 is not bright at present, so far as shipbuilding is concerned, but the basic factors, namely— efficiency of labor, the drop in price of materials, and the recognition by labor of the economic situation and willingness on their part to co-operate to overcome the adverse features, are such that a pessimistic cessation of effort to overcome these difficulties and to tide over the dull period is not justified, but a more vigorous attempt is necessary on the part of the Government and employers towards finding means by which the industry may be kept alive until exchange becomes normal and the Canadian shipbuilding industry has a real chance to continue on its own merits.

"In conclusion, in our opinion the present situation of the shipbuilding industry is such that, provided the difficulty of exchange which, after all, is an abnormal one due to the war, can be overcome by means of assistance from the Government or otherwise, there is every reason to believe that shipbuilding in Canada can continue as a profitable commercial business."

The Canadian Government Fleet

Our Canadian Government fleet has extended its activities during the year until now it is on every route in the British empire. Vessels are sailing from Halifax and St. John, N.B., in the winter, and Montreal and Quebec in the summer to Newfoundland, Liverpool, European Continental ports, British West Indies, and South America, while they are sailing to the East Indies and China from both the East and West coasts. The vessels are a fine type of ship, well found, with good accommodation for officers and crew, and in the later ships extra accommodation has been provided for three or four passengers. On the ships in the Australian service refrigerating space will be provided, as well as the domestic refrigeration that is fitted in most of them. The writer, who spent a good many years at sea, has never seen ships with such comfortable crew accommodation as these Canadian vessels, and it is a credit to their designers. The fleet, as we have had occasion to mention on several occasions, is managed exactly in the manner of a privately owned line, and on that account has managed to show a very good profit so far. The action of the management in linking up their fleet with such firms as the Cunard Line, and Alfred Holt's, have shown a foresightedness and absence of the usual bureaucratic spirit associated with Government undertakings that gives the fleet a hope of success in the lean time to come. Canada has received a wonderful advertisement in all parts of the world through this excellent fleet, and it would have been well worth while from that viewpoint alone. Apart from this the fleet has opened up trade routes for Canada, and we believe has opened up the possibilities of export trade to many who had not previously given it a thought. The fleet is pretty nearly completely manned by Canadians, and is about as thoroughly Canadian an institution as can be found in the country.

The not inconsiderable lake and river traffic carried on by the Canadian Lake transportation companies has been bigger than ever this year, and it is pleasing to note that a purely lake steamer was launched for a Canadian company from the yard at Midland during the early part of December. This is a business which Canadian lake yards are as well able to handle as any others, and they should find much profitable work of this kind in the future. The repair business is well catered for in Canada, with dry docks at Sydney, Halifax, on the St. Lawrence, at Port Arthur, Collingwood and Midland on the Great Lakes, and in the near future on the Canadian Pacific coast.

Engineering Development

It is interesting to look over the development that has taken place during the year in the engine equipment of the new vessels built. While not wishing to take the attitude of "I told you so," it is pleasing to note that the prophecy we made as to the adoption of the oil engine is coming true. There is a gradually increasing number of shipowners adopting this type of motor, and it is not in small ships by any means that it is being adopted. It is pleasing to note in this connection that the large engineering firms in Great Britain are keeping well ahead in the internal combustion engine field, and have produced several interesting designs which have proved very successful. The companies which have adopted the internal combustion engine for their ships include such well-known names as Elder, Dempster & Co., T. & J. Brocklebank and the Glen Line of Glasgow. The number of vessels fitted with internal combustion engines now represents near 2% of the number in Lloyd's Register. Those burning oil fuel under boilers number 16.3%, while 76½ use coal. The remainder are sailing vessels. There is little likelihood of the spread of the use of oil fuel to raise steam, as the price of oil is getting prohibitive for that purpose. It is not in the interests of conservation that oil should be largely used in this manner, as there is a large loss attending it. A good many vessels are fitted to burn either oil or coal under their boilers, and can change over from one to the other with little trouble, thus enabling them either to oil or coal their bunkers whichever is handiest, or cheapest.

The Geared Turbine in Favor

The geared turbine is also meeting with a large measure of favor, no less than 245 vessels of 1,286,046 tons being fitted with this class of machinery during the year. The reciprocating engine is still being put in a large number of ships, and there is no doubt that it will be in use for many years yet. Its record of reliability is something that will make many people cling to it till the newer forms have something of a record to show that can compare with the old three-legged job. The United States has experimented with ships fitted with the turbo-electric drive and have been sufficiently pleased with the results to decide on equipping a number of ships with this class of drive. It does not seem to have met with any favor in England or on the Continent. There seems to be a doubt if the extra weight and expense together with the complicated wiring, are justified in the ordinary cargo vessel.

There are, of course, certain advantages associated with the electric drive, notably in manoeuvring and in variable speeds, but for a cargo vessel these do not amount to much. The oil engine has developed to a point where it can be handled in manoeuvring as smartly as the triple expansion reciprocating, and nothing could be much handier than that. We do not anticipate the adoption of the electric drive in other countries than the United States.

There would seem to be little doubt but that shipping is in for a period of low

freights and hard times, and this will continue till the world settles down, if such a desirable consummation ever takes place. In circumstances like these, the only chance of survival is by practising the very bed rock of economy in the running of ships.

This means keeping personnel down to the minimum, quick turn round in every port, and the minimum of lost time for any cause. We believe the ship with oil engines will be far ahead of any of her rivals in this matter, but the number of them at present is so small that their effect on the general situation is small. However they are increasing in a very steady manner, and in the next ten years it would not be a matter of surprise to see the major portion of the carrying fleet fitted in this manner. The next

twelve months should bring some interesting developments in connection with the United States merchant fleet, which must be put on a commercial basis or go out of business. Just what method will be finally adopted to keep it afloat is hard to say, but we venture the opinion that enforcing the provisions of the Jones Act will do more harm to the new fleet than anything else that could be done.

The trouble is that the U. S. ship is costly to run, more so than any other nation ships, and there is no politician in the States with nerve enough to try and repeal the LaFollette Act. Even if there were, and he were successful, it is questionable if the ships could be manned with the special inducements removed. As it is, sailoring in United States ships is so desirable that numbers of men desert in U.S. ports from other country ships to get in on the good thing. There is no reason in the world why the United States should not have a merchant fleet,

if sufficient people are interested in it, but to try and create a merchant fleet by eliminating the ships of all other nations from their ports, is to say the least a mistaken idea. The practice of withdrawing tonnage for a time, and then throwing it on the market in an effort to wipe out competition has also resulted in wiping out American companies which was not quite within the intention of the gentlemen who were doing the manipulating.

France has been handicapped in her shipping activities to a great extent by the coal strike in England. A large number of the French vessels were of the small collier type, expressly for carrying coal from South Wales to French ports. Naturally they have not been able to do much in this line. The large French

transatlantic line is well under way again, after the cessation of the war. Italy has added considerably to her fleet, and has resumed all the services which she used to carry on. The Italians are experienced shipowners, and were possessed of some fine vessels before the war, and were among the heavy sufferers by the war. Japan has considerably increased her tonnage, but things are not too rosy in shipping circles in Nippon just now. Germany and Austria are of course still out of the reckoning as far as merchant shipping is concerned, although the arrangement made with the company of which Mr. Harriman is in control, points to the fact that the Hamburg-American line is not going to frozen out if they can help it.

This is not a time for prophesying, but more for hoping, and the hope is that the turmoil in Russia and other European countries will subside ere long and active trading commence between the various parts of the world. Only in that way

will there be cargoes for the ships of all nations, work for the people of all nations, and an equitable exchange rate which will once more put commodities in the reach of people of moderate means. In short, a return to the normal is what we hope, but with a higher standard of living than obtained before the war. This is quite within reach, once the peoples of the different countries settle down to working instead of fighting. Let it be soon.

REDUCING CORROSION

According to D. M. Buck, metallurgical engineer for the American Sheet & Tin Plate Co., Pittsburgh, there is an overwhelming mass of evidence to prove that by allowing with normal open-hearth or Bessemer steel a small copper content—0.15 to 0.25 per cent.—the corrosion produced by air and moisture is greatly reduced. The melting point of copper is about 700 deg. Fah: lower than the average tapping temperature, it diffuses readily, and once diffused does not segregate.

The manufacture of copper steel has heretofore been largely confined to sheet metal, and the product has been greatly improved. A conservative estimate indicates that the life of sheet metal is at least doubled by this treatment, and in speaking of this subject he advocated the use of copper in all iron and steel products called upon to resist the attack of air and moisture.

It was stated at a symposium on refractories for electric furnaces held by the Electric Furnace Association, of Columbus, U. S. A., that refining furnaces, handling metal and corrosive slag at temperatures limited only by the melting point of the furnace itself, have about reached the limit of their possible development. It would appear that a new, cheap, dependable refractory with a melting point higher than 1750 deg. Cent. is now required, a substance which is highly resistant to sudden temperature variation so as to permit intermittent operation, which has been calcined at electric furnace temperature, and which, therefore, possesses reasonable constancy of volume. It must resist the scour of slag—or if a roof brick, must withstand the action of fume, hot gases, and splatters of liquid. It must be a poor conductor of electricity, or leakage of current may be serious. Finally, it must have sufficient strength at high temperatures to meet the requirements for the arches and side walls of the furnace.

A proposal is being considered, says the Engineer, to preserve the French supply of gun-cotton and nitrocellulose explosives — about 90,000 tons — by placing it in tanks at the bottom of a lake in the Pyrenees. Stored away at a safe and constant temperature, it is believed that the material would neither deteriorate nor decompose.

LAUNCHING A CANADIAN GOVERNMENT STEAMER AT COLLINGWOOD—THE "CANADIAN ROVER."

WELDING AND CUTTING

The Welding of Locomotive Fire Boxes

Welding Door and Flue Sheets, Patches on Firebox Sheets, Patches on Mudring, Door Collars, Knuckle Cracks, Building Up Worn Edges of Fire Box.

IN December 16th issue of Canadian Machinery we commenced this article by telling how full welded fireboxes, semi-welded fireboxes, and full welded, half side sheets were prepared and completed. In this the concluding portion of the article, we will touch on the subjects mentioned in the heading.

Welding Full-Welded Door and Flue Sheets

The method to be followed in preparing to weld door and flue sheets is similar to that described for full-welded fireboxes, as shown in Fig. 25. In the full-welding of door and flue sheets, however, the work is done on the inside at all the joints except the one with the crown sheet. That part of the welding is to be done on the water side, unless the radial staybolts are in the door sheet, when the welding must be completed from the inside.

First fit and bolt the door or flue sheet to the mudring. Next bevel both edges of the vertical joints on the inside, leaving a 3-16 in. opening at the bottom of the vee, see Fig. 26. Now bevel the joint with the crown sheet on the outside. The door and flue sheets should be flanged, as shown in Figs. 27, 28 and 29.

Tack weld the joint for an inch at R1, Fig. 25. Then drop down about 10 inches and weld from R2 up to R1, from R3 to R2, so continuing to R5 and welding last at A on the mudring.

Assuming that the radial staybolts are not in the door sheet, both the door sheet and the flue sheet should be welded on the water side to the crown sheet, starting at c, about 10 inches from R1, Fig. 25, and welding back to R1, from d to c, e to d, on over to L1 and then welding on the inside of the firebox from L2 up to L1, thus welding along until the sheet is finally completed at B on the mudring.

Welding Patches on Firebox Sheets

Patches for Firebox Sheets should be triangular in shape whenever possible, the reason being that the sides of the triangular patch do not run parallel with the rows of holes, and weaken the joint, as would a rectangular patch.

To weld the patch shown at Fig. 30, proceed as follows: Bevel the edges, then tack weld at 1. Then, beginning about 10 inches from 1 at 2, weld up to 1, from 3 to 2, and from 4 to 3. Let the joint become cold. Then weld from 5 to 4, and 6 to 5.

Preheat as shown at xxx1, to allow for expansion. Weld from 8 to 1, 7 to 8, and 6 to 7, and complete the job by preheating at xxx6.

To weld the patches shown at Figures 31 and 32, first bolt the plate to the mudring. After bevelling the edges, tack weld at 1. Then weld from 2 to 1, 3 to 2 on down to the mudring at 5. A, should be done last. When the weld is cold, preheat at xxx1, and weld from 6 to 1, 7 to 6 on down to the mudring at A.

In welding the patch shown at Fig. 33, proceed as follows: Preheat along xxx1. Then tack at 1 and weld over about 10 inches from 2 to 1 and 3 to 2. Let the plate get cold. Preheat again at xxx1 and weld from 4 to 1. Then preheat at xxx5 and weld from 5 to 4. Let the

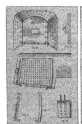

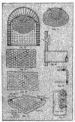

THIS VIEW TAKES IN FIGS. 25 TO FIG. 49.

sheet get cold. Preheat at xxx3 and weld from 6 to 3. Again preheat at xxx5 and weld from 5 to 6.

Figure 34 shows the proper way to round the corners of the patches, using a radius of about ¾ in. In preparing all of the joints, use a bevel of about 45° and leave the vee open about 3-16 in. at the bottom, as shown in Figure 35. In using the diamond patch at Figure 36, the welding should be done in the same way as that described for the triangular patches. Expansion must be provided for by the proper preheating and reheating at the ends of the line of weld.

Welding Patches on Flue Sheets

The triangular shaped patch should be used for flue sheets wherever practicable. The method of applying the patches may be described as follows, the welding being done on the water side of the firebox:

Welding on Patch of Figure 37

First cut the flue sheet as shown in the drawing. Next weld the bridges at 1, Figure 37. Then weld from 2 back to 1 and from 3 to 1.

Weld the knuckle at 6 and then from 10 to 6. Then weld the knuckle at 7 and from 11 to 7. Complete the job by welding to the flange at 12 and 13.

The next patch, Figure 38, is for a deep job and requires less welding to the flange. In this case prepare the sheet as described for Figure 37. Weld the bridges at 1 and from 2 to 1. Then weld from 3 to 1, 4 to 2, 5 to 3 and so on until knuckles are welded at 8 and 9. Next weld the patch to the flange at 10 and 11.

To weld patch Figure 39, proceed as follows: Weld successively bridges 1, 2, 3, 4, 5, and 6. Preheat at ends of rows,

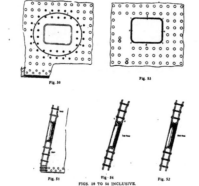

Fig. 50　　　Fig. 53

Fig. 51　　　Fig. 54　　　Fig. 52

FIGS. 50 TO 54 INCLUSIVE.

and weld bridges 7, 8, 9 and so on until all are welded. All welds should be butt-jointed as shown in Figure 40.

To weld cracked bridges such as shown at Figure 42, first preheat bridges marked x at 1 and 2, then weld at these points, and reheat at x.

Welding Patches of Figures 43 and 44

First weld patch to crown sheet. Then weld flue sheet to patch as shown at 2

and 3 of Figure 41. Next weld flue sheet to crown sheet at 6 and 7 (see Fig. 43), using butt joint 8 and 9, Figure 41. Now weld to the flange at 12 and 13. A patch may now be welded to the crown sheet as shown in Figure 44.

Welding Patches on Mudring

The following three methods of patching mudrings have proved to be most serviceable. Make sure that all rivets are left out adjacent to the weld. To weld the patch at Figure 45, proceed as follows: Weld from A to 1. Then weld from 2 to 1, and from 3 to 2. Next weld patch to flange at 3, Figure 45, as shown in detail at c and d, Figure 49. Then weld from 4 to 3 and on to B.

To weld patch shown at Figure 46, first weld from A to 1 on the mudring. Then back weld from 2 to 1, and allow weld to cool before welding from 3 to 2 and so along to 10. Complete the weld at B.

To weld on patch shown at Figure 47, proceed as follows: Remove the rivets adjacent to the weld on the mudring and side sheets, also remove the stay-bolts that would be affected by the heat. Start to weld at A. Then weld from 2 to 1, a distance of about 10 inches apart, and let the metal cool. Weld from 3 to 2, also welding to flange near 3. Then from 4 to 3 and B last. All weld joints should be beveled at 45 degrees and opened 3-16 in. at the bottom, as shown in Figure 48.

Welding Door Collars

In welding locomotive door collars the door sheet should be made large enough

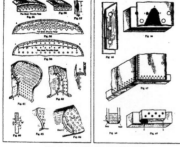

FIGS. 55 TO 69—SEE TEXT FOR DESCRIPTION

to include a row of staybolts within its area, as shown in Figure 50. First bolt the patch in place by screwing in some of the staybolts, then bevel the edges of the weld joint at 45 degrees on the inside of the firebox.

Next tack weld for about an inch at 1 and 5, Figure 50. Now advance about 10 inches to 2 and back weld to 1. Weld from 3 to 2, 4 to 3 and 6 to 4. Next weld from 8 to 1, 7 to 8, 6 to 7 and 5 to 6. Figure 51 gives an end sectional view of the welded joint of a door collar.

Welding Door Holes

In welding door holes for locomotive fireboxes the butt weld of Figure 52 should be used. The work should be done as follows: Bevel the edges to 45 degrees and leave bottom of weld open about 3-16 inch. Next tack at 1, Figure 53; weld from 2 to 1, 3 to 2 and so on to 5. Then weld from 8 to 1, 7 to 8, 6 to 7 and 5 to 6. A lap weld like that of Figure 54 is sometimes used for welding door holes, but it should not be resorted to if the butt weld of Figure 52 can be made.

Welding of Knuckle Cracks and Patches in Door and Flue Sheets

In the following work the welding should be done on the water side of the firebox wherever possible. To weld the flue sheet knuckle crack of Figure 55, proceed as follows:

Preheat at xxx1, and tack weld at 1. Then start at 2 and back weld to 1, from 3 to 2, and 4 to 3. Reheat at xxx4.

In welding the flue sheet knuckle crack of Figure 56, first weld from flue sheet over knuckle to crown sheet on the water side of the firebox, and complete weld from the inside. This weld should be reinforced about 1-8 in. and widened to about 2 inches at the top of the vee, as shown in Figure 57.

To weld the flue sheet knuckle crack of Figure 58, first start to weld the flange at A, Figure 58, working on the inside of the firebox. Then weld on the water side from 1 to A, back weld from 2 to 1, 3 to 2 and so on to the flange at B.

In welding the door sheet knuckle crack of Figure 59, commence working from inside of firebox, and weld from 1 to A, back weld from 2 to 1, 3 to 2 and on across to 8 and B. The distance between tack points of weld should be about 10 inches. Figure 60 gives the form of vee for the butt weld. In welding the flue sheet patch of Figure 61, first weld the flange at A. Then back weld from 2 to 1, and on to B.

When welding a flue and side sheet patch such as shown at Figure 62, first cut the side sheet to outside of the rivet holes, as shown in drawing c. Weld the flange at A, and then back weld down to the flange at C. Now weld from 6 to 5, 7 to 6 and so on to B.

The flue sheet weld of Figure 63 should be performed as described for Figure 55, and the crack C of Figure 64 should be welded the same as that of Figure 55, while patch D may be welded in place by the method shown in Figure 61.

Building-up Worn Edges of Firebox

The edges of fireboxes become worn, and cause leaky joints, from excessive caulking, erosion from escaping steam and corrosion. The restoration of these edges by building-up may be accomplished as follows:

First refer to Figure 65. Brush the rust off the edge and build up from 1 to 2. Light hammering of the weld while red hot will improve the joint. At Figure 66, a piece rusted out as shown in this figure may be built up by first removing the rivets 1, 2 and 3, and then adding metal from A down to the bottom of the sheet to the mudring. The last example, Figure 67, is somewhat different. In welding sheets to a mudring build up the joint of the two sheets for about 5 inches above the bottom. Bevel the bottom of the sheets as shown in figures 68 and 69.

Welding Full-Welded Half Door Sheets

There are two methods used to weld in half door sheets of fireboxes. That shown in Figure 70 is sometimes resorted to to save time. It is not the best way to do the work, however. This method leaves the old riveted flange on the side sheets. The welding is done by starting at 1, back welding from 2 up to 1, and so on around, ending at B' on the mudring. The most serviceable way to do the work is as shown in Figure 71.

The side sheet should be cut to the inner edge of the rivet holes, fitting the door sheet to the edge of the side sheet, and bevel the joints as shown in sketch at left of Figure 70.

Screw in the second row of staybolt holes, and bolt the door sheet to the mudring. Using a suitable tip and welding blowpipe with correct welding rod, start to weld at 1, Figure 71. Then weld the door sheet to the side sheet by welding from a point about 10 inches down at 2 up to 1, and from 3 to 2 and ending at A on the mudring. Back weld 10 inches from 6 to 1, from 7 to 6 and on around to B at the mudring. The sketch at the left of Figure 71 shows the style of joint between the door sheet and the side sheet.

Continued on Page 633

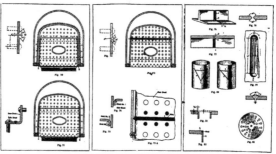

FIGS. 70 TO 83.—THIS CONCLUDES THE SKETCHES.

WHAT OUR READERS THINK AND DO

REDESIGNING SINGLE PURPOSE MACHINES

By G. Barrett

During the past two years we have heard a lot of talk regarding increased production, but are manufacturers taking the same interest in the development of jigs and fixtures that they did during the war period? There are a lot of old machines around the country at the present time that could be redesigned or modified for useful work. The accompanying sketch shows a special grinder constructed from an old single purpose shell machine. The headstock of this machine was removed and a knee bracket A bolted to the bed with the headstock on top. The vertical face of the knee bracket is provided with a dovetail slide to carry the movable tables B, which carries the compound rest C and D, the latter being operated by means of a rack and pinion. Fixtures of different design may be used for a variety of work.

The sketch shows the machine adopted for the grinding of pulleys. The jig consists of a cast iron block E, shaped at an angle of 2 degrees, with a stud set in at right angles to the inclined face. The stud F is fitted with ball bearings to facilitate revolving when grinding. A cup wheel G is used and numerous sizes of pulleys can be economically produced in this way.

In some instances it might be advisable to finish turn after grinding. This could be accomplished by form turning, as shown in Fig. 2, where two or more pulleys are located on the one arbor and turned at the same time.

Sketch Fig. 3 shows a speed cone bracket used on a drill press. These brackets were originally planed, which proved both long and expensive. The fixture for holding these brackets while grinding is shown in Fig. 3, and is simple and costs little to make when compared to the production of brackets, which is about six per hour. A similar fixture is used for grinding the base of the pump bearing shown in Fig. 4. The piece is held to the locating pin, insuring the base being square with the bearing. Fig. 5 shows a fixture for the grinding of the yokes for clutches. This is made of cast iron with a V-block to support the lug on the yoke. Two small pins are provided to keep the face of the yoke in alignment with the wheel. These are only a few of many jobs that could be economically performed on a machine of this description, and it would be well for many firms to give a little attention to machines that have probably been relegated to the scrap heap, the remodeling of which would undoubtedly prove a profitable undertaking.

Editor's Note.—We believe this article raises a point of real merit, namely the possibility of modifying war time special machinery for peace time prod-

ucts. No doubt many readers have seen various installations of such nature, and we would welcome sketches and descriptions of the same.

STEEL CUTTING EDGES

By W. S. Standiford

Probably one of the most useful and money saving metals that the machinery manufacturer receives the greatest benefits from is high-speed steel. By having the tools for production work made out of this material a greater output has been secured at a lower cost per piece, in spite of the fact that this grade of steel is a very high priced one. The fine qualities of high speed steel are getting to be more generally known among the rank and file of business men, and in the last few years its price has risen to such an extent that it has become a matter of great importance to make a pound of steel produce as many tools as possible.

To use a solid tool made of high priced steel is very expensive, and in cases of this kind great satisfaction can be obtained by having a blacksmith forge a shank of ordinary machine steel, and weld or braze a small piece of the high speed metal on top of the shank to be used as a cutting tip. The great drawback to either welding or brazing work of this character, exists in the fact that it is a very difficult matter to secure a perfect union between the relatively soft

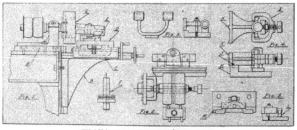

FIGS. 1 TO 5. SHOWING DETAILS OF THE SPECIAL GRINDER.

machine steel shank and piece of the high-speed metal so that it will withstand the heaviest roughing cuts that the machinery will allow. This is a necessity, as the greatest speed in doing work is obtained by cutting the roughing time down to its lowest possible limits; the production costs also being reduced by

HOW THE CUTTING EDGE IS PREPARED.

this method. By following the method outlined below, it will be found feasible to make a perfect union between the soft steel shank and the hard steel cutting tip that will give good satisfaction to the user of tool.

First forge the shank to its proper size and shape, and cut a slot in its top as shown in Fig. 1 of sufficient size as will allow of the insertion of a small piece of the extra hard metal to occupy this space, having the piece project out about 3-16 inch from the front and also 1-16 inch out from the sides of shank in order to allow for grinding after it is finished. It is also a good plan when making the shank to taper both front and sides at their bottoms. This facilitates quick grinding of the tool after hardening as a much smaller amount of metal has to be removed. The back end of both slot in shank and tool tip should be tapered to match. This makes it have a close contact when brazed and helps to keep the piece in position when a heavy strain is put on the front of the tool during the roughing cuts. Next cut a small piece of sheet copper about 1-32 inch thick to a slightly greater width and length than the bottom of slot that is shown in Fig. 1 and bend one end to fit the angle closely at the back of bar.

Brighten both the slot in tool shank and also copper strip with emery cloth or paper and treat the tip that is to be inserted the same way. Then coat both copper strip and the slot in machine steel bar with borax. Place tip over the copper piece and tie it tightly with wire so as to hold it in place during the heating operation. Fig. 2 depicts the tip placed over the sheet copper strip, no binding wires being shown in the picture. The

tool is now put into a heating furnace and heated to the proper temperature recommended by the makers of the steel. When this is effected, remove tool and with quick blows of a hammer used on front of tip, drive the beveled part of tip into its seat, then quickly strike the high-speed steel piece a couple of moderately heavy blows. This distributes the melted copper and borax uniformly over the joint and at the same time causes the two metals to unite. Quickly plunge tool into the quenching bath and keep it moving until it is cold. Grind it to shape and it is ready for use in the lathe or planer. The combination of a machine steel shank with a piece of extra high grade and priced steel, produces a cheap, but efficient tool that will give good service if the brazing has been very carefully done. It is possible by this method to successfully unite various brands of high speed steels to machine steel shanks, although the writer has not tried whether "stellite" (which is said to contain no steel whatever) could be welded or brazed to shanks of a cheaper metal.

A NEAT REPAIR
By Tyke

Some time ago the frame of a heavy duty power hack saw was broken at point "X" Fig. 1, due to an accidental obstruction of the stroke. The frame was of cast iron, and inverted U section. Due to this peculiar section and the small dimension between the inside of the U, it is doubtful if a really good welding job could have been made. In any case, it was decided to make a purely mechanical repair and the interesting part is that neither bolts nor rivets were used.

A forging of the proper shape and thickness was quickly made and simply

ground on the high spots so that it could be tapped lightly into place, though no pretence at fitting was made. When the two pieces of the frame butted closely together, the frame and forging were clamped and drilled, then reamed for No. 8 taper pins which are approximately 1-2-in. diameter.

It will be noticed that these pins were not driven in from the same side, but were alternated, see Fig. 2. Furthermore, the holes were slightly angled in order to give the pins proper key draft and ensure proportional stresses on the pins. This precaution proved to be a wise one, as the draft not only ensured the broken ends of the frame butting properly, but increased the holding properties of the pins themselves.

After the pins were driven home, babbitt metal was run into the frame about the forged piece in order to fill up any spaces that might exist due to any inequality of forging or casting. When cool, the pins were again lightly tapped to make sure they were tight and the frame assembled.

After many months' night and day work, the break was carefully examined and the break was hardly perceptible and certainly no more so than the day it was repaired.

TRANSFERRING HOLES
By Harry Moore

Transferring of holes in various classes of work is of such common occurrence in the life of the average mechanic that a brief description of several well tried methods would probably be appreciated by many readers. The use of the dividers, square and other tools, in the laying off of holes, does not always give the best results, as the chances of error are so numerous that it is frequently impossible to maintain the accuracy required. It is, of course, in the repair shop that these operations are mostly met with; new parts being fitted to old ones, old parts being replaced, etc. Fig. A shows a new shaft being fitted to an old collar, the latter containing the holes previous-

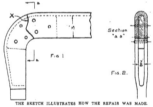

THE SKETCH ILLUSTRATES HOW THE REPAIR WAS MADE.

ly used. This looks like a very simple job, yet if the hole is drilled out of line with that already in the collar, the insertion of the pin becomes a neat little operation of hanging it around several corners until it finally peeps through the other side—a sorry looking pin, indeed.

A good way to avoid all this is to provide a V-block with a hole drilled through the center. It is then a simple matter to turn a piece of stock on the end to fit the hole in the collar and insert this in the hole drilled in the V-block. It will then be obvious that the collar and the shaft will be lined up properly and the drilling of the hole in the shaft will be done with a great degree of accuracy. A reverse condition to that just cited is shown at B, where the new collar requires drilling to fit the hole in the old shaft. The shaft, with the collar on the end, is placed in the V-block and adjusted so that the drill will enter the hole freely. A small parallel is then clamped to the shaft, with its side touching the V-block and the end resting on the drill table. The sleeve is now pushed into position and the shaft clamped down, first noting that the parallel is touching both surfaces as before. In starting a drill on a convex surface it is well to bring the drill lightly to the work until it has sufficient depth to guide it.

A convenient center punch for transferring the centers of holes is shown at C. It is made in three pieces, the punch being made of tool steel. All moving parts are made a sliding fit and the outer ring knurled to facilitate handling. Two small screws serve to keep the part together when not in use.

It is often required to fit a new cover to an old flange, where the latter has blind tapped holes. A case of this kind is shown at D. The old cover not being at hand, the work was set up as shown, a V-block being clamped on end and the flange pushed up against it, then lined up until the drill enters one of the holes. The new cover is now placed on top, clamped, and the hole drilled, a pin pushed into the hole, and the others drilled in the same way.

At E is shown a set up for drilling the holes in the periphery of a collar. Both old and new pieces are bolted to the angle iron and the latter position on the table controlled by two parallels fastened to the table, at right angles. Lateral movement of the angle to obtain the correct position for the new holes is obtained by means of a small cylindrical piece turned to the diameter equalling the center distance of the two collars. This method of drilling holes to conform to others is practically that of jig drilling. One big advantage with this method is that any inaccuracy in the initial piece is duplicated in the new piece, thus assuring a better result than if uniformity had been adhered to. There are very few cases where holes cannot be transferred direct by adopting a few simple principles as here outlined.

The gauge shown at F is one that may be used to good advantage in the locating of a center between two parallel slots or the sides of work. In operation the end pieces are adjusted to fit the slots when the bushing will be automatically centered; a center punch turned to fit the bushing is used to make the center mark.

GENEVA MOVEMENT PROBLEM
By G. Barrett

Relative to the Geneva movement problem appearing on page 398 of the October 28th issue, I submit the following as my method of manufacturing this piece. The material used would be cold rolled bar stock of the required diameter. The first operation would be performed in a small turret lathe, using two box tools, one for roughing and one for finishing the boss on the movement 19/64 inch in diameter, drilling the 7/32 inch hole in the center, and parting off to the desired thickness. The box tool, which is of the hollow mill design, is shown at A, Fig. 1. These tools would give good results provided ample chip clearance was allowed for. The output from this operation should be approxi-

mately 30 per hour, or about 2 cents apiece.

The next operation would be the milling of the slots. The jig or fixture for this work would be similar to that shown in Fig. 2, which is of simple design, made from machine steel and hardened. The upright portion is bored in the center to fit the boss of the movement which is held to the face by means of the clamp B. As the slot being milled is 7/64 inch wide, I would use a 3/32 inch saw cutter; this would leave a few thousandths for filing. This particular operation would constitute a milling of one of the slots in all of the pieces, and would cost about 40 cents per hundred.

The fixture used on the succeeding operation would be identical with that shown in Fig. 2, but would have a small locating block as indicated at C. This method would eliminate the necessity of having an indexing attachment on the jig, and would be far more accurate, as there would be no working parts to wear and produce errors. Completing this milling of the slots would approximate a cost of 1¼ cents per piece.

We now come to the stage where the 29/64 inch radius would be formed. This would be accomplished in a punch press using a punch and die similar to that shown in Fig. 3. The die is slotted at D to receive the boss of the movement, and also to receive the locating pins E and F. These pins would be placed to bring the movement in the proper position for punching. The punch is cut away in front so that the back portion will act as a guide before the punch part G comes in contact with the work. The cutting face of the punch is ground at an angle of about 8 degrees to provide for proper shear. The cost of this operation would be approximately 40 cents per hundred.

The final operation would be finish filing to gauge. The slots would be filed to fit the steel projection H, Fig. 4.

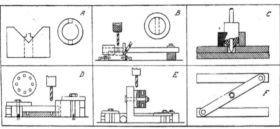

SIX HANDY HINTS. READ THE TEXT WHICH EXPLAINS THE STEPS CLEARLY.

The final gauging would be done on the gauge Fig. 4, which is made of a flat piece of steel with a hardened and ground pin to fit the bore of the movement, and another piece I hardened and ground and located in the block at the required center distance from the center pin. This final fitting would also serve to check up all other operations. This concluding operation would be done at a cost of about 8 cents apiece.

DOUBLE PARTING TOOL
By Clyde D. Thorburn

Some time ago we illustrated a double parting tool for use in cutting piston rings from sleeve casting, and the illustration herewith is that of another double tool for the same purpose. This tool was designed so as to have a longer life and would always be standard width between cutters. It is ground only on top of cutting edge, its design allowing the necessary clearances required. Body of tool is made from mild steel forging shaped on all flat surfaces and two grooves sawn in head to receive the high speed circular cutters. The cutters are turned slightly concave to insure continuous clearance and bevel on cutting face to make a clean cut-off.

The top tool is turned ¼ inch larger in diameter than bottom tool. This is to allow for sufficient strength of metal to hold on the lower ring when parting off upper ring. Six ⅜-inch holes are drilled through the cutters and hole to correspond through jaws being tapped through one side of jaw to take the retaining pin. This allows for the moving of cutters as they are ground back through wear. The tool can be used on the side head of a Bullard vertical or ordinary vertical boring mill. We are using it there and find it a distinct advantage over previous tools.

WELDING OF LOCOMOTIVE FIRE BOXES
Continued from Page 629

Welding Half Door Sheets Riveted to Side Sheets

In preparing to weld half door sheets above or below the doorhole, the sheet should be cut not nearer the doorhole than between the first and second rows of staybolts as shown in Figures 72 and 72A. All the staybolts, except those adjacent to the weld, may be screwed in place. The door sheet should be riveted to the side sheet and mudring before welding, and the weld vee should be formed as shown in Figure 73.

Next weld the flange at 1, Figures 72 and 74, then at a distance of about 10 inches from 1, back weld from 2 to 1, from 3 to 2 and continue to 8. Figure 75 gives details of the end of the weld.

Patches on Cylindrical Shells

Sketches A and B, Figure 80 show how patches should be applied to cylindrical shells and tanks. The base of the triangular patch should run with the circumference of the shell. A weld like that of sketch A should be started at 1 and go to 3, and be allowed to cool. Then from 3 to 5 and cool. Finally from 5 to 1. The weld of sketch B should be run from 1 to 3, and allowed to cool. Then from 3 to 5 and cool, and from 5 to 1.

Locomotive Flues

To weld locomotive flues, first countersink the fluehole about 3-16 in. deep, as shown in Figure 81. Extend the flue to within 1-8 in. of the outer surface of the sheet and 1-16 in. beyond the bottom of the countersinking. Leave out the copper ferrules, and expand the flues just enough to fill the flueholes. Start to weld at the bottom of the hole, and weld around to the top on one side, and then the same way on the other half. Weld the fillet flush with the inside of the flue and the face of the flue sheet. Figure 83 shows the order in which the flues should be welded so that unequal expansion may be allowed for. Weld the row of flues 1, 2, 3 and then 4, 5, 6, then 7, 8, 9 and so on to completion.

Johnny—"Mother, why did you marry my dad?"

Mother—"Johnny, I married your father because he once saved me from drowning."

Johnny—"I'll bet that's why he won't teach me how to swim."

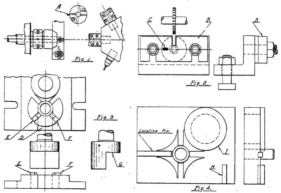

THIS IS HOW MR. BARRETT WOULD HANDLE THE JOB.

DEVELOPMENTS IN SHOP EQUIPMENT

NO. 4 PLAIN MILLING MACHINE

The Ford-Smith Machine Co., Ltd., Hamilton, Canada, have placed on the market a new machine to be known as their No. 4 plain milling machine.

This machine is of new design, and contains various features worthy of special attention. For example, special care has been given to centralized control of all movements of the knee and table, and rapid power is provided in all directions. This makes for speedy production, and the range and coarse feeds are said to be equal to those provided on geared millers.

Various Details

The column is of box section, with heavy wall and bars for supporting the knee and bearings. It is provided with door and shelves for holding necessary tools, etc. The spindle is a crucible steel forging, having a hole through its entire length. The front is threaded to receive a chuck or large cutter head, and this thread is protected by a cap when not in use. The front bearing is tapered, the back bearing parallel, the wear in the latter being adjusted by drawing up a bush which fits solidly into a tapered hole in the column.

The spindle is boxed for a B. & S. standard taper, and front of spindle is slotted, thus providing a positive solid drive. Each arbor support is provided with a wedge key, operated by a ball handle, this locking and locating the arbor support positively in line with the spindle.

The knee is designed to give maximum rigidity under the bending and twisting stress of the cutter. It is a box casting of the solid top type, being well braced internally with short ribs. There are no holes in top or sides of knee. It is locked on the column by means of a handle located at the front of the knee. A narrow guide with an adjustable taper gib is provided for the saddle slide. The cross screw is located near the centre of the guide. The knee has power elevating feed through a telescopic screw with a ball thrust. This avoids the cutting of hole in floor to clear the screw. Graduated collars reading to one thousandth are provided on table, cross, and elevating screws.

The table is of very rugged construction with wide bearing and great depth.

It is provided with oiling arrangements from the front, also felt wipers. The feed is autometic in both directions, and has a positive knock out.

The feed box js n_{01} carried on the column in the orthodox manner, but is bolted to an extension of the saddle on the knee. This new departure makes possible the arrangement of a dial feed hand wheel, handy to operator at working position in front of knee. Twelve changes of feed are obtained from the dial, which, in conjunction with the coarse feed handle at the front of column, gives twenty-four changes when the back gear is in.

All feed gears are hardened steel, ground in the bore. The feed shafts are also hardened and ground. A cone miller is often handicapped on large face milling work on account of the feed dropping in proportion to spindle speed, this, of course, causing lower production on work which must run at slow spindle speed. To overcome this trouble a feed

change is provided next to the spindle. For ordinary work the feed is driven off the feed pinion keyed to the spindle, but when the back gears are engaged, the feed may be driven off the cone pinion which speeds up the whole feed mechanism 3 to 1, or 10.8 to 1, according to the ratio of back gear engaged, thus making possible the very coarse feeds provided. The range of feed is from .006" per rev. of spindle, to 1.56" per rev.

The feed reversing mechanism is carried in the feed box on the knee, giving a reverse to all feeds. The rapid power control levers have to be held in position by hand while the rapid power is engaged, making accidental or thoughtless operation impossible. Should the feed run over at any time, a safety trip device prevents any breakage or damage. This machine is said to be the only cone driven machine having rapid power to knee as an integral part of the design. The countershaft is of two-speed friction

GENERAL VIEW OF NO. 4 PLAIN MILLER.

type with self-oiling bearings. Friction clutches are of improved expanding type, carefully balanced, with means for taking up wear. Machine is, of course, supplied with all necessary wrenches, oil pot, cutter arbor, etc. Following are the principal specifications:

Table working surface, 71″ x 14½″; T. slots, three ¾″; Longitudinal traverse, 42″; Cross traverse; 12″; Vertical traverse, 20″; Face of column to brace, 30″; Dia. of spindle in front bearing, 2½″; Size taper hole (B & S), No. 11; Size hole through, 13/16″; Spindle speeds No., 18; Spindle speeds range, 14 to 415 R.P.M.; Overarm dia., 4¾″. Distance centre of arbor to underside overarm, 7¾″; Large cone dia., 13″; Cone belt width, 3½″; Ratio back gear, 3 to 1, 10, 8, to 1; No. of feeds, 24 with back gear in, 12 without back gear in. Range of feeds per rev. of spindle, .002 to .39″ cross and elevating, .006 to 1.56″ table screw; Rapid power traverse, 12″ per min. elevating, 32″ per min. cross, 130″ per min. table traverse; vise plain, 8½″ wide, 2¾″ deep, 7″ opening; Countershaft pulleys, 16″ dia., 3″ face; Countershaft speeds, 180 and 220 R.P.M.

BALL BEARING DRILLS

Output, convenience, durability. These are the features much to be desired in any machine tool. The Alfred Herbert, Limited, Coventry, England, claim all three for their varied line of ball bearing drill presses. The first of the photographs (see Fig. 1) depicts a form spindle machine with three plain and one geared spindle. All spindles have auto-

FIG. 2—TWO-SPINDLE PLAIN LEVER FEED MACHINE.

matic, self engaging, self releasing feed mechanism: Fig. 2 shows a two spindle plain lever feed machine.

Every bearing in these machines is a ball, bearing, and we illustrate a sectional view of the spindle showing the spindle pulley bearing see Fig. 3. The spindle bearing construction, is uni-

FIG. 1—FOUR-SPINDLE MACHINE WITH THREE PLAIN AND ONE GEARED SPINDLE.

que in design and is patented. The only running bearing on the spindle is the one near the bottom of the spindle which takes all journal load, and upward and downward thrust. The reason of the success of this bearing under combined journal load and thrust is that on a vertical drilling spindle the end thrust is always greatly in excess of the side pressure, thus insuring that the load is continuously and equally shared by all the balls, and no ball ever has to sustain more than a very small fraction of the whole load.

The spindle passes through the feed sleeve without touching it, and is provided with a dust cap. The illustration of the spindle pulley clearly shows how the upper end of the spindle passes through but does not touch a fixed bush, the outside of which carries the two annular ball bearings of the spindle pulley. The spindle is splined and is driven by a key and centralized by the part of the pulley which projects above the fixed bush. This construction has proved of high mechanical efficiency, avoids all necessity for "floating" drives or other devices for preventing the spindle from binding in its bearings, and is simple and free from the possibility of faulty adjustment. The spindle is balanced by a flat helical spring acting on the pinion shaft, which eliminates the more common chain and balance weight and works with much less friction.

The self engaging, self releasing, automatic feed feature is worthy of note as it is said to be a great aid to production. The operator merely pulls down the feed lever until the drill touches the works, upon which the machine automatically feeds the drill, stops feeding at the required point, and then instantly

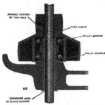

FIG. 3—DETAIL OF SPINDLE PULLEY BEARING.

withdraws the drill. Three rates of feed are provided with the automatic feature. Briefly, here are some other points of interest. These machines have a belt system by which six speeds to each spindle are obtained with the belt properly tracked for each speed, the convenience of the belt tensioning arrangement, the exceptional table area, the convenience of the two T slots in the tables, the con-

venience of the elevating and lowering motion to the table by rack and pinion. The general rigidity of the machine is worth notice, also the convenience of the pump and fitting arrangements, which have no easily choked flexible pipe connections.

These drilling machines are supplied in one to four spindles with any combination of spindles, i.e., with plain lever feed spindles, plain spindles with automatic feed, or geared spindles with automatic feeds. The maximum capacity in steel of the plain spindle is one inch, and of the geared spindle one and one-quarter inch. The range of six speeds on the plain spindle is 230, 530, 600, 900, 1,370, and 2,050 r.p.m., and on the geared spindle 77, 147, 170, 194, 321, and 424 r.p.m. This concern have their Canadian office and showroom at 1-3 Jarvis Street, Toronto.

LARGE TRIMMING PRESS

The accompanying photograph illustrates what is said to be the largest single crank trimming press, with side shear, that has ever been built. The press has a number of special features which are worthy of note. The frame is of the four-piece steel tie rod type of construction, which consists of a base, two uprights and a crown, held together by four large steel tie rods which are shrunk into place. These four rods take the stress. The uprights are extra heavy to withstand any lateral stresses which might occur when the slide is unevenly loaded. Presses with this type of construction are capable of economically operating dies for the heaviest kind of work that can be done in presses. The twin gears on the main shaft are an original feature in a trimming press of this character, and it will be noted that the design is such as to make an entirely self-contained machine without any outboard bearings and with the main gears close up to the uprights of the press. The gearing, moreover, is mounted high on the frame and in no way interferes with the operator at the back of the press.

The outside shearing or cutting off slide is particularly massive and substantial. It is operated by an eccentric shaft, which extends through the bearing in an auxiliary housing, secured to the outside slide housing by steel tie rods. The press is controlled by a powerful hand actuated friction clutch which enables the operator to start or stop the machine at any point of the stroke. The slide is raised and lowered to the proper adjustment by a special power driven elevating device. This saves considerable time when changing dies of varying heights. All the main bearings are lubricated from the floor by a centralized forced feed lubricating system. The press is driven by a direct connected 50 H. P. motor mounted on a bracket at the rear of the machine.

This press is specially adapted for trimming crankshafts of twelve cylinder aeroplane engines, and it will readily trim other work of larger dimensions which is forged in 15,000-lb. hammers or even larger.

An idea of the size of the press may be had from the following dimensions: weight about 185,000 lbs.; area of slide flange F. to B. x R. to L. 68 x 40"; area of bed F. to B. x R. to L. 72 x 48"; standard stroke of slide 16"; distance bed to slide stroke down and adjustment up 30"; stroke of outer slide 6"; ratio of gearing 44:1; number of strokes per minute 7"; extreme height over floor 229'; floor space over all F. to B. x R. to L. 137 x 166".

The above press has been designed

and built by the Toledo Machine and Tool Company, Toledo, Ohio.

HEAVY DUTY BORING MACHINE

As far back as 1916 the Rockford Drilling Machine Co., Rockford, Ill., became convinced there was a demand for a single purpose heavy duty horizontal drilling and boring machine, which could be highly standardized, and still be flexible enough to meet practically any requirement. The result of this belief is shown in the two photographs herewith. The machine illustrated is one of

THIS VIEW GIVES A GOOD IDEA OF THE SIZE OF THE PRESS.

three built for a large motor works, for boring crank shaft and cam shaft bearings. This machine is provided with two duplicate fixtures mounted on an index-

ing table. The loading station is at the rear of the machine, and the operating station at the front.

As a nucleus, the head of the Rockford vertical heavy duty drilling machine, complete with drive gears, spindle and feed mechanism, was selected. This head can be shortened to give as little as 12-inch spindle travel or increased to give 40-in of travel. In combination with the different spindle designs and multi-spindle drive heads, it gives the means of rotating the bars and tools.

The main drive may be by either motor or countershaft. In the countershaft drive, a three step cone is provided and a Hyatt roller bearing tight and loose pulley countershaft. The motor is directly connected through gears and without the use of flexible couplings or chains. Interchangeable pick off gears are provided for speed changes with both types of drive. Feed changes are four in number and are incorporated in a simple feed box with drive-key. The feeds available are varied to suit the work but are

spindle and the finished job around to the loading station. This method of handling makes the operation almost continuous.

This machine bores two crank shaft bearings 2 9-16 ins. diameter by 2 3-4 ins. and 2 15-16 ins. long and three cam shaft bearings 2 3-16 ins., 1 57-64 ins. and 1 55-64 ins. by 3 1-4 ins., 2 9-16 ins. and 3-4 in. long respectively, simultaneously in aluminum at a cutting speed of two hundred feet per minute and .010 in. feed per revolution of the spindle.

The operations are divided into rough boring, facing and finish boring, a machine being provided for each operation. All these operations with the addition of counter boring, drilling, etc., may be combined on one machine if the production does not warrant more equipment.

Care is taken in designing all fixtures to provide adequate clamping devices of such form that the case will be held firmly but not sprung out of shape. In this particular case, the lower surface of the crank case is clamped against hard-

iron and aluminum. They are rated at 2 1-2 in. diameter drilling from the solid in steel and have demonstrated their ability to withstand continuous production over a long period of time.

Following is a partial list of work for which these machines have been adapted: crankcase boring for tractors, trucks, and passenger cars; transmission case boring for the same; cylinder boring for tractors; front axle (king pin hole) drilling and reaming; rear axle drilling and boring; gun carriage work (77 m.m. and 150 m.m.); centering of large forgings; milling pads on crankcases; facing and counterboring bearings in crank cases and for double and drilling.

DUPLEX TURNING AND BORING MILL

An improved type of duplex turning and boring mill is shown in the accompanying illustration. This machine is an example of the new range of turning and

GENERAL VIEW OF THE MACHINE, SHOWING FRONT AND REAR VIEWS.

always in geometric ratio with a factor of 1.5. A forward and reverse feed mechanism can also be provided. All bearings on both drive and feed are bronze bushed. The base is of heavy box section, well ribbed and is bolted together in such manner as to facilitate the changes in design without many pattern changes. The feed and drive mechanism are completely mounted on one base section and the fixtures are mounted on the other section.

While one fixture is being unloaded and loaded, and the bars and cutters changed at the loading station, the other fixture is at the operating station and the piece it holds is being bored. When these operations are complete, the operator disconnects the bars from the drive head and steps on the foot treadle which projects through the front side of the base. This treadle pulls the index pin and lifts the table slightly off its bearing on the base and on to a large ball. The table is then rotated 180 degrees, bringing the new job in line with the

ened steel plates mounted on a vertical surface. This facilitates changing cutters and keeps the locating surfaces clean. Location of the case is by means of hardened dowels entering reamed holes in the lower surface of the crank case. These dowels are relieved of all cutting strain by means of spring plungers which are locked after the case is in position.

The equipment for the above includes a complete set of Kelly Reamer Co.'s bars and cutters. A floating drive is used between the bars and spindles so that the accuracy of the work depends upon the fixture alone. Liner bushings are provided in all cases so that it will be very easy to maintain alignment of the fixture. The bars are provided with wearing strips which may be packed up and re-ground to keep the bars up to the standard size.

These machines are adaptable to a wide range of work and are easily arranged to take care of holes from 1 inch in diameter in steel to 12 inches in cast

boring mills being manufactured by Webster & Bennett Ltd., Coventry, England.

It is a duplex machine with 42-inch diameter work tables, each forming a separately driven and operated unit. The other sizes of machine, which constitute the full range, are 20 and 30-inch duplex machines, and a 30-inch single machine. The machines in each case are built on the unit principle. The machine illustrated embodies some new and distinctive features making for high productivity with a minimum expenditure of labor. All controls are centralized, rapid power traverse is applied to all slide movements, and special provision is made whereby the turret heads may be quickly and easily rotated irrespective of lack of balance, due to heavy overhanging tools.

The distribution of metal to give maximum strength has received special attention, and the main castings, such as the base, columns, cross slides, saddles and turret slides, are heavily ribbed to enable them to withstand the heaviest duties to

which, by recent advances in cutting tools, the machines will be submitted.

Evidence of the consideration in this direction is seen in the method employed in mounting the turret-head cross slide, which incorporates the narrow guide principle. It will be noted that the cross slide is supported by square and V-shaped gibs at the bottom of the slide. This design also gives a direct thrust to the cross feed screw, which is located in the lower portion of the slide way.

The machine is entirely self-contained, power being supplied through all-geared speed boxes, driven by two separate motors—one to each unit. These are mounted on platforms seen immediately behind the turret heads. Correct tension is maintained on the belt by arranging the platforms on a rocker, and by means of screw-operating mechanism the platform and the motor contained thereon may be tilted over until the desired belt tension is obtained. The gear boxes are of the sliding gear type, and give a wide range of table speeds, rising in geometrical progression. The speeds may be quickly varied by simple lever movements, and in order to prevent gear breakages through careless operating the friction driving clutch becomes disengaged before the sliding members can be moved. The left-hand table in the duplex machines is equipped with a reversing motion, which enables the operator, when turning and surfacing, to work with the tools adjacent to the tool control levers. For drilling and boring or reaming, standard tools are used and the table runs in the reverse direction, whereby a heavy outlay in left-hand tools is avoided. The feed changes are effected by hand levers through the different combination of sliding gears contained in boxes at the foot of the columns.

On the change levers are numbered plates which indicate at any time the actual feed engaged. All the driving gears are made of steel or phosphor bronze, and those in the speed box are of chrome nickel steel, heat-treated for maximum strength. A special form of tooth is used throughout in these gears which, together with the high tensile steel of which they are made, makes practicable the use of fine pitches and small diameters, thus strength is maintained and high peripheral speeds are eliminated. Lubrication to the gear boxes is effected by means of filtered oil through siphon tubes, and oil emerging at the end of the bearings is automatically returned into the box.

Control Features

The levers seen at the side of each table control all speed changes. The feed changes are made by means of levers on the top of the feed box, the direction of their application, however, i.e. to vertical or longitudinal feeding, is determined by the lever directly under handwheel at right hand side of photograph. Rapid or slow power traverse movement, in either direction, for convenience in set-

ting the tool to the work, is actuated by means of the lever directly under the previous lever spoken of. By the above control arrangement, the operator seldom requires to move more than a few inches from the working position. Positive feeds may be instantly thrown out of engagement and rapid power traverse put into gear, thus enabling the turret to be quickly brought into any desired position, with the smallest expenditure of labor.

The latter lever has a compound movement, and actuates the positive feeds or the rapid power traverse motions in either direction. To operate the longitudinal traverse, the first lever is raised,

GENERAL APPEARANCE OF DUPLEX BORING MILL EQUIPPED WITH 42 INCH DIAMETER TABLES.

then by elevating or depressing the second lever the slow or fast feeds are engaged. By moving this lever either to the right or left, the saddle is operated in a corresponding direction.

An automatic trip device is embodied in the feed motion, whereby the feed may be automatically tripped at any point.

Arrangement of the Turret Head

Another feature of this machine is the arrangement of the turret heads. These, as will be noted, can be set over to operate at an angle, and are counterbalanced by means of a powerful spring, actuating through wire cable over a fusee, thus maintaining a constant pull against the weight of the slides and tools at all times and in all positions. The arrangement is safe and compact and in

no way restricts the movements of the operator. Within certain limits the displacement of the centre of gravity of a turret head, due as frequently happens to an unbalanced arrangement of the tools, is no disadvantage to the operator and no time is lost in bringing the turret into any required position. Beyond this a rotating mechanism becomes essential, both for the safety of the operator and as a waste time eliminator. On the 42-inch machine shown, arrangements are provided to neutralize the effects of overhang or heavy boring bars, tool holders and so forth, and the turret may be very easily rotated without the slightest risk.

To make a turret movement, it is first unclamped by raising the lever shown (which unlocks the turret from the slide), then by raising the lever on the left side of the turret, the indexing plunger in the turret is withdrawn, and by one revolution of the arm directly below this latter lever (which is geared by a pinion and internal gear to the turret) the next tool station is brought into the cutting position, after which it is locked by the two levers as before.

VARIETY SAW BENCH

The Oliver Machinery Co., Grand Rapids, Michigan, have placed on the market a new variety of saw bench. This machine is made in either motor-driven or belt-driven style. The machine is designed for the production of variety

wood work, wood patterns, furniture, automobile bodies, agricultural implements, etc. It will do ripping, cross cutting, and will cut mitres. In other words it is a wide range variety saw.

The construction is rugged throughout and all necessary attachments, such as mitre cut-off gauge, etc., are provided. Two types of tables can be furnished, plain and universal. The latter has a 15-inch rolling section to the left of saw, which rolls on ball bearing ways having vertical adjustment for alignment and wear. The rolling table can be moved four inches from saw, this permitting the use of dado saws and special heads. This table also permits doing accurate cross cutting, mitreing and grooving.

The table has a tilting device which consists of a hand wheel, worm and gear device, housed in a dust-proof casing attached to yoke with connecting link to table. This makes a positive and self-locking arrangement as a clamp screw is provided for positive clamping at any tilted position. The table can be tilted to 45 degrees to the left.

It has also a vertical adjustment, and if desired a plain table can be furnished. The universal table, however, is shown in the illustration.

It will be noted from Fig. 2 that the motor drive is not an afterthought, but has been built right into the machine; 5 horsepower is required when belt-driven, 3 to 4 horsepower when direct-motor driven. The motor on arbor machine is for 2 or 3 phase, 60 cycle alternating current. For other currents a standard motor running at about 1,800 r.p.m. is mounted on a sub base bolted directly to the main driving belt, thus eliminating the necessity of a countershaft. Motor pulley and belt are thoroughly guarded by suitable guard.

The saw arbor operates on ball bearings, and bearings are encased so completely that no dirt can come in contact with them. The arbor bearings are mounted in a one-piece rigidly constructed housing which is bolted and tongued to frame of machine. This housing is interchangeable as a unit with the motor head arbor when direct mounted motor is desired.

AUTOMATIC TURNING MACHINE

The machine depicted in Fig. 1 represents an entirely new development, manufactured by Thos. Ryder & Son, Ltd., Turner Bridge Works, Bolton, England. It is used for the turning of piston rings, and in fact is adapted specially to all motor manufacturing plants.

With the exception of checking the work, the machine is entirely automatic in action, and will simultaneously turn, bore, and part off bush castings into rings. Concentric or eccentric rings can be produced with equal facility. The machine can also be arranged to turn, form the ring and oil grooves in piston castings. In such cases the piston is located on a spigot bar, controlled by a hand wheel at the rear end of the main spindle. As the action of this machine is automatic, one operator can attend to a battery of machines, thus effecting a considerable saving.

When producing piston rings, the bush casting is secured to a face plate fixture. The machine is started and on engaging the feed the operation proceeds as follows. Boring tools carried in a bar mounted in the saddle, and turning tools, mounted in the front tool post, commence cutting simultaneously. Next the parting tools, which are carried in a holder mounted on a cross slide at back of bed, begin to part off the rings one by one.

GENERAL APPEARANCE OF AUTOMATIC TURNING MACHINE.

Operations continue until the turning and boring is completed, when the tools automatically stop.

The parting tools continue until the last few rings are cut off, and they also automatically stop. The machinery time for a normal 4-inch diameter ring is one minute.

The machine is of very rigid construction, and work produced from it is very uniform in character. The main frame is of a one piece box section and extends completely to the ground. Chutes are arranged to deposit the turnings away from the tools, slides, and moving parts. The latter are enclosed and protected from dust.

VIEW OF MACHINE WITH TABLE IN HORIZONTAL POSITION.

MACHINE WITH TABLE IN TILTED POSITION.

The drive is by fast and loose pulleys, through a geared head, giving a speed range between the ratios of 2 to 1. This allows for maximum cutting speeds. The main spindle runs in bearings of the firm's special cone-cum-ball bearing type, and splash lubrication is employed throughout. The slides carrying the turning, boring and parting tools are arranged with automatic feed and trip motions which are controlled by one handle only. When a complete cycle of operations is completed, one hand wheel winds the slides back to the original position, each slide keeping in

spigot face plate of machine. The arrangement is shown at Fig. 2. The spigot face plate, which is open at the sides to admit the gudgeon pin bosses as shown, is screwed on to the spindle nose, the bore of the latter being utilised to carry a support bush for the draw back bolt. It will be noticed that the outside roughing and finishing tools are followed by two roller steadies, these withstanding the thrust of the grooving and heading tools carried in the back slide. In the case of pistons with a skirt of full diameter, the steadies traverse out of the way as the sliding proceeds, allowing

HIGH SPEED MILLING ATTACHMENT

The accompanying illustrations show a new design high speed milling attachment made by the Brown & Sharpe Mfg. Co., Providence, R.I., for use on milling machines of their manufacture. As will be noted the mechanism is simple in construction and easily attached, no fixtures being required. The bracket and spindle support is a one-piece casting of substantial construction, so designed to protect the mechanism from dirt and injury.

The attachment is built in two sizes, the No. 1 for the smaller machines, and the No. 2 for the larger machines. The No. 2 size is adaptable by means of adjustable gib stops to the larger machines, having columns with different width of face, thus eliminating the necessity of a separate attachment for each different size of column on machine.

The positive means of locating the attachment is worth noting. To assure a positive vertical position, the attachment is provided with a locating segment which rests upon the spindle box of the machine. This spindle box projects beyond the face of the column and acts as a centering guide. The horizontal position is determined by first tightening the gib on the right-hand side of the attachment, thereby locating the attachment. The gib on the left-hand side securely clamps the attachment to the face of the column. This affords a large bearing surface, insuring rigidity and making the attachment practically an integral part of the machine.

Particular attention has been given to the drive that it may be positive, though simple, the only gearing being the large ring gear that fits on the taper nose spindle of the machine and the pinion on the attachment spindle. The large gear is made with an internal taper, ground to fit on the taper nose of the machine spindle and is held in position by the regular cutter driver. This gear is made of machinery steel and left soft to insure a smooth drive and eliminate chatter, also the objectionable "ring"

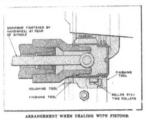

ARRANGEMENT WHEN DEALING WITH PISTONS.

register with the other automatically. The boring bar, as already mentioned, is carried on the saddle, where it is supported in a strong steady rest, and a pilot bush in the spindle nose engages the opposite end, thus ensuring great rigidity. Three boring tools are contained in the bar, which are of large section steel, sufficient to carry away heat. The foremost of these boring tools are circularly ground behind the cutting edge, and so act as effective following stays to the work.

When dealing with pistons, these are previously faced up and recessed at the bottom end, to act as a register on the

the finish-sizing tools to follow for sizing the top bands.

The maximum capacity of the machine is to turn pot castings up to 10 inches in length, and outside diameter of 8 3-16 ins. The maximum size of inside bore is 2 1-8 ins. Production times taken from actual practice have shown that pistons 4 1-4 inches diameter, 4 3-4 inches long, have been produced in 8 1-2 minutes each. It is said that one boy can feed three machines, owing to this machine's automatic action.

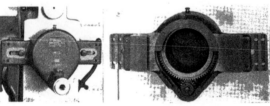

GENERAL APPEARANCE AND DETAIL VIEW OF HIGH-SPEED MILLING ATTACHMENT.

often produced by hardened gears at high speed. On the No. 2 attachment the pinion is heat treated and keyed to the attachment spindle, while on the No. 1 attachment the teeth of the pinion are cut directly on the spindle.

The spindle is hardened and ground and runs in a long phosphor bronze bearing, the bearing being tapered so that wear is taken up by tightening an adjustment nut, thereby forcing the spindle into the taper. Once having adjusted the spindle a small set screw securely clamps the adjusting nut in position. End thrust is taken by hardened steel and babbit washers located directly behind the shoulder on the front end of the spindle. The front end of the spindle has a taper hole to receive cutters, the smaller attachment having a No. 7 taper hole and the larger one a No. 9 taper hole. Oiling of the spindle bearing is effectively taken care of by means of an oil pocket on the front of the attachment containing wool which insures a constant and even distribution of the oil at all times.

The cutter is held in position by the taper in the spindle and is driven by the tenon. Cutters are readily removed by a releasing bolt which provides a positive means of ejection. This bolt is a part of the attachment and remains in the spindle.

CONCENTRIC CHUCKING

The importance of the chuck as a lathe accessory is too often imperfectly appreciated. The chuck is an indispensable link which connects the work to the source of power and it is obvious that the selection of an efficient chuck is of the greatest importance — otherwise one may have a costly and productive lathe stultified by its inefficiency. The desirable features of a chuck can be summarized as follows: Gripping power, strength, durability and adaptability.

The Coventry concentric chuck, illustrated, made by Alfred Herbert, Ltd., England, is said to embody all these features. Fig. 1 illustrates the chuck with one sector plate and jaw slide removed and from this, the principle of operation can be followed.

The work-gripping jaws are attached to radial jaw slides by square-headed collar screws, engaging nuts which slide in "tee" slots in the jaw slides. Both the base of the jaws and the front face of the jaw slides have fine pitch serrations which engage each other and positively prevent relative movement. The jaw slides are of "tee" section and are held in place by sector plates which fit inside the rim of the chuck body. The jaw slides are moved in and out by a scroll which is rotated through gearing. The scroll has three eccentric grooves in its face, each groove carrying a sliding block which pivots on the cylindrical portion on the jaw slide. The eccentric grooves are portions of a circle, and the sliding blocks are therefore in contact over the whole of the wearing surfaces, a result which is obtainable with spiral scrolls. Dimensioned circles on the face of the chuck facilitate the setting of the jaws.

The exceptional gripping power claimed by makers for this chuck is obtained by the high ratio of gearing used and the efficient design of scroll. The chuck is said to have exceptional strength, and on this point it is interesting to note the claim made by makers "that no Coventry chuck body has ever broken in use, and no body returned to them broken from any cause whatever."

Durability of this chuck is claimed by the makers on the grounds that all the component parts are of the highest class of material, and the refinements in design are such as to prevent rapid wear.

FIG. 1—THE CHUCK WITH ONE SECTOR PLATE AND JAW SLIDE REMOVED.

The makers' claim for adaptability is well illustrated in the accompanying illustrations, Figs. 2 and 3, showing respectively the chuck holding a gas engine cam, illustrating the convenience for gripping irregular or eccentric work, and a back axle casing, an example of the ease with which jaws can be applied for holding objects of irregular form. In this case the jaws shown are exceptionally long, two projecting 5 inches from the face of the chuck and the third 3 1-2 inches. They are shaped to suit the form of the work and arranged so that a five point grip is obtained, making it impossible for the work to rock. This concern carries a stock of soft jaws which customers can conveniently shape to their requirements.

The Coventry concentric chuck is made in six sizes, 9-inch, 12-inch, 15-inch, 18-inch, 21-inch and 25-inch. We have received from the Canadian office of Messrs. Herbert, Ltd., 1-3 Jarvis Street, Toronto, a copy of their booklet entitled "Concentric Chucking" which is something out of the ordinary in this line. Those interested in the question of chucks and chucking fixtures, can obtain a copy upon request.

FIG. 2—GRIPPING A GAS ENGINE CAM WITH PERFECT EASE

FIG. 3—NOTE THE CHUCK'S ADAPTABILITY ON THIS CLASS OF WORK.

The MacLean Publishing Company
LIMITED
(ESTABLISHED 1887)

JOHN BAYNE MACLEAN, President.   H. T. HUNTER, Vice-President
H. V. TYRRELL, General Manager.

PUBLISHERS OF

CANADIAN MACHINERY
～ MANUFACTURING NEWS ～

A weekly journal devoted to the machinery and manufacturing interests.
B. G. NEWTON, Manager.   A. R. KENNEDY, Managing Editor.
                    Associate Editors:
J. H. MOORE   T. H. FENNER   J. H. RODGERS (Montreal)
Office of Publication: 143-153 University Avenue, Toronto, Ontario.

VOL. XXIV.   TORONTO, DECEMBER 30, 1920   No. 27

PRINCIPAL CONTENTS

The Land of Achievement

WHEN one reads the phrase "The Land of Achievement" in a publication issued in New York, it would require no great stretch of imagination to surmise that United States was referred to. In this case it is not. "The Land of Achievement" is Canada, and the publication comes out under the auspices of the Bankers' Trust Company:

"As an evidence of Canada's phenomenal development the latest available figures indicate that the Dominion in relation to nine other industrial countries now stands first in area, second in potential water power, third in total railway mileage, fifth in total exports, sixth in pig iron production, total exports and foreign trade and eighth in population."

Canadians too often get the idea that other countries have a supply of super-men, while at home we have been served with a very average sort. After reviewing Canadian history since 1867, and paying a remarkably fine tribute to Canada's war record, the Bankers' Trust Company refers as follows to the men of Canada who have made this country what it is to-day:

"Needless to say, such a record of achievement as is afforded by the history of Canada during the fifty odd years which have elapsed since 'Dominion Day,' 1867, is not due to chance. Canada has been fortunate during this period in having in political life and in business a group of able, resourceful, morally strong, patriotic men who were not alone capable in action, but capable in leadership. They could not have achieved if there had not been working with them a splendid body of intelligent, determined and resourceful citizenry, men and women of moral force and courage, such as a new country alone seems to develop."

There is need right now in Canada for a decent appreciation of our own people, of our own institutions, resources and possibilities.

In our efforts to go into the markets of the world we need the confidence that comes from a knowledge that we CAN do things in this country.

We need to have developed in our national mind—and we must have a national mind—the idea that we have taken our place among the nations of the world.

Canada is suffering now because her own people simply assent to many of these things. They have never been seized by them, nor have they taken any of the inspiration in to board with them, either in their homes or in their business.

It is well that some enterprising outsider should now and then send out the fact that CANADA IS.

The way in which American settlers are coming to this country with their effects and their cash and the number of American firms already established here and coming over in 1921 is evidence that they have a lot of real dollars and cents faith in the future of this country. Were there any doubt in the matter, these shrewd people would not be staking millions of dollars in property and building here.

Get it, friend, get it. You are living in a real country, and if you are not plowing your field here, someone from the outside is going to see the chance and come in and ultimately reap the harvest.

No Sense of Responsibility

ONE of the lamentable traits that keeps coming to the surface in our industrial life is a lack of sense of individual responsibility for acts done in a collective way.

To illustrate:—A few days ago some forty boys, from 15 to 18 years of age, went on strike at the plant of the Canada Car and Foundry Co., at Fort William, because they were refused an increase from $5 and $5.50 per day to a schedule that would bring them $7 per day.

They promptly walked out, and as a result four hundred men in the plant were forced to quit work. These four hundred had no grievance with the firm, yet they were made idle because forty youngsters thought they could run the works.

It is not likely that any one of those forty boys would care to face those four hundred men and assume responsibility for forcing them out of work, and yet that is exactly what each one did.

The theory, too widely accepted, that there is no individual responsibility for acts done collectively, is a vicious thing, and yet it is surprising the number of people in the industrial world who cling tenaciously to it.

With stores shouting half-price sales and New York funds dangling around 18 per cent., a person has to have a pretty fair education to reckon what his week's pay is really worth.

Doctors in Paris believe that the skin may be so developed that it may perform the function of eyes. Thus will preference be given to the man who sprouts an eye in the back of his neck when a foreman is wanted for a gang of day workers.

The death of Horace E. Dodge following, within a few months the passing of his brother John F., removes from the mechanical world two brothers who started in a small way, faced the vicissitudes of business and conquered. The father of these boys was the owner of a small machine shop, in which the lads learned their trade and undoubtedly learned it well. They made typesetting machines, then bicycles. Following this they turned to automobile parts and then to the finished article, until to-day their plant covers over 100 acres of floor space.

TWO HOMELY LITTLE SERMONS FOR NEW YEAR

Of Course We Don't Refer to You

WERE someone to come around and suggest you were a knocker it may be that you would look highbrow and announce that you did not understand. Well, we'll tell you. Knocker is a pretty fair term—has some strength in it, and is part of the direct action vocabulary that is coming in every-day use when people want to say something when they speak.

Knocker means complainer, whiner, joy-killer, pessimist—a combination of all these things.

In high-up circles they call him a critic, or in a more high-up a critique. In other ultra-pink atmospheres he's labelled a connoisseur. But in honest-to-goodness language he's a knocker just the same.

If he agitates, he never seems to head his movement in the right direction. He uses up his energy to bust up things that his poor little fists could never construct again.

Then there's the different types. There's the chap who can't be chalked up as a noisy knocker. He doesn't exactly set to and heave the mallet in any particular direction.

But he does it in another way. If there's anything in the shop looking toward bettering conditions or securing better results, this bird does not get out on the steps and tell the sun what he thinks of it, neither does he call the boss on the 'phone at his house to tell him the dice are loaded and he won't play.

He's just a lukewarm oyster and about as palatable as a mouthful of soap.

He whispers to his neighbor in the shop, and his whisper puts the spike in any enthusiasm there might be in his corner of the works.

Perhaps he does not intend to be a brakesman on the train of progress, but he's certainly not tossing any fuel into the engine.

This chap is apt to be a poor prune at home. He's a mighty poor source of inspiration for the little woman who fell for his courtship guff, and as a father he's a blank for the children to copy. It would be the part of wisdom for Kith and Kin to forget him until he sprouts a little human enthusiasm and lets the shrub have a chance to grow a bit.

Say, chap, if you want a real tip and one that won't cost you the spending of one red cent—here 'tis—get enthusiastic.

Get in behind something worth while in the plant, in your community or at your own home. Put that little old hammer out on the porch where someone is sure to steal it and bear it away. There always will be some knockers, but for goodness sake don't get mixed up in the scramble for the available supply of hammers.

Just a few months ago the man around selling steel was grabbed as soon as he came on the premises and rushed into the combined hospitality of the general manager, the purchasing agent and the superintendent. To-day he sits on the bench and waits his turn.

In November, United States settlers entering Canada brought with them $17,519,033 in money and settlers' effects. There might well be a tear trickling down Uncle Sam's cheek as he waved farewell to this delegation.

Better Weigh in Now and Then

THE man whose job it was to clean out the premises of a big publishing house was making his rounds the other night carrying out the scrap paper accumulation of the day.

In one department he came to a big paper container that was so full that much had tumbled off on the floor. That basket looked for all the world as though it had done a good day's work.

It looked as though it had collected a tremendous pile of paper, so much in fact that it was absolutely unable to have another piece tossed at it or near it.

It had the premises around it all mussed up as though to bear witness to the fact that it was an overworked basket.

When the janitor came along he simply looked at it and grunted, and there wasn't even the suggestion of pity in the grunt.

He didn't even say that he'd have to get another basket to stand alongside of the one that seemed so overworked—nor did he say that this basket should only work a few hours and then be let off for the rest of the day.

Not a bit of it. He simply put his big, substantial foot in the basket and jammed that seeming overcrowded load into a little heap down at the bottom.

Then he gave the basket a shot and started it down the hall toward the baling machine.

Say, when that basket was all puffed up with scrap paper and wind and space, you might have been certain it would have made the baler puff to get away with the load. In the end all that frothy load made only a very small little filler in the corner of the baler of paper that was shipped out the next morning.

Did you ever run into any of those big, loose, puffed-up, overworked baskets in the shape of men?

Of course we don't mean you, but of course you know some person else that fits into this story like a fat woman into the average street car seat.

Always puffing around, looking wise, telling the world of the amount of overwork that is heaped on them, posing for the spotter who searches for the super-man, always keen to admit the boss doesn't bulge their pay envelope in keeping with their real weight in mental gold.

You know the rest. Some day they go up against the real thing. Some crisis or lesser event comes along and puts the foot and the pressure on them. They give and they quash, and they crumple up to next to nothing. When they are put in the baler there's not much left. When the wind and the froth and the boast are punctured there's mighty little left.

Better step up on the scales now and then to find out just how much is real and how much is just filling and padding.

The removal of the luxury tax leaves automobile dealers in a peculiar position. They have paid it to the makers and cannot collect it from the customer. If the Government refunds the amount it will defeat the aim of the tax. It goes to show that all matters of taxation and of tariff should be made as stable as possible in order that the business world may have a chance to study them and adjust their affairs accordingly.

MARKET
DEVELOPMENTS

Very Quiet Period in all the Markets

Steel Mills Are Down For Holidays, and the Independents May Cut Wages Before They Open Again—Corporation Mills Still Well Employed—Toronto and Montreal Doing Little Trading

THE end of the year finds the machine tool trade quiet in all departments. New York reports more inquiries, but no sales. Toronto and Montreal have somewhat the same experience and look for improvements early in the year.

The steel market is in the same position. Values are easing off a little more, bar material, both iron and steel, being marked down by fractions this week. An interesting situation is developing in the steel rolling district of United States. Most of the independent mills are down for the holiday week and it may be they are down because they have no orders to go on with. Will they reopen at the same wage scale, or will this be the chance to make a reduction? Several of the independent companies have already announced reductions as high as twenty per cent., and they are working under these conditions now. The Corporation,

on the other hand, with its books well filled with tonnage, is not likely to make any move in this direction for the present.

Talk of there being reduction made in steel prices has little sympathy now. It may be that the independents will announce their affirmation of the Corporation schedule. All the large mills have made money in recent months and are not in a position where they have to operate regardless of profit or loss. This being the case it seems unlikely that a runaway steel market can develop.

The scrap metal market ends the year in the dumps. There is nothing in sight to show where any betterment will come from. Even at the low levels quoted, nothing is being traded, and foundries and steel mills refuse to stock up, although it would be to their advantage to do so in some lines at least, as any sort of a buying movement will be sure to have the effect of stiffening prices.

LITTLE BUSINESS IS DONE IN MONTREAL AT THE YEAR'S END

Special to CANADIAN MACHINERY.

MONTREAL, Que., Dec. 30th. — The closing days of the year emphasize the dullness that has been more or less pronounced in many lines of activity during the past several weeks. The crowning event of the period of quiet trading has been the present holiday season, bringing forth an almost total absence of interest in all branches apart from that of retail domestic business. The closing of many industrial plants and the general tendency to shorter work days on the part of those factories continuing operations has still further added to the lull that has settled over many of the country's activities.

Plate and Sheet Prices Revised

There is little to feature trading in steel commodities. Buying is almost entirely confined to essential requirements, and the general trend to inactivity or curtailed operations in various lines of industry, only emphasizes the attitude of buyers. It may be truly said that this does not represent a healthy condition, but it is undoubtedly the reaction consequent on the abnormal state of world trade during the period of the war and the two succeeding years. As

one dealer stated—"it is not that industry is in a poor state of health, but rather that it has come to the point where the war paint must be washed off and replaced by more appropriate clothing suitable to the needs of the changed conditions."

"We are passing through a period," continued the dealer, "when we usually expect to experience a falling off of inquiry and demand, but there is no denying the fact that the present is more than ordinarily quiet, and this must be laid to the imperative need that is everywhere evident for establishing a firmer foundation upon which to build the industrial trade of the future, and the coming year in particular. We are more than hopeful that the early spring will prove the time for the laying of the corner stone of the redesigned business structure."

Local steel trading is at low tide this week, due to combined circumstances; holidays and festive season has diverted interest to domestic activities, and more than seasonable slackness in many plants and factories has added to the uncertainty that nearly always influences consumers towards a non-buying policy.

Dealers here anticipate a few weeks of quiet buying, but will not commit themselves further. Base quotations on iron and steel bars, from Montreal warehouses, have been revised downwards, the price now being 4½ cents, a decline of ½ cent per pound. Prices on plates are likewise lower, ¼ inch and up being quoted at 5 cents, and 3-16 inch at 5½ cents per pound. Black and blue annealed are easier, showing a drop of ¾ cent per pound, 28 gauge black being listed at $8.50 and blue annealed No. 10 are quoted at $7 per hundred. Quotations on Premier No. 28 are now 10 cents and on 10¾ oz. the price is 10½ cents, a decline of 1½ and one cent respectively.

Machinery Need Is Evident

The past week has been more than usually quiet in machine tool circles. The curtailed activity of many manufacturing plants has acted as a barometer to the purchasing departments, and in many instances the buying of equipment has been deferred to a more favorable time, or at least until the resumption of activity necessitates the installation of the additional tools, or the replacement of others. Inquiries coming in to the dealers would indicate that consumers and users on many lines of machinery would be in the market if they could assure themselves of an early return of industrial enterprise. The buy-

ing of supplies is lessening and future demand is comparatively quiet.

Dullness to Continue

The passing of the old year finds the old material market in the same stagnant condition that has featured its existence for several weeks back. That the early future holds little of interest is the opinion of many dealers, but some are looking forward to more active business as the winter is advanced. Prices which have obtained for the last month are still quoted but should be considered as nominal only, nearly every sale being made on its individual merits. If anything, the prices quoted are above those paid by the dealer.

WINDING UP THE YEAR'S AFFAIRS

Nearly All the Firms Have Had a Good Year, Though Hard to Get Required Material

TORONTO. — Many of the machine tool houses are busy just now getting their affairs of the year in readiness for the preparation of annual statements. They are finding out now whether they have bought wisely and turned over the stock quickly enough. One Toronto manager, speaking of this matter said this morning: "It is much harder running a business now than it was a few .months back, when everything was on a rising basis. Then we could go ahead and buy, knowing that before we would get rid of the lot the price would have gone up again. The makers in many cases were so busy that it was impossible to buy too heavily, as their capacity prevented that. Just now it is quite different, and we are working in a declining market, and supplies must be bought with that in mind. When we send out a requisition now we inquire about delivery on the lines mentioned before placing the order. If we find that the deliveries are all right and can be made from stock, as a general thing we cut the requisition. in two, knowing that we can order from stock at any time and get supplied. If we find deliveries are delayed we allow the requisition to stand. I don't suppose it means that we are doing any more or any less business, but it protects us in a market that is on the way down, and we feel safer for the present going along that way."

Have Had a Good Year

Machine tool men are not complaining about the business they have done during the year, neither are they willing to admit that it has been as profitable as the insistent nature of the demand might lead one to believe. Deliveries were very backward in the schedules of a good many makers, and in this way the dealers were not able to get

POINTS IN WEEK'S MARKETING NOTES

Steel and iron bars are quoted in Montreal warehouses at 4½ cents a pound, being a reduction of ½ cent from the last price.

—

Machine. tool men as a general thing have had a good year's business, the one drawback being. their inability to secure delivery when the selling was good. '

—

The steel warehouses are rather quiet this week and buying has narrowed down to immediate wants. The warehouses themselves are not taking in very much material.

—

The end of the year finds the scrap metal market in much the same condition that has prevailed for the last three months. Many of the yards are doing practically no business at all and more money is being lost than made.

—

The majority of the independent steel mills are idle for the holiday week and there is some speculation as to when they will open again.

—

A number of steel mills in United States outside of the Corporation have announced reductions in wages, running as high as 20 per cent. A wage reduction by the Corporation is not looked for at the present time.

—

There is less talk heard just now of any reduction being made in the price of steel and it seems likely that the schedule of the Corporation will be approved by the independents and there the matter will rest.

—

The selling of small tools is done only in small lots and users are not taking on anything in the way of surplus stock.

—

Representatives of Canadian firms returning from England report that German goods, especially all lines of cutlery, are being sold in London stores and are giving more competition to British goods than was thought possible a few months ago.

—

Very little pig iron is being sold in the United States or Canada at the present, and the market appears very weak. A requisition for a good sized tonnage would probably set a new low level.

—

through as much business as otherwise would have been the case. Neither is there any complaint made right now, although sales have been few

during the week. There is a certain feeling that business is going to be better, and that once the corner is turned the recovery will be quite brisk and satisfactory. Dealers are doing very little buying for stock, being content to place orders when they have a sale, knowing that in nearly every case they can se. cure tools out of stock now, something that has been quite out of the question for a good many months—yes, years in many lines.

The Steel Market

The year just closed has been a good one and yet a disappointment for many firms in the steel business. Good be. cause there has been plenty of business —disappointing because of their inability to secure material in sufficient quan. tities to take care of the business that was offering. It is quite correct that many yards have done a larger volume of business and made less money than many people imagine. Merchants, rather than let their stocks run out, were forced to buy where they could, and paid premiums that had the effect of shov. ing up the price. In this way they could not recover the same margin as when they were selling at normal figures, and more money has been required to finance the same volume of business.

Hard to Secure Material

There have been several' months in this year's business where firms in Canada which had to have steel in order to run their business have been put to no end of trouble to keep things moving in their direction and no small amount of enterprise has .been displayed in a. number of cases. For instance, the makers of agricultural machinery have kept men in the rolling mill districts, and at junction points and at any other spot where they felt it would be necessary to give attention to the movements of their shipments. This was especially so during the strike on · the railroads. One firm which depended on a supply of cold rolled kept their own crews in the mills, where they were being looked after, and their own trucks, taking the material as soon as it left the mills and trucking it some twenty miles to the nearest point where railway connection could be secured. It cost a lot of money to see that shipments were turned out and headed toward the Canadian border, but if the Canadian consumers did not take these precautions they were faced with the poor alternative of shutting down their plants until conditions should right themselves. All this activity cost money and plenty of it, and increased costs of manufacturing in Canada in a way that is not generally appreciated by those who did not have to face the conditions described above. '

Little Buying Now

The close of 1920 finds buyers in general standing pat in the matter of placing business. Many of them are full of

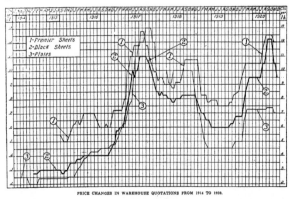

PRICE CHANGES IN WAREHOUSE QUOTATIONS FROM 1914 TO 1920.

ideas as to why others should rush into the market and buy heavily, but they cannot see for a minute why they should follow the same course themselves. Warehouses probably have sufficient to meet the demands of the present, which are fairly light.

Information from some of the large steel concerns warrants the belief that they are not looking forward to any extended period of depression. They claim that steel bought now is for actual and immediate consumption, and in this way none is being piled at the yards or at the plants of the consumers. Cancellations are not as serious as was anticipated, and many applications of users for cancellations have been converted into business again with delivery for the first quarter of 1921.

MORE INQUIRIES IN NEW YORK MARKET

But These Have Not Been Changed Into Real Businesses Yet—Next Year's Prospects

Special to CANADIAN MACHINERY.

NEW YORK, Dec. 27.—The year 1920 has been one of ups and downs in the machine-tool trade. The early part of the year was marked by very satisfactory buying, which began to taper off about the middle of the year, although some companies had good business up until September. The slump which hit the automobile industry was the begin-

nine of a downward trend in orders, which by the end of the year had reached a point of stagnation. Not in many years has the machine-tool industry of the United States known conditions so quiet as those through which it has passed in the last two months.

All eyes are naturally turned toward the new year, which is expected to bring forth some improvement. The tangible evidences of this prospective improvement are to be seen in a fairly good number of inquiries received in the past month or six weeks, most of which have

INDEPENDENTS MAY CUT WAGES BEFORE THEY START WORK AGAIN

Special to CANADIAN MACHINERY.

PITTSBURGH, December 30. — The great majority of the independent steel mills are practically idle this holiday week. Some had previously closed, but the bulk of the closing was done Christmas Eve. At some of the plants there were intimations that there might be resumption Monday, January 3, but it is very doubtful whether there will be much resumption at that time. A mill would not resume unless it had accumulated a fair volume of specifications during the period of idleness, as all the mills that closed had first completed all business on books. Then there is the matter of wages, as most of the independents are disposed to reduce the rate. In the past two weeks, prior to this week, operations by the independents, consid-

not been acted upon. In postponing purchases those companies which have made inquiry have announced that buying will not be done until after January 1. Whether this means immediately after the turn of the year will be determined only by events. There is a hopeful feeling however, that there will be a gradual resumption of business beginning with January 1 or at the latest by February 1. Christmas week was not productive of much new business or inquiries, nor is anything important expected to develop until after New Year's.

ering that some were closed entirely, probably averaged between 30 and 35 per cent. This week, with few plants running at all, and then only at low rates, the average of the whole is probably under 10 per cent.

The United States Steel Corporation has closed a plant or department here and there for the holiday week, not on account of lack of orders, but for the purpose of making necessary repairs or to allow one department to catch up with another. The major portions of the Pennsylvania tube works in Pittsburgh and National tube works at McKeesport are down for the week, one to make repairs and the other to allow finishing departments to work up stock, accumulated by the primary operations getting

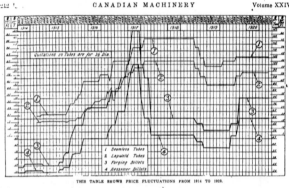

THIS TABLE SHOWS PRICE FLUCTUATIONS FROM 1914 TO 1920.

ahead. Apart from temporary closings this week the Steel Corporation is operating at over 90 per cent., or at a better rate than its average during the first ten months of the year.

Wages

It seems to be an established fact that the Steel Corporation will not reduce wages at this time, or until there has been a material change in conditions, particularly the corporation's own position, from that now existing. The independents are strongly disposed to reduce wages, but the Steel Corporation leadership, quite conspicuous in the various wage advances of the past five years, is absent. Furthermore, the independents do not confer together on any subject, either prices or wages, that having been considered undesirable for years on account of the state of public sentiment. Thus there is no opportunity for the rise of a general policy and each independent has to "go it alone." A few of the mills have taken action already. The Midvale Steel & Ordnance Company has made reductions at its Johnstown and Coatesville works, in Pennsylvania, the Wheeling Steel Corporation, operating chiefly in the Wheeling, W. Va., district, and a plant at Buffalo have lately announced reductions, to be effective this week, averaging about 20 per cent., with elimination of overtime payments in addition. Possibly some additional announcements will be made this week. A common guess is that many independents will announce wage reductions before they resume operations, the time of resumption being uncertain. If this proves to be the case it is probable that by about the middle of January the ma-

jority of independents will have announced wage reductions.

The problem is a difficult one for the independents. There is the lack of Steel Corporation leadership, already mentioned, on which the independents have so often depended, in various matters, in the past. Then there is the difficulty that while in some respects January 1 seems a good date for a wage reduction, the cost of living has been decreasing more or less steadily and further declines are expected in the future, while furthermore men would be more ready to accept a reduction after they had had a period of idleness. In particular, the cost of living depends largely on retail store prices and it is the feeling of all steel manufacturers that the retailers have not reduced prices as they should. A wage reduction, particularly in a town where the employees of the plant involved furnish the chief custom for the retail stores, would stimulate the retailers to reduce their prices. Thus, from one viewpoint the employer should wait for the cost of living to come down, while from another viewpoint he should make the reduction at once in order to help the cost of living to decline. One idea has been this, to make a small reduction now and another later, but a difficulty about this, in the minds of some of the steel makers, is that upon the appearance of the second reduction the buyers of steel would probably feel that they were entitled to lower prices. According to what practically all the independent mills say now, the Steel Corporation or Industrial Board prices are going to be maintained indefinitely, and thus anything that would encourage talk of lower steel prices would be very

objectionable. There are some observers, however, who insist that no matter how steel producers talk there will eventually be lower prices for steel before there is any general buying movement and before prices firm up and start advancing again. No definite predictions are made as to when such declines will come, whether in two months or many months.

Steel Prices Held

The finished steel market, as far as quoted prices go, is holding better than many people expected. There are some reports of price cutting, but not many, held. This may possibly be due to there having been so little enquiry that adequate temptation to cut prices has been lacking, and if so the turn of the year may see a more open market, for such a period of dullness as has prevailed in the past few weeks cannot be prolonged. There is another point, however. A mill cannot operate economically unless it has a fair tonnage. To cut prices and get a light operation would hardly pay, complete idleness involving less loss. As the steel mills are all in very comfortable position financially they have no incentive to operate except to make real profits.

The independent market on standard steel pipe continues at $7 to $10 above the Steel Corporation or Industrial Board list. There is a rumor that some or all of the independents will issue a new card January 1, conforming to the standard card. While the pipe mills have been running very well on orders they have had some cancellations and postponements and cannot run on orders

Continued on page 216

THIS WAY, PLEASE

Where quantity and exactness have to be considered, the Geometric Threading Machine is just the way that suits. These machines have been tested and proved over and over again on speed, accuracy and endurance.

Geometric Threading Machines are employed on a large class of small threaded parts that cannot be produced economically on the ordinary screw machine.

Made in three sizes—to cut ⅛ to ½ inch, ¼ to ¾ inch, and ¾ to 1½ inch diameter threads. The carriage is mounted on slides and on the largest size machine is moved back and forth by rack and pinion, and in the smaller sizes by hand.

Spindle speed changes readily made, adapting the machine to the diameter and material of the work. An adjustable stop assures accurate length of thread, and automatically opens the die head, permitting of drawing the work straight back.

A line from you brings full details regarding this machine. Tell us your threading requirements—let us recommend the proper Geometric Collapsing Tap or Self-opening Die.

THE GEOMETRIC TOOL COMPANY
NEW HAVEN CONNECTICUT

Canadian Agents:

Williams & Wilson, Ltd., Montreal. The A. R. Williams Machinery Co., Ltd., Toronto, Winnipeg, St. John, N.B., Halifax, N.S.
Canadian Fairbanks-Morse Co., Ltd., Manitoba, Saskatchewan, Alberta.

If interested tear out this page and place with letters to be answered.

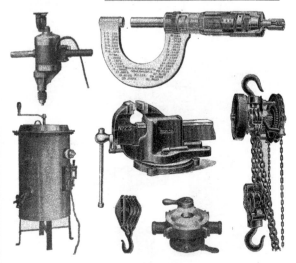

MAY CUT WAGES

Continued from page 212

for more than a very short time longer. However, it is the regular thing for pipe mills to carry stocks and ship from stock rather than current production. Hence the mills may hold their prices and make stock, as they have practically no stocks at present. The old practice may be resorted to of putting out a 30 or 60 days' guarantee, whereby if a new list should be issued, say February 1 or March 1, all shipments to jobbers' stocks for 30 or 60 days previously would receive an adjustment equal to the reduction.

A little foundry iron has changed hands at $35, valley, enough to show that the quotable market has not changed. Bessemer and basic have remained unchanged at $35 and $33 respectively, valley. The idea has become prevalent that some of the furnaces will formally announce price reductions January 1 or within a few days afterwards, possibly to about $30 for basic iron. The date has some significance, since several of the furnace interests will then get on to lower priced coke contracts and thus will have lower costs.

Some additional contract furnace coke business for the first half of the new year is reported as closed on the general basis of 5 to 1 against basic pig iron at valley furnaces. On $30 pig iron this would mean $6 for the coke. Altogether about 75,000 tons a month has been put under contract on substantially this basis.

NEW INDUSTRY AT PORT ARTHUR

In a large advertisement appearing in a recent edition of a Port Arthur paper, there is an announcement of the Port Arthur Shipbuilding Co., in which is recounted the falling off in the shipbuilding industry. The announcement goes on to state:

"Anticipating this condition of affairs, we have commenced the construction of pulp and paper machinery, and compressed air mine shovels, which work will, we hope, take up the slack caused by the falling off in shipbuilding. We have, however, in this new line, to go into competition with established manufacturers of long experience in this class of work. To succeed will mean not only continued prosperity for us, but the continuation of our quota of the prosperity of our city and the steady employment of the men who comprise our organization. Such success is what we are striving to attain for 1921, and we ask each and every one of our employees to unite with us in this purpose for the good of all concerned, and the welfare of the city in which we live. Of the citizens of Port Arthur, we ask a continuation of the co-operation and goodwill that they have given us in the past."

IRON, STEEL AND METALS SHOULD BE IN DEMAND DURING YEAR 1921

By W. S. LESLIE, of the A. C. Leslie & Co., Montreal

THE past year has been one of great changes in the metal market, some of these having proved disastrous to many manufacturers and dealers.

In iron and steel the scarcity prevailing at the end of last year was continued for the greater part of 1920, this condition being intensified by transportation difficulties, especially by the long drawn out strike of switchmen in the United States and also by the shortage of fuel. This shortage resulted in an abnormal advance in price of coal and coke, which forced makers who did not control their own supply to further advance their prices for iron and steel products, and, while the United States Steel Corporation steadily maintained their prices, they were unable to keep up with their deliveries' and even buyers who had contracts running at moderate prices were obliged to go into the market and pay heavy premiums for prompt or fairly prompt delivery. For the past few months, however, transportation conditions have been about normal and the fuel situation having been considerably relieved, makers have greatly improved their deliveries, and on this account the premiums commonly paid have been greatly reduced. In fact, the market is getting down to a point somewhere near the level established by the Steel Corporation.

This downward movement has been quickened and carried farther than would naturally have been the case by extensive cancellations and postponing of deliveries on the part of buyers, chiefly in the automobile and allied industries, and these conditions have naturally made buyers hold back and probably they are carrying this policy to an extreme.

British manufacturers were able during the year to regain some of their lost trade in Canada, being assisted by the scarcity of American material and the low rate of Sterling exchange. Great Britain also, and probably their products, will be more in evidence in the coming year. Already a considerable quantity of galvanized and a, smaller quantity of black sheets and tinplates have been imported and prices now being quoted are interesting for delivery at points where freight rate favors British imports.

The high price of fuel, however, in Great Britain and a shortage of pig iron have prevented any reduction in prices of the less manufactured forms of iron and steel that would make shipments to Canada possible.

The course of the non-ferrous metals has been most erratic. For the first five or six months of the year, tin, copper, lead, spelter and other metals were all very strong, though prices of some, more especially copper and spelter, were not abnormal compared with pre-war levels. A very serious slump, however, has since taken place in all these metals and values have apparently been carried lower than cost of production justifies, with the possible exception of lead, the supplies of which are comparatively small. Financial reasons, coupled with the fact that these metals are speculated in very heavily, account largely for the continued slump, and while there should undoubtedly be a good reaction in price of these metals, it is a question how soon this will take place. Certainly there must be more confidence than now exists before the market will be stabilized.

From what I have said the present would appear to be a time for buyers to continue a cautious policy in purchasing without allowing themselves to run out of supplies, but there is no reason why the demand for iron, steel and all other metals should not be good during the coming year, as there is a large amount of buying to be done when buyers have sufficient confidence, and are able to finance their requirements.

SELECTED MARKET QUOTATIONS

Being a record of prices current on raw and finished material entering into the manufacture of mechanical and general engineering products.

PIG IRON

Grey forge, Pittsburgh	$39 96
Lake Superior, charcoal, Chicago	53 50
Standard low phos., Philadelphia	44 79
Bessemer, Pittsburgh	41 96
Basic, Valley furnace	37 50
Toronto price:—	
Silicon, 2.25% to 2.75%	51 50
No. 2 Foundry, 1.75 to 2.25%	50 00

IRON AND STEEL

Per lb. to Large Buyers	Cents
Iron bars, base, Toronto	$ 4 75
Steel bars, base, Toronto	4 75
Iron bars, base, Montreal	4 50
Steel bars, base, Montreal	4 50
Reinforcing bars, base	5 50
Steel hoops	6 00
Tire steel	5 00
Spring steel	8 00
Band steel, No. 10 gauge and 3-16 in. base	5 50
Chequered floor plate, 3-16 in.	8 50
Chequered floor plate, ¼ in.	8 00
Bessemer rails, heavy, at mill	....
Steel bars, Pittsburgh	3 00-4 00
Tank plates, Pittsburgh	3 50
Structural shapes, Pittsburgh	3 00
Steel hoops, Pittsburgh	3 50-3 75
F.O.B., Toronto Warehouse	
Small shapes	5 50
F.O.B. Chicago Warehouse	
Steel bars	3 62
Structural shapes	3 72
Plates	3 67 to 5 50
Small shapes under 3"	3 62

FREIGHT RATES

Per 100 Pounds.

Pittsburgh to Following Points

	C.L.	L.C.L.
Montreal	58½	73
St. John, N.B.	84½	106½
Halifax	86	108
Toronto	38	54
Guelph	38	54
London	38	54
Windsor	35	50½

METALS

	Gross Montreal	Toronto
Lake copper	$19 00	$19 50
Electric copper	18 50	19 00
Castings, copper	18 00	19 00
Tin	44 00	46 00
Spelter	8 50	9 00
Lead	7 50	8 00
Antimony	8 00	9 00
Aluminum	34 00	35 00

Prices per 100 lbs.

PLATES

Plates, 3-16 in.	$ 5 50	$ 5 50
Plates, ¼ up	5 00	5 50

PIPE—WROUGHT

Standard Buttweld Pipe

Per 100 Ft.

	Steel Blk.	Galv.	Gen. Wrought Iron Blk.	Galv.
⅛	$ 6 50	$ 8 50	$.....	$.....
¼	6 81	7 41	6 91	8 01
⅜	5 31	7 41	5 91	8 01
½	7 10	8 63	7 95	9 48
¾	8 30	10 87	9 95	12 02
1	13 01	16 07	14 71	[illegible] 77
1¼	17 60	21 74	19 90	24 04
1½	21 04	25 99	23 79	28 74
2	28 31	34 97	32 01	38 67
2½	44 73	55 28	....	....
3	58 52	72 29	....	....
3½	74 06	90 62	....	....
4	87 75	107 37	....	....

Standard Lapweld Pipe

Per 100 Ft.

	Steel Blk.	Galv.	Gen. Wrought Iron Blk.	Galv.
2	$32 01	$ 38 67	$36 71	$42 37
2½	48 26	58 79	54 11	64 64
3	63 11	76 88	70 76	84 53
3½	75 90	92 46	85 10	101 66
4	89 93	107 55	100 83	120 45
4½	1 05	1 29	1 30	1 54
5	1 22	1 50	1 52	1 80
6	1 58	1 95	1 97	2 33
7	2 06	2 53	2 58	3 01
8L	2 16	2 66	2 66	3 18
8	2 49	3 07	3 07	3 64
9	2 98	3 67	3 67	4 36
10L	2 77	3 41	3 41	4 05
10	3 56	4 39	4 39	5 21

Prices—Ontario, Quebec and Maritime Provinces

WROUGHT NIPPLES

4" and under, 60%.

4½" and larger, 50%.

4" and under, running thread, 30%.

Standard couplings, 4-in. and under, 30%.

Do., 4½-in. and larger, 10%.

OLD MATERIAL

Dealers' Average Buying Prices.

Per 100 Pounds.	Montreal	Toronto
Copper, light	$10 50	$10 50
Copper, crucible	13 00	12 00
Copper, heavy	12 50	12 00
Copper wire	12 50	12 00
No. 1 machine composition	13 00	12 00
New brass cuttings	7 00	9 00
Red brass turnings	10 00	10 00
Yellow brass turnings	7 00	7 50
Light brass	5 00	5 00
Medium brass	6 50	6 00
Scrap zinc	5 00	5 50
Heavy lead	5 25	5 60
Tea lead	2 50	3 00
Aluminum	16 00	16 00

	Per Ton	Gross
Boiler plate	$11 00	$12 00
Heavy melting steel	18 00	23 00
Axles (wrought iron)	25 00	20 00
Rails (scrap)	18 00	18 00
Malleable scrap	20 00	25 00
No. 1 machine cast iron	32 00	33 00
Pipe, wrought	8 50	10 00
Car wheel	30 00	33 00
Steel axles	20 00	20 00
Mach. shop turnings	8 00	9 00
Stove plate	23 00	25 00
Cast boring	8 00	12 00

BOLTS, NUTS AND SCREWS

	Per Cent.
Carriage bolts, 7-16 and up	+10
Carriage bolts, ⅜-in. and less	Net
Coach and lag screws	—15
Stove bolts	55
Wrought washers	—25
Elevator bolts	+10
Machine bolts, 7-16 and over	+10
Machine bolts, ⅜-in. and less	+10
Blank bolts	Net
Bolt ends	Net
Machine screws, fl. and rd. hd., steel	27½
Machine screws, o. and fil. hd., steel	+25
Machine screws, fl. and rd. hd., brass	net
Machine screws, o. and fil. hd., brass	net
Nuts, square, blank .... +25 add	$2 00
Nuts, square, tapped ....add	2 25
Nuts, hex., blank ....add	2 50
Nuts, hex., tapped ....add	3 00
Copper rivets and burrs, list less	15
Burrs only, list plus	25
Iron rivets and burrs	40 and 5
Boiler rivets, base ¾" and larger	$8 50
Structural rivets, as above	8 40
Wood screws, O. & R., bright	75
Wood screws, flat, bright	77½
Wood screws, flat, brass	55
Wood screws, O. & R., brass	55½
Wood screws, flat, bronze	50
Wood screws, O. & R., bronze	47½

MILLED PRODUCTS

(Prices on unbroken packages)

	Per Cent
Set screws .... —20%, 25 and	5
Sq. and hex. hd. cap screws	12½
Rd. and fil. hd. cap screws ..plus	25
Flat but. hd. cap screws ....plus	50
Fin. and semi-fin. nuts up to 1-in.	12½
Fin. and Semi-fin. nuts, over 1 in., up to 1½-in.	—5
Fin. and Semi-fin. nuts over 1½ in., up to 2-in.	+12½
Studs	+5
Taper pins	—12½
Coupling bolts	+40
Planer head bolts, without fillet, list	+45
Planer head bolts, with fillet, list plus 10 and	+55
Planer head bolt nuts, same as finished nuts.	
Planer bolt washers	net
Hollow set screws	+60
Collar screws ......list plus 20,	30
Thumb screws	40
Thumb nuts	75
Patch bolts ....add	+85
Cold pressed nuts to 1½ in...add	$1 00
Cold pressed nuts over 1½ in..add	2 00

BILLETS

	Per gross ton
Bessemer billets	$60 00
Open-hearth billets	60 00
O.H. sheet bars	76 00
Forging billets	56 00-75 00
Wire rods	52 00-70 00

Government prices.

F.O.B. Pittsburgh.

NAILS AND SPIKES

Wire nails, base	$6 10
Cut nails, base	7 00
Miscellaneous wire nails	50%

ROPE AND PACKINGS

Plumbers' oakum, per lb.	0 10¼
Packing, square braided	0 38
Packing, No. 1 Italian	0 44
Packing, No. 2 Italian	0 36
Pure Manila rope	0 35½
British Manila rope	0 28
New Zealand hemp	0 28

POLISHED DRILL ROD

Discount off list, Montreal and Toronto .... net

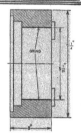

CAM
FOR DIE HEAD

Attached to the toolpost of the engine lathe on which this cam has
been machined, the Dumore Grinder
accurately grinds the hole without
necessitating another "set up."
The high efficiency standards of the
Geometric Tool Co. (New Haven
Conn.) make comment unnecessary
on this method of handling the job.
It is sufficient that the work of the
Dumore is so satisfactory, the finish so fine, that all cams this size
or larger, made in this shop are
ground in this manner.

The Average Shop Has Constant Use for Dumore Grinders

Dumore Grinders are made in types
and sizes that can be attached to all
classes of machines and used to successfully perform a wide variety of
grinding operations.

Equally valuable for the important
"odd jobs" that must be done in every

shop and on some manufacturing
work, they find a ready place in every
part of every metal working plant—
whether toolroom, factory or repair
shop.

*Ask us about DUMORE Grinders—full
details on request.*

Wisconsin Electric Company
2930 Sixteenth Street   Racine, Wisconsin

DUMORE HIGH SPEED GRINDERS

If interested tear out this page and place with letters to be answered.

MISCELLANEOUS

Solder, strictly	$ 0 28½
Solder, guaranteed	0 30½
Soldering coppers, lb.	0 62½
White lead, pure, cwt.	20 35
Red dry lead, 100-lb. kegs, per cwt.	15 00
Gasoline, per gal., bulk	0 42
Pure turp., single bbls., gal.	3 15
Linseed oil, raw, single bbls.	2 37
Linseed oil, boiled, single bbls.	2 40
Wood alcohol, per gal.	4 00
Whiting, plain, per 100 lbs.	3 00

CARBON DRILLS AND REAMERS

S.S. drills, wire size	40 and 5
Can. carbon cutters, plus	10
Standard drills, all sizes	40 and 5
3-fluted drills, plus	10
Jobbers' and letter sizes	40 and 5
Bit stock	50
Ratchet drills	10
S.S. drills for wood	40
Wood boring brace drills	25
Electricians' bits	30
Sockets	50
Sleeves	50
Taper pin reamers	25 off
Drills and countersinks	net
Bridge reamers, carbon	50
Centre reamers	5
Chucking reamers	net
Hand reamers	10
High speed drills, list net to plus	20
Can. high speed cutters, net to plus	10
American plus	40

COLD ROLLED STEEL

[At warehouse]

Rounds and squares	$7.50 base
Hexagons and flats	7.50 base

IRON PIPE FITTINGS

	Black	Galv.
Class A	70	85
Class B	30	40
Class C	20	30

Cast iron fittings, 5%; malleable bushings, 22½%; cast bushings, 22½%; unions, 37½%; plugs, 20% off list.

SHEETS

	Montreal	Toronto
Sheets, black, No. 28	$ 8 50	$ 9 00
Sheets, blue ann., No. 10	7 00	7 50
Canada plates, dull, 52 sheets	13 00	13 00
Can. plates, all bright	14 00	.....
Apollo brand, 10¾ oz. galvanized	....	....
Queen's Head, 28 B.W.G.	13 00	.....
Fleur-de-Lis, 28 B.W.G.	12 50	.....
Gorbal's Best, No. 28	....	....
Colborne Crown, No. 28	...	....
Premier, No. 28, U.S.	10 00	11 00
Premier, 10¾-oz.	10 50	11 40
Zinc sheets	16 50	20 00

PROOF COIL CHAIN

(Warehouse Price)

B

¼ in., $13.00; 5-16, $11.00; ⅜ in., $10.00; 7-16 in., $9.80; ½ in., $9.75; ⅝ in., $9.20; ¾ in., $9.30; ⅞ in., $9.50; 1 in., $9.10; Extra for B.B. Chain, $1.20; Extra for B.B.B. Chain, $1.80.

ELECTRIC WELD COIL CHAIN B.B.

⅛ in., $16.75; 3-16 in., $15.40; ¼ in., $13.00; 5-16 in., $11.00; ⅜ in., $10.00; 7-16 in., $9.80; ½ in., $9.75; ⅝ in., $9.50; ¾ in., $9.30.

Prices per 100 lbs.

FILES AND RASPS

	Per Cent.
Globe	50
Vulcan	50
P.H. and Imperial	50
Nicholson	32½
Black Diamond	27½
J. Barton Smith, Eagle	50
McClelland, Globe	50
Delta Files	20
Disston	40
Whitman & Barnes	50
Great Western-American	50
Kearney & Foot, Arcade	50

BOILER TUBES.

Size.	Seamless	Lapwelded
1 in.	$27 00	$.....
1¼ in.	29 50	.....
1½ in.	31 50	29 50
1¾ in.	31 50	30 00
2 in.	35 00	30 00
2¼ in.	35 00	29 00
2½ in.	42 00	37 00
3 in.	50 00	48 00
3¼ in.	.. ..	48 50
3½ in.	63 00	51 50
4 in.	85 00	65 50

Prices per 100 ft., Montreal and Toronto

OILS AND COMPOUNDS.

Castor oil, per lb.	
Royalite, per gal., bulk	29½
Palacine	32½
Machine oil, per gal.	65
Black oil, per gal.	34
Cylinder oil, Capital	1.01
Petroleum fuel oil, bbls., net	19

BELTING—No 1 OAK TANNED

Extra heavy, single and double		6½
Standard		6½
Cut leather lacing, No. 1		2 00
Leather in side	2 40	3 00

TAPES

Chesterman Metallic, 50 ft.	$2 00
Lufkin Metallic, 603, 50 ft.	2 00
Admiral Steel Tape, 50 ft.	2 75
Admiral Steel Tape, 100 ft.	4 45
Major Jun. Steel Tape, 50 ft.	3 50
Rival Steel Tape, 50 ft.	2 75
Rival Steel Tape, 100 ft.	4 45
Reliable Jun. Steel Tape, 50 ft.	3 50

PLATING SUPPLIES

Polishing wheels, felt	$4 50
Polishing wheels, bull-neck	2 00
Emery in kegs, Turkish	8¾
Pumice, ground	06
Emery glue	30
Tripoli composition	9½
Crocus composition	12
Emery composition	11
Rouge, silver	64
Rouge, powder, nickel	38

Prices per lb.

ARTIFICIAL CORUNDUM

Grits, 6 to 70 inclusive	.08½
Grits, 80 and finer	.6

BRASS—Warehouse Price

Brass rods, base ½ in. to 1 in. rod	0 30
Brass sheets, 24 gauge and heavier, base	0 38
Brass tubing, seamless	0 42
Copper tubing, seamless	0 44

WASTE

XXX Extra	.23	Atlas	.19
Peerless	.22	X Empire	.18½
Grand	.21½	Ideal	.18
Superior	.21½	X Press	.17
X L C R	.20		

Colored

Lion	.16	Popular	.12
Standard	.14	Keen	.10
No. 1	.14		

Wool Packing

Arrow	.35	Anvil	.22
Axle	.28	Anchor	.17

Washed Wipers

Select White .20 Dark colored .09

Mixed colored .10

This list subject to trade discount for quantity.

RUBBER BELTING

Standard ... 10% Best grades... 15%

ANODES

Nickel	.55 to .60
Copper	.38 to .40
Tin	.70 to .70
Zinc	.16 to .17

Prices per lb.

COPPER PRODUCTS

	Montreal	Toronto
Bars, ½ to 2 in.	$35 00	$37 00
Copper wire, list plus 10		
Plain sheets, 14 oz., 14x60 in.	40 00	44 00
Copper sheet, tinned, 14x60, 14 oz.	43 00	46 00
Copper sheet, planished, 16 oz. base	47 00	50 00
Braziers', in sheets, 6 x 4 base	39 00	42 00

LEAD SHEETS

	Montreal	Toronto
Sheets, 3 lbs. sq. ft.	$10 50	$14 50
Sheets, 3½ lbs. sq. ft.	10 25	14 00
Sheets, 4 to 6 lbs. sq. ft.	10 00	13 50

Cut sheets, ½c per lb. extra.

Cut sheets to size, 1c per lb. extra.

PLATING CHEMICALS

Acid, boracic	$ .23
Acid, hydrochloric	.04¾
Acid, nitric	.11
Acid, sulphuric	.04½
Ammonia, aqua	.15¾
Ammonium, carbonate	.23
Ammonium, chloride	.22
Ammonium hydrosulphuret	.75
Ammonium sulphate	.30
Arsenic, white	.16
Copper, carbonate, annhy	.41
Copper, sulphate	.13
Cobalt, sulphate	.20
Iron perchloride	.62
Lead acetate	.30
Nickel ammonium sulphate	.20
Nickel carbonate	.32
Nickel sulphate	.20
Potassium sulphide (substitute)	.40
Silver Chloride (per oz.)	1.15
Silver nitrate (per oz.)	1.10
Sodium bisulphate	.13
Sodium carbonate crystals	.04
Sodium cyanide, 127-130%	.39
Sodium hyposulphite per 100 lbs	9.00
Sodium phosphate	.15
Tin chloride	.30
Zinc chloride, C.P.	.30
Zinc sulphate	.08

Prices per lb. unless otherwise stated

GENERAL VIEW HAMILTON WORKS

Quality

THE
STEEL COMPANY
OF
CANADA
LIMITED
HAMILTON MONTREAL

Service

GENERAL VIEW OF NOTRE DAME WORKS MONTREAL

These Plants and the Organization that directs and operates them,
stand back of
and Guarantee Quality and Service at All Times

Machine Tool Designers Have Had a Busy Year

Brief Statements Touching on New Developments in Machine
Tools and Kindred Lines—Grinders, Presses, Shapers, Planers,
Drills, Lathes, and Miscellaneous Tools are Included

DURING the past year many new developments in the machine tool industry have taken place. These developments have been recorded from week to week in Canadian Machinery and the list to follow are only some that we have covered during the year. Of course, greater detail was given than that which follows, but this list is more of a handy reference for our readers rather than a description of the tool itself. Should a reader be interested further, he has merely to drop a line to the firm, in this way securing all the data.

Grinding Machines

The Fitchburg Grinding Machine Company, Pittsburgh, Pa.—A line of plain cylindrical grinding machines. Spindle is of tool steel, and spindle boxes of bronze, these boxes being provided with means of compensation for wear. Automatic cross feed is operated at each reversal of table and is self-releasing when work has been ground to size. These machines are built in six sizes.

The Badger Tool Co., Beloit, Wis.—A new line of grinding machines in which the spindle is fitted with both radial and thrust ball-bearings.

The Grand Rapids Machine Co., Grand Rapids, Mich.—A new line of tap grinding machines for grinding the taper and clearance on the ends of taps.

The Precision and Thread Grinding Mfg. Co., Philadelphia, Pa.—Several grinding attachments that are particularly adapted for thread grinding on lathes. Any style thread can be ground with these attachments. Grinding spindle is carried on ball-bearings, and is adjustable for radial wear and end thrust. Power transmission is accomplished by an endless belt which has a three-point contact with both the driving and driven pulleys. The belt then travels over a compensating two-pulley arrangement, which automatically keeps it at the proper tension to prevent slippage.

The H. F. Harris Engineering Co., Bridgeport, Conn.—An automatic hob grinding machine which embodies various interesting features.

The Lafayette Tool & Equipment Co., Philadelphia, Pa.—A universal grinder for precision tool room work. Both internal and external thread gauge grinding can be performed. The machine can also be used on the bench or as a grinding attachment for lathes, milling machines, shapers, planers, etc., etc.

The National Machine Tool Co., Racine, Wis.—A combination tool cutter and gauge grinding machine. Especially adapted for tool room or machine shop. Will grind cutters, arbors, reamers, mandrils, etc., between centres, or on large or small lathe. This machine can also be used for lapping small dies, bushings, also snap gauges.

The Badger Tool Co., Beloit, Wis.—A new double spindle type disc grinder. This machine is intended for finishing two opposite parallel faces of work at one time, such as drop-forged wrenches, piston rings, nuts, etc.

The Mummert-Dixon Company, Hanover, Pa.—A special design of portable grinding machine. This machine is self-contained and the wheel may be operated at any desired angle without interfering with the general alignment.

The Van Dorn Electric Tool Co., Cleveland, Ohio.—A small aerial grinder, which may be obtained for D.C. or A.C. current. Ball bearings are used throughout.

The Diamond Machine Co., Providence, R.I.—A heavy duty face grinding machine that is intended for grinding operations on heavy equipment that is usually accomplished on planing machines.

The Wordell Manufacturing Co., Cleveland, Ohio.—A new automatic metal cutting circular saw grinder. This machine will re-sharpen all screw-slotting, slitting and cold saws from 1½ inches to 18 inches diameter, and with teeth from 36 to the inch up to 2 inches from point to point. Saws are resharpened at a speed of 45 teeth per minute. The machine is fully automatic and once started requires no further attention.

The Bellevue Industrial Furnace Co., Detroit, Mich.—A line of automatic drill grinders. These grinders are said to sharpen a drill to a perfect angle. The machine is driven by an electric motor in the pedestal and is equipped with two-step pulleys so that two speeds are available.

A. B. Jardine Co., Hespeler, Ont.—A pedestal grinder for general grinding and polishing in machine shop, blacksmith shop, automobile shop, etc.

The Precision Trueing Machine & Tool Co., Cincinnati, Ohio.—A grinding wheel trueing machine. This machine can be applied to any style of grinder and is operated by either alternate or direct current.

The Beacon Engineering Co., Tipton, Eng.—A twist drill grinder for drills up to 1⅜", also a sensitive pillar treadle drill for holes up to ½".

The Bryant Chucking Grinder Co., Springfield, Vt.—A machine for the grinding of holes, also exterior surfaces. The machine is made in two types, single spindle and double.

Various Types of Presses

The Enterprise Machinery Co., Chicago, Ill.—A bench forming press that is intended for the rapid assembling of small parts and for such operations as seaming and riveting on hollow work.

The Lasalle Machine Works, Chicago, Ill.—A line of bench presses designed for accurate production of small parts, such as used by jewelers, typewriter manufacturers and for other kindred work.

The Streine Tool & Mfg. Co., New Bremen, Ohio.—A new toggle forming press for roofing, and special shapes, with a capacity up to 12 ft. in length. This machine is also provided with tables for die work.

The Toledo Machine & Tool Co., Toledo, Ohio.—These people have brought out a considerable number of very large presses during the past year.

Williams & White Co., Moline, Ill.—A new straight side press which has a working length of 16 ft. between the housings and is capable of exerting a pressure of 500 tons distributed load.

The Package Machine Co., Springfield, Mass.—A precision bench lathe on which the headstock is fitted with ball thrust bearings.

Shapers and Planers

The Columbia Machine Co., Hamilton, Ohio.—A varied line of heavy duty shapers ranging from 16" to 28" capacity.

The Simmons Machine Co., Albany, N.Y.—An openside planer that will take work 42 in. wide, 48 in. high and up to 12 ft. in length.

The Turbine Air Tool Co., Cleveland, Ohio.—A line of air planers which are used for shaving telegraph poles and is used on general construction work.

The Oliver Machinery Co., Grand Rapids, Mich.—A new direct-coupled, motor-driven, ball-bearing hand planer and jointer.

Alfred Herbert Co., Ltd., Toronto, are handling a special planer that is known as the Hilo-plane and which is electrically controlled.

The Universal Machine & Tool Co., Canton, Ohio.—A new 24-inch open side planer which is said to combine the

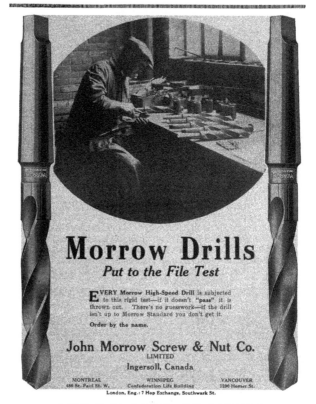

advantages of both shaper and planer. Variable stroke mechanism is used for driving the table, the speed of which is varied by means of shift gears.

Milling Machines

The Garvin Machine Tool Co.—A form miller of horizontal type. This machine is designed for cutting flat or cylindrical cams, and worm gears drive the work arbor. Power is transmitted through spur gearing, three changes of feed being given for the flat cam-cutting fixture. All gearing is housed, protecting it from damage or injury to the operator.

Marshalltown Manufacturing Co., Marshalltown, Iowa. —A plate milling machine for milling the edges of boiler plates. These machines are made in four sizes, having a range from 8 to 24 feet.

Tri State Milling Machine Co., Memphis, Tenn.—A milling machine attachment that has been adapted for use on shaping machines.

The Gabrielson Mfg. Corp., Syracuse, N.Y.—A new manufacturing type of milling machine. These machines are built in three sizes: 6, 9 and 12 inches between the spindle and table.

The Automatic Machine Co., Bridgeport, Conn.—A multiple spindle profile milling machine that is automatic in action. This machine is specially adapted for automobile work.

The Kempsmith Mfg. Co., Milwaukee, Wis.—A new vertical maximiller. This machine is of the usual type, but of very rigid design. Quick power traverse has been arranged, giving 100 inches per minute travel of the table in either direction, and 36 in. per min. on the vertical and transverse movement. The lubrication of the machine has been carefully gone into.

The Taylor & Fenn Co., Hartford, Conn.—A two-spindle spline machine that is automatic in operation and designed to simultaneously machine two spline grooves up to six inches in length on opposite sides of the same piece. Two cutters, operating simultaneously, reduce machining time on work where two spline grooves are cut. Work of any diameter up to 5 in. and of any length may be splined. The cutters can be made to advance either in unison or alternately and to knock off in like manner.

The Betts Machine Co., Rochester, N.Y.—A continuous milling machine that can be provided with whatever number of spindles desired.

The Toledo Milling Machine Co., Toledo, Ohio.—A new type of vertical milling machine. Unit construction has been adopted. Longitudinal table movement is 46 in., traverse 14 in. and vertical movement equal to the traverse. The spindle has a vertical movement of 6½ in.

The Ingersoll Milling Machine Co., Rockford, Ill.—A special type of reciprocating milling machine that may be designed to meet specific requirements. These machines are intended for working on duplicate parts.

The Ryerson-Conradson Co., Chicago, Ill.—A new helical drive high power miller. The proportions of this machine are adequate for heavy work, and as already stated, the drive is obtained through helical gears. Change gears are cut from solid, made of chrome nickel steel and are heat treated. Other gears are made of steel or bronze.

The Brown & Sharpe Mfg. Co., Providence, R.I.—A special automatic milling machine which is essentially a manufacturing machine and is used in plants where duplicate work is obtainable.

The Superior & Engineering Co., Detroit—A hand milling machine that is driven by a 1½ h.p. motor. Spindle has 12 changes of speed ranging from 100 to 600 r.p.m.

Various Types of Drills

The Joseph T. Ryerson & Sons, Chicago, Ill.—A new high power, plain, radial drilling machine.

The Arthur C. Mason, Inc., Hawthorne, N. J.—High speed, sensitive, ball-bearing drilling machine. Various

features have been incorporated, some of these being elimination of frictional bearings, absence of belt trouble, and direct drive. Machine is equipped with ball-bearings throughout, and all moving parts are protected. Four spindle speeds are arranged for, these varying from three thousand to ten thousand r. p. m. This machine is also made in bench type.

The Arthur C. Mason, Inc., Hamilton, N.J.—A heavy type of sensitive drilling machine. The spindle of this machine is equipped with ball bearings throughout.

National Automatic Tool Co., Richmond, Ind. — An inverted drilling machine which is said to be advantageous for the drilling of deep holes in cast iron.

The Barnes Drill Co., Rockford, Ill.—A gang drill built for heavy duty work. Spindles are exceptionally large and double spindled. This machine will drive a 2-in. twist drill at .041 in. feed per revolution of spindle, or at the rate of 6½ in. per minute in cast iron, without the gears.

The Niles-Bement-Pond Co., N.Y.—A new type of radial drilling machine. This machine is built in two sizes, 5 ft. and 6 ft. arms.

The Sibley Machine Co., South Bend, Ind.—An addition of two sizes to their regular line of drill presses.

Danber Kratsch Co., Oshkosh, Wis.—Upright drill press with all gear feed. Drive for machine consists of semi-steel and rawhide gears. There are no sliding gears in the gear box. Speed may be changed by single movement of clutch lever. Spindle has ball thrust bearings at lower end, and four changes of feed for each spindle speed is provided.

Joseph T. Ryerson and Sons Co., Chicago, Ill.—A high power, radial drilling machine. These machines are made in three sizes, four, five, and six feet arms. Ball bearings are installed.

Hilldrill Co., Philadelphia.—Gang drilling machine. Drill spindles are chain driven, and flexibility of chain drive makes it adaptable for gang drilling on any type of work. The cycle of operations is automatic, and the spacing of spindle is universal. Special mechanism clamps the work in place with a pressure of about ten tons. The drill heads are built as independent, detachable units.

The Morris Machine Tool Co., Cincinnati, Ohio. — Three sizes of radial drills, 2½ ft. 3 ft. and 3½ ft., respectively. Various features have been embodied, and with the exception of large spindle gear, all other gears are made of steel. Gears are fully inclosed and run in light grease or heavy oil. Six feeds are secured. Friction clutches in speed box and tapping attachment are of expanding ring type. Motor drive can be arranged if desired.

Genesee Mfg. Co. Rochester, N.Y.—A special tool for use in drilling machines, for the counter-boring of holes, facing, forming, and similar operations.

Lathes

The Canada Machinery Corp., Galt, Ont.—This concern has brought out a 16-inch by 20-inch shaper and a 42-inch engine lathe. The lathe is equipped with clutch control from the saddle, and an all gear head is driven by motor through silent chain drive. There is an internal gear face drive to the face plate.

R. A. Kelley Co., Xenia, Ohio.—A redesigned shaper which is said to be a considerable improvement over their old design. A special feed box is fitted to the ends of cross rails and feed is automatically disengaged at the end of the travel. Helical teeth are used on all gears and pinions.

Joseph T. Ryerson & Sons, Chicago, Ill.—A line of high power selective head engine lathes.

Alfred Herbert, Ltd., Coventry, England.—A universal turret lathe. A heavy duty combination turret machine, and heavy slotter. A new type of hexagon turret lathe that is suitable for both bar and combination turret lathe work. A new hydraulic turret lathe originally in-

THE MASSEY-HARRIS CO. HAVE A WELL-APPOINTED EMERGENCY STATION WELL FITTED UP

Healthy, Contented Workers Mean Greater Production

ACCIDENTS occur almost daily in Industrial plants; workers are incapacitated, time and money are lost, the men's spirits are demoralized. Safeguard against these possibilities by equipping a thoroughly modern, practical and efficient First Aid Department—a place where minor ills as well as serious accidents can be cared for. In this way the men will know you have their interests at heart—they will be more loyal, do better work and increase production.

Let us advise you regarding suitable first aid equipment. Our stock is very complete and consists of

STERILIZERS   INSTRUMENT CABINETS
PURE DRUGS   WASTE RECEPTACLES
              SURGICAL DRESSINGS

They are used and endorsed by the leading physicians.

Write for Our Complete Catalogue

INGRAM & BELL, Limited
256 McCaul St.   TORONTO, Ont.

If interested tear out this page and place with letters to be answered.

tended for use by disabled men. Chuck turret and cross slide are operated by means of hydraulic cylinders.

The Cincinnati Lathe and Tool Co., Oakley, Cincinnati.—A new line of gear head lathes. Seven sizes are made. These lathes are fitted with hexagon turret on bed, and power feed silent chain drive. As the speed variations are obtainable in the head, constant speed motors only are required. Twelve speed changes in geometric progression are secured on these lathes. This is accomplished through sliding gears and by shifting three levers.

The Seneca Falls Manufacturing Co., Inc., Seneca Falls, N.Y.—Two sizes of gap lathes, eleven inches by eighteen inches, and thirteen by twenty-one inches swing. These machines have wide range for screw cutting of all standard threads, from 3 to 72, including 11½ and 27 pipe threads.

O. R. Adams, Mfg. Co., Rochester, N.Y.—A lathe known as a short cut lathe. The head stock of this machine is designed to give 6 changes of speed with a range from 48 to 375 r.p.m. Speed changes can be made while lathe is in operation.

Miscellaneous Tools

Charles Engelhard, New York.—An automatic temperature regulator that opens or closes control valve at desired temperature.

Canadian Firefoam Co., Hamilton, Canada.—A fire extinguisher known as Foamite Firefoam. Various capacity engines, also small hand tanks are supplied, equipped with this particular extinguishing fluid.

The Ricker Instrument Co., Philadelphia, Pa.—A line of precision levels for machine shop use.

The Independent Pneumatic Tool Co., Chicago.—Pneumatic motor hoists up to two tons capacity, the speed lift is 32, 16 and 8 ft. respectively for the half, one, and two-ton sizes.

The Link-Belt Co., Chicago, Ill. — A new friction clutch which has a one-point adjustment. All moving parts are enclosed and parts are balanced. Parts are easily accessible, if repairs are necessary.

Canadian Reclaimers, Ltd.—A pulpwood reclaimer that is said to eliminate considerable labor when stacking or unstacking pulpwood.

The Pittsburg Instrument and Machine Co., Pittsburgh, Pa.—A sheet metal tester that is used to determine the drawing, stamping, compressive, and folding qualities of sheets of iron, steel, brass, copper, etc., etc.

Link-Belt Co., Chicago.—A hydraulic system for raising or lowering heavy doors such as are used on open hearth and heating furnace work.

The Roberts Mfg. Co., New Haven, Conn.—An improved friction clutch that is adapted to drilling, tapping, and stud setting.

The Kent Machine Co., Kent, Ohio.—A two-spindle pointing machine that is intended for the pointing of bolts, or rods up to three-quarter inch in diameter.

Black & Decker Mfg. Co., Baltimore, Md.—An electric air compressor that is made in two types, namely high and low pressure. The largest of these machines will work up to a pressure of 200 lbs. per square inch.

The Wayne Oil Tank & Pump Co., Fort Wayne, Ind. —An oil filtering cabinet which is intended to be placed under the machine from which the oil is flowing.

The Lapoint Machine Tool Co., Hudson, Mass. — A new type of broaching machine, known as the No. 3 Duplex. The power screws may be operated jointly, or as separate units as desired, and to any required length of broach.

D. MacKenzie Machinery Co., Guelph, Ont.—New saw table. This machine is adapted to electric or gasoline drive, and is designed not only for factories but for manual training schools.

The Scully-Jones, Chicago.—A new drill chuck that is designed to hold a straight shank drill in a taper hole of the drilling machine.

Frank G. Payson Co., Chicago, Ill.—A special line of air operated chucks, for use on double spindle turret lathes.

The Yoder Co., Cleveland, Ohio.—A specially designed slitting shear that is said to remove any irregularities in the strip after it has been cut. On metal up to one-eighth inch in thickness strips as narrow as 2 three-eighth inch, may be cut.

The Manhattan Machine and Tool Works, Grand Rapids, Mich.—A four-post screw press that is designed to suit all classes of arbor press work.

The Hoosier Drilling Machine Co., Goshen, Ind. — A new line of machine vises in which the jaws are of hardened steel, but may be modified to meet specific needs.

The Becker Milling Machine Co., Hyde Park, Boston, Mass.—A new machine known as Model D.1 die-sinker. Dies up to ten tons in weight can be accommodated. Centralized control is a feature of this machine. The spindle is back-geared and driven by a five-inch double belt. Thirty-two feeds are obtainable. Fourteen spindle speeds are provided. A seven and a half H. P. variable speed motor is used.

The Atlas Press Co., Kalamazoo, Mich.—Two new arbor presses, one 6-ton capacity, the other 15-ton. These presses are adapted for a variety of work, such as forcing arbors, straightening and broaching operations, etc.

The Cincinnati Gear Cutting Machine Co., Cincinnati, Ohio.—A new type of gear hobbing machine that has a maximum capacity of 17¼ inches diameter and 12-in face.

The Newton Machine Tool Works, Inc., Philadelphia, Pa.—A portable slotting machine that has a maximum stroke of 70 inches.

The Hercules Machine and Tool Co., Astoria, N.Y.—A gear hobbing machine which provides for exhaust adjustment in the centering of the hob. Automatic stops are provided for disengaging the feeds when machining the various classes of gears.

W. J. Pine Machine Co., Kenosha, Wis.—Straightening machine which combines a testing table and machine. Stock to be straightened is placed on upper testing table, then put in position under pressure screw. Operations performed on this machine are: offsetting connecting-rods and bars, bending irregular shapes in forms or dies, etc.

Volta Mfg. Co., Welland.—A furnace of the Heroult type. Furnace transformers have capacity of 1,500 K.V.A., and connection is made to the furnace catenaries by means of the usual copper busses. A tilting arrangement is provided. Furnace is balanced and a small-sized motor gives desired tilt.

The Stock Bridge Machine Co., Worcester, Mass. — A shaper attachment which will shape any irregular curve. Forming dies and various shaped blocks are possible with this attachment. A templet is carried in cross slide holder and is shaped to required outline of surface to be duplicated.

The Foster Machine Co., Elkhart, Ind. — A complete line of hand screw machines embracing five sizes. Several new features are noticed, but on the whole the machines do not represent any radical departure from established practice. Three different styles of cut-off units are supplied, being interchangeable. Power feed to turret is a separate unit, and machine can be built for power or hand feed.

W. S. Rockwell Co., New York.—New type of side-opening heat treating furnace. Furnace is 36 ft. wide, and doors can be raised or lowered independently, operated by compressed air and counter weight. Oil or gas can be used as fuel. Provision is made for convenient control of burners from above the rear of the furnace. Doors are controlled by air hoist mechanism.

Burton Engineering & Machinery Co., Cincinnati. — A line of gasoline and kerosene locomotives adaptable for any gauge track. These are built from three and a half to twelve tons, and power is transmitted through a spur friction, and power shaft.

REFUSE CHANGE
ASKED BY N. Y. C.

Step Preferred Instead of Pilot, But
Ottawa Does Not See It
That Way

OTTAWA.—Application of the New
York Central Railway for a modification
in the regulations regarding railway
safety appliance standards which would
permit the company to use step construction instead of pilot construction on
its freight engines operating into Canada has been refused by the Board of
Railway Commissioners. Dr. S. J. Mc-
Lean, in refusing the application, states
that "the board is asked to allow a
departure from its regulations and to
allow the substitution therefor of a practice depending entirely on tolerance."

Having in view Canadian conditions
and the careful consideration given before the regulations were adopted, he
advises refusal of the application.

The Railway Company represented that
the step had every advantage that a
pilot had, and that its use was allowed
under the rules of the Interstate Commerce Commission. The step equipment
was not, it was stated, objected to by
any of the Public Utilities Commissions
of the States through which the New
York Central operated, and there would
be difficulties in operation if a pilot was
required for the portion of the journey
in Canada.

PRODUCTION IS
UNDER WAR MARK

Figures Given Out Concerning Amount
of Iron and Steel Made in Canada

Production of steel and pig iron in
Canada in 1920 has shown an increase
over 1919, but the figures are considerably below the previous three years, when
war contracts so greatly speeded up the
output. According to a statement by the
Department of Mines at Ottawa, the
monthly production of steel in 1920 to
the end of September averaged 105,931
tons, compared with the monthly average in the previous four years, as follows: 1919, 86,157; 1918, 156,954; 1917,
145,494; 1916, 106,268.

The monthly production of pig iron
this year averaged 89,810 tons for the
first nine months, compared with an average of the following in previous years:
1919, 76,482; 1918, 99,629; 1917, 97,540;
1916, 97,438.

The blast furnace plants active during
the first nine months were those at Sydney and North Sydney, N.S., Hamilton,
Port Colborne and Sault Ste. Marie, Ontario. The blast furnace plants at Midland, Parry Sound, Deseronto and Port
Arthur, Ontario, were idle throughout
the period. At the end of September ten
stacks wer active and eight idle.

Pig iron was made from scrap iron and
steel at four electric furnace plants, lo-

cated at Hull, Montreal and Shawinigan
Falls, Quebec, and Orillia, Ontario.

PIG IRON TRADE

That some pig iron users are busy is
shown by purchases by rather important
companies. Further concessions in price
are noted. On the Pittsburg market No.
2 foundry may be obtained at $35 valley
furnace. The leading independent steel
company in this district has only about
one-third of its furnaces making iron at
present.

On the Chicago market the absence of
transactions of any size makes it difficult to name ruling prices. Southern
furnace prices are substantially the same
as a week ago.

On the Cleveland market the inquiry
is larger than in many weeks, but the
volume of sales shows little change.
Prices on foundry iron have settled down
to $35 base for No. 2. Some producers
believe there will be little resale iron
on the market after January 1st.

At Boston less resale iron is offered
than last week, presumably because of
the lack of demand. The feeling among
foundry owners is more hopeful,
although business is quiet.

The pig iron market at Birmingham
is at a standstill as there is practically
no business being done. With so many
furnaces out, a real buying movement
would find very little to go on.

Inquiries are showing improvement,
but sales are still light on the Buffalo
market. Five hundred tons of resale
foundry were sold at $34 to $35 base price
per ton. The total inquiries for the week
aggregate 3,000 tons, but little of this
developed into the actual order stage.

The market is very dull in the Cincinnati district but several inquiries for
first half of the year have appeared from
outside this territory. Furnace sales
continued light, there being no disposition to compete with resale prices on the
inquiries appearing.

Iron does not seem to be wanted at any
price is reported on the St. Louis market. Furnaces are making no prices,
and submission of offers is one of the
things more notable by absence than
presence in the market.

On the New York market it is extremely difficult to determine prices with
any accuracy as reports concerning re-
sales indicate a wide variation in quotations. Several eastern Pennsylvania
furnaces are expected to blow out or
be banked within the next few days.
Eastern Penn. No. 2 foundry 1.75 to
2.25 silicon delivered in New York
district is quoted at $36.52 to $37.52.

The Philadelphia market shows further declines in the price of pig iron.

Engineering

Schools are being planned by the Boards of Ste. Apelline, and St. Sylvestre, Que.

The United Farmers of Manitoba, in the district of St. Andrew's, are planning the erection of a hall there.

Victoria Hospital, London, Ont., is to have a new nurses' home. The estimated cost is $100,000. Erection will begin in the spring.

The Pigeon River Lumber Co., Port Arthur, Ont., are rebuilding their planing mill and machinery, destroyed by fire some months ago.

The town of Lauzon, Que., is considering the installation of a new water works system. The estimated cost is about $300,000.

W. Harris and Co. have plans prepared for the erection of a private garage and office building on Keating St., Toronto, at an estimated cost of $25,000.

The Hydro - Electric Commission, Brantford, Ont., is planning to duplicate its system at a cost of $750,000. The new work is projected for the coming year.

Ottawa, Ont., needs a new fire station, and a by-law will be submitted to the property owners at the coming elections to raise the sum of $200,000 for the purpose.

It has recently been announced that the Trans-Canada Theatres, Ltd., are contemplating the erection of theatres at Guelph, Ont., Regina, Sask., and Vancouver, B. C.

The Live Wire Co., Ltd., a new concern in Guelph, Ont., are producing from 30 to 35 thousand feet per day of rubber-covered insulated wire.

The ratepayers, Dutton, Ont., will be called on early in the year to sanction the expenditure of $18,000 for improvements to the community hall on Main street.

Operations are commencing at the new factory of the Canadian Detroit Twist Drill Co., at Walkerville. Their productions will comprise drills, reamers and milling cutters.

Good progress has been made on the 18-storey addition to the King Edward Hotel, Toronto. The steel work of the first 12 storeys is already up and the floors are being put in as the steel work goes up.

A new town is springing up around the paper mills of the Spruce Falls Pulp and Paper Company at Kapuskasing, Ont. There are about 1,250 men at present employed by the mills. Work on the mills is going ahead rapidly and it is expected the first pulpwood unit will be in operation by May next. The sum of $500,000 will be spent on the establishing of this townsite.

The Semet-Solvay Co., the largest concern of its kind in the world, is considering the erection of a by-product coke oven plant in Hamilton, costing $4,000,000. It is said this would produce all the artificial gas necessary to meet Hamilton's needs for several years.

Announcement has been made that plans have so far progressed that preliminary construction work on the international bridge to be built connecting Detroit and Windsor will be started early in the coming summer.

Classified Opportunities

The House of PETRIE and its Founder

50 Years of Progress

50 Years of Service

H.W. PETRIE.

THE history of the Metalworking and Woodworking industries of Canada is closely linked with the Machinery and Supply House of H. W. Petrie, Toronto and Hamilton. When H. W. Petrie entered his chosen field fifty years ago industrial Canada was in its infancy; but it grew and grew, and the house of Petrie grew up with it and became a by-word in every quarter of Canada where Machinery is mentioned. During its growth the house of Petrie endured and thrived through periodical trade depressions, political upheavals, wars and rumors of wars, for the good reason that a Petrie customer invariably comes back; he learns to expect quality, service and a square deal—and he is never disappointed!

It is interesting to note that the house of Petrie was founded at Brantford, Ont., where it had remarkable success. In 1890 the business was transferred to Toronto,

where conditions were more in keeping with the rapid growth of the concern.

Commenting upon this the Toronto Globe in its issue of December 27, 1890, referred to Mr. H. W. Petrie as one of the best known dealers of Machinery in the Dominion, who for many years conducted a business in Brantford, and owing to able management and thorough knowledge of the making and handling of machinery found his business had outgrown the confines of Brantford; as a consequence the proprietor found it expedient to remove his plant and erect at a great cost the present commodious depot at Toronto, situated immediately adjoining the tracks of the Canadian Pacific and Grand Trunk Railways.

In concluding its remarks the Globe on this occasion said: "This is one of the most extensive machinery establishments on this continent."

What the house of Petrie was to the machinery trade of the Dominion in the early nineties, it remains to-day—Canada's Premier Machinery and Supply House.

H. W. PETRIE LIMITED
TORONTO and HAMILTON

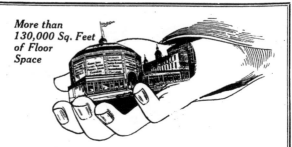

H.W. PETRIE LIMITED
TORONTO
HAMILTON
A Cincinnati Grinder
"DALTON SIX"
The Small Lathe for the Big Job
H. W. PETRIE
CINCINNATI SHAPERS
THE LODGE & SHIPLEY MANUFACTURING LATHE
The KEYSTONE to Satisfactory Service is Found in the Use of These Machine Tools ~ Sold by H.W. PETRIE Limited
Cincinnati Automatic Millers
CINCINNATI 16-INCH GEAR HOBBERS
For Accurate Gears and Quantity Production
CINCINNATI
COLBURN
BORING AND TURNING MILLS

MORROW
SCREWS, NUTS, STUDS

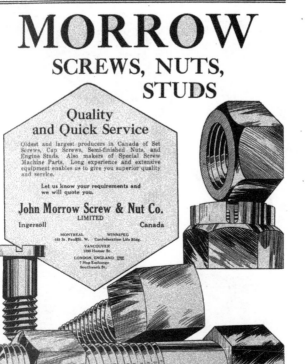

Quality and Quick Service

Oldest and largest producers in Canada of Set Screws, Cap Screws, Semi-finished Nuts, and Engine Studs. Also makers of Special Screw Machine Parts. Long experience and extensive equipment enables us to give you superior quality and service.

Let us know your requirements and we will quote you.

John Morrow Screw & Nut Co.
LIMITED
Ingersoll   Canada

MONTREAL   WINNIPEG
402 St. Paul St. W.   Confederation Life Bldg.
VANCOUVER
1290 Homer St.
LONDON, ENGLAND
7 Hop Exchange
Southwark St.

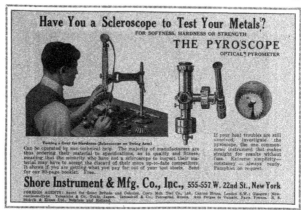

If interested tear out this page and place with letters to be answered.

"Production good, discipline good, everything running smoothly. Our Bonus System did it."

TIME is the heart, base and root of All Bonus Systems

A Bonus System will work wonders in connection with improving your output, keeping down overhead, getting full value from peak power loads, and in promoting discipline, contentment, and loyalty. It is doing this for hundreds of other firms, why not for you, too? In order to control the basic element of all bonus systems, which is TIME, you should use

International Time Recorders

An International makes every man his own time keeper. The time records are accurate, unchangeable—yet impersonal and impartial. Its constant urge to punctuality and discipline will cut down the daily total of lost working minutes that so often run into the hundreds of thousands and increase the overhead expense. Our Job Time Recorders take care of piece-work operations. Our In-and-Out Dial or Card Records insure punctuality at the daily shifts.

Send for our latest printed matter.

International Business Machines Co. Limited

F. E. MUTTON, Vice-President and General Manager

Head Office and Factory, 300-350 Campbell Avenue, Toronto

HALIFAX	ST. JOHN	QUEBEC	LONDON
44 Granville St.	18 Germain St.	501 Merger Bldg.	489 Richmond St.
OTTAWA	TORONTO	HAMILTON	
150 Queen St.	400½ Yonge St.	225 King St. E.	
WALKERVILLE	WINNIPEG	MONTREAL	VANCOUVER
44 Lincoln Rd.	227 McDermott Ave.	1 and 3 Notre Dame St. W.	110 Water St.

Also manufacturers of International Dayton Scales and International Electric Tabulators and Sorters

If interested tear out this page and place with letters to be answered.

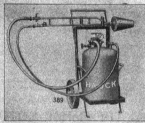

An Accessory Which Became a Principal

Jacobs Chucks, through their unusual accuracy of make and performance, have become a recognised principal part of the greater per cent. of drilling machines, and are used as a strong talking point by manufacturers, equippers, jobbers, and users alike.

Are you availing yourself of their super strength, durability, and accuracy?

The Jacobs Manufacturing Company
Hartford, Conn., U.S.A.

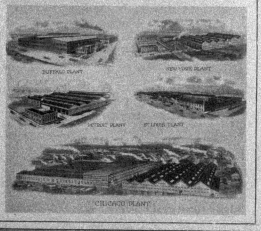

If what you need is not advertised, consult our Buyers' Directory and write advertisers listed under proper heading.

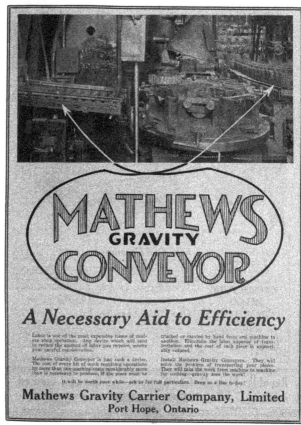

Republic

Flow Meters and

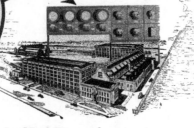

Republic

CO₂ Recorders

Point Out the Path to

Systematic Saving

THE Republic Flow Meter points out the path to systematic saving and tells you what your conditions are in the boiler room, the engine room, and in every department where steam is used.

It is a constant guide to saving, and, once the habit of saving has been acquired, the amount of savings is practically unlimited.

The Republic CO₂ Recorder has been developed to operate continuously under all boiler room conditions, the objectionable features found in all other CO₂ recorders having been eliminated. This enables the operator to observe the conditions of combustion at all times and to obtain the highest efficiency from his furnace—which means less work, less fuel, and less expense.

THE chart shown here demonstrates the relation between the percentage of CO₂ in the flue gases and the heat losses through the smokestack, showing the importance of observing and recording the CO₂ contents in the flue gases.

These curves were prepared by the Bureau of Mines at Washington, and we believe that they are absolutely reliable; the most up-to-date and exhaustive that it is possible to find.

By a reference to this chart it will be seen how the reduction of CO₂ shows the tremendous wastes that are going on in every plant to-day.

What is your percentage of CO₂? If you do not know, then you should know. If you do know, then refer to this chart, find out what your percentage of waste is, and the result, we feel sure, will be that the average CO₂ carried in your plant will be increased at once. Investigate Republic CO₂ Recorders now!

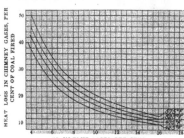

HEAT LOSS IN CHIMNEY GASES, PER CENT OF COAL FIRED

CO₂ IN FLUE GASES, PER CENT.

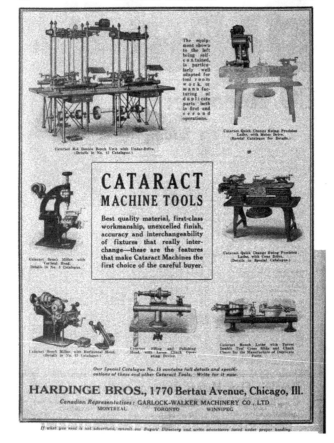

COLD DRAWN ALLOY STEEL BARS

Open Hearth and Electric Furnace

Chrome-Vanadium
Nickel—3½% and 5%
Chrome-Molybdenum
Chrome-Nickel
Carbon-Chrome
High Carbon-Special
Nickel-Chrome-Vanadium
Nickel-Chrome-Molybdenum

Our Steels are Melted—Rolled—Annealed—Heat Treated—Cold Drawn—in one large plant, and under supervision of Alloy Steel experts.

Our experience has been gained by specialization in the production of Alloy Steels.

UNITED ALLOY STEEL CORPORATION
CANTON, OHIO

Detroit	Philadelphia	Chicago
Syracuse	New York	San Francisco
Cleveland	Indianapolis	Portland

If interested tear out this page and place with letters to be answered.

The
Influential Four

For Self-Promotion and Sales
Promotion Use These Publications

CANADIAN MACHINERY
Weekly—$4.00 per year.
Covers the metal working field—
serving shop executives, purchasing agents and owners. First-class
mechanical paper and market
paper combined.

POWER HOUSE
Twice a month—$2.00 per year.
Serves the power plant engineer—
Steam, Electric Refrigeration, Hy-
draulic.

CANADIAN FOUNDRYMAN
Monthly—$2.00 per year.
A paper for the foundry owner,
superintendent and foreman. Claimed by its readers to be unexcelled.

MARINE ENGINEERING
Monthly—$2.00 per year.
Interprets marine engineering in
its broad sense, serving the ship
builders, navigation companies and
their officers. (The editor holds an
extra first-class B.O.T. certificate).

Published by

THE MACLEAN PUBLISHING COMPANY, LIMITED
143-153 UNIVERSITY AVE., TORONTO

A company whose idea of service has made it the largest concern of its kind in the British Empire.

Sample Copy and Advertising Rates sent
upon request.

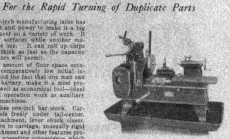

WALTHAM WHEEL SERVICE IN EASTERN CANADA

FOR three years the Standard Machinery & Supplies, Limited, have successfully represented the interests of the Waltham Grinding Wheel Company, of Waltham, Mass., for Eastern Canada. Its organization knows Waltham wheels and their application to grinding of all kinds.

It is natural that the Waltham Grinding Wheel Company of Canada, Limited, should place the sale of its products for Eastern Canada in the hands of the Standard Machinery & Supplies, Limited, who have so successfully built up their large business on the service they have rendered to grinding wheel users.

The combination of Waltham quality and Standard service for three years has meant much to grinders in Eastern Canada—it will mean even more to them in 1921.

ALOWALT  CARBOWALT

Sales Agents

STANDARD MACHINERY & SUPPLIES, LIMITED
261 Notre Dame Street West
MONTREAL

If interested tear out this page and place with letters to be answered.

If what you need is not advertised, consult our Buyers' Directory and write advertisers listed under proper heading.

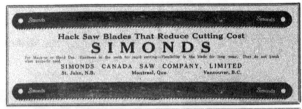

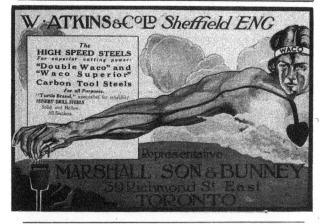

CRANES

ELECTRIC AND HAND OPERATED

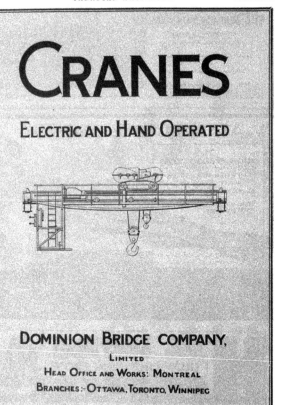

DOMINION BRIDGE COMPANY,
LIMITED
HEAD OFFICE AND WORKS: MONTREAL
BRANCHES:-OTTAWA, TORONTO, WINNIPEG

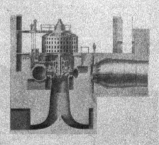

Mine Ventilating Drive.   Elevator.

Full Power Delivery

NO SLIPS   NO OIL BATHS   98.6 % EFFICIENT

By Service and Results the Morse Rocker Joint Silent Chain Drive has won recognition as the World's most efficient and durable power transmission. ..

Are you looking for improvement in power transmission, the saving of fuel, the saving of floor space by short centres, fixed speed ratios, improved quality, increased production, lower maintenance costs, longer life and dependability, the increase of profits now lost through slipping belts?

What more positive testimony of efficiency and durability of Morse Drives could be given than the fact that so many leaders of industry and engineers of prominence have specified Morse Rocker Joint Silent Chain for their Power Transmission?

BRUNO SLOTTING ATTACHMENT

For Shapers and Planers

Equips Your Shaper to do the
Work of a $1,000 Slotter

Used extensively in large plants on production as well as in small shops, tool rooms, etc., making possible intricate shaping on jigs, gauges, dies, key ways, slotting, etc.

It is bolted to the clapper in place of the tool post and can be applied as easily and as quickly as changing a tool.

Each attachment is equipped with two round shank cutting tools which may be turned to cut from any angle. Square shank cutting tools may be used if desired.

Specify size desired.

Sizes

No. 0 takes ⅛″ to ½″ round shank
No. 1 takes ½″ to ⅝″ round shank
No. 2 takes ⅝″ to ¾″ round shank

H. A. MOORE COMPANY
ROCHESTER, N.Y., U.S.A.

GALT SCREWS

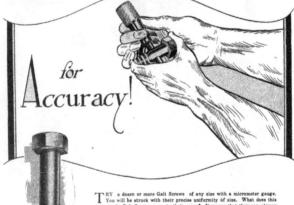

TRY a dozen or more Galt Screws of any size with a micrometer gauge. You will be struck with their precise uniformity of size. What does this accuracy in Galt Screws mean to their users? It means that they can always be depended upon to fit! Screws that always fit save a lot of time, trouble and money.

It is the dependable accuracy in Galt Screws that has brought them into wide demand for the building of machine tools throughout Canada.

TO THE MANUFACTURER:

We furnish accurate and therefore interchangeable parts, whether it be Cap and Set Screws, U.S.S. or Plain and Castellated S.A.E. Nuts, Taper Pins, or the more difficult automatic parts for engines, transmission and axles. Our facilities for doing secondary operations, such as drilling, milling, grinding, case-hardening and heat-treating are up-to-the-minute.

Submit Us Your Blueprints for Estimates

THE GALT MACHINE SCREW COMPANY, LIMITED
GALT, ONTARIO, CANADA

Eastern Representatives: F. Bacon & Co., 131 St. Paul St., Montreal, Quebec

MILLED SCREWS AND NUTS FOR EVERY PURPOSE

If what you need is not advertised, consult our Buyers' Directory and write advertisers listed under proper heading.

MODERN
Self-Opening Die Heads
(One Type - Two Styles - 17 Sizes)

Style
"H"

THEY HANDLE JOBS OTHERS CAN'T

Everywhere you find Moderns on difficult jobs. "Moderns" are almost daily supplanting other types because of efficiency on hard to handle work.

> ### WHAT'S THE SECRET?

The ability of Modern Self-Opening Die Heads to do work not possible for others is largely due to "Modern" distinctive design, although superfine workmanship is entitled to some of the credit.

Chasers in Modern Self-Opening Die Heads have ample and mechanically correct support. Users are not subjected to the annoyances of chasers shifting position during work. They will not "rock," "cock," or "bell at the mouth."

Based on record of accomplishment we've no hesitancy in claiming that Modern Self-Opening Die Heads will produce accurate, uniform threads the longest time without repairs.

Moderns have numerous mechanical advantages.

Send for complete technical bulletin, and study the reasons why Modern is preferred in the most exacting plants.

MODERN TOOL CO.   -   ERIE, PA.
Main Office and Works: State and Peach Streets

Branch Offices:

New York, N.Y., 3443 Lafayette Bldg.   Chicago, Ill., 32 N. Clinton St.
Detroit, Mich., 445 River Bldg.   Philadelphia, Pa., The Bourse.
Cleveland, Ohio, 433 Guardian Bldg.   Buffalo, N.Y., Associated Service Bldg.
Export Dept., Bishop Bldg., New York, N.Y.

Canadian Agents:
Rudel-Belnap Machinery Co., Ltd., Toronto and Montreal.

If interested tear out this page and place with letters to be answered.

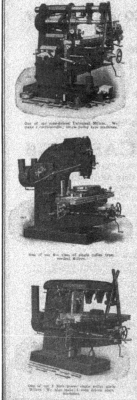

DEPENDABLE Steel Castings for Cement Mills and Mining Machinery, Stamp Mills, Crushing Plants, Excavating Outfits, Steel Car Wheels, Locomotive Driving Wheels and Frames, etc.

MACHINE-MOLDED GEARS

There's no pattern expense when you use "HISCO" Machine - Molded Gears. Made up to 18 feet in diameter.

They are right in quality, accurate, and low in price.

Write us for Estimates

HULL IRON & STEEL FOUNDRIES, LTD.
HULL, CANADA

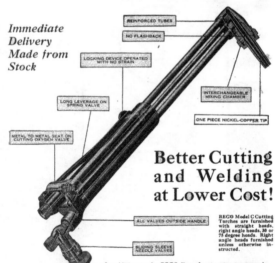
If what you need is not advertised, consult

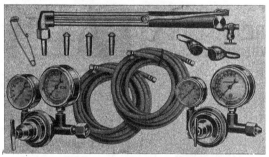
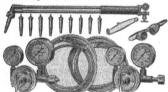

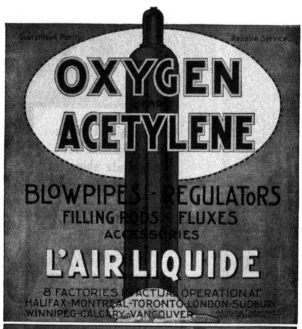

NAMCO Self-Opening
Dies—Capacity: 4½″

Acme Multiple Spindle Automatics—
Capacity: 4″ diam.

NAMCO Collapsing
Taps—Capacity: 7⅛″

NAMCO Automatic
Stud Threader—
Capacity: 1″ diam.

Gridley Multiple Spindle Automatics—
Capacity: 2⅝″ diam.

S.A.E. Plain and
Castellated Nuts

S.A.E. and U.S.S.
Cap and Set Screws

Gridley Automatic Turret Lathe—
Capacity: 5″ diam.

MOVE
YOUR GOODS
QUICKLY
SAFELY
ECONOMICALLY
on a
MORRIS RUNWAY

THE HERBERT MORRIS CRANE &
HOIST CO., LIMITED
NIAGARA FALLS   -   CANADA

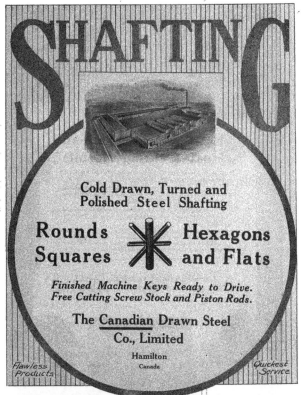

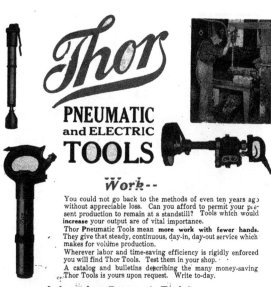

ANACONDA
BELT
Driving the Agitator in a Dissolving Tank

THE belt in the picture is driving the agitator in a dissolving tank in the recovery room of a kraft mill. The belt is subject to extreme heat; the only thing protecting it from the direct flame is a thin sheet of steel.

OTHER belts previously used on this machine only gave 50 per cent. of the service given by our ANACONDA, which is a heat, acid, and waterproof belt, and will withstand just such atmospheric conditions.

MAIN BELTING COMPANY OF CANADA LIMITED
MONTREAL: 10 St. Peter St.   TORONTO: 32 Front St. W.
WINNIPEG: W. W. Hicks, 567 Banning St.
CALGARY and EDMONTON: Gorman, Clancey & Grindley, Edmonton

If interested tear out this page and place with letters to be answered.

MUELLER
Heavy Duty Engine Lathe

*A machine that gives
Universal Satisfaction*

Give It Your Attention Before
Purchasing Another Lathe

ALL we have to say about the Heavy Duty MUELLER Engine Lathe is this: If you need an 18-in. Engine Lathe and fail to get a proposal on the MUELLER, you have missed an opportunity of a lifetime.

The next move is yours.

**MUELLER Radial Drills and Lathes
Sold in Canada Exclusively by**

——— FOR SALE IN CANADA BY ———

The **GEO. F. FOSS MACHINERY & SUPPLY COMPANY**
LIMITED
305 St. James Street, Montreal. Quebec

If what you need is not advertised, consult our Buyers' Directory and write advertisers listed under proper heading.

If what you need is not advertised, consult our Buyers' Directory and write advertisers listed under proper heading.

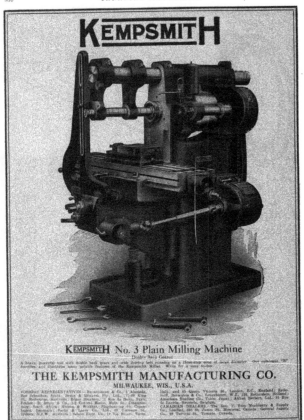

KEMPSMITH No. 3 Plain Milling Machine

Double Back Geared

A heavy, powerful tool with double back gears and wide driving belt running on a three-step cone of large diameter. Our catalogue "B" describes and illustrates many valuable features of the Kempsmith Miller. Write for a copy to-day.

THE KEMPSMITH MANUFACTURING CO.
MILWAUKEE, WIS., U.S.A.

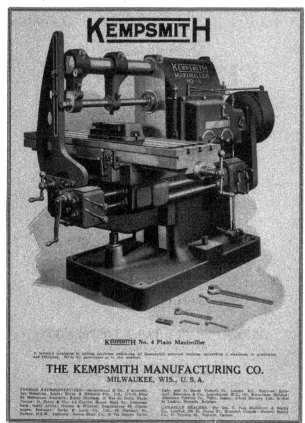

KEMPSMITH

KEMPSMITH No. 4 Plain Maximiller

A veritable maximum in milling machines embracing all Kempsmith patented features, permitting a maximum in production and efficiency. Write for particulars as to this machine.

THE KEMPSMITH MANUFACTURING CO.
MILWAUKEE, WIS., U.S.A.

If interested tear out this page and place with letters to be answered.

The OLIVER
No. 80 Variety Saw Bench

Either
Belt
or
Motor
Driven

View of
Self-contained
Motor-on-Arbor
Machine

This New Labor-Saver Commands Your Attention

IF YOU have a variety of work, the Oliver No. 80 is the machine you should have in your shop. It will save labor, time and expense, and the fact that it is an Oliver is proof of its general efficiency.

The machine is designed and built strictly to meet present-day requirements in the production of variety woodwork, wood patterns, furniture, automobile bodies, carriage bodies, agricultural implements, talking machine cabinets, and other high-grade woodwork. It will do ripping, cross-cutting and dadoing quickly and

efficiently. Cuts a perfect mitre and measures an angle instantly and accurately. It will cut off a length or rip to width. No calculating required to do this, neither does the operator have to refer to a rule. The construction is rugged throughout and all necessary attachments such as mitre cut-off gauge, etc., are provided. Two types of tables can be furnished, plain and universal.

There are many other features of the "Oliver" No. 80 you should know about. Send for complete description.

Before placing your next order for Pattern Shop Equipment, Woodworking Machinery, or Engine Lathes, you should investigate the "Oliver" line.

Oliver Machinery Co., Grand Rapids, Mich.

BRANCHES: New York, Chicago, St. Louis, Los Angeles,
San Francisco, Seattle, Salt Lake City, Denver, Phoenix,
Manchester, Eng.

*Beatty Bros., Limited, Fergus, Ontario,
with branches throughout Canada in a
recent issue of their weekly sales bulletin, "Facts," give a very clear idea of
the value of Hardware and Metal to the
Canadian trade. This is what "Facts"
has to say:—*

"Do You Read Hardware & Metal?"

"We do not often take space in FACTS to boost any particular paper, but we are making an exception in the case of Hardware and Metal.

"If we did not know that so many of our hardware salesmen subscribed to it, we should be tempted to republish in FACTS some of their recent articles on the price situation.

"Some ten or fifteen trade papers—as distinct from farm papers—come into the FACTS editorial sanctum. There is none with half the weight and breadth of information that is possessed by Hardware and Metal.

"During the easier days of the past few months,—when the main question was not what the Sales Force could sell but what the Factory could supply—we may have been tempted to let Hardware and Metal go by the board.

"You can't afford to neglect it in these days. It is just full of authoritative information on the price situation in particular and hardware merchandising in general.

"So read it carefully. It gives you just the dope you need. When you tell the merchant that prices are not going to slump—well, he may not quite believe you. He feels you may have some axe to grind.

"But when you can quote or indicate some article in Hardware and Metal to prove your point,—it convinces.

"Remember, Hardware and Metal carries a lot of weight. It is known to be thoroughly unprejudiced and reliable. There are very few stronger trade papers in America so far as their own particular fields are concerned.

"So don't neglect your weekly reading of Hardware and Metal. Clip out any specially good points and paste them in your notebook for easy reference.

"The first two or three items on page 40 of the October 23 issue are dandies—probably you noticed them. If not, turn up your copy and read them.

"We all have to be well posted in these days or we go under. This is no time for the half-informed salesman to be on the road."

HARDWARE AND METAL
"Canada's National Hardware Weekly"
143 University Avenue, Toronto, Canada
Montreal Branch, Southam Building   Winnipeg Branch, Union Trust Building

Published every Saturday since 1888. The only weekly hardware paper in Canada and the only hardware paper in Canada that gives you a circulation statement audited by the Audit Bureau of Circulations.

If interested tear out this page and place with letters to be answered.

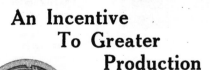

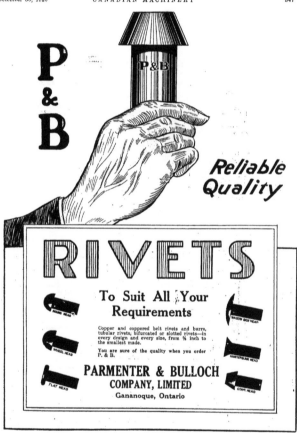

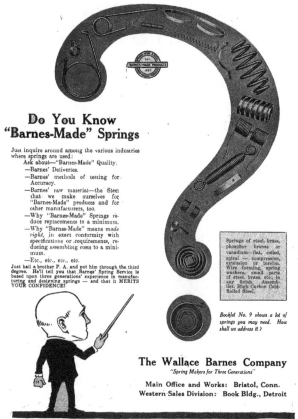

Do You Know "Barnes-Made" Springs

Just inquire around among the various industries where springs are used:

Ask about—"Barnes-Made" Quality.

—Barnes' Deliveries.

—Barnes' methods of testing for Accuracy.

—Barnes' raw material—the Steel that we make ourselves for "Barnes-Made" products and for other manufacturers, too.

—Why "Barnes-Made" Springs reduce replacements to a minimum.

—Why "Barnes-Made" means *made right*, in exact conformity with specifications or requirements, reducing assembling costs to a minimum.

—Etc., etc., etc., etc.

Just hail a brother P. A. and put him through the third degree. He'll tell you that Barnes' Spring Service is based upon three generations' experience in manufacturing and designing springs — and that it MERITS YOUR CONFIDENCE!

Springs of steel, brass, phosphor bronze or vanadium—flat, coiled, spiral — compression, extension or torsion. Wire forming, spring washers, small parts of steel, brass, etc., in any finish. Assemblies. High Carbon Cold-Rolled Steel.

Booklet No. 9 shows a lot of springs you may need. How shall we address it ?

The Wallace Barnes Company

"Spring Makers for Three Generations"

Main Office and Works: Bristol, Conn.

Western Sales Division: Book Bldg., Detroit

If interested tear out this page and place with letters to be answered.

Canadian Machinery
BUYERS DIRECTORY

If what you want is not here, write us, and we will tell you where to get it. Let us suggest that you consult also the advertisers' index—last page of book, after having secured advertisers' names from this directory. The information you desire may be found in the advertising pages. This department is maintained for the benefit and convenience of our readers. The insertion of our advertisers' names under proper headings is gladly undertaken, but does not become part of an advertising contract.

BUYERS' DIRECTORY

Castings, Aluminum
Canada Electric Castings Co., Ltd., Orillia.
Canada Metal Co., Ltd., Toronto, Ont.
Electric Steel & Engineering Co., Welland, Ont.
Tallman Brass & Metal Co., Hamilton, Ont.

Castings, Brass and Bronze
Algoma Steel Corp., Ltd., Sault Ste. Marie, Ont.
Canada Electric Castings Co., Ltd., Orillia.
Canada Metal Co., Ltd., Toronto, Ont.
Can. Driver-Harris Co., Walkerville, Ont.
Electric Steel & Engineering Co., Welland, Ont.
International Machinery & Supply Co., Montreal, Que.
Tallman Brass & Metal Co., Hamilton, Ont.

Castings, Marine
Can. Steel Foundries, Montreal, Que.

Castings, Die Molded
Electric Steel & Engineering Co., Welland, Ont.
Fisher Motor Co., Ltd., Orillia, Ont.
Franklin Die-Casting Corp., Syracuse, N.Y.
Katie Foundry Co., Galt, Ont.

Castings, Ferro-Alloy
Can. Steel Foundries, Montreal, Que.

Castings, Iron
Algoma Steel Corp., Ltd., Sault Ste. Marie, Ont.
Bernard Industrial Co., A., Fortierville, Que.
Bilton Machine Co., Bridgeport, Conn.
Brown, Boggs & Co., Ltd., Hamilton, Ont.
Canada Electric Castings Co., Ltd., Orillia.
Electric Steel & Engineering Co., Welland, Ont.
Fleck Ltd., Alexander, Ottawa, Ont.
Hanna & Co., M. A., Cleveland, Ohio.
Hepburn Ltd., John T., Toronto, Ont.
Katie Foundry Co., Galt, Ont.
Kennedy & Sons, Wm., Owen Sound, Ont.
McDougall Co., Ltd., R., Galt, Ont.
Victoria Foundry Co., Ltd., Ottawa, Ont.
Walker & Sons Metal Products, Ltd., Hiram, Walkerville, Ont.

Castings, Hyd. Press
Can. Steel Foundries, Montreal, Que.

Castings, Nichrome
Can. Driver-Harris Co., Walkerville, Ont.
Electric Steel & Engineering Co., Welland, Ont.
Hull Iron & Steel Foundries, Hull, Que.
Katie Foundry Co., Galt, Ont.
Walker & Sons Metal Products, Ltd., Hiram, Walkerville, Ont.

Castings, Semi-Steel
Davidson Mfg. Co., Thos., Montreal, Que.
Electric Steel & Engineering Co., Welland, Ont.
Halcomb Steel Co., Syracuse, N.Y.
Hull Iron & Steel Foundries, Hull, Que.
Katie Foundry Co., Galt, Ont.
Manitoba Steel Foundries, Ltd., Winnipeg, Man.

Castings, Steel
Can. Steel Foundries, Montreal, Que.
Kennedy & Sons, Wm., Owen Sound, Ont.
Swedish Crucible Steel Co. of Can., Ltd., Windsor, Ont.

Cements, Iron
Smooth Mfg. Co., Jersey City, N.J.

Centering Machines
Bertram & Son Co., Ltd., The John, Dundas, Ont.
Garlock-Walker Mchy. Co., Toronto, Ont.

Chains (See Sprockets and Chains)
Morris Crane & Hoist Co., Ltd., Niagara Falls, Ont.
Morse Chain Co., Ithaca, N.Y.
Philadelphia Gear Works, Philadelphia, Pa.
Renold (Hans) of Canada, Ltd., Montreal, Que.
Wright Mfg. Co., Lisbon, Ohio.

Chains, Driving
Can. Link-Belt Co., Toronto, Ont.
Greenfield Tap & Die Corp., Galt, Ont.
Jones & Glassco, Montreal, Que.
Morse Chain Co., Ithaca, N.Y.
Reading Chain & Block Co., Reading, Pa.
Renold (Hans) of Canada, Ltd., Montreal, Que.
Wright Mfg. Co., Lisbon, Ohio.

Chasers
Bertram & Son Co., Ltd., The John, Dundas, Ont.
Jones & Lamson Machine Co., Springfield, Vt.
Landis Machine Co., Inc., Waynesboro, Pa.
Pratt & Whitney Co., of Canada, Ltd., Dundas, Ont.

Chemists
Toronto Testing Laboratory, Toronto, Ont.

Chucking Machines
Acme Machine Tool Co., Cincinnati, Ohio
Brown & Sharpe Mfg. Co., Providence, R.I.
Gisholt Machine Co., Madison, Wis.
Pratt & Whitney Co., of Canada, Ltd., Dundas, Ont.
Steinle Turret Machine Co., Madison, Wis.
Warner & Swasey Co., Cleveland, Ohio.

Chucks, Drill
Jacobs Mfg. Co., Hartford, Conn.

Chucks, Drill and Tap
Aikenhead Hardware Ltd., Toronto, Ont.
Canadian SKF Co., Toronto, Ont.
Cushman Chuck Co., Hartford, Conn.
Dom. Steel Products Co., Brantford, Ont.
Goodell & Pratt Co., Greenfield, Mass.
Morrow Screw & Nut Co., Ltd., John, Ingersoll, Ont.
Morse Twist Drill & Machine Co., New Bedford, Mass.
McCroskey Tool Corp., Meadville, Pa.
Pratt & Whitney Co., of Canada, Ltd., Dundas, Ont.
Skinner Chuck Co., New Britain, Conn.
Union Mfg. Co., New Britain, Conn.
Williams & Wilson, Ltd., Montreal, Que.

Chucks, Lathe
Aikenhead Hardware Ltd., Toronto, Ont.
Bertram & Son Co., Ltd., The John, Dundas, Ont.
Cushman Chuck Co., Hartford, Conn.
Dom. Steel Products Co., Brantford, Ont.
Foss Machinery & Supply Co., Geo. F., Montreal, Que.
Geometric Tool Co., New Haven, Conn.
Gisholt Machine Co., Madison, Wis.
Ker & Goodwin Machine Co., Brantford, Ont.
Petrie, Ltd., H. W., Toronto, Ont.
Skinner Chuck Co., New Britain, Conn.
Union Mfg. Co., New Britain, Conn.
Williams & Wilson, Ltd., Montreal, Que.

Chucks, Magnetic
Pratt & Whitney Co., of Canada, Ltd., Dundas, Ont.

Chucks, Planer
Bertram & Son Co., Ltd., The John, Dundas, Ont.
Cushman Chuck Co., Hartford, Conn.
Skinner Chuck Co., New Britain, Conn.
Union Mfg. Co., New Britain, Conn.

Chucks, Vertical Boring Mill
Bertram & Son Co., Ltd., The John, Dundas, Ont.
Gisholt Machine Co., Madison, Wis.
Skinner Chuck Co., New Britain, Conn.
Union Mfg. Co., New Britain, Conn.

Clamps, Machinists'
Columbia Hdwe. Division, Cleveland, O.
Dickow, Fred C., Chicago, Ill.
Starrett Co., L. S., Athol, Mass.

Cleaners, Metal, Waste, General
Oakley Chemical Co., New York, N.Y.
Levis, The, Toronto, Canada.
Scythes & Company, Ltd., Toronto, Ont.

Clocks, Time
Gisholt Machine Co., Madison, Wis.
International Business Machines Co., Toronto, Ont.

Clutches, Friction
Bernard Industrial Co., A., Fortierville, Que.
Can. Link-Belt Co., Toronto, Ont.
Ford-Smith Machine Co., Hamilton, Ont.
Johnson Machine Co., Carlyle, Manchester, Conn.
Positive Clutch & Pulley Works, Toronto, Ont.

Coal and Ash Handling Machinery
Can. Ingersoll-Rand Co., Ltd., Sherbrooke, Que.
Can. Link-Belt Co., Toronto, Ont.
Morris Crane & Hoist Co., Ltd., Niagara Falls, Ont.

Coal-Storage Systems
Can. Link-Belt Co., Toronto, Ont.

Collars, Shaft or Set
Canada Foundries & Forgings Co., Welland, Ont.
Can. Link-Belt Co., Toronto, Ont.

Collets
Ackworthie, Ltd., John, Birmingham, Eng.
Butterfield & Co., Inc., Rock Island, Que.
Canada Machinery Corp., Galt, Ont.
Hendey Machine Co., Torrington, Conn.
Kearney & Trecker Co., Milwaukee, Wis.
Petrie, Ltd., H. W., Toronto, Ont.
Pratt & Whitney Co., of Canada, Ltd., Dundas, Ont.

Compounds, Carburizing, Case Hardening and Tempering
Cataract Refining Co., Toronto, Ont.

Compounds, Cleaning
Can. Hanson & Van Winkle Co., Ltd., Toronto, Ont.
Oakley Chemical Co., New York, N.Y.

Compounds, Cutting, Drilling, Grinding, Screw Cutting
Atkins & Co., Inc., E. C., Indianapolis, I.
Cataract Refining Co., Toronto, Ont.
Oakley Chemical Co., New York, N.Y.

Compressors, Air
Curtis Pneumatic Machinery Co., St. Louis, Mo.

Compressors, Air and Gas
Can. Ingersoll-Rand Co., Ltd., Sherbrooke, Que.
Garlock-Walker Mchy. Co., Toronto, Ont.
Holden Co., Ltd., Montreal, Que.
Petrie, Ltd., H. W., Toronto, Ont.

Cones, Friction
Norton Co. of Can., Ltd., Hamilton, Ont.

Connecting Rods and Straps
Canada Foundries & Forgings Co., Welland, Ont.

Contract Work
Ford-Smith Machine Co., Hamilton, Ont.
Victoria Foundry Co., Ltd., Ottawa, Ont.

Conveyors and Elevators (See Elevators)
Jones & Glassco, Montreal, Que.
Main Belting Co. of Can., Montreal, Que.
Mathews Gravity Carrier Co., Port Hope, Ont.

Copper
Brown's Copper & Brass Rolling Mills, Ltd., Toronto, Ont.

Cotter Pins
Morrow Screw & Nut Co., Ltd., John, Ingersoll, Ont.

Counterbores
Cleveland Twist Drill Co., Cleveland, O.
Eclipse Counterbore Co., Ltd., Walkerville, Ont.
Ingersoll Machine & Tool Co., Ltd., Ingersoll, Ont.
Pratt & Whitney Co., of Canada, Ltd., Dundas, Ont.

Counters, Revolution
Aikenhead Hardware Ltd., Toronto, Ont.
Starrett Co., L. S., Athol, Mass.

Countershafts
Bertram & Son Co., Ltd., The John, Dundas, Ont.
Canada Foundries & Forgings Co., Welland, Ont.
Ford-Smith Machine Co., Hamilton, Ont.
Johnson Machine Co., Carlyle, Manchester, Conn.
Kempsmith Mfg. Co., Milwaukee, Wis.
McDougall Co., Ltd., R., Galt, Ont.

Countersinks
Butterfield & Co., Inc., Rock Island, Que.
Eclipse Counterbore Co., Ltd., Walkerville, Ont.
Pratt & Whitney Co., of Canada, Ltd., Dundas, Ont.

Couplers, Car and Locomotive
Can. Steel Foundries, Montreal, Que.

Couplings, Flexible
Holden Co., Ltd., Montreal, Que.

Couplings, Rigid
Bernard Industrial Co., A., Fortierville, Que.

Couplings, Shaft
Bilton Machine Co., Bridgeport, Conn.
Can. Link-Belt Co., Toronto, Ont.

Cranes, Electric
Bertram & Son Co., Ltd., The John, Dundas, Ont.
Can. Link-Belt Co., Toronto, Ont.
Dominion Bridge Co., Ltd., Lachine, Que.
Hepburn Ltd., John T., Toronto, Ont.
Morris Crane & Hoist Co., Ltd., Niagara Falls, Ont.
Northern Crane Works, Walkerville, Ont.
Shepherd Electric Crane & Hoist Co., Montour Falls, N. Y.

Cranes, Hand (See Hoists, Hand)
Dominion Bridge Co., Ltd., Lachine, Que.
Hepburn Ltd., John T., Toronto, Ont.
Morris Crane & Hoist Co., Ltd., Niagara Falls, Ont.
Northern Crane Works, Walkerville, Ont.
Sheffield Engineering Supplies, Ltd., Montreal, Que.

Cranes, Locomotive
Can. Link-Belt Co., Toronto, Ont.
Holden Co., Ltd., Montreal, Que.

Cranes, Traveling
Bertram & Son Co., Ltd., The John, Dundas, Ont.
Can. Link-Belt Co., Toronto, Ont.
Dominion Bridge Co., Ltd., Lachine, Que.
Hepburn Ltd., John T., Toronto, Ont.
Morris Crane & Hoist Co., Ltd., Niagara Falls, Ont.
Northern Crane Works, Walkerville, Ont.
Reading Chain & Block Co., Reading, Pa.
Shepherd Electric Crane & Hoist Co., Montour Falls, N. Y.

Crank Pin Turning Machines
Garlock-Walker Mchy. Co., Toronto, Ont.
Herbert Ltd., Alfred, Toronto, Ont.
Underwood Corp., H. B., Philadelphia, Pa.

Cutters, Flue
Holden Co., Ltd., Montreal, Que.

Cutters, Gear
Armstrong Whitworth Co. of Can., Ltd., Montreal, Que.
Brown & Sharpe Mfg. Co., Providence, R.I.
Butterfield & Co., Inc., Rock Island, Que.
Pratt & Whitney Co., of Canada, Ltd., Dundas, Ont.

Cutters, High Speed
Atkins & Co., Inc., E. C., Indianapolis, I.
Bilton Machine Co., Bridgeport, Conn.
Butterfield & Co., Inc., Rock Island, Que.
Eclipse Counterbore Co., Ltd., Walkerville, Ont.
Ingersoll Machine & Tool Co., Ltd., Ingersoll, Ont.
Kearney & Trecker Co., Milwaukee, Wis.
Madison Mfg. Co., Muskegon, Mich.
Pilot Steel & Tool Co., Montreal, Que.
Pratt & Whitney Co., of Canada, Ltd., Dundas, Ont.

Cutters, Milling
Bilton Machine Co., Bridgeport, Conn.
Brown & Sharpe Mfg. Co., Providence, R.I.
Butterfield & Co., Inc., Rock Island, Que.
Cleveland Milling Machine Co., Cleveland.
Elliott & Whitehall Tool Co., Galt, Ont.
Ingersoll Machine & Tool Co., Ltd., Ingersoll, Ont.
Kearney & Trecker Co., Milwaukee, Wis.
Morse Twist Drill & Machine Co., New Bedford, Mass.
Pilot Steel & Tool Co., Montreal, Que.
Wilt Twist Drill Co. of Canada, Ltd., Walkerville, Ont.

Cutters, Stay Bolt
Acme Machinery Co., Cleveland, Ohio.
Landis Machine Co., Inc., Waynesboro, Pa.
Pratt & Whitney Co., of Canada, Ltd., Dundas, Ont.

Cutters, Thread
Butterfield & Co., Inc., Rock Island, Que.
Greenfield Tap & Die Corp., Galt, Ont.
Jones & Lamson Machine Co., Springfield, Vt.
Landis Machine Co., Inc., Waynesboro, Pa.

Cutting-Off Machines
Bertram & Son Co., Ltd., The John, Dundas, Ont.
Brown & Sharpe Mfg. Co., Providence, R.I.
Garlock-Walker Mchy. Co., Toronto, Ont.
Greenfield Tap & Die Corp., Galt, Ont.
Petrie, Ltd., H. W., Toronto, Ont.
Starrett Co., L. S., Athol, Mass.

Cutting-Off Machines, Pipe (See Pipe Cutting and Threading Machines)
Landis Machine Co., Inc., Waynesboro, Pa.
McDougall Co., Ltd., R., Galt, Ont.
Williams Tool Corp. of Can., Ltd., Brantford, Ont.

Cutting-Off Tools
Armstrong Bros. Tool Co., Chicago, Ill.
Pilot Steel & Tool Co., Montreal, Que.

Cutting Oil Filters (See Oil Filtering Systems)
Bowser & Co., Inc., S. F., Fort Wayne, I.
Cataract Refining Co., Toronto, Ont.

Cutting, Oxy-Acetylene
Carter Welding Co., Toronto, Ont.
Holden Co., Ltd., Montreal, Que.
Perdue, W. B., San Francisco, Calif.
Prest-O-Lite Co. of Can., Toronto, Ont.
Turner Brass Works, Sycamore, Ill.
Union Carbide Co. of Can., Welland, Ont.

Dealers, Machinery (See Searchlight Section)
Ford-Smith Machine Co., Hamilton, Ont.
Petrie, Ltd., H. W., Toronto, Ont.

Deckle Straps
Can. Consolidated Rubber Co., Ltd., Montreal, Que.

Diamonds, Black and Rough
Joyce-Koebel Co., Inc., New York, N.Y.

Diamond, Carbon and Bortz
Joyce-Koebel Co., Inc., New York, N.Y.

Diamond Tools
Aikenhead Hardware Ltd., Toronto, Ont.
Can. Desmond-Stephan Co., Hamilton, Ont.
Ford-Smith Machine Co., Hamilton, Ont.
Wheel Trueing Tool Co., Detroit, Mich.

Diamond Crossings
Can. Steel Foundries, Montreal, Que.

Die Sinking Machines, Automatic
Jones & Lamson Machine Co., Springfield, Vt.
Walcott Lathe Co., Jackson, Mich.

Dies, Screw and Thread Cutting
Ackworthie, Ltd., John, Birmingham, Eng.
Butterfield & Co., Inc., Rock Island, Que.
Greenfield Tap & Die Corp., Galt, Ont.
Jardine & Co., A. B., Hespeler, Ont.
Jones & Lamson Machine Co., Springfield, Vt.
Landis Machine Co., Inc., Waynesboro, Pa.
Murchey Machine & Tool Co., Detroit, Mich.
National Acme Co., Cleveland, Ohio.

Dies, Sheet-Metal and Sub-Press (See Tool Work)
Brown, Boggs & Co., Ltd., Hamilton, Ont.
Dyrab Steel (Consolidated) Ltd., Toronto, Ont.
Fisher Motor Co., Ltd., Orillia, Ont.
Ford-Smith Machine Co., Hamilton, Ont.
Toledo Machine & Tool Co., Toledo, Ohio

Dies, Forging
Brown, Boggs & Co., Ltd., Hamilton, Ont.
Canada Foundries & Forgings Co., Welland, Ont.
Canadian Atlas Crucible Steel Co., Ltd., Toronto, Ont.
Kimber & Hillier Mfg. Co., St. Catharines, Ont.

Dies, Self-Opening, Adjustable
Geometric Tool Co., New Haven, Conn.
Herbert Ltd., Alfred, Toronto, Ont.
Jones & Lamson Machine Co., Springfield, Vt.
Landis Machine Co., Inc., Waynesboro, Pa.
Murchey Machine & Tool Co., Detroit, Mich.
National Acme Co., Cleveland, Ohio.
Prest-O-Lite Co. of Can., Toronto, Ont.
Pratt & Whitney Co., of Canada, Ltd., Dundas, Ont.
Victor Tool Co., Waynesboro, Pa.

Dies, Threading-Opening
Jardine & Co., A. B., Hespeler, Ont.
Jones & Lamson Machine Co., Springfield, Vt.
Landis Machine Co., Inc., Waynesboro, Pa.
Morse Twist Drill & Machine Co., New Bedford, Mass.

Murchey Machine & Tool Co., Detroit, Mich.
National Acme Co., Cleveland, Ohio.
Pratt & Whitney Co., of Canada, Ltd., Dundas, Ont.
Rapid Tool & Machine Co., Lachine, Que.

Disc Cement
Ritchey Supply Co., Toronto, Ont.
Wausau Abrasives Co., Chicago, Ill.

Dividing Heads
Ackworthie, Ltd., John, Birmingham, Eng.
Ford-Smith Machine Co., Hamilton, Ont.
Hendey Machine Co., Torrington, Conn.
Kearney & Trecker Co., Milwaukee, Wis.
Petrie, Ltd., H. W., Toronto, Ont.

Dogs, Lathe and Milling Machine
Armstrong Bros. Tool Co., Chicago, Ill.

Drafting Boards and Tables
Darling Bros., Ltd., Montreal, Que.
Economy Drawing Table & Mfg. Co., Adrian, Mich.
Hughes Owens Co., Ltd., Montreal, Que.

Drafting Materials
American Lead Pencil Co., New York City, N.Y.
Darling Bros., Ltd., Montreal, Que.
Economy Drawing Table & Mfg. Co., Adrian, Mich.
Hughes Owens Co., Ltd., Montreal, Que.

Dressers, Grinding Wheel
Carborundum Co., Niagara Falls, N.Y.
Dom. Abrasive Wheel Co., Ltd., Mimico, Ont.
Ford-Smith Machine Co., Hamilton, Ont.
Joyce-Koebel Co., Inc., New York, N.Y.
Norton Co. of Can., Ltd., Hamilton, Ont.

Drill Holders
Armstrong Bros. Tool Co., Chicago, Ill.

Drill Rods
Aikenhead Hardware Ltd., Toronto, Ont.
Canadian Atlas Crucible Steel Co., Ltd., Toronto, Ont.

Drill Speeders
Canada Machinery Corp., Galt, Ont.

Drilling Machine Heads
Henry & Wright Mfg. Co., Hartford, Conn.
Hoefer Mfg. Co., Freeport, Ill.
United States Machine Tool Co., Cincinnati, Ohio.

Drilling Machines, Automatic
Hoosier Drilling Mach. Co., Goshen, Ind.
National Automatic Tool Co., Richmond, Ind.

Drilling Machines, Bench
Beacon Engineering Co., Tipton, England.
Can. Blower & Forge Co., Ltd., Kitchener.
Goodell & Pratt Co., Greenfield, Mass.
Henry & Wright Mfg. Co., Hartford, Conn.
High Speed Hammer Co., Rochester, N.Y.
Petrie, Ltd., H. W., Toronto, Ont.
Pratt & Whitney Co., of Canada, Ltd., Dundas, Ont.
Terry & Co., John C., Birmingham, Eng.
U.S. Electrical Tool Co., Cincinnati, O.
Wisconsin Electric Co., Racine, Wis.

Drilling Machines, Electric and Hand
Aikenhead Hardware Ltd., Toronto, Ont.
Cincinnati Electrical Tool Co., Cincinnati, Ohio.
Foss Machinery & Supply Co., Geo. F., Montreal, Que.
Garlock-Walker Mchy. Co., Toronto, Ont.
High Speed Hammer Co., Rochester, N.Y.
Holden Co., Ltd., Montreal, Que.
Independent Pneumatic Tool, Chicago, Ill.
Jardine & Co., A. B., Hespeler, Ont.
Wisconsin Electric Co., Racine, Wis.

Drilling Machines, Gang
Bertram & Son Co., Ltd., The John, Dundas, Ont.
Bilton Machine Co., Bridgeport, Conn.
Garlock-Walker Mchy. Co., Toronto, Ont.
Hoefer Mfg. Co., Freeport, Ill.
Pratt & Whitney Co., of Canada, Ltd., Dundas, Ont.

Drilling Machines, Heavy Duty
Bertram & Son Co., Ltd., The John, Dundas, Ont.
Canada Machinery Corp., Galt, Ont.
Garlock-Walker Mchy. Co., Toronto, Ont.
Hoosier Drilling Mach. Co., Goshen, Ind.
Rockford Lathe & Drill Co., Rockford, Ill.

Drilling Machines, Horizontal (See Boring, Drilling and Milling Machines, Horizontal)
Avery Drilling Machine Co., Cincinnati, Ohio.
Canada Machinery Corp., Galt, Ont.
Gisholt Machine Co., Madison, Wis.
Holly, R. B., Toronto, Ont.
Lucas Machine Tool Co., Cleveland, Ohio.
Rockford Drilling Machine Co., Rockford, Ill.
Rockford Lathe & Drill Co., Rockford, Ill.

Drilling Machines, Multiple Spindle
Beacon Engineering Co., Tipton, England.
Bertram & Son Co., Ltd., The John, Dundas, Ont.
Bilton Machine Co., Bridgeport, Conn.
Henry & Wright Mfg. Co., Hartford, Conn.
Hoefer Mfg. Co., Freeport, Ill.
National Acme Co., Cleveland, Ohio.
National Automatic Tool Co., Richmond, Ind.
Terry & Co., John C., Birmingham, Eng.

Drilling Machines, Pneumatic
Can. Ingersoll-Rand Co., Ltd., Sherbrooke, Que.
Cleveland Pneumatic Tool Co., Toronto, Ont.

BUYERS' DIRECTORY

Garlock-Walker Mchy. Co., Toronto, Ont.
Holden Co., Ltd., Montreal, Que.
Independent Pneumatic Tool, Chicago, Ill.

Drilling Machines, Portable
Haskins Co., R. G., Chicago, Ill.
Holden Co., Ltd., Montreal, Que.
Independent Pneumatic Tool, Chicago, Ill.
Jardine & Co., A. B., Hespeler, Ont.
Wisconsin Electric Co., Racine, Wis.

Drilling Machines, Radial
Bertram & Son Co., Ltd., The John, Dundas, Ont.
Canada Machinery Corp., Galt, Ont.
Foss Machinery & Supply Co., Geo. F., Montreal, Que.
Garlock-Walker Mchy. Co., Toronto, Ont.
Henry & Wright Mfg. Co., Hartford, Conn.
Herbert Ltd., Alfred, Toronto, Ont.
Mueller Machine & Tool Co., Cincinnati, Ohio.
Petrie, Ltd., H. W., Toronto, Ont.
Ryerson & Son, Jos. T., Chicago, Ill.
Toomey Inc., Frank, Philadelphia, Pa.
Williams Machinery Co., A. R., Toronto, Ont.
Williams Machinery & Supply Co., A. R., Montreal, Que.

Drilling Machines, Sensitive
Avery Drilling Machine Co., Cincinnati, Ohio.
Beacon Engineering Co., Tipton, England.
Bilton Machine Co., Bridgeport, Conn.
Henry & Wright Mfg. Co., Hartford, Conn.
Herbert Ltd., Alfred, Toronto, Ont.
High Speed Hammer Co., Rochester, N.Y.
Hoosier Drilling Mach. Co., Goshen, Ind.
Jones & Shipman, of Leicester, England. Toronto, Ont.
Pratt & Whitney Co., of Canada, Ltd., Dundas, Ont.
Rockford Drilling Machine Co., Rockford, Ill.
Terry & Co., John C., Birmingham, Eng.
United States Machine Tool Co., Cincinnati, Ohio.
Williams Machinery Co., A. R., Toronto, Ont.
Wisconsin Electric Co., Racine, Wis.

Drilling Machines, Turret
Gisholt Machine Co., Madison, Wis.
Steinle Turret Machine Co., Madison, Wis.
Williams Machinery Co., A. R., Toronto, Ont.

Drilling Machines, Vertical
Aurora Tool Works, Aurora, Ind.
Avery Drilling Machine Co., Cincinnati, Ohio.
Bertram & Son Co., Ltd., The John, Dundas, Ont.
Canada Machinery Corp., Galt, Ont.
Can. Blower & Forge Co., Ltd., Kitchener.
Garlock-Walker Mchy. Co., Toronto, Ont.
Herbert Ltd., Alfred, Toronto, Ont.
Hoefer Mfg. Co., Freeport, Ill.
Hoosier Drilling Mach. Co., Goshen, Ind.
McDougall Co., Ltd., R., Galt, Ont.
Petrie, Ltd., H. W., Toronto, Ont.
Rockford Drilling Machine Co., Rockford, Ill.
Rockford Lathe & Drill Co., Rockford, Ill.
Perfect Machine Co., Ltd., Galt, Ont.
Terry & Co., John C., Birmingham, Eng.
Strelinger Co. of Can., Ltd., Chas. A., Windsor, Ont.

Drills, Center
Butterfield & Co., Inc., Rock Island, Que.
Cleveland Twist Drill Co., Cleveland, O.
Ingersoll Machine & Tool Co., Ltd., Ingersoll, Ont.
Jones & Shipman, of Leicester, England. Toronto, Ont.
Morrow Screw & Nut Co., Ltd., John, Ingersoll, Ont.
Pratt & Whitney Co., of Canada, Ltd., Dundas, Ont.
Wilt Twist Drill Co. of Canada, Ltd., Walkerville, Ont.

Drills, High Speed Twist
Armstrong Whitworth Co. of Can., Ltd., Montreal, Que.
Butterfield & Co., Inc., Rock Island, Que.
Cleveland Twist Drill Co., Cleveland, O.
Can. Detroit Twist Drill Co., Walkerville, Ont.
Foss Machinery & Supply Co., Geo. F., Montreal, Que.
Garlock-Walker Mchy. Co., Toronto, Ont.
Ingersoll Machine & Tool Co., Ltd., Ingersoll, Ont.
International Machinery & Supply Co., Montreal, Que.
Lyman Tube & Supply Co., Montreal, Que.
Morrow Screw & Nut Co., Ltd., John, Ingersoll, Ont.
Morse Twist Drill & Machine Co., New Bedford, Mass.
Pilot Steel & Tool Co., Montreal, Que.
Pratt & Whitney Co., of Canada, Ltd., Dundas, Ont.
Sheffield Engineering Supplies, Ltd., Montreal, Que.
Sheffield Twist Drill & Steel Co., Sheffield, Eng.
Taylor Tool Co., J. A. M., Toronto, Ont.
Wilt Twist Drill Co. of Canada, Ltd., Walkerville, Ont.

Drills, Ratchet
Armstrong Bros. Tool Co., Chicago, Ill.
Butterfield & Co., Inc., Rock Island, Que.
Cleveland Twist Drill Co., Cleveland, O.

Morrow Screw & Nut Co., Ltd., John, Ingersoll, Ont.
Morse Twist Drill & Machine Co., New Bedford, Mass.

Drills, Twist and Flat
Butterfield & Co., Inc., Rock Island, Que.
Cleveland Twist Drill Co., Cleveland, O.
Can. Detroit Twist Drill Co., Walkerville, Ont.
Morrow Screw & Nut Co., Ltd., John, Ingersoll, Ont.
Pilot Steel & Tool Co., Montreal, Que.
Wilt Twist Drill Co. of Canada, Ltd., Walkerville, Ont.

Dust Handling Equipment
Can. Blower & Forge Co., Ltd., Kitchener.
Sturtevant Co., B. F., Boston, Mass.

Electrical Instruments
Bristol Co., Waterbury, Conn.
Northern Electric Co., Montreal, Que.

Electrical Supplies
Atkins & Co., Inc., E. C., Indianapolis, I.
Diamond State Fibre Co., Toronto, Ont.
Northern Electric Co., Montreal, Que.
U.S. Electrical Tool Co., Cincinnati, O.

Elevating Trucks (See Trucks)
Morris Crane & Hoist Co., Ltd., Niagara Falls, Ont.

Elevators and Conveyors
Can. Link-Belt Co., Toronto, Ont.
Jones & Glassco, Montreal, Que.
Lyman Tube & Supply Co., Montreal, Que.
Main Belting Co. of Can., Montreal, Que.
Mathews Gravity Carrier Co., Port Hope, Ont.

Emery Cloth
Wausau Abrasives Co., Chicago, Ill.

Emery Wheels (See Grinding Wheels)
Aikenhead Hardware Ltd., Toronto, Ont.
Atkins & Co., Inc., E. C., Indianapolis, I.
Carborundum Co., Niagara Falls, N.Y.
Dom. Abrasive Wheel Co., Ltd., Mimico, Ont.
Ford-Smith Machine Co., Hamilton, Ont.
International Machinery & Supply Co., Montreal, Que.
Norton Co. of Can., Ltd., Hamilton, Ont.

Engines, Capstan
Kennedy & Sons, Wm., Owen Sound, Ont.

Engineers, Mechanical
Ford-Smith Machine Co., Hamilton, Ont.
Gisholt Machine Co., Madison, Wis.
Hamilton Gear & Machine Co., Toronto, Ont.
Perdue, W. B., San Francisco, Calif.

Expanders, Tube
Garlock-Walker Mchy. Co., Toronto, Ont.
Holden Co., Ltd., Montreal, Que.
Jardine & Co., A. B., Hespeler, Ont.
Petrie, Ltd., H. W., Toronto, Ont.

Eyeglasses, Safety (See Goggles, Safety)
Prest-O-Lite Co. of Can., Toronto, Ont.

Fans, Electric
Can. Blower & Forge Co., Ltd., Kitchener.
Northern Electric Co., Montreal, Que.
Sturtevant Co., B. F., Boston, Mass.

Fans, Exhaust
Can. Blower & Forge Co., Ltd., Kitchener.
Petrie, Ltd., H. W., Toronto, Ont.
Sturtevant Co., B. F., Boston, Mass.

Fans, Ventilating
Can. Blower & Forge Co., Ltd., Kitchener.
Can. Ingersoll-Rand Co., Ltd., Sherbrooke, Que.
Petrie, Ltd., H. W., Toronto, Ont.
Sturtevant Co., B. F., Boston, Mass.

Fibre
Diamond State Fibre Co., Toronto, Ont.
Northern Electric Co., Montreal, Que.

File Handles
Ingersoll File Co., Ltd., Ingersoll, Ont.

Files and Rasps
Atkins & Co., Inc., E. C., Indianapolis, I.
Foss Machinery & Supply Co., Geo. F., Montreal, Que.
Ingersoll File Co., Ltd., Ingersoll, Ont.
International Machinery & Supply Co., Montreal, Que.
Morrow Screw & Nut Co., Ltd., John, Ingersoll, Ont.
Nicholson File Co., Port Hope, Ont.
Simonds Canada Saw Co., Montreal, Que.

Filing Machines
Garlock-Walker Mchy. Co., Toronto, Ont.
Oliver Machinery Co., Grand Rapids, Mich.
Williams Machinery & Supply Co., A. R., Montreal, Que.

Filler, Iron (See Cements, Iron)
Smooth Mfg. Co., Jersey City, N.J.

Fire Extinguishers
Can. Consolidated Rubber Co., Ltd., Montreal, Que.

Fittings, Pipe
International Malleable Iron Co., Guelph, Ont.

Flexible Shafts
Aikenhead Hardware Ltd., Toronto, Ont.
Haskins Co., R. G., Chicago, Ill.

Flux, Galvanizing
British [illegible] & Refining Co., Ltd., Montreal, Que.

Fluxes, Welding
L'Air Liquide Society, Toronto, Ont.

Forging Machinery
Acme Machinery Co., Cleveland, Ohio.
Bertram & Son Co., Ltd., The John, Dundas, Ont.
Brown, Boggs & Co., Ltd., Hamilton, Ont.
Canada Machinery Corp., Galt, Ont.
Garlock-Walker Mchy. Co., Toronto, Ont.
National Machinery Co., Tiffin, Ohio.
Stewart & Co., Duncan, Glasgow, Scot.

Forgings, Drop
Canada Foundries & Forgings Co., Welland, Ont.
Dominion Forge & Stamping Co., Ltd., Toronto, Ont.

Forgings, Hammer
Canada Foundries & Forgings Co., Welland, Ont.
Can. Atlas Crucible Steel Co., Ltd., Toronto, Ont.
Dominion Bridge Co., Ltd., Lachine, Que.
Dom. Foundries & Steel, Hamilton, Ont.
Hepburn Ltd., John T., Toronto, Ont.
N. S. Steel Co., Ltd., New Glasgow, N.S.

Foundry Equipment
Can. Ingersoll-Rand Co., Ltd., Sherbrooke, Que.
Can. Ingersoll-Rand Co., Ltd., Sherbrooke, Que.
Ford-Smith Machine Co., Hamilton, Ont.
Holden Co., Ltd., Montreal, Que.
McDougall Co., Ltd., R., Galt, Ont.
Petrie, Ltd., H. W., Toronto, Ont.
Rice Lewis & Son, Ltd., Toronto, Ont.

Foundry Supplies
Atkins & Co., Inc., E. C., Indianapolis, I.
Rice Lewis & Son, Ltd., Toronto, Ont.
Sturtevant Co., B. F., Boston, Mass.

Frogs, Spring or Rigid
Can. Steel Foundries, Montreal, Que.

Fuel Oil Burning System
Cole Ltd., George W., Toronto, Ont.
General Combustion Co. of Can., Ltd., Montreal, Que.

Furnaces, Electric
Electric Furnace Construction Co., Philadelphia, Pa.
General Combustion Co. of Can., Ltd., Montreal, Que.

Furnaces, Heat Treating Coal
General Combustion Co. of Can., Ltd., Montreal, Que.
Mechanical Engineering Co., Three Rivers, Que.
Rockwell Co., W. S., New York City.

Furnaces, Heat Treating Oil and Gas
Bellevue Industrial Furnace Co., Detroit.
Can. Ingersoll-Rand Co., Ltd., Sherbrooke, Que.
General Combustion Co. of Can., Ltd., Montreal, Que.
Mechanical Engineering Co., Three Rivers, Que.
Rockwell Co., W. S., New York City.
Walker & Sons Metal Products, Ltd., Hiram, Walkerville, Ont.

Furnaces and Ovens, Electric
Electric Furnace Construction Co., Philadelphia, Pa.
Petrie, Ltd., H. W., Toronto, Ont.
Volta Mfg. Co., Welland, Ont.
Walker & Sons Metal Products, Ltd., Hiram, Walkerville, Ont.

Furnaces, Tempering and Annealing
Brown & Sharpe Mfg. Co., Providence, R.I.
Electric Furnace Construction Co., Philadelphia, Pa.
Mechanical Engineering Co., Three Rivers, Que.
Rockwell Co., W. S., New York City.
Walker & Sons Metal Products, Ltd., Hiram, Walkerville, Ont.

Furniture, Machine Shop
Garlock-Walker Mchy. Co., Toronto, Ont.
Ministry of Munitions, London, Eng.
National Engineering Co., Sarnia, Ont.

Gages, Dial
Herbert Ltd., Alfred, Toronto, Ont.
Johansson Inc., C. E., Windsor, Ont.
Perdue, W. B., San Francisco, Calif.
Starrett Co., L. S., Athol, Mass.

Gages, Measuring (See Tool Work)
Chesterman & Co., Ltd., J., Sheffield, Eng.
Crescent Machine Co., Ltd., Montreal, Q.
Greenfield Tap & Die Corp., Galt, Ont.
Johansson Inc., C. E., Windsor, Ont.
Starrett Co., L. S., Athol, Mass.

Gages, Recording
Bristol Co., Waterbury, Conn.
Johansson Inc., C. E., Windsor, Ont.
Republic Flow Meters Co., Toronto, Ont.

Gages, Snap, Thread and Cylindrical
Ackworthie, Ltd., John, Birmingham, Eng.
Brown & Sharpe Mfg. Co., Providence, R.I.
Greenfield Tap & Die Corp., Galt, Ont.
Johansson Inc., C. E., Windsor, Ont.
Pratt & Whitney Co., of Canada, Ltd., Dundas, Ont.

Gages, Special Measuring (See Tool Work)
Greenfield Tap & Die Corp., Galt, Ont.
Pratt & Whitney Co., of Canada, Ltd., Dundas, Ont.

Gages, Standard
Armstrong Whitworth Co. of Can., Ltd., Montreal, Que.
Atkins & Co., Inc., E. C., Indianapolis, I.
Elliott & Whitehall Tool Co., Galt, Ont.
Johansson Inc., C. E., Windsor, Ont.

BUYERS' DIRECTORY

Gages, Thread
Ackworthle, Ltd., John, Birmingham, Eng.
Greenfield Tap & Die Corp., Galt, Ont.
Johansson Inc., C. E., Windsor, Ont.
Starrett Co., L. S., Athol, Mass.

Garnet, Emery and Flint Paper and Cloth
Ritchey Supply Co., Toronto, Ont.
Wausau Abrasives Co., Chicago, Ill.

Gas, Coal Compressed
L'Air Liquide Society, Toronto, Ont.

Gas, Compressed
Prest-O-Lite Co. of Can., Toronto, Ont.

Gaskets
Dunlop Tire & Rubber Goods Co., Ltd., Toronto, Ont.
Durabla Manufacturing Co., New York.
Goodyear Tire & Rubber Co. of Can., Ltd., Toronto, Ont.
Holden Co., Ltd., Montreal, Què.
Smooth Mfg. Co., Jersey City, N.J.
Voorhees Rubber Co., Jersey City, N.J.

Gear Blanks
Canada Foundries & Forgings Co., Welland, Ont.
Can. Steel Foundries, Montreal, Que.
Dom Foundries & Steel, Hamilton, Ont.
Philadelphia Gear Works, Philadelphia, Pa.

Gear-Cutting Machines
Bertram & Son Co., Ltd., The John, Dundas, Ont.
Bilton Machine Co., Bridgeport, Conn.
Brown & Sharpe Mfg. Co., Providence, R.I.
Fellows Gear Shaper Co., Springfield, Vt.
Petrie, Ltd., H. W., Toronto, Ont.
Whiton Machine Co., D. E., New London, Conn.

Gear Testing Machines
Brown & Sharpe Mfg. Co., Providence, R.I.

Gears, Cast
Can. Link-Belt Co., Toronto, Ont.
Can. Steel Foundries, Montreal, Que.
Dom. Foundries & Steel, Hamilton, Ont.
Fisher Motor Co., Ltd., Orillia, Ont.
Hull Iron & Steel Foundries, Hull, Que.

Gears, Cut
Brown & Sharpe Mfg. Co., Providence, R.I.
Canadian SKF Co., Toronto, Ont.
Crescent Machine Co., Ltd., Montreal, Q.
Diamond State Fibre Co., Toronto, Ont.
Dominion Bridge Co., Ltd., Lachine, Que.
Dom. Steel Products Co., Brantford, Ont.
Ford-Smith Machine Co., Hamilton, Ont.
Gardner & Son, Robt., Montreal, Que.
Hepburn Ltd., John T., Toronto, Ont.
Jardine & Co., A. B., Hespeler, Ont.
Jones & Glassco, Montreal, Que.
Lyman Tube & Supply Co., Montreal, Que.
McDougall Co., Ltd., R., Galt, Ont.
Philadelphia Gear Works, Philadelphia, Pa.
Henold (Hans) of Canada, Ltd., Montreal, Que.

Gears, Dressed
Kennedy & Sons, Wm., Owen Sound, Ont.

Gears, Forged
Canada Foundries & Forgings Co., Welland, Ont.
Lyman Tube & Supply Co., Montreal, Que.

Gears, Herringbone
Dom. Steel Products Co., Brantford, Ont.
Hamilton Gear & Machine Co., Toronto, Ont.
Philadelphia Gear Works, Philadelphia, Pa.

Gears, Machine Moulded
Can. Steel Foundries, Montreal, Que.

Gears, Rawhide (See Gears, Cut)
Philadelphia Gear Works, Philadelphia, Pa.

Gear, Silent Chain
Gardner & Son, Robt., Montreal, Que.
Morse Chain Co., Ithaca, N.Y.

Gears, Worm
Dom. Steel Products Co., Brantford, Ont.

Generators, Acetylene
L'Air Liquide Society, Toronto, Ont.

Generators, Electric
Holden Co., Ltd., Montreal, Que.
Northern Electric Co., Montreal, Que.
Petrie, Ltd., H. W., Toronto, Ont.
Sturtevant Co., B. F., Boston, Mass.

Goggles, Safety
Perdue, W. B., San Francisco, Calif.
Prest-O-Lite Co. of Can., Toronto, Ont.
Standard Optical Co., Geneva, N.Y.
Willson Goggles, Inc., Reading, Pa.

Grab Buckets
Can. Ingersoll-Rand Co., Ltd., Sherbrooke, Que.
Can. Link-Belt Co., Toronto, Ont.
Dominion Bridge Co., Ltd., Lachine, Que.
Morris Crane & Hoist Co., Ltd., Niagara Falls, Ont.

Grease Cups, Pressed Steel and Brass
Can. Winkley Co., Ltd., Windsor, Ont.

Greases, Lubricating
Canadian SKF Co., Toronto, Ont.
Cataract Refining Co., Toronto, Ont.

Grinding Discs
Ritchey Supply Co., Toronto, Ont.
Wausau Abrasives Co., Chicago, Ill.

Grinding Machines
Brown & Sharpe Mfg. Co., Providence, R.I.

Grinding Machines, Abrasive Belt
Beacon Engineering Co., Tipton, England.
Norton Co. of Can., Ltd., Hamilton, Ont.

Grinding Machines, Automatic
Pratt & Whitney Co., of Canada, Ltd., Dundas, Ont.
St. Louis Machine Co., St. Louis, Mo.

Grinding Machines, Bench
Aikenhead Hardware Ltd., Toronto, Ont.
Cowan & Company, of Galt, Ltd., Galt, Ont.
Ford-Smith Machine Co., Hamilton, Ont.
Foss Machinery & Supply Co., Geo. F., Montreal, Que.
Geometric Tool Co., New Haven, Conn.
Goodell & Pratt Co., Greenfield, Mass.
Holly, R. S., Toronto, Ont.
La Salle Tool Co., La Salle, Ill.
Landis Tool Co., Waynesboro, Pa.
Manhattan Machine & Tool Works, Grand Rapids, Mich.
Morse Twist Drill & Machine Co., New Bedford, Mass.
McDougall Co., Ltd., R., Galt, Ont.
Norton Co. of Can., Ltd., Hamilton, Ont.
Petrie, Ltd., H. W., Toronto, Ont.
Pratt & Whitney Co., of Canada, Ltd., Dundas, Ont.
Rockford Drilling Machine Co., Rockford, Ill.
Roelofson Machine & Tool Co., Toronto, Ont.
Waltham Grinding Wheel Co., Waltham, Mass.
Terry & Co., John C., Birmingham, Eng.
St. Louis Machine Co., St. Louis, Mo.
Strelinger Co. of Can., Ltd., Chas. A., Windsor, Ont.
Wilkinson & Kompass, Hamilton, Ont.
Williams Machinery & Supply Co., A. R., Montreal, Que.

Grinding Machines, Center
U.S. Electrical Tool Co., Cincinnati, O.
Wisconsin Electric Co., Racine, Wis.

Grinding Machines, Cutter and Reamer
Cincinnati Milling Machine Co., Cincinnati, Ohio.
Garlock-Walker Mchy. Co., Toronto, Ont.
Greenfield Machine Co., Greenfield, Mass.
Herbert Ltd., Alfred, Toronto, Ont.
Petrie, Ltd., H. W., Toronto, Ont.
Pratt & Whitney Co., of Canada, Ltd., Dundas, Ont.

Grinding Machines, Cylindrical
Garlock-Walker Mchy. Co., Toronto, Ont.
Greenfield Machine Co., Greenfield, Mass.
Manhattan Machine & Tool Works, Grand Rapids, Mich.
Pratt & Whitney Co., of Canada, Ltd., Dundas, Ont.

Grinding Machines, Die
Murchey Machine & Tool Co., Detroit, Mich.
National Acme Co., Cleveland, Ohio.
National Machinery Co., Tiffin, Ont.

Grinding Machines, Disc
Beacon Engineering Co., Tipton, England.
Ford-Smith Machine Co., Hamilton, Ont.

Grinding Machines, Drill
Beacon Engineering Co., Tipton, England.
Bertram & Son Co., Ltd., The John, Dundas, Ont.
Holden Co., Ltd., Montreal, Que.

Grinding Machines, Face
Ford-Smith Machine Co., Hamilton, Ont.

Grinding Machines, Floor and Tool
Beacon Engineering Co., Tipton, England.
Ford-Smith Machine Co., Hamilton, Ont.
Gisholt Machine Co., Madison, Wis.
Modern Tool Co., Erie, Pa.
National Acme Co., Cleveland, Ohio.
Petrie, Ltd., H. W., Toronto, Ont.
Terry & Co., John C., Birmingham, Eng.
Wilkinson & Kompass, Hamilton, Ont.

Grinding Machines, Internal
Garlock-Walker Mchy. Co., Toronto, Ont.
Holden Co., Ltd., Montreal, Que.
Manhattan Machine & Tool Works, Grand Rapids, Mich.

Grinding Machines, Portable
Can. Ingersoll-Rand Co., Ltd., Sherbrooke, Que.
Carborundum Co., Niagara Falls, N.Y.
Cincinnati Electrical Tool Co., Cincinnati, Ohio.
Cleveland Pneumatic Tool Co., Toronto, Ont.
Garlock-Walker Mchy. Co., Toronto, Ont.
Haskins Co., R. G., Chicago, Ill.
Holden Co., Ltd., Montreal, Que.
Independent Pneumatic Tool, Chicago, Ill.
Wilkinson & Kompass, Hamilton, Ont.
Wisconsin Electric Co., Racine, Wis.

Grinding Machines, Power Oscillating Tool
Herbert Ltd., Alfred, Toronto, Ont.

Grinding Machines, Ring Wheel
Ford-Smith Machine Co., Hamilton, Ont.

Grinding Machines, Snagging
Norton Co. of Can., Ltd., Hamilton, Ont.

Grinding Machines, Surface
Garlock-Walker Mchy. Co., Toronto, Ont.
La Salle Tool Co., Ltd., La Salle, Ill.
Perfect Machine Co., Ltd., Galt, Ont.
Petrie, Ltd., H. W., Toronto, Ont.
Pratt & Whitney Co., of Canada, Ltd., Dundas, Ont.

Grinding Machines, Thread
Pratt & Whitney Co., of Canada, Ltd., Dundas, Ont.

Grinding Machinery, Tool Post
Foss Machinery & Supply Co., Geo. F., Montreal, Que.
Gisholt Machine Co., Madison, Wis.
Wilkinson & Kompass, Hamilton, Ont.
Wisconsin Electric Co., Racine, Wis.

Grinding Machines, Universal
Foss Machinery & Supply Co., Geo. F., Montreal, Que.
Garlock-Walker Mchy. Co., Toronto, Ont.
Gisholt Machine Co., Madison, Wis.
Greenfield Machine Co., Greenfield, Mass.
Jones & Shipman, of Leicester, England, Toronto, Ont.
La Salle Tool Co., Ltd., La Salle, Ill.
Landis Tool Co., Waynesboro, Pa.
Manhattan Machine & Tool Works, Grand Rapids, Mich.
Modern Tool Co., Erie, Pa.
Morse Twist Drill & Machine Co., New Bedford, Mass.
Petrie, Ltd., H. W., Toronto, Ont.
Roelofson Machine & Tool Co., Toronto, Ont.
Waltham Grinding Wheel Co., Waltham, Mass.

Grinding Wheels
Aikenhead Hardware Ltd., Toronto, Ont.
Atkins & Co., Inc., E. C., Indianapolis, I.
Carborundum Co., Niagara Falls, N.Y.
Dom. Abrasive Wheel Co., Ltd., Mimico, Ont.
Ford-Smith Machine Co., Hamilton, Ont.
International Machinery & Supply Co., Montreal, Que.
Norton Co. of Can., Ltd., Hamilton, Ont.

Guards, Machinery and Window
Can. Wire & Iron Goods Co., Hamilton, Ont.

Gun-Barrel Machinery
Steinle Turret Machine Co., Madison, Wis.

Hack Saws, Power
Ackworthle, Ltd., John, Birmingham, Eng.
Aikenhead Hardware Ltd., Toronto, Ont.
Atkins & Co., Inc., E. C., Indianapolis, I.
Garlock-Walker Mchy. Co., Toronto, Ont.
Goodell & Pratt Co., Greenfield, Mass.
Lyman Tube & Supply Co., Montreal, Que.
McKenzie Machinery Co., Guelph, Ont.
Petrie, Ltd., H. W., Toronto, Ont.
Simonds Canada Saw Co., Montreal, Que.
Starrett Co., L. S., Athol, Mass.
Williams Machinery & Supply Co., A. R., Montreal, Que.

Hammers, Chipping
Cleveland Pneumatic Tool Co., Toronto, Ont.

Hammers, Drop
Bertram & Son Co., Ltd., The John, Dundas, Ont.
Bliss Co., E. W., Brooklyn, N.Y.
Brown, Boggs & Co., Ltd., Hamilton, Ont.
Canada Foundries & Forgings Co., Welland, Ont.
Canada Machinery Corp., Galt, Ont.

Hammers, Electric
Aikenhead Hardware Ltd., Toronto, Ont.
Brown, Boggs & Co., Ltd., Hamilton, Ont.
Holden Co., Ltd., Montreal, Que.

Hammers, Pneumatic
Can. Ingersoll-Rand Co., Ltd., Sherbrooke, Que.
Cleveland Pneumatic Tool Co., Toronto, Ont.
Garlock-Walker Mchy. Co., Toronto, Ont.
Holden Co., Ltd., Montreal, Que.
Independent Pneumatic Tool, Chicago, Ill.
Keller Pneumatic Tool Co., Grand Haven, Mich.
Ryerson & Son, Jos. T., Chicago, Ill.

Hammers, Power
Bertram & Son Co., Ltd., The John, Dundas, Ont.
Bradley & Son, Inc., C. C., Syracuse, N.Y.
Brown, Boggs & Co., Ltd., Hamilton, Ont.
High Speed Hammer Co., Rochester, N.Y.
Jardine & Co., A. B., Hespeler, Ont.
Petrie, Ltd., H. W., Toronto, Ont.
Ryerson & Son, Jos. T., Chicago, Ill.

Hammers, Rivetting
Cleveland Pneumatic Tool Co., Toronto, Ont.
Ryerson & Son, Jos. T., Chicago, Ill.

Hangers, Shafting
Can. Link-Belt Co., Toronto, Ont.
Canadian SKF Co., Toronto, Ont.
Chapman Double Ball Bearing Co., Toronto, Ont.
Ford-Smith Machine Co., Hamilton, Ont.
Foss Machinery & Supply Co., Geo. F., Montreal, Que.
Terry & Co., John C., Birmingham, Eng.
Williams Machinery & Supply Co., A. R., Montreal, Que.

Hardening, Case-Hardening and Tempering
Hamilton Gear & Machine Co., Toronto, Ont.

Hardness Testing Apparatus
Shore Instrument Co., Jamaica, N.Y.

Hobbing Machines
Barber-Colman Company, Rockford, Ill.
Herbert Ltd., Alfred, Toronto, Ont.
Petrie, Ltd., H. W., Toronto, Ont.

Hobs
Barber-Colman Company, Rockford, Ill.
Brown & Sharpe Mfg. Co., Providence, R.I.
Greenfield Tap & Die Corp., Galt, Ont.
Pratt & Whitney Co., of Canada, Ltd., Dundas, Ont.

Hoists, Electric
Can. Ingersoll-Rand Co., Ltd., Sherbrooke, Que.
Can. Link-Belt Co., Toronto, Ont.
Garlock-Walker Mchy. Co., Toronto, Ont.
Morris Crane & Hoist Co., Ltd., Niagara Falls, Ont.
Northern Crane Works, Walkerville, Ont.
Reading Chain & Block Co., Reading, Pa.
Shepherd Electric Crane & Hoist Co., Montour Falls, N. Y.
Volta Mfg. Co., Welland, Ont.

Hoists, Hand
Lyman Tube & Supply Co., Montreal, Que.
Morris Crane & Hoist Co., Ltd., Niagara Falls, Ont.
Wright Mfg. Co., Lisbon, Ohio.

Hoists, Pneumatic
Can. Ingersoll-Rand Co., Ltd., Sherbrooke, Que.
Curtis Pneumatic Machinery Co., St. Louis, Mo.
Garlock-Walker Mchy. Co., Toronto, Ont.
Holden Co., Ltd., Montreal, Que.
Independent Pneumatic Tool, Chicago, Ill.
Morris Crane & Hoist Co., Ltd., Niagara Falls, Ont.
Northern Crane Works, Walkerville, Ont.

Holders-On, Pneumatic
Can. Ingersoll-Rand Co., Ltd., Sherbrooke, Que.
Cleveland Pneumatic Tool Co., Toronto, Ont.
Holden Co., Ltd., Montreal, Que.
Independent Pneumatic Tool, Chicago, Ill.

Hose, All Kinds
Quaker City Rubber Co., Philadelphia, Pa.

Hose, Flexible
Gutta Percha & Rubber, Toronto, Ont.

Hose, Industrial
Dunlop Tire & Rubber Goods Co., Ltd., Toronto, Ont.
Goodyear Tire & Rubber Co. of Can., Ltd., Toronto, Ont.

Hose, Rubber
Can. Consolidated Rubber Co., Ltd., Montreal, Que.
Can. Foamite Firefoam Co., Hamilton, Ont.

Hose, Steam, Suction, Water
Quaker City Rubber Co., Philadelphia, Pa.

Hose, Steel
International Machinery & Supply Co., Montreal, Que.
Ontario Metal Products Co., Ltd., Toronto, Ont.

Hospital Supplies
Hartz, J. F., Co., Toronto, Ont.
Ingram & Bell, Ltd., Toronto, Ont.

Hydraulic Leather
Graton & Knight Mfg. Co., Worcester, Mass.

Hydraulic Machinery
Bertram & Son Co., Ltd., The John, Dundas, Ont.
Can. Ingersoll-Rand Co., Ltd., Sherbrooke, Que.
Garlock-Walker Mchy. Co., Toronto, Ont.
Stewart & Co., Duncan, Glasgow, Scot.

Igniters, Gas Engine
Canada Foundries & Forgings Co., Welland, Ont.

Indicators, Speed and Test
Aikenhead Hardware Ltd., Toronto, Ont.
Atkins & Co., Inc., E. C., Indianapolis, I.
Brown & Sharpe Mfg. Co., Providence, R.I.

Jacks, Hydraulic
International Machinery & Supply Co., Montreal, Que.
Norton, A. O., Boston, Mass.

Jacks, Planer
Armstrong Bros. Tool Co., Chicago, Ill.
Starrett Co., L. S., Athol, Mass.

Jigs and Fixtures (See Tool Work)
Bilton Machine Co., Bridgeport, Conn.
Burgess & Marchand, Montreal, Que.
Crescent Machine Co., Ltd., Montreal, Q.
Fisher Motor Co., Ltd., Orillia, Ont.
Ford-Smith Machine Co., Hamilton, Ont.
Gisholt Machine Co., Madison, Wis.
Hamilton Engineering Service, Ltd., Hamilton, Ont.
Rapid Tool & Machine Co., Lachine, Que.

Keyseating Machines
Bilton Machine Co., Bridgeport, Conn.
Elliott & Whitehall Tool Co., Galt, Ont.
Garlock-Walker Mchy. Co., Toronto, Ont.
Morton Mfg. Co., Muskegon, Mich.
Petrie, Ltd., H. W., Toronto, Ont.
Pratt & Whitney Co., of Canada, Ltd., Dundas, Ont.
Toomey Inc., Frank, Philadelphia, Pa.

Keys, Machine
Can. Drawn Steel Co., Hamilton, Ont.
Garlock-Walker Mchy. Co., Toronto, Ont.
Morton Mfg. Co., Muskegon, Mich.

Knives, Machine
Atkins & Co., Inc., E. C., Indianapolis, I.
Canada Machinery Corp., Galt, Ont.
Simonds Canada Saw Co., Montreal, Que.

Knurl Holders
Pratt & Whitney Co., of Canada, Ltd., Dundas, Ont.

Lacing Leather
Clipper Belt Lacer Co., Grand Rapids, Mich.
Main Belting Co. of Can., Montreal, Que.

BUYERS' DIRECTORY

BUYERS' DIRECTORY

Pipe Cutting and Threading Machines
- Crane Ltd., Montreal, Que.
- Greenfield Tap & Die Corp., Galt, Ont.
- Jardine & Co., A. B., Hespeler, Ont.
- Landis Machine Co., Inc., Waynesboro, Pa.
- Murchey Machine & Tool Co., Detroit, Mich.
- McDougall Co., Ltd., R., Galt, Ont.
- Petrie, Ltd., H. W., Toronto, Ont.
- Williams Tool Corp. of Can., Ltd., Brantford, Ont.

Pipe and Nipple Threading Machines
- Landis Machine Co., Inc., Waynesboro, Pa.

Pipe Fitters' Tools
- Aikenhead Hardware Ltd., Toronto, Ont.
- Crane Ltd., Montreal, Que.
- Rice Lewis & Son, Ltd., Toronto, Ont.

Pipe Threading Die Heads
- Landis Machine Co., Inc., Waynesboro, Pa.

Piston-Ring Machines
- National Acme Co., Cleveland, Ohio.
- Steinle Turret Machine Co., Madison, Wis.

Planers, Parallels
- L. & P. Mfg. Co., Niagara Falls, Ont.

Planing Machines
- Bertram & Son Co., Ltd., The John, Dundas, Ont.
- Canada Machinery Corp., Galt, Ont.
- Cowan & Company, of Galt, Ltd., Galt, Ont.
- Foss Machinery & Supply Co., Geo. F., Montreal, Que.
- Garlock-Walker Mchy. Co., Toronto, Ont.
- Hepburn Ltd., John T., Toronto, Ont.
- Herbert Ltd., Alfred, Toronto, Ont.
- L. & P. Mfg. Co., Niagara Falls, Ont.
- Morton Mfg. Co., Muskegon, Mich.
- Oliver Machinery Co., Grand Rapids, Mich.
- Toomey Inc., Frank, Philadelphia, Pa.
- Williams Machinery Co., A. R., Toronto, Ont.

Planing Machines, Rotary
- Bertram & Son Co., Ltd., The John, Dundas, Ont.
- Canada Machinery Corp., Galt, Ont.

Plate Rolls
- Bertram & Son Co., Ltd., The John, Dundas, Ont.

Pneumatic Tools
- Can. Ingersoll-Rand Co., Ltd., Sherbrooke, Que.
- Cleveland Pneumatic Tool Co., Toronto, Ont.
- Garlock-Walker Mchy. Co., Toronto, Ont.
- Holden Co., Ltd., Montreal, Que.
- Independent Pneumatic Tool, Chicago, Ill.
- International Machinery & Supply Co., Montreal, Que.
- Keller Pneumatic Tool Co., Grand Haven, Mich.

Polishing and Buffing Machines
- Ackworthie, Ltd., John, Birmingham, Eng.
- Archibald & Co., Chas. F., Montreal, Q.
- Brown & Sharpe Mfg. Co., Providence, R.I.
- Can. Hanson & Van Winkle Co., Ltd., Toronto, Ont.
- Ford-Smith Machine Co., Hamilton, Ont.
- Garlock-Walker Mchy. Co., Toronto, Ont.
- Terry & Co., John C., Birmingham, Eng.

Pressed Steel Parts
- Ackworthie, Ltd., John, Birmingham, Eng.
- American Pulley Co., Philadelphia, Pa.
- Fisher Motor Co., Ltd., Orillia, Ont.

Presses, Arbor
- Atlas Press Co., Kalamazoo, Mich.
- Lucas Machine Tool Co., Cleveland, Ohio.
- National Engineering Co., Sarnia, Ont.
- Petrie, Ltd., H. W., Toronto, Ont.
- Strelinger Co. of Can., Ltd., Chas. A., Windsor, Ont.

Presses, Banding

Presses, Broaching

Presses, Drop and Forging
- Brown, Boggs & Co., Ltd., Hamilton, Ont.
- Canada Foundries & Forgings Co., Welland, Ont.
- Toledo Machine & Tool Co., Toledo, Ohio.

Presses, Foot and Hand
- Brown, Boggs & Co., Ltd., Hamilton, Ont.
- Terry & Co., John C., Birmingham, Eng.

Presses, Forcing
- Atlas Press Co., Kalamazoo, Mich.
- Lucas Machine Tool Co., Cleveland, Ohio.
- Niagara Machine & Tool Works, Buffalo, N.Y.
- Stewart & Co., Duncan, Glasgow, Scot.

Presses, Hydraulic
- Baird Machine Co., Bridgeport, Conn.
- Bertram & Son Co., Ltd., The John, Dundas, Ont.
- Can. Ingersoll-Rand Co., Ltd., Sherbrooke, Que.
- Laurie Mfg. Co., Springfield, Ill.
- Niagara Machine & Tool Works, Buffalo, N.Y.
- Perrin Ltd., W. R., Toronto, Ont.
- Stewart & Co., Duncan, Glasgow, Scot.
- Williams Machinery Co., A. R., Toronto, Ont.

Presses, Power
- Bliss Co., E. W., Brooklyn, N.Y.
- Brown, Boggs & Co., Ltd., Hamilton, Ont.
- Canada Machinery Corp., Galt, Ont.
- Garlock-Walker Mchy. Co., Toronto, Ont.
- Hepburn Ltd., John T., Toronto, Ont.
- Henry & Wright Mfg. Co., Hartford, Conn.
- Lucas Machine Tool Co., Cleveland, Ohio.
- Niagara Machine & Tool Works, Buffalo, N.Y.
- Petrie, Ltd., H. W., Toronto, Ont.
- Stall Co., Inc., D. H., Buffalo, N.Y.
- Toledo Machine & Tool Co., Toledo, Ohio.

Presses, Screw
- Brown, Boggs & Co., Ltd., Hamilton, Ont.
- Petrie, Ltd., H. W., Toronto, Ont.

Profiling Machines
- Aikenhead Hardware Ltd., Toronto, Ont.
- Garlock-Walker Mchy. Co., Toronto, Ont.
- Pratt & Whitney Co., of Canada, Ltd., Dundas, Ont.

Protractors
- Brown & Sharpe Mfg. Co., Providence, R.I.

Propellers
- Kennedy & Sons, Wm., Owen Sound, Ont.

Pulleys, Cork Insert
- American Pulley Co., Philadelphia, Pa.
- Foss Machinery & Supply Co., Geo. F., Montreal, Que.
- Positive Clutch & Pulley Works, Toronto, Ont.

Pulleys, Metal
- American Pulley Co., Philadelphia, Pa.
- Bernard Industrial Co., A., Fortierville, Que.
- Can. Fairbanks-Morse Ltd., Montreal, Q.
- Canadian SKF Co., Toronto, Ont.
- Johnson Machine Co., Carlyle, Manchester, Conn.
- Kennedy & Sons, Wm., Owen Sound, Ont.
- Williams Machinery & Supply Co., A. R., Montreal, Que.

Pulp and Paper Mill Equipment
- MacKinnon Steel Co., Sherbrooke, Que.

Pumps, Barrel and Boiler-feed
- Trahern Pump Co., Rockford, Ill.

Pumps, Circulating and Coolant
- Trahern Pump Co., Rockford, Ill.

Pumps, Geared and Hand
- Trahern Pump Co., Rockford, Ill.

Pumps, Industrial
- Trahern Pump Co., Rockford, Ill.

Pumps, Hydraulic
- Can. Ingersoll-Rand Co., Ltd., Sherbrooke, Que.
- Electric Steel & Engineering Co., Welland, Ont.
- Hepburn Ltd., John T., Toronto, Ont.
- Holden Co., Ltd., Montreal, Que.
- Stewart & Co., Duncan, Glasgow, Scot.
- Trahern Pump Co., Rockford, Ill.

Pumps, Lubricant and Oil
- Bowser & Co., Inc., S. F., Fort Wayne, I.
- Can. Blower & Forge Co., Ltd., Kitchener.
- Hepburn Ltd., John T., Toronto, Ont.
- McDougall Co., Ltd., R., Galt, Ont.
- Trahern Pump Co., Rockford, Ill.

Pumps, Power
- Bowser & Co., Inc., S. F., Fort Wayne, I.
- Can. Blower & Forge Co., Ltd., Kitchener.
- Can. Fairbanks-Morse Ltd., Montreal, Q.
- Can. Ingersoll-Rand Co., Ltd., Sherbrooke, Que.
- Electric Steel & Engineering Co., Welland, Ont.
- Hepburn Ltd., John T., Toronto, Ont.
- Trahern Pump Co., Rockford, Ill.

Punches, Center
- Brown & Sharpe Mfg. Co., Providence, R.I.
- Pratt & Whitney Co., of Canada, Ltd., Dundas, Ont.
- Starrett Co., L. S., Athol, Mass.

Punches, Hand
- Brown, Boggs & Co., Ltd., Hamilton, Ont.
- Can. Blower & Forge Co., Ltd., Kitchener.
- Jardine & Co., A. B., Hespeler, Ont.
- Whitney Mfg. Co., W. A., Rockford, Ill.

Punches, Power
- Brown, Boggs & Co., Ltd., Hamilton, Ont.
- Canada Machinery Corp., Galt, Ont.
- Can. Blower & Forge Co., Ltd., Kitchener.
- Garlock-Walker Mchy. Co., Toronto, Ont.
- Petrie, Ltd., H. W., Toronto, Ont.
- Ryerson & Son, Jos. T., Chicago, Ill.
- Toledo Machine & Tool Co., Toledo, Ohio.

Punching Machines, Horizontal
- Bertrams Ltd., Edinburgh, Scotland.

Pyrometers, Electric
- Bristol Co., Waterbury, Conn.
- General Combustion Co. of Can., Ltd., Montreal, Que.
- Walker & Sons Metal Products, Ltd., Hiram, Walkerville, Ont.

Racks, Cut
- Ford-Smith Machine Co., Hamilton, Ont.
- Hamilton Gear & Machine Co., Toronto, Ont.

Racks, Storage (See Furniture, Machine Shop)
- Brantford Oven & Rack Co., Brantford, Ont.

Rammers, Foundry
- Holden Co., Ltd., Montreal, Que.

Reamer Holders
- Cleveland Twist Drill Co., Cleveland, O.
- Gisholt Machine Co., Madison, Wis.
- Victor Tool Co., Waynesboro, Pa.

Reamers, Expanding
- Aikenhead Hardware Ltd., Toronto, Ont.
- Can. Detroit Twist Drill Co., Walkerville, Ont.
- Cleveland Twist Drill Co., Cleveland, O.
- Gisholt Machine Co., Madison, Wis.
- Greenfield Tap & Die Corp., Galt, Ont.
- Ingersoll Machine & Tool Co., Ltd., Ingersoll, Ont.
- McCrosky Tool Corp., Meadville, Pa.
- Pratt & Whitney Co., of Canada, Ltd., Dundas, Ont.
- Wilt Twist Drill Co. of Canada, Ltd., Walkerville, Ont.

Reamers, Solid
- Armstrong Whitworth Co. of Can., Ltd., Montreal, Que.
- Butterfield & Co., Inc., Rock Island, Que.
- Can. Detroit Twist Drill Co., Walkerville, Ont.
- Cleveland Twist Drill Co., Cleveland, O.
- Foss Machinery & Supply Co., Geo. F., Montreal, Que.
- Greenfield Tap & Die Corp., Galt, Ont.
- Ingersoll Machine & Tool Co., Ltd., Ingersoll, Ont.
- International Machinery & Supply Co., Montreal, Que.
- Morse Twist Drill & Machine Co., New Bedford, Mass.
- Wilt Twist Drill Co. of Canada, Ltd., Walkerville, Ont.

Reamers, Taper
- Butterfield & Co., Inc., Rock Island, Que.
- Can. Detroit Twist Drill Co., Walkerville, Ont.
- Cleveland Twist Drill Co., Cleveland, O.
- Foss Machinery & Supply Co., Geo. F., Montreal, Que.
- Garlock-Walker Mchy. Co., Toronto, Ont.
- Gisholt Machine Co., Madison, Wis.
- Greenfield Tap & Die Corp., Galt, Ont.
- Ingersoll Machine & Tool Co., Ltd., Ingersoll, Ont.
- Morrow Screw & Nut Co., Ltd., John, Ingersoll, Ont.
- Pilot Steel & Tool Co., Montreal, Que.
- Pratt & Whitney Co., of Canada, Ltd., Dundas, Ont.
- Taylor Tool Co., J. A. M., Toronto, Ont.
- Wilt Twist Drill Co. of Canada, Ltd., Walkerville, Ont.

Recorders, Temperature
- Taylor Instrument Co., Rochester, N.Y.
- Walker & Sons Metal Products, Ltd., Hiram, Walkerville, Ont.

Recorders, Time
- Gisholt Machine Co., Madison, Wis.
- International Business Machines Co., Toronto, Ont.

Regulators, Automatic (for electric furnaces)
- Volta Mfg. Co., Welland, Ont.

Rheostats
- Northern Electric Co., Montreal, Que.

Resistance Materials
- Walker & Sons Metal Products, Ltd., Hiram, Walkerville, Ont.

Respirators
- Willson Goggles, Inc., Reading, Pa.

Rivets
- Parmenter & Bulloch Co., Gananoque, Ont.

Rivet Heaters
- Can. Ingersoll-Rand Co., Ltd., Sherbrooke, Que.
- General Combustion Co. of Can., Ltd., Montreal, Que.
- Volta Mfg. Co., Welland, Ont.

Rivet-Making Machinery
- Acme Machinery Co., Cleveland, Ohio.
- Bertram & Son Co., Ltd., The John, Dundas, Ont.
- National Machinery Co., Tiffin, Ohio.
- Ryerson & Son, Jos. T., Chicago, Ill.

Riveting Machines
- Bilton Machine Co., Bridgeport, Conn.
- Can. Ingersoll-Rand Co., Ltd., Sherbrooke, Que.
- High Speed Hammer Co., Rochester, N.Y.
- Holden Co., Ltd., Montreal, Que.
- Independent Pneumatic Tool, Chicago, Ill.
- Keller Pneumatic Tool Co., Grand Haven, Mich.
- Parmenter & Bulloch Co., Gananoque, Ont.
- Petrie, Ltd., H. W., Toronto, Ont.
- Ryerson & Son, Jos. T., Chicago, Ill.
- Schuster Co., F. B., New Haven, Conn.

Rolling Mill Equipment
- Stewart & Co., Duncan, Glasgow, Scot.

Rolls (Rubber Covered)
- Can. Consolidated Rubber Co., Ltd., Montreal, Que.

Rudder Frames, Steel
- Can. Steel Foundries, Montreal, Que.

Rubber Goods, Mechanical
- Quaker City Rubber Co., Philadelphia, Pa.

Rules, Steel
- Chesterman & Co., Ltd., J., Sheffield, Eng.

Rules, Steel and Wood
- Brown & Sharpe Mfg. Co., Providence, R.I.

Rust Preventatives
- Oakley Chemical Co., New York, N.Y.

Sand Paper
- Wausau Abrasives Co., Chicago, Ill.

Sand Equipment
- Can. Link-Belt Co., Toronto, Ont.

Sand Mills
- Frost Mfg. Co., Chicago, Ill.

Sand Rammers, Pneumatic
- Can. Ingersoll-Rand Co., Ltd., Sherbrooke, Que.
- Cleveland Pneumatic Tool Co., Toronto, Ont.
- Holden Co., Ltd., Montreal, Que.
- Independent Pneumatic Tool, Chicago, Ill.
- Keller Pneumatic Tool Co., Grand Haven, Mich.

Saw Frames and Blades, Hack
- Aikenhead Hardware Ltd., Toronto, Ont.
- Atkins & Co., Inc., E. C., Indianapolis, I.
- Clemson Bros., Inc., Hamilton, Ont.
- Diamond Saw & Stamping Works, Buffalo, N.Y.
- Foss Machinery & Supply Co., Geo. F., Montreal, Que.
- Rice Lewis & Son, Ltd., Toronto, Ont.
- Simonds Canada Saw Co., Montreal, Que.

Sawing Machines, Metal
- Atkins & Co., Inc., E. C., Indianapolis, I.
- Foss Machinery & Supply Co., Geo. F., Montreal, Que.
- Herbert Ltd., Alfred, Toronto, Ont.
- Lyman Tube & Supply Co., Montreal, Que.
- Ryerson & Son, Jos. T., Chicago, Ill.

Sawing Machines, Power Hack
- Ackworthie, Ltd., John, Birmingham, Eng.
- Atkins & Co., Inc., E. C., Indianapolis, I.
- Levis, The, Toronto, Canada.
- Perfect Machine Co., Ltd., Galt, Ont.
- Williams Machinery & Supply Co., A. R., Montreal, Que.

Saw Sharpening Machines
- Atkins & Co., Inc., E. C., Indianapolis, I.
- Oliver Machinery Co., Grand Rapids, Mich.

Saw Tables, Universal
- Atkins & Co., Inc., E. C., Indianapolis, I.
- Canada Machinery Corp., Galt, Ont.
- Cowan & Company, of Galt, Ltd., Galt, Ont.
- Garlock-Walker Mchy. Co., Toronto, Ont.
- McKenzie Machinery Co., Guelph, Ont.
- Oliver Machinery Co., Grand Rapids, Mich.
- Petrie, Ltd., H. W., Toronto, Ont.

Saws, Circular Metal
- Atkins & Co., Inc., E. C., Indianapolis, I.
- Cowan & Company, of Galt, Ltd., Galt, Ont.
- Simonds Canada Saw Co., Montreal, Que.
- Tabor Mfg. Co., Philadelphia, Pa.

Saws, Hand
- Aikenhead Hardware Ltd., Toronto, Ont.
- Atkins & Co., Inc., E. C., Indianapolis, I.
- Simonds Canada Saw Co., Montreal, Que.

Saws, Hot and Cold
- Atkins & Co., Inc., E. C., Indianapolis, I.
- Simonds Canada Saw Co., Montreal, Que.
- Stewart & Co., Duncan, Glasgow, Scot.

Saws, High Speed Steel
- Atkins & Co., Inc., E. C., Indianapolis, I.
- Butterfield & Co., Inc., Rock Island, Que.
- Pratt & Whitney Co., of Canada, Ltd., Dundas, Ont.
- Simonds Canada Saw Co., Montreal, Que.

Saws, Metal Band
- Atkins & Co., Inc., E. C., Indianapolis, I.
- Cowan & Company, of Galt, Ltd., Galt, Ont.
- Oliver Machinery Co., Grand Rapids, Mich.

Saws, Metal, Power
- Clemson Bros., Inc., Hamilton, Ont.

Saws, Metal Cutting
- Atkins & Co., Inc., E. C., Indianapolis, I.
- Brown & Sharpe Mfg. Co., Providence, R.I.
- Butterfield & Co., Inc., Rock Island, Que.
- Clemson Bros., Inc., Hamilton, Ont.
- Lyman Tube & Supply Co., Montreal, Que.
- Pratt & Whitney Co., of Canada, Ltd., Dundas, Ont.
- Simonds Canada Saw Co., Montreal, Que.
- Starrett Co., L. S., Athol, Mass.

Saws, Milling
- Atkins & Co., Inc., E. C., Indianapolis, I.
- Butterfield & Co., Inc., Rock Island, Que.
- Ingersoll Machine & Tool Co., Ltd., Ingersoll, Ont.
- Pratt & Whitney Co., of Canada, Ltd., Dundas, Ont.

Saws, Screw Slotting
- Atkins & Co., Inc., E. C., Indianapolis, I.
- Butterfield & Co., Inc., Rock Island, Que.
- Pratt & Whitney Co., of Canada, Ltd., Dundas, Ont.
- Simonds Canada Saw Co., Montreal, Que.

Saws, Swing Cut-off
- Oliver Machinery Co., Grand Rapids, Mich.

Scales
- Brown & Sharpe Mfg. Co., Providence, R.I.
- Can. Fairbanks-Morse Ltd., Montreal, Q.

Screens
- Can. Wire & Iron Goods Co., Hamilton, Ont.

Screw Driving Machine
- Canada Machinery Corp., Galt, Ont.
- Can. Ingersoll-Rand Co., Ltd., Sherbrooke, Que.
- Holden Co., Ltd., Montreal, Que.
- Independent Pneumatic Tool, Chicago, Ill.

Screw Extractors
- Cleveland Twist Drill Co., Cleveland, O.

Screw Machine Work
- Barnes Co., Wallace, Bristol, Conn.
- Cook Co., Asa S., Hartford, Conn.
- National Acme Co., Cleveland, Ohio.
- Tallman Brass & Metal Co., Hamilton, Ont.

Screw Machinery, Wood and Lag
Cook Co., Asa S., Hartford, Conn.

Screw Machines
Brown & Sharpe Mfg. Co., Providence, R.I.

Screw Machines, Automatic
Garlock-Walker Mchy. Co., Toronto, Ont.
Herbert Ltd., Alfred, Toronto, Ont.
National Acme Co., Cleveland, Ohio.

Screw Machines, Plain or Hand
Acme Machine Tool Co., Cincinnati, Ohio.
Greenfield Tap & Die Corp., Galt, Ont.
Herbert Ltd., Alfred, Toronto, Ont.
Pratt & Whitney Co., of Canada, Ltd., Dundas, Ont.
Warner & Swasey Co., Cleveland, Ohio.

Screw Plates
Aikenhead Hardware Ltd., Toronto, Ont.
Butterfield & Co., Inc., Rock Island, Que.
Greenfield Tap & Die Corp., Galt, Ont.
Jardine & Co., A. B., Hespeler, Ont.

Screws, Cap and Set
Galt Machine Screw Co., Galt, Ont.
Morrow Screw & Nut Co., Ltd., John, Ingersoll, Ont.
National Acme Co., Cleveland, Ohio.

Screws, Machine
Barnes Co., Wallace, Bristol, Conn.

Screws, Safety Set
Barnes Co., Wallace, Bristol, Conn.
Galt Machine Screw Co., Galt, Ont.
Morrow Screw & Nut Co., Ltd., John, Ingersoll, Ont.

Seamless Tubing. (See Tubing, Seamless Steel
Ontario Metal Products Co., Ltd., Toronto, Ont.

Second-Hand Machinery (See Searchlight Section)
Petrie, Ltd., H. W., Toronto, Ont.

Separators, Moisture and Oil
Bowser & Co., Inc., S. F., Fort Wayne, I.
Can. Ingersoll-Rand Co., Ltd., Sherbrooke, Que.

Separators, Oil and Waste
Bowser & Co., Inc., S. F., Fort Wayne, I.

Shafting
Canada Foundries & Forgings Co., Welland, Ont.
Can. Drawn Steel Co., Hamilton, Ont.
N.S. Steel Co., Ltd., New Glasgow, N.S.
Wilkinson & Kompass, Hamilton, Ont.
Williams Machinery Co., A. R., Toronto, Ont.
Williams Machinery & Supply Co., A. R., Montreal, Que.

Shapes, Cold-Drawn Special Steel
Union Drawn Steel Co., Hamilton, Ont.

Shaping Machines
Canada Machinery Corp., Galt, Ont.
Cowan & Company, of Galt, Ltd., Galt, Ont.
Foss Machinery & Supply Co., Geo. F., Montreal, Que.
Hendey Machine Co., Torrington, Conn.
Herbert Ltd., Alfred, Toronto, Ont.
Holly, R. E., Toronto, Ont.
Morton Mfg. Co., Muskegon, Mich.
McDougall Co., Ltd., R., Galt, Ont.
McKenzie Machinery Co., Guelph, Ont.
Roelofson Machine & Tool Co., Toronto, Ont.
Smith & Mills Co., Cincinnati, Ohio
Toomey Inc., Frank, Philadelphia, Pa.
Walcott Lathe Co., Jackson, Mich.
Williams Machinery Co., A. R., Toronto, Ont.

Shapers, Wood
Oliver Machinery Co., Grand Rapids, Mich.

Shears, Hand
Can. Blower & Forge Co., Ltd., Kitchener.
Whitney Mfg. Co., W. A., Rockford, Ill.

Shears, Power
Bliss Co., E. W., Brooklyn, N.Y.
Brown, Boggs & Co., Ltd., Hamilton, Ont.
Canada Machinery Corp., Galt, Ont.
Can. Blower & Forge Co., Ltd., Kitchener.
Niall Co., Inc., D. H., Buffalo, N.Y.
Stewart & Co., Duncan, Glasgow, Scot.
Toledo Machine & Tool Co., Toledo, Ohio.
Williams Machinery Co., A. R., Toronto, Ont.

Shearing Machines, Angle, Iron Bar and Gate
Bertrams Ltd., Edinburgh, Scotland.

Sheet Metal Working Machinery
Bliss Co., E. W., Brooklyn, N.Y.
Brown, Boggs & Co., Ltd., Hamilton, Ont.
Dominion Sheet Metal Corporation, Hamilton, Ont.
Garlock-Walker Mchy. Co., Toronto, Ont.
Herbert Ltd., Alfred, Toronto, Ont.
Niall Co., Inc., D. H., Buffalo, N.Y.
Terry & Co., John C., Birmingham, Eng.
Toledo Machine & Tool Co., Toledo, Ohio.

Sheets, Nickel, Resist, Alloy
International Nickel Co. of Can., Ltd., Toronto, Ont.

Side Frames, Locomotive
Can. Steel Foundries, Montreal, Que.

Slotting Attachments
Ford-Smith Machine Co., Ltd., Hamilton, Ont.
Kearney & Trecker Co., Milwaukee, Wis.
Kempsmith Mfg. Co., Milwaukee, Wis.
Moore H. A. Co., Rochester, N.Y.
National Acme Co., Cleveland, Ohio.

Slotting Machines
Bertram & Son Co., Ltd., The John, Dundas, Ont.
Canada Machinery Corp., Galt, Ont.

BUYERS' DIRECTORY

Ford-Smith Machine Co., Hamilton, Ont.
Herbert Ltd., Alfred, Toronto, Ont.
Moore H. A. Co., Rochester, N.Y.

Solders
British Smelting & Refining Co., Ltd., Montreal, Que.

Special Machinery and Tools
Brown Engineering Corp., Ltd., Toronto.
Can. Ingersoll-Rand Co., Ltd., Sherbrooke, Que.
Crescent Machine Co., Ltd., Montreal, Q.
Ford-Smith Machine Co., Hamilton, Ont.
Gisholt Machine Co., Madison, Wis.
Ingersoll Machine & Tool Co., Ltd., Ingersoll, Ont.
National Acme Co., Cleveland, Ohio.

Spectacles, Industrial
Willson Goggles, Inc., Reading, Pa.

Springs
Barnes Co., Wallace, Bristol, Conn.
Cleveland Wire Spring Co., Cleveland, O.
Dunbar Bros. Co., Bristol, Conn.
Dyrab Steel (Consolidated) Ltd., Toronto, Ont.
Steele Ltd., James, Guelph, Ont.

Sprockets and Chains
Can. Link-Belt Co., Toronto, Ont.
Jones & Glassco, Montreal, Que.
Lyman Tube & Supply Co., Montreal, Que.
Morse Chain Co., Ithaca, N.Y.
Renold (Hans) of Canada, Ltd., Montreal, Que.

Squares
Brown & Sharpe Mfg. Co., Providence, R.I.

Stamping, Metal
American Pulley Co., Philadelphia, Pa.
Barnes Co., Wallace, Bristol, Conn.
Diamond Saw & Stamping Works, Buffalo, N.Y.
Elliott & Whitehall Tool Co., Galt, Ont.
Fisher Motor Co., Ltd., Orillia, Ont.
Keller Pneumatic Tool Co., Grand Haven, Mich.
Parmenter & Bulloch Co., Gananoque, Ont.
Tallman Brass & Metal Co., Hamilton, Ont.

Stamps, Steel
Diamond Saw & Stamping Works, Buffalo, N.Y.

Stairways, Wrought Iron
Can. Wire & Iron Goods Co., Hamilton, Ont.

Steam Specialties
Crane Ltd., Montreal, Que.
Republic Flow Meters Co., Toronto, Ont.

Steel Plate
Hamilton Bridge Works Co., Ltd., Hamilton, Ont.
Dom. Foundries & Steel, Hamilton, Ont.

Steels, Tool
Vulcan Crucible Steel Co., Aliquippa, Pa.

Steel, Cold-Rolled Strip
Andrews Steel Co., Newport, Ky.
Barnes Co., Wallace, Bristol, Conn.
Can. Driver-Harris Co., Walkerville, Ont.
Firth & Sons, Ltd., Thos., Montreal, Q.
Ontario Metal Products Co., Ltd., Toronto, Ont.

Steel Castings
Dom. Foundries & Steel, Hamilton, Ont.

Steel Hardness Measuring Instruments
Halcomb Steel Co., Syracuse, N.Y.

Steel, Shafting and Free Cutting Screw
Barnes Co., Wallace, Bristol, Conn.
Can. Drawn Steel Co., Hamilton, Ont.
Union Drawn Steel Co., Hamilton, Ont.

Steel, Sheet
Andrews Steel Co., Newport, Ky.
Firth & Sons, Ltd., Thos., Montreal, Q.
Illingworth Steel Co., John, New York City, N.Y.
Rice Lewis & Son, Ltd., Toronto, Ont.
Steel Co. of Can., Ltd., Hamilton, Ont.
Toronto Iron Works, Toronto, Ont.

Steel, Tanks
Can. John Wood Mfg. Co., Toronto, Ont.

Steel, Stainless
Canadian Atlas Crucible Steel Co., Ltd., Toronto, Ont.
Steel Co. of Can., Ltd., Hamilton, Ont.
Vanadium Alloys Steel, Latrobe, Pa.

Steels, Alloy, Open Hearth and Electric
United Alloy Steel Corp., Canton, Ohio.

Steels, Alloy and Carbon
Algoma Steel Corp., Ltd., Sault Ste. Marie, Ont.
Andrews Steel Co., Newport, Ky.
Armstrong Whitworth Co. of Can., Ltd., Montreal, Que.
Atkins & Co., Ltd., Wm., Sheffield, Eng.
Barnes Co., Wallace, Bristol, Conn.
Canadian Atlas Crucible Steel Co., Ltd., Toronto, Ont.
Can. Driver-Harris Co., Walkerville, Ont.
Can. Steel Foundries, Montreal, Que.
Dom. Foundries & Steel, Hamilton, Ont.
Firth & Sons, Ltd., Thos., Montreal, Q.
Halcomb Steel Co., Syracuse, N.Y.
Illingworth Steel Co., John, New York City, N.Y.
Pilot Steel & Tool Co., Montreal, Que.
Rice Lewis & Son, Ltd., Toronto, Ont.
Steel Co. of Can., Ltd., Hamilton, Ont.
United Alloy Steel Corp., Canton, Ohio.
Vanadium Alloys Steel, Latrobe, Pa.
Vulcan Crucible Steel Co., Aliquippa, Pa.

Steels, High-Speed
Andrews Steel Co., Newport, Ky.
Armstrong Bros. Tool Co., Chicago, Ill.
Armstrong Whitworth Co. of Can., Ltd., Montreal, Que.
Atkins & Co., Ltd., Wm., Sheffield, Eng.
Baines & David, Ltd., Toronto, Ont.
Canadian Atlas Crucible Steel Co., Ltd., Toronto, Ont.
Drury Ltd., H. A., Montreal, Que.
Firth & Sons, Ltd., Thos., Montreal, Q.
Illingworth Steel Co., John, New York City, N.Y.
Pilot Steel & Tool Co., Montreal, Que.
Rice Lewis & Son, Ltd., Toronto, Ont.
Steel Co. of Can., Ltd., Hamilton, Ont.
Vanadium Alloys Steel, Latrobe, Pa.
Vulcan Crucible Steel Co., Aliquippa, Pa.

Steel, Magnet
Vanadium Alloys Steel, Latrobe, Pa.

Steel, Structural
MacKinnon Steel Co., Sherbrooke, Que.

Steel Tubing, Close Joint and Welded
Standard Tube & Fence Co., Ltd., Woodstock, Ont.

Stencil Machines
Diagraph Stencil Machine Corp'n, St. Louis, Mo.

Stencil Cutting Machines
Diagraph Stencil Machine Corp'n, St. Louis, Mo.

Stern Frames, Cast Steel
Can. Steel Foundries, Montreal, Que.

Stools, Shop
Ministry of Munitions, London, Eng.

Straightening Machinery
Bertrams Ltd., Edinburgh, Scotland.
Ryerson & Son, Jos. T., Chicago, Ill.

Studs
Galt Machine Screw Co., Galt, Ont.

Surface Plates
Bilton Machine Co., Bridgeport, Conn.

Swaging Machines
Atkins & Co., Inc., E. C., Indianapolis, I.

Switches, Railway
Can. Steel Foundries, Montreal, Que.

Switches and Switchboards
Northern Electric Co., Montreal, Que.

Tachometers
Aikenhead Hardware Ltd., Toronto, Ont.
Bristol Co., Waterbury, Conn.

Tanks, Steel
MacKinnon Steel Co., Sherbrooke, Que.
Can. John Wood Mfg. Co., Ltd., Toronto, Ont.

Tanks and Pumps, Oil
Bowser & Co., Inc., S. F., Fort Wayne, I.
Can. Ingersoll-Rand Co., Ltd., Sherbrooke, Que.
Toronto Iron Works, Toronto, Ont.

Tap Holders
Greenfield Tap & Die Corp., Galt, Ont.
Pratt & Whitney Co., of Canada, Ltd., Dundas, Ont.

Taper Pins
Galt Machine Screw Co., Galt, Ont.
Morrow Screw & Nut Co., Ltd., John, Ingersoll, Ont.
Pratt & Whitney Co., of Canada, Ltd., Dundas, Ont.

Tapes, Measuring
Chesterman & Co., Ltd., J., Sheffield, Eng.
Starrett Co., L. S., Athol, Mass.

Tapping Machines and Attachments
Ackworthie, Ltd., John, Birmingham, Eng.
Archibald & Co., Chas. P., Montreal, Q.
Burke Machine Tool Co., Conneaut, Ohio
Geometric Tool Co., New Haven, Conn.
Greenfield Tap & Die Corp., Galt, Ont.
Jardine & Co., A. B., Hespeler, Ont.
National Acme Co., Cleveland, Ohio.
Petrie, Ltd., H. W., Toronto, Ont.
Starrett Co., L. S., Athol, Mass.
St. Louis Machine Co., St. Louis, Mo.

Taps and Dies
Ackworthie, Ltd., John, Birmingham, Eng.
Butterfield & Co., Inc., Rock Island, Que.
Geometric Tool Co., New Haven, Conn.
Greenfield Tap & Die Corp., Galt, Ont.
International Machinery & Supply Co., Montreal, Que.
Jardine & Co., A. B., Hespeler, Ont.
Morse Twist Drill & Machine Co., New Bedford, Mass.
National Acme Co., Cleveland, Ohio.
Pratt & Whitney Co., of Canada, Ltd., Dundas, Ont.
Rickert Shafer Co., Erie, Pa.
Taylor Tool Co., J. A. M., Toronto, Ont.

Taps, Collapsing
Geometric Tool Co., New Haven, Conn.
Jardine & Co., A. B., Hespeler, Ont.
Murchey Machine & Tool Co., Detroit, Mich.
National Acme Co., Cleveland, Ohio.
Rickert Shafer Co., Erie, Pa.
Victor Tool Co., Waynesboro, Pa.

Teeth, Dredge Bucket
Kennedy & Sons, Wm., Owen Sound, Ont.

Testing Metals and Materials
Toronto Testing Laboratory, Toronto, Ont.

Thermometers
Bristol Co., Waterbury, Conn.

Thread-Cutting Tools
Butterfield & Co., Inc., Rock Island, Que.
Greenfield Tap & Die Corp., Galt, Ont.
Murchey Machine & Tool Co., Detroit, Mich.
National Acme Co., Cleveland, Ohio.
Pratt & Whitney Co., of Canada, Ltd., Dundas, Ont.
Victor Tool Co., Waynesboro, Pa.

Thread Cutting Machines
Landis Machine Co., Inc., Waynesboro, Pa.

Threading Machines
Acme Machinery Co., Cleveland, Ohio.
Geometric Tool Co., New Haven, Conn.
Greenfield Tap & Die Corp., Galt, Ont.
Murchey Machine & Tool Co., Detroit, Mich.
National Acme Co., Cleveland, Ohio.
National Machinery Co., Tiffin, Ohio.
Williams Tool Corp. of Can., Ltd., Brantford, Ont.

Thread Lead Testing Machines
Jones & Lamson Machine Co., Springfield, Vt.
Pratt & Whitney Co., of Canada, Ltd., Dundas, Ont.

Thread-Rolling Machines
Bliss Co., E. W., Brooklyn, N.Y.

Tongs
Hepburn Ltd., John T., Toronto, Ont.

Tool Cases
Rice Lewis & Son, Ltd., Toronto, Ont.

Tool Holders
Armstrong Bros. Tool Co., Chicago, Ill.
Bilton Machine Co., Bridgeport, Conn.
Gisholt Machine Co., Madison, Wis.
Williams & Co., J. H., Brooklyn, N.Y.

Tool Markers, Electric
Can. Ingersoll-Rand Co., Ltd., Sherbrooke, Que.

Tool Posts, Lathe
Bertram & Son Co., Ltd., The John, Dundas, Ont.
Canada Machinery Corp., Galt, Ont.
Williams & Co., J. H., Brooklyn, N.Y.

Tool Steels for all Purposes
Canadian Atlas Crucible Steel Co., Ltd., Toronto, Ont.

Tools, Small (See Machinists' Small Tools)
Armstrong Bros. Tool Co., Chicago, Ill.
Bertrams Ltd., Edinburgh, Scotland.
Burgess & Marchand, Montreal, Que.
Chesterman & Co., Ltd., J., Sheffield, Eng.
Dyrab Steel (Consolidated) Ltd., Toronto, Ont.
Elliott & Whitehall Tool Co., Galt, Ont.
Foss Machinery & Supply Co., Geo. F., Montreal, Que.
Geometric Tool Co., New Haven, Conn.
Greenfield Tap & Die Corp., Galt, Ont.
Hamilton Engineering Service, Ltd., Hamilton, Ont.
Jones & Shipman, of Leicester, England, Toronto, Ont.
Keller Pneumatic Tool Co., Grand Haven, Mich.
Kimber & Hillier Mfg. Co., St. Catharines, Ont.
Ministry of Munitions, London, Eng.
National Machine Tool Co., Racine, Wis.
Rapid Tool & Machine Co., Lachine, Que.
Rice Lewis & Son, Ltd., Toronto, Ont.
Rockford Milling Machine Co., Rockford, Ill.
Starrett Co., L. S., Athol, Mass.
Strelinger Co. of Can., Ltd., Chas. A., Windsor, Ont.
Wheel Trueing Tool Co., Detroit, Mich.
Williams Machinery Co., A. R., Toronto, Ont.
Williams Machinery & Supply Co., A. R., Montreal, Que.

Tool Work
Brown Engineering Corp., Ltd., Toronto.
Crescent Machine Co., Ltd., Montreal, Q.
Ford-Smith Machine Co., Hamilton, Ont.

Torches, Blow
Can. Hauck Burner Co., Port Hope, Ont.
International Machinery & Supply Co., Montreal, Que.
Perdue, W. R., San Francisco, Calif.
Prest-O-Lite Co. of Can., Toronto, Ont.
Rice Lewis & Son, Ltd., Toronto, Ont.

Trackwork, Railway
Can. Steel Foundries, Montreal, Que.

Trackwork, Manganese Steel
Can. Steel Foundries, Montreal, Que.

Transformers
Northern Electric Co., Montreal, Que.

Transmission Machinery
Bernard Industrial Co., A., Fortierville, Que.
Can. Link-Belt Co., Toronto, Ont.
Garlock-Walker Mchy. Co., Toronto, Ont.
Jones & Glassco, Montreal, Que.
Kennedy & Sons, Wm., Owen Sound, Ont.
Petrie, Ltd., H. W., Toronto, Ont.
Positive Clutch & Pulley Works, Toronto, Ont.
Renold (Hans) of Canada, Ltd., Montreal, Que.

Transportation Systems (See Trucks)
Mathews Gravity Carrier Co., Port Hope, Ont.

BUYERS' DIRECTORY

Treated Bits
Vanadium Alloys Steel, Latrobe, Pa.

Trolleys and Tramways
Can. Link-Belt Co., Toronto, Ont.
Morris Crane & Hoist Co., Ltd., Niagara Falls, Ont.
Northern Crane Works, Walkerville, Ont.
Reading Chain & Block Co., Reading, Pa.
Wright Mfg. Co., Lisbon, Ohio.

Trucks
Can. Fairbanks-Morse Ltd., Montreal, Q.
Cowan Truck Co. (R. B. Seuler), Toronto, Ont.
Diamond State Fibre Co., Toronto, Ont.
Hepburn Ltd., John T., Toronto, Ont.
Maple Leaf Mfg. Co., Montreal, Que.
Ministry of Munitions, London, Eng.
Morris Crane & Hoist Co., Ltd., Niagara Falls, Ont.
National Steel Car Corp., Ltd., Hamilton, Ont.

Trucks, Industrial Motor
Maple Leaf Mfg. Co., Montreal, Que.
Ministry of Munitions, London, Eng.
National Steel Car Corp., Ltd., Hamilton, Ont.

Tube, Products
Tube Co. of Canada, Toronto, Ont.

Tubing, Electric Welded or Oxy-Acety-lene Welded
Tube Co. of Canada, Toronto, Ont.

Tubing, Flexible
Dunlop Tire & Rubber Goods Co., Ltd., Toronto, Ont.
Goodyear Tire & Rubber Co. of Can., Ltd., Toronto, Ont.

Tubing, Seamless Steel
Tube Co. of Canada, Toronto, Ont.

Tubing, Seamless Steel, Brass and Copper
Dom. Steel Products Co., Brantford, Ont.
Lyman Tube & Supply Co., Montreal, Que.
Ontario Metal Products Co., Ltd., Toronto, Ont.
Tallman Brass & Metal Co., Hamilton, Ont.

Tubing, Welded
International Nickel Co. of Can., Ltd., Toronto, Ont.

Tubing, Welded Steel
Tube Co. of Canada, Toronto, Ont.

Turbines, Water
Kennedy & Sons, Wm., Owen Sound, Ont.

Turret Heads
Ackworths, Ltd., John, Birmingham, Eng.
Bertram & Son Co., Ltd., The John, Dundas, Ont.

Turret Machine (See Lathes, Horizontal Turret)
Acme Machine Tool Co., Cincinnati, Ohio.
Cook Co., Asa S., Hartford, Conn.
Gisholt Machine Co., Madison, Wis.
National Acme Co., Cleveland, Ohio.
Pratt & Whitney Co., of Canada, Ltd., Dundas, Ont.

Turret Posts
Gisholt Machine Co., Madison, Wis.
Warner & Swasey Co., Cleveland, Ohio.

Turrets, Tool Post
Gisholt Machine Co., Madison, Wis.
Motwood Tool Corp., Meadville, Pa.

Unions, Pipe
Crane Ltd., Montreal, Que.

Universal Joints
Ford-Smith Machine Co., Hamilton, Ont.
Holden Co., Ltd., Montreal, Que.

Valves
Can. Fairbanks-Morse Ltd., Montreal, Q.
Cleveland Pneumatic Tool Co., Toronto, Ont.
Crane Ltd., Montreal, Que.
Dunlop Tire & Rubber Goods Co., Ltd., Toronto, Ont.
Goodyear Tire & Rubber Co. of Can., Ltd., Toronto, Ont.

Valves, Rubber Pump
Quaker City Rubber Co., Philadelphia, Pa.

Vises, Drilling Machine
Hoosier Drilling Mach. Co., Goshen, Ind.
Kempsmith Mfg. Co., Milwaukee, Wis.
Reed Mfg. Co., Erie, Pa.

Vises, Metal Workers'
Brown & Sharpe Mfg. Co., Providence, R.I.
Columbia Hdwe. Division, Cleveland, O.

Vises, Milling Machine
Brown & Sharpe Mfg. Co., Providence, R.I.
Cincinnati Machine Co., Ltd., Montreal, Q.
Ford-Smith Machine Co., Hamilton, Ont.
Hendey Machine Co., Torrington, Conn.
Hoosier Drilling Mach. Co., Goshen, Ind.
Keeney & Trecker Co., Milwaukee, Wis.
Kempsmith Mfg. Co., Milwaukee, Wis.
Reed Mfg. Co., Erie, Pa.

Vises, Pipe
Columbia Hdwe. Division, Cleveland, O.
Greenfield Tap & Die Corp., Galt, Ont.

Vises, Planer and Shaper
Bertram & Son Co., Ltd., The John, Dundas, Ont.
Hendey Machine Co., Torrington, Conn.
Hoosier Drilling Mach. Co., Goshen, Ind.
Kempsmith Mfg. Co., Milwaukee, Wis.
McDougall Co., Ltd., R., Galt, Ont.
Reed Mfg. Co., Erie, Pa.

Vises, Universal Machine
Superior Machine Co., London, Ont.

Vises, Wood Workers'
Columbia Hdwe. Division, Cleveland, O.
Pipe Machinery & Supply Co., Cleve. F.
Montreal, Que.
Victor Tool Co., Waynesboro, Pa.

Voltmeters
Bristol Co., Waterbury, Conn.
Northern Electric Co., Montreal, Que.

Wagon Loaders
Link-Belt Co., Toronto, Ont.

Washers
Barnes Co., Wallace, Bristol, Conn.
Diamond State Fibre Co., Toronto, Ont.
Dunlop Tire & Rubber Goods Co., Ltd., Toronto, Ont.
Goodyear Tire & Rubber Co. of Can., Ltd., Toronto, Ont.

Washers, Rubber
Can. Ingersoll-Rand Co., Ltd., Sherbrooke, Que.

Welding Apparatus, Oxy-Acetylene
L'Air Liquide Society, Toronto, Ont.

Welding, Electric
Carter Welding Co., Toronto, Ont.
Lincoln Electric Co., Toronto, Ont.
National Electric Products, Toronto, Ont.

Welding Filler Rods
L'Air Liquide Society, Toronto, Ont.
Perdue, W. H., San Francisco, Calif.
Prest-O-Lite Co. of Can., Toronto, Ont.

Welding Machines, Oxy-Acetylene
Davis-Bournonville Co., Jersey City, N.J.
Holden Co., Ltd., Montreal, Que.
L'Air Liquide Society, Toronto, Ont.
Perdue, W. H., San Francisco, Calif.
Prest-O-Lite Co. of Can., Toronto, Ont.

Welding, Oxy-Acetylene
All-Weld Co., Toronto, Ont.
Carter Welding Co., Toronto, Ont.
Davis-Bournonville Co., Jersey City, N.J.
Lincoln Electric Co., Toronto, Ont.
National Electric Products, Toronto, Ont.
Turner Brass Works, Sycamore, Ill.
Union Carbide Co. of Can., Welland, Ont.

Welding Supplies
All-Weld Co., Toronto, Ont.
British Smelting & Refining Co., Ltd., Toronto, Ont.
Carter Welding Co., Toronto, Ont.
Davis-Bournonville Co., Jersey City, N.J.
L'Air Liquide Society, Toronto, Ont.
Lincoln Electric Co., Toronto, Ont.
National Electric Products, Toronto, Ont.
Perdue, W. H., San Francisco, Calif.
Prest-O-Lite Co. of Canada, Ltd., Toronto, Ont.
Turner Brass Works, Sycamore, Ill.
Union Carbide Co. of Can., Welland, Ont.

Wheels, Industrial
American Pulley Co., Philadelphia, Pa.
Bull Iron & Steel Foundries, Hull, Que.
Kennedy & Sons, Wm., Owen Sound, Ont.

Winches, Electric
Voile Mfg. Co., Welland, Ont.

Winches, Headgate
Kennedy & Sons, Wm., Owen Sound, Ont.

Winches, Stoping
Kennedy & Sons, Wm., Owen Sound, Ont.

Wire
Anglo-Canadian Wire Co., Montreal, Que.
Barnes Co., Wallace, Bristol, Conn.
Canada Metal Co., Ltd., Toronto, Ont.
Dennis Wire & Iron Works, London, Ont.
Greening Wire Co., B., Hamilton, Ont.
Northern Electric Co., Montreal, Que.

Wire Cloth
Can. Wire & Iron Goods Co., Hamilton, Ont.

Wire Rope
Can. Wire & Iron Goods Co., Hamilton, Ont.

Wire Straightening and Cutting Machinery
Baird Machine Co., Bridgeport, Conn.
Brown, Boggs & Co., Ltd., Hamilton, Ont.
Schuster Co., F. B., New Haven, Conn.

Wire, Welding
L'Air Liquide Society, Toronto, Ont.
Perdue, W. H., San Francisco, Calif.
Prest-O-Lite Co. of Can., Toronto, Ont.
Tallman Brass & Metal Co., Hamilton, Ont.

Wire, Special
Anglo-Canadian Wire Co., Montreal, Que.
Dennis Wire & Iron Works, London, Ont.
Greening Wire Co., B., Hamilton, Ont.
Walker & Sons Metal Products, Ltd., Hiram, Walkerville, Ont.

Woodworking Machinery
Canada Machinery Corp., Galt, Ont.
Cowan & Company, of Galt, Ltd., Galt, Ont.

Wrenches, Machinists'
Armstrong Bros. Tool Co., Chicago, Ill.
Canada Foundries & Forgings Co., Welland, Ont.

Wrenches, Drop Forged
Armstrong Bros. Tool Co., Chicago, Ill.
Canada Foundries & Forgings Co., Welland, Ont.

Wrenches, Pipe
Canada Foundries & Forgings Co., Welland, Ont.
Crane Ltd., Montreal, Que.
Greenfield Tap & Die Corp., Galt, Ont.

Wrenches, Tap
Butterfield & Co., Inc., Rock Island, Que.
Greenfield Tap & Die Corp., Galt, Ont.

INDEX TO ADVERTISERS